T0091748

Texts in Computer Science

Titles in this series now included in the Thomson Reuters Book Citation Index!

'Texts in Computer Science' (TCS) delivers high-quality instructional content for undergraduates and graduates in all areas of computing and information science, with a strong emphasis on core foundational and theoretical material but inclusive of some prominent applications-related content. TCS books should be reasonably self-contained and aim to provide students with modern and clear accounts of topics ranging across the computing curriculum. As a result, the books are ideal for semester courses or for individual self-study in cases where people need to expand their knowledge. All texts are authored by established experts in their fields, reviewed internally and by the series editors, and provide numerous examples, problems, and other pedagogical tools; many contain fully worked solutions.

The TCS series is comprised of high-quality, self-contained books that have broad and comprehensive coverage and are generally in hardback format and sometimes contain color. For undergraduate textbooks that are likely to be more brief and modular in their approach, require only black and white, and are under 275 pages, Springer offers the flexibly designed Undergraduate Topics in Computer Science series, to which we refer potential authors.

More information about this series at https://link.springer.com/bookseries/3191

Richard Szeliski

Computer Vision

Algorithms and Applications

Second Edition

Springer

Richard Szeliski
University of Washington
Seattle, WA, USA

ISSN 1868-0941 ISSN 1868-095X (electronic)
Texts in Computer Science
ISBN 978-3-030-34371-2 ISBN 978-3-030-34372-9 (eBook)
https://doi.org/10.1007/978-3-030-34372-9

Instance segmentation (delineating individual objects in a scene with pixel-accurate masks) using the Mask R-CNN system of He, Gkioxari et al. (2017), © 2017 IEEE (reproduced with permission).

This Springer imprint is published by the registered company Springer Nature Switzerland AG
The registered company address is: Gewerbestrasse 11, 6330 Cham, Switzerland

This book is dedicated to my parents,
Zdzisław and Jadwiga,
and my family,
Lyn, Anne, and Stephen.

1 Introduction 1

What is computer vision? • A brief history • Book overview • Sample syllabus • Notation

2 Image formation 27

Geometric primitives and transformations • Photometric image formation • The digital camera

3 Image processing 85

Point operators • Linear filtering • Non-linear filtering • Fourier transforms • Pyramids and wavelets • Geometric transformations

4 Model fitting and optimization 153

Scattered data interpolation • Variational methods and regularization • Markov random fields

5 Deep learning 187

Supervised learning • Unsupervised learning • Deep neural networks • Convolutional networks • More complex models

6 Recognition 273

Instance recognition • Image classification • Object detection • Semantic segmentation • Video understanding • Vision and language

7 Feature detection and matching 333

Points and patches • Edges and contours • Contour tracking • Lines and vanishing points • Segmentation

8 **Image alignment and stitching** 401

Pairwise alignment • Image stitching • Global alignment • Compositing

9 **Motion estimation** 443

Translational alignment • Parametric motion • Optical flow • Layered motion

10 **Computational photography** 483

Photometric calibration • High dynamic range imaging • Super-resolution, denoising, and blur removal • Image matting and compositing • Texture analysis and synthesis

11 **Structure from motion and SLAM** 543

Geometric intrinsic calibration • Pose estimation • Two-frame structure from motion • Multi-frame structure from motion • Simultaneous localization and mapping (SLAM)

12 **Depth estimation** 595

Epipolar geometry • Sparse correspondence • Dense correspondence • Local methods • Global optimization • Deep neural networks • Multi-view stereo • Monocular depth estimation

13 **3D reconstruction** 639

Shape from X • 3D scanning • Surface representations • Point-based representations • Volumetric representations • Model-based reconstruction • Recovering texture maps and albedos

14 **Image-based rendering** 681

View interpolation • Layered depth images • Light fields and Lumigraphs • Environment mattes • Video-based rendering • Neural rendering

Preface

The seeds for this book were first planted in 2001 when Steve Seitz at the University of Washington invited me to co-teach a course called "Computer Vision for Computer Graphics". At that time, computer vision techniques were increasingly being used in computer graphics to create image-based models of real-world objects, to create visual effects, and to merge real-world imagery using computational photography techniques. Our decision to focus on the applications of computer vision to fun problems such as image stitching and photo-based 3D modeling from personal photos seemed to resonate well with our students.

That initial course evolved into a more complete computer vision syllabus and project-oriented course structure that I used to co-teach general computer vision courses both at the University of Washington and at Stanford. (The latter was a course I co-taught with David Fleet in 2003.) Similar curricula were then adopted at a number of other universities and also incorporated into more specialized courses on computational photography. (For ideas on how to use this book in your own course, please see Table 1.1 in Section 1.4.)

This book also reflects my 40 years' experience doing computer vision research in corporate research labs, mostly at Digital Equipment Corporation's Cambridge Research Lab, Microsoft Research, and Facebook. In pursuing my work, I have mostly focused on problems and solution techniques (algorithms) that have practical real-world applications and that work well in practice. Thus, this book has more emphasis on basic techniques that work under real-world conditions and less on more esoteric mathematics that has intrinsic elegance but less practical applicability.

This book is suitable for teaching a senior-level undergraduate course in computer vision to students in both computer science and electrical engineering. I prefer students to have either an image processing or a computer graphics course as a prerequisite, so that they can spend less time learning general background mathematics and more time studying computer vision techniques. The book is also suitable for teaching graduate-level courses in computer vision, e.g., by delving into more specialized topics, and as a general reference to fundamental techniques and the recent research literature. To this end, I have attempted wherever possible to at least cite the newest research in each sub-field, even if the technical details are too complex to cover in the book itself.

In teaching our courses, we have found it useful for the students to attempt a number of small implementation projects, which often build on one another, in order to get them used to working with real-world images and the challenges that these present. The students are then asked to choose an individual topic for each of their small-group, final projects. (Sometimes these projects even turn into conference papers!) The exercises at the end of each chapter contain numerous suggestions for smaller mid-term projects, as well as more open-ended problems whose solutions are still active research topics. Wherever possible, I encourage students to try their algorithms on their own personal photographs, since this better motivates them, often leads to creative variants on the problems, and better acquaints them with the variety and complexity of real-world imagery.

In formulating and solving computer vision problems, I have often found it useful to draw inspiration from four high-level approaches:

- **Scientific:** build detailed models of the image formation process and develop mathematical techniques to invert these in order to recover the quantities of interest (where necessary, making simplifying assumptions to make the mathematics more tractable).

- **Statistical:** use probabilistic models to quantify the prior likelihood of your unknowns and the noisy measurement processes that produce the input images, then infer the best possible estimates of your desired quantities and analyze their resulting uncertainties. The inference algorithms used are often closely related to the optimization techniques used to invert the (scientific) image formation processes.

- **Engineering:** develop techniques that are simple to describe and implement but that are also known to work well in practice. Test these techniques to understand their limitation and failure modes, as well as their expected computational costs (run-time performance).

- **Data-driven:** collect a representative set of test data (ideally, with labels or ground-truth answers) and use these data to either tune or learn your model parameters, or at least to validate and quantify its performance.

These four approaches build on each other and are used throughout the book.

My personal research and development philosophy (and hence the exercises in the book) have a strong emphasis on *testing* algorithms. It's too easy in computer vision to develop an algorithm that does something *plausible* on a few images rather than something *correct*. The best way to validate your algorithms is to use a three-part strategy.

First, test your algorithm on clean synthetic data, for which the exact results are known. Second, add noise to the data and evaluate how the performance degrades as a function of noise level. Finally, test the algorithm on real-world data, preferably drawn from a wide variety of sources, such as photos found on the web. Only then can you truly know if your algorithm can deal with real-world complexity, i.e., images that do not fit some simplified model or assumptions.

In order to help students in this process, Appendix C includes pointers to commonly used datasets and software libraries that contain implementations of a wide variety of computer vision algorithms, which can enable you to tackle more ambitious projects (with your instructor's consent).

Notes on the Second Edition

The last decade has seen a truly dramatic explosion in the performance and applicability of computer vision algorithms, much of it engendered by the application of machine learning algorithms to large amounts of visual training data (Su and Crandall 2021).

Deep neural networks now play an essential role in so many vision algorithms that the new edition of this book introduces them early on as a fundamental technique that gets used extensively in subsequent chapters.

The most notable changes in the second edition include:

- Machine learning, deep learning, and deep neural networks are introduced early on in Chapter 5, as they play just as fundamental a role in vision algorithms as more classical techniques, such as image processing, graphical/probabilistic models, and energy minimization, which are introduced in the preceding two chapters.

- The recognition chapter has been moved earlier in the book to Chapter 6, since end-to-end deep learning systems no longer require the development of building blocks such as feature detection, matching, and segmentation. Many of the students taking vision classes are primarily interested in visual recognition, so presenting this material earlier in the course makes it easier for students to base their final project on these topics. This chapter also includes sections on semantic segmentation, video understanding, and vision and language.

- The application of neural networks and deep learning to myriad computer vision algorithms and applications, including flow and stereo, 3D shape modeling, and newly emerging fields such as neural rendering.

- New technologies such as SLAM (simultaneous localization and mapping) and VIO (visual inertial odometry) that now run reliably and are used in real-time applications such as augmented reality and autonomous navigation.

In addition to these larger changes, the book has been updated to reflect the latest state-of-the-art techniques such as internet-scale image search and phone-based computational photography. The new edition includes over 1500 new citations (papers) and has over 200 new figures.

Acknowledgements

I would like to gratefully acknowledge all of the people whose passion for research and inquiry as well as encouragement have helped me write this book.

Steve Zucker at McGill University first introduced me to computer vision, taught all of his students to question and debate research results and techniques, and encouraged me to pursue a graduate career in this area.

Takeo Kanade and Geoff Hinton, my PhD thesis advisors at Carnegie Mellon University, taught me the fundamentals of good research, writing, and presentation and mentored several generations of outstanding students and researchers. They fired up my interest in visual processing, 3D modeling, and statistical methods, while Larry Matthies introduced me to Kalman filtering and stereo matching. Geoff continues to inspire so many of us with this undiminished passion for trying to figure out "what makes the brain work". It's been a delight to see his pursuit of connectionist ideas bear so much fruit in this past decade.

Demetri Terzopoulos was my mentor at my first industrial research job and taught me the ropes of successful publishing. Yvan Leclerc and Pascal Fua, colleagues from my brief interlude at SRI International, gave me new perspectives on alternative approaches to computer vision.

During my six years of research at Digital Equipment Corporation's Cambridge Research Lab, I was fortunate to work with a great set of colleagues, including Ingrid Carlbom, Gudrun Klinker, Keith Waters, William Hsu, Richard Weiss, Stéphane Lavallée, and Sing Bing Kang, as well as to supervise the first of a long string of outstanding summer interns, including David Tonnesen, Sing Bing Kang, James Coughlan, and Harry Shum. This is also where I began my long-term collaboration with Daniel Scharstein.

At Microsoft Research, I had the outstanding fortune to work with some of the world's best researchers in computer vision and computer graphics, including Michael Cohen, Matt Uyttendaele, Sing Bing Kang, Harry Shum, Larry Zitnick, Sudipta Sinha, Drew Steedly, Simon Baker, Johannes Kopf, Neel Joshi, Krishnan Ramnath, Anandan, Phil Torr, Antonio Criminisi, Simon Winder, Matthew Brown, Michael Goesele, Richard Hartley, Hugues Hoppe, Stephen Gortler, Steve Shafer, Matthew Turk, Georg Petschnigg, Kentaro Toyama, Ramin Zabih, Shai Avidan, Patrice Simard, Chris Pal, Nebojsa Jojic, Patrick Baudisch, Dani Lischinski, Raanan Fattal, Eric Stollnitz, David Nistér, Blaise Aguera y Arcas, Andrew Fitzgibbon, Jamie Shotton, Wolf Kienzle, Piotr Dollar, and Ross Girshick. I was also lucky to have as interns such great students as Polina Golland, Simon

Baker, Mei Han, Arno Schödl, Ron Dror, Ashley Eden, Jonathan Shade, Jinxiang Chai, Rahul Swaminathan, Yanghai Tsin, Sam Hasinoff, Anat Levin, Matthew Brown, Eric Bennett, Vaibhav Vaish, Jan-Michael Frahm, James Diebel, Ce Liu, Josef Sivic, Grant Schindler, Colin Zheng, Neel Joshi, Sudipta Sinha, Zeev Farbman, Rahul Garg, Tim Cho, Yekeun Jeong, Richard Roberts, Varsha Hedau, Dilip Krishnan, Adarsh Kowdle, Edward Hsiao, Yong Seok Heo, Fabian Langguth, Andrew Owens, and Tianfan Xue. Working with such outstanding students also gave me the opportunity to collaborate with some of their amazing advisors, including Bill Freeman, Irfan Essa, Marc Pollefeys, Michael Black, Marc Levoy, and Andrew Zisserman.

Since moving to Facebook, I've had the pleasure to continue my collaborations with Michael Cohen, Matt Uyttendaele, Johannes Kopf, Wolf Kienzle, and Krishnan Ramnath, and also new colleagues including Kevin Matzen, Bryce Evans, Suhib Alsisan, Changil Kim, David Geraghty, Jan Herling, Nils Plath, Jan-Michael Frahm, True Price, Richard Newcombe, Thomas Whelan, Michael Goesele, Steven Lovegrove, Julian Straub, Simon Green, Brian Cabral, Michael Toksvig, Albert Para Pozzo, Laura Sevilla-Lara, Georgia Gkioxari, Justin Johnson, Chris Sweeney, and Vassileios Balntas. I've also had the pleasure to collaborate with some outstanding summer interns, including Tianfan Xue, Scott Wehrwein, Peter Hedman, Joel Janai, Aleksander Hołyński, Xuan Luo, Rui Wang, Olivia Wiles, and Yulun Tian. I'd like to thank in particular Michael Cohen, my mentor, colleague, and friend for the last 25 years for his unwavering support of my sprint to complete this second edition.

While working at Microsoft and Facebook, I've also had the opportunity to collaborate with wonderful colleagues at the University of Washington, where I hold an Affiliate Professor appointment. I'm indebted to Tony DeRose and David Salesin, who first encouraged me to get involved with the research going on at UW, my long-time collaborators Brian Curless, Steve Seitz, Maneesh Agrawala, Sameer Agarwal, and Yasu Furukawa, as well as the students I have had the privilege to supervise and interact with, including Fréderic Pighin, Yung-Yu Chuang, Doug Zongker, Colin Zheng, Aseem Agarwala, Dan Goldman, Noah Snavely, Ian Simon, Rahul Garg, Ryan Kaminsky, Juliet Fiss, Aleksander Hołyński, and Yifan Wang. As I mentioned at the beginning of this preface, this book owes its inception to the vision course that Steve Seitz invited me to co-teach, as well as to Steve's encouragement, course notes, and editorial input.

I'm also grateful to the many other computer vision researchers who have given me so many constructive suggestions about the book, including Sing Bing Kang, who was my informal book editor, Vladimir Kolmogorov, Daniel Scharstein, Richard Hartley, Simon Baker, Noah Snavely, Bill Freeman, Svetlana Lazebnik, Matthew Turk, Jitendra Malik, Alyosha Efros, Michael Black, Brian Curless, Sameer Agarwal, Li Zhang, Deva Ramanan, Olga Veksler, Yuri Boykov, Carsten Rother, Phil Torr, Bill Triggs, Bruce Maxwell, Rico Malvar, Jana Košecká, Eero Simoncelli, Aaron Hertzmann, Antonio Torralba, Tomaso Poggio, Theo Pavlidis, Baba Vemuri, Nando de Freitas, Chuck Dyer, Song Yi, Falk Schubert, Roman Pflugfelder, Marshall Tappen, James Coughlan, Sammy Rogmans, Klaus Strobel, Shanmuganathan, Andreas Siebert, Yongjun Wu, Fred Pighin, Juan Cockburn, Ronald Mallet, Tim Soper, Georgios Evangelidis, Dwight Fowler, Itzik Bayaz, Daniel O'Connor, Srikrishna Bhat, and Toru Tamaki, who wrote the Japanese translation and provided many useful errata.

For the second edition, I received significant help and advice from three key contributors. Daniel Scharstein helped me update the chapter on stereo, Matt Deitke contributed descriptions of the newest papers in deep learning, including the sections on transformers, variational autoencoders, and text-to-image synthesis, along with the exercises in Chapters 5 and 6 and some illustrations. Sing Bing Kang reviewed multiple drafts and provided useful suggestions. I'd also like to thank Andrew Glassner, whose book (Glassner 2018) and figures were a tremendous help, Justin Johnson,

Sean Bell, Ishan Misra, David Fouhey, Michael Brown, Abdelrahman Abdelhamed, Frank Dellaert, Xinlei Chen, Ross Girshick, Andreas Geiger, Dmytro Mishkin, Aleksander Hołyński, Joel Janai, Christoph Feichtenhofer, Yuandong Tian, Alyosha Efros, Pascal Fua, Torsten Sattler, Laura Leal-Taixé, Aljosa Osep, Qunjie Zhou, Jiří Matas, Eddy Ilg, Yann LeCun, Larry Jackel, Vasileios Balntas, Daniel DeTone, Zachary Teed, Junhwa Hur, Jun-Yan Zhu, Filip Radenović, Michael Zollhöfer, Matthias Nießner, Andrew Owens, Hervé Jégou, Luowei Zhou, Ricardo Martin Brualla, Pratul Srinivasan, Matteo Poggi, Fabio Tosi, Ahmed Osman, Dave Howell, Holger Heidrich, Howard Yen, Anton Papst, Syamprasad K. Rajagopalan, Abhishek Nagar, Vladimir Kuznetsov, Raphaël Fouque, Marian Ciobanu, Darko Simonovic, and Guilherme Schlinker.

In preparing the second edition, I taught some of the new material in two courses that I helped co-teach in 2020 at Facebook and UW. I'd like to thank my co-instructors Jan-Michael Frahm, Michael Goesele, Georgia Gkioxari, Ross Girshick, Jakob Julian Engel, Daniel Scharstein, Fernando de la Torre, Steve Seitz, and Harpreet Sawhney, from whom I learned a lot about the latest techniques that are included in the new edition. I'd also like to thank the TAs, including David Geraghty, True Price, Kevin Matzen, Akash Bapat, Aleksander Hołyński, Keunhong Park, and Svetoslav Kolev, for the wonderful job they did in creating and grading the assignments. I'd like to give a special thanks to Justin Johnson, whose excellent class slides (Johnson 2020), based on earlier slides from Stanford (Li, Johnson, and Yeung 2019), taught me the fundamentals of deep learning and which I used extensively in my own class and in preparing the new chapter on deep learning.

Shena Deuchers and Ian Kingston did a fantastic job copy-editing the first and second editions, respectively and suggesting many useful improvements, and Wayne Wheeler and Simon Rees at Springer were most helpful throughout the whole book publishing process. Keith Price's Annotated Computer Vision Bibliography was invaluable in tracking down references and related work.

If you have any suggestions for improving the book, please send me an e-mail, as I would like to keep the book as accurate, informative, and timely as possible.

The last year of writing this second edition took place during the worldwide COVID-19 pandemic. I would like to thank all of the first responders, medical and front-line workers, and everyone else who helped get us through these difficult and challenging times and to acknowledge the impact that this and other recent tragedies have had on all of us.

Lastly, this book would not have been possible or worthwhile without the incredible support and encouragement of my family. I dedicate this book to my parents, Zdzisław and Jadwiga, whose love, generosity, and accomplishments always inspired me; to my sister Basia for her lifelong friendship; and especially to Lyn, Anne, and Stephen, whose love and support in all matters (including my book projects) makes it all worthwhile.

Lake Wenatchee
May 2021

Contents

Preface **ix**

1 Introduction **1**
1.1 What is computer vision? 3
1.2 A brief history 9
1.3 Book overview 18
1.4 Sample syllabus 24
1.5 A note on notation 25
1.6 Additional reading 26

2 Image formation **27**
2.1 Geometric primitives and transformations 29
2.1.1 2D transformations 32
2.1.2 3D transformations 35
2.1.3 3D rotations 36
2.1.4 3D to 2D projections 41
2.1.5 Lens distortions 51
2.2 Photometric image formation 53
2.2.1 Lighting 53
2.2.2 Reflectance and shading 54
2.2.3 Optics 59
2.3 The digital camera 63
2.3.1 Sampling and aliasing 67
2.3.2 Color 69
2.3.3 Compression 78
2.4 Additional reading 80
2.5 Exercises 80

3 Image processing **85**
3.1 Point operators 87
3.1.1 Pixel transforms 89
3.1.2 Color transforms 90
3.1.3 Compositing and matting 91
3.1.4 Histogram equalization 92
3.1.5 *Application*: Tonal adjustment 95
3.2 Linear filtering 95
3.2.1 Separable filtering 99

3.2.2 Examples of linear filtering 100
3.2.3 Band-pass and steerable filters 101
3.3 More neighborhood operators 105
3.3.1 Non-linear filtering 105
3.3.2 Bilateral filtering 107
3.3.3 Binary image processing 110
3.4 Fourier transforms 113
3.4.1 Two-dimensional Fourier transforms 115
3.4.2 *Application*: Sharpening, blur, and noise removal 118
3.5 Pyramids and wavelets 119
3.5.1 Interpolation 119
3.5.2 Decimation 122
3.5.3 Multi-resolution representations 123
3.5.4 Wavelets 128
3.5.5 *Application*: Image blending 132
3.6 Geometric transformations 135
3.6.1 Parametric transformations 135
3.6.2 Mesh-based warping 140
3.6.3 *Application*: Feature-based morphing 142
3.7 Additional reading 143
3.8 Exercises 144

4 Model fitting and optimization 153
4.1 Scattered data interpolation 155
4.1.1 Radial basis functions 157
4.1.2 Overfitting and underfitting 159
4.1.3 Robust data fitting 162
4.2 Variational methods and regularization 163
4.2.1 Discrete energy minimization 166
4.2.2 Total variation 168
4.2.3 Bilateral solver 168
4.2.4 *Application*: Interactive colorization 169
4.3 Markov random fields 170
4.3.1 Conditional random fields 177
4.3.2 *Application*: Interactive segmentation 181
4.4 Additional reading 184
4.5 Exercises 185

5 Deep Learning 187
5.1 Supervised learning 191
5.1.1 Nearest neighbors 192
5.1.2 Bayesian classification 194
5.1.3 Logistic regression 198
5.1.4 Support vector machines 199
5.1.5 Decision trees and forests 202
5.2 Unsupervised learning 205
5.2.1 Clustering 205
5.2.2 K-means and Gaussians mixture models 206

5.2.3 Principal component analysis 209
5.2.4 Manifold learning 211
5.2.5 Semi-supervised learning 212
5.3 Deep neural networks 214
5.3.1 Weights and layers 215
5.3.2 Activation functions 217
5.3.3 Regularization and normalization 219
5.3.4 Loss functions 223
5.3.5 Backpropagation 226
5.3.6 Training and optimization 228
5.4 Convolutional neural networks 231
5.4.1 Pooling and unpooling 234
5.4.2 *Application*: Digit classification 237
5.4.3 Network architectures 238
5.4.4 Model zoos 242
5.4.5 Visualizing weights and activations 244
5.4.6 Adversarial examples 248
5.4.7 Self-supervised learning 249
5.5 More complex models 252
5.5.1 Three-dimensional CNNs 252
5.5.2 Recurrent neural networks 255
5.5.3 Transformers 257
5.5.4 Generative models 261
5.6 Additional reading 267
5.7 Exercises 268

6 Recognition **273**
6.1 Instance recognition 276
6.2 Image classification 278
6.2.1 Feature-based methods 278
6.2.2 Deep networks 285
6.2.3 *Application*: Visual similarity search 287
6.2.4 Face recognition 289
6.3 Object detection 295
6.3.1 Face detection 295
6.3.2 Pedestrian detection 299
6.3.3 General object detection 301
6.4 Semantic segmentation 307
6.4.1 *Application*: Medical image segmentation 310
6.4.2 Instance segmentation 311
6.4.3 Panoptic segmentation 312
6.4.4 *Application*: Intelligent photo editing 314
6.4.5 Pose estimation 315
6.5 Video understanding 316
6.6 Vision and language 319
6.7 Additional reading 326
6.8 Exercises 329

7 Feature detection and matching **333**

7.1 Points and patches . . . 335
7.1.1 Feature detectors . . . 337
7.1.2 Feature descriptors . . . 347
7.1.3 Feature matching . . . 352
7.1.4 Large-scale matching and retrieval . . . 358
7.1.5 Feature tracking . . . 361
7.1.6 *Application*: Performance-driven animation . . . 363
7.2 Edges and contours . . . 364
7.2.1 Edge detection . . . 364
7.2.2 Contour detection . . . 368
7.2.3 *Application*: Edge editing and enhancement . . . 372
7.3 Contour tracking . . . 373
7.3.1 Snakes and scissors . . . 373
7.3.2 Level Sets . . . 379
7.3.3 *Application*: Contour tracking and rotoscoping . . . 380
7.4 Lines and vanishing points . . . 381
7.4.1 Successive approximation . . . 381
7.4.2 Hough transforms . . . 381
7.4.3 Vanishing points . . . 384
7.5 Segmentation . . . 386
7.5.1 Graph-based segmentation . . . 388
7.5.2 Mean shift . . . 389
7.5.3 Normalized cuts . . . 391
7.6 Additional reading . . . 393
7.7 Exercises . . . 395

8 Image alignment and stitching **401**

8.1 Pairwise alignment . . . 403
8.1.1 2D alignment using least squares . . . 403
8.1.2 *Application*: Panography . . . 405
8.1.3 Iterative algorithms . . . 406
8.1.4 Robust least squares and RANSAC . . . 408
8.1.5 3D alignment . . . 410
8.2 Image stitching . . . 411
8.2.1 Parametric motion models . . . 412
8.2.2 *Application*: Whiteboard and document scanning . . . 414
8.2.3 Rotational panoramas . . . 414
8.2.4 Gap closing . . . 416
8.2.5 *Application*: Video summarization and compression . . . 417
8.2.6 Cylindrical and spherical coordinates . . . 418
8.3 Global alignment . . . 421
8.3.1 Bundle adjustment . . . 421
8.3.2 Parallax removal . . . 424
8.3.3 Recognizing panoramas . . . 425
8.4 Compositing . . . 426
8.4.1 Choosing a compositing surface . . . 426
8.4.2 Pixel selection and weighting (deghosting) . . . 430

8.4.3 *Application*: Photomontage . . . 435
8.4.4 Blending . . . 435
8.5 Additional reading . . . 437
8.6 Exercises . . . 438

9 Motion estimation **443**
9.1 Translational alignment . . . 445
9.1.1 Hierarchical motion estimation . . . 448
9.1.2 Fourier-based alignment . . . 449
9.1.3 Incremental refinement . . . 451
9.2 Parametric motion . . . 455
9.2.1 *Application*: Video stabilization . . . 457
9.2.2 Spline-based motion . . . 459
9.2.3 *Application*: Medical image registration . . . 461
9.3 Optical flow . . . 461
9.3.1 Deep learning approaches . . . 466
9.3.2 *Application*: Rolling shutter wobble removal . . . 468
9.3.3 Multi-frame motion estimation . . . 468
9.3.4 *Application*: Video denoising . . . 469
9.4 Layered motion . . . 470
9.4.1 *Application*: Frame interpolation . . . 473
9.4.2 Transparent layers and reflections . . . 474
9.4.3 Video object segmentation . . . 476
9.4.4 Video object tracking . . . 477
9.5 Additional reading . . . 478
9.6 Exercises . . . 479

10 Computational photography **483**
10.1 Photometric calibration . . . 486
10.1.1 Radiometric response function . . . 486
10.1.2 Noise level estimation . . . 488
10.1.3 Vignetting . . . 489
10.1.4 Optical blur (spatial response) estimation . . . 491
10.2 High dynamic range imaging . . . 494
10.2.1 Tone mapping . . . 499
10.2.2 *Application*: Flash photography . . . 506
10.3 Super-resolution, denoising, and blur removal . . . 508
10.3.1 Color image demosaicing . . . 515
10.3.2 Lens blur (bokeh) . . . 517
10.4 Image matting and compositing . . . 518
10.4.1 Blue screen matting . . . 519
10.4.2 Natural image matting . . . 521
10.4.3 Optimization-based matting . . . 524
10.4.4 Smoke, shadow, and flash matting . . . 527
10.4.5 Video matting . . . 528
10.5 Texture analysis and synthesis . . . 529
10.5.1 *Application*: Hole filling and inpainting . . . 531
10.5.2 *Application*: Non-photorealistic rendering . . . 532

10.5.3 Neural style transfer and semantic image synthesis 534
10.6 Additional reading 537
10.7 Exercises 538

11 Structure from motion and SLAM **543**
11.1 Geometric intrinsic calibration 545
11.1.1 Vanishing points 547
11.1.2 *Application*: Single view metrology 548
11.1.3 Rotational motion 549
11.1.4 Radial distortion 550
11.2 Pose estimation 552
11.2.1 Linear algorithms 552
11.2.2 Iterative non-linear algorithms 554
11.2.3 *Application*: Location recognition 555
11.2.4 Triangulation 558
11.3 Two-frame structure from motion 560
11.3.1 Eight, seven, and five-point algorithms 560
11.3.2 Special motions and structures 564
11.3.3 Projective (uncalibrated) reconstruction 565
11.3.4 Self-calibration 566
11.3.5 *Application*: View morphing 568
11.4 Multi-frame structure from motion 568
11.4.1 Factorization 568
11.4.2 Bundle adjustment 570
11.4.3 Exploiting sparsity 571
11.4.4 *Application*: Match move 574
11.4.5 Uncertainty and ambiguities 575
11.4.6 *Application*: Reconstruction from internet photos 576
11.4.7 Global structure from motion 578
11.4.8 Constrained structure and motion 580
11.5 Simultaneous localization and mapping (SLAM) 583
11.5.1 *Application*: Autonomous navigation 585
11.5.2 *Application*: Smartphone augmented reality 587
11.6 Additional reading 588
11.7 Exercises 590

12 Depth estimation **595**
12.1 Epipolar geometry 599
12.1.1 Rectification 600
12.1.2 Plane sweep 602
12.2 Sparse correspondence 604
12.2.1 3D curves and profiles 604
12.3 Dense correspondence 606
12.3.1 Similarity measures 607
12.4 Local methods 609
12.4.1 Sub-pixel estimation and uncertainty 610
12.4.2 *Application*: Stereo-based head tracking 611
12.5 Global optimization 612

12.5.1 Dynamic programming 614
12.5.2 Segmentation-based techniques 616
12.5.3 *Application*: Z-keying and background replacement 617
12.6 Deep neural networks 618
12.7 Multi-view stereo 620
12.7.1 Scene flow 624
12.7.2 Volumetric and 3D surface reconstruction 624
12.7.3 Shape from silhouettes 630
12.8 Monocular depth estimation 632
12.9 Additional reading 634
12.10 Exercises 635

13 3D reconstruction **639**
13.1 Shape from X 641
13.1.1 Shape from shading and photometric stereo 642
13.1.2 Shape from texture 645
13.1.3 Shape from focus 646
13.2 3D scanning 647
13.2.1 Range data merging 650
13.2.2 *Application*: Digital heritage 654
13.3 Surface representations 654
13.3.1 Surface interpolation 655
13.3.2 Surface simplification 656
13.3.3 Geometry images 656
13.4 Point-based representations 657
13.5 Volumetric representations 658
13.5.1 Implicit surfaces and level sets 658
13.6 Model-based reconstruction 660
13.6.1 Architecture 660
13.6.2 Facial modeling and tracking 663
13.6.3 *Application*: Facial animation 665
13.6.4 Human body modeling and tracking 668
13.7 Recovering texture maps and albedos 674
13.7.1 Estimating BRDFs 675
13.7.2 *Application*: 3D model capture 676
13.8 Additional reading 677
13.9 Exercises 679

14 Image-based rendering **681**
14.1 View interpolation 683
14.1.1 View-dependent texture maps 685
14.1.2 *Application*: Photo Tourism 686
14.2 Layered depth images 688
14.2.1 Impostors, sprites, and layers 688
14.2.2 *Application*: 3D photography 690
14.3 Light fields and Lumigraphs 693
14.3.1 Unstructured Lumigraph 696
14.3.2 Surface light fields 696

14.3.3 *Application*: Concentric mosaics . . . 698
14.3.4 *Application*: Synthetic re-focusing . . . 698
14.4 Environment mattes . . . 699
14.4.1 Higher-dimensional light fields . . . 700
14.4.2 The modeling to rendering continuum . . . 701
14.5 Video-based rendering . . . 701
14.5.1 Video-based animation . . . 702
14.5.2 Video textures . . . 703
14.5.3 *Application*: Animating pictures . . . 705
14.5.4 3D and free-viewpoint Video . . . 706
14.5.5 *Application*: Video-based walkthroughs . . . 708
14.6 Neural rendering . . . 711
14.7 Additional reading . . . 718
14.8 Exercises . . . 719

15 Conclusion **723**

A Linear algebra and numerical techniques **727**
A.1 Matrix decompositions . . . 728
A.1.1 Singular value decomposition . . . 728
A.1.2 Eigenvalue decomposition . . . 729
A.1.3 QR factorization . . . 731
A.1.4 Cholesky factorization . . . 732
A.2 Linear least squares . . . 733
A.2.1 Total least squares . . . 734
A.3 Non-linear least squares . . . 736
A.4 Direct sparse matrix techniques . . . 737
A.4.1 Variable reordering . . . 737
A.5 Iterative techniques . . . 738
A.5.1 Conjugate gradient . . . 739
A.5.2 Preconditioning . . . 740
A.5.3 Multigrid . . . 741

B Bayesian modeling and inference **743**
B.1 Estimation theory . . . 745
B.2 Maximum likelihood estimation and least squares . . . 747
B.3 Robust statistics . . . 748
B.4 Prior models and Bayesian inference . . . 750
B.5 Markov random fields . . . 751
B.6 Uncertainty estimation (error analysis) . . . 752

C Supplementary material **755**
C.1 Datasets and benchmarks . . . 756
C.2 Software . . . 761
C.3 Slides and lectures . . . 768

References **769**

Index **905**

Chapter 1

Introduction

1.1 What is computer vision? 3
1.2 A brief history 9
1.3 Book overview 18
1.4 Sample syllabus 24
1.5 A note on notation 25
1.6 Additional reading 26

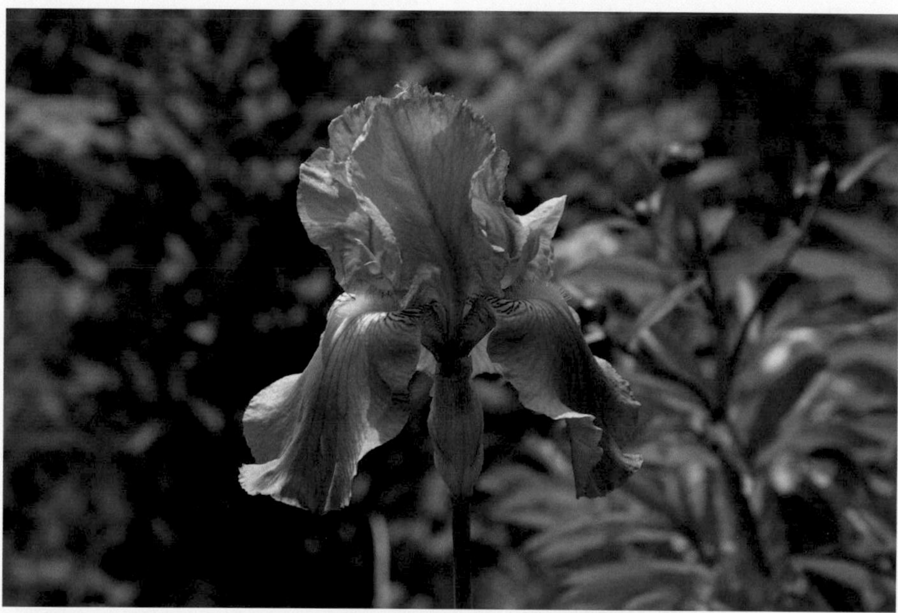

Figure 1.1 The human visual system has no problem interpreting the subtle variations in translucency and shading in this photograph and correctly segmenting the object from its background.

R. Szeliski, *Computer Vision*, Texts in Computer Science,
https://doi.org/10.1007/978-3-030-34372-9_1

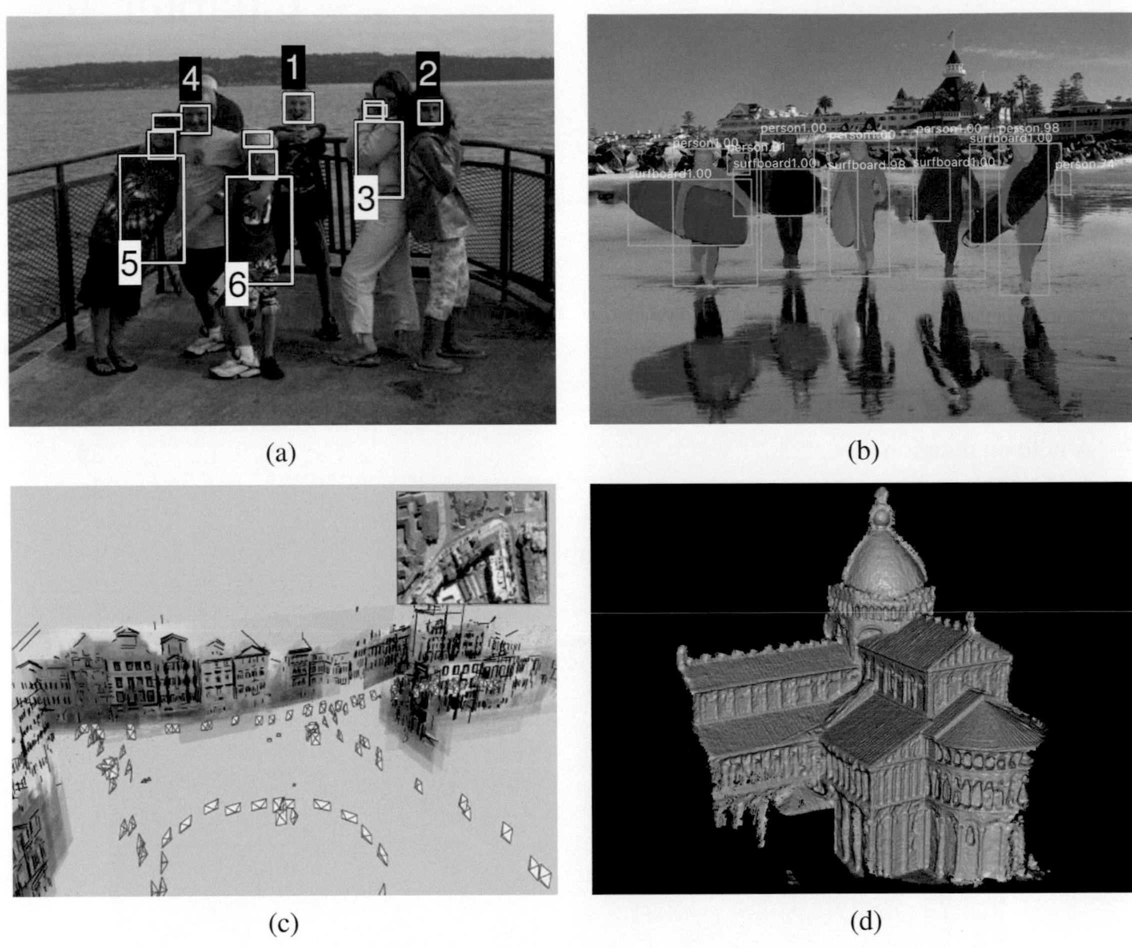

Figure 1.2 Some examples of computer vision algorithms and applications. (a) *Face detection* algorithms, coupled with color-based clothing and hair detection algorithms, can locate and recognize the individuals in this image (Sivic, Zitnick, and Szeliski 2006) © 2006 Springer. (b) *Object instance segmentation* can delineate each person and object in a complex scene (He, Gkioxari *et al.* 2017) © 2017 IEEE. (c) *Structure from motion* algorithms can reconstruct a sparse 3D point model of a large complex scene from hundreds of partially overlapping photographs (Snavely, Seitz, and Szeliski 2006) © 2006 ACM. (d) *Stereo matching* algorithms can build a detailed 3D model of a building façade from hundreds of differently exposed photographs taken from the internet (Goesele, Snavely *et al.* 2007) © 2007 IEEE.

1.1 What is computer vision?

As humans, we perceive the three-dimensional structure of the world around us with apparent ease. Think of how vivid the three-dimensional percept is when you look at a vase of flowers sitting on the table next to you. You can tell the shape and translucency of each petal through the subtle patterns of light and shading that play across its surface and effortlessly segment each flower from the background of the scene (Figure 1.1). Looking at a framed group portrait, you can easily count and name all of the people in the picture and even guess at their emotions from their facial expressions (Figure 1.2a). Perceptual psychologists have spent decades trying to understand how the visual system works and, even though they can devise optical illusions[1] to tease apart some of its principles (Figure 1.3), a complete solution to this puzzle remains elusive (Marr 1982; Wandell 1995; Palmer 1999; Livingstone 2008; Frisby and Stone 2010).

Researchers in computer vision have been developing, in parallel, mathematical techniques for recovering the three-dimensional shape and appearance of objects in imagery. Here, the progress in the last two decades has been rapid. We now have reliable techniques for accurately computing a 3D model of an environment from thousands of partially overlapping photographs (Figure 1.2c). Given a large enough set of views of a particular object or façade, we can create accurate dense 3D surface models using stereo matching (Figure 1.2d). We can even, with moderate success, delineate most of the people and objects in a photograph (Figure 1.2a). However, despite all of these advances, the dream of having a computer explain an image at the same level of detail and causality as a two-year old remains elusive.

Why is vision so difficult? In part, it is because it is an *inverse problem*, in which we seek to recover some unknowns given insufficient information to fully specify the solution. We must therefore resort to physics-based and probabilistic *models*, or machine learning from large sets of examples, to disambiguate between potential solutions. However, modeling the visual world in all of its rich complexity is far more difficult than, say, modeling the vocal tract that produces spoken sounds.

The *forward* models that we use in computer vision are usually developed in physics (radiometry, optics, and sensor design) and in computer graphics. Both of these fields model how objects move and animate, how light reflects off their surfaces, is scattered by the atmosphere, refracted through camera lenses (or human eyes), and finally projected onto a flat (or curved) image plane. While computer graphics are not yet perfect, in many domains, such as rendering a still scene composed of everyday objects or animating extinct creatures such as dinosaurs, the illusion of reality is essentially there.

In computer vision, we are trying to do the inverse, i.e., to describe the world that we see in one or more images and to reconstruct its properties, such as shape, illumination, and color distributions. It is amazing that humans and animals do this so effortlessly, while computer vision algorithms are so error prone. People who have not worked in the field often underestimate the difficulty of the problem. This misperception that vision should be easy dates back to the early days of artificial intelligence (see Section 1.2), when it was initially believed that the *cognitive* (logic proving and planning) parts of intelligence were intrinsically more difficult than the *perceptual* components (Boden 2006).

The good news is that computer vision *is* being used today in a wide variety of real-world applications, which include:

- **Optical character recognition (OCR)**: reading handwritten postal codes on letters (Fig

[1] Some fun pages with striking illusions include https://michaelbach.de/ot, https://www.illusionsindex.org, and http://www.ritsumei.ac.jp/~akitaoka/index-e.html.

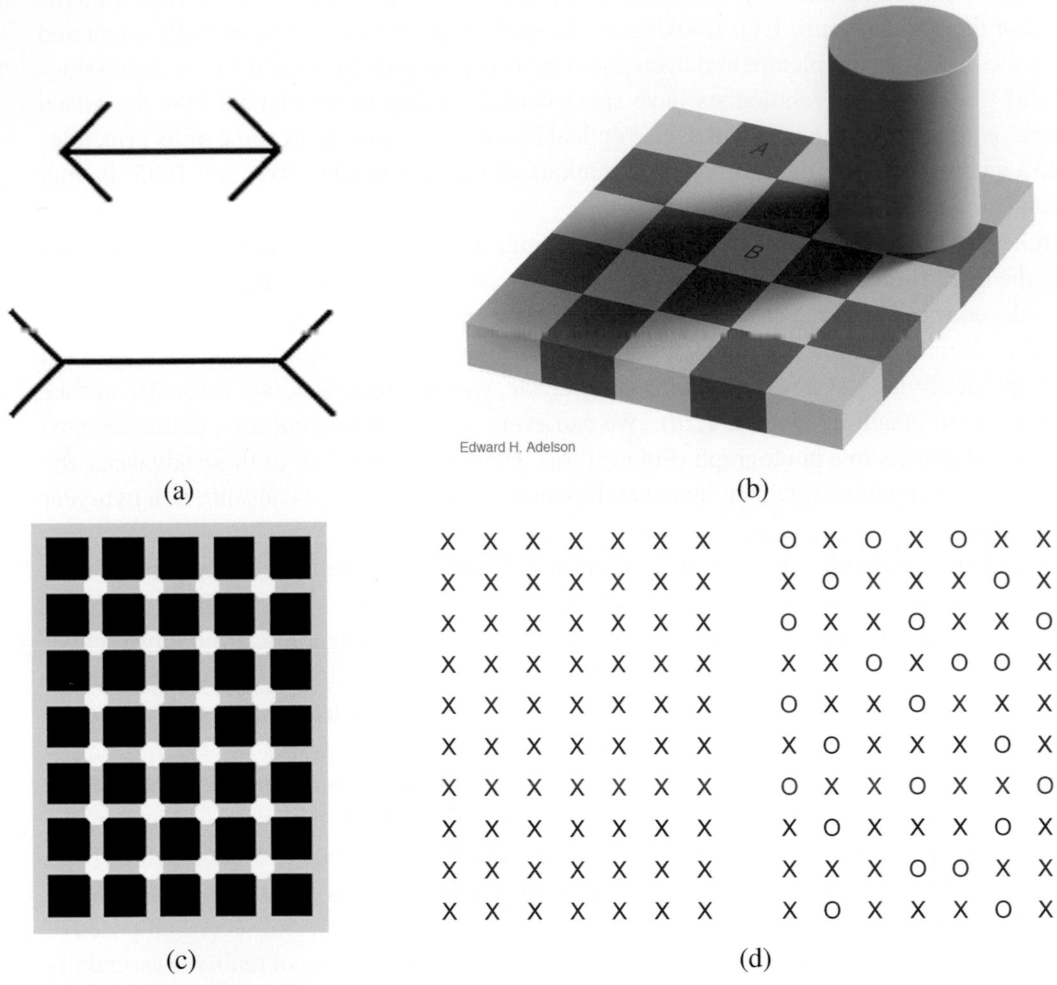

(a) (b) (c) (d)

Figure 1.3 Some common optical illusions and what they might tell us about the visual system: (a) The classic Müller-Lyer illusion, where the lengths of the two horizontal lines appear different, probably due to the imagined perspective effects. (b) The "white" square B in the shadow and the "black" square A in the light actually have the same absolute intensity value. The percept is due to *brightness constancy*, the visual system's attempt to discount illumination when interpreting colors. Image courtesy of Ted Adelson, http://persci.mit.edu/gallery/checkershadow. (c) A variation of the Hermann grid illusion, courtesy of Hany Farid. As you move your eyes over the figure, gray spots appear at the intersections. (d) Count the red Xs in the left half of the figure. Now count them in the right half. Is it significantly harder? The explanation has to do with a *pop-out* effect (Treisman 1985), which tells us about the operations of parallel perception and integration pathways in the brain.

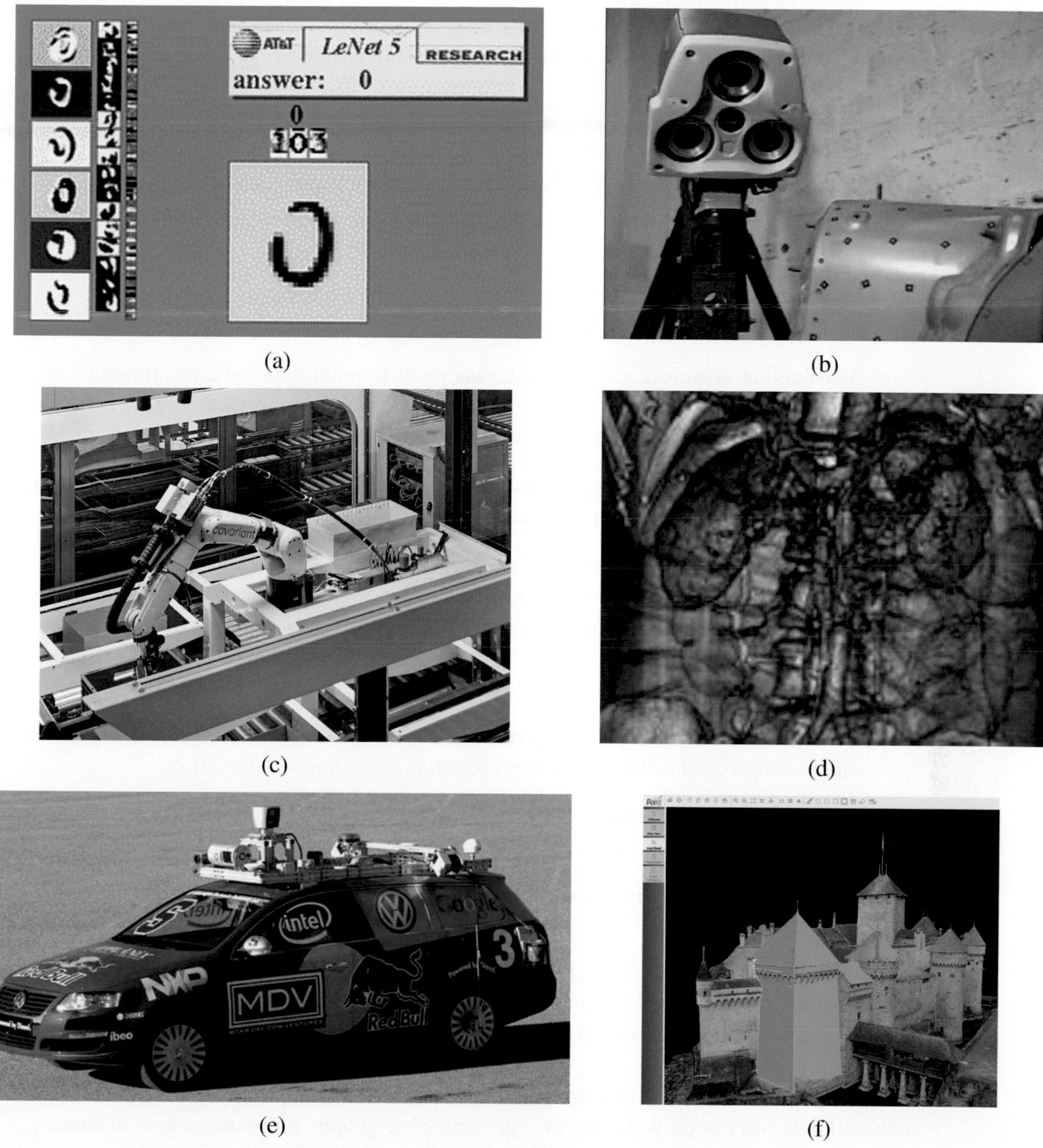

Figure 1.4 Some industrial applications of computer vision: (a) optical character recognition (OCR), http://yann.lecun.com/exdb/lenet; (b) mechanical inspection, http://www.cognitens.com; (c) warehouse picking, https://covariant.ai; (d) medical imaging, http://www.clarontech.com; (e) self-driving cars, (Montemerlo, Becker *et al.* 2008) © 2008 Wiley; (f) drone-based photogrammetry, https://www.pix4d.com/blog/mapping-chillon-castle-with-drone.

ure 1.4a) and automatic number plate recognition (ANPR);

- **Machine inspection:** rapid parts inspection for quality assurance using stereo vision with specialized illumination to measure tolerances on aircraft wings or auto body parts (Figure 1.4b) or looking for defects in steel castings using X-ray vision;
- **Retail:** object recognition for automated checkout lanes and fully automated stores (Wingfield 2019);
- **Warehouse logistics:** autonomous package delivery and pallet-carrying "drives" (Guizzo 2008; O'Brian 2019) and parts picking by robotic manipulators (Figure 1.4c; Ackerman 2020);
- **Medical imaging:** registering pre-operative and intra-operative imagery (Figure 1.4d) or performing long-term studies of people's brain morphology as they age;
- **Self-driving vehicles:** capable of driving point-to-point between cities (Figure 1.4e; Montemerlo, Becker *et al.* 2008; Urmson, Anhalt *et al.* 2008; Janai, Güney *et al.* 2020) as well as autonomous flight (Kaufmann, Gehrig *et al.* 2019);
- **3D model building (photogrammetry):** fully automated construction of 3D models from aerial and drone photographs (Figure 1.4f);
- **Match move:** merging computer-generated imagery (CGI) with live action footage by tracking feature points in the source video to estimate the 3D camera motion and shape of the environment. Such techniques are widely used in Hollywood, e.g., in movies such as Jurassic Park (Roble 1999; Roble and Zafar 2009); they also require the use of precise *matting* to insert new elements between foreground and background elements (Chuang, Agarwala *et al.* 2002).
- **Motion capture (mocap):** using retro-reflective markers viewed from multiple cameras or other vision-based techniques to capture actors for computer animation;
- **Surveillance:** monitoring for intruders, analyzing highway traffic and monitoring pools for drowning victims (e.g., https://swimeye.com);
- **Fingerprint recognition and biometrics:** for automatic access authentication as well as forensic applications.

David Lowe's website of industrial vision applications (http://www.cs.ubc.ca/spider/lowe/vision.html) lists many other interesting industrial applications of computer vision. While the above applications are all extremely important, they mostly pertain to fairly specialized kinds of imagery and narrow domains.

In addition to all of these industrial applications, there exist myriad *consumer-level* applications, such as things you can do with your own personal photographs and video. These include:

- **Stitching:** turning overlapping photos into a single seamlessly stitched panorama (Figure 1.5a), as described in Section 8.2;
- **Exposure bracketing:** merging multiple exposures taken under challenging lighting conditions (strong sunlight and shadows) into a single perfectly exposed image (Figure 1.5b), as described in Section 10.2;
- **Morphing:** turning a picture of one of your friends into another, using a seamless *morph* transition (Figure 1.5c);

- **3D modeling:** converting one or more snapshots into a 3D model of the object or person you are photographing (Figure 1.5d), as described in Section 13.6;
- **Video match move and stabilization:** inserting 2D pictures or 3D models into your videos by automatically tracking nearby reference points (see Section 11.4.4)[2] or using motion estimates to remove shake from your videos (see Section 9.2.1);
- **Photo-based walkthroughs:** navigating a large collection of photographs, such as the interior of your house, by flying between different photos in 3D (see Sections 14.1.2 and 14.5.5);
- **Face detection:** for improved camera focusing as well as more relevant image searching (see Section 6.3.1);
- **Visual authentication:** automatically logging family members onto your home computer as they sit down in front of the webcam (see Section 6.2.4).

The great thing about these applications is that they are already familiar to most students; they are, at least, technologies that students can immediately appreciate and use with their own personal media. Since computer vision is a challenging topic, given the wide range of mathematics being covered[3] and the intrinsically difficult nature of the problems being solved, having fun and relevant problems to work on can be highly motivating and inspiring.

The other major reason why this book has a strong focus on applications is that they can be used to *formulate* and *constrain* the potentially open-ended problems endemic in vision. Thus, it is better to think back from the problem at hand to suitable techniques, rather than to grab the first technique that you may have heard of. This kind of working back from problems to solutions is typical of an **engineering** approach to the study of vision and reflects my own background in the field.

First, I come up with a detailed problem definition and decide on the constraints and specifications for the problem. Then, I try to find out which techniques are known to work, implement a few of these, evaluate their performance, and finally make a selection. In order for this process to work, it is important to have realistic **test data**, both synthetic, which can be used to verify correctness and analyze noise sensitivity, and real-world data typical of the way the system will finally be used. If machine learning is being used, it is even more important to have representative unbiased **training data** in sufficient quantity to obtain good results on real-world inputs.

However, this book is not just an engineering text (a source of recipes). It also takes a **scientific** approach to basic vision problems. Here, I try to come up with the best possible models of the physics of the system at hand: how the scene is created, how light interacts with the scene and atmospheric effects, and how the sensors work, including sources of noise and uncertainty. The task is then to try to invert the acquisition process to come up with the best possible description of the scene.

The book often uses a **statistical** approach to formulating and solving computer vision problems. Where appropriate, probability distributions are used to model the scene and the noisy image acquisition process. The association of prior distributions with unknowns is often called *Bayesian modeling* (Appendix B). It is possible to associate a risk or loss function with misestimating the answer (Section B.2) and to set up your inference algorithm to minimize the expected risk. (Consider a robot trying to estimate the distance to an obstacle: it is usually safer to underestimate than to overestimate.) With statistical techniques, it often helps to gather lots of training data from which to

[2] For a fun student project on this topic, see the "PhotoBook" project at http://www.cc.gatech.edu/dvfx/videos/dvfx2005.html.

[3] These techniques include physics, Euclidean and projective geometry, statistics, and optimization. They make computer vision a fascinating field to study and a great way to learn techniques widely applicable in other fields.

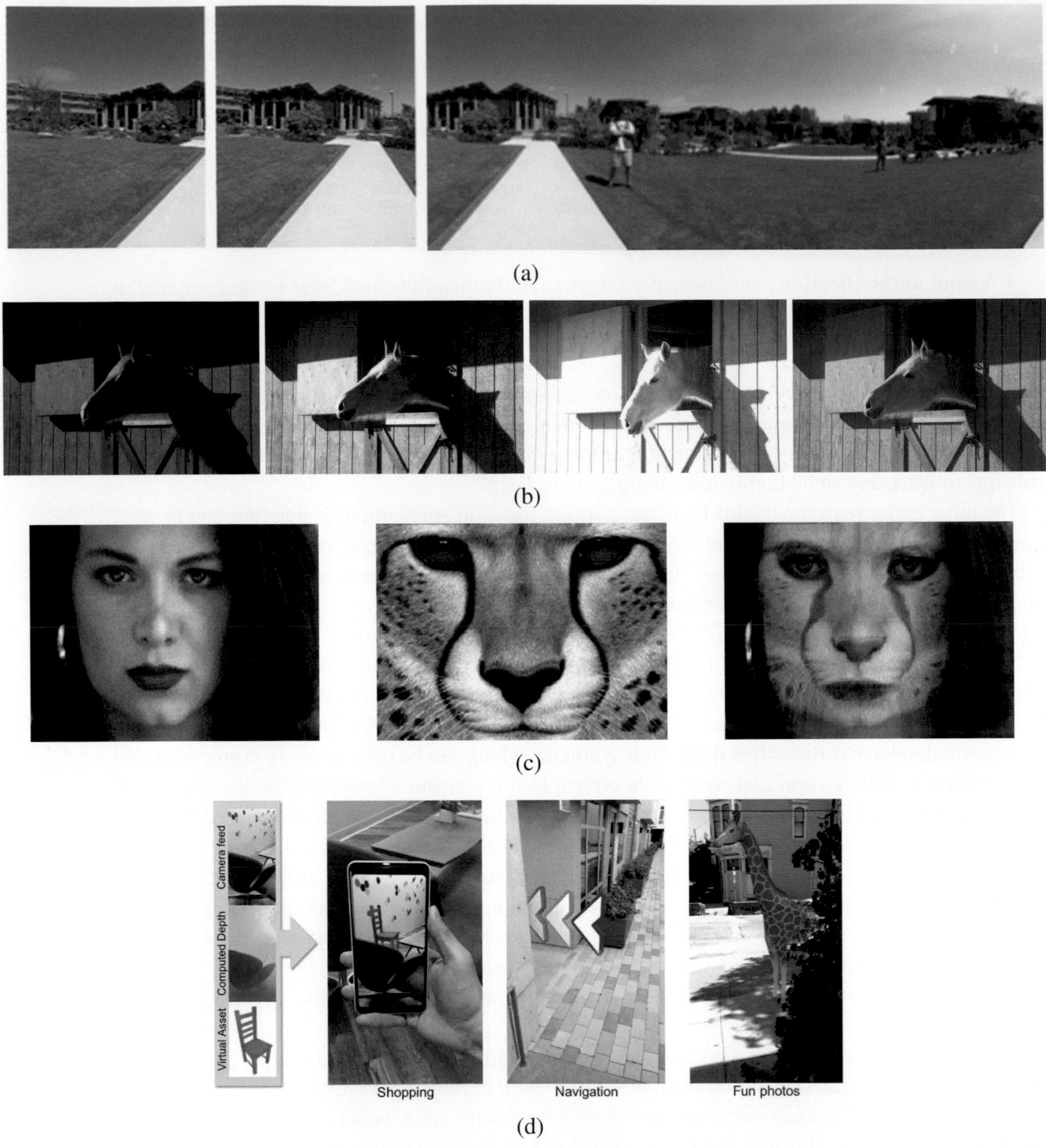

Figure 1.5 Some consumer applications of computer vision: (a) image stitching: merging different views (Szeliski and Shum 1997) © 1997 ACM; (b) exposure bracketing: merging different exposures; (c) morphing: blending between two photographs (Gomes, Darsa *et al.* 1999) © 1999 Morgan Kaufmann; (d) smartphone augmented reality showing real-time depth occlusion effects (Valentin, Kowdle *et al.* 2018) © 2018 ACM.

learn probabilistic models. Finally, statistical approaches enable you to use proven inference techniques to estimate the best answer (or distribution of answers) and to quantify the uncertainty in the resulting estimates.

Because so much of computer vision involves the solution of inverse problems or the estimation of unknown quantities, my book also has a heavy emphasis on **algorithms**, especially those that are known to work well in practice. For many vision problems, it is all too easy to come up with a mathematical description of the problem that either does not match realistic real-world conditions or does not lend itself to the stable estimation of the unknowns. What we need are algorithms that are both **robust** to noise and deviation from our models and reasonably **efficient** in terms of run-time resources and space. In this book, I go into these issues in detail, using Bayesian techniques, where applicable, to ensure robustness, and efficient search, minimization, and linear system solving algorithms to ensure efficiency.[4] Most of the algorithms described in this book are at a high level, being mostly a list of steps that have to be filled in by students or by reading more detailed descriptions elsewhere. In fact, many of the algorithms are sketched out in the exercises.

Now that I've described the goals of this book and the frameworks that I use, I devote the rest of this chapter to two additional topics. Section 1.2 is a brief synopsis of the history of computer vision. It can easily be skipped by those who want to get to "the meat" of the new material in this book and do not care as much about who invented what when.

The second is an overview of the book's contents, Section 1.3, which is useful reading for everyone who intends to make a study of this topic (or to jump in partway, since it describes chapter interdependencies). This outline is also useful for instructors looking to structure one or more courses around this topic, as it provides sample curricula based on the book's contents.

1.2 A brief history

In this section, I provide a brief personal synopsis of the main developments in computer vision over the last fifty years (Figure 1.6) with a focus on advances I find personally interesting and that have stood the test of time. Readers not interested in the provenance of various ideas and the evolution of this field should skip ahead to the book overview in Section 1.3.

1970s. When computer vision first started out in the early 1970s, it was viewed as the visual perception component of an ambitious agenda to mimic human intelligence and to endow robots with intelligent behavior. At the time, it was believed by some of the early pioneers of artificial intelligence and robotics (at places such as MIT, Stanford, and CMU) that solving the "visual input" problem would be an easy step along the path to solving more difficult problems such as higher-level reasoning and planning. According to one well-known story, in 1966, Marvin Minsky at MIT asked his undergraduate student Gerald Jay Sussman to "spend the summer linking a camera to a computer and getting the computer to describe what it saw" (Boden 2006, p. 781).[5] We now know that the problem is slightly more difficult than that.[6]

What distinguished computer vision from the already existing field of digital image processing (Rosenfeld and Pfaltz 1966; Rosenfeld and Kak 1976) was a desire to recover the three-dimensional

[4] In some cases, deep neural networks have also been shown to be an effective way to speed up algorithms that previously relied on iteration (Chen, Xu, and Koltun 2017).

[5] Boden (2006) cites (Crevier 1993) as the original source. The actual Vision Memo was authored by Seymour Papert (1966) and involved a whole cohort of students.

[6] To see how far robotic vision has come in the last six decades, have a look at some of the videos on the Boston Dynamics https://www.bostondynamics.com, Skydio https://www.skydio.com, and Covariant https://covariant.ai websites.

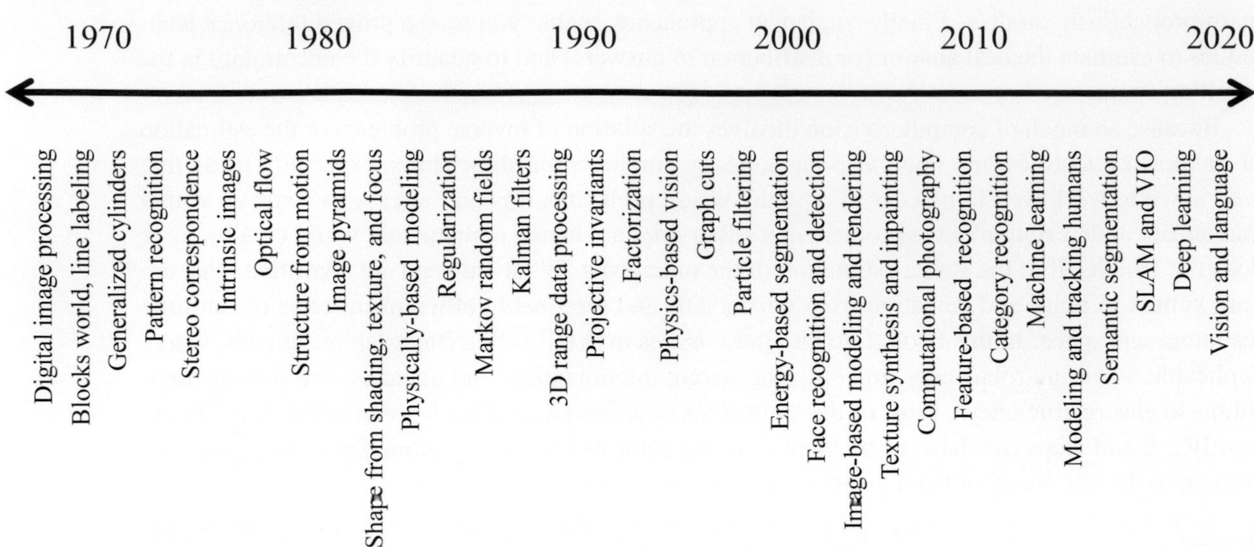

Figure 1.6 A rough timeline of some of the most active topics of research in computer vision.

structure of the world from images and to use this as a stepping stone towards full scene understanding. Winston (1975) and Hanson and Riseman (1978) provide two nice collections of classic papers from this early period.

Early attempts at scene understanding involved extracting edges and then inferring the 3D structure of an object or a "blocks world" from the topological structure of the 2D lines (Roberts 1965). Several *line labeling* algorithms (Figure 1.7a) were developed at that time (Huffman 1971; Clowes 1971; Waltz 1975; Rosenfeld, Hummel, and Zucker 1976; Kanade 1980). Nalwa (1993) gives a nice review of this area. The topic of edge detection was also an active area of research; a nice survey of contemporaneous work can be found in (Davis 1975).

Three-dimensional modeling of non-polyhedral objects was also being studied (Baumgart 1974; Baker 1977). One popular approach used *generalized cylinders*, i.e., solids of revolution and swept closed curves (Agin and Binford 1976; Nevatia and Binford 1977), often arranged into parts relationships[7] (Hinton 1977; Marr 1982) (Figure 1.7c). Fischler and Elschlager (1973) called such *elastic* arrangements of parts *pictorial structures* (Figure 1.7b).

A qualitative approach to understanding intensities and shading variations and explaining them by the effects of image formation phenomena, such as surface orientation and shadows, was championed by Barrow and Tenenbaum (1981) in their paper on *intrinsic images* (Figure 1.7d), along with the related *2½-D sketch* ideas of Marr (1982). This approach has seen periodic revivals, e.g., in the work of Tappen, Freeman, and Adelson (2005) and Barron and Malik (2012).

More quantitative approaches to computer vision were also developed at the time, including the first of many feature-based stereo correspondence algorithms (Figure 1.7e) (Dev 1974; Marr and Poggio 1976, 1979; Barnard and Fischler 1982; Ohta and Kanade 1985; Grimson 1985; Pollard, Mayhew, and Frisby 1985) and intensity-based optical flow algorithms (Figure 1.7f) (Horn and Schunck 1981; Huang 1981; Lucas and Kanade 1981; Nagel 1986). The early work in simultaneously recovering 3D structure and camera motion (see Chapter 11) also began around this time (Ullman 1979; Longuet-Higgins 1981).

[7] In robotics and computer animation, these linked-part graphs are often called *kinematic chains*.

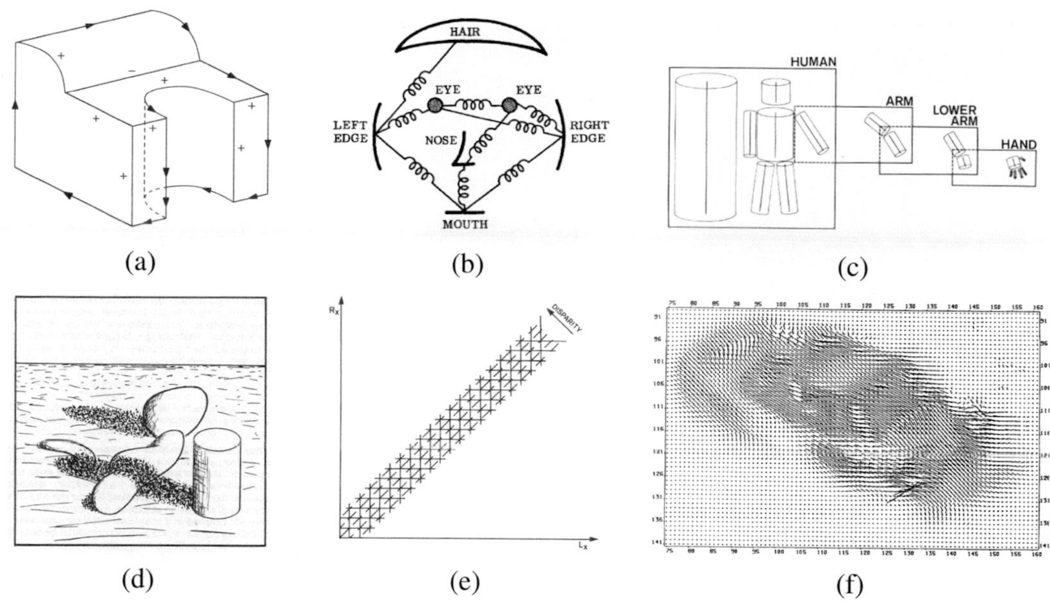

Figure 1.7 Some early (1970s) examples of computer vision algorithms: (a) line labeling (Nalwa 1993) © 1993 Addison-Wesley, (b) pictorial structures (Fischler and Elschlager 1973) © 1973 IEEE, (c) articulated body model (Marr 1982) © 1982 David Marr, (d) intrinsic images (Barrow and Tenenbaum 1981) © 1973 IEEE, (e) stereo correspondence (Marr 1982) © 1982 David Marr, (f) optical flow (Nagel and Enkelmann 1986) © 1986 IEEE.

A lot of the philosophy of how vision was believed to work at the time is summarized in David Marr's (1982) book.[8] In particular, Marr introduced his notion of the three levels of description of a (visual) information processing system. These three levels, very loosely paraphrased according to my own interpretation, are:

- **Computational theory:** What is the goal of the computation (task) and what are the constraints that are known or can be brought to bear on the problem?

- **Representations and algorithms:** How are the input, output, and intermediate information represented and which algorithms are used to calculate the desired result?

- **Hardware implementation:** How are the representations and algorithms mapped onto actual hardware, e.g., a biological vision system or a specialized piece of silicon? Conversely, how can hardware constraints be used to guide the choice of representation and algorithm? With the prevalent use of graphics chips (GPUs) and many-core architectures for computer vision, this question is again quite relevant.

As I mentioned earlier in this introduction, it is my conviction that a careful analysis of the problem specification and known constraints from image formation and priors (the scientific and statistical approaches) must be married with efficient and robust algorithms (the engineering approach) to design successful vision algorithms. Thus, it seems that Marr's philosophy is as good a guide to framing and solving problems in our field today as it was 25 years ago.

[8]More recent developments in visual perception theory are covered in (Wandell 1995; Palmer 1999; Livingstone 2008; Frisby and Stone 2010).

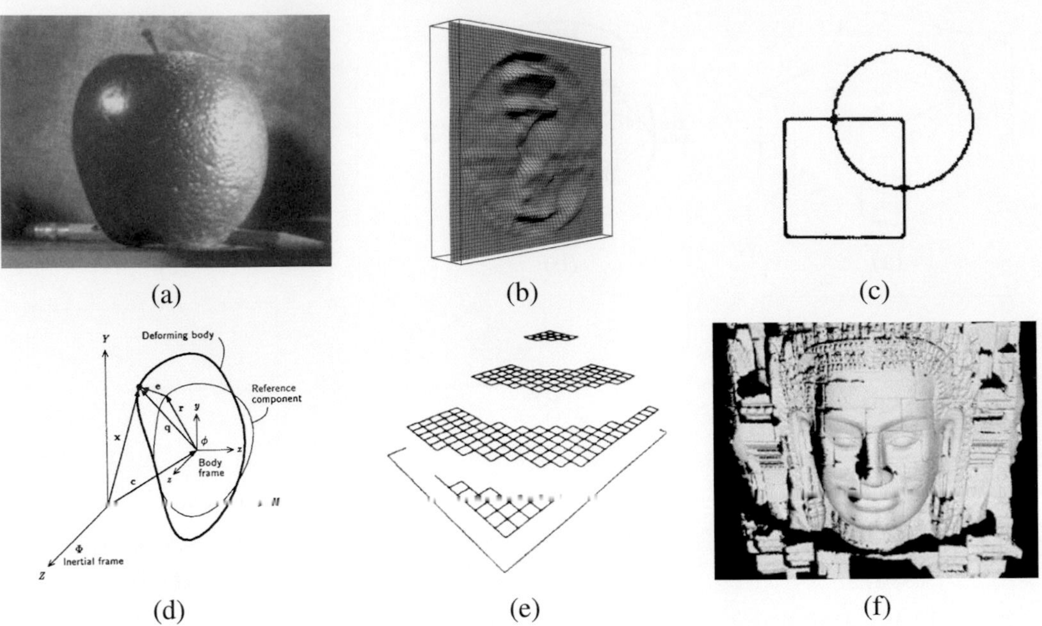

Figure 1.8 Examples of computer vision algorithms from the 1980s: (a) pyramid blending (Burt and Adelson 1983b) © 1983 ACM, (b) shape from shading (Freeman and Adelson 1991) © 1991 IEEE, (c) edge detection (Freeman and Adelson 1991) © 1991 IEEE, (d) physically based models (Terzopoulos and Witkin 1988) © 1988 IEEE, (e) regularization-based surface reconstruction (Terzopoulos 1988) © 1988 IEEE, (f) range data acquisition and merging (Banno, Masuda *et al.* 2008) © 2008 Springer.

1980s. In the 1980s, a lot of attention was focused on more sophisticated mathematical techniques for performing quantitative image and scene analysis.

Image pyramids (see Section 3.5) started being widely used to perform tasks such as image blending (Figure 1.8a) and coarse-to-fine correspondence search (Rosenfeld 1980; Burt and Adelson 1983b; Rosenfeld 1984; Quam 1984; Anandan 1989). Continuous versions of pyramids using the concept of *scale-space* processing were also developed (Witkin 1983; Witkin, Terzopoulos, and Kass 1986; Lindeberg 1990). In the late 1980s, wavelets (see Section 3.5.4) started displacing or augmenting regular image pyramids in some applications (Mallat 1989; Simoncelli and Adelson 1990a; Simoncelli, Freeman *et al.* 1992).

The use of stereo as a quantitative shape cue was extended by a wide variety of *shape-from-X* techniques, including shape from shading (Figure 1.8b) (see Section 13.1.1 and Horn 1975; Pentland 1984; Blake, Zisserman, and Knowles 1985; Horn and Brooks 1986, 1989), photometric stereo (see Section 13.1.1 and Woodham 1981), shape from texture (see Section 13.1.2 and Witkin 1981; Pentland 1984; Malik and Rosenholtz 1997), and shape from focus (see Section 13.1.3 and Nayar, Watanabe, and Noguchi 1995). Horn (1986) has a nice discussion of most of these techniques.

Research into better edge and contour detection (Figure 1.8c) (see Section 7.2) was also active during this period (Canny 1986; Nalwa and Binford 1986), including the introduction of dynamically evolving contour trackers (Section 7.3.1) such as *snakes* (Kass, Witkin, and Terzopoulos 1988), as well as three-dimensional *physically based models* (Figure 1.8d) (Terzopoulos, Witkin, and Kass 1987; Kass, Witkin, and Terzopoulos 1988; Terzopoulos and Fleischer 1988).

Researchers noticed that a lot of the stereo, flow, shape-from-X, and edge detection algorithms could be unified, or at least described, using the same mathematical framework if they were posed as variational optimization problems and made more robust (well-posed) using regularization (Fig-

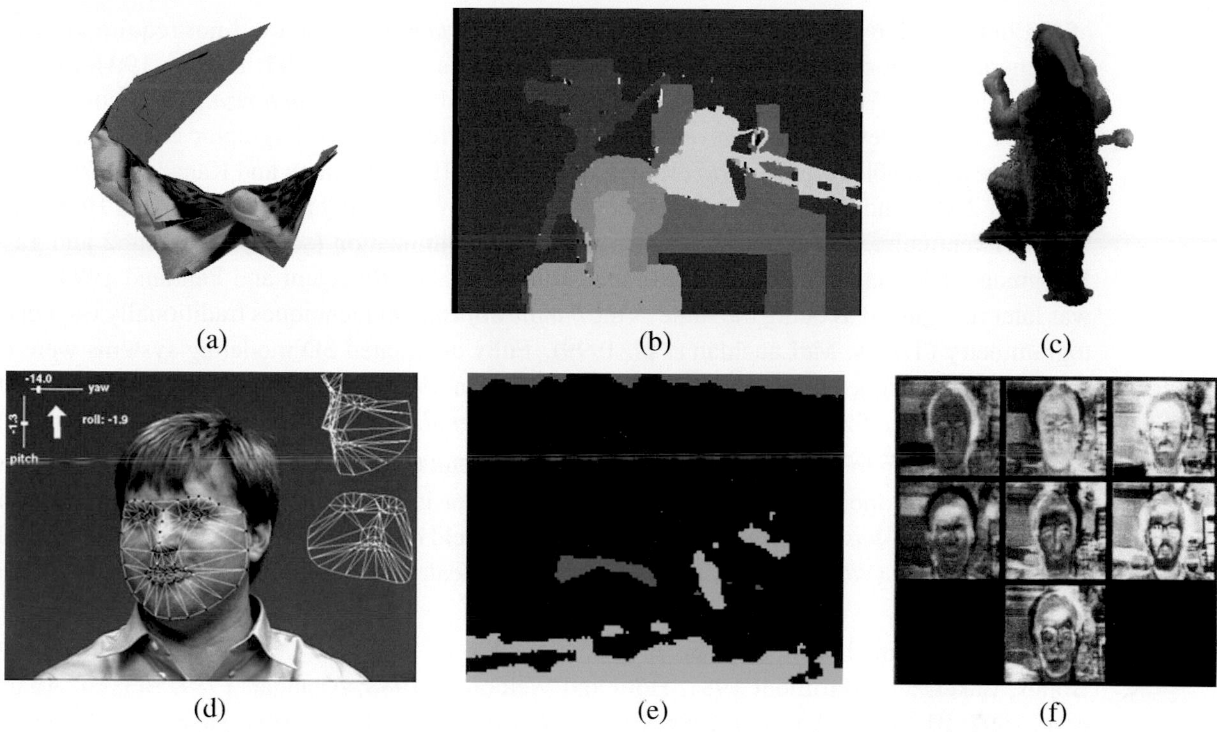

Figure 1.9 Examples of computer vision algorithms from the 1990s: (a) factorization-based structure from motion (Tomasi and Kanade 1992) © 1992 Springer, (b) dense stereo matching (Boykov, Veksler, and Zabih 2001), (c) multi-view reconstruction (Seitz and Dyer 1999) © 1999 Springer, (d) face tracking (Matthews, Xiao, and Baker 2007), (e) image segmentation (Belongie, Fowlkes *et al.* 2002) © 2002 Springer, (f) face recognition (Turk and Pentland 1991).

ure 1.8e) (see Section 4.2 and Terzopoulos 1983; Poggio, Torre, and Koch 1985; Terzopoulos 1986b; Blake and Zisserman 1987; Bertero, Poggio, and Torre 1988; Terzopoulos 1988). Around the same time, Geman and Geman (1984) pointed out that such problems could equally well be formulated using discrete *Markov random field* (MRF) models (see Section 4.3), which enabled the use of better (global) search and optimization algorithms, such as simulated annealing.

Online variants of MRF algorithms that modeled and updated uncertainties using the Kalman filter were introduced a little later (Dickmanns and Graefe 1988; Matthies, Kanade, and Szeliski 1989; Szeliski 1989). Attempts were also made to map both regularized and MRF algorithms onto parallel hardware (Poggio and Koch 1985; Poggio, Little *et al.* 1988; Fischler, Firschein *et al.* 1989). The book by Fischler and Firschein (1987) contains a nice collection of articles focusing on all of these topics (stereo, flow, regularization, MRFs, and even higher-level vision).

Three-dimensional range data processing (acquisition, merging, modeling, and recognition; see Figure 1.8f) continued being actively explored during this decade (Agin and Binford 1976; Besl and Jain 1985; Faugeras and Hebert 1987; Curless and Levoy 1996). The compilation by Kanade (1987) contains a lot of the interesting papers in this area.

1990s. While a lot of the previously mentioned topics continued to be explored, a few of them became significantly more active.

A burst of activity in using projective invariants for recognition (Mundy and Zisserman 1992) evolved into a concerted effort to solve the structure from motion problem (see Chapter 11). A lot of the initial activity was directed at *projective reconstructions*, which did not require knowledge of camera calibration (Faugeras 1992; Hartley, Gupta, and Chang 1992; Hartley 1994a; Faugeras and Luong 2001; Hartley and Zisserman 2004). Simultaneously, *factorization* techniques (Section 11.4.1) were developed to solve efficiently problems for which orthographic camera approximations were applicable (Figure 1.9a) (Tomasi and Kanade 1992; Poelman and Kanade 1997; Anandan and Irani 2002) and then later extended to the perspective case (Christy and Horaud 1996; Triggs 1996). Eventually, the field started using full global optimization (see Section 11.4.2 and Taylor, Kriegman, and Anandan 1991; Szeliski and Kang 1994; Azarbayejani and Pentland 1995), which was later recognized as being the same as the *bundle adjustment* techniques traditionally used in photogrammetry (Triggs, McLauchlan *et al.* 1999). Fully automated 3D modeling systems were built using such techniques (Beardsley, Torr, and Zisserman 1996; Schaffalitzky and Zisserman 2002; Snavely, Seitz, and Szeliski 2006; Agarwal, Furukawa *et al.* 2011; Frahm, Fite-Georgel *et al.* 2010).

Work begun in the 1980s on using detailed measurements of color and intensity combined with accurate physical models of radiance transport and color image formation created its own subfield known as *physics-based vision*. A good survey of the field can be found in the three-volume collection on this topic (Wolff, Shafer, and Healey 1992a; Healey and Shafer 1992; Shafer, Healey, and Wolff 1992).

Optical flow methods (see Chapter 9) continued to be improved (Nagel and Enkelmann 1986; Bolles, Baker, and Marimont 1987; Horn and Weldon Jr. 1988; Anandan 1989; Bergen, Anandan *et al.* 1992; Black and Anandan 1996; Bruhn, Weickert, and Schnörr 2005; Papenberg, Bruhn *et al.* 2006), with (Nagel 1986; Barron, Fleet, and Beauchemin 1994; Baker, Scharstein *et al.* 2011) being good surveys. Similarly, a lot of progress was made on dense stereo correspondence algorithms (see Chapter 12, Okutomi and Kanade (1993, 1994); Boykov, Veksler, and Zabih (1998); Birchfield and Tomasi (1999); Boykov, Veksler, and Zabih (2001), and the survey and comparison in Scharstein and Szeliski (2002)), with the biggest breakthrough being perhaps global optimization using *graph cut* techniques (Figure 1.9b) (Boykov, Veksler, and Zabih 2001).

Multi-view stereo algorithms (Figure 1.9c) that produce complete 3D surfaces (see Section 12.7) were also an active topic of research (Seitz and Dyer 1999; Kutulakos and Seitz 2000) that continues to be active today (Seitz, Curless *et al.* 2006; Schöps, Schönberger *et al.* 2017; Knapitsch, Park *et al.* 2017). Techniques for producing 3D volumetric descriptions from binary silhouettes (see Section 12.7.3) continued to be developed (Potmesil 1987; Srivasan, Liang, and Hackwood 1990; Szeliski 1993; Laurentini 1994), along with techniques based on tracking and reconstructing smooth occluding contours (see Section 12.2.1 and Cipolla and Blake 1992; Vaillant and Faugeras 1992; Zheng 1994; Boyer and Berger 1997; Szeliski and Weiss 1998; Cipolla and Giblin 2000).

Tracking algorithms also improved a lot, including contour tracking using *active contours* (see Section 7.3), such as *snakes* (Kass, Witkin, and Terzopoulos 1988), *particle filters* (Blake and Isard 1998), and *level sets* (Malladi, Sethian, and Vemuri 1995), as well as intensity-based (*direct*) techniques (Lucas and Kanade 1981; Shi and Tomasi 1994; Rehg and Kanade 1994), often applied to tracking faces (Figure 1.9d) (Lanitis, Taylor, and Cootes 1997; Matthews and Baker 2004; Matthews, Xiao, and Baker 2007) and whole bodies (Sidenbladh, Black, and Fleet 2000; Hilton, Fua, and Ronfard 2006; Moeslund, Hilton, and Krüger 2006).

Image segmentation (see Section 7.5) (Figure 1.9e), a topic which has been active since the earliest days of computer vision (Brice and Fennema 1970; Horowitz and Pavlidis 1976; Riseman and Arbib 1977; Rosenfeld and Davis 1979; Haralick and Shapiro 1985; Pavlidis and Liow 1990), was also an active topic of research, producing techniques based on minimum energy (Mumford and Shah 1989) and minimum description length (Leclerc 1989), *normalized cuts* (Shi and Malik 2000), and *mean shift* (Comaniciu and Meer 2002).

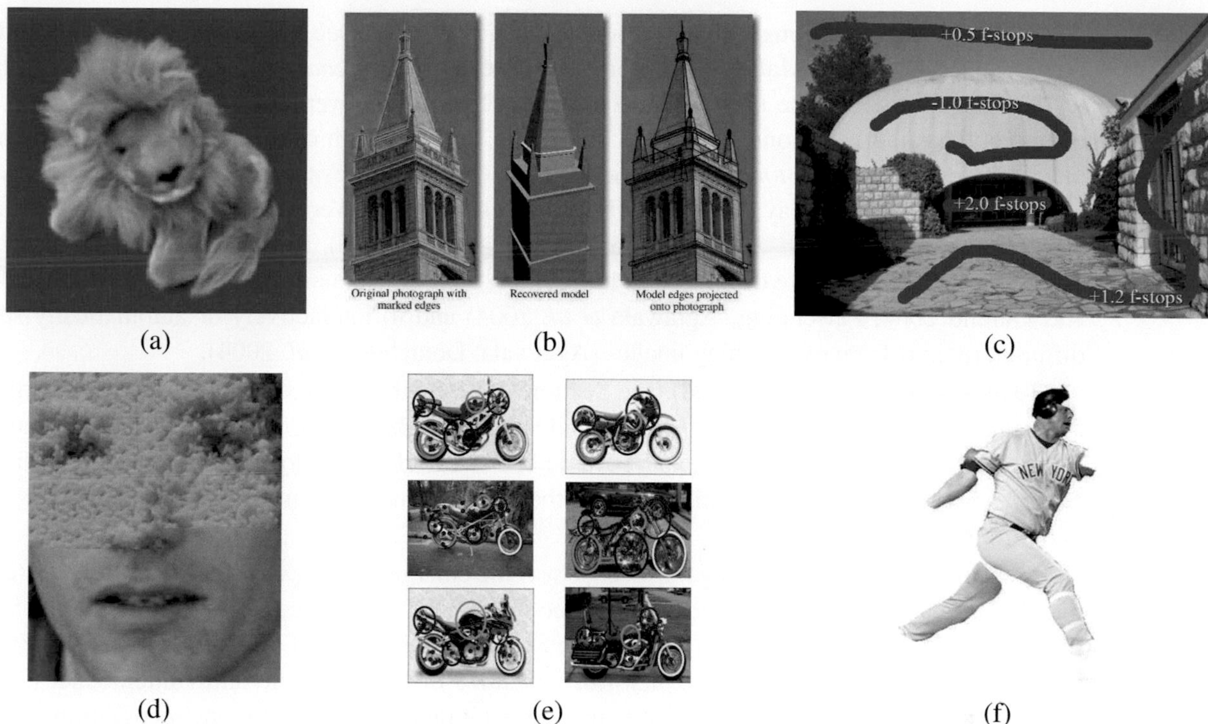

Figure 1.10 Examples of computer vision algorithms from the 2000s: (a) image-based rendering (Gortler, Grzeszczuk *et al.* 1996), (b) image-based modeling (Debevec, Taylor, and Malik 1996) © 1996 ACM, (c) interactive tone mapping (Lischinski, Farbman *et al.* 2006) (d) texture synthesis (Efros and Freeman 2001), (e) feature-based recognition (Fergus, Perona, and Zisserman 2007), (f) region-based recognition (Mori, Ren *et al.* 2004) © 2004 IEEE.

Statistical learning techniques started appearing, first in the application of principal component *eigenface* analysis to face recognition (Figure 1.9f) (see Section 5.2.3 and Turk and Pentland 1991) and linear dynamical systems for curve tracking (see Section 7.3.1 and Blake and Isard 1998).

Perhaps the most notable development in computer vision during this decade was the increased interaction with computer graphics (Seitz and Szeliski 1999), especially in the cross-disciplinary area of *image-based modeling and rendering* (see Chapter 14). The idea of manipulating real-world imagery directly to create new animations first came to prominence with *image morphing* techniques (Figure1.5c) (see Section 3.6.3 and Beier and Neely 1992) and was later applied to *view interpolation* (Chen and Williams 1993; Seitz and Dyer 1996), panoramic image stitching (Figure1.5a) (see Section 8.2 and Mann and Picard 1994; Chen 1995; Szeliski 1996; Szeliski and Shum 1997; Szeliski 2006a), and full light-field rendering (Figure 1.10a) (see Section 14.3 and Gortler, Grzeszczuk *et al.* 1996; Levoy and Hanrahan 1996; Shade, Gortler *et al.* 1998). At the same time, image-based modeling techniques (Figure 1.10b) for automatically creating realistic 3D models from collections of images were also being introduced (Beardsley, Torr, and Zisserman 1996; Debevec, Taylor, and Malik 1996; Taylor, Debevec, and Malik 1996).

2000s. This decade continued to deepen the interplay between the vision and graphics fields, but more importantly embraced data-driven and learning approaches as core components of vision. Many of the topics introduced under the rubric of image-based rendering, such as image stitching (see Section 8.2), light-field capture and rendering (see Section 14.3), and *high dynamic range*

(HDR) image capture through exposure bracketing (Figure1.5b) (see Section 10.2 and Mann and Picard 1995; Debevec and Malik 1997), were re-christened as *computational photography* (see Chapter 10) to acknowledge the increased use of such techniques in everyday digital photography. For example, the rapid adoption of exposure bracketing to create high dynamic range images necessitated the development of *tone mapping* algorithms (Figure 1.10c) (see Section 10.2.1) to convert such images back to displayable results (Fattal, Lischinski, and Werman 2002; Durand and Dorsey 2002; Reinhard, Stark *et al.* 2002; Lischinski, Farbman *et al.* 2006). In addition to merging multiple exposures, techniques were developed to merge flash images with non-flash counterparts (Eisemann and Durand 2004; Petschnigg, Agrawala *et al.* 2004) and to interactively or automatically select different regions from overlapping images (Agarwala, Dontcheva *et al.* 2004).

Texture synthesis (Figure 1.10d) (see Section 10.5), quilting (Efros and Leung 1999; Efros and Freeman 2001; Kwatra, Schödl *et al.* 2003), and inpainting (Bertalmio, Sapiro *et al.* 2000; Bertalmio, Vese *et al.* 2003; Criminisi, Pérez, and Toyama 2004) are additional topics that can be classified as computational photography techniques, since they re-combine input image samples to produce new photographs.

A second notable trend during this decade was the emergence of feature-based techniques (combined with learning) for object recognition (see Section 6.1 and Ponce, Hebert *et al.* 2006). Some of the notable papers in this area include the *constellation model* of Fergus, Perona, and Zisserman (2007) (Figure 1.10e) and the *pictorial structures* of Felzenszwalb and Huttenlocher (2005). Feature-based techniques also dominate other recognition tasks, such as scene recognition (Zhang, Marszalek *et al.* 2007) and panorama and location recognition (Brown and Lowe 2007; Schindler, Brown, and Szeliski 2007). And while *interest point* (patch-based) features tend to dominate current research, some groups are pursuing recognition based on contours (Belongie, Malik, and Puzicha 2002) and region segmentation (Figure 1.10f) (Mori, Ren *et al.* 2004).

Another significant trend from this decade was the development of more efficient algorithms for complex global optimization problems (see Chapter 4 and Appendix B.5 and Szeliski, Zabih *et al.* 2008; Blake, Kohli, and Rother 2011). While this trend began with work on graph cuts (Boykov, Veksler, and Zabih 2001; Kohli and Torr 2007), a lot of progress has also been made in message passing algorithms, such as *loopy belief propagation* (LBP) (Yedidia, Freeman, and Weiss 2001; Kumar and Torr 2006).

The most notable trend from this decade, which has by now completely taken over visual recognition and most other aspects of computer vision, was the application of sophisticated machine learning techniques to computer vision problems (see Chapters 5 and 6). This trend coincided with the increased availability of immense quantities of partially labeled data on the internet, as well as significant increases in computational power, which makes it more feasible to learn object categories without the use of careful human supervision.

2010s. The trend towards using large labeled (and also self-supervised) datasets to develop machine learning algorithms became a tidal wave that totally revolutionized the development of image recognition algorithms as well as other applications, such as denoising and optical flow, which previously used Bayesian and global optimization techniques.

This trend was enabled by the development of high-quality large-scale annotated datasets such as ImageNet (Deng, Dong *et al.* 2009; Russakovsky, Deng *et al.* 2015), Microsoft COCO (Common Objects in Context) (Lin, Maire *et al.* 2014), and LVIS (Gupta, Dollár, and Girshick 2019). These datasets provided not only reliable metrics for tracking the progress of recognition and semantic segmentation algorithms, but more importantly, sufficient labeled data to develop complete solutions based on machine learning.

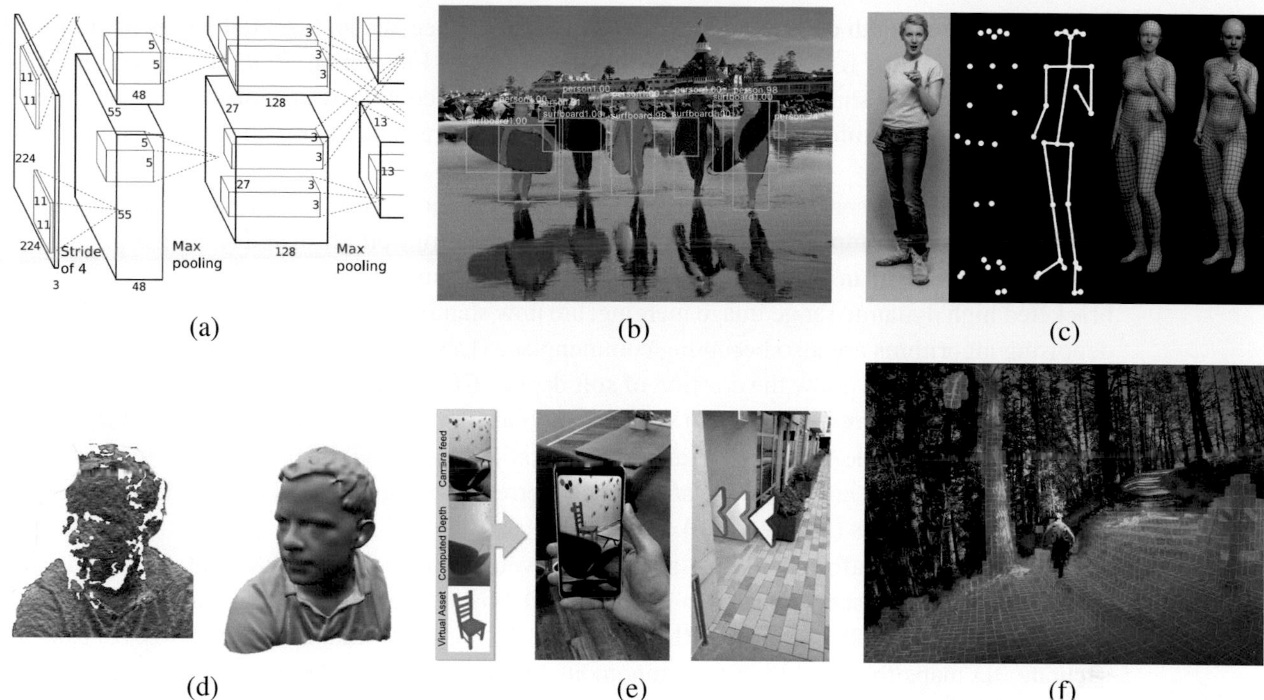

Figure 1.11 Examples of computer vision algorithms from the 2010s: (a) the SuperVision deep neural network © Krizhevsky, Sutskever, and Hinton (2012); (b) object instance segmentation (He, Gkioxari *et al.* 2017) © 2017 IEEE; (c) whole body, expression, and gesture fitting from a single image (Pavlakos, Choutas *et al.* 2019) © 2019 IEEE; (d) fusing multiple color depth images using the KinectFusion real-time system (Newcombe, Izadi *et al.* 2011) © 2011 IEEE; (e) smartphone augmented reality with real-time depth occlusion effects (Valentin, Kowdle *et al.* 2018) © 2018 ACM; (f) 3D map computed in real-time on a fully autonomous Skydio R1 drone (Cross 2019).

Another major trend was the dramatic increase in computational power available from the development of general purpose (data-parallel) algorithms on graphical processing units (GPGPU). The breakthrough SuperVision ("AlexNet") deep neural network (Figure 1.11a; Krizhevsky, Sutskever, and Hinton 2012), which was the first neural network to win the yearly ImageNet large-scale visual recognition challenge, relied on GPU training, as well as a number of technical advances, for its dramatic performance. After the publication of this paper, progress in using deep convolutional architectures accelerated dramatically, to the point where they are now the only architecture considered for recognition and semantic segmentation tasks (Figure 1.11b), as well as the preferred architecture for many other vision tasks (Chapter 5; LeCun, Bengio, and Hinton 2015), including optical flow (Sun, Yang *et al.* 2018)), denoising, and monocular depth inference (Li, Dekel *et al.* 2019).

Large datasets and GPU architectures, coupled with the rapid dissemination of ideas through timely publications on arXiv as well as the development of languages for deep learning and the open sourcing of neural network models, all contributed to an explosive growth in this area, both in rapid advances and capabilities, and also in the sheer number of publications and researchers now working on these topics. They also enabled the extension of image recognition approaches to video understanding tasks such as action recognition (Feichtenhofer, Fan *et al.* 2019), as well as structured regression tasks such as real-time multi-person body pose estimation (Cao, Simon *et al.* 2017).

Specialized sensors and hardware for computer vision tasks also continued to advance. The

Microsoft Kinect depth camera, released in 2010, quickly became an essential component of many 3D modeling (Figure 1.11d) and person tracking (Shotton, Fitzgibbon *et al.* 2011) systems. Over the decade, 3D body shape modeling and tracking systems continued to evolve, to the point where it is now possible to infer a person's 3D model with gestures and expression from a single image (Figure 1.11c).

And while depth sensors have not yet become ubiquitous (except for security applications on high-end phones), computational photography algorithms run on all of today's smartphones. Innovations introduced in the computer vision community, such as panoramic image stitching and bracketed high dynamic range image merging, are now standard features, and multi-image low-light denoising algorithms are also becoming commonplace (Liba, Murthy *et al.* 2019). Lightfield imaging algorithms, which allow the creation of soft depth-of-field effects, are now also becoming more available (Garg, Wadhwa *et al.* 2019). Finally, mobile augmented reality applications that perform real-time pose estimation and environment augmentation using combinations of feature tracking and inertial measurements are commonplace, and are currently being extended to include pixel-accurate depth occlusion effects (Figure 1.11e).

On higher-end platforms such as autonomous vehicles and drones, powerful real-time SLAM (simultaneous localization and mapping) and VIO (visual inertial odometry) algorithms (Engel, Schöps, and Cremers 2014; Forster, Zhang *et al.* 2017; Engel, Koltun, and Cremers 2018) can build accurate 3D maps that enable, e.g., autonomous flight through challenging scenes such as forests (Figure 1.11f).

In summary, this past decade has seen incredible advances in the performance and reliability of computer vision algorithms, brought in part by the shift to machine learning and training on very large sets of real-world data. It has also seen the application of vision algorithms in myriad commercial and consumer scenarios as well as new challenges engendered by their widespread use (Su and Crandall 2021).

1.3 Book overview

In the final part of this introduction, I give a brief tour of the material in this book, as well as a few notes on notation and some additional general references. Since computer vision is such a broad field, it is possible to study certain aspects of it, e.g., geometric image formation and 3D structure recovery, without requiring other parts, e.g., the modeling of reflectance and shading. Some of the chapters in this book are only loosely coupled with others, and it is not strictly necessary to read all of the material in sequence.

Figure 1.12 shows a rough layout of the contents of this book. Since computer vision involves going from images to both a semantic understanding as well as a 3D structural description of the scene, I have positioned the chapters horizontally in terms of where in this spectrum they land, in addition to vertically according to their dependence.[9]

Interspersed throughout the book are sample **applications**, which relate the algorithms and mathematical material being presented in various chapters to useful, real-world applications. Many of these applications are also presented in the exercises sections, so that students can write their own.

At the end of each section, I provide a set of **exercises** that the students can use to implement, test, and refine the algorithms and techniques presented in each section. Some of the exercises are suitable as written homework assignments, others as shorter one-week projects, and still others as

[9]For an interesting comparison with what is known about the human visual system, e.g., the largely parallel *what* and *where* pathways (Goodale and Milner 1992), see some textbooks on human perception (Palmer 1999; Livingstone 2008; Frisby and Stone 2010).

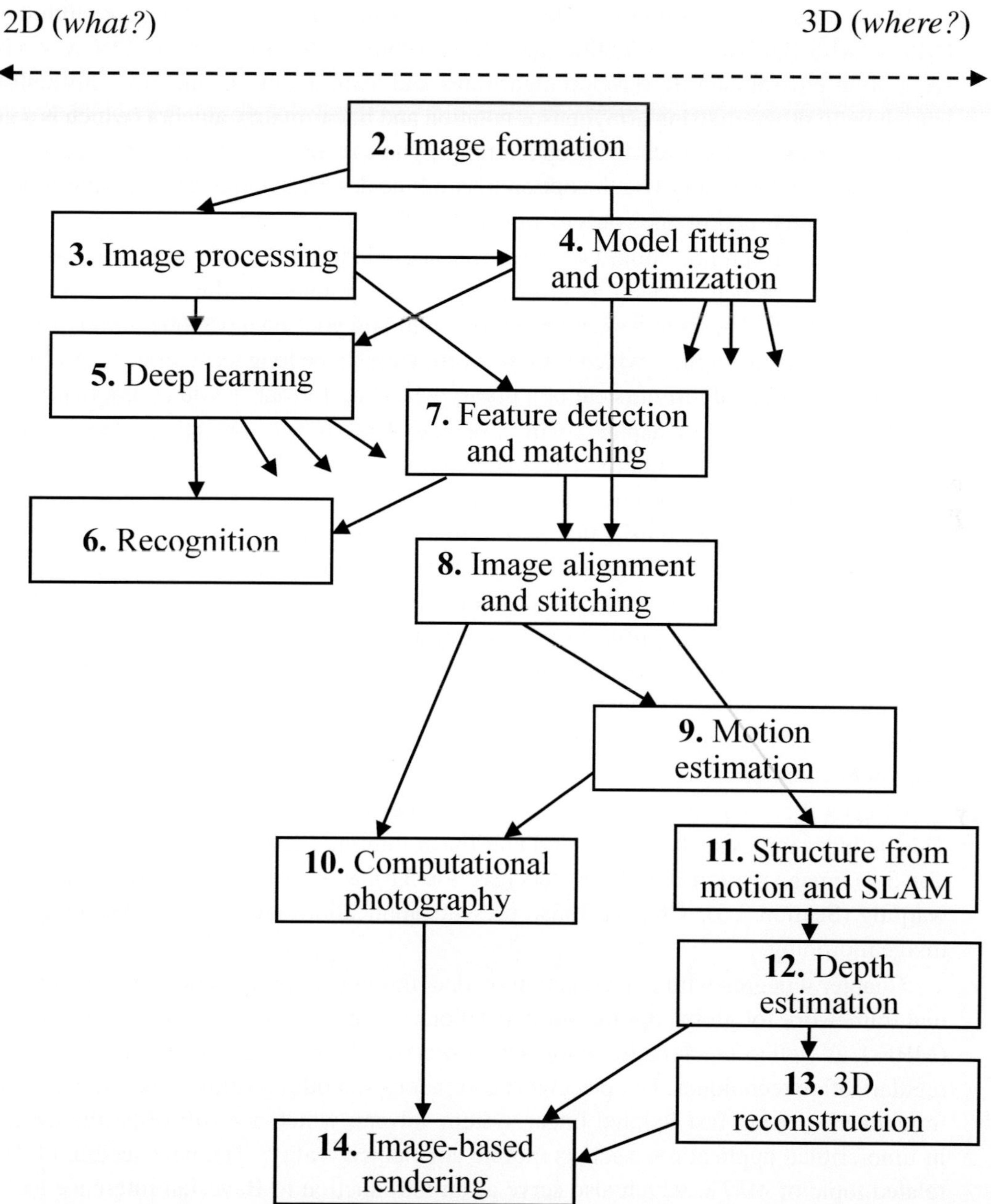

Figure 1.12 A taxonomy of the topics covered in this book, showing the (rough) dependencies between different chapters, which are roughly positioned along the left–right axis depending on whether they are more closely related to images (left) or 3D geometry (right) representations. The "what-where" along the top axis is a reference to separate visual pathways in the visual system (Goodale and Milner 1992), but should not be taken too seriously. Foundational techniques such as optimization and deep learning are widely used in subsequent chapters.

open-ended research problems that make for challenging final projects. Motivated students who implement a reasonable subset of these exercises will, by the end of the book, have a computer vision software library that can be used for a variety of interesting tasks and projects.

If the students or curriculum do not have a strong preference for programming languages, Python, with the NumPy scientific and array arithmetic library plus the OpenCV vision library, are a good environment to develop algorithms and learn about vision. Not only will the students learn how to program using array/tensor notation and linear/matrix algebra (which is a good foundation for later use of PyTorch for deep learning), you can also prepare classroom assignments using Jupyter notebooks, giving you the option to combine descriptive tutorials, sample code, and code to be extended/modified in one convenient location.[10]

As this is a reference book, I try wherever possible to discuss which techniques and algorithms work well in practice, as well as provide up-to-date pointers to the latest research results in the areas that I cover. The exercises can be used to build up your own personal library of self-tested and validated vision algorithms, which is more worthwhile in the long term (assuming you have the time) than simply pulling algorithms out of a library whose performance you do not really understand.

The book begins in Chapter 2 with a review of the image formation processes that create the images that we see and capture. Understanding this process is fundamental if you want to take a scientific (model-based) approach to computer vision. Students who are eager to just start implementing algorithms (or courses that have limited time) can skip ahead to the next chapter and dip into this material later. In Chapter 2, we break down image formation into three major components. Geometric image formation (Section 2.1) deals with points, lines, and planes, and how these are mapped onto images using *projective geometry* and other models (including radial lens distortion). Photometric image formation (Section 2.2) covers *radiometry*, which describes how light interacts with surfaces in the world, and *optics*, which projects light onto the sensor plane. Finally, Section 2.3 covers how sensors work, including topics such as sampling and aliasing, color sensing, and in-camera compression.

Chapter 3 covers image processing, which is needed in almost all computer vision applications. This includes topics such as linear and non-linear filtering (Section 3.3), the Fourier transform (Section 3.4), image pyramids and wavelets (Section 3.5), and geometric transformations such as image warping (Section 3.6). Chapter 3 also presents applications such as seamless image blending and image morphing.

Chapter 4 begins with a new section on data fitting and interpolation, which provides a conceptual framework for global optimization techniques such as *regularization* and *Markov random fields* (MRFs), as well as *machine learning*, which we cover in the next chapter. Section 4.2 covers classic regularization techniques, i.e., piecewise-continuous smoothing splines (aka *variational techniques*) implemented using fast iterated linear system solvers, which are still often the method of choice in time-critical applications such as mobile augmented reality. The next section (4.3) presents the related topic of *MRFs*, which also serve as an introduction to Bayesian inference techniques, covered at a more abstract level in Appendix B. The chapter also discusses applications to interactive colorization and segmentation.

Chapter 5 is a completely new chapter covering machine learning, deep learning, and deep neural networks. It begins in Section 5.1 with a review of classic *supervised machine learning* approaches, which are designed to classify images (or regress values) based on intermediate-level features. Section 5.2 looks at *unsupervised learning*, which is useful for both understanding unlabeled training data and providing models of real-world distributions. Section 5.3 presents the basic elements of

[10]You may also be able to run your notebooks and train your models using the Google Colab service at https://colab.research.google.com.

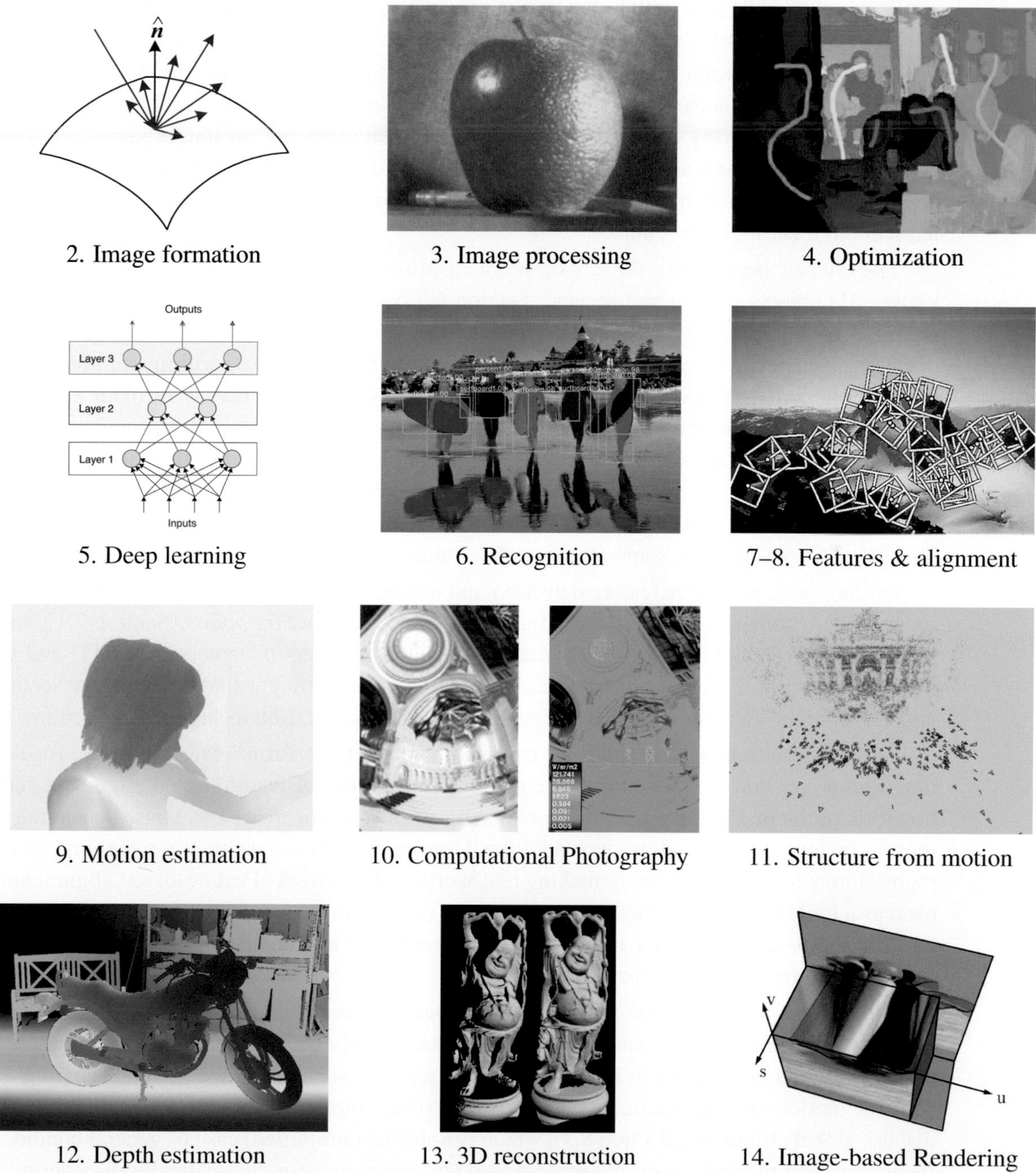

Figure 1.13 A pictorial summary of the chapter contents. Sources: Burt and Adelson (1983b); Agarwala, Dontcheva *et al.* (2004); Glassner (2018); He, Gkioxari *et al.* (2017); Brown, Szeliski, and Winder (2005); Butler, Wulff *et al.* (2012); Debevec and Malik (1997); Snavely, Seitz, and Szeliski (2006); Scharstein, Hirschmüller *et al.* (2014); Curless and Levoy (1996); Gortler, Grzeszczuk *et al.* (1996)—see the figures in the respective chapters for copyright information.

feedforward neural networks, including weights, layers, and activation functions, as well as methods for network training. Section 5.4 goes into more detail on convolutional networks and their applications to both recognition and image processing. The last section in the chapter discusses more complex networks, including 3D, spatio-temporal, recurrent, and generative networks.

Chapter 6 covers the topic of *recognition*. In the first edition of this book this chapter came last, since it built upon earlier methods such as segmentation and feature matching. With the advent of deep networks, many of these intermediate representations are no longer necessary, since the network can learn them as part of the training process. As so much of computer vision research is now devoted to various recognition topics, I decided to move this chapter up so that students can learn about it earlier in the course.

The chapter begins with the classic problem of *instance recognition*, i.e., finding instances of known 3D objects in cluttered scenes. Section 6.2 covers both traditional and deep network approaches to whole *image classification*, i.e., what used to be called *category recognition*. It also discusses the special case of facial recognition. Section 6.3 presents algorithms for *object detection* (drawing bounding boxes around recognized objects), with a brief review of older approaches to face and pedestrian detection. Section 6.4 covers various flavors of *semantic segmentation* (generating per-pixel labels), including *instance segmentation* (delineating separate objects), *pose estimation* (labeling pixels with body parts), and *panoptic segmentation* (labeling both things and stuff). In Section 6.5, we briefly look at some recent papers in *video understanding* and *action recognition*, while in Section 6.6 we mention some recent work in image captioning and visual question answering.

In Chapter 7, we cover feature detection and matching. A lot of current 3D reconstruction and recognition techniques are built on extracting and matching *feature points* (Section 7.1), so this is a fundamental technique required by many subsequent chapters (Chapters 8 and 11) and even in instance recognition (Section 6.1). We also cover edge and straight line detection in Sections 7.2 and 7.4, contour tracking in Section 7.3, and low-level segmentation techniques in Section 7.5.

Feature detection and matching are used in Chapter 8 to perform *image alignment* (or *registration*) and *image stitching*. We introduce the basic techniques of feature-based alignment and show how this problem can be solved using either linear or non-linear least squares, depending on the motion involved. We also introduce additional concepts, such as uncertainty weighting and robust regression, which are essential to making real-world systems work. Feature-based alignment is then used as a building block for both 2D applications such as image stitching (Section 8.2) and computational photography (Chapter 10), as well as 3D geometric alignment tasks such as pose estimation and structure from motion (Chapter 11).

The second part of Chapter 8 is devoted to *image stitching*, i.e., the construction of large panoramas and composites. While stitching is just one example of *computational photography* (see Chapter 10), there is enough depth here to warrant a separate section. We start by discussing various possible motion models (Section 8.2.1), including planar motion and pure camera rotation. We then discuss global alignment (Section 8.3), which is a special (simplified) case of general bundle adjustment, and then present *panorama recognition*, i.e., techniques for automatically discovering which images actually form overlapping panoramas. Finally, we cover the topics of *image compositing* and *blending* (Section 8.4), which involve both selecting which pixels from which images to use and blending them together so as to disguise exposure differences.

Image stitching is a wonderful application that ties together most of the material covered in earlier parts of this book. It also makes for a good mid-term course project that can build on previously developed techniques such as image warping and feature detection and matching. Sections 8.2–8.4 also present more specialized variants of stitching such as whiteboard and document scanning, video summarization, *panography*, full 360° spherical panoramas, and interactive photomontage for blending repeated action shots together.

In Chapter 9, we generalize the concept of feature-based image alignment to cover dense intensity-based motion estimation, i.e., *optical flow*. We start with the simplest possible motion models, translational motion (Section 9.1), and cover topics such as hierarchical (coarse-to-fine) motion estimation, Fourier-based techniques, and iterative refinement. We then present parametric motion models, which can be used to compensate for camera rotation and zooming, as well as affine or planar perspective motion (Section 9.2). This is then generalized to spline-based motion models (Section 9.2.2) and finally to general per-pixel optical flow (Section 9.3). We close the chapter in Section 9.4 with a discussion of layered and learned motion models as well as video object segmentation and tracking. Applications of motion estimation techniques include automated morphing, video denoising, and frame interpolation (slow motion).

Chapter 10 presents additional examples of *computational photography*, which is the process of creating new images from one or more input photographs, often based on the careful modeling and calibration of the image formation process (Section 10.1). Computational photography techniques include merging multiple exposures to create *high dynamic range* images (Section 10.2), increasing image resolution through blur removal and *super-resolution* (Section 10.3), and image editing and compositing operations (Section 10.4). We also cover the topics of texture analysis, synthesis, and *inpainting* (hole filling) in Section 10.5, as well as non-photorealistic rendering and style transfer.

Starting in Chapter 11, we delve more deeply into techniques for reconstructing 3D models from images. We begin by introducing methods for *intrinsic* camera calibration in Section 11.1 and *3D pose estimation*, i.e., *extrinsic* calibration, in Section 11.2. These sections also describe the applications of single-view reconstruction of building models and 3D *location recognition*. We then cover the topic of *triangulation* (Section 11.2.4), which is the 3D reconstruction of points from matched features when the camera positions are known.

Chapter 11 then moves on to the topic of *structure from motion*, which involves the simultaneous recovery of 3D camera motion and 3D scene structure from a collection of tracked 2D features. We begin with two-frame structure from motion (Section 11.3), for which algebraic techniques exist, as well as robust sampling techniques such as RANSAC that can discount erroneous feature matches. We then cover techniques for multi-frame structure from motion, including factorization (Section 11.4.1), bundle adjustment (Section 11.4.2), and constrained motion and structure models (Section 11.4.8). We present applications in visual effects (*match move*) and sparse 3D model construction for large (e.g., internet) photo collections. The final part of this chapter (Section 11.5) has a new section on *simultaneous localization and mapping* (SLAM) as well as its applications to autonomous navigation and mobile augmented reality (AR).

In Chapter 12, we turn to the topic of stereo correspondence, which can be thought of as a special case of motion estimation where the camera positions are already known (Section 12.1). This additional knowledge enables stereo algorithms to search over a much smaller space of correspondences to produce dense depth estimates using various combinations of matching criteria, optimization algorithm, and/or deep networks (Sections 12.3–12.6). We also cover *multi-view* stereo algorithms that build a true 3D surface representation instead of just a single depth map (Section 12.7), as well as *monocular depth inference* algorithms that hallucinate depth maps from just a single image (Section 12.8). Applications of stereo matching include head and gaze tracking, as well as depth-based background replacement (*Z-keying*).

Chapter 13 covers additional 3D shape and appearance modeling techniques. These include classic *shape-from-X* techniques such as shape from shading, shape from texture, and shape from focus (Section 13.1). An alternative to all of these *passive* computer vision techniques is to use *active rangefinding* (Section 13.2), i.e., to project patterned light onto scenes and recover the 3D geometry through triangulation. Processing all of these 3D representations often involves interpolating or simplifying the geometry (Section 13.3), or using alternative representations such as surface point sets (Section 13.4) or implicit functions (Section 13.5).

The collection of techniques for going from one or more images to partial or full 3D models is often called *image-based modeling* or *3D photography*. Section 13.6 examines three more specialized application areas (architecture, faces, and human bodies), which can use *model-based reconstruction* to fit parameterized models to the sensed data. Section 13.7 examines the topic of *appearance modeling*, i.e., techniques for estimating the texture maps, albedos, or even sometimes complete *bi-directional reflectance distribution functions* (BRDFs) that describe the appearance of 3D surfaces.

In Chapter 14, we discuss the large number of image-based rendering techniques that have been developed in the last three decades, including simpler techniques such as view interpolation (Section 14.1), layered depth images (Section 14.2), and sprites and layers (Section 14.2.1), as well as the more general framework of light fields and Lumigraphs (Section 14.3) and higher-order fields such as environment mattes (Section 14.4). Applications of these techniques include navigating 3D collections of photographs using *photo tourism*.

Next, we discuss video-based rendering, which is the temporal extension of image-based rendering. The topics we cover include video-based animation (Section 14.5.1), periodic video turned into *video textures* (Section 14.5.2), and 3D video constructed from multiple video streams (Section 14.5.4). Applications of these techniques include animating still images and creating home tours based on 360° video. We finish the chapter with an overview of the new emerging field of *neural rendering*.

To support the book's use as a textbook, the appendices and associated website contain more detailed mathematical topics and additional material. Appendix A covers linear algebra and numerical techniques, including matrix algebra, least squares, and iterative techniques. Appendix B covers Bayesian estimation theory, including maximum likelihood estimation, robust statistics, Markov random fields, and uncertainty modeling. Appendix C describes the supplementary material that can be used to complement this book, including images and datasets, pointers to software, and course slides.

1.4 Sample syllabus

Teaching all of the material covered in this book in a single quarter or semester course is a Herculean task and likely one not worth attempting.[11] It is better to simply pick and choose topics related to the lecturer's preferred emphasis and tailored to the set of mini-projects envisioned for the students.

Steve Seitz and I have successfully used a 10-week syllabus similar to the one shown in Table 1.1 as both an undergraduate and a graduate-level course in computer vision. The undergraduate course[12] tends to go lighter on the mathematics and takes more time reviewing basics, while the graduate-level course[13] dives more deeply into techniques and assumes the students already have a decent grounding in either vision or related mathematical techniques. Related courses have also been taught on the topics of 3D photography and computational photography. Appendix C.3 and the book's website list other courses that use this book to teach a similar curriculum.

[11] Some universities, such as Stanford (CS231A & 231N), Berkeley (CS194-26/294-26 & 280), and the University of Michigan (EECS 498/598 & 442), now split the material over two courses.

[12] http://www.cs.washington.edu/education/courses/455

[13] http://www.cs.washington.edu/education/courses/576

Week	Chapter	Topics
1.	Chapters 1–2	Introduction and image formation
2.	Chapter 3	Image processing
3.	Chapters 4–5	Optimization and learning
4.	Chapter 5	Deep learning
5.	Chapter 6	Recognition
6.	Chapter 7	Feature detection and matching
7.	Chapter 8	Image alignment and stitching
8.	Chapter 9	Motion estimation
9.	Chapter 10	Computational photography
10.	Chapter 11	Structure from motion
11.	Chapter 12	Depth estimation
12.	Chapter 13	3D reconstruction
13.	Chapter 14	Image-based rendering

Table 1.1 Sample syllabus for a one semester 13-week course. A 10-week quarter could go into lesser depth or omit some topics.

When Steve and I teach the course, we prefer to give the students several small programming assignments early in the course rather than focusing on written homework or quizzes. With a suitable choice of topics, it is possible for these projects to build on each other. For example, introducing feature matching early on can be used in a second assignment to do image alignment and stitching. Alternatively, direct (optical flow) techniques can be used to do the alignment and more focus can be put on either graph cut seam selection or multi-resolution blending techniques.

In the past, we have also asked the students to propose a final project (we provide a set of suggested topics for those who need ideas) by the middle of the course and reserved the last week of the class for student presentations. Sometimes, a few of these projects have actually turned into conference submissions!

No matter how you decide to structure the course or how you choose to use this book, I encourage you to try at least a few small programming tasks to get a feel for how vision techniques work and how they fail. Better yet, pick topics that are fun and can be used on your own photographs, and try to push your creative boundaries to come up with surprising results.

1.5 A note on notation

For better or worse, the notation found in computer vision and multi-view geometry textbooks tends to vary all over the map (Faugeras 1993; Hartley and Zisserman 2004; Girod, Greiner, and Niemann 2000; Faugeras and Luong 2001; Forsyth and Ponce 2003). In this book, I use the convention I first learned in my high school physics class (and later multi-variate calculus and computer graphics courses), which is that vectors $\mathbf{v}$ are lower case bold, matrices $\mathbf{M}$ are upper case bold, and scalars (T, s) are mixed case italic. Unless otherwise noted, vectors operate as column vectors, i.e., they post-multiply matrices, $\mathbf{Mv}$, although they are sometimes written as comma-separated parenthesized lists $\mathbf{x} = (x, y)$ instead of bracketed column vectors $\mathbf{x} = [x\ y]^T$. Some commonly used matrices are $\mathbf{R}$ for rotations, $\mathbf{K}$ for calibration matrices, and $\mathbf{I}$ for the identity matrix. Homogeneous coordinates (Section 2.1) are denoted with a tilde over the vector, e.g., $\tilde{\mathbf{x}} = (\tilde{x}, \tilde{y}, \tilde{w}) = \tilde{w}(x, y, 1) = \tilde{w}\bar{\mathbf{x}}$ in $\mathcal{P}^2$. The cross product operator in matrix form is denoted by $[\,]_\times$.

1.6 Additional reading

This book attempts to be self-contained, so that students can implement the basic assignments and algorithms described here without the need for outside references. However, it does presuppose a general familiarity with basic concepts in linear algebra and numerical techniques, which are reviewed in Appendix A, and image processing, which is reviewed in Chapter 3.

Students who want to delve more deeply into these topics can look in Golub and Van Loan (1996) for matrix algebra and Strang (1988) for linear algebra. In image processing, there are a number of popular textbooks, including Crane (1997), Gomes and Velho (1997), Jähne (1997), Pratt (2007), Russ (2007), Burger and Burge (2008), and Gonzalez and Woods (2017). For computer graphics, popular texts include Hughes, van Dam *et al.* (2013) and Marschner and Shirley (2015), with Glassner (1995) providing a more in-depth look at image formation and rendering. For statistics and machine learning, Chris Bishop's (2006) book is a wonderful and comprehensive introduction with a wealth of exercises, while Murphy (2012) provides a more recent take on the field and Hastie, Tibshirani, and Friedman (2009) a more classic treatment. A great introductory text to deep learning is Glassner (2018), while Goodfellow, Bengio, and Courville (2016) and Zhang, Lipton *et al.* (2021) provide more comprehensive treatments. Students may also want to look in other textbooks on computer vision for material that we do not cover here, as well as for additional project ideas (Nalwa 1993; Trucco and Verri 1998; Hartley and Zisserman 2004; Forsyth and Ponce 2011; Prince 2012; Davies 2017).

There is, however, no substitute for reading the latest research literature, both for the latest ideas and techniques and for the most up-to-date references to related literature.[14] In this book, I have attempted to cite the most recent work in each field so that students can read them directly and use them as inspiration for their own work. Browsing the last few years' conference proceedings from the major vision, graphics, and machine learning conferences, such as CVPR, ECCV, ICCV, SIGGRAPH, and NeurIPS, as well as keeping an eye out for the latest publications on arXiv, will provide a wealth of new ideas. The tutorials offered at these conferences, for which slides or notes are often available online, are also an invaluable resource.

[14]For a comprehensive bibliography and taxonomy of computer vision research, Keith Price's Annotated Computer Vision Bibliography https://www.visionbib.com/bibliography/contents.html is an invaluable resource.

Chapter 2

Image formation

2.1 Geometric primitives and transformations . . . 29
  2.1.1 2D transformations . . . 32
  2.1.2 3D transformations . . . 35
  2.1.3 3D rotations . . . 36
  2.1.4 3D to 2D projections . . . 41
  2.1.5 Lens distortions . . . 51
2.2 Photometric image formation . . . 53
  2.2.1 Lighting . . . 53
  2.2.2 Reflectance and shading . . . 54
  2.2.3 Optics . . . 59
2.3 The digital camera . . . 63
  2.3.1 Sampling and aliasing . . . 67
  2.3.2 Color . . . 69
  2.3.3 Compression . . . 78
2.4 Additional reading . . . 80
2.5 Exercises . . . 80

R. Szeliski, *Computer Vision*, Texts in Computer Science,
https://doi.org/10.1007/978-3-030-34372-9_2

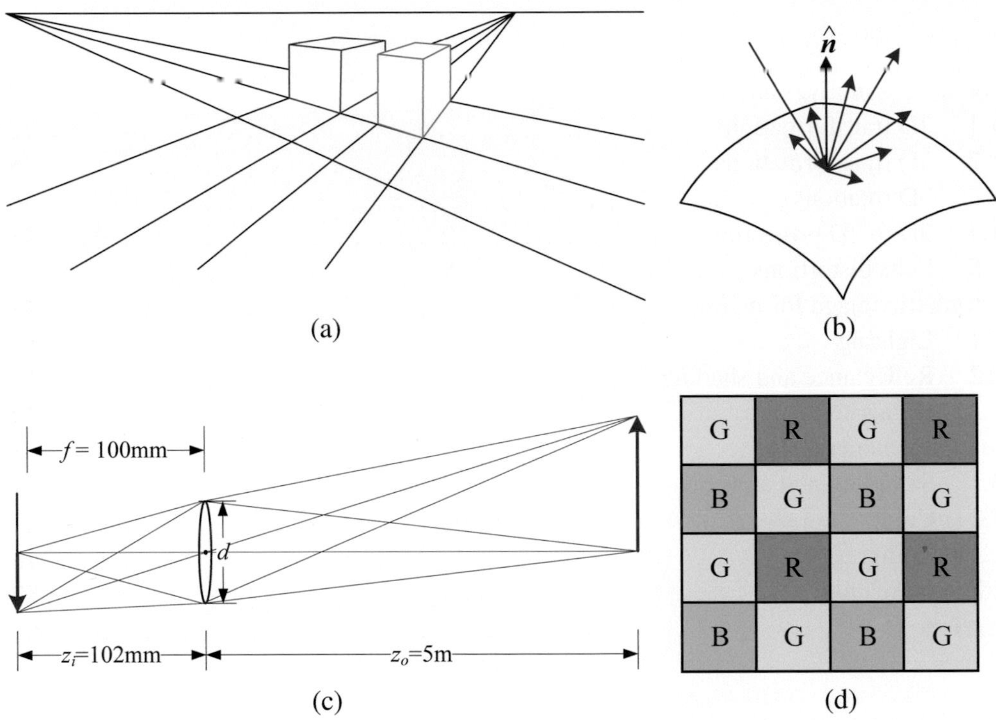

Figure 2.1 A few components of the image formation process: (a) perspective projection; (b) light scattering when hitting a surface; (c) lens optics; (d) Bayer color filter array.

Before we can analyze and manipulate images, we need to establish a vocabulary for describing the geometry of a scene. We also need to understand the image formation process that produced a particular image given a set of lighting conditions, scene geometry, surface properties, and camera optics. In this chapter, we present a simplified model of this image formation process.

Section 2.1 introduces the basic geometric primitives used throughout the book (points, lines, and planes) and the *geometric* transformations that project these 3D quantities into 2D image features (Figure 2.1a). Section 2.2 describes how lighting, surface properties (Figure 2.1b), and camera *optics* (Figure 2.1c) interact to produce the color values that fall onto the image sensor. Section 2.3 describes how continuous color images are turned into discrete digital *samples* inside the image sensor (Figure 2.1d) and how to avoid (or at least characterize) sampling deficiencies, such as aliasing.

The material covered in this chapter is but a brief summary of a very rich and deep set of topics, traditionally covered in a number of separate fields. A more thorough introduction to the geometry of points, lines, planes, and projections can be found in textbooks on multi-view geometry (Hartley and Zisserman 2004; Faugeras and Luong 2001) and computer graphics (Hughes, van Dam *et al.* 2013). The image formation (synthesis) process is traditionally taught as part of a computer graphics curriculum (Glassner 1995; Watt 1995; Hughes, van Dam *et al.* 2013; Marschner and Shirley 2015) but it is also studied in physics-based computer vision (Wolff, Shafer, and Healey 1992a). The behavior of camera lens systems is studied in optics (Möller 1988; Ray 2002; Hecht 2015). Some good books on color theory are Healey and Shafer (1992), Wandell (1995), and Wyszecki and Stiles (2000), with Livingstone (2008) providing a more fun and informal introduction to the topic of color perception. Topics relating to sampling and aliasing are covered in textbooks on signal and image processing (Crane 1997; Jähne 1997; Oppenheim and Schafer 1996; Oppenheim, Schafer, and Buck 1999; Pratt 2007; Russ 2007; Burger and Burge 2008; Gonzalez and Woods 2017). The recent book by Ikeuchi, Matsushita *et al.* (2020) also covers 3D geometry, photometry, and sensor models, with an emphasis on *active illumination* systems.

A note to students: If you have already studied computer graphics, you may want to skim the material in Section 2.1, although the sections on projective depth and object-centered projection near the end of Section 2.1.4 may be new to you. Similarly, physics students (as well as computer graphics students) will mostly be familiar with Section 2.2. Finally, students with a good background in image processing will already be familiar with sampling issues (Section 2.3) as well as some of the material in Chapter 3.

2.1 Geometric primitives and transformations

In this section, we introduce the basic 2D and 3D primitives used in this textbook, namely points, lines, and planes. We also describe how 3D features are projected into 2D features. More detailed descriptions of these topics (along with a gentler and more intuitive introduction) can be found in textbooks on multiple-view geometry (Hartley and Zisserman 2004; Faugeras and Luong 2001).

Geometric primitives form the basic building blocks used to describe three-dimensional shapes. In this section, we introduce points, lines, and planes. Later sections of the book discuss curves (Sections 7.3 and 12.2), surfaces (Section 13.3), and volumes (Section 13.5).

2D points. 2D points (pixel coordinates in an image) can be denoted using a pair of values, $\mathbf{x} = (x, y) \in \mathcal{R}^2$, or alternatively,

$$\mathbf{x} = \begin{bmatrix} x \\ y \end{bmatrix}. \tag{2.1}$$

(As stated in the introduction, we use the $(x_1, x_2, \ldots)$ notation to denote column vectors.)

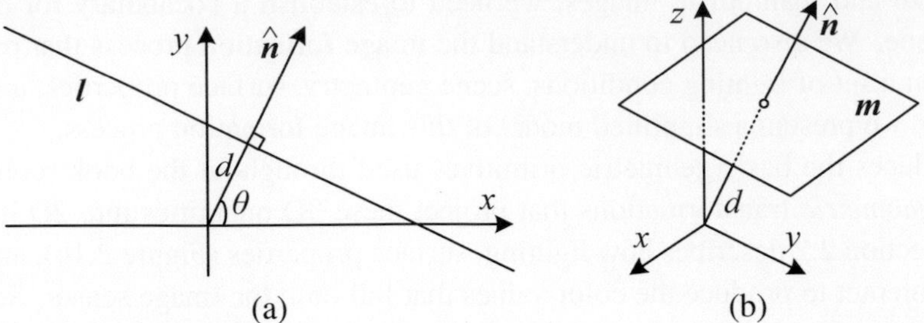

Figure 2.2 (a) 2D line equation and (b) 3D plane equation, expressed in terms of the normal $\hat{\mathbf{n}}$ and distance to the origin d.

2D points can also be represented using *homogeneous coordinates*, $\tilde{\mathbf{x}} = (\tilde{x}, \tilde{y}, \tilde{w}) \in \mathcal{P}^2$, where vectors that differ only by scale are considered to be equivalent. $\mathcal{P}^2 = \mathcal{R}^3 - (0,0,0)$ is called the 2D *projective space*.

A homogeneous vector $\tilde{\mathbf{x}}$ can be converted back into an *inhomogeneous* vector $\mathbf{x}$ by dividing through by the last element $\tilde{w}$, i.e.,

$$\tilde{\mathbf{x}} = (\tilde{x}, \tilde{y}, \tilde{w}) = \tilde{w}(x, y, 1) = \tilde{w}\bar{\mathbf{x}}, \tag{2.2}$$

where $\bar{\mathbf{x}} = (x, y, 1)$ is the *augmented vector*. Homogeneous points whose last element is $\tilde{w} = 0$ are called *ideal points* or *points at infinity* and do not have an equivalent inhomogeneous representation.

2D lines. 2D lines can also be represented using homogeneous coordinates $\tilde{\mathbf{l}} = (a, b, c)$. The corresponding *line equation* is

$$\bar{\mathbf{x}} \cdot \tilde{\mathbf{l}} = ax + by + c = 0. \tag{2.3}$$

We can normalize the line equation vector so that $\mathbf{l} = (\hat{n}_x, \hat{n}_y, d) = (\hat{\mathbf{n}}, d)$ with $\|\hat{\mathbf{n}}\| = 1$. In this case, $\hat{\mathbf{n}}$ is the *normal vector* perpendicular to the line and d is its distance to the origin (Figure 2.2). (The one exception to this normalization is the *line at infinity* $\tilde{\mathbf{l}} = (0, 0, 1)$, which includes all (ideal) points at infinity.)

We can also express $\hat{\mathbf{n}}$ as a function of rotation angle θ, $\hat{\mathbf{n}} = (\hat{n}_x, \hat{n}_y) = (\cos\theta, \sin\theta)$ (Figure 2.2a). This representation is commonly used in the *Hough transform* line-finding algorithm, which is discussed in Section 7.4.2. The combination (θ, d) is also known as *polar coordinates*.

When using homogeneous coordinates, we can compute the intersection of two lines as

$$\tilde{\mathbf{x}} = \tilde{\mathbf{l}}_1 \times \tilde{\mathbf{l}}_2, \tag{2.4}$$

where $\times$ is the cross product operator. Similarly, the line joining two points can be written as

$$\tilde{\mathbf{l}} = \tilde{\mathbf{x}}_1 \times \tilde{\mathbf{x}}_2. \tag{2.5}$$

When trying to fit an intersection point to multiple lines or, conversely, a line to multiple points, least squares techniques (Section 8.1.1 and Appendix A.2) can be used, as discussed in Exercise 2.1.

2D conics. There are other algebraic curves that can be expressed with simple polynomial homogeneous equations. For example, the *conic sections* (so called because they arise as the intersection of a plane and a 3D cone) can be written using a *quadric* equation

$$\tilde{\mathbf{x}}^T \mathbf{Q} \tilde{\mathbf{x}} = 0. \tag{2.6}$$

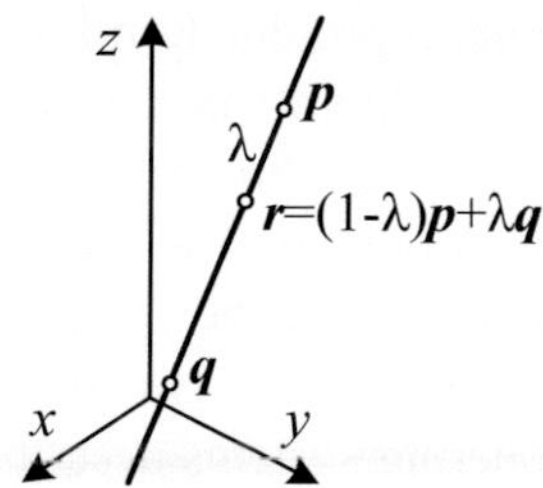

Figure 2.3 3D line equation, $\mathbf{r} = (1 - \lambda)\mathbf{p} + \lambda\mathbf{q}$.

Quadric equations play useful roles in the study of multi-view geometry and camera calibration (Hartley and Zisserman 2004; Faugeras and Luong 2001) but are not used extensively in this book.

3D points. Point coordinates in three dimensions can be written using inhomogeneous coordinates $\mathbf{x} = (x, y, z) \in \mathcal{R}^3$ or homogeneous coordinates $\tilde{\mathbf{x}} = (\tilde{x}, \tilde{y}, \tilde{z}, \tilde{w}) \in \mathcal{P}^3$. As before, it is sometimes useful to denote a 3D point using the augmented vector $\bar{\mathbf{x}} = (x, y, z, 1)$ with $\tilde{\mathbf{x}} = \tilde{w}\bar{\mathbf{x}}$.

3D planes. 3D planes can also be represented as homogeneous coordinates $\tilde{\mathbf{m}} = (a, b, c, d)$ with a corresponding plane equation

$$\bar{\mathbf{x}} \cdot \tilde{\mathbf{m}} = ax + by + cz + d = 0. \tag{2.7}$$

We can also normalize the plane equation as $\mathbf{m} = (\hat{n}_x, \hat{n}_y, \hat{n}_z, d) = (\hat{\mathbf{n}}, d)$ with $\|\hat{\mathbf{n}}\| = 1$. In this case, $\hat{\mathbf{n}}$ is the *normal vector* perpendicular to the plane and d is its distance to the origin (Figure 2.2b). As with the case of 2D lines, the *plane at infinity* $\tilde{\mathbf{m}} = (0, 0, 0, 1)$, which contains all the points at infinity, cannot be normalized (i.e., it does not have a unique normal or a finite distance).

We can express $\hat{\mathbf{n}}$ as a function of two angles (θ, ϕ),

$$\hat{\mathbf{n}} = (\cos\theta\cos\phi, \sin\theta\cos\phi, \sin\phi), \tag{2.8}$$

i.e., using *spherical coordinates*, but these are less commonly used than polar coordinates since they do not uniformly sample the space of possible normal vectors.

3D lines. Lines in 3D are less elegant than either lines in 2D or planes in 3D. One possible representation is to use two points on the line, $(\mathbf{p}, \mathbf{q})$. Any other point on the line can be expressed as a linear combination of these two points

$$\mathbf{r} = (1 - \lambda)\mathbf{p} + \lambda\mathbf{q}, \tag{2.9}$$

as shown in Figure 2.3. If we restrict $0 \le \lambda \le 1$, we get the *line segment* joining $\mathbf{p}$ and $\mathbf{q}$.

If we use homogeneous coordinates, we can write the line as

$$\tilde{\mathbf{r}} = \mu\tilde{\mathbf{p}} + \lambda\tilde{\mathbf{q}}. \tag{2.10}$$

A special case of this is when the second point is at infinity, i.e., $\tilde{\mathbf{q}} = (\hat{d}_x, \hat{d}_y, \hat{d}_z, 0) = (\hat{\mathbf{d}}, 0)$. Here, we see that $\hat{\mathbf{d}}$ is the *direction* of the line. We can then re-write the inhomogeneous 3D line equation as

$$\mathbf{r} = \mathbf{p} + \lambda\hat{\mathbf{d}}. \tag{2.11}$$

A disadvantage of the endpoint representation for 3D lines is that it has too many degrees of freedom, i.e., six (three for each endpoint) instead of the four degrees that a 3D line truly has. However, if we fix the two points on the line to lie in specific planes, we obtain a representation with four degrees of freedom. For example, if we are representing nearly vertical lines, then $z = 0$ and $z = 1$ form two suitable planes, i.e., the (x, y) coordinates in both planes provide the four coordinates describing the line. This kind of two-plane parameterization is used in the *light field* and *Lumigraph* image-based rendering systems described in Chapter 14 to represent the collection of rays seen by a camera as it moves in front of an object. The two-endpoint representation is also useful for representing line segments, even when their exact endpoints cannot be seen (only guessed at).

If we wish to represent all possible lines without bias towards any particular orientation, we can use *Plücker coordinates* (Hartley and Zisserman 2004, Section 3.2; Faugeras and Luong 2001, Chapter 3). These coordinates are the six independent non-zero entries in the 4×4 skew symmetric matrix

$$\mathbf{L} = \tilde{\mathbf{p}}\tilde{\mathbf{q}}^T - \tilde{\mathbf{q}}\tilde{\mathbf{p}}^T, \tag{2.12}$$

where $\tilde{\mathbf{p}}$ and $\tilde{\mathbf{q}}$ are *any* two (non-identical) points on the line. This representation has only four degrees of freedom, since $\mathbf{L}$ is homogeneous and also satisfies $|\mathbf{L}| = 0$, which results in a quadratic constraint on the Plücker coordinates.

In practice, the minimal representation is not essential for most applications. An adequate model of 3D lines can be obtained by estimating their direction (which may be known ahead of time, e.g., for architecture) and some point within the visible portion of the line (see Section 11.4.8) or by using the two endpoints, since lines are most often visible as finite line segments. However, if you are interested in more details about the topic of minimal line parameterizations, Förstner (2005) discusses various ways to infer and model 3D lines in projective geometry, as well as how to estimate the uncertainty in such fitted models.

3D quadrics. The 3D analog of a conic section is a quadric surface

$$\bar{\mathbf{x}}^T \mathbf{Q} \bar{\mathbf{x}} = 0 \tag{2.13}$$

(Hartley and Zisserman 2004, Chapter 3). Again, while quadric surfaces are useful in the study of multi-view geometry and can also serve as useful modeling primitives (spheres, ellipsoids, cylinders), we do not study them in great detail in this book.

2.1.1 2D transformations

Having defined our basic primitives, we can now turn our attention to how they can be transformed. The simplest transformations occur in the 2D plane are illustrated in Figure 2.4.

Translation. 2D translations can be written as $\mathbf{x}' = \mathbf{x} + \mathbf{t}$ or

$$\mathbf{x}' = \begin{bmatrix} \mathbf{I} & \mathbf{t} \end{bmatrix} \bar{\mathbf{x}}, \tag{2.14}$$

where $\mathbf{I}$ is the (2×2) identity matrix or

$$\bar{\mathbf{x}}' = \begin{bmatrix} \mathbf{I} & \mathbf{t} \\ \mathbf{0}^T & 1 \end{bmatrix} \bar{\mathbf{x}}, \tag{2.15}$$

where $\mathbf{0}$ is the zero vector. Using a 2×3 matrix results in a more compact notation, whereas using a full-rank 3×3 matrix (which can be obtained from the 2×3 matrix by appending a $[\mathbf{0}^T \; 1]$ row)

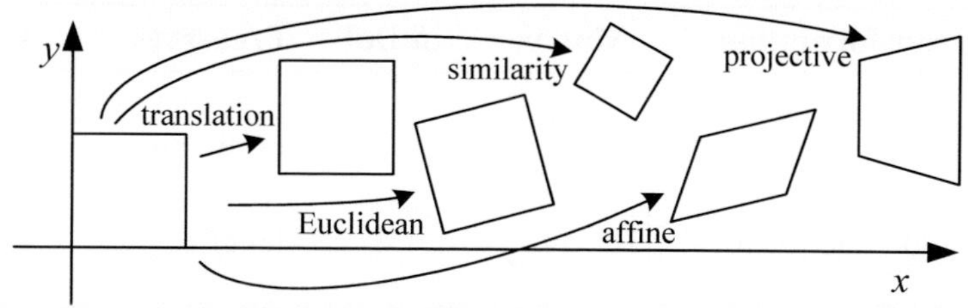

Figure 2.4 Basic set of 2D planar transformations.

makes it possible to chain transformations using matrix multiplication as well as to compute inverse transforms. Note that in any equation where an augmented vector such as $\bar{\mathbf{x}}$ appears on both sides, it can always be replaced with a full homogeneous vector $\tilde{\mathbf{x}}$.

Rotation + translation. This transformation is also known as *2D rigid body motion* or the *2D Euclidean transformation* (since Euclidean distances are preserved). It can be written as $\mathbf{x}' = \mathbf{R}\mathbf{x} + \mathbf{t}$ or

$$\mathbf{x}' = \begin{bmatrix} \mathbf{R} & \mathbf{t} \end{bmatrix} \bar{\mathbf{x}}. \tag{2.16}$$

where

$$\mathbf{R} = \begin{bmatrix} \cos\theta & -\sin\theta \\ \sin\theta & \cos\theta \end{bmatrix} \tag{2.17}$$

is an orthonormal rotation matrix with $\mathbf{R}\mathbf{R}^T = \mathbf{I}$ and $|\mathbf{R}| = 1$.

Scaled rotation. Also known as the *similarity transform*, this transformation can be expressed as $\mathbf{x}' = s\mathbf{R}\mathbf{x} + \mathbf{t}$, where s is an arbitrary scale factor. It can also be written as

$$\mathbf{x}' = \begin{bmatrix} s\mathbf{R} & \mathbf{t} \end{bmatrix} \bar{\mathbf{x}} = \begin{bmatrix} a & -b & t_x \\ b & a & t_y \end{bmatrix} \bar{\mathbf{x}}, \tag{2.18}$$

where we no longer require that $a^2 + b^2 = 1$. The similarity transform preserves angles between lines.

Affine. The affine transformation is written as $\mathbf{x}' = \mathbf{A}\bar{\mathbf{x}}$, where $\mathbf{A}$ is an arbitrary 2×3 matrix, i.e.,

$$\mathbf{x}' = \begin{bmatrix} a_{00} & a_{01} & a_{02} \\ a_{10} & a_{11} & a_{12} \end{bmatrix} \bar{\mathbf{x}}. \tag{2.19}$$

Parallel lines remain parallel under affine transformations.

Projective. This transformation, also known as a *perspective transform* or *homography*, operates on homogeneous coordinates,

$$\tilde{\mathbf{x}}' = \tilde{\mathbf{H}}\tilde{\mathbf{x}}, \tag{2.20}$$

where $\tilde{\mathbf{H}}$ is an arbitrary 3×3 matrix. Note that $\tilde{\mathbf{H}}$ is homogeneous, i.e., it is only defined up to a scale, and that two $\tilde{\mathbf{H}}$ matrices that differ only by scale are equivalent. The resulting homogeneous coordinate $\tilde{\mathbf{x}}'$ must be normalized in order to obtain an inhomogeneous result $\mathbf{x}$, i.e.,

$$x' = \frac{h_{00}x + h_{01}y + h_{02}}{h_{20}x + h_{21}y + h_{22}} \quad \text{and} \quad y' = \frac{h_{10}x + h_{11}y + h_{12}}{h_{20}x + h_{21}y + h_{22}}. \tag{2.21}$$

Transformation	Matrix	# DoF	Preserves	Icon
translation	$\begin{bmatrix} \mathbf{I} & \mathbf{t} \end{bmatrix}_{2\times 3}$	2	orientation	
rigid (Euclidean)	$\begin{bmatrix} \mathbf{R} & \mathbf{t} \end{bmatrix}_{2\times 3}$	3	lengths	
similarity	$\begin{bmatrix} s\mathbf{R} & \mathbf{t} \end{bmatrix}_{2\times 3}$	4	angles	
affine	$\begin{bmatrix} \mathbf{A} \end{bmatrix}_{2\times 3}$	6	parallelism	
projective	$\begin{bmatrix} \tilde{\mathbf{H}} \end{bmatrix}_{3\times 3}$	8	straight lines	

Table 2.1 Hierarchy of 2D coordinate transformations, listing the transformation name, its matrix form, the number of degrees of freedom, what geometric properties it preserves, and a mnemonic icon. Each transformation also preserves the properties listed in the rows below it, i.e., similarity preserves not only angles but also parallelism and straight lines. The 2×3 matrices are extended with a third $[\mathbf{0}^T \; 1]$ row to form a full 3×3 matrix for homogeneous coordinate transformations.

Perspective transformations preserve straight lines (i.e., they remain straight after the transformation).

Hierarchy of 2D transformations. The preceding set of transformations are illustrated in Figure 2.4 and summarized in Table 2.1. The easiest way to think of them is as a set of (potentially restricted) 3×3 matrices operating on 2D homogeneous coordinate vectors. Hartley and Zisserman (2004) contains a more detailed description of the hierarchy of 2D planar transformations.

The above transformations form a nested set of *groups*, i.e., they are closed under composition and have an inverse that is a member of the same group. (This will be important later when applying these transformations to images in Section 3.6.) Each (simpler) group is a subgroup of the more complex group below it. The mathematics of such *Lie groups* and their related algebras (tangent spaces at the origin) are discussed in a number of recent robotics tutorials (Dellaert and Kaess 2017; Blanco 2019; Solà, Deray, and Atchuthan 2019), where the 2D rotation and rigid transforms are called SO(2) and SE(2), which stand for the *special orthogonal* and *special Euclidean* groups.[1]

Co-vectors. While the above transformations can be used to transform points in a 2D plane, can they also be used directly to transform a line equation? Consider the homogeneous equation $\tilde{\mathbf{l}} \cdot \tilde{\mathbf{x}} = 0$. If we transform $\tilde{\mathbf{x}}' = \tilde{\mathbf{H}}\tilde{\mathbf{x}}$, we obtain

$$\tilde{\mathbf{l}}' \cdot \tilde{\mathbf{x}}' = \tilde{\mathbf{l}}'^T \tilde{\mathbf{H}}\tilde{\mathbf{x}} = (\tilde{\mathbf{H}}^T \tilde{\mathbf{l}}')^T \tilde{\mathbf{x}} = \tilde{\mathbf{l}} \cdot \tilde{\mathbf{x}} = 0, \tag{2.22}$$

i.e., $\tilde{\mathbf{l}}' = \tilde{\mathbf{H}}^{-T}\tilde{\mathbf{l}}$. Thus, the action of a projective transformation on a *co-vector* such as a 2D line or 3D normal can be represented by the transposed inverse of the matrix, which is equivalent to the *adjoint* of $\tilde{\mathbf{H}}$, since projective transformation matrices are homogeneous. Jim Blinn (1998) describes (in Chapters 9 and 10) the ins and outs of notating and manipulating co-vectors.

While the above transformations are the ones we use most extensively, a number of additional transformations are sometimes used.

[1]The term *special* refers to the desired condition of no reflection, i.e., $\det|\mathbf{R}| = 1$.

Stretch/squash. This transformation changes the aspect ratio of an image,

$$x' = s_x x + t_x$$
$$y' = s_y y + t_y,$$

and is a restricted form of an affine transformation. Unfortunately, it does not nest cleanly with the groups listed in Table 2.1.

Planar surface flow. This eight-parameter transformation (Horn 1986; Bergen, Anandan *et al.* 1992; Girod, Greiner, and Niemann 2000),

$$x' = a_0 + a_1 x + a_2 y + a_6 x^2 + a_7 xy$$
$$y' = a_3 + a_4 x + a_5 y + a_6 xy + a_7 y^2,$$

arises when a planar surface undergoes a small 3D motion. It can thus be thought of as a small motion approximation to a full homography. Its main attraction is that it is *linear* in the motion parameters, a_k, which are often the quantities being estimated.

Bilinear interpolant. This eight-parameter transform (Wolberg 1990),

$$x' = a_0 + a_1 x + a_2 y + a_6 xy$$
$$y' = a_3 + a_4 x + a_5 y + a_7 xy,$$

can be used to interpolate the deformation due to the motion of the four corner points of a square. (In fact, it can interpolate the motion of any four non-collinear points.) While the deformation is linear in the motion parameters, it does not generally preserve straight lines (only lines parallel to the square axes). However, it is often quite useful, e.g., in the interpolation of sparse grids using splines (Section 9.2.2).

2.1.2 3D transformations

The set of three-dimensional coordinate transformations is very similar to that available for 2D transformations and is summarized in Table 2.2. As in 2D, these transformations form a nested set of groups. Hartley and Zisserman (2004, Section 2.4) give a more detailed description of this hierarchy.

Translation. 3D translations can be written as $\mathbf{x}' = \mathbf{x} + \mathbf{t}$ or

$$\mathbf{x}' = \begin{bmatrix} \mathbf{I} & \mathbf{t} \end{bmatrix} \bar{\mathbf{x}}, \tag{2.23}$$

where $\mathbf{I}$ is the (3×3) identity matrix.

Rotation + translation. Also known as 3D *rigid body motion* or the 3D *Euclidean transformation* or SE(3), it can be written as $\mathbf{x}' = \mathbf{R}\mathbf{x} + \mathbf{t}$ or

$$\mathbf{x}' = \begin{bmatrix} \mathbf{R} & \mathbf{t} \end{bmatrix} \bar{\mathbf{x}}, \tag{2.24}$$

where $\mathbf{R}$ is a 3×3 orthonormal rotation matrix with $\mathbf{R}\mathbf{R}^T = \mathbf{I}$ and $|\mathbf{R}| = 1$. Note that sometimes it is more convenient to describe a rigid motion using

$$\mathbf{x}' = \mathbf{R}(\mathbf{x} - \mathbf{c}) = \mathbf{R}\mathbf{x} - \mathbf{R}\mathbf{c}, \tag{2.25}$$

Transformation	Matrix	# DoF	Preserves	Icon
translation	$\begin{bmatrix} \mathbf{I} & \mathbf{t} \end{bmatrix}_{3\times 4}$	3	orientation	
rigid (Euclidean)	$\begin{bmatrix} \mathbf{R} & \mathbf{t} \end{bmatrix}_{3\times 4}$	6	lengths	
similarity	$\begin{bmatrix} s\mathbf{R} & \mathbf{t} \end{bmatrix}_{3\times 4}$	7	angles	
affine	$\begin{bmatrix} \mathbf{A} \end{bmatrix}_{3\times 4}$	12	parallelism	
projective	$\begin{bmatrix} \tilde{\mathbf{H}} \end{bmatrix}_{4\times 4}$	15	straight lines	

Table 2.2 Hierarchy of 3D coordinate transformations. Each transformation also preserves the properties listed in the rows below it, i.e., similarity preserves not only angles but also parallelism and straight lines. The 3 $\times$ 4 matrices are extended with a fourth $[\mathbf{0}^T \; 1]$ row to form a full 4 $\times$ 4 matrix for homogeneous coordinate transformations. The mnemonic icons are drawn in 2D but are meant to suggest transformations occurring in a full 3D cube.

where $\mathbf{c}$ is the center of rotation (often the camera center).

Compactly parameterizing a 3D rotation is a non-trivial task, which we describe in more detail below.

Scaled rotation. The 3D *similarity transform* can be expressed as $\mathbf{x}' = s\mathbf{R}\mathbf{x} + \mathbf{t}$ where s is an arbitrary scale factor. It can also be written as

$$\mathbf{x}' = \begin{bmatrix} s\mathbf{R} & \mathbf{t} \end{bmatrix} \bar{\mathbf{x}}. \tag{2.26}$$

This transformation preserves angles between lines and planes.

Affine. The affine transform is written as $\mathbf{x}' = \mathbf{A}\bar{\mathbf{x}}$, where $\mathbf{A}$ is an arbitrary 3 $\times$ 4 matrix, i.e.,

$$\mathbf{x}' = \begin{bmatrix} a_{00} & a_{01} & a_{02} & a_{03} \\ a_{10} & a_{11} & a_{12} & a_{13} \\ a_{20} & a_{21} & a_{22} & a_{23} \end{bmatrix} \bar{\mathbf{x}}. \tag{2.27}$$

Parallel lines and planes remain parallel under affine transformations.

Projective. This transformation, variously known as a *3D perspective transform*, *homography*, or *collineation*, operates on homogeneous coordinates,

$$\tilde{\mathbf{x}}' = \tilde{\mathbf{H}}\tilde{\mathbf{x}}, \tag{2.28}$$

where $\tilde{\mathbf{H}}$ is an arbitrary 4 $\times$ 4 homogeneous matrix. As in 2D, the resulting homogeneous coordinate $\tilde{\mathbf{x}}'$ must be normalized in order to obtain an inhomogeneous result $\mathbf{x}$. Perspective transformations preserve straight lines (i.e., they remain straight after the transformation).

2.1.3 3D rotations

The biggest difference between 2D and 3D coordinate transformations is that the parameterization of the 3D rotation matrix $\mathbf{R}$ is not as straightforward, as several different possibilities exist.

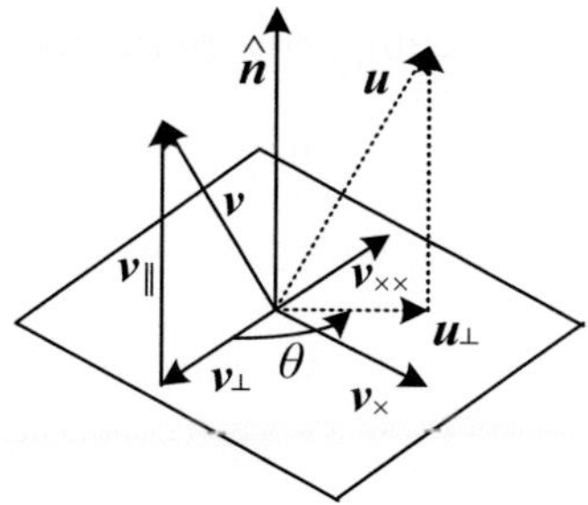

Figure 2.5 Rotation around an axis $\hat{\mathbf{n}}$ by an angle θ.

Euler angles

A rotation matrix can be formed as the product of three rotations around three cardinal axes, e.g., x, y, and z, or x, y, and x. This is generally a bad idea, as the result depends on the order in which the transforms are applied.[2] What is worse, it is not always possible to move smoothly in the parameter space, i.e., sometimes one or more of the Euler angles change dramatically in response to a small change in rotation.[3] For these reasons, we do not even give the formula for Euler angles in this book—interested readers can look in other textbooks or technical reports (Faugeras 1993; Diebel 2006). Note that, in some applications, if the rotations are known to be a set of uni-axial transforms, they can always be represented using an explicit set of rigid transformations.

Axis/angle (exponential twist)

A rotation can be represented by a rotation axis $\hat{\mathbf{n}}$ and an angle θ, or equivalently by a 3D vector $\boldsymbol{\omega} = \theta\hat{\mathbf{n}}$. Figure 2.5 shows how we can compute the equivalent rotation. First, we project the vector $\mathbf{v}$ onto the axis $\hat{\mathbf{n}}$ to obtain

$$\mathbf{v}_{\parallel} = \hat{\mathbf{n}}(\hat{\mathbf{n}} \cdot \mathbf{v}) = (\hat{\mathbf{n}}\hat{\mathbf{n}}^T)\mathbf{v}, \tag{2.29}$$

which is the component of $\mathbf{v}$ that is not affected by the rotation. Next, we compute the perpendicular residual of $\mathbf{v}$ from $\hat{\mathbf{n}}$,

$$\mathbf{v}_{\perp} = \mathbf{v} - \mathbf{v}_{\parallel} = (\mathbf{I} - \hat{\mathbf{n}}\hat{\mathbf{n}}^T)\mathbf{v}. \tag{2.30}$$

We can rotate this vector by 90° using the cross product,

$$\mathbf{v}_{\times} = \hat{\mathbf{n}} \times \mathbf{v}_{\perp} = \hat{\mathbf{n}} \times \mathbf{v} = [\hat{\mathbf{n}}]_{\times}\mathbf{v}, \tag{2.31}$$

where $[\hat{\mathbf{n}}]_{\times}$ is the matrix form of the cross product operator with the vector $\hat{\mathbf{n}} = (\hat{n}_x, \hat{n}_y, \hat{n}_z)$,

$$[\hat{\mathbf{n}}]_{\times} = \begin{bmatrix} 0 & -\hat{n}_z & \hat{n}_y \\ \hat{n}_z & 0 & -\hat{n}_x \\ -\hat{n}_y & \hat{n}_x & 0 \end{bmatrix}. \tag{2.32}$$

Note that rotating this vector by another 90° is equivalent to taking the cross product again,

$$\mathbf{v}_{\times\times} = \hat{\mathbf{n}} \times \mathbf{v}_{\times} = [\hat{\mathbf{n}}]^2_{\times}\mathbf{v} = -\mathbf{v}_{\perp},$$

and hence

$$\mathbf{v}_{\parallel} = \mathbf{v} - \mathbf{v}_{\perp} = \mathbf{v} + \mathbf{v}_{\times\times} = (\mathbf{I} + [\hat{\mathbf{n}}]^2_{\times})\mathbf{v}.$$

[2] However, in special situations, such as describing the motion of a pan-tilt head, these angles may be more intuitive.
[3] In robotics, this is sometimes referred to as *gimbal lock*.

We can now compute the in-plane component of the rotated vector $\mathbf{u}$ as

$$\mathbf{u}_\perp = \cos\theta\mathbf{v}_\perp + \sin\theta\mathbf{v}_\times = (\sin\theta[\hat{\mathbf{n}}]_\times - \cos\theta[\hat{\mathbf{n}}]_\times^2)\mathbf{v}.$$

Putting all these terms together, we obtain the final rotated vector as

$$\mathbf{u} = \mathbf{u}_\perp + \mathbf{v}_\parallel = (\mathbf{I} + \sin\theta[\hat{\mathbf{n}}]_\times + (1-\cos\theta)[\hat{\mathbf{n}}]_\times^2)\mathbf{v}. \tag{2.33}$$

We can therefore write the rotation matrix corresponding to a rotation by θ around an axis $\hat{\mathbf{n}}$ as

$$\mathbf{R}(\hat{\mathbf{n}}, \theta) = \mathbf{I} + \sin\theta[\hat{\mathbf{n}}]_\times + (1-\cos\theta)[\hat{\mathbf{n}}]_\times^2, \tag{2.34}$$

which is known as *Rodrigues' formula* (Ayache 1989).

The product of the axis $\hat{\mathbf{n}}$ and angle θ, $\boldsymbol{\omega} = \theta\hat{\mathbf{n}} = (\omega_x, \omega_y, \omega_z)$, is a minimal representation for a 3D rotation. Rotations through common angles such as multiples of 90° can be represented exactly (and converted to exact matrices) if θ is stored in degrees. Unfortunately, this representation is not unique, since we can always add a multiple of 360° (2π radians) to θ and get the same rotation matrix. As well, $(\hat{\mathbf{n}}, \theta)$ and $(-\hat{\mathbf{n}}, -\theta)$ represent the same rotation.

However, for small rotations (e.g., corrections to rotations), this is an excellent choice. In particular, for small (infinitesimal or instantaneous) rotations and θ expressed in radians, Rodrigues' formula simplifies to

$$\mathbf{R}(\boldsymbol{\omega}) \approx \mathbf{I} + \sin\theta[\hat{\mathbf{n}}]_\times \approx \mathbf{I} + [\theta\hat{\mathbf{n}}]_\times = \begin{bmatrix} 1 & -\omega_z & \omega_y \\ \omega_z & 1 & -\omega_x \\ -\omega_y & \omega_x & 1 \end{bmatrix}, \tag{2.35}$$

which gives a nice linearized relationship between the rotation parameters $\boldsymbol{\omega}$ and $\mathbf{R}$. We can also write $\mathbf{R}(\boldsymbol{\omega})\mathbf{v} \approx \mathbf{v} + \boldsymbol{\omega} \times \mathbf{v}$, which is handy when we want to compute the derivative of $\mathbf{R}\mathbf{v}$ with respect to $\boldsymbol{\omega}$,

$$\frac{\partial \mathbf{R}\mathbf{v}}{\partial \boldsymbol{\omega}^T} = -[\mathbf{v}]_\times = \begin{bmatrix} 0 & z & -y \\ -z & 0 & x \\ y & -x & 0 \end{bmatrix}. \tag{2.36}$$

Another way to derive a rotation through a finite angle is called the *exponential twist* (Murray, Li, and Sastry 1994). A rotation by an angle θ is equivalent to k rotations through θ/k. In the limit as $k \to \infty$, we obtain

$$\mathbf{R}(\hat{\mathbf{n}}, \theta) = \lim_{k\to\infty} (\mathbf{I} + \frac{1}{k}[\theta\hat{\mathbf{n}}]_\times)^k = \exp{[\boldsymbol{\omega}]_\times}. \tag{2.37}$$

If we expand the matrix exponential as a Taylor series (using the identity $[\hat{\mathbf{n}}]_\times^{k+2} = -[\hat{\mathbf{n}}]_\times^k$, $k > 0$, and again assuming θ is in radians),

$$\begin{aligned}
\exp{[\boldsymbol{\omega}]_\times} &= \mathbf{I} + \theta[\hat{\mathbf{n}}]_\times + \frac{\theta^2}{2}[\hat{\mathbf{n}}]_\times^2 + \frac{\theta^3}{3!}[\hat{\mathbf{n}}]_\times^3 + \cdots \\
&= \mathbf{I} + (\theta - \frac{\theta^3}{3!} + \cdots)[\hat{\mathbf{n}}]_\times + (\frac{\theta^2}{2} - \frac{\theta^4}{4!} + \cdots)[\hat{\mathbf{n}}]_\times^2 \\
&= \mathbf{I} + \sin\theta[\hat{\mathbf{n}}]_\times + (1-\cos\theta)[\hat{\mathbf{n}}]_\times^2,
\end{aligned} \tag{2.38}$$

which yields the familiar Rodrigues' formula.

In robotics (and group theory), rotations are called SO(3), i.e., the *special orthogonal* group in 3D. The incremental rotations $\boldsymbol{\omega}$ are associated with a Lie algebra se(3) and are the preferred way to formulate rotation derivatives and to model uncertainties in rotation estimates (Blanco 2019; Solà, Deray, and Atchuthan 2019).

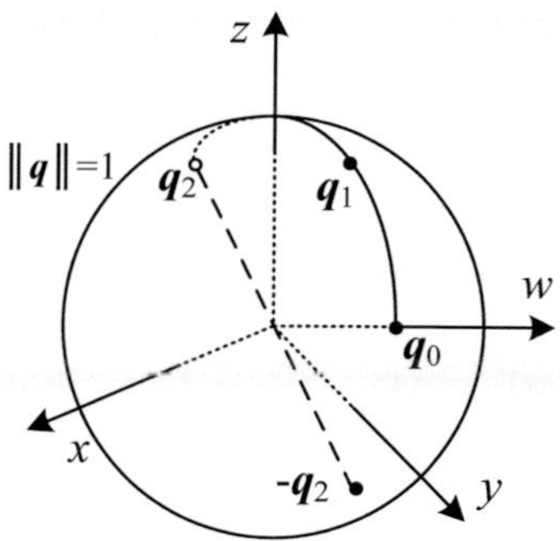

Figure 2.6 Unit quaternions live on the unit sphere $\|\mathbf{q}\| = 1$. This figure shows a smooth trajectory through the three quaternions $\mathbf{q}_0$, $\mathbf{q}_1$, and $\mathbf{q}_2$. The *antipodal* point to $\mathbf{q}_2$, namely $-\mathbf{q}_2$, represents the same rotation as $\mathbf{q}_2$.

Unit quaternions

The unit quaternion representation is closely related to the angle/axis representation. A unit quaternion is a unit length 4-vector whose components can be written as $\mathbf{q} = (q_x, q_y, q_z, q_w)$ or $\mathbf{q} = (x, y, z, w)$ for short. Unit quaternions live on the unit sphere $\|\mathbf{q}\| = 1$ and *antipodal* (opposite sign) quaternions, $\mathbf{q}$ and $-\mathbf{q}$, represent the same rotation (Figure 2.6). Other than this ambiguity (dual covering), the unit quaternion representation of a rotation is unique. Furthermore, the representation is *continuous*, i.e., as rotation matrices vary continuously, you can find a continuous quaternion representation, although the path on the quaternion sphere may wrap all the way around before returning to the "origin" $\mathbf{q}_o = (0, 0, 0, 1)$. For these and other reasons given below, quaternions are a very popular representation for pose and for pose interpolation in computer graphics (Shoemake 1985).

Quaternions can be derived from the axis/angle representation through the formula

$$\mathbf{q} = (\mathbf{v}, w) = (\sin\frac{\theta}{2}\hat{\mathbf{n}}, \cos\frac{\theta}{2}), \tag{2.39}$$

where $\hat{\mathbf{n}}$ and θ are the rotation axis and angle. Using the trigonometric identities $\sin\theta = 2\sin\frac{\theta}{2}\cos\frac{\theta}{2}$ and $(1 - \cos\theta) = 2\sin^2\frac{\theta}{2}$, Rodrigues' formula can be converted to

$$\begin{aligned} \mathbf{R}(\hat{\mathbf{n}}, \theta) &= \mathbf{I} + \sin\theta[\hat{\mathbf{n}}]_\times + (1 - \cos\theta)[\hat{\mathbf{n}}]^2_\times \\ &= \mathbf{I} + 2w[\mathbf{v}]_\times + 2[\mathbf{v}]^2_\times. \end{aligned} \tag{2.40}$$

This suggests a quick way to rotate a vector $\mathbf{v}$ by a quaternion using a series of cross products, scalings, and additions. To obtain a formula for $\mathbf{R}(\mathbf{q})$ as a function of (x, y, z, w), recall that

$$[\mathbf{v}]_\times = \begin{bmatrix} 0 & -z & y \\ z & 0 & -x \\ -y & x & 0 \end{bmatrix} \quad \text{and} \quad [\mathbf{v}]^2_\times = \begin{bmatrix} -y^2 - z^2 & xy & xz \\ xy & -x^2 - z^2 & yz \\ xz & yz & -x^2 - y^2 \end{bmatrix}.$$

We thus obtain

$$\mathbf{R}(\mathbf{q}) = \begin{bmatrix} 1 - 2(y^2 + z^2) & 2(xy - zw) & 2(xz + yw) \\ 2(xy + zw) & 1 - 2(x^2 + z^2) & 2(yz - xw) \\ 2(xz - yw) & 2(yz + xw) & 1 - 2(x^2 + y^2) \end{bmatrix}. \tag{2.41}$$

The diagonal terms can be made more symmetrical by replacing $1 - 2(y^2 + z^2)$ with $(x^2 + w^2 - y^2 - z^2)$, etc.

The nicest aspect of unit quaternions is that there is a simple algebra for composing rotations expressed as unit quaternions. Given two quaternions $\mathbf{q}_0 = (\mathbf{v}_0, w_0)$ and $\mathbf{q}_1 = (\mathbf{v}_1, w_1)$, the *quaternion multiply* operator is defined as

$$\mathbf{q}_2 = \mathbf{q}_0\mathbf{q}_1 = (\mathbf{v}_0 \times \mathbf{v}_1 + w_0\mathbf{v}_1 + w_1\mathbf{v}_0,\ w_0w_1 - \mathbf{v}_0 \cdot \mathbf{v}_1), \tag{2.42}$$

with the property that $\mathbf{R}(\mathbf{q}_2) = \mathbf{R}(\mathbf{q}_0)\mathbf{R}(\mathbf{q}_1)$. Note that quaternion multiplication is *not* commutative, just as 3D rotations and matrix multiplications are not.

Taking the inverse of a quaternion is easy: Just flip the sign of $\mathbf{v}$ or w (but not both!). (You can verify this has the desired effect of transposing the $\mathbf{R}$ matrix in (2.41).) Thus, we can also define *quaternion division* as

$$\mathbf{q}_2 = \mathbf{q}_0/\mathbf{q}_1 = \mathbf{q}_0\mathbf{q}_1^{-1} = (\mathbf{v}_0 \times \mathbf{v}_1 + w_0\mathbf{v}_1 - w_1\mathbf{v}_0,\ -w_0w_1 - \mathbf{v}_0 \cdot \mathbf{v}_1). \tag{2.43}$$

This is useful when the *incremental rotation* between two rotations is desired.

In particular, if we want to determine a rotation that is partway between two given rotations, we can compute the incremental rotation, take a fraction of the angle, and compute the new rotation. This procedure is called *spherical linear interpolation* or *slerp* for short (Shoemake 1985) and is given in Algorithm 2.1. Note that Shoemake presents two formulas other than the one given here. The first exponentiates $\mathbf{q}_r$ by alpha before multiplying the original quaternion,

$$\mathbf{q}_2 = \mathbf{q}_r^{\alpha}\mathbf{q}_0, \tag{2.44}$$

while the second treats the quaternions as 4-vectors on a sphere and uses

$$\mathbf{q}_2 = \frac{\sin(1-\alpha)\theta}{\sin\theta}\mathbf{q}_0 + \frac{\sin\alpha\theta}{\sin\theta}\mathbf{q}_1, \tag{2.45}$$

where $\theta = \cos^{-1}(\mathbf{q}_0 \cdot \mathbf{q}_1)$ and the dot product is directly between the quaternion 4-vectors. All of these formulas give comparable results, although care should be taken when $\mathbf{q}_0$ and $\mathbf{q}_1$ are close together, which is why I prefer to use an arctangent to establish the rotation angle.

Which rotation representation is better?

The choice of representation for 3D rotations depends partly on the application.

The axis/angle representation is minimal, and hence does not require any additional constraints on the parameters (no need to re-normalize after each update). If the angle is expressed in degrees, it is easier to understand the pose (say, 90° twist around x-axis), and also easier to express exact rotations. When the angle is in radians, the derivatives of $\mathbf{R}$ with respect to $\boldsymbol{\omega}$ can easily be computed (2.36).

Quaternions, on the other hand, are better if you want to keep track of a smoothly moving camera, since there are no discontinuities in the representation. It is also easier to interpolate between rotations and to chain rigid transformations (Murray, Li, and Sastry 1994; Bregler and Malik 1998).

My usual preference is to use quaternions, but to update their estimates using an incremental rotation, as described in Section 11.2.2.

procedure *slerp*$(\mathbf{q}_0, \mathbf{q}_1, \alpha)$:

1. $\mathbf{q}_r = \mathbf{q}_1/\mathbf{q}_0 = (\mathbf{v}_r, w_r)$
2. if $w_r < 0$ then $\mathbf{q}_r \leftarrow -\mathbf{q}_r$
3. $\theta_r = 2\tan^{-1}(\|\mathbf{v}_r\|/w_r)$
4. $\hat{\mathbf{n}}_r = \mathcal{N}(\mathbf{v}_r) = \mathbf{v}_r/\|\mathbf{v}_r\|$
5. $\theta_\alpha = \alpha\,\theta_r$
6. $\mathbf{q}_\alpha = (\sin\frac{\theta_\alpha}{2}\hat{\mathbf{n}}_r, \cos\frac{\theta_\alpha}{2})$
7. **return** $\mathbf{q}_2 = \mathbf{q}_\alpha \mathbf{q}_0$

Algorithm 2.1 Spherical linear interpolation (slerp). The axis and total angle are first computed from the quaternion ratio. (This computation can be lifted outside an inner loop that generates a set of interpolated position for animation.) An incremental quaternion is then computed and multiplied by the starting rotation quaternion.

2.1.4 3D to 2D projections

Now that we know how to represent 2D and 3D geometric primitives and how to transform them spatially, we need to specify how 3D primitives are projected onto the image plane. We can do this using a linear 3D to 2D projection matrix. The simplest model is orthography, which requires no division to get the final (inhomogeneous) result. The more commonly used model is perspective, since this more accurately models the behavior of real cameras.

Orthography and para-perspective

An orthographic projection simply drops the z component of the three-dimensional coordinate $\mathbf{p}$ to obtain the 2D point $\mathbf{x}$. (In this section, we use $\mathbf{p}$ to denote 3D points and $\mathbf{x}$ to denote 2D points.) This can be written as

$$\mathbf{x} = [\mathbf{I}_{2\times 2}|\mathbf{0}]\,\mathbf{p}. \tag{2.46}$$

If we are using homogeneous (projective) coordinates, we can write

$$\tilde{\mathbf{x}} = \begin{bmatrix} 1 & 0 & 0 & 0 \\ 0 & 1 & 0 & 0 \\ 0 & 0 & 0 & 1 \end{bmatrix} \tilde{\mathbf{p}}, \tag{2.47}$$

i.e., we drop the z component but keep the w component. Orthography is an approximate model for long focal length (telephoto) lenses and objects whose depth is *shallow* relative to their distance to the camera (Sawhney and Hanson 1991). It is exact only for *telecentric* lenses (Baker and Nayar 1999, 2001).

In practice, world coordinates (which may measure dimensions in meters) need to be scaled to fit onto an image sensor (physically measured in millimeters, but ultimately measured in pixels). For this reason, *scaled orthography* is actually more commonly used,

$$\mathbf{x} = [s\mathbf{I}_{2\times 2}|\mathbf{0}]\,\mathbf{p}. \tag{2.48}$$

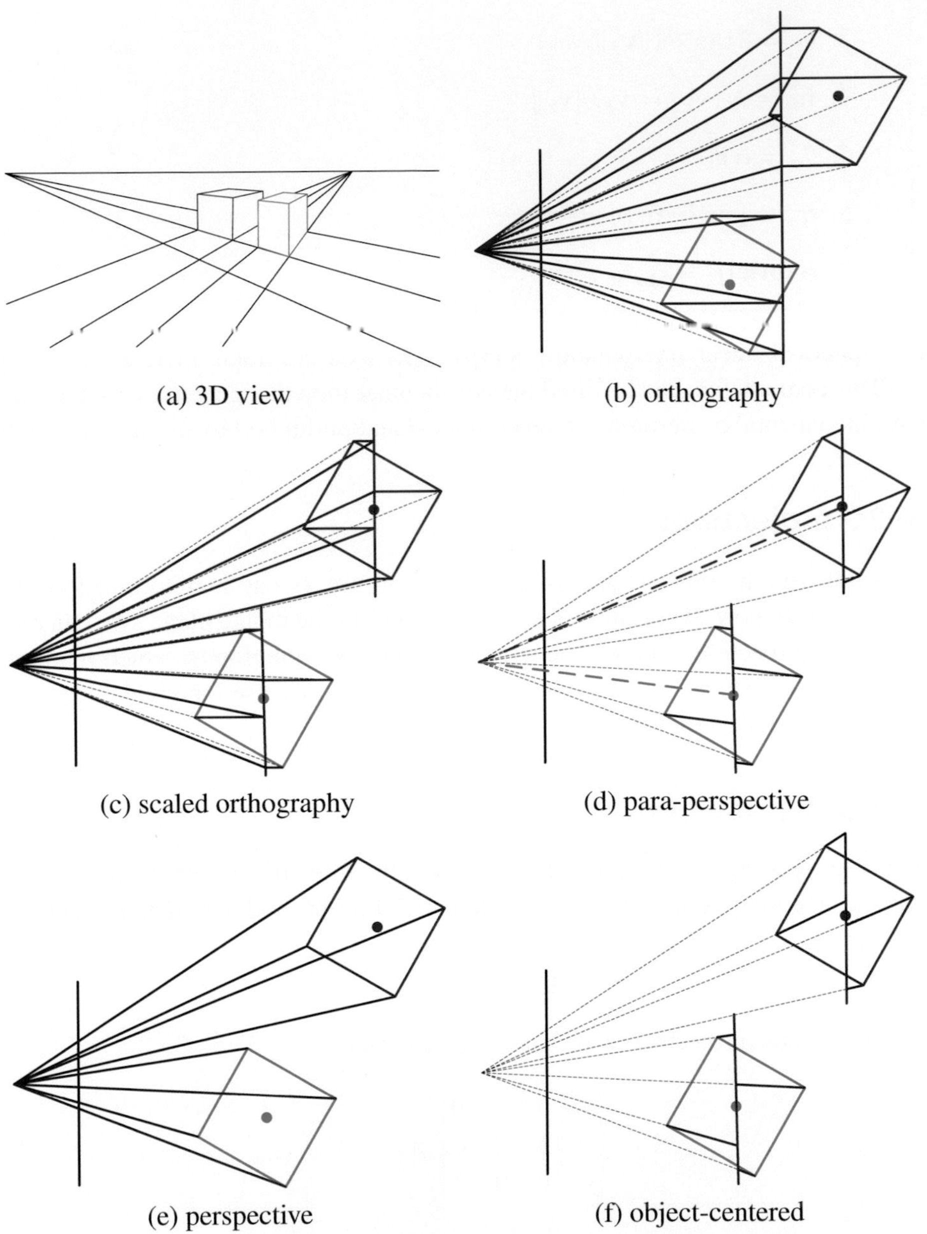

Figure 2.7 Commonly used projection models: (a) 3D view of world, (b) orthography, (c) scaled orthography, (d) para-perspective, (e) perspective, (f) object-centered. Each diagram shows a top-down view of the projection. Note how parallel lines on the ground plane and box sides remain parallel in the non-perspective projections.

This model is equivalent to first projecting the world points onto a local fronto-parallel image plane and then scaling this image using regular perspective projection. The scaling can be the same for all parts of the scene (Figure 2.7b) or it can be different for objects that are being modeled independently (Figure 2.7c). More importantly, the scaling can vary from frame to frame when estimating *structure from motion*, which can better model the scale change that occurs as an object approaches the camera.

Scaled orthography is a popular model for reconstructing the 3D shape of objects far away from the camera, since it greatly simplifies certain computations. For example, *pose* (camera orientation) can be estimated using simple least squares (Section 11.2.1). Under orthography, structure and motion can simultaneously be estimated using *factorization* (singular value decomposition), as discussed in Section 11.4.1 (Tomasi and Kanade 1992).

A closely related projection model is *para-perspective* (Aloimonos 1990; Poelman and Kanade 1997). In this model, object points are again first projected onto a local reference parallel to the image plane. However, rather than being projected orthogonally to this plane, they are projected *parallel* to the line of sight to the object center (Figure 2.7d). This is followed by the usual projection onto the final image plane, which again amounts to a scaling. The combination of these two projections is therefore *affine* and can be written as

$$\tilde{\mathbf{x}} = \begin{bmatrix} a_{00} & a_{01} & a_{02} & a_{03} \\ a_{10} & a_{11} & a_{12} & a_{13} \\ 0 & 0 & 0 & 1 \end{bmatrix} \tilde{\mathbf{p}}. \tag{2.49}$$

Note how parallel lines in 3D remain parallel after projection in Figure 2.7b–d. Para-perspective provides a more accurate projection model than scaled orthography, without incurring the added complexity of per-pixel perspective division, which invalidates traditional factorization methods (Poelman and Kanade 1997).

Perspective

The most commonly used projection in computer graphics and computer vision is true 3D *perspective* (Figure 2.7e). Here, points are projected onto the image plane by dividing them by their z component. Using inhomogeneous coordinates, this can be written as

$$\bar{\mathbf{x}} = \mathcal{P}_z(\mathbf{p}) = \begin{bmatrix} x/z \\ y/z \\ 1 \end{bmatrix}. \tag{2.50}$$

In homogeneous coordinates, the projection has a simple linear form,

$$\tilde{\mathbf{x}} = \begin{bmatrix} 1 & 0 & 0 & 0 \\ 0 & 1 & 0 & 0 \\ 0 & 0 & 1 & 0 \end{bmatrix} \tilde{\mathbf{p}}, \tag{2.51}$$

i.e., we drop the w component of $\mathbf{p}$. Thus, after projection, it is not possible to recover the *distance* of the 3D point from the image, which makes sense for a 2D imaging sensor.

A form often seen in computer graphics systems is a two-step projection that first projects 3D coordinates into *normalized device coordinates* $(x, y, z) \in [-1, 1] \times [-1, 1] \times [0, 1]$, and then rescales these coordinates to integer pixel coordinates using a *viewport* transformation (Watt 1995; OpenGL-

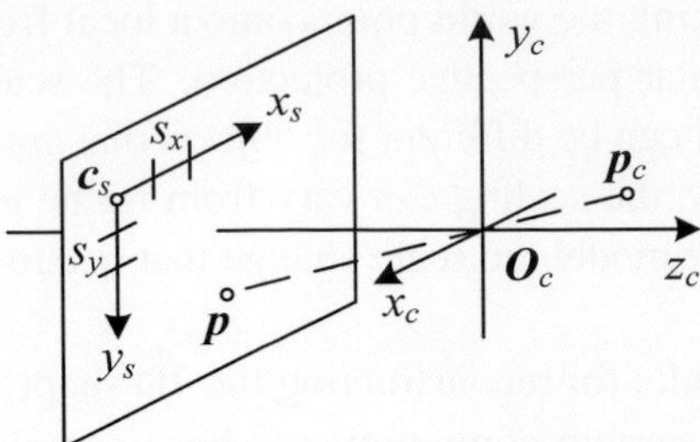

Figure 2.8 Projection of a 3D camera-centered point $\mathbf{p}_c$ onto the sensor planes at location $\mathbf{p}$. $\mathbf{O}_c$ is the optical center (nodal point), $\mathbf{c}_s$ is the 3D origin of the sensor plane coordinate system, and s_x and s_y are the pixel spacings.

ARB 1997). The (initial) perspective projection is then represented using a 4×4 matrix

$$\tilde{\mathbf{x}} = \begin{bmatrix} 1 & 0 & 0 & 0 \\ 0 & 1 & 0 & 0 \\ 0 & 0 & -z_{\text{far}}/z_{\text{range}} & z_{\text{near}} z_{\text{far}}/z_{\text{range}} \\ 0 & 0 & 1 & 0 \end{bmatrix} \tilde{\mathbf{p}}, \tag{2.52}$$

where z_{near} and z_{far} are the near and far z *clipping planes* and $z_{\text{range}} = z_{\text{far}} - z_{\text{near}}$. Note that the first two rows are actually scaled by the focal length and the aspect ratio so that visible rays are mapped to $(x, y, z) \in [-1, 1]^2$. The reason for keeping the third row, rather than dropping it, is that visibility operations, such as *z-buffering*, require a depth for every graphical element that is being rendered.

If we set $z_{\text{near}} = 1$, $z_{\text{far}} \to \infty$, and switch the sign of the third row, the third element of the normalized screen vector becomes the inverse depth, i.e., the *disparity* (Okutomi and Kanade 1993). This can be quite convenient in many cases since, for cameras moving around outdoors, the inverse depth to the camera is often a more well-conditioned parameterization than direct 3D distance.

While a regular 2D image sensor has no way of measuring distance to a surface point, *range sensors* (Section 13.2) and stereo matching algorithms (Chapter 12) can compute such values. It is then convenient to be able to map from a sensor-based depth or disparity value d directly back to a 3D location using the inverse of a 4×4 matrix (Section 2.1.4). We can do this if we represent perspective projection using a full-rank 4×4 matrix, as in (2.64).

Camera intrinsics

Once we have projected a 3D point through an ideal pinhole using a projection matrix, we must still transform the resulting coordinates according to the pixel sensor spacing and the relative position of the sensor plane to the origin. Figure 2.8 shows an illustration of the geometry involved. In this section, we first present a mapping from 2D pixel coordinates to 3D rays using a sensor homography $\mathbf{M}_s$, since this is easier to explain in terms of physically measurable quantities. We then relate these quantities to the more commonly used camera intrinsic matrix $\mathbf{K}$, which is used to map 3D camera-centered points $\mathbf{p}_c$ to 2D pixel coordinates $\tilde{\mathbf{x}}_s$.

Image sensors return pixel values indexed by integer *pixel coordinates* (x_s, y_s), often with the coordinates starting at the upper-left corner of the image and moving down and to the right. (This convention is not obeyed by all imaging libraries, but the adjustment for other coordinate systems is straightforward.) To map pixel centers to 3D coordinates, we first scale the (x_s, y_s) values by the pixel spacings (s_x, s_y) (sometimes expressed in microns for solid-state sensors) and then describe the orientation of the sensor array relative to the camera projection center $\mathbf{O}_c$ with an origin $\mathbf{c}_s$ and a 3D rotation $\mathbf{R}_s$ (Figure 2.8).

The combined 2D to 3D projection can then be written as

$$\mathbf{p} = \begin{bmatrix} \mathbf{R}_s & \mathbf{c}_s \end{bmatrix} \begin{bmatrix} s_x & 0 & 0 \\ 0 & s_y & 0 \\ 0 & 0 & 0 \\ 0 & 0 & 1 \end{bmatrix} \begin{bmatrix} x_s \\ y_s \\ 1 \end{bmatrix} = \mathbf{M}_s \bar{\mathbf{x}}_s. \tag{2.53}$$

The first two columns of the 3×3 matrix $\mathbf{M}_s$ are the 3D vectors corresponding to unit steps in the image pixel array along the x_s and y_s directions, while the third column is the 3D image array origin $\mathbf{c}_s$.

The matrix $\mathbf{M}_s$ is parameterized by eight unknowns: the three parameters describing the rotation $\mathbf{R}_s$, the three parameters describing the translation $\mathbf{c}_s$, and the two scale factors (s_x, s_y). Note that we ignore here the possibility of *skew* between the two axes on the image plane, since solid-state manufacturing techniques render this negligible. In practice, unless we have accurate external knowledge of the sensor spacing or sensor orientation, there are only seven degrees of freedom, since the distance of the sensor from the origin cannot be teased apart from the sensor spacing, based on external image measurement alone.

However, estimating a camera model $\mathbf{M}_s$ with the required seven degrees of freedom (i.e., where the first two columns are orthogonal after an appropriate re-scaling) is impractical, so most practitioners assume a general 3×3 homogeneous matrix form.

The relationship between the 3D pixel center $\mathbf{p}$ and the 3D camera-centered point $\mathbf{p}_c$ is given by an unknown scaling s, $\mathbf{p} = s\mathbf{p}_c$. We can therefore write the complete projection between $\mathbf{p}_c$ and a homogeneous version of the pixel address $\tilde{\mathbf{x}}_s$ as

$$\tilde{\mathbf{x}}_s = \alpha \mathbf{M}_s^{-1} \mathbf{p}_c = \mathbf{K}\mathbf{p}_c. \tag{2.54}$$

The 3×3 matrix $\mathbf{K}$ is called the *calibration matrix* and describes the camera *intrinsics* (as opposed to the camera's orientation in space, which are called the *extrinsics*).

From the above discussion, we see that $\mathbf{K}$ has seven degrees of freedom in theory and eight degrees of freedom (the full dimensionality of a 3×3 homogeneous matrix) in practice. Why, then, do most textbooks on 3D computer vision and multi-view geometry (Faugeras 1993; Hartley and Zisserman 2004; Faugeras and Luong 2001) treat $\mathbf{K}$ as an upper-triangular matrix with five degrees of freedom?

While this is usually not made explicit in these books, it is because we cannot recover the full $\mathbf{K}$ matrix based on external measurement alone. When calibrating a camera (Section 11.1) based on external 3D points or other measurements (Tsai 1987), we end up estimating the intrinsic ($\mathbf{K}$) and extrinsic $(\mathbf{R}, \mathbf{t})$ camera parameters simultaneously using a series of measurements,

$$\tilde{\mathbf{x}}_s = \mathbf{K} \begin{bmatrix} \mathbf{R} & \mathbf{t} \end{bmatrix} \mathbf{p}_w = \mathbf{P}\mathbf{p}_w, \tag{2.55}$$

where $\mathbf{p}_w$ are known 3D world coordinates and

$$\mathbf{P} = \mathbf{K}[\mathbf{R}|\mathbf{t}] \tag{2.56}$$

is known as the *camera matrix*. Inspecting this equation, we see that we can post-multiply $\mathbf{K}$ by $\mathbf{R}_1$ and pre-multiply $[\mathbf{R}|\mathbf{t}]$ by $\mathbf{R}_1^T$, and still end up with a valid calibration. Thus, it is impossible based on image measurements alone to know the true orientation of the sensor and the true camera intrinsics.

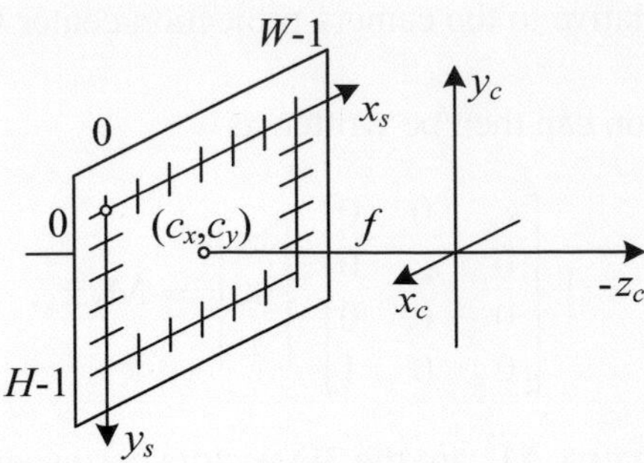

Figure 2.9 Simplified camera intrinsics showing the focal length f and the image center (c_x, c_y). The image width and height are W and H.

The choice of an upper-triangular form for $\mathbf{K}$ seems to be conventional. Given a full 3×4 camera matrix $\mathbf{P} = \mathbf{K}[\mathbf{R}|\mathbf{t}]$, we can compute an upper-triangular $\mathbf{K}$ matrix using QR factorization (Golub and Van Loan 1996). (Note the unfortunate clash of terminologies: In matrix algebra textbooks, $\mathbf{R}$ represents an upper-triangular (right of the diagonal) matrix; in computer vision, $\mathbf{R}$ is an orthogonal rotation.)

There are several ways to write the upper-triangular form of $\mathbf{K}$. One possibility is

$$\mathbf{K} = \begin{bmatrix} f_x & s & c_x \\ 0 & f_y & c_y \\ 0 & 0 & 1 \end{bmatrix}, \tag{2.57}$$

which uses independent *focal lengths* f_x and f_y for the sensor x and y dimensions. The entry s encodes any possible *skew* between the sensor axes due to the sensor not being mounted perpendicular to the optical axis and (c_x, c_y) denotes the *image center* expressed in pixel coordinates. The image center is also often called the *principal point* in the computer vision literature (Hartley and Zisserman 2004), although in optics, the principal points are 3D points usually inside the lens where the principal planes intersect the principal (optical) axis (Hecht 2015). Another possibility is

$$\mathbf{K} = \begin{bmatrix} f & s & c_x \\ 0 & af & c_y \\ 0 & 0 & 1 \end{bmatrix}, \tag{2.58}$$

where the *aspect ratio* a has been made explicit and a common focal length f is used.

In practice, for many applications an even simpler form can be obtained by setting $a = 1$ and $s = 0$,

$$\mathbf{K} = \begin{bmatrix} f & 0 & c_x \\ 0 & f & c_y \\ 0 & 0 & 1 \end{bmatrix}. \tag{2.59}$$

Often, setting the origin at roughly the center of the image, e.g., $(c_x, c_y) = (W/2, H/2)$, where W and H are the image width and height, respectively, can result in a perfectly usable camera model with a single unknown, i.e., the focal length f.

Figure 2.9 shows how these quantities can be visualized as part of a simplified imaging model. Note that now we have placed the image plane *in front* of the nodal point (projection center of the lens). The sense of the y-axis has also been flipped to get a coordinate system compatible with the way that most imaging libraries treat the vertical (row) coordinate.

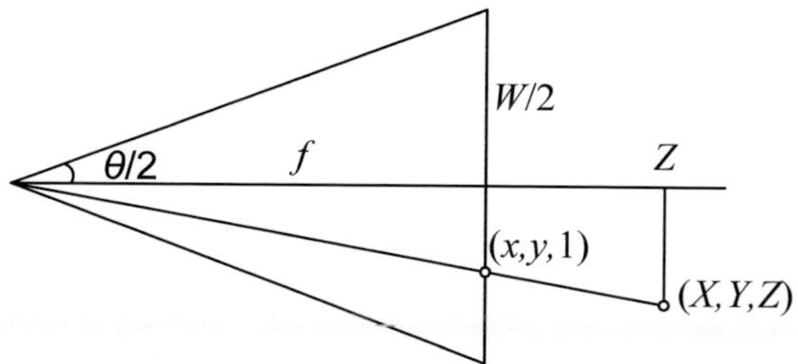

Figure 2.10 Central projection, showing the relationship between the 3D and 2D coordinates, $\mathbf{p}$ and $\mathbf{x}$, as well as the relationship between the focal length f, image width W, and the horizontal field of view θ_{H}.

A note on focal lengths

The issue of how to express focal lengths is one that often causes confusion in implementing computer vision algorithms and discussing their results. This is because the focal length depends on the units used to measure pixels.

If we number pixel coordinates using integer values, say $[0, W) \times [0, H)$, the focal length f and camera center (c_x, c_y) in (2.59) can be expressed as pixel values. How do these quantities relate to the more familiar focal lengths used by photographers?

Figure 2.10 illustrates the relationship between the focal length f, the sensor width W, and the horizontal field of view θ_{H}, which obey the formula

$$\tan \frac{\theta_{\mathrm{H}}}{2} = \frac{W}{2f} \quad \text{or} \quad f = \frac{W}{2}\left[\tan \frac{\theta_{\mathrm{H}}}{2}\right]^{-1}. \tag{2.60}$$

For a traditional 35mm film camera, whose active exposure area is 24mm $\times$ 36mm, we have $W =$ 36mm, and hence f is also expressed in millimeters.[4] For example, the "stock" lens that often comes with SLR (single lens reflex) cameras is 50mm, which is a good length, whereas 85mm is the standard for portrait photography. Since we work with digital images, however, it is more convenient to express W in pixels so that the focal length f can be used directly in the calibration matrix $\mathbf{K}$ as in (2.59).

Another possibility is to scale the pixel coordinates so that they go from $[-1, 1)$ along the longer image dimension and $[-a^{-1}, a^{-1})$ along the shorter axis, where $a \geq 1$ is the *image aspect ratio* (as opposed to the *sensor cell aspect ratio* introduced earlier). This can be accomplished using *modified normalized device coordinates,*

$$x'_s = (2x_s - W)/S \quad \text{and} \quad y'_s = (2y_s - H)/S, \qquad \text{where} \qquad S = \max(W, H). \tag{2.61}$$

This has the advantage that the focal length f and image center (c_x, c_y) become independent of the image resolution, which can be useful when using multi-resolution, image-processing algorithms, such as image pyramids (Section 3.5).[5] The use of S instead of W also makes the focal length the same for landscape (horizontal) and portrait (vertical) pictures, as is the case in 35mm photography. (In some computer graphics textbooks and systems, normalized device coordinates go from $[-1, 1] \times [-1, 1]$, which requires the use of two different focal lengths to describe the camera intrinsics (Watt

[4] 35mm denotes the width of the film strip, of which 24mm is used for exposing each frame and the remaining 11mm for perforation and frame numbering.

[5] To make the conversion truly accurate after a downsampling step in a pyramid, floating point values of W and H would have to be maintained, as they can become non-integer if they are ever odd at a larger resolution in the pyramid.

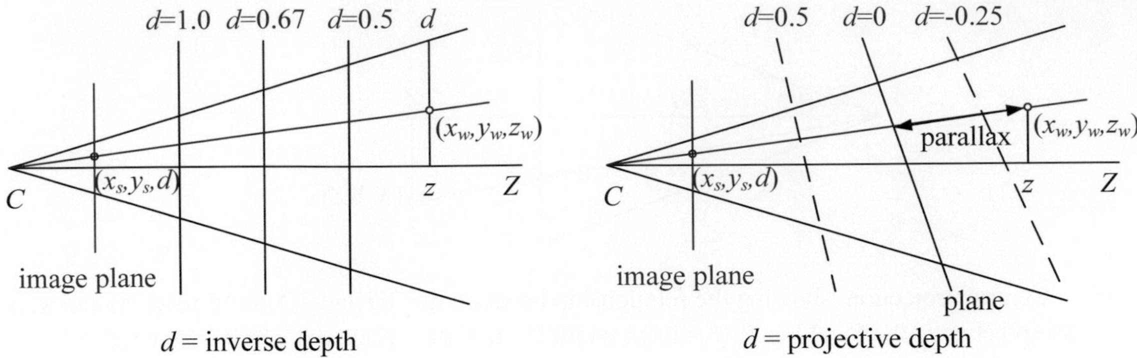

Figure 2.11 Regular disparity (inverse depth) and projective depth (parallax from a reference plane).

1995).) Setting $S = W = 2$ in (2.60), we obtain the simpler (unitless) relationship

$$f^{-1} = \tan \frac{\theta_{\rm H}}{2}. \tag{2.62}$$

The conversion between the various focal length representations is straightforward, e.g., to go from a unitless f to one expressed in pixels, multiply by $W/2$, while to convert from an f expressed in pixels to the equivalent 35mm focal length, multiply by 18mm.

Camera matrix

Now that we have shown how to parameterize the calibration matrix $\mathbf{K}$, we can put the camera intrinsics and extrinsics together to obtain a single 3×4 *camera matrix*

$$\mathbf{P} = \mathbf{K} \begin{bmatrix} \mathbf{R} & \mathbf{t} \end{bmatrix}. \tag{2.63}$$

It is sometimes preferable to use an invertible 4×4 matrix, which can be obtained by not dropping the last row in the $\mathbf{P}$ matrix,

$$\tilde{\mathbf{P}} = \begin{bmatrix} \mathbf{K} & \mathbf{0} \\ \mathbf{0}^T & 1 \end{bmatrix} \begin{bmatrix} \mathbf{R} & \mathbf{t} \\ \mathbf{0}^T & 1 \end{bmatrix} = \tilde{\mathbf{K}}\mathbf{E}, \tag{2.64}$$

where $\mathbf{E}$ is a 3D rigid-body (Euclidean) transformation and $\tilde{\mathbf{K}}$ is the full-rank calibration matrix. The 4×4 camera matrix $\tilde{\mathbf{P}}$ can be used to map directly from 3D world coordinates $\bar{\mathbf{p}}_w = (x_w, y_w, z_w, 1)$ to screen coordinates (plus disparity), $\mathbf{x}_s = (x_s, y_s, 1, d)$,

$$\mathbf{x}_s \sim \tilde{\mathbf{P}}\bar{\mathbf{p}}_w, \tag{2.65}$$

where $\sim$ indicates equality up to scale. Note that after multiplication by $\tilde{\mathbf{P}}$, the vector is divided by the *third* element of the vector to obtain the normalized form $\mathbf{x}_s = (x_s, y_s, 1, d)$.

Plane plus parallax (projective depth)

In general, when using the 4×4 matrix $\tilde{\mathbf{P}}$, we have the freedom to remap the last row to whatever suits our purpose (rather than just being the "standard" interpretation of disparity as inverse depth). Let us re-write the last row of $\tilde{\mathbf{P}}$ as $\mathbf{p}_3 = s_3[\hat{\mathbf{n}}_0|c_0]$, where $\|\hat{\mathbf{n}}_0\| = 1$. We then have the equation

$$d = \frac{s_3}{z}(\hat{\mathbf{n}}_0 \cdot \mathbf{p}_w + c_0), \tag{2.66}$$

where $z = \mathbf{p}_2 \cdot \bar{\mathbf{p}}_w = \mathbf{r}_z \cdot (\mathbf{p}_w - \mathbf{c})$ is the distance of $\mathbf{p}_w$ from the camera center C (2.25) along the optical axis Z (Figure 2.11). Thus, we can interpret d as the *projective disparity* or *projective depth* of a 3D scene point $\mathbf{p}_w$ from the *reference plane* $\hat{\mathbf{n}}_0 \cdot \mathbf{p}_w + c_0 = 0$ (Szeliski and Coughlan 1997; Szeliski and Golland 1999; Shade, Gortler *et al.* 1998; Baker, Szeliski, and Anandan 1998). (The projective depth is also sometimes called *parallax* in reconstruction algorithms that use the term *plane plus parallax* (Kumar, Anandan, and Hanna 1994; Sawhney 1994).) Setting $\hat{\mathbf{n}}_0 = \mathbf{0}$ and $c_0 = 1$, i.e., putting the reference plane at infinity, results in the more standard $d = 1/z$ version of disparity (Okutomi and Kanade 1993).

Another way to see this is to invert the $\tilde{\mathbf{P}}$ matrix so that we can map pixels plus disparity directly back to 3D points,

$$\tilde{\mathbf{p}}_w = \tilde{\mathbf{P}}^{-1}\mathbf{x}_s. \tag{2.67}$$

In general, we can choose $\tilde{\mathbf{P}}$ to have whatever form is convenient, i.e., to sample space using an arbitrary projection. This can come in particularly handy when setting up multi-view stereo reconstruction algorithms, since it allows us to sweep a series of planes (Section 12.1.2) through space with a variable (projective) sampling that best matches the sensed image motions (Collins 1996; Szeliski and Golland 1999; Saito and Kanade 1999).

Mapping from one camera to another

What happens when we take two images of a 3D scene from different camera positions or orientations (Figure 2.12a)? Using the full rank 4×4 camera matrix $\tilde{\mathbf{P}} = \tilde{\mathbf{K}}\mathbf{E}$ from (2.64), we can write the projection from world to screen coordinates as

$$\tilde{\mathbf{x}}_0 \sim \tilde{\mathbf{K}}_0\mathbf{E}_0\mathbf{p} = \tilde{\mathbf{P}}_0\mathbf{p}. \tag{2.68}$$

Assuming that we know the z-buffer or disparity value d_0 for a pixel in one image, we can compute the 3D point location $\mathbf{p}$ using

$$\mathbf{p} \sim \mathbf{E}_0^{-1}\tilde{\mathbf{K}}_0^{-1}\tilde{\mathbf{x}}_0 \tag{2.69}$$

and then project it into another image yielding

$$\tilde{\mathbf{x}}_1 \sim \tilde{\mathbf{K}}_1\mathbf{E}_1\mathbf{p} = \tilde{\mathbf{K}}_1\mathbf{E}_1\mathbf{E}_0^{-1}\tilde{\mathbf{K}}_0^{-1}\tilde{\mathbf{x}}_0 = \tilde{\mathbf{P}}_1\tilde{\mathbf{P}}_0^{-1}\tilde{\mathbf{x}}_0 = \mathbf{M}_{10}\tilde{\mathbf{x}}_0. \tag{2.70}$$

Unfortunately, we do not usually have access to the depth coordinates of pixels in a regular photographic image. However, for a *planar scene*, as discussed above in (2.66), we can replace the last row of $\mathbf{P}_0$ in (2.64) with a general *plane equation*, $\hat{\mathbf{n}}_0 \cdot \mathbf{p} + c_0$, that maps points on the plane to $d_0 = 0$ values (Figure 2.12b). Thus, if we set $d_0 = 0$, we can ignore the last column of $\mathbf{M}_{10}$ in (2.70) and also its last row, since we do not care about the final z-buffer depth. The mapping Equation (2.70) thus reduces to

$$\tilde{\mathbf{x}}_1 \sim \tilde{\mathbf{H}}_{10}\tilde{\mathbf{x}}_0, \tag{2.71}$$

where $\tilde{\mathbf{H}}_{10}$ is a general 3×3 homography matrix and $\tilde{\mathbf{x}}_1$ and $\tilde{\mathbf{x}}_0$ are now 2D homogeneous coordinates (i.e., 3-vectors) (Szeliski 1996). This justifies the use of the 8-parameter homography as a general alignment model for mosaics of planar scenes (Mann and Picard 1994; Szeliski 1996).

The other special case where we do not need to know depth to perform inter-camera mapping is when the camera is undergoing pure rotation (Section 8.2.3), i.e., when $\mathbf{t}_0 = \mathbf{t}_1$. In this case, we can write

$$\tilde{\mathbf{x}}_1 \sim \mathbf{K}_1\mathbf{R}_1\mathbf{R}_0^{-1}\mathbf{K}_0^{-1}\tilde{\mathbf{x}}_0 = \mathbf{K}_1\mathbf{R}_{10}\mathbf{K}_0^{-1}\tilde{\mathbf{x}}_0, \tag{2.72}$$

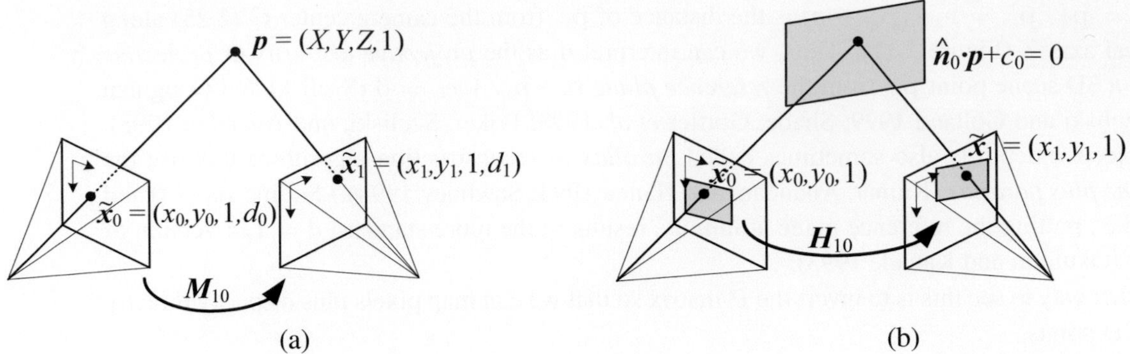

Figure 2.12 A point is projected into two images: (a) relationship between the 3D point coordinate $(X, Y, Z, 1)$ and the 2D projected point $(x, y, 1, d)$; (b) planar homography induced by points all lying on a common plane $\hat{\mathbf{n}}_0 \cdot \mathbf{p} + c_0 = 0$.

which again can be represented with a 3×3 homography. If we assume that the calibration matrices have known aspect ratios and centers of projection (2.59), this homography can be parameterized by the rotation amount and the two unknown focal lengths. This particular formulation is commonly used in image-stitching applications (Section 8.2.3).

Object-centered projection

When working with long focal length lenses, it often becomes difficult to reliably estimate the focal length from image measurements alone. This is because the focal length and the distance to the object are highly correlated and it becomes difficult to tease these two effects apart. For example, the change in scale of an object viewed through a zoom telephoto lens can either be due to a zoom change or to a motion towards the user. (This effect was put to dramatic use in some scenes of Alfred Hitchcock's film *Vertigo*, where the simultaneous change of zoom and camera motion produces a disquieting effect.)

This ambiguity becomes clearer if we write out the projection equation corresponding to the simple calibration matrix $\mathbf{K}$ (2.59),

$$x_s = f\frac{\mathbf{r}_x \cdot \mathbf{p} + t_x}{\mathbf{r}_z \cdot \mathbf{p} + t_z} + c_x \tag{2.73}$$

$$y_s = f\frac{\mathbf{r}_y \cdot \mathbf{p} + t_y}{\mathbf{r}_z \cdot \mathbf{p} + t_z} + c_y, \tag{2.74}$$

where $\mathbf{r}_x$, $\mathbf{r}_y$, and $\mathbf{r}_z$ are the three rows of $\mathbf{R}$. If the distance to the object center $t_z \gg \|\mathbf{p}\|$ (the size of the object), the denominator is approximately t_z and the overall scale of the projected object depends on the ratio of f to t_z. It therefore becomes difficult to disentangle these two quantities.

To see this more clearly, let $\eta_z = t_z^{-1}$ and $s = \eta_z f$. We can then re-write the above equations as

$$x_s = s\frac{\mathbf{r}_x \cdot \mathbf{p} + t_x}{1 + \eta_z \mathbf{r}_z \cdot \mathbf{p}} + c_x \tag{2.75}$$

$$y_s = s\frac{\mathbf{r}_y \cdot \mathbf{p} + t_y}{1 + \eta_z \mathbf{r}_z \cdot \mathbf{p}} + c_y \tag{2.76}$$

(Szeliski and Kang 1994; Pighin, Hecker *et al.* 1998). The scale of the projection s can be reliably estimated if we are looking at a known object (i.e., the 3D coordinates $\mathbf{p}$ are known). The inverse distance η_z is now mostly decoupled from the estimates of s and can be estimated from the amount of *foreshortening* as the object rotates. Furthermore, as the lens becomes longer, i.e., the projection model becomes orthographic, there is no need to replace a perspective imaging model with an orthographic one, since the same equation can be used, with $\eta_z \to 0$ (as opposed to f and t_z both going to infinity). This allows us to form a natural link between orthographic reconstruction techniques such as factorization and their projective/perspective counterparts (Section 11.4.1).

2.1.5 Lens distortions

The above imaging models all assume that cameras obey a *linear* projection model where straight lines in the world result in straight lines in the image. (This follows as a natural consequence of linear matrix operations being applied to homogeneous coordinates.) Unfortunately, many wide-angle lenses have noticeable *radial distortion*, which manifests itself as a visible curvature in the projection of straight lines. (See Section 2.2.3 for a more detailed discussion of lens optics, including chromatic aberration.) Unless this distortion is taken into account, it becomes impossible to create highly accurate photorealistic reconstructions. For example, image mosaics constructed without taking radial distortion into account will often exhibit blurring due to the misregistration of corresponding features before pixel blending (Section 8.2).

Fortunately, compensating for radial distortion is not that difficult in practice. For most lenses, a simple quartic model of distortion can produce good results. Let (x_c, y_c) be the pixel coordinates obtained *after* perspective division but *before* scaling by focal length f and shifting by the image center (c_x, c_y), i.e.,

$$
\begin{aligned}
x_c &= \frac{\mathbf{r}_x \cdot \mathbf{p} + t_x}{\mathbf{r}_z \cdot \mathbf{p} + t_z} \\
y_c &= \frac{\mathbf{r}_y \cdot \mathbf{p} + t_y}{\mathbf{r}_z \cdot \mathbf{p} + t_z}.
\end{aligned} \tag{2.77}
$$

The radial distortion model says that coordinates in the observed images are displaced towards (*barrel* distortion) or away (*pincushion* distortion) from the image center by an amount proportional to their radial distance (Figure 2.13a–b).[6] The simplest radial distortion models use low-order polynomials, e.g.,

$$
\begin{aligned}
\hat{x}_c &= x_c(1 + \kappa_1 r_c^2 + \kappa_2 r_c^4) \\
\hat{y}_c &= y_c(1 + \kappa_1 r_c^2 + \kappa_2 r_c^4),
\end{aligned} \tag{2.78}
$$

where $r_c^2 = x_c^2 + y_c^2$ and κ_1 and κ_2 are called the *radial distortion parameters*.[7] This model, which also includes a *tangential* component to account for lens decentering, was first proposed in the photogrammetry literature by Brown (1966), and so is sometimes called the *Brown* or *Brown–Conrady* model. However, the tangential components of the distortion are usually ignored because they can lead to less stable estimates (Zhang 2000).

After the radial distortion step, the final pixel coordinates can be computed using

$$
\begin{aligned}
x_s &= f\hat{x}_c + c_x \\
y_s &= f\hat{y}_c + c_y.
\end{aligned} \tag{2.79}
$$

[6] Anamorphic lenses, which are widely used in feature film production, do not follow this radial distortion model. Instead, they can be thought of, to a first approximation, as inducing different vertical and horizontal scaling, i.e., non-square pixels.

[7] Sometimes the relationship between x_c and $\hat{x}_c$ is expressed the other way around, i.e., $x_c = \hat{x}_c(1 + \kappa_1 \hat{r}_c^2 + \kappa_2 \hat{r}_c^4)$. This is convenient if we map image pixels into (warped) rays by dividing through by f. We can then undistort the rays and have true 3D rays in space.

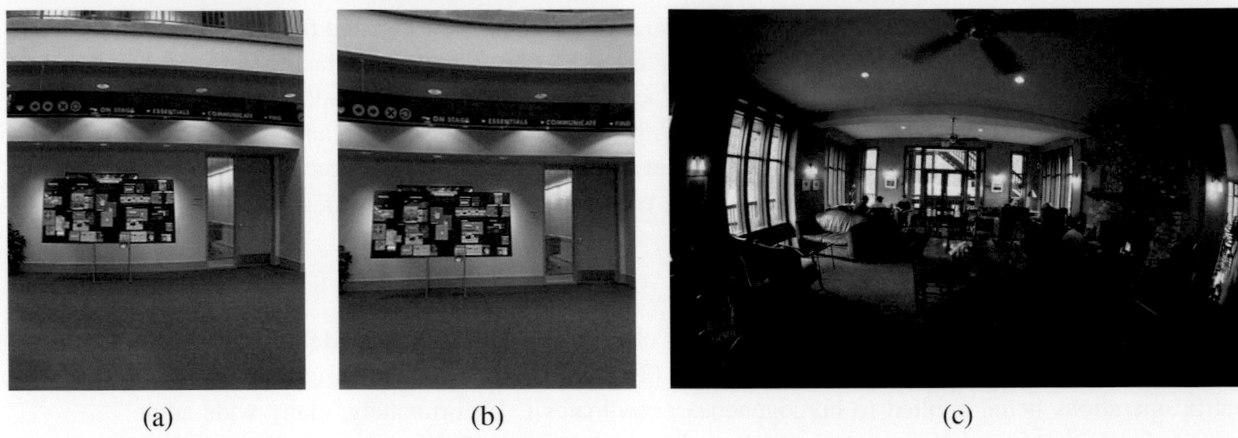

(a) (b) (c)

Figure 2.13 Radial lens distortions: (a) barrel, (b) pincushion, and (c) fisheye. The fisheye image spans almost 180° from side-to-side.

A variety of techniques can be used to estimate the radial distortion parameters for a given lens, as discussed in Section 11.1.4.

Sometimes the above simplified model does not model the true distortions produced by complex lenses accurately enough (especially at very wide angles). A more complete analytic model also includes *tangential distortions* and *decentering distortions* (Slama 1980).

Fisheye lenses (Figure 2.13c) require a model that differs from traditional polynomial models of radial distortion. Fisheye lenses behave, to a first approximation, as *equi-distance* projectors of angles away from the optical axis (Xiong and Turkowski 1997),

$$r = f\theta, \tag{2.80}$$

which is the same as the *polar projection* described by Equations (8.55–8.57). Because of the mostly linear mapping between distance from the center (pixels) and viewing angle, such lenses are sometimes called *f-theta lenses*, which is likely where the popular RICOH THETA 360° camera got its name. Xiong and Turkowski (1997) describe how this model can be extended with the addition of an extra quadratic correction in ϕ and how the unknown parameters (center of projection, scaling factor s, etc.) can be estimated from a set of overlapping fisheye images using a direct (intensity-based) non-linear minimization algorithm.

For even larger, less regular distortions, a parametric distortion model using splines may be necessary (Goshtasby 1989). If the lens does not have a single center of projection, it may become necessary to model the 3D *line* (as opposed to *direction*) corresponding to each pixel separately (Gremban, Thorpe, and Kanade 1988; Champleboux, Lavallée *et al.* 1992a; Grossberg and Nayar 2001; Sturm and Ramalingam 2004; Tardif, Sturm *et al.* 2009). Some of these techniques are described in more detail in Section 11.1.4, which discusses how to calibrate lens distortions.

There is one subtle issue associated with the simple radial distortion model that is often glossed over. We have introduced a non-linearity between the perspective projection and final sensor array projection steps. Therefore, we cannot, in general, post-multiply an arbitrary 3×3 matrix $\mathbf{K}$ with a rotation to put it into upper-triangular form and absorb this into the global rotation. However, this situation is not as bad as it may at first appear. For many applications, keeping the simplified diagonal form of (2.59) is still an adequate model. Furthermore, if we correct radial and other distortions to an accuracy where straight lines are preserved, we have essentially converted the sensor back into a linear imager and the previous decomposition still applies.

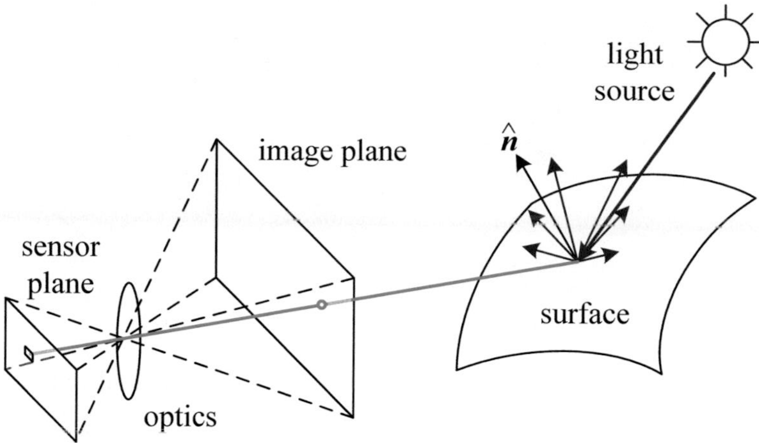

Figure 2.14 A simplified model of photometric image formation. Light is emitted by one or more light sources and is then reflected from an object's surface. A portion of this light is directed towards the camera. This simplified model ignores multiple reflections, which often occur in real-world scenes.

2.2 Photometric image formation

In modeling the image formation process, we have described how 3D geometric features in the world are projected into 2D features in an image. However, images are not composed of 2D features. Instead, they are made up of discrete color or intensity values. Where do these values come from? How do they relate to the lighting in the environment, surface properties and geometry, camera optics, and sensor properties (Figure 2.14)? In this section, we develop a set of models to describe these interactions and formulate a generative process of image formation. A more detailed treatment of these topics can be found in textbooks on computer graphics and image synthesis (Cohen and Wallace 1993; Sillion and Puech 1994; Watt 1995; Glassner 1995; Weyrich, Lawrence *et al.* 2009; Hughes, van Dam *et al.* 2013; Marschner and Shirley 2015).

2.2.1 Lighting

Images cannot exist without light. To produce an image, the scene must be illuminated with one or more light sources. (Certain modalities such as fluorescence microscopy and X-ray tomography do not fit this model, but we do not deal with them in this book.) Light sources can generally be divided into point and area light sources.

A point light source originates at a single location in space (e.g., a small light bulb), potentially at infinity (e.g., the Sun). (Note that for some applications such as modeling soft shadows (*penumbras*), the Sun may have to be treated as an area light source.) In addition to its location, a point light source has an intensity and a color spectrum, i.e., a distribution over wavelengths $L(\lambda)$. The intensity of a light source falls off with the square of the distance between the source and the object being lit, because the same light is being spread over a larger (spherical) area. A light source may also have a directional falloff (dependence), but we ignore this in our simplified model.

Area light sources are more complicated. A simple area light source such as a fluorescent ceiling light fixture with a diffuser can be modeled as a finite rectangular area emitting light equally in all directions (Cohen and Wallace 1993; Sillion and Puech 1994; Glassner 1995). When the distribution is strongly directional, a four-dimensional lightfield can be used instead (Ashdown 1993).

A more complex light distribution that approximates, say, the incident illumination on an object

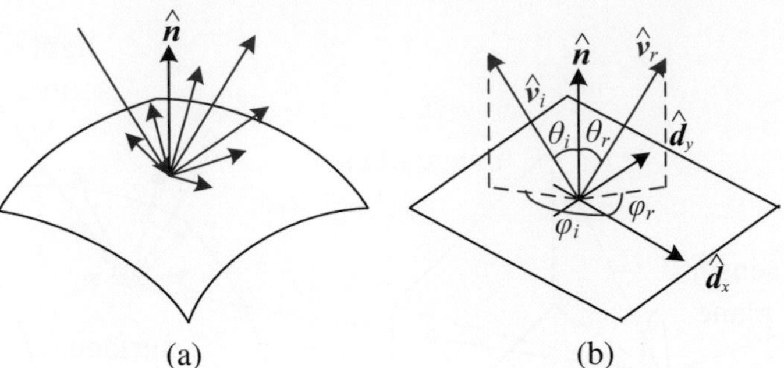

Figure 2.15 (a) Light scatters when it hits a surface. (b) The bidirectional reflectance distribution function (BRDF) $f(\theta_i, \phi_i, \theta_r, \phi_r)$ is parameterized by the angles that the incident, $\hat{\mathbf{v}}_i$, and reflected, $\hat{\mathbf{v}}_r$, light ray directions make with the local surface coordinate frame $(\hat{\mathbf{d}}_x, \hat{\mathbf{d}}_y, \hat{\mathbf{n}})$.

sitting in an outdoor courtyard, can often be represented using an *environment map* (Greene 1986) (originally called a *reflection map* (Blinn and Newell 1976)). This representation maps incident light directions $\hat{\mathbf{v}}$ to color values (or wavelengths, λ),

$$L(\hat{\mathbf{v}}; \lambda), \tag{2.81}$$

and is equivalent to assuming that all light sources are at infinity. Environment maps can be represented as a collection of cubical faces (Greene 1986), as a single longitude–latitude map (Blinn and Newell 1976), or as the image of a reflecting sphere (Watt 1995). A convenient way to get a rough model of a real-world environment map is to take an image of a reflective mirrored sphere (sometimes accompanied by a darker sphere to capture highlights) and to unwrap this image onto the desired environment map (Debevec 1998). Watt (1995) gives a nice discussion of environment mapping, including the formulas needed to map directions to pixels for the three most commonly used representations.

2.2.2 Reflectance and shading

When light hits an object's surface, it is scattered and reflected (Figure 2.15a). Many different models have been developed to describe this interaction. In this section, we first describe the most general form, the bidirectional reflectance distribution function, and then look at some more specialized models, including the diffuse, specular, and Phong shading models. We also discuss how these models can be used to compute the *global illumination* corresponding to a scene.

The Bidirectional Reflectance Distribution Function (BRDF)

The most general model of light scattering is the *bidirectional reflectance distribution function* (BRDF).[8] Relative to some local coordinate frame on the surface, the BRDF is a four-dimensional function that describes how much of each wavelength arriving at an *incident* direction $\hat{\mathbf{v}}_i$ is emitted in a *reflected* direction $\hat{\mathbf{v}}_r$ (Figure 2.15b). The function can be written in terms of the angles of the incident and reflected directions relative to the surface frame as

$$f_r(\theta_i, \phi_i, \theta_r, \phi_r; \lambda). \tag{2.82}$$

[8] Actually, even more general models of light transport exist, including some that model spatial variation along the surface, sub-surface scattering, and atmospheric effects—see Section 13.7.1—(Dorsey, Rushmeier, and Sillion 2007; Weyrich, Lawrence *et al.* 2009).

Figure 2.16 This close-up of a statue shows both diffuse (smooth shading) and specular (shiny highlight) reflection, as well as darkening in the grooves and creases due to reduced light visibility and interreflections. (Photo courtesy of the Caltech Vision Lab, http://www.vision.caltech.edu/archive.html.)

The BRDF is *reciprocal*, i.e., because of the physics of light transport, you can interchange the roles of $\hat{\mathbf{v}}_i$ and $\hat{\mathbf{v}}_r$ and still get the same answer (this is sometimes called *Helmholtz reciprocity*).

Most surfaces are *isotropic*, i.e., there are no preferred directions on the surface as far as light transport is concerned. (The exceptions are *anisotropic* surfaces such as brushed (scratched) aluminum, where the reflectance depends on the light orientation relative to the direction of the scratches.) For an isotropic material, we can simplify the BRDF to

$$f_r(\theta_i, \theta_r, |\phi_r - \phi_i|; \lambda) \quad \text{or} \quad f_r(\hat{\mathbf{v}}_i, \hat{\mathbf{v}}_r, \hat{\mathbf{n}}; \lambda), \tag{2.83}$$

as the quantities θ_i, θ_r, and $\phi_r - \phi_i$ can be computed from the directions $\hat{\mathbf{v}}_i$, $\hat{\mathbf{v}}_r$, and $\hat{\mathbf{n}}$.

To calculate the amount of light exiting a surface point $\mathbf{p}$ in a direction $\hat{\mathbf{v}}_r$ under a given lighting condition, we integrate the product of the incoming light $L_i(\hat{\mathbf{v}}_i; \lambda)$ with the BRDF (some authors call this step a *convolution*). Taking into account the *foreshortening* factor $\cos^+ \theta_i$, we obtain

$$L_r(\hat{\mathbf{v}}_r; \lambda) = \int L_i(\hat{\mathbf{v}}_i; \lambda) f_r(\hat{\mathbf{v}}_i, \hat{\mathbf{v}}_r, \hat{\mathbf{n}}; \lambda) \cos^+ \theta_i \, d\hat{\mathbf{v}}_i, \tag{2.84}$$

where

$$\cos^+ \theta_i = \max(0, \cos \theta_i). \tag{2.85}$$

If the light sources are discrete (a finite number of point light sources), we can replace the integral with a summation,

$$L_r(\hat{\mathbf{v}}_r; \lambda) = \sum_i L_i(\lambda) f_r(\hat{\mathbf{v}}_i, \hat{\mathbf{v}}_r, \hat{\mathbf{n}}; \lambda) \cos^+ \theta_i. \tag{2.86}$$

BRDFs for a given surface can be obtained through physical modeling (Torrance and Sparrow 1967; Cook and Torrance 1982; Glassner 1995), heuristic modeling (Phong 1975; Lafortune, Foo *et al.* 1997), or through empirical observation (Ward 1992; Westin, Arvo, and Torrance 1992; Dana, van Ginneken *et al.* 1999; Marschner, Westin *et al.* 2000; Matusik, Pfister *et al.* 2003; Dorsey, Rushmeier, and Sillion 2007; Weyrich, Lawrence *et al.* 2009; Shi, Mo *et al.* 2019).[9] Typical BRDFs can often be split into their *diffuse* and *specular* components, as described below.

[9] See http://www1.cs.columbia.edu/CAVE/software/curet for a database of some empirically sampled BRDFs.

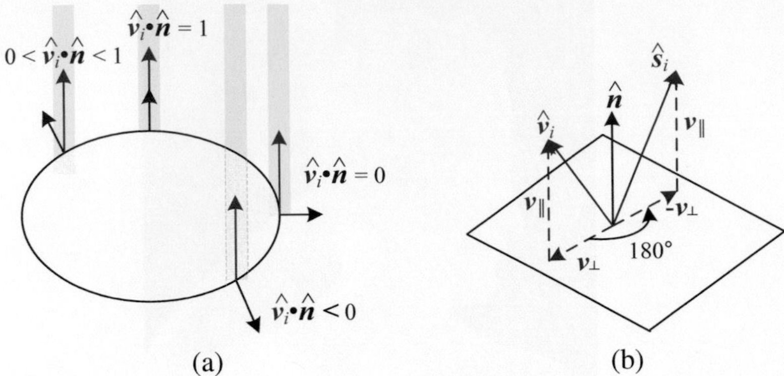

Figure 2.17 (a) The diminution of returned light caused by *foreshortening* depends on $\hat{\mathbf{v}}_i \cdot \hat{\mathbf{n}}$, the cosine of the angle between the incident light direction $\hat{\mathbf{v}}_i$ and the surface normal $\hat{\mathbf{n}}$. (b) Mirror (specular) reflection: The incident light ray direction $\hat{\mathbf{v}}_i$ is reflected onto the specular direction $\hat{\mathbf{s}}_i$ around the surface normal $\hat{\mathbf{n}}$.

Diffuse reflection

The diffuse component (also known as *Lambertian* or *matte* reflection) scatters light uniformly in all directions and is the phenomenon we most normally associate with *shading*, e.g., the smooth (non-shiny) variation of intensity with surface normal that is seen when observing a statue (Figure 2.16). Diffuse reflection also often imparts a strong *body color* to the light, as it is caused by selective absorption and re-emission of light inside the object's material (Shafer 1985; Glassner 1995).

While light is scattered uniformly in all directions, i.e., the BRDF is constant,

$$f_d(\hat{\mathbf{v}}_i, \hat{\mathbf{v}}_r, \hat{\mathbf{n}}; \lambda) = f_d(\lambda), \tag{2.87}$$

the amount of light depends on the angle between the incident light direction and the surface normal θ_i. This is because the surface area exposed to a given amount of light becomes larger at oblique angles, becoming completely self-shadowed as the outgoing surface normal points away from the light (Figure 2.17a). (Think about how you orient yourself towards the Sun or fireplace to get maximum warmth and how a flashlight projected obliquely against a wall is less bright than one pointing directly at it.) The *shading equation* for diffuse reflection can thus be written as

$$L_d(\hat{\mathbf{v}}_r; \lambda) = \sum_i L_i(\lambda) f_d(\lambda) \cos^+ \theta_i = \sum_i L_i(\lambda) f_d(\lambda) [\hat{\mathbf{v}}_i \cdot \hat{\mathbf{n}}]^+, \tag{2.88}$$

where

$$[\hat{\mathbf{v}}_i \cdot \hat{\mathbf{n}}]^+ = \max(0, \hat{\mathbf{v}}_i \cdot \hat{\mathbf{n}}). \tag{2.89}$$

Specular reflection

The second major component of a typical BRDF is *specular* (gloss or highlight) reflection, which depends strongly on the direction of the outgoing light. Consider light reflecting off a mirrored surface (Figure 2.17b). Incident light rays are reflected in a direction that is rotated by 180° around the surface normal $\hat{\mathbf{n}}$. Using the same notation as in Equations (2.29–2.30), we can compute the *specular reflection* direction $\hat{\mathbf{s}}_i$ as

$$\hat{\mathbf{s}}_i = \mathbf{v}_\| - \mathbf{v}_\perp = (2\hat{\mathbf{n}}\hat{\mathbf{n}}^T - \mathbf{I})\mathbf{v}_i. \tag{2.90}$$

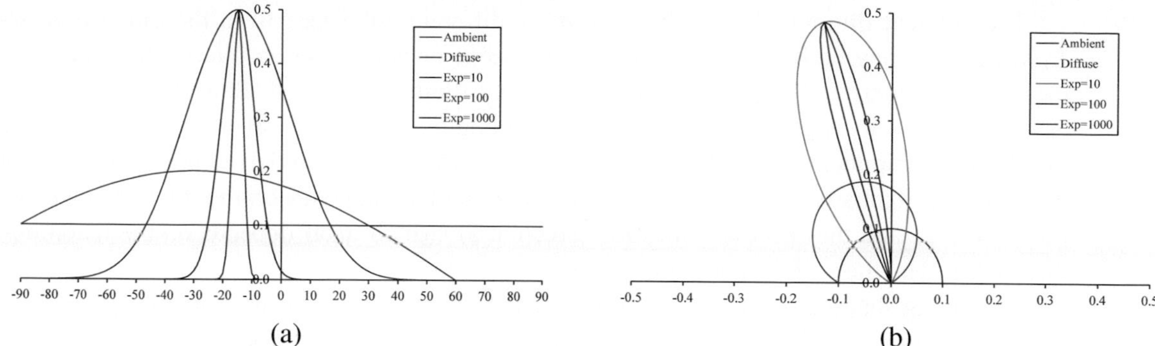

Figure 2.18 Cross-section through a Phong shading model BRDF for a fixed incident illumination direction: (a) component values as a function of angle away from surface normal; (b) polar plot. The value of the Phong exponent k_e is indicated by the "Exp" labels and the light source is at an angle of 30° away from the normal.

The amount of light reflected in a given direction $\hat{\mathbf{v}}_r$ thus depends on the angle $\theta_s = \cos^{-1}(\hat{\mathbf{v}}_r \cdot \hat{\mathbf{s}}_i)$ between the view direction $\hat{\mathbf{v}}_r$ and the specular direction $\hat{\mathbf{s}}_i$. For example, the Phong (1975) model uses a power of the cosine of the angle,

$$f_s(\theta_s; \lambda) = k_s(\lambda) \cos^{k_e} \theta_s, \tag{2.91}$$

while the Torrance and Sparrow (1967) micro-facet model uses a Gaussian,

$$f_s(\theta_s; \lambda) = k_s(\lambda) \exp(-c_s^2 \theta_s^2). \tag{2.92}$$

Larger exponents k_e (or inverse Gaussian widths c_s) correspond to more specular surfaces with distinct highlights, while smaller exponents better model materials with softer gloss.

Phong shading

Phong (1975) combined the diffuse and specular components of reflection with another term, which he called the *ambient illumination*. This term accounts for the fact that objects are generally illuminated not only by point light sources but also by a general diffuse illumination corresponding to inter-reflection (e.g., the walls in a room) or distant sources, such as the blue sky. In the Phong model, the ambient term does not depend on surface orientation, but depends on the color of both the ambient illumination $L_a(\lambda)$ and the object $k_a(\lambda)$,

$$f_a(\lambda) = k_a(\lambda) L_a(\lambda). \tag{2.93}$$

Putting all of these terms together, we arrive at the *Phong shading* model,

$$L_r(\hat{\mathbf{v}}_r; \lambda) = k_a(\lambda) L_a(\lambda) + k_d(\lambda) \sum_i L_i(\lambda) [\hat{\mathbf{v}}_i \cdot \hat{\mathbf{n}}]^+ + k_s(\lambda) \sum_i L_i(\lambda) (\hat{\mathbf{v}}_r \cdot \hat{\mathbf{s}}_i)^{k_e}. \tag{2.94}$$

Figure 2.18 shows a typical set of Phong shading model components as a function of the angle away from the surface normal (in a plane containing both the lighting direction and the viewer).

Typically, the ambient and diffuse reflection color distributions $k_a(\lambda)$ and $k_d(\lambda)$ are the same, since they are both due to sub-surface scattering (body reflection) inside the surface material (Shafer 1985). The specular reflection distribution $k_s(\lambda)$ is often uniform (white), since it is caused by interface reflections that do not change the light color. (The exception to this is emphmetallic materials, such as copper, as opposed to the more common *dielectric* materials, such as plastics.)

The ambient illumination $L_a(\lambda)$ often has a different color cast from the direct light sources $L_i(\lambda)$, e.g., it may be blue for a sunny outdoor scene or yellow for an interior lit with candles or incandescent lights. (The presence of ambient sky illumination in shadowed areas is what often causes shadows to appear bluer than the corresponding lit portions of a scene). Note also that the diffuse component of the Phong model (or of any shading model) depends on the angle of the *incoming* light source $\hat{\mathbf{v}}_i$, while the specular component depends on the relative angle between the viewer $\mathbf{v}_r$ and the specular reflection direction $\hat{\mathbf{s}}_i$ (which itself depends on the incoming light direction $\hat{\mathbf{v}}_i$ and the surface normal $\hat{\mathbf{n}}$).

The Phong shading model has been superseded in terms of physical accuracy by newer models in computer graphics, including the model developed by Cook and Torrance (1982) based on the original micro-facet model of Torrance and Sparrow (1967). While, initially, computer graphics hardware implemented the Phong model, the advent of programmable pixel shaders has made the use of more complex models feasible.

Di-chromatic reflection model

The Torrance and Sparrow (1967) model of reflection also forms the basis of Shafer's (1985) *di-chromatic reflection model*, which states that the apparent color of a uniform material lit from a single source depends on the sum of two terms,

$$L_r(\hat{\mathbf{v}}_r; \lambda) = L_i(\hat{\mathbf{v}}_r, \hat{\mathbf{v}}_i, \hat{\mathbf{n}}; \lambda) + L_b(\hat{\mathbf{v}}_r, \hat{\mathbf{v}}_i, \hat{\mathbf{n}}; \lambda) \quad (2.95)$$

$$= c_i(\lambda) m_i(\hat{\mathbf{v}}_r, \hat{\mathbf{v}}_i, \hat{\mathbf{n}}) + c_b(\lambda) m_b(\hat{\mathbf{v}}_r, \hat{\mathbf{v}}_i, \hat{\mathbf{n}}), \quad (2.96)$$

i.e., the radiance of the light reflected at the *interface*, L_i, and the radiance reflected at the *surface body*, L_b. Each of these, in turn, is a simple product between a relative power spectrum $c(\lambda)$, which depends only on wavelength, and a magnitude $m(\hat{\mathbf{v}}_r, \hat{\mathbf{v}}_i, \hat{\mathbf{n}})$, which depends only on geometry. (This model can easily be derived from a generalized version of Phong's model by assuming a single light source and no ambient illumination, and rearranging terms.) The di-chromatic model has been successfully used in computer vision to segment specular colored objects with large variations in shading (Klinker 1993) and has inspired local two-color models for applications such as Bayer pattern demosaicing (Bennett, Uyttendaele *et al.* 2006).

Global illumination (ray tracing and radiosity)

The simple shading model presented thus far assumes that light rays leave the light sources, bounce off surfaces visible to the camera, thereby changing in intensity or color, and arrive at the camera. In reality, light sources can be shadowed by occluders and rays can bounce multiple times around a scene while making their trip from a light source to the camera.

Two methods have traditionally been used to model such effects. If the scene is mostly specular (the classic example being scenes made of glass objects and mirrored or highly polished balls), the preferred approach is *ray tracing* or *path tracing* (Glassner 1995; Akenine-Möller and Haines 2002; Marschner and Shirley 2015), which follows individual rays from the camera across multiple bounces towards the light sources (or vice versa). If the scene is composed mostly of uniform albedo simple geometry illuminators and surfaces, *radiosity* (*global illumination*) techniques are preferred (Cohen and Wallace 1993; Sillion and Puech 1994; Glassner 1995). Combinations of the two techniques have also been developed (Wallace, Cohen, and Greenberg 1987), as well as more general *light transport* techniques for simulating effects such as the *caustics* cast by rippling water.

The basic ray tracing algorithm associates a light ray with each pixel in the camera image and finds its intersection with the nearest surface. A *primary* contribution can then be computed using

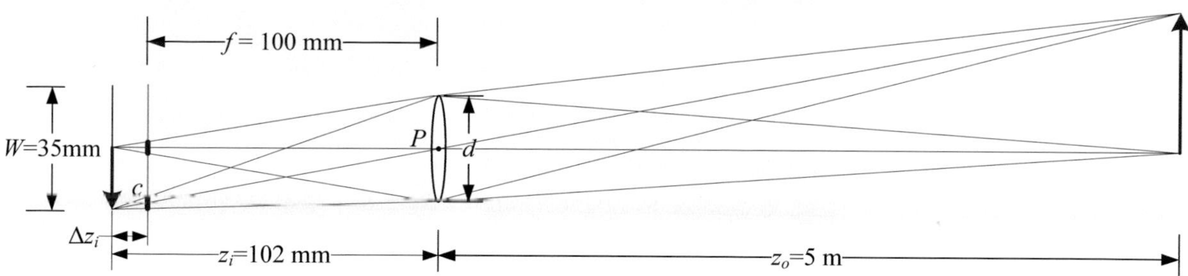

Figure 2.19 A thin lens of focal length f focuses the light from a plane at a distance z_o in front of the lens onto a plane at a distance z_i behind the lens, where $\frac{1}{z_o} + \frac{1}{z_i} = \frac{1}{f}$. If the focal plane (vertical gray line next to c) is moved forward, the images are no longer in focus and the *circle of confusion* c (small thick line segments) depends on the distance of the image plane motion Δz_i relative to the lens aperture diameter d. The field of view (f.o.v.) depends on the ratio between the sensor width W and the focal length f (or, more precisely, the focusing distance z_i, which is usually quite close to f).

the simple shading equations presented previously (e.g., Equation (2.94)) for all light sources that are visible for that surface element. (An alternative technique for computing which surfaces are illuminated by a light source is to compute a *shadow map*, or *shadow buffer*, i.e., a rendering of the scene from the light source's perspective, and then compare the depth of pixels being rendered with the map (Williams 1983; Akenine-Möller and Haines 2002).) Additional *secondary* rays can then be cast along the specular direction towards other objects in the scene, keeping track of any attenuation or color change that the specular reflection induces.

Radiosity works by associating lightness values with rectangular surface areas in the scene (including area light sources). The amount of light interchanged between any two (mutually visible) areas in the scene can be captured as a *form factor*, which depends on their relative orientation and surface reflectance properties, as well as the $1/r^2$ fall-off as light is distributed over a larger effective sphere the further away it is (Cohen and Wallace 1993; Sillion and Puech 1994; Glassner 1995). A large linear system can then be set up to solve for the final lightness of each area patch, using the light sources as the forcing function (right-hand side). Once the system has been solved, the scene can be rendered from any desired point of view. Under certain circumstances, it is possible to recover the global illumination in a scene from photographs using computer vision techniques (Yu, Debevec *et al.* 1999).

The basic radiosity algorithm does not take into account certain *near field* effects, such as the darkening inside corners and scratches, or the limited ambient illumination caused by partial shadowing from other surfaces. Such effects have been exploited in a number of computer vision algorithms (Nayar, Ikeuchi, and Kanade 1991; Langer and Zucker 1994).

While all of these global illumination effects can have a strong effect on the appearance of a scene, and hence its 3D interpretation, they are not covered in more detail in this book. (But see Section 13.7.1 for a discussion of recovering BRDFs from real scenes and objects.)

2.2.3 Optics

Once the light from a scene reaches the camera, it must still pass through the lens before reaching the analog or digital sensor. For many applications, it suffices to treat the lens as an ideal pinhole that simply projects all rays through a common center of projection (Figures 2.8 and 2.9).

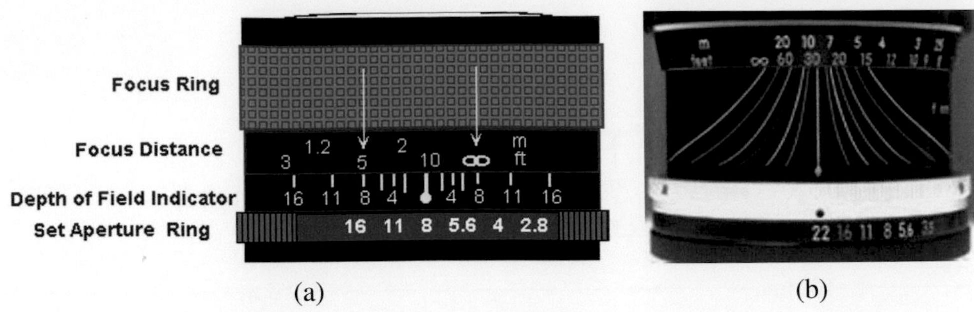

Figure 2.20 Regular and zoom lens depth of field indicators.

However, if we want to deal with issues such as focus, exposure, vignetting, and aberration, we need to develop a more sophisticated model, which is where the study of *optics* comes in (Möller 1988; Ray 2002; Hecht 2015).

Figure 2.19 shows a diagram of the most basic lens model, i.e., the *thin lens* composed of a single piece of glass with very low, equal curvature on both sides. According to the *lens law* (which can be derived using simple geometric arguments on light ray refraction), the relationship between the distance to an object z_o and the distance behind the lens at which a focused image is formed z_i can be expressed as

$$\frac{1}{z_o} + \frac{1}{z_i} = \frac{1}{f}, \tag{2.97}$$

where f is called the *focal length* of the lens. If we let $z_o \to \infty$, i.e., we adjust the lens (move the image plane) so that objects at infinity are in focus, we get $z_i = f$, which is why we can think of a lens of focal length f as being equivalent (to a first approximation) to a pinhole at a distance f from the focal plane (Figure 2.10), whose field of view is given by (2.60).

If the focal plane is moved away from its proper in-focus setting of z_i (e.g., by twisting the focus ring on the lens), objects at z_o are no longer in focus, as shown by the gray plane in Figure 2.19. The amount of misfocus is measured by the *circle of confusion* c (shown as short thick blue line segments on the gray plane).[10] The equation for the circle of confusion can be derived using similar triangles; it depends on the distance of travel in the focal plane Δz_i relative to the original focus distance z_i and the diameter of the aperture d (see Exercise 2.4).

The allowable depth variation in the scene that limits the circle of confusion to an acceptable number is commonly called the *depth of field* and is a function of both the focus distance and the aperture, as shown diagrammatically by many lens markings (Figure 2.20). Since this depth of field depends on the aperture diameter d, we also have to know how this varies with the commonly displayed *f-number*, which is usually denoted as $f/\#$ or N and is defined as

$$f/\# = N = \frac{f}{d}, \tag{2.98}$$

where the focal length f and the aperture diameter d are measured in the same unit (say, millimeters).

The usual way to write the f-number is to replace the $\#$ in $f/\#$ with the actual number, i.e., $f/1.4, f/2, f/2.8, \ldots, f/22$. (Alternatively, we can say $N = 1.4$, etc.) An easy way to interpret

[10] If the aperture is not completely circular, e.g., if it is caused by a hexagonal diaphragm, it is sometimes possible to see this effect in the actual blur function (Levin, Fergus *et al.* 2007; Joshi, Szeliski, and Kriegman 2008) or in the "glints" that are seen when shooting into the Sun.

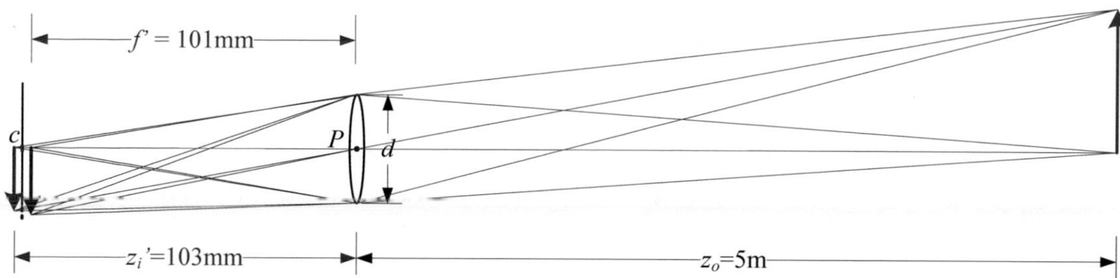

Figure 2.21 In a lens subject to *chromatic aberration*, light at different wavelengths (e.g., the red and blue arrows) is focused with a different focal length f' and hence a different depth z_i', resulting in both a geometric (in-plane) displacement and a loss of focus.

these numbers is to notice that dividing the focal length by the f-number gives us the diameter d, so these are just formulas for the aperture diameter.[11]

Notice that the usual progression for f-numbers is in *full stops*, which are multiples of $\sqrt{2}$, since this corresponds to doubling the area of the entrance pupil each time a smaller f-number is selected. (This doubling is also called changing the exposure by one *exposure value* or EV. It has the same effect on the amount of light reaching the sensor as doubling the exposure duration, e.g., from $^1/_{250}$ to $^1/_{125}$; see Exercise 2.5.)

Now that you know how to convert between f-numbers and aperture diameters, you can construct your own plots for the depth of field as a function of focal length f, circle of confusion c, and focus distance z_o, as explained in Exercise 2.4, and see how well these match what you observe on actual lenses, such as those shown in Figure 2.20.

Of course, real lenses are not infinitely thin and therefore suffer from geometric aberrations, unless compound elements are used to correct for them. The classic five *Seidel aberrations*, which arise when using *third-order optics*, include spherical aberration, coma, astigmatism, curvature of field, and distortion (Möller 1988; Ray 2002; Hecht 2015).

Chromatic aberration

Because the index of refraction for glass varies slightly as a function of wavelength, simple lenses suffer from *chromatic aberration*, which is the tendency for light of different colors to focus at slightly different distances (and hence also with slightly different magnification factors), as shown in Figure 2.21. The wavelength-dependent magnification factor, i.e., the *transverse chromatic aberration*, can be modeled as a per-color radial distortion (Section 2.1.5) and, hence, calibrated using the techniques described in Section 11.1.4. The wavelength-dependent blur caused by *longitudinal chromatic aberration* can be calibrated using techniques described in Section 10.1.4. Unfortunately, the blur induced by longitudinal aberration can be harder to undo, as higher frequencies can get strongly attenuated and hence hard to recover.

To reduce chromatic and other kinds of aberrations, most photographic lenses today are *compound lenses* made of different glass elements (with different coatings). Such lenses can no longer be modeled as having a single *nodal point* P through which all of the rays must pass (when approximating the lens with a pinhole model). Instead, these lenses have both a *front nodal point*, through which the rays enter the lens, and a *rear nodal point*, through which they leave on their way to the

[11] This also explains why, with zoom lenses, the f-number varies with the current zoom (focal length) setting.

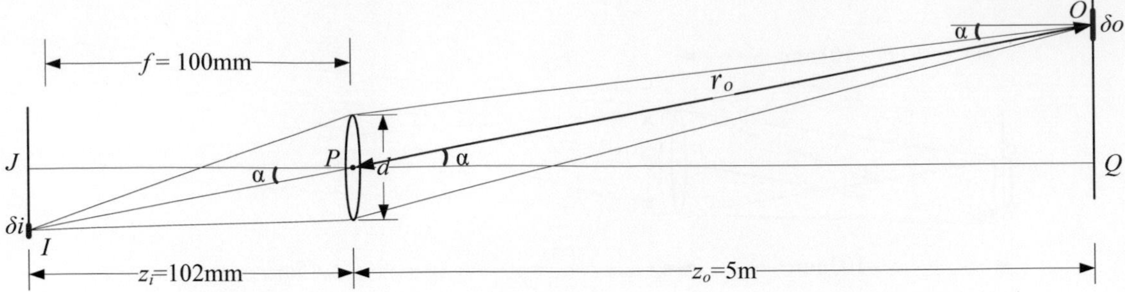

Figure 2.22 The amount of light hitting a pixel of surface area δi depends on the square of the ratio of the aperture diameter d to the focal length f, as well as the fourth power of the off-axis angle α cosine, $\cos^4 \alpha$.

sensor. In practice, only the location of the front nodal point is of interest when performing careful camera calibration, e.g., when determining the point around which to rotate to capture a parallax-free panorama (see Section 8.2.3 and Littlefield (2006) and Houghton (2013)).

Not all lenses, however, can be modeled as having a single nodal point. In particular, very wide-angle lenses such as fisheye lenses (Section 2.1.5) and certain *catadioptric* imaging systems consisting of lenses and curved mirrors (Baker and Nayar 1999) do not have a single point through which all of the acquired light rays pass. In such cases, it is preferable to explicitly construct a mapping function (look-up table) between pixel coordinates and 3D rays in space (Gremban, Thorpe, and Kanade 1988; Champleboux, Lavallée *et al.* 1992a; Grossberg and Nayar 2001; Sturm and Ramalingam 2004; Tardif, Sturm *et al.* 2009), as mentioned in Section 2.1.5.

Vignetting

Another property of real-world lenses is *vignetting*, which is the tendency for the brightness of the image to fall off towards the edge of the image.

Two kinds of phenomena usually contribute to this effect (Ray 2002). The first is called *natural vignetting* and is due to the foreshortening in the object surface, projected pixel, and lens aperture, as shown in Figure 2.22. Consider the light leaving the object surface patch of size δo located at an *off-axis angle* α. Because this patch is foreshortened with respect to the camera lens, the amount of light reaching the lens is reduced by a factor $\cos \alpha$. The amount of light reaching the lens is also subject to the usual $1/r^2$ fall-off; in this case, the distance $r_o = z_o / \cos \alpha$. The actual area of the aperture through which the light passes is foreshortened by an additional factor $\cos \alpha$, i.e., the aperture as seen from point O is an ellipse of dimensions $d \times d \cos \alpha$. Putting all of these factors together, we see that the amount of light leaving O and passing through the aperture on its way to the image pixel located at I is proportional to

$$\frac{\delta o \cos \alpha}{r_o^2} \pi \left(\frac{d}{2} \right)^2 \cos \alpha = \delta o \frac{\pi}{4} \frac{d^2}{z_o^2} \cos^4 \alpha. \tag{2.99}$$

Since triangles ΔOPQ and ΔIPJ are similar, the projected areas of the object surface δo and image pixel δi are in the same (squared) ratio as $z_o : z_i$,

$$\frac{\delta o}{\delta i} = \frac{z_o^2}{z_i^2}. \tag{2.100}$$

Putting these together, we obtain the final relationship between the amount of light reaching pixel i and the aperture diameter d, the focusing distance $z_i \approx f$, and the off-axis angle α,

$$\delta o \frac{\pi}{4} \frac{d^2}{z_o^2} \cos^4 \alpha = \delta i \frac{\pi}{4} \frac{d^2}{z_i^2} \cos^4 \alpha \approx \delta i \frac{\pi}{4} \left(\frac{d}{f} \right)^2 \cos^4 \alpha, \tag{2.101}$$

which is called the *fundamental radiometric relation* between the scene radiance L and the light (irradiance) E reaching the pixel sensor,

$$E = L \frac{\pi}{4} \left(\frac{d}{f} \right)^2 \cos^4 \alpha, \tag{2.102}$$

(Horn 1986; Nalwa 1993; Ray 2002; Hecht 2015). Notice in this equation how the amount of light depends on the pixel surface area (which is why the smaller sensors in point-and-shoot cameras are so much noisier than digital single lens reflex (SLR) cameras), the inverse square of the f-stop $N = f/d$ (2.98), and the fourth power of the $\cos^4 \alpha$ off-axis fall-off, which is the natural vignetting term.

The other major kind of vignetting, called *mechanical vignetting*, is caused by the internal occlusion of rays near the periphery of lens elements in a compound lens, and cannot easily be described mathematically without performing a full ray-tracing of the actual lens design.[12] However, unlike natural vignetting, mechanical vignetting can be decreased by reducing the camera aperture (increasing the f-number). It can also be calibrated (along with natural vignetting) using special devices such as integrating spheres, uniformly illuminated targets, or camera rotation, as discussed in Section 10.1.3.

2.3 The digital camera

After starting from one or more light sources, reflecting off one or more surfaces in the world, and passing through the camera's optics (lenses), light finally reaches the imaging sensor. How are the photons arriving at this sensor converted into the digital (R, G, B) values that we observe when we look at a digital image? In this section, we develop a simple model that accounts for the most important effects, such as exposure (gain and shutter speed), non-linear mappings, sampling and aliasing, and noise. Figure 2.23, which is based on camera models developed by Healey and Kondepudy (1994), Tsin, Ramesh, and Kanade (2001), and Liu, Szeliski *et al.* (2008), shows a simple version of the processing stages that occur in modern digital cameras. Chakrabarti, Scharstein, and Zickler (2009) developed a sophisticated 24-parameter model that is an even better match to the processing performed in digital cameras, while Kim, Lin *et al.* (2012), Hasinoff, Sharlet *et al.* (2016), and Karaimer and Brown (2016) provide more recent models of modern in-camera processing pipelines. Most recently, Brooks, Mildenhall *et al.* (2019) have developed detailed models of in-camera image processing pipelines to invert (*unprocess*) noisy JPEG images into their RAW originals, so that they can be better denoised, while Tseng, Yu *et al.* (2019) develop a tunable model of camera processing pipelines that can be used for image quality optimization.

Light falling on an imaging sensor is usually picked up by an *active sensing area*, integrated for the duration of the exposure (usually expressed as the shutter speed in a fraction of a second, e.g., $\frac{1}{125}, \frac{1}{60}, \frac{1}{30}$), and then passed to a set of *sense amplifiers*. The two main kinds of sensor used in digital still and video cameras today are charge-coupled device (CCD) and complementary metal oxide on silicon (CMOS).

[12]There are some empirical models that work well in practice (Kang and Weiss 2000; Zheng, Lin, and Kang 2006).

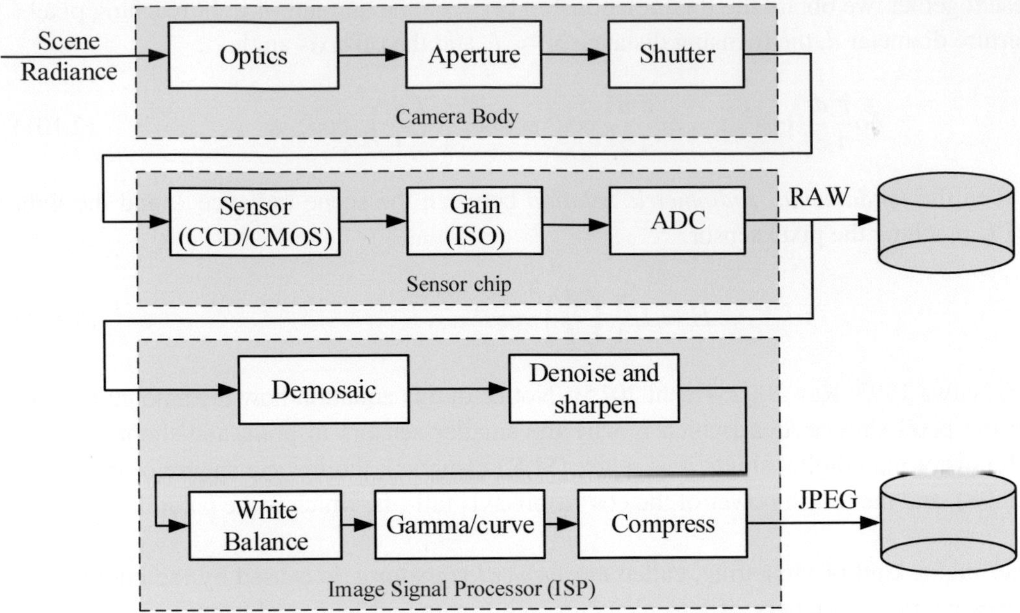

Figure 2.23 Image sensing pipeline, showing the various sources of noise as well as typical digital post-processing steps.

In a CCD, photons are accumulated in each active *well* during the exposure time. Then, in a *transfer* phase, the charges are transferred from well to well in a kind of "bucket brigade" until they are deposited at the sense amplifiers, which amplify the signal and pass it to an analog-to-digital converter (ADC).[13] Older CCD sensors were prone to *blooming*, when charges from one over-exposed pixel spilled into adjacent ones, but most newer CCDs have anti-blooming technology ("troughs" into which the excess charge can spill).

In CMOS, the photons hitting the sensor directly affect the conductivity (or gain) of a photodetector, which can be selectively gated to control exposure duration, and locally amplified before being read out using a multiplexing scheme. Traditionally, CCD sensors outperformed CMOS in quality-sensitive applications, such as digital SLRs, while CMOS was better for low-power applications, but today CMOS is used in most digital cameras.

The main factors affecting the performance of a digital image sensor are the shutter speed, sampling pitch, fill factor, chip size, analog gain, sensor noise, and the resolution (and quality) of the analog-to-digital converter. Many of the actual values for these parameters can be read from the EXIF tags embedded with digital images, while others can be obtained from the camera manufacturers' specification sheets or from camera review or calibration websites.[14]

Shutter speed. The shutter speed (exposure time) directly controls the amount of light reaching the sensor and hence determines if images are under- or over-exposed. (For bright scenes, where a large aperture or slow shutter speed is desired to get a shallow depth of field or motion blur, *neutral density filters* are sometimes used by photographers.) For dynamic scenes, the shutter speed

[13] In digital still cameras, a complete frame is captured and then read out sequentially at once. However, if video is being captured, a *rolling shutter*, which exposes and transfers each line separately, is often used. In older video cameras, the even fields (lines) were scanned first, followed by the odd fields, in a process that is called *interlacing*.

[14] http://www.clarkvision.com/imagedetail/digital.sensor.performance.summary

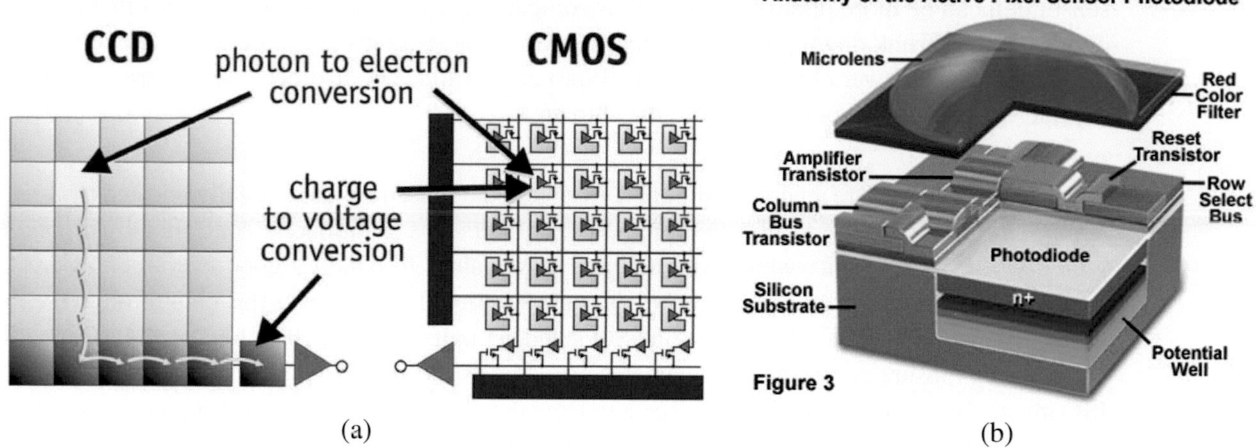

Figure 2.24 Digital imaging sensors: (a) CCDs move photogenerated charge from pixel to pixel and convert it to voltage at the output node; CMOS imagers convert charge to voltage inside each pixel (Litwiller 2005) © 2005 Photonics Spectra; (b) cutaway diagram of a CMOS pixel sensor, from https://micro.magnet.fsu.edu/primer/digitalimaging/cmosimagesensors.html.

also determines the amount of *motion blur* in the resulting picture. Usually, a higher shutter speed (less motion blur) makes subsequent analysis easier (see Section 10.3 for techniques to remove such blur). However, when video is being captured for display, some motion blur may be desirable to avoid stroboscopic effects.

Sampling pitch. The sampling pitch is the physical spacing between adjacent sensor cells on the imaging chip (Figure 2.24). A sensor with a smaller sampling pitch has a higher *sampling density* and hence provides a higher *resolution* (in terms of pixels) for a given active chip area. However, a smaller pitch also means that each sensor has a smaller area and cannot accumulate as many photons; this makes it not as *light sensitive* and more prone to noise.

Fill factor. The fill factor is the active sensing area size as a fraction of the theoretically available sensing area (the product of the horizontal and vertical sampling pitches). Higher fill factors are usually preferable, as they result in more light capture and less *aliasing* (see Section 2.3.1). While the fill factor was originally limited by the need to place additional electronics between the active sensing areas, modern *backside illumination* (or *back-illuminated*) sensors, coupled with efficient microlens designs, have largely removed this limitation (Fontaine 2015).[15] The fill factor of a camera can be determined empirically using a photometric camera calibration process (see Section 10.1.4).

Chip size. Video and point-and-shoot cameras have traditionally used small chip areas ($\frac{1}{4}$-inch to $\frac{1}{2}$-inch sensors[16]), while digital SLR cameras try to come closer to the traditional size of a 35mm film frame.[17] When overall device size is not important, having a larger chip size is preferable, since each

[15] https://en.wikipedia.org/wiki/Back-illuminated_sensor

[16] These numbers refer to the "tube diameter" of the old vidicon tubes used in video cameras. The 1/2.5" sensor on the Canon SD800 camera actually measures 5.76mm × 4.29mm, i.e., a sixth of the size (on side) of a 35mm full-frame (36mm × 24mm) DSLR sensor.

[17] When a DSLR chip does not fill the 35mm full frame, it results in a *multiplier effect* on the lens focal length. For example, a chip that is only 0.6 the dimension of a 35mm frame will make a 50mm lens image the same angular extent as a $50/0.6 = 50 \times 1.6 = 80$mm lens, as demonstrated in (2.60).

sensor cell can be more photo-sensitive. (For compact cameras, a smaller chip means that all of the optics can be shrunk down proportionately.) However, larger chips are more expensive to produce, not only because fewer chips can be packed into each wafer, but also because the probability of a chip defect goes up exponentially with the chip area.

Analog gain. Before analog-to-digital conversion, the sensed signal is usually boosted by a *sense amplifier*. In video cameras, the gain on these amplifiers was traditionally controlled by *automatic gain control* (AGC) logic, which would adjust these values to obtain a good overall exposure. In newer digital still cameras, the user now has some additional control over this gain through the *ISO setting*, which is typically expressed in ISO standard units such as 100, 200, or 400. Since the automated exposure control in most cameras also adjusts the aperture and shutter speed, setting the ISO manually removes one degree of freedom from the camera's control, just as manually specifying aperture and shutter speed does. In theory, a higher gain allows the camera to perform better under low light conditions (less motion blur due to long exposure times when the aperture is already maxed out). In practice, however, higher ISO settings usually amplify the *sensor noise*.

Sensor noise. Throughout the whole sensing process, noise is added from various sources, which may include *fixed pattern noise*, *dark current noise*, *shot noise*, *amplifier noise*, and *quantization noise* (Healey and Kondepudy 1994; Tsin, Ramesh, and Kanade 2001). The final amount of noise present in a sampled image depends on all of these quantities, as well as the incoming light (controlled by the scene radiance and aperture), the exposure time, and the sensor gain. Also, for low light conditions where the noise is due to low photon counts, a Poisson model of noise may be more appropriate than a Gaussian model (Alter, Matsushita, and Tang 2006; Matsushita and Lin 2007a; Wilburn, Xu, and Matsushita 2008; Takamatsu, Matsushita, and Ikeuchi 2008).

As discussed in more detail in Section 10.1.1, Liu, Szeliski *et al.* (2008) use this model, along with an empirical database of camera response functions (CRFs) obtained by Grossberg and Nayar (2004), to estimate the *noise level function* (NLF) for a given image, which predicts the overall noise variance at a given pixel as a function of its brightness (a separate NLF is estimated for each color channel). An alternative approach, when you have access to the camera before taking pictures, is to pre-calibrate the NLF by taking repeated shots of a scene containing a variety of colors and luminances, such as the Macbeth Color Chart shown in Figure 10.3b (McCamy, Marcus, and Davidson 1976). (When estimating the variance, be sure to throw away or downweight pixels with large gradients, as small shifts between exposures will affect the sensed values at such pixels.) Unfortunately, the pre-calibration process may have to be repeated for different exposure times and gain settings because of the complex interactions occurring within the sensing system.

In practice, most computer vision algorithms, such as image denoising, edge detection, and stereo matching, all benefit from at least a rudimentary estimate of the noise level. Barring the ability to pre-calibrate the camera or to take repeated shots of the same scene, the simplest approach is to look for regions of near-constant value and to estimate the noise variance in such regions (Liu, Szeliski *et al.* 2008).

ADC resolution. The final step in the analog processing chain occurring within an imaging sensor is the *analog to digital conversion* (ADC). While a variety of techniques can be used to implement this process, the two quantities of interest are the *resolution* of this process (how many bits it yields) and its noise level (how many of these bits are useful in practice). For most cameras, the number of bits quoted (eight bits for compressed JPEG images and a nominal 16 bits for the RAW formats provided by some DSLRs) exceeds the actual number of usable bits. The best way to

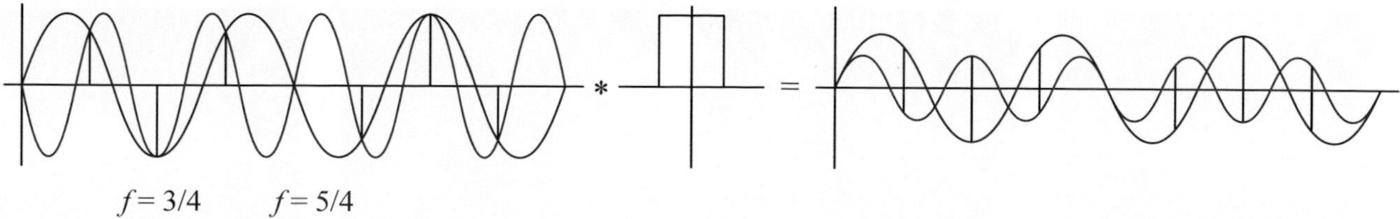

Figure 2.25 Aliasing of a one-dimensional signal: The blue sine wave at $f = 3/4$ and the red sine wave at $f = 5/4$ have the same digital samples, when sampled at $f = 2$. Even after convolution with a 100% fill factor box filter, the two signals, while no longer of the same magnitude, are still aliased in the sense that the sampled red signal looks like an inverted lower magnitude version of the blue signal. (The image on the right is scaled up for better visibility. The actual sine magnitudes are 30% and -18% of their original values.)

tell is to simply calibrate the noise of a given sensor, e.g., by taking repeated shots of the same scene and plotting the estimated noise as a function of brightness (Exercise 2.6).

Digital post-processing. Once the irradiance values arriving at the sensor have been converted to digital bits, most cameras perform a variety of *digital signal processing* (DSP) operations to enhance the image before compressing and storing the pixel values. These include color filter array (CFA) demosaicing, white point setting, and mapping of the luminance values through a *gamma function* to increase the perceived dynamic range of the signal. We cover these topics in Section 2.3.2 but, before we do, we return to the topic of aliasing, which was mentioned in connection with sensor array fill factors.

Newer imaging sensors. The capabilities of imaging sensor and related technologies such as depth sensors continue to evolve rapidly. Conferences that track these developments include the IS&T Symposium on Electronic Imaging Science and Technology sponsored by the Society for Imaging Science and Technology and the Image Sensors World blog.

2.3.1 Sampling and aliasing

What happens when a field of light impinging on the image sensor falls onto the active sense areas in the imaging chip? The photons arriving at each active cell are integrated and then digitized, as shown in Figure 2.24. However, if the fill factor on the chip is small and the signal is not otherwise *band-limited*, visually unpleasing aliasing can occur.

To explore the phenomenon of aliasing, let us first look at a one-dimensional signal (Figure 2.25), in which we have two sine waves, one at a frequency of $f = 3/4$ and the other at $f = 5/4$. If we sample these two signals at a frequency of $f = 2$, we see that they produce the same samples (shown in black), and so we say that they are *aliased*.[18] Why is this a bad effect? In essence, we can no longer reconstruct the original signal, since we do not know which of the two original frequencies was present.

In fact, Shannon's Sampling Theorem shows that the minimum sampling (Oppenheim and Schafer 1996; Oppenheim, Schafer, and Buck 1999) rate required to reconstruct a signal from its

[18] An alias is an alternate name for someone, so the sampled signal corresponds to two different *aliases*.

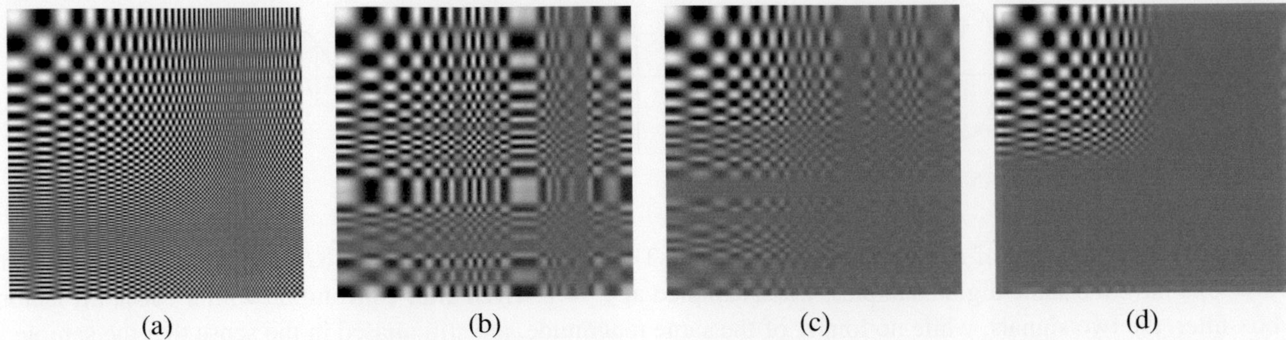

(a) (b) (c) (d)

Figure 2.26 Aliasing of a two-dimensional signal: (a) original full-resolution image; (b) downsampled 4 × with a 25% fill factor box filter; (c) downsampled 4 × with a 100% fill factor box filter; (d) downsampled 4 × with a high-quality 9-tap filter. Notice how the higher frequencies are aliased into visible frequencies with the lower quality filters, while the 9-tap filter completely removes these higher frequencies.

instantaneous samples must be at least twice the highest frequency,[19]

$$f_s \geq 2f_{\max}. \tag{2.103}$$

The maximum frequency in a signal is known as the *Nyquist frequency* and the inverse of the minimum sampling frequency $r_s = 1/f_s$ is known as the *Nyquist rate*.

However, you may ask, as an imaging chip actually *averages* the light field over a finite area, are the results on point sampling still applicable? Averaging over the sensor area does tend to attenuate some of the higher frequencies. However, even if the fill factor is 100%, as in the right image of Figure 2.25, frequencies above the Nyquist limit (half the sampling frequency) still produce an aliased signal, although with a smaller magnitude than the corresponding band-limited signals.

A more convincing argument as to why aliasing is bad can be seen by downsampling a signal using a poor quality filter such as a box (square) filter. Figure 2.26 shows a high-frequency *chirp* image (so called because the frequencies increase over time), along with the results of sampling it with a 25% fill-factor area sensor, a 100% fill-factor sensor, and a high-quality 9-tap filter. Additional examples of downsampling (*decimation*) filters can be found in Section 3.5.2 and Figure 3.29.

The best way to predict the amount of aliasing that an imaging system (or even an image processing algorithm) will produce is to estimate the *point spread function* (PSF), which represents the response of a particular pixel sensor to an ideal point light source. The PSF is a combination (convolution) of the blur induced by the optical system (lens) and the finite integration area of a chip sensor.[20]

If we know the blur function of the lens and the fill factor (sensor area shape and spacing) for the imaging chip (plus, optionally, the response of the anti-aliasing filter), we can convolve these (as described in Section 3.2) to obtain the PSF. Figure 2.27a shows the one-dimensional cross-section of a PSF for a lens whose blur function is assumed to be a disc with a radius equal to the pixel spacing s plus a sensing chip whose horizontal fill factor is 80%. Taking the Fourier transform of this PSF (Section 3.4), we obtain the *modulation transfer function* (MTF), from which we can estimate the amount of aliasing as the area of the Fourier magnitude outside the $f \leq f_s$ Nyquist frequency.[21]

[19] The actual theorem states that f_s must be at least twice the signal *bandwidth* but, as we are not dealing with modulated signals such as radio waves during image capture, the maximum frequency suffices.

[20] Imaging chips usually interpose an optical *anti-aliasing filter* just before the imaging chip to reduce or control the amount of aliasing.

[21] The complex Fourier transform of the PSF is actually called the *optical transfer function* (OTF) (Williams 1999). Its

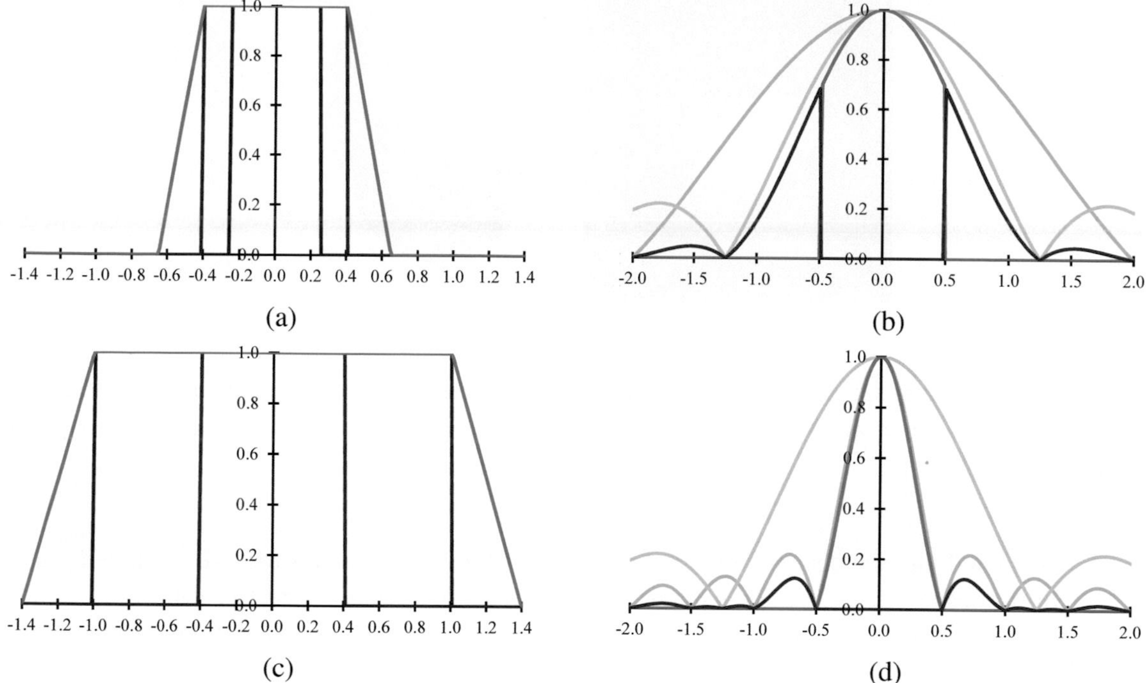

Figure 2.27 Sample point spread functions (PSF): The diameter of the blur disc (blue) in (a) is equal to half the pixel spacing, while the diameter in (c) is twice the pixel spacing. The horizontal fill factor of the sensing chip is 80% and is shown in brown. The convolution of these two kernels gives the point spread function, shown in green. The Fourier response of the PSF (the MTF) is plotted in (b) and (d). The area above the Nyquist frequency where aliasing occurs is shown in red.

If we defocus the lens so that the blur function has a radius of $2s$ (Figure 2.27c), we see that the amount of aliasing decreases significantly, but so does the amount of image detail (frequencies closer to $f = f_s$).

Under laboratory conditions, the PSF can be estimated (to pixel precision) by looking at a point light source such as a pinhole in a black piece of cardboard lit from behind. However, this PSF (the actual image of the pinhole) is only accurate to a pixel resolution and, while it can model larger blur (such as blur caused by defocus), it cannot model the sub-pixel shape of the PSF and predict the amount of aliasing. An alternative technique, described in Section 10.1.4, is to look at a calibration pattern (e.g., one consisting of slanted step edges (Reichenbach, Park, and Narayanswamy 1991; Williams and Burns 2001; Joshi, Szeliski, and Kriegman 2008)) whose ideal appearance can be re-synthesized to sub-pixel precision.

In addition to occurring during image acquisition, aliasing can also be introduced in various image processing operations, such as resampling, upsampling, and downsampling. Sections 3.4 and 3.5.2 discuss these issues and show how careful selection of filters can reduce the amount of aliasing.

2.3.2 Color

In Section 2.2, we saw how lighting and surface reflections are functions of wavelength. When the incoming light hits the imaging sensor, light from different parts of the spectrum is somehow

magnitude is called the *modulation transfer function* (MTF) and its phase is called the *phase transfer function* (PTF).

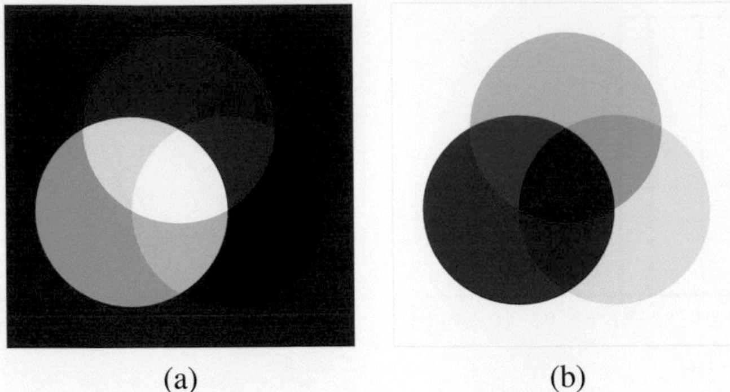

(a) (b)

Figure 2.28 Primary and secondary colors: (a) additive colors red, green, and blue can be mixed to produce cyan, magenta, yellow, and white; (b) subtractive colors cyan, magenta, and yellow can be mixed to produce red, green, blue, and black.

integrated into the discrete red, green, and blue (RGB) color values that we see in a digital image. How does this process work and how can we analyze and manipulate color values?

You probably recall from your childhood days the magical process of mixing paint colors to obtain new ones. You may recall that blue+yellow makes green, red+blue makes purple, and red+green makes brown. If you revisited this topic at a later age, you may have learned that the proper *subtractive* primaries are actually cyan (a light blue-green), magenta (pink), and yellow (Figure 2.28b), although black is also often used in four-color printing (CMYK).[22] If you ever subsequently took any painting classes, you learned that colors can have even more fanciful names, such as alizarin crimson, cerulean blue, and chartreuse. The subtractive colors are called subtractive because pigments in the paint absorb certain wavelengths in the color spectrum.

Later on, you may have learned about the *additive* primary colors (red, green, and blue) and how they can be added (with a slide projector or on a computer monitor) to produce cyan, magenta, yellow, white, and all the other colors we typically see on our TV sets and monitors (Figure 2.28a).

Through what process is it possible for two different colors, such as red and green, to interact to produce a third color like yellow? Are the wavelengths somehow mixed up to produce a new wavelength?

You probably know that the correct answer has nothing to do with physically mixing wavelengths. Instead, the existence of three primaries is a result of the *tri-stimulus* (or *tri-chromatic*) nature of the human visual system, since we have three different kinds of cells called cones, each of which responds selectively to a different portion of the color spectrum (Glassner 1995; Wandell 1995; Wyszecki and Stiles 2000; Livingstone 2008; Frisby and Stone 2010; Reinhard, Heidrich *et al.* 2010; Fairchild 2013).[23] Note that for machine vision applications, such as remote sensing and terrain classification, it is preferable to use many more wavelengths. Similarly, surveillance applications can often benefit from sensing in the near-infrared (NIR) range.

CIE RGB and XYZ

To test and quantify the tri-chromatic theory of perception, we can attempt to reproduce all *monochromatic* (single wavelength) colors as a mixture of three suitably chosen primaries. (Pure wavelength light can be obtained using either a prism or specially manufactured color filters.) In the

[22] It is possible to use additional inks such as orange, green, and violet to further extend the color gamut.

[23] See also Mark Fairchild's web page, http://markfairchild.org/WhyIsColor/books_links.html.

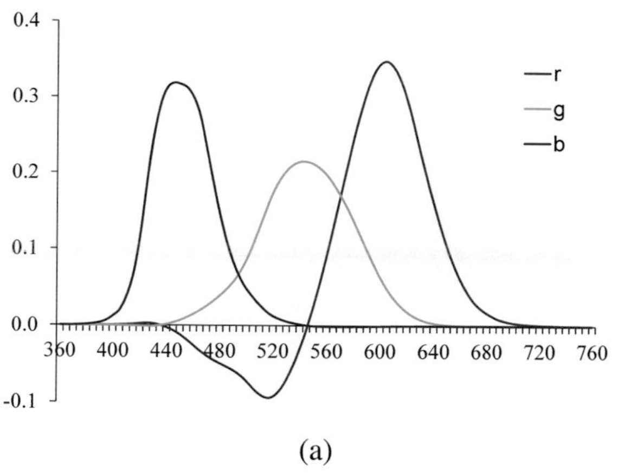

(a)

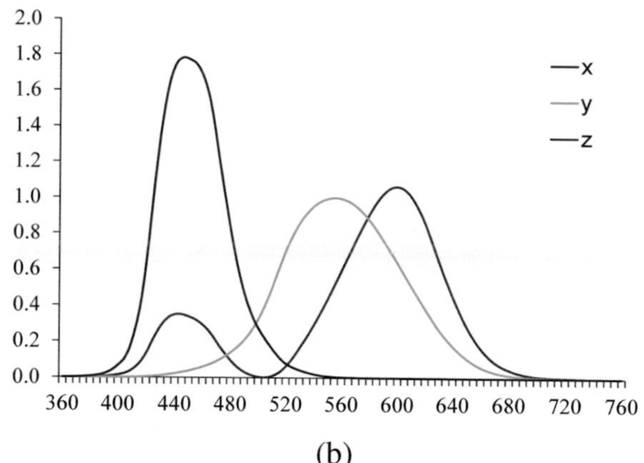

(b)

Figure 2.29 Standard CIE color matching functions: (a) $\bar{r}(\lambda), \bar{g}(\lambda), \bar{b}(\lambda)$ color spectra obtained from matching pure colors to the R=700.0nm, G=546.1nm, and B=435.8nm primaries; (b) $\bar{x}(\lambda), \bar{y}(\lambda), \bar{z}(\lambda)$ color matching functions, which are linear combinations of the $(\bar{r}(\lambda), \bar{g}(\lambda), \bar{b}(\lambda))$ spectra.

1930s, the Commission Internationale d'Eclairage (CIE) standardized the RGB representation by performing such *color matching* experiments using the primary colors of red (700.0nm wavelength), green (546.1nm), and blue (435.8nm).

Figure 2.29 shows the results of performing these experiments with a *standard observer*, i.e., averaging perceptual results over a large number of subjects.[24] You will notice that for certain pure spectra in the blue–green range, a *negative* amount of red light has to be added, i.e., a certain amount of red has to be added to the color being matched to get a color match. These results also provided a simple explanation for the existence of *metamers*, which are colors with different spectra that are perceptually indistinguishable. Note that two fabrics or paint colors that are metamers under one light may no longer be so under different lighting.

Because of the problem associated with mixing negative light, the CIE also developed a new color space called XYZ, which contains all of the pure spectral colors within its positive octant. (It also maps the Y axis to the *luminance*, i.e., perceived relative brightness, and maps pure white to a diagonal (equal-valued) vector.) The transformation from RGB to XYZ is given by

$$\begin{bmatrix} X \\ Y \\ Z \end{bmatrix} = \frac{1}{0.17697} \begin{bmatrix} 0.49 & 0.31 & 0.20 \\ 0.17697 & 0.81240 & 0.01063 \\ 0.00 & 0.01 & 0.99 \end{bmatrix} \begin{bmatrix} R \\ G \\ B \end{bmatrix}. \tag{2.104}$$

While the official definition of the CIE XYZ standard has the matrix normalized so that the Y value corresponding to pure red is 1, a more commonly used form is to omit the leading fraction, so that the second row adds up to one, i.e., the RGB triplet $(1, 1, 1)$ maps to a Y value of 1. Linearly blending the $(\bar{r}(\lambda), \bar{g}(\lambda), \bar{b}(\lambda))$ curves in Figure 2.29a according to (2.104), we obtain the resulting $(\bar{x}(\lambda), \bar{y}(\lambda), \bar{z}(\lambda))$ curves shown in Figure 2.29b. Notice how all three spectra (color matching functions) now have only positive values and how the $\bar{y}(\lambda)$ curve matches that of the luminance perceived by humans.

If we divide the XYZ values by the sum of X+Y+Z, we obtain the *chromaticity coordinates*

$$x = \frac{X}{X+Y+Z}, \quad y = \frac{Y}{X+Y+Z}, \quad z = \frac{Z}{X+Y+Z}, \tag{2.105}$$

[24]As Michael Brown notes in his tutorial on color (Brown 2019), the standard observer is actually an average taken over only 17 British subjects in the 1920s.

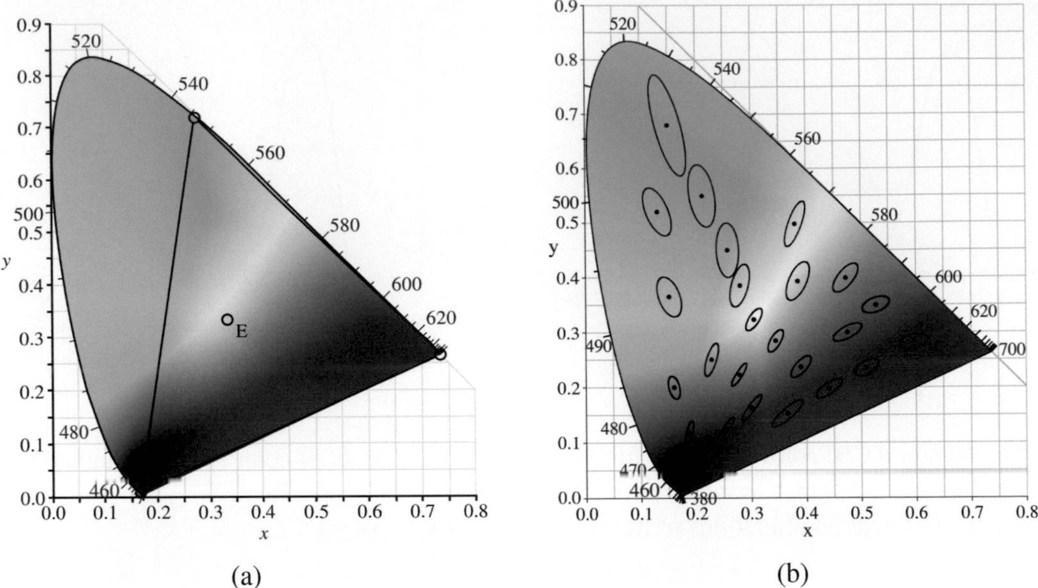

Figure 2.30 CIE chromaticity diagram, showing the pure single-wavelength spectral colors along the perimeter and the white point at E, plotted along their corresponding (x, y) values. (a) the red, green, and blue primaries do not span the complete gamut, so that negative amounts of red need to be added to span the blue–green range; (b) the MacAdam ellipses show color regions of equal discriminability, and form the basis of the Lab perceptual color space.

which sum to 1. The chromaticity coordinates discard the absolute intensity of a given color sample and just represent its pure color. If we sweep the monochromatic color λ parameter in Figure 2.29b from $\lambda = 380$nm to $\lambda = 800$nm, we obtain the familiar *chromaticity diagram* shown in Figure 2.30a. This figure shows the (x, y) value for every color value perceivable by most humans. (Of course, the CMYK reproduction process in this book does not actually span the whole gamut of perceivable colors.) The outer curved rim represents where all of the pure monochromatic color values map in (x, y) space, while the lower straight line, which connects the two endpoints, is known as the *purple line*. The inset triangle spans the red, green, and blue single-wavelength primaries used in the original color matching experiments, while E denotes the white point.

A convenient representation for color values, when we want to tease apart luminance and chromaticity, is therefore Yxy (luminance plus the two most distinctive chrominance components).

$L^*a^*b^*$ color space

While the XYZ color space has many convenient properties, including the ability to separate luminance from chrominance, it does not actually predict how well humans perceive *differences* in color or luminance.

Because the response of the human visual system is roughly logarithmic (we can perceive *relative* luminance differences of about 1%), the CIE defined a non-linear re-mapping of the XYZ space called L*a*b* (also sometimes called CIELAB), where differences in luminance or chrominance are more perceptually uniform, as shown in Figure 2.30b.[25]

[25] Another perceptually motivated color space called L*u*v* was developed and standardized simultaneously (Fairchild 2013).

The L* component of *lightness* is defined as

$$L^* = 116 f\left(\frac{Y}{Y_n}\right), \tag{2.106}$$

where Y_n is the luminance value for nominal white (Fairchild 2013) and

$$f(t) = \begin{cases} t^{1/3} & t > \delta^3 \\ t/(3\delta^2) + 2\delta/3 & \text{else}, \end{cases} \tag{2.107}$$

is a finite-slope approximation to the cube root with $\delta = 6/29$. The resulting 0...100 scale roughly measures equal amounts of lightness perceptibility.

In a similar fashion, the a* and b* components are defined as

$$a^* = 500\left[f\left(\frac{X}{X_n}\right) - f\left(\frac{Y}{Y_n}\right)\right] \quad \text{and} \quad b^* = 200\left[f\left(\frac{Y}{Y_n}\right) - f\left(\frac{Z}{Z_n}\right)\right], \tag{2.108}$$

where again, (X_n, Y_n, Z_n) is the measured white point. Figure 2.33i–k show the L*a*b* representation for a sample color image.

Color cameras

While the preceding discussion tells us how we can uniquely describe the perceived tri-stimulus description of any color (spectral distribution), it does not tell us how RGB still and video cameras actually work. Do they just measure the amount of light at the nominal wavelengths of red (700.0nm), green (546.1nm), and blue (435.8nm)? Do color monitors just emit exactly these wavelengths and, if so, how can they emit negative red light to reproduce colors in the cyan range?

In fact, the design of RGB video cameras has historically been based around the availability of colored phosphors that go into television sets. When standard-definition color television was invented (NTSC), a mapping was defined between the RGB values that would drive the three color guns in the cathode ray tube (CRT) and the XYZ values that unambiguously define perceived color (this standard was called ITU-R BT.601). With the advent of HDTV and newer monitors, a new standard called ITU-R BT.709 was created, which specifies the XYZ values of each of the color primaries,

$$\begin{bmatrix} X \\ Y \\ Z \end{bmatrix} = \begin{bmatrix} 0.412453 & 0.357580 & 0.180423 \\ 0.212671 & 0.715160 & 0.072169 \\ 0.019334 & 0.119193 & 0.950227 \end{bmatrix} \begin{bmatrix} R_{709} \\ G_{709} \\ B_{709} \end{bmatrix}. \tag{2.109}$$

In practice, each color camera integrates light according to the *spectral response function* of its red, green, and blue sensors,

$$\begin{aligned} R &= \int L(\lambda) S_R(\lambda) d\lambda, \\ G &= \int L(\lambda) S_G(\lambda) d\lambda, \\ B &= \int L(\lambda) S_B(\lambda) d\lambda, \end{aligned} \tag{2.110}$$

where $L(\lambda)$ is the incoming spectrum of light at a given pixel and $[S_R(\lambda), S_G(\lambda), S_B(\lambda)]$ are the red, green, and blue *spectral sensitivities* of the corresponding sensors.

Can we tell what spectral sensitivities the cameras actually have? Unless the camera manufacturer provides us with these data or we observe the response of the camera to a whole spectrum of

G	R	G	R
B	G	B	G
G	R	G	R
B	G	B	G

(a)

rGb	Rgb	rGb	Rgb
rgB	rGb	rgB	rGb
rGb	Rgb	rGb	Rgb
rgB	rGb	rgB	rGb

(b)

Figure 2.31 Bayer RGB pattern: (a) color filter array layout; (b) interpolated pixel values, with unknown (guessed) values shown as lower case.

monochromatic lights, these sensitivities are *not* specified by a standard such as BT.709. Instead, all that matters is that the tri-stimulus values for a given color produce the specified RGB values. The manufacturer is free to use sensors with sensitivities that do not match the standard XYZ definitions, so long as they can later be converted (through a linear transform) to the standard colors.

Similarly, while TV and computer monitors are supposed to produce RGB values as specified by Equation (2.109), there is no reason that they cannot use digital logic to transform the incoming RGB values into different signals to drive each of the color channels.[26] Properly calibrated monitors make this information available to software applications that perform *color management*, so that colors in real life, on the screen, and on the printer all match as closely as possible.

Color filter arrays

While early color TV cameras used three *vidicons* (tubes) to perform their sensing and later cameras used three separate RGB sensing chips, most of today's digital still and video cameras use a *color filter array* (CFA), where alternating sensors are covered by different colored filters (Figure 2.24).[27]

The most commonly used pattern in color cameras today is the *Bayer pattern* (Bayer 1976), which places green filters over half of the sensors (in a checkerboard pattern), and red and blue filters over the remaining ones (Figure 2.31). The reason that there are twice as many green filters as red and blue is because the luminance signal is mostly determined by green values and the visual system is much more sensitive to high-frequency detail in luminance than in chrominance (a fact that is exploited in color image compression—see Section 2.3.3). The process of *interpolating* the missing color values so that we have valid RGB values for all the pixels is known as *demosaicing* and is covered in detail in Section 10.3.1.

Similarly, color LCD monitors typically use alternating stripes of red, green, and blue filters placed in front of each liquid crystal active area to simulate the experience of a full color display. As before, because the visual system has higher resolution (acuity) in luminance than chrominance, it is possible to digitally prefilter RGB (and monochrome) images to enhance the perception of crispness (Betrisey, Blinn *et al.* 2000; Platt 2000b).

[26]The latest OLED TV monitors are now introducing higher dynamic range (HDR) and wide color gamut (WCG), https://www.cnet.com/how-to/what-is-wide-color-gamut-wcg.

[27]A chip design by Foveon stacked the red, green, and blue sensors beneath each other, but it never gained widespread adoption. Descriptions of alternative color filter arrays that have been proposed over the years can be found at https://en.wikipedia.org/wiki/Color_filter_array.

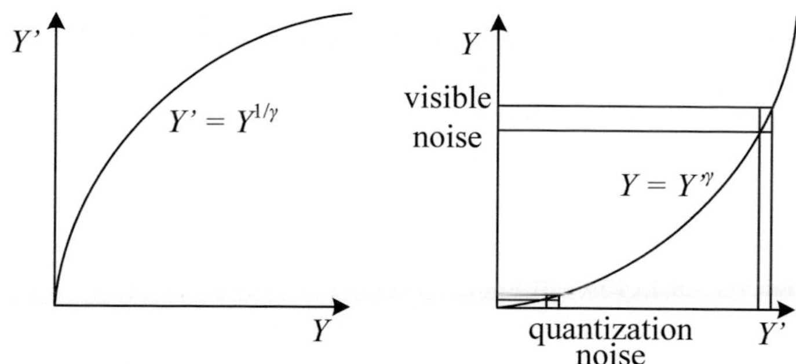

Figure 2.32 Gamma compression: (a) The relationship between the input signal luminance Y and the transmitted signal Y' is given by $Y' = Y^{1/\gamma}$. (b) At the receiver, the signal Y' is exponentiated by the factor γ, $\hat{Y} = Y'^{\gamma}$. Noise introduced during transmission is squashed in the dark regions, which corresponds to the more noise-sensitive region of the visual system.

Color balance

Before encoding the sensed RGB values, most cameras perform some kind of *color balancing* operation in an attempt to move the white point of a given image closer to pure white (equal RGB values). If the color system and the illumination are the same (the BT.709 system uses the daylight illuminant D_{65} as its reference white), the change may be minimal. However, if the illuminant is strongly colored, such as incandescent indoor lighting (which generally results in a yellow or orange hue), the compensation can be quite significant.

A simple way to perform color correction is to multiply each of the RGB values by a different factor (i.e., to apply a diagonal matrix transform to the RGB color space). More complicated transforms, which are sometimes the result of mapping to XYZ space and back, actually perform a *color twist*, i.e., they use a general 3×3 color transform matrix.[28] Exercise 2.8 has you explore some of these issues.

Gamma

In the early days of black and white television, the phosphors in the CRT used to display the TV signal responded non-linearly to their input voltage. The relationship between the voltage and the resulting brightness was characterized by a number called *gamma* (γ), since the formula was roughly

$$B = V^{\gamma}, \tag{2.111}$$

with a γ of about 2.2. To compensate for this effect, the electronics in the TV camera would pre-map the sensed luminance Y through an inverse gamma,

$$Y' = Y^{\frac{1}{\gamma}}, \tag{2.112}$$

with a typical value of $\frac{1}{\gamma} = 0.45$.

The mapping of the signal through this non-linearity before transmission had a beneficial side effect: noise added during transmission (remember, these were analog days!) would be reduced (after applying the gamma at the receiver) in the darker regions of the signal where it was more

[28] Those of you old enough to remember the early days of color television will naturally think of the *hue* adjustment knob on the television set, which could produce truly bizarre results.

visible (Figure 2.32).[29] (Remember that our visual system is roughly sensitive to relative differences in luminance.)

When color television was invented, it was decided to separately pass the red, green, and blue signals through the same gamma non-linearity before combining them for encoding. Today, even though we no longer have analog noise in our transmission systems, signals are still quantized during compression (see Section 2.3.3), so applying inverse gamma to sensed values remains useful.

Unfortunately, for both computer vision and computer graphics, the presence of gamma in images is often problematic. For example, the proper simulation of radiometric phenomena such as shading (see Section 2.2 and Equation (2.88)) occurs in a linear radiance space. Once all of the computations have been performed, the appropriate gamma should be applied before display. Unfortunately, many computer graphics systems (such as shading models) operate directly on RGB values and display these values directly. (Fortunately, newer color imaging standards such as the 16-bit scRGB use a linear space, which makes this less of a problem (Glassner 1995).)

In computer vision, the situation can be even more daunting. The accurate determination of surface normals, using a technique such as photometric stereo (Section 13.1.1) or even a simpler operation such as accurate image deblurring, require that the measurements be in a linear space of intensities. Therefore, it is imperative when performing detailed quantitative computations such as these to first undo the gamma and the per-image color re-balancing in the sensed color values. Chakrabarti, Scharstein, and Zickler (2009) develop a sophisticated 24-parameter model that is a good match to the processing performed by today's digital cameras; they also provide a database of color images you can use for your own testing.

For other vision applications, however, such as feature detection or the matching of signals in stereo and motion estimation, this linearization step is often not necessary. In fact, determining whether it is necessary to undo gamma can take some careful thinking, e.g., in the case of compensating for exposure variations in image stitching (see Exercise 2.7).

If all of these processing steps sound confusing to model, they are. Exercise 2.9 has you try to tease apart some of these phenomena using empirical investigation, i.e., taking pictures of color charts and comparing the RAW and JPEG compressed color values.

Other color spaces

While RGB and XYZ are the primary color spaces used to describe the spectral content (and hence tri-stimulus response) of color signals, a variety of other representations have been developed both in video and still image coding and in computer graphics.

The earliest color representation developed for video transmission was the YIQ standard developed for NTSC video in North America and the closely related YUV standard developed for PAL in Europe. In both of these cases, it was desired to have a *luma* channel Y (so called since it only roughly mimics true luminance) that would be comparable to the regular black-and-white TV signal, along with two lower frequency *chroma* channels.

In both systems, the Y signal (or more appropriately, the Y' luma signal since it is gamma compressed) is obtained from

$$Y'_{601} = 0.299R' + 0.587G' + 0.114B', \tag{2.113}$$

where R′G′B′ is the triplet of gamma-compressed color components. When using the newer color definitions for HDTV in BT.709, the formula is

$$Y'_{709} = 0.2125R' + 0.7154G' + 0.0721B'. \tag{2.114}$$

[29] A related technique called *companding* was the basis of the Dolby noise reduction systems used with audio tapes.

The UV components are derived from scaled versions of $(B' - Y')$ and $(R' - Y')$, namely,

$$U = 0.492111(B' - Y') \quad \text{and} \quad V = 0.877283(R' - Y'), \tag{2.115}$$

whereas the IQ components are the UV components rotated through an angle of 33°. In composite (NTSC and PAL) video, the chroma signals were then low-pass filtered horizontally before being modulated and superimposed on top of the Y' luma signal. Backward compatibility was achieved by having older black-and-white TV sets effectively ignore the high-frequency chroma signal (because of slow electronics) or, at worst, superimposing it as a high-frequency pattern on top of the main signal.

While these conversions were important in the early days of computer vision, when frame grabbers would directly digitize the composite TV signal, today all digital video and still image compression standards are based on the newer YCbCr conversion. YCbCr is closely related to YUV (the C_b and C_r signals carry the blue and red color difference signals and have more useful mnemonics than UV) but uses different scale factors to fit within the eight-bit range available with digital signals.

For video, the Y' signal is re-scaled to fit within the $[16 \ldots 235]$ range of values, while the Cb and Cr signals are scaled to fit within $[16 \ldots 240]$ (Gomes and Velho 1997; Fairchild 2013). For still images, the JPEG standard uses the full eight-bit range with no reserved values,

$$\begin{bmatrix} Y' \\ C_b \\ C_r \end{bmatrix} = \begin{bmatrix} 0.299 & 0.587 & 0.114 \\ -0.168736 & -0.331264 & 0.5 \\ 0.5 & -0.418688 & -0.081312 \end{bmatrix} \begin{bmatrix} R' \\ G' \\ B' \end{bmatrix} + \begin{bmatrix} 0 \\ 128 \\ 128 \end{bmatrix}, \tag{2.116}$$

where the R′G′B′ values are the eight-bit gamma-compressed color components (i.e., the actual RGB values we obtain when we open up or display a JPEG image). For most applications, this formula is not that important, since your image reading software will directly provide you with the eight-bit gamma-compressed R′G′B′ values. However, if you are trying to do careful image deblocking (Exercise 4.3), this information may be useful.

Another color space you may come across is *hue, saturation, value* (HSV), which is a projection of the RGB color cube onto a non-linear chroma angle, a radial saturation percentage, and a luminance-inspired value. In more detail, value is defined as either the mean or maximum color value, saturation is defined as scaled distance from the diagonal, and hue is defined as the direction around a color wheel (the exact formulas are described by Hall (1989), Hughes, van Dam *et al.* (2013), and Brown (2019)). Such a decomposition is quite natural in graphics applications such as color picking (it approximates the Munsell chart for color description). Figure 2.33l–n shows an HSV representation of a sample color image, where saturation is encoded using a gray scale (saturated = darker) and hue is depicted as a color.

If you want your computer vision algorithm to only affect the value (luminance) of an image and not its saturation or hue, a simpler solution is to use either the Yxy (luminance + chromaticity) coordinates defined in (2.105) or the even simpler *color ratios*,

$$r = \frac{R}{R+G+B}, \quad g = \frac{G}{R+G+B}, \quad b = \frac{B}{R+G+B} \tag{2.117}$$

(Figure 2.33e–h). After manipulating the luma (2.113), e.g., through the process of histogram equalization (Section 3.1.4), you can multiply each color ratio by the ratio of the new to old luma to obtain an adjusted RGB triplet.

While all of these color systems may sound confusing, in the end, it often may not matter that much which one you use. Poynton, in his *Color FAQ*, https://www.poynton.com/ColorFAQ.

(a) RGB (b) R (c) G (d) B

(e) rgb (f) r (g) g (h) b

(i) L* (j) a* (k) b*

(l) H (m) S (n) V

Figure 2.33 Color space transformations: (a–d) RGB; (e–h) rgb. (i–k) L*a*b*; (l–n) HSV. Note that the rgb, L*a*b*, and HSV values are all re-scaled to fit the dynamic range of the printed page.

html, notes that the perceptually motivated L*a*b* system is qualitatively similar to the gamma-compressed R′G′B′ system we mostly deal with, since both have a fractional power scaling (which approximates a logarithmic response) between the actual intensity values and the numbers being manipulated. As in all cases, think carefully about what you are trying to accomplish before deciding on a technique to use.

2.3.3 Compression

The last stage in a camera's processing pipeline is usually some form of image compression (unless you are using a lossless compression scheme such as camera RAW or PNG).

All color video and image compression algorithms start by converting the signal into YCbCr (or some closely related variant), so that they can compress the luminance signal with higher fidelity than the chrominance signal. (Recall that the human visual system has poorer frequency response to color than to luminance changes.) In video, it is common to subsample Cb and Cr by a factor

Figure 2.34 Image compressed with JPEG at three quality settings. Note how the amount of block artifact and high-frequency aliasing ("mosquito noise") increases from left to right.

of two horizontally; with still images (JPEG), the subsampling (averaging) occurs both horizontally and vertically.

Once the luminance and chrominance images have been appropriately subsampled and separated into individual images, they are then passed to a *block transform* stage. The most common technique used here is the *discrete cosine transform* (DCT), which is a real-valued variant of the discrete Fourier transform (DFT) (see Section 3.4.1). The DCT is a reasonable approximation to the Karhunen–Loève or eigenvalue decomposition of natural image patches, i.e., the decomposition that simultaneously packs the most energy into the first coefficients and diagonalizes the joint covariance matrix among the pixels (makes transform coefficients statistically independent). Both MPEG and JPEG use 8×8 DCT transforms (Wallace 1991; Le Gall 1991), although newer variants, including the new AV1 open standard,[30] use smaller 4×4 or even 2×2 blocks. Alternative transformations, such as wavelets (Taubman and Marcellin 2002) and lapped transforms (Malvar 1990, 1998, 2000) are used in compression standards such as JPEG 2000 and JPEG XR.

After transform coding, the coefficient values are quantized into a set of small integer values that can be coded using a variable bit length scheme such as a Huffman code or an arithmetic code (Wallace 1991; Marpe, Schwarz, and Wiegand 2003). (The DC (lowest frequency) coefficients are also adaptively predicted from the previous block's DC values. The term "DC" comes from "direct current", i.e., the non-sinusoidal or non-alternating part of a signal.) The step size in the quantization is the main variable controlled by the *quality* setting on the JPEG file (Figure 2.34).

With video, it is also usual to perform block-based *motion compensation*, i.e., to encode the difference between each block and a *predicted* set of pixel values obtained from a shifted block in the previous frame. (The exception is the *motion-JPEG* scheme used in older DV camcorders, which is nothing more than a series of individually JPEG compressed image frames.) While basic MPEG uses 16×16 motion compensation blocks with integer motion values (Le Gall 1991), newer standards use adaptively sized blocks, sub-pixel motions, and the ability to reference blocks from older frames (Sullivan, Ohm *et al.* 2012). In order to recover more gracefully from failures and to allow for random access to the video stream, predicted P frames are interleaved among independently coded I frames. (Bi-directional B frames are also sometimes used.)

The quality of a compression algorithm is usually reported using its *peak signal-to-noise ratio* (PSNR), which is derived from the average *mean square error*,

$$MSE = \frac{1}{n} \sum_{\mathbf{x}} \left[I(\mathbf{x}) - \hat{I}(\mathbf{x}) \right]^2, \tag{2.118}$$

[30] https://aomedia.org

where $I(\mathbf{x})$ is the original uncompressed image and $\hat{I}(\mathbf{x})$ is its compressed counterpart, or equivalently, the *root mean square error* (RMS error), which is defined as

$$RMS = \sqrt{MSE}. \tag{2.119}$$

The PSNR is defined as

$$PSNR = 10 \log_{10} \frac{I_{\max}^2}{MSE} = 20 \log_{10} \frac{I_{\max}}{RMS}, \tag{2.120}$$

where $I_{\max}$ is the maximum signal extent, e.g., 255 for eight-bit images.

While this is just a high-level sketch of how image compression works, it is useful to understand so that the artifacts introduced by such techniques can be compensated for in various computer vision applications. Note also that researchers are currently developing novel image and video compression algorithms based on deep neural networks, e.g., (Rippel and Bourdev 2017; Mentzer, Agustsson *et al.* 2019; Rippel, Nair *et al.* 2019) and https://www.compression.cc. It will be interesting to see what kinds of different artifacts these techniques produce.

2.4 Additional reading

As we mentioned at the beginning of this chapter, this book provides but a brief summary of a very rich and deep set of topics, traditionally covered in a number of separate fields.

A more thorough introduction to the geometry of points, lines, planes, and projections can be found in textbooks on multi-view geometry (Faugeras and Luong 2001; Hartley and Zisserman 2004) and computer graphics (Watt 1995; OpenGL-ARB 1997; Hughes, van Dam *et al.* 2013; Marschner and Shirley 2015). Topics covered in more depth include higher-order primitives such as quadrics, conics, and cubics, as well as three-view and multi-view geometry.

The image formation (synthesis) process is traditionally taught as part of a computer graphics curriculum (Glassner 1995; Watt 1995; Hughes, van Dam *et al.* 2013; Marschner and Shirley 2015) but it is also studied in physics-based computer vision (Wolff, Shafer, and Healey 1992a). The behavior of camera lens systems is studied in optics (Möller 1988; Ray 2002; Hecht 2015).

Some good books on color theory have been written by Healey and Shafer (1992), Wandell (1995), Wyszecki and Stiles (2000), and Fairchild (2013), with Livingstone (2008) providing a more fun and informal introduction to the topic of color perception. Mark Fairchild's page of color books and links[31] lists many other sources.

Topics relating to sampling and aliasing are covered in textbooks on signal and image processing (Crane 1997; Jähne 1997; Oppenheim and Schafer 1996; Oppenheim, Schafer, and Buck 1999; Pratt 2007; Russ 2007; Burger and Burge 2008; Gonzalez and Woods 2017).

Two courses that cover many of the above topics (image formation, lenses, color and sampling theory) in wonderful detail are Marc Levoy's Digital Photography course at Stanford (Levoy 2010) and Michael Brown's tutorial on the image processing pipeline at ICCV 2019 (Brown 2019). The recent book by Ikeuchi, Matsushita *et al.* (2020) also covers 3D geometry, photometry, and sensor models, but with an emphasis on *active illumination* systems.

2.5 Exercises

A note to students: This chapter is relatively light on exercises since it contains mostly background material and not that many usable techniques. If you really want to understand multi-view geometry

[31] http://markfairchild.org/WhyIsColor/books_links.html.

in a thorough way, I encourage you to read and do the exercises provided by Hartley and Zisserman (2004). Similarly, if you want some exercises related to the image formation process, Glassner's (1995) book is full of challenging problems.

Ex 2.1: Least squares intersection point and line fitting—advanced. Equation (2.4) shows how the intersection of two 2D lines can be expressed as their cross product, assuming the lines are expressed as homogeneous coordinates.

1. If you are given more than two lines and want to find a point $\tilde{\mathbf{x}}$ that minimizes the sum of squared distances to each line,
$$D = \sum_i (\tilde{\mathbf{x}} \cdot \tilde{\mathbf{l}}_i)^2, \tag{2.121}$$
how can you compute this quantity? (Hint: Write the dot product as $\tilde{\mathbf{x}}^T \tilde{\mathbf{l}}_i$ and turn the squared quantity into a *quadratic form*, $\tilde{\mathbf{x}}^T \mathbf{A} \tilde{\mathbf{x}}$.)

2. To fit a line to a bunch of points, you can compute the *centroid* (mean) of the points as well as the *covariance matrix* of the points around this mean. Show that the line passing through the centroid along the major axis of the covariance ellipsoid (largest eigenvector) minimizes the sum of squared distances to the points.

3. These two approaches are fundamentally different, even though projective duality tells us that points and lines are interchangeable. Why are these two algorithms so apparently different? Are they actually minimizing different objectives?

Ex 2.2: 2D transform editor. Write a program that lets you interactively create a set of rectangles and then modify their "pose" (2D transform). You should implement the following steps:

1. Open an empty window ("canvas").

2. Shift drag (rubber-band) to create a new rectangle.

3. Select the deformation mode (motion model): translation, rigid, similarity, affine, or perspective.

4. Drag any corner of the outline to change its transformation.

This exercise should be built on a set of pixel coordinate and transformation classes, either implemented by yourself or from a software library. Persistence of the created representation (save and load) should also be supported (for each rectangle, save its transformation).

Ex 2.3: 3D viewer. Write a simple viewer for 3D points, lines, and polygons. Import a set of point and line commands (primitives) as well as a viewing transform. Interactively modify the object or camera transform. This viewer can be an extension of the one you created in Exercise 2.2. Simply replace the viewing transformations with their 3D equivalents.

(Optional) Add a z-buffer to do hidden surface removal for polygons.

(Optional) Use a 3D drawing package and just write the viewer control.

Ex 2.4: Focus distance and depth of field. Figure out how the focus distance and depth of field indicators on a lens are determined.

1. Compute and plot the focus distance z_o as a function of the distance traveled from the focal length $\Delta z_i = f - z_i$ for a lens of focal length f (say, 100mm). Does this explain the hyperbolic progression of focus distances you see on a typical lens (Figure 2.20)?

2. Compute the depth of field (minimum and maximum focus distances) for a given focus setting z_o as a function of the circle of confusion diameter c (make it a fraction of the sensor width), the focal length f, and the f-stop number N (which relates to the aperture diameter d). Does this explain the usual depth of field markings on a lens that bracket the in-focus marker, as in Figure 2.20a?

3. Now consider a zoom lens with a varying focal length f. Assume that as you zoom, the lens stays in focus, i.e., the distance from the rear nodal point to the sensor plane z_i adjusts itself automatically for a fixed focus distance z_o. How do the depth of field indicators vary as a function of focal length? Can you reproduce a two-dimensional plot that mimics the curved depth of field lines seen on the lens in Figure 2.20b?

Ex 2.5: F-numbers and shutter speeds. List the common f-numbers and shutter speeds that your camera provides. On older model SLRs, they are visible on the lens and shutter speed dials. On newer cameras, you have to look at the electronic viewfinder (or LCD screen/indicator) as you manually adjust exposures.

1. Do these form geometric progressions; if so, what are the ratios? How do these relate to exposure values (EVs)?

2. If your camera has shutter speeds of $\frac{1}{60}$ and $\frac{1}{125}$, do you think that these two speeds are exactly a factor of two apart or a factor of 125/60 = 2.083 apart?

3. How accurate do you think these numbers are? Can you devise some way to measure exactly how the aperture affects how much light reaches the sensor and what the exact exposure times actually are?

Ex 2.6: Noise level calibration. Estimate the amount of noise in your camera by taking repeated shots of a scene with the camera mounted on a tripod. (Purchasing a remote shutter release is a good investment if you own a DSLR.) Alternatively, take a scene with constant color regions (such as a color checker chart) and estimate the variance by fitting a smooth function to each color region and then taking differences from the predicted function.

1. Plot your estimated variance as a function of level for each of your color channels separately.

2. Change the ISO setting on your camera; if you cannot do that, reduce the overall light in your scene (turn off lights, draw the curtains, wait until dusk). Does the amount of noise vary a lot with ISO/gain?

3. Compare your camera to another one at a different price point or year of make. Is there evidence to suggest that "you get what you pay for"? Does the quality of digital cameras seem to be improving over time?

Ex 2.7: Gamma correction in image stitching. Here's a relatively simple puzzle. Assume you are given two images that are part of a panorama that you want to stitch (see Section 8.2). The two images were taken with different exposures, so you want to adjust the RGB values so that they match along the seam line. Is it necessary to undo the gamma in the color values in order to achieve this?

Ex 2.8: White point balancing—tricky. A common (in-camera or post-processing) technique for performing white point adjustment is to take a picture of a white piece of paper and to adjust the RGB values of an image to make this a neutral color.

1. Describe how you would adjust the RGB values in an image given a sample "white color" of (R_w, G_w, B_w) to make this color neutral (without changing the exposure too much).

2. Does your transformation involve a simple (per-channel) scaling of the RGB values or do you need a full 3×3 color twist matrix (or something else)?

3. Convert your RGB values to XYZ. Does the appropriate correction now only depend on the XY (or xy) values? If so, when you convert back to RGB space, do you need a full 3×3 color twist matrix to achieve the same effect?

4. If you used pure diagonal scaling in the direct RGB mode but end up with a twist if you work in XYZ space, how do you explain this apparent dichotomy? Which approach is correct? (Or is it possible that neither approach is actually correct?)

If you want to find out what your camera *actually* does, continue on to the next exercise.

Ex 2.9: In-camera color processing—challenging. If your camera supports a RAW pixel mode, take a pair of RAW and JPEG images, and see if you can infer what the camera is doing when it converts the RAW pixel values to the final color-corrected and gamma-compressed eight-bit JPEG pixel values.

1. Deduce the pattern in your color filter array from the correspondence between co-located RAW and color-mapped pixel values. Use a color checker chart at this stage if it makes your life easier. You may find it helpful to split the RAW image into four separate images (subsampling even and odd columns and rows) and to treat each of these new images as a "virtual" sensor.

2. Evaluate the quality of the demosaicing algorithm by taking pictures of challenging scenes which contain strong color edges (such as those shown in in Section 10.3.1).

3. If you can take the same exact picture after changing the color balance values in your camera, compare how these settings affect this processing.

4. Compare your results against those presented in (Chakrabarti, Scharstein, and Zickler 2009), Kim, Lin *et al.* (2012), Hasinoff, Sharlet *et al.* (2016), Karaimer and Brown (2016), and Brooks, Mildenhall *et al.* (2019) or use the data available in their database of color images.

Chapter 3

Image processing

3.1 Point operators . . . 87
   3.1.1 Pixel transforms . . . 89
   3.1.2 Color transforms . . . 90
   3.1.3 Compositing and matting . . . 91
   3.1.4 Histogram equalization . . . 92
   3.1.5 *Application*: Tonal adjustment . . . 95
3.2 Linear filtering . . . 95
   3.2.1 Separable filtering . . . 99
   3.2.2 Examples of linear filtering . . . 100
   3.2.3 Band-pass and steerable filters . . . 101
3.3 More neighborhood operators . . . 105
   3.3.1 Non-linear filtering . . . 105
   3.3.2 Bilateral filtering . . . 107
   3.3.3 Binary image processing . . . 110
3.4 Fourier transforms . . . 113
   3.4.1 Two-dimensional Fourier transforms . . . 115
   3.4.2 *Application*: Sharpening, blur, and noise removal . . . 118
3.5 Pyramids and wavelets . . . 119
   3.5.1 Interpolation . . . 119
   3.5.2 Decimation . . . 122
   3.5.3 Multi-resolution representations . . . 123
   3.5.4 Wavelets . . . 128
   3.5.5 *Application*: Image blending . . . 132
3.6 Geometric transformations . . . 135
   3.6.1 Parametric transformations . . . 135
   3.6.2 Mesh-based warping . . . 140
   3.6.3 *Application*: Feature-based morphing . . . 142
3.7 Additional reading . . . 143
3.8 Exercises . . . 144

R. Szeliski, *Computer Vision*, Texts in Computer Science,
https://doi.org/10.1007/978-3-030-34372-9_3

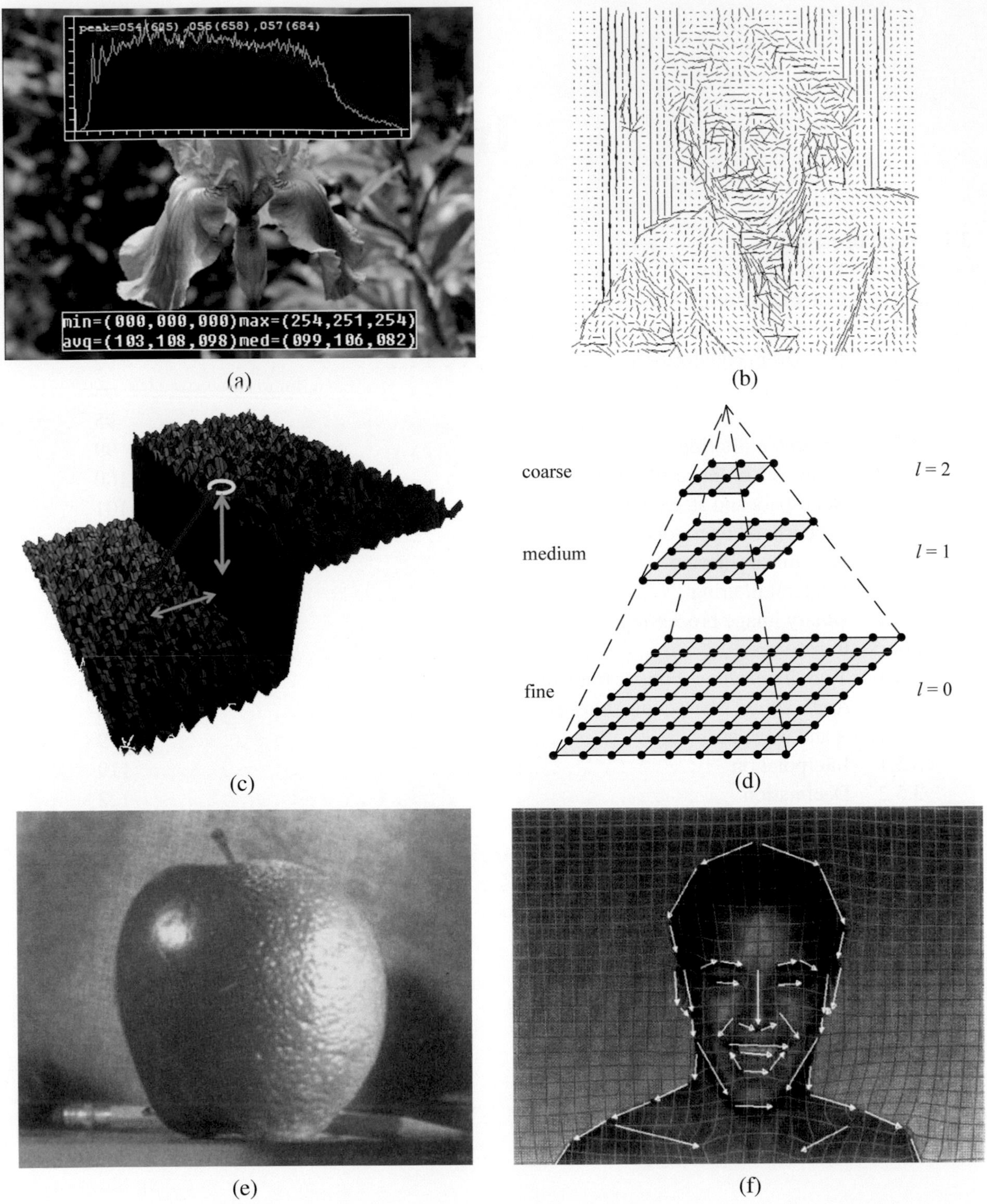

(a) (b) (c) (d) (e) (f)

Figure 3.1 Some common image processing operations: (a) partial histogram equalization; (b) orientation map computed from the second-order steerable filter (Freeman 1992) © 1992 IEEE; (c) bilateral filter (Durand and Dorsey 2002) © 2002 ACM; (d) image pyramid; (e) Laplacian pyramid blending (Burt and Adelson 1983b) © 1983 ACM; (f) line-based image warping (Beier and Neely 1992) © 1992 ACM.

Now that we have seen how images are formed through the interaction of 3D scene elements, lighting, and camera optics and sensors, let us look at the first stage in most computer vision algorithms, namely the use of image processing to preprocess the image and convert it into a form suitable for further analysis. Examples of such operations include exposure correction and color balancing, reducing image noise, increasing sharpness, or straightening the image by rotating it. Additional examples include image warping and image blending, which are often used for visual effects (Figures 3.1 and Section 3.6.3). While some may consider image processing to be outside the purview of computer vision, most computer vision applications, such as computational photography and even recognition, require care in designing the image processing stages to achieve acceptable results.

In this chapter, we review standard image processing operators that map pixel values from one image to another. Image processing is often taught in electrical engineering departments as a follow-on course to an introductory course in signal processing (Oppenheim and Schafer 1996; Oppenheim, Schafer, and Buck 1999). There are several popular textbooks for image processing, including Gomes and Velho (1997), Jähne (1997), Pratt (2007), Burger and Burge (2009), and Gonzalez and Woods (2017).

We begin this chapter with the simplest kind of image transforms, namely those that manipulate each pixel independently of its neighbors (Section 3.1). Such transforms are often called *point operators* or *point processes*. Next, we examine *neighborhood* (area-based) operators, where each new pixel's value depends on a small number of neighboring input values (Sections 3.2 and 3.3). A convenient tool to analyze (and sometimes accelerate) such neighborhood operations is the *Fourier Transform*, which we cover in Section 3.4. Neighborhood operators can be cascaded to form *image pyramids* and *wavelets*, which are useful for analyzing images at a variety of resolutions (scales) and for accelerating certain operations (Section 3.5). Another important class of global operators are *geometric transformations*, such as rotations, shears, and perspective deformations (Section 3.6).

While this chapter covers *classical* image processing techniques that consist mostly of linear and non-linear filtering operations, the next two chapters introduce energy-based and Bayesian graphical models, i.e., *Markov random fields* (Chapter 4), and then deep convolutional networks (Chapter 5), both of which are now widely used in image processing applications.

3.1 Point operators

The simplest kinds of image processing transforms are *point operators*, where each output pixel's value depends on only the corresponding input pixel value (plus, potentially, some globally collected information or parameters). Examples of such operators include brightness and contrast adjustments (Figure 3.2) as well as color correction and transformations. In the image processing literature, such operations are also known as *point processes* (Crane 1997).[1]

We begin this section with a quick review of simple point operators, such as brightness scaling and image addition. Next, we discuss how colors in images can be manipulated. We then present *image compositing* and *matting* operations, which play an important role in computational photography (Chapter 10) and computer graphics applications. Finally, we describe the more global process of *histogram equalization*. We close with an example application that manipulates *tonal values* (exposure and contrast) to improve image appearance.

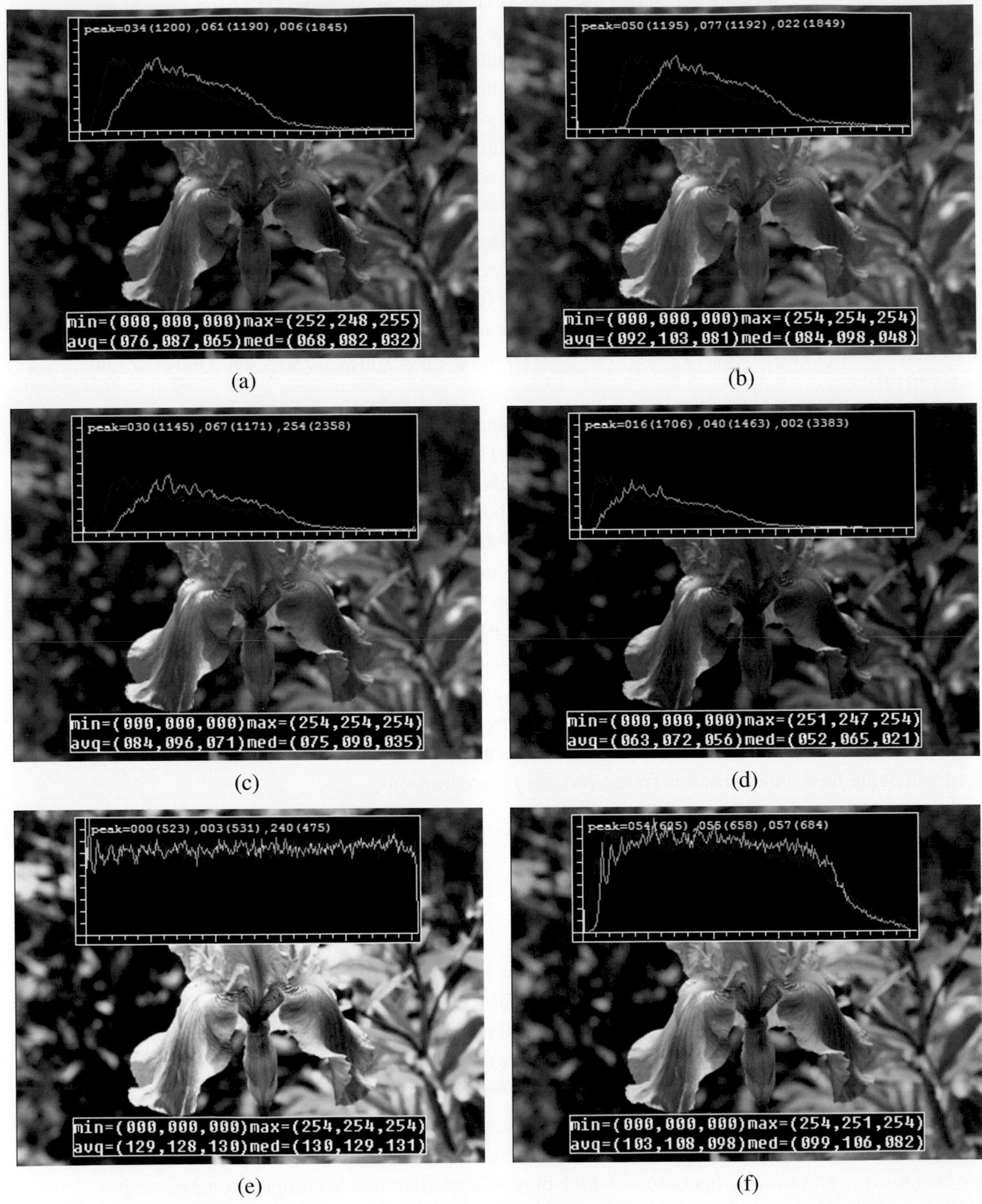

Figure 3.2 Some local image processing operations: (a) original image along with its three color (per-channel) histograms; (b) brightness increased (additive offset, $b = 16$); (c) contrast increased (multiplicative gain, $a = 1.1$); (d) gamma (partially) linearized ($\gamma = 1.2$); (e) full histogram equalization; (f) partial histogram equalization.

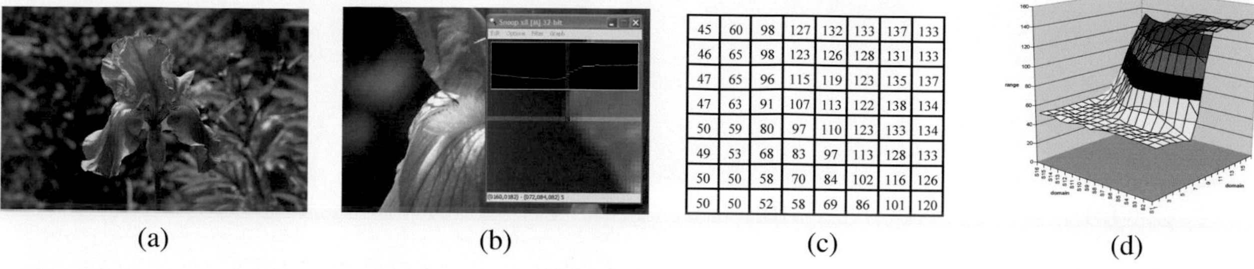

45	60	98	127	132	133	137	133
46	65	98	123	126	128	131	133
47	65	96	115	119	123	135	137
47	63	91	107	113	122	138	134
50	59	80	97	110	123	133	134
49	53	68	83	97	113	128	133
50	50	58	70	84	102	116	126
50	50	52	58	69	86	101	120

(a) (b) (c) (d)

Figure 3.3 Visualizing image data: (a) original image; (b) cropped portion and scanline plot using an image inspection tool; (c) grid of numbers; (d) surface plot. For figures (c)–(d), the image was first converted to grayscale.

3.1.1 Pixel transforms

A general image processing *operator* is a function that takes one or more input images and produces an output image. In the continuous domain, this can be denoted as

$$g(\mathbf{x}) = h(f(\mathbf{x})) \quad \text{or} \quad g(\mathbf{x}) = h(f_0(\mathbf{x}), \ldots, f_n(\mathbf{x})), \tag{3.1}$$

where $\mathbf{x}$ is in the D-dimensional (usually $D = 2$ for images) *domain* of the input and output functions f and g, which operate over some *range*, which can either be scalar or vector-valued, e.g., for color images or 2D motion. For discrete (sampled) images, the domain consists of a finite number of *pixel locations*, $\mathbf{x} = (i, j)$, and we can write

$$g(i,j) = h(f(i,j)). \tag{3.2}$$

Figure 3.3 shows how an image can be represented either by its color (appearance), as a grid of numbers, or as a two-dimensional function (surface plot).

Two commonly used point processes are multiplication and addition with a constant,

$$g(\mathbf{x}) = af(\mathbf{x}) + b. \tag{3.3}$$

The parameters $a > 0$ and b are often called the *gain* and *bias* parameters; sometimes these parameters are said to control *contrast* and *brightness*, respectively (Figures 3.2b–c).[2] The bias and gain parameters can also be spatially varying,

$$g(\mathbf{x}) = a(\mathbf{x})f(\mathbf{x}) + b(\mathbf{x}), \tag{3.4}$$

e.g., when simulating the *graded density filter* used by photographers to selectively darken the sky or when modeling vignetting in an optical system.

Multiplicative gain (both global and spatially varying) is a *linear* operation, as it obeys the *superposition principle*,

$$h(f_0 + f_1) = h(f_0) + h(f_1). \tag{3.5}$$

(We will have more to say about linear shift invariant operators in Section 3.2.) Operators such as image squaring (which is often used to get a local estimate of the *energy* in a band-pass filtered signal, see Section 3.5) are not linear.

[1] In convolutional neural networks (Section 5.4), such operations are sometimes called 1×1 convolutions.

[2] An image's luminance characteristics can also be summarized by its *key* (average luminance) and *range* (Kopf, Uyttendaele *et al.* 2007).

Figure 3.4 Image matting and compositing (Chuang, Curless *et al.* 2001) © 2001 IEEE: (a) source image; (b) extracted foreground object F; (c) alpha matte α shown in grayscale; (d) new composite C.

Another commonly used *dyadic* (two-input) operator is the *linear blend* operator,

$$g(\mathbf{x}) = (1-\alpha)f_0(\mathbf{x}) + \alpha f_1(\mathbf{x}). \tag{3.6}$$

By varying α from $0 \rightarrow 1$, this operator can be used to perform a temporal *cross-dissolve* between two images or videos, as seen in slide shows and film production, or as a component of image *morphing* algorithms (Section 3.6.3).

One highly used non-linear transform that is often applied to images before further processing is *gamma correction*, which is used to remove the non-linear mapping between input radiance and quantized pixel values (Section 2.3.2). To invert the gamma mapping applied by the sensor, we can use

$$g(\mathbf{x}) = [f(\mathbf{x})]^{1/\gamma}, \tag{3.7}$$

where a gamma value of $\gamma \approx 2.2$ is a reasonable fit for most digital cameras.

3.1.2 Color transforms

While color images can be treated as arbitrary vector-valued functions or collections of independent bands, it usually makes sense to think about them as highly correlated signals with strong connections to the image formation process (Section 2.2), sensor design (Section 2.3), and human perception (Section 2.3.2). Consider, for example, brightening a picture by adding a constant value to all three channels, as shown in Figure 3.2b. Can you tell if this achieves the desired effect of making the image look brighter? Can you see any undesirable side-effects or artifacts?

In fact, adding the same value to each color channel not only increases the apparent *intensity* of each pixel, it can also affect the pixel's *hue* and *saturation*. How can we define and manipulate such quantities in order to achieve the desired perceptual effects?

As discussed in Section 2.3.2, chromaticity coordinates (2.105) or even simpler color ratios (2.117) can first be computed and then used after manipulating (e.g., brightening) the luminance Y to re-compute a valid RGB image with the same hue and saturation. Figures 2.33f–h show some color ratio images multiplied by the middle gray value for better visualization.

Similarly, color balancing (e.g., to compensate for incandescent lighting) can be performed either by multiplying each channel with a different scale factor or by the more complex process of mapping to XYZ color space, changing the nominal white point, and mapping back to RGB, which can be written down using a linear 3×3 *color twist* transform matrix. Exercises 2.8 and 3.1 have you explore some of these issues.

Another fun project, best attempted after you have mastered the rest of the material in this chapter, is to take a picture with a rainbow in it and enhance the strength of the rainbow (Exercise 3.29).

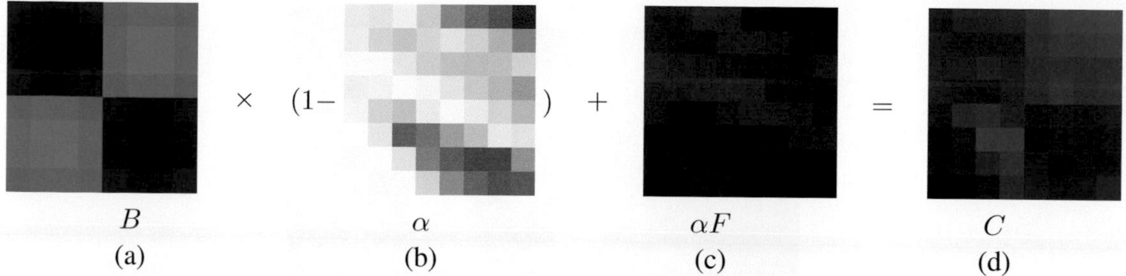

Figure 3.5 Compositing equation $C = (1 - \alpha)B + \alpha F$. The images are taken from a close-up of the region of the hair in the upper right part of the lion in Figure 3.4.

3.1.3 Compositing and matting

In many photo editing and visual effects applications, it is often desirable to cut a *foreground* object out of one scene and put it on top of a different *background* (Figure 3.4). The process of extracting the object from the original image is often called *matting* (Smith and Blinn 1996), while the process of inserting it into another image (without visible artifacts) is called *compositing* (Porter and Duff 1984; Blinn 1994a).

The intermediate representation used for the foreground object between these two stages is called an *alpha-matted color image* (Figure 3.4b–c). In addition to the three color RGB channels, an alpha-matted image contains a fourth *alpha* channel α (or A) that describes the relative amount of *opacity* or *fractional coverage* at each pixel (Figures 3.4c and 3.5b). The opacity is the opposite of the *transparency*. Pixels within the object are fully opaque ($\alpha = 1$), while pixels fully outside the object are transparent ($\alpha = 0$). Pixels on the boundary of the object vary smoothly between these two extremes, which hides the perceptual visible *jaggies* that occur if only binary opacities are used.

To composite a new (or foreground) image on top of an old (background) image, the *over operator*, first proposed by Porter and Duff (1984) and then studied extensively by Blinn (1994a; 1994b), is used:

$$C = (1 - \alpha)B + \alpha F. \tag{3.8}$$

This operator *attenuates* the influence of the background image B by a factor $(1 - \alpha)$ and then adds in the color (and opacity) values corresponding to the foreground layer F, as shown in Figure 3.5.

In many situations, it is convenient to represent the foreground colors in *pre-multiplied* form, i.e., to store (and manipulate) the αF values directly. As Blinn (1994b) shows, the pre-multiplied RGBA representation is preferred for several reasons, including the ability to blur or resample (e.g., rotate) alpha-matted images without any additional complications (just treating each RGBA band independently). However, when matting using local color consistency (Ruzon and Tomasi 2000; Chuang, Curless *et al.* 2001), the pure un-multiplied foreground colors F are used, since these remain constant (or vary slowly) in the vicinity of the object edge.

The over operation is not the only kind of compositing operation that can be used. Porter and Duff (1984) describe a number of additional operations that can be useful in photo editing and visual effects applications. In this book, we concern ourselves with only one additional commonly occurring case (but see Exercise 3.3).

When light reflects off clean transparent glass, the light passing through the glass and the light reflecting off the glass are simply added together (Figure 3.6). This model is useful in the analysis of *transparent motion* (Black and Anandan 1996; Szeliski, Avidan, and Anandan 2000), which occurs when such scenes are observed from a moving camera (see Section 9.4.2).

Figure 3.6 An example of light reflecting off the transparent glass of a picture frame (Black and Anandan 1996) © 1996 Elsevier. You can clearly see the woman's portrait inside the picture frame superimposed with the reflection of a man's face off the glass.

The actual process of *matting*, i.e., recovering the foreground, background, and alpha matte values from one or more images, has a rich history, which we study in Section 10.4. Smith and Blinn (1996) have a nice survey of traditional *blue-screen matting* techniques, while Toyama, Krumm *et al.* (1999) review *difference matting*. Since then, there has been a lot of activity in computational photography relating to *natural image matting* (Ruzon and Tomasi 2000; Chuang, Curless *et al.* 2001; Wang and Cohen 2009; Xu, Price *et al.* 2017), which attempts to extract the mattes from a single natural image (Figure 3.4a) or from extended video sequences (Chuang, Agarwala *et al.* 2002). All of these techniques are described in more detail in Section 10.4.

3.1.4 Histogram equalization

While the brightness and gain controls described in Section 3.1.1 can improve the appearance of an image, how can we automatically determine their best values? One approach might be to look at the darkest and brightest pixel values in an image and map them to pure black and pure white. Another approach might be to find the *average* value in the image, push it towards middle gray, and expand the *range* so that it more closely fills the displayable values (Kopf, Uyttendaele *et al.* 2007).

How can we visualize the set of lightness values in an image to test some of these heuristics? The answer is to plot the *histogram* of the individual color channels and luminance values, as shown in Figure 3.7b.[3] From this distribution, we can compute relevant statistics such as the minimum, maximum, and average intensity values. Notice that the image in Figure 3.7a has both an excess of dark values and light values, but that the mid-range values are largely under-populated. Would it not be better if we could simultaneously brighten some dark values and darken some light values, while still using the full extent of the available dynamic range? Can you think of a mapping that might do this?

One popular answer to this question is to perform *histogram equalization*, i.e., to find an intensity mapping function $f(I)$ such that the resulting histogram is flat. The trick to finding such a mapping is the same one that people use to generate random samples from a *probability density function*, which is to first compute the *cumulative distribution function* shown in Figure 3.7c.

[3] The histogram is simply the *count* of the number of pixels at each gray level value. For an eight-bit image, an accumulation table with 256 entries is needed. For higher bit depths, a table with the appropriate number of entries (probably fewer than the full number of gray levels) should be used.

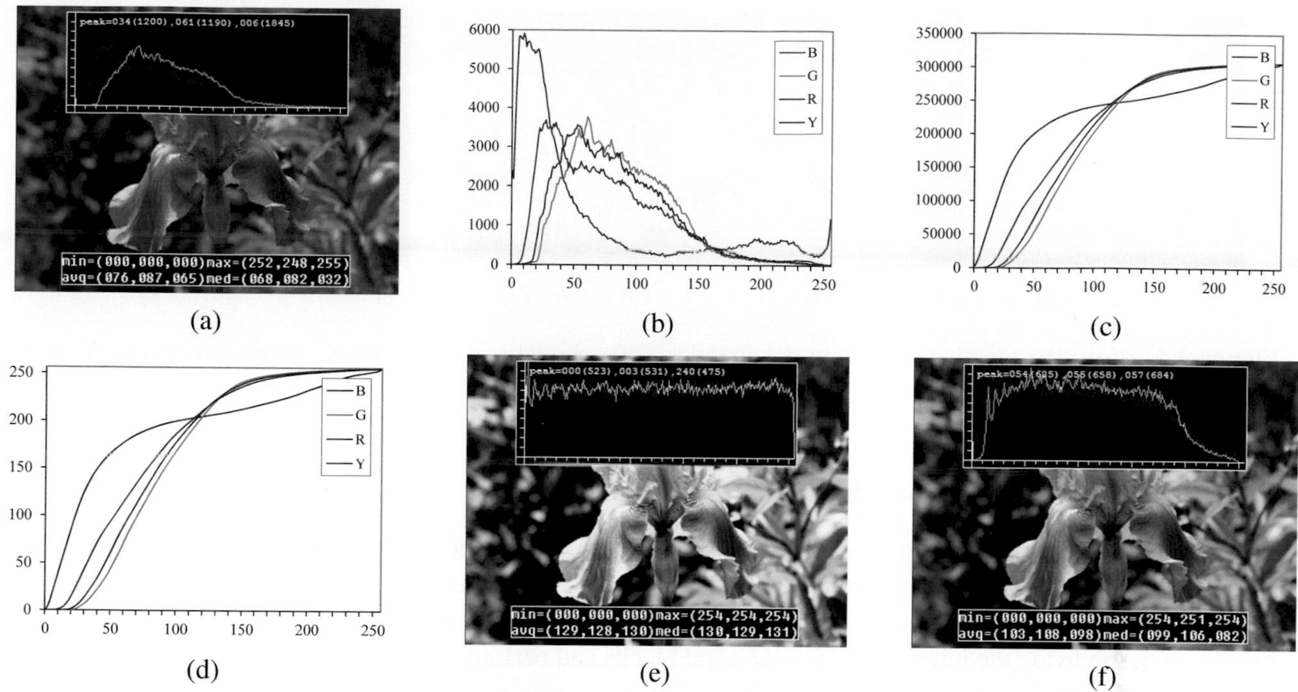

Figure 3.7 Histogram analysis and equalization: (a) original image; (b) color channel and intensity (luminance) histograms; (c) cumulative distribution functions; (d) equalization (transfer) functions; (e) full histogram equalization; (f) partial histogram equalization.

Think of the original histogram $h(I)$ as the distribution of grades in a class after some exam. How can we map a particular grade to its corresponding *percentile*, so that students at the 75% percentile range scored better than $^3/_4$ of their classmates? The answer is to integrate the distribution $h(I)$ to obtain the cumulative distribution $c(I)$,

$$c(I) = \frac{1}{N} \sum_{i=0}^{I} h(i) = c(I-1) + \frac{1}{N} h(I), \tag{3.9}$$

where N is the number of pixels in the image or students in the class. For any given grade or intensity, we can look up its corresponding percentile $c(I)$ and determine the final value that the pixel should take. When working with eight-bit pixel values, the I and c axes are rescaled from $[0, 255]$.

Figure 3.7e shows the result of applying $f(I) = c(I)$ to the original image. As we can see, the resulting histogram is flat; so is the resulting image (it is "flat" in the sense of a lack of contrast and being muddy looking). One way to compensate for this is to only *partially* compensate for the histogram unevenness, e.g., by using a mapping function $f(I) = \alpha c(I) + (1 - \alpha) I$, which is a linear blend between the cumulative distribution function and the identity transform (a straight line). As you can see in Figure 3.7f, the resulting image maintains more of its original grayscale distribution while having a more appealing balance.

Another potential problem with histogram equalization (or, in general, image brightening) is that noise in dark regions can be amplified and become more visible. Exercise 3.7 suggests some possible ways to mitigate this, as well as alternative techniques to maintain contrast and "punch" in the original images (Larson, Rushmeier, and Piatko 1997; Stark 2000).

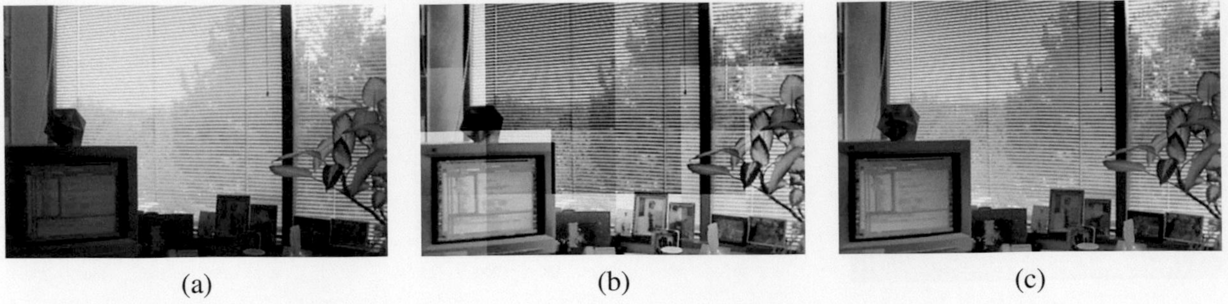

(a) (b) (c)

Figure 3.8 Locally adaptive histogram equalization: (a) original image; (b) block histogram equalization; (c) full locally adaptive equalization.

Locally adaptive histogram equalization

While global histogram equalization can be useful, for some images it might be preferable to apply different kinds of equalization in different regions. Consider for example the image in Figure 3.8a, which has a wide range of luminance values. Instead of computing a single curve, what if we were to subdivide the image into $M \times M$ pixel blocks and perform separate histogram equalization in each sub-block? As you can see in Figure 3.8b, the resulting image exhibits a lot of blocking artifacts, i.e., intensity discontinuities at block boundaries.

One way to eliminate blocking artifacts is to use a *moving window*, i.e., to recompute the histogram for every $M \times M$ block centered at each pixel. This process can be quite slow (M^2 operations per pixel), although with clever programming only the histogram entries corresponding to the pixels entering and leaving the block (in a raster scan across the image) need to be updated (M operations per pixel). Note that this operation is an example of the *non-linear neighborhood operations* we study in more detail in Section 3.3.1.

A more efficient approach is to compute non-overlapped block-based equalization functions as before, but to then smoothly interpolate the transfer functions as we move between blocks. This technique is known as *adaptive histogram equalization* (AHE) and its contrast-limited (gain-limited) version is known as CLAHE (Pizer, Amburn *et al.* 1987).[4] The weighting function for a given pixel (i, j) can be computed as a function of its horizontal and vertical position (s, t) within a block, as shown in Figure 3.9a. To blend the four lookup functions $\{f_{00}, \ldots, f_{11}\}$, a *bilinear* blending function,

$$f_{s,t}(I) = (1-s)(1-t)f_{00}(I) + s(1-t)f_{10}(I) + (1-s)tf_{01}(I) + stf_{11}(I) \tag{3.10}$$

can be used. (See Section 3.5.2 for higher-order generalizations of such *spline* functions.) Note that instead of blending the four lookup tables for each output pixel (which would be quite slow), we can instead blend the results of mapping a given pixel through the four neighboring lookups.

A variant on this algorithm is to place the lookup tables at the *corners* of each $M \times M$ block (see Figure 3.9b and Exercise 3.8). In addition to blending four lookups to compute the final value, we can also *distribute* each input pixel into four adjacent lookup tables during the histogram accumulation phase (notice that the gray arrows in Figure 3.9b point both ways), i.e.,

$$h_{k,l}(I(i,j)) \mathrel{+}= w(i,j,k,l), \tag{3.11}$$

where $w(i, j, k, l)$ is the bilinear weighting function between pixel (i, j) and lookup table (k, l).

[4] The CLAHE algorithm is part of OpenCV.

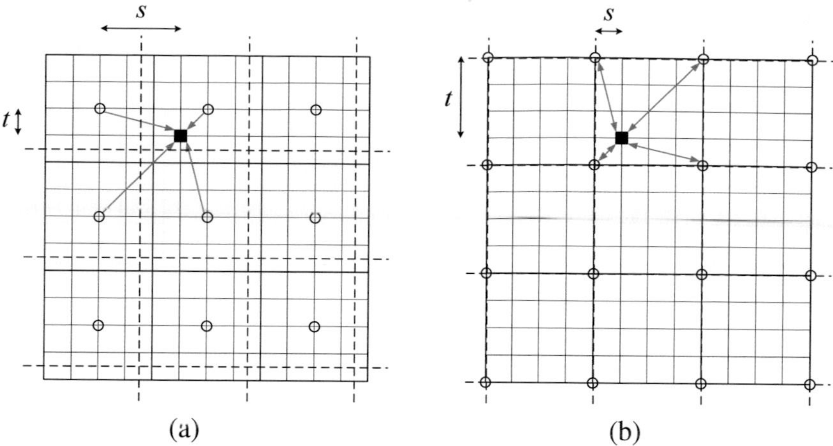

Figure 3.9 Local histogram interpolation using relative (s, t) coordinates: (a) block-based histograms, with block centers shown as circles; (b) corner-based "spline" histograms. Pixels are located on grid intersections. The black square pixel's transfer function is interpolated from the four adjacent lookup tables (gray arrows) using the computed (s, t) values. Block boundaries are shown as dashed lines.

This is an example of *soft histogramming*, which is used in a variety of other applications, including the construction of SIFT feature descriptors (Section 7.1.3) and vocabulary trees (Section 7.1.4).

3.1.5 *Application*: Tonal adjustment

One of the most widely used applications of point-wise image processing operators is the manipulation of contrast or *tone* in photographs, to make them look either more attractive or more interpretable. You can get a good sense of the range of operations possible by opening up any photo manipulation tool and trying out a variety of contrast, brightness, and color manipulation options, as shown in Figures 3.2 and 3.7.

Exercises 3.1, 3.6, and 3.7 have you implement some of these operations, to become familiar with basic image processing operators. More sophisticated techniques for tonal adjustment (Bae, Paris, and Durand 2006; Reinhard, Heidrich *et al.* 2010) are described in the section on high dynamic range tone mapping (Section 10.2.1).

3.2 Linear filtering

Locally adaptive histogram equalization is an example of a *neighborhood operator* or *local operator*, which uses a collection of pixel values in the vicinity of a given pixel to determine its final output value (Figure 3.10). In addition to performing local tone adjustment, neighborhood operators can be used to *filter* images to add soft blur, sharpen details, accentuate edges, or remove noise (Figure 3.11b–d). In this section, we look at *linear* filtering operators, which involve fixed weighted combinations of pixels in small neighborhoods. In Section 3.3, we look at non-linear operators such as morphological filters and distance transforms.

The most widely used type of neighborhood operator is a *linear filter*, where an output pixel's value is a weighted sum of pixel values within a small neighborhood $\mathcal{N}$ (Figure 3.10),

$$g(i, j) = \sum_{k,l} f(i + k, j + l)h(k, l). \tag{3.12}$$

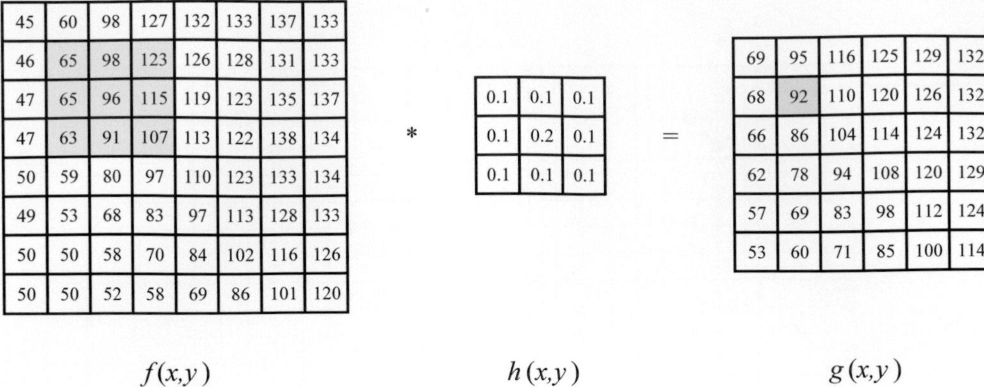

Figure 3.10 Neighborhood filtering (convolution): The image on the left is convolved with the filter in the middle to yield the image on the right. The light blue pixels indicate the source neighborhood for the light green destination pixel.

The entries in the weight *kernel* or *mask* $h(k, l)$ are often called the *filter coefficients*. The above *correlation* operator can be more compactly notated as

$$g = f \otimes h. \tag{3.13}$$

A common variant on this formula is

$$g(i, j) = \sum_{k,l} f(i - k, j - l)h(k, l) = \sum_{k,l} f(k, l)h(i - k, j - l), \tag{3.14}$$

where the sign of the offsets in f has been reversed, This is called the *convolution* operator,

$$g = f * h, \tag{3.15}$$

and h is then called the *impulse response function*.[5] The reason for this name is that the kernel function, h, convolved with an impulse signal, $\delta(i, j)$ (an image that is 0 everywhere except at the origin) reproduces itself, $h * \delta = h$, whereas correlation produces the reflected signal. (Try this yourself to verify that it is so.)

In fact, Equation (3.14) can be interpreted as the superposition (addition) of shifted impulse response functions $h(i - k, j - l)$ multiplied by the input pixel values $f(k, l)$. Convolution has additional nice properties, e.g., it is both commutative and associative. As well, the Fourier transform of two convolved images is the product of their individual Fourier transforms (Section 3.4).

Both correlation and convolution are *linear shift-invariant* (LSI) operators, which obey both the superposition principle (3.5),

$$h \circ (f_0 + f_1) = h \circ f_0 + h \circ f_1, \tag{3.16}$$

and the *shift invariance* principle,

$$g(i, j) = f(i + k, j + l) \quad \Leftrightarrow \quad (h \circ g)(i, j) = (h \circ f)(i + k, j + l), \tag{3.17}$$

which means that shifting a signal commutes with applying the operator ($\circ$ stands for the LSI operator). Another way to think of shift invariance is that the operator "behaves the same everywhere".

[5] The continuous version of convolution can be written as $g(\mathbf{x}) = \int f(\mathbf{x} - \mathbf{u})h(\mathbf{u})d\mathbf{u}$.

Figure 3.11 Some neighborhood operations: (a) original image; (b) blurred; (c) sharpened; (d) smoothed with edge-preserving filter; (e) binary image; (f) dilated; (g) distance transform; (h) connected components. For the dilation and connected components, black (ink) pixels are assumed to be active, i.e., to have a value of 1 in Equations (3.41–3.48).

$$\boxed{72 \mid 88 \mid 62 \mid 52 \mid 37} * \boxed{\tfrac{1}{4} \mid \tfrac{1}{2} \mid \tfrac{1}{4}} \quad \Leftrightarrow \quad \frac{1}{4}\begin{bmatrix} 2 & 1 & . & . & . \\ 1 & 2 & 1 & . & . \\ . & 1 & 2 & 1 & . \\ . & . & 1 & 2 & 1 \\ . & . & . & 1 & 2 \end{bmatrix} \begin{bmatrix} 72 \\ 88 \\ 62 \\ 52 \\ 37 \end{bmatrix}$$

Figure 3.12 One-dimensional signal convolution as a sparse matrix-vector multiplication, $\mathbf{g} = \mathbf{Hf}$.

Occasionally, a shift-variant version of correlation or convolution may be used, e.g.,

$$g(i, j) = \sum_{k,l} f(i - k, j - l)h(k, l; i, j), \tag{3.18}$$

where $h(k, l; i, j)$ is the convolution kernel at pixel (i, j). For example, such a spatially varying kernel can be used to model blur in an image due to variable depth-dependent defocus.

Correlation and convolution can both be written as a matrix-vector multiplication, if we first convert the two-dimensional images $f(i, j)$ and $g(i, j)$ into raster-ordered vectors $\mathbf{f}$ and $\mathbf{g}$,

$$\mathbf{g} = \mathbf{Hf}, \tag{3.19}$$

where the (sparse) $\mathbf{H}$ matrix contains the convolution kernels. Figure 3.12 shows how a one-dimensional convolution can be represented in matrix-vector form.

Padding (border effects)

The astute reader will notice that the correlation shown in Figure 3.10 produces a result that is smaller than the original image, which may not be desirable in many applications.[6] This is because the neighborhoods of typical correlation and convolution operations extend beyond the image boundaries near the edges, and so the filtered images suffer from *boundary effects*

To deal with this, a number of different *padding* or extension modes have been developed for neighborhood operations (Figure 3.13):

- *zero*: set all pixels outside the source image to 0 (a good choice for alpha-matted cutout images);
- *constant (border color)*: set all pixels outside the source image to a specified *border* value;
- *clamp (replicate or clamp to edge)*: repeat edge pixels indefinitely;
- *(cyclic) wrap (repeat or tile)*: loop “around” the image in a “toroidal” configuration;
- *mirror*: reflect pixels across the image edge;
- *extend*: extend the signal by subtracting the mirrored version of the signal from the edge pixel value.

In the computer graphics literature (Akenine-Möller and Haines 2002, p. 124), these mechanisms are known as the *wrapping mode* (OpenGL) or *texture addressing mode* (Direct3D). The formulas for these modes are left to the reader (Exercise 3.9).

[6] Note, however, that early convolutional networks such as LeNet (LeCun, Bottou *et al.* 1998) adopted this structure.

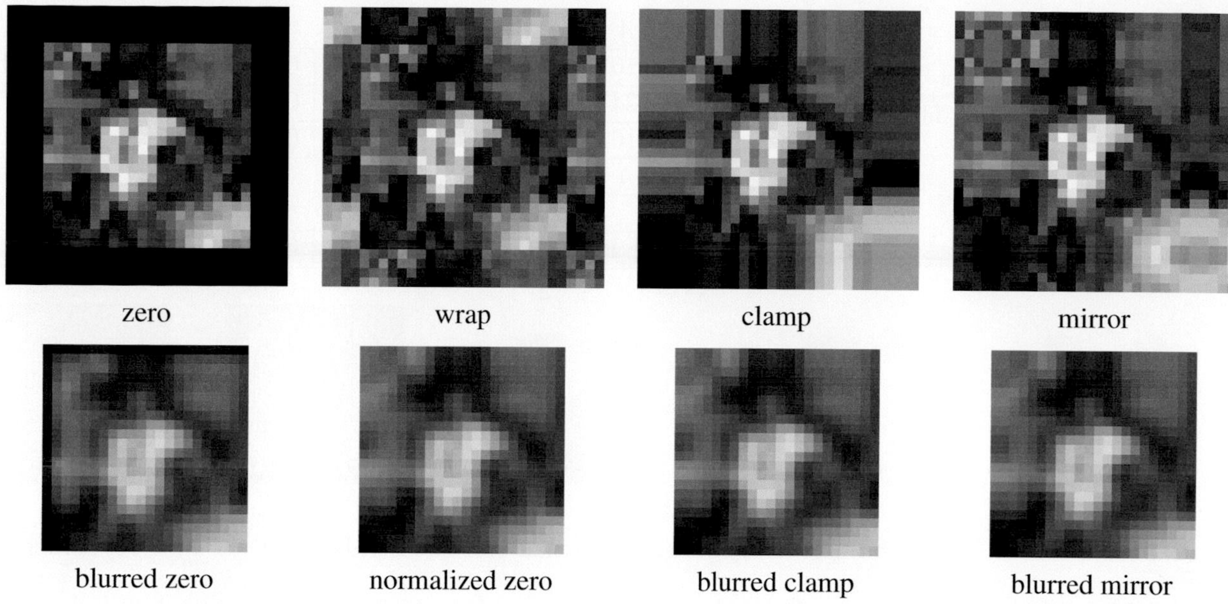

Figure 3.13 Border padding (top row) and the results of blurring the padded image (bottom row). The normalized zero image is the result of dividing (normalizing) the blurred zero-padded RGBA image by its corresponding soft alpha value.

Figure 3.13 shows the effects of padding an image with each of the above mechanisms and then blurring the resulting padded image. As you can see, zero padding darkens the edges, clamp (replication) padding propagates border values inward, mirror (reflection) padding preserves colors near the borders. Extension padding (not shown) keeps the border pixels fixed (during blur).

An alternative to padding is to blur the zero-padded RGBA image and to then divide the resulting image by its alpha value to remove the darkening effect. The results can be quite good, as seen in the normalized zero image in Figure 3.13.

3.2.1 Separable filtering

The process of performing a convolution requires K^2 (multiply-add) operations per pixel, where K is the size (width or height) of the convolution kernel, e.g., the box filter in Figure 3.14a. In many cases, this operation can be significantly sped up by first performing a one-dimensional horizontal convolution followed by a one-dimensional vertical convolution, which requires a total of 2K operations per pixel. A convolution kernel for which this is possible is said to be *separable*.

It is easy to show that the two-dimensional kernel $\mathbf{K}$ corresponding to successive convolution with a horizontal kernel $\mathbf{h}$ and a vertical kernel $\mathbf{v}$ is the *outer product* of the two kernels,

$$\mathbf{K} = \mathbf{v}\mathbf{h}^T \tag{3.20}$$

(see Figure 3.14 for some examples). Because of the increased efficiency, the design of convolution kernels for computer vision applications is often influenced by their separability.

How can we tell if a given kernel $\mathbf{K}$ is indeed separable? This can often be done by inspection or by looking at the analytic form of the kernel (Freeman and Adelson 1991). A more direct method is to treat the 2D kernel as a 2D matrix $\mathbf{K}$ and to take its singular value decomposition (SVD),

$$\mathbf{K} = \sum_i \sigma_i \mathbf{u}_i \mathbf{v}_i^T \tag{3.21}$$

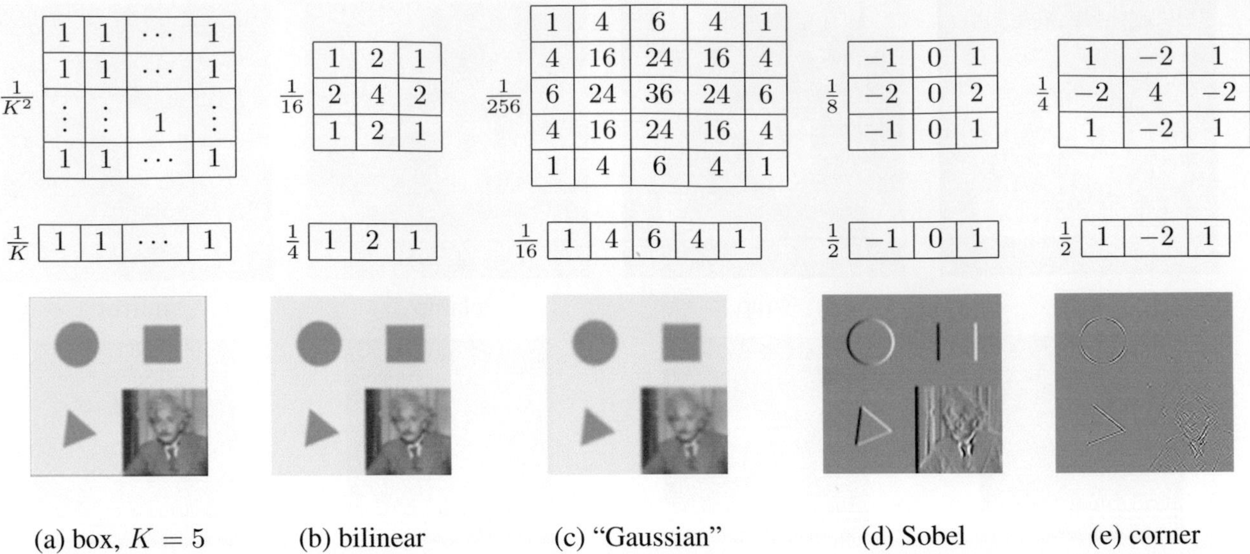

(a) box, $K = 5$ (b) bilinear (c) "Gaussian" (d) Sobel (e) corner

Figure 3.14 Separable linear filters: For each image (a)–(e), we show the 2D filter kernel (top), the corresponding horizontal 1D kernel (middle), and the filtered image (bottom). The filtered Sobel and corner images are signed, scaled up by $2\times$ and $4\times$, respectively, and added to a gray offset before display.

(see Appendix A.1.1 for the definition of the SVD). If only the first singular value σ_0 is non-zero, the kernel is separable and $\sqrt{\sigma_0}\mathbf{u}_0$ and $\sqrt{\sigma_0}\mathbf{v}_0^T$ provide the vertical and horizontal kernels (Perona 1995). For example, the Laplacian of Gaussian kernel (3.26 and 7.23) can be implemented as the sum of two separable filters (7.24) (Wiejak, Buxton, and Buxton 1985).

What if your kernel is not separable and yet you still want a faster way to implement it? Perona (1995), who first made the link between kernel separability and SVD, suggests using more terms in the (3.21) series, i.e., summing up a number of separable convolutions. Whether this is worth doing or not depends on the relative sizes of K and the number of significant singular values, as well as other considerations, such as cache coherency and memory locality.

3.2.2 Examples of linear filtering

Now that we have described the process for performing linear filtering, let us examine a number of frequently used filters.

The simplest filter to implement is the *moving average* or *box* filter, which simply averages the pixel values in a $K \times K$ window. This is equivalent to convolving the image with a kernel of all ones and then scaling (Figure 3.14a). For large kernels, a more efficient implementation is to slide a moving window across each scanline (in a separable filter) while adding the newest pixel and subtracting the oldest pixel from the running sum. This is related to the concept of *summed area tables*, which we describe shortly.

A smoother image can be obtained by separably convolving the image with a piecewise linear "tent" function (also known as a *Bartlett* filter). Figure 3.14b shows a 3×3 version of this filter, which is called the *bilinear* kernel, since it is the outer product of two linear (first-order) splines (see Section 3.5.2).

Convolving the linear tent function with itself yields the cubic approximating spline, which is called the "Gaussian" kernel (Figure 3.14c) in Burt and Adelson's (1983a) *Laplacian pyramid* representation (Section 3.5). Note that approximate Gaussian kernels can also be obtained by iterated convolution with box filters (Wells 1986). In applications where the filters really need to be rotationally symmetric, carefully tuned versions of sampled Gaussians should be used (Freeman and Adelson 1991) (Exercise 3.11).

The kernels we just discussed are all examples of blurring (smoothing) or *low-pass* kernels, since they pass through the lower frequencies while attenuating higher frequencies. How good are they at doing this? In Section 3.4, we use frequency-space Fourier analysis to examine the exact frequency response of these filters. We also introduce the *sinc* ($(\sin x)/x$) filter, which performs *ideal* low-pass filtering.

In practice, smoothing kernels are often used to reduce high-frequency noise. We have much more to say about using variants of smoothing to remove noise later (see Sections 3.3.1, 3.4, and as well as Chapters 4 and 5).

Surprisingly, smoothing kernels can also be used to *sharpen* images using a process called *unsharp masking*. Since blurring the image reduces high frequencies, adding some of the difference between the original and the blurred image makes it sharper,

$$g_{\text{sharp}} = f + \gamma(f - h_{\text{blur}} * f). \tag{3.22}$$

In fact, before the advent of digital photography, this was the standard way to sharpen images in the darkroom: create a blurred ("positive") negative from the original negative by misfocusing, then overlay the two negatives before printing the final image, which corresponds to

$$g_{\text{unsharp}} = f(1 - \gamma h_{\text{blur}} * f). \tag{3.23}$$

This is no longer a linear filter but it still works well.

Linear filtering can also be used as a pre-processing stage to edge extraction (Section 7.2) and interest point detection (Section 7.1) algorithms. Figure 3.14d shows a simple 3×3 edge extractor called the Sobel operator, which is a separable combination of a horizontal *central difference* (so called because the horizontal derivative is centered on the pixel) and a vertical tent filter (to smooth the results). As you can see in the image below the kernel, this filter effectively emphasizes vertical edges.

The simple corner detector (Figure 3.14e) looks for simultaneous horizontal and vertical second derivatives. As you can see, however, it responds not only to the corners of the square, but also along diagonal edges. Better corner detectors, or at least interest point detectors that are more rotationally invariant, are described in Section 7.1.

3.2.3 Band-pass and steerable filters

The Sobel and corner operators are simple examples of band-pass and oriented filters. More sophisticated kernels can be created by first smoothing the image with a (unit area) Gaussian filter,

$$G(x, y; \sigma) = \frac{1}{2\pi\sigma^2} e^{-\frac{x^2+y^2}{2\sigma^2}}, \tag{3.24}$$

and then taking the first or second derivatives (Marr 1982; Witkin 1983; Freeman and Adelson 1991). Such filters are known collectively as *band-pass filters*, since they filter out both low and high frequencies.

The (undirected) second derivative of a two-dimensional image,

$$\nabla^2 f = \frac{\partial^2 f}{\partial x^2} + \frac{\partial^2 f}{\partial y^2}, \tag{3.25}$$

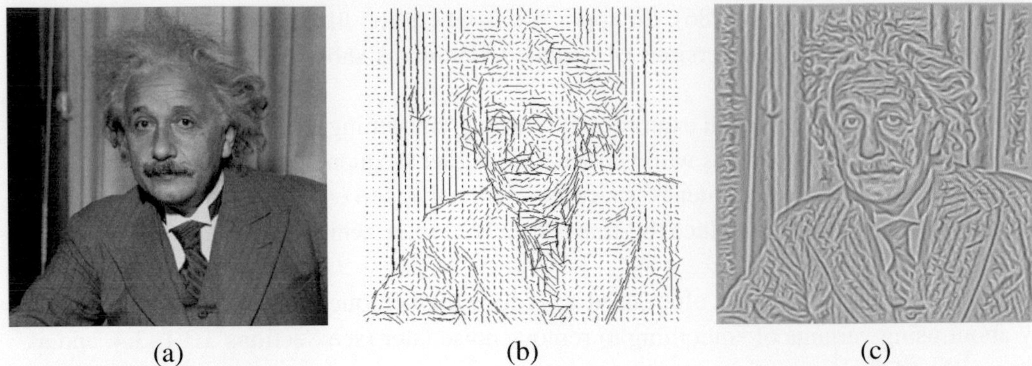

(a) (b) (c)

Figure 3.15 Second-order steerable filter (Freeman 1992) © 1992 IEEE: (a) original image of Einstein; (b) orientation map computed from the second-order oriented energy; (c) original image with oriented structures enhanced.

is known as the *Laplacian* operator. Blurring an image with a Gaussian and then taking its Laplacian is equivalent to convolving directly with the *Laplacian of Gaussian* (LoG) filter,

$$\nabla^2 G(x, y; \sigma) = \left(\frac{x^2 + y^2}{\sigma^4} - \frac{2}{\sigma^2} \right) G(x, y; \sigma), \tag{3.26}$$

which has certain nice *scale-space properties* (Witkin 1983; Witkin, Terzopoulos, and Kass 1986). The five-point Laplacian is just a compact approximation to this more sophisticated filter.

Likewise, the Sobel operator is a simple approximation to a *directional* or *oriented* filter, which can obtained by smoothing with a Gaussian (or some other filter) and then taking a *directional derivative* $\nabla_{\hat{\mathbf{u}}} = \frac{\partial}{\partial \hat{\mathbf{u}}}$, which is obtained by taking the dot product between the gradient field ∇ and a unit direction $\hat{\mathbf{u}} = (\cos\theta, \sin\theta)$,

$$\hat{\mathbf{u}} \cdot \nabla(G * f) = \nabla_{\hat{\mathbf{u}}}(G * f) = (\nabla_{\hat{\mathbf{u}}} G) * f. \tag{3.27}$$

The smoothed directional derivative filter,

$$G_{\hat{\mathbf{u}}} = uG_x + vG_y = u\frac{\partial G}{\partial x} + v\frac{\partial G}{\partial y}, \tag{3.28}$$

where $\hat{\mathbf{u}} = (u, v)$, is an example of a *steerable* filter, since the value of an image convolved with $G_{\hat{\mathbf{u}}}$ can be computed by first convolving with the pair of filters (G_x, G_y) and then *steering* the filter (potentially locally) by multiplying this gradient field with a unit vector $\hat{\mathbf{u}}$ (Freeman and Adelson 1991). The advantage of this approach is that a whole *family* of filters can be evaluated with very little cost.

How about steering a directional second derivative filter $\nabla_{\hat{\mathbf{u}}} \cdot \nabla_{\hat{\mathbf{u}}} G$, which is the result of taking a (smoothed) directional derivative and then taking the directional derivative again? For example, G_{xx} is the second directional derivative in the x direction.

At first glance, it would appear that the steering trick will not work, since for every direction $\hat{\mathbf{u}}$, we need to compute a different first directional derivative. Somewhat surprisingly, Freeman and Adelson (1991) showed that, for directional Gaussian derivatives, it is possible to steer *any* order of derivative with a relatively small number of basis functions. For example, only three basis functions are required for the second-order directional derivative,

$$G_{\hat{\mathbf{u}}\hat{\mathbf{u}}} = u^2 G_{xx} + 2uvG_{xy} + v^2 G_{yy}. \tag{3.29}$$

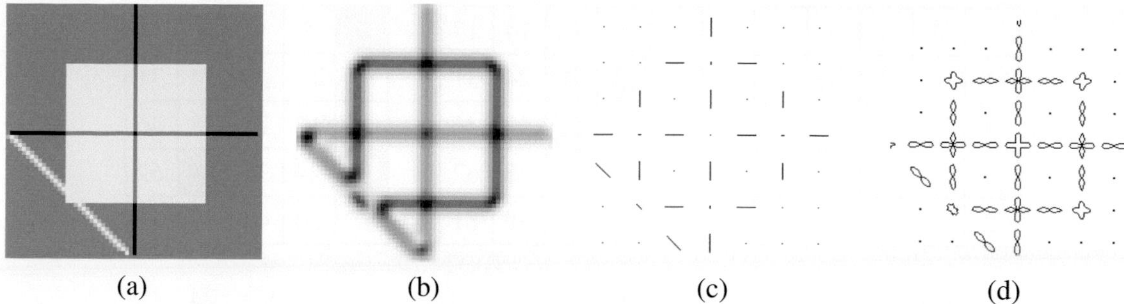

(a) (b) (c) (d)

Figure 3.16 Fourth-order steerable filter (Freeman and Adelson 1991) © 1991 IEEE: (a) test image containing bars (lines) and step edges at different orientations; (b) average oriented energy; (c) dominant orientation; (d) oriented energy as a function of angle (polar plot).

Furthermore, each of the basis filters, while not itself necessarily separable, can be computed using a linear combination of a small number of separable filters (Freeman and Adelson 1991).

This remarkable result makes it possible to construct directional derivative filters of increasingly greater *directional selectivity*, i.e., filters that only respond to edges that have strong local consistency in orientation (Figure 3.15). Furthermore, higher order steerable filters can respond to potentially more than a single edge orientation at a given location, and they can respond to both *bar* edges (thin lines) and the classic step edges (Figure 3.16). In order to do this, however, full *Hilbert transform pairs* need to be used for second-order and higher filters, as described in (Freeman and Adelson 1991).

Steerable filters are often used to construct both feature descriptors (Section 7.1.3) and edge detectors (Section 7.2). While the filters developed by Freeman and Adelson (1991) are best suited for detecting linear (edge-like) structures, more recent work by Koethe (2003) shows how a combined 2 × 2 boundary tensor can be used to encode both edge and junction ("corner") features. Exercise 3.13 has you implement such steerable filters and apply them to finding both edge and corner features.

Summed area table (integral image)

If an image is going to be repeatedly convolved with different box filters (and especially filters of different sizes at different locations), you can precompute the *summed area table* (Crow 1984), which is just the running sum of all the pixel values from the origin,

$$s(i,j) = \sum_{k=0}^{i} \sum_{l=0}^{j} f(k,l). \tag{3.30}$$

This can be efficiently computed using a recursive (raster-scan) algorithm,

$$s(i,j) = s(i-1,j) + s(i,j-1) - s(i-1,j-1) + f(i,j). \tag{3.31}$$

The image $s(i,j)$ is also often called an *integral image* (see Figure 3.17) and can actually be computed using only two additions per pixel if separate row sums are used (Viola and Jones 2004). To find the summed area (integral) inside a rectangle $[i_0, i_1] \times [j_0, j_1]$, we simply combine four samples from the summed area table,

$$S(i_0 \ldots i_1, j_0 \ldots j_1) = s(i_1, j_1) - s(i_1, j_0 - 1) - s(i_0 - 1, j_1) + s(i_0 - 1, j_0 - 1). \tag{3.32}$$

3	2	7	2	3
1	5	1	3	4
5	1	**3**	5	1
4	3	2	1	6
2	4	1	4	8

(a) S = 24

3	5	12	14	17
4	*11*	**19**	24	31
9	**17**	28	38	46
13	24	37	48	62
15	30	44	59	81

(b) s = 28

3	5	12	*14*	17
4	11	19	24	31
9	17	28	38	46
13	24	37	**48**	62
15	30	44	59	81

(c) S = 24

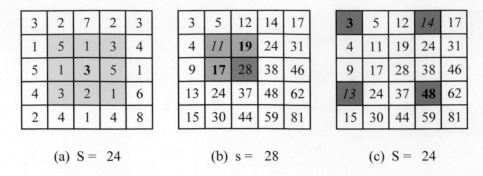

Figure 3.17 Summed area tables: (a) original image; (b) summed area table; (c) computation of area sum. Each value in the summed area table $s(i, j)$ (red) is computed recursively from its three adjacent (blue) neighbors (3.31). Area sums S (green) are computed by combining the four values at the rectangle corners (purple) (3.32). Positive values are shown in **bold** and negative values in *italics*.

A potential disadvantage of summed area tables is that they require $\log M + \log N$ extra bits in the accumulation image compared to the original image, where M and N are the image width and height. Extensions of summed area tables can also be used to approximate other convolution kernels (Wolberg (1990, Section 6.5.2) contains a review).

In computer vision, summed area tables have been used in face detection (Viola and Jones 2004) to compute simple multi-scale low-level features. Such features, which consist of adjacent rectangles of positive and negative values, are also known as *boxlets* (Simard, Bottou *et al.* 1998). In principle, summed area tables could also be used to compute the sums in the sum of squared differences (SSD) stereo and motion algorithms (Section 12.4). In practice, separable moving average filters are usually preferred (Kanade, Yoshida *et al.* 1996), unless many different window shapes and sizes are being considered (Veksler 2003).

Recursive filtering

The incremental formula (3.31) for the summed area is an example of a *recursive filter*, i.e., one whose values depends on previous filter outputs. In the signal processing literature, such filters are known as *infinite impulse response* (IIR), since the output of the filter to an impulse (single non-zero value) goes on forever. For example, for a summed area table, an impulse generates an infinite rectangle of 1s below and to the right of the impulse. The filters we have previously studied in this chapter, which involve the image with a finite extent kernel, are known as *finite impulse response* (FIR).

Two-dimensional IIR filters and recursive formulas are sometimes used to compute quantities that involve large area interactions, such as two-dimensional distance functions (Section 3.3.3) and connected components (Section 3.3.3).

More commonly, however, IIR filters are used inside one-dimensional separable filtering stages to compute large-extent smoothing kernels, such as efficient approximations to Gaussians and edge filters (Deriche 1990; Nielsen, Florack, and Deriche 1997). Pyramid-based algorithms (Section 3.5) can also be used to perform such large-area smoothing computations.

Figure 3.18 Median and bilateral filtering: (a) original image with Gaussian noise; (b) Gaussian filtered; (c) median filtered; (d) bilaterally filtered; (e) original image with shot noise; (f) Gaussian filtered; (g) median filtered; (h) bilaterally filtered. Note that the bilateral filter fails to remove the shot noise because the noisy pixels are too different from their neighbors.

3.3 More neighborhood operators

As we have just seen, linear filters can perform a wide variety of image transformations. However non-linear filters, such as edge-preserving median or bilateral filters, can sometimes perform even better. Other examples of neighborhood operators include *morphological* operators that operate on binary images, as well as *semi-global* operators that compute *distance transforms* and find *connected components* in binary images (Figure 3.11f–h).

3.3.1 Non-linear filtering

The filters we have looked at so far have all been *linear*, i.e., their response to a sum of two signals is the same as the sum of the individual responses. This is equivalent to saying that each output pixel is a weighted summation of some number of input pixels (3.19). Linear filters are easier to compose and are amenable to frequency response analysis (Section 3.4).

In many cases, however, better performance can be obtained by using a *non-linear* combination of neighboring pixels. Consider for example the image in Figure 3.18e, where the noise, rather than being Gaussian, is *shot noise*, i.e., it occasionally has very large values. In this case, regular blurring with a Gaussian filter fails to remove the noisy pixels and instead turns them into softer (but still visible) spots (Figure 3.18f).

Median filtering

A better filter to use in this case is the *median* filter, which selects the median value from each pixel's neighborhood (Figure 3.19a). Median values can be computed in expected linear time using a randomized select algorithm (Cormen 2001) and incremental variants have also been developed (Tomasi and Manduchi 1998; Bovik 2000, Section 3.2), as well as a constant time algorithm that

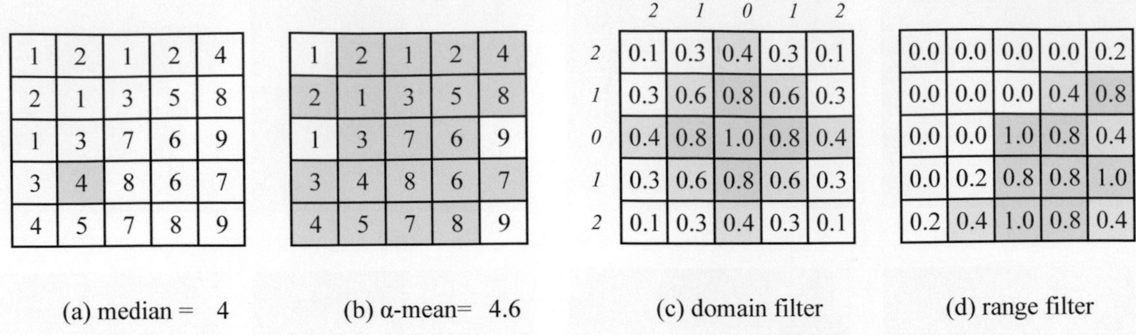

1	2	1	2	4
2	1	3	5	8
1	3	7	6	9
3	4	8	6	7
4	5	7	8	9

(a) median = 4

1	2	1	2	4
2	1	3	5	8
1	3	7	6	9
3	4	8	6	7
4	5	7	8	9

(b) α-mean= 4.6

	2	1	0	1	2
2	0.1	0.3	0.4	0.3	0.1
1	0.3	0.6	0.8	0.6	0.3
0	0.4	0.8	1.0	0.8	0.4
1	0.3	0.6	0.8	0.6	0.3
2	0.1	0.3	0.4	0.3	0.1

(c) domain filter

0.0	0.0	0.0	0.0	0.2
0.0	0.0	0.0	0.4	0.8
0.0	0.0	1.0	0.8	0.4
0.0	0.2	0.8	0.8	1.0
0.2	0.4	1.0	0.8	0.4

(d) range filter

Figure 3.19 Median and bilateral filtering: (a) median pixel (green); (b) selected α-trimmed mean pixels; (c) domain filter (numbers along edge are pixel distances); (d) range filter.

is independent of window size (Perreault and Hébert 2007). Since the shot noise value usually lies well outside the true values in the neighborhood, the median filter is able to filter away such bad pixels (Figure 3.18g).

One downside of the median filter, in addition to its moderate computational cost, is that because it selects only one input pixel value to replace each output pixel, it is not as *efficient* at averaging away regular Gaussian noise (Huber 1981; Hampel, Ronchetti *et al.* 1986; Stewart 1999). A better choice may be the α-trimmed mean (Lee and Redner 1990; Crane 1997, p. 109), which averages together all of the pixels except for the α fraction that are the smallest and the largest (Figure 3.19b).

Another possibility is to compute a *weighted median*, in which each pixel is used a number of times depending on its distance from the center. This turns out to be equivalent to minimizing the weighted objective function

$$\sum_{k,l} w(k,l)|f(i+k,j+l) - g(i,j)|^p, \tag{3.33}$$

where $g(i,j)$ is the desired output value and $p = 1$ for the weighted median. The value $p = 2$ is the usual *weighted mean*, which is equivalent to correlation (3.12) after normalizing by the sum of the weights (Haralick and Shapiro 1992, Section 7.2.6; Bovik 2000, Section 3.2). The weighted mean also has deep connections to other methods in robust statistics (see Appendix B.3), such as influence functions (Huber 1981; Hampel, Ronchetti *et al.* 1986).

Non-linear smoothing has another, perhaps even more important property, especially as shot noise is rare in today's cameras. Such filtering is more *edge preserving*, i.e., it has less tendency to soften edges while filtering away high-frequency noise.

Consider the noisy image in Figure 3.18a. In order to remove most of the noise, the Gaussian filter is forced to smooth away high-frequency detail, which is most noticeable near strong edges. Median filtering does better but, as mentioned before, does not do as well at smoothing away from discontinuities. See Tomasi and Manduchi (1998) for some additional references to edge-preserving smoothing techniques.

While we could try to use the α-trimmed mean or weighted median, these techniques still have a tendency to round sharp corners, since the majority of pixels in the smoothing area come from the background distribution.

3.3.2 Bilateral filtering

What if we were to combine the idea of a weighted filter kernel with a better version of outlier rejection? What if instead of rejecting a fixed percentage α, we simply reject (in a soft way) pixels whose *values* differ too much from the central pixel value? This is the essential idea in *bilateral filtering*, which was first popularized in the computer vision community by Tomasi and Manduchi (1998), although it had been proposed earlier by Aurich and Weule (1995) and Smith and Brady (1997). Paris, Kornprobst *et al.* (2008) provide a nice review of work in this area as well as myriad applications in computer vision, graphics, and computational photography.

In the bilateral filter, the output pixel value depends on a weighted combination of neighboring pixel values

$$\mathbf{g}(i,j) = \frac{\sum_{k,l} \mathbf{f}(k,l) w(i,j,k,l)}{\sum_{k,l} w(i,j,k,l)}. \tag{3.34}$$

The weighting coefficient $w(i,j,k,l)$ depends on the product of a *domain kernel*, (Figure 3.19c),

$$d(i,j,k,l) = \exp\left(-\frac{(i-k)^2 + (j-l)^2}{2\sigma_d^2}\right), \tag{3.35}$$

and a data-dependent *range kernel* (Figure 3.19d),

$$r(i,j,k,l) = \exp\left(-\frac{\|\mathbf{f}(i,j) - \mathbf{f}(k,l)\|^2}{2\sigma_r^2}\right). \tag{3.36}$$

When multiplied together, these yield the data-dependent *bilateral weight function*

$$w(i,j,k,l) = \exp\left(-\frac{(i-k)^2 + (j-l)^2}{2\sigma_d^2} - \frac{\|\mathbf{f}(i,j) - \mathbf{f}(k,l)\|^2}{2\sigma_r^2}\right). \tag{3.37}$$

Figure 3.20 shows an example of the bilateral filtering of a noisy step edge. Note how the domain kernel is the usual Gaussian, the range kernel measures appearance (intensity) similarity to the center pixel, and the bilateral filter kernel is a product of these two.

Notice that for color images, the range filter (3.36) uses the *vector distance* between the center and the neighboring pixel. This is important in color images, since an edge in any *one* of the color bands signals a change in material and hence the need to downweight a pixel's influence.[7]

Since bilateral filtering is quite slow compared to regular separable filtering, a number of acceleration techniques have been developed, as discussed in Durand and Dorsey (2002), Paris and Durand (2009), Chen, Paris, and Durand (2007), and Paris, Kornprobst *et al.* (2008). In particular, the *bilateral grid* (Chen, Paris, and Durand 2007), which subsamples the higher-dimensional color/position space on a uniform grid, continues to be widely used, including the application of the *bilateral solver* (Section 4.2.3 and Barron and Poole (2016)). An even faster implementation of bilateral filtering can be obtained using the *permutohedral lattice* approach developed by Adams, Baek, and Davis (2010).

Iterated adaptive smoothing and anisotropic diffusion

Bilateral (and other) filters can also be applied in an iterative fashion, especially if an appearance more like a "cartoon" is desired (Tomasi and Manduchi 1998). When iterated filtering is applied, a much smaller neighborhood can often be used.

[7] Tomasi and Manduchi (1998) show that using the vector distance (as opposed to filtering each color band separately) reduces color fringing effects. They also recommend taking the color difference in the more perceptually uniform CIELAB color space (see Section 2.3.2).

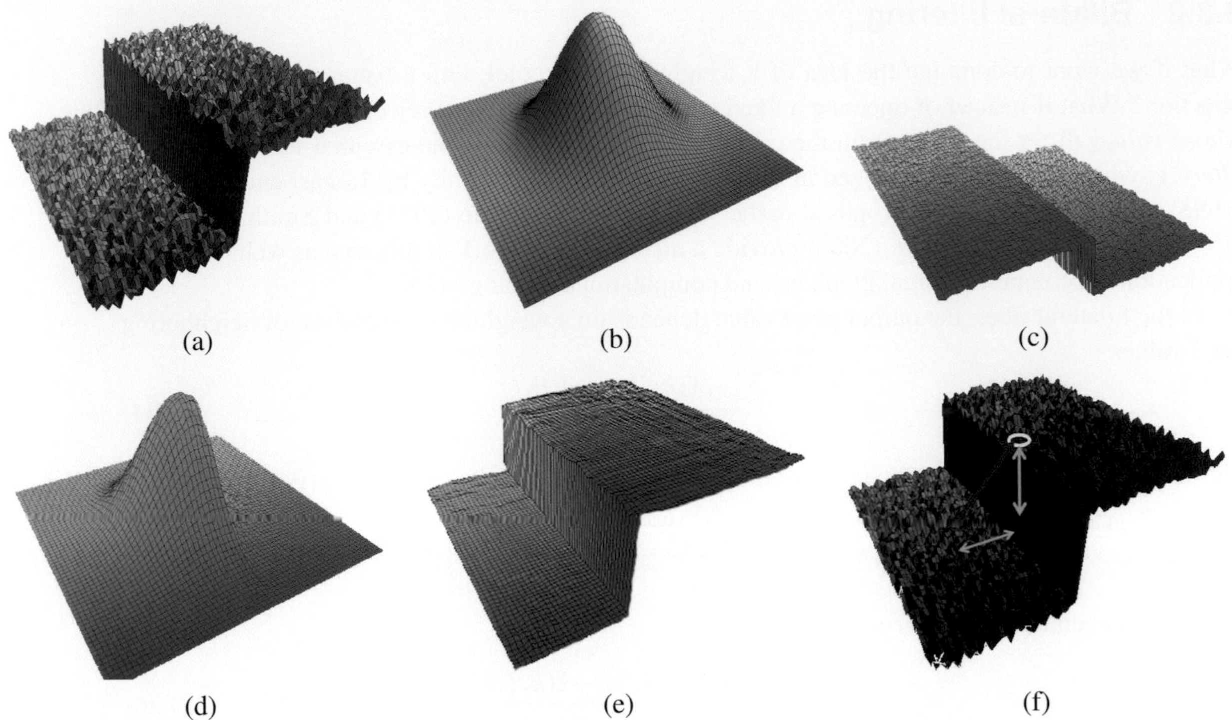

Figure 3.20 Bilateral filtering (Durand and Dorsey 2002) © 2002 ACM: (a) noisy step edge input; (b) domain filter (Gaussian); (c) range filter (similarity to center pixel value); (d) bilateral filter; (e) filtered step edge output; (f) 3D distance between pixels.

Consider, for example, using only the four nearest neighbors, i.e., restricting $|k-i|+|l-j| \leq 1$ in (3.34). Observe that

$$d(i,j,k,l) = \exp\left(-\frac{(i-k)^2+(j-l)^2}{2\sigma_d^2}\right) \tag{3.38}$$

$$= \begin{cases} 1, & |k-i|+|l-j|=0, \\ e^{-1/2\sigma_d^2}, & |k-i|+|l-j|=1. \end{cases} \tag{3.39}$$

We can thus re-write (3.34) as

$$\begin{aligned} f^{(t+1)}(i,j) &= \frac{f^{(t)}(i,j)+\eta\sum_{k,l} f^{(t)}(k,l)r(i,j,k,l)}{1+\eta\sum_{k,l} r(i,j,k,l)} \\ &= f^{(t)}(i,j) + \frac{\eta}{1+\eta R}\sum_{k,l} r(i,j,k,l)[f^{(t)}(k,l)-f^{(t)}(i,j)], \end{aligned} \tag{3.40}$$

where $R=\sum_{(k,l)} r(i,j,k,l)$, (k,l) are the $\mathcal{N}_4$ (nearest four) neighbors of (i,j), and we have made the iterative nature of the filtering explicit.

As Barash (2002) notes, (3.40) is the same as the discrete *anisotropic diffusion* equation first proposed by Perona and Malik (1990b).[8] Since its original introduction, anisotropic diffusion has been extended and applied to a wide range of problems (Nielsen, Florack, and Deriche 1997; Black, Sapiro *et al.* 1998; Weickert, ter Haar Romeny, and Viergever 1998; Weickert 1998). It has also

[8]The $1/(1+\eta R)$ factor is not present in anisotropic diffusion but becomes negligible as $\eta \to 0$.

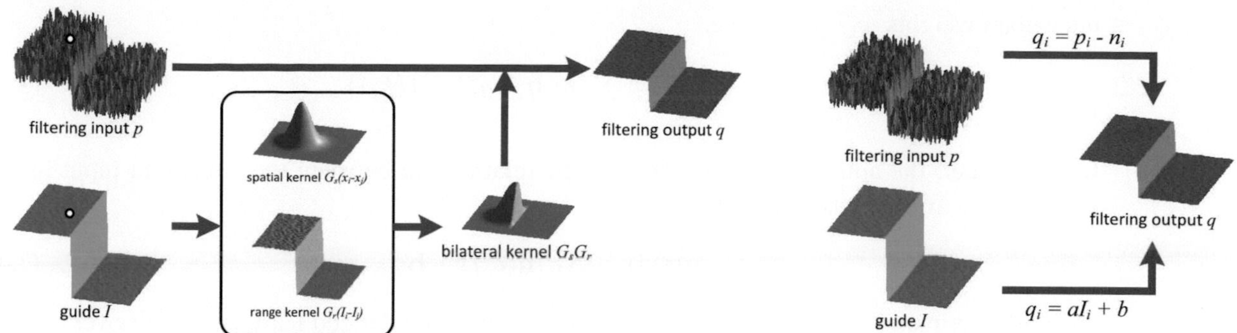

Figure 3.21 Guided image filtering (He, Sun, and Tang 2013) © 2013 IEEE. Unlike joint bilateral filtering, shown on the left, which computes a per pixel weight mask from the guide image (shown as I in the figure, but $\mathbf{h}$ in the text), the guided image filter models the output value (shown as q_i in the figure, but denoted as $\mathbf{g}(i, j)$ in the text) as a local affine transformation of the guide pixels.

been shown to be closely related to other *adaptive smoothing* techniques (Saint-Marc, Chen, and Medioni 1991; Barash 2002; Barash and Comaniciu 2004) as well as Bayesian regularization with a non-linear smoothness term that can be derived from image statistics (Scharr, Black, and Haussecker 2003).

In its general form, the range kernel $r(i, j, k, l) = r(\|f(i, j) - f(k, l)\|)$, which is usually called the *gain* or *edge-stopping* function, or diffusion coefficient, can be any monotonically increasing function with $r'(x) \to 0$ as $x \to \infty$. Black, Sapiro *et al.* (1998) show how anisotropic diffusion is equivalent to minimizing a robust penalty function on the image gradients, which we discuss in Sections 4.2 and 4.3. Scharr, Black, and Haussecker (2003) show how the edge-stopping function can be derived in a principled manner from local image statistics. They also extend the diffusion neighborhood from $\mathcal{N}_4$ to $\mathcal{N}_8$, which allows them to create a diffusion operator that is both rotationally invariant and incorporates information about the eigenvalues of the local structure tensor.

Note that, without a bias term towards the original image, anisotropic diffusion and iterative adaptive smoothing converge to a constant image. Unless a small number of iterations is used (e.g., for speed), it is usually preferable to formulate the smoothing problem as a joint minimization of a smoothness term and a data fidelity term, as discussed in Sections 4.2 and 4.3 and by Scharr, Black, and Haussecker (2003), which introduce such a bias in a principled manner.

Guided image filtering

While so far we have discussed techniques for filtering an image to obtain an improved version, e.g., one with less noise or sharper edges, it is also possible to use a different *guide* image to adaptively filter a noisy input (Eisemann and Durand 2004; Petschnigg, Agrawala *et al.* 2004; He, Sun, and Tang 2013). An example of this is using a flash image, which has strong edges but poor color, to adaptively filter a low-light non-flash color image, which has large amounts of noise, as described in Section 10.2.2. In their papers, where they apply the range filter (3.36) to a different guide image $\mathbf{h}()$, Eisemann and Durand (2004) call their approach a *cross-bilateral filter*, while Petschnigg, Agrawala *et al.* (2004) call it *joint bilateral filtering*.

He, Sun, and Tang (2013) point out that these papers are just two examples of the more general concept of *guided image filtering*, where the guide image $\mathbf{h}()$ is used to compute the locally adapted inter-pixel weights $w(i,j,k,l)$, i.e.,

$$\mathbf{g}(i,j) = \sum_{k,l} w(\mathbf{h}; i,j,k,l)\mathbf{f}(k,l). \tag{3.41}$$

In their paper, the authors suggest modeling the relationship between the guide and input images using a local affine transformation,

$$\mathbf{g}(i,j) = \mathbf{A}_{k,l}\mathbf{h}(i,j) + \mathbf{b}_{k,l}, \tag{3.42}$$

where the estimates for $\mathbf{A}_{k,l}$ and $\mathbf{b}_{k,l}$ are obtained from a regularized least squares fit over a square neighborhood centered around pixel (k,l), i.e., minimizing

$$\sum_{(i,j)\in\mathcal{N}_{k,l}} \|\mathbf{A}_{k,l}\mathbf{h}(i,j) + \mathbf{b}_{k,l} - \mathbf{f}(i,j)\|^2 + \lambda\|\mathbf{A}\|^2. \tag{3.43}$$

These kinds of regularized least squares problems are called *ridge regression* (Section 4.1). The concept behind this algorithm is illustrated in Figure 3.21.

Instead of just taking the predicted value of the filtered pixel $\mathbf{g}(i,j)$ from the window centered on that pixel, an average across all windows that cover the pixel is used. The resulting algorithm (He, Sun, and Tang 2013, Algorithm 1) consists of a series of local mean image and image moment filters, a per-pixel linear system solve (which reduces to a division if the guide image is scalar), and another set of filtering steps. The authors describe how this fast and simple process has been applied to a wide variety of computer vision problems, including image matting (Section 10.4.3), high dynamic range image tone mapping (Section 10.2.1), stereo matching (Hosni, Rhemann *et al.* 2013), and image denoising.

3.3.3 Binary image processing

While non-linear filters are often used to enhance grayscale and color images, they are also used extensively to process binary images. Such images often occur after a *thresholding* operation,

$$\theta(f,t) = \begin{cases} 1 & \text{if } f \geq t, \\ 0 & \text{else,} \end{cases} \tag{3.44}$$

e.g., converting a scanned grayscale document into a binary image for further processing, such as *optical character recognition*.

Morphology

The most common binary image operations are called *morphological operations*, because they change the *shape* of the underlying binary objects (Ritter and Wilson 2000, Chapter 7). To perform such an operation, we first convolve the binary image with a binary *structuring element* and then select a binary output value depending on the thresholded result of the convolution. (This is not the usual way in which these operations are described, but I find it a nice simple way to unify the processes.) The structuring element can be any shape, from a simple 3×3 box filter, to more complicated disc structures. It can even correspond to a particular shape that is being sought for in the image.

Figure 3.22 shows a close-up of the convolution of a binary image f with a 3×3 structuring element s and the resulting images for the operations described below. Let

$$c = f \otimes s \tag{3.45}$$

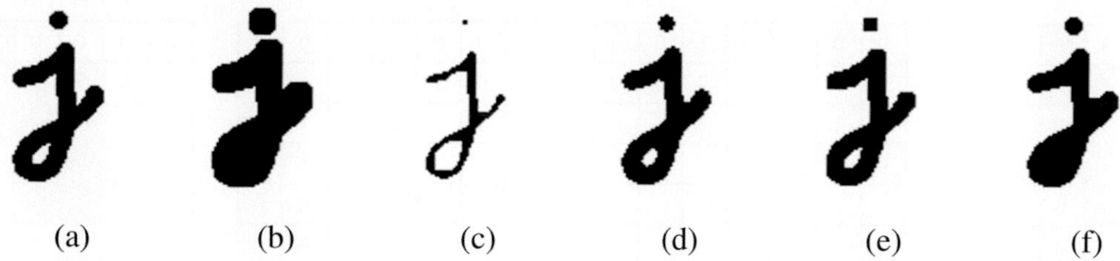

(a) (b) (c) (d) (e) (f)

Figure 3.22 Binary image morphology: (a) original image; (b) dilation; (c) erosion; (d) majority; (e) opening; (f) closing. The structuring element for all examples is a 5×5 square. The effects of majority are a subtle rounding of sharp corners. Opening fails to eliminate the dot, as it is not wide enough.

be the integer-valued *count* of the number of 1s inside each structuring element as it is scanned over the image and S be the size of the structuring element (number of pixels). The standard operations used in binary morphology include:

- **dilation**: $\text{dilate}(f, s) = \theta(c, 1)$;
- **erosion**: $\text{erode}(f, s) = \theta(c, S)$;
- **majority**: $\text{maj}(f, s) = \theta(c, S/2)$;
- **opening**: $\text{open}(f, s) = \text{dilate}(\text{erode}(f, s), s)$;
- **closing**: $\text{close}(f, s) = \text{erode}(\text{dilate}(f, s), s)$.

As we can see from Figure 3.22, dilation grows (thickens) objects consisting of 1s, while erosion shrinks (thins) them. The opening and closing operations tend to leave large regions and smooth boundaries unaffected, while removing small objects or holes and smoothing boundaries.

While we will not use mathematical morphology much in the rest of this book, it is a handy tool to have around whenever you need to clean up some thresholded images. You can find additional details on morphology in other textbooks on computer vision and image processing (Haralick and Shapiro 1992, Section 5.2; Bovik 2000, Section 2.2; Ritter and Wilson 2000, Section 7) as well as articles and books specifically on this topic (Serra 1982; Serra and Vincent 1992; Yuille, Vincent, and Geiger 1992; Soille 2006).

Distance transforms

The distance transform is useful in quickly precomputing the distance to a curve or set of points using a two-pass raster algorithm (Rosenfeld and Pfaltz 1966; Danielsson 1980; Borgefors 1986; Paglieroni 1992; Breu, Gil *et al.* 1995; Felzenszwalb and Huttenlocher 2012; Fabbri, Costa *et al.* 2008). It has many applications, including level sets (Section 7.3.2), fast *chamfer matching* (binary image alignment) (Huttenlocher, Klanderman, and Rucklidge 1993), feathering in image stitching and blending (Section 8.4.2), and nearest point alignment (Section 13.2.1).

The distance transform $D(i, j)$ of a binary image $b(i, j)$ is defined as follows. Let $d(k, l)$ be some *distance metric* between pixel offsets. Two commonly used metrics include the *city block* or *Manhattan* distance

$$d_1(k, l) = |k| + |l| \tag{3.46}$$

and the *Euclidean* distance

$$d_2(k, l) = \sqrt{k^2 + l^2}. \tag{3.47}$$

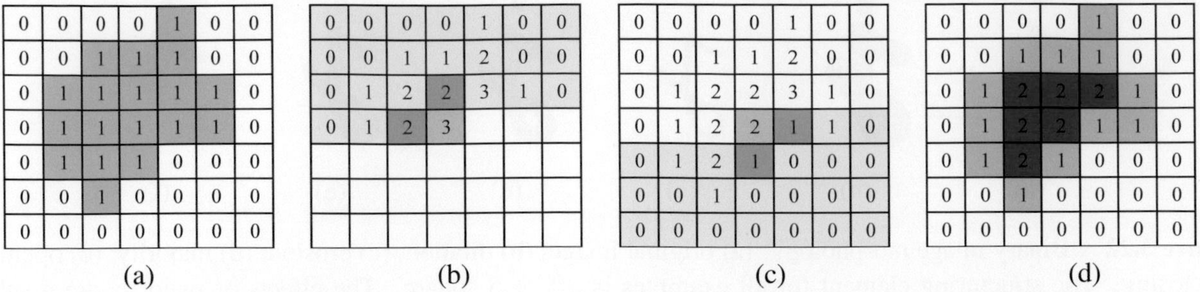

Figure 3.23 City block distance transform: (a) original binary image; (b) top to bottom (forward) raster sweep: green values are used to compute the orange value; (c) bottom to top (backward) raster sweep: green values are merged with old orange value; (d) final distance transform.

The distance transform is then defined as

$$D(i, j) = \min_{k,l:b(k,l)=0} d(i-k, j-l), \tag{3.48}$$

i.e., it is the distance to the *nearest* background pixel whose value is 0.

The D_1 city block distance transform can be efficiently computed using a forward and backward pass of a simple raster-scan algorithm, as shown in Figure 3.23. During the forward pass, each non-zero pixel in b is replaced by the minimum of 1 + the distance of its north or west neighbor. During the backward pass, the same occurs, except that the minimum is both over the current value D and 1 + the distance of the south and east neighbors (Figure 3.23).

Efficiently computing the Euclidean distance transform is more complicated (Danielsson 1980; Borgefors 1986). Here, just keeping the minimum scalar distance to the boundary during the two passes is not sufficient. Instead, a *vector-valued* distance consisting of both the x and y coordinates of the distance to the boundary must be kept and compared using the squared distance (hypotenuse) rule. As well, larger search regions need to be used to obtain reasonable results.

Figure 3.11g shows a distance transform computed from a binary image. Notice how the values grow away from the black (ink) regions and form ridges in the white area of the original image. Because of this linear growth from the starting boundary pixels, the distance transform is also sometimes known as the *grassfire transform*, since it describes the time at which a fire starting inside the black region would consume any given pixel, or a *chamfer*, because it resembles similar shapes used in woodworking and industrial design. The ridges in the distance transform become the *skeleton* (or *medial axis transform (MAT)*) of the region where the transform is computed, and consist of pixels that are of equal distance to two (or more) boundaries (Tek and Kimia 2003; Sebastian and Kimia 2005).

A useful extension of the basic distance transform is the *signed distance transform*, which computes distances to boundary pixels for *all* the pixels (Lavallée and Szeliski 1995). The simplest way to create this is to compute the distance transforms for both the original binary image and its complement and to negate one of them before combining. Because such distance fields tend to be smooth, it is possible to store them more compactly (with minimal loss in *relative* accuracy) using a spline defined over a quadtree or octree data structure (Lavallée and Szeliski 1995; Szeliski and Lavallée 1996; Frisken, Perry *et al.* 2000). Such precomputed signed distance transforms can be extremely useful in efficiently aligning and merging 2D curves and 3D surfaces (Huttenlocher, Klanderman, and Rucklidge 1993; Szeliski and Lavallée 1996; Curless and Levoy 1996), especially if the *vectorial* version of the distance transform, i.e., a pointer from each pixel or voxel to the nearest boundary or surface element, is stored and interpolated. Signed distance fields are also an essential component of level set evolution (Section 7.3.2), where they are called *characteristic functions*.

Connected components

Another useful semi-global image operation is finding *connected components*, which are defined as regions of adjacent pixels that have the same input value or label. Pixels are said to be $\mathcal{N}_4$ adjacent if they are immediately horizontally or vertically adjacent, and $\mathcal{N}_8$ if they can also be diagonally adjacent. Both variants of connected components are widely used in a variety of applications, such as finding individual letters in a scanned document or finding objects (say, cells) in a thresholded image and computing their area statistics. Over the years, a wide variety of efficient algorithms have been developed to find such components, including the ones described in Haralick and Shapiro (1992, Section 2.3) and He, Ren *et al.* (2017). Such algorithms are usually included in image processing libraries such as OpenCV.

Once a binary or multi-valued image has been segmented into its connected components, it is often useful to compute the area statistics for each individual region $\mathcal{R}$. Such statistics include:

- the area (number of pixels);
- the perimeter (number of boundary pixels);
- the centroid (average x and y values);
- the second moments,

$$\mathbf{M} = \sum_{(x,y)\in\mathcal{R}} \begin{bmatrix} x-\overline{x} \\ y-\overline{y} \end{bmatrix} \begin{bmatrix} x-\overline{x} & y-\overline{y} \end{bmatrix}, \tag{3.49}$$

  from which the major and minor axis orientation and lengths can be computed using eigenvalue analysis.

These statistics can then be used for further processing, e.g., for sorting the regions by the area size (to consider the largest regions first) or for preliminary matching of regions in different images.

3.4 Fourier transforms

In Section 3.2, we mentioned that Fourier analysis could be used to analyze the frequency characteristics of various filters. In this section, we explain both how Fourier analysis lets us determine these characteristics (i.e., the frequency *content* of an image) and how using the Fast Fourier Transform (FFT) lets us perform large-kernel convolutions in time that is independent of the kernel's size. More comprehensive introductions to Fourier transforms are provided by Bracewell (1986), Glassner (1995), Oppenheim and Schafer (1996), and Oppenheim, Schafer, and Buck (1999).

How can we analyze what a given filter does to high, medium, and low frequencies? The answer is to simply pass a sinusoid of known frequency through the filter and to observe by how much it is attenuated. Let

$$s(x) = \sin(2\pi f x + \phi_i) = \sin(\omega x + \phi_i) \tag{3.50}$$

be the input sinusoid whose *frequency* is f, *angular frequency* is $\omega = 2\pi f$, and *phase* is ϕ_i. Note that in this section, we use the variables x and y to denote the spatial coordinates of an image, rather than i and j as in the previous sections. This is both because the letters i and j are used for the *imaginary* number (the usage depends on whether you are reading complex variables or electrical

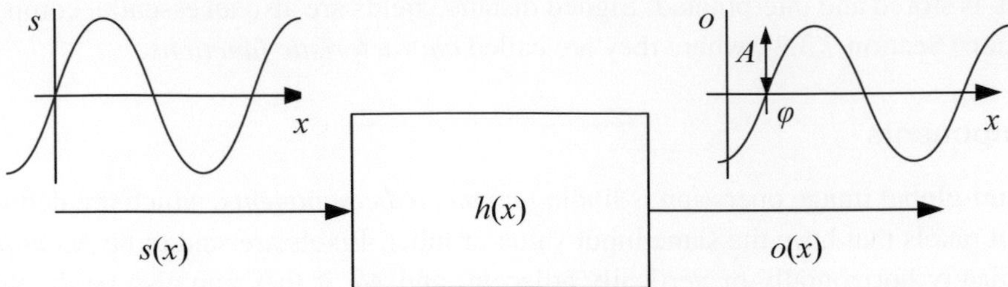

Figure 3.24 The Fourier Transform as the response of a filter $h(x)$ to an input sinusoid $s(x) = e^{j\omega x}$ yielding an output sinusoid $o(x) = h(x) * s(x) = Ae^{j(\omega x + \phi)}$.

engineering literature) and because it is clearer how to distinguish the horizontal (x) and vertical (y) components in frequency space. In this section, we use the letter j for the imaginary number, since that is the form more commonly found in the signal processing literature (Bracewell 1986; Oppenheim and Schafer 1996; Oppenheim, Schafer, and Buck 1999).

If we convolve the sinusoidal signal $s(x)$ with a filter whose impulse response is $h(x)$, we get another sinusoid of the same frequency but different magnitude A and phase ϕ_o,

$$o(x) = h(x) * s(x) = A \sin(\omega x + \phi_o), \tag{3.51}$$

as shown in Figure 3.24. To see that this is the case, remember that a convolution can be expressed as a weighted summation of shifted input signals (3.14) and that the summation of a bunch of shifted sinusoids of the same frequency is just a single sinusoid at that frequency.[9] The new magnitude A is called the *gain* or *magnitude* of the filter, while the phase difference $\Delta\phi = \phi_o - \phi_i$ is called the *shift* or *phase*.

In fact, a more compact notation is to use the complex-valued sinusoid

$$s(x) = e^{j\omega x} = \cos \omega x + j \sin \omega x. \tag{3.52}$$

In that case, we can simply write,

$$o(x) = h(x) * s(x) = Ae^{j(\omega x + \phi)}. \tag{3.53}$$

The *Fourier transform* is simply a tabulation of the magnitude and phase response at each frequency,

$$H(\omega) = \mathcal{F}\{h(x)\} = Ae^{j\phi}, \tag{3.54}$$

i.e., it is the response to a complex sinusoid of frequency ω passed through the filter $h(x)$. The Fourier transform pair is also often written as

$$h(x) \stackrel{\mathcal{F}}{\leftrightarrow} H(\omega). \tag{3.55}$$

Unfortunately, (3.54) does not give an actual *formula* for computing the Fourier transform. Instead, it gives a *recipe*, i.e., convolve the filter with a sinusoid, observe the magnitude and phase

[9] If h is a general (non-linear) transform, additional *harmonic* frequencies are introduced. This was traditionally the bane of audiophiles, who insisted on equipment with no *harmonic distortion*. Now that digital audio has introduced pure distortion-free sound, some audiophiles are buying retro tube amplifiers or digital signal processors that simulate such distortions because of their "warmer sound".

shift, repeat. Fortunately, closed form equations for the Fourier transform exist both in the continuous domain,

$$H(\omega) = \int_{-\infty}^{\infty} h(x) e^{-j\omega x} dx, \tag{3.56}$$

and in the discrete domain,

$$H(k) = \frac{1}{N} \sum_{x=0}^{N-1} h(x) e^{-j\frac{2\pi k x}{N}}, \tag{3.57}$$

where N is the length of the signal or region of analysis. These formulas apply both to filters, such as $h(x)$, and to signals or images, such as $s(x)$ or $g(x)$.

The discrete form of the Fourier transform (3.57) is known as the *Discrete Fourier Transform* (DFT). Note that while (3.57) can be evaluated for any value of k, it only makes sense for values in the range $k \in [-\frac{N}{2}, \frac{N}{2}]$. This is because larger values of k *alias* with lower frequencies and hence provide no additional information, as explained in the discussion on aliasing in Section 2.3.1.

At face value, the DFT takes $O(N^2)$ operations (multiply-adds) to evaluate. Fortunately, there exists a faster algorithm called the *Fast Fourier Transform* (FFT), which requires only $O(N \log_2 N)$ operations (Bracewell 1986; Oppenheim, Schafer, and Buck 1999). We do not explain the details of the algorithm here, except to say that it involves a series of $\log_2 N$ stages, where each stage performs small 2×2 transforms (matrix multiplications with known coefficients) followed by some semi-global permutations. (You will often see the term *butterfly* applied to these stages because of the pictorial shape of the signal processing graphs involved.) Implementations for the FFT can be found in most numerical and signal processing libraries.

The Fourier transform comes with a set of extremely useful properties relating original signals and their Fourier transforms, including superposition, shifting, reversal, convolution, correlation, multiplication, differentiation, domain scaling (stretching), and energy preservation (Parseval's Theorem). To make room for all of the new material in this second edition, I have removed all of these details, as well as a discussion of commonly used Fourier transform pairs. Interested readers should refer to (Szeliski 2010, Section 3.1, Tables 3.1–3.3) or standard textbooks on signal processing and Fourier transforms (Bracewell 1986; Glassner 1995; Oppenheim and Schafer 1996; Oppenheim, Schafer, and Buck 1999).

We can also compute the Fourier transforms for the small discrete kernels shown in Figure 3.14 (see Table 3.1). Notice how the moving average filters do not uniformly dampen higher frequencies and hence can lead to ringing artifacts. The binomial filter (Gomes and Velho 1997) used as the "Gaussian" in Burt and Adelson's (1983a) Laplacian pyramid (see Section 3.5), does a decent job of separating the high and low frequencies, but still leaves a fair amount of high-frequency detail, which can lead to aliasing after downsampling. The Sobel edge detector at first linearly accentuates frequencies, but then decays at higher frequencies, and hence has trouble detecting fine-scale edges, e.g., adjacent black and white columns. We look at additional examples of small kernel Fourier transforms in Section 3.5.2, where we study better kernels for prefiltering before decimation (size reduction).

3.4.1 Two-dimensional Fourier transforms

The formulas and insights we have developed for one-dimensional signals and their transforms translate directly to two-dimensional images. Here, instead of just specifying a horizontal or vertical frequency ω_x or ω_y, we can create an oriented sinusoid of frequency (ω_x, ω_y),

$$s(x, y) = \sin(\omega_x x + \omega_y y). \tag{3.58}$$

Name	Kernel	Transform	Plot
box-3	$\frac{1}{3}$ | 1 | 1 | 1 |	$\frac{1}{3}(1+2\cos\omega)$	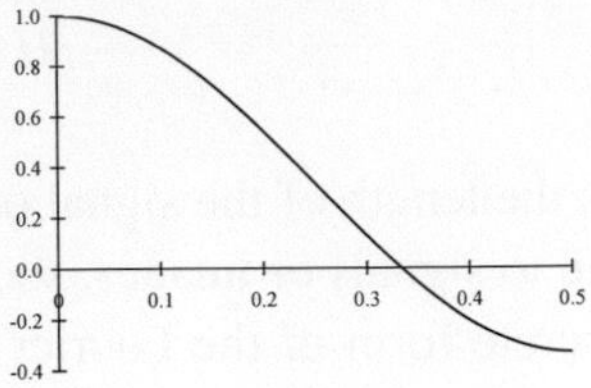
box-5	$\frac{1}{5}$ | 1 | 1 | 1 | 1 | 1 |	$\frac{1}{5}(1+2\cos\omega+2\cos 2\omega)$	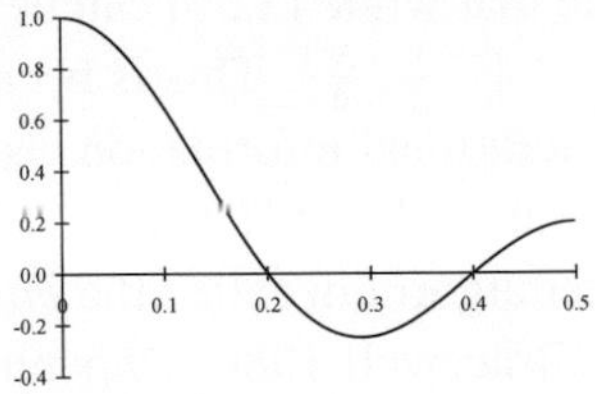
linear	$\frac{1}{4}$ | 1 | 2 | 1 |	$\frac{1}{2}(1+\cos\omega)$	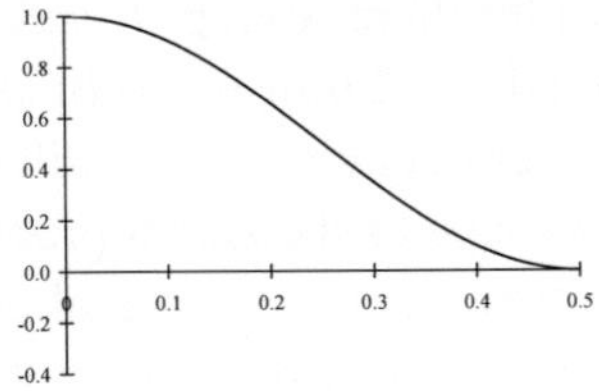
binomial	$\frac{1}{16}$ | 1 | 4 | 6 | 4 | 1 |	$\frac{1}{4}(1+\cos\omega)^2$	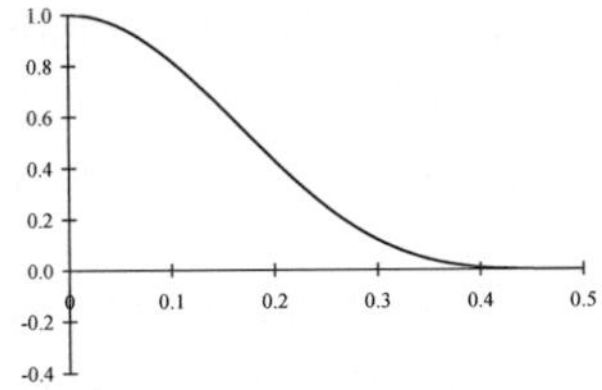
Sobel	$\frac{1}{2}$ | −1 | 0 | 1 |	$\sin\omega$	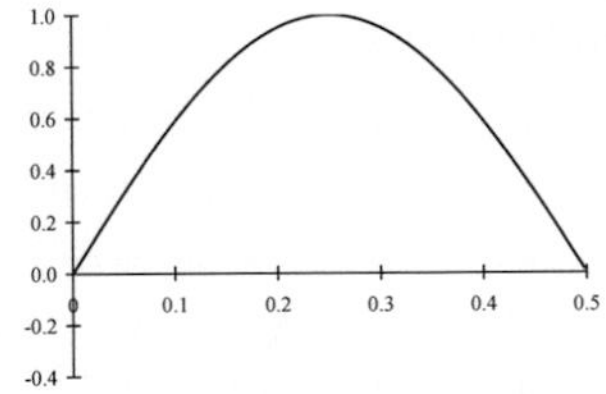
corner	$\frac{1}{2}$ | −1 | 2 | −1 |	$\frac{1}{2}(1-\cos\omega)$	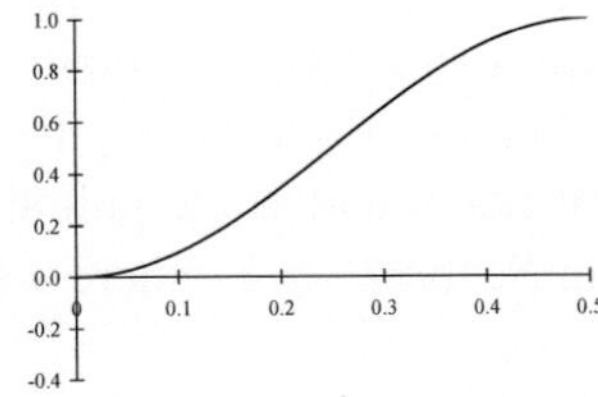

Table 3.1 Fourier transforms of the separable kernels shown in Figure 3.14, obtained by evaluating $\sum_k h(k)e^{-jk\omega}$.

The corresponding two-dimensional Fourier transforms are then

$$H(\omega_x, \omega_y) = \int_{-\infty}^{\infty} \int_{-\infty}^{\infty} h(x, y) e^{-j(\omega_x x + \omega_y y)} dx\, dy, \tag{3.59}$$

and in the discrete domain,

$$H(k_x, k_y) = \frac{1}{MN} \sum_{x=0}^{M-1} \sum_{y=0}^{N-1} h(x, y) e^{-j2\pi(k_x x/M + k_y y/N)} \tag{3.60}$$

where M and N are the width and height of the image.

All of the Fourier transform properties from 1D carry over to two dimensions if we replace the scalar variables x, ω, x_0 and a, with their 2D vector counterparts $\mathbf{x} = (x, y)$, $\boldsymbol{\omega} = (\omega_x, \omega_y)$, $\mathbf{x}_0 = (x_0, y_0)$, and $\mathbf{a} = (a_x, a_y)$, and use vector inner products instead of multiplications.

Wiener filtering

While the Fourier transform is a useful tool for analyzing the frequency characteristics of a filter kernel or image, it can also be used to analyze the frequency spectrum of a whole *class* of images.

A simple model for images is to assume that they are random noise fields whose expected magnitude at each frequency is given by this *power spectrum* $P_s(\omega_x, \omega_y)$, i.e.,

$$\left\langle [S(\omega_x, \omega_y)]^2 \right\rangle = P_s(\omega_x, \omega_y), \tag{3.61}$$

where the angle brackets $\langle\cdot\rangle$ denote the expected (mean) value of a random variable.[10] To generate such an image, we simply create a random Gaussian noise image $S(\omega_x, \omega_y)$ where each "pixel" is a zero-mean Gaussian of variance $P_s(\omega_x, \omega_y)$ and then take its inverse FFT.

The observation that signal spectra capture a first-order description of spatial statistics is widely used in signal and image processing. In particular, assuming that an image is a sample from a correlated Gaussian random noise field combined with a statistical model of the measurement process yields an optimum restoration filter known as the *Wiener filter*.

The first edition of this book contains a derivation of the Wiener filter (Szeliski 2010, Section 3.4.3), but I've decided to remove this from the current edition, since it is almost never used in practice any more, having been replaced with better-performing non-linear filters.

Discrete cosine transform

The *discrete cosine transform* (DCT) is a variant of the Fourier transform particularly well-suited to compressing images in a block-wise fashion. The one-dimensional DCT is computed by taking the dot product of each N-wide block of pixels with a set of cosines of different frequencies,

$$F(k) = \sum_{i=0}^{N-1} \cos\left(\frac{\pi}{N}(i + \frac{1}{2})k\right) f(i), \tag{3.62}$$

where k is the coefficient (frequency) index and the $1/2$-pixel offset is used to make the basis coefficients symmetric (Wallace 1991). Some of the discrete cosine basis functions are shown in Figure 3.25. As you can see, the first basis function (the straight blue line) encodes the average DC value in the block of pixels, while the second encodes a slightly curvy version of the slope.

[10] The notation $E[\cdot]$ is also commonly used.

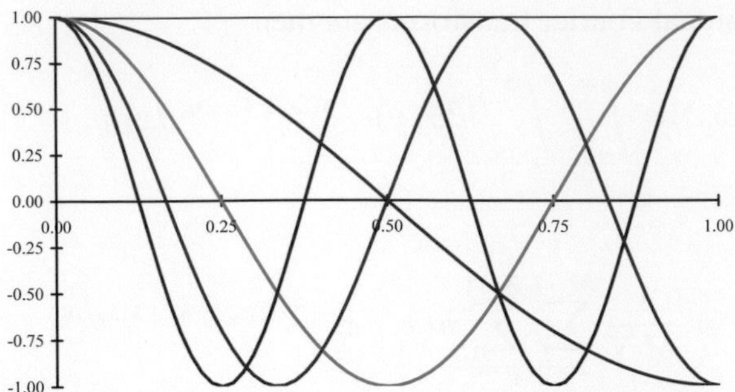

Figure 3.25 Discrete cosine transform (DCT) basis functions: The first DC (i.e., constant) basis is the horizontal blue line, the second is the brown half-cycle waveform, etc. These bases are widely used in image and video compression standards such as JPEG.

It turns out that the DCT is a good approximation to the optimal Karhunen–Loève decomposition of natural image statistics over small patches, which can be obtained by performing a principal component analysis (PCA) of images, as described in Section 5.2.3. The KL-transform decorrelates the signal optimally (assuming the signal is described by its spectrum) and thus, theoretically, leads to optimal compression.

The two-dimensional version of the DCT is defined similarly,

$$F(k,l) = \sum_{i=0}^{N-1} \sum_{j=0}^{N-1} \cos\left(\frac{\pi}{N}(i+\frac{1}{2})k\right) \cos\left(\frac{\pi}{N}(j+\frac{1}{2})l\right) f(i,j). \tag{3.63}$$

Like the 2D Fast Fourier Transform, the 2D DCT can be implemented separably, i.e., first computing the DCT of each line in the block and then computing the DCT of each resulting column. Like the FFT, each of the DCTs can also be computed in $O(N \log N)$ time.

As we mentioned in Section 2.3.3, the DCT is widely used in today's image and video compression algorithms, although alternatives such as wavelet transforms (Simoncelli and Adelson 1990b; Taubman and Marcellin 2002), discussed in Section 3.5.4, and overlapped variants of the DCT (Malvar 1990, 1998, 2000), are used in the JPEG2000 and JPEG XR standards. These newer algorithms suffer less from the *blocking artifacts* (visible edge-aligned discontinuities) that result from the pixels in each block (typically 8×8) being transformed and quantized independently. See Exercise 4.3 for ideas on how to remove blocking artifacts from compressed JPEG images.

3.4.2 *Application*: Sharpening, blur, and noise removal

Another common application of image processing is the enhancement of images through the use of sharpening and noise removal operations, which require some kind of neighborhood processing. Traditionally, these kinds of operations were performed using linear filtering (see Sections 3.2 and Section 3.4.1). Today, it is more common to use non-linear filters (Section 3.3.1), such as the weighted median or bilateral filter (3.34–3.37), anisotropic diffusion (3.39–3.40), or non-local means (Buades, Coll, and Morel 2008). Variational methods (Section 4.2), especially those using non-quadratic (robust) norms such as the L_1 norm (which is called *total variation*), are also often used. Most recently, deep neural networks have taken over the denoising community (Section 10.3). Figure 3.19 shows some examples of linear and non-linear filters being used to remove noise.

When measuring the effectiveness of image denoising algorithms, it is common to report the results as a *peak signal-to-noise ratio (PSNR)* measurement (2.120), where $I(\mathbf{x})$ is the original (noise-free) image and $\hat{I}(\mathbf{x})$ is the image after denoising; this is for the case where the noisy image has been synthetically generated, so that the clean image is known. A better way to measure the quality is to use a perceptually based similarity metric, such as the structural similarity (SSIM) index (Wang, Bovik *et al.* 2004; Wang, Bovik, and Simoncelli 2005) or FLIP image difference evaluator (Andersson, Nilsson *et al.* 2020). More recently, people have started measuring similarity using neural "perceptual" similarity metrics (Johnson, Alahi, and Fei-Fei 2016; Dosovitskiy and Brox 2016; Zhang, Isola *et al.* 2018; Tariq, Tursun *et al.* 2020; Czolbe, Krause *et al.* 2020), which, unlike L_2 (PSNR) or L_1 metrics, which encourage smooth or flat average results, prefer images with similar amounts of texture (Cho, Joshi *et al.* 2012). When the clean image is not available, it is also possible to assess the quality of an image using *no-reference image quality assessment* (Mittal, Moorthy, and Bovik 2012; Talebi and Milanfar 2018).

Exercises 3.12, 3.21, and 3.28 have you implement some of these operations and compare their effectiveness. More sophisticated techniques for blur removal and the related task of super-resolution are discussed in Section 10.3.

3.5 Pyramids and wavelets

So far in this chapter, all of the image transformations we have studied produce output images of the same size as the inputs. Often, however, we may wish to change the resolution of an image before proceeding further. For example, we may need to interpolate a small image to make its resolution match that of the output printer or computer screen. Alternatively, we may want to reduce the size of an image to speed up the execution of an algorithm or to save on storage space or transmission time.

Sometimes, we do not even know what the appropriate resolution for the image should be. Consider, for example, the task of finding a face in an image (Section 6.3.1). Since we do not know the scale at which the face will appear, we need to generate a whole *pyramid* of differently sized images and scan each one for possible faces. (Biological visual systems also operate on a hierarchy of scales (Marr 1982).) Such a pyramid can also be very helpful in accelerating the search for an object by first finding a smaller instance of that object at a coarser level of the pyramid and then looking for the full resolution object only in the vicinity of coarse-level detections (Section 9.1.1). Finally, image pyramids are extremely useful for performing multi-scale editing operations such as blending images while maintaining details.

In this section, we first discuss good filters for changing image resolution, i.e., upsampling (*interpolation*, Section 3.5.1) and downsampling (*decimation*, Section 3.5.2). We then present the concept of multi-resolution pyramids, which can be used to create a complete hierarchy of differently sized images and to enable a variety of applications (Section 3.5.3). A closely related concept is that of *wavelets*, which are a special kind of pyramid with higher frequency selectivity and other useful properties (Section 3.5.4). Finally, we present a useful application of pyramids, namely the blending of different images in a way that hides the seams between the image boundaries (Section 3.5.5).

3.5.1 Interpolation

In order to *interpolate* (or *upsample*) an image to a higher resolution, we need to select some interpolation kernel with which to convolve the image,

$$g(i,j) = \sum_{k,l} f(k,l)h(i-rk, j-rl). \tag{3.64}$$

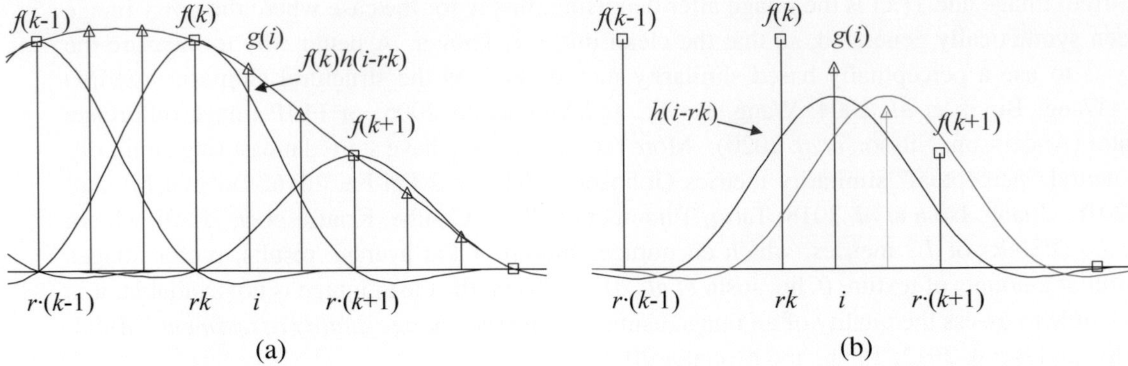

Figure 3.26 Signal interpolation, $g(i) = \sum_k f(k)h(i - rk)$: (a) weighted summation of input values; (b) polyphase filter interpretation.

This formula is related to the discrete convolution formula (3.14), except that we replace k and l in $h()$ with rk and rl, where r is the upsampling rate. Figure 3.26a shows how to think of this process as the superposition of sample weighted interpolation kernels, one centered at each input sample k. An alternative mental model is shown in Figure 3.26b, where the kernel is centered at the output pixel value i (the two forms are equivalent). The latter form is sometimes called the *polyphase filter* form, since the kernel values $h(i)$ can be stored as r separate kernels, each of which is selected for convolution with the input samples depending on the *phase* of i relative to the upsampled grid.

What kinds of kernel make good interpolators? The answer depends on the application and the computation time involved. Any of the smoothing kernels shown in Table 3.1 can be used after appropriate re-scaling.[11] The *linear* interpolator (corresponding to the tent kernel) produces interpolating piecewise linear curves, which result in unappealing *creases* when applied to images (Figure 3.27a). The cubic B-spline, whose discrete $1/2$-pixel sampling appears as the *binomial kernel* in Table 3.1, is an *approximating* kernel (the interpolated image does not pass through the input data points) that produces soft images with reduced high-frequency detail. The equation for the cubic B-spline is easiest to derive by convolving the tent function (linear B-spline) with itself.

While most graphics cards use the bilinear kernel (optionally combined with a MIP-map—see Section 3.5.3), most photo editing packages use *bicubic* interpolation. The cubic interpolant is a C^1 (derivative-continuous) piecewise-cubic *spline* (the term "spline" is synonymous with "piecewise-polynomial")[12] whose equation is

$$h(x) = \begin{cases} 1-(a+3)x^2+(a+2)|x|^3 & \text{if} \quad |x|<1 \\ a(|x|-1)(|x|-2)^2 & \text{if} \quad 1 \le |x| < 2 \\ 0 & \text{otherwise,} \end{cases} \tag{3.65}$$

where a specifies the derivative at $x = 1$ (Parker, Kenyon, and Troxel 1983). The value of a is often set to -1, since this best matches the frequency characteristics of a sinc function (Figure 3.28). It also introduces a small amount of sharpening, which can be visually appealing. Unfortunately, this choice does not linearly interpolate straight lines (intensity ramps), so some visible ringing may occur. A better choice for large amounts of interpolation is probably $a = -0.5$, which produces a

[11] The smoothing kernels in Table 3.1 have a unit area. To turn them into interpolating kernels, we simply scale them up by the interpolation rate r.

[12] The term "spline" comes from the draughtsman's workshop, where it was the name of a flexible piece of wood or metal used to draw smooth curves.

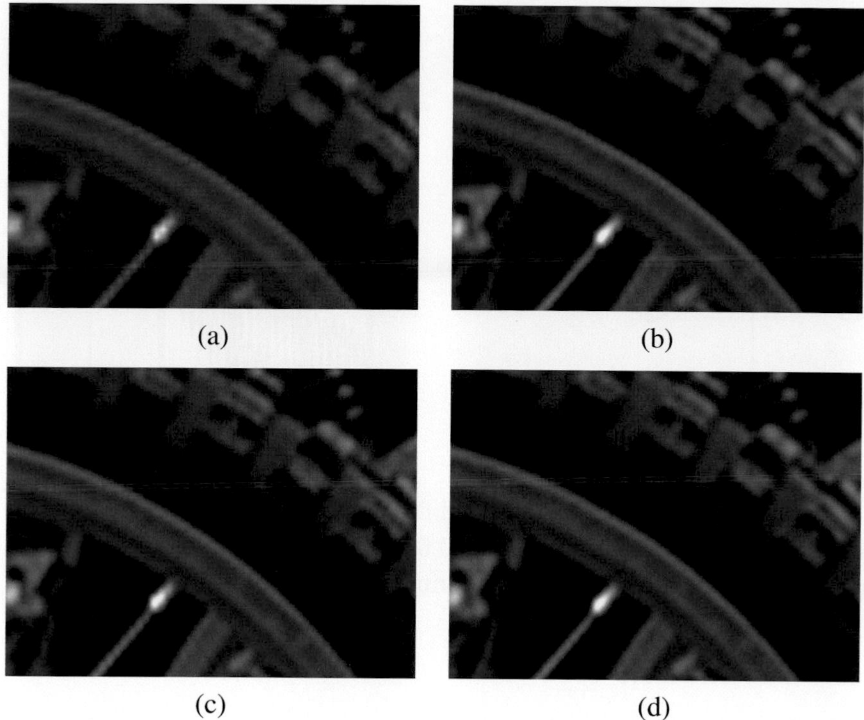

Figure 3.27 Two-dimensional image interpolation: (a) bilinear; (b) bicubic ($a = -1$); (c) bicubic ($a = -0.5$); (d) windowed sinc (nine taps).

quadratic reproducing spline; it interpolates linear and quadratic functions exactly (Wolberg 1990, Section 5.4.3). Figure 3.28 shows the $a = -1$ and $a = -0.5$ cubic interpolating kernel along with their Fourier transforms; Figure 3.27b and c shows them being applied to two-dimensional interpolation.

Splines have long been used for function and data value interpolation because of the ability to precisely specify derivatives at control points and efficient *incremental* algorithms for their evaluation (Bartels, Beatty, and Barsky 1987; Farin 1992, 2002). Splines are widely used in geometric modeling and computer-aided design (CAD) applications, although they have started being displaced by subdivision surfaces (Zorin, Schröder, and Sweldens 1996; Peters and Reif 2008). In computer vision, splines are often used for elastic image deformations (Section 3.6.2), scattered data interpolation (Section 4.1), motion estimation (Section 9.2.2), and surface interpolation (Section 13.3). In fact, it is possible to carry out most image processing operations by representing images as splines and manipulating them in a multi-resolution framework (Unser 1999; Nehab and Hoppe 2014).

The highest quality interpolator is generally believed to be the windowed sinc function because it both preserves details in the lower resolution image and avoids aliasing. (It is also possible to construct a C^1 piecewise-cubic approximation to the windowed sinc by matching its derivatives at zero crossing (Szeliski and Ito 1986).) However, some people object to the excessive *ringing* that can be introduced by the windowed sinc and to the repetitive nature of the ringing frequencies (see Figure 3.27d). For this reason, some photographers prefer to repeatedly interpolate images by a small fractional amount (this tends to decorrelate the original pixel grid with the final image). Additional possibilities include using the bilateral filter as an interpolator (Kopf, Cohen *et al.* 2007), using global optimization (Section 3.6) or hallucinating details (Section 10.3).

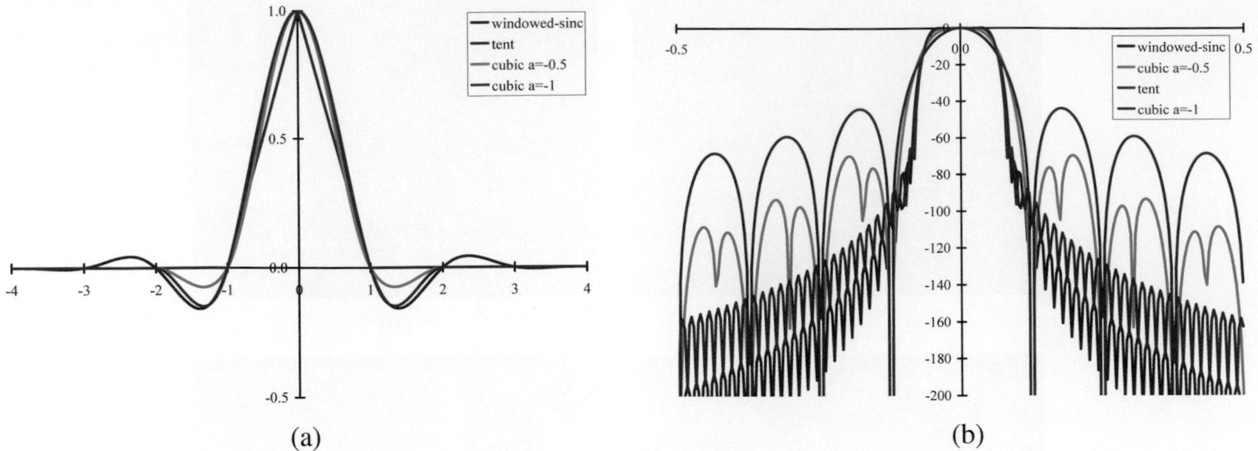

Figure 3.28 (a) Some windowed sinc functions and (b) their log Fourier transforms: raised-cosine windowed sinc in blue, cubic interpolators ($a = -1$ and $a = -0.5$) in green and purple, and tent function in brown. They are often used to perform high-accuracy low-pass filtering operations.

3.5.2 Decimation

While interpolation can be used to increase the resolution of an image, decimation (downsampling) is required to reduce the resolution.[13] To perform decimation, we first (conceptually) convolve the image with a low-pass filter (to avoid aliasing) and then keep every rth sample. In practice, we usually only evaluate the convolution at every rth sample,

$$g(i,j) = \sum_{k,l} f(k,l)h(ri-k, rj-l), \tag{3.66}$$

as shown in Figure 3.29. Note that the smoothing kernel $h(k,l)$, in this case, is often a stretched and re-scaled version of an interpolation kernel. Alternatively, we can write

$$g(i,j) = \frac{1}{r}\sum_{k,l} f(k,l)h(i-k/r, j-l/r) \tag{3.67}$$

and keep the same kernel $h(k,l)$ for both interpolation and decimation.

One commonly used ($r = 2$) decimation filter is the *binomial* filter introduced by Burt and Adelson (1983a). As shown in Table 3.1, this kernel does a decent job of separating the high and low frequencies, but still leaves a fair amount of high-frequency detail, which can lead to aliasing after downsampling. However, for applications such as image blending (discussed later in this section), this aliasing is of little concern.

If, however, the downsampled images will be displayed directly to the user or, perhaps, blended with other resolutions (as in MIP-mapping, Section 3.5.3), a higher-quality filter is desired. For high downsampling rates, the windowed sinc prefilter is a good choice (Figure 3.28). However, for small downsampling rates, e.g., $r = 2$, more careful filter design is required.

Table 3.2 shows a number of commonly used $r = 2$ downsampling filters, while Figure 3.30 shows their corresponding frequency responses. These filters include:

- the linear $[1, 2, 1]$ filter gives a relatively poor response;

[13]The term "decimation" has a gruesome etymology relating to the practice of killing every tenth soldier in a Roman unit guilty of cowardice. It is generally used in signal processing to mean any downsampling or rate reduction operation.

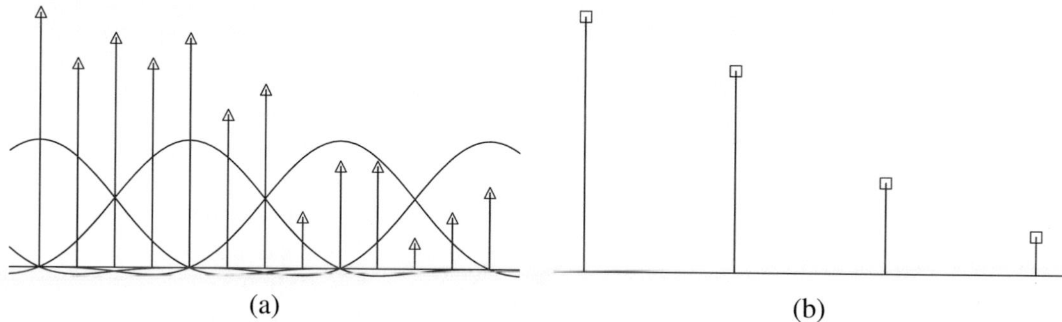

Figure 3.29 Signal decimation: (a) the original samples are (b) convolved with a low-pass filter before being downsampled.

- the binomial $[1, 4, 6, 4, 1]$ filter cuts off a lot of frequencies but is useful for computer vision analysis pyramids;
- the cubic filters from (3.65); the $a = -1$ filter has a sharper fall-off than the $a = -0.5$ filter (Figure 3.30);
- a cosine-windowed sinc function;
- the QMF-9 filter of Simoncelli and Adelson (1990b) is used for wavelet denoising and aliases a fair amount (note that the original filter coefficients are normalized to $\sqrt{2}$ gain so they can be "self-inverting");
- the 9/7 analysis filter from JPEG 2000 (Taubman and Marcellin 2002).

Please see the original papers for the full-precision values of some of these coefficients.

3.5.3 Multi-resolution representations

Now that we have described interpolation and decimation algorithms, we can build a complete image pyramid (Figure 3.31). As we mentioned before, pyramids can be used to accelerate coarse-to-fine search algorithms, to look for objects or patterns at different scales, and to perform multi-resolution blending operations. They are also widely used in computer graphics hardware and software to perform fractional-level decimation using the MIP-map, which we discuss in Section 3.6.

$\|n\|$	Linear	Binomial	Cubic $a = -1$	Cubic $a = -0.5$	Windowed sinc	QMF-9	JPEG 2000
0	0.50	0.3750	0.5000	0.50000	0.4939	0.5638	0.6029
1	0.25	0.2500	0.3125	0.28125	0.2684	0.2932	0.2669
2		0.0625	0.0000	0.00000	0.0000	-0.0519	-0.0782
3			-0.0625	-0.03125	-0.0153	-0.0431	-0.0169
4					0.0000	0.0198	0.0267

Table 3.2 Filter coefficients for $2 \times$ decimation. These filters are of odd length, are symmetric, and are normalized to have unit DC gain (sum up to 1). See Figure 3.30 for their associated frequency responses.

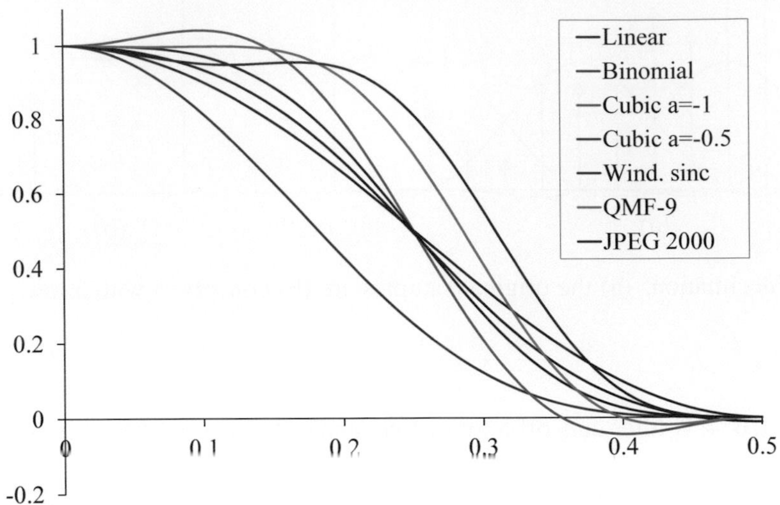

Figure 3.30 Frequency response for some 2 × decimation filters. The cubic $a = -1$ filter has the sharpest fall-off but also a bit of ringing; the wavelet analysis filters (QMF-9 and JPEG 2000), while useful for compression, have more aliasing.

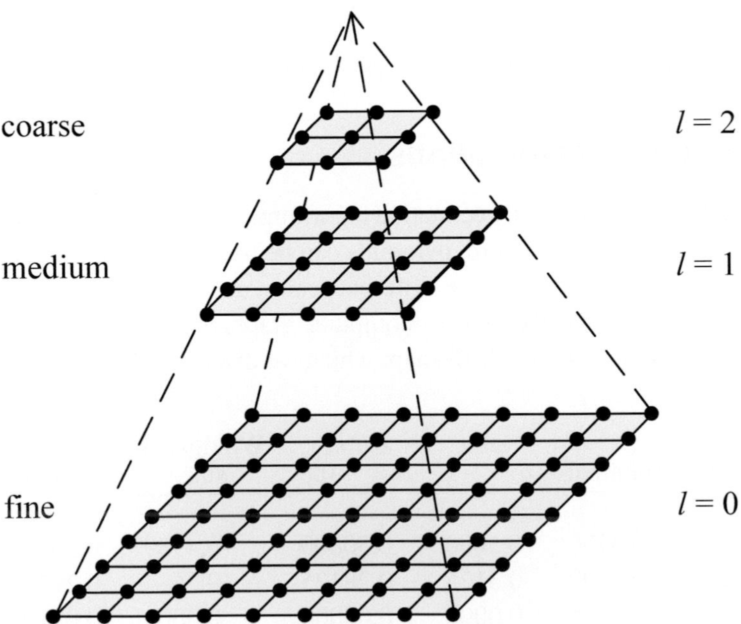

Figure 3.31 A traditional image pyramid: each level has half the resolution (width and height), and hence a quarter of the pixels, of its parent level.

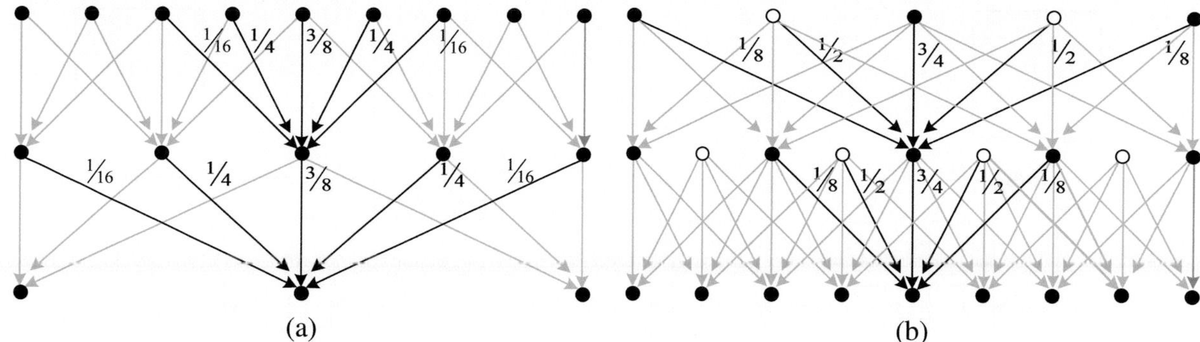

Figure 3.32 The Gaussian pyramid shown as a signal processing diagram: The (a) analysis and (b) re-synthesis stages are shown as using similar computations. The white circles indicate zero values inserted by the $\uparrow 2$ upsampling operation. Notice how the reconstruction filter coefficients are twice the analysis coefficients. The computation is shown as flowing down the page, regardless of whether we are going from coarse to fine or *vice versa*.

The best known (and probably most widely used) pyramid in computer vision is Burt and Adelson's (1983a) Laplacian pyramid. To construct the pyramid, we first blur and subsample the original image by a factor of two and store this in the next level of the pyramid (Figures 3.31 and 3.32). Because adjacent levels in the pyramid are related by a sampling rate $r = 2$, this kind of pyramid is known as an *octave pyramid*. Burt and Adelson originally proposed a five-tap kernel of the form

$$\boxed{c \mid b \mid a \mid b \mid c}, \tag{3.68}$$

with $b = 1/4$ and $c = 1/4 - a/2$. In practice, they and everyone else uses $a = 3/8$, which results in the familiar binomial kernel,

$$\frac{1}{16} \boxed{1 \mid 4 \mid 6 \mid 4 \mid 1}, \tag{3.69}$$

which is particularly easy to implement using shifts and adds. (This was important in the days when multipliers were expensive.) The reason they call their resulting pyramid a *Gaussian* pyramid is that repeated convolutions of the binomial kernel converge to a Gaussian.[14]

To compute the *Laplacian* pyramid, Burt and Adelson first interpolate a lower resolution image to obtain a *reconstructed* low-pass version of the original image (Figure 3.33). They then subtract this low-pass version from the original to yield the band-pass "Laplacian" image, which can be stored away for further processing. The resulting pyramid has *perfect reconstruction*, i.e., the Laplacian images plus the base-level Gaussian (L_2 in Figure 3.33) are sufficient to exactly reconstruct the original image. Figure 3.32 shows the same computation in one dimension as a signal processing diagram, which completely captures the computations being performed during the analysis and re-synthesis stages.

Burt and Adelson also describe a variant of the Laplacian pyramid, where the low-pass image is taken from the original blurred image rather than the reconstructed pyramid (piping the output of the L box directly to the subtraction in Figure 3.33). This variant has less aliasing, since it avoids one downsampling and upsampling round-trip, but it is not self-inverting, since the Laplacian images are no longer adequate to reproduce the original image.

As with the Gaussian pyramid, the term Laplacian is a bit of a misnomer, since their band-pass

[14] Then again, this is true for any smoothing kernel (Wells 1986).

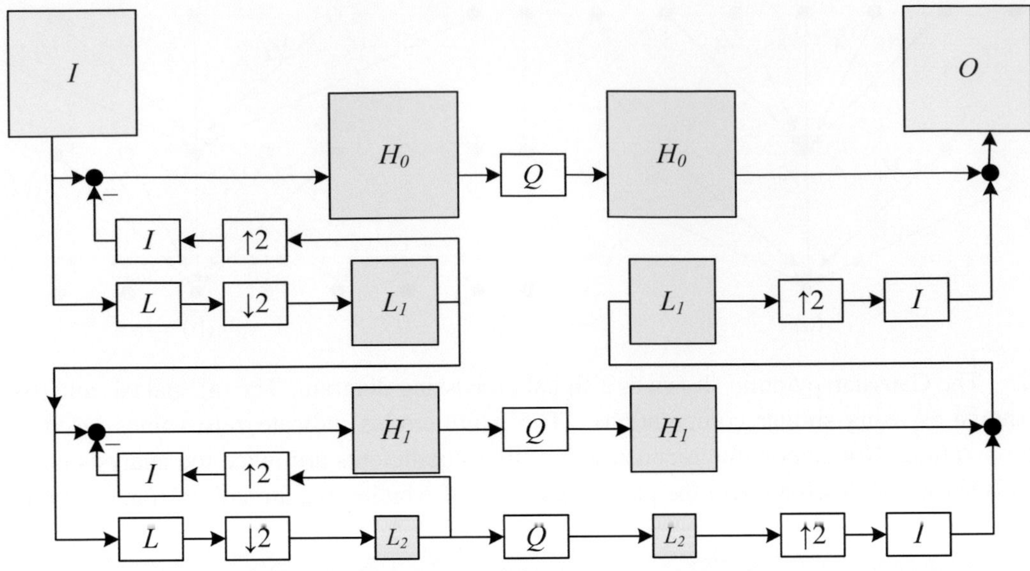

Figure 3.33 The Laplacian pyramid. The yellow images form the *Gaussian* pyramid, which is obtained by successively low-pass filtering and downsampling the input image. The blue images, together with the smallest low-pass image, which is needed for reconstruction, form the *Laplacian* pyramid. Each band-pass (blue) image is computed by upsampling and interpolating the lower-resolution Gaussian pyramid image, resulting in a blurred version of that level's low-pass image, which is subtracted from the low-pass to yield the blue band-pass image. During reconstruction, the interpolated images and the (optionally filtered) high-pass images are added back together starting with the coarsest level. The Q box indicates quantization or some other pyramid processing, e.g., noise removal by *coring* (setting small wavelet values to 0).

images are really differences of (approximate) Gaussians, or DoGs,

$$\text{DoG}\{I;\sigma_1,\sigma_2\} = G_{\sigma_1} * I - G_{\sigma_2} * I = (G_{\sigma_1} - G_{\sigma_2}) * I. \tag{3.70}$$

A Laplacian of Gaussian (which we saw in (3.26)) is actually its second derivative,

$$\text{LoG}\{I;\sigma\} = \nabla^2(G_\sigma * I) = (\nabla^2 G_\sigma) * I, \tag{3.71}$$

where

$$\nabla^2 = \frac{\partial^2}{\partial x^2} + \frac{\partial^2}{\partial y^2} \tag{3.72}$$

is the Laplacian (operator) of a function. Figure 3.34 shows how the Differences of Gaussian and Laplacians of Gaussian look in both space and frequency.

Laplacians of Gaussian have elegant mathematical properties, which have been widely studied in the *scale-space* community (Witkin 1983; Witkin, Terzopoulos, and Kass 1986; Lindeberg 1990; Nielsen, Florack, and Deriche 1997) and can be used for a variety of applications including edge detection (Marr and Hildreth 1980; Perona and Malik 1990b), stereo matching (Witkin, Terzopoulos, and Kass 1987), and image enhancement (Nielsen, Florack, and Deriche 1997).

One particularly useful application of the Laplacian pyramid is in the manipulation of local contrast as well as the tone mapping of high dynamic range images (Section 10.2.1). Paris, Hasinoff, and Kautz (2011) present a technique they call *local Laplacian filters*, which uses local range clipping in the construction of a modified Laplacian pyramid, as well as different accentuation and attenuation curves for small and large details, to implement edge-preserving filtering and tone mapping. Aubry,

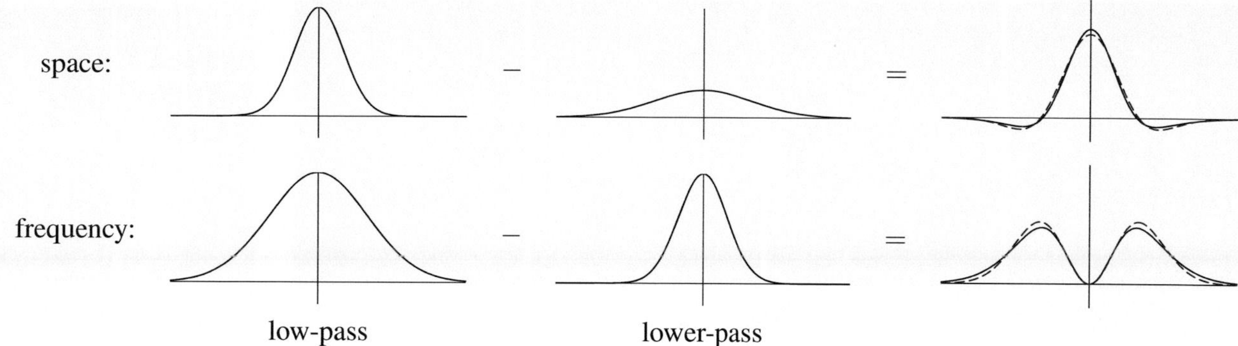

Figure 3.34 The difference of two low-pass filters results in a band-pass filter. The dashed blue lines show the close fit to a half-octave Laplacian of Gaussian.

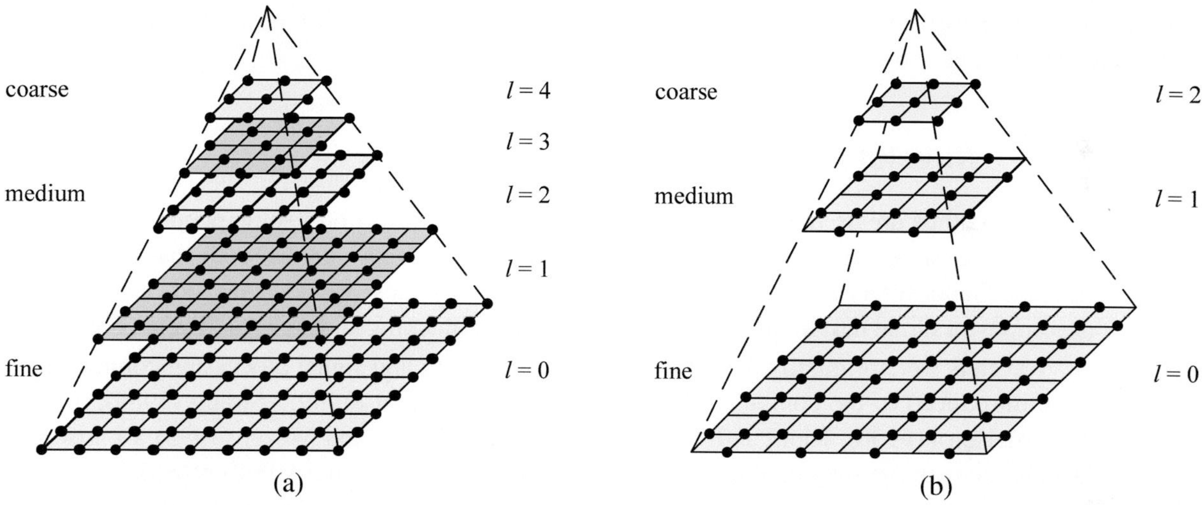

Figure 3.35 Multiresolution pyramids: (a) pyramid with half-octave (*quincunx*) sampling (odd levels are colored gray for clarity). (b) wavelet pyramid—each wavelet level stores 3/4 of the original pixels (usually the horizontal, vertical, and mixed gradients), so that the total number of wavelet coefficients and original pixels is the same.

Paris *et al.* (2014) discuss how to accelerate this processing for monotone (single channel) images and also show style transfer applications.

A less widely used variant is *half-octave pyramids*, shown in Figure 3.35a. These were first introduced to the vision community by Crowley and Stern (1984), who call them *Difference of Low-Pass* (DOLP) transforms. Because of the small scale change between adjacent levels, the authors claim that coarse-to-fine algorithms perform better. In the image-processing community, half-octave pyramids combined with checkerboard sampling grids are known as *quincunx* sampling (Feilner, Van De Ville, and Unser 2005). In detecting multi-scale features (Section 7.1.1), it is often common to use half-octave or even quarter-octave pyramids (Lowe 2004; Triggs 2004). However, in this case, the subsampling only occurs at every octave level, i.e., the image is repeatedly blurred with wider Gaussians until a full octave of resolution change has been achieved (Figure 7.11).

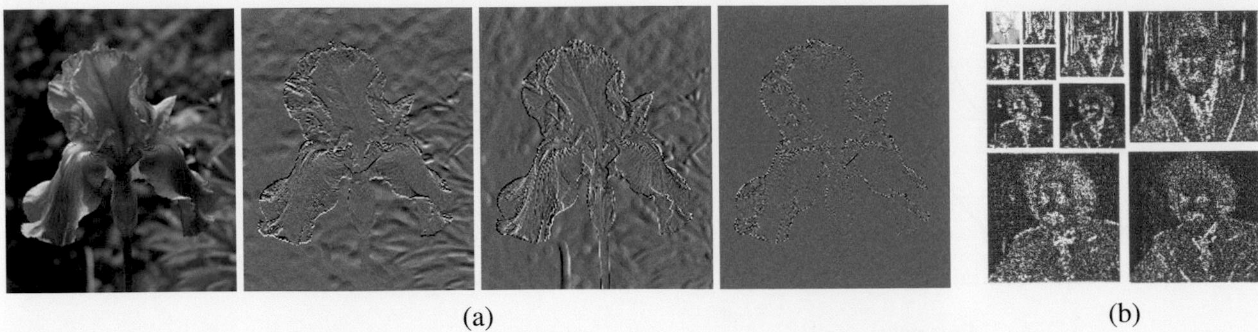

(a) (b)

Figure 3.36 A wavelet decomposition of an image: (a) single level decomposition with horizontal, vertical, and diagonal detail wavelets constructed using PyWavelet code (https://pywavelets.readthedocs.io); (b) coefficient magnitudes of a multi-level decomposition, with the high–high components in the lower right corner and the base in the upper left (Buccigrossi and Simoncelli 1999) © 1999 IEEE. Notice how the low–high and high–low components accentuate horizontal and vertical edges and gradients, while the high-high components store the less frequent mixed derivatives.

3.5.4 Wavelets

While pyramids are used extensively in computer vision applications, some people use *wavelet* decompositions as an alternative. Wavelets are filters that localize a signal in both space and frequency (like the Gabor filter) and are defined over a hierarchy of scales. Wavelets provide a smooth way to decompose a signal into frequency components without blocking and are closely related to pyramids.

Wavelets were originally developed in the applied math and signal processing communities and were introduced to the computer vision community by Mallat (1989). Strang (1989), Simoncelli and Adelson (1990b), Rioul and Vetterli (1991), Chui (1992), and Meyer (1993) all provide nice introductions to the subject along with historical reviews, while Chui (1992) provides a more comprehensive review and survey of applications. Sweldens (1997) describes the *lifting* approach to wavelets that we discuss shortly.

Wavelets are widely used in the computer graphics community to perform multi-resolution geometric processing (Stollnitz, DeRose, and Salesin 1996) and have also been used in computer vision for similar applications (Szeliski 1990b; Pentland 1994; Gortler and Cohen 1995; Yaou and Chang 1994; Lai and Vemuri 1997; Szeliski 2006b; Krishnan and Szeliski 2011; Krishnan, Fattal, and Szeliski 2013), as well as for multi-scale oriented filtering (Simoncelli, Freeman *et al.* 1992) and denoising (Portilla, Strela *et al.* 2003).

As both image pyramids and wavelets decompose an image into multi-resolution descriptions that are localized in both space and frequency, how do they differ? The usual answer is that traditional pyramids are *overcomplete*, i.e., they use more pixels than the original image to represent the decomposition, whereas wavelets provide a *tight frame*, i.e., they keep the size of the decomposition the same as the image (Figure 3.35b). However, some wavelet families *are*, in fact, overcomplete in order to provide better shiftability or steering in orientation (Simoncelli, Freeman *et al.* 1992). A better distinction, therefore, might be that wavelets are more orientation selective than regular band-pass pyramids.

How are two-dimensional wavelets constructed? Figure 3.37a shows a high-level diagram of one stage of the (recursive) coarse-to-fine construction (analysis) pipeline alongside the complementary re-construction (synthesis) stage. In this diagram, the high-pass filter followed by decimation keeps $^3/_4$ of the original pixels, while $^1/_4$ of the low-frequency coefficients are passed on to the next

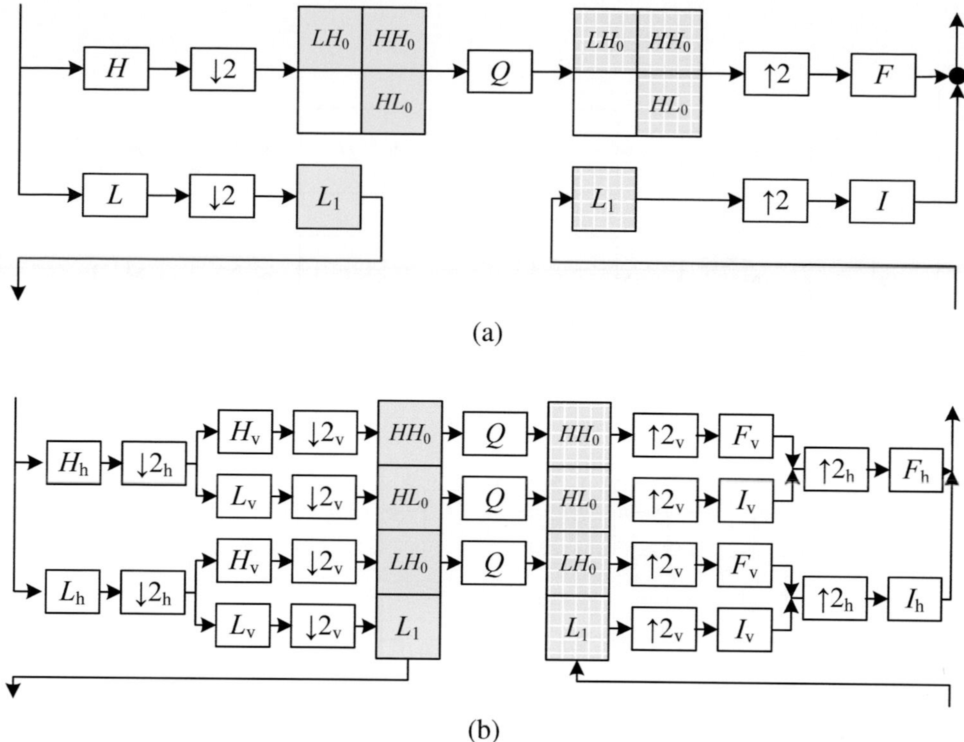

Figure 3.37 Two-dimensional wavelet decomposition: (a) high-level diagram showing the low-pass and high-pass transforms as single boxes; (b) separable implementation, which involves first performing the wavelet transform horizontally and then vertically. The I and F boxes are the interpolation and filtering boxes required to re-synthesize the image from its wavelet components.

stage for further analysis. In practice, the filtering is usually broken down into two separable sub-stages, as shown in Figure 3.37b. The resulting three wavelet images are sometimes called the high–high (HH), high–low (HL), and low–high (LH) images. The high–low and low–high images accentuate the horizontal and vertical edges and gradients, while the high–high image contains the less frequently occurring mixed derivatives (Figure 3.36).

How are the high-pass H and low-pass L filters shown in Figure 3.37b chosen and how can the corresponding reconstruction filters I and F be computed? Can filters be designed that all have finite impulse responses? This topic has been the main subject of study in the wavelet community for over two decades. The answer depends largely on the intended application, e.g., whether the wavelets are being used for compression, image analysis (feature finding), or denoising. Simoncelli and Adelson (1990b) show (in Table 4.1) some good odd-length quadrature mirror filter (QMF) coefficients that seem to work well in practice.

Since the design of wavelet filters is such a tricky art, is there perhaps a better way? Indeed, a simpler procedure is to split the signal into its even and odd components and then perform trivially reversible filtering operations on each sequence to produce what are called *lifted wavelets* (Figures 3.38 and 3.39). Sweldens (1996) gives a wonderfully understandable introduction to the *lifting scheme* for *second-generation wavelets*, followed by a comprehensive review (Sweldens 1997).

As Figure 3.38 demonstrates, rather than first filtering the whole input sequence (image) with high-pass and low-pass filters and then keeping the odd and even sub-sequences, the lifting scheme first splits the sequence into its even and odd sub-components. Filtering the even sequence with a

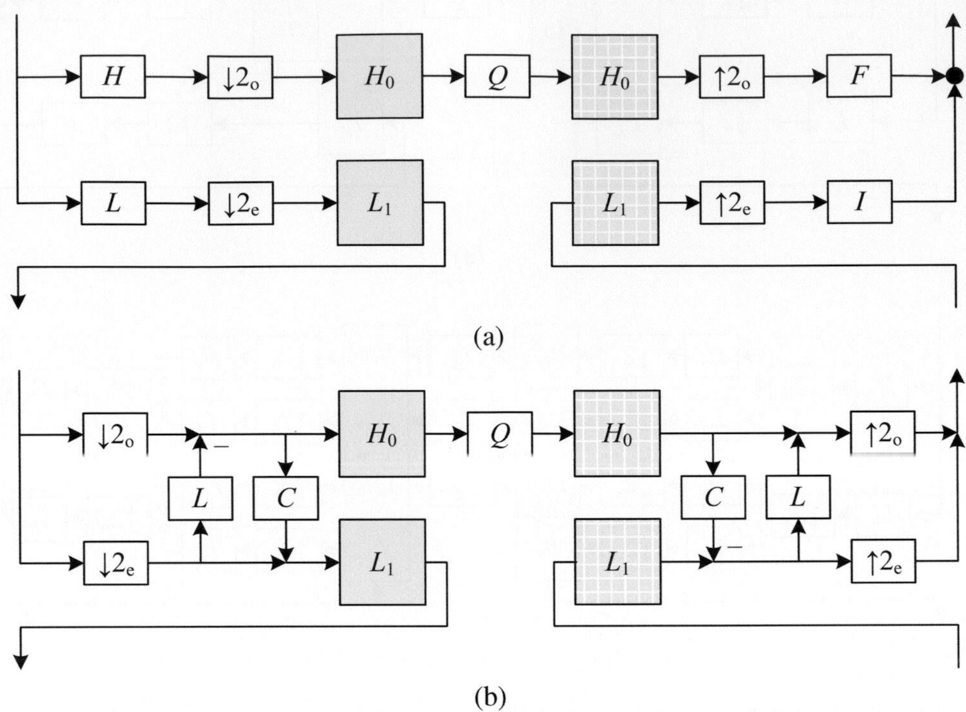

Figure 3.38 One-dimensional wavelet transform: (a) usual high-pass + low-pass filters followed by odd ($\downarrow 2_o$) and even ($\downarrow 2_e$) downsampling; (b) lifted version, which first selects the odd and even subsequences and then applies a low-pass prediction stage L and a high-pass correction stage C in an easily reversible manner.

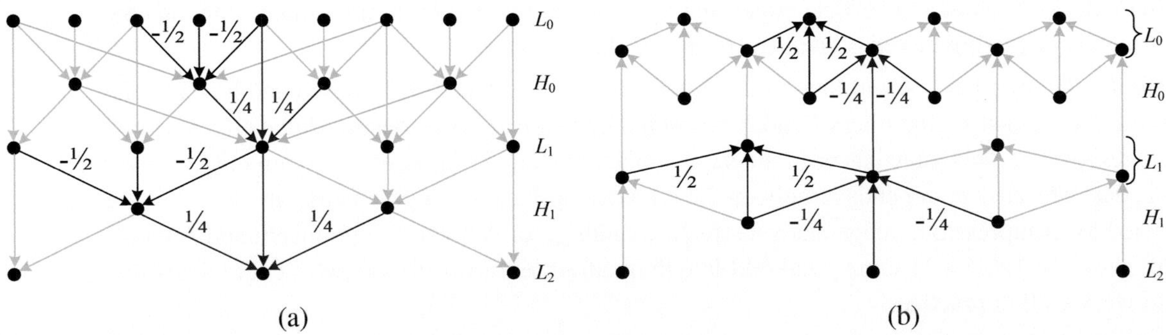

Figure 3.39 Lifted transform shown as a signal processing diagram: (a) The analysis stage first predicts the odd value from its even neighbors, stores the difference wavelet, and then compensates the coarser even value by adding in a fraction of the wavelet. (b) The synthesis stage simply reverses the flow of computation and the signs of some of the filters and operations. The light blue lines show what happens if we use four taps for the prediction and correction instead of just two.

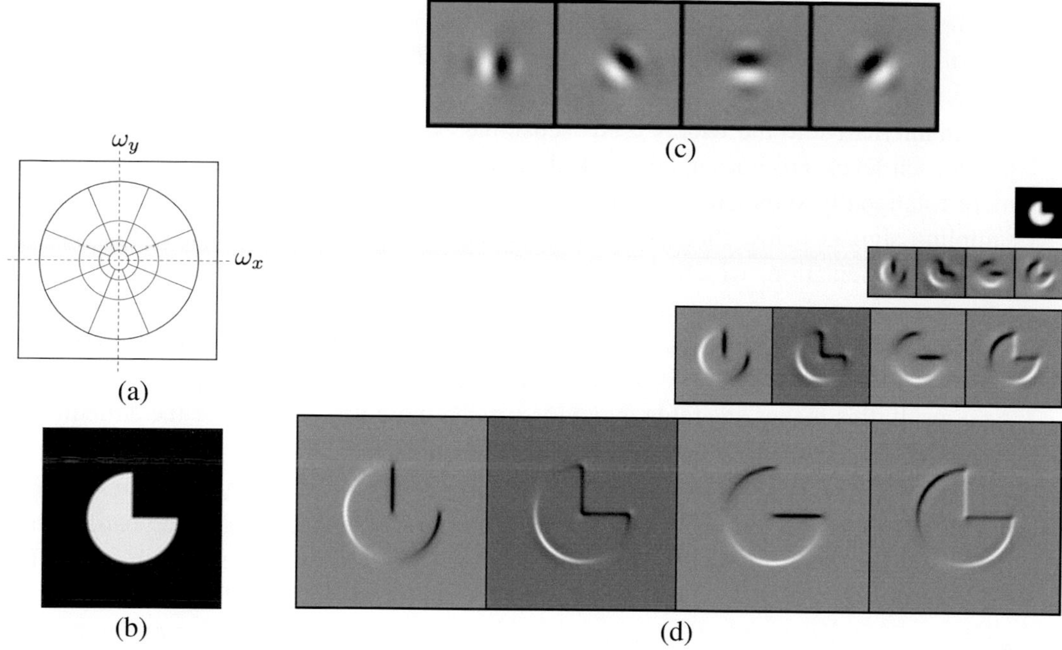

Figure 3.40 Steerable shiftable multiscale transforms (Simoncelli, Freeman *et al.* 1992) © 1992 IEEE: (a) radial multi-scale frequency domain decomposition; (b) original image; (c) a set of four steerable filters; (d) the radial multi-scale wavelet decomposition.

low-pass filter L and subtracting the result from the odd sequence is trivially reversible: simply perform the same filtering and then add the result back in. Furthermore, this operation can be performed in place, resulting in significant space savings. The same applies to filtering the difference signal with the correction filter C, which is used to ensure that the even sequence is low-pass. A series of such *lifting* steps can be used to create more complex filter responses with low computational cost and guaranteed reversibility.

This process can be more easily understood by considering the signal processing diagram in Figure 3.39. During analysis, the average of the even values is subtracted from the odd value to obtain a high-pass wavelet coefficient. However, the even samples still contain an aliased sample of the low-frequency signal. To compensate for this, a small amount of the high-pass wavelet is added back to the even sequence so that it is properly low-pass filtered. (It is easy to show that the effective low-pass filter is $[-1/8, 1/4, 3/4, 1/4, -1/8]$, which is indeed a low-pass filter.) During synthesis, the same operations are reversed with a judicious change in sign.

Of course, we need not restrict ourselves to two-tap filters. Figure 3.39 shows as light blue arrows additional filter coefficients that could optionally be added to the lifting scheme without affecting its reversibility. In fact, the low-pass and high-pass filtering operations can be interchanged, e.g., we could use a five-tap cubic low-pass filter on the odd sequence (plus center value) first, followed by a four-tap cubic low-pass predictor to estimate the wavelet, although I have not seen this scheme written down.

Lifted wavelets are called *second-generation wavelets* because they can easily adapt to non-regular sampling topologies, e.g., those that arise in computer graphics applications such as multi-resolution surface manipulation (Schröder and Sweldens 1995). It also turns out that lifted *weighted wavelets*, i.e., wavelets whose coefficients adapt to the underlying problem being solved (Fattal 2009), can be extremely effective for low-level image manipulation tasks and also for preconditioning the kinds of sparse linear systems that arise in the optimization-based approaches to vision algorithms that we discuss in Chapter 4 (Szeliski 2006b; Krishnan and Szeliski 2011; Krishnan, Fattal, and Szeliski 2013).

An alternative to the widely used "separable" approach to wavelet construction, which decomposes each level into horizontal, vertical, and "cross" sub-bands, is to use a representation that is more rotationally symmetric and orientationally selective and also avoids the aliasing inherent in sampling signals below their Nyquist frequency.[15] Simoncelli, Freeman *et al.* (1992) introduce such a representation, which they call a *pyramidal radial frequency implementation* of *shiftable multi-scale transforms* or, more succinctly, *steerable pyramids*. Their representation is not only overcomplete (which eliminates the aliasing problem) but is also orientationally selective and has identical analysis and synthesis basis functions, i.e., it is *self-inverting*, just like "regular" wavelets. As a result, this makes steerable pyramids a much more useful basis for the structural analysis and matching tasks commonly used in computer vision.

Figure 3.40a shows how such a decomposition looks in frequency space. Instead of recursively dividing the frequency domain into 2×2 squares, which results in checkerboard high frequencies, radial arcs are used instead. Figure 3.40d illustrates the resulting pyramid sub-bands. Even through the representation is *overcomplete*, i.e., there are more wavelet coefficients than input pixels, the additional frequency and orientation selectivity makes this representation preferable for tasks such as texture analysis and synthesis (Portilla and Simoncelli 2000) and image denoising (Portilla, Strela *et al.* 2003; Lyu and Simoncelli 2009).

3.5.5 *Application*: Image blending

One of the most engaging and fun applications of the Laplacian pyramid presented in Section 3.5.3 is the creation of blended composite images, as shown in Figure 3.41 (Burt and Adelson 1983b). While splicing the apple and orange images together along the midline produces a noticeable cut, *splining* them together (as Burt and Adelson (1983b) called their procedure) creates a beautiful illusion of a truly hybrid fruit. The key to their approach is that the low-frequency color variations between the red apple and the orange are smoothly blended, while the higher-frequency textures on each fruit are blended more quickly to avoid "ghosting" effects when two textures are overlaid.

To create the blended image, each source image is first decomposed into its own Laplacian pyramid (Figure 3.42, left and middle columns). Each band is then multiplied by a smooth weighting function whose extent is proportional to the pyramid level. The simplest and most general way to create these weights is to take a binary mask image (Figure 3.41g) and to construct a *Gaussian* pyramid from this mask. Each Laplacian pyramid image is then multiplied by its corresponding Gaussian mask and the sum of these two weighted pyramids is then used to construct the final image (Figure 3.42, right column).

Figure 3.41e–h shows that this process can be applied to arbitrary mask images with surprising results. It is also straightforward to extend the pyramid blend to an arbitrary number of images whose pixel provenance is indicated by an integer-valued label image (see Exercise 3.18). This is particularly useful in image stitching and compositing applications, where the exposures may vary between different images, as described in Section 8.4.4, where we also present more recent variants such as Poisson and gradient-domain blending (Pérez, Gangnet, and Blake 2003; Levin, Zomet *et al.* 2004).

[15] Such aliasing can often be seen as the signal content moving between bands as the original signal is slowly shifted.

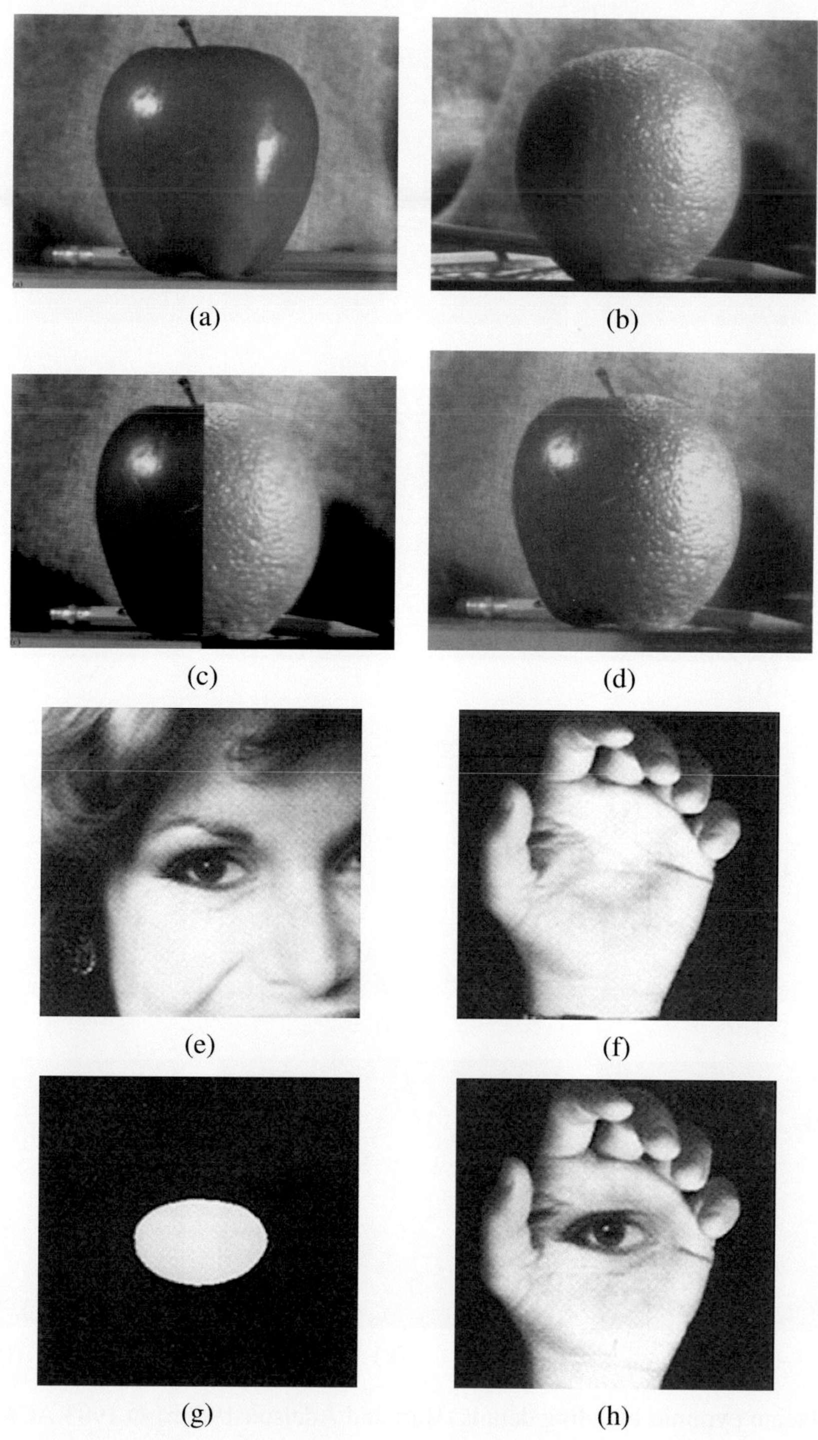

Figure 3.41 Laplacian pyramid blending (Burt and Adelson 1983b) © 1983 ACM: (a) original image of apple, (b) original image of orange, (c) regular splice, (d) pyramid blend. A masked blend of two images: (e) first input image, (f) second input image, (g) region mask, (h) blended image.

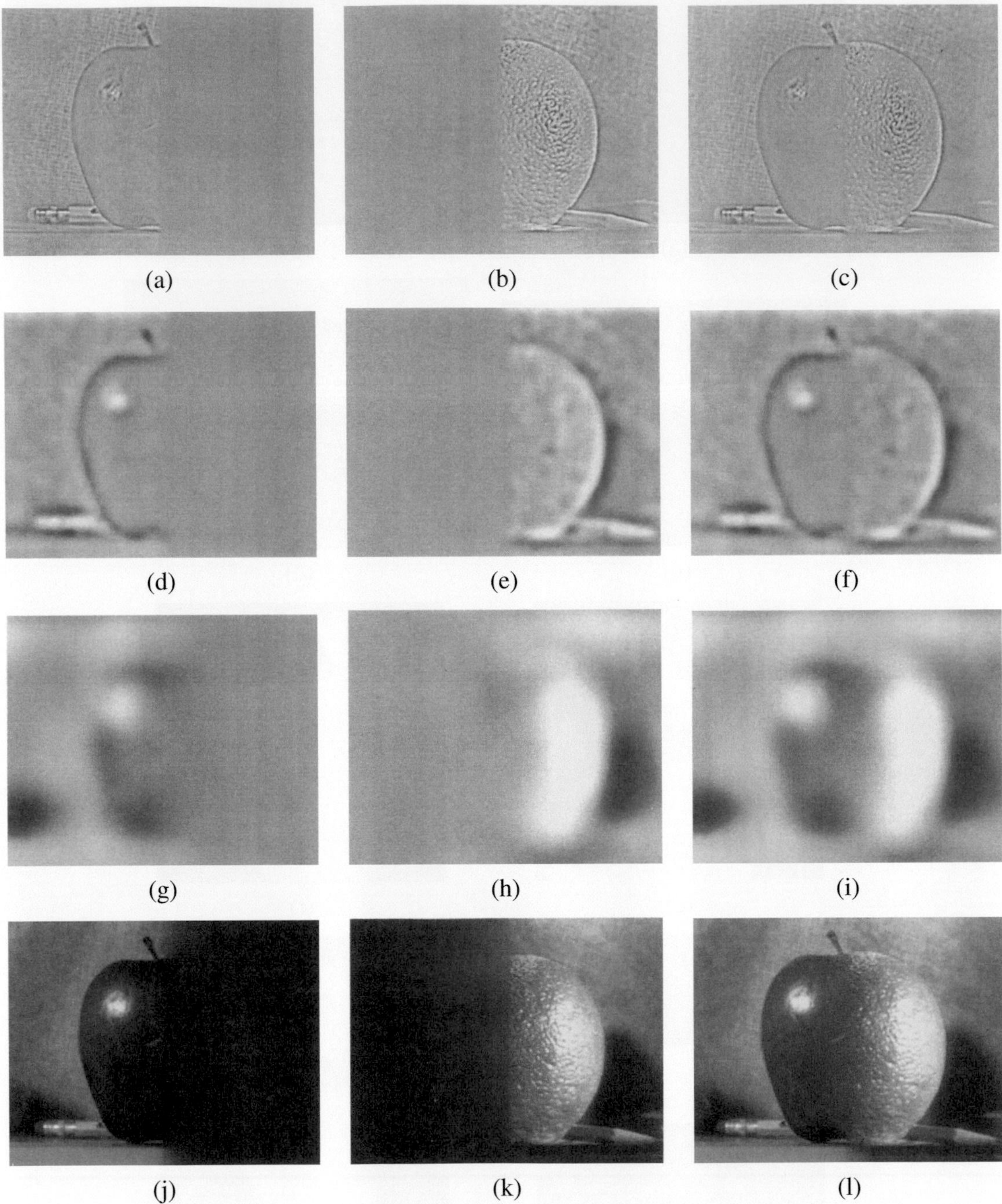

Figure 3.42 Laplacian pyramid blending details (Burt and Adelson 1983b) © 1983 ACM. The first three rows show the high, medium, and low-frequency parts of the Laplacian pyramid (taken from levels 0, 2, and 4). The left and middle columns show the original apple and orange images weighted by the smooth interpolation functions, while the right column shows the averaged contributions.

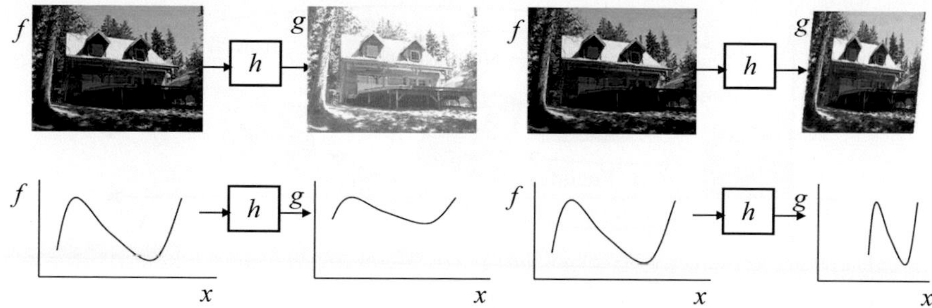

Figure 3.43 Image warping involves modifying the *domain* of an image function rather than its *range*.

3.6 Geometric transformations

In the previous sections, we saw how interpolation and decimation could be used to change the *resolution* of an image. In this section, we look at how to perform more general transformations, such as image rotations or general warps. In contrast to the point processes we saw in Section 3.1, where the function applied to an image transforms the *range* of the image,

$$g(\mathbf{x}) = h(f(\mathbf{x})), \tag{3.73}$$

here we look at functions that transform the *domain*,

$$g(\mathbf{x}) = f(\mathbf{h}(\mathbf{x})), \tag{3.74}$$

as shown in Figure 3.43.

We begin by studying the global *parametric* 2D transformation first introduced in Section 2.1.1. (Such a transformation is called parametric because it is controlled by a small number of parameters.) We then turn our attention to more local general deformations such as those defined on meshes (Section 3.6.2). Finally, we show in Section 3.6.3 how image warps can be combined with cross-dissolves to create interesting *morphs* (in-between animations). For readers interested in more details on these topics, there is an excellent survey by Heckbert (1986) as well as very accessible textbooks by Wolberg (1990), Gomes, Darsa *et al.* (1999) and Akenine-Möller and Haines (2002). Note that Heckbert's survey is on *texture mapping*, which is how the computer graphics community refers to the topic of warping images onto surfaces.

3.6.1 Parametric transformations

Parametric transformations apply a global deformation to an image, where the behavior of the transformation is controlled by a small number of parameters. Figure 3.44 shows a few examples of such transformations, which are based on the 2D geometric transformations shown in Figure 2.4. The formulas for these transformations were originally given in Table 2.1 and are reproduced here in Table 3.3 for ease of reference.

In general, given a transformation specified by a formula $\mathbf{x}' = \mathbf{h}(\mathbf{x})$ and a source image $f(\mathbf{x})$, how do we compute the values of the pixels in the new image $g(\mathbf{x})$, as given in (3.74)? Think about this for a minute before proceeding and see if you can figure it out.

If you are like most people, you will come up with an algorithm that looks something like Algorithm 3.1. This process is called *forward warping* or *forward mapping* and is shown in Figure 3.45a. Can you think of any problems with this approach?

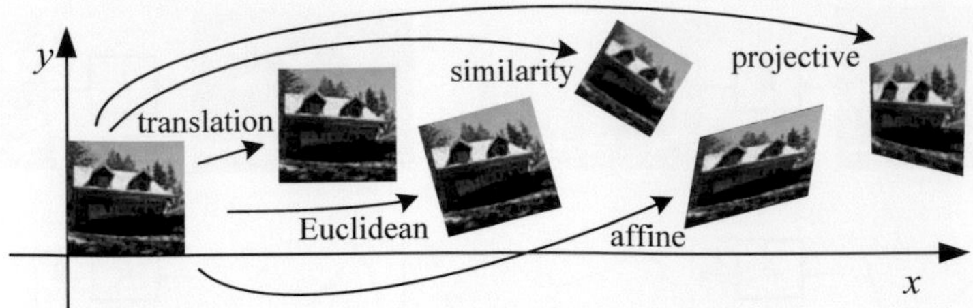

Figure 3.44 Basic set of 2D geometric image transformations.

Transformation	Matrix	# DoF	Preserves	Icon
translation	$\begin{bmatrix} \mathbf{I} & \mathbf{t} \end{bmatrix}_{2\times 3}$	2	orientation	
rigid (Euclidean)	$\begin{bmatrix} \mathbf{R} & \mathbf{t} \end{bmatrix}_{2\times 3}$	3	lengths	
similarity	$\begin{bmatrix} s\mathbf{R} & \mathbf{t} \end{bmatrix}_{2\times 3}$	4	angles	
affine	$\begin{bmatrix} \mathbf{A} \end{bmatrix}_{2\times 3}$	6	parallelism	
projective	$\begin{bmatrix} \tilde{\mathbf{H}} \end{bmatrix}_{3\times 3}$	8	straight lines	

Table 3.3 Hierarchy of 2D coordinate transformations. Each transformation also preserves the properties listed in the rows below it, i.e., similarity preserves not only angles but also parallelism and straight lines. The 2 $\times$ 3 matrices are extended with a third $[\mathbf{0}^T \; 1]$ row to form a full 3 $\times$ 3 matrix for homogeneous coordinate transformations.

In fact, this approach suffers from several limitations. The process of copying a pixel $f(\mathbf{x})$ to a location $\mathbf{x}'$ in g is not well defined when $\mathbf{x}'$ has a non-integer value. What do we do in such a case? What would you do?

You can round the value of $\mathbf{x}'$ to the nearest integer coordinate and copy the pixel there, but the resulting image has severe aliasing and pixels that jump around a lot when animating the transformation. You can also "distribute" the value among its four nearest neighbors in a weighted (bilinear) fashion, keeping track of the per-pixel weights and normalizing at the end. This technique is called *splatting* and is sometimes used for volume rendering in the graphics community (Levoy and Whitted 1985; Levoy 1988; Westover 1989; Rusinkiewicz and Levoy 2000). Unfortunately, it suffers from both moderate amounts of aliasing and a fair amount of blur (loss of high-resolution detail).

The second major problem with forward warping is the appearance of cracks and holes, especially when magnifying an image. Filling such holes with their nearby neighbors can lead to further aliasing and blurring.

What can we do instead? A preferable solution is to use *inverse warping* (Algorithm 3.2), where each pixel in the destination image $g(\mathbf{x}')$ is sampled from the original image $f(\mathbf{x})$ (Figure 3.46).

How does this differ from the forward warping algorithm? For one thing, since $\hat{\mathbf{h}}(\mathbf{x}')$ is (presumably) defined for all pixels in $g(\mathbf{x}')$, we no longer have holes. More importantly, resampling an image at non-integer locations is a well-studied problem (general image interpolation, see Section 3.5.2) and high-quality filters that control aliasing can be used.

Figure 3.45 Forward warping algorithm: (a) a pixel $f(\mathbf{x})$ is copied to its corresponding location $\mathbf{x}' = \mathbf{h}(\mathbf{x})$ in image $g(\mathbf{x}')$; (b) detail of the source and destination pixel locations.

procedure *forwardWarp*$(f, \mathbf{h}, \mathbf{out}\ g)$:

For every pixel $\mathbf{x}$ in $f(\mathbf{x})$

1. Compute the destination location $\mathbf{x}' = \mathbf{h}(\mathbf{x})$.
2. Copy the pixel $f(\mathbf{x})$ to $g(\mathbf{x}')$.

Algorithm 3.1 Forward warping algorithm for transforming an image $f(\mathbf{x})$ into an image $g(\mathbf{x}')$ through the parametric transform $\mathbf{x}' = \mathbf{h}(\mathbf{x})$.

Figure 3.46 Inverse warping algorithm: (a) a pixel $g(\mathbf{x}')$ is sampled from its corresponding location $\mathbf{x} = \hat{\mathbf{h}}(\mathbf{x}')$ in image $f(\mathbf{x})$; (b) detail of the source and destination pixel locations.

procedure *inverseWarp*$(f, \mathbf{h}, \mathbf{out}\ g)$:

For every pixel $\mathbf{x}'$ in $g(\mathbf{x}')$

1. Compute the source location $\mathbf{x} = \hat{\mathbf{h}}(\mathbf{x}')$
2. Resample $f(\mathbf{x})$ at location $\mathbf{x}$ and copy to $g(\mathbf{x}')$

Algorithm 3.2 Inverse warping algorithm for creating an image $g(\mathbf{x}')$ from an image $f(\mathbf{x})$ using the parametric transform $\mathbf{x}' = \mathbf{h}(\mathbf{x})$.

Where does the function $\hat{\mathbf{h}}(\mathbf{x}')$ come from? Quite often, it can simply be computed as the inverse of $\mathbf{h}(\mathbf{x})$. In fact, all of the parametric transforms listed in Table 3.3 have closed form solutions for the inverse transform: simply take the inverse of the 3×3 matrix specifying the transform.

In other cases, it is preferable to formulate the problem of image warping as that of resampling a source image $f(\mathbf{x})$ given a mapping $\mathbf{x} = \hat{\mathbf{h}}(\mathbf{x}')$ from destination pixels $\mathbf{x}'$ to source pixels $\mathbf{x}$. For example, in optical flow (Section 9.3), we estimate the flow field as the location of the *source* pixel that produced the current pixel whose flow is being estimated, as opposed to computing the *destination* pixel to which it is going. Similarly, when correcting for radial distortion (Section 2.1.5), we calibrate the lens by computing for each pixel in the final (undistorted) image the corresponding pixel location in the original (distorted) image.

What kinds of interpolation filter are suitable for the resampling process? Any of the filters we studied in Section 3.5.2 can be used, including nearest neighbor, bilinear, bicubic, and windowed sinc functions. While bilinear is often used for speed (e.g., inside the inner loop of a patch-tracking algorithm, see Section 9.1.3), bicubic, and windowed sinc are preferable where visual quality is important.

To compute the value of $f(\mathbf{x})$ at a non-integer location $\mathbf{x}$, we simply apply our usual FIR resampling filter,

$$g(x, y) = \sum_{k,l} f(k, l)h(x - k, y - l), \tag{3.75}$$

where (x, y) are the sub-pixel coordinate values and $h(x, y)$ is some interpolating or smoothing kernel. Recall from Section 3.5.2 that when decimation is being performed, the smoothing kernel is stretched and re-scaled according to the downsampling rate r.

Unfortunately, for a general (non-zoom) image transformation, the resampling rate r is not well defined. Consider a transformation that stretches the x dimensions while squashing the y dimensions. The resampling kernel should be performing regular interpolation along the x dimension and smoothing (to anti-alias the blurred image) in the y direction. This gets even more complicated for the case of general affine or perspective transforms.

What can we do? Fortunately, Fourier analysis can help. The two-dimensional generalization of the one-dimensional *domain scaling* law is

$$g(\mathbf{A}\mathbf{x}) \Leftrightarrow |\mathbf{A}|^{-1} G(\mathbf{A}^{-T}\mathbf{f}). \tag{3.76}$$

For all of the transforms in Table 3.3 except perspective, the matrix $\mathbf{A}$ is already defined. For perspective transformations, the matrix $\mathbf{A}$ is the linearized *derivative* of the perspective transformation (Figure 3.47a), i.e., the local affine approximation to the stretching induced by the projection (Heckbert 1986; Wolberg 1990; Gomes, Darsa *et al.* 1999; Akenine-Möller and Haines 2002).

To prevent aliasing, we need to prefilter the image $f(\mathbf{x})$ with a filter whose frequency response is the projection of the final desired spectrum through the $\mathbf{A}^{-T}$ transform (Szeliski, Winder, and Uyttendaele 2010). In general (for non-zoom transforms), this filter is non-separable and hence is very slow to compute. Therefore, a number of approximations to this filter are used in practice, include MIP-mapping, elliptically weighted Gaussian averaging, and anisotropic filtering (Akenine-Möller and Haines 2002).

MIP-mapping

MIP-mapping was first proposed by Williams (1983) as a means to rapidly prefilter images being used for *texture mapping* in computer graphics. A MIP-map[16] is a standard image pyramid

[16] The term "MIP" stands for *multi in parvo*, meaning "many in one".

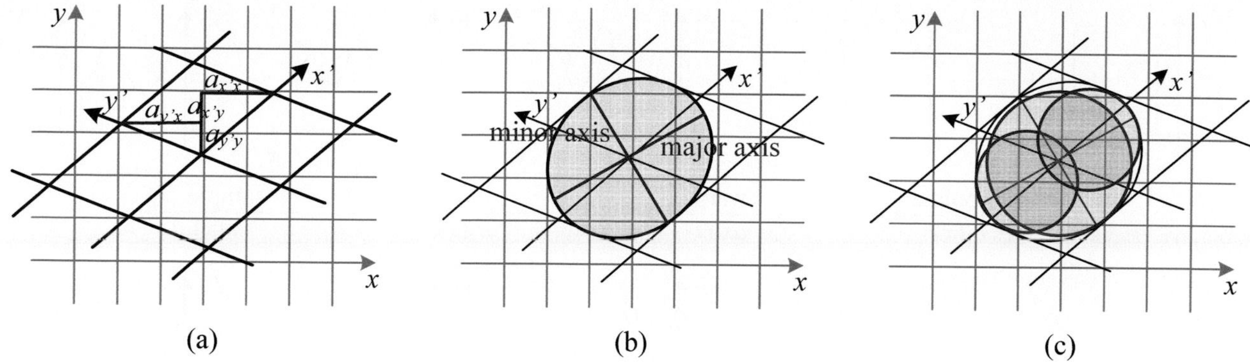

Figure 3.47 Anisotropic texture filtering: (a) Jacobian of transform $\mathbf{A}$ and the induced horizontal and vertical resampling rates $\{a_{x'x}, a_{x'y}, a_{y'x}, a_{y'y}\}$; (b) elliptical footprint of an EWA smoothing kernel; (c) anisotropic filtering using multiple samples along the major axis. Image pixels lie at line intersections.

(Figure 3.31), where each level is prefiltered with a high-quality filter rather than a poorer quality approximation, such as Burt and Adelson's (1983b) five-tap binomial. To resample an image from a MIP-map, a scalar estimate of the resampling rate r is first computed. For example, r can be the maximum of the absolute values in $\mathbf{A}$ (which suppresses aliasing) or it can be the minimum (which reduces blurring). Akenine-Möller and Haines (2002) discuss these issues in more detail.

Once a resampling rate has been specified, a *fractional* pyramid level is computed using the base 2 logarithm,

$$l = \log_2 r. \tag{3.77}$$

One simple solution is to resample the texture from the next higher or lower pyramid level, depending on whether it is preferable to reduce aliasing or blur. A better solution is to resample *both* images and blend them linearly using the fractional component of l. Since most MIP-map implementations use bilinear resampling within each level, this approach is usually called *trilinear MIP-mapping*. Computer graphics rendering APIs, such as OpenGL and Direct3D, have parameters that can be used to select which variant of MIP-mapping (and of the sampling rate r computation) should be used, depending on the desired tradeoff between speed and quality. Exercise 3.22 has you examine some of these tradeoffs in more detail.

Elliptical Weighted Average

The Elliptical Weighted Average (EWA) filter invented by Greene and Heckbert (1986) is based on the observation that the affine mapping $\mathbf{x} = \mathbf{A}\mathbf{x}'$ defines a skewed two-dimensional coordinate system in the vicinity of each source pixel $\mathbf{x}$ (Figure 3.47a). For every destination pixel $\mathbf{x}'$, the ellipsoidal projection of a small pixel grid in $\mathbf{x}'$ onto $\mathbf{x}$ is computed (Figure 3.47b). This is then used to filter the source image $g(\mathbf{x})$ with a Gaussian whose inverse covariance matrix is this ellipsoid.

Despite its reputation as a high-quality filter (Akenine-Möller and Haines 2002), we have found in our work (Szeliski, Winder, and Uyttendaele 2010) that because a Gaussian kernel is used, the technique suffers simultaneously from both blurring and aliasing, compared to higher-quality filters. The EWA is also quite slow, although faster variants based on MIP mapping have been proposed, as described in (Szeliski, Winder, and Uyttendaele 2010).

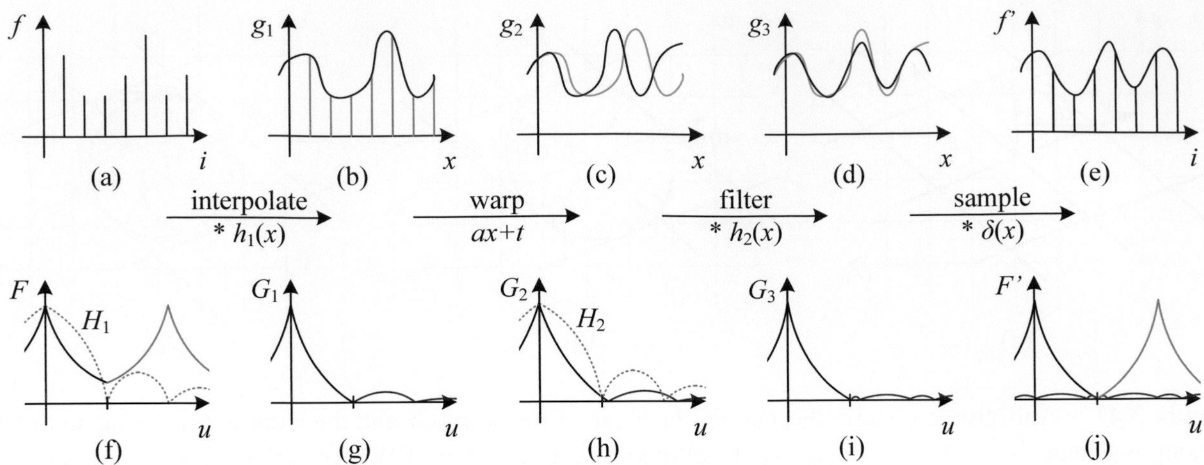

Figure 3.48 One-dimensional signal resampling (Szeliski, Winder, and Uyttendaele 2010): (a) original sampled signal $f(i)$; (b) interpolated signal $g_1(x)$; (c) warped signal $g_2(x)$; (d) filtered signal $g_3(x)$; (e) sampled signal $f'(i)$. The corresponding spectra are shown below the signals, with the aliased portions shown in red.

Anisotropic filtering

An alternative approach to filtering oriented textures, which is sometimes implemented in graphics hardware (GPUs), is to use anisotropic filtering (Barkans 1997; Akenine-Möller and Haines 2002). In this approach, several samples at different resolutions (fractional levels in the MIP-map) are combined along the major axis of the EWA Gaussian (Figure 3.47c).

Multi-pass transforms

The optimal approach to warping images without excessive blurring or aliasing is to adaptively prefilter the source image at each pixel using an ideal low-pass filter, i.e., an oriented skewed sinc or low-order (e.g., cubic) approximation (Figure 3.47a). Figure 3.48 shows how this works in one dimension. The signal is first (theoretically) interpolated to a continuous waveform, (ideally) low-pass filtered to below the new Nyquist rate, and then re-sampled to the final desired resolution. In practice, the interpolation and decimation steps are concatenated into a single *polyphase* digital filtering operation (Szeliski, Winder, and Uyttendaele 2010).

For parametric transforms, the oriented two-dimensional filtering and resampling operations can be approximated using a series of one-dimensional resampling and shearing transforms (Catmull and Smith 1980; Heckbert 1989; Wolberg 1990; Gomes, Darsa *et al.* 1999; Szeliski, Winder, and Uyttendaele 2010). The advantage of using a series of one-dimensional transforms is that they are much more efficient (in terms of basic arithmetic operations) than large, non-separable, two-dimensional filter kernels. In order to prevent aliasing, however, it may be necessary to upsample in the opposite direction before applying a shearing transformation (Szeliski, Winder, and Uyttendaele 2010).

3.6.2 Mesh-based warping

While parametric transforms specified by a small number of global parameters have many uses, *local* deformations with more degrees of freedom are often required.

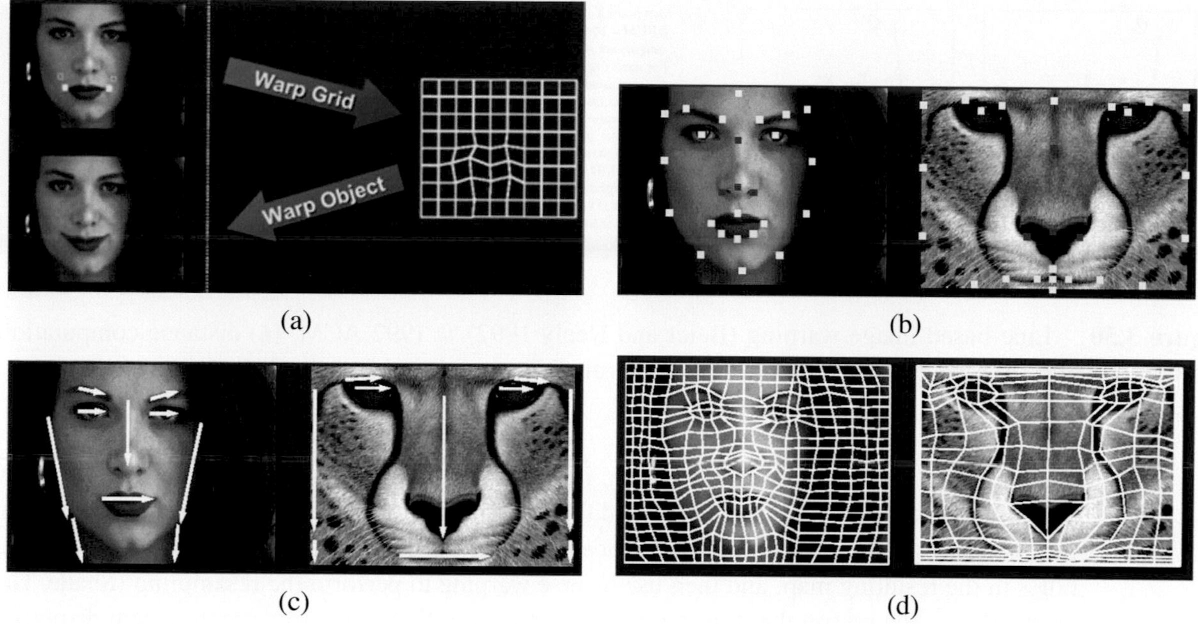

(a) (b) (c) (d)

Figure 3.49 Image warping alternatives (Gomes, Darsa *et al.* 1999) © 1999 Morgan Kaufmann: (a) sparse control points ⟶ deformation grid; (b) denser set of control point correspondences; (c) oriented line correspondences; (d) uniform quadrilateral grid.

Consider, for example, changing the appearance of a face from a frown to a smile (Figure 3.49a). What is needed in this case is to curve the corners of the mouth upwards while leaving the rest of the face intact.[17] To perform such a transformation, different amounts of motion are required in different parts of the image. Figure 3.49 shows some of the commonly used approaches.

The first approach, shown in Figure 3.49a–b, is to specify a *sparse* set of corresponding points. The displacement of these points can then be interpolated to a dense *displacement field* (Chapter 9) using a variety of techniques, which are described in more detail in Section 4.1 on scattered data interpolation. One possibility is to *triangulate* the set of points in one image (de Berg, Cheong *et al.* 2006; Litwinowicz and Williams 1994; Buck, Finkelstein *et al.* 2000) and to use an *affine* motion model (Table 3.3), specified by the three triangle vertices, inside each triangle. If the destination image is triangulated according to the new vertex locations, an inverse warping algorithm (Figure 3.46) can be used. If the source image is triangulated and used as a *texture map*, computer graphics rendering algorithms can be used to draw the new image (but care must be taken along triangle edges to avoid potential aliasing).

Alternative methods for interpolating a sparse set of displacements include moving nearby quadrilateral mesh vertices, as shown in Figure 3.49a, using *variational* (energy minimizing) interpolants such as regularization (Litwinowicz and Williams 1994), see Section 4.2, or using locally weighted (*radial basis function*) combinations of displacements (Section 4.1.1). (See Section 4.1 for additional *scattered data interpolation* techniques.) If quadrilateral meshes are used, it may be desirable to interpolate displacements down to individual pixel values using a smooth interpolant such as a quadratic B-spline (Farin 2002; Lee, Wolberg *et al.* 1996).

In some cases, e.g., if a dense depth map has been estimated for an image (Shade, Gortler *et al.* 1998), we only know the forward displacement for each pixel. As mentioned before, drawing

[17]See Section 6.2.4 on active appearance models for more sophisticated examples of changing facial expression and appearance.

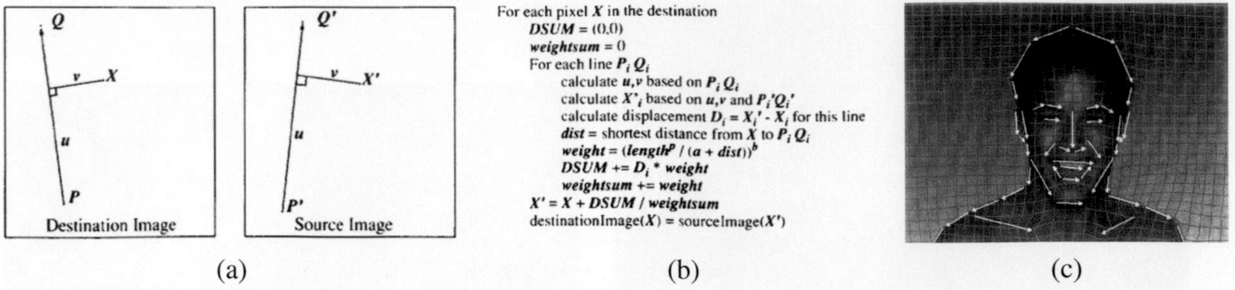

Q
v X
u
P
Destination Image
Q'
v X'
u
P'
Source Image
For each pixel X in the destination
DSUM = (0,0)
weightsum = 0
For each line P_i Q_i
calculate u,v based on P_i Q_i
calculate X'_i based on u,v and P_i'Q_i'
calculate displacement D_i = X_i' - X_i for this line
dist = shortest distance from X to P_i Q_i
weight = (length^p / (a + dist))^b
DSUM += D_i * weight
weightsum += weight
X' = X + DSUM / weightsum
destinationImage(X) = sourceImage(X')
(a)
(b)
(c)

Figure 3.50 Line-based image warping (Beier and Neely 1992) © 1992 ACM: (a) distance computation and position transfer; (b) rendering algorithm; (c) two intermediate warps used for morphing.

source pixels at their destination location, i.e., forward warping (Figure 3.45), suffers from several potential problems, including aliasing and the appearance of small cracks. An alternative technique in this case is to forward warp the *displacement field* (or depth map) to its new location, fill small holes in the resulting map, and then use inverse warping to perform the resampling (Shade, Gortler *et al.* 1998). The reason that this generally works better than forward warping is that displacement fields tend to be much smoother than images, so the aliasing introduced during the forward warping of the displacement field is much less noticeable.

A second approach to specifying displacements for local deformations is to use corresponding *oriented line segments* (Beier and Neely 1992), as shown in Figures 3.49c and 3.50. Pixels along each line segment are transferred from source to destination exactly as specified, and other pixels are warped using a smooth interpolation of these displacements. Each line segment correspondence specifies a translation, rotation, and scaling, i.e., a *similarity transform* (Table 3.3), for pixels in its vicinity, as shown in Figure 3.50a. Line segments influence the overall displacement of the image using a weighting function that depends on the minimum distance to the line segment (v in Figure 3.50a if $u \in [0, 1]$, else the shorter of the two distances to P and Q).

One final possibility for specifying displacement fields is to use a mesh specifically *adapted* to the underlying image content, as shown in Figure 3.49d. Specifying such meshes by hand can involve a fair amount of work; Gomes, Darsa *et al.* (1999) describe an interactive system for doing this. Once the two meshes have been specified, intermediate warps can be generated using linear interpolation and the displacements at mesh nodes can be interpolated using splines.

3.6.3 *Application*: Feature-based morphing

While warps can be used to change the appearance of or to animate a *single* image, even more powerful effects can be obtained by warping and blending two or more images using a process now commonly known as *morphing* (Beier and Neely 1992; Lee, Wolberg *et al.* 1996; Gomes, Darsa *et al.* 1999).

Figure 3.51 shows the essence of image morphing. Instead of simply cross-dissolving between two images, which leads to ghosting as shown in the top row, each image is warped toward the other image before blending, as shown in the bottom row. If the correspondences have been set up well (using any of the techniques shown in Figure 3.49), corresponding features are aligned and no ghosting results.

The above process is repeated for each intermediate frame being generated during a morph, using different blends (and amounts of deformation) at each interval. Let $t \in [0, 1]$ be the time parameter that describes the sequence of interpolated frames. The weighting functions for the two

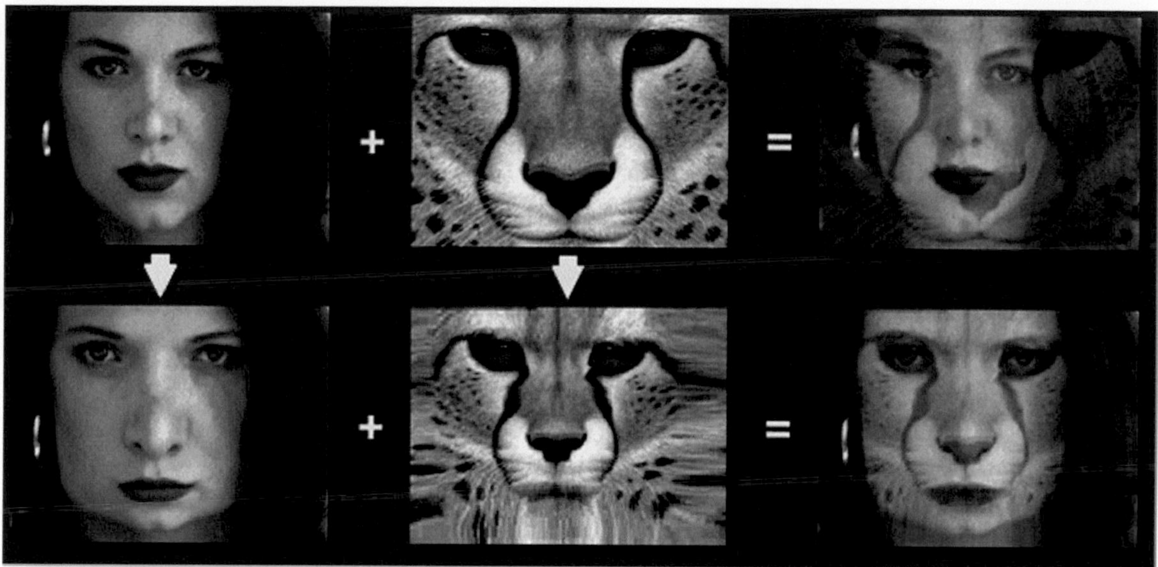

Figure 3.51 Image morphing (Gomes, Darsa *et al.* 1999) © 1999 Morgan Kaufmann. Top row: if the two images are just blended, visible ghosting results. Bottom row: both images are first warped to the same intermediate location (e.g., halfway towards the other image) and the resulting warped images are then blended resulting in a seamless morph.

warped images in the blend are $(1 - t)$ and t and the movements of the pixels specified by the correspondences are also linearly interpolated. Some care must be taken in defining what it means to partially warp an image towards a destination, especially if the desired motion is far from linear (Sederberg, Gao *et al.* 1993). Exercise 3.25 has you implement a morphing algorithm and test it out under such challenging conditions.

3.7 Additional reading

If you are interested in exploring the topic of image processing in more depth, some popular textbooks have been written by Gomes and Velho (1997), Jähne (1997), Pratt (2007), Burger and Burge (2009), and Gonzalez and Woods (2017). The pre-eminent conference and journal in this field are the IEEE International Conference on Image Processing and the IEEE Transactions on Image Processing.

For image compositing operators, the seminal reference is by Porter and Duff (1984) while Blinn (1994a,b) provides a more detailed tutorial. For image compositing, Smith and Blinn (1996) were the first to bring this topic to the attention of the graphics community, while Wang and Cohen (2009) provide a good in-depth survey.

In the realm of linear filtering, Freeman and Adelson (1991) provide a great introduction to separable and steerable oriented band-pass filters, while Perona (1995) shows how to approximate any filter as a sum of separable components.

The literature on non-linear filtering is quite wide and varied; it includes such topics as bilateral filtering (Tomasi and Manduchi 1998; Durand and Dorsey 2002; Chen, Paris, and Durand 2007; Paris and Durand 2009; Paris, Kornprobst *et al.* 2008), related iterative algorithms (Saint-Marc, Chen, and Medioni 1991; Nielsen, Florack, and Deriche 1997; Black, Sapiro *et al.* 1998; Weickert, ter Haar Romeny, and Viergever 1998; Weickert 1998; Barash 2002; Scharr, Black, and

Haussecker 2003; Barash and Comaniciu 2004) and variational approaches (Chan, Osher, and Shen 2001; Tschumperlé and Deriche 2005; Tschumperlé 2006; Kaftory, Schechner, and Zeevi 2007), and guided filtering (Eisemann and Durand 2004; Petschnigg, Agrawala *et al.* 2004; He, Sun, and Tang 2013).

Good references to image morphology include Haralick and Shapiro (1992, Section 5.2), Bovik (2000, Section 2.2), Ritter and Wilson (2000, Section 7) Serra (1982), Serra and Vincent (1992), Yuille, Vincent, and Geiger (1992), and Soille (2006).

The classic papers for image pyramids and pyramid blending are by Burt and Adelson (1983a,b). Wavelets were first introduced to the computer vision community by Mallat (1989) and good tutorial and review papers and books are available (Strang 1989; Simoncelli and Adelson 1990b; Rioul and Vetterli 1991; Chui 1992; Meyer 1993; Sweldens 1997). Wavelets are widely used in the computer graphics community to perform multi-resolution geometric processing (Stollnitz, DeRose, and Salesin 1996) and have been used in computer vision for similar applications (Szeliski 1990b; Pentland 1994; Gortler and Cohen 1995; Yaou and Chang 1994; Lai and Vemuri 1997; Szeliski 2006b; Krishnan and Szeliski 2011; Krishnan, Fattal, and Szeliski 2013), as well as for multi-scale oriented filtering (Simoncelli, Freeman *et al.* 1992) and denoising (Portilla, Strela *et al.* 2003).

While image pyramids (Section 3.5.3) are usually constructed using linear filtering operators, more recent work uses non-linear filters, since these can better preserve details and other salient features. Some representative papers in the computer vision literature are by Gluckman (2006a,b); Lyu and Simoncelli (2008) and in computational photography by Bae, Paris, and Durand (2006), Farbman, Fattal *et al.* (2008), and Fattal (2009).

High-quality algorithms for image warping and resampling are covered both in the image processing literature (Wolberg 1990; Dodgson 1992; Gomes, Darsa *et al.* 1999; Szeliski, Winder, and Uyttendaele 2010) and in computer graphics (Williams 1983; Heckbert 1986; Barkans 1997; Weinhaus and Devarajan 1997; Akenine-Möller and Haines 2002), where they go under the name of *texture mapping*. Combinations of image warping and image blending techniques are used to enable *morphing* between images, which is covered in a series of seminal papers and books (Beier and Neely 1992; Gomes, Darsa *et al.* 1999).

3.8 Exercises

Ex 3.1: Color balance. Write a simple application to change the color balance of an image by multiplying each color value by a different user-specified constant. If you want to get fancy, you can make this application interactive, with sliders.

1. Do you get different results if you take out the gamma transformation before or after doing the multiplication? Why or why not?

2. Take the same picture with your digital camera using different color balance settings (most cameras control the color balance from one of the menus). Can you recover what the color balance ratios are between the different settings? You may need to put your camera on a tripod and align the images manually or automatically to make this work. Alternatively, use a color checker chart (Figure 10.3b), as discussed in Sections 2.3 and 10.1.1.

3. Can you think of any reason why you might want to perform a color twist (Section 3.1.2) on the images? See also Exercise 2.8 for some related ideas.

Ex 3.2: Demosaicing. If you have access to the RAW image for the camera, perform the demosaicing yourself (Section 10.3.1). If not, just subsample an RGB image in a Bayer mosaic pattern.

Instead of just bilinear interpolation, try one of the more advanced techniques described in Section 10.3.1. Compare your result to the one produced by the camera. Does your camera perform a simple linear mapping between RAW values and the color-balanced values in a JPEG? Some high-end cameras have a RAW+JPEG mode, which makes this comparison much easier.

Ex 3.3: Compositing and reflections. Section 3.1.3 describes the process of compositing an alpha-matted image on top of another. Answer the following questions and optionally validate them experimentally:

1. Most captured images have gamma correction applied to them. Does this invalidate the basic compositing equation (3.8); if so, how should it be fixed?

2. The additive (pure reflection) model may have limitations. What happens if the glass is tinted, especially to a non-gray hue? How about if the glass is dirty or smudged? How could you model wavy glass or other kinds of refractive objects?

Ex 3.4: Blue screen matting. Set up a blue or green background, e.g., by buying a large piece of colored posterboard. Take a picture of the empty background, and then of the background with a new object in front of it. *Pull the matte* using the difference between each colored pixel and its assumed corresponding background pixel, using one of the techniques described in Section 3.1.3 or by Smith and Blinn (1996).

Ex 3.5: Difference keying. Implement a difference keying algorithm (see Section 3.1.3) (Toyama, Krumm *et al.* 1999), consisting of the following steps:

1. Compute the mean and variance (or median and robust variance) at each pixel in an "empty" video sequence.

2. For each new frame, classify each pixel as foreground or background (set the background pixels to RGBA=0).

3. (Optional) Compute the alpha channel and composite over a new background.

4. (Optional) Clean up the image using morphology (Section 3.3.1), label the connected components (Section 3.3.3), compute their centroids, and track them from frame to frame. Use this to build a "people counter".

Ex 3.6: Photo effects. Write a variety of photo enhancement or effects filters: contrast, solarization (quantization), etc. Which ones are useful (perform sensible corrections) and which ones are more creative (create unusual images)?

Ex 3.7: Histogram equalization. Compute the gray level (luminance) histogram for an image and equalize it so that the tones look better (and the image is less sensitive to exposure settings). You may want to use the following steps:

1. Convert the color image to luminance (Section 3.1.2).

2. Compute the histogram, the cumulative distribution, and the compensation transfer function (Section 3.1.4).

3. (Optional) Try to increase the "punch" in the image by ensuring that a certain fraction of pixels (say, 5%) are mapped to pure black and white.

4. (Optional) Limit the local *gain* $f'(I)$ in the transfer function. One way to do this is to limit $f(I) < \gamma I$ or $f'(I) < \gamma$ while performing the accumulation (3.9), keeping any unaccumulated values "in reserve". (I'll let you figure out the exact details.)

5. Compensate the luminance channel through the lookup table and re-generate the color image using color ratios (2.117).

6. (Optional) Color values that are *clipped* in the original image, i.e., have one or more saturated color channels, may appear unnatural when remapped to a non-clipped value. Extend your algorithm to handle this case in some useful way.

Ex 3.8: Local histogram equalization. Compute the gray level (luminance) histograms for each patch, but add to vertices based on distance (a spline).

1. Build on Exercise 3.7 (luminance computation).

2. Distribute values (counts) to adjacent vertices (bilinear).

3. Convert to CDF (look-up functions).

4. (Optional) Use low-pass filtering of CDFs.

5. Interpolate adjacent CDFs for final lookup.

Ex 3.9: Padding for neighborhood operations. Write down the formulas for computing the padded pixel values $\tilde{f}(i,j)$ as a function of the original pixel values $f(k,l)$ and the image width and height (M,N) for *each* of the padding modes shown in Figure 3.13. For example, for replication (clamping),

$$\tilde{f}(i,j) = f(k,l), \quad \begin{array}{l} k = \max(0, \min(M-1, i)), \\ l = \max(0, \min(N-1, j)), \end{array}$$

(Hint: you may want to use the min, max, mod, and absolute value operators in addition to the regular arithmetic operators.)

- Describe in more detail the advantages and disadvantages of these various modes.

- (Optional) Check what your graphics card does by drawing a texture-mapped rectangle where the texture coordinates lie beyond the $[0.0, 1.0]$ range and using different texture clamping modes.

Ex 3.10: Separable filters. Implement convolution with a separable kernel. The input should be a grayscale or color image along with the horizontal and vertical kernels. Make sure you support the padding mechanisms developed in the previous exercise. You will need this functionality for some of the later exercises. If you already have access to separable filtering in an image processing package you are using (such as IPL), skip this exercise.

- (Optional) Use Pietro Perona's (1995) technique to approximate convolution as a sum of a number of separable kernels. Let the user specify the number of kernels and report back some sensible metric of the approximation fidelity.

Ex 3.11: Discrete Gaussian filters. Discuss the following issues with implementing a discrete Gaussian filter:

- If you just sample the equation of a continuous Gaussian filter at discrete locations, will you get the desired properties, e.g., will the coefficients sum up to 1? Similarly, if you sample a derivative of a Gaussian, do the samples sum up to 0 or have vanishing higher-order moments?
- Would it be preferable to take the original signal, interpolate it with a sinc, blur with a continuous Gaussian, then prefilter with a sinc before re-sampling? Is there a simpler way to do this in the frequency domain?
- Would it make more sense to produce a Gaussian frequency response in the Fourier domain and to then take an inverse FFT to obtain a discrete filter?
- How does truncation of the filter change its frequency response? Does it introduce any additional artifacts?
- Are the resulting two-dimensional filters as rotationally invariant as their continuous analogs? Is there some way to improve this? In fact, can any 2D discrete (separable or non-separable) filter be truly rotationally invariant?

Ex 3.12: Sharpening, blur, and noise removal. Implement some softening, sharpening, and non-linear diffusion (selective sharpening or noise removal) filters, such as Gaussian, median, and bilateral (Section 3.3.1), as discussed in Section 3.4.2.

Take blurry or noisy images (shooting in low light is a good way to get both) and try to improve their appearance and legibility.

Ex 3.13: Steerable filters. Implement Freeman and Adelson's (1991) steerable filter algorithm. The input should be a grayscale or color image and the output should be a multi-banded image consisting of $G_1^{0^\circ}$ and $G_1^{90^\circ}$. The coefficients for the filters can be found in the paper by Freeman and Adelson (1991).

Test the various order filters on a number of images of your choice and see if you can reliably find corner and intersection features. These filters will be quite useful later to detect elongated structures, such as lines (Section 7.4).

Ex 3.14: Bilateral and guided image filters. Implement or download code for bilateral and/or guided image filtering and use this to implement some image enhancement or processing application, such as those described in Section 3.3.2

Ex 3.15: Fourier transform. Prove the properties of the Fourier transform listed in Szeliski (2010, Table 3.1) and derive the formulas for the Fourier transforms pairs listed in Szeliski (2010, Table 3.2) and Table 3.1. These exercises are very useful if you want to become comfortable working with Fourier transforms, which is a very useful skill when analyzing and designing the behavior and efficiency of many computer vision algorithms.

Ex 3.16: High-quality image resampling. Implement several of the low-pass filters presented in Section 3.5.2 and also the windowed sinc shown in Figure 3.28. Feel free to implement other filters (Wolberg 1990; Unser 1999).

Apply your filters to continuously resize an image, both magnifying (interpolating) and minifying (decimating) it; compare the resulting animations for several filters. Use both a synthetic chirp image (Figure 3.52a) and natural images with lots of high-frequency detail (Figure 3.52b–c).

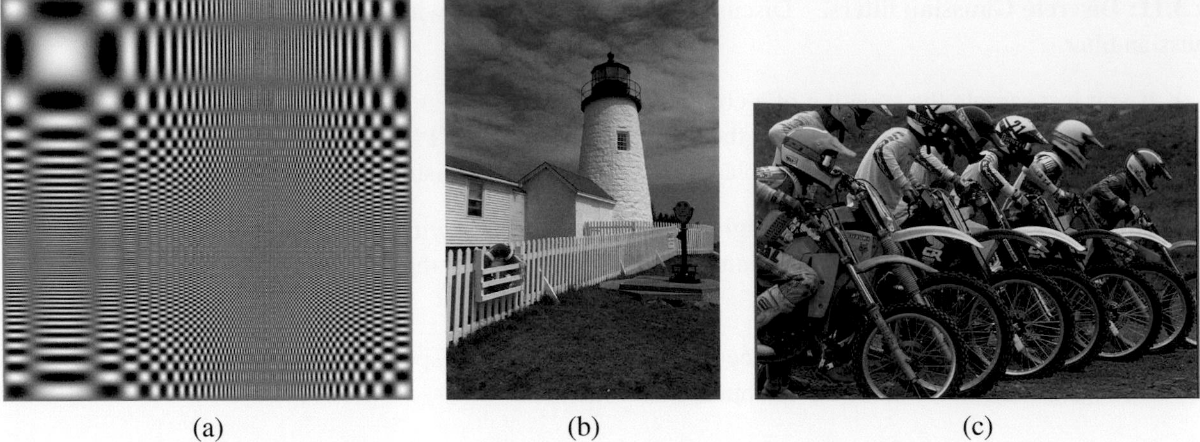

Figure 3.52 Sample images for testing the quality of resampling algorithms: (a) a synthetic chirp; (b) and (c) some high-frequency images from the image compression community.

You may find it helpful to write a simple visualization program that continuously plays the animations for two or more filters at once and that let you "blink" between different results.

Discuss the merits and deficiencies of each filter, as well as the tradeoff between speed and quality.

Ex 3.17: Pyramids. Construct an image pyramid. The inputs should be a grayscale or color image, a separable filter kernel, and the number of desired levels. Implement at least the following kernels:

- 2×2 block filtering;
- Burt and Adelson's binomial kernel $^1/_{16}(1, 4, 6, 4, 1)$ (Burt and Adelson 1983a);
- a high-quality seven- or nine-tap filter.

Compare the visual quality of the various decimation filters. Also, shift your input image by 1 to 4 pixels and compare the resulting decimated (quarter size) image sequence.

Ex 3.18: Pyramid blending. Write a program that takes as input two color images and a binary mask image and produces the Laplacian pyramid blend of the two images.

1. Construct the Laplacian pyramid for each image.
2. Construct the Gaussian pyramid for the two mask images (the input image and its complement).
3. Multiply each Laplacian image by its corresponding mask and sum the images (see Figure 3.41).
4. Reconstruct the final image from the blended Laplacian pyramid.

Generalize your algorithm to input n images and a label image with values $1 \ldots n$ (the value 0 can be reserved for "no input"). Discuss whether the weighted summation stage (step 3) needs to keep track of the total weight for renormalization, or whether the math just works out. Use your algorithm either to blend two differently exposed image (to avoid under- and over-exposed regions) or to make a creative blend of two different scenes.

Ex 3.19: Pyramid blending in PyTorch. Re-write your pyramid blending exercise in PyTorch.

1. PyTorch has support for all of the primitives you need, i.e., fixed size convolutions (make sure they filter each channel separately), downsampling, upsampling, and addition, subtraction, and multiplication (although the latter is rarely used).

2. The goal of this exercise is *not* to train the convolution weights, but just to become familiar with the DNN primitives available in PyTorch.

3. Compare your results to the ones using a standard Python or C++ computer vision library. They should be identical.

4. Discuss whether you like this API better or worse for these kinds of fixed pipeline imaging tasks.

Ex 3.20: Local Laplacian—challenging. Implement the local Laplacian contrast manipulation technique (Paris, Hasinoff, and Kautz 2011; Aubry, Paris *et al.* 2014) and use this to implement edge-preserving filtering and tone manipulation.

Ex 3.21: Wavelet construction and applications. Implement one of the wavelet families described in Section 3.5.4 or by Simoncelli and Adelson (1990b), as well as the basic Laplacian pyramid (Exercise 3.17). Apply the resulting representations to one of the following two tasks:

- **Compression:** Compute the entropy in each band for the different wavelet implementations, assuming a given quantization level (say, 1/4 gray level, to keep the rounding error acceptable). Quantize the wavelet coefficients and reconstruct the original images. Which technique performs better? (See Simoncelli and Adelson (1990b) or any of the multitude of wavelet compression papers for some typical results.)

- **Denoising.** After computing the wavelets, suppress small values using *coring*, i.e., set small values to zero using a piecewise linear or other C^0 function. Compare the results of your denoising using different wavelet and pyramid representations.

Ex 3.22: Parametric image warping. Write the code to do affine and perspective image warps (optionally bilinear as well). Try a variety of interpolants and report on their visual quality. In particular, discuss the following:

- In a MIP-map, selecting only the coarser level adjacent to the computed fractional level will produce a blurrier image, while selecting the finer level will lead to aliasing. Explain why this is so and discuss whether blending an aliased and a blurred image (tri-linear MIP-mapping) is a good idea.

- When the ratio of the horizontal and vertical resampling rates becomes very different (anisotropic), the MIP-map performs even worse. Suggest some approaches to reduce such problems.

Ex 3.23: Local image warping. Open an image and deform its appearance in one of the following ways:

1. Click on a number of pixels and move (drag) them to new locations. Interpolate the resulting sparse displacement field to obtain a dense motion field (Sections 3.6.2 and 3.5.1).

2. Draw a number of lines in the image. Move the endpoints of the lines to specify their new positions and use the Beier–Neely interpolation algorithm (Beier and Neely 1992), discussed in Section 3.6.2, to get a dense motion field.
3. Overlay a spline control grid and move one grid point at a time (optionally select the level of the deformation).
4. Have a dense per-pixel flow field and use a soft "paintbrush" to design a horizontal and vertical velocity field.
5. (Optional): Prove whether the Beier–Neely warp does or does not reduce to a sparse point-based deformation as the line segments become shorter (reduce to points).

Ex 3.24: Forward warping. Given a displacement field from the previous exercise, write a forward warping algorithm:

1. Write a forward warper using splatting, either nearest neighbor or soft accumulation (Section 3.6.1).
2. Write a two-pass algorithm that forward warps the displacement field, fills in small holes, and then uses inverse warping (Shade, Gortler *et al.* 1998).
3. Compare the quality of these two algorithms.

Ex 3.25: Feature-based morphing. Extend the warping code you wrote in Exercise 3.23 to import two different images and specify correspondences (point, line, or mesh-based) between the two images.

1. Create a morph by partially warping the images towards each other and cross-dissolving (Section 3.6.3).
2. Try using your morphing algorithm to perform an image rotation and discuss whether it behaves the way you want it to.

Ex 3.26: 2D image editor. Extend the program you wrote in Exercise 2.2 to import images and let you create a "collage" of pictures. You should implement the following steps:

1. Open up a new image (in a separate window).
2. Shift drag (rubber-band) to crop a subregion (or select whole image).
3. Paste into the current canvas.
4. Select the deformation mode (motion model): translation, rigid, similarity, affine, or perspective.
5. Drag any corner of the outline to change its transformation.
6. (Optional) Change the relative ordering of the images and which image is currently being manipulated.

The user should see the composition of the various images' pieces on top of each other.

This exercise should be built on the image transformation classes supported in the software library. Persistence of the created representation (save and load) should also be supported (for each image, save its transformation).

Figure 3.53 There is a faint image of a rainbow visible in the right-hand side of this picture. Can you think of a way to enhance it (Exercise 3.29)?

Ex 3.27: 3D texture-mapped viewer. Extend the viewer you created in Exercise 2.3 to include texture-mapped polygon rendering. Augment each polygon with (u, v, w) coordinates into an image.

Ex 3.28: Image denoising. Implement at least two of the various image denoising techniques described in this chapter and compare them on both synthetically noised image sequences and real-world (low-light) sequences. Does the performance of the algorithm depend on the correct choice of noise level estimate? Can you draw any conclusions as to which techniques work better?

Ex 3.29: Rainbow enhancer—challenging. Take a picture containing a rainbow, such as Figure 3.53, and enhance the strength (saturation) of the rainbow.

1. Draw an arc in the image delineating the extent of the rainbow.
2. Fit an *additive* rainbow function (explain why it is additive) to this arc (it is best to work with linearized pixel values), using the spectrum as the cross-section, and estimating the width of the arc and the amount of color being added. This is the trickiest part of the problem, as you need to tease apart the (low-frequency) rainbow pattern and the natural image hiding behind it.
3. Amplify the rainbow signal and add it back into the image, re-applying the gamma function if necessary to produce the final image.

Chapter 4

Model fitting and optimization

4.1 Scattered data interpolation 155
   4.1.1 Radial basis functions 157
   4.1.2 Overfitting and underfitting 159
   4.1.3 Robust data fitting 162
4.2 Variational methods and regularization 163
   4.2.1 Discrete energy minimization 166
   4.2.2 Total variation 168
   4.2.3 Bilateral solver 168
   4.2.4 *Application*: Interactive colorization 169
4.3 Markov random fields 170
   4.3.1 Conditional random fields 177
   4.3.2 *Application*: Interactive segmentation 181
4.4 Additional reading 184
4.5 Exercises 185

R. Szeliski, *Computer Vision*, Texts in Computer Science,
https://doi.org/10.1007/978-3-030-34372-9_4

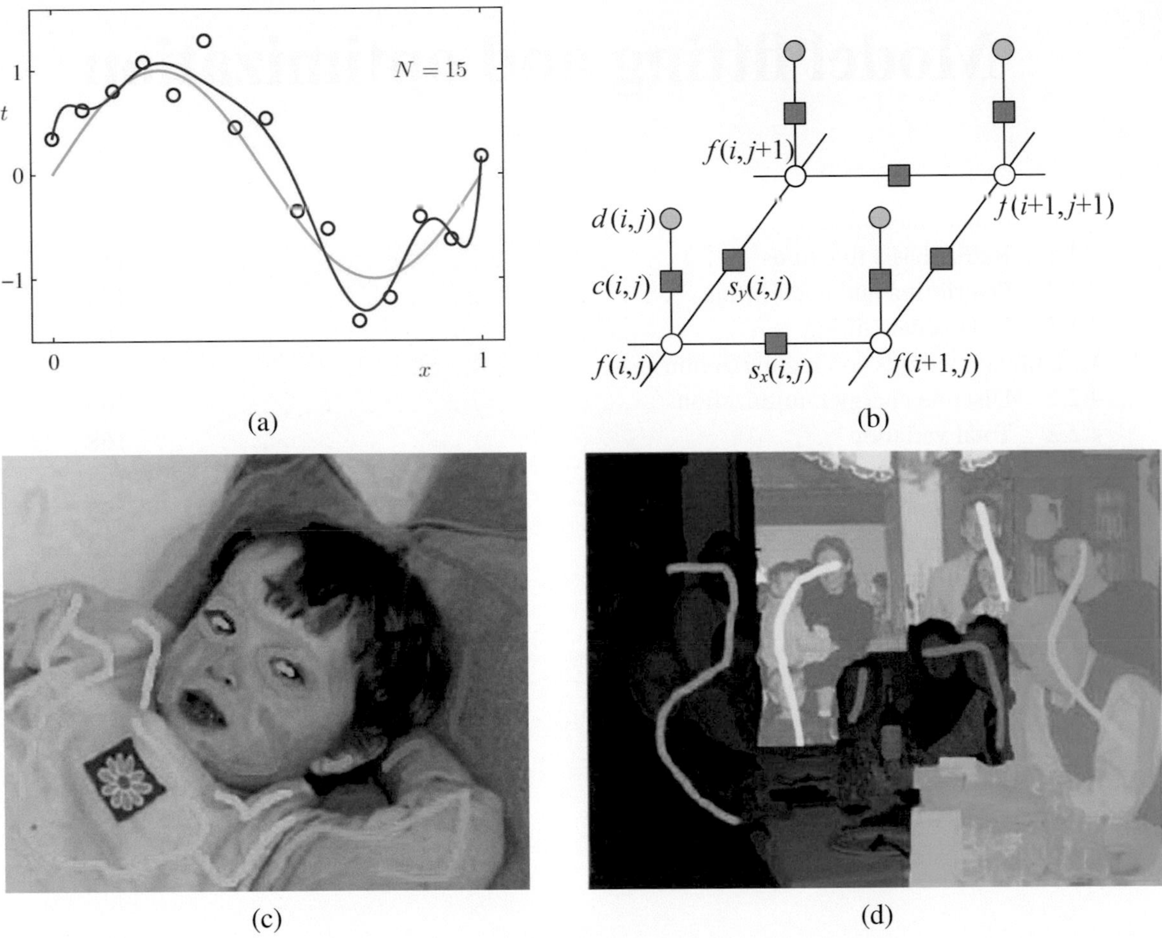

Figure 4.1 Examples of data interpolation and global optimization: (a) scattered data interpolation (curve fitting) (Bishop 2006) © 2006 Springer; (b) graphical model interpretation of first-order regularization; (c) colorization using optimization (Levin, Lischinski, and Weiss 2004) © 2004 ACM; (d) multi-image photomontage formulated as an unordered label MRF (Agarwala, Dontcheva *et al.* 2004) © 2004 ACM.

In the previous chapter, we covered a large number of image processing operators that take as input one or more images and produce some filtered or transformed version of these images. In many situations, however, we are given *incomplete* data as input, such as depths at a sparse number of locations, or user scribbles suggesting how an image should be colorized or segmented (Figure 4.1c–d).

The problem of interpolating a complete image (or more generally a *function* or *field*) from incomplete or varying quality data is often called *scattered data interpolation*. We begin this chapter with a review of techniques in this area, since in addition to being widely used in computer vision, they also form the basis of most machine learning algorithms, which we will study in the next chapter.

Instead of doing an exhaustive survey, we present in Section 4.1 some easy-to-use techniques, such as triangulation, spline interpolation, and radial basis functions. While these techniques are widely used, they cannot easily be modified to provide *controlled continuity*, i.e., to produce the kinds of piecewise continuous reconstructions we expect when estimating depth maps, label maps, or even color images.

For this reason, we introduce in Section 4.2 *variational methods*, which formulate the interpolation problem as the recovery of a piecewise smooth function subject to exact or approximate data constraints. Because the smoothness is controlled using penalties formulated as norms of the function, this class of techniques are often called *regularization* or *energy-based* approaches. To find the minimum-energy solutions to these problems, we discretize them (typically on a pixel grid), resulting in a discrete energy, which can then be minimized using sparse linear systems or related iterative techniques.

In the last part of this chapter, Section 4.3, we show how such energy-based formulations are related to Bayesian inference techniques formulated as *Markov random fields*, which are a special case of general probabilistic *graphical models*. In these formulations, data constraints can be interpreted as noisy and/or incomplete measurements, and piecewise smoothness constraints as *prior assumptions* or *models* over the solution space. Such formulations are also often called *generative models*, since we can, in principle, generate random samples from the prior distribution to see if they conform with our expectations. Because the prior models can be more complex than simple smoothness constraints, and because the solution space can have multiple local minima, more sophisticated optimization techniques have been developed, which we discuss in this section.

4.1 Scattered data interpolation

The goal of *scattered data interpolation* is to produce a (usually continuous and smooth) function $\mathbf{f}(\mathbf{x})$ that passes *through* a set of data points $\mathbf{d}_k$ placed at locations $\mathbf{x}_k$ such that

$$\mathbf{f}(\mathbf{x}_k) = \mathbf{d}_k. \tag{4.1}$$

The related problem of *scattered data approximation* only requires the function to pass *near* the data points (Amidror 2002; Wendland 2004; Anjyo, Lewis, and Pighin 2014). This is usually formulated using a penalty function such as

$$E_{\text{D}} = \sum_k \|\mathbf{f}(\mathbf{x}_k) - \mathbf{d}_k\|^2, \tag{4.2}$$

with the squared norm in the above formula sometimes replaced by a different norm or robust function (Section 4.1.3). In statistics and machine learning, the problem of predicting an output function

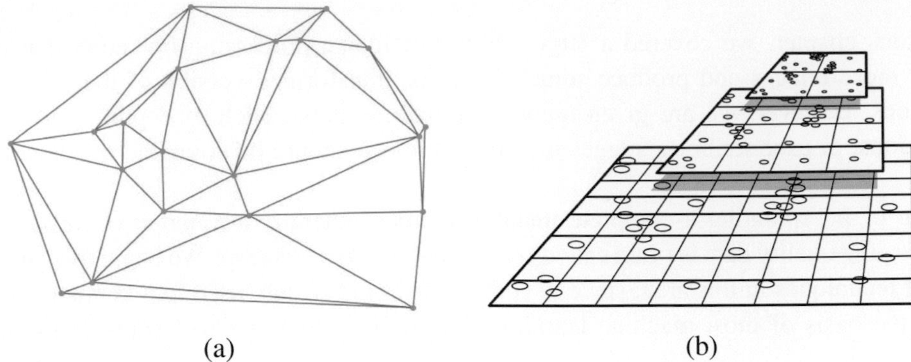

(a) (b)

Figure 4.2 Some simple scattered data interpolation and approximation algorithms: (a) a Delaunay triangulation defined over a set of data point locations; (b) data structure and intermediate results for the *pull-push* algorithm (Gortler, Grzeszczuk *et al.* 1996) © 1996 ACM.

given a finite number of samples is called *regression* (Section 5.1). The $\mathbf{x}$ vectors are called the *inputs* and the outputs $\mathbf{y}$ are called the *targets*. Figure 4.1a shows an example of one-dimensional scattered data interpolation, while Figures 4.2 and 4.8 show some two-dimensional examples.

At first glance, scattered data interpolation seems closely related to *image interpolation*, which we studied in Section 3.5.1. However, unlike images, which are regularly gridded, the data points in scattered data interpolation are irregularly placed throughout the domain, as shown in Figure 4.2. This requires some adjustments to the interpolation methods we use.

If the domain $\mathbf{x}$ is two-dimensional, as is the case with images, one simple approach is to *triangulate* the domain $\mathbf{x}$ using the data locations $\mathbf{x}_k$ as the triangle vertices. The resulting triangular network, shown in Figure 4.2a, is called a *triangular irregular network* (TIN), and was one of the early techniques used to produce elevation maps from scattered field measurements collected by surveys.

The triangulation in Figure 4.2a was produced using a *Delaunay triangulation*, which is the most widely used planar triangulation technique due to its attractive computational properties, such as the avoidance of long skinny triangles. Algorithms for efficiently computing such triangulation are readily available[1] and covered in textbooks on computational geometry (Preparata and Shamos 1985; de Berg, Cheong *et al.* 2008). The Delaunay triangulation can be extended to higher-dimensional domains using the property of circumscribing spheres, i.e., the requirement that all selected simplices (triangles, tetrahedra, etc.) have no other vertices inside their circumscribing spheres.

Once the triangulation has been defined, it is straightforward to define a piecewise-linear interpolant over each triangle, resulting in an interpolant that is C_0 but not generally C_1 continuous. The formulas for the function inside each triangle are usually derived using *barycentric coordinates*, which attain their maximal values at the vertices and sum up to one (Farin 2002; Amidror 2002).

If a smoother surface is desired as the interpolant, we can replace the piecewise linear functions on each triangle with higher-order *splines*, much as we did for image interpolation (Section 3.5.1). However, since these splines are now defined over irregular triangulations, more sophisticated techniques must be used (Farin 2002; Amidror 2002). Other, more recent interpolators based on geometric modeling techniques in computer graphics include subdivision surfaces (Peters and Reif 2008).

An alternative to triangulating the data points is to use a regular n-dimensional grid, as shown in

[1]For example, https://docs.scipy.org/doc/scipy/reference/tutorial/spatial.html

Figure 4.2b. Splines defined on such domains are often called *tensor product splines* and have been used to interpolate scattered data (Lee, Wolberg, and Shin 1997).

An even faster, but less accurate, approach is called the *pull-push* algorithm and was originally developed for interpolating missing 4D lightfield samples in a Lumigraph (Gortler, Grzeszczuk *et al.* 1996). The algorithm proceeds in three phases, as schematically illustrated in Figure 4.2b.

First, the irregular data samples are *splatted* onto (i.e., spread across) the nearest grid vertices, using the same approach we discussed in Section 3.6.1 on parametric image transformations. The splatting operations accumulate both values and weights at nearby vertices. In the second, *pull*, phase, values and weights are computed at a hierarchical set of lower resolution grids by combining the coefficient values from the higher resolution grids. In the lower resolution grids, the gaps (regions where the weights are low) become smaller. In the third, *push*, phase, information from each lower resolution grid is combined with the next higher resolution grid, filling in the gaps while not unduly blurring the higher resolution information already computed. Details of these three stages can be found in (Gortler, Grzeszczuk *et al.* 1996).

The pull-push algorithm is very fast, since it is essentially linear in the number of input data points and fine-level grid samples.

4.1.1 Radial basis functions

While the mesh-based representations I have just described can provide good-quality interpolants, they are typically limited to low-dimensional domains, because the size of the mesh grows combinatorially with the dimensionality of the domain. In higher dimensions, it is common to use *mesh-free* approaches that define the desired interpolant as a weighted sum of basis functions, similar to the formulation used in image interpolation (3.64). In machine learning, such approaches are often called *kernel functions* or *kernel regression* (Bishop 2006, Chapter 6; Murphy 2012, Chapter 14; Schölkopf and Smola 2001).

In more detail, the interpolated function f is a weighted sum (or *superposition*) of basis functions centered at each input data point

$$\mathbf{f}(\mathbf{x}) = \sum_k \mathbf{w}_k \phi(\|\mathbf{x} - \mathbf{x}_k\|), \tag{4.3}$$

where the $\mathbf{x}_k$ are the locations of the scattered data points, the ϕs are the *radial basis functions* (or kernels), and $\mathbf{w}_k$ are the local *weights* associated with each kernel. The basis functions $\phi()$ are called *radial* because they are applied to the radial distance between a data sample $\mathbf{x}_k$ and an evaluation point $\mathbf{x}$. The choice of ϕ determines the smoothness properties of the interpolant, while the choice of weights w_k determines how closely the function approximates the input.

Some commonly used basis functions (Anjyo, Lewis, and Pighin 2014) include

$$\text{Gaussian} \qquad \phi(r) = \exp(-r^2/c^2) \tag{4.4}$$

$$\text{Hardy multiquadric} \qquad \phi(r) = \sqrt{(r^2 + c^2)} \tag{4.5}$$

$$\text{Inverse multiquadric} \qquad \phi(r) = 1/\sqrt{(r^2 + c^2)} \tag{4.6}$$

$$\text{Thin plate spline} \qquad \phi(r) = r^2 \log r. \tag{4.7}$$

In these equations, r is the radial distance and c is a scale parameter that controls the size (radial falloff) of the basis functions, and hence its smoothness (more compact bases lead to "peakier" solutions). The thin plate spline equation holds for two dimensions (the general n-dimensional spline is called the *polyharmonic spline* and is given in (Anjyo, Lewis, and Pighin 2014)) and is the analytic solution to the second degree variational spline derived in (4.19).

If we want our function to exactly interpolate the data values, we solve the linear system of equations (4.1), i.e.,

$$\mathbf{f}(\mathbf{x}_k) = \sum_l \mathbf{w}_l \phi(\|\mathbf{x}_k - \mathbf{x}_l\|) = \mathbf{d}_k, \tag{4.8}$$

to obtain the desired set of weights $\mathbf{w}_k$. Note that for large amounts of basis function overlap (large values of c), these equations may be quite *ill-conditioned*, i.e., small changes in data values or locations can result in large changes in the interpolated function. Note also that the solution of such a system of equations is in general $O(m^3)$, where m is the number of data points (unless we use basis functions with finite extent to obtain a sparse set of equations).

A more prudent approach is to solve the *regularized data approximation problem*, which involves minimizing the data constraint energy (4.2) together with a weight penalty (*regularizer*) of the form

$$E_{\mathrm{W}} = \sum_k \|\mathbf{w}_k\|^p, \tag{4.9}$$

and to then minimize the regularized least squares problem

$$E(\{w_k\}) = E_{\mathrm{D}} + \lambda E_{\mathrm{W}} \tag{4.10}$$

$$= \sum_k \|\sum_l \mathbf{w}_l \phi(\|\mathbf{x}_k - \mathbf{x}_l\|) - \mathbf{d}_k\|^2 + \lambda \sum_k \|\mathbf{w}_k\|^p. \tag{4.11}$$

When $p = 2$ (quadratic weight penalty), the resulting energy is a pure least squares problem, and can be solved using the *normal equations* (Appendix A.2), where the λ value gets added along the diagonal to stabilize the system of equations.

In statistics and machine learning, the quadratic (regularized least squares) problem is called *ridge regression*. In neural networks, adding a quadratic penalty on the weights is called *weight decay*, because it encourages weights to decay towards zero (Section 5.3.3). When $p = 1$, the technique is called *lasso* (least absolute shrinkage and selection operator), since for sufficiently large values of λ, many of the weights $\mathbf{w}_k$ get driven to zero (Tibshirani 1996; Bishop 2006; Murphy 2012; Deisenroth, Faisal, and Ong 2020). This results in a *sparse* set of basis functions being used in the interpolant, which can greatly speed up the computation of new values of $\mathbf{f}(\mathbf{x})$. We will have more to say on sparse kernel techniques in the section on Support Vector Machines (Section 5.1.4).

An alternative to solving a set of equations to determine the weights $\mathbf{w}_k$ is to simply set them to the input data values $\mathbf{d}_k$. However, this fails to interpolate the data, and instead produces higher values in higher density regions. This can be useful if we are trying to estimate a probability density function from a set of samples. In this case, the resulting density function, obtained after normalizing the sum of sample-weighted basis functions to have a unit integral, is called the *Parzen window* or *kernel* approach to probability density estimation (Duda, Hart, and Stork 2001, Section 4.3; Bishop 2006, Section 2.5.1). Such probability densities can be used, among other things, for (spatially) clustering color values together for image segmentation in what is known as the *mean shift* approach (Comaniciu and Meer 2002) (Section 7.5.2).

If, instead of just estimating a density, we wish to actually interpolate a set of data values $\mathbf{d}_k$, we can use a related technique known as *kernel regression* or the *Nadaraya-Watson* model, in which we divide the data-weighted summed basis functions by the sum of all the basis functions,

$$\mathbf{f}(\mathbf{x}) = \frac{\sum_k \mathbf{d}_k \phi(\|\mathbf{x} - \mathbf{x}_k\|)}{\sum_l \phi(\|\mathbf{x} - \mathbf{x}_l\|)}. \tag{4.12}$$

Note how this operation is similar, in concept, to the *splatting* method for forward rendering we discussed in Section 3.6.1, except that here, the bases can be much wider than the nearest-neighbor bilinear bases used in graphics (Takeda, Farsiu, and Milanfar 2007).

Kernel regression is equivalent to creating a new set of spatially varying normalized shifted basis functions

$$\phi'_k(\mathbf{x}) = \frac{\phi(\|\mathbf{x} - \mathbf{x}_k\|)}{\sum_l \phi(\|\mathbf{x} - \mathbf{x}_l\|)}, \tag{4.13}$$

which form a *partition of unity*, i.e., sum up to 1 at every location (Anjyo, Lewis, and Pighin 2014). While the resulting interpolant can now be written more succinctly as

$$\mathbf{f}(\mathbf{x}) = \sum_k \mathbf{d}_k \phi'_k(\|\mathbf{x} - \mathbf{x}_k\|), \tag{4.14}$$

in most cases, it is more expensive to precompute and store the K ϕ'_k functions than to evaluate (4.12).

While not that widely used in computer vision, kernel regression techniques have been applied by Takeda, Farsiu, and Milanfar (2007) to a number of low-level image processing operations, including state-of-the-art handheld multi-frame super-resolution (Wronski, Garcia-Dorado *et al.* 2019).

One last scattered data interpolation technique worth mentioning is *moving least squares*, where a weighted subset of nearby points is used to compute a local smooth surface. Such techniques are mostly widely used in 3D computer graphics, especially for point-based surface modeling, as discussed in Section 13.4 and (Alexa, Behr *et al.* 2003; Pauly, Keiser *et al.* 2003; Anjyo, Lewis, and Pighin 2014).

4.1.2 Overfitting and underfitting

When we introduced weight regularization in (4.9), we said that it was usually preferable to approximate the data but we did not explain why. In most data fitting problems, the samples $\mathbf{d}_k$ (and sometimes even their locations $\mathbf{x}_k$) are noisy, so that fitting them exactly makes no sense. In fact, doing so can introduce a lot of spurious wiggles, when the true solution is likely to be smoother.

To delve into this phenomenon, let us start with a simple polynomial fitting example taken from (Bishop 2006, Chapter 1.1). Figure 4.3 shows a number of polynomial curves of different orders M fit to the blue circles, which are noisy samples from the underlying green sine curve. Notice how the low-order ($M = 0$ and $M = 1$) polynomials severely *underfit* the underlying data, resulting in curves that are too flat, while the $M = 9$ polynomial, which exactly fits the data, exhibits far more wiggle than is likely.

How can we quantify this amount of underfitting and overfitting, and how can we get just the right amount? This topic is widely studied in machine learning and covered in a number of texts, including Bishop (2006, Chapter 1.1), Glassner (2018, Chapter 9), Deisenroth, Faisal, and Ong (2020, Chapter 8), and Zhang, Lipton *et al.* (2021, Section 4.4.3).

One approach is to use regularized least squares, introduced in (4.11). Figure 4.4 shows an $M =$ 9th degree polynomial fit obtained by minimizing (4.11) with the polynomial basis functions $\phi_k(x) = x^k$ for two different values of λ. The left plot shows a reasonable amount of regularization, resulting in a plausible fit, while the larger value of λ on the right causes underfitting. Note that the $M = 9$ interpolant shown in the lower right quadrant of Figure 4.3 corresponds to the unregularized $\lambda = 0$ case.

If we were to now measure the difference between the red (estimated) and green (noise-free) curves, we see that choosing a good intermediate value of λ will produce the best result. In practice, however, we never have access to samples from the noise-free data.

Instead, if we are given a set of samples to interpolate, we can save some in a *validation* set in order to see if the function we compute is underfitting or overfitting. When we vary a parameter

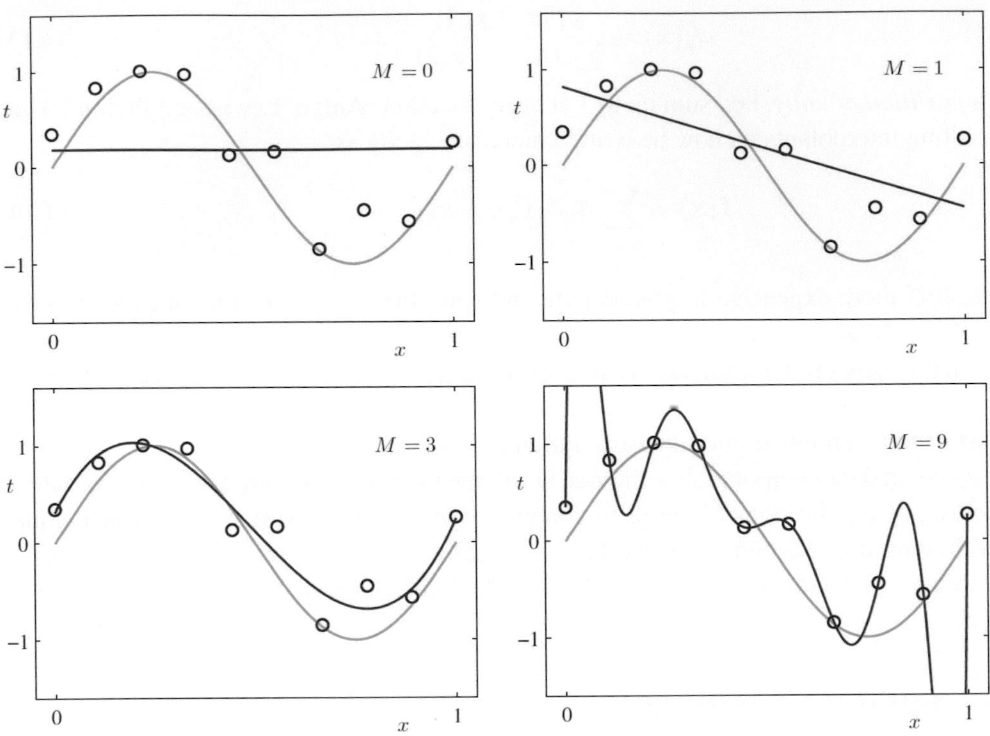

Figure 4.3 Polynomial curve fitting to the blue circles, which are noisy samples from the green sine curve (Bishop 2006) © 2006 Springer. The four plots show the 0th order constant function, the first order linear fit, the $M = 3$ cubic polynomial, and the 9th degree polynomial. Notice how the first two curves exhibit *underfitting*, while the last curve exhibits *overfitting*, i.e., excessive wiggle.

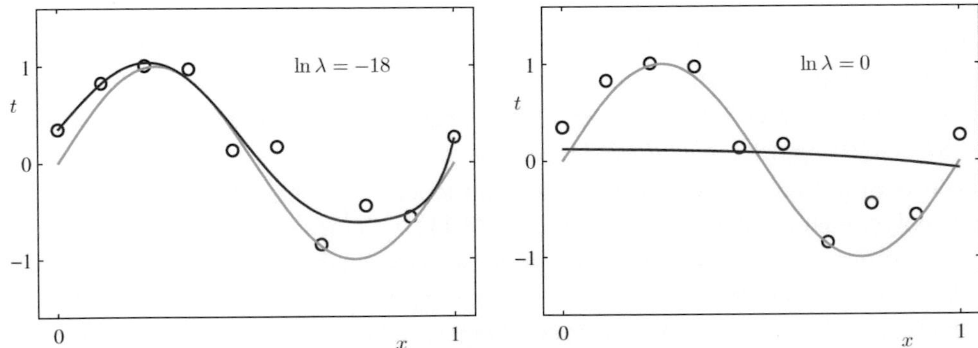

Figure 4.4 Regularized $M = 9$ polynomial fitting for two different values of λ (Bishop 2006) © 2006 Springer. The left plot shows a reasonable amount of regularization, resulting in a plausible fit, while the larger value of λ on the right causes underfitting.

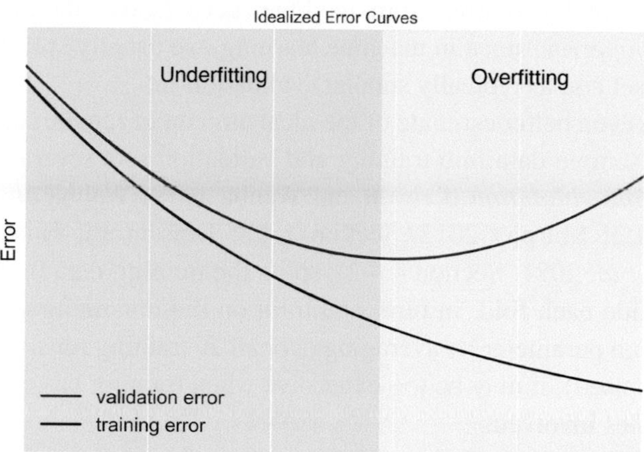

Figure 4.5 Fitting (training) and validation errors as a function of the amount of regularization or smoothing © Glassner (2018). The less regularized solutions on the right, while exhibiting lower fitting error, perform less well on the validation data.

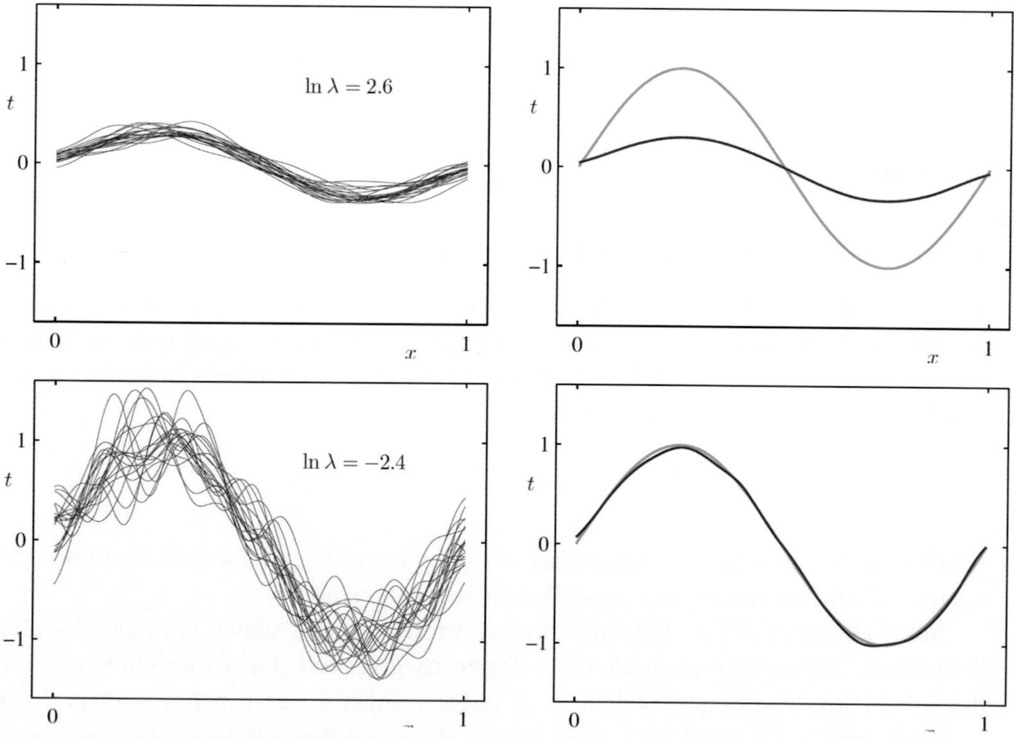

Figure 4.6 The more heavily regularized solution $\log \lambda = 2.6$ exhibits higher bias (deviation from original curve) than the less heavily regularized version ($\log \lambda = -2.4$), which has much higher variance (Bishop 2006) © 2006 Springer. The red curves on the left are $M = 24$ Gaussian basis fits to 25 randomly sampled points on the green curve. The red curve on the right is their mean.

such as λ (or use some other measure to control smoothness), we typically obtain a curve such as the one shown in Figure 4.5. In this figure, the blue curve denotes the fitting error, which in this case is called the *training error*, since in machine learning, we usually split the given data into a (typically larger) training set and a (typically smaller) validation set.

To obtain an even better estimate of the ideal amount of regularization, we can repeat the process of splitting our sample data into training and validation sets several times. One well-known technique, called *cross-validation* (Craven and Wahba 1979; Wahba and Wendelberger 1980; Bishop 2006, Section 1.3; Murphy 2012, Section 1.4.8; Deisenroth, Faisal, and Ong 2020, Chapter 8; Zhang, Lipton *et al.* 2021, Section 4.4.2), splits the training data into K *folds* (equal sized pieces). You then put aside each fold, in turn, and train on the remaining data. You can then estimate the best regularization parameter by averaging over all K training runs. While this generally works well ($K = 5$ is often used), it may be too expensive when training large neural networks because of the long training times involved.

Cross-validation is just one example of a class of *model selection* techniques that estimate *hyperparameters* in a training algorithm to achieve good performance. Additional methods include *information criteria* such as the Bayesian information criterion (BIC) (Torr 2002) and the Akaike information criterion (AIC) (Kanatani 1998), and Bayesian modeling approaches (Szeliski 1989; Bishop 2006; Murphy 2012).

One last topic worth mention with regard to data fitting, since it comes up often in discussions of statistical machine learning techniques, is the *bias-variance tradeoff* (Bishop 2006, Section 3.2). As you can see in Figure 4.6, using a large amount of regularization (top row) results in much lower variance between different random sample solutions, but much higher bias away from the true solution. Using insufficient regularization increases the variance dramatically, although an average over a large number of samples has low bias. The trick is to determine a reasonable compromise in terms of regularization so that any individual solution has a good expectation of being close to the *ground truth* (original clean continuous) data.

4.1.3 Robust data fitting

When we added a regularizer on the weights in (4.9), we noted that it did not have to be a quadratic penalty and could, instead, be a lower-order monomial that encouraged *sparsity* in the weights.

This same idea can be applied to data terms such as (4.2), where, instead of using a quadratic penalty, we can use a *robust loss function* $\rho()$,

$$E_{\rm R} = \sum_k \rho(\|\mathbf{r}_k\|), \quad \text{with } \mathbf{r}_k = \mathbf{f}(\mathbf{x}_k) - \mathbf{d}_k, \tag{4.15}$$

which gives lower weights to larger data fitting errors, which are more likely to be outlier measurements. (The fitting error term $\mathbf{r}_k$ is called the *residual error*.)

Some examples of loss functions from (Barron 2019) are shown in Figure 4.7 along with their derivatives. The regular quadratic ($\alpha = 2$) penalty gives full (linear) weight to each error, whereas the $\alpha = 1$ loss gives equal weight to all larger residuals, i.e., it behaves as an L_1 loss for large residuals, and L_2 for small ones. Even larger values of α discount large errors (outliers) even more, although they result in optimization problems that are *non-convex*, i.e., that can have multiple local minima. We will discuss techniques for finding good initial guesses for such problems later on in Section 8.1.4.

In statistics, minimizing non-quadratic loss functions to deal with potential outlier measurements is known as *M-estimation* (Huber 1981; Hampel, Ronchetti *et al.* 1986; Black and Rangarajan 1996; Stewart 1999). Such estimation problems are often solved using *iteratively reweighted least squares*,

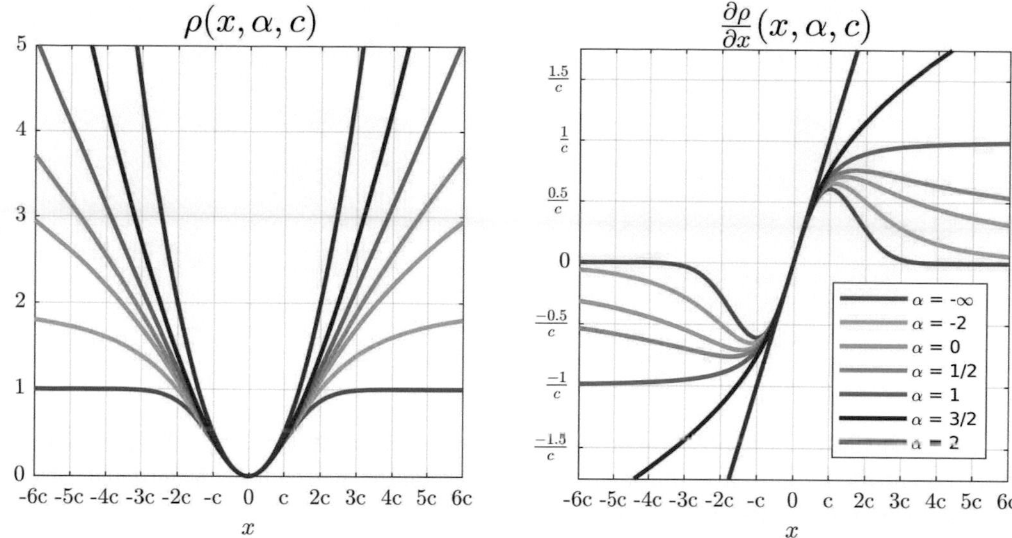

Figure 4.7 A general and adaptive loss function (left) and its gradient (right) for different values of its shape parameter α (Barron 2019) © 2019 IEEE. Several values of α reproduce existing loss functions: L_2 loss ($\alpha = 2$), Charbonnier loss ($\alpha = 1$), Cauchy loss ($\alpha = 0$), Geman-McClure loss ($\alpha = -2$), and Welsch loss ($\alpha = -1$).

which we discuss in more detail in Section 8.1.4 and Appendix B.3. The Appendix also discusses the relationship between robust statistics and non-Gaussian probabilistic models.

The generalized loss function introduced by Barron (2019) has two free parameters. The first one, α, controls how drastically outlier residuals are downweighted. The second (scale) parameter c controls the width of the quadratic well near the minimum, i.e., what range of residual values roughly corresponds to inliers. Traditionally, the choice of α, which corresponds to a variety of previously published loss functions, was determined heuristically, based on the expected shape of the outlier distribution and computational considerations (e.g., whether a convex loss was desired). The scale parameter c could be estimated using a robust measure of variance, as discussed in Appendix B.3.

In his paper, Barron (2019) discusses how both parameters can be determined at run time by maximizing the likelihood (or equivalently, minimizing the negative log-likelihood) of the given residuals, making such an algorithm self-tuning to a wide variety of noise levels and outlier distributions.

4.2 Variational methods and regularization

The theory of regularization we introduced in the previous section was first developed by statisticians trying to fit models to data that severely underconstrained the solution space (Tikhonov and Arsenin 1977; Engl, Hanke, and Neubauer 1996). Consider, for example, finding a smooth surface that passes through (or near) a set of measured data points (Figure 4.8). Such a problem is described as *ill-posed* because many possible surfaces can fit this data. Since small changes in the input can sometimes lead to large changes in the fit (e.g., if we use polynomial interpolation), such problems are also often *ill-conditioned*. Since we are trying to recover the unknown function $f(x, y)$ from which the data points $d(x_i, y_i)$ were sampled, such problems are also often called *inverse problems*. Many computer vision tasks can be viewed as inverse problems, since we are trying to recover a full description of the 3D world from a limited set of images.

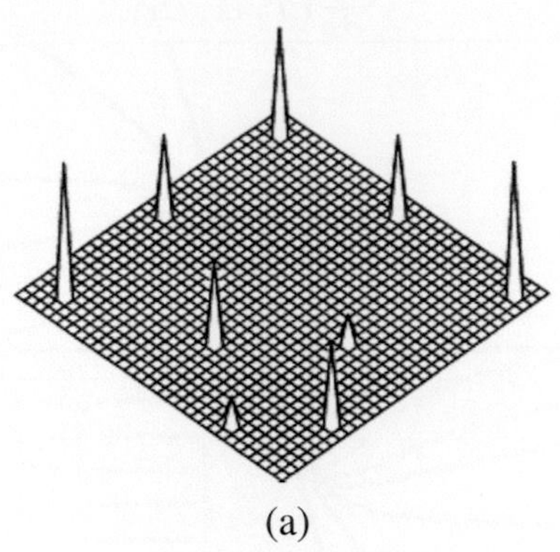

(a)

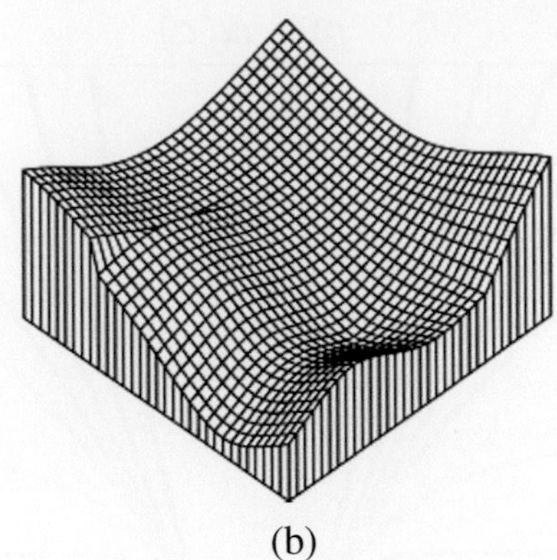

(b)

Figure 4.8 A simple surface interpolation problem: (a) nine data points of various heights scattered on a grid; (b) second-order, controlled-continuity, thin-plate spline interpolator, with a tear along its left edge and a crease along its right (Szeliski 1989) © 1989 Springer.

In the previous section, we attacked this problem using basis functions placed at the data points, or other heuristics such as the pull-push algorithm. While such techniques can provide reasonable solutions, they do not let us directly *quantify* and hence *optimize* the amount of smoothness in the solution, nor do they give us local control over where the solution should be discontinuous (Figure 4.8).

To do this, we use norms (measures) on function derivatives (described below) to formulate the problem and then find minimal energy solutions to these norms. Such techniques are often called *energy-based* or *optimization-based* approaches to computer vision. They are also often called *variational*, since we can use the *calculus of variations* to find the optimal solutions. Variational methods have been widely used in computer vision since the early 1980s to pose and solve a number of fundamental problems, including optical flow (Horn and Schunck 1981; Black and Anandan 1993; Brox, Bruhn *et al.* 2004; Werlberger, Pock, and Bischof 2010), segmentation (Kass, Witkin, and Terzopoulos 1988; Mumford and Shah 1989; Chan and Vese 2001), denoising (Rudin, Osher, and Fatemi 1992; Chan, Osher, and Shen 2001; Chan and Shen 2005), and multi-view stereo (Faugeras and Keriven 1998; Pons, Keriven, and Faugeras 2007; Kolev, Klodt *et al.* 2009). A more detailed list of relevant papers can be found in the Additional Reading section at the end of this chapter.

In order to quantify what it means to find a smooth solution, we can define a norm on the solution space. For one-dimensional functions $f(x)$, we can integrate the squared first derivative of the function,

$$\mathcal{E}_1 = \int f_x^2(x)\, dx \tag{4.16}$$

or perhaps integrate the squared second derivative,

$$\mathcal{E}_2 = \int f_{xx}^2(x)\, dx. \tag{4.17}$$

(Here, we use subscripts to denote differentiation.) Such energy measures are examples of *functionals*, which are operators that map functions to scalar values. They are also often called *variational methods*, because they measure the variation (non-smoothness) in a function.

In two dimensions (e.g., for images, flow fields, or surfaces), the corresponding smoothness functionals are

$$\mathcal{E}_1 = \int f_x^2(x,y) + f_y^2(x,y)\, dx\, dy = \int \|\nabla f(x,y)\|^2\, dx\, dy \tag{4.18}$$

and

$$\mathcal{E}_2 = \int f_{xx}^2(x,y) + 2f_{xy}^2(x,y) + f_{yy}^2(x,y)\,dx\,dy, \tag{4.19}$$

where the mixed $2f_{xy}^2$ term is needed to make the measure rotationally invariant (Grimson 1983).

The first derivative norm is often called the *membrane*, since interpolating a set of data points using this measure results in a tent-like structure. (In fact, this formula is a small-deflection approximation to the surface area, which is what soap bubbles minimize.) The second-order norm is called the *thin-plate spline*, since it approximates the behavior of thin plates (e.g., flexible steel) under small deformations. A blend of the two is called the *thin-plate spline under tension* (Terzopoulos 1986b).

The regularizers (smoothness functions) we have just described force the solution to be smooth and C_0 and/or C_1 continuous everywhere. In most computer vision applications, however, the fields we are trying to model or recover are only piecewise continuous, e.g., depth maps and optical flow fields jump at object discontinuities. Color images are even more discontinuous, since they also change appearance at albedo (surface color) and shading discontinuities.

To better model such functions, Terzopoulos (1986b) introduced *controlled-continuity splines*, where each derivative term is multiplied by a local weighting function,

$$\begin{aligned}\mathcal{E}_{\mathrm{CC}} = \int \rho(x,y)\{&[1-\tau(x,y)][f_x^2(x,y)+f_y^2(x,y)] \\ &+ \tau(x,y)[f_{xx}^2(x,y) + 2f_{xy}^2(x,y) + f_{yy}^2(x,y)]\}\,dx\,dy.\end{aligned} \tag{4.20}$$

Here, $\rho(x,y) \in [0,1]$ controls the *continuity* of the surface and $\tau(x,y) \in [0,1]$ controls the local *tension*, i.e., how flat the surface wants to be. Figure 4.8 shows a simple example of a controlled-continuity interpolator fit to nine scattered data points. In practice, it is more common to find first-order smoothness terms used with images and flow fields (Section 9.3) and second-order smoothness associated with surfaces (Section 13.3.1).

In addition to the smoothness term, variational problems also require a data term (or *data penalty*). For scattered data interpolation (Nielson 1993), the data term measures the distance between the function $f(x,y)$ and a set of data points $d_i = d(x_i, y_i)$,

$$\mathcal{E}_{\mathrm{D}} = \sum_i [f(x_i,y_i) - d_i]^2. \tag{4.21}$$

For a problem like noise removal, a continuous version of this measure can be used,

$$\mathcal{E}_{\mathrm{D}} = \int [f(x,y) - d(x,y)]^2\,dx\,dy. \tag{4.22}$$

To obtain a global energy that can be minimized, the two energy terms are usually added together,

$$\mathcal{E} = \mathcal{E}_{\mathrm{D}} + \lambda\mathcal{E}_{\mathrm{S}}, \tag{4.23}$$

where $\mathcal{E}_{\mathrm{S}}$ is the *smoothness penalty* ($\mathcal{E}_1$, $\mathcal{E}_2$ or some weighted blend such as $\mathcal{E}_{\mathrm{CC}}$) and λ is the *regularization parameter*, which controls the smoothness of the solution. As we saw in Section 4.1.2, good values for the regularization parameter can be estimated using techniques such as cross-validation.

4.2.1 Discrete energy minimization

In order to find the minimum of this continuous problem, the function $f(x, y)$ is usually first discretized on a regular grid.[2] The most principled way to perform this discretization is to use *finite element analysis*, i.e., to approximate the function with a piecewise continuous spline, and then perform the analytic integration (Bathe 2007).

Fortunately, for both the first-order and second-order smoothness functionals, the judicious selection of appropriate finite elements results in particularly simple discrete forms (Terzopoulos 1983). The corresponding *discrete* smoothness energy functions become

$$\begin{aligned} E_1 = \sum_{i,j} & s_x(i,j)[f(i+1,j) - f(i,j) - g_x(i,j)]^2 \\ & + s_y(i,j)[f(i,j+1) - f(i,j) - g_y(i,j)]^2 \end{aligned} \tag{4.24}$$

and

$$\begin{aligned} E_2 = h^{-2} \sum_{i,j} & c_x(i,j)[f(i+1,j) - 2f(i,j) + f(i-1,j)]^2 \\ & + 2c_m(i,j)[f(i+1,j+1) - f(i+1,j) - f(i,j+1) + f(i,j)]^2 \\ & + c_y(i,j)[f(i,j+1) - 2f(i,j) + f(i,j-1)]^2, \end{aligned} \tag{4.25}$$

where h is the size of the finite element grid. The h factor is only important if the energy is being discretized at a variety of resolutions, as in coarse-to-fine or multigrid techniques.

The optional smoothness weights $s_x(i,j)$ and $s_y(i,j)$ control the location of horizontal and vertical tears (or weaknesses) in the surface. For other problems, such as colorization (Levin, Lischinski, and Weiss 2004) and interactive tone mapping (Lischinski, Farbman *et al.* 2006), they control the smoothness in the interpolated chroma or exposure field and are often set inversely proportional to the local luminance gradient strength. For second-order problems, the crease variables $c_x(i,j)$, $c_m(i,j)$, and $c_y(i,j)$ control the locations of creases in the surface (Terzopoulos 1988; Szeliski 1990a).

The data values $g_x(i,j)$ and $g_y(i,j)$ are gradient data terms (constraints) used by algorithms, such as photometric stereo (Section 13.1.1), HDR tone mapping (Section 10.2.1) (Fattal, Lischinski, and Werman 2002), Poisson blending (Section 8.4.4) (Pérez, Gangnet, and Blake 2003), gradient-domain blending (Section 8.4.4) (Levin, Zomet *et al.* 2004), and Poisson surface reconstruction (Section 13.5.1) (Kazhdan, Bolitho, and Hoppe 2006; Kazhdan and Hoppe 2013). They are set to zero when just discretizing the conventional first-order smoothness functional (4.18). Note how separate smoothness and curvature terms can be imposed in the x, y, and mixed directions to produce local tears or creases (Terzopoulos 1988; Szeliski 1990a).

The two-dimensional discrete data energy is written as

$$E_\text{D} = \sum_{i,j} c(i,j)[f(i,j) - d(i,j)]^2, \tag{4.26}$$

where the local confidence weights $c(i,j)$ control how strongly the data constraint is enforced. These values are set to zero where there is no data and can be set to the inverse variance of the data measurements when there is data (as discussed by Szeliski (1989) and in Section 4.3).

The total energy of the discretized problem can now be written as a *quadratic form*

$$E = E_\text{D} + \lambda E_\text{S} = \mathbf{x}^T\mathbf{A}\mathbf{x} - 2\mathbf{x}^T\mathbf{b} + c, \tag{4.27}$$

[2]The alternative of using *kernel basis functions* centered on the data points (Boult and Kender 1986; Nielson 1993) is discussed in more detail in Section 13.3.1.

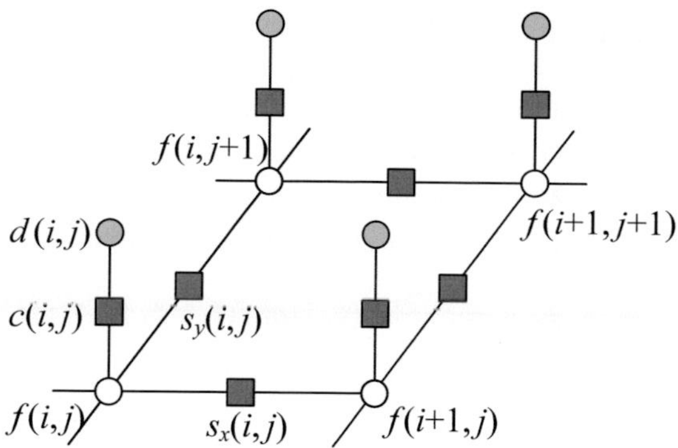

Figure 4.9 Graphical model interpretation of first-order regularization. The white circles are the unknowns $f(i,j)$ while the dark circles are the input data $d(i,j)$. In the resistive grid interpretation, the d and f values encode input and output voltages and the black squares denote resistors whose *conductance* is set to $s_x(i,j)$, $s_y(i,j)$, and $c(i,j)$. In the spring-mass system analogy, the circles denote elevations and the black squares denote springs. The same graphical model can be used to depict a first-order Markov random field (Figure 4.12).

where $\mathbf{x} = [f(0,0)\ldots f(m-1,n-1)]$ is called the *state vector.*[3]

The sparse symmetric positive-definite matrix $\mathbf{A}$ is called the *Hessian* since it encodes the second derivative of the energy function.[4] For the one-dimensional, first-order problem, $\mathbf{A}$ is tridiagonal; for the two-dimensional, first-order problem, it is multi-banded with five non-zero entries per row. We call $\mathbf{b}$ the *weighted data vector.* Minimizing the above quadratic form is equivalent to solving the sparse linear system

$$\mathbf{Ax} = \mathbf{b}, \tag{4.28}$$

which can be done using a variety of sparse matrix techniques, such as multigrid (Briggs, Henson, and McCormick 2000) and hierarchical preconditioners (Szeliski 2006b; Krishnan and Szeliski 2011; Krishnan, Fattal, and Szeliski 2013), as described in Appendix A.5 and illustrated in Figure 4.11. Using such techniques is essential to obtaining reasonable run-times, since properly preconditioned sparse linear systems have convergence times that are *linear* in the number of pixels.

While regularization was first introduced to the vision community by Poggio, Torre, and Koch (1985) and Terzopoulos (1986b) for problems such as surface interpolation, it was quickly adopted by other vision researchers for such varied problems as edge detection (Section 7.2), optical flow (Section 9.3), and shape from shading (Section 13.1) (Poggio, Torre, and Koch 1985; Horn and Brooks 1986; Terzopoulos 1986b; Bertero, Poggio, and Torre 1988; Brox, Bruhn *et al.* 2004). Poggio, Torre, and Koch (1985) also showed how the discrete energy defined by Equations (4.24–4.26) could be implemented in a resistive grid, as shown in Figure 4.9. In computational photography (Chapter 10), regularization and its variants are commonly used to solve problems such as high-dynamic range tone mapping (Fattal, Lischinski, and Werman 2002; Lischinski, Farbman *et al.* 2006), Poisson and gradient-domain blending (Pérez, Gangnet, and Blake 2003; Levin, Zomet *et al.* 2004; Agarwala, Dontcheva *et al.* 2004), colorization (Levin, Lischinski, and Weiss 2004), and natural image matting (Levin, Lischinski, and Weiss 2008).

[3] We use $\mathbf{x}$ instead of $\mathbf{f}$ because this is the more common form in the numerical analysis literature (Golub and Van Loan 1996).

[4] In numerical analysis, $\mathbf{A}$ is called the *coefficient* matrix (Saad 2003); in finite element analysis (Bathe 2007), it is called the *stiffness* matrix.

Robust regularization

While regularization is most commonly formulated using quadratic (L_2) norms, i.e., the squared derivatives in (4.16–4.19) and squared differences in (4.24–4.25), it can also be formulated using the non-quadratic *robust* penalty functions first introduced in Section 4.1.3 and discussed in more detail in Appendix B.3. For example, (4.24) can be generalized to

$$\begin{aligned} E_{1\text{R}} = \sum_{i,j} & s_x(i,j)\rho(f(i+1,j) - f(i,j)) \\ & + s_y(i,j)\rho(f(i,j+1) - f(i,j)), \end{aligned} \tag{4.29}$$

where $\rho(x)$ is some monotonically increasing penalty function. For example, the family of norms $\rho(x) = |x|^p$ is called p-norms. When $p < 2$, the resulting smoothness terms become more piecewise continuous than totally smooth, which can better model the discontinuous nature of images, flow fields, and 3D surfaces.

An early example of robust regularization is the *graduated non-convexity* (GNC) algorithm of Blake and Zisserman (1987). Here, the norms on the data and derivatives are clamped,

$$\rho(x) = \min(x^2, V). \tag{4.30}$$

Because the resulting problem is highly non-convex (it has many local minima), a *continuation* method is proposed, where a quadratic norm (which is convex) is gradually replaced by the non-convex robust norm (Allgower and Georg 2003). (Around the same time, Terzopoulos (1988) was also using continuation to infer the tear and crease variables in his surface interpolation problems.)

4.2.2 Total variation

Today, many regularized problems are formulated using the L_1 ($p = 1$) norm, which is often called *total variation* (Rudin, Osher, and Fatemi 1992; Chan, Osher, and Shen 2001; Chambolle 2004; Chan and Shen 2005; Tschumperlé and Deriche 2005; Tschumperlé 2006; Cremers 2007; Kaftory, Schechner, and Zeevi 2007; Kolev, Klodt *et al.* 2009; Werlberger, Pock, and Bischof 2010). The advantage of this norm is that it tends to better preserve discontinuities, but still results in a convex problem that has a globally unique solution. Other norms, for which the *influence* (derivative) more quickly decays to zero, are presented by Black and Rangarajan (1996), Black, Sapiro *et al.* (1998), and Barron (2019) and discussed in Section 4.1.3 and Appendix B.3.

Even more recently, *hyper-Laplacian* norms with $p < 1$ have gained popularity, based on the observation that the log-likelihood distribution of image derivatives follows a $p \approx 0.5 - 0.8$ slope and is therefore a hyper-Laplacian distribution (Simoncelli 1999; Levin and Weiss 2007; Weiss and Freeman 2007; Krishnan and Fergus 2009). Such norms have an even stronger tendency to prefer large discontinuities over small ones. See the related discussion in Section 4.3 (4.43).

While least squares regularized problems using L_2 norms can be solved using linear systems, other p-norms require different iterative techniques, such as iteratively reweighted least squares (IRLS), Levenberg–Marquardt, alternation between local non-linear subproblems and global quadratic regularization (Krishnan and Fergus 2009), or primal-dual algorithms (Chambolle and Pock 2011). Such techniques are discussed in Section 8.1.3 and Appendices A.3 and B.3.

4.2.3 Bilateral solver

In our discussion of variational methods, we have focused on energy minimization problems based on gradients and higher-order derivatives, which in the discrete setting involves evaluating weighted

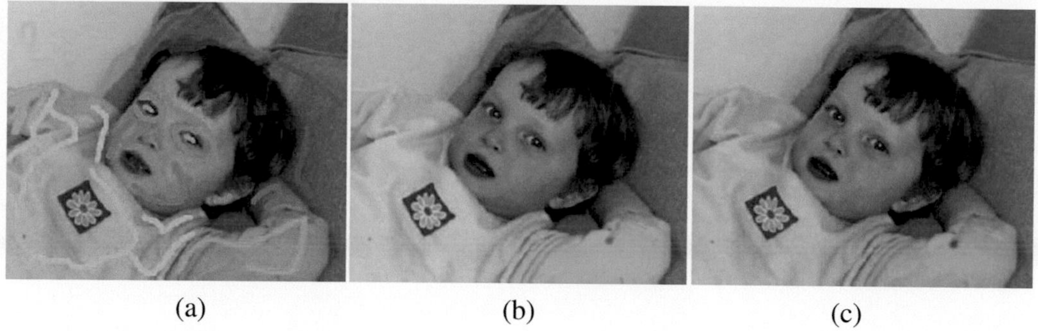

(a) (b) (c)

Figure 4.10 Colorization using optimization (Levin, Lischinski, and Weiss 2004) © 2004 ACM: (a) grayscale image with some color scribbles overlaid; (b) resulting colorized image; (c) original color image from which the grayscale image and the chrominance values for the scribbles were derived. Original photograph by Rotem Weiss.

errors between neighboring pixels. As we saw previously in our discussion of bilateral filtering in Section 3.3.2, we can often get better results by looking at a larger spatial neighborhood and combining pixels with similar colors or grayscale values. To extend this idea to a variational (energy minimization) setting, Barron and Poole (2016) propose replacing the usual first-order nearest-neighbor smoothness penalty (4.24) with a wider-neighborhood, bilaterally weighted version

$$E_{\rm B} = \sum_{i,j} \sum_{k,l} \hat{w}(i,j,k,l)[f(k,l) - f(i,j)]^2, \tag{4.31}$$

where

$$\hat{w}(i,j,k,l) = \frac{w(i,j,k,l)}{\sum_{m,n} w(i,j,m,n)}, \tag{4.32}$$

is the *bistochastized* (normalized) version of the *bilateral weight function* given in (3.37), which may depend on an input guide image, but not on the estimated values of f.[5]

To efficiently solve the resulting set of equations (which are much denser than nearest-neighbor versions), the authors use the same approach originally used to accelerate bilateral filtering, i.e., solving a related problem on a (spatially coarser) bilateral grid. The sequence of operations resembles those used for bilateral filtering, except that after splatting and before slicing, an iterative least squares solver is used instead of a multi-dimensional Gaussian blur. To further speed up the conjugate gradient solver, Barron and Poole (2016) use a multi-level preconditioner inspired by previous work on image-adapted preconditioners (Szeliski 2006b; Krishnan, Fattal, and Szeliski 2013).

Since its introduction, the bilateral solver has been used in a number of video processing and 3D reconstruction applications, including the stitching of binocular omnidirectional panoramic videos (Anderson, Gallup *et al.* 2016). The smartphone AR system developed by Valentin, Kowdle *et al.* (2018) extends the bilateral solver to have local *planar* models and uses a hardware-friendly real-time implementation (Mazumdar, Alaghi *et al.* 2017) to produce dense occlusion effects.

4.2.4 *Application*: Interactive colorization

A good use of edge-aware interpolation techniques is in *colorization*, i.e., manually adding colors to a "black and white" (grayscale) image. In most applications of colorization, the user draws some scribbles indicating the desired colors in certain regions (Figure 4.10a) and the system interpolates

[5]Note that in their paper, Barron and Poole (2016) use different σ_r values for the luminance and chrominance components of pixel color differences.

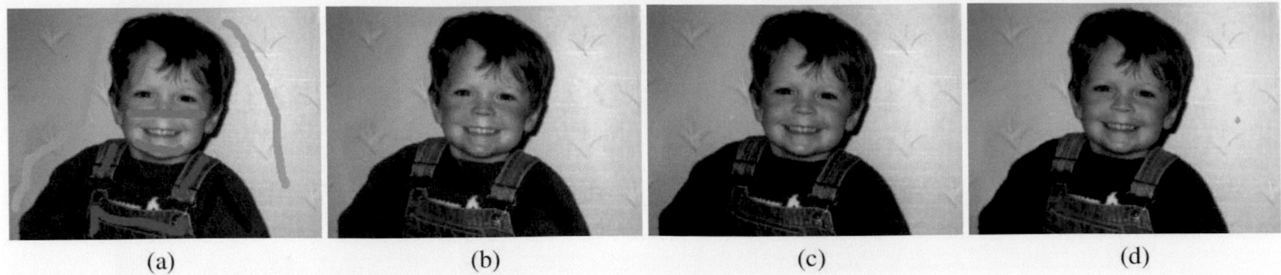

(a) (b) (c) (d)

Figure 4.11 Speeding up the inhomogeneous least squares colorization solver using locally adapted hierarchical basis preconditioning (Szeliski 2006b) © 2006 ACM: (a) input gray image with color strokes overlaid; (b) solution after 20 iterations of conjugate gradient; (c) using one iteration of hierarchical basis function preconditioning; (d) using one iteration of locally adapted hierarchical basis functions.

the specified chrominance (u, v) values to the whole image, which are then re-combined with the input luminance channel to produce a final colorized image, as shown in Figure 4.10b. In the system developed by Levin, Lischinski, and Weiss (2004), the interpolation is performed using locally weighted regularization (4.24), where the local smoothness weights are inversely proportional to luminance gradients. This approach to locally weighted regularization has inspired later algorithms for high dynamic range tone mapping (Lischinski, Farbman *et al.* 2006)(Section 10.2.1, as well as other applications of the weighted least squares (WLS) formulation (Farbman, Fattal *et al.* 2008). These techniques have benefitted greatly from image-adapted regularization techniques, such as those developed in Szeliski (2006b), Krishnan and Szeliski (2011), Krishnan, Fattal, and Szeliski (2013), and Barron and Poole (2016), as shown in Figure 4.11. An alternative approach to performing the sparse chrominance interpolation based on geodesic (edge-aware) distance functions has been developed by Yatziv and Sapiro (2006). Neural networks can also be used to implement *deep priors* for image colorization (Zhang, Zhu *et al.* 2017).

4.3 Markov random fields

As we have just seen, regularization, which involves the minimization of energy functionals defined over (piecewise) continuous functions, can be used to formulate and solve a variety of low-level computer vision problems. An alternative technique is to formulate a *Bayesian* or *generative* model, which separately models the noisy image formation (*measurement*) process, as well as assuming a statistical *prior* model over the solution space (Bishop 2006, Section 1.5.4). In this section, we look at priors based on Markov random fields, whose log-likelihood can be described using local neighborhood interaction (or penalty) terms (Kindermann and Snell 1980; Geman and Geman 1984; Marroquin, Mitter, and Poggio 1987; Li 1995; Szeliski, Zabih *et al.* 2008; Blake, Kohli, and Rother 2011).

The use of Bayesian modeling has several potential advantages over regularization (see also Appendix B). The ability to model measurement processes statistically enables us to extract the maximum information possible from each measurement, rather than just guessing what weighting to give the data. Similarly, the parameters of the prior distribution can often be *learned* by observing samples from the class we are modeling (Roth and Black 2007a; Tappen 2007; Li and Huttenlocher 2008). Furthermore, because our model is probabilistic, it is possible to estimate (in principle) complete probability *distributions* over the unknowns being recovered and, in particular, to model the *uncertainty* in the solution, which can be useful in later processing stages. Finally, Markov random field models can be defined over *discrete* variables, such as image labels (where the variables have no proper ordering), for which regularization does not apply.

According to Bayes' rule (Appendix B.4), the *posterior* distribution $p(\mathbf{x}|\mathbf{y})$ over the unknowns $\mathbf{x}$ given the measurements $\mathbf{y}$ can be obtained by multiplying the measurement likelihood $p(\mathbf{y}|\mathbf{x})$ by the prior distribution $p(\mathbf{x})$ and normalizing,

$$p(\mathbf{x}|\mathbf{y}) = \frac{p(\mathbf{y}|\mathbf{x})p(\mathbf{x})}{p(\mathbf{y})}, \tag{4.33}$$

where $p(\mathbf{y}) = \int_{\mathbf{x}} p(\mathbf{y}|\mathbf{x})p(\mathbf{x})$ is a normalizing constant used to make the $p(\mathbf{x}|\mathbf{y})$ distribution *proper* (integrate to 1). Taking the negative logarithm of both sides of (4.33), we get

$$-\log p(\mathbf{x}|\mathbf{y}) = -\log p(\mathbf{y}|\mathbf{x}) - \log p(\mathbf{x}) + C, \tag{4.34}$$

which is the *negative posterior log likelihood*.

To find the most likely (*maximum a posteriori* or MAP) solution $\mathbf{x}$ given some measurements $\mathbf{y}$, we simply minimize this negative log likelihood, which can also be thought of as an *energy*,

$$E(\mathbf{x}, \mathbf{y}) = E_{\mathrm{D}}(\mathbf{x}, \mathbf{y}) + E_{\mathrm{P}}(\mathbf{x}). \tag{4.35}$$

(We drop the constant C because its value does not matter during energy minimization.) The first term $E_{\mathrm{D}}(\mathbf{x}, \mathbf{y})$ is the *data energy* or *data penalty*; it measures the negative log likelihood that the data were observed given the unknown state $\mathbf{x}$. The second term $E_{\mathrm{P}}(\mathbf{x})$ is the *prior energy*; it plays a role analogous to the smoothness energy in regularization. Note that the MAP estimate may not always be desirable, as it selects the "peak" in the posterior distribution rather than some more stable statistic—see the discussion in Appendix B.2 and by Levin, Weiss *et al.* (2009).

For the remainder of this section, we focus on Markov random fields, which are probabilistic models defined over two or three-dimensional pixel or voxel grids. Before we dive into this, however, we should mention that MRFs are just one special case of the more general family of *graphical models* (Bishop 2006, Chapter 8; Koller and Friedman 2009; Nowozin and Lampert 2011; Murphy 2012, Chapters 10, 17, 19), which have sparse interactions between variables that can be captured in a *factor graph* (Dellaert and Kaess 2017; Dellaert 2021), such as the one shown in Figure 4.12. Graphical models come in a wide variety of topologies, including chains (used for audio and speech processing), trees (often used for modeling kinematic chains in tracking people (e.g., Felzenszwalb and Huttenlocher 2005)), stars (simplified models for people; Dalal and Triggs 2005; Felzenszwalb, Girshick *et al.* 2010, and constellations (Fergus, Perona, and Zisserman 2007). Such models were widely used for part-based recognition, as discussed in Section 6.2.1. For graphs that are acyclic, efficient linear-time inference algorithms based on dynamic programming can be used.

For image processing applications, the unknowns $\mathbf{x}$ are the set of output pixels

$$\mathbf{x} = [f(0,0) \ldots f(m-1, n-1)], \tag{4.36}$$

and the data are (in the simplest case) the input pixels

$$\mathbf{y} = [d(0,0) \ldots d(m-1, n-1)] \tag{4.37}$$

as shown in Figure 4.12.

For a Markov random field, the probability $p(\mathbf{x})$ is a *Gibbs* or *Boltzmann distribution*, whose negative log likelihood (according to the Hammersley–Clifford theorem) can be written as a sum of pairwise interaction potentials,

$$E_{\mathrm{P}}(\mathbf{x}) = \sum_{\{(i,j),(k,l)\} \in \mathcal{N}(i,j)} V_{i,j,k,l}(f(i,j), f(k,l)), \tag{4.38}$$

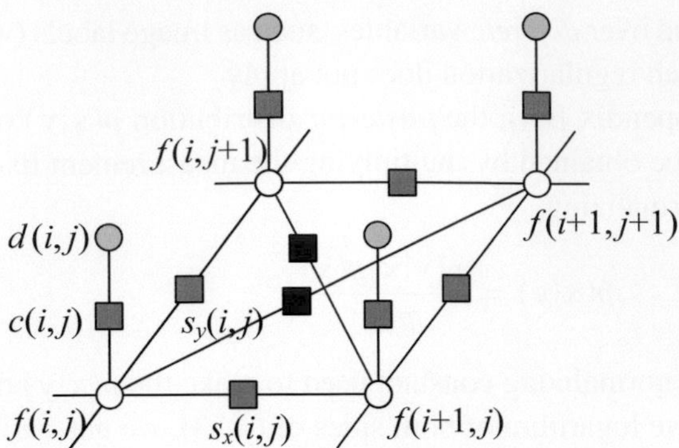

Figure 4.12 Graphical model for an $\mathcal{N}_4$ neighborhood Markov random field. (The blue edges are added for an $\mathcal{N}_8$ neighborhood.) The white circles are the unknowns $f(i,j)$, while the dark circles are the input data $d(i,j)$. The $s_x(i,j)$ and $s_y(i,j)$ black boxes denote arbitrary *interaction potentials* between adjacent nodes in the random field, and the $c(i,j)$ denote the *data penalty* functions. The same graphical model can be used to depict a discrete version of a first-order regularization problem (Figure 4.9).

where $\mathcal{N}(i,j)$ denotes the *neighbors* of pixel (i,j). In fact, the general version of the theorem says that the energy may have to be evaluated over a larger set of *cliques*, which depend on the *order* of the Markov random field (Kindermann and Snell 1980; Geman and Geman 1984; Bishop 2006; Kohli, Ladický, and Torr 2009; Kohli, Kumar, and Torr 2009).

The most commonly used neighborhood in Markov random field modeling is the $\mathcal{N}_4$ neighborhood, where each pixel in the field $f(i,j)$ interacts only with its immediate neighbors. The model in Figure 4.12, which we previously used in Figure 4.9 to illustrate the discrete version of first-order regularization, shows an $\mathcal{N}_4$ MRF. The $s_x(i,j)$ and $s_y(i,j)$ black boxes denote arbitrary *interaction potentials* between adjacent nodes in the random field and the $c(i,j)$ denote the data penalty functions. These square nodes can also be interpreted as *factors* in a *factor graph* version of the (undirected) graphical model (Bishop 2006; Dellaert and Kaess 2017; Dellaert 2021), which is another name for interaction potentials. (Strictly speaking, the factors are (improper) probability functions whose product is the (un-normalized) posterior distribution.)

As we will see in (4.41–4.42), there is a close relationship between these interaction potentials and the discretized versions of regularized image restoration problems. Thus, to a first approximation, we can view energy minimization being performed when solving a regularized problem and the maximum *a posteriori* inference being performed in an MRF as equivalent.

While $\mathcal{N}_4$ neighborhoods are most commonly used, in some applications $\mathcal{N}_8$ (or even higher order) neighborhoods perform better at tasks such as image segmentation because they can better model discontinuities at different orientations (Boykov and Kolmogorov 2003; Rother, Kohli *et al.* 2009; Kohli, Ladický, and Torr 2009; Kohli, Kumar, and Torr 2009).

Binary MRFs

The simplest possible example of a Markov random field is a binary field. Examples of such fields include 1-bit (black and white) scanned document images as well as images segmented into foreground and background regions.

To denoise a scanned image, we set the data penalty to reflect the agreement between the scanned and final images,

$$E_D(i,j) = w\delta(f(i,j), d(i,j)) \tag{4.39}$$

and the smoothness penalty to reflect the agreement between neighboring pixels

$$E_{\mathrm{P}}(i,j) = s\delta(f(i,j), f(i+1,j)) + s\delta(f(i,j), f(i,j+1)). \tag{4.40}$$

Once we have formulated the energy, how do we minimize it? The simplest approach is to perform gradient descent, flipping one state at a time if it produces a lower energy. This approach is known as *contextual classification* (Kittler and Föglein 1984), *iterated conditional modes* (ICM) (Besag 1986), or *highest confidence first* (HCF) (Chou and Brown 1990) if the pixel with the largest energy decrease is selected first.

Unfortunately, these downhill methods tend to get easily stuck in local minima. An alternative approach is to add some randomness to the process, which is known as *stochastic gradient descent* (Metropolis, Rosenbluth *et al.* 1953; Geman and Geman 1984). When the amount of noise is decreased over time, this technique is known as *simulated annealing* (Kirkpatrick, Gelatt, and Vecchi 1983; Carnevali, Coletti, and Patarnello 1985; Wolberg and Pavlidis 1985; Swendsen and Wang 1987) and was first popularized in computer vision by Geman and Geman (1984) and later applied to stereo matching by Barnard (1989), among others.

Even this technique, however, does not perform that well (Boykov, Veksler, and Zabih 2001). For binary images, a much better technique, introduced to the computer vision community by Boykov, Veksler, and Zabih (2001) is to re-formulate the energy minimization as a *max-flow/min-cut* graph optimization problem (Greig, Porteous, and Seheult 1989). This technique has informally come to be known as *graph cuts* in the computer vision community (Boykov and Kolmogorov 2011). For simple energy functions, e.g., those where the penalty for non-identical neighboring pixels is a constant, this algorithm is guaranteed to produce the *global minimum*. Kolmogorov and Zabih (2004) formally characterize the class of binary energy potentials (*regularity conditions*) for which these results hold, while newer work by Komodakis, Tziritas, and Paragios (2008) and Rother, Kolmogorov *et al.* (2007) provide good algorithms for the cases when they do not, i.e., for energy functions that are not *regular* or *sub-modular*.

In addition to the above mentioned techniques, a number of other optimization approaches have been developed for MRF energy minimization, such as (loopy) belief propagation and dynamic programming (for one-dimensional problems). These are discussed in more detail in Appendix B.5 as well as the comparative survey papers by Szeliski, Zabih *et al.* (2008) and Kappes, Andres *et al.* (2015), which have associated benchmarks and code at https://vision.middlebury.edu/MRF and http://hciweb2.iwr.uni-heidelberg.de/opengm.

Ordinal-valued MRFs

In addition to binary images, Markov random fields can be applied to ordinal-valued labels such as grayscale images or depth maps. The term "ordinal" indicates that the labels have an implied ordering, e.g., that higher values are lighter pixels. In the next section, we look at unordered labels, such as source image labels for image compositing.

In many cases, it is common to extend the binary data and smoothness prior terms as

$$E_{\mathrm{D}}(i,j) = c(i,j)\rho_d(f(i,j) - d(i,j)) \tag{4.41}$$

and

$$E_{\mathrm{P}}(i,j) = s_x(i,j)\rho_p(f(i,j) - f(i+1,j)) + s_y(i,j)\rho_p(f(i,j) - f(i,j+1)), \tag{4.42}$$

which are robust generalizations of the quadratic penalty terms (4.26) and (4.24), first introduced in (4.29). As before, the $c(i,j)$, $s_x(i,j)$, and $s_y(i,j)$ weights can be used to locally control the data

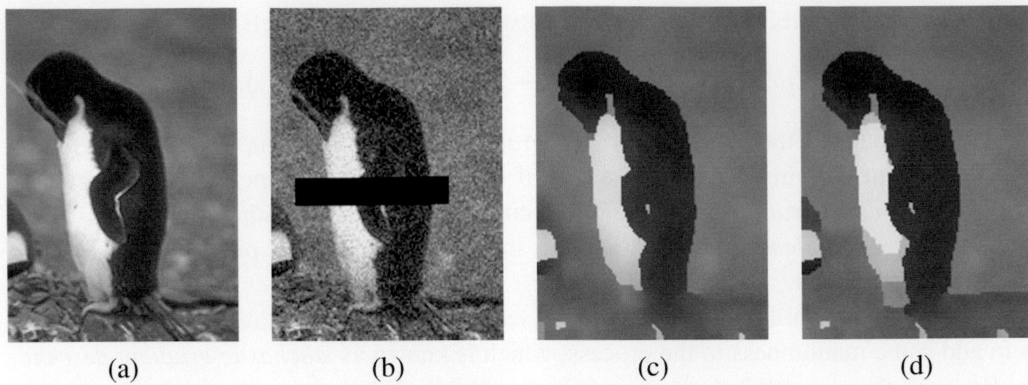

(a) (b) (c) (d)

Figure 4.13 Grayscale image denoising and inpainting: (a) original image; (b) image corrupted by noise and with missing data (black bar); (c) image restored using loopy belief propagation; (d) image restored using expansion move graph cuts. Images are from https://vision.middlebury.edu/MRF/results (Szeliski, Zabih *et al.* 2008).

weighting and the horizontal and vertical smoothness. Instead of using a quadratic penalty, however, a general monotonically increasing penalty function $\rho()$ is used. (Different functions can be used for the data and smoothness terms.) For example, ρ_p can be a hyper-Laplacian penalty

$$\rho_p(d) = |d|^p, \quad p < 1, \tag{4.43}$$

which better encodes the distribution of gradients (mainly edges) in an image than either a quadratic or linear (total variation) penalty.[6] Levin and Weiss (2007) use such a penalty to separate a transmitted and reflected image (Figure 9.16) by encouraging gradients to lie in one or the other image, but not both. Levin, Fergus *et al.* (2007) use the hyper-Laplacian as a prior for image deconvolution (deblurring) and Krishnan and Fergus (2009) develop a faster algorithm for solving such problems. For the data penalty, ρ_d can be quadratic (to model Gaussian noise) or the log of a *contaminated Gaussian* (Appendix B.3).

When ρ_p is a quadratic function, the resulting Markov random field is called a Gaussian Markov random field (GMRF) and its minimum can be found by sparse linear system solving (4.28). When the weighting functions are uniform, the GMRF becomes a special case of Wiener filtering (Section 3.4.1). Allowing the weighting functions to depend on the input image (a special kind of conditional random field, which we describe below) enables quite sophisticated image processing algorithms to be performed, including colorization (Levin, Lischinski, and Weiss 2004), interactive tone mapping (Lischinski, Farbman *et al.* 2006), natural image matting (Levin, Lischinski, and Weiss 2008), and image restoration (Tappen, Liu *et al.* 2007).

When ρ_d or ρ_p are non-quadratic functions, gradient descent techniques such as non-linear least squares or iteratively re-weighted least squares can sometimes be used (Appendix A.3). However, if the search space has lots of local minima, as is the case for stereo matching (Barnard 1989; Boykov, Veksler, and Zabih 2001), more sophisticated techniques are required.

The extension of graph cut techniques to multi-valued problems was first proposed by Boykov, Veksler, and Zabih (2001). In their paper, they develop two different algorithms, called the *swap move* and the *expansion move*, which iterate among a series of binary labeling sub-problems to

[6] Note that, unlike a quadratic penalty, the sum of the horizontal and vertical derivative p-norms is not rotationally invariant. A better approach may be to locally estimate the gradient direction and to impose different norms on the perpendicular and parallel components, which Roth and Black (2007b) call a *steerable random field*.

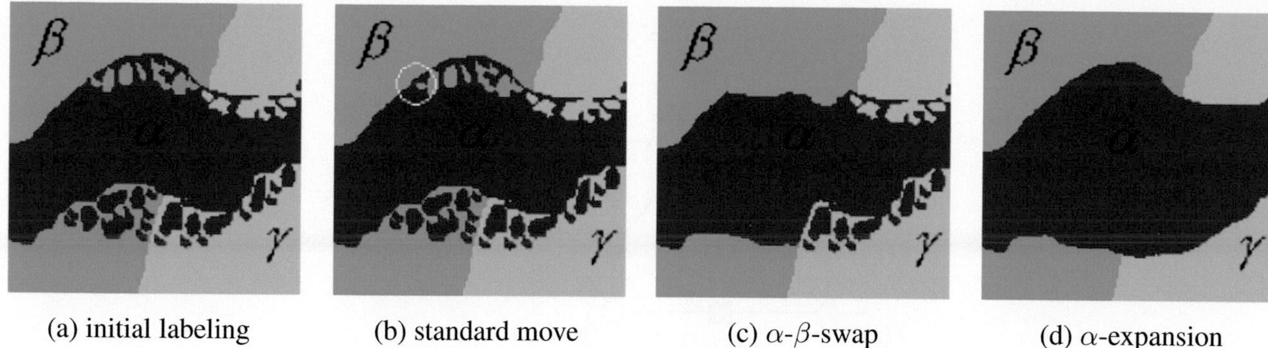

(a) initial labeling (b) standard move (c) α-β-swap (d) α-expansion

Figure 4.14 Multi-level graph optimization from Boykov, Veksler, and Zabih (2001) © 2001 IEEE: (a) initial problem configuration; (b) the standard move only changes one pixel; (c) the α-β-swap optimally exchanges all α and β-labeled pixels; (d) the α-expansion move optimally selects among current pixel values and the α label.

find a good solution (Figure 4.14). Note that a global solution is generally not achievable, as the problem is provably NP-hard for general energy functions. Because both these algorithms use a binary MRF optimization inside their inner loop, they are subject to the kind of constraints on the energy functions that occur in the binary labeling case (Kolmogorov and Zabih 2004).

Another MRF inference technique is *belief propagation* (BP). While belief propagation was originally developed for inference over trees, where it is exact (Pearl 1988), it has more recently been applied to graphs with loops such as Markov random fields (Freeman, Pasztor, and Carmichael 2000; Yedidia, Freeman, and Weiss 2001). In fact, some of the better performing stereo-matching algorithms use loopy belief propagation (LBP) to perform their inference (Sun, Zheng, and Shum 2003). LBP is discussed in more detail in comparative survey papera on MRF optimization (Szeliski, Zabih *et al.* 2008; Kappes, Andres *et al.* 2015).

Figure 4.13 shows an example of image denoising and inpainting (hole filling) using a non-quadratic energy function (non-Gaussian MRF). The original image has been corrupted by noise and a portion of the data has been removed (the black bar). In this case, the loopy belief propagation algorithm computes a slightly lower energy and also a smoother image than the alpha-expansion graph cut algorithm.

Of course, the above formula (4.42) for the smoothness term $E_{\mathrm{P}}(i, j)$ just shows the simplest case. In follow-on work, Roth and Black (2009) propose a *Field of Experts* (FoE) model, which sums up a large number of exponentiated local filter outputs to arrive at the smoothness penalty. Weiss and Freeman (2007) analyze this approach and compare it to the simpler hyper-Laplacian model of natural image statistics. Lyu and Simoncelli (2009) use *Gaussian Scale Mixtures* (GSMs) to construct an inhomogeneous multi-scale MRF, with one (positive exponential) GMRF modulating the variance (amplitude) of another Gaussian MRF.

It is also possible to extend the *measurement* model to make the sampled (noise-corrupted) input pixels correspond to blends of unknown (latent) image pixels, as in Figure 4.15. This is the commonly occurring case when trying to deblur an image. While this kind of a model is still a traditional generative Markov random field, i.e., we can in principle generate random samples from the prior distribution, finding an optimal solution can be difficult because the clique sizes get larger. In such situations, gradient descent techniques, such as iteratively reweighted least squares, can be used (Joshi, Zitnick *et al.* 2009). Exercise 4.4 has you explore some of these issues.

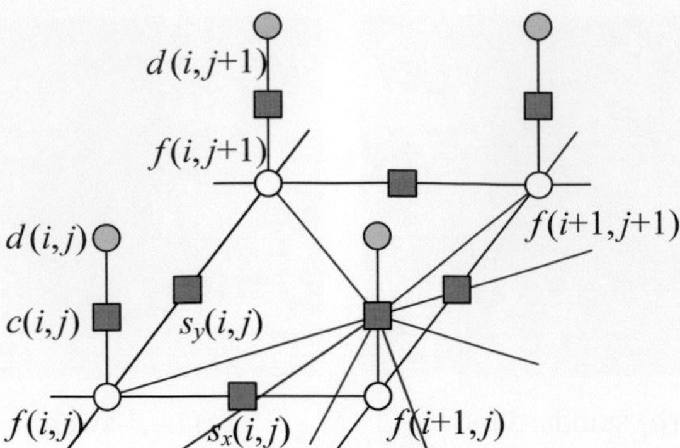

Figure 4.15 Graphical model for a Markov random field with a more complex measurement model. The additional colored edges show how combinations of unknown values (say, in a sharp image) produce the measured values (a noisy blurred image). The resulting graphical model is still a classic MRF and is just as easy to sample from, but some inference algorithms (e.g., those based on graph cuts) may not be applicable because of the increased network complexity, since state changes during the inference become more entangled and the posterior MRF has much larger cliques.

Unordered labels

Another case with multi-valued labels where Markov random fields are often applied is that of *unordered labels*, i.e., labels where there is no semantic meaning to the numerical difference between the values of two labels. For example, if we are classifying terrain from aerial imagery, it makes no sense to take the numerical difference between the labels assigned to forest, field, water, and pavement. In fact, the adjacencies of these various kinds of terrain each have different likelihoods, so it makes more sense to use a prior of the form

$$E_{\mathrm{P}}(i,j) = s_x(i,j)V(l(i,j),l(i+1,j)) + s_y(i,j)V(l(i,j),l(i,j+1)), \tag{4.44}$$

where $V(l_0, l_1)$ is a general *compatibility* or *potential* function. (Note that we have also replaced $f(i,j)$ with $l(i,j)$ to make it clearer that these are labels rather than function samples.) An alternative way to write this prior energy (Boykov, Veksler, and Zabih 2001; Szeliski, Zabih *et al.* 2008) is

$$E_{\mathrm{P}} = \sum_{(p,q)\in\mathcal{N}} V_{p,q}(l_p, l_q), \tag{4.45}$$

where the (p,q) are neighboring pixels and a spatially varying potential function $V_{p,q}$ is evaluated for each neighboring pair.

An important application of unordered MRF labeling is seam finding in image compositing (Davis 1998; Agarwala, Dontcheva *et al.* 2004) (see Figure 4.16, which is explained in more detail in Section 8.4.2). Here, the compatibility $V_{p,q}(l_p, l_q)$ measures the quality of the visual appearance that would result from placing a pixel p from image l_p next to a pixel q from image l_q. As with most MRFs, we assume that $V_{p,q}(l,l) = 0$. For different labels, however, the compatibility $V_{p,q}(l_p, l_q)$ may depend on the values of the underlying pixels $I_{l_p}(p)$ and $I_{l_q}(q)$.

Consider, for example, where one image I_0 is all sky blue, i.e., $I_0(p) = I_0(q) = B$, while the other image I_1 has a transition from sky blue, $I_1(p) = B$, to forest green, $I_1(q) = G$.

I_0 : | p | q | | p | q | : I_1

In this case, $V_{p,q}(1,0) = 0$ (the colors agree), while $V_{p,q}(0,1) > 0$ (the colors disagree).

Figure 4.16 An unordered label MRF (Agarwala, Dontcheva *et al.* 2004) © 2004 ACM: Strokes in each of the source images on the left are used as constraints on an MRF optimization, which is solved using graph cuts. The resulting multi-valued label field is shown as a color overlay in the middle image, and the final composite is shown on the right.

4.3.1 Conditional random fields

In a classic Bayesian model (4.33–4.35),

$$p(\mathbf{x}|\mathbf{y}) \propto p(\mathbf{y}|\mathbf{x})p(\mathbf{x}), \tag{4.46}$$

the prior distribution $p(\mathbf{x})$ is independent of the observations $\mathbf{y}$. Sometimes, however, it is useful to modify our prior assumptions, say about the smoothness of the field we are trying to estimate, in response to the sensed data. Whether this makes sense from a probability viewpoint is something we discuss once we have explained the new model.

Consider an interactive image segmentation system such as the one described in Boykov and Funka-Lea (2006). In this application, the user draws foreground and background strokes, and the system then solves a binary MRF labeling problem to estimate the extent of the foreground object. In addition to minimizing a data term, which measures the pointwise similarity between pixel colors and the inferred region distributions (Section 4.3.2), the MRF is modified so that the smoothness terms $s_x(x, y)$ and $s_y(x, y)$ in Figure 4.12 and (4.42) depend on the magnitude of the gradient between adjacent pixels.[7]

Since the smoothness term now depends on the data, Bayes' rule (4.46) no longer applies. Instead, we use a direct model for the posterior distribution $p(\mathbf{x}|\mathbf{y})$, whose negative log likelihood can be written as

$$\begin{aligned} E(\mathbf{x}|\mathbf{y}) &= E_{\mathrm{D}}(\mathbf{x}, \mathbf{y}) + E_{\mathrm{S}}(\mathbf{x}, \mathbf{y}) \\ &= \sum_p V_p(x_p, \mathbf{y}) + \sum_{(p,q)\in\mathcal{N}} V_{p,q}(x_p, x_q, \mathbf{y}), \end{aligned} \tag{4.47}$$

using the notation introduced in (4.45). The resulting probability distribution is called a *conditional random field* (CRF) and was first introduced to the computer vision field by Kumar and Hebert (2003), based on earlier work in text modeling by Lafferty, McCallum, and Pereira (2001).

Figure 4.17 shows a graphical model where the smoothness terms depend on the data values. In this particular model, each smoothness term depends only on its adjacent pair of data values, i.e., terms are of the form $V_{p,q}(x_p, x_q, y_p, y_q)$ in (4.47).

[7] An alternative formulation that also uses detected edges to modulate the smoothness of a depth or motion field and hence to integrate multiple lower level vision modules is presented by Poggio, Gamble, and Little (1988).

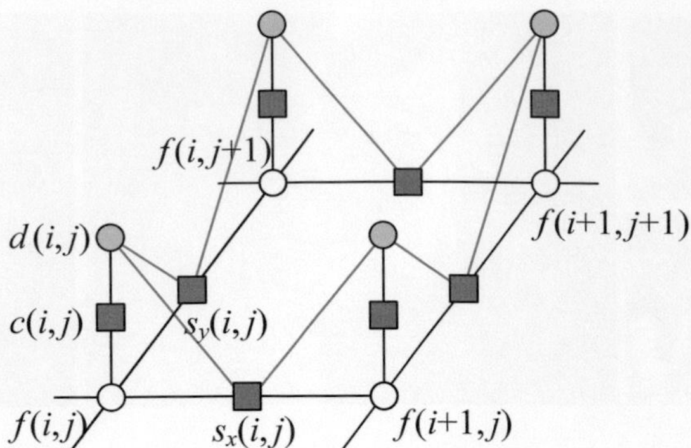

Figure 4.17 Graphical model for a conditional random field (CRF). The additional green edges show how combinations of sensed data influence the smoothness in the underlying MRF prior model, i.e., $s_x(i,j)$ and $s_y(i,j)$ in (4.42) depend on adjacent $d(i,j)$ values. These additional links (factors) enable the smoothness to depend on the input data. However, they make sampling from this MRF more complex.

The idea of modifying smoothness terms in response to input data is not new. For example, Boykov and Jolly (2001) used this idea for interactive segmentation, and it is now widely used in image segmentation (Section 4.3.2) (Blake, Rother *et al.* 2004; Rother, Kolmogorov, and Blake 2004), denoising (Tappen, Liu *et al.* 2007), and object recognition (Section 6.4) (Winn and Shotton 2006; Shotton, Winn *et al.* 2009).

In stereo matching, the idea of encouraging disparity discontinuities to coincide with intensity edges goes back even further to the early days of optimization and MRF-based algorithms (Poggio, Gamble, and Little 1988; Fua 1993; Bobick and Intille 1999; Boykov, Veksler, and Zabih 2001) and is discussed in more detail in (Section 12.5).

In addition to using smoothness terms that adapt to the input data, Kumar and Hebert (2003) also compute a neighborhood function over the input data for each $V_p(x_p, \mathbf{y})$ term, as illustrated in Figure 4.18, instead of using the classic unary MRF data term $V_p(x_p, y_p)$ shown in Figure 4.12.[8] Because such neighborhood functions can be thought of as *discriminant* functions (a term widely used in machine learning (Bishop 2006)), they call the resulting graphical model a *discriminative random field* (DRF). In their paper, Kumar and Hebert (2006) show that DRFs outperform similar CRFs on a number of applications, such as structure detection and binary image denoising.

Here again, one could argue that previous stereo correspondence algorithms also look at a neighborhood of input data, either explicitly, because they compute correlation measures (Criminisi, Cross *et al.* 2006) as data terms, or implicitly, because even pixel-wise disparity costs look at several pixels in either the left or right image (Barnard 1989; Boykov, Veksler, and Zabih 2001).

What then are the advantages and disadvantages of using conditional or discriminative random fields instead of MRFs?

Classic Bayesian inference (MRF) assumes that the prior distribution of the data is independent of the measurements. This makes a lot of sense: if you see a pair of sixes when you first throw a pair of dice, it would be unwise to assume that they will always show up thereafter. However, if after playing for a long time you detect a statistically significant bias, you may want to adjust your prior. What CRFs do, in essence, is to select or modify the prior model based on observed data. This can be viewed as making a partial inference over additional hidden variables or correlations between the unknowns (say, a label, depth, or clean image) and the knowns (observed images).

[8] Kumar and Hebert (2006) call the unary potentials $V_p(x_p, \mathbf{y})$ *association potentials* and the pairwise potentials $V_{p,q}(x_p, y_q, \mathbf{y})$ *interaction potentials*.

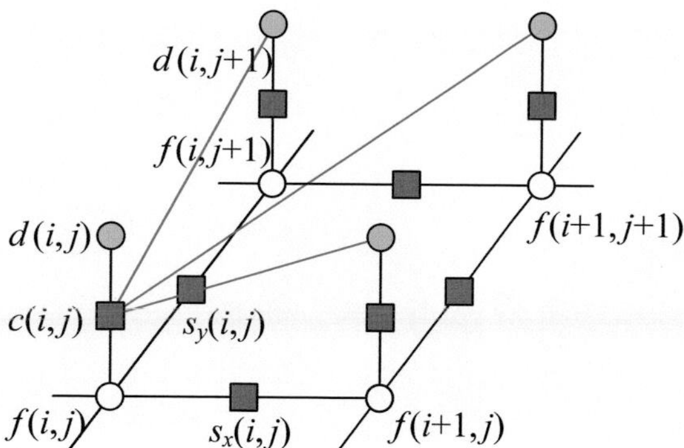

Figure 4.18 Graphical model for a discriminative random field (DRF). The additional green edges show how combinations of sensed data, e.g., $d(i, j+1)$, influence the data term for $f(i, j)$. The generative model is therefore more complex, i.e., we cannot just apply a simple function to the unknown variables and add noise.

In some cases, the CRF approach makes a lot of sense and is, in fact, the only plausible way to proceed. For example, in grayscale image colorization (Section 4.2.4) (Levin, Lischinski, and Weiss 2004), a commonly used way to transfer the continuity information from the input grayscale image to the unknown color image is to modify the local smoothness constraints. Similarly, for simultaneous segmentation and recognition (Winn and Shotton 2006; Shotton, Winn *et al.* 2009), it makes a lot of sense to permit strong color edges to increase the likelihood of semantic image label discontinuities.

In other cases, such as image denoising, the situation is more subtle. Using a non-quadratic (robust) smoothness term as in (4.42) plays a qualitatively similar role to setting the smoothness based on local gradient information in a Gaussian MRF (GMRF) (Tappen, Liu *et al.* 2007; Tanaka and Okutomi 2008). The advantage of Gaussian MRFs, when the smoothness can be correctly inferred, is that the resulting quadratic energy can be minimized in a single step, i.e., by solving a sparse set of linear equations. However, for situations where the discontinuities are not self-evident in the input data, such as for piecewise-smooth sparse data interpolation (Blake and Zisserman 1987; Terzopoulos 1988), classic robust smoothness energy minimization may be preferable. Thus, as with most computer vision algorithms, a careful analysis of the problem at hand and desired robustness and computation constraints may be required to choose the best technique.

Perhaps the biggest advantage of CRFs and DRFs, as argued by Kumar and Hebert (2006), Tappen, Liu *et al.* (2007), and Blake, Rother *et al.* (2004), is that learning the model parameters is more principled and sometimes easier. While learning parameters in MRFs and their variants is not a topic that we cover in this book, interested readers can find more details in publications by Kumar and Hebert (2006), Roth and Black (2007a), Tappen, Liu *et al.* (2007), Tappen (2007), and Li and Huttenlocher (2008).

Dense Conditional Random Fields (CRFs)

As with regular Markov random fields, conditional random fields (CRFs) are normally defined over small neighborhoods, e.g., the $\mathcal{N}_4$ neighborhood shown in Figure 4.17. However, images often contain longer-range interactions, e.g., pixels of similar colors may belong to related classes (Figure 4.19). In order to model such longer-range interactions, Krähenbühl and Koltun (2011) introduced what they call a *fully connected CRF*, which many people now call a *dense CRF*.

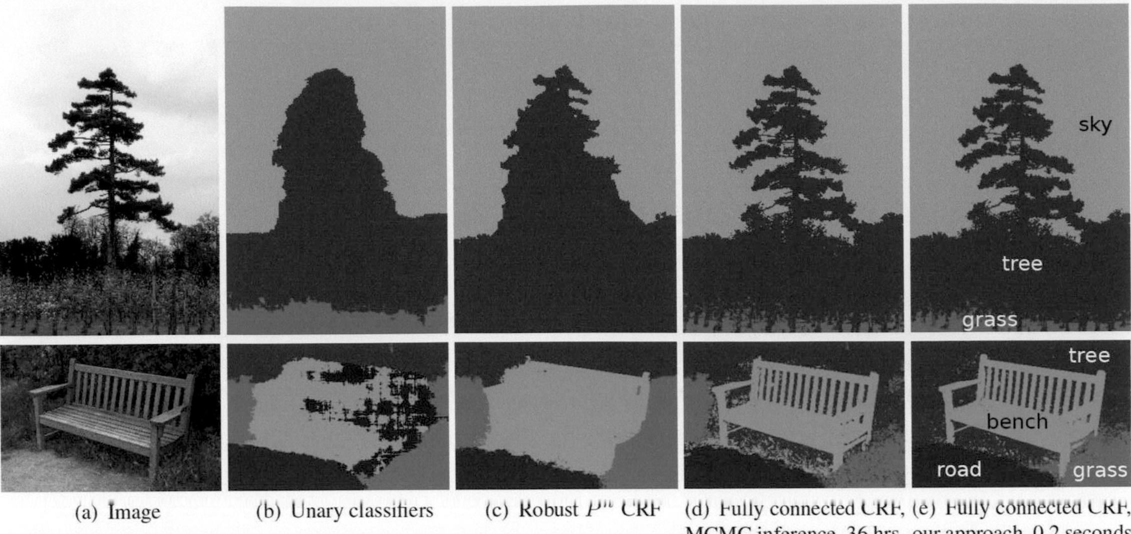

(a) Image (b) Unary classifiers (c) Robust P^n CRF (d) Fully connected CRF, MCMC inference, 36 hrs (e) Fully connected CRF, our approach, 0.2 seconds

Figure 4.19 Pixel-level classification with a fully connected CRF, from © Krähenbühl and Koltun (2011). The labels in each column describe the image or algorithm being run, which include a robust P^n CRF (Kohli, Ladický, and Torr 2009) and a very slow MCMC optimization algorithm.

As with traditional conditional random fields (4.47), their energy function consists of both unary terms and pairwise terms

$$E(\mathbf{x}|\mathbf{y}) = \sum_p V_p(x_p, \mathbf{y}) + \sum_{(p,q)} V_{p,q}(x_p, x_q, y_p, y_q), \tag{4.48}$$

where the (p, q) summation is now taken over *all* pairs of pixels, and not just adjacent ones.[9] The $\mathbf{y}$ denotes the input (guide) image over which the random field is conditioned. The pairwise interaction potentials have a restricted form

$$V_{p,q}(x_p, x_q, y_p, y_q) = \mu(x_p, x_q) \sum_{m=1}^{M} s_m w_m(p, q) \tag{4.49}$$

that is the product of a spatially invariant *label compatibility function* $\mu(x_p, x_q)$ and a sum of M Gaussian kernels of the same form (3.37) as is used in bilateral filtering and the bilateral solver. In their seminal paper, Krähenbühl and Koltun (2011) use two kernels, the first of which is an *appearance kernel* similar to (3.37) and the second is a spatial-only *smoothness kernel*.

Because of the special form of the long-range interaction potentials, which encapsulate all spatial and color similarity terms into a bilateral form, higher-dimensional filtering algorithms similar to those used in fast bilateral filters and solvers (Adams, Baek, and Davis 2010) can be used to efficiently compute a *mean field approximation* to the posterior conditional distribution (Krähenbühl and Koltun 2011). Figure 4.19 shows a comparison of their results (rightmost column) with previous approaches, including using simple unary terms, a robust CRF (Kohli, Ladický, and Torr 2009), and a very slow MCMC (Markov chain Monte Carlo) inference algorithm. As you can see, the fully connected CRF with a mean field solver produces dramatically better results in a very short time.

[9] In practice, as with bilateral filtering and the bilateral solver, the spatial extent may be over a large but finite region.

Since the publication of this paper, provably convergent and more efficient inference algorithms have been developed both by the original authors (Krähenbühl and Koltun 2013) and others (Vineet, Warrell, and Torr 2014; Desmaison, Bunel *et al.* 2016). Dense CRFs have seen widespread use in image segmentation problems and also as a "clean-up" stage for deep neural networks, as in the widely cited DeepLab paper by Chen, Papandreou *et al.* (2018).

4.3.2 *Application*: Interactive segmentation

The goal of image segmentation algorithms is to group pixels that have similar appearance (statistics) and to have the boundaries between pixels in different regions be of short length and across visible discontinuities. If we restrict the boundary measurements to be between immediate neighbors and compute region membership statistics by summing over pixels, we can formulate this as a classic pixel-based energy function using either a *variational formulation* (Section 4.2) or as a binary Markov random field (Section 4.3).

Examples of the continuous approach include Mumford and Shah (1989), Chan and Vese (2001), Zhu and Yuille (1996), and Tabb and Ahuja (1997) along with the level set approaches discussed in Section 7.3.2. An early example of a discrete labeling problem that combines both region-based and boundary-based energy terms is the work of Leclerc (1989), who used minimum description length (MDL) coding to derive the energy function being minimized. Boykov and Funka-Lea (2006) present a wonderful survey of various energy-based techniques for binary object segmentation, some of which we discuss below.

As we saw earlier in this chapter, the energy corresponding to a segmentation problem can be written (c.f. Equations (4.24) and (4.35–4.42)) as

$$E(f) = \sum_{i,j} E_{\mathrm{R}}(i,j) + E_{\mathrm{P}}(i,j), \tag{4.50}$$

where the region term

$$E_{\mathrm{R}}(i,j) = C(I(i,j); R(f(i,j))) \tag{4.51}$$

is the negative log likelihood that pixel intensity (or color) $I(i,j)$ is consistent with the statistics of region $R(f(i,j))$ and the boundary term

$$E_{\mathrm{P}}(i,j) = s_x(i,j)\delta(f(i,j), f(i+1,j)) + s_y(i,j)\delta(f(i,j), f(i,j+1)) \tag{4.52}$$

measures the inconsistency between $\mathcal{N}_4$ neighbors modulated by local horizontal and vertical smoothness terms $s_x(i,j)$ and $s_y(i,j)$.

Region statistics can be something as simple as the mean gray level or color (Leclerc 1989), in which case

$$C(I; \mu_k) = \|I - \mu_k\|^2. \tag{4.53}$$

Alternatively, they can be more complex, such as region intensity histograms (Boykov and Jolly 2001) or color Gaussian mixture models (Rother, Kolmogorov, and Blake 2004). For smoothness (boundary) terms, it is common to make the strength of the smoothness $s_x(i,j)$ inversely proportional to the local edge strength (Boykov, Veksler, and Zabih 2001).

Originally, energy-based segmentation problems were optimized using iterative gradient descent techniques, which were slow and prone to getting trapped in local minima. Boykov and Jolly (2001) were the first to apply the binary MRF optimization algorithm developed by Greig, Porteous, and Seheult (1989) to binary object segmentation.

In this approach, the user first delineates pixels in the background and foreground regions using a few strokes of an image brush. These pixels then become the *seeds* that tie nodes in the *S–T graph*

(a)

(b)

(c)

Figure 4.20 GrabCut image segmentation (Rother, Kolmogorov, and Blake 2004) © 2004 ACM: (a) the user draws a bounding box in red; (b) the algorithm guesses color distributions for the object and background and performs a binary segmentation; (c) the process is repeated with better region statistics.

to the source and sink labels S and T. Seed pixels can also be used to estimate foreground and background region statistics (intensity or color histograms).

The capacities of the other edges in the graph are derived from the region and boundary energy terms, i.e., pixels that are more compatible with the foreground or background region get stronger connections to the respective source or sink; adjacent pixels with greater smoothness also get stronger links. Once the minimum-cut/maximum-flow problem has been solved using a polynomial time algorithm (Goldberg and Tarjan 1988; Boykov and Kolmogorov 2004), pixels on either side of the computed cut are labeled according to the source or sink to which they remain connected. While graph cuts is just one of several known techniques for MRF energy minimization, it is still the one most commonly used for solving binary MRF problems.

The basic binary segmentation algorithm of Boykov and Jolly (2001) has been extended in a number of directions. The *GrabCut* system of Rother, Kolmogorov, and Blake (2004) iteratively re-estimates the region statistics, which are modeled as a mixtures of Gaussians in color space. This allows their system to operate given minimal user input, such as a single bounding box (Figure 4.20a)—the background color model is initialized from a strip of pixels around the box outline. (The foreground color model is initialized from the interior pixels, but quickly converges to a better estimate of the object.) The user can also place additional strokes to refine the segmentation as the solution progresses. Cui, Yang *et al.* (2008) use color and edge models derived from previous segmentations of similar objects to improve the local models used in GrabCut. Graph cut algorithms and other variants of Markov and conditional random fields have been applied to the *semantic segmentation* problem (Shotton, Winn *et al.* 2009; Krähenbühl and Koltun 2011), an example of which is shown in Figure 4.19 and which we study in more detail in Section 6.4.

Another major extension to the original binary segmentation formulation is the addition of *directed edges*, which allows boundary regions to be oriented, e.g., to prefer light to dark transitions or *vice versa* (Kolmogorov and Boykov 2005). Figure 4.21 shows an example where the directed graph cut correctly segments the light gray liver from its dark gray surround. The same approach can be used to measure the *flux* exiting a region, i.e., the signed gradient projected normal to the region boundary. Combining oriented graphs with larger neighborhoods enables approximating continuous problems such as those traditionally solved using level sets in the globally optimal graph cut framework (Boykov and Kolmogorov 2003; Kolmogorov and Boykov 2005).

More recent developments in graph cut-based segmentation techniques include the addition of connectivity priors to force the foreground to be in a single piece (Vicente, Kolmogorov, and Rother 2008) and shape priors to use knowledge about an object's shape during the segmentation process (Lempitsky and Boykov 2007; Lempitsky, Blake, and Rother 2008).

While optimizing the binary MRF energy (4.50) requires the use of combinatorial optimiza-

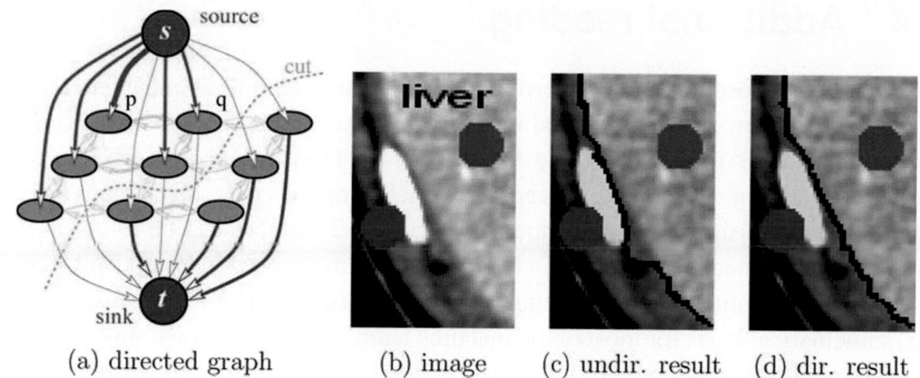

Figure 4.21 Segmentation with a directed graph cut (Boykov and Funka-Lea 2006) © 2006 Springer: (a) directed graph; (b) image with seed points; (c) the undirected graph incorrectly continues the boundary along the bright object; (d) the directed graph correctly segments the light gray region from its darker surround.

tion techniques, such as maximum flow, an approximate solution can be obtained by converting the binary energy terms into quadratic energy terms defined over a continuous $[0, 1]$ random field, which then becomes a classical membrane-based regularization problem (4.24–4.27). The resulting quadratic energy function can then be solved using standard linear system solvers (4.27–4.28), although if speed is an issue, you should use multigrid or one of its variants (Appendix A.5). Once the continuous solution has been computed, it can be thresholded at 0.5 to yield a binary segmentation.

The $[0, 1]$ continuous optimization problem can also be interpreted as computing the probability at each pixel that a *random walker* starting at that pixel ends up at one of the labeled seed pixels, which is also equivalent to computing the potential in a resistive grid where the resistors are equal to the edge weights (Grady 2006; Sinop and Grady 2007). K-way segmentations can also be computed by iterating through the seed labels, using a binary problem with one label set to 1 and all the others set to 0 to compute the relative membership probabilities for each pixel. In follow-on work, Grady and Ali (2008) use a precomputation of the eigenvectors of the linear system to make the solution with a novel set of seeds faster, which is related to the Laplacian matting problem presented in Section 10.4.3 (Levin, Acha, and Lischinski 2008). Couprie, Grady *et al.* (2009) relate the random walker to watersheds and other segmentation techniques. Singaraju, Grady, and Vidal (2008) add directed-edge constraints in order to support flux, which makes the energy piecewise quadratic and hence not solvable as a single linear system. The random walker algorithm can also be used to solve the Mumford–Shah segmentation problem (Grady and Alvino 2008) and to compute fast multigrid solutions (Grady 2008). A nice review of these techniques is given by Singaraju, Grady *et al.* (2011).

An even faster way to compute a continuous $[0, 1]$ approximate segmentation is to compute *weighted geodesic distances* between the 0 and 1 seed regions (Bai and Sapiro 2009), which can also be used to estimate soft alpha mattes (Section 10.4.3). A related approach by Criminisi, Sharp, and Blake (2008) can be used to find fast approximate solutions to general binary Markov random field optimization problems.

4.4 Additional reading

Scattered data interpolation and approximation techniques are fundamental to many different branches of applied mathematics. Some good introductory texts and articles include Amidror (2002), Wendland (2004), and Anjyo, Lewis, and Pighin (2014). These techniques are also related to geometric modeling techniques in computer graphics, which continues to be a very active research area. A nice introduction to basic spline techniques for curves and surfaces can be found in Farin (2002), while more recent approaches using subdivision surfaces are covered in Peters and Reif (2008).

Data interpolation and approximation also lie at the heart of *regression techniques*, which form the mathematical basis for most of the machine learning techniques we study in the next chapter. You can find good introductions to this topic (as well as underfitting, overfitting, and model selection) in texts on classic machine learning (Bishop 2006; Hastie, Tibshirani, and Friedman 2009; Murphy 2012; Deisenroth, Faisal, and Ong 2020) and deep learning (Goodfellow, Bengio, and Courville 2016; Glassner 2018; Zhang, Lipton *et al.* 2021).

Robust data fitting is also central to most computer vision problems. While introduced in this chapter, it is also revisited in Appendix B.3. Classic textbooks and articles on robust fitting and statistics include Huber (1981), Hampel, Ronchetti *et al.* (1986), Black and Rangarajan (1996), Rousseeuw and Leroy (1987), and Stewart (1999). The recent paper by Barron (2019) unifies many of the commonly used robust potential functions and shows how they can be used in machine learning applications.

The regularization approach to computer vision problems was first introduced to the vision community by Poggio, Torre, and Koch (1985) and Terzopoulos (1986a,b, 1988) and continues to be a popular framework for formulating and solving low-level vision problems (Ju, Black, and Jepson 1996; Nielsen, Florack, and Deriche 1997; Nordström 1990; Brox, Bruhn *et al.* 2004; Levin, Lischinski, and Weiss 2008). More detailed mathematical treatment and additional applications can be found in the applied mathematics and statistics literature (Tikhonov and Arsenin 1977; Engl, Hanke, and Neubauer 1996).

Variational formulations have been extensively used in low-level computer vision tasks, including optical flow (Horn and Schunck 1981; Nagel and Enkelmann 1986; Black and Anandan 1993; Alvarez, Weickert, and Sánchez 2000; Brox, Bruhn *et al.* 2004; Zach, Pock, and Bischof 2007a; Wedel, Cremers *et al.* 2009; Werlberger, Pock, and Bischof 2010), segmentation (Kass, Witkin, and Terzopoulos 1988; Mumford and Shah 1989; Caselles, Kimmel, and Sapiro 1997; Paragios and Deriche 2000; Chan and Vese 2001; Osher and Paragios 2003; Cremers 2007), denoising (Rudin, Osher, and Fatemi 1992), stereo (Pock, Schoenemann *et al.* 2008), multi-view stereo (Faugeras and Keriven 1998; Yezzi and Soatto 2003; Pons, Keriven, and Faugeras 2007; Labatut, Pons, and Keriven 2007; Kolev, Klodt *et al.* 2009), and scene flow (Wedel, Brox *et al.* 2011).

The literature on Markov random fields is truly immense, with publications in related fields such as optimization and control theory of which few vision practitioners are even aware. A good guide to the latest techniques is the book edited by Blake, Kohli, and Rother (2011). Other articles that contain nice literature reviews or experimental comparisons include Boykov and Funka-Lea (2006), Szeliski, Zabih *et al.* (2008), Kumar, Veksler, and Torr (2011), and Kappes, Andres *et al.* (2015). MRFs are just one version of the more general topic of graphical models, which is covered in several textbooks and survey, including Bishop (2006, Chapter 8), Koller and Friedman (2009), Nowozin and Lampert (2011), and Murphy (2012, Chapters 10, 17, 19)).

The seminal paper on Markov random fields is the work of Geman and Geman (1984), who introduced this formalism to computer vision researchers and also introduced the notion of *line processes*, additional binary variables that control whether smoothness penalties are enforced or not. Black and Rangarajan (1996) showed how independent line processes could be replaced with robust pairwise potentials; Boykov, Veksler, and Zabih (2001) developed iterative binary graph cut algorithms for optimizing multi-label MRFs; Kolmogorov and Zabih (2004) characterized the class of binary energy potentials required for these techniques to work; and Freeman, Pasztor, and Carmichael (2000) popularized the use of loopy belief propagation for MRF inference. Many more additional references can be found in Sections 4.3 and 4.3.2, and Appendix B.5.

Continuous-energy-based (variational) approaches to interactive segmentation include Leclerc (1989), Mumford and Shah (1989), Chan and Vese (2001), Zhu and Yuille (1996), and Tabb and Ahuja (1997). Discrete variants of such problems are usually optimized using binary graph cuts or other combinatorial energy minimization methods (Boykov and Jolly 2001; Boykov and Kolmogorov 2003; Rother, Kolmogorov, and Blake 2004; Kolmogorov and Boykov 2005; Cui, Yang *et al.* 2008; Vicente, Kolmogorov, and Rother 2008; Lempitsky and Boykov 2007; Lempitsky, Blake, and Rother 2008), although continuous optimization techniques followed by thresholding can also be used (Grady 2006; Grady and Ali 2008; Singaraju, Grady, and Vidal 2008; Criminisi, Sharp, and Blake 2008; Grady 2008; Bai and Sapiro 2009; Couprie, Grady *et al.* 2009). Boykov and Funka-Lea (2006) present a good survey of various energy-based techniques for binary object segmentation.

4.5 Exercises

Ex 4.1: Data fitting (scattered data interpolation). Generate some random samples from a smoothly varying function and then implement and evaluate one or more data interpolation techniques.

1. Generate a "random" 1-D or 2-D function by adding together a small number of sinusoids or Gaussians of random amplitudes and frequencies or scales.

2. Sample this function at a few dozen random locations.

3. Fit a function to these data points using one or more of the scattered data interpolation techniques described in Section 4.1.

4. Measure the fitting error between the estimated and original functions at some set of location, e.g., on a regular grid or at different random points.

5. Manually adjust any parameters your fitting algorithm may have to minimize the output sample fitting error, or use an automated technique such as cross-validation.

6. Repeat this exercise with a new set of random input sample and output sample locations. Does the optimal parameter change, and if so, by how much?

7. (Optional) Generate a piecewise-smooth test function by using different random parameters in different parts of of your image. How much more difficult does the data fitting problem become? Can you think of ways you might mitigate this?

Try to implement your algorithm in NumPy (or Matlab) using only array operations, in order to become more familiar with data-parallel programming and the linear algebra operators built into these systems. Use data visualization techniques such as those in Figures 4.3–4.6 to debug your algorithms and illustrate your results.

Ex 4.2: Graphical model optimization. Download and test out the software on the OpenGM2 library and benchmarks web site http://hciweb2.iwr.uni-heidelberg.de/opengm (Kappes, Andres *et al.* 2015). Try applying these algorithms to your own problems of interest (segmentation, de-noising, etc.). Which algorithms are more suitable for which problems? How does the quality compare to deep learning based approaches, which we study in the next chapter?

Ex 4.3: Image deblocking—challenging. Now that you have some good techniques to distinguish signal from noise, develop a technique to remove the *blocking artifacts* that occur with JPEG at high compression settings (Section 2.3.3). Your technique can be as simple as looking for unexpected edges along block boundaries, or looking at the quantization step as a projection of a convex region of the transform coefficient space onto the corresponding quantized values.

1. Does the knowledge of the compression factor, which is available in the JPEG header information, help you perform better deblocking? See Ehrlich, Lim *et al.* (2020) for a recent paper on this topic.

2. Because the quantization occurs in the DCT transformed YCbCr space (2.116), it may be preferable to perform the analysis in this space. On the other hand, image priors make more sense in an RGB space (or do they?). Decide how you will approach this dichotomy and discuss your choice.

3. While you are at it, since the YCbCr conversion is followed by a chrominance subsampling stage (before the DCT), see if you can restore some of the lost high-frequency chrominance signal using one of the better restoration techniques discussed in this chapter.

4. If your camera has a RAW + JPEG mode, how close can you come to the noise-free true pixel values? (This suggestion may not be that useful, since cameras generally use reasonably high quality settings for their RAW + JPEG models.)

Ex 4.4: Inference in deblurring—challenging. Write down the graphical model corresponding to Figure 4.15 for a non-blind image deblurring problem, i.e., one where the blur kernel is known ahead of time.

What kind of efficient inference (optimization) algorithms can you think of for solving such problems?

Chapter 5

Deep Learning

5.1 Supervised learning . . . 191
  5.1.1 Nearest neighbors . . . 192
  5.1.2 Bayesian classification . . . 194
  5.1.3 Logistic regression . . . 198
  5.1.4 Support vector machines . . . 199
  5.1.5 Decision trees and forests . . . 202
5.2 Unsupervised learning . . . 205
  5.2.1 Clustering . . . 205
  5.2.2 K-means and Gaussians mixture models . . . 206
  5.2.3 Principal component analysis . . . 209
  5.2.4 Manifold learning . . . 211
  5.2.5 Semi-supervised learning . . . 212
5.3 Deep neural networks . . . 214
  5.3.1 Weights and layers . . . 215
  5.3.2 Activation functions . . . 217
  5.3.3 Regularization and normalization . . . 219
  5.3.4 Loss functions . . . 223
  5.3.5 Backpropagation . . . 226
  5.3.6 Training and optimization . . . 228
5.4 Convolutional neural networks . . . 231
  5.4.1 Pooling and unpooling . . . 234
  5.4.2 *Application*: Digit classification . . . 237
  5.4.3 Network architectures . . . 238
  5.4.4 Model zoos . . . 242
  5.4.5 Visualizing weights and activations . . . 244
  5.4.6 Adversarial examples . . . 248
  5.4.7 Self-supervised learning . . . 249
5.5 More complex models . . . 252
  5.5.1 Three-dimensional CNNs . . . 252
  5.5.2 Recurrent neural networks . . . 255
  5.5.3 Transformers . . . 257
  5.5.4 Generative models . . . 261
5.6 Additional reading . . . 267
5.7 Exercises . . . 268

R. Szeliski, *Computer Vision*, Texts in Computer Science,
https://doi.org/10.1007/978-3-030-34372-9_5

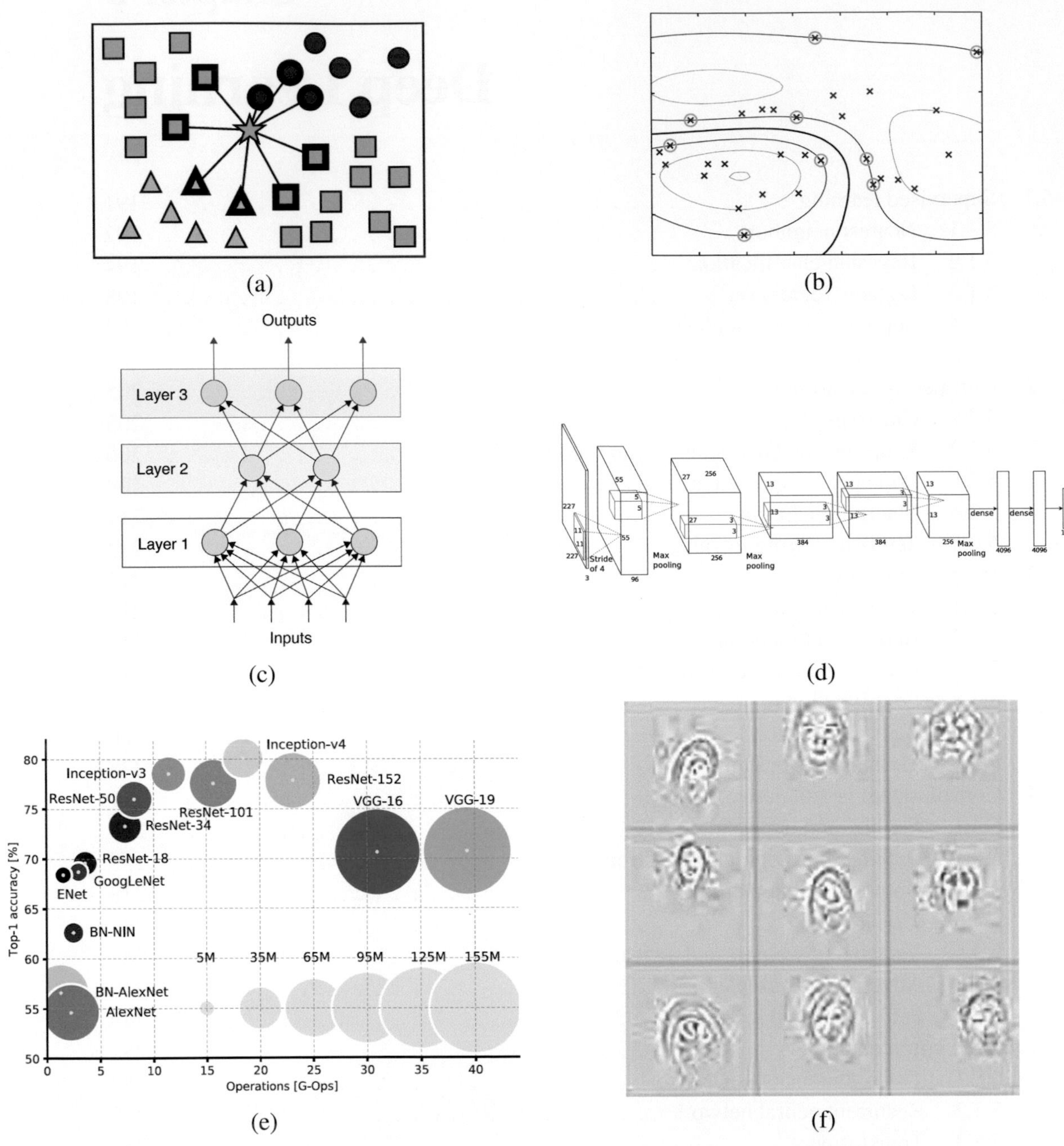

Figure 5.1 Machine learning and deep neural networks: (a) nearest neighbor classification © Glassner (2018); (b) Gaussian kernel support vector machine (Bishop 2006) © 2006 Springer; (c) a simple three-layer network © Glassner (2018); (d) the SuperVision deep neural network, courtesy of Matt Deitke after (Krizhevsky, Sutskever, and Hinton 2012); (e) network accuracy vs. size and operation counts (Canziani, Culurciello, and Paszke 2017) © 2017 IEEE; (f) visualizing network features (Zeiler and Fergus 2014) © 2014 Springer.

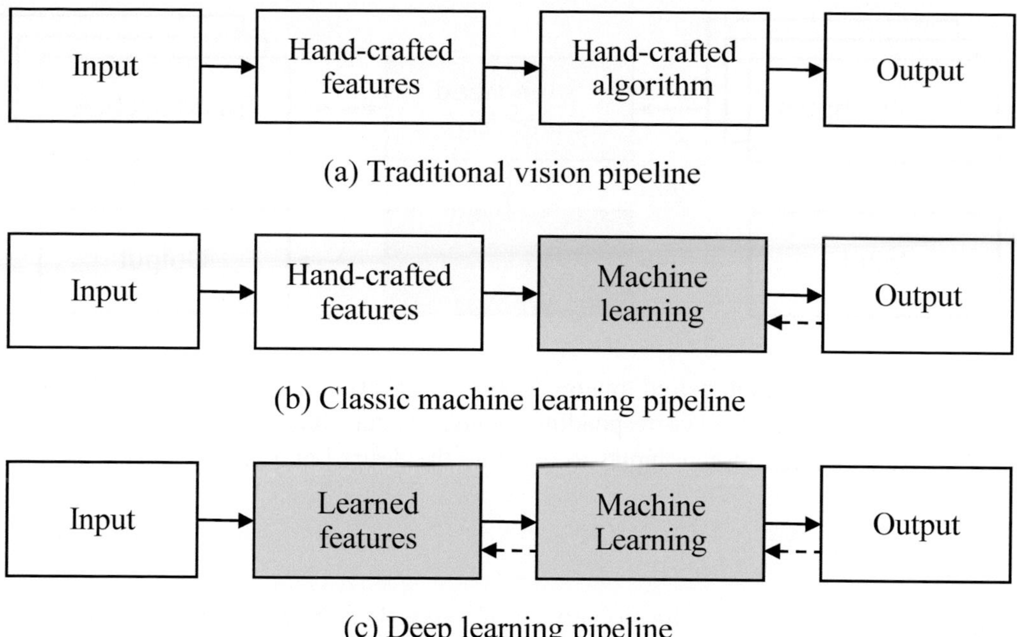

Figure 5.2 Traditional, machine learning, and deep learning pipelines, inspired by Goodfellow, Bengio, and Courville (2016, Figure 1.5). In a classic vision pipeline such as structure from motion, both the features and the algorithm were traditionally designed by hand (although learning techniques could be used, e.g., to design more repeatable features). Classic machine learning approaches take extracted features and use machine learning to build a classifier. Deep learning pipelines learn the whole pipeline, starting from pixels all the way to outputs, using end-to-end training (indicated by the backward dashed arrows) to fine-tune the model parameters.

Machine learning techniques have always played an important and often central role in the development of computer vision algorithms. Computer vision in the 1970s grew out of the fields of artificial intelligence, digital image processing, and pattern recognition (now called machine learning), and one of the premier journals in our field (*IEEE Transactions on Pattern Analysis and Machine Intelligence*) still bears testament to this heritage.

The image processing, scattered data interpolation, variational energy minimization, and graphical model techniques introduced in the previous two chapters have been essential tools in computer vision over the last five decades. While elements of machine learning and pattern recognition have also been widely used, e.g., for fine-tuning algorithm parameters, they really came into their own with the availability of large-scale labeled image datasets, such as ImageNet (Deng, Dong *et al.* 2009; Russakovsky, Deng *et al.* 2015), COCO (Lin, Maire *et al.* 2014), and LVIS (Gupta, Dollár, and Girshick 2019). Currently, deep neural networks are the most popular and widely used machine learning models in computer vision, not just for semantic classification and segmentation, but even for lower-level tasks such as image enhancement, motion estimation, and depth recovery (Bengio, LeCun, and Hinton 2021).

Figure 5.2 shows the main distinctions between traditional computer vision techniques, in which all of the processing stages were designed by hand, machine learning algorithms, in which hand crafted features were passed on to a machine learning stage, and deep networks, in which all of the algorithm components, including mid-level representations, are learned directly from the training data.

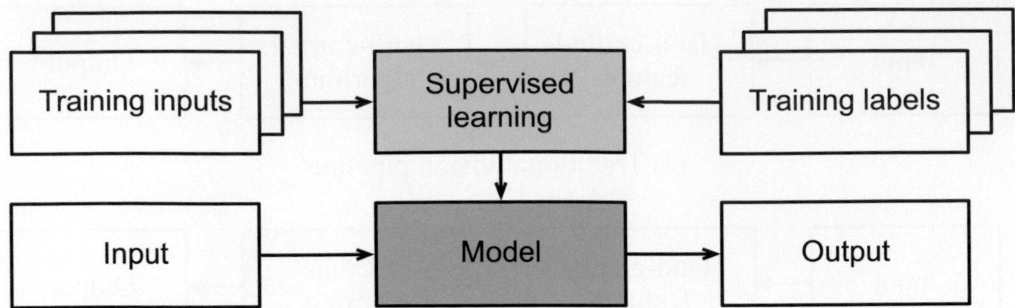

Figure 5.3 In supervised learning, paired training inputs and labels are used to estimate the model parameters that best predict the labels from their corresponding inputs. At run time, the model parameters are (usually) frozen, and the model is applied to new inputs to generate the desired outputs. © Zhang, Lipton *et al.* (2021, Figure 1.3)

We begin this chapter with an overview of classical machine learning approaches, such as nearest neighbors, logistic regression, support vector machines, and decision forests. This is a broad and deep subject, and we only provide a brief summary of the main popular approaches. More details on these techniques can be found in textbooks on this subject, which include Bishop (2006), Hastie, Tibshirani, and Friedman (2009), Murphy (2012), Criminisi and Shotton (2013), and Deisenroth, Faisal, and Ong (2020).

The machine learning part of the chapter focuses mostly on *supervised learning* for *classification* tasks, in which we are given a collection of inputs $\{\mathbf{x}_i\}$, which may be features derived from input images, paired with their corresponding class labels (or targets) $\{t_i\}$, which come from a set of classes $\{\mathcal{C}_k\}$. Most of the techniques described for supervised classification can easily be extended to *regression*, i.e., associating inputs $\{\mathbf{x}_i\}$ with real-valued scalar or vector outputs $\{\mathbf{y}_i\}$, which we have already studied in Section 4.1. We also look at some examples of *unsupervised learning* (Section 5.2), where there are no labels or outputs, as well as *semi-supervised learning*, in which labels or targets are only provided for a subset of the samples.

The second half of this chapter focuses on *deep neural networks*, which, over the last decade, have become the method of choice for most computer vision recognition and lower-level vision tasks. We begin with the elements that make up deep neural networks, including weights and activations, regularization terms, and training using backpropagation and stochastic gradient descents. Next, we introduce convolutional layers, review some of the classic architectures, and talk about how to pre-train networks and visualize their performance. Finally, we briefly touch on more advanced networks, such as three-dimensional and spatio-temporal models, as well as recurrent and generative adversarial networks.

Because machine learning and deep learning are such rich and deep topics, this chapter just briefly summarizes some of the main concepts and techniques. Comprehensive texts on classic machine learning include Bishop (2006), Hastie, Tibshirani, and Friedman (2009), Murphy (2012), and Deisenroth, Faisal, and Ong (2020) while textbooks focusing on deep learning include Goodfellow, Bengio, and Courville (2016), Glassner (2018), Glassner (2021), and Zhang, Lipton *et al.* (2021).

5.1 Supervised learning

Machine learning algorithms are usually categorized as either *supervised*, where paired inputs and outputs are given to the learning algorithm (Figure 5.3), or *unsupervised*, where statistical samples are provided without any corresponding labeled outputs (Section 5.2).

As shown in Figure 5.3, supervised learning involves feeding pairs of inputs $\{\mathbf{x}_i\}$ and their corresponding *target* output values $\{t_i\}$ into a learning algorithm, which adjusts the model's parameters so as to maximize the agreement between the model's predictions and the target outputs. The outputs can either be discrete labels that come from a set of classes $\{\mathcal{C}_k\}$, or they can be a set of continuous, potentially vector-valued *values*, which we denote by $\mathbf{y}_i$ to make the distinction between the two cases clearer. The first task is called *classification*, since we are trying to predict class membership, while the second is called *regression*, since historically, fitting a trend to data was called by that name (Section 4.1).[1]

After a *training phase* during which all of the *training data* (labeled input-output pairs) have been processed (often by iterating over them many times), the trained model can now be used to predict new output values for previously unseen inputs. This phase is often called the *test phase*, although this sometimes fools people into focusing excessively on performance on a given test set, rather than building a system that works robustly for any plausible inputs that might arise.

In this section, we focus more on classification, since we've already covered some of the simpler (linear and kernel) methods for regression in the previous chapter. One of the most common applications of classification in computer vision is semantic *image classification*, where we wish to label a complete image (or predetermined portion) with its most likely semantic category, e.g., horse, cat, or car (Section 6.2). This is the main application for which deep networks (Sections 5.3–5.4) were originally developed. More recently, however, such networks have also been applied to continuous pixel labeling tasks such as semantic segmentation, image denoising, and depth and motion estimation. More sophisticated tasks, such as object detection and instance segmentation, will be covered in Chapter 6.

Before we begin our review of traditional supervised learning techniques, we should define a little more formally what the system is trying to learn, i.e., what we meant by "maximize the agreement between the model's predictions and the target outputs." Ultimately, like any other computer algorithm that will occasionally make mistakes under uncertain, noisy, and/or incomplete data, we would like to maximize its expected utility, or conversely, minimize its expected *loss* or *risk*. This is the subject of *decision theory*, which is explained in more detail in textbooks on machine learning (Bishop 2006, Section 1.5; Hastie, Tibshirani, and Friedman 2009, Section 2.4; Murphy 2012, Section 6.5; Deisenroth, Faisal, and Ong 2020, Section 8.2).

We usually do not have access to the true probability distribution over the inputs, let alone the joint distribution over inputs and corresponding outputs. For this reason, we often use the training data distribution as a proxy for the real-world distribution. This approximation is known as *empirical risk minimization* (see above citations on decision theory), where the expected risk can be estimated with

$$E_{\text{Risk}}(\mathbf{w}) = \frac{1}{N}\sum L(\mathbf{y}_i, \mathbf{f}(\mathbf{x}_i;\mathbf{w})). \tag{5.1}$$

The loss function L measures the "cost" of predicting an output $\mathbf{f}(\mathbf{x}_i;\mathbf{w})$ for input $\mathbf{x}_i$ and model parameters $\mathbf{w}$ when the corresponding target is $\mathbf{y}_i$.[2]

[1] Note that in software engineering, a *regression* sometimes means a change in the code that results in degraded performance. That is not the kind of regression we will be studying here.

[2] In the machine learning literature, it is more common to write the loss using the letter L. But since we have used the letter E for energy (or summed error) in the previous chapter, we will stick to that notation throughout the book.

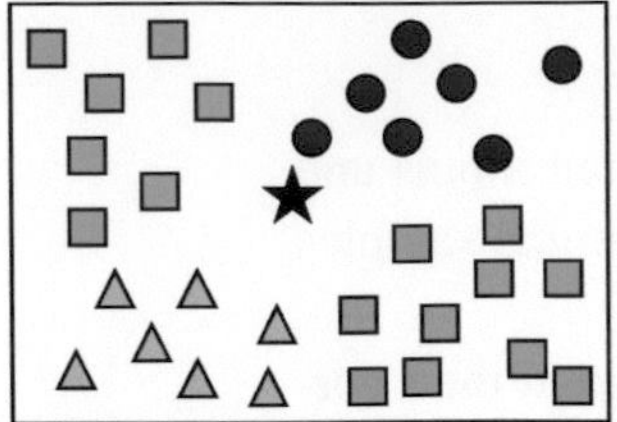
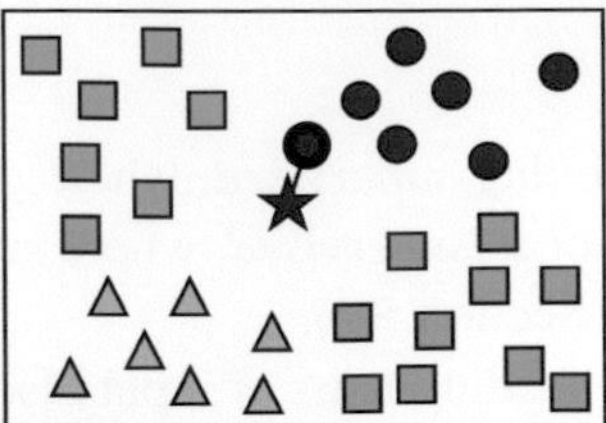
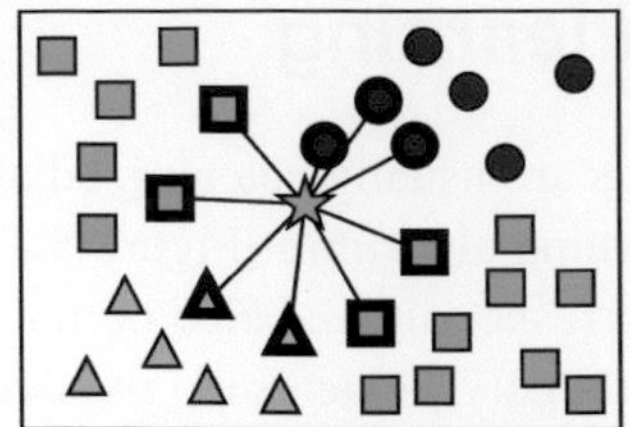
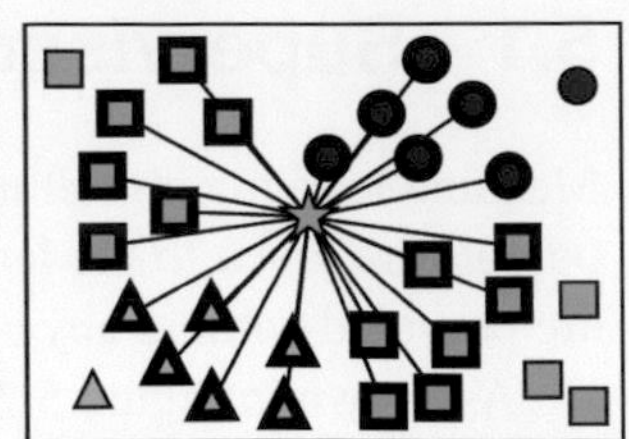

Figure 5.4 Nearest neighbor classification. To determine the class of the star (★) test sample, we find the k nearest neighbors and select the most popular class. This figure shows the results for $k = 1, 9$, and 25 samples. © Glassner (2018)

This formula should by now be quite familiar, since it is the same one we introduced in the previous chapter (4.2; 4.15) for *regression*. In those cases, the cost (penalty) is a simple quadratic or robust function of the difference between the target output $\mathbf{y}_i$ and the output predicted by the model $f(\mathbf{x}_i; \mathbf{w})$. In some situations, we may want the loss to model specific asymmetries in misprediction. For example, in autonomous navigation, it is usually more costly to over-estimate the distance to the nearest obstacle, potentially resulting in a collision, than to more conservatively under-estimate. We will see more examples of loss functions later on in this chapter, including Section 5.1.3 on Bayesian classification (5.19–5.24) and Section 5.3.4 on neural network loss (5.54–5.56).

In classification tasks, it is common to minimize the *misclassification rate*, i.e., penalizing all class prediction errors equally using a class-agnostic delta function (Bishop 2006, Sections 1.5.1–1.5.2). However, asymmetries often exist. For example, the cost of producing a *false negative* diagnosis in medicine, which may result in an untreated illness, is often greater than that of a *false positive*, which may suggest further tests. We will discuss true and false positives and negatives, along with error rates, in more detail in Section 7.1.3.

Data preprocessing

Before we start our review of widely used machine learning techniques, we should mention that it is usually a good idea to *center*, *standardize*, and if possible, *whiten* the input data (Glassner 2018, Section 10.5; Bishop 2006, Section 12.1.3). *Centering* the feature vectors means subtracting their mean value, while *standardizing* means also re-scaling each component so that its variance (average squared distance from the mean) is 1.

Whitening is a more computationally expensive process, which involves computing the covariance matrix of the inputs, taking its SVD, and then rotating the coordinate system so that the final dimensions are uncorrelated and have unit variance (under a Gaussian model). While this may be quite practical and helpful for low-dimension inputs, it can become prohibitively expensive for large sets of images. (But see the discussion in Section 5.2.3 on principal component analysis, where it can be feasible and useful.)

With this background in place, we now turn our attention to some widely used supervised learning techniques, namely nearest neighbors, Bayesian classification, logistic regression, support vector machines, and decision trees and forests.

5.1.1 Nearest neighbors

Nearest neighbors is a very simple *non-parametric* technique, i.e., one that does not involve a low-parameter analytic form for the underlying distribution. Instead, the training examples are all re-

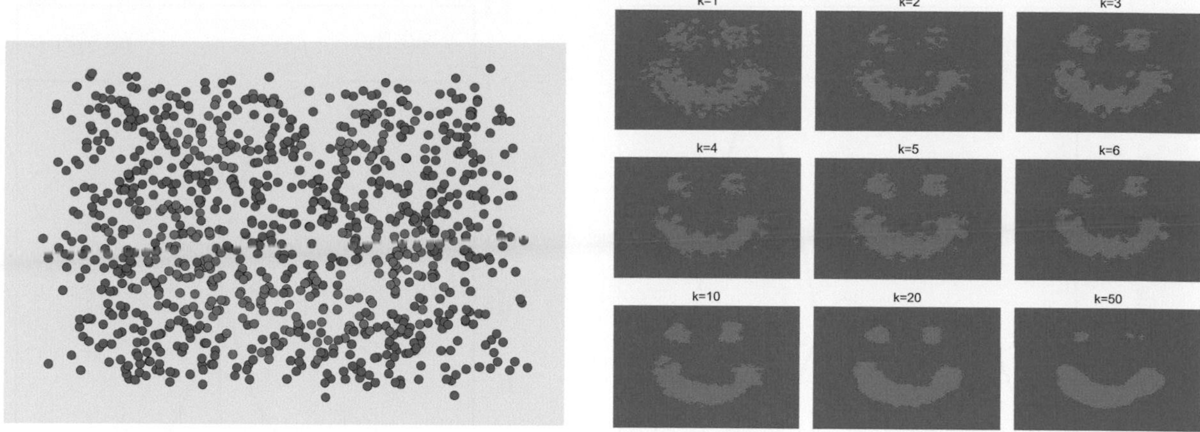

Figure 5.5 For noisy (intermingled) data, selecting too small a value of k results in irregular decision surfaces. Selecting too large a value can cause small regions to shrink or disappear. © Glassner (2018)

tained, and at evaluation time the "nearest" k neighbors are found and then averaged to produce the output.[3]

Figure 5.4 shows a simple graphical example for various values of k, i.e., from using the $k = 1$ nearest neighbor all the way to finding the $k = 25$ nearest neighbors and selecting the class with the highest count as the output label. As you can see, changing the number of neighbors affects the final class label, which changes from red to blue.

Figure 5.5 shows the effect of varying the number of neighbors in another way. The left half of the figure shows the initial samples, which fall into either blue or orange categories. As you can see, the training samples are highly *intermingled*, i.e., there is no clear (plausible) boundary that will correctly label all of the samples. The right side of this figure shows the *decision boundaries* for a k-NN classifier as we vary the values of k from 1 to 50. When k is too small, the classifier acts in a very random way, i.e., it is *overfitting* to the training data (Section 4.1.2). As k gets larger, the classifier *underfits* (over-smooths) the data, resulting in the shrinkage of the two smaller regions. The optimal number of nearest neighbors to use k is a *hyperparameter* for this algorithm. Techniques for determining a good value include cross-validation, which we discussed in Section 4.1.2.

While nearest neighbors is a rather brute-force machine learning technique (although Cover and Hart (1967) showed that it is statistically optimal in the large sample limit), but it can still be useful in many computer vision applications, such as large-scale matching and indexing (Section 7.1.4). As the number of samples gets large, however, efficient techniques must be used to find the (exact or approximate) nearest neighbors. Good algorithms for finding nearest neighbors have been developed in both the general computer science and more specialized computer vision communities.

Muja and Lowe (2014) developed a Fast Library for Approximate Nearest Neighbors (FLANN), which collects a number of previously developed algorithms and is incorporated as part of OpenCV. The library implements several powerful approximate nearest neighbor algorithms, including randomized k-d trees (Silpa-Anan and Hartley 2008), priority search k-means trees, approximate nearest neighbors (Friedman, Bentley, and Finkel 1977), and locality sensitive hashing (LSH) (Andoni and Indyk 2006). Their library can empirically determine which algorithm and parameters to use based on the characteristics of the data being indexed.

[3] The reason I put "nearest" in quotations is that standardizing and/or whitening the data will affect distances between vectors, and is usually helpful.

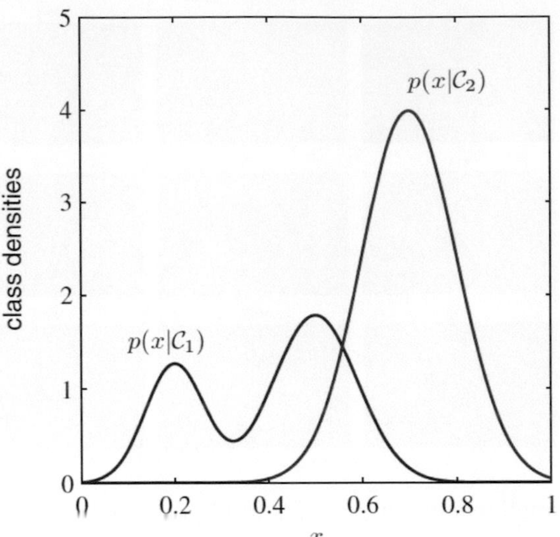

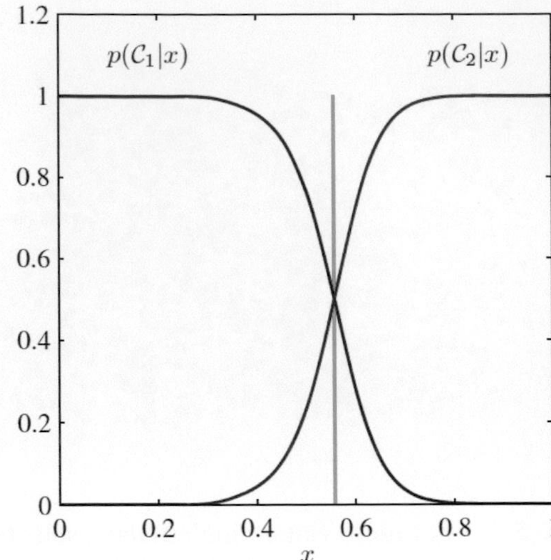

Figure 5.6 An example with two class conditional densities $p(x|\mathcal{C}_k)$ along with the corresponding posterior class probabilities $p(\mathcal{C}_k|x)$, which can be obtained using Bayes' rule, i.e., by dividing by the sum of the two curves (Bishop 2006) © 2006 Springer. The vertical green line is the optimal decision boundary for minimizing the misclassification rate.

More recently, Johnson, Douze, and Jégou (2021) developed the GPU-enabled Faiss library[4] for scaling similarity search (Section 6.2.3) to billions of vectors. The library is based on product quantization (Jégou, Douze, and Schmid 2010), which had been shown by the authors to perform better than LSH (Gordo, Perronnin *et al.* 2013) on the kinds of large-scale datasets the Faiss library was developed for.

5.1.2 Bayesian classification

For some simple machine learning problems, e.g., if we have an analytic model of feature construction and noising, or if we can gather enough samples, we can determine the probability distributions of the feature vectors for each class $p(\mathbf{x}|\mathcal{C}_k)$ as well as the prior class likelihoods $p(\mathcal{C}_k)$.[5] According to Bayes' rule (4.33), the likelihood of class $\mathcal{C}_k$ given a feature vector $\mathbf{x}$ (Figure 5.6) is given by

$$p_k = p(\mathcal{C}_k|\mathbf{x}) = \frac{p(\mathbf{x}|\mathcal{C}_k)p(\mathcal{C}_k)}{\sum_j p(\mathbf{x}|\mathcal{C}_j)p(\mathcal{C}_j)} \tag{5.2}$$

$$= \frac{\exp l_k}{\sum_j \exp l_j}, \tag{5.3}$$

where the second form (using the exp functions) is known as the *normalized exponential* or *softmax* function.[6] The quantity

$$l_k = \log p(\mathbf{x}|\mathcal{C}_k) + \log p(\mathcal{C}_k) \tag{5.4}$$

[4]https://github.com/facebookresearch/faiss

[5]The following notation and equations are adapted from Bishop (2006, Section 4.2), which describes *probabilistic generative classification.*

[6]For better numerical stability, it is common to subtract the largest value of l_j from all of the input values so that the exponentials are in the range $(0, 1]$ and there is less chance of roundoff error.

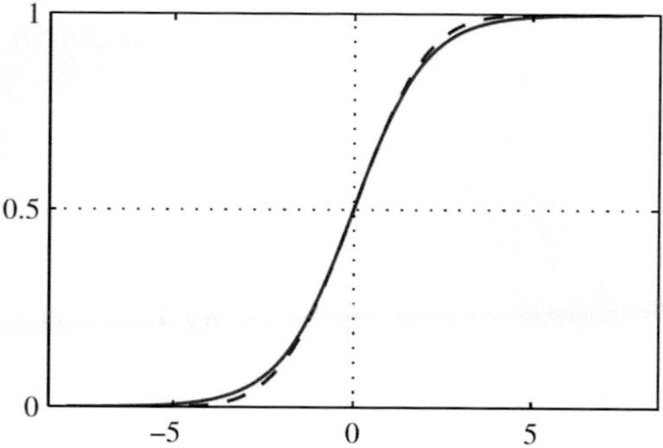

Figure 5.7 The logistic sigmoid function $\sigma(l)$, shown in red, along with a scaled error function, shown in dashed blue (Bishop 2006) © 2006 Springer.

is the *log-likelihood* of sample $\mathbf{x}$ being from class $\mathcal{C}_k$.[7] It is sometimes convenient to denote the softmax function (5.3) as a vector-to-vector valued function,

$$\mathbf{p} = \text{softmax}(\mathbf{l}). \tag{5.5}$$

The softmax function can be viewed as a soft version of a maximum indicator function, which returns 1 for the largest value of l_k whenever it dominates the other values. It is widely used in machine learning and statistics, including its frequent use as the final non-linearity in deep neural classification networks (Figure 5.27).

The process of using formula (5.2) to determine the likelihood of a class $\mathcal{C}_k$ given a feature vector $\mathbf{x}$ is known as *Bayesian classification*, since it combines a conditional feature likelihood $p(\mathbf{x}|\mathcal{C}_k)$ with a prior distribution over classes $p(\mathcal{C}_k)$ using Bayes' rule to determine the posterior class probabilities. In the case where the components of the feature vector are generated independently, i.e.,

$$p(\mathbf{x}|\mathcal{C}_k) = \prod_i p(x_i|\mathcal{C}_k), \tag{5.6}$$

the resulting technique is called a *naïve Bayes classifier*.

For the binary (two class) classification task, we can re-write (5.3) as

$$p(\mathcal{C}_0|\mathbf{x}) = \frac{1}{1+\exp(-l)} = \sigma(l), \tag{5.7}$$

where $l = l_0 - l_1$ is the difference between the two class log likelihood and is known as the *log odds* or *logit*.

The $\sigma(l)$ function is called the *logistic sigmoid function* (or simply the *logistic function* or *logistic curve*), where *sigmoid* means an S-shaped curve (Figure 5.7). The sigmoid was a popular *activation function* in earlier neural networks, although it has now been replaced by functions, as discussed in Section 5.3.2.

Linear and quadratic discriminant analysis

While probabilistic generative classification based on the normalized exponential and sigmoid can be applied to any set of log likelihoods, the formulas become much simpler when the distributions are multi-dimensional Gaussians.

[7] Some authors (e.g., Zhang, Lipton *et al.* 2021) use the term *logit* for the log-likelihood, although it is more commonly used to denote the *log odds*, discussed below, or the softmax function itself.

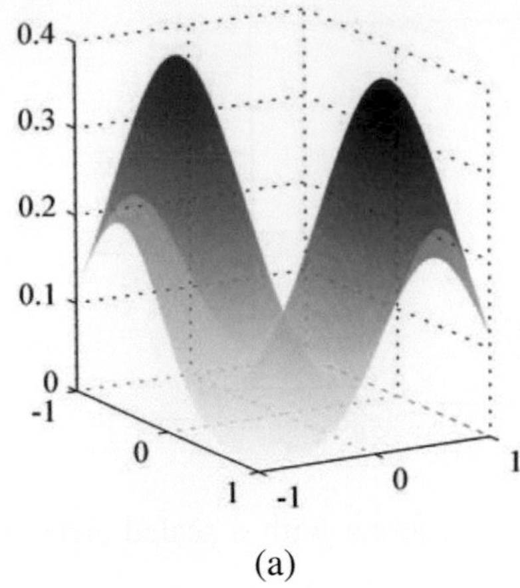

(a)

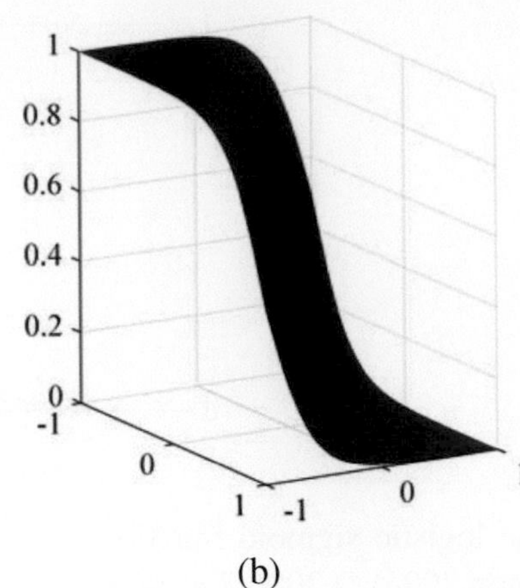

(b)

Figure 5.8 Logistic regression for two identically distributed Gaussian classes (Bishop 2006) © 2006 Springer: (a) two Gaussian distributions shown in red and blue; (b) the posterior probability $p(\mathcal{C}_0|\mathbf{x})$, shown as both the height of the function and the proportion of red ink.

For Gaussians with identical covariance matrices $\mathbf{\Sigma}$, we have

$$p(\mathbf{x}|\mathcal{C}_k) = \frac{1}{(2\pi)^{D/2}} \frac{1}{\|\mathbf{\Sigma}\|^{1/2}} \exp\left\{-\frac{1}{2}(\mathbf{x}-\boldsymbol{\mu}_k)^T \mathbf{\Sigma}^{-1}(\mathbf{x}-\boldsymbol{\mu}_k).\right\} \tag{5.8}$$

In the case of two classes (binary classification), we obtain (Bishop 2006, Section 4.2.1)

$$p(\mathcal{C}_0|\mathbf{x}) = \sigma(\mathbf{w}^T\mathbf{x} + b), \tag{5.9}$$

with

$$\mathbf{w} = \mathbf{\Sigma}^{-1}(\boldsymbol{\mu}_0 - \boldsymbol{\mu}_1), \quad \text{and} \tag{5.10}$$

$$b = \frac{1}{2}\boldsymbol{\mu}_0^T\mathbf{\Sigma}^{-1}\boldsymbol{\mu}_0 + \frac{1}{2}\boldsymbol{\mu}_1^T\mathbf{\Sigma}^{-1}\boldsymbol{\mu}_1 + \log\frac{p(\mathcal{C}_0)}{p(\mathcal{C}_1)}. \tag{5.11}$$

Equation (5.9), which we will revisit shortly in the context of non-generative (discriminative) classification (5.18), is called *logistic regression*, since we pass the output of a linear regression formula

$$l(\mathbf{x}) = \mathbf{w}^T\mathbf{x} + b \tag{5.12}$$

through the logistic function to obtain a class probability. Figure 5.8 illustrates this in two dimensions, there the posterior likelihood of the red class $p(\mathcal{C}_0|\mathbf{x})$ is shown on the right side.

In linear regression (5.12), $\mathbf{w}$ plays the role of the *weight* vector along which we project the feature vector $\mathbf{x}$, and b plays the role of the *bias*, which determines where to set the classification boundary. Note that the weight direction (5.10) aligns with the vector joining the distribution means (after rotating the coordinates by the inverse covariance $\mathbf{\Sigma}^{-1}$), while the bias term is proportional to the mean squared moments and the log class prior ratio $\log(p(\mathcal{C}_0)/p(\mathcal{C}_1))$.

For $K > 2$ classes, the softmax function (5.3) can be applied to the linear regression log likelihoods,

$$l_k(\mathbf{x}) = \mathbf{w}_k^T\mathbf{x} + b_k, \tag{5.13}$$

with

$$\mathbf{w}_k = \mathbf{\Sigma}^{-1}\boldsymbol{\mu}_k, \quad \text{and} \tag{5.14}$$

$$b_k = -\frac{1}{2}\boldsymbol{\mu}_k^T\mathbf{\Sigma}^{-1}\boldsymbol{\mu}_k + \log p(\mathcal{C}_k). \tag{5.15}$$

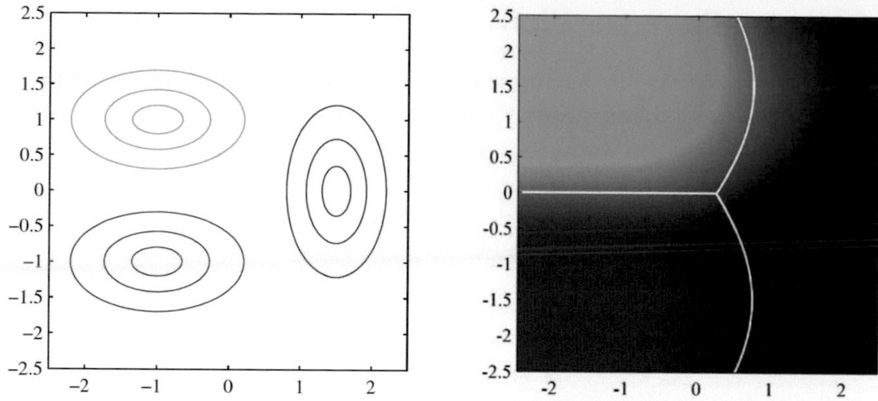

Figure 5.9 Quadratic discriminant analysis (Bishop 2006) © 2006 Springer. When the class covariances Σ_k are different, the decision surfaces between Gaussian distributions become quadratic surfaces.

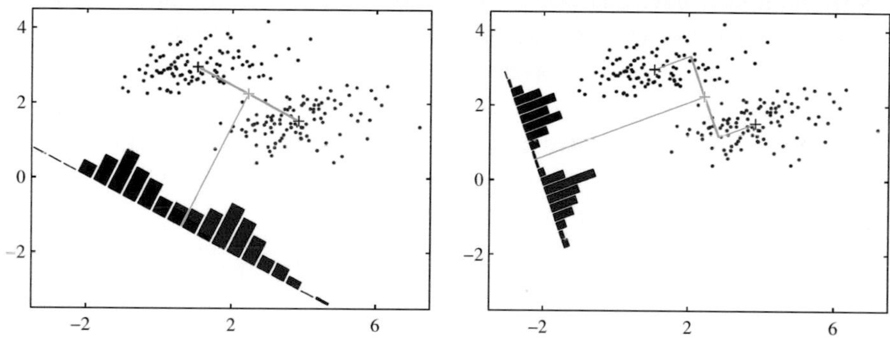

Figure 5.10 Fisher linear discriminant (Bishop 2006) © 2006 Springer. To find the projection direction to best separate two classes, we compute the sum of the two class covariances and then use its inverse to rotate the vector between the two class means.

Because the decision boundaries along which the classification switches from one class to another are linear,

$$\mathbf{w}_k \mathbf{x} + b_k > \mathbf{w}_l \mathbf{x} + b_l, \tag{5.16}$$

the technique of classifying examples using such criteria is known as *linear discriminant analysis* (Bishop 2006, Section 4.1; Murphy 2012, Section 4.2.2).[8]

Thus far, we have looked at the case where all of the class covariance matrices Σ_k are identical. When they vary between classes, the decision surfaces are no longer linear and they become quadratic (Figure 5.9). The derivation of these quadratic decision surfaces is known as *quadratic discriminant analysis* (Murphy 2012, Section 4.2.1).

In the case where Gaussian class distributions are not available, we can still find the best discriminant direction using Fisher discriminant analysis (Bishop 2006, Section 4.1.4; Murphy 2012, Section 8.6.3), as shown in Figure 5.10. Such analysis can be useful in separately modeling variability within different classes, e.g., the appearance variation of different people (Section 5.2.3).

[8]The acronym LDA is commonly used with linear discriminant analysis, but is sometimes also used for *latent Dirichlet allocation* in graphical models.

5.1.3 Logistic regression

In the previous section, we derived classification rules based on posterior probabilities applied to multivariate Gaussian distributions. Quite often, however, Gaussians are not appropriate models of our class distributions and we must resort to alternative techniques.

One of the simplest among these is *logistic regression*, which applies the same ideas as in the previous section, i.e., a linear projection onto a weight vector,

$$l_i = \mathbf{w} \cdot \mathbf{x}_i + b \tag{5.17}$$

followed by a logistic function

$$p_i = p(\mathcal{C}_0|\mathbf{x}_i) = \sigma(l_i) = \sigma(\mathbf{w}^T\mathbf{x}_i + b) \tag{5.18}$$

to obtain (binary) class probabilities. Logistic regression is a simple example of a *discriminative model*, since it does not construct or assume a prior distribution over unknowns, and hence is not *generative*, i.e., we cannot generate random samples from the class (Bishop 2006, Section 1.5.4).

As we no longer have analytic estimates for the class means and covariances (or they are poor models of the class distributions), we need some other method to determine the weights $\mathbf{w}$ and bias b. We do this by maximizing the posterior log likelihoods of the correct labels.

For the binary classification task, let $t_i \in \{0, 1\}$ be the class label for each training sample $\mathbf{x}_i$ and $p_i = p(\mathcal{C}_0|\mathbf{x})$ be the estimated likelihood predicted by (5.18) for a given weight and bias $(\mathbf{w}, b)$. We can maximize the likelihood of the correct labels being predicted by minimizing the negative log likelihood, i.e., the *cross-entropy loss* or error function,

$$E_{\text{CE}}(\mathbf{w}, b) = -\sum_i \{t_i \log p_i + (1 - t_i) \log(1 - p_i)\} \tag{5.19}$$

(Bishop 2006, Section 4.3.2).[9] Note how whenever the label $t_i = 0$, we want $p_i = p(\mathcal{C}_0|\mathbf{x}_i)$ to be high, and vice versa.

This formula can easily be extended to a multi-class loss by again defining the posterior probabilities as normalized exponentials over per-class linear regressions, as in (5.3) and (5.13),

$$p_{ik} = p(\mathcal{C}_k|\mathbf{x}_i) = \frac{\exp l_{ik}}{\sum_j \exp l_{ij}} = \frac{1}{Z_i} \exp l_{ik}, \tag{5.20}$$

with

$$l_{ik} = \mathbf{w}_k^T \mathbf{x}_i + b_k. \tag{5.21}$$

The term $Z_i = \sum_j \exp l_{ij}$ can be a useful shorthand in derivations and is sometimes called the *partition function*. After some manipulation (Bishop 2006, Section 4.3.4), the corresponding *multi-class cross-entropy loss* (a.k.a. *multinomial logistic regression objective*) becomes

$$E_{\text{MCCE}}(\{\mathbf{w}_k, b_k\}) = -\sum_i \sum_k \tilde{t}_{ik} \log p_{ik}, \tag{5.22}$$

where the 1-of-K (or *one-hot*) encoding has $\tilde{t}_{ik} = 1$ if sample i belongs to class k (and 0 otherwise).[10] It is more common to simply use the integer class value t_i as the target, in which case we

---

[9] Note, however, that since this derivation is based on the assumption of Gaussian noise, it may not perform well if there are outliers, e.g., errors in the labels. In such a case, a more robust measure such as mean absolute error (MAE) may be preferable (Ghosh, Kumar, and Sastry 2017) or it may be necessary to re-weight the training samples (Ren, Zeng *et al.* 2018).

[10] This kind of representation can be useful if we wish the target classes to be a mixture, e.g., in the *mixup* data augmentation technique of Zhang, Cisse *et al.* (2018).

can re-write this even more succinctly as

$$E(\{\mathbf{w}_k, b_k\}) = -\sum_i \log p_{it_i}, \tag{5.23}$$

i.e., we simply sum up the log likelihoods of the correct class for each training sample. Substituting the softmax formula (5.20) into this loss, we can re-write it as

$$E(\{\mathbf{w}_k, b_k\}) = \sum_i \left(\log Z_i - l_{it_i}\right). \tag{5.24}$$

To determine the best set of weights and biases, $\{\mathbf{w}_k, b_k\}$, we can use gradient descent, i.e., update their values using a Newton-Raphson second-order optimization scheme (Bishop 2006, Section 4.3.3),

$$\mathbf{w} \leftarrow \mathbf{w} - \mathbf{H}^{-1}\nabla E(\mathbf{w}), \tag{5.25}$$

where ∇E is the gradient of the loss function E with respect to the weight variables $\mathbf{w}$, and $\mathbf{H}$ is the Hessian matrix of second derivatives of E. Because the cross-entropy functions are not linear in the unknown weights, we need to iteratively solve this equation a few times to arrive at a good solution. Since the elements in $\mathbf{H}$ are updated after each iteration, this technique is also known as *iteratively reweighted least squares*, which we will study in more detail in Section 8.1.4. While many non-linear optimization problems have multiple local minima, the cross-entropy functions described in this section do not, so we are guaranteed to arrive at a unique solution.

Logistic regression does have some limitations, which is why it is often used for only the simplest classification tasks. If the classes in feature space are not linearly separable, using simple projections onto weight vectors may not produce adequate decision surfaces. In this case, *kernel methods* (Sections 4.1.1 and 5.1.4; Bishop 2006, Chapter 6; Murphy 2012, Chapter 14), which measure the distances between new (test) feature vectors and select training examples, can often provide good solutions.

Another problem with logistic regression is that if the classes actually are separable (either in the original feature space, or the lifted kernel space), there can be more than a single unique separating plane, as illustrated in Figure 5.11a. Furthermore, unless regularized, the weights $\mathbf{w}$ will continue to grow larger, as larger values of $\mathbf{w}_k$ lead to larger p_{ik} values (once a separating plane has been found) and hence a smaller overall loss.

For this reason, techniques that place the decision surfaces in a way that maximizes their separation to labeled examples have been developed, as we discuss next.

5.1.4 Support vector machines

As we have just mentioned, in some applications of logistic regression we cannot determine a single optimal decision surface (choice of weight and bias vectors $\{\mathbf{w}_k, b_k\}$ in (5.21)) because there are *gaps* in the feature space where any number of planes could be introduced. Consider Figure 5.11a, where the two classes are denoted in cyan and magenta colors. In addition to the two dashed lines and the solid line, there are infinitely many other lines that will also cleanly separate the two classes, including a swath of horizontal lines. Since the classification error for any of these lines is zero, how can we choose the best decision surface, keeping in mind that we only have a limited number of training examples, and that actual run-time examples may fall somewhere in between?

The answer to this problem is to use *maximum margin classifiers* (Bishop 2006, Section 7.1), as shown in Figure 5.11a, where the dashed lines indicate two parallel decision surfaces that have the

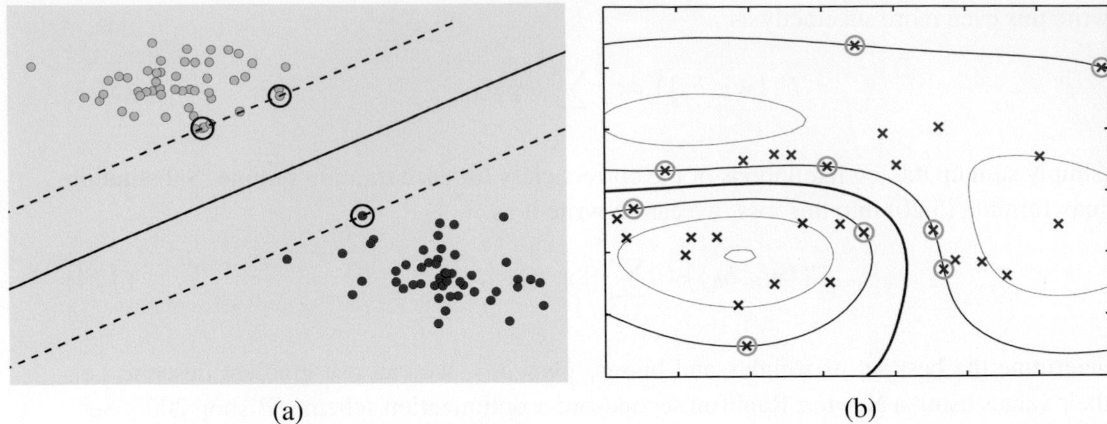

(a) (b)

Figure 5.11 (a) A support vector machine (SVM) finds the linear decision surface (hyper-plane) that maximizes the *margin* to the nearest training examples, which are called the *support vectors* © Glassner (2018). (b) A two-dimensional two class example of a Gaussian kernel support vector machine (Bishop 2006) © 2006 Springer. The red and blue ×s indicate the training samples, and the samples circled in green are the support vectors. The black lines indicate iso-contours of the kernel regression function, with the contours containing the blue and red support vectors indicating the ±1 contours and the dark contour in between being the decision surface.

maximum margin, i.e., the largest perpendicular distance between them. The solid line, which represents the hyperplane half-way between the dashed hyperplanes, is the maximum margin classifier.

Why is this a good idea? There are several potential derivations (Bishop 2006, Section 7.1), but a fairly intuitive explanation is that there may be real-world examples coming from the cyan and magenta classes that we have not yet seen. Under certain assumptions, the maximum margin classifier provides our best bet for correctly classifying as many of these unseen examples as possible.

To determine the maximum margin classifier, we need to find a weight-bias pair $(\mathbf{w}, b)$ for which all regression values $l_i = \mathbf{w} \cdot \mathbf{x}_i + b$ (5.17) have an absolute value of at least 1 as well as the correct sign. To denote this more compactly, let

$$\hat{t}_i = 2t_i - 1, \quad \hat{t}_i \in \{-1, 1\} \tag{5.26}$$

be the *signed class label*. We can now re-write the inequality condition as

$$\hat{t}_i(\mathbf{w} \cdot \mathbf{x}_i + b) \geq 1. \tag{5.27}$$

To maximize the margin, we simply find the smallest norm weight vector $\mathbf{w}$ that satisfies (5.27), i.e., we solve the optimization problem

$$\arg\min_{\mathbf{w},b} \|\mathbf{w}\|^2 \tag{5.28}$$

subject to (5.27). This is a classic *quadratic programming problem*, which can be solved using the method of Lagrange multipliers, as described in Bishop (2006, Section 7.1).

The inequality constraints are exactly satisfied, i.e., they turn into equalities, along the two dashed lines in Figure 5.11a, where we have $l_i = \mathbf{w}\mathbf{x}_i + b = \pm 1$. The circled points that touch the dashed lines are called the *support vectors*.[11] For a simple linear classifier, which can be denoted

[11] While the cyan and magenta dots may just look like points, they are, of course, schematic representations of higher-dimensional *vectors* lying in feature space.

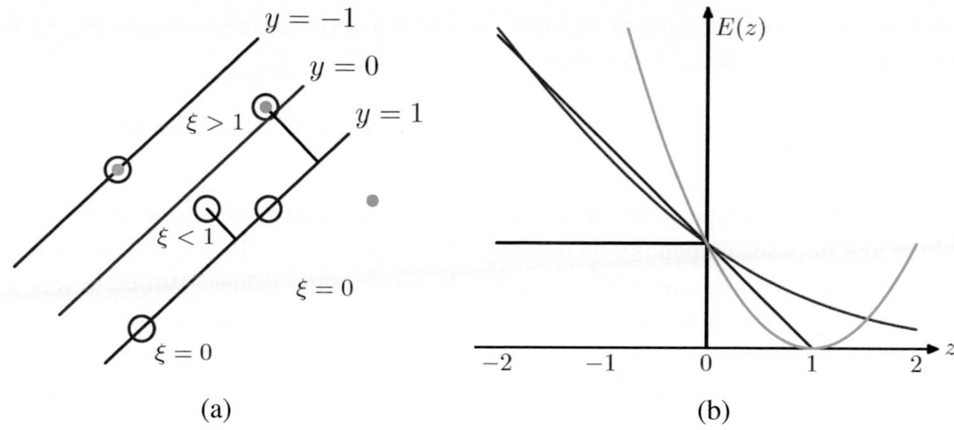

Figure 5.12 Support vector machine for overlapping class distributions (Bishop 2006) © 2006 Springer. (a) The green circled point is on the wrong side of the $y = 1$ decision contour and has a penalty of $\xi = 1 - y > 0$. (b) The "hinge" loss used in support vector machines is shown in blue, along with a rescaled version of the logistic regression loss function, shown in red, the misclassification error in black, and the squared error in green.

with a single weight and bias pair $(\mathbf{w}, b)$, there is no real advantage to computing the support vectors, except that they help us estimate the decision surface. However, as we will shortly see, when we apply kernel regression, having a small number of support vectors is a huge advantage.

What happens if the two classes are not *linearly* separable, and in fact require a complex curved surface to correctly classify samples, as in Figure 5.11b? In this case, we can replace linear regression with *kernel regression* (4.3), which we introduced in Section 4.1.1. Rather than multiplying the weight vector $\mathbf{w}$ with the feature vector $\mathbf{x}$, we instead multiply it with the value of K kernel functions centered at the data point locations $\mathbf{x}_k$,

$$l_i = f(\mathbf{x}_i; \mathbf{w}, b) = \sum_k w_k \phi(\|\mathbf{x}_i - \mathbf{x}_k\|) + b. \tag{5.29}$$

This is where the power of support vector machines truly comes in.

Instead of requiring the summation over all training samples $\mathbf{x}_k$, once we solve for the maximum margin classifier only a small subset of support vectors needs to be retained, as shown by the circled crosses in Figure 5.11b. As you can see in this figure, the decision boundary denoted by the dark black line nicely separates the red and blue class samples. Note that as with other applications of kernel regression, the width of the radial basis functions is still a free hyperparameter that must be reasonably tuned to avoid underfitting and overfitting.

Hinge loss. So far, we have focused on classification problems that are separable, i.e., for which a decision boundary exists that correctly classifies all the training examples. Support vector machines can also be applied to overlapping (mixed) class distributions (Figure 5.12a), which we previously approached using logistic regression. In this case, we replace the inequality conditions (5.27), i.e., $\hat{t}_i l_i \geq 1$, with a *hinge loss* penalty

$$E_{\text{HL}}(l_i, \hat{t}_i) = [1 - \hat{t}_i l_i]_+, [1 - \hat{t}_i l_i]_+, \tag{5.30}$$

where $[\cdot]_+$ denotes the positive part, i.e. $[x]+ = \max(0, x)$. The hinge loss penalty, shown in blue in Figure 5.12b, is 0 whenever the (previous) inequality is satisfied and ramps up linearly depending on how much the inequality is violated. To find the optimal weight values $(\mathbf{w}, b)$, we minimize the regularized sum of hinge loss values,

$$E_{\text{SV}}(\mathbf{w}, b) = \sum_i E_{\text{HL}}(l_i(\mathbf{x}_i; \mathbf{w}, b), \hat{t}_i) + \lambda \|\mathbf{w}\|^2. \tag{5.31}$$

Figure 5.12b compares the hinge loss to the logistic regression (cross-entropy) loss in (5.19). The hinge loss imposes no penalty on training samples that are on the correct side of the $|l_i| > 1$ boundary, whereas the cross-entropy loss prefers larger absolute values. While, in this section, we have focused on the two-class version of support vector machines, Bishop (2006, Chapter 7) describes the extension to multiple classes as well as efficient optimization algorithms such as *sequential minimal optimization* (SMO) (Platt 1989). There's also a nice online tutorial on the scikit-learn website.[12] A survey of SVMs and other kernel methods applied to computer vision can be found in Lampert (2008).

5.1.5 Decision trees and forests

In contrast to most of the supervised learning techniques we have studied so far in this chapter, which process complete feature vectors all at once (with either linear projections or distances to training examples), *decision trees* perform a sequence of simpler operations, often just looking at individual feature elements before deciding which element to look at next (Hastie, Tibshirani, and Friedman 2009, Chapter 17; Glassner 2018, Section 14.5; Criminisi, Shotton, and Konukoglu 2012; Criminisi and Shotton 2013). (Note that the *boosting approaches* we study in Section 6.3.1 also use similar simple *decision stumps*.) While decision trees have been used in statistical machine learning for several decades (Breiman, Friedman *et al.* 1984), the application of their more powerful extension, namely *decision forests*, only started gaining traction in computer vision a little over a decade ago (Lepetit and Fua 2006; Shotton, Johnson, and Cipolla 2008; Shotton, Girshick *et al.* 2013). Decision trees, like support vector machines, are discriminative classifiers (or regressors), since they never explicitly form a probabilistic (generative) model of the data they are classifying.

Figure 5.13 illustrates the basic concepts behind decision trees and random forests. In this example, training samples come from four different classes, each shown in a different color (a). A decision tree (b) is constructed top-to-bottom by selecting decisions at each node that split the training samples that have made it to that node into more specific (lower entropy) distributions. The thickness of each link shows the number of samples that get classified along that path, and the color of the link is the blend of the class colors that flow through that link. The color histograms show the class distributions at a few of the interior nodes.

A random forest (c) is created by building a set of decision trees, each of which makes slightly different decisions. At test (classification) time, a new sample is classified by each of the trees in the random forest, and the class distributions at the final leaf nodes are averaged to provide an answer that is more accurate than could be obtained with a single tree (with a given depth).

Random forests have several design parameters, which can be used to tailor their accuracy, generalization, and run-time and space complexity. These parameters include:

- the depth of each tree D,
- the number of trees T, and
- the number of samples examined at node construction time ρ.

[12] https://scikit-learn.org/stable/modules/svm.html#svm-classification

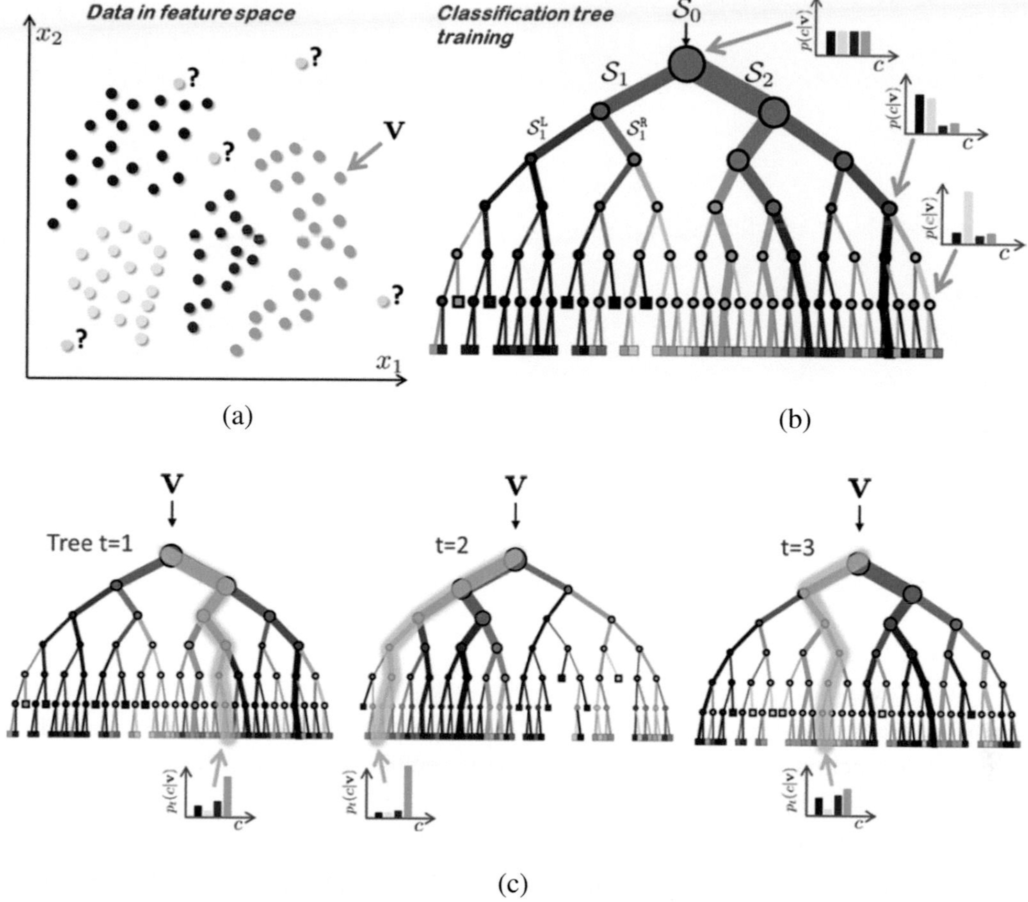

Figure 5.13 Decision trees and forests (Criminisi and Shotton 2013) © 2013 Springer. The top left figure (a) shows a set of training samples tags with four different class colors. The top right (b) shows a single decision tree with a distribution of classes at each node (the root node has the same distribution as the entire training set). During testing (c), each new example (feature vector) is tested at the root node, and depending on this test result (e.g., the comparison of some element to a threshold), a decision is made to walk down the tree to one of its children. This continues until a leaf node with a particular class distribution is reached. During training (b), decisions are selected such that they reduce the entropy (increase class specificity) at the node's children. The bottom diagram (c) shows an *ensemble* of three trees. After a particular test example has been classified by each tree, the class distributions of the leaf nodes of all the constituent trees are averaged.

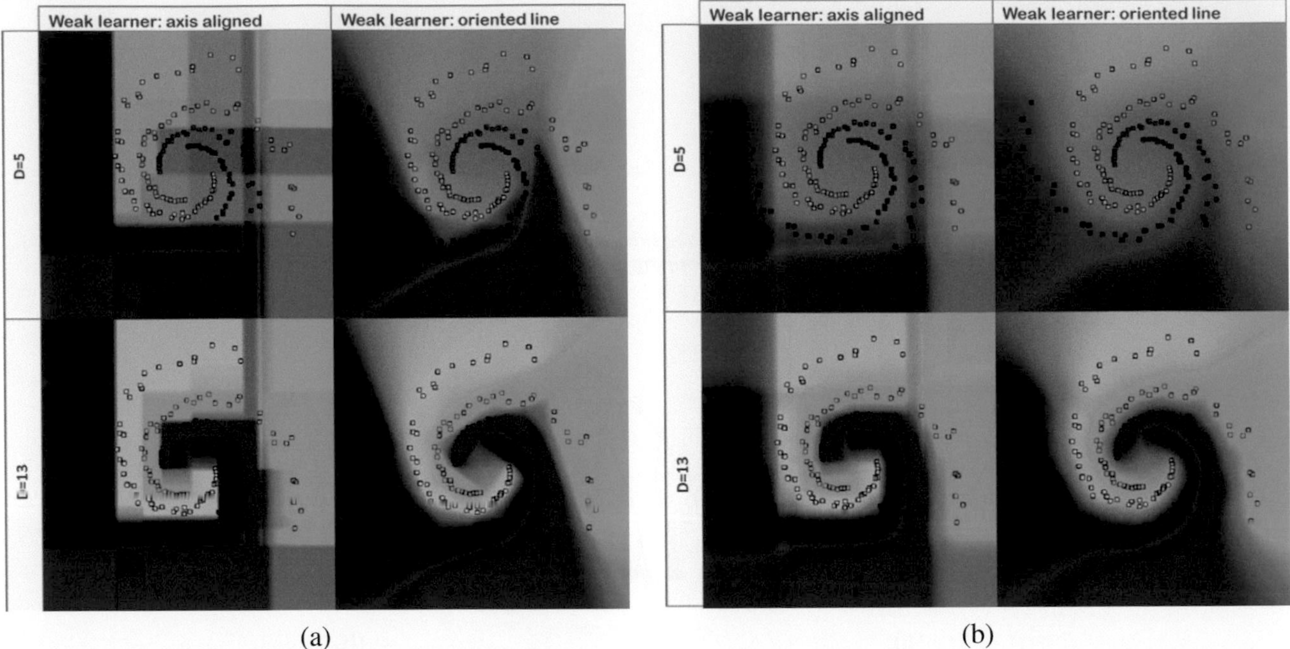

Figure 5.14 Random forest decision surfaces (Criminisi and Shotton 2013) © 2013 Springer. Figures (a) and (b) show smaller and larger amounts of "noise" between the $T = 400$ tree forests obtained by using $\rho = 500$ and $\rho = 5$ random hypotheses at each split node. Withing each figure, the two rows show trees of different depths ($D = 5$ and 13), while the columns show the effects of using axis-aligned or linear decision surfaces ("weak learners").

By only looking at a random subset ρ of all the training examples, each tree ends up having different decision functions at each node, so that the ensemble of trees can be averaged to produce softer decision boundaries.

Figure 5.14 shows the effects of some of these parameters on a simple four-class two-dimensional spiral dataset. In this figure, the number of trees has been fixed to $T = 400$. Criminisi and Shotton (2013, Chapter 4) have additional figures showing the effect of varying more parameters. The left (a) and right (b) halves of this figure show the effects of having less randomness ($\rho = 500$) and more randomness ($\rho = 5$) at the decision nodes. Less random trees produce sharper decision surfaces but may not generalize as well. Within each 2×2 grid of images, the top row shows a shallower $D = 5$ tree, while the bottom row shows a deeper $D = 13$ tree, which leads to finer details in the decision boundary. (As with all machine learning, better performance on training data may not lead to better generalization because of overfitting.) Finally, the right column shows what happens if axis-aligned (single element) decisions are replaced with linear combinations of feature elements.

When applied to computer vision, decision trees first made an impact in keypoint recognition (Lepetit and Fua 2006) and image segmentation (Shotton, Johnson, and Cipolla 2008). They were one of the key ingredients (along with massive amounts of synthetic training data) in the breakthrough success of human pose estimation from Kinect depth images (Shotton, Girshick *et al.* 2013). They also led to state-of-the-art medical image segmentation systems (Criminisi, Robertson *et al.* 2013), although these have now been supplanted by deep neural networks (Kamnitsas, Ferrante *et al.* 2016). Most of these applications, along with additional ones, are reviewed in the book edited by Criminisi and Shotton (2013).

5.2 Unsupervised learning

Thus far in this chapter, we have focused on *supervised learning* techniques where we are given training data consisting of paired input and target examples. In some applications, however, we are only given a set of data, which we wish to characterize, e.g., to see if there are any patterns, regularities, or typical distributions. This is typically the realm of classical statistics. In the machine learning community, this scenario is usually called *unsupervised learning*, since the sample data comes without labels. Examples of applications in computer vision include image segmentation (Section 7.5) and face and body recognition and reconstruction (Sections 13.6.2).

In this section, we look at some of the more widely used techniques in computer vision, namely clustering and mixture modeling (e.g., for segmentation) and principal component analysis (for appearance and shape modeling). Many other techniques are available, and are covered in textbooks on machine learning, such as Bishop (2006, Chapter 9), Hastie, Tibshirani, and Friedman (2009, Chapter 14), and Murphy (2012, Section 1.3).

5.2.1 Clustering

One of the simplest things you can do with your sample data is to group it into sets based on similarities (e.g., vector distances). In statistics, this problem is known as *cluster analysis* and is a widely studied area with hundreds of different algorithms (Jain and Dubes 1988; Kaufman and Rousseeuw 1990; Jain, Duin, and Mao 2000; Jain, Topchy *et al.* 2004). Murphy (2012, Chapter 25) has a nice exposition on clustering algorithms, including affinity propagation, spectral clustering, graph Laplacian, hierarchical, agglomerative, and divisive clustering. The survey by Xu and Wunsch (2005) is even more comprehensive, covering almost 300 different papers and such topics as similarity measures, vector quantization, mixture modeling, kernel methods, combinatorial and neural network algorithms, and visualization. Figure 5.15 shows some of the algorithms implemented in the https://scikit-learn.org cluster analysis package applied to some simple two-dimensional examples.

Splitting an image into successively finer regions (divisive clustering) is one of the oldest techniques in computer vision. Ohlander, Price, and Reddy (1978) present such a technique, which first computes a histogram for the whole image and then finds a threshold that best separates the large peaks in the histogram. This process is repeated until regions are either fairly uniform or below a certain size. More recent splitting algorithms often optimize some metric of intra-region similarity and inter-region dissimilarity. These are covered in Sections 7.5.3 and 4.3.2.

Region merging techniques also date back to the beginnings of computer vision. Brice and Fennema (1970) use a dual grid for representing boundaries between pixels and merge regions based on their relative boundary lengths and the strength of the visible edges at these boundaries.

In data clustering, algorithms can link clusters together based on the distance between their closest points (single-link clustering), their farthest points (complete-link clustering), or something in between (Jain, Topchy *et al.* 2004). Kamvar, Klein, and Manning (2002) provide a probabilistic interpretation of these algorithms and show how additional models can be incorporated within this framework. Applications of such agglomerative clustering (region merging) algorithms to image segmentation are discussed in Section 7.5.

Mean-shift (Section 7.5.2) and mode finding techniques, such as k-means and mixtures of Gaussians, model the feature vectors associated with each pixel (e.g., color and position) as samples from an unknown probability density function and then try to find clusters (modes) in this distribution.

Consider the color image shown in Figure 7.53a. How would you segment this image based on color alone? Figure 7.53b shows the distribution of pixels in L*u*v* space, which is equivalent to what a vision algorithm that ignores spatial location would see. To make the visualization simpler,

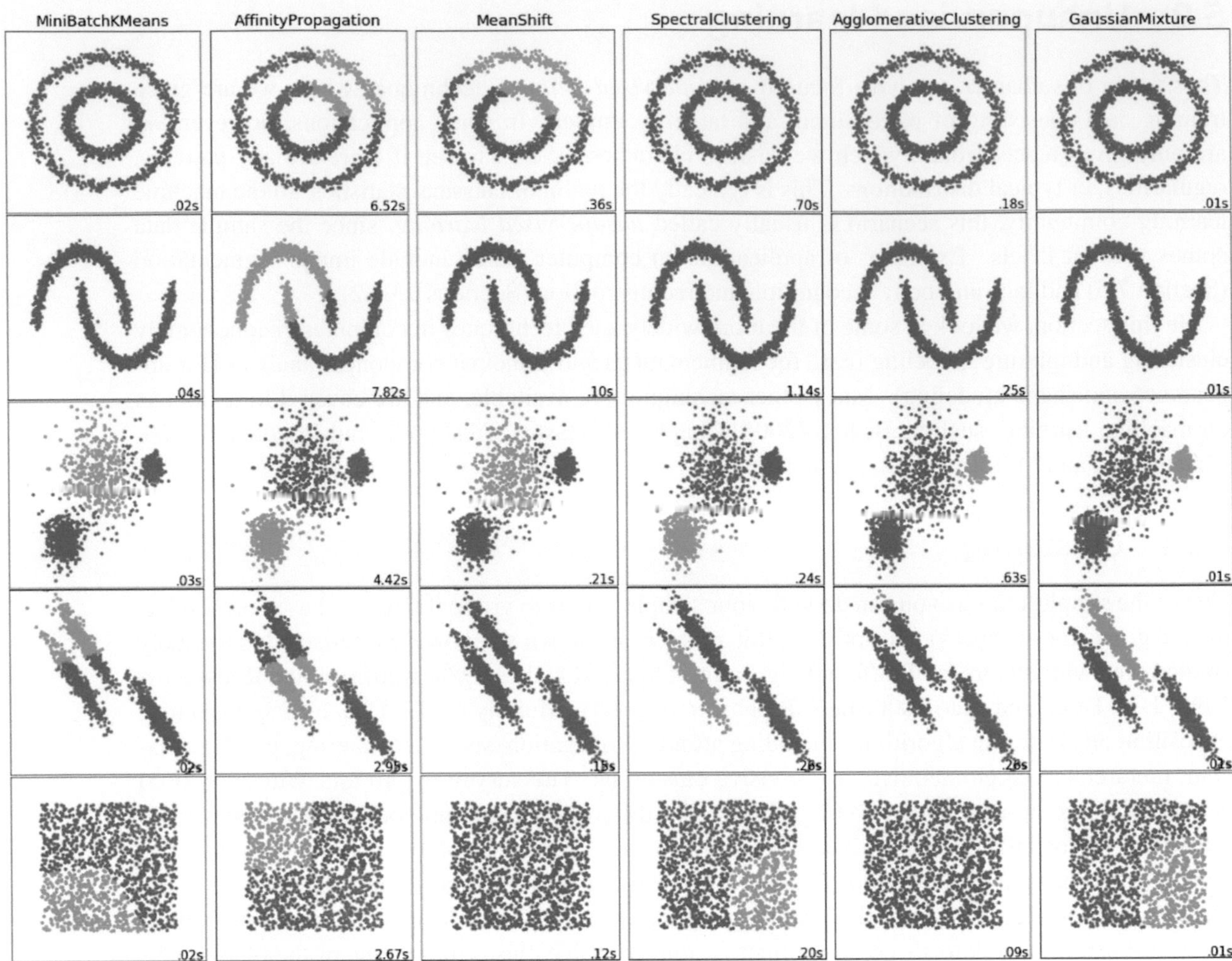

Figure 5.15 Comparison of different clustering algorithms on some toy datasets, generated using a simplified version of https://scikit-learn.org/stable/auto_examples/cluster/plot_cluster_comparison.html#sphx-glr-auto-examples-cluster-plot-cluster-comparison-py.

let us only consider the L*u* coordinates, as shown in Figure 7.53c. How many obvious (elongated) clusters do you see? How would you go about finding these clusters?

The k-means and mixtures of Gaussians techniques use a *parametric* model of the density function to answer this question, i.e., they assume the density is the superposition of a small number of simpler distributions (e.g., Gaussians) whose locations (centers) and shape (covariance) can be estimated. Mean shift, on the other hand, smoothes the distribution and finds its peaks as well as the regions of feature space that correspond to each peak. Since a complete density is being modeled, this approach is called *non-parametric* (Bishop 2006).

5.2.2 K-means and Gaussians mixture models

K-means implicitly model the probability density as a superposition of spherically symmetric distributions and does not require any probabilistic reasoning or modeling (Bishop 2006). Instead, the algorithm is given the number of clusters k it is supposed to find and is initialized by randomly

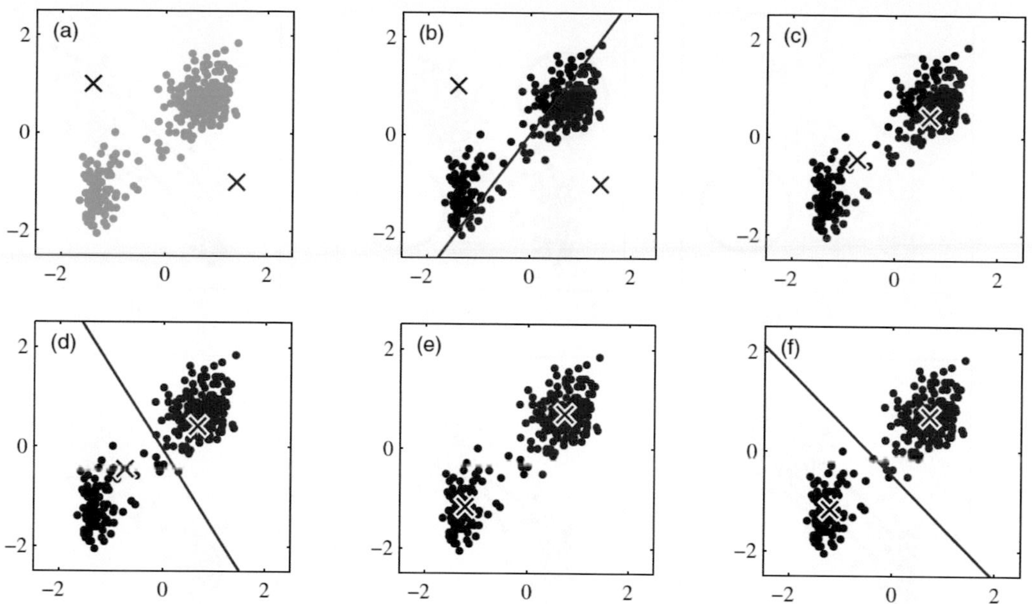

Figure 5.16 The k-means algorithm starts with a set of samples and the number of desired clusters (in this case, $k = 2$) (Bishop 2006) © 2006 Springer. It iteratively assigns samples to the nearest mean, and then re-computes the mean center until convergence.

sampling k centers from the input feature vectors. It then iteratively updates the cluster center location based on the samples that are closest to each center (Figure 5.16). Techniques have also been developed for splitting or merging cluster centers based on their statistics, and for accelerating the process of finding the nearest mean center (Bishop 2006).

In mixtures of Gaussians, each cluster center is augmented by a covariance matrix whose values are re-estimated from the corresponding samples (Figure 5.17). Instead of using nearest neighbors to associate input samples with cluster centers, a *Mahalanobis distance* (Appendix B.1) is used:

$$d(\mathbf{x}_i, \boldsymbol{\mu}_k; \boldsymbol{\Sigma}_k) = \|\mathbf{x}_i - \boldsymbol{\mu}_k\|_{\boldsymbol{\Sigma}_k^{-1}} = (\mathbf{x}_i - \boldsymbol{\mu}_k)^T \boldsymbol{\Sigma}_k^{-1} (\mathbf{x}_i - \boldsymbol{\mu}_k) \tag{5.32}$$

where $\mathbf{x}_i$ are the input samples, $\boldsymbol{\mu}_k$ are the cluster centers, and $\boldsymbol{\Sigma}_k$ are their covariance estimates. Samples can be associated with the nearest cluster center (a *hard assignment* of membership) or can be *softly assigned* to several nearby clusters.

This latter, more commonly used, approach corresponds to iteratively re-estimating the parameters for a Gaussians mixture model,

$$p(\mathbf{x}|\{\pi_k, \boldsymbol{\mu}_k, \boldsymbol{\Sigma}_k\}) = \sum_k \pi_k \mathcal{N}(\mathbf{x}|\boldsymbol{\mu}_k, \boldsymbol{\Sigma}_k), \tag{5.33}$$

where π_k are the *mixing coefficients*, $\boldsymbol{\mu}_k$ and $\boldsymbol{\Sigma}_k$ are the Gaussian means and covariances, and

$$\mathcal{N}(\mathbf{x}|\boldsymbol{\mu}_k, \boldsymbol{\Sigma}_k) = \frac{1}{|\boldsymbol{\Sigma}_k|} e^{-d(\mathbf{x}, \boldsymbol{\mu}_k; \boldsymbol{\Sigma}_k)} \tag{5.34}$$

is the *normal* (Gaussian) distribution (Bishop 2006).

To iteratively compute (a local) maximum likely estimate for the unknown mixture parameters $\{\pi_k, \boldsymbol{\mu}_k, \boldsymbol{\Sigma}_k\}$, the *expectation maximization* (EM) algorithm (Shlezinger 1968; Dempster, Laird, and Rubin 1977) proceeds in two alternating stages:

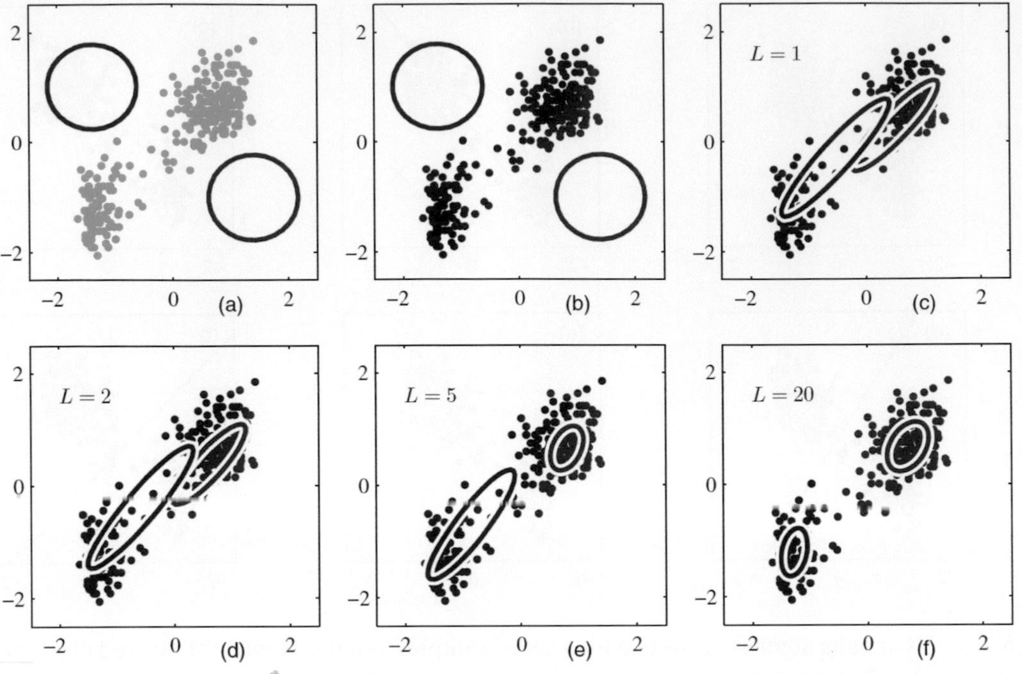

Figure 5.17 Gaussian mixture modeling (GMM) using expectation maximization (EM) (Bishop 2006) © 2006 Springer. Samples are softly assigned to cluster centers based on their *Mahalanobis distance* (inverse covariance weighted distance), and the new means and covariances are recomputed based on these weighted assignments.

1. The *expectation* stage (E step) estimates the *responsibilities*

$$z_{ik} = \frac{1}{Z_i} \pi_k \mathcal{N}(\mathbf{x}|\boldsymbol{\mu}_k, \boldsymbol{\Sigma}_k) \quad \text{with} \quad \sum_k z_{ik} = 1, \tag{5.35}$$

which are the estimates of how likely a sample $\mathbf{x}_i$ was generated from the kth Gaussian cluster.

2. The *maximization* stage (M step) updates the parameter values

$$\boldsymbol{\mu}_k = \frac{1}{N_k} \sum_i z_{ik} \mathbf{x}_i, \tag{5.36}$$

$$\boldsymbol{\Sigma}_k = \frac{1}{N_k} \sum_i z_{ik} (\mathbf{x}_i - \boldsymbol{\mu}_k)(\mathbf{x}_i - \boldsymbol{\mu}_k)^T, \tag{5.37}$$

$$\pi_k = \frac{N_k}{N}, \tag{5.38}$$

where

$$N_k = \sum_i z_{ik}. \tag{5.39}$$

is an estimate of the number of sample points assigned to each cluster.

Bishop (2006) has a wonderful exposition of both mixture of Gaussians estimation and the more general topic of expectation maximization.

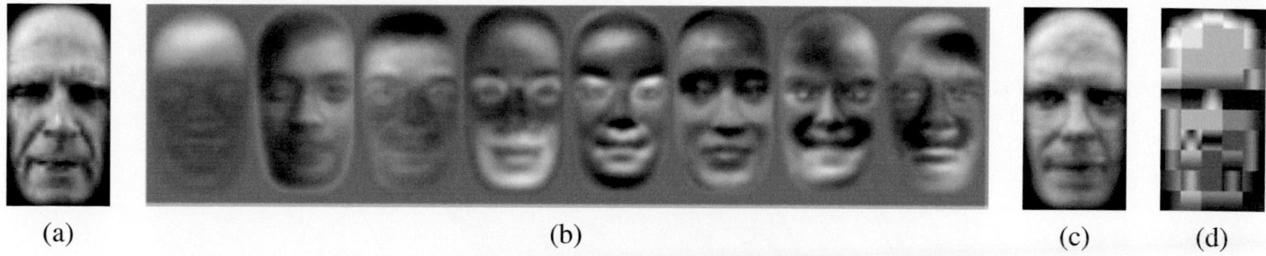

(a) (b) (c) (d)

Figure 5.18 Face modeling and compression using eigenfaces (Moghaddam and Pentland 1997) © 1997 IEEE: (a) input image; (b) the first eight eigenfaces; (c) image reconstructed by projecting onto this basis and compressing the image to 85 bytes; (d) image reconstructed using JPEG (530 bytes).

In the context of image segmentation, Ma, Derksen *et al.* (2007) present a nice review of segmentation using mixtures of Gaussians and develop their own extension based on Minimum Description Length (MDL) coding, which they show produces good results on the Berkeley segmentation dataset.

5.2.3 Principal component analysis

As we just saw in mixture analysis, modeling the samples within a cluster with a multi-variate Gaussian can be a powerful way to capture their distribution. Unfortunately, as the dimensionality of our sample space increases, estimating the full covariance quickly becomes infeasible.

Consider, for example, the space of all frontal faces (Figure 5.18). For an image consisting of P pixels, the covariance matrix has a size of $P \times P$. Fortunately, the full covariance normally does not have to be modeled, since a lower-rank approximation can be estimated using *principal component analysis*, as described in Appendix A.1.2.

PCA was originally used in computer vision for modeling faces, i.e., *eigenfaces*, initially for gray-scale images (Kirby and Sirovich 1990; Turk and Pentland 1991), and then for 3D models (Blanz and Vetter 1999; Egger, Smith *et al.* 2020) (Section 13.6.2) and active appearance models (Section 6.2.4), where they were also used to model facial shape deformations (Rowland and Perrett 1995; Cootes, Edwards, and Taylor 2001; Matthews, Xiao, and Baker 2007).

Eigenfaces. Eigenfaces rely on the observation first made by Kirby and Sirovich (1990) that an arbitrary face image $\mathbf{x}$ can be compressed and reconstructed by starting with a mean image $\mathbf{m}$ (Figure 6.1b) and adding a small number of scaled signed images $\mathbf{u}_i$,

$$\tilde{\mathbf{x}} = \mathbf{m} + \sum_{i=0}^{M-1} a_i \mathbf{u}_i, \tag{5.40}$$

where the signed basis images (Figure 5.18b) can be derived from an ensemble of training images using *principal component analysis* (also known as *eigenvalue analysis* or the *Karhunen–Loève transform*). Turk and Pentland (1991) recognized that the coefficients a_i in the eigenface expansion could themselves be used to construct a fast image matching algorithm.

In more detail, we start with a collection of *training images* $\{\mathbf{x}_j\}$, from which we compute the mean image $\mathbf{m}$ and a *scatter* or *covariance* matrix

$$\mathbf{C} = \frac{1}{N} \sum_{j=0}^{N-1} (\mathbf{x}_j - \mathbf{m})(\mathbf{x}_j - \mathbf{m})^T. \tag{5.41}$$

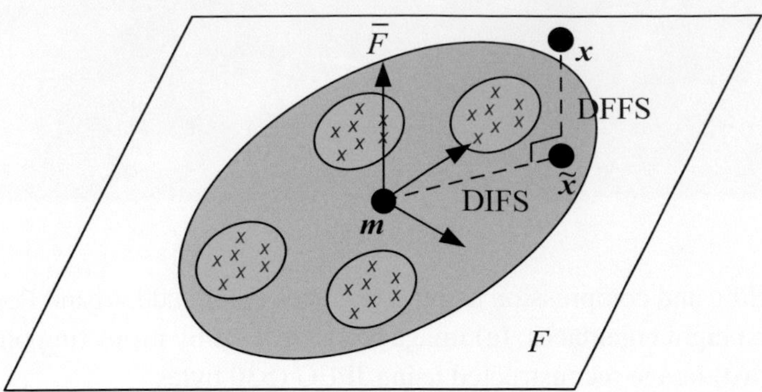

Figure 5.19 Projection onto the linear subspace spanned by the eigenface images (Moghaddam and Pentland 1997) © 1997 IEEE. The distance from face space (DFFS) is the orthogonal distance to the plane, while the distance in face space (DIFS) is the distance along the plane from the mean image. Both distances can be turned into Mahalanobis distances and given probabilistic interpretations.

We can apply the eigenvalue decomposition (A.6) to represent this matrix as

$$\mathbf{C} = \mathbf{U}\mathbf{\Lambda}\mathbf{U}^T = \sum_{i=0}^{N-1} \lambda_i \mathbf{u}_i \mathbf{u}_i^T, \tag{5.42}$$

where the λ_i are the eigenvalues of $\mathbf{C}$ and the $\mathbf{u}_i$ are the *eigenvectors*. For general images, Kirby and Sirovich (1990) call these vectors *eigenpictures*; for faces, Turk and Pentland (1991) call them *eigenfaces* (Figure 5.18b).[13]

Two important properties of the eigenvalue decomposition are that the optimal (best approximation) coefficients a_i for any new image $\mathbf{x}$ can be computed as

$$a_i = (\mathbf{x} - \mathbf{m}) \cdot \mathbf{u}_i, \tag{5.43}$$

and that, assuming the eigenvalues $\{\lambda_i\}$ are sorted in decreasing order, truncating the approximation given in (5.40) at any point M gives the best possible approximation (least error) between $\tilde{\mathbf{x}}$ and $\mathbf{x}$. Figure 5.18c shows the resulting approximation corresponding to Figure 5.18a and shows how much better it is at compressing a face image than JPEG.

Truncating the eigenface decomposition of a face image (5.40) after M components is equivalent to projecting the image onto a linear subspace F, which we can call the *face space* (Figure 5.19). Because the eigenvectors (eigenfaces) are orthogonal and of unit norm, the distance of a projected face $\tilde{\mathbf{x}}$ to the mean face $\mathbf{m}$ can be written as

$$\text{DIFS} = \|\tilde{\mathbf{x}} - \mathbf{m}\| = \left[\sum_{i=0}^{M-1} a_i^2\right]^{1/2}, \tag{5.44}$$

where DIFS stands for *distance in face space* (Moghaddam and Pentland 1997). The remaining distance between the original image $\mathbf{x}$ and its projection onto face space $\tilde{\mathbf{x}}$, i.e., the *distance from face space* (DFFS), can be computed directly in pixel space and represents the "faceness" of a

[13] In actual practice, the full $P \times P$ scatter matrix (5.41) is never computed. Instead, a smaller $N \times N$ matrix consisting of the inner products between all the signed deviations $(\mathbf{x}_i - \mathbf{m})$ is accumulated instead. See Appendix A.1.2 (A.13–A.14) for details.

particular image. It is also possible to measure the distance between two different faces in face space by taking the norm of their eigenface coefficients difference.

Computing such distances in Euclidean vector space, however, does not exploit the additional information that the eigenvalue decomposition of the covariance matrix (5.42) provides. To properly weight the distance based on the measured covariance, we can use the *Mahalanobis distance* (5.32) (Appendix B.1). A similar analysis can be performed for computing a sensible difference from face space (DFFS) (Moghaddam and Pentland 1997) and the two terms can be combined to produce an estimate of the likelihood of being a true face, which can be useful in doing face detection (Section 6.3.1). More detailed explanations of probabilistic and Bayesian PCA can be found in textbooks on statistical learning (Bishop 2006; Hastie, Tibshirani, and Friedman 2009; Murphy 2012), which also discuss techniques for selecting the optimum number of components M to use in modeling a distribution.

The original work on eigenfaces for recognition (Turk and Pentland 1991) was extended in Moghaddam and Pentland (1997), Heisele, Ho *et al.* (2003), and Heisele, Serre, and Poggio (2007) to include *modular eigenenspaces* for separately modeling the appearance of different facial components such as the eyes, nose, and mouth, as well as *view-based eigenspaces* to separately model different views of a face. It was also extended by Belhumeur, Hespanha, and Kriegman (1997) to handle appearance variation due to illumination, modeling *intrapersonal* and *extrapersonal* variability separately, and using Fisher linear discriminant analysis (Figure 5.10) to perform recognition. A Bayesian extension of this work was subsequently developed by Moghaddam, Jebara, and Pentland (2000). These extensions are described in more detail in the cited papers, as well as the first edition of this book (Szeliski 2010, Section 14.2).

It is also possible to generalize the bilinear factorization implicit in PCA and SVD approaches to multilinear (tensor) formulations that can model several interacting factors simultaneously (Vasilescu and Terzopoulos 2007). These ideas are related to additional topics in machine learning such as *subspace learning* (Cai, He *et al.* 2007), *local distance functions* (Frome, Singer *et al.* 2007; Ramanan and Baker 2009), and *metric learning* (Kulis 2013).

5.2.4 Manifold learning

In many cases, the data we are analyzing does not reside in a globally linear subspace, but does live on a lower-dimensional manifold. In this case, non-linear dimensionality reduction can be used (Lee and Verleysen 2007). Since these systems extract lower-dimensional manifolds in a higher-dimensional space, they are also known as *manifold learning* techniques (Zheng and Xue 2009). Figure 5.20 shows some examples of two-dimensional manifolds extracted from the three-dimensional S-shaped ribbon using the scikit-learn manifold learning package.[14]

These results are just a small sample from the large number of algorithms that have been developed, which include multidimensional scaling (Kruskal 1964a,b), Isomap (Tenenbaum, De Silva, and Langford 2000), Local Linear Embedding (Roweis and Saul 2000), Hessian Eigenmaps (Donoho and Grimes 2003), Laplacian Eigenmaps (Belkin and Niyogi 2003), local tangent space alignment (Zhang and Zha 2004), Dimensionality Reduction by Learning an Invariant Mapping (Hadsell, Chopra, and LeCun 2006), Modified LLE (Zhang and Wang 2007), t-distributed Stochastic Neighbor Embedding (t-SNE) (van der Maaten and Hinton 2008; van der Maaten 2014), and UMAP (McInnes, Healy, and Melville 2018). Many of these algorithms are reviewed in Lee and Verleysen (2007), Zheng and Xue (2009), and on Wikipedia.[15] Bengio, Paiement *et al.* (2004) describe

[14]https://scikit-learn.org/stable/modules/manifold.html

[15]https://en.wikipedia.org/wiki/Nonlinear_dimensionality_reduction

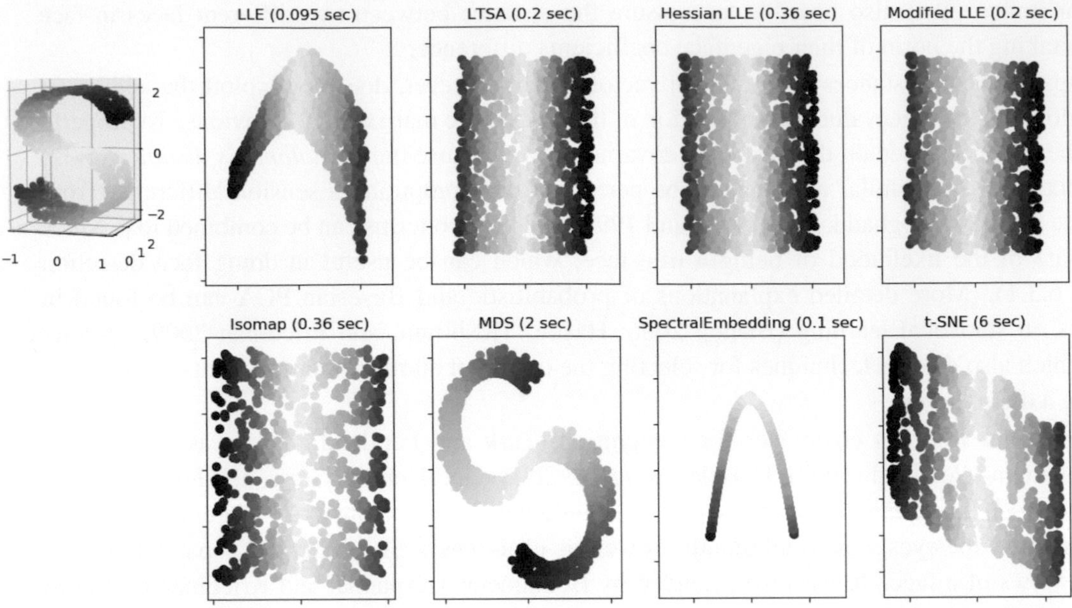

Figure 5.20 Examples of manifold learning, i.e., non-linear dimensionality reduction, applied to 1,000 points with 10 neighbors each, from https://scikit-learn.org/stable/modules/manifold.html. The eight sample outputs were produced by eight different embedding algorithms, as described in the scikit-learn manifold learning documentation page.

a method for extending such algorithms to compute the embedding of new ("out-of-sample") data points. McQueen, Meila *et al.* (2016) describe their megaman software package, which can efficiently solve embedding problems with millions of data points.

In addition to dimensionality reduction, which can be useful for regularizing data and accelerating similarity search, manifold learning algorithms can be used for visualizing input data distributions or neural network layer activations. Figure 5.21 show an example of applying two such algorithms (UMAP and t-SNE) to three different computer vision datasets.

5.2.5 Semi-supervised learning

In many machine learning settings, we have a modest amount of accurately labeled data and a far larger set of unlabeled or less accurate data. For example, an image classification dataset such as ImageNet may only contain one million labeled images, but the total number of images that can be found on the web is orders of magnitudes larger. Can we use this larger dataset, which still captures characteristics of our expect future inputs, to construct a better classifier or predictor?

Consider the simple diagrams in Figure 5.22. Even if only a small number of examples are labeled with the correct class (in this case, indicated by red and blue circles or dots), we can still imagine extending these labels (inductively) to nearby samples and therefore not only labeling all of the data, but also constructing appropriate decision surfaces for future inputs.

This area of study is called *semi-supervised learning* (Zhu and Goldberg 2009; Subramanya and Talukdar 2014). In general, it comes in two varieties. In *transductive learning*, the goal is to classify all of the unlabeled inputs that are given as one batch at the same time as the labeled examples, i.e., all of the dots and circles shown in Figure 5.22. In *inductive learning*, we train a machine learning system that will classify all future inputs, i.e., all the regions in the input space. The second form

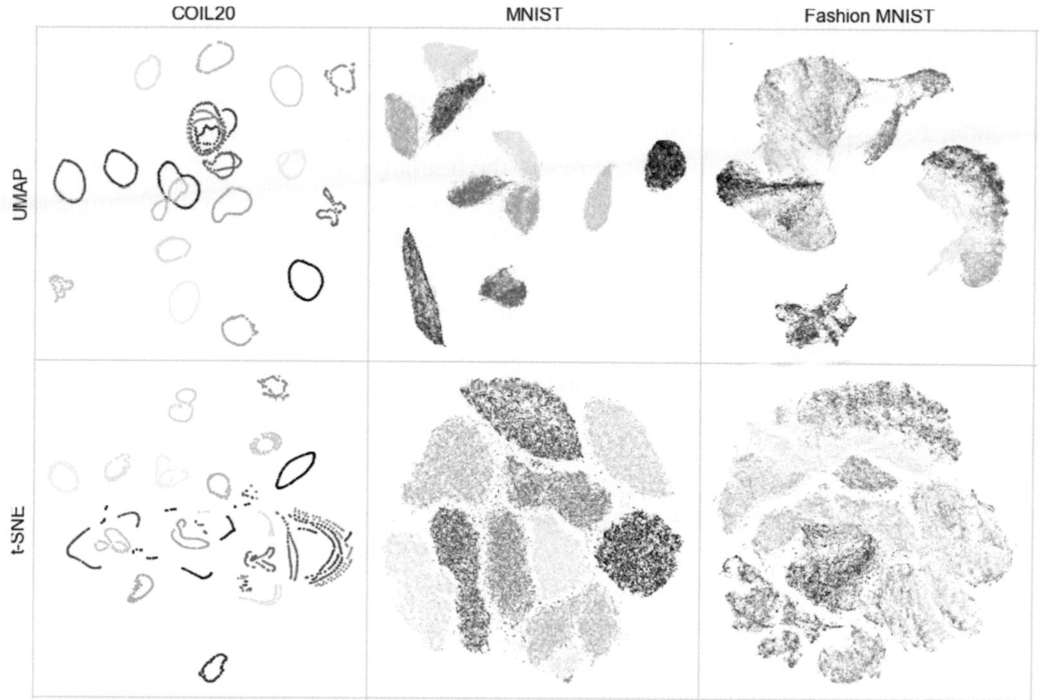

Figure 5.21 Comparison of UMAP and t-SNE manifold learning algorithms © McInnes, Healy, and Melville (2018) on three different computer vision learning recognition tasks: COIL (Nene, Nayar, and Murase 1996), MNIST (LeCun, Cortes, and Burges 1998), and Fashion MNIST (Xiao, Rasul, and Vollgraf 2017).

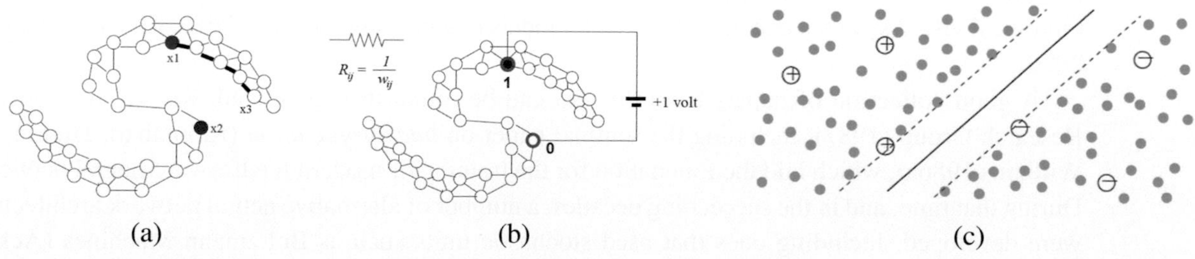

Figure 5.22 Examples of semi-supervised learning (Zhu and Goldberg 2009) © 2009 Morgan & Claypool: (a) two labeled samples and a graph connecting all of the samples; (b) solving binary labeling with harmonic functions, interpreted as a resistive electrical network; (c) using semi-supervised support vector machine (S3VM),

is much more widely used, since in practice, most machine learning systems are used for online applications such as autonomous driving or new content classification.

Semi-supervised learning is a subset of the larger class of *weakly supervised learning* problems, where the training data may not only be missing labels, but also have labels of questionable accuracy (Zhou 2018). Some early examples from computer vision (Torresani 2014) include building whole image classifiers from image labels found on the internet (Fergus, Perona, and Zisserman 2004; Fergus, Weiss, and Torralba 2009) and object detection and/or segmentation (localization) with missing or very rough delineations in the training data (Nguyen, Torresani *et al.* 2009; Deselaers, Alexe, and Ferrari 2012). In the deep learning era, weakly supervised learning continues to be widely used (Pathak, Krahenbuhl, and Darrell 2015; Bilen and Vedaldi 2016; Arandjelovic, Gronat *et al.* 2016; Khoreva, Benenson *et al.* 2017; Novotny, Larlus, and Vedaldi 2017; Zhai, Oliver *et al.* 2019). A recent example of weakly supervised learning being applied to billions of noisily labeled images is pre-training deep neural networks on Instagram images with hashtags (Mahajan, Girshick *et al.* 2018). We will look at weakly and self-supervised learning techniques for *pre-training* neural networks in Section 5.4.7.

5.3 Deep neural networks

As we saw in the introduction to this chapter (Figure 5.2), deep learning pipelines take an end-to-end approach to machine learning, optimizing every stage of the processing by searching for parameters that minimize the training loss. In order for such search to be feasible, it helps if the loss is a differentiable function of all these parameters. Deep neural networks provide a uniform, differentiable computation architecture, while also automatically discovering useful internal representations.

Interest in building computing systems that mimic neural (biological) computation has waxed and waned since the late 1950s, when Rosenblatt (1958) developed the *perceptron* and Widrow and Hoff (1960) derived the weight adaptation *delta rule*. Research into these topics was revitalized in the late 1970s by researchers who called themselves *connectionists*, organizing a series of meetings around this topic, which resulted in the foundation of the Neural Information Processing Systems (NeurIPS) conference in 1987. The recent book by Sejnowski (2018) has a nice historical review of this field's development, as do the introductions in Goodfellow, Bengio, and Courville (2016) and Zhang, Lipton *et al.* (2021), the review paper by Rawat and Wang (2017), and the Turing Award lecture by Bengio, LeCun, and Hinton (2021). And while most of the deep learning community has moved away from biologically plausible models, some research still studies the connection between biological visual systems and neural network models (Yamins and DiCarlo 2016; Zhuang, Yan *et al.* 2020).

A good collection of papers from this era can be found in McClelland, Rumelhart, and PDP Research Group (1987), including the seminal paper on *backpropagation* (Rumelhart, Hinton, and Williams 1986a), which laid the foundation for the training of modern feedforward neural networks. During that time, and in the succeeding decades, a number of alternative neural network architectures were developed, including ones that used stochastic units such as Boltzmann Machines (Ackley, Hinton, and Sejnowski 1985) and Restricted Boltzmann Machines (Hinton and Salakhutdinov 2006; Salakhutdinov and Hinton 2009). The survey by Bengio (2009) has a review of some of these earlier approaches to deep learning. Many of these architectures are examples of the *generative graphical models* we saw in Section 4.3.

Today's most popular deep neural networks are deterministic *discriminative feedforward networks* with real-valued activations, trained using gradient descent, i.e., the the backpropagation training rule (Rumelhart, Hinton, and Williams 1986b). When combined with ideas from convo-

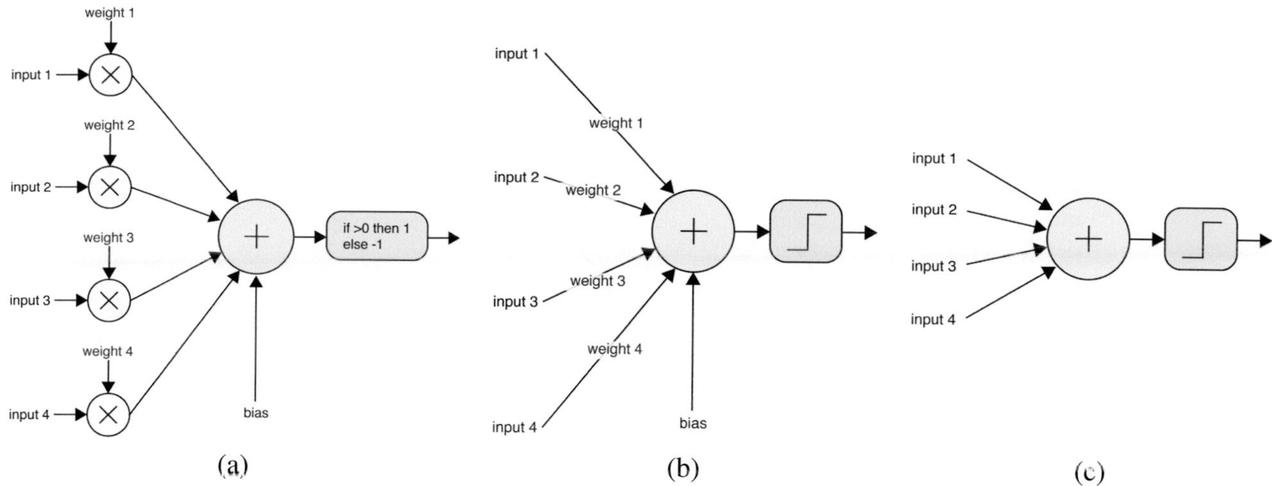

Figure 5.23 A perceptron unit (a) explicitly showing the weights being multiplied by the inputs, (b) with the weights written on the input connections, and (c) the most common form, with the weights and bias omitted. A non-linear activation function follows the weighted summation. © Glassner (2018)

lutional networks (Fukushima 1980; LeCun, Bottou *et al.* 1998), deep multi-layer neural networks produced the breakthroughs in speech recognition (Hinton, Deng *et al.* 2012) and visual recognition (Krizhevsky, Sutskever, and Hinton 2012; Simonyan and Zisserman 2014b) seen in the early 2010s. Zhang, Lipton *et al.* (2021, Chapter 7) have a nice description of the components that went into these breakthroughs and the rapid evolution in deep networks that has occurred since then, as does the earlier review paper by (Rawat and Wang 2017).

Compared to other machine learning techniques, which normally rely on several pre-processing stages to extract features on which classifiers can be built, deep learning approaches are usually trained *end-to-end*, going directly from raw pixels to final desired outputs (be they classifications or other images). In the next few sections, we describe the basic components that go into constructing and training such neural networks. More detailed explanations on each topic can be found in textbooks on deep learning (Nielsen 2015; Goodfellow, Bengio, and Courville 2016; Glassner 2018, 2021; Zhang, Lipton *et al.* 2021) as well as the excellent course notes by Li, Johnson, and Yeung (2019) and Johnson (2020).

5.3.1 Weights and layers

Deep neural networks (DNNs) are *feedforward* computation graphs composed of thousands of simple interconnected "neurons" (*units*), which, much like logistic regression (5.18), perform weighted sums of their inputs

$$s_i = \mathbf{w}_i^T \mathbf{x}_i + b_i \tag{5.45}$$

followed by a non-linear *activation function* re-mapping,

$$y_i = h(s_i), \tag{5.46}$$

as illustrated in Figure 5.23. The $\mathbf{x}_i$ are the inputs to the ith unit, $\mathbf{w}_i$ and b_i are its learnable *weights* and *bias*, s_i is the output of the weighted linear sum, and y_i is the final output after s_i is fed through the activation function h.[16] The outputs of each stage, which are often called the *activations*, are

[16]Note that we have switched to using s_i for the weighted summations, since we will want to use l to index neural network layers.

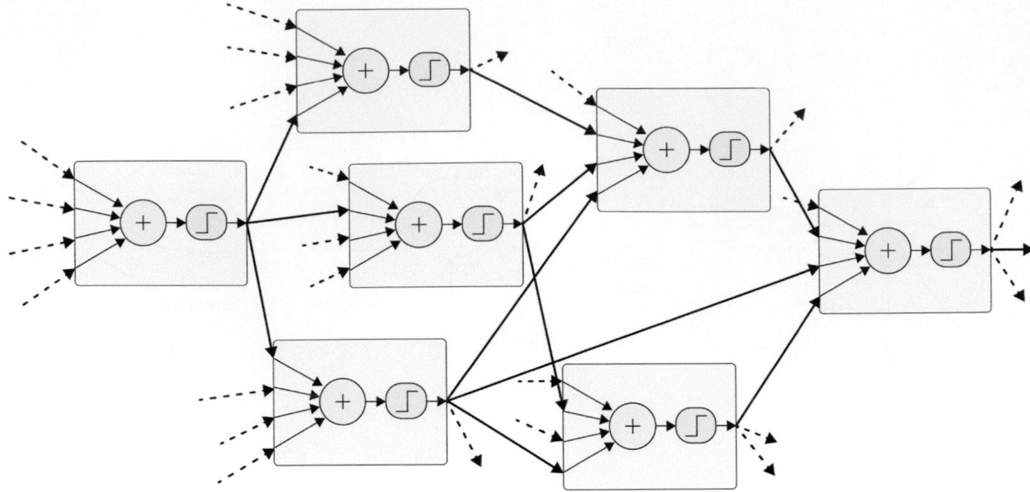

Figure 5.24 A multi-layer network, showing how the outputs of one unit are fed into additional units. © Glassner (2018)

then fed into units in later stages, as shown in Figure 5.24.[17]

The earliest such units were called *perceptrons* (Rosenblatt 1958) and were diagramed as shown in Figure 5.23a. Note that in this first diagram, the weights, which are optimized during the learning phase (Section 5.3.5), are shown explicitly along with the element-wise multiplications. Figure 5.23b shows a form in which the weights are written on top of the *connections* (arrows between units, although the arrowheads are often omitted). It is even more common to diagram nets as in Figure 5.23c, in which the weights (and bias) are completely omitted and assumed to be present.

Instead of being connected into an irregular computation graph as in Figure 5.24, neural networks are usually organized into consecutive *layers*, as shown in Figure 5.25. We can now think of all the units within a layer as being a vector, with the corresponding linear combinations written as

$$\mathbf{s}_l = \mathbf{W}_l \mathbf{x}_l, \tag{5.47}$$

where $\mathbf{x}_l$ are the inputs to layer l, $\mathbf{W}_l$ is a weight matrix, and $\mathbf{s}_l$ is the weighted sum, to which an element-wise non-linearity is applied using a set of activation functions,

$$\mathbf{x}_{l+1} = \mathbf{y}_l = \mathbf{h}(\mathbf{s}_l). \tag{5.48}$$

A layer in which a full (dense) weight matrix is used for the linear combination is called a *fully connected* (FC) layer, since all of the inputs to one layer are connected to all of its outputs. As we will see in Section 5.4, when processing pixels (or other signals), early stages of processing use convolutions instead of dense connections for both spatial invariance and better efficiency.[18] A network that consists only of fully connected (and no convolutional) layers is now often called a *multi-layer perceptron* (MLP).

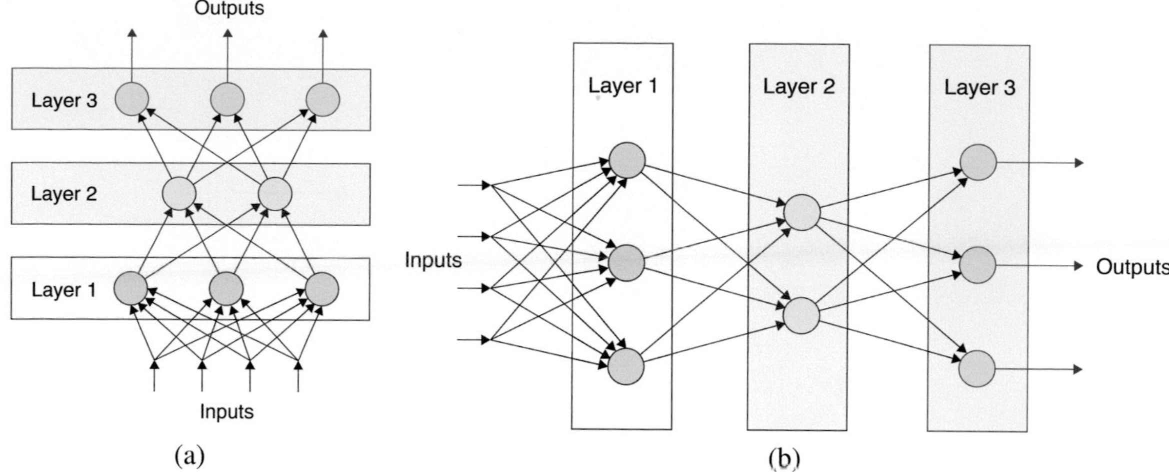

Figure 5.25 Two different ways to draw neural networks: (a) inputs at bottom, outputs at top, (b) inputs at left, outputs at right. © Glassner (2018)

5.3.2 Activation functions

Most early neural networks (Rumelhart, Hinton, and Williams 1986b; LeCun, Bottou *et al.* 1998) used sigmoidal functions similar to the ones used in logistic regression. Newer networks, starting with Nair and Hinton (2010) and Krizhevsky, Sutskever, and Hinton (2012), use Rectified Linear Units (ReLU) or variants. The ReLU activation function is defined as

$$h(y) = \max(0, y) \tag{5.49}$$

and is shown in the upper-left corner of Figure 5.26, along with some other popular functions, whose definitions can be found in a variety of publications (e.g., Goodfellow, Bengio, and Courville 2016, Section 6.3; Clevert, Unterthiner, and Hochreiter 2015; He, Zhang *et al.* 2015) and the Machine Learning Cheatsheet.[19]

While the ReLU is currently the most popular activation function, a widely cited observation in the CS231N course notes (Li, Johnson, and Yeung 2019) attributed to Andrej Karpathy warns that[20]

> *Unfortunately, ReLU units can be fragile during training and can "die". For example, a large gradient flowing through a ReLU neuron could cause the weights to update in such a way that the neuron will never activate on any datapoint again. If this happens, then the gradient flowing through the unit will forever be zero from that point on. That is, the ReLU units can irreversibly die during training since they can get knocked off the data manifold. ... With a proper setting of the learning rate this is less frequently an issue.*

The CS231n course notes advocate trying some alternative non-clipping activation functions if this problem arises.

[17]Note that while almost all feedforward neural networks use linear weighted summations of their inputs, the Neocognitron (Fukushima 1980) also included a divisive normalization stage inspired by the behavior of biological neurons. Some of the latest DNNs also support multiplicative interactions between activations using *conditional batch norm* (Section 5.3.3).

[18]Heads up for more confusing abbreviations: While a fully connected (dense) layer is often abbreviated as FC, a fully convolutional network, which is the opposite, i.e., sparsely connected with shared weights, is often abbreviated as FCN.

[19]https://ml-cheatsheet.readthedocs.io/en/latest/activation_functions.html

[20]http://cs231n.github.io/neural-networks-1/#actfun

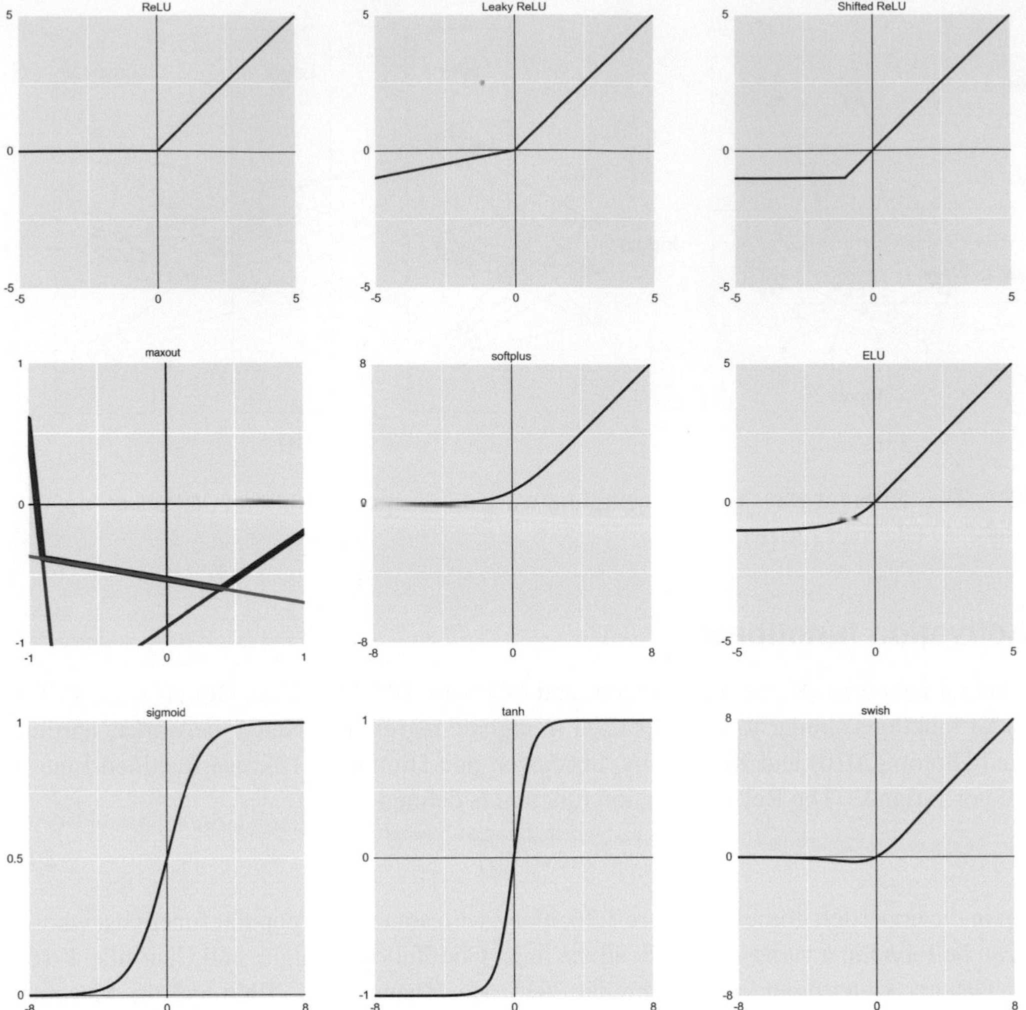

Figure 5.26 Some popular non-linear activation functions from © Glassner (2018): From top-left to bottom-right: ReLU, leaky ReLU, shifted ReLU, maxout, softplus, ELU, sigmoid, tanh, swish.

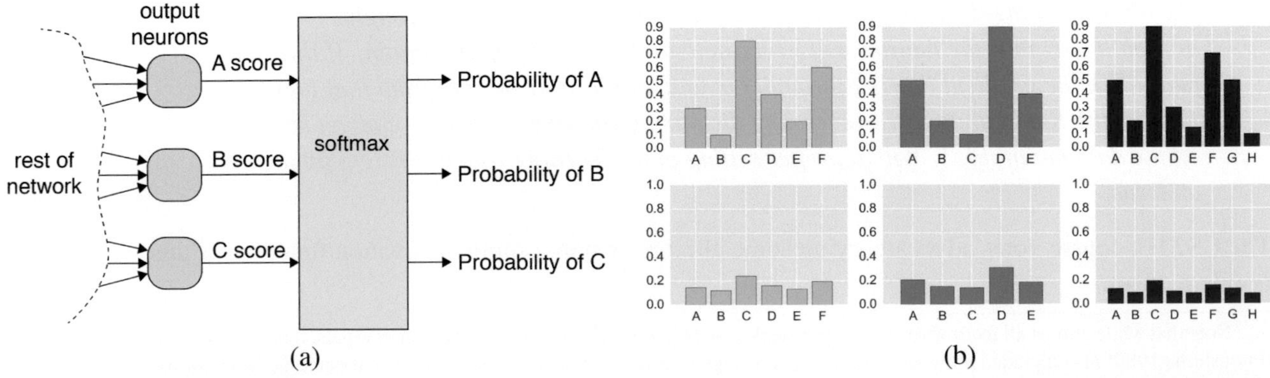

Figure 5.27 (a) A softmax layer used to convert from neural network activations ("score") to class likelihoods (b) The top row shows the activations, while the bottom shows the result of running the scores through softmax to obtain properly normalized likelihoods. © Glassner (2018).

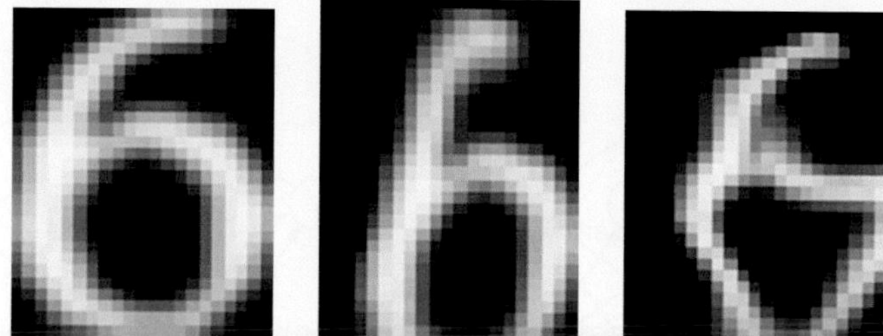

Figure 5.28 An original "6" digit from the MNIST database and two elastically distorted versions (Simard, Steinkraus, and Platt 2003) © 2003 IEEE.

For the final layer in networks used for classification, the softmax function (5.3) is normally used to convert from real-valued activations to class likelihoods, as shown in Figure 5.27. We can thus think of the penultimate set of neurons as determining directions in activation space that most closely match the log likelihoods of their corresponding class, while minimizing the log likelihoods of alternative classes. Since the inputs flow forward to the final output classes and probabilities, feedforward networks are discriminative, i.e., they have no statistical model of the classes they are outputting, nor any straightforward way to generate samples from such classes (but see Section 5.5.4 for techniques to do this).

5.3.3 Regularization and normalization

As with other forms of machine learning, regularization and other techniques can be used to prevent neural networks from overfitting so they can better generalize to unseen data. In this section, we discuss traditional methods such as regularization and data augmentation that can be applied to most machine learning systems, as well as techniques such as dropout and batch normalization, which are specific to neural networks.

Regularization and weight decay

As we saw in Section 4.1.1, quadratic or p-norm penalties on the weights (4.9) can be used to improve the conditioning of the system and to reduce overfitting. Setting $p = 2$ results in the usual L_2 regularization and makes large weights smaller, whereas using $p = 1$ is called *lasso* (least absolute shrinkage and selection operator) and can drive some weights all the way to zero. As the weights are being optimized inside a neural network, these terms make the weights smaller, so this kind of regularization is also known as *weight decay* (Bishop 2006, Section 3.1.4; Goodfellow, Bengio, and Courville 2016, Section 7.1; Zhang, Lipton *et al.* 2021, Section 4.5).[21] Note that for more complex optimization algorithms such as Adam, L_2 regularization and weight decay are *not* equivalent, but the desirable properties of weight decay can be restored using a modified algorithm (Loshchilov and Hutter 2019).

Dataset augmentation

Another powerful technique to reduce over-fitting is to add more training samples by perturbing the inputs and/or outputs of the samples that have already been collected. This technique is known as

[21]From a Bayesian perspective, we can also think of this penalty as a Gaussian prior on the weight distribution (Appendix B.4).

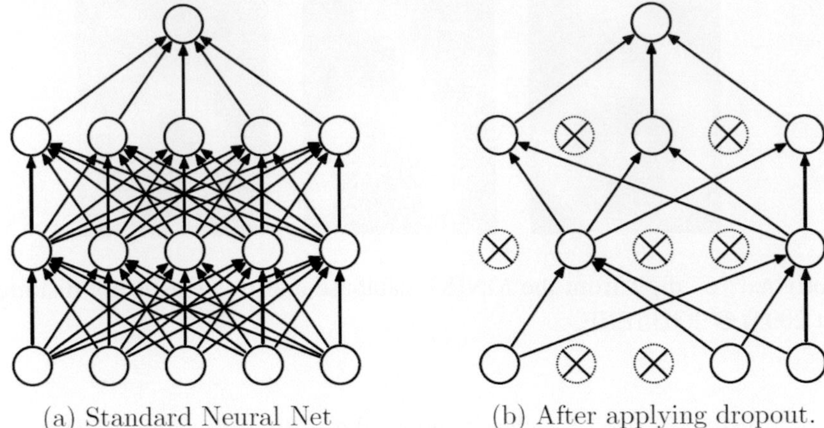

Figure 5.29 When using *dropout*, during training some fraction of units p is removed from the network (or, equivalently, clamped to zero) © Srivastava, Hinton *et al.* (2014). Doing this randomly for each mini-batch injects noise into the training process (at all levels of the network) and prevents the network from overly relying on particular units.

dataset augmentation (Zhang, Lipton *et al.* 2021, Section 13.1) and can be particularly effective on image classification tasks, since it is expensive to obtain labeled examples, and also since image classes should not change under small local perturbations.

An early example of such work applied to a neural network classification task is the *elastic distortion* technique proposed by Simard, Steinkraus, and Platt (2003). In their approach, random low-frequency displacement (warp) fields are synthetically generated for each training example and applied to the inputs during training (Figure 5.28). Note how such distortions are *not* the same as simply adding pixel noise to the inputs. Instead, distortions move pixels around, and therefore introduce much larger changes in the input vector space, while still preserving the semantic meaning of the examples (in this case, MNIST digits (LeCun, Cortes, and Burges 1998)).

Dropout

Dropout is a regularization technique introduced by Srivastava, Hinton *et al.* (2014), where at each mini-batch during training (Section 5.3.6), some percentage p (say 50%) of the units in each layer are clamped to zero, as shown in Figure 5.29. Randomly setting units to zero injects noise into the training process and also prevents the network from overly specializing units to particular samples or tasks, both of which can help reduce overfitting and improve generalization.

Because dropping (zeroing out) p of the units reduces the expected value of any sum the unit contributes to by a fraction $(1-p)$, the weighted sums s_i in each layer (5.45) are multiplied (during training) by $(1-p)^{-1}$. At test time, the network is run with no dropout and no compensation on the sums. A more detailed description of dropout can be found in Zhang, Lipton *et al.* (2021, Section 4.6) and Johnson (2020, Lecture 10).

Batch normalization

Optimizing the weights in a deep neural network, which we discuss in more detail in Section 5.3.6, is a tricky process and may be slow to converge.

One of the classic problems with iterative optimization techniques is poor *conditioning*, where the components of the gradient vary greatly in magnitude. While it is sometimes possible to reduce these effects with preconditioning techniques that scale individual elements in a gradient before taking a step (Section 5.3.6 and Appendix A.5.2), it is usually preferable to control the condition number of the system during the problem formulation.

In deep networks, one way in which poor conditioning can manifest itself is if the sizes of the weights or activations in successive layers become imbalanced. Say we take a given network and scale all of the weights in one layer by $100\times$ and scale down the weights in the next layer by the same amount. Because the ReLU activation function is linear in both of its domains, the outputs of the second layer will still be the same, although the activations at the output of the first layer with be 100 times larger. During the gradient descent step, the derivatives with respect to the weights will be vastly different after this rescaling, and will in fact be opposite in magnitude to the weights themselves, requiring tiny gradient descent steps to prevent overshooting (see Exercise 5.4).[22]

The idea behind batch normalization (Ioffe and Szegedy 2015) is to re-scale (and re-center) the activations at a given unit so that they have unit variance and zero mean (which, for a ReLU activation function, means that the unit will be active half the time). We perform this normalization by considering all of the training samples n in a given minibatch $\mathcal{B}$ (5.71) and computing the mean and variance statistics for unit i as

$$\mu_i = \frac{1}{|\mathcal{B}|} \sum_{n \in \mathcal{B}} s_i^{(n)} \tag{5.50}$$

$$\sigma_i^2 = \frac{1}{|\mathcal{B}|} \sum_{n \in \mathcal{B}} (s_i^{(n)} - \mu_i)^2 \tag{5.51}$$

$$\hat{s}_i^{(n)} = \frac{s_i^{(n)} - \mu_i}{\sqrt{\sigma_i^2 + \epsilon}}, \tag{5.52}$$

where s_i^n is the weighted sum of unit i for training sample n, $\hat{s}_i^{(n)}$ is the corresponding batch normalized sum, and ϵ (often 10^{-5}) is a small constant to prevent division by zero.

After batch normalization, the $\hat{s}_i^{(n)}$ activations now have zero mean and unit variance. However, this normalization may run at cross-purpose to the minimization of the loss function during training. For this reason, Ioffe and Szegedy (2015) add an extra gain γ_i and bias β_i parameter to each unit i and define the output of a batch normalization stage to be

$$y_i = \gamma_i \hat{s}_i + \beta_i. \tag{5.53}$$

These parameters act just like regular weights, i.e., they are modified using gradient descent during training to reduce the overall training loss.[23]

One subtlety with batch normalization is that the μ_i and σ_i^2 quantities depend analytically on all of the activation for a given unit in a minibatch. For gradient descent to be properly defined, the derivatives of the loss function with respect to these variables, and the derivatives of the quantities $\hat{s}_i$ and y_i with respect to these variables, must be computed as part of the gradient computation step,

[22] This motivating paragraph is my own explanation of why batch normalization might be a good idea, and is related to the idea that batch normalization reduces *internal covariate shift*, used by (Ioffe and Szegedy 2015) to justify their technique. This hypothesis is now being questioned and alternative theories are being developed (Bjorck, Gomes *et al.* 2018; Santurkar, Tsipras *et al.* 2018; Kohler, Daneshmand *et al.* 2019).

[23] There is a trick used by those in the know, which relies on the observation that any bias term b_i in the original summation s_i (5.45) shows up in the mean μ_i and gets subtracted out. For this reason, the bias term is often omitted when using batch (or other kinds of) normalization.

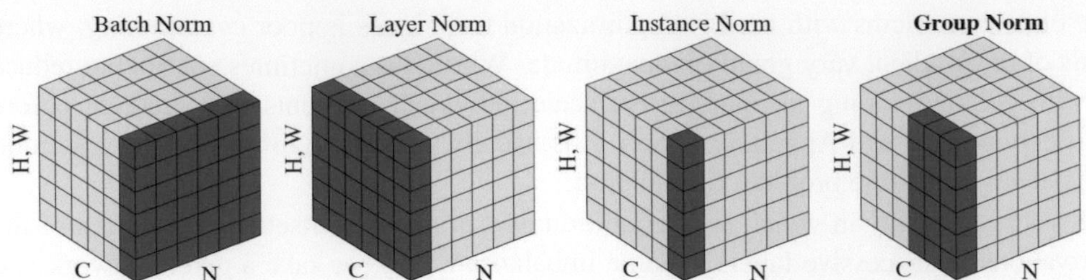

Figure 5.30 Batch norm, layer norm, instance norm, and group norm, from Wu and He (2018) © 2018 Springer. The (H, W) dimension denotes pixels, C denotes channels, and N denotes training samples in a minibatch. The pixels in blue are normalized by the same mean and variance.

using similar chain rule computations as the original backpropagation algorithm (5.65–5.68). These derivations can be found in Ioffe and Szegedy (2015) as well as several blogs.[24]

When batch normalization is applied to convolutional layers (Section 5.4), one could in principle compute a normalization separately for each pixel, but this would add a tremendous number of extra learnable bias and gain parameters (β_i, γ_i). Instead, batch normalization is usually implemented by computing the statistics as sums over all the pixels with the same convolution kernel, and then adding a single bias and gain parameter for each convolution kernel (Ioffe and Szegedy 2015; Johnson 2020, Lecture 10; Zhang, Lipton *et al.* 2021, Section 7.5).

Having described how batch normalization operates during training, we still need to decide what to do at test or inference time, i.e., when applying the trained network to new data. We cannot simply skip this stage, as the network was trained while removing common mean and variance estimates. For this reason, the mean and variance estimates are usually recomputed over the *whole* training set, or some running average of the per-batch statistics are used. Because of the linear form of (5.45) and (5.52–5.53), it is possible to fold the μ_i and σ_i estimates and learned (β_i, γ_i) parameters into the original weight and bias terms in (5.45).

Since the publication of the seminal paper by Ioffe and Szegedy (2015), a number of variants have been developed, some of which are illustrated in Figure 5.30. Instead of accumulating statistics over the samples in a minibatch $\mathcal{B}$, we can compute them over different subsets of activations in a layer. These subsets include:

- all the activations in a layer, which is called *layer normalization* (Ba, Kiros, and Hinton 2016);
- all the activations in a given convolutional output channel (see Section 5.4), which is called *instance normalization* (Ulyanov, Vedaldi, and Lempitsky 2017);
- different sub-groups of output channels, which is called *group normalization* (Wu and He 2018).

The paper by Wu and He (2018) describes each of these in more detail and also compares them experimentally. More recent work by Qiao, Wang *et al.* (2019a) and Qiao, Wang *et al.* (2019b) discusses some of the disadvantages of these newer variants and proposes two new techniques called *weight standardization* and *batch channel normalization* to mitigate these problems.

Instead of modifying the activations in a layer using their statistics, it is also possible to modify the weights in a layer to explicitly make the weight norm and weight vector direction separate

[24] https://kratzert.github.io/2016/02/12/understanding-the-gradient-flow-through-the-batch-normalization-layer.html, https://kevinzakka.github.io/2016/09/14/batch_normalization, https://deepnotes.io/batchnorm

parameters, which is called *weight normalization* (Salimans and Kingma 2016). A related technique called *spectral normalization* (Miyato, Kataoka *et al.* 2018) constrains the largest singular value of the weight matrix in each layer to be 1.

The bias and gain parameters (β_i, γ_i) may also depend on the activations in some other layer in the network, e.g., derived from a guide image.[25] Such techniques are referred to as *conditional batch normalization* and have been used to select between different artistic styles (Dumoulin, Shlens, and Kudlur 2017) and to enable local semantic guidance in image synthesis (Park, Liu *et al.* 2019). Related techniques and applications are discussed in more detail in Section 14.6 on neural rendering.

The reasons why batch and other kinds of normalization help deep networks converge faster and generalize better are still being debated. Some recent papers on this topic include Bjorck, Gomes *et al.* (2018), Hoffer, Banner *et al.* (2018), Santurkar, Tsipras *et al.* (2018), and Kohler, Daneshmand *et al.* (2019).

5.3.4 Loss functions

In order to optimize the weights in a neural network, we need to first define a *loss function* that we minimize over the training examples. We have already seen the main loss functions used in machine learning in previous parts of this chapter.

For classification, most neural networks use a final softmax layer (5.3), as shown in Figure 5.27. Since the outputs are meant to be class probabilities that sum up to 1, it is natural to use the cross-entropy loss given in (5.19) or (5.23–5.24) as the function to minimize during training. Since in our description of the feedforward networks we have used indices i and j to denote neural units, we will, in this section, use n to index a particular training example.

The multi-class cross-entropy loss can thus be re-written as

$$E(\mathbf{w}) = \sum_n E_n(\mathbf{w}) = -\sum_n \log p_{nt_n}, \tag{5.54}$$

where $\mathbf{w}$ is the vector of all weights, biases, and other model parameters, and p_{nk} is the network's current estimate of the probability of class k for sample n, and t_n is the integer denoting the correct class. Substituting the definition of p_{nk} from (5.20) with the appropriate replacement of l_{ik} with s_{nk} (the notation we use for neural nets), we get

$$E_n(\mathbf{w}) = \log Z_n - s_{nt_n} \tag{5.55}$$

with $Z_n = \sum_j \exp s_{nj}$. Gómez (2018) has a nice discussion of some of the losses widely used in deep learning.

For networks that perform regression, i.e., generate one or more continuous variables such as depth maps or denoised images, it is common to use an L_2 loss,

$$E(\mathbf{w}) = \sum_n E_n(\mathbf{w}) = -\sum_n \|\mathbf{y}_n - \mathbf{t}_n\|^2, \tag{5.56}$$

where $\mathbf{y}_n$ is the network output for sample n and $\mathbf{t}_n$ is the corresponding training (target) value, since this is a natural measure of error between continuous variables. However, if we believe there may be outliers in the training data, or if gross errors are not so harmful as to merit a quadratic penalty, more robust norms such as L_1 can be used (Barron 2019; Ranftl, Lasinger *et al.* 2020). (It

[25] Note that this gives neural networks the ability to multiply two layers in a network, which we used previously to perform locally (Section 3.5.5).

is also possible to use robust norms for classification, e.g., adding an outlier probability to the class labels.)

As it is common to interpret the final outputs of a network as a probability distribution, we need to ask whether it is wise to use such probabilities as a measure of confidence in a particular answer. If a network is properly trained and predicting answers with good accuracy, it is tempting to make this assumption. The training losses we have presented so far, however, only encourage the network to maximize the probability-weighted correct answers, and do not, in fact, encourage the network outputs to be properly *confidence calibrated.* Guo, Pleiss *et al.* (2017) discuss this issue, and present some simple measures, such as multiplying the log-likelihoods by a *temperature* (Platt 2000a), to improve the match between classifier probabilities and true reliability. The GrokNet image recognition system (Bell, Liu *et al.* 2020), which we discuss in Section 6.2.3, uses calibration to obtain better attribute probability estimates.

For networks that hallucinate new images, e.g., when introducing missing high-frequency details (Section 10.3) or doing image transfer tasks (Section 14.6), we may want to use a *perceptual loss* (Johnson, Alahi, and Fei-Fei 2016; Dosovitskiy and Brox 2016; Zhang, Isola *et al.* 2018), which uses intermediate layer neural network responses as the basis of comparison between target and output images. It is also possible to train a separate *discriminator* network to evaluate the quality (and plausibility) of synthesized images, as discussed in Section 5.5.4 More details on the application of loss functions to image synthesis can be found in Section 14.6 on neural rendering.

While loss functions are traditionally applied to supervised learning tasks, where the correct label or target value $\mathbf{t}_n$ is given for each input, it is also possible to use loss functions in an unsupervised setting. An early example of this was the *contrastive loss* function proposed by Hadsell, Chopra, and LeCun (2006) to cluster samples that are similar together while spreading dissimilar samples further apart. More formally, we are given a set of inputs $\{\mathbf{x}_i\}$ and pairwise indicator variables $\{t_{ij}\}$ that indicate whether two inputs are similar.[26] The goal is now to compute an embedding $\mathbf{v}_i$ for each input $\mathbf{x}_i$ such that similar input pairs have similar embeddings (low distances), while dissimilar inputs have large embedding distances. Finding mappings or embeddings that create useful distances between samples is known as *(distance) metric learning* (Köstinger, Hirzer *et al.* 2012; Kulis 2013) and is a commonly used tool in machine learning. The losses used to encourage the creation of such meaningful distances are collectively known as *ranking losses* (Gómez 2019) and can be used to relate features from different domains such as text and images (Karpathy, Joulin, and Fei-Fei 2014).

The contrastive loss from (Hadsell, Chopra, and LeCun 2006) is defined as

$$E_{\mathrm{CL}} = \sum_{(i,j)\in\mathcal{P}} \{t_{ij} \log L_{\mathrm{S}}(d_{ij}) + (1 - t_{ij}) \log L_{\mathrm{D}}(d_{ij})\}, \tag{5.57}$$

where $\mathcal{P}$ is the set of all labeled input pairs, L_{S} and L_{D} are the similar and dissimilar loss functions, and $d_{ij} = \|\mathbf{v}_i - \mathbf{v}_j\|$ are the pairwise distance between paired embeddings.[27] This has a form similar to the cross-entropy loss given in (5.19), except that we measure squared distances between encodings $\mathbf{v}_i$ and $\mathbf{v}_j$. In their paper, Hadsell, Chopra, and LeCun (2006) suggest using a quadratic function for L_{S} and a quadratic hinge loss (c.f. (5.30)) $L_{\mathrm{D}} = [m - d_{ij}]_+^2$ for dissimilarity, where m is a margin beyond which there is no penalty.

To train with a contrastive loss, you can run both pairs of inputs through the neural network, compute the loss, and then backpropagate the gradients through both instantiations (activations) of the network. This can also be thought of as constructing a *Siamese network* consisting of two copies with shared weights (Bromley, Guyon *et al.* 1994; Chopra, Hadsell, and LeCun 2005). It is

[26] Indicator variables are often denoted as y_{ij}, but we will stick to the t_{ij} notation to be consistent with Section 5.1.3.

[27] In metric learning, the embeddings are very often normalized to unit length.

also possible to construct a *triplet loss* that takes as input a pair of matching samples and a third non-matching sample and ensures that the distance between non-matching samples is greater than the distance between matches plus some margin (Weinberger and Saul 2009; Weston, Bengio, and Usunier 2011; Schroff, Kalenichenko, and Philbin 2015; Rawat and Wang 2017).

Both pairwise contrastive and triplet losses can be used to learn embeddings for visual similarity search (Bell and Bala 2015; Wu, Manmatha *et al.* 2017; Bell, Liu *et al.* 2020), as discussed in more detail in Section 6.2.3. They have also been recently used for unsupervised pre-training of neural networks (Wu, Xiong *et al.* 2018; He, Fan *et al.* 2020; Chen, Kornblith *et al.* 2020), which we discuss in Section 5.4.7. In this case, it is more common to use a different contrastive loss function, inspired by softmax (5.3) and multi-class cross-entropy (5.20–5.22), which was first proposed by (Sohn 2016). Before computing the loss, the embeddings are all normalized to unit norm, $\|\hat{\mathbf{v}}_i\|^2 = 1$. Then, the following loss is summed over all matching embeddings,

$$l_{ij} = -\log \frac{\exp(\hat{\mathbf{v}}_i \cdot \hat{\mathbf{v}}_j/\tau)}{\sum_k \exp(\hat{\mathbf{v}}_i \cdot \hat{\mathbf{v}}_k/\tau)}, \tag{5.58}$$

with the denominator summed over non-matches as well. The τ variable denotes the "temperature" and controls how tight the clusters will be; it is sometimes replaced with an s multiplier parameterizing the hyper-sphere radius (Deng, Guo *et al.* 2019). The exact details of how the matches are computed vary by exact implementation.

This loss goes by several names, including InfoNCE (Oord, Li, and Vinyals 2018), and NT-Xent (normalized temperature cross-entropy loss) in Chen, Kornblith *et al.* (2020). Generalized versions of this loss called SphereFace, CosFace, and ArcFace are discussed and compared in the ArcFace paper (Deng, Guo *et al.* 2019) and used by Bell, Liu *et al.* (2020) as part of their visual similarity search system. The *smoothed average precision* loss recently proposed by Brown, Xie *et al.* (2020) can sometimes be used as an alternative to the metric losses discussed in this section. Some recent papers that compare and discuss deep metric learning approaches include (Jacob, Picard *et al.* 2019; Musgrave, Belongie, and Lim 2020).

Weight initialization

Before we can start optimizing the weights in our network, we must first initialize them. Early neural networks used small random weights to break the symmetry, i.e., to make sure that all of the gradients were not zero. It was observed, however, that in deeper layers, the activations would get progressively smaller.

To maintain a comparable variance in the activations of successive layers, we must take into account the *fan-in* of each layer, i.e., the number of incoming connections where activations get multiplied by weights. Glorot and Bengio (2010) did an initial analysis of this issue, and came up with a recommendation to set the random initial weight variance as the inverse of the fan-in. Their analysis, however, assumed a linear activation function (at least around the origin), such as a tanh function.

Since most modern deep neural networks use the ReLU activation function (5.49), He, Zhang *et al.* (2015) updated this analysis to take into account this asymmetric non-linearity. If we initialize the weights to have zero mean and variance V_l for layer l and set the original biases to zero, the linear summation in (5.45) will have a variance of

$$Var[s_l] = n_l V_l E[x_l^2], \tag{5.59}$$

where n_l is the number of incoming activations/weights and $E[x_l^2]$ is the expectation of the squared incoming activations. When the summations s_l, which have zero mean, are fed through the ReLU, the negative ones will get clamped to zero, so the expectation of the squared output $E[y_l^2]$ is half the variance of s_l, $Var[s_l]$.

In order to avoid decaying or increasing average activations in deeper layers, we want the magnitude of the activations in successive layers to stay about the same. Since we have

$$E[y_l^2] = \frac{1}{2}Var[s_l] = \frac{1}{2}n_l V_l E[x_l^2], \tag{5.60}$$

we conclude that the variance in the initial weights V_l should be set to

$$V_l = \frac{2}{n_l}, \tag{5.61}$$

i.e., the inverse of half the fan-in of a given unit or layer. This weight initialization rule is commonly called *He initialization*.

Neural network initialization continues to be an active research area, with publications that include Krähenbühl, Doersch *et al.* (2016), Mishkin and Matas (2016), Frankle and Carbin (2019), and Zhang, Dauphin, and Ma (2019)

5.3.5 Backpropagation

Once we have set up our neural network by deciding on the number of layers, their widths and depths, added some regularization terms, defined the loss function, and initialized the weights, we are ready to train the network with our sample data. To do this, we use gradient descent or one of its variants to iteratively modify the weights until the network has converged to a good set of values, i.e., an acceptable level of performance on the training and validation data.

To do this, we compute the derivatives (gradients) of the loss function E_n for training sample n with respect to the weights $\mathbf{w}$ using the chain rule, starting with the outputs and working our way back through the network towards the inputs, as shown in Figure 5.31. This procedure is known as *backpropagation* (Rumelhart, Hinton, and Williams 1986b) and stands for *backward propagation of errors*. You can find alternative descriptions of this technique in textbooks and course notes on deep learning, including Bishop (2006, Section 5.3.1), Goodfellow, Bengio, and Courville (2016, Section 6.5), Glassner (2018, Chapter 18), Johnson (2020, Lecture 6), and Zhang, Lipton *et al.* (2021).

Recall that in the forward (evaluation) pass of a neural network, activations (layer outputs) are computed layer-by-layer, starting with the first layer and finishing at the last. We will see in the next section that many newer DNNs have an acyclic graph structure, as shown in Figures 5.42–5.43, rather than just a single linear pipeline. In this case, any breadth-first traversal of the graph can be used. The reason for this evaluation order is computational efficiency. Activations need only be computed once for each input sample and can be re-used in succeeding stages of computation.

During backpropagation, we perform a similar breadth-first traversal of the reverse graph. However, instead of computing activations, we compute derivatives of the loss with respect to the weights and inputs, which we call *errors*. Let us look at this in more detail, starting with the loss function.

The derivative of the cross-entropy loss E_n (5.54) with respect to the output probability p_{nk} is simply $-\delta_{nt_n}/p_{nk}$. What is more interesting is the derivative of the loss with respect to the *scores* s_{nk} going into the softmax layer (5.55) shown in Figure 5.27,

$$\frac{\partial E_n}{\partial s_{nk}} = -\delta_{nt_n} + \frac{1}{Z_n}\exp s_{nk} = p_{nk} - \delta_{nt_n} = p_{nk} - \tilde{t}_{nk}. \tag{5.62}$$

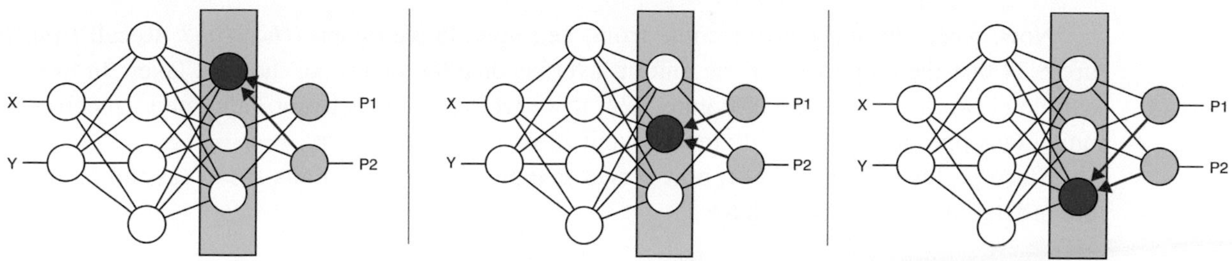

Figure 5.31 Backpropagating the derivatives (errors) through an intermediate layer of the deep network © Glassner (2018). The derivatives of the loss function applied to a single training example with respect to each of the pink unit inputs are summed together and the process is repeated chaining backward through the network.

(The last form is useful if we are using one-hot encoding or the targets have non-binary probabilities.) This has a satisfyingly intuitive explanation as the difference between the predicted class probability p_{nk} and the true class identity t_{nk}.

For the L_2 loss in (5.56), we get a similar result,

$$\frac{\partial E_n}{\partial y_{nk}} = y_{nk} - t_{nk}, \tag{5.63}$$

which in this case denotes the real-valued difference between the predicted and target values.

In the rest of this section, we drop the sample index n from the activations x_{in} and y_{in}, since the derivatives for each sample n can typically be computed independently from other samples.[28]

To compute the partial derivatives of the loss term with respect to earlier weights and activations, we work our way back through the network, as shown in Figure 5.31. Recall from (5.45–5.46) that we first compute a weighted sum s_i by taking a dot product between the input activations $\mathbf{x}_i$ and the unit's weight vector $\mathbf{w}_i$,

$$s_i = \mathbf{w}_i^T \mathbf{x}_i + b_i = \sum_j w_{ij} x_{ij} + b_i. \tag{5.64}$$

We then pass this weighted sum through an activation function h to obtain $y_i = h(s_i)$.

To compute the derivative of the loss E_n with respect to the weights, bias, and input activations, we use the chain rule,

$$e_i = \frac{\partial E_n}{\partial s_i} = h'(s_i)\frac{\partial E_n}{\partial y_i}, \tag{5.65}$$

$$\frac{\partial E_n}{\partial w_{ij}} = x_{ij}\frac{\partial E_n}{\partial s_i} = x_{ij} e_i, \tag{5.66}$$

$$\frac{\partial E_n}{\partial b_i} = \frac{\partial E_n}{\partial s_i} = e_i, \quad \text{and} \tag{5.67}$$

$$\frac{\partial E_n}{\partial x_{ij}} = w_{ij}\frac{\partial E_n}{\partial s_i} = w_{ij} e_i. \tag{5.68}$$

We call the term $e_i = \partial E_n / \partial s_i$, i.e., the partial derivative of the loss E_n with respect to the summed activation s_i, the *error*, as it gets propagated backward through the network.

[28] This is not the case if batch or other kinds of normalization (Section 5.3.3) are being used. For batch normalization, we have to accumulate the statistics across all the samples in the batch and then take their derivatives with respect to each weight (Ioffe and Szegedy 2015). For instance and group norm, we compute the statistics across all the pixels in a given channel or group, and then have to compute these additional derivatives as well.

Now, where do these errors come from, i.e., how do we obtain $\partial E_n / \partial y_i$? Recall from Figure 5.24 that the outputs from one unit or layer become the inputs for the next layer. In fact, for a simple network like the one in Figure 5.24, if we let x_{ij} be the activation that unit i receives from unit j (as opposed to just the jth input to unit i), we can simply set $x_{ij} = y_j$.

Since y_i, the output of unit i, now serves as input for the other units $k > i$ (assuming the units are ordered breadth first), we have

$$\frac{\partial E_n}{\partial y_i} = \sum_{k>i} \frac{\partial E_n}{\partial x_{ki}} = \sum_{k>i} w_{ki} e_k \tag{5.69}$$

and

$$e_i = h'(s_i) \frac{\partial E_n}{\partial y_i} = h'(s_i) \sum_{k>i} w_{ki} e_k. \tag{5.70}$$

In other words, to compute a unit's (backpropagation) error, we compute a weighted sum of the errors coming from the units it feeds into and then multiply this by the derivative of the current activation function $h'(s_i)$. This backward flow of errors is shown in Figure 5.31, where the errors for the three units in the shaded box are computed using weighted sums of the errors coming from later in the network.

This backpropagation rule has a very intuitive explanation. The error (derivative of the loss) for a given unit depends on the errors of the units that it feeds multiplied by the weights that couple them together. This is a simple application of the chain rule. The slope of the activation function $h'(s_i)$ modulates this interaction. If the unit's output is clamped to zero or small, e.g., with a negative-input ReLU or the "flat" part of a sigmoidal response, the unit's error is itself zero or small. The gradient of the weight, i.e., how much the weight should be perturbed to reduce the loss, is a signed product of the incoming activation and the unit's error, $x_{ij} e_i$. This is closely related to the Hebbian update rule (Hebb 1949), which observes that synaptic efficiency in biological neurons increases with correlated firing in the presynaptic and postsynaptic cells. An easier way to remember this rule is "neurons wire together if they fire together" (Lowel and Singer 1992).

There are, of course, other computational elements in modern neural networks, including convolutions and pooling, which we cover in the next section. The derivatives and error propagation through such other units follows the same procedure as we sketched here, i.e., recursively apply the chain rule, taking analytic derivatives of the functions being applied, until you have the derivatives of the loss function with respect to all the parameters being optimized, i.e., the gradient of the loss.

As you may have noticed, the computation of the gradients with respect to the weights requires the unit activations computed in the forward pass. A typical implementation of neural network training stores the activations for a given sample and uses these during the backprop (backward error propagation) stage to compute the weight derivatives. Modern neural networks, however, may have millions of units and hence activations (Figure 5.44). The number of activations that need to be stored can be reduced by only storing them at certain layers and then re-computing the rest as needed, which goes under the name *gradient checkpointing* (Griewank and Walther 2000; Chen, Xu *et al.* 2016; Bulatov 2018).[29] A more extensive review of low-memory training can be found in the technical report by Sohoni, Aberger *et al.* (2019).

5.3.6 Training and optimization

At this point, we have all of the elements needed to train a neural network. We have defined the network's topology in terms of the sizes and depths of each layer, specified our activation functions,

[29]This name seems a little weird, since it's actually the activations that are saved instead of the gradients.

added regularization terms, specified our loss function, and initialized the weights. We have even described how to compute the gradients, i.e., the derivatives of the regularized loss with respect to all of our weights. What we need at this point is some algorithm to turn these gradients into weight updates that will optimize the loss function and produce a network that generalizes well to new, unseen data.

In most computer vision algorithms such as optical flow (Section 9.1.3), 3D reconstruction using bundle adjustment (Section 11.4.2), and even in smaller-scale machine learning problems such as logistic regression (Section 5.1.3), the method of choice is linearized least squares (Appendix A.3). The optimization is performed using a second-order method such as Gauss-Newton, in which we evaluate all of the terms in our loss function and then take an optimally-sized downhill step using a direction derived from the gradients and the Hessian of the energy function.

Unfortunately, deep learning problems are far too large (in terms of number of parameters and training samples; see Figure 5.44) to make this approach practical. Instead, practitioners have developed a series of optimization algorithms based on extensions to *stochastic gradient descent* (SGD) (Zhang, Lipton *et al.* 2021, Chapter 11). In SGD, instead of evaluating the loss function by summing over all the training samples, as in (5.54) or (5.56), we instead just evaluate a single training sample n and compute the derivatives of the associated loss $E_n(\mathbf{w})$. We then take a tiny downhill step along the direction of this gradient.

In practice, the directions obtained from just a single sample are incredibly noisy estimates of a good descent direction, so the losses and gradients are usually summed over a small subset of the training data,

$$E_{\mathcal{B}}(\mathbf{w}) = \sum_{n \in \mathcal{B}} E_n(\mathbf{w}), \tag{5.71}$$

where each subset $\mathcal{B}$ is called a *minibatch*. Before we start to train, we randomly assign the training samples into a fixed set of minibatches, each of which has a fixed size that commonly ranges from 32 at the low end to 8k at the higher end (Goyal, Dollár *et al.* 2017). The resulting algorithm is called *minibatch stochastic gradient descent*, although in practice, most people just call it SGD (omitting the reference to minibatches).[30]

After evaluating the gradients $\mathbf{g} = \nabla_{\mathbf{w}} E_{\mathcal{B}}$ by summing over the samples in the minibatch, it is time to update the weights. The simplest way to do this is to take a small step in the gradient direction,

$$\mathbf{w} \leftarrow \mathbf{w} - \alpha \mathbf{g} \quad \text{or} \tag{5.72}$$

$$\mathbf{w}_{t+i} = \mathbf{w}_t - \alpha_t \mathbf{g}_t \tag{5.73}$$

where the first variant looks more like an assignment statement (see, e.g., Zhang, Lipton *et al.* 2021, Chapter 11; Loshchilov and Hutter 2019), while the second makes the temporal dependence explicit, using t to denote each successive step in the gradient descent.[31]

The step size parameter α is often called the *learning rate* and must be carefully adjusted to ensure good progress while avoiding overshooting and exploding gradients. In practice, it is common to start with a larger (but still small) learning rate α_t and to decrease it over time so that the optimization settles into a good minimum (Johnson 2020, Lecture 11; Zhang, Lipton *et al.* 2021, Chapter 11).

[30] In the deep learning community, classic algorithms that sum over all the measurements are called *batch gradient descent*, although this term is not widely used elsewhere, as it is assumed that using all measurement at once is the preferred approach. In large-scale problems such as bundle adjustment, it's possible that using minibatches may result in better performance, but this has so far not been explored.

[31] I use the index k in discussing iterative algorithms in Appendix A.5.

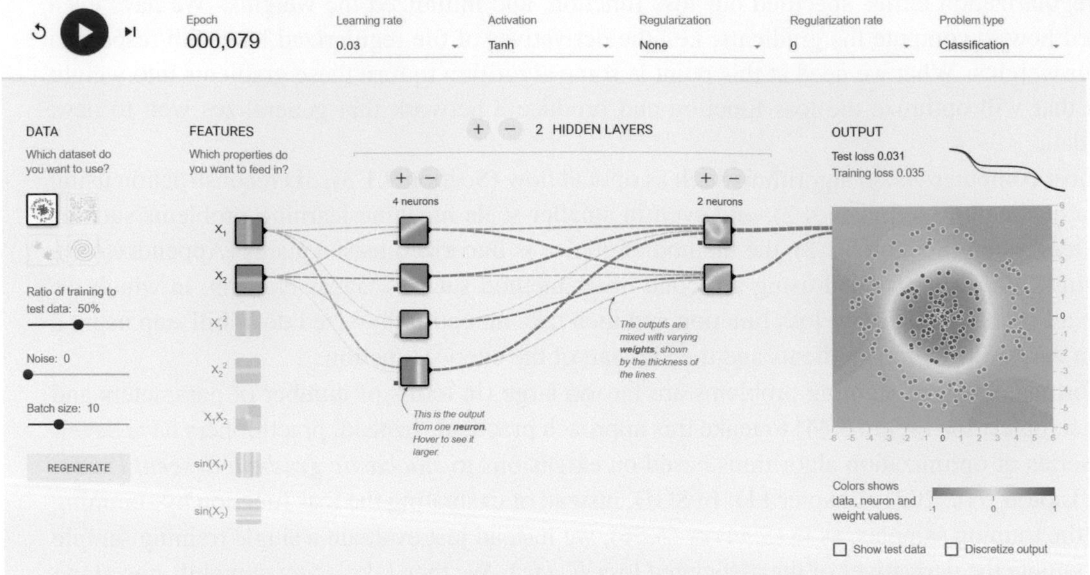

Figure 5.32 Screenshot from http://playground.tensorflow.org, where you can build and train your own small network in your web browser. Because the input space is two-dimensional, you can visualize the responses to all 2D inputs at each unit in the network.

Regular gradient descent is prone to stalling when the current solution reaches a "flat spot" in the search space, and stochastic gradient descent only pays attention to the errors in the current minibatch. For these reasons, the SGD algorithms may use the concept of *momentum*, where an exponentially decaying ("leaky") running average of the gradients is accumulated and used as the update direction,

$$\mathbf{v}_{t+i} = \rho \mathbf{v}_t + \mathbf{g}_t \tag{5.74}$$

$$\mathbf{w}_{t+i} = \mathbf{w}_t - \alpha_t \mathbf{v}_t. \tag{5.75}$$

A relatively large value of $\rho \in [0.9, 0.99]$ is used to give the algorithm good memory, effectively averaging gradients over more batches.[32]

Over the last decade, a number of more sophisticated optimization techniques have been applied to deep network training, as described in more detail in Johnson (2020, Lecture 11) and Zhang, Lipton *et al.* (2021, Chapter 11)). These algorithms include:

- *Nesterov momentum*, where the gradient is (effectively) computed at the state predicted from the velocity update;
- *AdaGrad* (Adaptive Gradient), where each component in the gradient is divided by the square root of the per-component summed squared gradients (Duchi, Hazan, and Singer 2011);
- *RMSProp*, where the running sum of squared gradients is replaced with a leaky (decaying) sum (Hinton 2012);
- *Adadelta*, which augments RMSProp with a leaky sum of the actual per-component changes in the parameters and uses these in the gradient re-scaling equation (Zeiler 2012);

[32]Note that a recursive formula such as (5.74), which is the same as a temporal infinite impulse response filter (3.2.3) converges in the limit to a value of $\mathbf{g}/(1-\rho)$, so α needs to be correspondingly adjusted.

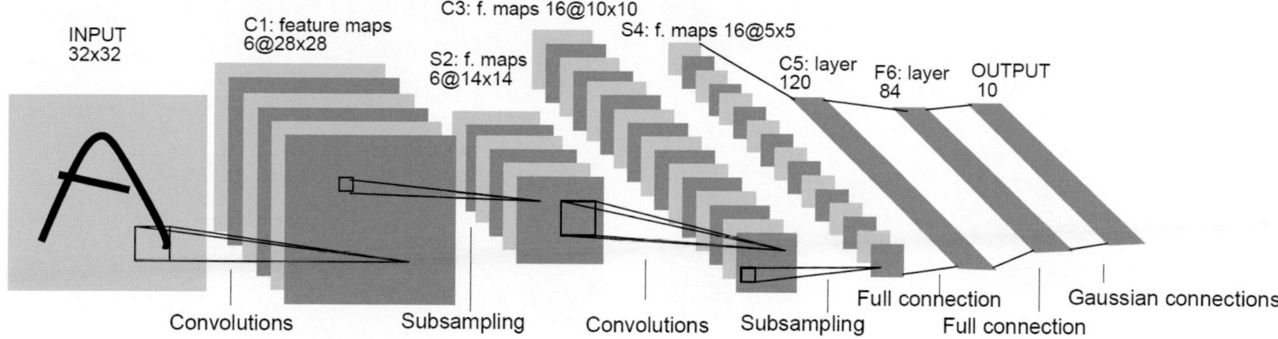

Figure 5.33 Architecture of LeNet-5, a convolutional neural network for digit recognition (LeCun, Bottou *et al.* 1998) © 1998 IEEE. This network uses multiple channels in each layer and alternates multi-channel convolutions with downsampling operations, followed by some fully connected layers that produce one activation for each of the 10 digits being classified.

- *Adam*, which combines elements of all the previous ideas into a single framework and also de-biases the initial leaky estimates (Kingma and Ba 2015); and
- *AdamW*, which is Adam with decoupled weight decay (Loshchilov and Hutter 2019).

Adam and AdamW are currently the most popular optimizers for deep networks, although even with all their sophistication, learning rates need to be set carefully (and probably decayed over time) to achieve good results. Setting the right *hyperparameters*, such as the learning rate initial value and decay rate, momentum terms such as ρ, and amount of regularization, so that the network achieves good performance within a reasonable training time is itself an open research area. The lecture notes by Johnson (2020, Lecture 11) provide some guidance, although in many cases, people perform a search over hyperparameters to find which ones produce the best performing network.

A simple two-input example

A great way to get some intuition on how deep networks update the weights and carve out a solution space during training is to play with the interactive visualization at http://playground.tensorflow.org.[33] As shown in Figure 5.32, just click the "run" (▷) button to get started, then reset the network to a new start (button to the left of run) and try single-stepping the network, using different numbers of units per hidden layer and different activation functions. Especially when using ReLUs, you can see how the network carves out different parts of the input space and then combines these sub-pieces together. Section 5.4.5 discusses visualization tools to get insights into the behavior of larger, deeper networks.

5.4 Convolutional neural networks

The previous sections on deep learning have covered all of the essential elements of constructing and training deep networks. However, they have omitted what is likely the most crucial component of deep networks for image processing and computer vision, which is the use of trainable multi-layer

[33] Additional informative interactive demonstrations can be found at https://cs.stanford.edu/people/karpathy/convnetjs.

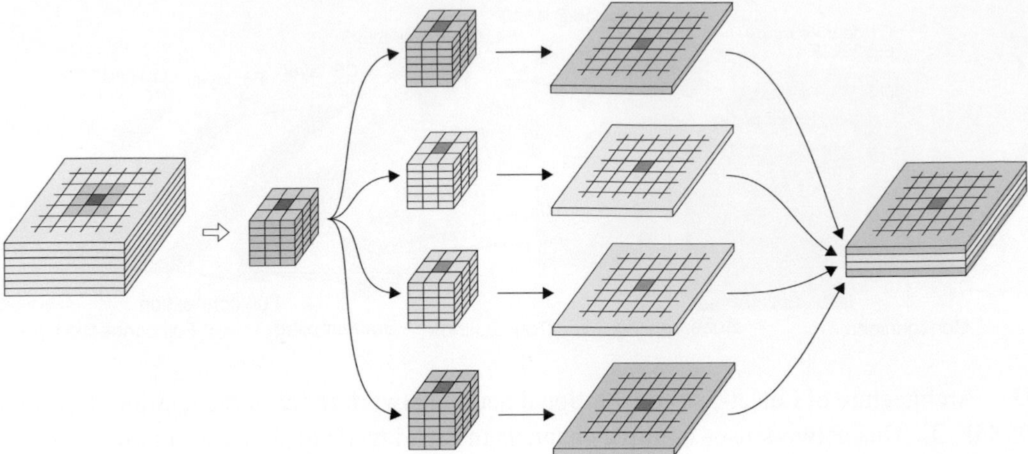

Figure 5.34 2D convolution with multiple input and output channels © Glassner (2018). Each 2D convolution kernel takes as input all of the C_1 channels in the preceding layer, windowed to a small area, and produces the values (after the activation function non-linearity) in one of the C_2 channels in the next layer. For each of the output channels, we have $S^2 \times C_1$ kernel weights, so the total number of learnable parameters in each convolutional layer is $S^2 \times C_1 \times C_2$. In this figure, we have $C_1 = 6$ input channels and $C_2 = 4$ output channels, with an $S = 3$ convolution window, for a total of $9 \times 6 \times 4$ learnable weights, shown in the middle column of the figure. Since the convolution is applied at each of the $W \times H$ pixels in a given layer, the amount of computation (multiply-adds) in each forward and backward pass over one sample in a given layer is $WHS^2C_1C_2$.

convolutions. The idea of convolutional neural networks was popularized by LeCun, Bottou *et al.* (1998), where they introduced the LeNet-5 network for digit recognition shown in Figure 5.33.[34]

Instead of connecting all of the units in a layer to all the units in a preceding layer, convolutional networks organize each layer into *feature maps* (LeCun, Bottou *et al.* 1998), which you can think of as parallel planes or *channels*, as shown in Figure 5.33. In a convolutional layer, the weighted sums are only performed within a small local window, and weights are identical for all pixels, just as in regular shift-invariant image convolution and correlation (3.12–3.15).

Unlike image convolution, however, where the same filter is applied to each (color) channel, neural network convolutions typically linearly combine the activations from each of the C_1 input channels in a previous layer and use different convolution kernels for each of the C_2 output channels, as shown in Figures 5.34–5.35.[35] This makes sense, as the main task in convolutional neural network layers is to construct local features (Figure 3.40c) and to then combine them in different ways to produce more discriminative and semantically meaningful features.[36] Visualizations of the kinds of features that deep networks extract are shown in Figure 5.47 in Section 5.4.5.

With these intuitions in place, we can write the weighted linear sums (5.45) performed in a convolutional layer as

$$s(i,j,c_2) = \sum_{c_1 \in \{C_1\}} \sum_{(k,l) \in \mathcal{N}} w(k,l,c_1,c_2) x(i+k, j+l, c_1) + b(c_2), \tag{5.76}$$

[34]A similar convolutional architecture, but without the gradient descent training procedure, was earlier proposed by Fukushima (1980).

[35]The number of channels in a given network layer is sometimes called its *depth*, but the number of layers in a deep network is also called its depth. So, be careful when reading network descriptions.

[36]Note that pixels in different input and output channels (within the convolution window size) are fully connected, unless grouped convolutions, discussed below, are used.

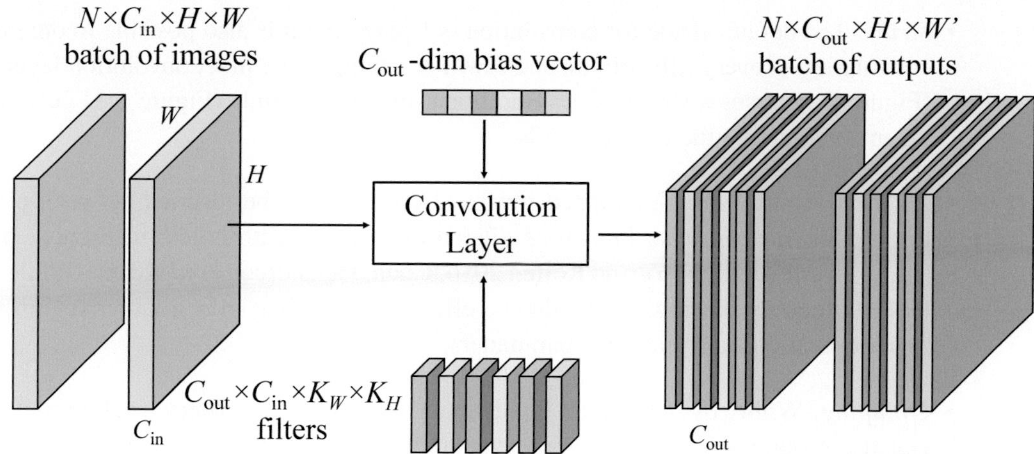

Figure 5.35 2D convolution with multiple batches, input, and output channels, © Johnson (2020). When doing mini-batch gradient descent, a whole batch of training images or features is passed into a convolutional layer, which takes as input all of the C_{in} channels in the preceding layer, windowed to a small area, and produces the values (after the activation function non-linearity) in one of the C_{out} channels in the next layer. As before, for each of the output channels, we have $K_w \times K_h \times C_{\text{in}}$ kernel weights, so the total number of learnable parameters in each convolutional layer is $K_w \times K_h \times C_{\text{in}} \times C_{\text{in}}$. In this figure, we have $C_{\text{in}} = 3$ input channels and $C_{\text{out}} = 6$ output channels.

where the $x(i, j, c_1)$ are the activations in the previous layer, just as in (5.45), $\mathcal{N}$ are the S^2 signed offsets in the 2D spatial kernel, and the notation $c_1 \in \{C_1\}$ denotes $c_1 \in [0, C_1)$. Note that because the offsets (k, l) are added to (instead of subtracted from) the (i, j) pixel coordinates, this operation is actually a *correlation* (3.13), but this distinction is usually glossed over.[37]

In neural network diagrams such as those shown in Figures 5.33 and 5.39–5.43, it is common to indicate the convolution kernel size S and the number of channels in a layer C, and only sometimes to show the image dimensions, as in Figures 5.33 and 5.39. Note that some neural networks such as the Inception module in GoogLeNet (Szegedy, Liu *et al.* 2015) shown in Figure 5.42 use 1×1 convolutions, which do not actually perform convolutions but rather combine various channels on a per-pixel basis, often with the goal of reducing the dimensionality of the feature space.

Because the weights in a convolution kernel are the same for all of the pixels within a given layer and channel, these weights are actually *shared* across what would result if we drew all of the connections between different pixels in different layers. This means that there are many fewer weights to learn than in fully connected layers. It also means that during backpropagation, kernel weight updates are summed over all of the pixels in a given layer/channel.

To fully determine the behavior of a convolutional layer, we still need to specify a few additional parameters.[38] These include:

- *Padding.* Early networks such as LeNet-5 did not pad the image, which therefore shrank after each convolution. Modern networks can optionally specify a padding width and mode, using one of the choices used with traditional image processing, such as zero padding or pixel replication, as shown in Figure 3.13.

[37] Since the weights in a neural network are learned, this reversal does not really matter.

[38] Most of the neural network building blocks we present in this chapter have corresponding functions in widely used deep learning frameworks, where you can get more detailed information about their operation. For example, the 2D convolution operator is called *Conv2d* in PyTorch and is documented at https://pytorch.org/docs/stable/nn.html#convolution-layers.

- *Stride.* The default stride for convolution is 1 pixel, but it is also possible to only evaluate the convolution at every nth column and row. For example, the first convolution layer in AlexNet (Figure 5.39) uses a stride of 4. Traditional image pyramids (Figure 3.31) use a stride of 2 when constructing the coarser levels.

- *Dilation.* Extra "space" (skipped rows and column) can be inserted between pixel samples during convolution, also known as dilated or *à trous* (with holes, in French, or often just "atrous") convolution (Yu and Koltun 2016; Chen, Papandreou *et al.* 2018). While in principle this can lead to aliasing, it can also be effective at pooling over a larger region while using fewer operations and learnable parameters.

- *Grouping.* While, by default, all input channels are used to produce each output channel, we can also group the input and output layers into G separate groups, each of which is convolved separately (Xie, Girshick *et al.* 2017). $G = 1$ corresponds to regular convolution, while $G = C_1$ means that each corresponding input channel is convolved independently from the others, which is known as *depthwise* or *channel-separated* convolution (Howard, Zhu *et al.* 2017; Tran, Wang *et al.* 2019).

A nice animation of the effects of these different parameters created by Vincent Dumoulin can be found at https://github.com/vdumoulin/conv_arithmetic as well as Dumoulin and Visin (2016).

In certain applications such as image inpainting (Section 10.5.1), the input image may come with an associated binary *mask*, indicating which pixels are valid and which need to be filled in. This is similar to the concept of alpha-matted images we studied in Section 3.1.3. In this case, one can use *partial convolutions* (Liu, Reda *et al.* 2018), where the input pixels are multiplied by the mask pixels and then normalized by the count of non-zero mask pixels. The mask channel output is set to 1 if any input mask pixels are non-zero. This resembles the pull-push algorithm of Gortler, Grzeszczuk *et al.* (1996) that we presented in Figure 4.2, except that the convolution weights are learned.

A more sophisticated version of partial convolutions is *gated convolutions* (Yu, Lin *et al.* 2019; Chang, Liu *et al.* 2019), where the per-pixel masks are derived from the previous layer using a learned convolution followed by a sigmoid non-linearity. This enables the network not only to learn a better measure of per-pixel confidence (weighting), but also to incorporate additional features such as user-drawn sketches or derived semantic information.

5.4.1 Pooling and unpooling

As we just saw in the discussion of convolution, strides of greater than 1 can be used to reduce the resolution of a given layer, as in the first convolutional layer of AlexNet (Figure 5.39). When the weights inside the convolution kernel are identical and sum up to 1, this is called *average pooling* and is typically applied in a channel-wise manner.

A widely used variant is to compute the maximum response within a square window, which is called *max pooling*. Common strides and window sizes for max pooling are a stride of 2 and 2×2 non-overlapping windows or 3×3 overlapping windows. Max pooling layers can be thought of as a "logical or", since they only require one of the units in the pooling region to be turned on. They are also supposed to provide some shift invariance over the inputs. However, most deep networks are not all that shift-invariant, which degrades their performance. The paper by Zhang (2019) has a nice discussion of this issue and some simple suggestions to mitigate this problem.

One issue that commonly comes up is how to backpropagate through a max pooling layer. The max pool operator acts like a "switch" that shunts (connects) one of the input units to the output

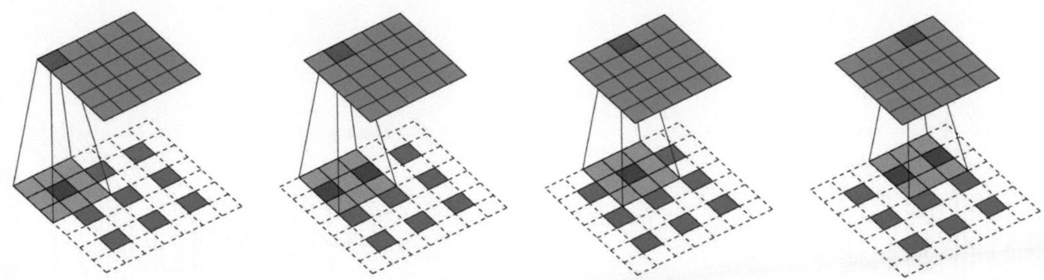

Figure 5.36 Transposed convolution (© Dumoulin and Visin (2016)) can be used to upsample (increase the size of) an image. Before applying the convolution operator, $(s - 1)$ extra rows and columns of zeros are inserted between the input samples, where s is the upsampling stride.

unit. Therefore, during backpropagation, we only need to pass the error and derivatives down to this maximally active unit, as long as we have remembered which unit has this response.

This same *max unpooling* mechanism can be used to create a "deconvolution network" when searching for the stimulus (Figure 5.47) that most strongly activates a particular unit (Zeiler and Fergus 2014).

If we want a more continuous behavior, we could construct a pooling unit that computes an L_p norm over its inputs, since the $L_{p \to \infty}$ effectively computes a maximum over its components (Springenberg, Dosovitskiy *et al.* 2015). However, such a unit requires more computation, so it is not widely used in practice, except sometimes at the final layer, where it is known as *generalized mean (GeM) pooling* (Dollár, Tu *et al.* 2009; Tolias, Sicre, and Jégou 2016; Gordo, Almazán *et al.* 2017; Radenović, Tolias, and Chum 2019) or *dynamic mean (DAME) pooling* (Yang, Kien Nguyen *et al.* 2019). In their paper, Springenberg, Dosovitskiy *et al.* (2015) also show that using strided convolution instead of max pooling can produce competitive results.

While unpooling can be used to (approximately) reverse the effect of max pooling operation, if we want to reverse a convolutional layer, we can look at learned variants of the interpolation operator we studied in Sections 3.5.1 and 3.5.3. The easiest way to visualize this operation is to add extra rows and columns of zeros between the pixels in the input layer, and to then run a regular convolution (Figure 5.36). This operation is sometimes called *backward convolution* with a *fractional stride* (Long, Shelhamer, and Darrell 2015), although it is more commonly known as *transposed convolution* (Dumoulin and Visin 2016), because when convolutions are written in matrix form, this operation is a multiplication with a transposed sparse weight matrix. Just as with regular convolution, padding, stride, dilation, and grouping parameters can be specified. However, in this case, the stride specifies the factor by which the image will be upsampled instead of downsampled.

U-Nets and Feature Pyramid Networks

When discussing the Laplacian pyramid in Section 3.5.3, we saw how image downsampling and upsampling can be combined to achieve a variety of multi-resolution image processing tasks (Figure 3.33). The same kinds of combinations can be used in deep convolutional networks, in particular, when we want the output to be a full-resolution image. Examples of such applications include pixel-wise semantic labeling (Section 6.4), image denoising and super-resolution (Section 10.3), monocular depth inference (Section 12.8), and neural style transfer (Section 14.6). The idea of reducing the resolution of a network and then expanding it again is sometimes called a *bottleneck* and is related to earlier self-supervised network training using *autoencoders* (Hinton and Zemel 1994;

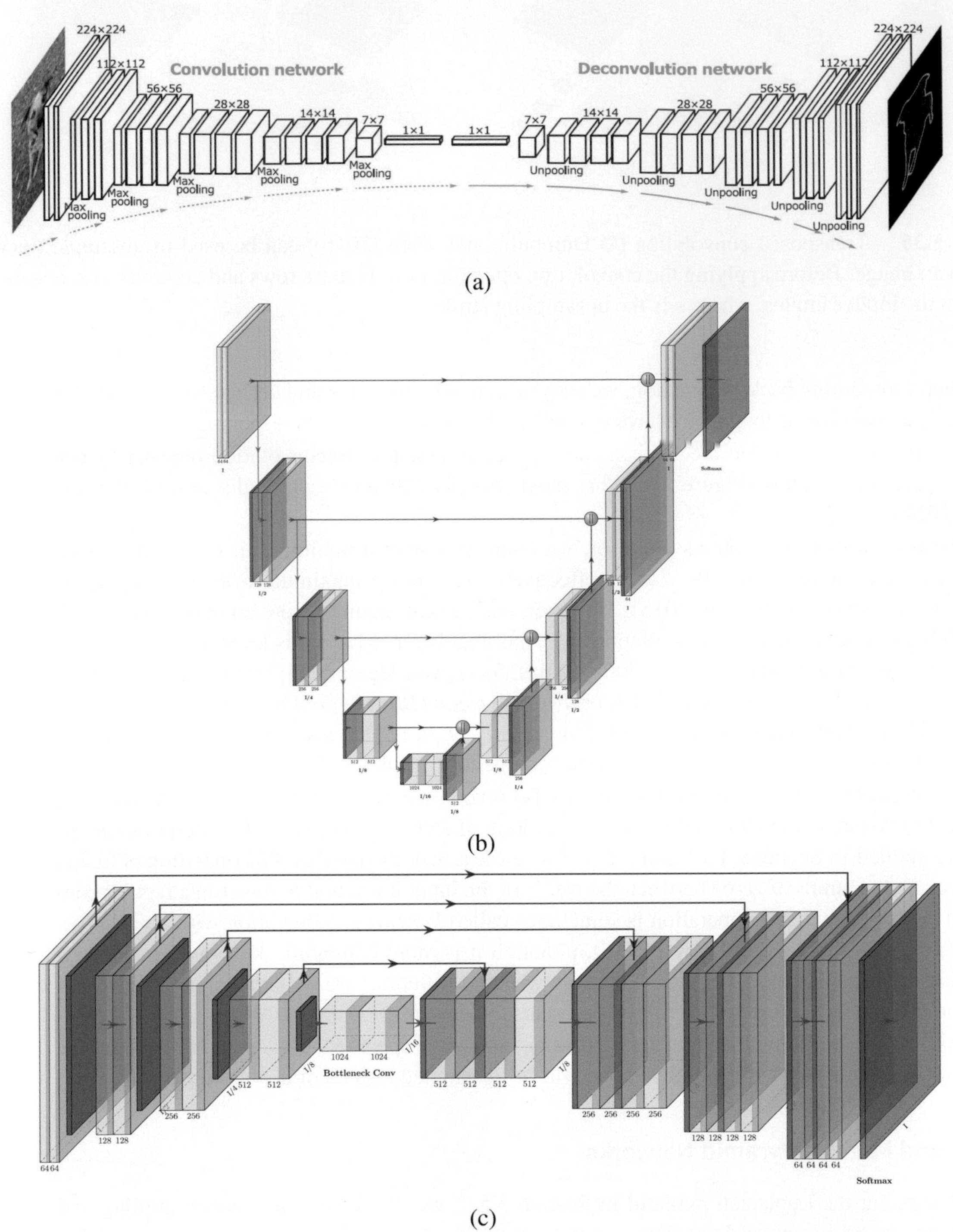

Figure 5.37 (a) The deconvolution network of Noh, Hong, and Han (2015) © 2015 IEEE and (b–c) the U-Net of Ronneberger, Fischer, and Brox (2015), drawn using the PlotNeuralNet LaTeX package. In addition to the fine-to-coarse-to-fine bottleneck used in (a), the U-Net also has skip connections between encoding and decoding layers at the same resolution.

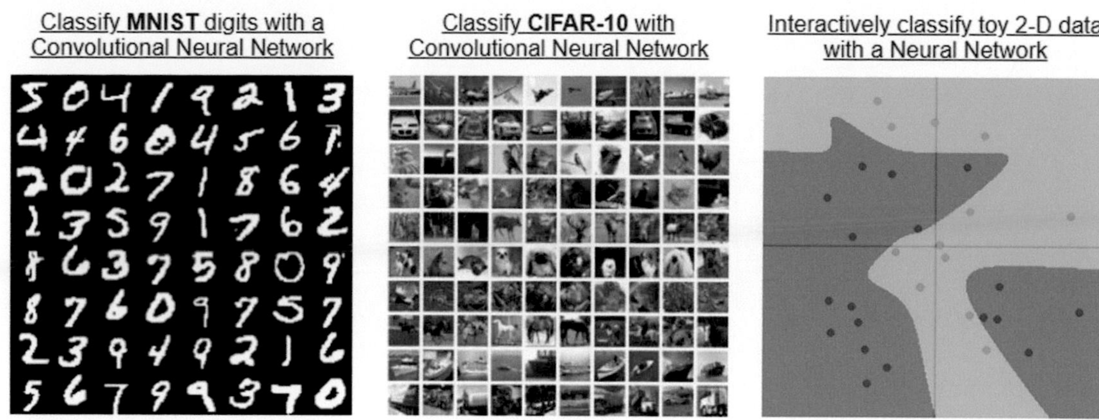

Figure 5.38 Screenshot from Andrej Karpathy's web browser demos at https://cs.stanford.edu/people/karpathy/convnetjs, where you can run a number of small neural networks, including CNNs for digit and tiny image classification.

Goodfellow, Bengio, and Courville 2016, Chapter 14).

One of the earliest applications of this idea was the *fully convolutional network* developed by Long, Shelhamer, and Darrell (2015). This paper inspired myriad follow-on architectures, including the hourglass-shaped "deconvolution" network of Noh, Hong, and Han (2015), the U-Net of Ronneberger, Fischer, and Brox (2015), the atrous convolution network with CRF refinement layer of Chen, Papandreou *et al.* (2018), and the panoptic feature pyramid networks of Kirillov, Girshick *et al.* (2019). Figure 5.37 shows the general layout of two of these networks, which are discussed in more detail in Section 6.4 on semantic segmentation. We will see other uses of these kinds of *backbone networks* (He, Gkioxari *et al.* 2017) in later sections on image denoising and super-resolution (Section 10.3), monocular depth inference (Section 12.8), and neural style transfer (Section 14.6).

5.4.2 *Application*: Digit classification

One of the earliest commercial application of convolutional neural networks was the LeNet-5 system created by LeCun, Bottou *et al.* (1998) whose architecture is shown in Figure 5.33. This network contained most of the elements of modern CNNs, although it used sigmoid non-linearities, average pooling, and Gaussian RBF units instead of softmax at its output. If you want to experiment with this simple digit recognition CNN, you can visit the interactive JavaScript demo created by Andrej Karpathy at https://cs.stanford.edu/people/karpathy/convnetjs (Figure 5.38).

The network was initially deployed around 1995 by AT&T to automatically read checks deposited in NCR ATM machines to verify that the written and keyed check amounts were the same. The system was then incorporated into NCR's high-speed check reading systems, which at some point were processing somewhere between 10% and 20% of all the checks in the US.[39]

Today, variants of the LeNet-5 architecture (Figure 5.33) are commonly used as the first convolutional neural network introduced in courses and tutorials on the subject.[40] Although the MNIST dataset (LeCun, Cortes, and Burges 1998) originally used to train LeNet-5 is still sometimes used,

[39]This information courtesy of Yann LeCun and Larry Jackel, who were two of the principals in the development of this system.

[40]See, e.g., https://pytorch.org/tutorials/beginner/blitz/cifar10_tutorial.html.

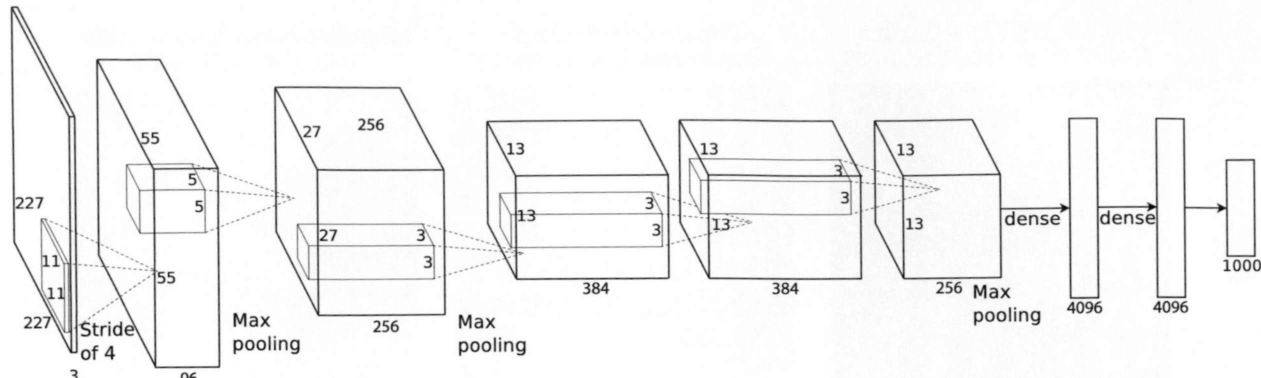

Figure 5.39 Architecture of the SuperVision deep neural network (more commonly known as "AlexNet"), courtesy of Matt Deitke (redrawn from (Krizhevsky, Sutskever, and Hinton 2012)). The network consists of multiple convolutional layers with ReLU activations, max pooling, some fully connected layers, and a softmax to produce the final class probabilities.

it is more common to use the more challenging CIFAR-10 (Krizhevsky 2009) or Fashion MNIST (Xiao, Rasul, and Vollgraf 2017) as datasets for training and testing.

5.4.3 Network architectures

While modern convolutional neural networks were first developed and deployed in the late 1990s, it was not until the breakthrough publication by Krizhevsky, Sutskever, and Hinton (2012) that they started outperforming more traditional techniques on natural image classification (Figure 5.40). As you can see in this figure, the AlexNet system (the more widely used name for their SuperVision network) led to a dramatic drop in error rates from 25.8% to 16.4%. This was rapidly followed in the next few years with additional dramatic performance improvements, due to further developments as well as the use of deeper networks, e.g., from the original 8-layer AlexNet to a 152-layer ResNet.

Figure 5.39 shows the architecture of the SuperVision network, which contains a series of convolutional layers with ReLU (rectified linear) non-linearities, max pooling, some fully connected layers, and a final softmax layer, which is fed into a multi-class cross-entropy loss. Krizhevsky, Sutskever, and Hinton (2012) also used dropout (Figure 5.29), small translation and color manipulation for data augmentation, momentum, and weight decay (L_2 weight penalties).

The next few years after the publication of this paper saw dramatic improvement in the classification performance on the ImageNet Large Scale Visual Recognition Challenge (Russakovsky, Deng *et al.* 2015), as shown in Figure 5.40. A nice description of the innovations in these various networks, as well as their capacities and computational cost, can be found in the lecture slides by Justin Johnson (2020, Lecture 8).

The winning entry from 2013 by Zeiler and Fergus (2014) used a larger version of AlexNet with more channels in the convolution stages and lowered the error rate by about 30%. The 2014 Oxford Visual Geometry Group (VGG) winning entry by Simonyan and Zisserman (2014b) used repeated 3×3 convolution/ReLU blocks interspersed with 2×2 max pooling and channel doubling (Figure 5.41), followed by some fully connected layers, to produce 16–19 layer networks that further reduced the error by 40%. However, as shown in Figure 5.44, this increased performance came at a greatly increased amount of computation.

The 2015 GoogLeNet of Szegedy, Liu *et al.* (2015) focused instead on efficiency. GoogLeNet

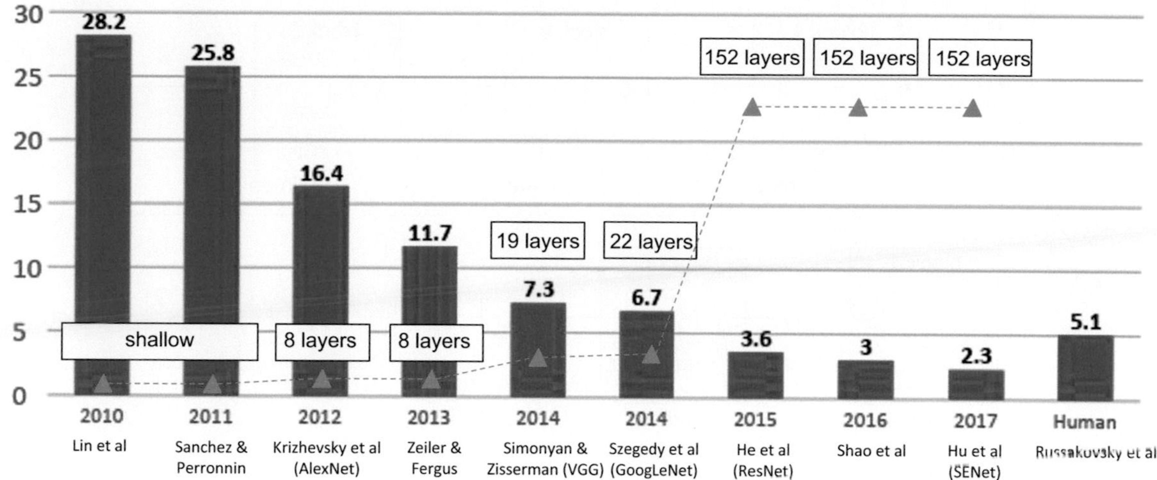

Figure 5.40 Top-5 error rate and network depths of winning entries from the ImageNet Large Scale Visual Recognition Challenge (ILSVRC) © Li, Johnson, and Yeung (2019).

begins with an aggressive stem network that uses a series of strided and regular convolutions and max pool layers to quickly reduce the image resolutions from 224^2 to 28^2. It then uses a number of *Inception modules* (Figure 5.42), each of which is a small branching neural network whose features get concatenated at the end. One of the important characteristics of this module is that it uses 1×1 "bottleneck" convolutions to reduce the number of channels before performing larger 3×3 and 5×5 convolutions, thereby saving a significant amount of computation. This kind of projection followed by an additional convolution is similar in spirit to the approximation of filters as a sum of separable convolutions proposed by Perona (1995). GoogLeNet also removed the fully connected (MLP) layers at the end, relying instead on global average pooling followed by one linear layer before the softmax. Its performance was similar to that of VGG but at dramatically lower computation and model size costs (Figure 5.44).

The following year saw the introduction of Residual Networks (He, Zhang *et al.* 2016a), which dramatically expanded the number of layers that could be successfully trained (Figure 5.40). The main technical innovation was the introduction of *skip connections* (originally called "shortcut connections"), which allow information (and gradients) to flow around a set of convolutional layers, as shown in Figure 5.43. The networks are called *residual networks* because they allow the network to learn the residuals (differences) between a set of incoming and outgoing activations. A variant on the basic residual block is the "bottleneck block" shown on the right side of Figure 5.43, which reduces the number of channels before performing the 3×3 convolutional layer. A further extension, described in (He, Zhang *et al.* 2016b), moves the ReLU non-linearity to before the residual summation, thereby allowing true identity mappings to be modeled at no cost.

To build a ResNet, various residual blocks are interspersed with strided convolutions and channel doubling to achieve the desired number of layers. (Similar downsampling stems and average pooled softmax layers as in GoogLeNet are used at the beginning and end.) By combining various numbers of residual blocks, ResNets consisting of 18, 34, 50, 101, and 152 layers have been constructed and evaluated. The deeper networks have higher accuracy but more computational cost (Figure 5.44). In 2015, ResNet not only took first place in the ILSVRC (ImageNet) classification, detection, and localization challenges, but also took first place in the detection and segmentation challenges on the newer COCO dataset and benchmark (Lin, Maire *et al.* 2014).

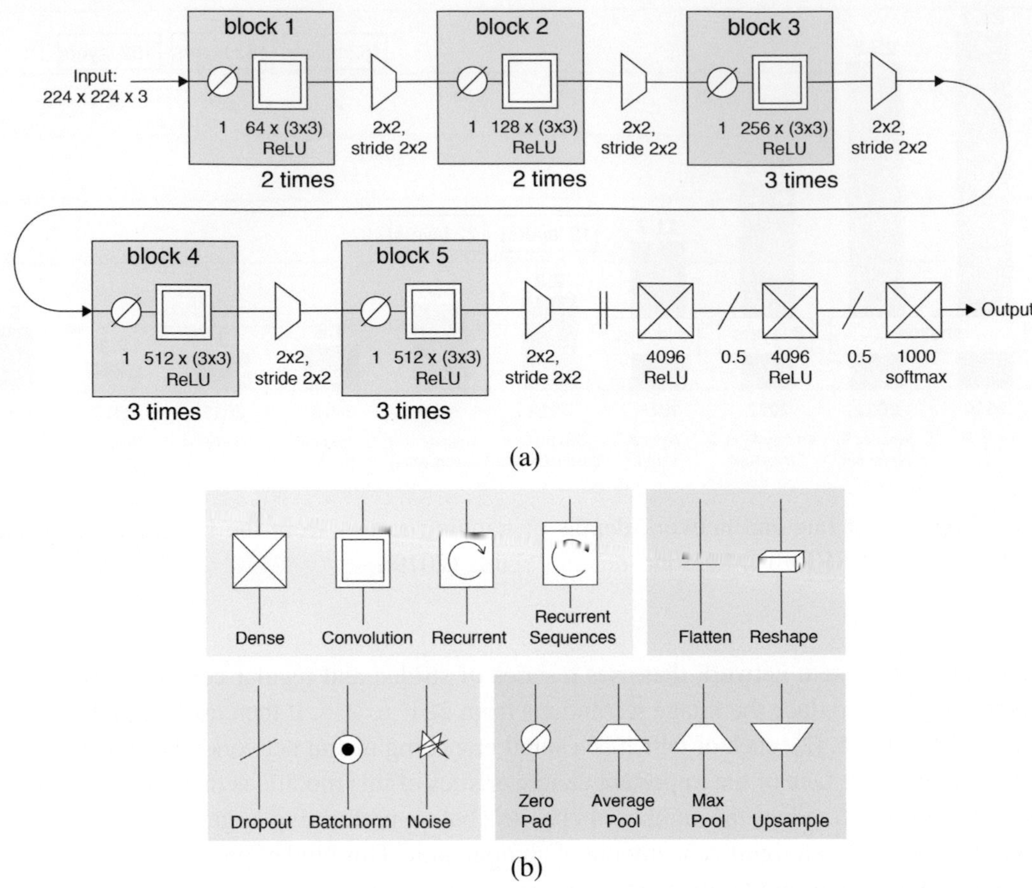

Figure 5.41 The VGG16 network of Simonyan and Zisserman (2014b) © Glassner (2018). (a) The network consists of repeated zero-pad, 3×3 convolution, ReLU blocks interspersed with 2×2 max pooling and a doubling in the number of channels. This is followed by some fully connected and dropout layers, with a final softmax into the 1,000 ImagetNet categories. (b) Some of the schematic neural network symbols used by Glassner (2018).

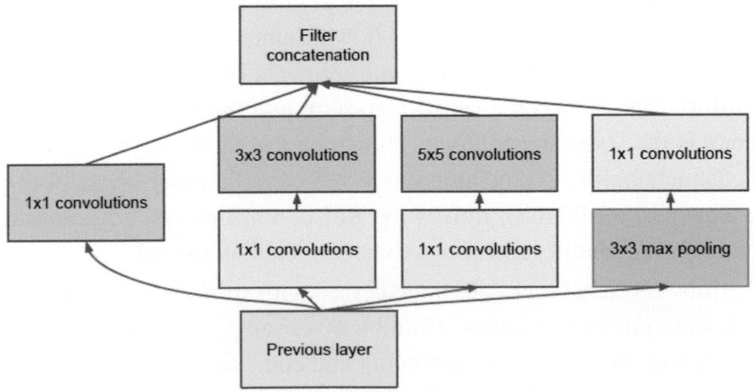

Figure 5.42 An Inception module from (Szegedy, Liu *et al.* 2015) © 2015 IEEE, which combines dimensionality reduction, multiple convolution sizes, and max pooling as different channels that get stacked together into a final feature map.

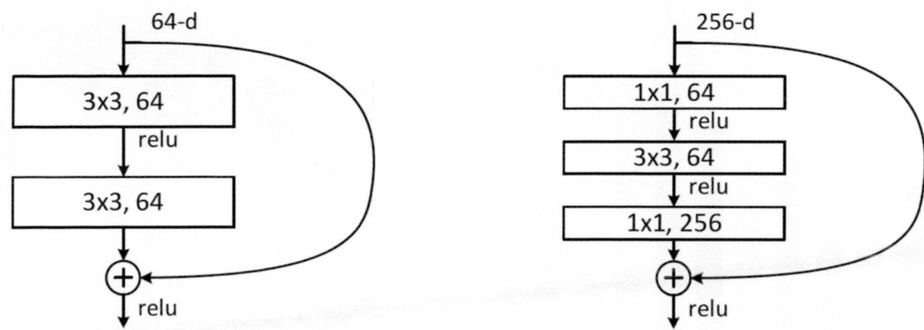

Figure 5.43 ResNet residual networks (He, Zhang *et al.* 2016a) © 2016 IEEE, showing skip connections going around a series of convolutional layers. The figure on the right uses a bottleneck to reduce the number of channels before the convolution. Having direct connections that shortcut the convolutional layer allows gradients to more easily flow backward through the network during training.

Since then, myriad extensions and variants have been constructed and evaluated. The ResNeXt system from Xie, Girshick *et al.* (2017) used grouped convolutions to slightly improve accuracy. Denseley connected CNNs (Huang, Liu *et al.* 2017) added skip connections between non-adjacent convolution and/or pool blocks. Finally, the Squeeze-and-Excitation network (SENet) by Hu, Shen, and Sun (2018) added global context (via global pooling) to each layer to obtain a noticeable increase in accuracy. More information about these and other CNN architectures can be found in both the original papers as well as class notes on this topic (Li, Johnson, and Yeung 2019; Johnson 2020).

Mobile networks

As deep neural networks were getting deeper and larger, a countervailing trend emerged in the construction of smaller, less computationally expensive networks that could be used in mobile and embedded applications. One of the earliest networks tailored for lighter-weight execution was MobileNets (Howard, Zhu *et al.* 2017), which used *depthwise convolutions*, a special case of grouped convolutions where the number of groups equals the number of channels. By varying two hyperparameters, namely a *width multiplier* and a *resolution multiplier*, the network architecture could be tuned along an accuracy vs. size vs. computational efficiency tradeoff. The follow-on MobileNetV2 system (Sandler, Howard *et al.* 2018) added an "inverted residual structure", where the shortcut connections were between the bottleneck layers. ShuffleNet (Zhang, Zhou *et al.* 2018) added a "shuffle" stage between grouped convolutions to enable channels in different groups to co-mingle. ShuffleNet V2 (Ma, Zhang *et al.* 2018) added a channel split operator and tuned the network architectures using end-to-end performance measures. Two additional networks designed for computational efficiency are ESPNet (Mehta, Rastegari *et al.* 2018) and ESPNetv2 (Mehta, Rastegari *et al.* 2019), which use pyramids of (depth-wide) dilated separable convolutions.

The concepts of grouped, depthwise, and channel-separated convolutions continue to be a widely used tool for managing computational efficiency and model size (Choudhary, Mishra *et al.* 2020), not only in mobile networks, but also in video classification (Tran, Wang *et al.* 2019), which we discuss in more detail in Section 5.5.2.

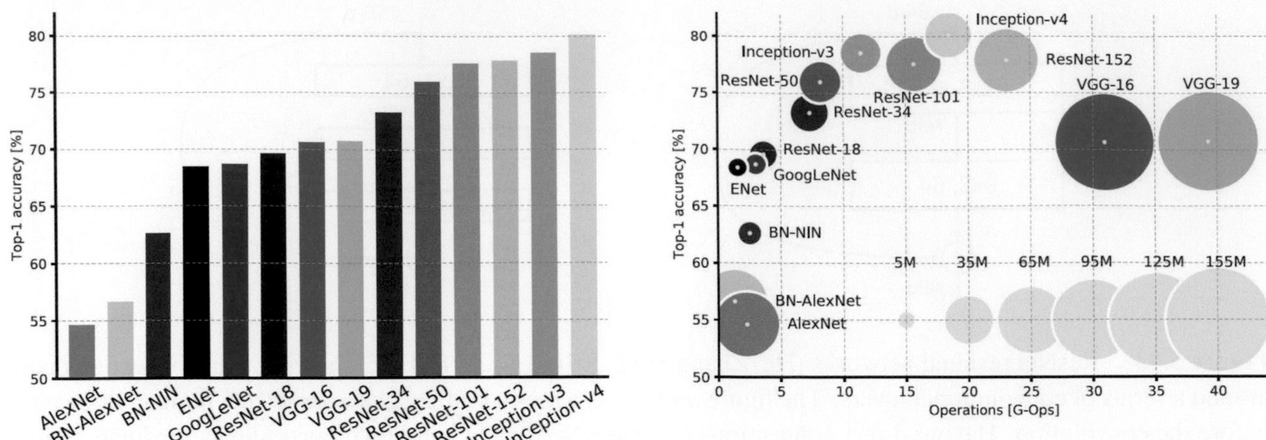

Figure 5.44 Network accuracy vs. size and operation counts (Canziani, Culurciello, and Paszke 2017) © 2017 IEEE: In the right figure, the network accuracy is plotted against operation count (1–40 G-Ops), while the size of the circle indicates the number of parameters (10–155 M). The initials BN indicate a batch normalized version of a network.

5.4.4 Model zoos

A great way to experiment with these various CNN architectures is to download pre-trained models from a *model zoo*[41] such as the TorchVision library at https://github.com/pytorch/vision. If you look in the torchvision/models folder, you will find implementations of AlexNet, VGG, GoogleNet, Inception, ResNet, DenseNet, MobileNet, and ShuffleNet, along with other models for classification, object detection, and image segmentation. Even more recent models, some of which are discussed in the upcoming sections, can be found in the PyTorch Image Models library (timm), https://github.com/rwightman/pytorch-image-models. Similar collections of pre-trained models exist for other languages, e.g., https://www.tensorflow.org/lite/models for efficient (mobile) TensorFlow models.

While people often download and use pre-trained neural networks for their applications, it is more common to at least *fine-tune* such networks on data more characteristic of the application (as opposed to the public benchmark data on which most zoo models are trained).[42] It is also quite common to replace the last few layers, i.e., the *head* of the network (so called because it lies at the top of a layer diagram when the layers are stacked bottom-to-top) while leaving the *backbone* intact. The terms backbone and head(s) are widely used and were popularized by the Mask-RCNN paper (He, Gkioxari *et al.* 2017). Some more recent papers refer to the backbone and head as the *trunk* and its *branches* (Ding and Tao 2018; Kirillov, Girshick *et al.* 2019; Bell, Liu *et al.* 2020), with the term *neck* also being occasionally used (Chen, Wang *et al.* 2019).[43]

When adding a new head, its parameters can be trained using the new data specific to the intended application. Depending on the amount and quality of new training data available, the head can be as simple as a linear model such as an SVM or logistic regression/softmax (Donahue, Jia *et al.* 2014; Sharif Razavian, Azizpour *et al.* 2014), or as complex as a fully connected or convolutional network (Xiao, Liu *et al.* 2018). Fine-tuning some of the layers in the backbone is also an option,

[41]The name "model zoo" itself is a fanciful invention of Evan Shelhamer, lead developer on Caffe (Jia, Shelhamer *et al.* 2014), who first used it on https://caffe.berkeleyvision.org/model_zoo.html to describe a collection of various pre-trained DNN models (personal communication).

[42]See, e.g., https://classyvision.ai/tutorials/fine_tuning and (Zhang, Lipton *et al.* 2021, Section 13.2).

[43]Classy Vision uses a trunk and heads terminology, https://classyvision.ai/tutorials/classy_model.

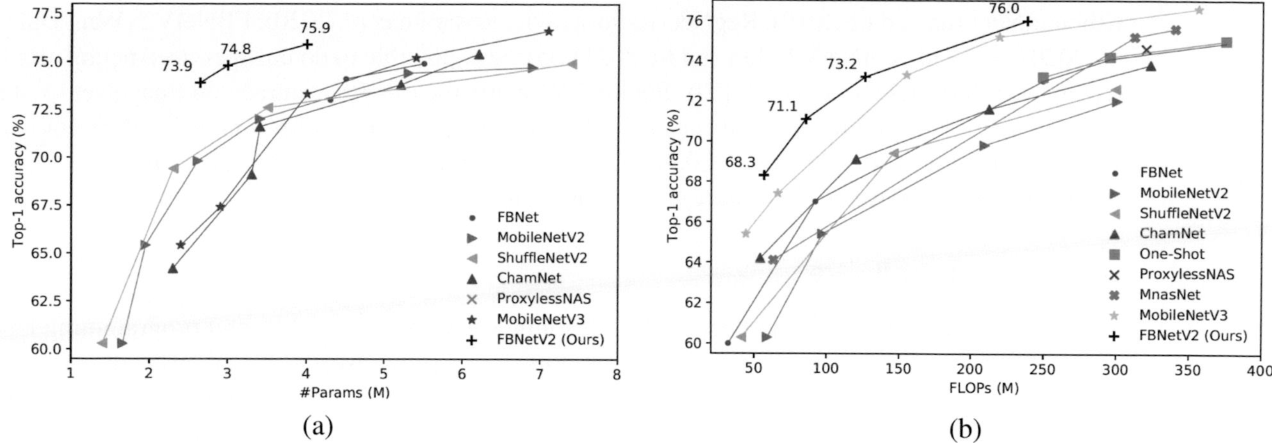

Figure 5.45 ImageNet accuracy vs. (a) size (# of parameters) and (b) operation counts for a number of recent efficient networks (Wan, Dai *et al.* 2020) © 2020 IEEE.

but requires sufficient data and a slower learning rate so that the benefits of the pre-training are not lost. The process of pre-training a machine learning system on one dataset and then applying it to another domain is called *transfer learning* (Pan and Yang 2009). We will take a closer look at transfer learning in Section 5.4.7 on self-supervised learning.

Model size and efficiency

As you can tell from the previous discussion, neural network models come in a large variety of sizes (typically measured in number of parameters, i.e., weights and biases) and computational loads (often measured in FLOPs per forward inference pass). The evaluation by Canziani, Culurciello, and Paszke (2017), summarized in Figure 5.44, gives a snapshot of the performance (accuracy and size+operations) of the top-performing networks on the ImageNet challenge from 2012–2017. In addition to the networks we have already discussed, the study includes Inception-v3 (Szegedy, Vanhoucke *et al.* 2016) and Inception-v4 (Szegedy, Ioffe *et al.* 2017).

Because deep neural networks can be so memory- and compute-intensive, a number of researchers have investigated methods to reduce both, using lower precision (e.g., fixed-point) arithmetic and weight compression (Han, Mao, and Dally 2016; Iandola, Han *et al.* 2016). The XNOR-Net paper by Rastegari, Ordonez *et al.* (2016) investigates using binary weights (on-off connections) and optionally binary activations. It also has a nice review of previous binary networks and other compression techniques, as do more recent survey papers (Sze, Chen *et al.* 2017; Gu, Wang *et al.* 2018; Choudhary, Mishra *et al.* 2020).

Neural Architecture Search (NAS)

One of the most recent trends in neural network design is the use of *Neural Architecture Search* (NAS) algorithms to try different network topologies and parameterizations (Zoph and Le 2017; Zoph, Vasudevan *et al.* 2018; Liu, Zoph *et al.* 2018; Pham, Guan *et al.* 2018; Liu, Simonyan, and Yang 2019; Hutter, Kotthoff, and Vanschoren 2019). This process is more efficient (in terms of a researcher's time) than the trial-and-error approach that characterized earlier network design. Elsken, Metzen, and Hutter (2019) survey these and additional papers on this rapidly evolving topic. More recent publications include FBNet (Wu, Dai *et al.* 2019), RandomNets (Xie, Kirillov *et al.* 2019) ,

EfficientNet (Tan and Le 2019), RegNet (Radosavovic, Kosaraju *et al.* 2020), FBNetV2 (Wan, Dai *et al.* 2020), and EfficientNetV2 (Tan and Le 2021). It is also possible to do unsupervised neural architecture search (Liu, Dollár *et al.* 2020). Figure 5.45 shows the top-1% accuracy on ImageNet vs. the network size (# of parameters) and forward inference operation counts for a number of recent network architectures (Wan, Dai *et al.* 2020). Compared to the earlier networks shown in Figure 5.44, the newer networks use 10× (or more) fewer parameters.

Deep learning software

Over the last decade, a large number of deep learning software frameworks and programming language extensions have been developed. The Wikipedia entry on deep learning software lists over twenty such frameworks, about a half of which are still being actively developed.[44] While Caffe (Jia, Shelhamer *et al.* 2014) was one of the first to be developed and used for computer vision applications, it has mostly been supplanted by PyTorch and TensorFlow, at least if we judge by the open source implementations that now accompany most computer vision research papers.

Andrew Glassner's (2018) introductory deep learning book uses the Keras library because of its simplicity. The Dive into Deep Learning book (Zhang, Lipton *et al.* 2021) and associated course (Smola and Li 2019) use MXNet for all the examples in the text, but they have recently released PyTorch and TensorFlow code samples as well. Stanford's CS231n (Li, Johnson, and Yeung 2019) and Johnson (2020) include a lecture on the fundamentals of PyTorch and TensorFlow. Some classes also use simplified frameworks that require the students to implement more components, such as the Educational Framework (EDF) developed by McAllester (2020) and used in Geiger (2021).

In addition to software frameworks and libraries, deep learning code development usually benefits from good visualization libraries such as TensorBoard[45] and Visdom.[46] And in addition to the model zoos mentioned earlier in this section, there are even higher-level packages such as Classy Vision,[47] which allow you to train or fine-tune your own classifier with no or minimal programming. Andrej Karpathy also provides a useful guide for training neural networks at http://karpathy.github.io/2019/04/25/recipe, which may help avoid common issues.

5.4.5 Visualizing weights and activations

Visualizing intermediate and final results has always been an integral part of computer vision algorithm development (e.g., Figures 1.7–1.11) and is an excellent way to develop intuitions and debug or refine results. In this chapter, we have already seen examples of tools for simple two-input neural network visualizations, e.g., the TensorFlow Playground in Figure 5.32 and ConvNetJS in Figure 5.38. In this section, we discuss tools for visualizing network weights and, more importantly, the *response functions* of different units or layers in a network.

For a simple small network such as the one shown in Figure 5.32, we can indicate the strengths of connections using line widths and colors. What about networks with more units? A clever way to do this, called *Hinton diagrams* in honor of its inventor, is to indicate the strengths of the incoming and outgoing weights as black or white boxes of different sizes, as shown in Figure 5.46 (Ackley, Hinton, and Sejnowski 1985; Rumelhart, Hinton, and Williams 1986b).[48]

If we wish to display the set of activations in a given layer, e.g., the response of the final 10-category layer in MNIST or CIFAR-10, across some or all of the inputs, we can use non-linear

[44] https://en.wikipedia.org/wiki/Comparison_of_deep-learning_software

[45] https://www.tensorflow.org/tensorboard

[46] https://github.com/fossasia/visdom

[47] https://classyvision.ai

[48] In the early days of neural networks, bit-mapped displays and printers only supported 1-bit black and white images.

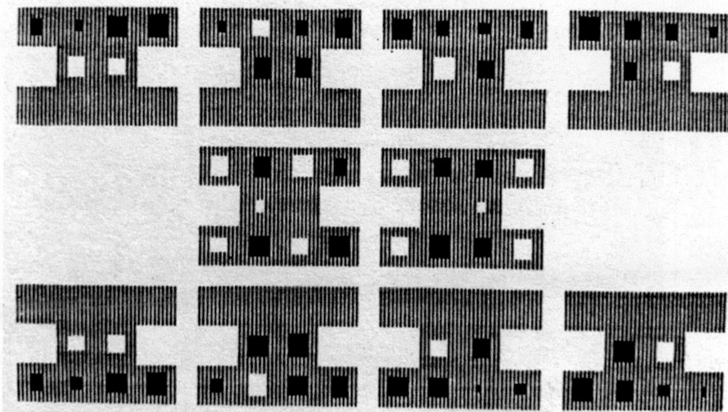

Figure 5.46 A Hinton diagram showing the weights connecting the units in a a three layer neural network, courtesy of Geoffrey Hinton. The size of each small box indicates the magnitude of each weight and its color (black or white) indicates the sign.

dimensionality reduction techniques such as t-SNE and UMap discussed in Section 5.2.4 and Figure 5.21.

How can we visualize what individual units ("neurons") in a deep network respond to? For the first layer in a network (Figure 5.47, upper left corner), the response can be read directly from the incoming weights (grayish images) for a given channel. We can also find the patches in the validation set that produce the largest responses across the units in a given channel (colorful patches in the upper left corner of Figure 5.47). (Remember that in a convolutional neural network, different units in a particular channel respond similarly to shifted versions of the input, ignoring boundary and aliasing effects.)

For deeper layers in a network, we can again find maximally activating patches in the input images. Once these are found, Zeiler and Fergus (2014) pair a *deconvolution network* with the original network to backpropagate feature activations all the way back to the image patch, which results in the grayish images in layers 2–5 in Figure 5.47. A related technique called *guided backpropagation* developed by Springenberg, Dosovitskiy *et al.* (2015) produces slightly higher contrast results.

Another way to probe a CNN feature map is to determine how strongly parts of an input image activate units in a given channel. Zeiler and Fergus (2014) do this by masking sub-regions of the input image with a gray square, which not only produces activation maps, but can also show the most likely labels associated with each image region (Figure 5.48). Simonyan, Vedaldi, and Zisserman (2013) describe a related technique they call *saliency maps*, Nguyen, Yosinski, and Clune (2016) call their related technique *activation maximization*, and Selvaraju, Cogswell *et al.* (2017) call their visualization technique *gradient-weighted class activation mapping* (Grad-CAM).

Many more techniques for visualizing neural network responses and behaviors have been described in various papers and blogs (Mordvintsev, Olah, and Tyka 2015; Zhou, Khosla *et al.* 2015; Nguyen, Yosinski, and Clune 2016; Bau, Zhou *et al.* 2017; Olah, Mordvintsev, and Schubert 2017; Olah, Satyanarayan *et al.* 2018; Cammarata, Carter *et al.* 2020), as well as the extensive lecture slides by Johnson (2020, Lecture 14). Figure 5.49 shows one example, visualizing different layers in a pre-trained GoogLeNet. OpenAI also recently released a great interactive tool called Microscope,[49] which allows people to visualize the significance of every neuron in many common neural networks.

[49] https://microscope.openai.com/models

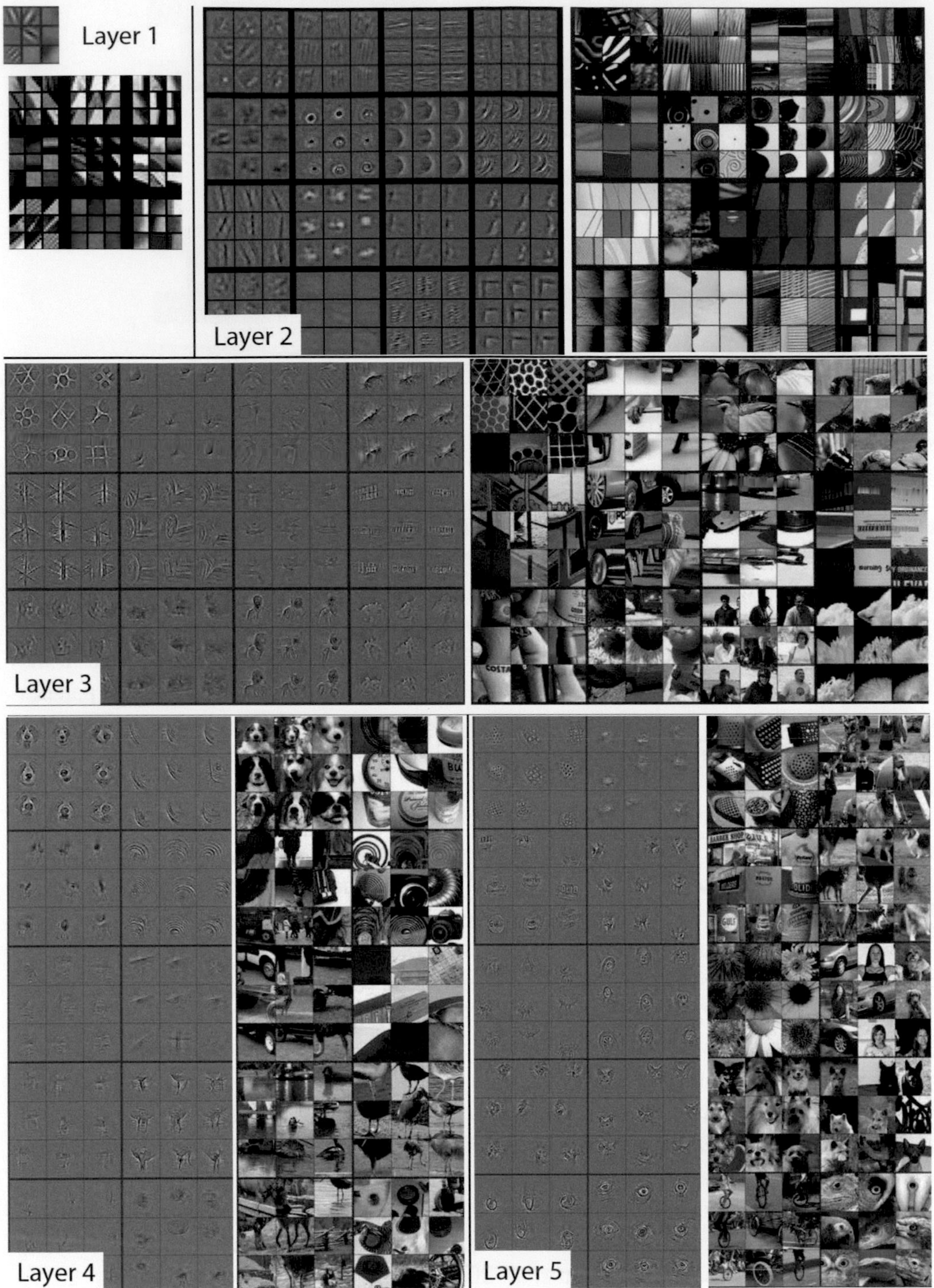

Figure 5.47 Visualizing network weights and features (Zeiler and Fergus 2014) © 2014 Springer. Each visualized convolutional layer is taken from a network adapted from the SuperVision net of Krizhevsky, Sutskever, and Hinton (2012). The 3×3 subimages denote the top nine responses in one feature map (channel in a given layer) projected back into pixel space (higher layers project to larger pixel patches), with the color images on the right showing the most responsive image patches from the validation set, and the grayish signed images on the left showing the corresponding maximum stimulus pre-images.

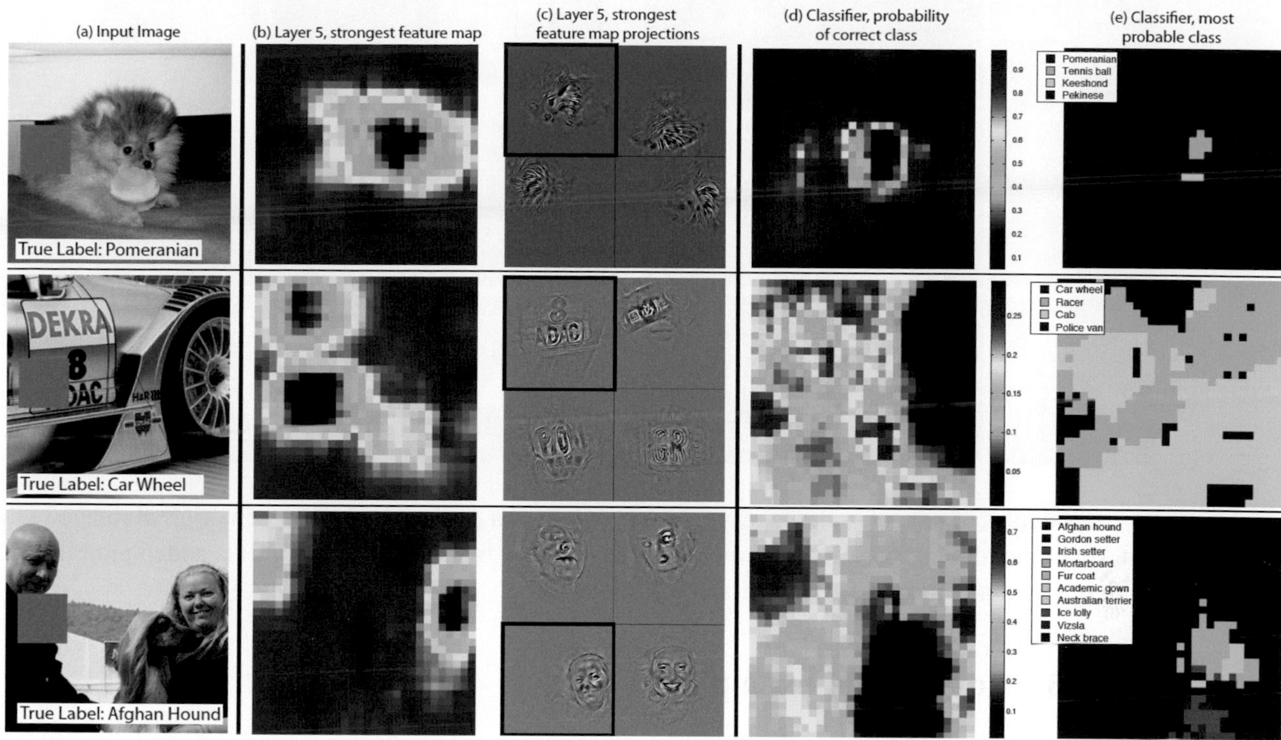

Figure 5.48 Heat map visualization from Zeiler and Fergus (2014) © 2014 Springer. By covering up portions of the input image with a small gray square, the response of a highly active channel in layer 5 can be visualized (second column), as can the feature map projections (third column), the likelihood of the correct class, and the most likely class per pixel.

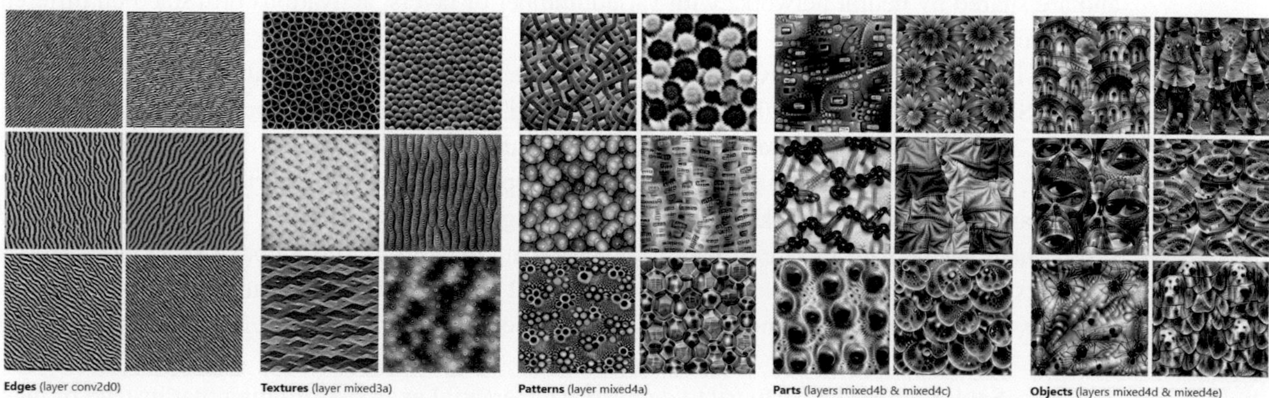

Figure 5.49 Feature visualization of how GoogLeNet (Szegedy, Liu *et al.* 2015) trained on ImageNet builds up its representations over different layers, from Olah, Mordvintsev, and Schubert (2017).

Figure 5.50 Examples of adversarial images © Szegedy, Zaremba *et al.* (2013). For each original image in the left column, a small random perturbation (shown magnified by 10× in the middle column) is added to obtain the image in the right column, which is always classified as an ostrich.

5.4.6 Adversarial examples

While techniques such as guided backpropagation can help us better visualize neural network responses, they can also be used to "trick" deep networks into misclassifying inputs by subtly perturbing them, as shown in Figure 5.50. The key to creating such images is to take a set of final activations and to then backpropagate a gradient in the direction of the "fake" class, updating the input image until the fake class becomes the dominant activation. Szegedy, Zaremba *et al.* (2013) call such perturbed images *adversarial examples*.

Running this backpropagation requires access to the network and its weights, which means that this is a *white box attack*, as opposed to a *black box attack*, where nothing is known about the network. Surprisingly, however, the authors find "... that adversarial examples are relatively robust, and are shared by neural networks with varied number of layers, activations or trained on different subsets of the training data."

The initial discovery of adversarial attacks spurred a flurry of additional investigations (Goodfellow, Shlens, and Szegedy 2015; Nguyen, Yosinski, and Clune 2015; Kurakin, Goodfellow, and Bengio 2016; Moosavi-Dezfooli, Fawzi, and Frossard 2016; Goodfellow, Papernot *et al.* 2017). Eykholt, Evtimov *et al.* (2018) show how adding simple stickers to real world objects (such as stop signs) can cause neural networks to misclassify photographs of such objects. Hendrycks, Zhao *et al.* (2021) have created a database of natural images that consistently fool popular deep classification networks trained on ImageNet. And while adversarial examples are mostly used to demonstrate the weaknesses of deep learning models, they can also be used to improve recognition (Xie, Tan *et al.* 2020).

Ilyas, Santurkar *et al.* (2019) try to demystify adversarial examples, finding that instead of making the anticipated large-scale perturbations that affect a human label, they are performing a type of shortcut learning (Lapuschkin, Wäldchen *et al.* 2019; Geirhos, Jacobsen *et al.* 2020). They find that optimizers are exploiting the *non-robust features* for an image label; that is, non-random correlations for an image class that exist in the dataset, but are not easily detected by humans. These non-robust features look merely like noise to a human observer, leaving images perturbed by them predominantly the same. Their claim is supported by training classifiers solely on non-robust features and finding that they correlate with image classification performance.

Are there ways to guard against adversarial attacks? The `cleverhans` software library (Papernot, Faghri *et al.* 2018) provides implementations of adversarial example construction techniques and adversarial training. There's also an associated http://www.cleverhans.io blog on security and privacy in machine learning. Madry, Makelov *et al.* (2018) show how to train a network that is robust to bounded additive perturbations in known test images. There's also recent work on detecting (Qin, Frosst *et al.* 2020b) and *deflecting* adversarial attacks (Qin, Frosst *et al.* 2020a) by forcing the perturbed images to visually resemble their (false) target class. This continues to be an evolving area, with profound implications for the robustness and safety of machine learning-based applications, as is the issue of *dataset bias* (Torralba and Efros 2011), which can be guarded against, to some extent, by testing *cross-dataset transfer* performance (Ranftl, Lasinger *et al.* 2020).

5.4.7 Self-supervised learning

As we mentioned previously, it is quite common to *pre-train* a *backbone* (or *trunk*) network for one task, e.g., whole image classification, and to then replace the final (few) layers with a new *head* (or one or more *branches*), which are then trained for a different task, e.g., semantic image segmentation (He, Gkioxari *et al.* 2017). Optionally, the last few layers of the original backbone network can be *fine-tuned.*

The idea of training on one task and then using the learning on another is called *transfer learning*, while the process of modifying the final network to its intended application and statistics is called *domain adaptation*. While this idea was originally applied to backbones trained on labeled datasets such as ImageNet, i.e., in a *fully supervised* manner, the possibility of pre-training on the immensely larger set of unlabeled real-world images always remained a tantalizing possibility.

The central idea in *self-supervised* learning is to create a supervised *pretext task* where the labels can be automatically derived from unlabeled images, e.g., by asking the network to predict a subset of the information from the rest. Once pre-trained, the network can then be modified and fine-tuned on the final intended *downstream task*. Weng (2019) has a wonderful introductory blog post on this topic, and Zisserman (2018) has a great lecture, where the term *proxy task* is used. Additional good introductions can be found in the survey by Jing and Tian (2020) and the bibliography by Ren (2020).

Figure 5.51 shows some examples of pretext tasks that have been proposed for pre-training image classification networks. These include:

- Context prediction (Doersch, Gupta, and Efros 2015): take nearby image patches and predict their relative positions.
- Context encoders (Pathak, Krahenbuhl *et al.* 2016): inpaint one or more missing regions in an image.
- 9-tile jigsaw puzzle (Noroozi and Favaro 2016): rearrange the tiles into their correct positions.
- Colorizing black and white images (Zhang, Isola, and Efros 2016).
- Rotating images by multiples of 90° to make them upright (Gidaris, Singh, and Komodakis 2018). The paper compares itself against 11 previous self-supervised techniques.

In addition to using single-image pretext tasks, many researchers have used video clips, since successive frames contain semantically related content. One way to use this information is to order

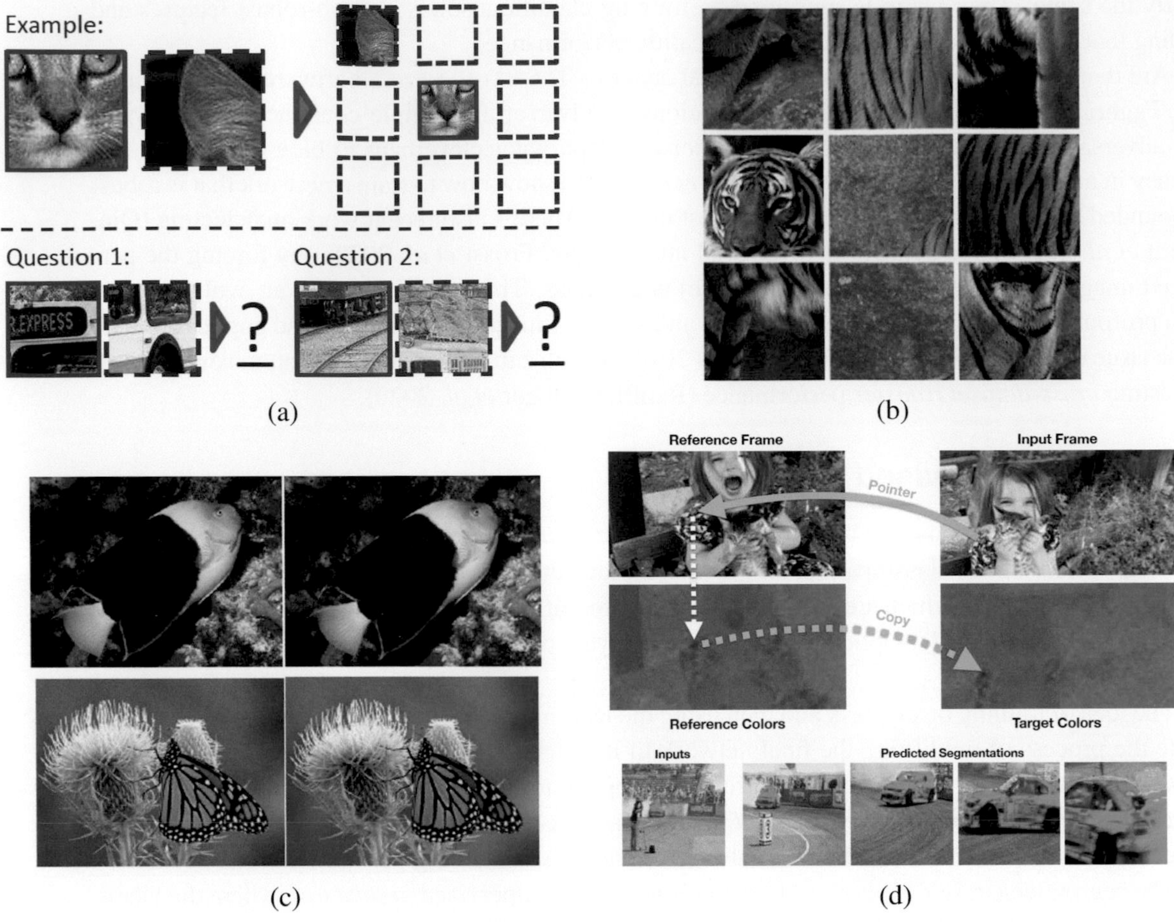

Figure 5.51 Examples of self-supervised learning tasks: (a) guessing the relative positions of image patches—can you guess the answers to Q1 and Q2? (Doersch, Gupta, and Efros 2015) © 2015 IEEE; (b) solving a nine-tile jigsaw puzzle (Noroozi and Favaro 2016) © 2016 Springer; (c) image colorization (Zhang, Isola, and Efros 2016) © 2016 Springer; (d) video color transfer for tracking (Vondrick, Shrivastava *et al.* 2018) © 2016 Springer.

video frames correctly in time, i.e., to use a temporal version of context prediction and jigsaw puzzles (Misra, Zitnick, and Hebert 2016; Wei, Lim *et al.* 2018). Another is to extend colorization to video, with the colors in the first frame given (Vondrick, Shrivastava *et al.* 2018), which encourages the network to learn semantic categories and correspondences. And since videos usually come with sounds, these can be used as additional cues in self-supervision, e.g., by asking a network to align visual and audio signals (Chung and Zisserman 2016; Arandjelovic and Zisserman 2018; Owens and Efros 2018), or in an unsupervised (contrastive) framework (Alwassel, Mahajan *et al.* 2020; Patrick, Asano *et al.* 2020).

Since self-supervised learning shows such great promise, an open question is whether such techniques could at some point surpass the performance of fully-supervised networks trained on smaller fully-labeled datasets.[50] Some impressive results have been shown using *semi-supervised (weak) learning* (Section 5.2.5) on very large (300M–3.5B) partially or noisily labeled datasets such as JFT-300M (Sun, Shrivastava *et al.* 2017) and Instagram hashtags (Mahajan, Girshick *et al.* 2018). Other researchers have tried simultaneously using supervised learning on labeled data and self-supervised

[50]https://people.eecs.berkeley.edu/~efros/gelato_bet.html

pretext task learning on unlabeled data (Zhai, Oliver *et al.* 2019; Sun, Tzeng *et al.* 2019). It turns out that getting the most out of such approaches requires careful attention to dataset size, model architecture and capacity, and the extract details (and difficulty) of the pretext tasks (Kolesnikov, Zhai, and Beyer 2019; Goyal, Mahajan *et al.* 2019; Misra and Maaten 2020). At the same time, others are investigating how much real benefit pre-training actually gives in downstream tasks (He, Girshick, and Dollár 2019; Newell and Deng 2020; Feichtenhofer, Fan *et al.* 2021).

Semi-supervised training systems automatically generate ground truth labels for pretext tasks so that these can be used in a supervised manner (e.g, by minimizing classification errors). An alternative is to use unsupervised learning with a contrastive loss (Section 5.3.4) or other ranking loss (Gómez 2019) to encourage semantically similar inputs to produce similar encodings while spreading dissimilar inputs further apart. This is commonly now called *contrastive (metric) learning*.

Wu, Xiong *et al.* (2018) train a network to produce a separate embedding for each instance (training example), which they store in a moving average *memory bank* as new samples are fed through the neural network being trained. They then classify new images using nearest neighbors in the embedding space. Momentum Contrast (MoCo) replaces the memory bank with a fixed-length queue of encoded samples fed through a temporally adapted momentum encoder, which is separate from the actual network being trained (He, Fan *et al.* 2020). Pretext-invariant representation learning (PIRL) uses pretext tasks and "multi-crop" data augmentation, but then compares their outputs using a memory bank and contrastive loss (Misra and Maaten 2020). SimCLR (simple framework for contrastive learning) uses fixed mini-batches and applies a contrastive loss (normalized temperature cross-entropy, similar to (5.58)) between each sample in the batch and all the other samples, along with aggressive data augmentation (Chen, Kornblith *et al.* 2020). MoCo v2 combines ideas from MoCo and SimCLR to obtain even better results (Chen, Fan *et al.* 2020). Rather than directly comparing the generated embeddings, a fully connected network (MLP) is first applied.

Contrastive losses are a useful tool in metric learning, since they encourage distances in an embedding space to be small for semantically related inputs. An alternative is to use deep clustering to similarly encourage related inputs to produce similar outputs (Caron, Bojanowski *et al.* 2018; Ji, Henriques, and Vedaldi 2019; Asano, Rupprecht, and Vedaldi 2020; Gidaris, Bursuc *et al.* 2020; Yan, Misra *et al.* 2020). Some of the latest results using clustering for unsupervised learning now produce results competitive with contrastive metric learning and also suggest that the kinds of data augmentation being used are even more important than the actual losses that are chosen (Caron, Misra *et al.* 2020; Tian, Chen, and Ganguli 2021). In the context of vision and language (Section 6.6), CLIP (Radford, Kim *et al.* 2021) has achieved remarkable generalization for image classification using contrastive learning and "natural-language supervision." With a dataset of 400 million text and image pairs, their task is to take in a single image and a random sample of 32,768 text snippets and predict which text snippet is truly paired with the image.

Interestingly, it has recently been discovered that representation learning that *only* enforces similarity between semantically similar inputs also works well. This seems counter-intuitive, because without negative pairs as in contrastive learning, the representation can easily collapse to trivial solutions by predicting a constant for any input and maximizing similarity. To avoid this collapse, careful attention is often paid to the network design. Bootstrap Your Own Latent (BYOL) (Grill, Strub *et al.* 2020) shows that with a momentum encoder, an extra predictor MLP on the online network side, and a stop-gradient operation on the target network side, one can successfully remove the negatives from MoCo v2 training. SimSiam (Chen and He 2021) further shown that even the momentum encoder is not required and only a stop-gradient operation is sufficient for the network to learn meaningful representations. While both systems jointly train the predictor MLP and the encoder with gradient updates, it has been even more recently shown that the predictor weights can be directly set using statistics of the input right before the predictor layer (Tian, Chen, and Ganguli 2021). Feichtenhofer, Fan *et al.* (2021) compare a number of these unsupervised representation learning techniques on a variety of video understanding tasks and find that the learned spatiotemporal representations generalize well to different tasks.

Contrastive learning and related work rely on compositions of data augmentations (e.g. color jitters, random crops, etc.) to learn representations that are *invariant* to such changes (Chen and He 2021). An alternative attempt is to use generative modeling (Chen, Radford *et al.* 2020), where the representations are pre-trained by predicting pixels either in an auto-regressive (GPT- or other language model) manner or a de-noising (BERT-, masked auto-encoder) manner. Generative modeling has the potential to bridge self-supervised learning across domains from vision to NLP, where scalable pre-trained models are now dominant.

One final variant on self-supervised learning is using a student-teacher model, where the teacher network is used to provide training examples to a student network. These kinds of architectures were originally called *model compression* (Bucilă, Caruana, and Niculescu-Mizil 2006) and *knowledge distillation* (Hinton, Vinyals, and Dean 2015) and were used to produce smaller models. However, when coupled with additional data and larger capacity networks, they can also be used to improve performance. Xie, Luong *et al.* (2020) train an EfficientNet (Tan and Le 2019) on the labeled ImageNet training set, and then use this network to label an additional 300M unlabeled images. The true labels and pseudo-labeled images are then used to train a higher-capacity "student", using regularization (e.g., dropout) and data augmentation to improve generalization. The process is then repeated to yield further improvements.

In all, self-supervised learning is currently one of the most exciting sub-areas in deep learning,[51] and many leading researchers believe that it may hold the key to even better deep learning (LeCun and Bengio 2020). To explore implementations further, VISSL provides open-source PyTorch implementations of many state-of-the-art self-supervised learning models (with weights) that were described in this section.[52]

5.5 More complex models

While deep neural networks started off being used in 2D image understanding and processing applications, they are now also widely used for 3D data such as medical images and video sequences. We can also chain a series of deep networks together in time by feeding the results from one time frame to the next (or even forward-backward). In this section, we look first at three-dimensional convolutions and then at recurrent models that propagate information forward or bi-directionally in time. We also look at *generative models* that can synthesize completely new images from semantic or related inputs.

5.5.1 Three-dimensional CNNs

As we just mentioned, deep neural networks in computer vision started off being used in the processing of regular two-dimensional images. However, as the amount of video being shared and analyzed increases, deep networks are also being applied to video understanding, which we discuss in more detail Section 6.5. We are also seeing applications in three-dimensional volumetric models such as occupancy maps created from range data (Section 13.5) and volumetric medical images (Section 6.4.1).

[51] https://sites.google.com/view/self-supervised-icml2019
[52] https://github.com/facebookresearch/vissl

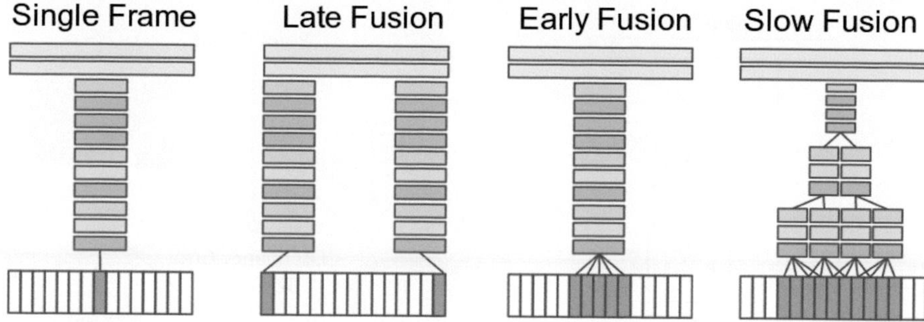

Figure 5.52 Alternative approaches to information fusion over the temporal dimensions (Karpathy, Toderici *et al.* 2014) © 2014 IEEE.

It may appear, at first glance, that the convolutional networks we have already studied, such as the ones illustrated in Figures 5.33, 5.34, and 5.39 already perform 3D convolutions, since their input receptive fields are 3D boxes in (x, y, c), where c is the (feature) *channel* dimension. So, we could in principle fit a sliding window (say in time, or elevation) into a 2D network and be done. Or, we could use something like *grouped convolutions*. However, it's more convenient to operate on a complete 3D volume all at once, and to have *weight sharing* across the third dimension for all kernels, as well as multiple input and output feature channels at each layer.

One of the earliest applications of 3D convolutions was in the processing of video data to classify human actions (Kim, Lee, and Yang 2007; Ji, Xu *et al.* 2013; Baccouche, Mamalet *et al.* 2011). Karpathy, Toderici *et al.* (2014) describe a number of alternative architectures for fusing temporal information, as illustrated in Figure 5.52. The single frame approach classifies each frame independently, depending purely on that frame's content. Late fusion takes features generated from each frame and makes a per-clip classification. Early fusion groups small sets of adjacent frames into multiple channels in a 2D CNN. As mentioned before, the interactions across time do not have the convolutional aspects of weight sharing and temporal shift invariance. Finally, 3D CNNs (Ji, Xu *et al.* 2013) (not shown in this figure) learn 3D space and time-invariance kernels that are run over spatio-temporal windows and fused into a final score.

Tran, Bourdev *et al.* (2015) show how very simple $3 \times 3 \times 3$ convolutions combined with pooling in a deep network can be used to obtain even better performance. Their C3D network can be thought of as the "VGG of 3D CNNs" (Johnson 2020, Lecture 18). Carreira and Zisserman (2017) compare this architecture to alternatives that include *two-stream* models built by analyzing pixels and optical flows in parallel pathways (Figure 6.44b). Section 6.5 on video understanding discusses these and other architectures used for such problems, which have also been attacked using sequential models such as recurrent neural networks (RNNs) and LSTM, which we discuss in Section 5.5.2. Lecture 18 on video understanding by Johnson (2020) has a nice review of all these video understanding architectures.

In addition to video processing, 3D convolutional neural networks have been applied to volumetric image processing. Two examples of shape modeling and recognition from range data, i.e., 3D ShapeNets (Wu, Song *et al.* 2015) and VoxNet (Maturana and Scherer 2015) are shown in Figure 5.53. Examples of their application to medical image segmentation (Kamnitsas, Ferrante *et al.* 2016; Kamnitsas, Ledig *et al.* 2017) are discussed in Section 6.4.1. We discuss neural network approaches to 3D modeling in more detail in Sections 13.5.1 and 14.6.

Like regular 2D CNNs, 3D CNN architectures can exploit different spatial and temporal resolutions, striding, and channel depths, but they can be very computation and memory intensive. To

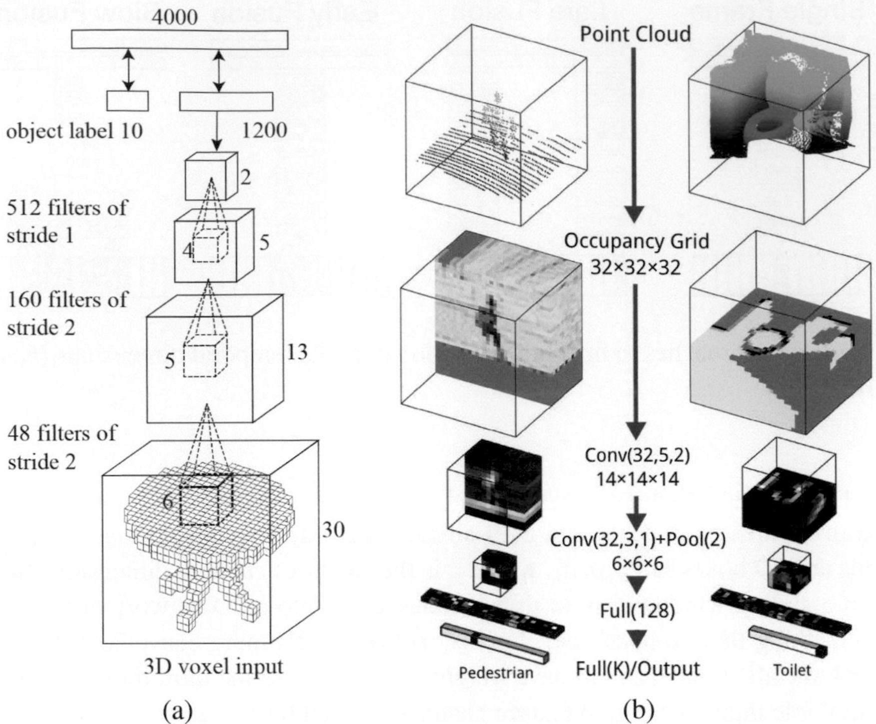

Figure 5.53 3D convolutional networks applied to volumetric data for object detection: (a) 3D ShapeNets (Wu, Song *et al.* 2015) © 2015 IEEE; (b) VoxNet (Maturana and Scherer 2015) © 2015 IEEE.

counteract this, Feichtenhofer, Fan *et al.* (2019) develop a two-stream SlowFast architecture, where a slow pathway operates at a lower frame rate and is combined with features from a fast pathway with higher temporal sampling but fewer channels (Figure 6.44c). Video processing networks can also be made more efficient using channel-separated convolutions (Tran, Wang *et al.* 2019) and neural architecture search (Feichtenhofer 2020). Multigrid techniques (Appendix A.5.3) can also be used to accelerate the training of video recognition models (Wu, Girshick *et al.* 2020).

3D point clouds and meshes

In addition to processing 3D gridded data such as volumetric density, implicit distance functions, and video sequences, neural networks can be used to infer 3D models from single images. One approach is to predict per-pixel depth, which we study in Section 12.8. Another is to reconstruct full 3D models represented using volumetric density (Choy, Xu *et al.* 2016), which we study in Sections 13.5.1 and 14.6. Some more recent experiments, however, suggest that some of these 3D inference networks (Tatarchenko, Dosovitskiy, and Brox 2017; Groueix, Fisher *et al.* 2018; Richter and Roth 2018) may just be recognizing the general object category and doing a small amount of fitting (Tatarchenko, Richter *et al.* 2019).

Generating and processing 3D point clouds has also been extensively studied (Fan, Su, and Guibas 2017; Qi, Su *et al.* 2017; Wang, Sun *et al.* 2019). Guo, Wang *et al.* (2020) provide a comprehensive survey that reviews over 200 publications in this area.

A final alternative is to infer 3D triangulated meshes from either RGB-D (Wang, Zhang *et al.* 2018) or regular RGB (Gkioxari, Malik, and Johnson 2019; Wickramasinghe, Fua, and Knott 2021) images. Figure 5.54 illustrates the components of the Mesh R-CNN system, which detects images

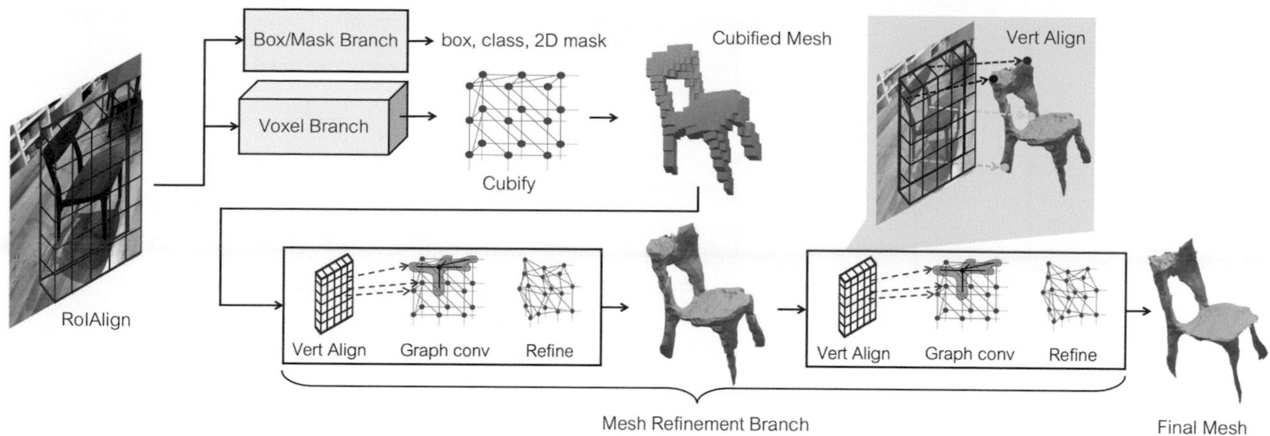

Figure 5.54 Overview of the Mesh R-CNN system (Gkioxari, Malik, and Johnson 2019) © 2019 IEEE. A Mask R-CNN backbone is augmented with two 3D shape inference branches. The voxel branch predicts a coarse shape for each detected object, which is further deformed with a sequence of refinement stages in the mesh refinement branch.

of 3D objects and turns each one into a triangulated mesh after first reconstructing a volumetric model. The primitive operations and representations needed to process such meshes using deep neural networks can be found in the PyTorch3D library.[53]

5.5.2 Recurrent neural networks

While 2D and 3D convolutional networks are a good fit for images and volumes, sometimes we wish to process a *sequence* of images, audio signals, or text. A good way to exploit previously seen information is to pass features detected at one time instant (e.g., video frame) as input to the next frame's processing. Such architectures are called Recurrent Neural Networks (RNNs) and are described in more detail in Goodfellow, Bengio, and Courville (2016, Chapter 10) and Zhang, Lipton *et al.* (2021, Chapter 8). Figure 5.55 shows a schematic sketch of such an architecture. Deep network layers not only pass information on to subsequent layers (and an eventual output), but also feed some of their information as input to the layer processing the next frame of data. Individual layers share weights across time (a bit like 3D convolution kernels), and backpropagation requires computing derivatives for all of the "unrolled" units (time instances) and summing these derivatives to obtain weight updates.

Because gradients can propagate for a long distance backward in time, and can therefore vanish or explode (just as in deep networks before the advent of residual networks), it is also possible to add extra *gating* units to modulate how information flows between frames. Such architectures are called *Gated Recurrent Units* (GRUs) and *Long short-term memory* (LSTM) (Hochreiter and Schmidhuber 1997; Zhang, Lipton *et al.* 2021, Chapter 9).

RNNs and LSTMs are often used for video processing, since they can fuse information over time and model temporal dependencies (Baccouche, Mamalet *et al.* 2011; Donahue, Hendricks *et al.* 2015; Ng, Hausknecht *et al.* 2015; Srivastava, Mansimov, and Salakhudinov 2015; Ballas, Yao *et al.* 2016), as well as language modeling, image captioning, and visual question answering. We discuss these topics in more detail in Sections 6.5 and 6.6. They have also occasionally been used to merge multi-view information in stereo (Yao, Luo *et al.* 2019; Riegler and Koltun 2020a) and to

[53]https://github.com/facebookresearch/pytorch3d

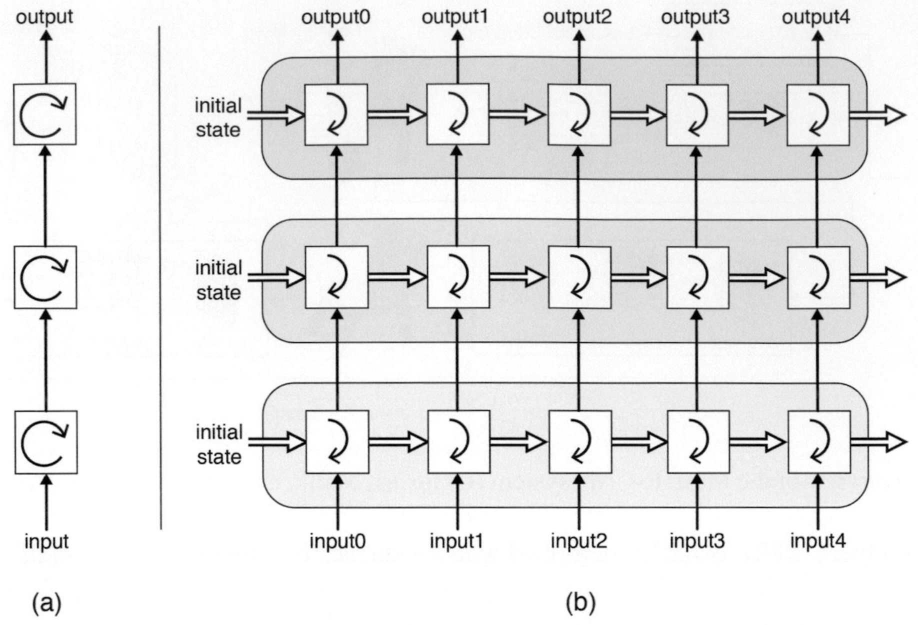

Figure 5.55 A deep recurrent neural network (RNN) uses multiple stages to process sequential data, with the output of one stage feeding the input of the next © Glassner (2018). Each stage maintains its own state and backpropagates its own gradients, although weights are shared between all stages. Column (a) shows a more compact rolled-up diagram, while column (b) shows the corresponding unrolled version.

simulate iterative flow algorithms in a fully differentiable (and hence trainable) manner (Hur and Roth 2019; Teed and Deng 2020b).

To propagate information forward in time, RNNs, GRUs, and LSTMs need to encode *all* of the potentially useful previous information in the hidden state being passed between time steps. In some situations, it is useful for a sequence modeling network to look further back (or even forward) in time. This kind of capability is often called *attention* and is described in more detail in Zhang, Lipton *et al.* (2021, Chapter 10), Johnson (2020, Lecture 12), and Section 5.5.3 on transformers. In brief, networks with attention store lists of *keys* and *values*, which can be probed with a *query* to return a weighted blend of values depending on the alignment between the query and each key. In this sense, they are similar to *kernel regression* (4.12–4.14), which we studied in Section 4.1.1, except that the query and the keys are multiplied (with appropriate weights) before being passed through a softmax to determine the blending weights.

Attention can either be used to look backward at the hidden states in previous time instances (which is called *self-attention*), or to look at different parts of the image (*visual attention*, as illustrated in Figure 6.46). We discuss these topics in more detail in Section 6.6 on vision and language. When recognizing or generating sequences, such as the words in a sentence, attention modules often used to work in tandem with sequential models such as RNNs or LSTMs. However, more recent works have made it possible to apply attention to the entire input sequence in one parallel step, as described in Section 5.5.3 on transformers.

The brief descriptions in this section just barely skim the broad topic of deep sequence modeling, which is usually covered in several lectures in courses on deep learning (e.g., Johnson 2020, Lectures 12–13) and several chapters in deep learning textbooks (Zhang, Lipton *et al.* 2021, Chapters 8–10). Interested readers should consult these sources for more detailed information.

5.5.3 Transformers

Transformers, which are a novel architecture that adds attention mechanisms (which we describe below) to deep neural networks, were first introduced by Vaswani, Shazeer *et al.* (2017) in the context of neural machine translation, where the task consists of translating text from one language to another (Mikolov, Sutskever *et al.* 2013). In contrast to RNNs and their variants (Section 5.5.2), which process input tokens one at a time, transformers can to operate on the entire input sequence at once. In the years after first being introduced, transformers became the dominant paradigm for many tasks in natural language processing (NLP), enabling the impressive results produced by BERT (Devlin, Chang *et al.* 2018), RoBERTa (Liu, Ott *et al.* 2019), and GPT-3 (Brown, Mann *et al.* 2020), among many others. Transformers then began seeing success when processing the natural language component and later layers of many vision and language tasks (Section 6.6). More recently, they have gained traction in pure computer vision tasks, even outperforming CNNs on several popular benchmarks.

The motivation for applying transformers to computer vision is different than that of applying it to NLP. Whereas RNNs suffer from sequentially processing the input, convolutions do not have this problem, as their operations are already inherently parallel. Instead, the problem with convolutions has to do with their *inductive biases*, i.e., the default assumptions encoded into convolutional models.

A convolution operation assumes that nearby pixels are more important than far away pixels. Only after several convolutional layers are stacked together does the receptive field grow large enough to attend to the entire image (Araujo, Norris, and Sim 2019), unless the network is endowed with non-local operations (Wang, Girshick *et al.* 2018) similar to those used in some image denoising algorithms (Buades, Coll, and Morel 2005a). As we have seen in this chapter, convolution's spatial locality bias has led to remarkable success across many aspects of computer vision. But as datasets, models, and computational power grow by orders of magnitude, these inductive biases may become a factor inhibiting further progress.[54]

The fundamental component of a transformer is *self-attention*, which is itself built out of applying *attention* to each of N unit activations in a given layer in the network.[55] Attention is often described using an analogy to the concept of *associative maps* or *dictionaries* found as data structures in programming languages and databases. Given a set of *key-value pairs*, $\{(\mathbf{k}_i, \mathbf{v}_i)\}$ and a query $\mathbf{q}$, a dictionary returns the value $\mathbf{v}_i$ corresponding to the key $\mathbf{k}_i$ that exactly matches the query. In neural networks, the key and query values are real-valued vectors (e.g., linear projections of activations), so the corresponding operation returns a weighted sum of values where the weights depend on the pairwise distances between a query and the set of keys. This is basically the same as *scattered data interpolation*, which we studied in Section 4.1.1, as pointed out in Zhang, Lipton *et al.* (2021, Section 10.2). However, instead of using radial distances as in (4.14), attention mechanisms in neural networks more commonly use *scaled dot-product attention* (Vaswani, Shazeer *et al.* 2017; Zhang, Lipton *et al.* 2021, Section 10.3.3), which involves taking the dot product between the query and key vectors, scaling down by the square root of the dimension of these embeddings D,[56] and then applying the softmax function of (5.5), i.e.,

$$\mathbf{y} = \sum_i \alpha(\mathbf{q} \cdot \mathbf{k}_i / D)\mathbf{v}_i = \text{softmax}(\mathbf{q}^T \mathbf{K}/D)^T \mathbf{V}, \tag{5.77}$$

[54] Rich Sutton, a pioneer in reinforcement learning, believes that learning to leverage computation, instead of encoding human knowledge, is *the bitter lesson* to learn from the history of AI research (Sutton 2019). Others disagree with this view, believing that it is essential to be able to learn from small amounts of data (Lake, Salakhutdinov, and Tenenbaum 2015; Marcus 2020).

[55] N may indicate the number of words in a sentence or patches in an image

[56] We divide the dot product by $\sqrt{D}$ so that the variance of the scaled dot product does not increase for larger embedding dimensions, which could result in vanishing gradients.

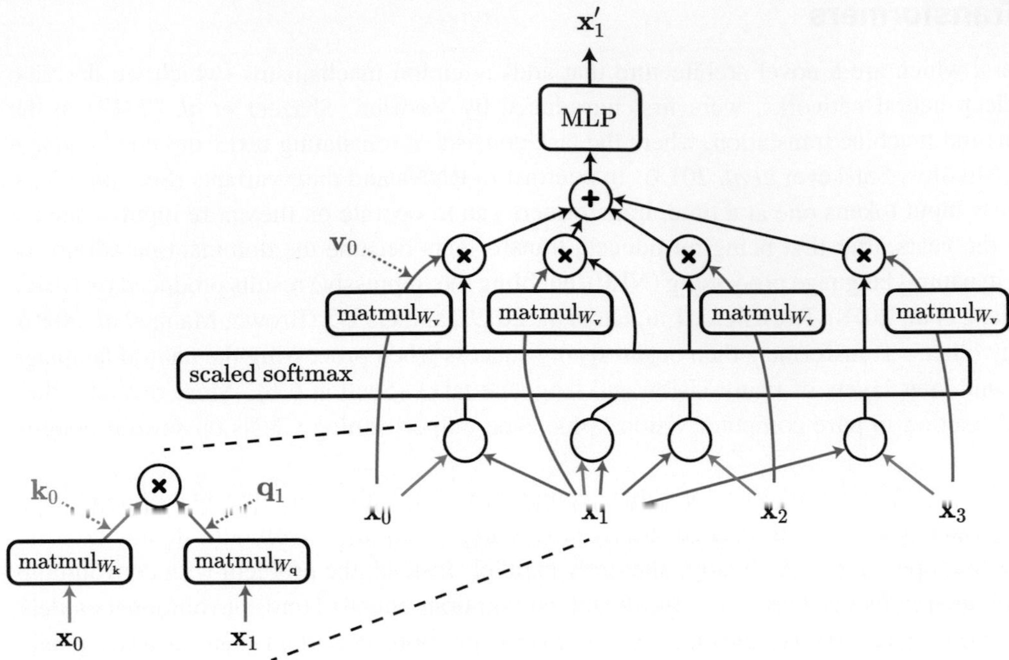

Figure 5.56 The self-attention computation graph to compute a single output vector $\mathbf{x}'_2$, courtesy of Matt Deitke, adapted from Vaswani, Huang, and Manning (2019). Note that the full self-attention operation also computes outputs for $\mathbf{x}'_1$, $\mathbf{x}'_3$, and $\mathbf{x}'_4$ by shifting the input to the query ($\mathbf{x}_2$ in this case) between $\mathbf{x}_1$, $\mathbf{x}_3$, and $\mathbf{x}_4$, respectively. For each of matmul$_V$, matmul$_K$, and matmul$_Q$, there is a single matrix of weights that gets reused with each call.

where $\mathbf{K}$ and $\mathbf{V}$ are the row-stacked matrices composed of the key and value vectors, respectively, and $\mathbf{y}$ is the output of the attention operator.[57]

Given a set of input vectors $\{\mathbf{x}_0, \mathbf{x}_1, \ldots, \mathbf{x}_{N-1}\}$, the self-attention operation produces a set of output vectors $\{\mathbf{x}'_0, \mathbf{x}'_1, \ldots, \mathbf{x}'_{N-1}\}$. Figure 5.56 shows the case for $N = 4$, where the self-attention computation graph is used to obtain a single output vector $\mathbf{x}'_2$. As pictured, self-attention uses three learned weight matrices, $\mathbf{W_q}$, $\mathbf{W_k}$, and $\mathbf{W_v}$, which determine the

$$\mathbf{q}_i = \mathbf{W_q}\mathbf{x}_i, \; \mathbf{k}_i = \mathbf{W_k}\mathbf{x}_i, \; \text{and } \mathbf{v}_i = \mathbf{W_v}\mathbf{x}_i \tag{5.78}$$

per-unit query, key, and value vectors going into each attention block. The weighted sum of values is then optionally passed through a multi-layer perceptron (MLP) to produce $\mathbf{x}'_2$.

In comparison to a fully connected or convolutional layer, self-attention computes each output (e.g., $\mathbf{x}'_i$) based on all of the input vectors $\{\mathbf{x}_0, \mathbf{x}_1, \ldots, \mathbf{x}_{N-1}\}$. In that sense, it is often compared to a fully connected layer, but instead of the weights being fixed for each input, the weights are adapted on the spot, based on the input (Khan, Naseer *et al.* 2021). Compared to convolutions, self-attention is able to attend to every part of the input from the start, instead of constraining itself to local regions of the input, which may help it introduce the kind of context information needed to disambiguate the objects shown in Figure 6.8.

There are several components that are combined with self-attention to produce a transformer block, as described in Vaswani, Shazeer *et al.* (2017). The full transformer consists of both an

[57] The partition of unity function α notation is borrowed from Zhang, Lipton *et al.* (2021, Section 10.3).

encoder and a decoder block, although both share many of the same components. In many applications, an encoder can be used without a decoder (Devlin, Chang *et al.* 2018; Dosovitskiy, Beyer *et al.* 2021) and vice versa (Razavi, van den Oord, and Vinyals 2019).

The right side of Figure 5.57 shows an example of a transformer encoder block. For both the encoder and decoder:

- Instead of modeling set-to-set operations, we can model sequence-to-sequence operations by adding a **positional encoding** to each input vector (Gehring, Auli *et al.* 2017). The positional encoding typically consists of a set of temporally shifted sine waves from which position information can be decoded. (Such position encodings have also recently been added to implicit neural shape representations, which we study in Sections 13.5.1 and 14.6.)

- In lieu of applying a single self-attention operation to the input, multiple self-attention operations, with different learned weight matrices to build different keys, values, and queries, are often joined together to form **multi-headed self-attention** (Vaswani, Shazeer *et al.* 2017). The result of each head is then concatenated together before everything is passed through an MLP.

- **Layer normalization** (Ba, Kiros, and Hinton 2016) is then applied to the output of the MLP. Each vector may then independently be passed through another MLP with shared weights before layer normalization is applied again.

- **Residual connections** (He, Zhang *et al.* 2016a) are employed after multi-headed attention and after the final MLP.

During training, the biggest difference in the decoder is that some of the input vectors to self-attention may be masked out, which helps support parallel training in autoregressive prediction tasks. Further exposition of the details and implementation of the transformer architecture is provided in Vaswani, Shazeer *et al.* (2017) and in the additional reading (Section 5.6).

A key challenge of applying transformers to the image domain has to do with the size of image input (Vaswani, Shazeer *et al.* 2017). Let N denote the length of the input, D denote the number of dimensions for each input entry, and K denote a convolution's (on side) kernel size.[58] The number of floating point operations (FLOPs) required for self-attention is on the order of $O(N^2D)$, whereas the FLOPs for a convolution operation is on the order of $O(ND^2K^2)$. For instance, with an ImageNet image scaled to size $224 \times 224 \times 3$, if each pixel is treated independently, $N = 224 \times 224 = 50176$ and $D = 3$. Here, a convolution is *significantly* more efficient than self-attention. In contrast, applications like neural machine translation may only have N as the number of words in a sentence and D as the dimension for each word embedding (Mikolov, Sutskever *et al.* 2013), which makes self-attention much more efficient.

The Image Transformer (Parmar, Vaswani *et al.* 2018) was the first attempt at applying the full transformer model to the image domain, with many of the same authors that introduced the transformer. It used both an encoder and decoder to try and build an autoregressive generative model that predicts the next pixel, given a sequence of input pixels and all the previously predicted pixels. (The earlier work on non-local networks by Wang, Girshick *et al.* (2018) also used ideas inspired by transformers, but with a simpler attention block and a fully two-dimensional setup.) Each vector input to the transformer corresponded to a single pixel, which ultimately constrained them to generate small images (i.e., 32×32), since the quadratic cost of self-attention was too expensive otherwise.

[58]In Section 5.4 on convolutional architectures, we use C to denote the number of channels instead of D to denote the embedding dimensions.

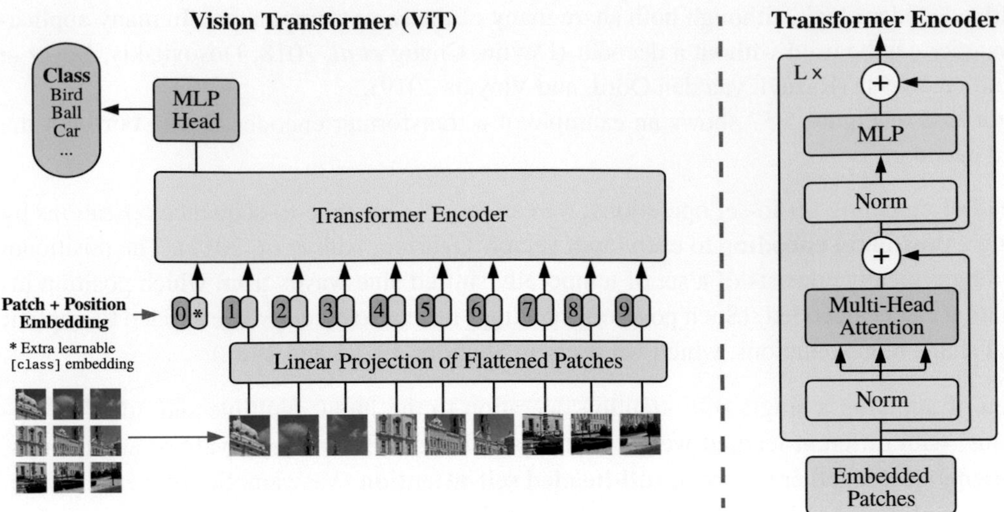

Figure 5.57 The Vision Transformer (ViT) model from (Dosovitskiy, Beyer *et al.* 2021) breaks an image into a 16 × 16 grid of patches. Each patch is then flattened, passed through a shared embedding matrix, and combined with a positional encoding vector. These inputs are then passed through a transformer encoder (right) several times before predicting an image's class.

Dosovitskiy, Beyer *et al.* (2021) had a breakthrough that allowed transformers to process much larger images. Figure 5.57 shows the diagram of the model, named the Vision Transformer (ViT). For the task of image recognition, instead of treating each pixel as a separate input vector to the transformer, they divide an image (of size 224 × 224) into 196 distinct 16 × 16 gridded image patches. Each patch is then flattened, and passed through a shared embedding matrix, which is equivalent to a strided 16 × 16 convolution, and the results are combined with a positional encoding vector and then passed to the transformer. Earlier work from Cordonnier, Loukas, and Jaggi (2019) introduced a similar patching approach, but on a smaller scale with 2 × 2 patches.

ViT was only able to outperform their convolutional baseline BiT (Kolesnikov, Beyer *et al.* 2020) when using over 100 million training images from JFT-300M (Sun, Shrivastava *et al.* 2017). When using ImageNet alone, or a random subset of 10 or 30 million training samples from JPT-300, the ViT model typically performed much worse than the BiT baseline. Their results suggest that in low-data domains, the inductive biases present in convolutions are typically quite useful. But, with orders of magnitude of more data, a transformer model might discover even better representations that are not representable with a CNN.

Some works have also gone into combining the inductive biases of convolutions with transformers (Srinivas, Lin *et al.* 2021; Wu, Xiao *et al.* 2021; Lu, Batra *et al.* 2019; Yuan, Guo *et al.* 2021). An influential example of such a network is DETR (Carion, Massa *et al.* 2020), which is applied to the task of object detection. It first processes the image with a ResNet backbone, with the output getting passed to a transformer encoder-decoder architecture. They find that the addition of a transformer improves the ability to detect large objects, which is believed to be because of its ability to reason globally about correspondences between inputted encoding vectors.

The application and usefulness of transformers in the realm of computer vision is still being widely researched. Already, however, they have achieved impressive performance on a wide range of tasks, with new papers being published rapidly.[59] Some more notable applications include image

[59] https://github.com/dk-liang/Awesome-Visual-Transformer

classification (Liu, Lin *et al.* 2021; Touvron, Cord *et al.* 2020), object detection (Dai, Cai *et al.* 2020; Liu, Lin *et al.* 2021), image pre-training (Chen, Radford *et al.* 2020), semantic segmentation (Zheng, Lu *et al.* 2020), pose recognition (Li, Wang *et al.* 2021), super-resolution (Zeng, Fu, and Chao 2020), colorization (Kumar, Weissenborn, and Kalchbrenner 2021), generative modeling (Jiang, Chang, and Wang 2021; Hudson and Zitnick 2021), and video classification (Arnab, Dehghani *et al.* 2021; Fan, Xiong *et al.* 2021; Li, Zhang *et al.* 2021). Recent works have also found success extending ViT's patch embedding to pure MLP vision architectures (Tolstikhin, Houlsby *et al.* 2021; Liu, Dai *et al.* 2021; Touvron, Bojanowski *et al.* 2021). Applications to vision and language are discussed in Section 6.6.

5.5.4 Generative models

Throughout this chapter, I have mentioned that machine learning algorithms such as logistic regression, support vector machines, random trees, and feedforward deep neural networks are all examples of *discriminative* systems that never form an explicit *generative* model of the quantities they are trying to estimate (Bishop 2006, Section 1.5; Murphy 2012, Section 8.6). In addition to the potential benefits of generative models discussed in these two textbooks, Goodfellow (2016) and Kingma and Welling (2019) list some additional ones, such as the ability to visualize our assumptions about our unknowns, training with missing or incompletely labeled data, and the ability to generate multiple, alternative, results.

In computer graphics, which is sometimes called *image synthesis* (as opposed to the *image understanding* or *image analysis* we do in computer vision), the ability to easily generate realistic random images and models has long been an essential tool. Examples of such algorithms include texture synthesis and style transfer, which we study in more detail in Section 10.5, as well as fractal terrain (Fournier, Fussel, and Carpenter 1982) and tree generation (Prusinkiewicz and Lindenmayer 1996). Examples of deep neural networks being used to generate such novel images, often under user control, are shown in Figures 5.60 and 10.58. Related techniques are also used in the nascent field of *neural rendering*, which we discuss in Section 14.6.

How can we unlock the demonstrated power of deep neural networks to capture semantics in order to visualize sample images and generate new ones? One approach could be to use the visualization techniques introduced in Section 5.4.5. But as you can see from Figure 5.49, while such techniques can give us insights into individual units, they fail to create fully realistic images.

Another approach might be to construct a *decoder* network to undo the classification performed by the original (*encoder*) network. This kind of "bottleneck" architecture is widely used, as shown in Figure 5.37a, to derive semantic per-pixel labels from images. Can we use a similar idea to generate realistic looking images?

Variational autoencoders

A network that encodes an image into small compact codes and then attempts to decode it back into the same image is called an *autoencoder*. The compact codes are typically represented as a vector, which is often called the *latent* vector to emphasize that it is hidden and unknown. Autoencoders have a long history of use in neural networks, even predating today's feedforward networks (Kingma and Welling 2019). It was once believed that this might be a good way to pre-train networks, but the more challenging proxy tasks we studied in Section 5.4.7 have proven to be more effective.

At a high level, to train an autoencoder on a dataset of images, we can use an unsupervised objective that tries to have the output image of the decoder match the training image input to the encoder. To generate a new image, we can then randomly sample a latent vector and hope that from

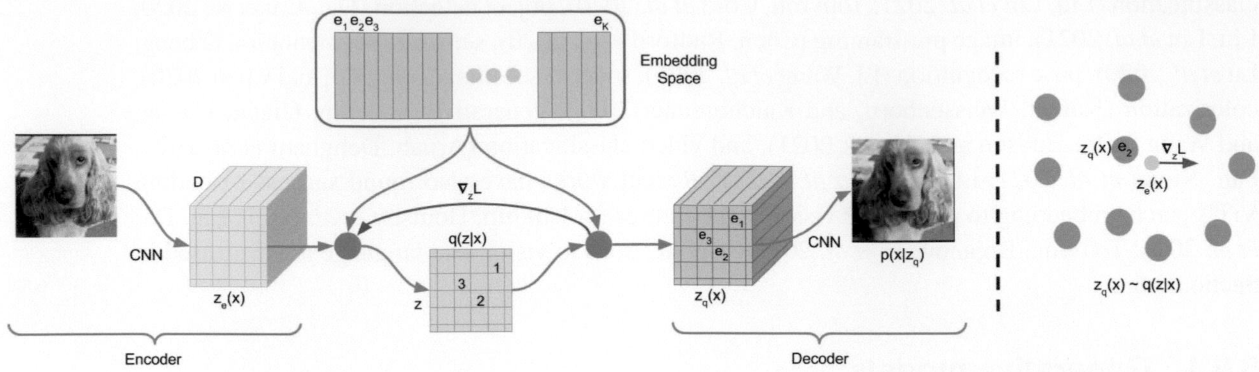

Figure 5.58 The VQ-VAE model. On the left, $z_e(x)$ represents the output of the encoder, the embedding space on top represents the codebook of K embedding vectors, and $q(z \mid x)$ represents the process of replacing each spatial (i.e., channel-wise) vector in the output of the encoder with its nearest vector in the codebook. On the right, we see how a $z_e(x)$ vector (green) may be rounded to e_2, and that the gradient in the encoder network (red) may push the vector away from e_2 during backpropagation. © van den Oord, Vinyals, and Kavukcuoglu (2017)

that vector, the decoder can generate a new image that looks like it came from the distribution of training images in our dataset.

With an autoencoder, there is a deterministic, one-to-one mapping from each input to its latent vector. Hence, the number of latent vectors that are generated exactly matches the number of input data points. If the encoder's objective is to produce a latent vector that makes it easy to decode, one possible solution would be for every latent vector to be extremely far away from every other latent vector. Here, the decoder can overfit all the latent vectors it has seen since they would all be unique with little overlap. However, as our goal is to randomly generate latent vectors that can be passed to the decoder to generate realistic images, we want the latent space to both be well explored and to encode some meaning, such as nearby vectors being semantically similar. Ghosh, Sajjadi *et al.* (2019) propose one potential solution, where they inject noise into the latent vector and empirically find that it works quite well.

Another extension of the autoencoder is the *variational autoencoder* (VAE) (Kingma and Welling 2013; Rezende, Mohamed, and Wierstra 2014; Kingma and Welling 2019). Instead of generating a single latent vector for each input, it generates the mean and covariance that define a chosen distribution of latent vectors. The distribution can then be sampled from to produce a single latent vector, which gets passed into the decoder. To avoid having the covariance matrix become the zero matrix, making the sampling process deterministic, the objective function often includes a regularization term to penalize the distribution if it is far from some chosen (e.g., Gaussian) distribution. Due to their probabilistic nature, VAEs can explore the space of possible latent vectors significantly better than autoencoders, making it harder for the decoder to overfit the training data.

Motivated by how natural language is discrete and by how images can typically be described in language (Section 6.6), the vector quantized VAE (VQ-VAE) of van den Oord, Vinyals, and Kavukcuoglu (2017) takes the approach of modeling the latent space with categorical variables Figure 5.58 shows an outline of the VQ-VAE architecture. The encoder and decoder operate like a normal VAE, where the encoder predicts some latent representation from the input, and the decoder generates an image from the latent representation. However, in contrast to the normal VAE, the VQ-VAE replaces each spatial dimension of the predicted latent representation with its nearest vector from a discrete set of vectors (named the *codebook*). The discretized latent representation is then

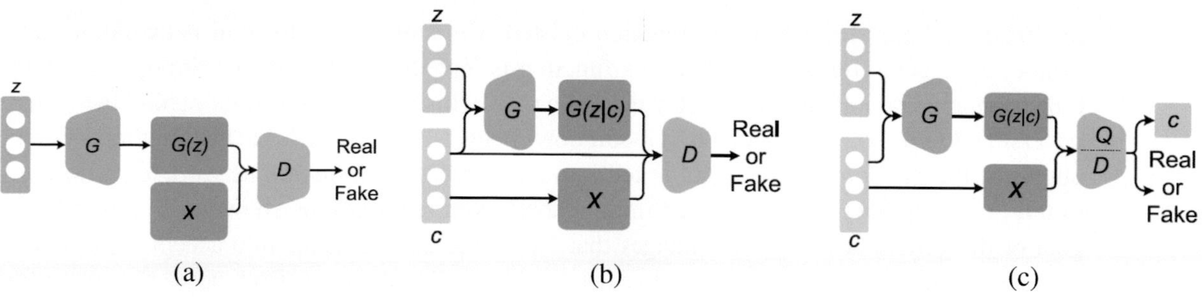

Figure 5.59 Generative adversarial network (GAN) architectures from Pan, Yu *et al.* (2019) © 2019 IEEE. (a) In a regular GAN, random "latent" noise vectors z are fed into a generator network G, which produces synthetic "fake" images $x' = G(z)$. The job of the discriminator D is to tell the fake images apart from real samples x. (b) In a conditional GAN (cGAN), the network iterates (during training) over all the classes that we wish to synthesize. The generator G gets both a class id c and a random noise vector z as input, and the discriminator D gets the class id as well and needs to determine if its input is a real member of the given class. (c) The discriminator in an InfoGAN does not have access to the class id, but must instead infer it from the samples it is given.

passed to the decoder. The vectors in the codebook are trained simultaneously with the VAE's encoder and decoder. Here, the codebook vectors are optimized to move closer to the spatial vectors outputted by the encoder.

Although a VQ-VAE uses a discrete codebook of vectors, the number of possible images it can represent is still monstrously large. In some of their image experiments, they set the size of the codebook to $K = 512$ vectors and set the size of the latent variable to be $z = 32 \times 32 \times 1$. Here, they can represent $512^{32 \cdot 32 \cdot 1}$ possible images.

Compared to a VAE, which typically assumes a Gaussian latent distribution, the latent distribution of a VQ-VAE is not as clearly defined, so a separate generative model is trained to sample latent variables z. The model is trained on the final latent variables outputted from the trained VQ-VAE encoder across the training data. For images, entries in z are often spatially dependent, e.g., an object may be encoded over many neighboring entries. With entries being chosen from a discrete codebook of vectors, we can use a PixelCNN (van den Oord, Kalchbrenner *et al.* 2016) to autoregressively sample new entries in the latent variable based on previously sampled neighboring entries. The PixelCNN can also be conditionally trained, which enables the ability to sample latent variables corresponding to a particular image class or feature.

A follow-up to the VQ-VAE model, named VQ-VAE-2 (Razavi, van den Oord, and Vinyals 2019), uses a two-level approach to decoding images, where with both a small and large latent vector, they can get much higher fidelity reconstructed and generated images. Section 6.6 discusses Dall·E (Ramesh, Pavlov *et al.* 2021), a model that applies VQ-VAE-2 to text-to-image generation and achieves remarkable results.

Generative adversarial networks

Another possibility for image synthesis is to use the multi-resolution features computed by pre-trained networks to match the statistics of a given texture or style image, as described in Figure 10.57. While such networks are useful for matching the *style* of a given artist and the high-level *content* (layout) of a photograph, they are not sufficient to generate completely photorealistic images.

In order to create truly photorealistic synthetic images, we want to determine if an image is real(istic) or fake. If such a loss function existed, we could use it to train networks to generate synthetic images. But, since such a loss function is incredibly difficult to write by hand, why not train a separate neural network to play the critic role? This is the main insight behind the *generative adversarial networks* introduced by Goodfellow, Pouget-Abadie *et al.* (2014). In their system, the output of the *generator* network G is fed into a separate *discriminator* network D, whose task is to tell "fake" synthetically generated images apart from real ones, as shown in Figure 5.59a. The goal of the generator is to create images that "fool" the discriminator into accepting them as real, while the goal of the discriminator is to catch the "forger" in their act. Both networks are co-trained simultaneously, using a blend of loss functions that encourage each network to do its job. The joint loss function can be written as

$$E_{\text{GAN}}(\mathbf{w}_G, \mathbf{w}_D) = \sum_n \log D(\mathbf{x}_n) + \log\left(1 - D(G(\mathbf{z}_n))\right), \tag{5.79}$$

where the $\{\mathbf{x}_n\}$ are the real-world training images, $\{\mathbf{z}_n\}$ are random vectors, which are passed through the generator G to produce synthetic images $\mathbf{x}'_n$, and the $\{\mathbf{w}_G, \mathbf{w}_D\}$ are the weights (parameters) in the generator and discriminator.

Instead of minimizing this loss, we adjust the weights of the generator to minimize the second term (they do not affect the first), and adjust the weights of the discriminator to *maximize* both terms, i.e., minimize the discriminator's error. This process is often called a *minimax game*.[60] More details about the formulation and how to optimize it can be found in the original paper by Goodfellow, Pouget-Abadie *et al.* (2014), as well as deep learning textbooks (Zhang, Lipton *et al.* 2021, Chapter 17), lectures (Johnson 2020, Lecture 20), tutorials (Goodfellow, Isola *et al.* 2018), and review articles (Creswell, White *et al.* 2018; Pan, Yu *et al.* 2019).

The original paper by Goodfellow, Pouget-Abadie *et al.* (2014) used a small, fully connected network to demonstrate the basic idea, so it could only generate 32×32 images such as MNIST digits and low-resolution faces. The *Deep Convolutional GAN* (DCGAN) introduced by Radford, Metz, and Chintala (2015) uses the second half of the deconvolution network shown in Figure 5.37a to map from the random latent vectors $\mathbf{z}$ to arbitrary size images and can therefore generate a much wider variety of outputs, while LAPGAN uses a Laplacian pyramid of adversarial networks (Denton, Chintala *et al.* 2015). Blending between different latent vectors (or perturbing them in certain directions) generates in-between synthetic images.

GANs and DCGANs can be trained to generate new samples from a given class, but it is even more useful to generate samples from different classes using the same trained network. The *conditional GAN* (cGAN) proposed by Mirza and Osindero (2014) achieves this by feeding a class vector into both the generator, which *conditions* its output on this second input, as well as the discriminator, as shown in Figure 5.59b. It is also possible to make the discriminator predict classes that correlate with the class vector using an extra mutual information term, as shown in Figure 5.59c (Chen, Duan *et al.* 2016). This allows the resulting InfoGAN network to learn disentangled representations, such as the digit shapes and writing styles in MNIST, or pose and lighting.

While generating random images can have many useful graphics applications, such as generating textures, filling holes, and stylizing photographs, as discussed in Section 10.5, it becomes even more useful when it can be done under a person's artistic control (Lee, Zitnick, and Cohen 2011). The iGAN interactive image editing system developed by Zhu, Krähenbühl *et al.* (2016) does this by learning a manifold of photorealistic images using a generative adversarial network and then constraining user edits (or even sketches) to produce images that lie on this manifold.

[60] Note that the term *adversarial* in GANs refers to this adversarial game between the generator and the discriminator, which helps the generator create better pictures. This is distinct from the *adversarial examples* we discussed in Section 5.4.6, which are images designed to fool recognition systems.

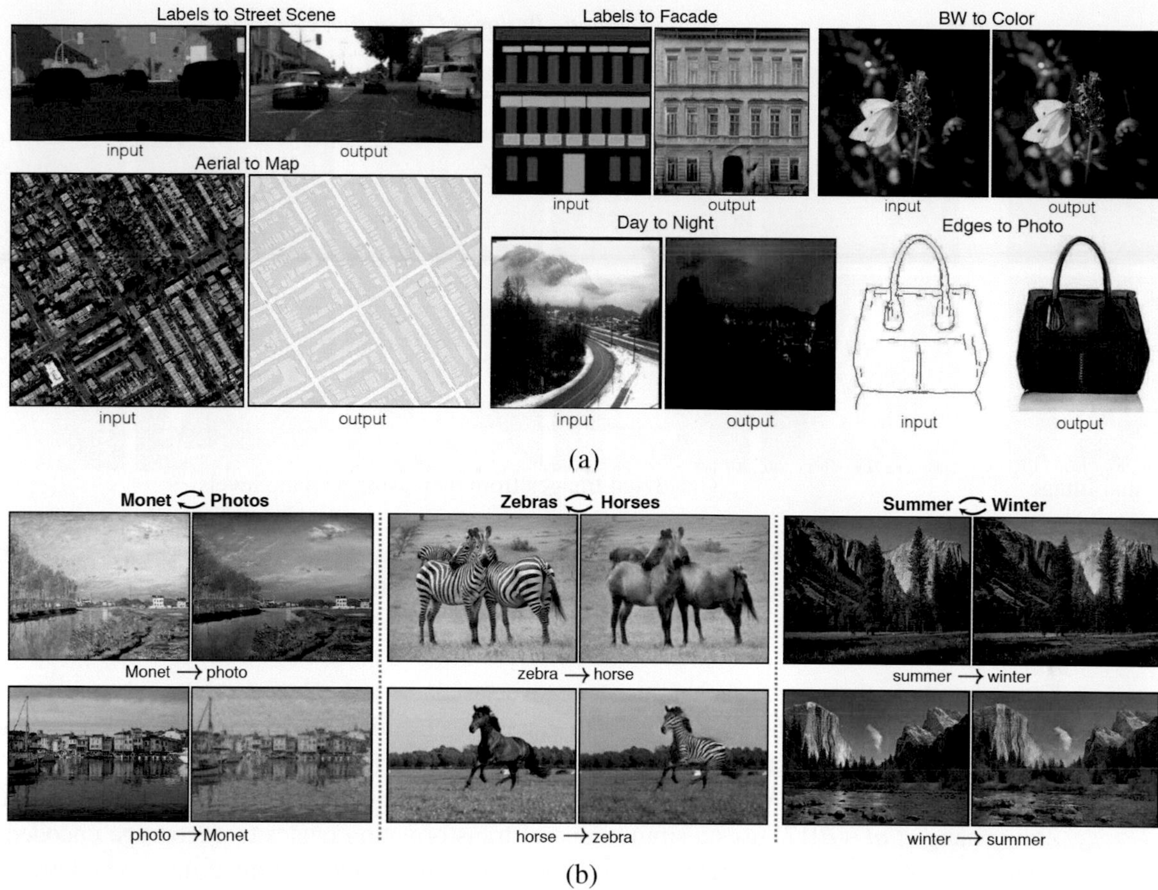

Figure 5.60 Image-to-image translation. (a) Given paired training images, the original `pix2pix` system learns how to turn sketches into photos, semantic maps to images, and other pixel remapping tasks (Isola, Zhu *et al.* 2017) © 2017 IEEE. (b) CycleGAN does not require paired training images, just collections coming from different sources, such as painting and photographs or horses and zebras (Zhu, Park *et al.* 2017) © 2017 IEEE.

This approach was generalized by Isola, Zhu *et al.* (2017) to all kinds of other image-to-image translation tasks, as shown in Figure 5.60a. In their `pix2pix` system, images, which can just be sketches or semantic labels, are fed into a modified U-Net, which converts them to images with different semantic meanings or styles (e.g., photographs or road maps). When the input is a semantic label map and the output is a photorealistic image, this process is often called *semantic image synthesis*. The translation network is trained with a conditional GAN, which takes paired images from the two domains at training time and has the discriminator decide if the synthesized (translated) image together with the input image are a real or fake pair. Referring back to Figure 5.59b, the class c is now a complete image, which is fed into both G and the discriminator D, along with its paired or synthesized output. Instead of making a decision for the whole image, the discriminator looks at overlapping patches and makes decisions on a patch-by-patch basis, which requires fewer parameters and provides more training data and more discriminative feedback. In their implementation, there is no random vector z; instead, dropout is used during both training and "test" (translation) time, which is equivalent to injecting noise at different levels in the network.

In many situations, paired images are not available, e.g., when you have collections of paintings and photographs from different locations, or pictures of animals in two different classes, as shown in

Figure 5.61 The Semantic Image Pyramid can be used to choose which semantic level in a deep network to modify when editing an image (Shocher, Gandelsman *et al.* 2020) © 2020 IEEE.

Figure 5.60b. In this case, a cycle-consistent adversarial network (CycleGAN) can be used to require the mappings between the two domains to encourage identity, while also ensuring that generated images are perceptually similar to the training images (Zhu, Park *et al.* 2017). DualGAN (Yi, Zhang *et al.* 2017) and DiscoGAN (Kim, Cha *et al.* 2017) use related ideas. The BicycleGAN system of Zhu, Zhang *et al.* (2017) uses a similar idea of transformation cycles to encourage encoded latent vectors to correspond to different modes in the outputs for better interpretability and control.

Since the publication of the original GAN paper, the number of extensions, applications, and follow-on papers has exploded. The GAN Zoo website[61] lists over 500 GAN papers published between 2014 and mid-2018, at which point it stopped being updated. Large number of papers continue to appear each year in vision, machine learning, and graphics conferences.

Some of the more important papers since 2017 include Wasserstein GANs (Arjovsky, Chintala, and Bottou 2017), Progressive GANs (Karras, Aila *et al.* 2018), UNIT (Liu, Breuel, and Kautz 2017) and MUNIT (Huang, Liu *et al.* 2018), spectral normalization (Miyato, Kataoka *et al.* 2018), SAGAN (Zhang, Goodfellow *et al.* 2019), BigGAN (Brock, Donahue, and Simonyan 2019), StarGAN (Choi, Choi *et al.* 2018) and StyleGAN (Karras, Laine, and Aila 2019) and follow-on papers (Choi, Uh *et al.* 2020; Karras, Laine *et al.* 2020; Viazovetskyi, Ivashkin, and Kashin 2020), SPADE (Park, Liu *et al.* 2019), GANSpace (Härkönen, Hertzmann *et al.* 2020), and VQGAN (Esser, Rombach, and Ommer 2020). You can find more detailed explanations and references to many more papers in the lectures by Johnson (2020, Lecture 20), tutorials by Goodfellow, Isola *et al.* (2018), and review articles by Creswell, White *et al.* (2018), Pan, Yu *et al.* (2019), and Tewari, Fried *et al.* (2020).

In summary, generative adversarial networks and their myriad extensions continue to be an extremely vibrant and useful research area, with applications such as image super-resolution (Section 10.3), photorealistic image synthesis (Section 10.5.3), image-to-image translation, and interactive image editing. Two very recent examples of this last application are the Semantic Pyramid for Image Generation by Shocher, Gandelsman *et al.* (2020), in which the semantic manipulation level can be controlled (from small texture changes to higher-level layout changes), as shown in Figure 5.61, and the Swapping Autoencoder by Park, Zhu *et al.* (2020), where structure and texture can be independently edited.

[61] https://github.com/hindupuravinash/the-gan-zoo

5.6 Additional reading

Machine learning and deep learning are rich, broad subjects which properly deserve their own course of study to master. Fortunately, there are a large number of good textbooks and online courses available to learn this material.

My own favorite for machine learning is the book by Bishop (2006), since it provides a broad treatment with a Bayesian flavor and excellent figures, which I have re-used in this book. The books by Glassner (2018, 2021) provide an even gentler introduction to both classic machine learning and deep learning, as well as additional figures I reference in this book. Two additional widely used textbooks for machine learning are Hastie, Tibshirani, and Friedman (2009) and Murphy (2012). Deisenroth, Faisal, and Ong (2020) provide a nice compact treatment of mathematics for machine learning, including linear and matrix algebra, probability theory, model fitting, regression, PCA, and SVMs, with a more in-depth exposition than the terse summaries I provide in this book. The book on Automated Machine Learning edited by Hutter, Kotthoff, and Vanschoren (2019) surveys automated techniques for designing and optimizing machine learning algorithms.

For deep learning, Goodfellow, Bengio, and Courville (2016) were the first to provide a comprehensive treatment, but it has not recently been revised. Glassner (2018, 2021) provides a wonderful introduction to deep learning, with lots of figures and no equations. I recommend it even to experienced practitioners since it helps develop and solidify intuitions about how learning works. An up-to-date reference on deep learning is the *Dive into Deep Learning* online textbook by Zhang, Lipton *et al.* (2021), which comes with interactive Python notebooks sprinkled throughout the text, as well as an associated course (Smola and Li 2019). Some introductory courses to deep learning use Charniak (2019).

Rawat and Wang (2017) provide a nice review article on deep learning, including a history of early and later neural networks, as well in-depth discussion of many deep learning components, such as pooling, activation functions, losses, regularization, and optimization. Additional surveys related to advances in deep learning include Sze, Chen *et al.* (2017), Elsken, Metzen, and Hutter (2019), Gu, Wang *et al.* (2018), and Choudhary, Mishra *et al.* (2020). Sejnowski (2018) provides an in-depth history of the early days of neural networks.

The Deep Learning for Computer Vision course slides by Johnson (2020) are an outstanding reference and a great way to learn the material, both for the depth of their information and how up-to-date the presentations are kept. They are based on Stanford's CS231n course (Li, Johnson, and Yeung 2019), which is also a great up-to-date source. Additional classes on deep learning with slides and/or video lectures include Grosse and Ba (2019), McAllester (2020), Leal-Taixé and Nießner (2020), Leal-Taixé and Nießner (2021), and Geiger (2021)

For transformers, Bloem (2019) provides a nice starting tutorial on implementing the standard transformer encoder and decoder block in PyTorch, from scratch. More comprehensive surveys of transformers applied to computer vision include Khan, Naseer *et al.* (2021) and Han, Wang *et al.* (2020). Tay, Dehghani *et al.* (2020) provides an overview of many attempts to reduce the quadratic cost of self-attention.Wightman (2021) makes available a fantastic collection of computer vision transformer implementations in PyTorch, with pre-trained weights and great documentation. Additional course lectures introducing transformers with videos and slides include Johnson (2020, Lecture 13), Vaswani, Huang, and Manning (2019, Lecture 14) and LeCun and Canziani (2020, Week 12).

For GANs, the new deep learning textbook by Zhang, Lipton *et al.* (2021, Chapter 17), lectures by Johnson (2020, Lecture 20), tutorials by Goodfellow, Isola *et al.* (2018), and review articles by Creswell, White *et al.* (2018), Pan, Yu *et al.* (2019), and Tewari, Fried *et al.* (2020) are all good sources. For a survey of the latest visual recognition techniques, the tutorials presented at ICCV

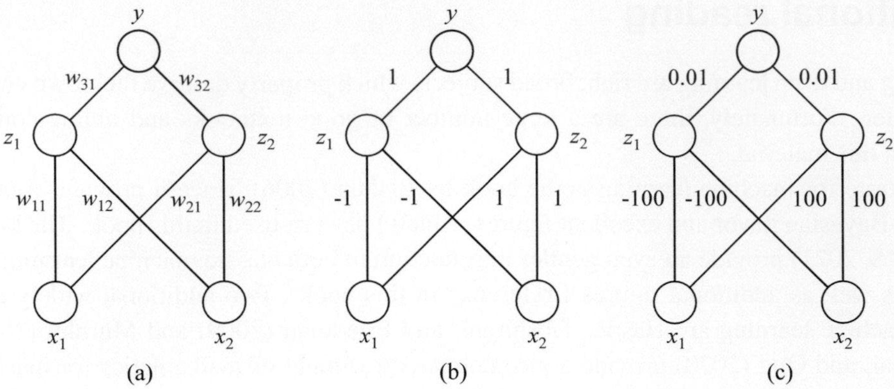

Figure 5.62 Simple two hidden unit network with a ReLU activation function and no bias parameters for regressing the function $y = |x_1 + 1.1x_2|$: (a) can you guess a set of weights that would fit this function?; (b) a reasonable set of starting weights; (c) a poorly scaled set of weights.

(Xie, Girshick *et al.* 2019), CVPR (Girshick, Kirillov *et al.* 2020), and ECCV (Xie, Girshick *et al.* 2020) are excellent up-to-date sources.

5.7 Exercises

Ex 5.1: Backpropagation and weight updates. Implement the forward activation, backward gradient and error propagation, and weight update steps in a simple neural network. You can find examples of such code in HW3 of the 2020 UW CSE 576 class[62] or the Educational Framework (EDF) developed by McAllester (2020) and used in Geiger (2021).

Ex 5.2: LeNet. Download, train, and test a simple "LeNet" (LeCun, Bottou *et al.* 1998) convolutional neural network on the CIFAR-10 (Krizhevsky 2009) or Fashion MNIST (Xiao, Rasul, and Vollgraf 2017) datasets. You can find such code in numerous places on the web, including HW4 of the 2020 UW CSE 576 class or the PyTorch beginner tutorial on Neural Networks.[63]

Modify the network to remove the non-linearities. How does the performance change? Can you improve the performance of the original network by increasing the number of channels, layers, or convolution sizes? Do the training and testing accuracies move in the same or different directions as you modify your network?

Ex 5.3: Deep learning textbooks. Both the *Deep Learning: From Basics to Practice* book by Glassner (2018, Chapters 15, 23, and 24) and the *Dive into Deep Learning* book by Zhang, Lipton *et al.* (2021) contain myriad graded exercises with code samples to develop your understanding of deep neural networks. If you have the time, try to work through most of these.

Ex 5.4: Activation and weight scaling. Consider the two hidden unit network shown in Figure 5.62, which uses ReLU activation functions and has no additive bias parameters. Your task is to find a set of weights that will fit the function

$$y = |x_1 + 1.1x_2|. \tag{5.80}$$

[62] https://courses.cs.washington.edu/courses/cse576/20sp/calendar/

[63] https://pytorch.org/tutorials/beginner/blitz/neural_networks_tutorial.html

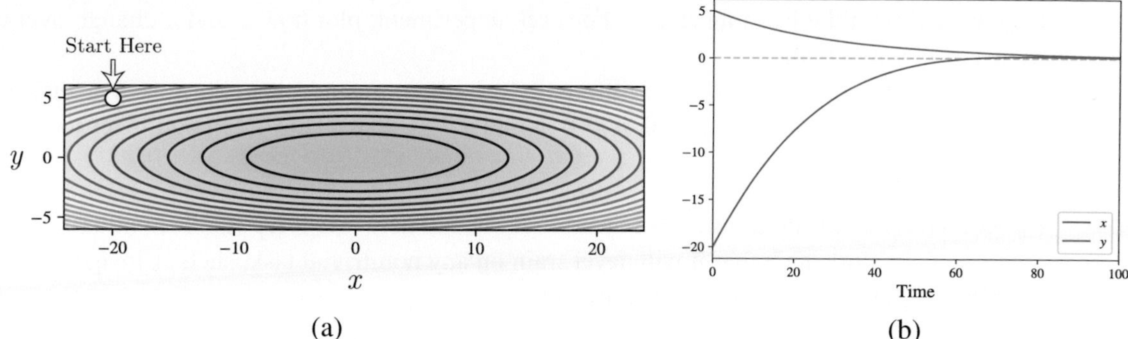

Figure 5.63 Function optimization: (a) the contour plot of $f(x, y) = x^2 + 20y^2$ with the function being minimized at $(0, 0)$; (b) ideal gradient descent optimization that quickly converges towards the minimum at $x = 0$, $y = 0$.

1. Can you guess a set of weights that will fit this function?
2. Starting with the weights shown in column b, compute the activations for the hidden and final units as well as the regression loss for the nine input values $(x_1, x_2) \in \{-1, 0, 1\} \times \{-1, 0, 1\}$.
3. Now compute the gradients of the squared loss with respect to all six weights using the backpropagation chain rule equations (5.65–5.68) and sum them up across the training samples to get a final gradient.
4. What step size should you take in the gradient direction, and what would your update squared loss become?
5. Repeat this exercise for the initial weights in column (c) of Figure 5.62.
6. Given this new set of weights, how much worse is your error decrease, and how many iterations would you expect it to take to achieve a reasonable solution?
7. Would batch normalization help in this case?

Note: the following exercises were suggested by Matt Deitke.

Ex 5.5: Function optimization. Consider the function $f(x, y) = x^2+20y^2$ shown in Figure 5.63a. Begin by solving for the following:

1. Calculate ∇f, i.e., the gradient of f.
2. Evaluate the gradient at $x = -20$, $y = 5$.

Implement some of the common gradient descent optimizers, which should take you from the starting point $x = -20$, $y = 5$ to near the minimum at $x = 0$, $y = 0$. Try each of the following optimizers:

1. Standard gradient descent.
2. Gradient descent with momentum, starting with the momentum term as $\rho = 0.99$.
3. Adam, starting with decay rates of $\beta_1 = 0.9$ and $b_2 = 0.999$.

Play around with the learning rate α. For each experiment, plot how x and y change over time, as shown in Figure 5.63b.

How do the optimizers behave differently? Is there a single learning rate that makes all the optimizers converge towards $x = 0, y = 0$ in under 200 steps? Does each optimizer monotonically trend towards $x = 0, y = 0$?

Ex 5.6: Weight initialization. For an arbitrary neural network, is it possible to initialize the weights of a neural network such that it will never train on any non-trivial task, such as image classification or object detection? Explain why or why not.

Ex 5.7: Convolutions. Consider convolving a $256 \times 256 \times 3$ image with 64 separate convolution kernels. For kernels with heights and widths of $\{(3 \times 3), (5 \times 5), (7 \times 7), \text{and } (9 \times 9)\}$, answer each of the following:

1. How many parameters (i.e., weights) make up the convolution operation?

2. What is the output size after convolving the image with the kernels?

Ex 5.8: Data augmentation. The figure below shows image augmentations that translate and scale an image.

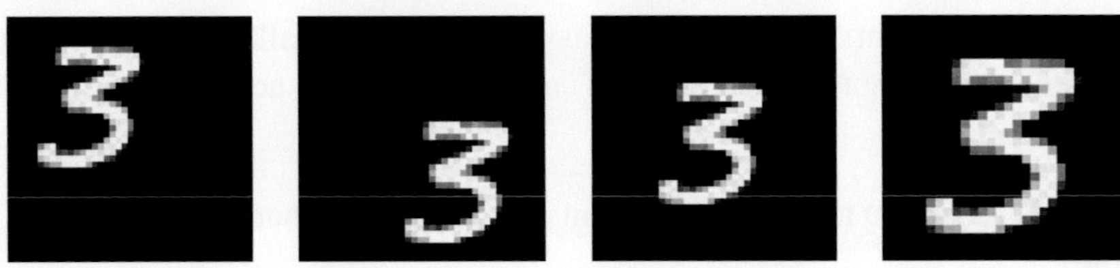

Let CONV denote a convolution operation, f denote an arbitrary function (such as scaling or translating an image), and IMAGE denote the input image. A function f has invariance, with respect to a convolution, when $\text{CONV}(\text{IMAGE}) = \text{CONV}(f(\text{IMAGE}))$, and equivariance when $\text{CONV}(f(\text{IMAGE})) = f(\text{CONV}(\text{IMAGE}))$. Answer and explain each of the following:

1. Are convolutions translation invariant?

2. Are convolutions translation equivariant?

3. Are convolutions scale invariant?

4. Are convolutions scale equivariant?

Ex 5.9: Training vs. validation. Suppose your model is performing significantly better on the training data than it is on the validation data. What changes might be made to the loss function, training data, and network architecture to prevent such overfitting?

Ex 5.10: Cascaded convolutions. With only a single matrix multiplication, how can multiple convolutional kernel's convolve over an entire input image? Here, let the input image be of size $256 \times 256 \times 3$ and each of the 64 kernels be of size $3 \times 3 \times 3$.[64]

Ex 5.11: Pooling vs. 1 × 1 convolutions. Pooling layers and 1×1 convolutions are both commonly used to shrink the size of the proceeding layer. When would you use one over the other?

[64] *Hint*: You will need to reshape the input and each convolution's kernel size.

Ex 5.12: Inception. Why is an inception module more efficient than a residual block? What are the comparative disadvantages of using an inception module?

Ex 5.13: ResNets. Why is it easier to train a ResNet with 100 layers than a VGG network with 100 layers?

Ex 5.14: U-Nets. An alternative to the U-Net architecture is to not change the size of the height and width intermediate activations throughout the network. The final layer would then be able to output the same transformed pixel-wise representation of the input image. What is the disadvantage of this approach?

Ex 5.15: Early vs. late fusion in video processing. What are two advantages of early fusion compared to late fusion?

Ex 5.16: Video-to-video translation. Independently pass each frame in a video through a `pix2pix` model. For instance, if the video is of the day, then the output might be each frame at night. Stitch the output frames together to form a video. What do you notice? Does the video look plausible?

Ex 5.17: Vision Transformer. Using a Vision Transformer (ViT) model, pass several images through it and create a histogram of the activations after each layer normalization operation. Do the histograms tend to form of a normal distribution?

Ex 5.18: GAN training. In the GAN loss formulation, suppose the discriminator D is near-perfect, such that it correctly outputs near 1 for real images $\mathbf{x}_n$ and near 0 for synthetically generated images $G(\mathbf{z}_n)$.

1. For both the discriminator and the generator, compute its approximate loss with

$$\mathcal{L}_{\text{GAN}}(\mathbf{x}_n, \mathbf{z}_n) = \log D(\mathbf{x}_n) + \log(1 - D(G(\mathbf{z}_n))), \tag{5.81}$$

   where the discriminator tries to minimize $\mathcal{L}_{\text{GAN}}$ and the generator tries to maximize $\mathcal{L}_{\text{GAN}}$.

2. How well can this discriminator be used to train the generator?

3. Can you modify the generator's loss function, $\min \log(1 - D(G(\mathbf{z}_n)))$, such that it is easier to train with both a great discriminator and a discriminator that is no better than random?[65]

Ex 5.19: Colorization. Even though large amounts of unsupervised data can be collected for image colorization, it often does not train well using a pixel-wise regression loss between an image's predicted colors and its true colors. Why is that? Is there another loss function that may be better suited for the problem?

[65] *Hint:* The loss function should suggest a relatively large change to fool a great discriminator and a relatively small change with a discriminator that is no better than random.

Chapter 6

Recognition

6.1 Instance recognition . . . 276
6.2 Image classification . . . 278
 6.2.1 Feature-based methods . . . 278
 6.2.2 Deep networks . . . 285
 6.2.3 *Application*: Visual similarity search . . . 287
 6.2.4 Face recognition . . . 289
6.3 Object detection . . . 295
 6.3.1 Face detection . . . 295
 6.3.2 Pedestrian detection . . . 299
 6.3.3 General object detection . . . 301
6.4 Semantic segmentation . . . 307
 6.4.1 *Application*: Medical image segmentation . . . 310
 6.4.2 Instance segmentation . . . 311
 6.4.3 Panoptic segmentation . . . 312
 6.4.4 *Application*: Intelligent photo editing . . . 314
 6.4.5 Pose estimation . . . 315
6.5 Video understanding . . . 316
6.6 Vision and language . . . 319
6.7 Additional reading . . . 326
6.8 Exercises . . . 329

R. Szeliski, *Computer Vision*, Texts in Computer Science,
https://doi.org/10.1007/978-3-030-34372-9_6

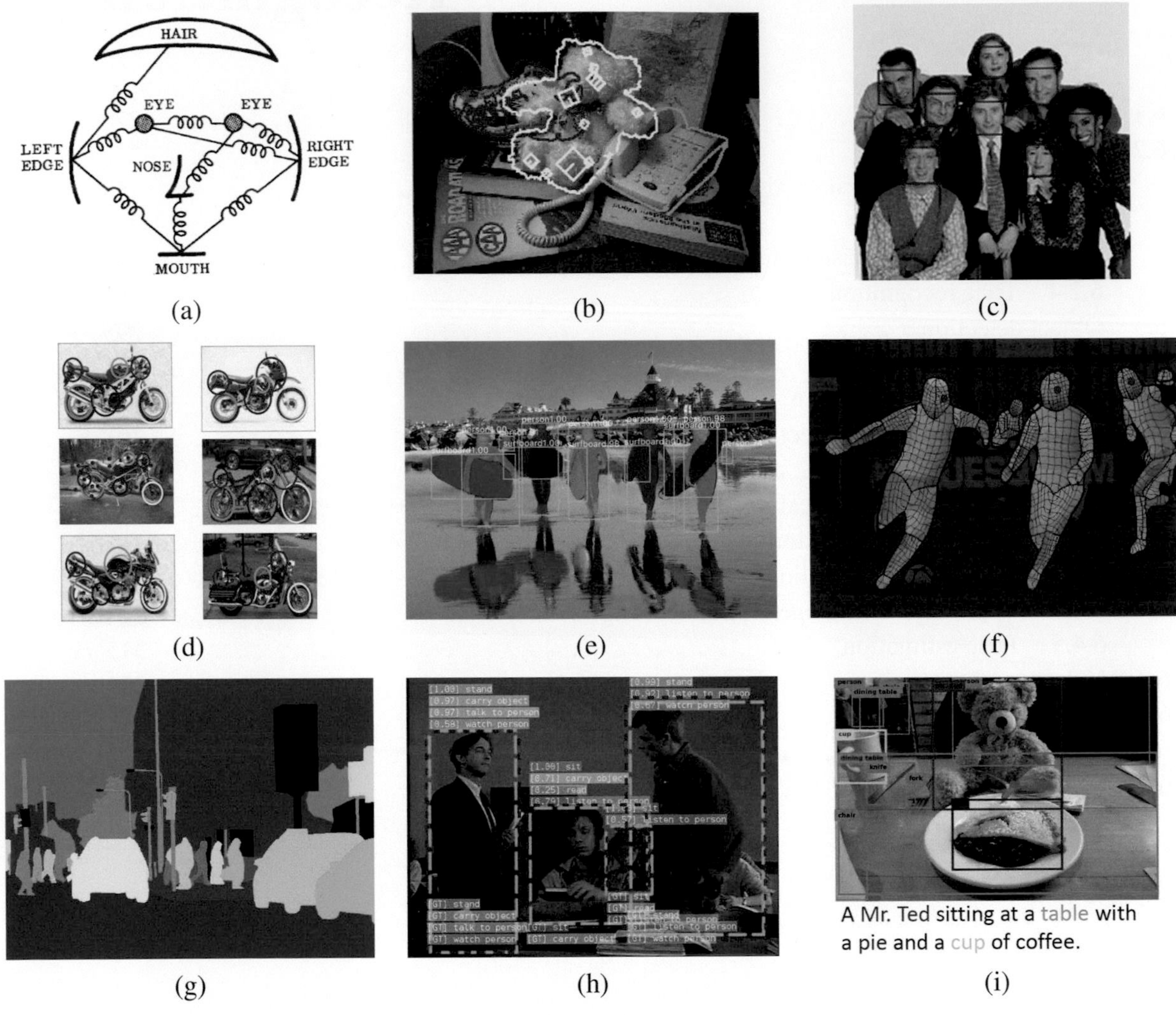

Figure 6.1 Various kinds of recognition: (a) face recognition with pictorial structures (Fischler and Elschlager 1973) © 1973 IEEE; (b) instance (known object) recognition (Lowe 1999) © 1999 IEEE; (c) real-time face detection (Viola and Jones 2004) © 2004 Springer; (d) feature-based recognition (Fergus, Perona, and Zisserman 2007) © 2007 Springer; (e) instance segmentation using Mask R-CNN (He, Gkioxari *et al.* 2017) © 2017 IEEE; (f) pose estimation (Güler, Neverova, and Kokkinos 2018) © 2018 IEEE; (g) panoptic segmentation (Kirillov, He *et al.* 2019) © 2019 IEEE; (h) video action recognition (Feichtenhofer, Fan *et al.* 2019); (i) image captioning (Lu, Yang *et al.* 2018) © 2018 IEEE.

Of all the computer vision topics covered in this book, visual recognition has undergone the largest changes and fastest development in the last decade, due in part to the availability of much larger labeled datasets as well as breakthroughs in deep learning (Figure 5.40). In the first edition of this book (Szeliski 2010), recognition was the last chapter, since it was considered a "high-level task" to be layered on top of lower-level components such as feature detection and matching. In fact, many introductory vision courses still teach recognition at the end, often covering "classic" (non-learning) vision algorithms and applications first, and then shifting to deep learning and recognition.

As I mentioned in the preface and introduction, I have now moved machine and deep learning to early in the book, since it is foundational technology widely used in other parts of computer vision. I also decided to move the recognition chapter right after deep learning, since most of the modern techniques for recognition are natural applications of deep neural networks. The majority of the old recognition chapter has been replaced with newer deep learning techniques, so you will sometimes find terse descriptions of classical recognition techniques along with pointers to the first edition and relevant surveys or seminal papers.

A good example of the classic approach is *instance recognition*, where we are trying to find exemplars of a particular manufactured object such as a stop sign or sneaker (Figure 6.1b). (An even earlier example is face recognition using relative feature locations, as shown in Figure 6.1a.) The general approach of finding distinctive features while dealing with local appearance variation (Section 7.1.2), and then checking for their co-occurrence and relative positions in an image, is still widely used for manufactured 3D object detection (Figure 6.3), 3D structure and pose recovery (Chapter 11), and location recognition (Section 11.2.3). Highly accurate and widely used feature-based approaches to instance recognition were developed in the 2000s (Figure 7.27) and, despite more recent deep learning-based alternatives, are often still the preferred method (Sattler, Zhou *et al.* 2019). We review instance recognition in Section 6.1, although some of the needed components, such as feature detection, description, and matching (Chapter 7), as well as 3D pose estimation and verification (Chapter 11), will not be introduced until later.

The more difficult problem of *category* or *class recognition* (e.g., recognizing members of highly variable categories such as cats, dogs, or motorcycles) was also initially attacked using feature-based approaches and relative locations (*part-based models*), such as the one depicted in Figure 6.1d. We begin our discussion of *image classification* (another name for whole-image category recognition) in Section 6.2 with a review of such "classic" (though now rarely used) techniques. We then show how the deep neural networks described in the previous chapter are ideally suited to these kinds of classification problems. Next, we cover visual similarity search, where instead of categorizing an image into a predefined number of categories, we retrieve other images that are semantically similar. Finally, we focus on face recognition, which is one of the longest studied topics in computer vision.

In Section 6.3, we turn to the topic of *object detection*, where we categorize not just whole images but delineate (with bounding boxes) where various objects are located. This topic includes more specialized variants such as *face detection* and *pedestrian detection*, as well as the detection of objects in generic categories. In Section 6.4, we study *semantic segmentation*, where the task is now to delineate various objects and materials in a pixel-accurate manner, i.e., to label each pixel with an object identity and class. Variants on this include *instance segmentation*, where each separate object gets a unique label, *panoptic segmentation*, where both objects and stuff (e.g., grass, sky) get labeled, and *pose estimation*, where pixels get labeled with people's body parts and orientations. The last two sections of this chapter briefly touch on *video understanding* (Section 6.5) and *vision and language* (Section 6.6).

Before starting to describe individual recognition algorithms and variants, I should briefly mention the critical role that large-scale datasets and benchmarks have played in the rapid advancement

Figure 6.2 Recognizing objects in a cluttered scene (Lowe 2004) © 2004 Springer. Two of the training images in the database are shown on the left. They are matched to the cluttered scene in the middle using SIFT features, shown as small squares in the right image. The affine warp of each recognized database image onto the scene is shown as a larger parallelogram in the right image.

of recognition systems. While small datasets such as Xerox 10 (Csurka, Dance *et al.* 2006) and Caltech-101 (Fei-Fei, Fergus, and Perona 2006) played an early role in evaluating object recognition systems, the PASCAL Visual Object Class (VOC) challenge (Everingham, Van Gool *et al.* 2010; Everingham, Eslami *et al.* 2015) was the first dataset large and challenging enough to significantly propel the field forward. However, PASCAL VOC only contained 20 classes. The introduction of the ImageNet dataset (Deng, Dong *et al.* 2009; Russakovsky, Deng *et al.* 2015), which had 1,000 classes and over one million labeled images, finally provided enough data to enable end-to-end learning systems to break through. The Microsoft COCO (Common Objects in Context) dataset spurred further development (Lin, Maire *et al.* 2014), especially in accurate per-object segmentation, which we study in Section 6.4. A nice review of crowdsourcing methods to construct such datasets is presented in (Kovashka, Russakovsky *et al.* 2016). We will mention additional, sometimes more specialized, datasets throughout this chapter. A listing of the most popular and active datasets and benchmarks is provided in Tables 6.1–6.4.

6.1 Instance recognition

General object recognition falls into two broad categories, namely *instance recognition* and *class recognition*. The former involves re-recognizing a known 2D or 3D rigid object, potentially being viewed from a novel viewpoint, against a cluttered background, and with partial occlusions.[1] The latter, which is also known as *category-level* or *generic* object recognition (Ponce, Hebert *et al.* 2006), is the much more challenging problem of recognizing any instance of a particular general class, such as "cat", "car", or "bicycle".

Over the years, many different algorithms have been developed for instance recognition. Mundy (2006) surveys earlier approaches, which focused on extracting lines, contours, or 3D surfaces from images and matching them to known 3D object models. Another popular approach was to acquire images from a large set of viewpoints and illuminations and to represent them using an eigenspace

[1]The Microsoft COCO dataset paper (Lin, Maire *et al.* 2014) introduced the newer concept of *instance segmentation*, which is the pixel-accurate delineation of different objects drawn from a set of generic classes (Section 6.4.2). This now sometimes leads to confusion, unless you look at these two terms (instance recognition vs. segmentation) carefully.

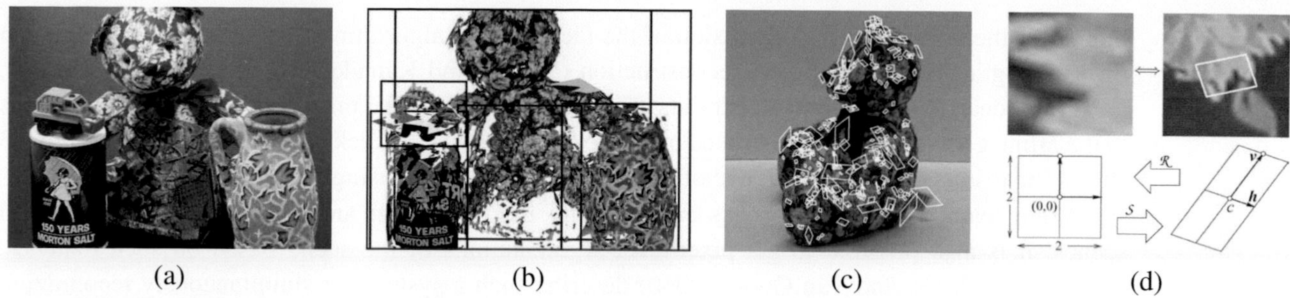

(a) (b) (c) (d)

Figure 6.3 3D object recognition with affine regions (Rothganger, Lazebnik *et al.* 2006) © 2006 Springer: (a) sample input image; (b) five of the recognized (reprojected) objects along with their bounding boxes; (c) a few of the local affine regions; (d) local affine region (patch) reprojected into a canonical (square) frame, along with its geometric affine transformations.

decomposition (Murase and Nayar 1995). More recent approaches (Lowe 2004; Lepetit and Fua 2005; Rothganger, Lazebnik *et al.* 2006; Ferrari, Tuytelaars, and Van Gool 2006b; Gordon and Lowe 2006; Obdržálek and Matas 2006; Sivic and Zisserman 2009; Zheng, Yang, and Tian 2018) tend to use viewpoint-invariant 2D features, such as those we will discuss in Section 7.1.2. After extracting informative sparse 2D features from both the new image and the images in the database, image features are matched against the object database, using one of the sparse feature matching strategies described in Section 7.1.3. Whenever a sufficient number of matches have been found, they are verified by finding a geometric transformation that aligns the two sets of features (Figure 6.2).

Geometric alignment

To recognize one or more instances of some known objects, such as those shown in the left column of Figure 6.2, the recognition system first extracts a set of interest points in each database image and stores the associated descriptors (and original positions) in an indexing structure such as a search tree (Section 7.1.3). At recognition time, features are extracted from the new image and compared against the stored object features. Whenever a sufficient number of matching features (say, three or more) are found for a given object, the system then invokes a *match verification* stage, whose job is to determine whether the spatial arrangement of matching features is consistent with those in the database image.

Because images can be highly cluttered and similar features may belong to several objects, the original set of feature matches can have a large number of outliers. For this reason, Lowe (2004) suggests using a Hough transform (Section 7.4.2) to accumulate votes for likely geometric transformations. In his system, he uses an affine transformation between the database object and the collection of scene features, which works well for objects that are mostly planar, or where at least several corresponding features share a quasi-planar geometry.[2]

Another system that uses local affine frames is the one developed by Rothganger, Lazebnik *et al.* (2006). In their system, the affine region detector of Mikolajczyk and Schmid (2004) is used to rectify local image patches (Figure 6.3d), from which both a SIFT descriptor and a 10×10 UV color histogram are computed and used for matching and recognition. Corresponding patches in different views of the same object, along with their local affine deformations, are used to compute a 3D affine

[2]When a larger number of features is available, a full fundamental matrix can be used (Brown and Lowe 2002; Gordon and Lowe 2006). When image stitching is being performed (Brown and Lowe 2007), the motion models discussed in Section 8.2.1 can be used instead.

model for the object using an extension of the factorization algorithm of Section 11.4.1, which can then be upgraded to a Euclidean reconstruction (Tomasi and Kanade 1992). At recognition time, local Euclidean neighborhood constraints are used to filter potential matches, in a manner analogous to the affine geometric constraints used by Lowe (2004) and Obdržálek and Matas (2006). Figure 6.3 shows the results of recognizing five objects in a cluttered scene using this approach.

While feature-based approaches are normally used to detect and localize known objects in scenes, it is also possible to get pixel-level segmentations of the scene based on such matches. Ferrari, Tuytelaars, and Van Gool (2006b) describe such a system for simultaneously recognizing objects and segmenting scenes, while Kannala, Rahtu *et al.* (2008) extend this approach to non-rigid deformations. Section 6.4 re-visits this topic of joint recognition and segmentation in the context of generic class (category) recognition.

While instance recognition in the early to mid-2000s focused on the problem of locating a known 3D object in an image, as shown in Figures 6.2–6.3, attention shifted to the more challenging problem of *instance retrieval* (also known as *content-based image retrieval*), in which the number of images being searched can be very large. Section 7.1.4 reviews such techniques, a snapshot of which can be seen in Figure 7.27 and the survey by Zheng, Yang, and Tian (2018). This topic is also related to visual similarity search (Section 6.2.3 and 3D pose estimation (Section 11.2).

6.2 Image classification

While instance recognition techniques are relatively mature and are used in commercial applications such as traffic sign recognition (Stallkamp, Schlipsing *et al.* 2012), generic category (class) recognition is still a rapidly evolving research area. Consider for example the set of photographs in Figure 6.4a, which shows objects taken from 10 different visual categories. (I'll leave it up to you to name each of the categories.) How would you go about writing a program to categorize each of these images into the appropriate class, especially if you were also given the choice "none of the above"?

As you can tell from this example, visual category recognition is an extremely challenging problem. However, the progress in the field has been quite dramatic, if judged by how much better today's algorithms are compared to those of a decade ago.

In this section, we review the main classes of algorithms used for whole-image classification. We begin with classic feature-based approaches that rely on handcrafted features and their statistics, optionally using machine learning to do the final classification (Figure 5.2b). Since such techniques are no longer widely used, we present a fairly terse description of the most important techniques. More details can be found in the first edition of this book (Szeliski 2010, Chapter 14) and in the cited journal papers and surveys. Next, we describe modern image classification systems, which are based on the deep neural networks we introduced in the previous chapter. We then describe visual similarity search, where the task is to find visually and semantically similar images, rather than classification into a fixed set of categories. Finally, we look at face recognition, since this topic has its own long history and set of techniques.

6.2.1 Feature-based methods

In this section, we review "classic" feature-based approaches to category recognition (image classification). While, historically, *part-based* representations and recognition algorithms (Section 6.2.1) were the preferred approach (Fischler and Elschlager 1973; Felzenszwalb and Huttenlocher 2005; Fergus, Perona, and Zisserman 2007), we begin by describing simpler *bag-of-features* approaches

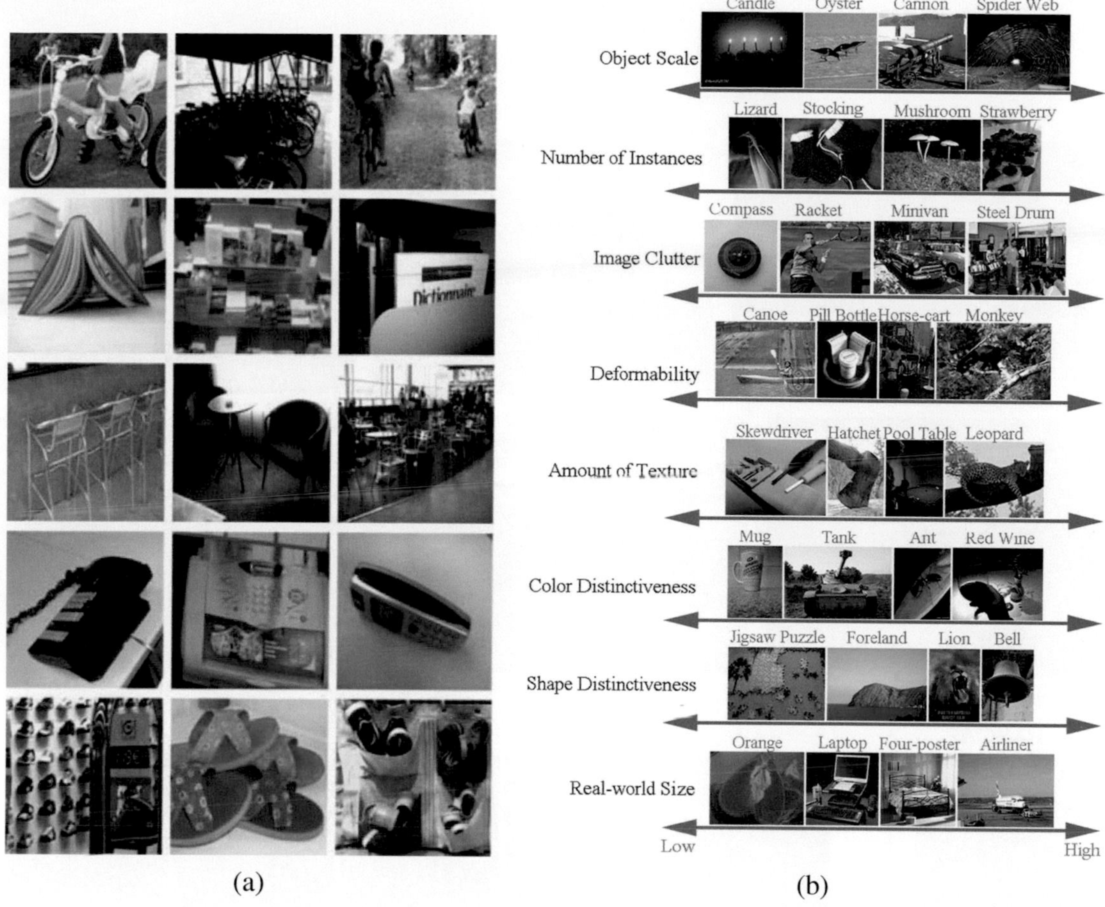

Figure 6.4 Challenges in image recognition: (a) sample images from the Xerox 10 class dataset (Csurka, Dance *et al.* 2006) © 2007 Springer; (b) axes of difficulty and variation from the ImageNet dataset (Russakovsky, Deng *et al.* 2015) © 2015 Springer.

that represent objects and images as unordered collections of feature descriptors. We then review more complex systems constructed with part-based models, and then look at how context and scene understanding, as well as machine learning, can improve overall recognition results. Additional details on the techniques presented in this section can be found in older survey articles, paper collections, and courses (Pinz 2005; Ponce, Hebert *et al.* 2006; Dickinson, Leonardis *et al.* 2007; Fei-Fei, Fergus, and Torralba 2009), as well as two review articles on the PASCAL and ImageNet recognition challenges (Everingham, Van Gool *et al.* 2010; Everingham, Eslami *et al.* 2015; Russakovsky, Deng *et al.* 2015) and the first edition of this book (Szeliski 2010, Chapter 14).

Bag of words

One of the simplest algorithms for category recognition is the *bag of words* (also known as *bag of features* or *bag of keypoints*) approach (Csurka, Dance *et al.* 2004; Lazebnik, Schmid, and Ponce 2006; Csurka, Dance *et al.* 2006; Zhang, Marszalek *et al.* 2007). As shown in Figure 6.6, this algorithm simply computes the distribution (histogram) of visual words found in the query image and compares this distribution to those found in the training images. We will give more details of this approach in Section 7.1.4. The biggest difference from instance recognition is the absence of

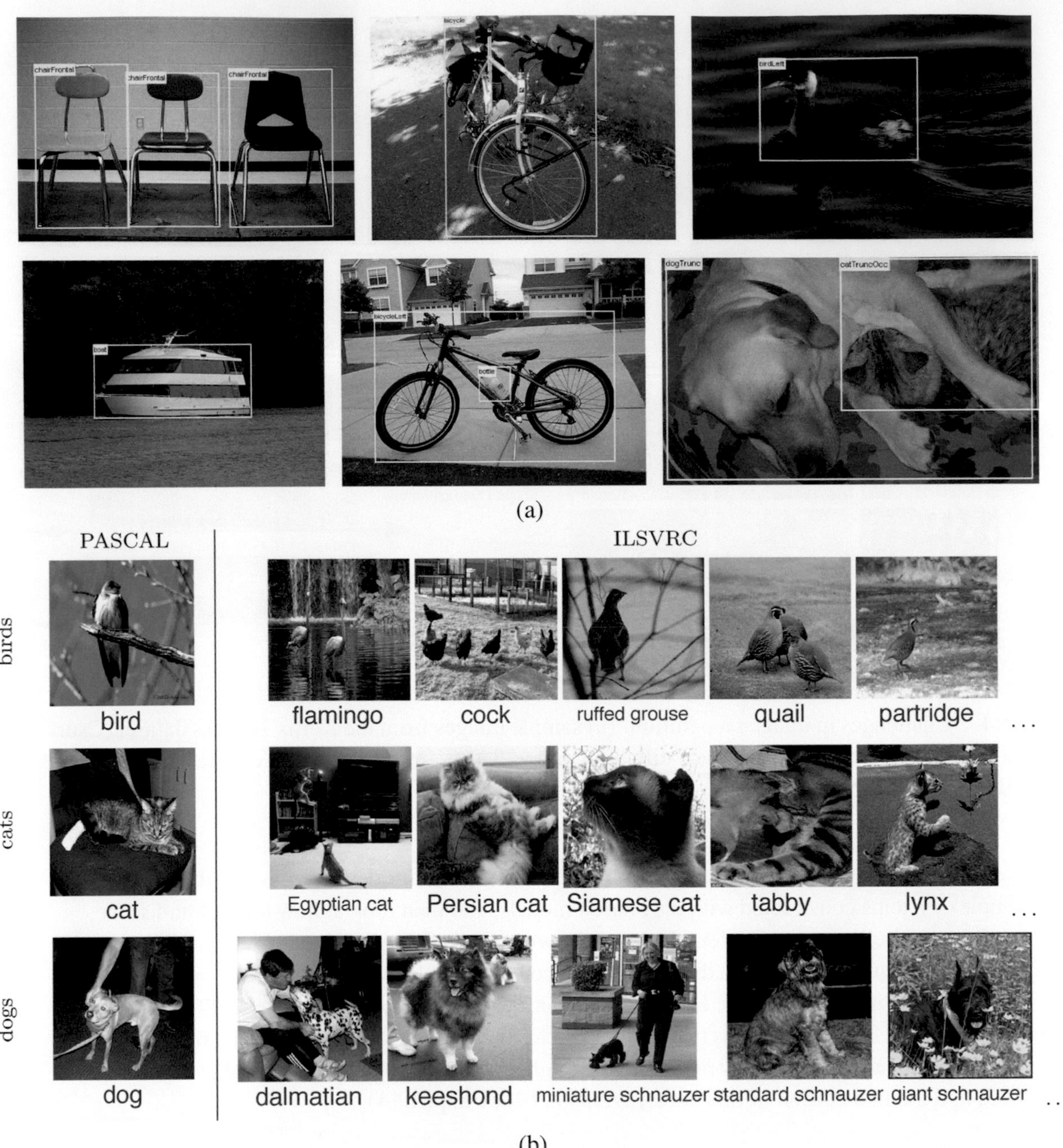

Figure 6.5 Sample images from two widely used image classification datasets: (a) Pascal Visual Object Categories (VOC) (Everingham, Eslami *et al.* 2015) © 2015 Springer; (b) ImageNet (Russakovsky, Deng *et al.* 2015) © 2015 Springer.

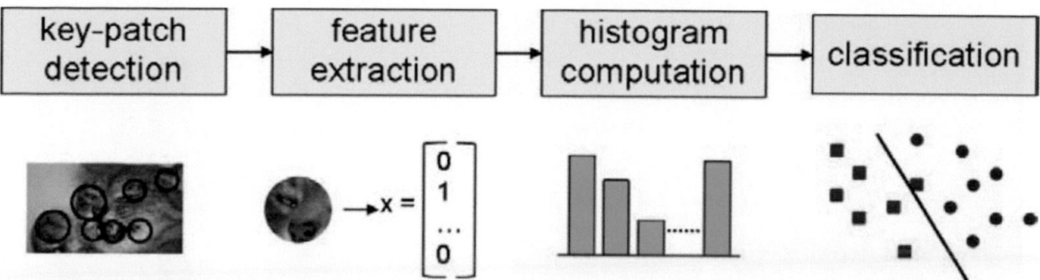

Figure 6.6 A typical processing pipeline for a bag-of-words category recognition system (Csurka, Dance *et al.* 2006) © 2007 Springer. Features are first extracted at keypoints and then quantized to get a distribution (histogram) over the learned *visual words* (feature cluster centers). The feature distribution histogram is used to learn a decision surface using a classification algorithm, such as a support vector machine.

a geometric verification stage (Section 6.1), since individual instances of generic visual categories, such as those shown in Figure 6.4a, have relatively little spatial coherence to their features (but see the work by Lazebnik, Schmid, and Ponce (2006)).

Csurka, Dance *et al.* (2004) were the first to use the term *bag of keypoints* to describe such approaches and among the first to demonstrate the utility of frequency-based techniques for category recognition. Their original system used affine covariant regions and SIFT descriptors, k-means visual vocabulary construction, and both a naïve Bayesian classifier and support vector machines for classification. (The latter was found to perform better.) Their newer system (Csurka, Dance *et al.* 2006) uses regular (non-affine) SIFT patches and boosting instead of SVMs and incorporates a small amount of geometric consistency information.

Zhang, Marszalek *et al.* (2007) perform a more detailed study of such bag of features systems. They compare a number of feature detectors (Harris–Laplace (Mikolajczyk and Schmid 2004) and Laplacian (Lindeberg 1998b)), descriptors (SIFT, RIFT, and SPIN (Lazebnik, Schmid, and Ponce 2005)), and SVM kernel functions.

Instead of quantizing feature vectors to visual words, Grauman and Darrell (2007b) develop a technique for directly computing an approximate distance between two variably sized collections of feature vectors. Their approach is to bin the feature vectors into a multi-resolution pyramid defined in feature space and count the number of features that land in corresponding bins B_{il} and B'_{il}. The distance between the two sets of feature vectors (which can be thought of as points in a high-dimensional space) is computed using histogram intersection between corresponding bins, while discounting matches already found at finer levels and weighting finer matches more heavily. In follow-on work, Grauman and Darrell (2007a) show how an explicit construction of the pyramid can be avoided using hashing techniques.

Inspired by this work, Lazebnik, Schmid, and Ponce (2006) show how a similar idea can be employed to augment bags of keypoints with loose notions of 2D spatial location analogous to the pooling performed by SIFT (Lowe 2004) and "gist" (Torralba, Murphy *et al.* 2003). In their work, they extract affine region descriptors (Lazebnik, Schmid, and Ponce 2005) and quantize them into visual words. (Based on previous results by Fei-Fei and Perona (2005), the feature descriptors are extracted densely (on a regular grid) over the image, which can be helpful in describing textureless regions such as the sky.) They then form a spatial pyramid of bins containing word counts (histograms) and use a similar pyramid match kernel to combine histogram intersection counts in a hierarchical fashion.

The debate about whether to use quantized feature descriptors or continuous descriptors and

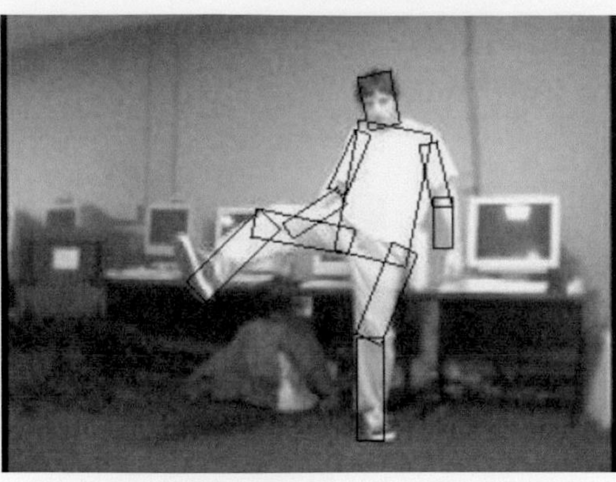
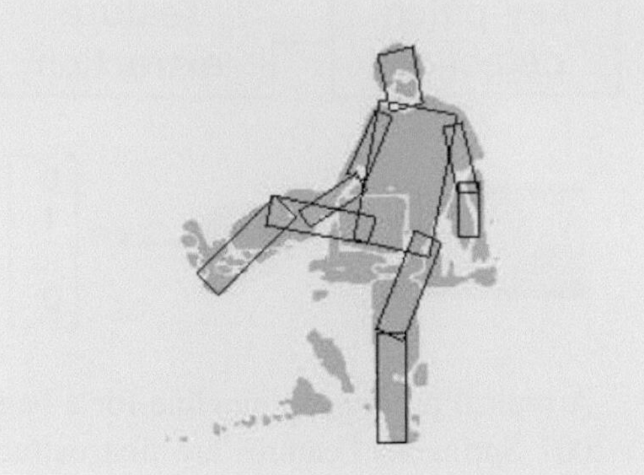

Figure 6.7 Using pictorial structures to locate and track a person (Felzenszwalb and Huttenlocher 2005) © 2005 Springer. The structure consists of articulated rectangular body parts (torso, head, and limbs) connected in a tree topology that encodes relative part positions and orientations. To fit a pictorial structure model, a binary silhouette image is first computed using background subtraction.

also whether to use sparse or dense features went on for many years. Boiman, Shechtman, and Irani (2008) show that if query images are compared to *all* the features representing a given class, rather than just each class image individually, nearest-neighbor matching followed by a naïve Bayes classifier outperforms quantized visual words. Instead of using generic feature detectors and descriptors, some authors have been investigating *learning* class-specific features (Ferencz, Learned-Miller, and Malik 2008), often using randomized forests (Philbin, Chum *et al.* 2007; Moosmann, Nowak, and Jurie 2008; Shotton, Johnson, and Cipolla 2008) or combining the feature generation and image classification stages (Yang, Jin *et al.* 2008). Others, such as Serre, Wolf, and Poggio (2005) and Mutch and Lowe (2008) use hierarchies of dense feature transforms inspired by biological (visual cortical) processing combined with SVMs for final classification.

Part-based models

Recognizing an object by finding its constituent parts and measuring their geometric relationships is one of the oldest approaches to object recognition (Fischler and Elschlager 1973; Kanade 1977; Yuille 1991). Part-based approaches were often used for face recognition (Moghaddam and Pentland 1997; Heisele, Ho *et al.* 2003; Heisele, Serre, and Poggio 2007) and continue being used for pedestrian detection (Figure 6.24) (Felzenszwalb, McAllester, and Ramanan 2008) and pose estimation (Güler, Neverova, and Kokkinos 2018).

In this overview, we discuss some of the central issues in part-based recognition, namely, the representation of geometric relationships, the representation of individual parts, and algorithms for learning such descriptions and recognizing them at run time. More details on part-based models for recognition can be found in the course notes by Fergus (2009).

The earliest approaches to representing geometric relationships were dubbed *pictorial structures* by Fischler and Elschlager (1973) and consisted of spring-like connections between different feature locations (Figure 6.1a). To fit a pictorial structure to an image, an energy function of the form

$$E = \sum_i V_i(\mathbf{l}_i) + \sum_{ij \in E} V_{ij}(\mathbf{l}_i, \mathbf{l}_j) \tag{6.1}$$

is minimized over all potential part locations or poses $\{\mathbf{l}_i\}$ and pairs of parts (i, j) for which an edge (geometric relationship) exists in E. Note how this energy is closely related to that used with

Markov random fields (4.35–4.38), which can be used to embed pictorial structures in a probabilistic framework that makes parameter learning easier (Felzenszwalb and Huttenlocher 2005).

Part-based models can have different topologies for the geometric connections between the parts (Carneiro and Lowe 2006). For example, Felzenszwalb and Huttenlocher (2005) restrict the connections to a tree, which makes learning and inference more tractable. A tree topology enables the use of a recursive Viterbi (dynamic programming) algorithm (Pearl 1988; Bishop 2006), in which leaf nodes are first optimized as a function of their parents, and the resulting values are then plugged in and eliminated from the energy function, To further increase the efficiency of the inference algorithm, Felzenszwalb and Huttenlocher (2005) restrict the pairwise energy functions $V_{ij}(\mathbf{l}_i, \mathbf{l}_j)$ to be Mahalanobis distances on functions of location variables and then use fast distance transform algorithms to minimize each pairwise interaction in time that is closer to linear in N.

Figure 6.7 shows the results of using their pictorial structures algorithm to fit an articulated body model to a binary image obtained by background segmentation. In this application of pictorial structures, parts are parameterized by the locations, sizes, and orientations of their approximating rectangles. Unary matching potentials $V_i(\mathbf{l}_i)$ are determined by counting the percentage of foreground and background pixels inside and just outside the tilted rectangle representing each part.

A large number of different graphical models have been proposed for part-based recognition. Carneiro and Lowe (2006) discuss a number of these models and propose one of their own, which they call a *sparse flexible model*; it involves ordering the parts and having each part's location depend on at most k of its ancestor locations.

The simplest models are bags of words, where there are no geometric relationships between different parts or features. While such models can be very efficient, they have a very limited capacity to express the spatial arrangement of parts. Trees and stars (a special case of trees where all leaf nodes are directly connected to a common root) are the most efficient in terms of inference and hence also learning (Felzenszwalb and Huttenlocher 2005; Fergus, Perona, and Zisserman 2005; Felzenszwalb, McAllester, and Ramanan 2008). Directed acyclic graphs come next in terms of complexity and can still support efficient inference, although at the cost of imposing a causal structure on the part model (Bouchard and Triggs 2005; Carneiro and Lowe 2006). k-fans, in which a clique of size k forms the root of a star-shaped model have inference complexity $O(N^{k+1})$, although with distance transforms and Gaussian priors, this can be lowered to $O(N^k)$ (Crandall, Felzenszwalb, and Huttenlocher 2005; Crandall and Huttenlocher 2006). Finally, fully connected *constellation* models are the most general, but the assignment of features to parts becomes intractable for moderate numbers of parts P, since the complexity of such an assignment is $O(N^P)$ (Fergus, Perona, and Zisserman 2007).

The original constellation model was developed by Burl, Weber, and Perona (1998) and consists of a number of parts whose relative positions are encoded by their mean locations and a full covariance matrix, which is used to denote not only positional uncertainty but also potential correlations between different parts. Weber, Welling, and Perona (2000) extended this technique to a weakly supervised setting, where both the appearance of each part and its locations are automatically learned given whole image labels. Fergus, Perona, and Zisserman (2007) further extend this approach to simultaneous learning of appearance and shape models from scale-invariant keypoint detections.

The part-based approach to recognition has also been extended to learning new categories from small numbers of examples, building on recognition components developed for other classes (Fei-Fei, Fergus, and Perona 2006). More complex hierarchical part-based models can be developed using the concept of grammars (Bouchard and Triggs 2005; Zhu and Mumford 2006). A simpler way to use parts is to have keypoints that are recognized as being part of a class vote for the estimated part locations (Leibe, Leonardis, and Schiele 2008). Parts can also be a useful component of fine-grained category recognition systems, as shown in Figure 6.9.

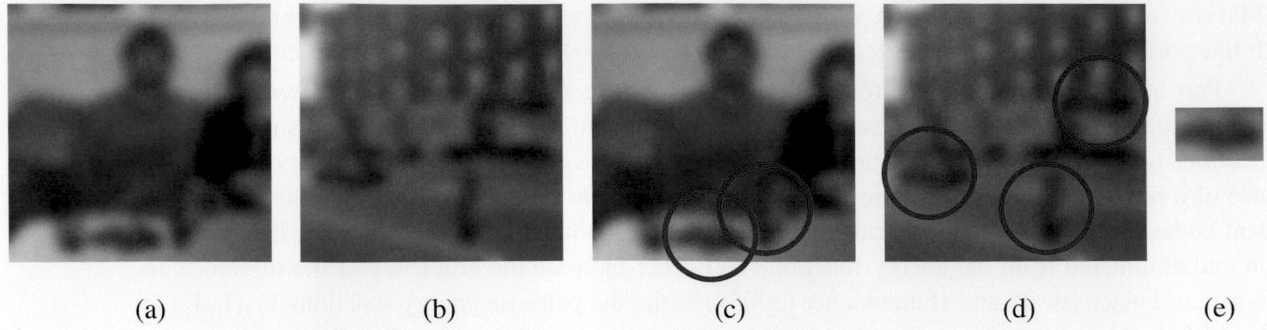

(a) (b) (c) (d) (e)

Figure 6.8 The importance of context (images courtesy of Antonio Torralba). Can you name all of the objects in images (a–b), especially those that are circled in (c–d). Look carefully at the circled objects. Did you notice that they all have the same shape (after being rotated), as shown in column (e)?

Context and scene understanding

Thus far, we have mostly considered the task of recognizing and localizing objects in isolation from that of understanding the scene (context) in which the object occur. This is a big limitation, as context plays a very important role in human object recognition (Oliva and Torralba 2007). Context can greatly improve the performance of object recognition algorithms (Divvala, Hoiem *et al.* 2009), as well as providing useful semantic clues for general scene understanding (Torralba 2008).

Consider the two photographs in Figure 6.8a–b. Can you name all of the objects, especially those circled in images (c–d)? Now have a closer look at the circled objects. Do see any similarity in their shapes? In fact, if you rotate them by 90°, they are all the same as the "blob" shown in Figure 6.8e. So much for our ability to recognize object by their shape!

Even though we have not addressed context explicitly earlier in this chapter, we have already seen several instances of this general idea being used. A simple way to incorporate spatial information into a recognition algorithm is to compute feature statistics over different regions, as in the spatial pyramid system of Lazebnik, Schmid, and Ponce (2006). Part-based models (Figure 6.7) use a kind of local context, where various parts need to be arranged in a proper geometric relationship to constitute an object.

The biggest difference between part-based and context models is that the latter combine objects into scenes and the number of constituent objects from each class is not known in advance. In fact, it is possible to combine part-based and context models into the same recognition architecture (Murphy, Torralba, and Freeman 2003; Sudderth, Torralba *et al.* 2008; Crandall and Huttenlocher 2007).

Consider an image database consisting of street and office scenes. If we have enough training images with labeled regions, such as buildings, cars, and roads, or monitors, keyboards, and mice, we can develop a geometric model for describing their relative positions. Sudderth, Torralba *et al.* (2008) develop such a model, which can be thought of as a two-level constellation model. At the top level, the distributions of objects relative to each other (say, buildings with respect to cars) is modeled as a Gaussian. At the bottom level, the distribution of parts (affine covariant features) with respect to the object center is modeled using a mixture of Gaussians. However, since the number of objects in the scene and parts in each object are unknown, a *latent Dirichlet process* (LDP) is used to model object and part creation in a generative framework. The distributions for all of the objects and parts are learned from a large labeled database and then later used during inference (recognition) to label the elements of a scene.

Another example of context is in simultaneous segmentation and recognition (Section 6.4 and Figure 6.33), where the arrangements of various objects in a scene are used as part of the labeling process. Torralba, Murphy, and Freeman (2004) describe a conditional random field where the estimated locations of building and roads influence the detection of cars, and where boosting is used to learn the structure of the CRF. Rabinovich, Vedaldi *et al.* (2007) use context to improve the results of CRF segmentation by noting that certain adjacencies (relationships) are more likely than others, e.g., a person is more likely to be on a horse than on a dog. Galleguillos and Belongie (2010) review various approaches proposed for adding context to object categorization, while Yao and Fei-Fei (2012) study human-object interactions. (For a more recent take on this problem, see Gkioxari, Girshick *et al.* (2018).)

Context also plays an important role in 3D inference from single images (Figure 6.41), using computer vision techniques for labeling pixels as belonging to the ground, vertical surfaces, or sky (Hoiem, Efros, and Hebert 2005a). This line of work has been extended to a more holistic approach that simultaneously reasons about object identity, location, surface orientations, occlusions, and camera viewing parameters (Hoiem, Efros, and Hebert 2008).

A number of approaches use the *gist* of a scene (Torralba 2003; Torralba, Murphy *et al.* 2003) to determine where instances of particular objects are likely to occur. For example, Murphy, Torralba, and Freeman (2003) train a regressor to predict the vertical locations of objects such as pedestrians, cars, and buildings (or screens and keyboards for indoor office scenes) based on the gist of an image. These location distributions are then used with classic object detectors to improve the performance of the detectors. Gists can also be used to directly match complete images, as we saw in the scene completion work of Hays and Efros (2007).

Finally, some of the work in scene understanding exploits the existence of large numbers of labeled (or even unlabeled) images to perform matching directly against whole images, where the images themselves implicitly encode the expected relationships between objects (Russell, Torralba *et al.* 2007; Malisiewicz and Efros 2008; Galleguillos and Belongie 2010). This, of course, is one of the central benefits of using deep neural networks, which we discuss in the next section.

6.2.2 Deep networks

As we saw in Section 5.4.3, deep networks started outperforming "shallow" learning-based approaches on the ImageNet Large Scale Visual Recognition Challenge (ILSVRC) with the introduction of the "AlexNet" SuperVision system of Krizhevsky, Sutskever, and Hinton (2012). Since that time, recognition accuracy has continued to improve dramatically (Figure 5.40) driven to a large degree by deeper networks and better training algorithms. More recently, more efficient networks have become the focus of research (Figure 5.45) as well as larger (unlabeled) training datasets (Section 5.4.7). There are now open-source frameworks such as Classy Vision[3] for training and fine tuning your own image and video classification models. Users can also upload custom images on the web to the Computer Vision Explorer[4] to see how well many popular computer vision models perform on their own images.

In addition to recognizing commonly occurring categories such as those found in the ImageNet and COCO datasets, researchers have studied the problem of *fine-grained* category recognition (Duan, Parikh *et al.* 2012; Zhang, Donahue *et al.* 2014; Krause, Jin *et al.* 2015), where the differences between sub-categories can be subtle and the number of exemplars is quite low (Figure 6.9). Examples of categories with fine-grained sub classes include flowers (Nilsback and Zisserman 2006), cats and dogs (Parkhi, Vedaldi *et al.* 2012), birds (Wah, Branson *et al.* 2011; Van Horn,

[3]https://classyvision.ai

[4]https://vision-explorer.allenai.org

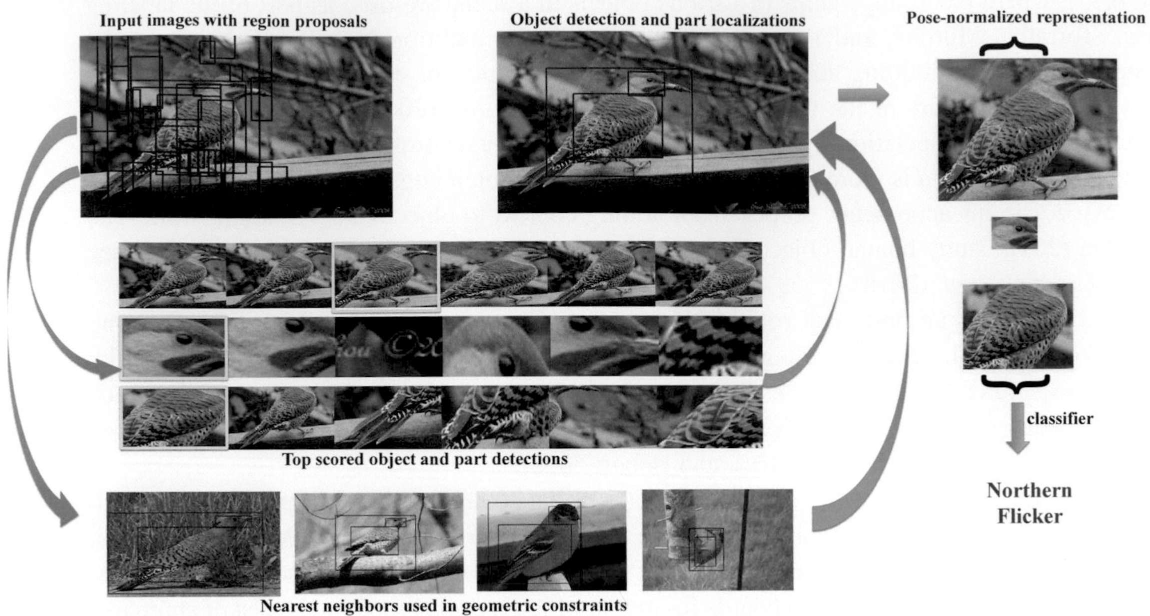

Figure 6.9 Fine-grained category recognition using parts (Zhang, Donahue *et al.* 2014) © 2014 Springer. Deep neural network object and part detectors are trained and their outputs are combined using geometric constraints. A classifier trained on features from the extracted parts is used for the final categorization.

Figure 6.10 Fine-grained category recognition. (a) The iNaturalist website and app allows citizen scientists to collect and classify images on their phones (Van Horn, Mac Aodha *et al.* 2018) © 2018 IEEE. (b) Attributes can be used for fine-grained categorization and zero-shot learning (Lampert, Nickisch, and Harmeling 2014) © 2014 Springer. These images are part of the *Animals with Attributes* dataset.

Branson *et al.* 2015), and cars (Yang, Luo *et al.* 2015). A recent example of fine-grained categorization is the iNaturalist system (Van Horn, Mac Aodha *et al.* 2018),[5] which allows both specialists and citizen scientists to photograph and label biological species, using a fine-grained category recognition system to label new images (Figure 6.10a).

Fine-grained categorization is often attacked using *attributes* of images and classes (Lampert, Nickisch, and Harmeling 2009; Parikh and Grauman 2011; Lampert, Nickisch, and Harmeling 2014), as shown in Figure 6.10b. Extracting attributes can enable *zero-shot learning* (Xian, Lampert *et al.* 2019), where previously unseen categories can be described using combinations of such attributes. However, some caution must be used in order not to learn spurious correlations between different attributes (Jayaraman, Sha, and Grauman 2014) or between objects and their common contexts (Singh, Mahajan *et al.* 2020). Fine-grained recognition can also be tackled using metric learning (Wu, Manmatha *et al.* 2017) or nearest-neighbor visual similarity search (Touvron, Sablayrolles *et al.* 2020), which we discuss next.

6.2.3 *Application*: Visual similarity search

Automatically classifying images into categories and tagging them with attributes using computer vision algorithms makes it easier to find them in catalogues and on the web. This is commonly used in *image search* or *image retrieval* engines, which find likely images based on keywords, just as regular web search engines find relevant documents and pages.

Sometimes, however, it's easier to find the information you need from an image, i.e., using *visual search*. Examples of this include fine-grained categorization, which we have just seen, as well as instance retrieval, i.e., finding the exact same object (Section 6.1) or location (Section 11.2.3). Another variant is finding visually similar images (often called *visual similarity search* or *reverse image search*), which is useful when the search intent cannot be succinctly captured in words.[6]

The topic of searching by visual similarity has a long history and goes by a variety of names, including query by image content (QBIC) (Flickner, Sawhney *et al.* 1995) and content-based image retrieval (CBIR) (Smeulders, Worring *et al.* 2000; Lew, Sebe *et al.* 2006; Vasconcelos 2007; Datta, Joshi *et al.* 2008). Early publications in these fields were based primarily on simple whole-image similarity metrics, such as color and texture (Swain and Ballard 1991; Jacobs, Finkelstein, and Salesin 1995; Manjunathi and Ma 1996).

Later architectures, such as that by Fergus, Perona, and Zisserman (2004), use a feature-based learning and recognition algorithm to re-rank the outputs from a traditional keyword-based image search engine. In follow-on work, Fergus, Fei-Fei *et al.* (2005) cluster the results returned by image search using an extension of probabilistic latest semantic analysis (PLSA) (Hofmann 1999) and then select the clusters associated with the highest ranked results as the representative images for that category. Other approaches rely on carefully annotated image databases such as LabelMe (Russell, Torralba *et al.* 2008). For example, Malisiewicz and Efros (2008) describe a system that, given a query image, can find similar LabelMe images, whereas Liu, Yuen, and Torralba (2009) combine feature-based correspondence algorithms with the labeled database to perform simultaneous recognition and segmentation.

Newer approaches to visual similarity search use whole-image descriptors such as Fisher kernels and the Vector of Locally Aggregated Descriptors (VLAD) (Jégou, Perronnin *et al.* 2012) or pooled CNN activations (Babenko and Lempitsky 2015a; Tolias, Sicre, and Jégou 2016; Cao, Araujo, and Sim 2020; Ng, Balntas *et al.* 2020; Tolias, Jenicek, and Chum 2020) combined with metric learning

[5] https://www.inaturalist.org

[6] Some authors use the term image retrieval to denote visual similarity search, (e.g., Jégou, Perronnin *et al.* 2012; Radenović, Tolias, and Chum 2019).

Figure 6.11 The GrokNet product recognition service is used for product tagging, visual search, and recommendations © Bell, Liu *et al.* (2020): (a) recognizing all the products in a photo; (b) automatically sourcing data for metric learning using weakly supervised data augmentation.

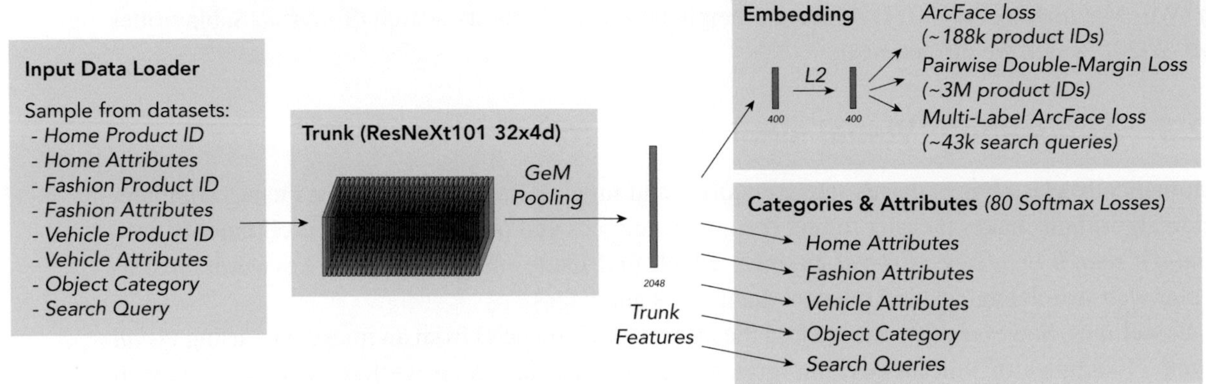

Figure 6.12 The GrokNet training architecture uses seven datasets, a common DNN trunk, two branches, and 83 loss functions (80 categorical losses + 3 embedding losses) © Bell, Liu *et al.* (2020).

(Bell and Bala 2015; Song, Xiang *et al.* 2016; Gordo, Almazán *et al.* 2017; Wu, Manmatha *et al.* 2017; Berman, Jégou *et al.* 2019) to represent each image with a compact descriptor that can be used to measure similarity in large databases (Johnson, Douze, and Jégou 2021). It is also possible to combine several techniques, such as deep networks with VLAD (Arandjelovic, Gronat *et al.* 2016), generalized mean (GeM) pooling (Radenović, Tolias, and Chum 2019), or dynamic mean (DAME) pooling (Yang, Kien Nguyen *et al.* 2019) into complete systems that are end-to-end tunable. Gordo, Almazán *et al.* (2017) provide a comprehensive review and experimental comparison of many of these techniques, which we also discuss in Section 7.1.4 on large-scale matching and retrieval. Some of the latest techniques for image retrieval use combinations of local and global descriptors to obtain state-of-the art performance on the landmark recognition tasks (Cao, Araujo, and Sim 2020; Ng, Balntas *et al.* 2020; Tolias, Jenicek, and Chum 2020). The ECCV 2020 Workshop on Instance-Level Recognition[7] has pointers to some of the latest work in this area, while the upcoming NeurIPS'21 Image Similarity Challenge[8] has new datasets for detecting content manipulation.

A recent example of a commercial system that uses visual similarity search, in addition to category recognition, is the GrokNet product recognition service described by Bell, Liu *et al.* (2020). GrokNet takes as input user images and shopping queries and returns indexed items similar to the ones in the query image (Figure 6.11a). The reason for needing a similarity search component is that the world contains too many "long-tail" items such as "a fur sink, an electric dog polisher, or a

[7]https://ilr-workshop.github.io/ECCVW2020
[8]https://www.drivendata.org/competitions/79/

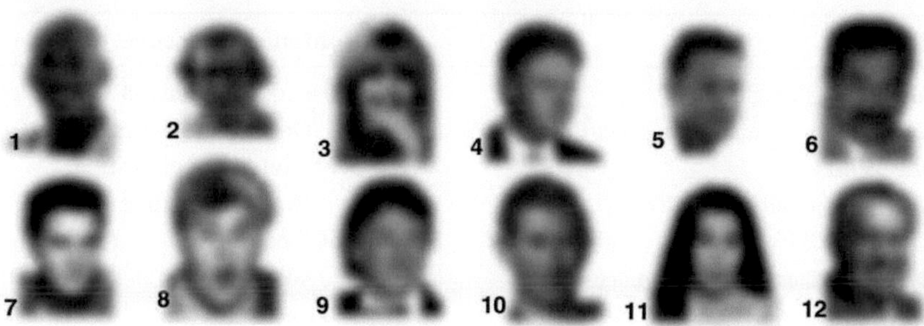

Figure 6.13 Humans can recognize low-resolution faces of familiar people (Sinha, Balas *et al.* 2006) © 2006 IEEE.

gasoline powered turtleneck sweater",[9] to make full categorization practical.

At training time, GrokNet takes both weakly labeled images, with category and/or attribute labels, and unlabeled images, where features in objects are detected and then used for metric learning, using a modification of ArcFace loss (Deng, Guo *et al.* 2019) and a novel pairwise margin loss (Figure 6.11b). The overall system takes in large collections of unlabeled and weakly labeled images and trains a ResNeXt101 trunk using a combination of category and attribute softmax losses and three different metric losses on the embeddings (Figure 6.12). GrokNet is just one example of a large number of commercial visual product search systems that have recently been developed. Others include systems from Amazon (Wu, Manmatha *et al.* 2017), Pinterest (Zhai, Wu *et al.* 2019), and Facebook (Tang, Borisyuk *et al.* 2019). In addition to helping people find items they may with to purchase, large-scale similarity search can also speed the search for harmful content on the web, as exemplified in Facebook's SimSearchNet.[10]

6.2.4 Face recognition

Among the various recognition tasks that computers are asked to perform, face recognition is the one where they have arguably had the most success.[11] While even people cannot readily distinguish between similar people with whom they are not familiar (O'Toole, Jiang *et al.* 2006; O'Toole, Phillips *et al.* 2009), computers' ability to distinguish among a small number of family members and friends has found its way into consumer-level photo applications. Face recognition can be used in a variety of additional applications, including human–computer interaction (HCI), identity verification (Kirovski, Jojic, and Jancke 2004), desktop login, parental controls, and patient monitoring (Zhao, Chellappa *et al.* 2003), but it also has the potential for misuse (Chokshi 2019; Ovide 2020).

Face recognizers work best when they are given images of faces under a wide variety of pose, illumination, and expression (PIE) conditions (Phillips, Moon *et al.* 2000; Sim, Baker, and Bsat 2003; Gross, Shi, and Cohn 2005; Huang, Ramesh *et al.* 2007; Phillips, Scruggs *et al.* 2010). More recent widely used datasets include labeled Faces in the Wild (LFW) (Huang, Ramesh *et al.* 2007; Learned-Miller, Huang *et al.* 2016), YouTube Faces (YTF) (Wolf, Hassner, and Maoz 2011), MegaFace (Kemelmacher-Shlizerman, Seitz *et al.* 2016; Nech and Kemelmacher-Shlizerman 2017), and the IARPA Janus Benchmark (IJB) (Klare, Klein *et al.* 2015; Maze, Adams *et al.* 2018), as tabulated in

[9] https://www.google.com/search?q=gasoline+powered+turtleneck+sweater

[10] https://ai.facebook.com/blog/using-ai-to-detect-covid-19-misinformation-and-exploitative-content

[11] Instance recognition, i.e., the re-recognition of known objects such as locations or planar objects, is the other most successful application of general image recognition. In the general domain of *biometrics*, i.e., identity recognition, specialized images such as irises and fingerprints perform even better (Jain, Bolle, and Pankanti 1999; Daugman 2004).

Name/URL	Contents/Reference
CMU Multi-PIE database http://www.cs.cmu.edu/afs/cs/project/PIE/MultiPie	337 people's faces in various poses Gross, Matthews *et al.* (2010)
Faces in the Wild http://vis-www.cs.umass.edu/lfw	5,749 internet celebrities Huang, Ramesh *et al.* (2007)
YouTube Faces (YTF) https://www.cs.tau.ac.il/~wolf/ytfaces	1,595 people in 3,425 YouTube videos Wolf, Hassner, and Maoz (2011)
MegaFace https://megaface.cs.washington.edu	1M internet faces Nech and Kemelmacher-Shlizerman (2017)
IARPA Janus Benchmark (IJB) https://www.nist.gov/programs-projects/face-challenges	31,334 faces of 3,531 people in videos Maze, Adams *et al.* (2018)
WIDER FACE http://shuoyang1213.me/WIDERFACE	32,203 images for face *detection* Yang, Luo *et al.* (2016)

Table 6.1 Face recognition and detection datasets, adapted from Maze, Adams *et al.* (2018).

Table 6.1. (See Masi, Wu *et al.* (2018) for additional datasets used for training.)

Some of the earliest approaches to face recognition involved finding the locations of distinctive image features, such as the eyes, nose, and mouth, and measuring the distances between these feature locations (Fischler and Elschlager 1973; Kanade 1977; Yuille 1991). Other approaches relied on comparing gray-level images projected onto lower dimensional subspaces called *eigenfaces* (Section 5.2.3) and jointly modeling shape and appearance variations (while discounting pose variations) using *active appearance models* (Section 6.2.4). Descriptions of "classic" (pre-DNN) face recognition systems can be found in a number of surveys and books on this topic (Chellappa, Wilson, and Sirohey 1995; Zhao, Chellappa *et al.* 2003; Li and Jain 2005) as well as the Face Recognition website.[12] The survey on face recognition by humans by Sinha, Balas *et al.* (2006) is also well worth reading; it includes a number of surprising results, such as humans' ability to recognize low-resolution images of familiar faces (Figure 6.13) and the importance of eyebrows in recognition. Researchers have also studied the automatic recognition of facial expressions. See Chang, Hu *et al.* (2006), Shan, Gong, and McOwan (2009), and Li and Deng (2020) for some representative papers.

Active appearance and 3D shape models

The need to use modular or view-based eigenspaces for face recognition, which we discussed in Section 5.2.3, is symptomatic of a more general observation, i.e., that facial appearance and identifiability depend as much on *shape* as they do on color or texture (which is what eigenfaces capture). Furthermore, when dealing with 3D head rotations, the *pose* of a person's head should be discounted when performing recognition.

In fact, the earliest face recognition systems, such as those by Fischler and Elschlager (1973), Kanade (1977), and Yuille (1991), found distinctive feature points on facial images and performed recognition on the basis of their relative positions or distances. Later techniques such as *local feature analysis* (Penev and Atick 1996) and *elastic bunch graph matching* (Wiskott, Fellous *et al.* 1997) combined local filter responses (jets) at distinctive feature locations together with shape models to perform recognition.

A visually compelling example of why both shape and texture are important is the work of Rowland and Perrett (1995), who manually traced the contours of facial features and then used

[12] https://www.face-rec.org

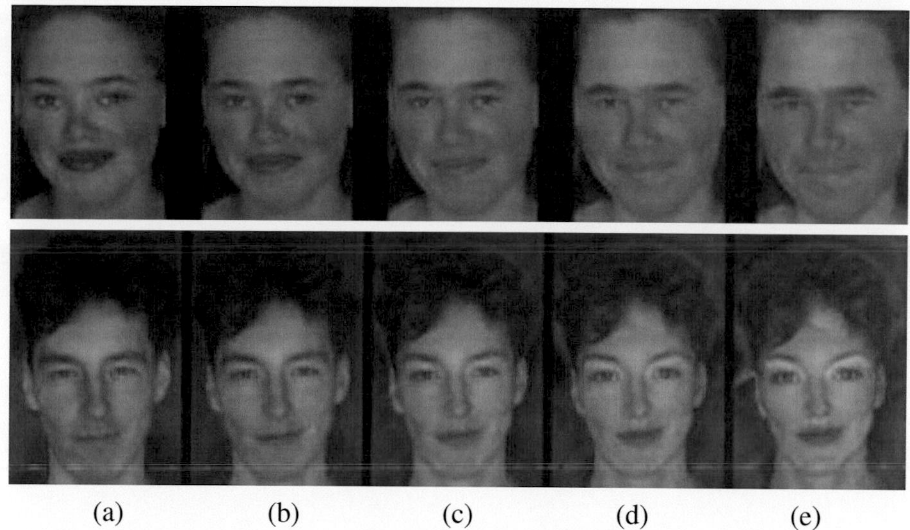

(a) (b) (c) (d) (e)

Figure 6.14 Manipulating facial appearance through shape and color (Rowland and Perrett 1995) © 1995 IEEE. By adding or subtracting gender-specific shape and color characteristics to an input image (b), different amounts of gender variation can be induced. The amounts added (from the mean) are: (a) +50% (gender enhancement), (b) 0% (original image), (c) –50% (near "androgyny"), (d) –100% (gender switched), and (e) –150% (opposite gender attributes enhanced).

these contours to normalize (warp) each image to a canonical shape. After analyzing both the shape and color images for deviations from the mean, they were able to associate certain shape and color deformations with personal characteristics such as age and gender (Figure 6.14). Their work demonstrates that both shape and color have an important influence on the perception of such characteristics.

Around the same time, researchers in computer vision were beginning to use simultaneous shape deformations and texture interpolation to model the variability in facial appearance caused by identity or expression (Beymer 1996; Vetter and Poggio 1997), developing techniques such as Active Shape Models (Lanitis, Taylor, and Cootes 1997), 3D Morphable Models (Blanz and Vetter 1999; Egger, Smith *et al.* 2020), and Elastic Bunch Graph Matching (Wiskott, Fellous *et al.* 1997).[13]

The *active appearance models* (AAMs) of Cootes, Edwards, and Taylor (2001) model both the variation in the shape of an image $\mathbf{s}$, which is normally encoded by the location of key feature points on the image, as well as the variation in texture $\mathbf{t}$, which is normalized to a canonical shape before being analyzed. Both shape and texture are represented as deviations from a mean shape $\bar{\mathbf{s}}$ and texture $\bar{\mathbf{t}}$,

$$\mathbf{s} = \bar{\mathbf{s}} + \mathbf{U}_s \mathbf{a} \tag{6.2}$$

$$\mathbf{t} = \bar{\mathbf{t}} + \mathbf{U}_t \mathbf{a}, \tag{6.3}$$

where the eigenvectors in $\mathbf{U}_s$ and $\mathbf{U}_t$ have been pre-scaled (whitened) so that unit vectors in $\mathbf{a}$ represent one standard deviation of variation observed in the training data. In addition to these principal deformations, the shape parameters are transformed by a global similarity to match the location, size, and orientation of a given face. Similarly, the texture image contains a scale and offset to best match novel illumination conditions.

[13]We will look at the application of PCA to 3D head and face modeling and animation in Section 13.6.3.

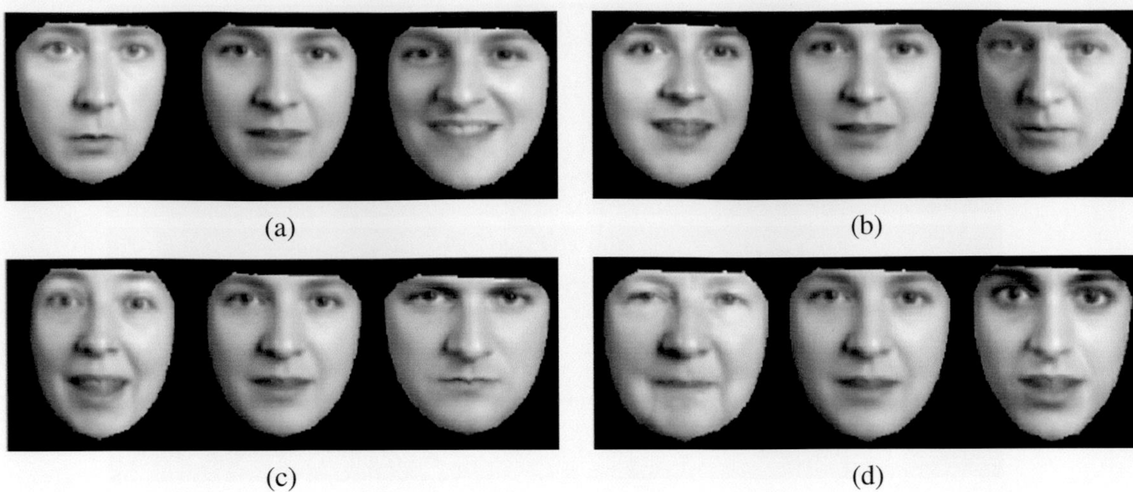

Figure 6.15 Principal modes of variation in active appearance models (Cootes, Edwards, and Taylor 2001) © 2001 IEEE. The four images show the effects of simultaneously changing the first four modes of variation in both shape and texture by $\pm\sigma$ from the mean. You can clearly see how the shape of the face and the shading are simultaneously affected.

As you can see, the same appearance parameters $\mathbf{a}$ in (6.2–6.3) simultaneously control both the shape and texture deformations from the mean, which makes sense if we believe them to be correlated. Figure 6.15 shows how moving three standard deviations along each of the first four principal directions ends up changing several correlated factors in a person's appearance, including expression, gender, age, and identity.

Although active appearance models are primarily designed to accurately capture the variability in appearance and deformation that are characteristic of faces, they can be adapted to face recognition by computing an identity subspace that separates variation in identity from other sources of variability such as lighting, pose, and expression (Costen, Cootes *et al.* 1999). The basic idea, which is modeled after similar work in eigenfaces (Belhumeur, Hespanha, and Kriegman 1997; Moghaddam, Jebara, and Pentland 2000), is to compute separate statistics for intrapersonal and extrapersonal variation and then find discriminating directions in these subspaces. While AAMs have sometimes been used directly for recognition (Blanz and Vetter 2003), their main use in the context of recognition is to align faces into a canonical pose (Liang, Xiao *et al.* 2008; Ren, Cao *et al.* 2014) so that more traditional methods of face recognition (Penev and Atick 1996; Wiskott, Fellous *et al.* 1997; Ahonen, Hadid, and Pietikäinen 2006; Zhao and Pietikäinen 2007; Cao, Yin *et al.* 2010) can be used.

Active appearance models have been extended to deal with illumination and viewpoint variation (Gross, Baker *et al.* 2005) as well as occlusions (Gross, Matthews, and Baker 2006). One of the most significant extensions is to construct 3D models of shape (Matthews, Xiao, and Baker 2007), which are much better at capturing and explaining the full variability of facial appearance across wide changes in pose. Such models can be constructed either from monocular video sequences (Matthews, Xiao, and Baker 2007), as shown in Figure 6.16a, or from multi-view video sequences (Ramnath, Koterba *et al.* 2008), which provide even greater reliability and accuracy in reconstruction and tracking (Murphy-Chutorian and Trivedi 2009).

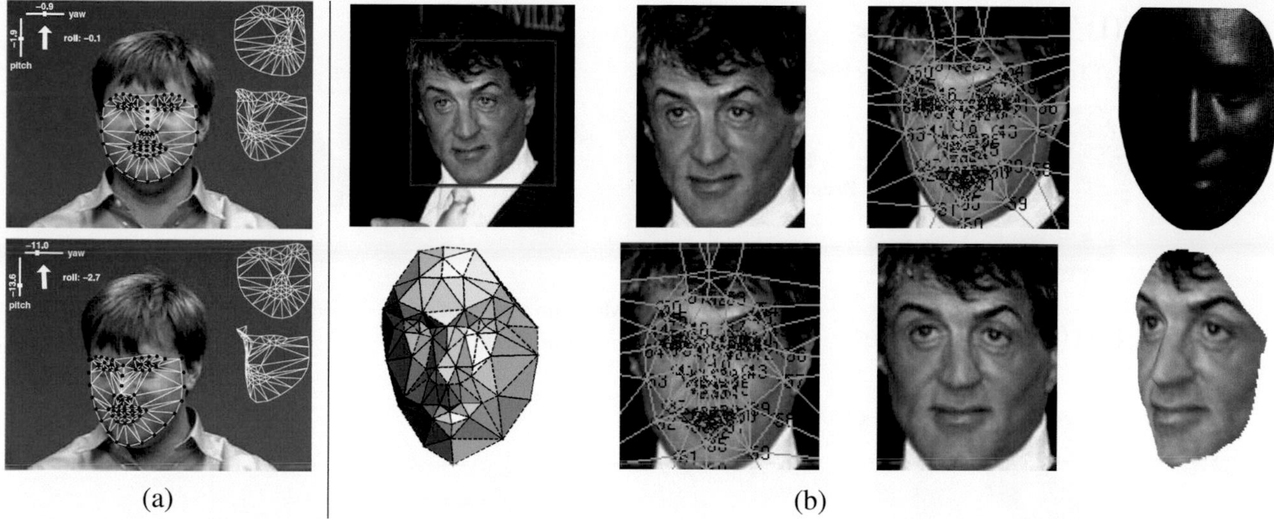

Figure 6.16 Head tracking and frontalization: (a) using 3D active appearance models (AAMs) (Matthews, Xiao, and Baker 2007) © 2007 Springer, showing video frames along with the estimated yaw, pitch, and roll parameters and the fitted 3D deformable mesh; (b) using six and then 67 fiducial points in the DeepFace system (Taigman, Yang *et al.* 2014) © 2014 IEEE, used to frontalize the face image (bottom row).

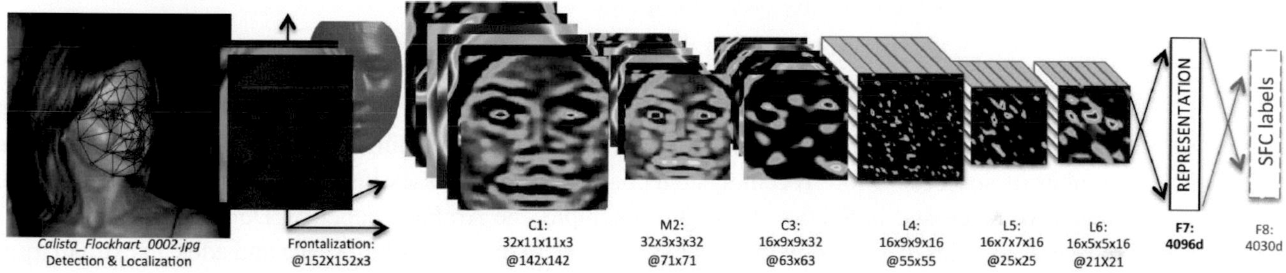

Figure 6.17 The DeepFace architecture (Taigman, Yang *et al.* 2014) © 2014 IEEE, starts with a frontalization stage, followed by several locally connected (non-convolutional) layers, and then two fully connected layers with a K-class softmax.

Facial recognition using deep learning

Prompted by the dramatic success of deep networks in whole-image categorization, face recognition researchers started using deep neural network backbones as part of their systems. Figures 6.16b–6.17 shows two stages in the DeepFace system of Taigman, Yang *et al.* (2014), which was one of the first systems to realize large gains using deep networks. In their system, a landmark-based pre-processing *frontalization* step is used to convert the original color image into a well-cropped front-looking face. Then, a deep locally connected network (where the convolution kernels can vary spatially) is fed into two final fully connected layers before classification.

Some of the more recent deep face recognizers omit the frontalization stage and instead use data augmentation (Section 5.3.3) to create synthetic inputs with a larger variety of poses (Schroff, Kalenichenko, and Philbin 2015; Parkhi, Vedaldi, and Zisserman 2015). Masi, Wu *et al.* (2018) provide an excellent tutorial and survey on deep face recognition, including a list of widely used training and testing datasets, a discussion of frontalization and dataset augmentation, and a section on training losses (Figure 6.18). This last topic is central to the ability to scale to larger and larger

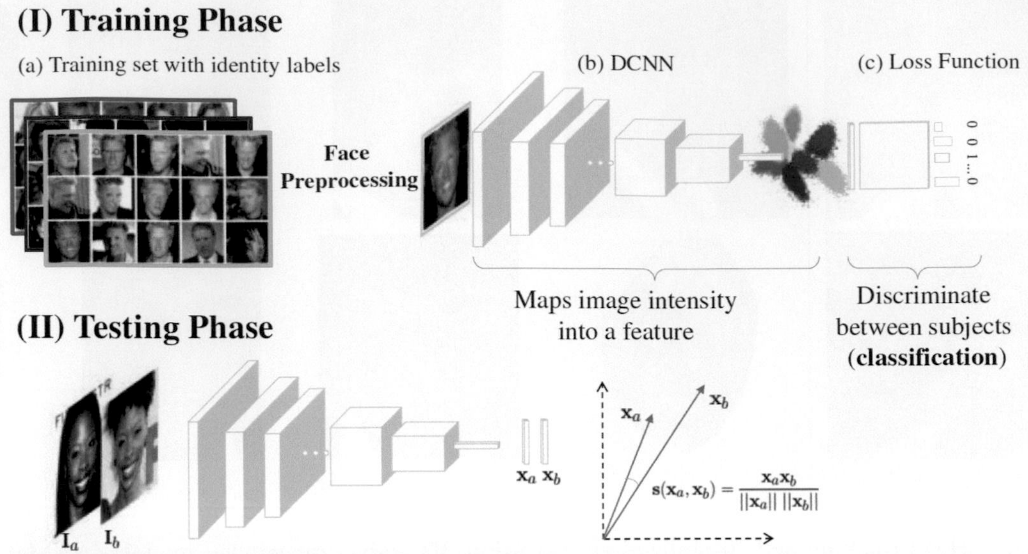

Figure 6.18 A typical modern deep face recognition architecture, from the survey by Masi, Wu *et al.* (2018) © 2018 IEEE. At training time, a huge labeled face set (a) is used to constrain the weights of a DCNN (b), optimizing a loss function (c) for a classification task. At test time, the classification layer is often discarded, and the DCNN is used as a feature extractor for comparing face descriptors.

numbers of people. Schroff, Kalenichenko, and Philbin (2015) and Parkhi, Vedaldi, and Zisserman (2015) use triplet losses to construct a low-dimensional embedding space that is independent of the number of subjects. More recent systems use contrastive losses inspired by the softmax function, which we discussed in Section 5.3.4. For example, the ArcFace paper by Deng, Guo *et al.* (2019) measures angular distances on the unit hypersphere in the embedding space and adds an extra margin to get identities to clump together. This idea has been further extended for visual similarity search (Bell, Liu *et al.* 2020) and face recognition (Huang, Shen *et al.* 2020; Deng, Guo *et al.* 2020a).

Personal photo collections

In addition to digital cameras automatically finding faces to aid in auto-focusing and video cameras finding faces in video conferencing to center on the speaker (either mechanically or digitally), face detection has found its way into most consumer-level photo organization packages and photo sharing sites. Finding faces and allowing users to tag them makes it easier to find photos of selected people at a later date or to automatically share them with friends. In fact, the ability to tag friends in photos is one of the more popular features on Facebook.

Sometimes, however, faces can be hard to find and recognize, especially if they are small, turned away from the camera, or otherwise occluded. In such cases, combining face recognition with person detection and clothes recognition can be very effective, as illustrated in Figure 6.19 (Sivic, Zitnick, and Szeliski 2006). Combining person recognition with other kinds of context, such as location recognition (Section 11.2.3) or activity or event recognition, can also help boost performance (Lin, Kapoor *et al.* 2010).

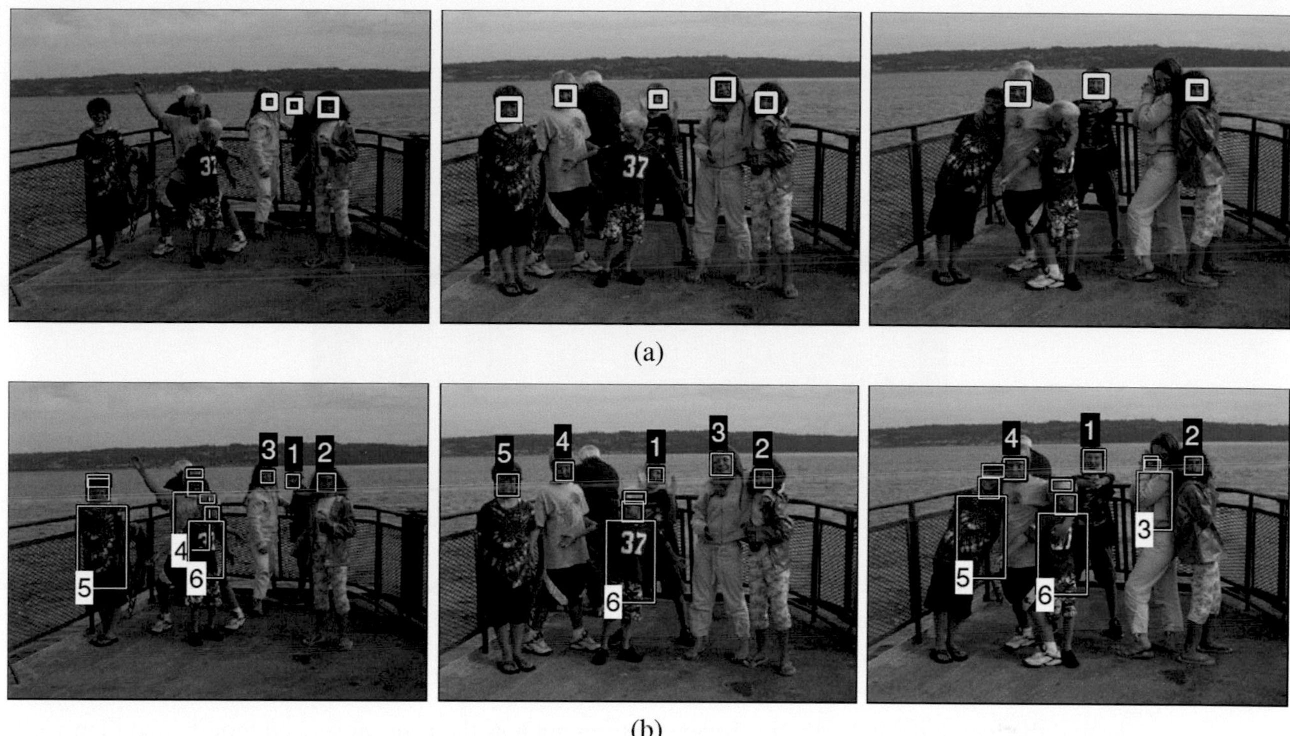

Figure 6.19 Person detection and re-recognition using a combined face, hair, and torso model (Sivic, Zitnick, and Szeliski 2006) © 2006 Springer. (a) Using face detection alone, several of the heads are missed. (b) The combined face and clothing model successfully re-finds all the people.

6.3 Object detection

If we are given an image to analyze, such as the group portrait in Figure 6.20, we could try to apply a recognition algorithm to every possible sub-window in this image. Such algorithms are likely to be both slow and error-prone. Instead, it is more effective to construct special-purpose *detectors*, whose job it is to rapidly find likely regions where particular objects might occur.

We begin this section with face detectors, which were some of the earliest successful examples of recognition. Such algorithms are built into most of today's digital cameras to enhance auto-focus and into video conferencing systems to control panning and zooming. We then look at pedestrian detectors, as an example of more general methods for object detection. Finally, we turn to the problem of multi-class object detection, which today is solved using deep neural networks.

6.3.1 Face detection

Before face recognition can be applied to a general image, the locations and sizes of any faces must first be found (Figures 6.1c and 6.20). In principle, we could apply a face recognition algorithm at every pixel and scale (Moghaddam and Pentland 1997) but such a process would be too slow in practice.

Over the last four decades, a wide variety of fast face detection algorithms have been developed. Yang, Kriegman, and Ahuja (2002) and Zhao, Chellappa *et al.* (2003) provide comprehensive surveys of earlier work in this field. According to their taxonomy, face detection techniques can be classified as feature-based, template-based, or appearance-based. Feature-based techniques attempt

Figure 6.20 Face detection results produced by Rowley, Baluja, and Kanade (1998) © 1998 IEEE. Can you find the one false positive (a box around a non-face) among the 57 true positive results?

to find the locations of distinctive image features such as the eyes, nose, and mouth, and then verify whether these features are in a plausible geometrical arrangement. These techniques include some of the early approaches to face recognition (Fischler and Elschlager 1973; Kanade 1977; Yuille 1991), as well as later approaches based on modular eigenspaces (Moghaddam and Pentland 1997), local filter jets (Leung, Burl, and Perona 1995; Penev and Atick 1996; Wiskott, Fellous *et al.* 1997), support vector machines (Heisele, Ho *et al.* 2003; Heisele, Serre, and Poggio 2007), and boosting (Schneiderman and Kanade 2004).

Template-based approaches, such as active appearance models (AAMs) (Section 6.2.4), can deal with a wide range of pose and expression variability. Typically, they require good initialization near a real face and are therefore not suitable as fast face detectors.

Appearance-based approaches scan over small overlapping rectangular patches of the image searching for likely face candidates, which can then be refined using a *cascade* of more expensive but selective detection algorithms (Sung and Poggio 1998; Rowley, Baluja, and Kanade 1998; Romdhani, Torr *et al.* 2001; Fleuret and Geman 2001; Viola and Jones 2004). To deal with scale variation, the image is usually converted into a sub-octave pyramid and a separate scan is performed on each level. Most appearance-based approaches rely heavily on training classifiers using sets of labeled face and non-face patches.

Sung and Poggio (1998) and Rowley, Baluja, and Kanade (1998) present two of the earliest appearance-based face detectors and introduce a number of innovations that are widely used in later work by others. To start with, both systems collect a set of labeled face patches (Figure 6.20) as well as a set of patches taken from images that are known not to contain faces, such as aerial images or vegetation. The collected face images are augmented by artificially mirroring, rotating, scaling, and translating the images by small amounts to make the face detectors less sensitive to such effects.

The next few paragraphs provide quick reviews of a number of early appearance-based face detectors, keyed by the machine algorithms they are based on. These systems provide an interesting glimpse into the gradual adoption and evolution of machine learning in computer vision. More detailed descriptions can be found in the original papers, as well as the first edition of this book (Szeliski 2010).

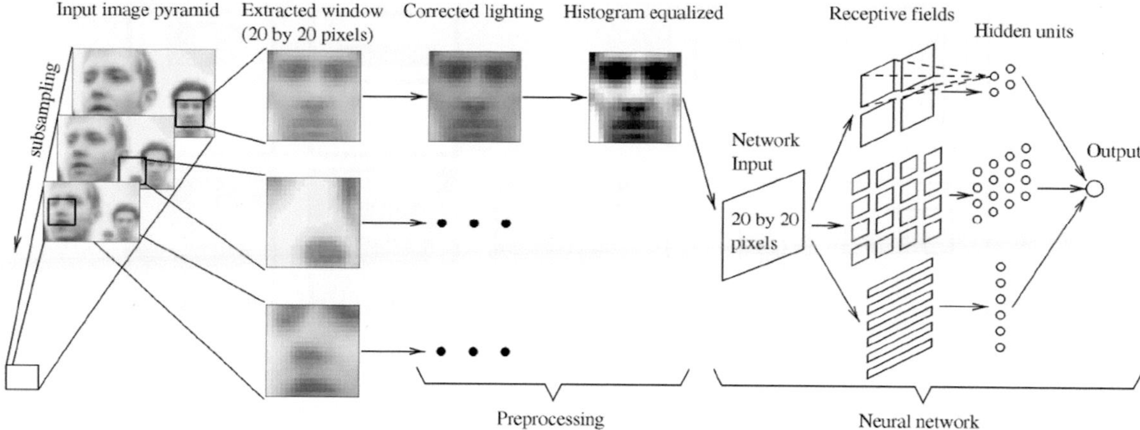

Figure 6.21 A neural network for face detection (Rowley, Baluja, and Kanade 1998) © 1998 IEEE. Overlapping patches are extracted from different levels of a pyramid and then pre-processed. A three-layer neural network is then used to detect likely face locations.

Clustering and PCA. Once the face and non-face patterns have been pre-processed, Sung and Poggio (1998) cluster each of these datasets into six separate clusters using k-means and then fit PCA subspaces to each of the resulting 12 clusters. At detection time, the DIFS and DFFS metrics first developed by Moghaddam and Pentland (1997) are used to produce 24 Mahalanobis distance measurements (two per cluster). The resulting 24 measurements are input to a multi-layer perceptron (MLP), i.e., a fully connected neural network.

Neural networks. Instead of first clustering the data and computing Mahalanobis distances to the cluster centers, Rowley, Baluja, and Kanade (1998) apply a neural network (MLP) directly to the 20×20 pixel patches of gray-level intensities, using a variety of differently sized hand-crafted "receptive fields" to capture both large-scale and smaller scale structure (Figure 6.21). The resulting neural network directly outputs the likelihood of a face at the center of every overlapping patch in a multi-resolution pyramid. Since several overlapping patches (in both space and resolution) may fire near a face, an additional merging network is used to merge overlapping detections. The authors also experiment with training several networks and merging their outputs. Figure 6.20 shows a sample result from their face detector.

Support vector machines. Instead of using a neural network to classify patches, Osuna, Freund, and Girosi (1997) use *support vector machines* (SVMs), which we discussed in Section 5.1.4, to classify the same preprocessed patches as Sung and Poggio (1998). An SVM searches for a series of *maximum margin* separating planes in feature space between different classes (in this case, face and non-face patches). In those cases where linear classification boundaries are insufficient, the feature space can be lifted into higher-dimensional features using *kernels* (5.29). SVMs have been used by other researchers for both face detection and face recognition (Heisele, Ho *et al.* 2003; Heisele, Serre, and Poggio 2007) as well as general object recognition (Lampert 2008).

Boosting. Of all the face detectors developed in the 2000s, the one introduced by Viola and Jones (2004) is probably the best known. Their technique was the first to introduce the concept of *boosting* to the computer vision community, which involves training a series of increasingly discriminating

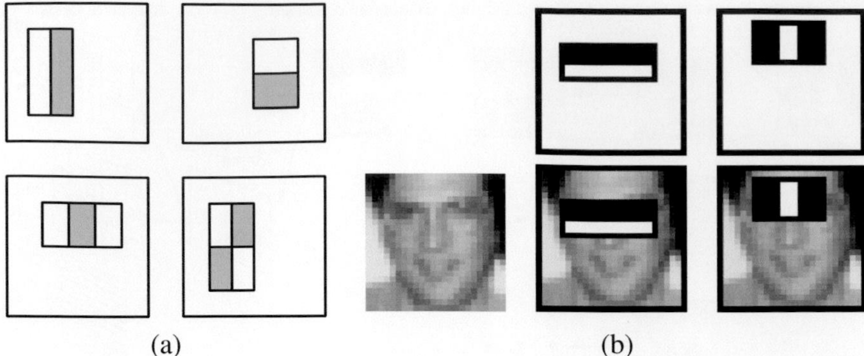

Figure 6.22 Simple features used in boosting-based face detector (Viola and Jones 2004) © 2004 Springer: (a) difference of rectangle feature composed of 2–4 different rectangles (pixels inside the white rectangles are subtracted from the gray ones); (b) the first and second features selected by AdaBoost. The first feature measures the differences in intensity between the eyes and the cheeks, the second one between the eyes and the bridge of the nose.

simple classifiers and then blending their outputs (Bishop 2006, Section 14.3; Hastie, Tibshirani, and Friedman 2009, Chapter 10; Murphy 2012, Section 16.4; Glassner 2018, Section 14.7).

In more detail, boosting involves constructing a *classifier* $h(\mathbf{x})$ as a sum of simple *weak learners*,

$$h(\mathbf{x}) = \text{sign}\left[\sum_{j=0}^{m-1} \alpha_j h_j(\mathbf{x})\right], \tag{6.4}$$

where each of the weak learners $h_j(\mathbf{x})$ is an extremely simple function of the input, and hence is not expected to contribute much (in isolation) to the classification performance.

In most variants of boosting, the weak learners are threshold functions,

$$h_j(\mathbf{x}) = a_j[f_j < \theta_j] + b_j[f_j \geq \theta_j] = \begin{cases} a_j & \text{if} \quad f_j < \theta_j \\ b_j & \text{otherwise,} \end{cases} \tag{6.5}$$

which are also known as *decision stumps* (basically, the simplest possible version of *decision trees*). In most cases, it is also traditional (and simpler) to set a_j and b_j to ± 1, i.e., $a_j = -s_j$, $b_j = +s_j$, so that only the feature f_j, the threshold value θ_j, and the polarity of the threshold $s_j \in \pm 1$ need to be selected.[14]

In many applications of boosting, the features are simply coordinate axes x_k, i.e., the boosting algorithm selects one of the input vector components as the best one to threshold. In Viola and Jones' face detector, the features are differences of rectangular regions in the input patch, as shown in Figure 6.22. The advantage of using these features is that, while they are more discriminating than single pixels, they are extremely fast to compute once a summed area table has been precomputed, as described in Section 3.2.3 (3.31–3.32). Essentially, for the cost of an $O(N)$ precomputation phase (where N is the number of pixels in the image), subsequent differences of rectangles can be computed in $4r$ additions or subtractions, where $r \in \{2, 3, 4\}$ is the number of rectangles in the feature.

The key to the success of boosting is the method for incrementally selecting the weak learners and for re-weighting the training examples after each stage. The AdaBoost (Adaptive Boosting)

[14] Some variants, such as that of Viola and Jones (2004), use $(a_j, b_j) \in [0, 1]$ and adjust the learning algorithm accordingly.

Figure 6.23 Pedestrian detection using histograms of oriented gradients (Dalal and Triggs 2005) © 2005 IEEE: (a) the average gradient image over the training examples; (b) each "pixel" shows the maximum positive SVM weight in the block centered on the pixel; (c) likewise, for the negative SVM weights; (d) a test image; (e) the computed R-HOG (rectangular histogram of gradients) descriptor; (f) the R-HOG descriptor weighted by the positive SVM weights; (g) the R-HOG descriptor weighted by the negative SVM weights.

algorithm (Bishop 2006; Hastie, Tibshirani, and Friedman 2009; Murphy 2012) does this by re-weighting each sample as a function of whether it is correctly classified at each stage, and using the stage-wise average classification error to determine the final weightings α_j among the weak classifiers.

To further increase the speed of the detector, it is possible to create a *cascade* of classifiers, where each classifier uses a small number of tests (say, a two-term AdaBoost classifier) to reject a large fraction of non-faces while trying to pass through all potential face candidates (Fleuret and Geman 2001; Viola and Jones 2004; Brubaker, Wu *et al.* 2008).

Deep networks. Since the initial burst of face detection research in the early 2000s, face detection algorithms have continued to evolve and improve (Zafeiriou, Zhang, and Zhang 2015). Researchers have proposed using cascades of features (Li and Zhang 2013), deformable parts models (Mathias, Benenson *et al.* 2014), aggregated channel features (Yang, Yan *et al.* 2014), and neural networks (Li, Lin *et al.* 2015; Yang, Luo *et al.* 2015). The WIDER FACE benchmark[15,16] (Yang, Luo *et al.* 2016) contains results from, and pointers to, more recent papers, including RetinaFace (Deng, Guo *et al.* 2020b), which combines ideas from other recent neural networks and object detectors such as Feature Pyramid Networks (Lin, Dollár *et al.* 2017) and RetinaNet (Lin, Goyal *et al.* 2017), and also has a nice review of other recent face detectors.

6.3.2 Pedestrian detection

While a lot of the early research on object detection focused on faces, the detection of other objects, such as pedestrians and cars, has also received widespread attention (Gavrila and Philomin 1999; Gavrila 1999; Papageorgiou and Poggio 2000; Mohan, Papageorgiou, and Poggio 2001; Schneiderman and Kanade 2004). Some of these techniques maintained the same focus as face detection on speed and efficiency. Others, however, focused on accuracy, viewing detection as a more challenging

[15] http://shuoyang1213.me/WIDERFACE

[16] The WIDER FACE benchmark has expanded to a larger set of detection challenges and workshops: https://wider-challenge.org/2019.html.

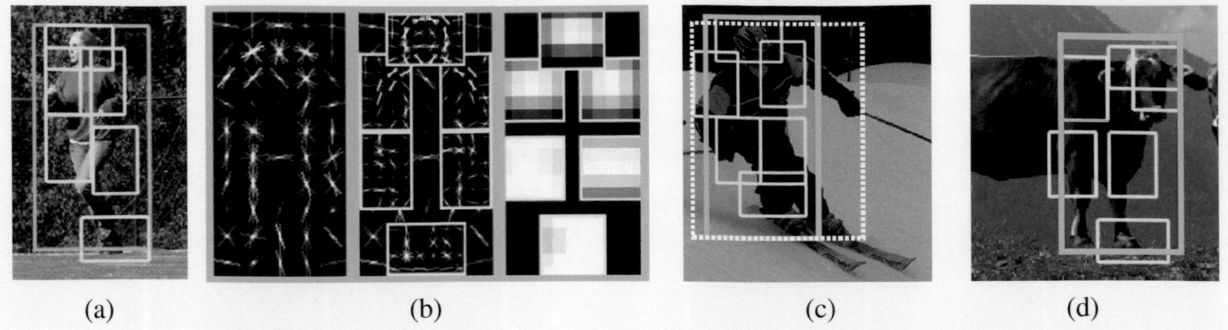

(a) (b) (c) (d)

Figure 6.24 Part-based object detection (Felzenszwalb, McAllester, and Ramanan 2008) © 2008 IEEE: (a) An input photograph and its associated person (blue) and part (yellow) detection results. (b) The detection model is defined by a coarse template, several higher resolution part templates, and a spatial model for the location of each part. (c) True positive detection of a skier and (d) false positive detection of a cow (labeled as a person).

variant of generic class recognition (Section 6.3.3) in which the locations and extents of objects are to be determined as accurately as possible (Everingham, Van Gool *et al.* 2010; Everingham, Eslami *et al.* 2015; Lin, Maire *et al.* 2014).

An example of a well-known pedestrian detector is the algorithm developed by Dalal and Triggs (2005), who use a set of overlapping *histogram of oriented gradients* (HOG) descriptors fed into a support vector machine (Figure 6.23). Each HOG has cells to accumulate magnitude-weighted votes for gradients at particular orientations, just as in the scale invariant feature transform (SIFT) developed by Lowe (2004), which we will describe in Section 7.1.2 and Figure 7.16. Unlike SIFT, however, which is only evaluated at interest point locations, HOGs are evaluated on a regular overlapping grid and their descriptor magnitudes are normalized using an even coarser grid; they are only computed at a single scale and a fixed orientation. To capture the subtle variations in orientation around a person's outline, a large number of orientation bins are used and no smoothing is performed in the central difference gradient computation—see Dalal and Triggs (2005) for more implementation details. Figure 6.23d shows a sample input image, while Figure 6.23e shows the associated HOG descriptors.

Once the descriptors have been computed, a support vector machine (SVM) is trained on the resulting high-dimensional continuous descriptor vectors. Figures 6.23b–c show a diagram of the (most) positive and negative SVM weights in each block, while Figures 6.23f–g show the corresponding weighted HOG responses for the central input image. As you can see, there are a fair number of positive responses around the head, torso, and feet of the person, and relatively few negative responses (mainly around the middle and the neck of the sweater).

Much like face detection, the fields of pedestrian and general object detection continued to advance rapidly in the 2000s (Belongie, Malik, and Puzicha 2002; Mikolajczyk, Schmid, and Zisserman 2004; Dalal and Triggs 2005; Leibe, Seemann, and Schiele 2005; Opelt, Pinz, and Zisserman 2006; Torralba 2007; Andriluka, Roth, and Schiele 2009; Maji and Berg 2009; Andriluka, Roth, and Schiele 2010; Dollár, Belongie, and Perona 2010).

A significant advance in the field of person detection was the work of Felzenszwalb, McAllester, and Ramanan (2008), who extend the histogram of oriented gradients person detector to incorporate flexible parts models (Section 6.2.1). Each part is trained and detected on HOGs evaluated at two pyramid levels below the overall object model and the locations of the parts relative to the parent node (the overall bounding box) are also learned and used during recognition (Figure 6.24b). To compensate for inaccuracies or inconsistencies in the training example bounding boxes (dashed

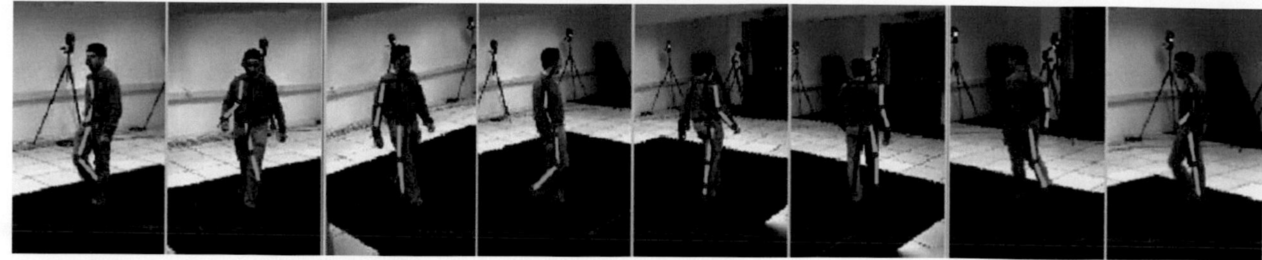

Figure 6.25 Pose detection using random forests (Rogez, Rihan *et al.* 2008) © 2008 IEEE. The estimated pose (state of the kinematic model) is drawn over each input frame.

white lines in Figure 6.24c), the "true" location of the parent (blue) bounding box is considered a latent (hidden) variable and is inferred during both training and recognition. Since the locations of the parts are also latent, the system can be trained in a semi-supervised fashion, without needing part labels in the training data. An extension to this system (Felzenszwalb, Girshick *et al.* 2010), which includes among its improvements a simple contextual model, was among the two best object detection systems in the 2008 Visual Object Classes detection challenge (Everingham, Van Gool *et al.* 2010). Improvements to part-based person detection and pose estimation include work by Andriluka, Roth, and Schiele (2009) and Kumar, Zisserman, and Torr (2009).

An even more accurate estimate of a person's pose and location is presented by Rogez, Rihan *et al.* (2008), who compute both the phase of a person in a walk cycle and the locations of individual joints, using random forests built on top of HOGs (Figure 6.25). Since their system produces full 3D pose information, it is closer in its application domain to 3D person trackers (Sidenbladh, Black, and Fleet 2000; Andriluka, Roth, and Schiele 2010), which we will discussed in Section 13.6.4. When video sequences are available, the additional information present in the optical flow and motion discontinuities can greatly aid in the detection task, as discussed by Efros, Berg *et al.* (2003), Viola, Jones, and Snow (2003), and Dalal, Triggs, and Schmid (2006).

Since the 2000s, pedestrian and general person detection have continued to be actively developed, often in the context of more general multi-class object detection (Everingham, Van Gool *et al.* 2010; Everingham, Eslami *et al.* 2015; Lin, Maire *et al.* 2014). The Caltech pedestrian detection benchmark[17] and survey by Dollár, Belongie, and Perona (2010) introduces a new dataset and provides a nice review of algorithms through 2012, including Integral Channel Features (Dollár, Tu *et al.* 2009), the Fastest Pedestrian Detector in the West (Dollár, Belongie, and Perona 2010), and 3D pose estimation algorithms such as Poselets (Bourdev and Malik 2009). Since its original construction, this benchmark continues to tabulate and evaluate more recent detectors, including Dollár, Appel, and Kienzle (2012), Dollár, Appel *et al.* (2014), and more recent algorithms based on deep neural networks (Sermanet, Kavukcuoglu *et al.* 2013; Ouyang and Wang 2013; Tian, Luo *et al.* 2015; Zhang, Lin *et al.* 2016). The CityPersons dataset (Zhang, Benenson, and Schiele 2017) and WIDER Face and Person Challenge[18] also report results on recent algorithms.

6.3.3 General object detection

While face and pedestrian detection algorithms were the earliest to be extensively studied, computer vision has always been interested in solving the general object detection and labeling problem, in

[17] http://www.vision.caltech.edu/Image_Datasets/CaltechPedestrians

[18] https://wider-challenge.org/2019.html

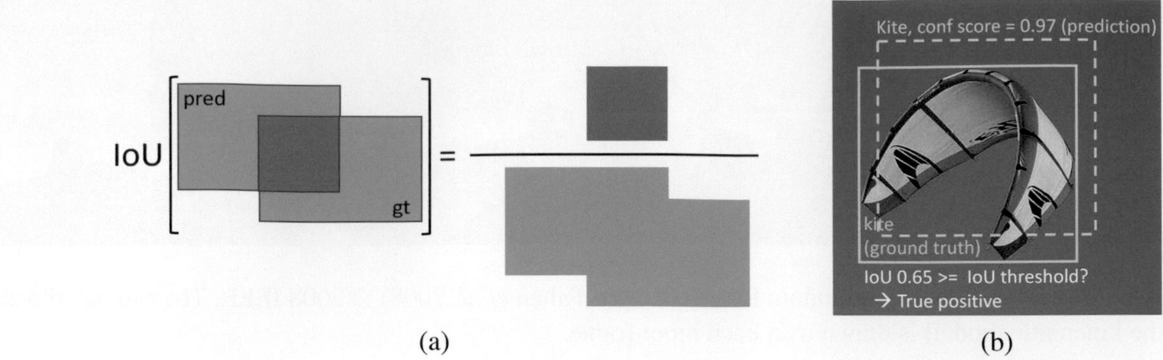

Figure 6.26 Intersection over union (IoU): (a) schematic formula, (b) real-world example © 2020 Ross Girshick.

addition to whole-image classification. The PASCAL Visual Object Classes (VOC) Challenge (Everingham, Van Gool *et al.* 2010), which contained 20 classes, had both classification and detection challenges. Early entries that did well on the detection challenge include a feature-based detector and spatial pyramid matching SVM classifier by Chum and Zisserman (2007), a star-topology deformable part model by Felzenszwalb, McAllester, and Ramanan (2008), and a sliding window SVM classifier by Lampert, Blaschko, and Hofmann (2008). The competition was re-run annually, with the two top entries in the 2012 detection challenge (Everingham, Eslami *et al.* 2015) using a sliding window spatial pyramid matching (SPM) SVM (de Sande, Uijlings *et al.* 2011) and a University of Oxford re-implementation of a deformable parts model (Felzenszwalb, Girshick *et al.* 2010).

The ImageNet Large Scale Visual Recognition Challenge (ILSVRC), released in 2010, scaled up the dataset from around 20 thousand images in PASCAL VOC 2010 to over 1.4 million in ILSVRC 2010, and from 20 object classes to 1,000 object classes (Russakovsky, Deng *et al.* 2015). Like PASCAL, it also had an object detection task, but it contained a much wider range of challenging images (Figure 6.4). The Microsoft COCO (Common Objects in Context) dataset (Lin, Maire *et al.* 2014) contained even more objects per image, as well as pixel-accurate segmentations of multiple objects, enabling the study of not only *semantic segmentation* (Section 6.4), but also individual object *instance segmentation* (Section 6.4.2). Table 6.2 list some of the datasets used for training and testing general object detection algorithms.

The release of COCO coincided with a wholescale shift to deep networks for image classification, object detection, and segmentation (Jiao, Zhang *et al.* 2019; Zhao, Zheng *et al.* 2019). Figure 6.29 shows the rapid improvements in average precision (AP) on the COCO object detection task, which correlates strongly with advances in deep neural network architectures (Figure 5.40).

Precision vs. recall

Before we describe the elements of modern object detectors, we should first discuss what metrics they are trying to optimize. The main task in object detection, as illustrated in Figures 6.5a and 6.26b, is to put accurate bounding boxes around all the objects of interest and to correctly label such objects. To measure the accuracy of each bounding box (not too small and not too big), the common metric is *intersection over union* (IoU), which is also known as the *Jaccard index* or *Jaccard similarity coefficient* (Rezatofighi, Tsoi *et al.* 2019). The IoU is computed by taking the predicted and ground truth bounding boxes $B_{\rm pr}$ and $B_{\rm gt}$ for an object and computing the ratio of their area of

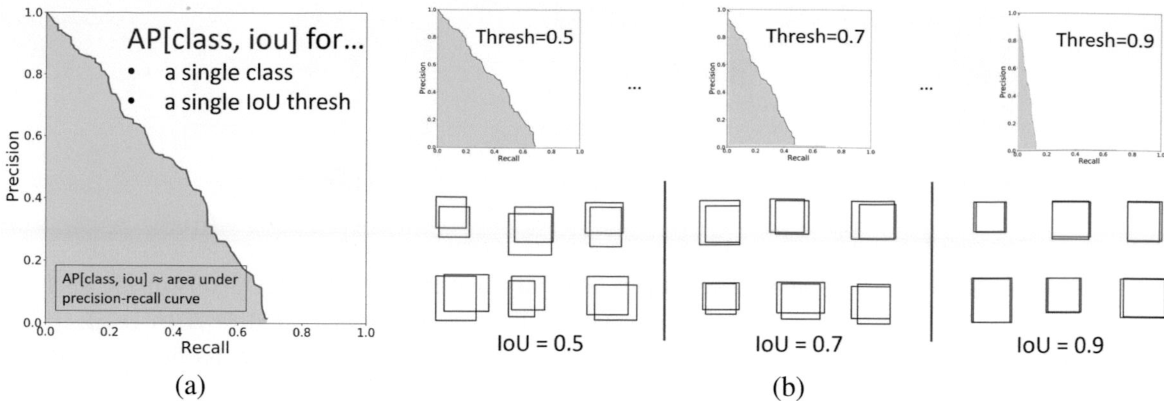

Figure 6.27 Object detector average precision © 2020 Ross Girshick: (a) a precision-recall curve for a single class and IoU threshold, with the AP being the area under the P-R curve; (b) average precision averaged over several IoU thresholds (from looser to tighter).

intersection and their area of union,

$$IoU = \frac{B_{\text{pr}} \cap B_{\text{gt}}}{B_{\text{pr}} \cup B_{\text{gt}}}, \tag{6.6}$$

as shown in Figure 6.26a.

As we will shortly see, object detectors operate by first proposing a number of plausible rectangular regions (detections) and then classifying each detection while also producing a confidence score (Figure 6.26b). These regions are then run through some kind of *non-maximal suppression* (NMS) stage, which removes weaker detections that have too much overlap with stronger detections, using a greedy most-confident-first algorithm.

To evaluate the performance of an object detector, we run through all of the detections, from most confident to least, and classify them as *true positive* TP (correct label and sufficiently high IoU) or *false positive* FP (incorrect label or ground truth object already matched). For each new decreasing confidence threshold, we can compute the *precision* and *recall* as

$$\text{precision} = \frac{\text{TP}}{\text{TP+FP}} \tag{6.7}$$

$$\text{recall} = \frac{\text{TP}}{\text{P}}, \tag{6.8}$$

where P is the number of positive examples, i.e., the number of labeled ground truth detections in the test image.[19] (See Section 7.1.3 on feature matching for additional terms that are often used in measuring and describing error rates.)

Computing the precision and recall at every confidence threshold allows us to populate a precision-recall curve, such as the one in Figure 6.27a. The area under this curve is called *average precision* (AP). A separate AP score can be computed for each class being detected, and the results averaged to produce a *mean average precision* (mAP). Another widely used measure if the While earlier benchmarks such as PASCAL VOC determined the mAP using a single IoU threshold of 0.5 (Everingham, Eslami *et al.* 2015), the COCO benchmark (Lin, Maire *et al.* 2014) averages the mAP over a set of IoU thresholds, IoU $\in \{0.50, 0.55, \ldots, 0.95\}$, as shown in Figure 6.27a. While this AP score

[19] Another widely reported measure is the *F-score*, which is the harmonic mean of the precision and recall.

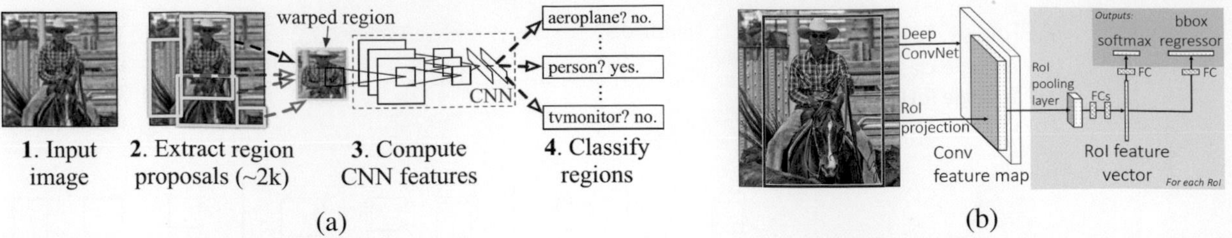

Figure 6.28 The R-CNN and Fast R-CNN object detectors. (a) R-CNN rescales pixels inside each proposal region and performs a CNN + SVM classification (Girshick, Donahue *et al.* 2015) © 2015 IEEE. (b) Fast R-CNN resamples convolutional features and uses fully connected layers to perform classification and bounding box regression (Girshick 2015) © 2015 IEEE.

continues to be widely used, an alternative *probability-based detection quality* (PDQ) score has recently been proposed (Hall, Dayoub *et al.* 2020). A smoother version of average precision called Smooth-AP has also been proposed and shown to have benefits on large-scale image retrieval tasks (Brown, Xie *et al.* 2020).

Modern object detectors

The first stage in detecting objects in an image is to propose a set of plausible rectangular regions in which to run a classifier. The development of such *region proposal* algorithms was an active research area in the early 2000s (Alexe, Deselaers, and Ferrari 2012; Uijlings, Van De Sande *et al.* 2013; Cheng, Zhang *et al.* 2014; Zitnick and Dollár 2014).

One of the earliest object detectors based on neural networks is R-CNN, the Region-based Convolutional Network developed by Girshick, Donahue *et al.* (2014). As illustrated in Figure 6.28a, this detector starts by extracting about 2,000 region proposals using the selective search algorithm of Uijlings, Van De Sande *et al.* (2013). Each proposed regions is then rescaled (warped) to a 224 square image and passed through an AlexNet or VGG neural network with a support vector machine (SVM) final classifier.

The follow-on Fast R-CNN paper by Girshick (2015) interchanges the convolutional neural network and region extraction stages and replaces the SVM with some fully connected (FC) layers, which compute both an object class and a bounding box refinement (Figure 6.28b). This reuses the CNN computations and leads to much faster training and test times, as well as dramatically better accuracy compared to previous networks (Figure 6.29). As you can see from Figure 6.28b, Fast R-CNN is an example of a deep network with a shared *backbone* and two separate *heads*, and hence two different loss functions, although these terms were not introduced until the Mask R-CNN paper by He, Gkioxari *et al.* (2017).

The Faster R-CNN system, introduced a few month later by Ren, He *et al.* (2015), replaces the relatively slow selective search stage with a convolutional region proposal network (RPN), resulting in much faster inference. After computing convolutional features, the RPN suggests at each coarse location a number of potential *anchor boxes*, which vary in shape and size to accommodate different potential objects. Each proposal is then classified and refined by an instance of the Fast R-CNN heads and the final detections are ranked and merged using non-maximal suppression.

R-CNN, Fast R-CNN, and Faster R-CNN all operate on a single resolution convolutional feature map (Figure 6.30b). To obtain better scale invariance, it would be preferable to operate on a range of resolutions, e.g, by computing a feature map at each image pyramid level, as shown in Figure 6.30a, but this is computationally expensive. We could, instead, simply start with the various

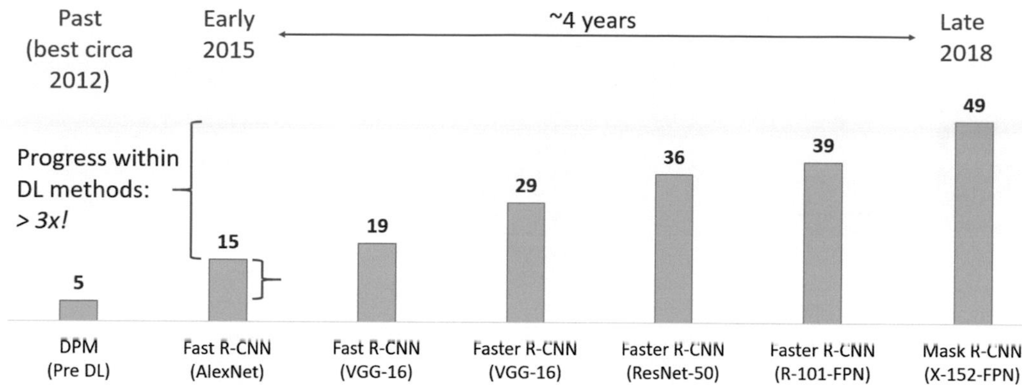

Figure 6.29 Best average precision (AP) results by year on the COCO object detection task (Lin, Maire *et al.* 2014) © 2020 Ross Girshick.

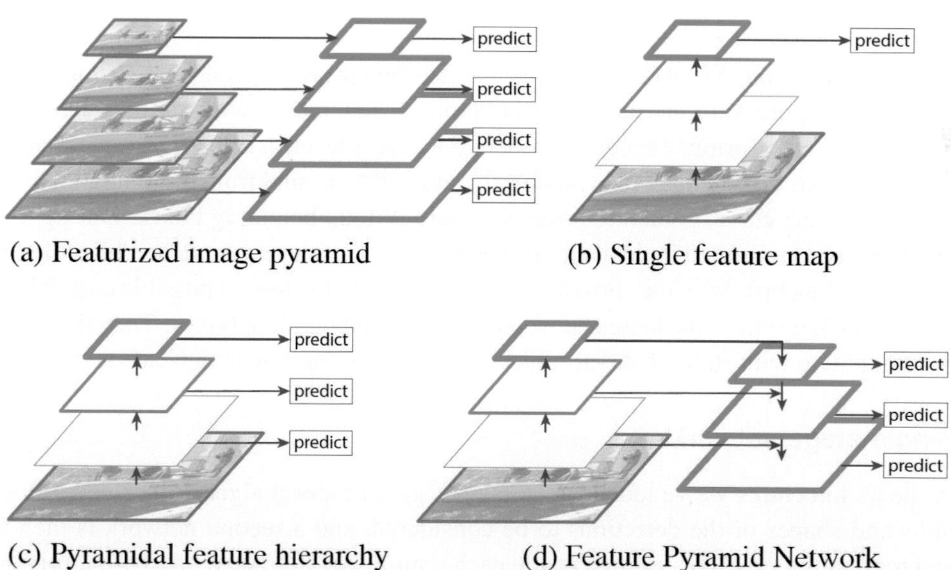

Figure 6.30 A Feature Pyramid Network and its precursors (Lin, Dollár *et al.* 2017) © 2017 IEEE: (a) deep features extracted at each level in an image pyramid; (b) a single low-resolution feature map; (c) a deep feature pyramid, with higher levels having greater abstraction; (d) a Feature Pyramid Network, with top-down context for all levels.

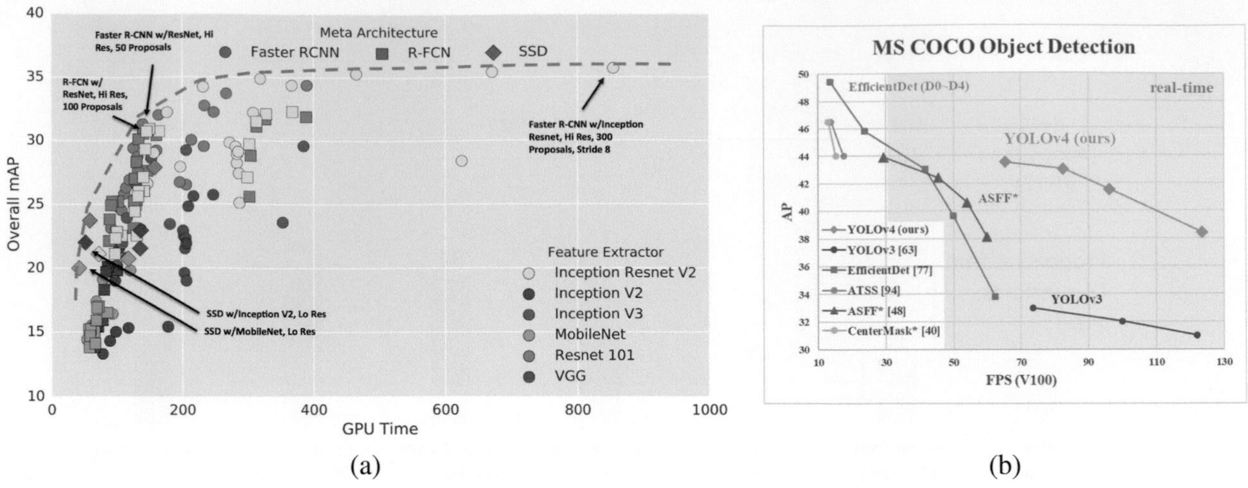

(a) (b)

Figure 6.31 Speed/accuracy trade-offs for convolutional object detectors: (a) (Huang, Rathod *et al.* 2017) © 2017 IEEE; (b) YOLOv4 © Bochkovskiy, Wang, and Liao (2020).

levels inside the convolutional network (Figure 6.30c), but these levels have different degrees of semantic abstraction, i.e., higher/smaller levels are attuned to more abstract constructs. The best solution is to construct a *Feature Pyramid Network* (FPN), as shown in Figure 6.30d, where top-down connections are used to endow higher-resolution (lower) pyramid levels with the semantics inferred at higher levels (Lin, Dollár *et al.* 2017).[20] This additional information significantly enhances the performance of object detectors (and other downstream tasks) and makes their behavior much less sensitive to object size.

DETR (Carion, Massa *et al.* 2020) uses a simpler architecture that eliminates the use of non-maximum suppression and anchor generation. Their model consists of a ResNet backbone that feeds into a transformer encoder-decoder. At a high level, it makes N bounding box predictions, some of which may include the "no object class". The ground truth bounding boxes are also padded with "no object class" bounding boxes to obtain N total bounding boxes. During training, *bipartite matching* is then used to build a one-to-one mapping from every predicted bounding box to a ground truth bounding box, with the chosen mapping leading to the lowest possible cost. The overall training loss is then the sum of the losses between the matched bounding boxes. They find that their approach is competitive with state-of-the-art object detection performance on COCO.

Single-stage networks

In the architectures we've looked at so far, a region proposal algorithm or network selects the locations and shapes of the detections to be considered, and a second network is then used to classify and regress the pixels or features inside each region. An alternative is to use a *single-stage network*, which uses a single neural network to output detections at a variety of locations. Two examples of such detectors are SSD (Single Shot MultiBox Detector) from Liu, Anguelov *et al.* (2016) and the family of YOLO (You Only Look Once) detectors described in Redmon, Divvala *et al.* (2016),Redmon and Farhadi (2017), and Redmon and Farhadi (2018). RetinaNet (Lin, Goyal *et al.* 2017) is also a single-stage detector built on top of a feature pyramid network. It uses a *focal loss* to focus the training on hard examples by downweighting the loss on well-classified samples, thus preventing

[20]It's interesting to note that the human visual system is full of such re-entrant or feedback pathways (Gilbert and Li 2013), although the extent to which cognition influences perception is still being debated (Firestone and Scholl 2016).

(a) (b) (c) (d)

Figure 6.32 Examples of image segmentation (Kirillov, He *et al.* 2019) © 2019 IEEE: (a) original image; (b) semantic segmentation (per-pixel classification); (c) instance segmentation (delineate each object); (d) panoptic segmentation (label all things and stuff).

the larger number of easy negatives from overwhelming the training. These and more recent convolutional object detectors are described in the recent survey by Jiao, Zhang *et al.* (2019). Figure 6.31 shows the speed and accuracy of detectors published up through early 2017.

The latest in the family of YOLO detectors is YOLOv4 by Bochkovskiy, Wang, and Liao (2020). In addition to outperforming other recent fast detectors such as EfficientDet (Tan, Pang, and Le 2020), as shown in Figure 6.31b, the paper breaks the processing pipeline into several stages, including a *neck*, which performs the top-down feature enhancement found in the feature pyramid network. The paper also evaluates many different components, which they categorize into a "bag of freebies" that can be used during training and a "bag of specials" that can be used at detection time with minimal additional cost.

While most bounding box object detectors continue to evaluate their results on the COCO dataset (Lin, Maire *et al.* 2014),[21] newer datasets such as Open Images (Kuznetsova, Rom *et al.* 2020), and LVIS: Large Vocabulary Instance Segmentation (Gupta, Dollár, and Girshick 2019) are now also being used (see Table 6.2). Two recent workshops that highlight the latest results using these datasets are Zendel *et al.* (2020) and Kirillov, Lin *et al.* (2020) and also have challenges related to instance segmentation, panoptic segmentation, keypoint estimation, and dense pose estimation, which are topics we discuss later in this chapter. Open-source frameworks for training and fine-tuning object detectors include the TensorFlow Object Detection API[22] and PyTorch's Detectron2.[23]

6.4 Semantic segmentation

A challenging version of general object recognition and scene understanding is to simultaneously perform recognition and accurate boundary segmentation (Fergus 2007). In this section, we examine a number of related problems, namely *semantic segmentation* (per-pixel class labeling), *instance segmentation* (accurately delineating each separate object), *panoptic segmentation* (labeling both objects and stuff), and dense pose estimation (labeling pixels belonging to people and their body parts). Figures 6.32 and 6.43 show some of these kinds of segmentations.

The basic approach to simultaneous recognition and segmentation is to formulate the problem as one of labeling every pixel in an image with its class membership. Older approaches often did this using energy minimization or Bayesian inference techniques, i.e., conditional random fields (Section 4.3.1). The TextonBoost system of Shotton, Winn *et al.* (2009) uses unary (pixel-wise) potentials based on image-specific color distributions (Section 4.3.2), location information (e.g.,

[21] See https://codalab.org for the latest competitions and leaderboards.

[22] https://github.com/tensorflow/models/tree/master/research/object_detection

[23] https://github.com/facebookresearch/detectron2

Name/URL	Extents	Contents/Reference
Object recognition		
Oxford buildings dataset	Pictures of buildings	5,062 images
https://www.robots.ox.ac.uk/~vgg/data/oxbuildings		Philbin, Chum *et al.* (2007)
INRIA Holidays	Holiday scenes	1,491 images
https://lear.inrialpes.fr/people/jegou/data.php		Jégou, Douze, and Schmid (2008)
PASCAL	Segmentations, boxes	11k images (2.9k with segmentations)
http://host.robots.ox.ac.uk/pascal/VOC		Everingham, Eslami *et al.* (2015)
ImageNet	Complete images	21k (WordNet) classes, 14M images
https://www.image-net.org		Deng, Dong *et al.* (2009)
Fashion MNIST	Complete images	70k fashion products
https://github.com/zalandoresearch/fashion-mnist		Xiao, Rasul, and Vollgraf (2017)
Object detection and segmentation		
Caltech Pedestrian Dataset	Bounding boxes	Pedestrians
http://www.vision.caltech.edu/Image_Datasets/CaltechPedestrians		Dollár, Wojek *et al.* (2009)
MSR Cambridge	Per-pixel segmentations	23 classes
https://www.microsoft.com/en-us/research/project/image-understanding		Shotton, Winn *et al.* (2009)
LabelMe dataset	Polygonal boundaries	>500 categories
http://labelme.csail.mit.edu		Russell, Torralba *et al.* (2008)
Microsoft COCO	Segmentations, boxes	330k images
https://cocodataset.org		Lin, Maire *et al.* (2014)
Cityscapes	Polygonal boundaries	30 classes, 25,000 images
https://www.cityscapes-dataset.com		Cordts, Omran *et al.* (2016)
Broden	Segmentation masks	A variety of visual concepts
http://netdissect.csail.mit.edu		Bau, Zhou *et al.* (2017)
Broden+	Segmentation masks	A variety of visual concepts
https://github.com/CSAILVision/unifiedparsing		Xiao, Liu *et al.* (2018)
LVIS	Instance segmentations	1,000 categories, 2.2M images
https://www.lvisdataset.org		Gupta, Dollár, and Girshick (2019)
Open Images	Segs., relationships	478k images, 3M relationships
https://g.co/dataset/openimages		Kuznetsova, Rom *et al.* (2020)

Table 6.2 Image databases for classification, detection, and localization.

foreground objects are more likely to be in the middle of the image, sky is likely to be higher, and road is likely to be lower), and novel texture-layout classifiers trained using shared boosting. It also uses traditional pairwise potentials that look at image color gradients. The texton-layout features first filter the image with a series of 17 oriented filter banks and then cluster the responses to classify each pixel into 30 different texton classes (Malik, Belongie *et al.* 2001). The responses are then filtered using offset rectangular regions trained with joint boosting (Viola and Jones 2004) to produce the texton-layout features used as unary potentials. Figure 6.33 shows some examples of images successfully labeled and segmented using TextonBoost

The TextonBoost conditional random field framework has been extended to LayoutCRFs by Winn and Shotton (2006), who incorporate additional constraints to recognize multiple object instances and deal with occlusions, and by Hoiem, Rother, and Winn (2007) to incorporate full 3D models. Conditional random fields continued to be widely used and extended for simultaneous recognition and segmentation applications, as described in the first edition of this book (Szeliski

Figure 6.33 Simultaneous recognition and segmentation using TextonBoost (Shotton, Winn *et al.* 2009) © 2009 Springer.

2010, Section 14.4.3), along with approaches that first performed low-level or hierarchical segmentations (Section 7.5).

The development of fully convolutional networks (Long, Shelhamer, and Darrell 2015), which we described in Section 5.4.1, enabled per-pixel semantic labeling using a single neural network. While the first networks suffered from poor resolution (very loose boundaries), the addition of conditional random fields at a final stage (Chen, Papandreou *et al.* 2018; Zheng, Jayasumana *et al.* 2015), deconvolutional upsampling (Noh, Hong, and Han 2015), and fine-level connections in U-nets (Ronneberger, Fischer, and Brox 2015), all helped improve accuracy and resolution.

Modern semantic segmentation systems are often built on architectures such as the feature pyramid network (Lin, Dollár *et al.* 2017), which have top-down connections to help percolate semantic information down to higher-resolution maps. For example, the Pyramid Scene Parsing Network (PSPNet) of Zhao, Shi *et al.* (2017) uses spatial pyramid pooling (He, Zhang *et al.* 2015) to aggregate features at various resolution levels. The Unified Perceptual Parsing network (UPerNet) of Xiao, Liu *et al.* (2018) uses both a feature pyramid network and a pyramid pooling module to label image pixels not only with object categories but also materials, parts, and textures, as shown in Figure 6.34. HRNet (Wang, Sun *et al.* 2020) keeps high-resolution versions of feature maps throughout the pipeline with occasional interchange of information between channels at different resolution layers. Such networks can also be used to estimate surface normals and depths in an image (Huang, Zhou *et al.* 2019; Wang, Geraghty *et al.* 2020).

Semantic segmentation algorithms were initially trained and tested on datasets such as MSRC (Shotton, Winn *et al.* 2009) and PASCAL VOC (Everingham, Eslami *et al.* 2015). More recent datasets include the Cityscapes dataset for urban scene understanding (Cordts, Omran *et al.* 2016) and ADE20K (Zhou, Zhao *et al.* 2019), which labels pixels in a wider variety of indoor and outdoor scenes with 150 different category and part labels. The Broadly and Densely Labeled Dataset (Broden) created by Bau, Zhou *et al.* (2017) federates a number of such densely labeled datasets, including ADE20K, Pascal-Context, Pascal-Part, OpenSurfaces, and Describable Textures to obtain

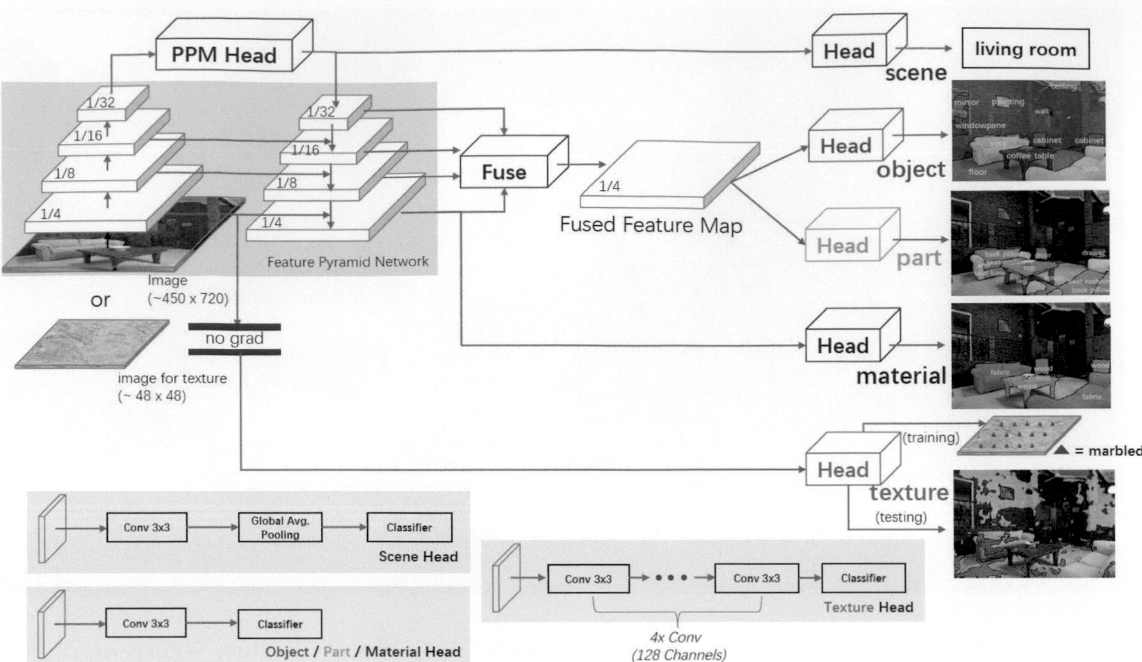

Figure 6.34 The UPerNet framework for Unified Perceptual Parsing (Xiao, Liu *et al.* 2018) © 2018 Springer. A Feature Pyramid Network (FPN) backbone is appended with a Pyramid Pooling Module (PPM) before feeding it into the top-down branch of the FPN. before feeding it into the top-down branch of the FPN. Various layers of the FPN and/or PPM are fed into different heads, including a scene head for image classification, object and part heads from the fused FPN features, a material head operating on the finest level of the FPN, and a texture head that does not participate in the FPN fine tuning. The bottom gray squares give more details into some of the heads.

a wide range of labels such as materials and textures in addition to basic object semantics. While this dataset was originally developed to aid in the interpretability of deep networks, it has also proven useful (with extensions) for training unified multi-task labeling systems such as UPerNet (Xiao, Liu *et al.* 2018). Table 6.2 list some of the datasets used for training and testing semantic segmentation algorithms.

One final note. While semantic image segmentation and labeling have widespread applications in image understanding, the converse problem of going from a semantic sketch or painting of a scene to a photorealistic image has also received widespread attention (Johnson, Gupta, and Fei-Fei 2018; Park, Liu *et al.* 2019; Bau, Strobelt *et al.* 2019; Ntavelis, Romero *et al.* 2020b). We look at this topic in more detail in Section 10.5.3 on semantic image synthesis.

6.4.1 *Application*: Medical image segmentation

One of the most promising applications of image segmentation is in the medical imaging domain, where it can be used to segment anatomical tissues for later quantitative analysis. Figure 4.21 shows a binary graph cut with directed edges being used to segment the liver tissue (light gray) from its surrounding bone (white) and muscle (dark gray) tissue. Figure 6.35 shows the segmentation of a brain scan for the detection of brain tumors. Before the development of the mature optimization and deep learning techniques used in modern image segmentation algorithms, such processing required much more laborious manual tracing of individual X-ray slices.

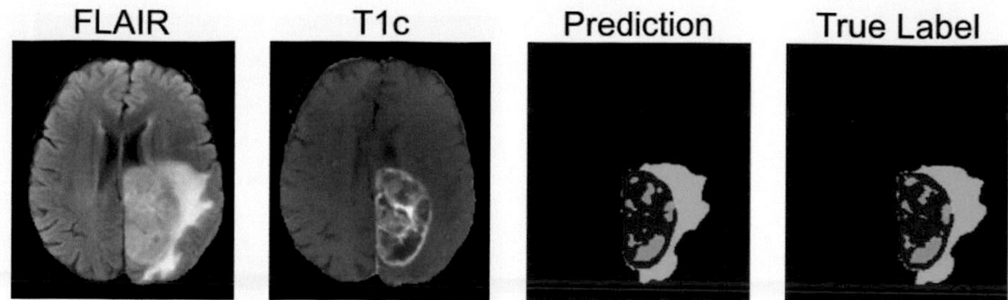

Figure 6.35 3D volumetric medical image segmentation using a deep network (Kamnitsas, Ferrante *et al.* 2016) © 2016 Springer.

Initially, optimization techniques such as Markov random fields (Section 4.3.2) and discriminative classifiers such as random forests (Section 5.1.5) were used for medical image segmentation (Criminisi, Robertson *et al.* 2013). More recently, the field has shifted to deep learning approaches (Kamnitsas, Ferrante *et al.* 2016; Kamnitsas, Ledig *et al.* 2017; Havaei, Davy *et al.* 2017).

The fields of medical image segmentation (McInerney and Terzopoulos 1996) and medical image registration (Kybic and Unser 2003) (Section 9.2.3) are rich research fields with their own specialized conferences, such as *Medical Imaging Computing and Computer Assisted Intervention (MICCAI)*, and journals, such as *Medical Image Analysis* and *IEEE Transactions on Medical Imaging*. These can be great sources of references and ideas for research in this area.

6.4.2 Instance segmentation

Instance segmentation is the task of finding all of the relevant objects in an image and producing pixel-accurate masks for their visible regions (Figure 6.36b). One potential approach to this task is to perform known object instance recognition (Section 6.1) and to then backproject the object model into the scene (Lowe 2004), as shown in Figure 6.1d, or matching portions of the new scene to pre-learned (segmented) object models (Ferrari, Tuytelaars, and Van Gool 2006b; Kannala, Rahtu *et al.* 2008). However, this approach only works for known rigid 3D models.

For more complex (flexible) object models, such as those for humans, a different approach is to pre-segment the image into larger or smaller pieces (Section 7.5) and to then match such pieces to portions of the model (Mori, Ren *et al.* 2004; Mori 2005; He, Zemel, and Ray 2006; Gu, Lim *et al.* 2009). For general highly variable classes, a related approach is to vote for potential object locations and scales based on feature correspondences and to then infer the object extents (Leibe, Leonardis, and Schiele 2008).

With the advent of deep learning, researchers started combining region proposals or image pre-segmentations with convolutional second stages to infer the final instance segmentations (Hariharan, Arbeláez *et al.* 2014; Hariharan, Arbeláez *et al.* 2015; Dai, He, and Sun 2015; Pinheiro, Lin *et al.* 2016; Dai, He, and Sun 2016; Li, Qi *et al.* 2017).

A breakthrough in instance segmentation came with the introduction of Mask R-CNN (He, Gkioxari *et al.* 2017). As shown in Figure 6.36a, Mask R-CNN uses the same region proposal network as Faster R-CNN (Ren, He *et al.* 2015), but then adds an additional *branch* for predicting the object *mask*, in addition to the existing branch for bounding box refinement and classification.[24] As with other networks that have multiple branches (or heads) and outputs, the training losses corre-

[24]Mask R-CNN was the first paper to introduce the terms *backbone* and *head* to describe the common deep convolutional feature extraction front end and the specialized back end branches.

Figure 6.36 Instance segmentation using Mask R-CNN (He, Gkioxari *et al.* 2017) © 2017 IEEE: (a) system architecture, with an additional segmentation branch; (b) sample results.

Figure 6.37 Person keypoint detection and segmentation using Mask R-CNN (He, Gkioxari *et al.* 2017) © 2017 IEEE

sponding to each supervised output need to be carefully balanced. It is also possible to add additional branches, e.g., branches trained to detect human keypoint locations (implemented as per-keypoint mask images), as shown in Figure 6.37.

Since its introduction, the performance of Mask R-CNN and its extensions has continued to improve with advances in backbone architectures (Liu, Qi *et al.* 2018; Chen, Pang *et al.* 2019). Two recent workshops that highlight the latest results in this area are the COCO + LVIS Joint Recognition Challenge (Kirillov, Lin *et al.* 2020) and the Robust Vision Challenge (Zendel *et al.* 2020).[25] It is also possible to replace the pixel masks produced by most instance segmentation techniques with time-evolving closed contours, i.e., "snakes" (Section 7.3.1), as in Peng, Jiang *et al.* (2020). In order to encourage higher-quality segmentation boundaries, Cheng, Girshick *et al.* (2021) propose a new Boundary Intersection-over-Union (Boundary IoU) metric to replace the commonly used Mask IoU metric.

6.4.3 Panoptic segmentation

As we have seen, semantic segmentation classifies each pixel in an image into its semantic category, i.e., what *stuff* does each pixel correspond to. Instance segmentation associates pixels with individ-

[25]You can find the leaderboards for instance segmentation and other COCO recognition tasks at https://cocodataset.org.

Figure 6.38 Panoptic segmentation results produced using a Panoptic Feature Pyramid Network (Kirillov, Girshick *et al.* 2019) © 2019 IEEE.

Figure 6.39 Detectron2 panoptic segmentation results on some of my personal photos. (Click on the "Colab Notebook" link at https://github.com/facebookresearch/detectron2 and then edit the input image URL to try your own.)

ual objects, i.e., how many *objects* are there and what are their extents (Figure 6.32). Putting both of these systems together has long been a goal of semantic scene understanding (Yao, Fidler, and Urtasun 2012; Tighe and Lazebnik 2013; Tu, Chen *et al.* 2005). Doing this on a per-pixel level results in a *panoptic segmentation* of the scene, where all of the objects are correctly segmented and the remaining stuff is correctly labeled (Kirillov, He *et al.* 2019). Producing a sensible *panoptic quality* (PQ) metric that simultaneously balances the accuracy on both tasks takes some careful design. In their paper, Kirillov, He *et al.* (2019) describe their proposed metric and analyze the performance of both humans (in terms of consistency) and recent algorithms on three different datasets.

The COCO dataset has now been extended to include a panoptic segmentation task, on which some recent results can be found in the ECCV 2020 workshop on this topic (Kirillov, Lin *et al.* 2020). Figure 6.38 show some segmentations produced by the panoptic feature pyramid network described by Kirillov, Girshick *et al.* (2019), which adds two branches for instance segmentation and semantic segmentation to a feature pyramid network.

(a) (b) (c) (d)

Figure 6.40 Scene completion using millions of photographs (Hays and Efros 2007) © 2007 ACM: (a) original image; (b) after unwanted foreground removal; (c) plausible scene matches, with the one the user selected highlighted in red; (d) output image after replacement and blending.

(a) (b) (c) (d) (e)

Figure 6.41 Automatic photo pop-up (Hoiem, Efros, and Hebert 2005a) © 2005 ACM: (a) input image; (b) superpixels are grouped into (c) multiple regions; (d) labels indicating ground (green), vertical (red), and sky (blue); (e) novel view of resulting piecewise-planar 3D model.

6.4.4 *Application*: Intelligent photo editing

Advances in object recognition and scene understanding have greatly increased the power of intelligent (semi-automated) photo editing applications. One example is the Photo Clip Art system of Lalonde, Hoiem *et al.* (2007), which recognizes and segments objects of interest, such as pedestrians, in internet photo collections and then allows users to paste them into their own photos. Another is the scene completion system of Hays and Efros (2007), which tackles the same *inpainting* problem we will study in Section 10.5. Given an image in which we wish to erase and fill in a large section (Figure 6.40a–b), where do you get the pixels to fill in the gaps in the edited image? Traditional approaches either use smooth continuation (Bertalmio, Sapiro *et al.* 2000) or borrow pixels from other parts of the image (Efros and Leung 1999; Criminisi, Pérez, and Toyama 2004; Efros and Freeman 2001). With the availability of huge numbers of images on the web, it often makes more sense to find a *different* image to serve as the source of the missing pixels.

In their system, Hays and Efros (2007) compute the *gist* of each image (Oliva and Torralba 2001; Torralba, Murphy *et al.* 2003) to find images with similar colors and composition. They then run a graph cut algorithm that minimizes image gradient differences and composite the new replacement piece into the original image using Poisson image blending (Section 8.4.4) (Pérez, Gangnet, and Blake 2003). Figure 6.40d shows the resulting image with the erased foreground rooftops region replaced with sailboats. Additional examples of photo editing and computational photography applications enabled by what has been dubbed "internet computer vision" can be found in the special journal issue edited by Avidan, Baker, and Shan (2010).

A different application of image recognition and segmentation is to infer 3D structure from a single photo by recognizing certain scene structures. For example, Criminisi, Reid, and Zisserman

Figure 6.42 OpenPose real-time multi-person 2D pose estimation (Cao, Simon *et al.* 2017) © 2017 IEEE.

(2000) detect vanishing points and have the user draw basic structures, such as walls, to infer the 3D geometry (Section 11.1.2). Hoiem, Efros, and Hebert (2005a), on the other hand, work with more "organic" scenes such as the one shown in Figure 6.41. Their system uses a variety of classifiers and statistics learned from labeled images to classify each pixel as either ground, vertical, or sky (Figure 6.41d). To do this, they begin by computing superpixels (Figure 6.41b) and then group them into plausible regions that are likely to share similar geometric labels (Figure 6.41c). After all the pixels have been labeled, the boundaries between the vertical and ground pixels can be used to infer 3D lines along which the image can be folded into a "pop-up" (after removing the sky pixels), as shown in Figure 6.41e. In related work, Saxena, Sun, and Ng (2009) develop a system that directly infers the depth and orientation of each pixel instead of using just three geometric class labels. We will examine techniques to infer depth from single images in more detail in Section 12.8.

6.4.5 Pose estimation

The inference of human pose (head, body, and limb locations and attitude) from a single images can be viewed as yet another kind of segmentation task. We have already discussed some pose estimation techniques in Section 6.3.2 on pedestrian detection section, as shown in Figure 6.25. Starting with the seminal work by Felzenszwalb and Huttenlocher (2005), 2D and 3D pose detection and estimation rapidly developed as an active research area, with important advances and datasets (Sigal and Black 2006a; Rogez, Rihan *et al.* 2008; Andriluka, Roth, and Schiele 2009; Bourdev and Malik 2009; Johnson and Everingham 2011; Yang and Ramanan 2011; Pishchulin, Andriluka *et al.* 2013; Sapp and Taskar 2013; Andriluka, Pishchulin *et al.* 2014).

More recently, deep networks have become the preferred technique to identify human body keypoints in order to convert these into pose estimates (Tompson, Jain *et al.* 2014; Toshev and

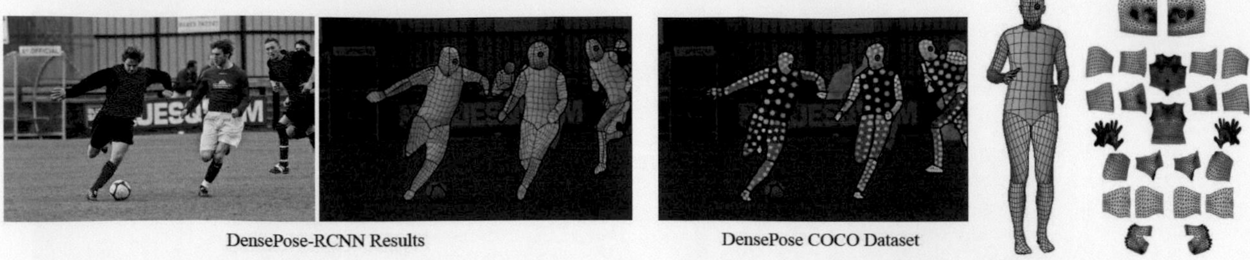

Figure 6.43 Dense pose estimation aims at mapping all human pixels of an RGB image to the 3D surface of the human body (Güler, Neverova, and Kokkinos 2018) © 2018 IEEE. The paper describes DensePose-COCO, a large-scale ground-truth dataset containing manually annotated image-to-surface correspondences for 50K persons and a DensePose-RCNN trained to densely regress UV coordinates at multiple frames per second.

Szegedy 2014; Pishchulin, Insafutdinov *et al.* 2016; Wei, Ramakrishna *et al.* 2016; Cao, Simon *et al.* 2017; He, Gkioxari *et al.* 2017; Hidalgo, Raaj *et al.* 2019; Huang, Zhu *et al.* 2020).[26] Figure 6.42 shows some of the impressive real-time multi-person 2D pose estimation results produced by the OpenPose system (Cao, Hidalgo *et al.* 2019).

The latest, most challenging, task in human pose estimation is the DensePose task introduced by Güler, Neverova, and Kokkinos (2018), where the task is to associate each pixel in RGB images of people with 3D points on a surface-based model, as shown in Figure 6.43. The authors provide dense annotations for 50,000 people appearing in COCO images and evaluate a number of correspondence networks, including their own DensePose-RCNN with several extensions. A more in-depth discussion on 3D human body modeling and tracking can be found in Section 13.6.4.

6.5 Video understanding

As we've seen in the previous sections of this chapter, image understanding mostly concerns itself with naming and delineating the objects and stuff in an image, although the relationships between objects and people are also sometimes inferred (Yao and Fei-Fei 2012; Gupta and Malik 2015; Yatskar, Zettlemoyer, and Farhadi 2016; Gkioxari, Girshick *et al.* 2018). (We will look at the topic of describing complete images in the next section on vision and language.)

What, then, is video understanding? For many researchers, it starts with the detection and description of human actions, which are taken as the basic atomic units of videos. Of course, just as with images, these basic primitives can be chained into more complete descriptions of longer video sequences.

Human activity recognition began being studied in the 1990s, along with related topics such as human motion tracking, which we discuss in Sections 9.4.4 and 13.6.4. Aggarwal and Cai (1999) provide a comprehensive review of these two areas, which they call *human motion analysis*. Some of the techniques they survey use point and mesh tracking, as well as spatio-temporal signatures.

In the 2000s, attention shifted to spatio-temporal features, such as the clever use of optical flow in small patches to recognize sports activities (Efros, Berg *et al.* 2003) or spatio-temporal feature detectors for classifying actions in movies (Laptev, Marszalek *et al.* 2008), later combined with image context (Marszalek, Laptev, and Schmid 2009) and tracked feature trajectories (Wang and Schmid 2013). Poppe (2010), Aggarwal and Ryoo (2011), and Weinland, Ronfard, and Boyer (2011)

[26] You can find the leaderboards for human keypoint detection at https://cocodataset.org.

Name/URL	Metadata	Contents/Reference
Charades	Actions, objects, descriptions	9.8k videos
https://prior.allenai.org/projects/charades		Sigurdsson, Varol *et al.* (2016)
YouTube8M	Entities	4.8k visual entities, 8M videos
https://research.google.com/youtube8m		Abu-El-Haija, Kothari *et al.* (2016)
Kinetics	Action classes	700 action classes, 650k videos
https://deepmind.com/research/open-source/kinetics		Carreira and Zisserman (2017)
"Something-something"	Actions with objects	174 actions, 220k videos
https://20bn.com/datasets/something-something		Goyal, Kahou *et al.* (2017)
AVA	Actions	80 actions in 430 15-minute videos
https://research.google.com/ava		Gu, Sun *et al.* (2018)
EPIC-KITCHENS	Actions and objects	100 hours of egocentric videos
https://epic-kitchens.github.io		Damen, Doughty *et al.* (2018)

Table 6.3 Datasets for video understanding and action recognition.

provide surveys of algorithms from this decade. Some of the datasets used in this research include the KTH human motion dataset (Schüldt, Laptev, and Caputo 2004), the UCF sports action dataset (Rodriguez, Ahmed, and Shah 2008), the Hollywood human action dataset (Marszalek, Laptev, and Schmid 2009), UCF-101 (Soomro, Zamir, and Shah 2012), and the HMDB human motion database (Kuehne, Jhuang *et al.* 2011).

In the last decade, video understanding techniques have shifted to using deep networks (Ji, Xu *et al.* 2013; Karpathy, Toderici *et al.* 2014; Simonyan and Zisserman 2014a; Tran, Bourdev *et al.* 2015; Feichtenhofer, Pinz, and Zisserman 2016; Carreira and Zisserman 2017; Varol, Laptev, and Schmid 2017; Wang, Xiong *et al.* 2019; Zhu, Li *et al.* 2020), sometimes combined with temporal models such as LSTMs (Baccouche, Mamalet *et al.* 2011; Donahue, Hendricks *et al.* 2015; Ng, Hausknecht *et al.* 2015; Srivastava, Mansimov, and Salakhudinov 2015).

While it is possible to apply these networks directly to the pixels in the video stream, e.g., using 3D convolutions (Section 5.5.1), researchers have also investigated using optical flow (Chapter 9.3) as an additional input. The resulting *two-stream architecture* was proposed by Simonyan and Zisserman (2014a) and is shown in Figure 6.44a. A later paper by Carreira and Zisserman (2017) compares this architecture to alternatives such as 3D convolutions on the pixel stream as well as hybrids of two streams and 3D convolutions (Figure 6.44b).

The latest architectures for video understanding have gone back to using 3D convolutions on the raw pixel stream (Tran, Wang *et al.* 2018, 2019; Kumawat, Verma *et al.* 2021). Wu, Feichtenhofer *et al.* (2019) store 3D CNN features into what they call a *long-term feature bank* to give a broader temporal context for action recognition. Feichtenhofer, Fan *et al.* (2019) propose a two-stream SlowFast architecture, where a slow pathway operates at a lower frame rate and is combined with features from a fast pathway with higher temporal sampling but fewer channels (Figure 6.44c). Some widely used datasets used for evaluating these algorithms are summarized in Table 6.3. They include Charades (Sigurdsson, Varol *et al.* 2016), YouTube8M (Abu-El-Haija, Kothari *et al.* 2016), Kinetics (Carreira and Zisserman 2017), "Something-something" (Goyal, Kahou *et al.* 2017), AVA (Gu, Sun *et al.* 2018), EPIC-KITCHENS (Damen, Doughty *et al.* 2018), and AVA-Kinetics (Li, Thotakuri *et al.* 2020). A nice exposition of these and other video understanding algorithms can be found in Johnson (2020, Lecture 18).

As with image recognition, researchers have also started using self-supervised algorithms to train video understanding systems. Unlike images, video clips are usually *multi-modal*, i.e., they contain

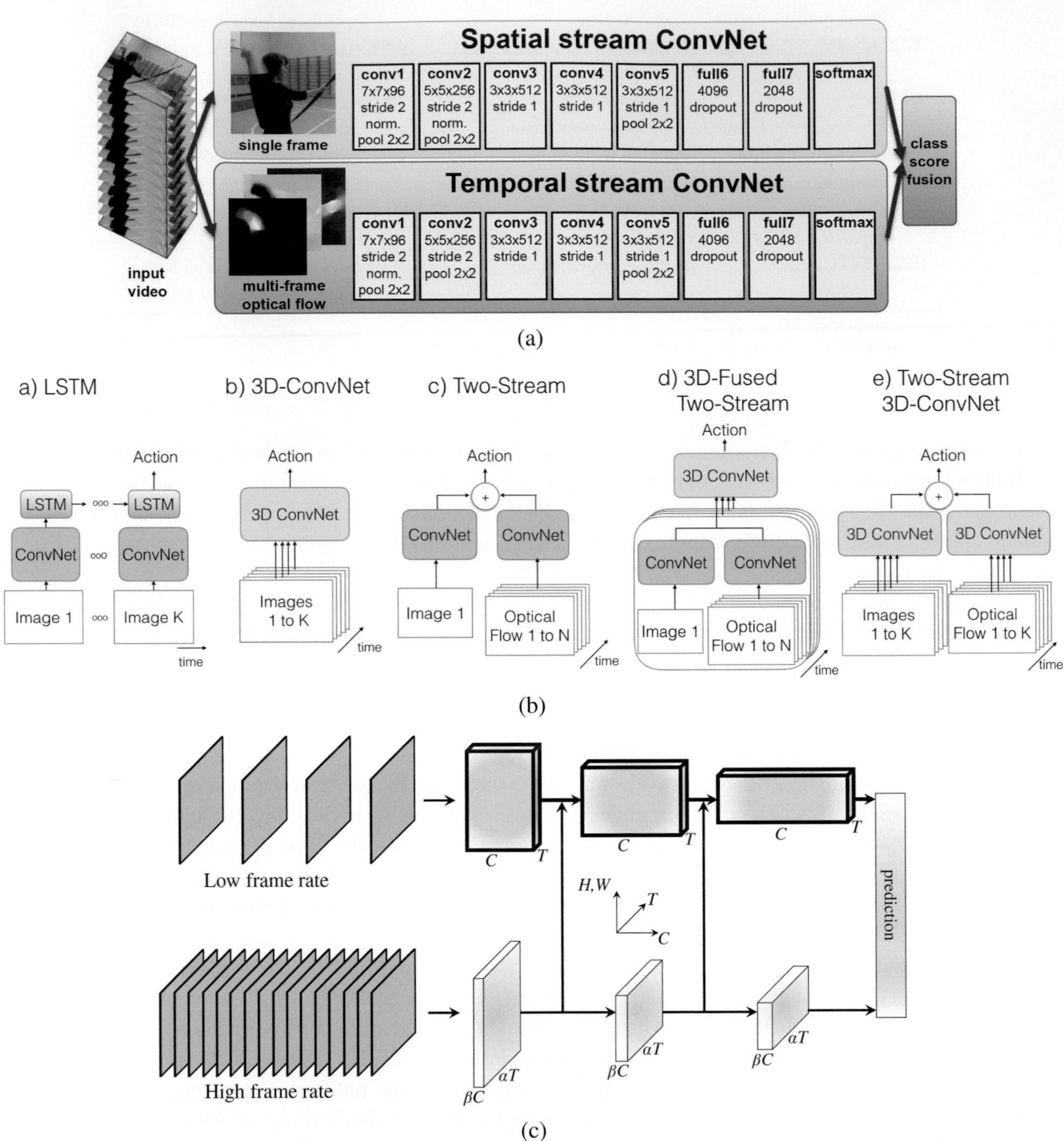

Figure 6.44 Video understanding using neural networks: (a) two-stream architecture for video classification © Simonyan and Zisserman (2014a); (b) some alternative video processing architectures (Carreira and Zisserman 2017) © 2017 IEEE; (c) a SlowFast network with a low frame rate, low temporal resolution Slow pathway and a high frame rate, higher temporal resolution Fast pathway (Feichtenhofer, Fan *et al.* 2019) © 2019 IEEE.

audio tracks in addition to the pixels, which can be an excellent source of unlabeled supervisory signals (Alwassel, Mahajan *et al.* 2020; Patrick, Asano *et al.* 2020). When available at inference time, audio signals can improve the accuracy of such systems (Xiao, Lee *et al.* 2020).

Finally, while action recognition is the main focus of most recent video understanding work, it is also possible to classify videos into different scene categories such as "beach", "fireworks", or "snowing." This problem is called *dynamic scene recognition* and can be addressed using spatio-temporal CNNs (Feichtenhofer, Pinz, and Wildes 2017).

6.6 Vision and language

The ultimate goal of much of computer vision research is not just to solve simpler tasks such as building 3D models of the world or finding relevant images, but to become an essential component of *artificial general intelligence* (AGI). This requires vision to integrate with other components of artificial intelligence such as speech and language understanding and synthesis, logical inference, and commonsense and specialized knowledge representation and reasoning.

Advances in speech and language processing have enabled the widespread deployment of speech-based intelligent virtual assistants such as Siri, Google Assistant, and Alexa. Earlier in this chapter, we've seen how computer vision systems can name individual objects in images and find similar images by appearance or keywords. The next natural step of integration with other AI components is to merge vision and language, i.e., *natural language processing* (NLP).

While this area has been studied for a long time (Duygulu, Barnard *et al.* 2002; Farhadi, Hejrati *et al.* 2010), the last decade has seen a rapid increase in performance and capabilities (Mogadala, Kalimuthu, and Klakow 2021; Gan, Yu *et al.* 2020). An example of this is the BabyTalk system developed by Kulkarni, Premraj *et al.* (2013), which first detects objects, their attributes, and their positional relationships, then infers a likely compatible labeling of these objects, and finally generates an image caption, as shown in Figure 6.45a.

Visual captioning

The next few years brought a veritable explosion of papers on the topic of image captioning and description, including (Chen and Lawrence Zitnick 2015; Donahue, Hendricks *et al.* 2015; Fang, Gupta *et al.* 2015; Karpathy and Fei-Fei 2015; Vinyals, Toshev *et al.* 2015; Xu, Ba *et al.* 2015; Johnson, Karpathy, and Fei-Fei 2016; Yang, He *et al.* 2016; You, Jin *et al.* 2016). Many of these systems combine CNN-based image understanding components (mostly object and human action detectors) with RNNs or LSTMs to generate the description, often in conjunction with other techniques such as multiple instance learning, maximum entropy language models, and visual attention. One somewhat surprising early result was that nearest-neighbor techniques, i.e., finding sets of similar looking images with captions and then creating a consensus caption, work surprisingly well (Devlin, Gupta *et al.* 2015).

Over the last few years, attention-based systems have continued to be essential components of image captioning systems (Lu, Xiong *et al.* 2017; Anderson, He *et al.* 2018; Lu, Yang *et al.* 2018). Figure 6.46 shows examples from two such papers, where each word in the generated caption is *grounded* with a corresponding image region. The CVPR 2020 tutorial by (Zhou 2020) summarizes over two dozen related papers from the last five years, including papers that use transformers (Section 5.5.3) to do the captioning. It also covers video description and dense video captioning (Aafaq, Mian *et al.* 2019; Zhou, Kalantidis *et al.* 2019) and vision-language pre-training (Sun, Myers *et al.* 2019; Zhou, Palangi *et al.* 2020; Li, Yin *et al.* 2020). The tutorial also has lectures on

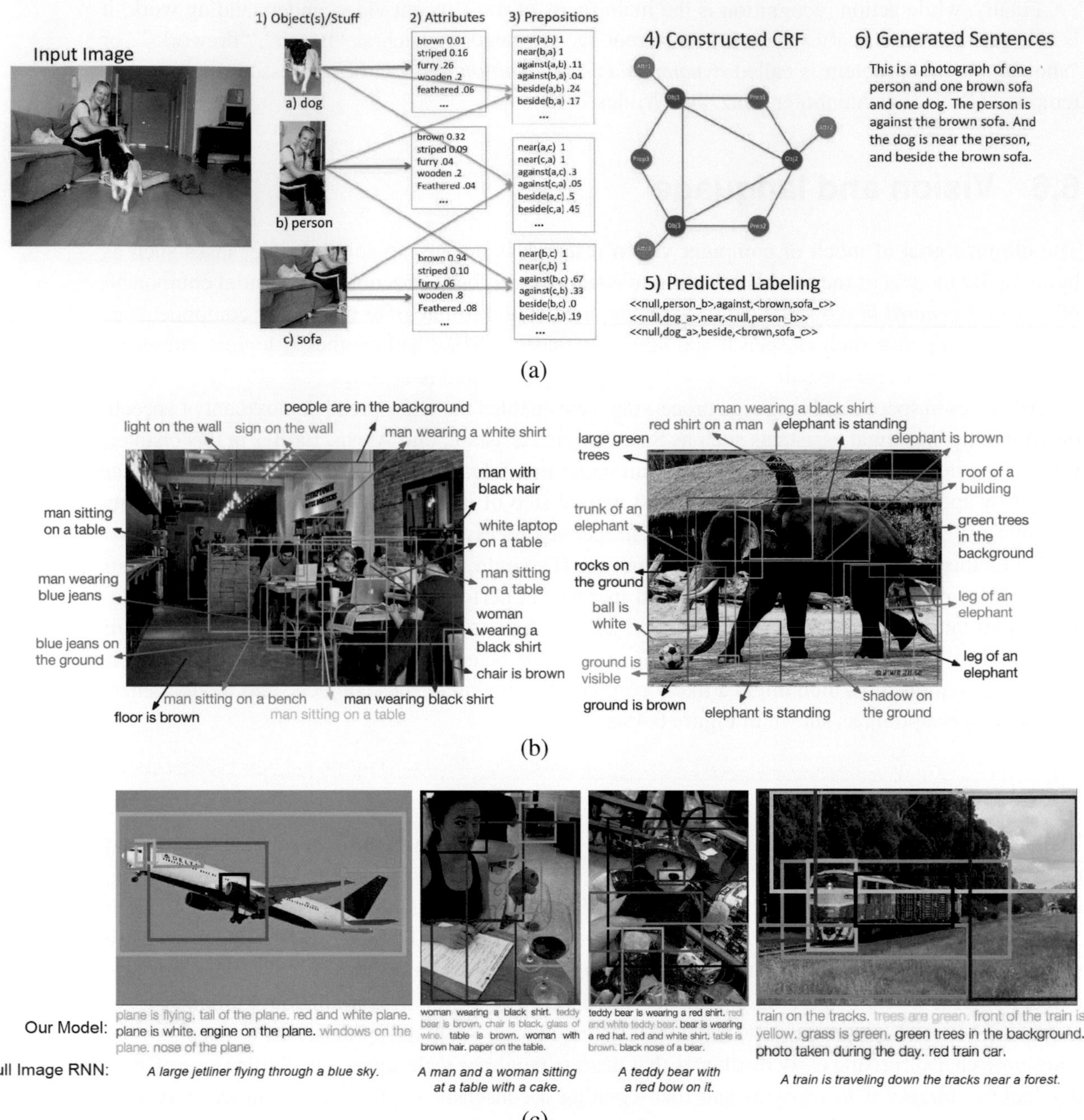

Figure 6.45 Image captioning systems: (a) BabyTalk detects objects, attributes, and positional relationships and composes these into image captions (Kulkarni, Premraj *et al.* 2013) © 2013 IEEE; (b–c) DenseCap associates word phrases with regions and then uses an RNN to construct plausible sentences (Johnson, Karpathy, and Fei-Fei 2016) © 2016 IEEE.

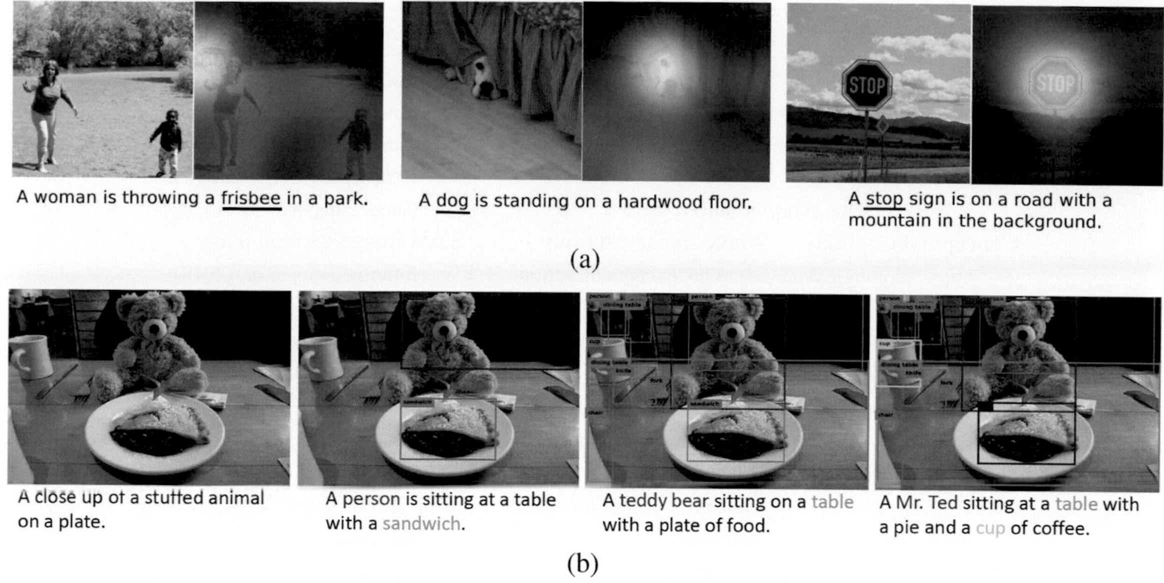

Figure 6.46 Image captioning with attention: (a) The "Show, Attend, and Tell" system, which uses hard attention to align generated words with image regions © Xu, Ba *et al.* (2015); (b) Neural Baby Talk captions generated using different detectors, showing the association between words and grounding regions (Lu, Yang *et al.* 2018) © 2018 IEEE.

	NO LABEL	LABELED "IPOD"	LABELED "LIBRARY"	LABELED "PIZZA"
Granny Smith	85.61%	0.13%	1.14%	0.89%
iPod	0.42%	99.68%	0.08%	0%
library	0%	0%	90.53%	0%
pizza	0%	0%	0%	65.35%
rifle	0%	0%	0%	0%
toaster	0%	0%	0%	0%

Figure 6.47 An adversarial typographic attack used against CLIP (Radford, Kim *et al.* 2021) discovered by ©Goh, Cammarata *et al.* (2021). Instead of predicting the object that exists in the scene, CLIP predicts the output based on the adversarial handwritten label.

visual question answering and reasoning (Gan 2020), text-to-image synthesis (Cheng 2020), and vision-language pre-training (Yu, Chen, and Li 2020).

For the task of image classification (Section 6.2), one of the major restrictions is that a model can only predict a label from the discrete pre-defined set of labels it trained on. CLIP (Radford, Kim *et al.* 2021) proposes an alternative approach that relies on image captions to enable zero-shot transfer to any possible set of labels. Given an image with a set of labels (e.g., $\{$dog, cat, $\ldots$, house$\}$), CLIP predicts the label that maximizes the probability that the image is captioned with a prompt similar to "A photo of a $\{$label$\}$". Section 5.4.7 discusses the training aspect of CLIP, which collects 400 million text-image pairs and uses contrastive learning to determine how likely it is for an image to be paired with a caption.

Remarkably, without having seen or fine-tuned to many popular image classification benchmarks (e.g., ImageNet, Caltech 101), CLIP can outperform independently fine-tuned ResNet-50 models supervised on each specific dataset. Moreover, compared to state-of-the-art classification models, CLIP's zero-shot generalization is significantly more robust to dataset distribution shifts, performing well on each of ImageNet Sketch (Wang, Ge *et al.* 2019), ImageNetV2 (Recht, Roelofs *et al.* 2019), and ImageNet-R (Hendrycks, Basart *et al.* 2020), without being specifically trained on any of them.

Name/URL	Metadata	Contents/Reference
Flickr30k (Entities)	Image captions (grounded)	30k images (+ bounding boxes)
https://shannon.cs.illinois.edu/DenotationGraph		Young, Lai *et al.* (2014)
http://bryanplummer.com/Flickr30kEntities		Plummer, Wang *et al.* (2017)
COCO Captions	Whole image captions	1.5M captions, 330k images
https://cocodataset.org/#captions-2015		Chen, Fang *et al.* (2015)
Conceptual Captions	Whole image captions	3.3M image caption pairs
https://ai.google.com/research/ConceptualCaptions		Sharma, Ding *et al.* (2018)
YFCC100M	Flickr metadata	100M images with metadata
http://projects.dfki.uni-kl.de/yfcc100m		Thomee, Shamma *et al.* (2016)
Visual Genome	Dense annotations	108k images with region graphs
https://visualgenome.org		Krishna, Zhu *et al.* (2017)
VQA v2.0	Question/answer pairs	265k images
https://visualqa.org		Goyal, Khot *et al.* (2017)
VCR	Multiple choice questions	110k movie clips, 290k QAs
https://visualcommonsense.com		Zellers, Bisk *et al.* (2019)
GQA	Compositional QA	22M questions on Visual Genome
https://visualreasoning.net		Hudson and Manning (2019)
VisDial	Dialogs for chatbot	120k COCO images + dialogs
https://visualdialog.org		Das, Kottur *et al.* (2017)

Table 6.4 Image datasets for vision and language research.

In fact, Goh, Cammarata *et al.* (2021) found that CLIP units responded similarly with concepts presented in different modalities (e.g., an image of Spiderman, text of the word spider, and a drawing of Spiderman). Figure 6.47 shows the adversarial typographic attack they discovered that could fool CLIP. By simply placing a handwritten class label (e.g., iPod) on a real-world object (e.g., Apple), CLIP often predicted the class written on the label.

As with other areas of visual recognition and learning-based systems, datasets have played an important role in the development of vision and language systems. Some widely used datasets of images with captions include Conceptual Captions (Sharma, Ding *et al.* 2018), the UIUC Pascal Sentence Dataset (Farhadi, Hejrati *et al.* 2010), the SBU Captioned Photo Dataset (Ordonez, Kulkarni, and Berg 2011), Flickr30k (Young, Lai *et al.* 2014), COCO Captions (Chen, Fang *et al.* 2015), and their extensions to 50 sentences per image (Vedantam, Lawrence Zitnick, and Parikh 2015) (see Table 6.4). More densely annotated datasets such as Visual Genome (Krishna, Zhu *et al.* 2017) describe different sub-regions of an image with their own phrases, i.e., provide *dense captioning*, as shown in Figure 6.48. YFCC100M (Thomee, Shamma *et al.* 2016) contains around 100M images from Flickr, but it only includes the raw user uploaded metadata for each image, such as the title, time of upload, description, tags, and (optionally) the location of the image.

Metrics for measuring sentence similarity also play an important role in the development of image captioning and other vision and language systems. Some widely used metrics include BLEU: BiLingual Evaluation Understudy (Papineni, Roukos *et al.* 2002), ROUGE: Recall Oriented Understudy of Gisting Evaluation (Lin 2004), METEOR: Metric for Evaluation of Translation with Explicit ORdering (Banerjee and Lavie 2005), CIDEr: Consensus-based Image Description Evaluation (Vedantam, Lawrence Zitnick, and Parikh 2015), and SPICE: Semantic Propositional Image Caption Evaluation (Anderson, Fernando *et al.* 2016).[27]

[27]See https://www.cs.toronto.edu/~fidler/slides/2017/CSC2539/Kaustav_slides.pdf.

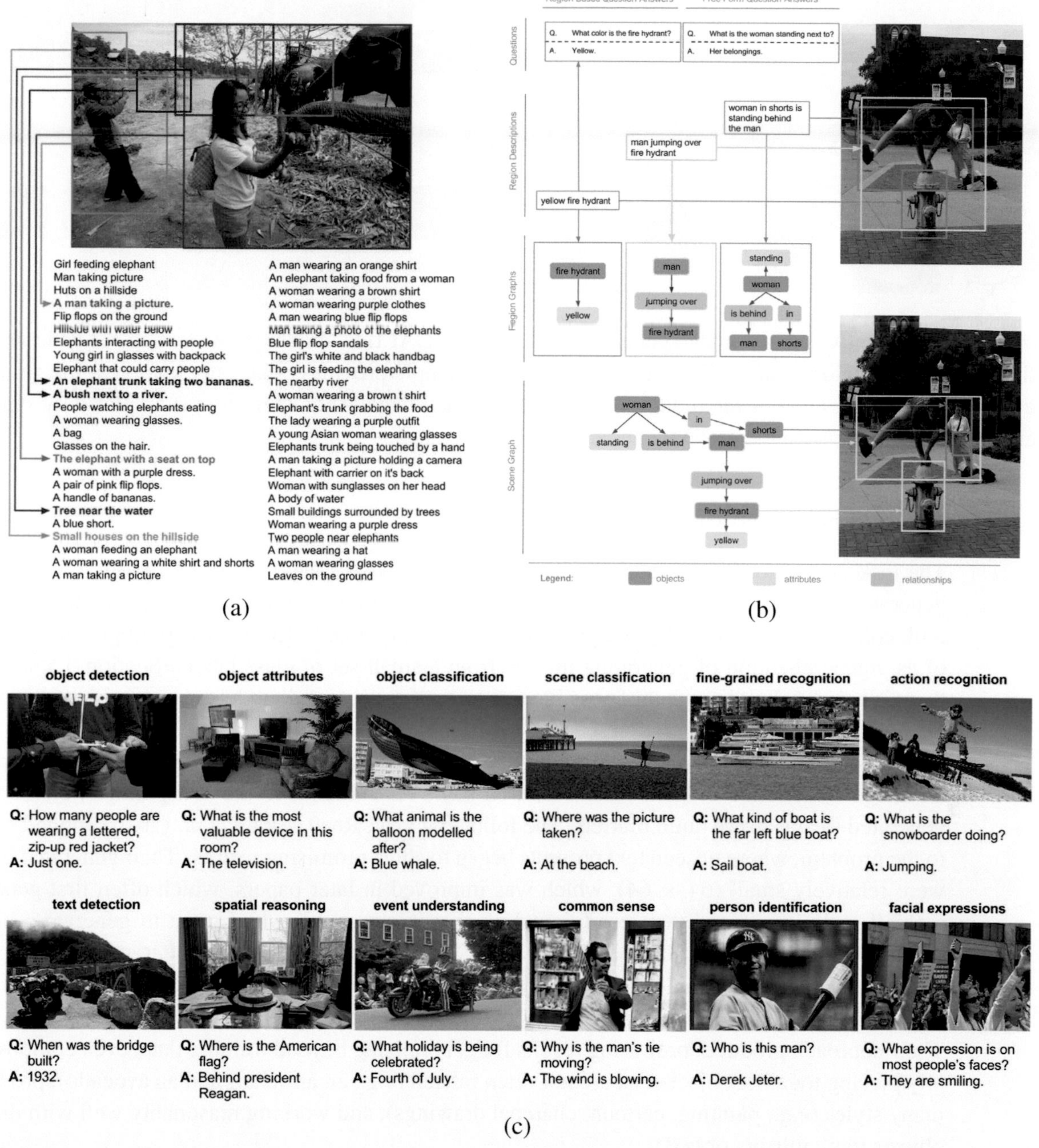

Figure 6.48 Images and data from the Visual Genome dataset (Krishna, Zhu *et al.* 2017) © 2017 Springer. (a) An example image with its region descriptors. (b) Each region has a graph representation of objects, attributes, and pairwise relationships, which are combined into a scene graph where all the objects are grounded to the image, and also associated questions and answers. (c) Some sample question and answer pairs, which cover a spectrum of visual tasks from recognition to high-level reasoning.

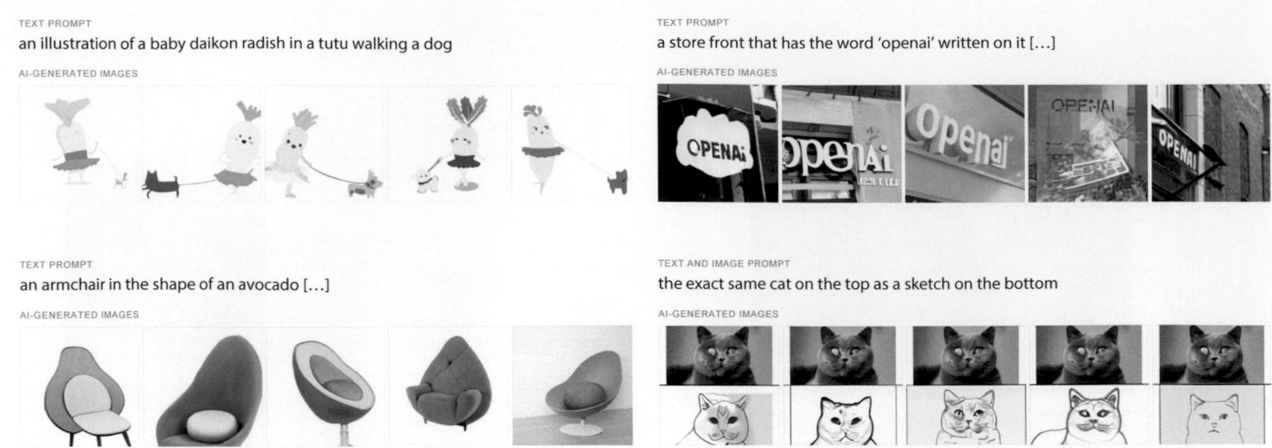

Figure 6.49 Qualitative text-to-image generation results from DALL·E, showing a wide range of generalization abilities ©Ramesh, Pavlov *et al.* (2021). The bottom right example provides a partially complete image prompt of a cat, along with text, and has the model fill in the rest of the image. The other three examples only start with the text prompt as input, with the model generating the entire image.

Text-to-image generation

The task of text-to-image generation is the inverse of visual captioning, i.e., given a text prompt, generate the image. Since images are represented in such high dimensionality, generating them to look coherent has historically been difficult. Generating images from a text prompt can be thought of as a generalization of generating images from a small set of class labels (Section 5.5.4). Since there is a near-infinite number of possible text prompts, successful models must be able to generalize from the relatively small fraction seen during training.

Early work on this task from Mansimov, Parisotto *et al.* (2016) used an RNN to iteratively draw an image from scratch. Their results showed some resemblance to the text prompts, although the generated images were quite blurred. The following year, Reed, Akata *et al.* (2016) applied a GAN to the problem, where unseen text prompts began to show promising results. Their generated images were relatively small (64×64), which was improved in later papers, which often first generated a small-scale image and then conditioned on that image and the text input to generate a higher-resolution image (Zhang, Xu *et al.* 2017, 2018; Xu, Zhang *et al.* 2018; Li, Qi *et al.* 2019).

DALL·E (Ramesh, Pavlov *et al.* 2021) uses orders of magnitude of more data (250 million text-image pairs on the internet) and compute to achieve astonishing qualitative results (Figure 6.49).[28] Their approach produces promising results for generalizing beyond training data, even compositionally piecing together objects that are not often related (e.g., an armchair and an avocado), producing many styles (e.g., painting, cartoon, charcoal drawings), and working reasonably well with difficult objects (e.g., mirrors or text).

The model for DALL·E consists of two components: a VQ-VAE-2 (Section 5.5.4) and a decoder transformer (Section 5.5.3). The text is tokenized into 256 tokens, each of which is one of 16,384 possible vectors using a BPE-encoding (Sennrich, Haddow, and Birch 2015). The VQ-VAE-2 uses a codebook of size 8,192 (significantly larger than the codebook of size 512 used in the original VQ-VAE-2 paper) to compress images as a 32×32 grid of vector tokens. At inference time, DALL·E uses a transformer decoder, which starts with the 256 text tokens to autoregressively predict the 32×32 grid of image tokens. Given such a grid, the VQ-VAE-2 is able to use its decoder to generate

[28] Play with the results at https://openai.com/blog/dall-e.

the final RGB image of size 256 $\times$ 256. To achieve better empirical results, DALL·E generates 512 image candidates and reranks them using CLIP (Radford, Kim *et al.* 2021), which determines how likely a given caption is associated with a given image.

An intriguing extension of DALL·E is to use the VQ-VAE-2 encoder to predict a subset of the compressed image tokens. For instance, suppose we are given a text input and an image. The text input can be tokenized into its 256 tokens, and one can obtain the 32 $\times$ 32 image tokens using the VQ-VAE-2 encoder. If we then discard the bottom half of the image tokens, the transformer decoder can be used to autoregressively predict which tokens might be there. These tokens, along with the non-discarded ones from the original image, can be passed into the VQ-VAE-2 decoder to produce a completed image. Figure 6.49 (bottom right) shows how such a text and partial image prompt can be used for applications such as image-to-image translation (Section 5.5.4).

Visual Question Answering and Reasoning

Image and video captioning are useful tasks that bring us closer to building artificially intelligent systems, as they demonstrate the ability to put together visual cues such as object identities, attributes, and actions. However, it remains unclear if the system has understood the scene at a deeper level and if it can reason about the constituent pieces and how they fit together.

To address these concerns, researchers have been building *visual question answering* (VQA) systems, which require the vision algorithm to answer open-ended questions about the image, such as the ones shown in Figure 6.48c. A lot of this work started with the creation of the Visual Question Answering (VQA) dataset (Antol, Agrawal *et al.* 2015), which spurred a large amount of subsequent research. The following year, VQA v2.0 improved this dataset by creating a *balanced* set of image pairs, where each question had different answers in the two images (Goyal, Khot *et al.* 2017).[29] This dataset was further extended to reduce the influence of prior assumptions and data distributions and to encourage answers to be grounded in the images (Agrawal, Batra *et al.* 2018).

Since then, many additional VQA datasets have been created. These include the VCR dataset for visual commonsense reasoning (Zellers, Bisk *et al.* 2019) and the GQA dataset and metrics for evaluating visual reasoning and compositional question answering (Hudson and Manning 2019), which is built on top of the information about objects, attributes, and relations provided through the Visual Genome scene graphs (Krishna, Zhu *et al.* 2017). A discussion of these and other datasets for VQA can be found in the CVPR 2020 tutorial by Gan (2020), including datasets that test visual grounding and referring expression comprehension, visual entailment, using external knowledge, reading text, answering sub-questions, and using logic. Some of these datasets are summarized in Table 6.4.

As with image and video captioning, VQA systems use various flavors of attention to associate pixel regions with semantic concepts (Yang, He *et al.* 2016). However, instead of using sequence models such as RNNs, LSTMs, or transformers to generate text, the natural language question is first parsed to produce an encoding that is then fused with the image embedding to generate the desired answer.

The image semantic features can either be computed on a coarse grid, or a "bottom-up" object detector can be combined with a "top-down" attention mechanism to provide feature weightings (Anderson, He *et al.* 2018). In recent years, the pendulum has swung back and forth between techniques that use bottom-up regions and gridded feature descriptors, with two of the recent best-performing algorithms going back to the simpler (and much faster) gridded approach (Jiang, Misra *et al.* 2020; Huang, Zeng *et al.* 2020). The CVPR 2020 tutorial by Gan (2020) discusses these and

[29] https://visualqa.org

dozens of other VQA systems as well as their subcomponents, such as multimodal fusion variants (bilinear pooling, alignment, relational reasoning), neural module networks, robust VQA, and multimodal pre-training, The survey by Mogadala, Kalimuthu, and Klakow (2021) and the annual VQA Challeng workshop (Shrivastava, Hudson *et al.* 2020) are also excellent sources of additional information. And if you would like to test out the current state of VQA systems, you can upload your own image to https://vqa.cloudcv.org and ask the system your own questions.

Visual Dialog. An even more challenging version of VQA is *visual dialog*, where a chatbot is given an image and asked to answer open-ended questions about the image while also referring to previous elements of the conversation. The VisDial dataset was the earliest to be widely used for this task (Das, Kottur *et al.* 2017).[30] You can find pointers to systems that have been developed for this task at the Visual Dialog workshop and challenge (Shrivastava, Hudson *et al.* 2020). There's also a chatbot at https://visualchatbot.cloudcv.org where you can upload your own image and start a conversation, which can sometimes lead to humorous (or weird) outcomes (Shane 2019).

Vision-language pre-training. As with many other recognition tasks, pre-training has had some dramatic success in the last few years, with systems such as ViLBERT (Lu, Batra *et al.* 2019), Oscar (Li, Yin *et al.* 2020), and many other systems described in the CVPR 2020 tutorial on self-supervised learning for vision-and-language (Yu, Chen, and Li 2020).

6.7 Additional reading

Unlike machine learning or deep learning, there are no recent textbooks or surveys devoted specifically to the general topics of image recognition and scene understanding. Some earlier surveys (Pinz 2005; Andreopoulos and Tsotsos 2013) and collections of papers (Ponce, Hebert *et al.* 2006; Dickinson, Leonardis *et al.* 2007) review the "classic" (pre-deep learning) approaches, but given the tremendous changes in the last decade, many of these techniques are no longer used. Currently, some of the best sources for the latest material, in addition to this chapter and university computer vision courses, are tutorials at the major vision conferences such as ICCV (Xie, Girshick *et al.* 2019), CVPR (Girshick, Kirillov *et al.* 2020), and ECCV (Xie, Girshick *et al.* 2020). Image recognition datasets such as those listed in Tables 6.1–6.4 that maintain active leaderboards can also be a good source for recent papers.

Algorithms for instance recognition, i.e., the detection of static manufactured objects that only vary slightly in appearance but may vary in 3D pose, are still often based on detecting 2D points of interest and describing them using viewpoint-invariant descriptors, as discussed in Chapter 7 and (Lowe 2004), Rothganger, Lazebnik *et al.* (2006), and Gordon and Lowe (2006). In more recent years, attention has shifted to the more challenging problem of *instance retrieval* (also known as *content-based image retrieval*), in which the number of images being searched can be very large (Sivic and Zisserman 2009). Section 7.1.4 in the next chapter reviews such techniques, as does the survey in (Zheng, Yang, and Tian 2018). This topic is also related to visual similarity search (Bell and Bala 2015; Arandjelovic, Gronat *et al.* 2016; Song, Xiang *et al.* 2016; Gordo, Almazán *et al.* 2017; Rawat and Wang 2017; Bell, Liu *et al.* 2020), which was covered in Section 6.2.3.

A number of surveys, collections of papers, and course notes have been written on the topic of feature-based whole image (single-object) category recognition (Pinz 2005; Ponce, Hebert *et al.* 2006; Dickinson, Leonardis *et al.* 2007; Fei-Fei, Fergus, and Torralba 2009). Some of these papers

[30] https://visualdialog.org

use a bag of words or keypoints (Csurka, Dance *et al.* 2004; Lazebnik, Schmid, and Ponce 2006; Csurka, Dance *et al.* 2006; Grauman and Darrell 2007b; Zhang, Marszalek *et al.* 2007; Boiman, Shechtman, and Irani 2008; Ferencz, Learned-Miller, and Malik 2008). Other papers recognize objects based on their contours, e.g., using shape contexts (Belongie, Malik, and Puzicha 2002) or other techniques (Shotton, Blake, and Cipolla 2005; Opelt, Pinz, and Zisserman 2006; Ferrari, Tuytelaars, and Van Gool 2006a).

Many object recognition algorithms use part-based decompositions to provide greater invariance to articulation and pose. Early algorithms focused on the relative positions of the parts (Fischler and Elschlager 1973; Kanade 1977; Yuille 1991) while later algorithms used more sophisticated models of appearance (Felzenszwalb and Huttenlocher 2005; Fergus, Perona, and Zisserman 2007; Felzenszwalb, McAllester, and Ramanan 2008). Good overviews on part-based models for recognition can be found in the course notes by Fergus (2009). Carneiro and Lowe (2006) discuss a number of graphical models used for part-based recognition, which include trees and stars, k-fans, and constellations.

Classical recognition algorithms often used scene context as part of their recognition strategy. Representative papers in this area include Torralba (2003), Torralba, Murphy *et al.* (2003), Rabinovich, Vedaldi *et al.* (2007), Russell, Torralba *et al.* (2007), Sudderth, Torralba *et al.* (2008), and Divvala, Hoiem *et al.* (2009). Machine learning also became a key component of classical object detection and recognition algorithms (Felzenszwalb, McAllester, and Ramanan 2008; Sivic, Russell *et al.* 2008), as did exploiting large human-labeled databases (Russell, Torralba *et al.* 2007; Torralba, Freeman, and Fergus 2008).

The breakthrough success of the "AlexNet" SuperVision system of Krizhevsky, Sutskever, and Hinton (2012) shifted the focus in category recognition research from feature-based approaches to deep neural networks. The rapid improvement in recognition accuracy, captured in Figure 5.40 and described in more detail in Section 5.4.3 has been driven to a large degree by deeper networks and better training algorithms, and also in part by larger (unlabeled) training datasets (Section 5.4.7).

More specialized recognition systems such as those for recognizing faces underwent a similar evolution. While some of the earliest approaches to face recognition involved finding the distinctive image features and measuring the distances between them (Fischler and Elschlager 1973; Kanade 1977; Yuille 1991), later approaches relied on comparing gray-level images, often projected onto lower dimensional subspaces (Turk and Pentland 1991; Belhumeur, Hespanha, and Kriegman 1997; Heisele, Ho *et al.* 2003) or local binary patterns (Ahonen, Hadid, and Pietikäinen 2006). A variety of shape and pose deformation models were also developed (Beymer 1996; Vetter and Poggio 1997), including Active Shape Models (Cootes, Cooper *et al.* 1995), 3D Morphable Models (Blanz and Vetter 1999; Egger, Smith *et al.* 2020), and Active Appearance Models (Cootes, Edwards, and Taylor 2001; Matthews and Baker 2004; Ramnath, Koterba *et al.* 2008). Additional information about classic face recognition algorithms can be found in a number of surveys and books on this topic (Chellappa, Wilson, and Sirohey 1995; Zhao, Chellappa *et al.* 2003; Li and Jain 2005).

The concept of shape models for *frontalization* continued to be used as the community shifted to deep neural network approaches (Taigman, Yang *et al.* 2014). Some more recent deep face recognizers, however, omit the frontalization stage and instead use data augmentation to create synthetic inputs with a larger variety of poses (Schroff, Kalenichenko, and Philbin 2015; Parkhi, Vedaldi, and Zisserman 2015). Masi, Wu *et al.* (2018) provide an excellent tutorial and survey on deep face recognition, including a list of widely used training and testing datasets, a discussion of frontalization and dataset augmentation, and a section on training losses.

As the problem of whole-image (single object) category recognition became more "solved", attention shifted to multiple object delineation and labeling, i.e., object detection. Object detection was originally studied in the context of specific categories such as faces, pedestrians, cars, etc. Seminal papers in face detection include those by Osuna, Freund, and Girosi (1997); Sung and Poggio (1998); Rowley, Baluja, and Kanade (1998); Viola and Jones (2004); Heisele, Ho *et al.* (2003), with Yang, Kriegman, and Ahuja (2002) providing a comprehensive survey of early work in this field. Early work in pedestrian and car detection was carried out by Gavrila and Philomin (1999); Gavrila (1999); Papageorgiou and Poggio (2000); Schneiderman and Kanade (2004). Subsequent papers include (Mikolajczyk, Schmid, and Zisserman 2004; Dalal and Triggs 2005; Leibe, Seemann, and Schiele 2005; Andriluka, Roth, and Schiele 2009; Dollár, Belongie, and Perona 2010; Felzenszwalb, Girshick *et al.* 2010).

Modern generic object detectors are typically constructed using a region proposal algorithm (Uijlings, Van De Sande *et al.* 2013; Zitnick and Dollár 2014) that then feeds selected regions of the image (either as pixels or precomputed neural features) into a multi-way classifier, resulting in architectures such as R-CNN (Girshick, Donahue *et al.* 2014), Fast R-CNN (Girshick 2015), Faster R-CCNN (Ren, He *et al.* 2015), and FPN (Lin, Dollár *et al.* 2017). An alternative to this two-stage approach is a *single-stage network*, which uses a single network to output detections at a variety of locations. Examples of such architectures include SSD (Liu, Anguelov *et al.* 2016), RetinaNet (Lin, Goyal *et al.* 2017), and YOLO (Redmon, Divvala *et al.* 2016; Redmon and Farhadi 2017, 2018; Bochkovskiy, Wang, and Liao 2020). These and more recent convolutional object detectors are described in the recent survey by Jiao, Zhang *et al.* (2019).

While object detection can be sufficient in many computer vision applications such as counting cars or pedestrians or even describing images, a detailed pixel-accurate labeling can be potentially even more useful, e.g., for photo editing. This kind of labeling comes in several flavors, including semantic segmentation (what stuff is this?), instance segmentation (which countable object is this?), panoptic segmentation (what stuff or object is it?). One early approach to this problem was to pre-segment the image into pieces and then match these pieces to portions of the model (Mori, Ren *et al.* 2004; Russell, Efros *et al.* 2006; Borenstein and Ullman 2008; Gu, Lim *et al.* 2009). Another popular approach was to use conditional random fields (Kumar and Hebert 2006; He, Zemel, and Carreira-Perpiñán 2004; Winn and Shotton 2006; Rabinovich, Vedaldi *et al.* 2007; Shotton, Winn *et al.* 2009). which at that time produced some of the best results on the PASCAL VOC segmentation challenge. Modern semantic segmentation algorithms use pyramidal fully-convolutional architectures to map input pixels to class labels (Long, Shelhamer, and Darrell 2015; Zhao, Shi *et al.* 2017; Xiao, Liu *et al.* 2018; Wang, Sun *et al.* 2020).

The more challenging task of instance segmentation, where each distinct object gets its own unique label, is usually tackled using a combination of object detectors and per-object segmentation, as exemplified in the seminal Mask R-CNN paper by He, Gkioxari *et al.* (2017). Follow-on work uses more sophisticated backbone architectures (Liu, Qi *et al.* 2018; Chen, Pang *et al.* 2019). Two recent workshops that highlight the latest results in this area are the COCO + LVIS Joint Recognition Challenge (Kirillov, Lin *et al.* 2020) and the Robust Vision Challenge (Zendel *et al.* 2020).

Putting semantic and instance segmentation together has long been a goal of semantic scene understanding (Yao, Fidler, and Urtasun 2012; Tighe and Lazebnik 2013; Tu, Chen *et al.* 2005). Doing this on a per-pixel level results in a *panoptic segmentation* of the scene, where all of the objects are correctly segmented and the remaining stuff is correctly labeled (Kirillov, He *et al.* 2019; Kirillov, Girshick *et al.* 2019). The COCO dataset has now been extended to include a panoptic segmentation task, on which some recent results can be found in the ECCV 2020 workshop on this topic (Kirillov, Lin *et al.* 2020).

Research in video understanding, or more specifically human activity recognition, dates back to the 1990s; some good surveys include (Aggarwal and Cai 1999; Poppe 2010; Aggarwal and Ryoo

2011; Weinland, Ronfard, and Boyer 2011). In the last decade, video understanding techniques shifted to using deep networks (Ji, Xu *et al.* 2013; Karpathy, Toderici *et al.* 2014; Simonyan and Zisserman 2014a; Donahue, Hendricks *et al.* 2015; Tran, Bourdev *et al.* 2015; Feichtenhofer, Pinz, and Zisserman 2016; Carreira and Zisserman 2017; Tran, Wang *et al.* 2019; Wu, Feichtenhofer *et al.* 2019; Feichtenhofer, Fan *et al.* 2019). Some widely used datasets used for evaluating these algorithms are summarized in Table 6.3.

While associating words with images has been studied for a while (Duygulu, Barnard *et al.* 2002), sustained research into describing images with captions and complete sentences started in the early 2010s (Farhadi, Hejrati *et al.* 2010; Kulkarni, Premraj *et al.* 2013). The last decade has seen a rapid increase in performance and capabilities of such systems (Mogadala, Kalimuthu, and Klakow 2021; Gan, Yu *et al.* 2020). The first sub-problem to be widely studied was image captioning (Donahue, Hendricks *et al.* 2015; Fang, Gupta *et al.* 2015; Karpathy and Fei-Fei 2015; Vinyals, Toshev *et al.* 2015; Xu, Ba *et al.* 2015; Devlin, Gupta *et al.* 2015), with later systems using attention mechanisms (Anderson, He *et al.* 2018; Lu, Yang *et al.* 2018). More recently, researchers have developed systems for visual question answering (Antol, Agrawal *et al.* 2015) and visual commonsense reasoning (Zellers, Bisk *et al.* 2019).

The CVPR 2020 tutorial on recent advances in visual captioning (Zhou 2020) summarizes over two dozen related papers from the last five years, including papers that use Transformers to do the captioning. It also covers video description and dense video captioning (Aafaq, Mian *et al.* 2019; Zhou, Kalantidis *et al.* 2019) and vision-language pre-training (Sun, Myers *et al.* 2019; Zhou, Palangi *et al.* 2020; Li, Yin *et al.* 2020). The tutorial also has lectures on visual question answering and reasoning (Gan 2020), text-to-image synthesis (Cheng 2020), and vision-language pre-training (Yu, Chen, and Li 2020).

6.8 Exercises

Ex 6.1: Pre-trained recognition networks. Find a pre-trained network for image classification, segmentation, or some other task such as face recognition or pedestrian detection.

After running the network, can you characterize the most common kinds of errors the network is making? Create a "confusion matrix" indicating which categories get classified as other categories. Now try the network on your own data, either from a web search or from your personal photo collection. Are there surprising results?

My own favorite code to try is Detectron2,[31] which I used to generate the panoptic segmentation results shown in Figure 6.39.

Ex 6.2: Re-training recognition networks. After analyzing the performance of your pre-trained network, try re-training it on the original dataset on which it was trained, but with modified parameters (numbers of layers, channels, training parameters) or with additional examples. Can you get the network to perform more to you liking?

Many of the online tutorials, such as the Detectron2 Collab notebook mentioned above, come with instructions on how to re-train the network from scratch on a different dataset. Can you create your own dataset, e.g., using a web search and figure out how to label the examples? A low effort (but not very accurate) way is to trust the results of the web search. Russakovsky, Deng *et al.* (2015), Kovashka, Russakovsky *et al.* (2016), and other papers on image datasets discuss the challenges in obtaining accurate labels.

[31] Click on the "Colab Notebook" link at https://github.com/facebookresearch/detectron2 and then edit the input image URL to try your own.

Train your network, try to optimize its architecture, and report on the challenges you faced and discoveries you made.

Note: the following exercises were suggested by Matt Deitke.

Ex 6.3: Image perturbations. Download either ImageNet or Imagenette.[32] Now, perturb each image by adding a small square to the top left of the image, where the color of the square is unique for each label, as shown in the following figure:

(a) cassette player (b) golf ball (c) English Springer

Using any image classification model,[33] e.g., ResNet, EfficientNet, or ViT, train the model from scratch on the perturbed images. Does the model overfit to the color of the square and ignore the rest of the image? When evaluating the model on the training and validation data, try adversarially swapping colors between different labels.

Ex 6.4: Image normalization. Using the same dataset downloaded for the previous exercise, take a ViT model and remove all the intermediate layer normalization operations. Are you able to train the network? Using techniques in Li, Xu *et al.* (2018), how do the plots of the loss landscape appear with and without the intermediate layer normalization operations?

Ex 6.5: Semantic segmentation. Explain the differences between instance segmentation, semantic segmentation, and panoptic segmentation. For each type of segmentation, can it be post-processed to obtain the other kinds of segmentation?

Ex 6.6: Class encoding. Categorical inputs to a neural network, such as a word or object, can be encoded with one-hot encoded vector.[34] However, it is common to pass the one-hot encoded vector through an embedding matrix, where the output is then passed into the neural network loss function. What are the advantages of vector embedding over using one-hot encoding?

Ex 6.7: Object detection. For object detection, how do the number of parameters for DETR, Faster-RCNN, and YOLOv4 compare? Try training each of them on MS COCO. Which one tends to train the slowest? How long does it take each model to evaluate a single image at inference time?

Ex 6.8: Image classification vs. description. For image classification, list at least two significant differences between using categorical labels and natural language descriptions.

Ex 6.9: ImageNet Sketch. Try taking several pre-trained models on ImageNet and evaluating them, without any fine-tuning, on ImageNet Sketch (Wang, Ge *et al.* 2019). For each of these models, to what extent does the performance drop due to the shift in distribution?

Ex 6.10: Self-supervised learning. Provide examples of self-supervised learning pretext tasks for each of the following data types: static images, videos, and vision-and-language.

[32] Imagenette, https://github.com/fastai/imagenette, is a smaller 10-class subset of ImageNet that is easier to use with limited computing resources. .

[33] You may find the PyTorch Image Models at https://github.com/rwightman/pytorch-image-models useful.

[34] With a categorical variable, one-hot encoding is used to represent which label is chosen, i.e., when a label is chosen, its entry in the vector is 1 with all other entries being 0.

Ex 6.11: Video understanding. For many video understanding tasks, we may be interested in tracking an object through time. Why might this be preferred to making predictions independently for each frame? Assume that inference speed is not a problem.

Ex 6.12: Fine-tuning a new head. Take the backbone of a network trained for object classification and fine-tune it for object detection with a variant of YOLO. Why might it be desirable to freeze the early layers of the network?

Ex 6.13: Movie understanding. Currently, most video understanding networks, such as those discussed in this chapter, tend to only deal with short video clips as input. What modifications might be necessary in order to operate over longer sequences such as an entire movie?

Chapter 7

Feature detection and matching

7.1 Points and patches 335
7.1.1 Feature detectors 337
7.1.2 Feature descriptors 347
7.1.3 Feature matching 352
7.1.4 Large-scale matching and retrieval 358
7.1.5 Feature tracking 361
7.1.6 *Application*: Performance-driven animation 363
7.2 Edges and contours 364
7.2.1 Edge detection 364
7.2.2 Contour detection 368
7.2.3 *Application*: Edge editing and enhancement 372
7.3 Contour tracking 373
7.3.1 Snakes and scissors 373
7.3.2 Level Sets 379
7.3.3 *Application*: Contour tracking and rotoscoping 380
7.4 Lines and vanishing points 381
7.4.1 Successive approximation 381
7.4.2 Hough transforms 381
7.4.3 Vanishing points 384
7.5 Segmentation 386
7.5.1 Graph-based segmentation 388
7.5.2 Mean shift 389
7.5.3 Normalized cuts 391
7.6 Additional reading 393
7.7 Exercises 395

R. Szeliski, *Computer Vision*, Texts in Computer Science,
https://doi.org/10.1007/978-3-030-34372-9_7

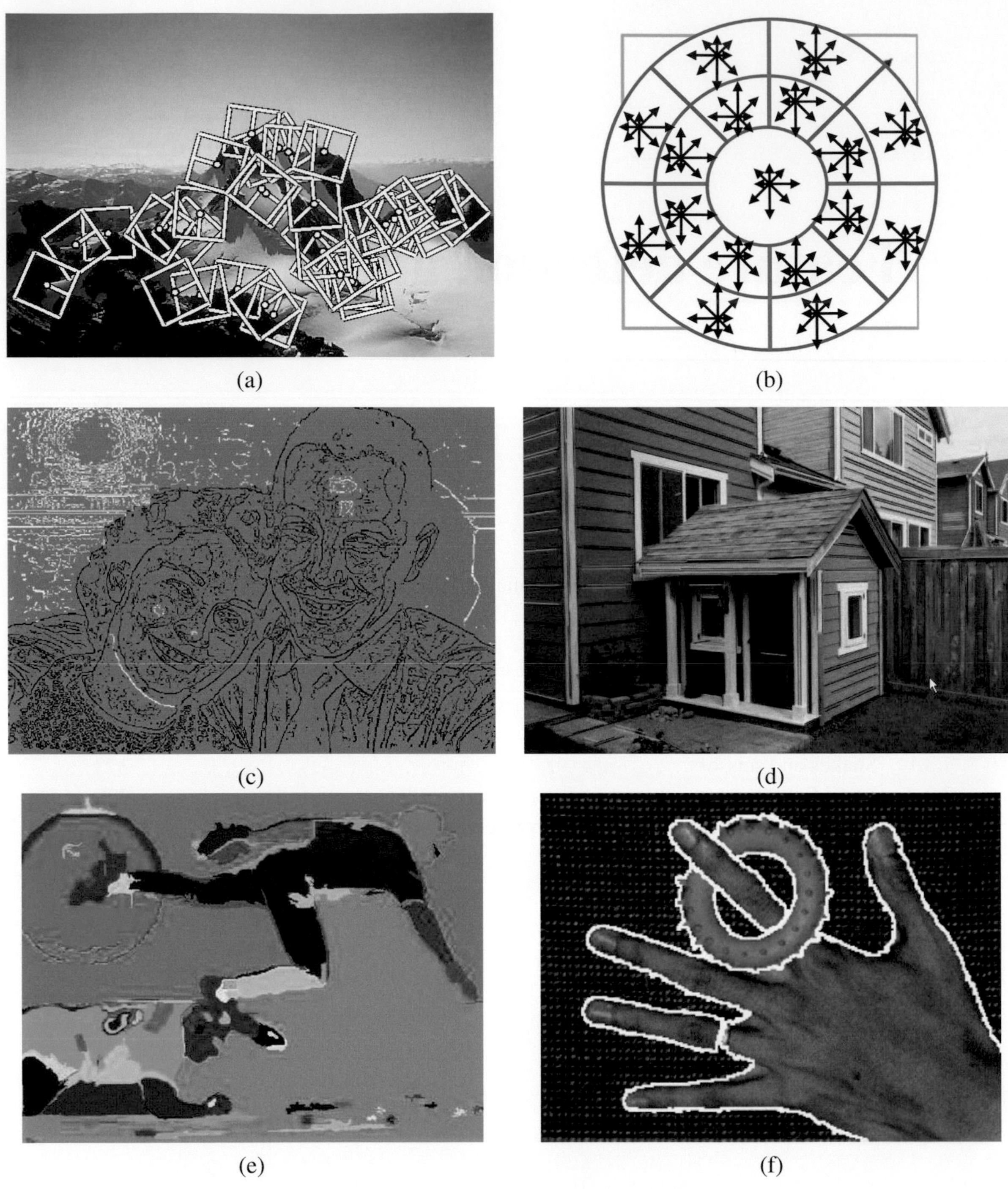

Figure 7.1 Feature detectors and descriptors can be used to analyze, describe and match images: (a) point-like interest operators (Brown, Szeliski, and Winder 2005) © 2005 IEEE; (b) GLOH descriptor (Mikolajczyk and Schmid 2005); (c) edges (Elder and Goldberg 2001) © 2001 IEEE; (d) straight lines (Sinha, Steedly *et al.* 2008) © 2008 ACM; (e) graph-based merging (Felzenszwalb and Huttenlocher 2004) © 2004 Springer; (f) mean shift (Comaniciu and Meer 2002) © 2002 IEEE.

Feature detection and matching are an essential component of many computer vision applications. Consider the two pairs of images shown in Figure 7.2. For the first pair, we may wish to *align* the two images so that they can be seamlessly stitched into a composite mosaic (Section 8.2). For the second pair, we may wish to establish a dense set of *correspondences* so that a 3D model can be constructed or an in-between view can be generated (Chapter 12). In either case, what kinds of *features* should you detect and then match to establish such an alignment or set of correspondences? Think about this for a few moments before reading on.

The first kind of feature that you may notice are specific locations in the images, such as mountain peaks, building corners, doorways, or interestingly shaped patches of snow. These kinds of localized features are often called *keypoint features* or *interest points* (or even *corners*) and are often described by the appearance of pixel patches surrounding the point location (Section 7.1). Another class of important features are *edges*, e.g., the profile of mountains against the sky (Section 7.2). These kinds of features can be matched based on their orientation and local appearance (edge profiles) and can also be good indicators of object boundaries and *occlusion* events in image sequences. Edges can be grouped into longer *curves* and *contours*, which can then be tracked (Section 7.3). They can also be grouped into *straight line segments*, which can be directly matched or analyzed to find *vanishing points* and hence internal and external camera parameters (Section 7.4).

In this chapter, we describe some practical approaches to detecting such features and also discuss how feature correspondences can be established across different images. Point features are now used in such a wide variety of applications that it is good practice to read and implement some of the algorithms from Section 7.1. Edges and lines provide information that is complementary to both keypoint and region-based descriptors and are well suited to describing the boundaries of manufactured objects. These alternative descriptors, while extremely useful, can be skipped in a short introductory course.

The last part of this chapter (Section 7.5) discusses bottom-up non-semantic segmentation techniques. While these were once widely used as essential components of both recognition and matching algorithms, they have mostly been supplanted by the *semantic segmentation* techniques we studied in Section 6.4. They are still used occasionally to group pixels together for faster or more reliable matching.

7.1 Points and patches

Point features can be used to find a sparse set of corresponding locations in different images, often as a precursor to computing camera pose (Chapter 11), which is a prerequisite for computing a denser set of correspondences using stereo matching (Chapter 12). Such correspondences can also be used to align different images, e.g., when stitching image mosaics (Section 8.2) or high dynamic range images (Section 10.2), or performing video stabilization (Section 9.2.1). They are also used extensively to perform object instance recognition (Section 6.1). A key advantage of keypoints is that they permit matching even in the presence of clutter (occlusion) and large scale and orientation changes.

Feature-based correspondence techniques have been used since the early days of stereo matching (Hannah 1974; Moravec 1983; Hannah 1988) and subsequently gained popularity for image-stitching applications (Zoghlami, Faugeras, and Deriche 1997; Brown and Lowe 2007) as well as fully automated 3D modeling (Beardsley, Torr, and Zisserman 1996; Schaffalitzky and Zisserman 2002; Brown and Lowe 2005; Snavely, Seitz, and Szeliski 2006).

There are two main approaches to finding feature points and their correspondences. The first is to find features in one image that can be accurately *tracked* using a local search technique, such

Figure 7.2 Two pairs of images to be matched. What kinds of features might one use to establish a set of *correspondences* between these images?

as correlation or least squares (Section 7.1.5). The second is to independently detect features in all the images under consideration and then *match* features based on their local appearance (Section 7.1.3). The former approach is more suitable when images are taken from nearby viewpoints or in rapid succession (e.g., video sequences), while the latter is more suitable when a large amount of motion or appearance change is expected, e.g., in stitching together panoramas (Brown and Lowe 2007), establishing correspondences in *wide baseline stereo* (Schaffalitzky and Zisserman 2002), or performing object recognition (Fergus, Perona, and Zisserman 2007).

In this section, we split the keypoint detection and matching pipeline into four separate stages. During the *feature detection* (extraction) stage (Section 7.1.1), each image is searched for locations that are likely to match well in other images. In the *feature description* stage (Section 7.1.2), each region around detected keypoint locations is converted into a more compact and stable (invariant) *descriptor* that can be matched against other descriptors. The *feature matching* stage (Sections 7.1.3 and 7.1.4) efficiently searches for likely matching candidates in other images. The *feature tracking* stage (Section 7.1.5) is an alternative to the third stage that only searches a small neighborhood around each detected feature and is therefore more suitable for video processing.

A wonderful example of all of these stages can be found in David Lowe's (2004) paper, which describes the development and refinement of his *Scale Invariant Feature Transform* (SIFT). Comprehensive descriptions of alternative techniques can be found in a series of survey and evaluation papers covering both feature detection (Schmid, Mohr, and Bauckhage 2000; Mikolajczyk, Tuytelaars *et al.* 2005; Tuytelaars and Mikolajczyk 2008) and feature descriptors (Mikolajczyk and Schmid 2005; Balntas, Lenc *et al.* 2020). Shi and Tomasi (1994) and Triggs (2004) also provide nice reviews of classic (pre-neural network) feature detection techniques.

Figure 7.3 Image pairs with extracted patches below. Notice how some patches can be localized or matched with higher accuracy than others.

7.1.1 Feature detectors

How can we find image locations where we can reliably find correspondences with other images, i.e., what are good features to track (Shi and Tomasi 1994; Triggs 2004)? Look again at the image pair shown in Figure 7.3 and at the three sample *patches* to see how well they might be matched or tracked. As you may notice, textureless patches are nearly impossible to localize. Patches with large contrast changes (gradients) are easier to localize, although straight line segments at a single orientation suffer from the *aperture problem* (Horn and Schunck 1981; Lucas and Kanade 1981; Anandan 1989), i.e., it is only possible to align the patches along the direction *normal* to the edge direction (Figure 7.4b). Patches with gradients in at least two (significantly) different orientations are the easiest to localize, as shown schematically in Figure 7.4a.

These intuitions can be formalized by looking at the simplest possible matching criterion for comparing two image patches, i.e., their (weighted) summed square difference,

$$E_{\text{WSSD}}(\mathbf{u}) = \sum_i w(\mathbf{x}_i)[I_1(\mathbf{x}_i + \mathbf{u}) - I_0(\mathbf{x}_i)]^2, \tag{7.1}$$

where I_0 and I_1 are the two images being compared, $\mathbf{u} = (u, v)$ is the *displacement* vector, $w(\mathbf{x})$ is a spatially varying weighting (or window) function, and the summation i is over all the pixels in the patch. Note that this is the same formulation we later use to estimate motion between complete images (Section 9.1).

When performing feature detection, we do not know which other image locations the feature will end up being matched against. Therefore, we can only compute how stable this metric is with respect to small variations in position $\Delta\mathbf{u}$ by comparing an image patch against itself, which is known as an *auto-correlation function* or *surface*

$$E_{\text{AC}}(\Delta\mathbf{u}) = \sum_i w(\mathbf{x}_i)[I_0(\mathbf{x}_i + \Delta\mathbf{u}) - I_0(\mathbf{x}_i)]^2 \tag{7.2}$$

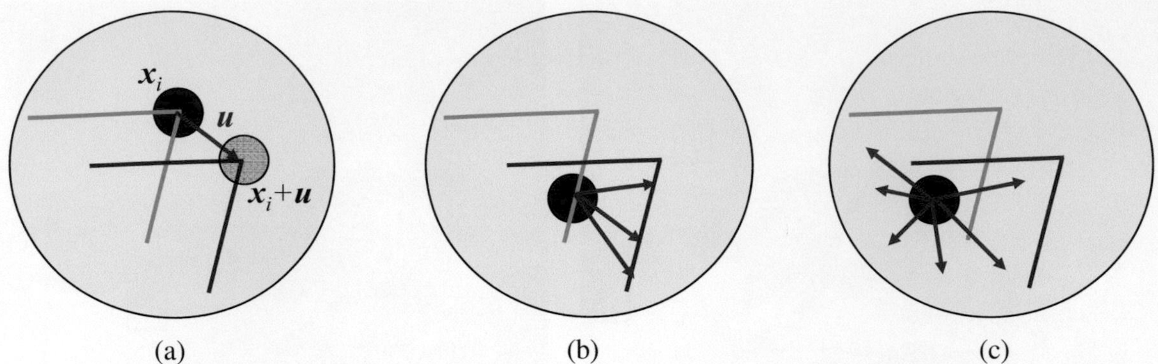

Figure 7.4 Aperture problems for different image patches: (a) stable ("corner-like") flow; (b) classic aperture problem (barber-pole illusion); (c) textureless region. The two images I_0 (yellow) and I_1 (red) are overlaid. The red vector $\mathbf{u}$ indicates the displacement between the patch centers and the $w(\mathbf{x}_i)$ weighting function (patch window) is shown as a dark circle.

(Figure 7.5).[1] Note how the auto-correlation surface for the textured flower bed (Figure 7.5b and the red cross in the lower right quadrant of Figure 7.5a) exhibits a strong minimum, indicating that it can be well localized. The correlation surface corresponding to the roof edge (Figure 7.5c) has a strong ambiguity along one direction, while the correlation surface corresponding to the cloud region (Figure 7.5d) has no stable minimum.

Using a Taylor Series expansion of the image function $I_0(\mathbf{x}_i + \Delta\mathbf{u}) \approx I_0(\mathbf{x}_i) + \nabla I_0(\mathbf{x}_i) \cdot \Delta\mathbf{u}$ (Lucas and Kanade 1981; Shi and Tomasi 1994), we can approximate the auto-correlation surface as

$$E_{\mathrm{AC}}(\Delta\mathbf{u}) = \sum_i w(\mathbf{x}_i)[I_0(\mathbf{x}_i + \Delta\mathbf{u}) - I_0(\mathbf{x}_i)]^2 \tag{7.3}$$

$$\approx \sum_i w(\mathbf{x}_i)[I_0(\mathbf{x}_i) + \nabla I_0(\mathbf{x}_i) \cdot \Delta\mathbf{u} - I_0(\mathbf{x}_i)]^2 \tag{7.4}$$

$$= \sum_i w(\mathbf{x}_i)[\nabla I_0(\mathbf{x}_i) \cdot \Delta\mathbf{u}]^2 \tag{7.5}$$

$$= \Delta\mathbf{u}^T \mathbf{A} \Delta\mathbf{u}, \tag{7.6}$$

where

$$\nabla I_0(\mathbf{x}_i) = (\frac{\partial I_0}{\partial x}, \frac{\partial I_0}{\partial y})(\mathbf{x}_i) \tag{7.7}$$

is the *image gradient* at $\mathbf{x}_i$. This gradient can be computed using a variety of techniques (Schmid, Mohr, and Bauckhage 2000). The classic "Harris" detector (Harris and Stephens 1988) uses a [–2 –1 0 1 2] filter, but more modern variants (Schmid, Mohr, and Bauckhage 2000; Triggs 2004) convolve the image with horizontal and vertical derivatives of a Gaussian (typically with $\sigma = 1$).

The auto-correlation matrix $\mathbf{A}$ can be written as

$$\mathbf{A} = w * \begin{bmatrix} I_x^2 & I_x I_y \\ I_x I_y & I_y^2 \end{bmatrix}, \tag{7.8}$$

where we have replaced the weighted summations with discrete convolutions with the weighting kernel w. This matrix can be interpreted as a tensor (multiband) image, where the outer products of

[1] Strictly speaking, a correlation is the *product* of two patches (3.12); I'm using the term here in a more qualitative sense. The weighted sum of squared differences is often called an *SSD surface* (Section 9.1).

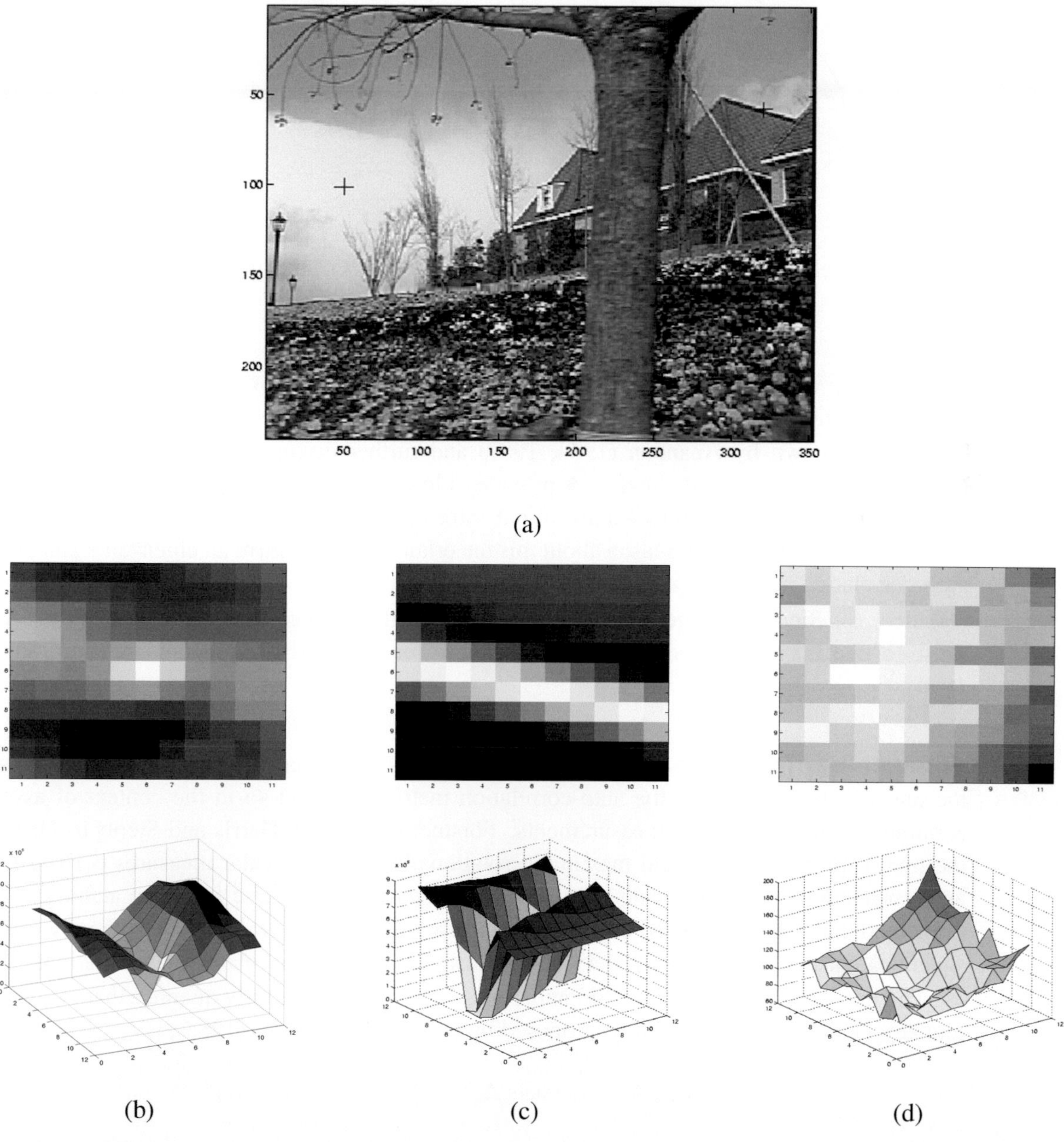

Figure 7.5 Three auto-correlation surfaces $E_{\text{AC}}(\Delta\mathbf{u})$ shown as both grayscale images and surface plots: (a) The original image is marked with three red crosses to denote where the auto-correlation surfaces were computed; (b) this patch is from the flower bed (good unique minimum); (c) this patch is from the roof edge (one-dimensional aperture problem); and (d) this patch is from the cloud (no good peak). Each grid point in figures b–d is one value of $\Delta\mathbf{u}$.

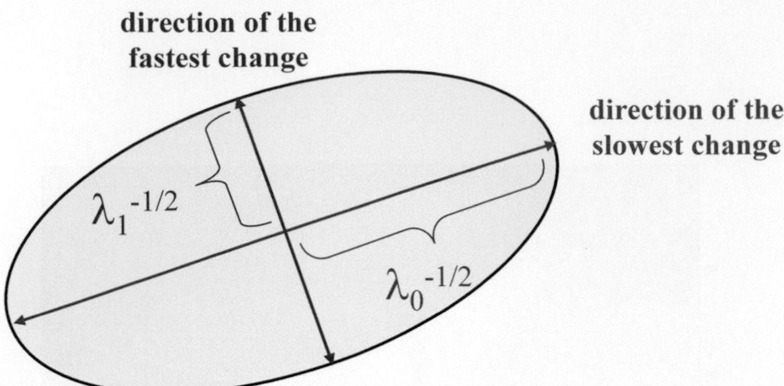

Figure 7.6 Uncertainty ellipse corresponding to an eigenvalue analysis of the auto-correlation matrix $\mathbf{A}$.

the gradients ∇I are convolved with a weighting function w to provide a per-pixel estimate of the local (quadratic) shape of the auto-correlation function.

As first shown by Anandan (1984; 1989) and further discussed in Section 9.1.3 and Equation (9.37), the inverse of the matrix $\mathbf{A}$ provides a lower bound on the uncertainty in the location of a matching patch. It is therefore a useful indicator of which patches can be reliably matched. The easiest way to visualize and reason about this uncertainty is to perform an eigenvalue analysis of the auto-correlation matrix $\mathbf{A}$, which produces two eigenvalues (λ_0, λ_1) and two eigenvector directions (Figure 7.6). Since the larger uncertainty depends on the smaller eigenvalue, i.e., $\lambda_0^{-1/2}$, it makes sense to find maxima in the smaller eigenvalue to locate good features to track (Shi and Tomasi 1994).

Förstner–Harris. While Anandan (1984) and Lucas and Kanade (1981) were the first to analyze the uncertainty structure of the auto-correlation matrix, they did so in the context of associating certainties with optical flow measurements. Förstner (1986) and Harris and Stephens (1988) were the first to propose using local maxima in rotationally invariant scalar measures derived from the auto-correlation matrix to locate keypoints for the purpose of sparse feature matching.[2] Both of these techniques also proposed using a Gaussian weighting window instead of the previously used square patches, which makes the detector response insensitive to in-plane image rotations.

The minimum eigenvalue λ_0 (Shi and Tomasi 1994) is not the only quantity that can be used to find keypoints. A simpler quantity, proposed by Harris and Stephens (1988), is

$$\det(\mathbf{A}) - \alpha\,\mathrm{trace}(\mathbf{A})^2 = \lambda_0\lambda_1 - \alpha(\lambda_0 + \lambda_1)^2 \tag{7.9}$$

with $\alpha = 0.06$. Unlike eigenvalue analysis, this quantity does not require the use of square roots and yet is still rotationally invariant and also downweights edge-like features where $\lambda_1 \gg \lambda_0$. Triggs (2004) suggests using the quantity

$$\lambda_0 - \alpha\lambda_1 \tag{7.10}$$

(say, with $\alpha = 0.05$), which also reduces the response at 1D edges, where aliasing errors sometimes inflate the smaller eigenvalue. He also shows how the basic 2×2 Hessian can be extended to

[2] Schmid, Mohr, and Bauckhage (2000) and Triggs (2004) give more detailed historical reviews of feature detection algorithms.

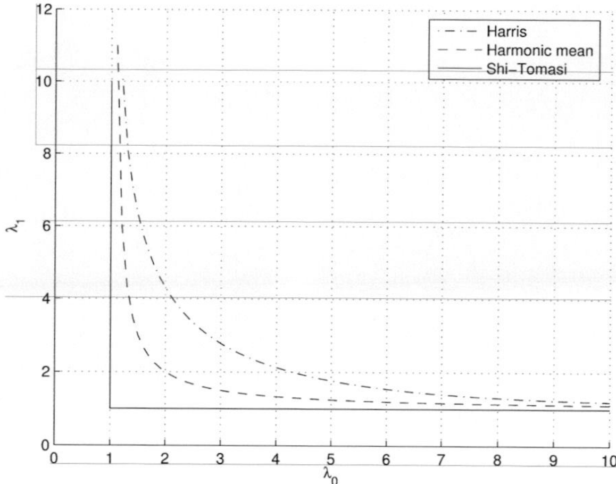

Figure 7.7 Isocontours of popular keypoint detection functions (Brown, Szeliski, and Winder 2004). Each detector looks for points where the eigenvalues λ_0, λ_1 of $\mathbf{A} = w * \nabla I \nabla I^T$ are both large.

(a) (b) (c)

Figure 7.8 Interest operator responses: (a) Sample image, (b) Harris response, and (c) DoG response. The circle sizes and colors indicate the scale at which each interest point was detected. Notice how the two detectors tend to respond at complementary locations.

parametric motions to detect points that are also accurately localizable in scale and rotation. Brown, Szeliski, and Winder (2005), on the other hand, use the harmonic mean,

$$\frac{\det \mathbf{A}}{\operatorname{tr} \mathbf{A}} = \frac{\lambda_0 \lambda_1}{\lambda_0 + \lambda_1}, \tag{7.11}$$

which is a smoother function in the region where $\lambda_0 \approx \lambda_1$. Figure 7.7 shows isocontours of the various interest point operators, from which we can see how the two eigenvalues are blended to determine the final interest value. Figure 7.8 shows the resulting interest operator responses for the classic Harris detector as well as the difference of Gaussian (DoG) detector discussed below.

Adaptive non-maximal suppression (ANMS). While most feature detectors simply look for local maxima in the interest function, this can lead to an uneven distribution of feature points across the image, e.g., points will be denser in regions of higher contrast. To mitigate this problem, Brown, Szeliski, and Winder (2005) only detect features that are both local maxima and whose response value is significantly (10%) greater than that of all of its neighbors within a radius r (Figure 7.9c–d). They devise an efficient way to associate suppression radii with all local maxima by first sorting

(a) Strongest 250 (b) Strongest 500

(c) ANMS 250, $r = 24$ (d) ANMS 500, $r = 16$

Figure 7.9 Adaptive non-maximal suppression (ANMS) (Brown, Szeliski, and Winder 2005) © 2005 IEEE: The upper two images show the strongest 250 and 500 interest points, while the lower two images show the interest points selected with adaptive non-maximal suppression, along with the corresponding suppression radius r. Note how the latter features have a much more uniform spatial distribution across the image.

them by their response strength and then creating a second list sorted by decreasing suppression radius (Brown, Szeliski, and Winder 2005). Figure 7.9 shows a qualitative comparison of selecting the top n features and using ANMS. Note that non-maximal suppression is now also an essential component of DNN-based object detectors, as discussed in Section 6.3.3.

Measuring repeatability. Given the large number of feature detectors that have been developed in computer vision, how can we decide which ones to use? Schmid, Mohr, and Bauckhage (2000) were the first to propose measuring the *repeatability* of feature detectors, which they define as the frequency with which keypoints detected in one image are found within ϵ (say, $\epsilon = 1.5$) pixels of the corresponding location in a transformed image. In their paper, they transform their planar images by applying rotations, scale changes, illumination changes, viewpoint changes, and adding noise. They also measure the *information content* available at each detected feature point, which they define as the entropy of a set of rotationally invariant local grayscale descriptors. Among the techniques they survey, they find that the improved (Gaussian derivative) version of the Harris operator with $\sigma_d = 1$ (scale of the derivative Gaussian) and $\sigma_i = 2$ (scale of the integration Gaussian) works best.

Figure 7.10 Multi-scale oriented patches (MOPS) extracted at five pyramid levels (Brown, Szeliski, and Winder 2005) © 2005 IEEE. The boxes show the feature orientation and the region from which the descriptor vectors are sampled.

Scale invariance

In many situations, detecting features at the finest stable scale possible may not be appropriate. For example, when matching images with little high-frequency detail (e.g., clouds), fine-scale features may not exist.

One solution to the problem is to extract features at a variety of scales, e.g., by performing the same operations at multiple resolutions in a pyramid and then matching features at the same level. This kind of approach is suitable when the images being matched do not undergo large scale changes, e.g., when matching successive aerial images taken from an airplane or stitching panoramas taken with a fixed-focal-length camera. Figure 7.10 shows the output of one such approach: the multi-scale oriented patch detector of Brown, Szeliski, and Winder (2005), for which responses at five different scales are shown.

However, for most object recognition applications, the scale of the object in the image is unknown. Instead of extracting features at many different scales and then matching all of them, it is more efficient to extract features that are stable in both location *and* scale (Lowe 2004; Mikolajczyk and Schmid 2004).

Early investigations into scale selection were performed by Lindeberg (1993; 1998b), who first proposed using extrema in the Laplacian of Gaussian (LoG) function as interest point locations. Based on this work, Lowe (2004) proposed computing a set of sub-octave Difference of Gaussian filters (Figure 7.11a), looking for 3D (space+scale) maxima in the resulting structure (Figure 7.11b), and then computing a sub-pixel space+scale location using a quadratic fit (Brown and Lowe 2002). The number of sub-octave levels was determined, after careful empirical investigation, to be three, which corresponds to a quarter-octave pyramid, which is the same as used by Triggs (2001).

As with the Harris operator, pixels where there is strong asymmetry in the local curvature of the indicator function (in this case, the DoG) are rejected. This is implemented by first computing the

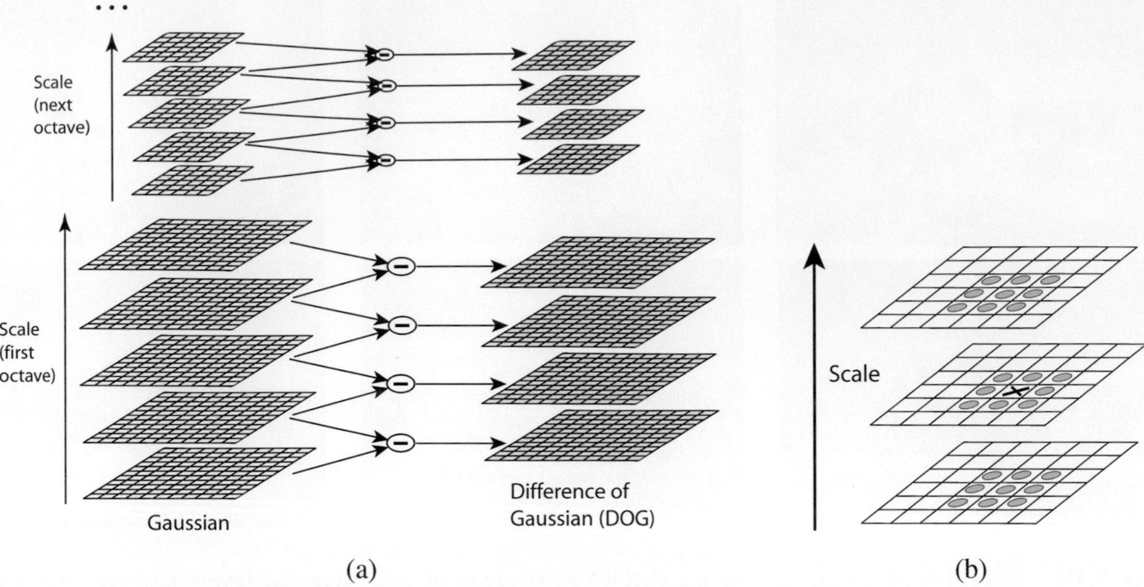

Figure 7.11 Scale-space feature detection using a sub-octave Difference of Gaussian pyramid (Lowe 2004) © 2004 Springer: (a) Adjacent levels of a sub-octave Gaussian pyramid are subtracted to produce Difference of Gaussian images; (b) extrema (maxima and minima) in the resulting 3D volume are detected by comparing a pixel to its 26 neighbors.

local Hessian of the difference image D,

$$\mathbf{H} = \begin{bmatrix} D_{xx} & D_{xy} \\ D_{xy} & D_{yy} \end{bmatrix}, \tag{7.12}$$

and then rejecting keypoints for which

$$\frac{\mathrm{Tr}(\mathbf{H})^2}{\mathrm{Det}(\mathbf{H})} > 10. \tag{7.13}$$

While Lowe's Scale Invariant Feature Transform (SIFT) performs well in practice, it is not based on the same theoretical foundation of maximum spatial stability as the auto-correlation-based detectors. (In fact, its detection locations are often complementary to those produced by such techniques and can therefore be used in conjunction with these other approaches.) In order to add a scale selection mechanism to the Harris corner detector, Mikolajczyk and Schmid (2004) evaluate the Laplacian of Gaussian function at each detected Harris point (in a multi-scale pyramid) and keep only those points for which the Laplacian is extremal (larger or smaller than both its coarser and finer-level values). An optional iterative refinement for both scale and position is also proposed and evaluated. Additional examples of scale-invariant region detectors are discussed by Mikolajczyk, Tuytelaars *et al.* (2005) and Tuytelaars and Mikolajczyk (2008).

Rotational invariance and orientation estimation

In addition to dealing with scale changes, most image matching and object recognition algorithms need to deal with (at least) in-plane image rotation. One way to deal with this problem is to design descriptors that are rotationally invariant (Schmid and Mohr 1997), but such descriptors have poor discriminability, i.e. they map different looking patches to the same descriptor.

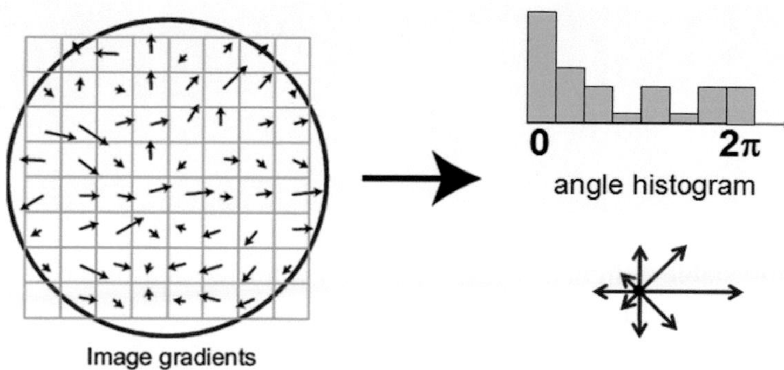

Figure 7.12 A dominant orientation estimate can be computed by creating a histogram of all the gradient orientations (weighted by their magnitudes or after thresholding out small gradients) and then finding the significant peaks in this distribution (Lowe 2004) © 2004 Springer.

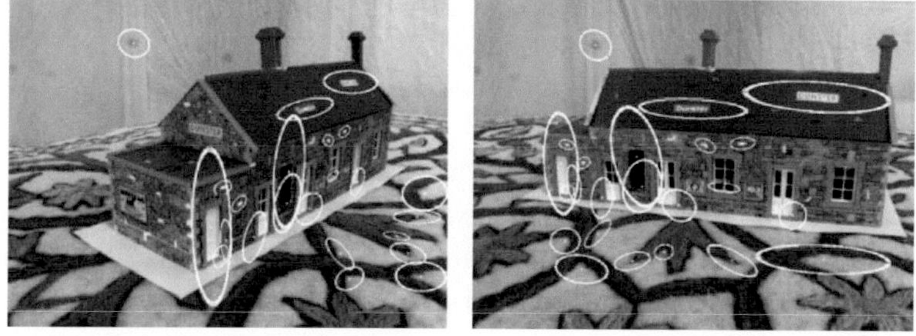

Figure 7.13 Affine region detectors used to match two images taken from dramatically different viewpoints (Mikolajczyk and Schmid 2004) © 2004 Springer.

A better method is to estimate a *dominant orientation* at each detected keypoint. Once the local orientation and scale of a keypoint have been estimated, a scaled and oriented patch around the detected point can be extracted and used to form a feature descriptor (Figures 7.10 and 7.15).

The simplest possible orientation estimate is the average gradient within a region around the keypoint. If a Gaussian weighting function is used (Brown, Szeliski, and Winder 2005), this average gradient is equivalent to a first-order steerable filter (Section 3.2.3), i.e., it can be computed using an image convolution with the horizontal and vertical derivatives of Gaussian filter (Freeman and Adelson 1991). To make this estimate more reliable, it is usually preferable to use a larger aggregation window (Gaussian kernel size) than detection window (Brown, Szeliski, and Winder 2005). The orientations of the square boxes shown in Figure 7.10 were computed using this technique.

Sometimes, however, the averaged (signed) gradient in a region can be small and therefore an unreliable indicator of orientation. A more reliable technique is to look at the *histogram* of orientations computed around the keypoint. Lowe (2004) computes a 36-bin histogram of edge orientations weighted by both gradient magnitude and Gaussian distance to the center, finds all peaks within 80% of the global maximum, and then computes a more accurate orientation estimate using a three-bin parabolic fit (Figure 7.12).

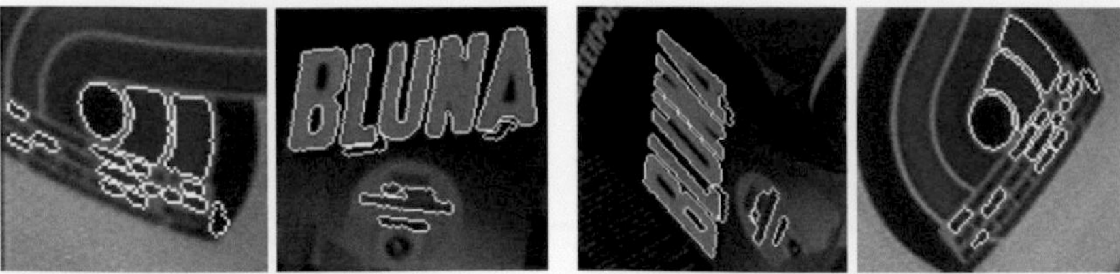

Figure 7.14 Maximally stable extremal regions (MSERs) extracted and matched from a number of images (Matas, Chum *et al.* 2004) © 2004 Elsevier.

Affine invariance

While scale and rotation invariance are highly desirable, for many applications such as *wide baseline stereo matching* (Pritchett and Zisserman 1998; Schaffalitzky and Zisserman 2002) or location recognition (Chum, Philbin *et al.* 2007), full affine invariance is preferred. Affine-invariant detectors not only respond at consistent locations after scale and orientation changes, they also respond consistently across affine deformations such as (local) perspective foreshortening (Figure 7.13). In fact, for a small enough patch, any continuous image warping can be well approximated by an affine deformation.

To introduce affine invariance, several authors have proposed fitting an ellipse to the autocorrelation or Hessian matrix (using eigenvalue analysis) and then using the principal axes and ratios of this fit as the affine coordinate frame (Lindeberg and Gårding 1997; Baumberg 2000; Mikolajczyk and Schmid 2004; Mikolajczyk, Tuytelaars *et al.* 2005; Tuytelaars and Mikolajczyk 2008).

Another important affine invariant region detector is the maximally stable extremal region (MSER) detector developed by Matas, Chum *et al.* (2004). To detect MSERs, binary regions are computed by thresholding the image at all possible gray levels (the technique therefore only works for grayscale images). This operation can be performed efficiently by first sorting all pixels by gray value and then incrementally adding pixels to each connected component as the threshold is changed (Nistér and Stewénius 2008). As the threshold is changed, the area of each component (region) is monitored; regions whose rate of change of area with respect to the threshold is minimal are defined as *maximally stable*. Such regions are therefore invariant to both affine geometric and photometric (linear bias-gain or smooth monotonic) transformations (Figure 7.14). If desired, an affine coordinate frame can be fit to each detected region using its moment matrix.

The area of feature point detection continues to be very active, with papers appearing every year at major computer vision conferences. Mikolajczyk, Tuytelaars *et al.* (2005) and Tuytelaars and Mikolajczyk (2008) survey a number of popular (pre-DNN) affine region detectors and provide experimental comparisons of their invariance to common image transformations such as scaling, rotations, noise, and blur.

More recent papers published in the last decade include:

- SURF (Bay, Ess *et al.* 2008), which uses integral images for faster convolutions;
- FAST and FASTER (Rosten, Porter, and Drummond 2010), one of the first learned detectors;
- BRISK (Leutenegger, Chli, and Siegwart 2011), which uses a scale-space FAST detector together with a bit-string descriptor;
- ORB (Rublee, Rabaud *et al.* 2011), which adds orientation to FAST; and

- KAZE (Alcantarilla, Bartoli, and Davison 2012) and Accelerated-KAZE (Alcantarilla, Nuevo, and Bartoli 2013), which use non-linear diffusion to select the scale for feature detection.

While FAST introduced the idea of machine learning for feature detectors, more recent papers use convolutional neural networks to perform the detection. These include:

- Learning covariant feature detectors (Lenc and Vedaldi 2016);
- Learning to assign orientations to feature points (Yi, Verdie *et al.* 2016);
- LIFT, learned invariant feature transforms (Yi, Trulls *et al.* 2016), SuperPoint, self-supervised interest point detection and description (DeTone, Malisiewicz, and Rabinovich 2018), and LF-Net, learning local features from images (Ono, Trulls *et al.* 2018), all three of which jointly optimize the detectors and descriptors in a single (multi-head) pipeline;
- AffNet (Mishkin, Radenovic, and Matas 2018), which detects matchable affine-covariant regions;
- Key.Net (Barroso-Laguna, Riba *et al.* 2019), which uses a combination of handcrafted and learned CNN features; and
- D2-Net (Dusmanu, Rocco *et al.* 2019), R2D2 (Revaud, Weinzaepfel *et al.* 2019), and D2D (Tian, Balntas *et al.* 2020), which all extract dense local feature descriptors and then keeps the ones that have high saliency or repeatability.

These last two papers also contains a nice review of other recent feature detectors, as does the paper by Balntas, Lenc *et al.* (2020).

Of course, keypoints are not the only features that can be used for registering images. Zoghlami, Faugeras, and Deriche (1997) use line segments as well as point-like features to estimate homographies between pairs of images, whereas Bartoli, Coquerelle, and Sturm (2004) use line segments with local correspondences along the edges to extract 3D structure and motion. Tuytelaars and Van Gool (2004) use affine invariant regions to detect correspondences for wide baseline stereo matching, whereas Kadir, Zisserman, and Brady (2004) detect salient regions where patch entropy and its rate of change with scale are locally maximal. Corso and Hager (2005) use a related technique to fit 2D oriented Gaussian kernels to homogeneous regions. More details on techniques for finding and matching curves, lines, and regions can be found later in this chapter.

7.1.2 Feature descriptors

After detecting keypoint features, we must *match* them, i.e., we must determine which features come from corresponding locations in different images. In some situations, e.g., for video sequences (Shi and Tomasi 1994) or for stereo pairs that have been *rectified* (Zhang, Deriche *et al.* 1995; Loop and Zhang 1999; Scharstein and Szeliski 2002), the local motion around each feature point may be mostly translational. In this case, simple error metrics, such as the *sum of squared differences* or *normalized cross-correlation*, described in Section 9.1, can be used to directly compare the intensities in small patches around each feature point. (The comparative study by Mikolajczyk and Schmid (2005), discussed below, uses cross-correlation.) Because feature points may not be exactly located, a more accurate matching score can be computed by performing incremental motion refinement as described in Section 9.1.3, but this can be time-consuming and can sometimes even decrease performance (Brown, Szeliski, and Winder 2005).

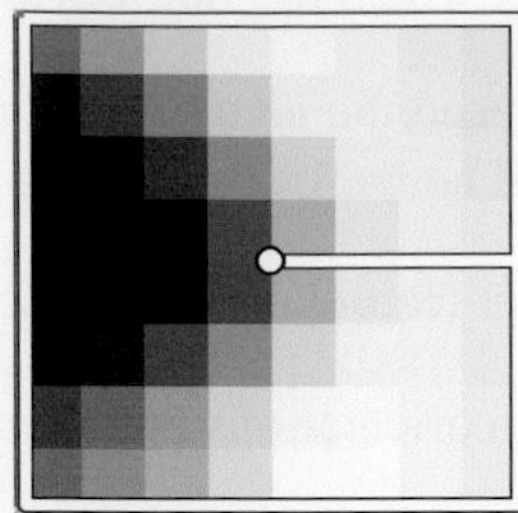

Figure 7.15 Once a local scale and orientation estimate has been determined, MOPS descriptors are formed using an 8×8 sampling of bias and gain normalized intensity values, with a sample spacing of five pixels relative to the detection scale (Brown, Szeliski, and Winder 2005) © 2005 IEEE. This low frequency sampling gives the features some robustness to interest point location error and is achieved by sampling at a higher pyramid level than the detection scale.

In most cases, however, the local appearance of features will change in orientation and scale, and sometimes even undergo affine deformations. Extracting a local scale, orientation, or affine frame estimate and then using this to resample the patch before forming the feature descriptor is thus usually preferable (Figure 7.15).

Even after compensating for these changes, the local appearance of image patches will usually still vary from image to image. How can we make image descriptors more invariant to such changes, while still preserving discriminability between different (non-corresponding) patches? Mikolajczyk and Schmid (2005) review a number of view-invariant local image descriptors and experimentally compare their performance. More recently, Balntas, Lenc *et al.* (2020) and Jin, Mishkin *et al.* (2021) compare the large number of learned feature descriptors developed in the prior decade.[3] Below, we describe a few of these descriptors in more detail.

Bias and gain normalization (MOPS). For tasks that do not exhibit large amounts of foreshortening, such as image stitching, simple normalized intensity patches perform reasonably well and are simple to implement (Brown, Szeliski, and Winder 2005) (Figure 7.15). To compensate for slight inaccuracies in the feature point detector (location, orientation, and scale), multi-scale oriented patches (MOPS) are sampled at a spacing of five pixels relative to the detection scale, using a coarser level of the image pyramid to avoid aliasing. To compensate for affine photometric variations (linear exposure changes or bias and gain, (3.3)), patch intensities are re-scaled so that their mean is zero and their variance is one.

Scale invariant feature transform (SIFT). SIFT features (Lowe 2004) are formed by computing the gradient at each pixel in a 16×16 window around the detected keypoint, using the appropriate level of the Gaussian pyramid at which the keypoint was detected. The gradient magnitudes are downweighted by a Gaussian fall-off function (shown as a blue circle in Figure 7.16a) to reduce the influence of gradients far from the center, as these are more affected by small misregistrations.

In each 4×4 quadrant, a gradient orientation histogram is formed by (conceptually) adding the gradient values weighted by the Gaussian fall-off function to one of eight orientation histogram bins. To reduce the effects of location and dominant orientation misestimation, each of the original 256

[3]Many recent publications such as Tian, Yu *et al.* (2019) use their HPatches dataset to compare their performance against previous approaches.

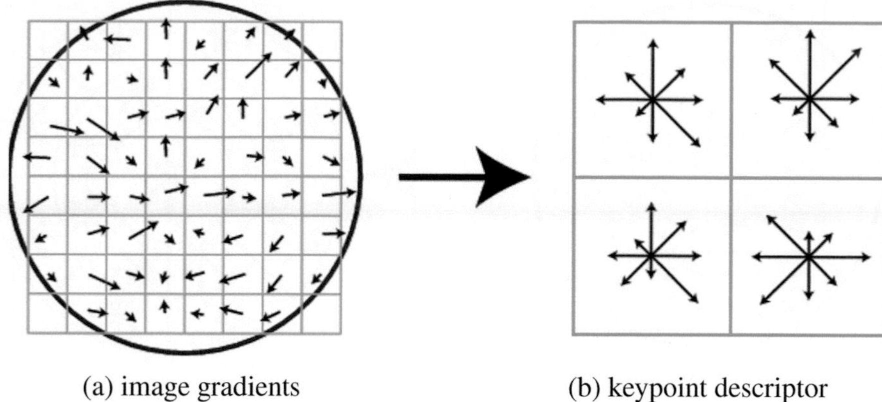

(a) image gradients (b) keypoint descriptor

Figure 7.16 A schematic representation of Lowe's (2004) scale invariant feature transform (SIFT): (a) Gradient orientations and magnitudes are computed at each pixel and weighted by a Gaussian fall-off function (blue circle). (b) A weighted gradient orientation histogram is then computed in each subregion, using trilinear interpolation. While this figure shows an 8×8 pixel patch and a 2×2 descriptor array, Lowe's actual implementation uses 16×16 patches and a 4×4 array of eight-bin histograms.

weighted gradient magnitudes is softly added to $2 \times 2 \times 2$ adjacent histogram bins in the (x, y, θ) space using trilinear interpolation. Softly distributing values to adjacent histogram bins is generally a good idea in any application where histograms are being computed, e.g., for Hough transforms (Section 7.4.2) or local histogram equalization (Section 3.1.4).

The 4x4 array of eight-bin histogram yields 128 non-negative values form a raw version of the SIFT descriptor vector. To reduce the effects of contrast or gain (additive variations are already removed by the gradient), the 128-D vector is normalized to unit length. To further make the descriptor robust to other photometric variations, values are clipped to 0.2 and the resulting vector is once again renormalized to unit length.

PCA-SIFT. Ke and Sukthankar (2004) propose a simpler way to compute descriptors inspired by SIFT; it computes the x and y (gradient) derivatives over a 39×39 patch and then reduces the resulting 3042-dimensional vector to 36 using principal component analysis (PCA) (Section 5.2.3 and Appendix A.1.2). Another popular variant of SIFT is SURF (Bay, Ess *et al.* 2008), which uses box filters to approximate the derivatives and integrals used in SIFT.

RootSIFT. Arandjelović and Zisserman (2012) observe that by simply re-normalizing SIFT descriptors using an L_1 measure and then taking the square root of each component, a dramatic increase in performance (discriminability) can be obtained.

Gradient location-orientation histogram (GLOH). This descriptor, developed by Mikolajczyk and Schmid (2005), is a variant of SIFT that uses a log-polar binning structure instead of the four quadrants used by Lowe (2004) (Figure 7.17). The spatial bins extend over the radii 0...6, 6...11, and 11...15, with eight angular bins (except for the single central region), for a total of 17 spatial bins and GLOH uses 16 orientation bins instead of the 8 used in SIFT. The 272-dimensional histogram is then projected onto a 128-dimensional descriptor using PCA trained on a large database. In their evaluation, Mikolajczyk and Schmid (2005) found that GLOH, which has the best performance overall, outperforms SIFT by a small margin.

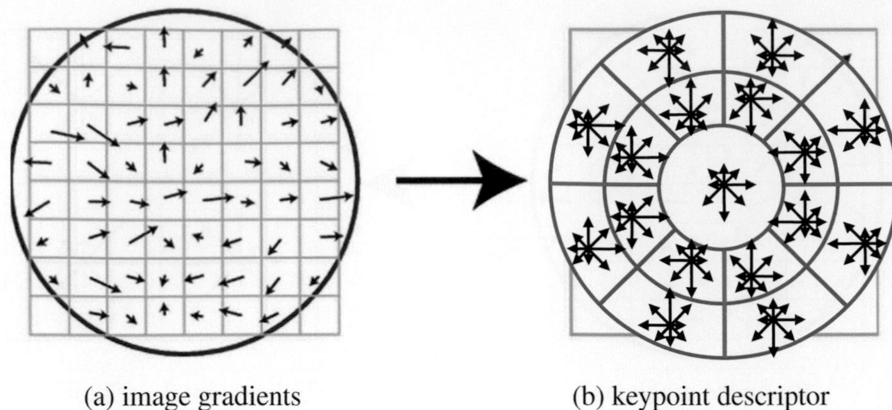

(a) image gradients (b) keypoint descriptor

Figure 7.17 The gradient location-orientation histogram (GLOH) descriptor uses log-polar bins instead of square bins to compute orientation histograms (Mikolajczyk and Schmid 2005). GLOH uses 16 gradient orientations inside each bin, although this figure only shows 8 to appear less cluttered.

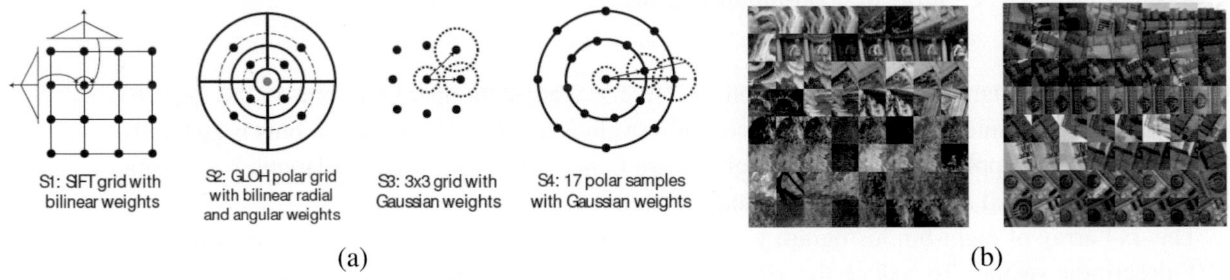

Figure 7.18 Spatial summation blocks for SIFT, GLOH, and some related feature descriptors (Winder and Brown 2007) © 2007 IEEE: (a) The parameters for the features, e.g., their Gaussian weights, are learned from a training database of (b) matched real-world image patches obtained from robust structure from motion applied to internet photo collections (Hua, Brown, and Winder 2007).

Steerable filters. Steerable filters (Section 3.2.3) are combinations of derivative of Gaussian filters that permit the rapid computation of even and odd (symmetric and anti-symmetric) edge-like and corner-like features at all possible orientations (Freeman and Adelson 1991). Because they use reasonably broad Gaussians, they too are somewhat insensitive to localization and orientation errors.

Performance of local descriptors. Among the local descriptors that Mikolajczyk and Schmid (2005) compared, they found that GLOH performed best, followed closely by SIFT. They also present results for many other descriptors not covered in this book.

The field of feature descriptors continued to advance rapidly, with some techniques looking at local color information (van de Weijer and Schmid 2006; Abdel-Hakim and Farag 2006). Winder and Brown (2007) develop a multi-stage framework for feature descriptor computation that subsumes both SIFT and GLOH (Figure 7.18a) and also allows them to learn optimal parameters for newer descriptors that outperform previous hand-tuned descriptors. Hua, Brown, and Winder (2007) extend this work by learning lower-dimensional projections of higher-dimensional descriptors that have the best discriminative power, and Brown, Hua, and Winder (2011) further extend it by learning the optimal placement of the pooling regions. All of these papers use a database of real-world image patches (Figure 7.18b) obtained by sampling images at locations that were reliably matched

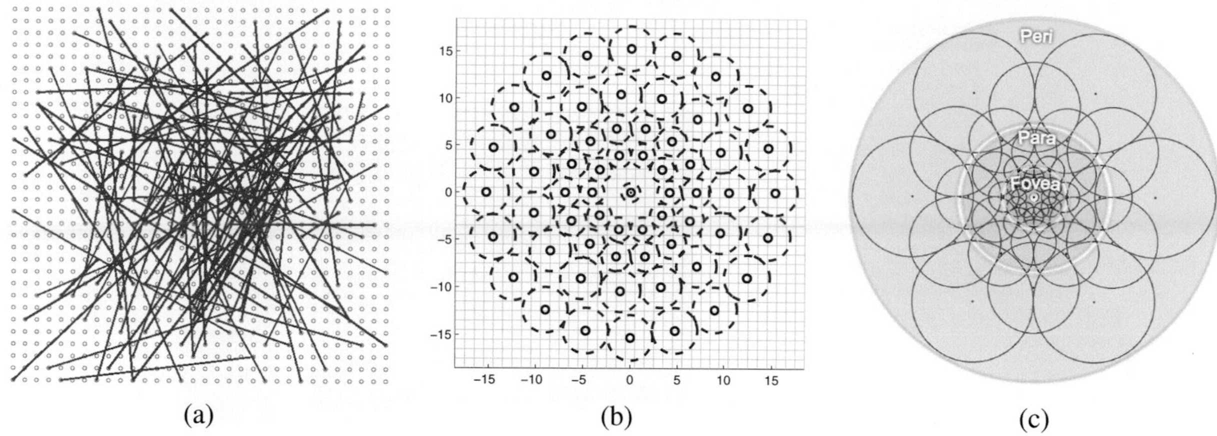

(a) (b) (c)

Figure 7.19 Binary bit-string feature descriptors: (a) the BRIEF descriptor compares 128 pairs of pixel values (denoted by line segments) and stores the comparison results in a 128-bit vector (Calonder, Lepetit *et al.* 2010) © 2010 Springer; (b) BRISK sampling pattern and Gaussian blur radii; (Leutenegger, Chli, and Siegwart 2011) © 2011 IEEE; (c) FREAK retinal sampling pattern (Alahi, Ortiz, and Vandergheynst 2012) © 2012 IEEE.

using a robust structure-from-motion algorithm applied to internet photo collections (Snavely, Seitz, and Szeliski 2006; Goesele, Snavely *et al.* 2007). In concurrent work, Tola, Lepetit, and Fua (2010) developed a similar DAISY descriptor for dense stereo matching and optimized its parameters based on ground truth stereo data.

While these techniques construct feature detectors that optimize for repeatability across *all* object classes, it is also possible to develop class- or instance-specific feature detectors that maximize *discriminability* from other classes (Ferencz, Learned-Miller, and Malik 2008). If planar surface orientations can be determined in the images being matched, it is also possible to extract viewpoint-invariant patches (Wu, Clipp *et al.* 2008).

A more recent trend has been the development of binary bit-string feature descriptors, which can take advantage of fast Hamming distance operators in modern computer architectures. The BRIEF descriptor (Calonder, Lepetit *et al.* 2010) compares 128 different pairs of pixel values (denoted as line segments in Figure 7.19a) scattered around the keypoint location to obtain a 128-bit vector. ORB (Rublee, Rabaud *et al.* 2011) adds an orientation component to the FAST detector before computing oriented BRIEF descriptors. BRISK (Leutenegger, Chli, and Siegwart 2011) adds scale-space analysis to the FAST detector and a radially symmetric sampling pattern (Figure 7.19b) to produce the binary descriptor. FREAK (Alahi, Ortiz, and Vandergheynst 2012) uses a more pronounced "retinal" (log-polar) sampling pattern paired with a cascade of bit comparisons for even greater speed and efficiency. The survey and evaluation by Mukherjee, Wu, and Wang (2015) compares all of these "classic" feature detectors and descriptors.

Since 2015 or so, most of the new feature descriptors are constructed using deep learning techniques, as surveyed in Balntas, Lenc *et al.* (2020) and Jin, Mishkin *et al.* (2021). Some of these descriptors, such as LIFT (Yi, Trulls *et al.* 2016), TFeat (Balntas, Riba *et al.* 2016), HPatches (Balntas, Lenc *et al.* 2020), L2-Net (Tian, Fan, and Wu 2017), HardNet (Mishchuk, Mishkin *et al.* 2017), Geodesc (Luo, Shen *et al.* 2018), LF-Net (Ono, Trulls *et al.* 2018), SOSNet (Tian, Yu *et al.* 2019), and Key.Net (Barroso-Laguna, Riba *et al.* 2019) operate on patches, much like the classical SIFT approach. They hence require an initial local feature detector to determine the center of the patch and use a predetermined patch size when constructing the input to the network.

In contrast, approaches such as DELF (Noh, Araujo *et al.* 2017), SuperPoint (DeTone, Mal-

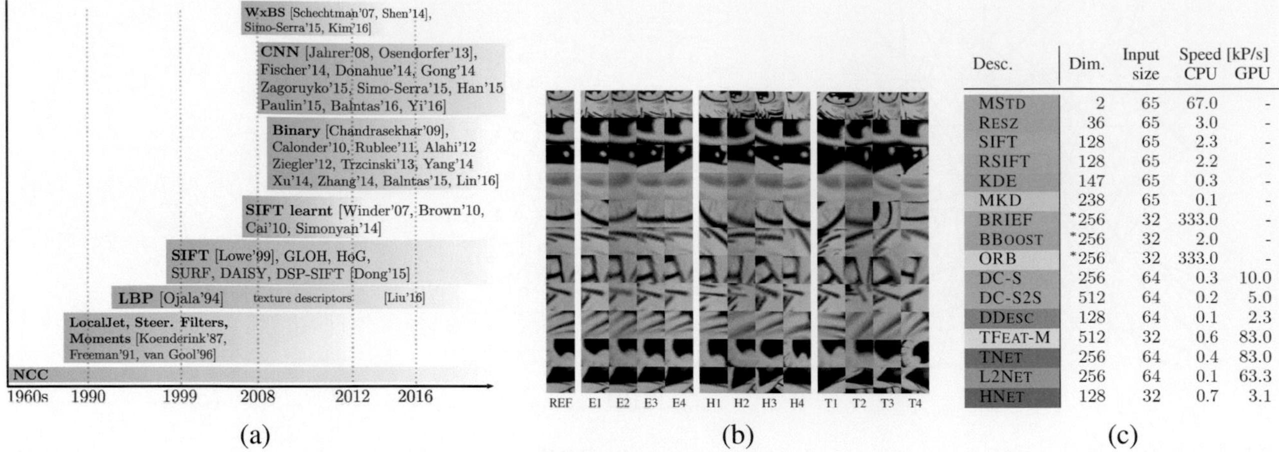

Desc.	Dim.	Input size	Speed [kP/s] CPU	GPU
MSTD	2	65	67.0	-
RESZ	36	65	3.0	-
SIFT	128	65	2.3	-
RSIFT	128	65	2.2	-
KDE	147	65	0.3	-
MKD	238	65	0.1	-
BRIEF	*256	32	333.0	-
BBOOST	*256	32	2.0	-
ORB	*256	32	333.0	-
DC-S	256	64	0.3	10.0
DC-S2S	512	64	0.2	5.0
DDESC	128	64	0.1	2.3
TFEAT-M	512	32	0.6	83.0
TNET	256	64	0.4	83.0
L2NET	256	64	0.1	63.3
HNET	128	32	0.7	3.1

Figure 7.20 HPatches local descriptors benchmark (Balntas, Lenc *et al.* 2020) © 2019 IEEE: (a) chronology of feature descriptors; (b) typical patches in the dataset (grouped by Easy, Hard, and Tough); (c) size and speed of different descriptors.

isiewicz, and Rabinovich 2018), D2-Net (Dusmanu, Rocco *et al.* 2019), ContextDesc (Luo, Shen *et al.* 2019), R2D2 (Revaud, Weinzaepfel *et al.* 2019), ASLFeat (Luo, Zhou *et al.* 2020), and CAPS (Wang, Zhou *et al.* 2020) use the entire image as the input to the descriptor computation. This has the added benefit that the receptive field used to compute the descriptor can be learned from the data and does not require specifying a patch size. Theoretically, these CNN models can learn receptive fields that use all of the pixels in the image, although in practice they tend to use Gaussian-like receptive fields (Zhou, Khosla *et al.* 2015; Luo, Li *et al.* 2016; Selvaraju, Cogswell *et al.* 2017).

In the HPatches benchmark (Figure 7.20) for evaluating patch matching by Balntas, Lenc *et al.* (2020), HardNet and L2-net performed the best on average. Another paper (Wang, Zhou *et al.* 2020) shows CAPS and R2D2 as the best performers, while S2DNet (Germain, Bourmaud, and Lepetit 2020) and LISRD (Pautrat, Larsson *et al.* 2020) also claim state-of-the-art performance, while the WISW benchmark (Bellavia and Colombo 2020) shows that traditional descriptors such as SIFT enhanced with more recent ideas do the best. On the wide baseline image matching benchmark by Jin, Mishkin *et al.* (2021),[4] HardNet, Key.Net, and D2-Net were top performers (e.g., D2-Net had the highest number of landmarks), although the results were quite task-dependent and the Difference of Gaussian detector was still the best. The performance of these descriptors on matching features across large illumination differences (day-night) has also been studied (Radenović, Schönberger *et al.* 2016; Zhou, Sattler, and Jacobs 2016; Mishkin 2021).

The most recent trend in wide-baseline matching has been to densely extract features without a detector stage and to then match and refine the set of correspondences (Jiang, Trulls *et al.* 2021; Sarlin, Unagar *et al.* 2021; Sun, Shen *et al.* 2021; Truong, Danelljan *et al.* 2021; Zhou, Sattler, and Leal-Taixé 2021). Some of these more recent techniques have been evaluated by Mishkin (2021).

7.1.3 Feature matching

Once we have extracted features and their descriptors from two or more images, the next step is to establish some preliminary feature matches between these images. The approach we take depends

[4]The benchmark is associated with the CVPR Workshop on Image Matching: Local Features & Beyond: https://image-matching-workshop.github.io.

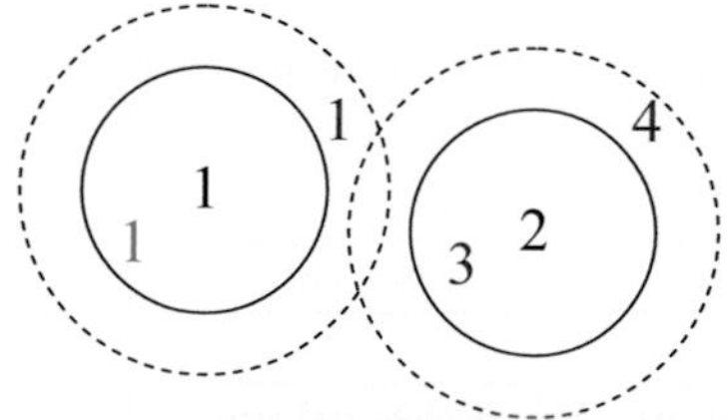

Figure 7.21 False positives and negatives: The black digits 1 and 2 are features being matched against a database of features in other images. At the current threshold setting (the solid circles), the green 1 is a *true positive* (good match), the blue 1 is a *false negative* (failure to match), and the red 3 is a *false positive* (incorrect match). If we set the threshold higher (the dashed circles), the blue 1 becomes a true positive but the brown 4 becomes an additional false positive.

partially on the application, e.g., different strategies may be preferable for matching images that are known to overlap (e.g., in image stitching) vs. images that may have no correspondence whatsoever (e.g., when trying to recognize objects from a database).

In this section, we divide this problem into two separate components. The first is to select a *matching strategy*, which determines which correspondences are passed on to the next stage for further processing. The second is to devise efficient *data structures* and *algorithms* to perform this matching as quickly as possible, which we expand on in Section 7.1.4.

Matching strategy and error rates

Determining which feature matches are reasonable to process further depends on the context in which the matching is being performed. Say we are given two images that overlap to a fair amount (e.g., for image stitching or for tracking objects in a video). We know that most features in one image are likely to match the other image, although some may not match because they are occluded or their appearance has changed too much.

On the other hand, if we are trying to recognize how many known objects appear in a cluttered scene (Figure 6.2), most of the features may not match. Furthermore, a large number of potentially matching objects must be searched, which requires more efficient strategies, as described below.

To begin with, we assume that the feature descriptors have been designed so that Euclidean (vector magnitude) distances in feature space can be directly used for ranking potential matches. If it turns out that certain parameters (axes) in a descriptor are more reliable than others, it is usually preferable to re-scale these axes ahead of time, e.g., by determining how much they vary when compared against other known good matches (Hua, Brown, and Winder 2007). A more general process, which involves transforming feature vectors into a new scaled basis, is called *whitening* and is discussed in more detail in the context of eigenface-based face recognition (Section 5.2.3).

Given a Euclidean distance metric, the simplest matching strategy is to set a threshold (maximum distance) and to return all matches from other images within this threshold. Setting the threshold too high results in too many *false positives*, i.e., incorrect matches being returned. Setting the threshold too low results in too many *false negatives*, i.e., too many correct matches being missed (Figure 7.21).

We can quantify the performance of a matching algorithm at a particular threshold by first counting the number of true and false matches and match failures, using the following definitions (Fawcett 2006), which we already discussed in Section 6.3.3:

- TP: true positives, i.e., number of correct matches;

	True matches	True non-matches		
Predicted matches	TP = 18	FP = 4	P' = 22	PPV = 0.82
Predicted non-matches	FN = 2	TN = 76	N' = 78	
	P = 20	N = 80	Total = 100	
	TPR = 0.90	FPR = 0.05		ACC = 0.94

Table 7.1 The number of matches correctly and incorrectly estimated by a feature matching algorithm, showing the number of true positives (TP), false positives (FP), false negatives (FN), and true negatives (TN). The columns sum up to the actual number of positives (P) and negatives (N), while the rows sum up to the predicted number of positives (P′) and negatives (N′). The formulas for the true positive rate (TPR), the false positive rate (FPR), the positive predictive value (PPV), and the accuracy (ACC) are given in the text.

- FN: false negatives, matches that were not correctly detected;
- FP: false positives, proposed matches that are incorrect;
- TN: true negatives, non-matches that were correctly rejected.

Table 7.1 shows a sample *confusion matrix* (contingency table) containing such numbers.

We can convert these numbers into *unit rates* by defining the following quantities (Fawcett 2006):

- true positive rate (TPR),

$$\mathrm{TPR} = \frac{\mathrm{TP}}{\mathrm{TP{+}FN}} = \frac{\mathrm{TP}}{\mathrm{P}}; \tag{7.14}$$

- false positive rate (FPR),

$$\mathrm{FPR} = \frac{\mathrm{FP}}{\mathrm{FP{+}TN}} = \frac{\mathrm{FP}}{\mathrm{N}}; \tag{7.15}$$

- positive predictive value (PPV),

$$\mathrm{PPV} = \frac{\mathrm{TP}}{\mathrm{TP{+}FP}} = \frac{\mathrm{TP}}{\mathrm{P}'}; \tag{7.16}$$

- accuracy (ACC),

$$\mathrm{ACC} = \frac{\mathrm{TP{+}TN}}{\mathrm{P{+}N}}. \tag{7.17}$$

In the *information retrieval* (or document retrieval) literature (Baeza-Yates and Ribeiro-Neto 1999; Manning, Raghavan, and Schütze 2008), the term *precision* (how many returned documents are relevant) is used instead of PPV and *recall* (what fraction of relevant documents was found) is used instead of TPR (see also Section 6.3.3). The precision and recall can be combined into a single measure called the *F-score*, which is their harmonic mean. This single measure is often used to rank vision algorithms (Knapitsch, Park *et al.* 2017).

Any particular matching strategy (at a particular threshold or parameter setting) can be rated by the TPR and FPR numbers; ideally, the true positive rate will be close to 1 and the false positive rate close to 0. As we vary the matching threshold, we obtain a family of such points, which are collectively known as the *receiver operating characteristic (ROC) curve* (Fawcett 2006) (Figure 7.22a). The closer this curve lies to the upper left corner, i.e., the larger the area under the curve (AUC), the better its performance. Figure 7.22b shows how we can plot the number of matches and non-matches as a function of inter-feature distance d. These curves can then be used to plot an ROC

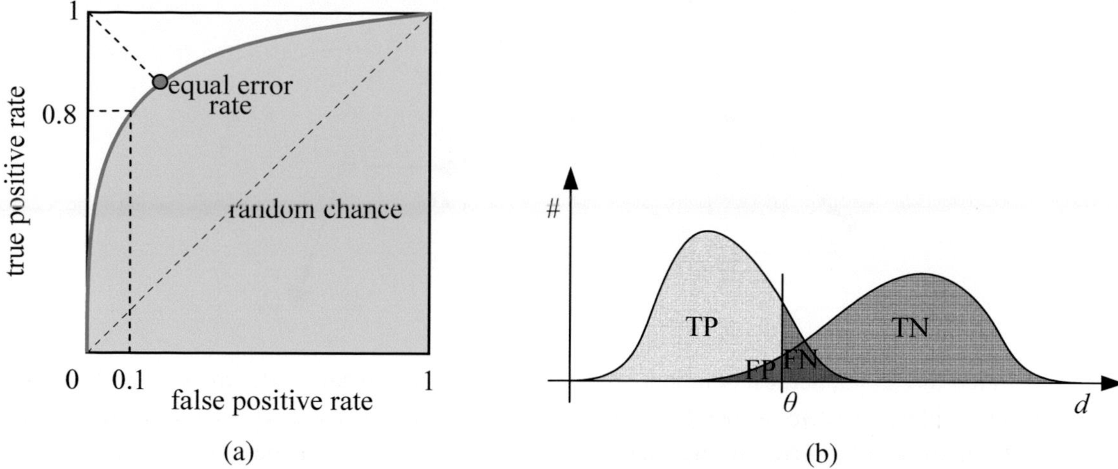

Figure 7.22 ROC curve and its related rates: (a) The ROC curve plots the true positive rate against the false positive rate for a particular combination of feature extraction and matching algorithms. Ideally, the true positive rate should be close to 1, while the false positive rate is close to 0. The area under the ROC curve (AUC) is often used as a single (scalar) measure of algorithm performance. Alternatively, the equal error rate is sometimes used. (b) The distribution of positives (matches) and negatives (non-matches) as a function of inter-feature distance d. As the threshold θ is increased, the number of true positives (TP) and false positives (FP) increases.

curve (Exercise 7.3). The ROC curve can also be used to calculate the *mean average precision*, which is the average precision (PPV) as you vary the threshold to select the best results, then the two top results, etc. (see Section 6.3.3 and Figure 6.27).

The problem with using a fixed threshold is that it is difficult to set; the useful range of thresholds can vary a lot as we move to different parts of the feature space (Lowe 2004; Mikolajczyk and Schmid 2005). A better strategy in such cases is to simply match the *nearest neighbor* in feature space. Since some features may have no matches (e.g., they may be part of background clutter in object recognition or they may be occluded in the other image), a threshold is still used to reduce the number of false positives.

Ideally, this threshold itself will adapt to different regions of the feature space. If sufficient training data is available (Hua, Brown, and Winder 2007), it is sometimes possible to learn different thresholds for different features. Often, however, we are simply given a collection of images to match, e.g., when stitching images or constructing 3D models from unordered photo collections (Brown and Lowe 2007, 2005; Snavely, Seitz, and Szeliski 2006). In this case, a useful heuristic can be to compare the nearest neighbor distance to that of the second nearest neighbor, preferably taken from an image that is known not to match the target (e.g., a different object in the database) (Brown and Lowe 2002; Lowe 2004; Mishkin, Matas, and Perdoch 2015). We can define this *nearest neighbor distance ratio* (Mikolajczyk and Schmid 2005) as

$$\text{NNDR} = \frac{d_1}{d_2} = \frac{\|D_A - D_B\|}{\|D_A - D_C\|}, \tag{7.18}$$

where d_1 and d_2 are the nearest and second nearest neighbor distances, D_A is the target descriptor, and D_B and D_C are its closest two neighbors (Figure 7.23). Recent work has shown that mutual NNDR (or, at least NNDR with cross-consistency check) work noticeably better than one-way NNDR (Bellavia and Colombo 2020; Jin, Mishkin *et al.* 2021).

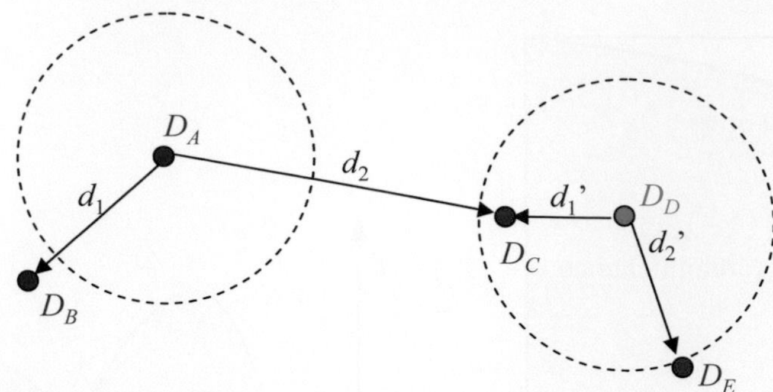

Figure 7.23 Fixed threshold, nearest neighbor, and nearest neighbor distance ratio matching. At a fixed distance threshold (dashed circles), descriptor D_A fails to match D_B and D_D incorrectly matches D_C and D_E. If we pick the nearest neighbor, D_A correctly matches D_B but D_D incorrectly matches D_C. Using nearest neighbor distance ratio (NNDR) matching, the small NNDR d_1/d_2 correctly matches D_A with D_B, and the large NNDR d_1'/d_2' correctly rejects matches for D_D.

Efficient matching

Once we have decided on a matching strategy, we still need to efficiently search for potential candidates. The simplest way to find all corresponding feature points is to compare all features against all other features in each pair of potentially matching images. While traditionally this has been too computationally expensive, modern GPUs have enabled such comparisons.

A more efficient approach is to devise an *indexing structure*, such as a multi-dimensional search tree or a hash table, to rapidly search for features near a given feature. Such indexing structures can either be built for each image independently (which is useful if we want to only consider certain potential matches, e.g., searching for a particular object) or globally for all the images in a given database, which can potentially be faster, since it removes the need to iterate over each image. For extremely large databases (millions of images or more), even more efficient structures based on ideas from document retrieval, e.g., *vocabulary trees* (Nistér and Stewénius 2006), *product quantization* (Jégou, Douze, and Schmid 2010; Johnson, Douze, and Jégou 2021), or an *inverted multi-index* (Babenko and Lempitsky 2015b) can be used, as discussed in Section 7.1.4.

One of the simpler techniques to implement is multi-dimensional hashing, which maps descriptors into fixed size buckets based on some function applied to each descriptor vector. At matching time, each new feature is hashed into a bucket, and a search of nearby buckets is used to return potential candidates, which can then be sorted or graded to determine which are valid matches.

A simple example of hashing is the Haar wavelets used by Brown, Szeliski, and Winder (2005) in their MOPS paper. During the matching structure construction, each 8×8 scaled, oriented, and normalized MOPS patch is converted into a three-element index by performing sums over different quadrants of the patch. The resulting three values are normalized by their expected standard deviations and then mapped to the two (of $b = 10$) nearest 1D bins. The three-dimensional indices formed by concatenating the three quantized values are used to index the $2^3 = 8$ bins where the feature is stored (added). At query time, only the primary (closest) indices are used, so only a single three-dimensional bin needs to be examined. The coefficients in the bin can then be used to select k approximate nearest neighbors for further processing (such as computing the NNDR).

A more complex, but more widely applicable, version of hashing is called *locality sensitive hashing*, which uses unions of independently computed hashing functions to index the features (Gionis, Indyk, and Motwani 1999; Shakhnarovich, Darrell, and Indyk 2006). Shakhnarovich, Vi-

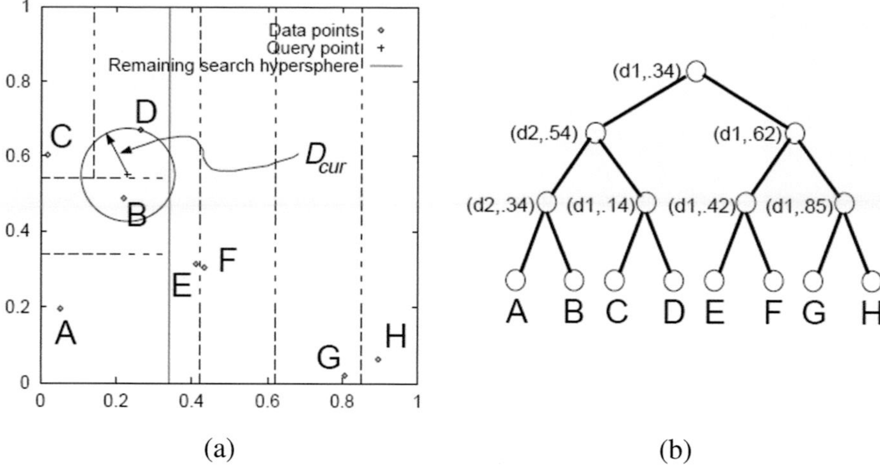

(a) (b)

Figure 7.24 K-d tree and best bin first (BBF) search (Beis and Lowe 1999) © 1999 IEEE: (a) The spatial arrangement of the axis-aligned cutting planes is shown using dashed lines. Individual data points are shown as small diamonds. (b) The same subdivision can be represented as a tree, where each interior node represents an axis-aligned cutting plane (e.g., the top node cuts along dimension d1 at value .34) and each leaf node is a data point. During a BBF search, a query point (denoted by "+") first looks in its containing bin (D) and then in its nearest adjacent bin (B), rather than its closest neighbor in the tree (C).

ola, and Darrell (2003) extend this technique to be more sensitive to the distribution of points in parameter space, which they call *parameter-sensitive hashing*. More recent work converts high-dimensional descriptor vectors into binary codes that can be compared using Hamming distances (Torralba, Weiss, and Fergus 2008; Weiss, Torralba, and Fergus 2008) or that can accommodate arbitrary kernel functions (Kulis and Grauman 2009; Raginsky and Lazebnik 2009).

Another widely used class of indexing structures are multi-dimensional search trees. The best known of these are *k-d trees*, also often written as kd-trees, which divide the multi-dimensional feature space along alternating axis-aligned hyperplanes, choosing the threshold along each axis so as to maximize some criterion, such as the search tree balance (Samet 1989). Figure 7.24 shows an example of a two-dimensional k-d tree. Here, eight different data points A–H are shown as small diamonds arranged on a two-dimensional plane. The k-d tree recursively splits this plane along axis-aligned (horizontal or vertical) cutting planes. Each split can be denoted using the dimension number and split value (Figure 7.24b). The splits are arranged so as to try to balance the tree, i.e., to keep its maximum depth as small as possible. At query time, a classic k-d tree search first locates the query point (+) in its appropriate bin (D), and then searches nearby leaves in the tree (C, B, . . .) until it can guarantee that the nearest neighbor has been found. The best bin first (BBF) search (Beis and Lowe 1999) searches bins in order of their spatial proximity to the query point and is therefore usually more efficient.

Many additional data structures have been developed for solving exact and approximate nearest neighbor problems (Arya, Mount *et al.* 1998; Liang, Liu *et al.* 2001; Hjaltason and Samet 2003). For example, Nene and Nayar (1997) developed a technique they call *slicing* that uses a series of 1D binary searches on the point list sorted along different dimensions to efficiently cull down a list of candidate points that lie within a hypercube of the query point. Grauman and Darrell (2005) reweight the matches at different levels of an indexing tree, which allows their technique to be less sensitive to discretization errors in the tree construction. Nistér and Stewénius (2006) use a *metric tree*, which compares feature descriptors to a small number of prototypes at each level in a hierarchy. The

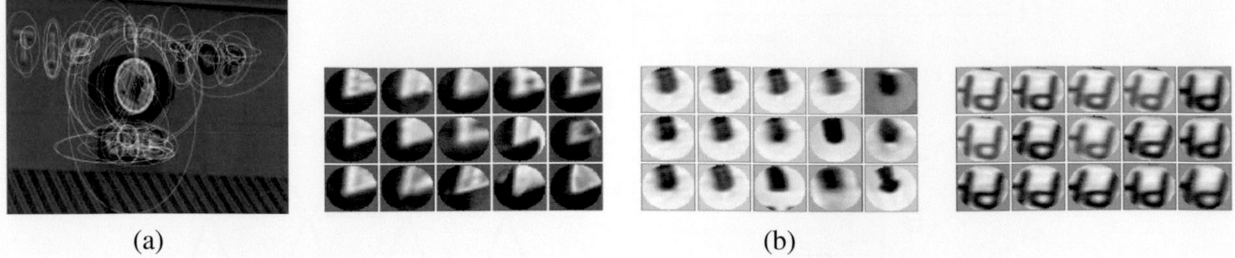

Figure 7.25 Visual words obtained from elliptical normalized affine regions (Sivic and Zisserman 2009) © 2009 IEEE. (a) Affine covariant regions are extracted from each frame and clustered into visual words using k-means clustering on SIFT descriptors with a learned Mahalanobis distance. (b) The central patch in each grid shows the query and the surrounding patches show the nearest neighbors.

resulting quantized *visual words* can then be used with classical information retrieval (document relevance) techniques to quickly winnow down a set of potential candidates from a database of millions of images (Section 7.1.4). Muja and Lowe (2009) compare a number of these approaches, introduce a new one of their own (priority search on hierarchical k-means trees), and conclude that multiple randomized k-d trees often provide the best performance. Modern libraries for computing approximate nearest neighbors include FLANN (Muja and Lowe 2014) and Faiss (Johnson, Douze, and Jégou 2021), which are discussed in Section 5.1.1 and Appendix C.2.

Feature match verification and densification

Once we have some candidate matches, we can use geometric alignment (Section 8.1) to verify which matches are *inliers* and which ones are *outliers*. For example, if we expect the whole image to be translated or rotated in the matching view, we can fit a global geometric transform and keep only those feature matches that are sufficiently close to this estimated transformation. The process of selecting a small set of seed matches and then verifying a larger set is often called *random sampling* or RANSAC (Section 8.1.4). Once an initial set of correspondences has been established, some systems look for additional matches, e.g., by looking for additional correspondences along epipolar lines (Section 12.1) or in the vicinity of estimated locations based on the global transform. It is also possible to use deep neural networks to perform feature matching and filtering, as in the SuperGlue system of Sarlin, DeTone *et al.* (2020). These topics are discussed further in Sections 8.1 and 12.2.

7.1.4 Large-scale matching and retrieval

As the number of objects in the database starts to grow (say, billions of objects or video frames), the time it takes to match a new image against each database image can become prohibitive. Instead of comparing the images one at a time, techniques are needed to quickly narrow down the search to a few likely images, which can then be compared using a more conservative verification stage.

The problem of quickly finding partial matches between documents is one of the central problems in *information retrieval* (IR) (Baeza-Yates and Ribeiro-Neto 1999; Manning, Raghavan, and Schütze 2008). In computer vision, the problem of finding a particular object in a large collection is called *content-based image retrieval (CBIR)* (Smeulders, Worring *et al.* 2000; Lew, Sebe *et al.* 2006; Vasconcelos 2007; Datta, Joshi *et al.* 2008) or *instance retrieval* (Zheng, Yang, and Tian 2018). The basic approach in fast document retrieval algorithms is to precompute an *inverted index* between individual words and the documents (or web pages or news stories) where they occur. More precisely,

Figure 7.26 Location or building recognition using randomized trees (Philbin, Chum *et al.* 2007) © 2007 IEEE. The left image is the query, the other images are the highest-ranked results.

the *frequency* of occurrence of particular words in a document is used to quickly find documents that match a particular query.

Sivic and Zisserman (2009) were the first to adapt IR techniques to visual search. In their Video Google system, affine invariant features are first detected in all the video frames they are indexing using both *shape adapted* regions around Harris feature points (Schaffalitzky and Zisserman 2002; Mikolajczyk and Schmid 2004) and maximally stable extremal regions (Matas, Chum *et al.* 2004; Section 7.1.1), as shown in Figure 7.25a. Next, 128-dimensional SIFT descriptors are computed from each normalized region (i.e., the patches shown in Figure 7.25b). Then, an average covariance matrix for these descriptors is estimated by accumulating statistics for features tracked from frame to frame. The feature descriptor covariance $\mathbf{\Sigma}$ is then used to define a Mahalanobis distance (5.32) between feature descriptors. In practice, feature descriptors are *whitened* by pre-multiplying them by $\mathbf{\Sigma}^{-1/2}$ so that Euclidean distances can be used.[5]

To apply fast information retrieval techniques to images, the high-dimensional feature descriptors that occur in each image must first be mapped into discrete *visual words*. Sivic and Zisserman (2003) perform this mapping using k-means clustering, while some of the later methods (Nistér and Stewénius 2006; Philbin, Chum *et al.* 2007) use alternative techniques, such as vocabulary trees or randomized forests. To keep the clustering time manageable, only a few hundred video frames are used to learn the cluster centers, which still involves estimating several thousand clusters from about 300,000 descriptors, although subsequent work has greatly extended this capacity (Nistér and Stewénius 2006; Philbin, Chum *et al.* 2007; Mikulik, Perdoch *et al.* 2013). At visual query time, each feature in a new query region (e.g., Figure 7.25a, which is a cropped region from a larger video frame) is mapped to its corresponding visual word. To keep very common patterns from contaminating the results, a *stop list* of the most common visual words is created and such words are dropped from further consideration.

Once a query image or region has been mapped into its constituent visual words, likely matching images must then be retrieved from the database. The exact details of how this is done can be found in Sivic and Zisserman (2009), Nistér and Stewénius (2006), Philbin, Chum *et al.* (2007), Chum, Philbin *et al.* (2007), Philbin, Chum *et al.* (2008), and also in the first edition of this book (Szeliski 2010, Section 14.3.2). Because of the high efficiency in both quantizing and scoring features, the vocabulary-tree-based recognition system built by Nistér and Stewénius (2006) was able to process incoming images in real time against a database of 40,000 CD covers and at 1Hz when matching a database of one million frames taken from six feature-length movies.

[5]Note that the computation of feature covariances from matched feature points is much more sensible than simply performing a PCA on the descriptor space (Winder and Brown 2007). This corresponds roughly to the *within-class* scatter matrix we studied in Section 5.2.3.

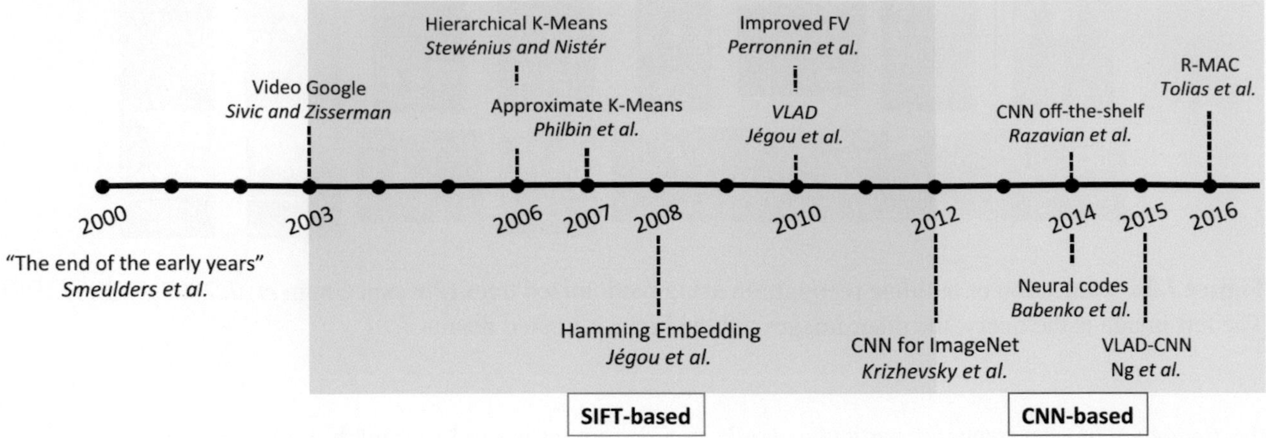

Figure 7.27 Milestones in instance retrieval (Zheng, Yang, and Tian 2018) © 2018 IEEE, showing the shift from hand-crafted feature-based retrieval to CNN-based approaches.

Instance recognition systems continued to improve rapidly in the 2000s. Philbin, Chum *et al.* (2007) showed that randomized forest of k-d trees perform better than vocabulary trees on a large location recognition task (Figure 7.26). They also compared the effects of using different 2D motion models (Section 2.1.1) in the verification stage. In follow-on work, Chum, Philbin *et al.* (2007) applied another idea from information retrieval, namely *query expansion*, which involves re-submitting top-ranked images from the initial query as additional queries to generate additional candidate results.[6] Philbin, Chum *et al.* (2008) showed how to mitigate quantization problems in visual words selection using *soft assignment*, where each feature descriptor is mapped to a number of nearby visual words, which is similar to the multiple assignment idea proposed earlier by Jégou, Harzallah, and Schmid (2007). However, such techniques tend to reduce the sparsity of visual word vectors and increase the memory and computation costs. Jégou, Douze, and Schmid (2008) incorporated partial geometrical information and an explicit matching scheme between local descriptors in the initial large-scale image ranking stage. Taken together, these algorithms helped instance recognition algorithms perform Web-scale retrieval, matching, 3D reconstruction tasks (Agarwal, Furukawa *et al.* 2010, 2011; Frahm, Fite-Georgel *et al.* 2010; Snavely, Simon *et al.* 2010).

Since the "deep learning revolution" in 2012, researchers have started developing neural feature detectors and descriptors (Sections 7.1.1 and 7.1.2) and sometimes combining them into end-to-end matching systems.[7] Figure 7.27 shows some of the major milestones in instance retrieval, while Figure 7.28 shows the variety of different classic and CNN-based retrieval architectures that have been considered. The survey paper by Zheng, Yang, and Tian (2018) describes and contrasts these various algorithms in more detail and also provides an experimental comparison of some of these algorithms on image retrieval datasets. You can also find more details on related techniques and systems in Section 6.2.3 on visual similarity search, which discusses global descriptors that represent an image with a single vector (Arandjelovic, Gronat *et al.* 2016; Radenović, Tolias, and Chum 2019; Yang, Kien Nguyen *et al.* 2019; Cao, Araujo, and Sim 2020; Ng, Balntas *et al.* 2020; Tolias, Jenicek, and Chum 2020) as alternatives to bags of local features, Section 11.2.3 on location recognition, and Section 11.4.6 on large-scale 3D reconstruction from community (internet) photos.

[6] An alternative to query expansion is *database-side augmentation* (Arandjelović and Zisserman 2012).

[7] But note that some popular open-source large-scale reconstruction systems such as COLMAP still use traditional features and indexing schemes (Schönberger and Frahm 2016).

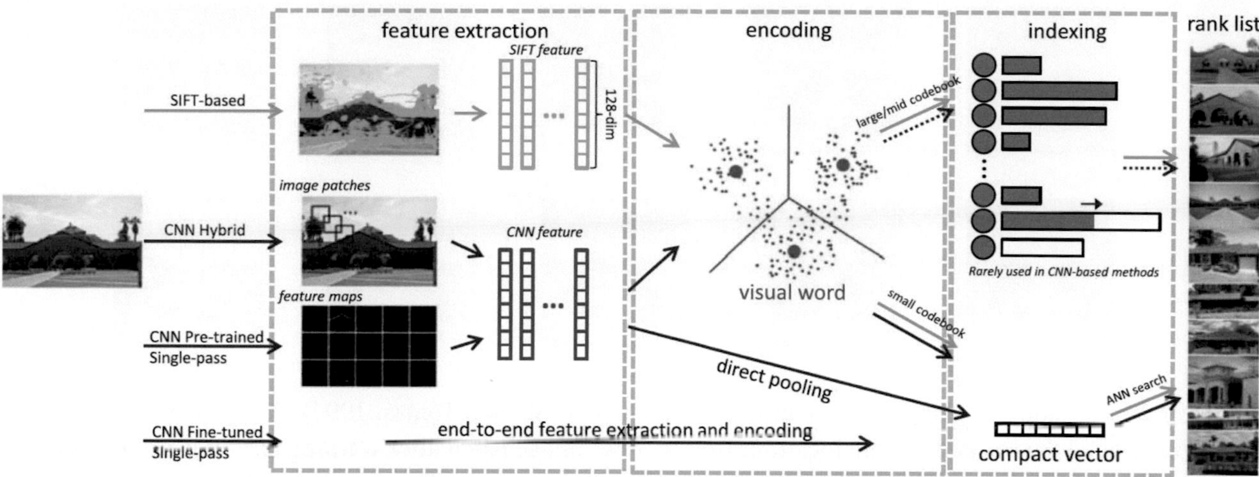

Figure 7.28 Typical pipeline for feature-based instance retrieval (Zheng, Yang, and Tian 2018) © 2018 IEEE, showing the feature extraction, encoding, and indexing portions, which are often collapsed when using a deep learning framework.

7.1.5 Feature tracking

An alternative to independently finding features in all candidate images and then matching them is to find a set of likely feature locations in a first image and to then *search* for their corresponding locations in subsequent images. This kind of *detect then track* approach is more widely used for video tracking applications, where the expected amount of motion and appearance deformation between adjacent frames is expected to be small.

The process of selecting good features to track is closely related to selecting good features for more general recognition applications. In practice, regions containing high gradients in both directions, i.e., which have high eigenvalues in the auto-correlation matrix (7.8), provide stable locations at which to find correspondences (Shi and Tomasi 1994).

In subsequent frames, searching for locations where the corresponding patch has low squared difference (7.1) often works well enough. However, if the images are undergoing brightness change, explicitly compensating for such variations (9.9) or using *normalized cross-correlation* (9.11) may be preferable. If the search range is large, it is also often more efficient to use a *hierarchical* search strategy, which uses matches in lower-resolution images to provide better initial guesses and hence speed up the search (Section 9.1.1). Alternatives to this strategy involve learning what the appearance of the patch being tracked should be and then searching for it in the vicinity of its predicted position (Avidan 2001; Jurie and Dhome 2002; Williams, Blake, and Cipolla 2003). These topics are all covered in more detail in Section 9.1.3.

If features are being tracked over longer image sequences, their appearance can undergo larger changes. You then have to decide whether to continue matching against the originally detected patch (feature) or to re-sample each subsequent frame at the matching location. The former strategy is prone to failure, as the original patch can undergo appearance changes such as foreshortening. The latter runs the risk of the feature drifting from its original location to some other location in the image (Shi and Tomasi 1994). (Mathematically, small misregistration errors compound to create a *Markov random walk*, which leads to larger drift over time.)

A preferable solution is to compare the original patch to later image locations using an *affine*

Figure 7.29 Feature tracking using an affine motion model (Shi and Tomasi 1994) © 1994 IEEE, Top row: image patch around the tracked feature location. Bottom row: image patch after warping back toward the first frame using an affine deformation. Even though the speed sign gets larger from frame to frame, the affine transformation maintains a good resemblance between the original and subsequent tracked frames.

motion model (Section 9.2). Shi and Tomasi (1994) first compare patches in neighboring frames using a translational model and then use the location estimates produced by this step to initialize an affine registration between the patch in the current frame and the base frame where a feature was first detected (Figure 7.29). In their system, features are only detected infrequently, i.e., only in regions where tracking has failed. In the usual case, an area around the current *predicted* location of the feature is searched with an incremental registration algorithm (Section 9.1.3). The resulting tracker is often called the Kanade–Lucas–Tomasi (KLT) tracker.

Since their original work on feature tracking, Shi and Tomasi's approach has generated a plethora of follow-on papers and applications. Beardsley, Torr, and Zisserman (1996) use extended feature tracking combined with structure from motion (Chapter 11) to incrementally build up sparse 3D models from video sequences. Kang, Szeliski, and Shum (1997) tie together the corners of adjacent (regularly gridded) patches to provide some additional stability to the tracking, at the cost of poorer handling of occlusions. Tommasini, Fusiello *et al.* (1998) provide a better spurious match rejection criterion for the basic Shi and Tomasi algorithm, Collins and Liu (2003) provide improved mechanisms for feature selection and dealing with larger appearance changes over time, and Shafique and Shah (2005) develop algorithms for feature matching (data association) for videos with large numbers of moving objects or points. Lepetit and Fua (2005) and Yilmaz, Javed, and Shah (2006) survey the larger field of object tracking, which includes not only feature-based techniques but also alternative techniques based on contour and region (Section 7.3).

A more recent development in feature tracking is the use of learning algorithms to build special-purpose recognizers to rapidly search for matching features anywhere in an image (Lepetit, Pilet, and Fua 2006; Hinterstoisser, Benhimane *et al.* 2008; Rogez, Rihan *et al.* 2008; Özuysal, Calonder *et al.* 2010). By taking the time to train classifiers on sample patches and their affine deformations, extremely fast and reliable feature detectors can be constructed, which enables much faster motions to be supported (Figure 7.30). Coupling such features to deformable models (Pilet, Lepetit, and Fua 2008) or structure-from-motion algorithms (Klein and Murray 2008) can result in even higher stability.

While feature-based tracking is still widely used in real-time applications such as SLAM, autonomous navigation, and augmented reality (Section 11.5), a lot of current work on tracking is focused on whole *object tracking* (Chellappa, Sankaranarayanan *et al.* 2010; Smeulders, Chu *et al.* 2014), which we study in more detail in Section 9.4.4.

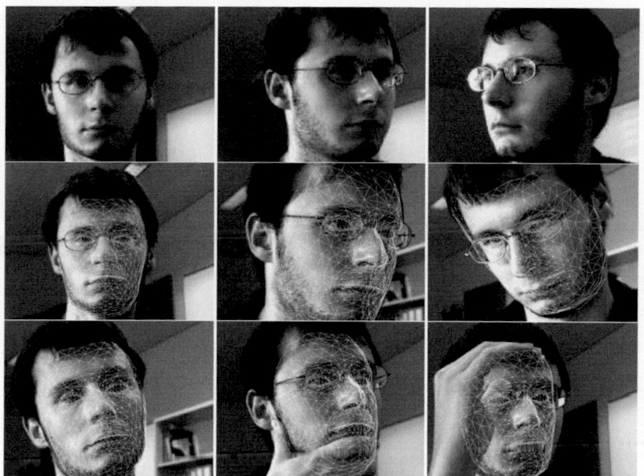

Figure 7.30 Real-time head tracking using fast trained classifiers (Lepetit, Pilet, and Fua 2004) © 2004 IEEE.

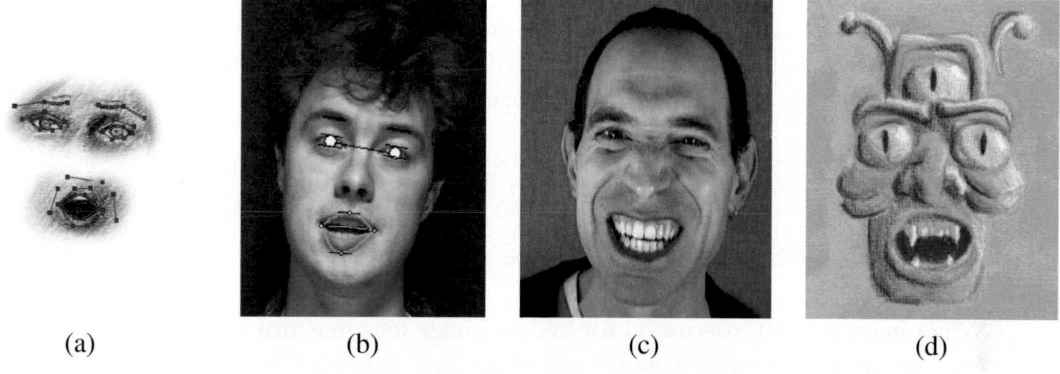

(a) (b) (c) (d)

Figure 7.31 Performance-driven, hand-drawn animation (Buck, Finkelstein *et al.* 2000) © 2000 ACM: (a) eye and mouth portions of hand-drawn sketch with their overlaid control lines; (b) an input video frame with the tracked features overlaid; (c) a different input video frame along with its (d) corresponding hand-drawn animation.

7.1.6 *Application*: Performance-driven animation

One of the most compelling applications of fast feature tracking is *performance-driven animation*, i.e., the interactive deformation of a 3D graphics model based on tracking a user's motions (Williams 1990; Litwinowicz and Williams 1994; Lepetit, Pilet, and Fua 2004).

Buck, Finkelstein *et al.* (2000) present a system that tracks a user's facial expressions and head motions and then uses them to morph among a series of hand-drawn sketches. An animator first extracts the eye and mouth regions of each sketch and draws control lines over each image (Figure 7.31a). At run time, a face-tracking system (Toyama 1998) determines the current location of these features (Figure 7.31b). The animation system decides which input images to morph based on nearest neighbor feature appearance matching and triangular barycentric interpolation. It also computes the global location and orientation of the head from the tracked features. The resulting morphed eye and mouth regions are then composited back into the overall head model to yield a frame of hand-drawn animation (Figure 7.31d).

In more recent work, Barnes, Jacobs *et al.* (2008) watch users animate paper cutouts on a desk

Figure 7.32 Human boundary detection (Martin, Fowlkes, and Malik 2004) © 2004 IEEE. The darkness of the edges corresponds to how many human subjects marked an object boundary at that location.

and then turn the resulting motions and drawings into seamless 2D animations. Feature-based facial trackers continue to be widely used (Zollhöfer, Thies *et al.* 2018), both in the visual effects industry, as well as for real-time smartphone augmented reality effects such as Facebook's Spark AR Face Masks.

7.2 Edges and contours

While interest points are useful for finding image locations that can be accurately matched in 2D, edge points are far more plentiful and often carry important semantic associations. For example, the boundaries of objects, which also correspond to occlusion events in 3D, are usually delineated by visible contours. Other kinds of edges correspond to shadow boundaries or crease edges, where surface orientation changes rapidly. Isolated edge points can also be grouped into longer *curves* or *contours*, as well as *straight line segments* (Section 7.4). It is interesting that even young children have no difficulty in recognizing familiar objects or animals from such simple line drawings.

7.2.1 Edge detection

Given an image, how can we find the salient edges? Consider the color images in Figure 7.32. If someone asked you to point out the most "salient" or "strongest" edges or the object boundaries, which ones would you trace? How closely do your perceptions match the edge images shown in Figure 7.32?

Qualitatively, edges occur at boundaries between regions of different color, intensity, or texture (Martin, Fowlkes, and Malik 2004; Arbeláez, Maire *et al.* 2011; Pont-Tuset, Arbeláez *et al.* 2017). Unfortunately, segmenting an image into coherent regions is a difficult task, which we address in Section 7.5. Often, it is preferable to detect edges using only purely local information.

Under such conditions, a reasonable approach is to define an edge as a location of *rapid intensity or color variation*. Think of an image as a height field. On such a surface, edges occur at locations of *steep slopes*, or equivalently, in regions of closely packed contour lines (on a topographic map).

A mathematical way to define the slope and direction of a surface is through its gradient,

$$\mathbf{J}(\mathbf{x}) = \nabla I(\mathbf{x}) = \left(\frac{\partial I}{\partial x}, \frac{\partial I}{\partial y}\right)(\mathbf{x}). \tag{7.19}$$

The local gradient vector **J** points in the direction of *steepest ascent* in the intensity function. Its magnitude is an indication of the slope or strength of the variation, while its orientation points in a direction *perpendicular* to the local contour.

Unfortunately, taking image derivatives accentuates high frequencies and hence amplifies noise, as the proportion of noise to signal is larger at high frequencies. It is therefore prudent to smooth the image with a low-pass filter prior to computing the gradient. Because we would like the response of our edge detector to be independent of orientation, a circularly symmetric smoothing filter is desirable. As we saw in Section 3.2, the Gaussian is the only separable circularly symmetric filter, so it is used in most edge detection algorithms. Canny (1986) discusses alternative filters and a number of researchers review alternative edge detection algorithms and compare their performance (Davis 1975; Nalwa and Binford 1986; Nalwa 1987; Deriche 1987; Freeman and Adelson 1991; Nalwa 1993; Heath, Sarkar *et al.* 1998; Crane 1997; Ritter and Wilson 2000; Bowyer, Kranenburg, and Dougherty 2001; Arbeláez, Maire *et al.* 2011; Pont-Tuset, Arbeláez *et al.* 2017).

Because differentiation is a linear operation, it commutes with other linear filtering operations. The gradient of the smoothed image can therefore be written as

$$\mathbf{J}_\sigma(\mathbf{x}) = \nabla[G_\sigma(\mathbf{x}) * I(\mathbf{x})] = [\nabla G_\sigma](\mathbf{x}) * I(\mathbf{x}), \tag{7.20}$$

i.e., we can convolve the image with the horizontal and vertical derivatives of the Gaussian kernel function,

$$\nabla G_\sigma(\mathbf{x}) = \left(\frac{\partial G_\sigma}{\partial x}, \frac{\partial G_\sigma}{\partial y}\right)(\mathbf{x}) = [-x \;\; -y]\frac{1}{\sigma^2}\exp\left(-\frac{x^2+y^2}{2\sigma^2}\right), \tag{7.21}$$

where the parameter σ indicates the width of the Gaussian. This is the same computation that is performed by Freeman and Adelson's (1991) first-order steerable filter, which we have already covered in Section 3.2.3.

For many applications, however, we wish to thin such a continuous gradient image to return isolated edges only, i.e., as single pixels at discrete locations along the edge contours. This can be achieved by looking for *maxima* in the edge strength (gradient magnitude) in a direction *perpendicular* to the edge orientation, i.e., along the gradient direction.

Finding this maximum corresponds to taking a directional derivative of the strength field in the direction of the gradient and then looking for zero crossings. The desired directional derivative is equivalent to the dot product between a second gradient operator and the results of the first,

$$S_\sigma(\mathbf{x}) = \nabla \cdot \mathbf{J}_\sigma(\mathbf{x}) = [\nabla^2 G_\sigma](\mathbf{x}) * I(\mathbf{x}). \tag{7.22}$$

The gradient operator dot product with the gradient is called the *Laplacian*. The convolution kernel

$$\nabla^2 G_\sigma(\mathbf{x}) = \left(\frac{x^2+y^2}{\sigma^4} - \frac{2}{\sigma^2}\right) G_\sigma(\mathbf{x}), \tag{7.23}$$

is therefore called the *Laplacian of Gaussian* (LoG) kernel (Marr and Hildreth 1980). This kernel can be split into two separable parts,

$$\nabla^2 G_\sigma(\mathbf{x}) = \left(\frac{x^2}{2\sigma^4} - \frac{1}{\sigma^2}\right) G_\sigma(x)G_\sigma(y) + \left(\frac{y^2}{2\sigma^4} - \frac{1}{\sigma^2}\right) G_\sigma(y)G_\sigma(x) \tag{7.24}$$

(Wiejak, Buxton, and Buxton 1985), which allows for a much more efficient implementation using separable filtering (Section 3.2.1).

In practice, it is quite common to replace the Laplacian of Gaussian convolution with a difference of Gaussian (DoG) computation, since the kernel shapes are qualitatively similar (Figure 3.34). This is especially convenient if a "Laplacian pyramid" (Section 3.5) has already been computed.[8]

In fact, it is not strictly necessary to take differences between adjacent levels when computing the edge field. Think about what a zero crossing in a "generalized" difference of Gaussians image represents. The finer (smaller kernel) Gaussian is a noise-reduced version of the original image. The coarser (larger kernel) Gaussian is an estimate of the average intensity over a larger region. Thus, whenever the DoG image changes sign, this corresponds to the (slightly blurred) image going from relatively darker to relatively lighter, as compared to the average intensity in that neighborhood.

Once we have computed the sign function $S(\mathbf{x})$, we must find its *zero crossings* and convert these into edge elements (*edgels*). An easy way to detect and represent zero crossings is to look for adjacent pixel locations $\mathbf{x}_i$ and $\mathbf{x}_j$ where the sign changes value, i.e., $[S(\mathbf{x}_i) > 0] \neq [S(\mathbf{x}_j) > 0]$.

The sub-pixel location of this crossing can be obtained by computing the "x-intercept" of the "line" connecting $S(\mathbf{x}_i)$ and $S(\mathbf{x}_j)$,

$$\mathbf{x}_z = \frac{\mathbf{x}_i S(\mathbf{x}_j) - \mathbf{x}_j S(\mathbf{x}_i)}{S(\mathbf{x}_j) - S(\mathbf{x}_i)}. \tag{7.25}$$

The orientation and strength of such edgels can be obtained by linearly interpolating the gradient values computed on the original pixel grid.

An alternative edgel representation can be obtained by linking adjacent edgels on the dual grid to form edgels that live *inside* each square formed by four adjacent pixels in the original pixel grid.[9] The advantage of this representation is that the edgels now live on a grid offset by half a pixel from the original pixel grid and are thus easier to store and access. As before, the orientations and strengths of the edges can be computed by interpolating the gradient field or estimating these values from the difference of Gaussian image (see Exercise 7.7).

In applications where the accuracy of the edge orientation is more important, higher-order steerable filters can be used (Freeman and Adelson 1991) (see Section 3.2.3). Such filters are more selective for more elongated edges and also have the possibility of better modeling curve intersections because they can represent multiple orientations at the same pixel (Figure 3.16). Their disadvantage is that they are more expensive to compute and the directional derivative of the edge strength does not have a simple closed form solution.[10]

Scale selection and blur estimation

As we mentioned before, the derivative, Laplacian, and Difference of Gaussian filters (7.20–7.23) all require the selection of a spatial scale parameter σ. If we are only interested in detecting sharp edges, the width of the filter can be determined from image noise characteristics (Canny 1986; Elder and Zucker 1998). However, if we want to detect edges that occur at different resolutions (Figures 7.33b–c), a *scale-space* approach that detects and then selects edges at different scales may be necessary (Witkin 1983; Lindeberg 1994, 1998a; Nielsen, Florack, and Deriche 1997).

Elder and Zucker (1998) present a principled approach to solving this problem. Given a known image noise level, their technique computes, for every pixel, the minimum scale at which an edge

[8]Recall that Burt and Adelson's (1983a) "Laplacian pyramid" actually computes differences of Gaussian-filtered levels.

[9]This algorithm is a 2D version of the 3D *marching cubes* isosurface extraction algorithm (Lorensen and Cline 1987).

[10]In fact, the edge orientation can have a 180° ambiguity for "bar edges", which makes the computation of zero crossings in the derivative more tricky.

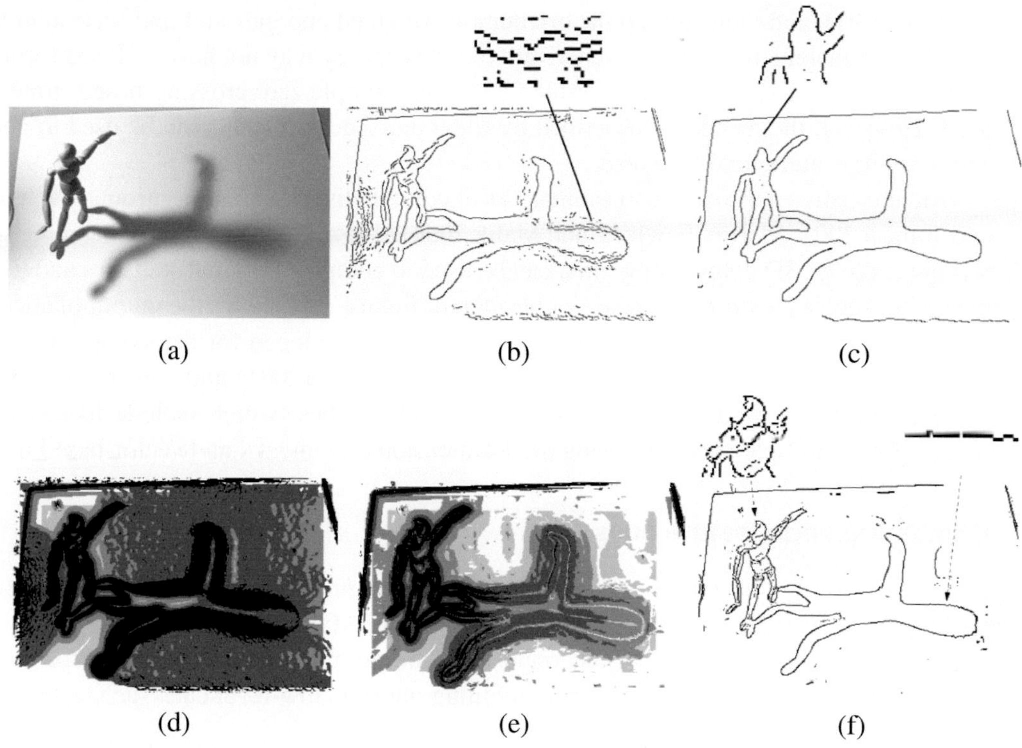

Figure 7.33 Scale selection for edge detection (Elder and Zucker 1998) © 1998 IEEE: (a) original image; (b–c) Canny/Deriche edge detector tuned to the finer (mannequin) and coarser (shadow) scales; (d) minimum reliable scale for gradient estimation; (e) minimum reliable scale for second derivative estimation; (f) final detected edges.

can be reliably detected (Figure 7.33d). Their approach first computes gradients densely over an image by selecting among gradient estimates computed at different scales, based on their gradient magnitudes. It then performs a similar estimate of minimum scale for directed second derivatives and uses zero crossings of this latter quantity to robustly select edges (Figures 7.33e–f). As an optional final step, the blur width of each edge can be computed from the distance between extrema in the second derivative response minus the width of the Gaussian filter.

Color edge detection

While most edge detection techniques have been developed for grayscale images, color images can provide additional information. For example, noticeable edges between *iso-luminant* colors (colors that have the same luminance) are useful cues but fail to be detected by grayscale edge operators.

One simple approach is to combine the outputs of grayscale detectors run on each color band separately.[11] However, some care must be taken. For example, if we simply sum up the gradients in each of the color bands, the signed gradients may actually cancel each other! (Consider, for example a pure red-to-green edge.) We could also detect edges independently in each band and then take the union of these, but this might lead to thickened or doubled edges that are hard to link.

A better approach is to compute the *oriented energy* in each band (Morrone and Burr 1988; Perona and Malik 1990a), e.g., using a second-order steerable filter (Section 3.2.3) (Freeman and

[11] Instead of using the raw RGB space, a more perceptually uniform color space such as L*a*b* (see Section 2.3.2) can be used instead. When trying to match human performance (Martin, Fowlkes, and Malik 2004), this makes sense. However, in terms of the physics of the underlying image formation and sensing, it may be a questionable strategy.

Adelson 1991), and then sum up the orientation-weighted energies and find their joint best orientation. Unfortunately, the directional derivative of this energy may not have a closed form solution (as in the case of signed first-order steerable filters), so a simple zero crossing-based strategy cannot be used. However, the technique described by Elder and Zucker (1998) can be used to compute these zero crossings numerically instead.

An alternative approach is to estimate local color statistics in regions around each pixel (Ruzon and Tomasi 2001; Martin, Fowlkes, and Malik 2004). This has the advantage that more sophisticated techniques (e.g., 3D color histograms) can be used to compare regional statistics and that additional measures, such as texture, can also be considered. Figure 7.34 shows the output of such detectors.

Over the years, many other approaches have been developed for detecting color edges, dating back to early work by Nevatia (1977). Ruzon and Tomasi (2001) and Gevers, van de Weijer, and Stokman (2006) provide good reviews of these approaches, which include ideas such as fusing outputs from multiple channels, using multidimensional gradients, and vector-based methods.

Combining edge feature cues

If the goal of edge detection is to match human *boundary detection* performance (Bowyer, Kranenburg, and Dougherty 2001; Martin, Fowlkes, and Malik 2004; Arbeláez, Maire *et al.* 2011; Pont-Tuset, Arbeláez *et al.* 2017), as opposed to simply finding stable features for matching, even better detectors can be constructed by combining multiple low-level cues such as brightness, color, and texture.

Martin, Fowlkes, and Malik (2004) describe a system that combines brightness, color, and texture edges to produce state-of-the-art performance on a database of hand-segmented natural color images (Martin, Fowlkes *et al.* 2001). First, they construct and train separate oriented half-disc detectors for measuring significant differences in brightness (luminance), color (a* and b* channels, summed responses), and texture (un-normalized filter bank responses from the work of Malik, Belongie *et al.* (2001)). Some of the responses are then sharpened using a soft non-maximal suppression technique. Finally, the outputs of the three detectors are combined using a variety of machine-learning techniques, from which logistic regression is found to have the best tradeoff between speed, space, and accuracy . The resulting system (see Figure 7.34 for some examples) is shown to outperform previously developed techniques. Maire, Arbelaez *et al.* (2008) improve on these results by combining the detector based on local appearance with a *spectral* (segmentation-based) detector (Belongie and Malik 1998). In follow-on work, Arbeláez, Maire *et al.* (2011) build a hierarchical segmentation on top of this edge detector using a variant of the watershed algorithm.

7.2.2 Contour detection

While isolated edges can be useful for a variety of applications, such as line detection (Section 7.4) and sparse stereo matching (Section 12.2), they become even more useful when linked into continuous contours.

If the edges have been detected using zero crossings of some function, linking them up is straightforward, since adjacent edgels share common endpoints. Linking the edgels into chains involves picking up an unlinked edgel and following its neighbors in both directions. Either a sorted list of edgels (sorted first by x coordinates and then by y coordinates, for example) or a 2D array can be used to accelerate the neighbor finding. If edges were not detected using zero crossings, finding the continuation of an edgel can be tricky. In this case, comparing the orientation (and, optionally, phase) of adjacent edgels can be used for disambiguation. Ideas from connected component computation can also sometimes be used to make the edge linking process even faster (see Exercise 7.8).

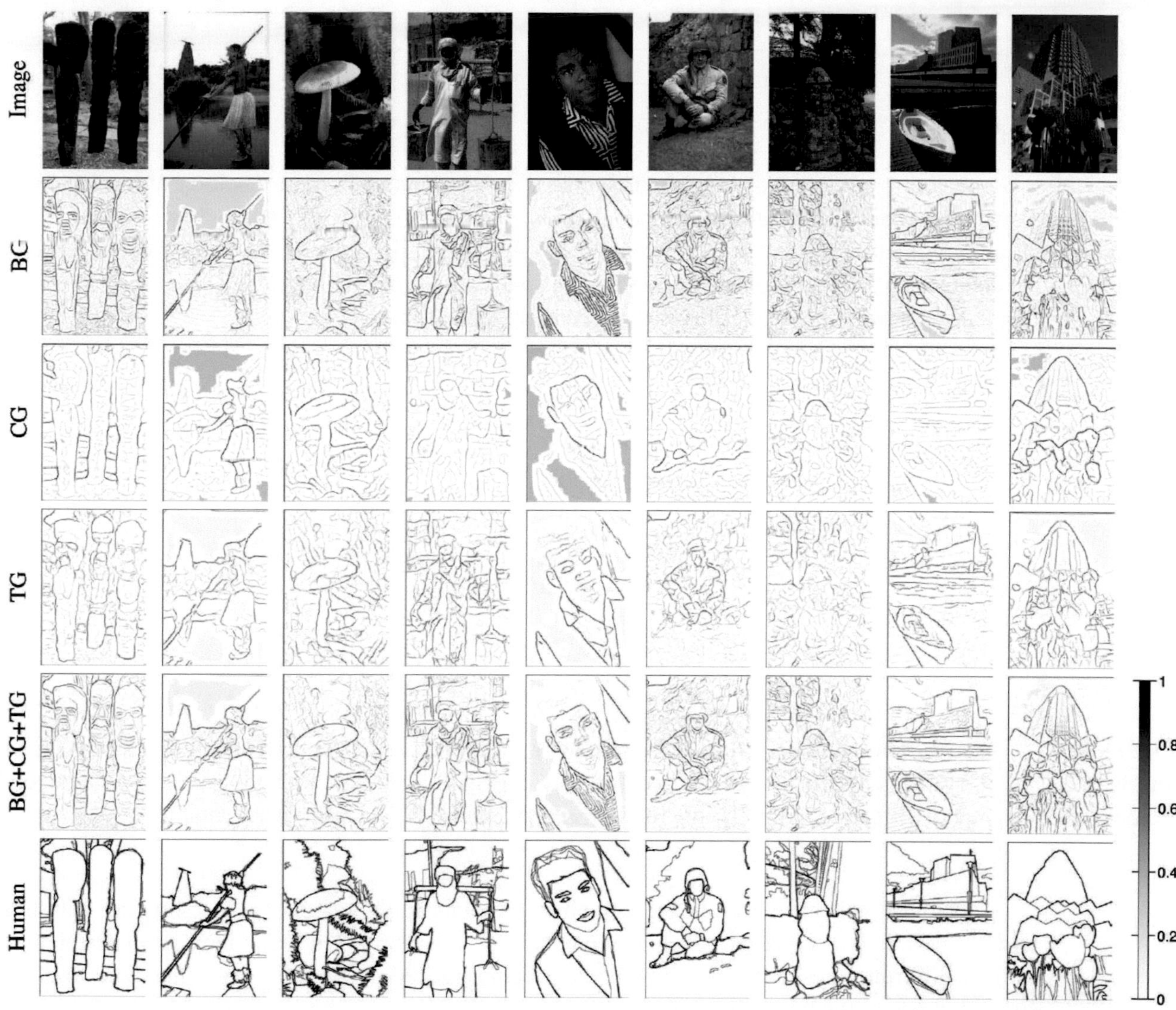

Figure 7.34 Combined brightness, color, texture boundary detector (Martin, Fowlkes, and Malik 2004) © 2004 IEEE. Successive rows show the outputs of the brightness gradient (BG), color gradient (CG), texture gradient (TG), and combined (BG+CG+TG) detectors. The final row shows human-labeled boundaries derived from a database of hand-segmented images (Martin, Fowlkes *et al.* 2001).

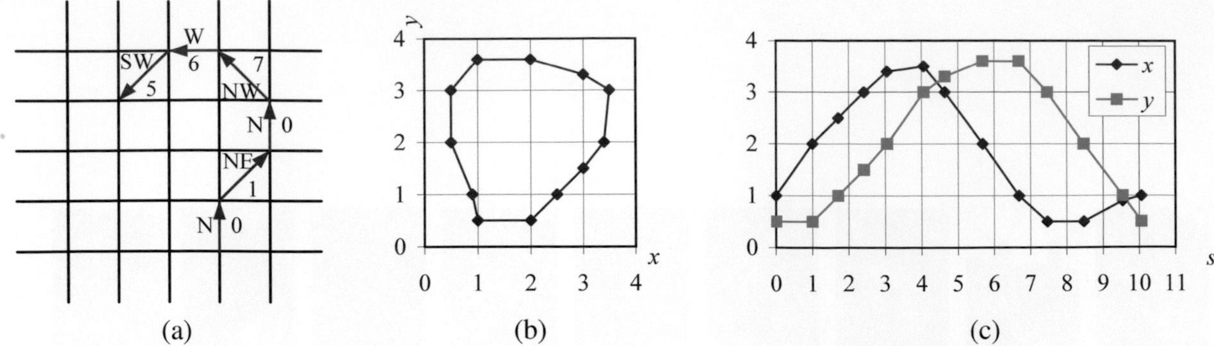

Figure 7.35 Some coding alternatives for linked contours. (a) A chain code representation of a grid-aligned linked edge chain. The code is represented as a series of direction codes, e.g., 0 1 0 7 6 5, which can further be compressed using predictive and run-length coding. (b–c) Arc-length parameterization of a contour. Discrete points along the contour (b) are first transcribed as (c) (x, y) pairs along the arc length s. This curve can then be regularly re-sampled or converted into alternative (e.g., Fourier) representations.

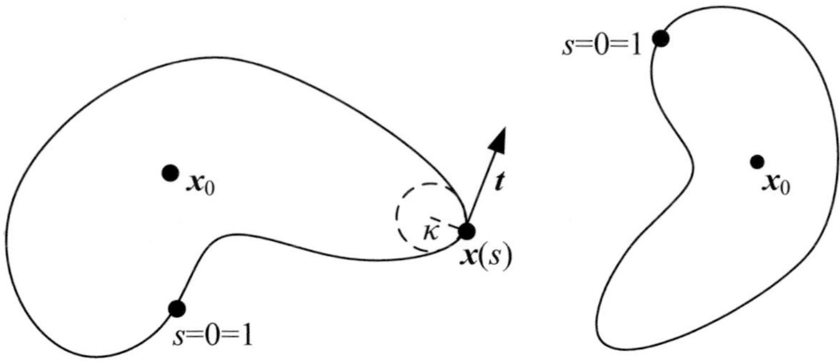

Figure 7.36 Matching two contours using their arc-length parameterization. If both curves are normalized to unit length, $s \in [0, 1]$ and centered around their centroid $\mathbf{x}_0$, they will have the same descriptor up to an overall "temporal" shift (due to different starting points for $s = 0$) and a phase (x-y) shift (due to rotation).

Once the edgels have been linked into chains, we can apply an optional thresholding with hysteresis to remove low-strength contour segments (Canny 1986). The basic idea of hysteresis is to set two different thresholds and allow a curve being tracked above the higher threshold to dip in strength down to the lower threshold.

Linked edgel lists can be encoded more compactly using a variety of alternative representations. A *chain code* encodes a list of connected points lying on an $\mathcal{N}_8$ grid using a three-bit code corresponding to the eight cardinal directions (N, NE, E, SE, S, SW, W, NW) between a point and its successor (Figure 7.35a). While this representation is more compact than the original edgel list (especially if predictive variable-length coding is used), it is not very suitable for further processing.

A more useful representation is the *arc length parameterization* of a contour, $\mathbf{x}(s)$, where s denotes the arc length along a curve. Consider the linked set of edgels shown in Figure 7.35b. We start at one point (the dot at $(1.0, 0.5)$ in Figure 7.35c) and plot it at coordinate $s = 0$ (Figure 7.35c). The next point at $(2.0, 0.5)$ gets plotted at $s = 1$, and the next point at $(2.5, 1.0)$ gets plotted at $s = 1.7071$, i.e., we increment s by the length of each edge segment. The resulting plot can be resampled on a regular (say, integral) s grid before further processing.

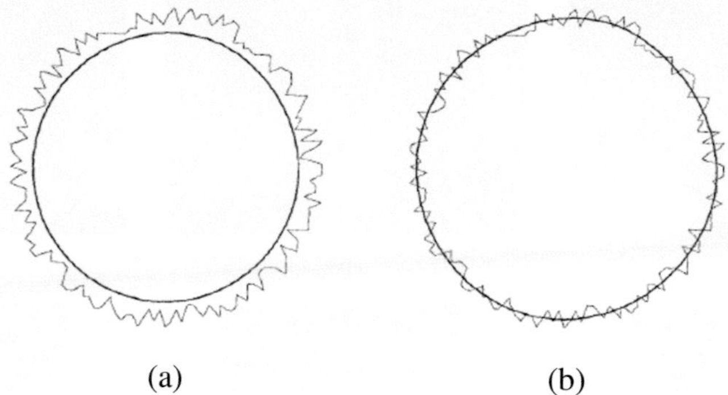

(a) (b)

Figure 7.37 Curve smoothing with a Gaussian kernel (Lowe 1988) © 1998 IEEE: (a) without a shrinkage correction term; (b) with a shrinkage correction term.

Figure 7.38 Changing the character of a curve without affecting its sweep (Finkelstein and Salesin 1994) © 1994 ACM: higher frequency wavelets can be replaced with exemplars from a style library to effect different local appearances.

The advantage of the arc-length parameterization is that it makes matching and processing (e.g., smoothing) operations much easier. Consider the two curves describing similar shapes shown in Figure 7.36. To compare the curves, we first subtract the average values $\mathbf{x}_0 = \int_s \mathbf{x}(s)$ from each descriptor. Next, we rescale each descriptor so that s goes from 0 to 1 instead of 0 to S, i.e., we divide $\mathbf{x}(s)$ by S. Finally, we take the Fourier transform of each normalized descriptor, treating each $\mathbf{x} = (x, y)$ value as a complex number. If the original curves are the same (up to an unknown scale and rotation), the resulting Fourier transforms should differ only by a scale change in magnitude plus a constant complex phase shift, due to rotation, and a linear phase shift in the domain, due to different starting points for s (see Exercise 7.9).

Arc-length parameterization can also be used to smooth curves to remove digitization noise. However, if we just apply a regular smoothing filter, the curve tends to shrink on itself (Figure 7.37a). Lowe (1989) and Taubin (1995) describe techniques that compensate for this shrinkage by adding an offset term based on second derivative estimates or a larger smoothing kernel (Figure 7.37b). An alternative approach, based on selectively modifying different frequencies in a wavelet decomposition, is presented by Finkelstein and Salesin (1994). In addition to controlling shrinkage without affecting its "sweep", wavelets allow the "character" of a curve to be interactively modified, as shown in Figure 7.38.

The evolution of curves as they are smoothed and simplified is related to "grassfire" (distance)

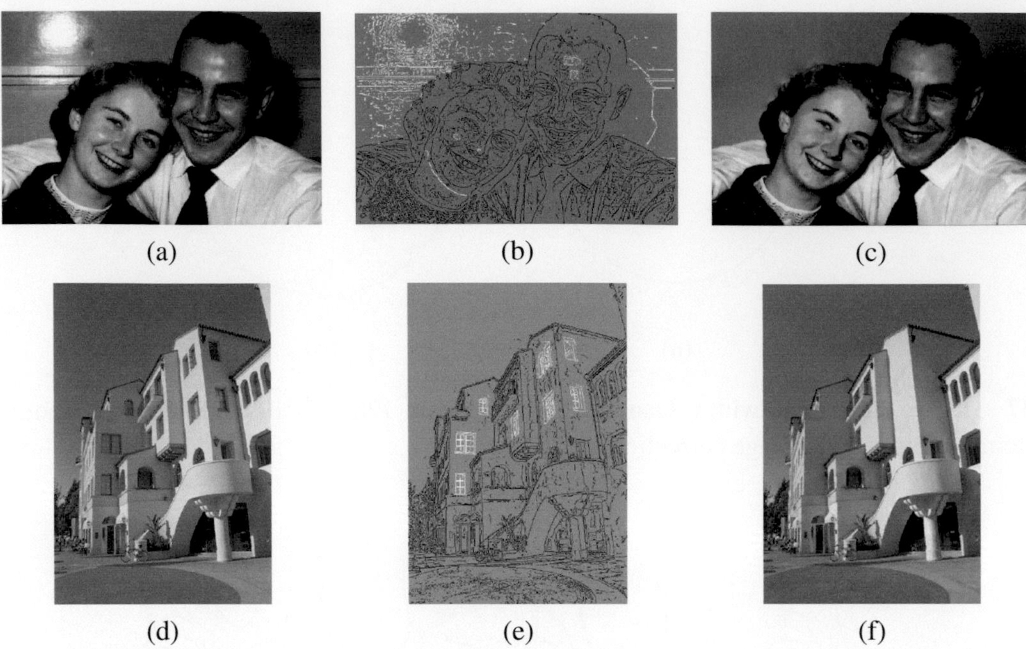

Figure 7.39 Image editing in the contour domain (Elder and Goldberg 2001) © 2001 IEEE: (a) and (d) original images; (b) and (e) extracted edges (edges to be deleted are marked in white); (c) and (f) reconstructed edited images.

transforms and region skeletons (Section 3.3.3) (Tek and Kimia 2003), and can be used to recognize objects based on their contour shape (Sebastian and Kimia 2005). More local descriptors of curve shape such as *shape contexts* (Belongie, Malik, and Puzicha 2002) can also be used for recognition and are potentially more robust to missing parts due to occlusions.

The field of contour detection and linking continues to evolve rapidly and now includes techniques for global contour grouping, boundary completion, and junction detection (Maire, Arbelaez *et al.* 2008), as well as grouping contours into likely regions (Arbeláez, Maire *et al.* 2011) and wide-baseline correspondence (Meltzer and Soatto 2008). Some additional papers that address contour detection include Xiaofeng and Bo (2012), Lim, Zitnick, and Dollár (2013), Dollár and Zitnick (2015), Xie and Tu (2015), and Pont-Tuset, Arbeláez *et al.* (2017).

7.2.3 *Application*: Edge editing and enhancement

While edges can serve as components for object recognition or features for matching, they can also be used directly for image editing.

In fact, if the edge magnitude and blur estimate are kept along with each edge, a visually similar image can be reconstructed from this information (Elder 1999). Based on this principle, Elder and Goldberg (2001) propose a system for "image editing in the contour domain". Their system allows users to selectively remove edges corresponding to unwanted features such as specularities, shadows, or distracting visual elements. After reconstructing the image from the remaining edges, the undesirable visual features have been removed (Figure 7.39).

Another potential application is to enhance perceptually salient edges while simplifying the underlying image to produce a cartoon-like or "pen-and-ink" stylized image (DeCarlo and Santella 2002). This application is discussed in more detail in Section 10.5.2.

7.3 Contour tracking

While lines, vanishing points, and rectangles are commonplace in the human-made world, curves corresponding to object boundaries are even more common, especially in the natural environment. In this section, we describe some approaches to locating such boundary curves in images.

The first, originally called *snakes* by its inventors (Kass, Witkin, and Terzopoulos 1988) (Section 7.3.1), is an energy-minimizing, two-dimensional spline curve that evolves (moves) towards image features such as strong edges. The second, *intelligent scissors* (Mortensen and Barrett 1995) (Section 7.3.1), allows the user to sketch in real time a curve that clings to object boundaries. Finally, *level set* techniques (Section 7.3.2) evolve the curve as the zero-set of a *characteristic function*, which allows them to easily change topology and incorporate region-based statistics.

All three of these are examples of *active contours* (Blake and Isard 1998; Mortensen 1999), since these boundary detectors iteratively move towards their final solution under the combination of image and optional user-guidance forces. The presentation below is heavily shortened from that presented in the first edition of this book (Szeliski 2010, Section 5.1), where interested readers can find more details.

7.3.1 Snakes and scissors

Snakes are a two-dimensional generalization of the 1D energy-minimizing splines first introduced in Section 4.2,

$$\mathcal{E}_{\text{int}} = \int \alpha(s)\|\mathbf{f}_s(s)\|^2 + \beta(s)\|\mathbf{f}_{ss}(s)\|^2\, ds, \tag{7.26}$$

where s is the arc-length along the curve $\mathbf{f}(s) = (x(s), y(s))$ and $\alpha(s)$ and $\beta(s)$ are first- and second-order continuity weighting functions analogous to the $s(x,y)$ and $c(x,y)$ terms introduced in (4.24–4.25). We can discretize this energy by sampling the initial curve position evenly along its length (Figure 7.35c) to obtain

$$\begin{aligned} E_{\text{int}} = \sum_i & \alpha(i)\|f(i+1) - f(i)\|^2/h^2 \\ & + \beta(i)\|f(i+1) - 2f(i) + f(i-1)\|^2/h^4, \end{aligned} \tag{7.27}$$

where h is the step size, which can be neglected if we resample the curve along its arc-length after each iteration.

In addition to this *internal* spline energy, a snake simultaneously minimizes external image-based and constraint-based potentials. The image-based potentials are the sum of several terms

$$\mathcal{E}_{\text{image}} = w_{\text{line}}\mathcal{E}_{\text{line}} + w_{\text{edge}}\mathcal{E}_{\text{edge}} + w_{\text{term}}\mathcal{E}_{\text{term}}, \tag{7.28}$$

where the *line* term attracts the snake to dark ridges, the *edge* term attracts it to strong gradients (edges), and the *term* term attracts it to line terminations. As the snakes evolve by minimizing their energy, they often "wiggle" and "slither", which accounts for their popular name.

Because regular snakes have a tendency to shrink, it is usually better to initialize them by drawing the snake outside the object of interest to be tracked. Alternatively, an expansion *ballooning* force can be added to the dynamics (Cohen and Cohen 1993), essentially moving each point outwards along its normal. It is also possible to replace the energy-minimizing variational evolution equations with a deep neural network to significantly improve performance (Peng, Jiang *et al.* 2020).

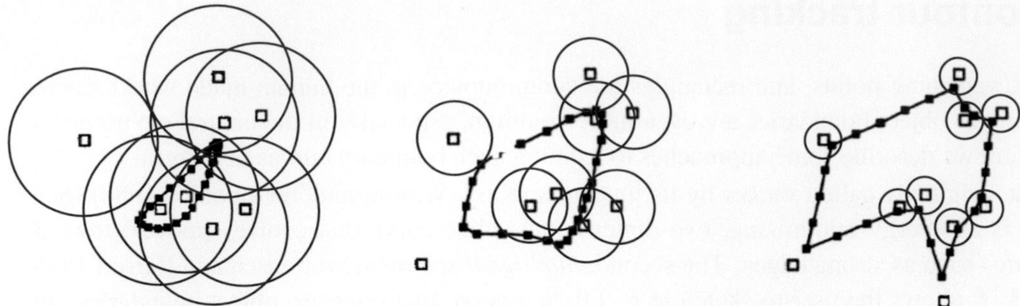

Figure 7.40 Elastic net: The open squares indicate the cities and the closed squares linked by straight line segments are the tour points. The blue circles indicate the approximate extent of the attraction force of each city, which is reduced over time. Under the Bayesian interpretation of the elastic net, the blue circles correspond to one standard deviation of the circular Gaussian that generates each city from some unknown tour point.

Elastic nets and slippery springs

An interesting variant on snakes, first proposed by Durbin and Willshaw (1987) and later re-formulated in an energy-minimizing framework by Durbin, Szeliski, and Yuille (1989), is the *elastic net* formulation of the Traveling Salesman Problem (TSP). Recall that in a TSP, the salesman must visit each city once while minimizing the total distance traversed. A snake that is constrained to pass through each city could solve this problem (without any optimality guarantees) but it is impossible to tell ahead of time which snake control point should be associated with each city.

Instead of having a fixed constraint between snake nodes and cities, a city is assumed to pass near *some* point along the tour (Figure 7.40). In a probabilistic interpretation, each city is generated as a *mixture* of Gaussians centered at each tour point,

$$p(\mathbf{d}(j)) = \sum_i p_{ij} \quad \text{with} \quad p_{ij} = e^{-d_{ij}^2/(2\sigma^2)}, \tag{7.29}$$

where σ is the standard deviation of the Gaussian and

$$d_{ij} = \|\mathbf{f}(i) - \mathbf{d}(j)\| \tag{7.30}$$

is the Euclidean distance between a tour point $\mathbf{f}(i)$ and a city location $\mathbf{d}(j)$. The corresponding data fitting energy (negative log likelihood) is

$$E_{\text{slippery}} = -\sum_j \log p(\mathbf{d}(j)) = -\sum_j \log \left[\sum e^{-\|\mathbf{f}(i)-\mathbf{d}(j)\|^2/2\sigma^2} \right]. \tag{7.31}$$

This energy derives its name from the fact that, unlike a regular spring, which couples a given snake point to a given constraint, this alternative energy defines a *slippery spring* that allows the association between constraints (cities) and curve (tour) points to evolve over time (Szeliski 1989). Note that this is a soft variant of the popular *iterative closest point* data constraint that is often used in fitting or aligning surfaces to data points or to each other (Section 13.2.1) (Besl and McKay 1992; Chen and Medioni 1992; Zhang 1994).

To compute a good solution to the TSP, the slippery spring data association energy is combined with a regular first-order internal smoothness energy (7.27) to define the cost of a tour. The tour $\mathbf{f}(s)$ is initialized as a small circle around the mean of the city points and σ is progressively lowered

(Figure 7.40). For large σ values, the tour tries to stay near the centroid of the points but as σ decreases each city pulls more and more strongly on its closest tour points (Durbin, Szeliski, and Yuille 1989). In the limit as $\sigma \to 0$, each city is guaranteed to capture at least one tour point and the tours between subsequent cites become straight lines.

Splines and shape priors

While snakes can be very good at capturing the fine and irregular detail in many real-world contours, they sometimes exhibit too many degrees of freedom, making it more likely that they can get trapped in local minima during their evolution.

One solution to this problem is to control the snake with fewer degrees of freedom through the use of B-spline approximations (Menet, Saint-Marc, and Medioni 1990b,a; Cipolla and Blake 1990). The resulting *B-snake* can be written as

$$\mathbf{f}(s) = \sum_k B_k(s)\mathbf{x}_k. \tag{7.32}$$

If the object being tracked or recognized has large variations in location, scale, or orientation, these can be modeled as an additional transformation on the control points, e.g., $\mathbf{x}'_k = s\mathbf{R}\mathbf{x}_k + \mathbf{t}$ (2.18), which can be estimated at the same time as the values of the control points. Alternatively, separate *detection* and *alignment* stages can be run to first localize and orient the objects of interest (Cootes, Cooper *et al.* 1995).

In a B-snake, because the snake is controlled by fewer degrees of freedom, there is less need for the internal smoothness forces used with the original snakes, although these can still be derived and implemented using finite element analysis, i.e., taking derivatives and integrals of the B-spline basis functions (Terzopoulos 1983; Bathe 2007).

In practice, it is more common to estimate a set of *shape priors* on the typical distribution of the control points $\{\mathbf{x}_k\}$ (Cootes, Cooper *et al.* 1995). One potential way of describing this distribution would be by the location $\bar{\mathbf{x}}_k$ and 2D covariance $\mathbf{C}_k$ of each individual point $\mathbf{x}_k$. These could then be turned into a quadratic penalty (prior energy) on the point location. In practice, however, the variation in point locations is usually highly correlated.

A preferable approach is to estimate the joint covariance of all the points simultaneously. First, concatenate all of the point locations $\{\mathbf{x}_k\}$ into a single vector $\mathbf{x}$, e.g., by interleaving the x and y locations of each point. The distribution of these vectors across all training examples can be described with a mean $\bar{\mathbf{x}}$ and a covariance

$$\mathbf{C} = \frac{1}{P}\sum_p (\mathbf{x}_p - \bar{\mathbf{x}})(\mathbf{x}_p - \bar{\mathbf{x}})^T, \tag{7.33}$$

where $\mathbf{x}_p$ are the P training examples. Using *eigenvalue analysis* (Appendix A.1.2), which is also known as *principal component analysis* (PCA) (Section 5.2.3 and Appendix B.1), the covariance matrix can be written as,

$$\mathbf{C} = \mathbf{\Phi}\ \mathrm{diag}(\lambda_0 \ldots \lambda_{K-1})\ \mathbf{\Phi}^T. \tag{7.34}$$

In most cases, the likely appearance of the points can be modeled using only a few eigenvectors with the largest eigenvalues. The resulting *point distribution model* (Cootes, Taylor *et al.* 1993; Cootes, Cooper *et al.* 1995) can be written as

$$\mathbf{x} = \bar{\mathbf{x}} + \hat{\mathbf{\Phi}}\ \mathbf{b}, \tag{7.35}$$

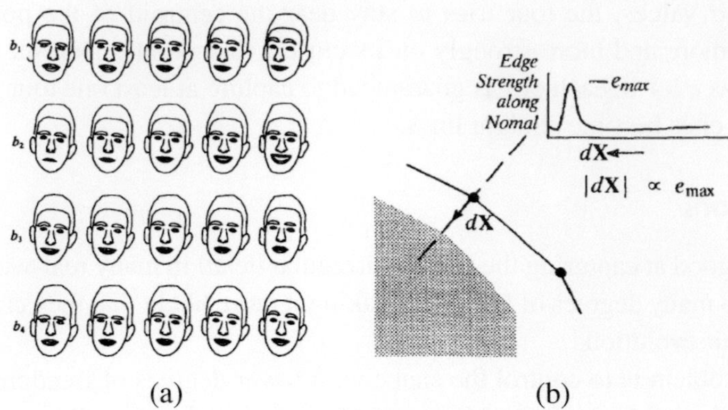

Figure 7.41 Active Shape Model (ASM): (a) the effect of varying the first four shape parameters for a set of faces (Cootes, Taylor *et al.* 1993) © 1993 IEEE; (b) searching for the strongest gradient along the normal to each control point (Cootes, Cooper *et al.* 1995) © 1995 Elsevier.

where $\mathbf{b}$ is an $M \ll K$ element *shape parameter* vector and $\hat{\mathbf{\Phi}}$ are the first m columns of $\mathbf{\Phi}$. To constrain the shape parameters to reasonable values, we can use a quadratic penalty of the form

$$E_{\text{shape}} = \frac{1}{2}\mathbf{b}^T \operatorname{diag}(\lambda_0 \ldots \lambda_{M-1})\, \mathbf{b} = \sum_m b_m^2 / 2\lambda_m. \tag{7.36}$$

Alternatively, the range of allowable b_m values can be limited to some range, e.g., $|b_m| \le 3\sqrt{\lambda_m}$ (Cootes, Cooper *et al.* 1995). Alternative approaches for deriving a set of shape vectors are reviewed by Isard and Blake (1998). Varying the individual shape parameters b_m over the range $-2\sqrt{\lambda_m} \le 2\sqrt{\lambda_m}$ can give a good indication of the expected variation in appearance, as shown in Figure 7.41a.

To align a point distribution model with an image, each control point searches in a direction normal to the contour to find the most likely corresponding image edge point (Figure 7.41b). These individual measurements can be combined with priors on the shape parameters (and, if desired, position, scale, and orientation parameters) to estimate a new set of parameters. The resulting *active shape model* (ASM) can be iteratively minimized to fit images to non-rigidly deforming objects, such as medical images, or body parts, such as hands (Cootes, Cooper *et al.* 1995). The ASM can also be combined with a PCA analysis of the underlying gray-level distribution to create an *active appearance model* (AAM) (Cootes, Edwards, and Taylor 2001), which we discussed in more detail in Section 6.2.4.

Dynamic snakes and CONDENSATION

In many applications of active contours, the object of interest is being tracked from frame to frame as it deforms and evolves. In this case, it makes sense to use estimates from the previous frame to predict and constrain the new estimates.

One way to do this is to use Kalman filtering, which results in a formulation called *Kalman snakes* (Terzopoulos and Szeliski 1992; Blake, Curwen, and Zisserman 1993). The Kalman filter is based on a linear dynamic model of shape parameter evolution,

$$\mathbf{x}_t = \mathbf{A}\mathbf{x}_{t-1} + \mathbf{w}_t, \tag{7.37}$$

where $\mathbf{x}_t$ and $\mathbf{x}_{t-1}$ are the current and previous state variables, $\mathbf{A}$ is the linear *transition matrix*, and $\mathbf{w}$ is a noise (perturbation) vector, which is often modeled as a Gaussian (Gelb 1974). The

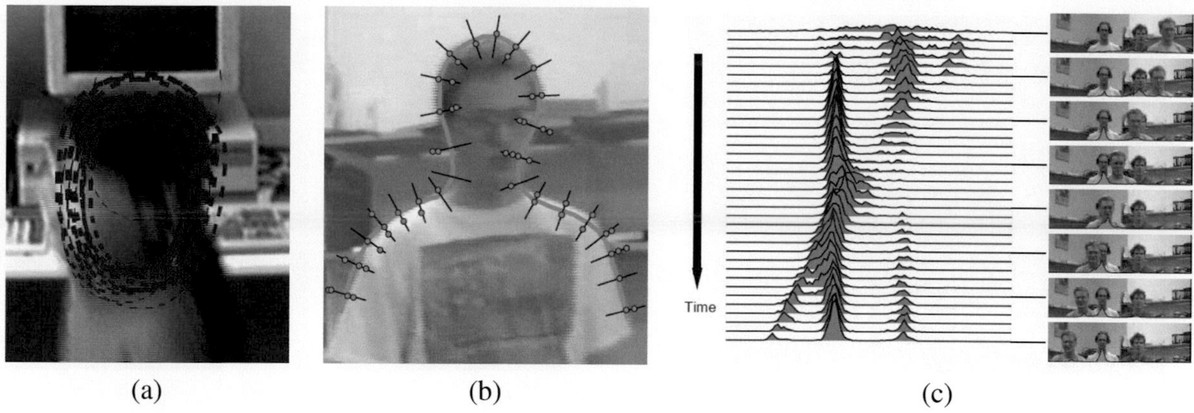

Figure 7.42 Head tracking using CONDENSATION (Isard and Blake 1998) © 1998 Springer: (a) sample set representation of head estimate distribution; (b) multiple measurements at each control vertex location; (c) multi-hypothesis tracking over time.

matrices **A** and the noise covariance can be learned ahead of time by observing typical sequences of the object being tracked (Blake and Isard 1998).

In many situations, however, such as when tracking in clutter, a better estimate for the contour can be obtained if we remove the assumptions that the distributions are Gaussian, which is what the Kalman filter requires. In this case, a general multi-modal distribution is propagated. To model such multi-modal distributions, Isard and Blake (1998) introduced the use of *particle filtering* to the computer vision community.[12] Particle filtering techniques represent a probability distribution using a collection of weighted point samples (Andrieu, de Freitas *et al.* 2003; Bishop 2006; Koller and Friedman 2009).

To update the locations of the samples according to the linear dynamics (deterministic drift), the centers of the samples are updated and multiple samples are generated for each point. These are then perturbed to account for the stochastic diffusion, i.e., their locations are moved by random vectors taken from the distribution of **w**.[13] Finally, the weights of these samples are multiplied by the measurement probability density, i.e., we take each sample and measure its likelihood given the current (new) measurements. Because the point samples represent and propagate conditional estimates of the multi-modal density, Isard and Blake (1998) dubbed their algorithm CONditional DENSity propagATION or CONDENSATION.

Figure 7.42a shows what a factored sample of a head tracker might look like, drawing a red B-spline contour for each of (a subset of) the particles being tracked. Figure 7.42b shows why the measurement density itself is often multi-modal: the locations of the edges perpendicular to the spline curve can have multiple local maxima due to background clutter. Finally, Figure 7.42c shows the temporal evolution of the conditional density (x coordinate of the head and shoulder tracker centroid) as it tracks several people over time.

Scissors

Active contours allow a user to roughly specify a boundary of interest and have the system evolve the contour towards a more accurate location as well as track it over time. The results of this curve

[12] Alternatives to modeling multi-modal distributions include *mixtures of Gaussians* (Section 5.2.2) and *multiple hypothesis tracking* (Bar-Shalom and Fortmann 1988; Cham and Rehg 1999).

[13] Note that because of the structure of these steps, non-linear dynamics and non-Gaussian noise can be used.

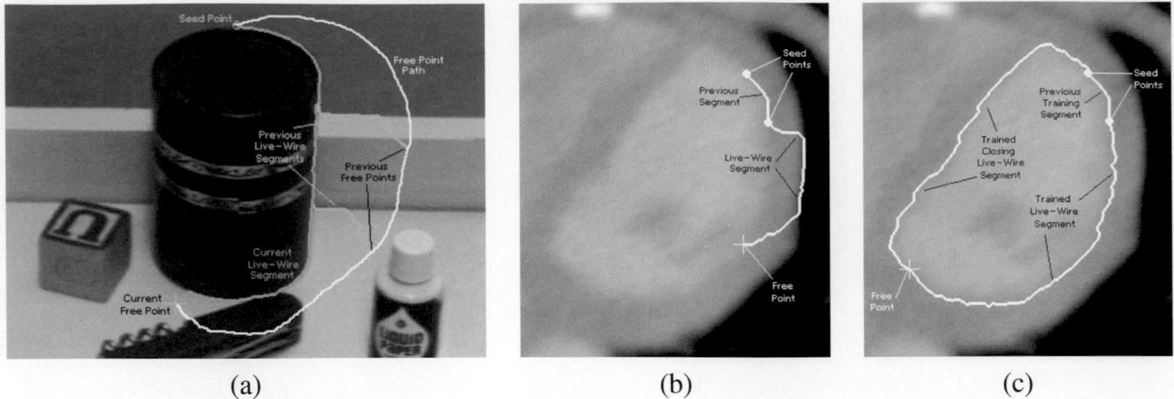

Figure 7.43 Intelligent scissors: (a) as the mouse traces the white path, the scissors follow the orange path along the object boundary (the green curves show intermediate positions) (Mortensen and Barrett 1995) © 1995 ACM; (b) regular scissors can sometimes jump to a stronger (incorrect) boundary; (c) after training to the previous segment, similar edge profiles are preferred (Mortensen and Barrett 1998) © 1995 Elsevier.

evolution, however, may be unpredictable and may require additional user-based hints to achieve the desired result.

An alternative approach is to have the system optimize the contour in real time as the user is drawing (Mortensen 1999). The *intelligent scissors* system developed by Mortensen and Barrett (1995) does just that. As the user draws a rough outline (the white curve in Figure 7.43a), the system computes and draws a better curve that clings to high-contrast edges (the orange curve).

To compute the optimal curve path (*live-wire*), the image is first pre-processed to associate low costs with edges (links between neighboring horizontal, vertical, and diagonal, i.e., $\mathcal{N}_8$ neighbors) that are likely to be boundary elements. Their system uses a combination of zero-crossing, gradient magnitudes, and gradient orientations to compute these costs.

Next, as the user traces a rough curve, the system continuously recomputes the lowest-cost path between the starting *seed point* and the current mouse location using Dijkstra's algorithm, a breadth-first dynamic programming algorithm that terminates at the current target location.

In order to keep the system from jumping around unpredictably, the system will "freeze" the curve to date (reset the seed point) after a period of inactivity. To prevent the live wire from jumping onto adjacent higher-contrast contours, the system also "learns" the intensity profile under the current optimized curve, and uses this to preferentially keep the wire moving along the same (or a similar looking) boundary (Figure 7.43b–c).

Several extensions have been proposed to the basic algorithm, which works remarkably well even in its original form. Mortensen and Barrett (1999) use *tobogganing*, which is a simple form of watershed region segmentation, to pre-segment the image into regions whose boundaries become candidates for optimized curve paths. The resulting region boundaries are turned into a much smaller graph, where nodes are located wherever three or four regions meet. The Dijkstra algorithm is then run on this reduced graph, resulting in much faster (and often more stable) performance. Another extension to intelligent scissors is to use a probabilistic framework that takes into account the current trajectory of the boundary, resulting in a system called JetStream (Pérez, Blake, and Gangnet 2001).

Instead of re-computing an optimal curve at each time instant, a simpler system can be developed by simply "snapping" the current mouse position to the nearest likely boundary point (Gleicher 1995). Applications of these boundary extraction techniques to image cutting and pasting are presented in Section 10.4.

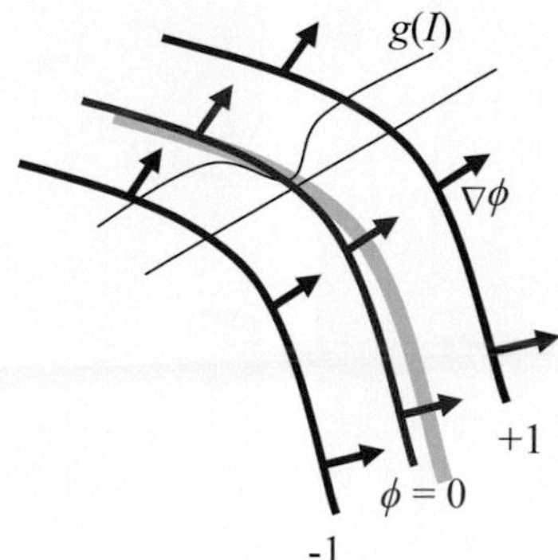

Figure 7.44 Level set evolution for a geodesic active contour. The embedding function ϕ is updated based on the curvature of the underlying surface modulated by the edge/speed function $g(I)$, as well as the gradient of $g(I)$, thereby attracting it to strong edges.

7.3.2 Level Sets

A limitation of active contours based on parametric curves of the form $\mathbf{f}(s)$, e.g., snakes, B-snakes, and CONDENSATION, is that it is challenging to change the topology of the curve as it evolves (McInerney and Terzopoulos 1999, 2000). Furthermore, if the shape changes dramatically, curve reparameterization may also be required.

An alternative representation for such closed contours is to use a *level set*, where the *zero-crossing(s)* of a *characteristic* (or signed distance (Section 3.3.3)) function define the curve. Level sets evolve to fit and track objects of interest by modifying the underlying *embedding function* (another name for this 2D function) $\phi(x, y)$ instead of the curve $\mathbf{f}(s)$ (Malladi, Sethian, and Vemuri 1995; Sethian 1999; Sapiro 2001; Osher and Paragios 2003). To reduce the amount of computation required, only a small strip (frontier) around the locations of the current zero-crossing needs to updated at each step, which results in what are called *fast marching methods* (Sethian 1999).

An example of an evolution equation is the *geodesic active contour* proposed by Caselles, Kimmel, and Sapiro (1997) and Yezzi, Kichenassamy *et al.* (1997),

$$\begin{aligned}\frac{d\phi}{dt} &= |\nabla\phi|\text{div}\left(g(I)\frac{\nabla\phi}{|\nabla\phi|}\right)\\ &= g(I)|\nabla\phi|\text{div}\left(\frac{\nabla\phi}{|\nabla\phi|}\right) + \nabla g(I)\cdot\nabla\phi,\end{aligned} \tag{7.38}$$

where $g(I)$ is a generalized version of the snake edge potential. To get an intuitive sense of the curve's behavior, assume that the embedding function ϕ is a signed distance function away from the curve (Figure 7.44), in which case $|\phi| = 1$. The first term in Equation (7.38) moves the curve in the direction of its curvature, i.e., it acts to straighten the curve, under the influence of the modulation function $g(I)$. The second term moves the curve down the gradient of $g(I)$, encouraging the curve to migrate towards minima of $g(I)$.

While this level-set formulation can readily change topology, it is still susceptible to local minima, since it is based on local measurements such as image gradients. An alternative approach is to re-cast the problem in a segmentation framework, where the energy measures the consistency of the image statistics (e.g., color, texture, motion) inside and outside the segmented regions (Cremers, Rousson, and Deriche 2007; Rousson and Paragios 2008; Houhou, Thiran, and Bresson 2008). These approaches build on earlier energy-based segmentation frameworks introduced by Leclerc

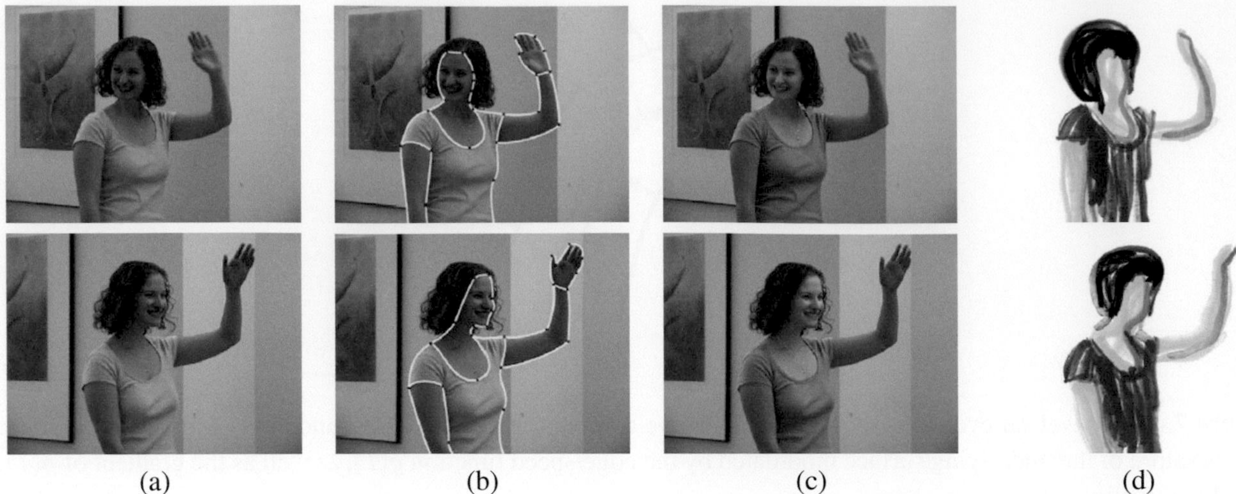

Figure 7.45 Keyframe-based rotoscoping (Agarwala, Hertzmann *et al.* 2004) © 2004 ACM: (a) original frames; (b) rotoscoped contours; (c) re-colored blouse; (d) rotoscoped hand-drawn animation.

(1989), Mumford and Shah (1989), and Chan and Vese (2001), which are discussed in more detail in Section 4.3.2.

For more information on level sets and their applications, please see the collection of papers edited by Osher and Paragios (2003) as well as the series of Workshops on Variational and Level Set Methods in Computer Vision (Paragios, Faugeras *et al.* 2005) and Special Issues on Scale Space and Variational Methods in Computer Vision (Paragios and Sgallari 2009).

7.3.3 *Application*: Contour tracking and rotoscoping

Active contours can be used in a wide variety of object-tracking applications (Blake and Isard 1998; Yilmaz, Javed, and Shah 2006). For example, they can be used to track facial features for performance-driven animation (Terzopoulos and Waters 1990; Lee, Terzopoulos, and Waters 1995; Parke and Waters 1996; Bregler, Covell, and Slaney 1997). They can also be used to track heads and people, as shown in Figure 7.42, as well as moving vehicles (Paragios and Deriche 2000). Additional applications include medical image segmentation, where contours can be tracked from slice to slice in computed tomography (Cootes and Taylor 2001), or over time, as in ultrasound scans.

An interesting application that is closer to computer animation and visual effects is *rotoscoping*, which uses the tracked contours to deform a set of hand-drawn animations (or to modify or replace the original video frames).[14] Agarwala, Hertzmann *et al.* (2004) present a system based on tracking hand-drawn B-spline contours drawn at selected keyframes, using a combination of geometric and appearance-based criteria (Figure 7.45). They also provide an excellent review of previous rotoscoping and image-based, contour-tracking systems.

Additional applications of rotoscoping (object contour detection and segmentation), such as cutting and pasting objects from one photograph into another, are presented in Section 10.4.

[14] The term comes from a device (a rotoscope) that projected frames of a live-action film underneath an acetate so that artists could draw animations directly over the actors' shapes.

7.4 Lines and vanishing points

While edges and general curves are suitable for describing the contours of natural objects, the human-made world is full of straight lines. Detecting and matching these lines can be useful in a variety of applications, including architectural modeling, pose estimation in urban environments, and the analysis of printed document layouts.

In this section, we present some techniques for extracting *piecewise linear* descriptions from the curves computed in the previous section. We begin with some algorithms for approximating a curve as a piecewise-linear polyline. We then describe the *Hough transform*, which can be used to group edgels into line segments even across gaps and occlusions. Finally, we describe how 3D lines with common *vanishing points* can be grouped together. These vanishing points can be used to calibrate a camera and to determine its orientation relative to a rectahedral scene, as described in Section 11.1.1.

7.4.1 Successive approximation

As we saw in Section 7.2.2, describing a curve as a series of 2D locations $\mathbf{x}_i = \mathbf{x}(s_i)$ provides a general representation suitable for matching and further processing. In many applications, however, it is preferable to approximate such a curve with a simpler representation, e.g., as a piecewise-linear polyline or as a B-spline curve (Farin 2002).

Many techniques have been developed over the years to perform this approximation, which is also known as *line simplification*. One of the oldest, and simplest, is the one proposed by Ramer (1972) and Douglas and Peucker (1973), who recursively subdivide the curve at the point furthest away from the line joining the two endpoints (or the current coarse polyline approximation). Hershberger and Snoeyink (1992) provide a more efficient implementation and also cite some of the other related work in this area.

Once the line simplification has been computed, it can be used to approximate the original curve. If a smoother representation or visualization is desired, either approximating or interpolating splines or curves can be used (Sections 3.5.1 and 7.3.1) (Szeliski and Ito 1986; Bartels, Beatty, and Barsky 1987; Farin 2002).

7.4.2 Hough transforms

While curve approximation with polylines can often lead to successful line extraction, lines in the real world are sometimes broken up into disconnected components or made up of many collinear line segments. In many cases, it is desirable to group such collinear segments into extended lines. At a further processing stage (described in Section 7.4.3), we can then group such lines into collections with common vanishing points.

The Hough transform, named after its original inventor (Hough 1962), is a well-known technique for having edges "vote" for plausible line locations (Duda and Hart 1972; Ballard 1981; Illingworth and Kittler 1988). In its original formulation (Figure 7.46), each edge point votes for *all* possible lines passing through it, and lines corresponding to high *accumulator* or *bin* values are examined for potential line fits.[15] Unless the points on a line are truly punctate, a better approach is to use the local orientation information at each edgel to vote for a *single* accumulator cell (Figure 7.47), as described below. A hybrid strategy, where each edgel votes for a number of possible orientation or location pairs centered around the estimate orientation, may be desirable in some cases.

[15]The Hough transform can also be *generalized* to look for other geometric features, such as circles (Ballard 1981).

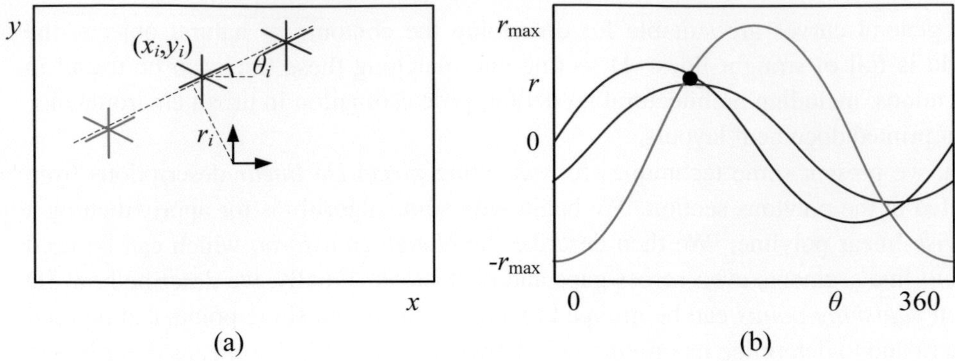

Figure 7.46 Original Hough transform: (a) each point votes for a complete family of potential lines $r_i(\theta) = x_i \cos\theta + y_i \sin\theta$; (b) each pencil of lines sweeps out a sinusoid in (r, θ); their intersection provides the desired line equation.

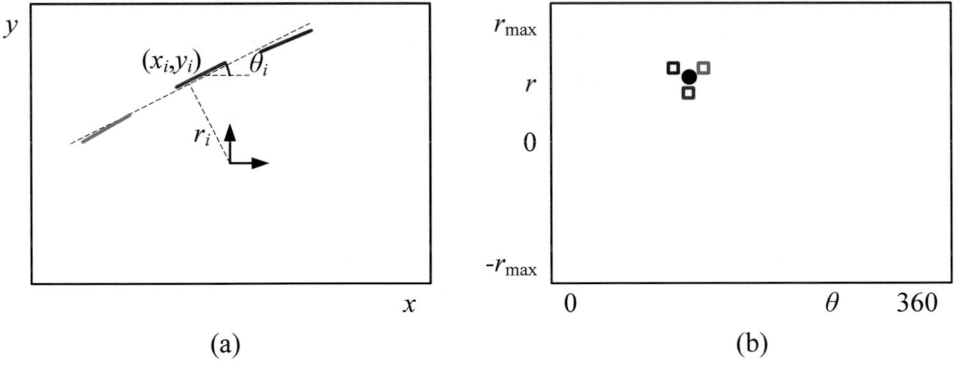

Figure 7.47 Oriented Hough transform: (a) an edgel re-parameterized in polar (r, θ) coordinates, with $\hat{\mathbf{n}}_i = (\cos\theta_i, \sin\theta_i)$ and $r_i = \hat{\mathbf{n}}_i \cdot \mathbf{x}_i$; (b) (r, θ) accumulator array, showing the votes for the three edgels marked in red, green, and blue.

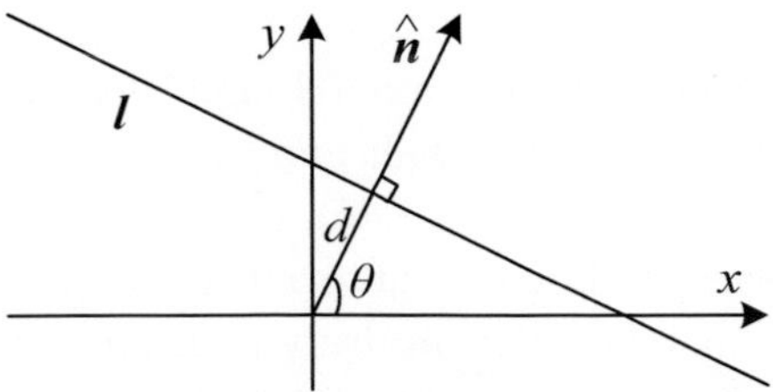

Figure 7.48 2D line equation expressed in terms of the normal $\hat{\mathbf{n}}$ and distance to the origin d.

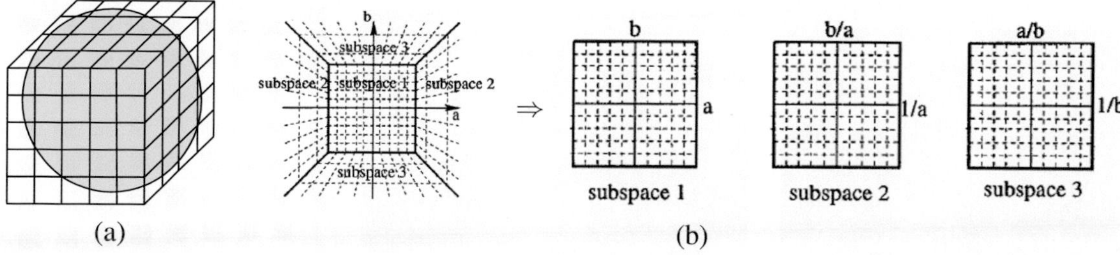

Figure 7.49 Cube map representation for line equations and vanishing points: (a) a cube map surrounding the unit sphere; (b) projecting the half-cube onto three subspaces (Tuytelaars, Van Gool, and Proesmans 1997) © 1997 IEEE.

Before we can vote for line hypotheses, we must first choose a suitable representation. Figure 7.48 (copied from Figure 2.2a) shows the normal-distance $(\hat{\mathbf{n}}, d)$ parameterization for a line. Since lines are made up of edge segments, we adopt the convention that the line normal $\hat{\mathbf{n}}$ points in the same direction (i.e., has the same sign) as the image gradient $\mathbf{J}(\mathbf{x}) = \nabla I(\mathbf{x})$ (7.19). To obtain a minimal two-parameter representation for lines, we convert the normal vector into an angle

$$\theta = \tan^{-1} n_y/n_x, \tag{7.39}$$

as shown in Figure 7.48. The range of possible (θ, d) values is $[-180°, 180°] \times [-\sqrt{2}, \sqrt{2}]$, assuming that we are using normalized pixel coordinates (2.61) that lie in $[-1, 1]$. The number of bins to use along each axis depends on the accuracy of the position and orientation estimate available at each edgel and the expected line density, and is best set experimentally with some test runs on sample imagery.

There are a lot of details in getting the Hough transform to work well, including using edge segment lengths or strengths during the voting process, keeping a list of constituent edgels in the accumulator array for easier post-processing, and optionally combining edges of different "polarity" into the same line segments. These are best worked out by writing an implementation and testing it out on sample data.

An alternative to the 2D polar (θ, d) representation for lines is to use the full 3D $\mathbf{m} = (\hat{\mathbf{n}}, d)$ line equation, projected onto the unit sphere. While the sphere can be parameterized using spherical coordinates (2.8),

$$\hat{\mathbf{m}} = (\cos\theta\cos\phi, \sin\theta\cos\phi, \sin\phi), \tag{7.40}$$

this does not uniformly sample the sphere and still requires the use of trigonometry.

An alternative representation can be obtained by using a *cube map*, i.e., projecting $\mathbf{m}$ onto the face of a unit cube (Figure 7.49a). To compute the cube map coordinate of a 3D vector $\mathbf{m}$, first find the largest (absolute value) component of $\mathbf{m}$, i.e., $m = \pm\max(|n_x|, |n_y|, |d|)$, and use this to select one of the six cube faces. Divide the remaining two coordinates by m and use these as indices into the cube face. While this avoids the use of trigonometry, it does require some decision logic.

One advantage of using the cube map, first pointed out by Tuytelaars, Van Gool, and Proesmans (1997), is that all of the lines passing through a point correspond to line segments on the cube faces, which is useful if the original (full voting) variant of the Hough transform is being used. In their work, they represent the line equation as $ax + b + y = 0$, which does not treat the x and y axes symmetrically. Note that if we restrict $d \geq 0$ by ignoring the polarity of the edge orientation (gradient sign), we can use a half-cube instead, which can be represented using only three cube faces, as shown in Figure 7.49b (Tuytelaars, Van Gool, and Proesmans 1997).

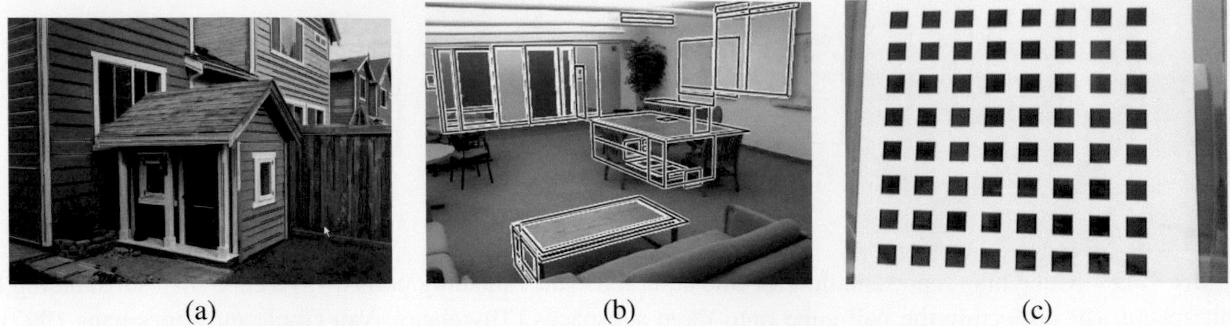

(a) (b) (c)

Figure 7.50 Real-world vanishing points: (a) architecture (Sinha, Steedly *et al.* 2008), (b) furniture (Mičušík, Wildenauer, and Košecká 2008) © 2008 IEEE, and (c) calibration patterns (Zhang 2000).

RANSAC-based line detection. Another alternative to the Hough transform is the RANdom SAmple Consensus (RANSAC) algorithm described in more detail in Section 8.1.4. In brief, RANSAC randomly chooses pairs of edgels to form a line hypothesis and then tests how many other edgels fall onto this line. (If the edge orientations are accurate enough, a single edgel can produce this hypothesis.) Lines with sufficiently large numbers of *inliers* (matching edgels) are then selected as the desired line segments.

An advantage of RANSAC is that no accumulator array is needed, so the algorithm can be more space efficient and potentially less prone to the choice of bin size. The disadvantage is that many more hypotheses may need to be generated and tested than those obtained by finding peaks in the accumulator array.

Bottom-up grouping. Yet another approach to line segment detection is to iteratively group edgels with similar orientations into oriented rectangular *line-support regions* (Burns, Hanson, and Riseman 1986). The validity of such regions can then be determined using a statistical analysis, as described in the LSD paper by Grompone von Gioi, Jakubowicz *et al.* (2008). The resulting algorithm is quite fast, does a good job of distinguishing line segments from texture, and is widely used in practice because of its performance and open source availability. Recently, deep neural network algorithms have been developed to simultaneously extract line segments and their junctions (Huang, Wang *et al.* 2018; Zhang, Li *et al.* 2019; Huang, Qin *et al.* 2020; Lin, Pintea, and van Gemert 2020).

In general, there is no clear consensus on which line estimation technique performs best. It is therefore a good idea to think carefully about the problem at hand and to implement several approaches (successive approximation, Hough, and RANSAC) to determine the one that works best for your application.

7.4.3 Vanishing points

In many scenes, structurally important lines have the same vanishing point because they are parallel in 3D. Examples of such lines are horizontal and vertical building edges, zebra crossings, railway tracks, the edges of furniture such as tables and dressers, and of course, the ubiquitous calibration pattern (Figure 7.50). Finding the vanishing points common to such line sets can help refine their position in the image and, in certain cases, help determine the intrinsic and extrinsic orientation of the camera (Section 11.1.1).

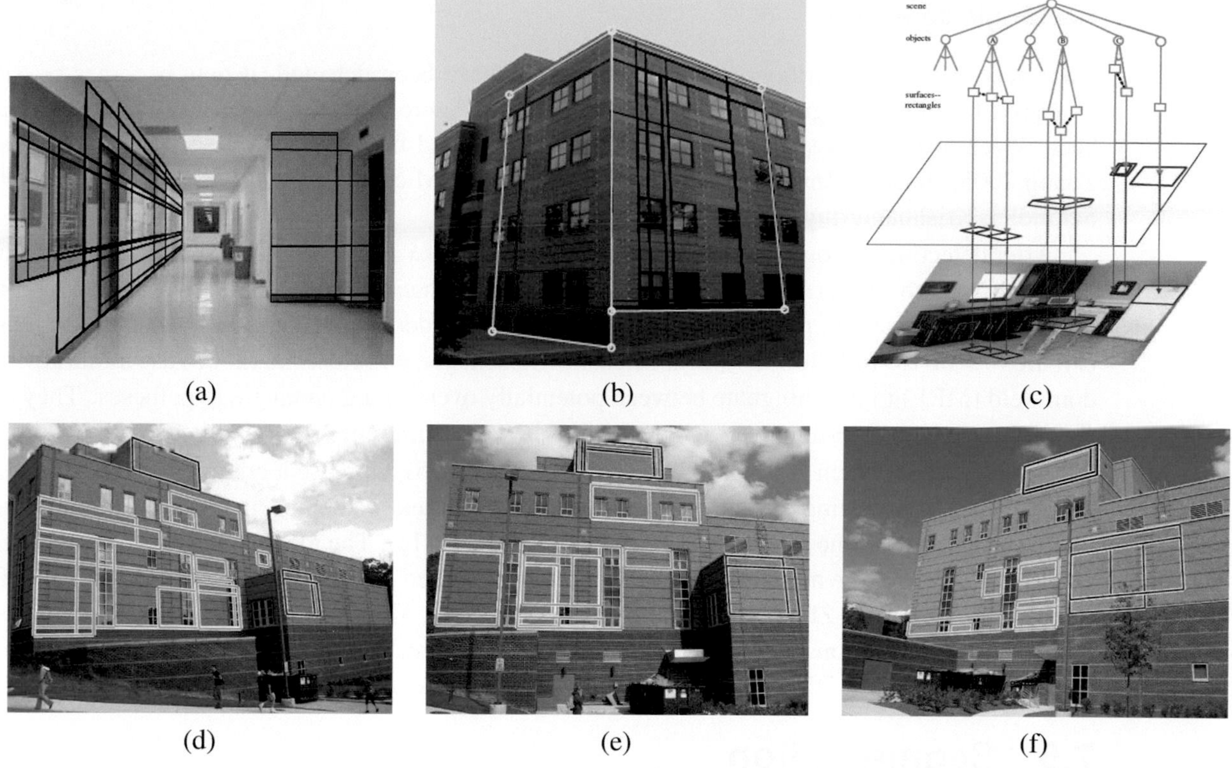

Figure 7.51 Rectangle detection: (a) indoor corridor and (b) building exterior with grouped facades (Košecká and Zhang 2005) © 2005 Elsevier; (c) grammar-based recognition (Han and Zhu 2005) © 2005 IEEE; (d–f) rectangle matching using a plane sweep algorithm (Mičušík, Wildenauer, and Košecká 2008) © 2008 IEEE.

Over the years, a large number of techniques have been developed for finding vanishing points (Quan and Mohr 1989; Collins and Weiss 1990; Brillaut-O'Mahoney 1991; McLean and Kotturi 1995; Becker and Bove 1995; Shufelt 1999; Tuytelaars, Van Gool, and Proesmans 1997; Schaffalitzky and Zisserman 2000; Antone and Teller 2002; Rother 2002; Košecká and Zhang 2005; Denis, Elder, and Estrada 2008; Pflugfelder 2008; Tardif 2009; Bazin, Seo *et al.* 2012; Antunes and Barreto 2013; Kluger, Ackermann *et al.* 2017; Zhou, Qi *et al.* 2019a)—see some of the more recent papers for additional references and alternative approaches.

In the first edition of this book (Szeliski 2010, Section 4.3.3), I presented a simple Hough technique based on having line pairs vote for potential vanishing point locations, followed by a robust least squares fitting stage. While my technique proceeds in two discrete stages, better results may be obtained by alternating between assigning lines to vanishing points and refitting the vanishing point locations (Antone and Teller 2002; Košecká and Zhang 2005; Pflugfelder 2008). The results of detecting individual vanishing points can also be made more robust by simultaneously searching for pairs or triplets of mutually orthogonal vanishing points (Shufelt 1999; Antone and Teller 2002; Rother 2002; Sinha, Steedly *et al.* 2008; Li, Kim *et al.* 2020). Some results of such vanishing point detection algorithms can be seen in Figure 7.50. It is also possible to simultaneously detect line segments and their junctions using a neural network (Zhang, Li *et al.* 2019) and to then use these to construct complete 3D wireframe models (Zhou, Qi, and Ma 2019; Zhou, Qi *et al.* 2019b).

Rectangle detection

Once sets of mutually orthogonal vanishing points have been detected, it now becomes possible to search for 3D rectangular structures in the image (Figure 7.51). A variety of techniques have been developed to find such rectangles, primarily focused on architectural scenes (Košecká and Zhang 2005; Han and Zhu 2005; Shaw and Barnes 2006; Mičušík, Wildenauer, and Košecká 2008; Schindler, Krishnamurthy *et al.* 2008).

After detecting orthogonal vanishing directions, Košecká and Zhang (2005) refine the fitted line equations, search for corners near line intersections, and then verify rectangle hypotheses by rectifying the corresponding patches and looking for a preponderance of horizontal and vertical edges (Figures 7.51a–b). In follow-on work, Mičušík, Wildenauer, and Košecká (2008) use a Markov random field (MRF) to disambiguate between potentially overlapping rectangle hypotheses. They also use a plane sweep algorithm to match rectangles between different views (Figures 7.51d–f).

A different approach is proposed by Han and Zhu (2005), who use a grammar of potential rectangle shapes and nesting structures (between rectangles and vanishing points) to infer the most likely assignment of line segments to rectangles (Figure 7.51c). The idea of using regular, repetitive structures as part of the modeling process is now being called *holistic 3D reconstruction* (Zhou, Furukawa, and Ma 2019; Zhou, Furukawa *et al.* 2020; Pintore, Mura *et al.* 2020) and will be discussed in more detail in Section 13.6.1 on modeling 3D architecture.

7.5 Segmentation

Image segmentation is the task of finding groups of pixels that "go together". In statistics and machine learning, this problem is known as *cluster analysis* or more simply *clustering* and is a widely studied area with hundreds of different algorithms (Jain and Dubes 1988; Kaufman and Rousseeuw 1990; Jain, Duin, and Mao 2000; Jain, Topchy *et al.* 2004; Xu and Wunsch 2005). We've already discussed general vector-space clustering algorithms in Section 5.2.1. The main difference between clustering and segmentation is that the former usually ignores pixel layout and neighborhoods, while the latter relies heavily on spatial cues and constraints.

In computer vision, image segmentation is one of the oldest and most widely studied problems (Brice and Fennema 1970; Pavlidis 1977; Riseman and Arbib 1977; Ohlander, Price, and Reddy 1978; Rosenfeld and Davis 1979; Haralick and Shapiro 1985). Early techniques often used region splitting or merging (Brice and Fennema 1970; Horowitz and Pavlidis 1976; Ohlander, Price, and Reddy 1978; Pavlidis and Liow 1990), which correspond to *divisive* and *agglomerative* algorithms (Jain, Topchy *et al.* 2004; Xu and Wunsch 2005), which we introduced in Section 5.2.1. More recent algorithms typically optimize some global criterion, such as intra-region consistency and inter-region boundary lengths or dissimilarity (Leclerc 1989; Mumford and Shah 1989; Shi and Malik 2000; Comaniciu and Meer 2002; Felzenszwalb and Huttenlocher 2004; Cremers, Rousson, and Deriche 2007; Pont-Tuset, Arbeláez *et al.* 2017).

We have already seen examples of image segmentation using image morphology (Section 3.3.3), Markov random fields (Section 4.3), active contours (Section 7.3), and level sets (Section 7.3.2). In the recognition chapter (Section 6.4), we studied *semantic segmentation*, whose goal is to break the image up into semantically labeled regions such as sky, grass, and individual people and animals. In this section, we review some additional techniques for bottom-up general (non-semantic) image segmentation. These include algorithms based on region splitting and merging, graph-based segmentation, and probabilistic aggregation (Section 7.5.1), *mean shift* mode finding (Section 7.5.2), and *normalized cuts* splitting based on pixel similarity metrics (Section 7.5.3). Since many of these algorithms are no longer widely used, a lot of the descriptions have been considerably shortened from those found in the first edition of this book (Szeliski 2010, Chapter 5), where you can find longer descriptions.

Since the literature on image segmentation is so vast, a good way to get a handle on some of the better performing algorithms is to look at experimental comparisons on human-labeled databases (Arbeláez, Maire *et al.* 2011; Pont-Tuset, Arbeláez *et al.* 2017). The best known of these is the Berkeley Segmentation Dataset and Benchmark (Martin, Fowlkes *et al.* 2001), which consists of 1,000 images from a Corel image dataset that were hand-labeled by 30 human subjects, for which Unnikrishnan, Pantofaru, and Hebert (2007) propose new metrics for comparing segmentation algorithms, while Estrada and Jepson (2009) compare four well-known segmentation algorithms. A newer database of foreground and background segmentations, used by Alpert, Galun *et al.* (2007), is also available.

As mentioned in Section 3.3.3, the simplest possible technique for segmenting a grayscale image is to select a threshold and then compute connected components. Unfortunately, a single threshold is rarely sufficient for the whole image because of lighting and intra-object statistical variations.

Region splitting (divisive clustering). Splitting the image into successively finer regions is one of the oldest techniques in computer vision. Ohlander, Price, and Reddy (1978) present such a technique, which first computes a histogram for the whole image and then finds a threshold that best separates the large peaks in the histogram. This process is repeated until regions are either fairly uniform or below a certain size. More recent splitting algorithms often optimize some metric of intra-region similarity and inter-region dissimilarity. These are covered in Sections 4.3.2 and Sections 7.5.3.

Region merging (agglomerative clustering). Region merging techniques also date back to the beginnings of computer vision. Brice and Fennema (1970) use a dual grid for representing boundaries between pixels and merge regions based on their relative boundary lengths and the strength of the visible edges at these boundaries.

A very simple version of pixel-based merging combines adjacent regions whose average color difference is below a threshold or whose regions are too small. Segmenting the image into such *superpixels* (Mori, Ren *et al.* 2004), which are not semantically meaningful, can be a useful preprocessing stage to make higher-level algorithms such as stereo matching (Zitnick, Kang *et al.* 2004; Taguchi, Wilburn, and Zitnick 2008), optical flow (Zitnick, Jojic, and Kang 2005; Brox, Bregler, and Malik 2009), and recognition (Mori, Ren *et al.* 2004; Mori 2005; Gu, Lim *et al.* 2009; Lim, Arbeláez *et al.* 2009) both faster and more robust. It is also possible to combine both splitting and merging by starting with a medium-grain segmentation (in a quadtree representation) and then allowing both merging and splitting operations (Horowitz and Pavlidis 1976; Pavlidis and Liow 1990).

Watershed. A technique related to thresholding, since it operates on a grayscale image, is *watershed* computation (Vincent and Soille 1991). This technique segments an image into several *catchment basins*, which are the regions of an image (interpreted as a height field or landscape) where rain would flow into the same lake. An efficient way to compute such regions is to start flooding the landscape at all of the local minima and to label ridges wherever differently evolving components meet. The whole algorithm can be implemented using a priority queue of pixels and breadth-first search (Vincent and Soille 1991).[16]

[16]A related algorithm can be used to compute maximally stable extremal regions (MSERs) efficiently (Section 7.1.1) (Nistér and Stewénius 2008).

(a) (b) (c)

Figure 7.52 Graph-based merging segmentation (Felzenszwalb and Huttenlocher 2004) © 2004 Springer: (a) input grayscale image that is successfully segmented into three regions even though the variation inside the smaller rectangle is larger than the variation across the middle edge; (b) input grayscale image; (c) resulting segmentation using an $\mathcal{N}_8$ pixel neighborhood.

Since images rarely have dark regions separated by lighter ridges, watershed segmentation is usually applied to a smoothed version of the gradient magnitude image, which also makes it usable with color images. As an alternative, the maximum oriented energy in a steerable filter (3.28–3.29) (Freeman and Adelson 1991) can be used as the basis of the *oriented watershed transform* developed by Arbeláez, Maire *et al.* (2011). Such techniques end up finding smooth regions separated by visible (higher gradient) boundaries. Since such boundaries are what active contours usually follow, active contour algorithms (Mortensen and Barrett 1999; Li, Sun *et al.* 2004) often precompute such a segmentation using either the watershed or the related *tobogganing* technique (Section 7.3.1).

7.5.1 Graph-based segmentation

While many merging algorithms simply apply a fixed rule that groups pixels and regions together, Felzenszwalb and Huttenlocher (2004) present a merging algorithm that uses *relative dissimilarities* between regions to determine which ones should be merged; it produces an algorithm that provably optimizes a global grouping metric. They start with a pixel-to-pixel dissimilarity measure $w(e)$ that measures, for example, intensity differences between $\mathcal{N}_8$ neighbors. Alternatively, they can use the *joint feature space* distances introduced by Comaniciu and Meer (2002), which we discuss in Sections 7.5.2 and 7.5.3. Figure 7.52 shows two examples of images segmented using their technique.

Probabilistic aggregation

Alpert, Galun *et al.* (2007) develop a probabilistic merging algorithm based on two cues, namely gray-level similarity and texture similarity. The gray-level similarity between regions R_i and R_j is based on the *minimal external difference* from other neighboring regions, which is compared to the *average intensity difference* to compute the likelihoods p_{ij} that two regions should be merged. Merging proceeds in a hierarchical fashion inspired by algebraic multigrid techniques (Brandt 1986; Briggs, Henson, and McCormick 2000) and previously used by Alpert, Galun *et al.* (2007) in their segmentation by weighted aggregation (SWA) algorithm (Sharon, Galun *et al.* 2006). Figure 7.56 shows the segmentations produced by this algorithm compared to other popular segmentation algorithms.

7.5.2 Mean shift

Mean-shift and mode finding techniques, such as k-means and mixtures of Gaussians, model the feature vectors associated with each pixel (e.g., color and position) as samples from an unknown probability density function and then try to find clusters (modes) in this distribution.

Consider the color image shown in Figure 7.53a. How would you segment this image based on color alone? Figure 7.53b shows the distribution of pixels in L*u*v* space, which is equivalent to what a vision algorithm that ignores spatial location would see. To make the visualization simpler, let us only consider the L*u* coordinates, as shown in Figure 7.53c. How many obvious (elongated) clusters do you see? How would you go about finding these clusters?

The k-means and mixtures of Gaussians techniques we studied in Section 5.2.2 use a *parametric* model of the density function to answer this question, i.e., they assume the density is the superposition of a small number of simpler distributions (e.g., Gaussians) whose locations (centers) and shape (covariance) can be estimated. Mean shift, on the other hand, smoothes the distribution and finds its peaks as well as the regions of feature space that correspond to each peak. Since a complete density is being modeled, this approach is called *non-parametric* (Bishop 2006).

The key to mean shift is a technique for efficiently finding peaks in this high-dimensional data distribution without ever computing the complete function explicitly (Fukunaga and Hostetler 1975; Cheng 1995; Comaniciu and Meer 2002). Consider once again the data points shown in Figure 7.53c, which can be thought of as having been drawn from some probability density function. If we could compute this density function, as visualized in Figure 7.53e, we could find its major peaks (*modes*) and identify regions of the input space that climb to the same peak as being part of the same region. This is the inverse of the *watershed* algorithm described in Section 7.5, which climbs downhill to find *basins of attraction*.

The first question, then, is how to estimate the density function given a sparse set of samples. One of the simplest approaches is to just smooth the data, e.g., by convolving it with a fixed kernel of width h, which, as we saw in Section 4.1.1, is the *Parzen window* approach to density estimation (Duda, Hart, and Stork 2001, Section 4.3; Bishop 2006, Section 2.5.1). Once we have computed $f(\mathbf{x})$, as shown in Figure 7.53e, we can find its local maxima using gradient ascent or some other optimization technique.

The problem with this "brute force" approach is that, for higher dimensions, it becomes computationally prohibitive to evaluate $f(\mathbf{x})$ over the complete search space. Instead, mean shift uses a variant of what is known in the optimization literature as *multiple restart gradient descent*. Starting at some guess for a local maximum, $\mathbf{y}_k$, which can be a random input data point $\mathbf{x}_i$, mean shift computes the gradient of the density estimate $f(\mathbf{x})$ at $\mathbf{y}_k$ and takes an uphill step in that direction. Details on how this can be done efficiently can be found in papers on mean shift (Comaniciu and Meer 2002; Paris and Durand 2007) as well as the first edition of this book (Szeliski 2010, Section 5.3.2).

The color-based segmentation shown in Figure 7.53 only looks at pixel colors when determining the best clustering. It may therefore cluster together small isolated pixels that happen to have the same color, which may not correspond to a semantically meaningful segmentation of the image. Better results can usually be obtained by clustering in the *joint domain* of color and location. In this approach, the spatial coordinates of the image $\mathbf{x}_s = (x, y)$, which are called the *spatial domain*, are concatenated with the color values $\mathbf{x}_r$, which are known as the *range domain*, and mean-shift clustering is applied in this five-dimensional space $\mathbf{x}_j$. Since location and color may have different scales, the kernels are adjusted separately, just as in the bilateral filter kernel (3.34–3.37) discussed in Section 3.3.2. The difference between mean shift and bilateral filtering, however, is that in mean shift, the spatial coordinates of each pixel are adjusted along with its color values, so that the pixel migrates more quickly towards other pixels with similar colors, and can therefore later be used for

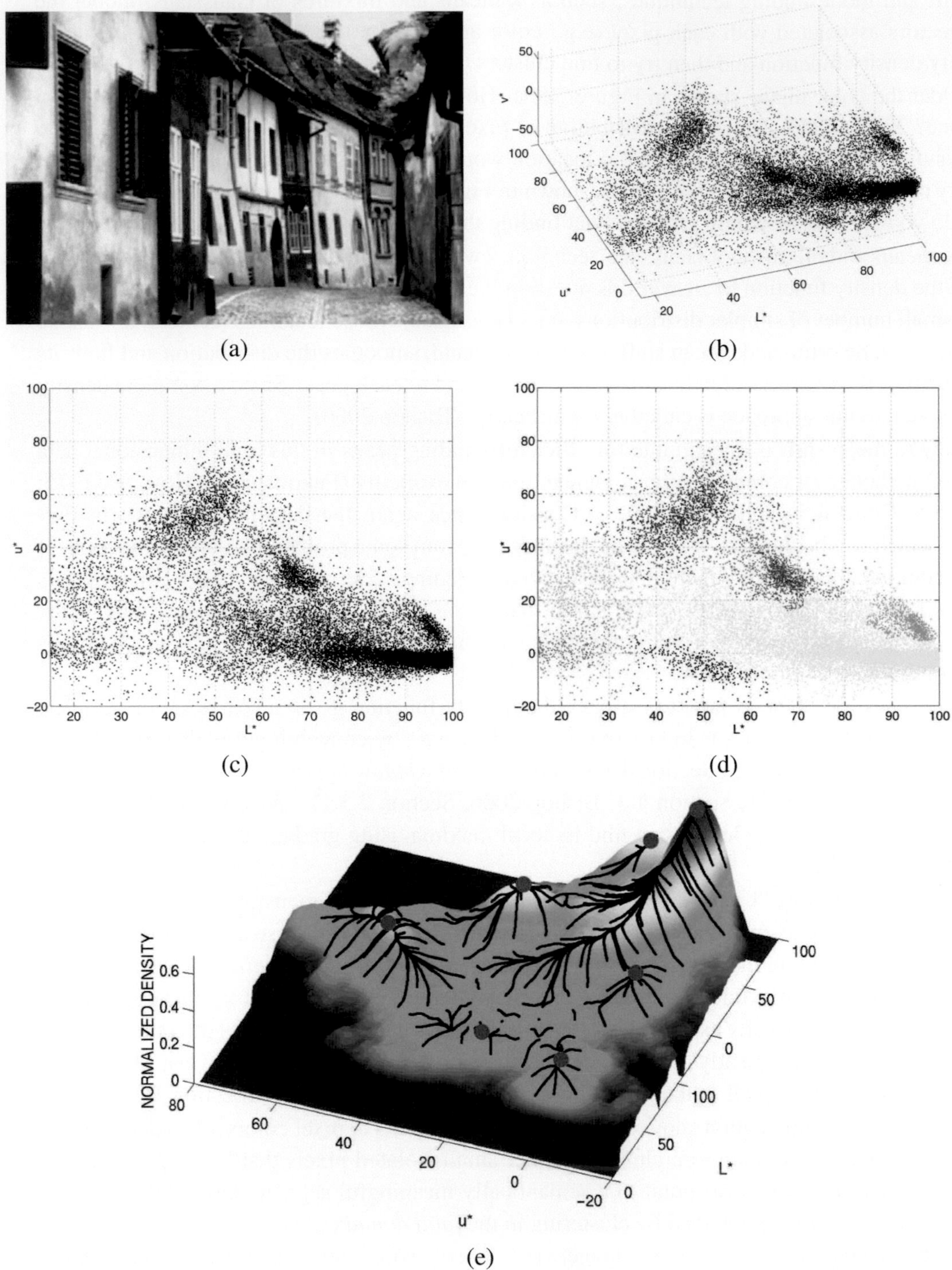

Figure 7.53 Mean-shift image segmentation (Comaniciu and Meer 2002) © 2002 IEEE: (a) input color image; (b) pixels plotted in L*u*v* space; (c) L*u* space distribution; (d) clustered results after 159 mean-shift procedures; (e) corresponding trajectories with peaks marked as red dots.

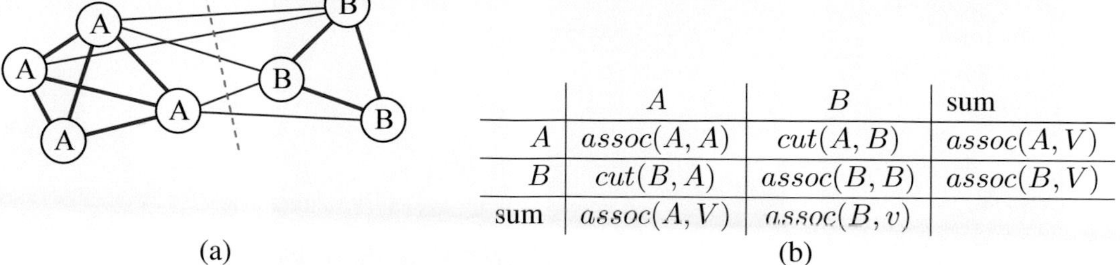

	A	B	sum
A	$assoc(A,A)$	$cut(A,B)$	$assoc(A,V)$
B	$cut(B,A)$	$assoc(B,B)$	$assoc(B,V)$
sum	$assoc(A,V)$	$assoc(B,v)$	

(a) (b)

Figure 7.54 Sample weighted graph and its normalized cut: (a) a small sample graph and its smallest normalized cut; (b) tabular form of the associations and cuts for this graph. The *assoc* and *cut* entries are computed as area sums of the associated weight matrix $\mathbf{W}$. Normalizing the table entries by the row or column sums produces normalized associations and cuts $Nassoc$ and $Ncut$.

clustering and segmentation.

Mean shift has been applied to a number of different problems in computer vision, including face tracking, 2D shape extraction, and texture segmentation (Comaniciu and Meer 2002), stereo matching (Wei and Quan 2004), non-photorealistic rendering (Section 10.5.2) (DeCarlo and Santella 2002), and video editing (Section 10.4.5) (Wang, Bhat *et al.* 2005). Paris and Durand (2007) provide a nice review of such applications, as well as techniques for more efficiently solving the mean-shift equations and producing hierarchical segmentations.

7.5.3 Normalized cuts

While bottom-up merging techniques aggregate regions into coherent wholes and mean-shift techniques try to find clusters of similar pixels using mode finding, the normalized cuts technique introduced by Shi and Malik (2000) examines the *affinities* (similarities) between nearby pixels and tries to separate groups that are connected by weak affinities.

Consider the simple graph shown in Figure 7.54a. The pixels in group A are all strongly connected with high affinities, shown as thick red lines, as are the pixels in group B. The connections between these two groups, shown as thinner blue lines, are much weaker. A *normalized cut* between the two groups, shown as a dashed line, separates them into two clusters.

The cut between two groups A and B is defined as the sum of all the weights being cut, where the weights between two pixels (or regions) i and j measure their similarity. Using a minimum cut as a segmentation criterion, however, does not result in reasonable clusters, since the smallest cuts usually involve isolating a single pixel.

A better measure of segmentation is the normalized cut, which is defined as

$$Ncut(A,B) = \frac{cut(A,B)}{assoc(A,V)} + \frac{cut(A,B)}{assoc(B,V)}, \tag{7.41}$$

where $assoc(A,A) = \sum_{i \in A, j \in A} w_{ij}$ is the *association* (sum of all the weights) within a cluster and $assoc(A,V) = assoc(A,A) + cut(A,B)$ is the sum of *all* the weights associated with nodes in A. Figure 7.54b shows how the cuts and associations can be thought of as area sums in the weight matrix $\mathbf{W} = [w_{ij}]$, where the entries of the matrix have been arranged so that the nodes in A come first and the nodes in B come second. Dividing each of these areas by the corresponding row sum (the rightmost column of Figure 7.54b) results in the normalized cut and association values. These normalized values better reflect the fitness of a particular segmentation, since they look for

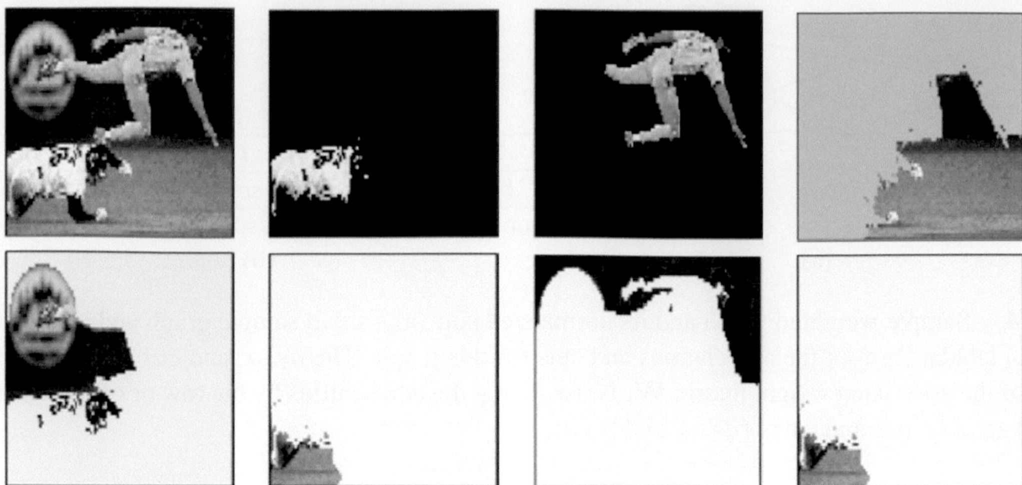

Figure 7.55 Normalized cuts segmentation (Shi and Malik 2000) © 2000 IEEE: The input image and the components returned by the normalized cuts algorithm.

collections of edges that are weak relative to all of the edges both inside and emanating from a particular region.

Unfortunately, computing the optimal normalized cut is NP-complete. Instead, Shi and Malik (2000) suggest computing a real-valued assignment of nodes to groups, using a generalized eigenvalue analysis of the *normalized* affinity matrix (Weiss 1999), as described in more detail in the normalized cuts paper and (Szeliski 2010, Section 5.4). Because these eigenvectors can be interpreted as the large modes of vibration in a spring-mass system, normalized cuts is an example of a *spectral method* for image segmentation. After the real-valued eigenvector is computed, the variables corresponding to positive and negative eigenvector values are associated with the two cut components. This process can be further repeated to hierarchically subdivide an image, as shown in Figure 7.55.

The original algorithm proposed by Shi and Malik (2000) used spatial position and image feature differences to compute the pixel-wise affinities. In subsequent work, Malik, Belongie *et al.* (2001) look for *intervening contours* between pixels i and j to define intervening contour weights and then multiply these weights with a texton-based texture similarity metric. They then use an initial over-segmentation based purely on local pixel-wise features to re-estimate intervening contours and texture statistics in a region-based manner. Figure 7.56 shows the results of running this improved algorithm on a number of test images.

Because it requires the solution of large sparse eigenvalue problems, normalized cuts can be quite slow. Sharon, Galun *et al.* (2006) present a way to accelerate the computation of the normalized cuts using an approach inspired by algebraic multigrid (Brandt 1986; Briggs, Henson, and McCormick 2000).

An example of the segmentation produced by weighted aggregation (SWA) is shown in Figure 7.56, along with the most recent probabilistic bottom-up merging algorithm by Alpert, Galun *et al.* (2007). In more recent work, Pont-Tuset, Arbeláez *et al.* (2017) speed up normalized cuts and extend it to multiple scales to obtain state-of-the-art results on both the Berkeley Segmentation Dataset as well as (at the time) object proposals on the VOC and COCO datasets.

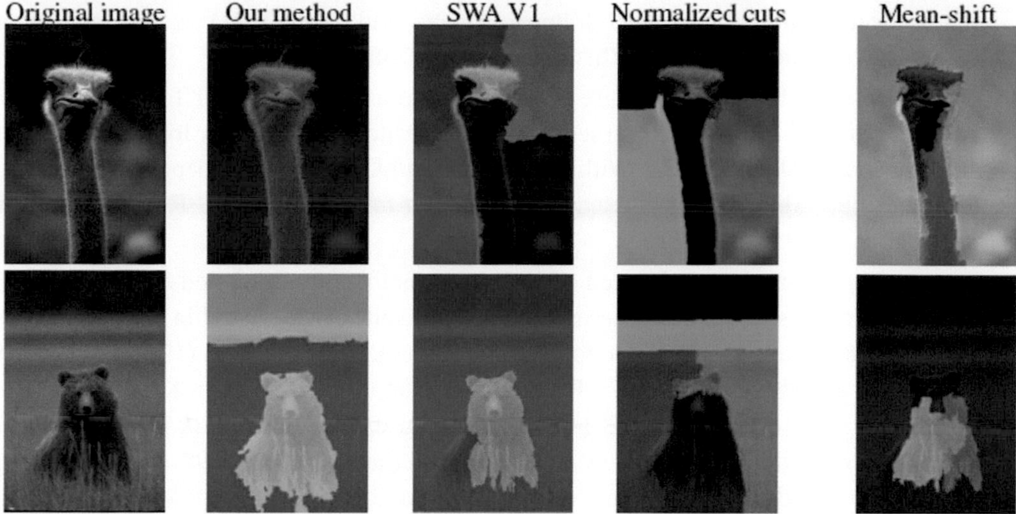

Figure 7.56 Comparative segmentation results (Alpert, Galun *et al.* 2007) © 2007 IEEE. "Our method" refers to the probabilistic bottom-up merging algorithm developed by Alpert *et al.*

7.6 Additional reading

One of the seminal papers on feature detection, description, and matching is by Lowe (2004). Comprehensive surveys and evaluations of such techniques have been made by Schmid, Mohr, and Bauckhage (2000), Mikolajczyk and Schmid (2005), Mikolajczyk, Tuytelaars *et al.* (2005), and Tuytelaars and Mikolajczyk (2008), while Shi and Tomasi (1994) and Triggs (2004) also provide nice reviews.

In the area of feature detectors (Mikolajczyk, Tuytelaars *et al.* 2005), in addition to such classic approaches as Förstner–Harris (Förstner 1986; Harris and Stephens 1988) and difference of Gaussians (Lindeberg 1993, 1998b; Lowe 2004), maximally stable extremal regions (MSERs) are widely used for applications that require affine invariance (Matas, Chum *et al.* 2004; Nistér and Stewénius 2008). More recent interest point detectors are discussed by Xiao and Shah (2003), Koethe (2003), Carneiro and Jepson (2005), Kenney, Zuliani, and Manjunath (2005), Bay, Ess *et al.* (2008), Platel, Balmachnova *et al.* (2006), and Rosten, Porter, and Drummond (2010), as are techniques based on line matching (Zoghlami, Faugeras, and Deriche 1997; Bartoli, Coquerelle, and Sturm 2004) and region detection (Kadir, Zisserman, and Brady 2004; Matas, Chum *et al.* 2004; Tuytelaars and Van Gool 2004; Corso and Hager 2005). Three recent papers with nice reviews of DNN-based feature detectors are Balntas, Lenc *et al.* (2020), Barroso-Laguna, Riba *et al.* (2019), and Tian, Balntas *et al.* (2020).

A variety of local feature descriptors (and matching heuristics) are surveyed and compared by Mikolajczyk and Schmid (2005). More recent publications in this area include those by van de Weijer and Schmid (2006), Abdel-Hakim and Farag (2006), Winder and Brown (2007), and Hua, Brown, and Winder (2007) and the recent evaluations by Balntas, Lenc *et al.* (2020) and Jin, Mishkin *et al.* (2021). Techniques for efficiently matching features include k-d trees (Beis and Lowe 1999; Lowe 2004; Muja and Lowe 2009), pyramid matching kernels (Grauman and Darrell 2005), metric (vocabulary) trees (Nistér and Stewénius 2006), variety of multi-dimensional hashing techniques (Shakhnarovich, Viola, and Darrell 2003; Torralba, Weiss, and Fergus 2008; Weiss, Torralba, and Fergus 2008; Kulis and Grauman 2009; Raginsky and Lazebnik 2009), and product quantization

(Jégou, Douze, and Schmid 2010; Johnson, Douze, and Jégou 2021). A good review of large-scale systems for instance retrieval is Zheng, Yang, and Tian (2018).

The classic reference on feature detection and tracking is Shi and Tomasi (1994). More recent work in this field has focused on learning better matching functions for specific features (Avidan 2001; Jurie and Dhome 2002; Williams, Blake, and Cipolla 2003; Lepetit and Fua 2005; Lepetit, Pilet, and Fua 2006; Hinterstoisser, Benhimane *et al.* 2008; Rogez, Rihan *et al.* 2008; Özuysal, Calonder *et al.* 2010).

A highly cited and widely used edge detector is the one developed by Canny (1986). Alternative edge detectors as well as experimental comparisons can be found in publications by Nalwa and Binford (1986), Nalwa (1987), Deriche (1987), Freeman and Adelson (1991), Nalwa (1993), Heath, Sarkar *et al.* (1998), Crane (1997), Ritter and Wilson (2000), Bowyer, Kranenburg, and Dougherty (2001), Arbeláez, Maire *et al.* (2011), and Pont-Tuset, Arbeláez *et al.* (2017). The topic of scale selection in edge detection is nicely treated by Elder and Zucker (1998), while approaches to color and texture edge detection can be found in Ruzon and Tomasi (2001), Martin, Fowlkes, and Malik (2004), and Gevers, van de Weijer, and Stokman (2006). Edge detectors have also been combined with region segmentation techniques to further improve the detection of semantically salient boundaries (Maire, Arbelaez *et al.* 2008; Arbeláez, Maire *et al.* 2011; Xiaofeng and Bo 2012; Pont-Tuset, Arbeláez *et al.* 2017). Edges linked into contours can be smoothed and manipulated for artistic effect (Lowe 1989; Finkelstein and Salesin 1994; Taubin 1995) and used for recognition (Belongie, Malik, and Puzicha 2002; Tek and Kimia 2003; Sebastian and Kimia 2005).

The topic of active contours has a long history, beginning with the seminal work on snakes and other energy-minimizing variational methods (Kass, Witkin, and Terzopoulos 1988; Cootes, Cooper *et al.* 1995; Blake and Isard 1998), continuing through techniques such as intelligent scissors (Mortensen and Barrett 1995, 1999; Pérez, Blake, and Gangnet 2001), and culminating in level sets (Malladi, Sethian, and Vemuri 1995; Caselles, Kimmel, and Sapiro 1997; Sethian 1999; Paragios and Deriche 2000; Sapiro 2001; Osher and Paragios 2003; Paragios, Faugeras *et al.* 2005; Cremers, Rousson, and Deriche 2007; Rousson and Paragios 2008; Paragios and Sgallari 2009), which are currently the most widely used active contour methods.

An early, well-regarded paper on straight line extraction in images was written by Burns, Hanson, and Riseman (1986). Their idea of bottom-up *line-support regions* was extended by Grompone von Gioi, Jakubowicz *et al.* (2008) to construct the popular LSD line segment detector. The literature on vanishing point detection is quite vast and still evolving (Quan and Mohr 1989; Collins and Weiss 1990; Brillaut-O'Mahoney 1991; McLean and Kotturi 1995; Becker and Bove 1995; Shufelt 1999; Tuytelaars, Van Gool, and Proesmans 1997; Schaffalitzky and Zisserman 2000; Antone and Teller 2002; Rother 2002; Košecká and Zhang 2005; Denis, Elder, and Estrada 2008; Pflugfelder 2008; Tardif 2009; Bazin, Seo *et al.* 2012; Antunes and Barreto 2013; Zhou, Qi *et al.* 2019a). Simultaneous line and junction detection techniques have also been developed (Huang, Wang *et al.* 2018; Zhang, Li *et al.* 2019).

The topic of image segmentation is closely related to clustering techniques, which are treated in a number of monographs and review articles (Jain and Dubes 1988; Kaufman and Rousseeuw 1990; Jain, Duin, and Mao 2000; Jain, Topchy *et al.* 2004). Some early segmentation techniques include those described by Brice and Fennema (1970), Pavlidis (1977), Riseman and Arbib (1977), Ohlander, Price, and Reddy (1978), Rosenfeld and Davis (1979), and Haralick and Shapiro (1985), while examples of newer techniques are developed by Leclerc (1989), Mumford and Shah (1989), Shi and Malik (2000), and Felzenszwalb and Huttenlocher (2004).

Arbeláez, Maire *et al.* (2011) and Pont-Tuset, Arbeláez *et al.* (2017) provide good reviews of automatic segmentation techniques and compare their performance on the Berkeley Segmentation Dataset and Benchmark (Martin, Fowlkes *et al.* 2001).[17] Additional comparison papers and databases include those by Unnikrishnan, Pantofaru, and Hebert (2007), Alpert, Galun *et al.* (2007), and Estrada and Jepson (2009).

Techniques for segmenting images based on local pixel similarities combined with aggregation or splitting methods include watersheds (Vincent and Soille 1991; Beare 2006; Arbeláez, Maire *et al.* 2011), region splitting (Ohlander, Price, and Reddy 1978), region merging (Brice and Fennema 1970; Pavlidis and Liow 1990; Jain, Topchy *et al.* 2004), as well as graph-based and probabilistic multi-scale approaches (Felzenszwalb and Huttenlocher 2004; Alpert, Galun *et al.* 2007).

Mean-shift algorithms, which find modes (peaks) in a density function representation of the pixels, are presented by Comaniciu and Meer (2002) and Paris and Durand (2007). Parametric mixtures of Gaussians can also be used to represent and segment such pixel densities (Bishop 2006; Ma, Derksen *et al.* 2007).

The seminal work on spectral (eigenvalue) methods for image segmentation is the *normalized cut* algorithm of Shi and Malik (2000). Related work includes that by Weiss (1999), Meilă and Shi (2000), Meilă and Shi (2001), Malik, Belongie *et al.* (2001), Ng, Jordan, and Weiss (2001), Yu and Shi (2003), Cour, Bénézit, and Shi (2005), Sharon, Galun *et al.* (2006), Tolliver and Miller (2006), and Wang and Oliensis (2010).

7.7 Exercises

Ex 7.1: Interest point detector. Implement one or more keypoint detectors and compare their performance (with your own or with a classmate's detector).

Possible detectors:

- Laplacian or Difference of Gaussian;
- Förstner–Harris Hessian (try different formula variants given in (7.9–7.11));
- oriented/steerable filter, looking for either second-order high second response or two edges in a window (Koethe 2003), as discussed in Section 7.1.1.
- any of the newer DNN-based detectors.

Other detectors are described in Mikolajczyk, Tuytelaars *et al.* (2005), Tuytelaars and Mikolajczyk (2008), and Balntas, Lenc *et al.* (2020). Additional optional steps could include:

1. Compute the detections on a sub-octave pyramid and find 3D maxima.
2. Find local orientation estimates using steerable filter responses or a gradient histogramming method.
3. Implement non-maximal suppression, such as the adaptive technique of Brown, Szeliski, and Winder (2005).
4. Vary the window shape and size (prefilter and aggregation).

To test for repeatability, download the code from https://www.robots.ox.ac.uk/~vgg/research/affine (Mikolajczyk, Tuytelaars *et al.* 2005; Tuytelaars and Mikolajczyk 2008) or simply rotate or shear your own test images. (Pick a domain you may want to use later, e.g., for outdoor stitching.)

Be sure to measure and report the stability of your scale and orientation estimates.

[17] http://www.eecs.berkeley.edu/Research/Projects/CS/vision/grouping/segbench.

Ex 7.2: Interest point descriptor. Implement two or more descriptors from Section 7.1.2 (steered to local scale and orientation estimates, if appropriate) and compare their performance on some images of your own choosing.

You can either use the evaluation methodologies (and optionally software) described in Mikolajczyk and Schmid (2005), Balntas, Lenc *et al.* (2020), or Jin, Mishkin *et al.* (2021).

Ex 7.3: ROC curve computation. Given a pair of curves (histograms) plotting the number of matching and non-matching features as a function of Euclidean distance d as shown in Figure 7.22b, derive an algorithm for plotting a ROC curve (Figure 7.22a). In particular, let $t(d)$ be the distribution of true matches and $f(d)$ be the distribution of (false) non-matches. Write down the equations for the ROC, i.e., TPR(FPR), and the AUC.

(Hint: Plot the cumulative distributions $T(d) = \int t(d)$ and $F(d) = \int f(d)$ and see if these help you derive the TPR and FPR at a given threshold θ.)

Ex 7.4: Feature matcher. After extracting features from a collection of overlapping or distorted images,[18] match them up by their descriptors either using nearest neighbor matching or a more efficient matching strategy such as a k-d tree.

See whether you can improve the accuracy of your matches using techniques such as the nearest neighbor distance ratio.

Ex 7.5: Feature tracker. Instead of finding feature points independently in multiple images and then matching them, find features in the first image of a video or image sequence and then re-locate the corresponding points in the next frames using either search and gradient descent (Shi and Tomasi 1994) or learned feature detectors (Lepetit, Pilet, and Fua 2006; Fossati, Dimitrijevic *et al.* 2007). When the number of tracked points drops below a threshold or new regions in the image become visible, find additional points to track.

(Optional) Winnow out incorrect matches by estimating a homography (8.19–8.23) or fundamental matrix (Section 11.3.3).

(Optional) Refine the accuracy of your matches using the iterative registration algorithm described in Section 9.2 and Exercise 9.2.

Ex 7.6: Facial feature tracker. Apply your feature tracker to tracking points on a person's face, either manually initialized to interesting locations such as eye corners or automatically initialized at interest points.

(Optional) Match features between two people and use these features to perform image morphing (Exercise 3.25).

Ex 7.7: Edge detector. Implement an edge detector of your choice. Compare its performance to that of your classmates' detectors or code downloaded from the internet.

A simple but well-performing sub-pixel edge detector can be created as follows:

1. Blur the input image a little,

$$B_\sigma(\mathbf{x}) = G_\sigma(\mathbf{x}) * I(\mathbf{x}).$$

2. Construct a Gaussian pyramid (Exercise 3.17),

$$P = \text{Pyramid}\{B_\sigma(\mathbf{x})\}$$

[18] https://www.robots.ox.ac.uk/~vgg/research/affine.

3. Subtract an interpolated coarser-level pyramid image from the original resolution blurred image,

$$S(\mathbf{x}) = B_\sigma(\mathbf{x}) - P.\text{InterpolatedLevel}(L).$$

4. For each quad of pixels, $\{(i,j),(i+1,j),(i,j+1),(i+1,j+1)\}$, count the number of zero crossings along the four edges.

5. When there are exactly two zero crossings, compute their locations using (7.25) and store these edgel endpoints along with the midpoint in the edgel structure.

6. For each edgel, compute the local gradient by taking the horizontal and vertical differences between the values of S along the zero crossing edges.

7. Store the magnitude of this gradient as the edge strength and either its orientation or that of the segment joining the edgel endpoints as the edge orientation.

8. Add the edgel to a list of edgels or store it in a 2D array of edgels (addressed by pixel coordinates).

Ex 7.8: Edge linking and thresholding. Link up the edges computed in the previous exercise into chains and optionally perform thresholding with hysteresis.

The steps may include:

1. Store the edgels either in a 2D array (say, an integer image with indices into the edgel list) or pre-sort the edgel list first by (integer) x coordinates and then y coordinates, for faster neighbor finding.

2. Pick up an edgel from the list of unlinked edgels and find its neighbors in both directions until no neighbor is found or a closed contour is obtained. Flag edgels as linked as you visit them and push them onto your list of linked edgels.

3. (Optional) Perform hysteresis-based thresholding (Canny 1986). Use two thresholds "hi" and "lo" for the edge strength. A candidate edgel is considered an edge if either its strength is above the "hi" threshold or its strength is above the "lo" threshold and it is (recursively) connected to a previously detected edge.

4. (Optional) Link together contours that have small gaps but whose endpoints have similar orientations.

5. (Optional) Find junctions between adjacent contours, e.g., using some of the ideas (or references) from Maire, Arbelaez *et al.* (2008).

Ex 7.9: Contour matching. Convert a closed contour (linked edgel list) into its arc-length parameterization and use this to match object outlines.

The steps may include:

1. Walk along the contour and create a list of (x_i, y_i, s_i) triplets, using the arc-length formula

$$s_{i+1} = s_i + \|\mathbf{x}_{i+1} - \mathbf{x}_i\|. \tag{7.42}$$

2. Resample this list onto a regular set of (x_j, y_j, j) samples using linear interpolation of each segment.

3. Compute the average values of x and y, i.e., $\overline{x}$ and $\overline{y}$ and subtract them from your sampled curve points.

4. Resample the original (x_i, y_i, s_i) piecewise-linear function onto a length-independent set of samples, say $j \in [0, 1023]$. (Using a length which is a power of two makes subsequent Fourier transforms more convenient.)

5. Compute the Fourier transform of the curve, treating each (x, y) pair as a complex number.

6. To compare two curves, fit a linear equation to the phase difference between the two curves. (Careful: phase wraps around at 360°. Also, you may wish to weight samples by their Fourier spectrum magnitude—see Section 9.1.2.)

7. (Optional) Prove that the constant phase component corresponds to the temporal shift in s, while the linear component corresponds to rotation.

Of course, feel free to try any other curve descriptor and matching technique from the computer vision literature (Tek and Kimia 2003; Sebastian and Kimia 2005).

Ex 7.10: Jigsaw puzzle solver—challenging. Write a program to automatically solve a jigsaw puzzle from a set of scanned puzzle pieces. Your software may include the following components:

1. Scan the pieces (either face up or face down) on a flatbed scanner with a distinctively colored background.

2. (Optional) Scan in the box top to use as a low-resolution reference image.

3. Use color-based thresholding to isolate the pieces.

4. Extract the contour of each piece using edge finding and linking.

5. (Optional) Re-represent each contour using an arc-length or some other re-parameterization. Break up the contours into meaningful matchable pieces. (Is this hard?)

6. (Optional) Associate color values with each contour to help in the matching.

7. (Optional) Match pieces to the reference image using some rotationally invariant feature descriptors.

8. Solve a global optimization or (backtracking) search problem to snap pieces together and place them in the correct location relative to the reference image.

9. Test your algorithm on a succession of more difficult puzzles and compare your results with those of others.

For some additional ideas, have a look at Cho, Avidan, and Freeman (2010).

Ex 7.11: Successive approximation line detector. Implement a line simplification algorithm (Section 7.4.1) (Ramer 1972; Douglas and Peucker 1973) to convert a hand-drawn curve (or linked edge image) into a small set of polylines.

(Optional) Re-render this curve using either an approximating or interpolating spline or Bezier curve (Szeliski and Ito 1986; Bartels, Beatty, and Barsky 1987; Farin 2002).

Ex 7.12: Line fitting uncertainty. Estimate the uncertainty (covariance) in your line fit using uncertainty analysis.

1. After determining which edgels belong to the line segment (using either successive approximation or Hough transform), re-fit the line segment using total least squares (Van Huffel and Vandewalle 1991; Van Huffel and Lemmerling 2002), i.e., find the mean or centroid of the edgels and then use eigenvalue analysis to find the dominant orientation.

2. Compute the perpendicular errors (deviations) to the line and robustly estimate the variance of the fitting noise using an estimator such as MAD (Appendix B.3).

3. (Optional) re-fit the line parameters by throwing away outliers or using a robust norm or influence function.

4. Estimate the error in the perpendicular location of the line segment and its orientation.

Ex 7.13: Vanishing points. Compute the vanishing points in an image using one of the techniques described in Section 7.4.3 and optionally refine the original line equations associated with each vanishing point. Your results can be used later to track a target or reconstruct architecture (Section 13.6.1).

Ex 7.14: Vanishing point uncertainty. Perform an uncertainty analysis on your estimated vanishing points. You will need to decide how to represent your vanishing point, e.g., homogeneous coordinates on a sphere, to handle vanishing points near infinity.

See the discussion of Bingham distributions by Collins and Weiss (1990) for some ideas.

Ex 7.15: Region segmentation. Implement one of the region segmentation algorithms described in this chapter. Some popular segmentation algorithms include:

- k-means (Section 5.2.2);
- mixtures of Gaussians (Section 5.2.2);
- mean shift (Section 7.5.2);
- normalized cuts (Section 7.5.3);
- similarity graph-based segmentation (Section 7.5.1);
- binary Markov random fields solved using graph cuts (Section 4.3.2).

Apply your region segmentation to a video sequence and use it to track moving regions from frame to frame.

Alternatively, test out your segmentation algorithm on the Berkeley segmentation database (Martin, Fowlkes *et al.* 2001).

Chapter 8

Image alignment and stitching

8.1 Pairwise alignment . . . 403
- 8.1.1 2D alignment using least squares . . . 403
- 8.1.2 *Application*: Panography . . . 405
- 8.1.3 Iterative algorithms . . . 406
- 8.1.4 Robust least squares and RANSAC . . . 408
- 8.1.5 3D alignment . . . 410

8.2 Image stitching . . . 411
- 8.2.1 Parametric motion models . . . 412
- 8.2.2 *Application*: Whiteboard and document scanning . . . 414
- 8.2.3 Rotational panoramas . . . 414
- 8.2.4 Gap closing . . . 416
- 8.2.5 *Application*: Video summarization and compression . . . 417
- 8.2.6 Cylindrical and spherical coordinates . . . 418

8.3 Global alignment . . . 421
- 8.3.1 Bundle adjustment . . . 421
- 8.3.2 Parallax removal . . . 424
- 8.3.3 Recognizing panoramas . . . 425

8.4 Compositing . . . 426
- 8.4.1 Choosing a compositing surface . . . 426
- 8.4.2 Pixel selection and weighting (deghosting) . . . 430
- 8.4.3 *Application*: Photomontage . . . 435
- 8.4.4 Blending . . . 435

8.5 Additional reading . . . 437

8.6 Exercises . . . 438

R. Szeliski, *Computer Vision*, Texts in Computer Science,
https://doi.org/10.1007/978-3-030-34372-9_8

Figure 8.1 Image stitching: (a) geometric alignment of 2D images for stitching (Szeliski and Shum 1997) © 1997 ACM; (b) a spherical panorama constructed from 54 photographs (Szeliski and Shum 1997) © 1997 ACM; (c) a multi-image panorama automatically assembled from an unordered photo collection; a multi-image stitch (d) without and (e) with moving object removal (Uyttendaele, Eden, and Szeliski 2001) © 2001 IEEE.

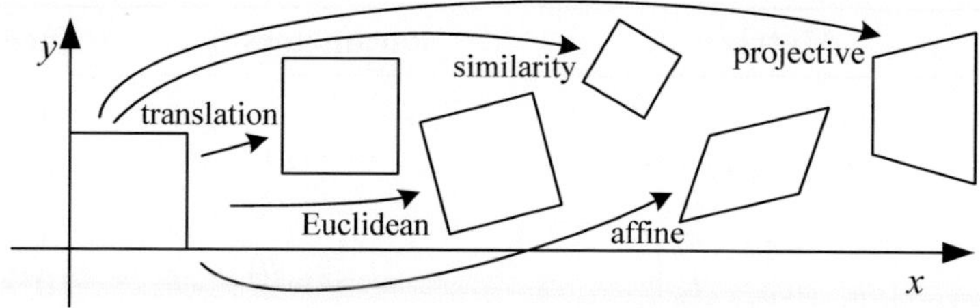

Figure 8.2 Basic set of 2D planar transformations

Once we have extracted features from images, the next stage in many vision algorithms is to match these features across different images (Section 7.1.3). An important component of this matching is to verify whether the set of matching features is geometrically consistent, e.g., whether the feature displacements can be described by a simple 2D or 3D geometric transformation. The computed motions can then be used in other applications such as image stitching (Section 8.2) or augmented reality (Section 11.2.2).

In this chapter, we look at the topic of geometric image registration, i.e., the computation of 2D and 3D transformations that map features in one image to another (Section 8.1). In Chapter 11, we look at the related problems of *pose estimation*, which is determining a camera's position relative to a known 3D object or scene, and *structure from motion*, i.e., how to simultaneously estimate 3D geometry and camera motion.

8.1 Pairwise alignment

Feature-based alignment is the problem of estimating the motion between two or more sets of matched 2D or 3D points. In this section, we restrict ourselves to global *parametric* transformations, such as those described in Section 2.1.1 and shown in Table 2.1 and Figure 8.2, or higher order transformation for curved surfaces (Shashua and Toelg 1997; Can, Stewart *et al.* 2002). Applications to non-rigid or elastic deformations (Bookstein 1989; Kambhamettu, Goldgof *et al.* 1994; Szeliski and Lavallée 1996; Torresani, Hertzmann, and Bregler 2008) are examined in Sections 9.2.2 and 13.6.4.

8.1.1 2D alignment using least squares

Given a set of matched feature points $\{(\mathbf{x}_i, \mathbf{x}'_i)\}$ and a planar parametric transformation[1] of the form

$$\mathbf{x}' = \mathbf{f}(\mathbf{x}; \mathbf{p}), \tag{8.1}$$

how can we produce the best estimate of the motion parameters $\mathbf{p}$? The usual way to do this is to use least squares, i.e., to minimize the sum of squared residuals

$$E_{\text{LS}} = \sum_i \|\mathbf{r}_i\|^2 = \sum_i \|\mathbf{f}(\mathbf{x}_i; \mathbf{p}) - \mathbf{x}'_i\|^2, \tag{8.2}$$

where

$$\mathbf{r}_i = \mathbf{x}'_i - \mathbf{f}(\mathbf{x}_i; \mathbf{p}) = \hat{\mathbf{x}}'_i - \tilde{\mathbf{x}}'_i \tag{8.3}$$

[1]For examples of non-planar parametric models, such as quadrics, see the work of Shashua and Toelg (1997) and Shashua and Wexler (2001).

Transform	Matrix	Parameters $\mathbf{p}$	Jacobian $\mathbf{J}$
translation	$\begin{bmatrix} 1 & 0 & t_x \\ 0 & 1 & t_y \end{bmatrix}$	(t_x, t_y)	$\begin{bmatrix} 1 & 0 \\ 0 & 1 \end{bmatrix}$
Euclidean	$\begin{bmatrix} c_\theta & -s_\theta & t_x \\ s_\theta & c_\theta & t_y \end{bmatrix}$	(t_x, t_y, θ)	$\begin{bmatrix} 1 & 0 & -s_\theta x - c_\theta y \\ 0 & 1 & c_\theta x - s_\theta y \end{bmatrix}$
similarity	$\begin{bmatrix} 1+a & -b & t_x \\ b & 1+a & t_y \end{bmatrix}$	(t_x, t_y, a, b)	$\begin{bmatrix} 1 & 0 & x & -y \\ 0 & 1 & y & x \end{bmatrix}$
affine	$\begin{bmatrix} 1+a_{00} & a_{01} & t_x \\ a_{10} & 1+a_{11} & t_y \end{bmatrix}$	$(t_x, t_y, a_{00}, a_{01}, a_{10}, a_{11})$	$\begin{bmatrix} 1 & 0 & x & y & 0 & 0 \\ 0 & 1 & 0 & 0 & x & y \end{bmatrix}$
projective	$\begin{bmatrix} 1+h_{00} & h_{01} & h_{02} \\ h_{10} & 1+h_{11} & h_{12} \\ h_{20} & h_{21} & 1 \end{bmatrix}$	$(h_{00}, h_{01}, \ldots, h_{21})$	(see Section 8.1.3)

Table 8.1 Jacobians of the 2D coordinate transformations $\mathbf{x}' = \mathbf{f}(\mathbf{x}; \mathbf{p})$ shown in Table 2.1, where we have re-parameterized the motions so that they are identity for $\mathbf{p} = 0$.

is the *residual* between the measured location $\hat{\mathbf{x}}'_i$ and its corresponding current *predicted* location $\tilde{\mathbf{x}}'_i = \mathbf{f}(\mathbf{x}_i; \mathbf{p})$. (See Appendix A.2 for more on least squares and Appendix B.2 for a statistical justification.)

Many of the motion models presented in Section 2.1.1 and Table 2.1, i.e., translation, similarity, and affine, have a *linear* relationship between the amount of motion $\Delta\mathbf{x} = \mathbf{x}' - \mathbf{x}$ and the unknown parameters $\mathbf{p}$,

$$\Delta\mathbf{x} = \mathbf{x}' - \mathbf{x} = \mathbf{J}(\mathbf{x})\mathbf{p}, \tag{8.4}$$

where $\mathbf{J} = \partial\mathbf{f}/\partial\mathbf{p}$ is the *Jacobian* of the transformation $\mathbf{f}$ with respect to the motion parameters $\mathbf{p}$ (see Table 8.1). In this case, a simple *linear* regression (linear least squares problem) can be formulated as

$$E_{\text{LLS}} = \sum_i \|\mathbf{J}(\mathbf{x}_i)\mathbf{p} - \Delta\mathbf{x}_i\|^2 \tag{8.5}$$

$$= \mathbf{p}^T \left[\sum_i \mathbf{J}^T(\mathbf{x}_i)\mathbf{J}(\mathbf{x}_i)\right] \mathbf{p} - 2\mathbf{p}^T \left[\sum_i \mathbf{J}^T(\mathbf{x}_i)\Delta\mathbf{x}_i\right] + \sum_i \|\Delta\mathbf{x}_i\|^2 \tag{8.6}$$

$$= \mathbf{p}^T\mathbf{A}\mathbf{p} - 2\mathbf{p}^T\mathbf{b} + c. \tag{8.7}$$

The minimum can be found by solving the symmetric positive definite (SPD) system of *normal equations*[2]

$$\mathbf{A}\mathbf{p} = \mathbf{b}, \tag{8.8}$$

where

$$\mathbf{A} = \sum_i \mathbf{J}^T(\mathbf{x}_i)\mathbf{J}(\mathbf{x}_i) \tag{8.9}$$

[2] For poorly conditioned problems, it is better to use QR decomposition on the set of linear equations $\mathbf{J}(\mathbf{x}_i)\mathbf{p} = \Delta\mathbf{x}_i$ instead of the normal equations (Björck 1996; Golub and Van Loan 1996). However, such conditions rarely arise in image registration.

Figure 8.3 A simple panograph consisting of three images automatically aligned with a translational model and then averaged together.

is called the *Hessian* and $\mathbf{b} = \sum_i \mathbf{J}^T(\mathbf{x}_i)\Delta\mathbf{x}_i$. For the case of pure translation, the resulting equations have a particularly simple form, i.e., the translation is the average translation between corresponding points or, equivalently, the translation of the point centroids.

Uncertainty weighting. The above least squares formulation assumes that all feature points are matched with the same accuracy. This is often not the case, since certain points may fall into more textured regions than others. If we associate a scalar variance estimate σ_i^2 with each correspondence, we can minimize the *weighted least squares* problem instead,[3]

$$E_{\text{WLS}} = \sum_i \sigma_i^{-2} \|\mathbf{r}_i\|^2. \tag{8.10}$$

As shown in Section 9.1.3, a covariance estimate for patch-based matching can be obtained by multiplying the inverse of the *patch Hessian* $\mathbf{A}_i$ (9.48) with the per-pixel noise covariance σ_n^2 (9.37). Weighting each squared residual by its inverse covariance $\mathbf{\Sigma}_i^{-1} = \sigma_n^{-2}\mathbf{A}_i$ (which is called the *information matrix*), we obtain

$$E_{\text{CWLS}} = \sum_i \|\mathbf{r}_i\|^2_{\mathbf{\Sigma}_i^{-1}} = \sum_i \mathbf{r}_i^T \mathbf{\Sigma}_i^{-1} \mathbf{r}_i = \sum_i \sigma_n^{-2} \mathbf{r}_i^T \mathbf{A}_i \mathbf{r}_i. \tag{8.11}$$

8.1.2 *Application*: Panography

One of the simplest (and most fun) applications of image alignment is a special form of image stitching called *panography*. In a panograph, images are translated and optionally rotated and scaled before being blended with simple averaging (Figure 8.3). This process mimics the photographic collages created by artist David Hockney, although his compositions use an opaque overlay model, being created out of regular photographs.

In most of the examples seen on the web, the images are aligned by hand for best artistic effect.[4] However, it is also possible to use feature matching and alignment techniques to perform the registration automatically (Nomura, Zhang, and Nayar 2007; Zelnik-Manor and Perona 2007).

[3] Problems where each measurement can have a different variance or uncertainty are called *heteroscedastic models*.

[4] https://www.flickr.com/groups/panography.

Consider a simple translational model. We want all the corresponding features in different images to line up as best as possible. Let $\mathbf{t}_j$ be the location of the jth image coordinate frame in the global composite frame and $\mathbf{x}_{ij}$ be the location of the ith matched feature in the jth image. In order to align the images, we wish to minimize the least squares error

$$E_{\mathrm{PLS}} = \sum_{ij} \|(\mathbf{t}_j + \mathbf{x}_{ij}) - \mathbf{x}_i\|^2, \tag{8.12}$$

where $\mathbf{x}_i$ is the consensus (average) position of feature i in the global coordinate frame. (An alternative approach is to register each pair of overlapping images separately and then compute a consensus location for each frame—see Exercise 8.2.)

The above least squares problem is indeterminate (you can add a constant offset to all the frame and point locations $\mathbf{t}_j$ and $\mathbf{x}_i$). To fix this, either pick one frame as being at the origin or add a constraint to make the average frame offsets be 0.

The formulas for adding rotation and scale transformations are straightforward and are left as an exercise (Exercise 8.2). See if you can create some collages that you would be happy to share with others on the web.

8.1.3 Iterative algorithms

While linear least squares is the simplest method for estimating parameters, most problems in computer vision do not have a simple linear relationship between the measurements and the unknowns. In this case, the resulting problem is called *non-linear least squares* or *non-linear regression*.

Consider, for example, the problem of estimating a rigid Euclidean 2D transformation (translation plus rotation) between two sets of points. If we parameterize this transformation by the translation amount (t_x, t_y) and the rotation angle θ, as in Table 2.1, the Jacobian of this transformation, given in Table 8.1, depends on the current value of θ. Notice how in Table 8.1, we have re-parameterized the motion matrices so that they are always the identity at the origin $\mathbf{p} = 0$, which makes it easier to initialize the motion parameters.

To minimize the non-linear least squares problem, we iteratively find an update $\Delta\mathbf{p}$ to the current parameter estimate $\mathbf{p}$ by minimizing

$$E_{\mathrm{NLS}}(\Delta\mathbf{p}) = \sum_i \|\mathbf{f}(\mathbf{x}_i; \mathbf{p} + \Delta\mathbf{p}) - \mathbf{x}'_i\|^2 \tag{8.13}$$

$$\approx \sum_i \|\mathbf{J}(\mathbf{x}_i; \mathbf{p})\Delta\mathbf{p} - \mathbf{r}_i\|^2 \tag{8.14}$$

$$= \Delta\mathbf{p}^T \left[\sum_i \mathbf{J}^T\mathbf{J}\right] \Delta\mathbf{p} - 2\Delta\mathbf{p}^T \left[\sum_i \mathbf{J}^T\mathbf{r}_i\right] + \sum_i \|\mathbf{r}_i\|^2 \tag{8.15}$$

$$= \Delta\mathbf{p}^T\mathbf{A}\Delta\mathbf{p} - 2\Delta\mathbf{p}^T\mathbf{b} + c, \tag{8.16}$$

where the “Hessian”[5] $\mathbf{A}$ is the same as Equation (8.9) and the right-hand side vector

$$\mathbf{b} = \sum_i \mathbf{J}^T(\mathbf{x}_i)\mathbf{r}_i \tag{8.17}$$

is now a Jacobian-weighted sum of residual vectors. This makes intuitive sense, as the parameters are pulled in the direction of the prediction error with a strength proportional to the Jacobian.

[5]The “Hessian” $\mathbf{A}$ is not the true Hessian (second derivative) of the non-linear least squares problem (8.13). Instead, it is the approximate Hessian, which neglects second (and higher) order derivatives of $\mathbf{f}(\mathbf{x}_i; \mathbf{p} + \Delta\mathbf{p})$.

Once $\mathbf{A}$ and $\mathbf{b}$ have been computed, we solve for $\Delta\mathbf{p}$ using

$$(\mathbf{A} + \lambda \text{diag}(\mathbf{A}))\Delta\mathbf{p} = \mathbf{b}, \tag{8.18}$$

and update the parameter vector $\mathbf{p} \leftarrow \mathbf{p} + \Delta\mathbf{p}$ accordingly. The parameter λ is an additional damping parameter used to ensure that the system takes a "downhill" step in energy (squared error) and is an essential component of the Levenberg–Marquardt algorithm (described in more detail in Appendix A.3). In many applications, it can be set to 0 if the system is successfully converging.

For the case of our 2D translation+rotation, we end up with a 3×3 set of normal equations in the unknowns $(\delta t_x, \delta t_y, \delta\theta)$. An initial guess for (t_x, t_y, θ) can be obtained by fitting a four-parameter similarity transform in (t_x, t_y, c, s) and then setting $\theta = \tan^{-1}(s/c)$. An alternative approach is to estimate the translation parameters using the centroids of the 2D points and to then estimate the rotation angle using polar coordinates (Exercise 8.3).

For the other 2D motion models, the derivatives in Table 8.1 are all fairly straightforward, except for the projective 2D motion (homography), which arises in image-stitching applications (Section 8.2). These equations can be re-written from (2.21) in their new parametric form as

$$x' = \frac{(1 + h_{00})x + h_{01}y + h_{02}}{h_{20}x + h_{21}y + 1} \quad \text{and} \quad y' = \frac{h_{10}x + (1 + h_{11})y + h_{12}}{h_{20}x + h_{21}y + 1}. \tag{8.19}$$

The Jacobian is therefore

$$\mathbf{J} = \frac{\partial \mathbf{f}}{\partial \mathbf{p}} = \frac{1}{D} \begin{bmatrix} x & y & 1 & 0 & 0 & 0 & -x'x & -x'y \\ 0 & 0 & 0 & x & y & 1 & -y'x & -y'y \end{bmatrix}, \tag{8.20}$$

where $D = h_{20}x + h_{21}y + 1$ is the denominator in (8.19), which depends on the current parameter settings (as do x' and y').

An initial guess for the eight unknowns $\{h_{00}, h_{01}, \ldots, h_{21}\}$ can be obtained by multiplying both sides of the equations in (8.19) through by the denominator, which yields the linear set of equations,

$$\begin{bmatrix} \hat{x}' - x \\ \hat{y}' - y \end{bmatrix} = \begin{bmatrix} x & y & 1 & 0 & 0 & 0 & -\hat{x}'x & -\hat{x}'y \\ 0 & 0 & 0 & x & y & 1 & -\hat{y}'x & -\hat{y}'y \end{bmatrix} \begin{bmatrix} h_{00} \\ \vdots \\ h_{21} \end{bmatrix}. \tag{8.21}$$

However, this is not optimal from a statistical point of view, since the denominator D, which was used to multiply each equation, can vary quite a bit from point to point.[6]

One way to compensate for this is to *reweight* each equation by the inverse of the current estimate of the denominator, D,

$$\frac{1}{D}\begin{bmatrix} \hat{x}' - x \\ \hat{y}' - y \end{bmatrix} = \frac{1}{D}\begin{bmatrix} x & y & 1 & 0 & 0 & 0 & -\hat{x}'x & -\hat{x}'y \\ 0 & 0 & 0 & x & y & 1 & -\hat{y}'x & -\hat{y}'y \end{bmatrix} \begin{bmatrix} h_{00} \\ \vdots \\ h_{21} \end{bmatrix}. \tag{8.22}$$

While this may at first seem to be the exact same set of equations as (8.21), because least squares is being used to solve the over-determined set of equations, the weightings *do* matter and produce a different set of normal equations that performs better in practice.

[6]Hartley and Zisserman (2004) call this strategy of forming linear equations from rational equations the *direct linear transform*, but that term is more commonly associated with pose estimation (Section 11.2). Note also that our definition of the h_{ij} parameters differs from that used in their book, since we define h_{ii} to be the *difference* from unity and we do not leave h_{22} as a free parameter, which means that we cannot handle certain extreme homographies.

The most principled way to do the estimation, however, is to directly minimize the squared residual Equations (8.13) using the Gauss–Newton approximation, i.e., performing a first-order Taylor series expansion in $\mathbf{p}$, as shown in (8.14), which yields the set of equations

$$\begin{bmatrix}\hat{x}' - \tilde{x}' \\ \hat{y}' - \tilde{y}'\end{bmatrix} = \frac{1}{D}\begin{bmatrix} x & y & 1 & 0 & 0 & 0 & -\tilde{x}'x & -\tilde{x}'y \\ 0 & 0 & 0 & x & y & 1 & -\tilde{y}'x & -\tilde{y}'y \end{bmatrix}\begin{bmatrix}\Delta h_{00} \\ \vdots \\ \Delta h_{21}\end{bmatrix}. \tag{8.23}$$

While these look similar to (8.22), they differ in two important respects. First, the left-hand side consists of unweighted *prediction errors* rather than point displacements and the solution vector is a *perturbation* to the parameter vector $\mathbf{p}$. Second, the quantities inside $\mathbf{J}$ involve *predicted* feature locations $(\tilde{x}', \tilde{y}')$ instead of *sensed* feature locations $(\hat{x}', \hat{y}')$. Both of these differences are subtle and yet they lead to an algorithm that, when combined with proper checking for downhill steps (as in the Levenberg–Marquardt algorithm), will converge to a local minimum. Note that iterating Equations (8.22) is not guaranteed to converge, since it is not minimizing a well-defined energy function.

Equation (8.23) is analogous to the *additive* algorithm for direct intensity-based registration (Section 9.2), since the change to the full transformation is being computed. If we prepend an incremental homography to the current homography instead, i.e., we use a *compositional* algorithm (described in Section 9.2), we get $D = 1$ (since $\mathbf{p} = 0$) and the above formula simplifies to

$$\begin{bmatrix}\hat{x}' - x \\ \hat{y}' - y\end{bmatrix} = \begin{bmatrix} x & y & 1 & 0 & 0 & 0 & -x^2 & -xy \\ 0 & 0 & 0 & x & y & 1 & -xy & -y^2 \end{bmatrix}\begin{bmatrix}\Delta h_{00} \\ \vdots \\ \Delta h_{21}\end{bmatrix}, \tag{8.24}$$

where we have replaced $(\tilde{x}', \tilde{y}')$ with (x, y) for conciseness.

8.1.4 Robust least squares and RANSAC

While regular least squares is the method of choice for measurements where the noise follows a normal (Gaussian) distribution, more robust versions of least squares are required when there are outliers among the correspondences (as there almost always are). In this case, it is preferable to use an *M-estimator* (Huber 1981; Hampel, Ronchetti *et al.* 1986; Black and Rangarajan 1996; Stewart 1999), which involves applying a robust penalty function $\rho(r)$ to the residuals

$$E_{\text{RLS}}(\Delta\mathbf{p}) = \sum_i \rho(\|\mathbf{r}_i\|) \tag{8.25}$$

instead of squaring them.[7]

We can take the derivative of this function with respect to $\mathbf{p}$ and set it to 0,

$$\sum_i \psi(\|\mathbf{r}_i\|)\frac{\partial\|\mathbf{r}_i\|}{\partial\mathbf{p}} = \sum_i \frac{\psi(\|\mathbf{r}_i\|)}{\|\mathbf{r}_i\|}\mathbf{r}_i^T\frac{\partial\mathbf{r}_i}{\partial\mathbf{p}} = 0, \tag{8.26}$$

where $\psi(r) = \rho'(r)$ is the derivative of ρ and is called the *influence function*. If we introduce a *weight function*, $w(r) = \psi(r)/r$, we observe that finding the stationary point of (8.25) using (8.26) is equivalent to minimizing the *iteratively reweighted least squares* (IRLS) problem

$$E_{\text{IRLS}} = \sum_i w(\|\mathbf{r}_i\|)\|\mathbf{r}_i\|^2, \tag{8.27}$$

[7] The plots for some commonly used robust penalty functions ρ can be found in Figure 4.7.

where the $w(\|\mathbf{r}_i\|)$ play the same local weighting role as σ_i^{-2} in (8.10). The IRLS algorithm alternates between computing the influence functions $w(\|\mathbf{r}_i\|)$ and solving the resulting weighted least squares problem (with fixed w values). Other incremental robust least squares algorithms can be found in the work of Sawhney and Ayer (1996), Black and Anandan (1996), Black and Rangarajan (1996), and Baker, Gross *et al.* (2003) and in textbooks and tutorials on robust statistics (Huber 1981; Hampel, Ronchetti *et al.* 1986; Rousseeuw and Leroy 1987; Stewart 1999).

While M-estimators can definitely help reduce the influence of outliers, in some cases, starting with too many outliers will prevent IRLS (or other gradient descent algorithms) from converging to the global optimum. A better approach is often to find a starting set of *inlier* correspondences, i.e., points that are consistent with a dominant motion estimate.[8]

Two widely used approaches to this problem are called RANdom SAmple Consensus, or RANSAC for short (Fischler and Bolles 1981), and *least median of squares* (LMS) (Rousseeuw 1984). Both techniques start by selecting (at random) a subset of k correspondences, which is then used to compute an initial estimate for $\mathbf{p}$. The *residuals* of the full set of correspondences are then computed as

$$\mathbf{r}_i = \tilde{\mathbf{x}}_i'(\mathbf{x}_i; \mathbf{p}) - \hat{\mathbf{x}}_i', \tag{8.28}$$

where $\tilde{\mathbf{x}}_i'$ are the *estimated* (mapped) locations and $\hat{\mathbf{x}}_i'$ are the sensed (detected) feature point locations.[9]

The RANSAC technique then counts the number of *inliers* that are within ϵ of their predicted location, i.e., whose $\|\mathbf{r}_i\| \leq \epsilon$. (The ϵ value is application dependent but is often around 1–3 pixels.) Least median of squares finds the median value of the $\|\mathbf{r}_i\|^2$ values. The random selection process is repeated S times and the sample set with the largest number of inliers (or with the smallest median residual) is kept as the final solution. Either the initial parameter guess $\mathbf{p}$ or the full set of computed inliers is then passed on to the next data fitting stage.

When the number of measurements is quite large, it may be preferable to only score a subset of the measurements in an initial round that selects the most plausible hypotheses for additional scoring and selection. This modification of RANSAC, which can significantly speed up its performance, is called *Preemptive RANSAC* (Nistér 2003). In another variant on RANSAC called PROSAC (PROgressive SAmple Consensus), random samples are initially added from the most "confident" matches, thereby speeding up the process of finding a (statistically) likely good set of inliers (Chum and Matas 2005). Raguram, Chum *et al.* (2012) provide a unified framework from which most of these techniques can be derived as well as a nice experimental comparison.

Additional variants on RANSAC include MLESAC (Torr and Zisserman 2000), DSAC (Brachmann, Krull *et al.* 2017), Graph-Cut RANSAC (Barath and Matas 2018), MAGSAC (Barath, Matas, and Noskova 2019), and ESAC (Brachmann and Rother 2019). Some of these algorithms, such as DSAC (Differentiable RANSAC), are designed to be differentiable so they can be used in end-to-end training of feature detection and matching pipelines (Section 7.1). The MAGSAC++ paper by Barath, Noskova *et al.* (2020) compares many of these variants. Yang, Antonante *et al.* (2020) claim that using a robust penalty function with a decreasing outlier parameter, i.e., *graduated non-convexity* (Blake and Zisserman 1987; Barron 2019), can outperform RANSAC in many geometric correspondence and pose estimation problems. To ensure that the random sampling has a good chance of finding a true set of inliers, a sufficient number of trials S must be evaluated. Let p be the probability that any given correspondence is valid and P be the probability of success after S trials. The likelihood in one trial that all k random samples are inliers is p^k. Therefore, the likelihood that

[8]For pixel-based alignment methods (Section 9.1.1), hierarchical (coarse-to-fine) techniques are often used to lock onto the *dominant motion* in a scene.

[9]For problems such as epipolar geometry estimation, the residual may be the distance between a point and a line.

k	p	S
3	0.5	35
6	0.6	97
6	0.5	293

Table 8.2 Number of trials S to attain a 99% probability of success (Stewart 1999).

S such trials will all fail is

$$1 - P = (1 - p^k)^S \tag{8.29}$$

and the required minimum number of trials is

$$S = \frac{\log(1 - P)}{\log(1 - p^k)}. \tag{8.30}$$

Stewart (1999) gives examples of the required number of trials S to attain a 99% probability of success. As you can see from Table 8.2, the number of trials grows quickly with the number of sample points used. This provides a strong incentive to use the *minimum* number of sample points k possible for any given trial, which is how RANSAC is normally used in practice.

Uncertainty modeling

In addition to robustly computing a good alignment, some applications require the computation of uncertainty (see Appendix B.6). For linear problems, this estimate can be obtained by inverting the Hessian matrix (8.9) and multiplying it by the feature position noise, if these have not already been used to weight the individual measurements, as in Equations (8.10) and (8.11). In statistics, the Hessian, which is the inverse covariance, is sometimes called the (Fisher) *information matrix* (Appendix B.1).

When the problem involves non-linear least squares, the inverse of the Hessian matrix provides the *Cramer–Rao lower bound* on the covariance matrix, i.e., it provides the *minimum* amount of covariance in a given solution, which can actually have a wider spread ("longer tails") if the energy flattens out away from the local minimum where the optimal solution is found.

8.1.5 3D alignment

Instead of aligning 2D sets of image features, many computer vision applications require the alignment of 3D points. In the case where the 3D transformations are linear in the motion parameters, e.g., for translation, similarity, and affine, regular least squares (8.5) can be used.

The case of rigid (Euclidean) motion,

$$E_{\text{R3D}} = \sum_i \|\mathbf{x}'_i - \mathbf{R}\mathbf{x}_i - \mathbf{t}\|^2, \tag{8.31}$$

which arises more frequently and is often called the *absolute orientation* problem (Horn 1987), requires slightly different techniques. If only scalar weightings are being used (as opposed to full 3D per-point anisotropic covariance estimates), the weighted centroids of the two point clouds $\mathbf{c}$ and $\mathbf{c}'$ can be used to estimate the translation $\mathbf{t} = \mathbf{c}' - \mathbf{R}\mathbf{c}$.[10] We are then left with the problem of

[10] When full covariances are used, they are transformed by the rotation, so a closed-form solution for translation is not possible.

estimating the rotation between two sets of points $\{\hat{\mathbf{x}}_i = \mathbf{x}_i - \mathbf{c}\}$ and $\{\hat{\mathbf{x}}'_i = \mathbf{x}'_i - \mathbf{c}'\}$ that are both centered at the origin.

One commonly used technique is called the *orthogonal Procrustes algorithm* (Golub and Van Loan 1996, p. 601) and involves computing the singular value decomposition (SVD) of the 3 × 3 correlation matrix

$$\mathbf{C} = \sum_i \hat{\mathbf{x}}'\hat{\mathbf{x}}^T = \mathbf{U}\boldsymbol{\Sigma}\mathbf{V}^T. \tag{8.32}$$

The rotation matrix is then obtained as $\mathbf{R} = \mathbf{U}\mathbf{V}^T$. (Verify this for yourself when $\hat{\mathbf{x}}' = \mathbf{R}\hat{\mathbf{x}}$.)

Another technique is the absolute orientation algorithm (Horn 1987) for estimating the unit quaternion corresponding to the rotation matrix $\mathbf{R}$, which involves forming a 4 × 4 matrix from the entries in $\mathbf{C}$ and then finding the eigenvector associated with its largest positive eigenvalue.

Lorusso, Eggert, and Fisher (1995) experimentally compare these two techniques to two additional techniques proposed in the literature, but find that the difference in accuracy is negligible (well below the effects of measurement noise).

In situations where these closed-form algorithms are not applicable, e.g., when full 3D covariances are being used or when the 3D alignment is part of some larger optimization, the incremental rotation update introduced in Section 2.1.3 (2.35–2.36), which is parameterized by an instantaneous rotation vector $\boldsymbol{\omega}$, can be used (See Section 8.2.3 for an application to image stitching.)

In some situations, e.g., when merging range data maps, the correspondence between data points is not known *a priori*. In this case, iterative algorithms that start by matching nearby points and then update the most likely correspondence can be used (Besl and McKay 1992; Zhang 1994; Szeliski and Lavallée 1996; Gold, Rangarajan *et al.* 1998; David, DeMenthon *et al.* 2004; Li and Hartley 2007; Enqvist, Josephson, and Kahl 2009). These techniques are discussed in more detail in Section 13.2.1.

8.2 Image stitching

Algorithms for aligning images and stitching them into seamless photo-mosaics are among the oldest and most widely used in computer vision (Milgram 1975; Peleg 1981). Image stitching algorithms create the high-resolution photo-mosaics used to produce today's digital maps and satellite photos. They are also now a standard mode in smartphone cameras and can be used to create beautiful ultra wide-angle panoramas.

Image stitching originated in the photogrammetry community, where more manually intensive methods based on surveyed *ground control points* or manually registered *tie points* have long been used to register aerial photos into large-scale photo-mosaics (Slama 1980). One of the key advances in this community was the development of *bundle adjustment* algorithms (Section 11.4.2), which could simultaneously solve for the locations of all of the camera positions, thus yielding globally consistent solutions (Triggs, McLauchlan *et al.* 1999). Another recurring problem in creating photo-mosaics is the elimination of visible seams, for which a variety of techniques have been developed over the years (Milgram 1975, 1977; Peleg 1981; Davis 1998; Agarwala, Dontcheva *et al.* 2004)

In film photography, special cameras were developed in the 1990s to take ultra-wide-angle panoramas, often by exposing the film through a vertical slit as the camera rotated on its axis (Meehan 1990). In the mid-1990s, image alignment techniques started being applied to the construction of wide-angle seamless panoramas from regular hand-held cameras (Mann and Picard 1994; Chen 1995; Szeliski 1996). Subsequent algorithms addressed the need to compute globally consistent alignments (Szeliski and Shum 1997; Sawhney and Kumar 1999; Shum and Szeliski 2000), to remove "ghosts" due to parallax and object movement (Davis 1998; Shum and Szeliski 2000; Uyttendaele, Eden, and Szeliski 2001; Agarwala, Dontcheva *et al.* 2004), and to deal with varying exposures (Mann and Picard 1994; Uyttendaele, Eden, and Szeliski 2001; Levin, Zomet *et al.* 2004; Eden, Uyttendaele, and Szeliski 2006; Kopf, Uyttendaele *et al.* 2007).[11]

While early techniques worked by directly minimizing pixel-to-pixel dissimilarities, today's algorithms extract a sparse set of features and match them to each other, as described in Chapter 7. Such feature-based approaches (Zoghlami, Faugeras, and Deriche 1997; Capel and Zisserman 1998; Cham and Cipolla 1998; Badra, Qumsieh, and Dudek 1998; McLauchlan and Jaenicke 2002; Brown and Lowe 2007) have the advantage of being more robust against scene movement and are usually faster,[12] Their biggest advantage, however, is the ability to "recognize panoramas", i.e., to automatically discover the adjacency (overlap) relationships among an unordered set of images, which makes them ideally suited for fully automated stitching of panoramas taken by casual users (Brown and Lowe 2007).

What, then, are the essential problems in image stitching? As with image alignment, we must first determine the appropriate mathematical model relating pixel coordinates in one image to pixel coordinates in another; Section 8.2.1 reviews the basic models we have studied and presents some new motion models related specifically to panoramic image stitching. Next, we must somehow estimate the correct alignments relating various pairs (or collections) of images. Chapter 7 discusses how distinctive features can be found in each image and then efficiently matched to rapidly establish correspondences between pairs of images. Chapter 9 discusses how direct pixel-to-pixel comparisons combined with gradient descent (and other optimization techniques) can also be used to estimate these parameters. When multiple images exist in a panorama, global optimization techniques can be used to compute a globally consistent set of alignments and to efficiently discover which images overlap one another. In Section 8.3, we look at how each of these previously developed techniques can be modified to take advantage of the imaging setups commonly used to create panoramas.

Once we have aligned the images, we must choose a final compositing surface for warping the aligned images (Section 8.4.1). We also need algorithms to seamlessly cut and blend overlapping images, even in the presence of parallax, lens distortion, scene motion, and exposure differences (Section 8.4.2–8.4.4).

8.2.1 Parametric motion models

Before we can register and align images, we need to establish the mathematical relationships that map pixel coordinates from one image to another. A variety of such *parametric motion models* are possible, from simple 2D transforms, to planar perspective models, 3D camera rotations, lens distortions, and mapping to non-planar (e.g., cylindrical) surfaces.

We already covered several of these models in Sections 2.1 and 8.1. In particular, we saw in Section 2.1.4 how the parametric motion describing the deformation of a planar surface as viewed from different positions can be described with an eight-parameter homography (2.71) (Mann and Picard 1994; Szeliski 1996). We also saw how a camera undergoing a pure rotation induces a different kind of homography (2.72).

In this section, we review both of these models and show how they can be applied to different stitching situations. We also introduce spherical and cylindrical compositing surfaces and show how, under favorable circumstances, they can be used to perform alignment using pure translations (Section 8.2.6). Deciding which alignment model is most appropriate for a given situation or set of

[11] A collection of some of these papers was compiled by Benosman and Kang (2001) and they are surveyed by Szeliski (2006a).

[12] See a discussion of the pros and cons of direct vs. feature-based techniques in (Triggs, Zisserman, and Szeliski 2000) and in the first edition of this book (Szeliski 2010, Section 8.3.4).

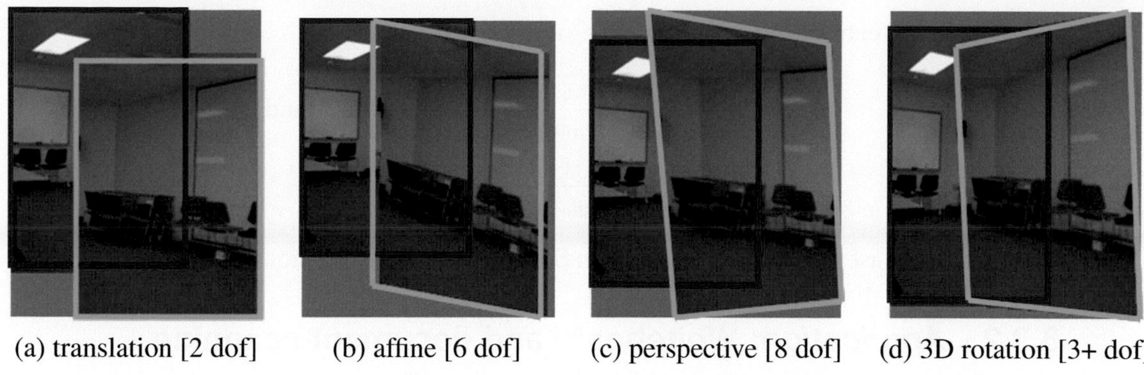

(a) translation [2 dof] (b) affine [6 dof] (c) perspective [8 dof] (d) 3D rotation [3+ dof]

Figure 8.4 Two-dimensional motion models and how they can be used for image stitching.

data is a *model selection* problem (Torr 2002; Bishop 2006; Robert 2007; Hastie, Tibshirani, and Friedman 2009; Murphy 2012), an important topic we do not cover in this book.

Planar perspective motion

The simplest possible motion model to use when aligning images is to simply translate and rotate them in 2D (Figure 8.4a). This is exactly the same kind of motion that you would use if you had overlapping photographic prints. It is also the kind of technique favored by David Hockney to create the collages that he calls *joiners* (Zelnik-Manor and Perona 2007; Nomura, Zhang, and Nayar 2007). Creating such collages, which show visible seams and inconsistencies that add to the artistic effect, is popular on websites such as Flickr, where they more commonly go under the name *panography* (Section 8.1.2). Translation and rotation are also usually adequate motion models to compensate for small camera motions in applications such as photo and video stabilization and merging (Exercise 8.1 and Section 9.2.1).

In Section 2.1.4, we saw how the mapping between two cameras viewing a common plane can be described using a 3×3 homography (2.71). Consider the matrix $\mathbf{M}_{10}$ that arises when mapping a pixel in one image to a 3D point and then back onto a second image,

$$\tilde{\mathbf{x}}_1 \sim \tilde{\mathbf{P}}_1 \tilde{\mathbf{P}}_0^{-1} \tilde{\mathbf{x}}_0 = \mathbf{M}_{10} \tilde{\mathbf{x}}_0. \tag{8.33}$$

When the last row of the $\mathbf{P}_0$ matrix is replaced with a plane equation $\hat{\mathbf{n}}_0 \cdot \mathbf{p} + c_0$ and points are assumed to lie on this plane, i.e., their disparity is $d_0 = 0$, we can ignore the last column of $\mathbf{M}_{10}$ and also its last row, since we do not care about the final z-buffer depth. The resulting homography matrix $\tilde{\mathbf{H}}_{10}$ (the upper left 3×3 sub-matrix of $\mathbf{M}_{10}$) describes the mapping between pixels in the two images,

$$\tilde{\mathbf{x}}_1 \sim \tilde{\mathbf{H}}_{10} \tilde{\mathbf{x}}_0. \tag{8.34}$$

This observation formed the basis of some of the earliest automated image stitching algorithms (Mann and Picard 1994; Szeliski 1994, 1996). Because reliable feature matching techniques had not yet been developed, these algorithms used direct pixel value matching, i.e., direct parametric motion estimation, as described in Section 9.2 and Equations (8.19–8.20).

More recent stitching algorithms first extract features and then match them up, often using robust techniques such as RANSAC (Section 8.1.4) to compute a good set of inliers. The final computation of the homography (8.34), i.e., the solution of the least squares fitting problem given pairs of corresponding features,

$$x_1 = \frac{(1+h_{00})x_0 + h_{01}y_0 + h_{02}}{h_{20}x_0 + h_{21}y_0 + 1} \quad \text{and} \tag{8.35}$$

$$y_1 = \frac{h_{10}x_0 + (1+h_{11})y_0 + h_{12}}{h_{20}x_0 + h_{21}y_0 + 1}, \tag{8.36}$$

uses iterative least squares, as described in Section 8.1.3 and Equations (8.21–8.23).

8.2.2 *Application*: Whiteboard and document scanning

The simplest image-stitching application is to stitch together a number of image scans taken on a flatbed scanner. Say you have a large map, or a piece of child's artwork, that is too large to fit on your scanner. Simply take multiple scans of the document, making sure to overlap the scans by a large enough amount to ensure that there are enough common features. Next, take successive pairs of images that you know overlap, extract features, match them up, and estimate the 2D rigid transform (2.16),

$$\mathbf{x}_{k+1} = \mathbf{R}_k \mathbf{x}_k + \mathbf{t}_k, \tag{8.37}$$

that best matches the features, using two-point RANSAC, if necessary, to find a good set of inliers. Then, on a final compositing surface (aligned with the first scan, for example), resample your images (Section 3.6.1) and average them together. Can you see any potential problems with this scheme?

One complication is that a 2D rigid transformation is non-linear in the rotation angle θ, so you will have to either use non-linear least squares or constrain $\mathbf{R}$ to be orthonormal, as described in Section 8.1.3.

A bigger problem lies in the pairwise alignment process. As you align more and more pairs, the solution may drift so that it is no longer globally consistent. In this case, a global optimization procedure, as described in Section 8.3, may be required. Such global optimization often requires a large system of non-linear equations to be solved, although in some cases, such as linearized homographies (Section 8.2.3) or similarity transforms (Section 8.1.2), regular least squares may be an option.

A slightly more complex scenario is when you take multiple overlapping handheld pictures of a whiteboard or other large planar object (He and Zhang 2005; Zhang and He 2007). Here, the natural motion model to use is a homography, although a more complex model that estimates the 3D rigid motion relative to the plane (plus the focal length, if unknown), could in principle be used.

8.2.3 Rotational panoramas

The most typical case for panoramic image stitching is when the camera undergoes a pure rotation. Think of standing at the rim of the Grand Canyon. Relative to the distant geometry in the scene, as you snap away, the camera is undergoing a pure rotation, which is equivalent to assuming that all points are very far from the camera, i.e., on the *plane at infinity* (Figure 8.5).[13] Setting $\mathbf{t}_0 = \mathbf{t}_1 = 0$, we get the simplified 3×3 homography

$$\tilde{\mathbf{H}}_{10} = \mathbf{K}_1 \mathbf{R}_1 \mathbf{R}_0^{-1} \mathbf{K}_0^{-1} = \mathbf{K}_1 \mathbf{R}_{10} \mathbf{K}_0^{-1}, \tag{8.38}$$

[13] In a more general (e.g., indoor) scene, if we want to ensure that there is no *parallax* (visible relative movement between objects at different depths), we need to rotate the camera around the lens's front *no-parallax point* (Littlefield 2006). This can be achieved by using a specialized panoramic rotation head with a built-in translation stage (Houghton 2013) or by determining the front nodal point using observations of collinear points—see Debevec, Wenger *et al.* (2002) and Szeliski (2010, Figure 6.7).

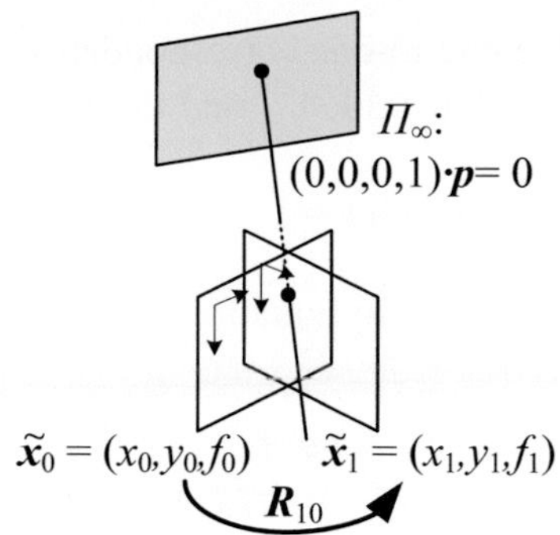

Figure 8.5 Pure 3D camera rotation. The form of the homography (mapping) is particularly simple and depends only on the 3D rotation matrix and focal lengths.

where $\mathbf{K}_k = \operatorname{diag}(f_k, f_k, 1)$ is the simplified camera intrinsic matrix (2.59), assuming that $c_x = c_y = 0$, i.e., we are indexing the pixels starting from the image center (Szeliski 1996). This can also be re-written as

$$\begin{bmatrix} x_1 \\ y_1 \\ 1 \end{bmatrix} \sim \begin{bmatrix} f_1 & & \\ & f_1 & \\ & & 1 \end{bmatrix} \mathbf{R}_{10} \begin{bmatrix} f_0^{-1} & & \\ & f_0^{-1} & \\ & & 1 \end{bmatrix} \begin{bmatrix} x_0 \\ y_0 \\ 1 \end{bmatrix} \tag{8.39}$$

or

$$\begin{bmatrix} x_1 \\ y_1 \\ f_1 \end{bmatrix} \sim \mathbf{R}_{10} \begin{bmatrix} x_0 \\ y_0 \\ f_0 \end{bmatrix}, \tag{8.40}$$

which reveals the simplicity of the mapping equations and makes all of the motion parameters explicit. Thus, instead of the general eight-parameter homography relating a pair of images, we get the three-, four-, or five-parameter *3D rotation* motion models corresponding to the cases where the focal length f is known, fixed, or variable (Szeliski and Shum 1997).[14] Estimating the 3D rotation matrix (and, optionally, focal length) associated with each image is intrinsically more stable than estimating a homography with a full eight degrees of freedom, which makes this the method of choice for large-scale image stitching algorithms (Szeliski and Shum 1997; Shum and Szeliski 2000; Brown and Lowe 2007).

Given this representation, how do we update the rotation matrices to best align two overlapping images? Given a current estimate for the homography $\tilde{\mathbf{H}}_{10}$ in (8.38), the best way to update $\mathbf{R}_{10}$ is to prepend an *incremental* rotation matrix $\mathbf{R}(\boldsymbol{\omega})$ to the current estimate $\mathbf{R}_{10}$ (Szeliski and Shum 1997; Shum and Szeliski 2000),

$$\tilde{\mathbf{H}}(\boldsymbol{\omega}) = \mathbf{K}_1 \mathbf{R}(\boldsymbol{\omega}) \mathbf{R}_{10} \mathbf{K}_0^{-1} = [\mathbf{K}_1 \mathbf{R}(\boldsymbol{\omega}) \mathbf{K}_1^{-1}][\mathbf{K}_1 \mathbf{R}_{10} \mathbf{K}_0^{-1}] = \mathbf{D}\tilde{\mathbf{H}}_{10}. \tag{8.41}$$

Note that here we have written the update rule in the *compositional* form, where the incremental update $\mathbf{D}$ is *prepended* to the current homography $\tilde{\mathbf{H}}_{10}$. Using the small-angle approximation to $\mathbf{R}(\boldsymbol{\omega})$ given in (2.35), we can write the incremental update matrix as

$$\mathbf{D} = \mathbf{K}_1 \mathbf{R}(\boldsymbol{\omega}) \mathbf{K}_1^{-1} \approx \mathbf{K}_1 (\mathbf{I} + [\boldsymbol{\omega}]_\times) \mathbf{K}_1^{-1} = \begin{bmatrix} 1 & -\omega_z & f_1 \omega_y \\ \omega_z & 1 & -f_1 \omega_x \\ -\omega_y / f_1 & \omega_x / f_1 & 1 \end{bmatrix} \tag{8.42}$$

[14] An initial estimate of the focal lengths can be obtained using the intrinsic calibration techniques described in Section 11.1.3 or from EXIF tags.

Notice how there is now a nice one-to-one correspondence between the entries in the $\mathbf{D}$ matrix and the $h_{00}, \ldots, h_{21}$ parameters used in Table 8.1 and Equation (8.19), i.e.,

$$(h_{00}, h_{01}, h_{02}, h_{00}, h_{11}, h_{12}, h_{20}, h_{21}) = (0, -\omega_z, f_1\omega_y, \omega_z, 0, -f_1\omega_x, -\omega_y/f_1, \omega_x/f_1). \quad (8.43)$$

We can therefore apply the chain rule to Equations (8.24 and 8.43) to obtain

$$\begin{bmatrix} \hat{x}' - x \\ \hat{y}' - y \end{bmatrix} = \begin{bmatrix} -xy/f_1 & f_1 + x^2/f_1 & -y \\ -(f_1 + y^2/f_1) & xy/f_1 & x \end{bmatrix} \begin{bmatrix} \omega_x \\ \omega_y \\ \omega_z \end{bmatrix}, \quad (8.44)$$

which give us the linearized update equations needed to estimate $\boldsymbol{\omega} = (\omega_x, \omega_y, \omega_z)$.[15] Notice that this update rule depends on the focal length f_1 of the *target* view and is independent of the focal length f_0 of the *template* view. This is because the compositional algorithm essentially makes small perturbations to the target. Once the incremental rotation vector $\boldsymbol{\omega}$ has been computed, the $\mathbf{R}_1$ rotation matrix can be updated using $\mathbf{R}_1 \leftarrow \mathbf{R}(\boldsymbol{\omega})\mathbf{R}_1$.

The formulas for updating the focal length estimates are a little more involved and are given in Shum and Szeliski (2000). We will not repeat them here, since an alternative update rule, based on minimizing the difference between back-projected 3D rays, is given in Section 8.3.1. Figure 8.1a shows the alignment of four images under the 3D rotation motion model.

8.2.4 Gap closing

The techniques presented in this section can be used to estimate a series of rotation matrices and focal lengths, which can be chained together to create large panoramas. Unfortunately, because of accumulated errors, this approach will rarely produce a closed 360° panorama. Instead, there will invariably be either a gap or an overlap (Figure 8.6).

We can solve this problem by matching the first image in the sequence with the last one. The difference between the two rotation matrix estimates associated with the repeated first image indicates the amount of misregistration. This error can be distributed evenly across the whole sequence by taking the quotient of the two quaternions associated with these rotations and dividing this "error quaternion" by the number of images in the sequence (Szeliski and Shum 1997). We can also update the estimated focal length based on the amount of misregistration. To do this, we first convert the error quaternion into a *gap angle*, θ_g and then update the focal length using the equation $f' = f(1 - \theta_g/360°)$.

Figure 8.6a shows the end of registered image sequence and the first image. There is a big gap between the last image and the first, which are in fact the same image. The gap is 32° because the wrong estimate of focal length ($f = 510$) was used. Figure 8.6b shows the registration after closing the gap with the correct focal length ($f = 468$). Notice that both mosaics show very little visual misregistration (except at the gap), yet Figure 8.6a has been computed using a focal length that has 9% error. Related approaches have been developed by Hartley (1994b), McMillan and Bishop (1995), Stein (1995), and Kang and Weiss (1997) to solve the focal length estimation problem using pure panning motion and cylindrical images.

Unfortunately, this gap-closing heuristic only works for the kind of "one-dimensional" panorama where the camera is continuously turning in the same direction. In Section 8.3, we describe a different approach to removing gaps and overlaps that works for arbitrary camera motions.

[15] This is the same as the rotational component of instantaneous rigid flow (Bergen, Anandan *et al.* 1992) and the update equations given by Szeliski and Shum (1997) and Shum and Szeliski (2000).

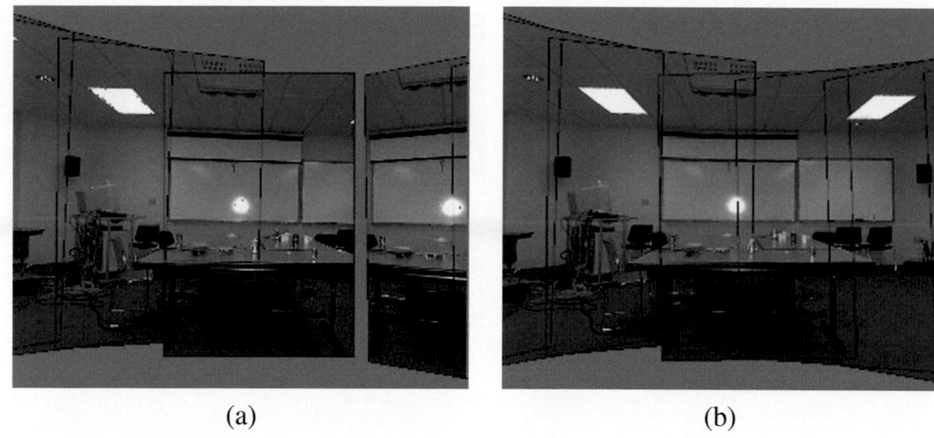

(a) (b)

Figure 8.6 Gap closing (Szeliski and Shum 1997) © 1997 ACM: (a) A gap is visible when the focal length is wrong ($f = 510$). (b) No gap is visible for the correct focal length ($f = 468$).

8.2.5 *Application*: Video summarization and compression

An interesting application of image stitching is the ability to summarize and compress videos taken with a panning camera. This application was first suggested by Teodosio and Bender (1993), who called their mosaic-based summaries *salient stills*. These ideas were then extended by Irani, Hsu, and Anandan (1995) and Irani and Anandan (1998) to additional applications, such as video compression and video indexing. While these early approaches used affine motion models and were therefore restricted to long focal lengths, the techniques were generalized by Lee, Chen *et al.* (1997) to full eight-parameter homographies and incorporated into the MPEG-4 video compression standard, where the stitched background layers were called *video sprites* (Figure 8.7).

While video stitching is in many ways a straightforward generalization of multiple-image stitching (Steedly, Pal, and Szeliski 2005; Baudisch, Tan *et al.* 2006), the potential presence of large amounts of independent motion, camera zoom, and the desire to visualize dynamic events impose additional challenges. For example, moving foreground objects can often be removed using *median filtering*. Alternatively, foreground objects can be extracted into a separate layer (Sawhney and Ayer 1996) and later composited back into the stitched panoramas, sometimes as multiple instances to give the impressions of a "Chronophotograph" (Massey and Bender 1996) and sometimes as video overlays (Irani and Anandan 1998). Videos can also be used to create animated *panoramic video textures* (Section 14.5.2), in which different portions of a panoramic scene are animated with independently moving video loops (Agarwala, Zheng *et al.* 2005; Rav-Acha, Pritch *et al.* 2005; Joshi, Mehta *et al.* 2012; Yan, Liu, and Furukawa 2017; He, Liao *et al.* 2017; Oh, Joo *et al.* 2017), or to shine "video flashlights" onto a composite mosaic of a scene (Sawhney, Arpa *et al.* 2002).

Video can also provide an interesting source of content for creating panoramas taken from moving cameras. While this invalidates the usual assumption of a single point of view (optical center), interesting results can still be obtained. For example, the VideoBrush system of Sawhney, Kumar *et al.* (1998) uses thin strips taken from the center of the image to create a panorama taken from a horizontally moving camera. This idea can be generalized to other camera motions and compositing surfaces using the concept of mosaics on an adaptive manifold (Peleg, Rousso *et al.* 2000), and also used to generate panoramic stereograms (Ishiguro, Yamamoto, and Tsuji 1992; Peleg, Ben-Ezra, and Pritch 2001).[16] Related ideas have been used to create panoramic matte paintings for multi-

[16]A similar technique was likely used in the Google Cardboard Camera, https://blog.google/products/google-vr/

Figure 8.7 Video stitching the background scene to create a single *sprite* image that can be transmitted and used to re-create the background in each frame (Lee, Chen *et al.* 1997) © 1997 IEEE.

plane cel animation (Wood, Finkelstein *et al.* 1997), for creating stitched images of scenes with parallax (Kumar, Anandan *et al.* 1995), and as 3D representations of more complex scenes using *multiple-center-of-projection images* (Rademacher and Bishop 1998) and *multi-perspective panoramas* (Román, Garg, and Levoy 2004; Román and Lensch 2006; Agarwala, Agrawala *et al.* 2006; Kopf, Chen *et al.* 2010).

Another interesting variant on video-based panoramas is *concentric mosaics* (Section 14.3.3) (Shum and He 1999). Here, rather than trying to produce a single panoramic image, the complete original video is kept and used to re-synthesize views (from different camera origins) using ray remapping (light field rendering), thus endowing the panorama with a sense of 3D depth. The same dataset can also be used to explicitly reconstruct the depth using multi-baseline stereo (Ishiguro, Yamamoto, and Tsuji 1992; Peleg, Ben-Ezra, and Pritch 2001; Li, Shum *et al.* 2004; Zheng, Kang *et al.* 2007).

8.2.6 Cylindrical and spherical coordinates

An alternative to using homographies or 3D motions to align images is to first warp the images into *cylindrical* coordinates and then use a pure translational model to align them (Chen 1995; Szeliski 1996). Unfortunately, this only works if the images are all taken with a level camera or with a known tilt angle.

Assume for now that the camera is in its canonical position, i.e., its rotation matrix is the identity, $\mathbf{R} = \mathbf{I}$, so that the optical axis is aligned with the z-axis and the y-axis is aligned vertically. The 3D ray corresponding to an (x, y) pixel is therefore (x, y, f).

We wish to project this image onto a *cylindrical surface* of unit radius (Szeliski 1996). Points on this surface are parameterized by an angle θ and a height h, with the 3D cylindrical coordinates corresponding to (θ, h) given by

$$(\sin\theta, h, \cos\theta) \propto (x, y, f), \tag{8.45}$$

cardboard-camera-ios.

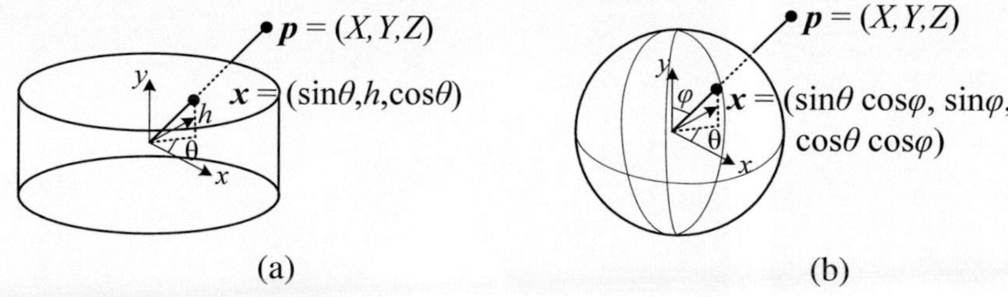

Figure 8.8 Projection from 3D to (a) cylindrical and (b) spherical coordinates.

as shown in Figure 8.8a. From this correspondence, we can compute the formula for the *warped* or *mapped* coordinates (Szeliski and Shum 1997),

$$x' = s\theta = s \tan^{-1} \frac{x}{f}, \tag{8.46}$$

$$y' = sh = s \frac{y}{\sqrt{x^2 + f^2}}, \tag{8.47}$$

where s is an arbitrary scaling factor (sometimes called the *radius* of the cylinder) that can be set to $s = f$ to minimize the distortion (scaling) near the center of the image.[17] The inverse of this mapping equation is given by

$$x = f \tan \theta = f \tan \frac{x'}{s}, \tag{8.48}$$

$$y = h\sqrt{x^2 + f^2} = \frac{y'}{s} f \sqrt{1 + \tan^2 x'/s} = f \frac{y'}{s} \sec \frac{x'}{s}. \tag{8.49}$$

Images can also be projected onto a *spherical surface* (Szeliski and Shum 1997), which is useful if the final panorama includes a full sphere or hemisphere of views, instead of just a cylindrical strip. In this case, the sphere is parameterized by two angles (θ, ϕ), with 3D spherical coordinates given by

$$(\sin\theta\cos\phi, \sin\phi, \cos\theta\cos\phi) \propto (x, y, f), \tag{8.50}$$

as shown in Figure 8.8b.[18] The correspondence between coordinates is now given by (Szeliski and Shum 1997):

$$x' = s\theta = s \tan^{-1} \frac{x}{f}, \tag{8.51}$$

$$y' = s\phi = s \tan^{-1} \frac{y}{\sqrt{x^2 + f^2}}, \tag{8.52}$$

while the inverse is given by

$$x = f \tan \theta = f \tan \frac{x'}{s}, \tag{8.53}$$

$$y = \sqrt{x^2 + f^2} \tan \phi = \tan \frac{y'}{s} f \sqrt{1 + \tan^2 x'/s} = f \tan \frac{y'}{s} \sec \frac{x'}{s}. \tag{8.54}$$

[17] The scale can also be set to a larger or smaller value for the final compositing surface, depending on the desired output panorama resolution—see Section 8.4.

[18] Note that these are not the usual spherical coordinates, first presented in Equation (2.8). Here, the y-axis points at the north pole instead of the z-axis, since we are used to viewing images taken horizontally, i.e., with the y-axis pointing in the direction of the gravity vector.

(a)

(b)

Figure 8.9 A cylindrical panorama (Szeliski and Shum 1997) © 1997 ACM: (a) two cylindrically warped images related by a horizontal translation; (b) part of a cylindrical panorama composited from a sequence of images.

Note that it may be simpler to generate a scaled (x, y, z) direction from Equation (8.50) followed by a perspective division by z and a scaling by f.

Cylindrical image stitching algorithms are most commonly used when the camera is known to be level and only rotating around its vertical axis (Chen 1995). Under these conditions, images at different rotations are related by a pure horizontal translation.[19] This makes it attractive as an initial class project in an introductory computer vision course, since the full complexity of the perspective alignment algorithm (Sections 8.1, 9.2, and 8.2.3) can be avoided. Figure 8.9 shows how two cylindrically warped images from a leveled rotational panorama are related by a pure translation (Szeliski and Shum 1997).

Professional panoramic photographers often use pan-tilt heads that make it easy to control the tilt and to stop at specific *detents* in the rotation angle. Motorized rotation heads are also sometimes used for the acquisition of larger panoramas (Kopf, Uyttendaele *et al.* 2007).[20] Not only do they ensure a uniform coverage of the visual field with a desired amount of image overlap but they also make it possible to stitch the images using cylindrical or spherical coordinates and pure translations. In this case, pixel coordinates (x, y, f) must first be rotated using the known tilt and panning angles before being projected into cylindrical or spherical coordinates (Chen 1995). Having a roughly known panning angle also makes it easier to compute the alignment, as the rough relative positioning of all the input images is known ahead of time, enabling a reduced search range for alignment. Figure 8.1b shows a full 3D rotational panorama unwrapped onto the surface of a sphere (Szeliski and Shum 1997).

One final coordinate mapping worth mentioning is the *polar* mapping, where the north pole lies along the optical axis rather than the vertical axis,

$$(\cos\theta\sin\phi, \sin\theta\sin\phi, \cos\phi) = s\,(x, y, z). \tag{8.55}$$

In this case, the mapping equations become

$$x' = s\phi\cos\theta = s\frac{x}{r}\tan^{-1}\frac{r}{z}, \tag{8.56}$$

$$y' = s\phi\sin\theta = s\frac{y}{r}\tan^{-1}\frac{r}{z}, \tag{8.57}$$

[19]Small vertical tilts can sometimes be compensated for with vertical translations.

[20]See also https://gigapan.org.

where $r = \sqrt{x^2 + y^2}$ is the *radial distance* in the (x, y) plane and $s\phi$ plays a similar role in the (x', y') plane. This mapping provides an attractive visualization surface for certain kinds of wide-angle panoramas and is also a good model for the distortion induced by *fisheye lenses*, as discussed in Section 2.1.5. Note how for small values of (x, y), the mapping equations reduce to $x' \approx sx/z$, which suggests that s plays a role similar to the focal length f.

8.3 Global alignment

So far, we have discussed how to register pairs of images using a variety of motion models. In most applications, we are given more than a single pair of images to register. The goal is then to find a globally consistent set of alignment parameters that minimize the misregistration between all pairs of images (Szeliski and Shum 1997; Shum and Szeliski 2000; Sawhney and Kumar 1999; Coorg and Teller 2000).

In this section, we extend the pairwise matching criteria (8.2, 9.1, and 9.43) to a global energy function that involves all of the per-image pose parameters (Section 8.3.1). Once we have computed the global alignment, we often need to perform *local adjustments*, such as *parallax removal*, to reduce double images and blurring due to local misregistrations (Section 8.3.2). Finally, if we are given an unordered set of images to register, we need to discover which images go together to form one or more panoramas. This process of *panorama recognition* is described in Section 8.3.3.

8.3.1 Bundle adjustment

One way to register a large number of images is to add new images to the panorama one at a time, aligning the most recent image with the previous ones already in the collection (Szeliski and Shum 1997) and discovering, if necessary, which images it overlaps (Sawhney and Kumar 1999). In the case of 360° panoramas, accumulated error may lead to the presence of a gap (or excessive overlap) between the two ends of the panorama, which can be fixed by stretching the alignment of all the images using a process called *gap closing* (Section 8.2.4). However, a better alternative is to simultaneously align all the images using a least-squares framework to correctly distribute any misregistration errors.

The process of simultaneously adjusting pose parameters and 3D point locations for a large collection of overlapping images is called *bundle adjustment* in the photogrammetry community (Triggs, McLauchlan *et al.* 1999). In computer vision, it was first applied to the general structure from motion problem (Szeliski and Kang 1994) and then later specialized for panoramic image stitching (Shum and Szeliski 2000; Sawhney and Kumar 1999; Coorg and Teller 2000).

In this section, we formulate the problem of global alignment using a feature-based approach, since this results in a simpler system. An equivalent direct approach can be obtained either by dividing images into patches and creating a virtual feature correspondence for each one (Shum and Szeliski 2000) or by replacing the per-feature error metrics with per-pixel metrics (Irani and Anandan 1999).

Before we describe this in more details, we should mention that a simpler, although less accurate, approach is to compute pairwise rotation estimates between overlapping images, and to then use a *rotation averaging* approach to estimate a global rotation for each camera (Hartley, Trumpf *et al.* 2013). However, since the measurement errors in each feature point location are not being counted correctly, as is the case in bundle adjustment, the solution will not have the same theoretical optimality.

Consider the feature-based alignment problem given in Equation (8.2), i.e.,

$$E_{\text{pairwise-LS}} = \sum_i \|\mathbf{r}_i\|^2 = \|\tilde{\mathbf{x}}_i'(\mathbf{x}_i; \mathbf{p}) - \hat{\mathbf{x}}_i'\|^2. \tag{8.58}$$

For multi-image alignment, instead of having a single collection of pairwise feature correspondences, $\{(\mathbf{x}_i, \hat{\mathbf{x}}_i')\}$, we have a collection of n features, with the location of the ith feature point in the jth image denoted by $\mathbf{x}_{ij}$ and its scalar confidence (i.e., inverse variance) denoted by c_{ij}.[21] Each image also has some associated pose parameters.

In this section, we assume that this pose consists of a rotation matrix $\mathbf{R}_j$ and a focal length f_j, although formulations in terms of homographies are also possible (Szeliski and Shum 1997; Sawhney and Kumar 1999). The equation mapping a 3D point $\mathbf{x}_i$ into a point $\mathbf{x}_{ij}$ in frame j can be re-written from Equations (2.68) and (8.38) as

$$\tilde{\mathbf{x}}_{ij} \sim \mathbf{K}_j \mathbf{R}_j \mathbf{x}_i \quad \text{and} \quad \mathbf{x}_i \sim \mathbf{R}_j^{-1} \mathbf{K}_j^{-1} \tilde{\mathbf{x}}_{ij}, \tag{8.59}$$

where $\mathbf{K}_j = \text{diag}(f_j, f_j, 1)$ is the simplified form of the calibration matrix. The motion mapping a point $\mathbf{x}_{ij}$ from frame j into a point $\mathbf{x}_{ik}$ in frame k is similarly given by

$$\tilde{\mathbf{x}}_{ik} \sim \tilde{\mathbf{H}}_{kj} \tilde{\mathbf{x}}_{ij} = \mathbf{K}_k \mathbf{R}_k \mathbf{R}_j^{-1} \mathbf{K}_j^{-1} \tilde{\mathbf{x}}_{ij}. \tag{8.60}$$

Given an initial set of $\{(\mathbf{R}_j, f_j)\}$ estimates obtained from chaining pairwise alignments, how do we refine these estimates?

One approach is to directly extend the pairwise energy $E_{\text{pairwise-LS}}$ (8.58) to a multiview formulation,

$$E_{\text{all-pairs-2D}} = \sum_i \sum_{jk} c_{ij} c_{ik} \|\tilde{\mathbf{x}}_{ik}(\hat{\mathbf{x}}_{ij}; \mathbf{R}_j, f_j, \mathbf{R}_k, f_k) - \hat{\mathbf{x}}_{ik}\|^2, \tag{8.61}$$

where the $\tilde{\mathbf{x}}_{ik}$ function is the *predicted* location of feature i in frame k given by (8.60), $\hat{\mathbf{x}}_{ij}$ is the *observed* location, and the "2D" in the subscript indicates that an image-plane error is being minimized (Shum and Szeliski 2000). Note that since $\tilde{\mathbf{x}}_{ik}$ depends on the $\hat{\mathbf{x}}_{ij}$ observed value, we actually have an *errors-in-variable* problem, which in principle requires more sophisticated techniques than least squares to solve (Van Huffel and Lemmerling 2002; Matei and Meer 2006). However, in practice, if we have enough features, we can directly minimize the above quantity using regular non-linear least squares and obtain an accurate multi-frame alignment.

While this approach works pretty well, it suffers from two potential disadvantages. First, because a summation is taken over all pairs with corresponding features, features that are observed many times are overweighted in the final solution. (In effect, a feature observed m times gets counted $\binom{m}{2}$ times instead of m times.) Second, the derivatives of $\tilde{\mathbf{x}}_{ik}$ with respect to the $\{(\mathbf{R}_j, f_j)\}$ are a little cumbersome, although using the incremental correction to $\mathbf{R}_j$ introduced in Section 8.2.3 makes this more tractable.

An alternative way to formulate the optimization is to use true bundle adjustment, i.e., to solve not only for the pose parameters $\{(\mathbf{R}_j, f_j)\}$ but also for the 3D point positions $\{\mathbf{x}_i\}$,

$$E_{\text{BA-2D}} = \sum_i \sum_j c_{ij} \|\tilde{\mathbf{x}}_{ij}(\mathbf{x}_i; \mathbf{R}_j, f_j) - \hat{\mathbf{x}}_{ij}\|^2, \tag{8.62}$$

where $\tilde{\mathbf{x}}_{ij}(\mathbf{x}_i; \mathbf{R}_j, f_j)$ is given by (8.59). The disadvantage of full bundle adjustment is that there are more variables to solve for, so each iteration and also the overall convergence may be slower.

[21] Features that are not seen in image j have $c_{ij} = 0$. We can also use 2×2 inverse covariance matrices Σ_{ij}^{-1} in place of c_{ij}, as shown in Equation (8.11).

(Imagine how the 3D points need to "shift" each time some rotation matrices are updated.) However, the computational complexity of each linearized Gauss–Newton step can be reduced using sparse matrix techniques (Section 11.4.3) (Szeliski and Kang 1994; Triggs, McLauchlan *et al.* 1999; Hartley and Zisserman 2004).

An alternative formulation is to minimize the error in 3D projected ray directions (Shum and Szeliski 2000), i.e.,

$$E_{\text{BA}-\text{3D}} = \sum_i \sum_j c_{ij} \|\tilde{\mathbf{x}}_i(\hat{\mathbf{x}}_{ij}; \mathbf{R}_j, f_j) - \mathbf{x}_i\|^2, \tag{8.63}$$

where $\tilde{\mathbf{x}}_i(\mathbf{x}_{ij}; \mathbf{R}_j, f_j)$ is given by the second half of (8.59). This has no particular advantage over (8.62). In fact, since errors are being minimized in 3D ray space, there is a bias towards estimating longer focal lengths, since the angles between rays become smaller as f increases.

However, if we eliminate the 3D rays $\mathbf{x}_i$, we can derive a pairwise energy formulated in 3D ray space (Shum and Szeliski 2000),

$$E_{\text{all}-\text{pairs}-\text{3D}} = \sum_i \sum_{jk} c_{ij} c_{ik} \|\tilde{\mathbf{x}}_i(\hat{\mathbf{x}}_{ij}; \mathbf{R}_j, f_j) - \tilde{\mathbf{x}}_i(\hat{\mathbf{x}}_{ik}; \mathbf{R}_k, f_k)\|^2. \tag{8.64}$$

This results in the simplest set of update equations (Shum and Szeliski 2000), since the f_k can be folded into the creation of the homogeneous coordinate vector as in Equation (8.40). Thus, even though this formula over-weights features that occur more frequently, it is the method used by Shum and Szeliski (2000) and Brown, Szeliski, and Winder (2005). To reduce the bias towards longer focal lengths, we multiply each residual (3D error) by $\sqrt{f_j f_k}$, which is similar to projecting the 3D rays into a "virtual camera" of intermediate focal length.

Up vector selection. As mentioned above, there exists a global ambiguity in the pose of the 3D cameras computed by the above methods. While this may not appear to matter, people prefer that the final stitched image is "upright" rather than twisted or tilted. More concretely, people are used to seeing photographs displayed so that the vertical (gravity) axis points straight up in the image. Consider how you usually shoot photographs: while you may pan and tilt the camera any which way, you usually keep the horizontal edge of your camera (its x-axis) parallel to the ground plane (perpendicular to the world gravity direction).

Mathematically, this constraint on the rotation matrices can be expressed as follows. Recall from Equation (8.59) that the 3D to 2D projection is given by

$$\tilde{\mathbf{x}}_{ik} \sim \mathbf{K}_k \mathbf{R}_k \mathbf{x}_i. \tag{8.65}$$

We wish to post-multiply each rotation matrix $\mathbf{R}_k$ by a global rotation $\mathbf{R}_\text{G}$ such that the projection of the global y-axis, $\hat{\jmath} = (0, 1, 0)$ is perpendicular to the image x-axis, $\hat{\imath} = (1, 0, 0)$.[22]

This constraint can be written as

$$\hat{\imath}^T \mathbf{R}_k \mathbf{R}_\text{G} \hat{\jmath} = 0 \tag{8.66}$$

(note that the scaling by the calibration matrix is irrelevant here). This is equivalent to requiring that the first row of $\mathbf{R}_k$, $\mathbf{r}_{k0} = \hat{\imath}^T \mathbf{R}_k$ be perpendicular to the second column of $\mathbf{R}_\text{G}$, $\mathbf{r}_{\text{G}1} = \mathbf{R}_\text{G}\hat{\jmath}$. This set of constraints (one per input image) can be written as a least squares problem,

$$\mathbf{r}_{\text{G}1} = \arg\min_{\mathbf{r}} \sum_k (\mathbf{r}^T \mathbf{r}_{k0})^2 = \arg\min_{\mathbf{r}} \mathbf{r}^T \left[\sum_k \mathbf{r}_{k0} \mathbf{r}_{k0}^T \right] \mathbf{r}. \tag{8.67}$$

[22] Note that here we use the convention common in computer graphics that the vertical world axis corresponds to y. This is a natural choice if we wish the rotation matrix associated with a "regular" image taken horizontally to be the identity, rather than a 90° rotation around the x-axis.

Thus, $\mathbf{r}_{G1}$ is the smallest eigenvector of the *scatter* or *moment* matrix spanned by the individual camera rotation x-vectors, which should generally be of the form $(c, 0, s)$ when the cameras are upright.

To fully specify the $\mathbf{R}_G$ global rotation, we need to specify one additional constraint. This is related to the *view selection* problem discussed in Section 8.4.1. One simple heuristic is to prefer the average z-axis of the individual rotation matrices, $\overline{\mathbf{k}} = \sum_k \hat{\mathbf{k}}^T \mathbf{R}_k$ to be close to the world z-axis, $\mathbf{r}_{G2} = \mathbf{R}_G \hat{\mathbf{k}}$. We can therefore compute the full rotation matrix $\mathbf{R}_G$ in three steps:

1. $\mathbf{r}_{G1} = \min \text{ eigenvector } (\sum_k \mathbf{r}_{k0} \mathbf{r}_{k0}^T)$;

2. $\mathbf{r}_{G0} = \mathcal{N}((\sum_k \mathbf{r}_{k2}) \times \mathbf{r}_{G1})$;

3. $\mathbf{r}_{G2} = \mathbf{r}_{G0} \times \mathbf{r}_{G1}$,

where $\mathcal{N}(\mathbf{v}) = \mathbf{v}/\|\mathbf{v}\|$ normalizes a vector $\mathbf{v}$.

8.3.2 Parallax removal

Once we have optimized the global orientations and focal lengths of our cameras, we may find that the images are still not perfectly aligned, i.e., the resulting stitched image looks blurry or ghosted in some places. This can be caused by a variety of factors, including unmodeled radial distortion, 3D parallax (failure to rotate the camera around its front nodal point), small scene motions such as waving tree branches, and large-scale scene motions such as people moving in and out of pictures.

Each of these problems can be treated with a different approach. Radial distortion can be estimated (potentially ahead of time) using one of the techniques discussed in Section 2.1.5. For example, the *plumb-line method* (Brown 1971; Kang 2001; El-Melegy and Farag 2003) adjusts radial distortion parameters until slightly curved lines become straight, while mosaic-based approaches adjust them until misregistration is reduced in image overlap areas (Stein 1997; Sawhney and Kumar 1999).

3D parallax can be handled by doing a full 3D bundle adjustment, i.e., by replacing the projection Equation (8.59) used in Equation (8.62) with Equation (2.68), which models camera translations. The 3D positions of the matched feature points and cameras can then be simultaneously recovered, although this can be significantly more expensive than parallax-free image registration. Once the 3D structure has been recovered, the scene could (in theory) be projected to a single (central) viewpoint that contains no parallax. However, to do this, dense *stereo* correspondence needs to be performed (Section 12.3) (Li, Shum *et al.* 2004; Zheng, Kang *et al.* 2007), which may not be possible if the images contain only partial overlap. In that case, it may be necessary to correct for parallax only in the overlap areas, which can be accomplished using a *multi-perspective plane sweep* (MPPS) algorithm (Kang, Szeliski, and Uyttendaele 2004; Uyttendaele, Criminisi *et al.* 2004).

When the motion in the scene is very large, i.e., when objects appear and disappear completely, a sensible solution is to simply *select* pixels from only one image at a time as the source for the final composite (Milgram 1977; Davis 1998; Agarwala, Dontcheva *et al.* 2004), as discussed in Section 8.4.2. However, when the motion is reasonably small (on the order of a few pixels), general 2D motion estimation (optical flow) can be used to perform an appropriate correction before blending using a process called *local alignment* (Shum and Szeliski 2000; Kang, Uyttendaele *et al.* 2003). This same process can also be used to compensate for radial distortion and 3D parallax, although it uses a weaker motion model than explicitly modeling the source of error and may, therefore, fail more often or introduce unwanted distortions.

(a) (b) (c)

Figure 8.10 Deghosting a mosaic with motion parallax (Shum and Szeliski 2000) © 2000 IEEE: (a) composite with parallax; (b) after a single deghosting step (patch size 32); (c) after multiple steps (sizes 32, 16 and 8).

The local alignment technique introduced by Shum and Szeliski (2000) starts with the global bundle adjustment (8.64) used to optimize the camera poses. Once these have been estimated, the *desired* location of a 3D point $\mathbf{x}_i$ can be estimated as the *average* of the back-projected 3D locations,

$$\bar{\mathbf{x}}_i \sim \sum_j c_{ij} \tilde{\mathbf{x}}_i(\hat{\mathbf{x}}_{ij}; \mathbf{R}_j, f_j) \bigg/ \sum_j c_{ij}\,, \tag{8.68}$$

which can be projected into each image j to obtain a *target location* $\bar{\mathbf{x}}_{ij}$. The difference between the target locations $\bar{\mathbf{x}}_{ij}$ and the original features $\mathbf{x}_{ij}$ provide a set of local motion estimates

$$\mathbf{u}_{ij} = \bar{\mathbf{x}}_{ij} - \mathbf{x}_{ij}, \tag{8.69}$$

which can be interpolated to form a dense correction field $\mathbf{u}_j(\mathbf{x}_j)$. In their system, Shum and Szeliski (2000) use an *inverse warping* algorithm where the sparse $-\mathbf{u}_{ij}$ values are placed at the new target locations $\bar{\mathbf{x}}_{ij}$, interpolated using bilinear kernel functions (Nielson 1993) and then added to the original pixel coordinates when computing the warped (corrected) image. To get a reasonably dense set of features to interpolate, Shum and Szeliski (2000) place a feature point at the center of each patch (the patch size controls the smoothness in the local alignment stage), rather than relying on features extracted using an interest operator (Figure 8.10).

An alternative approach to motion-based deghosting was proposed by Kang, Uyttendaele *et al.* (2003), who estimate dense optical flow between each input image and a central *reference* image. The accuracy of the flow vector is checked using a photo-consistency measure before a given warped pixel is considered valid and is used to compute a high dynamic range radiance estimate, which is the goal of their overall algorithm. The requirement for a reference image makes their approach less applicable to general image mosaicing, although an extension to this case could certainly be envisaged.

The idea of combining *global* parametric warps with *local* mesh-based warps or multiple motion models to compensate for parallax has been refined in a number of more recent papers (Zaragoza, Chin *et al.* 2013; Zhang and Liu 2014; Lin, Pankanti *et al.* 2015; Lin, Jiang *et al.* 2016; Herrmann, Wang *et al.* 2018b; Lee and Sim 2020). Some of these papers use *content-preserving warps* (Liu, Gleicher *et al.* 2009) for their local deformations, while others include a rolling shutter model (Zhuang and Tran 2020).

8.3.3 Recognizing panoramas

The final piece needed to perform fully automated image stitching is a technique to recognize which images actually go together, which Brown and Lowe (2007) call *recognizing panoramas*. If the user takes images in sequence so that each image overlaps its predecessor and also specifies the first and last images to be stitched, bundle adjustment combined with the process of *topology inference* can be used to automatically assemble a panorama (Sawhney and Kumar 1999). However, users often jump around when taking panoramas, e.g., they may start a new row on top of a previous one, jump back to take a repeat shot, or create 360° panoramas where end-to-end overlaps need to be discovered. Furthermore, the ability to discover multiple panoramas taken by a user over an extended period of time can be a big convenience.

To recognize panoramas, Brown and Lowe (2007) first find all pairwise image overlaps using a feature-based method and then find connected components in the overlap graph to "recognize" individual panoramas (Figure 8.11). The feature-based matching stage first extracts scale invariant feature transform (SIFT) feature locations and feature descriptors (Lowe 2004) from all the input images and places them in an indexing structure, as described in Section 7.1.3. For each image pair under consideration, the nearest matching neighbor is found for each feature in the first image, using the indexing structure to rapidly find candidates and then comparing feature descriptors to find the best match. RANSAC is used to find a set of *inlier* matches; pairs of matches are used to hypothesize similarity motion models that are then used to count the number of inliers. A RANSAC algorithm tailored specifically for rotational panoramas is described by Brown, Hartley, and Nistér (2007).

In practice, the most difficult part of getting a fully automated stitching algorithm to work is deciding which pairs of images actually correspond to the same parts of the scene. Repeated structures such as windows (Figure 8.12) can lead to false matches when using a feature-based approach. One way to mitigate this problem is to perform a direct pixel-based comparison between the registered images to determine if they actually are different views of the same scene. Unfortunately, this heuristic may fail if there are moving objects in the scene (Figure 8.13). While there is no magic bullet for this problem, short of full scene understanding, further improvements can likely be made by applying domain-specific heuristics, such as priors on typical camera motions as well as machine learning techniques applied to the problem of match validation.

8.4 Compositing

Once we have registered all of the input images with respect to each other, we need to decide how to produce the final stitched mosaic image. This involves selecting a final compositing surface (flat, cylindrical, spherical, etc.) and view (reference image). It also involves selecting which pixels contribute to the final composite and how to optimally blend these pixels to minimize visible seams, blur, and ghosting.

In this section, we review techniques that address these problems, namely compositing surface parameterization, pixel and seam selection, blending, and exposure compensation. Our emphasis is on fully automated approaches to the problem. Since the creation of high-quality panoramas and composites is as much an artistic endeavor as a computational one, various interactive tools have been developed to assist this process (Agarwala, Dontcheva *et al.* 2004; Li, Sun *et al.* 2004; Rother, Kolmogorov, and Blake 2004). Some of these are covered in more detail in Section 10.4.

8.4.1 Choosing a compositing surface

The first choice to be made is how to represent the final image. If only a few images are stitched together, a natural approach is to select one of the images as the *reference* and to then warp all of the other images into its reference coordinate system. The resulting composite is sometimes called

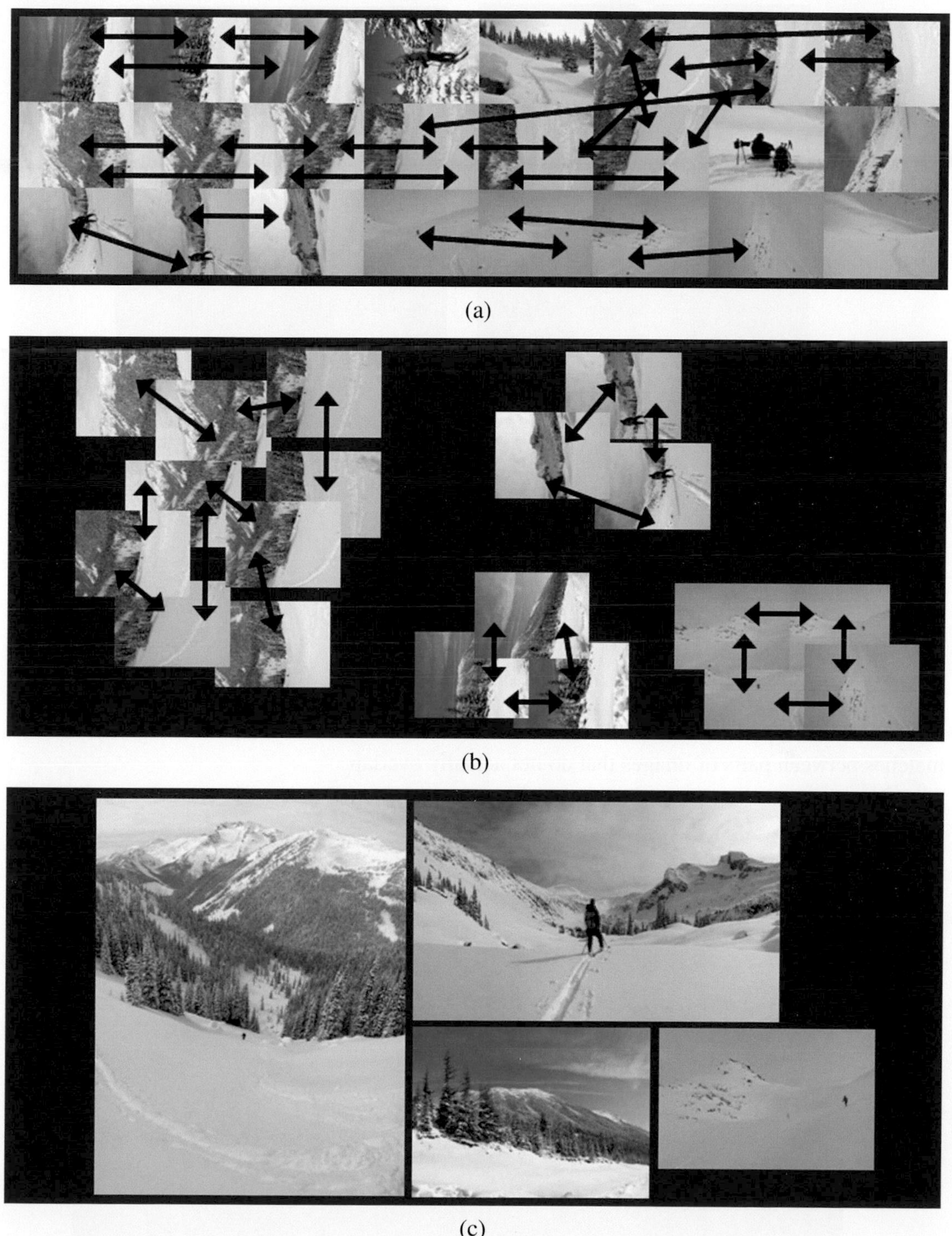

Figure 8.11 Recognizing panoramas (Brown, Szeliski, and Winder 2005), figures courtesy of Matthew Brown: (a) input images with pairwise matches; (b) images grouped into connected components (panoramas); (c) individual panoramas registered and blended into stitched composites.

Figure 8.12 Matching errors (Brown, Szeliski, and Winder 2004): accidental matching of several features can lead to matches between pairs of images that do not actually overlap.

Figure 8.13 Validation of image matches by direct pixel error comparison can fail when the scene contains moving objects (Uyttendaele, Eden, and Szeliski 2001) © 2001 IEEE.

a *flat* panorama, since the projection onto the final surface is still a perspective projection, and hence straight lines remain straight (which is often a desirable attribute).[23]

For larger fields of view, however, we cannot maintain a flat representation without excessively stretching pixels near the border of the image. (In practice, flat panoramas start to look severely distorted once the field of view exceeds 90° or so.) The usual choice for compositing larger panoramas is to use a cylindrical (Chen 1995; Szeliski 1996) or spherical (Szeliski and Shum 1997) projection, as described in Section 8.2.6. In fact, any surface used for *environment mapping* in computer graphics can be used, including a *cube map*, which represents the full viewing sphere with the six square faces of a cube (Greene 1986; Szeliski and Shum 1997). Cartographers have also developed a number of alternative methods for representing the globe (Bugayevskiy and Snyder 1995).

The choice of parameterization is somewhat application-dependent and involves a tradeoff between keeping the local appearance undistorted (e.g., keeping straight lines straight) and providing a reasonably uniform sampling of the environment. Automatically making this selection and smoothly transitioning between representations based on the extent of the panorama is discussed in Kopf, Uyttendaele *et al.* (2007). A recent trend in panoramic photography has been the use of stereographic projections looking down at the ground (in an outdoor scene) to create "little planet" renderings.[24]

View selection. Once we have chosen the output parameterization, we still need to determine which part of the scene will be *centered* in the final view. As mentioned above, for a flat composite, we can choose one of the images as a reference. Often, a reasonable choice is the one that is geometrically most central. For example, for rotational panoramas represented as a collection of 3D rotation matrices, we can choose the image whose z-axis is closest to the average z-axis (assuming a reasonable field of view). Alternatively, we can use the average z-axis (or quaternion, but this is trickier) to define the reference rotation matrix.

For larger, e.g., cylindrical or spherical, panoramas, we can use the same heuristic if a subset of the viewing sphere has been imaged. In the case of full 360° panoramas, a better choice is to choose the middle image from the sequence of inputs, or sometimes the first image, assuming this contains the object of greatest interest. In all of these cases, having the user control the final view is often highly desirable. If the "up vector" computation described in Section 8.3.1 is working correctly, this can be as simple as panning over the image or setting a vertical "center line" for the final panorama.

Coordinate transformations. After selecting the parameterization and reference view, we still need to compute the mappings between the input and output pixels coordinates.

If the final compositing surface is flat (e.g., a single plane or the face of a cube map) and the input images have no radial distortion, the coordinate transformation is the simple homography described by Equation (8.38). This kind of warping can be performed in graphics hardware by appropriately setting texture mapping coordinates and rendering a single quadrilateral.

If the final composite surface has some other analytic form (e.g., cylindrical or spherical), we need to convert every pixel in the final panorama into a viewing ray (3D point) and then map it back into each image according to the projection (and optionally radial distortion) equations. This process can be made more efficient by precomputing some lookup tables, e.g., the partial trigonometric functions needed to map cylindrical or spherical coordinates to 3D coordinates or the radial distortion field at each pixel. It is also possible to accelerate this process by computing exact pixel mappings on a coarser grid and then interpolating these values.

[23] Techniques have also been developed to straighten curved lines in cylindrical and spherical panoramas (Carroll, Agrawala, and Agarwala 2009; Kopf, Lischinski *et al.* 2009; Carroll, Agarwala, and Agrawala 2010).

[24] These are inspired by *The Little Prince* by Antoine De Saint-Exupery. Go to https://www.flickr.com and search for "little planet projection".

When the final compositing surface is a texture-mapped polyhedron, a slightly more sophisticated algorithm must be used. Not only do the 3D and texture map coordinates have to be properly handled, but a small amount of *overdraw* outside the triangle footprints in the texture map is necessary, to ensure that the texture pixels being interpolated during 3D rendering have valid values (Szeliski and Shum 1997).

Sampling issues. While the above computations can yield the correct (fractional) pixel addresses in each input image, we still need to pay attention to sampling issues. For example, if the final panorama has a lower resolution than the input images, prefiltering the input images is necessary to avoid aliasing. These issues have been extensively studied in both the image processing and computer graphics communities. The basic problem is to compute the appropriate prefilter, which depends on the distance (and arrangement) between neighboring samples in a source image. As discussed in Sections 3.5.2 and 3.6.1, various approximate solutions, such as MIP mapping (Williams 1983) or elliptically weighted Gaussian averaging (Greene and Heckbert 1986) have been developed in the graphics community. For highest visual quality, a higher order (e.g., cubic) interpolator combined with a spatially adaptive prefilter may be necessary (Wang, Kang *et al.* 2001). Under certain conditions, it may also be possible to produce images with a higher resolution than the input images using the process of *super-resolution* (Section 10.3).

8.4.2 Pixel selection and weighting (deghosting)

Once the source pixels have been mapped onto the final composite surface, we must still decide how to blend them in order to create an attractive-looking panorama. If all of the images are in perfect registration and identically exposed, this is an easy problem, i.e., any pixel or combination will do. However, for real images, visible seams (due to exposure differences), blurring (due to misregistration), or ghosting (due to moving objects) can occur.

Creating clean, pleasing-looking panoramas involves both deciding which pixels to use and how to weight or blend them. The distinction between these two stages is a little fluid, since per-pixel weighting can be thought of as a combination of selection and blending. In this section, we discuss spatially varying weighting, pixel selection (seam placement), and then more sophisticated blending.

Feathering and center-weighting. The simplest way to create a final composite is to simply take an *average* value at each pixel,

$$C(\mathbf{x}) = \sum_k w_k(\mathbf{x}) \tilde{I}_k(\mathbf{x}) \bigg/ \sum_k w_k(\mathbf{x}) , \tag{8.70}$$

where $\tilde{I}_k(\mathbf{x})$ are the *warped* (re-sampled) images and $w_k(\mathbf{x})$ is 1 at valid pixels and 0 elsewhere. On computer graphics hardware, this kind of summation can be performed in an *accumulation buffer* (using the A channel as the weight).

Simple averaging usually does not work very well, since exposure differences, misregistrations, and scene movement are all very visible (Figure 8.14a). If rapidly moving objects are the only problem, taking a *median* filter (which is a kind of pixel selection operator) can often be used to remove them (Figure 8.14b) (Irani and Anandan 1998). Conversely, center-weighting (discussed below) and *minimum likelihood* selection (Agarwala, Dontcheva *et al.* 2004) can sometimes be used to retain multiple copies of a moving object (Figure 8.17).

A better approach to averaging is to weight pixels near the center of the image more heavily and to down-weight pixels near the edges. When an image has some cutout regions, down-weighting

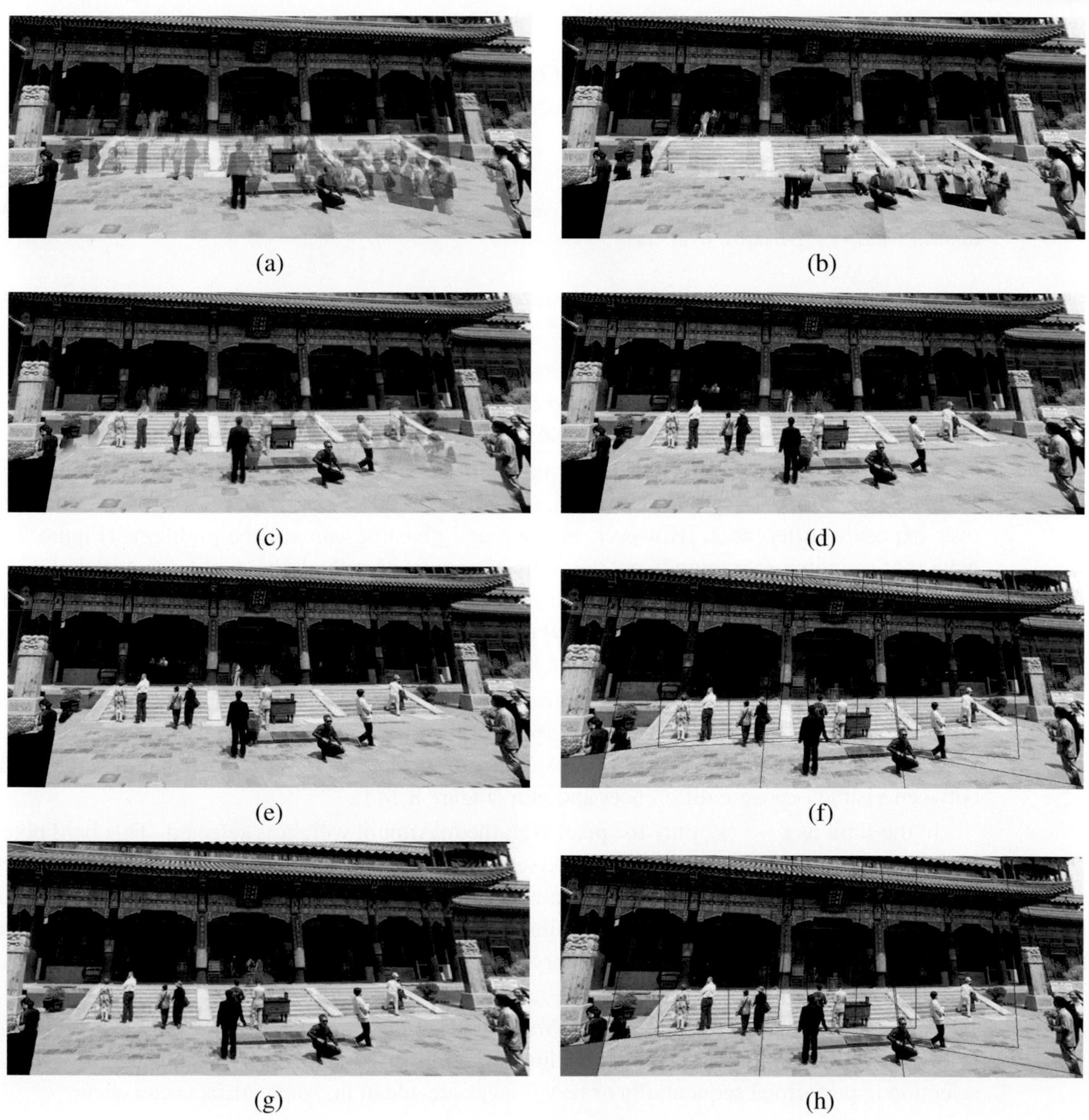

Figure 8.14 Final composites computed by a variety of algorithms (Szeliski 2006a): (a) average, (b) median, (c) feathered average, (d) *p-norm* $p = 10$, (e) Voronoi, (f) weighted ROD vertex cover with feathering, (g) graph cut seams with Poisson blending, and (h) with pyramid blending.

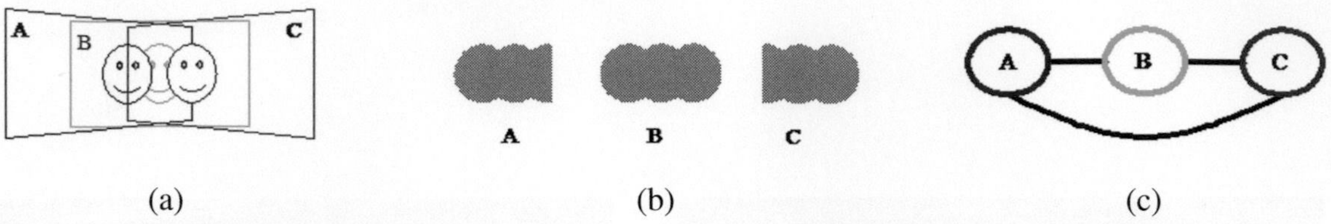

(a) (b) (c)

Figure 8.15 Computation of regions of difference (RODs) (Uyttendaele, Eden, and Szeliski 2001) © 2001 IEEE: (a) three overlapping images with a moving face; (b) corresponding RODs; (c) graph of coincident RODs.

pixels near the edges of both cutouts and the image is preferable. This can be done by computing a *distance map* or *grassfire transform*,

$$w_k(\mathbf{x}) = \arg\min_{\mathbf{y}} \{ \|\mathbf{y}\| \mid \tilde{I}_k(\mathbf{x} + \mathbf{y}) \text{ is invalid } \}, \tag{8.71}$$

where each valid pixel is tagged with its Euclidean distance to the nearest invalid pixel (Section 3.3.3). The Euclidean distance map can be efficiently computed using a two-pass raster algorithm (Danielsson 1980; Borgefors 1986).

Weighted averaging with a distance map is often called *feathering* (Szeliski and Shum 1997; Chen and Klette 1999; Uyttendaele, Eden, and Szeliski 2001) and does a reasonable job of blending over exposure differences. However, blurring and ghosting can still be problems (Figure 8.14c). Note that weighted averaging is *not* the same as compositing the individual images with the classic *over* operation (Porter and Duff 1984; Blinn 1994a), even when using the weight values (normalized to sum up to one) as *alpha* (translucency) channels. This is because the over operation attenuates the values from more distant surfaces and, hence, is not equivalent to a direct sum.

One way to improve feathering is to raise the distance map values to some large power, i.e., to use $w_k^p(\mathbf{x})$ in Equation (8.70). The weighted averages then become dominated by the larger values, i.e., they act somewhat like a *p-norm*. The resulting composite can often provide a reasonable tradeoff between visible exposure differences and blur (Figure 8.14d).

In the limit as $p \rightarrow \infty$, only the pixel with the maximum weight is selected. This hard pixel selection process produces a visibility mask-sensitive variant of the familiar *Voronoi diagram*, which assigns each pixel to the nearest image center in the set (Wood, Finkelstein *et al.* 1997; Peleg, Rousso *et al.* 2000). The resulting composite, while useful for artistic guidance and in high-overlap panoramas (*manifold mosaics*) tends to have very hard edges with noticeable seams when the exposures vary (Figure 8.14e).

Xiong and Turkowski (1998) use this Voronoi idea (local maximum of the grassfire transform) to select seams for Laplacian pyramid blending (which is discussed below). However, since the seam selection is performed sequentially as new images are added in, some artifacts can occur.

Optimal seam selection. Computing the Voronoi diagram is one way to select the *seams* between regions where different images contribute to the final composite. However, Voronoi images totally ignore the local image structure underlying the seam. A better approach is to place the seams in regions where the images agree, so that transitions from one source to another are not visible. In this way, the algorithm avoids "cutting through" moving objects where a seam would look unnatural (Davis 1998). For a pair of images, this process can be formulated as a simple dynamic program starting from one edge of the overlap region and ending at the other (Milgram 1975, 1977; Davis 1998; Efros and Freeman 2001).

Figure 8.16 Photomontage (Agarwala, Dontcheva *et al.* 2004) © 2004 ACM. From a set of five source images (of which four are shown on the left), Photomontage quickly creates a composite family portrait in which everyone is smiling and looking at the camera (right). Users simply flip through the stack and coarsely draw strokes using the designated source image objective over the people they wish to add to the composite. The user-applied strokes and computed regions (middle) are color-coded by the borders of the source images on the left.

When multiple images are being composited, the dynamic program idea does not readily generalize. (For square texture tiles being composited sequentially, Efros and Freeman (2001) run a dynamic program along each of the four tile sides.)

To overcome this problem, Uyttendaele, Eden, and Szeliski (2001) observed that, for well-registered images, moving objects produce the most visible artifacts, namely translucent looking *ghosts*. Their system therefore decides which objects to keep and which ones to erase. First, the algorithm compares all overlapping input image pairs to determine *regions of difference* (RODs) where the images disagree. Next, a graph is constructed with the RODs as vertices and edges representing ROD pairs that overlap in the final composite (Figure 8.15). Since the presence of an edge indicates an area of disagreement, vertices (regions) must be removed from the final composite until no edge spans a pair of remaining vertices. The smallest such set can be computed using a *vertex cover* algorithm. Since several such covers may exist, a *weighted vertex cover* is used instead, where the vertex weights are computed by summing the feather weights in the ROD (Uyttendaele, Eden, and Szeliski 2001). The algorithm therefore prefers removing regions that are near the edge of the image, which reduces the likelihood that partially visible objects will appear in the final composite. (It is also possible to infer which object in a region of difference is the foreground object by the "edginess" (pixel differences) across the ROD boundary, which should be higher when an object is present (Herley 2005).) Once the desired excess regions of difference have been removed, the final composite can be created by feathering (Figure 8.14f).

A different approach to pixel selection and seam placement is described by Agarwala, Dontcheva *et al.* (2004). Their system computes the label assignment that optimizes the sum of two objective functions. The first is a per-pixel *image objective* that determines which pixels are likely to produce good composites,

$$E_{\rm D} = \sum_{\mathbf{x}} D(\mathbf{x}, l(\mathbf{x})), \tag{8.72}$$

where $D(\mathbf{x}, l)$ is the *data penalty* associated with choosing image l at pixel $\mathbf{x}$. In their system, users can select which pixels to use by "painting" over an image with the desired object or appearance, which sets $D(\mathbf{x}, l)$ to a large value for all labels l other than the one selected by the user (Figure 8.16). Alternatively, automated selection criteria can be used, such as *maximum likelihood*, which prefers pixels that occur repeatedly in the background (for object removal), or *minimum likelihood* for objects that occur infrequently, i.e., for moving object retention. Using a more traditional

Figure 8.17 Set of five photos tracking a snowboarder's jump stitched together into a seamless composite. Because the algorithm prefers pixels near the center of the image, multiple copies of the boarder are retained.

center-weighted data term tends to favor objects that are centered in the input images (Figure 8.17).

The second term is a *seam objective* that penalizes differences in labels between adjacent images,

$$E_{\text{S}} = \sum_{(\mathbf{x},\mathbf{y})\in\mathcal{N}} S(\mathbf{x}, \mathbf{y}, l(\mathbf{x}), l(\mathbf{y})), \tag{8.73}$$

where $S(\mathbf{x}, \mathbf{y}, l_x, l_y)$ is the image-dependent *interaction penalty* or *seam cost* of placing a seam between pixels $\mathbf{x}$ and $\mathbf{y}$, and $\mathcal{N}$ is the set of $\mathcal{N}_4$ neighboring pixels. For example, the simple color-based seam penalty used in Kwatra, Schödl *et al.* (2003) and Agarwala, Dontcheva *et al.* (2004) can be written as

$$S(\mathbf{x}, \mathbf{y}, l_x, l_y) = \|\tilde{I}_{l_x}(\mathbf{x}) - \tilde{I}_{l_y}(\mathbf{x})\| + \|\tilde{I}_{l_x}(\mathbf{y}) - \tilde{I}_{l_y}(\mathbf{y})\|. \tag{8.74}$$

More sophisticated seam penalties can also look at image gradients or the presence of image edges (Agarwala, Dontcheva *et al.* 2004). Seam penalties are widely used in other computer vision applications such as stereo matching (Boykov, Veksler, and Zabih 2001) to give the labeling function its *coherence* or *smoothness*. An alternative approach, which places seams along strong consistent edges in overlapping images using a watershed computation is described by Soille (2006).

The sum of these two objective functions gives rise to a *Markov random field* (MRF), for which good optimization algorithms are described in Sections 4.3 and 4.3.2 and Appendix B.5. For label computations of this kind, the *α-expansion* algorithm developed by Boykov, Veksler, and Zabih (2001) works particularly well (Szeliski, Zabih *et al.* 2008).

For the result shown in Figure 8.14g, Agarwala, Dontcheva *et al.* (2004) use a large data penalty for invalid pixels and 0 for valid pixels. Notice how the seam placement algorithm avoids regions of difference, including those that border the image and that might result in objects being cut off. Graph cuts (Agarwala, Dontcheva *et al.* 2004) and vertex cover (Uyttendaele, Eden, and Szeliski 2001) often produce similar looking results, although the former is significantly slower since it optimizes over all pixels, while the latter is more sensitive to the thresholds used to determine regions of difference. More recent approaches to seam selection include SEAGULL (Lin, Jiang *et al.* 2016), which jointly optimizes local alignment and seam selection, and object-centered image stitching (Herrmann, Wang *et al.* 2018a), which uses an off-the-shelf object detector to avoid cutting through objects.

8.4.3 *Application*: Photomontage

While image stitching is normally used to composite partially overlapping photographs, it can also be used to composite repeated shots of a scene taken with the aim of obtaining the best possible composition and appearance of each element.

Figure 8.16 shows the *Photomontage* system developed by Agarwala, Dontcheva *et al.* (2004), where users draw strokes over a set of pre-aligned images to indicate which regions they wish to keep from each image. Once the system solves the resulting multi-label graph cut (8.72–8.73), the various pieces taken from each source photo are blended together using a variant of Poisson image blending (8.75–8.77). Their system can also be used to automatically composite an all-focus image from a series of bracketed focus images (Hasinoff, Kutulakos *et al.* 2009) or to remove wires and other unwanted elements from sets of photographs. Exercise 8.14 has you implement this system and try out some of its variants.

8.4.4 Blending

Once the seams between images have been determined and unwanted objects removed, we still need to blend the images to compensate for exposure differences and other misalignments. The spatially varying weighting (feathering) previously discussed can often be used to accomplish this. However, it is difficult in practice to achieve a pleasing balance between smoothing out low-frequency exposure variations and retaining sharp enough transitions to prevent blurring (although using a high exponent in feathering can help).

Laplacian pyramid blending. An attractive solution to this problem is the Laplacian pyramid blending technique developed by Burt and Adelson (1983b), which we discussed in Section 3.5.5. Instead of using a single transition width, a frequency-adaptive width is used by creating a band-pass (Laplacian) pyramid and making the transition widths within each level a function of the level, i.e., the same width in pixels. In practice, a small number of levels, i.e., as few as two (Brown and Lowe 2007), may be adequate to compensate for differences in exposure. The result of applying this pyramid blending is shown in Figure 8.14h.

Gradient domain blending. An alternative approach to multi-band image blending is to perform the operations in the *gradient domain*. Reconstructing images from their gradient fields has a long history in computer vision (Horn 1986), starting originally with work in brightness constancy (Horn 1974), shape from shading (Horn and Brooks 1989), and photometric stereo (Woodham 1981). Related ideas have also been used for reconstructing images from their edges (Elder and Goldberg 2001), removing shadows from images (Weiss 2001), separating reflections from a single image (Levin, Zomet, and Weiss 2004; Levin and Weiss 2007), and *tone mapping* high dynamic range images by reducing the magnitude of image edges (gradients) (Fattal, Lischinski, and Werman 2002).

Pérez, Gangnet, and Blake (2003) show how gradient domain reconstruction can be used to do seamless object insertion in image editing applications (Figure 8.18). Rather than copying pixels, the *gradients* of the new image fragment are copied instead. The actual pixel values for the copied area are then computed by solving a *Poisson equation* that locally matches the gradients while obeying the fixed *Dirichlet* (exact matching) conditions at the seam boundary. Pérez, Gangnet, and Blake (2003) show that this is equivalent to computing an additive *membrane* interpolant of the mismatch

Figure 8.18 Poisson image editing (Pérez, Gangnet, and Blake 2003) © 2003 ACM: (a) The dog and the two children are chosen as source images to be pasted into the destination swimming pool. (b) Simple pasting fails to match the colors at the boundaries, whereas (c) Poisson image blending masks these differences.

between the source and destination images along the boundary.[25] In earlier work, Peleg (1981) also proposed adding a smooth function to enforce consistency along the seam curve.

Agarwala, Dontcheva *et al.* (2004) extended this idea to a multi-source formulation, where it no longer makes sense to talk of a destination image whose exact pixel values must be matched at the seam. Instead, *each* source image contributes its own gradient field and the Poisson equation is solved using *Neumann* boundary conditions, i.e., dropping any equations that involve pixels outside the boundary of the image.

Rather than solving the Poisson partial differential equations, Agarwala, Dontcheva *et al.* (2004) directly minimize a *variational problem,*

$$\min_{C(\mathbf{x})} \|\nabla C(\mathbf{x}) - \nabla \tilde{I}_{l(\mathbf{x})}(\mathbf{x})\|^2. \tag{8.75}$$

The discretized form of this equation is a set of gradient constraint equations

$$C(\mathbf{x}+\hat{\imath}) - C(\mathbf{x}) = \tilde{I}_{l(\mathbf{x})}(\mathbf{x}+\hat{\imath}) - \tilde{I}_{l(\mathbf{x})}(\mathbf{x}) \quad \text{and} \tag{8.76}$$

$$C(\mathbf{x}+\hat{\jmath}) - C(\mathbf{x}) = \tilde{I}_{l(\mathbf{x})}(\mathbf{x}+\hat{\jmath}) - \tilde{I}_{l(\mathbf{x})}(\mathbf{x}), \tag{8.77}$$

where $\hat{\imath} = (1,0)$ and $\hat{\jmath} = (0,1)$ are unit vectors in the x and y directions.[26] They then solve the associated sparse least squares problem. Since this system of equations is only defined up to an additive constraint, Agarwala, Dontcheva *et al.* (2004) ask the user to select the value of one pixel. In practice, a better choice might be to weakly bias the solution towards reproducing the original color values.

In order to accelerate the solution of this sparse linear system, Fattal, Lischinski, and Werman (2002) use multigrid, whereas Agarwala, Dontcheva *et al.* (2004) use hierarchical basis preconditioned conjugate gradient descent (Szeliski 1990b, 2006b; Krishnan and Szeliski 2011; Krishnan, Fattal, and Szeliski 2013) (Appendix A.5). In subsequent work, Agarwala (2007) shows how using a quadtree representation for the solution can further accelerate the computation with minimal loss

[25] The membrane interpolant is known to have nicer interpolation properties for arbitrary-shaped constraints than frequency-domain interpolants (Nielson 1993).

[26] At seam locations, the right-hand side is replaced by the average of the gradients in the two source images.

in accuracy, while Szeliski, Uyttendaele, and Steedly (2008) show how representing the per-image offset fields using coarser splines is even faster. This latter work also argues that blending in the log domain, i.e., using multiplicative rather than additive offsets, is preferable, as it more closely matches texture contrasts across seam boundaries. The resulting seam blending works very well in practice (Figure 8.14h), although care must be taken when copying large gradient values near seams so that a "double edge" is not introduced.

Copying gradients directly from the source images after seam placement is just one approach to gradient domain blending. The paper by Levin, Zomet *et al.* (2004) examines several different variants of this approach, which they call *Gradient-domain Image STitching* (GIST). The techniques they examine include feathering (blending) the gradients from the source images, as well as using an L_1norm in performing the reconstruction of the image from the gradient field, rather than using an L_2norm as in Equation (8.75). Their preferred technique is the L_1 optimization of a feathered (blended) cost function on the original image gradients (which they call GIST1-l_1). Since L_1optimization using linear programming can be slow, they develop a faster iterative median-based algorithm in a multigrid framework. Visual comparisons between their preferred approach and what they call *optimal seam on the gradients* (which is equivalent to the approach of Agarwala, Dontcheva *et al.* (2004)) show similar results, while significantly improving on pyramid blending and feathering algorithms.

Exposure compensation. Pyramid and gradient domain blending can do a good job of compensating for moderate amounts of exposure differences between images. However, when the exposure differences become large, alternative approaches may be necessary.

Uyttendaele, Eden, and Szeliski (2001) iteratively estimate a local correction between each source image and a blended composite. First, a block-based quadratic transfer function is fit between each source image and an initial feathered composite. Next, transfer functions are averaged with their neighbors to get a smoother mapping and per-pixel transfer functions are computed by *splining* (interpolating) between neighboring block values. Once each source image has been smoothly adjusted, a new feathered composite is computed and the process is repeated (typically three times). The results shown by Uyttendaele, Eden, and Szeliski (2001) demonstrate that this does a better job of exposure compensation than simple feathering and can handle local variations in exposure due to effects such as lens vignetting.

Ultimately, however, the most principled way to deal with exposure differences is to stitch images in the radiance domain, i.e., to convert each image into a radiance image using its exposure value and then create a stitched, high dynamic range image, as discussed in Section 10.2 and Eden, Uyttendaele, and Szeliski (2006).

8.5 Additional reading

Hartley and Zisserman (2004) provide a wonderful introduction to the topics of feature-based alignment and optimal motion estimation. Techniques for robust estimation are discussed in more detail in Appendix B.3 and in monographs and review articles on this topic (Huber 1981; Hampel, Ronchetti *et al.* 1986; Rousseeuw and Leroy 1987; Black and Rangarajan 1996; Stewart 1999). The most commonly used robust initialization technique in computer vision is RANdom SAmple Consensus (RANSAC) (Fischler and Bolles 1981), which has spawned a series of more efficient variants (Torr and Zisserman 2000; Nistér 2003; Chum and Matas 2005; Raguram, Chum *et al.* 2012; Brachmann, Krull *et al.* 2017; Barath and Matas 2018; Barath, Matas, and Noskova 2019; Brachmann and

Rother 2019). The MAGSAC++ paper by Barath, Noskova *et al.* (2020) compares many of these variants.

The literature on image stitching dates back to work in the photogrammetry community in the 1970s (Milgram 1975, 1977; Slama 1980). In computer vision, papers started appearing in the early 1980s (Peleg 1981), while the development of fully automated techniques came about a decade later (Mann and Picard 1994; Chen 1995; Szeliski 1996; Szeliski and Shum 1997; Sawhney and Kumar 1999; Shum and Szeliski 2000). Those techniques used direct pixel-based alignment but feature-based approaches are now the norm (Zoghlami, Faugeras, and Deriche 1997; Capel and Zisserman 1998; Cham and Cipolla 1998; Badra, Qumsieh, and Dudek 1998; McLauchlan and Jaenicke 2002; Brown and Lowe 2007). A collection of some of these papers can be found in the book by Benosman and Kang (2001). Szeliski (2006a) provides a comprehensive survey of image stitching, on which the material in this chapter is based. More recent publications include Zaragoza, Chin *et al.* (2013), Zhang and Liu (2014), Lin, Pankanti *et al.* (2015), Lin, Jiang *et al.* (2016), Herrmann, Wang *et al.* (2018b), Lee and Sim (2020), and Zhuang and Tran (2020).

High-quality techniques for optimal seam finding and blending are another important component of image stitching systems. Important developments in this field include work by Milgram (1977), Burt and Adelson (1983b), Davis (1998), Uyttendaele, Eden, and Szeliski (2001), Pérez, Gangnet, and Blake (2003), Levin, Zomet *et al.* (2004), Agarwala, Dontcheva *et al.* (2004), Eden, Uyttendaele, and Szeliski (2006), Kopf, Uyttendaele *et al.* (2007), Lin, Jiang *et al.* (2016), and Herrmann, Wang *et al.* (2018a).

In addition to the merging of multiple overlapping photographs taken for aerial or terrestrial panoramic image creation, stitching techniques can be used for automated whiteboard scanning (He and Zhang 2005; Zhang and He 2007), scanning with a mouse (Nakao, Kashitani, and Kaneyoshi 1998), and retinal image mosaics (Can, Stewart *et al.* 2002). They can also be applied to video sequences (Teodosio and Bender 1993; Irani, Hsu, and Anandan 1995; Kumar, Anandan *et al.* 1995; Sawhney and Ayer 1996; Massey and Bender 1996; Irani and Anandan 1998; Sawhney, Arpa *et al.* 2002; Agarwala, Zheng *et al.* 2005; Rav-Acha, Pritch *et al.* 2005; Steedly, Pal, and Szeliski 2005; Baudisch, Tan *et al.* 2006) and can even be used for video compression (Lee, Chen *et al.* 1997).

8.6 Exercises

Ex 8.1: Feature-based image alignment for flip-book animations. Take a set of photos of an action scene or portrait (preferably in burst shooting mode) and align them to make a composite or flip-book animation.

1. Extract features and feature descriptors using some of the techniques described in Sections 7.1.1–7.1.2.

2. Match your features using nearest neighbor matching with a nearest neighbor distance ratio test (7.18).

3. Compute an optimal 2D translation and rotation between the first image and all subsequent images, using least squares (Section 8.1.1) with optional RANSAC for robustness (Section 8.1.4).

4. Resample all of the images onto the first image's coordinate frame (Section 3.6.1) using either bilinear or bicubic resampling and optionally crop them to their common area.

5. Convert the resulting images into an animated GIF (using software available from the web) or optionally implement cross-dissolves to turn them into a "slo-mo" video.

6. (Optional) Combine this technique with feature-based (Exercise 3.25) morphing.

Ex 8.2: Panography. Create the kind of panograph discussed in Section 8.1.2 and commonly found on the web.

1. Take a series of interesting overlapping photos.
2. Use the feature detector, descriptor, and matcher developed in Exercises 7.1–7.4 (or existing software) to match features among the images.
3. Turn each connected component of matching features into a *track*, i.e., assign a unique index i to each track, discarding any tracks that are inconsistent (contain two different features in the same image).
4. Compute a global translation for each image using Equation (8.12).
5. Since your matches probably contain errors, turn the above least square metric into a robust metric (8.25) and re-solve your system using iteratively reweighted least squares.
6. Compute the size of the resulting composite canvas and resample each image into its final position on the canvas. (Keeping track of bounding boxes will make this more efficient.)
7. Average all of the images, or choose some kind of ordering and implement translucent *over* compositing (3.8).
8. (Optional) Extend your parametric motion model to include rotations and scale, i.e., the similarity transform given in Table 8.1. Discuss how you could handle the case of translations and rotations only (no scale).
9. (Optional) Write a simple tool to let the user adjust the ordering and opacity, and add or remove images.
10. (Optional) Write down a different least squares problem that involves pairwise matching of images. Discuss why this might be better or worse than the global matching formula given in (8.12).

Ex 8.3: 2D rigid/Euclidean matching. Several alternative approaches are given in Section 8.1.3 for estimating a 2D rigid (Euclidean) alignment.

1. Implement the various alternatives and compare their accuracy on synthetic data, i.e., random 2D point clouds with noisy feature positions.
2. One approach is to estimate the translations from the centroids and then estimate rotation in polar coordinates. Do you need to weight the angles obtained from a polar decomposition in some way to get the statistically correct estimate?
3. How can you modify your techniques to take into account either scalar (8.10) or full two-dimensional point covariance weightings (8.11)? Do all of the previously developed "short-cuts" still work or does full weighting require iterative optimization?

Ex 8.4: 2D match move/augmented reality. Replace a picture in a magazine or a book with a different image or video.

1. Take a picture of a magazine or book page.

2. Outline a figure or picture on the page with a rectangle, i.e., draw over the four sides as they appear in the image.
3. Match features in this area with each new image frame.
4. Replace the original image with an "advertising" insert, warping the new image with the appropriate homography.
5. Try your approach on a clip from a sporting event (e.g., indoor or outdoor soccer) to implement a billboard replacement.

Ex 8.5: Direct pixel-based alignment. Take a pair of images, compute a coarse-to-fine affine alignment (Exercise 9.2) and then blend them using either averaging (Exercise 8.2) or a Laplacian pyramid (Exercise 3.18). Extend your motion model from affine to perspective (homography) to better deal with rotational mosaics and planar surfaces seen under arbitrary motion.

Ex 8.6: Featured-based stitching. Extend your feature-based alignment technique from Exercise 8.2 to use a full perspective model and then blend the resulting mosaic using either averaging or more sophisticated distance-based feathering (Exercise 8.13).

Ex 8.7: Cylindrical strip panoramas. To generate cylindrical or spherical panoramas from a horizontally panning (rotating) camera, it is best to use a tripod. Set your camera up to take a series of 50% overlapped photos and then use the following steps to create your panorama:

1. Estimate the amount of radial distortion by taking some pictures with lots of long straight lines near the edges of the image and then using the plumb-line method from Exercise 11.5.
2. Compute the focal length either by using a ruler and paper (Debevec, Wenger *et al.* 2002) or by rotating your camera on the tripod, overlapping the images by exactly 0% and counting the number of images it takes to make a 360° panorama.
3. Convert each of your images to cylindrical coordinates using (8.45–8.49).
4. Line up the images with a translational motion model using either a direct pixel-based technique, such as coarse-to-fine incremental or an FFT, or a feature-based technique.
5. (Optional) If doing a complete 360° panorama, align the first and last images. Compute the amount of accumulated vertical misregistration and re-distribute this among the images.
6. Blend the resulting images using feathering or some other technique.

Ex 8.8: Coarse alignment. Use FFT or phase correlation (Section 9.1.2) to estimate the initial alignment between successive images. How well does this work? Over what range of overlaps? If it does not work, does aligning sub-sections (e.g., quarters) do better?

Ex 8.9: Automated mosaicing. Use feature-based alignment with four-point RANSAC for homographies (Section 8.1.3, Equations (8.19–8.23)) or three-point RANSAC for rotational motions (Brown, Hartley, and Nistér 2007) to match up all pairs of overlapping images.

Merge these pairwise estimates together by finding a spanning tree of pairwise relations. Visualize the resulting global alignment, e.g., by displaying a blend of each image with all other images that overlap it.

For greater robustness, try multiple spanning trees (perhaps randomly sampled based on the confidence in pairwise alignments) to see if you can recover from bad pairwise matches (Zach, Klopschitz, and Pollefeys 2010). As a measure of fitness, count how many pairwise estimates are consistent with the global alignment.

Ex 8.10: Global optimization. Use the initialization from the previous algorithm to perform a full bundle adjustment over all of the camera rotations and focal lengths, as described in Section 11.4.2 and by Shum and Szeliski (2000). Optionally, estimate radial distortion parameters as well or support fisheye lenses (Section 2.1.5).

As in the previous exercise, visualize the quality of your registration by creating composites of each input image with its neighbors, optionally blinking between the original image and the composite to better see misalignment artifacts.

Ex 8.11: Deghosting. Use the results of the previous bundle adjustment to predict the location of each feature in a consensus geometry. Use the difference between the predicted and actual feature locations to correct for small misregistrations, as described in Section 8.3.2 (Shum and Szeliski 2000).

Ex 8.12: Compositing surface. Choose a compositing surface (Section 8.4.1), e.g., a single reference image extended to a larger plane, a sphere represented using cylindrical or spherical coordinates, a stereographic "little planet" projection, or a cube map.

Project all of your images onto this surface and blend them with equal weighting, for now (just to see where the original image seams are).

Ex 8.13: Feathering and blending. Compute a feather (distance) map for each warped source image and use these maps to blend the warped images.

Alternatively, use Laplacian pyramid blending (Exercise 3.18) or gradient domain blending.

Ex 8.14: Photomontage and object removal. Implement a "Photomontage" system in which users can indicate desired or unwanted regions in pre-registered images using strokes or other primitives (such as bounding boxes).

(Optional) Devise an automatic moving objects remover (or "keeper") by analyzing which inconsistent regions are more or less typical given some consensus (e.g., median filtering) of the aligned images. Figure 8.17 shows an example where the moving object was kept. Try to make this work for sequences with large amounts of overlaps and consider averaging the images to make the moving object look more ghosted.

Chapter 9

Motion estimation

9.1 Translational alignment 445
9.1.1 Hierarchical motion estimation 448
9.1.2 Fourier-based alignment 449
9.1.3 Incremental refinement 451
9.2 Parametric motion 455
9.2.1 *Application*: Video stabilization 457
9.2.2 Spline-based motion 459
9.2.3 *Application*: Medical image registration 461
9.3 Optical flow 461
9.3.1 Deep learning approaches 466
9.3.2 *Application*: Rolling shutter wobble removal 468
9.3.3 Multi-frame motion estimation 468
9.3.4 *Application*: Video denoising 469
9.4 Layered motion 470
9.4.1 *Application*: Frame interpolation 473
9.4.2 Transparent layers and reflections 474
9.4.3 Video object segmentation 476
9.4.4 Video object tracking 477
9.5 Additional reading 478
9.6 Exercises 479

R. Szeliski, *Computer Vision*, Texts in Computer Science,
https://doi.org/10.1007/978-3-030-34372-9_9

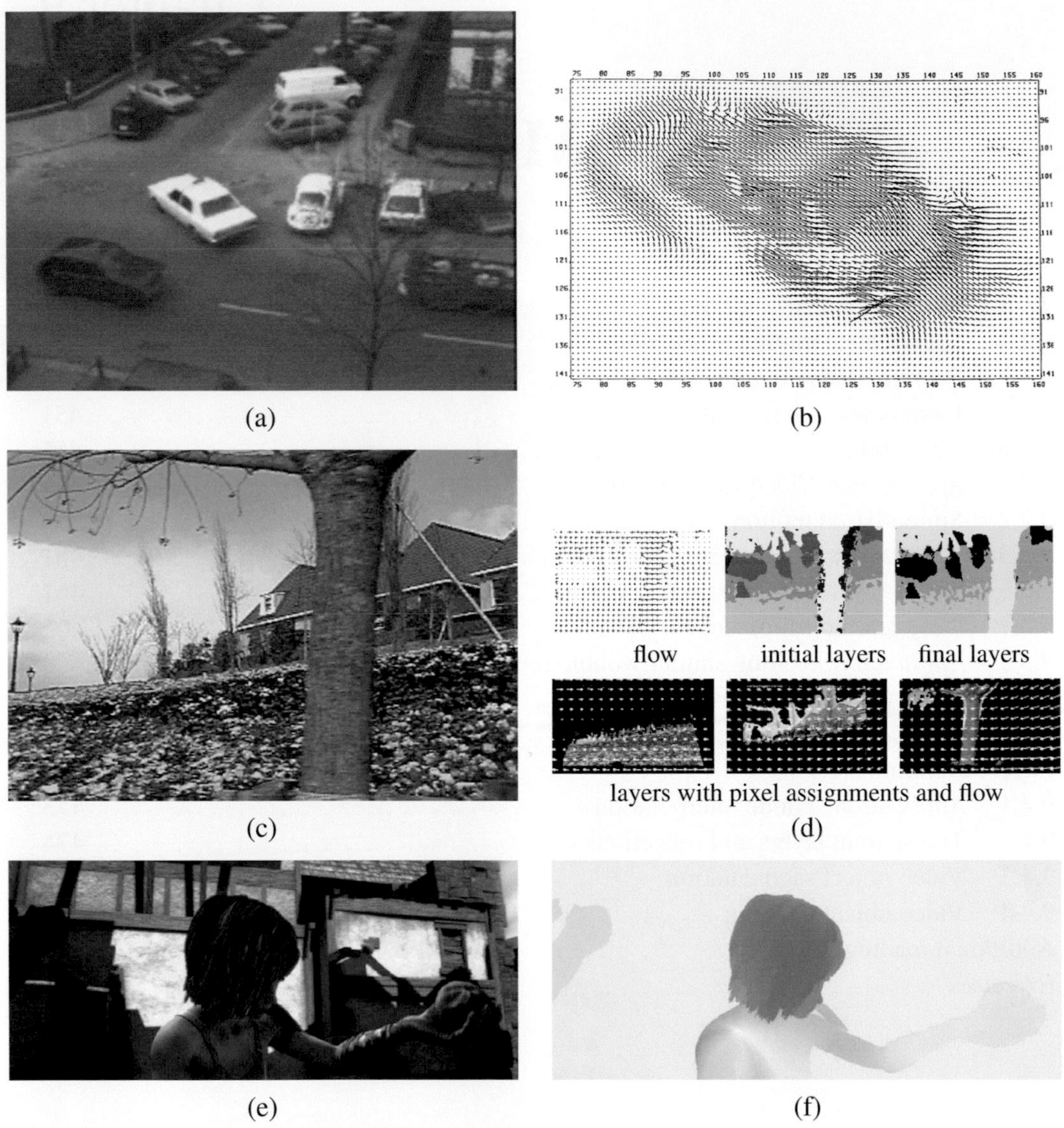

Figure 9.1 Motion estimation: (a–b) regularization-based optical flow (Nagel and Enkelmann 1986) © 1986 IEEE; (c–d) layered motion estimation (Wang and Adelson 1994) © 1994 IEEE; (e–f) sample image and ground truth flow from evaluation database (Butler, Wulff *et al.* 2012) © 2012 Springer.

Algorithms for aligning images and estimating motion in video sequences are among the most widely used in computer vision. For example, frame-rate image alignment is widely used in digital cameras to implement their image stabilization (IS) feature.

An early example of a widely used image registration algorithm is the patch-based translational alignment (optical flow) technique developed by Lucas and Kanade (1981). Variants of this algorithm are used in almost all motion-compensated video compression schemes such as MPEG/H.263 (Le Gall 1991) and HEVC/H.265 (Sullivan, Ohm *et al.* 2012). Similar parametric motion estimation algorithms have found a wide variety of applications, including video summarization (Teodosio and Bender 1993; Irani and Anandan 1998), video stabilization (Hansen, Anandan *et al.* 1994; Srinivasan, Chellappa *et al.* 2005; Matsushita, Ofek *et al.* 2006), and video compression (Irani, Hsu, and Anandan 1995; Lee, Chen *et al.* 1997). More sophisticated image registration algorithms have also been developed for medical imaging and remote sensing. Image registration techniques are surveyed by Brown (1992), Zitov'aa and Flusser (2003), Goshtasby (2005), and Szeliski (2006a).

To estimate the motion between two or more images, a suitable *error metric* must first be chosen to compare the images (Section 9.1). Once this has been established, a suitable *search* technique must be devised. The simplest technique is to exhaustively try all possible alignments, i.e., to do a *full search*. In practice, this may be too slow, so *hierarchical* coarse-to-fine techniques (Section 9.1.1) based on image pyramids are normally used. Alternatively, Fourier transforms (Section 9.1.2) can be used to speed up the computation.

To get sub-pixel precision in the alignment, *incremental* methods (Section 9.1.3) based on a Taylor series expansion of the image function are often used. These can also be applied to *parametric motion models* (Section 9.2), which model global image transformations such as rotation or shearing. Motion estimation can be made more reliable by *learning* the typical dynamics or motion statistics of the scenes or objects being tracked, e.g., the natural gait of walking people (Section 9.2). For more complex motions, piecewise parametric *spline motion models* (Section 9.2.2) can be used.

In the presence of multiple independent (and perhaps non-rigid) motions, general-purpose *optical flow* (or *optic flow*) techniques need to be used, as described in Section 9.3. In recent years, the best-performing techniques have started using deep neural networks (Section 9.3.1). For even more complex motions that include a lot of occlusions, *layered motion models* (Section 9.4), which decompose the scene into coherently moving layers, can work well. Such representations can also be used to perform video object segmentation (Section 9.4.3) and object tracking (Section 9.4.4).

In this chapter, we describe each of these techniques in more detail. Additional details can be found in review and comparative evaluation papers on motion estimation (Barron, Fleet, and Beauchemin 1994; Mitiche and Bouthemy 1996; Stiller and Konrad 1999; Szeliski 2006a; Baker, Scharstein *et al.* 2011; Sun, Yang *et al.* 2018; Janai, Güney *et al.* 2020; Hur and Roth 2020).

9.1 Translational alignment

The simplest way to establish an alignment between two images or image patches is to shift one image relative to the other. Given a *template* image $I_0(\mathbf{x})$ sampled at discrete pixel locations $\{\mathbf{x}_i = (x_i, y_i)\}$, we wish to find where it is located in image $I_1(\mathbf{x})$. A least squares solution to this problem is to find the minimum of the *sum of squared differences* (SSD) function

$$E_{\text{SSD}}(\mathbf{u}) = \sum_i [I_1(\mathbf{x}_i + \mathbf{u}) - I_0(\mathbf{x}_i)]^2 = \sum_i e_i^2, \tag{9.1}$$

where $\mathbf{u} = (u, v)$ is the *displacement* and $e_i = I_1(\mathbf{x}_i + \mathbf{u}) - I_0(\mathbf{x}_i)$ is called the *residual error* (or the *displaced frame difference* in the video coding literature).[1] (We ignore for the moment the possibility that parts of I_0 may lie outside the boundaries of I_1 or be otherwise not visible.) The assumption that corresponding pixel values remain the same in the two images is often called the *brightness constancy constraint*.[2]

In general, the displacement $\mathbf{u}$ can be fractional, so a suitable interpolation function must be applied to image $I_1(\mathbf{x})$. In practice, a bilinear interpolant is often used, but bicubic interpolation can yield slightly better results (Szeliski and Scharstein 2004). Color images can be processed by summing differences across all three color channels, although it is also possible to first transform the images into a different color space or to only use the luminance (which is often done in video encoders).

Robust error metrics. We can make the above error metric more robust to outliers by replacing the squared error terms with a robust function $\rho(e_i)$ (Huber 1981; Hampel, Ronchetti *et al.* 1986; Black and Anandan 1996; Stewart 1999) to obtain

$$E_{\text{SRD}}(\mathbf{u}) = \sum_i \rho(I_1(\mathbf{x}_i + \mathbf{u}) - I_0(\mathbf{x}_i)) = \sum_i \rho(e_i). \tag{9.2}$$

The robust norm $\rho(e)$ is a function that grows less quickly than the quadratic penalty associated with least squares. One such function, sometimes used in motion estimation for video coding because of its speed, is the *sum of absolute differences* (SAD) metric[3] or L_1 norm, i.e.,

$$E_{\text{SAD}}(\mathbf{u}) = \sum_i |I_1(\mathbf{x}_i + \mathbf{u}) - I_0(\mathbf{x}_i)| = \sum_i |e_i|. \tag{9.3}$$

However, because this function is not differentiable at the origin, it is not well suited to gradient-descent approaches such as the ones presented in Section 9.1.3.

Instead, a smoothly varying function that is quadratic for small values but grows more slowly away from the origin is often used. Black and Rangarajan (1996) discuss a variety of such functions, including the *Geman–McClure* function,

$$\rho_{\text{GM}}(x) = \frac{x^2}{1 + x^2/a^2}, \tag{9.4}$$

where a is a constant that can be thought of as an *outlier threshold*. An appropriate value for the threshold can itself be derived using robust statistics (Huber 1981; Hampel, Ronchetti *et al.* 1986; Rousseeuw and Leroy 1987), e.g., by computing the *median absolute deviation*, $MAD = \text{med}_i|e_i|$, and multiplying it by 1.4 to obtain a robust estimate of the standard deviation of the inlier noise process (Stewart 1999). Barron (2019) proposes a generalized robust loss function that can model various outlier distributions and thresholds, as discussed in more detail in Sections 4.1.3 and Appendix B.3, and also has a Bayesian method for estimating the loss function parameters.

[1] The usual justification for using least squares is that it is the optimal estimate with respect to Gaussian noise. See the discussion below on robust error metrics as well as Appendix B.3.

[2] Brightness constancy (Horn 1974) is the tendency for objects to maintain their perceived brightness under varying illumination conditions.

[3] In video compression, e.g., the H.264 standard (https://www.itu.int/rec/T-REC-H.264), the sum of absolute transformed differences (SATD), which measures the differences in a frequency transform space, e.g., using a Hadamard transform, is often used, as it more accurately predicts quality (Richardson 2003).

Spatially varying weights. The error metrics above ignore that fact that for a given alignment, some of the pixels being compared may lie outside the original image boundaries. Furthermore, we may want to partially or completely downweight the contributions of certain pixels. For example, we may want to selectively "erase" some parts of an image from consideration when stitching a mosaic where unwanted foreground objects have been cut out. For applications such as background stabilization, we may want to downweight the middle part of the image, which often contains independently moving objects being tracked by the camera.

All of these tasks can be accomplished by associating a spatially varying per-pixel weight with each of the two images being matched. The error metric then becomes the weighted (or *windowed*) SSD function,

$$E_{\text{WSSD}}(\mathbf{u}) = \sum_i w_0(\mathbf{x}_i) w_1(\mathbf{x}_i + \mathbf{u})[I_1(\mathbf{x}_i + \mathbf{u}) - I_0(\mathbf{x}_i)]^2, \tag{9.5}$$

where the weighting functions w_0 and w_1 are zero outside the image boundaries.

If a large range of potential motions is allowed, the above metric can have a bias towards smaller overlap solutions. To counteract this bias, the windowed SSD score can be divided by the overlap area

$$A = \sum_i w_0(\mathbf{x}_i) w_1(\mathbf{x}_i + \mathbf{u}) \tag{9.6}$$

to compute a *per-pixel* (or mean) squared pixel error E_{WSSD}/A. The square root of this quantity is the *root mean square* intensity error

$$RMS = \sqrt{E_{\text{WSSD}}/A} \tag{9.7}$$

often reported in comparative studies.

Bias and gain (exposure differences). Often, the two images being aligned were not taken with the same exposure. A simple model of linear (affine) intensity variation between the two images is the *bias and gain* model,

$$I_1(\mathbf{x} + \mathbf{u}) = (1 + \alpha) I_0(\mathbf{x}) + \beta, \tag{9.8}$$

where β is the *bias* and α is the *gain* (Lucas and Kanade 1981; Gennert 1988; Fuh and Maragos 1991; Baker, Gross, and Matthews 2003; Evangelidis and Psarakis 2008). The least squares formulation then becomes

$$E_{\text{BG}}(\mathbf{u}) = \sum_i [I_1(\mathbf{x}_i + \mathbf{u}) - (1 + \alpha) I_0(\mathbf{x}_i) - \beta]^2 = \sum_i [\alpha I_0(\mathbf{x}_i) + \beta - e_i]^2. \tag{9.9}$$

Rather than taking a simple squared difference between corresponding patches, it becomes necessary to perform a *linear regression* (Appendix A.2), which is somewhat more costly. Note that for color images, it may be necessary to estimate a different bias and gain for each color channel to compensate for the automatic *color correction* performed by some digital cameras (Section 2.3.2). Bias and gain compensation are also used in video codecs, where they are known as *weighted prediction* (Richardson 2003).

A more general (spatially varying, non-parametric) model of intensity variation, which is computed as part of the registration process, is used in Negahdaripour (1998), Jia and Tang (2003), and Seitz and Baker (2009). This can be useful for dealing with local variations such as the *vignetting* caused by wide-angle lenses, wide apertures, or lens housings. It is also possible to pre-process the images before comparing their values, e.g., using band-pass filtered images (Anandan 1989;

Bergen, Anandan *et al.* 1992), or gradients (Scharstein 1994; Papenberg, Bruhn *et al.* 2006), or using other local transformations such as histograms or rank transforms (Cox, Roy, and Hingorani 1995; Zabih and Woodfill 1994), or to maximize *mutual information* (Viola and Wells III 1997; Kim, Kolmogorov, and Zabih 2003). Hirschmüller and Scharstein (2009) compare a number of these approaches and report on their relative performance in scenes with exposure differences.

Correlation. An alternative to taking intensity differences is to perform *correlation*, i.e., to maximize the *product* (or *cross-correlation*) of the two aligned images,

$$E_{\mathrm{CC}}(\mathbf{u}) = \sum_i I_0(\mathbf{x}_i) I_1(\mathbf{x}_i + \mathbf{u}). \tag{9.10}$$

At first glance, this may appear to make bias and gain modeling unnecessary, since the images will prefer to line up regardless of their relative scales and offsets. However, this is actually not true. If a very bright patch exists in $I_1(\mathbf{x})$, the maximum product may actually lie in that area.

For this reason, *normalized cross-correlation* is more commonly used,

$$E_{\mathrm{NCC}}(\mathbf{u}) = \frac{\sum_i [I_0(\mathbf{x}_i) - \overline{I_0}]\,[I_1(\mathbf{x}_i + \mathbf{u}) - \overline{I_1}]}{\sqrt{\sum_i [I_0(\mathbf{x}_i) - \overline{I_0}]^2} \sqrt{\sum_i [I_1(\mathbf{x}_i + \mathbf{u}) - \overline{I_1}]^2}}, \tag{9.11}$$

where

$$\overline{I_0} = \frac{1}{N} \sum_i I_0(\mathbf{x}_i) \quad \text{and} \tag{9.12}$$

$$\overline{I_1} = \frac{1}{N} \sum_i I_1(\mathbf{x}_i + \mathbf{u}) \tag{9.13}$$

are the *mean images* of the corresponding patches and N is the number of pixels in the patch. The normalized cross-correlation score is always guaranteed to be in the range $[-1, 1]$, which makes it easier to handle in some higher-level applications, such as deciding which patches truly match. Normalized correlation works well when matching images taken with different exposures, e.g., when creating high dynamic range images (Section 10.2). Note, however, that the NCC score is undefined if either of the two patches has zero variance (and, in fact, its performance degrades for noisy low-contrast regions).

A variant on NCC, which is related to the bias–gain regression implicit in the matching score (9.9), is the *normalized SSD* score

$$E_{\mathrm{NSSD}}(\mathbf{u}) = \frac{1}{2} \frac{\sum_i \left[[I_0(\mathbf{x}_i) - \overline{I_0}] - [I_1(\mathbf{x}_i + \mathbf{u}) - \overline{I_1}] \right]^2}{\sqrt{\sum_i [I_0(\mathbf{x}_i) - \overline{I_0}]^2 + [I_1(\mathbf{x}_i + \mathbf{u}) - \overline{I_1}]^2}} \tag{9.14}$$

proposed by Criminisi, Shotton *et al.* (2007). In their experiments, they find that it produces comparable results to NCC, but is more efficient when applied to a large number of overlapping patches using a moving average technique (Section 3.2.2).

9.1.1 Hierarchical motion estimation

Now that we have a well-defined alignment cost function to optimize, how can we find its minimum? The simplest solution is to do a *full search* over some range of shifts, using either integer or sub-pixel steps. This is often the approach used for *block matching* in *motion compensated video compression*, where a range of possible motions (say, ± 16 pixels) is explored.[4]

To accelerate this search process, *hierarchical motion estimation* is often used: an image pyramid (Section 3.5) is constructed and a search over a smaller number of discrete pixels (corresponding to the same range of motion) is first performed at coarser levels (Quam 1984; Anandan 1989; Bergen, Anandan *et al.* 1992). The motion estimate from one level of the pyramid is then used to initialize a smaller *local* search at the next finer level. Alternatively, several seeds (good solutions) from the coarse level can be used to initialize the fine-level search. While this is not guaranteed to produce the same result as a full search, it usually works almost as well and is much faster.

More formally, let

$$I_k^{(l)}(\mathbf{x}_j) \leftarrow \tilde{I}_k^{(l-1)}(2\mathbf{x}_j) \tag{9.15}$$

be the *decimated* image at level l obtained by subsampling (*downsampling*) a smoothed version of the image at level $l-1$. See Section 3.5 for how to perform the required downsampling (pyramid construction) without introducing too much aliasing.

At the coarsest level, we search for the best displacement $\mathbf{u}^{(l)}$ that minimizes the difference between images $I_0^{(l)}$ and $I_1^{(l)}$. This is usually done using a full search over some range of displacements $\mathbf{u}^{(l)} \in 2^{-l}[-S, S]^2$, where S is the desired *search range* at the finest (original) resolution level, optionally followed by the incremental refinement step described in Section 9.1.3.

Once a suitable motion vector has been estimated, it is used to *predict* a likely displacement

$$\hat{\mathbf{u}}^{(l-1)} \leftarrow 2\mathbf{u}^{(l)} \tag{9.16}$$

for the next finer level.[5] The search over displacements is then repeated at the finer level over a much narrower range of displacements, say $\hat{\mathbf{u}}^{(l-1)} \pm 1$, again optionally combined with an incremental refinement step (Anandan 1989). Alternatively, one of the images can be *warped* (resampled) by the current motion estimate, in which case only small incremental motions need to be computed at the finer level. A nice description of the whole process, extended to parametric motion estimation (Section 9.2), is provided by Bergen, Anandan *et al.* (1992).

9.1.2 Fourier-based alignment

When the search range corresponds to a significant fraction of the larger image (as is the case in image stitching, see Section 8.2), the hierarchical approach may not work that well, as it is often not possible to coarsen the representation too much before significant features are blurred away. In this case, a Fourier-based approach may be preferable.

Fourier-based alignment relies on the fact that the Fourier transform of a shifted signal has the same magnitude as the original signal, but a linearly varying phase (Section 3.4), i.e.,

$$\mathcal{F}\{I_1(\mathbf{x}+\mathbf{u})\} = \mathcal{F}\{I_1(\mathbf{x})\}\, e^{-j\mathbf{u}\cdot\boldsymbol{\omega}} = \mathcal{I}_1(\boldsymbol{\omega})e^{-j\mathbf{u}\cdot\boldsymbol{\omega}}, \tag{9.17}$$

where $\boldsymbol{\omega}$ is the vector-valued angular frequency of the Fourier transform and we use calligraphic notation $\mathcal{I}_1(\boldsymbol{\omega}) = \mathcal{F}\{I_1(\mathbf{x})\}$ to denote the Fourier transform of a signal (Section 3.4).

[4] In stereo matching (Section 12.1.2), an explicit search over all possible disparities (i.e., a *plane sweep*) is almost always performed, as the number of search hypotheses is much smaller due to the 1D nature of the potential displacements.

[5] This doubling of displacements is only necessary if displacements are defined in integer *pixel* coordinates, which is the usual case in the literature (Bergen, Anandan *et al.* 1992). If *normalized device coordinates* (Section 2.1.4) are used instead, the displacements (and search ranges) need not change from level to level, although the step sizes will need to be adjusted, to keep search steps of roughly one pixel or less.

Another useful property of Fourier transforms is that convolution in the spatial domain corresponds to multiplication in the Fourier domain (Section 3.4).[6] The Fourier transform of the cross-correlation function $E_{\rm CC}$ can thus be written as

$$\mathcal{F}\{E_{\rm CC}(\mathbf{u})\} = \mathcal{F}\left\{\sum_i I_0(\mathbf{x}_i) I_1(\mathbf{x}_i + \mathbf{u})\right\} = \mathcal{F}\{I_0(\mathbf{u})\bar{*} I_1(\mathbf{u})\} = \mathcal{I}_0(\boldsymbol{\omega})\mathcal{I}_1^*(\boldsymbol{\omega}), \tag{9.18}$$

where

$$f(\mathbf{u})\bar{*}g(\mathbf{u}) = \sum_i f(\mathbf{x}_i) g(\mathbf{x}_i + \mathbf{u}) \tag{9.19}$$

is the *correlation* function, i.e., the convolution of one signal with the reverse of the other, and $\mathcal{I}_1^*(\boldsymbol{\omega})$ is the *complex conjugate* of $\mathcal{I}_1(\boldsymbol{\omega})$. This is because convolution is defined as the summation of one signal with the reverse of the other (Section 3.4).

To efficiently evaluate $E_{\rm CC}$ over the range of all possible values of $\mathbf{u}$, we take the Fourier transforms of both images $I_0(\mathbf{x})$ and $I_1(\mathbf{x})$, multiply both transforms together (after conjugating the second one), and take the inverse transform of the result. The Fast Fourier Transform algorithm can compute the transform of an $N \times M$ image in $\mathrm{O}(NM \log NM)$ operations (Bracewell 1986). This can be significantly faster than the $\mathrm{O}(N^2M^2)$ operations required to do a full search when the full range of image overlaps is considered.

While Fourier-based convolution is often used to accelerate the computation of image correlations, it can also be used to accelerate the sum of squared differences function (and its variants). Consider the SSD formula given in (9.1). Its Fourier transform can be written as

$$\begin{aligned}\mathcal{F}\{E_{\rm SSD}(\mathbf{u})\} &= \mathcal{F}\left\{\sum_i [I_1(\mathbf{x}_i + \mathbf{u}) - I_0(\mathbf{x}_i)]^2\right\} \\ &= \delta(\boldsymbol{\omega}) \sum_i [I_0^2(\mathbf{x}_i) + I_1^2(\mathbf{x}_i)] - 2\mathcal{I}_0(\boldsymbol{\omega})\mathcal{I}_1^*(\boldsymbol{\omega}).\end{aligned} \tag{9.20}$$

Thus, the SSD function can be computed by taking twice the correlation function and subtracting it from the sum of the energies in the two images (or patches).

Windowed correlation. Unfortunately, the Fourier convolution theorem only applies when the summation over $\mathbf{x}_i$ is performed over *all* the pixels in both images, using a circular shift of the image when accessing pixels outside the original boundaries. While this is acceptable for small shifts and comparably sized images, it makes no sense when the images overlap by a small amount or one image is a small subset of the other.

In that case, the cross-correlation function should be replaced with a *windowed* (weighted) cross-correlation function,

$$E_{\rm WCC}(\mathbf{u}) = \sum_i w_0(\mathbf{x}_i) I_0(\mathbf{x}_i)\, w_1(\mathbf{x}_i + \mathbf{u}) I_1(\mathbf{x}_i + \mathbf{u}), \tag{9.21}$$

$$= [w_0(\mathbf{x}) I_0(\mathbf{x})]\bar{*}[w_1(\mathbf{x}) I_1(\mathbf{x})] \tag{9.22}$$

where the weighting functions w_0 and w_1 are zero outside the valid ranges of the images and both images are padded so that circular shifts return 0 values outside the original image boundaries.

[6] In fact, the Fourier shift property (9.17) derives from the convolution theorem by observing that shifting is equivalent to convolution with a displaced delta function $\delta(\mathbf{x} - \mathbf{u})$.

An even more interesting case is the computation of the *weighted* SSD function introduced in Equation (9.5),

$$E_{\text{WSSD}}(\mathbf{u}) = \sum_i w_0(\mathbf{x}_i) w_1(\mathbf{x}_i + \mathbf{u}) [I_1(\mathbf{x}_i + \mathbf{u}) - I_0(\mathbf{x}_i)]^2. \tag{9.23}$$

Expanding this as a sum of correlations and deriving the appropriate set of Fourier transforms is left for Exercise 9.1.

The same kind of derivation can also be applied to the bias–gain corrected sum of squared difference function E_{BG} (9.9). Again, Fourier transforms can be used to efficiently compute all the correlations needed to perform the linear regression in the bias and gain parameters in order to estimate the exposure-compensated difference for each potential shift (Exercise 9.1). It is also possible to use Fourier transforms to estimate the rotation and scale between two patches that are centered on the same pixel, as described in De Castro and Morandi (1987) and Szeliski (2010, Section 8.1.2).

Phase correlation. A variant of regular correlation (9.18) that is sometimes used for motion estimation is *phase correlation* (Kuglin and Hines 1975; Brown 1992). Here, the spectrum of the two signals being matched is *whitened* by dividing each per-frequency product in (9.18) by the magnitudes of the Fourier transforms,

$$\mathcal{F}\{E_{\text{PC}}(\mathbf{u})\} = \frac{\mathcal{I}_0(\boldsymbol{\omega})\mathcal{I}_1^*(\boldsymbol{\omega})}{\|\mathcal{I}_0(\boldsymbol{\omega})\|\|\mathcal{I}_1(\boldsymbol{\omega})\|} \tag{9.24}$$

before taking the final inverse Fourier transform. In the case of noiseless signals with perfect (cyclic) shift, we have $I_1(\mathbf{x} + \mathbf{u}) = I_0(\mathbf{x})$ and hence, from Equation (9.17), we obtain

$$\begin{aligned} \mathcal{F}\{I_1(\mathbf{x} + \mathbf{u})\} &= \mathcal{I}_1(\boldsymbol{\omega}) e^{-2\pi j \mathbf{u} \cdot \boldsymbol{\omega}} = \mathcal{I}_0(\boldsymbol{\omega}) \quad \text{and} \\ \mathcal{F}\{E_{\text{PC}}(\mathbf{u})\} &= e^{-2\pi j \mathbf{u} \cdot \boldsymbol{\omega}}. \end{aligned} \tag{9.25}$$

The output of phase correlation (under ideal conditions) is therefore a single spike (impulse) located at the correct value of $\mathbf{u}$, which (in principle) makes it easier to find the correct estimate.

Phase correlation has a reputation in some quarters of outperforming regular correlation, but this behavior depends on the characteristics of the signals and noise. If the original images are contaminated by noise in a narrow frequency band (e.g., low-frequency noise or peaked frequency "hum"), the whitening process effectively de-emphasizes the noise in these regions. However, if the original signals have very low signal-to-noise ratio at some frequencies (say, two blurry or low-textured images with lots of high-frequency noise), the whitening process can actually decrease performance (see Exercise 9.1).

Gradient cross-correlation has emerged as a promising alternative to phase correlation (Argyriou and Vlachos 2003), although further systematic studies are probably warranted. Phase correlation has also been studied by Fleet and Jepson (1990) as a method for estimating general optical flow and stereo disparity.

9.1.3 Incremental refinement

The techniques described up till now can estimate alignment to the nearest pixel (or potentially fractional pixel if smaller search steps are used). In general, image stabilization and stitching applications require much higher accuracies to obtain acceptable results.

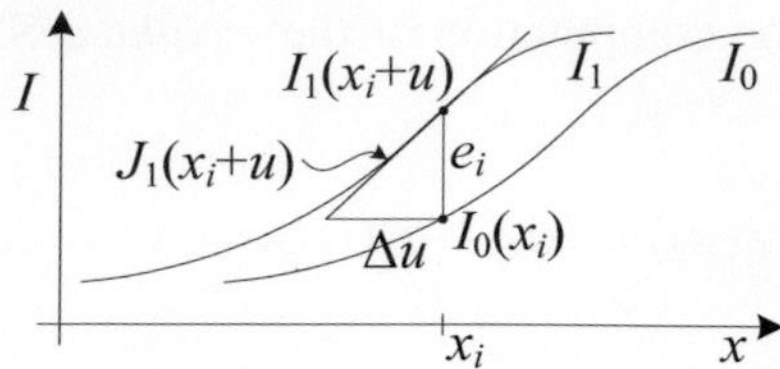

Figure 9.2 Taylor series approximation of a function and the incremental computation of the optical flow correction amount. $\mathbf{J}_1(\mathbf{x}_i + \mathbf{u})$ is the image gradient at $(\mathbf{x}_i + \mathbf{u})$ and e_i is the current intensity difference.

To obtain better *sub-pixel* estimates, we can use one of several techniques described by Tian and Huhns (1986). One possibility is to evaluate several discrete (integer or fractional) values of (u, v) around the best value found so far and to *interpolate* the matching score to find an analytic minimum (Szeliski and Scharstein 2004).

A more commonly used approach, first proposed by Lucas and Kanade (1981), is to perform *gradient descent* on the SSD energy function (9.1), using a Taylor series expansion of the image function (Figure 9.2),

$$E_{\text{LK-SSD}}(\mathbf{u} + \Delta\mathbf{u}) = \sum_i [I_1(\mathbf{x}_i + \mathbf{u} + \Delta\mathbf{u}) - I_0(\mathbf{x}_i)]^2 \tag{9.26}$$

$$\approx \sum_i [I_1(\mathbf{x}_i + \mathbf{u}) + \mathbf{J}_1(\mathbf{x}_i + \mathbf{u})\Delta\mathbf{u} - I_0(\mathbf{x}_i)]^2 \tag{9.27}$$

$$= \sum_i [\mathbf{J}_1(\mathbf{x}_i + \mathbf{u})\Delta\mathbf{u} + e_i]^2, \tag{9.28}$$

where

$$\mathbf{J}_1(\mathbf{x}_i + \mathbf{u}) = \nabla I_1(\mathbf{x}_i + \mathbf{u}) = \left(\frac{\partial I_1}{\partial x}, \frac{\partial I_1}{\partial y}\right)(\mathbf{x}_i + \mathbf{u}) \tag{9.29}$$

is the *image gradient* or *Jacobian* at $(\mathbf{x}_i + \mathbf{u})$ and

$$e_i = I_1(\mathbf{x}_i + \mathbf{u}) - I_0(\mathbf{x}_i), \tag{9.30}$$

first introduced in (9.1), is the current intensity error.[7] The gradient at a particular sub-pixel location $(\mathbf{x}_i + \mathbf{u})$ can be computed using a variety of techniques, the simplest of which is simply to take the horizontal and vertical differences between pixels $\mathbf{x}$ and $\mathbf{x} + (1, 0)$ or $\mathbf{x} + (0, 1)$. More sophisticated derivatives can sometimes lead to noticeable performance improvements.

The linearized form of the incremental update to the SSD error (9.28) is often called the *optical flow constraint* or *brightness constancy constraint* equation (Horn and Schunck 1981)

$$I_x u + I_y v + I_t = 0, \tag{9.31}$$

where the subscripts in I_x and I_y denote spatial derivatives, and I_t is called the *temporal derivative*, which makes sense if we are computing instantaneous velocity in a video sequence. When squared and summed or integrated over a region, it can be used to compute optical flow (Horn and Schunck 1981).

The above least squares problem (9.28) can be minimized by solving the associated *normal equations* (Appendix A.2),

$$\mathbf{A}\Delta\mathbf{u} = \mathbf{b} \tag{9.32}$$

[7] We follow the convention, commonly used in robotics and by Baker and Matthews (2004), that derivatives with respect to (column) vectors result in row vectors, so that fewer transposes are needed in the formulas.

where

$$\mathbf{A} = \sum_i \mathbf{J}_1^T(\mathbf{x}_i + \mathbf{u})\mathbf{J}_1(\mathbf{x}_i + \mathbf{u}) \tag{9.33}$$

and

$$\mathbf{b} = -\sum_i e_i \mathbf{J}_1^T(\mathbf{x}_i + \mathbf{u}) \tag{9.34}$$

are called the (Gauss–Newton approximation of the) *Hessian* and *gradient-weighted residual vector*, respectively.[8] These matrices are also often written as

$$\mathbf{A} = \begin{bmatrix} \sum I_x^2 & \sum I_x I_y \\ \sum I_x I_y & \sum I_y^2 \end{bmatrix} \quad \text{and} \quad \mathbf{b} = -\begin{bmatrix} \sum I_x I_t \\ \sum I_y I_t \end{bmatrix}. \tag{9.35}$$

The gradients required for $\mathbf{J}_1(\mathbf{x}_i + \mathbf{u})$ can be evaluated at the same time as the image warps required to estimate $I_1(\mathbf{x}_i + \mathbf{u})$ (Section 3.6.1 (3.75)) and, in fact, are often computed as a side-product of image interpolation. If efficiency is a concern, these gradients can be replaced by the gradients in the *template* image,

$$\mathbf{J}_1(\mathbf{x}_i + \mathbf{u}) \approx \mathbf{J}_0(\mathbf{x}_i), \tag{9.36}$$

because near the correct alignment, the template and displaced target images should look similar. This has the advantage of allowing the precomputation of the Hessian and Jacobian images, which can result in significant computational savings (Hager and Belhumeur 1998; Baker and Matthews 2004). A further reduction in computation can be obtained by writing the warped image $I_1(\mathbf{x}_i + \mathbf{u})$ used to compute e_i in (9.30) as a convolution of a sub-pixel interpolation filter with the discrete samples in I_1 (Peleg and Rav-Acha 2006). Precomputing the inner product between the gradient field and shifted version of I_1 allows the iterative re-computation of e_i to be performed in constant time (independent of the number of pixels).

The effectiveness of the above incremental update rule relies on the quality of the Taylor series approximation. When far away from the true displacement (say, 1–2 pixels), several iterations may be needed. It is possible, however, to estimate a value for $\mathbf{J}_1$ using a least squares fit to a series of larger displacements to increase the range of convergence (Jurie and Dhome 2002) or to "learn" a special-purpose recognizer for a given patch (Avidan 2001; Williams, Blake, and Cipolla 2003; Lepetit, Pilet, and Fua 2006; Hinterstoisser, Benhimane *et al.* 2008; Özuysal, Calonder *et al.* 2010) as discussed in Section 7.1.5.

A commonly used stopping criterion for incremental updating is to monitor the magnitude of the displacement correction $\|\mathbf{u}\|$ and to stop when it drops below a certain threshold (say, $^1/_{10}$ of a pixel). For larger motions, it is usual to combine the incremental update rule with a hierarchical coarse-to-fine search strategy, as described in Section 9.1.1.

Conditioning and aperture problems. Sometimes, the inversion of the linear system (9.32) can be poorly conditioned because of lack of two-dimensional texture in the patch being aligned. A commonly occurring example of this is the *aperture problem*, first identified in some of the early papers on optical flow (Horn and Schunck 1981) and then studied more extensively by Anandan (1989). Consider an image patch that consists of a slanted edge moving to the right (Figure 7.4). Only the *normal* component of the velocity (displacement) can be reliably recovered in this case.

[8] The true Hessian is the full second derivative of the error function E, which may not be positive definite—see Section 8.1.3 and Appendix A.3.

This manifests itself in (9.32) as a *rank-deficient* matrix $\mathbf{A}$, i.e., one whose smaller eigenvalue is very close to zero.[9]

When Equation (9.32) is solved, the component of the displacement along the edge is very poorly conditioned and can result in wild guesses under small noise perturbations. One way to mitigate this problem is to add a *prior* (soft constraint) on the expected range of motions (Simoncelli, Adelson, and Heeger 1991; Baker, Gross, and Matthews 2004; Govindu 2006). This can be accomplished by adding a small value to the diagonal of $\mathbf{A}$, which essentially biases the solution towards smaller $\Delta\mathbf{u}$ values that still (mostly) minimize the squared error.

However, the pure Gaussian model assumed when using a simple (fixed) quadratic prior, as in Simoncelli, Adelson, and Heeger (1991), does not always hold in practice, e.g., because of aliasing along strong edges (Triggs 2004). For this reason, it may be prudent to add some small fraction (say, 5%) of the larger eigenvalue to the smaller one before doing the matrix inversion.

Uncertainty modeling. The reliability of a particular patch-based motion estimate can be captured more formally with an *uncertainty model*. The simplest such model is a *covariance matrix*, which captures the expected variance in the motion estimate in all possible directions. As discussed in Section 8.1.4 and Appendix B.6, under small amounts of additive Gaussian noise, it can be shown that the covariance matrix $\mathbf{\Sigma_u}$ is proportional to the inverse of the Hessian $\mathbf{A}$,

$$\mathbf{\Sigma_u} = \sigma_n^2 \mathbf{A}^{-1}, \tag{9.37}$$

where σ_n^2 is the variance of the additive Gaussian noise (Anandan 1989; Matthies, Kanade, and Szeliski 1989; Szeliski 1989).

For larger amounts of noise, the linearization performed by the Lucas–Kanade algorithm in (9.28) is only approximate, so the above quantity becomes a *Cramer–Rao lower bound* on the true covariance. Thus, the minimum and maximum eigenvalues of the Hessian $\mathbf{A}$ can now be interpreted as the (scaled) inverse variances in the least-certain and most-certain directions of motion. (A more detailed analysis using a more realistic model of image noise is given by Steele and Jaynes (2005).) Figure 7.5 shows the local SSD surfaces for three different pixel locations in an image. As you can see, the surface has a clear minimum in the highly textured region and suffers from the aperture problem near the strong edge.

Bias and gain, weighting, and robust error metrics. The Lucas–Kanade update rule can also be applied to the bias–gain equation (9.9) to obtain

$$E_{\text{LK-BG}}(\mathbf{u} + \Delta\mathbf{u}) = \sum_i [\mathbf{J}_1(\mathbf{x}_i + \mathbf{u})\Delta\mathbf{u} + e_i - \alpha I_0(\mathbf{x}_i) - \beta]^2 \tag{9.38}$$

(Lucas and Kanade 1981; Gennert 1988; Fuh and Maragos 1991; Baker, Gross, and Matthews 2003). The resulting 4×4 system of equations can be solved to simultaneously estimate the translational displacement update $\Delta\mathbf{u}$ and the bias and gain parameters β and α.

A similar formulation can be derived for images (templates) that have a *linear appearance variation*,

$$I_1(\mathbf{x} + \mathbf{u}) \approx I_0(\mathbf{x}) + \sum_j \lambda_j B_j(\mathbf{x}), \tag{9.39}$$

[9] The matrix $\mathbf{A}$ is by construction always guaranteed to be symmetric positive semi-definite, i.e., it has real non-negative eigenvalues.

where the $B_j(\mathbf{x})$ are the *basis images* and the λ_j are the unknown coefficients (Hager and Belhumeur 1998; Baker, Gross *et al.* 2003; Baker, Gross, and Matthews 2003). Potential linear appearance variations include illumination changes (Hager and Belhumeur 1998) and small non-rigid deformations (Black and Jepson 1998; Kambhamettu, Goldgof *et al.* 2003).

A weighted (windowed) version of the Lucas–Kanade algorithm is also possible:

$$E_{\text{LK-WSSD}}(\mathbf{u}+\Delta\mathbf{u}) = \sum_i w_0(\mathbf{x}_i)w_1(\mathbf{x}_i+\mathbf{u})[\mathbf{J}_1(\mathbf{x}_i+\mathbf{u})\Delta\mathbf{u}+e_i]^2. \tag{9.40}$$

Note that here, in deriving the Lucas–Kanade update from the original weighted SSD function (9.5), we have neglected taking the derivative of the $w_1(\mathbf{x}_i+\mathbf{u})$ weighting function with respect to $\mathbf{u}$, which is usually acceptable in practice, especially if the weighting function is a binary mask with relatively few transitions.

Baker, Gross *et al.* (2003) only use the $w_0(\mathbf{x})$ term, which is reasonable if the two images have the same extent and no (independent) cutouts in the overlap region. They also discuss the idea of making the weighting proportional to $\nabla I(\mathbf{x})$, which helps for very noisy images, where the gradient itself is noisy. Similar observations, formulated in terms of *total least squares* (Van Huffel and Vandewalle 1991; Van Huffel and Lemmerling 2002), have been made by other researchers studying optical flow (Weber and Malik 1995; Bab-Hadiashar and Suter 1998b; Mühlich and Mester 1998). Baker, Gross *et al.* (2003) show how evaluating Equation (9.40) at just the *most reliable* (highest gradient) pixels does not significantly reduce performance for large enough images, even if only 5–10% of the pixels are used. (This idea was originally proposed by Dellaert and Collins (1999), who used a more sophisticated selection criterion.)

The Lucas–Kanade incremental refinement step can also be applied to the robust error metric introduced in Section 9.1,

$$E_{\text{LK-SRD}}(\mathbf{u}+\Delta\mathbf{u}) = \sum_i \rho(\mathbf{J}_1(\mathbf{x}_i+\mathbf{u})\Delta\mathbf{u}+e_i), \tag{9.41}$$

which can be solved using the *iteratively reweighted least squares* technique described in Section 8.1.4.

9.2 Parametric motion

Many image alignment tasks, for example image stitching with handheld cameras, require the use of more sophisticated motion models, as described in Section 2.1.1. As these models, e.g., affine deformations, typically have more parameters than pure translation, a full search over the possible range of values is impractical. Instead, the incremental Lucas–Kanade algorithm can be generalized to parametric motion models and used in conjunction with a hierarchical search algorithm (Lucas and Kanade 1981; Rehg and Witkin 1991; Fuh and Maragos 1991; Bergen, Anandan *et al.* 1992; Shashua and Toelg 1997; Shashua and Wexler 2001; Baker and Matthews 2004).

For parametric motion, instead of using a single constant translation vector $\mathbf{u}$, we use a spatially varying *motion field* or *correspondence map*, $\mathbf{x}'(\mathbf{x};\mathbf{p})$, parameterized by a low-dimensional vector $\mathbf{p}$, where $\mathbf{x}'$ can be any of the motion models presented in Section 2.1.1. The parametric incremental motion update rule now becomes

$$E_{\text{LK-PM}}(\mathbf{p} + \Delta\mathbf{p}) = \sum_i [I_1(\mathbf{x}'(\mathbf{x}_i; \mathbf{p} + \Delta\mathbf{p})) - I_0(\mathbf{x}_i)]^2 \tag{9.42}$$

$$\approx \sum_i [I_1(\mathbf{x}'_i) + \mathbf{J}_1(\mathbf{x}'_i)\Delta\mathbf{p} - I_0(\mathbf{x}_i)]^2 \tag{9.43}$$

$$= \sum_i [\mathbf{J}_1(\mathbf{x}'_i)\Delta\mathbf{p} + e_i]^2, \tag{9.44}$$

where the Jacobian is now

$$\mathbf{J}_1(\mathbf{x}'_i) = \frac{\partial I_1}{\partial \mathbf{p}} = \nabla I_1(\mathbf{x}'_i)\frac{\partial \mathbf{x}'}{\partial \mathbf{p}}(\mathbf{x}_i), \tag{9.45}$$

i.e., the product of the image gradient ∇I_1 with the Jacobian of the correspondence field, $\mathbf{J}_{x'} = \partial\mathbf{x}'/\partial\mathbf{p}$.

The motion Jacobians $\mathbf{J}_{\mathbf{x}'}$ for the 2D planar transformations introduced in Section 2.1.1 and Table 2.1 are given in Table 8.1. Note how we have re-parameterized the motion matrices so that they are always the identity at the origin $\mathbf{p} = 0$. This becomes useful later, when we talk about the compositional and inverse compositional algorithms. (It also makes it easier to impose priors on the motions.)

For parametric motion, the (Gauss–Newton) *Hessian* and *gradient-weighted residual vector* become

$$\mathbf{A} = \sum_i \mathbf{J}^T_{\mathbf{x}'}(\mathbf{x}_i)[\nabla I_1^T(\mathbf{x}'_i)\nabla I_1(\mathbf{x}'_i)]\mathbf{J}_{\mathbf{x}'}(\mathbf{x}_i) \tag{9.46}$$

and

$$\mathbf{b} = -\sum_i \mathbf{J}^T_{\mathbf{x}'}(\mathbf{x}_i)[e_i \nabla I_1^T(\mathbf{x}'_i)]. \tag{9.47}$$

Note how the expressions inside the square brackets are the same ones evaluated for the simpler translational motion case (9.33–9.34).

Patch-based approximation. The computation of the Hessian and residual vectors for parametric motion can be significantly more expensive than for the translational case. For parametric motion with n parameters and N pixels, the accumulation of $\mathbf{A}$ and $\mathbf{b}$ takes $O(n^2N)$ operations (Baker and Matthews 2004). One way to reduce this by a significant amount is to divide the image up into smaller sub-blocks (patches) P_j and to only accumulate the simpler 2×2 quantities inside the square brackets at the pixel level (Shum and Szeliski 2000),

$$\mathbf{A}_j = \sum_{i \in P_j} \nabla I_1^T(\mathbf{x}'_i)\nabla I_1(\mathbf{x}'_i) \tag{9.48}$$

$$\mathbf{b}_j = \sum_{i \in P_j} e_i \nabla I_1^T(\mathbf{x}'_i). \tag{9.49}$$

The full Hessian and residual can then be approximated as

$$\mathbf{A} \approx \sum_j \mathbf{J}^T_{\mathbf{x}'}(\hat{\mathbf{x}}_j)[\sum_{i \in P_j} \nabla I_1^T(\mathbf{x}'_i)\nabla I_1(\mathbf{x}'_i)]\mathbf{J}_{\mathbf{x}'}(\hat{\mathbf{x}}_j) = \sum_j \mathbf{J}^T_{\mathbf{x}'}(\hat{\mathbf{x}}_j)\mathbf{A}_j\mathbf{J}_{\mathbf{x}'}(\hat{\mathbf{x}}_j) \tag{9.50}$$

and

$$\mathbf{b} \approx -\sum_j \mathbf{J}^T_{\mathbf{x}'}(\hat{\mathbf{x}}_j)[\sum_{i \in P_j} e_i \nabla I_1^T(\mathbf{x}'_i)] = -\sum_j \mathbf{J}^T_{\mathbf{x}'}(\hat{\mathbf{x}}_j)\mathbf{b}_j, \tag{9.51}$$

where $\hat{\mathbf{x}}_j$ is the *center* of each patch P_j (Shum and Szeliski 2000). This is equivalent to replacing the true motion Jacobian with a piecewise-constant approximation. In practice, this works quite well.

Compositional approach. For a complex parametric motion such as a homography, the computation of the motion Jacobian becomes complicated and may involve a per-pixel division. Szeliski and Shum (1997) observed that this can be simplified by first warping the target image I_1 according to the current motion estimate $\mathbf{x}'(\mathbf{x}; \mathbf{p})$,

$$\tilde{I}_1(\mathbf{x}) = I_1(\mathbf{x}'(\mathbf{x}; \mathbf{p})), \tag{9.52}$$

and then comparing this *warped* image against the template $I_0(\mathbf{x})$. Subsequently Hager and Belhumeur (1998) suggested replacing the gradient of $\tilde{I}_1(\mathbf{x})$ with the gradient of $I_0(\mathbf{x})$, as described previously in (9.36), which allows the precomputation (and inversion) of the Hessian matrix $\mathbf{A}$ given in (9.46). The residual vector $\mathbf{b}$ (9.47) can also be partially precomputed, i.e., the *steepest descent* images $\nabla I_0(\mathbf{x})\mathbf{J}_{\tilde{\mathbf{x}}}(\mathbf{x})$ can be precomputed and stored for later multiplication with the $e(\mathbf{x}) = \tilde{I}_1(\mathbf{x}) - I_0(\mathbf{x})$ error images, as described in (Szeliski 2010, Section 8.2) and (Baker and Matthews 2004), where this is called the *inverse additive* scheme. Baker and Matthews (2004) also introduce one more variant they call the *inverse compositional* algorithm where they warp the template image $I_0(\mathbf{x})$ and precompute the inverse Hessian and the steepest descent images, which makes it the preferred approach. They also discuss the advantage of using Gauss–Newton iteration (i.e., the first-order expansion of the least squares, as above) compared to other approaches such as steepest descent and Levenberg–Marquardt.

Subsequent parts of the series (Baker, Gross *et al.* 2003; Baker, Gross, and Matthews 2003, 2004) discuss more advanced topics such as per-pixel weighting, pixel selection for efficiency, a more in-depth discussion of robust metrics and algorithms, linear appearance variations, and priors on parameters. They make for invaluable reading for anyone interested in implementing a highly tuned implementation of incremental image registration and have been widely used as components of subsequent object trackers, which are discussed in Section 9.4.4. Evangelidis and Psarakis (2008) provide some detailed experimental evaluations of these and other related approaches.

Learned motion models

An alternative to parameterizing the motion field with a geometric deformation such as an affine transform is to learn a set of basis functions tailored to a particular application (Black, Yacoob *et al.* 1997). First, a set of dense motion fields (Section 9.3) is computed from a set of training videos. Next, singular value decomposition (SVD) is applied to the stack of motion fields $\mathbf{u}_t(\mathbf{x})$ to compute the first few singular vectors $\mathbf{v}_k(\mathbf{x})$. Finally, for a new test sequence, a novel flow field is computed using a coarse-to-fine algorithm that estimates the unknown coefficient a_k in the parameterized flow field

$$\mathbf{u}(\mathbf{x}) = \sum_k a_k \mathbf{v}_k(\mathbf{x}). \tag{9.53}$$

Figure 9.3a shows a set of basis fields learned by observing videos of walking motions. Figure 9.3b shows the temporal evolution of the basis coefficients as well as a few of the recovered parametric motion fields. Note that similar ideas can also be applied to feature tracks (Torresani, Hertzmann, and Bregler 2008), which is a topic we discuss in more detail in Sections 7.1.5 and 13.6.4, as well as video stabilization (Yu and Ramamoorthi 2020).

9.2.1 *Application*: Video stabilization

Video stabilization is one of the most widely used applications of parametric motion estimation (Hansen, Anandan *et al.* 1994; Irani, Rousso, and Peleg 1997; Morimoto and Chellappa 1997; Srinivasan, Chellappa *et al.* 2005; Grundmann, Kwatra, and Essa 2011). Algorithms for stabilization

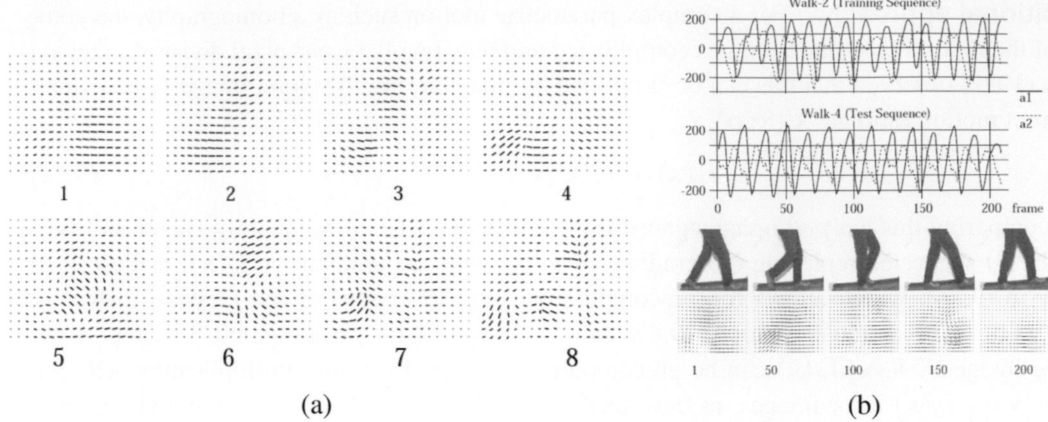

Figure 9.3 Learned parameterized motion fields for a walking sequence (Black, Yacoob *et al.* 1997) © 1997 IEEE: (a) learned basis flow fields; (b) plots of motion coefficients over time and corresponding estimated motion fields.

run inside both hardware devices, such as camcorders and still cameras, and software packages for improving the visual quality of shaky videos.

In their paper on full-frame video stabilization, Matsushita, Ofek *et al.* (2006) give a nice overview of the three major stages of stabilization, namely motion estimation, motion smoothing, and image warping. Motion estimation algorithms often use a similarity transform to handle camera translations, rotations, and zooming. The tricky part is getting these algorithms to lock onto the background motion, which is a result of the camera movement, without getting distracted by independently moving foreground objects (Yu and Ramamoorthi 2018, 2020; Yu, Ramamoorthi *et al.* 2021). Motion smoothing algorithms recover the low-frequency (slowly varying) part of the motion and then estimate the high-frequency shake component that needs to be removed. While quadratic penalties on motion derivatives are commonly used, more realistic virtual camera motions (locked and linear) can be obtained using L_1 minimization of derivatives (Grundmann, Kwatra, and Essa 2011). Finally, image warping algorithms apply the high-frequency correction to render the original frames as if the camera had undergone only the smooth motion.

The resulting stabilization algorithms can greatly improve the appearance of shaky videos but they often still contain visual artifacts. For example, image warping can result in missing borders around the image, which must be cropped, filled using information from other frames, or hallucinated using inpainting techniques (Section 10.5.1). Furthermore, video frames captured during fast motion are often blurry. Their appearance can be improved either by using deblurring techniques (Section 10.3) or by stealing sharper pixels from other frames with less motion or better focus (Matsushita, Ofek *et al.* 2006). Exercise 9.3 has you implement and test some of these ideas.

In situations where the camera is translating a lot in 3D, e.g., when the videographer is walking, an even better approach is to compute a full structure from motion reconstruction of the camera motion and 3D scene. One or more smooth 3D camera paths can then be computed and the original video re-rendered using view interpolation with the interpolated 3D point cloud serving as the proxy geometry while preserving salient features in what is sometimes called *content preserving warps* (Liu, Gleicher *et al.* 2009, 2011; Liu, Yuan *et al.* 2013; Kopf, Cohen, and Szeliski 2014). If you have access to a camera array instead of a single video camera, you can do even better using a light field rendering approach (Section 14.3) (Smith, Zhang *et al.* 2009).

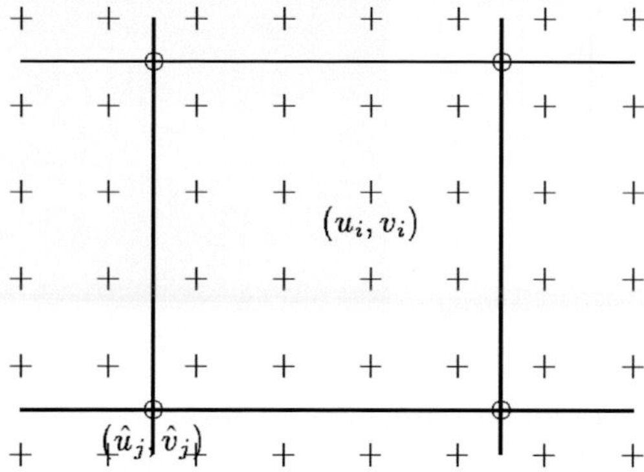

Figure 9.4 Spline motion field: the displacement vectors $\mathbf{u}_i = (u_i, v_i)$ are shown as pluses (+) and are controlled by the smaller number of control vertices $\hat{\mathbf{u}}_j = (\hat{u}_i, \hat{v}_j)$, which are shown as circles (∘).

9.2.2 Spline-based motion

While parametric motion models are useful in a wide variety of applications (such as video stabilization and mapping onto planar surfaces), most image motion is too complicated to be captured by such low-dimensional models.

Traditionally, optical flow algorithms (Section 9.3) compute an independent motion estimate for each pixel, i.e., the number of flow vectors computed is equal to the number of input pixels. The general optical flow analog to Equation (9.1) can thus be written as

$$E_{\text{SSD-OF}}(\{\mathbf{u}_i\}) = \sum_i [I_1(\mathbf{x}_i + \mathbf{u}_i) - I_0(\mathbf{x}_i)]^2. \tag{9.54}$$

Notice how in the above equation, the number of variables $\{\mathbf{u}_i\}$ is twice the number of measurements, so the problem is underconstrained.

The two classic approaches to this problem, which we study in Section 9.3, are to perform the summation over overlapping regions (the *patch-based* or *window-based* approach) or to add smoothness terms on the $\{\mathbf{u}_i\}$ field using *regularization* or *Markov random fields* (Chapter 4). In this section, we describe an alternative approach that lies somewhere between general optical flow (independent flow at each pixel) and parametric flow (a small number of global parameters). The approach is to represent the motion field as a two-dimensional *spline* controlled by a smaller number of *control vertices* $\{\hat{\mathbf{u}}_j\}$ (Figure 9.4),

$$\mathbf{u}_i = \sum_j \hat{\mathbf{u}}_j B_j(\mathbf{x}_i) = \sum_j \hat{\mathbf{u}}_j w_{i,j}, \tag{9.55}$$

where the $B_j(\mathbf{x}_i)$ are called the *basis functions* and are only non-zero over a small *finite support* interval (Szeliski and Coughlan 1997). We call the $w_{ij} = B_j(\mathbf{x}_i)$ *weights* to emphasize that the $\{\mathbf{u}_i\}$ are known linear combinations of the $\{\hat{\mathbf{u}}_j\}$.

Substituting the formula for the individual per-pixel flow vectors $\mathbf{u}_i$ (9.55) into the SSD error metric (9.54) yields a parametric motion formula similar to Equation (9.43). The biggest difference is that the Jacobian $\mathbf{J}_1(\mathbf{x}'_i)$ (9.45) now consists of the sparse entries in the weight matrix $\mathbf{W} = [w_{ij}]$.

In situations where we know something more about the motion field, e.g., when the motion is due to a camera moving in a static scene, we can use more specialized motion models. For example, the *plane plus parallax* model (Section 2.1.4) can be naturally combined with a spline-based motion

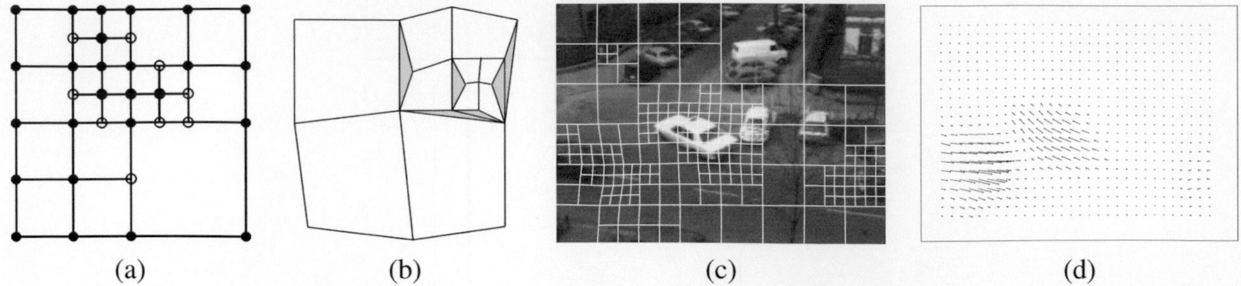

Figure 9.5 Quadtree spline-based motion estimation (Szeliski and Shum 1996) © 1996 IEEE: (a) quadtree spline representation, (b) which can lead to *cracks*, unless the white nodes are constrained to depend on their parents; (c) deformed quadtree spline mesh overlaid on grayscale image; (d) flow field visualized as a needle diagram.

representation, where the in-plane motion is represented by a homography (8.19) and the out-of-plane parallax d is represented by a scalar variable at each spline control point (Szeliski and Kang 1995; Szeliski and Coughlan 1997).

In many cases, the small number of spline vertices results in a motion estimation problem that is well conditioned. However, if large textureless regions (or elongated edges subject to the aperture problem) persist across several spline patches, it may be necessary to add a *regularization* term to make the problem well posed (Section 4.2). The simplest way to do this is to directly add squared difference penalties between adjacent vertices in the spline control mesh $\{\hat{\mathbf{u}}_j\}$, as in (4.24). If a multi-resolution (coarse-to-fine) strategy is being used, it is important to re-scale these smoothness terms while going from level to level.

The linear system corresponding to the spline-based motion estimator is sparse and regular. Because it is usually of moderate size, it can often be solved using direct techniques such as Cholesky decomposition (Appendix A.4). Alternatively, if the problem becomes too large and subject to excessive fill-in, iterative techniques such as hierarchically preconditioned conjugate gradient (Szeliski 1990b, 2006b; Krishnan and Szeliski 2011; Krishnan, Fattal, and Szeliski 2013) can be used instead (Appendix A.5).

Because of its robustness, spline-based motion estimation has been used for a number of applications, including visual effects (Roble 1999) and medical image registration (Section 9.2.3) (Szeliski and Lavallée 1996; Kybic and Unser 2003).

One disadvantage of the basic technique, however, is that the model does a poor job near motion discontinuities, unless an excessive number of nodes are used. To remedy this situation, Szeliski and Shum (1996) propose using a *quadtree* representation embedded in the spline control grid (Figure 9.5a). Large cells are used to present regions of smooth motion, while smaller cells are added in regions of motion discontinuities (Figure 9.5c).

To estimate the motion, a coarse-to-fine strategy is used. Starting with a regular spline imposed over a lower-resolution image, an initial motion estimate is obtained. Spline patches where the motion is inconsistent, i.e., the squared residual (9.54) is above a threshold, are subdivided into smaller patches. To avoid *cracks* in the resulting motion field (Figure 9.5b), the values of certain nodes in the refined mesh, i.e., those adjacent to larger cells, need to be *restricted* so that they depend on their parent values. This is most easily accomplished using a hierarchical basis representation for the quadtree spline (Szeliski 1990b) and selectively setting some of the hierarchical basis functions to 0, as described in (Szeliski and Shum 1996).

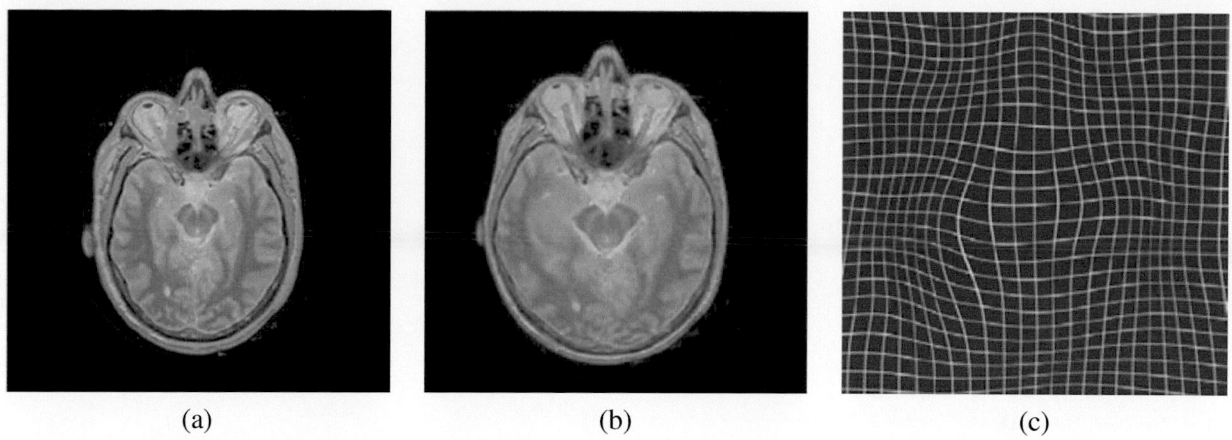

(a) (b) (c)

Figure 9.6 Elastic brain registration (Kybic and Unser 2003) © 2003 IEEE: (a) original brain atlas and patient MRI images overlaid in red–green; (b) after elastic registration with eight user-specified landmarks (not shown); (c) a cubic B-spline deformation field, shown as a deformed grid.

9.2.3 *Application*: Medical image registration

Because they excel at representing smooth *elastic* deformation fields, spline-based motion models have found widespread use in medical image registration (Bajcsy and Kovacic 1989; Szeliski and Lavallée 1996; Christensen, Joshi, and Miller 1997).[10] Registration techniques can be used both to track an individual patient's development or progress over time (a *longitudinal* study) or to match different patient images together to find commonalities and detect variations or pathologies (*cross-sectional* studies). When different imaging *modalities* are being registered, e.g., computed tomography (CT) scans and magnetic resonance images (MRI), *mutual information* measures of similarity are often necessary (Viola and Wells III 1997; Maes, Collignon *et al.* 1997).

Kybic and Unser (2003) provide a nice literature review and describe a complete working system based on representing both the images and the deformation fields as multi-resolution splines. Figure 9.6 shows an example of the Kybic and Unser system being used to register a patient's brain MRI with a labeled brain atlas image. The system can be run in a fully automatic mode but more accurate results can be obtained by locating a few key *landmarks*. More recent papers on deformable medical image registration, including performance evaluations, include Klein, Staring, and Pluim (2007), Glocker, Komodakis *et al.* (2008), and the survey by Sotiras, Davatzikos, and Paragios (2013).

As with other applications, regular volumetric splines can be enhanced using selective refinement. In the case of 3D volumetric image or surface registration, these are known as *octree splines* (Szeliski and Lavallée 1996) and have been used to register medical surface models such as vertebrae and faces from different patients (Figure 9.7).

9.3 Optical flow

The most general (and challenging) version of motion estimation is to compute an independent estimate of motion at *each* pixel, which is generally known as *optical* (or *optic*) *flow*. As we mentioned in the previous section, this generally involves minimizing the brightness or color difference between

[10]In computer graphics, such elastic volumetric deformations are known as *free-form deformations* (Sederberg and Parry 1986; Coquillart 1990; Celniker and Gossard 1991).

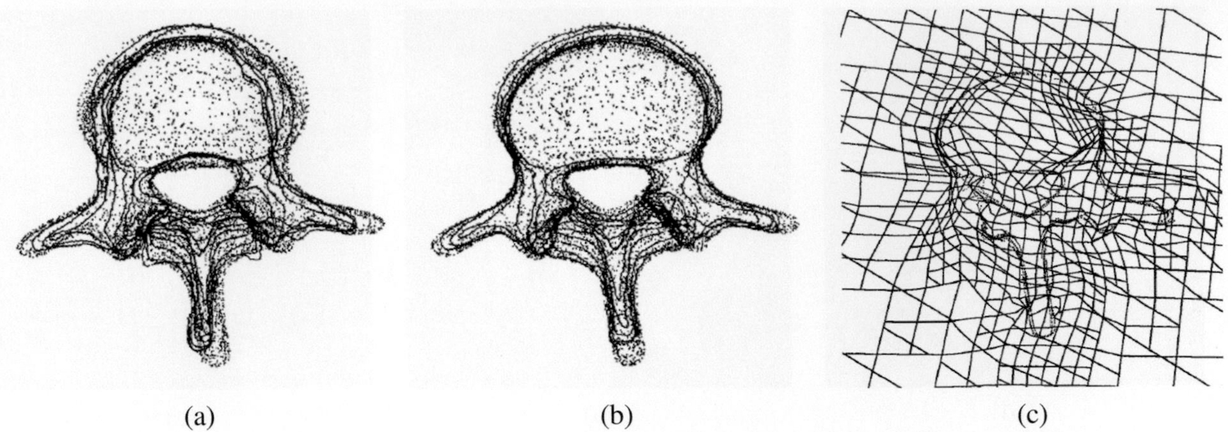

(a) (b) (c)

Figure 9.7 Octree spline-based image registration of two vertebral surface models (Szeliski and Lavallée 1996) © 1996 Springer: (a) after initial rigid alignment; (b) after elastic alignment; (c) a cross-section through the adapted octree spline deformation field.

corresponding pixels summed over the image,

$$E_{\text{SSD-OF}}(\{\mathbf{u}_i\}) = \sum_i [I_1(\mathbf{x}_i + \mathbf{u}_i) - I_0(\mathbf{x}_i)]^2. \tag{9.56}$$

Because the number of variables $\{\mathbf{u}_i\}$ is twice the number of measurements, the problem is underconstrained. The two classic approaches to this problem are to perform the summation *locally* over overlapping regions (the *patch-based* or *window-based* approach) or to add smoothness terms on the $\{\mathbf{u}_i\}$ field using regularization or Markov random fields (Chapter 4) and to search for a global minimum. Good overviews of recent optical flow algorithms can be found in Baker, Scharstein *et al.* (2011), Sun, Yang *et al.* (2018), Janai, Güney *et al.* (2020), and Hur and Roth (2020).

The patch-based approach usually involves using a Taylor series expansion of the displaced image function (9.28) to obtain sub-pixel estimates (Lucas and Kanade 1981). Anandan (1989) shows how a series of local discrete search steps can be interleaved with Lucas–Kanade incremental refinement steps in a coarse-to-fine pyramid scheme, which allows the estimation of large motions, as described in Section 9.1.1. He also analyzes how the *uncertainty* in local motion estimates is related to the eigenvalues of the local Hessian matrix $\mathbf{A}_i$ (9.37), as shown in Figures 7.4 and 7.5.

Bergen, Anandan *et al.* (1992) develop a unified framework for describing both parametric (Section 9.2) and patch-based optical flow algorithms and provide a nice introduction to this topic. After each iteration of optical flow estimation in a coarse-to-fine pyramid, they re-warp one of the images so that only incremental flow estimates are computed (Section 9.1.1). When overlapping patches are used, an efficient implementation is to first compute the outer products of the gradients and intensity errors (9.33–9.34) at every pixel and then perform the overlapping window sums using a moving average filter.[11]

Instead of solving for each motion (or motion update) independently, Horn and Schunck (1981) develop a regularization-based framework where (9.56) is simultaneously minimized over all flow vectors $\{\mathbf{u}_i\}$. To constrain the problem, smoothness constraints, i.e., squared penalties on flow derivatives, are added to the basic per-pixel error metric. Because the technique was originally developed for small motions in a variational (continuous function) framework, the linearized *brightness constancy constraint* corresponding to (9.28), i.e., (9.31), is more commonly written as an

[11] Other smoothing or aggregation filters can also be used at this stage (Bruhn, Weickert, and Schnörr 2005).

analytic integral

$$E_{\text{HS}} = \int (I_x u + I_y v + I_t)^2 \, dx \, dy, \tag{9.57}$$

where $(I_x, I_y) = \nabla I_1 = \mathbf{J}_1$, $I_t = e_i$ is the *temporal derivative*, i.e., the brightness change between images, and $u(x, y)$ and $v(x, y)$ are the 2D optical flow functions. The Horn and Schunck model can also be viewed as the limiting case of spline-based motion estimation as the splines become 1×1 pixel patches.

It is also possible to combine ideas from local and global flow estimation into a single framework by using a locally aggregated (as opposed to single-pixel) Hessian as the brightness constancy term (Bruhn, Weickert, and Schnörr 2005). Consider the discrete analog (9.28) to the analytic global energy (9.57),

$$E_{\text{HSD}} = \sum_i \mathbf{u}_i^T [\mathbf{J}_i \mathbf{J}_i^T] \mathbf{u}_i + 2 e_i \mathbf{J}_i^T \mathbf{u}_i + e_i^2. \tag{9.58}$$

If we replace the per-pixel (rank 1) Hessians $\mathbf{A}_i = [\mathbf{J}_i \mathbf{J}_i^T]$ and residuals $\mathbf{b}_i = \mathbf{J}_i e_i$ with area-aggregated versions (9.33–9.34), we obtain a global minimization algorithm where region-based brightness constraints are used.

Another extension to the basic optical flow model is to use a combination of global (parametric) and local motion models. For example, if we know that the motion is due to a camera moving in a static scene (rigid motion), we can re-formulate the problem as the estimation of a per-pixel depth along with the parameters of the global camera motion (Adiv 1989; Hanna 1991; Bergen, Anandan *et al.* 1992; Szeliski and Coughlan 1997; Nir, Bruckstein, and Kimmel 2008; Wedel, Cremers *et al.* 2009). Such techniques are closely related to stereo matching (Chapter 12). Alternatively, we can estimate either per-image or per-segment affine motion models combined with per-pixel *residual* corrections (Black and Jepson 1996; Ju, Black, and Jepson 1996; Chang, Tekalp, and Sezan 1997; Mémin and Pérez 2002). We revisit this topic in Section 9.4.

Of course, image brightness may not always be an appropriate metric for measuring appearance consistency, e.g., when the lighting in an image is varying. As discussed in Section 9.1, matching gradients, filtered images, or other metrics such as image Hessians (second derivative measures) may be more appropriate. It is also possible to locally compute the *phase* of steerable filters in the image, which is insensitive to both bias and gain transformations (Fleet and Jepson 1990). Papenberg, Bruhn *et al.* (2006) review and explore such constraints and also provide a detailed analysis and justification for iteratively re-warping images during incremental flow computation.

Because the brightness constancy constraint is evaluated at each pixel independently, rather than being summed over patches where the constant flow assumption may be violated, global optimization approaches tend to perform better near motion discontinuities. This is especially true if robust metrics are used in the smoothness constraint (Black and Anandan 1996; Bab-Hadiashar and Suter 1998a).[12] One popular choice for robust metrics is the L_1 norm, also known as *total variation* (TV), which results in a convex energy whose global minimum can be found (Bruhn, Weickert, and Schnörr 2005; Papenberg, Bruhn *et al.* 2006; Zach, Pock, and Bischof 2007b; Zimmer, Bruhn, and Weickert 2011). Anisotropic smoothness priors, which apply a different smoothness in the directions parallel and perpendicular to the image gradient, are another popular choice (Nagel and Enkelmann 1986; Sun, Roth *et al.* 2008; Werlberger, Trobin *et al.* 2009; Werlberger, Pock, and Bischof 2010). It is also possible to learn a set of better smoothness constraints (derivative filters and robust functions) from a set of paired flow and intensity images (Sun, Roth *et al.* 2008). Many of these techniques are discussed in more detail by Baker, Scharstein *et al.* (2011) and Sun, Roth, and Black (2014).

[12] Robust brightness metrics (Section 9.1, (9.2)) can also help improve the performance of window-based approaches (Black and Anandan 1996).

Optical flow evaluation results **Statistics:** Average SD R0.5 R1.0 R2.0 A50 A75 A95
Error type: endpoint angle interpolation normalized interpolation

Average endpoint error	avg.	Army (Hidden texture) GT im0 im1			Mequon (Hidden texture) GT im0 im1			Schefflera (Hidden texture) GT im0 im1			Wooden (Hidden texture) GT im0 im1			Grove (Synthetic) GT im0 im1			Urban (Synthetic) GT im0 im1			Yosemite (Synthetic) GT im0 im1			Teddy (Stereo) GT im0 im1		
	rank	all	disc	untext	all	disc	untext	all	disc	untext	all	disc	untext	all	disc	untext	all	disc	untext	all	disc	untext	all	disc	untext
Adaptive [20]	4.4	**0.09** 1	**0.26** 1	**0.06** 1	0.23 5	0.78 4	[illegible] 5	0.54 8	[illegible] 10	0.21 3	**0.18** 1	0.91 3	**0.10** 1	0.88 3	1.25 3	0.73 5	0.50 3	1.28 3	0.31 3	0.14 10	0.16 12	0.22 10	0.65 3	1.37 3	0.79 4
Complementary OF [21]	5.7	0.11 5	0.28 3	0.10 9	**0.18** 1	**0.63** [illegible]	**0.12** 1	0.31 3	0.75 [illegible]	**0.18** 1	0.19 2	0.97 5	0.12 3	0.97 10	1.31 6	1.00 11	1.78 20	1.73 7	0.87 14	0.11 4	0.12 2	0.22 10	0.68 4	1.48 4	0.95 8
Aniso. Huber-L1 [22]	5.8	0.10 3	0.28 3	0.08 3	0.31 11	0.88 [illegible]	[illegible]	[illegible] 10	[illegible]	0.29 12	0.20 4	0.92 4	0.13 5	0.84 2	1.20 2	0.70 2	**0.39** 1	**1.23** 1	**0.28** 1	0.17 15	0.15 9	0.27 16	0.64 2	1.36 2	0.79 4
DPOF [18]	6.1	0.13 12	0.35 12	0.09 4	0.25 6	0.79 5	[illegible]	[illegible]	[illegible] 1	0.21 3	0.19 2	**0.62** 1	0.15 11	**0.74** 1	**1.09** 1	**0.49** 1	0.66 7	1.80 10	0.63 8	0.19 17	0.17 14	0.35 20	**0.50** 1	**1.08** 1	**0.55** 1
TV-L1-improved [17]	7.2	**0.09** 1	**0.26** 1	0.07 2	0.20 3	0.71 3	0.16 2	0.53 7	1.18 9	0.22 5	0.21 7	1.24 11	0.11 2	0.90 4	1.31 6	0.72 3	1.51 14	1.93 11	0.84 11	0.18 16	0.17 14	0.31 17	0.73 8	1.62 9	0.87 7
CBF [12]	7.8	0.10 3	0.28 3	0.09 4	0.34 12	0.80 6	0.37 13	0.43 5	0.95 5	0.26 8	0.21 7	1.14 8	0.13 5	0.90 4	1.27 4	0.82 7	0.41 2	**1.23** 1	0.30 2	0.23 22	0.19 20	0.39 21	0.76 9	1.56 6	1.02 9
Brox et al. [5]	8.4	0.11 5	0.32 8	0.11 12	0.27 9	0.93 10	0.22 9	0.39 4	0.94 4	0.24 7	0.24 9	1.25 12	0.13 5	1.10 13	1.39 12	1.43 17	0.89 8	1.77 8	0.55 7	0.10 2	0.13 4	**0.11** 1	0.91 11	1.83 12	1.13 12
Rannacher [23]	8.5	0.11 5	0.31 6	0.09 4	0.25 6	0.84 7	0.21 8	0.57 12	1.27 15	0.26 8	0.24 9	1.32 14	0.13 5	0.91 7	1.33 8	0.72 3	1.49 13	1.95 13	0.78 9	0.15 12	0.14 7	0.26 13	0.69 6	1.58 8	0.86 6
F-TV-L1 [15]	8.8	0.14 13	0.35 12	0.14 15	0.34 12	0.98 12	0.26 11	0.59 14	1.19 10	0.26 8	0.27 13	1.36 15	0.16 12	0.90 4	1.30 5	0.76 6	0.54 4	1.62 6	0.36 4	0.13 6	0.15 9	0.20 9	0.68 4	1.56 6	0.66 2
Second-order prior [8]	9.0	0.11 5	0.31 6	0.09 4	0.26 8	0.93 10	0.20 7	0.57 12	1.25 14	0.26 8	0.20 4	1.04 6	0.12 3	0.94 8	1.34 9	0.83 9	0.61 6	1.93 11	0.47 6	0.20 18	0.16 12	0.34 19	0.77 10	1.64 10	1.07 10
Fusion [6]	9.4	0.11 5	0.34 10	0.10 9	0.19 2	0.69 2	0.16 2	0.29 2	0.66 2	0.23 6	0.20 4	1.19 10	0.14 9	1.07 11	1.42 13	1.22 13	1.35 10	1.49 5	0.86 13	0.20 18	0.20 21	0.26 13	1.07 14	2.07 16	1.39 16
Dynamic MRF [7]	11.1	0.12 11	0.34 10	0.11 12	0.22 4	0.89 9	0.16 2	0.44 6	1.13 7	0.20 2	0.24 9	1.29 13	0.14 9	1.11 14	1.52 17	1.13 12	1.54 15	2.37 20	0.93 15	0.13 6	0.12 2	0.31 17	1.27 18	2.33 20	1.66 17
SegOF [10]	11.7	0.15 14	0.36 14	0.10 9	0.57 15	1.16 15	0.59 19	0.68 15	1.24 12	0.64 14	0.32 15	0.86 2	0.26 15	1.18 17	1.50 16	1.47 18	1.63 18	2.09 14	0.96 16	**0.08** 1	0.13 4	0.12 2	0.70 7	1.50 5	0.69 3
Learning Flow [11]	13.3	0.11 5	0.32 8	0.09 4	0.29 10	0.99 13	0.23 10	0.55 9	1.24 12	0.29 12	0.36 16	1.56 17	0.25 14	1.25 19	1.64 21	1.41 16	1.55 17	2.32 19	0.85 12	0.14 10	0.18 18	0.24 12	1.09 15	2.09 18	1.27 13
Filter Flow [19]	14.3	0.17 16	0.39 16	0.13 14	0.43 14	1.09 14	0.38 14	0.75 16	1.34 16	0.78 19	0.70 19	1.54 16	0.68 19	1.13 16	1.38 11	1.51 19	0.57 5	1.32 4	0.44 5	0.22 20	0.23 23	0.26 13	0.96 12	1.66 11	1.12 11
GraphCuts [14]	14.5	0.16 15	0.38 15	0.14 15	0.59 18	1.36 19	0.46 15	0.56 10	1.07 6	0.64 14	0.26 12	1.14 8	0.17 13	0.96 9	1.35 10	0.84 10	2.25 23	1.79 9	1.22 21	0.22 20	0.17 14	0.43 22	1.22 17	2.05 15	1.78 19
Black & Anandan [4]	15.0	0.18 17	0.42 17	0.19 18	0.58 17	1.31 17	0.50 16	0.95 19	1.58 18	0.70 16	0.49 17	1.59 18	0.45 17	1.08 12	1.42 13	1.22 13	1.43 11	2.28 17	0.83 10	0.15 12	0.17 14	0.17 6	1.11 16	1.98 14	1.30 14
SPSA-learn [13]	15.7	0.18 17	0.45 18	0.17 17	0.57 15	1.32 18	0.51 17	0.84 17	1.50 17	0.72 17	0.52 18	1.64 19	0.49 18	1.12 15	1.42 13	1.39 15	1.75 19	2.14 15	1.06 20	0.13 6	0.13 4	0.19 7	1.32 19	2.08 17	1.73 18
GroupFlow [9]	15.9	0.21 19	0.51 19	0.21 19	0.79 21	1.69 21	0.72 21	0.86 18	1.64 19	0.74 18	0.30 14	1.07 7	0.26 15	1.29 22	1.81 22	0.82 7	1.94 21	2.30 18	1.36 22	0.11 4	0.14 7	0.19 7	1.06 13	1.96 13	1.35 15
2D-CLG [1]	17.4	0.28 21	0.62 22	0.21 19	0.67 20	1.21 16	0.70 20	1.12 21	1.80 21	0.99 22	1.07 22	2.06 21	1.12 22	1.23 18	1.52 17	1.62 22	1.54 15	2.15 16	0.96 16	0.10 2	**0.11** 1	0.16 4	1.38 20	2.26 19	1.83 20
Horn & Schunck [3]	18.6	0.22 20	0.55 20	0.22 21	0.61 19	1.53 20	0.52 18	1.01 20	1.73 20	0.80 20	0.78 20	2.02 20	0.77 20	1.26 20	1.58 19	1.55 20	1.43 11	2.59 22	1.00 18	0.16 14	0.18 18	0.15 3	1.51 21	2.50 21	1.88 21
TI-DOFE [24]	19.6	0.38 23	0.64 23	0.47 23	1.16 22	1.72 22	1.26 22	1.39 23	2.06 24	1.17 23	1.29 23	2.21 23	1.41 23	1.27 21	1.61 20	1.57 21	1.28 9	2.57 21	1.01 19	0.13 6	0.15 9	0.16 4	1.87 22	2.71 22	2.53 22
FOLKI [16]	22.6	0.29 22	0.73 24	0.33 22	1.52 23	1.96 24	1.80 23	1.23 22	2.04 23	0.95 21	0.99 21	2.20 22	1.08 21	1.53 23	1.85 23	2.07 23	2.14 22	3.23 24	1.60 23	0.26 23	0.21 22	0.68 23	2.67 23	3.27 23	4.32 23
Pyramid LK [2]	23.7	0.39 24	0.61 21	0.61 24	1.67 24	1.78 23	2.00 24	1.50 24	1.97 22	1.38 24	1.57 24	2.39 24	1.78 24	2.94 24	3.72 24	2.98 24	3.33 24	2.74 23	2.43 24	0.30 24	0.24 24	0.73 24	3.80 24	5.08 24	4.88 24

Move the mouse over the numbers in the table to see the corresponding images. Click to compare with the ground truth.

Figure 9.8 Evaluation of the results of 24 optical flow algorithms, October 2009, https://vision.middlebury.edu/flow, (Baker, Scharstein *et al.* 2009). By moving the mouse pointer over an underlined performance score, the user can interactively view the corresponding flow and error maps. Clicking on a score toggles between the computed and ground truth flows. Next to each score, the corresponding rank in the current column is indicated by a smaller blue number. The minimum (best) score in each column is shown in boldface. The table is sorted by the average rank (computed over all 24 columns, three region masks for each of the eight sequences). The average rank serves as an *approximate* measure of performance *under the selected metric/statistic*.

Because of the large, two-dimensional search space in estimating flow, most algorithms use variations of gradient descent and coarse-to-fine continuation methods to minimize the global energy function. This contrasts starkly with stereo matching, which is an "easier" one-dimensional disparity estimation problem, where combinatorial optimization techniques were the method of choice until the advent of deep neural networks.[13] One way to deal with this complexity is to start with efficient patch-based correspondences (Kroeger, Timofte *et al.* 2016). Another way to deal with the large two-dimensional search space is to integrate sparse feature matches into a variational formulation, as was initially proposed by Brox and Malik (2010a). This approach was later extended by several authors, including Weinzaepfel, Revaud *et al.* (2013), whose DeepFlow system use a hand-crafted (non-learnt) convolutional network to compute initial quasi-dense correspondences, and Revaud, Weinzaepfel *et al.* (2015), whose EpicFlow system added an edge and occlusion-aware interpolation step before the variational optimization.

Combinatorial optimization methods based on Markov random fields were among the better-performing methods on the optical flow database of Baker, Scharstein *et al.* (2011)[14] when it was originally released, but have now been overtaken by deep neural networks. Examples of such techniques include the one developed by Glocker, Paragios *et al.* (2008), who use a coarse-to-fine strategy with per-pixel 2D uncertainty estimates, which are then used to guide the refinement and search at the next finer level. Lempitsky, Roth, and Rother (2008) use fusion moves (Lempitsky, Rother, and Blake 2007) over proposals generated from basic flow algorithms (Horn and Schunck 1981; Lucas and Kanade 1981) to find good solutions.

A careful empirical analysis of these kinds of "classic" coarse-to-fine energy-minimization approaches is provided in the meticulously executed paper by Sun, Roth, and Black (2014).[15] Figure 9.9a shows the main components of the framework they examine, including an initial warping based on the previous level's flow (or a grid search at the coarsest level), followed by energy minimizing flow updates, and then an optional post-processing step. In their paper, the authors not only review dozens of variational (energy-minimization) approaches developed from the 1980s (Horn and Schunck 1981) through to 2013, but also show that algorithmic details such as median filtering post-processing, often glossed over by previous authors, have a strong influence on the results. In addition to performing their analysis on the Middlebury Flow dataset (Baker, Scharstein *et al.* 2011), they also evaluate on the newer Sintel dataset (Butler, Wulff *et al.* 2012).[16]

The field of accurate motion estimation continues to evolve at a rapid pace, with significant advances in performance occurring every year. While the Middlebury optical flow website (Figure 9.8) continues to be a good source of pointers to high-performing algorithms, more recent publications tend to focus (both training and evaluation) on the MPI Sintel dataset developed by Butler, Wulff *et al.* (2012), some samples of which are shown in Figure 9.1e–f. Some algorithms also train and test on the KITTI flow benchmark (Geiger, Lenz, and Urtasun 2012), although that dataset focuses on video acquired from a driving vehicle. In general, it appears that learning-based algorithms trained on one dataset still have trouble when applied to a different dataset.[17]

[13]Some exceptions to this trend of not exploring the full 4D cost volume can be found in Xu, Ranftl, and Koltun (2017) and Teed and Deng (2020b).

[14]https://vision.middlebury.edu/flow

[15]The earlier conference version of this paper had the eye-catching title of "Secrets of optical flow estimation and their principles" (Sun, Roth, and Black 2010).

[16]http://sintel.is.tue.mpg.de

[17]http://www.robustvision.net

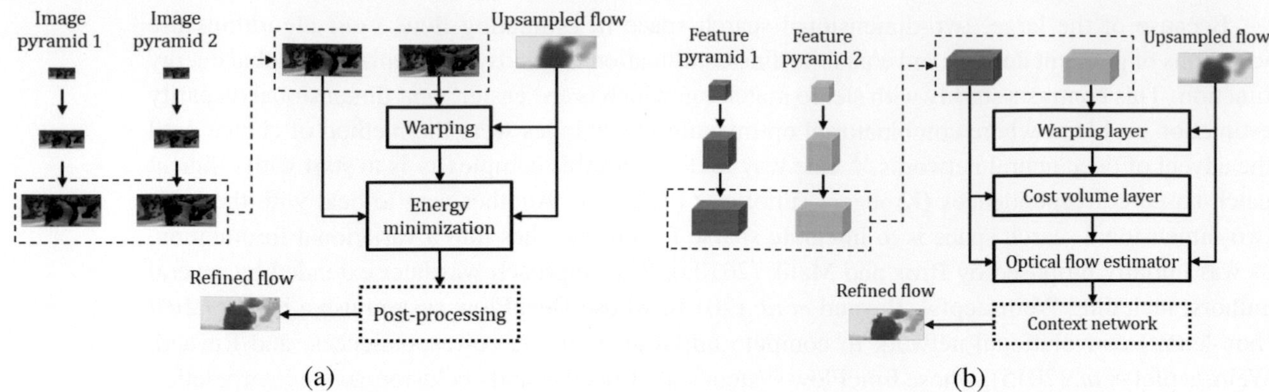

Figure 9.9 Iterative coarse-to-fine optical flow estimation (Sun, Yang *et al.* 2018) © 2018 IEEE: (a) "classic" variational (energy minimization) approach (Sun, Roth, and Black 2014); (b) newer neural network approach trained with end-to-end deep learning (Sun, Yang *et al.* 2018). Both figures show the processing at a *single* level of the coarse-to-fine pyramid, taking as input the flow computed by the previous (coarser) level and passing the refined flow onto the finer level below.

9.3.1 Deep learning approaches

Over the last decade, deep neural networks have become an essential component of all highly-performant optical flow algorithms, as described in the survey articles by Janai, Güney *et al.* (2020, Chapter 11) and Hur and Roth (2020). An early approach to use non-linear aggregation inspired by deep convolutional networks is the DeepFlow system of Weinzaepfel, Revaud *et al.* (2013), which uses a hand-crafted (non-learned) convolutions and pooling to compute multi-level response maps (matching costs), which are then optimized using a classic energy-minimizing variational framework.

The first system to use full deep end-to-end learning in an encoder-decoder network was FlowNetS (Dosovitskiy, Fischer *et al.* 2015), which was trained on the authors' synthetic FlyingChairs dataset. The paper also introduced FlowNetC, which uses a correlation network (local cost volume). The follow-on FlowNet 2.0 system uses the initial flow estimates to warp the images and then refines the flow estimates using cascaded encoder-decoder networks (Ilg, Mayer *et al.* 2017), while subsequent papers also deal with occlusions and uncertainty modeling (Ilg, Saikia *et al.* 2018; Ilg, Çiçek *et al.* 2018).

An alternative to stacking full-resolution networks in series is to use image and flow pyramids together with coarse-to-fine warping and refinement, as first explored in the SPyNet paper by Ranjan and Black (2017). The more recent PWC-Net of Sun, Yang *et al.* (2018, 2019) shown in Figure 9.9b extends this idea by first computing a feature pyramid from each frame, warping the second set of features by the flow interpolated from the previous resolution level, and then computing a cost volume by correlating these features using a dot product between feature maps shifted by up to $d = \pm 4$ pixels. The refined optical flow estimates at the current level are produced using a multi-layer CNN whose inputs are the cost volume, the image features, and the interpolated flow from the previous level. A final context network takes as input the flow estimate and features from the second to last level and uses dilated convolutions to endow the network with a broader context. If you compare Figures 9.9a–b, you will see a pleasing correspondence between the various processing stages of classic and deep coarse-to-fine flow estimation algorithms.[18]

[18]Note that as with other coarse-to-fine warping approaches, these algorithms struggle with fast-moving fine structures that may not be visible at coarser levels (Brox and Malik 2010a).

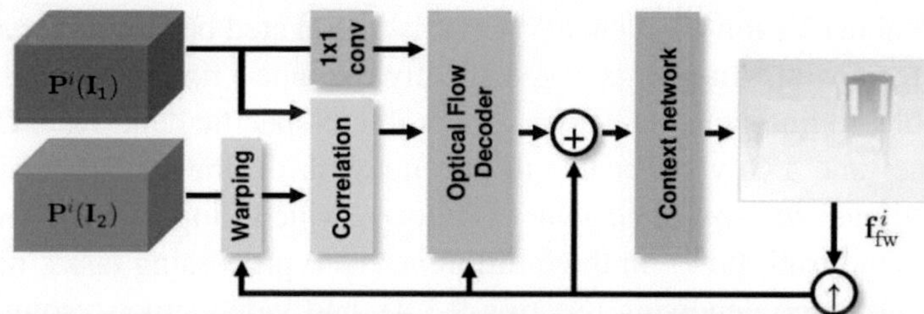

Figure 9.10 Iterative residual refinement optical flow estimation (Hur and Roth 2019) © 2019 IEEE. The coarse-to-fine cascade of Sun, Yang *et al.* (2018) in Figure 9.9b is replaced with a recurrent neural network (RNN) that cycles interpolated coarser level flow estimates as warping inputs to the next finer level but uses the same convolutional weights at each level.

A variant on the coarse-to-fine PWC-Net developed by Hur and Roth (2019) is the Iterative Residual Refinement network shown in Figure 9.10. Instead of cascading a set of different deep networks as in FlowNet 2.0 and PWC-Net, IRRs re-use the same structure and convolution weights at each layer, which allows the network to be re-drawn in the "rolled up" version, as shown in this figure. The network can thus be thought of as a simple recurrent neural network (RNN) that upsamples the output flow estimates after each stage. In addition to having fewer parameters, this weight sharing also improves accuracy. In their paper, the authors also show how this network can be extended (doubled) to simultaneously compute forward and backward flows as well as occlusions.

In more recent work, Jonschkowski, Stone *et al.* (2020) take PWC-Net as their basic architecture and systematically study all of the components involved in training the flow estimator in an *unsupervised* manner, i.e., using regular real-world videos with no ground truth flow, which can enable much larger training sets to be used (Ahmadi and Patras 2016; Meister, Hur, and Roth 2018). In their paper, Jonschkowski *et al.* systematically compare photometric losses, occlusion estimation, self-supervision, and smoothness constraints, and analyze the effect of other choices, such as pre-training, image resolution, data augmentation, and batch size. They also propose four improvements to these key components, including cost volume normalization, gradient stopping for occlusion estimation, applying smoothness at the native flow resolution, and image resizing for self-supervision. Another recent paper that explicitly deals with occlusions is Jiang, Campbell *et al.* (2021).

Another recent trend has been to model the uncertainty that arises in flow field estimation due to homogeneous and occluded regions (Ilg, Çiçek *et al.* 2018). The HD^3 network developed by Yin, Darrell, and Yu (2019) models correspondence distributions across multiple resolution levels, while the LiteFlowNet3 network of Hui and Loy (2020) extends their small and fast LiteFlowNet2 network (Hui, Tang, and Loy 2021) with cost volume modulation and flow field deformation modules to significantly improve accuracy at minimal cost. In concurrent work, Hofinger, Rota Bulò *et al.* (2020) introduce novel components such as replacing warping by sampling, smart gradient blocking, and knowledge distillation, which not only improve the quality of their flow estimates but can also be used in other applications such as stereo matching. Teed and Deng (2020b) build on the idea of a recurrent network (Hur and Roth 2019), but instead of warping feature maps, they precompute a full $(W \times H)^2$ multi-resolution correlation volume (Recurrent All-Pairs Field Transforms or RAFT), which is accessed at each iteration based on the current flow estimates. Computing a sparse correlation volume storing only the k closest matches for each reference image feature can further accelerate the computation (Jiang, Lu *et al.* 2021).

Given the rapid evolution in optical flow techniques, which is the best one to use? The answer is highly problem-dependent. One way to assess this is to look across a number of datasets, as is done in the Robust Vision Challenge.[19] On this aggregated benchmark, variants of RAFT, IRR, and PWC all perform well. Another is to specifically evaluate a flow algorithm based on its indented use, and, if possible, to fine-tune the network on problem-specific data. Xue, Chen *et al.* (2019) describe how they fine-tune a SPyNet coarse-to-fine network on their synthetically degraded Vimeo-90K dataset to estimate *task-oriented flow* (TOFlow), which outperforms "higher accuracy" networks (and even ground truth flow) on three different video processing tasks, namely frame interpolation (Section 9.4.1), video denoising (Section 9.3.4), and video super-resolution. It is also possible to significantly improve the performance of learning-based flow algorithms by tailoring the synthetic training data to a target dataset (Sun, Vlasic *et al.* 2021).

9.3.2 *Application*: Rolling shutter wobble removal

To save on silicon circuitry and enable greater photo sensitivity or fill factors, many CMOS imaging sensors such as those found in mobile phones use a *rolling shutter*, where different rows or columns are exposed in sequence. When photographing or filming a scene with fast scene or camera motions, this can result in straight lines becoming slanted or curved (e.g., the propeller blades on a plane or helicopter) or rigid parts of the scene wobbling (also known as the *jello effect*), e.g., when the camera is rapidly vibrating during action photography.

To compensate for these distortion, which are caused by different exposure times for different scanlines, accurate per-pixel optical flow must be estimated, as opposed to the whole-frame parametric motion that can sometimes be used for slower-motion video stabilization (Section 9.2.1). Baker, Bennett *et al.* (2010) and Forssén and Ringaby (2010) were among the first computer vision researchers to study this problem. In their paper, Baker, Bennett *et al.* (2010) recover a high-frequency motion field from the lower-frequency inter-frame motions and use this to resample each output scanline. Forssén and Ringaby (2010) perform similar computations using models of camera rotation, which require intrinsic lens calibration. Grundmann, Kwatra *et al.* (2012) remove the need for such calibration using mixtures of homographies to model the camera and scene motions, while Liu, Gleicher *et al.* (2011) use subspace constraints. Accurate rolling shutter correction is also required to produce high-quality image stitching results (Zhuang and Tran 2020).

While in some modern imaging systems such as action cameras, inertial measurements units (IMUs) can provide high-frequency estimates of camera motion, they cannot directly provide estimates of depth-dependent parallax and independent object motions. For this reason, the best in-camera image stabilizers use a combination of IMU data and sophisticated image processing.[20] Modeling rolling shutter is also important to obtain accurate pose estimates in structure from motion (Hedborg, Forssén *et al.* 2012; Kukelova, Albl *et al.* 2018; Albl, Kukelova *et al.* 2020; Kukelova, Albl *et al.* 2020) and visual-inertial fusion in SLAM (Patron-Perez, Lovegrove, and Sibley 2015; Schubert, Demmel *et al.* 2018), which are discussed in Sections 11.4.2 and 11.5.

9.3.3 Multi-frame motion estimation

So far, we have looked at motion estimation as a two-frame problem, where the goal is to compute a motion field that aligns pixels from one image with those in another. In practice, motion estimation is usually applied to video, where a whole sequence of frames is available to perform this task.

One classic approach to multi-frame motion is to *filter* the spatio-temporal volume using oriented or steerable filters (Heeger 1988), in a manner analogous to oriented edge detection (Section 3.2.3). Figure 9.11 shows two frames from the commonly used *flower garden* sequence, as well

[19] http://www.robustvision.net/leaderboard.php?benchmark=flow

[20] https://gopro.com/en/us/news/hero7-black-hypersmooth-technology

Figure 9.11 Slice through a spatio-temporal volume (Szeliski 1999a) © 1999 IEEE: (a–b) two frames from the *flower garden* sequence; (c) a horizontal slice through the complete spatio-temporal volume, with the arrows indicating locations of potential key frames where flow is estimated. Note that the colors for the flower garden sequence are incorrect; the correct colors (yellow flowers) are shown in Figure 9.13.

as a horizontal slice through the spatio-temporal volume, i.e., the 3D volume created by stacking all of the video frames together. Because the pixel motion is mostly horizontal, the slopes of individual (textured) pixel tracks, which correspond to their horizontal velocities, can clearly be seen. Spatio-temporal filtering uses a 3D volume around each pixel to determine the best orientation in space–time, which corresponds directly to a pixel's velocity.

Unfortunately, to obtain reasonably accurate velocity estimates everywhere in an image, spatio-temporal filters have moderately large extents, which severely degrades the quality of their estimates near motion discontinuities. (This same problem is endemic in 2D window-based motion estimators.) An alternative to full spatio-temporal filtering is to estimate more local spatio-temporal derivatives and use them inside a global optimization framework to fill in textureless regions (Bruhn, Weickert, and Schnörr 2005; Govindu 2006).

Another alternative is to simultaneously estimate multiple motion estimates, while also optionally reasoning about occlusion relationships (Szeliski 1999a). Figure 9.11c shows schematically one potential approach to this problem. The horizontal arrows show the locations of keyframes s where motion is estimated, while other slices indicate video frames t whose colors are matched with those predicted by interpolating between the keyframes. Motion estimation can be cast as a global energy minimization problem that simultaneously minimizes brightness compatibility and flow compatibility terms between keyframes and other frames, in addition to using robust smoothness terms.

The multi-view framework is potentially even more appropriate for rigid scene motion (multi-view stereo) (Section 12.7), where the unknowns at each pixel are disparities and occlusion relationships can be determined directly from pixel depths (Szeliski 1999a; Kolmogorov and Zabih 2002). However, it is also applicable to general motion, with the addition of models for occlusion relationships, as in the MirrorFlow system of Hur and Roth (2017) as well as multi-frame versions (Janai, Guney *et al.* 2018; Neoral, Šochman, and Matas 2018; Ren, Gallo *et al.* 2019).

9.3.4 *Application*: Video denoising

Video denoising is the process of removing noise and other artifacts such as scratches from film and video (Kokaram 2004; Gai and Kang 2009; Liu and Freeman 2010). Unlike single image denoising, where the only information available is in the current picture, video denoisers can average or borrow information from adjacent frames. However, to do this without introducing blur or jitter (irregular motion), they need accurate per-pixel motion estimates. One way to do this is to use task-oriented flow, where the flow network is specifically tuned end-to-end to provide the best denoising performance (Xue, Chen *et al.* 2019).

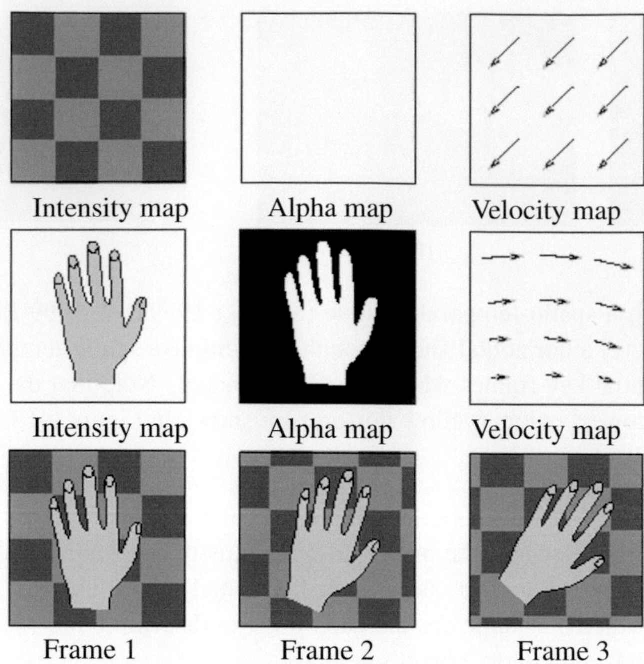

Figure 9.12 Layered motion estimation framework (Wang and Adelson 1994) © 1994 IEEE: The top two rows describe the two layers, each of which consists of an intensity (color) image, an alpha mask (black=transparent), and a parametric motion field. The layers are composited with different amounts of motion to recreate the video sequence.

Exercise 9.6 lists some of the steps required, which include the ability to determine if the current motion estimate is accurate enough to permit averaging with other frames. And while some recent papers continue to estimate flow as part of the multi-frame denoising pipeline (Tassano, Delon, and Veit 2019; Xue, Chen *et al.* 2019), others either concatenate similar patches from different frames (Maggioni, Boracchi *et al.* 2012) or concatenate small subsets of frames into a deep network that never explicitly estimates a motion representation (Claus and van Gemert 2019; Tassano, Delon, and Veit 2020). A more general form of video enhancement and restoration called *video quality mapping* has also recently started being investigated (Fuoli, Huang *et al.* 2020).

9.4 Layered motion

In many situations, visual motion is caused by the movement of a small number of objects at different depths in the scene. In such situations, the pixel motions can be described more succinctly (and estimated more reliably) if pixels are grouped into appropriate objects or *layers* (Wang and Adelson 1994).

Figure 9.12 shows this approach schematically. The motion in this sequence is caused by the translational motion of the checkered background and the rotation of the foreground hand. The complete motion sequence can be reconstructed from the appearance of the foreground and background elements, which can be represented as alpha-matted images (*sprites* or *video objects*) and the parametric motion corresponding to each layer. Displacing and compositing these layers in back to front order (Section 3.1.3) recreates the original video sequence.

Layered motion representations not only lead to compact representations (Wang and Adelson

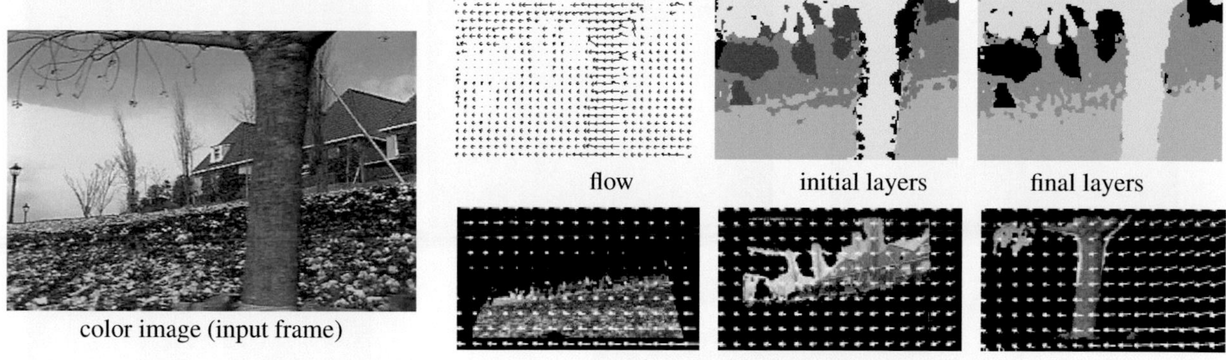

Figure 9.13 Layered motion estimation results (Wang and Adelson 1994) © 1994 IEEE.

1994; Lee, Chen *et al.* 1997), but they also exploit the information available in multiple video frames, as well as accurately modeling the appearance of pixels near motion discontinuities. This makes them particularly suited as a representation for image-based rendering (Section 14.2.1) (Shade, Gortler *et al.* 1998; Zitnick, Kang *et al.* 2004) as well as object-level video editing.

To compute a layered representation of a video sequence, Wang and Adelson (1994) first estimate affine motion models over a collection of non-overlapping patches and then cluster these estimates using k-means. They then alternate between assigning pixels to layers and recomputing motion estimates for each layer using the assigned pixels, using a technique first proposed by Darrell and Pentland (1991). Once the parametric motions and pixel-wise layer assignments have been computed for each frame independently, layers are constructed by warping and merging the various layer pieces from all of the frames together. Median filtering is used to produce sharp composite layers that are robust to small intensity variations, as well as to infer occlusion relationships between the layers. Figure 9.13 shows the results of this process on the *flower garden* sequence. You can see both the initial and final layer assignments for one of the frames, as well as the composite flow and the alpha-matted layers with their corresponding flow vectors overlaid.

In follow-on work, Weiss and Adelson (1996) use a formal probabilistic mixture model to infer both the optimal number of layers and the per-pixel layer assignments. Weiss (1997) further generalizes this approach by replacing the per-layer affine motion models with smooth regularized per-pixel motion estimates, which allows the system to better handle curved and undulating layers, such as those seen in most real-world sequences.

The above approaches, however, still make a distinction between estimating the motions and layer assignments and then later estimating the layer colors. In the system described by Baker, Szeliski, and Anandan (1998), the generative model is generalized to account for real-world rigid motion scenes. The motion of each frame is described using a 3D camera model and the motion of each layer is described using a 3D plane equation plus per-pixel residual depth offsets (the *plane plus parallax* representation (Section 2.1.4)). The initial layer estimation proceeds in a manner similar to that of Wang and Adelson (1994), except that rigid planar motions (homographies) are used instead of affine motion models. The final model refinement, however, jointly re-optimizes the layer pixel color and opacity values L_l and the 3D depth, plane, and motion parameters z_l, $\mathbf{n}_l$, and $\mathbf{P}_t$ by minimizing the discrepancy between the re-synthesized and observed motion sequences (Baker, Szeliski, and Anandan 1998).

Figure 9.14 shows the final results obtained with this algorithm. As you can see, the motion

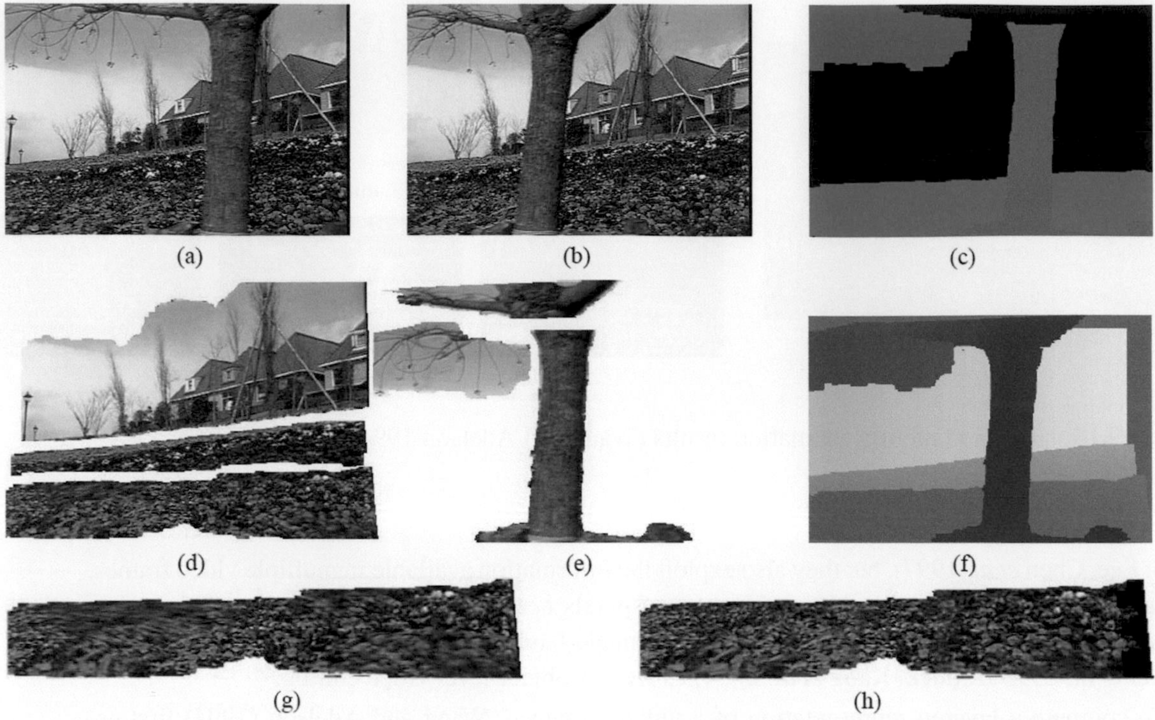

Figure 9.14 Layered stereo reconstruction (Baker, Szeliski, and Anandan 1998) © 1998 IEEE: (a) first and (b) last input images; (c) initial segmentation into six layers; (d) and (e) the six layer sprites; (f) depth map for planar sprites (darker denotes closer); front layer (g) before and (h) after residual depth estimation. Note that the colors for the flower garden sequence are incorrect; the correct colors (yellow flowers) are shown in Figure 9.13.

boundaries and layer assignments are much crisper than those in Figure 9.13. Because of the per-pixel depth offsets, the individual layer color values are also sharper than those obtained with affine or planar motion models. While the original system of Baker, Szeliski, and Anandan (1998) required a rough initial assignment of pixels to layers, Torr, Szeliski, and Anandan (2001) describe automated Bayesian techniques for initializing this system and determining the optimal number of layers.

Layered motion estimation continues to be an active area of research. Representative papers from the 2000s include (Sawhney and Ayer 1996; Jojic and Frey 2001; Xiao and Shah 2005; Kumar, Torr, and Zisserman 2008; Thayananthan, Iwasaki, and Cipolla 2008; Schoenemann and Cremers 2008), while more recent papers include (Sun, Sudderth, and Black 2012; Sun, Wulff *et al.* 2013; Sun, Liu, and Pfister 2014; Wulff and Black 2015) and (Sevilla-Lara, Sun *et al.* 2016), which jointly performs semantic segmentation and motion estimation.

Layers are not the only way to introduce segmentation into motion estimation. A large number of algorithms have been developed that alternate between estimating optical flow vectors and segmenting them into coherent regions (Black and Jepson 1996; Ju, Black, and Jepson 1996; Chang, Tekalp, and Sezan 1997; Mémin and Pérez 2002; Cremers and Soatto 2005). Some of these techniques rely on first segmenting the input color images and then estimating per-segment motions that produce a coherent motion field while also modeling occlusions (Zitnick, Kang *et al.* 2004; Zitnick, Jojic, and Kang 2005; Stein, Hoiem, and Hebert 2007; Thayananthan, Iwasaki, and Cipolla 2008). In fact, the segmentation of videos into coherently moving parts has evolved into its own topic, namely *video object segmentation*, which we study in Section 9.4.3.

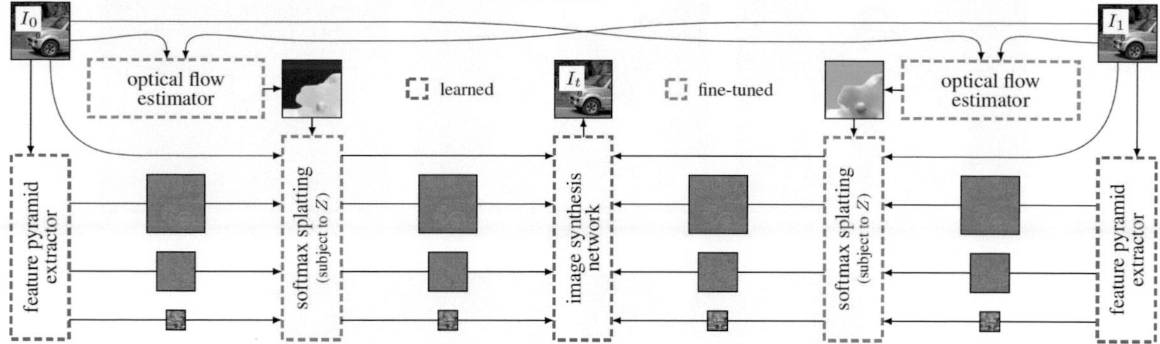

Figure 9.15 Deep feature video interpolation network (Niklaus and Liu 2020) © 2020 IEEE. This multi-stage network first computes bi directional flow, encodes each frame using feature pyramids, and then warps and com bines these features using softmax splatting. The combined features are then fed into a final image synthesis network (decoder).

9.4.1 *Application*: Frame interpolation

Frame interpolation is a widely used application of motion estimation, often implemented in hardware to match an incoming video to a monitor's actual refresh rate, where information in novel in-between frames needs to be interpolated from preceding and subsequent frames. The best results can be obtained if an accurate motion estimate can be computed at each unknown pixel's location. However, in addition to computing the motion, occlusion information is critical to prevent colors from being contaminated by moving foreground objects that might obscure a particular pixel in a preceding or subsequent frame.

In a little more detail, consider Figure 9.11c and assume that the arrows denote keyframes between which we wish to interpolate additional images. The orientations of the streaks in this figure encode the velocities of individual pixels. If the same motion estimate $\mathbf{u}_0$ is obtained at location $\mathbf{x}_0$ in image I_0 as is obtained at location $\mathbf{x}_0 + \mathbf{u}_0$ in image I_1, the flow vectors are said to be *consistent*. This motion estimate can be transferred to location $\mathbf{x}_0 + t\mathbf{u}_0$ in the image I_t being generated, where $t \in (0, 1)$ is the time of interpolation. The final color value at pixel $\mathbf{x}_0 + t\mathbf{u}_0$ can be computed as a linear blend,

$$I_t(\mathbf{x}_0 + t\mathbf{u}_0) = (1 - t)I_0(\mathbf{x}_0) + tI_1(\mathbf{x}_0 + \mathbf{u}_0). \tag{9.59}$$

If, however, the motion vectors are different at corresponding locations, some method must be used to determine which is correct and which image contains colors that are occluded. The actual reasoning is even more subtle than this. One example of such an interpolation algorithm, based on earlier work in depth map interpolation by Shade, Gortler *et al.* (1998) and Zitnick, Kang *et al.* (2004), is the one used in the flow evaluation paper of Baker, Scharstein *et al.* (2011). An even higher-quality frame interpolation algorithm, which uses gradient-based reconstruction, is presented by Mahajan, Huang *et al.* (2009). Accuracy on frame interpolation tasks is also sometimes used to gauge the quality of motion estimation algorithms (Szeliski 1999b; Baker, Scharstein *et al.* 2011).

More recent frame interpolation techniques use deep neural networks as part of their architectures. Some approaches use spatio-temporal convolutions (Niklaus, Mai, and Liu 2017), while others use DNNs to compute bi-directional optical flow (Xue, Chen *et al.* 2019) and then combine the contributions from the two original frames using either context features (Niklaus and Liu 2018) or soft visibility maps (Jiang, Sun *et al.* 2018). The system by Niklaus and Liu (2020) encodes the input frames as deep multi-resolution neural features, forward warps these using bi-directional

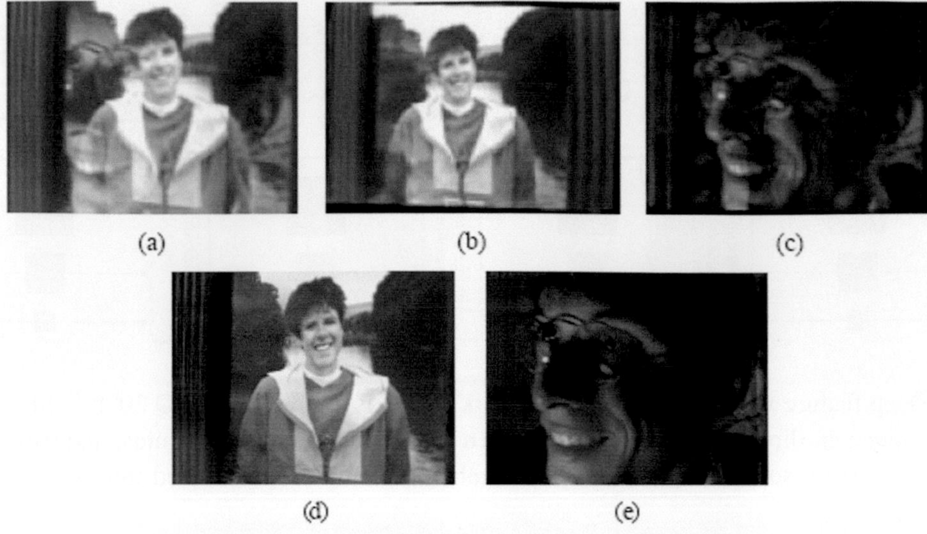

Figure 9.16 Light reflecting off the transparent glass of a picture frame: (a) first image from the input sequence; (b) dominant motion layer *min-composite*; (c) secondary motion residual layer *max-composite*; (d–e) final estimated picture and reflection layers The original images are from Black and Anandan (1996), while the separated layers are from Szeliski, Avidan, and Anandan (2000) © 2000 IEEE.

flow, combines these features using softmax splatting, and then uses a final deep network to decode these combined features, as shown in Figure 9.15. A similar architecture can also be used to create temporally textured looping videos from a single still image (Holynski, Curless *et al.* 2021). Other recently developed frame interpolation networks include Choi, Choi *et al.* (2020), Lee, Kim *et al.* (2020), Kang, Jo *et al.* (2020), and Park, Ko *et al.* (2020).

9.4.2 Transparent layers and reflections

A special case of layered motion that occurs quite often is transparent motion, which is usually caused by reflections seen in windows and picture frames (Figures 9.16 and 9.17).

Some of the early work in this area handles transparent motion by either just estimating the component motions (Shizawa and Mase 1991; Bergen, Burt *et al.* 1992; Darrell and Simoncelli 1993; Irani, Rousso, and Peleg 1994) or by assigning individual pixels to competing motion layers (Darrell and Pentland 1995; Black and Anandan 1996; Ju, Black, and Jepson 1996), which is appropriate for scenes partially seen through a fine occluder (e.g., foliage). However, to accurately separate truly transparent layers, a better model for motion due to reflections is required. Because of the way that light is both reflected from and transmitted through a glass surface, the correct model for reflections is an *additive* one, where each moving layer contributes some intensity to the final image (Szeliski, Avidan, and Anandan 2000).

If the motions of the individual layers are known, the recovery of the individual layers is a simple constrained least squares problem, with the individual layer images are constrained to be positive and saturated pixels provide an inequality constraint on the summed values. However, this problem can suffer from extended low-frequency ambiguities, especially if either of the layers lacks dark (black) pixels or the motion is uni-directional. In their paper, Szeliski, Avidan, and Anandan (2000) show that the simultaneous estimation of the motions and layer values can be obtained by alternating between robustly computing the motion layers and then making conservative (upper- or

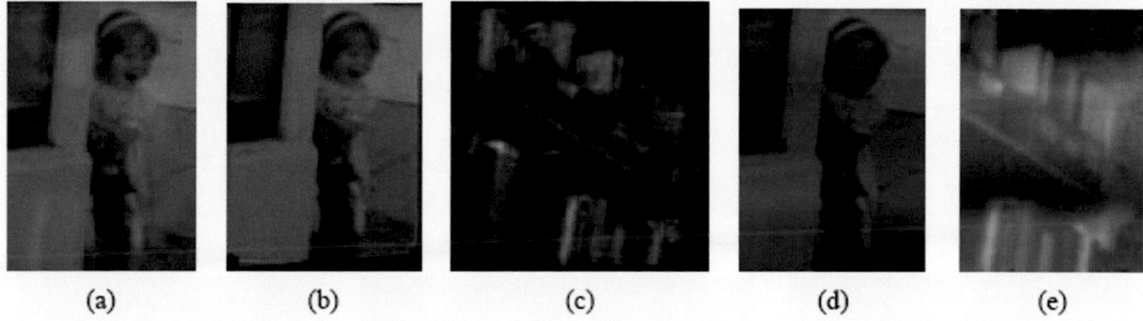

Figure 9.17 Transparent motion separation (Szeliski, Avidan, and Anandan 2000) © 2000 IEEE: (a) first image from input sequence; (b) dominant motion layer *min-composite*; (c) secondary motion residual layer *max-composite*; (d–e) final estimated picture and reflection layers. Note that the reflected layers in (c) and (e) are doubled in intensity to better show their structure.

lower-bound) estimates of the layer intensities. The final motion and layer estimates can then be polished using gradient descent on a joint constrained least squares formulation similar to Baker, Szeliski, and Anandan (1998), where the *over* compositing operator is replaced with addition.

Figures 9.16 and 9.17 show the results of applying these techniques to two different picture frames with reflections. Notice how, in the second sequence, the amount of reflected light is quite low compared to the transmitted light (the picture of the girl) and yet the algorithm is still able to recover both layers.

Unfortunately, the simple parametric motion models used in Szeliski, Avidan, and Anandan (2000) are only valid for planar reflectors and scenes with shallow depth. The extension of these techniques to curved reflectors and scenes with significant depth has also been studied (Swaminathan, Kang *et al.* 2002; Criminisi, Kang *et al.* 2005; Jacquet, Hane *et al.* 2013), as has the extension to scenes with more complex 3D depth (Tsin, Kang, and Szeliski 2006). While motion sequences used to evaluate optical flow techniques have also started to include reflection and transparency (Baker, Scharstein *et al.* 2011; Butler, Wulff *et al.* 2012), the ground truth flow estimates they provide and use for evaluation only include the dominant motion at each pixel, e.g., ignoring mist and reflections.

In more recent work, Sinha, Kopf *et al.* (2012) model 3D scenes with reflections captured from a moving camera using two layers with varying depth and reflectivity and then use these to produce image-based renderings (novel view synthesis), which we discuss in more detail in Section 14.2.1. Kopf, Langguth *et al.* (2013) extend the modeling and rendering component of this system to recover colored image *gradients* for each layer and then use gradient-domain rendering to reconstruct the novel views. Xue, Rubinstein *et al.* (2015) extend these models with a gradient sparsity prior to enable *obstruction-free photography* when looking through windows and fences. More recent papers on this topic include Yang, Li *et al.* (2016), Nandoriya, Elgharib *et al.* (2017), and Liu, Lai *et al.* (2020a). The advent of dual-pixel imaging sensors, originally designed to provide fast focusing, can also be used to remove reflections by separating gradients into different depth planes (Punnappurath and Brown 2019).

While all of these techniques are useful for separating or eliminating reflections that appear as coherent images, more complex 3D geometries often give rise to spatially distributed *specularities* (Section 2.2.2) that are not amenable to layer-based representation. In such cases, lightfield representations such as surface lightfields (Section 14.3.2 and Figure 14.13) and neural light fields (Section 14.6 and Figure 14.24b) may be more appropriate.

(a)

(b)

Figure 9.18 Sample sequences from the Densely Annotated VIdeo Segmentation (DAVIS) datasets © Pont-Tuset, Perazzi *et al.* (2017). The DAVIS 2016 dataset (a) only contains foreground-background segmentations (red regions), while the DAVIS 2017 dataset (b) contains multiple annotated objects in each sequence (brightly colored regions).

9.4.3 Video object segmentation

As we have seen throughout this chapter, the accurate estimation of motion usually requires the segmentation of a video into coherently moving regions or objects as well as the correct modeling of occlusions. Segmenting a video clip into coherent objects is the temporal analog to still image segmentation, which we studied in Section 7.5. In addition to providing more accurate motion estimates, video object segmentation supports a variety of editing tasks, such as object removal and insertion (Section 10.4.5) as well as video understanding and interpretation.

While the segmentation of foreground and background layers has been studied for a long time (Bergen, Anandan *et al.* 1992; Wang and Adelson 1994; Gorelick, Blank *et al.* 2007; Lee and Grauman 2010; Brox and Malik 2010b; Lee, Kim, and Grauman 2011; Fragkiadaki, Zhang, and Shi 2012; Papazoglou and Ferrari 2013; Wang, Shen, and Porikli 2015; Perazzi, Wang *et al.* 2015), the introduction of DAVIS (Densely Annotated VIdeo Segmentation) by Perazzi, Pont-Tuset *et al.* (2016) greatly accelerated research in this area. Figure 9.18a shows some frames from the original DAVIS 2016 dataset, where the first frame is annotated with a foreground pixel mask (shown in red) and the task is to estimate foreground masks for the remaining frames. The DAVIS 2017 dataset (Pont-Tuset, Perazzi *et al.* 2017) increased the number of video clips from 50 to 150, added more challenging elements such as motion blur and foreground occlusions, and most importantly, added more than one annotated object per sequence (Figure 9.18b).

Algorithm for video object segmentation such as OSVOS (Caelles, Maninis *et al.* 2017), FusionSeg (Jain, Xiong, and Grauman 2017), MaskTrack (Perazzi, Khoreva *et al.* 2017), and SegFlow (Cheng, Tsai *et al.* 2017), usually consist of a deep per-frame segmentation network as well as a motion estimation algorithm, which is used to link and refine the segmentations. Some approaches (Caelles, Maninis *et al.* 2017; Khoreva, Benenson *et al.* 2019) also fine-tune the segmentation networks based on the first frame annotations. More recent approaches have focused on increasing the computational efficiency of the pipelines (Chen, Pont-Tuset *et al.* 2018; Cheng, Tsai *et al.* 2018;

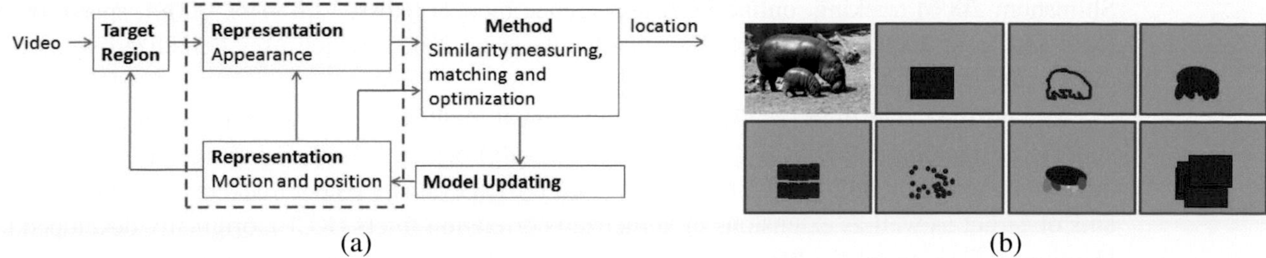

Figure 9.19 Visual object tracking (Smeulders, Chu *et al.* 2014) ©2014 IEEE: (a) high-level model showing main tracker components; (b) some tracked region representations, including a single bounding box, contour, blob, patch-based, sparse features, parts, and multiple bounding boxes.

Wug Oh, Lee *et al.* 2018; Wang, Zhang *et al.* 2019; Meinhardt and Leal-Taixé 2020).

Since 2017, an annual challenge and workshop on the DAVIS dataset have been held in conjunction with CVPR. More recent additions to the challenges have been segmentation with weaker annotations/scribbles (Caelles, Montes *et al.* 2018) or completely unsupervised segmentation, where the algorithms compute temporally linked segmentations of the video frames (Caelles, Pont-Tuset *et al.* 2019). There is also a newer, larger, dataset called YouTube-VOS (Xu, Yang *et al.* 2018) with its own associated set of challenges and leaderboards. The number of papers published on the topic continues to be high. The best sources for recent work are the challenge leaderboards at https://davischallenge.org and https://youtube-vos.org, which are accompanied by short papers describing the techniques, as well as the large number of conference papers, which usually have "Video Object Segmentation" in their titles.

9.4.4 Video object tracking

One of the most widely used applications of computer vision to video analysis is *video object tracking*. These applications include surveillance (Benfold and Reid 2011), animal and cell tracking (Khan, Balch, and Dellaert 2005), sports player tracking (Lu, Ting *et al.* 2013), and automotive safety (Janai, Güney *et al.* 2020, Chapter 6).

We have already discussed simpler examples of tracking in previous chapters, including feature (patch) tracking in Section 7.1.5 and contour tracking in Section 7.3. Surveys and experimental evaluation of such techniques include Lepetit and Fua (2005), Yilmaz, Javed, and Shah (2006), Wu, Lim, and Yang (2013), and Janai, Güney *et al.* (2020, Chapter 6).

A great starting point for learning more about tracking is the survey and tutorial by Smeulders, Chu *et al.* (2014), which was also one of the first large-scale tracking datasets, with over 300 video clips, ranging from a few seconds to a few minutes. Figure 9.19a shows some of the main components usually present in an online tracking system, which include choosing representations for shape, motion, position, and appearance, as well as similarity measures, optimization, and optional model updating. Figure 9.19b shows some of the choices for representing shapes and appearance, including a single bounding box, contours, patches, features, and parts.

The paper includes a discussion of previous surveys and techniques, as well as datasets, evaluation measures, and the above-mentioned model choices. It then categorizes a selection of well-known and more recent algorithms into a taxonomy that includes simple matching with fixed templates, extended and constrained (sparse) appearance models, discriminative classifiers, and tracking by detection. The algorithms discussed and evaluated include KLT, as implemented by Baker and Matthews (2004), mean-shift (Comaniciu and Meer 2002) and fragments-based (Adam, Rivlin, and

Shimshoni 2006) tracking, online PCA appearance models (Ross, Lim *et al.* 2008), sparse bases (Mei and Ling 2009), and Struct (Hare, Golodetz *et al.* 2015), which uses kernelized structured output support vector machine.

Around the same time (2013), a series of annual challenges and workshops on single-target short-term tracking called VOT (visual object tracking) began.[21] In their journal paper describing the evaluation methodology, Kristan, Matas *et al.* (2016) evaluate recent trackers and find that variants of Struct as well as extensions of kernelized correlation filters (KCF), originally developed by Henriques, Caseiro *et al.* (2014), performed the best. Other highly influential papers from this era include (Bertinetto, Valmadre *et al.* 2016a,b; Danelljan, Robinson *et al.* 2016). Since that time, deep networks have played an essential role in visual object tracking, often using Siamese networks (Section 5.3.4; Bromley, Guyon *et al.* 1994; Chopra, Hadsell, and LeCun 2005) to map regions being tracked into neural embeddings. Lists and descriptions of more recent tracking algorithms can be found in the annual reports that accompany the VOT challenges and workshops, the most recent of which is Kristan, Leonardis *et al.* (2020).

In parallel with the single-object VOT challenges and workshops, a multiple object tracking was introduced as part of the KITTI vision benchmark (Geiger, Lenz, and Urtasun 2012) and a separate benchmark was developed by Leal-Taixé, Milan *et al.* (2015) along with a series of challenges, with the most recent results described in Dendorfer, Ošep *et al.* (2021).[22] A survey of multiple object tracking papers through 2016 can be found in Luo, Xing *et al.* (2021). Simple and fast multiple object trackers include Bergmann, Meinhardt, and Leal-Taixé (2019) and Zhou, Koltun, and Krähenbühl (2020). Until recently, however, tracking datasets have focused mostly on people, vehicles, and animals. To expand the range of objects that can be tracked, Dave, Khurana *et al.* (2020) created the TAO (tracking any object) dataset, consisting of 2,907 videos, which were annotated "bottom-up" by first having users tag anything that moves and then classifying such objects into 833 categories.

While in this section, we have focused mostly on object tracking, the primary goal of which is to locate an object in contiguous video frames, it is also possible to simultaneously track and segment (Voigtlaender, Krause *et al.* 2019; Wang, Zhang *et al.* 2019) or to track non-rigidly deforming objects such as T-shirts with deformable models from either video (Kambhamettu, Goldgof *et al.* 2003; White, Crane, and Forsyth 2007; Pilet, Lepetit, and Fua 2008; Furukawa and Ponce 2008; Salzmann and Fua 2010) or RGB-D streams (Božič, Zollhöfer *et al.* 2020; Božič, Palafox *et al.* 2020, 2021). The recent TrackFormer paper by Meinhardt, Kirillov *et al.* (2021) includes a nice review of recent work of multi-object tracking and segmentation.

9.5 Additional reading

Some of the earliest algorithms for motion estimation were developed for motion-compensated video coding (Netravali and Robbins 1979) and such techniques continue to be used in modern coding standards such as MPEG, H.263, and H.264 (Le Gall 1991; Richardson 2003).[23] In computer vision, this field was originally called *image sequence analysis* (Huang 1981). Some of the early seminal papers include the variational approaches developed by Horn and Schunck (1981) and Nagel and Enkelmann (1986), and the patch-based translational alignment technique developed by Lucas and Kanade (1981). Hierarchical (coarse-to-fine) versions of such algorithms were developed by Quam

[21] https://www.votchallenge.net

[22] https://motchallenge.net

[23] https://www.itu.int/rec/T-REC-H.264.

(1984), Anandan (1989), and Bergen, Anandan *et al.* (1992), although they have also long been used in motion estimation for video coding.

Translational motion models were generalized to affine motion by Rehg and Witkin (1991), Fuh and Maragos (1991), and Bergen, Anandan *et al.* (1992) and to quadric reference surfaces by Shashua and Toelg (1997) and Shashua and Wexler (2001)—see Baker and Matthews (2004) for a nice review. Such parametric motion estimation algorithms have found widespread application in video summarization (Teodosio and Bender 1993; Irani and Anandan 1998), video stabilization (Hansen, Anandan *et al.* 1994; Srinivasan, Chellappa *et al.* 2005; Matsushita, Ofek *et al.* 2006), and video compression (Irani, Hsu, and Anandan 1995; Lee, Chen *et al.* 1997). Surveys of parametric image registration include those by Brown (1992), Zitov'aa and Flusser (2003), Goshtasby (2005), and Szeliski (2006a).

Good general surveys and comparisons of optical flow algorithms include those by Aggarwal and Nandhakumar (1988), Barron, Fleet, and Beauchemin (1994), Otte and Nagel (1994), Mitiche and Bouthemy (1996), Stiller and Konrad (1999), McCane, Novins *et al.* (2001), Szeliski (2006a), and Baker, Scharstein *et al.* (2011), Sun, Yang *et al.* (2018), Janai, Güney *et al.* (2020), and Hur and Roth (2020). The topic of matching primitives, i.e., pre-transforming images using filtering or other techniques before matching, is treated in a number of papers (Anandan 1989; Bergen, Anandan *et al.* 1992; Scharstein 1994; Zabih and Woodfill 1994; Cox, Roy, and Hingorani 1995; Viola and Wells III 1997; Negahdaripour 1998; Kim, Kolmogorov, and Zabih 2003; Jia and Tang 2003; Papenberg, Bruhn *et al.* 2006; Seitz and Baker 2009). Hirschmüller and Scharstein (2009) compare a number of these approaches and report on their relative performance in scenes with exposure differences.

The publication of the first large benchmark for evaluating optical flow algorithms by Baker, Scharstein *et al.* (2011) led to rapid advances in the quality of estimation algorithms. While most of the best performing algorithms used robust data and smoothness norms such as L_1 or TV and continuous variational optimization techniques, some algorithms used discrete optimization or segmentation (Papenberg, Bruhn *et al.* 2006; Trobin, Pock *et al.* 2008; Xu, Chen, and Jia 2008; Lempitsky, Roth, and Rother 2008; Werlberger, Trobin *et al.* 2009; Lei and Yang 2009; Wedel, Cremers *et al.* 2009).

The creation of the Sintel (Butler, Wulff *et al.* 2012) and KITTI (Geiger, Lenz, and Urtasun 2012) datasets further accelerated progress in optical flow algorithms. Significant papers from this past decade include Weinzaepfel, Revaud *et al.* (2013), Sun, Roth, and Black (2014), Revaud, Weinzaepfel *et al.* (2015), Ilg, Mayer *et al.* (2017), Xu, Ranftl, and Koltun (2017), Sun, Yang *et al.* (2018, 2019), Hur and Roth (2019), and Teed and Deng (2020b). Good review of flow papers from the last decade can be found in Sun, Yang *et al.* (2018), Janai, Güney *et al.* (2020), and Hur and Roth (2020).

Good starting places to read about video object segmentation and video object tracking are recent workshops associated with the main datasets and challenges on these topics (Pont-Tuset, Perazzi *et al.* 2017; Xu, Yang *et al.* 2018; Kristan, Leonardis *et al.* 2020; Dave, Khurana *et al.* 2020; Dendorfer, Ošep *et al.* 2021).

9.6 Exercises

Ex 9.1: Correlation. Implement and compare the performance of the following correlation algorithms:

- sum of squared differences (9.1)
- sum of robust differences (9.2)

- sum of absolute differences (9.3)
- bias–gain compensated squared differences (9.9)
- normalized cross-correlation (9.11)
- windowed versions of the above (9.22–9.23)
- Fourier-based implementations of the above measures (9.18–9.20)
- phase correlation (9.24)
- gradient cross-correlation (Argyriou and Vlachos 2003).

Compare a few of your algorithms on different motion sequences with different amounts of noise, exposure variation, occlusion, and frequency variations (e.g., high-frequency textures, such as sand or cloth, and low-frequency images, such as clouds or motion-blurred video). Some datasets with illumination variation and ground truth correspondences (horizontal motion) can be found at https://vision.middlebury.edu/stereo/data (the 2005 and 2006 datasets).

Some additional ideas, variants, and questions:

1. When do you think that phase correlation will outperform regular correlation or SSD? Can you show this experimentally or justify it analytically?
2. For the Fourier-based masked or windowed correlation and sum of squared differences, the results should be the same as the direct implementations. Note that you will have to expand (9.5) into a sum of pairwise correlations, just as in (9.22). (This is part of the exercise.)
3. For the bias–gain corrected variant of squared differences (9.9), you will also have to expand the terms to end up with a 3×3 (least squares) system of equations. If implementing the Fast Fourier Transform version, you will need to figure out how all of these entries can be evaluated in the Fourier domain.
4. (Optional) Implement some of the additional techniques studied by Hirschmüller and Scharstein (2009) and see if your results agree with theirs.

Ex 9.2: Affine registration. Implement a coarse-to-fine direct method for affine and projective image alignment.

1. Does it help to use lower-order (simpler) models at coarser levels of the pyramid (Bergen, Anandan *et al.* 1992)?
2. (Optional) Implement patch-based acceleration (Shum and Szeliski 2000; Baker and Matthews 2004).
3. See the Baker and Matthews (2004) survey for more comparisons and ideas.

Ex 9.3: Stabilization. Write a program to stabilize an input video sequence. You could implement the following steps, as described in Section 9.2.1:

1. Compute the translation (and, optionally, rotation) between successive frames with robust outlier rejection.
2. Perform temporal high-pass filtering on the motion parameters to remove the low-frequency component (smooth the motion).

3. Compensate for the high-frequency motion, zooming in slightly (a user-specified amount) to avoid missing edge pixels.

4. (Optional) Do not zoom in, but instead borrow pixels from previous or subsequent frames to fill in.

5. (Optional) Compensate for images that are blurry because of fast motion by "stealing" higher frequencies from adjacent frames.

Ex 9.4: Optical flow. Compute optical flow (spline-based or per-pixel) between two images, using one or more of the techniques described in this chapter.

1. Test your algorithms on the motion sequences available at https://vision.middlebury.edu/flow or http://sintel.is.tue.mpg.de and compare your results (visually) to those available on these websites. If you think your algorithm is competitive with the best, consider submitting it for formal evaluation.

2. Visualize the quality of your results by generating in-between images using frame interpolation (Exercise 9.5).

3. What can you say about the relative efficiency (speed) of your approach?

Ex 9.5: Automated morphing and frame interpolation. Write a program to automatically morph between pairs of images. Implement the following steps, as sketched out in Section 9.4.1 and by Baker, Scharstein *et al.* (2011):

1. Compute the flow both ways (previous exercise). Consider using a multi-frame ($n > 2$) technique to better deal with occluded regions.

2. For each intermediate (morphed) image, compute a set of flow vectors and which images should be used in the final composition.

3. Blend (cross-dissolve) the images and view with a sequence viewer.

Try this out on images of your friends and colleagues and see what kinds of morphs you get. Alternatively, take a video sequence and do a high-quality slow-motion effect. Compare your algorithm with simple cross-fading.

Ex 9.6: Video denoising. Implement the algorithm sketched in Application 9.3.4. Your algorithm should contain the following steps:

1. Compute accurate per-pixel flow.

2. Determine which pixels in the reference image have good matches with other frames.

3. Either average all of the matched pixels or choose the sharpest image, if trying to compensate for blur. Don't forget to use regular single-frame denoising techniques as part of your solution, (see Section 3.4.2 and Exercise 3.12).

4. Devise a fall-back strategy for areas where you don't think the flow estimates are accurate enough.

Ex 9.7: Layered motion estimation. Decompose into separate layers (Section 9.4) a video sequence of a scene taken with a moving camera:

1. Find the set of dominant (affine or planar perspective) motions, either by computing them in blocks or by finding a robust estimate and then iteratively re-fitting outliers.
2. Determine which pixels go with each motion.
3. Construct the layers by blending pixels from different frames.
4. (Optional) Add per-pixel residual flows or depths.
5. (Optional) Refine your estimates using an iterative global optimization technique.
6. (Optional) Write an interactive renderer to generate in-between frames or view the scene from different viewpoints (Shade, Gortler *et al.* 1998).
7. (Optional) Construct an *unwrap mosaic* from a more complex scene and use this to do some video editing (Rav-Acha, Kohli *et al.* 2008).

Ex 9.8: Transparent motion and reflection estimation. Take a video sequence looking through a window (or picture frame) and see if you can remove the reflection to better see what is inside.

The steps are described in Section 9.4.2 and by Szeliski, Avidan, and Anandan (2000). Alternative approaches can be found in work by Shizawa and Mase (1991), Bergen, Burt *et al.* (1992), Darrell and Simoncelli (1993), Darrell and Pentland (1995), Irani, Rousso, and Peleg (1994), Black and Anandan (1996), and Ju, Black, and Jepson (1996).

Ex 9.9: Motion segmentation. Write a program to segment an image into separately moving regions or to reliably find motion boundaries.

Use the DAVIS motion segmentation database (Pont-Tuset, Perazzi *et al.* 2017) as some of your test data.

Ex 9.10: Video object tracking. Write an object tracker and test it out on one of the latest video object tracking datasets (Leal-Taixé, Milan *et al.* 2015; Kristan, Matas *et al.* 2016; Dave, Khurana *et al.* 2020; Kristan, Leonardis *et al.* 2020; Dendorfer, Ošep *et al.* 2021).

Chapter 10

Computational photography

10.1 Photometric calibration 486
10.1.1 Radiometric response function 486
10.1.2 Noise level estimation 488
10.1.3 Vignetting 489
10.1.4 Optical blur (spatial response) estimation 491
10.2 High dynamic range imaging 494
10.2.1 Tone mapping 499
10.2.2 *Application*: Flash photography 506
10.3 Super-resolution, denoising, and blur removal 508
10.3.1 Color image demosaicing 515
10.3.2 Lens blur (bokeh) 517
10.4 Image matting and compositing 518
10.4.1 Blue screen matting 519
10.4.2 Natural image matting 521
10.4.3 Optimization-based matting 524
10.4.4 Smoke, shadow, and flash matting 527
10.4.5 Video matting 528
10.5 Texture analysis and synthesis 529
10.5.1 *Application*: Hole filling and inpainting 531
10.5.2 *Application*: Non-photorealistic rendering 532
10.5.3 Neural style transfer and semantic image synthesis 534
10.6 Additional reading 537
10.7 Exercises 538

R. Szeliski, *Computer Vision*, Texts in Computer Science,
https://doi.org/10.1007/978-3-030-34372-9_10

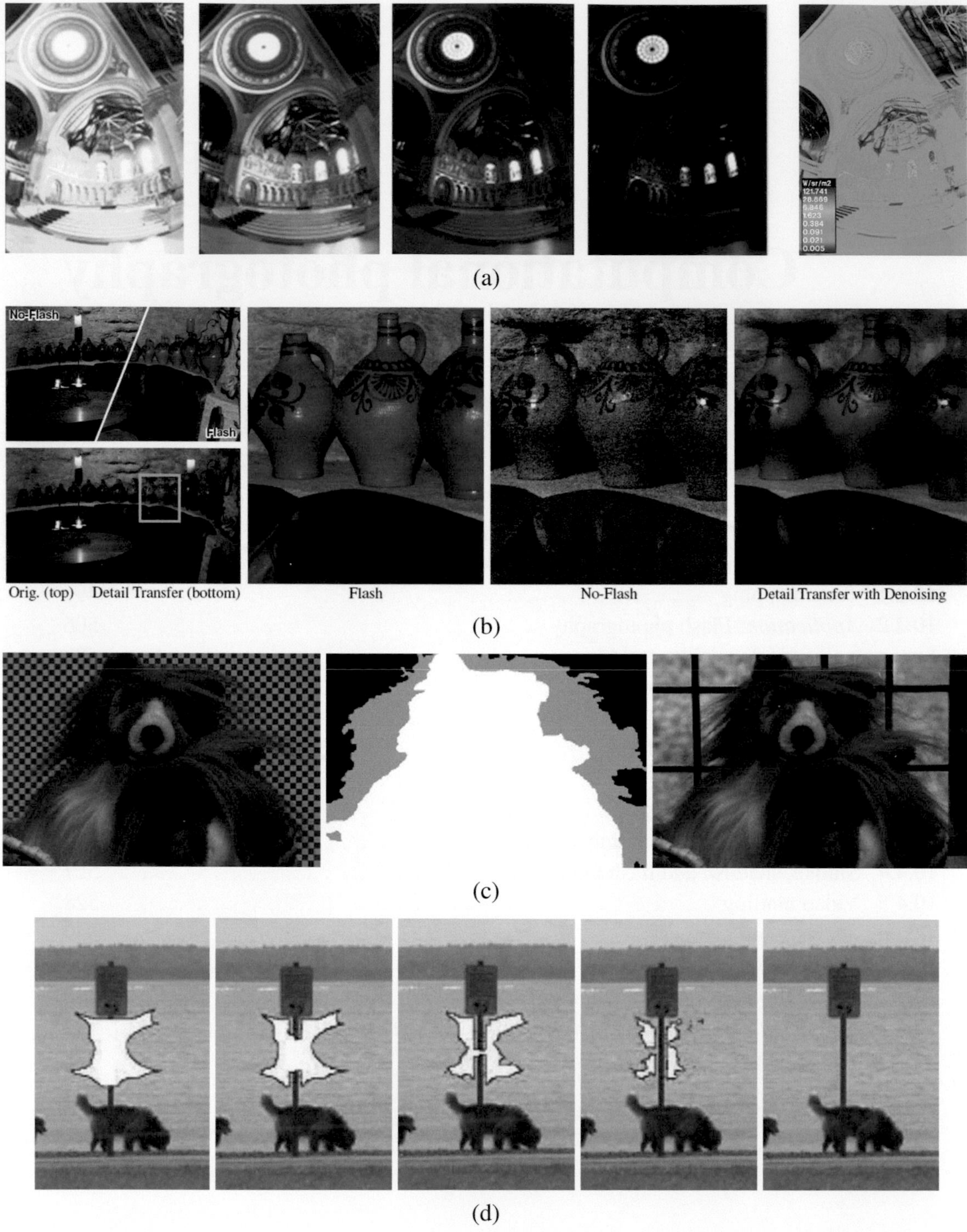

Figure 10.1 Computational photography: (a) merging multiple exposures to create high dynamic range images (Debevec and Malik 1997) © 1997 ACM; (b) merging flash and non-flash photographs; (Petschnigg, Agrawala *et al.* 2004) © 2004 ACM; (c) image matting and compositing; (Chuang, Curless *et al.* 2001) © 2001 IEEE; (d) hole filling with inpainting (Criminisi, Pérez, and Toyama 2004) © 2004 IEEE.

Of all the advances in computer vision in the last decade, computational photography has arguably had the most widespread commercial impact. In 2010, the seminal *Frankencamera* paper by Adams, Talvala *et al.* (2010) had just been released, as had one of the first widely used in-camera panoramic image stitching apps.[1] Fast forward to 2020, and every smartphone now has built-in panoramic stitching, high dynamic range (HDR) exposure merging, and multi-image denoising and super-resolution (Hasinoff, Sharlet *et al.* 2016; Wronski, Garcia-Dorado *et al.* 2019; Liba, Murthy *et al.* 2019), and the newest phones are also simulating shallow depth of field (bokeh) with multiple lenses or dual pixels (Barron, Adams *et al.* 2015; Wadhwa, Garg *et al.* 2018; Garg, Wadhwa *et al.* 2019; Zhang, Wadhwa *et al.* 2020).

In Section 8.2, we described how to stitch multiple images into wide field of view panoramas, allowing us to create photographs that could not be captured with a regular camera. This is just one instance of *computational photography*, where image analysis and processing algorithms are applied to one or more photographs to create images that go beyond the capabilities of traditional imaging systems.

In this chapter, we cover a number of additional computational photography algorithms. We begin with a review of photometric image calibration (Section 10.1), i.e., the measurement of camera and lens responses, which is a prerequisite for many of the algorithms we describe later. We then discuss *high dynamic range imaging* (Section 10.2), which captures the full range of brightness in a scene through the use of multiple exposures (Figure 10.1a). We also discuss *tone mapping operators*, which map wide-gamut images back into regular display devices such as screens and printers, as well as algorithms that merge flash and regular images to obtain better exposures (Figure 10.1b).

Next, we discuss how the resolution and visual quality of images can be improved either by merging multiple photographs together or using sophisticated image priors or deep networks (Section 10.3). This includes algorithms for extracting full-color images from the patterned Bayer mosaics present in most cameras.

In Section 10.4, we discuss algorithms for cutting pieces of images from one photograph and pasting them into others (Figure 10.1c). In Section 10.5, we describe how to generate novel textures from real-world samples for applications such as filling holes in images (Figure 10.1d). We close with a brief overview of *non-photorealistic rendering* (Section 10.5.2), which can turn regular photographs into artistic renderings that resemble traditional drawings and paintings, and a discussion of neural network approaches to style transfer and semantic image synthesis (Section 10.5.3.

One topic that we do not cover extensively in this book is novel computational sensors, optics, and cameras. A nice survey can be found in an article by Nayar (2006), the book by Raskar and Tumblin (2010), and research papers such as Levin, Fergus *et al.* (2007). Some related discussion can also be found in Sections 10.2 and 14.3.

A good general-audience introduction to computational photography can be found in the article by Hayes (2008) as well as survey papers by Nayar (2006), Cohen and Szeliski (2006), Levoy (2006), and Debevec (2006).[2] Raskar and Tumblin (2010) give extensive coverage of topics in this area, with particular emphasis on computational cameras and sensors. The sub-field of high dynamic range imaging has its own book discussing research in this area (Reinhard, Heidrich *et al.* 2010), as well as a wonderful book aimed more at professional photographers (Freeman 2008).[3] A good survey of image matting is provided by Wang and Cohen (2009).

There are also several courses on computational photography where the instructors have provided extensive online materials, e.g., Yannis Gkioulekas' class at Carnegie Mellon,[4] Alyosha Efros'

[1] https://en.wikipedia.org/wiki/Photosynth#Mobile_apps

[2] See also the two special issue journals edited by Bimber (2006) and Durand and Szeliski (2007).

[3] Gulbins and Gulbins (2009) discuss related photographic techniques.

[4] CMU 15-463, http://graphics.cs.cmu.edu/courses/15-463

class at Berkeley,[5] Frédo Durand's Computation Photography course at MIT,[6] Marc Levoy's class at Stanford,[7] and a series of SIGGRAPH courses on Computational Photography.[8]

10.1 Photometric calibration

Before we can successfully merge multiple photographs, we need to characterize the functions that map incoming irradiance into pixel values and also the amount of noise present in each image. In this section, we examine three components of the imaging pipeline (Figure 10.2) that affect this mapping. For a more comprehensive, tunable model of modern digital camera processing pipelines, see the recent paper by Tseng, Yu *et al.* (2019).

The first is the *radiometric response function* (Mitsunaga and Nayar 1999), which maps photons arriving at the lens into digital values stored in the image file (Section 10.1.1). The second is *vignetting*, which darkens pixel values near the periphery of images, especially at large apertures (Section 10.1.3). The third is the *point spread function*, which characterizes the blur induced by the lens, anti-aliasing filters, and finite sensor areas (Section 10.1.4).[9] The material in this section builds on the image formation processes described in Sections 2.2.3 and 2.3.3, so if it has been a while since you looked at those sections, please go back and review them.

10.1.1 Radiometric response function

As we can see in Figure 10.2, a number of factors affect how the intensity of light arriving at the lens ends up being mapped into stored digital values. Let us ignore for now any non-uniform attenuation that may occur inside the lens, which we cover in Section 10.1.3.

The first factors to affect this mapping are the aperture and shutter speed (Section 2.3), which can be modeled as global multipliers on the incoming light, most conveniently measured in *exposure values* ($\log_2$ brightness ratios). Next, the analog to digital (A/D) converter on the sensing chip applies an electronic gain, usually controlled by the ISO setting on your camera. While in theory this gain is linear, as with any electronics non-linearities may be present (either unintentionally or by design). Ignoring, for now, photon noise, on-chip noise, amplifier noise, and quantization noise, which we discuss shortly, you can often assume that the mapping between incoming light and the values stored in a RAW camera file (if your camera supports this) is roughly linear.

If images are being stored in the more common JPEG format, the camera's image signal processor (ISP) next performs Bayer pattern demosaicing (Sections 2.3.2 and 10.3.1), which is a mostly linear (but often non-stationary) process. Some sharpening is also often applied at this stage. Next, the color values are multiplied by different constants (or sometimes a 3×3 color twist matrix) to perform color balancing, i.e., to move the white point closer to pure white. Finally, a standard gamma is applied to the intensities in each color channel and the colors are converted into YCbCr format before being transformed by a DCT, quantized, and then compressed into the JPEG format (Section 2.3.3). Figure 10.2 shows all of these steps in pictorial form.

Given the complexity of all of this processing, it is difficult to model the camera response function (Figure 10.3a), i.e., the mapping between incoming irradiance and digital RGB values, from

[5] Berkeley CS194-26/294-26, https://inst.eecs.berkeley.edu/~cs194-26/fa20

[6] MIT 6.815/6.865, https://stellar.mit.edu/S/course/6/sp15/6.815

[7] Stanford CS 448A, https://graphics.stanford.edu/courses/cs448a-10

[8] https://web.media.mit.edu/~raskar/photo.

[9] Additional photometric camera and lens effects include sensor glare, blooming, and chromatic aberration, which can also be thought of as a spectrally varying form of geometric aberration (Section 2.2.3).

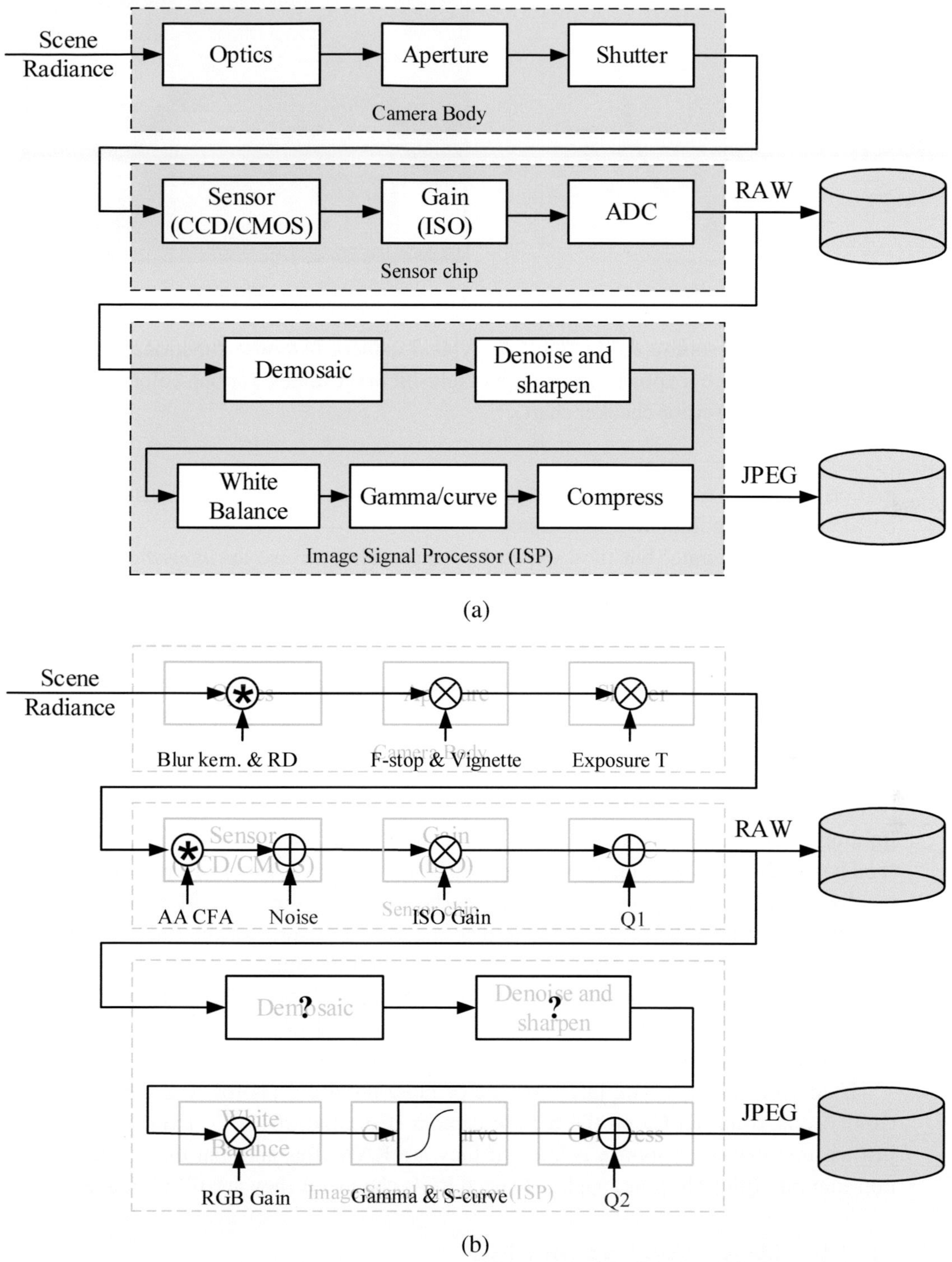

Figure 10.2 Image sensing pipeline: (a) block diagram showing the various sources of noise as well as the typical digital post-processing steps; (b) equivalent signal transforms, including convolution, gain, and noise injection. The abbreviations are: RD = radial distortion, AA = anti-aliasing filter, CFA = color filter array, Q1 and Q2 = quantization noise.

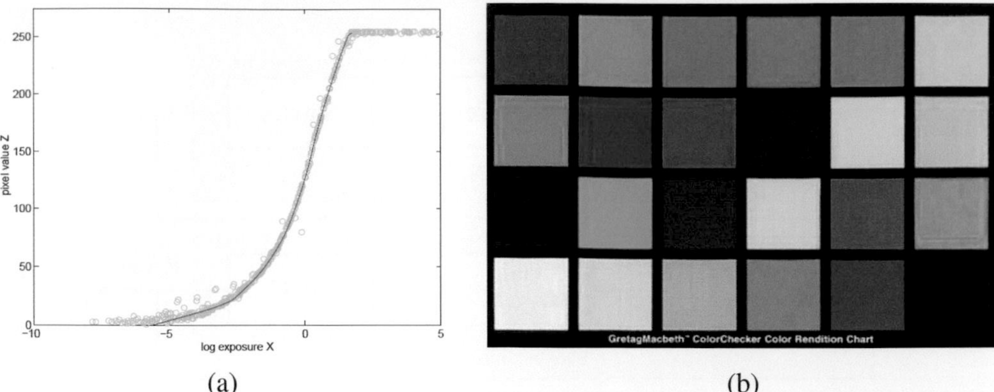

(a) (b)

Figure 10.3 Radiometric response calibration: (a) typical camera response function, showing the mapping between incoming log irradiance (exposure) and output eight-bit pixel values, for one color channel (Debevec and Malik 1997) © 1997 ACM; (b) color checker chart.

first principles. A more practical approach is to calibrate the camera by measuring correspondences between incoming light and final values.

The most accurate, but most expensive, approach is to use an *integrating sphere*, which is a large (typically 1m diameter) sphere carefully painted on the inside with white matte paint. An accurately calibrated light at the top controls the amount of radiance inside the sphere (which is constant everywhere because of the sphere's radiometry) and a small opening at the side allows for a camera/lens combination to be mounted. By slowly varying the current going into the light, an accurate correspondence can be established between incoming radiance and measured pixel values. The vignetting and noise characteristics of the camera can also be simultaneously determined.

A more practical alternative is to use a calibration chart (Figure 10.3b) such as the Macbeth or Munsell ColorChecker Chart.[10] The biggest problem with this approach is to ensure uniform lighting. One approach is to use a large dark room with a high-quality light source far away from (and perpendicular to) the chart. Another is to place the chart outdoors away from any shadows. (The results will differ under these two conditions, because the color of the illuminant will be different.)

The easiest approach is probably to take multiple exposures of the same scene while the camera is on a tripod and to recover the response function by simultaneously estimating the incoming irradiance at each pixel and the response curve (Mann and Picard 1995; Debevec and Malik 1997; Mitsunaga and Nayar 1999). This approach is discussed in more detail in Section 10.2 on high dynamic range imaging.

If all else fails, i.e., you just have one or more unrelated photos, you can use an International Color Consortium (ICC) profile for the camera (Fairchild 2013).[11] Even more simply, you can just assume that the response is linear if they are RAW files and that the images have a $\gamma = 2.2$ non-linearity (plus clipping) applied to each RGB channel if they are JPEG images.

10.1.2 Noise level estimation

In addition to knowing the camera response function, it is also often important to know the amount of noise being injected under a particular camera setting (e.g., ISO/gain level). The simplest characterization of noise is a single standard deviation, usually measured in gray levels, independent of

[10] https://www.xrite.com.

[11] See also the ICC *Information on Profiles*, https://www.color.org/info_profiles2.xalter.

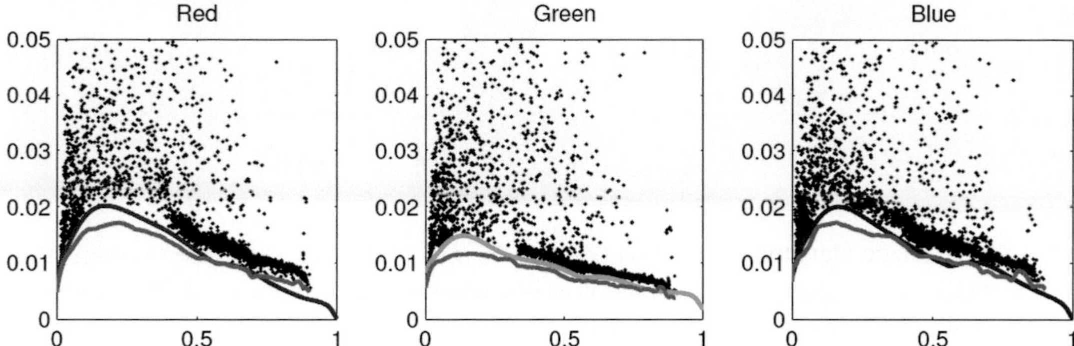

Figure 10.4 Noise level function estimates obtained from a single color photograph (Liu, Szeliski *et al.* 2008) © 2008 IEEE. The colored curves are the estimated NLF fit as the probabilistic lower envelope of the measured deviations between the noisy piecewise-smooth images. The ground truth NLFs obtained by averaging 29 images are shown in gray.

pixel value. A more accurate model can be obtained by estimating the noise level as a function of pixel value (Figure 10.4), which is known as the *noise level function* (Liu, Szeliski *et al.* 2008).

As with the camera response function, the simplest way to estimate these quantities is in the lab, using either an integrating sphere or a calibration chart. The noise can be estimated either at each pixel independently, by taking repeated exposures and computing the temporal variance in the measurements (Healey and Kondepudy 1994), or over regions, by assuming that pixel values should all be the same within some region (e.g., inside a color checker square) and computing a spatial variance.

This approach can be generalized to photos where there are regions of constant or slowly varying intensity (Liu, Szeliski *et al.* 2008). First, segment the image into such regions and fit a constant or linear function inside each region. Next, measure the (spatial) standard deviation of the differences between the noisy input pixels and the smooth fitted function away from large gradients and region boundaries. Plot these as a function of output level for each color channel, as shown in Figure 10.4. Finally, fit a lower envelope to this distribution to ignore pixels or deviations that are outliers. A fully Bayesian approach to this problem that models the statistical distribution of each quantity is presented by Liu, Szeliski *et al.* (2008). A simpler approach, which should produce useful results in most cases, is to fit a low-dimensional function (e.g., positive valued B-spline) to the lower envelope (see Exercise 10.2).

Matsushita and Lin (2007b) present a technique for simultaneously estimating a camera's response and noise level functions based on skew (asymmetries) in level-dependent noise distributions. Their paper also contains extensive references to previous work in these areas.

10.1.3 Vignetting

A common problem with using wide-angle and wide-aperture lenses is that the image tends to darken in the corners (Figure 10.5a). This problem is generally known as *vignetting* and comes in several different forms, including natural, optical, and mechanical vignetting (Section 2.2.3) (Ray 2002). As with radiometric response function calibration, the most accurate way to calibrate vignetting is to use an integrating sphere or a picture of a uniformly colored and illuminated blank wall.

An alternative approach is to stitch a panoramic scene and to assume that the true radiance at

Figure 10.5 Single image vignetting correction (Zheng, Yu *et al.* 2008) © 2008 IEEE: (a) original image with strong visible vignetting; (b) vignetting compensation as described by Zheng, Zhou *et al.* (2006); (c–d) vignetting compensation as described by Zheng, Yu *et al.* (2008).

Figure 10.6 Simultaneous estimation of vignetting, exposure, and radiometric response (Goldman 2010) © 2011 IEEE: (a) original average of the input images; (b) after compensating for vignetting; (c) using gradient domain blending only (note the remaining mottled look); (d) after both vignetting compensation and blending.

each pixel comes from the central portion of each input image. This is easier to do if the radiometric response function is already known (e.g., by shooting in RAW mode) and if the exposure is kept constant. If the response function, image exposures, and vignetting function are unknown, they can still be recovered by optimizing a large least squares fitting problem (Litvinov and Schechner 2005; Goldman 2010). Figure 10.6 shows an example of simultaneously estimating the vignetting, exposure, and radiometric response function from a set of overlapping photographs (Goldman 2010). Note that unless vignetting is modeled and compensated, regular gradient-domain image blending (Section 8.4.4) will not create an attractive image.

If only a single image is available, vignetting can be estimated by looking for slow consistent intensity variations in the radial direction. The original algorithm proposed by Zheng, Lin, and Kang (2006) first pre-segmented the image into smoothly varying regions and then performed an analysis inside each region. Instead of pre-segmenting the image, Zheng, Yu *et al.* (2008) compute the radial gradients at all the pixels and use the asymmetry in this distribution (because gradients away from the center are, on average, slightly negative) to estimate the vignetting. Figure 10.5 shows the results of applying each of these algorithms to an image with a large amount of vignetting. Exercise 10.3 has you implement some of the above techniques.

Figure 10.7 Calibration pattern with edges equally distributed at all orientations that can be used for PSF and radial distortion estimation (Joshi, Szeliski, and Kriegman 2008) © 2008 IEEE. A portion of an actual sensed image is shown in the middle and a close-up of the ideal pattern is on the right.

10.1.4 Optical blur (spatial response) estimation

One final characteristic of imaging systems that you should calibrate is the spatial response function, which encodes the optical blur that gets convolved with the incoming image to produce the point-sampled image. The shape of the convolution kernel, which is also known as the *point spread function (PSF)* or *optical transfer function*, depends on several factors, including lens blur and radial distortion (Section 2.2.3), anti-aliasing filters in front of the sensor, and the shape and extent of each active pixel area (Section 2.3) (Figure 10.2). A good estimate of this function is required for applications such as multi-image super-resolution and deblurring (Section 10.3).

In theory, one could estimate the PSF by simply observing an infinitely small point light source everywhere in the image. Creating an array of samples by drilling through a dark plate and back-lighting with a very bright light source is difficult in practice.

A more practical approach is to observe an image composed of long straight lines or bars, as these can be fitted to arbitrary precision. Because the location of a horizontal or vertical edge can be *aliased* during acquisition, slightly slanted edges are preferred. The profile and locations of such edges can be estimated to sub-pixel precision, which makes it possible to estimate the PSF at sub-pixel resolutions (Reichenbach, Park, and Narayanswamy 1991; Burns and Williams 1999; Williams and Burns 2001; Goesele, Fuchs, and Seidel 2003). The thesis by Murphy (2005) contains a nice survey of all aspects of camera calibration, including the spatial frequency response (SFR), spatial uniformity, tone reproduction, color reproduction, noise, dynamic range, color channel registration, and depth of field. It also includes a description of a slant-edge calibration algorithm called `sfrmat2`.

The slant-edge technique can be used to recover a 1D projection of the 2D PSF, e.g., slightly vertical edges are used to recover the horizontal *line spread function* (LSF) (Williams 1999). The LSF is then often converted into the Fourier domain and its magnitude plotted as a one-dimensional *modulation transfer function* (MTF), which indicates which image frequencies are lost (blurred) and aliased during the acquisition process (Section 2.3.1). For most computational photography applications, it is preferable to directly estimate the full 2D PSF, as it can be hard to recover from its projections (Williams 1999).

Figure 10.7 shows a pattern containing edges at all orientations, which can be used to directly recover a two-dimensional PSF. First, corners in the pattern are located by extracting edges in the sensed image, linking them, and finding the intersections of the circular arcs. Next, the ideal pattern, whose analytic form is known, is warped (using a homography) to fit the central portion of the input

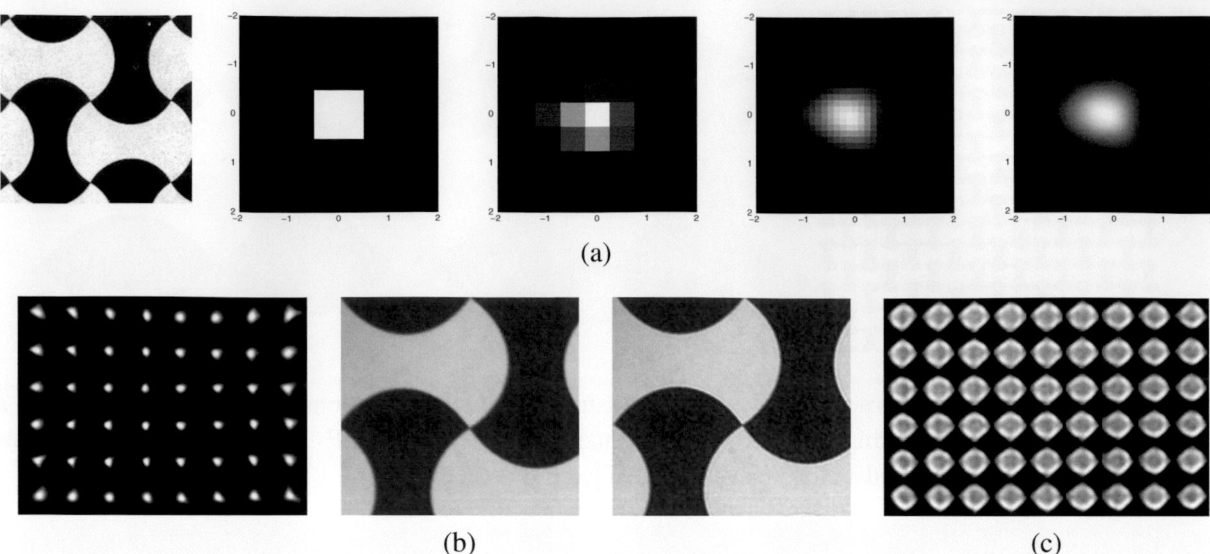

Figure 10.8 Point spread function estimation using a calibration target (Joshi, Szeliski, and Kriegman 2008) © 2008 IEEE. (a) Sub-pixel PSFs at successively higher resolutions (note the interaction between the square sensing area and the circular lens blur). (b) The radial distortion and chromatic aberration can also be estimated and removed. (c) PSF for a misfocused (blurred) lens showing some diffraction and vignetting effects in the corners.

image and its intensities are adjusted to fit the ones in the sensed image. If desired, the pattern can be rendered at a higher resolution than the input image, which enables the estimation of the PSF to sub-pixel resolution (Figure 10.8a). Finally a large linear least squares system is solved to recover the unknown PSF kernel K,

$$K = \arg\min_{K} \|B - D(I * K)\|^2, \tag{10.1}$$

where B is the sensed (blurred) image, I is the predicted (sharp) image, and D is an optional downsampling operator that matches the resolution of the ideal and sensed images (Joshi, Szeliski, and Kriegman 2008). An alternative solution technique is to estimate 1D PSF profiles first and to then combine them using a Radon transform (Cho, Paris *et al.* 2011).

If the process of estimating the PSF is done locally in overlapping patches of the image, it can also be used to estimate the radial distortion and chromatic aberration induced by the lens (Figure 10.8b). Because the homography mapping the ideal target to the sensed image is estimated in the central (undistorted) part of the image, any (per-channel) shifts induced by the optics manifest themselves as a displacement in the PSF centers.[12] Compensating for these shifts eliminates both the achromatic radial distortion and the inter-channel shifts that result in visible chromatic aberration. The color-dependent blurring caused by chromatic aberration (Figure 2.21) can also be removed using the deblurring techniques discussed in Section 10.3. Figure 10.8b shows how the radial distortion and chromatic aberration manifest themselves as elongated and displaced PSFs, along with the result of removing these effects in a region of the calibration target.

The local 2D PSF estimation technique can also be used to estimate vignetting. Figure 10.8c shows how the mechanical vignetting manifests itself as clipping of the PSF in the corners of the

[12] This process confounds the distinction between geometric and photometric calibration. In principle, any geometric distortion could be modeled by spatially varying displaced PSFs. In practice, it is easier to fold any large shifts into the geometric correction component.

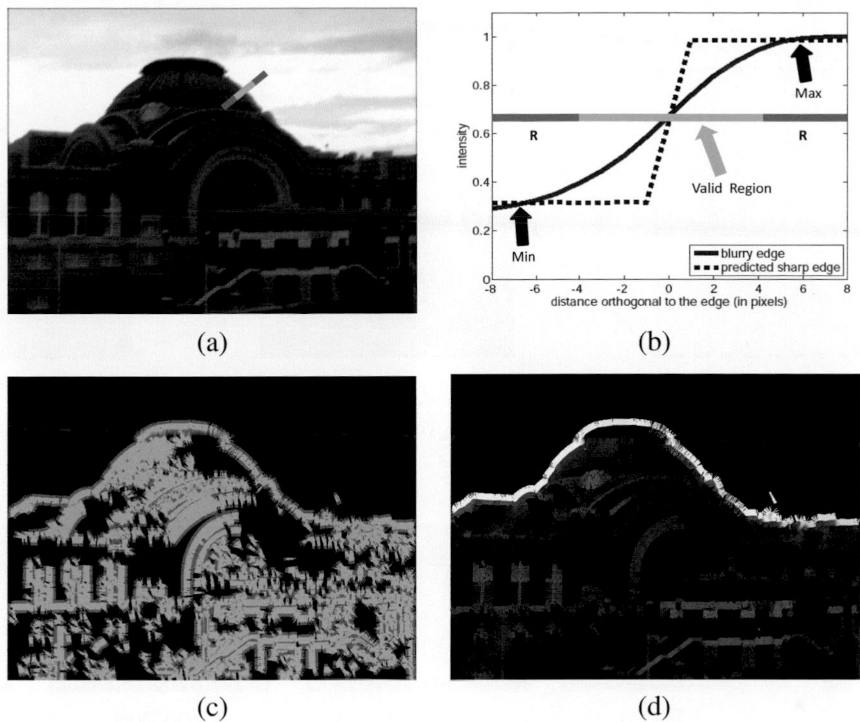

Figure 10.9 Estimating the PSF without using a calibration pattern (Joshi, Szeliski, and Kriegman 2008) © 2008 IEEE: (a) Input image with blue cross-section (profile) location, (b) Profile of sensed and predicted step edges, (c–d) Locations and values of the predicted colors near the edge locations.

image. For the overall dimming associated with vignetting to be properly captured, the modified intensities of the ideal pattern need to be extrapolated from the center, which is best done with a uniformly illuminated target.

When working with RAW Bayer-pattern images, the correct way to estimate the PSF is to only evaluate the least squares terms in (10.1) at sensed pixel values, while interpolating the ideal image to all values. For JPEG images, you should linearize your intensities first, e.g., remove the gamma and any other non-linearities in your estimated radiometric response function.

What if you have an image that was taken with an uncalibrated camera? Can you still recover the PSF an use it to correct the image? In fact, with a slight modification, the previous algorithms still work.

Instead of assuming a known calibration image, you can detect strong elongated edges and fit ideal step edges in such regions (Figure 10.9b), resulting in the sharp image shown in Figure 10.9d. For every pixel that is surrounded by a complete set of valid estimated neighbors (green pixels in Figure 10.9c), apply the least squares formula (10.1) to estimate the kernel K. The resulting locally estimated PSFs can be used to correct for chromatic aberration (because the relative displacements between per-channel PSFs can be computed), as shown by Joshi, Szeliski, and Kriegman (2008).

Exercise 10.4 provides some more detailed instructions for implementing and testing edge-based PSF estimation algorithms. An alternative approach, which does not require the explicit detection of edges but uses image statistics (gradient distributions) instead, is presented by Fergus, Singh *et al.* (2006).

Figure 10.10 Sample indoor image where the areas outside the window are overexposed and inside the room are too dark.

Figure 10.11 Relative brightness of different scenes, ranging from 1 inside a dark room lit by a monitor to 2,000,000 looking at the Sun. Photos courtesy of Paul Debevec.

10.2 High dynamic range imaging

As we mentioned earlier in this chapter, registered images taken at different exposures can be used to calibrate the radiometric response function of a camera. More importantly, they can help you create well-exposed photographs under challenging conditions, such as brightly lit scenes where any single exposure contains saturated (overexposed) and dark (underexposed) regions (Figure 10.10). This problem is quite common, because the natural world contains a range of radiance values that is far greater than can be captured with any photographic sensor or film (Figure 10.11). Taking a set of *bracketed exposures* (exposures taken by a camera in automatic exposure bracketing (AEB) mode to deliberately under- and over-expose the image) gives you the material from which to create a properly exposed photograph, as shown in Figure 10.12 (Freeman 2008; Gulbins and Gulbins 2009; Hasinoff, Durand, and Freeman 2010; Reinhard, Heidrich *et al.* 2010).

While it is possible to combine pixels from different exposures directly into a final composite (Burt and Kolczynski 1993; Mertens, Kautz, and Reeth 2007), this approach runs the risk of creating contrast reversals and halos. Instead, the more common approach is to proceed in three stages:

1. Estimate the radiometric response function from the aligned images.

2. Estimate a *radiance map* by selecting or blending pixels from different exposures.

3. Tone map the resulting high dynamic range (HDR) image back into a displayable gamut.

The idea behind estimating the radiometric response function is relatively straightforward (Mann and Picard 1995; Debevec and Malik 1997; Mitsunaga and Nayar 1999; Reinhard, Heidrich *et al.*

Figure 10.12 A bracketed set of shots (using the camera's automatic exposure bracketing (AEB) mode) and the resulting high dynamic range (HDR) composite.

2010). Suppose you take three sets of images at different exposures (shutter speeds), say at ± 2 exposure values.[13] If we were able to determine the irradiance (exposure) E_i at each pixel (2.102), we could plot it against the measured pixel value z_{ij} for each exposure time t_j, as shown in Figure 10.13.

Unfortunately, we do not know the irradiance values E_i, so these have to be estimated at the same time as the radiometric response function f, which can be written (Debevec and Malik 1997) as

$$z_{ij} = f(E_i \, t_j), \tag{10.2}$$

where t_j is the exposure time for the jth image. The inverse response curve f^{-1} is given by

$$f^{-1}(z_{ij}) = E_i \, t_j. \tag{10.3}$$

Taking logarithms of both sides (base 2 is convenient, as we can now measure quantities in EVs), we obtain

$$g(z_{ij}) = \log f^{-1}(z_{ij}) = \log E_i + \log t_j, \tag{10.4}$$

where $g = \log f^{-1}$ (which maps pixel values z_{ij} into log irradiance) is the curve we are estimating (Figure 10.13 turned on its side).

Debevec and Malik (1997) assume that the exposure times t_j are known. (Recall that these can be obtained from a camera's EXIF tags, but that they actually follow a power of 2 progression ..., $^1/_{128}$, $^1/_{64}$, $^1/_{32}$, $^1/_{16}$, $^1/_{8}$, ... instead of the marked ..., $^1/_{125}$, $^1/_{60}$, $^1/_{30}$, $^1/_{15}$, $^1/_{8}$, ... values—see Exercise 2.5.) The unknowns are therefore the per-pixel exposures E_i and the response values $g_k = g(k)$, where g can be discretized according to the 256 pixel values commonly observed in eight-bit images. (The response curves are calibrated separately for each color channel.)

In order to make the response curve smooth, Debevec and Malik (1997) add a second-order smoothness constraint

$$\lambda \sum_k g''(k)^2 = \lambda \sum [g(k-1) - 2g(k) + g(k+1)]^2, \tag{10.5}$$

which is similar to the one used in snakes (7.27). Because pixel values are more reliable in the middle of their range (and the g function becomes singular near saturation values), they also add a weighting (hat) function $w(k)$ that decays to zero at both ends of the pixel value range,

$$w(z) = \begin{cases} z - z_{\min} & z \le (z_{\min} + z_{\max})/2 \\ z_{\max} - z & z > (z_{\min} + z_{\max})/2. \end{cases} \tag{10.6}$$

[13] Changing the shutter speed is preferable to changing the aperture, as the latter can modify the vignetting and focus. Using ± 2 "f-stops" (technically, exposure values, or EVs, as f-stops refer to apertures) is usually the right compromise between capturing a good dynamic range and having properly exposed pixels everywhere.

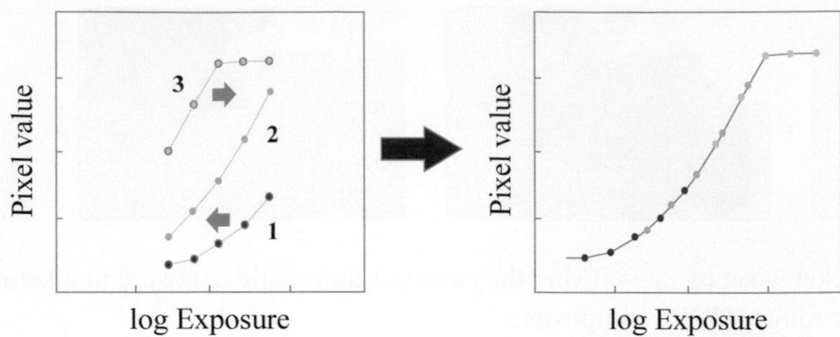

Figure 10.13 Radiometric calibration using multiple exposures (Debevec and Malik 1997). Corresponding pixel values are plotted as functions of log exposures (irradiance). The curves on the left are shifted to account for each pixel's unknown radiance until they all line up into a single smooth curve.

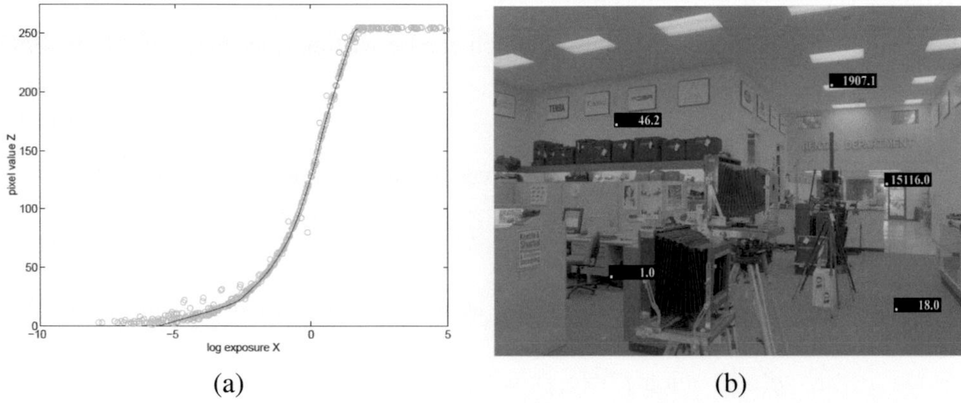

(a) (b)

Figure 10.14 Recovered response function and radiance image for a real digital camera (DCS460) (Debevec and Malik 1997) © 1997 ACM.

Putting all of these terms together, they obtain a least squares problem in the unknowns $\{g_k\}$ and $\{E_i\}$,

$$E = \sum_i \sum_j w(z_{i,j})[g(z_{i,j}) - \log E_i - \log t_j]^2 + \lambda \sum_k w(k) g''(k)^2. \tag{10.7}$$

(To remove the overall shift ambiguity in the response curve and irradiance values, the middle of the response curve is set to 0.) Debevec and Malik (1997) show how this can be implemented in 21 lines of MATLAB code, which partially accounts for the popularity of their technique.

While Debevec and Malik (1997) assume that the exposure times t_j are known exactly, there is no reason why these additional variables cannot be thrown into the least squares problem, constraining their final estimated values to lie close to their nominal values $\hat{t}_j$ with an extra term $\eta \sum_j (t_j - \hat{t}_j)^2$.

Figure 10.14 shows the recovered radiometric response function for a digital camera along with select (relative) radiance values in the overall radiance map. Figure 10.15 shows the bracketed input images captured on color film and the corresponding radiance map. Note that while most research on high dynamic range imaging assumes that the radiometric (or camera) response function is independent of exposure, this is not actually the case. Rodríguez, Vazquez-Corral, and Bertalmío (2019) describe how to take this into account to get improved results.

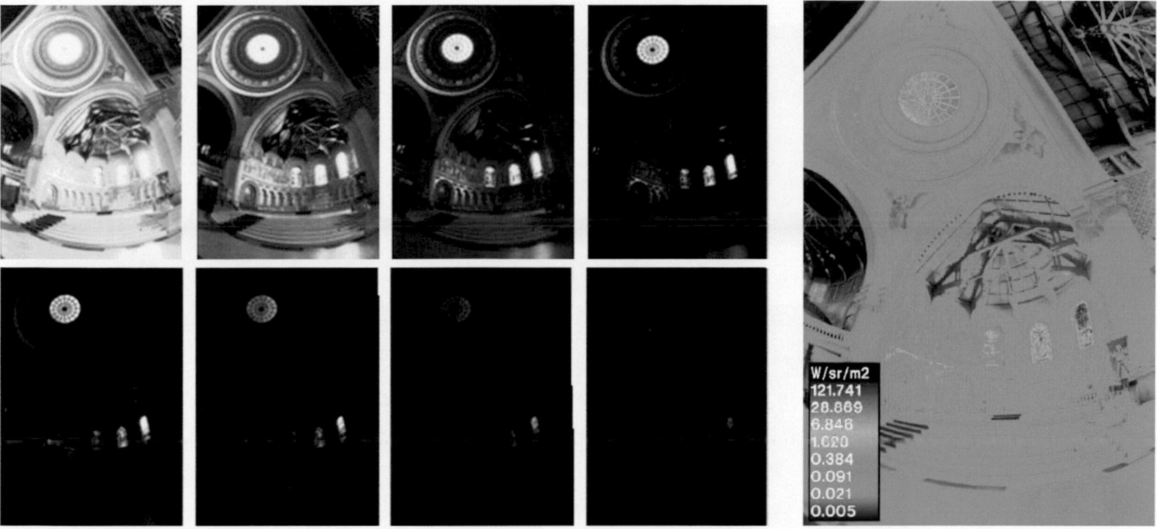

Figure 10.15 Bracketed set of exposures captured with a film camera and the resulting radiance image displayed in pseudocolor (Debevec and Malik 1997) © 1997 ACM.

While Debevec and Malik (1997) use a general second-order smooth curve g to parameterize their response curve, Mann and Picard (1995) use a three-parameter function

$$f(E) = \alpha + \beta E^{\gamma}, \tag{10.8}$$

while Mitsunaga and Nayar (1999) use a low-order ($N \leq 10$) polynomial for the inverse response function g. Pal, Szeliski *et al.* (2004) derive a Bayesian model that estimates an independent smooth response function for each image, which can better model the more sophisticated (and hence less predictable) automatic contrast and tone adjustment performed in today's digital cameras.

Once the response function has been estimated, the second step in creating high dynamic range photographs is to merge the input images into a composite *radiance map*. If the response function and images were known exactly, i.e., if they were noise free, you could use any non-saturated pixel value to estimate the corresponding radiance by mapping it through the inverse response curve $E = g(z)$.

Unfortunately, pixels are noisy, especially under low-light conditions when fewer photons arrive at the sensor. To compensate for this, Mann and Picard (1995) use the derivative of the response function as a weight in determining the final radiance estimate, because "flatter" regions of the curve tell us less about the incoming irradiance. Debevec and Malik (1997) use a hat function (10.6) which accentuates mid-tone pixels while avoiding saturated values. Mitsunaga and Nayar (1999) show that to maximize the signal-to-noise ratio (SNR), the weighting function must emphasize both higher pixel values and larger gradients in the transfer function, i.e.,

$$w(z) = g(z)/g'(z), \tag{10.9}$$

where the weights w are used to form the final irradiance estimate

$$\log E_i = \frac{\sum_j w(z_{ij})[g(z_{ij}) - \log t_j]}{\sum_j w(z_{ij})}. \tag{10.10}$$

Exercise 10.1 has you implement one of the radiometric response function calibration techniques and then use it to create radiance maps.

Figure 10.16 Merging multiple exposures to create a high dynamic range composite (Kang, Uyttendaele *et al.* 2003): (a–c) three different exposures; (d) merging the exposures using classic algorithms (note the ghosting due to the horse's head movement); (e) merging the exposures with motion compensation.

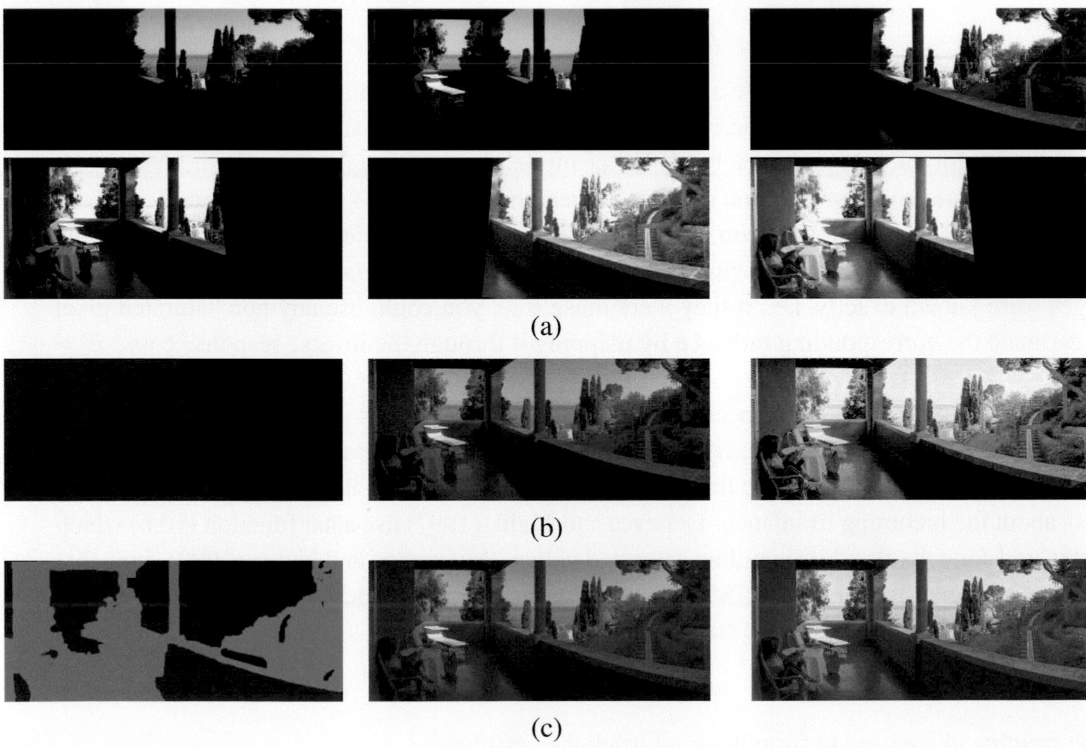

Figure 10.17 HDR merging with large amounts of motion (Eden, Uyttendaele, and Szeliski 2006) © 2006 IEEE: (a) registered bracketed input images; (b) results after the first pass of image selection: reference labels, image, and tone-mapped image; (c) results after the second pass of image selection: final labels, compressed HDR image, and tone-mapped image

Under real-world conditions, casually acquired images may not be perfectly registered and may contain moving objects. Ward (2003) uses a global (parametric) transform to align the input images, while Kang, Uyttendaele *et al.* (2003) present an algorithm that combines global registration with local motion estimation (optical flow) to accurately align the images before blending their radiance estimates (Figure 10.16). Because the images may have widely different exposures, care must be taken when estimating the motions, which must themselves be checked for consistency to avoid the creation of ghosts and object fragments.

Even this approach, however, may not work when the camera is simultaneously undergoing large panning motions and exposure changes, which is a common occurrence in casually acquired panoramas. Under such conditions, different parts of the image may be seen at one or more exposures. Devising a method to blend all of these different sources while avoiding sharp transitions and dealing with scene motion is a challenging problem. One approach is to first find a consensus mosaic and to then selectively compute radiances in under- and over-exposed regions (Eden, Uyttendaele, and Szeliski 2006), as shown in Figure 10.17. Additional techniques for constructing and displaying high dynamic range video are discussed in Myszkowski, Mantiuk, and Krawczyk (2008), Tocci, Kiser *et al.* (2011), Sen, Kalantari *et al.* (2012), Dufaux, Le Callet *et al.* (2016), Banterle, Artusi *et al.* (2017), and Kalantari and Ramamoorthi (2017). Another approach is to use deep learning techniques to infer the high dynamic range radiance image from a single low dynamic range image (Liu, Lai *et al.* 2020b).

Some cameras, such as the Sony α550 and Pentax K-7, have started integrating multiple exposure merging and tone mapping directly into the camera body. In the future, the need to compute high dynamic range images from multiple exposures may be eliminated by advances in camera sensor technology (Yang, El Gamal *et al.* 1999; Nayar and Mitsunaga 2000; Nayar and Branzoi 2003; Kang, Uyttendaele *et al.* 2003; Narasimhan and Nayar 2005; Tumblin, Agrawal, and Raskar 2005). However, the need to blend such images and to tone map them to lower-gamut displays is likely to remain.

HDR image formats. Before we discuss techniques for mapping HDR images back to a displayable gamut, we should discuss the commonly used formats for storing HDR images.

If storage space is not an issue, storing each of the R, G, and B values as a 32-bit IEEE float is the best solution. The commonly used Portable PixMap (.ppm) format, which supports both uncompressed ASCII and raw binary encodings of values, can be extended to a Portable FloatMap (.pfm) format by modifying the header. TIFF also supports full floating point values.

A more compact representation is the Radiance format (.pic, .hdr) (Ward 1994), which uses a single common exponent and per-channel mantissas. An intermediate encoding, OpenEXR from ILM,[14] uses 16-bit floats for each channel, which is a format supported natively on most modern GPUs. Ward (2004) describes these and other data formats such as LogLuv (Larson 1998) in more detail, as do the books by Freeman (2008) and Reinhard, Heidrich *et al.* (2010). An even more recent HDR image format is the JPEG XR standard.

10.2.1 Tone mapping

Once a radiance map has been computed, it is usually necessary to display it on a lower gamut (i.e., eight-bit) screen or printer. A variety of *tone mapping* techniques has been developed for this purpose, which involve either computing spatially varying transfer functions or reducing image gradients to fit the available dynamic range (Reinhard, Heidrich *et al.* 2010).

[14] https://www.openexr.net.

(a) (b) (c)

Figure 10.18 Global tone mapping: (a) input HDR image, linearly mapped; (b) gamma applied to each color channel independently; (c) gamma applied to intensity (colors are less washed out). Original HDR image courtesy of Paul Debevec, https://www.pauldebevec.com/Research/HDR. Processed images courtesy of Frédo Durand, MIT 6.815/6.865 course on Computational Photography.

The simplest way to compress a high dynamic range radiance image into a low dynamic range gamut is to use a global transfer curve (Larson, Rushmeier, and Piatko 1997). Figure 10.18 shows one such example, where a gamma curve is used to map an HDR image back into a displayable gamut. If gamma is applied separately to each channel (Figure 10.18b), the colors become muted (less saturated), as higher-valued color channels contribute less (proportionately) to the final color. Extracting the luminance channel from the color image using (2.104), applying the global mapping to the luminance channel, and then reconstituting the color image using (10.19) works better (Figure 10.18c).

Unfortunately, when the image has a really wide range of exposures, this global approach still fails to preserve details in regions with widely varying exposures. What is needed, instead, is something akin to the dodging and burning performed by photographers in the darkroom. Mathematically, this is similar to dividing each pixel by the *average* brightness in a region around that pixel.

Figure 10.19 shows how this process works. As before, the image is split into its luminance and chrominance channels. The log luminance image

$$H(x,y) = \log L(x,y) \tag{10.11}$$

is then low-pass filtered to produce a *base layer*

$$H_{\mathrm{L}}(x,y) = B(x,y) * H(x,y), \tag{10.12}$$

and a high-pass *detail layer*

$$H_{\mathrm{H}}(x,y) = H(x,y) - H_{\mathrm{L}}(x,y). \tag{10.13}$$

The base layer is then contrast reduced by scaling to the desired log-luminance range,

$$H'_{\mathrm{H}}(x,y) = s\, H_{\mathrm{H}}(x,y) \tag{10.14}$$

and added to the detail layer to produce the new log-luminance image

$$I(x,y) = H'_{\mathrm{H}}(x,y) + H_{\mathrm{L}}(x,y), \tag{10.15}$$

which can then be exponentiated to produce the tone-mapped (compressed) luminance image. Note that this process is equivalent to dividing each luminance value by (a monotonic mapping of) the average log-luminance value in a region around that pixel.

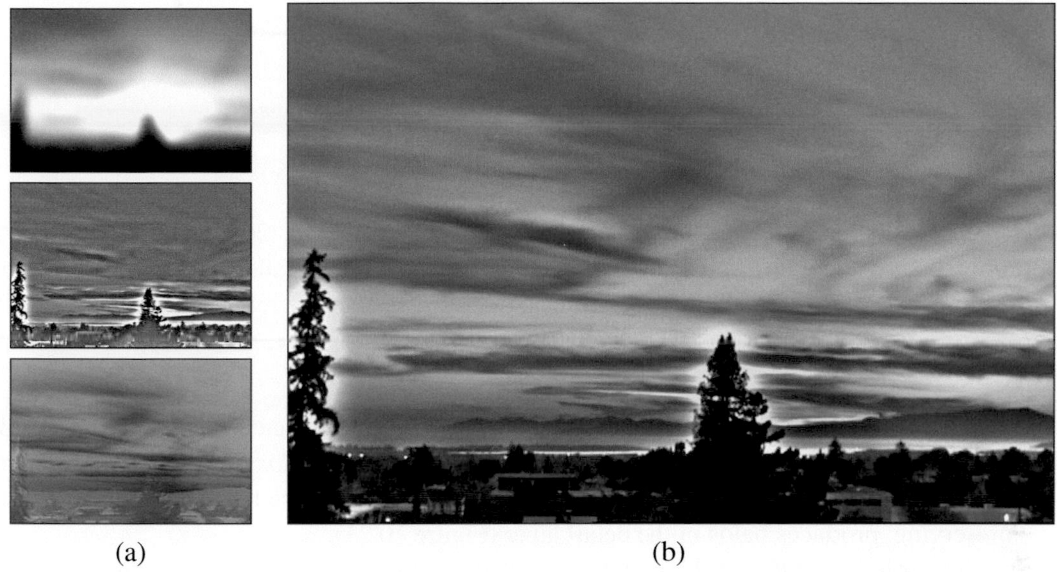

(a) (b)

Figure 10.19 Local tone mapping using linear filters: (a) low-pass and high-pass filtered log luminance images and color (chrominance) image; (b) resulting tone-mapped image (after attenuating the low-pass log luminance image) shows visible halos around the trees. Processed images courtesy of Frédo Durand, MIT 6.815/6.865 course on Computational Photography.

(a) (b)

Figure 10.20 Local tone mapping using a bilateral filter (Durand and Dorsey 2002): (a) low-pass and high-pass bilateral filtered log luminance images and color (chrominance) image; (b) resulting tone-mapped image (after attenuating the low-pass log luminance image) shows no halos. Processed images courtesy of Frédo Durand, MIT 6.815/6.865 course on Computational Photography.

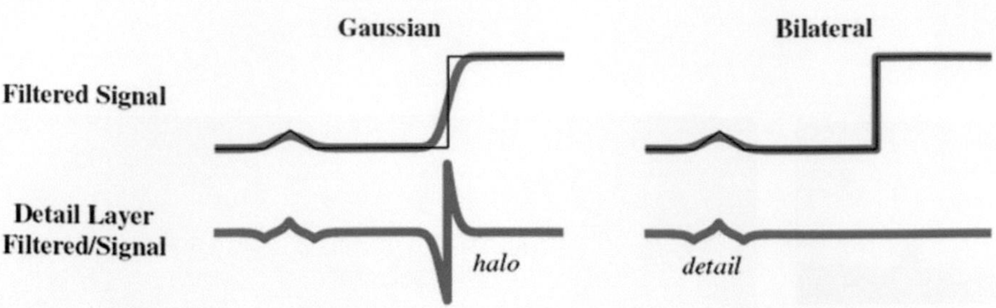

Figure 10.21 Gaussian vs. bilateral filtering (Petschnigg, Agrawala *et al.* 2004) © 2004 ACM: A Gaussian low-pass filter blurs across all edges and therefore creates strong peaks and valleys in the detail image that cause halos. The bilateral filter does not smooth across strong edges and thereby reduces halos while still capturing detail.

Figure 10.19 shows the low-pass and high-pass log luminance image and the resulting tone-mapped color image. Note how the detail layer has visible *halos* around the high-contrast edges, which are visible in the final tone-mapped image. This is because linear filtering, which is not edge preserving, produces halos in the detail layer (Figure 10.21).

The solution to this problem is to use an edge-preserving filter to create the base layer. Durand and Dorsey (2002) study a number of such edge-preserving filters, including anisotropic and robust anisotropic diffusion, and select bilateral filtering (Section 3.3.1) as their edge-preserving filter. (The paper by Farbman, Fattal *et al.* (2008) argues in favor of using a weighted least squares (WLF) filter as an alternative to the bilateral filter and Paris, Kornprobst *et al.* (2008) reviews bilateral filtering and its applications in computer vision and computational photography.) Figure 10.20 shows how replacing the linear low-pass filter with a bilateral filter produces tone-mapped images with no visible halos. Figure 10.22 summarizes the complete information flow in this process, starting with the decomposition into log luminance and chrominance images, bilateral filtering, contrast reduction, and re-composition into the final output image.

An alternative to compressing the base layer is to compress its *derivatives*, i.e., the gradient of the log-luminance image (Fattal, Lischinski, and Werman 2002). Figure 10.23 illustrates this process. The log-luminance image is differentiated to obtain a gradient image

$$H'(x, y) = \nabla H(x, y). \tag{10.16}$$

This gradient image is then attenuated by a spatially varying attenuation function $\Phi(x, y)$,

$$G(x, y) = H'(x, y)\, \Phi(x, y). \tag{10.17}$$

The attenuation function $I(x, y)$ is designed to attenuate large-scale brightness changes (Figure 10.24a) and is designed to take into account gradients at different spatial scales (Fattal, Lischinski, and Werman 2002).

After attenuation, the resulting gradient field is re-integrated by solving a first-order variational (least squares) problem,

$$\min \int\int \|\nabla I(x, y) - G(x, y)\|^2 dx\, dy \tag{10.18}$$

to obtain the compressed log-luminance image $I(x, y)$. This least squares problem is the same that was used for Poisson blending (Section 8.4.4) and was first introduced in our study of regularization (Section 4.2, 4.24). It can efficiently be solved using techniques such as multigrid and hierarchical basis preconditioning (Fattal, Lischinski, and Werman 2002; Szeliski 2006b; Farbman, Fattal *et al.*

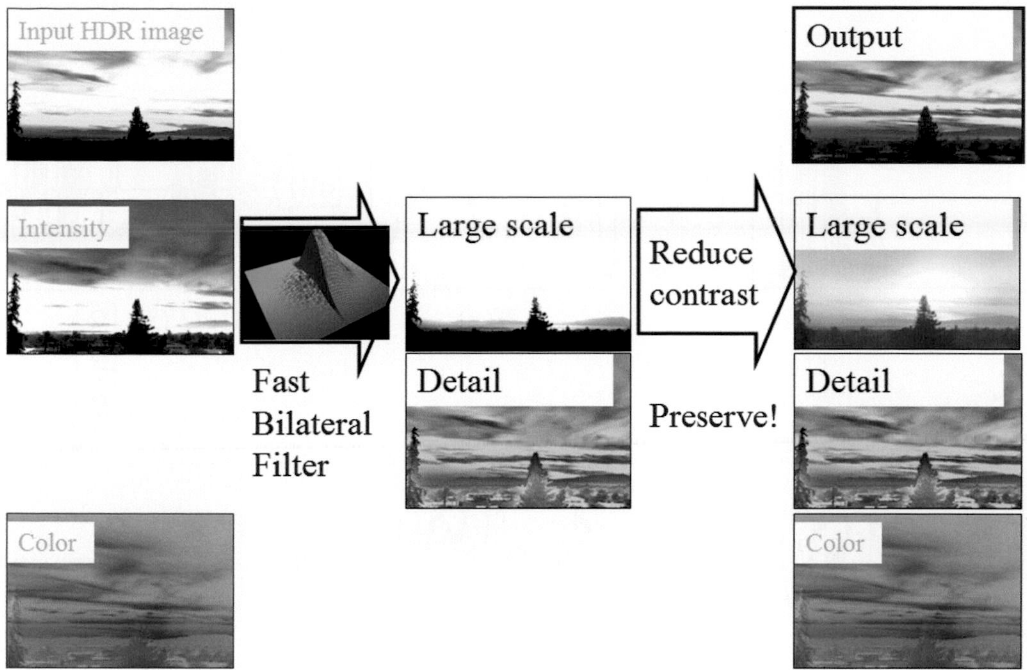

Figure 10.22 Local tone mapping using a bilateral filter (Durand and Dorsey 2002): summary of algorithm workflow. Images courtesy of Frédo Durand, MIT 6.815/6.865 course on Computational Photography.

2008; Krishnan and Szeliski 2011; Krishnan, Fattal, and Szeliski 2013). Once the new luminance image has been computed, it is combined with the original color image using

$$C_{\text{out}} = \left(\frac{C_{\text{in}}}{L_{\text{in}}} \right)^s L_{\text{out}}, \tag{10.19}$$

where $C = (R, G, B)$ and L_{in} and L_{out} are the original and compressed luminance images. The exponent s controls the saturation of the colors and is typically in the range $s \in [0.4, 0.6]$ (Fattal, Lischinski, and Werman 2002). Figure 10.24b shows the final tone-mapped color image, which shows no visible halos despite the extremely large variation in input radiance values.

Yet another alternative to these two approaches is to perform the local dodging and burning using a locally scale-selective operator (Reinhard, Stark *et al.* 2002). Figure 10.25 shows how such a scale selection operator can determine a radius (scale) that only includes similar color values within the inner circle while avoiding much brighter values in the surrounding circle. In practice, a difference of Gaussians normalized by the inner Gaussian response is evaluated over a range of scales, and the largest scale whose metric is below a threshold is selected (Reinhard, Stark *et al.* 2002).

Another recently developed approach to tone mapping based on multi-resolution decomposition is the Local Laplacian Filter (Paris, Hasinoff, and Kautz 2011), which we introduced in Section 3.5.3. Coefficients in a Laplacian pyramid are constructed from locally contrast-adjusted patches, which enables the technique to not only tone map HDR images, but also to enhance local details and do style transfer (Aubry, Paris *et al.* 2014).

What all of these techniques have in common is that they adaptively attenuate or brighten different regions of the image so that they can be displayed in a limited gamut without loss of contrast. Lischinski, Farbman *et al.* (2006) introduce an *interactive* technique that performs this operation by interpolating a set of sparse user-drawn adjustments (strokes and associated exposure value cor-

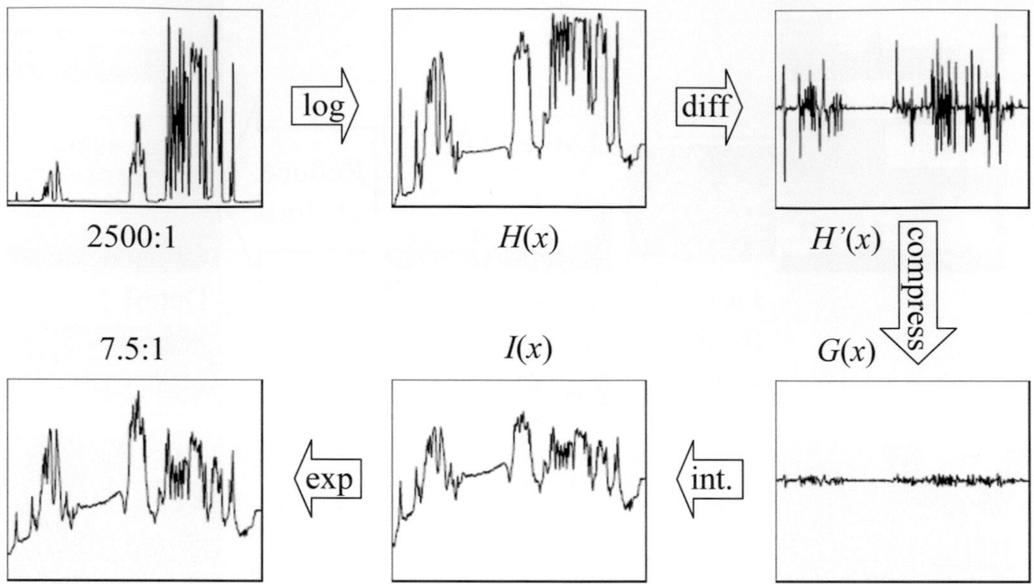

Figure 10.23 Gradient domain tone mapping (Fattal, Lischinski, and Werman 2002) © 2002 ACM. The original image with a dynamic range of 2415:1 is first converted into the log domain, $H(x)$, and its gradients are computed, $H'(x)$. These are attenuated (compressed) based on local contrast, $G(x)$, and integrated to produce the new logarithmic exposure image $I(x)$, which is exponentiated to produce the final intensity image, whose dynamic range is 7.5:1.

(a) (b)

Figure 10.24 Gradient domain tone mapping (Fattal, Lischinski, and Werman 2002) © 2002 ACM: (a) attenuation map, with darker values corresponding to more attenuation; (b) final tone-mapped image.

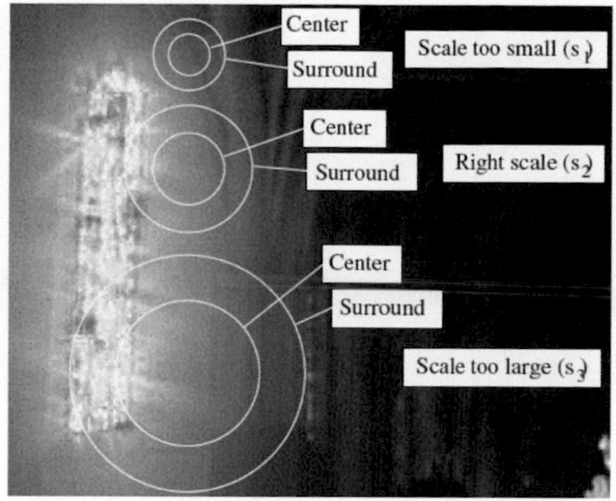

Figure 10.25 Scale selection for tone mapping (Reinhard, Stark *et al.* 2002) © 2002 ACM.

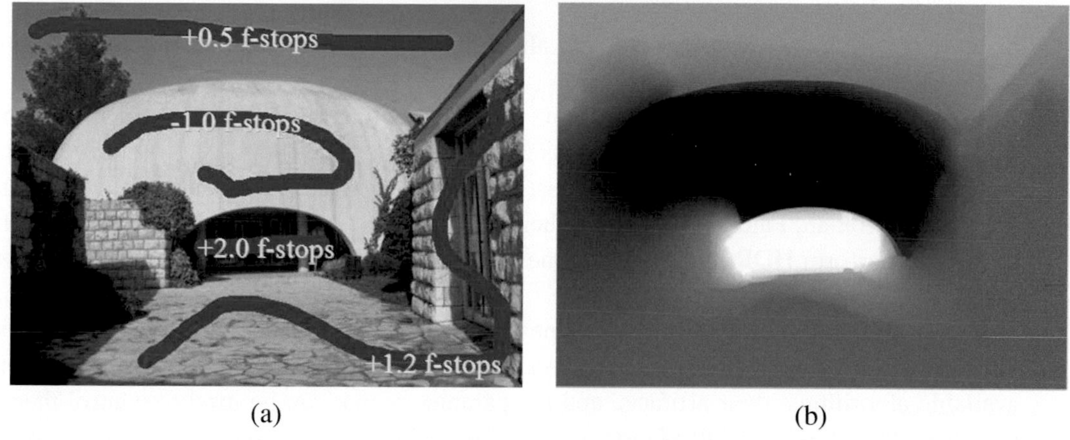

(a) (b)

Figure 10.26 Interactive local tone mapping (Lischinski, Farbman *et al.* 2006) © 2006 ACM: (a) user-drawn strokes with associated exposure values $g(x, y)$; (b) corresponding piecewise-smooth exposure adjustment map $f(x, y)$.

rections) to a piecewise-continuous exposure correction map (Figure 10.26). The interpolation is performed by minimizing a locally weighted least squares (WLS) variational problem,

$$\min \int\int w_{\text{d}}(x,y)\|f(x,y) - g(x,y)\|^2 dx\, dy + \lambda \int\int w_{\text{s}}(x,y)\|\nabla f(x,y)\|^2 dx\, dy, \quad (10.20)$$

where $g(x, y)$ and $f(x, y)$ are the input and output log exposure (attenuation) maps (Figure 10.26). The data weighting term $w_{\text{d}}(x, y)$ is 1 at stroke locations and 0 elsewhere. The smoothness weighting term $w_{\text{s}}(x, y)$ is inversely proportional to the log-luminance gradient,

$$w_{\text{s}} = \frac{1}{\|\nabla H\|^{\alpha} + \epsilon} \quad (10.21)$$

and hence encourages the $f(x, y)$ map to be smoother in low-gradient areas than along high gradient discontinuities.[15] The same approach can also be used for fully automated tone mapping by setting

[15]In practice, the x and y discrete derivatives are weighted separately (Lischinski, Farbman *et al.* 2006). Their default parameter settings are $\lambda = 0.2$, $\alpha = 1$, and $\epsilon = 0.0001$.

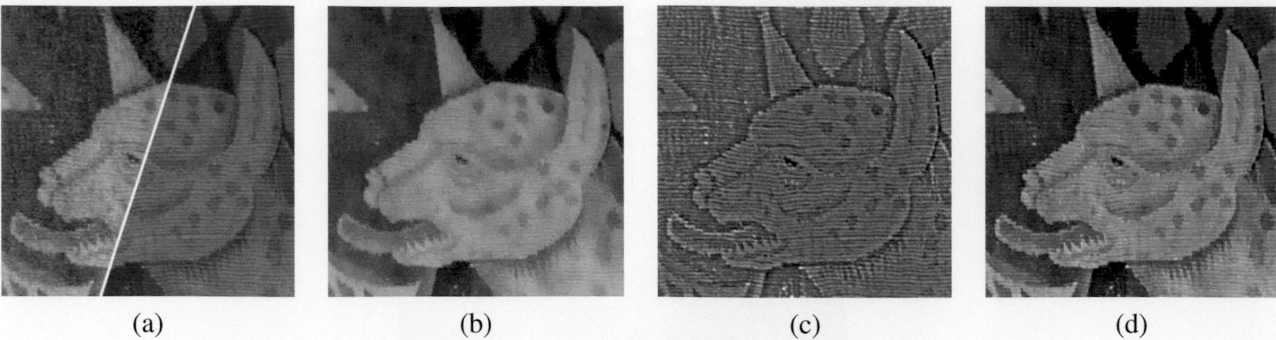

Figure 10.27 Detail transfer in flash/no-flash photography (Petschnigg, Agrawala *et al.* 2004) © 2004 ACM: (a) details of input ambient A and flash F images; (b) joint bilaterally filtered no-flash image A^{NR}; (c) detail layer F^{Detail} computed from the flash image F; (d) final merged image A^{Final}.

target exposure values at each pixel and allowing the weighted least squares to convert these into piecewise smooth adjustment maps.

The weighted least squares algorithm, which was originally developed for image colorization applications (Levin, Lischinski, and Weiss 2004), has since been applied to general edge-preserving smoothing in applications such as contrast enhancement (Bae, Paris, and Durand 2006) and tone mapping (Farbman, Fattal *et al.* 2008) where the bilateral filtering was previously used. It can also be used to perform HDR merging and tone mapping simultaneously (Raman and Chaudhuri 2007, 2009).

Given the wide range of locally adaptive tone mapping algorithms that have been developed, which ones should be used in practice? Freeman (2008) provides a great discussion of commercially available algorithms, their artifacts, and the parameters that can be used to control them. He also has a wealth of tips for HDR photography and workflow. I highly recommend his book for anyone contemplating additional research (or personal photography) in this area.

10.2.2 *Application*: Flash photography

While high dynamic range imaging combines images of a scene taken at different exposures, it is also possible to combine flash and non-flash images to achieve better exposure and color balance and to reduce noise (Eisemann and Durand 2004; Petschnigg, Agrawala *et al.* 2004).

The problem with flash images is that the color is often unnatural (it fails to capture the ambient illumination), there may be strong shadows or specularities, and there is a radial falloff in brightness away from the camera (Figures 10.1b and 10.27a). Non-flash photos taken under low light conditions often suffer from excessive noise (because of the high ISO gains and low photon counts) and blur (due to longer exposures). Is there some way to combine a non-flash photo taken just before the flash goes off with the flash photo to produce an image with good color values, sharpness, and low noise? In fact, the discontinued FujiFilm FinePix F40fd camera takes a pair of flash and no flash images in quick succession; however, it only lets you decide to keep one of them.

Petschnigg, Agrawala *et al.* (2004) approach this problem by first filtering the no-flash (ambient) image A with a variant of the bilateral filter called the *joint bilateral filter*[16] in which the range kernel (3.36)

$$r(i,j,k,l) = \exp\left(-\frac{\|f(i,j) - f(k,l)\|^2}{2\sigma_r^2}\right) \tag{10.22}$$

[16]Eisemann and Durand (2004) call this the *cross bilateral filter*.

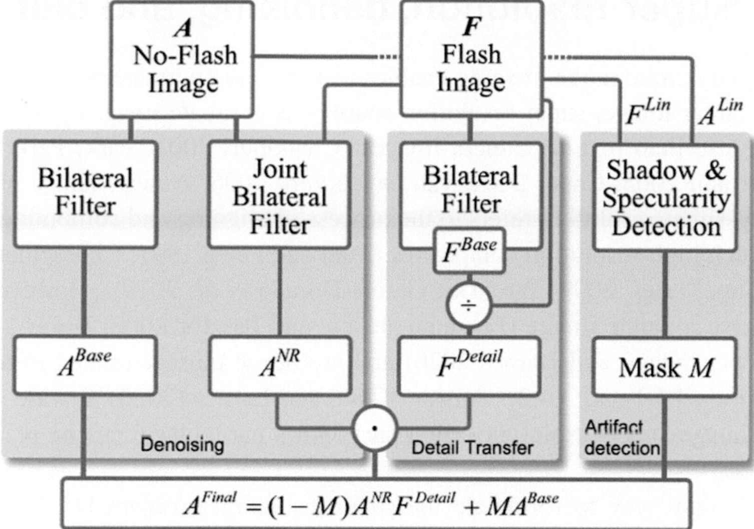

Figure 10.28 Flash/no-flash photography algorithm (Petschnigg, Agrawala *et al.* 2004) © 2004 ACM. The ambient (no-flash) image A is filtered with a regular bilateral filter to produce A^{Base}, which is used in shadow and specularity regions, and a joint bilaterally filtered noise reduced image A^{NR}. The flash image F is bilaterally filtered to produce a base image F^{Base} and a detail (ratio) image F^{Detail}, which is used to modulate the denoised ambient image. The shadow/specularity mask M is computed by comparing linearized versions of the flash and no-flash images.

is evaluated on the flash image F instead of the ambient image A, as the flash image is less noisy and hence has more reliable edges (Figure 10.27b). Because the contents of the flash image can be unreliable inside and at the boundaries of shadows and specularities, these are detected and a regular bilaterally filtered image A^{Base} is used instead (Figure 10.28).

The second stage of their algorithm computes a flash detail image

$$F^{Detail} = \frac{F + \epsilon}{F^{\text{Base}} + \epsilon}, \tag{10.23}$$

where F^{Base} is a bilaterally filtered version of the flash image F and $\epsilon = 0.02$. This detail image (Figure 10.27c) encodes details that may have been filtered away from the noise-reduced no-flash image A^{NR}, as well as additional details created by the flash camera, which often add crispness. The detail image is used to modulate the noise-reduced ambient image A^{NR} to produce the final results

$$A^{Final} = (1 - M)A^{NR}F^{Detail} + MA^{Base} \tag{10.24}$$

shown in Figures 10.1b and 10.27d.

Eisemann and Durand (2004) present an alternative algorithm that shares some of the same basic concepts. Both papers are well worth reading and contrasting (Exercise 10.6).

Flash images can also be used for a variety of additional applications such as extracting more reliable foreground mattes of objects (Raskar, Tan *et al.* 2004; Sun, Li *et al.* 2006). Given a large enough training set, it is also possible to decompose single flash images into their ambient and flash illumination components, which can be used to adjust their appearance (Aksoy, Kim *et al.* 2018). Flash photography is just one instance of the more general topic of *active illumination*, which is discussed in more detail by Raskar and Tumblin (2010) and Ikeuchi, Matsushita *et al.* (2020).

10.3 Super-resolution, denoising, and blur removal

While high dynamic range imaging enables us to obtain an image with a larger dynamic range than a single regular image, super-resolution enables us to create images with higher *spatial* resolution and less noise than regular camera images (Chaudhuri 2001; Park, Park, and Kang 2003; Capel and Zisserman 2003; Capel 2004; van Ouwerkerk 2006; Anwar, Khan, and Barnes 2020). Most commonly, super-resolution refers to the process of aligning and combining several input images to produce such high-resolution composites (Irani and Peleg 1991; Cheeseman, Kanefsky *et al.* 1993; Pickup, Capel *et al.* 2009; Wronski, Garcia-Dorado *et al.* 2019). However, some techniques can super-resolve a single image (Freeman, Jones, and Pasztor 2002; Baker and Kanade 2002; Fattal 2007; Anwar, Khan, and Barnes 2020) and are hence closely related to techniques for removing blur (Sections 3.4.1 and 3.4.2). Anwar, Khan, and Barnes (2020) provide a comprehensive review of single image super-resolution techniques with a particular focus on recent deep learning-based approaches.

A traditional way to formulate the super-resolution problem is to write down the stochastic image formation equations and image priors and to then use Bayesian inference to recover the super-resolved (original) sharp image. We can do this by generalizing the image formation equations used for image deblurring (Section 3.4.1), which we also used for blur kernel (PSF) estimation (Section 10.1.4). In this case, we have several observed images $\{o_k(\mathbf{x})\}$, as well as an image warping function $\hat{\mathbf{h}}_k(\mathbf{x})$ for each observed image (Figure 3.46). Combining all of these elements, we get the (noisy) observation equations[17]

$$o_k(\mathbf{x}) = D\{b(\mathbf{x}) * s(\hat{\mathbf{h}}_k(\mathbf{x}))\} + n_k(\mathbf{x}), \tag{10.25}$$

where D is the downsampling operator, which operates *after* the super-resolved (sharp) warped image $s(\hat{\mathbf{h}}_k(\mathbf{x}))$ has been convolved with the blur kernel $b(\mathbf{x})$. The above image formation equations lead to the following least squares problem,

$$\sum_k \|o_k(\mathbf{x}) - D\{b_k(\mathbf{x}) * s(\hat{\mathbf{h}}_k(\mathbf{x}))\}\|^2. \tag{10.26}$$

In most super-resolution algorithms, the alignment (warping) $\hat{\mathbf{h}}_k$ is estimated using one of the input frames as the *reference frame*; either feature-based (Section 8.1.3) or direct (image-based) (Section 9.2) parametric alignment techniques can be used. (A few algorithms, such as those described by Schultz and Stevenson (1996), Capel (2004), and Wronski, Garcia-Dorado *et al.* (2019) use dense (per-pixel flow) estimates.) A better approach is to re-compute the alignment by directly minimizing (10.26) once an initial estimate of $s(\mathbf{x})$ has been computed (Hardie, Barnard, and Armstrong 1997) or to *marginalize* out the motion parameters altogether (Pickup, Capel *et al.* 2007).

The point spread function (blur kernel) b_k is either inferred from knowledge of the image formation process (e.g., the amount of motion or defocus blur and the camera sensor optics) or calibrated from a test image or the observed images $\{o_k\}$ using one of the techniques described in Section 10.1.4. The problem of simultaneously inferring the blur kernel and the sharp image is known as *blind image deconvolution* (Kundur and Hatzinakos 1996; Levin 2006; Levin, Weiss *et al.* 2011; Campisi and Egiazarian 2017).[18]

[17] It is also possible to add an unknown bias–gain term to each observation (Capel 2004), as was done for motion estimation in (9.8).

[18] Notice that there is a chicken-and-egg problem if both the blur kernel and the super-resolved image are unknown. This can be "broken" either using structural assumptions about the sharp image, e.g., the presence of edges (Joshi, Szeliski, and Kriegman 2008) or prior models for the image, such as edge sparsity (Fergus, Singh *et al.* 2006).

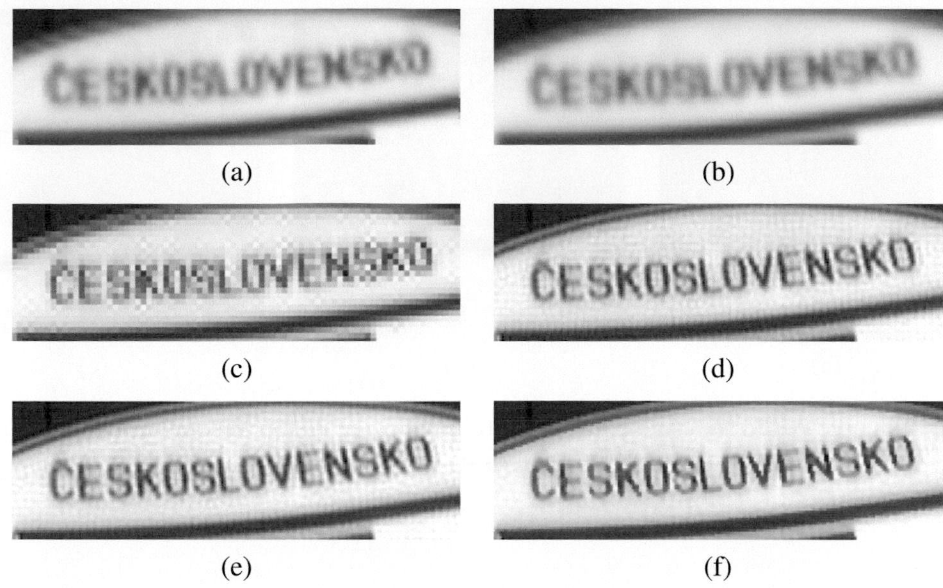

Figure 10.29 Super-resolution results using a variety of image priors (Capel 2001): (a) Low-res ROI (bicubic $3\times$ zoom); (b) average image; (c) MLE @ $1.25\times$ pixel-zoom; (d) simple $\|x\|^2$ prior ($\lambda = 0.004$); (e) GMRF ($\lambda = 0.003$); (f) HMRF ($\lambda = 0.01$, $\alpha = 0.04$). 10 images are used as input and a $3\times$ super-resolved image is produced in each case, except for the MLE result in (c).

Given an estimate of $\hat{\mathbf{h}}_k$ and $b_k(\mathbf{x})$, (10.26) can be re-written using matrix/vector notation as a large sparse least squares problem in the unknown values of the super-resolved pixels $\mathbf{s}$,

$$\sum_k \|\mathbf{o}_k - \mathbf{D}\mathbf{B}_k\mathbf{W}_k\mathbf{s}\|^2. \tag{10.27}$$

(Recall from (3.75) that once the warping function $\hat{\mathbf{h}}_k$ is known, values of $s(\hat{\mathbf{h}}_k(\mathbf{x}))$ depend linearly on those in $s(\mathbf{x})$.) An efficient way to solve this least squares problem is to use preconditioned conjugate gradient descent (Capel 2004), although some earlier algorithms, such as the one developed by Irani and Peleg (1991), used regular gradient descent (also known as iterative back projection (IBP) in the computed tomography literature).

The above formulation assumes that warping can be expressed as a simple (sinc or bicubic) interpolated resampling of the super-resolved sharp image, followed by a stationary (spatially invariant) blurring (PSF) and area integration process. However, if the surface is severely foreshortened, we have to take into account the spatially varying filtering that occurs during the image warping (Section 3.6.1), before we can then model the PSF induced by the optics and camera sensor (Wang, Kang *et al.* 2001; Capel 2004).

How well does this least squares (MLE) approach to super-resolution work? In practice, this depends a lot on the amount of blur and aliasing in the camera optics, as well as the accuracy in the motion and PSF estimates (Baker and Kanade 2002; Jiang, Wong, and Bao 2003; Capel 2004). Less blurring and more aliasing means that there is more (aliased) high frequency information available to be recovered. However, because the least squares (maximum likelihood) formulation uses no image prior, a lot of high-frequency noise can be introduced into the solution (Figure 10.29c).

For this reason, classic super-resolution algorithms assume some form of image prior. The simplest of these is to place a penalty on the image derivatives similar to Equations (4.29) and

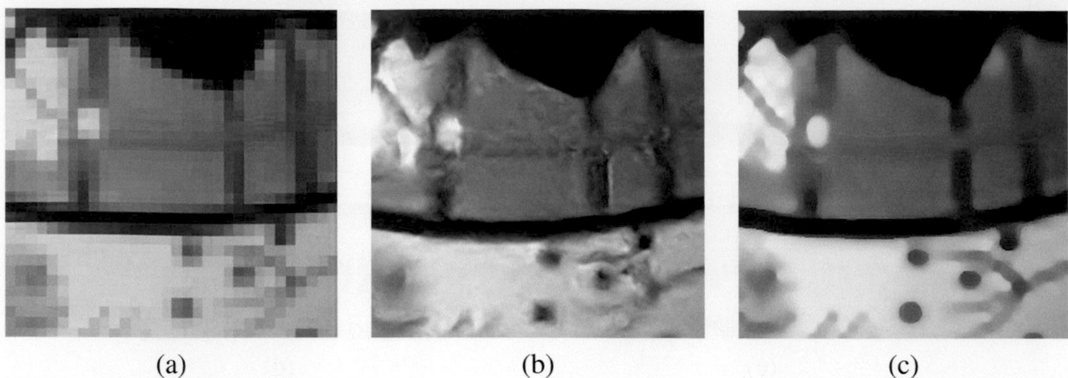

(a) (b) (c)

Figure 10.30 Example-based super-resolution: (a) original 32 × 32 low-resolution image; (b) example-based super-resolved 256 × 256 image (Freeman, Jones, and Pasztor 2002) © 2002 IEEE; (c) upsampling via imposed edge statistics (Fattal 2007) © 2007 ACM.

(4.42), e.g.,

$$\sum_{(i,j)} \rho_p(s(i,j) - s(i+1,j)) + \rho_p(s(i,j) - s(i,j+1)). \tag{10.28}$$

As discussed in Section 4.3, when ρ_p is quadratic, this is a form of Tikhonov regularization (Section 4.2), and the overall problem is still linear least squares. The resulting prior image model is a Gaussian Markov random field (GMRF), which can be extended to other (e.g., diagonal) differences, as in Capel (2004) and Figure 10.29.

Unfortunately, GMRFs tend to produce solutions with visible ripples, which can also be interpreted as increased noise sensitivity in middle frequencies. A better image prior is a robust prior that encourages piecewise continuous solutions (Black and Rangarajan 1996), see Appendix B.3. Examples of such priors include the Huber potential (Schultz and Stevenson 1996; Capel and Zisserman 2003), which is a blend of a Gaussian with a longer-tailed Laplacian, and the even sparser (heavier-tailed) hyper-Laplacians used by Levin, Fergus *et al.* (2007) and Krishnan and Fergus (2009). It is also possible to learn the parameters for such priors using cross-validation (Capel 2004; Pickup 2007).

While sparse (robust) derivative priors can reduce rippling effects and increase edge sharpness, they cannot *hallucinate* higher-frequency texture or details. To do this, a training set of sample images can be used to find plausible mappings between low-frequency originals and the missing higher frequencies. Inspired by some of the example-based texture synthesis algorithms we discuss in Section 10.5, the *example-based super-resolution* algorithm developed by Freeman, Jones, and Pasztor (2002) uses training images to *learn* the mapping between local texture patches and missing higher-frequency details. To ensure that overlapping patches are similar in appearance, a Markov random field is used and optimized using either belief propagation (Freeman, Pasztor, and Carmichael 2000) or a raster-scan deterministic variant (Freeman, Jones, and Pasztor 2002). Figure 10.30 shows the results of hallucinating missing details using this approach and compares these results to a more recent algorithm by Fattal (2007). This latter algorithm learns to predict oriented gradient magnitudes in the finer resolution image based on a pixel's location relative to the nearest detected edge along with the corresponding edge statistics (magnitude and width). It is also possible to combine sparse (robust) derivative priors with example-based super-resolution, as shown by Tappen, Russell, and Freeman (2003).

An alternative (but closely related) form of hallucination is to *recognize* the parts of a training

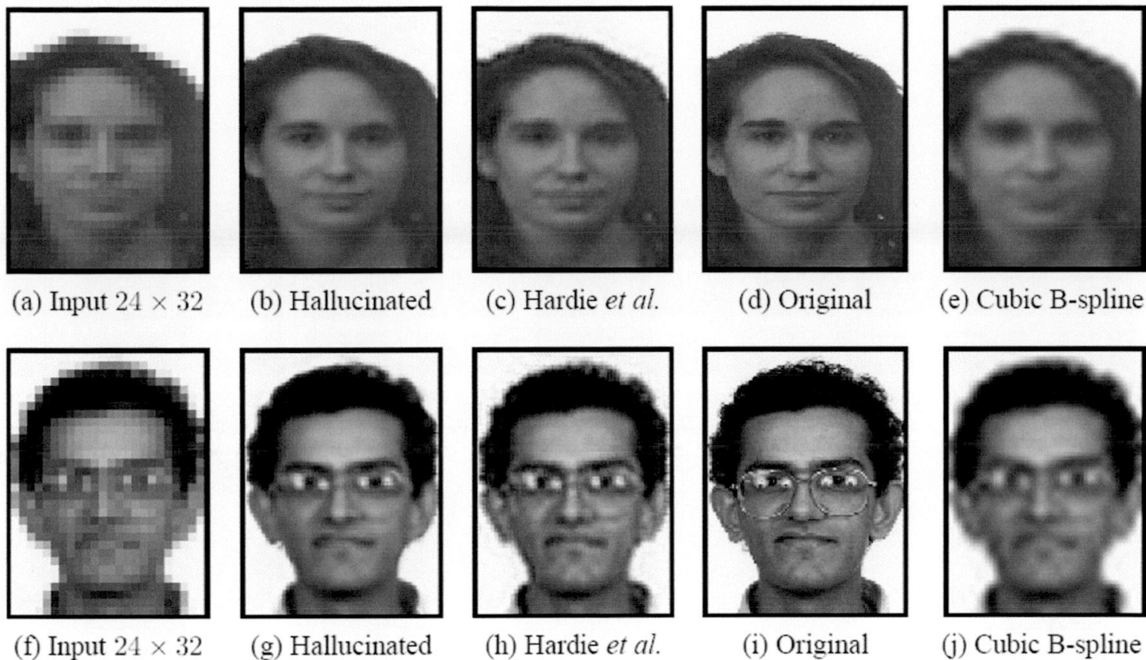

Figure 10.31 Recognition-based super-resolution (Baker and Kanade 2002) © 2002 IEEE. The *Hallucinated* column shows the results of the recognition-based algorithm compared to the regularization-based approach of Hardie, Barnard, and Armstrong (1997).

database of images to which a low-resolution pixel might correspond. In their work, Baker and Kanade (2002) use local derivative-of-Gaussian filter responses as features and then match *parent structure* vectors in a manner similar to De Bonet (1997).[19] The high-frequency gradient at each recognized training image location is then used as a constraint on the super-resolved image, along with the usual reconstruction (prediction) Equation (10.26). Figure 10.31 shows the result of hallucinating higher-resolution faces from lower-resolution inputs; Baker and Kanade (2002) also show examples of super-resolving known-font text. Exercise 10.7 gives more details on how to implement and test one or more of these super-resolution techniques.

The latest trend in super-resolution has been the use of deep neural networks to directly predict super-resolved images. This approach, which began with the seminal work of Dong, Loy *et al.* (2016), has generated dozens of different DNNs and architectures, including the Deep Learning Super Sampling hardware embedded in the latest NVIDIA graphics cards (Burnes 2020). The recent survey on single-image super-resolution by Anwar, Khan, and Barnes (2020) categorizes these algorithms into a taxonomy (Figure 10.32a), provides a pictorial summary network architectures (Figure 10.32b), and compares the super-resolution results both numerically and visually on noise-free known bicubic-kernel decimation image datasets. While the results shown in Figure 10.33 show dramatic differences between algorithms, it is not clear how well these algorithms generalize to real-world noisy input with unknown blur kernels. The RealSR real-world super-resolution dataset developed by (Cai, Zeng *et al.* 2019), shot using a zoom lens on a digital camera, provides a means to test (and train) algorithms on real imaging degradations. This dataset forms the basis

[19]For face super-resolution, where all the images are pre-aligned, only corresponding pixels in different images are examined.

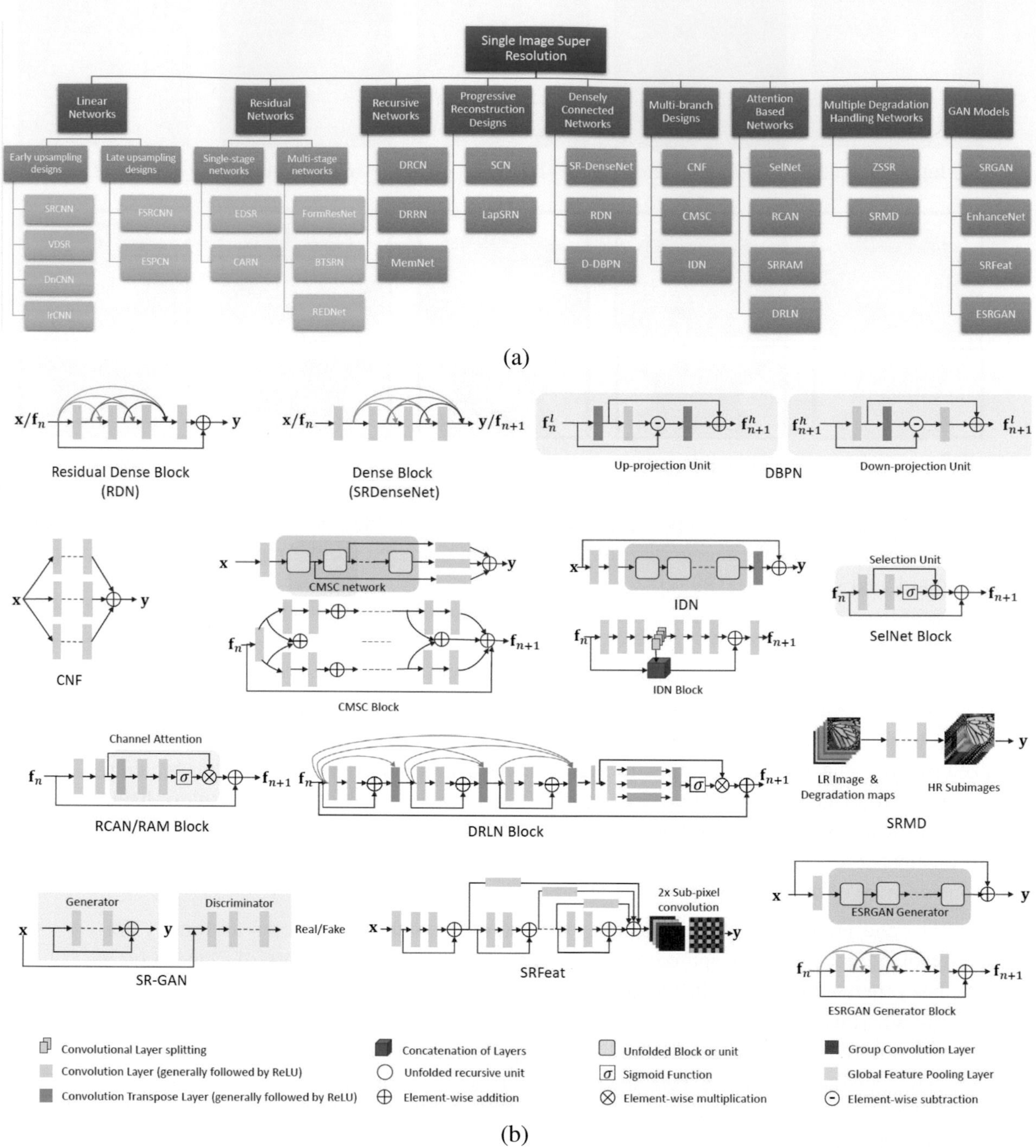

Figure 10.32 Recent deep neural network algorithms for single image super-resolution (Anwar, Khan, and Barnes 2020) © 2020 ACM: (a) a taxonomy of the algorithms based on their general approach; (b) schematic architectures for a subset of the algorithms.

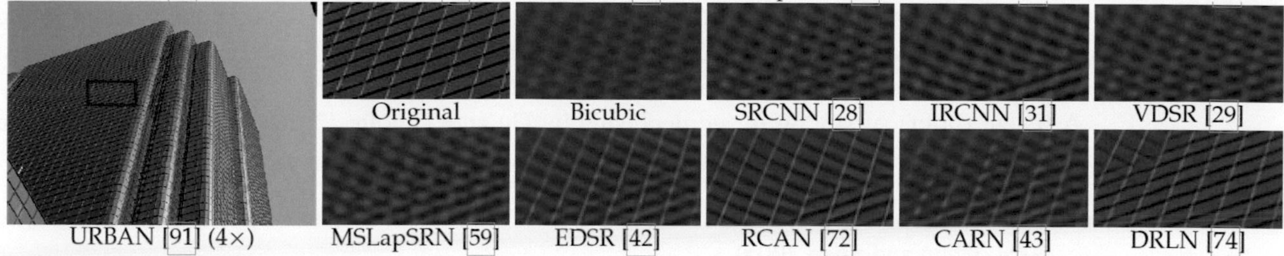

Figure 10.33 Visual comparison of some super-resolution algorithms (Anwar, Khan, and Barnes 2020) © 2020 ACM.

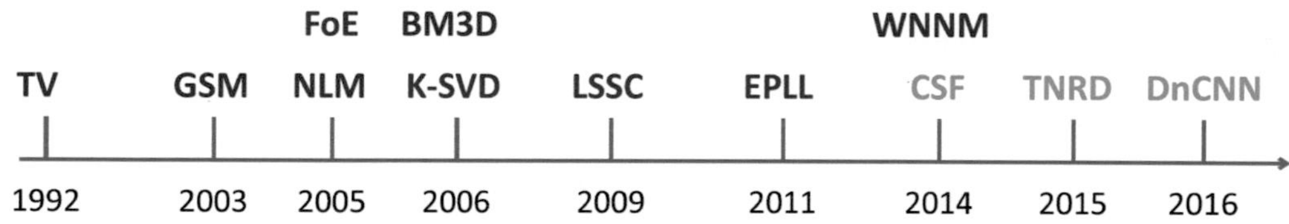

Figure 10.34 Timeline of denoising algorithms from Gu and Timofte (2019) © 2019 Springer.

for the NTIRE challenges on real image super-resolution (Cai, Gu *et al.* 2019),[20] which provide empirical comparisons of recent deep network-based algorithms.

While single-image super-resolution is interesting, much more impressive (and practical) results can be obtained by building a multi-frame super-resolution algorithm directly into a smartphone camera, where the processing can be done jointly with the image demosaicing. We discuss recent work by Wronski, Garcia-Dorado *et al.* (2019) in Section 10.3.1 and Figure 10.38 on color image demosaicing. It is also possible to upsample videos temporally using frame interpolation (Section 9.4.1), spatially using video super-resolution (Liu and Sun 2013; Kappeler, Yoo *et al.* 2016; Shi, Caballero *et al.* 2016; Tao, Gao *et al.* 2017; Nah, Timofte *et al.* 2019; Isobe, Jia *et al.* 2020; Li, Tao *et al.* 2020), or simultaneously in both the spatial and temporal dimensions (Kang, Jo *et al.* 2020).

Single and multi-frame denoising

Image denoising is one of the classic problems in image processing and computer vision (Perona and Malik 1990b; Rudin, Osher, and Fatemi 1992; Buades, Coll, and Morel 2005b). Over the last four decades, hundreds of algorithms have been developed, and the field continues to be actively studied, with recent algorithms all being based on deep neural networks.

The latest benchmark for comparing image denoising algorithms, the NTIRE 2020 Challenge on Real Image Denoising (Abdelhamed, Afifi *et al.* 2020), is based on a smartphone image denoising dataset (SIDD) (Abdelhamed, Lin, and Brown 2018), where the noise-free ground truth images

[20] https://data.vision.ee.ethz.ch/cvl/ntire20/, https://data.vision.ee.ethz.ch/cvl/aim20/

were obtained by averaging sets of 150 noisy images. This provides much more realistic and varied real-world noise and image processing models than the synthetically noised images used in most previous benchmarks (with the exception of (Plötz and Roth 2017)).

A recent (brief) survey on image denoising by Gu and Timofte (2019) includes the following seminal denoising papers[21] (see Figure 10.34 for a timeline):

- total variation (TV) (Rudin, Osher, and Fatemi 1992; Chan, Osher, and Shen 2001; Chambolle 2004; Chan and Shen 2005),
- Gaussian scale mixtures (GSMs) (Lyu and Simoncelli 2009),
- Field of Experts (FoE) (Roth and Black 2009),
- non-local means (NLM) (Buades, Coll, and Morel 2005a,b),
- BM3D (Dabov, Foi *et al.* 2007),
- sparse overcomplete dictionaries (K-SVD) (Aharon, Elad, and Bruckstein 2006),
- expected patch log likelihood (EPLL) (Zoran and Weiss 2011),
- an MLP denoiser (Burger, Schuler, and Harmeling 2012),
- weighted nuclear norm minimization (WNNM) (Gu, Zhang *et al.* 2014),
- shrinkage fields (CSF) (Schmidt and Roth 2014),
- Trainable Nonlinear Reaction Diffusion (TNRD) (Chen and Pock 2016),
- a cross-channel noise model for color images (Nam, Hwang *et al.* 2016),
- a denoising residual CNN (DnCNN) (Zhang, Zuo *et al.* 2017), which is now considered the baseline for DNN denoising, and
- learning to see in the dark (Chen, Chen *et al.* 2018).

While these results show dramatic improvement over time, today's imaging sensors for the most part produce relatively clean images, except in low-light situations, where the ISO camera gain must be increased and the read and photon noise become comparable to the signal strength. In this regime, it is preferable, if possible, to take a rapid burst of images at low ISO (gain) and then combine these to obtain a denoised image (Hasinoff, Kutulakos *et al.* 2009; Hasinoff, Durand, and Freeman 2010; Liu, Yuan *et al.* 2014). This approach was generalized and applied to low-light photography in the HDR+ system of Hasinoff, Sharlet *et al.* (2016). More recent work along these lines, some of which combines low-light photography, demosaicing, and in some cases super-resolution, includes papers by Godard, Matzen, and Uyttendaele (2018), Chen, Chen *et al.* (2018), Mildenhall, Barron *et al.* (2018), Wronski, Garcia-Dorado *et al.* (2019), and (Rong, Demandolx *et al.* 2020). Liba, Murthy *et al.* (2019) describe the technology that underlies Google's *Night Sight* feature, which not only robustly aligns and merges different moving regions together under noisy conditions, but also introduces the concept of "motion metering" to determine the optimal number of frames and exposure times.

[21] I have added a few more papers from the ICCV tutorial by Brown (2019) and a few additional recommendations from Abdelrahman Abdelhamed.

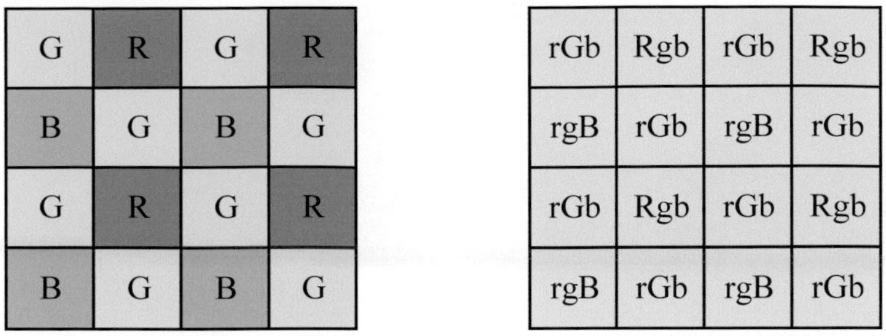

G	R	G	R
B	G	B	G
G	R	G	R
B	G	B	G

(a)

rGb	Rgb	rGb	Rgb
rgB	rGb	rgB	rGb
rGb	Rgb	rGb	Rgb
rgB	rGb	rgB	rGb

(b)

Figure 10.35 Bayer RGB pattern: (a) color filter array layout; (b) interpolated pixel values, with unknown (guessed) values shown as lower case.

Blur removal

Under favorable conditions, super-resolution and related upsampling techniques can increase the resolution of a well-photographed image or image collection. When the input images are blurry to start with, the best one can often hope for is to reduce the amount of blur. This problem is closely related to super-resolution, with the biggest differences being that the blur kernel b is usually much larger (and unknown) and the downsampling factor D is unity.

A large literature on image deblurring exists; some publications with nice literature reviews include those by Fergus, Singh *et al.* (2006), Yuan, Sun *et al.* (2008), and Joshi, Zitnick *et al.* (2009). It is also possible to reduce blur by combining sharp (but noisy) images with blurrier (but cleaner) images (Yuan, Sun *et al.* 2007), take lots of quick exposures (Hasinoff and Kutulakos 2011; Hasinoff, Kutulakos *et al.* 2009; Hasinoff, Durand, and Freeman 2010), or use *coded aperture* techniques to simultaneously estimate depth and reduce blur (Levin, Fergus *et al.* 2007; Zhou, Lin, and Nayar 2009). When available, data from on-board IMUs (inertial measurement units) can be used for blur kernel determination (Joshi, Kang *et al.* 2010). It is also possible to use information from dual-pixel sensors to aid the deblurring of misfocused images (Abuolaim and Brown 2020).

The past decade has seen the introductions of a large number of new learning-based deblurring algorithms (Sun, Cao *et al.* 2015; Schuler, Hirsch *et al.* 2016; Nah, Hyun Kim, and Mu Lee 2017; Kupyn, Budzan *et al.* 2018; Tao, Gao *et al.* 2018; Zhang, Dai *et al.* 2019; Kupyn, Martyniuk *et al.* 2019). There has also been some work on artificially re-introducing texture in deblurred images to better match the expected image statistics (Cho, Joshi *et al.* 2012), i.e., what is now commonly called *perceptual loss* (Section 5.3.4).

10.3.1 Color image demosaicing

A special case of super-resolution, which is used daily in most digital still cameras, is the process of *demosaicing* samples from a color filter array (CFA) into a full-color RGB image. Figure 10.35 shows the most commonly used CFA known as the *Bayer pattern*, which has twice as many green (G) sensors as red and blue sensors.

The process of going from the known CFA pixels values to the full RGB image is quite challenging. Unlike regular super-resolution, where small errors in guessing unknown values usually show up as blur or aliasing, demosaicing artifacts often produce spurious colors or high-frequency patterned *zippering*, which are quite visible to the eye (Figure 10.36b).

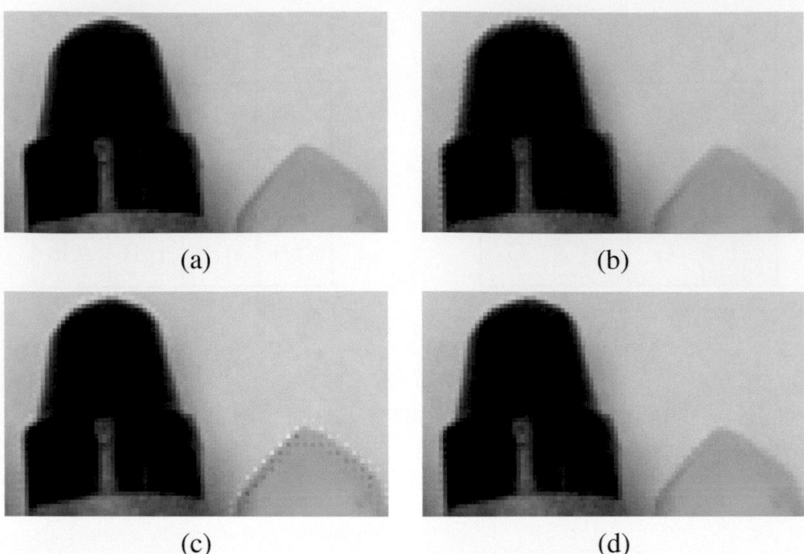

Figure 10.36 CFA demosaicing results (Bennett, Uyttendaele *et al.* 2006) © 2006 Springer: (a) original full-resolution image (a color subsampled version is used as the input to the algorithms); (b) bilinear interpolation results, showing color fringing near the tip of the blue crayon and zippering near its left (vertical) edge; (c) the high-quality linear interpolation results of Malvar, He, and Cutler (2004) (note the strong halo/checkerboard artifacts on the yellow crayon); (d) using the local two-color prior of Bennett, Uyttendaele *et al.* (2006).

Over the years, a variety of techniques have been developed for image demosaicing (Kimmel 1999). Longere, Delahunt *et al.* (2002), Tappen, Russell, and Freeman (2003), and Li, Gunturk, and Zhang (2008) provide surveys of the field as well as comparisons of previously developed techniques using perceptually motivated metrics. To reduce the zippering effect, most techniques use the edge or gradient information from the green channel, which is more reliable because it is sampled more densely, to infer plausible values for the red and blue channels, which are more sparsely sampled.

To reduce color fringing, some techniques perform a color space analysis, e.g., using median filtering on color opponent channels (Longere, Delahunt *et al.* 2002). The approach of Bennett, Uyttendaele *et al.* (2006) computes local two-color models from an initial demosaicing result, using a moving 5×5 window to find the two dominant colors (Figure 10.37).[22]

Once the local color model has been estimated at each pixel, a Bayesian approach is then used to encourage pixel values to lie along each color line and to cluster around the dominant color values, which reduces halos (Figure 10.36d). The Bayesian approach also supports the simultaneous application of demosaicing, denoising, and super-resolution, i.e., multiple CFA inputs can be merged into a higher-quality full-color image. More recent work that combines demosaicing and denoising includes papers by Chatterjee, Joshi *et al.* (2011) and Gharbi, Chaurasia *et al.* (2016). The NTIRE 2020 Challenge on Real Image Denoising (Abdelhamed, Afifi *et al.* 2020) includes a track on denoising RAW (i.e., color filter array) images. There's also an interesting paper by Jin, Facciolo, and Morel (2020) studying whether denoising should be applied before or after demosaicing.

As we mentioned before, burst photography (Cohen and Szeliski 2006; Hasinoff, Kutulakos *et al.* 2009; Hasinoff and Kutulakos 2011), i.e., the combination of rapidly acquired sequences of images, is becoming ubiquitous in smartphone cameras. A wonderful example of a recent system

[22]Previous work on locally linear color models (Klinker, Shafer, and Kanade 1990; Omer and Werman 2004) focuses on color and illumination variation within a single material, whereas Bennett, Uyttendaele *et al.* (2006) use the two-color model to describe variations across color (material) edges.

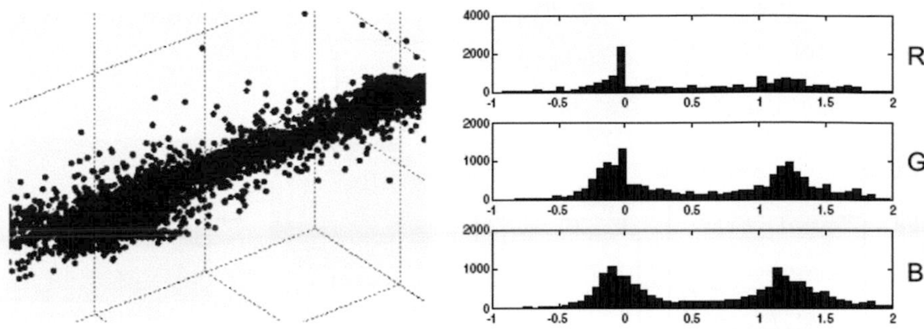

Figure 10.37 Two-color model computed from a collection of local 5 × 5 neighborhoods (Bennett, Uyttendaele *et al.* 2006) © 2006 Springer. After two-means clustering and reprojection along the line joining the two dominant colors (red dots), the majority of the pixels fall near the fitted line. The distribution along the line, projected along the RGB axes, is peaked at 0 and 1, the two dominant colors.

that performs joint demosaicing and multi-frame super-resolutions, based on locally adapted kernel functions (Figure 10.38), is the paper by Wronski, Garcia-Dorado *et al.* (2019), which underlies the *Super Res Zoom* feature in Google's Pixel smartphones.

10.3.2 Lens blur (bokeh)

The ability to create a shallow depth-of-field photograph using a large aperture (Section 2.2.3) has always been one of the advantages of large-format, e.g., single lens reflex (SLR), cameras. The desire to artificially simulate refocusable, shallow depth-of-field cameras was one of the driving impetuses behind computational photography (Levoy 2006) and led to the development of lightfield cameras (Ng, Levoy *et al.* 2005), which we discuss in Section 14.3.4. Although some commercial models, such as the Lytro, were produced, the ability to create such images with smartphone cameras has only recently become widespread.[23]

The Apple iPhone 7 Plus with its dual (wide/telephoto) lens was the first smartphone to introduce this feature, which they called the *Portrait mode*. Although the technical details behind this feature have never been published, the algorithm that estimates the depth image (which can be read out of the metadata in the portrait images) probably uses some combination of stereo matching and deep learning. A little later, Google released its own Portrait Mode, which uses the dual pixels, originally designed for focusing the camera optics, along with person segmentation to compute a depth map, as described in the paper by Wadhwa, Garg *et al.* (2018). Once the depth map has been estimated, a fast approximation to a back-to-front blurred *over* compositing operator is used to correctly blur the background without including foreground colors. More recently Garg, Wadhwa *et al.* (2019) have improved the quality of the depth estimation using a deep network, and also used two lenses (along with dual pixels) to produce even higher-quality depth maps (Zhang, Wadhwa *et al.* 2020).

One final word on *bokeh*, which is the term photographers use to describe the shape of the glints or highlights that appear in an image. This shape is determined by the configuration of the *aperture blades* that control how much light enters the lens (on larger-format cameras). Traditionally, these were made with straight metal leaves, which resulted in polygonal apertures, but they were then mostly replaced by curved leaves to produce a more circular shape. When using computational

[23] An earlier feature called Google Lens Blur, which required moving the camera in a pattern, https://ai.googleblog.com/2014/04/lens-blur-in-new-google-camera-app.html, was never widely used.

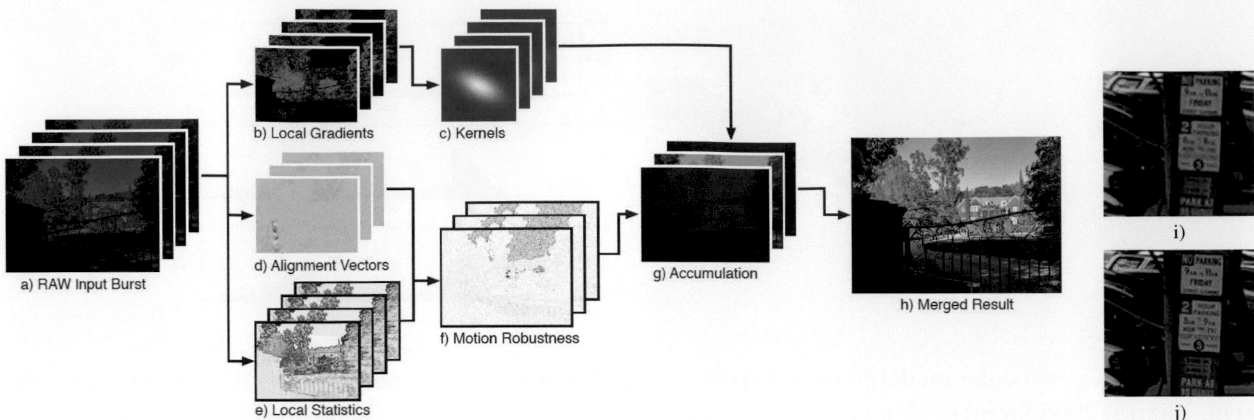

Figure 10.38 Hand-held multi-frame super-resolution (Wronski, Garcia-Dorado *et al.* 2019) © 2019 ACM. Processing pipeline, showing: (a) the captured burst of raw (Bayer CFA) images; (b) local gradients used to compute oriented kernels (c); (d) motion estimates, combined with local statistics (e) to compute blend weights (f). Results from (i) the previous method of Hasinoff, Sharlet *et al.* (2016) and (j) Wronski, Garcia-Dorado *et al.* (2019).

photography, we can use whatever shape is pleasing to the photographer, but preferably *not* a Gaussian blur, which does not correspond to any real aperture and produces indistinct highlights. The paper by Wadhwa, Garg *et al.* (2018) uses a circular bokeh for their depth-of-field effect and a more recent version performs the computations in the HDR (radiance) space to produce more accurate highlights.[24]

10.4 Image matting and compositing

Image matting and compositing is the process of cutting a foreground object out of one image and pasting it against a new background (Smith and Blinn 1996; Wang and Cohen 2009). It is commonly used in television and film production to composite a live actor in front of computer-generated imagery such as weather maps or 3D virtual characters and scenery (Wright 2006; Brinkmann 2008), and it has recently become a popular feature in video conferencing systems.

We have already seen a number of tools for interactively segmenting objects in an image, including snakes (Section 7.3.1), scissors (Section 7.3.1), and GrabCut segmentation (Section 4.3.2). While these techniques can generate reasonable pixel-accurate segmentations, they fail to capture the subtle interplay of foreground and background colors at *mixed pixels* along the boundary (Szeliski and Golland 1999) (Figure 10.39a).

To successfully copy a foreground object from one image to another without visible discretization artifacts, we need to *pull a matte*, i.e., to estimate a soft opacity channel α and the uncontaminated foreground colors F from the input composite image C. Recall from Section 3.1.3 (Figure 3.4) that the compositing equation (3.8) can be written as

$$C = (1 - \alpha)B + \alpha F. \tag{10.29}$$

This operator attenuates the influence of the background image B by a factor $(1 - \alpha)$ and then adds in the (partial) color values corresponding to the foreground element F.

[24]https://ai.googleblog.com/2019/12/improvements-to-portrait-mode-on-google.html

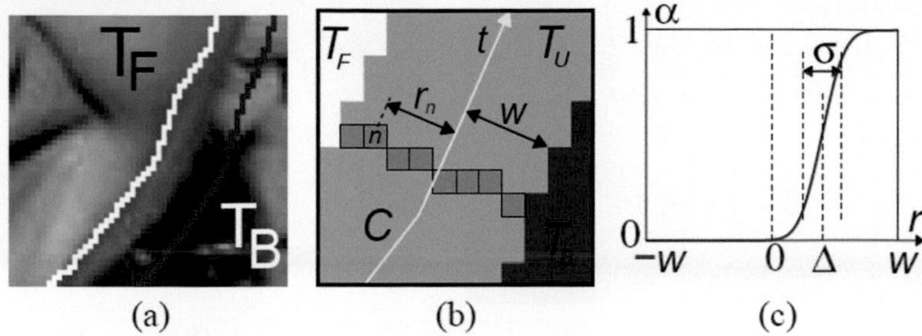

Figure 10.39 Softening a hard segmentation boundary (border matting) (Rother, Kolmogorov, and Blake 2004) © 2004 ACM: (a) the region surrounding a segmentation boundary where pixels of mixed foreground and background colors are visible; (b) pixel values along the boundary are used to compute a soft alpha matte; (c) at each point along the curve t, a displacement Δ and a width σ are estimated.

While the compositing operation is easy to implement, the reverse *matting* operation of estimating F, α, and B given an input image C is much more challenging (Figure 10.40). To see why, observe that while the composite pixel color C provides three measurements, the F, α, and B unknowns have a total of seven degrees of freedom. Devising techniques to estimate these unknowns despite the underconstrained nature of the problem is the essence of image matting.

In this section, we review a number of image matting techniques. We begin with *blue screen matting*, which assumes that the background is a constant known color, and discuss its variants, two-screen matting (when multiple backgrounds can be used) and difference matting (where the known background is arbitrary). We then discuss local variants of *natural image matting*, where both the foreground and background are unknown. In these applications, it is usual to first specify a *trimap*, i.e., a three-way labeling of the image into foreground, background, and unknown regions (Figure 10.40b). Next, we present some global optimization approaches to natural image matting. Finally, we discuss variants on the matting problem, including shadow matting, flash matting, and environment matting.

10.4.1 Blue screen matting

Blue screen matting involves filming an actor (or object) in front of a constant colored background. While originally bright blue was the preferred color, bright green is now more commonly used (Wright 2006; Brinkmann 2008). Smith and Blinn (1996) discuss a number of techniques for blue screen matting, which are mostly described in patents rather than in the open research literature. Early techniques used linear combinations of object color channels with user-tuned parameters to estimate the opacity α.

Chuang, Curless *et al.* (2001) describe a newer technique called Mishima's algorithm, which involves fitting two polyhedral surfaces (centered at the mean background color), separating the foreground and background color distributions, and then measuring the relative distance of a novel color to these surfaces to estimate α (Figure 10.41e). While this technique works well in many studio settings, it can still suffer from *blue spill*, where translucent pixels around the edges of an object acquire some of the background blue coloration.

Two-screen matting. In their paper, Smith and Blinn (1996) also introduce an algorithm called *triangulation matting* that uses more than one known background color to over-constrain the equations required to estimate the opacity α and foreground color F.

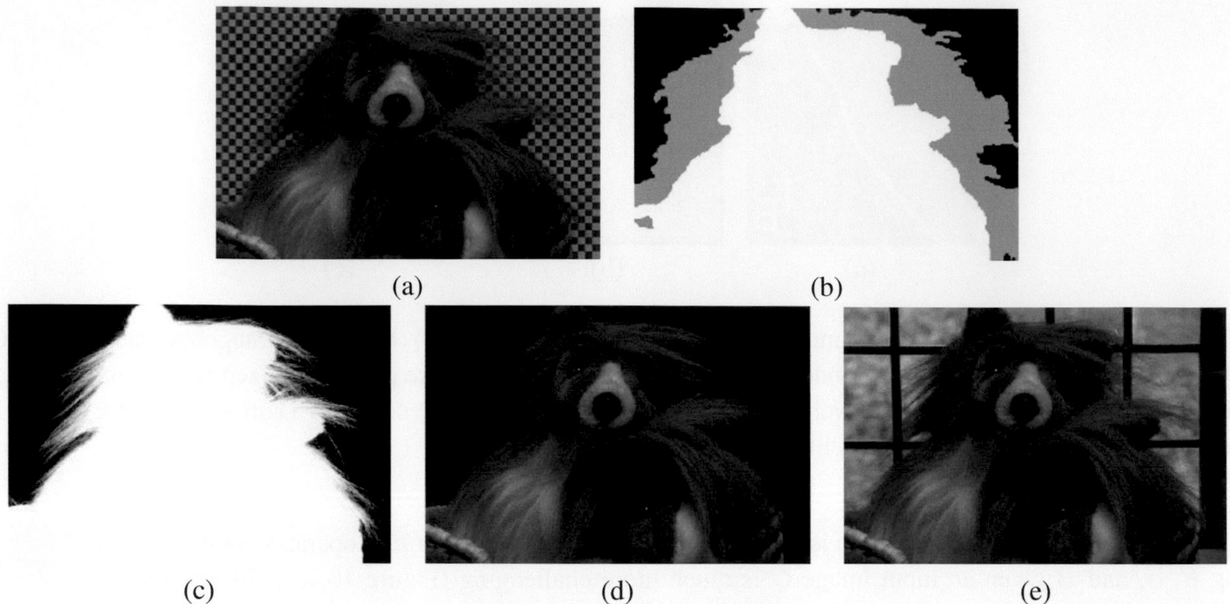

Figure 10.40 Natural image matting (Chuang, Curless *et al.* 2001) © 2001 IEEE: (a) input image with a "natural" (non-constant) background; (b) hand-drawn trimap—gray indicates unknown regions; (c) extracted alpha map; (d) extracted (premultiplied) foreground colors; (e) composite over a new background.

For example, consider in the compositing equation (10.29) setting the background color to black, i.e., $B = 0$. The resulting composite image C is therefore equal to αF. Replacing the background color with a different known non-zero value B now results in

$$C - \alpha F = (1 - \alpha)B, \tag{10.30}$$

which is an overconstrained set of (color) equations for estimating α. In practice, B should be chosen so as not to saturate C and, for best accuracy, several values of B should be used. It is also important that colors be linearized before processing, which is the case for *all* image matting algorithms. Papers that generate ground truth alpha mattes for evaluation purposes normally use these techniques to obtain accurate matte estimates (Chuang, Curless *et al.* 2001; Wang and Cohen 2007a; Levin, Acha, and Lischinski 2008; Rhemann, Rother *et al.* 2008, 2009).[25] Exercise 10.8 has you do this as well.

Difference matting. A related approach when the background is irregular but known is called *difference matting* (Wright 2006; Brinkmann 2008). It is most commonly used when the actor or object is filmed against a static background, e.g., for office video conferencing, person tracking applications (Toyama, Krumm *et al.* 1999), or to produce silhouettes for volumetric 3D reconstruction techniques (Section 12.7.3) (Szeliski 1993; Seitz and Dyer 1997; Seitz, Curless *et al.* 2006). It can also be used with a panning camera where the background is composited from frames where the foreground has been removed using a *garbage matte* (Section 10.4.5) (Chuang, Agarwala *et al.* 2002). Another application is the detection of visual continuity errors in films, i.e., differences in the background when a shot is re-taken at a later time (Pickup and Zisserman 2009).

In the case where the foreground and background motions can both be specified with parametric transforms, high-quality mattes can be extracted using a generalization of triangulation matting

[25] See the alpha matting evaluation website at http://alphamatting.com.

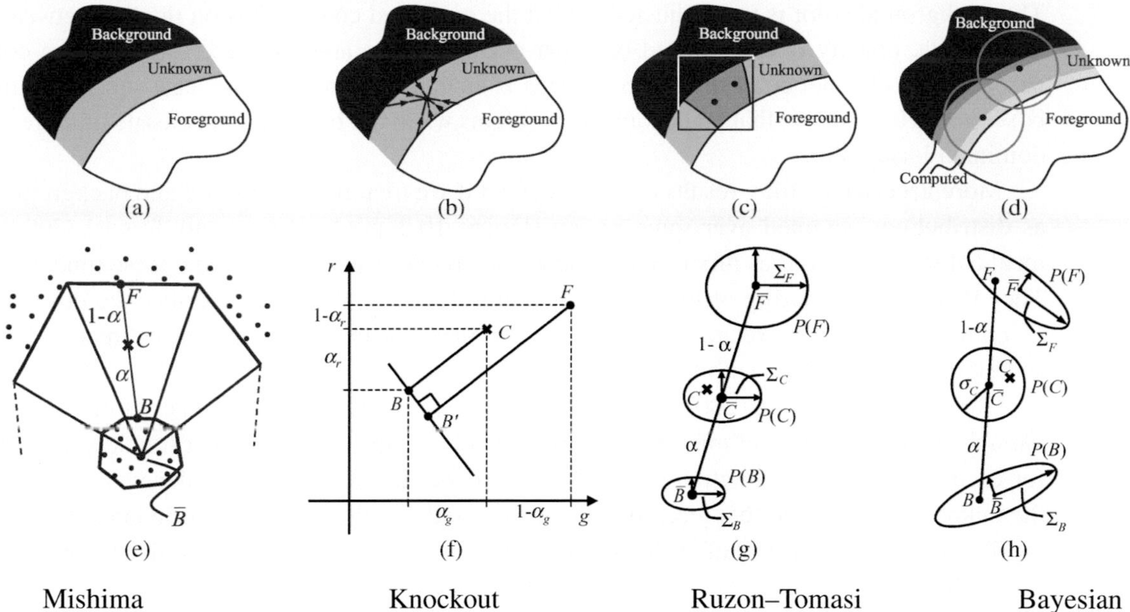

Figure 10.41 Image matting algorithms (Chuang, Curless *et al.* 2001) © 2001 IEEE. Mishima's algorithm models global foreground and background color distribution as polyhedral surfaces centered around the mean background (blue) color. Knockout uses a local color estimate of foreground and background for each pixel and computes α along each color axis. Ruzon and Tomasi's algorithm locally models foreground and background colors and variances. Chuang *et al.*'s Bayesian matting approach computes a MAP estimate of (fractional) foreground color and opacity given the local foreground and background distributions.

(Wexler, Fitzgibbon, and Zisserman 2002). When frames need to be processed independently, however, the results are often of poor quality (Figure 10.42). In such cases, using a pair of stereo cameras as input can dramatically improve the quality of the results (Criminisi, Cross *et al.* 2006; Yin, Criminisi *et al.* 2007).

10.4.2 Natural image matting

The most general version of image matting is when nothing is known about the background except, perhaps, for a rough segmentation of the scene into foreground, background, and unknown regions, which is known as the *trimap* (Figure 10.40b). Some techniques, however, relax this requirement and allow the user to just draw a few strokes or scribbles in the image: see Figures 10.45 and 10.46 (Wang and Cohen 2005; Wang, Agrawala, and Cohen 2007; Levin, Lischinski, and Weiss 2008; Rhemann, Rother *et al.* 2008; Rhemann, Rother, and Gelautz 2008). Fully automated single image matting results have also been reported (Levin, Acha, and Lischinski 2008; Singaraju, Rother, and Rhemann 2009). The survey paper by Wang and Cohen (2009) has detailed descriptions and comparisons of all of these techniques, a selection of which are described briefly below, while the website http://alphamatting.com has up-to-date lists and numerical comparisons of the most recent algorithms.

A relatively simple algorithm for performing natural image matting is Knockout, as described by Chuang, Curless *et al.* (2001) and illustrated in Figure 10.41f. In this algorithm, the nearest known foreground and background pixels (in image space) are determined and then blended with neighboring known pixels to produce a per-pixel foreground F and background B color estimate.

The background color is then adjusted so that the measured color C lies on the line between F and B. Finally, opacity α is estimated on a per-channel basis, and the three estimates are combined based on per-channel color differences. (This is an approximation to the least squares solution for α.) Figure 10.42 shows that Knockout has problems when the background consists of more than one dominant local color.

More accurate matting results can be obtained if we treat the foreground and background colors as distributions sampled over some region (Figure 10.41g–h). Ruzon and Tomasi (2000) model local color distributions as mixtures of (uncorrelated) Gaussians and compute these models in strips. They then find the pairing of mixture components F and B that best describes the observed color C, compute the α as the relative distance between these means, and adjust the estimates of F and B so that they are collinear with C.

Chuang, Curless *et al.* (2001) and Hillman, Hannah, and Renshaw (2001) use full 3×3 color covariance matrices to model mixtures of correlated Gaussians, and compute estimates independently for each pixel. Matte extraction proceeds in strips starting from known color values growing into the unknown regions, so that recently computed F and B colors can be used in later stages.

To estimate the most likely value of an unknown pixel's opacity and (unmixed) foreground and background colors, Chuang *et al.* use a fully Bayesian formulation that maximizes

$$P(F, B, \alpha|C) = P(C|F, B, \alpha)P(F)P(B)P(\alpha)/P(C). \tag{10.31}$$

This is equivalent to minimizing the negative log likelihood

$$L(F, B, \alpha|C) = L(C|F, B, \alpha) + L(F) + L(B) + L(\alpha) \tag{10.32}$$

(dropping the $L(C)$ term because it is constant).

Let us examine each of these terms in turn. The first, $L(C|F, B, \alpha)$, is the likelihood that pixel color C was observed given values for the unknowns (F, B, α). If we assume Gaussian noise in our observation with variance σ_C^2, this negative log likelihood (data term) is

$$L(C) = {}^1\!/_2\|C - [\alpha F + (1 - \alpha)B]\|^2/\sigma_C^2, \tag{10.33}$$

as illustrated in Figure 10.41h.

The second term, $L(F)$, corresponds to the likelihood that a particular foreground color F comes from the Gaussian mixture model. After partitioning the sample foreground colors into clusters, a weighted mean $\overline{F}$ and covariance Σ_F are computed, where the weights are proportional to a given foreground pixel's opacity and distance from the unknown pixel.[26] The negative log likelihood for each cluster is thus given by

$$L(F) = (F - \overline{F})^T \Sigma_F^{-1} (F - \overline{F}). \tag{10.34}$$

A similar method is used to estimate unknown background color distributions. If the background is already known, i.e., for blue screen or difference matting applications, its measured color value and variance are used instead.

An alternative to modeling the foreground and background color distributions as mixtures of Gaussians is to keep around the original color samples and to compute the most likely pairings that explain the observed color C (Wang and Cohen 2005, 2007a). These techniques are described in more detail in (Wang and Cohen 2009).

[26] Note that in this whole chapter, we mostly use upper-case italics to denote images or pixel values, even when they are color vectors. The covariance Σ_F is a 3×3 matrix for each foreground cluster.

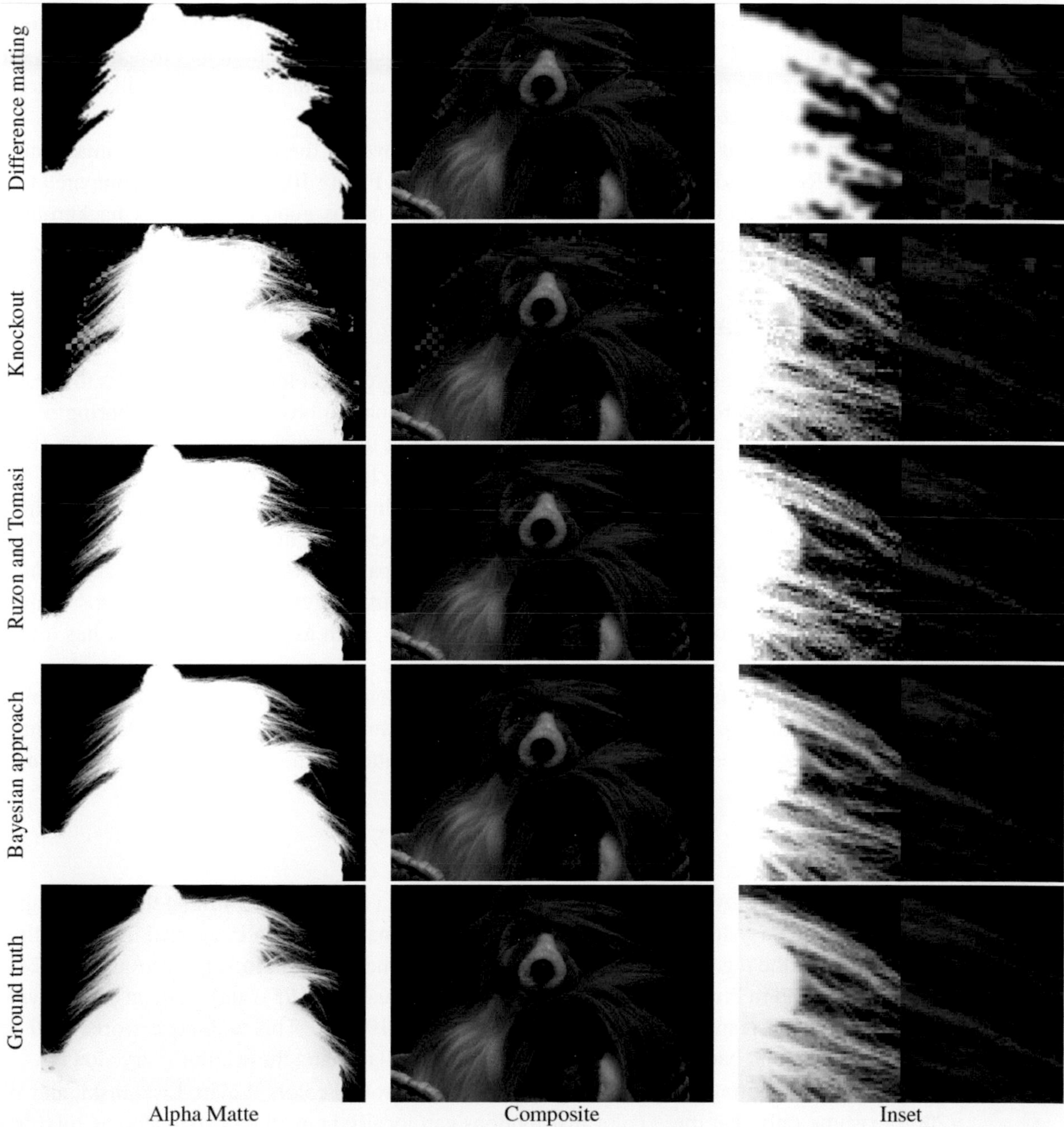

Figure 10.42 Natural image matting results (Chuang, Curless *et al.* 2001) © 2001 IEEE. Difference matting and Knockout both perform poorly on this kind of background, while the newer natural image matting techniques perform well. Chuang *et al.*'s results are slightly smoother and closer to the ground truth.

In their Bayesian matting paper, Chuang, Curless *et al.* (2001) assume a constant (non-informative) distribution for $L(\alpha)$. Follow-on papers assume this distribution to be more peaked around 0 and 1, or sometimes use Markov random fields (MRFs) to define a global correlated prior on $P(\alpha)$ (Wang and Cohen 2009).

To compute the most likely estimates for (F, B, α), the Bayesian matting algorithm alternates between computing (F, B) and α, as each of these problems is quadratic and hence can be solved as a small linear system. When several color clusters are estimated, the most likely pairing of foreground and background color clusters is used.

Bayesian image matting produces results that improve on the original natural image matting algorithm by Ruzon and Tomasi (2000), as can be seen in Figure 10.42. However, compared to later techniques (Wang and Cohen 2009), its performance is not as good for complex backgrounds or inaccurate trimaps (Figure 10.44).

10.4.3 Optimization-based matting

An alternative to estimating each pixel's opacity and foreground color independently is to use global optimization to compute a matte that takes into account correlations between neighboring α values. Two examples of this are border matting in the GrabCut interactive segmentation system (Rother, Kolmogorov, and Blake 2004) and Poisson Matting (Sun, Jia *et al.* 2004).

Border matting first dilates the region around the binary segmentation produced by GrabCut (Section 4.3.2) and then solves for a sub-pixel boundary location Δ and a blur width σ for every point along the boundary (Figure 10.39). Smoothness in these parameters along the boundary is enforced using regularization and the optimization is performed using dynamic programming. While this technique can obtain good results for smooth boundaries, such as a person's face, it has difficulty with fine details, such as hair.

Poisson matting (Sun, Jia *et al.* 2004) assumes a known foreground and background color for each pixel in the trimap (as with Bayesian matting). However, instead of independently estimating each α value, it assumes that the gradient of the alpha matte and the gradient of the color image are related by

$$\nabla \alpha = \frac{F - B}{\|F - B\|^2} \cdot \nabla C, \tag{10.35}$$

which can be derived by taking gradients of both sides of (10.29) and assuming that the foreground and background vary slowly. The per-pixel gradient estimates are then integrated into a continuous $\alpha(\mathbf{x})$ field using the regularization (least squares) technique first described in Section 4.2 (4.24) and subsequently used in Poisson blending (Section 8.4.4, Equation (8.75)) and gradient-based dynamic range compression mapping (Section 10.2.1, Equation (10.18)). This technique works well when good foreground and background color estimates are available and these colors vary slowly.

Instead of computing per-pixel foreground and background colors, Levin, Lischinski, and Weiss (2008) assume only that these color distributions can locally be well approximated as mixtures of two colors, which is known as the *color line model* (Figure 10.43a–c). Under this assumption, a closed-form estimate for α at each pixel i in a (say, 3×3) window W_k is given by

$$\alpha_i = \mathbf{a}_k \cdot (\mathbf{C}_i - \mathbf{B}_0) = \mathbf{a}_k \cdot \mathbf{C} + b_k, \tag{10.36}$$

where $\mathbf{C}_i$ is the pixel color treated as a three-vector, $\mathbf{B}_0$ is any pixel along the background color line, and $\mathbf{a}_k$ is the vector joining the two closest points on the foreground and background color lines, as shown in Figure 10.43c. (Note that the geometric derivation shown in this figure is an alternative

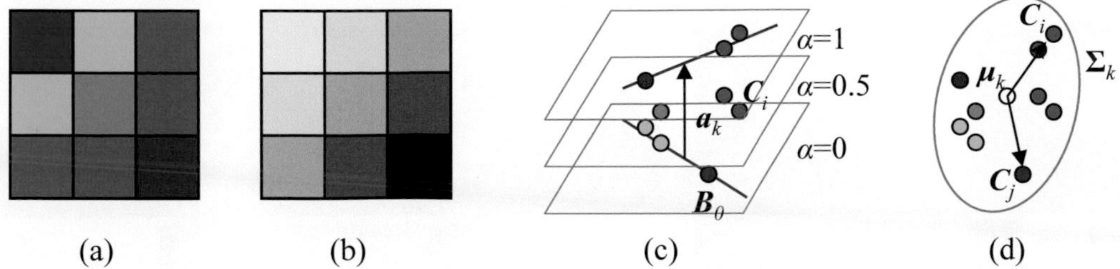

Figure 10.43 Color line matting (Levin, Lischinski, and Weiss 2008): (a) local 3 × 3 patch of colors; (b) potential assignment of α values; (c) foreground and background color lines, the vector $\mathbf{a}_k$ joining their closest points of intersection, and the family of parallel planes of constant α values, $\alpha_i = \mathbf{a}_k \cdot (\mathbf{C}_i - \mathbf{B}_0)$; (d) a scatter plot of sample colors and the deviations from the mean μ_k for two sample colors $\mathbf{C}_i$ and $\mathbf{C}_j$.

to the algebraic derivation presented by Levin, Lischinski, and Weiss (2008).) Minimizing the deviations of the alpha values α_i from their respective color line models (10.36) over all overlapping windows W_k in the image gives rise to the cost

$$E_\alpha = \sum_k \left(\sum_{i \in W_k} (\alpha_i - \mathbf{a}_k \cdot \mathbf{C}_i - b_k)^2 + \epsilon \|\mathbf{a}_k\| \right), \tag{10.37}$$

where the ϵ term is used to regularize the value of $\mathbf{a}_k$ in the case where the two color distributions overlap (i.e., in constant α regions).

Because this formula is quadratic in the unknowns $\{(\mathbf{a}_k, b_k)\}$, they can be eliminated inside each window W_k, leading to a final energy

$$E_\alpha = \alpha^T \mathbf{L} \alpha, \tag{10.38}$$

where the entries in the $\mathbf{L}$ matrix are given by

$$L_{ij} = \sum_{k: i \in W_k \wedge j \in W_k} \left(\delta_{ij} - \frac{1}{M} \left(1 + (\mathbf{C}_i - \mu_k)^T \hat{\mathbf{\Sigma}}_k^{-1} (\mathbf{C}_j - \mu_k) \right) \right), \tag{10.39}$$

where $M = |W_k|$ is the number of pixels in each (overlapping) window, μ_k is the mean color of the pixels in window W_k, and $\hat{\mathbf{\Sigma}}_k$ is the 3 × 3 covariance of the pixel colors plus $\epsilon/M \mathbf{I}$.

Figure 10.43d shows the intuition behind the entries in this affinity matrix, which is called the *matting Laplacian*. Note how when two pixels $\mathbf{C}_i$ and $\mathbf{C}_j$ in W_k point in opposite directions away from the mean μ_k, their weighted dot product is close to -1, and so their affinity becomes close to 0. Pixels close to each other in color space (and hence with similar expected α values) will have affinities close to $-2/M$.

Minimizing the quadratic energy (10.38) constrained by the known values of $\alpha = \{0, 1\}$ at scribbles only requires the solution of a sparse set of linear equations, which is why the authors call their technique a *closed-form solution* to natural image matting. Once α has been computed, the foreground and background colors are estimated using a least squares minimization of the compositing equation (10.29) regularized with a spatially varying first-order smoothness,

$$E = \sum_i \|C_i - [\alpha + F_i + (1 - \alpha_i) B_i]\|^2 + \lambda |\nabla \alpha_i| (\|\nabla F_i\|^2 + \|\nabla B_i\|^2), \tag{10.40}$$

where the $|\nabla \alpha_i|$ weight is applied separately for the x and y components of the F and B derivatives (Levin, Lischinski, and Weiss 2008).

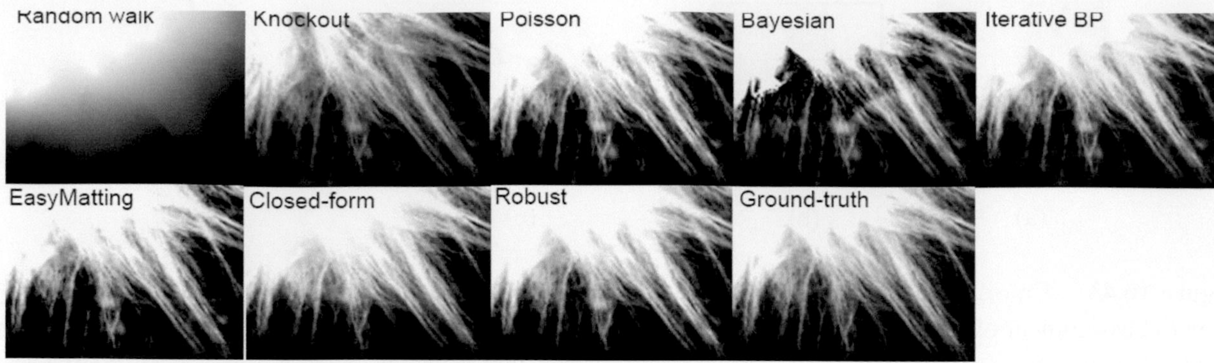

Figure 10.44 Comparative matting results for a medium accuracy trimap. Wang and Cohen (2009) describe the individual techniques being compared.

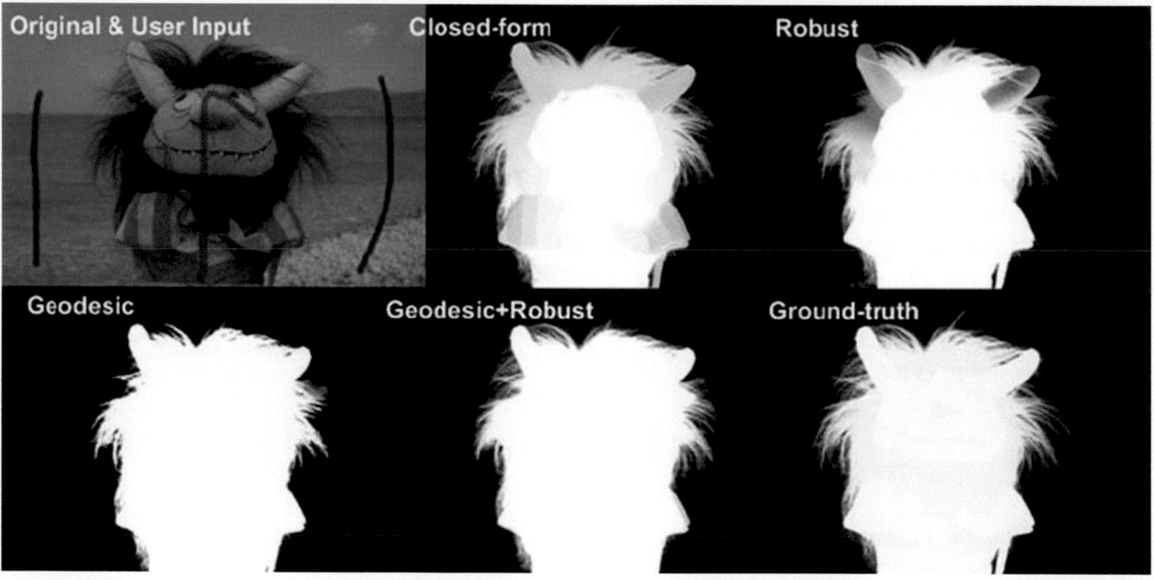

Figure 10.45 Comparative matting results with scribble-based inputs. Wang and Cohen (2009) describe the individual techniques being compared.

Figure 10.46 Stroke-based segmentation result (Rhemann, Rother *et al.* 2008) © 2008 IEEE.

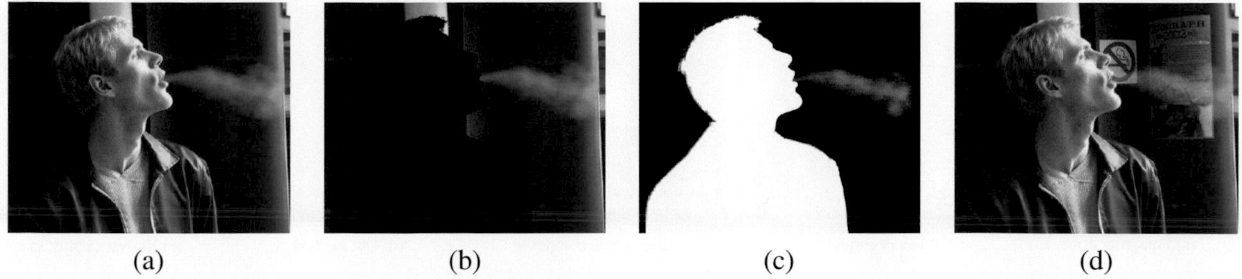

(a) (b) (c) (d)

Figure 10.47 Smoke matting (Chuang, Agarwala *et al.* 2002) © 2002 ACM: (a) input video frame; (b) after removing the foreground object; (c) estimated alpha matte; (d) insertion of new objects into the background.

Laplacian (closed-form) matting is just one of many optimization-based techniques surveyed and compared by Wang and Cohen (2009). Some of these techniques use alternative formulations for the affinities or smoothness terms on the α matte, alternative estimation techniques such as belief propagation, or alternative representations (e.g., local histograms) for modeling local foreground and background color distributions (Wang and Cohen 2005, 2007a,b). Some of these techniques also provide real-time results as the user draws a contour line or sparse set of scribbles (Wang, Agrawala, and Cohen 2007; Rhemann, Rother *et al.* 2008) or even pre-segment the image into a small number of mattes that the user can select with simple clicks (Levin, Acha, and Lischinski 2008).

Figure 10.44 shows the results of running a number of the surveyed algorithms on a region of toy animal fur where a trimap has been specified, while Figure 10.45 shows results for techniques that can produce mattes with only a few scribbles as input. Figure 10.46 shows a result for an even more recent algorithm (Rhemann, Rother *et al.* 2008) that claims to outperform all of the techniques surveyed by Wang and Cohen (2009).

The latest results on natural image matting can be found on the http://alphamatting.com website created by Rhemann, Rother *et al.* (2009). It currently lists over 60 different algorithms, with most of the more recent algorithms using deep neural networks. The Deep Image Matting paper by Xu, Price *et al.* (2017) provides a larger database of 49,300 training images and 1,000 test images constructed by overlaying manually created color foreground mattes over a variety of backgrounds.[27]

Pasting. Once a matte has been pulled from an image, it is usually composited directly over the new background, unless the seams between the cutout and background regions are to be hidden, in which case Poisson blending (Pérez, Gangnet, and Blake 2003) can be used (Section 8.4.4).

In the latter case, it is helpful if the matte boundary passes through regions that either have little texture or look similar in the old and new images. Papers by Jia, Sun *et al.* (2006) and Wang and Cohen (2007b) explain how to do this.

10.4.4 Smoke, shadow, and flash matting

In addition to matting out solid objects with fractional boundaries, it is also possible to matte out translucent media such as smoke (Chuang, Agarwala *et al.* 2002). Starting with a video sequence, each pixel is modeled as a linear combination of its (unknown) background color and a constant foreground (smoke) color that is common to all pixels. Voting in color space is used to estimate this

[27] https://sites.google.com/view/deepimagematting

(a) Foreground scene (b) Background scene (c) Blue screen composite (d) Our method (e) Reference photograph

Figure 10.48 Shadow matting (Chuang, Goldman *et al.* 2003) © 2003 ACM. Instead of simply darkening the new scene with the shadow (c), shadow matting correctly dims the lit scene with the new shadow and drapes the shadow over 3D geometry (d).

foreground color and the distance along each color line is used to estimate the per-pixel temporally varying alpha (Figure 10.47).

Extracting and re-inserting shadows is also possible using a related technique (Chuang, Goldman *et al.* 2003; Wang, Curless, and Seitz 2020). Here, instead of assuming a constant foreground color, each pixel is assumed to vary between its fully lit and fully shadowed colors, which can be estimated by taking (robust) minimum and maximum values over time as a shadow passes over the scene (Exercise 10.9). The resulting fractional *shadow matte* can be used to re-project the shadow into a new scene. If the destination scene has a non-planar geometry, it can be scanned by waving a straight stick shadow across the scene. The new shadow matte can then be warped with the computed deformation field to have it drape correctly over the new scene (Figure 10.48). Shadows can also be extracted from video streams by extending video object segmentation algorithms (Section 9.4.3) to include shadows and other effects such as smoke (Lu, Cole *et al.* 2021). An example of useful shadow manipulation in photographs is the removal or softening of harsh shadows in people's portraits (Sun, Barron *et al.* 2019; Zhou, Hadap *et al.* 2019; Zhang, Barron *et al.* 2020), which is available as the Portrait Light feature in Google Photos.[28]

The quality and reliability of matting algorithms can also be enhanced using more sophisticated acquisition systems. For example, taking a flash and non-flash image pair supports the reliable extraction of foreground mattes, which show up as regions of large illumination change between the two images (Sun, Li *et al.* 2006). Taking simultaneous video streams focused at different distances (McGuire, Matusik *et al.* 2005) or using multi-camera arrays (Joshi, Matusik, and Avidan 2006) are also good approaches to producing high-quality mattes. These techniques are described in more detail in (Wang and Cohen 2009).

Lastly, photographing a refractive object in front of a number of patterned backgrounds allows the object to be placed in novel 3D environments. These environment matting techniques (Zongker, Werner *et al.* 1999; Chuang, Zongker *et al.* 2000) are discussed in Section 14.4.

10.4.5 Video matting

While regular single-frame matting techniques such as blue or green screen matting (Smith and Blinn 1996; Wright 2006; Brinkmann 2008) can be applied to video sequences, the presence of moving objects can sometimes make the matting process easier, as portions of the background may get revealed in preceding or subsequent frames.

[28] https://blog.google/products/photos/new-helpful-editor

Chuang, Agarwala *et al.* (2002) describe a nice approach to this *video matting* problem, where foreground objects are first removed using a conservative *garbage matte* and the resulting *background plates* are aligned and composited to yield a high-quality background estimate. They also describe how trimaps drawn at sparse keyframes can be interpolated to in-between frames using bi-direction optical flow. Alternative approaches to video matting, such as rotoscoping, which involves drawing curves or strokes in video sequence keyframes (Agarwala, Hertzmann *et al.* 2004; Wang, Bhat *et al.* 2005), are discussed in the matting survey paper by Wang and Cohen (2009). There is also a newer dataset of carefully matted stop-motion animation videos created by Erofeev, Gitman *et al.* (2015).[29]

Since the original development of video matting techniques, improved algorithms have been developed for both interactive and fully automated *video object segmentation*, as discussed in Section 9.4.3. The paper by Sengupta, Jayaram *et al.* (2020) uses deep learning and adversarial loss, as well as a motion prior, to provide high-quality mattes from small-motion handheld videos where a *clean plate* of the background has also been captured. Wang, Curless, and Seitz (2020) describe a system where shadows and occlusions can be determined by observing people walking around a scene, enabling the insertion of new people at correct scales and lighting. In follow-up work Lin, Ryabtsev *et al.* (2021) describe a high-resolution real-time video matting system along with two new video and image matting datasets. Finally, Lu, Cole *et al.* (2021) describe how to extract shadows, reflections, and other effects associated with objects being tracked and segmented in videos.

10.5 Texture analysis and synthesis

While texture analysis and synthesis may not at first seem like computational photography techniques, they are, in fact, widely used to repair defects, such as small holes, in images or to create non-photorealistic painterly renderings from regular photographs.

The problem of texture synthesis can be formulated as follows: given a small sample of a "texture" (Figure 10.49a), generate a larger similar-looking image (Figure 10.49b). As you can imagine, for certain sample textures, this problem can be quite challenging.

Traditional approaches to texture analysis and synthesis try to match the spectrum of the source image while generating shaped noise. Matching the frequency characteristics, which is equivalent to matching spatial correlations, is in itself not sufficient. The distributions of the responses at different frequencies must also match. Heeger and Bergen (1995) develop an algorithm that alternates between matching the histograms of multi-scale (steerable pyramid) responses and matching the final image histogram. Portilla and Simoncelli (2000) improve on this technique by also matching pairwise statistics across scale and orientations. De Bonet (1997) uses a coarse-to-fine strategy to find locations in the source texture with a similar *parent structure*, i.e., similar multi-scale oriented filter responses, and then randomly chooses one of these matching locations as the current sample value. Gatys, Ecker, and Bethge (2015) also use a pyramidal fine-to-coarse-to-fine algorithm, but using deep networks trained for object recognition. At each level in the deep network, they gather correlation statistics between various features. During generation, they iteratively update the random image until these more perceptually motivated statistic (Zhang, Isola *et al.* 2018) are matched. We give more details on this and other neural approaches to texture synthesis, such as Shaham, Dekel, and Michaeli (2019), in Section 10.5.3 on neural style transfer.

Exemplar-based texture synthesis algorithms sequentially generate texture pixels by looking for neighborhoods in the source texture that are similar to the currently synthesized image (Efros and Leung 1999). Consider the (as yet) unknown pixel $\mathbf{p}$ in the partially constructed texture on the left

[29] https://videomatting.com

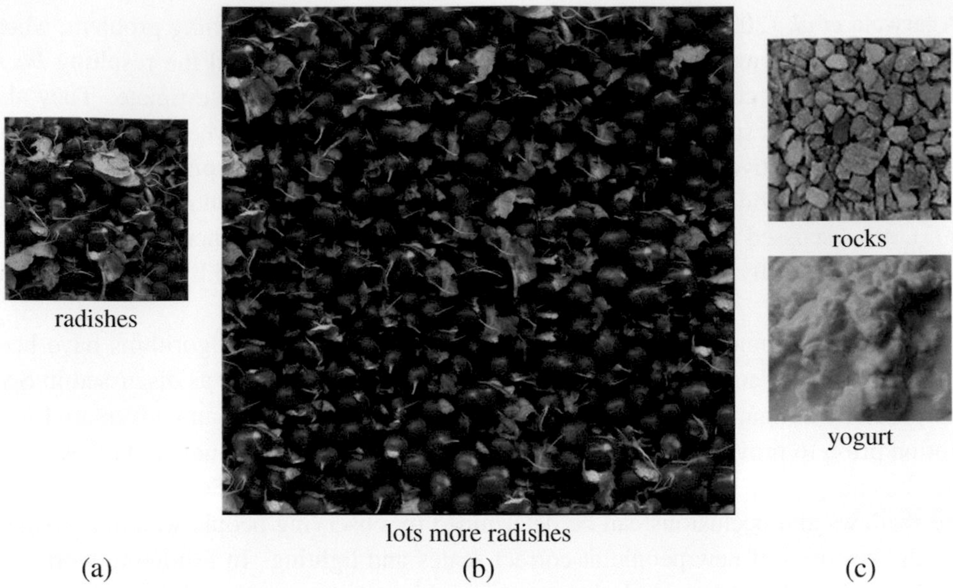

Figure 10.49 Texture synthesis: (a) given a small patch of texture, the task is to synthesize (b) a similar-looking larger patch; (c) other semi-structured textures that are challenging to synthesize. (Images courtesy of Alyosha Efros.)

side of Figure 10.50. As some of its neighboring pixels have been already been synthesized, we can look for similar partial neighborhoods in the sample texture image on the right and randomly select one of these as the new value of $\mathbf{p}$. This process can be repeated down the new image either in a raster fashion or by scanning around the periphery ("onion peeling") when filling holes, as discussed in (Section 10.5.1). In their actual implementation, Efros and Leung (1999) find the most similar neighborhood and then include all other neighborhoods within a $d = (1 + \epsilon)$ distance, with $\epsilon = 0.1$. They also optionally weight the random pixel selections by the similarity metric d.

To accelerate this process and improve its visual quality, Wei and Levoy (2000) extend this technique using a coarse-to-fine generation process, where coarser levels of the pyramid, which have already been synthesized, are also considered during the matching (De Bonet 1997). To accelerate the nearest neighbor finding, tree-structured vector quantization is used. A much faster version of such nearest neighbor search is the widely used randomized PatchMatch iterative update algorithm developed by Barnes, Shechtman *et al.* (2009).

Efros and Freeman (2001) propose an alternative acceleration and visual quality improvement technique. Instead of synthesizing a single pixel at a time, overlapping square blocks are selected using similarity with previously synthesized regions (Figure 10.51). Once the appropriate blocks have been selected, the seam between newly overlapping blocks is determined using dynamic programming. (Full graph cut seam selection is not required, because only one seam location per row is needed for a vertical boundary.) Because this process involves selecting small patches and them stitching them together, Efros and Freeman (2001) call their system *image quilting*. Komodakis and Tziritas (2007) present an MRF-based version of this block synthesis algorithm that uses a new, efficient version of loopy belief propagation they call "Priority-BP". Wei, Lefebvre *et al.* (2009) present a comprehensive survey of work in exemplar-based texture synthesis through 2009.

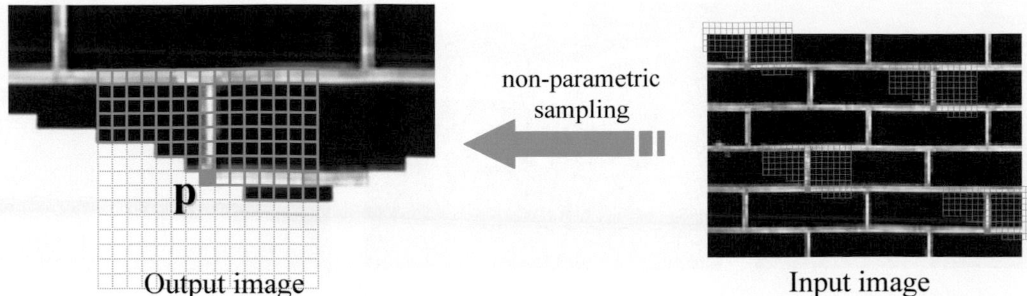

Figure 10.50 Texture synthesis using non-parametric sampling (Efros and Leung 1999). The value of the newest pixel $\mathbf{p}$ is randomly chosen from similar local (partial) patches in the source texture (input image). (Figure courtesy of Alyosha Efros.)

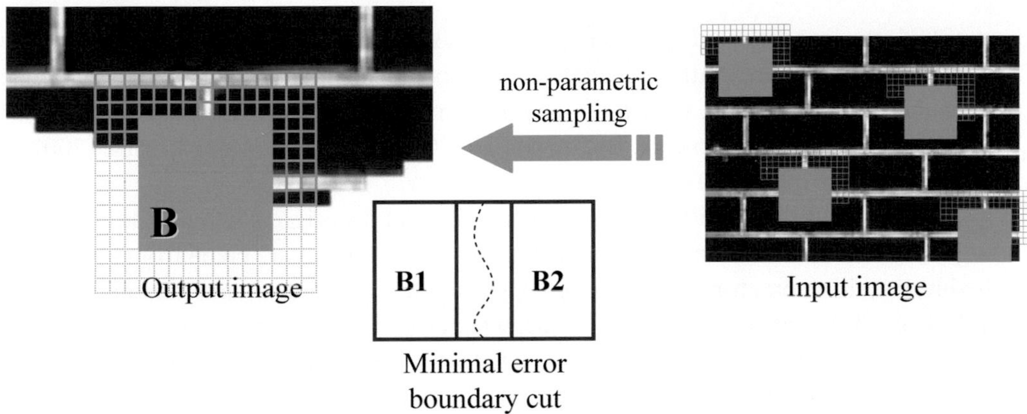

Figure 10.51 Texture synthesis by image quilting (Efros and Freeman 2001). Instead of generating a single pixel at a time, larger blocks are copied from the source texture. The transitions in the overlap regions between the selected blocks are then optimized using dynamic programming. (Figure courtesy of Alyosha Efros.)

10.5.1 *Application*: Hole filling and inpainting

Filling holes left behind when objects or defects are excised from photographs, which is known as *inpainting*, is one of the most common applications of texture synthesis. Such techniques are used not only to remove unwanted people or interlopers from photographs (King 1997) but also to fix small defects in old photos and movies (*scratch removal*) or to remove wires holding props or actors in mid-air during filming (*wire removal*). Bertalmio, Sapiro *et al.* (2000) solve the problem by propagating pixel values along isophote (constant-value) directions interleaved with some anisotropic diffusion steps (Figure 10.52a–b). Telea (2004) develops a faster technique that uses the fast marching method from level sets (Section 7.3.2). However, these techniques will not hallucinate texture in the missing regions. Bertalmio, Vese *et al.* (2003) augment their earlier technique by adding synthetic texture to the infilled regions.

The example-based (non-parametric) texture generation techniques discussed in the previous section can also be used by filling the holes from the outside in (the "onion-peel" ordering). However, this approach may fail to propagate strong oriented structures. Criminisi, Pérez, and Toyama (2004) use exemplar-based texture synthesis where the order of synthesis is determined by the strength of the gradient along the region boundary (Figures 10.1d and 10.52c–d). Sun, Yuan *et*

(a) (b) (c) (d)

Figure 10.52 Image inpainting (hole filling): (a–b) propagation along isophote directions (Bertalmio, Sapiro *et al.* 2000) © 2000 ACM; (c–d) exemplar-based inpainting with confidence-based filling order (Criminisi, Pérez, and Toyama 2004).

al. (2004) present a related approach where the user draws interactive lines to indicate where structures should be preferentially propagated. Additional techniques related to these approaches include those developed by Drori, Cohen-Or, and Yeshurun (2003), Kwatra, Schödl *et al.* (2003), Kwatra, Essa *et al.* (2005), Wilczkowiak, Brostow *et al.* (2005), Komodakis and Tziritas (2007), and Wexler, Shechtman, and Irani (2007).

Most hole filling algorithms borrow small pieces of the original image to fill in the holes. When a large database of source images is available, e.g., when images are taken from a photo sharing site or the internet, it is sometimes possible to copy a single contiguous image region to fill the hole. Hays and Efros (2007) present such a technique, which uses image context and boundary compatibility to select the source image, which is then blended with the original (holey) image using graph cuts and Poisson blending. This technique is discussed in more detail in Section 6.4.4 and Figure 6.40.

As with other areas of image processing, deep neural networks are used in all of the latest techniques (Yang, Lu *et al.* 2017; Yu, Lin *et al.* 2018; Liu, Reda *et al.* 2018; Zeng, Fu *et al.* 2019; Yu, Lin *et al.* 2019; Chang, Liu *et al.* 2019; Nazeri, Ng *et al.* 2019; Ren, Yu *et al.* 2019; Shih, Su *et al.* 2020; Yi, Tang *et al.* 2020). Some of these papers have introduced interesting new extensions to neural network architectures, such as *partial convolutions* (Liu, Reda *et al.* 2018) and *partial convolutions* (Yu, Lin *et al.* 2019), the propagation of edge structures (Nazeri, Ng *et al.* 2019; Ren, Yu *et al.* 2019), multi-resolution attention and residuals (Yi, Tang *et al.* 2020), and iterative confidence feedback (Zeng, Lin *et al.* 2020). Inpainting has also been applied to video sequences (e.g., Gao, Saraf *et al.* 2020). Results on recent challenges on image inpainting can be found in the AIM 2020 Workshop and Challenges on this topic (Ntavelis, Romero *et al.* 2020a).

10.5.2 *Application*: Non-photorealistic rendering

Two more applications of the exemplar-based texture synthesis ideas are texture transfer (Efros and Freeman 2001) and image analogies (Hertzmann, Jacobs *et al.* 2001), which are both examples of non-photorealistic rendering (Gooch and Gooch 2001).

In addition to using a source texture image, texture transfer also takes a reference (or target) image, and tries to match certain characteristics of the target image with the newly synthesized image. For example, the new image being rendered in Figure 10.53c not only tries to satisfy the usual similarity constraints with the source texture in Figure 10.53b, but it also tries to match the luminance characteristics of the reference image. Efros and Freeman (2001) mention that blurred image intensities or local image orientation angles are alternative quantities that could be matched.

Hertzmann, Jacobs *et al.* (2001) formulate the following problem:

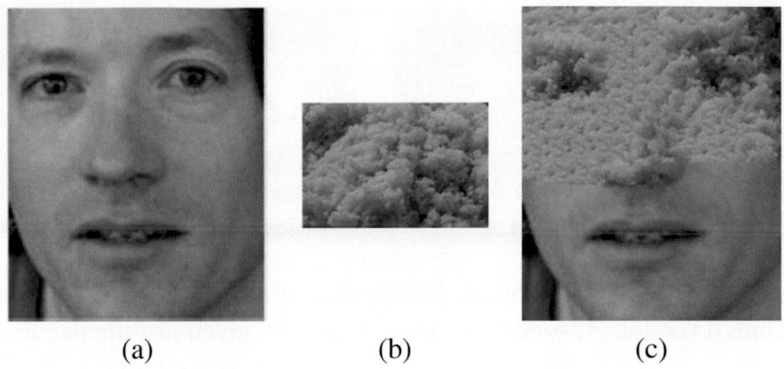

Figure 10.53 Texture transfer (Efros and Freeman 2001) © 2001 ACM: (a) reference (target) image; (b) source texture, (c) image (partially) rendered using the texture.

Figure 10.54 Image analogies (Hertzmann, Jacobs *et al.* 2001) © 2001 ACM. Given an example pair of a source image A and its rendered (filtered) version A', generate the rendered version B' from another unfiltered source image B.

> Given a pair of images A and A' (the unfiltered and filtered source images, respectively), along with some additional unfiltered target image B, synthesize a new filtered target image B' such that
>
> $$A : A' :: B : B'.$$

Instead of having the user program a certain non-photorealistic rendering effect, it is sufficient to supply the system with examples of before and after images, and let the system synthesize the novel image using exemplar-based synthesis, as shown in Figure 10.54.

The algorithm used to solve image analogies proceeds in a manner analogous to the texture synthesis algorithms of Efros and Leung (1999) and Wei and Levoy (2000). Once Gaussian pyramids have been computed for all of the source and reference images, the algorithm looks for neighborhoods in the source filtered pyramids generated from A' that are similar to the partially constructed neighborhood in B', while at the same time having similar multi-resolution appearances at corresponding locations in A and B. As with texture transfer, appearance characteristics can include not only (blurred) color or luminance values but also orientations.

This general framework allows image analogies to be applied to a variety of rendering tasks. In addition to exemplar-based non-photorealistic rendering, image analogies can be used for traditional texture synthesis, super-resolution, and texture transfer (using the same textured image for both A and A'). If only the filtered (rendered) image A' is available, as is the case with paintings, the missing reference image A can be hallucinated using a smart (edge preserving) blur operator. Finally, it is possible to train a system to perform *texture-by-numbers* by manually painting over a natural image with pseudocolors corresponding to pixels' semantic meanings, e.g., water, trees, and

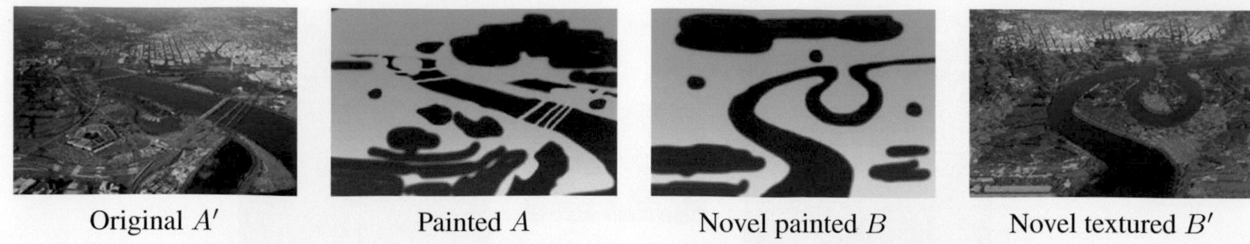

Figure 10.55 Texture-by-numbers (Hertzmann, Jacobs *et al.* 2001) © 2001 ACM. Given a textured image A' and a hand-labeled (painted) version A, synthesize a new image B' given just the painted version B.

Figure 10.56 Non-photorealistic abstraction of photographs: (a) (DeCarlo and Santella 2002) © 2002 ACM and (b) (Farbman, Fattal *et al.* 2008) © 2008 ACM.

grass (Figure 10.55a–b). The resulting system can then convert a novel sketch into a fully rendered synthetic photograph (Figure 10.55c–d). In more recent work, Cheng, Vishwanathan, and Zhang (2008) add ideas from image quilting (Efros and Freeman 2001) and MRF inference (Komodakis, Tziritas, and Paragios 2008) to the basic image analogies algorithm, while Ramanarayanan and Bala (2007) recast this process as energy minimization, which means it can be viewed as a conditional random field (Section 4.3.1), and devise an efficient algorithm to find a good minimum.

More traditional filtering and feature detection techniques can also be used for non-photorealistic rendering.[30] For example, pen-and-ink illustration (Winkenbach and Salesin 1994) and painterly rendering techniques (Litwinowicz 1997) use local color, intensity, and orientation estimates as an input to their procedural rendering algorithms. Techniques for stylizing and simplifying photographs and video (DeCarlo and Santella 2002; Winnemöller, Olsen, and Gooch 2006; Farbman, Fattal *et al.* 2008), as in Figure 10.56, use combinations of edge-preserving blurring (Section 3.3.1) and edge detection and enhancement (Section 7.2.3).

10.5.3 Neural style transfer and semantic image synthesis

With the advent of deep learning, image-guided exemplar-based texture synthesis has mostly been replaced with statistics matching in deep networks (Gatys, Ecker, and Bethge 2015). Figure 10.57 illustrates the basic idea used in neural style transfer networks. In the original work of Gatys, Ecker,

[30] For a good selection of papers, see the Symposia on Non-Photorealistic Animation and Rendering (NPAR).

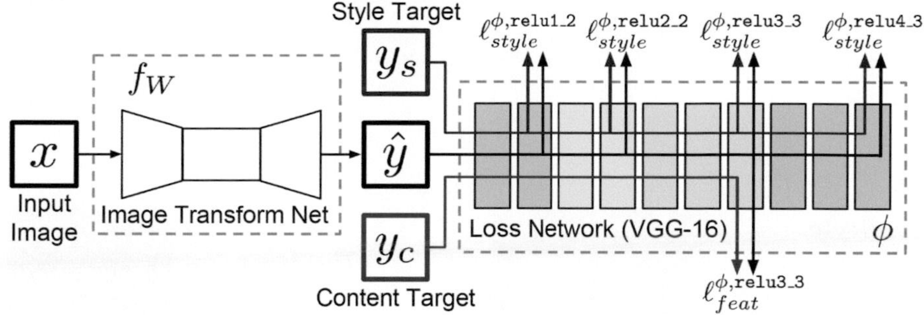

Figure 10.57 Network architecture for neural style transfer, which learns to transform images in one particular style (Johnson, Alahi, and Fei-Fei 2016) © 2016 Springer. During training, the content target image y_c is fed into the image transformation network as an input x, along with a style image y_s, and the network weights are updated so as to minimize the perceptual losses, i.e., the style reconstruction loss l_{style} and the feature reconstruction loss l_{feat}. The earlier network by Gatys, Ecker, and Bethge (2015) did not have an image transformation network, and instead used the losses to optimize the transformed image $\hat{y}$.

and Bethge (2016), a style image y_s and a content image y_c (see Figure 10.58 for examples) are input to a *loss network*, which compares features derived from the style and target images with those derived from the image $\hat{y}$ being synthesized. These losses are normally a combination of a *perceptual loss*. The gradients of these losses are used to adjust the generated image $\hat{y}$ in an iterative fashion, which makes this process quite slow.

To accelerate this, Johnson, Alahi, and Fei-Fei (2016) train a feedforward *image transformation network* with a fixed style image and many different content targets, adjusting the network weights so that the stylized image $\hat{y}$ resulting from a target y_c matches the desired statistics. When a new image x is presented to be stylized, it is simply run through the image transformation network. Figure 10.58a shows some comparisons between Gatys, Ecker, and Bethge (2016) and Johnson, Alahi, and Fei-Fei (2016).

Perceptual loss has now become a standard component of image synthesis systems (Dosovitskiy and Brox 2016), often as an additional component to the generative adversarial loss (Section 5.5.4). They are also sometimes used as an alternative to older image quality metrics such as SSIM (Zhang, Isola *et al.* 2018; Talebi and Milanfar 2018; Tariq, Tursun *et al.* 2020; Czolbe, Krause *et al.* 2020).

The basic architecture in Johnson, Alahi, and Fei-Fei (2016) was extended by Ulyanov, Vedaldi, and Lempitsky (2017), who show that using *instance normalization* instead of batch normalization significantly improves the results. Dumoulin, Shlens, and Kudlur (2017) and Huang and Belongie (2017) further extended these ideas to train one network to mimic different styles, using *conditional instance normalization* and *adaptive instance normalization* to select among the pre-trained styles (or in-between blends), as shown in Figure 10.58b.

Neural style transfer continues to be an actively studied area, with related approaches working on more generalized *image-to-image translation* (Isola, Zhu *et al.* 2017) and *semantic photo synthesis* (Chen and Koltun 2017; Park, Liu *et al.* 2019; Bau, Strobelt *et al.* 2019; Ntavelis, Romero *et al.* 2020b) applications—see Tewari, Fried *et al.* (2020, Section 6.1) for a recent survey. Most of the newer architectures use generative adversarial networks (GANs) (Kotovenko, Sanakoyeu *et al.* 2019; Shaham, Dekel, and Michaeli 2019; Yang, Wang *et al.* 2019; Svoboda, Anoosheh *et al.* 2020, Wang, Li *et al.* 2020; Xia, Zhang *et al.* 2020; Härkönen, Hertzmann *et al.* 2020), which we discussed in Section 5.5.4. There's also a recent course on the more general topic of learning-based image synthesis (Zhu 2021).

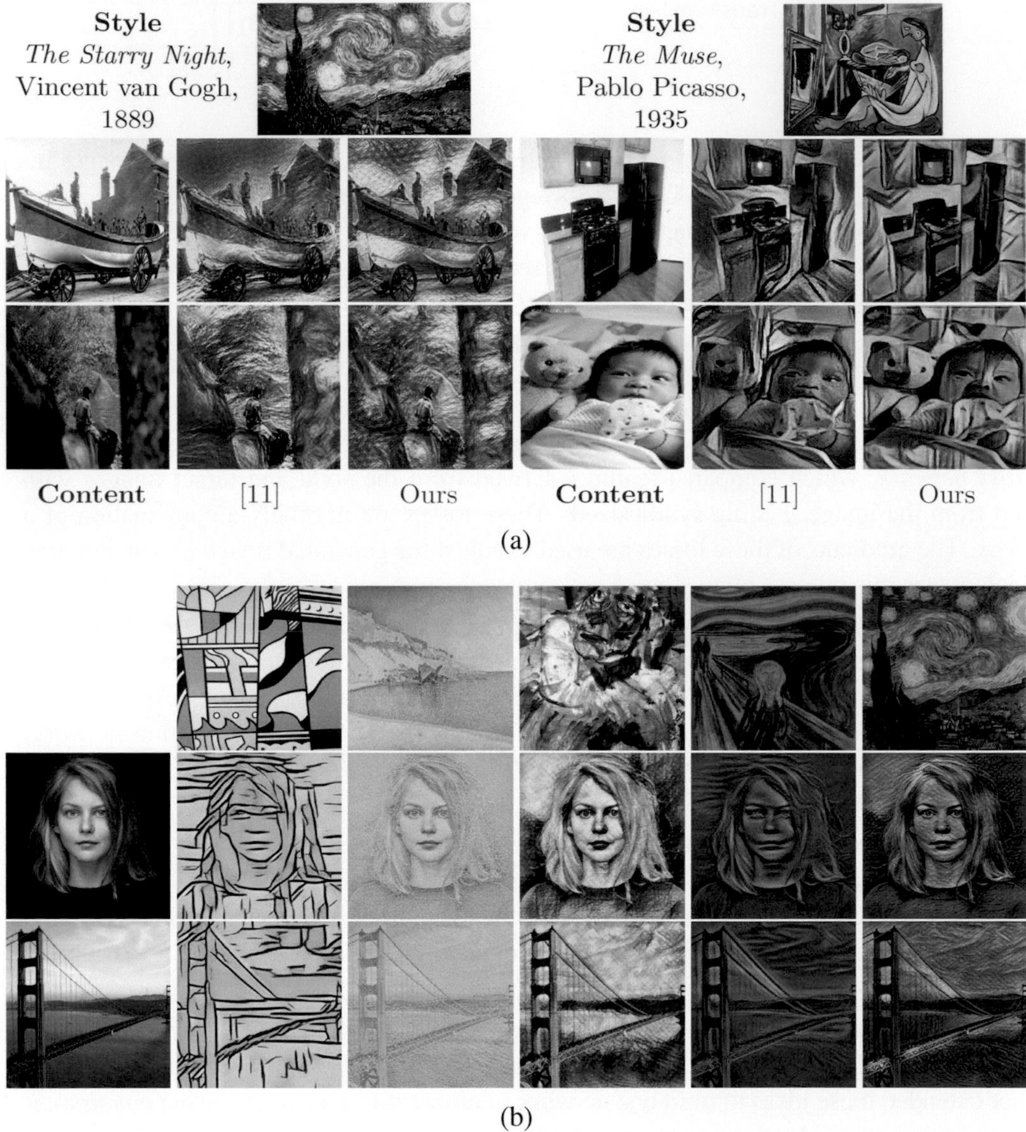

Figure 10.58 Two examples of neural style transfer: (a) the pre-trained network of Johnson, Alahi, and Fei-Fei (2016) © 2016 Springer (labeled "Ours") vs. (Gatys, Ecker, and Bethge 2016) (labeled "[11]");, (b) a network that uses conditional instance normalization to mimic different styles (top row) applied to various content (left column) © (Dumoulin, Shlens, and Kudlur 2017).

10.6 Additional reading

Good overviews of the first decade of computational photography can be found in the book by Raskar and Tumblin (2010) and survey articles by Nayar (2006), Cohen and Szeliski (2006), Levoy (2006), Debevec (2006), and Hayes (2008), as well as two special journal issues edited by Bimber (2006) and Durand and Szeliski (2007). Notes from the courses on computational photography mentioned at the beginning of this chapter are another great source for more recent material and references.[31]

The sub-field of high dynamic range imaging has its own book discussing research in this area (Reinhard, Heidrich *et al.* 2010), as well as some books describing related photographic techniques (Freeman 2008; Gulbins and Gulbins 2009). Algorithms for calibrating the radiometric response function of a camera can be found in articles by Mann and Picard (1995), Debevec and Malik (1997), and Mitsunaga and Nayar (1999).

The subject of tone mapping is treated extensively in (Reinhard, Heidrich *et al.* 2010). Representative papers from the large volume of literature on this topic include (Tumblin and Rushmeier 1993; Larson, Rushmeier, and Piatko 1997; Pattanaik, Ferwerda *et al.* 1998; Tumblin and Turk 1999; Durand and Dorsey 2002; Fattal, Lischinski, and Werman 2002; Reinhard, Stark *et al.* 2002; Lischinski, Farbman *et al.* 2006; Farbman, Fattal *et al.* 2008; Paris, Hasinoff, and Kautz 2011; Aubry, Paris *et al.* 2014).

The literature on super-resolution is quite extensive (Chaudhuri 2001; Park, Park, and Kang 2003; Capel and Zisserman 2003; Capel 2004; van Ouwerkerk 2006). The term super-resolution usually describes techniques for aligning and merging multiple images to produce higher-resolution composites (Keren, Peleg, and Brada 1988; Irani and Peleg 1991; Cheeseman, Kanefsky *et al.* 1993; Mann and Picard 1994; Chiang and Boult 1996; Bascle, Blake, and Zisserman 1996; Capel and Zisserman 1998; Smelyanskiy, Cheeseman *et al.* 2000; Capel and Zisserman 2000; Pickup, Capel *et al.* 2009; Gulbins and Gulbins 2009; Hasinoff, Sharlet *et al.* 2016; Wronski, Garcia-Dorado *et al.* 2019). However, single-image super-resolution techniques have also been developed (Freeman, Jones, and Pasztor 2002; Baker and Kanade 2002; Fattal 2007; Dong, Loy *et al.* 2016; Cai, Gu *et al.* 2019; Anwar, Khan, and Barnes 2020). Such techniques are closely related to denoising (Zhang, Zuo *et al.* 2017; Brown 2019; Liba, Murthy *et al.* 2019; Gu and Timofte 2019), deblurring and blind image deconvolution (Campisi and Egiazarian 2017; Zhang, Dai *et al.* 2019; Kupyn, Martyniuk *et al.* 2019), and demosaicing (Chatterjee, Joshi *et al.* 2011; Gharbi, Chaurasia *et al.* 2016; Abdelhamed, Afifi *et al.* 2020).

A good survey on image matting is given by Wang and Cohen (2009). Representative papers, which include extensive comparisons with previous work, include (Chuang, Curless *et al.* 2001; Wang and Cohen 2007a; Levin, Acha, and Lischinski 2008; Rhemann, Rother *et al.* 2008, 2009; Xu, Price *et al.* 2017). You can find pointers to recent papers and results on the http://alphamatting.com website created by Rhemann, Rother *et al.* (2009). Matting ideas can also be applied to manipulate shadows (Chuang, Goldman *et al.* 2003; Sun, Barron *et al.* 2019; Zhou, Hadap *et al.* 2019; Zhang, Barron *et al.* 2020; Wang, Curless, and Seitz 2020) and videos (Chuang, Agarwala *et al.* 2002; Wang, Bhat *et al.* 2005; Erofeev, Gitman *et al.* 2015; Sengupta, Jayaram *et al.* 2020; Lin, Ryabtsev *et al.* 2021).

The literature on texture synthesis and hole filling includes traditional approaches to texture synthesis, which try to match image statistics between source and destination images (Heeger and

[31] CMU 15-463, http://graphics.cs.cmu.edu/courses/15-463/2008_fall, Berkeley CS194-26/294-26, https://inst.eecs.berkeley.edu/~cs194-26/fa20, MIT 6.815/6.865, https://stellar.mit.edu/S/course/6/sp08/6.815/materials.html, Stanford CS 448A, https://graphics.stanford.edu/courses/cs448a-08-spring, CMU 16-726, https://learning-image-synthesis.github.io, and SIGGRAPH courses, https://web.media.mit.edu/~raskar/photo.

Bergen 1995; De Bonet 1997; Portilla and Simoncelli 2000), as well as approaches that search for matching neighborhoods or patches inside the source sample (Efros and Leung 1999; Wei and Levoy 2000; Efros and Freeman 2001; Wei, Lefebvre *et al.* 2009) or use neural networks (Gatys, Ecker, and Bethge 2015; Shaham, Dekel, and Michaeli 2019). In a similar vein, traditional approaches to hole filling involve the solution of local variational (smooth continuation) problems (Bertalmio, Sapiro *et al.* 2000; Bertalmio, Vese *et al.* 2003; Telea 2004). The next wave of techniques use data-driven texture synthesis approaches (Drori, Cohen-Or, and Yeshurun 2003; Kwatra, Schödl *et al.* 2003; Criminisi, Pérez, and Toyama 2004; Sun, Yuan *et al.* 2004; Kwatra, Essa *et al.* 2005; Wilczkowiak, Brostow *et al.* 2005; Komodakis and Tziritas 2007; Wexler, Shechtman, and Irani 2007). The most recent algorithms for image and video inpainting use deep neural networks (Yang, Lu *et al.* 2017; Yu, Lin *et al.* 2018; Liu, Reda *et al.* 2018; Shih, Su *et al.* 2020; Yi, Tang *et al.* 2020; Gao, Saraf *et al.* 2020; Ntavelis, Romero *et al.* 2020a). In addition to generating isolated patches of texture or inpainting missing region, related techniques can also be used to transfer the style of an image or painting to another one (Efros and Freeman 2001; Hertzmann, Jacobs *et al.* 2001; Gatys, Ecker, and Bethge 2016; Johnson, Alahi, and Fei-Fei 2016; Dumoulin, Shlens, and Kudlur 2017; Huang and Belongie 2017; Shaham, Dekel, and Michaeli 2019).

10.7 Exercises

Ex 10.1: Radiometric calibration. Implement one of the multi-exposure radiometric calibration algorithms described in Section 10.2 (Debevec and Malik 1997; Mitsunaga and Nayar 1999; Reinhard, Heidrich *et al.* 2010). This calibration will be useful in a number of different applications, such as stitching images or stereo matching with different exposures and shape from shading.

1. Take a series of bracketed images with your camera on a tripod. If your camera has an automatic exposure bracketing (AEB) modes, taking three images may be sufficient to calibrate most of your camera's dynamic range, especially if your scene has a lot of bright and dark regions. (Shooting outdoors or through a window on a sunny day is best.)

2. If your images are not taken on a tripod, first perform a global alignment.

3. Estimate the radiometric response function using one of the techniques cited above.

4. Estimate the high dynamic range radiance image by selecting or blending pixels from different exposures (Debevec and Malik 1997; Mitsunaga and Nayar 1999; Eden, Uyttendaele, and Szeliski 2006).

5. Repeat your calibration experiments under different conditions, e.g., indoors under incandescent light, to get a sense for the range of color balancing effects that your camera imposes.

6. If your camera supports RAW and JPEG mode, calibrate both sets of images simultaneously and to each other (the radiance at each pixel will correspond). See if you can come up with a model for what your camera does, e.g., whether it treats color balance as a diagonal or full 3 $\times$ 3 matrix multiply, whether it uses non-linearities in addition to gamma, whether it sharpens the image while "developing" the JPEG image, etc.

7. Develop an interactive viewer to change the exposure of an image based on the average exposure of a region around the mouse. (One variant is to show the adjusted image inside a window around the mouse. Another is to adjust the complete image based on the mouse position.)

8. Implement a tone mapping operator (Exercise 10.5) and use this to map your radiance image to a displayable gamut.

Ex 10.2: Noise level function. Determine your camera's noise level function using either multiple shots or by analyzing smooth regions.

1. Set up your camera on a tripod looking at a calibration target or a static scene with a good variation in input levels and colors. (Check your camera's histogram to ensure that all values are being sampled.)
2. Take repeated images of the same scene (ideally with a remote shutter release) and average them to compute the variance at each pixel. Discarding pixels near high gradients (which are affected by camera motion), plot for each color channel the standard deviation at each pixel as a function of its output value.
3. Fit a lower envelope to these measurements and use this as your noise level function. How much variation do you see in the noise as a function of input level? How much of this is significant, i.e., away from flat regions in your camera response function where you do not want to be sampling anyway?
4. (Optional) Using the same images, develop a technique that segments the image into near-constant regions (Liu, Szeliski *et al.* 2008). (This is easier if you are photographing a calibration chart.) Compute the deviations for each region from a *single* image and use them to estimate the NLF. How does this compare to the multi-image technique, and how stable are your estimates from image to image?

Ex 10.3: Vignetting. Estimate the amount of vignetting in some of your lenses using one of the following three techniques (or devise one of your choosing):

1. Take an image of a large uniform intensity region (well-illuminated wall or blue sky—but be careful of brightness gradients) and fit a radial polynomial curve to estimate the vignetting.
2. Construct a center-weighted panorama and compare these pixel values to the input image values to estimate the vignetting function. Weight pixels in slowly varying regions more highly, as small misalignments will give large errors at high gradients. Optionally estimate the radiometric response function as well (Litvinov and Schechner 2005; Goldman 2010).
3. Analyze the radial gradients (especially in low-gradient regions) and fit the robust means of these gradients to the derivative of the vignetting function, as described by Zheng, Yu *et al.* (2008).

For the parametric form of your vignetting function, you can either use a simple radial function, e.g.,

$$f(r) = 1 + \alpha_1 r + \alpha_2 r^2 + \cdots \tag{10.41}$$

or one of the specialized equations developed by Kang and Weiss (2000) and Zheng, Lin, and Kang (2006).

In all of these cases, be sure that you are using linearized intensity measurements, by using either RAW images or images linearized through a radiometric response function, or at least images where the gamma curve has been removed.

(Optional) What happens if you forget to undo the gamma before fitting a (multiplicative) vignetting function?

Ex 10.4: Optical blur (PSF) estimation. Compute the optical PSF either using a known target (Figure 10.7) or by detecting and fitting step edges (Section 10.1.4) (Joshi, Szeliski, and Kriegman 2008; Cho, Paris *et al.* 2011).

1. Detect strong edges to sub-pixel precision.

2. Fit a local profile to each oriented edge and fill these pixels into an ideal target image, either at image resolution or at a higher resolution (Figure 10.9c–d).

3. Use least squares (10.1) at valid pixels to estimate the PSF kernel K, either globally or in locally overlapping sub-regions of the image.

4. Visualize the recovered PSFs and use them to remove chromatic aberration or deblur the image.

Ex 10.5: Tone mapping. Implement one of the tone mapping algorithms discussed in Section 10.2.1 (Durand and Dorsey 2002; Fattal, Lischinski, and Werman 2002; Reinhard, Stark *et al.* 2002; Lischinski, Farbman *et al.* 2006) or any of the numerous additional algorithms discussed by Reinhard, Heidrich *et al.* (2010) and https://stellar.mit.edu/S/course/6/sp08/6.815/materials.html.

(Optional) Compare your algorithm to local histogram equalization (Section 3.1.4).

Ex 10.6: Flash enhancement. Develop an algorithm to combine flash and non-flash photographs to best effect. You can use ideas from Eisemann and Durand (2004) and Petschnigg, Agrawala *et al.* (2004) or anything else you think might work well.

Ex 10.7: Super-resolution. Implement one or more super-resolution algorithms and compare their performance.

1. Take a set of photographs of the same scene using a hand-held camera (to ensure that there is some jitter between the photographs).

2. Determine the PSF for the images you are trying to super-resolve using one of the techniques in Exercise 10.4.

3. Alternatively, simulate a collection of lower-resolution images by taking a high-quality photograph (avoid those with compression artifacts) and applying your own prefilter kernel and downsampling.

4. Estimate the relative motion between the images using a parametric translation and rotation motion estimation algorithm (Sections 8.1.3 or 9.2).

5. Implement a basic least squares super-resolution algorithm by minimizing the difference between the observed and downsampled images (10.26–10.27).

6. Add in a gradient image prior, either as another least squares term or as a robust term that can be minimized using iteratively reweighted least squares (Appendix A.3).

7. (Optional) Implement one of the example-based super-resolution techniques, where matching against a set of exemplar images is used either to infer higher-frequency information to be added to the reconstruction (Freeman, Jones, and Pasztor 2002) or higher-frequency gradients to be matched in the super-resolved image (Baker and Kanade 2002).

8. (Optional) Use local edge statistic information to improve the quality of the super-resolved image (Fattal 2007).

9. (Optional) Try some of the newest DNN-based super-resolution algorithms.

Ex 10.8: Image matting. Develop an algorithm for pulling a foreground matte from natural images, as described in Section 10.4.

1. Make sure that the images you are taking are linearized (Exercise 10.1 and Section 10.1) and that your camera exposure is fixed (full manual mode), at least when taking multiple shots of the same scene.

2. To acquire ground truth data, place your object in front of a computer monitor and display a variety of solid background colors as well as some natural imagery.

3. Remove your object and re-display the same images to acquire known background colors.

4. Use triangulation matting (Smith and Blinn 1996) to estimate the ground truth opacities α and pre-multiplied foreground colors αF for your objects.

5. Implement one or more of the natural image matting algorithms described in Section 10.4 and compare your results to the ground truth values you computed. Alternatively, use the matting test images published on http://alphamatting.com.

6. (Optional) Run your algorithms on other images taken with the same calibrated camera (or other images you find interesting).

Ex 10.9: Smoke and shadow matting. Extract smoke or shadow mattes from one scene and insert them into another (Chuang, Agarwala *et al.* 2002; Chuang, Goldman *et al.* 2003).

1. Take a still or video sequence of images with and without some intermittent smoke and shadows. (Remember to linearize your images before proceeding with any computations.)

2. For each pixel, fit a line to the observed color values.

3. If performing smoke matting, robustly compute the intersection of these lines to obtain the smoke color estimate. Then, estimate the background color as the other extremum (unless you have already taken a smoke-free background image).

   If performing shadow matting, compute robust shadow (minimum) and lit (maximum) values for each pixel.

4. Extract the smoke or shadow mattes from each frame as the fraction between these two values (background and smoke or shadowed and lit).

5. Scan a new (destination) scene or modify the original background with an image editor.

6. Re-insert the smoke or shadow matte, along with any other foreground objects you may have extracted.

7. (Optional) Using a series of cast stick shadows, estimate the deformation field for the destination scene to correctly warp (drape) the shadows across the new geometry. (This is related to the shadow scanning technique developed by Bouguet and Perona (1999) and implemented in Exercise 13.2.)

8. (Optional) Chuang, Goldman *et al.* (2003) only demonstrated their technique for planar source geometries. Can you extend their technique to capture shadows acquired from an irregular source geometry?

9. (Optional) Can you change the direction of the shadow, i.e., simulate the effect of changing the light source direction?

10. (Optional) Re-implement the facial shadow removal algorithm of Zhang, Barron *et al.* (2020) and try applying it to other domains.

Ex 10.10: Texture synthesis. Implement one of the texture synthesis or hole filling algorithms presented in Section 10.5. Here is one possible procedure:

1. Implement the basic Efros and Leung (1999) algorithm, i.e., starting from the outside (for hole filling) or in raster order (for texture synthesis), search for a similar neighborhood in the source texture image, and copy that pixel.

2. Add in the Wei and Levoy (2000) extension of generating the pixels in a coarse-to-fine fashion, i.e., generate a lower-resolution synthetic texture (or filled image), and use this as a guide for matching regions in the finer resolution version.

3. Add in the Criminisi, Pérez, and Toyama (2004) idea of prioritizing pixels to be filled by some function of the local structure (gradient or orientation strength).

4. Extend any of the above algorithms by selecting sub-blocks in the source texture and using optimization to determine the seam between the new block and the existing image that it overlaps (Efros and Freeman 2001).

5. (Optional) Implement one of the isophote (smooth continuation) inpainting algorithms (Bertalmio, Sapiro *et al.* 2000; Telea 2004).

6. (Optional) Add the ability to supply a target (reference) image (Efros and Freeman 2001) or to provide sample filtered or unfiltered (reference and rendered) images (Hertzmann, Jacobs *et al.* 2001), see Section 10.5.2.

7. (Optional) Try some of the newer DNN-based inpainting algorithms described at the end of Section 10.5.1.

Ex 10.11: Colorization. Implement the Levin, Lischinski, and Weiss (2004) colorization algorithm that is sketched out in Section 4.2.4 and Figure 4.10. If you prefer, you can implement this as a neural network (Zhang, Zhu *et al.* 2017). Find some historic monochrome photographs and some modern color ones. Write an interactive tool that lets you "pick" colors from a modern photo and paint over the old one. Tune the algorithm parameters to give you good results. Are you pleased with the results? Can you think of ways to make them look more "antique", e.g., with softer (less saturated and edgy) colors?

(Alternative) Implement or test out one of the newer "automatic colorization" algorithms such as Zhang, Isola, and Efros (2016) or (Vondrick, Shrivastava *et al.* 2018).

Ex 10.12: Style transfer. Try some of the non-photorealistic rendering or style transfer algorithms from Sections 10.5.2–10.5.3 on your own images. Can you come up with surprising results? How about failure cases?

Chapter 11

Structure from motion and SLAM

11.1 Geometric intrinsic calibration 545
  11.1.1 Vanishing points 547
  11.1.2 *Application*: Single view metrology 548
  11.1.3 Rotational motion 549
  11.1.4 Radial distortion 550
11.2 Pose estimation 552
  11.2.1 Linear algorithms 552
  11.2.2 Iterative non-linear algorithms 554
  11.2.3 *Application*: Location recognition 555
  11.2.4 Triangulation 558
11.3 Two-frame structure from motion 560
  11.3.1 Eight, seven, and five-point algorithms 560
  11.3.2 Special motions and structures 564
  11.3.3 Projective (uncalibrated) reconstruction 565
  11.3.4 Self-calibration 566
  11.3.5 *Application*: View morphing 568
11.4 Multi-frame structure from motion 568
  11.4.1 Factorization 568
  11.4.2 Bundle adjustment 570
  11.4.3 Exploiting sparsity 571
  11.4.4 *Application*: Match move 574
  11.4.5 Uncertainty and ambiguities 575
  11.4.6 *Application*: Reconstruction from internet photos 576
  11.4.7 Global structure from motion 578
  11.4.8 Constrained structure and motion 580
11.5 Simultaneous localization and mapping (SLAM) 583
  11.5.1 *Application*: Autonomous navigation 585
  11.5.2 *Application*: Smartphone augmented reality 587
11.6 Additional reading 588
11.7 Exercises 590

R. Szeliski, *Computer Vision*, Texts in Computer Science,
https://doi.org/10.1007/978-3-030-34372-9_11

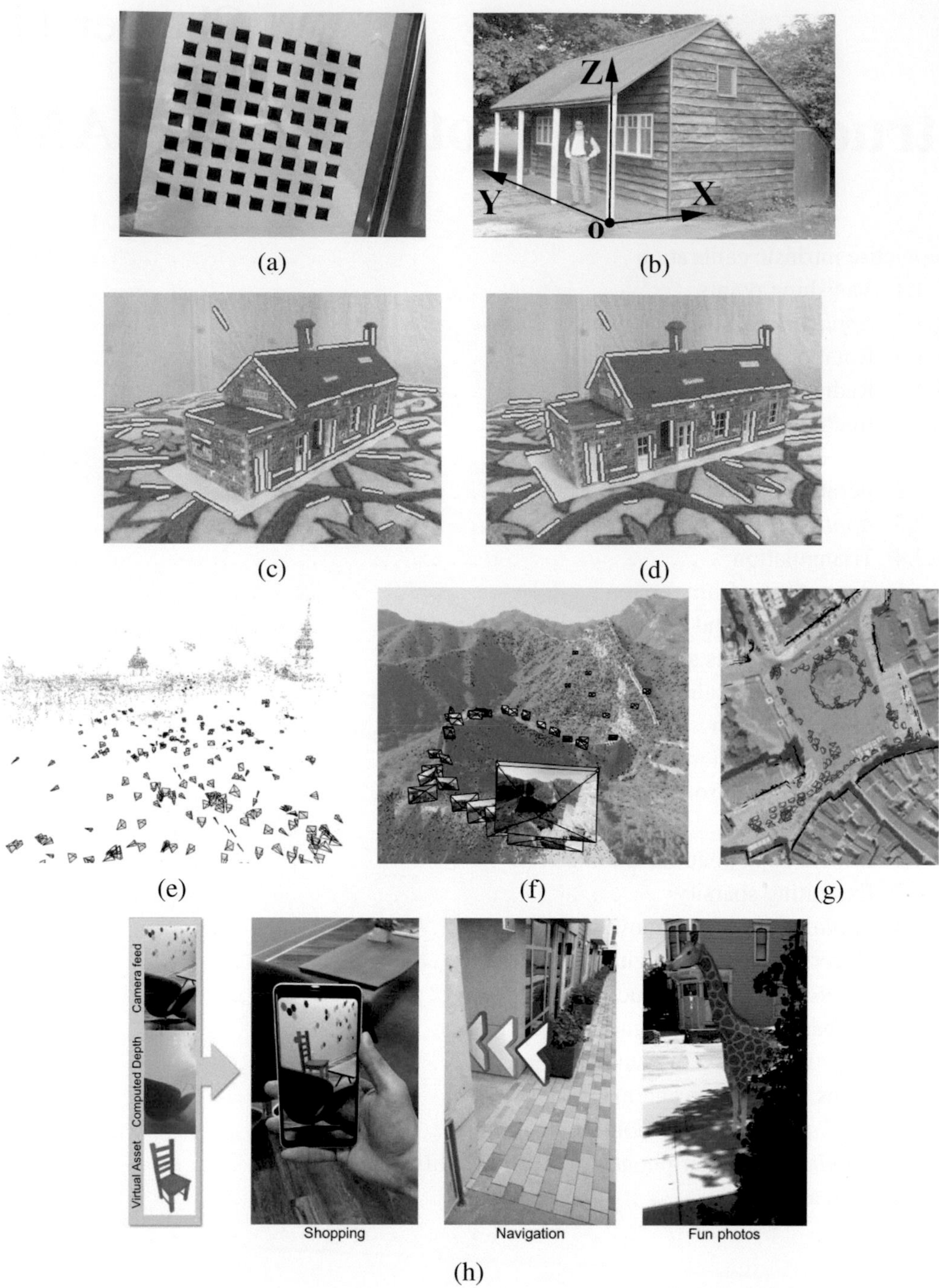

Figure 11.1 Structure from motion examples: (a) a two-dimensional calibration target (Zhang 2000) © 2000 IEEE; (b) single view metrology (Criminisi, Reid, and Zisserman 2000) © 2000 Springer. (c–d) line matching (Schmid and Zisserman 1997) © 1997 IEEE; (e–g) 3D reconstructions of Trafalgar Square, Great Wall of China, and Prague Old Town Square (Snavely, Seitz, and Szeliski 2006) © 2006 ACM; (h) smartphone augmented reality showing real-time depth occlusion effects (Valentin, Kowdle *et al.* 2018) © 2018 ACM.

The reconstruction of 3D models from images has been one of the central topics in computer vision since its inception (Figure 1.7). In fact, it was then believed that the construction of 3D models was a prerequisite for scene understanding and recognition (Marr 1982), although work in the last few decades has proven otherwise. However, 3D modeling has also proven to be immensely useful in applications such as virtual tourism (Section 11.4.6), autonomous navigation (Section 11.5.1), and augmented reality (Section 11.5.2).

In the last three chapters, we focused on techniques for establishing correspondences between 2D images and using these in a variety of applications such as image stitching, video enhancement, and computational photography. In this chapter, we turn to the topic of using such correspondences to build sparse 3D models of a scene and to re-localize cameras with respect to such models. While this process often involves simultaneously estimating both 3D geometry (structure) and camera pose (motion), it is commonly known (for historical reasons) as *structure from motion* (Ullman 1979).

The topics of projective geometry and structure from motion are extremely rich and some excellent textbooks and surveys have been written on them (Faugeras and Luong 2001; Hartley and Zisserman 2004; Moons, Van Gool, and Vergauwen 2010; Ma, Soatto *et al.* 2012). This chapter skips over a lot of the richer material available in these books, such as the trifocal tensor and algebraic techniques for full self-calibration, and concentrates instead on the basics that we have found useful in large-scale, image-based reconstruction problems (Snavely, Seitz, and Szeliski 2006).

We begin this chapter in Section 11.1 with a review of commonly used techniques for calibrating the camera *intrinsics*, e.g., the focal length and radial distortion parameters we introduced in Sections 2.1.4–2.1.5. Next, we discuss how to estimate the *extrinsic pose* of a camera from 3D to 2D point correspondences (Section 11.2) as well as how to *triangulate* a set of 2D correspondences to estimate a point's 3D location. We then look at the two-frame structure from motion problem (Section 11.3), which involves the determination of the *epipolar geometry* between two cameras and which can also be used to recover certain information about the camera intrinsics using self-calibration (Section 11.3.4). Section 11.4.1 looks at *factorization* approaches to simultaneously estimating structure and motion from large numbers of point tracks using orthographic approximations to the projection model. We then develop a more general and useful approach to structure from motion, namely the simultaneous *bundle adjustment* of all the camera and 3D structure parameters (Section 11.4.2). We also look at special cases that arise when there are higher-level structures, such as lines and planes, in the scene (Section 11.4.8). In the last part of this chapter (Section 11.5), we look at real-time systems for simultaneous localization and mapping (SLAM), which reconstruct a 3D world model while moving through an environment, and can be applied to both visual navigation and augmented reality.

11.1 Geometric intrinsic calibration

As we discuss in the next section (Equations (11.14–11.15)), the computation of the internal (intrinsic) camera calibration parameters can occur simultaneously with the estimation of the (extrinsic) pose of the camera with respect to a known calibration target. This, indeed, is the "classic" approach to camera calibration used in both the photogrammetry (Slama 1980) and the computer vision (Tsai 1987) communities. In this section, we look at simpler alternative formulations that may not involve the full solution of a non-linear regression problem, the use of alternative calibration targets, and the estimation of the non-linear part of camera optics such as radial distortion. In some applications, you can use the EXIF tags associated with a JPEG image to obtain a rough estimate of a camera's focal length and hence to initialize iterative estimation algorithms; but this technique should be used with caution as the results are often inaccurate.

(a) (b)

Figure 11.2 Calibration patterns: (a) a three-dimensional target (Quan and Lan 1999) © 1999 IEEE; (b) a two-dimensional target (Zhang 2000) © 2000 IEEE. Note that radial distortion needs to be removed from such images before the feature points can be used for calibration.

Calibration patterns

The use of a calibration pattern or set of markers is one of the more reliable ways to estimate a camera's intrinsic parameters. In photogrammetry, it is common to set up a camera in a large field looking at distant calibration targets whose exact location has been precomputed using surveying equipment (Slama 1980; Atkinson 1996; Kraus 1997). In this case, the translational component of the pose becomes irrelevant and only the camera rotation and intrinsic parameters need to be recovered.

If a smaller calibration rig needs to be used, e.g., for indoor robotics applications or for mobile robots that carry their own calibration target, it is best if the calibration object can span as much of the workspace as possible (Figure 11.2a), as planar targets often fail to accurately predict the components of the pose that lie far away from the plane. A good way to determine if the calibration has been successfully performed is to estimate the covariance in the parameters (Section 8.1.4) and then project 3D points from various points in the workspace into the image in order to estimate their 2D positional uncertainty.

If no calibration pattern is available, it is also possible to perform calibration simultaneously with structure and pose recovery (Sections 11.1.3 and 11.4.2), which is known as *self-calibration* (Faugeras, Luong, and Maybank 1992; Pollefeys, Koch, and Van Gool 1999; Hartley and Zisserman 2004; Moons, Van Gool, and Vergauwen 2010). However, such an approach requires a large amount of imagery to be accurate.

Planar calibration patterns

When a finite workspace is being used and accurate machining and motion control platforms are available, a good way to perform calibration is to move a planar calibration target through the workspace volume and use the known 3D point locations for calibration. This approach is sometimes called the *N-planes* calibration approach (Gremban, Thorpe, and Kanade 1988; Champleboux, Lavallée *et al.* 1992b; Grossberg and Nayar 2001) and has the advantage that each camera pixel can be mapped to a unique 3D ray in space, which takes care of both linear effects modeled by the calibration matrix $\mathbf{K}$ and non-linear effects such as radial distortion (Section 11.1.4).

A less cumbersome but also less accurate calibration can be obtained by waving a planar calibra-

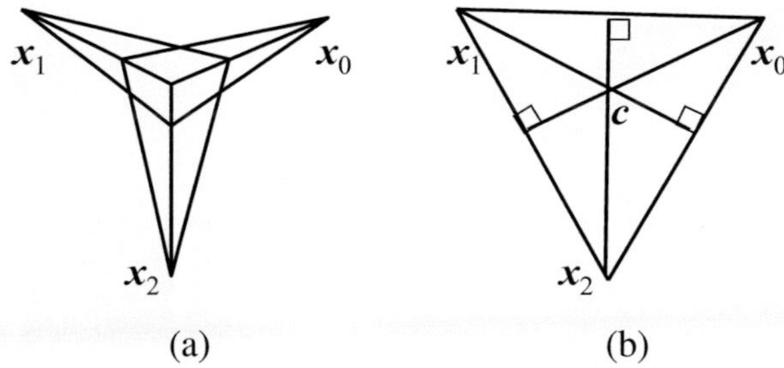

Figure 11.3 Calibration from vanishing points: (a) any pair of finite vanishing points $(\hat{\mathbf{x}}_i, \hat{\mathbf{x}}_j)$ can be used to estimate the focal length; (b) the orthocenter of the vanishing point triangle gives the image center of the image **c**.

tion pattern in front of a camera (Figure 11.2b) In this case, the pattern's pose has to be recovered in conjunction with the intrinsics. In this technique, each input image is used to compute a separate homography (8.19–8.23) $\tilde{\mathbf{H}}$ mapping the plane's calibration points $(X_i, Y_i, 1)$ into image coordinates (x_i, y_i),

$$\mathbf{x}_i = \begin{bmatrix} x_i \\ y_i \\ 1 \end{bmatrix} \sim \mathbf{K} \begin{bmatrix} \mathbf{r}_0 & \mathbf{r}_1 & \mathbf{t} \end{bmatrix} \begin{bmatrix} X_i \\ Y_i \\ 1 \end{bmatrix} \sim \tilde{\mathbf{H}} \mathbf{p}_i, \tag{11.1}$$

where the $\mathbf{r}_i$ are the first two columns of $\mathbf{R}$ and $\sim$ indicates equality up to scale. From these, Zhang (2000) shows how to form linear constraints on the nine entries in the $\mathbf{B} = \mathbf{K}^{-T}\mathbf{K}^{-1}$ matrix, from which the calibration matrix $\mathbf{K}$ can be recovered using a matrix square root and inversion. The matrix $\mathbf{B}$ is known as the *image of the absolute conic* (IAC) in projective geometry and is commonly used for camera calibration (Hartley and Zisserman 2004, Section 8.5). If only the focal length is being recovered, the even simpler approach of using vanishing points described below can be used instead.

11.1.1 Vanishing points

A common case for calibration that occurs often in practice is when the camera is looking at a manufactured or architectural scene with long extended rectangular patterns such as boxes or building walls. In this case, we can intersect the 2D lines corresponding to 3D parallel lines to compute their *vanishing points*, as described in Section 7.4.3, and use these to determine the intrinsic and extrinsic calibration parameters (Caprile and Torre 1990; Becker and Bove 1995; Liebowitz and Zisserman 1998; Cipolla, Drummond, and Robertson 1999; Antone and Teller 2002; Criminisi, Reid, and Zisserman 2000; Hartley and Zisserman 2004; Pflugfelder 2008).

Let us assume that we have detected two or more orthogonal vanishing points, all of which are *finite*, i.e., they are not obtained from lines that appear to be parallel in the image plane (Figure 11.3a). Let us also assume a simplified form for the calibration matrix $\mathbf{K}$ where only the focal length is unknown (2.59). It is often safe for rough 3D modeling to assume that the optical center is at the center of the image, that the aspect ratio is 1, and that there is no skew. In this case, the projection equation for the vanishing points can be written as

$$\hat{\mathbf{x}}_i = \begin{bmatrix} x_i - c_x \\ y_i - c_y \\ f \end{bmatrix} \sim \mathbf{R}\mathbf{p}_i = \mathbf{r}_i, \tag{11.2}$$

where $\mathbf{p}_i$ corresponds to one of the cardinal directions $(1, 0, 0)$, $(0, 1, 0)$, or $(0, 0, 1)$, and $\mathbf{r}_i$ is the ith column of the rotation matrix $\mathbf{R}$.

(a) (b)

Figure 11.4 Single view metrology (Criminisi, Reid, and Zisserman 2000) © 2000 Springer: (a) input image showing the three coordinate axes computed from the two horizontal vanishing points (which can be determined from the sidings on the shed); (b) a new view of the 3D reconstruction.

From the orthogonality between columns of the rotation matrix, we have

$$\mathbf{r}_i \cdot \mathbf{r}_j \sim (x_i - c_x)(x_j - c_x) + (y_i - c_y)(y_j - c_y) + f^2 = 0, \quad i \neq j \tag{11.3}$$

from which we can obtain an estimate for f^2. Note that the accuracy of this estimate increases as the vanishing points move closer to the center of the image. In other words, it is best to tilt the calibration pattern a decent amount around the 45° axis, as in Figure 11.3a. Once the focal length f has been determined, the individual columns of $\mathbf{R}$ can be estimated by normalizing the left-hand side of (11.2) and taking cross products. Alternatively, the orthogonal Procrustes algorithm (8.32) can be used.

If all three vanishing points are visible and finite in the same image, it is also possible to estimate the image center as the orthocenter of the triangle formed by the three vanishing points (Caprile and Torre 1990; Hartley and Zisserman 2004, Section 8.6) (Figure 11.3b). In practice, however, it is more accurate to re-estimate any unknown intrinsic calibration parameters using non-linear least squares (11.14).

11.1.2 *Application*: Single view metrology

A fun application of vanishing point estimation and camera calibration is the *single view metrology* system developed by Criminisi, Reid, and Zisserman (2000). Their system allows people to interactively measure heights and other dimensions as well as to build piecewise-planar 3D models, as shown in Figure 11.4.

The first step in their system is to identify two orthogonal vanishing points on the ground plane and the vanishing point for the vertical direction, which can be done by drawing some parallel sets of lines in the image. Alternatively, automated techniques such as those discussed in Section 7.4.3 or by Schaffalitzky and Zisserman (2000) could be used. The user then marks a few dimensions in the image, such as the height of a reference object, and the system can automatically compute the height of another object. Walls and other planar impostors (geometry) can also be sketched and reconstructed.

In the formulation originally developed by Criminisi, Reid, and Zisserman (2000), the system produces an *affine* reconstruction, i.e., one that is only known up to a set of independent scaling

Figure 11.5 Four images taken with a hand-held camera registered using a 3D rotation motion model, which can be used to estimate the focal length of the camera (Szeliski and Shum 1997) © 2000 ACM.

factors along each axis. A potentially more useful system can be constructed by assuming that the camera is calibrated up to an unknown focal length, which can be recovered from orthogonal (finite) vanishing directions, as we have just described in Section 11.1.1. Once this is done, the user can indicate an origin on the ground plane and another point a known distance away. From this, points on the ground plane can be directly projected into 3D, and points above the ground plane, when paired with their ground plane projections, can also be recovered. A fully metric reconstruction of the scene then becomes possible.

Exercise 11.4 has you implement such a system and then use it to model some simple 3D scenes. Section 13.6.1 describes other, potentially multi-view, approaches to architectural reconstruction, including an interactive piecewise-planar modeling system that uses vanishing points to establish 3D line directions and plane normals (Sinha, Steedly *et al.* 2008).

11.1.3 Rotational motion

When no calibration targets or known structures are available but you can rotate the camera around its front nodal point (or, equivalently, work in a large open environment where all objects are distant), the camera can be calibrated from a set of overlapping images by assuming that it is undergoing pure rotational motion, as shown in Figure 11.5 (Stein 1995; Hartley 1997b; Hartley, Hayman *et al.* 2000; de Agapito, Hayman, and Reid 2001; Kang and Weiss 1999; Shum and Szeliski 2000; Frahm and Koch 2003). When a full 360° motion is used to perform this calibration, a very accurate estimate of the focal length f can be obtained, as the accuracy in this estimate is proportional to the total number of pixels in the resulting cylindrical panorama (Section 8.2.6) (Stein 1995; Shum and Szeliski 2000).

To use this technique, we first compute the homographies $\tilde{\mathbf{H}}_{ij}$ between all overlapping pairs of images, as explained in Equations (8.19–8.23). Then, we use the observation, first made in Equation (2.72) and explored in more detail in Equation (8.38), that each homography is related to the inter-camera rotation $\mathbf{R}_{ij}$ through the (unknown) calibration matrices $\mathbf{K}_i$ and $\mathbf{K}_j$,

$$\tilde{\mathbf{H}}_{ij} = \mathbf{K}_i \mathbf{R}_i \mathbf{R}_j^{-1} \mathbf{K}_j^{-1} = \mathbf{K}_i \mathbf{R}_{ij} \mathbf{K}_j^{-1}. \tag{11.4}$$

The simplest way to obtain the calibration is to use the simplified form of the calibration matrix (2.59), where we assume that the pixels are square and the image center lies at the geometric center of the 2D pixel array, i.e., $\mathbf{K}_k = \mathrm{diag}(f_k, f_k, 1)$. We subtract half the width and height from the original pixel coordinates to that the pixel $(x, y) = (0, 0)$ lies at the center of the image. We can then rewrite Equation (11.4) as

$$\mathbf{R}_{10} \sim \mathbf{K}_1^{-1}\tilde{\mathbf{H}}_{10}\mathbf{K}_0 \sim \begin{bmatrix} h_{00} & h_{01} & f_0^{-1}h_{02} \\ h_{10} & h_{11} & f_0^{-1}h_{12} \\ f_1 h_{20} & f_1 h_{21} & f_0^{-1}f_1 h_{22} \end{bmatrix}, \tag{11.5}$$

where h_{ij} are the elements of $\tilde{\mathbf{H}}_{10}$.

Using the orthonormality properties of the rotation matrix $\mathbf{R}_{10}$ and the fact that the right-hand side of (11.5) is known only up to a scale, we obtain

$$h_{00}^2 + h_{01}^2 + f_0^{-2}h_{02}^2 = h_{10}^2 + h_{11}^2 + f_0^{-2}h_{12}^2 \tag{11.6}$$

and

$$h_{00}h_{10} + h_{01}h_{11} + f_0^{-2}h_{02}h_{12} = 0. \tag{11.7}$$

From this, we can compute estimates for f_0 of

$$f_0^2 = \frac{h_{12}^2 - h_{02}^2}{h_{00}^2 + h_{01}^2 - h_{10}^2 - h_{11}^2} \quad \text{if} \quad \mathrm{h}_{00}^2 + \mathrm{h}_{01}^2 \neq \mathrm{h}_{10}^2 + \mathrm{h}_{11}^2 \tag{11.8}$$

or

$$f_0^2 = -\frac{h_{02}h_{12}}{h_{00}h_{10} + h_{01}h_{11}} \quad \text{if} \quad \mathrm{h}_{00}\mathrm{h}_{10} \neq -\mathrm{h}_{01}\mathrm{h}_{11}. \tag{11.9}$$

If neither of these conditions holds, we can also take the dot products between the first (or second) row and the third one. Similar results can be obtained for f_1 as well, by analyzing the columns of $\tilde{\mathbf{H}}_{10}$. If the focal length is the same for both images, we can take the geometric mean of f_0 and f_1 as the estimated focal length $f = \sqrt{f_1 f_0}$. When multiple estimates of f are available, e.g., from different homographies, the median value can be used as the final estimate. A more general (upper-triangular) estimate of $\mathbf{K}$ can be obtained in the case of a fixed-parameter camera $\mathbf{K}_i = \mathbf{K}$ using the technique of Hartley (1997b). Extensions to the cases of temporally varying calibration parameters and non-stationary cameras are discussed by Hartley, Hayman *et al.* (2000) and de Agapito, Hayman, and Reid (2001).

The quality of the intrinsic camera parameters can be greatly increased by constructing a full 360° panorama, as mis-estimating the focal length will result in a gap (or excessive overlap) when the first image in the sequence is stitched to itself (Figure 8.6). The resulting misalignment can be used to improve the estimate of the focal length and to re-adjust the rotation estimates, as described in Section 8.2.4. Rotating the camera by 90° around its optical axis and re-shooting the panorama is a good way to check for aspect ratio and skew pixel problems, as is generating a full hemi-spherical panorama when there is sufficient texture.

Ultimately, however, the most accurate estimate of the calibration parameters (including radial distortion) can be obtained using a full simultaneous non-linear minimization of the intrinsic and extrinsic (rotation) parameters, as described in Section 11.2.2.

11.1.4 Radial distortion

When images are taken with wide-angle lenses, it is often necessary to model *lens distortions* such as *radial distortion*. As discussed in Section 2.1.5, the radial distortion model says that coordinates in the observed images are displaced towards (*barrel* distortion) or away (*pincushion* distortion) from the image center by an amount proportional to their radial distance (Figure 2.13a–b). The simplest radial distortion models use low-order polynomials (c.f. Equation (2.78)),

$$\begin{aligned} \hat{x} &= x(1 + \kappa_1 r^2 + \kappa_2 r^4) \\ \hat{y} &= y(1 + \kappa_1 r^2 + \kappa_2 r^4), \end{aligned} \tag{11.10}$$

where $(x, y) = (0, 0)$ at the radial distortion center (2.77), $r^2 = x^2 + y^2$, and κ_1 and κ_2 are called the *radial distortion parameters* (Brown 1971; Slama 1980).[1]

A variety of techniques can be used to estimate the radial distortion parameters for a given lens, if the digital camera has not already done this in its capture software. One of the simplest and most useful is to take an image of a scene with a lot of straight lines, especially lines aligned with and near the edges of the image. The radial distortion parameters can then be adjusted until all of the lines in the image are straight, which is commonly called the *plumb-line method* (Brown 1971; Kang 2001; El-Melegy and Farag 2003). Exercise 11.5 gives some more details on how to implement such a technique.

Another approach is to use several overlapping images and to combine the estimation of the radial distortion parameters with the image alignment process, i.e., by extending the pipeline used for stitching in Section 8.3.1. Sawhney and Kumar (1999) use a hierarchy of motion models (translation, affine, projective) in a coarse-to-fine strategy coupled with a quadratic radial distortion correction term. They use direct (intensity-based) minimization to compute the alignment. Stein (1997) uses a feature-based approach combined with a general 3D motion model (and quadratic radial distortion), which requires more matches than a parallax-free rotational panorama but is potentially more general. More recent approaches sometimes simultaneously compute both the unknown intrinsic parameters and the radial distortion coefficients, which may include higher-order terms or more complex rational or non-parametric forms (Claus and Fitzgibbon 2005; Sturm 2005; Thirthala and Pollefeys 2005; Barreto and Daniilidis 2005; Hartley and Kang 2005; Steele and Jaynes 2006; Tardif, Sturm *et al.* 2009).

When a known calibration target is being used (Figure 11.2), the radial distortion estimation can be folded into the estimation of the other intrinsic and extrinsic parameters (Zhang 2000; Hartley and Kang 2007; Tardif, Sturm *et al.* 2009). This can be viewed as adding another stage to the general non-linear minimization pipeline shown in Figure 11.7 between the intrinsic parameter multiplication box $\mathbf{f}_\mathrm{C}$ and the perspective division box $\mathbf{f}_\mathrm{P}$. (See Exercise 11.6 on more details for the case of a planar calibration target.)

Of course, as discussed in Section 2.1.5, more general models of lens distortion, such as fisheye and non-central projection, may sometimes be required. While the parameterization of such lenses may be more complicated (Section 2.1.5), the general approach of either using calibration rigs with known 3D positions or self-calibration through the use of multiple overlapping images of a scene can both be used (Hartley and Kang 2007; Tardif, Sturm, and Roy 2007). The same techniques used to calibrate for radial distortion can also be used to reduce the amount of chromatic aberration by separately calibrating each color channel and then warping the channels to put them back into alignment (Exercise 11.7).

[1] Sometimes the relationship between x and $\hat{x}$ is expressed the other way around, i.e., using primed (final) coordinates on the right-hand side, $x = \hat{x}(1 + \kappa_1 \hat{r}^2 + \kappa_2 \hat{r}^4)$. This is convenient if we map image pixels into (warped) rays and then undistort the rays to obtain 3D rays in space, i.e., if we are using inverse warping.

11.2 Pose estimation

A particular instance of feature-based alignment, which occurs very often, is estimating an object's 3D pose from a set of 2D point projections. This *pose estimation* problem is also known as *extrinsic* calibration, as opposed to the *intrinsic* calibration of internal camera parameters such as focal length, which we discuss in Section 11.1. The problem of recovering pose from three correspondences, which is the minimal amount of information necessary, is known as the *perspective-3-point-problem* (P3P),[2] with extensions to larger numbers of points collectively known as PnP (Haralick, Lee *et al.* 1994; Quan and Lan 1999; Gao, Hou *et al.* 2003; Moreno-Noguer, Lepetit, and Fua 2007; Persson and Nordberg 2018).

In this section, we look at some of the techniques that have been developed to solve such problems, starting with the *direct linear transform* (DLT), which recovers a 3×4 camera matrix, followed by other "linear" algorithms, and then looking at statistically optimal iterative algorithms.

11.2.1 Linear algorithms

The simplest way to recover the pose of the camera is to form a set of rational linear equations analogous to those used for 2D motion estimation (8.19) from the camera matrix form of perspective projection (2.55–2.56),

$$x_i = \frac{p_{00}X_i + p_{01}Y_i + p_{02}Z_i + p_{03}}{p_{20}X_i + p_{21}Y_i + p_{22}Z_i + p_{23}} \tag{11.11}$$

$$y_i = \frac{p_{10}X_i + p_{11}Y_i + p_{12}Z_i + p_{13}}{p_{20}X_i + p_{21}Y_i + p_{22}Z_i + p_{23}}, \tag{11.12}$$

where (x_i, y_i) are the measured 2D feature locations and (X_i, Y_i, Z_i) are the known 3D feature locations (Figure 11.6). As with (8.21), this system of equations can be solved in a linear fashion for the unknowns in the camera matrix $\mathbf{P}$ by multiplying the denominator on both sides of the equation.Because $\mathbf{P}$ is unknown up to a scale, we can either fix one of the entries, e.g., $p_{23} = 1$, or find the smallest singular vector of the set of linear equations. The resulting algorithm is called the *direct linear transform* (DLT) and is commonly attributed to Sutherland (1974). (For a more in-depth discussion, see Hartley and Zisserman (2004).) To compute the 12 (or 11) unknowns in $\mathbf{P}$, at least six correspondences between 3D and 2D locations must be known.

As with the case of estimating homographies (8.21–8.23), more accurate results for the entries in $\mathbf{P}$ can be obtained by directly minimizing the set of Equations (11.11–11.12) using non-linear least squares with a small number of iterations. Note that instead of taking the ratios of the X/Z and Y/Z values as in (11.11–11.12), it is also possible to take a cross product of the 3-vector $(x_i, y_i, 1)$ image measurement and the 3-D ray (X, Y, Z) and set the three elements of this cross-product to 0. The resulting three equations, when interpreted as a set of least squares constraints, in effect compute the squared sine of the angle between the two rays.

Once the entries in $\mathbf{P}$ have been recovered, it is possible to recover both the intrinsic calibration matrix $\mathbf{K}$ and the rigid transformation $(\mathbf{R}, \mathbf{t})$ by observing from Equation (2.56) that

$$\mathbf{P} = \mathbf{K}[\mathbf{R}|\mathbf{t}]. \tag{11.13}$$

Because $\mathbf{K}$ is upper-triangular (see the discussion in Section 2.1.4), both $\mathbf{K}$ and $\mathbf{R}$ can be obtained from the front 3×3 sub-matrix of $\mathbf{P}$ using RQ factorization (Golub and Van Loan 1996).[3]

[2] The "3-point" algorithms actually require a 4th point to resolve a 4-way ambiguity.

[3] Note the unfortunate clash of terminologies: In matrix algebra textbooks, $\mathbf{R}$ represents an upper-triangular matrix; in computer vision, $\mathbf{R}$ is an orthogonal rotation.

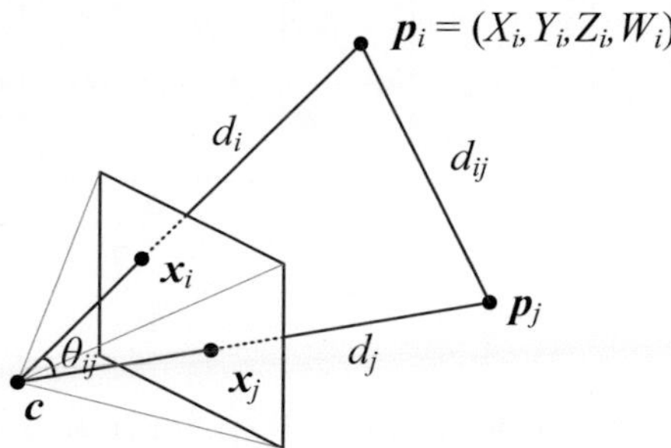

Figure 11.6 Pose estimation by the direct linear transform and by measuring visual angles and distances between pairs of points.

In most applications, however, we have some prior knowledge about the intrinsic calibration matrix $\mathbf{K}$, e.g., that the pixels are square, the skew is very small, and the image center is near the geometric center of the image (2.57–2.59). Such constraints can be incorporated into a non-linear minimization of the parameters in $\mathbf{K}$ and $(\mathbf{R}, \mathbf{t})$, as described in Section 11.2.2.

In the case where the camera is already calibrated, i.e., the matrix $\mathbf{K}$ is known (Section 11.1), we can perform pose estimation using as few as three points (Fischler and Bolles 1981; Haralick, Lee *et al.* 1994; Quan and Lan 1999). The basic observation that these *linear PnP* (*perspective n-point*) algorithms employ is that the visual angle between any pair of 2D points $\hat{\mathbf{x}}_i$ and $\hat{\mathbf{x}}_j$ must be the same as the angle between their corresponding 3D points $\mathbf{p}_i$ and $\mathbf{p}_j$ (Figure 11.6).

A full derivation of this approach can be found in the first edition of this book (Szeliski 2010, Section 6.2.1) and also in (Quan and Lan 1999), where the authors provide accuracy results for this and other techniques, which use fewer points but require more complicated algebraic manipulations. The paper by Moreno-Noguer, Lepetit, and Fua (2007) reviews other alternatives and also gives a lower complexity algorithm that typically produces more accurate results. An even more recent paper by Terzakis and Lourakis (2020) reviews papers published in the last decade.

Unfortunately, because minimal PnP solutions can be quite noise sensitive and also suffer from *bas-relief ambiguities* (e.g., depth reversals) (Section 11.4.5), it is prudent to optimize the initial estimates from PnP using the iterative technique described in Section 11.2.2. An alternative pose estimation algorithm involves starting with a scaled orthographic projection model and then iteratively refining this initial estimate using a more accurate perspective projection model (DeMenthon and Davis 1995). The attraction of this model, as stated in the paper's title, is that it can be implemented "in 25 lines of [Mathematica] code".

CNN-based pose estimation

As with other areas on computer vision, deep neural networks have also been applied to pose estimation. Some representative papers include Xiang, Schmidt *et al.* (2018), Oberweger, Rad, and Lepetit (2018), Hu, Hugonot *et al.* (2019), Peng, Liu *et al.* (2019), and (Hu, Fua *et al.* 2020) for object pose estimation, and papers such as Kendall and Cipolla (2017) and Kim, Dunn, and Frahm (2017) discussed in Section 11.2.3 on location recognition. There is also a very active community around estimating pose from RGB-D images, with the most recent papers (Hagelskjær and Buch 2020; Labbé, Carpentier *et al.* 2020) evaluated on the BOP (Benchmark for 6DOF Object Pose) (Hodaň, Michel *et al.* 2018).[4]

[4]https://bop.felk.cvut.cz/challenges/bop-challenge-2020, https://cmp.felk.cvut.cz/sixd/workshop_2020

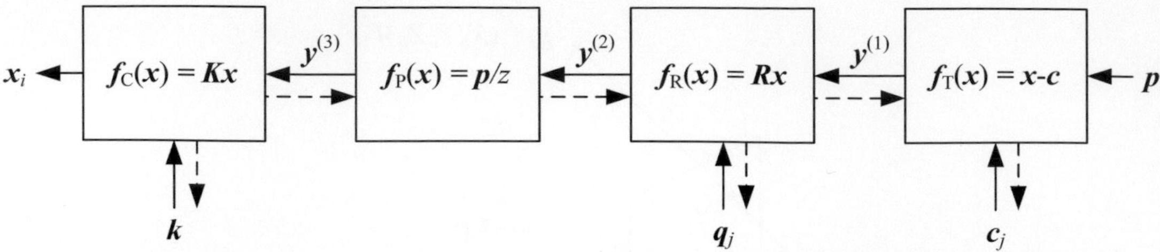

Figure 11.7 A set of chained transforms for projecting a 3D point $\mathbf{p}_i$ to a 2D measurement $\mathbf{x}_i$ through a series of transformations $\mathbf{f}^{(k)}$, each of which is controlled by its own set of parameters. The dashed lines indicate the flow of information as partial derivatives are computed during a backward pass.

11.2.2 Iterative non-linear algorithms

The most accurate and flexible way to estimate pose is to directly minimize the squared (or robust) reprojection error for the 2D points as a function of the unknown pose parameters in $(\mathbf{R}, \mathbf{t})$ and optionally $\mathbf{K}$ using non-linear least squares (Tsai 1987; Bogart 1991; Gleicher and Witkin 1992). We can write the projection equations as

$$\mathbf{x}_i = \mathbf{f}(\mathbf{p}_i; \mathbf{R}, \mathbf{t}, \mathbf{K}) \tag{11.14}$$

and iteratively minimize the robustified linearized reprojection errors

$$E_{\text{NLP}} = \sum_i \rho\left(\frac{\partial \mathbf{f}}{\partial \mathbf{R}}\Delta\mathbf{R} + \frac{\partial \mathbf{f}}{\partial \mathbf{t}}\Delta\mathbf{t} + \frac{\partial \mathbf{f}}{\partial \mathbf{K}}\Delta\mathbf{K} - \mathbf{r}_i\right), \tag{11.15}$$

where $\mathbf{r}_i = \tilde{\mathbf{x}}_i - \hat{\mathbf{x}}_i$ is the current residual vector (2D error in predicted position) and the partial derivatives are with respect to the unknown pose parameters (rotation, translation, and optionally calibration). The *robust loss function* ρ, which we first introduced in (4.15) in Section 4.1.3, is used to reduce the influence of outlier correspondences. Note that if full 2D covariance estimates are available for the 2D feature locations, the above squared norm can be weighted by the inverse point covariance matrix, as in Equation (8.11).

An easier to understand (and implement) version of the above non-linear regression problem can be constructed by re-writing the projection equations as a concatenation of simpler steps, each of which transforms a 4D homogeneous coordinate $\mathbf{p}_i$ by a simple transformation such as translation, rotation, or perspective division (Figure 11.7). The resulting projection equations can be written as

$$\mathbf{y}^{(1)} = \mathbf{f}_{\text{T}}(\mathbf{p}_i; \mathbf{c}_j) = \mathbf{p}_i - \mathbf{c}_j, \tag{11.16}$$

$$\mathbf{y}^{(2)} = \mathbf{f}_{\text{R}}(\mathbf{y}^{(1)}; \mathbf{q}_j) = \mathbf{R}(\mathbf{q}_j)\,\mathbf{y}^{(1)}, \tag{11.17}$$

$$\mathbf{y}^{(3)} = \mathbf{f}_{\text{P}}(\mathbf{y}^{(2)}) = \frac{\mathbf{y}^{(2)}}{z^{(2)}}, \tag{11.18}$$

$$\mathbf{x}_i = \mathbf{f}_{\text{C}}(\mathbf{y}^{(3)}; \mathbf{k}) = \mathbf{K}(\mathbf{k})\,\mathbf{y}^{(3)}. \tag{11.19}$$

Note that in these equations, we have indexed the camera centers $\mathbf{c}_j$ and camera rotation quaternions $\mathbf{q}_j$ by an index j, in case more than one pose of the calibration object is being used (see also Section 11.4.2.) We are also using the camera center $\mathbf{c}_j$ instead of the world translation $\mathbf{t}_j$, as this is a more natural parameter to estimate.

The advantage of this chained set of transformations is that each one has a simple partial derivative with respect both to its parameters and to its input. Thus, once the predicted value of $\tilde{\mathbf{x}}_i$ has been computed based on the 3D point location $\mathbf{p}_i$ and the current values of the pose parameters $(\mathbf{c}_j, \mathbf{q}_j, \mathbf{k})$, we can obtain all of the required partial derivatives using the chain rule

$$\frac{\partial \mathbf{r}_i}{\partial \mathbf{p}^{(k)}} = \frac{\partial \mathbf{r}_i}{\partial \mathbf{y}^{(k)}} \frac{\partial \mathbf{y}^{(k)}}{\partial \mathbf{p}^{(k)}}, \tag{11.20}$$

where $\mathbf{p}^{(k)}$ indicates one of the parameter vectors that is being optimized. (This same "trick" is used in neural networks as part of *backpropagation*, which we presented in Figure 5.31.)

The one special case in this formulation that can be considerably simplified is the computation of the rotation update. Instead of directly computing the derivatives of the 3×3 rotation matrix $\mathbf{R}(\mathbf{q})$ as a function of the unit quaternion entries, you can prepend the incremental rotation matrix $\Delta\mathbf{R}(\boldsymbol{\omega})$ given in Equation (2.35) to the current rotation matrix and compute the partial derivative of the transform with respect to these parameters, which results in a simple cross product of the backward chaining partial derivative and the outgoing 3D vector, as explained in Equation (2.36).

Target-based augmented reality

A widely used application of pose estimation is *augmented reality*, where virtual 3D images or annotations are superimposed on top of a live video feed, either through the use of see-through glasses (a head-mounted display) or on a regular computer or mobile device screen (Azuma, Baillot *et al.* 2001; Haller, Billinghurst, and Thomas 2007; Billinghurst, Clark, and Lee 2015). In some applications, a special pattern printed on cards or in a book is tracked to perform the augmentation (Kato, Billinghurst *et al.* 2000; Billinghurst, Kato, and Poupyrev 2001). For a desktop application, a grid of dots printed on a mouse pad can be tracked by a camera embedded in an augmented mouse to give the user control of a full six degrees of freedom over their position and orientation in a 3D space (Hinckley, Sinclair *et al.* 1999). Today, tracking known targets such as movie posters is used in some phone-based augmented reality systems such as Facebook's Spark AR.[5]

Sometimes, the scene itself provides a convenient object to track, such as the rectangle defining a desktop used in *through-the-lens camera control* (Gleicher and Witkin 1992). In outdoor locations, such as film sets, it is more common to place special markers such as brightly colored balls in the scene to make it easier to find and track them (Bogart 1991). In older applications, surveying techniques were used to determine the locations of these balls before filming. Today, it is more common to apply structure-from-motion directly to the film footage itself (Section 11.5.2).

Exercise 8.4 has you implement a tracking and pose estimation system for augmented-reality applications.

11.2.3 *Application*: Location recognition

One of the most exciting applications of pose estimation is in the area of location recognition, which can be used both in desktop applications (*"Where did I take this holiday snap?"*) and in mobile smartphone applications. The latter case includes not only finding out your current location based on a cell-phone image, but also providing you with navigation directions or annotating your images with useful information, such as building names and restaurant reviews (i.e., a pocketable form of *augmented reality*). This problem is also often called *visual (or image-based) localization* (Se, Lowe, and Little 2002; Zhang and Kosecka 2006; Janai, Güney *et al.* 2020, Section 13.3) or *visual place recognition* (Lowry, Sünderhauf *et al.* 2015).

[5] https://sparkar.facebook.com/ar-studio

Figure 11.8 Feature-based location recognition (Schindler, Brown, and Szeliski 2007) © 2007 IEEE: (a) three typical series of overlapping street photos; (b) handheld camera shots and (c) their corresponding database photos.

Some approaches to location recognition assume that the photos consist of architectural scenes for which vanishing directions can be used to pre-rectify the images for easier matching (Robertson and Cipolla 2004). Other approaches use general affine covariant interest points to perform *wide baseline matching* (Schaffalitzky and Zisserman 2002), with the winning entry on the ICCV 2005 Computer Vision Contest (Szeliski 2005) using this approach (Zhang and Kosecka 2006). The Photo Tourism system of Snavely, Seitz, and Szeliski (2006) (Section 14.1.2) was the first to apply these kinds of ideas to large-scale image matching and (implicit) location recognition from internet photo collections taken under a wide variety of viewing conditions.

The main difficulty in location recognition is in dealing with the extremely large community (user-generated) photo collections on websites such as Flickr (Philbin, Chum *et al.* 2007; Chum, Philbin *et al.* 2007; Philbin, Chum *et al.* 2008; Irschara, Zach *et al.* 2009; Turcot and Lowe 2009; Sattler, Leibe, and Kobbelt 2011, 2017) or commercially captured databases (Schindler, Brown, and Szeliski 2007; Klingner, Martin, and Roseborough 2013; Torii, Arandjelović *et al.* 2018). The prevalence of commonly appearing elements such as foliage, signs, and common architectural elements further complicates the task (Schindler, Brown, and Szeliski 2007; Jegou, Douze, and Schmid 2009; Chum and Matas 2010b; Knopp, Sivic, and Pajdla 2010; Torii, Sivic *et al.* 2013; Sattler, Havlena *et al.* 2016). Figure 7.26 shows some results on location recognition from community photo collections, while Figure 11.8 shows sample results from denser commercially acquired datasets. In the latter case, the overlap between adjacent database images can be used to verify and prune potential matches using "temporal" filtering, i.e., requiring the query image to match nearby overlapping database images before accepting the match. Similar ideas have been used to improve location recognition from panoramic video sequences (Levin and Szeliski 2004; Samano, Zhou, and Calway 2020) and to combine local SLAM reconstructions from image sequences with matching against a precomputed map for higher reliability (Stenborg, Sattler, and Hammarstrand 2020). Recognizing indoor locations inside buildings and shopping malls poses its own set of challenges, including textureless areas and repeated elements (Levin and Szeliski 2004; Wang, Fidler, and Urtasun 2015; Sun, Xie *et al.* 2017; Taira, Okutomi *et al.* 2018; Taira, Rocco *et al.* 2019; Lee, Ryu *et al.* 2021). The matching of ground-level to aerial images has also been studied (Kaminsky, Snavely *et al.* 2009; Shan, Wu *et al.* 2014).

Some of the initial research on location recognition was organized around the Oxford 5k and Paris 6k datasets (Philbin, Chum *et al.* 2007, 2008; Radenović, Iscen *et al.* 2018), as well as the Vienna (Irschara, Zach *et al.* 2009) and Photo Tourism (Li, Snavely, and Huttenlocher 2010) datasets, and later around the 7 scenes indoor RGB-D dataset (Shotton, Glocker *et al.* 2013) and Cambridge Landmarks (Kendall, Grimes, and Cipolla 2015). The NetVLAD paper (Arandjelovic, Gronat *et al.* 2016) was tested on Google Street View Time Machine data. Currently, the most widely used visual localization datasets are collected at the Long-Term Visual Localization Benchmark[6] and include such datasets as Aachen Day-Night (Sattler, Maddern *et al.* 2018) and InLoc (Taira, Okutomi *et al.* 2018). And while most localization systems work from collections of ground-level images, it is also possible to re-localize based on textured digital elevation (terrain) models for outdoor (non-city) applications (Baatz, Saurer *et al.* 2012; Brejcha, Lukáč *et al.* 2020).

Some of the most recent approaches to localization use deep networks to generate feature descriptors (Arandjelovic, Gronat *et al.* 2016; Kim, Dunn, and Frahm 2017; Torii, Arandjelović *et al.* 2018; Radenović, Tolias, and Chum 2019; Yang, Kien Nguyen *et al.* 2019; Sarlin, Unagar *et al.* 2021), perform large-scale instance retrieval (Radenović, Tolias, and Chum 2019; Cao, Araujo, and Sim 2020; Ng, Balntas *et al.* 2020; Tolias, Jenicek, and Chum 2020; Pion, Humenberger *et al.* 2020 and Section 6.2.3), map images to 3D scene coordinates (Brachmann and Rother 2018), or perform end-to-end scene coordinate regression (Shotton, Glocker *et al.* 2013), absolute pose regression (APR) (Kendall, Grimes, and Cipolla 2015; Kendall and Cipolla 2017), or relative pose regression (RPR) (Melekhov, Ylioinas *et al.* 2017; Balntas, Li, and Prisacariu 2018). Recent evaluations of these techniques have shown that classical approaches based on feature matching followed by geometric pose optimization typically outperform pose regression approaches in terms of accuracy and generalization (Sattler, Zhou *et al.* 2019; Zhou, Sattler *et al.* 2019; Ding, Wang *et al.* 2019; Lee, Ryu *et al.* 2021; Sarlin, Unagar *et al.* 2021).

The Long-Term Visual Localization benchmark has a leaderboard listing the best-performing localization systems. In the CVPR 2020 workshop and challenge, some of the winning entries were based on recent detectors, descriptors, and matchers such as SuperGlue (Sarlin, DeTone *et al.* 2020), ASLFeat (Luo, Zhou *et al.* 2020), and R2D2 (Revaud, Weinzaepfel *et al.* 2019). Other systems that did well include HF-Net (Sarlin, Cadena *et al.* 2019), ONavi (Fan, Zhou *et al.* 2020), and D2-Net (Dusmanu, Rocco *et al.* 2019). An even more recent trend is to use DNNs or transformers to establish dense coarse-to-fine matches (Jiang, Trulls *et al.* 2021; Sun, Shen *et al.* 2021).

Another variant on location recognition is the automatic discovery of *landmarks*, i.e., frequently photographed objects and locations. Simon, Snavely, and Seitz (2007) show how these kinds of objects can be discovered simply by analyzing the matching graph constructed as part of the 3D modeling process in Photo Tourism. More recent work has extended this approach to larger datasets using efficient clustering techniques (Philbin and Zisserman 2008; Li, Wu *et al.* 2008; Chum, Philbin, and Zisserman 2008; Chum and Matas 2010a; Arandjelović and Zisserman 2012), combining metadata such as GPS and textual tags with visual search (Quack, Leibe, and Van Gool 2008; Crandall, Backstrom *et al.* 2009; Li, Snavely *et al.* 2012), and using multiple descriptors to obtain real-time performance in micro aerial vehicle navigation (Lim, Sinha *et al.* 2012). It is now even possible to automatically associate object tags with images based on their co-occurrence in multiple loosely tagged images (Simon and Seitz 2008; Gammeter, Bossard *et al.* 2009).

The concept of organizing the world's photo collections by location has even been recently extended to organizing all of the universe's (astronomical) photos in an application called *astrometry*,[7] The technique used to match any two star fields is to take quadruplets of nearby stars (a pair of

[6] https://www.visuallocalization.net

[7] https://astrometry.net

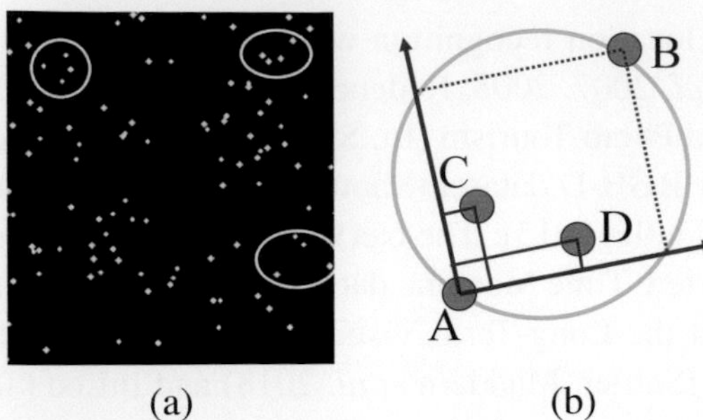

Figure 11.9 Locating star fields using astrometry, https://astrometry.net. (a) Input star field and some selected star quads. (b) The 2D coordinates of stars C and D are encoded relative to the unit square defined by A and B.

stars and another pair inside their diameter) to form a 30-bit *geometric hash* by encoding the relative positions of the second pair of points using the inscribed square as the reference frame, as shown in Figure 11.9. Traditional information retrieval techniques (k-d trees built for different parts of a sky atlas) are then used to find matching quads as potential star field location hypotheses, which can then be verified using a similarity transform.

11.2.4 Triangulation

The problem of determining a point's 3D position from a set of corresponding image locations and known camera positions is known as *triangulation*. This problem is the converse of the pose estimation problem we studied in Section 11.2.

One of the simplest ways to solve this problem is to find the 3D point $\mathbf{p}$ that lies closest to all of the 3D rays corresponding to the 2D matching feature locations $\{\mathbf{x}_j\}$ observed by cameras $\{\mathbf{P}_j = \mathbf{K}_j[\mathbf{R}_j|\mathbf{t}_j]\}$, where $\mathbf{t}_j = -\mathbf{R}_j\mathbf{c}_j$ and $\mathbf{c}_j$ is the jth camera center (2.55–2.56). As you can see in Figure 11.10, these rays originate at $\mathbf{c}_j$ in a direction $\hat{\mathbf{v}}_j = \mathcal{N}(\mathbf{R}_j^{-1}\mathbf{K}_j^{-1}\mathbf{x}_j)$, where $\mathcal{N}(\mathbf{v})$ normalizes a vector $\mathbf{v}$ to unit length. The nearest point to $\mathbf{p}$ on this ray, which we denote as $\mathbf{q}_j = \mathbf{c}_j + d_j\hat{\mathbf{v}}_j$, minimizes the distance

$$\|\mathbf{q}_j - \mathbf{p}\|^2 = \|\mathbf{c}_j + d_j\hat{\mathbf{v}}_j - \mathbf{p}\|^2, \tag{11.21}$$

which has a minimum at $d_j = \hat{\mathbf{v}}_j \cdot (\mathbf{p} - \mathbf{c}_j)$. Hence,

$$\mathbf{q}_j = \mathbf{c}_j + (\hat{\mathbf{v}}_j\hat{\mathbf{v}}_j^T)(\mathbf{p} - \mathbf{c}_j) = \mathbf{c}_j + (\mathbf{p} - \mathbf{c}_j)_{\|}, \tag{11.22}$$

in the notation of Equation (2.29), and the squared distance between $\mathbf{p}$ and $\mathbf{q}_j$ is

$$r_j^2 = \|(\mathbf{I} - \hat{\mathbf{v}}_j\hat{\mathbf{v}}_j^T)(\mathbf{p} - \mathbf{c}_j)\|^2 = \|(\mathbf{p} - \mathbf{c}_j)_\perp\|^2. \tag{11.23}$$

The optimal value for $\mathbf{p}$, which lies closest to all of the rays, can be computed as a regular least squares problem by summing over all the r_j^2 and finding the optimal value of $\mathbf{p}$,

$$\mathbf{p} = \left[\sum_j(\mathbf{I} - \hat{\mathbf{v}}_j\hat{\mathbf{v}}_j^T)\right]^{-1}\left[\sum_j(\mathbf{I} - \hat{\mathbf{v}}_j\hat{\mathbf{v}}_j^T)\mathbf{c}_j\right]. \tag{11.24}$$

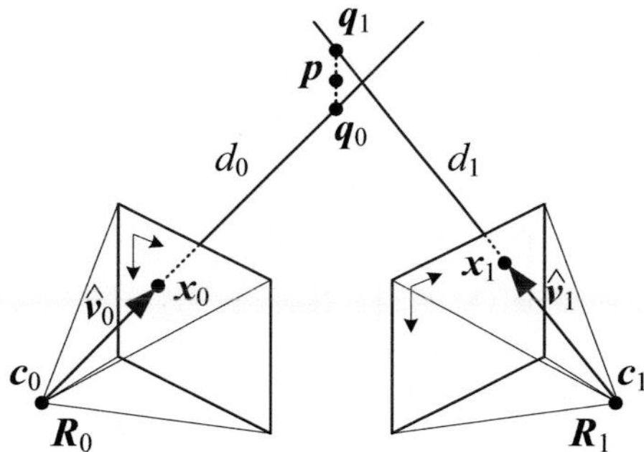

Figure 11.10 3D point triangulation by finding the point $\mathbf{p}$ that lies nearest to all of the optical rays $\mathbf{c}_j + d_j\hat{\mathbf{v}}_j$.

An alternative formulation, which is more statistically optimal and which can produce significantly better estimates if some of the cameras are closer to the 3D point than others, is to minimize the residual in the measurement equations

$$x_j = \frac{p_{00}^{(j)}X + p_{01}^{(j)}Y + p_{02}^{(j)}Z + p_{03}^{(j)}W}{p_{20}^{(j)}X + p_{21}^{(j)}Y + p_{22}^{(j)}Z + p_{23}^{(j)}W} \tag{11.25}$$

$$y_j = \frac{p_{10}^{(j)}X + p_{11}^{(j)}Y + p_{12}^{(j)}Z + p_{13}^{(j)}W}{p_{20}^{(j)}X + p_{21}^{(j)}Y + p_{22}^{(j)}Z + p_{23}^{(j)}W}, \tag{11.26}$$

where (x_j, y_j) are the measured 2D feature locations and $\{p_{00}^{(j)} \ldots p_{23}^{(j)}\}$ are the known entries in camera matrix $\mathbf{P}_j$ (Sutherland 1974).

As with Equations (8.21, 11.11, and 11.12), this set of non-linear equations can be converted into a linear least squares problem by multiplying both sides of the denominator, again resulting in the direct linear transform (DLT) formulation. Note that if we use homogeneous coordinates $\mathbf{p} = (X, Y, Z, W)$, the resulting set of equations is homogeneous and is best solved as a singular value decomposition (SVD) or eigenvalue problem (looking for the smallest singular vector or eigenvector). If we set $W = 1$, we can use regular linear least squares, but the resulting system may be singular or poorly conditioned, i.e., if all of the viewing rays are parallel, as occurs for points far away from the camera.

For this reason, it is generally preferable to parameterize 3D points using homogeneous coordinates, especially if we know that there are likely to be points at greatly varying distances from the cameras. Of course, minimizing the set of observations (11.25–11.26) using non-linear least squares, as described in (8.14 and 8.23), is preferable to using linear least squares, regardless of the representation chosen.

For the case of two observations, it turns out that the location of the point $\mathbf{p}$ that exactly minimizes the true reprojection error (11.25–11.26) can be computed using the solution of degree six polynomial equations (Hartley and Sturm 1997). Another problem to watch out for with triangulation is the issue of *cheirality*, i.e., ensuring that the reconstructed points lie in front of all the cameras (Hartley 1998). While this cannot always be guaranteed, a useful heuristic is to take the points that lie behind the cameras because their rays are diverging (imagine Figure 11.10 where the rays were pointing *away* from each other) and to place them on the plane at infinity by setting their W values to 0.

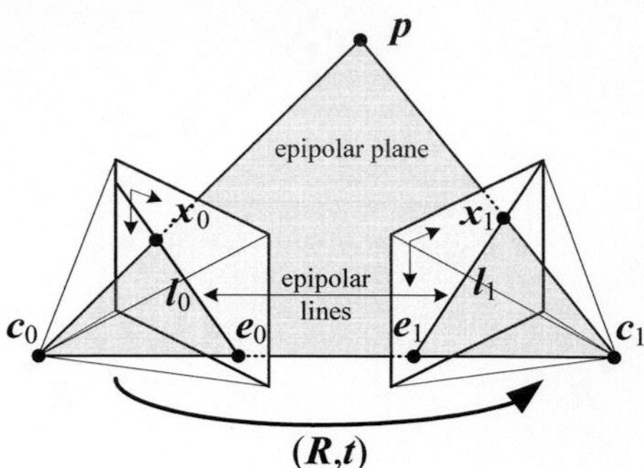

Figure 11.11 Epipolar geometry: The vectors $\mathbf{t} = \mathbf{c}_1 - \mathbf{c}_0$, $\mathbf{p} - \mathbf{c}_0$ and $\mathbf{p} - \mathbf{c}_1$ are co-planar and define the basic epipolar constraint expressed in terms of the pixel measurements $\mathbf{x}_0$ and $\mathbf{x}_1$.

11.3 Two-frame structure from motion

So far in our study of 3D reconstruction, we have always assumed that either the 3D point positions or the 3D camera poses are known in advance. In this section, we take our first look at *structure from motion*, which is the simultaneous recovery of 3D structure and pose from image correspondences. In particular, we examine techniques that operate on just two frames with point correspondences. We divide this section into the study of classic "n-point" algorithms, special (degenerate) cases, projective (uncalibrated) reconstruction, and self-calibration for cameras whose intrinsic calibrations are unknown.

11.3.1 Eight, seven, and five-point algorithms

Consider Figure 11.11, which shows a 3D point $\mathbf{p}$ being viewed from two cameras whose relative position can be encoded by a rotation $\mathbf{R}$ and a translation $\mathbf{t}$. As we do not know anything about the camera positions, without loss of generality, we can set the first camera at the origin $\mathbf{c}_0 = \mathbf{0}$ and at a canonical orientation $\mathbf{R}_0 = \mathbf{I}$.

The 3D point $\mathbf{p}_0 = d_0\hat{\mathbf{x}}_0$ observed in the first image at location $\hat{\mathbf{x}}_0$ and at a z distance of d_0 is mapped into the second image by the transformation

$$d_1\hat{\mathbf{x}}_1 = \mathbf{p}_1 = \mathbf{R}\mathbf{p}_0 + \mathbf{t} = \mathbf{R}(d_0\hat{\mathbf{x}}_0) + \mathbf{t}, \tag{11.27}$$

where $\hat{\mathbf{x}}_j = \mathbf{K}_j^{-1}\mathbf{x}_j$ are the (local) ray direction vectors. Taking the cross product of the two (interchanged) sides with $\mathbf{t}$ in order to annihilate it on the right-hand side yields[8]

$$d_1[\mathbf{t}]_\times\hat{\mathbf{x}}_1 = d_0[\mathbf{t}]_\times\mathbf{R}\hat{\mathbf{x}}_0. \tag{11.28}$$

Taking the dot product of both sides with $\hat{\mathbf{x}}_1$ yields

$$d_0\hat{\mathbf{x}}_1^T([\mathbf{t}]_\times\mathbf{R})\hat{\mathbf{x}}_0 = d_1\hat{\mathbf{x}}_1^T[\mathbf{t}]_\times\hat{\mathbf{x}}_1 = 0, \tag{11.29}$$

because the right-hand side is a triple product with two identical entries. (Another way to say this is that the cross product matrix $[\mathbf{t}]_\times$ is skew symmetric and returns 0 when pre- and post-multiplied by the same vector.)

[8]The cross-product operator $[\,]_\times$ was introduced in (2.32).

We therefore arrive at the basic *epipolar constraint*

$$\hat{\mathbf{x}}_1^T \mathbf{E}\, \hat{\mathbf{x}}_0 = 0, \tag{11.30}$$

where

$$\mathbf{E} = [\mathbf{t}]_\times \mathbf{R} \tag{11.31}$$

is called the *essential matrix* (Longuet-Higgins 1981).

An alternative way to derive the epipolar constraint is to notice that, for the cameras to be oriented so that the rays $\hat{\mathbf{x}}_0$ and $\hat{\mathbf{x}}_1$ intersect in 3D at point $\mathbf{p}$, the vectors connecting the two camera centers $\mathbf{c}_1 - \mathbf{c}_0 = -\mathbf{R}_1^{-1}\mathbf{t}$ and the rays corresponding to pixels $\mathbf{x}_0$ and $\mathbf{x}_1$, namely $\mathbf{R}_j^{-1}\hat{\mathbf{x}}_j$, must be co-planar. This requires that the triple product

$$(\hat{\mathbf{x}}_0, \mathbf{R}^{-1}\hat{\mathbf{x}}_1, -\mathbf{R}^{-1}\mathbf{t}) = (\mathbf{R}\hat{\mathbf{x}}_0, \hat{\mathbf{x}}_1, -\mathbf{t}) = \hat{\mathbf{x}}_1 \cdot (\mathbf{t} \times \mathbf{R}\hat{\mathbf{x}}_0) = \hat{\mathbf{x}}_1^T([\mathbf{t}]_\times \mathbf{R})\hat{\mathbf{x}}_0 = 0. \tag{11.32}$$

Notice that the essential matrix $\mathbf{E}$ maps a point $\hat{\mathbf{x}}_0$ in image 0 into a line $\mathbf{l}_1 = \mathbf{E}\hat{\mathbf{x}}_0 = [\mathbf{t}]_\times \mathbf{R}\hat{\mathbf{x}}_0$ in image 1, because $\hat{\mathbf{x}}_1^T \mathbf{l}_1 = 0$ (Figure 11.11). All such lines must pass through the second *epipole* $\mathbf{e}_1$, which is therefore defined as the left singular vector of $\mathbf{E}$ with a 0 singular value, or, equivalently, the projection of the vector $\mathbf{t}$ into image 1. The dual (transpose) of these relationships gives us the epipolar line in the first image as $\mathbf{l}_0 = \mathbf{E}^T \hat{\mathbf{x}}_1$ and $\mathbf{e}_0$ as the zero-value right singular vector of $\mathbf{E}$.

Eight-point algorithm. Given this fundamental relationship (11.30), how can we use it to recover the camera motion encoded in the essential matrix $\mathbf{E}$? If we have N corresponding measurements $\{(\mathbf{x}_{i0}, \mathbf{x}_{i1})\}$, we can form N homogeneous equations in the nine elements of $\mathbf{E} = \{e_{00} \ldots e_{22}\}$,

$$\begin{array}{cccc} x_{i0}x_{i1}e_{00} & +y_{i0}x_{i1}e_{01} & +x_{i1}e_{02} & + \\ x_{i0}y_{i1}e_{00} & +y_{i0}y_{i1}e_{11} & +y_{i1}e_{12} & + \\ x_{i0}e_{20} & +y_{i0}e_{21} & +e_{22} & =0 \end{array} \tag{11.33}$$

where $\mathbf{x}_{ij} = (x_{ij}, y_{ij}, 1)$. This can be written more compactly as

$$[\mathbf{x}_{i1}\, \mathbf{x}_{i0}^T] \otimes \mathbf{E} = \mathbf{Z}_i \otimes \mathbf{E} = \mathbf{z}_i \cdot \mathbf{f} = 0, \tag{11.34}$$

where $\otimes$ indicates an element-wise multiplication and summation of matrix elements, and $\mathbf{z}_i$ and $\mathbf{f}$ are the vectorized forms of the $\mathbf{Z}_i = \hat{\mathbf{x}}_{i1}\hat{\mathbf{x}}_{i0}^T$ and $\mathbf{E}$ matrices.[9] Given $N \geq 8$ such equations, we can compute an estimate (up to scale) for the entries in $\mathbf{E}$ using an SVD.

In the presence of noisy measurements, how close is this estimate to being statistically optimal? If you look at the entries in (11.33), you can see that some entries are the products of image measurements such as $x_{i0}y_{i1}$ and others are direct image measurements (or even the identity). If the measurements have comparable noise, the terms that are products of measurements have their noise amplified by the other element in the product, which can lead to very poor scaling, e.g., an inordinately large influence of points with large coordinates (far away from the image center).

To counteract this trend, Hartley (1997a) suggests that the point coordinates should be translated and scaled so that their centroid lies at the origin and their variance is unity, i.e.,

$$\tilde{x}_i = s(x_i - \mu_x) \tag{11.35}$$

$$\tilde{y}_i = s(y_i - \mu_y) \tag{11.36}$$

[9] We use $\mathbf{f}$ instead of $\mathbf{e}$ to denote the vectorized form of $\mathbf{E}$ to avoid confusion with the epipoles $\mathbf{e}_j$.

such that $\sum_i \tilde{x}_i = \sum_i \tilde{y}_i = 0$ and $\sum_i \tilde{x}_i^2 + \sum_i \tilde{y}_i^2 = 2n$, where n is the number of points.[10]

Once the essential matrix $\tilde{\mathbf{E}}$ has been computed from the transformed coordinates $\{(\tilde{\mathbf{x}}_{i0}, \tilde{\mathbf{x}}_{i1})\}$, where $\tilde{\mathbf{x}}_{ij} = \mathbf{T}_j \hat{\mathbf{x}}_{ij}$ and $\mathbf{T}_j$ is the 3×3 matrix that implements the shift and scale operations in (11.35–11.36), the original essential matrix $\mathbf{E}$ can be recovered as

$$\mathbf{E} = \mathbf{T}_1^T \tilde{\mathbf{E}} \mathbf{T}_0. \tag{11.37}$$

In his paper, Hartley (1997a) compares the improvement due to his re-normalization strategy to alternative distance measures proposed by others such as Zhang (1998a,b) and concludes that his simple re-normalization in most cases is as effective as (or better than) alternative techniques. Torr and Fitzgibbon (2004) recommend a variant on this algorithm where the norm of the upper 2×2 sub-matrix of $\mathbf{E}$ is set to 1 and show that it has even better stability with respect to 2D coordinate transformations.

7-point algorithm. Because $\mathbf{E}$ is rank-deficient, it turns out that we actually only need seven correspondences of the form of Equation (11.34) instead of eight to estimate this matrix (Hartley 1994a; Torr and Murray 1997; Hartley and Zisserman 2004). The advantage of using fewer correspondences inside a RANSAC robust fitting stage is that fewer random samples need to be generated. From this set of seven homogeneous equations (which we can stack into a 7×9 matrix for SVD analysis), we can find two independent vectors, say $\mathbf{f}_0$ and $\mathbf{f}_1$ such that $\mathbf{z}_i \cdot \mathbf{f}_j = 0$. These two vectors can be converted back into 3×3 matrices $\mathbf{E}_0$ and $\mathbf{E}_1$, which span the solution space for

$$\mathbf{E} = \alpha \mathbf{E}_0 + (1 - \alpha)\mathbf{E}_1. \tag{11.38}$$

To find the correct value of α, we observe that $\mathbf{E}$ has a zero determinant, as it is rank deficient, and hence

$$|\alpha \mathbf{E}_0 + (1 - \alpha)\mathbf{E}_1| = 0. \tag{11.39}$$

This gives us a cubic equation in α, which has either one or three solutions (roots). Substituting these values into (11.38) to obtain $\mathbf{E}$, we can test this essential matrix against other unused feature correspondences to select the correct one.

The normalized "eight-point algorithm" (Hartley 1997a) and seven-point algorithm described above are not the only way to estimate the camera motion from correspondences. Additional variants include a five-point algorithm that requires finding the roots of a 10th degree polynomial (Nistér 2004) as well as variants that handle special (restricted) motions or scene structures, as discussed later on in this section. Because such algorithms use fewer points to compute their estimates, they are less sensitive to outliers when used as part of a random sampling (RANSAC) strategy.[11]

Recovering $\mathbf{t}$ and $\mathbf{R}$. Once an estimate for the essential matrix $\mathbf{E}$ has been recovered, the direction of the translation vector $\mathbf{t}$ can be estimated. Note that the absolute distance between the two cameras can never be recovered from pure image measurements alone, regardless of how many cameras or points are used. Knowledge about absolute camera and point positions or distances, often called *ground control points* in photogrammetry, is always required to establish the final scale, position, and orientation.

[10] More precisely, Hartley (1997a) suggests scaling the points "so that the average distance from the origin is equal to $\sqrt{2}$" but the heuristic of unit variance is faster to compute (does not require per-point square roots) and should yield comparable improvements.

[11] You can find an experimental comparison of a number of RANSAC variants at https://opencv.org/evaluating-opencvs-new-ransacs/.

To estimate this direction $\hat{\mathbf{t}}$, observe that under ideal noise-free conditions, the essential matrix $\mathbf{E}$ is singular, i.e., $\hat{\mathbf{t}}^T\mathbf{E} = 0$. This singularity shows up as a singular value of 0 when an SVD of $\mathbf{E}$ is performed,

$$\mathbf{E} = [\hat{\mathbf{t}}]_\times \mathbf{R} = \mathbf{U}\boldsymbol{\Sigma}\mathbf{V}^T = \begin{bmatrix} \mathbf{u}_0 & \mathbf{u}_1 & \hat{\mathbf{t}} \end{bmatrix} \begin{bmatrix} 1 & & \\ & 1 & \\ & & 0 \end{bmatrix} \begin{bmatrix} \mathbf{v}_0^T \\ \mathbf{v}_1^T \\ \mathbf{v}_2^T \end{bmatrix}. \tag{11.40}$$

When $\mathbf{E}$ is computed from noisy measurements, the singular vector associated with the smallest singular value gives us $\hat{\mathbf{t}}$. (The other two singular values should be similar but are not, in general, equal to 1 because $\mathbf{E}$ is only computed up to an unknown scale.)

Once $\hat{\mathbf{t}}$ has been recovered, how can we estimate the corresponding rotation matrix $\mathbf{R}$? Recall that the cross-product operator $[\hat{\mathbf{t}}]_\times$ (2.32) projects a vector onto a set of orthogonal basis vectors that include $\hat{\mathbf{t}}$, zeros out the $\hat{\mathbf{t}}$ component, and rotates the other two by 90°,

$$[\hat{\mathbf{t}}]_\times = \mathbf{S}\mathbf{Z}\mathbf{R}_{90^\circ}\mathbf{S}^T = \begin{bmatrix} \mathbf{s}_0 & \mathbf{s}_1 & \hat{\mathbf{t}} \end{bmatrix} \begin{bmatrix} 1 & & \\ & 1 & \\ & & 0 \end{bmatrix} \begin{bmatrix} 0 & -1 & \\ 1 & 0 & \\ & & 1 \end{bmatrix} \begin{bmatrix} \mathbf{s}_0^T \\ \mathbf{s}_1^T \\ \hat{\mathbf{t}}^T \end{bmatrix}, \tag{11.41}$$

where $\hat{\mathbf{t}} = \mathbf{s}_0 \times \mathbf{s}_1$. From Equations (11.40 and 11.41), we get

$$\mathbf{E} = [\hat{\mathbf{t}}]_\times \mathbf{R} = \mathbf{S}\mathbf{Z}\mathbf{R}_{90^\circ}\mathbf{S}^T\mathbf{R} = \mathbf{U}\boldsymbol{\Sigma}\mathbf{V}^T, \tag{11.42}$$

from which we can conclude that $\mathbf{S} = \mathbf{U}$. Recall that for a noise-free essential matrix, $(\boldsymbol{\Sigma} = \mathbf{Z})$, and hence

$$\mathbf{R}_{90^\circ}\mathbf{U}^T\mathbf{R} = \mathbf{V}^T \tag{11.43}$$

and

$$\mathbf{R} = \mathbf{U}\mathbf{R}_{90^\circ}^T\mathbf{V}^T. \tag{11.44}$$

Unfortunately, we only know both $\mathbf{E}$ and $\hat{\mathbf{t}}$ up to a sign. Furthermore, the matrices $\mathbf{U}$ and $\mathbf{V}$ are not guaranteed to be rotations (you can flip both their signs and still get a valid SVD). For this reason, we have to generate all four possible rotation matrices

$$\mathbf{R} = \pm\mathbf{U}\mathbf{R}_{\pm 90^\circ}^T\mathbf{V}^T \tag{11.45}$$

and keep the two whose determinant $|\mathbf{R}| = 1$. To disambiguate between the remaining pair of potential rotations, which form a *twisted pair* (Hartley and Zisserman 2004, p. 259), we need to pair them with both possible signs of the translation direction $\pm\hat{\mathbf{t}}$ and select the combination for which the largest number of points is seen in front of both cameras.[12]

The property that points must lie in front of the camera, i.e., at a positive distance along the viewing rays emanating from the camera, is known as *cheirality* (Hartley 1998). In addition to determining the signs of the rotation and translation, as described above, the cheirality (sign of the distances) of the points in a reconstruction can be used inside a RANSAC procedure (along with the reprojection errors) to distinguish between likely and unlikely configurations.[13] cheirality can also be used to transform projective reconstructions (Sections 11.3.3 and 11.3.4) into *quasi-affine* reconstructions (Hartley 1998).

[12] In the noise-free case, a single point suffices. It is safer, however, to test all or a sufficient subset of points, downweighting the ones that lie close to the plane at infinity, for which it is easy to get depth reversals.

[13] Note that as points get further away from a camera, i.e., closer toward the plane at infinity, errors in cheirality become more likely.

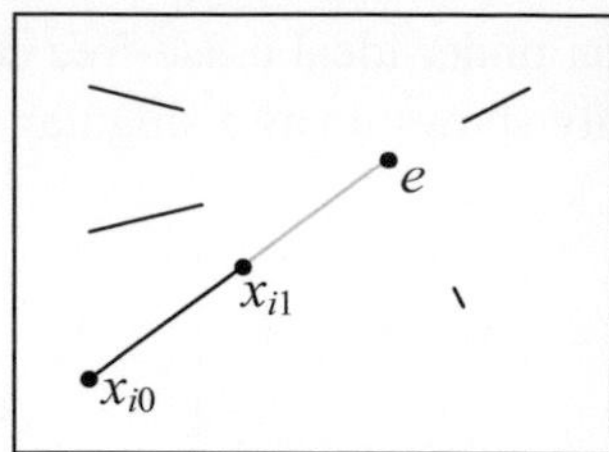

Figure 11.12 Pure translational camera motion results in visual motion where all the points move towards (or away from) a common *focus of expansion* (FOE) $\mathbf{e}$. They therefore satisfy the triple product condition $(\mathbf{x}_0, \mathbf{x}_1, \mathbf{e}) = \mathbf{e} \cdot (\mathbf{x}_0 \times \mathbf{x}_1) = 0$.

11.3.2 Special motions and structures

In certain situations, specially tailored algorithms can take advantage of known (or guessed) camera arrangements or 3D structures.

Pure translation (known rotation). In the case where we know the rotation, we can pre-rotate the points in the second image to match the viewing direction of the first. The resulting set of 3D points all move towards (or away from) the *focus of expansion* (FOE), as shown in Figure 11.12.[14] The resulting essential matrix $\mathbf{E}$ is (in the noise-free case) skew symmetric and so can be estimated more directly by setting $e_{ij} = -e_{ji}$ and $e_{ii} = 0$ in (11.33). Two points with non-zero parallax now suffice to estimate the FOE.

A more direct derivation of the FOE estimate can be obtained by minimizing the triple product

$$\sum_i (\mathbf{x}_{i0}, \mathbf{x}_{i1}, \mathbf{e})^2 = \sum_i ((\mathbf{x}_{i0} \times \mathbf{x}_{i1}) \cdot \mathbf{e})^2, \tag{11.46}$$

which is equivalent to finding the null space for the set of equations

$$(y_{i0} - y_{i1})e_0 + (x_{i1} - x_{i0})e_1 + (x_{i0}y_{i1} - y_{i0}x_{i1})e_2 = 0. \tag{11.47}$$

Note that, as in the eight-point algorithm, it is advisable to normalize the 2D points to have unit variance before computing this estimate.

In situations where a large number of points at infinity are available, e.g., when shooting outdoor scenes or when the camera motion is small compared to distant objects, this suggests an alternative RANSAC strategy for estimating the camera motion. First, pick a pair of points to estimate a rotation, hoping that both of the points lie at infinity (very far from the camera). Then, compute the FOE and check whether the residual error is small (indicating agreement with this rotation hypothesis) and whether the motions towards or away from the epipole (FOE) are all in the same direction (ignoring very small motions, which may be noise-contaminated).

Pure rotation. The case of pure rotation results in a degenerate estimate of the essential matrix $\mathbf{E}$ and of the translation direction $\hat{\mathbf{t}}$. Consider first the case of the rotation matrix being known. The estimates for the FOE will be degenerate, because $\mathbf{x}_{i0} \approx \mathbf{x}_{i1}$, and hence (11.47), is degenerate. A similar argument shows that the equations for the essential matrix (11.33) are also rank-deficient.

This suggests that it might be prudent before computing a full essential matrix to first compute a rotation estimate $\mathbf{R}$ using (8.32), potentially with just a small number of points, and then compute the residuals after rotating the points before proceeding with a full $\mathbf{E}$ computation.

[14] Fans of *Star Trek* and *Star Wars* will recognize this as the "jump to hyperdrive" visual effect.

Dominant planar structure. When a dominant plane is present in the scene, DEGENSAC, which tests whether too many correspondences are co-planar, can be used to recover the fundamental matrix more reliably than the seven-point algorithm (Chum, Werner, and Matas 2005).

As you can tell from the previous special cases, there exist many different specialized cases of two-frame structure-from-motion as well as many alternative appropriate techniques. The OpenGV library developed by Kneip and Furgale (2014) contains open-source implementations of many of these algorithms.[15]

11.3.3 Projective (uncalibrated) reconstruction

In many cases, such as when trying to build a 3D model from internet or legacy photos taken by unknown cameras without any EXIF tags, we do not know ahead of time the intrinsic calibration parameters associated with the input images. In such situations, we can still estimate a two-frame reconstruction, although the true metric structure may not be available, e.g., orthogonal lines or planes in the world may not end up being reconstructed as orthogonal.

Consider the derivations we used to estimate the essential matrix $\mathbf{E}$ (11.30–11.32). In the uncalibrated case, we do not know the calibration matrices $\mathbf{K}_j$, so we cannot use the normalized ray directions $\hat{\mathbf{x}}_j = \mathbf{K}_j^{-1}\mathbf{x}_j$. Instead, we have access only to the image coordinates $\mathbf{x}_j$, and so the essential matrix equation (11.30) becomes

$$\hat{\mathbf{x}}_1^T\mathbf{E}\hat{\mathbf{x}}_1 = \mathbf{x}_1^T\mathbf{K}_1^{-T}\mathbf{E}\mathbf{K}_0^{-1}\mathbf{x}_0 = \mathbf{x}_1^T\mathbf{F}\mathbf{x}_0 = 0, \tag{11.48}$$

where

$$\mathbf{F} = \mathbf{K}_1^{-T}\mathbf{E}\mathbf{K}_0^{-1} \tag{11.49}$$

is called the *fundamental matrix* (Faugeras 1992; Hartley, Gupta, and Chang 1992; Hartley and Zisserman 2004).

Like the essential matrix, the fundamental matrix is (in principle) rank two,

$$\mathbf{F} = \mathbf{U}\boldsymbol{\Sigma}\mathbf{V}^T = \begin{bmatrix}\mathbf{u}_0 & \mathbf{u}_1 & \mathbf{e}_1\end{bmatrix}\begin{bmatrix}\sigma_0 & & \\ & \sigma_1 & \\ & & 0\end{bmatrix}\begin{bmatrix}\mathbf{v}_0^T \\ \mathbf{v}_1^T \\ \mathbf{e}_0^T\end{bmatrix}. \tag{11.50}$$

Its smallest left singular vector indicates the epipole $\mathbf{e}_1$ in the image 1 and its smallest right singular vector is $\mathbf{e}_0$ (Figure 11.11). The fundamental matrix can be factored into a skew-symmetric cross product matrix $[\mathbf{e}]_\times$ and a homography $\tilde{\mathbf{H}}$,

$$\mathbf{F} = [\mathbf{e}]_\times\tilde{\mathbf{H}}. \tag{11.51}$$

The homography $\tilde{\mathbf{H}}$, which in principle from (11.49) should equal

$$\tilde{\mathbf{H}} = \mathbf{K}_1^{-T}\mathbf{R}\mathbf{K}_0^{-1}, \tag{11.52}$$

cannot be uniquely recovered from $\mathbf{F}$, as any homography of the form $\tilde{\mathbf{H}}' = \tilde{\mathbf{H}} + \mathbf{e}\mathbf{v}^T$ results in the same $\mathbf{F}$ matrix. (Note that $[\mathbf{e}]_\times$ annihilates any multiple of $\mathbf{e}$.)

Any one of these valid homographies $\tilde{\mathbf{H}}$ maps some plane in the scene from one image to the other. It is not possible to tell in advance which one it is without either selecting four or more co-planar correspondences to compute $\tilde{\mathbf{H}}$ as part of the $\mathbf{F}$ estimation process (in a manner analogous to guessing a rotation for $\mathbf{E}$) or mapping all points in one image through $\tilde{\mathbf{H}}$ and seeing which ones

[15] https://laurentkneip.github.io/opengv

line up with their corresponding locations in the other. The resulting representation is often referred to as *plane plus parallax* (Kumar, Anandan, and Hanna 1994; Sawhney 1994) and is described in more detail in Section 2.1.4.

To create a *projective* reconstruction of the scene, we can pick any valid homography $\tilde{\mathbf{H}}$ that satisfies Equation (11.49). For example, following a technique analogous to Equations (11.40–11.44), we get

$$\mathbf{F} = [\mathbf{e}]_\times \tilde{\mathbf{H}} = \mathbf{S}\mathbf{Z}\mathbf{R}_{90^\circ}\mathbf{S}^T\tilde{\mathbf{H}} = \mathbf{U}\boldsymbol{\Sigma}\mathbf{V}^T \tag{11.53}$$

and hence

$$\tilde{\mathbf{H}} = \mathbf{U}\mathbf{R}_{90^\circ}^T\hat{\boldsymbol{\Sigma}}\mathbf{V}^T, \tag{11.54}$$

where $\hat{\boldsymbol{\Sigma}}$ is the singular value matrix with the smallest value replaced by a reasonable alternative (say, the middle value).[16] We can then form a pair of camera matrices

$$\mathbf{P}_0 = [\mathbf{I}|\mathbf{0}] \qquad \text{and} \qquad \mathbf{P}_0 = [\tilde{\mathbf{H}}|\mathbf{e}], \tag{11.55}$$

from which a projective reconstruction of the scene can be computed using triangulation (Section 11.2.4).

While the projective reconstruction may not be useful on its own, it can often be *upgraded* to an affine or metric reconstruction, as described below. Even without this step, however, the fundamental matrix $\mathbf{F}$ can be very useful in finding additional correspondences, as they must all lie on corresponding epipolar lines, i.e., any feature $\mathbf{x}_0$ in image 0 must have its correspondence lying on the associated epipolar line $\mathbf{l}_1 = \mathbf{F}\mathbf{x}_0$ in image 1, assuming that the point motions are due to a rigid transformation.

11.3.4 Self-calibration

The results of structure from motion computation are much more useful if a *metric* reconstruction is obtained, i.e., one in which parallel lines are parallel, orthogonal walls are at right angles, and the reconstructed model is a scaled version of reality. Over the years, a large number of *self-calibration* (or *auto-calibration*) techniques have been developed for converting a projective reconstruction into a metric one, which is equivalent to recovering the unknown calibration matrices $\mathbf{K}_j$ associated with each image (Hartley and Zisserman 2004; Moons, Van Gool, and Vergauwen 2010).

In situations where additional information is known about the scene, different methods may be employed. For example, if there are parallel lines in the scene, three or more vanishing points, which are the images of points at infinity, can be used to establish the homography for the plane at infinity, from which focal lengths and rotations can be recovered. If two or more finite *orthogonal* vanishing points have been observed, the single-image calibration method based on vanishing points (Section 11.1.1) can be used instead.

In the absence of such external information, it is not possible to recover a fully parameterized independent calibration matrix $\mathbf{K}_j$ for each image from correspondences alone. To see this, consider the set of all camera matrices $\mathbf{P}_j = \mathbf{K}_j[\mathbf{R}_j|\mathbf{t}_j]$ projecting world coordinates $\mathbf{p}_i = (X_i, Y_i, Z_i, W_i)$ into screen coordinates $\mathbf{x}_{ij} \sim \mathbf{P}_j\mathbf{p}_i$. Now consider transforming the 3D scene $\{\mathbf{p}_i\}$ through an arbitrary 4×4 projective transformation $\tilde{\mathbf{H}}$, yielding a new model consisting of points $\mathbf{p}'_i = \tilde{\mathbf{H}}\mathbf{p}_i$. Post-multiplying each $\mathbf{P}_j$ matrix by $\tilde{\mathbf{H}}^{-1}$ still produces the same screen coordinates and a new set calibration matrices can be computed by applying RQ decomposition to the new camera matrix $\mathbf{P}'_j = \mathbf{P}_j\tilde{\mathbf{H}}^{-1}$.

[16] Hartley and Zisserman (2004, p. 256) recommend using $\tilde{\mathbf{H}} = [\mathbf{e}]_\times\mathbf{F}$ (Luong and Viéville 1996), which places the camera on the plane at infinity.

For this reason, all self-calibration methods assume some restricted form of the calibration matrix, either by setting or equating some of their elements or by assuming that they do not vary over time. While most of the techniques discussed by Hartley and Zisserman (2004); Moons, Van Gool, and Vergauwen (2010) require three or more frames, in this section we present a simple technique that can recover the focal lengths (f_0, f_1) of both images from the fundamental matrix $\mathbf{F}$ in a two-frame reconstruction (Hartley and Zisserman 2004, p. 472).

To accomplish this, we assume that the camera has zero skew, a known aspect ratio (usually set to 1), and a known image center, as in Equation (2.59). How reasonable is this assumption in practice? The answer, as with many questions, is "it depends".

If absolute metric accuracy is required, as in photogrammetry applications, it is imperative to pre-calibrate the cameras using one of the techniques from Section 11.1 and to use ground control points to pin down the reconstruction. If instead, we simply wish to reconstruct the world for visualization or image based rendering applications, as in the Photo Tourism system of Snavely, Seitz, and Szeliski (2006), this assumption is quite reasonable in practice.

Most cameras today have square pixels and an image center near the middle of the image, and are much more likely to deviate from a simple camera model due to radial distortion (Section 11.1.4), which should be compensated for whenever possible. The biggest problems occur when images have been cropped off-center, in which case the image center will no longer be in the middle, or when perspective pictures have been taken of a different picture, in which case a general camera matrix becomes necessary.[17]

Given these caveats, the two-frame focal length estimation algorithm based on the Kruppa equations developed by Hartley and Zisserman (2004, p. 456) proceeds as follows. Take the left and right singular vectors $\{\mathbf{u}_0, \mathbf{u}_1, \mathbf{v}_0, \mathbf{v}_1\}$ of the fundamental matrix $\mathbf{F}$ (11.50) and their associated singular values $\{\sigma_0, \sigma_1\}$ and form the following set of equations:

$$\frac{\mathbf{u}_1^T \mathbf{D}_0 \mathbf{u}_1}{\sigma_0^2 \mathbf{v}_0^T \mathbf{D}_1 \mathbf{v}_0} = -\frac{\mathbf{u}_0^T \mathbf{D}_0 \mathbf{u}_1}{\sigma_0 \sigma_1 \mathbf{v}_0^T \mathbf{D}_1 \mathbf{v}_1} = \frac{\mathbf{u}_0^T \mathbf{D}_0 \mathbf{u}_0}{\sigma_1^2 \mathbf{v}_1^T \mathbf{D}_1 \mathbf{v}_1}, \tag{11.56}$$

where the two matrices

$$\mathbf{D}_j = \mathbf{K}_j \mathbf{K}_j^T = \text{diag}(f_j^2, f_j^2, 1) = \begin{bmatrix} f_j^2 & & \\ & f_j^2 & \\ & & 1 \end{bmatrix} \tag{11.57}$$

encode the unknown focal lengths. For simplicity, let us rewrite each of the numerators and denominators in (11.56) as

$$e_{ij0}(f_0^2) = \mathbf{u}_i^T \mathbf{D}_0 \mathbf{u}_j = a_{ij} + b_{ij} f_0^2, \tag{11.58}$$

$$e_{ij1}(f_1^2) = \sigma_i \sigma_j \mathbf{v}_i^T \mathbf{D}_1 \mathbf{v}_j = c_{ij} + d_{ij} f_1^2. \tag{11.59}$$

Notice that each of these is affine (linear plus constant) in either f_0^2 or f_1^2. Hence, we can cross-multiply these equations to obtain quadratic equations in f_j^2, which can readily be solved. (See also the work by Bougnoux (1998) and Kanatani and Matsunaga (2000) for some alternative formulations.)

An alternative solution technique is to observe that we have a set of three equations related by an unknown scalar λ, i.e.,

$$e_{ij0}(f_0^2) = \lambda e_{ij1}(f_1^2) \tag{11.60}$$

[17] In Photo Tourism, our system registered photographs of an information sign outside Notre Dame with real pictures of the cathedral.

(Richard Hartley, personal communication, July 2009). These can readily be solved to yield $(f_0^2, \lambda f_1^2, \lambda)$ and hence (f_0, f_1).

How well does this approach work in practice? There are certain degenerate configurations, such as when there is no rotation or when the optical axes intersect, when it does not work at all. (In such a situation, you can vary the focal lengths of the cameras and obtain a deeper or shallower reconstruction, which is an example of a *bas-relief ambiguity* (Section 11.4.5).) Hartley and Zisserman (2004) recommend using techniques based on three or more frames. However, if you find two images for which the estimates of $(f_0^2, \lambda f_1^2, \lambda)$ are well conditioned, they can be used to initialize a more complete bundle adjustment of all the parameters (Section 11.4.2). An alternative, which is often used in systems such as Photo Tourism, is to use camera EXIF tags or generic default values to initialize focal length estimates and refine them as part of bundle adjustment.

11.3.5 *Application*: View morphing

An interesting application of basic two-frame structure from motion is *view morphing* (also known as *view interpolation*, see Section 14.1), which can be used to generate a smooth 3D animation from one view of a 3D scene to another (Chen and Williams 1993; Seitz and Dyer 1996).

To create such a transition, you must first smoothly interpolate the camera matrices, i.e., the camera positions, orientations, and focal lengths. While simple linear interpolation can be used (representing rotations as quaternions (Section 2.1.3)), a more pleasing effect is obtained by *easing in* and *easing out* the camera parameters, e.g., using a raised cosine, as well as moving the camera along a more circular trajectory (Snavely, Seitz, and Szeliski 2006).

To generate in-between frames, either a full set of 3D correspondences needs to be established (Section 12.3) or 3D models (proxies) must be created for each reference view. Section 14.1 describes several widely used approaches to this problem. One of the simplest is to just triangulate the set of matched feature points in each image, e.g., using Delaunay triangulation. As the 3D points are re-projected into their intermediate views, pixels can be mapped from their original source images to their new views using affine or projective mapping (Szeliski and Shum 1997). The final image is then composited using a linear blend of the two reference images, as with usual morphing (Section 3.6.3).

11.4 Multi-frame structure from motion

While two-frame techniques are useful for reconstructing sparse geometry from stereo image pairs and for initializing larger-scale 3D reconstructions, most applications can benefit from the much larger number of images that are usually available in photo collections and videos of scenes.

In this section, we briefly review an older technique called *factorization*, which can provide useful solutions for short video sequences, and then turn to the more commonly used *bundle adjustment* approach, which uses non-linear least squares to obtain optimal solutions under general camera configurations.

11.4.1 Factorization

When processing video sequences, we often get extended *feature tracks* (Section 7.1.5) from which it is possible to recover the structure and motion using a process called *factorization*. Consider the tracks generated by a rotating ping pong ball, which has been marked with dots to make its shape

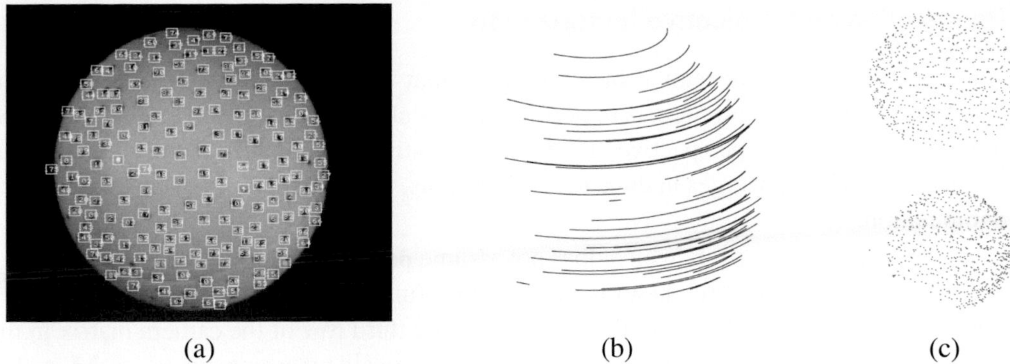

(a) (b) (c)

Figure 11.13 3D reconstruction of a rotating ping pong ball using factorization (Tomasi and Kanade 1992) © 1992 Springer: (a) sample image with tracked features overlaid; (b) subsampled feature motion stream; (c) two views of the reconstructed 3D model.

and motion more discernable (Figure 11.13). We can readily see from the shape of the tracks that the moving object must be a sphere, but how can we infer this mathematically?

It turns out that, under orthography or related models we discuss below, the shape and motion can be recovered simultaneously using a singular value decomposition (Tomasi and Kanade 1992). The details of how to do this are presented in the paper by Tomasi and Kanade (1992) and also in the first edition of this book (Szeliski 2010, Section 7.3).

Once the rotation matrices and 3D point locations have been recovered, there still exists a bas-relief ambiguity, i.e., we can never be sure if the object is rotating left to right or if its depth reversed version is moving the other way. (This can be seen in the classic rotating Necker Cube visual illusion.) Additional cues, such as the appearance and disappearance of points, or perspective effects, both of which are discussed below, can be used to remove this ambiguity.

For motion models other than pure orthography, e.g., for scaled orthography or para-perspective, the approach above must be extended in the appropriate manner. Such techniques are relatively straightforward to derive from first principles; more details can be found in papers that extend the basic factorization approach to these more flexible models (Poelman and Kanade 1997). Additional extensions of the original factorization algorithm include multi-body rigid motion (Costeira and Kanade 1995), sequential updates to the factorization (Morita and Kanade 1997), the addition of lines and planes (Morris and Kanade 1998), and re-scaling the measurements to incorporate individual location uncertainties (Anandan and Irani 2002).

A disadvantage of factorization approaches is that they require a complete set of tracks, i.e., each point must be visible in each frame, for the factorization approach to work. Tomasi and Kanade (1992) deal with this problem by first applying factorization to smaller denser subsets and then using known camera (motion) or point (structure) estimates to *hallucinate* additional missing values, which allows them to incrementally incorporate more features and cameras. Huynh, Hartley, and Heyden (2003) extend this approach to view missing data as special cases of outliers. Buchanan and Fitzgibbon (2005) develop fast iterative algorithms for performing large matrix factorizations with missing data. The general topic of principal component analysis (PCA) with missing data also appears in other computer vision problems (Shum, Ikeuchi, and Reddy 1995; De la Torre and Black 2003; Gross, Matthews, and Baker 2006; Torresani, Hertzmann, and Bregler 2008; Vidal, Ma, and Sastry 2016).

Perspective and projective factorization

Another disadvantage of regular factorization is that it cannot deal with perspective cameras. One way to get around this problem is to perform an initial affine (e.g., orthographic) reconstruction and to then correct for the perspective effects in an iterative manner (Christy and Horaud 1996). This algorithm usually converges in three to five iterations, with the majority of the time spent in the SVD computation.

An alternative approach, which does not assume partially calibrated cameras (known image center, square pixels, and zero skew) is to perform a fully *projective* factorization (Sturm and Triggs 1996; Triggs 1996). In this case, the inclusion of the third row of the camera matrix in the measurement matrix is equivalent to multiplying each reconstructed measurement $\mathbf{x}_{ji} = \mathbf{M}_j \mathbf{p}_i$ by its inverse (projective) depth $\eta_{ji} = d_{ji}^{-1} = 1/(\mathbf{P}_{j2}\mathbf{p}_i)$ or, equivalently, multiplying each measured position by its projective depth d_{ji}. In the original paper by Sturm and Triggs (1996), the projective depths d_{ji} are obtained from two-frame reconstructions, while in later work (Triggs 1996; Oliensis and Hartley 2007), they are initialized to $d_{ji} = 1$ and updated after each iteration. Oliensis and Hartley (2007) present an update formula that is guaranteed to converge to a fixed point. None of these authors suggest actually estimating the third row of $\mathbf{P}_j$ as part of the projective depth computations. In any case, it is unclear when a fully projective reconstruction would be preferable to a partially calibrated one, especially if they are being used to initialize a full bundle adjustment of all the parameters.

One of the attractions of factorization methods is that they provide a "closed form" (sometimes called a "linear") method to initialize iterative techniques such as bundle adjustment. An alternative initialization technique is to estimate the homographies corresponding to some common plane seen by all the cameras (Rother and Carlsson 2002). In a calibrated camera setting, this can correspond to estimating consistent rotations for all of the cameras, for example, using matched vanishing points (Antone and Teller 2002). Once these have been recovered, the camera positions can then be obtained by solving a linear system (Antone and Teller 2002; Rother and Carlsson 2002; Rother 2003).

11.4.2 Bundle adjustment

As we have mentioned several times before, the most accurate way to recover structure and motion is to perform robust non-linear minimization of the measurement (re-projection) errors, which is commonly known in the photogrammetry (and now computer vision) communities as *bundle adjustment*.[18] Triggs, McLauchlan *et al.* (1999) provide an excellent overview of this topic, including its historical development, pointers to the photogrammetry literature (Slama 1980; Atkinson 1996; Kraus 1997), and subtle issues with gauge ambiguities. The topic is also treated in depth in textbooks and surveys on multi-view geometry (Faugeras and Luong 2001; Hartley and Zisserman 2004; Moons, Van Gool, and Vergauwen 2010).

We have already introduced the elements of bundle adjustment in our discussion on iterative pose estimation (Section 11.2.2), i.e., Equations (11.14–11.20) and Figure 11.7. The biggest difference between these formulas and full bundle adjustment is that our feature location measurements $\mathbf{x}_{ij}$ now depend not only on the point (track) index i but also on the camera pose index j,

$$\mathbf{x}_{ij} = \mathbf{f}(\mathbf{p}_i, \mathbf{R}_j, \mathbf{c}_j, \mathbf{K}_j), \tag{11.61}$$

[18] The term "bundle" refers to the bundles of rays connecting camera centers to 3D points and the term "adjustment" refers to the iterative minimization of re-projection error. Alternative terms for this in the vision community include *optimal motion estimation* (Weng, Ahuja, and Huang 1993) and *non-linear least squares* (Appendix A.3) (Taylor, Kriegman, and Anandan 1991; Szeliski and Kang 1994).

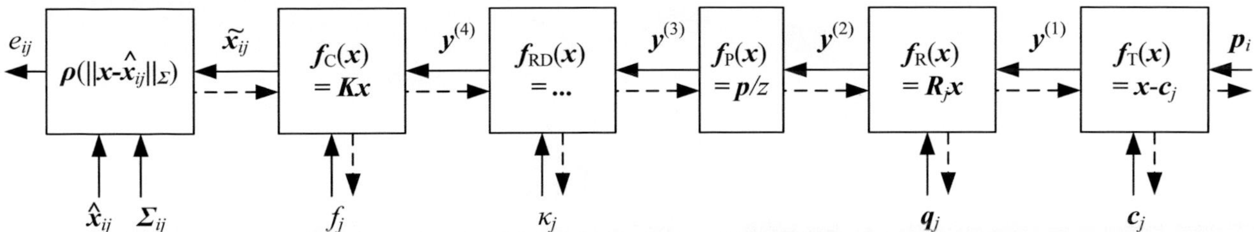

Figure 11.14 A set of chained transforms for projecting a 3D point $\mathbf{p}_i$ into a 2D measurement $\mathbf{x}_{ij}$ through a series of transformations $\mathbf{f}^{(k)}$, each of which is controlled by its own set of parameters. The dashed lines indicate the flow of information as partial derivatives are computed during a backward pass. The formula for the radial distortion function is $\mathbf{f}_{\text{RD}}(\mathbf{x}) = (1 + \kappa_1 r^2 + \kappa_2 r^4)\mathbf{x}$.

and that the 3D point positions $\mathbf{p}_i$ are also being simultaneously updated. In addition, it is common to add a stage for radial distortion parameter estimation (2.78),

$$\mathbf{f}_{\text{RD}}(\mathbf{x}) = (1 + \kappa_1 r^2 + \kappa_2 r^4)\mathbf{x}, \tag{11.62}$$

if the cameras being used have not been pre-calibrated, as shown in Figure 11.14.

While most of the boxes (transforms) in Figure 11.14 have previously been explained (11.19), the leftmost box has not. This box performs a robust comparison of the predicted and measured 2D locations $\hat{\mathbf{x}}_{ij}$ and $\tilde{\mathbf{x}}_{ij}$ after re-scaling by the measurement noise covariance $\boldsymbol{\Sigma}_{ij}$. In more detail, this operation can be written as

$$\mathbf{r}_{ij} = \tilde{\mathbf{x}}_{ij} - \hat{\mathbf{x}}_{ij}, \tag{11.63}$$

$$s_{ij}^2 = \mathbf{r}_{ij}^T \boldsymbol{\Sigma}_{ij}^{-1} \mathbf{r}_{ij}, \tag{11.64}$$

$$e_{ij} = \hat{\rho}(s_{ij}^2), \tag{11.65}$$

where $\hat{\rho}(r^2) = \rho(r)$. The corresponding Jacobians (partial derivatives) can be written as

$$\frac{\partial e_{ij}}{\partial s_{ij}^2} = \hat{\rho}'(s_{ij}^2), \tag{11.66}$$

$$\frac{\partial s_{ij}^2}{\partial \tilde{\mathbf{x}}_{ij}} = \boldsymbol{\Sigma}_{ij}^{-1} \mathbf{r}_{ij}. \tag{11.67}$$

The advantage of the chained representation introduced above is that it not only makes the computations of the partial derivatives and Jacobians simpler but it can also be adapted to any camera configuration. Consider for example a pair of cameras mounted on a robot that is moving around in the world, as shown in Figure 11.15a. By replacing the rightmost two transformations in Figure 11.14 with the transformations shown in Figure 11.15b, we can simultaneously recover the position of the robot at each time and the calibration of each camera with respect to the rig, in addition to the 3D structure of the world.

11.4.3 Exploiting sparsity

Large bundle adjustment problems, such as those involving reconstructing 3D scenes from thousands of internet photographs (Snavely, Seitz, and Szeliski 2008b; Agarwal, Furukawa *et al.* 2010, 2011; Snavely, Simon *et al.* 2010), can require solving non-linear least squares problems with millions of

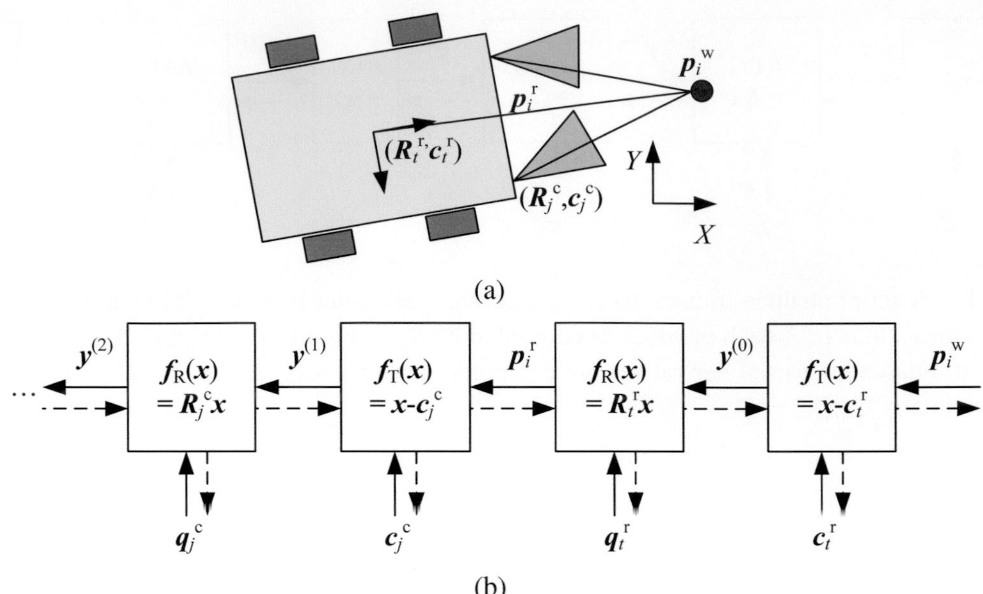

Figure 11.15 A camera rig and its associated transform chain. (a) As the mobile rig (robot) moves around in the world, its pose with respect to the world at time t is captured by $(\mathbf{R}_t^r, \mathbf{c}_t^r)$. Each camera's pose with respect to the rig is captured by $(\mathbf{R}_j^c, \mathbf{c}_j^c)$. (b) A 3D point with world coordinates $\mathbf{p}_i^w$ is first transformed into rig coordinates $\mathbf{p}_i^r$, and then through the rest of the camera-specific chain, as shown in Figure 11.14.

measurements (feature matches) and tens of thousands of unknown parameters (3D point positions and camera poses). Unless some care is taken, these kinds of problem can become intractable, because the (direct) solution of dense least squares problems is cubic in the number of unknowns.

Fortunately, structure from motion is a *bipartite* problem in structure and motion. Each feature point $\mathbf{x}_{ij}$ in a given image depends on one 3D point position $\mathbf{p}_i$ and one 3D camera pose $(\mathbf{R}_j, \mathbf{c}_j)$. This is illustrated in Figure 11.16a, where each circle (1–9) indicates a 3D point, each square (A–D) indicates a camera, and lines (edges) indicate which points are visible in which cameras (2D features). If the values for all the points are known or fixed, the equations for all the cameras become independent, and vice versa.

If we order the structure variables before the motion variables in the Hessian matrix $\mathbf{A}$ (and hence also the right-hand side vector $\mathbf{b}$), we obtain a structure for the Hessian shown in Figure 11.16c.[19] When such a system is solved using sparse Cholesky factorization (see Appendix A.4) (Björck 1996; Golub and Van Loan 1996), the *fill-in* occurs in the smaller motion Hessian A_{cc} (Szeliski and Kang 1994; Triggs, McLauchlan *et al.* 1999; Hartley and Zisserman 2004; Lourakis and Argyros 2009; Engels, Stewénius, and Nistér 2006). More recent papers (Byröd and Åström 2009; Jeong, Nistér *et al.* 2010; Agarwal, Snavely *et al.* 2010; Jeong, Nistér *et al.* 2012) explore the use of iterative (conjugate gradient) techniques for the solution of bundle adjustment problems. Other papers explore the use of parallel multicore algorithms (Wu, Agarwal *et al.* 2011).

In more detail, the *reduced* motion Hessian is computed using the *Schur complement*,

$$\mathbf{A}'_{\text{CC}} = \mathbf{A}_{\text{CC}} - \mathbf{A}_{\text{PC}}^T \mathbf{A}_{\text{PP}}^{-1} \mathbf{A}_{\text{pc}}, \tag{11.68}$$

where $\mathbf{A}_{\text{PP}}$ is the point (structure) Hessian (the top left block of Figure 11.16c), $\mathbf{A}_{\text{PC}}$ is the point-

[19] This ordering is preferable when there are fewer cameras than 3D points, which is the usual case. The exception is when we are tracking a small number of points through many video frames, in which case this ordering should be reversed.

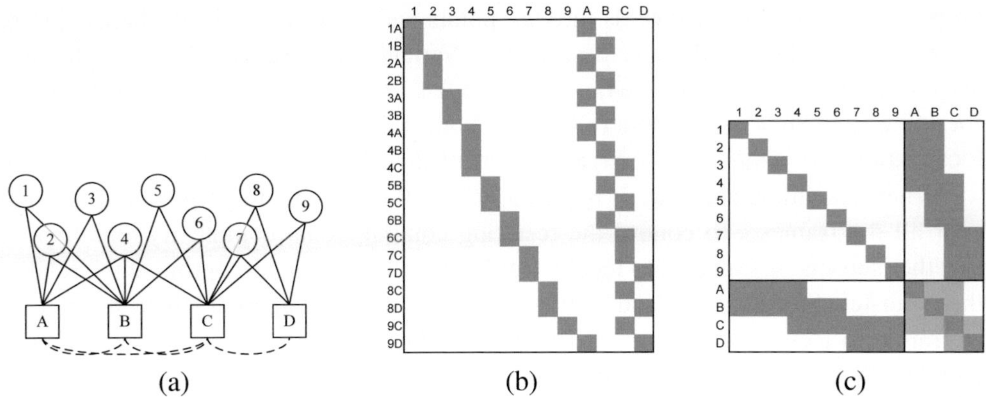

Figure 11.16 (a) Bipartite graph for a toy structure from motion problem and (b) its associated Jacobian $\mathbf{J}$ and (c) Hessian $\mathbf{A}$. Numbers indicate 3D points and letters indicate cameras. The dashed arcs and light blue squares indicate the fill-in that occurs when the structure (point) variables are eliminated.

camera Hessian (the top right block), and $\mathbf{A}_{\mathrm{CC}}$ and $\mathbf{A}'_{\mathrm{CC}}$ are the motion Hessians before and after the point variable elimination (the bottom right block of Figure 11.16c). Notice that $\mathbf{A}'_{\mathrm{CC}}$ has a non-zero entry between two cameras if they see any 3D point in common. This is indicated with dashed arcs in Figure 11.16a and light blue squares in Figure 11.16c.

Whenever there are global parameters present in the reconstruction algorithm, such as camera intrinsics that are common to all of the cameras, or camera rig calibration parameters such as those shown in Figure 11.15, they should be ordered last (placed along the right and bottom edges of $\mathbf{A}$) to reduce fill-in.

Engels, Stewénius, and Nistér (2006) provide a nice recipe for sparse bundle adjustment, including all the steps needed to initialize the iterations, as well as typical computation times for a system that uses a fixed number of backward-looking frames in a real-time setting. They also recommend using homogeneous coordinates for the structure parameters $\mathbf{p}_i$, which is a good idea, as it avoids numerical instabilities for points near infinity.

Bundle adjustment is now the standard method of choice for most structure-from-motion problems and is commonly applied to problems with hundreds of weakly calibrated images and tens of thousands of points. (Much larger problems are commonly solved in photogrammetry and aerial imagery, but these are usually carefully calibrated and make use of surveyed ground control points.) However, as the problems become larger, it becomes impractical to re-solve full bundle adjustment problems at each iteration.

One approach to dealing with this problem is to use an incremental algorithm, where new cameras are added over time. (This makes particular sense if the data is being acquired from a video camera or moving vehicle (Nistér, Naroditsky, and Bergen 2006; Pollefeys, Nistér *et al.* 2008).) A Kalman filter can be used to incrementally update estimates as new information is acquired. Unfortunately, such sequential updating is only statistically optimal for linear least squares problems.

For non-linear problems such as structure from motion, an extended Kalman filter, which linearizes measurement and update equations around the current estimate, needs to be used (Gelb 1974; Viéville and Faugeras 1990). To overcome this limitation, several passes can be made through the data (Azarbayejani and Pentland 1995). Because points disappear from view (and old cameras become irrelevant), a *variable state dimension filter* (VSDF) can be used to adjust the set of state variables over time, for example, by keeping only cameras and point tracks seen in the last k frames (McLauchlan 2000). A more flexible approach to using a fixed number of frames is to propagate corrections backwards through points and cameras until the changes on parameters are below a threshold (Steedly and Essa 2001). Variants of these techniques, including methods that use a fixed window for bundle adjustment (Engels, Stewénius, and Nistér 2006) or select keyframes for doing full bundle adjustment (Klein and Murray 2008) are now commonly used in simultaneous localization and mapping (SLAM) and augmented-reality applications, as discussed in Section 11.5.

When maximum accuracy is required, it is still preferable to perform a full bundle adjustment over all the frames. To control the resulting computational complexity, one approach is to lock together subsets of frames into locally rigid configurations and to optimize the relative positions of these cluster (Steedly, Essa, and Dellaert 2003). A different approach is to select a smaller number of frames to form a *skeletal set* that still spans the whole dataset and produces reconstructions of comparable accuracy (Snavely, Seitz, and Szeliski 2008b). We describe this latter technique in more detail in Section 11.4.6, where we discuss applications of structure from motion to large image sets. Additional techniques for efficiently solving large structure from motion and SLAM systems can be found in the survey by Dellaert and Kaess (2017); Dellaert (2021).

While bundle adjustment and other robust non-linear least squares techniques are the methods of choice for most structure-from-motion problems, they suffer from initialization problems, i.e., they can get stuck in local energy minima if not started sufficiently close to the global optimum. Many systems try to mitigate this by being conservative in what reconstruction they perform early on and which cameras and points they add to the solution (Section 11.4.6). An alternative, however, is to re-formulate the problem using a norm that supports the computation of global optima.

Kahl and Hartley (2008) describe techniques for using L_∞ norms in geometric reconstruction problems. The advantage of such norms is that globally optimal solutions can be efficiently computed using second-order cone programming (SOCP). The disadvantage is that L_∞ norms are particularly sensitive to outliers and so must be combined with good outlier rejection techniques before they can be used.

A large number of high-quality open source bundle adjustment packages have been developed, including the Ceres Solver,[20] Multicore Bundle Adjustment (Wu, Agarwal *et al.* 2011),[21] the Sparse Levenberg-Marquardt based non-linear least squares optimizer and bundle adjuster,[22] and OpenSfM.[23] You can find more pointers to open-source software in Appendix Appendix C.2 and reviews of open-source and commercial photogrammetry software[24] as well as examples of their application[25] on the web.

11.4.4 *Application*: Match move

One of the neatest applications of structure from motion is to estimate the 3D motion of a video or film camera, along with the geometry of a 3D scene, in order to superimpose 3D graphics or computer-generated images (CGI) on the scene. In the visual effects industry, this is known as the *match move* problem (Roble 1999), as the motion of the synthetic 3D camera used to render the graphics must be *matched* to that of the real-world camera. For very small motions, or motions involving pure camera rotations, one or two tracked points can suffice to compute the necessary visual motion. For planar surfaces moving in 3D, four points are needed to compute the homography,

[20] http://ceres-solver.org

[21] https://grail.cs.washington.edu/projects/mcba

[22] https://github.com/chzach/SSBA

[23] https://www.opensfm.org

[24] https://peterfalkingham.com/2020/07/10/free-and-commercial-photogrammetry-software-review-2020

[25] https://beforesandafters.com/2020/07/06/tales-from-on-set-lidar-scanning-for-joker-and-john-wick-3, https://rd.nytimes.com/projects/reconstructing-journalistic-scenes-in-3d

which can then be used to insert planar overlays, e.g., to replace the contents of advertising billboards during sporting events.

The general version of this problem requires the estimation of the full 3D camera pose along with the focal length (zoom) of the lens and potentially its radial distortion parameters (Roble 1999). When the 3D structure of the scene is known ahead of time, pose estimation techniques such as *view correlation* (Bogart 1991) or *through-the-lens camera control* (Gleicher and Witkin 1992) can be used, as described in Section 11.4.4.

For more complex scenes, it is usually preferable to recover the 3D structure simultaneously with the camera motion using structure-from-motion techniques. The trick with using such techniques is that to prevent any visible jitter between the synthetic graphics and the actual scene, features must be tracked to very high accuracy and ample feature tracks must be available in the vicinity of the insertion location. Some of today's best known match move software packages, such as the *boujou* package from 2d3, which won an Emmy award in 2002, originated in structure-from-motion research in the computer vision community (Fitzgibbon and Zisserman 1998).

11.4.5 Uncertainty and ambiguities

Because structure from motion involves the estimation of so many highly coupled parameters, often with no known "ground truth" components, the estimates produced by structure from motion algorithms can often exhibit large amounts of uncertainty (Szeliski and Kang 1997; Wilson and Wehrwein 2020). An example of this is the classic *bas-relief ambiguity*, which makes it hard to simultaneously estimate the 3D depth of a scene and the amount of camera motion (Oliensis 2005).[26]

As mentioned before, a unique coordinate frame and scale for a reconstructed scene cannot be recovered from monocular visual measurements alone. (When a stereo rig is used, the scale can be recovered if we know the distance (baseline) between the cameras.) This seven-degree-of-freedom (coordinate frame and scale) *gauge ambiguity* makes it tricky to compute the covariance matrix associated with a 3D reconstruction (Triggs, McLauchlan *et al.* 1999; Kanatani and Morris 2001). A simple way to compute a covariance matrix that ignores the gauge freedom (indeterminacy) is to throw away the seven smallest eigenvalues of the information matrix (inverse covariance), whose values are equivalent to the problem Hessian $\mathbf{A}$ up to noise scaling (see Section 8.1.4 and Appendix B.6). After we do this, the resulting matrix can be inverted to obtain an estimate of the parameter covariance.

Szeliski and Kang (1997) use this approach to visualize the largest directions of variation in typical structure from motion problems. Not surprisingly, they find that, ignoring the gauge freedoms, the greatest uncertainties for problems such as observing an object from a small number of nearby viewpoints are in the depths of the 3D structure relative to the extent of the camera motion.[27]

It is also possible to estimate *local* or *marginal* uncertainties for individual parameters, which corresponds simply to taking block sub-matrices from the full covariance matrix. Under certain conditions, such as when the camera poses are relatively certain compared to 3D point locations, such uncertainty estimates can be meaningful. However, in many cases, individual uncertainty measures can mask the extent to which reconstruction errors are correlated, which is why looking at the first few modes of greatest joint variation can be helpful.

[26]Bas-relief refers to a kind of sculpture in which objects, often on ornamental friezes, are sculpted with less depth than they actually occupy. When lit from above by sunlight, they appear to have true 3D depth because of the ambiguity between relative depth and the angle of the illuminant (Section 13.1.1).

[27]A good way to minimize the amount of such ambiguities is to use wide field of view cameras (Antone and Teller 2002; Levin and Szeliski 2006).

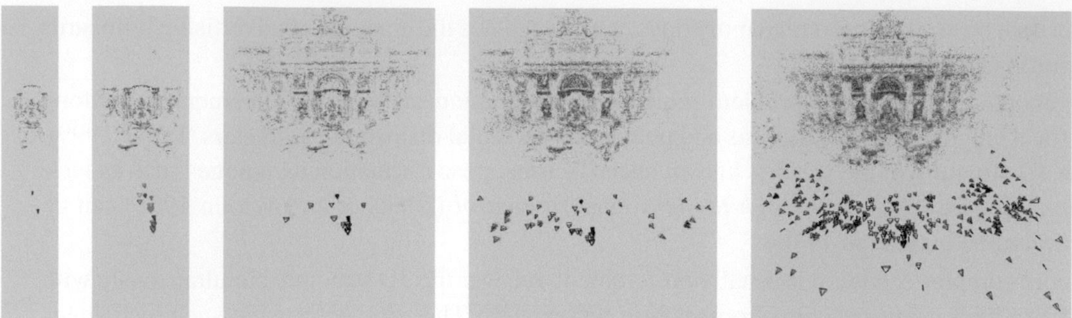

Figure 11.17 Incremental structure from motion (Snavely, Seitz, and Szeliski 2006) © 2006 ACM. Starting with an initial two-frame reconstruction of Trevi Fountain, batches of images are added using pose estimation, and their positions (along with the 3D model) are refined using bundle adjustment.

The other way in which gauge ambiguities affect structure from motion and, in particular, bundle adjustment is that they make the system Hessian matrix $\mathbf{A}$ rank-deficient and hence impossible to invert. A number of techniques have been proposed to mitigate this problem (Triggs, McLauchlan *et al.* 1999; Bartoli 2003). In practice, however, it appears that simply adding a small amount of the Hessian diagonal $\lambda \text{diag}(\mathbf{A})$ to the Hessian $\mathbf{A}$ itself, as is done in the Levenberg–Marquardt non-linear least squares algorithm (Appendix A.3), usually works well.

11.4.6 *Application*: Reconstruction from internet photos

The most widely used application of structure from motion is in the reconstruction of 3D objects and scenes from video sequences and collections of images (Pollefeys and Van Gool 2002). The last two decades have seen an explosion of techniques for performing this task automatically without the need for any manual correspondence or pre-surveyed ground control points. A lot of these techniques assume that the scene is taken with the same camera and hence the images all have the same intrinsics (Fitzgibbon and Zisserman 1998; Koch, Pollefeys, and Van Gool 2000; Schaffalitzky and Zisserman 2002; Tuytelaars and Van Gool 2004; Pollefeys, Nistér *et al.* 2008; Moons, Van Gool, and Vergauwen 2010). Many of these techniques take the results of the sparse feature matching and structure from motion computation and then compute dense 3D surface models using multi-view stereo techniques (Section 12.7) (Koch, Pollefeys, and Van Gool 2000; Pollefeys and Van Gool 2002; Pollefeys, Nistér *et al.* 2008; Moons, Van Gool, and Vergauwen 2010; Schönberger, Zheng *et al.* 2016).

An exciting innovation in this space has been the application of structure from motion and multi-view stereo techniques to thousands of images taken from the internet, where very little is known about the cameras taking the photographs (Snavely, Seitz, and Szeliski 2008a). Before the structure from motion computation can begin, it is first necessary to establish sparse correspondences between different pairs of images and to then link such correspondences into *feature tracks*, which associate individual 2D image features with global 3D points. Because the $O(N^2)$ comparison of all pairs of images can be very slow, a number of techniques have been developed in the recognition community to make this process faster (Section 7.1.4) (Nistér and Stewénius 2006; Philbin, Chum *et al.* 2008; Li, Wu *et al.* 2008; Chum, Philbin, and Zisserman 2008; Chum and Matas 2010a; Arandjelović and Zisserman 2012).

To begin the reconstruction process, it is important to select a good pair of images, where there are both a large number of consistent matches (to lower the likelihood of incorrect correspondences)

(a) (b) (c)

Figure 11.18 3D reconstructions produced by the incremental structure from motion algorithm developed by Snavely, Seitz, and Szeliski (2006) © 2006 ACM: (a) cameras and point cloud from Trafalgar Square; (b) cameras and points overlaid on an image from the Great Wall of China; (c) overhead view of a reconstruction of the Old Town Square in Prague registered to an aerial photograph.

and a significant amount of out-of-plane parallax,[28] to ensure that a stable reconstruction can be obtained (Snavely, Seitz, and Szeliski 2006). The EXIF tags associated with the photographs can be used to get good initial estimates for camera focal lengths, although this is not always strictly necessary, because these parameters are re-adjusted as part of the bundle adjustment process.

Once an initial pair has been reconstructed, the pose of cameras that see a sufficient number of the resulting 3D points can be estimated (Section 11.2) and the complete set of cameras and feature correspondences can be used to perform another round of bundle adjustment. Figure 11.17 shows the progression of the incremental bundle adjustment algorithm, where sets of cameras are added after each successive round of bundle adjustment, while Figure 11.18 shows some additional results. An alternative to this kind of *seed and grow* approach is to first reconstruct triplets of images and then hierarchically merge them into larger collections (Fitzgibbon and Zisserman 1998).

Unfortunately, as the incremental structure from motion algorithm continues to add more cameras and points, it can become extremely slow. The direct solution of a dense system of $O(N)$ equations for the camera pose updates can take $O(N^3)$ time; while structure from motion problems are rarely dense, scenes such as city squares have a high percentage of cameras that see points in common. Re-running the bundle adjustment algorithm after every few camera additions results in a quartic scaling of the run time with the number of images in the dataset. One approach to solving this problem is to select a smaller number of images for the original scene reconstruction and to fold in the remaining images at the very end.

Snavely, Seitz, and Szeliski (2008b) develop an algorithm for computing such a *skeletal set* of images, which is guaranteed to produce a reconstruction whose error is within a bounded factor of the optimal reconstruction accuracy. Their algorithm first evaluates all pairwise uncertainties (position covariances) between overlapping images and then chains them together to estimate a lower bound for the relative uncertainty of any distant pair. The skeletal set is constructed so that the maximal uncertainty between any pair grows by no more than a constant factor. Figure 11.19 shows an example of the skeletal set computed for 784 images of the Pantheon in Rome. As you can see, even though the skeletal set contains just a fraction of the original images, the shapes of the skeletal set and full bundle adjusted reconstructions are virtually indistinguishable.

Since the initial publication on large-scale internet photo reconstruction by Snavely, Seitz, and

[28] A simple way to compute this is to robustly fit a homography to the correspondences and measure reprojection errors.

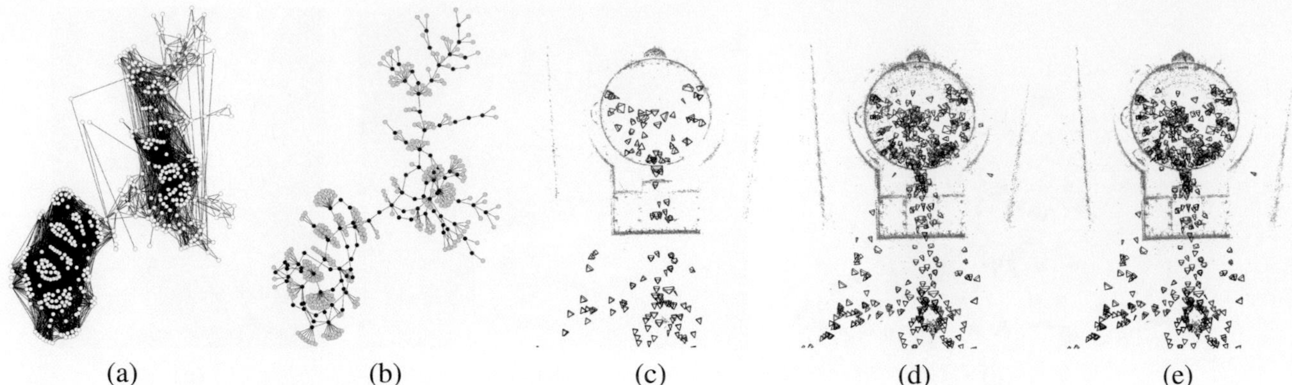

Figure 11.19 Large-scale structure from motion using skeletal sets (Snavely, Seitz, and Szeliski 2008b) © 2008 IEEE: (a) original match graph for 784 images; (b) skeletal set containing 101 images; (c) top-down view of scene (Pantheon) reconstructed from the skeletal set; (d) reconstruction after adding in the remaining images using pose estimation; (e) final bundle adjusted reconstruction, which is almost identical.

Szeliski (2008a,b), there have been a large number of follow-on papers exploring even larger datasets and more efficient algorithms (Agarwal, Furukawa *et al.* 2010, 2011; Frahm, Fite-Georgel *et al.* 2010; Wu 2013; Heinly, Schönberger *et al.* 2015; Schönberger and Frahm 2016). Among these, the COLMAP open source structure from motion and multi-view stereo system is currently one of the most widely used, as it can reconstruct extremely large scenes, such as the one shown in Figure 11.20 (Schönberger and Frahm 2016).[29]

The ability to automatically reconstruct 3D models from large, unstructured image collections has also brought to light subtle problems with traditional structure from motion algorithms, including the need to deal with repetitive and duplicate structures (Wu, Frahm, and Pollefeys 2010; Roberts, Sinha *et al.* 2011; Wilson and Snavely 2013; Heinly, Dunn, and Frahm 2014) as well as dynamic visual objects such as people (Ji, Dunn, and Frahm 2014; Zheng, Wang *et al.* 2014). It has also opened up a wide variety of additional applications, including the ability to automatically find and label locations and regions of interest (Simon, Snavely, and Seitz 2007; Simon and Seitz 2008; Gammeter, Bossard *et al.* 2009) and to cluster large image collections so that they can be automatically labeled (Li, Wu *et al.* 2008; Quack, Leibe, and Van Gool 2008). Some additional applications related to image-based rendering are discussed in more detail in Section 14.1.2.

11.4.7 Global structure from motion

While incremental bundle adjustment algorithms are still the most commonly used approaches for large-scale reconstruction (Schönberger and Frahm 2016), they can be quite slow because of the need to successively solve increasing larger optimization problems. An alternative to iteratively growing the solution is to solve for all of the structure and motion unknowns in a single global step, once the feature correspondences have been established.

One approach to this is to set up a linear system of equations that relate all of the camera centers and 3D point, line, and plane equations to the known 2D feature or line positions (Kaucic, Hartley, and Dano 2001; Rother 2003). However, these approaches require a reference plane (e.g., building wall) to be visible and matched in all images, and are also sensitive to distant points, which must first be discarded. These approaches, while theoretically interesting, are not widely used.

[29] https://colmap.github.io

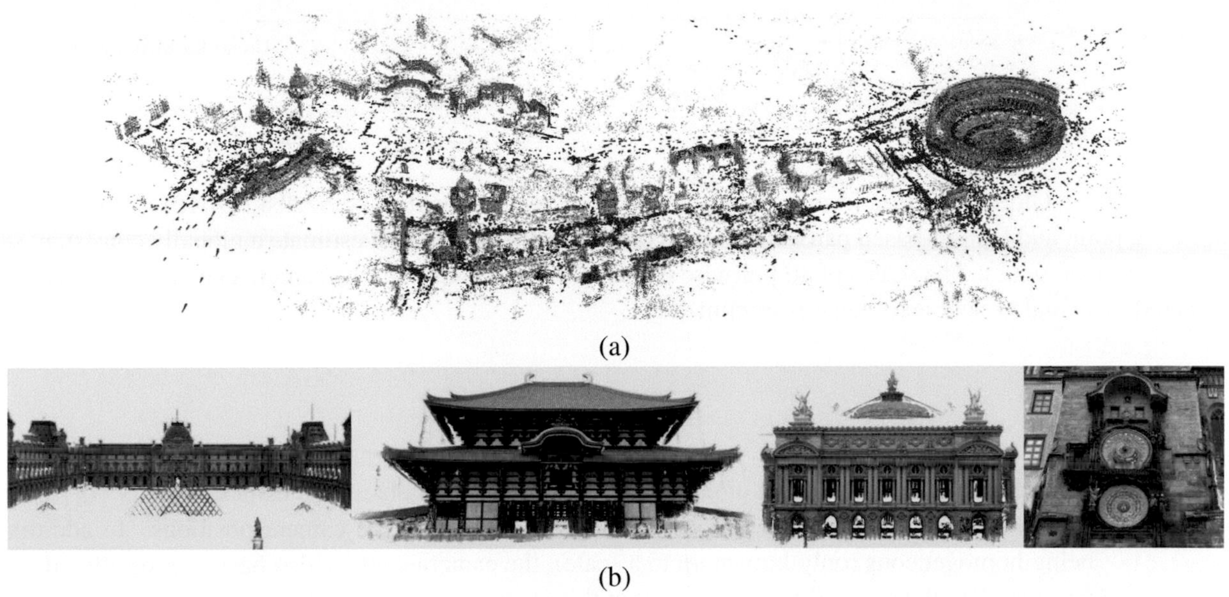

(a)

(b)

Figure 11.20 Large-scale reconstructions created with the COLMAP structure from motion and multi-view stereo system: (a) sparse model of central Rome constructed from 21K photos (Schönberger and Frahm 2016) © 2016 IEEE; (b) dense models of several landmarks produced with the MVS pipeline (Schönberger, Zheng *et al.* 2016) © 2016 Springer.

A second approach, first proposed by Govindu (2001), starts by computing pairwise Euclidean structure and motion reconstructions using the techniques discussed in Section 11.3.[30] Pairwise rotation estimates are then used to compute a globally consistent orientation estimate for each camera, using a process known as *rotation averaging* (Govindu 2001; Martinec and Pajdla 2007; Chatterjee and Govindu 2013; Hartley, Trumpf *et al.* 2013; Dellaert, Rosen *et al.* 2020).[31] In a final step, the camera positions are determined by scaling each of the local camera translations, after they have been rotated into a global coordinate system (Govindu 2001, 2004; Martinec and Pajdla 2007; Sinha, Steedly, and Szeliski 2010). In the robotics (SLAM) community, this last step is called *pose graph optimization* (Carlone, Tron *et al.* 2015).

Figure 11.21 shows a more recent pipeline implementing this concept, which includes the initial feature point extraction, matching, and two-view reconstruction, followed by global rotation estimation, and then a final solve for the camera centers. The pipeline developed by Sinha, Steedly, and Szeliski (2010) also matches vanishing points, when these can be found, in order to eliminate rotational drift in the global orientation estimates.

While there are several alternative algorithms for estimating the global rotations, an even wider variety of algorithms exists for estimating the camera centers. After rotating all of the cameras by their global rotation estimate, we can compute globally oriented local translation direction in each reconstructed pair ij and denote this as $\hat{\mathbf{t}}_{ij}$. The fundamental relationship between the unknown camera centers $\{\mathbf{c}_i\}$ and the translation directions can be written as

$$\mathbf{c}_j - \mathbf{c}_i = s_{ij}\hat{\mathbf{t}}_{ij} \tag{11.69}$$

[30]While almost of all of these techniques assume known calibration (focal lengths) for each image, Sweeney, Kneip *et al.* (2015) estimate focal lengths from refined fundamental matrices.

[31]We have already introduced the concept of rotation averaging when we discussed global registration of panoramas in Section 8.3.1.

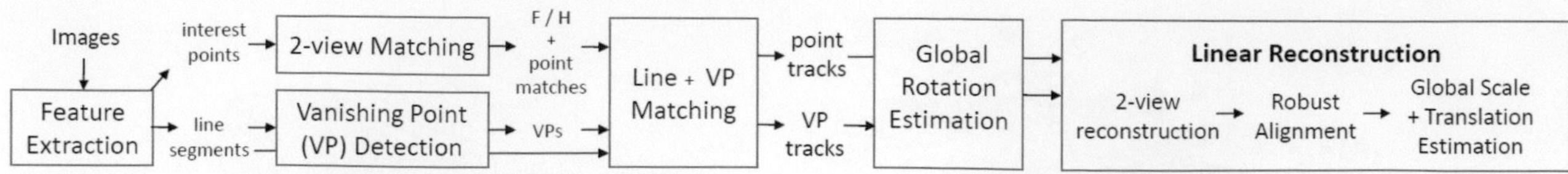

Figure 11.21 Global structure from motion pipeline from Sinha, Steedly, and Szeliski (2010) © 2010 Springer. Vanishing point and feature-based pairwise rotation estimates are used to first estimate a globally consistent set of orientations (rotations). The scales of all pairwise reconstructions along with the camera center positions are then estimated in a single linear least squares minimization.

or

$$\hat{\mathbf{t}}_{ij} \times (\mathbf{c}_j - \mathbf{c}_i) = 0 \tag{11.70}$$

(Govindu 2001). The first set of equations can be solved to obtain the camera centers $\{\mathbf{c}_i\}$ and the scale variables s_{ij}, while the second directly produces only the camera positions. In addition to being homogeneous (only known up to a scale), the camera centers also have a translational *gauge freedom*, i.e., they can all be translated (but this is always the case with structure from motion).

Because these equations minimize the algebraic alignment between local translation directions and global camera center differences, they do not correctly weight reconstructions with different baselines. Several alternatives have been proposed to remediate this (Govindu 2004; Sinha, Steedly, and Szeliski 2010; Jiang, Cui, and Tan 2013; Moulon, Monasse, and Marlet 2013; Wilson and Snavely 2014; Cui and Tan 2015; Özyeşil and Singer 2015; Holynski, Geraghty *et al.* 2020). Some of these techniques also cannot handle collinear cameras, as in the original formulation, as well as some more recent ones, we can shift cameras along a collinear segment and still satisfy the directional constraints.

For community photo collections taken over a large area such as a plaza, this is not a crucial problem (Wilson and Snavely 2014). However, for reconstructions from video or walks around or through a building, the collinear camera problem is a real issue. Sinha, Steedly, and Szeliski (2010) handle this by estimating the relative scales of pairwise reconstructions that share a common camera and then use these relative scales to constraint all of the global scales.

Two open-source structure from motion pipelines that include some of these global techniques are Theia[32] (Sweeney, Hollerer, and Turk 2015) and OpenMVG[33] (Moulon, Monasse *et al.* 2016). The papers have nice reviews of the related algorithms.

11.4.8 Constrained structure and motion

The most general algorithms for structure from motion make no prior assumptions about the objects or scenes that they are reconstructing. In many cases, however, the scene contains higher-level geometric primitives, such as lines and planes. These can provide information complementary to interest points and also serve as useful building blocks for 3D modeling and visualization. Furthermore, these primitives are often arranged in particular relationships, i.e., many lines and planes are either parallel or orthogonal to each other (Zhou, Furukawa, and Ma 2019; Zhou, Furukawa *et al.* 2020). This is particularly true of architectural scenes and models, which we study in more detail in Section 13.6.1.

Sometimes, instead of exploiting regularity in the scene structure, it is possible to take advantage of a constrained motion model. For example, if the object of interest is rotating on a turntable

[32] http://www.theia-sfm.org
[33] https://github.com/openMVG/openMVG

Figure 11.22 Two images of a toy house along with their matched 3D line segments (Schmid and Zisserman 1997) © 1997 Springer.

(Szeliski 1991b), i.e., around a fixed but unknown axis, specialized techniques can be used to recover this motion (Fitzgibbon, Cross, and Zisserman 1998). In other situations, the camera itself may be moving in a fixed arc around some center of rotation (Shum and He 1999). Specialized capture setups, such as mobile stereo camera rigs or moving vehicles equipped with multiple fixed cameras, can also take advantage of the knowledge that individual cameras are (mostly) fixed with respect to the capture rig, as shown in Figure 11.15.[34]

Line-based techniques

It is well known that pairwise epipolar geometry cannot be recovered from line matches alone, even if the cameras are calibrated. To see this, think of projecting the set of lines in each image into a set of 3D planes in space. You can move the two cameras around into any configuration you like and still obtain a valid reconstruction for 3D lines.

When lines are visible in three or more views, the trifocal tensor can be used to transfer lines from one pair of images to another (Hartley and Zisserman 2004). The trifocal tensor can also be computed on the basis of line matches alone.

Schmid and Zisserman (1997) describe a widely used technique for matching 2D lines based on the average of 15×15 pixel correlation scores evaluated at all pixels along their common line segment intersection.[35] In their system, the epipolar geometry is assumed to be known, e.g., computed from point matches. For wide baselines, all possible homographies corresponding to planes passing through the 3D line are used to warp pixels and the maximum correlation score is used. For triplets of images, the trifocal tensor is used to verify that the lines are in geometric correspondence before evaluating the correlations between line segments. Figure 11.22 shows the results of using their system.

Bartoli and Sturm (2003) describe a complete system for extending three view relations (trifocal tensors) computed from manual line correspondences to a full bundle adjustment of all the line and camera parameters. The key to their approach is to use the Plücker coordinates (2.12) to parameterize lines and to directly minimize reprojection errors. It is also possible to represent 3D line segments by their endpoints and to measure either the reprojection error perpendicular to the detected 2D line segments in each image or the 2D errors using an elongated uncertainty ellipse aligned with the line segment direction (Szeliski and Kang 1994).

[34] Because of mechanical compliance and jitter, it may be prudent to allow for a small amount of individual camera rotation around a nominal position.

[35] Because lines often occur at depth or orientation discontinuities, it may be preferable to compute correlation scores (or to match color histograms (Bay, Ferrari, and Van Gool 2005)) separately on each side of the line.

Instead of reconstructing 3D lines, Bay, Ferrari, and Van Gool (2005) use RANSAC to group lines into likely coplanar subsets. Four lines are chosen at random to compute a homography, which is then verified for these and other plausible line segment matches by evaluating color histogram-based correlation scores. The 2D intersection points of lines belonging to the same plane are then used as virtual measurements to estimate the epipolar geometry, which is more accurate than using the homographies directly.

An alternative to grouping lines into coplanar subsets is to group lines by parallelism. Whenever three or more 2D lines share a common vanishing point, there is a good likelihood that they are parallel in 3D. By finding multiple vanishing points in an image (Section 7.4.3) and establishing correspondences between such vanishing points in different images, the relative rotations between the various images (and often the camera intrinsics) can be directly estimated (Section 11.1.1). Finding an orthogonal set of vanishing points and using these to establish a global orientation is often called invoking the *Manhattan world assumption* (Coughlan and Yuille 1999). A generalized version where streets can meet at non-orthogonal angles was called the *Atlanta world* by Schindler and Dellaert (2004).

Shum, Han, and Szeliski (1998) describe a 3D modeling system that constructs calibrated panoramas from multiple images (Section 11.4.2) and then has the user draw vertical and horizontal lines in the image to demarcate the boundaries of planar regions. The lines are used to establish an absolute rotation for each panorama and are then used (along with the inferred vertices and planes) to build a 3D structure, which can be recovered up to scale from one or more images (Figure 13.20).

A fully automated approach to line-based structure from motion is presented by Werner and Zisserman (2002). In their system, they first find lines and group them by common vanishing points in each image (Section 7.4.3). The vanishing points are then used to calibrate the camera, i.e., to perform a "metric upgrade" (Section 11.1.1). Lines corresponding to common vanishing points are then matched using both appearance (Schmid and Zisserman 1997) and trifocal tensors. These lines are then used to infer planes and a block-structured model for the scene, as described in more detail in Section 13.6.1. More recent work using deep neural networks can also be used to construct 3D wireframe models from one or more images.

Plane-based techniques

In scenes that are rich in planar structures, e.g., in architecture, it is possible to directly estimate homographies between different planes, using either feature-based or intensity-based methods. In principle, this information can be used to simultaneously infer the camera poses and the plane equations, i.e., to compute plane-based structure from motion.

Luong and Faugeras (1996) show how a fundamental matrix can be directly computed from two or more homographies using algebraic manipulations and least squares. Unfortunately, this approach often performs poorly, because the algebraic errors do not correspond to meaningful reprojection errors (Szeliski and Torr 1998).

A better approach is to *hallucinate* virtual point correspondences within the areas from which each homography was computed and to feed them into a standard structure from motion algorithm (Szeliski and Torr 1998). An even better approach is to use full bundle adjustment with explicit plane equations, as well as additional constraints to force reconstructed co-planar features to lie exactly on their corresponding planes. (A principled way to do this is to establish a coordinate frame for each plane, e.g., at one of the feature points, and to use 2D in-plane parameterizations for the other points.) The system developed by Shum, Han, and Szeliski (1998) shows an example of such an approach, where the directions of lines and normals for planes in the scene are prespecified by the user. In more recent work, Micusik and Wildenauer (2017) use planes as additional constraints

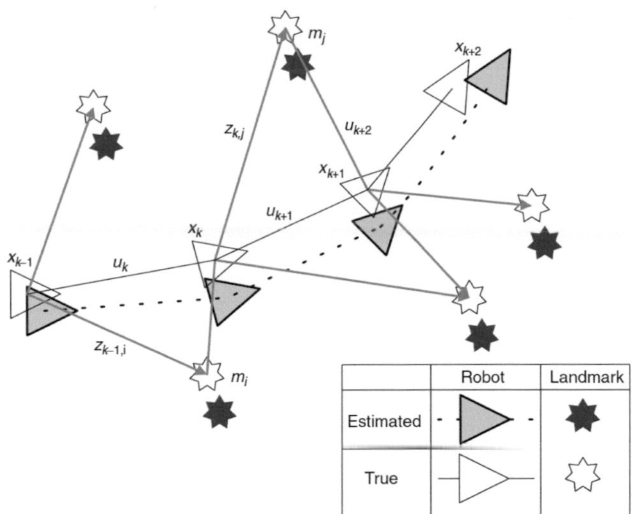

Figure 11.23 In simultaneous localization and mapping (SLAM), the system simultaneously estimates the positions of a robot and its nearby landmarks (Durrant-Whyte and Bailey 2006) © 2006 IEEE.

inside a bundle adjustment formulation. Other recent papers that use combinations of lines and/or planes to reduce drift in 3D reconstructions include (Zhou, Zou *et al.* 2015), Li, Yao *et al.* (2018), Yang and Scherer (2019), and Holynski, Geraghty *et al.* (2020).

11.5 Simultaneous localization and mapping (SLAM)

While the computer vision community has been studying structure from motion, i.e., the reconstruction of sparse 3D models from multiple images and videos, since the early 1980s (Longuet-Higgins 1981), the mobile robotics community has in parallel been studying the automatic construction of 3D maps from moving robots.[36] In robotics, the problem was formulated as the simultaneous estimation of 3D robot and *landmark* poses (Figure 11.23), and was known as *probabilistic mapping* (Thrun, Burgard, and Fox 2005) and *simultaneous localization and mapping* (SLAM) (Durrant-Whyte and Bailey 2006; Bailey and Durrant-Whyte 2006; Cadena, Carlone *et al.* 2016). In the computer vision community, the problem was originally called *visual odometry* (Levin and Szeliski 2004; Nistér, Naroditsky, and Bergen 2006; Maimone, Cheng, and Matthies 2007), although that term is now usually reserved for shorter-range motion estimation that does not involve building a global map with *loop closing* (Cadena, Carlone *et al.* 2016).

Early versions of such algorithms used range-sensing techniques, such as ultrasound, laser range finders, or stereo matching, to estimate local 3D geometry, which could then be fused into a 3D model. Newer techniques can perform the same task based purely on visual feature tracking from a monocular camera (Davison, Reid *et al.* 2007). Good introductory tutorials can be found in Durrant-Whyte and Bailey (2006) and Bailey and Durrant-Whyte (2006), while more comprehensive surveys of more recent techniques are presented in (Fuentes-Pacheco, Ruiz-Ascencio, and Rendón-Mancha 2015) and Cadena, Carlone *et al.* (2016).

SLAM differs from bundle adjustment in two fundamental aspects. First, it allows for a variety

[36]In the 1980s, the vision and robotics communities were essentially the same set of researchers working in these two sub-fields of artificial intelligence.

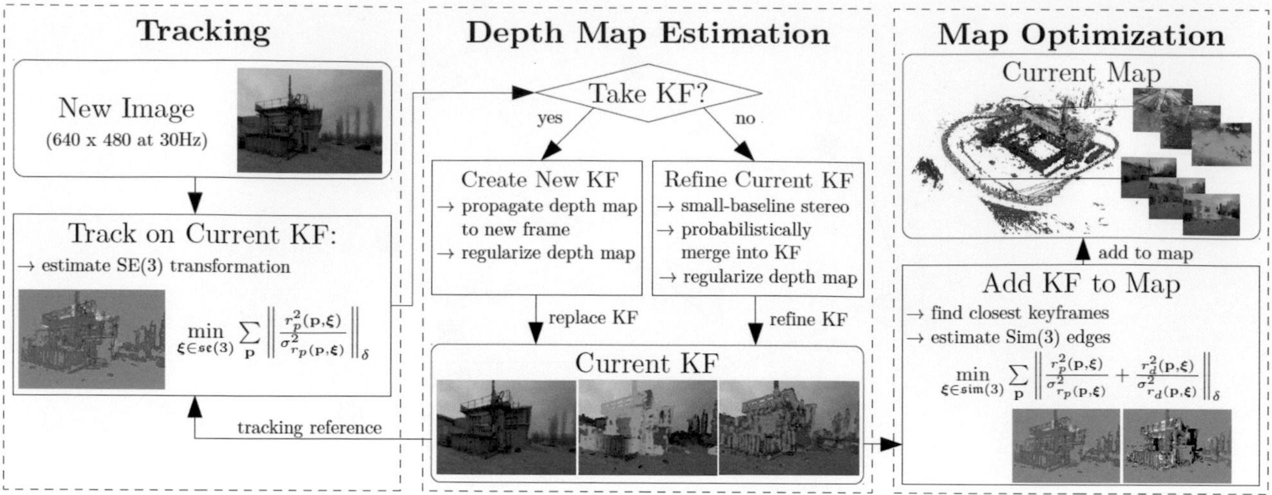

Figure 11.24 The architecture of the LSD-SLAM system (Engel, Schöps, and Cremers 2014) © 2014 Springer, showing the front end, which does the tracking, data association, and local 3D pose and structure (depth map) updating, and the back end, which does global map optimization.

of sensing devices, instead of just being restricted to tracked or matched feature points. Second, it solves the localization problem *online*, i.e., with no or very little lag in providing the current sensor pose. This makes it the method of choice for both time-critical robotics applications such as autonomous navigation (Section 11.5.1) and real-time augmented reality (Section 11.5.2).

Some of the important milestones in SLAM include:

- the application of SLAM to monocular cameras (MonoSLAM) (Davison, Reid *et al.* 2007);
- parallel tracking and mapping (PTAM) (Klein and Murray 2007), which split the front end (tracking) and back end (mapping) processes (Figure 11.24) onto two separate threads running at different rates (Figure 11.27) and then implemented the whole process on a camera phone (Klein and Murray 2009);
- adaptive relative bundle adjustment (Sibley, Mei *et al.* 2009, 2010), which maintains collections of local reconstructions anchored at different keyframes;
- incremental smoothing and mapping (iSAM) (Kaess, Ranganathan, and Dellaert 2008; Kaess, Johannsson *et al.* 2012) and other applications of factor graphs to handle the speed-accuracy-delay tradeoff (Dellaert and Kaess 2017; Dellaert 2021);
- dense tracking and mapping (DTAM) (Newcombe, Lovegrove, and Davison 2011), which estimates and updates a dense depth map for every frame;
- ORB-SLAM (Mur-Artal, Montiel, and Tardos 2015) and ORB-SLAM2 (Mur-Artal and Tardós 2017), which handle monocular, stereo, and RGB-D cameras as well as loop closures;
- SVO (semi-direct visual odometry) (Forster, Zhang *et al.* 2017), which combines patch-based tracking with classic bundle adjustment; and
- LSD-SLAM (large-scale direct SLAM) (Engel, Schöps, and Cremers 2014) and DSO (direct sparse odometry) (Engel, Koltun, and Cremers 2018), which only keep depth estimates at strong gradient locations (Figure 11.24).

- BAD SLAM (bundle adjusted direct RGB-D SLAM) (Schöps, Sattler, and Pollefeys 2019a).

Many of these systems have open source implementations. Some widely used benchmarks include a benchmark for RGB-D SLAM systems (Sturm, Engelhard *et al.* 2012), the KITTI Visual Odometry / SLAM benchmark (Geiger, Lenz *et al.* 2013), the synthetic ICL-NUIM dataset (Handa, Whelan *et al.* 2014), the TUM monoVO dataset (Engel, Usenko, and Cremers 2016), the EuRoC MAV dataset (Burri, Nikolic *et al.* 2016), the ETH3D SLAM benchmark (Schöps, Sattler, and Pollefeys 2019a), and the GSLAM general SLAM benchmark (Zhao, Xu *et al.* 2019).

The most recent trend in SLAM has been the integration with visual-inertial odometry (VIO) algorithms (Mourikis and Roumeliotis 2007; Li and Mourikis 2013; Forster, Carlone *et al.* 2016), which combine higher-frequency inertial measurement unit (IMU) measurements with visual tracks, which serve to remove low-frequency drift. Because IMUs are now commonplace in consumer devices such as cell phones and action cameras, VIO-enhanced SLAM systems serve as the foundation for widely used mobile augmented reality frameworks such as ARKit and ARCore (Section 11.5.2). A dataset and evaluation of open-source VIO systems can be found at Schubert, Goll *et al.* (2018).

As you can tell from this very brief overview, SLAM is an incredibly rich and rapidly evolving field of research, full of challenging robust optimization and real-time performance problems. A good source for finding a list of the most recent papers and algorithms is the KITTI Visual Odometry/SLAM Evaluation[37] (Geiger, Lenz, and Urtasun 2012) and the recent survey paper on computer vision for autonomous driving (Janai, Güney *et al.* 2020, Section 13.2).

11.5.1 *Application*: Autonomous navigation

Since the early days of artificial intelligence and robotics, computer vision has been used to enable manipulation for dextrous robots and navigation for autonomous robots (Janai, Güney *et al.* 2020; Kubota 2019). Some of the earliest vision-based navigation systems include the Stanford Cart (Figure 11.25a) and CMU Rover (Moravec 1980, 1983), the Terregator (Wallace, Stentz *et al.* 1985), and the CMU Nablab (Thorpe, Hebert *et al.* 1988), which originally could only advance 4m every 10 sec ($<$ 1 mph), and which was also the first system to use a neural network for driving (Pomerleau 1989).

The early algorithms and technologies advanced rapidly, with the VaMoRs system of Dickmanns and Mysliwetz (1992) operating a 25Hz Kalman filter loop and driving with good lane markings at 100 km/h. By the mid 2000s, when DARPA introduced their Grand Challenge and Urban Challenge, vehicles equipped with both range-finding lidar cameras and stereo cameras were able to traverse rough outdoor terrain and navigate city streets at regular human driving speeds (Urmson, Anhalt *et al.* 2008; Montemerlo, Becker *et al.* 2008).[38] These systems led to the formation of industrial research projects at companies such as Google and Tesla,[39] as well numerous startups, many of which exhibit their vehicles at computer vision conferences (Figure 11.25c–d).

A comprehensive review of computer vision technologies for autonomous vehicles can be found in the survey by Janai, Güney *et al.* (2020), which also comes with a useful on-line visualization tool of relevant papers.[40] The survey contains chapters on the large number of vision algorithms and components that go into autonomous navigation, which include datasets and benchmarks, sensors, object detection and tracking, segmentation, stereo, flow and scene flow, SLAM, scene understanding, and end-to-end learning of autonomous driving behaviors.

[37] http://www.cvlibs.net/datasets/kitti/eval_odometry.php

[38] Algorithms that use range data as part of their map building and localization are commonly called *RGB-D SLAM* systems (Sturm, Engelhard *et al.* 2012).

[39] You can find a number of talks about Tesla's efforts on Andrej Karpathy's web page, https://karpathy.ai.

[40] http://www.cvlibs.net/projects/autonomous_vision_survey

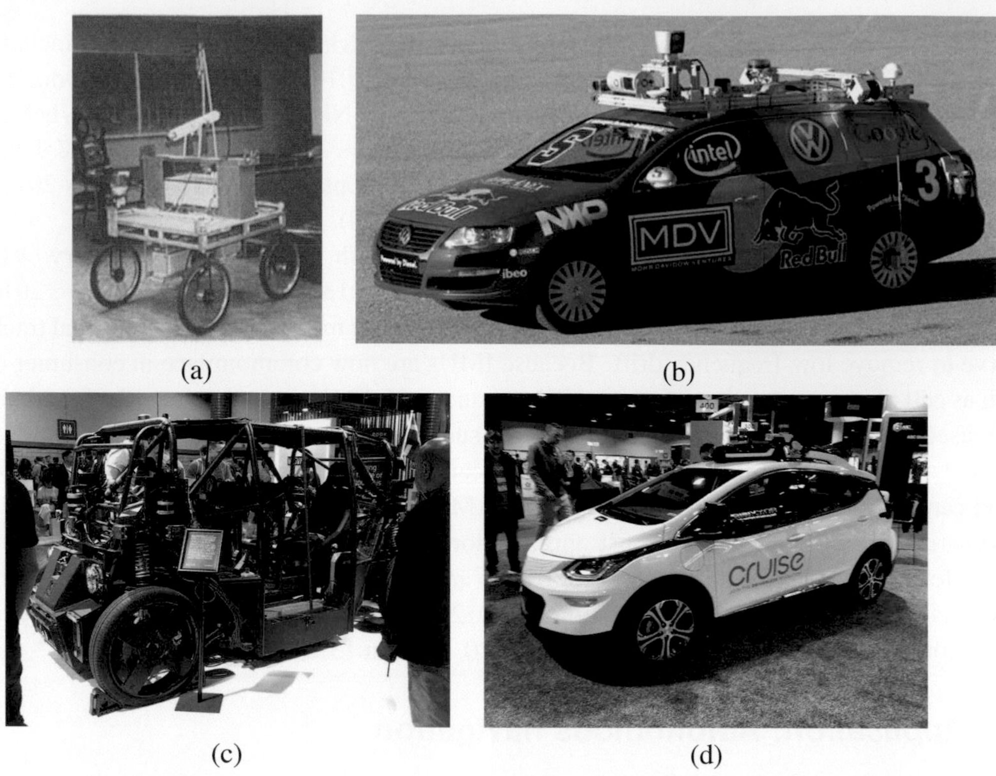

(a) (b)

(c) (d)

Figure 11.25 Autonomous vehicles: (a) the Stanford Cart (Moravec 1983) ©1983 IEEE; (b) Junior: The Stanford entry in the Urban Challenge (Montemerlo, Becker *et al.* 2008) © 2008 Wiley; (c–d) self-driving car prototypes from the CVPR 2019 exhibit floor.

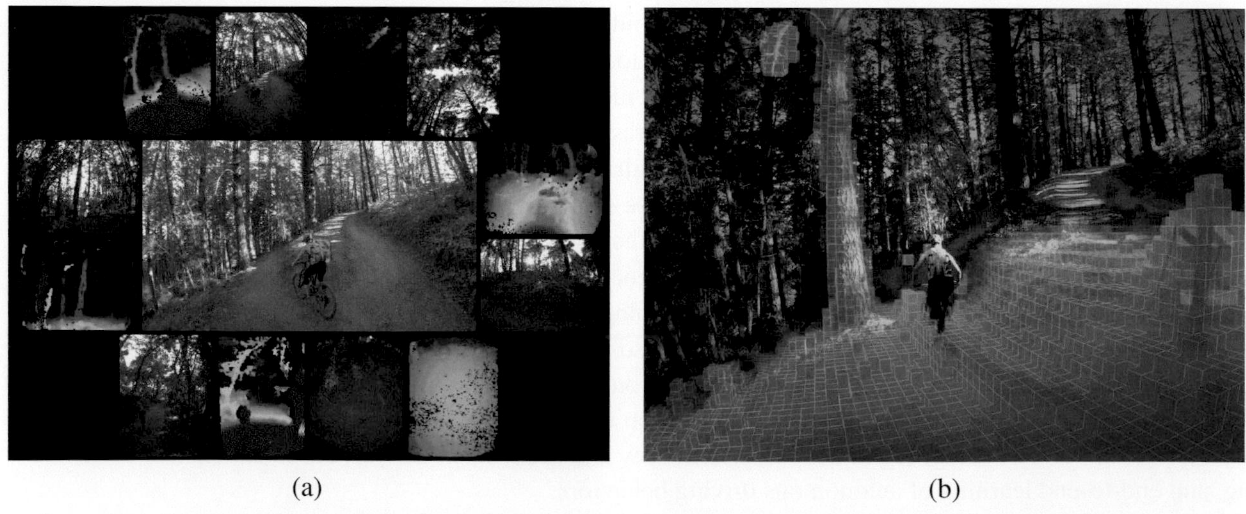

(a) (b)

Figure 11.26 Fully autonomous Skydio R1 drone flying in the wild © 2019 Skydio: (a) multiple input images and depth maps; (b) fully integrated 3D map (Cross 2019).

(a) (b)

Figure 11.27 3D augmented reality: (a) Darth Vader and a horde of Ewoks battle it out on a table-top recovered using real-time, keyframe-based structure from motion (Klein and Murray 2007) © 2007 IEEE; (b) a virtual teapot is fixed to the top of a real-world coffee cup, whose pose is re-recognized at each time frame (Gordon and Lowe 2006) © 2007 Springer.

In addition to autonomous navigation for wheeled (and legged) robots and vehicles, computer vision algorithms are widely used in the control of autonomous drones for both recreational applications (Ackerman 2019) (Figure 11.26) and drone racing (Jung, Hwang *et al.* 2018; Kaufmann, Gehrig *et al.* 2019). A great talk describing Skydio's approach to visual autonomous navigation by Gareth Cross (2019) can be found in the ICRA 2019 Workshop on Algorithms and Architectures for Learning In-The-Loop Systems in Autonomous Flight[41] as well as Lecture 23 in Pieter Abbeel's (2019) class on Advanced Robotics, which has dozens of other interesting related lectures.

11.5.2 *Application*: Smartphone augmented reality

Another closely related application is *augmented reality*, where 3D objects are inserted into a video feed in real time, often to annotate or help users understand a scene (Azuma, Baillot *et al.* 2001; Feiner 2002; Billinghurst, Clark, and Lee 2015). While traditional systems require prior knowledge about the scene or object being visually tracked (Rosten and Drummond 2005), newer systems can simultaneously build up a model of the 3D environment and then track it so that graphics can be superimposed (Reitmayr and Drummond 2006; Wagner, Reitmayr *et al.* 2008).

Klein and Murray (2007) describe a *parallel tracking and mapping* (PTAM) system, which simultaneously applies full bundle adjustment to keyframes selected from a video stream, while performing robust real-time pose estimation on intermediate frames (Figure 11.27a). Once an initial 3D scene has been reconstructed, a dominant plane is estimated (in this case, the table-top) and 3D animated characters are virtually inserted. Klein and Murray (2008) extend this system to handle even faster camera motion by adding edge features, which can still be tracked even when interest points become too blurred. They also use a direct (intensity-based) rotation estimation algorithm for even faster motions.

Instead of modeling the whole scene as one rigid reference frame, Gordon and Lowe (2006) first build a 3D model of an individual object using feature matching and structure from motion. Once the system has been initialized, for every new frame they find the object and its pose using a 3D

[41] https://uav-learning-icra.github.io/2019

Figure 11.28 Smartphone augmented reality showing real-time depth occlusion effects (Valentin, Kowdle *et al.* 2018) © 2018 ACM.

instance recognition algorithm, and then superimpose a graphical object onto that model, as shown in Figure 11.27b.

While reliably tracking such objects and environments is now a well-solved problem, with frameworks such as ARKit,[42] ARCore,[43] and Spark AR[44] being widely used for mobile AR application development, determining which pixels should be occluded by foreground scene elements (Chuang, Agarwala *et al.* 2002; Wang and Cohen 2009) still remains an active research area.

One recent example of such work is the Smartphone AR system developed by Valentin, Kowdle *et al.* (2018) shown in Figure 11.28. The system proceeds by generating a semi-dense depth map by matching the current frame to a previous keyframe using a CRF followed by a filtering step. This map is then interpolated to full resolution using a novel planar bilateral solver, and the resulting depth map used for occlusion effects. As accurate per-pixel depth is such an essential component of augmented reality effects, we are likely to see rapid progress in this area, using both active and passive depth sensing technologies.

11.6 Additional reading

Camera calibration was first studied in photogrammetry (Brown 1971; Slama 1980; Atkinson 1996; Kraus 1997) but it has also been widely studied in computer vision (Tsai 1987; Gremban, Thorpe, and Kanade 1988; Champleboux, Lavallée *et al.* 1992b; Zhang 2000; Grossberg and Nayar 2001). Vanishing points observed either from rectahedral calibration objects or architecture are often used to perform rudimentary calibration (Caprile and Torre 1990; Becker and Bove 1995; Liebowitz and Zisserman 1998; Cipolla, Drummond, and Robertson 1999; Antone and Teller 2002; Criminisi, Reid, and Zisserman 2000; Hartley and Zisserman 2004; Pflugfelder 2008). Performing camera calibration without using known targets is known as *self-calibration* and is discussed in textbooks and surveys on structure from motion (Faugeras, Luong, and Maybank 1992; Hartley and Zisserman

[42] https://developer.apple.com/augmented-reality

[43] https://developers.google.com/ar

[44] https://sparkar.facebook.com/ar-studio

2004; Moons, Van Gool, and Vergauwen 2010). One popular subset of such techniques uses pure rotational motion (Stein 1995; Hartley 1997b; Hartley, Hayman *et al.* 2000; de Agapito, Hayman, and Reid 2001; Kang and Weiss 1999; Shum and Szeliski 2000; Frahm and Koch 2003).

The topic of registering 3D point datasets is called *absolute orientation* (Horn 1987) and *3D pose estimation* (Lorusso, Eggert, and Fisher 1995). A variety of techniques has been developed for simultaneously computing 3D point correspondences and their corresponding rigid transformations (Besl and McKay 1992; Zhang 1994; Szeliski and Lavallée 1996; Gold, Rangarajan *et al.* 1998; David, DeMenthon *et al.* 2004; Li and Hartley 2007; Enqvist, Josephson, and Kahl 2009). When only 2D observations are available, a variety of algorithms for the *linear PnP* (*perspective n-point*) have been developed (DeMenthon and Davis 1995; Quan and Lan 1999; Moreno-Noguer, Lepetit, and Fua 2007; Terzakis and Lourakis 2020). More recent approaches to pose estimation use deep networks (Arandjelovic, Gronat *et al.* 2016; Brachmann, Krull *et al.* 2017; Xiang, Schmidt *et al.* 2018, Oberweger, Rad, and Lepetit 2018; Hu, Hugonot *et al.* 2019; Peng, Liu *et al.* 2019). Estimating pose from RGB-D images is also very active (Drost, Ulrich *et al.* 2010; Brachmann, Michel *et al.* 2016; Labbé, Carpentier *et al.* 2020). In addition to recognizing object pose for robotics tasks, pose estimation is widely used in location recognition (Sattler, Zhou *et al.* 2019; Revaud, Weinzaepfel *et al.* 2019; Zhou, Sattler *et al.* 2019; Sarlin, DeTone *et al.* 2020; Luo, Zhou *et al.* 2020).

The topic of structure from motion is extensively covered in books and review articles on multi-view geometry (Faugeras and Luong 2001; Hartley and Zisserman 2004; Moons, Van Gool, and Vergauwen 2010) with survey of more recent developments in Özyeşil, Voroninski *et al.* (2017). For two-frame reconstruction, Hartley (1997a) wrote a highly cited paper on the "eight-point algorithm" for computing an essential or fundamental matrix with reasonable point normalization. When the cameras are calibrated, the five-point algorithm of Nistér (2004) can be used in conjunction with RANSAC to obtain initial reconstructions from the minimum number of points. When the cameras are uncalibrated, various self-calibration techniques can be found in work by Hartley and Zisserman (2004) and Moons, Van Gool, and Vergauwen (2010).

Triggs, McLauchlan *et al.* (1999) provide a good tutorial and survey on bundle adjustment, while Lourakis and Argyros (2009) and Engels, Stewénius, and Nistér (2006) provide tips on implementation and effective practices. Bundle adjustment is also covered in textbooks and surveys on multi-view geometry (Faugeras and Luong 2001; Hartley and Zisserman 2004; Moons, Van Gool, and Vergauwen 2010). Techniques for handling larger problems are described by Snavely, Seitz, and Szeliski (2008b), Agarwal, Snavely *et al.* (2009), Agarwal, Snavely *et al.* (2010), Jeong, Nistér *et al.* (2012), Wu (2013), Heinly, Schönberger *et al.* (2015), Schönberger and Frahm (2016), and Dellaert and Kaess (2017). While bundle adjustment is often called as an inner loop inside incremental reconstruction algorithms (Snavely, Seitz, and Szeliski 2006), hierarchical (Fitzgibbon and Zisserman 1998; Farenzena, Fusiello, and Gherardi 2009) and global (Rother and Carlsson 2002; Martinec and Pajdla 2007; Sinha, Steedly, and Szeliski 2010; Jiang, Cui, and Tan 2013; Moulon, Monasse, and Marlet 2013; Wilson and Snavely 2014; Cui and Tan 2015; Özyeşil and Singer 2015; Holynski, Geraghty *et al.* 2020) approaches for initialization are also possible and perhaps even preferable.

In the robotics community, techniques for reconstructing a 3D environment from a moving robot are called *simultaneous localization and mapping* (SLAM) (Thrun, Burgard, and Fox 2005; Durrant-Whyte and Bailey 2006; Bailey and Durrant-Whyte 2006; Fuentes-Pacheco, Ruiz-Ascencio, and Rendón-Mancha 2015; Cadena, Carlone *et al.* 2016). SLAM differs from bundle adjustment in that it allows for a variety of sensing devices and that it solves the localization problem *online*. This makes it the method of choice for both time-critical robotics applications such as autonomous navigation (Janai, Güney *et al.* 2020) and real-time augmented reality (Valentin, Kowdle *et al.* 2018). Important papers in this field include (Davison, Reid *et al.* 2007; Klein and Murray 2007, 2009; Newcombe,

Lovegrove, and Davison 2011; Kaess, Johannsson *et al.* 2012; Engel, Schöps, and Cremers 2014; Mur-Artal and Tardós 2017; Forster, Zhang *et al.* 2017; Dellaert and Kaess 2017; Engel, Koltun, and Cremers 2018; Schöps, Sattler, and Pollefeys 2019a) as well as papers that integrate SLAM with IMUs to obtain *visual inertial odometry* (VIO) (Mourikis and Roumeliotis 2007; Li and Mourikis 2013; Forster, Carlone *et al.* 2016; Schubert, Goll *et al.* 2018).

11.7 Exercises

Ex 11.1: Rotation-based calibration. Take an outdoor or indoor sequence from a rotating camera with very little parallax and use it to calibrate the focal length of your camera using the techniques described in Section 11.1.3 or Sections 8.2.3–8.3.1.

1. Take out any radial distortion in the images using one of the techniques from Exercises 11.5–11.6 or using parameters supplied for a given camera by your instructor.

2. Detect and match feature points across neighboring frames and chain them into feature tracks.

3. Compute homographies between overlapping frames and use Equations (11.8–11.9) to get an estimate of the focal length.

4. Compute a full 360° panorama and update your focal length estimate to close the gap (Section 8.2.4).

5. (Optional) Perform a complete bundle adjustment in the rotation matrices and focal length to obtain the highest quality estimate (Section 8.3.1).

Ex 11.2: Target-based calibration. Use a three-dimensional target to calibrate your camera.

1. Construct a three-dimensional calibration pattern with known 3D locations. It is not easy to get high accuracy unless you use a machine shop, but you can get close using heavy plywood and printed patterns.

2. Find the corners, e.g, using a line finder and intersecting the lines.

3. Implement one of the iterative calibration and pose estimation algorithms described in Tsai (1987), Bogart (1991), or Gleicher and Witkin (1992) or the system described in Section 11.2.2.

4. Take many pictures at different distances and orientations relative to the calibration target and report on both your re-projection errors and accuracy. (To do the latter, you may need to use simulated data.)

Ex 11.3: Calibration accuracy. Compare the three calibration techniques (plane-based, rotation-based, and 3D-target-based).

One approach is to have a different student implement each one and to compare the results. Another approach is to use synthetic data, potentially re-using the software you developed for Exercise 2.3. The advantage of using synthetic data is that you know the ground truth for the calibration and pose parameters, you can easily run lots of experiments, and you can synthetically vary the noise in your measurements.

Here are some possible guidelines for constructing your test sets:

1. Assume a medium-wide focal length (say, 50° field of view).

2. For the plane-based technique, generate a 2D grid target and project it at different inclinations.

3. For a 3D target, create an inner cube corner and position it so that it fills most of field of view.

4. For the rotation technique, scatter points uniformly on a sphere until you get a similar number of points as for other techniques.

Before comparing your techniques, predict which one will be the most accurate (normalize your results by the square root of the number of points used).

Add varying amounts of noise to your measurements and describe the noise sensitivity of your various techniques.

Ex 11.4: Single view metrology. Implement a system to measure dimensions and reconstruct a 3D model from a single image of an architectural scene using visible vanishing directions (Section 11.1.2) (Criminisi, Reid, and Zisserman 2000).

1. Find the three orthogonal vanishing points from parallel lines and use them to establish the three coordinate axes (rotation matrix $\mathbf{R}$ of the camera relative to the scene). If two of the vanishing points are finite (not at infinity), use them to compute the focal length, assuming a known image center. Otherwise, find some other way to calibrate your camera; you could use some of the techniques described by Schaffalitzky and Zisserman (2000).

2. Click on a ground plane point to establish your origin and click on a point a known distance away to establish the scene scale. This lets you compute the translation $\mathbf{t}$ between the camera and the scene. As an alternative, click on a pair of points, one on the ground plane and one above it, and use the known height to establish the scene scale.

3. Write a user interface that lets you click on ground plane points to recover their 3D locations. (Hint: you already know the camera matrix, so knowledge of a point's z value is sufficient to recover its 3D location.) Click on pairs of points (one on the ground plane, one above it) to measure vertical heights.

4. Extend your system to let you draw quadrilaterals in the scene that correspond to axis-aligned rectangles in the world, using some of the techniques described by Sinha, Steedly *et al.* (2008). Export your 3D rectangles to a VRML or PLY[45] file.

5. (Optional) Warp the pixels enclosed by the quadrilateral using the correct homography to produce a texture map for each planar polygon.

Ex 11.5: Radial distortion with plumb lines. Implement a plumb-line algorithm to determine the radial distortion parameters.

1. Take some images of scenes with lots of straight lines, e.g., hallways in your home or office, and try to get some of the lines as close to the edges of the image as possible.

2. Extract the edges and link them into curves, as described in Section 7.2.2 and Exercise 7.8.

3. Fit quadratic or elliptic curves to the linked edges using a generalization of the successive line approximation algorithm described in Section 7.4.1 and Exercise 7.11 and keep the curves that fit this form well.

[45] https://meshlab.net.

4. For each curved segment, fit a straight line and minimize the perpendicular distance between the curve and the line while adjusting the radial distortion parameters.

5. Alternate between re-fitting the straight line and adjusting the radial distortion parameters until convergence.

Ex 11.6: Radial distortion with a calibration target. Use a grid calibration target to determine the radial distortion parameters.

1. Print out a planar calibration target, mount it on a stiff board, and get it to fill your field of view.

2. Detect the squares, lines, or dots in your calibration target.

3. Estimate the homography mapping the target to the camera from the central portion of the image that does not have any radial distortion.

4. Predict the positions of the remaining targets and use the differences between the observed and predicted positions to estimate the radial distortion.

5. (Optional) Fit a general spline model (for severe distortion) instead of the quartic distortion model.

6. (Optional) Extend your technique to calibrate a fisheye lens.

Ex 11.7: Chromatic aberration. Use the radial distortion estimates for each color channel computed in the previous exercise to clean up wide-angle lens images by warping all of the channels into alignment. (Optional) Straighten out the images at the same time.

Can you think of any reasons why this warping strategy may not always work?

Ex 11.8: Triangulation. Use the calibration pattern you built and tested in Exercise 11.2 to test your triangulation accuracy. As an alternative, generate synthetic 3D points and cameras and add noise to the 2D point measurements.

1. Assume that you know the camera pose, i.e., the camera matrices. Use the 3D distance to rays (11.24) or linearized versions of Equations (11.25–11.26) to compute an initial set of 3D locations. Compare these to your known ground truth locations.

2. Use iterative non-linear minimization to improve your initial estimates and report on the improvement in accuracy.

3. (Optional) Use the technique described by Hartley and Sturm (1997) to perform two-frame triangulation.

4. See if any of the failure modes reported by Hartley and Sturm (1997) or Hartley (1998) occur in practice.

Ex 11.9: Essential and fundamental matrix. Implement the two-frame $\mathbf{E}$ and $\mathbf{F}$ matrix estimation techniques presented in Section 11.3, with suitable re-scaling for better noise immunity.

1. Use the data from Exercise 11.8 to validate your algorithms and to report on their accuracy.

2. (Optional) Implement one of the improved **F** or **E** estimation algorithms, e.g., using renormalization (Zhang 1998b; Torr and Fitzgibbon 2004; Hartley and Zisserman 2004), RANSAC (Torr and Murray 1997), least median of squares (LMS), or the five-point algorithm developed by Nistér (2004).

Ex 11.10: View morphing and interpolation. Implement automatic view morphing, i.e., compute two-frame structure from motion and then use these results to generate a smooth animation from one image to the next (Section 11.3.5).

1. Decide how to represent your 3D scene, e.g., compute a Delaunay triangulation of the matched point and decide what to do with the triangles near the border. (Hint: try fitting a plane to the scene, e.g., behind most of the points.)

2. Compute your in-between camera positions and orientations.

3. Warp each triangle to its new location, preferably using the correct perspective projection (Szeliski and Shum 1997).

4. (Optional) If you have a denser 3D model (e.g., from stereo), decide what to do at the "cracks".

5. (Optional) For a non-rigid scene, e.g., two pictures of a face with different expressions, not all of your matched points will obey the epipolar geometry. Decide how to handle them to achieve the best effect.

Ex 11.11: Bundle adjuster. Implement a full bundle adjuster. This may sound daunting, but it really is not.

1. Devise the internal data structures and external file representations to hold your camera parameters (position, orientation, and focal length), 3D point locations (Euclidean or homogeneous), and 2D point tracks (frame and point identifier as well as 2D locations).

2. Use some other technique, such as factorization, to initialize the 3D point and camera locations from your 2D tracks (e.g., a subset of points that appears in all frames).

3. Implement the code corresponding to the forward transformations in Figure 11.14, i.e., for each 2D point measurement, take the corresponding 3D point, map it through the camera transformations (including perspective projection and focal length scaling), and compare it to the 2D point measurement to get a residual error.

4. Take the residual error and compute its derivatives with respect to all the unknown motion and structure parameters, using backward chaining, as shown, e.g., in Figure 11.14 and Equation (11.19). This gives you the sparse Jacobian **J** used in Equations (8.13–8.17) and Equation (11.15).

5. Use a sparse least squares or linear system solver, e.g., MATLAB, SparseSuite, or SPARSKIT (see Appendix A.4 and A.5), to solve the corresponding linearized system, adding a small amount of diagonal preconditioning, as in Levenberg–Marquardt.

6. Update your parameters, make sure your rotation matrices are still orthonormal (e.g., by re-computing them from your quaternions), and continue iterating while monitoring your residual error.

7. (Optional) Use the "Schur complement trick" (11.68) to reduce the size of the system being solved (Triggs, McLauchlan *et al.* 1999; Hartley and Zisserman 2004; Lourakis and Argyros 2009; Engels, Stewénius, and Nistér 2006).

8. (Optional) Implement your own iterative sparse solver, e.g., conjugate gradient, and compare its performance to a direct method.

9. (Optional) Make your bundle adjuster robust to outliers, or try adding some of the other improvements discussed in (Engels, Stewénius, and Nistér 2006). Can you think of any other ways to make your algorithm even faster or more robust?

Ex 11.12: Match move and augmented reality. Use the results of the previous exercise to superimpose a rendered 3D model on top of video. See Section 11.4.4 for more details and ideas. Check for how "locked down" the objects are.

Ex 11.13: Line-based reconstruction. Augment the previously developed bundle adjuster to include lines, possibly with known 3D orientations.

Optionally, use co-planar sets of points and lines to hypothesize planes and to enforce coplanarity (Schaffalitzky and Zisserman 2002; Robertson and Cipolla 2002).

Ex 11.14: Flexible bundle adjuster. Design a bundle adjuster that allows for arbitrary chains of transformations and prior knowledge about the unknowns, as suggested in Figures 11.14–11.15.

Ex 11.15: Unordered image matching. Compute the camera pose and 3D structure of a scene from an arbitrary collection of photographs (Brown and Lowe 2005; Snavely, Seitz, and Szeliski 2006).

Ex 11.16: Augmented reality toolkits. Write a simple mobile AR app based on one of the widely used augmented reality frameworks such as ARKit or ARCore. What fun effects can you create? What are the conditions that make your AR system lose track? Can you move a large distance and come back to your original location without too much drift?

Chapter 12

Depth estimation

12.1 Epipolar geometry . . . 599
12.1.1 Rectification . . . 600
12.1.2 Plane sweep . . . 602
12.2 Sparse correspondence . . . 604
12.2.1 3D curves and profiles . . . 604
12.3 Dense correspondence . . . 606
12.3.1 Similarity measures . . . 607
12.4 Local methods . . . 609
12.4.1 Sub-pixel estimation and uncertainty . . . 610
12.4.2 *Application*: Stereo-based head tracking . . . 611
12.5 Global optimization . . . 612
12.5.1 Dynamic programming . . . 614
12.5.2 Segmentation-based techniques . . . 616
12.5.3 *Application*: Z-keying and background replacement . . . 617
12.6 Deep neural networks . . . 618
12.7 Multi-view stereo . . . 620
12.7.1 Scene flow . . . 624
12.7.2 Volumetric and 3D surface reconstruction . . . 624
12.7.3 Shape from silhouettes . . . 630
12.8 Monocular depth estimation . . . 632
12.9 Additional reading . . . 634
12.10 Exercises . . . 635

R. Szeliski, *Computer Vision*, Texts in Computer Science,
https://doi.org/10.1007/978-3-030-34372-9_12

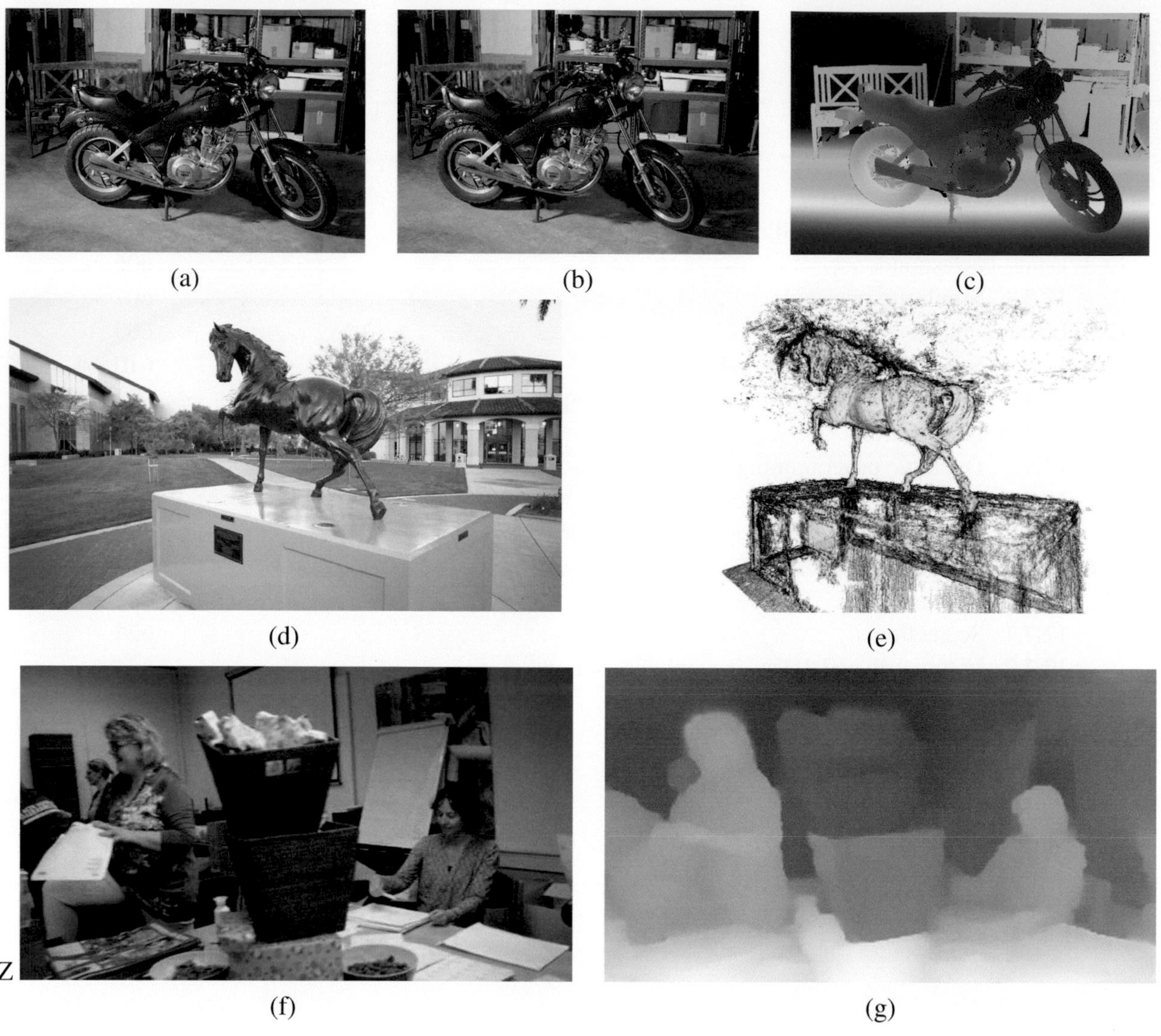

Figure 12.1 Depth estimation algorithms can convert a pair of color images (a–b) into a depth map (c) (Scharstein, Hirschmüller *et al.* 2014) © 2014 Springer, a sequence of images (d) into a 3D model (e) (Knapitsch, Park *et al.* 2017) © 2017 ACM, or a single image (f) into a depth map (g) (Li, Dekel *et al.* 2019) © 2019 IEEE.

Stereo matching is the process of taking two or more images and building a 3D model of the scene by finding matching pixels in the images and converting their 2D positions into 3D depths. In Chapter 11, we described techniques for recovering camera positions and building sparse 3D models of scenes or objects. In this chapter, we address the question of how to build a more complete 3D model, e.g., a sparse or dense *depth map* that assigns relative depths to pixels in the input images. We also look at the topic of *multi-view stereo* algorithms that produce complete 3D volumetric or surface-based object models, as well as *monocular* depth recovery algorithms that infer plausible depths from just a single image.

Why are people interested in depth estimation and stereo matching? From the earliest inquiries into visual perception, it was known that we perceive depth based on the differences in appearance between the left and right eye.[1] As a simple experiment, hold your finger vertically in front of your eyes and close each eye alternately. You will notice that the finger jumps left and right relative to the background of the scene. The same phenomenon is visible in the image pair shown in Figure 12.1a–b, in which the foreground objects shift left and right relative to the background.

As we will shortly see, under simple imaging configurations (both eyes or cameras looking straight ahead), the amount of horizontal motion or *disparity* is inversely proportional to the distance from the observer. While the basic physics and geometry relating visual disparity to scene structure are well understood (Section 12.1), automatically measuring this disparity by establishing dense and accurate inter-image *correspondences* is a challenging task.

The earliest stereo matching algorithms were developed in the field of *photogrammetry* for automatically constructing topographic elevation maps from overlapping aerial images. Prior to this, operators would use photogrammetric stereo plotters, which displayed shifted versions of such images to each eye and allowed the operator to float a dot cursor around constant elevation contours. The development of fully automated stereo matching algorithms was a major advance in this field, enabling much more rapid and less expensive processing of aerial imagery (Hannah 1974; Hsieh, McKeown, and Perlant 1992).

In computer vision, the topic of stereo matching has been one of the most widely studied and fundamental problems (Marr and Poggio 1976; Barnard and Fischler 1982; Dhond and Aggarwal 1989; Scharstein and Szeliski 2002; Brown, Burschka, and Hager 2003; Seitz, Curless *et al.* 2006), and continues to be one of the most active research areas (Poggi, Tosi *et al.* 2021). While photogrammetric matching concentrated mainly on aerial imagery, computer vision applications include modeling the human visual system (Marr 1982), robotic navigation and manipulation (Moravec 1983; Konolige 1997; Thrun, Montemerlo *et al.* 2006; Janai, Güney *et al.* 2020) and Figures 12.2j and 11.26, as well as view interpolation and image-based rendering (Figure 12.2a–d), 3D model building (Figure 12.2e–f and h–i), mixing live action with computer-generated imagery (Figure 12.2g), and augmented reality (Valentin, Kowdle *et al.* 2018; Chaurasia, Nieuwoudt *et al.* 2020) and Figure 11.28.

In this chapter, we describe the fundamental principles behind stereo matching, following the general taxonomy proposed by Scharstein and Szeliski (2002). We begin in Section 12.1 with a review of the *geometry* of stereo image matching, i.e., how to compute for a given pixel in one image the range of possible locations the pixel might appear at in the other image, i.e., its *epipolar line*. We describe how to pre-warp images so that corresponding epipolar lines are coincident (*rectification*). We also describe a general resampling algorithm called *plane sweep* that can be used to perform multi-image stereo matching with arbitrary camera configurations.

Next, we briefly survey techniques for the *sparse* stereo matching of interest points and edge-like features (Section 12.2). We then turn to the main topic of this chapter, namely the estimation

[1]The word *stereo* comes from the Greek for *solid*; stereo vision is how we perceive solid shape (Koenderink 1990).

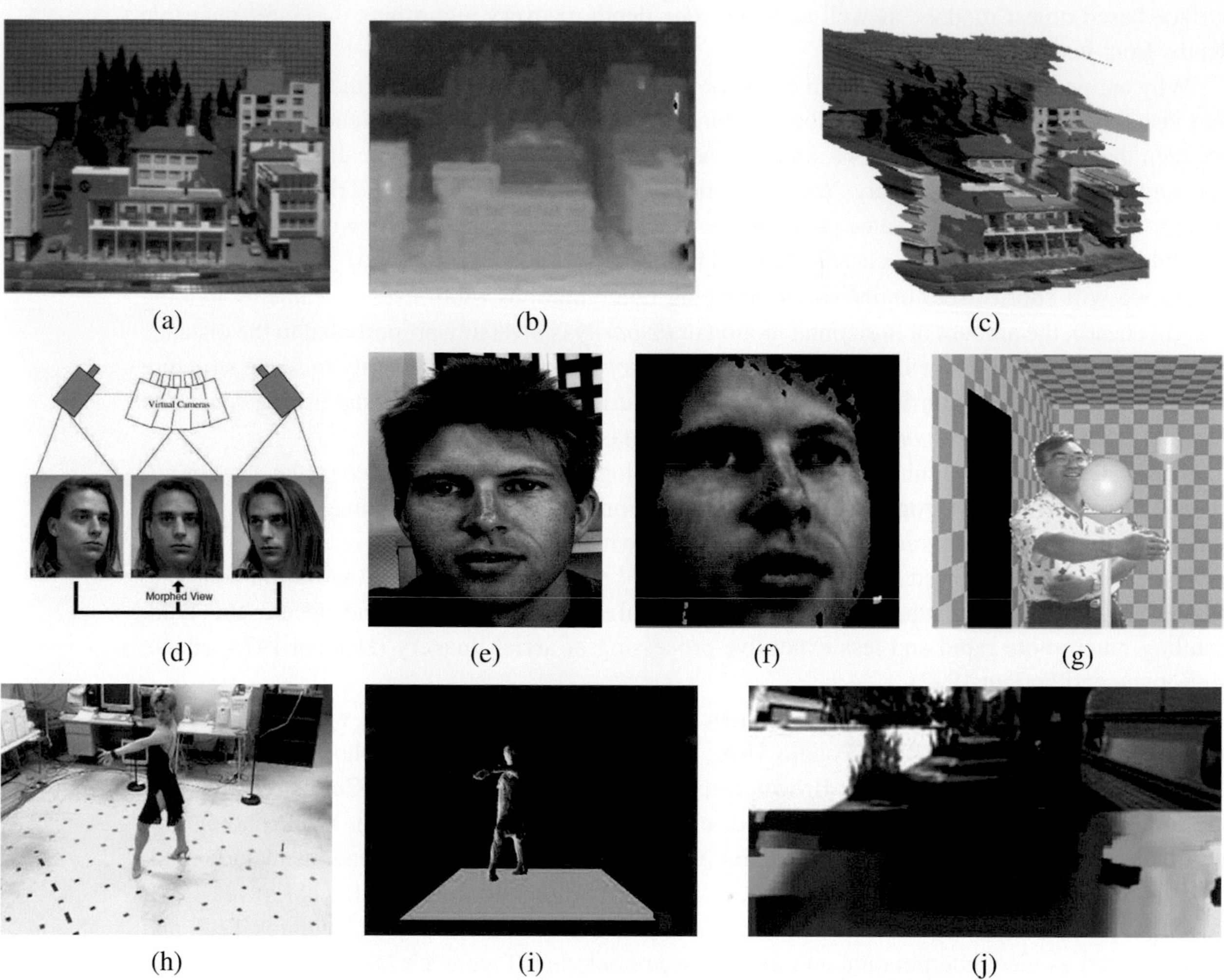

Figure 12.2 Applications of stereo vision: (a) input image, (b) computed depth map, and (c) new view generation from multi-view stereo (Matthies, Kanade, and Szeliski 1989) © 1989 Springer; (d) view morphing between two images (Seitz and Dyer 1996) © 1996 ACM; (e–f) 3D face modeling (images courtesy of Frédéric Devernay); (g) *z-keying* live and computer-generated imagery (Kanade, Yoshida *et al.* 1996) © 1996 IEEE; (h–i) building 3D surface models from multiple video streams in Virtualized Reality (Kanade, Rander, and Narayanan 1997) © 1997 IEEE; (j) computing depth maps for autonomous navigation (Geiger, Lenz, and Urtasun 2012) © 2012 IEEE.

Name/URL	Contents/Reference
Middlebury stereo https://vision.middlebury.edu/stereo	33 high-resolution stereo pairs (Scharstein, Hirschmüller *et al.* 2014)
Middlebury multi-view https://vision.middlebury.edu/mview	6 3D objects scanned from 300+ views (Seitz, Curless *et al.* 2006)
EPFL (no longer active)	6 outdoor multi-view sets of images (Strecha, von Hansen *et al.* 2008)
KITTI 2015 http://www.cvlibs.net/datasets/kitti/eval_stereo_flow.php	200 train + 200 test stereo pairs (Menze and Geiger 2015)
DTU https://roboimagedata.compute.dtu.dk/?page_id=36	124 toy scenes with 49–64 images each (Jensen, Dahl *et al.* 2014)
Freiburg Scene Flow https://lmb.informatik.uni-freiburg.de/resources/datasets	39k synthetic stereo pairs (Mayer, Ilg *et al.* 2018)
ETH3D https://www.eth3d.net	13 training + 12 test high-res scenes (Schöps, Schönberger *et al.* 2017)
Tanks and Temples https://www.tanksandtemples.org	7 training + 14 test 4K video scenes (Knapitsch, Park *et al.* 2017)
BlendedMVS https://github.com/YoYo000/BlendedMVS	17k MVS images covering 113 scenes (Yao, Luo *et al.* 2020)

Table 12.1 Widely used stereo datasets and benchmarks.

of a *dense* set of pixel-wise correspondences in the form of a *disparity map* (Figure 12.1c). This involves first selecting a pixel matching criterion (Section 12.3) and then using either local area-based aggregation (Section 12.4), global optimization (Section 12.5), or deep networks (Section 12.6), to help disambiguate potential matches. In Section 12.7, we discuss *multi-view stereo* that use more than pairs of images in order to produce higher-quality depth maps or complete 3D object or scene models (Figure 12.1d–e). Finally, in Section 12.8 we present algorithms for inferring depth from just a single image, which has now become possible using machine learning and deep networks.

Throughout this chapter, we will often refer to datasets and benchmarks that have been used to develop depth inference algorithms and gauge their performance. Of these, the most widely used and influential include the Middlebury stereo and multi-view datasets benchmarks, which were among the first to keep up-to-date leaderboards, the EPFL multi-view dataset, the KITTI benchmarks for autonomous driving (stereo, flow, scene flow, and others), the DTU dataset, ETH3D benchmark, Tanks and Temples benchmark, and BlendedMVS dataset, which are all summarized in Table 12.1. Pointers to additional datasets can be found in Mayer, Ilg *et al.* (2018), Janai, Güney *et al.* (2020), Laga, Jospin *et al.* (2020), and Poggi, Tosi *et al.* (2021).

12.1 Epipolar geometry

Given a pixel in one image, how can we compute its correspondence in the other image? In Chapter 9, we saw that a variety of search techniques can be used to match pixels based on their local appearance as well as the motions of neighboring pixels. In the case of stereo matching, however, we have some additional information available, namely the positions and calibration data for the cameras that took the pictures of the same static scene (Section 11.3).

How can we exploit this information to reduce the number of potential correspondences, and hence both speed up the matching and increase its reliability? Figure 12.3a shows how a pixel in

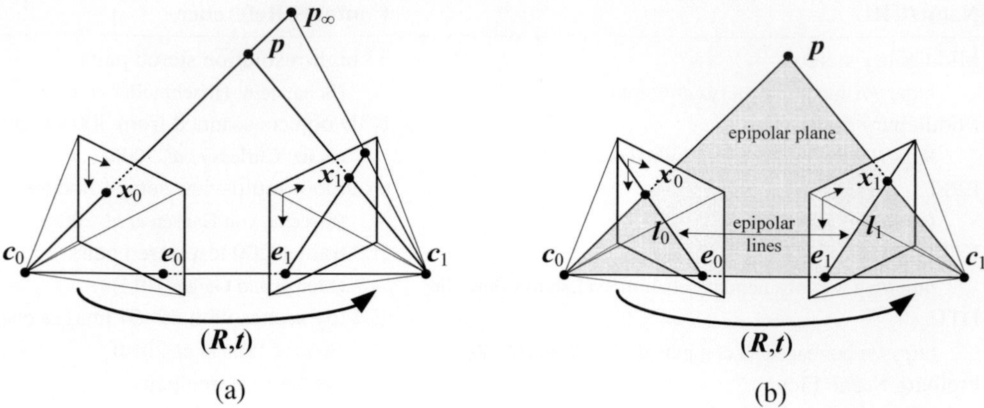

Figure 12.3 Epipolar geometry: (a) epipolar line segment corresponding to one ray; (b) corresponding set of epipolar lines and their epipolar plane.

one image $\mathbf{x}_0$ projects to an *epipolar line segment* in the other image. The segment is bounded at one end by the projection of the original viewing ray at infinity $\mathbf{p}_\infty$ and at the other end by the projection of the original camera center $\mathbf{c}_0$ into the second camera, which is known as the *epipole* $\mathbf{e}_1$. If we project the epipolar line in the second image back into the first, we get another line (segment), this time bounded by the other corresponding epipole $\mathbf{e}_0$. Extending both line segments to infinity, we get a pair of corresponding *epipolar lines* (Figure 12.3b), which are the intersection of the two image planes with the *epipolar plane* that passes through both camera centers $\mathbf{c}_0$ and $\mathbf{c}_1$ as well as the point of interest $\mathbf{p}$ (Faugeras and Luong 2001; Hartley and Zisserman 2004).

12.1.1 Rectification

As we saw in Section 11.3, the epipolar geometry for a pair of cameras is implicit in the relative pose and calibrations of the cameras, and can easily be computed from seven or more point matches using the fundamental matrix (or five or more points for the calibrated essential matrix) (Zhang 1998a,b; Faugeras and Luong 2001; Hartley and Zisserman 2004). Once this geometry has been computed, we can use the epipolar line corresponding to a pixel in one image to constrain the search for corresponding pixels in the other image. One way to do this is to use a general correspondence algorithm, such as optical flow (Section 9.3), but to only consider locations along the epipolar line (or to project any flow vectors that fall off back onto the line).

A more efficient algorithm can be obtained by first *rectifying* (i.e., warping) the input images so that corresponding horizontal scanlines are epipolar lines (Loop and Zhang 1999; Faugeras and Luong 2001; Hartley and Zisserman 2004).[2] Afterwards, it is possible to match horizontal scanlines independently or to shift images horizontally while computing matching scores (Figure 12.4).

A simple way to rectify the two images is to first rotate both cameras so that they are looking perpendicular to the line joining the camera centers $\mathbf{c}_0$ and $\mathbf{c}_1$. As there is a degree of freedom in the *tilt*, the smallest rotations that achieve this should be used. Next, to determine the desired twist around the optical axes, make the *up vector* (the camera y-axis) perpendicular to the camera center line. This ensures that corresponding epipolar lines are horizontal and that the disparity for points

[2]This makes most sense if the cameras are next to each other, although by rotating the cameras, rectification can be performed on any pair that is not *verged* too much or has too much of a scale change. In those latter cases, using plane sweep (below) or hypothesizing small planar patch locations in 3D (Goesele, Snavely *et al.* 2007) may be preferable.

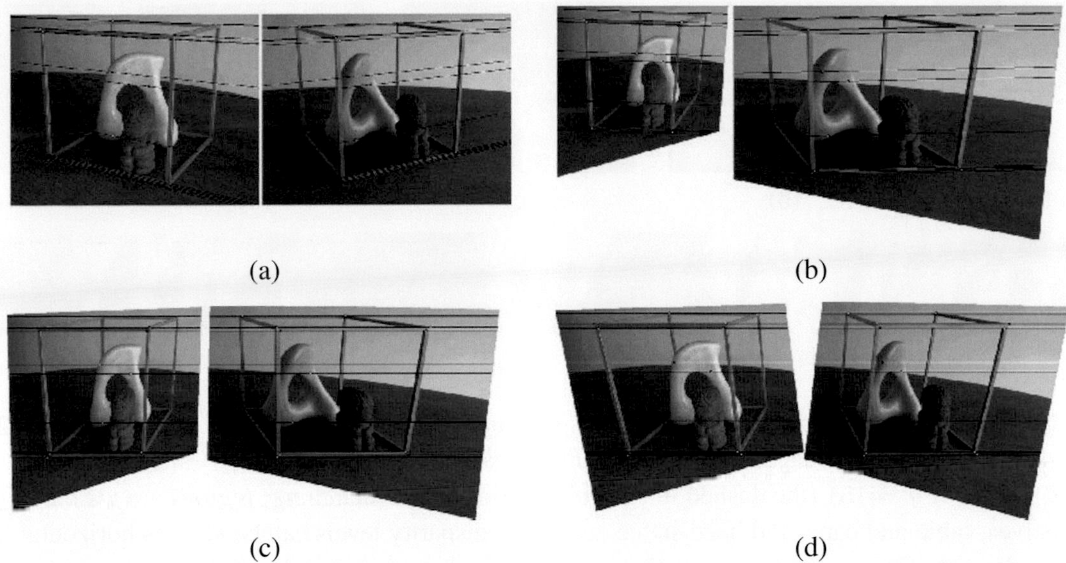

Figure 12.4 The multi-stage stereo rectification algorithm of Loop and Zhang (1999) © 1999 IEEE. (a) Original image pair overlaid with several epipolar lines; (b) images transformed so that epipolar lines are parallel; (c) images rectified so that epipolar lines are horizontal and in vertical correspondence; (d) final rectification that minimizes horizontal distortions.

at infinity is 0. Finally, re-scale the images, if necessary, to account for different focal lengths, magnifying the smaller image to avoid aliasing. (The full details of this procedure can be found in Fusiello, Trucco, and Verri (2000) and Exercise 12.1.) When additional information about the imaging process is available, e.g., that the images were formed on co-planar photographic plates, more specialized and accurate algorithms can be developed (Luo, Kong *et al.* 2020). Note that in general, it is not possible to rectify an arbitrary collection of images simultaneously unless their optical centers are collinear, although rotating the cameras so that they all point in the same direction reduces the inter-camera pixel movements to scalings and translations.

The resulting *standard rectified geometry* is employed in a lot of stereo camera setups and stereo algorithms, and leads to a very simple inverse relationship between 3D depths Z and disparities d,

$$d = f\frac{B}{Z}, \tag{12.1}$$

where f is the focal length (measured in pixels), B is the baseline, and

$$x' = x + d(x, y), \quad y' = y \tag{12.2}$$

describes the relationship between corresponding pixel coordinates in the left and right images (Bolles, Baker, and Marimont 1987; Okutomi and Kanade 1993; Scharstein and Szeliski 2002).[3] The task of extracting depth from a set of images then becomes one of estimating the *disparity map* $d(x, y)$.

After rectification, we can easily compare the similarity of pixels at corresponding locations (x, y) and $(x', y') = (x + d, y)$ and store them in a *disparity space image* (DSI) $C(x, y, d)$ for

[3] The term *disparity* was first introduced in the human vision literature to describe the difference in location of corresponding features seen by the left and right eyes (Marr 1982). Horizontal disparity is the most commonly studied phenomenon, but vertical disparity is possible if the eyes are verged.

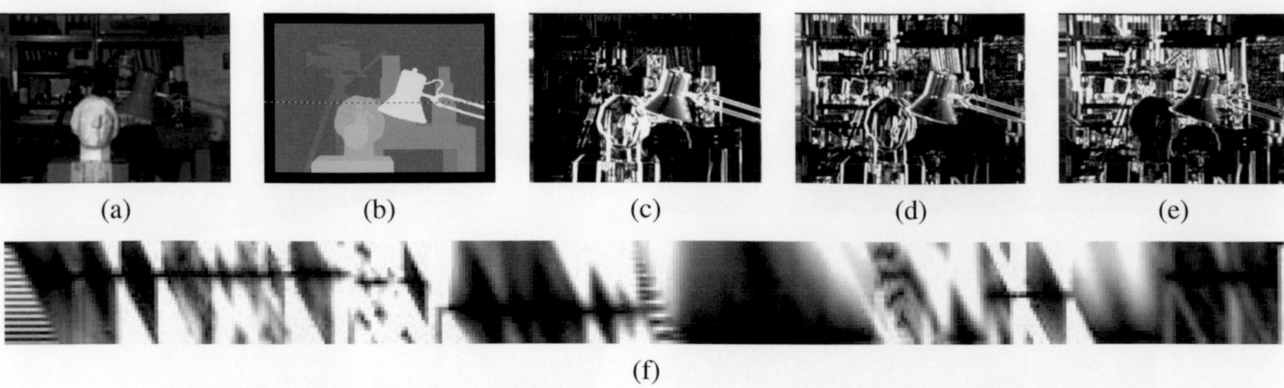

(a) (b) (c) (d) (e)

(f)

Figure 12.5 Slices through a typical disparity space image (DSI) (Scharstein and Szeliski 2002) © 2002 Springer: (a) original color image; (b) ground truth disparities; (c–e) three (x, y) slices for $d = 10, 16, 21$; (f) an (x, d) slice for $y = 151$ (the dashed line in (b)). Various dark (matching) regions are visible in (c–e), e.g., the bookshelves, table and cans, and head statue, and three disparity levels can be seen as horizontal lines in (f). The dark bands in the DSIs indicate regions that match at this disparity. (Smaller dark regions are often the result of textureless regions.) Additional examples of DSIs are discussed by Bobick and Intille (1999).

further processing (Figure 12.5). The concept of the disparity space (x, y, d) dates back to early work in stereo matching (Marr and Poggio 1976), while the concept of a disparity space image (volume) is generally associated with Yang, Yuille, and Lu (1993) and Intille and Bobick (1994).

12.1.2 Plane sweep

An alternative to pre-rectifying the images before matching is to sweep a set of planes through the scene and to measure the *photoconsistency* of different images as they are re-projected onto these planes (Figure 12.6). This process is commonly known as the *plane sweep* algorithm (Collins 1996; Szeliski and Golland 1999; Saito and Kanade 1999).

As we saw in Section 2.1.4, where we introduced projective depth (also known as *plane plus parallax* (Kumar, Anandan, and Hanna 1994; Sawhney 1994; Szeliski and Coughlan 1997)), the last row of a full-rank 4×4 projection matrix $\tilde{\mathbf{P}}$ can be set to an arbitrary plane equation $\mathbf{p}_3 = s_3[\hat{\mathbf{n}}_0|c_0]$. The resulting four-dimensional projective transform (*collineation*) (2.68) maps 3D world points $\mathbf{p} = (X, Y, Z, 1)$ into screen coordinates $\mathbf{x}_s = (x_s, y_s, 1, d)$, where the *projective depth* (or *parallax*) d (2.66) is 0 on the reference plane (Figure 2.11).

Sweeping d through a series of disparity hypotheses, as shown in Figure 12.6a, corresponds to mapping each input image into the *virtual camera* $\tilde{\mathbf{P}}$ defining the disparity space through a series of homographies (2.68–2.71),

$$\tilde{\mathbf{x}}_k \sim \tilde{\mathbf{P}}_k \tilde{\mathbf{P}}^{-1} \mathbf{x}_s = \tilde{\mathbf{H}}_k \tilde{\mathbf{x}} + \mathbf{t}_k d = (\tilde{\mathbf{H}}_k + \mathbf{t}_k [0\ 0\ d]) \tilde{\mathbf{x}}, \tag{12.3}$$

as shown in Figure 2.12b, where $\tilde{\mathbf{x}}_k$ and $\tilde{\mathbf{x}}$ are the homogeneous pixel coordinates in the source and virtual (reference) images (Szeliski and Golland 1999). The members of the family of homographies $\tilde{\mathbf{H}}_k(d) = \tilde{\mathbf{H}}_k + \mathbf{t}_k[0\ 0\ d]$, which are parameterized by the addition of a rank-1 matrix, are related to each other through a *planar homology* (Hartley and Zisserman 2004, A5.2).

The choice of virtual camera and parameterization is application dependent and is what gives this framework a lot of its flexibility. In many applications, one of the input cameras (the *reference* camera) is used, thus computing a depth map that is registered with one of the input images and

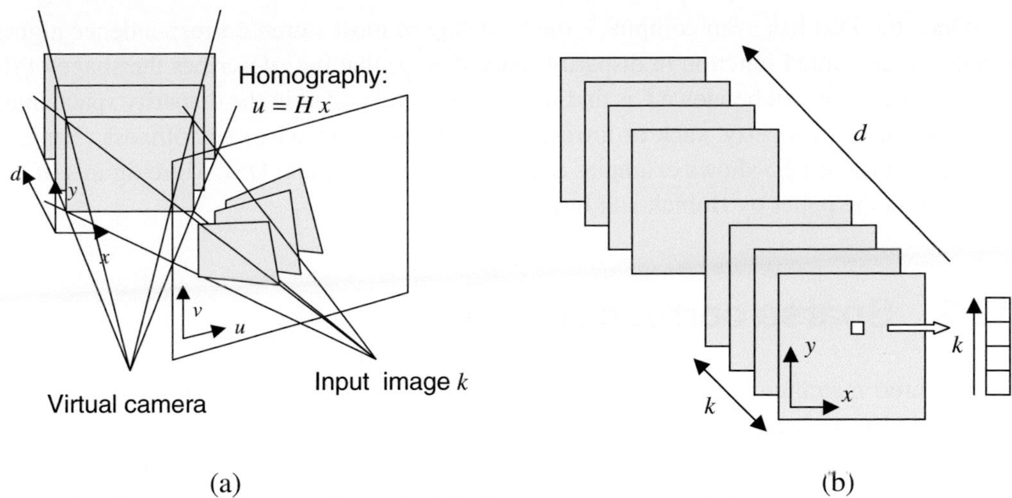

Figure 12.6 Sweeping a set of planes through a scene (Szeliski and Golland 1999) © 1999 Springer: (a) The set of planes seen from a virtual camera induces a set of homographies in any other source (input) camera image. (b) The warped images from all the other cameras can be stacked into a generalized disparity space volume $\tilde{I}(x, y, d, k)$ indexed by pixel location (x, y), disparity d, and camera k.

which can later be used for image-based rendering (Sections 14.1 and 14.2). In other applications, such as view interpolation for gaze correction in video-conferencing (Section 12.4.2) (Ott, Lewis, and Cox 1993; Criminisi, Shotton *et al.* 2003), a camera centrally located between the two input cameras is preferable, because it provides the needed per-pixel disparities to hallucinate the virtual middle image.

The choice of disparity sampling, i.e., the setting of the zero parallax plane and the scaling of integer disparities, is also application dependent, and is usually set to bracket the range of interest, i.e., the *working volume*, while scaling disparities to sample the image in pixel (or sub-pixel) shifts. For example, when using stereo vision for obstacle avoidance in robot navigation, it is most convenient to set up disparity to measure per-pixel elevation above the ground (Ivanchenko, Shen, and Coughlan 2009).

As each input image is warped onto the current planes parameterized by disparity d, it can be stacked into a *generalized disparity space image* $\tilde{I}(x, y, d, k)$ for further processing (Figure 12.6b) (Szeliski and Golland 1999). In most stereo algorithms, the photoconsistency (e.g., sum of squared or robust differences) with respect to the reference image I_r is calculated and stored in the DSI

$$C(x, y, d) = \sum_k \rho(\tilde{I}(x, y, d, k) - I_r(x, y)). \tag{12.4}$$

However, it is also possible to compute alternative statistics such as robust variance, focus, or entropy (Section 12.3.1) (Vaish, Szeliski *et al.* 2006) or to use this representation to reason about occlusions (Szeliski and Golland 1999; Kang and Szeliski 2004). The generalized DSI will come in particularly handy when we come back to the topic of multi-view stereo in Section 12.7.2.

Of course, planes are not the only surfaces that can be used to define a 3D sweep through the space of interest. Cylindrical surfaces, especially when coupled with panoramic photography (Section 8.2), are often used (Ishiguro, Yamamoto, and Tsuji 1992; Kang and Szeliski 1997; Shum and Szeliski 1999; Li, Shum *et al.* 2004; Zheng, Kang *et al.* 2007). It is also possible to define other manifold topologies, e.g., ones where the camera rotates around a fixed axis (Seitz 2001).

Once the DSI has been computed, the next step in most stereo correspondence algorithms is to produce a univalued function in disparity space $d(x, y)$ that best describes the shape of the surfaces in the scene. This can be viewed as finding a surface embedded in the disparity space image that has some optimality property, such as lowest cost and best (piecewise) smoothness (Yang, Yuille, and Lu 1993). Figure 12.5 shows examples of slices through a typical DSI. More figures of this kind can be found in the paper by Bobick and Intille (1999).

12.2 Sparse correspondence

Early stereo matching algorithms were *feature-based*, i.e., they first extracted a set of potentially matchable image locations, using either interest operators or edge detectors, and then searched for corresponding locations in other images using a patch-based metric (Hannah 1974; Marr and Poggio 1979; Mayhew and Frisby 1980; Baker and Binford 1981; Arnold 1983; Grimson 1985; Ohta and Kanade 1985; Bolles, Baker, and Marimont 1987; Matthies, Kanade, and Szeliski 1989; Hsieh, McKeown, and Perlant 1992; Bolles, Baker, and Hannah 1993). This limitation to sparse correspondences was partially due to computational resource limitations, but was also driven by a desire to limit the answers produced by stereo algorithms to matches with high certainty. In some applications, there was also a desire to match scenes with potentially very different illuminations, where edges might be the only stable features (Collins 1996). Such sparse 3D reconstructions could later be interpolated using surface fitting algorithms such as those discussed in Sections 4.2 and 13.3.1.

More recent work in this area has focused on first extracting highly reliable features and then using these as *seeds* to grow additional matches (Zhang and Shan 2000; Lhuillier and Quan 2002; Čech and Šára 2007) or as inputs to a dense per-pixel depth solver (Valentin, Kowdle *et al.* 2018). Similar approaches have also been extended to wide baseline multi-view stereo problems and combined with 3D surface reconstruction (Lhuillier and Quan 2005; Strecha, Tuytelaars, and Van Gool 2003; Goesele, Snavely *et al.* 2007) or free-space reasoning (Taylor 2003), as described in more detail in Section 12.7.

12.2.1 3D curves and profiles

Another example of sparse correspondence is the matching of *profile curves* (or *occluding contours*), which occur at the boundaries of objects (Figure 12.7) and at interior self occlusions, where the surface curves away from the camera viewpoint.

The difficulty in matching profile curves is that in general, the locations of profile curves vary as a function of camera viewpoint. Therefore, matching curves directly in two images and then triangulating these matches can lead to erroneous shape measurements. Fortunately, if three or more closely spaced frames are available, it is possible to fit a local circular arc to the locations of corresponding edgels (Figure 12.7a) and therefore obtain semi-dense curved surface meshes directly from the matches (Figures 12.7c and g). Another advantage of matching such curves is that they can be used to reconstruct surface shape for untextured surfaces, so long as there is a visible difference between foreground and background colors.

Over the years, a number of different techniques have been developed for reconstructing surface shape from profile curves (Giblin and Weiss 1987; Cipolla and Blake 1992; Vaillant and Faugeras 1992; Zheng 1994; Boyer and Berger 1997; Szeliski and Weiss 1998). Cipolla and Giblin (2000) describe many of these techniques, as well as related topics such as inferring camera motion from profile curve sequences. Below, we summarize the approach developed by Szeliski and Weiss (1998),

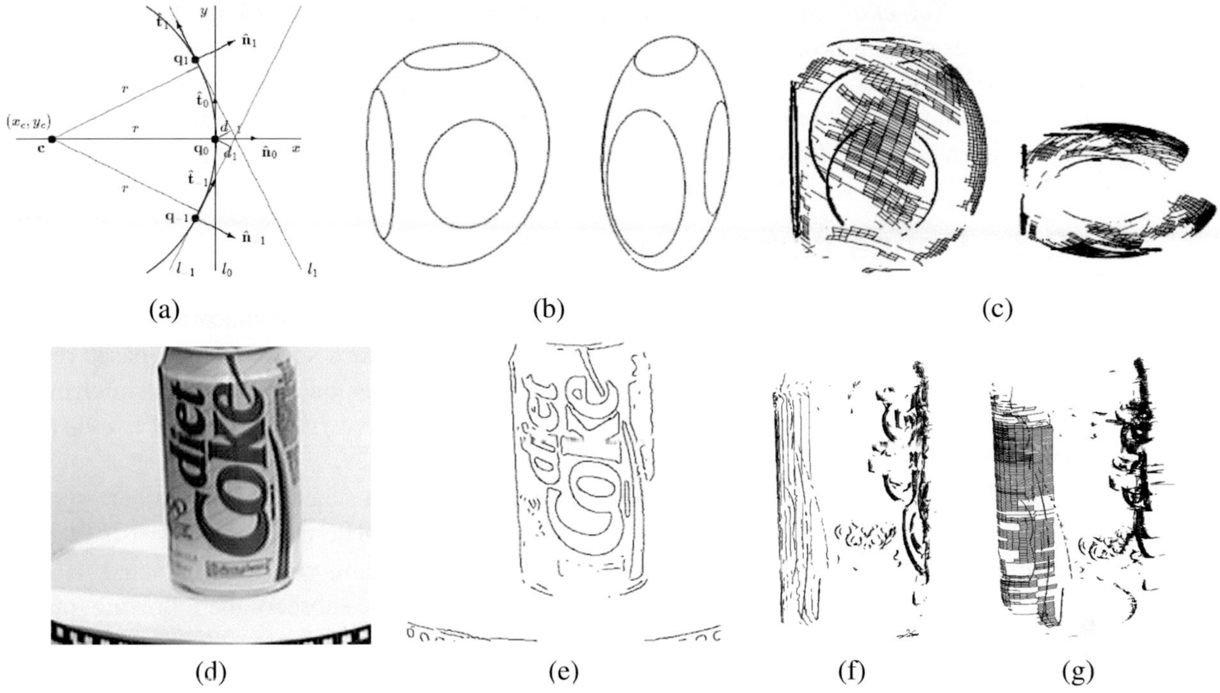

Figure 12.7 Surface reconstruction from occluding contours (Szeliski and Weiss 1998) © 2002 Springer: (a) circular arc fitting in the epipolar plane; (b) synthetic example of an ellipsoid with a truncated side and elliptic surface markings; (c) partially reconstructed surface mesh seen from an oblique and top-down view; (d) real-world image sequence of a soda can on a turntable; (e) extracted edges; (f) partially reconstructed profile curves; (g) partially reconstructed surface mesh. (Partial reconstructions are shown so as not to clutter the images.)

which assumes a discrete set of images, rather than formulating the problem in a continuous differential framework.

Let us assume that the camera is moving smoothly enough that the local epipolar geometry varies slowly, i.e., the epipolar planes induced by the successive camera centers and an edgel under consideration are nearly co-planar. The first step in the processing pipeline is to extract and link edges in each of the input images (Figures 12.7b and e). Next, edgels in successive images are matched using pairwise epipolar geometry, proximity and (optionally) appearance. This provides a linked set of edges in the spatio-temporal volume, which is sometimes called the *weaving wall* (Baker 1989).

To reconstruct the 3D location of an individual edgel, along with its local in-plane normal and curvature, we project the viewing rays corresponding to its neighbors onto the instantaneous epipolar plane defined by the camera center, the viewing ray, and the camera velocity, as shown in Figure 12.7a. We then fit an *osculating circle* to the projected lines, from which we can compute a 3D point position (Szeliski and Weiss 1998).

The resulting set of 3D points, along with their spatial (in-image) and temporal (between-image) neighbors, form a 3D surface mesh with local normal and curvature estimates (Figures 12.7c and g). Note that whenever a curve is due to a surface marking or a sharp crease edge, rather than a smooth surface profile curve, this shows up as a 0 or small radius of curvature. Such curves result in isolated 3D space curves, rather than elements of smooth surface meshes, but can still be incorporated into the 3D surface model during a later stage of surface interpolation (Section 13.3.1).

More recent examples of 3D curve reconstruction from sequences of RGB and RGB-D images include (Li, Yao *et al.* 2018; Liu, Chen *et al.* 2018; Wang, Liu *et al.* 2020), the latest of which can even recover camera pose with untextured backgrounds. When the thin structures being modeled are planar manifolds, such as leaves or paper, as opposed to true 3D curves such as wires, specially tailored mesh representations may be more appropriate (Kim, Zimmer *et al.* 2013; Yücer, Kim *et al.* 2016; Yücer, Sorkine-Hornung *et al.* 2016), as discussed in more detail in Sections 12.7.2 and 14.3.

12.3 Dense correspondence

While sparse matching algorithms are still occasionally used, most stereo matching algorithms today focus on dense correspondence, as this is required for applications such as image-based rendering or modeling. This problem is more challenging than sparse correspondence, because inferring depth values in textureless regions requires a certain amount of guesswork. (Think of a solid colored background seen through a picket fence. What depth should it be?)

In this section, we review the taxonomy and categorization scheme for dense correspondence algorithms first proposed by Scharstein and Szeliski (2002). The taxonomy consists of a set of algorithmic "building blocks" from which a large set of algorithms can be constructed. It is based on the observation that stereo algorithms generally perform some subset of the following four steps:

1. matching cost computation;
2. cost (support) aggregation;
3. disparity computation and optimization; and
4. disparity refinement.

For example, *local* (window-based) algorithms (Section 12.4), where the disparity computation at a given point depends only on intensity values within a finite window, usually make implicit smoothness assumptions by aggregating support. Some of these algorithms can cleanly be broken down into steps 1, 2, 3. For example, the traditional sum-of-squared-differences (SSD) algorithm can be described as:

1. The matching cost is the squared difference of intensity values at a given disparity.
2. Aggregation is done by summing the matching cost over square windows with constant disparity.
3. Disparities are computed by selecting the minimal (winning) aggregated value at each pixel.

Some local algorithms, however, combine steps 1 and 2 and use a matching cost that is based on a support region, e.g., normalized cross-correlation (Hannah 1974; Bolles, Baker, and Hannah 1993) and the rank transform (Zabih and Woodfill 1994) and other ordinal measures (Bhat and Nayar 1998). (This can also be viewed as a preprocessing step; see Section 12.3.1.)

Global algorithms, on the other hand, make explicit smoothness assumptions and then solve a global optimization problem (Section 12.5). Such algorithms typically do not perform an aggregation step, but rather seek a disparity assignment (step 3) that minimizes a global cost function that consists of data (step 1) terms and smoothness terms. The main distinction among these algorithms is the minimization procedure used, e.g., simulated annealing (Marroquin, Mitter, and Poggio 1987; Barnard 1989), probabilistic (mean-field) diffusion (Scharstein and Szeliski 1998), expectation maximization (EM) (Birchfield, Natarajan, and Tomasi 2007), graph cuts (Boykov, Veksler, and Zabih 2001), or loopy belief propagation (Sun, Zheng, and Shum 2003), to name just a few.

In between these two broad classes are certain iterative algorithms that do not explicitly specify a global function to be minimized, but whose behavior mimics closely that of iterative optimization algorithms (Marr and Poggio 1976; Zitnick and Kanade 2000). Hierarchical (coarse-to-fine) algorithms resemble such iterative algorithms, but typically operate on an image pyramid where results from coarser levels are used to constrain a more local search at finer levels (Witkin, Terzopoulos, and Kass 1987; Quam 1984; Bergen, Anandan *et al.* 1992). Also situated between local and global methods is *semi-global-matching* (SGM) (Hirschmüller 2008), which approximates minimizing a 2D cost function via 1D optimization (see Section 12.5.1), as well as methods that avoid exploring the whole search space, e.g., PatchMatch stereo (Bleyer, Rhemann, and Rother 2011) and local plane sweeps (LPS) (Sinha, Scharstein, and Szeliski 2014). A large number of neural network algorithms have also been developed for stereo matching, which we review in Section 12.6.

While most stereo matching algorithms produce a single disparity map with respect to a reference input image, or a path through the disparity space that encodes a continuous surface (Figure 12.13), a few algorithms compute fractional opacity values along with depths and colors for each pixel (Szeliski and Golland 1999; Zhou, Tucker *et al.* 2018; Flynn, Broxton *et al.* 2019). As these are closely related to volumetric reconstruction techniques, we discuss them in Section 12.7.2 as well as Section 14.2.1 on image-based rendering with layers.

12.3.1 Similarity measures

The first component of any dense stereo matching algorithm is a similarity measure that compares pixel values in order to determine how likely they are to be in correspondence. In this section, we briefly review the similarity measures introduced in Section 9.1 and mention a few others that have been developed specifically for stereo matching (Scharstein and Szeliski 2002; Hirschmüller and Scharstein 2009).

The most common pixel-based matching costs include sums of *squared intensity differences* (SSD) (Hannah 1974) and *absolute intensity differences* (SAD) (Kanade 1994). In the video processing community, these matching criteria are referred to as the *mean-squared error* (MSE) and *mean absolute difference* (MAD) measures; the term *displaced frame difference* is also often used (Tekalp 1995).

More recently, robust measures (9.2), including truncated quadratics and contaminated Gaussians, have been proposed (Black and Anandan 1996; Black and Rangarajan 1996; Scharstein and Szeliski 1998; Barron 2019). These measures are useful because they limit the influence of mismatches during aggregation. Vaish, Szeliski *et al.* (2006) compare a number of such robust measures, including a new one based on the entropy of the pixel values at each disparity hypothesis (Zitnick, Kang *et al.* 2004), which is particularly useful in multi-view stereo.

Other traditional matching costs include normalized cross-correlation (9.11) (Hannah 1974; Bolles, Baker, and Hannah 1993; Evangelidis and Psarakis 2008), which behaves similarly to sum-of-squared-differences (SSD), and binary matching costs (i.e., match or no match) (Marr and Poggio 1976), based on binary features such as edges (Baker and Binford 1981; Grimson 1985) or the sign of the Laplacian (Nishihara 1984). Because of their poor discriminability, simple binary matching costs are no longer used in dense stereo matching.

Some costs are insensitive to differences in camera gain or bias, for example gradient-based measures (Seitz 1989; Scharstein 1994), phase and filter-bank responses (Marr and Poggio 1979; Kass 1988; Jenkin, Jepson, and Tsotsos 1991; Jones and Malik 1992), filters that remove regular or robust (bilaterally filtered) means (Ansar, Castano, and Matthies 2004; Hirschmüller and Scharstein 2009), dense feature descriptor (Tola, Lepetit, and Fua 2010), and non-parametric measures such as rank and census transforms (Zabih and Woodfill 1994), ordinal measures (Bhat and Nayar 1998),

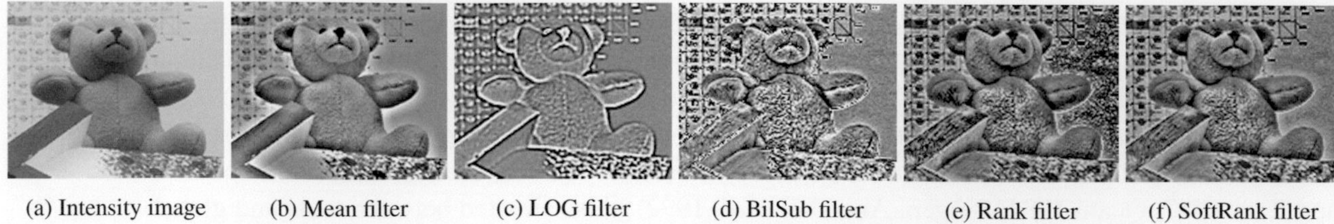

(a) Intensity image (b) Mean filter (c) LOG filter (d) BilSub filter (e) Rank filter (f) SoftRank filter

Figure 12.8 Various similarity measures (pre-processing filters) studied in (Hirschmüller and Scharstein 2009) © 2009 IEEE. The contrast of (b)–(d) has been increased for better visualization.

or entropy (Zitnick, Kang *et al.* 2004; Zitnick and Kang 2007). The census transform, which converts each pixel inside a moving window into a bit vector representing which neighbors are above or below the central pixel, was found by Hirschmüller and Scharstein (2009) to be quite robust against large-scale, non-stationary exposure and illumination changes. Figure 12.8 shows a few of the transformations that can be applied to images to improve their similarity across illumination variations.

It is also possible to correct for differing global camera characteristics by performing a preprocessing or iterative refinement step that estimates inter-image bias–gain variations using global regression (Gennert 1988), histogram equalization (Cox, Roy, and Hingorani 1995), or mutual information (Kim, Kolmogorov, and Zabih 2003; Hirschmüller 2008). Local, smoothly varying compensation fields have also been proposed (Strecha, Tuytelaars, and Van Gool 2003; Zhang, McMillan, and Yu 2006).

To compensate for sampling issues, i.e., dramatically different pixel values in high-frequency areas, Birchfield and Tomasi (1998) proposed a matching cost that is less sensitive to shifts in image sampling. Rather than just comparing pixel values shifted by integral amounts (which may miss a valid match), they compare each pixel in the reference image against a linearly interpolated function of the other image. More detailed studies of these and additional matching costs are explored in Szeliski and Scharstein (2004) and Hirschmüller and Scharstein (2009). In particular, if you expect there to be significant exposure or appearance variation between images that you are matching, some of the more robust measures that performed well in the evaluation by Hirschmüller and Scharstein (2009), such as the census transform (Zabih and Woodfill 1994), ordinal measures (Bhat and Nayar 1998), bilateral subtraction (Ansar, Castano, and Matthies 2004), or hierarchical mutual information (Hirschmüller 2008), should be used. Interestingly, color information does not appear to help when utilized in matching costs (Bleyer and Chambon 2010), although it is important for aggregation (discussed in next section). When matching more than pairs of images, more sophisticated variants of similarity (photoconsistency) measures can be used, as discussed in Section 12.7 and (Furukawa and Hernández 2015, Chapter 2).

More recently, one of the first successes of deep learning for stereo was the learning of matching costs. Žbontar and LeCun (2016) trained a neural network to compare image patches, trained on data extracted from the Middlebury (Scharstein, Hirschmüller *et al.* 2014) and KITTI (Geiger, Lenz, and Urtasun 2012) datasets. This matching cost is still widely used in top-performing methods on these two benchmarks.

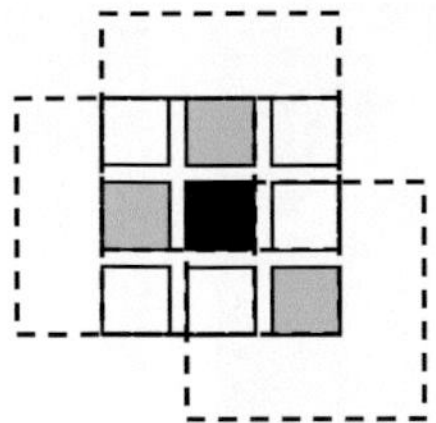

Figure 12.9 Shiftable window (Scharstein and Szeliski 2002) © 2002 Springer. The effect of trying all 3 × 3 shifted windows around the black pixel is the same as taking the minimum matching score across all *centered* (non-shifted) windows in the same neighborhood. (For clarity, only three of the neighboring shifted windows are shown here.)

12.4 Local methods

Local and window-based methods aggregate the matching cost by summing or averaging over a *support region* in the DSI $C(x, y, d)$.[4] A support region can be either two-dimensional at a fixed disparity (favoring fronto-parallel surfaces), or three-dimensional in x-y-d space (supporting slanted surfaces). Two-dimensional evidence aggregation has been implemented using square windows or Gaussian convolution (traditional), multiple windows anchored at different points, i.e., shiftable windows (Arnold 1983; Fusiello, Roberto, and Trucco 1997; Bobick and Intille 1999), windows with adaptive sizes (Okutomi and Kanade 1992; Kanade and Okutomi 1994; Kang, Szeliski, and Chai 2001; Veksler 2001, 2003), windows based on connected components of constant disparity (Boykov, Veksler, and Zabih 1998), the results of color-based segmentation (Yoon and Kweon 2006; Tombari, Mattoccia *et al.* 2008), or with a guided filter (Hosni, Rhemann *et al.* 2013). Three-dimensional support functions that have been proposed include limited disparity difference (Grimson 1985), limited disparity gradient (Pollard, Mayhew, and Frisby 1985), Prazdny's coherence principle (Prazdny 1985), and the work by Zitnick and Kanade (2000), which includes visibility and occlusion reasoning. PatchMatch stereo (Bleyer, Rhemann, and Rother 2011), discussed in more detail below, also does aggregation in 3D via slanted support windows.

Aggregation with a fixed support region can be performed using 2D or 3D convolution,

$$C(x, y, d) = w(x, y, d) * C_0(x, y, d), \tag{12.5}$$

or, in the case of rectangular windows, using efficient moving average box-filters (Section 3.2.2) (Kanade, Yoshida *et al.* 1996; Kimura, Shinbo *et al.* 1999). Shiftable windows can also be implemented efficiently using a separable sliding min-filter (Figure 12.9) (Scharstein and Szeliski 2002, Section 4.2). Selecting among windows of different shapes and sizes can be performed more efficiently by first computing a *summed area table* (Section 3.2.3, 3.30–3.32) (Veksler 2003). Selecting the right window is important, because windows must be large enough to contain sufficient texture and yet small enough so that they do not straddle depth discontinuities (Figure 12.10). An alternative method for aggregation is *iterative diffusion*, i.e., repeatedly adding to each pixel's cost the weighted values of its neighboring pixels' costs (Szeliski and Hinton 1985; Shah 1993; Scharstein and Szeliski 1998).

Of the local aggregation methods compared by Gong, Yang *et al.* (2007) and Tombari, Mattoccia *et al.* (2008), the fast variable window approach of Veksler (2003) and the locally weighting approach developed by Yoon and Kweon (2006) consistently stood out as having the best tradeoff

[4]For two surveys and comparisons of such techniques, please see the work of Gong, Yang *et al.* (2007) and Tombari, Mattoccia *et al.* (2008).

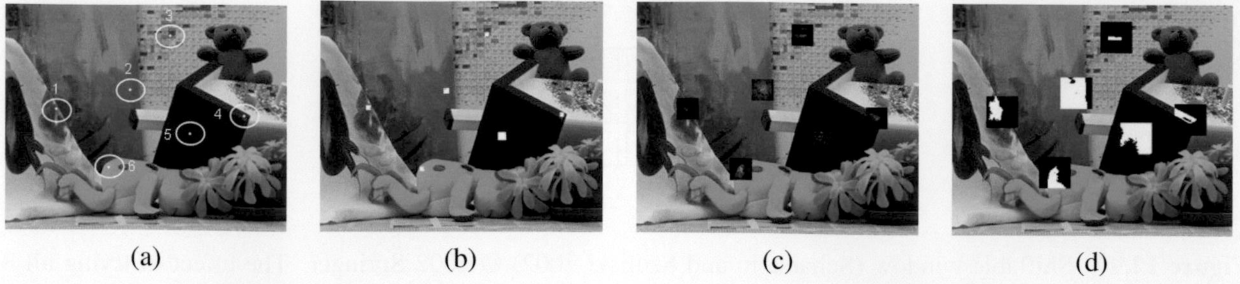

Figure 12.10 Aggregation window sizes and weights adapted to image content (Tombari, Mattoccia *et al.* 2008) © 2008 IEEE: (a) original image with selected evaluation points; (b) variable windows (Veksler 2003); (c) adaptive weights (Yoon and Kweon 2006); (d) segmentation-based (Tombari, Mattoccia, and Di Stefano 2007). Notice how the adaptive weights and segmentation-based techniques adapt their support to similarly colored pixels.

between performance and speed.[5] The local weighting technique, in particular, is interesting because, instead of using square windows with uniform weighting, each pixel within an aggregation window influences the final matching cost based on its color similarity and spatial distance, just as in bilateral filtering (Figure 12.10c). (In fact, their aggregation step is closely related to doing a joint bilateral filter on the color/disparity image, except that it is done symmetrically in both reference and target images.) The segmentation-based aggregation method of Tombari, Mattoccia, and Di Stefano (2007) did even better, although a fast implementation of this algorithm does not yet exist. Another approach to aggregation is to aggregate along one or more minimum spanning trees based on pixel similarities (Yang 2015; Li, Yu *et al.* 2017).

In local methods, the emphasis is on the matching cost computation and cost aggregation steps. Computing the final disparities is trivial: simply choose at each pixel the disparity associated with the minimum cost value. Thus, these methods perform a local "winner-take-all" (WTA) optimization at each pixel. A limitation of this approach (and many other correspondence algorithms) is that uniqueness of matches is only enforced for one image (the *reference image*), while points in the other image might match multiple points, unless cross-checking and subsequent hole filling is used (Fua 1993; Hirschmüller and Scharstein 2009).

12.4.1 Sub-pixel estimation and uncertainty

Most stereo correspondence algorithms compute a set of disparity estimates in some discretized space, e.g., for integer disparities (exceptions include continuous optimization techniques such as optical flow (Bergen, Anandan *et al.* 1992) or splines (Szeliski and Coughlan 1997)). For applications such as robot navigation or people tracking, these may be perfectly adequate. However for image-based rendering, such quantized maps lead to very unappealing view synthesis results, i.e., the scene appears to be made up of many thin shearing layers. To remedy this situation, many algorithms apply a sub-pixel refinement stage after the initial discrete correspondence stage. (An alternative is to simply start with more discrete disparity levels (Szeliski and Scharstein 2004).)

Sub-pixel disparity estimates can be computed in a variety of ways, including iterative gradient descent and fitting a curve to the matching costs at discrete disparity levels (Ryan, Gray, and Hunt 1980; Lucas and Kanade 1981; Tian and Huhns 1986; Matthies, Kanade, and Szeliski 1989; Kanade and Okutomi 1994). This provides an easy way to increase the resolution of a stereo algorithm with little additional computation. However, to work well, the intensities being matched must vary

[5]Extensive results from Tombari, Mattoccia *et al.* (2008) can be found at http://www.vision.deis.unibo.it/spe.

(a) (b) (c)

Figure 12.11 Uncertainty in stereo depth estimation (Szeliski 1991b): (a) input image; (b) estimated depth map (blue is closer); (c) estimated confidence(red is higher). As you can see, more textured areas have higher confidence.

smoothly, and the regions over which these estimates are computed must be on the same (correct) surface.

Some questions have been raised about the advisability of fitting correlation curves to integer-sampled matching costs (Shimizu and Okutomi 2001). This situation may even be worse when sampling-insensitive dissimilarity measures are used (Birchfield and Tomasi 1998). These issues are explored in more depth by Szeliski and Scharstein (2004) and Haller and Nedevschi (2012).

Besides sub-pixel computations, there are other ways of post-processing the computed disparities. Occluded areas can be detected using cross-checking, i.e., comparing left-to-right and right-to-left disparity maps (Fua 1993). A median filter can be applied to clean up spurious mismatches, and holes due to occlusion can be filled by surface fitting or by distributing neighboring disparity estimates (Birchfield and Tomasi 1999; Scharstein 1999; Hirschmüller and Scharstein 2009).

Another kind of post-processing, which can be useful in later processing stages, is to associate *confidences* with per-pixel depth estimates (Figure 12.11), which can be done by looking at the curvature of the correlation surface, i.e., how strong the minimum in the DSI image is at the winning disparity. Matthies, Kanade, and Szeliski (1989) show that under the assumption of small noise, photometrically calibrated images, and densely sampled disparities, the variance of a local depth estimate can be estimated as

$$Var(d) = \frac{\sigma_I^2}{a}, \tag{12.6}$$

where a is the curvature of the DSI as a function of d, which can be measured using a local parabolic fit or by squaring all the horizontal gradients in the window, and σ_I^2 is the variance of the image noise, which can be estimated from the minimum SSD score. (See also Section 8.1.4, (9.37), and Appendix B.6.) Over the years, a variety of stereo confidence measures have been proposed. Hu and Mordohai (2012) and Poggi, Kim *et al.* (2021) provide thorough surveys of this topic.

12.4.2 *Application*: Stereo-based head tracking

A common application of real-time stereo algorithms is for tracking the position of a user interacting with a computer or game system. The use of stereo can dramatically improve the reliability of such a system compared to trying to use monocular color and intensity information (Darrell, Gordon *et al.* 2000). Once recovered, this information can be used in a variety of applications, including controlling a virtual environment or game, correcting the apparent gaze during video conferencing, and background replacement. We discuss the first two applications below and defer the discussion of background replacement to Section 12.5.3.

The use of head tracking to control a user's virtual viewpoint while viewing a 3D object or environment on a computer monitor is sometimes called *fish tank virtual reality*, as the user is observing a 3D world as if it were contained inside a fish tank (Ware, Arthur, and Booth 1993). Early versions of these systems used mechanical head tracking devices and stereo glasses. Today, such systems can be controlled using stereo-based head tracking and stereo glasses can be replaced with autostereoscopic displays. Head tracking can also be used to construct a "virtual mirror", where the user's head can be modified in real time using a variety of visual effects (Darrell, Baker *et al.* 1997).

Another application of stereo head tracking and 3D reconstruction is in gaze correction (Ott, Lewis, and Cox 1993). When a user participates in a desktop video conference or video chat, the camera is usually placed on top of the monitor. Because the person is gazing at a window somewhere on the screen, it appears as if they are looking down and away from the other participants, instead of straight at them. Replacing the single camera with two or more cameras enables a virtual view to be constructed right at the position where they are looking, resulting in virtual eye contact. Real-time stereo matching is used to construct an accurate 3D head model and view interpolation (Section 14.1) is used to synthesize the novel in-between view (Criminisi, Shotton *et al.* 2003). More recent publications on gaze correction in video conferencing include Kuster, Popa *et al.* (2012) and Kononenko and Lempitsky (2015), and the technology has been deployed in several commercial video conferencing systems.[6]

12.5 Global optimization

Global stereo matching methods perform some optimization or iteration steps after the disparity computation phase and often skip the aggregation step altogether, because the global smoothness constraints perform a similar function. Many global methods are formulated in an energy-minimization framework, where, as we saw in Chapters 4 (4.24–4.27) and 9, the objective is to find a solution d that minimizes a global energy,

$$E(d) = E_{\text{D}}(d) + \lambda E_{\text{S}}(d). \tag{12.7}$$

The data term, $E_{\text{D}}(d)$, measures how well the disparity function d agrees with the input image pair. Using our previously defined disparity space image, we define this energy as

$$E_{\text{D}}(d) = \sum_{(x,y)} C(x, y, d(x, y)), \tag{12.8}$$

where C is the (initial or aggregated) matching cost DSI.

The smoothness term $E_{\text{S}}(d)$ encodes the smoothness assumptions made by the algorithm. To make the optimization computationally tractable, the smoothness term is often restricted to measuring only the differences between neighboring pixels' disparities,

$$E_{\text{S}}(d) = \sum_{(x,y)} \rho(d(x, y) - d(x + 1, y)) + \rho(d(x, y) - d(x, y + 1)), \tag{12.9}$$

[6] https://venturebeat.com/2019/10/03/microsofts-ai-powered-eye-gaze-tech-is-exclusive-to-the-surface-pro-x

where ρ is some monotonically increasing function of disparity difference. It is also possible to use larger neighborhoods, such as $\mathcal{N}_8$, which can lead to better boundaries (Boykov and Kolmogorov 2003), or to use second-order smoothness terms (Woodford, Reid *et al.* 2008), but such terms require more complex optimization techniques. An alternative to smoothness functionals is to use a lower-dimensional representation, such as splines (Szeliski and Coughlan 1997).

In standard regularization (Section 4.2), ρ is a quadratic function, which makes d smooth everywhere and may lead to poor results at object boundaries. Energy functions that do not have this problem are called *discontinuity-preserving* and are based on robust ρ functions (Terzopoulos 1986b; Black and Rangarajan 1996). The seminal paper by Geman and Geman (1984) gave a Bayesian interpretation of these kinds of energy functions and proposed a discontinuity-preserving energy function based on Markov random fields (MRFs) and additional *line processes*, which are additional binary variables that control whether smoothness penalties are enforced or not. Black and Rangarajan (1996) show how independent line process variables can be replaced by robust pairwise disparity terms.

The terms in E_S can also be made to depend on the intensity differences, e.g.,

$$\rho_D(d(x,y) - d(x+1,y)) \cdot \rho_I(\|I(x,y) - I(x+1,y)\|), \tag{12.10}$$

where ρ_I is some monotonically decreasing function of intensity differences that lowers smoothness costs at high-intensity gradients. This idea (Gamble and Poggio 1987; Fua 1993; Bobick and Intille 1999; Boykov, Veksler, and Zabih 2001) encourages disparity discontinuities to coincide with intensity or color edges and appears to account for some of the good performance of global optimization approaches. While most researchers set these functions heuristically, Pal, Weinman *et al.* (2012) show how the free parameters in such *conditional random fields* (Section 4.3, (4.47)) can be learned from ground truth disparity maps.

Once the global energy has been defined, a variety of algorithms can be used to find a (local) minimum. Traditional approaches associated with regularization and Markov random fields include continuation (Blake and Zisserman 1987), simulated annealing (Geman and Geman 1984; Marroquin, Mitter, and Poggio 1987; Barnard 1989), highest confidence first (Chou and Brown 1990), and mean-field annealing (Geiger and Girosi 1991).

Max-flow and *graph cut* methods have been proposed to solve a special class of global optimization problems (Roy and Cox 1998; Boykov, Veksler, and Zabih 2001; Ishikawa 2003). Such methods are more efficient than simulated annealing and have produced good results, as have techniques based on loopy belief propagation (Sun, Zheng, and Shum 2003; Tappen and Freeman 2003). Appendix B.5 and survey papers on MRF inference (Szeliski, Zabih *et al.* 2008; Blake, Kohli, and Rother 2011; Kappes, Andres *et al.* 2015) discuss and compare such techniques in more detail.

While global optimization techniques have largely been displaced by deep learning approaches (Section 12.6) for datasets such as KITI with large amounts of training images and high overlap with the test distributions, they still perform the best on challenging stereo pairs with fine details such as the high-resolution Middlebury pairs (Scharstein, Hirschmüller *et al.* 2014). One example of such an approach is the local expansion moves algorithm developed by Taniai, Matsushita *et al.* (2018). Below, we describe some related techniques that are of historical interest, run faster, or are tailored to handle specific situations.

Cooperative algorithms. Cooperative algorithms, inspired by computational models of human stereo vision, were among the earliest methods proposed for disparity computation (Dev 1974; Marr and Poggio 1976; Marroquin 1983; Szeliski and Hinton 1985; Zitnick and Kanade 2000). Such algorithms iteratively update disparity estimates using non-linear operations based on neighboring

(a) (b) (c) (d)

Figure 12.12 Stereo matching using local plane sweeps (Sinha, Scharstein, and Szeliski 2014) © 2014 IEEE: (a) input image; (b) initial sparse matches; (c) matches grouped by slanted planes; (d) 3D visualization of planes and grouped features.

disparity and matching values and result in an overall behavior similar to global optimization algorithms. In fact, for some of these algorithms, it is possible to explicitly state a global function that is being minimized (Scharstein and Szeliski 1998). There are also iterative algorithms that look at a larger neighborhood in the image, such as PatchMatch Stereo (Bleyer, Rhemann, and Rother 2011), which estimates a local 3D plane at each pixel and uses the non-local PatchMatch algorithm (Barnes, Shechtman *et al.* 2009) to quickly find approximate nearest neighbors in plane space. This approach has recently been applied to the multi-view stereo setting to produce an extremely time and space-efficient high-quality algorithm (Wang, Galliani *et al.* 2021).

Coarse-to-fine and incremental warping. Most of today's best algorithms first enumerate all possible matches at all possible disparities and then select the best set of matches in some way. Faster approaches can sometimes be obtained using methods inspired by classic (infinitesimal) optical flow computation. Here, images are successively warped and disparity estimates incrementally updated until a satisfactory registration is achieved. These techniques are most often implemented within a coarse-to-fine hierarchical refinement framework (Quam 1984; Bergen, Anandan *et al.* 1992; Barron, Fleet, and Beauchemin 1994; Szeliski and Coughlan 1997). Recently, coarse-to-fine or *pyramid* approaches have been having a renaissance in modern deep networks, applied both to optical flow (Ranjan and Black 2017; Sun, Yang *et al.* 2018) and stereo (Chang and Chen 2018).

Local plane sweeps. Instead of sweeping planes perpendicular to the viewing direction, it is also possible to model the scene using a collection of slanted planes, which is beneficial if the scene contains highly slanted planar surfaces such as floors or walls, as shown in Figure 12.12 (Sinha, Scharstein, and Szeliski 2014). Once such planes have been estimated and pixels assigned to each plane, it is then possible to estimate per-pixel out-of-plane displacements to better model curved surfaces. Slanted planes were also used earlier in the the PatchMatch stereo algorithm (Bleyer, Rhemann, and Rother 2011), and have also been used more recently in the *planar bilateral solver* used for smartphone AR (Valentin, Kowdle *et al.* 2018).

12.5.1 Dynamic programming

A different class of global optimization algorithm is based on *dynamic programming*. While the 2D optimization of Equation (12.7) can be shown to be NP-hard for common classes of smoothness functions (Veksler 1999), dynamic programming can find the global minimum for independent scanlines in polynomial time. Dynamic programming was first used for stereo vision in sparse, edge-based methods (Baker and Binford 1981; Ohta and Kanade 1985). More recent approaches

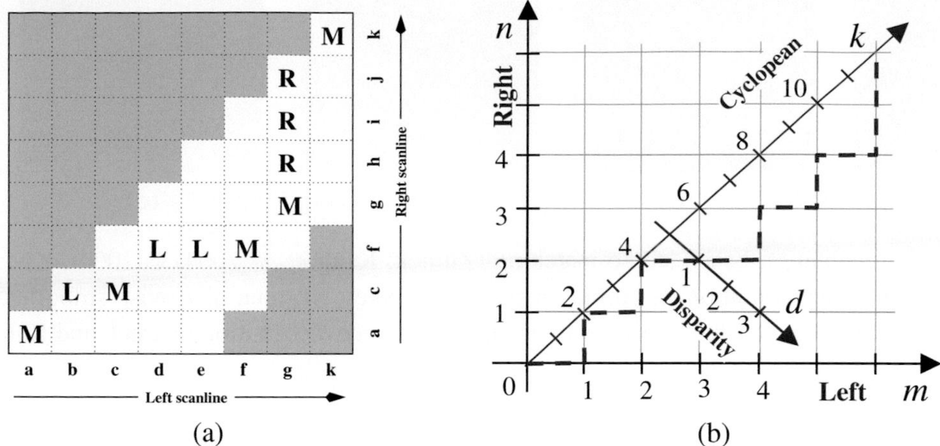

Figure 12.13 Stereo matching using dynamic programming, as illustrated by (a) Scharstein and Szeliski (2002) © 2002 Springer and (b) Kolmogorov, Criminisi *et al.* (2006) © 2006 IEEE. For each pair of corresponding scanlines, a minimizing path through the matrix of all pairwise matching costs (DSI) is selected. Lowercase letters (a–k) symbolize the intensities along each scanline. Uppercase letters represent the selected path through the matrix. Matches are indicated by M, while partially occluded points (which have a fixed cost) are indicated by L or R, corresponding to points only visible in the left or right images, respectively. Usually, only a limited disparity range is considered (0–4 in the figure, indicated by the non-shaded squares). The representation in (a) allows for diagonal moves while the representation in (b) does not. Note that these diagrams, which use the *Cyclopean* representation of depth, i.e., depth relative to a camera between the two input cameras, show an "unskewed" x-d slice through the DSI.

have focused on the dense (intensity-based) scanline matching problem (Belhumeur 1996; Geiger, Ladendorf, and Yuille 1992; Cox, Hingorani *et al.* 1996; Bobick and Intille 1999; Birchfield and Tomasi 1999). These approaches work by computing the minimum-cost path through the matrix of all pairwise matching costs between two corresponding scanlines, i.e., through a horizontal slice of the DSI. Partial occlusion is handled explicitly by assigning a group of pixels in one image to a single pixel in the other image. Figure 12.13 schematically shows how DP works, while Figure 12.5f shows a real DSI slice over which the DP is applied.

To implement dynamic programming for a scanline y, each entry (state) in a 2D cost matrix $D(m, n)$ is computed by combining its DSI matching cost value with one of its predecessor cost values while also including a fixed penalty for occluded pixels. The aggregation rules corresponding to Figure 12.13b are given by Kolmogorov, Criminisi *et al.* (2006), who also use a two-state foreground–background model for bi-layer segmentation.

Problems with dynamic programming stereo include the selection of the right cost for occluded pixels and the difficulty of enforcing inter-scanline consistency, although several methods propose ways of addressing the latter (Ohta and Kanade 1985; Belhumeur 1996; Cox, Hingorani *et al.* 1996; Bobick and Intille 1999; Birchfield and Tomasi 1999; Kolmogorov, Criminisi *et al.* 2006). Another problem is that the dynamic programming approach requires enforcing the *monotonicity* or *ordering constraint* (Yuille and Poggio 1984). This constraint requires that the relative ordering of pixels on a scanline remain the same between the two views, which may not be the case in scenes containing narrow foreground objects.

An alternative to traditional dynamic programming, introduced by Scharstein and Szeliski (2002), is to neglect the vertical smoothness constraints in (12.9) and simply optimize independent scanlines

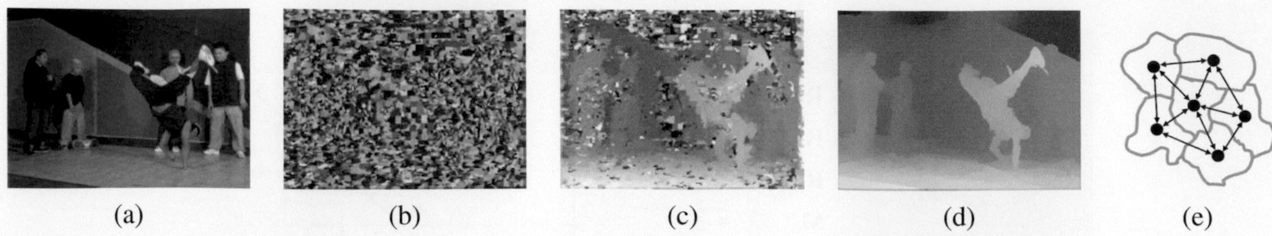

(a) (b) (c) (d) (e)

Figure 12.14 Segmentation-based stereo matching (Zitnick, Kang *et al.* 2004) © 2004 ACM: (a) input color image; (b) color-based segmentation; (c) initial disparity estimates; (d) final piecewise-smoothed disparities; (e) MRF neighborhood defined over the segments in the disparity space distribution (Zitnick and Kang 2007) © 2007 Springer.

in the global energy function (12.7). The advantage of this *scanline optimization* algorithm is that it computes the same representation and minimizes a reduced version of the same energy function as the full 2D energy function (12.7). Unfortunately, it still suffers from the same streaking artifacts as dynamic programming. Dynamic programming is also possible on tree structures, which can ameliorate the streaking (Veksler 2005).

Much higher quality results can be obtained by summing up the cumulative cost function from multiple directions, e.g, from the eight cardinal directions, N, E, W, S, NE, SE, SW, NW (Hirschmüller 2008). The resulting *semi-global matching* (SGM) algorithm performs quite well and is extremely efficient, enabling real-time low-power implementations (Gehrig, Eberli, and Meyer 2009). Drory, Haubold *et al.* (2014) show that SGM is equivalent to early stopping for a particular variant of belief propagation. Semi-global matching has also been extended using learned components, e.g., SGM-Net (Seki and Pollefeys 2017), which uses a CNN to adjust transition costs, and SGM-Forest (Schönberger, Sinha, and Pollefeys 2018), which uses a random-forest classifier to fuse disparity proposals from different directions.

12.5.2 Segmentation-based techniques

While most stereo matching algorithms perform their computations on a per-pixel basis, some techniques first segment the images into regions and then try to label each region with a disparity.

For example, Tao, Sawhney, and Kumar (2001) segment the reference image, estimate per-pixel disparities using a local technique, and then do local plane fits inside each segment before applying smoothness constraints between neighboring segments. Zitnick, Kang *et al.* (2004) and Zitnick and Kang (2007) use over-segmentation to mitigate initial bad segmentations. After a set of initial cost values for each segment has been stored into a *disparity space distribution* (DSD), iterative relaxation (or loopy belief propagation, in the more recent work of Zitnick and Kang (2007)) is used to adjust the disparity estimates for each segment, as shown in Figure 12.14. Taguchi, Wilburn, and Zitnick (2008) refine the segment shapes as part of the optimization process, which leads to much improved results, as shown in Figure 12.15.

Even more accurate results are obtained by Klaus, Sormann, and Karner (2006), who first segment the reference image using mean shift, run a small (3×3) SAD plus gradient SAD (weighted by cross-checking) to get initial disparity estimates, fit local planes, re-fit with global planes, and then run a final MRF on plane assignments with loopy belief propagation. When the algorithm was first introduced in 2006, it was the top ranked algorithm on the existing Middlebury benchmark.

The algorithm by Wang and Zheng (2008) follows a similar approach of segmenting the image, doing local plane fits, and then performing cooperative optimization of neighboring plane fit pa-

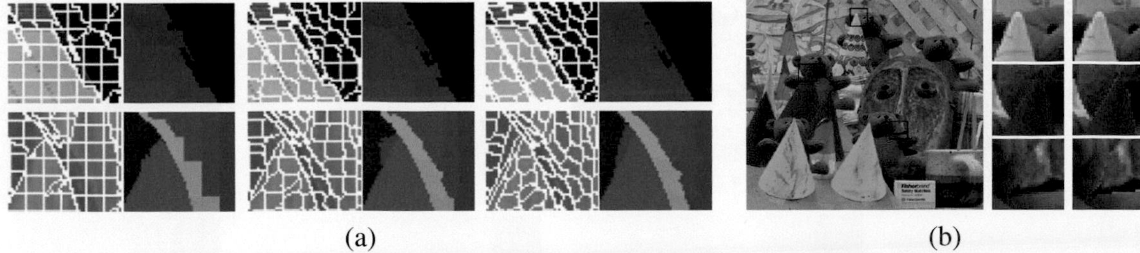

(a) (b)

Figure 12.15 Stereo matching with adaptive over-segmentation and matting (Taguchi, Wilburn, and Zitnick 2008) © 2008 IEEE: (a) segment boundaries are refined during the optimization, leading to more accurate results (e.g., the thin green leaf in the bottom row); (b) alpha mattes are extracted at segment boundaries, which leads to visually better compositing results (middle column).

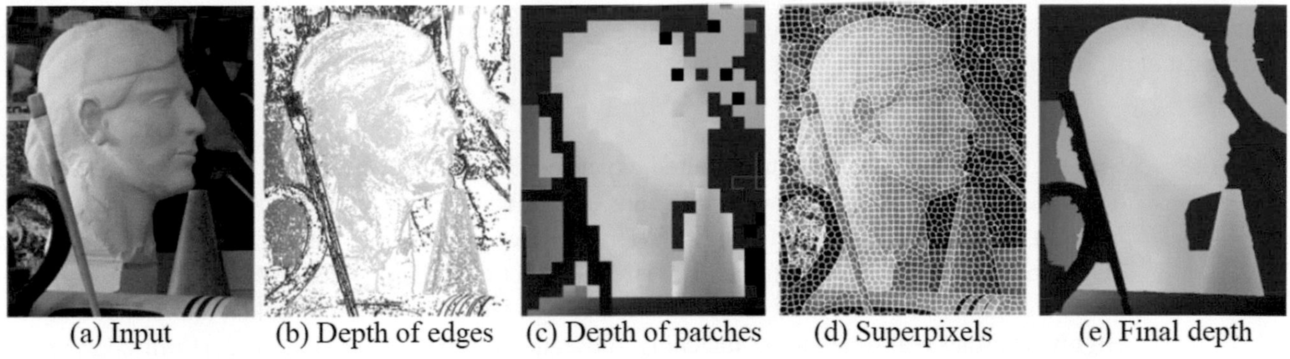

(a) Input (b) Depth of edges (c) Depth of patches (d) Superpixels (e) Final depth

Figure 12.16 Multiframe matching using edges, planes, and superpixels (Xue, Owens *et al.* 2019) © 2019 Elsevier.

rameters. The algorithm by Yang, Wang *et al.* (2009), uses the color correlation approach of Yoon and Kweon (2006) and hierarchical belief propagation to obtain an initial set of disparity estimates. Gallup, Frahm, and Pollefeys (2010) segment the image into planar and non-planar regions and use different representations for these two classes of surfaces.

More recently, Xue, Owens *et al.* (2019) start by matching edges across a multi-frame stereo sequence and then fit overlapping square patches to obtain local plane hypotheses. These are then refined using superpixels and a final edge-aware relaxation to get continuous depth maps.

Another important ability of segmentation-based stereo algorithms, which they share with algorithms that use explicit layers (Baker, Szeliski, and Anandan 1998; Szeliski and Golland 1999) or boundary extraction (Hasinoff, Kang, and Szeliski 2006), is the ability to extract fractional pixel alpha mattes at depth discontinuities (Bleyer, Gelautz *et al.* 2009). This ability is crucial when attempting to create virtual view interpolation without clinging boundary or tearing artifacts (Zitnick, Kang *et al.* 2004) and also to seamlessly insert virtual objects (Taguchi, Wilburn, and Zitnick 2008), as shown in Figure 12.15b.

12.5.3 *Application*: Z-keying and background replacement

Another application of real time stereo matching is *Z-keying*, which is the process of segmenting a foreground actor from the background using depth information, usually for the purpose of replacing the background with some computer-generated imagery, as shown in Figure 12.2g.

Figure 12.17 Background replacement using z-keying with a bi-layer segmentation algorithm (Kolmogorov, Criminisi *et al.* 2006) © 2006 IEEE.

Originally, z-keying systems required expensive custom-built hardware to produce the desired depth maps in real time and were, therefore, restricted to broadcast studio applications (Kanade, Yoshida *et al.* 1996; Iddan and Yahav 2001). Off-line systems were also developed for estimating 3D multi-viewpoint geometry from video streams (Section 14.5.4) (Kanade, Rander, and Narayanan 1997; Carranza, Theobalt *et al.* 2003; Zitnick, Kang *et al.* 2004; Vedula, Baker, and Kanade 2005). Highly accurate real-time stereo matching subsequently made it possible to perform z-keying on regular PCs, enabling desktop video conferencing applications such as those shown in Figure 12.17 (Kolmogorov, Criminisi *et al.* 2006), but these have mostly been replaced with deep networks for background replacement (Sengupta, Jayaram *et al.* 2020) and real-time 3D phone-based reconstruction algorithms for augmented reality (Figure 11.28 and Valentin, Kowdle *et al.* 2018).

12.6 Deep neural networks

As with other areas of computer vision, deep neural networks and end-to-end learning have had a large impact on stereo matching. In this section, we briefly review how DNNs have been used in stereo correspondence algorithms. We follow the same structure as the two recent surveys by Poggi, Tosi *et al.* (2021) and Laga, Jospin *et al.* (2020), which classify techniques into three categories, namely,

1. learning in the stereo pipeline,
2. end-to-end learning with 2D architectures, and
3. end-to-end learning with 3D architectures.

We briefly discuss a few papers in each group and refer the reader to the full surveys for more details (Janai, Güney *et al.* 2020; Poggi, Tosi *et al.* 2021; Laga, Jospin *et al.* 2020).

Learning in the stereo pipeline

Even before the advent of deep learning, several authors proposed learning components of the traditional stereo pipeline, e.g., to learn hyperparameters of MRF and CRF stereo models (Zhang and

Seitz 2007; Pal, Weinman *et al.* 2012). Žbontar and LeCun (2016) were the first to bring deep learning to stereo by training features to optimize a pairwise matching cost. These learned matching costs are still widely used in top-performing methods on the Middlebury stereo evaluation. Many other authors have since proposed CNNs for matching cost computation and aggregation (Luo, Schwing, and Urtasun 2016; Park and Lee 2017; Zhang, Prisacariu *et al.* 2019).

Learning has also been used to improve traditional optimization techniques, in particular the widely used SGM algorithm of Hirschmüller (2008). This includes SGM-Net (Seki and Pollefeys 2017), which uses a CNN to adjust transition costs, and SGM-Forest (Schönberger, Sinha, and Pollefeys 2018), which uses a random-forest classifier to select among disparity values from multiple incident directions. CNNs have also been used in the refinement stage, replacing earlier techniques such as bilateral filtering (Gidaris and Komodakis 2017; Batsos and Mordohai 2018; Knöbelreiter and Pock 2019).

End-to-end learning with 2D architectures

The availability of large synthetic datasets with ground truth disparities, in particular the Freiburg SceneFlow dataset (Mayer, Ilg *et al.* 2016, 2018) enabled the end-to-end training of stereo networks and resulted in a proliferation of new methods. These methods work well on benchmarks that provide enough training data so that the network can be tuned to the domain, notably KITTI (Geiger, Lenz, and Urtasun 2012; Geiger, Lenz *et al.* 2013; Menze and Geiger 2015), where deep-learning based methods started to dominate the leaderboards in 2016.

The first deep learning architectures for stereo were similar to those designed for dense regression tasks such as semantic segmentation (Chen, Zhu *et al.* 2018). These *2D architectures* typically employ an encoder-decoder design inspired by U-Net (Ronneberger, Fischer, and Brox 2015). The first such model was DispNet-C, introduced in the seminal paper by Mayer, Ilg *et al.* (2016), utilizing a *correlation layer* (Dosovitskiy, Fischer *et al.* 2015) to compute the similarity between image layers.

Subsequent improvements to 2D architectures included the idea of residual networks that apply residual corrections to the original disparities (Pang, Sun *et al.* 2017), which can also be done in an iterative fashion (Liang, Feng *et al.* 2018). Coarse-to-fine processing can be used (Tonioni, Tosi *et al.* 2019; Yin, Darrell, and Yu 2019), and networks can estimate occlusions and depth boundaries (Ilg, Saikia *et al.* 2018; Song, Zhao *et al.* 2020) or use neural architecture search (NAS) to improve performance (Saikia, Marrakchi *et al.* 2019). HITNet incorporates several of these ideas and produces efficient state-of-the-art results using local slanted plane hypotheses and iterative refinement (Tankovich, Hane *et al.* 2021).

The 2D architecture developed by Knöbelreiter, Reinbacher *et al.* (2017) uses a joint CNN and Conditional Random Field (CRF) model to infer dense disparity maps. Another promising approach is multi-task learning, for instance, jointly estimating disparities and semantic segmentation (Yang, Zhao *et al.* 2018; Jiang, Sun *et al.* 2019). It is also possible to increase the apparent resolution of the output depth map and reduce over-smoothing by representing the output as a bimodal mixture distribution (Tosi, Liao *et al.* 2021).

End-to-end learning with 3D architectures

An alternative approach is to use *3D architectures*, which explicitly encode geometry by processing features over a 3D volume, where the third dimension corresponds to the disparity search range. In other words, such architectures explicitly represent the disparity space image (DSI), while still

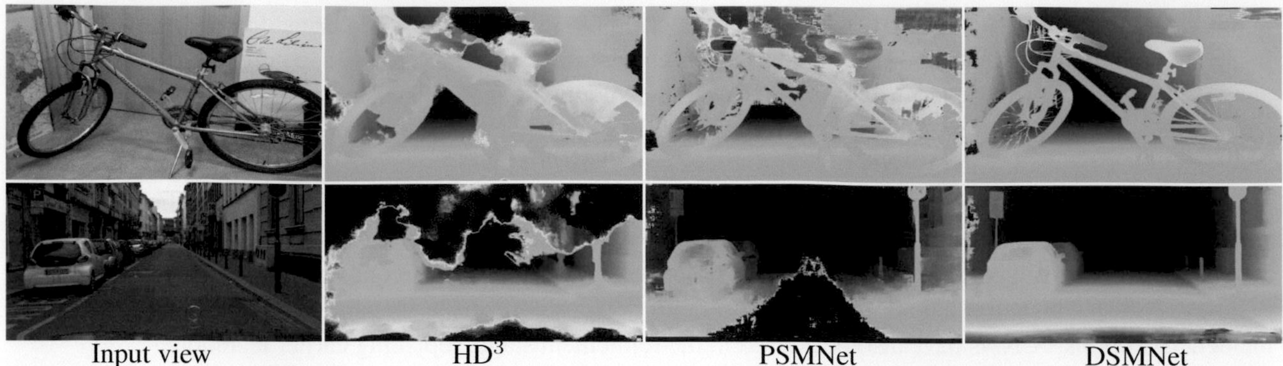

Figure 12.18 Disparity maps computed by three different DNN stereo matchers trained on synthetic data and applied to real-world image pairs (Zhang, Qi *et al.* 2020) © 2020 Springer.

keeping multiple feature channels instead of just scalar cost values. Compared to 2D architectures, they incur much higher memory requirements and runtimes.

The first examples of such architectures include GC-Net (Kendall, Martirosyan *et al.* 2017) and PSMNet (Chang and Chen 2018). 3D architectures also allow the integration of traditional local aggregation methods (Zhang, Prisacariu *et al.* 2019) and methods to avoid geometric inconsistencies (Chabra, Straub *et al.* 2019). While resource constraints often mean that 3D DNN-based stereo methods operate at fairly low resolutions, the Hierarchical Stereo Matching (HSM) network (Yang, Manela *et al.* 2019) uses a pyramid approach that selectively restricts the search space at higher resolutions and enables *anytime on-demand inference*, i.e., stopping the processing early for higher frame rates. Duggal, Wang *et al.* (2019) address limited resources by developing a differentiable version of PatchMatch (Bleyer, Rhemann, and Rother 2011) in a recurrent neural net. Cheng, Zhong *et al.* (2020) use neural architecture search (NAS) to create a state-of-the-art 3D architecture.

While supervised deep learning approaches have come to dominate individual benchmarks that include dedicated training sets such as KITTI, they do not yet generalize well across domains (Zendel *et al.* 2020). On the Middlebury benchmark, which features high-resolution images and only provides very limited training data, deep learning methods are still notably absent. Poggi, Tosi *et al.* (2021) identify the following two major challenges that remain open: (1) generalization across different domains, and (2) applicability on high-resolution images. For cross-domain generalization, Poggi, Tosi *et al.* (2021) describe techniques for both offline and online self-supervised adaptation and guided deep learning, while Laga, Jospin *et al.* (2020) discuss both fine-tuning and data transformation. A recent example of domain generalization is the domain-invariant stereo matching network (DSMNet) of Zhang, Qi *et al.* (2020), which compares favorably with alternative state-of-the-art models such as HD^3 (Yin, Darrell, and Yu 2019) and PSMNet (Chang and Chen 2018), as shown in Figure 12.18. Another example of domain adaptation is AdaStereo (Song, Yang *et al.* 2021). For high-resolution images, techniques have been developed to increase resolution in a coarse-to-fine manner (Khamis, Fanello *et al.* 2018; Chabra, Straub *et al.* 2019).

12.7 Multi-view stereo

While matching pairs of images is a useful way of obtaining depth information, using more images can significantly improve results. In this section, we review not only techniques for creating complete 3D object models, but also simpler techniques for improving the quality of depth maps using

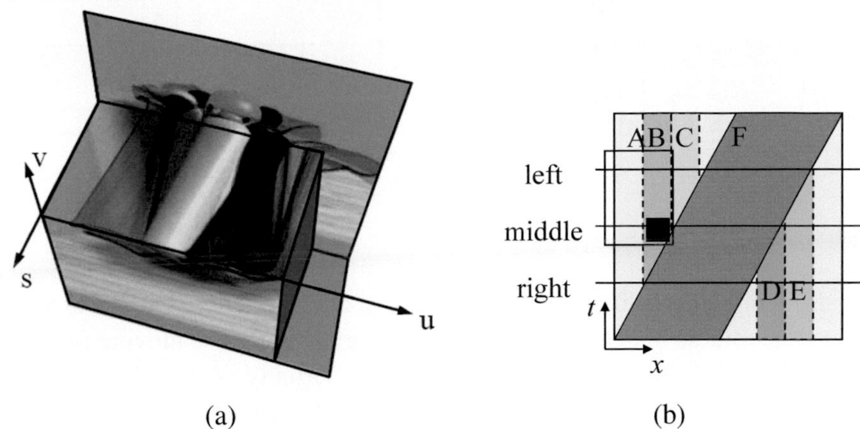

Figure 12.19 Epipolar plane image (EPI) (Gortler, Grzeszczuk *et al.* 1996) © 1996 ACM and a schematic EPI (Kang, Szeliski, and Chai 2001) © 2001 IEEE. (a) The Lumigraph (light field) (Section 14.3) is the 4D space of all light rays passing through a volume of space. Taking a 2D slice results in all of the light rays embedded in a plane and is equivalent to a scanline taken from a stacked EPI volume. Objects at different depths move sideways with velocities (slopes) proportional to their inverse depth. Occlusion (and translucency) effects can easily be seen in this representation. (b) The EPI corresponding to Figure 12.20 showing the three images (middle, left, and right) as slices through the EPI volume. The spatially and temporally shifted window around the black pixel is indicated by the rectangle, showing that the right image is not being used in matching.

multiple source images. A good survey of techniques developed up through 2015 can be found in Furukawa and Hernández (2015) and a more recent review in Janai, Güney *et al.* (2020, Chapter 10).

As we saw in our discussion of plane sweep (Section 12.1.2), it is possible to resample all neighboring k images at each disparity hypothesis d into a generalized disparity space volume $\tilde{I}(x, y, d, k)$. The simplest way to take advantage of these additional images is to sum up their differences from the reference image I_r as in (12.4),

$$C(x, y, d) = \sum_k \rho(\tilde{I}(x, y, d, k) - I_r(x, y)). \tag{12.11}$$

This is the basis of the well-known sum of summed-squared-difference (SSSD) and SSAD approaches (Okutomi and Kanade 1993; Kang, Webb *et al.* 1995), which can be extended to reason about likely patterns of occlusion (Nakamura, Matsuura *et al.* 1996). More recent work by Gallup, Frahm *et al.* (2008) shows how to adapt the baselines used to the expected depth to get the best tradeoff between geometric accuracy (wide baseline) and robustness to occlusion (narrow baseline). Alternative multi-view cost metrics include measures such as synthetic focus sharpness and the entropy of the pixel color distribution (Vaish, Szeliski *et al.* 2006).

A useful way to visualize the multi-frame stereo estimation problem is to examine the *epipolar plane image* (EPI) formed by stacking corresponding scanlines from all the images, as shown in Figures 9.11c and 12.19 (Bolles, Baker, and Marimont 1987; Baker and Bolles 1989; Baker 1989). As you can see in Figure 12.19, as a camera translates horizontally (in a standard horizontally rectified geometry), objects at different depths move sideways at a rate inversely proportional to their depth (12.1).[7] Foreground objects occlude background objects, which can be seen as *EPI-strips* (Criminisi, Kang *et al.* 2005) occluding other strips in the EPI. If we are given a dense enough

[7]The four-dimensional generalization of the EPI is the *light field*, which we study in Section 14.3.

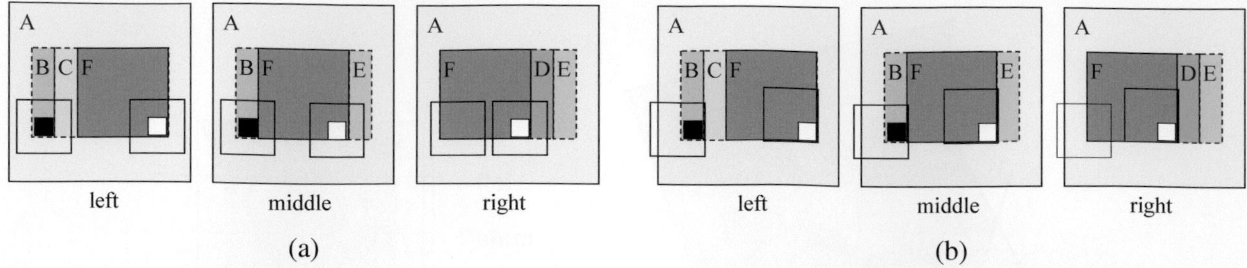

Figure 12.20 Spatio-temporally shiftable windows (Kang, Szeliski, and Chai 2001) © 2001 IEEE: A simple three-image sequence (the middle image is the reference image), which has a moving frontal gray square (marked F) and a stationary background. Regions B, C, D, and E are partially occluded. (a) A regular SSD algorithm will make mistakes when matching pixels in these regions (e.g., the window centered on the black pixel in region B) and in windows straddling depth discontinuities (the window centered on the white pixel in region F). (b) Shiftable windows help mitigate the problems in partially occluded regions and near depth discontinuities. The shifted window centered on the white pixel in region F matches correctly in all frames. The shifted window centered on the black pixel in region B matches correctly in the left image, but requires temporal selection to disable matching the right image. Figure 12.19b shows an EPI corresponding to this sequence and describes in more detail how temporal selection works.

set of images, we can find such strips and reason about their relationships to both reconstruct the 3D scene and make inferences about translucent objects (Tsin, Kang, and Szeliski 2006) and specular reflections (Swaminathan, Kang *et al.* 2002; Criminisi, Kang *et al.* 2005). Alternatively, we can treat the series of images as a set of sequential observations and merge them using Kalman filtering (Matthies, Kanade, and Szeliski 1989) or maximum likelihood inference (Cox 1994).

When fewer images are available, it becomes necessary to fall back on aggregation techniques, such as sliding windows or global optimization. With additional input images, however, the likelihood of occlusions increases. It is therefore prudent to adjust not only the best window locations using a shiftable window approach, as shown in Figure 12.20a, but also to optionally select a subset of neighboring frames to discount those images where the region of interest is occluded, as shown in Figure 12.20b (Kang, Szeliski, and Chai 2001). Figure12.19b shows how such spatio-temporal selection or shifting of windows corresponds to selecting the most likely un-occluded volumetric region in the epipolar plane image volume.

The results of applying these techniques to the multi-frame *flower garden* image sequence are shown in Figure 12.21, which compares the results of using regular (non-shifted) SSSD with spatially shifted windows and full spatio-temporal window selection. (The task of applying stereo to a rigid scene filmed with a moving camera is sometimes called *motion stereo*). Similar improvements from using spatio-temporal selection are reported by Kang and Szeliski (2004) and are evident even when local measurements are combined with global optimization.

While computing a depth map from multiple inputs outperforms pairwise stereo matching, even more dramatic improvements can be obtained by estimating multiple depth maps simultaneously (Szeliski 1999a; Kang and Szeliski 2004). The existence of multiple depth maps enables more accurate reasoning about occlusions, as regions that are occluded in one image may be visible (and matchable) in others. The multi-view reconstruction problem can be formulated as the simultaneous estimation of depth maps at key frames (Figure 9.11c) while maximizing not only photoconsistency and piecewise disparity smoothness, but also the consistency between disparity estimates at different frames. While Szeliski (1999a) and Kang and Szeliski (2004) use soft (penalty-based) constraints to encourage multiple disparity maps to be consistent, Kolmogorov and Zabih (2002) show how such

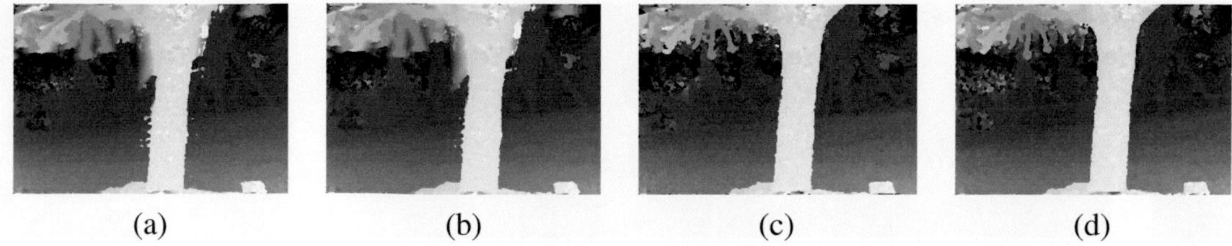

Figure 12.21 Local (5 × 5 window-based) matching results (Kang, Szeliski, and Chai 2001) © 2001 IEEE: (a) window that is not spatially perturbed (centered); (b) spatially perturbed window; (c) using the best five of 10 neighboring frames; (d) using the better half sequence. Notice how the results near the tree trunk are improved using temporal selection.

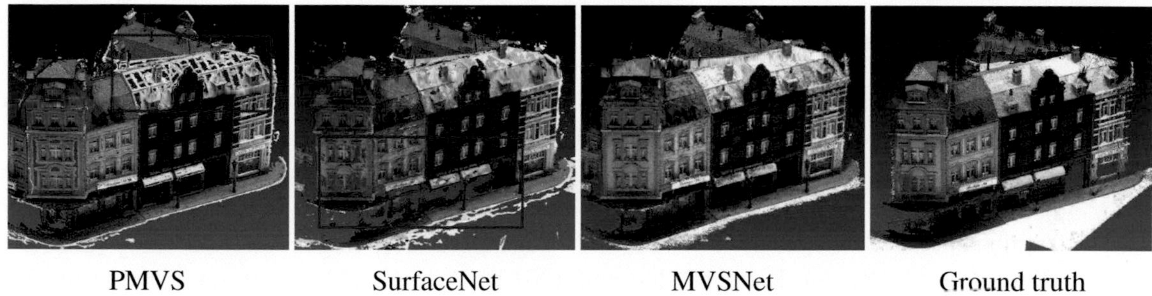

Figure 12.22 Depth maps computed using three different multi-view stereo algorithms shown as colored point clouds (Yao, Luo *et al.* 2018) © 2018 Springer. The red boxes indicate problem areas where MVSNet does better.

consistency measures can be encoded as hard constraints, which guarantee that the multiple depth maps are not only similar but actually identical in overlapping regions. Additional algorithms that simultaneously estimate multiple disparity maps include those of Maitre, Shinagawa, and Do (2008) and Zhang, Jia *et al.* (2008) and the widely used COLMAP algorithm (Schönberger, Zheng *et al.* 2016), which uses view selection and geometric consistency between multiple depth maps to filter matches, as shown in Figure 12.26b.

The latest multi-view stereo algorithms use deep neural networks to compute matching (cost) volumes and to fuse these into disparity maps. The DeepMVS system computes pairwise matching costs between a reference image and neighboring images and then fuses them together with max pooling, followed by a dense CRF refinement (Huang, Matzen *et al.* 2018). MVSNet computes the variance between all encoded images warped onto each sweep plane, uses a 3D U-Net to regularize the costs, and then a soft argmin and depth refinement network to produce good results on the DTU and Tanks and Temples datasets (Yao, Luo *et al.* 2018), as shown in Figure 12.22.

More recent variants on such networks include P-MVSNet (Luo, Guan *et al.* 2019), which uses a patch-wise matching confidence aggregator, and CasMVSNet (Gu, Fan *et al.* 2020) and CVP-MVSNet (Yang, Mao *et al.* 2020), both of which use coarse-to-fine pyramid processing. Four even more recent papers that all score well on the DTU, ETH3D, Tanks and Temples, and/or Blended MVS datasets are Vis-MVSNet (Zhang, Yao *et al.* 2020), D^{2}HC-RMVSNet (Yan, Wei *et al.* 2020), DeepC-MVS (Kuhn, Sormann *et al.* 2020), and PatchmatchNet (Wang, Galliani *et al.* 2021). These algorithms use various combinations of visibility and occlusion reasoning, confidence or uncertainty maps, and geometric consistency checks, and efficient propagation schemes to achieve good results. As so many new multi-view stereo papers continue to get published, the ETH3D and Tanks and Temples leaderboards (Table12.1) are good places to look for the latest results.

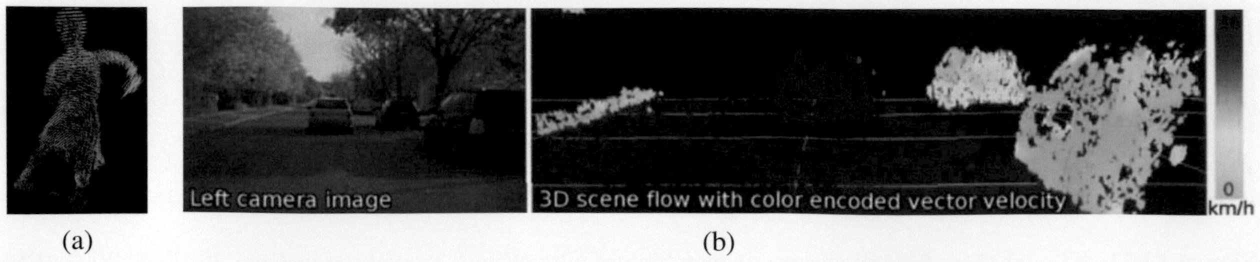

(a) (b)

Figure 12.23 Three-dimensional scene flow: (a) computed from a multi-camera dome surrounding the dancer shown in Figure 12.2h–j (Vedula, Baker *et al.* 2005) © 2005 IEEE; (b) computed from stereo cameras mounted on a moving vehicle (Wedel, Rabe *et al.* 2008) © 2008 Springer.

12.7.1 Scene flow

A closely related topic to multi-frame stereo estimation is *scene flow*, in which multiple cameras are used to capture a dynamic scene. The task is then to simultaneously recover the 3D shape of the object at every instant in time and to estimate the full 3D motion of every surface point between frames. Representative papers in this area include Vedula, Baker *et al.* (2005), Zhang and Kambhamettu (2003), Pons, Keriven, and Faugeras (2007), Huguet and Devernay (2007), Wedel, Rabe *et al.* (2008), and Rabe, Müller *et al.* (2010). Figure 12.23a shows an image of the 3D scene flow for the tango dancer shown in Figure 12.2h–j, while Figure 12.23b shows 3D scene flows captured from a moving vehicle for the purpose of obstacle avoidance. In addition to supporting mensuration and safety applications, scene flow can be used to support both spatial and temporal view interpolation (Section 14.5.4), as demonstrated by Vedula, Baker, and Kanade (2005).

The creation of the KITTI scene flow dataset (Geiger, Lenz, and Urtasun 2012) as well as increased interest in autonomous driving have accelerated research into scene flow algorithms (Janai, Güney *et al.* 2020, Chapter 12). One way to help regularize the problem is to adopt a piecewise planar representation (Vogel, Schindler, and Roth 2015). Another is to decompose the scene into rigid separately moving objects such as vehicles (Menze and Geiger 2015), using semantic segmentation (Behl, Hosseini Jafari *et al.* 2017), as well as to use other segmentation cues (Ilg, Saikia *et al.* 2018; Ma, Wang *et al.* 2019; Jiang, Sun *et al.* 2019). The more widespread availability of 3D sensors has enabled the extension of scene flow algorithms to use this modality as an additional input (Sun, Sudderth, and Pfister 2015; Behl, Paschalidou *et al.* 2019).

12.7.2 Volumetric and 3D surface reconstruction

The most challenging but also most useful variant of multi-view stereo reconstruction is the construction of globally consistent 3D models (Seitz, Curless *et al.* 2006). This topic has a long history in computer vision, starting with surface mesh reconstruction techniques such as the one developed by Fua and Leclerc (1995) (Figure 12.24a). A variety of approaches and representations have been used to solve this problem, including 3D voxels (Seitz and Dyer 1999; Szeliski and Golland 1999; De Bonet and Viola 1999; Kutulakos and Seitz 2000; Eisert, Steinbach, and Girod 2000; Slabaugh, Culbertson *et al.* 2004; Sinha and Pollefeys 2005; Vogiatzis, Hernández *et al.* 2007; Hiep, Keriven *et al.* 2009), level sets (Faugeras and Keriven 1998; Pons, Keriven, and Faugeras 2007), polygonal meshes (Fua and Leclerc 1995; Narayanan, Rander, and Kanade 1998; Hernández and Schmitt 2004; Furukawa and Ponce 2009), and multiple depth maps (Kolmogorov and Zabih 2002). Figure 12.24 shows representative examples of 3D object models reconstructed using some of these techniques.

To organize and compare all these techniques, Seitz, Curless *et al.* (2006) developed a six-point

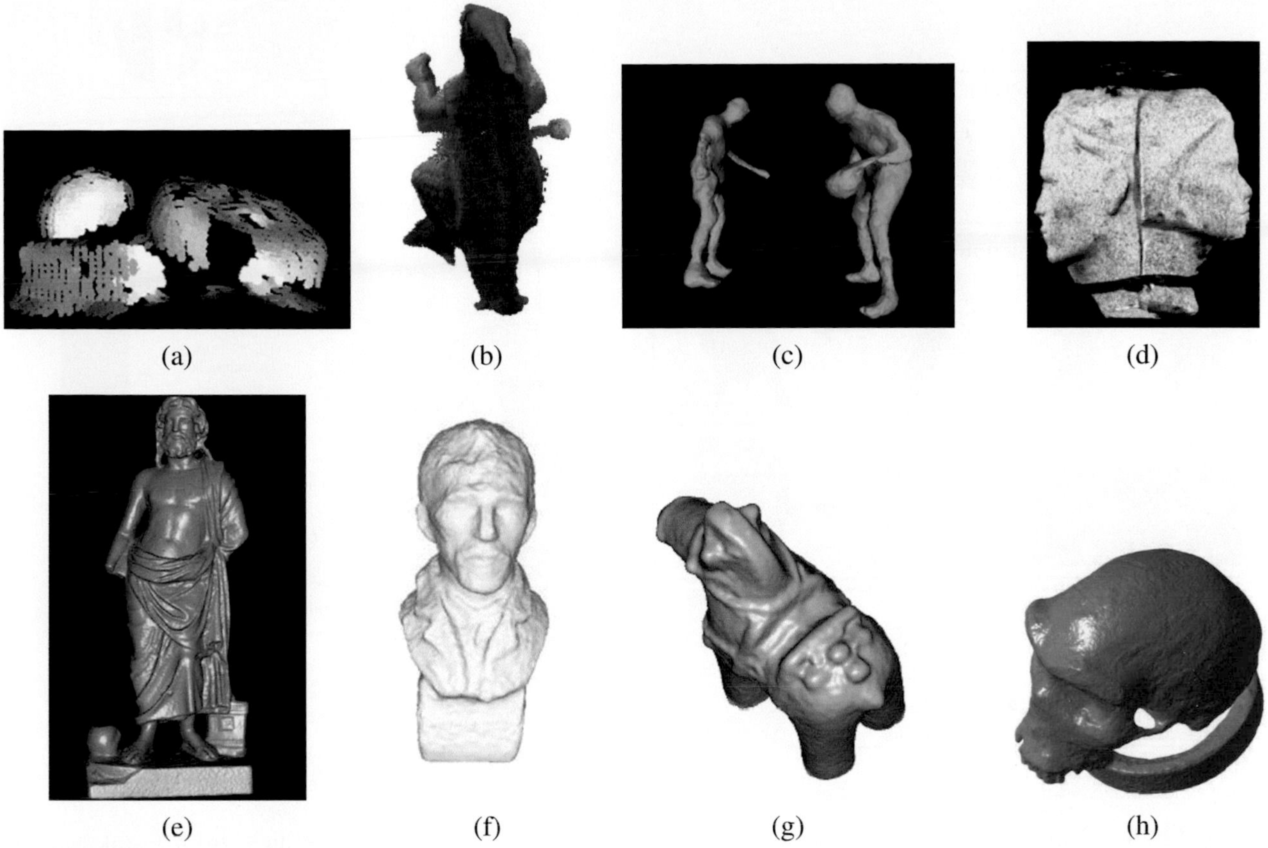

Figure 12.24 Multi-view stereo algorithms: (a) surface-based stereo (Fua and Leclerc 1995) © 1995 Springer; (b) voxel coloring (Seitz and Dyer 1999) © 1999 Springer; (c) depth map merging (Narayanan, Rander, and Kanade 1998) © 1998 IEEE; (d) level set evolution (Faugeras and Keriven 1998) © 1998 IEEE; (e) silhouette and stereo fusion (Hernández and Schmitt 2004) © 2004 Elsevier; (f) multi-view image matching (Pons, Keriven, and Faugeras 2005) © 2005 IEEE; (g) volumetric graph cut (Vogiatzis, Torr, and Cipolla 2005) © 2005 IEEE; (h) carved visual hulls (Furukawa and Ponce 2009) © 2009 Springer.

taxonomy that can help classify algorithms according to the *scene representation*, *photoconsistency measure*, *visibility model*, *shape priors*, *reconstruction algorithm*, and *initialization requirements* they use. Below, we summarize some of these choices and list a few representative papers. For more details, please consult the full survey paper (Seitz, Curless *et al.* 2006) as well as more recent surveys by Furukawa and Ponce (2010) and Janai, Güney *et al.* (2020, Chapter 10). The ETH3D and Tanks and Temples leaderboards list the most up-to-date results and pointers to recent papers.

Scene representation. According to the taxonomy proposed by Furukawa and Ponce (2010), multi-view stereo algorithms primarily use four scene representations, namely *depth maps*, *point clouds*, *volumetric fields*, and *3D meshes*, as shown in Figure 12.25a. These are often combined into a complete pipeline that includes camera pose estimation, per-image depth map or point cloud computation, volumetric fusion, and surface mesh refinement (Pollefeys, Nistér *et al.* 2008), as shown in Figure 12.25b.

We have already discussed multi-view depth map estimation earlier in this section. An example of a point cloud representation is the patch-based multi-view stereo (PMVS) algorithm developed

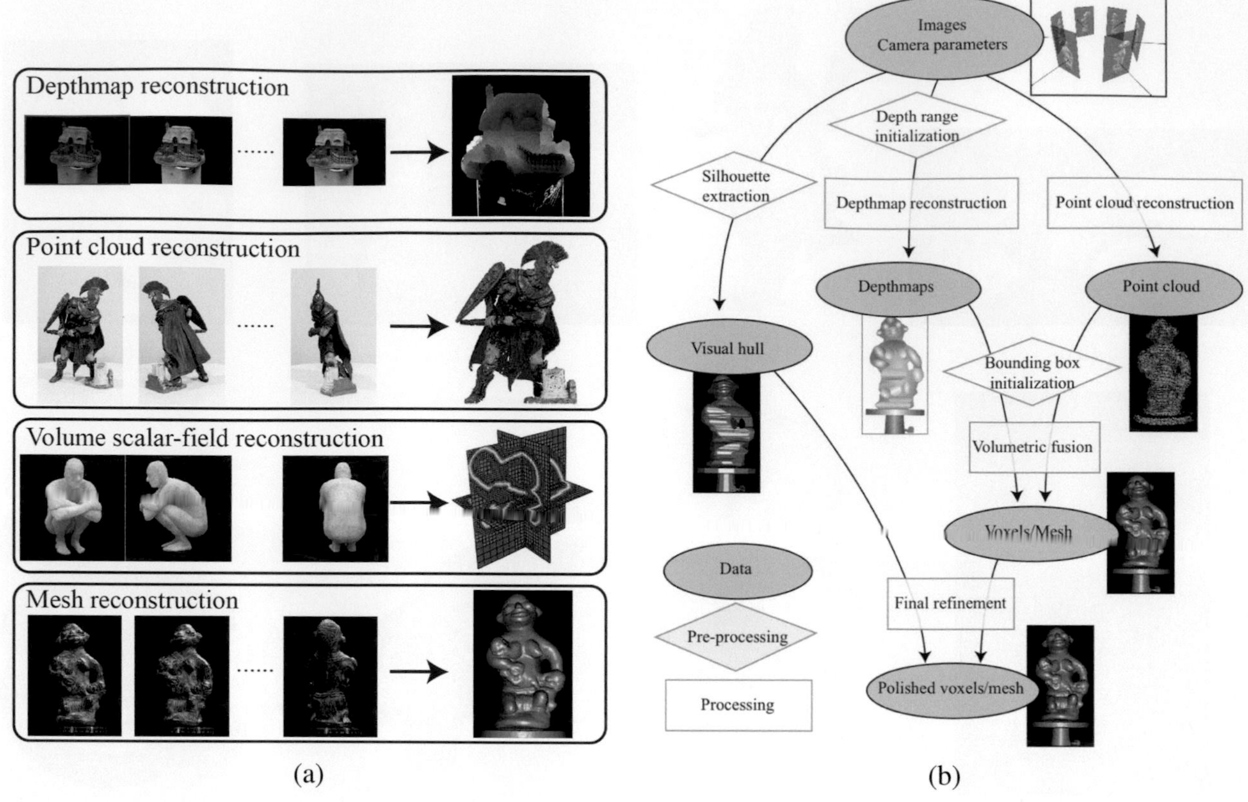

Figure 12.25 Multi-view stereo (a) scene representations and (b) processing pipelines, from Furukawa and Hernández (2015) © 2015 now publishers.

by Furukawa and Ponce (2010), which starts with sparse 3D points reconstructed using structure from motion, then optimizes and densifies local oriented patches or *surfels* (Szeliski and Tonnesen 1992; Section 13.4) while taking into account visibility constraints, as shown in Figure 12.26a. This representation can then be turned into a mesh for final optimization. Since then, improved techniques have been developed for view selection and filtering as well as normal estimation, as exemplified in the systems developed by Fuhrmann and Goesele (2014) and Schönberger, Zheng *et al.* (2016), the latter of which (shown in Figures 11.20b and 12.26b) provides the dense multi-view stereo component of the popular COLMAP open-source reconstruction system (Schönberger and Frahm 2016). When highly sampled video sequences are available, reconstructing point clouds from tracked edges may be more appropriate, as discussed in Section 12.2.1, Kim, Zimmer *et al.* (2013) and Yücer, Kim *et al.* (2016) and illustrated in Figure 12.26c.

One of the more popular 3D representations is a uniform grid of 3D voxels,[8] which can be reconstructed using a variety of carving techniques (Seitz and Dyer 1999; Kutulakos and Seitz 2000) or optimization (Sinha and Pollefeys 2005; Vogiatzis, Hernández *et al.* 2007; Hiep, Keriven *et al.* 2009; Jancosek and Pajdla 2011; Vu, Labatut *et al.* 2012), some of which are illustrated in Figure 12.24. Level set techniques (Section 7.3.2) also operate on a uniform grid but, instead of representing a binary occupancy map, they represent the signed distance to the surface (Faugeras and Keriven 1998; Pons, Keriven, and Faugeras 2007), which can encode a finer level of detail and also be used to

[8]For outdoor scenes that go to infinity, an inverted gridding of space may be preferable (Slabaugh, Culbertson *et al.* 2004; Zhang, Riegler *et al.* 2020).

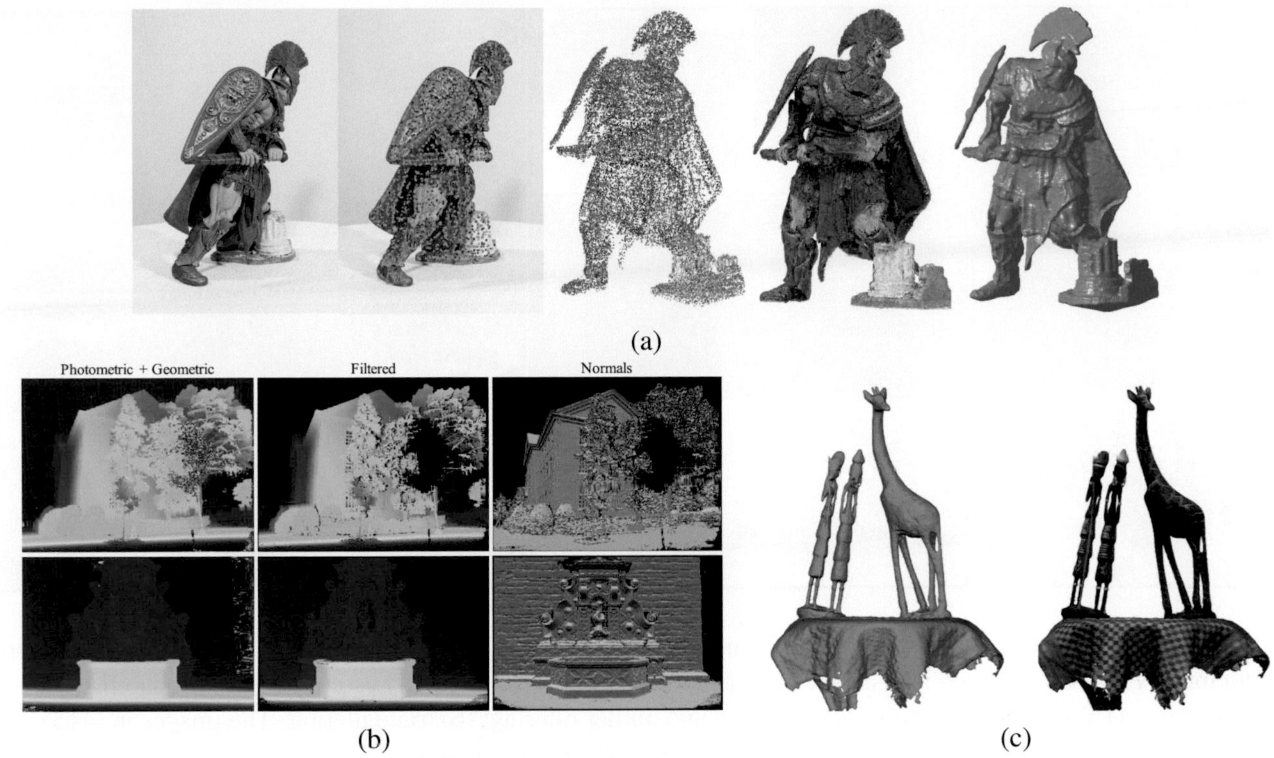

Figure 12.26 Point cloud reconstruction: (a) the PMVS pipeline, showing a sample input image, detected features, initial reconstructed patches, patches after expansion and filtering, and the final mesh model (Furukawa and Ponce 2010) © 2010 IEEE; (b) depth maps and surface normals from two stages of the COLMAP multi-view stereo pipeline (Schönberger, Zheng *et al.* 2016) © 2016 Springer; (c) thin structures recovered from gradients in a dense orbiting camera light field (Yücer, Kim *et al.* 2016) © 2016 IEEE.

merge multiple point clouds or range data scans, as discussed extensively in Section 13.2.1. Instead of using a uniformly sampled volume, which works best for compact 3D objects, it is also possible to create a view frustum corresponding to one of the input images and to sample the z dimension as inverse depths, i.e., uniform disparities for a set of co-planar cameras (Figure 14.7). This kind of representation is called a *stack of acetates* in Szeliski and Golland (1999) and *multiplane images* in Zhou, Tucker *et al.* (2018).

Polygonal meshes are another popular representation (Fua and Leclerc 1995; Narayanan, Rander, and Kanade 1998; Isidoro and Sclaroff 2003; Hernández and Schmitt 2004; Furukawa and Ponce 2009; Hiep, Keriven *et al.* 2009). Meshes are the standard representation used in computer graphics and also readily support the computation of visibility and occlusions. Finally, as we discussed in the previous section, multiple depth maps can also be used (Szeliski 1999a; Kolmogorov and Zabih 2002; Kang and Szeliski 2004). Many algorithms also use more than a single representation, e.g., they may start by computing multiple depth maps and then merge them into a 3D object model (Narayanan, Rander, and Kanade 1998; Furukawa and Ponce 2009; Goesele, Curless, and Seitz 2006; Goesele, Snavely *et al.* 2007; Pollefeys, Nistér *et al.* 2008; Furukawa, Curless *et al.* 2010; Furukawa and Ponce 2010; Vu, Labatut *et al.* 2012; Schönberger, Zheng *et al.* 2016), as illustrated in Figure 12.25b.

An example of a recent system that combines several representations into a scalable distributed approach that can handle datasets with hundreds of high-resolution images is the LTVRE multi-

Figure 12.27 3D reconstruction pipeline from Kuhn, Hirschmüller *et al.* (2017) © 2017 Springer: (0) structure from motion; (1) stereo matching using semi-global matching; (2) depth quality estimation; (3) probabilistic space occupancy; (4+5) probabilistic point optimization and outlier filtering; (6) triangulation. The images in (4+5) and (6) are half texture-mapped and half flat shaded to show more surface detail.

view stereo system by Kuhn, Hirschmüller *et al.* (2017). The system starts from pairwise disparity maps computed with SGM (Hirschmüller 2008). These depth estimates are fused with a probabilistic multi-scale approach using a learned stereo error model, using an octree to handle variable resolution, followed by filtering of conflicting points based on visibility constraints, and finally triangulation. Figure 12.27 shows an illustration of the processing pipeline.

Photoconsistency measure. As we discussed in (Section 12.3.1), a variety of similarity measures can be used to compare pixel values in different images, including measures that try to discount illumination effects or be less sensitive to outliers. In multi-view stereo, algorithms have a choice of computing these measures directly on the surface of the model, i.e., in *scene space*, or projecting pixel values from one image (or from a textured model) back into another image, i.e., in *image space*. (The latter corresponds more closely to a Bayesian approach, because input images are noisy measurements of the colored 3D model.) The geometry of the object, i.e., its distance to each camera and its local surface normal, when available, can be used to adjust the matching windows used in the computation to account for foreshortening and scale change (Goesele, Snavely *et al.* 2007).

Visibility model. A big advantage that multi-view stereo algorithms have over single-depth-map approaches is their ability to reason in a principled manner about visibility and occlusions. Techniques that use the current state of the 3D model to predict which surface pixels are visible in each image (Kutulakos and Seitz 2000; Faugeras and Keriven 1998; Vogiatzis, Hernández *et al.* 2007; Hiep, Keriven *et al.* 2009; Furukawa and Ponce 2010; Schönberger, Zheng *et al.* 2016) are classified as using *geometric visibility models* in the taxonomy of Seitz, Curless *et al.* (2006). Techniques that select a neighboring subset of image to match are called *quasi-geometric* (Narayanan, Rander, and Kanade 1998; Kang and Szeliski 2004; Hernández and Schmitt 2004), while techniques that use traditional robust similarity measures are called *outlier-based*. While full geometric reasoning is the most principled and accurate approach, it can be very slow to evaluate and depends on the evolving quality of the current surface estimate to predict visibility, which can be a bit of a chicken-and-egg problem, unless conservative assumptions are used, as they are by Kutulakos and Seitz (2000).

Shape priors. Because stereo matching is often underconstrained, especially in textureless regions, most matching algorithms adopt (either explicitly or implicitly) some form of prior model for the expected shape. Many of the techniques that rely on optimization use a 3D smoothness or area-based photoconsistency constraint, which, because of the natural tendency of smooth surfaces to shrink inwards, often results in a *minimal surface* prior (Faugeras and Keriven 1998; Sinha and Pollefeys 2005; Vogiatzis, Hernández *et al.* 2007). Approaches that carve away the volume of space often stop once a photoconsistent solution is found (Seitz and Dyer 1999; Kutulakos and Seitz 2000), which corresponds to a *maximal surface* bias, i.e., these techniques tend to over-estimate the true shape. Finally, multiple depth map approaches often adopt traditional *image-based* smoothness (regularization) constraints.

Higher-level shape priors can also be used, such as Manhattan world assumptions that assume most surfaces of interest are axis-aligned (Furukawa, Curless *et al.* 2009a,b) or at related orientations such as slanted roofs (Sinha, Steedly, and Szeliski 2009; Osman Ulusoy, Black, and Geiger 2017). These kinds of *architectural priors* are discussed in more detail in Section 13.6.1. It is also possible to use 2D semantic segmentation in images, e.g., into wall, ground, and foliage classes, to apply different kinds of regularization and surface normal priors in different regions of the model (Häne, Zach *et al.* 2013).

Reconstruction algorithm. The details of how the actual reconstruction algorithm proceeds is where the largest variety—and greatest innovation—in multi-view stereo algorithms can be found.

Some approaches use global optimization defined over a three-dimensional photoconsistency volume to recover a complete surface. Approaches based on graph cuts use polynomial complexity binary segmentation algorithms to recover the object model defined on the voxel grid (Sinha and Pollefeys 2005; Vogiatzis, Hernández *et al.* 2007; Hiep, Keriven *et al.* 2009; Jancosek and Pajdla 2011; Vu, Labatut *et al.* 2012). Level set approaches use a continuous surface evolution to find a good minimum in the configuration space of potential surfaces and therefore require a reasonably good initialization (Faugeras and Keriven 1998; Pons, Keriven, and Faugeras 2007). For the photoconsistency volume to be meaningful, matching costs need to be computed in some robust fashion, e.g., using sets of limited views or by aggregating multiple depth maps.

An alternative approach to global optimization is to sweep through the 3D volume while computing both photoconsistency and visibility simultaneously. The *voxel coloring* algorithm of Seitz and Dyer (1999) performs a front-to-back plane sweep. On every plane, any voxels that are sufficiently photoconsistent are labeled as part of the object. The corresponding pixels in the source images can then be "erased", as they are already accounted for, and therefore do not contribute to further photoconsistency computations. (A similar approach, albeit without the front-to-back sweep order, is used by Szeliski and Golland (1999).) The resulting 3D volume, under noise- and resampling-free conditions, is guaranteed to produce both a photoconsistent 3D model and to enclose whatever true 3D object model generated the images (Figure 12.28a–b).

Unfortunately, voxel coloring is only guaranteed to work if all of the cameras lie on the same side of the sweep planes, which is not possible in general ring configurations of cameras. Kutulakos and Seitz (2000) generalize voxel coloring to *space carving*, where subsets of cameras that satisfy the voxel coloring constraint are iteratively selected and the 3D voxel grid is alternately carved away

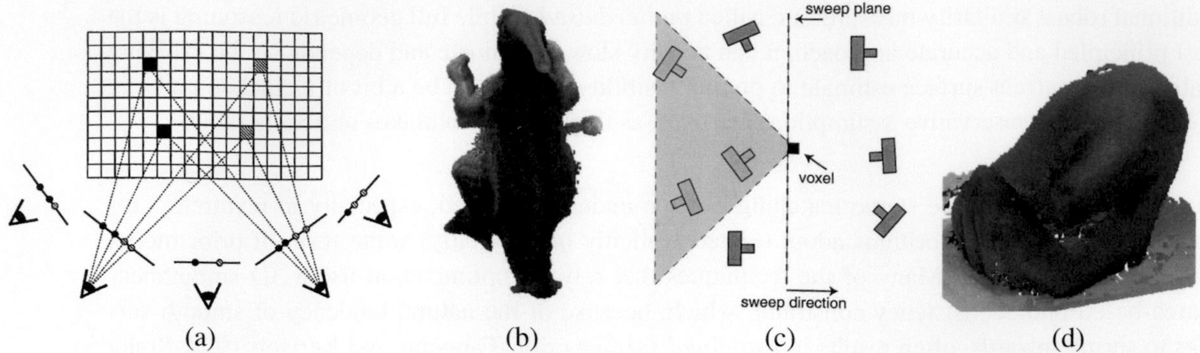

Figure 12.28 Voxel coloring (Seitz and Dyer 1999) © 1999 Springer and space carving (Kutulakos and Seitz 2000) © 2000 Springer. (a–b): voxel coloring sweeps a plane through the scene in a front-to-back manner with respect to the cameras. (c–d): space carving uses multiple sweep directions to deal with more general camera configurations.

along different axes (Figure 12.28c–d).

Another popular approach to multi-view stereo is to first independently compute multiple depth maps and then merge these partial maps into a complete 3D model. Approaches to depth map merging, which are discussed in more detail in Section 13.2.1, include signed distance functions (Curless and Levoy 1996), used by Goesele, Curless, and Seitz (2006), and Poisson surface reconstruction (Kazhdan, Bolitho, and Hoppe 2006), used by Goesele, Snavely *et al.* (2007).

Initialization requirements. One final element discussed by Seitz, Curless *et al.* (2006) is the varying degrees of initialization required by different algorithms. Because some algorithms refine or evolve a rough 3D model, they require a reasonably accurate (or overcomplete) initial model, which can often be obtained by reconstructing a volume from object silhouettes, as discussed in Section 12.7.3. However, if the algorithm performs a global optimization (Kolev, Klodt *et al.* 2009; Kolev and Cremers 2009), this dependence on initialization is not an issue.

Empirical evaluation. The widespread adoption of datasets and benchmarks has led to the rapid advances in multi-view reconstruction over the last two decades. Table 12.1 lists some of the most widely used and influential ones, with sample images and/or results shown in Figures 12.1, 12.22, and 12.26. Pointers to additional datasets can be found in Mayer, Ilg *et al.* (2018), Janai, Güney *et al.* (2020), Laga, Jospin *et al.* (2020), and Poggi, Tosi *et al.* (2021). Pointers to the most recent algorithms can usually be found on the leaderboards of the ETH3D and Tanks and Temples benchmarks.

12.7.3 Shape from silhouettes

In many situations, performing a foreground–background segmentation of the object of interest is a good way to initialize or fit a 3D model (Grauman, Shakhnarovich, and Darrell 2003; Vlasic, Baran *et al.* 2008) or to impose a convex set of constraints on multi-view stereo (Kolev and Cremers 2008). Over the years, a number of techniques have been developed to reconstruct a 3D volumetric model from the intersection of the binary silhouettes projected into 3D. The resulting model is called a *visual hull* (or sometimes a *line hull*), analogous with the convex hull of a set of points, because the volume is maximal with respect to the visual silhouettes and surface elements are tangent to

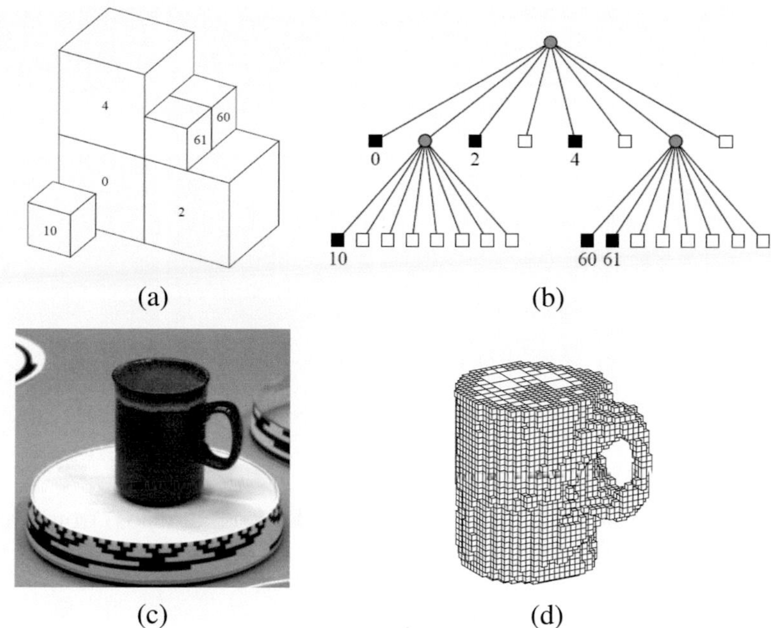

Figure 12.29 Volumetric octree reconstruction from binary silhouettes (Szeliski 1993) © 1993 Elsevier: (a) octree representation and its corresponding (b) tree structure; (c) input image of an object on a turntable; (d) computed 3D volumetric octree model.

the viewing rays (lines) along the silhouette boundaries (Laurentini 1994). It is also possible to carve away a more accurate reconstruction using multi-view stereo (Sinha and Pollefeys 2005) or by analyzing cast shadows (Savarese, Andreetto *et al.* 2007).

Some techniques first approximate each silhouette with a polygonal representation and then intersect the resulting faceted conical regions in three-space to produce polyhedral models (Baumgart 1974; Martin and Aggarwal 1983; Matusik, Buehler, and McMillan 2001), which can later be refined using triangular splines (Sullivan and Ponce 1998). Other approaches use voxel-based representations, usually encoded as octrees (Samet 1989), because of the resulting space–time efficiency. Figures 12.29a–b show an example of a 3D octree model and its associated colored tree, where black nodes are interior to the model, white nodes are exterior, and gray nodes are of mixed occupancy. Examples of octree-based reconstruction approaches include Potmesil (1987), Noborio, Fukada, and Arimoto (1988), Srivasan, Liang, and Hackwood (1990), and Szeliski (1993).

The approach of Szeliski (1993) first converts each binary silhouette into a one-sided variant of a distance map, where each pixel in the map indicates the largest square that is completely inside (or outside) the silhouette. This makes it fast to project an octree cell into the silhouette to confirm whether it is completely inside or outside the object, so that it can be colored black, or white, or left as gray (mixed) for further refinement on a smaller grid. The octree construction algorithm proceeds in a coarse-to-fine manner, first building an octree at a relatively coarse resolution, and then refining it by revisiting and subdividing all the input images for the gray (mixed) cells whose occupancy has not yet been determined. Figure 12.29d shows the resulting octree model computed from a coffee cup rotating on a turntable.

More recent work on visual hull computation borrows ideas from image-based rendering, and is hence called an *image based visual hull* (Matusik, Buehler *et al.* 2000). Instead of precomputing a global 3D model, an image-based visual hull is recomputed for each new viewpoint, by successively intersecting viewing ray segments with the binary silhouettes in each image. This not only leads to a

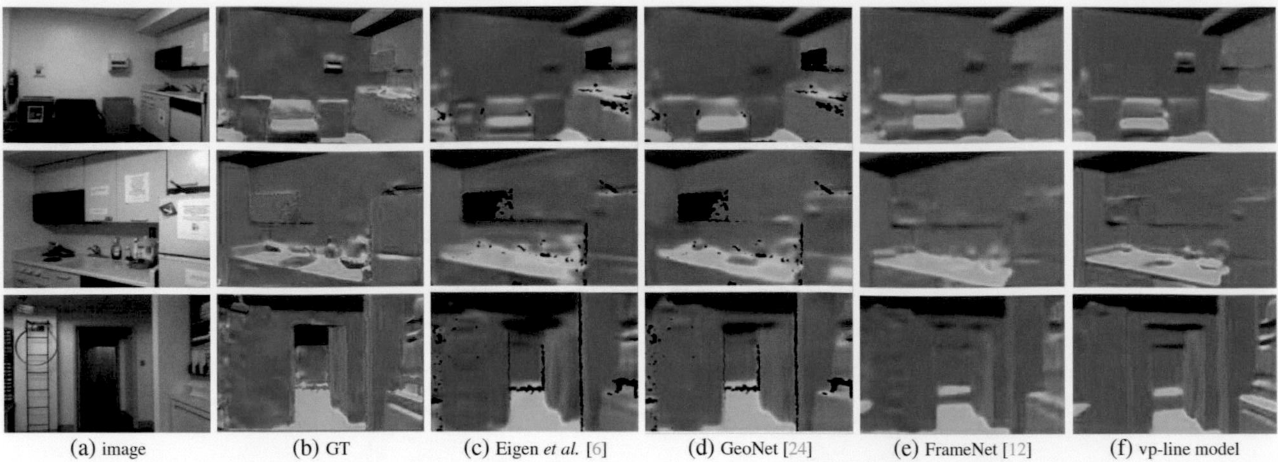

Figure 12.30 Monocular depth inference (shown as color-coded normal maps) from images in the NYU Depth Dataset V2 (Wang, Geraghty *et al.* 2020) © 2020 IEEE.

fast computation algorithm but also enables fast texturing of the recovered model with color values from the input images. This approach can also be combined with high-quality deformable templates to capture and re-animate whole body motion (Vlasic, Baran *et al.* 2008).

While the methods described above start with a binary foreground/background silhouette image, it is also possible to extract silhouette curves, usually to sub-pixel precision, and to reconstruct partial surface meshes from tracking these, as discussed in Section 12.2.1. Such silhouette curves can also be combined with regular image edges to construct complete surface models (Yücer, Kim *et al.* 2016), such as the ones shown in Figure 12.26c.

12.8 Monocular depth estimation

The ability to infer (or hallucinate?) depth maps from single images opens up all kinds of creative possibilities, such as displaying them in 3D (Figure 6.41 and Kopf, Matzen *et al.* 2020), creating soft focus effects (Section 10.3.2), and potentially to aid scene understanding. It can also be used in robotics applications such as autonomous navigation (Figure 12.31), although most (autonomous and regular) vehicles have more than one camera or range sensors, if equipped with computer vision.

We already saw in Section 6.4.4 how the automatic photo pop-up system can use image segmentation and classification to create "cardboard cut-out" versions of a photo (Hoiem, Efros, and Hebert 2005a; Saxena, Sun, and Ng 2009). More recent systems to infer depth from single images use deep neural networks. These are described in two recent surveys (Poggi, Tosi *et al.* 2021, Section 7; Zhao, Sun *et al.* 2020), which discuss 20 and over 50 related techniques, respectively, and benchmark them on the KITTI dataset (Geiger, Lenz, and Urtasun 2012) shown in Figure 12.31.

One of the first papers to use a neural network to compute a depth map was the system developed by Eigen, Puhrsch, and Fergus (2014), which was subsequently extended to also infer surface normals and semantic labels (Eigen and Fergus 2015). These systems were trained and tested on the NYU Depth Dataset V2 (Silberman, Hoiem *et al.* 2012), shown in Figure 12.30, and the KITTI dataset. Most of the subsequent work in this area trains and tests on these two fairly restricted datasets (indoor apartments or outdoor street scenes), although authors sometimes use Make3D (Saxena, Sun, and Ng 2009) or Cityscapes (Cordts, Omran *et al.* 2016), which are both outdoor

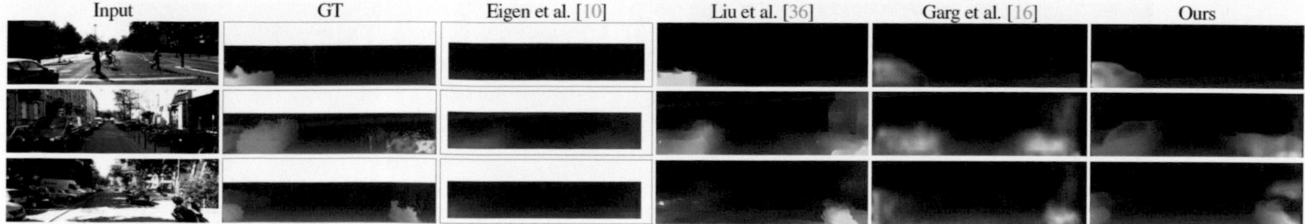

Figure 12.31 Monocular depth map estimates from images in the KITTI dataset (Godard, Mac Aodha, and Brostow 2017) © 2017 IEEE.

street scenes, or ScanNet (Dai, Chang *et al.* 2017), which has indoor scenes. The danger in training and testing on such "closed world" datasets is that the network can learn shortcuts, such as inferring depth based on the location along the ground plane or failing to "pop up" objects that are not in commonly occurring classes (van Dijk and de Croon 2019).

Early systems were trained on the depth images that came with datasets such as NYU Depth, KITTI, and ScanNet, where it turns out that adding soft constraints such as co-planarity can improve performance (Wang, Shen *et al.* 2016; Yin, Liu *et al.* 2019). Later, "unsupervised" techniques were introduced that use photometric consistency between warped stereo pairs of images (Godard, Mac Aodha, and Brostow 2017; Xian, Shen *et al.* 2018) or image pairs in video sequences (Zhou, Brown *et al.* 2017). It is also possible to train on 3D reconstructions of famous landmarks (Li and Snavely 2018), image sets containing people posing in a "Mannequin Challenge" (Li, Dekel *et al.* 2019), or to take more diverse "images in the wild" and have them labeled with relative depths (Chen, Fu *et al.* 2016).

A recent paper that federates several such datasets is the MiDaS system developed by Ranftl, Lasinger *et al.* (2020), who not only use a number of existing "in the wild" datasets to train a network based on Xian, Shen *et al.* (2018), but also use thousands of stereo image pairs from over a dozen 3D movies as additional training, validation, and test data. In their paper, they not only show that their system produces superior results to previous approaches (Figure 12.32), but also argue that their *zero-shot cross-dataset transfer* protocol, i.e., testing on data sets separate from training sets, rather than using random train and test subsets, produces a system that works better on real-world images and avoids unintended dataset bias (Torralba and Efros 2011).

An alternative to inferring depth maps from single images is to infer complete closed 3D shapes, using either volumetric (Choy, Xu *et al.* 2016; Girdhar, Fouhey *et al.* 2016; Tulsiani, Gupta *et al.* 2018) or mesh-based (Gkioxari, Malik, and Johnson 2019) representations (Han, Laga, and Bennamoun 2021). Instead of applying deep networks to just a single color image, it is also possible to augment such networks with additional cues and representations, such as oriented lines and planes (Wang, Geraghty *et al.* 2020), which serve as higher-level *shape priors* (Sections 12.7.2 and 13.6.1). Neural rendering can also be used to create novel views (Tucker and Snavely 2020; Wiles, Gkioxari *et al.* 2020; Figure 14.22d), and to make the monocular depth predictions consistent over time (Luo, Huang *et al.* 2020; Teed and Deng 2020a; Kopf, Rong, and Huang 2021). An example of a consumer application of monocular depth inference is One Shot 3D Photography (Kopf, Matzen *et al.* 2020), where the system, implemented on a mobile phone using a compact and efficient DNN, first infers a depth map, then converts this to multiple layers, inpaints the background, creates a mesh and texture atlas, and then provides real-time interactive viewing on the phone, as shown in Figure 14.10c.

Figure 12.32 Monocular depth map estimates and novel views from images in COCO dataset (Ranftl, Lasinger *et al.* 2020) © 2020 IEEE.

12.9 Additional reading

The field of stereo correspondence and depth estimation is one of the oldest and most widely studied topics in computer vision. A number of good surveys have been written over the years (Marr and Poggio 1976; Barnard and Fischler 1982; Dhond and Aggarwal 1989; Scharstein and Szeliski 2002; Brown, Burschka, and Hager 2003; Seitz, Curless *et al.* 2006; Furukawa and Hernández 2015; Janai, Güney *et al.* 2020; Laga, Jospin *et al.* 2020; Poggi, Tosi *et al.* 2021) and they can serve as good guides to this extensive literature.

Because of computational limitations and the desire to find appearance-invariant correspondences, early algorithms often focused on finding *sparse correspondences* (Hannah 1974; Marr and Poggio 1979; Mayhew and Frisby 1980; Ohta and Kanade 1985; Bolles, Baker, and Marimont 1987; Matthies, Kanade, and Szeliski 1989).

The topic of computing epipolar geometry and pre-rectifying images is covered in Sections 11.3 and 12.1 and is also treated in textbooks on multi-view geometry (Faugeras and Luong 2001; Hartley and Zisserman 2004) and articles specifically on this topic (Torr and Murray 1997; Zhang 1998a,b). The concepts of the *disparity space* and *disparity space image* are often associated with the seminal work by Marr (1982) and the papers of Yang, Yuille, and Lu (1993) and Intille and Bobick (1994). The plane sweep algorithm was first popularized by Collins (1996) and then generalized to a full arbitrary projective setting by Szeliski and Golland (1999) and Saito and Kanade (1999). Plane sweeps can also be formulated using cylindrical surfaces (Ishiguro, Yamamoto, and Tsuji 1992; Kang and Szeliski 1997; Shum and Szeliski 1999; Li, Shum *et al.* 2004; Zheng, Kang *et al.* 2007) or even more general topologies (Seitz 2001).

Once the topology for the cost volume or DSI has been set up, we need to compute the actual photoconsistency measures for each pixel and potential depth. A wide range of such measures have been proposed, as discussed in Section 12.3.1. Some of these are compared in recent surveys and evaluations of matching costs (Scharstein and Szeliski 2002; Hirschmüller and Scharstein 2009).

To compute an actual depth map from these costs, some form of optimization or selection criterion must be used. The simplest of these are sliding windows of various kinds, which are discussed in Section 12.4 and surveyed by Gong, Yang *et al.* (2007) and Tombari, Mattoccia *et al.* (2008). Global optimization frameworks are often used to compute the best disparity field, as described in Section 12.5. These techniques include dynamic programming and truly global optimization algorithms, such as graph cuts and loopy belief propagation. More recently, deep neural networks have become popular for computing correspondence and disparity maps, as discussed in Section 12.6 and surveyed in Laga, Jospin *et al.* (2020) and Poggi, Tosi *et al.* (2021). A good place to find pointers to the latest results in this field is the list of benchmarks in Table 12.1.

Algorithms for multi-view stereo typically fall into two categories (Furukawa and Hernández 2015). The first include algorithms that compute traditional depth maps using several images for computing photoconsistency measures (Okutomi and Kanade 1993; Kang, Webb *et al.* 1995; Szeliski and Golland 1999; Vaish, Szeliski *et al.* 2006; Gallup, Frahm *et al.* 2008; Huang, Matzen *et al.* 2018; Yao, Luo *et al.* 2018). Some of these techniques compute multiple depth maps and use additional constraints to encourage the different depth maps to be consistent (Szeliski 1999a; Kolmogorov and Zabih 2002; Kang and Szeliski 2004; Maitre, Shinagawa, and Do 2008; Zhang, Jia *et al.* 2008; Yan, Wei *et al.* 2020; Zhang, Yao *et al.* 2020).

The second category consists of papers that compute true 3D volumetric or surface-based object models. Again, because of the large number of papers published on this topic, rather than citing them here, we refer you to the material in Section 12.7.2, the surveys by Seitz, Curless *et al.* (2006), Furukawa and Hernández (2015), and Janai, Güney *et al.* (2020), and the online evaluation websites listed in Table 12.1.

The topic of monocular depth inference is currently very active. Good places to start, in addition to Section 12.8, are the recent surveys by Poggi, Tosi *et al.* (2021, Section 7) and Zhao, Sun *et al.* (2020).

12.10 Exercises

Ex 12.1: Stereo pair rectification. Implement the following simple algorithm (Section 12.1.1):

1. Rotate both cameras so that they are looking perpendicular to the line joining the two camera centers $\mathbf{c}_0$ and $\mathbf{c}_1$. The smallest rotation can be computed from the cross product between the original and desired optical axes.

2. Twist the optical axes so that the horizontal axis of each camera looks in the direction of the other camera. (Again, the cross product between the current x-axis after the first rotation and the line joining the cameras gives the rotation.)

3. If needed, scale up the smaller (less detailed) image so that it has the same resolution (and hence line-to-line correspondence) as the other image.

Now compare your results to the algorithm proposed by Loop and Zhang (1999). Can you think of situations where their approach may be preferable?

Ex 12.2: Rigid direct alignment. Modify your spline-based or optical flow motion estimator from Exercise 9.4 to use epipolar geometry, i.e. to only estimate disparity.

(Optional) Extend your algorithm to simultaneously estimate the epipolar geometry (without first using point correspondences) by estimating a base homography corresponding to a reference plane for the dominant motion and then an epipole for the residual parallax (motion).

Ex 12.3: Shape from profiles. Reconstruct a surface model from a series of edge images (Section 12.2.1).

1. Extract edges and link them (Exercises 7.7–7.8).
2. Based on previously computed epipolar geometry, match up edges in triplets (or longer sets) of images.
3. Reconstruct the 3D locations of the curves using osculating circles (Szeliski and Weiss 1998).
4. Render the resulting 3D surface model as a sparse mesh, i.e., drawing the reconstructed 3D profile curves and links between 3D points in neighboring images with similar osculating circles.

Ex 12.4: Plane sweep. Implement a plane sweep algorithm (Section 12.1.2).

If the images are already pre-rectified, this consists simply of shifting images relative to each other and comparing pixels. If the images are not pre-rectified, compute the homography that resamples the target image into the reference image's coordinate system for each plane.

Evaluate a subset of the following similarity measures (Section 12.3.1) and compare their performance by visualizing the disparity space image (DSI), which should be dark for pixels at correct depths:

- squared difference (SD);
- absolute difference (AD);
- truncated or robust measures;
- gradient differences;
- rank or census transform (the latter usually performs better);
- mutual information from a precomputed joint density function.

Consider using the Birchfield and Tomasi (1998) technique of comparing ranges between neighboring pixels (different shifted or warped images). Also, try pre-compensating images for bias or gain variations using one or more of the techniques discussed in Section 12.3.1.

Ex 12.5: Aggregation and window-based stereo. Implement one or more of the matching cost aggregation strategies described in Section 12.4:

- convolution with a box or Gaussian kernel;
- shifting window locations by applying a min filter (Scharstein and Szeliski 2002);
- picking a window that maximizes some match-reliability metric (Veksler 2001, 2003);
- weighting pixels by their similarity to the central pixel (Yoon and Kweon 2006).

Once you have aggregated the costs in the DSI, pick the winner at each pixel (winner-take-all), and then optionally perform one or more of the following post-processing steps:

1. compute matches both ways and pick only the reliable matches (draw the others in another color);

2. tag matches that are unsure (whose confidence is too low);
3. fill in the matches that are unsure from neighboring values;
4. refine your matches to sub-pixel disparity by either fitting a parabola to the DSI values around the winner or by using an iteration of Lukas–Kanade.

Ex 12.6: Optimization-based stereo. Compute the disparity space image (DSI) volume using one of the techniques you implemented in Exercise 12.4 and then implement one (or more) of the global optimization techniques described in Section 12.5 to compute the depth map. Potential choices include:

- dynamic programming or scanline optimization (relatively easy);
- semi-global optimization (Hirschmüller 2008), which is a simple extension of scanline optimization and performs well;
- graph cuts using alpha expansions (Boykov, Veksler, and Zabih 2001), for which you will need to find a max-flow or min-cut algorithm (https://vision.middlebury.edu/stereo);
- loopy belief propagation (Freeman, Pasztor, and Carmichael 2000);
- deep neural networks, as described in Section 12.6.

Evaluate your algorithm by running it on the Middlebury stereo datasets.

How well does your algorithm do against local aggregation (Yoon and Kweon 2006)? Can you think of some extensions or modifications to make it even better?

Ex 12.7: View interpolation, revisited. Compute a dense depth map using one of the techniques you developed above and use it (or, better yet, a depth map for each source image) to generate smooth in-between views from a stereo dataset.

Compare your results against using the ground truth depth data (if available).

What kinds of artifacts do you see? Can you think of ways to reduce them?

More details on implementing such algorithms can be found in Section 14.1 and Exercises 14.1–14.4.

Ex 12.8: Multi-frame stereo. Extend one of your previous techniques to use multiple input frames (Section 12.7) and try to improve the results you obtained with just two views.

If helpful, try using temporal selection (Kang and Szeliski 2004) to deal with the increased number of occlusions in multi-frame datasets.

You can also try to simultaneously estimate multiple depth maps and make them consistent (Kolmogorov and Zabih 2002; Kang and Szeliski 2004).

Or just use one of the latest DNN-based multi-view stereo algorithms.

Test your algorithms out on some standard multi-view datasets.

Ex 12.9: Volumetric stereo. Implement voxel coloring (Seitz and Dyer 1999) as a simple extension to the plane sweep algorithm you implemented in Exercise 12.4.

1. Instead of computing the complete DSI all at once, evaluate each plane one at a time from front to back.
2. Tag every voxel whose photoconsistency is below a certain threshold as being part of the object and remember its average (or robust) color (Seitz and Dyer 1999; Eisert, Steinbach, and Girod 2000; Kutulakos 2000; Slabaugh, Culbertson *et al.* 2004).

3. Erase the input pixels corresponding to tagged voxels in the input images, e.g., by setting their alpha value to 0 (or to some reduced number, depending on occupancy).

4. As you evaluate the next plane, use the source image alpha values to modify your photoconsistency score, e.g., only consider pixels that have full alpha or weight pixels by their alpha values.

5. If the cameras are not all on the same side of your plane sweeps, use space carving (Kutulakos and Seitz 2000) to cycle through different subsets of source images while carving away the volume from different directions.

Ex 12.10: Depth map merging. Use the technique you developed for multi-frame stereo in Exercise 12.8 or a different technique, such as the one described by Goesele, Snavely *et al.* (2007), to compute a depth map for every input image.

Merge these depth maps into a coherent 3D model, e.g., using Poisson surface reconstruction (Kazhdan, Bolitho, and Hoppe 2006).

Ex 12.11: Monocular depth estimation. Test out of the recent monocular depth inference algorithms on your own images. Can you create a "3D photo" effect where wiggling your camera or moving your mouse makes the photo move in 3D. Tabulate the failure cases of the depth inference and conjecture some possible reasons and/or avenues for improvement.

Chapter 13

3D reconstruction

13.1 Shape from X . . . 641
    13.1.1 Shape from shading and photometric stereo . . . 642
    13.1.2 Shape from texture . . . 645
    13.1.3 Shape from focus . . . 646
13.2 3D scanning . . . 647
    13.2.1 Range data merging . . . 650
    13.2.2 *Application*: Digital heritage . . . 654
13.3 Surface representations . . . 654
    13.3.1 Surface interpolation . . . 655
    13.3.2 Surface simplification . . . 656
    13.3.3 Geometry images . . . 656
13.4 Point-based representations . . . 657
13.5 Volumetric representations . . . 658
    13.5.1 Implicit surfaces and level sets . . . 658
13.6 Model-based reconstruction . . . 660
    13.6.1 Architecture . . . 660
    13.6.2 Facial modeling and tracking . . . 663
    13.6.3 *Application*: Facial animation . . . 665
    13.6.4 Human body modeling and tracking . . . 668
13.7 Recovering texture maps and albedos . . . 674
    13.7.1 Estimating BRDFs . . . 675
    13.7.2 *Application*: 3D model capture . . . 676
13.8 Additional reading . . . 677
13.9 Exercises . . . 679

R. Szeliski, *Computer Vision*, Texts in Computer Science,
https://doi.org/10.1007/978-3-030-34372-9_13

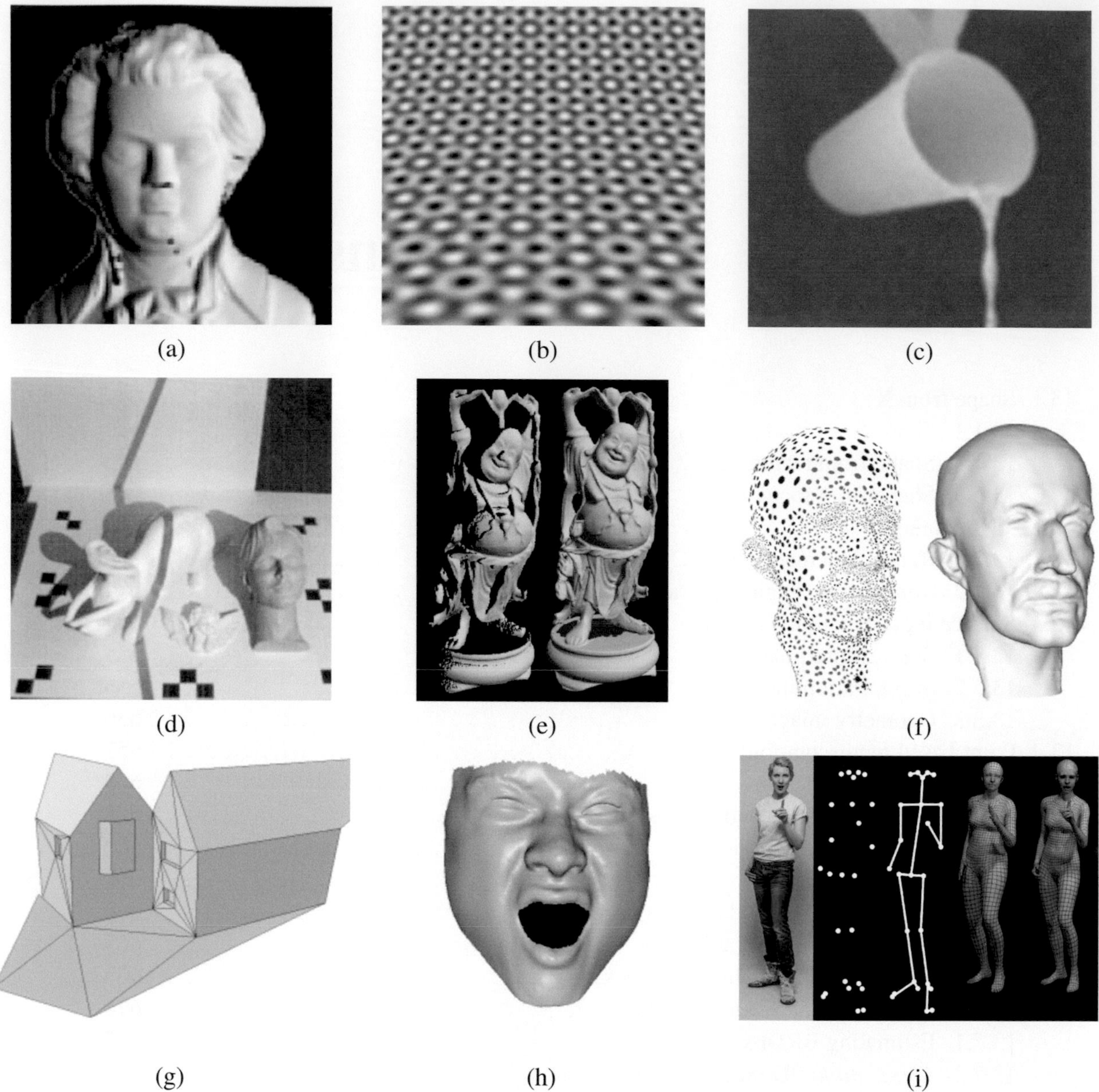

Figure 13.1 3D shape acquisition and modeling techniques: (a) shaded image (Zhang, Tsai *et al.* 1999) © 1999 IEEE; (b) texture gradient (Gårding 1992) © 1992 Springer; (c) real-time depth from focus (Nayar, Watanabe, and Noguchi 1996) © 1996 IEEE; (d) scanning a scene with a stick shadow (Bouguet and Perona 1999) © 1999 Springer; (e) merging range maps into a 3D model (Curless and Levoy 1996) © 1996 ACM; (f) point-based surface modeling (Pauly, Keiser *et al.* 2003) © 2003 ACM; (g) automated modeling of a 3D building using lines and planes (Werner and Zisserman 2002) © 2002 Springer; (h) 3D face model from spacetime stereo (Zhang, Snavely *et al.* 2004) © 2004 ACM; (i) whole body, expression, and gesture fitting from a single image (Pavlakos, Choutas *et al.* 2019) © 2019 IEEE.

As we saw in the previous chapter, many stereo matching techniques have been developed to reconstruct high-quality 3D models from two or more images. However, stereo is just one of the many potential cues that can be used to infer shape from images. In this chapter, we investigate a number of such techniques, which include not only visual cues such as shading and focus, but also techniques for merging multiple range or depth images into 3D models, as well as techniques for reconstructing specialized models, such as heads, bodies, or architecture.

Among the various cues that can be used to infer shape, the shading on a surface (Figure 13.1a) can provide a lot of information about local surface orientations and hence overall surface shape (Section 13.1.1). This approach becomes even more powerful when lights shining from different directions can be turned on and off separately (*photometric stereo*). Texture gradients (Figure 13.1b), i.e., the foreshortening of regular patterns as the surface slants or bends away from the camera, can provide similar cues on local surface orientation (Section 13.1.2). Focus is another powerful cue to scene depth, especially when two or more images with different focus settings are used (Section 13.1.3).

3D shape can also be estimated using active illumination techniques such as light stripes (Figure 13.1d) or time of flight range finders (Section 13.2). The partial surface models obtained using such techniques (or passive image-based stereo) can then be merged into more coherent 3D surface models (Figure 13.1e), as discussed in Section 13.2.1. Such techniques have been used to construct highly detailed and accurate models of cultural heritage such as historic sites (Section 13.2.2). The resulting surface models can then be simplified to support viewing at different resolutions and streaming across the web (Section 13.3.2). An alternative to working with continuous surfaces is to represent 3D surfaces as dense collections of 3D oriented points (Section 13.4) or as volumetric primitives (Section 13.5).

3D modeling can be more efficient and effective if we know something about the objects we are trying to reconstruct. In Section 13.6, we look at three specialized but commonly occurring examples, namely architecture (Figure 13.1g), heads and faces (Figure 13.1h), and whole bodies (Figure 13.1i). In addition to modeling people, we also discuss techniques for tracking them.

The last stage of shape and appearance modeling is to extract some colored textures to paint onto our 3D models (Section 13.7). Some techniques go beyond this and actually estimate full BRDFs (Section 13.7.1), although if there is no desire to re-light the scene, Surface Light Fields may be easier to acquire (Section 14.3.2).

Because there exists such a large variety of techniques to perform 3D modeling, this chapter does not go into detail on any one of these. Readers are encouraged to find more information in the cited references and recent computer vision conferences, as well as more specialized conferences devoted to these topics, e.g., the International Conference on 3D Vision (3DV) and the IEEE International Conference on Automatic Face and Gesture Recognition (FG).

13.1 Shape from X

In addition to binocular disparity, shading, texture, and focus all play a role in how we perceive shape. The study of how shape can be inferred from such cues is sometimes called *shape from X*, because the individual instances are called *shape from shading*, *shape from texture*, and *shape from focus*.[1] In this section, we look at these three cues and how they can be used to reconstruct 3D geometry. A good overview of all these topics can be found in the collection of papers on physics-based shape inference edited by Wolff, Shafer, and Healey (1992b), the survey by Ackermann and Goesele (2015) and the book by Ikeuchi, Matsushita *et al.* (2020).

[1] We have already seen examples of shape from stereo, shape from profiles, and shape from silhouettes in Chapter 12.

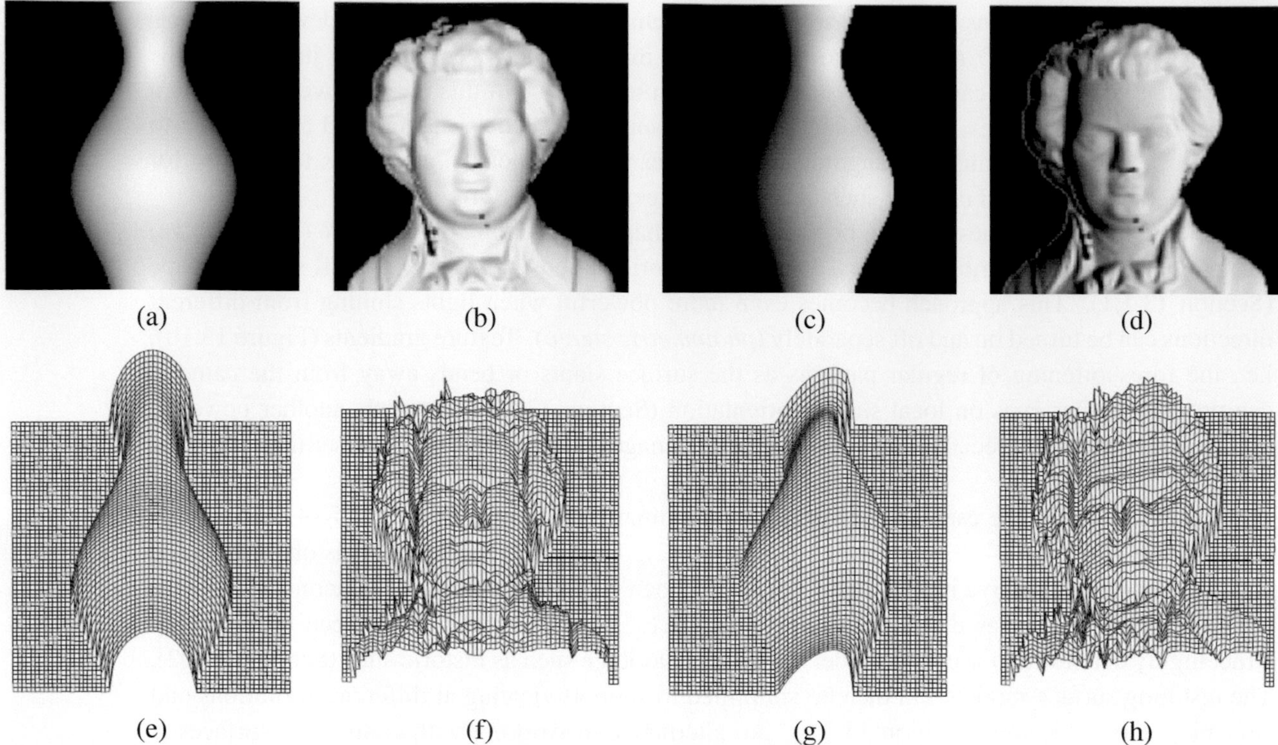

Figure 13.2 Synthetic shape from shading (Zhang, Tsai *et al.* 1999) © 1999 IEEE: shaded images, (a–b) with light from in front $(0, 0, 1)$ and (c–d) with light from the front right $(1, 0, 1)$; (e–f) corresponding shape from shading reconstructions using the technique of Tsai and Shah (1994).

13.1.1 Shape from shading and photometric stereo

When you look at images of smooth shaded objects, such as the ones shown in Figure 13.2, you can clearly see the shape of the object from just the shading variation. How is this possible? The answer is that as the surface normal changes across the object, the apparent brightness changes as a function of the angle between the local surface orientation and the incident illumination, as shown in Figure 2.15 (Section 2.2.2).

The problem of recovering the shape of a surface from this intensity variation is known as *shape from shading* and is one of the classic problems in computer vision (Horn 1975). The collection of papers edited by Horn and Brooks (1989) is a great source of information on this topic, especially the chapter on variational approaches. The survey by Zhang, Tsai *et al.* (1999) not only reviews more recent techniques, but also provides some comparative results.

Most shape from shading algorithms assume that the surface under consideration is of a uniform albedo and reflectance, and that the light source directions are either known or can be calibrated by the use of a reference object. Under the assumptions of distant light sources and observer, the variation in intensity (*irradiance equation*) becomes purely a function of the local surface orientation,

$$I(x, y) = R(p(x, y), q(x, y)), \tag{13.1}$$

where $(p, q) = (z_x, z_y)$ are the depth map derivatives and $R(p, q)$ is called the *reflectance map*. For example, a diffuse (Lambertian) surface has a reflectance map that is the (non-negative) dot product (2.89) between the surface normal $\hat{\mathbf{n}} = (p, q, 1)/\sqrt{1 + p^2 + q^2}$ and the light source direction $\mathbf{v} =$

(v_x, v_y, v_z),

$$R(p,q) = \max\left(0, \rho \frac{pv_x + qv_y + v_z}{\sqrt{1+p^2+q^2}}\right), \tag{13.2}$$

where ρ is the surface reflectance factor (albedo).

In principle, Equations (13.1–13.2) can be used to estimate (p, q) using non-linear least squares or some other method. Unfortunately, unless additional constraints are imposed, there are more unknowns per pixel (p, q) than there are measurements (I). One commonly used constraint is the smoothness constraint,

$$\mathcal{E}_s = \int p_x^2 + p_y^2 + q_x^2 + q_y^2 \, dx\, dy = \int \|\nabla p\|^2 + \|\nabla q\|^2 \, dx\, dy, \tag{13.3}$$

which we have already seen in Section 4.2 (4.18). The other is the *integrability constraint*,

$$\mathcal{E}_i = \int (p_y - q_x)^2 \, dx\, dy, \tag{13.4}$$

which arises naturally, because for a valid depth map $z(x, y)$ with $(p, q) = (z_x, z_y)$, we have $p_y = z_{xy} = z_{yx} = q_x$.

Instead of first recovering the orientation fields (p, q) and integrating them to obtain a surface, it is also possible to directly minimize the discrepancy in the image formation equation (13.1) while finding the optimal depth map $z(x, y)$ (Horn 1990). Unfortunately, shape from shading is susceptible to local minima in the search space and, like other variational problems that involve the simultaneous estimation of many variables, can also suffer from slow convergence. Using multi-resolution techniques (Szeliski 1991a) can help accelerate the convergence, while using more sophisticated optimization techniques (Dupuis and Oliensis 1994) can help avoid local minima.

In practice, surfaces other than plaster casts are rarely of a single uniform albedo. Shape from shading therefore needs to be combined with some other technique or extended in some way to make it useful. One way to do this is to combine it with stereo matching (Fua and Leclerc 1995; Logothetis, Mecca, and Cipolla 2019) or known texture (surface patterns) (White and Forsyth 2006). The stereo and texture components provide information in textured regions, while shape from shading helps fill in the information across uniformly colored regions and also provides finer information about surface shape.

Photometric stereo. Another way to make shape from shading more reliable is to use multiple light sources that can be selectively turned on and off. This technique is called *photometric stereo*, as the light sources play a role analogous to the cameras located at different locations in traditional stereo (Woodham 1981).[2] For each light source, we have a different reflectance map, $R_1(p, q)$, $R_2(p, q)$, etc. Given the corresponding intensities I_1, I_2, etc. at a pixel, we can in principle recover both an unknown albedo ρ and a surface orientation estimate (p, q).

For diffuse surfaces (13.2), if we parameterize the local orientation by $\hat{\mathbf{n}}$, we get (for non-shadowed pixels) a set of linear equations of the form

$$I_k = \rho \hat{\mathbf{n}} \cdot \mathbf{v}_k, \tag{13.5}$$

from which we can recover $\rho\hat{\mathbf{n}}$ using linear least squares. These equations are well conditioned as long as the (three or more) vectors $\mathbf{v}_k$ are linearly independent, i.e., they are not along the same azimuth (direction away from the viewer).

[2] An alternative to turning lights on-and-off is to use three colored lights (Woodham 1994; Hernandez, Vogiatzis *et al.* 2007; Hernández and Vogiatzis 2010).

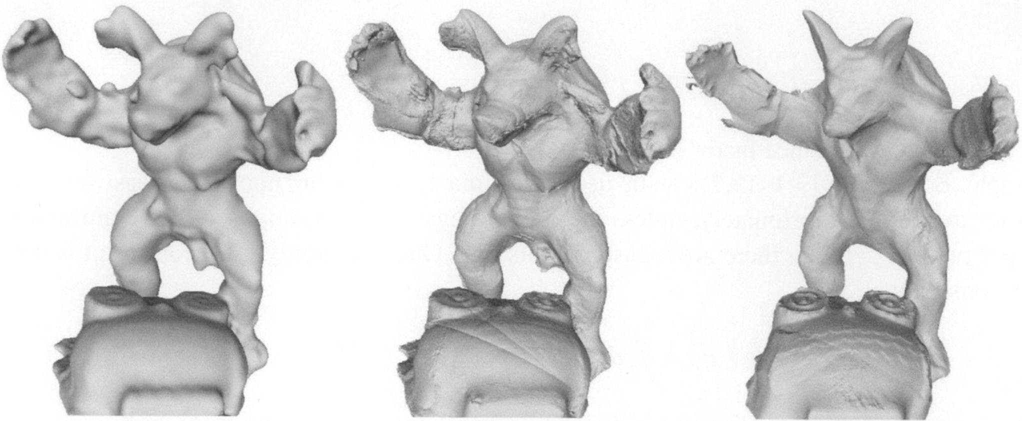

Figure 13.3 Multi-view photometric stereo (Logothetis, Mecca, and Cipolla 2019) © 2019 IEEE: initial COLMAP multi-view stereo reconstruction; refined with (Park, Sinha *et al.* 2017); and (Logothetis, Mecca, and Cipolla 2019).

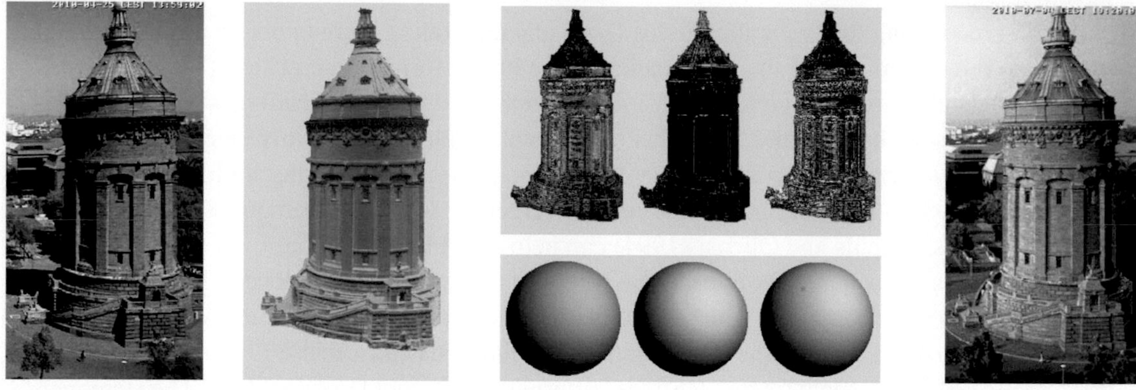

Figure 13.4 Webcam-based outdoor photometric stereo (Ackermann, Langguth *et al.* 2012) © 2012 IEEE: an input image, the recovered normal map, three basis BRDFs below their respective material maps, and a synthetic rendering from a new sun position.

Once the surface normals or gradients have been recovered at each pixel, they can be integrated into a depth map using a variant of regularized surface fitting (4.24). Nehab, Rusinkiewicz *et al.* (2005) and Harker and O'Leary (2008) discuss more sophisticated techniques for doing this. The combination of multi-view stereo for coarse shape and photometric stereo for fine detail continues to be an active area of research (Hernández, Vogiatzis, and Cipolla 2008; Wu, Liu *et al.* 2010; Park, Sinha *et al.* 2017). Logothetis, Mecca, and Cipolla (2019) describe such a system that can produce very high-quality scans (Figure 13.3), although it requires a sophisticated laboratory setup. A more practical setup that only requires a stereo camera and a flash to produce a flash/non-flash pair is describe by Cao, Waechter *et al.* (2020). It is also possible to apply photometric stereo to outdoor web camera sequences (Figure 13.4), using the trajectory of the Sun as a variable direction illuminator (Ackermann, Langguth *et al.* 2012).

When surfaces are specular, more than three light directions may be required. In fact, the irradiance equation given in (13.1) not only requires that the light sources and camera be distant from the surface, it also neglects inter-reflections, which can be a significant source of the shading ob-

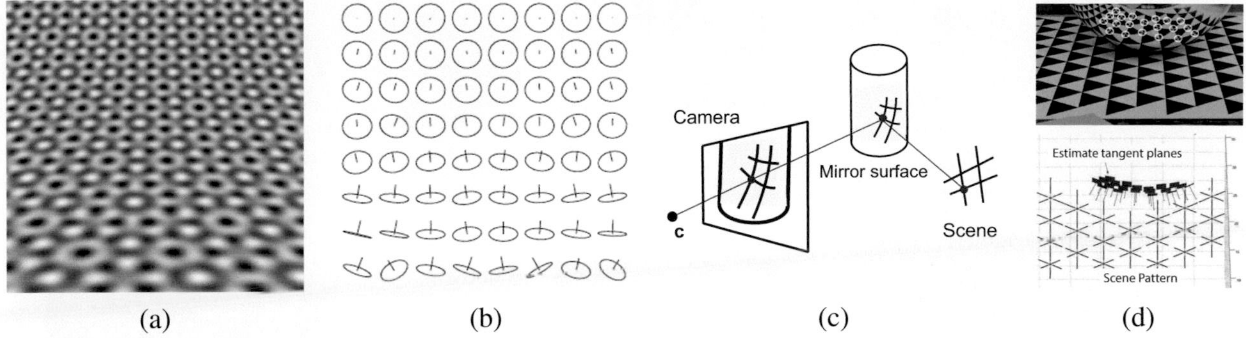

(a) (b) (c) (d)

Figure 13.5 Synthetic shape from texture (Gårding 1992) © 1992 Springer: (a) regular texture wrapped onto a curved surface and (b) the corresponding surface normal estimates. Shape from mirror reflections (Savarese, Chen, and Perona 2005) © 2005 Springer: (c) a regular pattern reflecting off a curved mirror gives rise to (d) curved lines, from which 3D point locations and normals can be inferred.

served on object surfaces, e.g., the darkening seen inside concave structures such as grooves and crevasses (Nayar, Ikeuchi, and Kanade 1991). However, if one can control the placements of lights and cameras so that they are *reciprocal*, i.e., the position of lights and cameras can be (conceptually) switched, it is possible to recover constraints on surface depths and normals using a procedure known as *Helmholtz stereopsis* (Zickler, Belhumeur, and Kriegman 2002).

While earlier work on photometric stereo assumed known illuminant directions and reflectance (BRDF) functions, more recent work aims to loosen these constraints. Ackermann and Goesele (2015) provide an extensive survey of such techniques, while Shi, Mo *et al.* (2019) describe their DiLiGenT dataset and benchmark for evaluating non-Lambertian photometric stereo and cite over 100 related papers. As with other areas of computer vision, deep networks and end-to-end learning are now commonly used to to recover shape and illuminant direction from photometrics stereo. Some recent papers include Chen, Han *et al.* (2019), Li, Robles-Kelly *et al.* (2019), Haefner, Ye *et al.* (2019), Chen, Waechter *et al.* (2020), and Santo, Waechter, and Matsushita (2020).

13.1.2 Shape from texture

The variation in foreshortening observed in regular textures can also provide useful information about local surface orientation. Figure 13.5 shows an example of such a pattern, along with the estimated local surface orientations. Shape from texture algorithms require a number of processing steps, including the extraction of repeated patterns or the measurement of local frequencies to compute local affine deformations, and a subsequent stage to infer local surface orientation. Details on these various stages can be found in the research literature (Witkin 1981; Ikeuchi 1981; Blostein and Ahuja 1987; Gårding 1992; Malik and Rosenholtz 1997; Lobay and Forsyth 2006). A more recent paper uses a generative model to represent the repetitive appearance of textures and jointly optimizes the model along with the local surface orientations at every pixel (Verbin and Zickler 2020).

When the original pattern is regular, it is possible to fit a regular but slightly deformed grid to the image and use this grid for a variety of image replacement or analysis tasks (Liu, Collins, and Tsin 2004; Liu, Lin, and Hays 2004; Hays, Leordeanu *et al.* 2006; Lin, Hays *et al.* 2006; Park, Brocklehurst *et al.* 2009). This process becomes even easier if specially printed textured cloth patterns are used (White and Forsyth 2006; White, Crane, and Forsyth 2007).

The deformations induced in a regular pattern when it is viewed in the reflection of a curved mirror, as shown in Figure 13.5c–d, can be used to recover the shape of the surface (Savarese, Chen,

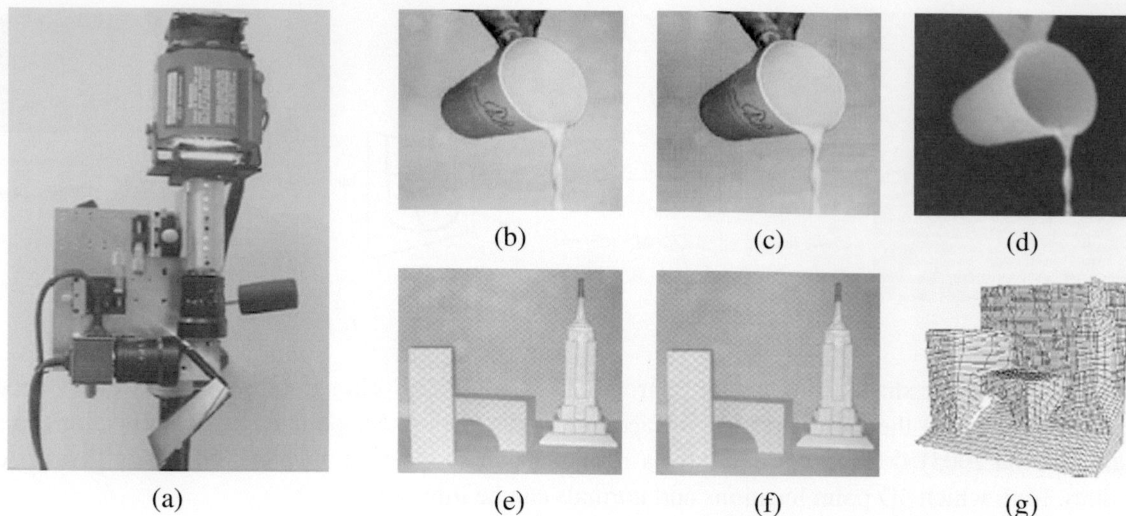

Figure 13.6 Real-time depth from defocus (Nayar, Watanabe, and Noguchi 1996) © 1996 IEEE: (a) the real-time focus range sensor, which includes a half-silvered mirror between the two telecentric lenses (lower right), a prism that splits the image into two CCD sensors (lower left), and an edged checkerboard pattern illuminated by a Xenon lamp (top); (b–c) input video frames from the two cameras along with (d) the corresponding depth map; (e–f) two frames (you can see the texture if you zoom in) and (g) the corresponding 3D mesh model.

and Perona 2005; Rozenfeld, Shimshoni, and Lindenbaum 2011). It is also possible to infer local shape information from *specular flow*, i.e., the motion of specularities when viewed from a moving camera (Oren and Nayar 1997; Zisserman, Giblin, and Blake 1989; Swaminathan, Kang *et al.* 2002).

13.1.3 Shape from focus

A strong cue for object depth is the amount of blur, which increases as the object's surface moves away from the camera's focusing distance. As shown in Figure 2.19, moving the object surface away from the focus plane increases the circle of confusion, according to a formula that is easy to establish using similar triangles (Exercise 2.4).

A number of techniques have been developed to estimate depth from the amount of defocus (*depth from defocus*) (Pentland 1987; Nayar and Nakagawa 1994; Nayar, Watanabe, and Noguchi 1996; Watanabe and Nayar 1998; Chaudhuri and Rajagopalan 1999; Favaro and Soatto 2006). To make such a technique practical, a number of issues need to be addressed:

- The amount of blur increase in *both* directions as you move away from the focus plane. Therefore, it is necessary to use two or more images captured with different focus distance settings (Pentland 1987; Nayar, Watanabe, and Noguchi 1996) or to translate the object in depth and look for the point of maximum sharpness (Nayar and Nakagawa 1994).

- The magnification of the object can vary as the focus distance is changed or the object is moved. This can be modeled either explicitly (making correspondence more difficult) or using *telecentric optics*, which approximate an orthographic camera and require an aperture in front of the lens (Nayar, Watanabe, and Noguchi 1996).

- The amount of defocus must be reliably estimated. A simple approach is to average the squared gradient in a region, but this suffers from several problems, including the image mag-

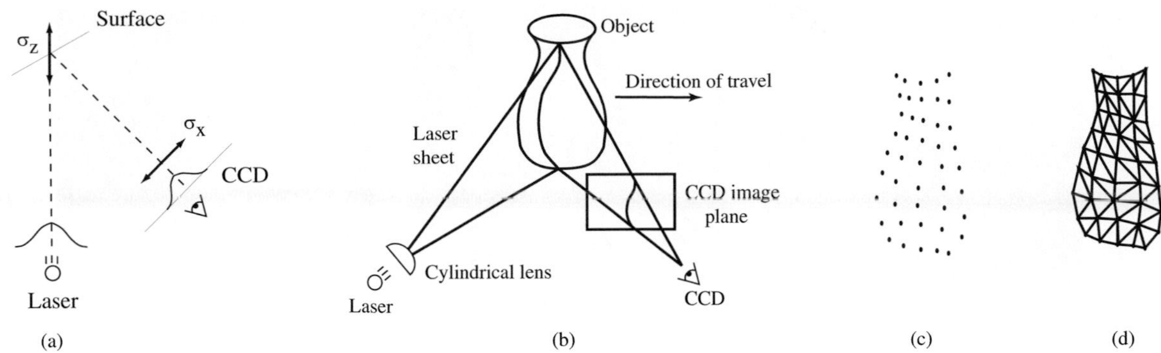

Figure 13.7 Range data scanning (Curless and Levoy 1996) © 1996 ACM: (a) a laser dot on a surface is imaged by a CCD sensor; (b) a laser stripe (sheet) is imaged by the sensor (the deformation of the stripe encodes the distance to the object); (c) the resulting set of 3D points are turned into (d) a triangulated mesh.

nification problem mentioned above. A better solution is to use carefully designed *rational filters* (Watanabe and Nayar 1998).

Figure 13.6 shows an example of a real-time depth from defocus sensor, which employs two imaging chips at slightly different depths sharing a common optical path, as well as an active illumination system that projects a checkerboard pattern from the same direction. As you can see in Figure 13.6b–g, the system produces high-accuracy real-time depth maps for both static and dynamic scenes.

13.2 3D scanning

As we have seen in the previous section, actively lighting a scene, whether for the purpose of estimating normals using photometric stereo, or for adding artificial texture for shape from defocus, can greatly improve the performance of vision systems. This kind of *active illumination* has been used from the earliest days of machine vision to construct highly precise sensors for estimating 3D depth images using a variety of *rangefinding* (or *range sensing*) techniques (Besl 1989; Curless 1999; Hebert 2000; Zhang 2018).[3] While rangefinders such as lidar (Light Detection and Ranging) and laser-based 3D scanners were once limited to commercial and laboratory applications, the development of low-cost *depth cameras* such as the Microsoft Kinect (Zhang 2012) have revolutionized many aspects of computer vision. It is now common to refer to the registered color and depth frames produced by such cameras as *RGB-D* (or *RGBD*) images (Silberman, Hoiem *et al.* 2012).

One of the early active illumination sensors used in computer vision and computer graphics was a laser or light stripe sensor, which sweeps a plane of light across the scene or object while observing it from an offset viewpoint, as shown in Figure 13.7b (Rioux and Bird 1993; Curless and Levoy 1995). As the stripe falls across the object, it deforms its shape according to the shape of the surface it is illuminating. It is then a simple matter of using *optical triangulation* to estimate the 3D locations of all the points seen in a particular stripe. In more detail, knowledge of the 3D plane equation of the light stripe allows us to infer the 3D location corresponding to each illuminated pixel, as previously discussed in (2.70–2.71). The accuracy of light striping techniques can be improved by finding the exact temporal peak in illumination for each pixel (Curless and Levoy 1995). The final accuracy of

[3] Rangefinding is an old-fashioned word for measuring distance, often using passive or active optical means.

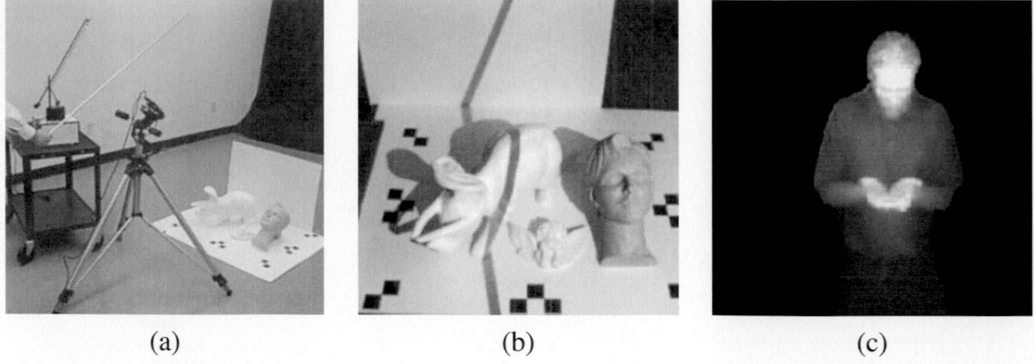

(a) (b) (c)

Figure 13.8 Shape scanning using cast shadows (Bouguet and Perona 1999) © 1999 Springer: (a) camera setup with a point light source (a desk lamp without its reflector), a hand-held stick casting a shadow, and (b) the objects being scanned in front of two planar backgrounds. (c) Real-time depth map using a pulsed illumination system (Iddan and Yahav 2001) © 2001 SPIE.

a scanner can be determined using slant edge modulation techniques, i.e., by imaging sharp creases in a calibration object (Goesele, Fuchs, and Seidel 2003).

An interesting variant on light stripe rangefinding is presented by Bouguet and Perona (1999). Instead of projecting a light stripe, they simply wave a stick casting a shadow over a scene or object illuminated by a point light source such as a lamp or the Sun (Figure 13.8a). As the shadow falls across two background planes whose orientation relative to the camera is known (or inferred during pre-calibration), the plane equation for each stripe can be inferred from the two projected lines, whose 3D equations are known (Figure 13.8b). The deformation of the shadow as it crosses the object being scanned then reveals its 3D shape, as with regular light stripe rangefinding (Exercise 13.2). This technique can also be used to estimate the 3D geometry of a background scene and how its appearance varies as it moves into shadow, to cast new shadows onto the scene (Chuang, Goldman *et al.* 2003) (Section 10.4.3).

The time it takes to scan an object using a light stripe technique is proportional to the number of depth planes used, which is usually comparable to the number of pixels across an image. A much faster scanner can be constructed by turning different projector pixels on and off in a structured manner, e.g., using a binary or Gray code (Besl 1989). For example, let us assume that the LCD projector we are using has 1,024 columns of pixels. Taking the 10-bit binary code corresponding to each column's address (0...1,023), we project the first bit, then the second, etc. After 10 projections (e.g., a third of a second for a synchronized 30Hz camera-projector system), each pixel in the camera knows which of the 1,024 columns of projector light it is seeing. A similar approach can also be used to estimate the refractive properties of an object by placing a monitor behind the object (Zongker, Werner *et al.* 1999; Chuang, Zongker *et al.* 2000) (Section 14.4). Very fast scanners can also be constructed with a single laser beam, i.e., a real-time *flying spot* optical triangulation scanner (Rioux, Bechthold *et al.* 1987).

If even faster, i.e., frame-rate, scanning is required, we can project a single textured pattern into the scene. Proesmans, Van Gool, and Defoort (1998) describe a system where a checkerboard grid is projected onto an object and the deformation of the grid is used to infer 3D shape. Unfortunately, such a technique only works if the surface is continuous enough to link all of the grid points. Instead of projecting a grid, it is also possible to project one or more sinusoidal *fringe patterns* and to then recover deformations in the surface from the relative phase displacements using a process called *fringe projection profilometry* (Su and Zhang 2010; Zuo, Huang *et al.* 2016; Zhang 2018).

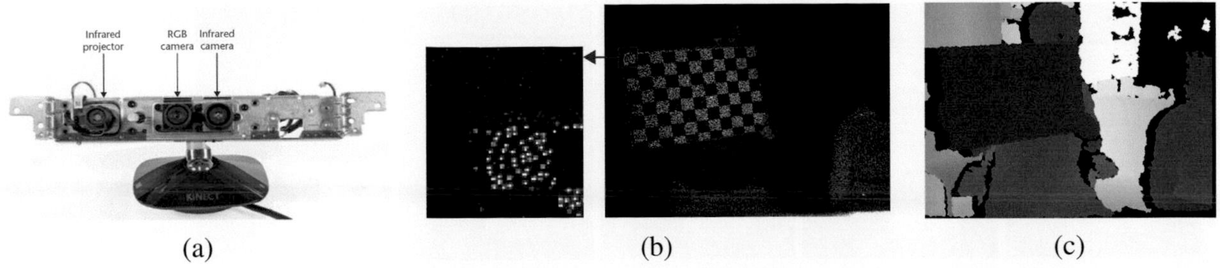

Figure 13.9 The Microsoft Kinect depth camera (Zhang 2012) © 2012 IEEE: (a) the hardware, comprising an infrared (IR) speckle pattern projector and a color and IR camera pair; (b) close-up of a sample infrared image, showing the projected dots; (c) the final depth map, which has black "shadows" in the areas not illuminated by the projector.

The Microsoft Kinect (Zhang 2012) depth camera uses a variant of this technique, projecting an infrared (IR) *speckle pattern*, which looks like a bunch of random dots, but which in fact consists of a known calibrated pseudo-random pattern (Figure 13.9). By measuring the horizontal displacement (parallax) between the dots seen in the IR camera and their expected locations, a depth map can be computed, interpolating over the pixels not illuminated by the dots (Fanello, Rhemann *et al.* 2016; Fanello, Valentin *et al.* 2017b). Since its release, the Kinect camera has been widely used in computer vision research (Zhang 2012; Han, Shao *et al.* 2013), as well as applications such as 3D body tracking (Section 13.6.4) and object scanning and home interior reconstruction (Section 13.2.1). Kinect sensors were used to create the first widely used dataset for 3D semantic scene understanding (Silberman, Hoiem *et al.* 2012), although larger 3D scanned datasets have since been created (Dai, Chang *et al.* 2017).

A higher resolution system can be constructed using high-speed custom illumination and sensing hardware. Iddan and Yahav (2001) describe the construction of their 3DV Zcam video-rate depth sensing camera, which projects a pulsed plane of light onto the scene and then integrates the returning light for a short interval, essentially obtaining time-of-flight measurement for the distance to individual pixels in the scene. A good description of earlier time-of-flight systems, including amplitude and frequency modulation schemes for lidar, can be found in (Besl 1989), and a more recent description can be found in the book by Hansard, Lee *et al.* (2012). While the initial version of the Microsoft Kinect depth camera used a speckle pattern structured light system (Zhang 2012), the newer Kinect V2 uses a time-of-flight (ToF) sensor that uses phase measurements of amplitude-modulated light signals (Bamji, O'Connor *et al.* 2014). Traditional multi-frequency phase unwrapping techniques can be used to estimate absolute depth, but more accurate depths for dynamic scenes can be obtained by simultaneously modeling depths and object velocities (Stühmer, Nowozin *et al.* 2015).

Instead of using a single camera, it is also possible to construct an active illumination range sensor using stereo imaging setups, resulting in a system that is often called *active (illumination) stereo*. The simplest way to do this is to just project random stripe patterns onto the scene to create synthetic texture, which helps match textureless surfaces (Kang, Webb *et al.* 1995). Projecting a known series of stripes, just as in coded pattern single camera rangefinding, makes the correspondence between pixels unambiguous and allows for the recovery of depth estimates at pixels only seen in a single camera (Scharstein and Szeliski 2003). This technique has been used to produce large numbers of highly accurate registered multi-image stereo pairs and depth maps for the purpose of evaluating stereo correspondence algorithms (Scharstein and Szeliski 2002; Hirschmüller and Scharstein 2009; Scharstein, Hirschmüller *et al.* 2014) and learning depth map priors and parameters (Pal, Wein-

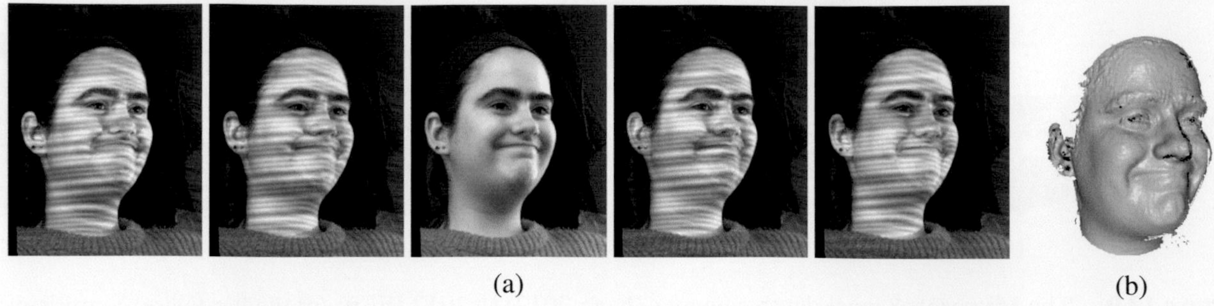

(a) (b)

Figure 13.10 Real-time dense 3D face capture using spacetime stereo (Zhang, Snavely *et al.* 2004) © 2004 ACM: (a) set of five consecutive video frames from one of two stereo cameras (every fifth frame is free of stripe patterns, in order to extract texture); (b) resulting high-quality 3D surface model (depth map visualized as a shaded rendering).

man *et al.* 2012). Carefully designed algorithms can perform local matching of patterns at 500Hz (Fanello, Valentin *et al.* 2017a,b).

While projecting multiple patterns usually requires the scene or object to remain still, additional processing can enable the production of real-time depth maps for dynamic scenes. The basic idea (Davis, Ramamoorthi, and Rusinkiewicz 2003; Zhang, Curless, and Seitz 2003) is to assume that depth is nearly constant within a 3D space–time window around each pixel and to use the 3D window for matching and reconstruction. Depending on the surface shape and motion, this assumption may be error-prone, as shown in Davis, Nahab *et al.* (2005). To model shapes more accurately, Zhang, Curless, and Seitz (2003) model the linear disparity variation within the space–time window and show that better results can be obtained by globally optimizing disparity and disparity gradient estimates over video volumes (Zhang, Snavely *et al.* 2004). Figure 13.10 shows the results of applying this system to a person's face; the frame-rate 3D surface model can then be used for further model-based fitting and computer graphics manipulation (Section 13.6.2). As mentioned previously, motion modeling can also be applied to phase-based time-of-flight sensors (Stühmer, Nowozin *et al.* 2015).

One word of caution about active range sensing. When the surfaces being scanned are too reflective, the camera may see a reflection off the object's surface and assume that this *virtual image* is the true scene. For surfaces with moderate amounts of reflection, such as the ceramic models in Wood, Azuma *et al.* (2000) or the Corn Cho puffs in Park, Newcombe, and Seitz (2018), there is still sufficient diffuse reflection under the specular layer to obtain a 3D range map. (The specular part can then be recovered separately to produce a *surface light field*, as described in Section 14.3.2.) However, for true mirrors, active range scanners will invariably capture the virtual 3D model seen reflected in the mirror, so that additional techniques such as looking for a reflection of the scanning device must be used (Whelan, Goesele *et al.* 2018).

13.2.1 Range data merging

While individual range images can be useful for applications such as real-time z-keying or facial motion capture, they are often used as building blocks for more complete 3D object modeling. In such applications, the next two steps in processing are the registration (alignment) of partial 3D surface models and their integration into coherent 3D surfaces (Curless 1999). If desired, this can be followed by a model fitting stage using either parametric representations such as generalized cylinders (Agin and Binford 1976; Nevatia and Binford 1977; Marr and Nishihara 1978; Brooks

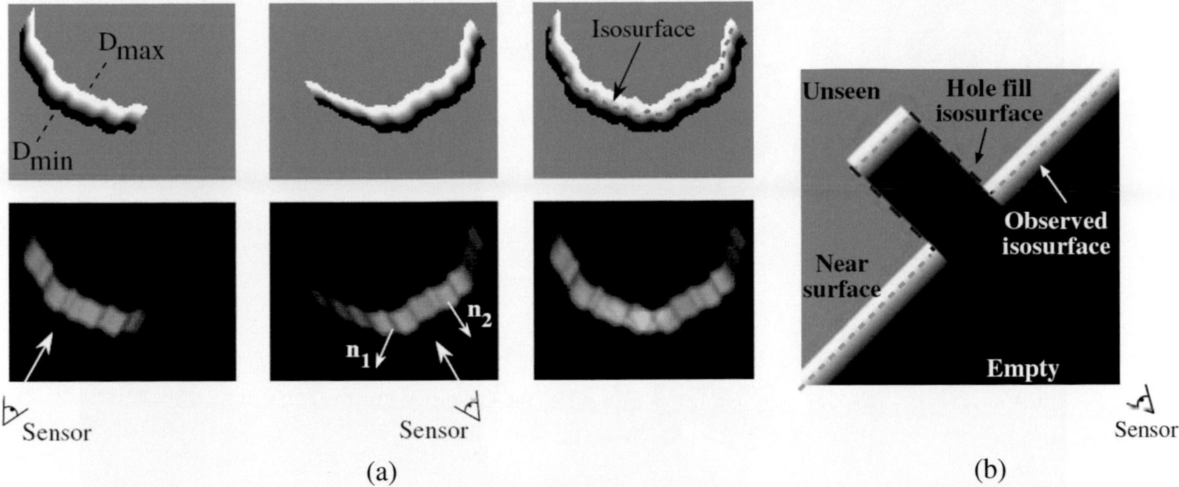

Figure 13.11 Range data merging (Curless and Levoy 1996) © 1996 ACM: (a) two signed distance functions (top left) are merged with their (weights) bottom left to produce a combined set of functions (right column) from which an isosurface can be extracted (green dashed line); (b) the signed distance functions are combined with empty and unseen space labels to fill holes in the isosurface.

1981), superquadrics (Pentland 1986; Solina and Bajcsy 1990; Terzopoulos and Metaxas 1991), or non-parametric models such as triangular meshes (Boissonat 1984) or physically based models (Terzopoulos, Witkin, and Kass 1988; Delingette, Hebert, and Ikeuichi 1992; Terzopoulos and Metaxas 1991; McInerney and Terzopoulos 1993; Terzopoulos 1999). A number of techniques have also been developed for segmenting range images into simpler constituent surfaces (Hoover, Jean-Baptiste *et al.* 1996).

The most widely used 3D registration technique is the *iterative closest point* (ICP) algorithm, which alternates between finding the closest point matches between the two surfaces being aligned and then solving a 3D *absolute orientation* problem (Section 8.1.5, (8.31–8.32) (Besl and McKay 1992; Chen and Medioni 1992; Zhang 1994; Szeliski and Lavallée 1996; Gold, Rangarajan *et al.* 1998; David, DeMenthon *et al.* 2004; Li and Hartley 2007; Enqvist, Josephson, and Kahl 2009). Some techniques, such as the one developed by Chen and Medioni (1992), use local surface tangent planes to make this computation more accurate and to accelerate convergence. More recently, Rusinkiewicz (2019) proposed a symmetric oriented point distance similar to the energy terms used in *oriented particles* (Szeliski and Tonnesen 1992). A nice review of ICP and its related variants can be found in the papers by Tam, Cheng *et al.* (2012) and Pomerleau, Colas, and Siegwart (2015).

As the two surfaces being aligned usually only have partial overlap and may also have outliers, robust matching criteria (Section 8.1.4 and Appendix B.3) are typically used. To speed up the determination of the closest point, and also to make the distance-to-surface computation more accurate, one of the two point sets (e.g., the current merged model) can be converted into a *signed distance function*, optionally represented using an *octree spline* for compactness (Lavallée and Szeliski 1995). Variants on the basic ICP algorithm can be used to register 3D point sets under non-rigid deformations, e.g., for medical applications (Feldmar and Ayache 1996; Szeliski and Lavallée 1996). Color values associated with the points or range measurements can also be used as part of the registration process to improve robustness (Johnson and Kang 1997; Pulli 1999).

Unfortunately, the ICP algorithm and its variants can only find a locally optimal alignment between 3D surfaces. If this is not known *a priori*, more global correspondence or search techniques,

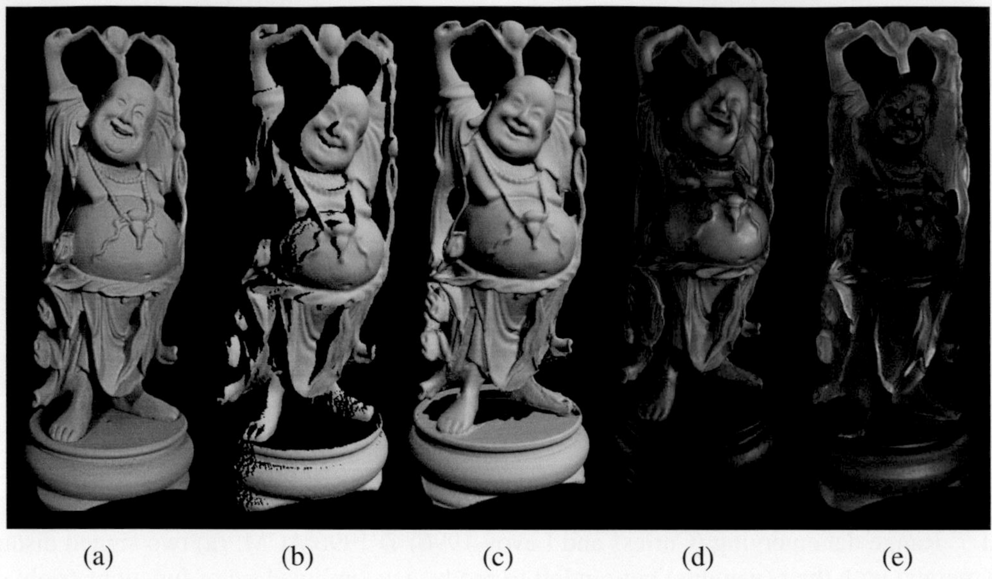

Figure 13.12 Reconstruction and hardcopy of the "Happy Buddha" statuette (Curless and Levoy 1996) © 1996 ACM: (a) photograph of the original statue after spray painting with matte gray; (b) partial range scan; (c) merged range scans; (d) colored rendering of the reconstructed model; (e) hardcopy of the model constructed using stereolithography.

based on local descriptors invariant to 3D rigid transformations, need to be used. An example of such a descriptor is the *spin image*, which is a local circular projection of a 3D surface patch around the local normal axis (Johnson and Hebert 1999). Another (earlier) example is the *splash* representation introduced by Stein and Medioni (1992). More recent work along these lines studies the problem of *pose estimation* (Section 11.2) from RGB-D images, which is essentially the same problem as aligning a range map to a 3D model. Recent papers on this topic (Drost, Ulrich *et al.* 2010; Brachmann, Michel *et al.* 2016; Vidal, Lin *et al.* 2018) typically evaluate themselves on the Benchmark for 6DOF Object Pose Estimation,[4] which also hosts a series of yearly workshops on this topic.

Once two or more 3D surfaces have been aligned, they can be merged into a single model. One approach is to represent each surface using a triangulated mesh and combine these meshes using a process that is sometimes called *zippering* (Soucy and Laurendeau 1992; Turk and Levoy 1994). Another, now more widely used, approach is to compute a (truncated) signed distance function that fits all of the 3D data points (Hoppe, DeRose *et al.* 1992; Curless and Levoy 1996; Hilton, Stoddart *et al.* 1996; Wheeler, Sato, and Ikeuchi 1998).

Figure 13.11 shows one such approach, the *volumetric range image processing* (VRIP) technique developed by Curless and Levoy (1996), which first computes a weighted signed distance function from each range image and then merges them using a weighted averaging process. To make the representation more compact, run-length coding is used to encode the empty, seen, and varying (signed distance) voxels, and only the signed distance values near each surface are stored.[5] Once the merged signed distance function has been computed, a zero-crossing surface extraction algorithm, such as *marching cubes* (Lorensen and Cline 1987), can be used to recover a meshed surface model. Figure 13.12 shows an example of the complete range data merging and isosurface extraction pipeline. Rusinkiewicz, Hall-Holt, and Levoy (2002) present a real-time system that combines fast ICP and

[4]https://bop.felk.cvut.cz/home

[5]An alternative, even more compact, representation could be to use octrees (Lavallée and Szeliski 1995).

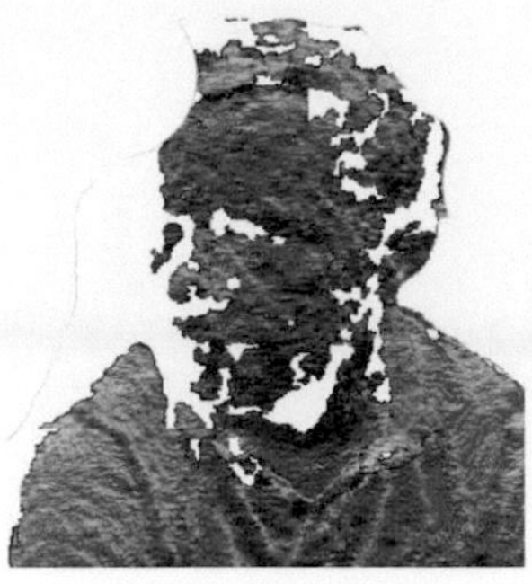

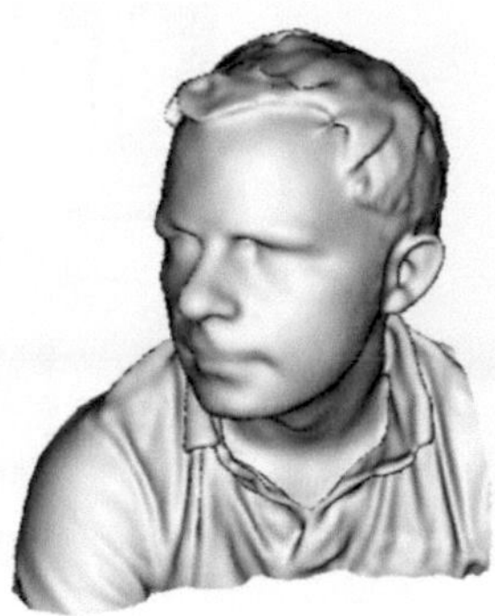

Figure 13.13 Fusing multiple depth images using the KinectFusion real-time system (Newcombe, Izadi *et al.* 2011) © 2011 IEEE. The three images show an original (noisy) range scan, rendered as a colored normal map, and the fused 3D model, rendered as both a normal map and Phong-shaded.

point-based merging and rendering.

The advent of consumer-level RGB-D cameras such as Kinect created renewed interest in large-scale range data registration and merging (Zhang 2012; Han, Shao *et al.* 2013). An influential paper in this area is Kinect Fusion (Izadi, Kim *et al.* 2011; Newcombe, Izadi *et al.* 2011), which combines an ICP-like SLAM technique called DTAM (Newcombe, Lovegrove, and Davison 2011) with real-time TSDF (truncated signed distance function) volumetric integration, which is described in more detail in Section 13.5.1. Follow-on papers include Elastic Fragments for non-rigid alignment (Zhou, Miller, and Koltun 2013), Octomap (Hornung, Wurm *et al.* 2013), which uses an octree and probabilistic occupancy, and Voxel Hashing (Nießner, Zollhöfer *et al.* 2013) and Chisel (Klingensmith, Dryanovski *et al.* 2015), both of which uses spatial hashing to compress the TSDF. KinectFusion has also been extended to handle highly variable scanning resolution (Fuhrmann and Goesele 2011, 2014), dynamic scenes (DynamicFusion (Newcombe, Fox, and Seitz 2015), VolumeDeform (Innmann, Zollhöfer *et al.* 2016), and Motion2Fusion (Dou, Davidson *et al.* 2017)), to use non-rigid surface deformations for global model refinement (ElasticFusion: Whelan, Salas-Moreno *et al.* (2016)), to produce a globally consistent BundleFusion model (Dai, Nießner *et al.* 2017), and to use a deep network to perform the non-rigid matching (Božič, Zollhöfer *et al.* 2020). More details on these and other techniques for constructing 3D models from RGB-D scans can be found in the survey by Zollhöfer, Stotko *et al.* (2018).

Some of the most recent work in range data merging uses neural networks to represent the TSDF (Park, Florence *et al.* 2019), update a TSDF with incoming range data scans (Weder, Schonberger *et al.* 2020, 2021), or provide local priors (Chabra, Lenssen *et al.* 2020). Range data merging techniques are often used for both 3D object scanning and for visual map building and navigation (RGB-D SLAM), which we discussed in Section 11.5. And now that depth sensing (aka lidar) technology is starting to appear in mobile phones, it can be used to build complete texture-mapped 3D room models, e.g., using Occipital's Canvas app (Stein 2020).[6]

Volumetric range data merging techniques based on signed distance or characteristic (inside–outside) functions are also widely used to extract smooth well behaved surfaces from oriented or unoriented point sets (Hoppe, DeRose *et al.* 1992, Ohtake, Belyaev *et al.* 2003; Kazhdan, Bolitho, and Hoppe 2006; Lempitsky and Boykov 2007; Zach, Pock, and Bischof 2007b; Zach 2008), as discussed in more detail in Section 13.5.1 and the survey paper by Berger, Tagliasacchi *et al.* (2017).

[6] https://canvas.io

(a) (b) (c)

Figure 13.14 Laser range modeling of the Bayon temple at Angkor-Thom (Banno, Masuda *et al.* 2008) © 2008 Springer: (a) sample photograph from the site; (b) a detailed head model scanned from the ground; (c) final merged 3D model of the temple scanned using a laser range sensor mounted on a balloon.

13.2.2 *Application*: Digital heritage

Active rangefinding technologies, combined with surface modeling and appearance modeling techniques (Section 13.7), are widely used in the fields of archaeological and historical preservation, which often also goes under the name *digital heritage* (MacDonald 2006). In such applications, detailed 3D models of cultural objects are acquired and later used for applications such as analysis, preservation, restoration, and the production of duplicate artwork (Rioux and Bird 1993).

A notable example of such an endeavor is the Digital Michelangelo project of Levoy, Pulli *et al.* (2000), which used Cyberware laser stripe scanners and high-quality digital SLR cameras mounted on a large gantry to obtain detailed scans of Michelangelo's David and other sculptures in Florence. The project also took scans of the *Forma Urbis Romae*, an ancient stone map of Rome that had shattered into pieces, for which new matches were obtained using digital techniques. The whole process, from initial planning, to software development, acquisition, and post-processing, took several years (and many volunteers), and produced a wealth of 3D shape and appearance modeling techniques as a result.

Even larger-scale projects have since been attempted, for example, the scanning of complete temple sites such as Angkor-Thom (Ikeuchi and Sato 2001; Ikeuchi and Miyazaki 2007; Banno, Masuda *et al.* 2008). Figure 13.14 shows details from this project, including a sample photograph, a detailed 3D (sculptural) head model scanned from ground level, and an aerial overview of the final merged 3D site model, which was acquired using a balloon.

13.3 Surface representations

In previous sections, we have seen different representations being used to integrate 3D range scans. We now look at several of these representations in more detail. Explicit surface representations, such as triangle meshes, splines (Farin 1992, 2002), and subdivision surfaces (Stollnitz, DeRose, and Salesin 1996; Zorin, Schröder, and Sweldens 1996; Warren and Weimer 2001; Peters and Reif 2008), enable not only the creation of highly detailed models but also processing operations, such as interpolation (Section 13.3.1), fairing or smoothing, and decimation and simplification (Section 13.3.2). We also examine discrete point-based representations (Section 13.4) and volumetric representations (Section 13.5).

13.3.1 Surface interpolation

One of the most common operations on surfaces is their reconstruction from a set of sparse data constraints, i.e., *scattered data interpolation*, which we covered in Section 4.1. When formulating such problems, surfaces may be parameterized as height fields $f(\mathbf{x})$, as 3D parametric surfaces $\mathbf{f}(\mathbf{x})$, or as non parametric models such as collections of triangles.

In Section 4.2, we saw how two-dimensional function interpolation and approximation problems $\{d_i\} \rightarrow f(\mathbf{x})$ could be cast as energy minimization problems using regularization (4.18–4.23). Such problems can also specify the locations of discontinuities in the surface as well as local orientation constraints (Terzopoulos 1986b; Zhang, Dugas-Phocion *et al.* 2002).

One approach to solving such problems is to discretize both the surface and the energy on a discrete grid or mesh using finite element analysis (4.24–4.27) (Terzopoulos 1986b). Such problems can then be solved using sparse system solving techniques, such as multigrid (Briggs, Henson, and McCormick 2000) or hierarchically preconditioned conjugate gradient (Szeliski 2006b; Krishnan and Szeliski 2011; Krishnan, Fattal, and Szeliski 2013). The surface can also be represented using a hierarchical combination of multilevel B-splines (Lee, Wolberg, and Shin 1997).

An alternative approach is to use *radial basis* (or *kernel*) functions (Boult and Kender 1986; Nielson 1993), which we covered in Section 4.1.1. As we mentioned in that section, if we want the function $\mathbf{f}(\mathbf{x})$ to exactly interpolate the data points, a dense linear system must be solved to determine the magnitude associated with each basis function (Boult and Kender 1986). It turns out that, for certain regularized problems, e.g., (4.18–4.21), there exist radial basis functions (kernels) that give the same results as a full analytical solution (Boult and Kender 1986). Unfortunately, because the dense system solving is cubic in the number of data points, basis function approaches can only be used for small problems such as feature-based image morphing (Beier and Neely 1992).

When a three-dimensional *parametric surface* is being modeled, the vector-valued function $\mathbf{f}$ in (4.18–4.27) encodes 3D coordinates (x, y, z) on the surface and the domain $\mathbf{x} = (s, t)$ encodes the surface parameterization. One example of such surfaces are symmetry-seeking parametric models, which are elastically deformable versions of *generalized cylinders*[7] (Terzopoulos, Witkin, and Kass 1987). In these models, s is the parameter *along* the spine of the deformable tube and t is the parameter *around* the tube. A variety of smoothness and radial symmetry forces are used to constrain the model while it is fitted to image-based silhouette curves.

It is also possible to define *non-parametric* surface models, such as general triangulated meshes, and to equip such meshes (using finite element analysis) with both internal smoothness metrics and external data fitting metrics (Sander and Zucker 1990; Fua and Sander 1992; Delingette, Hebert, and Ikeuichi 1992; McInerney and Terzopoulos 1993). While most of these approaches assume a standard *elastic* deformation model, which uses quadratic internal smoothness terms, it is also possible to use sub-linear energy models to better preserve surface creases (Diebel, Thrun, and Brünig 2006) or to use graph-convolutional neural networks (GCNNs) as an alternative to the update equations, as in Deep Active Surface Models (Wickramasinghe, Fua, and Knott 2021). Triangle meshes can also be augmented with either spline elements (Sullivan and Ponce 1998) or subdivision surfaces (Stollnitz, DeRose, and Salesin 1996; Zorin, Schröder, and Sweldens 1996; Warren and Weimer 2001; Peters and Reif 2008) to produce surfaces with better smoothness control.

Both parametric and non-parametric surface models assume that the topology of the surface is known and fixed ahead of time. For more flexible surface modeling, we can either represent the surface as a collection of oriented points (Section 13.4) or use 3D implicit functions (Section 13.5.1),

[7] A generalized cylinder (Brooks 1981) is a *solid of revolution*, i.e., the result of rotating a (usually smooth) curve around an axis. It can also be generated by sweeping a slowly varying circular cross-section along the axis. (These two interpretations are equivalent.)

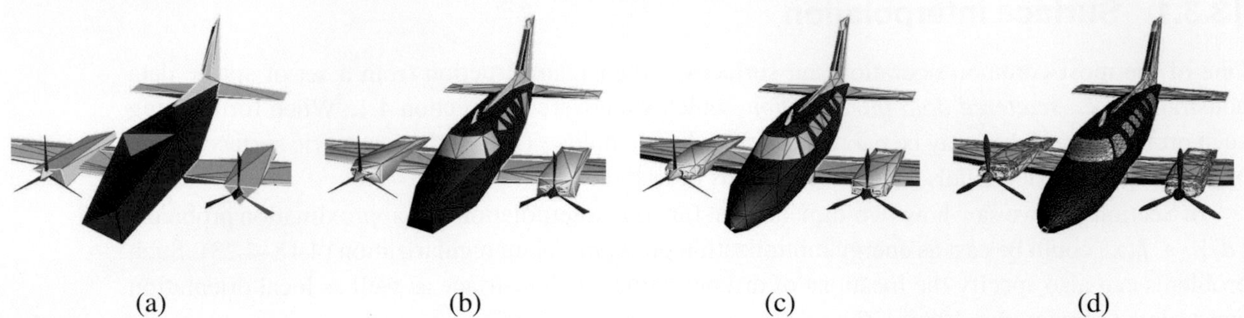

(a) (b) (c) (d)

Figure 13.15 Progressive mesh representation of an airplane model (Hoppe 1996) © 1996 ACM: (a) base mesh M^0 (150 faces); (b) mesh M^{175} (500 faces); (c) mesh M^{425} (1,000 faces); (d) original mesh $M = M^n$ (13,546 faces).

which can also be combined with elastic 3D surface models (McInerney and Terzopoulos 1993).

The field of surface reconstruction from unorganized point samples continues to advance rapidly, with more recent work addressing issues with data imperfections, as described in the survey by Berger, Tagliasacchi *et al.* (2017) .

13.3.2 Surface simplification

Once a triangle mesh has been created from 3D data, it is often desirable to create a hierarchy of mesh models, for example, to control the displayed *level of detail* (LOD) in a computer graphics application. (In essence, this is a 3D analog to image pyramids (Section 3.5).) One approach to doing this is to approximate a given mesh with one that has *subdivision connectivity*, over which a set of triangular wavelet coefficients can then be computed (Eck, DeRose *et al.* 1995). A more continuous approach is to use sequential *edge collapse* operations to go from the original fine-resolution mesh to a coarse base-level mesh (Hoppe 1996; Lee, Sweldens *et al.* 1998). The resulting *progressive mesh* (PM) representation can be used to render the 3D model at arbitrary levels of detail, as shown in Figure 13.15. More recent papers on multiresolution geometric modeling can be found in the survey by Floater and Hormann (2005) and the collection of papers edited by Dodgson, Floater, and Sabin (2005).

13.3.3 Geometry images

While multi-resolution surface representations such as Eck, DeRose *et al.* (1995), Hoppe (1996), and Lee, Sweldens *et al.* (1998) support level of detail operations, they still consist of an irregular collection of triangles, which makes them more difficult to compress and store in a cache-efficient manner.[8]

To make the triangulation completely regular (uniform and gridded), Gu, Gortler, and Hoppe (2002) describe how to create *geometry images* by cutting surface meshes along well-chosen lines and "flattening" the resulting representation into a square. Figure 13.16a shows the resulting (x, y, z) values of the surface mesh mapped over the unit square, while Figure 13.16b shows the associated (n_x, n_y, n_z) *normal map*, i.e., the surface normals associated with each mesh vertex, which can be used to compensate for loss in visual fidelity if the original geometry image is heavily compressed.

[8] Subdivision triangulations, such as those in Eck, DeRose *et al.* (1995), are *semi-regular*, i.e., regular (ordered and nested) within each subdivided base triangle.

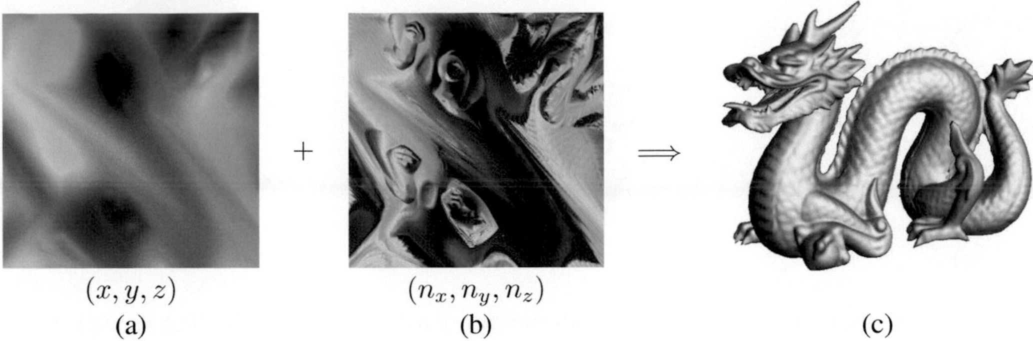

Figure 13.16 Geometry images (Gu, Gortler, and Hoppe 2002) © 2002 ACM: (a) the 257 × 257 geometry image defines a mesh over the surface; (b) the 512 × 512 normal map defines vertex normals; (c) final lit 3D model.

13.4 Point-based representations

As we mentioned previously, triangle-based surface models assume that the topology (and often the rough shape) of the 3D model is known ahead of time. While it is possible to re-mesh a model as it is being deformed or fitted, a simpler solution is to dispense with an explicit triangle mesh altogether and to have triangle vertices behave as *oriented points*, or particles, or *surface elements* (surfels) (Szeliski and Tonnesen 1992).

To endow the resulting particle system with internal smoothness constraints, pairwise interaction potentials can be defined that approximate the equivalent elastic bending energies that would be obtained using local finite-element analysis.[9] Instead of defining the finite element neighborhood for each particle (vertex) ahead of time, a soft influence function is used to couple nearby particles. The resulting 3D model can change both topology and particle density as it evolves and can therefore be used to interpolate partial 3D data with holes (Szeliski, Tonnesen, and Terzopoulos 1993b). Discontinuities in both the surface orientation and crease curves can also be modeled (Szeliski, Tonnesen, and Terzopoulos 1993a).

To render the particle system as a continuous surface, local dynamic triangulation heuristics (Szeliski and Tonnesen 1992) or direct surface element *splatting* (Pfister, Zwicker *et al.* 2000) can be used. Another alternative is to first convert the point cloud into an implicit signed distance or inside–outside function, using either minimum signed distances to the oriented points (Hoppe, DeRose *et al.* 1992) or by interpolating a characteristic (inside–outside) function using radial basis functions (Turk and O'Brien 2002; Dinh, Turk, and Slabaugh 2002). Even greater precision over the implicit function fitting, including the ability to handle irregular point densities, can be obtained by computing a *moving least squares* (MLS) estimate of the signed distance function (Alexa, Behr *et al.* 2003; Pauly, Keiser *et al.* 2003), as shown in Figure 13.17. Further improvements can be obtained using local sphere fitting (Guennebaud and Gross 2007), faster and more accurate re-sampling (Guennebaud, Germann, and Gross 2008), and kernel regression to better tolerate outliers (Oztireli, Guennebaud, and Gross 2008).

The survey by Berger, Tagliasacchi *et al.* (2017) discusses more recent work on reconstructing smooth complete surfaces from point clouds. The SurfelMeshing paper by Schöps, Sattler, and Pollefeys (2020) presents an RGB-D SLAM system based on a variable-resolution surfel represen-

[9] As mentioned before, an alternative is to use *sub-linear* interaction potentials, which encourage the preservation of surface creases (Diebel, Thrun, and Brünig 2006).

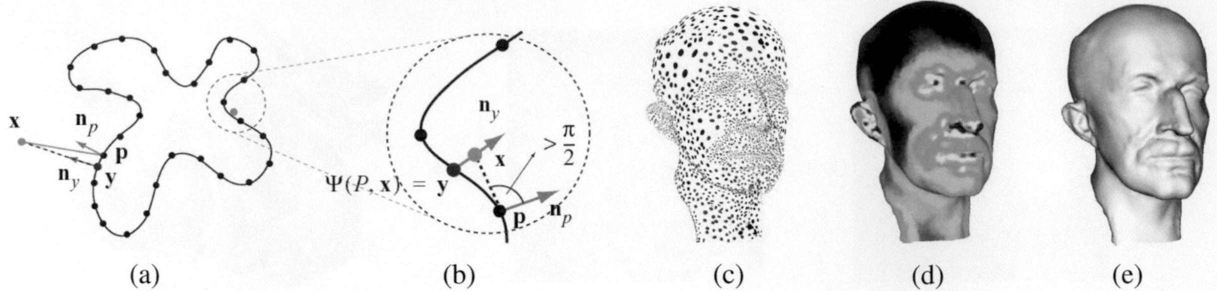

Figure 13.17 Point-based surface modeling with moving least squares (MLS) (Pauly, Keiser *et al.* 2003) © 2003 ACM: (a) a set of points (black dots) is turned into an implicit inside–outside function (black curve); (b) the signed distance to the nearest oriented point can serve as an approximation to the inside–outside distance; (c) a set of oriented points with variable sampling density representing a 3D surface (head model); (d) local estimate of sampling density, which is used in the moving least squares; (e) reconstructed continuous 3D surface.

tation that gets re-triangulated as more scans are integrated. Other recent approaches to 3D point clouds that use deep learning, mentioned previously in Section 5.5.1, are discussed in the survey by Guo, Wang *et al.* (2020). Even more recent algorithms to estimate better normals in 3D models are presented in Ben-Shabat and Gould (2020) and Zhu and Smith (2020).

13.5 Volumetric representations

A third alternative for modeling 3D surfaces is to construct 3D volumetric inside–outside functions. We have already seen examples of this in Section 12.7.2, where we looked at voxel coloring (Seitz and Dyer 1999), space carving (Kutulakos and Seitz 2000), and level set (Pons, Keriven, and Faugeras 2007) techniques for stereo matching, and Section 12.7.3, where we discussed using binary silhouette images to reconstruct volumes.

In this section, we look at continuous *implicit* (inside–outside) functions to represent 3D shape.

13.5.1 Implicit surfaces and level sets

While polyhedral and voxel-based representations can represent three-dimensional shapes to an arbitrary precision, they lack some of the intrinsic smoothness properties available with continuous implicit surfaces, which use an *indicator function* (or *characteristic function*) $F(x, y, z)$ to indicate which 3D points are inside $F(x, y, z) < 0$ or outside $F(x, y, z) > 0$ the object.

An early example of using implicit functions to model 3D objects in computer vision were superquadrics (Pentland 1986; Solina and Bajcsy 1990; Waithe and Ferrie 1991; Leonardis, Jaklič, and Solina 1997). To model a wider variety of shapes, superquadrics are usually combined with either rigid or non-rigid deformations (Terzopoulos and Metaxas 1991; Metaxas and Terzopoulos 2002). Superquadric models can either be fitted to range data or used directly for stereo matching.

A different kind of implicit shape model can be constructed by defining a *signed distance function* over a regular three-dimensional grid, optionally using an octree spline to represent this function more coarsely away from its surface (zero-set) (Lavallée and Szeliski 1995; Szeliski and Lavallée 1996; Frisken, Perry *et al.* 2000; Ohtake, Belyaev *et al.* 2003). We have already seen examples of signed distance functions being used to represent distance transforms (Section 3.3.3), level sets for 2D contour fitting and tracking (Section 7.3.2), volumetric stereo (Section 12.7.2), range data

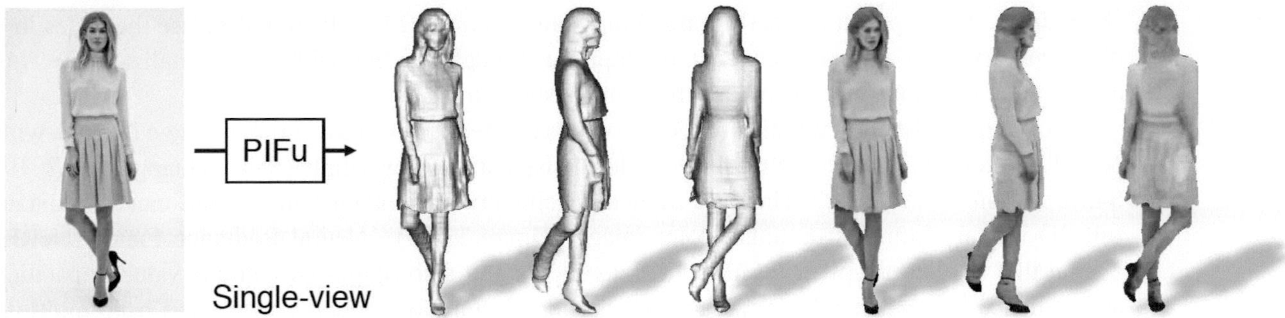

Figure 13.18 A Pixel-aligned Implicit Function (PIFu) network can recover a high-resolution 3D textured model of a clothed human from a single input image (Saito, Huang *et al.* 2019) © 2019 IEEE.

merging (Section 13.2.1), and point-based modeling (Section 13.4). The advantage of representing such functions directly on a grid is that it is quick and easy to look up distance function values for any (x, y, z) location and also easy to extract the isosurface using the marching cubes algorithm (Lorensen and Cline 1987). The work of Ohtake, Belyaev *et al.* (2003) is particularly notable, as it allows for several distance functions to be used simultaneously and then combined locally to produce sharp features such as creases.

Poisson surface reconstruction (Kazhdan, Bolitho, and Hoppe 2006; Kazhdan and Hoppe 2013) uses a closely related volumetric function, namely a smoothed 0/1 inside–outside (characteristic or occupancy) function, which can be thought of as a clipped signed distance function. The gradients for this function are set to lie along oriented surface normals near known surface points and 0 elsewhere. The function itself is represented using a quadratic tensor-product B-spline over an octree, which provides a compact representation with larger cells away from the surface or in regions of lower point density, and also admits the efficient solution of the related Poisson equations (4.24–4.27), e.g., Section 8.4.4 and Pérez, Gangnet, and Blake (2003).

It is also possible to replace the quadratic penalties used in the Poisson equations with L_1 (total variation) constraints and still obtain a convex optimization problem, which can be solved using either continuous (Zach, Pock, and Bischof 2007b; Zach 2008) or discrete graph cut (Lempitsky and Boykov 2007) techniques.

Signed distance functions also play an integral role in level-set evolution equations (Sections 7.3.2 and 12.7.2), where the values of distance transforms on the mesh are updated as the surface evolves to fit multi-view stereo photoconsistency measures (Faugeras and Keriven 1998).

As with many other areas of computer vision, deep neural networks have started being applied to the construction and modeling of volumetric object representations. Some neural networks construct 3D surface or volumetric occupancy grid models from single images (Choy, Xu *et al.* 2016; Tatarchenko, Dosovitskiy, and Brox 2017; Groueix, Fisher *et al.* 2018; Richter and Roth 2018), although more recent experiments suggest that these networks may just be recognizing the general object category and doing a small amount of fitting (Tatarchenko, Richter *et al.* 2019). DeepSDFs (Park, Florence *et al.* 2019), IM-NET (Chen and Zhang 2019), Occupancy Networks (Mescheder, Oechsle *et al.* 2019), Deep Implicit Surface (DISN) networks (Xu, Wang *et al.* 2019), and UCLID-Net (Guillard, Remelli, and Fua 2020) train networks to transform continuous (x, y, z) inputs into signed distance or $[0, 1]$ occupancy values and sometimes combine convolutional image encoders with MLPs to represent color and surface details (Oechsle, Mescheder *et al.* 2019), while MeshSDF can continuously transform SDFs into deformable meshes (Remelli, Lukoianov *et al.* 2020). All of these networks use latent codes to represent individual instances from a generic class (e.g., car or chair) from the ShapeNet dataset (Chang, Funkhouser *et al.* 2015), although they use the codes in a different part of the network (either in the input or through conditional batch normalization). This allows them to reconstruct 3D models from just a single image.

Pixel-aligned Implicit function (PIFu) networks combine fully convolutional image features with neural implicit functions to better preserve local shape and color details (Saito, Huang *et al.* 2019; Saito, Simon *et al.* 2020). They are trained specifically on clothed humans and can hallucinate full 3D models from just a single color image (Figure 13.18). Neural Radiance Fields (NeRF) extend this to also use pixel ray directions as inputs and also output continuous valued opacities and radiance values, enabling ray-traced rendering of shiny 3D models constructed from multiple input images (Mildenhall, Srinivasan *et al.* 2020). This representation is related to Lumigraphs and Surface Light Fields, which we study in Section 14.3. Both of these systems are examples of *neural rendering* approaches to generating photorealistic novel views, which we discuss in more detail in Section 14.6.

To deal with larger (e.g., building-scale) scenes, Convolutional Occupancy Networks (Peng, Niemeyer *et al.* 2020) first retrieve local features from a 2D, multiplane, or 3D grid, and then use a trained MLP (fully connected network) to decode these into local occupancy volumes. Instead of modeling a complete 3D scene, Local Implicit Grid Representations (Jiang, Sud *et al.* 2020) model small local sub-volumes, allowing them to be used as a kind of prior for other shape reconstruction methods.

13.6 Model-based reconstruction

When we know something ahead of time about the objects we are trying to model, we can construct more detailed and reliable 3D models using specialized techniques and representations. For example, architecture is usually made up of large planar regions and other parametric forms (such as surfaces of revolution), usually oriented perpendicular to gravity and to each other (Section 13.6.1). Heads and faces can be represented using low-dimensional, non-rigid shape models, because the variability in shape and appearance of human faces, while extremely large, is still bounded (Section 13.6.2). Human bodies or parts, such as hands, form highly articulated structures, which can be represented using kinematic chains of piecewise rigid skeletal elements linked by joints (Section 13.6.4).

In this section, we highlight some of the main ideas, representations, and modeling algorithms used for these three cases. Additional details and references can be found in specialized conferences and workshops devoted to these topics, e.g., the International Conference on 3D Vision (3DV) and the IEEE International Conference on Automatic Face and Gesture Recognition (FG).

13.6.1 Architecture

Architectural modeling, especially from aerial photography, has been one of the longest studied problems in both photogrammetry and computer vision (Walker and Herman 1988). In the last two decades, the development of reliable image-based modeling techniques, as well as the prevalence of digital cameras and 3D computer games, has led to widespread deployment of such systems.

The work by Debevec, Taylor, and Malik (1996) was one of the earliest hybrid geometry- and image-based modeling and rendering systems. Their Façade system combines an interactive image-guided geometric modeling tool with model-based (local plane plus parallax) stereo matching and view-dependent texture mapping. During the interactive photogrammetric modeling phase, the user selects block elements and aligns their edges with visible edges in the input images (Figure 13.19a).

Figure 13.19 Interactive architectural modeling using the Façade system (Debevec, Taylor, and Malik 1996) © 1996 ACM: (a) input image with user-drawn edges shown in green; (b) shaded 3D solid model; (c) geometric primitives overlaid onto the input image; (d) final view-dependent, texture-mapped 3D model.

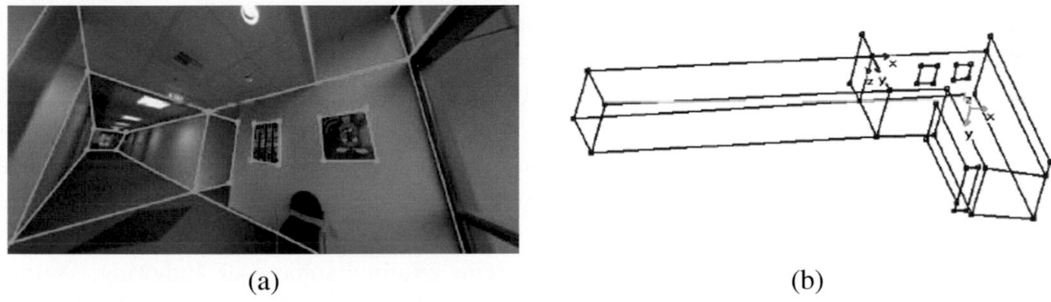

(a) (b)

Figure 13.20 Interactive 3D modeling from panoramas (Shum, Han, and Szeliski 1998) © 1998 IEEE: (a) wide-angle view of a panorama with user-drawn vertical and horizontal (axis-aligned) lines; (b) single-view reconstruction of the corridors.

The system then automatically computes the dimensions and locations of the blocks along with the camera positions using constrained optimization (Figure 13.19b–c). This approach is intrinsically more reliable than general feature-based structure from motion, because it exploits the strong geometry available in the block primitives. Related work by Becker and Bove (1995), Horry, Anjyo, and Arai (1997), Criminisi, Reid, and Zisserman (2000), and Holynski, Geraghty *et al.* (2020) exploits similar information available from vanishing points. In the interactive, image-based modeling system of Sinha, Steedly *et al.* (2008), vanishing point directions are used to guide the user drawing of polygons, which are then automatically fitted to sparse 3D points recovered using structure from motion.

Once the rough geometry has been estimated, more detailed offset maps can be computed for each planar face using a local plane sweep, which Debevec, Taylor, and Malik (1996) call *model-based stereo*. Finally, during rendering, images from different viewpoints are warped and blended together as the camera moves around the scene, using a process (related to light field and Lumigraph rendering; see Section 14.3) called *view-dependent texture mapping* (Figure 13.19d).

For interior modeling, instead of working with single pictures, it is more useful to work with panoramas, as you can see larger extents of walls and other structures. The 3D modeling system developed by Shum, Han, and Szeliski (1998) first constructs calibrated panoramas from multiple images (Section 11.4.2) and then has the user draw vertical and horizontal lines in the image to

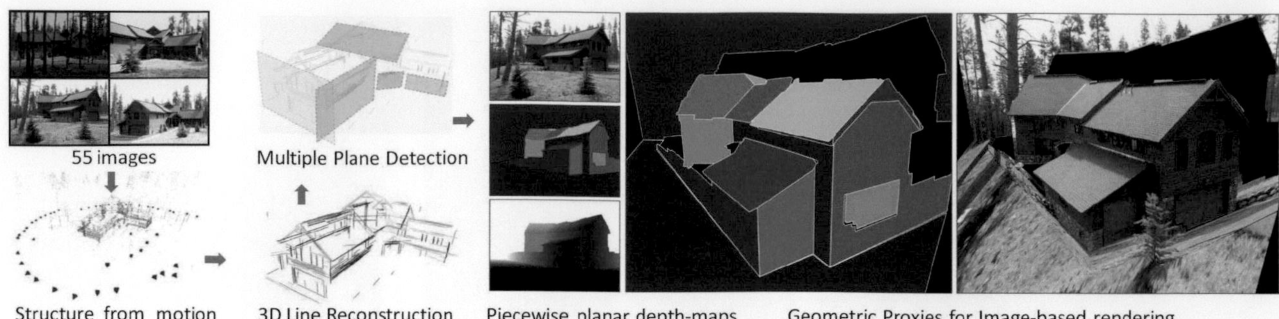

Figure 13.21 Automated architectural reconstruction using 3D lines and planes (Sinha, Steedly, and Szeliski 2009) © 2009 IEEE.

demarcate the boundaries of planar regions. The lines are initially used to establish an absolute rotation for each panorama and are later used (along with the inferred vertices and planes) to optimize the 3D structure, which can be recovered up to scale from one or more images (Figure 13.20). Recent advances in deep networks now make it possible to both automatically infer the lines and their junctions (Huang, Wang *et al.* 2018; Zhang, Li *et al.* 2019) and to build complete 3D wireframe models (Zhou, Qi, and Ma 2019; Zhou, Qi *et al.* 2019b). 360° high dynamic range panoramas can also be used for outdoor modeling, because they provide highly reliable estimates of relative camera orientations as well as vanishing point directions (Antone and Teller 2002; Teller, Antone *et al.* 2003).

While earlier image-based modeling systems required some user authoring, Werner and Zisserman (2002) present a fully automated line-based reconstruction system. As described in Section 11.4.8, they first detect lines and vanishing points and use them to calibrate the camera; then they establish line correspondences using both appearance matching and trifocal tensors, which enables them to reconstruct families of 3D line segments. They then generate plane hypotheses, using both co-planar 3D lines and a plane sweep (Section 12.1.2) based on cross-correlation scores evaluated at interest points. Intersections of planes are used to determine the extent of each plane, i.e., an initial coarse geometry, which is then refined with the addition of rectangular or wedge-shaped indentations and extrusions. Note that when top-down maps of the buildings being modeled are available, these can be used to further constrain the 3D modeling process (Robertson and Cipolla 2002, 2009). The idea of using matched 3D lines for estimating vanishing point directions and dominant planes is used in a number of fully automated image-based architectural modeling systems (Zebedin, Bauer *et al.* 2008; Mičušík and Košecká 2009; Furukawa, Curless *et al.* 2009b; Sinha, Steedly, and Szeliski 2009; Holynski, Geraghty *et al.* 2020) as well as SLAM systems (Zhou, Zou *et al.* 2015; Li, Yao *et al.* 2018; Yang and Scherer 2019). Figure 13.21 shows some of the processing stages in the system developed by Sinha, Steedly, and Szeliski (2009).

Another common characteristic of architecture is the repeated use of primitives such as windows, doors, and colonnades. Architectural modeling systems can be designed to search for such repeated elements and to use them as part of the structure inference process (Dick, Torr, and Cipolla 2004; Mueller, Zeng *et al.* 2007; Schindler, Krishnamurthy *et al.* 2008; Pauly, Mitra *et al.* 2008; Sinha, Steedly *et al.* 2008). The combination of structured elements such as parallel lines, junctions, and rectangles with full axis-aligned 3D models for the modeling of architectural environments has recently been called *holistic 3D reconstruction*. More details can be found in the recent tutorial by Zhou, Furukawa, and Ma (2019), workshop (Zhou, Furukawa *et al.* 2020), and state-of-the-art report by Pintore, Mura *et al.* (2020).

The combination of all these techniques now makes it possible to reconstruct the structure of large 3D scenes (Zhu and Kanade 2008). For example, the *Urbanscan* system of Pollefeys, Nistér *et al.* (2008) reconstructs texture-mapped 3D models of city streets from videos acquired with a GPS-equipped vehicle. To obtain real-time performance, they use both optimized online structure-from-motion algorithms, as well as GPU implementations of plane-sweep stereo aligned to dominant planes and depth map fusion. Cornelis, Leibe *et al.* (2008) present a related system that also uses plane-sweep stereo (aligned to vertical building façades) combined with object recognition and segmentation for vehicles. Mičušík and Košecká (2009) build on these results using omni-directional images and superpixel-based stereo matching along dominant plane orientations. Reconstruction directly from active range scanning data combined with color imagery that has been compensated for exposure and lighting variations is also possible (Chen and Chen 2008; Stamos, Liu *et al.* 2008; Troccoli and Allen 2008).

Numerous photogrammetric reconstruction systems that produce detailed texture-mapped 3D models have been developed based on these computer vision techniques.[10] Examples of commercial software that can be used to reconstruct large-scale 3D models from aerial drone and ground level photography include Pix4D,[11] Metashape,[12] and RealityCapture.[13] Another example is Occipital's Canvas mobile phone app[14] (Stein 2020), which appears to use a combination of photogrammetry (3D point and line matching and reconstruction, as discussed above) and depth map fusion.

13.6.2 Facial modeling and tracking

Another area in which specialized shape and appearance models are extremely helpful is in the modeling of heads and faces. Even though the appearance of people seems at first glance to be infinitely variable, the actual shape of a person's head and face can be described reasonably well using a few dozen parameters (Pighin, Hecker *et al.* 1998; Guenter, Grimm *et al.* 1998; DeCarlo, Metaxas, and Stone 1998; Blanz and Vetter 1999; Shan, Liu, and Zhang 2001; Zollhöfer, Thies *et al.* 2018; Egger, Smith *et al.* 2020).

Figure 13.22 shows an example of an image-based modeling system, where user-specified keypoints in several images are used to fit a generic head model to a person's face. As you can see in Figure 13.22c, after specifying just over 100 keypoints, the shape of the face has become quite adapted and recognizable. Extracting a texture map from the original images and then applying it to the head model results in an animatable model with striking visual fidelity (Figure 13.23a).

A more powerful system can be built by applying *principal component analysis* (PCA) to a collection of 3D scanned faces, which is a topic we discuss in Section 13.6.3. As you can see in Figure 13.25, it is then possible to fit morphable 3D models to single images and to use such models for a variety of animation and visual effects (Blanz and Vetter 1999; Egger, Smith *et al.* 2020). It is also possible to design stereo matching algorithms that optimize directly for the head model parameters (Shan, Liu, and Zhang 2001; Kang and Jones 2002) or to use the output of real-time stereo with active illumination (Zhang, Snavely *et al.* 2004) (Figures 13.10 and 13.23b).

As the sophistication of 3D facial capture systems evolved, so did the detail and realism in the reconstructed models. Modern systems can capture (in real-time) not only surface details such as wrinkles and creases, but also accurate models of skin reflection, translucency, and sub-surface scattering (Debevec, Hawkins *et al.* 2000, Weyrich, Matusik *et al.* 2006; Golovinskiy, Matusik *et al.*

[10] https://all3dp.com/1/best-photogrammetry-software

[11] https://www.pix4d.com

[12] https://www.agisoft.com/

[13] https://www.capturingreality.com/

[14] https://canvas.io

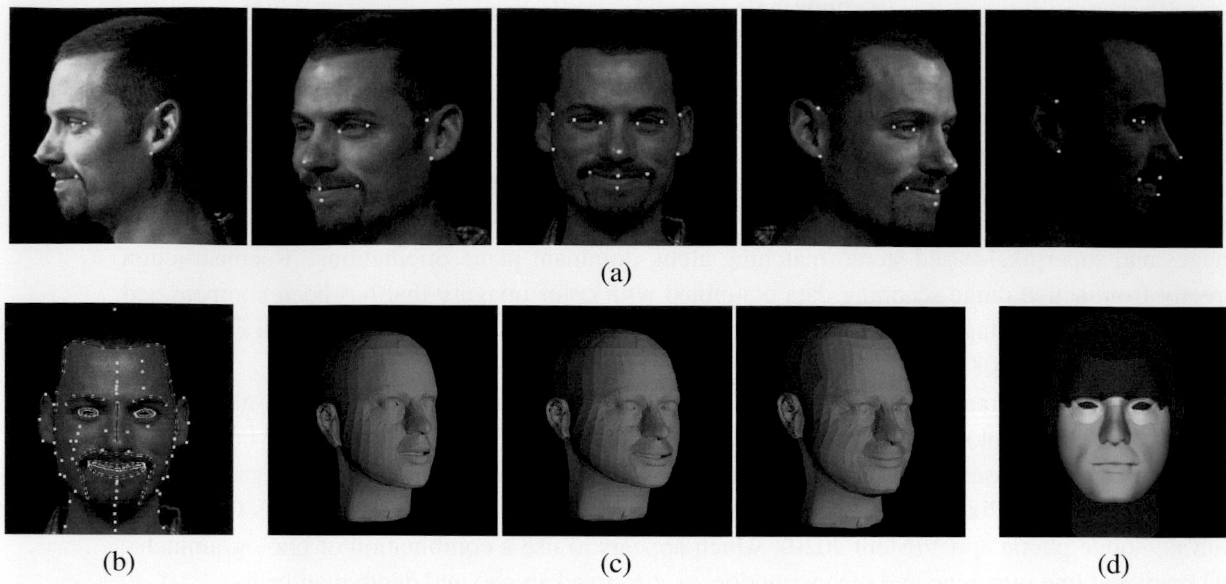

Figure 13.22 3D model fitting to a collection of images: (Pighin, Hecker *et al.* 1998) © 1998 ACM: (a) set of five input images along with user-selected keypoints; (b) the complete set of keypoints and curves; (c) three meshes—the original, adapted after 13 keypoints, and after an additional 99 keypoints; (d) the partition of the image into separately animatable regions.

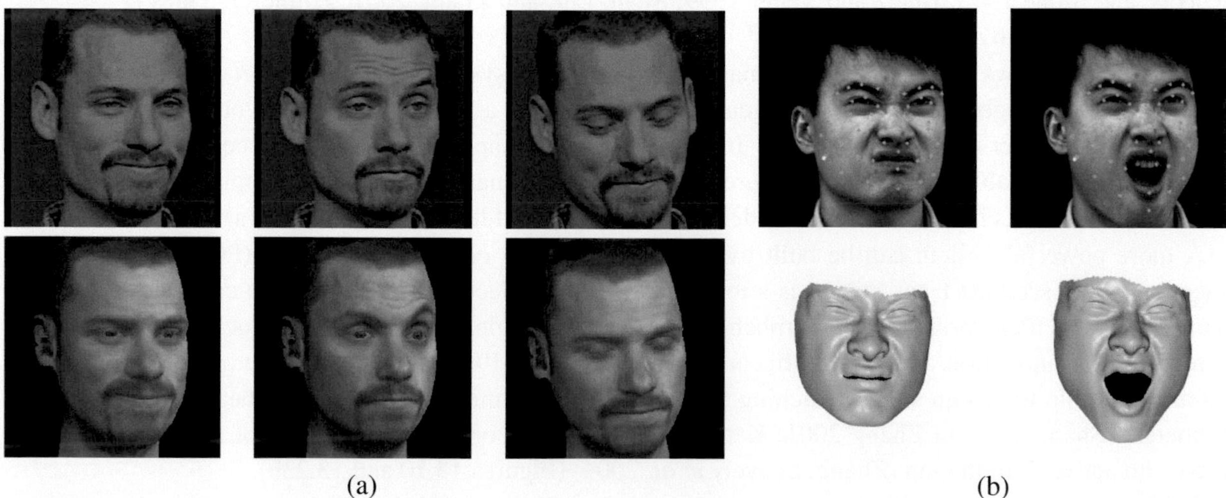

Figure 13.23 Head and expression tracking and re-animation using deformable 3D models. (a) Models fitted directly to five input video streams (Pighin, Szeliski, and Salesin 2002) © 2002 Springer: The bottom row shows the results of re-animating a synthetic texture-mapped 3D model with pose and expression parameters fitted to the input images in the top row. (b) Models fitted to frame-rate spacetime stereo surface models (Zhang, Snavely *et al.* 2004) © 2004 ACM: The top row shows the input images with synthetic green markers overlaid, while the bottom row shows the fitted 3D surface model.

Figure 13.24 Portrait shadow removal and manipulation (Zhang, Barron *et al.* 2020) © 2020 ACM. The top row shows the original photographs and the bottom row the corresponding enhanced photographs after more flattering lighting has been simulated.

2006; Bickel, Botsch *et al.* 2007; Igarashi, Nishino, and Nayar 2007; Meka, Haene *et al.* 2019).

Once a 3D head model has been constructed, it can be used in a variety of applications, such as head tracking (Toyama 1998; Lepetit, Pilet, and Fua 2004; Matthews, Xiao, and Baker 2007), as shown in Figures 7.30 and face transfer, i.e., replacing one person's face with another in a video (Bregler, Covell, and Slaney 1997; Vlasic, Brand *et al.* 2005). Additional applications include face beautification by warping face images toward a more attractive "standard" (Leyvand, Cohen-Or *et al.* 2008), face de-identification for privacy protection (Gross, Sweeney *et al.* 2008), and face swapping (Bitouk, Kumar *et al.* 2008).

More recent applications of 3D head models include photorealistic avatars for video conferencing (Chu, Ma *et al.* 2020), 3D unwarping for better selfies (Fried, Shechtman *et al.* 2016; Zhao, Huang *et al.* 2019; Ma, Lin *et al.* 2020), and single image portrait relighting (Sun, Barron *et al.* 2019; Zhou, Hadap *et al.* 2019; Zhang, Barron *et al.* 2020), an example of which is shown in Figure 13.24. This last application is available as the Portrait Light feature in Google Photos.[15] Additional applications can be found in the survey papers by Zollhöfer, Thies *et al.* (2018) and Egger, Smith *et al.* (2020).

13.6.3 *Application*: Facial animation

Perhaps the most widely used application of 3D head modeling is facial animation (Zollhöfer, Thies *et al.* 2018). Once a parameterized 3D model of the shape and appearance (surface texture) of a person's head has been constructed, it can be used directly to track a person's facial motions (Figure 13.23a) and to animate a different character with these same motions and expressions (Pighin, Szeliski, and Salesin 2002).

An improved version of such a system can be constructed by first applying principal component analysis (PCA) to the space of possible head shapes and facial appearances. Blanz and Vetter (1999) describe a system where they first capture a set of 200 colored range scans of faces (Figure 13.25a),

[15]https://blog.google/products/photos/new-helpful-editor

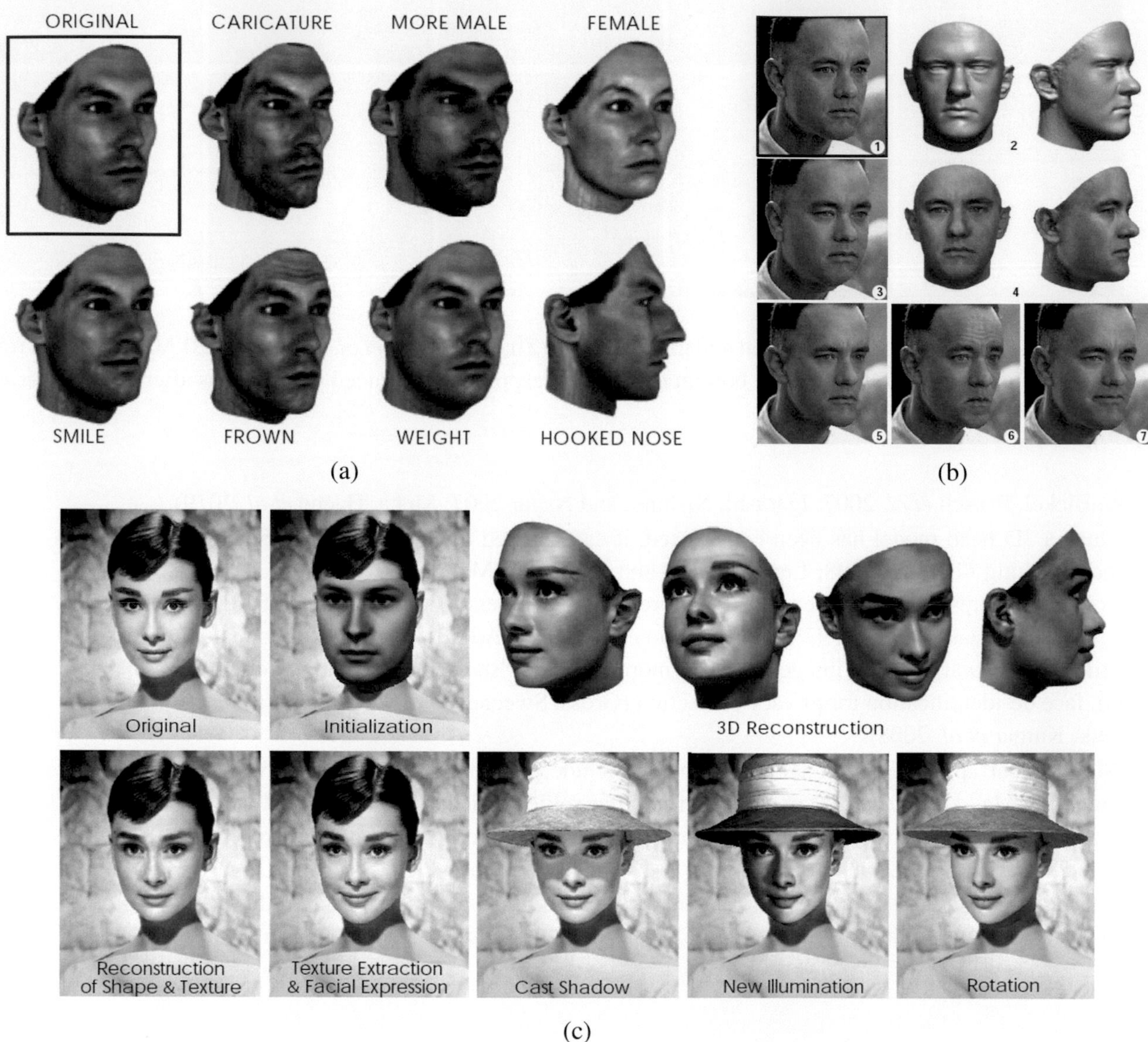

Figure 13.25 3D morphable face model (Blanz and Vetter 1999) © 1999 ACM: (a) original 3D face model with the addition of shape and texture variations in specific directions: deviation from the mean (caricature), gender, expression, weight, and nose shape; (b) a 3D morphable model is fitted to a single image, after which its weight or expression can be manipulated; (c) another example of a 3D reconstruction along with a different set of 3D manipulations, such as lighting and pose change.

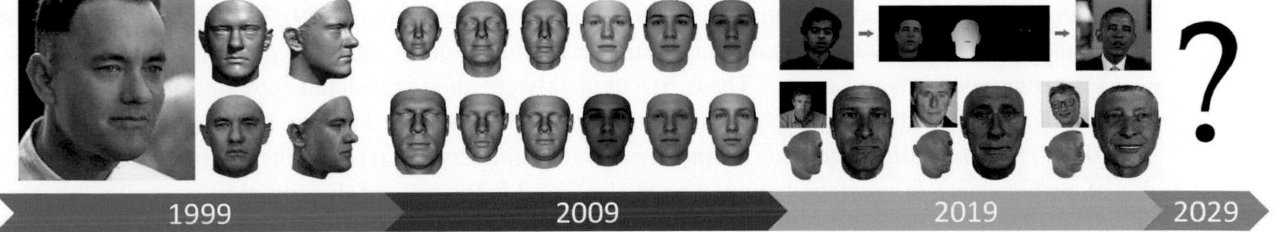

Figure 13.26 A timeline of twenty years of 3D morphable head models (Egger, Smith *et al.* 2020) © 2020 ACM, including results from the original paper by Blanz and Vetter (1999), the first publicly available morphable model (Paysan, Knothe *et al.* 2009), facial re-enactment results (Kim, Garrido *et al.* 2018), and GAN-based models (Gecer, Ploumpis *et al.* 2019).

which can be represented as a large collection of (X, Y, Z, R, G, B) samples (vertices).[16] For 3D morphing to be meaningful, corresponding vertices in different people's scans must first be put into correspondence (Pighin, Hecker *et al.* 1998). Once this is done, PCA can be applied to more naturally parameterize the 3D morphable model. The flexibility of this model can be increased by performing separate analyses in different subregions, such as the eyes, nose, and mouth, just as in modular eigenspaces (Moghaddam and Pentland 1997).

After computing a subspace representation, different directions in this space can be associated with different characteristics such as gender, facial expressions, or facial features (Figure 13.25a). As in the work of Rowland and Perrett (1995), faces can be turned into caricatures by exaggerating their displacement from the mean image.

3D morphable models can be fitted to a single image using gradient descent on the error between the input image and the re-synthesized model image, after an initial manual placement of the model in an approximately correct pose, scale, and location (Figures 13.25b–c). The efficiency of this fitting process can be increased using inverse compositional image alignment (Baker and Matthews 2004) as described by Romdhani and Vetter (2003).

The resulting texture-mapped 3D model can then be modified to produce a variety of visual effects, including changing a person's weight or expression, or three-dimensional effects such as re-lighting or 3D video-based animation (Section 14.5.1). Such models can also be used for video compression, e.g., by only transmitting a small number of facial expression and pose parameters to drive a synthetic avatar (Eisert, Wiegand, and Girod 2000; Gao, Chen *et al.* 2003; Lombardi, Saragih *et al.* 2018; Wei, Saragih *et al.* 2019) or to bring a still portrait image to life (Averbuch-Elor, Cohen-Or *et al.* 2017). The survey paper on 3D morphable face models by Egger, Smith *et al.* (2020) (Figure 13.26) discusses additional research and applications in this area.

3D facial animation is often matched to the performance of an actor, in what is known as *performance-driven animation* (Section 7.1.6) (Williams 1990). Traditional performance-driven animation systems use marker-based motion capture (Ma, Jones *et al.* 2008), while some newer systems use depth cameras or regular video to control the animation (Buck, Finkelstein *et al.* 2000; Pighin, Szeliski, and Salesin 2002; Zhang, Snavely *et al.* 2004; Vlasic, Brand *et al.* 2005; Weise, Bouaziz *et al.* 2011; Thies, Zollhofer *et al.* 2016; Thies, Zollhöfer *et al.* 2018).

An example of the latter approach is the system developed for the film *The Curious Case of Benjamin Button*, in which Digital Domain used the CONTOUR system from Mova[17] to capture

[16]A cylindrical coordinate system provides a natural two-dimensional embedding for this collection, but such an embedding is not necessary to perform PCA.

[17]http://www.mova.com.

actor Brad Pitt's facial motions and expressions (Roble and Zafar 2009). CONTOUR uses a combination of phosphorescent paint and multiple high-resolution video cameras to capture real-time 3D range scans of the actor. These 3D models were then translated into Facial Action Coding System (FACS) shape and expression parameters (Ekman and Friesen 1978) to drive a different (older) synthetically animated computer-generated imagery (CGI) character. More recent examples of performance-driven facial animation can be found in the state of the art report by Zollhöfer, Thies *et al.* (2018).

13.6.4 Human body modeling and tracking

The topics of tracking humans, modeling their shape and appearance, and recognizing their activities, are some of the most actively studied areas of computer vision. Annual conferences[18] and special journal issues (Hilton, Fua, and Ronfard 2006) are devoted to this subject, and two surveys (Forsyth, Arikan *et al.* 2006; Moeslund, Hilton, and Krüger 2006) each list over 400 papers devoted to these topics.[19] The HumanEva database of articulated human motions contains multi-view video sequences of human actions along with corresponding motion capture data, evaluation code, and a reference 3D tracker based on particle filtering. The companion paper by Sigal, Balan, and Black (2010) not only describes the database and evaluation but also has a nice survey of important work in this field. The more recent MPI FAUST dataset (Bogo, Romero *et al.* 2014) has 300 real, high-resolution human scans with automatically computed ground-truth correspondences, while the even newer AMASS dataset (Mahmood, Ghorbani *et al.* 2019) has more than 40 hours of motion data, spanning over 300 subjects and 11,000 motions.[20]

Given the breadth of this area, it is difficult to categorize all of this research, especially as different techniques usually build on each other. Moeslund, Hilton, and Krüger (2006) divide their survey into initialization, tracking (which includes background modeling and segmentation), pose estimation, and action (activity) recognition. Forsyth, Arikan *et al.* (2006) divide their survey into sections on tracking (background subtraction, deformable templates, flow, and probabilistic models), recovering 3D pose from 2D observations, and data association and body parts. They also include a section on motion synthesis, which is more widely studied in computer graphics (Arikan and Forsyth 2002; Kovar, Gleicher, and Pighin 2002; Lee, Chai *et al.* 2002; Li, Wang, and Shum 2002; Pullen and Bregler 2002): see Section 14.5.2. Another potential taxonomy for work in this field would be along the lines of whether 2D or 3D (or multi-view) images are used as input and whether 2D or 3D kinematic models are used.

In this section, we briefly review some of the more seminal and widely cited papers in the areas of background subtraction, initialization and detection, tracking with flow, 3D kinematic models, probabilistic models, adaptive shape modeling, and activity recognition. We refer the reader to the previously mentioned surveys for other topics and more details.

Background subtraction. One of the first steps in many human tracking systems is to model the background to extract the moving foreground objects (silhouettes) corresponding to people. Toyama, Krumm *et al.* (1999) review several *difference matting* and *background maintenance* (modeling) techniques and provide a good introduction to this topic. Stauffer and Grimson (1999) describe some techniques based on mixture models, while Sidenbladh and Black (2003) develop a

[18] International Conference on Automatic Face and Gesture Recognition (FG) and IEEE Workshop on Analysis and Modeling of Faces and Gestures (AMFG).

[19] Older surveys include those by Gavrila (1999) and Moeslund and Granum (2001). Some surveys on gesture recognition, which we do not cover in this book, include those by Pavlović, Sharma, and Huang (1997) and Yang, Ahuja, and Tabb (2002).

[20] Additional datasets from the MPI Perceiving Systems group can be found at https://ps.is.mpg.de/code.

more comprehensive treatment, which models not only the background image statistics but also the appearance of the foreground objects, e.g., their edge and motion (frame difference) statistics. More recent techniques for video background matting, such as those of Sengupta, Jayaram *et al.* (2020) and Lin, Ryabtsev *et al.* (2021) are discussed in Section 10.4.5 on video matting.

Once silhouettes have been extracted from one or more cameras, they can then be modeled using deformable templates or other contour models (Baumberg and Hogg 1996; Wren, Azarbayejani *et al.* 1997). Tracking such silhouettes over time supports the analysis of multiple people moving around a scene, including building shape and appearance models and detecting if they are carrying objects (Haritaoglu, Harwood, and Davis 2000; Mittal and Davis 2003; Dimitrijevic, Lepetit, and Fua 2006).

Initialization and detection. To track people in a fully automated manner, it is necessary to first detect (or re-acquire) their presence in individual video frames. This topic is closely related to *pedestrian detection*, which is often considered as a kind of object recognition (Mori, Ren *et al.* 2004; Felzenszwalb and Huttenlocher 2005; Felzenszwalb, McAllester, and Ramanan 2008; Dollár, Wojek *et al.* 2012; Dollár, Appel *et al.* 2014; Sermanet, Kavukcuoglu *et al.* 2013; Ouyang and Wang 2013; Tian, Luo *et al.* 2015; Zhang, Lin *et al.* 2016), and is therefore treated in more depth in Section 6.3.2. Additional techniques for initializing 3D trackers based on 2D images include those described by Howe, Leventon, and Freeman (2000), Rosales and Sclaroff (2000), Shakhnarovich, Viola, and Darrell (2003), Sminchisescu, Kanaujia *et al.* (2005), Agarwal and Triggs (2006), Lee and Cohen (2006), Sigal and Black (2006b), and Stenger, Thayananthan *et al.* (2006).

Single-frame human detection and pose estimation algorithms can be used by themselves to perform tracking (Ramanan, Forsyth, and Zisserman 2005; Rogez, Rihan *et al.* 2008; Bourdev and Malik 2009; Güler, Neverova, and Kokkinos 2018; Cao, Hidalgo *et al.* 2019), as described in Section 6.3.2 (Figure 6.25) and Section 6.4.5 (Figure 6.42–6.43). They are often combined, however, with frame-to-frame tracking techniques to provide better reliability (Fossati, Dimitrijevic *et al.* 2007; Andriluka, Roth, and Schiele 2008; Ferrari, Marin-Jimenez, and Zisserman 2008).

Tracking with flow. The tracking of people and their pose from frame to frame can be enhanced by computing optical flow or matching the appearance of their limbs from one frame to another. For example, the *cardboard people* model of Ju, Black, and Yacoob (1996) models the appearance of each leg portion (upper and lower) as a moving rectangle, and uses optical flow to estimate their location in each subsequent frame. Cham and Rehg (1999) and Sidenbladh, Black, and Fleet (2000) track limbs using optical flow and templates, along with techniques for dealing with multiple hypotheses and uncertainty. Bregler, Malik, and Pullen (2004) use a full 3D model of limb and body motion, as described below. It is also possible to match the estimated motion field itself to some prototypes in order to identify the particular phase of a running motion or to match two low-resolution video portions to perform video replacement (Efros, Berg *et al.* 2003). Flow-based tracking can also be used to track non-rigidly deforming objects such as T-shirts (White, Crane, and Forsyth 2007; Pilet, Lepetit, and Fua 2008; Furukawa and Ponce 2008; Salzmann and Fua 2010; Božič, Zollhöfer *et al.* 2020; Božič, Palafox *et al.* 2020, 2021). It is also possible to use inter-frame motion to estimate an evolving textured 3D mesh model of a moving person (de Aguiar, Stoll *et al.* 2008).

3D kinematic models. The effectiveness of human modeling and tracking can be greatly enhanced using a more accurate 3D model of a person's shape and motion. Underlying such representations, which are ubiquitous in 3D computer animation in games and special effects, is a *kinematic*

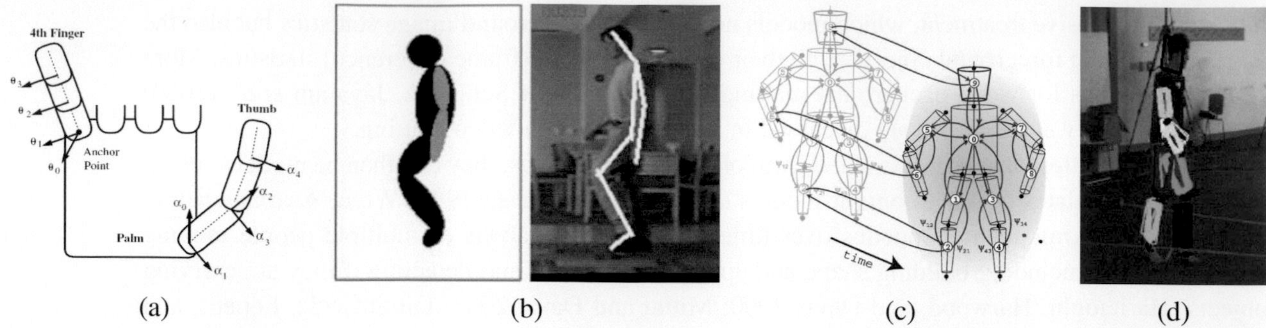

Figure 13.27 Tracking 3D human motion: (a) kinematic chain model for a human hand (Rehg, Morris, and Kanade 2003) © 2003, reprinted by permission of SAGE; (b) tracking a kinematic chain blob model in a video sequence (Bregler, Malik, and Pullen 2004) © 2004 Springer; (c–d) probabilistic loose-limbed collection of body parts (Sigal, Bhatia *et al.* 2004) © 2004 IEEE.

model or *kinematic chain*, which specifies the length of each limb in a skeleton as well as the 2D or 3D rotation angles between the limbs or segments (Figure 13.27a–b). Inferring the values of the joint angles from the locations of the visible surface points is called *inverse kinematics* (IK) and is widely studied in computer graphics.

Figure 13.27a shows the kinematic model for a human hand used by Rehg, Morris, and Kanade (2003) to track hand motion in a video. As you can see, the attachment points between the fingers and the thumb have two degrees of freedom, while the finger joints themselves have only one. Using this kind of model can greatly enhance the ability of an edge-based tracker to cope with rapid motion, ambiguities in 3D pose, and partial occlusions.

One of the biggest advances in reliable real-time hand tracking and modeling was the introduction of the Kinect consumer RGB-D camera (Sharp, Keskin *et al.* 2015; Taylor, Bordeaux *et al.* 2016), Since then, regular RGB tracking and modeling has also improved significantly, with newer techniques using neural networks for reliability and speed (Zimmermann and Brox 2017; Mueller, Bernard *et al.* 2018; Hasson, Varol *et al.* 2019; Shan, Geng *et al.* 2020; Moon, Shiratori, and Lee 2020; Moon, Yu *et al.* 2020; Spurr, Iqbal *et al.* 2020; Taheri, Ghorbani *et al.* 2020). Several systems also combine body and hand tracking to more accurately capture human expressions and activities (Romero, Tzionas, and Black 2017; Joo, Simon, and Sheikh 2018; Pavlakos, Choutas *et al.* 2019; Rong, Shiratori, and Joo 2020).

In addition to hands, kinematic chain models are even more widely used for whole body modeling and tracking (O'Rourke and Badler 1980; Hogg 1983; Rohr 1994). One popular approach is to associate an ellipsoid or superquadric with each rigid limb in the kinematic model, as shown in Figure 13.27b. This model can then be fitted to each frame in one or more video streams either by matching silhouettes extracted from known backgrounds or by matching and tracking the locations of occluding edges (Gavrila and Davis 1996; Kakadiaris and Metaxas 2000; Bregler, Malik, and Pullen 2004; Kehl and Van Gool 2006).

One of the big breakthroughs in real-time skeletal tracking was the introduction of the Kinect consumer depth camera for interactive video game control (Shotton, Fitzgibbon *et al.* 2011; Taylor, Shotton *et al.* 2012; Shotton, Girshick *et al.* 2013) as shown in Figure 13.28. In the current landscape of skeletal tracking, some techniques use 2D models coupled to 2D measurements, some use 3D measurements (range data or multi-view video) with 3D models (Baak, Mueller *et al.* 2011), and some use monocular video to infer and track 3D models directly (Mehta, Sridhar *et al.* 2017; Habermann, Xu *et al.* 2019).

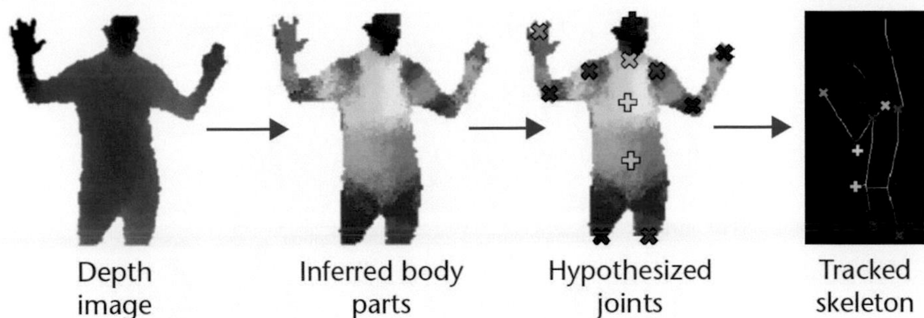

Figure 13.28 The Kinect skeletal tracking pipeline, which consists of per-pixel body-part classification, body joint hypotheses, and then mapping to a skeleton using temporal continuity and prior knowledge (Shotton, Girshick *et al.* 2013). This figure is taken from (Zhang 2012) © 2012 IEEE.

It is also possible to use temporal models to improve the tracking of periodic motions, such as walking, by analyzing the joint angles as functions of time (Polana and Nelson 1997; Seitz and Dyer 1997; Cutler and Davis 2000). The generality and applicability of such techniques can be improved by learning typical motion patterns using principal component analysis (Sidenbladh, Black, and Fleet 2000; Urtasun, Fleet, and Fua 2006).

Probabilistic models. Because tracking can be such a difficult task, sophisticated probabilistic inference techniques are often used to estimate the likely states of the person being tracked. One popular approach, called *particle filtering* (Isard and Blake 1998), was originally developed for tracking the outlines of people and hands, as described in Section 7.3.1. It was subsequently applied to whole-body tracking (Deutscher, Blake, and Reid 2000; Sidenbladh, Black, and Fleet 2000; Deutscher and Reid 2005) and continues to be used in modern trackers (Ong, Micilotta *et al.* 2006). Alternative approaches to handling the uncertainty inherent in tracking include multiple hypothesis tracking (Cham and Rehg 1999) and inflated covariances (Sminchisescu and Triggs 2001).

Figure 13.27c–d shows an example of a sophisticated spatio-temporal probabilistic graphical model called *loose-limbed people*, which models not only the geometric relationship between various limbs, but also their likely temporal dynamics (Sigal, Bhatia *et al.* 2004). The conditional probabilities relating various limbs and time instances are learned from training data, and particle filtering is used to perform the final pose inference.

Adaptive shape modeling. Another essential component of whole body modeling and tracking is the fitting of parameterized shape models to visual data. As we saw in Section 13.6.3 (Figure 13.25), the availability of large numbers of registered 3D range scans can be used to create *morphable models* of shape and appearance (Allen, Curless, and Popović 2003). Building on this work, Anguelov, Srinivasan *et al.* (2005) develop a sophisticated system called SCAPE (Shape Completion and Animation for PEople), which first acquires a large number of range scans of different people in varied poses, and then registers these scans using semi-automated marker placement. The registered datasets are used to model the variation in shape as a function of personal characteristics and skeletal pose, e.g., the bulging of muscles as certain joints are flexed (Figure 13.29, top row). The resulting system can then be used for *shape completion*, i.e., the recovery of a full 3D mesh model from a small number of captured markers, by finding the best model parameters in both shape and pose space that fit the measured data.

Because it is constructed completely from scans of people in close-fitting clothing and uses a

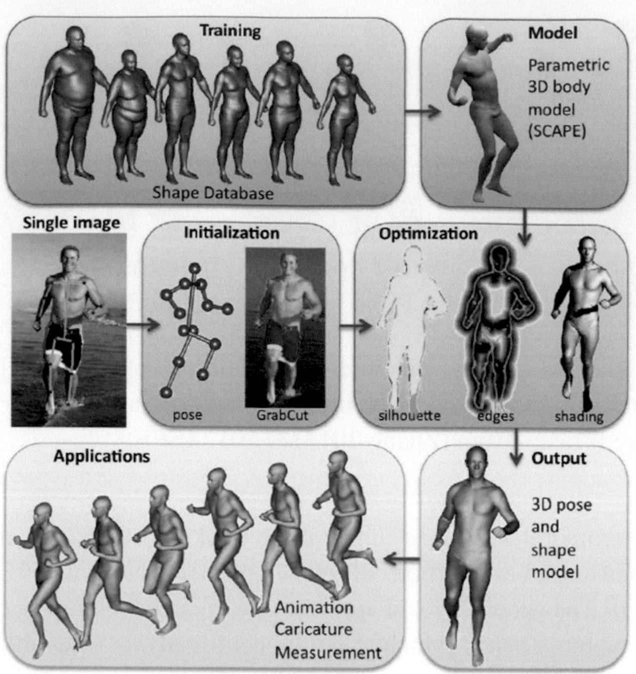

Figure 13.29 Estimating human shape and pose from a single image using a parametric 3D model (Guan, Weiss *et al.* 2009) © 2009 IEEE.

parametric shape model, the SCAPE system cannot cope with people wearing loose-fitting clothing. Bălan and Black (2008) overcome this limitation by estimating the body shape that fits within the visual hull of the same person observed in multiple poses, while Vlasic, Baran *et al.* (2008) adapt an initial surface mesh fitted with a parametric shape model to better match the visual hull.

While the preceding body fitting and pose estimation systems use multiple views to estimate body shape, Guan, Weiss *et al.* (2009) fit a human shape and pose model to a single image of a person on a natural background. Manual initialization is used to estimate a rough pose (skeleton) and height model, and this is then used to segment the person's outline using the Grab Cut segmentation algorithm (Section 4.3.2). The shape and pose estimate are then refined using a combination of silhouette edge cues and shading information (Figure 13.29). The resulting 3D model can be used to create novel animations.

While some of the original work on 3D body and pose fitting was done using the SCAPE and BlendSCAPE (Hirshberg, Loper *et al.* 2012) models, the Skinned Multi-Person Linear model (SMPL) developed by Loper, Mahmood *et al.* (2015) introduced a skinned vertex-based model that accurately represents a wide variety of body shapes in natural human poses. The model consists of a rest pose template, pose-dependent blend shapes, and identity-dependent blend shapes, and is built by training on a large collection of aligned 3D human scans. Bogo, Kanazawa *et al.* (2016) show how the parameters of this 3D model can be estimated from just a single image using their SMPLfy method.

In subsequent work Romero, Tzionas, and Black (2017) extend this model by adding a hand Model with Articulated and Non-rigid defOrmations (MANO). Joo, Simon, and Sheikh (2018) stitch together the SMPL body model with a face and a hand model to create the 3D Frank and Adam models that can track multiple people in a social setting. And Pavlakos, Choutas *et al.* (2019) use thousands of 3D scans to train a new, unified, 3D model of the human body (SMPL-X) that extends

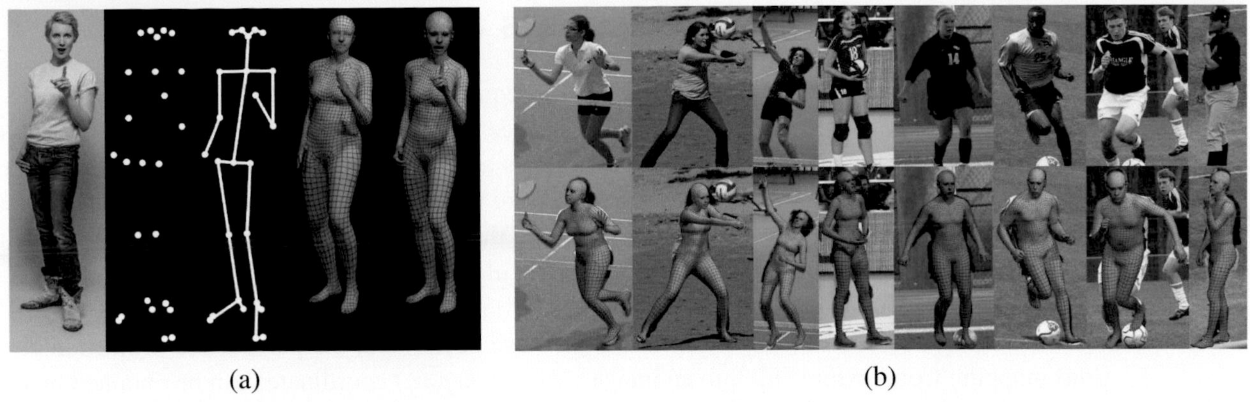

(a) (b)

Figure 13.30 Whole body, expression, and gesture fitting from a single image using the SMPL-X model from Pavlakos, Choutas *et al.* (2019) © 2019 IEEE: (a) estimating the major joints, skeleton, SMPL, and SMPL-X models from a single image; (b) qualitative results of SMPL-X for some in-the-wild images.

SMPL with gender-specific models and includes fully articulated hands and an expressive face, as shown in Figure 13.30. They also replace the mixture of Gaussians prior in SMPL with a variational autoencoder (VAE) and develop a new VPoser prior trained on the large-scale AMASS motion capture dataset collected by Mahmood, Ghorbani *et al.* (2019).

In more recent work, Kocabas, Athanasiou, and Black (2020) introduce VIBE, a system for video inference of human body pose and shape that makes use of AMASS. Choutas, Pavlakos *et al.* (2020) develop a system they call ExPose (EXpressive POse and Shape rEgression), which directly regresses the body, face, and hands SMPL-X parameters from an RGB image. The more recent STAR (Sparse Trained Articulated human body Regressor) model (Osman, Bolkart, and Black 2020), has many fewer parameters than SMPL and removes spurious long-range correlations between vertices. It also includes shape-dependent pose-corrective blend shapes that depend on both body pose and BMI and also models a much wider range of variation in the human population by training STAR with an additional 10,000 scans of male and female subjects. GHUM and GHUML (Xu, Bazavan *et al.* 2020) rely on non-linear shape spaces constructed from deep variational autoencoders for body and facial deformation and on normalizing flow representations for skeleton (body and hand) kinematics. Recent papers that continue to improve the accuracy and speed of single-image model fitting on the challenging 3D Poses in the Wild (3DPW) benchmark and dataset (von Marcard, Henschel *et al.* 2018) include Song, Chen, and Hilliges (2020), Joo, Neverova, and Vedaldi (2020), and Rong, Shiratori, and Joo (2020).

Activity recognition. The final widely studied topic in human modeling is motion, activity, and action recognition (Bobick 1997; Hu, Tan *et al.* 2004; Hilton, Fua, and Ronfard 2006). Examples of actions that are commonly recognized include walking and running, jumping, dancing, picking up objects, sitting down and standing up, and waving. Papers on these topics include Robertson and Reid (2006), Sminchisescu, Kanaujia, and Metaxas (2006), Weinland, Ronfard, and Boyer (2006), Yilmaz and Shah (2006), and Gorelick, Blank *et al.* (2007), as well as more recent video understanding papers such as the ones we covered in Section 6.5, e.g., Carreira and Zisserman (2017), Tran, Wang *et al.* (2018), Tran, Wang *et al.* (2019), Wu, Feichtenhofer *et al.* (2019), and Feichtenhofer, Fan *et al.* (2019).

13.7 Recovering texture maps and albedos

After a 3D model of an object or person has been acquired, the final step in modeling is usually to recover a *texture map* to describe the object's surface appearance. This first requires establishing a parameterization for the (u, v) texture coordinates as a function of 3D surface position.[21] One simple way to do this is to associate a separate texture map with each triangle (or pair of triangles). More space-efficient techniques involve unwrapping the surface onto one or more maps, e.g., using a subdivision mesh (Section 13.3.2) (Eck, DeRose *et al.* 1995) or a geometry image (Section 13.3.3) (Gu, Gortler, and Hoppe 2002).

Once the (u, v) coordinates for each triangle have been fixed, the perspective projection equations mapping from texture (u, v) to an image j's pixel (u_j, v_j) coordinates can be obtained by concatenating the affine $(u, v) \rightarrow (X, Y, Z)$ mapping with the perspective homography $(X, Y, Z) \rightarrow (u_j, v_j)$ (Szeliski and Shum 1997). The color values for the (u, v) texture map can then be resampled and stored, or the original image can itself be used as the texture source using projective texture mapping (OpenGL-ARB 1997).

The situation becomes more involved when more than one source image is available for appearance recovery, which is the usual case. One possibility is to use a *view-dependent texture map* (Section 14.1.1), in which a different source image (or combination of source images) is used for each polygonal face based on the angles between the virtual camera, the surface normals, and the source images (Debevec, Taylor, and Malik 1996; Pighin, Hecker *et al.* 1998). An alternative approach is to estimate a complete Surface Light Field for each surface point (Wood, Azuma *et al.* 2000), as described in Section 14.3.2.

In some situations, e.g., when using models in traditional 3D games, it is preferable to merge all of the source images into a single coherent texture map during pre-processing (Weinhaus and Devarajan 1997). Ideally, each surface triangle should select the source image where it is seen most directly (perpendicular to its normal) and at the resolution best matching the texture map resolution.[22] This can be posed as a graph cut optimization problem, where the smoothness term encourages adjacent triangles to use similar source images, followed by blending to compensate for exposure differences (Lempitsky and Ivanov 2007; Sinha, Steedly *et al.* 2008). Even better results can be obtained by explicitly modeling geometric and photometric misalignments between the source images (Shum and Szeliski 2000; Gal, Wexler *et al.* 2010; Waechter, Moehrle, and Goesele 2014; Zhou and Koltun 2014; Huang, Dai *et al.* 2017; Fu, Yan *et al.* 2018; Schöps, Sattler, and Pollefeys 2019b; Lee, Ha *et al.* 2020). "Neural" texture map representations can also be used as an alternative to RGB color fields (Oechsle, Mescheder *et al.* 2019; Mihajlovic, Weder *et al.* 2021). Zollhöfer, Stotko *et al.* (2018, Section 4.1) discuss related techniques in more detail.

These kinds of approaches produce good results when the lighting stays fixed with respect to the object, i.e., when the camera moves around the object or space. When the lighting is strongly directional, however, and the object is being moved relative to this lighting, strong shading effects or specularities may be present, which will interfere with the reliable recovery of a texture (albedo) map. In this case, it is preferable to explicitly undo the shading effects (Section 13.1) by modeling the light source directions and estimating the surface reflectance properties while recovering the texture map (Sato and Ikeuchi 1996; Sato, Wheeler, and Ikeuchi 1997; Yu and Malik 1998; Yu, Debevec *et al.* 1999). Figure 13.31 shows the results of one such approach, where the specularities are first removed while estimating the matte reflectance component (albedo) and then later re-introduced by

[21] Although a few recent papers have directly constructed a mapping from (x, y, z) to color values (Saito, Huang *et al.* 2019; Saito, Simon *et al.* 2020; Mildenhall, Srinivasan *et al.* 2020)—see Section 14.6.

[22] When surfaces are seen at oblique viewing angles, it may be necessary to blend different images together to obtain the best resolution (Wang, Kang *et al.* 2001).

(a) (b) (c)

Figure 13.31 Estimating the diffuse albedo and reflectance parameters for a scanned 3D model (Sato, Wheeler, and Ikeuchi 1997) © 1997 ACM: (a) set of input images projected onto the model; (b) the complete diffuse reflection (albedo) model; (c) rendering from the reflectance model including the specular component.

estimating the specular component k_s in a Torrance–Sparrow reflection model (2.92).

13.7.1 Estimating BRDFs

A more ambitious approach to the problem of view-dependent appearance modeling is to estimate a general bidirectional reflectance distribution function (BRDF) for each point on an object's surface. Dana, van Ginneken *et al.* (1999), Jensen, Marschner *et al.* (2001), and Lensch, Kautz *et al.* (2003) present different techniques for estimating such functions, while Dorsey, Rushmeier, and Sillion (2007) and Weyrich, Lawrence *et al.* (2009) provide surveys of the topics of BRDF modeling, recovery, and rendering.

As we saw in Section 2.2.2 (2.82), the BRDF can be written as

$$f_r(\theta_i, \phi_i, \theta_r, \phi_r; \lambda), \tag{13.6}$$

where (θ_i, ϕ_i) and (θ_r, ϕ_r) are the angles the incident $\hat{\mathbf{v}}_i$ and reflected $\hat{\mathbf{v}}_r$ light ray directions make with the local surface coordinate frame $(\hat{\mathbf{d}}_x, \hat{\mathbf{d}}_y, \hat{\mathbf{n}})$ shown in Figure 2.15. When modeling the appearance of an object, as opposed to the appearance of a patch of material, we need to estimate this function at every point (x, y) on the object's surface, which gives us the *spatially varying* BRDF, or SVBRDF (Weyrich, Lawrence *et al.* 2009),

$$f_v(x, y, \theta_i, \phi_i, \theta_r, \phi_r; \lambda). \tag{13.7}$$

If sub-surface scattering effects are being modeled, such as the long-range transmission of light through materials such as alabaster, the eight-dimensional bidirectional scattering-surface reflectance-distribution function (BSSRDF) is used instead,

$$f_e(x_i, y_i, \theta_i, \phi_i, x_e, y_e, \theta_e, \phi_e; \lambda), \tag{13.8}$$

where the e subscript now represents the *emitted* rather than the *reflected* light directions.

Weyrich, Lawrence *et al.* (2009) provide a nice survey of these and related topics, including basic photometry, BRDF models, traditional BRDF acquisition using *gonio reflectometry*, i.e., the precise measurement of visual angles and reflectances (Marschner, Westin *et al.* 2000; Dupuy and Jakob 2018), multiplexed illumination (Schechner, Nayar, and Belhumeur 2009), skin modeling (Debevec, Hawkins *et al.* 2000; Weyrich, Matusik *et al.* 2006), and image-based acquisition techniques, which simultaneously recover an object's 3D shape and reflectometry from multiple photographs.

A nice example of this latter approach is the system developed by Lensch, Kautz *et al.* (2003), who estimate locally varying BRDFs and refine their shape models using local estimates of surface

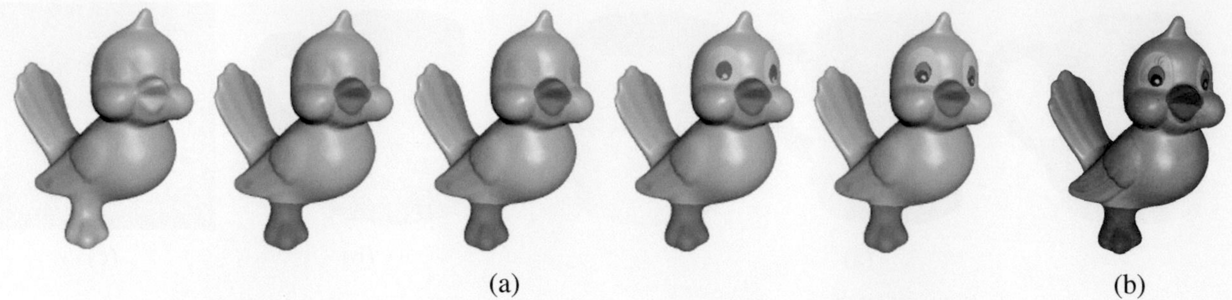

Figure 13.32 Image-based reconstruction of appearance and detailed geometry (Lensch, Kautz *et al.* 2003) © 2003 ACM. (a) Appearance models (BRDFs) are re-estimated using divisive clustering. (b) To model detailed spatially varying appearance, each lumitexel is projected onto the basis formed by the clustered materials.

normals. To build up their models, they first associate a *lumitexel*, which contains a 3D position, a surface normal, and a set of sparse radiance samples, with each surface point. Next, they cluster such lumitexels into materials that share common properties, using a Lafortune reflectance model (Lafortune, Foo *et al.* 1997) and a divisive clustering approach (Figure 13.32a). Finally, to model detailed spatially varying appearance, each lumitexel (surface point) is projected onto the basis of clustered appearance models (Figure 13.32b). A more accurate system for estimating normals can be obtained using polarized lighting, as described by Ma, Hawkins *et al.* (2007).

More recent approaches to recovering spatially varying BRDFs (SVBRDFs) either start with RGB-D scanners (Park, Newcombe, and Seitz 2018; Schmitt, Donne *et al.* 2020), flash/no-flash image pairs (Aittala, Weyrich, and Lehtinen 2015), or use deep learning approaches to simultaneously estimate surface normals and appearance models (Li, Sunkavalli, and Chandraker 2018; Li, Xu *et al.* 2018). Even more sophisticated systems can also estimate shape and environmental lighting from range scanner sequences (Park, Holynski, and Seitz 2020) or single monocular images (Boss, Jampani *et al.* 2020; Li, Shafiei *et al.* 2020; Chen, Nobuhara, and Nishino 2020) and even perform relighting on such scenes (Bi, Xu *et al.* 2020a,b; Sang and Chandraker 2020; Bi, Xu *et al.* 2020c). A more in-depth review of techniques for capturing the 3D shape and appearance of objects with RGB-D cameras can be found in the state of the art report by Zollhöfer, Stotko *et al.* (2018).

While most of the techniques discussed in this section require large numbers of views to estimate surface properties, an interesting challenge is to take these techniques out of the lab and into the real world, and to combine them with regular and internet photo image-based modeling approaches.

13.7.2 *Application*: 3D model capture

The techniques described in this chapter for building complete 3D models from multiple images and then recovering their surface appearance have opened up a whole new range of applications that often go under the name *3D photography*. Pollefeys and Van Gool (2002) and Pollefeys, Van Gool *et al.* (2004) provide nice introductions to such systems, including the processing steps of feature matching, structure from motion recovery, dense depth map estimation, 3D model building, and texture map recovery. A complete web-based system for automatically performing all of these tasks, called ARC3D, is described by Vergauwen and Van Gool (2006) and Moons, Van Gool, and Vergauwen (2010). The latter paper provides not only an in-depth survey of this whole field but also a detailed description of their complete end-to-end system.

An example of a more recent commercial photogrammetric modeling system that can be used for both object and scene capture is Pix4D, whose website shows a wonderful example of a 3D texture-mapped castle reconstructed from both regular and aerial drone photographs.[23] Examples of casual 3D photography enabled by the advent of smartphones include Hedman, Alsisan *et al.* (2017), Hedman and Kopf (2018), and Kopf, Matzen *et al.* (2020) and are described in more detail in Section 14.2.2.

An alternative to such fully automated systems is to put the user in the loop in what is sometimes called *interactive computer vision*. An early example of this was the Façade architectural modeling system developed by Debevec, Taylor, and Malik (1996). van den Hengel, Dick *et al.* (2007) describe their VideoTrace system, which performs automated point tracking and 3D structure recovery from video and then lets the user draw triangles and surfaces on top of the resulting point cloud, as well as interactively adjusting the locations of model vertices. Sinha, Steedly *et al.* (2008) describe a related system that uses matched vanishing points in multiple images (Figure 7.50) to infer 3D line orientations and plane normals. These are then used to guide the user drawing axis-aligned planes, which are automatically fitted to the recovered 3D point cloud. Fully automated variants on these ideas are described by Zebedin, Bauer *et al.* (2008), Furukawa, Curless *et al.* (2009a), Furukawa, Curless *et al.* (2009b), Mičušík and Košecká (2009), and Sinha, Steedly, and Szeliski (2009).

As the sophistication and reliability of these techniques continues to improve, we can expect to see even more user-friendly applications for photorealistic 3D modeling from images (Exercise 13.8).

13.8 Additional reading

Shape from shading is one of the classic problems in computer vision (Horn 1975). Some representative papers in this area include those by Horn (1977), Ikeuchi and Horn (1981), Pentland (1984), Horn and Brooks (1986), Horn (1990), Szeliski (1991a), Mancini and Wolff (1992), Dupuis and Oliensis (1994), and Fua and Leclerc (1995). The collection of papers edited by Horn and Brooks (1989) is a great source of information on this topic, especially the chapter on variational approaches. The survey by Zhang, Tsai *et al.* (1999) reviews such techniques and also provides some comparative results.

Woodham (1981) wrote the seminal paper of photometric stereo. Shape from texture techniques include those by Witkin (1981), Ikeuchi (1981), Blostein and Ahuja (1987), Gårding (1992), Malik and Rosenholtz (1997), Liu, Collins, and Tsin (2004), Liu, Lin, and Hays (2004), Hays, Leordeanu *et al.* (2006), Lin, Hays *et al.* (2006), Lobay and Forsyth (2006), White and Forsyth (2006), White, Crane, and Forsyth (2007), and Park, Brocklehurst *et al.* (2009). Good papers and books on depth from defocus have been written by Pentland (1987), Nayar and Nakagawa (1994), Nayar, Watanabe, and Noguchi (1996), Watanabe and Nayar (1998), Chaudhuri and Rajagopalan (1999), and Favaro and Soatto (2006). Additional techniques for recovering shape from various kinds of illumination effects, including inter-reflections (Nayar, Ikeuchi, and Kanade 1991), are discussed in the book on shape recovery edited by Wolff, Shafer, and Healey (1992b). A more recent survey on photometric stereo is Ackermann and Goesele (2015) and recent papers include Logothetis, Mecca, and Cipolla (2019), Haefner, Ye *et al.* (2019), and Santo, Waechter, and Matsushita (2020).

Active rangefinding systems, which use laser or natural light illumination projected into the scene, have been described by Besl (1989), Rioux and Bird (1993), Kang, Webb *et al.* (1995), Curless and Levoy (1995), Curless and Levoy (1996), Proesmans, Van Gool, and Defoort (1998), Bouguet and Perona (1999), Curless (1999), Hebert (2000), Iddan and Yahav (2001), Goesele, Fuchs, and Seidel (2003), Scharstein and Szeliski (2003), Davis, Ramamoorthi, and Rusinkiewicz (2003), Zhang, Curless, and Seitz (2003), Zhang, Snavely *et al.* (2004), and Moons, Van Gool, and

[23]https://www.pix4d.com/blog/mapping-chillon-castle-with-drone

Vergauwen (2010), and in the more recent reviews by Zhang (2018) and Ikeuchi, Matsushita *et al.* (2020). Individual range scans can be aligned using 3D correspondence and distance optimization techniques such as *iterative closest points* and its variants (Besl and McKay 1992; Zhang 1994; Szeliski and Lavallée 1996; Johnson and Kang 1997; Gold, Rangarajan *et al.* 1998; Johnson and Hebert 1999; Pulli 1999; David, DeMenthon *et al.* 2004; Li and Hartley 2007; Enqvist, Josephson, and Kahl 2009; Pomerleau, Colas, and Siegwart 2015; Rusinkiewicz 2019). Once they have been aligned, range scans can be merged using techniques that model the signed distance of surfaces to volumetric sample points (Hoppe, DeRose *et al.* 1992; Curless and Levoy 1996; Hilton, Stoddart *et al.* 1996; Wheeler, Sato, and Ikeuchi 1998; Kazhdan, Bolitho, and Hoppe 2006; Lempitsky and Boykov 2007; Zach, Pock, and Bischof 2007b; Zach 2008; Newcombe, Izadi *et al.* 2011; Zhou, Miller, and Koltun 2013; Newcombe, Fox, and Seitz 2015; Zollhöfer, Stotko *et al.* 2018).

Once constructed, 3D surfaces can be modeled and manipulated using a variety of three-dimensional representations, which include triangle meshes (Eck, DeRose *et al.* 1995; Hoppe 1996), splines (Farin 1992; Lee, Wolberg, and Shin 1997; Farin 2002), subdivision surfaces (Stollnitz, DeRose, and Salesin 1996; Zorin, Schröder, and Sweldens 1996; Warren and Weimer 2001; Peters and Reif 2008), and geometry images (Gu, Gortler, and Hoppe 2002). Alternatively, they can be represented as collections of point samples with local orientation estimates (Hoppe, DeRose *et al.* 1992; Szeliski and Tonnesen 1992; Turk and O'Brien 2002; Pfister, Zwicker *et al.* 2000; Alexa, Behr *et al.* 2003; Pauly, Keiser *et al.* 2003; Diebel, Thrun, and Brünig 2006; Guennebaud and Gross 2007; Guennebaud, Germann, and Gross 2008; Oztireli, Guennebaud, and Gross 2008; Berger, Tagliasacchi *et al.* 2017). They can also be modeled using implicit inside–outside characteristic or signed distance functions sampled on regular or irregular (octree) volumetric grids (Lavallée and Szeliski 1995; Szeliski and Lavallée 1996; Frisken, Perry *et al.* 2000; Dinh, Turk, and Slabaugh 2002; Kazhdan, Bolitho, and Hoppe 2006; Lempitsky and Boykov 2007; Zach, Pock, and Bischof 2007b; Zach 2008; Kazhdan and Hoppe 2013).

The literature on model-based 3D reconstruction is extensive. For modeling architecture and urban scenes, both interactive and fully automated systems have been developed. A special journal issue devoted to the reconstruction of large-scale 3D scenes (Zhu and Kanade 2008) is a good source of references and Robertson and Cipolla (2009) give a nice description of a complete system. Lots of additional references can be found in Section 13.6.1.

Face and whole body modeling and tracking is a very active sub-field of computer vision, with its own conferences and workshops, e.g., the International Conference on Automatic Face and Gesture Recognition (FG) and IEEE Workshop on Analysis and Modeling of Faces and Gestures (AMFG). Two recent survey papers on 3D face modeling and tracking are Zollhöfer, Thies *et al.* (2018) and Egger, Smith *et al.* (2020), while surveys on the topic of whole body modeling and tracking include Forsyth, Arikan *et al.* (2006), Moeslund, Hilton, and Krüger (2006), and Sigal, Balan, and Black (2010).

Some representative papers on recovering texture maps from multiple color and RGB-D images include Gal, Wexler *et al.* (2010), Waechter, Moehrle, and Goesele (2014), Zhou and Koltun (2014), and Lee, Ha *et al.* (2020) as well as Zollhöfer, Stotko *et al.* (2018, Section 4.1). The more complex process of recovering spatially varying BRDFs is covered in surveys by Dorsey, Rushmeier, and Sillion (2007) and Weyrich, Lawrence *et al.* (2009). More recent techniques that can do this using fewer images and RGB-D images include Aittala, Weyrich, and Lehtinen (2015), Li, Sunkavalli, and Chandraker (2018), Schmitt, Donne *et al.* (2020), and Boss, Jampani *et al.* (2020) and the survey by Zollhöfer, Stotko *et al.* (2018).

13.9 Exercises

Ex 13.1: Shape from focus. Grab a series of focused images with a digital SLR set to manual focus (or get one that allows for programmatic focus control) and recover the depth of an object.

1. Take some calibration images, e.g., of a checkerboard, so that you can compute a mapping between the amount of defocus and the focus setting.
2. Try both a fronto-parallel planar target and one which is slanted so that it covers the working range of the sensor. Which one works better?
3. Now put a real object in the scene and perform a similar focus sweep.
4. For each pixel, compute the local sharpness and fit a parabolic curve over focus settings to find the most in-focus setting.
5. Map these focus settings to depth and compare your result to ground truth. If you are using a known simple object, such as a sphere or cylinder (a ball or a soda can), it's easy to measure its true shape.
6. (Optional) See if you can recover the depth map from just two or three focus settings.
7. (Optional) Use an LCD projector to project artificial texture onto the scene. Use a pair of cameras to compare the accuracy of your shape from focus and shape from stereo techniques.
8. (Optional) Create an all-in-focus image using the technique of Agarwala, Dontcheva *et al.* (2004).

Ex 13.2: Shadow striping. Implement the handheld shadow striping system of Bouguet and Perona (1999). The basic steps include the following:

1. Set up two background planes behind the object of interest and calculate their orientation relative to the viewer, e.g., with fiducial marks.
2. Cast a moving shadow with a stick across the scene; record the video or capture the data with a webcam.
3. Estimate each light plane equation from the projections of the cast shadow against the two backgrounds.
4. Triangulate to the remaining points on each curve to get a 3D stripe and display the stripes using a 3D graphics engine.
5. (Optional) remove the requirement for a known second (vertical) plane and infer its location (or that of the light source) using the techniques described by Bouguet and Perona (1999). The techniques from Exercise 10.9 may also be helpful here.

Ex 13.3: Range data registration. Register two or more 3D datasets using either iterative closest points (ICP) (Besl and McKay 1992, Zhang 1994, Gold, Rangarajan *et al.* 1998) or octree signed distance fields (Szeliski and Lavallée 1996) (Section 13.2.1).

Apply your technique to narrow-baseline stereo pairs, e.g., obtained by moving a camera around an object, using structure from motion to recover the camera poses, and using a standard stereo matching algorithm.

Ex 13.4: Range data merging. Merge the datasets that you registered in the previous exercise using signed distance fields (Curless and Levoy 1996; Hilton, Stoddart *et al.* 1996) or one of their newer variants (Newcombe, Izadi *et al.* 2011; Hornung, Wurm *et al.* 2013; Nießner, Zollhöfer *et al.* 2013; Klingensmith, Dryanovski *et al.* 2015; Dai, Nießner *et al.* 2017; Zollhöfer, Stotko *et al.* 2018). Extract a meshed surface model from the signed distance field using marching cubes and display the resulting model.

Ex 13.5: Surface simplification. Use progressive meshes (Hoppe 1996) or some other technique from Section 13.3.2 to create a hierarchical simplification of your surface model.

Ex 13.6: Architectural modeler. Build a 3D interior or exterior model of some architectural structure, such as your house, from a series of handheld wide-angle photographs.

1. Extract lines and vanishing points (Exercises 7.11–7.14) to estimate the dominant directions in each image.

2. Use structure from motion to recover all of the camera poses and match up the vanishing points.

3. Let the user sketch the locations of the walls by drawing lines corresponding to wall bottoms, tops, and horizontal extents onto the images (Sinha, Steedly *et al.* 2008)—see also Exercise 11.4. Do something similar for openings (doors and windows) and simple furniture (tables and countertops).

4. Convert the resulting polygonal meshes into a 3D model (e.g., VRML) and optionally texture-map these surfaces from the images.

Ex 13.7: Body tracker. Download some human body movement sequences from one of the datasets such as HumanEva, MPI FAUST, or AMASS discussed in Section 13.6.4. Either implement a human motion tracker from scratch or extend existing code in some interesting way.

Ex 13.8: 3D photography. Combine all of your previously developed techniques to produce a system that takes a series of photographs or a video and constructs a photorealistic texture-mapped 3D model.

Chapter 14

Image-based rendering

14.1 View interpolation 683
14.1.1 View-dependent texture maps 685
14.1.2 *Application*: Photo Tourism 686
14.2 Layered depth images 688
14.2.1 Impostors, sprites, and layers 688
14.2.2 *Application*: 3D photography 690
14.3 Light fields and Lumigraphs 693
14.3.1 Unstructured Lumigraph 696
14.3.2 Surface light fields 696
14.3.3 *Application*: Concentric mosaics 698
14.3.4 *Application*: Synthetic re-focusing 698
14.4 Environment mattes 699
14.4.1 Higher-dimensional light fields 700
14.4.2 The modeling to rendering continuum 701
14.5 Video-based rendering 701
14.5.1 Video-based animation 702
14.5.2 Video textures 703
14.5.3 *Application*: Animating pictures 705
14.5.4 3D and free-viewpoint Video 706
14.5.5 *Application*: Video-based walkthroughs 708
14.6 Neural rendering 711
14.7 Additional reading 718
14.8 Exercises 719

R. Szeliski, *Computer Vision*, Texts in Computer Science,
https://doi.org/10.1007/978-3-030-34372-9_14

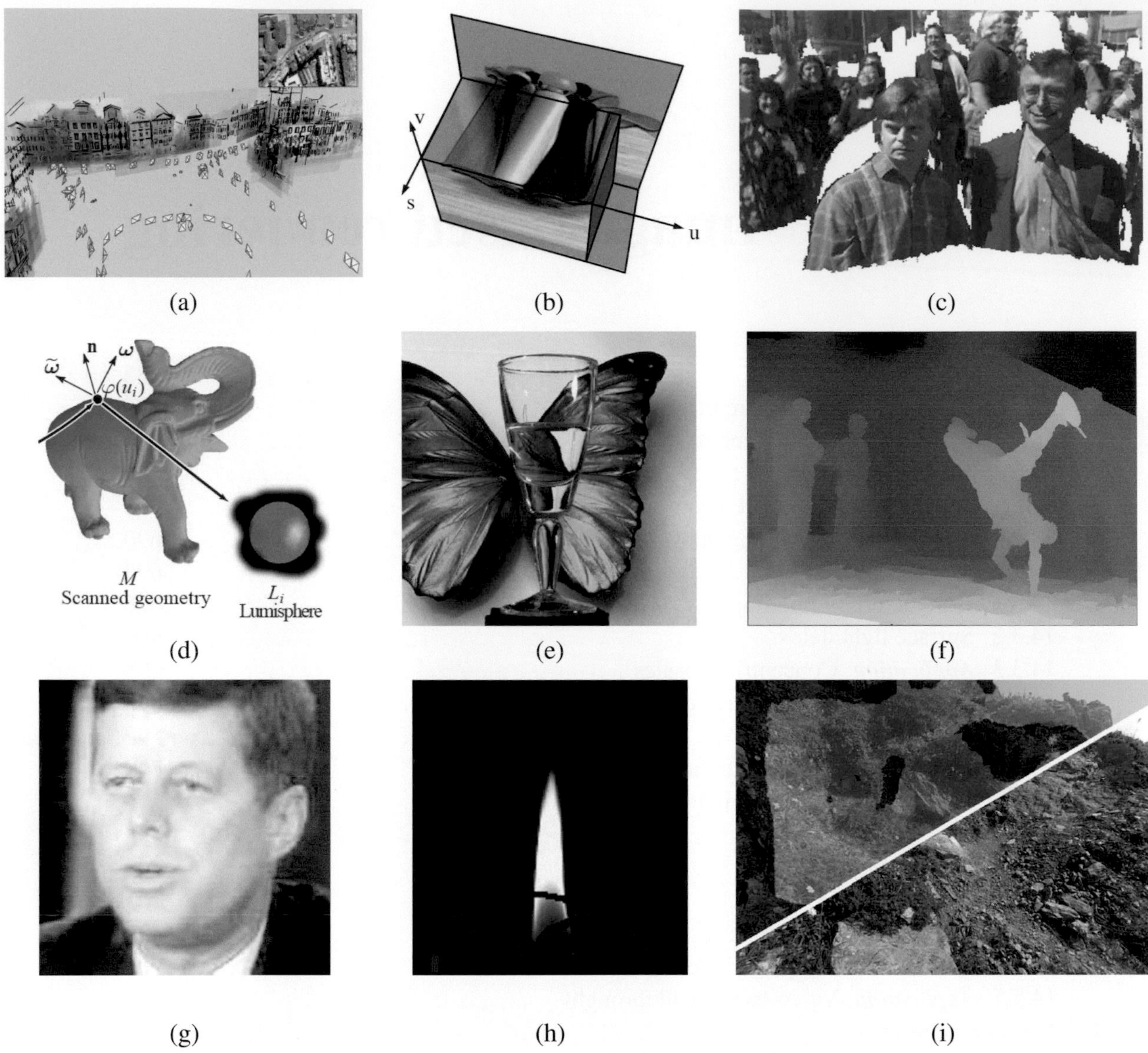

(a) (b) (c)

(d) (e) (f)

(g) (h) (i)

Figure 14.1 Image-based and video-based rendering: (a) a 3D view of a Photo Tourism reconstruction (Snavely, Seitz, and Szeliski 2006) © 2006 ACM; (b) a slice through a 4D light field (Gortler, Grzeszczuk *et al.* 1996) © 1996 ACM; (c) sprites with depth (Shade, Gortler *et al.* 1998) © 1998 ACM; (d) surface light field (Wood, Azuma *et al.* 2000) © 2000 ACM; (e) environment matte in front of a novel background (Zongker, Werner *et al.* 1999) © 1999 ACM; (f) video view interpolation (Zitnick, Kang *et al.* 2004) © 2004 ACM; (g) Video Rewrite used to re-animate old video (Bregler, Covell, and Slaney 1997) © 1997 ACM; (h) video texture of a candle flame (Schödl, Szeliski *et al.* 2000) © 2000 ACM; (i) hyperlapse video, stitching multiple frames with 3D proxies (Kopf, Cohen, and Szeliski 2014) © 2014 ACM.

Over the last few decades, image-based rendering has emerged as one of the most exciting applications of computer vision (Kang, Li *et al.* 2006; Shum, Chan, and Kang 2007; Gallo, Troccoli *et al.* 2020). In image-based rendering, 3D reconstruction techniques from computer vision are combined with computer graphics rendering techniques that use multiple views of a scene to create interactive photo-realistic experiences such as the Photo Tourism system shown in Figure 14.1a. Commercial versions of such systems include immersive street-level navigation in online mapping systems such as Google Maps and the creation of 3D Photosynths from large collections of casually acquired photographs.

In this chapter, we explore a variety of image-based rendering techniques, such as those illustrated in Figure 14.1. We begin with *view interpolation* (Section 14.1), which creates a seamless transition between a pair of reference images using one or more precomputed depth maps. Closely related to this idea are *view-dependent texture maps* (Section 14.1.1), which blend multiple texture maps on a 3D model's surface. The representations used for both the color imagery and the 3D geometry in view interpolation include a number of clever variants such as *layered depth images* (Section 14.2) and *sprites with depth* (Section 14.2.1).

We continue our exploration of image-based rendering with the *light field* and *Lumigraph* four-dimensional representations of a scene's appearance (Section 14.3), which can be used to render the scene from any arbitrary viewpoint. Variants on these representations include the *unstructured Lumigraph* (Section 14.3.1), *surface light fields* (Section 14.3.2), *concentric mosaics* (Section 14.3.3), and *environment mattes* (Section 14.4).

We then explore the topic of *video-based rendering*, which uses one or more videos To create novel video-based experiences (Section 14.5). The topics we cover include video-based facial animation (Section 14.5.1), as well as *video textures* (Section 14.5.2), in which short video clips can be seamlessly looped to create dynamic real-time video-based renderings of a scene.

We continue with a discussion of *3D videos* created from multiple video streams (Section 14.5.4), as well as *video-based walkthroughs* of environments (Section 14.5.5), which have found widespread application in immersive outdoor mapping and driving direction systems. We finish this chapter with a review of recent work in *neural rendering* (Section 14.6), where generative neural networks are used to create more realistic reconstructions of both static scenes and objects as well as people.

14.1 View interpolation

While the term *image-based rendering* first appeared in the papers by Chen (1995) and McMillan and Bishop (1995), the work on *view interpolation* by Chen and Williams (1993) is considered as the seminal paper in the field. In view interpolation, pairs of rendered images are combined with their precomputed depth maps to generate interpolated views that mimic what a virtual camera would see in between the two reference views. Since its original introduction, the whole field of *novel view synthesis* from captured images has continued to be a very active area. A good historical overview and recent results can be found in the CVPR tutorial on this topic (Gallo, Troccoli *et al.* 2020).

View interpolation combines two ideas that were previously used in computer vision and computer graphics. The first is the idea of pairing a recovered depth map with the reference image used in its computation and then using the resulting texture-mapped 3D model to generate novel views (Figure 12.1). The second is the idea of *morphing* (Section 3.6.3) (Figure 3.51), where correspondences between pairs of images are used to warp each reference image to an in-between location while simultaneously cross-dissolving between the two warped images.

Figure 14.2 illustrates this process in more detail. First, both source images are warped to the novel view, using both the knowledge of the reference and virtual 3D camera pose along with each

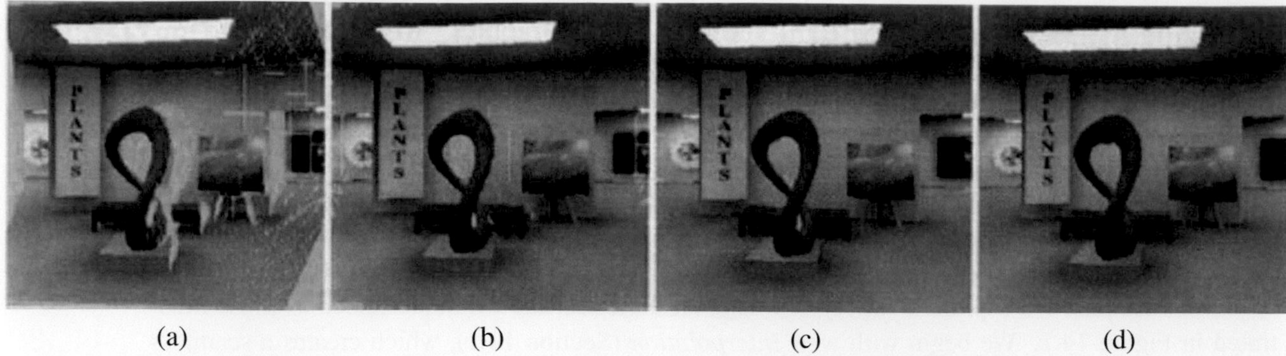

(a) (b) (c) (d)

Figure 14.2 View interpolation (Chen and Williams 1993) © 1993 ACM: (a) holes from one source image (shown in blue); (b) holes after combining two widely spaced images; (c) holes after combining two closely spaced images; (d) after interpolation (hole filling).

image's depth map (2.68–2.70). In the paper by Chen and Williams (1993), a *forward warping* algorithm (Algorithm 3.1 and Figure 3.45) is used. The depth maps are represented as quadtrees for both space and rendering time efficiency (Samet 1989).

During the forward warping process, multiple pixels (which occlude one another) may land on the same destination pixel. To resolve this conflict, either a *z-buffer* depth value can be associated with each destination pixel or the images can be warped in back-to-front order, which can be computed based on the knowledge of epipolar geometry (Chen and Williams 1993; Laveau and Faugeras 1994; McMillan and Bishop 1995).

Once the two reference images have been warped to the novel view (Figure 14.2a–b), they can be merged to create a coherent composite (Figure 14.2c). Whenever one of the images has a *hole* (illustrated as a cyan pixel), the other image is used as the final value. When both images have pixels to contribute, these can be blended as in usual morphing, i.e., according to the relative distances between the virtual and source cameras. Note that if the two images have very different exposures, which can happen when performing view interpolation on real images, the hole-filled regions and the blended regions will have different exposures, leading to subtle artifacts.

The final step in view interpolation (Figure 14.2d) is to fill any remaining holes or cracks due to the forward warping process or lack of source data (scene visibility). This can be done by copying pixels from the *further* pixels adjacent to the hole. (Otherwise, foreground objects are subject to a "fattening effect".)

The above process works well for rigid scenes, although its visual quality (lack of aliasing) can be improved using a two-pass, forward–backward algorithm (Section 14.2.1) (Shade, Gortler *et al.* 1998) or full 3D rendering (Zitnick, Kang *et al.* 2004). In the case where the two reference images are views of a non-rigid scene, e.g., a person smiling in one image and frowning in the other, *view morphing*, which combines ideas from view interpolation with regular morphing, can be used (Seitz and Dyer 1996). A depth map fitted to a face can also be used to synthesize a view from a longer distance, removing the enlarged nose and other facial features common to "selfie" photography (Fried, Shechtman *et al.* 2016).

While the original view interpolation paper describes how to generate novel views based on similar precomputed (linear perspective) images, the *plenoptic modeling* paper of McMillan and Bishop (1995) argues that cylindrical images should be used to store the precomputed rendering or real-world images. Chen (1995) also proposes using environment maps (cylindrical, cubic, or spherical) as source images for view interpolation.

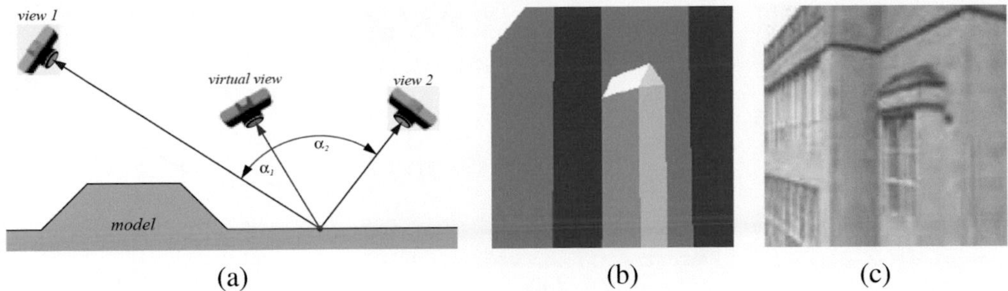

Figure 14.3 View-dependent texture mapping (Debevec, Taylor, and Malik 1996) © 1996 ACM. (a) The weighting given to each input view depends on the relative angles between the novel (virtual) view and the original views; (b) simplified 3D model geometry; (c) with view-dependent texture mapping, the geometry appears to have more detail (recessed windows).

14.1.1 View-dependent texture maps

View-dependent texture maps (Debevec, Taylor, and Malik 1996) are closely related to view interpolation. Instead of associating a separate depth map with each input image, a single 3D model is created for the scene, but different images are used as texture map sources depending on the virtual camera's current position (Figure 14.3a).[1]

In more detail, given a new virtual camera position, the similarity of this camera's view of each polygon (or pixel) is compared to that of potential source images. The images are then blended using a weighting that is inversely proportional to the angles α_i between the virtual view and the source views (Figure 14.3a).[2] Even though the geometric model can be fairly coarse (Figure 14.3b), blending different views gives a strong sense of more detailed geometry because of the visual motion between corresponding pixels. While the original paper performs the weighted blend computation separately at each pixel or coarsened polygon face, follow-on work by Debevec, Yu, and Borshukov (1998) presents a more efficient implementation based on precomputing contributions for various portions of viewing space and then using projective texture mapping (OpenGL-ARB 1997).

The idea of view-dependent texture mapping has been used in a large number of subsequent image-based rendering systems, including facial modeling and animation (Pighin, Hecker *et al.* 1998) and 3D scanning and visualization (Pulli, Abi-Rached *et al.* 1998). Closely related to view-dependent texture mapping is the idea of blending between light rays in 4D space, which forms the basis of the Lumigraph and unstructured Lumigraph systems (Section 14.3) (Gortler, Grzeszczuk *et al.* 1996; Buehler, Bosse *et al.* 2001).

To provide even more realism in their Façade system, Debevec, Taylor, and Malik (1996) also include a *model-based stereo* component, which computes an offset (parallax) map for each coarse planar facet of their model. They call the resulting analysis and rendering system a *hybrid geometry- and image-based* approach, as it uses traditional 3D geometric modeling to create the global 3D model, but then uses local depth offsets, along with view interpolation, to add visual realism. Instead of warping per-pixel depth maps or coarser triangulated geometry (as in unstructured Lumigraphs, Section 14.3.1), it is also possible to use super-pixels as the basic primitives being warped (Chaurasia, Duchene *et al.* 2013). Fixed rules for view-dependent blending can also be replaced with deep neural networks, as in the *deep blending* system by Hedman, Philip *et al.* (2018).

[1]The term *image-based modeling*, which is now commonly used to describe the creation of texture-mapped 3D models from multiple images, appears to have first been used by Debevec, Taylor, and Malik (1996), who also used the term *photogrammetric modeling* to describe the same process.

[2]More sophisticated blending weights are discussed in Section 14.3.1 on unstructured Lumigraph rendering.

Figure 14.4 Photo Tourism (Snavely, Seitz, and Szeliski 2006) © 2006 ACM: (a) a 3D overview of the scene, with translucent washes and lines painted onto the planar impostors; (b) once the user has selected a region of interest, a set of related thumbnails is displayed along the bottom; (c) planar proxy selection for optimal stabilization (Snavely, Garg *et al.* 2008) © 2008 ACM.

14.1.2 *Application*: Photo Tourism

While view interpolation was originally developed to accelerate the rendering of 3D scenes on low-powered processors and systems without graphics acceleration, it turns out that it can be applied directly to large collections of casually acquired photographs. The *Photo Tourism* system developed by Snavely, Seitz, and Szeliski (2006) uses structure from motion to compute the 3D locations and poses of all the cameras taking the images, along with a sparse 3D point-cloud model of the scene (Section 11.4.6, Figure 11.17).

To perform an image-based exploration of the resulting *sea of images* (Aliaga, Funkhouser *et al.* 2003), Photo Tourism first associates a 3D proxy with each image. While a triangulated mesh obtained from the point cloud can sometimes form a suitable proxy, e.g., for outdoor terrain models, a simple dominant plane fit to the 3D points visible in each image often performs better, because it does not contain any erroneous segments or connections that pop out as artifacts. As automated 3D modeling techniques continue to improve, however, the pendulum may swing back to more detailed 3D geometry (Goesele, Snavely *et al.* 2007; Sinha, Steedly, and Szeliski 2009). One example is the hybrid rendering system developed by Goesele, Ackermann *et al.* (2010), who use dense per-image depth maps for the well-reconstructed portions of each image and 3D colored point clouds for the less confident regions.

The resulting image-based navigation system lets users move from photo to photo, either by selecting cameras from a top-down view of the scene (Figure 14.4a) or by selecting regions of interest in an image, navigating to nearby views, or selecting related thumbnails (Figure 14.4b). To create a background for the 3D scene, e.g., when being viewed from above, non-photorealistic techniques (Section 10.5.2), such as translucent color washes or highlighted 3D line segments, can be used (Figure 14.4a). The system can also be used to annotate regions of images and to automatically propagate such annotations to other photographs.

The 3D planar proxies used in Photo Tourism and the related Photosynth system from Microsoft result in non-photorealistic transitions reminiscent of visual effects such as "page flips". Selecting a stable 3D axis for all the planes can reduce the amount of swimming and enhance the perception of 3D (Figure 14.4c) (Snavely, Garg *et al.* 2008). It is also possible to automatically detect objects in the scene that are seen from multiple views and create "orbits" of viewpoints around such objects. Furthermore, nearby images in both 3D position and viewing direction can be linked to create "virtual paths", which can then be used to navigate between arbitrary pairs of images, such as those you might take yourself while walking around a popular tourist site (Snavely, Garg *et al.* 2008). This idea

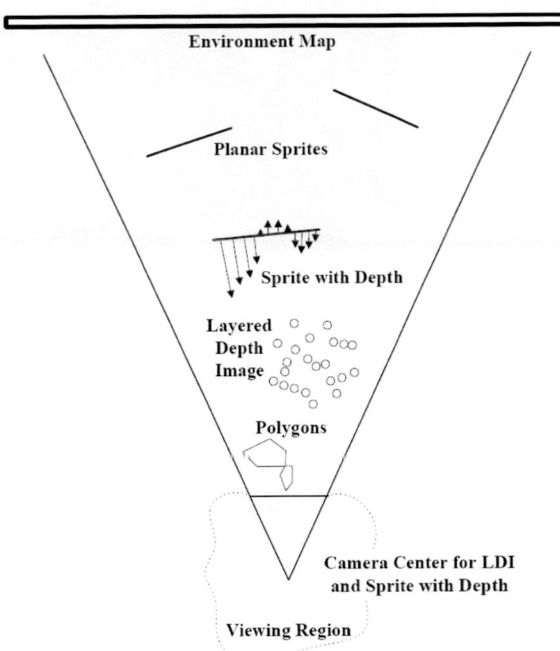

Figure 14.5 A variety of image-based rendering primitives, which can be used depending on the distance between the camera and the object of interest (Shade, Gortler *et al.* 1998) © 1998 ACM. Closer objects may require more detailed polygonal representations, while mid-level objects can use a layered depth image (LDI), and far-away objects can use sprites (potentially with depth) and environment maps.

has been further developed and released as a feature on Google Maps called Photo Tours (Kushal, Self *et al.* 2012).[3] The quality of such synthesized virtual views has become so accurate that Shan, Adams *et al.* (2013) propose a *visual Turing test* to distinguish between synthetic and real images. Waechter, Beljan *et al.* (2017) produce higher-resolution quality assessments of image-based modeling and rendering system using what they call *virtual rephotography*. Further improvements can be obtained using even more recent *neural rendering* techniques (Hedman, Philip *et al.* 2018; Meshry, Goldman *et al.* 2019; Li, Xian *et al.* 2020), which we discuss in Section 14.6.

The spatial matching of image features and regions performed by Photo Tourism can also be used to infer more information from large image collections. For example, Simon, Snavely, and Seitz (2007) show how the match graph between images of popular tourist sites can be used to find the most *iconic* (commonly photographed) objects in the collection, along with their related tags. In follow-on work, Simon and Seitz (2008) show how such tags can be propagated to sub-regions of each image, using an analysis of which 3D points appear in the central portions of photographs. Extensions of these techniques to *all* of the world's images, including the use of GPS tags where available, have been investigated as well (Li, Wu *et al.* 2008; Quack, Leibe, and Van Gool 2008; Crandall, Backstrom *et al.* 2009; Li, Crandall, and Huttenlocher 2009; Zheng, Zhao *et al.* 2009; Raguram, Wu *et al.* 2011).

[3]https://maps.googleblog.com/2012/04/visit-global-landmarks-with-photo-tours.html

Figure 14.6 Sprites with depth (Shade, Gortler *et al.* 1998) © 1998 ACM: (a) alpha-matted color sprite; (b) corresponding relative depth or parallax; (c) rendering without relative depth; (d) rendering with depth (note the curved object boundaries).

14.2 Layered depth images

Traditional view interpolation techniques associate a single depth map with each source or reference image. Unfortunately, when such a depth map is warped to a novel view, holes and cracks inevitably appear behind the foreground objects. One way to alleviate this problem is to keep several depth and color values (*depth pixels*) at every pixel in a reference image (or, at least for pixels near foreground–background transitions) (Figure 14.5). The resulting data structure, which is called a *layered depth image* (LDI), can be used to render new views using a back-to-front forward warping (splatting) algorithm (Shade, Gortler *et al.* 1998).

14.2.1 Impostors, sprites, and layers

An alternative to keeping lists of color-depth values at each pixel, as is done in the LDI, is to organize objects into different *layers* or *sprites*. The term sprite originates in the computer game industry, where it is used to designate flat animated characters in games such as Pac-Man or Mario Bros. When put into a 3D setting, such objects are often called *impostors*, because they use a piece of flat, alpha-matted geometry to represent simplified versions of 3D objects that are far away from the camera (Shade, Lischinski *et al.* 1996; Lengyel and Snyder 1997; Torborg and Kajiya 1996). In computer vision, such representations are usually called *layers* (Wang and Adelson 1994; Baker, Szeliski, and Anandan 1998; Torr, Szeliski, and Anandan 1999; Birchfield, Natarajan, and Tomasi 2007). Section 9.4.2 discusses the topics of transparent layers and reflections, which occur on specular and transparent surfaces such as glass.

While flat layers can often serve as an adequate representation of geometry and appearance for far-away objects, better geometric fidelity can be achieved by also modeling the per-pixel offsets relative to a base plane, as shown in Figures 14.5 and 14.6a–b. Such representations are called *plane plus parallax* in the computer vision literature (Kumar, Anandan, and Hanna 1994; Sawhney 1994; Szeliski and Coughlan 1997; Baker, Szeliski, and Anandan 1998), as discussed in Section 9.4 (Figure 9.14). In addition to fully automated stereo techniques, it is also possible to paint in depth layers (Kang 1998; Oh, Chen *et al.* 2001; Shum, Sun *et al.* 2004) or to infer their 3D structure from monocular image cues (Sections 6.4.4 and 12.8) (Hoiem, Efros, and Hebert 2005b; Saxena, Sun, and Ng 2009).

How can we render a sprite with depth from a novel viewpoint? One possibility, as with a regular depth map, is to just forward warp each pixel to its new location, which can cause aliasing and cracks. A better way, which we have already mentioned in Section 3.6.2, is to first warp the depth (or (u, v) displacement) map to the novel view, fill in the cracks, and then use higher-quality

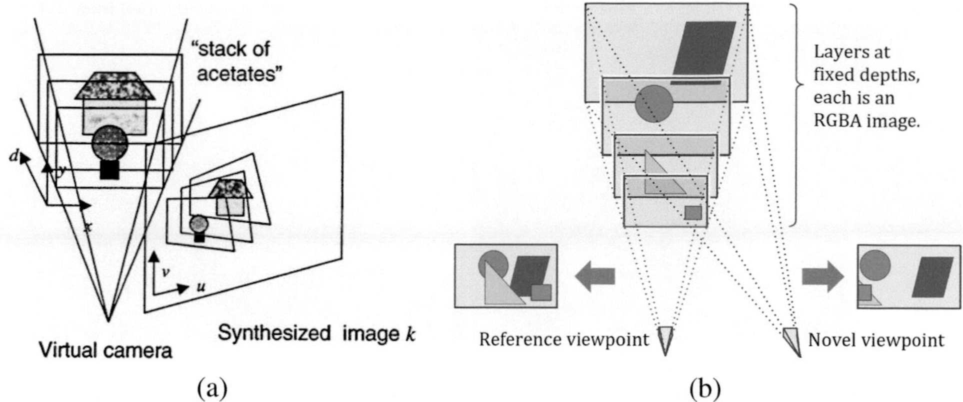

(a) (b)

Figure 14.7 Finely sliced fronto-parallel layers: (a) stack of acetates (Szeliski and Golland 1999) © 1999 Springer and (b) multiplane images (Zhou, Tucker *et al.* 2018) © 2018 ACM. These representations (which are equivalent) consist of a set of fronto-parallel planes at fixed depths from a reference camera coordinate frame, with each plane encoding an RGB image and an alpha map that capture the scene appearance at the corresponding depth.

inverse warping to resample the color image (Shade, Gortler *et al.* 1998). Figure 14.6d shows the results of applying such a two-pass rendering algorithm. From this still image, you can appreciate that the foreground sprites look more rounded; however, to fully appreciate the improvement in realism, you would have to look at the actual animated sequence.

Sprites with depth can also be rendered using conventional graphics hardware, as described in (Zitnick, Kang *et al.* 2004). Rogmans, Lu *et al.* (2009) describe GPU implementations of both real-time stereo matching and real-time forward and inverse rendering algorithms.

An alternative to constructing a small number of layers is to discretize the viewing frustum subtending a layered depth image into a large number of fronto-parallel planes, each of which contains RGBA values (Szeliski and Golland 1999), as shown in Figure 14.7. This is the same spatial representation we presented in Section 12.1.2 and Figure 12.6 on *plane sweep* approaches to stereo, except that here it is being used to represent a colored 3D scene instead of accumulating a matching cost volume. This representation is essentially a perspective variant of a volumetric representation containing RGB color and α opacity values (Sections 13.2.1 and 13.5).

This representation was recently rediscovered and now goes under the popular name of *multiplane images (MPI)* (Zhou, Tucker *et al.* 2018). Figure 14.8 shows an MPI representation derived from a stereo image pair along with a novel synthesized view. MPIs are easier to derive from pairs or collections of stereo images than true (minimal) layered representations because there is a 1:1 correspondence between pixels (actually, voxels) in a plane sweep cost volume (Figure 12.5) and an MPI. However, they are not as compact and can lead to tearing artifacts once the viewpoint exceeds a certain range. (We will talk about using inpainting to mitigate such holes in image-based representations in Section 14.2.2). MPIs are also related to the *soft 3D* volumetric representation proposed earlier by Penner and Zhang (2017).

Since their initial development for novel view extrapolation, i.e., "stereo magnification" (Zhou, Tucker *et al.* 2018), MPIs have found a wide range of applications in image-based rendering, including extension to multiple input images and faster inference (Flynn, Broxton *et al.* 2019), CNN refinement and better inpainting (Srinivasan, Tucker *et al.* 2019), interpolating between collections of MPIs (Mildenhall, Srinivasan *et al.* 2019), and large view extrapolations (Choi, Gallo *et al.* 2019). The planar MPI structure has also been generalized to curved surfaces for representing partial or

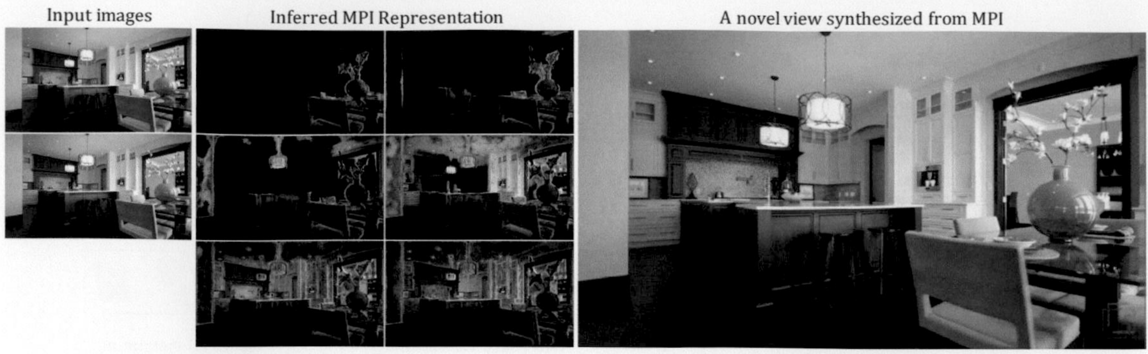

Figure 14.8 MPI representation constructed from a stereo pair of color images, along with a novel view reconstructed from the MPI (Zhou, Tucker *et al.* 2018) © 2018 ACM. Note how the planes slice the 3D scene into thin layers, each of which has colors and full or partial opacities in only a small region.

complete 3D panoramas (Broxton, Flynn *et al.* 2020; Attal, Ling *et al.* 2020; Lin, Xu *et al.* 2020).[4]

Another important application of layers is in the modeling of reflections. When the reflector (e.g., a glass pane) is planar, the reflection forms a *virtual image*, which can be modeled as a separate layer (Section 9.4.2 and Figures 9.16–9.17), so long as *additive* (instead of *over*) compositing is used to combine the reflected and transmitted images (Szeliski, Avidan, and Anandan 2000; Sinha, Kopf *et al.* 2012; Kopf, Langguth *et al.* 2013). Figure 14.9 shows an example of a two-layer decomposition reconstructed from a short video clip, which can be re-rendered from novel views by adding warped versions of the two layers (each of which has its own depth map). When the reflective surface is curved, a quasi-stable virtual image may still be available, although this depends on the local variations in principal curvatures (Swaminathan, Kang *et al.* 2002; Criminisi, Kang *et al.* 2005). The modeling of reflections is one of the advantages attributed to layered representations such as MPIs (Zhou, Tucker *et al.* 2018; Broxton, Flynn *et al.* 2020), although in these papers over compositing is still used, which results in plausible but not physically correct renderings.

14.2.2 *Application*: 3D photography

The desire to capture and view photographs of the world in 3D prompted the development of stereo cameras and viewers in the mid-1800s (Luo, Kong *et al.* 2020) and more recently the popularity of 3D movies.[5] It has also underpinned much of the research in 3D shape and appearance capture and modeling we studied in the previous chapter and more specifically Section 13.7.2. Until recently, however, while the required multiple images could be captured with hand-held cameras (Pollefeys, Van Gool *et al.* 2004; Snavely, Seitz, and Szeliski 2006), desktop or laptop computers were required to process and interactively view the images.

The ability to capture, construct, and widely share such 3D models has dramatically increased in the last few years and now goes under the name of *3D photography*. Hedman, Alsisan *et al.* (2017) describe their *Casual 3D Photography* system, which takes a sequence of overlapping images taken from a moving camera and then uses a combination of structure from motion, multi-view stereo, and 3D image warping and stitching to construct two-layer partial panoramas that can be viewed on a computer, as shown in Figure 14.10. The Instant 3D system of Hedman and Kopf (2018) builds a

[4]Exploring the interactive 3D videos on the authors' websites, e.g., https://augmentedperception.github.io/deepviewvideo, is a good way to get a sense of this new medium.

[5]It is interesting to note, however, that for now (at least), in-home 3D TV sets have failed to take off.

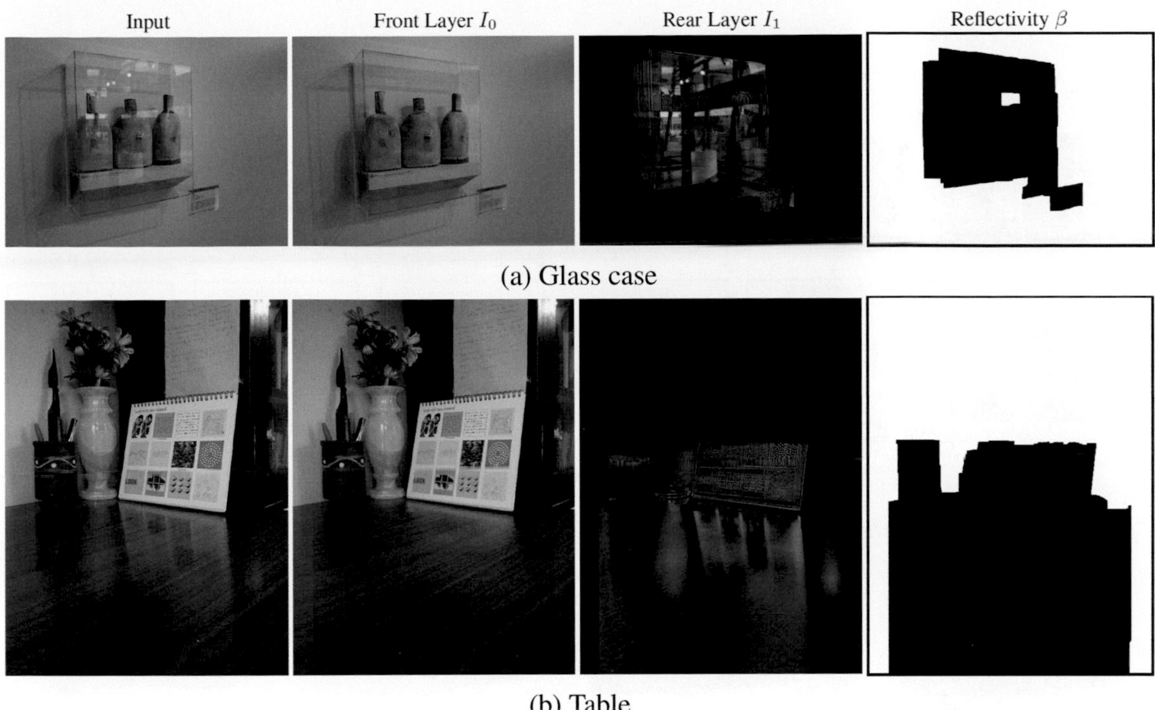

Figure 14.9 Image-based rendering of scenes with reflections using multiple additive layers (Sinha, Kopf *et al.* 2012) © 2012 ACM. The left column shows an image from the input sequence and the next two columns show the two separated layers (transmitted and reflected light). The last column is an estimate of which portions of the scene are reflective. As you can see, stray bits of reflections sometimes cling to the transmitted light layer. Note how in the table, the amount of reflected light (gloss) decreases towards the bottom of the image because of Fresnel reflection.

similar system, but starts with the depth images available from newer dual-camera smartphones to significantly speed up the process. Note, however, that the individual depth images are not *metric*, i.e., related to true depth with a single global scalar transformation, so must be deformably warped before being stitched together. A *texture atlas* is then constructed to compactly store the pixel color values while also supporting multiple layers.

While these systems produce beautiful wide 3D images that can create a true sense of immersion ("being there"), much more practical and fast solutions can be constructed using a single depth image. Kopf, Alsisan *et al.* (2019) describe their phone-based system, which takes a single dual-lens photograph with its estimated depth map and constructs a multi-layer 3D photograph with occluded pixels being *inpainted* from nearby background pixels (see Section 10.5.1 and Shih, Su *et al.* 2020).[6] To remove the requirement for depth maps being associated with the input images Kopf, Matzen *et al.* (2020) use a monocular depth inference network (Section 12.8) to estimate the depth, thereby enabling 3D photos to be produced from any photograph in a phone's camera roll, or even from historical photographs, as shown in Figure 14.10c.[7] When historic stereographs are available, these can be used to create even more accurate 3D photographs, as shown by Luo, Kong *et al.* (2020). It is also possible to create a "3D Ken Burns" effect, i.e., small looming video clips, from regular images

[6] Facebook rolled out 3D photographs for the iPhone in October 2018, https://facebook360.fb.com/2018/10/11/3d-photos-now-rolling-out-on-facebook-and-in-vr, along with the ability to post and interactively view the photos.

[7] In February 2020, Facebook released the ability to use regular photos, https://ai.facebook.com/blog/powered-by-ai-turning-any-2d-photo-into-3d-using-convolutional-neural-nets.

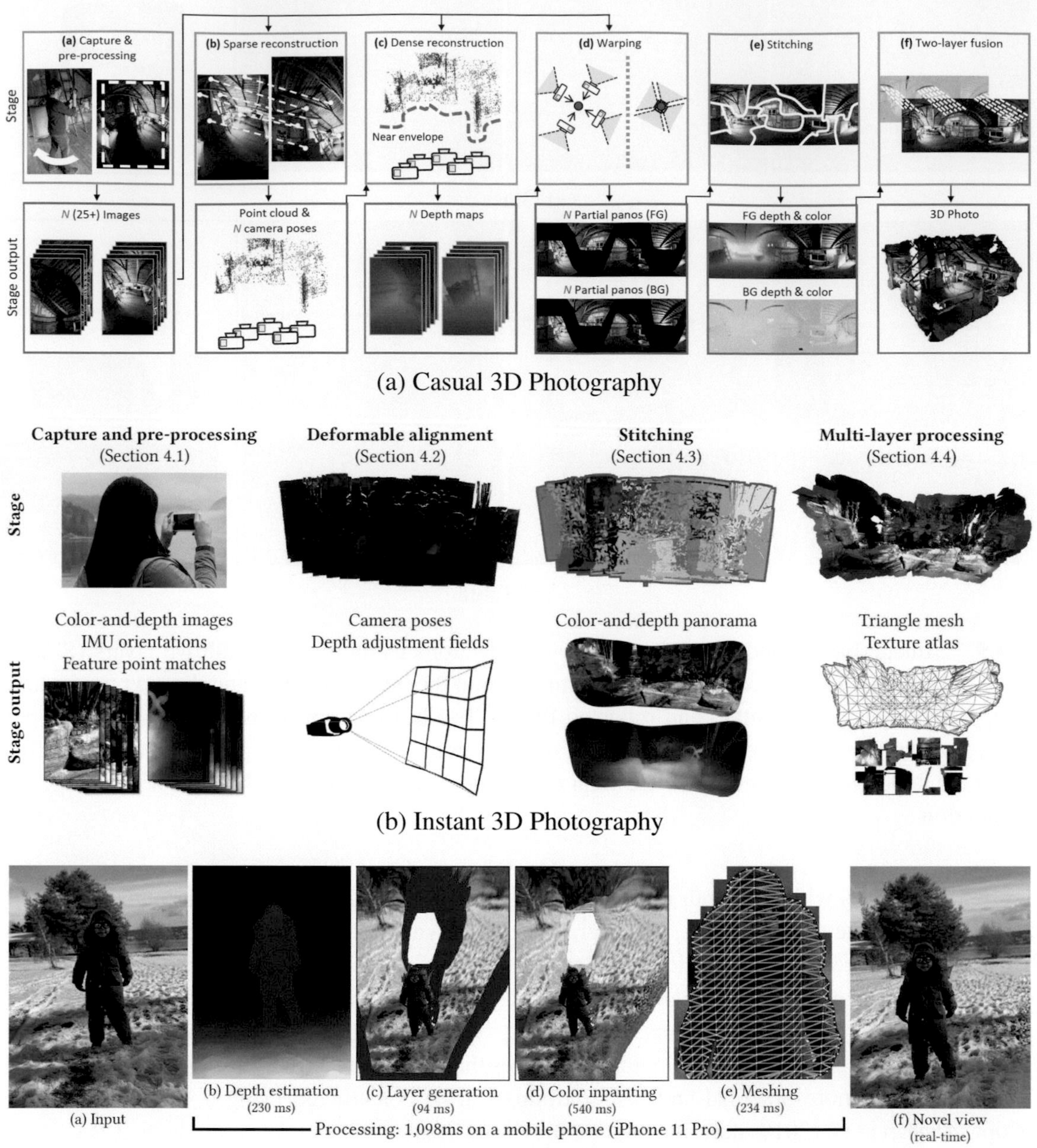

Figure 14.10 Systems for capturing and modeling 3D scenes from handheld photographs. (a) Casual 3D Photography takes a series of overlapping images and constructs per-image depth maps, which are then warped and blended together into a two-layer representation (Hedman, Alsisan *et al.* 2017) © 2017 ACM. (b) Instant 3D Photography starts with the depth maps produced by a dual-lens smartphone and warps and registers the depth maps to create a similar representation with far less computation (Hedman and Kopf 2018) © 2018 ACM. (c) One Shot 3D Photography starts with a single photo, performs monocular depth estimation, layer construction and inpainting, and mesh and atlas generation, enabling phone-based reconstruction and interactive viewing (Kopf, Matzen *et al.* 2020) © 2020 ACM.

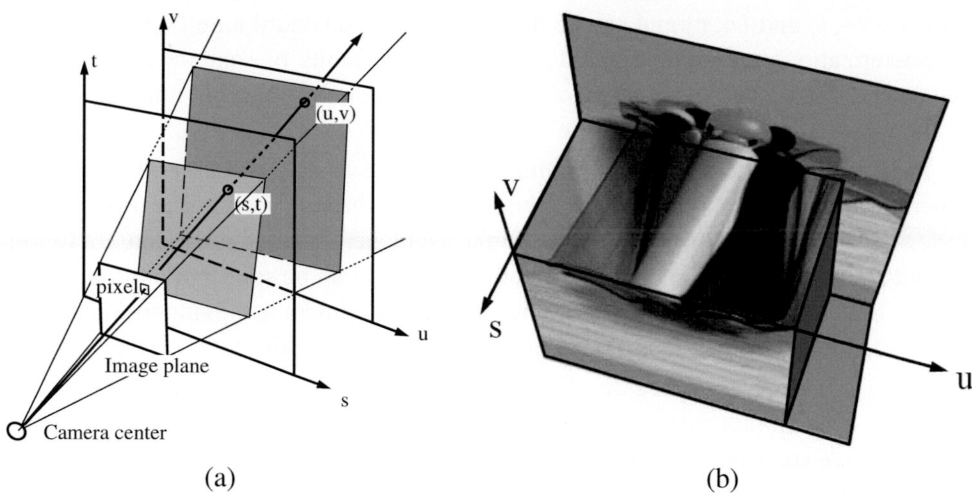

Figure 14.11 The Lumigraph (Gortler, Grzeszczuk *et al.* 1996) © 1996 ACM: (a) a ray is represented by its 4D two-plane parameters (s, t) and (u, v); (b) a slice through the 3D light field subset (u, v, s).

using monocular depth inference (Niklaus, Mai *et al.* 2019).[8]

14.3 Light fields and Lumigraphs

While image-based rendering approaches can synthesize scene renderings from novel viewpoints, they raise the following more general question:

> *Is is possible to capture and render the appearance of a scene from all possible viewpoints and, if so, what is the complexity of the resulting structure?*

Let us assume that we are looking at a static scene, i.e., one where the objects and illuminants are fixed, and only the observer is moving around. Under these conditions, we can describe each image by the location and orientation of the virtual camera (6 dof) as well as its intrinsics (e.g., its focal length). However, if we capture a two-dimensional *spherical* image around each possible camera location, we can re-render any view from this information.[9] Thus, taking the cross-product of the three-dimensional space of camera positions with the 2D space of spherical images, we obtain the 5D *plenoptic function* of Adelson and Bergen (1991), which forms the basis of the image-based rendering system of McMillan and Bishop (1995).

Notice, however, that when there is no light dispersion in the scene, i.e., no smoke or fog, all the coincident rays along a portion of free space (between solid or refractive objects) have the same color value. Under these conditions, we can reduce the 5D plenoptic function to the 4D *light field* of all possible rays (Gortler, Grzeszczuk *et al.* 1996; Levoy and Hanrahan 1996; Levoy 2006).[10]

To make the parameterization of this 4D function simpler, let us put two planes in the 3D scene roughly bounding the area of interest, as shown in Figure 14.11a. Any light ray terminating at a camera that lives in front of the st plane (assuming that this space is empty) passes through the two

[8] Google released a similar feature called Cinematic photos https://blog.google/products/photos/new-cinematic-photos-and-more-ways-relive-your-memories.

[9] As we are counting dimensions, we ignore for now any sampling or resolution issues.

[10] Levoy and Hanrahan (1996) borrowed the term *light field* from a paper by Gershun (1939). Another name for this representation is the *photic field* (Moon and Spencer 1981).

planes at (s, t) and (u, v) and can be described by its 4D coordinate (s, t, u, v). This diagram (and parameterization) can be interpreted as describing a family of cameras living on the st plane with their image planes being the uv plane. The uv plane can be placed at infinity, which corresponds to all the virtual cameras looking in the same direction.

In practice, if the planes are of finite extent, the finite *light slab* $L(s, t, u, v)$ can be used to generate any synthetic view that a camera would see through a (finite) *viewport* in the st plane with a view frustum that wholly intersects the far uv plane. To enable the camera to move all the way around an object, the 3D space surrounding the object can be split into multiple domains, each with its own light slab parameterization. Conversely, if the camera is moving inside a bounded volume of free space looking outward, multiple cube faces surrounding the camera can be used as (s, t) planes.

Thinking about 4D spaces is difficult, so let us drop our visualization by one dimension. If we fix the row value t and constrain our camera to move along the s axis while looking at the uv plane, we can stack all of the stabilized images the camera sees to get the (u, v, s) *epipolar volume*, which we discussed in Section 12.7. A "horizontal" cross-section through this volume is the well-known *epipolar plane image* (Bolles, Baker, and Marimont 1987), which is the us slice shown in Figure 14.11b.

As you can see in this slice, each color pixel moves along a linear track whose slope is related to its depth (parallax) from the uv plane. (Pixels exactly on the uv plane appear "vertical", i.e., they do not move as the camera moves along s.) Furthermore, pixel tracks occlude one another as their corresponding 3D surface elements occlude. Translucent pixels, however, composite *over* background pixels (Section 3.1.3 (3.8)) rather than occluding them. Thus, we can think of adjacent pixels sharing a similar planar geometry as *EPI strips* or *EPI tubes* (Criminisi, Kang *et al.* 2005). 3D lightfields taken from a camera slowly moving through a static scene can be an excellent source for high-accuracy 3D reconstruction, as demonstrated in the papers by Kim, Zimmer *et al.* (2013), Yücer, Kim *et al.* (2016), and Yücer, Sorkine-Hornung *et al.* (2016).

The equations mapping from pixels (x, y) in a virtual camera and the corresponding (s, t, u, v) coordinates are relatively straightforward to derive and are sketched out in Exercise 14.7. It is also possible to show that the set of pixels corresponding to a regular orthographic or perspective camera, i.e., one that has a linear projective relationship between 3D points and (x, y) pixels (2.63), lie along a two-dimensional hyperplane in the (s, t, u, v) light field (Exercise 14.7).

While a light field can be used to render a complex 3D scene from novel viewpoints, a much better rendering (with less ghosting) can be obtained if something is known about its 3D geometry. The Lumigraph system of Gortler, Grzeszczuk *et al.* (1996) extends the basic light field rendering approach by taking into account the 3D location of surface points corresponding to each 3D ray.

Consider the ray (s, u) corresponding to the dashed line in Figure 14.12, which intersects the object's surface at a distance z from the uv plane. When we look up the pixel's color in camera s_i (assuming that the light field is discretely sampled on a regular 4D (s, t, u, v) grid), the actual pixel coordinate is u', instead of the original u value specified by the (s, u) ray. Similarly, for camera s_{i+1} (where $s_i \leq s \leq s_{i+1}$), pixel address u'' is used. Thus, instead of using quadri-linear interpolation of the nearest sampled (s, t, u, v) values around a given ray to determine its color, the (u, v) values are modified for each discrete (s_i, t_i) camera.

Figure 14.12 also shows the same reasoning in *ray space*. Here, the original continuous-valued (s, u) ray is represented by a triangle and the nearby sampled discrete values are shown as circles. Instead of just blending the four nearest samples, as would be indicated by the vertical and horizontal dashed lines, the modified (s_i, u') and (s_{i+1}, u'') values are sampled instead and their values are then blended.

The resulting rendering system produces images of much better quality than a proxy-free light

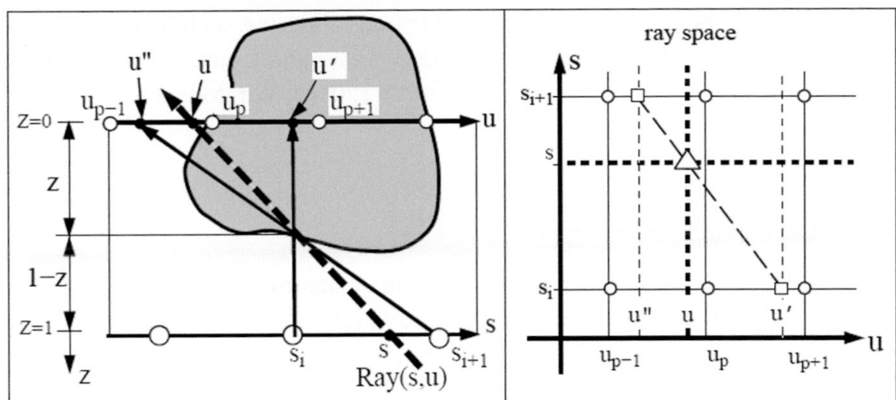

Figure 14.12 Depth compensation in the Lumigraph (Gortler, Grzeszczuk *et al.* 1996) © 1996 ACM. To resample the (s, u) dashed light ray, the u parameter corresponding to each discrete s_i camera location is modified according to the out-of-plane depth z to yield new coordinates u and u'; in (u, s) ray space, the original sample ($\triangle$) is resampled from the (s_i, u') and (s_{i+1}, u'') samples, which are themselves linear blends of their adjacent ($\circ$) samples.

field and is the method of choice whenever 3D geometry can be inferred. In subsequent work, Isaksen, McMillan, and Gortler (2000) show how a planar proxy for the scene, which is a simpler 3D model, can be used to simplify the resampling equations. They also describe how to create synthetic aperture photos, which mimic what might be seen by a wide-aperture lens, by blending more nearby samples (Levoy and Hanrahan 1996). A similar approach can be used to re-focus images taken with a plenoptic (microlens array) camera (Ng, Levoy *et al.* 2005; Ng 2005) or a light field microscope (Levoy, Ng *et al.* 2006). It can also be used to see through obstacles, using extremely large synthetic apertures focused on a background that can blur out foreground objects and make them appear translucent (Wilburn, Joshi *et al.* 2005; Vaish, Szeliski *et al.* 2006).

Now that we understand how to render new images from a light field, how do we go about capturing such datasets? One answer is to move a calibrated camera with a motion control rig or *gantry*.[11] Another approach is to take handheld photographs and to determine the pose and intrinsic calibration of each image using either a calibrated stage or structure from motion. In this case, the images need to be *rebinned* into a regular 4D (s, t, u, v) space before they can be used for rendering (Gortler, Grzeszczuk *et al.* 1996). Alternatively, the original images can be used directly using a process called the *unstructured Lumigraph*, which we describe below.

Because of the large number of images involved, light fields and Lumigraphs can be quite voluminous to store and transmit. Fortunately, as you can tell from Figure 14.11b, there is a tremendous amount of redundancy (coherence) in a light field, which can be made even more explicit by first computing a 3D model, as in the Lumigraph. A number of techniques have been developed to compress and progressively transmit such representations (Gortler, Grzeszczuk *et al.* 1996; Levoy and Hanrahan 1996; Rademacher and Bishop 1998; Magnor and Girod 2000; Wood, Azuma *et al.* 2000; Shum, Kang, and Chan 2003; Magnor, Ramanathan, and Girod 2003; Zhang and Chen 2004; Shum, Chan, and Kang 2007).

Since the original burst of research on lightfields in the mid-1990 and early 2000s, better tech-

[11] See http://lightfield.stanford.edu/acq.html for a description of some of the gantries and camera arrays built at the Stanford Computer Graphics Laboratory (Wilburn, Joshi *et al.* 2005). A more recent dataset was created by Honauer, Johannsen *et al.* (2016) and is available at https://lightfield-analysis.uni-konstanz.de Both websites provide light field datasets that are a great source of research and project material.

niques continue to be developed for analyzing and rendering such images. Some representative papers and datasets from the last decade include Wanner and Goldluecke (2014), Honauer, Johannsen *et al.* (2016), Kalantari, Wang, and Ramamoorthi (2016), Wu, Masia *et al.* (2017), and Shin, Jeon *et al.* (2018).

14.3.1 Unstructured Lumigraph

When the images in a Lumigraph are acquired in an unstructured (irregular) manner, it can be counterproductive to resample the resulting light rays into a regularly binned (s, t, u, v) data structure. This is both because resampling always introduces a certain amount of aliasing and because the resulting gridded light field can be populated very sparsely or irregularly.

The alternative is to render directly from the acquired images, by finding for each light ray in a virtual camera the closest pixels in the original images. The *unstructured Lumigraph* rendering (ULR) system of Buehler, Bosse *et al.* (2001) describes how to select such pixels by combining a number of fidelity criteria, including *epipole consistency* (distance of rays to a source camera's center), *angular deviation* (similar incidence direction on the surface), *resolution* (similar sampling density along the surface), *continuity* (to nearby pixels), and *consistency* (along the ray). These criteria can all be combined to determine a weighting function between each virtual camera's pixel and a number of candidate input cameras from which it can draw colors. To make the algorithm more efficient, the computations are performed by discretizing the virtual camera's image plane using a regular grid overlaid with the polyhedral object mesh model and the input camera centers of projection and interpolating the weighting functions between vertices.

The unstructured Lumigraph generalizes previous work in both image-based rendering and light field rendering. When the input cameras are gridded, the ULR behaves the same way as regular Lumigraph rendering. When fewer cameras are available but the geometry is accurate, the algorithm behaves similarly to view-dependent texture mapping (Section 14.1.1). If RGB-D depth images are available, these can be fused into lower-resolution proxies that can be combined with higher-resolution source images at rendering time (Hedman, Ritschel *et al.* 2016). And while the original ULR paper uses manually constructed rules for determining pixel weights, it is also possible to learn such blending weights using a deep neural network (Hedman, Philip *et al.* 2018; Riegler and Koltun 2020a).

14.3.2 Surface light fields

Of course, using a two-plane parameterization for a light field is not the only possible choice. (It is the one usually presented first, as the projection equations and visualizations are the easiest to draw and understand.) As we mentioned on the topic of light field compression, if we know the 3D shape of the object or scene whose light field is being modeled, we can effectively compress the field because nearby rays emanating from nearby surface elements have similar color values.

In fact, if the object is totally diffuse, ignoring occlusions, which can be handled using 3D graphics algorithms or z-buffering, all rays passing through a given surface point will have the same color value. Hence, the light field "collapses" to the usual 2D texture-map defined over an object's surface. Conversely, if the surface is totally specular (e.g., mirrored), each surface point reflects a miniature copy of the environment surrounding that point. In the absence of inter-reflections (e.g., a convex object in a large open space), each surface point simply reflects the far-field *environment map* (Section 2.2.1), which again is two-dimensional. Therefore, is seems that re-parameterizing the 4D light field to lie on the object's surface can be extremely beneficial.

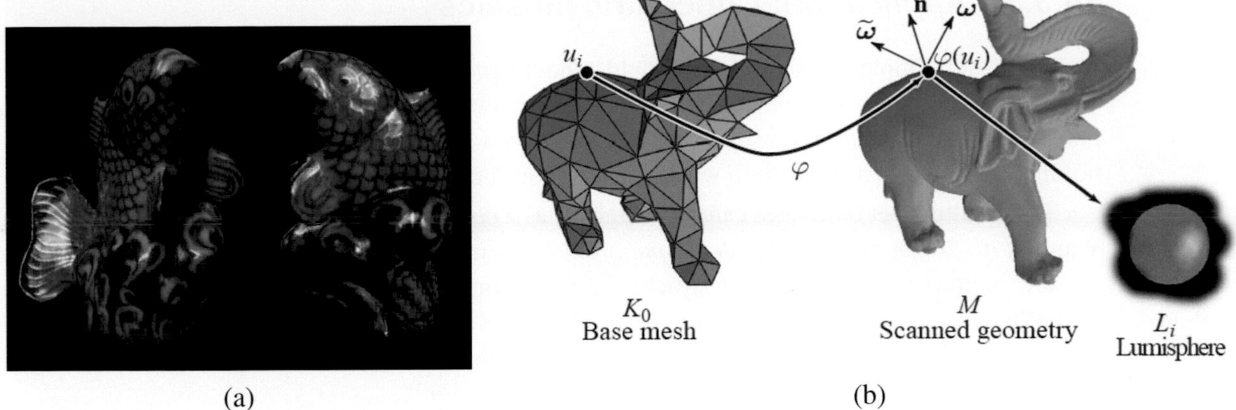

Figure 14.13 Surface light fields (Wood, Azuma *et al.* 2000) © 2000 ACM: (a) example of a highly specular object with strong inter-reflections; (b) the surface light field stores the light emanating from each surface point in all visible directions as a "Lumisphere".

These observations underlie the *surface light field* representation introduced by Wood, Azuma *et al.* (2000). In their system, an accurate 3D model is built of the object being represented. Then the *Lumisphere* of all rays emanating from each surface point is estimated or captured (Figure 14.13). Nearby Lumispheres will be highly correlated and hence amenable to both compression and manipulation.

To estimate the diffuse component of each Lumisphere, a median filtering over all visible exiting directions is first performed for each channel. Once this has been subtracted from the Lumisphere, the remaining values, which should consist mostly of the specular components, are *reflected* around the local surface normal (2.90), which turns each Lumisphere into a copy of the local environment around that point. Nearby Lumispheres can then be compressed using predictive coding, vector quantization, or principal component analysis.

The decomposition into a diffuse and specular component can also be used to perform editing or manipulation operations, such as re-painting the surface, changing the specular component of the reflection (e.g., by blurring or sharpening the specular Lumispheres), or even geometrically deforming the object while preserving detailed surface appearance.

In more recent work, Park, Newcombe, and Seitz (2018) use an RGB-D camera to acquire a 3D model and its diffuse reflectance layer using min compositing and iteratively reweighted least squares, as discussed in Section 9.4.2. They then estimate a simple piecewise-constant BRDF model to account for the specular components. In their follow-on Seeing the World in a Bag of Chips paper, Park, Holynski, and Seitz (2020) also estimate the *specular reflectance map*, which is a convolution of the environment map with the object's specular BRDF. Additional techniques to estimate spatially varying BRDFs are discussed in Section 13.7.1.

In summary, surface light fields are a good representation to add realism to scanned 3D object models by modeling their specular properties, thus avoiding the "cardboard" (matte) appearance of such models when their reflections are ignored. For larger scenes, especially those containing large planar reflectors such as glass windows or glossy tables, modeling the reflections as separate layers, as discussed in Sections 9.4.2 and 14.2.1, or as true mirror surfaces (Whelan, Goesele *et al.* 2018), may be more appropriate.

14.3.3 *Application*: Concentric mosaics

A useful and simple version of light field rendering is a panoramic image with parallax, i.e., a video or series of photographs taken from a camera swinging in front of some rotation point. Such panoramas can be captured by placing a camera on a boom on a tripod, or even more simply, by holding a camera at arm's length while rotating your body around a fixed axis.

The resulting set of images can be thought of as a *concentric mosaic* (Shum and He 1999; Shum, Wang *et al.* 2002) or a *layered depth panorama* (Zheng, Kang *et al.* 2007). The term "concentric mosaic" comes from a particular structure that can be used to re-bin all of the sampled rays, essentially associating each column of pixels with the "radius" of the concentric circle to which it is tangent (Ishiguro, Yamamoto, and Tsuji 1992; Shum and He 1999; Peleg, Ben-Ezra, and Pritch 2001).

Rendering from such data structures is fast and straightforward. If we assume that the scene is far enough away, for any virtual camera location, we can associate each column of pixels in the virtual camera with the nearest column of pixels in the input image set. (For a regularly captured set of images, this computation can be performed analytically.) If we have some rough knowledge of the depth of such pixels, columns can be stretched vertically to compensate for the change in depth between the two cameras. If we have an even more detailed depth map (Peleg, Ben-Ezra, and Pritch 2001; Li, Shum *et al.* 2004; Zheng, Kang *et al.* 2007), we can perform pixel-by-pixel depth corrections.

While the virtual camera's motion is constrained to lie in the plane of the original cameras and within the radius of the original capture ring, the resulting experience can exhibit complex rendering phenomena, such as reflections and translucencies, which cannot be captured using a texture-mapped 3D model of the world. Exercise 14.10 has you construct a concentric mosaic rendering system from a series of hand-held photos or video.

While concentric mosaics are captured by moving the camera on a (roughly) circular arc, it is also possible to construct *manifold projections* (Peleg and Herman 1997), *multiple-center-of-projection images* (Rademacher and Bishop 1998), and *multi-perspective panoramas* (Román, Garg, and Levoy 2004; Román and Lensch 2006; Agarwala, Agrawala *et al.* 2006; Kopf, Chen *et al.* 2010), which we discussed briefly in Section 8.2.5.

14.3.4 *Application*: Synthetic re-focusing

In addition to the interactive viewing of captured scenes and objects, light field rendering can be used to add synthetic depth of field effects to photographs (Levoy 2006). In the computational photography chapter (Section 10.3.2), we mentioned how the depth estimates produced by modern dual-lens and/or dual-pixel smartphones can be used to synthetically blur photographs (Wadhwa, Garg *et al.* 2018; Garg, Wadhwa *et al.* 2019; Zhang, Wadhwa *et al.* 2020).

When larger numbers of input images are available, e.g., when using microlens arrays, the images can be shifted and combined to simulate the effects of a larger aperture lens in what is known as *synthetic aperture photography* (Ng, Levoy *et al.* 2005; Ng 2005), which was the basis of the Lytro light field camera. Related ideas have been used for shallow depth of field in *light field microscopy* (Levoy, Chen *et al.* 2004; Levoy, Ng *et al.* 2006), obstruction removal (Wilburn, Joshi *et al.* 2005; Vaish, Szeliski *et al.* 2006; Xue, Rubinstein *et al.* 2015; Liu, Lai *et al.* 2020a), and *coded aperture photography* (Levin, Fergus *et al.* 2007; Zhou, Lin, and Nayar 2009).

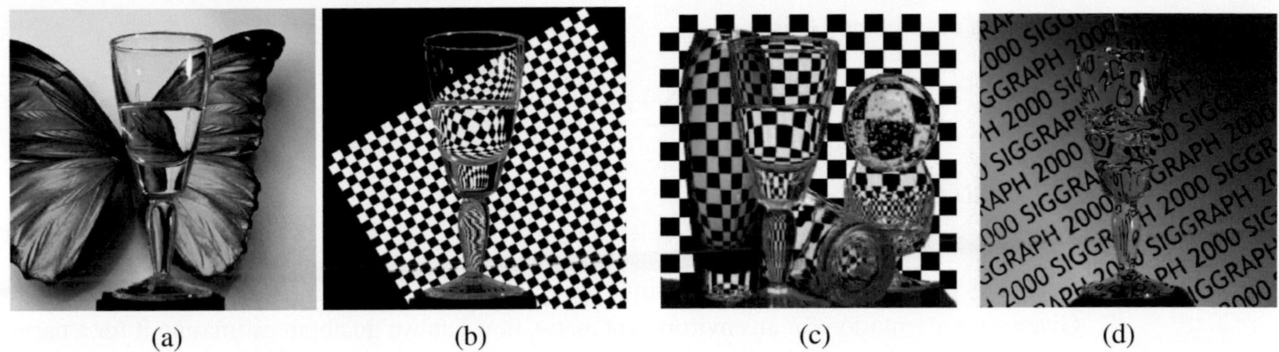

Figure 14.14 Environment mattes: (a–b) a refractive object can be placed in front of a series of backgrounds and their light patterns will be correctly refracted (Zongker, Werner *et al.* 1999) (c) multiple refractions can be handled using a Gaussian mixture model and (d) real-time mattes can be pulled using a single graded colored background (Chuang, Zongker *et al.* 2000) © 2000 ACM.

14.4 Environment mattes

So far in this chapter, we have dealt with view interpolation and light fields, which are techniques for modeling and rendering complex static scenes seen from different viewpoints.

What if, instead of moving around a virtual camera, we take a complex, refractive object, such as the water goblet shown in Figure 14.14, and place it in front of a new background? Instead of modeling the 4D space of rays emanating from a scene, we now need to model how each pixel in our view of this object refracts incident light coming from its environment.

What is the intrinsic dimensionality of such a representation and how do we go about capturing it? Let us assume that if we trace a light ray from the camera at pixel (x, y) toward the object, it is reflected or refracted back out toward its environment at an angle (ϕ, θ). If we assume that other objects and illuminants are sufficiently distant (the same assumption we made for surface light fields in Section 14.3.2), this 4D mapping $(x, y) \rightarrow (\phi, \theta)$ captures all the information between a refractive object and its environment. Zongker, Werner *et al.* (1999) call such a representation an *environment matte*, as it generalizes the process of object matting (Section 10.4) to not only cut and paste an object from one image into another but also take into account the subtle refractive or reflective interplay between the object and its environment.

Recall from Equations (3.8) and (10.29) that a foreground object can be represented by its pre-multiplied colors and opacities $(\alpha F, \alpha)$. Such a matte can then be composited onto a new background B using

$$C_i = \alpha_i F_i + (1 - \alpha_i) B_i, \tag{14.1}$$

where i is the pixel under consideration. In environment matting, we augment this equation with a reflective or refractive term to model indirect light paths between the environment and the camera. In the original work of Zongker, Werner *et al.* (1999), this indirect component I_i is modeled as

$$I_i = R_i \int A_i(\mathbf{x}) B(\mathbf{x}) d\mathbf{x}, \tag{14.2}$$

where A_i is the rectangular *area of support* for that pixel, R_i is the colored reflectance or transmittance (for colored glossy surfaces or glass), and $B(\mathbf{x})$ is the background (environment) image, which is integrated over the area $A_i(\mathbf{x})$. In follow-on work, Chuang, Zongker *et al.* (2000) use a superposition of oriented Gaussians,

$$I_i = \sum_j R_{ij} \int G_{ij}(\mathbf{x}) B(\mathbf{x}) d\mathbf{x}, \tag{14.3}$$

where each 2D Gaussian

$$G_{ij}(\mathbf{x}) = G_{2\mathrm{D}}(\mathbf{x}; \mathbf{c}_{ij}, \sigma_{ij}, \theta_{ij}) \tag{14.4}$$

is modeled by its center $\mathbf{c}_{ij}$, unrotated widths $\sigma_{ij} = (\sigma^x_{ij}, \sigma^y_{ij})$, and orientation θ_{ij}.

Given a representation for an environment matte, how can we go about estimating it for a particular object? The trick is to place the object in front of a monitor (or surrounded by a set of monitors), where we can change the illumination patterns $B(\mathbf{x})$ and observe the value of each composite pixel C_i.[12]

As with traditional two-screen matting (Section 10.4.1), we can use a variety of solid colored backgrounds to estimate each pixel's foreground color $\alpha_i F_i$ and partial coverage (opacity) α_i. To estimate the area of support A_i in (14.2), Zongker, Werner *et al.* (1999) use a series of periodic horizontal and vertical solid stripes at different frequencies and phases, which is reminiscent of the structured light patterns used in active rangefinding (Section 13.2). For the more sophisticated Gaussian mixture model (14.3), Chuang, Zongker *et al.* (2000) sweep a series of narrow Gaussian stripes at four different orientations (horizontal, vertical, and two diagonals), which enables them to estimate multiple oriented Gaussian responses at each pixel.

Once an environment matte has been "pulled", it is then a simple matter to replace the background with a new image $B(\mathbf{x})$ to obtain a novel composite of the object placed in a different environment (Figure 14.14a–c). The use of multiple backgrounds during the matting process, however, precludes the use of this technique with dynamic scenes, e.g., water pouring into a glass (Figure 14.14d). In this case, a single graded color background can be used to estimate a single 2D monochromatic displacement for each pixel (Chuang, Zongker *et al.* 2000).

14.4.1 Higher-dimensional light fields

As you can tell from the preceding discussion, an environment matte in principle maps every pixel (x, y) into a 4D distribution over light rays and is, hence, a six-dimensional representation. (In practice, each 2D pixel's response is parameterized using a dozen or so parameters, e.g., $\{F, \alpha, B, R, A\}$, instead of a full mapping.) What if we want to model an object's refractive properties from every potential point of view? In this case, we need a mapping from every incoming 4D light ray to every potential exiting 4D light ray, which is an 8D representation. If we use the same trick as with surface light fields, we can parameterize each surface point by its 4D BRDF to reduce this mapping back down to 6D, but this loses the ability to handle multiple refractive paths.

If we want to handle dynamic light fields, we need to add another temporal dimension. (Wenger, Gardner *et al.* (2005) gives a nice example of a dynamic appearance and illumination acquisition system.) Similarly, if we want a continuous distribution over wavelengths, this becomes another dimension.

These examples illustrate how modeling the full complexity of a visual scene through sampling can be extremely expensive. Fortunately, constructing specialized models, which exploit knowledge about the physics of light transport along with the natural coherence of real-world objects, can make these problems more tractable.

[12] If we relax the assumption that the environment is distant, the monitor can be placed at several depths to estimate a depth-dependent mapping function (Zongker, Werner *et al.* 1999).

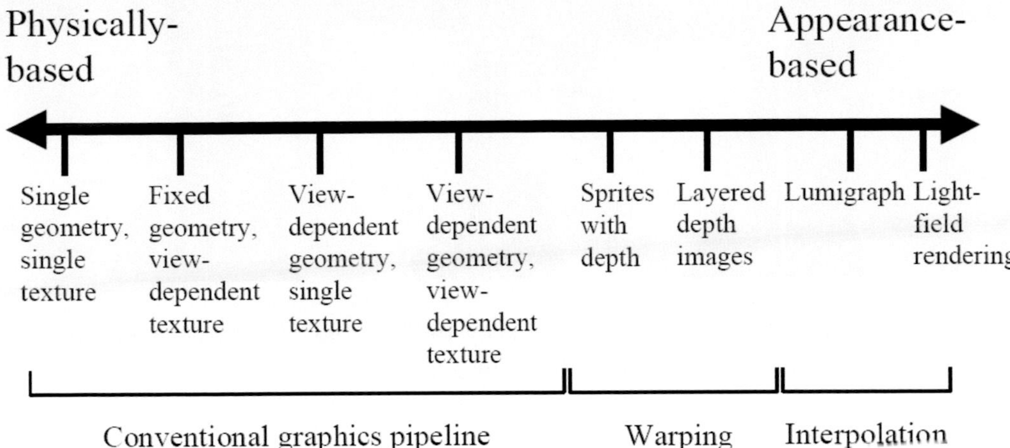

Figure 14.15 The geometry–image continuum in image-based rendering (Kang, Szeliski, and Anandan 2000) © 2000 IEEE. Representations at the left of the spectrum use more detailed geometry and simpler image representations, while representations and algorithms on the right use more images and less geometry.

14.4.2 The modeling to rendering continuum

The image-based rendering representations and algorithms we have studied in this chapter span a continuum ranging from classic 3D texture-mapped models all the way to pure sampled ray-based representations such as light fields (Figure 14.15). Representations such as view-dependent texture maps and Lumigraphs still use a single global geometric model, but select the colors to map onto these surfaces from nearby images. View-dependent geometry, e.g., multiple depth maps, sidestep the need for coherent 3D geometry, and can sometimes better model local non-rigid effects such as specular motion (Swaminathan, Kang *et al.* 2002; Criminisi, Kang *et al.* 2005). Sprites with depth and layered depth images use image-based representations of both color and geometry and can be efficiently rendered using warping operations rather than 3D geometric rasterization.

The best choice of representation and rendering algorithm depends on both the quantity and quality of the input imagery as well as the intended application. When nearby views are being rendered, image-based representations capture more of the visual fidelity of the real world because they directly sample its appearance. On the other hand, if only a few input images are available or the image-based models need to be manipulated, e.g., to change their shape or appearance, more abstract 3D representations such as geometric and local reflection models are a better fit. As we continue to capture and manipulate increasingly larger quantities of visual data, research into these aspects of image-based modeling and rendering will continue to evolve.

14.5 Video-based rendering

As multiple images can be used to render new images or interactive experiences, can something similar be done with video? In fact, a fair amount of work has been done in the area of *video-based rendering* and *video-based animation*, two terms first introduced by Schödl, Szeliski *et al.* (2000) to denote the process of generating new video sequences from captured video footage. An early example of such work is Video Rewrite (Bregler, Covell, and Slaney 1997), in which archival video footage is "re-animated" by having actors say new utterances (Figure 14.16). More recently, the term video-based rendering has been used by some researchers to denote the creation of virtual

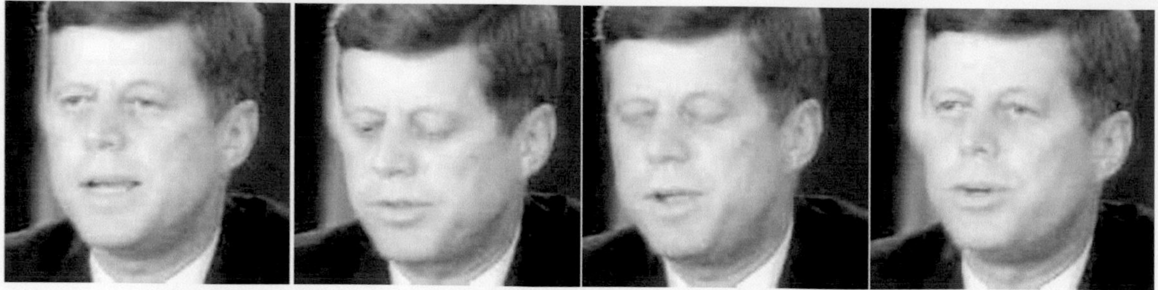

Figure 14.16 Video Rewrite (Bregler, Covell, and Slaney 1997) © 1997 ACM: the video frames are composed from bits and pieces of old video footage matched to a new audio track.

camera moves from a set of synchronized video cameras placed in a studio (Magnor 2005). (The terms *free-viewpoint video* and *3D video* are also sometimes used: see Section 14.5.4.)

In this section, we present a number of video-based rendering systems and applications. We start with *video-based animation* (Section 14.5.1), in which video footage is re-arranged or modified, e.g., in the capture and re-rendering of facial expressions. A special case of this is *video textures* (Section 14.5.2), in which source video is automatically cut into segments and re-looped to create infinitely long video animations. It is also possible to create such animations from still pictures or paintings, by segmenting the image into separately moving regions and animating them using stochastic motion fields (Section 14.5.3).

Next, we turn our attention to *3D video* (Section 14.5.4), in which multiple synchronized video cameras are used to film a scene from different directions. The source video frames can then be re-combined using image-based rendering techniques, such as view interpolation, to create virtual camera paths between the source cameras as part of a real-time viewing experience. Finally, we discuss capturing environments by driving or walking through them with panoramic video cameras to create interactive video-based walkthrough experiences (Section 14.5.5).

14.5.1 Video-based animation

As we mentioned above, an early example of video-based animation is Video Rewrite, in which frames from original video footage are rearranged to match them to novel spoken utterances, e.g., for movie dubbing (Figure 14.16). This is similar in spirit to the way that *concatenative speech synthesis* systems work (Taylor 2009).

In their system, Bregler, Covell, and Slaney (1997) first use speech recognition to extract phonemes from both the source video material and the novel audio stream. Phonemes are grouped into *triphones* (triplets of phonemes), as these better model the *coarticulation* effect present when people speak. Matching triphones are then found in the source footage and audio track. The mouth images corresponding to the selected video frames are then cut and pasted into the desired video footage being re-animated or dubbed, with appropriate geometric transformations to account for head motion. During the analysis phase, features corresponding to the lips, chin, and head are tracked using computer vision techniques. During synthesis, image morphing techniques are used to blend and stitch adjacent mouth shapes into a more coherent whole. In subsequent work, Ezzat, Geiger, and Poggio (2002) describe how to use a *multidimensional morphable model* (Section 13.6.2) combined with regularized trajectory synthesis to improve these results.

A more sophisticated version of this system, called *face transfer*, uses a novel source video, instead of just an audio track, to drive the animation of a previously captured video, i.e., to re-render a video of a talking head with the appropriate visual speech, expression, and head pose elements (Vlasic, Brand *et al.* 2005). This work is one of many *performance-driven animation* systems (Section 7.1.6), which are often used to animate 3D facial models (Figures 13.23–13.25). While traditional performance-driven animation systems use marker-based motion capture (Williams 1990; Litwinowicz and Williams 1994; Ma, Jones *et al.* 2008), video footage can now be used directly to control the animation (Buck, Finkelstein *et al.* 2000; Pighin, Szeliski, and Salesin 2002; Zhang, Snavely *et al.* 2004; Vlasic, Brand *et al.* 2005; Roble and Zafar 2009; Thies, Zollhofer *et al.* 2016; Thies, Zollhöfer *et al.* 2018; Zollhöfer, Thies *et al.* 2018; Fried, Tewari *et al.* 2019; Egger, Smith *et al.* 2020; Tewari, Fried *et al.* 2020). More details on related techniques can also be found in Section 13.6.3 on facial animation and Section 14.6 on neural rendering.

In addition to its most common application to facial animation, video-based animation can also be applied to whole body motion (Section 13.6.4), e.g., by matching the flow fields between two different source videos and using one to drive the other (Efros, Berg *et al.* 2003; Wang, Liu *et al.* 2018; Chan, Ginosar *et al.* 2019). Another approach to video-based rendering is to use flow or 3D modeling to *unwrap* surface textures into stabilized images, which can then be manipulated and re-rendered onto the original video (Pighin, Szeliski, and Salesin 2002; Rav-Acha, Kohli *et al.* 2008).

14.5.2 Video textures

Video-based animation is a powerful means of creating photo-realistic videos by re-purposing existing video footage to match some other desired activity or script. What if, instead of constructing a special animation or narrative, we simply want the video to continue playing in a plausible manner? For example, many websites use images or videos to highlight their destinations, e.g., to portray attractive beaches with surf and palm trees waving in the wind. Instead of using a static image or a video clip that has a discontinuity when it loops, can we transform the video clip into an infinite-length animation that plays forever?

This idea is the basis of *video textures*, in which a short video clip can be arbitrarily extended by re-arranging video frames while preserving visual continuity (Schödl, Szeliski *et al.* 2000). The basic problem in creating video textures is how to perform this re-arrangement without introducing visual artifacts. Can you think of how you might do this?

The simplest approach is to match frames by visual similarity (e.g., L_2 distance) and to jump between frames that appear similar. Unfortunately, if the motions in the two frames are different, a dramatic visual artifact will occur (the video will appear to "stutter"). For example, if we fail to match the motions of the clock pendulum in Figure 14.17a, it can suddenly change direction in mid-swing.

How can we extend our basic frame matching to also match motion? In principle, we could compute optical flow at each frame and match this. However, flow estimates are often unreliable (especially in textureless regions) and it is not clear how to weight the visual and motion similarities relative to each other. As an alternative, Schödl, Szeliski *et al.* (2000) suggest matching *triplets* or larger neighborhoods of adjacent video frames, much in the same way as Video Rewrite matches triphones. Once we have constructed an $n \times n$ similarity matrix between all video frames (where n is the number of frames), a simple finite impulse response (FIR) filtering of each match sequence can be used to emphasize subsequences that match well.

The results of this match computation gives us a *jump table* or, equivalently, a transition probability between any two frames in the original video. This is shown schematically as red arcs in Figure 14.17b, where the red bar indicates which video frame is currently being displayed, and arcs light up as a forward or backward transition is taken. We can view these transition probabilities as encoding the *hidden Markov model* (HMM) that underlies a stochastic video generation process.

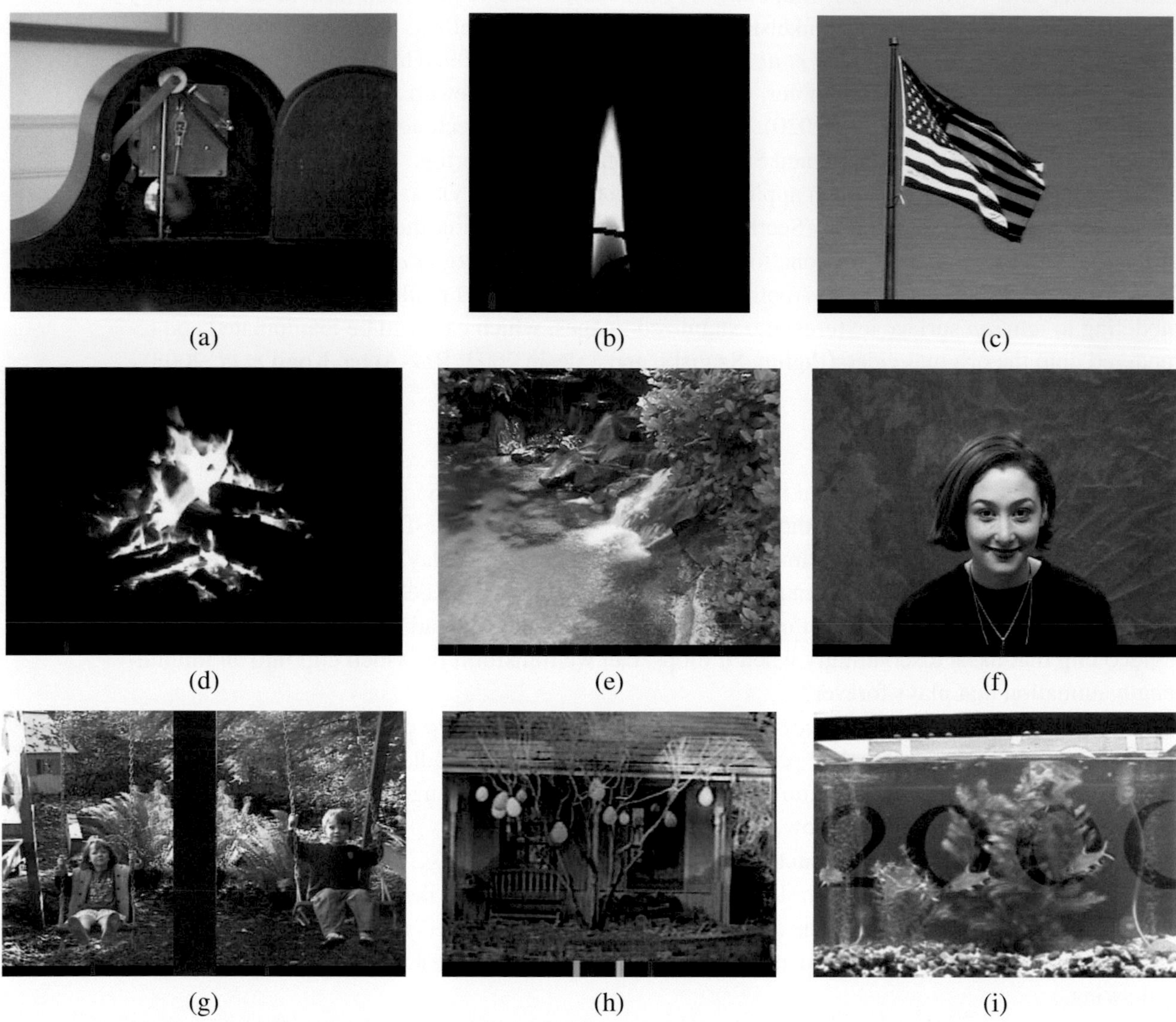

Figure 14.17 Video textures (Schödl, Szeliski *et al.* 2000) © 2000 ACM: (a) a clock pendulum, with correctly matched direction of motion; (b) a candle flame, showing temporal transition arcs; (c) the flag is generated using morphing at jumps; (d) a bonfire uses longer cross-dissolves; (e) a waterfall cross-dissolves several sequences at once; (f) a smiling animated face; (g) two swinging children are animated separately; (h) the balloons are automatically segmented into separate moving regions; (i) a synthetic fish tank consisting of bubbles, plants, and fish. Videos corresponding to these images can be found at https://www.cc.gatech.edu/gvu/perception/projects/videotexture.

Sometimes, it is not possible to find exactly matching subsequences in the original video. In this case, morphing, i.e., warping and blending frames during transitions (Section 3.6.3) can be used to hide the visual differences (Figure 14.17c). If the motion is chaotic enough, as in a bonfire or a waterfall (Figures 14.17d–e), simple blending (extended cross-dissolves) may be sufficient. Improved transitions can also be obtained by performing 3D graph cuts on the spatio-temporal volume around a transition (Kwatra, Schödl *et al.* 2003).

Video textures need not be restricted to chaotic random phenomena such as fire, wind, and water. Pleasing video textures can be created of people, e.g., a smiling face (as in Figure 14.17f) or someone running on a treadmill (Schödl, Szeliski *et al.* 2000). When multiple people or objects are moving independently, as in Figures 14.17g–h, we must first segment the video into independently moving regions and animate each region separately. It is also possible to create large panoramic video textures from a slowly panning camera (Agarwala, Zheng *et al.* 2005; He, Liao *et al.* 2017).

Instead of just playing back the original frames in a stochastic (random) manner, video textures can also be used to create scripted or interactive animations. If we extract individual elements, such as fish in a fishtank (Figure 14.17i) into separate *video sprites*, we can animate them along prespecified paths (by matching the path direction with the original sprite motion) to make our video elements move in a desired fashion (Schödl and Essa 2002). A more recent example of controlling video sprites is the Vid2Player system, which models the movements and shots of tennis players to create synthetic video-realistic games (Zhang, Sciutto *et al.* 2021). In fact, work on video textures inspired research on systems that re-synthesize new motion sequences from motion capture data, which some people refer to as "mocap soup" (Arikan and Forsyth 2002; Kovar, Gleicher, and Pighin 2002; Lee, Chai *et al.* 2002; Li, Wang, and Shum 2002; Pullen and Bregler 2002).

While video textures primarily analyze the video as a sequence of frames (or regions) that can be re-arranged in time, *temporal textures* (Szummer and Picard 1996; Bar-Joseph, El-Yaniv *et al.* 2001) and *dynamic textures* (Doretto, Chiuso *et al.* 2003; Yuan, Wen *et al.* 2004; Doretto and Soatto 2006) treat the video as a 3D spatio-temporal volume with textural properties, which can be described using auto-regressive temporal models and combined with layered representations (Chan and Vasconcelos 2009). In more recent work, video texture authoring systems have been extended to allow for control over the *dynamism* (amount of motion) in different regions (Joshi, Mehta *et al.* 2012; Liao, Joshi, and Hoppe 2013; Yan, Liu, and Furukawa 2017; He, Liao *et al.* 2017; Oh, Joo *et al.* 2017) and improved loop transitions (Liao, Finch, and Hoppe 2015).

14.5.3 *Application*: Animating pictures

While video textures can turn a short video clip into an infinitely long video, can the same thing be done with a single still image? The answer is yes, if you are willing to first segment the image into different layers and then animate each layer separately.

Chuang, Goldman *et al.* (2005) describe how an image can be decomposed into separate layers using interactive matting techniques. Each layer is then animated using a class-specific synthetic motion. As shown in Figure 14.18, boats rock back and forth, trees sway in the wind, clouds move horizontally, and water ripples, using a shaped noise displacement map. All of these effects can be tied to some global control parameters, such as the velocity and direction of a virtual wind. After being individually animated, the layers can be composited to create a final dynamic rendering.

In more recent work, Holynski, Curless *et al.* (2021) train a deep network to take a static photo, hallucinate a plausible motion field, encode the image as deep multi-resolution features, and then advect these features bi-directionally in time using Eulerian motion, using an architecture inspired by Niklaus and Liu (2020) and Wiles, Gkioxari *et al.* (2020). The resulting deep features are then decoded to produce a looping video clip with synthetic stochastic fluid motions.

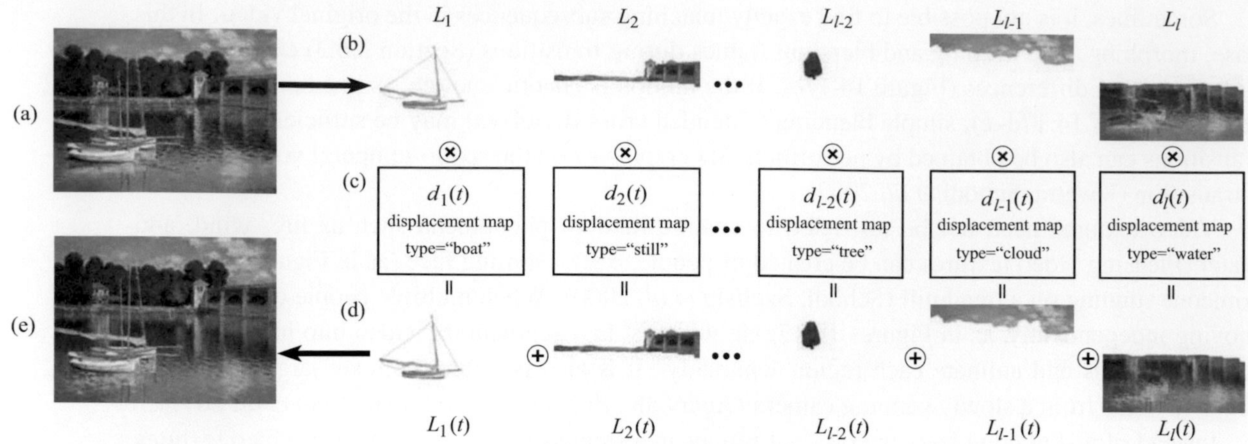

Figure 14.18 Animating still pictures (Chuang, Goldman *et al.* 2005) © 2005 ACM. (a) The input still image is manually segmented into (b) several layers. (c) Each layer is then animated with a different stochastic motion texture (d) The animated layers are then composited to produce (e) the final animation

14.5.4 3D and free-viewpoint Video

In the last decade, the 3D movies have become an established medium. Currently, such releases are filmed using stereoscopic camera rigs and displayed in theaters (or at home) to viewers wearing polarized glasses. In the future, however, home audiences may wish to view such movies with multi-zone auto-stereoscopic displays, where each person gets his or her own customized stereo stream and can move around a scene to see it from different perspectives.

The stereo matching techniques developed in the computer vision community along with image-based rendering (view interpolation) techniques from graphics are both essential components in such scenarios, which are sometimes called *free-viewpoint video* (Carranza, Theobalt *et al.* 2003) or *virtual viewpoint video* (Zitnick, Kang *et al.* 2004). In addition to solving a series of per-frame reconstruction and view interpolation problems, the depth maps or proxies produced by the analysis phase must be temporally consistent in order to avoid flickering artifacts. Neural rendering techniques (Tewari, Fried *et al.* 2020, Section 6.3) can also be used for both the reconstruction and rendering phases.

Shum, Chan, and Kang (2007) and Magnor (2005) present nice overviews of various video view interpolation techniques and systems. These include the Virtualized Reality system of Kanade, Rander, and Narayanan (1997) and Vedula, Baker, and Kanade (2005), Immersive Video (Moezzi, Katkere *et al.* 1996), Image-Based Visual Hulls (Matusik, Buehler *et al.* 2000; Matusik, Buehler, and McMillan 2001), and Free-Viewpoint Video (Carranza, Theobalt *et al.* 2003), which all use global 3D geometric models (surface-based (Section 13.3) or volumetric (Section 13.5)) as their proxies for rendering. The work of Vedula, Baker, and Kanade (2005) also computes *scene flow*, i.e., the 3D motion between corresponding surface elements, which can then be used to perform spatio-temporal interpolation of the multi-view video stream. A more recent variant of scene flow is the *occupancy flow* work of Niemeyer, Mescheder *et al.* (2019).

The Virtual Viewpoint Video system of Zitnick, Kang *et al.* (2004), on the other hand, associates a two-layer depth map with each input image, which allows them to accurately model occlusion effects such as the mixed pixels that occur at object boundaries. Their system, which consists of eight synchronized video cameras connected to a disk array (Figure 14.19a), first uses segmentation-based stereo to extract a depth map for each input image (Figure 14.19e). Near object boundaries

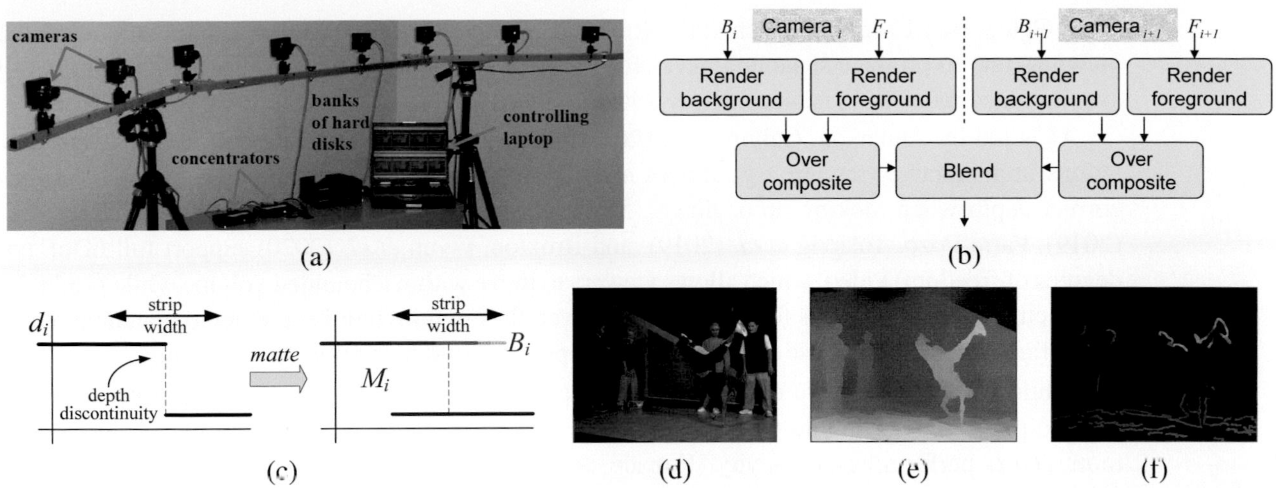

Figure 14.19 Video view interpolation (Zitnick, Kang *et al.* 2004) © 2004 ACM: (a) the capture hardware consists of eight synchronized cameras; (b) the background and foreground images from each camera are rendered and composited before blending; (c) the two-layer representation, before and after boundary matting; (d) background color estimates; (e) background depth estimates; (f) foreground color estimates.

(depth discontinuities), the background layer is extended along a strip behind the foreground object (Figure 14.19c) and its color is estimated from the neighboring images where it is not occluded (Figure 14.19d). Automated matting techniques (Section 10.4) are then used to estimate the fractional opacity and color of boundary pixels in the foreground layer (Figure 14.19f).

At render time, given a new virtual camera that lies between two of the original cameras, the layers in the neighboring cameras are rendered as texture-mapped triangles and the foreground layer (which may have fractional opacities) is then composited over the background layer (Figure 14.19b). The resulting two images are merged and blended by comparing their respective z-buffer values. (Whenever the two z-values are sufficiently close, a linear blend of the two colors is computed.) The interactive rendering system runs in real time using regular graphics hardware. It can therefore be used to change the observer's viewpoint while playing the video or to freeze the scene and explore it in 3D. Rogmans, Lu *et al.* (2009) subsequently developed GPU implementations of both real-time stereo matching and real-time rendering algorithms, which enable them to explore algorithmic alternatives in a real-time setting.

The depth maps computed from the eight stereo cameras using off-line stereo matching have been used in studies of 3D video compression (Smolic and Kauff 2005; Gotchev and Rosenhahn 2009; Tech, Chen *et al.* 2015). Active video-rate depth sensing cameras, such as the 3DV Zcam (Iddan and Yahav 2001), which we discussed in Section 13.2.1, are another potential source of such data.

When large numbers of closely spaced cameras are available, as in the Stanford Light Field Camera (Wilburn, Joshi *et al.* 2005), it may not always be necessary to compute explicit depth maps to create video-based rendering effects, although the results are usually of higher quality if you do (Vaish, Szeliski *et al.* 2006).

The last few years have seen a revival of research into 3D video, spurred in part by the wider availability of virtual reality headsets, which can be used to view such videos with a strong sense of immersion. The Jump virtual reality capture system from Google (Anderson, Gallup *et al.* 2016) uses 16 GoPro cameras arranged on a 28cm diameter ring to capture multiple videos, which are then stitched offline into a pair of omnidirectional stereo (ODS) videos (Ishiguro, Yamamoto, and Tsuji

1992; Peleg, Ben-Ezra, and Pritch 2001; Richardt, Pritch *et al.* 2013), which can then be warped at viewing time to produce separate images for each eye. A similar system, constructed from tightly synchronized industrial vision cameras, was introduced around the same time by Cabral (2016).

As noted by Anderson, Gallup *et al.* (2016), however, the ODS representation has severe limitations in interactive viewing, e.g., it does not support head tilt, or translational motion, or produce correct depth when looking up or down. More recent systems developed by Serrano, Kim *et al.* (2019), Parra Pozo, Toksvig *et al.* (2019), and Broxton, Flynn *et al.* (2020) support full 6DoF (six degrees of freedom) video, which allows viewers to move within a bounded volume while producing perspectively correct images for each eye. However, they require multi-view stereo matching during the offline construction phase to produce the 3D proxies need to support such viewing.

While these systems are designed to capture *inside out* experiences, where a user can watch a video unfolding all around them, pointing the cameras *outside in* can be used to capture one or more actors performing an activity (Kanade, Rander, and Narayanan 1997; Joo, Liu *et al.* 2015; Tang, Dou *et al.* 2018). Such setups are often called *free-viewpoint video* or *volumetric performance capture* systems. The most recent versions of such systems use deep networks to reconstruct, represent, compress, and/or render time-evolving volumetric scenes (Martin-Brualla, Pandey *et al.* 2018; Pandey, Tkach *et al.* 2019; Lombardi, Simon *et al.* 2019; Tang, Singh *et al.* 2020; Peng, Zhang *et al.* 2021), as summarized in the recent survey on neural rendering by Tewari, Fried *et al.* (2020, Section 6.3). And while most of these systems require custom-built multi-camera rigs, it is also possible to construct 3D videos from collections of handheld videos (Bansal, Vo *et al.* 2020) or even a single moving smartphone camera (Yoon, Kim *et al.* 2020; Luo, Huang *et al.* 2020).

14.5.5 *Application*: Video-based walkthroughs

Video camera arrays enable the simultaneous capture of 3D dynamic scenes from multiple viewpoints, which can then enable the viewer to explore the scene from viewpoints near the original capture locations. What if, instead we wish to capture an extended area, such as a home, a movie set, or even an entire city?

In this case, it makes more sense to move the camera through the environment and play back the video as an interactive video-based walkthrough. To allow the viewer to look around in all directions, it is preferable to use a panoramic video camera (Uyttendaele, Criminisi *et al.* 2004).[13]

One way to structure the acquisition process is to capture these images in a 2D horizontal plane, e.g., over a grid superimposed inside a room. The resulting *sea of images* (Aliaga, Funkhouser *et al.* 2003) can be used to enable continuous motion between the captured locations.[14] However, extending this idea to larger settings, e.g., beyond a single room, can become tedious and data-intensive.

Instead, a natural way to explore a space is often to just walk through it along some prespecified paths, just as museums or home tours guide users along a particular path, say down the middle of each room.[15] Similarly, city-level exploration can be achieved by driving down the middle of each street and allowing the user to branch at each intersection. This idea dates back to the Aspen MovieMap project (Lippman 1980), which recorded analog video taken from moving cars onto videodiscs for later interactive playback.

[13] See https://www.cis.upenn.edu/~kostas/omni.html for descriptions of panoramic (omnidirectional) vision systems and associated workshops.

[14] The Photo Tourism system of Snavely, Seitz, and Szeliski (2006) applies this idea to less structured collections.

[15] In computer games, restricting a player to forward and backward motion along predetermined paths is called *rail-based gaming*.

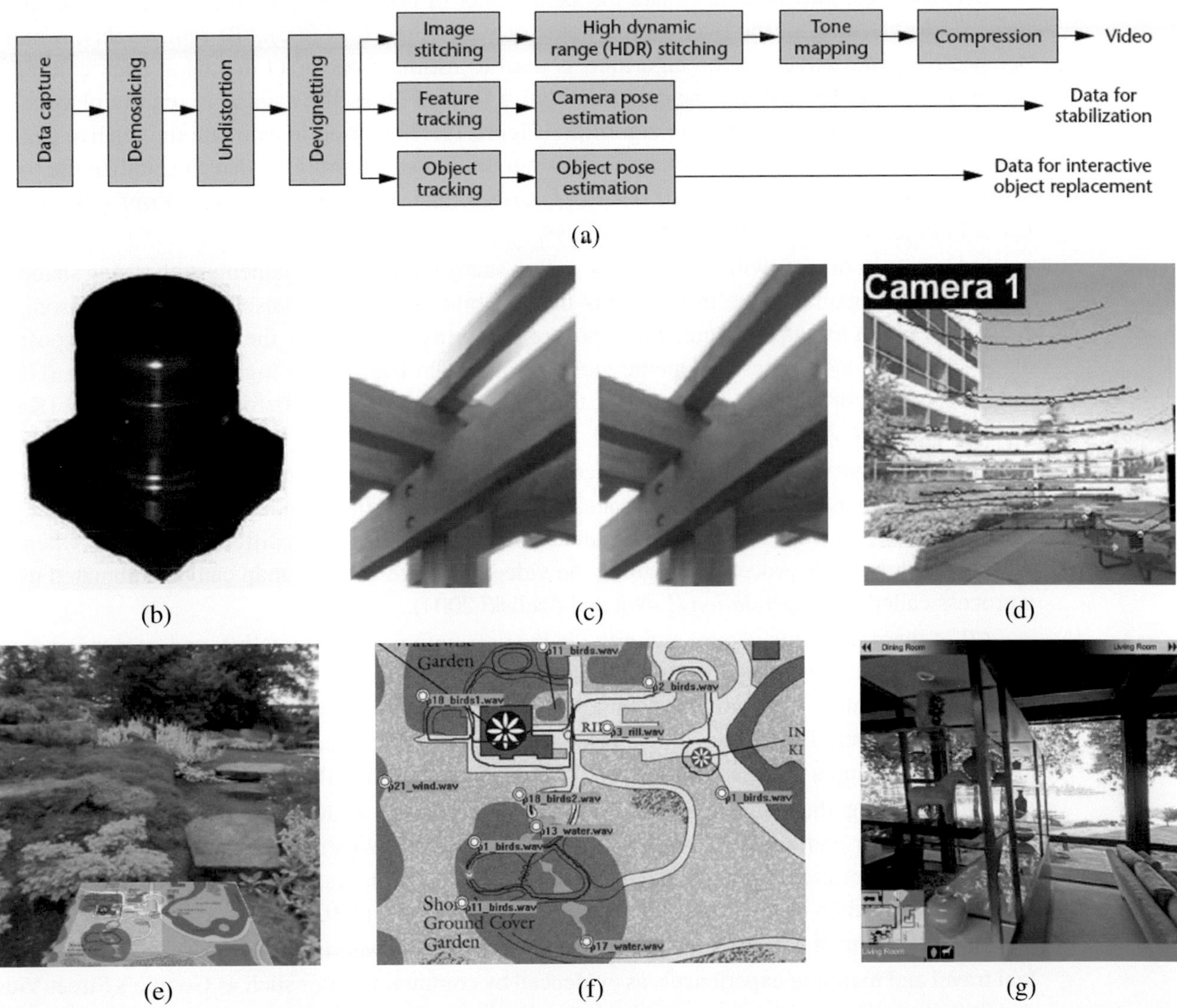

Figure 14.20 Video-based walkthroughs (Uyttendaele, Criminisi *et al.* 2004) © 2004 IEEE: (a) system diagram of video pre-processing; (b) the Point Grey Ladybug camera; (c) ghost removal using multi-perspective plane sweep; (d) point tracking, used both for calibration and stabilization; (e) interactive garden walkthrough with map below; (f) overhead map authoring and sound placement; (g) interactive home walkthrough with navigation bar (top) and icons of interest (bottom).

Improvements in video technology enabled the capture of panoramic (spherical) video using a small co-located array of cameras, such as the Point Grey Ladybug camera (Figure 14.20b) developed by Uyttendaele, Criminisi *et al.* (2004) for their interactive video-based walkthrough project. In their system, the synchronized video streams from the six cameras (Figure 14.20a) are stitched together into 360° panoramas using a variety of techniques developed specifically for this project.

Because the cameras do not share the same center of projection, parallax between the cameras can lead to ghosting in the overlapping fields of view (Figure 14.20c). To remove this, a multi-perspective plane sweep stereo algorithm is used to estimate per-pixel depths at each column in the overlap area. To calibrate the cameras relative to each other, the camera is spun in place and a constrained structure from motion algorithm (Figure 11.15) is used to estimate the relative camera poses and intrinsics. Feature tracking is then run on the walk-through video to stabilize the video sequence. Liu, Gleicher *et al.* (2009), Kopf, Cohen, and Szeliski (2014), and Kopf (2016) have carried out more recent work along these lines.

Indoor environments with windows, as well as sunny outdoor environments with strong shadows, often have a dynamic range that exceeds the capabilities of video sensors. For this reason, the Ladybug camera has a programmable exposure capability that enables the bracketing of exposures at subsequent video frames. To merge the resulting video frames into high dynamic range (HDR) video, pixels from adjacent frames need to be motion-compensated before being merged (Kang, Uyttendaele *et al.* 2003).

The interactive walk-through experience becomes much richer and more navigable if an overview map is available as part of the experience. In Figure 14.20f, the map has annotations, which can show up during the tour, and localized sound sources, which play (with different volumes) when the viewer is nearby. The process of aligning the video sequence with the map can be automated using a process called *map correlation* (Levin and Szeliski 2004).

All of these elements combine to provide the user with a rich, interactive, and immersive experience. Figure 14.20e shows a walk through the Bellevue Botanical Gardens, with an overview map in perspective below the live video window. Arrows on the ground are used to indicate potential directions of travel. The viewer simply orients their view towards one of the arrows (the experience can be driven using a game controller) and "walks" forward along the desired path.

Figure 14.20g shows an indoor home tour experience. In addition to a schematic map in the lower left corner and adjacent room names along the top navigation bar, icons appear along the bottom whenever items of interest, such as a homeowner's art pieces, are visible in the main window. These icons can then be clicked to provide more information and 3D views.

The development of interactive video tours spurred a renewed interest in 360° video-based virtual travel and mapping experiences, as evidenced by commercial sites such as Google's Street View and 360cities. The same videos can also be used to generate turn-by-turn driving directions, taking advantage of both expanded fields of view and image-based rendering to enhance the experience (Chen, Neubert *et al.* 2009).

While initially, 360° cameras were exotic and expensive, they have more recently become widely available consumer products, such as the popular RICOH THETA camera, first introduced in 2013, and the GoPro MAX action camera. When shooting 360° videos, it is possible to stabilize the video using algorithms tailored to such videos (Kopf 2016) or proprietary algorithms based on the camera's IMU readings. And while most of these cameras produce monocular photos and videos, VR180 cameras have two lenses and so can create wide field-of-view stereoscopic content. It is even possible to produce 3D 360° content by carefully stitching and transforming two 360° camera streams (Matzen, Cohen *et al.* 2017).

In addition to capturing immersive photos and videos of scenic locations and popular events,

(a) Scene reconstruction (b) Proxy geometry (c) Stitched & blended

Figure 14.21 First-person hyperlapse video creation (Kopf, Cohen, and Szeliski 2014) © 2014 ACM: (a) 3D camera path and point cloud recovery, followed by smooth path planning; (b) 3D per-camera proxy estimation; and (c) source frame and seam selection using an MRF and Poisson blending.

360° and regular action cameras can also be worn, moved through an environment, and then sped up to create *hyperlapse* videos (Kopf, Cohen, and Szeliski 2014). Because such videos may exhibit large amounts of translational motion and parallax when heavily sped up, it is insufficient to simply compensate for camera rotations or even to warp individual input frames, because the large amounts of compensating motion may force the virtual camera to look outside the video frames. Instead, after constructing a sparse 3D model and smoothing the camera path, keyframes are selected and 3D proxies are computed for each of these by interpolating the sparse 3D point cloud, as shown in Figure 14.21. These frames are then warped and stitched together (using Poisson blending) using a Markov random field to ensure as much smoothness and visual continuity as possible. This system combines many different previously developed 3D modeling, computational photography, and image-based rendering algorithms to produce remarkably smooth high-speed tours of large-scale environments (such as cities) and activities (such as rock climbing and skiing).

As we continue to capture more and more of our real world with large amounts of high-quality imagery and video, the interactive modeling, exploration, and rendering techniques described in this chapter will play an even bigger role in bringing virtual experiences based in remote areas of the world as well as re-living special memories closer to everyone.

14.6 Neural rendering

The most recent development in image-based rendering is the introduction of deep neural networks into both the modeling (construction) and viewing parts of image-based rendering pipelines. Neural rendering has been applied in a number of different domains, including style and texture manipulation and 2D semantic photo synthesis (Sections 5.5.4 and 10.5.3), 3D object shape and appearance modeling (Section 13.5.1), facial animation and reenactment (Section 13.6.3), 3D body capture and replay (Section 13.6.4), novel view synthesis (Section 14.1), free-viewpoint video (Section 14.5.4), and relighting (Duchêne, Riant *et al.* 2015; Meka, Haene *et al.* 2019; Philip, Gharbi *et al.* 2019; Sun, Barron *et al.* 2019; Zhou, Hadap *et al.* 2019; Zhang, Barron *et al.* 2020).

A comprehensive survey of all of these applications and techniques can be found in the state of the art report by Tewari, Fried *et al.* (2020), whose abstract states:

> *Neural rendering is a new and rapidly emerging field that combines generative machine learning techniques with physical knowledge from computer graphics, e.g., by the integration of differentiable rendering into network training. With a plethora of applications in computer graphics and vision, neural rendering is poised to become a new area in the graphics community...*

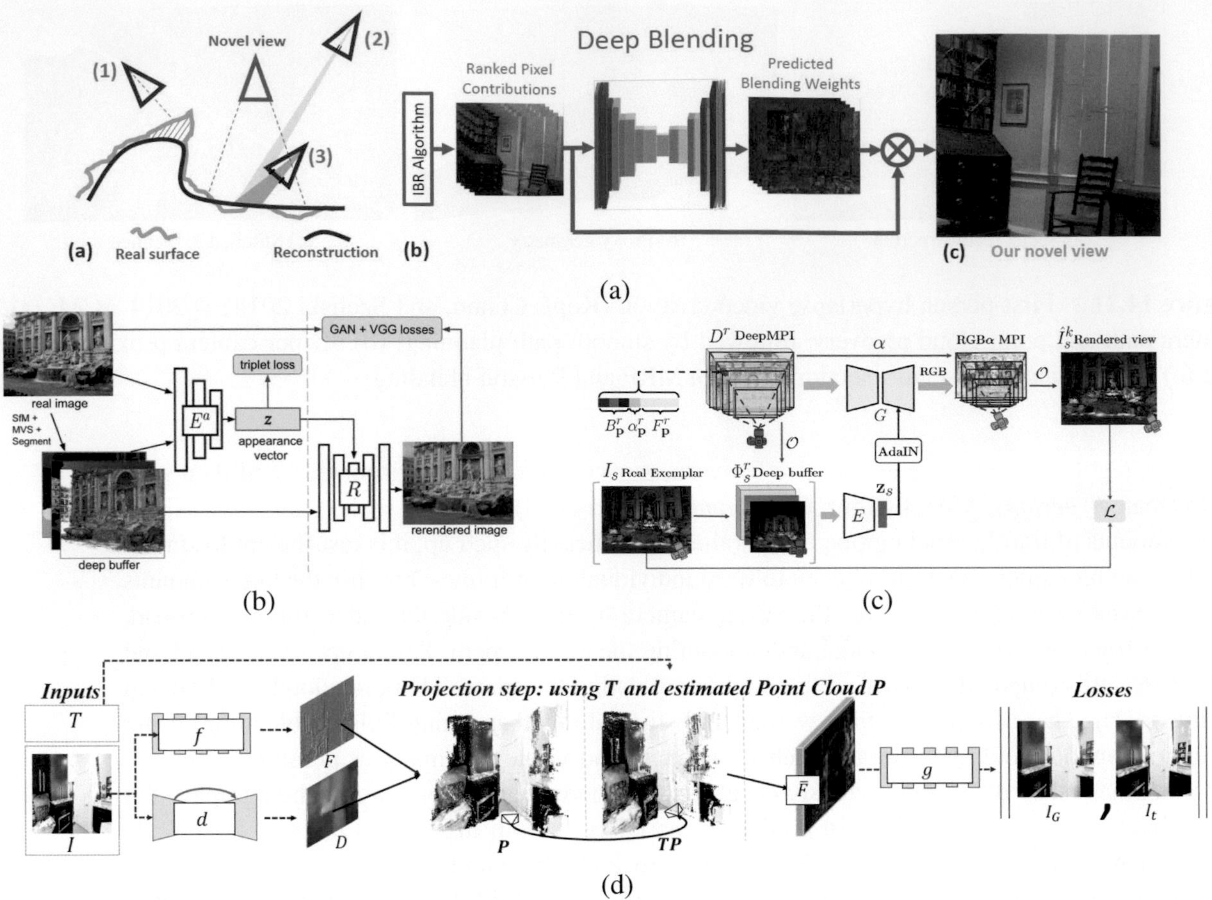

Figure 14.22 Examples of neural image-based rendering: (a) deep blending of depth-warped source images (Hedman, Philip *et al.* 2018) © 2018 ACM; (b) neural re-rendering in the wild with controllable view and lighting (Meshry, Goldman *et al.* 2019) © 2019 IEEE; (c) crowdsampling the plenoptic function with a deep MPI (Li, Xian *et al.* 2020) © 2020 Springer. (d) SynSin: novel view synthesis from a single image (Wiles, Gkioxari *et al.* 2020) © 2020 IEEE.

The survey contains over 230 references and highlights 46 representative papers, grouped into six general categories, namely semantic photo synthesis, novel view synthesis, free viewpoint video, relighting, facial reenactment, and body reenactment. As you can tell, these categories overlap with the sections of the book mentioned in the previous paragraph. A set of lectures based on this content can be found in the related CVPR tutorial on neural rendering (Tewari, Zollhöfer *et al.* 2020), and several of the lectures in the TUM AI Guest Lecture Series are also on neural rendering research.[16] The X-Fields paper by Bemana, Myszkowski *et al.* (2020, Table 1) also has a nice tabulation of related space, time, and illumination interpolation papers with an emphasis on deep methods, while the short bibliography by Dellaert and Yen-Chen (2021) summarizes even more recent techniques. Some neural rendering systems are implemented using differentiable rendering, which is surveyed by Kato, Beker *et al.* (2020).

As we have already seen many of these neural rendering techniques in the previous sections mentioned above, we focus here on their application to 3D image-based modeling and rendering. There are many ways to organize the last few years' worth of research in neural rendering. In this

[16]https://niessner.github.io/TUM-AI-Lecture-Series

section, I have chosen to use four broad categories of underlying 3D representations, which we have studied in the last two chapters, namely: texture-mapped meshes, depth images and layers, volumetric grids, and implicit functions.

Texture-mapped meshes. As described in Chapter 13, a convenient representation for modeling and rendering a 3D scene is a triangle mesh, which can be reconstructed from images using multi-view stereo. One of the earliest papers to use a neural network as part of the 3D rendering process was the deep blending system of Hedman, Philip *et al.* (2018), who augment an unstructured Lumigraph rendering pipeline (Buehler, Bosse *et al.* 2001) with a deep neural network that computes the per-pixel blending weights for the warped images selected for each novel view, as shown in Figure 14.22a. LookinGood (Martin-Brualla, Pandey *et al.* 2018) takes a single or multiple-image texture-mapped 3D upper or whole-body rendering and fills in the holes, denoises the appearance, and increases the resolution using a U-Net trained on held out views. Along a similar line, Deep Learning Super Sampling (DLSS) uses an encoder-decoder DNN implemented in GPU hardware to increase the resolution of rendered games in real time (Burnes 2020).

While these systems warp colored textures or images (i.e., view-dependent textures) and then apply a neural net post-process, it is also possible to first convert the images into a "neural" encoding and then warp and blend such representations. Free View Synthesis (Riegler and Koltun 2020a) starts by building a local 3D model for the novel view using multi-view stereo. It then encodes the source images as neural codes, reprojects these codes to the novel viewpoint, and composites them using a recurrent neural network and softmax. Instead of warping neural codes at render time and then blending and decoding them, the follow-on Stable View Synthesis system (Riegler and Koltun 2020b) collects neural codes from all incoming rays for every surface point and then combines these with an *on-surface aggregation* network to produce outgoing neural codes along the rays to the novel view camera. Deferred Neural Rendering (Thies, Zollhöfer, and Nießner 2019) uses a (u, v) parameterization over the 3D surface to learn and store a 2D texture map of neural codes, which can be sampled and decoded at rendering time.

Depth images and layers. To deal with images taken at different times of day and weather, i.e., "in the wild", Meshry, Goldman *et al.* (2019) use a DNN to compute a latent "appearance" vector for each input image and its associated depth image (computed using traditional multi-view stereo), as shown in Figure 14.22b. At render time, the appearance can be manipulated (in addition to the 3D viewpoint) to explore the range of conditions under which the images were taken. Li, Xian *et al.* (2020) develop a related pipeline (Figure 14.22c), which instead of storing a single "deep" color/depth/appearance image or buffer uses a multiplane image (MPI). As with the previous system, an encoder-decoder modulated with the appearance vector (using Adaptive Instance Normalization) is used to render the final image, in this case through an intermediate MPI that does the view warping and over compositing. Instead of using many parallel finely sliced planes, the GeLaTO (Generative Latent Textured Objects) system uses a small number of oriented planes ("billboards") with associated neural textures to model thin transparent objects such as eyeglasses (Martin-Brualla, Pandey *et al.* 2020). At render time, these textures are warped and then decoded and composited using a U-Net to produce a final RGBA sprite.

While all of these previous systems use multiple images to build a 3D neural representation, SynSin (Synthesis from a Single Image) (Wiles, Gkioxari *et al.* 2020) starts with just a single color image and uses a DNN to turn this image into a neural features F and depth D buffer pair, as shown in Figure 14.22d. At render time, the neural features are warped according to their associated depths and the camera view matrix, splatted with soft weights, and composited back-to-front to obtain a

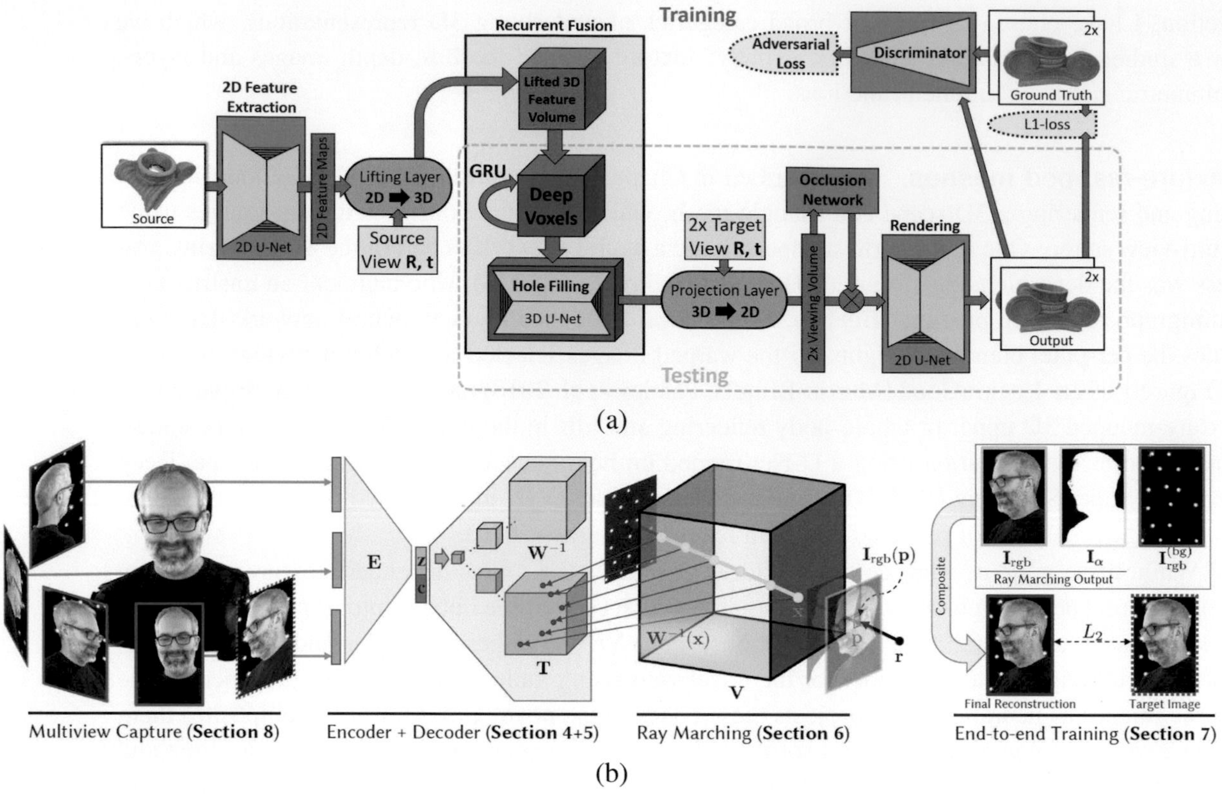

Figure 14.23 Examples of voxel grid neural rendering: (a) DeepVoxels (Sitzmann, Thies *et al.* 2019) © 2019 IEEE; (b) Neural Volumes (Lombardi, Simon *et al.* 2019) © 2019 ACM.

neural rendered frame $\tilde{F}$, which is then decoded into the final color novel view I_G. In Semantic View Synthesis Huang, Tseng *et al.* (2020) start with a semantic label map and use semantic image synthesis (Section 5.5.4) to convert this into a synthetic color image and depth map. These are then used to create a multiplane image from which novel views can be rendered. Holynski, Curless *et al.* (2021) train a deep network to take a static photo, hallucinate a plausible motion field, encode the image as deep features with soft blending weights, advect these features bi-directionally in time, and decode the rendered neural feature frames to produce a looping video clip with synthetic stochastic fluid motions, as discussed in Section 14.5.3.

Voxel representations. Another 3D representation that can be used for neural rendering is a 3D voxel grid. Figure 14.23 shows the modeling and rendering pipelines from two such papers. DeepVoxels (Sitzmann, Thies *et al.* 2019) learn a 3D embedding of neural codes for a given 3D object. At render time, these are projected into 2D view, filtered through an occlusion network (similar to back-to-front alpha compositing), and then decoded into a final image. Neural Volumes (Lombardi, Simon *et al.* 2019) use an encoder-decoder to convert a set of multi-view color images into a 3D RGBα volume and an associated volumetric warp field that can model facial expression variation. At render time, the color volume is warped and then ray marching is used to create a final 2D RGBα foreground image.[17] In more recent work, Weng, Curless, and Kemelmacher-Shlizerman

[17]Note that we mostly use RGBA in earlier parts of the book to denote three color channels with an opacity. In the remainder of this section, I use RGBα to be consistent with recent papers.

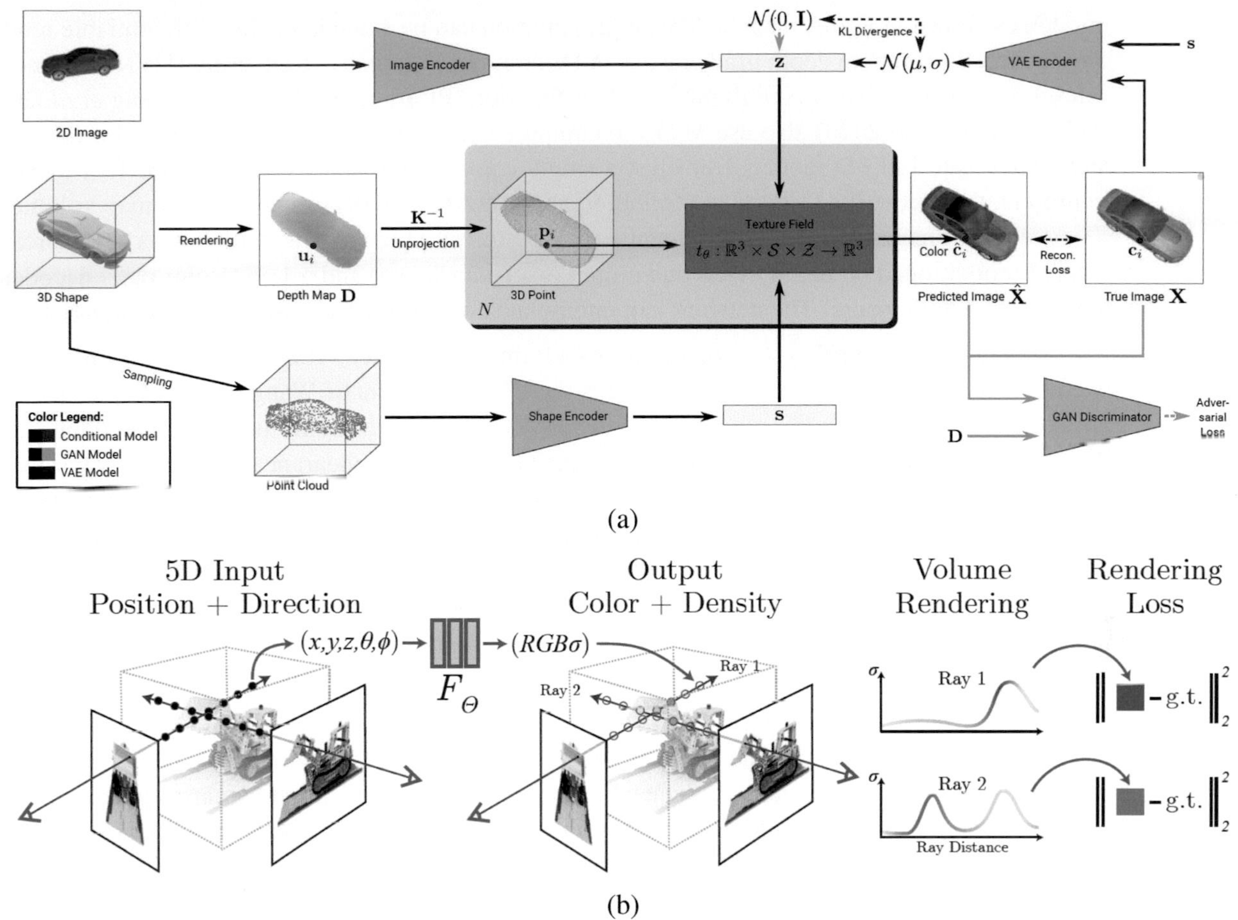

Figure 14.24 Examples of implicit function (MLP) neural rendering: (a) Texture Fields (Oechsle, Mescheder *et al.* 2019) © 2019 IEEE; (b) Neural Radiance Fields (Mildenhall, Srinivasan *et al.* 2020) © 2020 Springer.

(2020) show how deformable Neural Volumes can be constructed and animated from monocular videos of moving people, such as athletes.

Coordinate-based neural representations. The final representation we discuss in this section are implicit functions implemented using fully connected networks, which are now more commonly known as *multilayer perceptrons* or MLPs.[18] We have already seen the use of $[0, 1]$ occupancy and implicit signed distance functions for 3D shape modeling in Section 13.5.1, where we mentioned papers such as Occupancy Networks (Mescheder, Oechsle *et al.* 2019), IM-NET (Chen and Zhang 2019), DeepSDF (Park, Florence *et al.* 2019), and Convolutional Occupancy Networks (Peng, Niemeyer *et al.* 2020).

To render colored images, such representations also need to encode the appearance (e.g., color, texture, or light field) information at either the surface or throughout the volume. Texture Fields (Oechsle, Mescheder *et al.* 2019) train an MLP conditioned on both 3D shape and latent appearance (e.g., car color) to produce a 3D volumetric color field that can then be used to texture map a 3D

[18] As Jon Barron and others have pointed out, only signed distance functions actually encode "implicit functions" as level-sets of their volumetric values. The more general class of techniques that includes opacity models is often called *coordinate regression networks* or *coordinate-based MLPs*.

model, as shown in Figure 14.24a. This representation can be extended using differentiable rendering to directly compute depth gradients, as in Differential Volumetric Rendering (DVR) (Niemeyer, Mescheder *et al.* 2020). Pixel-aligned Implicit function (PIFu) networks (Saito, Huang *et al.* 2019; Saito, Simon *et al.* 2020) also use MLPs to compute volumetric inside/outside and color fields and can hallucinate full 3D models from just a single color image, as shown in Figure 13.18. Scene representation networks (Sitzmann, Zollhöfer, and Wetzstein 2019) use an MLP to map volumetric (x, y, z) coordinates to high-dimensional neural features, which are used by both a ray marching LSTM (conditioned on the 3D view and output pixel coordinate) and a 1×1 color pixel decoder to generate the final image. The network can interpolate both appearance and shape latent variables.

An interesting hybrid system that replaces a trained per-object MLP with on-the-fly multi-view stereo matching and image-based rendering is the IBRNet system of Wang, Wang *et al.* (2021). As with other volumetric neural renders, the network evaluates each ray in the novel viewpoint image by marching along the ray and computing a density and neural appearance feature at each sampled location. However, instead of looking up these values from a pre-trained MLP, it samples the neural features from a small number of adjacent input images, much like in Unstructured Lumigraph (Buehler, Bosse *et al.* 2001; Hedman, Philip *et al.* 2018) and Stable View Synthesis (Riegler and Koltun 2020b), which use a precomputed 3D surface model (which IBRNet does not). The opacity and appearance values along the ray are refined using a transformer architecture, which replaces the more traditional winner-take-all module in a stereo matcher, followed by a classic volumetric compositing of the colors and densities.

To model viewpoint dependent effects such as highlights on plastic objects, i.e., to model a full light field (Section 14.3), Neural Radiance Fields (NeRF) extend the implicit mapping from (x, y, z) spatial positions to also include a viewing direction (θ, ϕ) as inputs, as shown in Figure 14.24b (Mildenhall, Srinivasan *et al.* 2020). Each (x, y, z) query is first turned into a *positional encoding* that consists of sinusoidal waves at octave frequencies before going into a 256-channel MLP. These positional codes are also injected into the fifth layer, and an encoding of the viewing direction is injected at the ninth layer, which is where the opacities are computed (Mildenhall, Srinivasan *et al.* 2020, Figure 7). It turns out that these positional encodings are essential to enabling the MLP to represent fine details, as explored in more depth by Tancik, Srinivasan *et al.* (2020), as well as in the SIREN (Sinusoidal Representation Network) paper by Sitzmann, Martel *et al.* (2020), which uses periodic (sinusoidal) activation functions.

It is also possible to pre-train these neural networks, i.e., use *meta-learning*, on a wider class of objects to speed up the optimization task for new images (Sitzmann, Chan *et al.* 2020; Tancik, Mildenhall *et al.* 2021) and also to use cone tracing together with *integrated positional encoding* to reduce aliasing and handle multi-resolution inputs and output (Barron, Mildenhall *et al.* 2021). The NeRF++ paper by Zhang, Riegler *et al.* (2020) extends the original NeRF representation to handle unbounded 3D scenes by adding an "inside-out" $1/r$ *inverted sphere parameterization*, while Neural Sparse Voxel Fields build an octree with implicit neural functions inside each non-empty cell (Liu, Gu *et al.* 2020).

Instead of modeling opacities, the Implicit Differentiable Renderer (IDR) developed by Yariv, Kasten *et al.* (2020) models a signed distance function, which enables them at rendering time to extract a level-set surface with analytic normals, which are then passed to the neural renderer, which models viewpoint-dependent effects. The system also automatically adjusts input camera positions using differentiable rendering. Neural Lumigraph Rendering uses sinusoidal representation networks to produce more compact representations (Kellnhofer, Jebe *et al.* 2021). They can also export a 3D mesh for much faster view-dependent Lumigraph rendering. Takikawa, Litalien *et al.* (2021) also construct an implicit signed distance field, but instead of using a single MLP, they build a sparse octree structure that stores neural features in cells (much like neural sparse voxel fields) and supports both level of detail and fast sphere tracing. Neural Implicit Surfaces (NeuS) also use a signed distance representation but use a rendering formula that better handles surface occlusions (Wang, Liu *et al.* 2021).

While NeRF, IDR, and NSVF require a large number of images of a static object taken under controlled (uniform lighting) conditions, NeRF in the Wild (Martin-Brualla, Radwan *et al.* 2021) takes an unstructured set of images from a landmark tourist location and not only models appearance changes such as weather and time of day but also removes transient occluders such as tourists. NeRFs can also be constructed from a single or small number of images by conditioning a class-specific neural radiance field on such inputs as in pixelNeRF (Yu, Ye *et al.* 2020). Deformable neural radiance fields or "nerfies" (Park, Sinha *et al.* 2020), Neural Scene Flow Fields (Li, Niklaus *et al.* 2021), Dynamic Neural Radiance Fields (Pumarola, Corona *et al.* 2021), Space-time Neural Irradiance Fields (Xian, Huang *et al.* 2021), and HyperNeRF (Park, Sinha *et al.* 2021) all take as input hand-held videos taken around a person or moving through a scene. They model both the viewpoint variation and volumetric non-rigid deformations such as head or body movements and expression changes, either using a learned deformation field, adding time as an extra input variable, or embedding the representation in a higher dimension.

It is also possible to extend NeRFs to model not only the opacities and view-dependent colors of 3D coordinates, but also their interactions with potential illuminants. Neural Reflectance and Visibility Fields (NeRV) do this by also returning for each query 3D coordinate a surface normal and parametric BRDF as well as the environment visibility and expected termination depth for outgoing rays at that point (Srinivasan, Deng *et al.* 2021). Neural Reflection Decomposition (NeRD) models densities and colors using an implicit MLP that also returns an appearance vector, which is decoded into a parametric BRDF (Boss, Braun *et al.* 2020). It then uses the environmental illumination, approximated using spherical Gaussians, along with the density normal and BRDF, to render the final color sample at that voxel. PhySG uses a similar approach, using a signed distance field to represent the shape and a mixture of spherical Gaussian to represent the BRDF (Zhang, Luan *et al.* 2021).

Most of the neural rendering techniques that include view-dependent effects are quite slow to render, since they require sampling a volumetric space along each ray, using expensive MLPs to perform each location/direction lookup. To achieve real-time rendering while modeling view-dependent effects, a number of recent papers use efficient spatial data structures (octrees, sparse grids, or multiplane images) to store opacities and base colors (or potentially small MLPs) and then use factored approximations of the radiance field to model view-dependent effects (Wizadwongsa, Phongthawee *et al.* 2021; Garbin, Kowalski *et al.* 2021; Reiser, Peng *et al.* 2021; Yu, Li *et al.* 2021; Hedman, Srinivasan *et al.* 2021). While the exact details of the representations used in the various stages vary amongst these papers, they all start with high-fidelity view-dependent models related to the original NeRF paper or its extensions and then "bake" or "distill" these into faster to evaluate spatial data structures and simplified (but still accurate) view-dependent models. The resulting systems produce the same high fidelity renderings as full Neural Radiance Fields while running often 1000x faster than pure MLP-based representations.

As you can tell from the brief discussion in this section, neural rendering is an extremely active research area with new architectures being proposed every few months (Dellaert and Yen-Chen 2021). The best place to find the latest developments, as with other topics in computer vision, is to look on arXiv and in the leading computer vision, graphics, and machine learning conferences.

14.7 Additional reading

Two good surveys of image-based rendering are by Kang, Li *et al.* (2006) and Shum, Chan, and Kang (2007), with earlier surveys available from Kang (1999), McMillan and Gortler (1999), and Debevec (1999). Today, the field often goes under the name of *novel view synthesis* (NVS), with a recent tutorial at CVPR (Gallo, Troccoli *et al.* 2020) providing a good overview of historical and current techniques.

The term *image-based rendering* was introduced by McMillan and Bishop (1995), although the seminal paper in the field is the view interpolation paper by Chen and Williams (1993). Debevec, Taylor, and Malik (1996) describe their Façade system, which not only created a variety of image-based modeling tools but also introduced the widely used technique of *view-dependent texture mapping*. Early work on planar impostors and layers was carried out by Shade, Lischinski *et al.* (1996), Lengyel and Snyder (1997), and Torborg and Kajiya (1996), while newer work based on *sprites with depth* is described by Shade, Gortler *et al.* (1998). Using a large number of parallel planes with RGBA colors and opacities (originally dubbed the "stack of acetates" model by Szeliski and Golland (1999)) was rediscovered by Zhou, Tucker *et al.* (2018) and now goes by the name of multiplane images (MPI). This representation is widely used in recent 3D capture and rendering pipelines (Mildenhall, Srinivasan *et al.* 2019; Choi, Gallo *et al.* 2019; Broxton, Flynn *et al.* 2020; Attal, Ling *et al.* 2020; Lin, Xu *et al.* 2020). To accurately model reflections, the alpha-compositing operator used in MPIs needs to be replaced with an additive model, as in Sinha, Kopf *et al.* (2012) and Kopf, Langguth *et al.* (2013).

The two foundational papers in image-based rendering are *Light field rendering* by Levoy and Hanrahan (1996) and *The Lumigraph* by Gortler, Grzeszczuk *et al.* (1996). Buehler, Bosse *et al.* (2001) generalize the Lumigraph approach to irregularly spaced collections of images, while Levoy (2006) provides a survey and more gentle introduction to the topic of light field and image-based rendering. Wu, Masia *et al.* (2017) provide a more recent survey of this topic. More recently, neural rendering techniques have been used to improve the blending heuristics used in the Unstructured Lumigraph (Hedman, Philip *et al.* 2018; Riegler and Koltun 2020a).

Surface light fields (Wood, Azuma *et al.* 2000; Park, Newcombe, and Seitz 2018; Yariv, Kasten *et al.* 2020) provide an alternative parameterization for light fields with accurately known surface geometry and support both better compression and the possibility of editing surface properties. Concentric mosaics (Shum and He 1999; Shum, Wang *et al.* 2002) and panoramas with depth (Peleg, Ben-Ezra, and Pritch 2001; Li, Shum *et al.* 2004; Zheng, Kang *et al.* 2007), provide useful parameterizations for light fields captured with panning cameras. Multi-perspective images (Rademacher and Bishop 1998) and manifold projections (Peleg and Herman 1997), although not true light fields, are also closely related to these ideas.

Among the possible extensions of light fields to higher-dimensional structures, environment mattes (Zongker, Werner *et al.* 1999; Chuang, Zongker *et al.* 2000) are the most useful, especially for placing captured objects into new scenes.

Video-based rendering, i.e., the re-use of video to create new animations or virtual experiences, started with the seminal work of Szummer and Picard (1996), Bregler, Covell, and Slaney (1997), and Schödl, Szeliski *et al.* (2000). Important follow-on work to these basic re-targeting approaches includes Schödl and Essa (2002), Kwatra, Schödl *et al.* (2003), Doretto, Chiuso *et al.* (2003), Wang and Zhu (2003), Zhong and Sclaroff (2003), Yuan, Wen *et al.* (2004), Doretto and Soatto (2006), Zhao and Pietikäinen (2007), Chan and Vasconcelos (2009), Joshi, Mehta *et al.* (2012), Liao, Joshi, and Hoppe (2013), Liao, Finch, and Hoppe (2015), Yan, Liu, and Furukawa (2017), He, Liao *et al.* (2017), and Oh, Joo *et al.* (2017). Related techniques have also been used for performance driven video animation (Zollhöfer, Thies *et al.* 2018; Fried, Tewari *et al.* 2019; Chan, Ginosar *et al.* 2019;

Egger, Smith *et al.* 2020).

Systems that allow users to change their 3D viewpoint based on multiple synchronized video streams include Moezzi, Katkere *et al.* (1996), Kanade, Rander, and Narayanan (1997), Matusik, Buehler *et al.* (2000), Matusik, Buehler, and McMillan (2001), Carranza, Theobalt *et al.* (2003), Zitnick, Kang *et al.* (2004), Magnor (2005), Vedula, Baker, and Kanade (2005), Joo, Liu *et al.* (2015), Anderson, Gallup *et al.* (2016), Tang, Dou *et al.* (2018), Serrano, Kim *et al.* (2019), Parra Pozo, Toksvig *et al.* (2019), Bansal, Vo *et al.* (2020), Broxton, Flynn *et al.* (2020), and Tewari, Fried *et al.* (2020). 3D (multi-view) video coding and compression is also an active area of research (Smolic and Kauff 2005; Gotchev and Rosenhahn 2009), and is used in 3D Blu-Ray discs and multi-view video coding (MVC) extensions to the High Efficientcy Video Coding (HEVC) standard (Tech, Chen *et al.* 2015).

The whole field of neural rendering is quite recent, with initial publications focusing on 2D image synthesis (Zhu, Krähenbühl *et al.* 2016; Isola, Zhu *et al.* 2017) and only more recently being applied to 3D novel view synthesis (Hedman, Philip *et al.* 2018; Martin-Brualla, Pandey *et al.* 2018). Tewari, Fried *et al.* (2020) provide an excellent survey of this area, with 230 references and 46 highlighted papers. Additional overviews include the related CVPR tutorial on neural rendering (Tewari, Zollhöfer *et al.* 2020), several of the lectures in the TUM AI Guest Lecture Series, the X-Fields paper by Bemana, Myszkowski *et al.* (2020, Table 1), and a recent bibliography by Dellaert and Yen-Chen (2021).

14.8 Exercises

Ex 14.1: Depth image rendering. Develop a "view extrapolation" algorithm to re-render a previously computed stereo depth map coupled with its corresponding reference color image.

1. Use a 3D graphics mesh rendering system such as OpenGL with two triangles per pixel quad and perspective (projective) texture mapping (Debevec, Yu, and Borshukov 1998).

2. Alternatively, use the one- or two-pass forward warper you constructed in Exercise 3.24, extended using (2.68–2.70) to convert from disparities or depths into displacements.

3. (Optional) Kinks in straight lines introduced during view interpolation or extrapolation are visually noticeable, which is one reason why image morphing systems let you specify line correspondences (Beier and Neely 1992). Modify your depth estimation algorithm to match and estimate the geometry of straight lines and incorporate it into your image-based rendering algorithm.

Ex 14.2: View interpolation. Extend the system you created in the previous exercise to render two reference views and then blend the images using a combination of z-buffering, hole filing, and blending (morphing) to create the final image (Section 14.1).

1. (Optional) If the two source images have very different exposures, the hole-filled regions and the blended regions will have different exposures. Can you extend your algorithm to mitigate this?

2. (Optional) Extend your algorithm to perform three-way (trilinear) interpolation between neighboring views. You can triangulate the reference camera poses and use barycentric coordinates for the virtual camera to determine the blending weights.

Ex 14.3: View morphing. Modify your view interpolation algorithm to perform morphs between views of a non-rigid object, such as a person changing expressions.

1. Instead of using a pure stereo algorithm, use a general flow algorithm to compute displacements, but separate them into a rigid displacement due to camera motion and a non-rigid deformation.

2. At render time, use the rigid geometry to determine the new pixel location but then add a fraction of the non-rigid displacement as well.

3. (Optional) Take a single image, such as the Mona Lisa or a friend's picture, and create an animated 3D view morph (Seitz and Dyer 1996).

   (a) Find the vertical axis of symmetry in the image and reflect your reference image to provide a virtual pair (assuming the person's hairstyle is somewhat symmetric).

   (b) Use structure from motion to determine the relative camera pose of the pair.

   (c) Use dense stereo matching to estimate the 3D shape.

   (d) Use view morphing to create a 3D animation.

Ex 14.4: View dependent texture mapping. Use a 3D model you created along with the original images to implement a view-dependent texture mapping system.

1. Use one of the 3D reconstruction techniques you developed in Exercises 11.10, 12.9, 12.10, or 13.8 to build a triangulated 3D image-based model from multiple photographs.

2. Extract textures for each model face from your photographs, either by performing the appropriate resampling or by figuring out how to use the texture mapping software to directly access the source images.

3. For each new camera view, select the best source image for each visible model face.

4. Extend this to blend between the top two or three textures. This is trickier, because it involves the use of texture blending or pixel shading (Debevec, Taylor, and Malik 1996; Debevec, Yu, and Borshukov 1998; Pighin, Hecker *et al.* 1998).

Ex 14.5: Layered depth images. Extend your view interpolation algorithm (Exercise 14.2) to store more than one depth or color value per pixel (Shade, Gortler *et al.* 1998), i.e., a layered depth image (LDI). Modify your rendering algorithm accordingly. For your data, you can use synthetic ray tracing, a layered reconstructed model, or a volumetric reconstruction.

Ex 14.6: Rendering from sprites or layers. Extend your view interpolation algorithm to handle multiple planes or sprites (Section 14.2.1) (Shade, Gortler *et al.* 1998).

1. Extract your layers using the technique you developed in Exercise 9.7.

2. Alternatively, use an interactive painting and 3D placement system to extract your layers (Kang 1998; Oh, Chen *et al.* 2001; Shum, Sun *et al.* 2004).

3. Determine a back-to-front order based on expected visibility or add a z-buffer to your rendering algorithm to handle occlusions.

4. Render and composite all of the resulting layers, with optional alpha matting to handle the edges of layers and sprites.

5. Try one of the newer multiplane image (MPI) techniques (Zhou, Tucker *et al.* 2018).

Ex 14.7: Light field transformations. Derive the equations relating regular images to 4D light field coordinates.

1. Determine the mapping between the far plane (u, v) coordinates and a virtual camera's (x, y) coordinates.
   (a) Start by parameterizing a 3D point on the uv plane in terms of its (u, v) coordinates.
   (b) Project the resulting 3D point to the camera pixels $(x, y, 1)$ using the usual 3×4 camera matrix $\mathbf{P}$ (2.63).
   (c) Derive the 2D homography relating (u, v) and (x, y) coordinates.
2. Write down a similar transformation for (s, t) to (x, y) coordinates.
3. Prove that if the virtual camera is actually on the (s, t) plane, the (s, t) value depends only on the camera's image center and is independent of (x, y).
4. Prove that an image taken by a regular orthographic or perspective camera, i.e., one that has a linear projective relationship between 3D points and (x, y) pixels (2.63), samples the (s, t, u, v) light field along a two-dimensional hyperplane.

Ex 14.8: Light field and Lumigraph rendering. Implement a light field or Lumigraph rendering system:

1. Download one of the light field datasets from http://lightfield.stanford.edu or https://lightfield-analysis.uni-konstanz.de.
2. Write an algorithm to synthesize a new view from this light field, using quadri-linear interpolation of (s, t, u, v) ray samples.
3. Try varying the focal plane corresponding to your desired view (Isaksen, McMillan, and Gortler 2000) and see if the resulting image looks sharper.
4. Determine a 3D proxy for the objects in your scene. You can do this by running multi-view stereo over one of your light fields to obtain a depth map per image.
5. Implement the Lumigraph rendering algorithm, which modifies the sampling of rays according to the 3D location of each surface element.
6. Collect a set of images yourself and determine their pose using structure from motion.
7. Implement the unstructured Lumigraph rendering algorithm from Buehler, Bosse *et al.* (2001).

Ex 14.9: Surface light fields. Construct a surface light field (Wood, Azuma *et al.* 2000) and see how well you can compress it.

1. Acquire an interesting light field of a specular scene or object, or download one from http://lightfield.stanford.edu or https://lightfield-analysis.uni-konstanz.de.
2. Build a 3D model of the object using a multi-view stereo algorithm that is robust to outliers due to specularities.
3. Estimate the Lumisphere for each surface point on the object.

4. Estimate its diffuse components. Is the median the best way to do this? Why not use the minimum color value? What happens if there is Lambertian shading on the diffuse component?
5. Model and compress the remaining portion of the Lumisphere using one of the techniques suggested by Wood, Azuma *et al.* (2000) or invent one of your own.
6. Study how well your compression algorithm works and what artifacts it produces.
7. (Optional) Develop a system to edit and manipulate your surface light field.

Ex 14.10: Handheld concentric mosaics. Develop a system to navigate a handheld concentric mosaic.

1. Stand in the middle of a room with a camcorder held at arm's length in front of you and spin in a circle.
2. Use a structure from motion system to determine the camera pose and sparse 3D structure for each input frame.
3. (Optional) Re-bin your image pixels into a more regular concentric mosaic structure.
4. At view time, determine from the new camera's view (which should be near the plane of your original capture) which source pixels to display. You can simplify your computations to determine a source column (and scaling) for each output column.
5. (Optional) Use your sparse 3D structure, interpolated to a dense depth map, to improve your rendering (Zheng, Kang *et al.* 2007).

Ex 14.11: Video textures. Capture some videos of natural phenomena, such as a water fountain, fire, or smiling face, and loop the video seamlessly into an infinite length video (Schödl, Szeliski *et al.* 2000).

1. Compare all the frames in the original clip using an L_2 (sum of square difference) metric. (This assumes the videos were shot on a tripod or have already been stabilized.)
2. Filter the comparison table temporally to accentuate temporal sub-sequences that match well together.
3. Convert your similarity table into a jump probability table through some exponential distribution. Be sure to modify transitions near the end so you do not get "stuck" in the last frame.
4. Starting with the first frame, use your transition table to decide whether to jump forward, backward, or continue to the next frame.
5. (Optional) Add any of the other extensions to the original video textures idea, such as multiple moving regions, interactive control, or graph cut spatio-temporal texture seaming.

Ex 14.12: Neural rendering. Most of the recent neural rendering papers come with open source code as well as carefully acquired datasets.

Try downloading more than one of these and run different algorithms on different datasets. Compare the quality of the renderings you obtain and list the visual artifacts you detect and how you might improve them.

Try capturing your own dataset, if this is feasible, and describe additional breaking points of the current algorithms.

Chapter 15

Conclusion

In this book, we have covered a broad range of computer vision topics. We started with a review of basic geometry and optics, as well as mathematical tools such as image and signal processing, continuous and discrete optimization, statistical modeling, and machine learning. We then used these to develop computer vision algorithms such as image enhancement and segmentation, object detection and classification, motion estimation, and 3D shape reconstruction. These components, in turn, enabled us to build more complex applications, such as large-scale image retrieval, converting images to descriptions, stitching multiple images into wider and higher dynamic range composites, tracking people and objects, navigating in new unseen environments, and augmenting video with embedded 3D overlays.

In the decade since the publication of the first edition of this book, the computer vision field has exploded, both in the maturity and reliability of vision algorithms, as well as the number of practitioners and commercial applications. The most notable advance has been in deep learning, which now enables visual recognition at a level of performance that has eclipsed what we could do ten years ago. Deep learning has also found widespread application in basic vision algorithms such as image enhancement, motion estimation, and 3D shape recovery. Other advances, such as reliable real-time tracking and reconstruction have enabled applications such as autonomous navigation and phone-based augmented reality. And advances in sophisticated image processing have produced computational photography algorithms that run in every mobile phone producing images that surpass the quality available with much more expensive traditional photographic equipment.

You may ask: Why is our field so broad and aren't there any unifying principles that can be used to simplify our study? Part of the answer lies in the expansive definition of computer vision, which is the capture, analysis, and interpretation of our 3D environment using images and video, as well as the incredible complexity inherent in the formation of visual imagery. In some ways, our field is as complex as the study of automotive engineering, which requires an understanding of internal combustion, mechanics, aerodynamics, ergonomics, electrical circuitry, and control systems, among other topics. Computer vision similarly draws on a wide variety of sub-disciplines, which makes it challenging to cover in a one-semester course, or even to achieve mastery during a course of graduate studies. Conversely, the incredible breadth and technical complexity of computer vision is what draws many people to this research field.

Because of this richness and the difficulty in making and measuring progress, I attempt to instill in my students, and hopefully in the readers of this book, a discipline founded on principles from engineering, science, statistics, and machine learning.

The engineering approach to problem solving is to first carefully define the overall problem

R. Szeliski, *Computer Vision*, Texts in Computer Science,
https://doi.org/10.1007/978-3-030-34372-9_15

being tackled and to question the basic assumptions and goals inherent in this process. Once this has been done, a number of alternative solutions or approaches are implemented and carefully tested, paying attention to issues such as reliability and computational cost. Finally, one or more solutions are deployed and evaluated in real-world settings. For this reason, this book contains different alternatives for solving vision problems, many of which are sketched out in the exercises for students to implement and test on their own.

The scientific approach builds upon a basic understanding of physical principles. In the case of computer vision, this includes the physics of natural and artificial structures, image formation, including lighting and atmospheric effects, optics, and noisy sensors. The task is to then invert this formation using stable and efficient algorithms to obtain reliable descriptions of the scene and other quantities of interest. The scientific approach also encourages us to formulate and test hypotheses, which is similar to the extensive testing and evaluation inherent in engineering disciplines.

Because so much about the image formation process is inherently uncertain and ambiguous, a statistical approach, which models both the uncertainty and prior distributions in the world, as well as the degradations in the image formation process, is often essential. Bayesian inference techniques can then be used to combine prior and measurement models to estimate the unknowns and to model their uncertainty. Efficient learning and inference algorithms, such as dynamic programming, graph cuts, and belief propagation, often play a crucial role in this process.

Finally, machine learning techniques, driven by large amounts of training data—both labeled (supervised) and unlabeled (unsupervised)—enable the development of models that can discover hard to describe regularities and patterns in the world, which can make inference more reliable. However, despite the incredible advances enabled by learning techniques, we must still remain cautious about the inherent limitations of learning-based approaches, and not just slough off problems due to "insufficient or biased training data" as someone else's problem.

Along these lines, I was inspired by a segment from Shree Nayar's *First Principles of Computer Vision* online lecture series:[1]

> Since deep learning is very popular today, you may be wondering if it is worth knowing the first principles of vision, or for that matter, the first principles of any field. Given a task, why not just train a neural network with tons of data to solve the task?
>
> Indeed, there are applications where such an approach may suffice. But there are several reasons to embrace the basics. First, it would be laborious and unnecessary to train a network to learn a phenomenon that can be precisely and concisely described using first principles. Second, when a network does not perform well enough, first principles are your only hope for understanding why. Third, collecting data to train a network can be tedious, and sometimes even impractical. In such cases, models based on first principles can be used to synthesize the data, instead of collecting it. And finally, the most compelling reason to learn the first principles of any field is curiosity. What makes humans unique is that innate desire to know why things work the way they do.

Given the breadth of material we have covered in this book, what new developments are we likely to see in the future? It seems fairly obvious from the tremendous advances in the last decade that machine learning, including the ability to fine-tune architectures and algorithm to optimize continuous criteria and metrics, will continue to evolve and produce significant improvements. The current dominance of feedforward convolutional architectures, mostly using weighted linear summation and simple non-linearities, is likely to evolve to include more complex architectures with attention and top-down feedback, as we are already starting to see. Sophisticated application-specific imaging

[1] https://fpcv.cs.columbia.edu, Introduction:Overview video

sensors will likely start being used more often, displacing and enhancing the use of visible light imaging sensors originally developed for photography. Integration with additional sensors, such as IMUs and potentially active sensing (where power permits) will make classic problems such as real-time localization and 3D reconstruction much more reliable and ubiquitous.

The most challenging applications of computer vision will likely remain in the realm of artificial general intelligence (AGI), which aims to create systems that exhibit the same range of understanding and behaviors as people. Since progress here depends on concurrent progress in many other aspects of artificial intelligence, it will be interesting to see how these different AI modalities and capabilities leverage each other for improved performance.

Whatever the outcome of these research endeavors, computer vision is already having a tremendous impact in many areas, including digital photography, visual effects, medical imaging, safety and surveillance, image search, product recommendations, and aids for the visually impaired. The breadth of the problems and techniques inherent in this field, combined with the richness of the mathematics and the utility of the resulting algorithms, will ensure that this remains an exciting area of study for years to come.

Appendix A

Linear algebra and numerical techniques

A.1 Matrix decompositions 728
A.1.1 Singular value decomposition 728
A.1.2 Eigenvalue decomposition 729
A.1.3 QR factorization 731
A.1.4 Cholesky factorization 732
A.2 Linear least squares 733
A.2.1 Total least squares 734
A.3 Non-linear least squares 736
A.4 Direct sparse matrix techniques 737
A.4.1 Variable reordering 737
A.5 Iterative techniques 738
A.5.1 Conjugate gradient 739
A.5.2 Preconditioning 740
A.5.3 Multigrid 741

R. Szeliski, *Computer Vision*, Texts in Computer Science,
https://doi.org/10.1007/978-3-030-34372-9

In this appendix, we introduce some elements of linear algebra and numerical techniques that are used elsewhere in the book. We start with some basic decompositions in matrix algebra, including the singular value decomposition (SVD), eigenvalue decompositions, and other matrix decompositions (factorizations). Next, we look at the problem of linear least squares, which can be solved using either the QR decomposition or normal equations. This is followed by non-linear least squares, which arise when the measurement equations are not linear in the unknowns or when robust error functions are used. Such problems require iteration to find a solution. Next, we look at direct solution (factorization) techniques for sparse problems, where the ordering of the variables may have a large influence on the computation and memory requirements. Finally, we discuss iterative techniques for solving large linear (or linearized) least squares problems. Good general references for much of this material include books by Björck (1996), Golub and Van Loan (1996), Trefethen and Bau (1997), Meyer (2000), Nocedal and Wright (2006), Björck and Dahlquist (2010), and Deisenroth, Faisal, and Ong (2020) and the collection of matrix formulas compiled by (Petersen and Pedersen 2012).

A note on vector and matrix indexing. To be consistent with the rest of the book and with the general usage in the computer science and computer vision communities, I adopt a 0-based indexing scheme for vector and matrix element indexing. Please note that most mathematical textbooks and papers use 1-based indexing, so you need to be aware of the differences when you read this book.

A.1 Matrix decompositions

To better understand the structure of matrices and more stably perform operations such as inversion and system solving, a number of decompositions (or factorizations) can be used. In this section, we review singular value decomposition (SVD), eigenvalue decomposition, QR factorization, and Cholesky factorization.

A.1.1 Singular value decomposition

One of the most useful decompositions in matrix algebra is the *singular value decomposition* (SVD), which states that any real-valued $m \times n$ matrix $\mathbf{A}$ can be written as

$$\begin{aligned} \mathbf{A}_{m\times n} &= \mathbf{U}_{m\times p}\mathbf{\Sigma}_{p\times p}\mathbf{V}^T_{p\times n} \qquad \text{(A.1)} \\ &= \begin{bmatrix} \\ \mathbf{u}_0 & \cdots & \mathbf{u}_{p-1} \\ \\ \end{bmatrix} \begin{bmatrix} \sigma_0 & & \\ & \ddots & \\ & & \sigma_{p-1} \end{bmatrix} \begin{bmatrix} \mathbf{v}_0^T \\ \cdots \\ \mathbf{v}_{p-1}^T \end{bmatrix}, \end{aligned}$$

where $p = \min(m, n)$. The matrices $\mathbf{U}$ and $\mathbf{V}$ are orthonormal, i.e., $\mathbf{U}^T\mathbf{U} = \mathbf{I}$ and $\mathbf{V}^T\mathbf{V} = \mathbf{I}$, and so are their column vectors,

$$\mathbf{u}_i \cdot \mathbf{u}_j = \mathbf{v}_i \cdot \mathbf{v}_j = \delta_{ij}. \qquad \text{(A.2)}$$

The singular values are all non-negative and can be ordered in decreasing order

$$\sigma_0 \geq \sigma_1 \geq \cdots \geq \sigma_{p-1} \geq 0. \qquad \text{(A.3)}$$

A geometric intuition for the SVD of a matrix $\mathbf{A}$ can be obtained by re-writing $\mathbf{A} = \mathbf{U\Sigma V}^T$ in (A.1) as

$$\mathbf{AV} = \mathbf{U\Sigma} \qquad \text{or} \qquad \mathbf{A}\mathbf{v}_j = \sigma_j\mathbf{u}_j. \qquad \text{(A.4)}$$

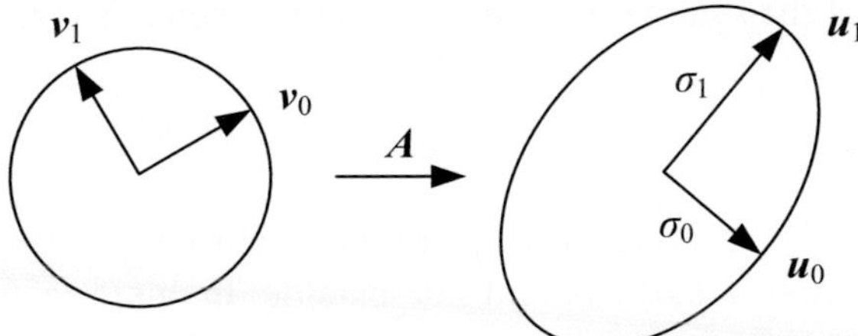

Figure A.1 The action of a matrix $\mathbf{A}$ can be visualized by thinking of the domain as being spanned by a set of orthonormal vectors $\mathbf{v}_j$, each of which is transformed to a new orthogonal vector $\mathbf{u}_j$ with a length σ_j. When $\mathbf{A}$ is interpreted as a covariance matrix and its eigenvalue decomposition is performed, each of the $\mathbf{u}_j$ axes denote a principal direction (component) and each σ_j denotes one standard deviation along that direction.

This formula says that the matrix $\mathbf{A}$ takes any basis vector $\mathbf{v}_j$ and maps it to a direction $\mathbf{u}_j$ with length σ_j, as shown in Figure A.1

If only the first r singular values are positive, the matrix $\mathbf{A}$ is of *rank* r and the index p in the SVD decomposition (A.1) can be replaced by r. (In other words, we can drop the last $p-r$ columns of $\mathbf{U}$ and $\mathbf{V}$.)

An important property of the singular value decomposition of a matrix (also true for the eigenvalue decomposition of a real symmetric non-negative definite matrix) is that if we truncate the expansion

$$\mathbf{A} = \sum_{j=0}^{t} \sigma_j \mathbf{u}_j \mathbf{v}_j^T, \tag{A.5}$$

we obtain the best possible least squares approximation to the original matrix $\mathbf{A}$. This is used both in eigenface-based face recognition systems (Section 5.2.3) and in the separable approximation of convolution kernels (3.21).

A.1.2 Eigenvalue decomposition

If the matrix $\mathbf{C}$ is symmetric ($m = n$),[1] it can be written as an eigenvalue decomposition,

$$\mathbf{C} = \mathbf{U}\boldsymbol{\Lambda}\mathbf{U}^T = \begin{bmatrix} \mathbf{u}_0 & \cdots & \mathbf{u}_{n-1} \end{bmatrix} \begin{bmatrix} \lambda_0 & & \\ & \ddots & \\ & & \lambda_{n-1} \end{bmatrix} \begin{bmatrix} \mathbf{u}_0^T \\ \cdots \\ \mathbf{u}_{n-1}^T \end{bmatrix}$$

$$= \sum_{i=0}^{n-1} \lambda_i \mathbf{u}_i \mathbf{u}_i^T. \tag{A.6}$$

(The eigenvector matrix $\mathbf{U}$ is sometimes written as $\boldsymbol{\Phi}$ and the eigenvectors $\mathbf{u}$ as ϕ.) In this case, the eigenvalues

$$\lambda_0 \geq \lambda_1 \geq \cdots \geq \lambda_{n-1} \tag{A.7}$$

can be both positive and negative.[2]

[1] In this appendix, we denote symmetric matrices using $\mathbf{C}$ and general rectangular matrices using $\mathbf{A}$.

[2] Eigenvalue decompositions can be computed for non-symmetric matrices, but the eigenvalues and eigenvectors can have complex entries in that case.

A special case of the symmetric matrix $\mathbf{C}$ occurs when it is constructed as the sum of a number of outer products

$$\mathbf{C} = \sum_i \mathbf{a}_i \mathbf{a}_i^T = \mathbf{A}\mathbf{A}^T, \tag{A.8}$$

which often occurs when solving least squares problems (Appendix A.2), where the matrix $\mathbf{A}$ consists of all the $\mathbf{a}_i$ column vectors stacked side-by-side. In this case, we are guaranteed that all of the eigenvalues λ_i are non-negative. The associated matrix $\mathbf{C}$ is *positive semi-definite*

$$\mathbf{x}^T \mathbf{C} \mathbf{x} \geq 0, \quad \forall \mathbf{x}. \tag{A.9}$$

If the matrix $\mathbf{C}$ is of full rank, the eigenvalues are all positive and the matrix is called *symmetric positive definite* (SPD).

Symmetric positive semi-definite matrices also arise in the statistical analysis of data, as they represent the *covariance* of a set of $\{\mathbf{x}_i\}$ points around their mean $\bar{\mathbf{x}}$,

$$\mathbf{C} = \frac{1}{n} \sum_i (\mathbf{x}_i - \bar{\mathbf{x}})(\mathbf{x}_i - \bar{\mathbf{x}})^T. \tag{A.10}$$

In this case, performing the eigenvalue decomposition is known as *principal component analysis* (PCA), because it models the principal directions (and magnitudes) of variation of the point distribution around their mean, as shown in Section 7.3.1, Section 5.2.3 (5.41), and Appendix B.1 (B.10). Figure A.1 shows how the principal components of the covariance matrix $\mathbf{C}$ denote the principal axes $\mathbf{u}_j$ of the uncertainty ellipsoid corresponding to this point distribution and how the $\sigma_j = \sqrt{\lambda_j}$ denote the standard deviations along each axis.

The eigenvalues and eigenvectors of $\mathbf{C}$ and the singular values and singular vectors of $\mathbf{A}$ are closely related. Given

$$\mathbf{A} = \mathbf{U}\boldsymbol{\Sigma}\mathbf{V}^T, \tag{A.11}$$

we get

$$\mathbf{C} = \mathbf{A}\mathbf{A}^T = \mathbf{U}\boldsymbol{\Sigma}\mathbf{V}^T\mathbf{V}\boldsymbol{\Sigma}\mathbf{U}^T = \mathbf{U}\boldsymbol{\Lambda}\mathbf{U}^T. \tag{A.12}$$

From this, we see that $\lambda_i = \sigma_i^2$ and that the left singular vectors of $\mathbf{A}$ are the eigenvectors of $\mathbf{C}$.

This relationship gives us an efficient method for computing the eigenvalue decomposition of large matrices that are rank deficient, such as the scatter matrices observed in computing eigenfaces (Section 5.2.3). Observe that the covariance matrix $\mathbf{C}$ in (5.41) is exactly the same as $\mathbf{C}$ in (A.8). Note also that the individual difference-from-mean images $\mathbf{a}_i = \mathbf{x}_i - \bar{\mathbf{x}}$ are long vectors of length P (the number of pixels in the image), while the total number of exemplars N (the number of faces in the training database) is much smaller. Instead of forming $\mathbf{C} = \mathbf{A}\mathbf{A}^T$, which is $P \times P$, we form the matrix

$$\hat{\mathbf{C}} = \mathbf{A}^T\mathbf{A}, \tag{A.13}$$

which is $N \times N$. (This involves taking the dot product between every pair of difference images $\mathbf{a}_i$ and $\mathbf{a}_j$.) The eigenvalues of $\hat{\mathbf{C}}$ are the squared singular values of $\mathbf{A}$, namely $\boldsymbol{\Sigma}^2$, and are hence also the eigenvalues of $\mathbf{C}$. The eigenvectors of $\hat{\mathbf{C}}$ are the right singular vectors $\mathbf{V}$ of $\mathbf{A}$, from which the desired eigenfaces $\mathbf{U}$, which are the left singular vectors of $\mathbf{A}$, can be computed as

$$\mathbf{U} = \mathbf{A}\mathbf{V}\boldsymbol{\Sigma}^{-1}. \tag{A.14}$$

This final step is essentially computing the eigenfaces as linear combinations of the difference images (Turk and Pentland 1991). If you have access to a high-quality linear algebra package such as LAPACK, routines for efficiently computing a small number of the left singular vectors and singular values of rectangular matrices such as $\mathbf{A}$ are usually provided (Appendix C.2). However, if storing all of the images in memory is prohibitive, the construction of $\hat{\mathbf{C}}$ in (A.13) can be used instead.

How can eigenvalue and singular value decompositions actually be computed? Notice that an eigenvector is defined by the equation

$$\lambda_i \mathbf{u}_i = \mathbf{C}\mathbf{u}_i \quad \text{or} \quad (\lambda_i \mathbf{I} - \mathbf{C})\mathbf{u}_i = 0. \tag{A.15}$$

(This can be derived from (A.6) by post-multiplying both sides by $\mathbf{u}_i$.) Because the latter equation is *homogeneous*, i.e., it has a zero right-hand-side, it can only have a non-zero (non-trivial) solution for $\mathbf{u}_i$ if the system is rank deficient, i.e.,

$$|(\lambda \mathbf{I} - \mathbf{C})| = 0. \tag{A.16}$$

Evaluating this determinant yields a *characteristic* polynomial equation in λ, which can be solved for small problems, e.g., 2×2 or 3×3 matrices, in closed form.

For larger matrices, iterative algorithms that first reduce the matrix $\mathbf{C}$ to a real symmetric tridiagonal form using orthogonal transforms and then perform QR iterations are normally used (Golub and Van Loan 1996; Trefethen and Bau 1997; Björck and Dahlquist 2010). As these techniques are rather involved, it is best to use a linear algebra package such as LAPACK (Anderson, Bai *et al.* 1999)—see Appendix C.2.

Factorization with missing data requires different kinds of iterative algorithms, which often involve either hallucinating the missing terms or minimizing some weighted reconstruction metric, which is intrinsically much more challenging than regular factorization. This area has been widely studied in computer vision (Shum, Ikeuchi, and Reddy 1995; De la Torre and Black 2003; Huynh, Hartley, and Heyden 2003; Buchanan and Fitzgibbon 2005; Gross, Matthews, and Baker 2006; Torresani, Hertzmann, and Bregler 2008) and is sometimes called *generalized PCA*. However, this term is also sometimes used to denote algebraic subspace clustering techniques, which is the subject of the monograph by Vidal, Ma, and Sastry (2016).

A.1.3 QR factorization

A widely used technique for stably solving poorly conditioned least squares problems (Björck 1996), and the basis of more complex algorithms, such as computing the SVD and eigenvalue decompositions, is the QR factorization,

$$\mathbf{A} = \mathbf{Q}\mathbf{R}, \tag{A.17}$$

where $\mathbf{Q}$ is an *orthonormal* (or *unitary*) matrix $\mathbf{Q}\mathbf{Q}^T = \mathbf{I}$ and $\mathbf{R}$ is upper triangular.[3] In computer vision, QR can be used to convert a camera matrix into a rotation matrix and an upper-triangular calibration matrix (11.13) and also in various self-calibration algorithms (Section 11.3.4). The most common algorithms for computing QR decompositions (modified Gram–Schmidt, Householder transformations, and Givens rotations) are described by Golub and Van Loan (1996), Trefethen and Bau (1997), and Björck and Dahlquist (2010) and are also found in LAPACK. Unlike the SVD and eigenvalue decompositions, QR factorization does not require iteration and can be computed exactly in $O(MN^2 + N^3)$ operations, where M is the number of rows and N is the number of columns (for a tall matrix).

[3]The term "R" comes from the German name for the lower–upper (LU) decomposition, which is LR for "links" and "rechts" (left and right of the diagonal).

procedure *Cholesky*$(\mathbf{C}, \mathbf{R})$:

$\mathbf{R} = \mathbf{C}$

for $i = 0 \ldots n-1$

for $j = i+1 \ldots n-1$

$\mathbf{R}_{j,j:n-1} = \mathbf{R}_{j,j:n-1} - r_{ij} r_{ii}^{-1} \mathbf{R}_{i,j:n-1}$

$\mathbf{R}_{i,i:n-1} = r_{ii}^{-1/2} \mathbf{R}_{i,i:n-1}$

Algorithm A.1 Cholesky decomposition of the matrix $\mathbf{C}$ into its upper triangular form $\mathbf{R}$.

A.1.4 Cholesky factorization

Cholesky factorization can be applied to any symmetric positive definite matrix $\mathbf{C}$ to convert it into a product of symmetric lower and upper triangular matrices,

$$\mathbf{C} = \mathbf{L}\mathbf{L}^T = \mathbf{R}^T\mathbf{R}, \tag{A.18}$$

where $\mathbf{L}$ is a lower-triangular matrix and $\mathbf{R}$ is an upper-triangular matrix. Unlike Gaussian elimination, which may require pivoting (row and column reordering) or may become unstable (sensitive to roundoff errors or reordering), Cholesky factorization remains stable for positive definite matrices, such as those that arise from normal equations in least squares problems (Appendix A.2). Because of the form of (A.18), the matrices $\mathbf{L}$ and $\mathbf{R}$ are sometimes called *matrix square roots*.[4]

The algorithm to compute an upper triangular Cholesky decomposition of $\mathbf{C}$ is a straightforward symmetric generalization of Gaussian elimination and is based on the decomposition (Björck 1996; Golub and Van Loan 1996)

$$\mathbf{C} = \begin{bmatrix} \gamma & \mathbf{c}^T \\ \mathbf{c} & \mathbf{C}_{11} \end{bmatrix} \tag{A.19}$$

$$= \begin{bmatrix} \gamma^{1/2} & \mathbf{0}^T \\ \mathbf{c}\gamma^{-1/2} & \mathbf{I} \end{bmatrix} \begin{bmatrix} 1 & \mathbf{0}^T \\ \mathbf{0} & \mathbf{C}_{11} - \mathbf{c}\gamma^{-1}\mathbf{c}^T \end{bmatrix} \begin{bmatrix} \gamma^{1/2} & \gamma^{-1/2}\mathbf{c}^T \\ \mathbf{0} & \mathbf{I} \end{bmatrix} \tag{A.20}$$

$$= \mathbf{R}_0^T \mathbf{C}_1 \mathbf{R}_0, \tag{A.21}$$

which, through recursion, can be turned into

$$\mathbf{C} = \mathbf{R}_0^T \ldots \mathbf{R}_{n-1}^T \mathbf{R}_{n-1} \ldots \mathbf{R}_0 = \mathbf{R}^T\mathbf{R}. \tag{A.22}$$

Algorithm A.1 provides a more procedural definition, which can store the upper-triangular matrix $\mathbf{R}$ in the same space as $\mathbf{C}$, if desired. The total operation count for Cholesky factorization is $O(N^3)$ for a dense matrix but can be significantly lower for sparse matrices with low fill-in (Appendix A.4).

Note that Cholesky decomposition can also be applied to block-structured matrices, where the term γ in (A.19) is now a square block sub-matrix and $\mathbf{c}$ is a rectangular matrix (Golub and Van Loan 1996). The computation of square roots can be avoided by leaving the γ on the diagonal of the middle factor in (A.20), which results in the $\mathbf{C} = \mathbf{L}\mathbf{D}\mathbf{L}^T$ factorization, where $\mathbf{D}$ is a diagonal matrix. However, as square roots are relatively fast on modern computers, this is not worth the bother and Cholesky factorization is usually preferred.

[4] In fact, there exists a whole family of matrix square roots. Any matrix of the form $\mathbf{L}\mathbf{Q}$ or $\mathbf{Q}\mathbf{R}$, where $\mathbf{Q}$ is a unitary matrix, is a square root of $\mathbf{C}$.

A.2 Linear least squares

Least squares fitting problems are pervasive in computer vision. For example, the alignment of images based on matching feature points involves the minimization of a squared distance objective function (8.2),

$$E_{\mathrm{LS}} = \sum_i \|\mathbf{r}_i\|^2 = \sum_i \|\mathbf{f}(\mathbf{x}_i;\mathbf{p}) - \mathbf{x}'_i\|^2, \tag{A.23}$$

where

$$\mathbf{r}_i = \mathbf{x}'_i - \mathbf{f}(\mathbf{x}_i;\mathbf{p}) = \hat{\mathbf{x}}'_i - \tilde{\mathbf{x}}'_i \tag{A.24}$$

is the *residual* between the measured location $\hat{\mathbf{x}}'_i$ and its corresponding current *predicted* location $\tilde{\mathbf{x}}'_i = \mathbf{f}(\mathbf{x}_i;\mathbf{p})$. More complex versions of least squares problems, such as large-scale structure from motion (Section 11.4.2), may involve the minimization of functions of thousands of variables. Even problems such as image filtering (Section 3.4.1) and regularization (Section 4.2) may involve the minimization of sums of squared errors.

Figure A.2a shows an example of a simple least squares line fitting problem, where the quantities being estimated are the line equation parameters (m, b). When the sampled vertical values y_i are assumed to be noisy versions of points on the line $y = mx + b$, the optimal estimates for (m, b) can be found by minimizing the squared vertical residuals

$$E_{\mathrm{VLS}} = \sum_i |y_i - (mx_i + b)|^2. \tag{A.25}$$

Note that the function being fitted need not itself be linear to use linear least squares. All that is required is that the function be linear in the unknown parameters. For example, polynomial fitting can be written as

$$E_{\mathrm{PLS}} = \sum_i |y_i - (\sum_{j=0}^{p} a_j x_i^j)|^2, \tag{A.26}$$

while sinusoid fitting with unknown amplitude A and phase ϕ (but known frequency f) can be written as

$$E_{\mathrm{SLS}} = \sum_i |y_i - A\sin(2\pi f x_i + \phi)|^2 = \sum_i |y_i - (B\sin 2\pi f x_i + C\cos 2\pi f x_i)|^2, \tag{A.27}$$

which is linear in (B, C).

In general, it is more common to denote the unknown parameters using $\mathbf{x}$ and to write the general form of linear least squares as[5]

$$E_{\mathrm{LLS}} = \sum_i |\mathbf{a}_i\mathbf{x} - b_i|^2 = \|\mathbf{A}\mathbf{x} - \mathbf{b}\|^2. \tag{A.28}$$

Expanding the above equation gives us

$$E_{\mathrm{LLS}} = \mathbf{x}^T(\mathbf{A}^T\mathbf{A})\mathbf{x} - 2\mathbf{x}^T(\mathbf{A}^T\mathbf{b}) + \|\mathbf{b}\|^2, \tag{A.29}$$

whose minimum value for $\mathbf{x}$ can be found by solving the associated *normal equations* (Björck 1996; Golub and Van Loan 1996)

$$(\mathbf{A}^T\mathbf{A})\mathbf{x} = \mathbf{A}^T\mathbf{b}. \tag{A.30}$$

[5] Be extra careful in interpreting the variable names here. In the 2D line-fitting example, x is used to denote the horizontal axis, but in the general least squares problem, $\mathbf{x} = (m, b)$ denotes the unknown parameter vector.

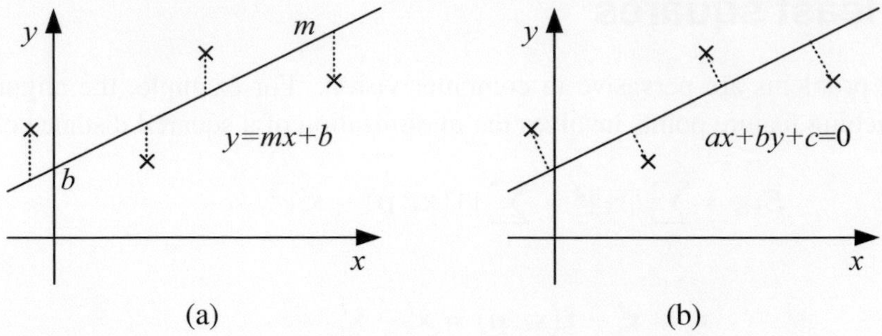

Figure A.2 Least squares regression. (a) The line $y = mx + b$ is fitted to the four noisy data points, $\{(x_i, y_i)\}$, denoted by $\times$, by minimizing the squared vertical residuals between the data points and the line, $\sum_i \|y_i - (mx_i + b)\|^2$. (b) When the measurements $\{(x_i, y_i)\}$ are assumed to have noise in all directions, the sum of orthogonal squared distances to the line $\sum_i \|ax_i + by_i + c\|^2$ is minimized using total least squares.

The preferred way to solve the normal equations is to use Cholesky factorization. Let

$$\mathbf{C} = \mathbf{A}^T\mathbf{A} = \mathbf{R}^T\mathbf{R}, \tag{A.31}$$

where $\mathbf{R}$ is the upper-triangular Cholesky factor of the Hessian $\mathbf{C}$, and

$$\mathbf{d} = \mathbf{A}^T\mathbf{b}. \tag{A.32}$$

After factorization, the solution for $\mathbf{x}$ can be obtained as

$$\mathbf{R}^T\mathbf{z} = \mathbf{d}, \qquad \mathbf{R}\mathbf{x} = \mathbf{z}, \tag{A.33}$$

which involves the solution of two triangular systems, i.e., forward and backward substitution (Björck 1996).

In cases where the least squares problem is numerically poorly conditioned (which should generally be avoided by adding sufficient regularization or prior knowledge about the parameters (Appendix A.3)), it is possible to use QR factorization or SVD directly on the matrix $\mathbf{A}$ (Björck 1996; Golub and Van Loan 1996; Trefethen and Bau 1997; Nocedal and Wright 2006; Björck and Dahlquist 2010), e.g.,

$$\mathbf{A}\mathbf{x} = \mathbf{Q}\mathbf{R}\mathbf{x} = \mathbf{b} \qquad \longrightarrow \qquad \mathbf{R}\mathbf{x} = \mathbf{Q}^T\mathbf{b}. \tag{A.34}$$

Note that the upper triangular matrices $\mathbf{R}$ produced by the Cholesky factorization of $\mathbf{C} = \mathbf{A}^T\mathbf{A}$ and the QR factorization of $\mathbf{A}$ are the same, but that solving (A.34) is generally more stable (less sensitive to roundoff error) but slower (by a constant factor).

A.2.1 Total least squares

In some problems, e.g., when performing geometric line fitting in 2D images or 3D plane fitting to point cloud data, instead of having measurement error along one particular axis, the measured points have uncertainty in all directions, which is known as the *errors-in-variables* model (Van Huffel and Lemmerling 2002; Matei and Meer 2006). In this case, it makes more sense to minimize a set of homogeneous squared errors of the form

$$E_{\text{TLS}} = \sum_i (\mathbf{a}_i\mathbf{x})^2 = \|\mathbf{A}\mathbf{x}\|^2, \tag{A.35}$$

which is known as *total least squares* (TLS) (Van Huffel and Vandewalle 1991; Björck 1996; Golub and Van Loan 1996; Van Huffel and Lemmerling 2002).

The above error metric has a trivial minimum solution at $\mathbf{x} = 0$ and is, in fact, homogeneous in $\mathbf{x}$. For this reason, we augment this minimization problem with the requirement that $\|\mathbf{x}\|^2 = 1$. which results in the eigenvalue problem

$$\mathbf{x} = \arg\min_{\mathbf{x}} \mathbf{x}^T(\mathbf{A}^T\mathbf{A})\mathbf{x} \qquad \text{such that} \qquad \|\mathbf{x}\|^2 = 1. \tag{A.36}$$

The value of $\mathbf{x}$ that minimizes this constrained problem is the eigenvector associated with the smallest eigenvalue of $\mathbf{A}^T\mathbf{A}$. This is the same as the last right singular vector of $\mathbf{A}$, because

$$\mathbf{A} = \mathbf{U}\boldsymbol{\Sigma}\mathbf{V}^T, \tag{A.37}$$

$$\mathbf{A}^T\mathbf{A} = \mathbf{V}\boldsymbol{\Sigma}^2\mathbf{V}^T, \tag{A.38}$$

$$\mathbf{A}^T\mathbf{A}\mathbf{v}_k = \sigma_k^2\mathbf{v}_k, \tag{A.39}$$

which is minimized by selecting the smallest σ_k value.

Figure A.2b shows a line-fitting problem where, in this case, the measurement errors are assumed to be isotropic in (x, y). The solution for the best line equation $ax + by + c = 0$ is found by minimizing

$$E_{\text{TLS-2D}} = \sum_i (ax_i + by_i + c)^2, \tag{A.40}$$

i.e., finding the eigenvector associated with the smallest eigenvalue of[6]

$$\mathbf{C} = \mathbf{A}^T\mathbf{A} = \sum_i \begin{bmatrix} x_i \\ y_i \\ 1 \end{bmatrix} \begin{bmatrix} x_i & y_i & 1 \end{bmatrix}. \tag{A.41}$$

Notice, however, that minimizing $\sum_i(\mathbf{a}_i\mathbf{x})^2$ in (A.35) is only statistically optimal (Appendix B.1) if all of the measured terms in the $\mathbf{a}_i$, e.g., the $(x_i, y_i, 1)$ measurements, have equal noise. This is definitely not the case in the line-fitting example of Figure A.2b (A.40), as the 1 values are noise-free. To mitigate this, we first subtract the mean x and y values from all the measured points

$$\hat{x}_i = x_i - \bar{x} \tag{A.42}$$

$$\hat{y}_i = y_i - \bar{y} \tag{A.43}$$

and then fit the 2D line equation $a(x - \bar{x}) + b(y - \bar{y}) = 0$ by minimizing

$$E_{\text{TLS-2Dm}} = \sum_i (a\hat{x}_i + b\hat{y}_i)^2. \tag{A.44}$$

The more general case where each individual measurement component can have different noise level, as is the case in estimating essential and fundamental matrices (Section 11.3), is called the *heteroscedastic* errors-in-variable (HEIV) model and is discussed by Matei and Meer (2006).

[6] Again, be careful with the variable names here. The measurement equation is $\mathbf{a}_i = (x_i, y_i, 1)$ and the unknown parameters are $\mathbf{x} = (a, b, c)$.

A.3 Non-linear least squares

In many vision problems, such as structure from motion, the least squares problem formulated in (A.23) involves functions $\mathbf{f}(\mathbf{x}_i;\mathbf{p})$ that are *not* linear in the unknown parameters $\mathbf{p}$. This problem is known as *non-linear least squares* or *non-linear regression* (Björck 1996; Madsen, Nielsen, and Tingleff 2004; Nocedal and Wright 2006). It is usually solved by iteratively re-linearizing (A.23) around the current estimate of $\mathbf{p}$ using the gradient derivative (Jacobian) $\mathbf{J} = \partial\mathbf{f}/\partial\mathbf{p}$ and computing an incremental improvement $\Delta\mathbf{p}$.

As shown in Equations (8.13–8.17), this results in

$$E_{\text{NLS}}(\Delta\mathbf{p}) = \sum_i \|\mathbf{f}(\mathbf{x}_i;\mathbf{p}+\Delta\mathbf{p}) - \mathbf{x}'_i\|^2 \tag{A.45}$$

$$\approx \sum_i \|\mathbf{J}(\mathbf{x}_i;\mathbf{p})\Delta\mathbf{p} - \mathbf{r}_i\|^2, \tag{A.46}$$

where the Jacobians $\mathbf{J}(\mathbf{x}_i;\mathbf{p})$ and residual vectors $\mathbf{r}_i$ play the same role in forming the normal equations as $\mathbf{a}_i$ and b_i in (A.28).

Because the above approximation only holds near a local minimum or for small values of $\Delta\mathbf{p}$, the update $\mathbf{p} \leftarrow \mathbf{p} + \Delta\mathbf{p}$ may not always decrease the summed square residual error (A.45). One way to mitigate this problem is to take a smaller step,

$$\mathbf{p} \leftarrow \mathbf{p} + \alpha\Delta\mathbf{p}, \qquad 0 < \alpha \leq 1. \tag{A.47}$$

A simple way to determine a reasonable value of α is to start with 1 and successively halve the value, which is a simple form of *line search* (Al-Baali and Fletcher 1986; Björck 1996; Nocedal and Wright 2006).

Another approach to ensuring a downhill step in error is to add a diagonal damping term to the approximate Hessian

$$\mathbf{C} = \sum_i \mathbf{J}^T(\mathbf{x}_i)\mathbf{J}(\mathbf{x}_i), \tag{A.48}$$

i.e., to solve

$$[\mathbf{C} + \lambda\,\text{diag}(\mathbf{C})]\Delta\mathbf{p} = \mathbf{d}, \tag{A.49}$$

where

$$\mathbf{d} = \sum_i \mathbf{J}^T(\mathbf{x}_i)\mathbf{r}_i, \tag{A.50}$$

which is called a *damped Gauss–Newton* method. The damping parameter λ is increased if the squared residual is not decreasing as fast as expected, i.e., as predicted by (A.46), and is decreased if the expected decrease is obtained (Madsen, Nielsen, and Tingleff 2004). The combination of the Newton (first-order Taylor series) approximation (A.46) and the adaptive damping parameter λ is commonly known as the Levenberg–Marquardt algorithm (Levenberg 1944; Marquardt 1963) and is an example of more general *trust region methods*, which are discussed in more detail in Björck (1996), Conn, Gould, and Toint (2000), Madsen, Nielsen, and Tingleff (2004), and Nocedal and Wright (2006).

When the initial solution is far away from its quadratic region of convergence around a local minimum, *large residual methods*, e.g., *Newton-type methods*, which add a second-order term to the Taylor series expansion in (A.46), may converge faster. Quasi-Newton methods such as BFGS, which require only gradient evaluations, can also be useful if memory size is an issue. Such techniques are discussed in textbooks and papers on numerical optimization (Toint 1987; Björck 1996; Conn, Gould, and Toint 2000; Nocedal and Wright 2006).

A.4 Direct sparse matrix techniques

Many optimization problems in computer vision, such as bundle adjustment (Szeliski and Kang 1994; Triggs, McLauchlan *et al.* 1999; Hartley and Zisserman 2004; Snavely, Seitz, and Szeliski 2008b; Agarwal, Snavely *et al.* 2009) have Jacobian and (approximate) Hessian matrices that are extremely sparse (Section 11.4.3). For example, Figure 11.16a shows the *bipartite* model typical of structure from motion problems, in which most points are only observed by a subset of the cameras, which results in the sparsity patterns for the Jacobian and Hessian shown in Figure 11.16b–c.

Whenever the Hessian matrix is sparse enough, it is more efficient to use sparse Cholesky factorization instead of regular Cholesky factorization. In such sparse direct techniques, the Hessian matrix $\mathbf{C}$ and its associated Cholesky factor $\mathbf{R}$ are stored in *compressed form*, in which the amount of storage is proportional to the number of (potentially) non-zero entries (Björck 1996; Davis 2006).[7] Algorithms for computing the non zero elements in $\mathbf{C}$ and $\mathbf{R}$ from the sparsity pattern of the Jacobian matrix $\mathbf{J}$ are given by Björck (1996, Section 6.4), and algorithms for computing the numerical Cholesky and QR decompositions (once the sparsity pattern has been computed and storage allocated) are discussed by Björck (1996, Section 6.5). More recent publications on direct sparse techniques which discuss supernodal and multifrontal algorithms for large sparse systems include Davis (2006) and Davis, Rajamanickam, and Sid-Lakhdar (2016).

A.4.1 Variable reordering

The key to efficiently solving sparse problems using direct (non-iterative) techniques is to determine an efficient *ordering* for the variables, which reduces the amount of *fill-in*, i.e., the number of non-zero entries in $\mathbf{R}$ that were zero in the original $\mathbf{C}$ matrix. We have already seen in Section 11.4.3 how storing the more numerous 3D point parameters before the camera parameters and using the Schur complement (11.68) results in a more efficient algorithm. Similarly, sorting parameters by time in video-based reconstruction problems usually results in lower fill-in. Furthermore, any problem whose adjacency graph (the graph corresponding to the sparsity pattern) is a tree can be solved in linear time with an appropriate reordering of the variables (putting all the children before their parents). All of these are examples of good reordering techniques.

In the general case of unstructured data, there are many heuristics available to find good reorderings (Björck 1996; Davis 2006).[8] For general adjacency (sparsity) graphs, *minimum degree orderings* generally produce good results. For planar graphs, which often arise on image or spline grids (Section 9.2.2), *nested dissection*, which recursively splits the graph into two equal halves along a *frontier* (or boundary) of small size, generally works well. Such *domain decomposition* (or *multi-frontal*) techniques also enable the use of parallel processing, as independent sub-graphs can be processed in parallel on separate processors (Davis 2011).

The overall set of steps used to perform the direct solution of sparse least squares problems is summarized in Algorithm A.2, which is a modified version of Algorithm 6.6.1 by Björck (1996, Section 6.6)). If a series of related least squares problems is being solved, as is the case in iterative non-linear least squares (Appendix A.3), steps 1–3 can be performed ahead of time and reused for each new invocation with different $\mathbf{C}$ and $\mathbf{d}$ values. When the problem is block-structured, as is the case in structure from motion where point (structure) variables have dense 3×3 sub-entries in

[7] For example, you can store a list of (i, j, c_{ij}) triples. One example of such a scheme is *compressed sparse row (CSR)* storage. An alternative storage method called *skyline*, which stores adjacent vertical spans of non-zero elements (Bathe 2007), is sometimes used in finite element analysis. Banded systems such as snakes (7.27) can store just the non-zero band elements (Björck 1996, Section 6.2) and can be solved in $O(nb^2)$, where n is the number of variables and b is the bandwidth.

[8] Finding the optimal reordering with minimal fill-in is provably NP-hard.

procedure *SparseCholeskySolve*$(\mathbf{C}, \mathbf{d})$:

1. Determine symbolically the structure of $\mathbf{C}$, i.e., the adjacency graph.
2. (Optional) Compute a reordering for the variables, taking into account any block structure inherent in the problem.
3. Determine the fill-in pattern for $\mathbf{R}$ and allocate the compressed storage for $\mathbf{R}$ as well as storage for the permuted right-hand side $\hat{\mathbf{d}}$.
4. Copy the elements of $\mathbf{C}$ and $\mathbf{d}$ into $\mathbf{R}$ and $\hat{\mathbf{d}}$, permuting the values according to the computed ordering.
5. Perform the numerical factorization of $\mathbf{R}$ using Algorithm A.1.
6. Solve the factored system (A.33), i.e.,

$$\mathbf{R}^T\mathbf{z} = \hat{\mathbf{d}}, \qquad \mathbf{R}\mathbf{x} = \mathbf{z}.$$

7. Return the solution $\mathbf{x}$, after undoing the permutation.

Algorithm A.2 Sparse least squares using a sparse Cholesky decomposition of the matrix $\mathbf{C}$.

$\mathbf{C}$ and cameras have 6×6 (or larger) entries, the cost of performing the reordering computation is small compared to the actual numerical factorization, which can benefit from block-structured matrix operations (Golub and Van Loan 1996). It is also possible to apply sparse reordering and multifrontal techniques to QR factorization (Davis 2011), which may be preferable when the least squares problems are poorly conditioned.

A.5 Iterative techniques

When problems become large, the amount of memory required to store the Hessian matrix $\mathbf{C}$ and its factor $\mathbf{R}$, and the amount of time it takes to compute the factorization, can become prohibitively large, especially when there are large amounts of fill-in. This is often the case with image processing problems defined on pixel grids, because, even with the optimal reordering (nested dissection) the amount of fill can still be large.

A preferable approach to solving such linear systems is to use iterative techniques, which compute a series of estimates that converge to the final solution, e.g., by taking a series of downhill steps in an energy function such as (A.29).

A large number of iterative techniques have been developed over the years, including such well-known algorithms as successive overrelaxation and multi-grid. These are described in specialized textbooks on iterative solution techniques (Axelsson 1996; Saad 2003) as well as in more general books on numerical linear algebra and least squares techniques (Björck 1996; Golub and Van Loan 1996; Trefethen and Bau 1997; Nocedal and Wright 2006; Björck and Dahlquist 2010).

ConjugateGradient$(\mathbf{C}, \mathbf{d}, \mathbf{x}_0)$

1. $\mathbf{r}_0 = \mathbf{d} - \mathbf{C}\mathbf{x}_0$
2. $\mathbf{p}_0 = \mathbf{r}_0$
3. **for** $k = 0 \ldots$
4. $\quad \mathbf{w}_k = \mathbf{C}\mathbf{p}_k$
5. $\quad \alpha_k = \|\mathbf{r}_k\|^2 / (\mathbf{p}_k \cdot \mathbf{w}_k)$
6. $\quad \mathbf{x}_{k+1} = \mathbf{x}_k + \alpha_k \mathbf{p}_k$
7. $\quad \mathbf{r}_{k+1} = \mathbf{r}_k - \alpha_k \mathbf{w}_k$
8.
9. $\quad \beta_{k+1} = \|\mathbf{r}_{k+1}\|^2 / \|\mathbf{r}_k\|^2$
10. $\quad \mathbf{p}_{k+1} = \mathbf{r}_{k+1} + \beta_{k+1}\mathbf{p}_k$

ConjugateGradientLS$(\mathbf{A}, \mathbf{b}, \mathbf{x}_0)$

1. $\mathbf{q}_0 = \mathbf{b} - \mathbf{A}\mathbf{x}_0, \quad \mathbf{r}_0 = \mathbf{A}^T\mathbf{q}_0$
2. $\mathbf{p}_0 = \mathbf{r}_0$
3. **for** $k = 0 \ldots$
4. $\quad \mathbf{v}_k = \mathbf{A}\mathbf{p}_k$
5. $\quad \alpha_k = \|\mathbf{r}_k\|^2 / \|\mathbf{v}_k\|^2$
6. $\quad \mathbf{x}_{k+1} = \mathbf{x}_k + \alpha_k \mathbf{p}_k$
7. $\quad \mathbf{q}_{k+1} = \mathbf{q}_k - \alpha_k \mathbf{v}_k$
8. $\quad \mathbf{r}_{k+1} = \mathbf{A}^T\mathbf{q}_{k+1}$
9. $\quad \beta_{k+1} = \|\mathbf{r}_{k+1}\|^2 / \|\mathbf{r}_k\|^2$
10. $\quad \mathbf{p}_{k+1} = \mathbf{r}_{k+1} + \beta_{k+1}\mathbf{p}_k$

Algorithm A.3 Conjugate gradient and conjugate gradient least squares algorithms. The algorithms are described in more detail in the text, but in brief, they choose descent directions $\mathbf{p}_k$ that are conjugate to each other with respect to $\mathbf{C}$ by computing a factor β by which to discount the previous search direction $\mathbf{p}_{k-1}$. They then find the optimal step size α and take a downhill step by an amount $\alpha_k\mathbf{p}_k$.

A.5.1 Conjugate gradient

The iterative solution technique that often performs best is conjugate gradient descent, which takes a series of downhill steps that are *conjugate* to each other with respect to the $\mathbf{C}$ matrix, i.e., if the $\mathbf{u}$ and $\mathbf{v}$ descent directions satisfy $\mathbf{u}^T\mathbf{C}\mathbf{v} = 0$. In practice, conjugate gradient descent outperforms other kinds of gradient descent algorithm because its convergence rate is proportional to the square root of the *condition number* of $\mathbf{C}$ instead of the condition number itself.[9] Shewchuk (1994) provides a nice introduction to this topic, with clear intuitive explanations of the reasoning behind the conjugate gradient algorithm and its performance.

Algorithm A.3 describes the conjugate gradient algorithm and its related least squares counterpart, which can be used when the original set of least squares linear equations is available in the form of $\mathbf{A}\mathbf{x} = \mathbf{b}$ (A.28). While it is easy to convince yourself that the two forms are mathematically equivalent, the least squares form is preferable if rounding errors start to affect the results because of poor conditioning. It may also be preferable if, due to the sparsity structure of $\mathbf{A}$, multiplies with the original $\mathbf{A}$ matrix are faster or more space efficient than multiplies with $\mathbf{C}$.

The conjugate gradient algorithm starts by computing the current residual $\mathbf{r}_0 = \mathbf{d} - \mathbf{C}\mathbf{x}_0$, which is the direction of steepest descent of the energy function (A.28). It sets the original descent direction $\mathbf{p}_0 = \mathbf{r}_0$. Next, it multiplies the descent direction by the quadratic form (Hessian) matrix $\mathbf{C}$ and combines this with the residual to estimate the optimal step size α_k. The solution vector $\mathbf{x}_k$ and the residual vector $\mathbf{r}_k$ are then updated using this step size. (Notice how the least squares variant of the conjugate gradient algorithm splits the multiplication by the $\mathbf{C} = \mathbf{A}^T\mathbf{A}$ matrix across steps 4 and

[9]The condition number $\kappa(\mathbf{C})$ is the ratio of the largest and smallest eigenvalues of $\mathbf{C}$. The actual convergence rate depends on the clustering of the eigenvalues, as discussed in the references cited in this section.

8.) Finally, a new search direction is calculated by first computing a factor β as the ratio of current to previous residual magnitudes. The new search direction $\mathbf{p}_{k+1}$ is then set to the residual plus β times the old search direction $\mathbf{p}_k$, which keeps the directions conjugate with respect to $\mathbf{C}$.

It turns out that conjugate gradient descent can also be directly applied to non-quadratic energy functions, e.g., those arising from non-linear least squares (Appendix A.3). Instead of explicitly forming a local quadratic approximation $\mathbf{C}$ and then computing residuals $\mathbf{r}_k$, non-linear conjugate gradient descent computes the gradient of the energy function E (A.45) directly inside each iteration and uses it to set the search direction (Nocedal and Wright 2006). Because the quadratic approximation to the energy function may not exist or may be inaccurate, line search is often used to determine the step size α_k. Furthermore, to compensate for errors in finding the true function minimum, alternative formulas for β_{k+1}, such as Polak–Ribière,

$$\beta_{k+1} = \frac{\nabla E(\mathbf{x}_{k+1})[\nabla E(\mathbf{x}_{k+1}) - \nabla E(\mathbf{x}_k)]}{\|\nabla E(\mathbf{x}_k)\|^2} \tag{A.51}$$

are often used (Nocedal and Wright 2006).

A.5.2 Preconditioning

As we mentioned previously, the rate of convergence of the conjugate gradient algorithm is governed in large part by the condition number $\kappa(\mathbf{C})$. Its effectiveness can therefore be increased dramatically by reducing this number, e.g., by rescaling elements in $\mathbf{x}$, which corresponds to rescaling rows and columns in $\mathbf{C}$.

In general, preconditioning is usually thought of as a change of basis from the vector $\mathbf{x}$ to a new vector

$$\hat{x} = \mathbf{S}\mathbf{x}. \tag{A.52}$$

The corresponding linear system being solved then becomes

$$\mathbf{A}\mathbf{S}^{-1}\hat{x} = \mathbf{S}^{-1}\mathbf{b} \qquad \text{or} \qquad \hat{\mathbf{A}}\hat{x} = \hat{\mathbf{b}}, \tag{A.53}$$

with a corresponding least squares energy (A.29) of the form

$$E_{\text{PLS}} = \hat{x}^T(\mathbf{S}^{-T}\mathbf{C}\mathbf{S}^{-1})\hat{x} - 2\hat{x}^T(\mathbf{S}^{-T}\mathbf{d}) + \|\hat{\mathbf{b}}\|^2. \tag{A.54}$$

The actual preconditioned matrix $\hat{\mathbf{C}} = \mathbf{S}^{-T}\mathbf{C}\mathbf{S}^{-1}$ is usually not explicitly computed. Instead, Algorithm A.3 is extended to insert $\mathbf{S}^{-T}$ and $\mathbf{S}^T$ operations at the appropriate places (Björck 1996; Golub and Van Loan 1996; Trefethen and Bau 1997; Saad 2003; Nocedal and Wright 2006).

A good preconditioner $\mathbf{S}$ is easy and cheap to compute, but is also a decent approximation to a square root of $\mathbf{C}$, so that $\kappa(\mathbf{S}^{-T}\mathbf{C}\mathbf{S}^{-1})$ is closer to 1. The simplest such choice is the square root of the diagonal matrix $\mathbf{S} = \mathbf{D}^{1/2}$, with $\mathbf{D} = \text{diag}(\mathbf{C})$. This has the advantage that any scalar change in variables (e.g., using radians instead of degrees for angular measurements) has no effect on the range of convergence of the iterative technique. For problems that are naturally block-structured, e.g., for structure from motion, where 3D point positions or 6D camera poses are being estimated, a block diagonal preconditioner is often a good choice.

A wide variety of more sophisticated preconditioners have been developed over the years (Björck 1996; Golub and Van Loan 1996; Trefethen and Bau 1997; Saad 2003; Nocedal and Wright 2006), many of which can be directly applied to problems in computer vision (Byröd and Åström 2009; Agarwal, Snavely *et al.* 2010; Jeong, Nistér *et al.* 2012). Some of these are based on an *incomplete Cholesky* factorization of $\mathbf{C}$, i.e., one in which the amount of fill-in in $\mathbf{R}$ is strictly limited, e.g., to just the original non-zero elements in $\mathbf{C}$.[10] Other preconditioners are based on a sparsified, e.g., tree-based or clustered, approximation to $\mathbf{C}$ (Koutis 2007; Koutis and Miller 2008; Grady 2008; Koutis, Miller, and Tolliver 2009), as these are known to have efficient inversion properties.

For grid-based image-processing applications, *parallel* or *hierarchical* preconditioners often perform extremely well (Yserentant 1986; Szeliski 1990b; Pentland 1994; Saad 2003; Szeliski 2006b; Krishnan and Szeliski 2011; Krishnan, Fattal, and Szeliski 2013). These approaches use a change of basis transformation $\mathbf{S}$ that resembles the pyramidal or wavelet representations discussed in Section 3.5, and are hence amenable to parallel and GPU-based implementations (Figure 3.35b). Coarser elements in the new representation quickly converge to the low-frequency components in the solution, while finer-level elements encode the higher-frequency components. Some of the relationships between hierarchical preconditioners, incomplete Cholesky factorization, and multigrid techniques are explored by Saad (2003) and Szeliski (2006b), Krishnan and Szeliski (2011), and Krishnan, Fattal, and Szeliski (2013).

A.5.3 Multigrid

One other class of iterative techniques widely used in computer vision is *multigrid* techniques (Briggs, Henson, and McCormick 2000; Trottenberg, Oosterlee, and Schuller 2000), which have been applied to problems such as surface interpolation (Terzopoulos 1986a), optical flow (Terzopoulos 1986a; Bruhn, Weickert *et al.* 2006), high dynamic range tone mapping (Fattal, Lischinski, and Werman 2002), colorization (Levin, Lischinski, and Weiss 2004), natural image matting (Levin, Lischinski, and Weiss 2008), and segmentation (Grady 2008).

The main idea behind multigrid is to form coarser (lower-resolution) versions of the problems and use them to compute the low-frequency components of the solution. However, unlike simple coarse-to-fine techniques, which use the coarse solutions to initialize the fine solution, multigrid techniques only *correct* the low-frequency component of the current solution and use multiple rounds of coarsening and refinement (in what are often called "V" and "W" patterns of motion across the pyramid) to obtain rapid convergence.

On certain simple homogeneous problems (such as solving Poisson equations), multigrid techniques can achieve optimal performance, i.e., computation times linear in the number of variables. However, for more inhomogeneous problems or problems on irregular grids, variants on these techniques, such as *algebraic multigrid* (AMG) approaches, which look at the structure of $\mathbf{C}$ to derive coarse level problems, may be preferable. Saad (2003) has a nice discussion of the relationship between multigrid and parallel preconditioners and on the relative merits of using multigrid or conjugate gradient approaches.

[10] If a complete Cholesky factorization $\mathbf{C} = \mathbf{R}^T\mathbf{R}$ is used, we get $\hat{\mathbf{C}} = \mathbf{R}^{-T}\mathbf{C}\mathbf{R}^{-1} = \mathbf{I}$ and all iterative algorithms converge in a single step, thereby obviating the need to use them, but the complete factorization is often too expensive. Note that incomplete factorization can also benefit from reordering.

Appendix B

Bayesian modeling and inference

B.1 Estimation theory 745
B.2 Maximum likelihood estimation and least squares 747
B.3 Robust statistics 748
B.4 Prior models and Bayesian inference 750
B.5 Markov random fields 751
B.6 Uncertainty estimation (error analysis) 752

R. Szeliski, *Computer Vision*, Texts in Computer Science,
https://doi.org/10.1007/978-3-030-34372-9

As you may have noticed, the following problem commonly recurs in computer vision applications. Given a number of measurements (images, feature positions, etc.), estimate the values of some unknown structure or parameters (camera positions, object shape, etc.). These kinds of problems are in general called *inverse* problems because they involve estimating unknown model parameters instead of simulating the forward formation equations.[1] Computer graphics is a classic forward modeling problem (given some objects, cameras, and lighting, simulate the images that would result), while computer vision problems are usually of the inverse kind (given one or more images, recover the scene that gave rise to these images).

Given an instance of an inverse problem, there are, in general, several ways to proceed. For instance, through clever (or sometimes straightforward) algebraic manipulation, a closed form solution for the unknowns can sometimes be derived. Consider, for example, the *camera matrix calibration* problem (Section 11.2.1): given an image of a calibration pattern consisting of known 3D point positions, compute the 3×4 camera matrix $\mathbf{P}$ that maps these points onto the image plane.

In more detail, we can write this problem as (11.11–11.12)

$$x_i = \frac{p_{00}X_i + p_{01}Y_i + p_{02}Z_i + p_{03}}{p_{20}X_i + p_{21}Y_i + p_{22}Z_i + p_{23}} \tag{B.1}$$

$$y_i = \frac{p_{10}X_i + p_{11}Y_i + p_{12}Z_i + p_{13}}{p_{20}X_i + p_{21}Y_i + p_{22}Z_i + p_{23}}, \tag{B.2}$$

where (x_i, y_i) is the feature position of the ith point measured in the image plane, (X_i, Y_i, Z_i) is the corresponding 3D point position, and the p_{ij} are the unknown entries of the camera matrix $\mathbf{P}$. Moving the denominator over to the left-hand side, we end up with a set of simultaneous linear equations,

$$x_i(p_{20}X_i + p_{21}Y_i + p_{22}Z_i + p_{23}) = p_{00}X_i + p_{01}Y_i + p_{02}Z_i + p_{03}, \tag{B.3}$$

$$y_i(p_{20}X_i + p_{21}Y_i + p_{22}Z_i + p_{23}) = p_{10}X_i + p_{11}Y_i + p_{12}Z_i + p_{13}, \tag{B.4}$$

which we can solve using linear least squares (Appendix A.2) to obtain an estimate of $\mathbf{P}$.

The question then arises: Is this set of equations the right ones to be solving? If the measurements are totally noise-free or we do not care about getting the best possible answer, then the answer is yes. However, in general, we cannot be sure that we have a reasonable algorithm unless we make a model of the likely sources of error and devise an algorithm that performs as well as possible given these potential errors.

In the rest of this appendix, we provide a brief tutorial on the fundamentals of Bayesian modeling and inference. We start with estimation theory (how to build forward models that account for noise) and show how to model likelihoods under Gaussian noise. We then show how when the measurements are linear, these result in least squares regression. In Appendix B.3, we review robust estimation techniques designed to deal with measurement outliers (gross errors). Appendices B.4 and B.5 discuss Bayesian prior models and Markov random fields, which are compact local priors suitable for image processing. We also describe a number of widely used *inference* algorithms for finding good solutions to MRF problems. Finally, Appendix B.6 describes how we can model the posterior *uncertainty* in our estimates.

[1] As we saw in Chapters 4 and 5, these problems are called *regression problems*, because we are trying to estimate a *continuous* quantity from noisy inputs, as opposed to a discrete *classification* task (Bishop 2006).

B.1 Estimation theory

The study of inverse inference problems from noisy data is often called *estimation theory* (Gelb 1974), and its extension to problems where we explicitly choose a loss function is called *statistical decision theory* (Berger 1993; MacKay 2003; Bishop 2006; Robert 2007; Hastie, Tibshirani, and Friedman 2009; Murphy 2012; Deisenroth, Faisal, and Ong 2020). We first start by writing down the forward process that leads from our unknowns (and knowns) to a set of noise-corrupted measurements. We then devise an algorithm that will give us an estimate (or set of estimates) that are both insensitive to the noise (as best they can be) and also quantify the reliability of these estimates. In this Appendix, I provide a very condensed overview of this topic, including an introduction to basic probability and Bayesian inference. Much more detailed and informative treatment can be found in the books by Bishop (2006), Hastie, Tibshirani, and Friedman (2009), and (Murphy 2012) and Deisenroth, Faisal, and Ong (2020)).

The perspective projection equations above are just a particular instance of a more general set of *measurement equations*,

$$\mathbf{y}_i = \mathbf{f}_i(\mathbf{x}) + \mathbf{n}_i. \tag{B.5}$$

Here, the $\mathbf{y}_i$ are the noise-corrupted *measurements*, e.g., (x_i, y_i) in Equations (B.1–B.2) and $\mathbf{x}$ is the unknown *state vector*.[2]

Each measurement comes with its associated *measurement model* $\mathbf{f}_i(\mathbf{x})$, which maps the unknown into that particular measurement. Note that the use of the $\mathbf{f}_i(\mathbf{x})$ form makes it straightforward to have measurements of different dimensions, which becomes useful when we start adding in prior information (Appendix B.4).

Each measurement is also contaminated with some noise $\mathbf{n}_i$. In Equation (B.7) we specify that $\mathbf{n}_i$ is a zero-mean normal (Gaussian) random variable with a covariance matrix $\mathbf{\Sigma}_i$. In general, the noise need not be Gaussian and, in fact, it is usually prudent to assume that some measurements may be outliers. However, we defer this discussion to Appendix B.3, after we have explored the simpler Gaussian noise case more fully. We also assume that the noise vectors $\mathbf{n}_i$ are independent. In the case where they are not (e.g., when some constant gain or offset contaminates all of the pixels in a given image), we can add this effect as a *nuisance parameter* to our state vector $\mathbf{x}$ and later estimate its value (and discard it, if so desired).

Likelihood for multivariate Gaussian noise

Given all of the noisy measurements $\mathbf{y} = \{\mathbf{y}_i\}$, we would like to infer a probability distribution on the unknown $\mathbf{x}$ vector. We can write the *likelihood* of having observed the $\{\mathbf{y}_i\}$ given a particular value of $\mathbf{x}$ as

$$L = p(\mathbf{y}|\mathbf{x}) = \prod_i p(\mathbf{y}_i|\mathbf{x}) = \prod_i p(\mathbf{y}_i|\mathbf{f}_i(\mathbf{x})) = \prod_i p(\mathbf{n}_i). \tag{B.6}$$

When each noise vector $\mathbf{n}_i$ is a multivariate Gaussian with covariance $\mathbf{\Sigma}_i$,

$$\mathbf{n}_i \sim \mathcal{N}(0, \mathbf{\Sigma}_i), \tag{B.7}$$

we can write this likelihood as

$$\begin{aligned} L &= \prod_i |2\pi\mathbf{\Sigma}_i|^{-1/2} \exp\left(-\frac{1}{2}(\mathbf{y}_i - \mathbf{f}_i(\mathbf{x}))^T \mathbf{\Sigma}_i^{-1} (\mathbf{y}_i - \mathbf{f}_i(\mathbf{x}))\right) \\ &= \prod_i |2\pi\mathbf{\Sigma}_i|^{-1/2} \exp\left(-\frac{1}{2}\|\mathbf{y}_i - \mathbf{f}_i(\mathbf{x})\|^2_{\mathbf{\Sigma}_i^{-1}}\right), \end{aligned} \tag{B.8}$$

[2] In the Kalman filtering literature (Gelb 1974), it is more common to use $\mathbf{z}$ instead of $\mathbf{y}$ to denote measurements.

where the matrix norm $\|\mathbf{x}\|^2_{\mathbf{A}}$ is a shorthand notation for $\mathbf{x}^T\mathbf{A}\mathbf{x}$.

The norm $\|\mathbf{y}_i - \bar{\mathbf{y}}_i\|_{\mathbf{\Sigma}_i^{-1}}$ is often called the *Mahalanobis distance*, which we introduced in (5.32), and is used to measure the distance between a measurement and the mean of a multivariate Gaussian distribution (Bishop 2006, Section 2.3; Hartley and Zisserman 2004, Appendix 2). Contours of equal Mahalanobis distance are equi-probability contours (Figure 5.9). Note that when the measurement covariance is isotropic (the same in all directions), i.e., when $\mathbf{\Sigma}_i = \sigma_i^2\mathbf{I}$, the likelihood can be written as

$$L = \prod_i (2\pi\sigma_i^2)^{-N_i/2} \exp\left(-\frac{1}{2\sigma_i^2}\|\mathbf{y}_i - \mathbf{f}_i(\mathbf{x})\|^2\right), \tag{B.9}$$

where N_i is the length of the ith measurement vector $\mathbf{y}_i$.

We can more easily visualize the structure of the covariance matrix and the corresponding Mahalanobis distance if we first perform an *eigenvalue* or *principal component* analysis (PCA) of the covariance matrix (A.6),

$$\mathbf{\Sigma}_i = \mathbf{\Phi}\ \mathrm{diag}(\lambda_0 \ldots \lambda_{N-1})\ \mathbf{\Phi}^T. \tag{B.10}$$

Equal-probability contours of the corresponding multi-variate Gaussian, which are also equi-distance contours in the Mahalanobis distance (Figure 5.19), are multi-dimensional ellipsoids whose axis directions are given by the columns of $\mathbf{\Phi}$ (the *eigenvectors*) and whose lengths are given by the $\sigma_j = \sqrt{\lambda_j}$ (Figure A.1).

It is usually more convenient to work with the negative log likelihood, which we can think of as a *cost* or *energy*

$$E = -\log L = \frac{1}{2}\sum_i (\mathbf{y}_i - \mathbf{f}_i(\mathbf{x}))^T\mathbf{\Sigma}_i^{-1}(\mathbf{y}_i - \mathbf{f}_i(\mathbf{x})) + k \tag{B.11}$$

$$= \frac{1}{2}\sum_i \|\mathbf{y}_i - \mathbf{f}_i(\mathbf{x})\|^2_{\mathbf{\Sigma}_i^{-1}} + k, \tag{B.12}$$

where $k = \sum_i \log|2\pi\mathbf{\Sigma}_i|$ is a constant that depends on the measurement variances, but is independent of $\mathbf{x}$.

Notice that the inverse covariance $\mathbf{C}_i = \mathbf{\Sigma}_i^{-1}$ plays the role of a *weight* on each of the measurement error *residuals*, i.e., the difference between the contaminated measurement $\mathbf{y}_i$ and its uncontaminated (predicted) value $\mathbf{f}_i(\mathbf{x})$. In fact, the inverse covariance is often called the (Fisher) *information matrix* (Bishop 2006), because it tells us how much information is contained in a given measurement, i.e., how well it constrains the final estimate. We can also think of this matrix as denoting the amount of *confidence* to associate with each measurement (hence the letter $\mathbf{C}$).

In this formulation, it is quite acceptable for some information matrices to be singular (of degenerate rank) or even zero (if the measurement is missing altogether). Rank-deficient measurements often occur, for example, when using a line feature or edge to measure a 3D edge-like feature, as its exact position along the edge is unknown (or of infinite or extremely large variance) (Section 9.1.3).

To make the distinction between the noise contaminated measurement and its expected value for a particular setting of $\mathbf{x}$ more explicit, we adopt the notation $\tilde{\mathbf{y}}$ for the former (think of the tilde as the approximate or noisy value) and $\hat{\mathbf{y}} = \mathbf{f}_i(\mathbf{x})$ for the latter (think of the hat as the predicted or expected value). We can then write the negative log likelihood as

$$E = -\log L = \frac{1}{2}\sum_i \|\tilde{\mathbf{y}}_i - \hat{\mathbf{y}}_i\|^2_{\mathbf{\Sigma}_i^{-1}} + k. \tag{B.13}$$

B.2 Maximum likelihood estimation and least squares

Now that we have presented the likelihood and log likelihood functions, how can we find the optimal value for our state estimate $\mathbf{x}$? One plausible choice might be to select the value of $\mathbf{x}$ that maximizes $L = p(\mathbf{y}|\mathbf{x})$. In fact, in the absence of any prior model for $\mathbf{x}$ (Appendix B.4), we have

$$L = p(\mathbf{y}|\mathbf{x}) = p(\mathbf{y}, \mathbf{x}) = p(\mathbf{x}|\mathbf{y}). \tag{B.14}$$

Therefore, choosing the value of $\mathbf{x}$ that maximizes the likelihood is equivalent to choosing the maximum of our probability density estimate for $\mathbf{x}$.

When might this be a good idea? If the data (measurements) constrain the possible values of $\mathbf{x}$ so that they all cluster tightly around one value (e.g., if the distribution $p(\mathbf{x}|\mathbf{y})$ is a unimodal Gaussian), the maximum likelihood estimate is the optimal one in that it is both unbiased and has the least possible variance. In many other cases, e.g., if a single estimate is all that is required, it is still often the best estimate.[3]

However, if the probability is multi-modal, i.e., it has several local minima in the log likelihood, much more care may be required. In particular, it might be necessary to defer certain decisions (such as the ultimate position of an object being tracked) until more measurements have been taken. The CONDENSATION algorithm presented in Section 7.3.1 is one possible method for modeling and updating such multi-modal distributions but is just one example of more general *particle filtering* and *Markov Chain Monte Carlo* (MCMC) techniques (Andrieu, de Freitas *et al.* 2003; Bishop 2006; Koller and Friedman 2009).

Another possible way to choose the best estimate is to maximize the *expected utility* (or, conversely, to minimize the expected risk or loss) associated with obtaining the correct estimate, i.e., by minimizing

$$E_{\text{loss}}(\mathbf{x}, \mathbf{y}) = \int l(\mathbf{x} - \mathbf{z}) p(\mathbf{z}|\mathbf{y}) d\mathbf{z}. \tag{B.15}$$

For example, if a robot wants to avoid hitting a wall at all costs, the loss function will be high whenever the estimate underestimates the true distance to the wall. When $l(\mathbf{x} - \mathbf{y}) = \delta(\mathbf{x} - \mathbf{y})$, we obtain the maximum likelihood estimate, whereas when $l(\mathbf{x} - \mathbf{y}) = \|\mathbf{x} - \mathbf{y}\|^2$, we obtain the *mean square error* (MSE) or *expected value* estimate. The explicit modeling of a utility or loss function is what characterizes *statistical decision theory* (Berger 1993; MacKay 2003; Bishop 2006; Robert 2007; Hastie, Tibshirani, and Friedman 2009; Murphy 2012; Deisenroth, Faisal, and Ong 2020) and the minimization of expected risk (in machine learning) is called *empirical risk minimization*, which we discussed in Section 5.1, Equation (5.1).

How do we find the maximum likelihood estimate? If the measurement noise is Gaussian, we can minimize the quadratic objective function (B.13). This becomes even simpler if the measurement equations are linear, i.e.,

$$\mathbf{f}_i(\mathbf{x}) = \mathbf{H}_i \mathbf{x}, \tag{B.16}$$

where $\mathbf{H}$ is the *measurement matrix* relating unknown state variables $\mathbf{x}$ to measurements $\tilde{\mathbf{y}}$. In this case, (B.13) becomes

$$E = \sum_i \|\tilde{\mathbf{y}}_i - \mathbf{H}_i \mathbf{x}\|_{\mathbf{\Sigma}_i^{-1}} = \sum_i (\tilde{\mathbf{y}}_i - \mathbf{H}_i \mathbf{x})^T \mathbf{C}_i (\tilde{\mathbf{y}}_i - \mathbf{H}_i \mathbf{x}), \tag{B.17}$$

[3] According to the Gauss-Markov theorem, least squares produces the best linear unbiased estimator (BLUE) for a linear measurement model regardless of the actual noise distribution, assuming that the noise is zero mean and uncorrelated.

which is a simple quadratic form in $\mathbf{x}$, which can be solved using linear least squares (Appendix A.2) to obtain the minimum energy (maximum likelihood) solution

$$\mathbf{x} = \left(\sum_i \mathbf{H}_i^T \mathbf{C}_i \mathbf{H}_i\right)^{-1} \left(\sum_i \mathbf{H}_i^T \mathbf{C}_i \tilde{\mathbf{y}}_i\right) \tag{B.18}$$

with a corresponding posterior covariance of

$$\mathbf{\Sigma} = \mathbf{C}^{-1} = \left(\sum_i \mathbf{H}_i^T \mathbf{C}_i \mathbf{H}_i\right)^{-1}. \tag{B.19}$$

When $\mathbf{H}_i = \mathbf{I}$, i.e., when we are just taking an average of covariance-weighted measurements, we obtain the even simpler formula

$$\mathbf{x} = \left(\sum_i \mathbf{C}_i\right)^{-1} \left(\sum_i \mathbf{C}_i \tilde{\mathbf{y}}_i\right), \tag{B.20}$$

which is a simple information-weighted mean, with a final covariance (uncertainty) of $\mathbf{\Sigma} = \left(\sum_i \mathbf{C}_i\right)^{-1}$.

When the measurements are non-linear, the system must be solved iteratively using non-linear least squares (Appendix A.3). In this case, we can compute a *Cramer–Rao lower bound* (CRLB) on the posterior covariance using the same covariance formula as before (B.19) except that we use the Jacobians $\mathbf{J}(\mathbf{x}_i; \mathbf{p})$ from (A.46) are used instead of the measurement matrices $\mathbf{H}_i$.

B.3 Robust statistics

In Appendix B.1, we assumed that the noise being added to each measurement (B.5) was multivariate Gaussian (B.7). This is an appropriate model if the noise is the result of lots of tiny errors being added together, e.g., from thermal noise in a silicon imager. In most cases, however, measurements can be contaminated with larger *outliers*, i.e., gross failures in the measurement process. Examples of such outliers include bad feature matches (Section 8.1.4), occlusions in stereo matching (Chapter 12), and discontinuities in an otherwise smooth image, depth map, or label image (Sections 4.2.1 and 4.3).

In such cases, it makes more sense to model the measurement noise with a long-tailed *contaminated* noise model, such as a Laplacian. The negative log likelihood in this case, rather than being quadratic in the measurement residuals (B.12–B.17), has a slower growth in the penalty function to account for the increased likelihood of large errors.

This formulation of the inference problem is called an *M-estimator* in the robust statistics literature (Huber 1981; Hampel, Ronchetti *et al.* 1986; Black and Rangarajan 1996; Stewart 1999; Barron 2019) and involves applying a robust penalty function $\rho(r)$ to the residuals

$$E_{\mathrm{RLS}}(\Delta\mathbf{p}) = \sum_i \rho(\|\mathbf{r}_i\|) \tag{B.21}$$

instead of squaring them. Over the years, a variety of robust loss functions have been developed, as discussed in the above references. Recently, Barron (2019) unified a number of these under a two-parameter loss function, which we introduced in Section 4.1.3. This loss function, shown in Figure 4.7, can be written as

$$\rho(x; \alpha, c) = \frac{|\alpha - 2|}{2}\left(\left(\frac{(x/c)^2}{|\alpha - 2|} + 1\right)^{\alpha/2} - 1\right), \tag{B.22}$$

where α is a shape parameter that controls the robustness of the loss and $c > 0$ is a scale parameter that controls the size of the loss's quadratic bowl near $x = 0$. In his paper, Barron (2019) discusses how both parameters can be determined at run time by maximizing the likelihood (or equivalently, minimizing the negative log-likelihood) of the given residuals, making such an algorithm self-tuning to a wide variety of noise levels and outlier distributions.

As we mentioned in Section 8.1.4, we can take the derivative of this function with respect to the unknown parameters $\mathbf{p}$ we are estimating and set it to 0,

$$\sum_i \psi(\|\mathbf{r}_i\|)\frac{\partial\|\mathbf{r}_i\|}{\partial\mathbf{p}} = \sum_i \frac{\psi(\|\mathbf{r}_i\|)}{\|\mathbf{r}_i\|}\mathbf{r}_i^T\frac{\partial\mathbf{r}_i}{\partial\mathbf{p}} = 0, \tag{B.23}$$

where $\psi(r) = \rho'(r)$ is the derivative of ρ and is called the *influence function*. If we introduce a *weight function*, $w(r) = \Psi(r)/r$, we observe that finding the stationary point of (B.21) using (B.23) is equivalent to minimizing the *iteratively re-weighted least squares* (IRLS) problem

$$E_{\text{IRLS}} = \sum_i w(\|\mathbf{r}_i\|)\|\mathbf{r}_i\|^2, \tag{B.24}$$

where the $w(\|\mathbf{r}_i\|)$ play the same local weighting role as $\mathbf{C}_i = \mathbf{\Sigma}_i^{-1}$ in (B.12). Black and Anandan (1996) describe a variety of robust penalty functions and their corresponding influence and weighting function.

The IRLS algorithm alternates between computing the influence functions $w(\|\mathbf{r}_i\|)$ and solving the resulting weighted least squares problem (with fixed w values). Alternative incremental robust least squares algorithms can be found in the work of Sawhney and Ayer (1996), Black and Anandan (1996), Black and Rangarajan (1996), and Baker, Gross *et al.* (2003) and textbooks and tutorials on robust statistics (Huber 1981; Hampel, Ronchetti *et al.* 1986; Rousseeuw and Leroy 1987; Stewart 1999). It is also possible to apply general optimization techniques (Appendix A.3) directly to the non-linear cost function given in Equation (B.24), which may sometimes have better convergence properties.

Most robust penalty functions involve a scale parameter, which should typically be set to the variance (or standard deviation, depending on the formulation) of the non-contaminated (inlier) noise. Estimating such noise levels directly from the measurements or their residuals, however, can be problematic, as such estimates themselves become contaminated by outliers. The robust statistics literature contains a variety of techniques to estimate such parameters. One of the simplest and most effective is the *median absolute deviation* (MAD),

$$MAD = \text{med}_i\|\mathbf{r}_i\|, \tag{B.25}$$

which, when multiplied by 1.4, provides a robust estimate of the standard deviation of the inlier noise process.

As mentioned in Section 8.1.4, it is often better to start iterative non-linear minimization techniques, such as IRLS, in the vicinity of a good solution by first randomly selecting small subsets of measurements until a good set of inliers is found. The best known of these techniques is RANdom SAmple Consensus (RANSAC) (Fischler and Bolles 1981), although even better variants such as Preemptive RANSAC (Nistér 2003), PROgressive SAmple Consensus (PROSAC) (Chum and Matas 2005), USAC (Raguram, Chum *et al.* 2012), and Latent RANSAC (Korman and Litman 2018) have since been developed. The paper by Raguram, Chum *et al.* (2012) provides a nice experimental comparison of most of these techniques.

Additional variants on RANSAC include MLESAC (Torr and Zisserman 2000), DSAC (Brachmann, Krull *et al.* 2017), Graph-Cut RANSAC (Barath and Matas 2018), MAGSAC (Barath, Matas, and Noskova 2019), and ESAC (Brachmann and Rother 2019). The MAGSAC++ paper by Barath, Noskova *et al.* (2020) compares many of these variants. Yang, Antonante *et al.* (2020) claim that using a robust penalty function with a decreasing outlier parameter, i.e., *graduated non-convexity* (Blake and Zisserman 1987; Barron 2019), can outperform RANSAC in many geometric correspondence and pose estimation problems.

B.4 Prior models and Bayesian inference

While maximum likelihood estimation can often lead to good solutions, in some cases the range of possible solutions consistent with the measurements is too large to be useful. For example, consider the problem of image denoising (Section 3.4.2). If we estimate each pixel separately based on just its noisy version, we cannot make any progress, as there are a large number of values that could lead to each noisy measurement.[4] Instead, we need to rely on typical properties of images, e.g., that they tend to be piecewise smooth (Section 4.2.1).

The propensity of images to be piecewise smooth can be encoded in a *prior distribution* $p(\mathbf{x})$, which measures the likelihood of an image being a natural image. Statistical models where we construct or estimate a prior distribution over the unknowns we are trying to recover are known as *generative models*. As the prior distribution is known, we can *generate* random samples and see if they conform to our expected appearance or distribution, although sometimes the sampling process may itself involve a lot of computation. For example, to encode piecewise smoothness, we can use a *Markov random field* model (4.38 and B.29) whose negative log likelihood is proportional to a robustified measure of image smoothness (gradient magnitudes).

Prior models need not be restricted to image processing applications. For example, we may have some external knowledge about the rough dimensions of an object being scanned, the focal length of a lens being calibrated, or the likelihood that a particular object might appear in an image. All of these are examples of prior distributions or probabilities and they can be used to produce more reliable estimates.

As we have already seen in (4.33), Bayes' rule states that a *posterior* distribution $p(\mathbf{x}|\mathbf{y})$ over the unknowns $\mathbf{x}$ given the measurements $\mathbf{y}$ can be obtained by multiplying the measurement likelihood $p(\mathbf{y}|\mathbf{x})$ by the prior distribution $p(\mathbf{x})$ and normalizing,

$$p(\mathbf{x}|\mathbf{y}) = \frac{p(\mathbf{y}|\mathbf{x})p(\mathbf{x})}{p(\mathbf{y})}, \tag{B.26}$$

where $p(\mathbf{y}) = \int_{\mathbf{x}} p(\mathbf{y}|\mathbf{x})p(\mathbf{x})$ is a normalizing constant used to make the $p(\mathbf{x}|\mathbf{y})$ distribution *proper* (integrate to 1). Taking the negative logarithm of both sides of Equation (B.26), we get

$$-\log p(\mathbf{x}|\mathbf{y}) = -\log p(\mathbf{y}|\mathbf{x}) - \log p(\mathbf{x}) + \log p(\mathbf{y}), \tag{B.27}$$

which is the *negative posterior log likelihood*. It is common to drop the constant $\log p(\mathbf{y})$ because its value does not matter during energy minimization. However, if the prior distribution $p(\mathbf{x})$ depends on some unknown parameters, we may wish to keep $\log p(\mathbf{y})$ in order to compute the most likely value of these parameters using *Occam's razor*, i.e., by maximizing the likelihood of the observations, or to select the correct number of free parameters using *model selection* (Torr 2002; Bishop 2006; Robert 2007; Hastie, Tibshirani, and Friedman 2009).

To find the most likely (*maximum a posteriori* or MAP) solution $\mathbf{x}$ given some measurements $\mathbf{y}$, we simply minimize this negative log likelihood, which can also be thought of as an *energy*,

$$E(\mathbf{x}, \mathbf{y}) = E_d(\mathbf{x}, \mathbf{y}) + E_p(\mathbf{x}). \tag{B.28}$$

[4] In fact, the maximum likelihood estimate is just the noisy image itself.

The first term $E_d(\mathbf{x}, \mathbf{y})$ is the *data energy* or *data penalty* and measures the negative log likelihood that the measurements $\mathbf{y}$ were observed given the unknown state $\mathbf{x}$. The second term $E_p(\mathbf{x})$ is the *prior energy* and it plays a role analogous to the smoothness energy in regularization. Note that the MAP estimate may not always be desirable, because it selects the "peak" in the posterior distribution rather than some more stable statistic such as MSE—see the discussion in Appendix B.2 about loss functions and decision theory.

B.5 Markov random fields

Markov random fields (Blake, Kohli, and Rother 2011) are the most popular types of prior model for gridded image-like data, which include not only regular natural images (Section 4.3) but also two-dimensional fields such as optical flow (Chapter 9) or depth maps (Chapter 12), as well as binary fields, such as segmentations (Section 4.3.2).[5]

As we discussed in Section 4.3, the prior probability $p(\mathbf{x})$ for a Markov random field is a *Gibbs* or *Boltzmann distribution*, whose negative log likelihood (according to the Hammersley–Clifford Theorem) can be written as a sum of pairwise *interaction potentials*,

$$E_{\mathrm{P}}(\mathbf{x}) = \sum_{\{(i,j),(k,l)\} \in \mathcal{N}} V_{i,j,k,l}(f(i,j), f(k,l)), \tag{B.29}$$

where $\mathcal{N}(i,j)$ denotes the *neighbors* of pixel (i,j). In the more general case, MRFs can also contain unary potentials, as well as *higher-order potentials* defined over larger cardinality *cliques* (Kindermann and Snell 1980; Geman and Geman 1984; Bishop 2006; Potetz and Lee 2008; Kohli, Kumar, and Torr 2009; Kohli, Ladický, and Torr 2009; Rother, Kohli *et al.* 2009; Alahari, Kohli, and Torr 2010). They can also contain *line processes*, i.e., additional binary variables that mediate discontinuities between adjacent elements (Geman and Geman 1984). Black and Rangarajan (1996) show how independent line process variables can be eliminated and incorporated into regular MRFs using robust pairwise penalty functions.

The most commonly used neighborhood in Markov random field modeling is the $\mathcal{N}_4$ neighborhood, where each pixel in the field $f(i,j)$ interacts only with its immediate neighbors; Figure 4.12 shows such an $\mathcal{N}_4$ MRF. The $s_x(i,j)$ and $s_y(i,j)$ black boxes denote arbitrary interaction potentials between adjacent nodes in the random field and the $w(i,j)$ denote the elemental data penalty terms in E_d (B.28). These square nodes can also be interpreted as *factors* in a *factor graph* version of the undirected graphical model (Bishop 2006; Wainwright and Jordan 2008; Koller and Friedman 2009; Dellaert and Kaess 2017; Dellaert 2021), which is another name for interaction potentials. (Strictly speaking, the factors are improper probability functions whose product is the un-normalized posterior distribution.)

More complex and higher-dimensional interaction models and neighborhoods are also possible. For example, 2D grids can be enhanced with the addition of diagonal connections (an $\mathcal{N}_8$ neighborhood) or even larger numbers of pairwise terms (Boykov and Kolmogorov 2003; Rother, Kolmogorov *et al.* 2007). 3D grids can be used to compute globally optimal segmentations in 3D volumetric medical images (Boykov and Funka-Lea 2006) (Section 6.4.1). Higher-order cliques can also be used to develop more sophisticated models (Potetz and Lee 2008; Kohli, Ladický, and Torr 2009; Kohli, Kumar, and Torr 2009).

[5] Alternative formulations include power spectra (Section 3.4.1) and non-local means (Buades, Coll, and Morel 2008). Many people would argue that deep neural networks provide *learned* priors over the output distributions, although these are not strictly Bayesian priors that can be additively combined with measurements in a log likelihood domain.

One of the biggest challenges in using MRF models is to develop efficient *inference algorithms* that will find low-energy solutions (Veksler 1999; Boykov, Veksler, and Zabih 2001; Kohli 2007; Kumar 2008). Over the years, a large variety of such algorithms have been developed, including simulated annealing, graph cuts, and loopy belief propagation. The choice of inference technique can greatly affect the overall performance of a vision system. For example, most of the top-performing algorithms on the Middlebury Stereo Evaluation page use either belief propagation or graph cuts.

The first edition of this book (Szeliski 2010, Appendix B.5) had more detailed explanations of the most widely used MRF inference techniques, including gradient descent and simulated annealing, dynamic programming, belief propagation, graph cuts, and linear programming, which are a subset of the methods evaluated by Kappes, Andres *et al.* (2015) and shown in Figure B.1. However, since MRFs have now largely been replaced with deep neural networks in most applications, I have omitted these descriptions from this new edition. Instead, interested readers should look in the first edition and also the book on advanced MRF techniques by Blake, Kohli, and Rother (2011). Experimental comparisons, along with test datasets and reference software, can be found in the papers by Szeliski, Zabih *et al.* (2008)[6] and Kappes, Andres *et al.* (2015).[7]

B.6 Uncertainty estimation (error analysis)

In addition to computing the most likely estimate, many applications require an estimate for the *uncertainty* in this estimate.[8] The most general way to do this is to compute a complete probability distribution over all of the unknowns, but this is generally intractable. The one special case where it is easy to obtain a simple description for this distribution is linear estimation problems with Gaussian noise, where the joint energy function (negative log likelihood of the posterior estimate) is a quadratic. In this case, the posterior distribution is a multi-variate Gaussian and its covariance $\mathbf{\Sigma}$ can be computed directly from the inverse of the noise-weighted problem Hessian, as shown in (B.19. (Another name for the inverse covariance matrix, which is equal to the Hessian in such simple cases, is the *information matrix*.)

Even here, however, the full covariance matrix may be too large to compute and store. For example, in large structure from motion problems, a large sparse Hessian normally results in a full dense covariance matrix. In such cases, it is often considered acceptable to report only the variance in the estimated quantities or simple covariance estimates on individual parameters, such as 3D point positions or camera pose estimates (Szeliski 1990a). More insight into the problem, e.g., the dominant *modes* of uncertainty, can be obtained using eigenvalue analysis (Szeliski and Kang 1997).

For problems where the posterior energy is non-quadratic, e.g., in non-linear or robustified least squares, it is still often possible to obtain an estimate of the Hessian in the vicinity of the optimal solution. In this case, the *Cramer–Rao lower bound* on the uncertainty (covariance) can be computed as the inverse of the Hessian. Another way of saying this is that while the local Hessian can underestimate how "wide" the energy function can be, the covariance can never be smaller than the estimate based on this local quadratic approximation. It is also possible to estimate a different kind of uncertainty (min-marginal energies) in general MRFs where the MAP inference is performed using graph cuts (Kohli and Torr 2008).

While many computer vision applications ignore uncertainty modeling, it is often useful to compute these estimates just to get an intuitive feeling for the reliability of the estimates. Certain ap-

[6] https://vision.middlebury.edu/MRF.

[7] http://hciweb2.iwr.uni-heidelberg.de/opengm

[8] This is particularly true of classic photogrammetry applications, where the reporting of precision is almost always considered mandatory (Förstner 2005).

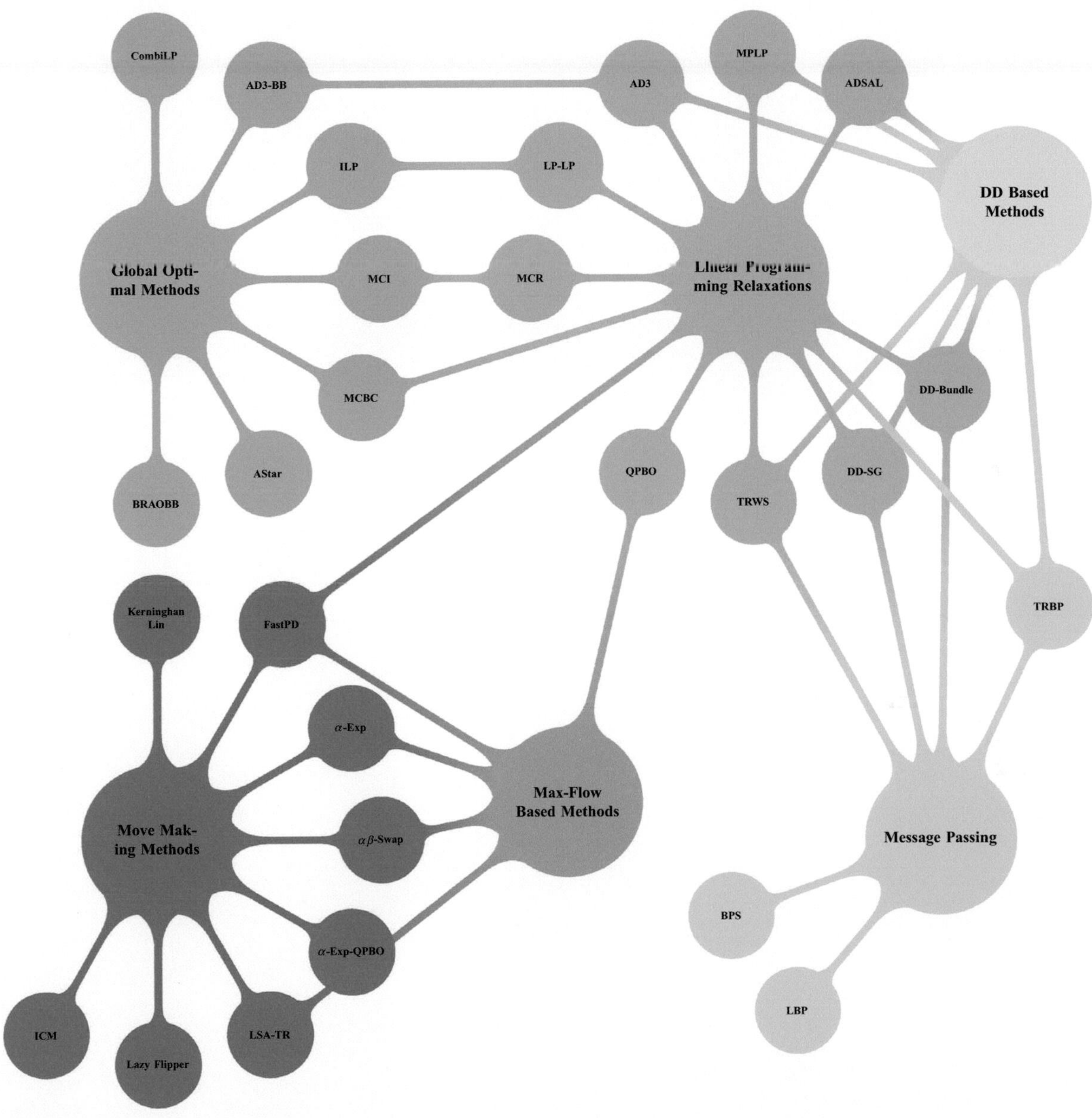

Figure B.1 Schematic taxonomy of the inference methods evaluated in the benchmark study by Kappes, Andres *et al.* (2015) © 2015 Springer.

plications, such as Kalman filtering, require the computation of this uncertainty (either explicitly as posterior covariances or implicitly as inverse covariances) to optimally integrate new measurements with previously computed estimates (Dickmanns and Graefe 1988; Matthies, Kanade, and Szeliski 1989; Szeliski 1989).

Appendix C

Supplementary material

C.1 Datasets and benchmarks 756
C.2 Software 761
C.3 Slides and lectures 768

R. Szeliski, *Computer Vision*, Texts in Computer Science,
https://doi.org/10.1007/978-3-030-34372-9

In this final appendix, I summarize some of the supplementary materials that may be useful to students, instructors, and researchers. The book's website at https://szeliski.org/Book contains updated lists of related courses, so please check there as well.

C.1 Datasets and benchmarks

As I mentioned in the introduction, one of the keys to developing reliable vision algorithms is to test your procedures on challenging and representative datasets. When ground truth or other people's results are available, such test can be even more informative (and quantitative).

Over the years, a large number of datasets have been developed for testing and evaluating computer vision algorithms, e.g., Middlebury stereo (Scharstein and Szeliski 2002), PASCAL (Everingham, Van Gool *et al.* 2010), ImageNet (Russakovsky, Deng *et al.* 2015), KITTI (Geiger, Lenz, and Urtasun 2012), Sintel (Butler, Wulff *et al.* 2012), and COCO (Lin, Maire *et al.* 2014).

Many of these datasets come with associated benchmarks where the results (and often pointers to code) for the latest algorithms can be found. I have already mentioned (and in some cases tabulated) many of these datasets in previous chapters of the book. In this appendix, I provide a summary of these datasets. You can also find older, less frequently used datasets in the first edition of this book (Szeliski 2010, Appendix C.1) and an up-to-date list on VisionBib.Com (http://datasets.visionbib.com), which has been curated and maintained by Keith Price since 1994.

Below, I list some of the more popular datasets, grouped by the book chapters to which they most closely correspond.

Chapter 2: Image formation

- CUReT: Columbia-Utrecht Reflectance and Texture Database, https://www1.cs.columbia.edu/CAVE/software/curet (Dana, van Ginneken *et al.* 1999).

- Middlebury Color Datasets: registered color images taken by different cameras to study how they transform gamuts and colors, https://vision.middlebury.edu/color/data (Chakrabarti, Scharstein, and Zickler 2009).

Chapter 4: Model fitting and optimization

- Middlebury test datasets for evaluating MRF minimization/inference algorithms, https://vision.middlebury.edu/MRF/results (Szeliski, Zabih *et al.* 2008).

- The OpenGM2 library and benchmarks for discrete factor graph models, http://hciweb2.iwr.uni-heidelberg.de/opengm (Kappes, Andres *et al.* 2015).

Chapter 5: Deep learning

- Small-scale datasets suitable for training a simple CNN as a useful teaching tool:[1] MNIST (LeCun, Cortes, and Burges 1998), CIFAR-100 (Krizhevsky 2009), and Fashion MNIST (Xiao, Rasul, and Vollgraf 2017).

- PyTorch TorchVision provides a great way to easily download some of the popular computer vision datasets, https://pytorch.org/vision/stable/datasets.html. TensorFlow also provides similar support with TensorFlow Datasets, https://www.tensorflow.org/datasets.

[1] See, e.g., https://pytorch.org/tutorials/beginner/blitz/cifar10_tutorial.html.

- Widely used recognition, detection, and segmentation datasets and benchmarks, as listed in Tables 6.1–6.4; separate datasets for other tasks such as image enhancement, motion estimation, and stereo, are discussed in later sections.

Chapter 6: Recognition

- The face recognition and detection datasets listed in Table 6.1 and Masi, Wu *et al.* (2018).
- The Caltech pedestrian detection benchmark (Dollár, Belongie, and Perona 2010) and person detection subtasks in datasets such as KITTI, http://www.cvlibs.net/datasets/kitti (Geiger, Lenz, and Urtasun 2012) and Cityscapes, https://www.cityscapes-dataset.com (Cordts, Omran *et al.* 2016)
- Table 6.2 lists datasets and benchmarks for image classification, general object detection, and segmentation. Two recent workshops that highlight the latest results on these datasets are the Robust Vision Challenge Zendel *et al.* (2020) (see Table C.1) and the COCO + LVIS Joint Recognition Challenge Kirillov, Lin *et al.* (2020).
- Datasets and benchmarks for fine-grained category recognition can be found at the CVPR Workshop on Fine-Grained Visual Categorization, https://sites.google.com/view/fgvc8 as well as some of the papers on this topic discussed in Section 6.2.2.
- Table 6.3 lists some datasets for video understanding and action recognition.
- Table 6.4 lists some widely used datasets for vision and language research, which includes image captioning, dense annotation, visual question answering, and visual dialog.

Chapter 7: Feature detection and matching

- The HPatches dataset and benchmark (Balntas, Lenc *et al.* 2020) is often used to evaluate new feature detectors and descriptors.
- The Image Matching Benchmark (Jin, Mishkin *et al.* 2021) is also widely used and has associated workshops.
- Visual localization datasets such as Aachen Day-Night (Sattler, Maddern *et al.* 2018) are also often used.
- Pointers to datasets for evaluating instance retrieval algorithms can be found in Zheng, Yang, and Tian (2018).
- Non-semantic image segmentation (splitting an image into "reasonable pieces" without labeling their content) is not widely studied any more. Pointers to classic datasets such as the Berkeley Segmentation Dataset and Benchmark (Martin, Fowlkes *et al.* 2001) can be found in the first edition of this book (Szeliski 2010, Appendix C.1).

Chapter 9: Motion estimation

- The Middlebury optical flow evaluation website, https://vision.middlebury.edu/flow (Baker, Scharstein *et al.* 2011) continues to be used for evaluation, since it contains a variety of short real-world sequences.

- Most optical flow algorithms are evaluated on the Sintel dataset, http://sintel.is.tue.mpg.de (Butler, Wulff *et al.* 2012), since it contains both training and test subsets and an active leaderboard, although the videos are stylized computer animations.
- Many algorithms also train and test on the KITTI flow benchmark (Geiger, Lenz, and Urtasun 2012), although it only contains videos acquired from a driving vehicle. The computer-generated sequences in the VIsual PERception (VIPER) benchmark (Richter, Hayder, and Koltun 2017) also contain driving sequences. Mayer, Ilg *et al.* (2018, Table 1) tabulates widely-used datasets for optical flow and depth estimation and shows some sample images in Figure 1.
- A comparison of flow algorithm performance across different datasets (listed in Table C.1) can be found in the Robust Vision Challenge workshop (http://www.robustvision.net).
- For video object segmentation, the Densely Annotated VIdeo Segmentation (DAVIS) dataset Pont-Tuset, Perazzi *et al.* (2017) contains a set of widely-used evaluation video clips with ground-truth segmentation data. There is also a newer, larger, dataset called YouTube-VOS (Xu, Yang *et al.* 2018) with its own associated set of challenges and leaderboards.
- Datasets for video object tracking (VOT) and multiple object tracking (MOT) can be found at the associated workshops (Kristan, Leonardis *et al.* 2020; Dendorfer, Ošep *et al.* 2021). A wider range of objects to track can be found in the Track Any Object (TAO) dataset by Dave, Khurana *et al.* (2020).

Chapter 10: Computational photography

- The High Dynamic Range radiance maps captured by Debevec and Malik (1997) at https://www.debevec.org/Research/HDR are still the go-to place to find high-quality HDR images.
- The RealSR real-world super-resolution dataset developed by Cai, Zeng *et al.* (2019) can be used to train and test SR algorithms on real imaging degradations. This dataset forms the basis for the NTIRE challenges on real image super-resolution (Cai, Gu *et al.* 2019), which provide empirical comparisons of recent deep network-based algorithms.
- The latest benchmark for comparing image denoising algorithms, the NTIRE 2020 Challenge on Real Image Denoising (Abdelhamed, Afifi *et al.* 2020), is based on a smartphone image denoising dataset (SIDD) (Abdelhamed, Lin, and Brown 2018) created by averaging sets of real-world noisy images.
- Thea alpha matting evaluation website, http://alphamatting.com (Rhemann, Rother *et al.* 2009) provides a standard set of test images and a leaderboard.
- The video matting dataset at https://videomatting.com (Erofeev, Gitman *et al.* 2015) provides stop-motion animation videos created by carefully hand-matting each frame.
- Lin, Ryabtsev *et al.* (2021) describe a high-resolution real-time video matting system along with two new video and image matting datasets.
- The AIM 2020 Workshop and Challenges on image inpainting (Ntavelis, Romero *et al.* 2020a) provides datasets for evaluating such algorithms.

Chapter 11: Structure from motion and SLAM

- The Benchmark for 6DOF Object Pose (BOP) developed by Hodaň, Michel *et al.* (2018) has results from the recent challenge and workshop at https://bop.felk.cvut.cz/challenges/bop-challenge-2020 and http://cmp.felk.cvut.cz/sixd/workshop_2020.

- The Long-Term Visual Localization Benchmark, https://www.visuallocalization.net, includes datasets such as Aachen Day-Night (Sattler, Maddern *et al.* 2018) and InLoc (Taira, Okutomi *et al.* 2018) along with an associated set of challenges and workshop held at ECCV 2020.

- The 1DSfM collection of landmark images created by Wilson and Snavely (2014) (https://www.cs.cornell.edu/projects/1dsfm), which is an extension of the Photo Tourism dataset created by Snavely, Seitz, and Szeliski (2008a), is widely used to test large-scale structure from motion algorithms. The poses provided with this dataset, which were obtained using the software in Wilson and Snavely (2014), are generally considered as "ground truth" when testing more efficient algorithms, although they have never been geo-registered. The ETH3D, https://www.eth3d.net (Schöps, Schönberger *et al.* 2017) and Tanks and Temples, https://www.tanksandtemples.org (Knapitsch, Park *et al.* 2017) datasets are also occasionally used.

- Some widely used benchmarks for SLAM systems include a benchmark for RGB-D SLAM systems (Sturm, Engelhard *et al.* 2012), the KITTI Visual Odometry / SLAM benchmark (Geiger, Lenz *et al.* 2013), the synthetic ICL-NUIM dataset (Handa, Whelan *et al.* 2014), the TUM monoVO dataset (Engel, Usenko, and Cremers 2016), the EuRoC MAV dataset (Burri, Nikolic *et al.* 2016), the ETH3D SLAM benchmark (Schöps, Sattler, and Pollefeys 2019a), and the GSLAM general SLAM framework and benchmark (Zhao, Xu *et al.* 2019). Many of these are surveyed and categorized in the paper by Ye, Zhao, and Vela (2019), which was presented at the ICRA 2019 Workshop on Dataset Generation and Benchmarking of SLAM Algorithms for Robotics and VR/AR, https://sites.google.com/view/icra-2019-workshop/home.

Chapter 12: Depth estimation

- The most widely used datasets and benchmarks for two-frame and multi-view stereo are listed in Tables 12.1 and C.1. Among these, Middlebury stereo, KITTI, and ETH3D maintain active leaderboards tabulating the performance of two-frame stereo algorithms. For multi-view stereo, ETH3D and Tanks and Temples have leaderboards, and DTU is widely used and self-reported in papers.

- Many algorithms that train and test on the same dataset (e.g., KITTI) do not perform as well when tested on different datasets (Zendel *et al.* 2020). Song, Yang *et al.* (2021) discuss this issue and domain adaptation techniques that can reduce this problem.

- KeystoneDepth has a large set of rectified historical image pairs, but without ground truth depth (Luo, Kong *et al.* 2020).

- For monocular depth inference, many algorithms train and test on the KITTI outdoor driving image sequences. The MiDaS system developed by Ranftl, Lasinger *et al.* (2020) federates a number of monocular depth inference datasets and also adds thousands of stereo image pairs from 3D movies for training, validation, and testing.

	Stereo	Flow	Depth	Obj. Det.	Semantic	Instance	Panoptic
ADE20K[1]					X		
COCO[2]				X	X	X	X
Cityscapes[3]					X	X	X
ETH3D[4]	X						
HD1K[5]		X					
KITTI[6]	X	X	X		X	X	X
MVD[7]				X	X	X	X
Middlebury[8]	X	X					
MPI Sintel[9]		X	X				
Objects365[10]				X			
OID[11]				X		X	
rabbitai[12]			X				
ScanNet[13]					X	X	
VIPER[14]		X	X		X	X	X
WildDash[15]					X	X	X

[1] http://sceneparsing.csail.mit.edu (Zhou, Zhao *et al.* 2019)
[2] http://cocodataset.org (Lin, Maire *et al.* 2014)
[3] https://www.cityscapes-dataset.com (Cordts, Omran *et al.* 2016)
[4] https://www.eth3d.net (Schöps, Schönberger *et al.* 2017)
[5] http://hci-benchmark.org (Kondermann, Nair *et al.* 2016)
[6] http://www.cvlibs.net/datasets/kitti (Menze and Geiger 2015)
[7] http://mapillary.com/dataset/vistas (Neuhold, Ollmann *et al.* 2017)
[8] http://vision.middlebury.edu (Scharstein, Hirschmüller *et al.* 2014)
[9] http://sintel.is.tue.mpg.de (Butler, Wulff *et al.* 2012)
[10] https://www.objects365.org (Shao, Li *et al.* 2019)
[11] https://storage.googleapis.com/openimages/web/index.html (Kuznetsova, Rom *et al.* 2020)
[12] https://rabbitai.de/benchmark (Schilling, Gutsche *et al.* 2020)
[13] http://kaldir.vc.in.tum.de/scannet_benchmark (Dai, Chang *et al.* 2017)
[14] https://playing-for-benchmarks.org (Richter, Hayder, and Koltun 2017)
[15] https://www.wilddash.cc (Zendel, Honauer *et al.* 2018)

Table C.1 The list of seven challenges (one per column) in the Robust Vision Challenge 2020 (http://www.robustvision.net) along with the datasets and benchmarks that are included in each challenge.

Chapter 13: 3D reconstruction

- The DiLiGenT photometric stereo dataset provides images taken under calibrated directional lighting and objects with general reflectance along with ground truth shapes (Shi, Mo *et al.* 2019). It also provides a taxonomy and evaluation of photometric stereo methods for general non-Lambertian materials and unknown lighting.

- NYU3D (Silberman, Hoiem *et al.* 2012) and ScanNet (Dai, Chang *et al.* 2017) were some of the early 3D indoor scene datasets used to study 3D reconstruction and range fusion algorithms. More recent algorithms such as Chabra, Lenssen *et al.* (2020) or Weder, Schonberger *et al.* (2021) use some combination of 3D Scenes (Zhou and Koltun 2013), ICL-NUIM (Handa, Whelan *et al.* 2014), ShapeNet (Chang, Funkhouser *et al.* 2015), and Tanks and Temples (Knapitsch, Park *et al.* 2017). Reviews of RGB-D datasets can be found in Firman (2016) and Zollhöfer, Stotko *et al.* (2018).

- Over the years, a number of 3D human body and motion datasets have been captured, including HumanEva (Sigal, Balan, and Black 2010), MPI FAUST (Bogo, Romero *et al.* 2014), Panoptic Studio (Joo, Simon *et al.* 2019), EHF (Pavlakos, Choutas *et al.* 2019), AMASS (Mahmood, Ghorbani *et al.* 2019), and 3D Poses in the Wild (3DPW) (von Marcard, Henschel *et al.* 2018).[2]

- In parallel with these datasets, 3D human body models and fitting algorithms have been developed, including SCAPE (Anguelov, Srinivasan *et al.* 2005), BlendSCAPE (Hirshberg, Loper *et al.* 2012). SMPL (Loper, Mahmood *et al.* 2015), MANO (Joo, Simon, and Sheikh 2018), SMPL-X (Pavlakos, Choutas *et al.* 2019), VIBE (Kocabas, Athanasiou, and Black 2020), ExPose (Choutas, Pavlakos *et al.* 2020), STAR (Osman, Bolkart, and Black 2020), Learned Gradient Descent (Song, Chen, and Hilliges 2020), and FrankMoCap (Rong, Shiratori, and Joo 2020). These are described in more detail in Section 13.6.4.

Chapter 14: Image-based rendering

- The original Photo Tourism dataset created by Snavely, Seitz, and Szeliski (2008a) was extended by Wilson and Snavely (2014) to the much larger 1DSfM collection of landmark images at https://www.cs.cornell.edu/projects/1dsfm.

- The Stanford Light Field Archive, http://lightfield.stanford.edu (Wilburn, Joshi *et al.* 2005) and the 4D Light Field Dataset, https://lightfield-analysis.uni-konstanz.de (Honauer, Johannsen *et al.* 2016) both provide high-quality light fields for research and projects.

- The Virtual Viewpoint Video multi-viewpoint video with per-frame depth maps, https://www.microsoft.com/en-us/research/group/interactive-visual-media/#!downloads (Zitnick, Kang *et al.* 2004) continues to be widely used for research into 3D and multi-view video compression. Newer multi-view video datasets include Facebook Surround 360, https://github.com/facebook/Surround360 (Parra Pozo, Toksvig *et al.* 2019) and Deep View Video https://augmentedperception.github.io/deepviewvideo (Broxton, Flynn *et al.* 2020).

- Most of the recent Neural Rendering papers discussed in Section 14.6 either provide their own multi-view datasets or re-use datasets from previously published papers.

C.2 Software

Since the publication of the first edition of this book, when high quality open source computer vision software was still scarce, the last decade has seen an explosion in such software. Most research papers today come with open source software implementation, often tested on well-known datasets. The web site Papers with Code (https://paperswithcode.com) lists many of the latest machine learning research papers along with pointers to their implementations.

When getting started in computer vision, many students either dive into using and extending such code, or work through tutorials on deep learning frameworks such as PyTorch (https://pytorch.org/tutorials) or TensorFlow (https://www.tensorflow.org/tutorials). The Dive into Deep Learning book and web site (Zhang, Lipton *et al.* 2021) has associated Python Notebooks, based on the Apache MXNet machine learning framework, which can be downloaded and run as students are working through the material.

[2] Additional datasets can be found on the MPI Perceiving Systems https://ps.is.mpg.de/code and Virtual Humans group https://virtualhumans.mpi-inf.mpg.de/software.html web pages.

For “classic” computer vision algorithms not based on deep learning, one of the best sources continues to be the Open Source Computer Vision (OpenCV) library (https://opencv.org), which was originally developed by Gary Bradski and his colleagues at Intel (Bradsky and Kaehler 2008; Kaehler and Bradski 2017). The library has more than 2500 optimized algorithms, which includes both classic and state-of-the-art computer vision and machine learning algorithms, with C++, Python, Java and MATLAB interfaces.

For most of my research career, I did my software development in C++, since I liked its run-time efficiency, strong type checking, and object-oriented framework. In the last few years, however, I’ve shifted to Python. Having an interactive environment that does not require re-compilation and linking is a big plus. Even better, the NumPy (https://numpy.org/) multidimensional array (tensor) library, when used in the right way, introduces developers to array-based (matrix) arithmetic and (hopefully) dissuades them from writing pixel-iteration loops that are slow to write and error-prone. A big advantage of writing in this fashion is that it maps closely to the abstractions used in the deep learning frameworks such as PyTorch and TensorFlow. It also often results in highly optimized code that can be run on both CPUs and GPUs with minimal changes.[3]

In the rest of this section, I list some additional software packages and libraries that students may find useful. You can also find pointers to older (currently less used) software packages in the first edition of this book (Szeliski 2010, Appendix C.2).

Chapter 3: Image processing

- Before diving into OpenCV, I would encourage you to write some simple image processing functions in NumPy using the built-in multidimensional array notation. It’s fine to use OpenCV for image input/output and to use Matplotlib for visualization. There are also other high-level packages for image processing, such as scikit-image and PIL/Pillow. A more recently developed computer vision library is MMCV (https://openmmlab.com/codebase#MMCV).

- As a warm-up exercise, before diving into machine learning but after doing the basic PyTorch or TensorFlow tutorials, try porting your NumPy code into one of these languages.

- Another language that supports array-level functional programming is Halide (https://halide-lang.org) (Ragan-Kelley, Barnes *et al.* 2013), which provides optimized compilation onto a large number of targets, including CPUs, GPUs, mobile processors, and DSPs such as the Qualcomm Hexagon.

- For wavelets, PyWavelets (https://pywavelets.readthedocs.io) has a nice extensive set of variants.

- I have always found it helpful to have an image viewer where I can quickly flip between aligned images to look for differences, which show up much better than when viewing images side-by-side.

Chapter 4: Model fitting and optimization

- Scikit-learn (https://scikit-learn.org) implements a number of algorithms for regression, i.e., scattered data interpolation.

[3] See, e.g., https://cupy.dev or https://devopedia.org/numpy.

- OpenGM (http://hciweb2.iwr.uni-heidelberg.de/opengm) is a C++ template library for discrete factor graph models and distributive operations on these models. It includes state-of-the-art optimization and inference algorithms beyond message passing.

Chapter 5: Deep learning

- Scikit-learn (https://scikit-learn.org) includes a large number of traditional machine learning algorithms and tutorials. Glassner (2018, Chapter 15) has a nice review of these algorithms along with some exercises.

- Over the last decade, a large number of deep learning software frameworks and programming language extensions have been developed. The Wikipedia entry on deep learning software lists over twenty such frameworks.[4]

- The Dive into Deep Learning book (Zhang, Lipton *et al.* 2021) and associated course (Smola and Li 2019) use MXNet for all the examples in the text, but they have recently released PyTorch and TensorFlow code samples as well. Stanford's CS231n (Li, Johnson, and Yeung 2019) and Johnson (2020) include a lecture on the fundamentals of PyTorch and TensorFlow.

- Some classes also use simplified frameworks that require the students to implement more components, such as the Educational Framework (EDF) developed by McAllester (2020) and used in Geiger (2021).

- PyTorch (https://pytorch.org) and TensorFlow (https://www.tensorflow.org) are currently the most widely used deep learning frameworks. Compared to NumPy, they enable much faster numerical computing by leveraging a GPU.

- Tensor Processing Units (TPUs) are specialized hardware optimized specifically for deep learning and can offer speed improvements over GPUs. TPUs are only available through Google Cloud. While they are still less popular than GPUs, many of the new papers using TPUs find it most effective to use JAX (https://github.com/google/jax).

- Even though deep learning frameworks provide some support for image augmentation, the `imgaug` library (https://github.com/aleju/imgaug) provides a much wider range of augmentation possibilities.

- VISSL (https://vissl.ai) is an extendable self-supervised learning framework written in PyTorch. It provides many benchmarks, model implementations, and weights.

- Google Colab (https://colab.research.google.com) is often used as a free cloud computing platform for the assignments in computer vision courses that can benefit from a GPU. It provides access to a GPU and memory to download datasets. The programming environment uses Jupyter interactive notebooks, which makes code easy to share and reproduce.

- Kaggle (https://www.kaggle.com), a Google subsidiary, provides a platform to compete with your own models on many popular computer vision datasets. The vast majority of winning models now using deep learning, with many of the challenges providing lively discussions about how different people attempted the problem and explored the data.

[4] https://en.wikipedia.org/wiki/Comparison_of_deep-learning_software

- Variants of the LeNet-5 architecture (Figure 5.33) are commonly used as the first convolutional neural network introduced in courses and tutorials on the subject.[5] Although the MNIST dataset (LeCun, Cortes, and Burges 1998) originally used to train LeNet-5 is still sometimes used, it is more common to use the more challenging CIFAR-10 (Krizhevsky 2009) or Fashion MNIST (Xiao, Rasul, and Vollgraf 2017).

- Andrej Karpathy provides a useful guide for training neural networks at https://karpathy.github.io/2019/04/25/recipe, which may help avoid common issues.

- A great way to experiment with various CNN architectures is to download pre-trained models from a *model zoo* such as the TorchVision library (https://github.com/pytorch/vision). If you look in the torchvision/models folder, you will find implementations of AlexNet, VGG, GoogleNet, Inception, ResNet, DenseNet, MobileNet, and ShuffleNet, along with other models for classification, object detection, and image segmentation. Even more recent models can be found in the PyTorch Image Models library (timm), https://github.com/rwightman/pytorch-image-models. Similar collections of pre-trained models exist for other languages, e.g., https://www.tensorflow.org/lite/models for efficient (mobile) TensorFlow models.

- In addition to software frameworks and libraries, deep learning code development usually benefits from good visualization libraries such as TensorBoard (https://www.tensorflow.org/tensorboard) and Visdom (https://github.com/fossasia/visdom). A great way to get some intuition on how deep networks update the weights and carve out a solution space during training is to play with the interactive visualization at https://playground.tensorflow.org, as shown in Figure 5.32.[6] OpenAI also recently released a great interactive tool called Microscope (https://microscope.openai.com/models), which allows people to visualize the significance of every neuron in a network.

- The PyTorch3D library (https://github.com/facebookresearch/pytorch3d) provides representations and functions to process 3D volumes and 3D meshes using deep neural networks.

Chapter 6: Recognition

- For large-scale similarity search and clustering, the GPU-enabled Faiss library (https://github.com/facebookresearch/faiss) developed by Johnson, Douze, and Jégou (2021) can scale to very large datasets.

- There are many open-source frameworks such as Classy Vision (https://classyvision.ai), TensorFlow Core (https://www.tensorflow.org/tutorials/images/classification), and MMClassification (https://openmmlab.com) for training and fine tuning image and video classification models. You can also upload your images to the Computer Vision Explorer (https://vision-explorer.allenai.org) to see how well popular computer vision models perform on them.

- Open-source frameworks for training and fine-tuning object detectors include the TensorFlow Object Detection API (https://github.com/tensorflow/models/tree/master/research/object_detection), PyTorch's Detectron2 (https://github.com/facebookresearch/detectron2), and OpenMMLab's MMDetection (https://openmmlab.com/codebase#MMDetection) (Chen, Wang *et al.* 2019).

---

[5] See, e.g., https://pytorch.org/tutorials/beginner/blitz/cifar10_tutorial.html.

[6] Additional interactive demonstrations can be found at https://cs.stanford.edu/people/karpathy/convnetjs.

- Detectron2 also includes semantic and panoptic segmentation, which can also be found in TensorFlow Core (https://www.tensorflow.org/tutorials/images/segmentation) and many other libraries.
- OpenPose (Cao, Hidalgo *et al.* 2019) and DensePose (Güler, Neverova, and Kokkinos 2018) are two popular software packages for determining "stick figure" and dense pixel-labeled 3D pose from 2D images.
- Pointers to software for more specialized tasks such as face detection and recognition, pedestrian detection, video understanding, and vision and language can usually be found alongside the latest papers discussed in Chapter 6.

Chapter 7: Feature detection and matching

- Implementations of many of the "classic" feature detectors and descriptors can be found in the OpenCV Features2D class and sub-classes.
- Implementations of newer DNN-based detectors and descriptors can be found associated with the papers discussed in Chapter 7 and the datasets discussed in Appendix C.1.

Chapter 9: Motion estimation

- The leaderboards (evaluation results) for the Middlebury (https://vision.middlebury.edu/flow/eval/results/results-e1.php), Sintel (http://sintel.is.tue.mpg.de/results), and KITTI (http://www.cvlibs.net/datasets/kitti/eval_scene_flow.php?benchmark=flow) datasets contain pointers to the latest optical flow papers and code.

Chapter 10: Computational photography

- Pointers to papers and algorithms for a variety of computational photography tasks such as super-resolution, image denoising, image and video matting, and inpainting can be found at the benchmarks and workshops associated with these topics, as discussed in Chapter 10 and the list of datasets in Appendix C.1.

Chapter 11: Structure from motion and SLAM

- OpenCV implements a number of widely used camera calibration and pose estimation algorithm in the calib3d module, as does OpenGV (https://laurentkneip.github.io/opengv) (Kneip and Furgale 2014) and OpenMVG (https://github.com/openMVG/openMVG) (Moulon, Monasse *et al.* 2016).
- You can find an experimental comparison of a number of RANSAC variants at https://opencv.org/evaluating-opencvs-new-ransacs/.
- A large number of open-source bundle adjustment algorithms designed to handle unordered photo collections have been developed over the years, including:
  - SBA: sparse bundle adjustment (https://www.ics.forth.gr/~lourakis/sba) (Lourakis and Argyros 2009).
  - Simple sparse bundle adjustment (SSBA) (https://github.com/chzach/SSBA).

- Bundler, structure from motion for unordered image collections (https://phototour.cs.washington.edu/bundler) (Snavely, Seitz, and Szeliski 2006).
- The Ceres Solver for bundle adjustment and general non-linear least squares (http://ceres-solver.org).
- MCBA (Multicore Bundle Adjustment) (https://grail.cs.washington.edu/projects/mcba) (Wu, Agarwal *et al.* 2011).
- Visual SfM (http://ccwu.me/vsfm), which wraps a GUI around several reconstruction algorithms (Wu, Agarwal *et al.* 2011; Wu 2013).
- MVE (https://www.gcc.tu-darmstadt.de/home/proj/mve), a complete SfM pipeline with densification, meshing, and texturing (Fuhrmann, Langguth *et al.* 2015).
- The Theia global structure from motion library (http://www.theia-sfm.org) (Sweeney, Hollerer, and Turk 2015).
- OpenMVG (Open Multiple View Geometry) https://github.com/openMVG/openMVG (Moulon, Monasse *et al.* 2016).
- COLMAP (https://github.com/colmap/colmap), which includes both a large-scale structure from motion system (Schönberger and Frahm 2016) and a multi-view stereo pipeline (Schönberger, Zheng *et al.* 2016).
- Square Root Bundle Adjustment (https://vision.in.tum.de/research/vslam/rootba) (Demmel, Sommer *et al.* 2021).

Among these, COLMAP appears to be the most often used today in other research projects, e.g., for image-based rendering systems.

- Popular open-source packages for Simultaneous Localization and Mapping (SLAM) and Visual Odometry (VO or VIO) include
  - LSD-SLAM (large-scale direct SLAM) (Engel, Schöps, and Cremers 2014),
  - ORB-SLAM (Mur-Artal, Montiel, and Tardos 2015) and ORB-SLAM2 (Mur-Artal and Tardós 2017),
  - SVO (semi-direct visual odometry) (Forster, Zhang *et al.* 2017),
  - GTSAM (Dellaert and Kaess 2017; Dellaert 2021),
  - DSO (direct sparse odometry) (Engel, Koltun, and Cremers 2018),
  - BAD SLAM (bundle adjusted direct RGB-D SLAM) (Schöps, Sattler, and Pollefeys 2019a), and
  - GSLAM (a general SLAM framework and benchmark) (Zhao, Xu *et al.* 2019).
- There are also highly-optimized SLAM/VIO libraries available on mobile platforms, such as iOS (ARKit), Android (ARCore), and Facebook (Spark AR Studio), designed for easy integration into mobile augmented reality applications.

Chapter 12: Stereo correspondence

- Open-source software for the latest stereo matching, multi-view, and monocular depth inference algorithms usually accompanies recently published papers. Lists of the most recent and best performing algorithms can be found on the leaderboards associated with the most popular benchmarks such as Middlebury, KITTI, ETH3D, and Tanks and Temples, which are discussed in Appendix C.1 and Tables 12.1 and C.1. algorithm

- Both MVE (https://www.gcc.tu-darmstadt.de/home/proj/mve) (Fuhrmann, Langguth *et al.* 2015) and COLMAP (https://github.com/colmap/colmap) (Schönberger, Zheng *et al.* 2016) provide complete 3D reconstruction pipelines that include structure from motion, multi-view stereo densification, mesh generation, and texturing. A review of earlier packages can be found in Furukawa and Hernández (2015).
- A number of high-quality commercial photogrammetry packages such as CapturingReality, ContextCapture, Metashape, and Pix4D, which grew out of computer vision research labs, provide similar functionality.[7]

Chapter 13: 3D reconstruction

- The Scanalyze package (https://graphics.stanford.edu/software/scanalyze) developed at the Stanford Graphics lab contains a number of algorithms for aligning, registering, and fusing range images and 3D meshes.
- Open3D (http://www.open3d.org) is a more recent package with similar registration and volumetric merging capabilities (Zhou, Park, and Koltun 2018).
- MeshLab (https://www.meshlab.net) is a widely used package for processing, editing, and viewing 3D triangular meshes (Cignoni, Callieri *et al.* 2008).
- X3D is an XML-based format for representing 3D geometry and is an updated version of the original VRML (.wrl) format. A number of high-quality interactive viewers can be found on the web.
- The Point Cloud Library (PCL) at https://pointclouds.org is a library for point cloud processing and includes functions for feature detection, registration, segmentation, and visualization.
- As mentioned previously, both MVE and COLMAP have functions to generate 3D texture-mapped meshes (Fuhrmann, Langguth *et al.* 2015; Schönberger, Zheng *et al.* 2016).
- Canvas (https://canvas.io) is a phone-based 3D capture app that merges depth data from the phone's lidar sensor to produce complete textured 3D meshes.

Chapter 14: Image-based rendering

- As with other areas of computer vision, most recently published image-based rendering and neural rendering papers now come with open source implementations.

Appendix A: Linear algebra and numerical techniques

- The first edition of this book (Szeliski 2010, Appendix C.2) lists a number of widely used linear algebra and non-linear least squares packages such as BLAS, LAPACK, ATLAS, MKL, MINPACK, PARADISO, TAUCS, HSL, and ITSOL. Most of these are now integrated into larger packages such as Python's NumPy and GPU machine learning frameworks such as PyTorch and TensorFlow.
- If you are interested in sparse linear least squares solvers, it is worth looking at SuiteSparse (https://people.engr.tamu.edu/davis/suitesparse.html), since it contains a wide range of algorithms and associated publications (Davis 2006, 2011).

[7] See also https://peterfalkingham.com/2020/07/10/free-and-commercial-photogrammetry-software-review-2020 and https://all3dp.com/1/best-photogrammetry-software.

Appendix B: Bayesian modeling and inference

- The Middlebury benchmark for MRF minimization, https://vision.middlebury.edu/MRF/code, contains implementations of basic MRF inference algorithms (Szeliski, Zabih *et al.* 2008).
- The OpenGM2 library and benchmarks for discrete factor graph models, http://hciweb2.iwr.uni-heidelberg.de/opengm, contains a more extensive and up-to-date set of algorithms. (Kappes, Andres *et al.* 2015).

C.3 Slides and lectures

While there are no official slide sets to go with this book, its content largely parallels that of the courses I have co-taught at the University of Washington, https://www.cs.washington.edu/education/courses/cse576.

Related computer vision and deep learning classes include:

- Noah Snavely's Introduction to Computer Vision class at Cornell Tech, https://www.cs.cornell.edu/courses/cs5670/2021sp/
- Alyosha Efros' Intro to Computer Vision and Computational Photography class at Berkeley https://inst.eecs.berkeley.edu/~cs194-26/fa20.
- David Fouhey's and Justin Johnson's Computer Vision class at the University of Michigan, https://web.eecs.umich.edu/~justincj/teaching/eecs442.
- Bill Freeman, Antonio Torralba, and Phillip Isola's Advances in Computer Vision class at MIT http://6.869.csail.mit.edu/sp21.
- Justin Johnson's Deep Learning for Computer Vision class at the University of Michigan, https://web.eecs.umich.edu/~justincj/teaching/eecs498.
- Yann LeCun and Alfredo Canziani's Deep Learning class at NYU, https://atcold.github.io/NYU-DLSP21.
- UC Berkeley's class on Deep Unsupervised Learning, https://sites.google.com/view/berkeley-cs294-158-sp20.

You can find a more comprehensive list of such courses on the book's web site, https://szeliski.org/Book/default.htm#Slides.

There are also some great online lectures series, including:

- The 2004 UW-MSR Course on Vision Algorithms, http://www.cs.washington.edu/education/courses/cse577/04sp/index.htm.
- The 2020-2021 TUM AI Guest Lecture Series, https://niessner.github.io/TUM-AI-Lecture-Series.
- The 2020-2021 3DGV virtual seminar series on Geometry Processing and 3D Computer Vision, https://3dgv.github.io.

References

Aafaq, N., Mian, A., Liu, W., Gilani, S. Z., and Shah, M. (2019). Video description: A survey of methods, datasets, and evaluation metrics. *ACM Computing Surveys*, 52(6):1–37.

Abbeel, P. (2019). University of California at Berkeley CS 287 class: Advanced robotics. Slides and video recordings available at https://people.eecs.berkeley.edu/~pabbeel/cs287-fa19.

Abdel-Hakim, A. E. and Farag, A. A. (2006). CSIFT: A SIFT descriptor with color invariant characterstics. In *IEEE Computer Society Conference on Computer Vision and Pattern Recognition (CVPR)*, pp. 1978–1983.

Abdelhamed, A., Lin, S., and Brown, M. S. (2018). A high-quality denoising dataset for smartphone cameras. In *IEEE Conference on Computer Vision and Pattern Recognition (CVPR)*.

Abdelhamed, A., Afifi, M., Timofte, R., and Brown, M. S. (2020). NTIRE 2020 challenge on real image denoising: Dataset, methods and results. In *IEEE/CVF Conference on Computer Vision and Pattern Recognition (CVPR) Workshops*.

Abu-El-Haija, S., Kothari, N., Lee, J., Natsev, P., Toderici, G., Varadarajan, B., and Vijayanarasimhan, S. (2016). YouTube-8M: A large-scale video classification benchmark. *arXiv preprint arXiv:1609.08675*.

Abuolaim, A. and Brown, M. S. (2020). Defocus deblurring using dual-pixel data. In *European Conference on Computer Vision (ECCV)*.

Ackerman, E. (2019). Skydio's new drone is smaller, even smarter, and (almost) affordable. *IEEE Spectrum*. https://spectrum.ieee.org/automaton/robotics/drones/skydios-new-drone-is-smaller-even-smarter-and-almost-affordable.

Ackerman, E. (2020). Covariant uses simple robot and gigantic neural net to automate warehouse picking. *IEEE Spectrum*. https://spectrum.ieee.org/automaton/robotics/industrial-robots/covariant-ai-gigantic-neural-network-to-automate-warehouse-picking.

Ackermann, J. and Goesele, M. (2015). A survey of photometric stereo techniques. *Foundations and Trends® in Computer Graphics and Vision*, 9(3-4):149–254.

Ackermann, J., Langguth, F., Fuhrmann, S., and Goesele, M. (2012). Photometric stereo for outdoor webcams. In *IEEE Computer Society Conference on Computer Vision and Pattern Recognition (CVPR)*.

Ackley, D. H., Hinton, G. E., and Sejnowski, T. J. (1985). A learning algorithm for Boltzmann Machines. *Cognitive Science*, 9:147–169.

Adam, A., Rivlin, E., and Shimshoni, I. (2006). Robust fragments-based tracking using the integral histogram. In *IEEE Computer Society Conference on Computer Vision and Pattern Recognition (CVPR)*, pp. 798–805.

Adams, A., Talvala, E.-V., Park, S. H., Jacobs, D. E., Ajdin, B., Gelfand, N., Dolson, J., Vaquero, D., Baek, J., Tico, M., Lensch, H. P. A., Matusik, W., Pulli, K., Horowitz, M., and Levoy, M. (2010). The Frankencamera: An experimental platform for computational photography. *ACM Transactions on Graphics*, 29(4):29.

Adams, A., Baek, J., and Davis, M. A. (2010). Fast high-dimensional filtering using the permutohedral lattice. In *Computer Graphics Forum (Eurographics)*, pp. 753–762.

Adelson, E. H. and Bergen, J. (1991). The plenoptic function and the elements of early vision. In *Computational Models of Visual Processing*, pp. 3–20.

Adiv, G. (1989). Inherent ambiguities in recovering 3-D motion and structure from a noisy flow field. *IEEE Transactions on Pattern Analysis and Machine Intelligence*, 11(5):477–490.

Agarwal, A. and Triggs, B. (2006). Recovering 3D human pose from monocular images. *IEEE Transactions on Pattern Analysis and Machine Intelligence*, 28(1):44–58.

R. Szeliski, *Computer Vision*, Texts in Computer Science,
https://doi.org/10.1007/978-3-030-34372-9

Agarwal, S., Snavely, N., Seitz, S. M., and Szeliski, R. (2010). Bundle adjustment in the large. In *European Conference on Computer Vision (ECCV)*, pp. 29–42.

Agarwal, S., Snavely, N., Simon, I., Seitz, S. M., and Szeliski, R. (2009). Building Rome in a day. In *IEEE International Conference on Computer Vision (ICCV)*.

Agarwal, S., Furukawa, Y., Snavely, N., Curless, B., Seitz, S. M., and Szeliski, R. (2010). Reconstructing Rome. *Computer*, 43(6):40–47.

Agarwal, S., Furukawa, Y., Snavely, N., Simon, I., Curless, B., Seitz, S. M., and Szeliski, R. (2011). Building Rome in a day. *Comuminications of the ACM*, 54(10):105–112.

Agarwala, A. (2007). Efficient gradient-domain compositing using quadtrees. *ACM Transactions on Graphics*, 26(3).

Agarwala, A., Hertzmann, A., Seitz, S., and Salesin, D. (2004). Keyframe-based tracking for rotoscoping and animation. *ACM Transactions on Graphics (Proc. SIGGRAPH)*, 23(3):584–591.

Agarwala, A., Agrawala, M., Cohen, M., Salesin, D., and Szeliski, R. (2006). Photographing long scenes with multi-viewpoint panoramas. *ACM Transactions on Graphics (Proc. SIGGRAPH)*, 25(3):853–861.

Agarwala, A., Dontcheva, M., Agrawala, M., Drucker, S., Colburn, A., Curless, B., Salesin, D. H., and Cohen, M. F. (2004). Interactive digital photomontage. *ACM Transactions on Graphics (Proc. SIGGRAPH)*, 23(3):292–300.

Agarwala, A., Zheng, K. C., Pal, C., Agrawala, M., Cohen, M., Curless, B., Salesin, D., and Szeliski, R. (2005). Panoramic video textures. *ACM Transactions on Graphics (Proc. SIGGRAPH)*, 24(3):821–827.

Aggarwal, J. K. and Cai, Q. (1999). Human motion analysis: A review. *Computer Vision and Image Understanding*, 73(3):428–440.

Aggarwal, J. K. and Nandhakumar, N. (1988). On the computation of motion from sequences of images—a review. *Proceedings of the IEEE*, 76(8):917–935.

Aggarwal, J. K. and Ryoo, M. S. (2011). Human activity analysis: A review. *ACM Computing Surveys*, 43(3):1–43.

Agin, G. J. and Binford, T. O. (1976). Computer description of curved objects. *IEEE Transactions on Computers*, C-25(4):439–449.

Agrawal, A., Batra, D., Parikh, D., and Kembhavi, A. (2018). Don't just assume; look and answer: Overcoming priors for visual question answering. In *IEEE Conference on Computer Vision and Pattern Recognition (CVPR)*.

Aharon, M., Elad, M., and Bruckstein, A. (2006). K-SVD: An algorithm for designing overcomplete dictionaries for sparse representation. *IEEE Transactions on Signal Processing*, 54(11):4311–4322.

Ahmadi, A. and Patras, I. (2016). Unsupervised convolutional neural networks for motion estimation. In *2016 IEEE International Conference on Image Processing (ICIP)*, pp. 1629–1633.

Ahonen, T., Hadid, A., and Pietikäinen, M. (2006). Face description with local binary patterns: Application to face recognition. *IEEE Transactions on Pattern Analysis and Machine Intelligence*, 28(12):2037–2041.

Aittala, M., Weyrich, T., and Lehtinen, J. (2015). Two-shot SVBRDF capture for stationary materials. *ACM Transactions On Graphics (Proc. SIGGRAPH)*, 34(4):110–1.

Akenine-Möller, T. and Haines, E. (2002). *Real-Time Rendering*. A K Peters, Wellesley, Massachusetts, 2nd edition.

Aksoy, Y., Kim, C., Kellnhofer, P., Paris, S., Elgharib, M., Pollefeys, M., and Matusik, W. (2018). A dataset of flash and ambient illumination pairs from the crowd. In *European Conference on Computer Vision (ECCV)*.

Al-Baali, M. and Fletcher, R. (1986). An efficient line search for nonlinear least squares. *Journal Journal of Optimization Theory and Applications*, 48(3):359–377.

Alahari, K., Kohli, P., and Torr, P. (2010). Dynamic hybrid algorithms for MAP inference in discrete MRFs. *IEEE Transactions on Pattern Analysis and Machine Intelligence*, 32(10):1846–1857.

Alahi, A., Ortiz, R., and Vandergheynst, P. (2012). Freak: Fast retina keypoint. In *IEEE Computer Society Conference on Computer Vision and Pattern Recognition (CVPR)*.

Albl, C., Kukelova, Z., Larsson, V., Polic, M., Pajdla, T., and Schindler, K. (2020). From two rolling shutters to one global shutter. In *IEEE/CVF Conference on Computer Vision and Pattern Recognition (CVPR)*.

Alcantarilla, P. F., Bartoli, A., and Davison, A. J. (2012). KAZE features. In *European Conference on Computer Vision (ECCV)*, pp. 214–227.

Alcantarilla, P. F., Nuevo, J., and Bartoli, A. (2013). Fast explicit diffusion for accelerated features in nonlinear scale spaces. In *British Machine Vision Conference (BMVC)*.

Alexa, M., Behr, J., Cohen-Or, D., Fleishman, S., Levin, D., and Silva, C. T. (2003). Computing and rendering point set surfaces. *IEEE Transactions on Visualization and Computer Graphics*, 9(1):3–15.

Alexe, B., Deselaers, T., and Ferrari, V. (2012). Measuring the objectness of image windows. *IEEE Transactions on Pattern Analysis and Machine Intelligence*, 34(11):2189–2202.

Aliaga, D. G., Funkhouser, T., Yanovsky, D., and Carlbom, I. (2003). Sea of images. *IEEE Computer Graphics and Applications*, 23(6):22–30.

Allen, B., Curless, B., and Popović, Z. (2003). The space of human body shapes: reconstruction and parameterization from range scans. *ACM Transactions on Graphics (Proc. SIGGRAPH)*, 22(3):587–594.

Allgower, E. L. and Georg, K. (2003). *Introduction to Numerical Continuation Methods*. Society for Industrial and Applied Mathematics.

Aloimonos, J. (1990). Perspective approximations. *Image and Vision Computing*, 8:177–192.

Alpert, S., Galun, M., Basri, R., and Brandt, A. (2007). Image segmentation by probabilistic bottom-up aggregation and cue integration. In *IEEE Computer Society Conference on Computer Vision and Pattern Recognition (CVPR)*.

Alter, F., Matsushita, Y., and Tang, X. (2006). An intensity similarity measure in low-light conditions. In *European Conference on Computer Vision (ECCV)*, pp. 267–280.

Alvarez, L., Weickert, J., and Sánchez, J. (2000). Reliable estimation of dense optical flow fields with large displacements. *International Journal of Computer Vision*, 39(1):41–56.

Alwassel, H., Mahajan, D., Korbar, B., Torresani, L., Ghanem, B., and Tran, D. (2020). Self-supervised learning by cross-modal audio-video clustering. In *Advances in Neural Information Processing Systems (NeurIPS)*.

Amidror, I. (2002). Scattered data interpolation methods for electronic imaging systems: a survey. *Journal of Electronic Imaging*, 11(2):157–76.

Anandan, P. (1984). Computing dense displacement fields with confidence measures in scenes containing occlusion. In *Image Understanding Workshop*, pp. 236–246.

Anandan, P. (1989). A computational framework and an algorithm for the measurement of visual motion. *International Journal of Computer Vision*, 2(3):283–310.

Anandan, P. and Irani, M. (2002). Factorization with uncertainty. *International Journal of Computer Vision*, 49(2–3):101–116.

Anderson, E., Bai, Z., Bischof, C., Blackford, S., Demmel, J. W., Dongarra, J. J., Croz, J. D., Greenbaum, A., Hammarling, S., McKenney, A., and Sorensen, D. C. (1999). *LAPACK Users' Guide*. Society for Industrial and Applied Mathematics, 3rd edition.

Anderson, P., Fernando, B., Johnson, M., and Gould, S. (2016). SPICE: Semantic propositional image caption evaluation. In *European Conference on Computer Vision (ECCV)*, pp. 382–398.

Anderson, P., He, X., Buehler, C., Teney, D., Johnson, M., Gould, S., and Zhang, L. (2018). Bottom-up and top-down attention for image captioning and visual question answering. In *IEEE Conference on Computer Vision and Pattern Recognition (CVPR)*.

Anderson, R., Gallup, D., Barron, J. T., Kontkanen, J., Snavely, N., Hernández, C., Agarwal, S., and Seitz, S. M. (2016). Jump: virtual reality video. *ACM Transactions on Graphics (Proc. SIGGRAPH Asia)*, 35(6):198.

Andersson, P., Nilsson, J., Akenine-Möller, T., Oskarsson, M., Åström, K., and Fairchild, M. D. (2020). FLIP: A difference evaluator for alternating images. *Proceedings of the ACM on Computer Graphics and Interactive Techniques (High-Performance Graphics)*, 3(2):15.

Andoni, A. and Indyk, P. (2006). Near-optimal hashing algorithms for approximate nearest neighbor in high dimensions. In *IEEE Symposium on Foundations of Computer Science (FOCS)*, pp. 459–468.

Andreopoulos, A. and Tsotsos, J. K. (2013). 50 years of object recognition: Directions forward. *Computer Vision and Image Understanding*, 117(8):827–891.

Andrieu, C., de Freitas, N., Doucet, A., and Jordan, M. I. (2003). An introduction to MCMC for machine learning. *Machine Learning*, 50(1–2):5–43.

Andriluka, M., Roth, S., and Schiele, B. (2008). People-tracking-by-detection and people-detection-by-tracking. In *IEEE Computer Society Conference on Computer Vision and Pattern Recognition (CVPR)*.

Andriluka, M., Roth, S., and Schiele, B. (2009). Pictorial structures revisited: People detection and articulated pose estimation. In *IEEE Computer Society Conference on Computer Vision and Pattern Recognition (CVPR)*.

Andriluka, M., Roth, S., and Schiele, B. (2010). Monocular 3D pose estimation and tracking by detection. In *IEEE Computer Society Conference on Computer Vision and Pattern Recognition (CVPR)*.

Andriluka, M., Pishchulin, L., Gehler, P., and Schiele, B. (2014). 2D human pose estimation: New benchmark and state of the art analysis. In *IEEE Conference on Computer Vision and Pattern Recognition (CVPR)*.

Anguelov, D., Srinivasan, P., Koller, D., Thrun, S., Rodgers, J., and Davis, J. (2005). SCAPE: Shape completion and animation of people. *ACM Transactions on Graphics (Proc. SIGGRAPH)*, 24(3):408–416.

Anjyo, K., Lewis, J. P., and Pighin, F. (2014). Scattered data interpolation for computer graphics. In *ACM SIGGRAPH Course Notes*.

Ansar, A., Castano, A., and Matthies, L. (2004). Enhanced real-time stereo using bilateral filtering. In *International Symposium on 3D Data Processing, Visualization, and Transmission (3DPVT)*.

Antol, S., Agrawal, A., Lu, J., Mitchell, M., Batra, D., Zitnick, C. L., and Parikh, D. (2015). VQA: Visual question answering. In *IEEE International Conference on Computer Vision (ICCV)*.

Antone, M. and Teller, S. (2002). Scalable extrinsic calibration of omni-directional image networks. *International Journal of Computer Vision*, 49(2–3):143–174.

Antunes, M. and Barreto, J. P. (2013). A global approach for the detection of vanishing points and mutually orthogonal vanishing directions. In *IEEE Computer Society Conference on Computer Vision and Pattern Recognition (CVPR)*.

Anwar, S., Khan, S., and Barnes, N. (2020). A deep journey into super-resolution: A survey. *ACM Computing Surveys*, 53(3):60.

Arandjelović, R. and Zisserman, A. (2012). Three things everyone should know to improve object retrieval. In *IEEE Computer Society Conference on Computer Vision and Pattern Recognition (CVPR)*.

Arandjelovic, R. and Zisserman, A. (2018). Objects that sound. In *European Conference on Computer Vision (ECCV)*.

Arandjelovic, R., Gronat, P., Torii, A., Pajdla, T., and Sivic, J. (2016). NetVLAD: CNN architecture for weakly supervised place recognition. In *IEEE Conference on Computer Vision and Pattern Recognition (CVPR)*.

Araujo, A., Norris, W., and Sim, J. (2019). Computing receptive fields of convolutional neural networks. *Distill*, 4(11):e21.

Arbeláez, P., Maire, M., Fowlkes, C., and Malik, J. (2011). Contour detection and hierarchical image segmentation. *IEEE Transactions on Pattern Analysis and Machine Intelligence*, 33(5):898–916.

Argyriou, V. and Vlachos, T. (2003). Estimation of sub-pixel motion using gradient cross-correlation. *Electronic Letters*, 39(13):980–982.

Arikan, O. and Forsyth, D. A. (2002). Interactive motion generation from examples. *ACM Transactions on Graphics*, 21(3):483–490.

Arjovsky, M., Chintala, S., and Bottou, L. (2017). Wasserstein generative adversarial networks. In *International Conference on Machine Learning (ICML)*, pp. 214–223.

Arnab, A., Dehghani, M., Heigold, G., Sun, C., Lučić, M., and Schmid, C. (2021). ViViT: A video vision transformer. *arXiv preprint arXiv:2103.15691*.

Arnold, R. D. (1983). *Automated Stereo Perception*. Technical Report AIM-351, Artificial Intelligence Laboratory, Stanford University.

Arya, S., Mount, D. M., Netanyahu, N. S., Silverman, R., and Wu, A. Y. (1998). An optimal algorithm for approximate nearest neighbor searching in fixed dimensions. *Journal of the ACM*, 45(6):891–923.

Asano, Y. M., Rupprecht, C., and Vedaldi, A. (2020). Self-labelling via simultaneous clustering and representation learning. In *International Conference on Learning Representations (ICLR)*.

Ashdown, I. (1993). Near-field photometry: A new approach. *Journal of the Illuminating Engineering Society*, 22(1):163–180.

Atkinson, K. B. (1996). *Close Range Photogrammetry and Machine Vision*. Whittles Publishing, Scotland, UK.

Attal, B., Ling, S., Gokaslan, A., Richardt, C., and Tompkin, J. (2020). MatryODShka: Real-time 6dof video view synthesis using multi-sphere images. In *European Conference on Computer Vision (ECCV)*.

Aubry, M., Paris, S., Hasinoff, S. W., Kautz, J., and Durand, F. (2014). Fast local Laplacian filters: Theory and applications. *ACM Transactions on Graphics (Proc. SIGGRAPH)*, 33(5):167.

Aurich, V. and Weule, J. (1995). Non-linear Gaussian filters performing edge preserving diffusion. In *DAGM Symposium*, pp. 538–545.

Averbuch-Elor, H., Cohen-Or, D., Kopf, J., and Cohen, M. F. (2017). Bringing portraits to life. *Transactions on Graphics (Proc. SIGGRAPH Asia)*, 36(6):196.

Avidan, S. (2001). Support vector tracking. In *IEEE Computer Society Conference on Computer Vision and Pattern Recognition (CVPR)*, pp. 283–290.

Avidan, S., Baker, S., and Shan, Y. (2010). Special issue on Internet Vision. *Proceedings of the IEEE*, 98(8):1367–1369.

Axelsson, O. (1996). *Iterative Solution Methods*. Cambridge University Press, Cambridge.

Ayache, N. (1989). *Vision Stéréoscopique et Perception Multisensorielle*. InterEditions, Paris.

Azarbayejani, A. and Pentland, A. P. (1995). Recursive estimation of motion, structure, and focal length. *IEEE Transactions on Pattern Analysis and Machine Intelligence*, 17(6):562–575.

Azuma, R. T., Baillot, Y., Behringer, R., Feiner, S. K., Julier, S., and MacIntyre, B. (2001). Recent advances in augmented reality. *IEEE Computer Graphics and Applications*, 21(6):34–47.

Ba, J. L., Kiros, J. R., and Hinton, G. E. (2016). Layer normalization. *arXiv preprint arXiv:1607.06450*.

Baak, A., Mueller, M., Bharaj, G., Seidel, H.-P., and Theobalt, C. (2011). A data-driven approach for real-time full body pose reconstruction from a depth camera. In *International Conference on Computer Vision (ICCV)*.

Baatz, G., Saurer, O., Köser, K., and Pollefeys, M. (2012). Large scale visual geo-localization of images in mountainous terrain. In *European Conference on Computer Vision (ECCV)*, pp. 517–530.

Bab-Hadiashar, A. and Suter, D. (1998a). Robust optic flow computation. *International Journal of Computer Vision*, 29(1):59–77.

Bab-Hadiashar, A. and Suter, D. (1998b). Robust total least squares based optic flow computation. In *Asian Conference on Computer Vision (ACCV)*, pp. 566–573.

Babenko, A. and Lempitsky, V. (2015a). Aggregating local deep features for image retrieval. In *IEEE International Conference on Computer Vision (ICCV)*.

Babenko, A. and Lempitsky, V. (2015b). The inverted multi-index. *IEEE transactions on pattern analysis and machine intelligence*, 37(6):1247–1260.

Baccouche, M., Mamalet, F., Wolf, C., Garcia, C., and Baskurt, A. (2011). Sequential deep learning for human action recognition. In *International Workshop on Human Behavior Understanding*, pp. 29–39.

Badra, F., Qumsieh, A., and Dudek, G. (1998). Rotation and zooming in image mosaicing. In *IEEE Workshop on Applications of Computer Vision (WACV)*, pp. 50–55.

Bae, S., Paris, S., and Durand, F. (2006). Two-scale tone management for photographic look. *ACM Transactions on Graphics*, 25(3):637–645.

Baeza-Yates, R. and Ribeiro-Neto, B. (1999). *Modern Information Retrieval*. Addison Wesley.

Bai, X. and Sapiro, G. (2009). Geodesic matting: A framework for fast interactive image and video segmentation and matting. *International Journal of Computer Vision*, 82(2):113–132.

Bailey, T. and Durrant-Whyte, H. (2006). Simultaneous localization and mapping (SLAM): Part II. *IEEE Robotics & Automation Magazine*, 13(3):108–117.

Bajcsy, R. and Kovacic, S. (1989). Multiresolution elastic matching. *Computer Vision, Graphics, and Image Processing*, 46(1):1–21.

Baker, H. H. (1977). Three-dimensional modeling. In *International Joint Conference on Artificial Intelligence (IJCAI)*, pp. 649–655.

Baker, H. H. (1989). Building surfaces of evolution: The weaving wall. *International Journal of Computer Vision*, 3(1):50–71.

Baker, H. H. and Binford, T. O. (1981). Depth from edge and intensity based stereo. In *IJCAI*, pp. 631–636.

Baker, H. H. and Bolles, R. C. (1989). Generalizing epipolar-plane image analysis on the spatiotemporal surface. *International Journal of Computer Vision*, 3(1):33–49.

Baker, S. and Kanade, T. (2002). Limits on super-resolution and how to break them. *IEEE Transactions on Pattern Analysis and Machine Intelligence*, 24(9):1167–1183.

Baker, S. and Matthews, I. (2004). Lucas-Kanade 20 years on: A unifying framework: Part 1: The quantity approximated, the warp update rule, and the gradient descent approximation. *International Journal of Computer Vision*, 56(3):221–255.

Baker, S. and Nayar, S. (1999). A theory of single-viewpoint catadioptric image formation. *International Journal of Computer Vision*, 5(2):175–196.

Baker, S. and Nayar, S. K. (2001). Single viewpoint catadioptric cameras. In Benosman, R. and Kang, S. B. (eds), *Panoramic Vision: Sensors, Theory, and Applications*, pp. 39–71, Springer, New York.

Baker, S., Gross, R., and Matthews, I. (2003). *Lucas-Kanade 20 Years On: A Unifying Framework: Part 3*. Technical Report CMU-RI-TR-03-35, The Robotics Institute, Carnegie Mellon University.

Baker, S., Gross, R., and Matthews, I. (2004). *Lucas-Kanade 20 Years On: A Unifying Framework: Part 4*. Technical Report CMU-RI-TR-04-14, The Robotics Institute, Carnegie Mellon University.

Baker, S., Szeliski, R., and Anandan, P. (1998). A layered approach to stereo reconstruction. In *IEEE Computer Society Conference on Computer Vision and Pattern Recognition (CVPR)*, pp. 434–441.

Baker, S., Bennett, E., Kang, S. B., and Szeliski, R. (2010). Removing rolling shutter wobble. In *IEEE Computer Society Conference on Computer Vision and Pattern Recognition (CVPR)*.

Baker, S., Gross, R., Ishikawa, T., and Matthews, I. (2003). *Lucas-Kanade 20 Years On: A Unifying Framework: Part 2*. Technical Report CMU-RI-TR-03-01, The Robotics Institute, Carnegie Mellon University.

Baker, S., Scharstein, D., Lewis, J., Roth, S., Black, M. J., and Szeliski, R. (2009). *A Database and Evaluation Methodology for Optical Flow*. Technical Report MSR-TR-2009-179, Microsoft Research.

Baker, S., Scharstein, D., Lewis, J., Roth, S., Black, M. J., and Szeliski, R. (2011). A database and evaluation methodology for optical flow. *International Journal of Computer Vision*, 92(1):1–31.

Bălan, A. O. and Black, M. J. (2008). The naked truth: Estimating body shape under clothing. In *European Conference on Computer Vision (ECCV)*, pp. 15–29.

Ballard, D. H. (1981). Generalizing the Hough transform to detect arbitrary patterns. *Pattern Recognition*, 13(2):111–122.

Ballas, N., Yao, L., Pal, C., and Courville, A. (2016). Delving deeper into convolutional networks for learning video representations. In *International Conference on Learning Representations (ICLR)*.

Balntas, V., Li, S., and Prisacariu, V. (2018). RelocNet: Continuous metric learning relocalisation using neural nets. In *European Conference on Computer Vision (ECCV)*.

Balntas, V., Riba, E., Ponsa, D., and Mikolajczyk, K. (2016). Learning local feature descriptors with triplets and shallow convolutional neural networks. In *British Machine Vision Conference (BMVC)*, p. 3.

Balntas, V., Lenc, K., Vedaldi, A., Tuytelaars, T., Matas, J., and Mikolajczyk, K. (2020). $\mathbb{H}$Patches: A benchmark and evaluation of handcrafted and learned local descriptors. *IEEE Transactions on Pattern Analysis and Machine Intelligence*, 42(11):2825–2841.

Bamji, C. S., O'Connor, P., Elkhatib, T., Mehta, S., Thompson, B., Prather, L. A., Snow, D., Akkaya, O. C., Daniel, A., Payne, A. D., Perry, T., Fenton, M., and Chan, V.-H. (2014). A 0.13 μm CMOS system-on-chip for a 512 $\times$ 424 time-of-flight image sensor with multi-frequency photo-demodulation up to 130 MHz and 2 GS/s ADC. *IEEE Journal of Solid-State Circuits*, 50(1):303–319.

Banerjee, S. and Lavie, A. (2005). METEOR: An automatic metric for mt evaluation with improved correlation with human judgments. In *Proceedings of the ACL Workshop on Intrinsic and Extrinsic Evaluation Measures for Machine Translation and/or Summarization*, pp. 65–72.

Banno, A., Masuda, T., Oishi, T., and Ikeuchi, K. (2008). Flying laser range sensor for large-scale site-modeling and its applications in Bayon digital archival project. *International Journal of Computer Vision*, 78(2–3):207–222.

Bansal, A., Vo, M., Sheikh, Y., Ramanan, D., and Narasimhan, S. (2020). 4D visualization of dynamic events from unconstrained multi-view videos. In *IEEE/CVF Conference on Computer Vision and Pattern Recognition (CVPR)*.

Banterle, F., Artusi, A., Debattista, K., and Chalmers, A. (2017). *Advanced high dynamic range imaging*. CRC Press.

Bar-Joseph, Z., El-Yaniv, R., Lischinski, D., and Werman, M. (2001). Texture mixing and texture movie synthesis using statistical learning. *IEEE Transactions on Visualization and Computer Graphics*, 7(2):120–135.

Bar-Shalom, Y. and Fortmann, T. E. (1988). *Tracking and data association*. Academic Press, Boston.

Barash, D. (2002). A fundamental relationship between bilateral filtering, adaptive smoothing, and the nonlinear diffusion equation. *IEEE Transactions on Pattern Analysis and Machine Intelligence*, 24(6):844–847.

Barash, D. and Comaniciu, D. (2004). A common framework for nonlinear diffusion, adaptive smoothing, bilateral filtering and mean shift. *Image and Vision Computing*, 22(1):73–81.

Barath, D. and Matas, J. (2018). Graph-cut RANSAC. In *IEEE Conference on Computer Vision and Pattern Recognition (CVPR)*.

Barath, D., Matas, J., and Noskova, J. (2019). MAGSAC: Marginalizing sample consensus. In *IEEE/CVF Conference on Computer Vision and Pattern Recognition (CVPR)*.

Barath, D., Noskova, J., Ivashechkin, M., and Matas, J. (2020). MAGSAC++, a fast, reliable and accurate robust estimator. In *IEEE/CVF Conference on Computer Vision and Pattern Recognition (CVPR)*.

Barkans, A. C. (1997). High quality rendering using the Talisman architecture. In *Eurographics Workshop on Graphics Hardware*.

Barnard, S. T. (1989). Stochastic stereo matching over scale. *International Journal of Computer Vision*, 3(1):17–32.

Barnard, S. T. and Fischler, M. A. (1982). Computational stereo. *Computing Surveys*, 14(4):553–572.

Barnes, C., Shechtman, E., Finkelstein, A., and Goldman, D. (2009). PatchMatch: A randomized correspondence algorithm for structural image editing. *ACM Transactions on Graphics*, 28(3):24.

Barnes, C., Jacobs, D. E., Sanders, J., Goldman, D. B., Rusinkiewicz, S., Finkelstein, A., and Agrawala, M. (2008). Video puppetry: A performative interface for cutout animation. *ACM Transactions on Graphics*, 27(5):124.

Barreto, J. P. and Daniilidis, K. (2005). Fundamental matrix for cameras with radial distortion. In *International Conference on Computer Vision (ICCV)*, pp. 625–632.

Barron, J. L., Fleet, D. J., and Beauchemin, S. S. (1994). Performance of optical flow techniques. *International Journal of Computer Vision*, 12(1):43–77.

Barron, J. T. (2019). A general and adaptive robust loss function. In *IEEE/CVF Conference on Computer Vision and Pattern Recognition (CVPR)*.

Barron, J. T. and Malik, J. (2012). Shape, albedo, and illumination from a single image of an unknown object. In *IEEE Computer Society Conference on Computer Vision and Pattern Recognition (CVPR)*.

Barron, J. T. and Poole, B. (2016). The fast bilateral solver. In *European Conference on Computer Vision (ECCV)*.

Barron, J. T., Adams, A., Shih, Y., and Hernandez, C. (2015). Fast bilateral-space stereo for synthetic defocus. In *IEEE Conference on Computer Vision and Pattern Recognition (CVPR)*.

Barron, J. T., Mildenhall, B., Tancik, M., Hedman, P., Martin-Brualla, R., and Srinivasan, P. P. (2021). Mip-NeRF: A multiscale representation for anti-aliasing neural radiance fields. *arXiv preprint arXiv:2103.13415*.

Barroso-Laguna, A., Riba, E., Ponsa, D., and Mikolajczyk, K. (2019). Key.Net: Keypoint detection by handcrafted and learned CNN filters. In *IEEE/CVF International Conference on Computer Vision (ICCV)*.

Barrow, H. G. and Tenenbaum, J. M. (1981). Computational vision. *Proceedings of the IEEE*, 69(5):572–595.

Bartels, R. H., Beatty, J. C., and Barsky, B. A. (1987). *An Introduction to Splines for use in Computer Graphics and Geeometric Modeling*. Morgan Kaufmann Publishers, Los Altos.

Bartoli, A. (2003). Towards gauge invariant bundle adjustment: A solution based on gauge dependent damping. In *International Conference on Computer Vision (ICCV)*, pp. 760–765.

Bartoli, A. and Sturm, P. (2003). Multiple-view structure and motion from line correspondences. In *International Conference on Computer Vision (ICCV)*, pp. 207–212.

Bartoli, A., Coquerelle, M., and Sturm, P. (2004). A framework for pencil-of-points structure-from-motion. In *European Conference on Computer Vision (ECCV)*, pp. 28–40.

Bascle, B., Blake, A., and Zisserman, A. (1996). Motion deblurring and super-resolution from an image sequence. In *European Conference on Computer Vision (ECCV)*, pp. 573–582.

Bathe, K.-J. (2007). *Finite Element Procedures*. Prentice-Hall, Inc., Englewood Cliffs, New Jersey.

Batsos, K. and Mordohai, P. (2018). RecResNet: A recurrent residual CNN architecture for disparity map enhancement. In *International Conference on 3D Vision (3DV)*, pp. 238–247.

Bau, D., Zhou, B., Khosla, A., Oliva, A., and Torralba, A. (2017). Network dissection: Quantifying interpretability of deep visual representations. In *IEEE Conference on Computer Vision and Pattern Recognition (CVPR)*.

Bau, D., Strobelt, H., Peebles, W., Wulff, J., Zhou, B., Zhu, J.-Y., and Torralba, A. (2019). Semantic photo manipulation with a generative image prior. *ACM Transactions on Graphics (Proc. SIGGRAPH)*, 38(4):59:1–59:11.

Baudisch, P., Tan, D., Steedly, D., Rudolph, E., Uyttendaele, M., Pal, C., and Szeliski, R. (2006). An exploration of user interface designs for real-time panoramic photography. *Australian Journal of Information Systems*, 13(2).

Baumberg, A. (2000). Reliable feature matching across widely separated views. In *IEEE Computer Society Conference on Computer Vision and Pattern Recognition (CVPR)*, pp. 774–781.

Baumberg, A. M. and Hogg, D. C. (1996). Generating spatiotemporal models from examples. *Image and Vision Computing*, 14(8):525–532.

Baumgart, B. G. (1974). *Geometric Modeling for Computer Vision.* Technical Report AIM-249, Artificial Intelligence Laboratory, Stanford University.

Bay, H., Ferrari, V., and Van Gool, L. (2005). Wide-baseline stereo matching with line segments. In *IEEE Computer Society Conference on Computer Vision and Pattern Recognition (CVPR)*, pp. 329–336.

Bay, H., Ess, A., Tuytelaars, T., and Van Gool, L. (2008). Speeded-up robust features (SURF). *Computer Vision and Image Understanding*, 110:346–359.

Bayer, B. E. (1976). Color imaging array. US Patent No. 3,971,065.

Bazin, J.-C., Seo, Y. D., Demonceaux, C., Vasseur, P., Ikeuchi, K., Kweon, I. S., and Pollefeys, M. (2012). Globally optimal line clustering and vanishing point estimation in Manhattan world. In *IEEE Computer Society Conference on Computer Vision and Pattern Recognition (CVPR)*.

Beardsley, P., Torr, P., and Zisserman, A. (1996). 3D model acquisition from extended image sequences. In *European Conference on Computer Vision (ECCV)*, pp. 683–695.

Beare, R. (2006). A locally constrained watershed transform. *IEEE Transactions on Pattern Analysis and Machine Intelligence*, 28(7):1063–1074.

Becker, S. and Bove, V. M. (1995). Semiautomatic 3-D model extraction from uncalibrated 2-D camera views. In *SPIE Vol. 2410, Visual Data Exploration and Analysis II*, pp. 447–461.

Behl, A., Paschalidou, D., Donne, S., and Geiger, A. (2019). PointFlowNet: Learning representations for rigid motion estimation from point clouds. In *IEEE/CVF Conference on Computer Vision and Pattern Recognition (CVPR)*.

Behl, A., Hosseini Jafari, O., Karthik Mustikovela, S., Abu Alhaija, H., Rother, C., and Geiger, A. (2017). Bounding boxes, segmentations and object coordinates: How important is recognition for 3D scene flow estimation in autonomous driving scenarios? In *IEEE International Conference on Computer Vision (ICCV)*.

Beier, T. and Neely, S. (1992). Feature-based image metamorphosis. *Computer Graphics (SIGGRAPH)*, 26(2):35–42.

Beis, J. S. and Lowe, D. G. (1999). Indexing without invariants in 3D object recognition. *IEEE Transactions on Pattern Analysis and Machine Intelligence*, 21(10):1000–1015.

Belhumeur, P. N. (1996). A Bayesian approach to binocular stereopsis. *International Journal of Computer Vision*, 19(3):237–260.

Belhumeur, P. N., Hespanha, J. P., and Kriegman, D. J. (1997). Eigenfaces vs. Fisherfaces: Recognition using class specific linear projection. *IEEE Transactions on Pattern Analysis and Machine Intelligence*, 19(7):711–720.

Belkin, M. and Niyogi, P. (2003). Laplacian eigenmaps for dimensionality reduction and data representation. *Neural Computation*, 15(6):1373–1396.

Bell, S. and Bala, K. (2015). Learning visual similarity for product design with convolutional neural networks. *ACM Transactions on Graphics (Proc. SIGGRAPH)*, 34(4):98.

Bell, S., Liu, Y., Alsheikh, S., Tang, Y., Pizzi, E., Henning, M., Singh, K., Parkhi, O., and Borisyuk, F. (2020). GrokNet: Unified computer vision model trunk and embeddings for commerce. In *ACM SIGKDD International Conference on Knowledge Discovery & Data Mining*, New York, NY, USA.

Bellavia, F. and Colombo, C. (2020). Is there anything new to say about SIFT matching? *International Journal of Computer Vision*, 128(7):1847–1866.

Belongie, S. and Malik, J. (1998). Finding boundaries in natural images: a new method using point descriptors and area completion. In *European Conference on Computer Vision (ECCV)*, pp. 751–766.

Belongie, S., Malik, J., and Puzicha, J. (2002). Shape matching and object recognition using shape contexts. *IEEE Transactions on Pattern Analysis and Machine Intelligence*, 24(4):509–522.

Belongie, S., Fowlkes, C., Chung, F., and Malik, J. (2002). Spectral partitioning with indefinite kernels using the Nyström extension. In *European Conference on Computer Vision (ECCV)*, pp. 531–543.

Bemana, M., Myszkowski, K., Seidel, H.-P., and Ritschel, T. (2020). X-Fields: Implicit neural view-, light- and time-image interpolation. *ACM Transactions on Graphics (Proc. SIGGRAPH Asia)*, 39(6).

Ben-Shabat, Y. and Gould, S. (2020). DeepFit: 3D surface fitting via neural network weighted least squares. In *European Conference on Computer Vision (ECCV)*.

Benfold, B. and Reid, I. (2011). Stable multi-target tracking in real-time surveillance video. In *IEEE Computer Society Conference on Computer Vision and Pattern Recognition (CVPR)*.

Bengio, Y. (2009). Learning deep architectures for AI. *Foundations and Trends® in Machine Learning*, 2(1):1–127.

Bengio, Y., LeCun, Y., and Hinton, G. (2021). Deep learning for AI. *Communications of the ACM*, 64(7):58–65.

Bengio, Y., Paiement, J.-F., Vincent, P., Delalleau, O., Roux, N. L., and Ouimet, M. (2004). Out-of-Sample extensions for LLE, Isomap, MDS, Eigenmaps, and spectral clustering. In *Advances in Neural Information Processing Systems (NeurIPS)*, pp. 177–184.

Bennett, E., Uyttendaele, M., Zitnick, C. L., Szeliski, R., and Kang, S. B. (2006). Video and image Bayesian demosaicing with a two color image prior. In *European Conference on Computer Vision (ECCV)*, pp. 508–521.

Benosman, R. and Kang, S. B. (eds). (2001). *Panoramic Vision: Sensors, Theory, and Applications*, Springer, New York.

Bergen, J. R., Anandan, P., Hanna, K. J., and Hingorani, R. (1992). Hierarchical model-based motion estimation. In *European Conference on Computer Vision (ECCV)*, pp. 237–252.

Bergen, J. R., Burt, P. J., Hingorani, R., and Peleg, S. (1992). A three-frame algorithm for estimating two-component image motion. *IEEE Transactions on Pattern Analysis and Machine Intelligence*, 14(9):886–896.

Berger, J. O. (1993). *Statistical Decision Theory and Bayesian Analysis*. Springer, New York, 2nd edition.

Berger, M., Tagliasacchi, A., Seversky, L. M., Alliez, P., Guennebaud, G., Levine, J. A., Sharf, A., and Silva, C. T. (2017). A survey of surface reconstruction from point clouds. *Computer Graphics Forum*, 36(1):301–329.

Bergmann, P., Meinhardt, T., and Leal-Taixé, L. (2019). Tracking without bells and whistles. In *IEEE/CVF International Conference on Computer Vision (ICCV)*.

Berman, M., Jégou, H., Vedaldi, A., Kokkinos, I., and Douze, M. (2019). MultiGrain: a unified image embedding for classes and instances. *arXiv preprint arXiv:1902.05509*.

Bertalmio, M., Sapiro, G., Caselles, V., and Ballester, C. (2000). Image inpainting. In *ACM SIGGRAPH Conference Proceedings*, pp. 417–424.

Bertalmio, M., Vese, L., Sapiro, G., and Osher, S. (2003). Simultaneous structure and texture image inpainting. *IEEE Transactions on Image Processing*, 12(8):882–889.

Bertero, M., Poggio, T. A., and Torre, V. (1988). Ill-posed problems in early vision. *Proceedings of the IEEE*, 76(8):869–889.

Bertinetto, L., Valmadre, J., Golodetz, S., Miksik, O., and Torr, P. H. S. (2016a). Staple: Complementary learners for real-time tracking. In *IEEE Conference on Computer Vision and Pattern Recognition (CVPR)*.

Bertinetto, L., Valmadre, J., Henriques, J. F., Vedaldi, A., and Torr, P. H. S. (2016b). Fully-convolutional Siamese networks for object tracking. In *European Conference on Computer Vision (ECCV)*, pp. 850–865.

Besag, J. (1986). On the statistical analysis of dirty pictures. *Journal of the Royal Statistical Society B*, 48(3):259–302.

Besl, P. (1989). Active optical range imaging sensors. In Sanz, J. L. (ed.), *Advances in Machine Vision*, chapter 1, pp. 1–63, Springer-Verlag.

Besl, P. J. and Jain, R. C. (1985). Three-dimensional object recognition. *Computing Surveys*, 17(1):75–145.

Besl, P. J. and McKay, N. D. (1992). A method for registration of 3-D shapes. *IEEE Transactions on Pattern Analysis and Machine Intelligence*, 14(2):239–256.

Betrisey, C., Blinn, J. F., Dresevic, B., Hill, B., Hitchcock, G., Keely, B., Mitchell, D. P., Platt, J. C., and Whitted, T. (2000). Displaced filtering for patterned displays. In *Society for Information Display Symposium,*, pp. 296–299.

Beymer, D. (1996). Feature correspondence by interleaving shape and texture computations. In *IEEE Computer Society Conference on Computer Vision and Pattern Recognition (CVPR)*, pp. 921–928.

Bhat, D. N. and Nayar, S. K. (1998). Ordinal measures for image correspondence. *IEEE Transactions on Pattern Analysis and Machine Intelligence*, 20(4):415–423.

Bi, S., Xu, Z., Sunkavalli, K., Kriegman, D., and Ramamoorthi, R. (2020a). Deep 3D capture: Geometry and reflectance from sparse multi-view images. In *IEEE/CVF Conference on Computer Vision and Pattern Recognition (CVPR)*.

Bi, S., Xu, Z., Sunkavalli, K., Hašan, M., Hold-Geoffroy, Y., Kriegman, D., and Ramamoorthi, R. (2020b). Deep reflectance volumes: Relightable reconstructions from multi-view photometric images. In *European Conference on Computer Vision (ECCV)*.

Bi, S., Xu, Z., Srinivasan, P., Mildenhall, B., Sunkavalli, K., Hašan, M., Hold-Geoffroy, Y., Kriegman, D., and Ramamoorthi, R. (2020c). Neural reflectance fields for appearance acquisition. *arXiv preprint arXiv:2008.03824*.

Bickel, B., Botsch, M., Angst, R., Matusik, W., Otaduy, M., Pfister, H., and Gross, M. (2007). Multi-scale capture of facial geometry and motion. *ACM Transactions on Graphics*, 26(3).

Bilen, H. and Vedaldi, A. (2016). Weakly supervised deep detection networks. In *IEEE Conference on Computer Vision and Pattern Recognition (CVPR)*.

Billinghurst, M., Clark, A., and Lee, G. (2015). A survey of augmented reality. *Foundations and Trends® in Human-Computer Interaction*, 8(2–3):73–272.

Billinghurst, M., Kato, H., and Poupyrev, I. (2001). The MagicBook: a transitional AR interface. *Computers & Graphics*, 25:745–753.

Bimber, O. (2006). Computational photography—the next big step. *Computer*, 39(8):28–29.

Birchfield, S. and Tomasi, C. (1998). A pixel dissimilarity measure that is insensitive to image sampling. *IEEE Transactions on Pattern Analysis and Machine Intelligence*, 20(4):401–406.

Birchfield, S. and Tomasi, C. (1999). Depth discontinuities by pixel-to-pixel stereo. *International Journal of Computer Vision*, 35(3):269–293.

Birchfield, S. T., Natarajan, B., and Tomasi, C. (2007). Correspondence as energy-based segmentation. *Image and Vision Computing*, 25(8):1329–1340.

Bishop, C. M. (2006). *Pattern Recognition and Machine Learning*. Springer, New York, NY.

Bitouk, D., Kumar, N., Dhillon, S., Belhumeur, P., and Nayar, S. K. (2008). Face swapping: Automatically replacing faces in photographs. *ACM Transactions on Graphics*, 27(3):39.

Björck, A. (1996). *Numerical Methods for Least Squares Problems*. Society for Industrial and Applied Mathematics.

Björck, A. and Dahlquist, G. (2010). *Numerical Methods in Scientific Computing*. Volume II, Society for Industrial and Applied Mathematics.

Bjorck, N., Gomes, C. P., Selman, B., and Weinberger, K. Q. (2018). Understanding batch normalization. In *Advances in Neural Information Processing Systems (NeurIPS)*, pp. 7694–7705.

Black, M., Yacoob, Y., Jepson, A. D., and Fleet, D. J. (1997). Learning parameterized models of image motion. In *IEEE Computer Society Conference on Computer Vision and Pattern Recognition (CVPR)*, pp. 561–567.

Black, M. J. and Anandan, P. (1993). A framework for the robust estimation of optic flow. In *International Conference on Computer Vision (ICCV)*, pp. 231–236.

Black, M. J. and Anandan, P. (1996). The robust estimation of multiple motions: Parametric and piecewise-smooth flow fields. *Computer Vision and Image Understanding*, 63(1):75–104.

Black, M. J. and Jepson, A. D. (1996). Estimating optical flow in segmented images using variable-order parametric models with local deformations. *IEEE Transactions on Pattern Analysis and Machine Intelligence*, 18(10):972–986.

Black, M. J. and Jepson, A. D. (1998). EigenTracking: robust matching and tracking of articulated objects using a view-based representation. *International Journal of Computer Vision*, 26(1):63–84.

Black, M. J. and Rangarajan, A. (1996). On the unification of line processes, outlier rejection, and robust statistics with applications in early vision. *International Journal of Computer Vision*, 19(1):57–91.

Black, M. J., Sapiro, G., Marimont, D. H., and Heeger, D. (1998). Robust anisotropic diffusion. *IEEE Transactions on Image Processing*, 7(3):421–432.

Blake, A. and Isard, M. (1998). *Active Contours: The Application of Techniques from Graphics, Vision, Control Theory and Statistics to Visual Tracking of Shapes in Motion*. Springer Verlag, London.

Blake, A. and Zisserman, A. (1987). *Visual Reconstruction*. MIT Press, Cambridge, Massachusetts.

Blake, A., Curwen, R., and Zisserman, A. (1993). A framework for spatio-temporal control in the tracking of visual contour. *International Journal of Computer Vision*, 11(2):127–145.

Blake, A., Kohli, P., and Rother, C. (eds). (2011). *Markov Random Fields for Vision and Image Processing*, MIT Press.

Blake, A., Zisserman, A., and Knowles, G. (1985). Surface descriptions from stereo and shading. *Image and Vision Computing*, 3(4):183–191.

Blake, A., Rother, C., Brown, M., Perez, P., and Torr, P. (2004). Interactive image segmentation using an adaptive GMMRF model. In *European Conference on Computer Vision (ECCV)*, pp. 428–441.

Blanco, J.-L. (2019). *A tutorial on SE(3) transformation parameterizations and on-manifold optimization*. Technical Report, University of Malaga.

Blanz, V. and Vetter, T. (1999). A morphable model for the synthesis of 3D faces. In *ACM SIGGRAPH Conference Proceedings*, pp. 187–194.

Blanz, V. and Vetter, T. (2003). Face recognition based on fitting a 3D morphable model. *IEEE Transactions on Pattern Analysis and Machine Intelligence*, 25(9):1063–1074.

Bleyer, M. and Chambon, S. (2010). Does color really help in dense stereo matching? In *International Symposium on 3D Data Processing, Visualization and Transmission (3DPVT)*.

Bleyer, M., Rhemann, C., and Rother, C. (2011). PatchMatch stereo — stereo matching with slanted support windows. In *British Machine Vision Conference (BMVC)*.

Bleyer, M., Gelautz, M., Rother, C., and Rhemann, C. (2009). A stereo approach that handles the matting problem via image warping. In *IEEE Computer Society Conference on Computer Vision and Pattern Recognition (CVPR)*.

Blinn, J. (1998). *Dirty Pixels*. Morgan Kaufmann Publishers, San Francisco.

Blinn, J. F. (1994a). Jim Blinn's corner: Compositing, part 1: Theory. *IEEE Computer Graphics and Applications*, 14(5):83–87.

Blinn, J. F. (1994b). Jim Blinn's corner: Compositing, part 2: Practice. *IEEE Computer Graphics and Applications*, 14(6):78–82.

Blinn, J. F. and Newell, M. E. (1976). Texture and reflection in computer generated images. *Communications of the ACM*, 19(10):542–547.

Bloem, P. (2019). Transformers from scratch. http://peterbloem.nl/blog/transformers.

Blostein, D. and Ahuja, N. (1987). Shape from texture: Integrating texture-element extraction and surface estimation. *IEEE Transactions on Pattern Analysis and Machine Intelligence*, 11(12):1233–1251.

Bobick, A. F. (1997). Movement, activity and action: the role of knowledge in the perception of motion. *Proceedings of the Royal Society of London*, B 352:1257–1265.

Bobick, A. F. and Intille, S. S. (1999). Large occlusion stereo. *International Journal of Computer Vision*, 33(3):181–200.

Bochkovskiy, A., Wang, C.-Y., and Liao, H.-Y. M. (2020). YOLOv4: Optimal speed and accuracy of object detection. *arXiv preprint arXiv:2004.10934*.

Boden, M. A. (2006). *Mind As Machine: A History of Cognitive Science*. Oxford University Press, Oxford, England.

Bogart, R. G. (1991). View correlation. In Arvo, J. (ed.), *Graphics Gems II*, pp. 181–190, Academic Press, Boston.

Bogo, F., Romero, J., Loper, M., and Black, M. J. (2014). FAUST: Dataset and evaluation for 3D mesh registration. In *IEEE Conference on Computer Vision and Pattern Recognition (CVPR)*.

Bogo, F., Kanazawa, A., Lassner, C., Gehler, P., Romero, J., and Black, M. J. (2016). Keep it SMPL: Automatic estimation of 3D human pose and shape from a single image. In *European Conference on Computer Vision*, pp. 561–578.

Boiman, O., Shechtman, E., and Irani, M. (2008). In defense of nearest-neighbor based image classification. In *IEEE Computer Society Conference on Computer Vision and Pattern Recognition (CVPR)*.

Boissonat, J.-D. (1984). Representing 2D and 3D shapes with the Delaunay triangulation. In *International Conference on Pattern Recognition (ICPR)*, pp. 745–748.

Bolles, R. C., Baker, H. H., and Hannah, M. J. (1993). The JISCT stereo evaluation. In *Image Understanding Workshop*, pp. 263–274.

Bolles, R. C., Baker, H. H., and Marimont, D. H. (1987). Epipolar-plane image analysis: An approach to determining structure from motion. *International Journal of Computer Vision*, 1:7–55.

Bookstein, F. L. (1989). Principal warps: Thin-plate splines and the decomposition of deformations. *IEEE Transactions on Pattern Analysis and Machine Intelligence*, 11(6):567–585.

Borenstein, E. and Ullman, S. (2008). Combined top-down/bottom-up segmentation. *IEEE Transactions on Pattern Analysis and Machine Intelligence*, 30(12):2109–2125.

Borgefors, G. (1986). Distance transformations in digital images. *Computer Vision, Graphics and Image Processing*, 34(3):227–248.

Boss, M., Jampani, V., Kim, K., Lensch, H. P., and Kautz, J. (2020). Two-shot spatially-varying BRDF and shape estimation. In *IEEE/CVF Conference on Computer Vision and Pattern Recognition (CVPR)*.

Boss, M., Braun, R., Jampani, V., Barron, J. T., Liu, C., and Lensch, H. (2020). NeRD: Neural reflectance decomposition from image collections. *arXiv preprint arXiv:2012.03918*.

Bouchard, G. and Triggs, B. (2005). Hierarchical part-based visual object categorization. In *IEEE Computer Society Conference on Computer Vision and Pattern Recognition (CVPR)*, pp. 709–714.

Bougnoux, S. (1998). From projective to Euclidean space under any practical situation, a criticism of self-calibration. In *International Conference on Computer Vision (ICCV)*, pp. 790–798.

Bouguet, J.-Y. and Perona, P. (1999). 3D photography using shadows in dual-space geometry. *International Journal of Computer Vision*, 35(2):129–149.

Boult, T. E. and Kender, J. R. (1986). Visual surface reconstruction using sparse depth data. In *IEEE Computer Society Conference on Computer Vision and Pattern Recognition (CVPR)*, pp. 68–76.

Bourdev, L. and Malik, J. (2009). Poselets: Body part detectors trained using 3D human pose annotations. In *International Conference on Computer Vision (ICCV)*.

Bovik, A. (ed.). (2000). *Handbook of Image and Video Processing*, Academic Press, San Diego.

Bowyer, K. W., Kranenburg, C., and Dougherty, S. (2001). Edge detector evaluation using empirical ROC curves. *Computer Vision and Image Understanding*, 84(1):77–103.

Boyer, E. and Berger, M. O. (1997). 3D surface reconstruction using occluding contours. *International Journal of Computer Vision*, 22(3):219–233.

Boykov, Y. and Funka-Lea, G. (2006). Graph cuts and efficient N-D image segmentation. *International Journal of Computer Vision*, 70(2):109–131.

Boykov, Y. and Jolly, M.-P. (2001). Interactive graph cuts for optimal boundary and region segmentation of objects in N-D images. In *International Conference on Computer Vision (ICCV)*, pp. 105–112.

Boykov, Y. and Kolmogorov, V. (2003). Computing geodesics and minimal surfaces via graph cuts. In *International Conference on Computer Vision (ICCV)*, pp. 26–33.

Boykov, Y. and Kolmogorov, V. (2004). An experimental comparison of min-cut/max-flow algorithms for energy minimization in vision. *IEEE Transactions on Pattern Analysis and Machine Intelligence*, 26(9):1124–1137.

Boykov, Y. and Kolmogorov, V. (2011). Basic graph cut algorithms. In Blake, A., Kohli, P., and Rother, C. (eds), *Markov Random Fields for Vision and Image Processing*, pp. 31–50, MIT Press.

Boykov, Y., Veksler, O., and Zabih, R. (1998). A variable window approach to early vision. *IEEE Transactions on Pattern Analysis and Machine Intelligence*, 20(12):1283–1294.

Boykov, Y., Veksler, O., and Zabih, R. (2001). Fast approximate energy minimization via graph cuts. *IEEE Transactions on Pattern Analysis and Machine Intelligence*, 23(11):1222–1239.

Božič, A., Zollhöfer, M., Theobalt, C., and Nießner, M. (2020). DeepDeform: Learning non-rigid RGB-D reconstruction with semi-supervised data. In *IEEE/CVF Conference on Computer Vision and Pattern Recognition (CVPR)*.

Božič, A., Palafox, P., Zollhöfer, M., Dai, J. T. A., and Nießner, M. (2021). Neural deformation graphs for globally-consistent non-rigid reconstruction. In *IEEE/CVF Conference on Computer Vision and Pattern Recognition (CVPR)*.

Božič, A., Palafox, P., Zollhöfer, M., Dai, A., Thies, J., and Nießner, M. (2020). Neural non-rigid tracking. In *Advances in Neural Information Processing Systems (NeurIPS)*.

Bracewell, R. N. (1986). *The Fourier Transform and its Applications*. McGraw-Hill, New York, 2nd edition.

Brachmann, E. and Rother, C. (2018). Learning less is more — 6d camera localization via 3D surface regression. In *IEEE Conference on Computer Vision and Pattern Recognition (CVPR)*.

Brachmann, E. and Rother, C. (2019). Expert sample consensus applied to camera re-localization. In *IEEE/CVF International Conference on Computer Vision (ICCV)*.

Brachmann, E., Michel, F., Krull, A., Yang, M. Y., Gumhold, S., and Rother, C. (2016). Uncertainty-driven 6d pose estimation of objects and scenes from a single RGB image. In *IEEE Conference on Computer Vision and Pattern Recognition (CVPR)*.

Brachmann, E., Krull, A., Nowozin, S., Shotton, J., Michel, F., Gumhold, S., and Rother, C. (2017). DSAC — differentiable RANSAC for camera localization. In *IEEE Conference on Computer Vision and Pattern Recognition (CVPR)*.

Bradsky, G. and Kaehler, A. (2008). *Learning OpenCV: Computer Vision with the OpenCV Library*. O'Reilly, Sebastopol, CA.

Brandt, A. (1986). Algebraic multigrid theory: The symmetric case. *Applied Mathematics and Computation*, 19(1–4):23–56.

Bregler, C. and Malik, J. (1998). Tracking people with twists and exponential maps. In *IEEE Computer Society Conference on Computer Vision and Pattern Recognition (CVPR)*, pp. 8–15.

Bregler, C., Covell, M., and Slaney, M. (1997). Video rewrite: Driving visual speech with audio. In *ACM SIGGRAPH Conference Proceedings*, pp. 353–360.

Bregler, C., Malik, J., and Pullen, K. (2004). Twist based acquisition and tracking of animal and human kinematics. *International Journal of Computer Vision*, 56(3):179–194.

Breiman, L., Friedman, J., Stone, C. J., and Olshen, R. A. (1984). *Classification and regression trees*. CRC Press.

Brejcha, J., Lukáč, M., Hold-Geoffroy, Y., Wang, O., and Čadík, M. (2020). LandscapeAR: Large scale outdoor augmented reality by matching photographs with terrain models using learned descriptors. In *European Conference on Computer Vision (ECCV)*.

Breu, H., Gil, J., Kirkpatrick, D., and Werman, M. (1995). Linear time Euclidean distance transform algorithms. *IEEE Transactions on Pattern Analysis and Machine Intelligence*, 17(5):529–533.

Brice, C. R. and Fennema, C. L. (1970). Scene analysis using regions. *Artificial Intelligence*, 1(3–4):205–226.

Briggs, W. L., Henson, V. E., and McCormick, S. F. (2000). *A Multigrid Tutorial*. Society for Industrial and Applied Mathematics, Philadelphia, 2nd edition.

Brillaut-O'Mahoney, B. (1991). New method for vanishing point detection. *Computer Vision, Graphics, and Image Processing*, 54(2):289–300.

Brinkmann, R. (2008). *The Art and Science of Digital Compositing*. Morgan Kaufmann Publishers, San Francisco, 2nd edition.

Brock, A., Donahue, J., and Simonyan, K. (2019). Large scale GAN training for high fidelity natural image synthesis. In *International Conference on Learning Representations (ICLR)*.

Bromley, J., Guyon, I., LeCun, Y., Säckinger, E., and Shah, R. (1994). Signature verification using a "siamese" time delay neural network. In *Advances in Neural Information Processing Systems (NeurIPS)*, pp. 737–744.

Brooks, R. A. (1981). Symbolic reasoning among 3-D models and 2-D images. *Artificial Intelligence*, 17:285–348.

Brooks, T., Mildenhall, B., Xue, T., Chen, J., Sharlet, D., and Barron, J. T. (2019). Unprocessing images for learned raw denoising. In *IEEE/CVF Conference on Computer Vision and Pattern Recognition (CVPR)*.

Brown, A., Xie, W., Kalogeiton, V., and Zisserman, A. (2020). Smooth-AP: Smoothing the path towards large-scale image retrieval. In *European Conference on Computer Vision (ECCV)*.

Brown, D. C. (1966). Decentering distortion of lenses. *Photogrammetric Engineering and Remote Sensing*, 32(3):444–462.

Brown, D. C. (1971). Close-range camera calibration. *Photogrammetric Engineering*, 37(8):855–866.

Brown, L. G. (1992). A survey of image registration techniques. *Computing Surveys*, 24(4):325–376.

Brown, M. and Lowe, D. (2002). Invariant features from interest point groups. In *British Machine Vision Conference*, pp. 656–665.

Brown, M. and Lowe, D. (2005). Unsupervised 3D object recognition and reconstruction in unordered datasets. In *International Conference on 3D Imaging and Modelling*, pp. 1218–1225.

Brown, M. and Lowe, D. (2007). Automatic panoramic image stitching using invariant features. *International Journal of Computer Vision*, 74(1):59–73.

Brown, M., Hartley, R., and Nistér, D. (2007). Minimal solutions for panoramic stitching. In *IEEE Computer Society Conference on Computer Vision and Pattern Recognition (CVPR)*.

Brown, M., Hua, G., and Winder, S. (2011). Discriminative learning of local image descriptors. *IEEE Transactions on Pattern Analysis and Machine Intelligence*, 33(1):43–57.

Brown, M., Szeliski, R., and Winder, S. (2004). *Multi-Image Matching Using Multi-Scale Oriented Patches*. Technical Report MSR-TR-2004-133, Microsoft Research.

Brown, M., Szeliski, R., and Winder, S. (2005). Multi-image matching using multi-scale oriented patches. In *IEEE Computer Society Conference on Computer Vision and Pattern Recognition (CVPR)*, pp. 510–517.

Brown, M. S. (2019). ICCV 2019 tutorial on understanding color and the in-camera image processing pipeline for computer vision. https://www.eecs.yorku.ca/~mbrown/ICCV2019_Brown.html.

Brown, M. Z., Burschka, D., and Hager, G. D. (2003). Advances in computational stereo. *IEEE Transactions on Pattern Analysis and Machine Intelligence*, 25(8):993–1008.

Brown, T. B., Mann, B., Ryder, N., Subbiah, M., Kaplan, J., Dhariwal, P., Neelakantan, A., Shyam, P., Sastry, G., Askell, A. *et al.* (2020). Language models are few-shot learners. *arXiv preprint arXiv:2005.14165*.

Brox, T. and Malik, J. (2010a). Large displacement optical flow: descriptor matching in variational motion estimation. *IEEE Transactions on Pattern Analysis and Machine Intelligence*, 33(3):500–513.

Brox, T. and Malik, J. (2010b). Object segmentation by long term analysis of point trajectories. In *European Conference on Computer Vision (ECCV)*, pp. 282–295.

Brox, T., Bregler, C., and Malik, J. (2009). Large displacement optical flow. In *IEEE Computer Society Conference on Computer Vision and Pattern Recognition (CVPR)*.

Brox, T., Bruhn, A., Papenberg, N., and Weickert, J. (2004). High accuracy optical flow estimation based on a theory for warping. In *European Conference on Computer Vision (ECCV)*, pp. 25–36.

Broxton, M., Flynn, J., Overbeck, R., Erickson, D., Hedman, P., DuVall, M., Dourgarian, J., Busch, J., Whalen, M., and Debevec, P. (2020). Immersive light field video with a layered mesh representation. In *ACM Transactions on Graphics (Proc. SIGGRAPH)*, pp. 86:1–86:15.

Brubaker, S. C., Wu, J., Sun, J., Mullin, M. D., and Rehg, J. M. (2008). On the design of cascades of boosted ensembles for face detection. *International Journal of Computer Vision*, 77(1–3):65–86.

Bruhn, A., Weickert, J., and Schnörr, C. (2005). Lucas/Kanade meets Horn/Schunck: Combining local and global optic flow methods. *International Journal of Computer Vision*, 61(3):211–231.

Bruhn, A., Weickert, J., Kohlberger, T., and Schnörr, C. (2006). A multigrid platform for real-time motion computation with discontinuity-preserving variational methods. *International Journal of Computer Vision*, 70(3):257–277.

Buades, A., Coll, B., and Morel, J.-M. (2005a). A non-local algorithm for image denoising. In *IEEE Computer Society Conference on Computer Vision and Pattern Recognition (CVPR)*, pp. 60–65.

Buades, A., Coll, B., and Morel, J.-M. (2005b). A review of image denoising algorithms, with a new one. *Multiscale Modeling and Simulation*, 4(2):490–530.

Buades, A., Coll, B., and Morel, J.-M. (2008). Nonlocal image and movie denoising. *International Journal of Computer Vision*, 76(2):123–139.

Buccigrossi, R. W. and Simoncelli, E. P. (1999). Image compression via joint statistical characterization in the wavelet domain. *IEEE Transactions on Image Processing*, 8(12):1688–1701.

Buchanan, A. and Fitzgibbon, A. (2005). Damped Newton algorithms for matrix factorization with missing data. In *IEEE Computer Society Conference on Computer Vision and Pattern Recognition (CVPR)*, pp. 316–322.

Bucilă, C., Caruana, R., and Niculescu-Mizil, A. (2006). Model compression. In *ACM SIGKDD International Conference on Knowledge Discovery & Data Mining*, pp. 535–541.

Buck, I., Finkelstein, A., Jacobs, C., Klein, A., Salesin, D. H., Seims, J., Szeliski, R., and Toyama, K. (2000). Performance-driven hand-drawn animation. In *Symposium on Non Photorealistic Animation and Rendering*, pp. 101–108.

Buehler, C., Bosse, M., McMillan, L., Gortler, S. J., and Cohen, M. F. (2001). Unstructured Lumigraph rendering. In *ACM SIGGRAPH Conference Proceedings*, pp. 425–432.

Bugayevskiy, L. M. and Snyder, J. P. (1995). *Map Projections: A Reference Manual.* CRC Press.

Bulatov, Y. (2018). Fitting larger networks into memory. https://medium.com/tensorflow/fitting-larger-networks-into-memory-583e3c758ff9.

Burger, H. C., Schuler, C. J., and Harmeling, S. (2012). Image denoising: Can plain neural networks compete with BM3D? In *IEEE Computer Society Conference on Computer Vision and Pattern Recognition (CVPR)*.

Burger, W. and Burge, M. J. (2008). *Digital Image Processing: An Algorithmic Introduction Using Java.* Springer, New York, NY.

Burger, W. and Burge, M. J. (2009). *Principles of Digital Image Processing*. Springer, New York, NY.

Burl, M. C., Weber, M., and Perona, P. (1998). A probabilistic approach to object recognition using local photometry and global geometry. In *European Conference on Computer Vision (ECCV)*, pp. 628–641.

Burnes, A. (2020). NVIDIA DLSS 2.0: A big leap in AI rendering. https://www.nvidia.com/en-us/geforce/news/nvidia-dlss-2-0-a-big-leap-in-ai-rendering.

Burns, J. B., Hanson, A. R., and Riseman, E. M. (1986). Extracting straight lines. *IEEE Transactions on Pattern Analysis and Machine Intelligence*, PAMI-8(4):425–455.

Burns, P. D. and Williams, D. (1999). Using slanted edge analysis for color registration measurement. In *IS&T PICS Conference*, pp. 51–53.

Burri, M., Nikolic, J., Gohl, P., Schneider, T., Rehder, J., Omari, S., Achtelik, M. W., and Siegwart, R. (2016). The EuRoC micro aerial vehicle datasets. *International Journal of Robotics Research*, 35(10):1157–1163.

Burt, P. J. and Adelson, E. H. (1983a). The Laplacian pyramid as a compact image code. *IEEE Transactions on Communications*, COM-31(4):532–540.

Burt, P. J. and Adelson, E. H. (1983b). A multiresolution spline with applications to image mosaics. *ACM Transactions on Graphics*, 2(4):217–236.

Burt, P. J. and Kolczynski, R. J. (1993). Enhanced image capture through fusion. In *International Conference on Computer Vision (ICCV)*, pp. 173–182.

Butler, D. J., Wulff, J., Stanley, G. B., and Black, M. J. (2012). A naturalistic open source movie for optical flow evaluation. In *European Conference on Computer Vision (ECCV)*, pp. 611–625.

Byröd, M. and Åström, K. (2009). Bundle adjustment using conjugate gradients with multiscale preconditioning. In *British Machine Vision Conference (BMVC)*.

Cabral, B. K. (2016). Introducing Facebook Surround 360: An open, high-quality 3D-360 video capture system. https://engineering.fb.com/video-engineering/introducing-facebook-surround-360-an-open-high-quality-3d-360-video-capture-system.

Cadena, C., Carlone, L., Carrillo, H., Latif, Y., Scaramuzza, D., Neira, J., Reid, I., and Leonard, J. J. (2016). Past, present, and future of simultaneous localization and mapping: Toward the robust-perception age. *IEEE Transactions on Robotics*, 32(6):1309–1332.

Caelles, S., Maninis, K.-K., Pont-Tuset, J., Leal-Taixé, L., Cremers, D., and Van Gool, L. (2017). One-shot video object segmentation. In *IEEE Conference on Computer Vision and Pattern Recognition (CVPR)*.

Caelles, S., Pont-Tuset, J., Perazzi, F., Montes, A., Maninis, K.-K., and Van Gool, L. (2019). The 2019 DAVIS challenge on VOS: Unsupervised multi-object segmentation. *arXiv preprint arXiv:1905.00737*.

Caelles, S., Montes, A., Maninis, K.-K., Chen, Y., Van Gool, L., Perazzi, F., and Pont-Tuset, J. (2018). The 2018 DAVIS challenge on video object segmentation. *arXiv preprint arXiv:1803.00557*.

Cai, D., He, X., Hu, Y., Han, J., and Huang, T. (2007). Learning a spatially smooth subspace for face recognition. In *IEEE Computer Society Conference on Computer Vision and Pattern Recognition (CVPR)*.

Cai, J., Gu, S., Timofte, R., and Zhang, L. (2019). NTIRE 2019 challenge on real image super-resolution: Methods and results. In *IEEE Conference on Computer Vision and Pattern Recognition (CVPR) Workshops*.

Cai, J., Zeng, H., Yong, H., Cao, Z., and Zhang, L. (2019). Toward real-world single image super-resolution: A new benchmark and a new model. In *IEEE/CVF International Conference on Computer Vision (ICCV)*.

Calonder, M., Lepetit, V., Strecha, C., and Fua, P. (2010). BRIEF: Binary robust independent elementary features. In *European Conference on Computer Vision (ECCV)*, pp. 778–792.

Cammarata, N., Carter, S., Goh, G., Olah, C., Petrov, M., and Schubert, L. (2020). Thread: Circuits. *Distill*, 5(3):e24.

Campisi, P. and Egiazarian, K. (2017). *Blind image deconvolution: theory and applications*. CRC Press.

Can, A., Stewart, C., Roysam, B., and Tanenbaum, H. (2002). A feature-based, robust, hierarchical algorithm for registering pairs of images of the curved human retina. *IEEE Transactions on Pattern Analysis and Machine Intelligence*, 24(3):347–364.

Canny, J. (1986). A computational approach to edge detection. *IEEE Transactions on Pattern Analysis and Machine Intelligence*, PAMI-8(6):679–698.

Canziani, A., Culurciello, E., and Paszke, A. (2017). Evaluation of neural network architectures for embedded systems. In *IEEE International Symposium on Circuits and Systems (ISCAS)*, pp. 1–4.

Cao, B., Araujo, A., and Sim, J. (2020). Unifying deep local and global features for image search. In *European Conference on Computer Vision (ECCV)*.

Cao, X., Waechter, M., Shi, B., Gao, Y., Zheng, B., and Matsushita, Y. (2020). Stereoscopic flash and no-flash photography for shape and albedo recovery. In *IEEE/CVF Conference on Computer Vision and Pattern Recognition (CVPR)*.

Cao, Z., Simon, T., Wei, S.-E., and Sheikh, Y. (2017). Realtime multi-person 2D pose estimation using part affinity fields. In *IEEE Conference on Computer Vision and Pattern Recognition (CVPR)*.

Cao, Z., Yin, Q., Tang, X., and Sun, J. (2010). Face recognition with learning-based descriptor. In *IEEE Computer Society Conference on Computer Vision and Pattern Recognition (CVPR)*.

Cao, Z., Hidalgo, G., Simon, T., Wei, S.-E., and Sheikh, Y. (2019). OpenPose: Realtime multi-person 2D pose estimation using part affinity fields. *IEEE Transactions on Pattern Analysis and Machine Intelligence*, 43(1):172–186.

Capel, D. (2004). *Image Mosaicing and Super-resolution. Distinguished Dissertation Series, British Computer Society*, Springer-Verlag.

Capel, D. and Zisserman, A. (1998). Automated mosaicing with super-resolution zoom. In *IEEE Computer Society Conference on Computer Vision and Pattern Recognition (CVPR)*, pp. 885–891.

Capel, D. and Zisserman, A. (2000). Super-resolution enhancement of text image sequences. In *International Conference on Pattern Recognition (ICPR)*, pp. 600–605.

Capel, D. and Zisserman, A. (2003). Computer vision applied to super resolution. *IEEE Signal Processing Magazine*, 20(3):75–86.

Capel, D. P. (2001). *Super-resolution and Image Mosaicing*. Ph.D. thesis, University of Oxford.

Caprile, B. and Torre, V. (1990). Using vanishing points for camera calibration. *International Journal of Computer Vision*, 4(2):127–139.

Carion, N., Massa, F., Synnaeve, G., Usunier, N., Kirillov, A., and Zagoruyko, S. (2020). End-to-end object detection with transformers. In *European Conference on Computer Vision (ECCV)*.

Carlone, L., Tron, R., Daniilidis, K., and Dellaert, F. (2015). Initialization techniques for 3D SLAM: a survey on rotation estimation and its use in pose graph optimization. In *IEEE International Conference on Robotics and Automation (ICRA)*, pp. 4597–4604.

Carneiro, G. and Jepson, A. (2005). The distinctiveness, detectability, and robustness of local image features. In *IEEE Computer Society Conference on Computer Vision and Pattern Recognition (CVPR)*, pp. 296–301.

Carneiro, G. and Lowe, D. (2006). Sparse flexible models of local features. In *European Conference on Computer Vision (ECCV)*, pp. 29–43.

Carnevali, P., Coletti, L., and Patarnello, S. (1985). Image processing by simulated annealing. *IBM Journal of Research and Development*, 29(6):569–579.

Caron, M., Bojanowski, P., Joulin, A., and Douze, M. (2018). Deep clustering for unsupervised learning of visual features. In *European Conference on Computer Vision (ECCV)*.

Caron, M., Misra, I., Mairal, J., Goyal, P., Bojanowski, P., and Joulin, A. (2020). Unsupervised learning of visual features by contrasting cluster assignments. In *Advances in Neural Information Processing Systems (NeurIPS)*.

Carranza, J., Theobalt, C., Magnor, M. A., and Seidel, H.-P. (2003). Free-viewpoint video of human actors. *ACM Transactions on Graphics (Proc. SIGGRAPH)*, 22(3):569–577.

Carreira, J. and Zisserman, A. (2017). Quo vadis, action recognition? a new model and the kinetics dataset. In *IEEE Conference on Computer Vision and Pattern Recognition (CVPR)*.

Carroll, R., Agarwala, A., and Agrawala, M. (2010). Image warps for artistic perspective manipulation. *ACM Transactions on Graphics*, 29(4):127.

Carroll, R., Agrawala, M., and Agarwala, A. (2009). Optimizing content-preserving projections for wide-angle images. *ACM Transactions on Graphics*, 28(3):43.

Caselles, V., Kimmel, R., and Sapiro, G. (1997). Geodesic active contours. *International Journal of Computer Vision*, 21(1):61–79.

Catmull, E. and Smith, A. R. (1980). 3-D transformations of images in scanline order. *Computer Graphics (SIGGRAPH)*, 14(3):279–285.

Celniker, G. and Gossard, D. (1991). Deformable curve and surface finite-elements for free-form shape design. *Computer Graphics (SIGGRAPH)*, 25(4):257–266.

Chabra, R., Straub, J., Sweeney, C., Newcombe, R., and Fuchs, H. (2019). StereoDRNet: Dilated residual stereonet. In *IEEE/CVF Conference on Computer Vision and Pattern Recognition (CVPR)*.

Chabra, R., Lenssen, J. E., Ilg, E., Schmidt, T., Straub, J., Lovegrove, S., and Newcombe, R. (2020). Deep local shapes: Learning local SDF priors for detailed 3D reconstruction. In *European Conference on Computer Vision (ECCV)*.

Chakrabarti, A., Scharstein, D., and Zickler, T. (2009). An empirical camera model for internet color vision. In *British Machine Vision Conference (BMVC)*.

Cham, T. J. and Cipolla, R. (1998). A statistical framework for long-range feature matching in uncalibrated image mosaicing. In *IEEE Computer Society Conference on Computer Vision and Pattern Recognition (CVPR)*, pp. 442–447.

Cham, T.-J. and Rehg, J. M. (1999). A multiple hypothesis approach to figure tracking. In *IEEE Computer Society Conference on Computer Vision and Pattern Recognition (CVPR)*, pp. 239–245.

Chambolle, A. (2004). An algorithm for total variation minimization and applications. *Journal of Mathematical Imaging and Vision*, 20(1–2):89–97.

Chambolle, A. and Pock, T. (2011). A first-order primal-dual algorithm for convex problems with applications to imaging. *Journal of Mathematical Imaging and Vision*, 40(1):120–145.

Champleboux, G., Lavallée, S., Sautot, P., and Cinquin, P. (1992a). Accurate calibration of cameras and range imaging sensors, the NPBS method. In *IEEE International Conference on Robotics and Automation*, pp. 1552–1558.

Champleboux, G., Lavallée, S., Szeliski, R., and Brunie, L. (1992b). From accurate range imaging sensor calibration to accurate model-based 3-D object localization. In *IEEE Computer Society Conference on Computer Vision and Pattern Recognition (CVPR)*, pp. 83–89.

Chan, A. B. and Vasconcelos, N. (2009). Layered dynamic textures. *IEEE Transactions on Pattern Analysis and Machine Intelligence*, 31(10):1862–1879.

Chan, C., Ginosar, S., Zhou, T., and Efros, A. A. (2019). Everybody dance now. In *IEEE/CVF International Conference on Computer Vision (ICCV)*.

Chan, T. F. and Shen, J. J. (2005). *Image processing and analysis: variational, PDE, wavelet, and stochastic methods*. SIAM.

Chan, T. F. and Vese, L. A. (2001). Active contours without edges. *IEEE Transactions on Image Processing*, 10(2):266–277.

Chan, T. F., Osher, S., and Shen, J. (2001). The digital TV filter and nonlinear denoising. *IEEE Transactions on Image Processing*, 10(2):231–241.

Chang, A. X., Funkhouser, T., Guibas, L., Hanrahan, P., Huang, Q., Li, Z., Savarese, S., Savva, M., Song, S., Su, H., Xiao, J., Yi, L., and Yu, F. (2015). ShapeNet: An information-rich 3D model repository. *arXiv preprint arXiv:1512.03012*.

Chang, J.-R. and Chen, Y.-S. (2018). Pyramid stereo matching network. In *IEEE Conference on Computer Vision and Pattern Recognition (CVPR)*.

Chang, M. M., Tekalp, A. M., and Sezan, M. I. (1997). Simultaneous motion estimation and segmentation. *IEEE Transactions on Image Processing*, 6(9):1326–1333.

Chang, Y., Hu, C., Feris, R., and Turk, M. (2006). Manifold based analysis of facial expression. *Image and Vision Computing*, 24(6):605–614.

Chang, Y.-L., Liu, Z. Y., Lee, K.-Y., and Hsu, W. (2019). Free-form video inpainting with 3D gated convolution and temporal PatchGAN. In *IEEE/CVF International Conference on Computer Vision (ICCV)*.

Charniak, E. (2019). *Introduction to Deep Learning*. MIT Press.

Chatterjee, A. and Govindu, V. M. (2013). Efficient and robust large-scale rotation averaging. In *International Conference on Computer Vision (ICCV)*.

Chatterjee, P., Joshi, N., Kang, S. B., and Matsushita, Y. (2011). Noise suppression in low-light images through joint denoising and demosaicing. In *IEEE Computer Society Conference on Computer Vision and Pattern Recognition (CVPR)*.

Chaudhuri, S. (2001). *Super-Resolution Imaging*. Springer.

Chaudhuri, S. and Rajagopalan, A. N. (1999). *Depth from Defocus: A Real Aperture Imaging Approach*. Springer.

Chaurasia, G., Duchene, S., Sorkine-Hornung, O., and Drettakis, G. (2013). Depth synthesis and local warps for plausible image-based navigation. *ACM Transactions on Graphics (ToG)*, 32(3):1–12.

Chaurasia, G., Nieuwoudt, A., Ichim, A.-E., Szeliski, R., and Sorkine-Hornung, A. (2020). Passthrough+: Real-time stereoscopic view synthesis for mobile mixed reality. *Proceedings of the ACM on Computer Graphics and Interactive Techniques*, 3(1):7.

Cheeseman, P., Kanefsky, B., Hanson, R., and Stutz, J. (1993). *Super-Resolved Surface Reconstruction From Multiple Images*. Technical Report FIA-93-02, NASA Ames Research Center, Artificial Intelligence Branch.

Chellappa, R., Wilson, C., and Sirohey, S. (1995). Human and machine recognition of faces: A survey. *Proceedings of the IEEE*, 83(5):705–740.

Chellappa, R., Sankaranarayanan, A. C., Veeraraghavan, A., and Turaga, P. (2010). Statistical methods and models for video-based tracking, modeling, and recognition. *Foundations and Trends® in Signal Processing*, 3(1–2):1–151.

Chen, B., Neubert, B., Ofek, E., Deussen, O., and Cohen, M. F. (2009). Integrated videos and maps for driving directions. In *ACM Symposium on User Interface Software and Technology (UIST)*, pp. 223–232, New York, NY, USA.

Chen, C., Chen, Q., Xu, J., and Koltun, V. (2018). Learning to see in the dark. In *IEEE Conference on Computer Vision and Pattern Recognition (CVPR)*.

Chen, C.-Y. and Klette, R. (1999). Image stitching - comparisons and new techniques. In *Computer Analysis of Images and Patterns (CAIP)*, pp. 615–622.

Chen, G., Han, K., Shi, B., Matsushita, Y., and Wong, K.-Y. K. (2019). Self-calibrating deep photometric stereo networks. In *IEEE/CVF Conference on Computer Vision and Pattern Recognition (CVPR)*.

Chen, G., Waechter, M., Shi, B., Wong, K.-Y. K., and Matsushita, Y. (2020). What is learned in deep uncalibrated photometric stereo? In *European Conference on Computer Vision (ECCV)*.

Chen, J. and Chen, B. (2008). Architectural modeling from sparsely scanned range data. *International Journal of Computer Vision*, 78(2–3):223–236.

Chen, J., Paris, S., and Durand, F. (2007). Real-time edge-aware image processing with the bilateral grid. *ACM Transactions on Graphics*, 26(3).

Chen, K., Pang, J., Wang, J., Xiong, Y., Li, X., Sun, S., Feng, W., Liu, Z., Shi, J., Ouyang, W., Loy, C. C., and Lin, D. (2019). Hybrid task cascade for instance segmentation. In *IEEE/CVF Conference on Computer Vision and Pattern Recognition (CVPR)*.

Chen, K., Wang, J., Pang, J., Cao, Y., Xiong, Y., Li, X., Sun, S., Feng, W., Liu, Z., Xu, J., Zhang, Z., Cheng, D., Zhu, C., Cheng, T., Zhao, Q., Li, B., Lu, X., Zhu, R., Wu, Y., Dai, J., Wang, J., Shi, J., Ouyang, W., Loy, C. C., and Lin, D. (2019). MMDetection: Open MMLab detection toolbox and benchmark. *arXiv preprint arXiv:1906.07155*.

Chen, L.-C., Papandreou, G., Kokkinos, I., Murphy, K., and Yuille, A. L. (2018). DeepLab: Semantic image segmentation with deep convolutional nets, atrous convolution, and fully connected CRFs. *IEEE Transactions on Pattern Analysis and Machine Intelligence*, 40(4):834–848.

Chen, L.-C., Zhu, Y., Papandreou, G., Schroff, F., and Adam, H. (2018). Encoder-decoder with atrous separable convolution for semantic image segmentation. In *European Conference on Computer Vision (ECCV)*, pp. 833–851.

Chen, M., Radford, A., Child, R., Wu, J., Jun, H., Luan, D., and Sutskever, I. (2020). Generative pretraining from pixels. In *International Conference on Machine Learning (ICML)*, pp. 1691–1703.

Chen, Q. and Koltun, V. (2017). Photographic image synthesis with cascaded refinement networks. In *IEEE International Conference on Computer Vision (ICCV)*.

Chen, Q., Xu, J., and Koltun, V. (2017). Fast image processing with fully-convolutional networks. In *IEEE International Conference on Computer Vision (ICCV)*.

Chen, S. and Williams, L. (1993). View interpolation for image synthesis. In *ACM SIGGRAPH Conference Proceedings*, pp. 279–288.

Chen, S. E. (1995). QuickTime VR – an image-based approach to virtual environment navigation. In *ACM SIGGRAPH Conference Proceedings*, pp. 29–38.

Chen, T., Kornblith, S., Norouzi, M., and Hinton, G. (2020). A simple framework for contrastive learning of visual representations. In *International Conference on Machine Learning (ICML)*, pp. 1597–1607.

Chen, T., Xu, B., Zhang, C., and Guestrin, C. (2016). Training deep nets with sublinear memory cost. *arXiv preprint arXiv:1604.06174*.

Chen, W., Fu, Z., Yang, D., and Deng, J. (2016). Single-image depth perception in the wild. In *Advances in Neural Information Processing Systems (NeurIPS)*, pp. 730–738.

Chen, X. and He, K. (2021). Exploring simple siamese representation learning. In *IEEE/CVF Conference on Computer Vision and Pattern Recognition (CVPR)*.

Chen, X. and Lawrence Zitnick, C. (2015). Mind's eye: A recurrent visual representation for image caption generation. In *IEEE Conference on Computer Vision and Pattern Recognition (CVPR)*.

Chen, X., Fan, H., Girshick, R., and He, K. (2020). Improved baselines with momentum contrastive learning. *arXiv preprint arXiv:2003.04297*.

Chen, X., Duan, Y., Houthooft, R., Schulman, J., Sutskever, I., and Abbeel, P. (2016). InfoGAN: Interpretable representation learning by information maximizing generative adversarial nets. In *Advances in Neural Information Processing Systems (NeurIPS)*, pp. 2172–2180.

Chen, X., Fang, H., Lin, T.-Y., Vedantam, R., Gupta, S., Dollár, P., and Zitnick, C. L. (2015). Microsoft COCO captions: Data collection and evaluation server. *arXiv preprint arXiv:1504.00325*.

Chen, Y. and Medioni, G. (1992). Object modeling by registration of multiple range images. *Image and Vision Computing*, 10(3):145–155.

Chen, Y. and Pock, T. (2016). Trainable nonlinear reaction diffusion: A flexible framework for fast and effective image restoration. *IEEE Transactions on Pattern Analysis and Machine Intelligence*, 39(6):1256–1272.

Chen, Y., Pont-Tuset, J., Montes, A., and Van Gool, L. (2018). Blazingly fast video object segmentation with pixel-wise metric learning. In *IEEE Conference on Computer Vision and Pattern Recognition (CVPR)*.

Chen, Z. and Zhang, H. (2019). Learning implicit fields for generative shape modeling. In *IEEE/CVF Conference on Computer Vision and Pattern Recognition (CVPR)*.

Chen, Z., Nobuhara, S., and Nishino, K. (2020). Invertible neural BRDF for object inverse rendering. In *European Conference on Computer Vision (ECCV)*.

Cheng, B., Girshick, R., Dollar, P., Berg, A. C., and Kirillov, A. (2021). Boundary IoU: Improving object-centric image segmentation evaluation. In *IEEE/CVF Conference on Computer Vision and Pattern Recognition (CVPR)*.

Cheng, J., Tsai, Y.-H., Wang, S., and Yang, M.-H. (2017). SegFlow: Joint learning for video object segmentation and optical flow. In *IEEE International Conference on Computer Vision (ICCV)*.

Cheng, J., Tsai, Y.-H., Hung, W.-C., Wang, S., and Yang, M.-H. (2018). Fast and accurate online video object segmentation via tracking parts. In *IEEE Conference on Computer Vision and Pattern Recognition (CVPR)*.

Cheng, L., Vishwanathan, S. V. N., and Zhang, X. (2008). Consistent image analogies using semi-supervised learning. In *IEEE Computer Society Conference on Computer Vision and Pattern Recognition (CVPR)*.

Cheng, M.-M., Zhang, Z., Lin, W.-Y., and Torr, P. (2014). BING: Binarized normed gradients for objectness estimation at 300fps. In *IEEE Conference on Computer Vision and Pattern Recognition (CVPR)*.

Cheng, X., Zhong, Y., Harandi, M., Dai, Y., Chang, X., Li, H., Drummond, T., and Ge, Z. (2020). Hierarchical neural architecture search for deep stereo matching. In *Advances in Neural Information Processing Systems (NeurIPS)*.

Cheng, Y. (1995). Mean shift, mode seeking, and clustering. *IEEE Transactions on Pattern Analysis and Machine Intelligence*, 17(8):790–799.

Cheng, Y. (2020). CVPR 2020 tutorial on text-to-image generation. https://rohit497.github.io/Recent-Advances-in-Vision-and-Language-Research.

Chiang, M.-C. and Boult, T. E. (1996). Efficient image warping and super-resolution. In *IEEE Workshop on Applications of Computer Vision (WACV)*, pp. 56–61.

Cho, T. S., Avidan, S., and Freeman, W. T. (2010). A probabilistic image jigsaw puzzle solver. In *IEEE Conference on Computer Vision and Pattern Recognition (CVPR)*.

Cho, T. S., Paris, S., Freeman, B., and Horn, B. (2011). Blur kernel estimation using the radon transform. In *IEEE Computer Society Conference on Computer Vision and Pattern Recognition (CVPR)*.

Cho, T. S., Joshi, N., Zitnick, C. L., Kang, S. B., Szeliski, R., and Freeman, W. T. (2012). Image restoration by matching gradient distributions. *IEEE Transactions on Pattern Analysis and Machine Intelligence*, 34(4):683–694.

Choi, I., Gallo, O., Troccoli, A., Kim, M. H., and Kautz, J. (2019). Extreme view synthesis. In *IEEE/CVF International Conference on Computer Vision (ICCV)*.

Choi, M., Choi, J., Baik, S., Kim, T. H., and Lee, K. M. (2020). Scene-adaptive video frame interpolation via meta-learning. In *IEEE/CVF Conference on Computer Vision and Pattern Recognition (CVPR)*.

Choi, Y., Uh, Y., Yoo, J., and Ha, J.-W. (2020). Stargan v2: Diverse image synthesis for multiple domains. In *IEEE/CVF Conference on Computer Vision and Pattern Recognition (CVPR)*.

Choi, Y., Choi, M., Kim, M., Ha, J.-W., Kim, S., and Choo, J. (2018). StarGAN: Unified generative adversarial networks for multi-domain image-to-image translation. In *IEEE Conference on Computer Vision and Pattern Recognition (CVPR)*.

Chokshi, N. (2019). Facial recognition's many controversies, from stadium surveillance to racist software. *New York Times*.

Chopra, S., Hadsell, R., and LeCun, Y. (2005). Learning a similarity metric discriminatively, with application to face verification. In *IEEE Computer Society Conference on Computer Vision and Pattern Recognition (CVPR)*, pp. 539–546.

Chou, P. B. and Brown, C. M. (1990). The theory and practice of Bayesian image labeling. *International Journal of Computer Vision*, 4(3):185–210.

Choudhary, T., Mishra, V., Goswami, A., and Sarangapani, J. (2020). A comprehensive survey on model compression and acceleration. *Artificial Intelligence Review*, (53):5113–5155.

Choutas, V., Pavlakos, G., Bolkart, T., Tzionas, D., and Black, M. J. (2020). Monocular expressive body regression through body-driven attention. In *European Conference on Computer Vision (ECCV)*.

Choy, C. B., Xu, D., Gwak, J., Chen, K., and Savarese, S. (2016). 3D-R2N2: A unified approach for single and multi-view 3D object reconstruction. In *European Conference on Computer Vision (ECCV)*, pp. 628–644.

Christensen, G., Joshi, S., and Miller, M. (1997). Volumetric transformation of brain anatomy. *IEEE Transactions on Medical Imaging*, 16(6):864–877.

Christy, S. and Horaud, R. (1996). Euclidean shape and motion from multiple perspective views by affine iterations. *IEEE Transactions on Pattern Analysis and Machine Intelligence*, 18(11):1098–1104.

Chu, H., Ma, S., la Torre, F. D., Fidler, S., and Sheikh, Y. (2020). Expressive telepresence via modular codec avatars. In *European Conference on Computer Vision (ECCV)*.

Chuang, Y.-Y., Curless, B., Salesin, D. H., and Szeliski, R. (2001). A Bayesian approach to digital matting. In *IEEE Computer Society Conference on Computer Vision and Pattern Recognition (CVPR)*, pp. 264–271.

Chuang, Y.-Y., Agarwala, A., Curless, B., Salesin, D. H., and Szeliski, R. (2002). Video matting of complex scenes. *ACM Transactions on Graphics (Proc. SIGGRAPH)*, 21(3):243–248.

Chuang, Y.-Y., Goldman, D. B., Curless, B., Salesin, D. H., and Szeliski, R. (2003). Shadow matting. *ACM Transactions on Graphics (Proc. SIGGRAPH)*, 22(3):494–500.

Chuang, Y.-Y., Goldman, D. B., Zheng, K. C., Curless, B., Salesin, D. H., and Szeliski, R. (2005). Animating pictures with stochastic motion textures. *ACM Transactions on Graphics (Proc. SIGGRAPH)*, 24(3):853–860.

Chuang, Y.-Y., Zongker, D., Hindorff, J., Curless, B., Salesin, D. H., and Szeliski, R. (2000). Environment matting extensions: Towards higher accuracy and real-time capture. In *ACM SIGGRAPH Conference Proceedings*, pp. 121–130.

Chui, C. K. (1992). *Wavelet Analysis and Its Applications*. Academic Press, New York.

Chum, O. and Matas, J. (2005). Matching with PROSAC—progressive sample consensus. In *IEEE Computer Society Conference on Computer Vision and Pattern Recognition (CVPR)*, pp. 220–226.

Chum, O. and Matas, J. (2010a). Large-scale discovery of spatially related images. *IEEE Transactions on Pattern Analysis and Machine Intelligence*, 32(2):371–377.

Chum, O. and Matas, J. (2010b). Unsupervised discovery of co-occurrence in sparse high dimensional data. In *IEEE Computer Society Conference on Computer Vision and Pattern Recognition (CVPR)*.

Chum, O. and Zisserman, A. (2007). An exemplar model for learning object classes. In *IEEE Computer Society Conference on Computer Vision and Pattern Recognition (CVPR)*.

Chum, O., Philbin, J., and Zisserman, A. (2008). Near duplicate image detection: min-hash and tf-idf weighting. In *British Machine Vision Conference (BMVC)*.

Chum, O., Werner, T., and Matas, J. (2005). Two-view geometry estimation unaffected by a dominant plane. In *IEEE Computer Society Conference on Computer Vision and Pattern Recognition (CVPR)*, pp. 772–779.

Chum, O., Philbin, J., Sivic, J., Isard, M., and Zisserman, A. (2007). Total recall: Automatic query expansion with a generative feature model for object retrieval. In *International Conference on Computer Vision (ICCV)*.

Chung, J. S. and Zisserman, A. (2016). Out of time: automated lip sync in the wild. In *Asian Conference on Computer Vision (ACCV) Workshops*, pp. 251–263.

Cignoni, P., Callieri, M., Corsini, M., Dellepiane, M., Ganovelli, F., and Ranzuglia, G. (2008). MeshLab: an Open-Source Mesh Processing Tool. In *Eurographics Italian Chapter Conference*.

Cipolla, R. and Blake, A. (1990). The dynamic analysis of apparent contours. In *International Conference on Computer Vision (ICCV)*, pp. 616–623.

Cipolla, R. and Blake, A. (1992). Surface shape from the deformation of apparent contours. *International Journal of Computer Vision*, 9(2):83–112.

Cipolla, R. and Giblin, P. (2000). *Visual Motion of Curves and Surfaces*. Cambridge University Press, Cambridge.

Cipolla, R., Drummond, T., and Robertson, D. P. (1999). Camera calibration from vanishing points in images of architectural scenes. In *British Machine Vision Conference (BMVC)*.

Claus, D. and Fitzgibbon, A. (2005). A rational function lens distortion model for general cameras. In *IEEE Computer Society Conference on Computer Vision and Pattern Recognition (CVPR)*, pp. 213–219.

Claus, M. and van Gemert, J. (2019). ViDeNN: Deep blind video denoising. In *IEEE/CVF Conference on Computer Vision and Pattern Recognition Workshops (CVPRW)*.

Clevert, D.-A., Unterthiner, T., and Hochreiter, S. (2015). Fast and accurate deep network learning by exponential linear units (ELUs). *arXiv preprint arXiv:1511.07289*.

Clowes, M. B. (1971). On seeing things. *Artificial Intelligence*, 2:79–116.

Cohen, L. D. and Cohen, I. (1993). Finite-element methods for active contour models and balloons for 2-D and 3-D images. *IEEE Transactions on Pattern Analysis and Machine Intelligence*, 15(11):1131–1147.

Cohen, M. and Wallace, J. (1993). *Radiosity and Realistic Image Synthesis*. Morgan Kaufmann.

Cohen, M. F. and Szeliski, R. (2006). The Moment Camera. *Computer*, 39(8):40–45.

Collins, R. T. (1996). A space-sweep approach to true multi-image matching. In *IEEE Computer Society Conference on Computer Vision and Pattern Recognition (CVPR)*, pp. 358–363.

Collins, R. T. and Liu, Y. (2003). On-line selection of discriminative tracking features. In *International Conference on Computer Vision (ICCV)*, pp. 346–352.

Collins, R. T. and Weiss, R. S. (1990). Vanishing point calculation as a statistical inference on the unit sphere. In *International Conference on Computer Vision (ICCV)*, pp. 400–403.

Comaniciu, D. and Meer, P. (2002). Mean shift: A robust approach toward feature space analysis. *IEEE Transactions on Pattern Analysis and Machine Intelligence*, 24(5):603–619.

Conn, A. R., Gould, N. I. M., and Toint, P. L. (2000). *Trust-Region Methods*. Society for Industrial and Applied Mathematics, Philadephia.

Cook, R. L. and Torrance, K. E. (1982). A reflectance model for computer graphics. *ACM Transactions on Graphics*, 1(1):7–24.

Coorg, S. and Teller, S. (2000). Spherical mosaics with quaternions and dense correlation. *International Journal of Computer Vision*, 37(3):259–273.

Cootes, T., Edwards, G. J., and Taylor, C. J. (2001). Active appearance models. *IEEE Transactions on Pattern Analysis and Machine Intelligence*, 23(6):681–685.

Cootes, T., Cooper, D., Taylor, C., and Graham, J. (1995). Active shape models—their training and application. *Computer Vision and Image Understanding*, 61(1):38–59.

Cootes, T., Taylor, C., Lanitis, A., Cooper, D., and Graham, J. (1993). Building and using flexible models incorporating grey-level information. In *International Conference on Computer Vision (ICCV)*, pp. 242–246.

Cootes, T. F. and Taylor, C. J. (2001). Statistical models of appearance for medical image analysis and computer vision. In *Medical Imaging*.

Coquillart, S. (1990). Extended free-form deformations: A sculpturing tool for 3D geometric modeling. *Computer Graphics (SIGGRAPH)*, 24(4):187–196.

Cordonnier, J.-B., Loukas, A., and Jaggi, M. (2019). On the relationship between self-attention and convolutional layers. *arXiv preprint arXiv:1911.03584*.

Cordts, M., Omran, M., Ramos, S., Rehfeld, T., Enzweiler, M., Benenson, R., Franke, U., Roth, S., and Schiele, B. (2016). The Cityscapes dataset for semantic urban scene understanding. In *IEEE Conference on Computer Vision and Pattern Recognition (CVPR)*.

Cormen, T. H. (2001). *Introduction to Algorithms*. MIT Press, Cambridge, Massachusetts.

Cornelis, N., Leibe, B., Cornelis, K., and Van Gool, L. (2008). 3D urban scene modeling integrating recognition and reconstruction. *International Journal of Computer Vision*, 78(2–3):121–141.

Corso, J. and Hager, G. (2005). Coherent regions for concise and stable image description. In *IEEE Computer Society Conference on Computer Vision and Pattern Recognition (CVPR)*, pp. 184–190.

Costeira, J. and Kanade, T. (1995). A multi-body factorization method for motion analysis. In *International Conference on Computer Vision (ICCV)*, pp. 1071–1076.

Costen, N., Cootes, T. F., Edwards, G. J., and Taylor, C. J. (1999). Simultaneous extraction of functional face subspaces. In *IEEE Computer Society Conference on Computer Vision and Pattern Recognition (CVPR)*, pp. 492–497.

Coughlan, J. M. and Yuille, A. L. (1999). Manhattan world: Compass direction from a single image by Bayesian inference. In *IEEE International Conference on Computer Vision (ICCV)*, pp. 941–947.

Couprie, C., Grady, L., Najman, L., and Talbot, H. (2009). Power watersheds: A new image segmentation framework extending graph cuts, random walker and optimal spanning forest. In *International Conference on Computer Vision (ICCV)*.

Cour, T., Bénézit, F., and Shi, J. (2005). Spectral segmentation with multiscale graph decomposition. In *IEEE Computer Society Conference on Computer Vision and Pattern Recognition (CVPR)*, pp. 1123–1130.

Cover, T. and Hart, P. (1967). Nearest neighbor pattern classification. *IEEE Transactions on Information Theory*, 13(1):21–27.

Cox, I. J. (1994). A maximum likelihood N-camera stereo algorithm. In *IEEE Computer Society Conference on Computer Vision and Pattern Recognition (CVPR)*, pp. 733–739.

Cox, I. J., Roy, S., and Hingorani, S. L. (1995). Dynamic histogram warping of image pairs for constant image brightness. In *IEEE International Conference on Image Processing (ICIP)*, pp. 366–369.

Cox, I. J., Hingorani, S. L., Rao, S. B., and Maggs, B. M. (1996). A maximum likelihood stereo algorithm. *Computer Vision and Image Understanding*, 63(3):542–567.

Crandall, D. and Huttenlocher, D. (2007). Composite models of objects and scenes for category recognition. In *IEEE Computer Society Conference on Computer Vision and Pattern Recognition (CVPR)*.

Crandall, D., Felzenszwalb, P., and Huttenlocher, D. (2005). Spatial priors for part-based recognition using statistical models. In *IEEE Computer Society Conference on Computer Vision and Pattern Recognition (CVPR)*, pp. 10–17.

Crandall, D., Backstrom, L., Huttenlocher, D., and Kleinberg, J. (2009). Mapping the world's photos. In *International World Wide Web Conference*, pp. 761–770.

Crandall, D. J. and Huttenlocher, D. P. (2006). Weakly supervised learning of part-based spatial models for visual object recognition. In *European Conference on Computer Vision (ECCV)*, pp. 16–29.

Crane, R. (1997). *A Simplified Approach to Image Processing*. Prentice Hall, Upper Saddle River, NJ.

Craven, P. and Wahba, G. (1979). Smoothing noisy data with spline functions: Estimating the correct degree of smoothing by the method of generalized cross-validation. *Numerische Mathematik*, 31:377–403.

Cremers, D. (2007). Nonlinear dynamical shape priors for level set segmentation. In *IEEE Computer Society Conference on Computer Vision and Pattern Recognition (CVPR)*.

Cremers, D. and Soatto, S. (2005). Motion competition: A variational framework for piecewise parametric motion segmentation. *International Journal of Computer Vision*, 62(3):249–265.

Cremers, D., Rousson, M., and Deriche, R. (2007). A review of statistical approaches to level set segmentation: integrating color, texture, motion and shape. *International Journal of Computer Vision*, 72(2):195–215.

Creswell, A., White, T., Dumoulin, V., Arulkumaran, K., Sengupta, B., and Bharath, A. A. (2018). Generative adversarial networks: An overview. *IEEE Signal Processing Magazine*, 35(1):53–65.

Crevier, D. (1993). *AI: The Tumultuous Search for Artificial Intelligence*. BasicBooks, New York, NY.

Criminisi, A. and Shotton, J. (2013). *Decision forests for computer vision and medical image analysis*. Springer.

Criminisi, A., Pérez, P., and Toyama, K. (2004). Region filling and object removal by exemplar-based inpainting. *IEEE Transactions on Image Processing*, 13(9):1200–1212.

Criminisi, A., Reid, I., and Zisserman, A. (2000). Single view metrology. *International Journal of Computer Vision*, 40(2):123–148.

Criminisi, A., Sharp, T., and Blake, A. (2008). GeoS: Geodesic image segmentation. In *European Conference on Computer Vision (ECCV)*, pp. 99–112.

Criminisi, A., Shotton, J., and Konukoglu, E. (2012). Decision forests: A unified framework for classification, regression, density estimation, manifold learning and semi-supervised learning. *Foundations and Trends® in Computer Graphics and Vision*, 7(2–3):81–227.

Criminisi, A., Cross, G., Blake, A., and Kolmogorov, V. (2006). Bilayer segmentation of live video. In *IEEE Computer Society Conference on Computer Vision and Pattern Recognition (CVPR)*, pp. 53–60.

Criminisi, A., Shotton, J., Blake, A., and Torr, P. (2003). Gaze manipulation for one-to-one teleconferencing. In *International Conference on Computer Vision (ICCV)*, pp. 191–198.

Criminisi, A., Kang, S. B., Swaminathan, R., Szeliski, R., and Anandan, P. (2005). Extracting layers and analyzing their specular properties using epipolar-plane-image analysis. *Computer Vision and Image Understanding*, 97(1):51–85.

Criminisi, A., Shotton, J., Blake, A., Rother, C., and Torr, P. H. S. (2007). Efficient dense stereo with occlusion by four-state dynamic programming. *International Journal of Computer Vision*, 71(1):89–110.

Criminisi, A., Robertson, D., Konukoglu, E., Shotton, J., Pathak, S., White, S., and Siddiqui, K. (2013). Regression forests for efficient anatomy detection and localization in computed tomography scans. *Medical Image Analysis*, 17(8):1293–1303.

Cross, G. (2019). Fully autonomous flying robots in the wild - challenges and opportunities. Keynote talk at the ICRA 2019 Workshop on Algorithms And Architectures For Learning In-The-Loop Systems In Autonomous Flight, https://ieee-aerialrobotics-uavs.org/wp-content/uploads/2019/06/ICRA-workshop-talk-skydio.pdf.

Crow, F. C. (1984). Summed-area table for texture mapping. *Computer Graphics (SIGGRAPH)*, 18(3):207–212.

Crowley, J. L. and Stern, R. M. (1984). Fast computation of the difference of low-pass transform. *IEEE Transactions on Pattern Analysis and Machine Intelligence*, 6(2):212–222.

Csurka, G., Dance, C. R., Perronnin, F., and Willamowski, J. (2006). Generic visual categorization using weak geometry. In Ponce, J., Hebert, M., Schmid, C., and Zisserman, A. (eds), *Toward Category-Level Object Recognition*, pp. 207–224, Springer, New York.

Csurka, G., Dance, C. R., Fan, L., Willamowski, J., and Bray, C. (2004). Visual categorization with bags of keypoints. In *ECCV International Workshop on Statistical Learning in Computer Vision*.

Cui, J., Yang, Q., Wen, F., Wu, Q., Zhang, C., Van Gool, L., and Tang, X. (2008). Transductive object cutout. In *IEEE Computer Society Conference on Computer Vision and Pattern Recognition (CVPR)*.

Cui, Z. and Tan, P. (2015). Global structure-from-motion by similarity averaging. In *IEEE International Conference on Computer Vision (ICCV)*.

Curless, B. (1999). From range scans to 3D models. *Computer Graphics*, 33(4):38–41.

Curless, B. and Levoy, M. (1995). Better optical triangulation through spacetime analysis. In *International Conference on Computer Vision (ICCV)*, pp. 987–994.

Curless, B. and Levoy, M. (1996). A volumetric method for building complex models from range images. In *ACM SIGGRAPH Conference Proceedings*, pp. 303–312.

Cutler, R. and Davis, L. S. (2000). Robust real-time periodic motion detection, analysis, and applications. *IEEE Transactions on Pattern Analysis and Machine Intelligence*, 22(8):781–796.

Czolbe, S., Krause, O., Cox, I., and Igel, C. (2020). A loss function for generative neural networks based on Watson's perceptual model. In *Advances in Neural Information Processing Systems (NeurIPS)*.

Čech, J. and Šára, R. (2007). Efficient sampling of disparity space for fast and accurate matching. In *Towards Benchmarking Automated Calibration, Orientation and Surface Reconstruction from Images (BenCOS)*.

Dabov, K., Foi, A., Katkovnik, V., and Egiazarian, K. (2007). Image denoising by sparse 3-d transform-domain collaborative filtering. *IEEE Transactions on Image Processing*, 16(8):2080–2095.

Dai, A., Nießner, M., Zollhöfer, M., Izadi, S., and Theobalt, C. (2017). BundleFusion: Real-time globally consistent 3D reconstruction using on-the-fly surface reintegration. *ACM Transactions on Graphics*, 36(4):76a.

Dai, A., Chang, A. X., Savva, M., Halber, M., Funkhouser, T., and Nießner, M. (2017). ScanNet: Richly-annotated 3D reconstructions of indoor scenes. In *IEEE Conference on Computer Vision and Pattern Recognition (CVPR)*.

Dai, J., He, K., and Sun, J. (2015). Convolutional feature masking for joint object and stuff segmentation. In *IEEE Conference on Computer Vision and Pattern Recognition (CVPR)*.

Dai, J., He, K., and Sun, J. (2016). Instance-aware semantic segmentation via multi-task network cascades. In *IEEE Conference on Computer Vision and Pattern Recognition (CVPR)*.

Dai, Z., Cai, B., Lin, Y., and Chen, J. (2020). UP-DETR: Unsupervised pre-training for object detection with transformers. *arXiv preprint arXiv:2011.09094*.

Dalal, N. and Triggs, B. (2005). Histograms of oriented gradients for human detection. In *IEEE Computer Society Conference on Computer Vision and Pattern Recognition (CVPR)*, pp. 886–893.

Dalal, N., Triggs, B., and Schmid, C. (2006). Human detection using oriented histograms of flow and appearance. In *European Conference on Computer Vision (ECCV)*, pp. 428–441.

Damen, D., Doughty, H., Maria Farinella, G., Fidler, S., Furnari, A., Kazakos, E., Moltisanti, D., Munro, J., Perrett, T., Price, W., and Wray, M. (2018). Scaling egocentric vision: The EPIC-KITCHENS dataset. In *European Conference on Computer Vision (ECCV)*.

Dana, K. J., van Ginneken, B., Nayar, S. K., and Koenderink, J. J. (1999). Reflectance and texture of real world surfaces. *ACM Transactions on Graphics*, 18(1):1–34.

Danelljan, M., Robinson, A., Khan, F. S., and Felsberg, M. (2016). Beyond correlation filters: Learning continuous convolution operators for visual tracking. In *European Conference on Computer Vision*, pp. 472–488.

Danielsson, P. E. (1980). Euclidean distance mapping. *Computer Graphics and Image Processing*, 14(3):227–248.

Darrell, T. and Pentland, A. (1991). Robust estimation of a multi-layered motion representation. In *IEEE Workshop on Visual Motion*, pp. 173–178.

Darrell, T. and Pentland, A. (1995). Cooperative robust estimation using layers of support. *IEEE Transactions on Pattern Analysis and Machine Intelligence*, 17(5):474–487.

Darrell, T. and Simoncelli, E. (1993). "Nulling" filters and the separation of transparent motion. In *IEEE Computer Society Conference on Computer Vision and Pattern Recognition (CVPR)*, pp. 738–739.

Darrell, T., Gordon, G., Harville, M., and Woodfill, J. (2000). Integrated person tracking using stereo, color, and pattern detection. *International Journal of Computer Vision*, 37(2):175–185.

Darrell, T., Baker, H., Crow, F., Gordon, G., and Woodfill, J. (1997). Magic morphin mirror: face-sensitive distortion and exaggeration. In *ACM SIGGRAPH Visual Proceedings*.

Das, A., Kottur, S., Gupta, K., Singh, A., Yadav, D., Moura, J. M. F., Parikh, D., and Batra, D. (2017). Visual dialog. In *IEEE Conference on Computer Vision and Pattern Recognition (CVPR)*.

Datta, R., Joshi, D., Li, J., and Wang, J. Z. (2008). Image retrieval: Ideas, influences, and trends of the new age. *ACM Computing Surveys*, 40(2):5.

Daugman, J. (2004). How iris recognition works. *IEEE Transactions on Circuits and Systems for Video Technology*, 14(1):21–30.

Dave, A., Khurana, T., Tokmakov, P., Schmid, C., and Ramanan, D. (2020). TAO: A large-scale benchmark for tracking any object. In *European Conference on Computer Vision (ECCV)*.

David, P., DeMenthon, D., Duraiswami, R., and Samet, H. (2004). SoftPOSIT: Simultaneous pose and correspondence determination. *International Journal of Computer Vision*, 59(3):259–284.

Davies, E. R. (2017). *Computer vision: principles, algorithms, applications, learning*. Academic Press, 5th edition.

Davis, J. (1998). Mosaics of scenes with moving objects. In *IEEE Computer Society Conference on Computer Vision and Pattern Recognition (CVPR)*, pp. 354–360.

Davis, J., Ramamoorthi, R., and Rusinkiewicz, S. (2003). Spacetime stereo: A unifying framework for depth from triangulation. In *IEEE Computer Society Conference on Computer Vision and Pattern Recognition (CVPR)*, pp. 359–366.

Davis, J., Nahab, D., Ramamoorthi, R., and Rusinkiewicz, S. (2005). Spacetime stereo: A unifying framework for depth from triangulation. *IEEE Transactions on Pattern Analysis and Machine Intelligence*, 27(2):296–302.

Davis, L. (1975). A survey of edge detection techniques. *Computer Graphics and Image Processing*, 4(3):248–270.

Davis, T. A. (2006). *Direct Methods for Sparse Linear Systems*. SIAM.

Davis, T. A. (2011). SuiteSparseQR: Multifrontal multithreaded rank-revealing sparse QR factorization. *ACM Transactions on Mathematical Software*, 38(1):8:1–8:22.

Davis, T. A., Rajamanickam, S., and Sid-Lakhdar, W. M. (2016). A survey of direct methods for sparse linear systems. *Acta Numerica*, 25:383–566.

Davison, A., Reid, I., Molton, N. D., and Stasse, O. (2007). MonoSLAM: Real-time single camera SLAM. *IEEE Transactions on Pattern Analysis and Machine Intelligence*, 29(6):1052–1067.

de Agapito, L., Hayman, E., and Reid, I. (2001). Self-calibration of rotating and zooming cameras. *International Journal of Computer Vision*, 45(2):107–127.

de Aguiar, E., Stoll, C., Theobalt, C., Ahmed, N., Seidel, H.-P., and Thrun, S. (2008). Performance capture from sparse multi-view video. *ACM Transactions on Graphics*, 27(3):98.

de Berg, M., Cheong, O., van Kreveld, M., and Overmars, M. (2006). *Computational Geometry: Algorithms and Applications*. Springer, New York, NY, 3rd edition.

de Berg, M., Cheong, O., van Kreveld, M., and Overmars, M. (2008). *Computational Geometry: Algorithms and Applications*. Springer-Verlag.

De Bonet, J. (1997). Multiresolution sampling procedure for analysis and synthesis of texture images. In *ACM SIGGRAPH Conference Proceedings*, pp. 361–368.

De Bonet, J. S. and Viola, P. (1999). Poxels: Probabilistic voxelized volume reconstruction. In *International Conference on Computer Vision (ICCV)*, pp. 418–425.

De Castro, E. and Morandi, C. (1987). Registration of translated and rotated images using finite Fourier transforms. *IEEE Transactions on Pattern Analysis and Machine Intelligence*, PAMI-9(5):700–703.

De la Torre, F. and Black, M. J. (2003). A framework for robust subspace learning. *International Journal of Computer Vision*, 54(1/2/3):117–142.

de Sande, K. V., Uijlings, J., Gevers, T., and Smeulders, A. (2011). Segmentation as selective search for object recognition. In *International Conference on Computer Vision (ICCV)*.

Debevec, P. (1998). Rendering synthetic objects into real scenes: Bridging traditional and image-based graphics with global illumination and high dynamic range photography. In *ACM SIGGRAPH Conference Proceedings*, pp. 189–198.

Debevec, P. (2006). Virtual cinematography: Relighting through computation. *Computer*, 39(8):57–65.

Debevec, P., Hawkins, T., Tchou, C., Duiker, H.-P., Sarokin, W., and Sagar, M. (2000). Acquiring the reflectance field of a human face. In *ACM SIGGRAPH Conference Proceedings*, pp. 145–156.

Debevec, P., Wenger, A., Tchou, C., Gardner, A., Waese, J., and Hawkins, T. (2002). A lighting reproduction approach to live-action compositing. *ACM Transactions on Graphics (Proc. SIGGRAPH)*, 21(3):547–556.

Debevec, P. E. (1999). Image-based modeling and lighting. *Computer Graphics*, 33(4):46–50.

Debevec, P. E. and Malik, J. (1997). Recovering high dynamic range radiance maps from photographs. In *ACM SIGGRAPH Conference Proceedings*, pp. 369–378.

Debevec, P. E., Taylor, C. J., and Malik, J. (1996). Modeling and rendering architecture from photographs: A hybrid geometry- and image-based approach. In *ACM SIGGRAPH Conference Proceedings*, pp. 11–20.

Debevec, P. E., Yu, Y., and Borshukov, G. D. (1998). Efficient view-dependent image-based rendering with projective texture-mapping. In *Eurographics Rendering Workshop*, pp. 105–116.

DeCarlo, D. and Santella, A. (2002). Stylization and abstraction of photographs. *ACM Transactions on Graphics (Proc. SIGGRAPH)*, 21(3):769–776.

DeCarlo, D., Metaxas, D., and Stone, M. (1998). An anthropometric face model using variational techniques. In *ACM SIGGRAPH Conference Proceedings*, pp. 67–74.

Deisenroth, M. P., Faisal, A. A., and Ong, C. S. (2020). *Mathematics for Machine Learning*. Cambridge University Press. https://mml-book.github.io.

Delingette, H., Hebert, M., and Ikeuichi, K. (1992). Shape representation and image segmentation using deformable surfaces. *Image and Vision Computing*, 10(3):132–144.

Dellaert, F. (2021). Factor graphs: Exploiting structure in robotics. *Annual Review of Control, Robotics, and Autonomous Systems*, 4(1):141–166.

Dellaert, F. and Collins, R. (1999). Fast image-based tracking by selective pixel integration. In *ICCV Workshop on Frame-Rate Vision*, pp. 1–22.

Dellaert, F. and Kaess, M. (2017). Factor graphs for robot perception. *Foundations and Trends® in Robotics*, 6(1-2):1–139.

Dellaert, F. and Yen-Chen, L. (2021). Neural volume rendering: NeRF and beyond. *arXiv preprint arXiv:2101.05204*. See also https://dellaert.github.io/NeRF.

Dellaert, F., Rosen, D. M., Wu, J., Mahony, R., and Carlone, L. (2020). Shonan rotation averaging: Global optimality by surfing so(p)(n). In *European Conference on Computer Vision (ECCV)*.

DeMenthon, D. I. and Davis, L. S. (1995). Model-based object pose in 25 lines of code. *International Journal of Computer Vision*, 15(1–2):123–141.

Demmel, N., Sommer, C., Cremers, D., and Usenko, V. (2021). Square root bundle adjustment for large-scale reconstruction. In *IEEE/CVF Conference on Computer Vision and Pattern Recognition (CVPR)*.

Dempster, A., Laird, N. M., and Rubin, D. B. (1977). Maximum likelihood from incomplete data via the EM algorithm. *Journal of the Royal Statistical Society B*, 39(1):1–38.

Dendorfer, P., Ošep, A., Milan, A., Schindler, K., Cremers, D., Reid, I., Roth, S., and Leal-Taixé, L. (2021). MOTChallenge: A benchmark for single-camera multiple target tracking. *International Journal of Computer Vision*, 129(4):845–881.

Deng, J., Guo, J., Xue, N., and Zafeiriou, S. (2019). ArcFace: Additive angular margin loss for deep face recognition. In *IEEE/CVF Conference on Computer Vision and Pattern Recognition (CVPR)*.

Deng, J., Guo, J., Liu, T., Gong, M., and Zafeiriou, S. (2020a). Sub-center ArcFace: Boosting face recognition by large-scale noisy web faces. In *European Conference on Computer Vision (ECCV)*.

Deng, J., Guo, J., Ververas, E., Kotsia, I., and Zafeiriou, S. (2020b). RetinaFace: Single-shot multi-level face localisation in the wild. In *IEEE/CVF Conference on Computer Vision and Pattern Recognition (CVPR)*.

Deng, J., Dong, W., Socher, R., Li, L.-J., Li, K., and Fei-Fei, L. (2009). ImageNet: A large-scale hierarchical image database. In *IEEE Computer Society Conference on Computer Vision and Pattern Recognition (CVPR)*, pp. 248–255.

Denis, P., Elder, J. H., and Estrada, F. J. (2008). Efficient edge-based methods for estimating Manhattan frames in urban imagery. In *European Conference on Computer Vision (ECCV)*, pp. 197–210.

Denton, E., Chintala, S., Szlam, A., and Fergus, R. (2015). Deep generative image models using a Laplacian pyramid of adversarial networks. In *Advances in Neural Information Processing Systems (NeurIPS)*.

Deriche, R. (1987). Using Canny's criteria to derive a recursively implemented optimal edge detector. *International Journal of Computer Vision*, 1(2):167–187.

Deriche, R. (1990). Fast algorithms for low-level vision. *IEEE Transactions on Pattern Analysis and Machine Intelligence*, 12(1):78–87.

Deselaers, T., Alexe, B., and Ferrari, V. (2012). Weakly supervised localization and learning with generic knowledge. *International Journal of Computer Vision*, 100(3):275–293.

Desmaison, A., Bunel, R., Kohli, P., Torr, P. H. S., and Kumar, M. P. (2016). Efficient continuous relaxations for dense CRF. In *European Conference on Computer Vision (ECCV)*, pp. 818–833.

DeTone, D., Malisiewicz, T., and Rabinovich, A. (2018). SuperPoint: Self-supervised interest point detection and description. In *IEEE Conference on Computer Vision and Pattern Recognition Workshops*, pp. 224–236.

Deutscher, J. and Reid, I. (2005). Articulated body motion capture by stochastic search. *International Journal of Computer Vision*, 61(2):185–205.

Deutscher, J., Blake, A., and Reid, I. (2000). Articulated body motion capture by annealed particle filtering. In *IEEE Computer Society Conference on Computer Vision and Pattern Recognition (CVPR)*, pp. 126–133.

Dev, P. (1974). *Segmentation Processes in Visual Perception: A Cooperative Neural Model*. COINS Technical Report 74C-5, University of Massachusetts at Amherst.

Devlin, J., Chang, M.-W., Lee, K., and Toutanova, K. (2018). BERT: Pre-training of deep bidirectional transformers for language understanding. *arXiv preprint arXiv:1810.04805*.

Devlin, J., Gupta, S., Girshick, R., Mitchell, M., and Zitnick, C. L. (2015). Exploring nearest neighbor approaches for image captioning. *arXiv preprint arXiv:1505.04467*.

Dhond, U. R. and Aggarwal, J. K. (1989). Structure from stereo—a review. *IEEE Transactions on Systems, Man, and Cybernetics*, 19(6):1489–1510.

Dick, A., Torr, P. H. S., and Cipolla, R. (2004). Modelling and interpretation of architecture from several images. *International Journal of Computer Vision*, 60(2):111–134.

Dickinson, S., Leonardis, A., Schiele, B., and Tarr, M. J. (eds). (2007). *Object Categorization: Computer and Human Vision Perspectives*, Cambridge University Press, New York.

Dickmanns, E. D. and Graefe, V. (1988). Dynamic monocular machine vision. *Machine Vision and Applications*, 1:223–240.

Dickmanns, E. D. and Mysliwetz, B. D. (1992). Recursive 3-D road and relative ego-state recognition. *IEEE Transactions on Pattern Analysis and Machine Intelligence*, 14(2):199–213.

Diebel, J. (2006). *Representing Attitude: Euler Angles, Quaternions, and Rotation Vectors*. Technical Report, Stanford University.

Diebel, J. R., Thrun, S., and Brünig, M. (2006). A Bayesian method for probable surface reconstruction and decimation. *ACM Transactions on Graphics*, 25(1).

Dimitrijevic, M., Lepetit, V., and Fua, P. (2006). Human body pose detection using Bayesian spatio-temporal templates. *Computer Vision and Image Understanding*, 104(2–3):127–139.

Ding, C. and Tao, D. (2018). Trunk-branch ensemble convolutional neural networks for video-based face recognition. *IEEE Transactions on Pattern Analysis and Machine Intelligence*, 40(4):1002–1014.

Ding, M., Wang, Z., Sun, J., Shi, J., and Luo, P. (2019). CamNet: Coarse-to-fine retrieval for camera re-localization. In *IEEE/CVF International Conference on Computer Vision (ICCV)*.

Dinh, H. Q., Turk, G., and Slabaugh, G. (2002). Reconstructing surfaces by volumetric regularization using radial basis functions. *IEEE Transactions on Pattern Analysis and Machine Intelligence*, 24(10):1358–1371.

Divvala, S., Hoiem, D., Hays, J., Efros, A. A., and Hebert, M. (2009). An empirical study of context in object detection. In *IEEE Computer Society Conference on Computer Vision and Pattern Recognition (CVPR)*.

Dodgson, N. A. (1992). *Image Resampling*. Technical Report TR261, Wolfson College and Computer Laboratory, University of Cambridge.

Dodgson, N. A., Floater, M. S., and Sabin, M. A. (2005). *Advances in Multiresolution for Geometric Modelling*. Springer.

Doersch, C., Gupta, A., and Efros, A. A. (2015). Unsupervised visual representation learning by context prediction. In *IEEE International Conference on Computer Vision (ICCV)*.

Dollár, P. and Zitnick, C. L. (2015). Fast edge detection using structured forests. *IEEE Transactions on Pattern Analysis and Machine Intelligence*, 37(8):1558–1570.

Dollár, P., Appel, R., and Kienzle, W. (2012). Crosstalk cascades for frame-rate pedestrian detection. In *European Conference on Computer Vision (ECCV)*, pp. 645–659.

Dollár, P., Belongie, S., and Perona, P. (2010). The fastest pedestrian detector in the West. In *British Machine Vision Conference (BMVC)*.

Dollár, P., Appel, R., Belongie, S., and Perona, P. (2014). Fast feature pyramids for object detection. *IEEE Transactions on Pattern Analysis and Machine Intelligence*, 36(8):1532–1545.

Dollár, P., Tu, Z., Perona, P., and Belongie, S. (2009). Integral channel features. In *British Machine Vision Conference*.

Dollár, P., Wojek, C., Schiele, B., and Perona, P. (2009). Pedestrian detection: A benchmark. In *IEEE Computer Society Conference on Computer Vision and Pattern Recognition (CVPR)*.

Dollár, P., Wojek, C., Schiele, B., and Perona, P. (2012). Pedestrian detection: An evaluation of the state of the art. *IEEE Transactions on Pattern Analysis and Machine Intelligence*, 34(4):743–761.

Donahue, J., Hendricks, L. A., Guadarrama, S., Rohrbach, M., Venugopalan, S., Saenko, K., and Darrell, T. (2015). Long-term recurrent convolutional networks for visual recognition and description. In *IEEE Conference on Computer Vision and Pattern Recognition (CVPR)*.

Donahue, J., Jia, Y., Vinyals, O., Hoffman, J., Zhang, N., Tzeng, E., and Darrell, T. (2014). DeCAF: A deep convolutional activation feature for generic visual recognition. In *International Conference on Machine Learning (ICML)*, pp. 647–655.

Dong, C., Loy, C. C., He, K., and Tang, X. (2016). Image super-resolution using deep convolutional networks. *IEEE Transactions on Pattern Analysis and Machine Intelligence*, 38(2):295–307.

Donoho, D. L. and Grimes, C. (2003). Hessian Eigenmaps: Locally linear embedding techniques for high-dimensional data. *Proceedings of the National Academy of Sciences*, 100(10):5591–5596.

Doretto, G. and Soatto, S. (2006). Dynamic shape and appearance models. *IEEE Transactions on Pattern Analysis and Machine Intelligence*, 28(12):2006–2019.

Doretto, G., Chiuso, A., Wu, Y. N., and Soatto, S. (2003). Dynamic textures. *International Journal of Computer Vision*, 51(2):91–109.

Dorsey, J., Rushmeier, H., and Sillion, F. (2007). *Digital Modeling of Material Appearance*. Morgan Kaufmann, San Francisco.

Dosovitskiy, A., Beyer, L., Kolesnikov, A., Weissenborn, D., Zhai, X., Unterthiner, T., Dehghani, M., Minderer, M., Heigold, G., Gelly, S., Uszkoreit, J., and Houlsby, N. (2021). An image is worth 16x16 words: Transformers for image recognition at scale. In *International Conference on Learning Representations (ICLR)*. arXiv preprint arXiv:2010.11929.

Dosovitskiy, A. and Brox, T. (2016). Generating images with perceptual similarity metrics based on deep networks. In *Advances in Neural Information Processing Systems (NeurIPS)*, pp. 658–666.

Dosovitskiy, A., Fischer, P., Ilg, E., Hausser, P., Hazirbas, C., Golkov, V., van der Smagt, P., Cremers, D., and Brox, T. (2015). FlowNet: Learning optical flow with convolutional networks. In *IEEE International Conference on Computer Vision (ICCV)*, pp. 2758–2766.

Dou, M., Davidson, P., Fanello, S. R., Khamis, S., Kowdle, A., Rhemann, C., Tankovich, V., and Izadi, S. (2017). Motion2Fusion: Real-time volumetric performance capture. *ACM Transactions on Graphics (Proc. SIGGRAPH Asia)*, 36(6):1–16.

Douglas, D. H. and Peucker, T. K. (1973). Algorithms for the reduction of the number of points required to represent a digitized line or its caricature. *The Canadian Cartographer*, 10(2):112–122.

Drori, I., Cohen-Or, D., and Yeshurun, H. (2003). Fragment-based image completion. *ACM Transactions on Graphics (Proc. SIGGRAPH)*, 22(3):303–312.

Drory, A., Haubold, C., Avidan, S., and Hamprecht, F. A. (2014). Semi-global matching: a principled derivation in terms of message passing. In *German Conference on Pattern Recognition (GCPR)*, pp. 43–53.

Drost, B., Ulrich, M., Navab, N., and Ilic, S. (2010). Model globally, match locally: Efficient and robust 3D object recognition. In *IEEE Computer Society Conference on Computer Vision and Pattern Recognition (CVPR)*.

Duan, K., Parikh, D., Crandall, D., and Grauman, K. (2012). Discovering localized attributes for fine-grained recognition. In *IEEE Computer Society Conference on Computer Vision and Pattern Recognition (CVPR)*.

Duchêne, S., Riant, C., Chaurasia, G., Lopez-Moreno, J., Laffont, P.-Y., Popov, S., Bousseau, A., and Drettakis, G. (2015). Multi-view intrinsic images of outdoors scenes with an application to relighting. *ACM Transactions on Graphics (Proc. SIGGRAPH)*.

Duchi, J., Hazan, E., and Singer, Y. (2011). Adaptive subgradient methods for online learning and stochastic optimization. *Journal of Machine Learning Research*, 12(7).

Duda, R. O. and Hart, P. E. (1972). Use of the Hough transform to detect lines and curves in pictures. *Communications of the ACM*, 15(1):11–15.

Duda, R. O., Hart, P. E., and Stork, D. G. (2001). *Pattern Classification*. John Wiley & Sons, New York, 2nd edition.

Dufaux, F., Le Callet, P., Mantiuk, R., and Mrak, M. (2016). *High dynamic range video: from acquisition, to display and applications*. Academic Press.

Duggal, S., Wang, S., Ma, W.-C., Hu, R., and Urtasun, R. (2019). DeepPruner: Learning efficient stereo matching via differentiable PatchMatch. In *International Conference on Computer Vision (ICCV)*.

Dumoulin, V. and Visin, F. (2016). A guide to convolution arithmetic for deep learning. *arXiv preprint arXiv:1603.07285*.

Dumoulin, V., Shlens, J., and Kudlur, M. (2017). A learned representation for artistic style. In *International Conference on Learning Representations (ICLR)*.

Dupuis, P. and Oliensis, J. (1994). An optimal control formulation and related numerical methods for a problem in shape reconstruction. *Annals of Applied Probability*, 4(2):287–346.

Dupuy, J. and Jakob, W. (2018). An adaptive parameterization for efficient material acquisition and rendering. *ACM Transactions on Graphics*, 37(6):Article 274.

Durand, F. and Dorsey, J. (2002). Fast bilateral filtering for the display of high-dynamic-range images. *ACM Transactions on Graphics (Proc. SIGGRAPH)*, 21(3):257–266.

Durand, F. and Szeliski, R. (2007). Computational photography. *IEEE Computer Graphics and Applications*, 27(2):21–22. Guest Editors' Introduction to Special Issue.

Durbin, R. and Willshaw, D. (1987). An analogue approach to the traveling salesman problem using an elastic net method. *Nature*, 326:689–691.

Durbin, R., Szeliski, R., and Yuille, A. (1989). An analysis of the elastic net approach to the travelling salesman problem. *Neural Computation*, 1(3):348–358.

Durrant-Whyte, H. and Bailey, T. (2006). Simultaneous localization and mapping: part I. *IEEE Robotics & Automation Magazine*, 13(2):99–110.

Dusmanu, M., Rocco, I., Pajdla, T., Pollefeys, M., Sivic, J., Torii, A., and Sattler, T. (2019). D2-Net: A trainable CNN for joint description and detection of local features. In *IEEE/CVF Conference on Computer Vision and Pattern Recognition (CVPR)*.

Duygulu, P., Barnard, K., de Freitas, J. F. G., and Forsyth, D. A. (2002). Object recognition as machine translation: Learning a lexicon for a fixed image vocabulary. In *European Conference on Computer Vision*, pp. 97–112.

Eck, M., DeRose, T., Duchamp, T., Hoppe, H., Lounsbery, M., and Stuetzle, W. (1995). Multiresolution analysis of arbitrary meshes. In *ACM SIGGRAPH Conference Proceedings*, pp. 173–182.

Eden, A., Uyttendaele, M., and Szeliski, R. (2006). Seamless image stitching of scenes with large motions and exposure differences. In *IEEE Computer Society Conference on Computer Vision and Pattern Recognition (CVPR)*, pp. 2498–2505.

Efros, A. A. and Freeman, W. T. (2001). Image quilting for texture synthesis and transfer. In *ACM SIGGRAPH Conference Proceedings*, pp. 341–346.

Efros, A. A. and Leung, T. K. (1999). Texture synthesis by non-parametric sampling. In *International Conference on Computer Vision (ICCV)*, pp. 1033–1038.

Efros, A. A., Berg, A. C., Mori, G., and Malik, J. (2003). Recognizing action at a distance. In *International Conference on Computer Vision (ICCV)*, pp. 726–733.

Egger, B., Smith, W. A. P., Tewari, A., Wuhrer, S., Zollhöfer, M., Beeler, T., Bernard, F., Bolkart, T., Kortylewski, A., Romdhani, S., Theobalt, C., Blanz, V., and Vetter, T. (2020). 3D morphable face models—past, present, and future. *ACM Transactions on Graphics*, 39(5):157.

Ehrlich, M., Lim, S.-N., Davis, L., and Shrivastava, A. (2020). Quantization guided JPEG artifact correction. In *European Conference on Computer Vision (ECCV)*.

Eigen, D. and Fergus, R. (2015). Predicting depth, surface normals and semantic labels with a common multi-scale convolutional architecture. In *IEEE International Conference on Computer Vision (ICCV)*.

Eigen, D., Puhrsch, C., and Fergus, R. (2014). Depth map prediction from a single image using a multi-scale deep network. In *Advances in Neural Information Processing Systems (NeurIPS)*, p. 2366–2374.

Eisemann, E. and Durand, F. (2004). Flash photography enhancement via intrinsic relighting. *ACM Transactions on Graphics*, 23(3):673–678.

Eisert, P., Steinbach, E., and Girod, B. (2000). Automatic reconstruction of stationary 3-D objects from multiple uncalibrated camera views. *IEEE Transactions on Circuits and Systems for Video Technology*, 10(2):261–277.

Eisert, P., Wiegand, T., and Girod, B. (2000). Model-aided coding: a new approach to incorporate facial animation into motion-compensated video coding. *IEEE Transactions on Circuits and Systems for Video Technology*, 10(3):344–358.

Ekman, P. and Friesen, W. V. (1978). *Facial Action Coding System: A Technique for the Measurement of Facial Movement*. Consulting Psychologists Press, Palo Alto, CA.

El-Melegy, M. and Farag, A. (2003). Nonmetric lens distortion calibration: Closed-form solutions, robust estimation and model selection. In *International Conference on Computer Vision (ICCV)*, pp. 554–559.

Elder, J. H. (1999). Are edges incomplete? *International Journal of Computer Vision*, 34(2/3):97–122.

Elder, J. H. and Goldberg, R. M. (2001). Image editing in the contour domain. *IEEE Transactions on Pattern Analysis and Machine Intelligence*, 23(3):291–296.

Elder, J. H. and Zucker, S. W. (1998). Local scale control for edge detection and blur estimation. *IEEE Transactions on Pattern Analysis and Machine Intelligence*, 20(7):699–716.

Elsken, T., Metzen, J. H., and Hutter, F. (2019). Neural architecture search: A survey. *Journal of Machine Learning Research*, 20(55):1–21.

Engel, J., Koltun, V., and Cremers, D. (2018). Direct sparse odometry. *IEEE Transactions on Pattern Analysis and Machine Intelligence*, 40(3):611–625.

Engel, J., Schöps, T., and Cremers, D. (2014). LSD-SLAM: Large-scale direct monocular SLAM. In *European Conference on Computer Vision (ECCV)*, pp. 834–849.

Engel, J., Usenko, V., and Cremers, D. (2016). A photometrically calibrated benchmark for monocular visual odometry. *arXiv preprint arXiv:1607.02555*.

Engels, C., Stewénius, H., and Nistér, D. (2006). Bundle adjustment rules. In *Photogrammetric Computer Vision (PCV)*.

Engl, H. W., Hanke, M., and Neubauer, A. (1996). *Regularization of Inverse Problems*. Kluwer Academic Publishers, Dordrecht.

Enqvist, O., Josephson, K., and Kahl, F. (2009). Optimal correspondences from pairwise constraints. In *International Conference on Computer Vision (ICCV)*.

Erofeev, M., Gitman, Y., Vatolin, D., Fedorov, A., and Wang, J. (2015). Perceptually motivated benchmark for video matting. In *British Machine Vision Conference (BMVC)*, pp. 99.1–99.12.

Esser, P., Rombach, R., and Ommer, B. (2020). Taming transformers for high-resolution image synthesis. *arXiv preprint arXiv:2012.09841*.

Estrada, F. J. and Jepson, A. D. (2009). Benchmarking image segmentation algorithms. *International Journal of Computer Vision*, 85(2):167–181.

Evangelidis, G. D. and Psarakis, E. Z. (2008). Parametric image alignment using enhanced correlation coefficient maximization. *IEEE Transactions on Pattern Analysis and Machine Intelligence*, 30(10):1858–1865.

Everingham, M., Van Gool, L., Williams, C. K. I., Winn, J., and Zisserman, A. (2010). The PASCAL visual object classes (VOC) challenge. *International Journal of Computer Vision*, 88(2):147–168.

Everingham, M., Eslami, S. M. A., Van Gool, L., Williams, C. K. I., Winn, J., and Zisserman, A. (2015). The PASCAL visual object classes challenge: A retrospective. *International Journal of Computer Vision*, 111(1):98–136.

Eykholt, K., Evtimov, I., Fernandes, E., Li, B., Rahmati, A., Xiao, C., Prakash, A., Kohno, T., and Song, D. (2018). Robust physical-world attacks on deep learning visual classification. In *IEEE Conference on Computer Vision and Pattern Recognition (CVPR)*.

Ezzat, T., Geiger, G., and Poggio, T. (2002). Trainable videorealistic speech animation. *ACM Transactions on Graphics (Proc. SIGGRAPH)*, 21(3):388–398.

Fabbri, R., Costa, L. D. F., Torelli, J. C., and Bruno, O. M. (2008). 2D Euclidean distance transform algorithms: A comparative survey. *ACM Computing Surveys*, 40(1):2.

Fairchild, M. D. (2013). *Color Appearance Models*. Wiley, 3rd edition.

Fan, H., Su, H., and Guibas, L. J. (2017). A point set generation network for 3D object reconstruction from a single image. In *IEEE Conference on Computer Vision and Pattern Recognition (CVPR)*.

Fan, H., Zhou, Y., Li, A., Gao, S., Li, J., and Guo, Y. (2020). Visual localization using semantic segmentation and depth prediction. *arXiv preprint arXiv:2005.11922*.

Fan, H., Xiong, B., Mangalam, K., Li, Y., Yan, Z., Malik, J., and Feichtenhofer, C. (2021). Multiscale vision transformers. *arXiv preprint arXiv:2104.11227*.

Fanello, S. R., Rhemann, C., Tankovich, V., Kowdle, A., Escolano, S. O., Kim, D., and Izadi, S. (2016). HyperDepth: Learning depth from structured light without matching. In *IEEE Conference on Computer Vision and Pattern Recognition (CVPR)*.

Fanello, S. R., Valentin, J., Rhemann, C., Kowdle, A., Tankovich, V., Davidson, P., and Izadi, S. (2017a). UltraStereo: Efficient learning-based matching for active stereo systems. In *IEEE Conference on Computer Vision and Pattern Recognition (CVPR)*.

Fanello, S. R., Valentin, J., Kowdle, A., Rhemann, C., Tankovich, V., Ciliberto, C., Davidson, P., and Izadi, S. (2017b). Low compute and fully parallel computer vision with HashMatch. In *IEEE International Conference on Computer Vision (ICCV)*.

Fang, H., Gupta, S., Iandola, F., Srivastava, R. K., Deng, L., Dollár, P., Gao, J., He, X., Mitchell, M., Platt, J. C., Zitnick, C. L., and Zweig, G. (2015). From captions to visual concepts and back. In *IEEE Conference on Computer Vision and Pattern Recognition (CVPR)*.

Farbman, Z., Fattal, R., Lischinski, D., and Szeliski, R. (2008). Edge-preserving decompositions for multi-scale tone and detail manipulation. *ACM Transactions on Graphics (Proc. SIGGRAPH)*, 27(3):67.

Farenzena, M., Fusiello, A., and Gherardi, R. (2009). Structure-and-motion pipeline on a hierarchical cluster tree. In *IEEE International Workshop on 3D Digital Imaging and Modeling (3DIM)*.

Farhadi, A., Hejrati, M., Sadeghi, M. A., Young, P., Rashtchian, C., Hockenmaier, J., and Forsyth, D. (2010). Every picture tells a story: Generating sentences from images. In *European Conference on Computer Vision (ECCV)*, pp. 15–29.

Farin, G. (1992). From conics to NURBS: A tutorial and survey. *IEEE Computer Graphics and Applications*, 12(5):78–86.

Farin, G. E. (2002). *Curves and Surfaces for CAGD: A Practical Guide*. Academic Press, Boston, Massachusetts, 5th edition.

Fattal, R. (2007). Image upsampling via imposed edge statistics. *ACM Transactions on Graphics*, 26(3).

Fattal, R. (2009). Edge-avoiding wavelets and their applications. *ACM Transactions on Graphics*, 28(3):22.

Fattal, R., Lischinski, D., and Werman, M. (2002). Gradient domain high dynamic range compression. *ACM Transactions on Graphics (Proc. SIGGRAPH)*, 21(3):249–256.

Faugeras, O. (1993). *Three-dimensional computer vision: A geometric viewpoint*. MIT Press, Cambridge, Massachusetts.

Faugeras, O. and Keriven, R. (1998). Variational principles, surface evolution, PDEs, level set methods, and the stereo problem. *IEEE Transactions on Image Processing*, 7(3):336–344.

Faugeras, O. and Luong, Q.-T. (2001). *The Geometry of Multiple Images*. MIT Press, Cambridge, MA.

Faugeras, O. D. (1992). What can be seen in three dimensions with an uncalibrated stereo rig? In *European Conference on Computer Vision (ECCV)*, pp. 563–578.

Faugeras, O. D. and Hebert, M. (1987). The representation, recognition and positioning of 3-D shapes from range data. In Kanade, T. (ed.), *Three-Dimensional Machine Vision*, pp. 301–353, Kluwer Academic Publishers, Boston.

Faugeras, O. D., Luong, Q.-T., and Maybank, S. J. (1992). Camera self-calibration: Theory and experiments. In *European Conference on Computer Vision (ECCV)*, pp. 321–334.

Favaro, P. and Soatto, S. (2006). *3-D Shape Estimation and Image Restoration: Exploiting Defocus and Motion-Blur*. Springer.

Fawcett, T. (2006). An introduction to ROC analysis. *Pattern Recognition Letters*, 27(8):861–874.

Fei-Fei, L. and Perona, P. (2005). A Bayesian hierarchical model for learning natural scene categories. In *IEEE Computer Society Conference on Computer Vision and Pattern Recognition (CVPR)*, pp. 524–531.

Fei-Fei, L., Fergus, R., and Perona, P. (2006). One-shot learning of object categories. *IEEE Transactions on Pattern Analysis and Machine Intelligence*, 28(4):594–611.

Fei-Fei, L., Fergus, R., and Torralba, A. (2009). ICCV 2009 short course on recognizing and learning object categories. In *International Conference on Computer Vision (ICCV)*. https://people.csail.mit.edu/torralba/shortCourseRLOC.

Feichtenhofer, C. (2020). X3D: Expanding architectures for efficient video recognition. In *IEEE/CVF Conference on Computer Vision and Pattern Recognition (CVPR)*.

Feichtenhofer, C., Pinz, A., and Wildes, R. P. (2017). Temporal residual networks for dynamic scene recognition. In *IEEE Conference on Computer Vision and Pattern Recognition (CVPR)*.

Feichtenhofer, C., Pinz, A., and Zisserman, A. (2016). Convolutional two-stream network fusion for video action recognition. In *IEEE Conference on Computer Vision and Pattern Recognition (CVPR)*.

Feichtenhofer, C., Fan, H., Malik, J., and He, K. (2019). Slowfast networks for video recognition. In *IEEE/CVF International Conference on Computer Vision (ICCV)*.

Feichtenhofer, C., Fan, H., Xiong, B., Girshick, R., and He, K. (2021). A large-scale study on unsupervised spatiotemporal representation learning. In *IEEE/CVF Conference on Computer Vision and Pattern Recognition (CVPR)*.

Feilner, M., Van De Ville, D., and Unser, M. (2005). An orthogonal family of quincunx wavelets with continuously adjustable order. *IEEE Transactions on Image Processing*, 14(4):499–520.

Feiner, S. K. (2002). Augmented reality: A new way of seeing. *Scientific American*, 286(4):48–55.

Feldmar, J. and Ayache, N. (1996). Rigid, affine, and locally affine registration of free-form surfaces. *International Journal of Computer Vision*, 18(2):99–119.

Felzenszwalb, P., McAllester, D., and Ramanan, D. (2008). A discriminatively trained, multiscale, deformable part model. In *IEEE Computer Society Conference on Computer Vision and Pattern Recognition (CVPR)*.

Felzenszwalb, P. F. and Huttenlocher, D. P. (2004). Efficient graph-based image segmentation. *International Journal of Computer Vision*, 59(2):167–181.

Felzenszwalb, P. F. and Huttenlocher, D. P. (2005). Pictorial structures for object recognition. *International Journal of Computer Vision*, 61(1):55–79.

Felzenszwalb, P. F. and Huttenlocher, D. P. (2012). Distance transforms of sampled functions. *Theory of Computing*, 8(1):415–428.

Felzenszwalb, P. F., Girshick, R. B., McAllester, D., and Ramanan, D. (2010). Object detection with discriminatively trained part-based models. *IEEE Transactions on Pattern Analysis and Machine Intelligence*, 32(9):1627–1645.

Ferencz, A., Learned-Miller, E. G., and Malik, J. (2008). Learning to locate informative features for visual identification. *International Journal of Computer Vision*, 77(1–3):3–24.

Fergus, R. (2007). Combined segmentation and recognition. In *CVPR Short Course on Recognizing and Learning Object Categories*. https://people.csail.mit.edu/torralba/shortCourseRLOC.

Fergus, R. (2009). Classical methods for object recognition. In *ICCV 2009 Short Course on Recognizing and Learning Object Categories*. https://people.csail.mit.edu/torralba/shortCourseRLOC.

Fergus, R., Perona, P., and Zisserman, A. (2004). A visual category filter for Google images. In *European Conference on Computer Vision (ECCV)*, pp. 242–256.

Fergus, R., Perona, P., and Zisserman, A. (2005). A sparse object category model for efficient learning and exhaustive recognition. In *IEEE Computer Society Conference on Computer Vision and Pattern Recognition (CVPR)*, pp. 380–387.

Fergus, R., Perona, P., and Zisserman, A. (2007). Weakly supervised scale-invariant learning of models for visual recognition. *International Journal of Computer Vision*, 71(3):273–303.

Fergus, R., Weiss, Y., and Torralba, A. (2009). Semi-supervised learning in gigantic image collections. In *Advances in Neural Information Processing Systems (NeurIPS)*, pp. 522–530.

Fergus, R., Fei-Fei, L., Perona, P., and Zisserman, A. (2005). Learning object categories from Google's image search. In *International Conference on Computer Vision (ICCV)*, pp. 1816–1823.

Fergus, R., Singh, B., Hertzmann, A., Roweis, S. T., and Freeman, W. T. (2006). Removing camera shake from a single photograph. *ACM Transactions on Graphics*, 25(3):787–794.

Ferrari, V., Marin-Jimenez, M. J., and Zisserman, A. (2008). Progressive search space reduction for human pose estimation. In *IEEE Computer Society Conference on Computer Vision and Pattern Recognition (CVPR)*.

Ferrari, V., Tuytelaars, T., and Van Gool, L. (2006a). Object detection by contour segment networks. In *European Conference on Computer Vision (ECCV)*, pp. 14–28.

Ferrari, V., Tuytelaars, T., and Van Gool, L. (2006b). Simultaneous object recognition and segmentation from single or multiple model views. *International Journal of Computer Vision*, 67(2):159–188.

Finkelstein, A. and Salesin, D. H. (1994). Multiresolution curves. In *ACM SIGGRAPH Conference Proceedings*, pp. 261–268.

Firestone, C. and Scholl, B. J. (2016). Cognition does not affect perception: Evaluating the evidence for "top-down" effects. *Behavioral and Brain Sciences*, 39:E229.

Firman, M. (2016). RGBD datasets: Past, present and future. In *IEEE Conference on Computer Vision and Pattern Recognition (CVPR) Workshops*.

Fischler, M. A. and Bolles, R. C. (1981). Random sample consensus: A paradigm for model fitting with applications to image analysis and automated cartography. *Communications of the ACM*, 24(6):381–395.

Fischler, M. A. and Elschlager, R. A. (1973). The representation and matching of pictorial structures. *IEEE Transactions on Computers*, 22(1):67–92.

Fischler, M. A. and Firschein, O. (1987). *Readings in Computer Vision*. Morgan Kaufmann Publishers, Inc., Los Altos.

Fischler, M. A., Firschein, O., Barnard, S. T., Fua, P. V., and Leclerc, Y. (1989). *The Vision Problem: Exploiting Parallel Computation*. Technical Note 458, SRI International, Menlo Park.

Fitzgibbon, A. W. and Zisserman, A. (1998). Automatic camera recovery for closed and open image sequences. In *European Conference on Computer Vision (ECCV)*, pp. 311–326.

Fitzgibbon, A. W., Cross, G., and Zisserman, A. (1998). Automatic 3D model construction for turn-table sequences. In *European Workshop on 3D Structure from Multiple Images of Large-Scale Environments (SMILE)*, pp. 155–170.

Fleet, D. and Jepson, A. (1990). Computation of component image velocity from local phase information. *International Journal of Computer Vision*, 5(1):77–104.

Fleuret, F. and Geman, D. (2001). Coarse-to-fine face detection. *International Journal of Computer Vision*, 41(1/2):85–107.

Flickner, M., Sawhney, H., Niblack, W., Ashley, J., Huang, Q., Dom, B., Gorkani, M., Hafner, J., Lee, D., Petkovic, D., Steele, D., and Yanker, P. (1995). Query by image and video content: The QBIC system. *Computer*, 28(9):23–32.

Floater, M. S. and Hormann, K. (2005). Surface parameterization: a tutorial and survey. In Dodgson, Neil A.and Floater, M. S. and Sabin, M. A. (eds), *Advances in Multiresolution for Geometric Modelling*, pp. 157–186, Springer.

Flynn, J., Broxton, M., Debevec, P., DuVall, M., Fyffe, G., Overbeck, R., Snavely, N., and Tucker, R. (2019). DeepView: View synthesis with learned gradient descent. In *IEEE/CVF Conference on Computer Vision and Pattern Recognition (CVPR)*.

Fontaine, R. (2015). The state-of-the-art of mainstream cmos image sensors. In *Proceedings of the International Image Sensors Workshop*, pp. 6–12.

Forssén, P.-E. and Ringaby, E. (2010). Rectifying rolling shutter video from hand-held devices. In *IEEE Computer Society Conference on Computer Vision and Pattern Recognition (CVPR)*.

Forster, C., Carlone, L., Dellaert, F., and Scaramuzza, D. (2016). On-manifold preintegration for real-time visual–inertial odometry. *IEEE Transactions on Robotics*, 33(1):1–21.

Forster, C., Zhang, Z., Gassner, M., Werlberger, M., and Scaramuzza, D. (2017). SVO: Semidirect visual odometry for monocular and multicamera systems. *IEEE Transactions on Robotics*, 33(2):249–265.

Förstner, W. (1986). A feature-based correspondence algorithm for image matching. *Intl. Arch. Photogrammetry & Remote Sensing*, 26(3):150–166.

Förstner, W. (2005). Uncertainty and projective geometry. In Bayro-Corrochano, E. (ed.), *Handbook of Geometric Computing*, pp. 493–534, Springer, New York.

Forsyth, D. and Ponce, J. (2003). *Computer Vision: A Modern Approach*. Prentice Hall, Upper Saddle River, NJ.

Forsyth, D. and Ponce, J. (2011). *Computer Vision: A Modern Approach*. Pearson, 2nd edition.

Forsyth, D. A., Arikan, O., Ikemoto, L., O'Brien, J., and Ramanan, D. (2006). Computational studies of human motion: Part 1, tracking and motion synthesis. *Foundations and Trends® in Computer Graphics and Computer Vision*, 1(2/3):77–254.

Fossati, A., Dimitrijevic, M., Lepetit, V., and Fua, P. (2007). Bridging the gap between detection and tracking for 3D monocular video-based motion capture. In *IEEE Computer Society Conference on Computer Vision and Pattern Recognition (CVPR)*.

Fournier, A., Fussel, D., and Carpenter, L. (1982). Computer rendering of stochastic models. *Communications of the ACM*, 25(6):371–384.

Fragkiadaki, K., Zhang, G., and Shi, J. (2012). Video segmentation by tracing discontinuities in a trajectory embedding. In *IEEE Computer Society Conference on Computer Vision and Pattern Recognition (CVPR)*.

Frahm, J.-M., Fite-Georgel, P., Gallup, D., Johnson, T., Raguram, R., Wu, C., Jen, Y.-H., Dunn, E., Clipp, B., Lazebnik, S., and Pollefeys, M. (2010). Building Rome on a cloudless day. In *European Conference on Computer Vision (ECCV)*, pp. 368–381.

Frahm, J.-M. and Koch, R. (2003). Camera calibration with known rotation. In *International Conference on Computer Vision (ICCV)*, pp. 1418–1425.

Frankle, J. and Carbin, M. (2019). The lottery ticket hypothesis: Finding sparse, trainable neural networks. In *International Conference on Learning Representations (ICLR)*.

Freeman, M. (2008). *Mastering HDR Photography*. Amphoto Books, New York.

Freeman, W. T. (1992). *Steerable Filters and Local Analysis of Image Structure*. Ph.D. thesis, Massachusetts Institute of Technology.

Freeman, W. T. and Adelson, E. H. (1991). The design and use of steerable filters. *IEEE Transactions on Pattern Analysis and Machine Intelligence*, 13(9):891–906.

Freeman, W. T., Jones, T. R., and Pasztor, E. C. (2002). Example-based super-resolution. *IEEE Computer Graphics and Applications*, 22(2):56–65.

Freeman, W. T., Pasztor, E. C., and Carmichael, O. T. (2000). Learning low-level vision. *International Journal of Computer Vision*, 40(1):25–47.

Fried, O., Tewari, A., Zollhöfer, M., Finkelstein, A., Shechtman, E., Goldman, D. B., Genova, K., Jin, Z., Theobalt, C., and Agrawala, M. (2019). Text-based editing of talking-head video. *ACM Transactions on Graphics (Proc. SIGGRAPH)*, 38(4):68:1–68:14.

Fried, O., Shechtman, E., Goldman, D. B., and Finkelstein, A. (2016). Perspective-aware manipulation of portrait photos. *ACM Transactions on Graphics (Proc. SIGGRAPH)*, 35(4):1–10.

Friedman, J. H., Bentley, J. L., and Finkel, R. A. (1977). An algorithm for finding best matches in logarithmic expected time. *ACM Transactions on Mathematical Software*, 3(3):209–226.

Frisby, J. P. and Stone, J. V. (2010). *Seeing: The computational approach to biological vision*. MIT Press, 2nd edition.

Frisken, S. F., Perry, R. N., Rockwood, A. P., and Jones, T. R. (2000). Adaptively sampled distance fields: A general representation of shape for computer graphics. In *ACM SIGGRAPH Conference Proceedings*, pp. 249–254.

Frome, A., Singer, Y., Sha, F., and Malik, J. (2007). Learning globally-consistent local distance functions for shape-based image retrieval and classification. In *International Conference on Computer Vision (ICCV)*.

Fu, Y., Yan, Q., Yang, L., Liao, J., and Xiao, C. (2018). Texture mapping for 3D reconstruction with RGB-D sensor. In *IEEE Conference on Computer Vision and Pattern Recognition (CVPR)*.

Fua, P. (1993). A parallel stereo algorithm that produces dense depth maps and preserves image features. *Machine Vision and Applications*, 6(1):35–49.

Fua, P. and Leclerc, Y. G. (1995). Object-centered surface reconstruction: Combining multi-image stereo and shading. *International Journal of Computer Vision*, 16(1):35–56.

Fua, P. and Sander, P. (1992). Segmenting unstructured 3D points into surfaces. In *European Conference on Computer Vision (ECCV)*, pp. 676–680.

Fuentes-Pacheco, J., Ruiz-Ascencio, J., and Rendón-Mancha, J. M. (2015). Visual simultaneous localization and mapping: a survey. *Artificial Intelligence Review*, 43(1):55–81.

Fuh, C.-S. and Maragos, P. (1991). Motion displacement estimation using an affine model for image matching. *Optical Engineering*, 30(7):881–887.

Fuhrmann, S. and Goesele, M. (2011). Fusion of depth maps with multiple scales. In *ACM Transactions on Graphics (Proc. SIGGRAPH Asia)*.

Fuhrmann, S. and Goesele, M. (2014). Floating scale surface reconstruction. *ACM Transactions on Graphics (Proc. SIGGRAPH)*, 33(4):46.

Fuhrmann, S., Langguth, F., Moehrle, N., Waechter, M., and Goesele, M. (2015). MVE — an image-based reconstruction environment. *Computers & Graphics*, 53:44–53.

Fukunaga, K. and Hostetler, L. D. (1975). The estimation of the gradient of a density function, with applications in pattern recognition. *IEEE Transactions on Information Theory*, 21:32–40.

Fukushima, K. (1980). Neocognitron: A self-organizing neural network model for a mechanism of pattern recognition unaffected by shift in position. *Biological Cybernetics*, 36(4):193–202.

Fuoli, D., Huang, Z., Danelljan, M., and Timofte, R. (2020). NTIRE 2020 challenge on video quality mapping: Methods and results. In *IEEE/CVF Conference on Computer Vision and Pattern Recognition (CVPR) Workshops*.

Furukawa, Y. and Hernández, C. (2015). Multi-view stereo: A tutorial. *Foundations and Trends® in Computer Graphics and Vision*, 9(1-2):1–148.

Furukawa, Y. and Ponce, J. (2008). Dense 3D motion capture from synchronized video streams. In *IEEE Computer Society Conference on Computer Vision and Pattern Recognition (CVPR)*.

Furukawa, Y. and Ponce, J. (2009). Carved visual hulls for image-based modeling. *International Journal of Computer Vision*, 81(1):53–67.

Furukawa, Y. and Ponce, J. (2010). Accurate, dense, and robust multi-view stereopsis. *IEEE Transactions on Pattern Analysis and Machine Intelligence*, 32(8):1362–1376.

Furukawa, Y., Curless, B., Seitz, S. M., and Szeliski, R. (2009a). Manhattan-world stereo. In *IEEE Computer Society Conference on Computer Vision and Pattern Recognition (CVPR)*, pp. 1422–1429.

Furukawa, Y., Curless, B., Seitz, S. M., and Szeliski, R. (2009b). Reconstructing building interiors from images. In *IEEE International Conference on Computer Vision (ICCV)*.

Furukawa, Y., Curless, B., Seitz, S. M., and Szeliski, R. (2010). Towards internet-scale multi-view stereo. In *IEEE Computer Society Conference on Computer Vision and Pattern Recognition (CVPR)*.

Fusiello, A., Roberto, V., and Trucco, E. (1997). Efficient stereo with multiple windowing. In *IEEE Computer Society Conference on Computer Vision and Pattern Recognition (CVPR)*, pp. 858–863.

Fusiello, A., Trucco, E., and Verri, A. (2000). A compact algorithm for rectification of stereo pairs. *Machine Vision and Applications*, 12(1):16–22.

Gårding, J. (1992). Shape from texture for smooth curved surfaces in perspective projection. *Journal of Mathematical Imaging and Vision*, 2:329–352.

Gai, J. and Kang, S. B. (2009). Matte-based restoration of vintage video. *IEEE Transactions on Image Processing*, 18:2185–2197.

Gal, R., Wexler, Y., Ofek, E., Hoppe, H., and Cohen-Or, D. (2010). Seamless montage for texturing models. In *Eurographics*.

Galleguillos, C. and Belongie, S. (2010). Context based object categorization: A critical survey. *Computer Vision and Image Understanding*, 114(6):712–722.

Gallo, O., Troccoli, A., Jampani, V., Szeliski, R., Wiles, O., Tucker, R., Chaurasia, G., Kalantari, N., Srinivasan, P., and Yoon, J. S. (2020). CVPR 2020 tutorial on novel view synthesis: From depth-based warping to multi-plane images and beyond. https://nvlabs.github.io/nvs-tutorial-cvpr2020.

Gallup, D., Frahm, J.-M., and Pollefeys, M. (2010). Piecewise planar and non-planar stereo for urban scene reconstruction. In *IEEE Computer Society Conference on Computer Vision and Pattern Recognition (CVPR)*.

Gallup, D., Frahm, J.-M., Mordohai, P., and Pollefeys, M. (2008). Variable baseline/resolution stereo. In *IEEE Computer Society Conference on Computer Vision and Pattern Recognition (CVPR)*.

Gamble, E. and Poggio, T. (1987). *Visual integration and detection of discontinuities: the key role of intensity edges*. A. I. Memo 970, Artificial Intelligence Laboratory, Massachusetts Institute of Technology.

Gammeter, S., Bossard, L., Quack, T., and Van Gool, L. (2009). I know what you did last summer: Object-level auto-annotation of holiday snaps. In *International Conference on Computer Vision (ICCV)*.

Gan, Z. (2020). CVPR 2020 tutorial on visual question answering and visual reasoning. https://rohit497.github.io/Recent-Advances-in-Vision-and-Language-Research.

Gan, Z., Yu, L., Cheng, Y., Zhou, L., Li, L., Chen, Y.-C., Liu, J., and He, X. (2020). CVPR 2020 tutorial on recent advances in vision-and-language research. https://rohit497.github.io/Recent-Advances-in-Vision-and-Language-Research.

Gao, C., Saraf, A., Huang, J.-B., and Kopf, J. (2020). Flow-edge guided video completion. In *European Conference on Computer Vision (ECCV)*.

Gao, W., Chen, Y., Wang, R., Shan, S., and Jiang, D. (2003). Learning and synthesizing MPEG-4 compatible 3-D face animation from video sequence. *IEEE Transactions on Circuits and Systems for Video Technology*, 13(11):1119–1128.

Gao, X.-S., Hou, X.-R., Tang, J., and Cheng, H.-F. (2003). Complete solution classification for the perspective-three-point problem. *IEEE Transactions on Pattern Analysis and Machine Intelligence*, 25(8):930–943.

Garbin, S. J., Kowalski, M., Johnson, M., Shotton, J., and Valentin, J. (2021). FastNeRF: High-fidelity neural rendering at 200fps. *arXiv preprint arXiv:2103.10380*.

Garg, R., Wadhwa, N., Ansari, S., and Barron, J. T. (2019). Learning single camera depth estimation using dual-pixels. In *IEEE/CVF International Conference on Computer Vision (ICCV)*.

Gatys, L., Ecker, A. S., and Bethge, M. (2015). Texture synthesis using convolutional neural networks. In *Advances in Neural Information Processing Systems (NeurIPS)*, pp. 262–270.

Gatys, L. A., Ecker, A. S., and Bethge, M. (2016). Image style transfer using convolutional neural networks. In *IEEE Conference on Computer Vision and Pattern Recognition (CVPR)*.

Gavrila, D. M. (1999). The visual analysis of human movement: A survey. *Computer Vision and Image Understanding*, 73(1):82–98.

Gavrila, D. M. and Davis, L. S. (1996). 3D model-based tracking of humans in action: A multi-view approach. In *IEEE Computer Society Conference on Computer Vision and Pattern Recognition (CVPR)*, pp. 73–80.

Gavrila, D. M. and Philomin, V. (1999). Real-time object detection for smart vehicles. In *International Conference on Computer Vision (ICCV)*, pp. 87–93.

Gecer, B., Ploumpis, S., Kotsia, I., and Zafeiriou, S. (2019). GANFIT: Generative adversarial network fitting for high fidelity 3D face reconstruction. In *IEEE/CVF Conference on Computer Vision and Pattern Recognition (CVPR)*.

Gehrig, S. K., Eberli, F., and Meyer, T. (2009). A real-time low-power stereo vision engine using semi-global matching. In *International Conference on Computer Vision Systems*, pp. 134–143.

Gehring, J., Auli, M., Grangier, D., Yarats, D., and Dauphin, Y. N. (2017). Convolutional sequence to sequence learning. In *International Conference on Machine Learning (ICML)*, pp. 1243–1252.

Geiger, A. (2021). University of Tübingen ML-4103 course: Deep learning. Slides and lecture notes available at https://uni-tuebingen.de/de/203146.

Geiger, A., Lenz, P., and Urtasun, R. (2012). Are we ready for autonomous driving? The KITTI vision benchmark suite. In *IEEE Computer Society Conference on Computer Vision and Pattern Recognition (CVPR)*.

Geiger, A., Lenz, P., Stiller, C., and Urtasun, R. (2013). Vision meets robotics: The KITTI dataset. *International Journal of Robotics Research*, 32(11):1231–1237.

Geiger, D. and Girosi, F. (1991). Parallel and deterministic algorithms for MRFs: Surface reconstruction. *IEEE Transactions on Pattern Analysis and Machine Intelligence*, 13(5):401–412.

Geiger, D., Ladendorf, B., and Yuille, A. (1992). Occlusions and binocular stereo. In *European Conference on Computer Vision (ECCV)*, pp. 425–433.

Geirhos, R., Jacobsen, J.-H., Michaelis, C., Zemel, R., Brendel, W., Bethge, M., and Wichmann, F. A. (2020). Shortcut learning in deep neural networks. *Nature Machine Intelligence*, 2(11):665–673.

Gelb, A. (ed.). (1974). *Applied Optimal Estimation*. MIT Press, Cambridge, Massachusetts.

Geman, S. and Geman, D. (1984). Stochastic relaxation, Gibbs distribution, and the Bayesian restoration of images. *IEEE Transactions on Pattern Analysis and Machine Intelligence*, PAMI-6(6):721–741.

Gennert, M. A. (1988). Brightness-based stereo matching. In *International Conference on Computer Vision (ICCV)*, pp. 139–143.

Germain, H., Bourmaud, G., and Lepetit, V. (2020). S2DNet: Learning image features for accurate sparse-to-dense matching. In *European Conference on Computer Vision (ECCV)*.

Gershun, A. (1939). The light field. *Journal of Mathematics and Physics*, XVIII:51–151.

Gevers, T., van de Weijer, J., and Stokman, H. (2006). Color feature detection. In Lukac, R. and Plataniotis, K. N. (eds), *Color Image Processing: Methods and Applications*, CRC Press.

Gharbi, M., Chaurasia, G., Paris, S., and Durand, F. (2016). Deep joint demosaicking and denoising. *ACM Transactions on Graphics (Proc. SIGGRAPH Asia)*, 35(6):191.

Ghosh, A., Kumar, H., and Sastry, P. S. (2017). Robust loss functions under label noise for deep neural networks. In *AAAI Conference on Artificial Intelligence (AAAI)*, pp. 1919–1925.

Ghosh, P., Sajjadi, M. S. M., Vergari, A., Black, M., and Schölkopf, B. (2019). From variational to deterministic autoencoders. *arXiv preprint arXiv:1903.12436*.

Giblin, P. and Weiss, R. (1987). Reconstruction of surfaces from profiles. In *International Conference on Computer Vision (ICCV)*, pp. 136–144.

Gidaris, S. and Komodakis, N. (2017). Detect, replace, refine: Deep structured prediction for pixel wise labeling. In *IEEE Conference on Computer Vision and Pattern Recognition (CVPR)*.

Gidaris, S., Singh, P., and Komodakis, N. (2018). Unsupervised representation learning by predicting image rotations. In *International Conference on Learning Representations (ICLR)*.

Gidaris, S., Bursuc, A., Komodakis, N., Perez, P., and Cord, M. (2020). Learning representations by predicting bags of visual words. In *IEEE/CVF Conference on Computer Vision and Pattern Recognition (CVPR)*.

Gilbert, C. D. and Li, W. (2013). Top-down influences on visual processing. *Nature Reviews Neuroscience*, 14(5):350–363.

Gionis, A., Indyk, P., and Motwani, R. (1999). Similarity search in high dimensions via hashing. In *International Conference on Very Large Data Bases (VLDB)*, pp. 518–529.

Girdhar, R., Fouhey, D. F., Rodriguez, M., and Gupta, A. (2016). Learning a predictable and generative vector representation for objects. In *European Conference on Computer Vision (ECCV)*, pp. 484–499.

Girod, B., Greiner, G., and Niemann, H. (eds). (2000). *Principles of 3D Image Analysis and Synthesis*, Kluwer, Boston.

Girshick, R. (2015). Fast R-CNN. In *IEEE International Conference on Computer Vision (ICCV)*.

Girshick, R., Kirillov, A., Wu, Y., Feichtenhofer, C., Fan, H., Gkioxari, G., Johnson, J., Ravi, N., Dollár, P., Lo, W.-Y., and Xie, S. (2020). CVPR 2020 tutorial on visual recognition for images, video, and 3D. https://s9xie.github.io/Tutorials/CVPR2020.

Girshick, R., Donahue, J., Darrell, T., and Malik, J. (2014). Rich feature hierarchies for accurate object detection and semantic segmentation. In *IEEE Conference on Computer Vision and Pattern Recognition (CVPR)*.

Girshick, R., Donahue, J., Darrell, T., and Malik, J. (2015). Region-based convolutional networks for accurate object detection and segmentation. *IEEE Transactions on Pattern Analysis and Machine Intelligence*, 38(1):142–158.

Gkioxari, G., Malik, J., and Johnson, J. (2019). Mesh R-CNN. In *IEEE/CVF International Conference on Computer Vision (ICCV)*.

Gkioxari, G., Girshick, R., Dollár, P., and He, K. (2018). Detecting and recognizing human-object interactions. In *IEEE Conference on Computer Vision and Pattern Recognition (CVPR)*.

Glassner, A. (2018). *Deep Learning: From Basics to Practice*. The Imaginary Institute. https://www.glassner.com/portfolio/deep-learning-from-basics-to-practice.

Glassner, A. (2021). *Deep Learning: A Visual Approach*. no starch press. https://nostarch.com/deep-learning-visual-approach.

Glassner, A. S. (1995). *Principles of Digital Image Synthesis*. Morgan Kaufmann Publishers, San Francisco.

Gleicher, M. (1995). Image snapping. In *ACM SIGGRAPH Conference Proceedings*, pp. 183–190.

Gleicher, M. and Witkin, A. (1992). Through-the-lens camera control. *Computer Graphics (SIGGRAPH)*, 26(2):331–340.

Glocker, B., Komodakis, N., Tziritas, G., Navab, N., and Paragios, N. (2008). Dense image registration through MRFs and efficient linear programming. *Medical Image Analysis*, 12(6):731–741.

Glocker, B., Paragios, N., Komodakis, N., Tziritas, G., and Navab, N. (2008). Optical flow estimation with uncertainties through dynamic MRFs. In *IEEE Computer Society Conference on Computer Vision and Pattern Recognition (CVPR)*.

Glorot, X. and Bengio, Y. (2010). Understanding the difficulty of training deep feedforward neural networks. In *International Conference on Artificial Intelligence and Statistics*, pp. 249–256.

Gluckman, J. (2006a). Higher order image pyramids. In *European Conference on Computer Vision (ECCV)*, pp. 308–320.

Gluckman, J. (2006b). Scale variant image pyramids. In *IEEE Computer Society Conference on Computer Vision and Pattern Recognition (CVPR)*, pp. 1069–1075.

Godard, C., Mac Aodha, O., and Brostow, G. J. (2017). Unsupervised monocular depth estimation with left-right consistency. In *IEEE Conference on Computer Vision and Pattern Recognition (CVPR)*.

Godard, C., Matzen, K., and Uyttendaele, M. (2018). Deep burst denoising. In *European Conference on Computer Vision (ECCV)*.

Goesele, M., Curless, B., and Seitz, S. (2006). Multi-view stereo revisited. In *IEEE Computer Society Conference on Computer Vision and Pattern Recognition (CVPR)*, pp. 2402–2409.

Goesele, M., Fuchs, C., and Seidel, H.-P. (2003). Accuracy of 3D range scanners by measurement of the slanted edge modulation transfer function. In *International Conference on 3-D Digital Imaging and Modeling*.

Goesele, M., Snavely, N., Curless, B., Hoppe, H., and Seitz, S. M. (2007). Multi-view stereo for community photo collections. In *International Conference on Computer Vision (ICCV)*.

Goesele, M., Ackermann, J., Fuhrmann, S., Haubold, C., Klowsky, R., Steedly, D., and Szeliski, R. (2010). Ambient point clouds for view interpolation. *ACM Transactions on Graphics (Proc. SIGGRAPH)*, 29(4):95.

Goh, G., Cammarata, N., Voss, C., Carter, S., Petrov, M., Schubert, L., Radford, A., and Olah, C. (2021). Multimodal neurons in artificial neural networks. *Distill*. https://distill.pub/2021/multimodal-neurons.

Gold, S., Rangarajan, A., Lu, C., Pappu, S., and Mjolsness, E. (1998). New algorithms for 2D and 3D point matching: Pose estimation and correspondence. *Pattern Recognition*, 31(8):1019–1031.

Goldberg, A. V. and Tarjan, R. E. (1988). A new approach to the maximum-flow problem. *Journal of the ACMD*, 35(4):921–940.

Goldman, D. B. (2010). Vignette and exposure calibration and compensation. *IEEE Transactions on Pattern Analysis and Machine Intelligence*, 32(12).

Golovinskiy, A., Matusik, W., ster, H. P., Rusinkiewicz, S., and Funkhouser, T. (2006). A statistical model for synthesis of detailed facial geometry. *ACM Transactions on Graphics*, 25(3):1025–1034.

Golub, G. and Van Loan, C. F. (1996). *Matrix Computation*. The John Hopkins University Press, Baltimore and London, 3rd edition.

Gomes, J. and Velho, L. (1997). *Image Processing for Computer Graphics*. Springer-Verlag, New York.

Gomes, J., Darsa, L., Costa, B., and Velho, L. (1999). *Warping and Morphing of Graphical Objects*. Morgan Kaufmann Publishers, San Francisco.

Gómez, R. (2018). Understanding categorical cross-entropy loss, binary cross-entropy loss, softmax loss, logistic loss, focal loss and all those confusing names. https://gombru.github.io/2018/05/23/cross_entropy_loss.

Gómez, R. (2019). Understanding ranking loss, contrastive loss, margin loss, triplet loss, hinge loss and all those confusing names. https://gombru.github.io/2019/04/03/ranking_loss.

Gong, M., Yang, R., Wang, L., and Gong, M. (2007). A performance study on different cost aggregation approaches used in realtime stereo matching. *International Journal of Computer Vision*, 75(2):283–296.

Gonzalez, R. C. and Woods, R. E. (2017). *Digital Image Processing*. Prentice-Hall, Upper Saddle River, NJ, 4th edition.

Gooch, B. and Gooch, A. (2001). *Non-Photorealistic Rendering*. A K Peters, Ltd, Natick, Massachusetts.

Goodale, M. A. and Milner, A. D. (1992). Separate visual pathways for perception and action. *Trends in Neurosciences*, 15(1):20–25.

Goodfellow, I. (2016). Nips 2016 tutorial: Generative adversarial networks. *arXiv preprint arXiv:1701.00160*.

Goodfellow, I., Isola, P., Park, T., Zhu, J.-Y., Arora, S., Denton, E., Rosca, M., Liu, M.-Y., Hoffman, J., Wang, X., Ermon, S., Vondrick, C., and Efros, A. A. (2018). CVPR 2018 tutorial on GANs. https://sites.google.com/view/cvpr2018tutorialongans.

Goodfellow, I., Bengio, Y., and Courville, A. (2016). *Deep Learning*. MIT Press. https://www.deeplearningbook.org.

Goodfellow, I., Shlens, J., and Szegedy, C. (2015). Explaining and harnessing adversarial examples. In *International Conference on Learning Representations (ICLR)*.

Goodfellow, I., Papernot, N., Huang, S., Duan, R., Abbeel, P., and Clark, J. (2017). Attacking machine learning with adversarial examples. *OpenAI Blog*. https://openai.com/blog/adversarial-example-research.

Goodfellow, I., Pouget-Abadie, J., Mirza, M., Xu, B., Warde-Farley, D., Ozair, S., Courville, A., and Bengio, Y. (2014). Generative adversarial nets. In *Advances in Neural Information Processing Systems (NeurIPS)*, pp. 2672–2680.

Gordo, A., Almazán, J., Revaud, J., and Larlus, D. (2017). End-to-end learning of deep visual representations for image retrieval. *International Journal of Computer Vision*, 124(2):237–254.

Gordo, A., Perronnin, F., Gong, Y., and Lazebnik, S. (2013). Asymmetric distances for binary embeddings. *IEEE Transactions on Pattern Analysis and Machine Intelligence*, 36(1):33–47.

Gordon, I. and Lowe, D. G. (2006). What and where: 3D object recognition with accurate pose. In Ponce, J., Hebert, M., Schmid, C., and Zisserman, A. (eds), *Toward Category-Level Object Recognition*, pp. 67–82, Springer, New York.

Gorelick, L., Blank, M., Shechtman, E., Irani, M., and Basri, R. (2007). Actions as space-time shapes. *IEEE Transactions on Pattern Analysis and Machine Intelligence*, 29(12):2247–2253.

Gortler, S. J. and Cohen, M. F. (1995). Hierarchical and variational geometric modeling with wavelets. In *Symposium on Interactive 3D Graphics*, pp. 35–43.

Gortler, S. J., Grzeszczuk, R., Szeliski, R., and Cohen, M. F. (1996). The Lumigraph. In *ACM SIGGRAPH Conference Proceedings*, pp. 43–54.

Goshtasby, A. (1989). Correction of image deformation from lens distortion using Bézier patches. *Computer Vision, Graphics, and Image Processing*, 47(4):385–394.

Goshtasby, A. (2005). *2-D and 3-D Image Registration*. Wiley, New York.

Gotchev, A. and Rosenhahn, B. (eds). (2009). *Proceedings of the 3DTV Conference: The True Vision—Capture, Transmission and Display of 3D Video*, IEEE Computer Society Press.

Govindu, V. M. (2001). Combining two-view constraints for motion estimation. In *IEEE Computer Society Conference on Computer Vision and Pattern Recognition (CVPR)*, pp. 684–691.

Govindu, V. M. (2004). Lie-algebraic averaging for globally consistent motion estimation. In *IEEE Computer Society Conference on Computer Vision and Pattern Recognition (CVPR)*, pp. 684–691.

Govindu, V. M. (2006). Revisiting the brightness constraint: Probabilistic formulation and algorithms. In *European Conference on Computer Vision (ECCV)*, pp. 177–188.

Goyal, P., Mahajan, D., Gupta, A., and Misra, I. (2019). Scaling and benchmarking self-supervised visual representation learning. In *IEEE/CVF International Conference on Computer Vision (ICCV)*.

Goyal, P., Dollár, P., Girshick, R., Noordhuis, P., Wesolowski, L., Kyrola, A., Tulloch, A., Jia, Y., and He, K. (2017). Accurate, large minibatch SGD: Training ImageNet in 1 hour. *arXiv preprint arXiv:1706.02677*.

Goyal, R., Kahou, E. S., Michalski, V., Materzyńska, J., Westphal, S., Kim, H., Haenel, V., Fruend, I., Yianilos, P., Müller-Freitag, M., Hoppe, F., Thurau, C., Bax, I., and Memisevic, R. (2017). The "something something" video database for learning and evaluating visual common sense. In *IEEE International Conference on Computer Vision (ICCV)*.

Goyal, Y., Khot, T., Summers-Stay, D., Batra, D., and Parikh, D. (2017). Making the V in VQA matter: Elevating the role of image understanding in visual question answering. In *IEEE Conference on Computer Vision and Pattern Recognition (CVPR)*.

Grady, L. (2006). Random walks for image segmentation. *IEEE Transactions on Pattern Analysis and Machine Intelligence*, 28(11):1768–1783.

Grady, L. (2008). A lattice-preserving multigrid method for solving the inhomogeneous Poisson equations used in image analysis. In *European Conference on Computer Vision (ECCV)*, pp. 252–264.

Grady, L. and Ali, S. (2008). Fast approximate random walker segmentation using eigenvector precomputation. In *IEEE Computer Society Conference on Computer Vision and Pattern Recognition (CVPR)*.

Grady, L. and Alvino, C. (2008). Reformulating and optimizing the Mumford–Shah functional on a graph — a faster, lower energy solution. In *European Conference on Computer Vision (ECCV)*, pp. 248–261.

Grauman, K. and Darrell, T. (2005). Efficient image matching with distributions of local invariant features. In *IEEE Computer Society Conference on Computer Vision and Pattern Recognition (CVPR)*, pp. 627–634.

Grauman, K. and Darrell, T. (2007a). Pyramid match hashing: Sub-linear time indexing over partial correspondences. In *IEEE Computer Society Conference on Computer Vision and Pattern Recognition (CVPR)*.

Grauman, K. and Darrell, T. (2007b). The pyramid match kernel: Efficient learning with sets of features. *Journal of Machine Learning Research*, 8:725–760.

Grauman, K., Shakhnarovich, G., and Darrell, T. (2003). Inferring 3D structure with a statistical image-based shape model. In *International Conference on Computer Vision (ICCV)*, pp. 641–648.

Greene, N. (1986). Environment mapping and other applications of world projections. *IEEE Computer Graphics and Applications*, 6(11):21–29.

Greene, N. and Heckbert, P. (1986). Creating raster Omnimax images from multiple perspective views using the elliptical weighted average filter. *IEEE Computer Graphics and Applications*, 6(6):21–27.

Greig, D., Porteous, B., and Seheult, A. (1989). Exact maximum a posteriori estimation for binary images. *Journal of the Royal Statistical Society, Series B*, 51(2):271–279.

Gremban, K. D., Thorpe, C. E., and Kanade, T. (1988). Geometric camera calibration using systems of linear equations. In *IEEE International Conference on Robotics and Automation*, pp. 562–567.

Griewank, A. and Walther, A. (2000). Algorithm 799: revolve: an implementation of checkpointing for the reverse or adjoint mode of computational differentiation. *ACM Transactions on Mathematical Software*, 26(1):19–45.

Grill, J.-B., Strub, F., Altché, F., Tallec, C., Richemond, P., Buchatskaya, E., Doersch, C., Avila Pires, B., Guo, Z., Gheshlaghi Azar, M., Piot, B., Kavukcuoglu, K., Munos, R., and Valko, M. (2020). Bootstrap your own latent - a new approach to self-supervised learning. In *Advances in Neural Information Processing Systems (NeurIPS)*.

Grimson, W. E. L. (1983). An implementation of a computational theory of visual surface interpolation. *Computer Vision, Graphics, and Image Processing*, 22:39–69.

Grimson, W. E. L. (1985). Computational experiments with a feature based stereo algorithm. *IEEE Transactions on Pattern Analysis and Machine Intelligence*, PAMI-7(1):17–34.

Grompone von Gioi, R., Jakubowicz, J., Morel, J.-M., and Randall, G. (2008). LSD: A fast line segment detector with a false detection control. *IEEE Transactions on Pattern Analysis and Machine Intelligence*, 32(4):722–732.

Gross, R., Matthews, I., and Baker, S. (2006). Active appearance models with occlusion. *Image and Vision Computing*, 24(6):593–604.

Gross, R., Shi, J., and Cohn, J. F. (2005). Quo vadis face recognition? In *IEEE Workshop on Empirical Evaluation Methods in Computer Vision*.

Gross, R., Baker, S., Matthews, I., and Kanade, T. (2005). Face recognition across pose and illumination. In Li, S. Z. and Jain, A. K. (eds), *Handbook of Face Recognition*, Springer.

Gross, R., Sweeney, L., De la Torre, F., and Baker, S. (2008). Semi-supervised learning of multi-factor models for face de-identification. In *IEEE Computer Society Conference on Computer Vision and Pattern Recognition (CVPR)*.

Gross, R., Matthews, I., Cohn, J., Kanade, T., and Baker, S. (2010). Multi-PIE. *Image and Vision Computing*, 28(5):807–813.

Grossberg, M. D. and Nayar, S. K. (2001). A general imaging model and a method for finding its parameters. In *International Conference on Computer Vision (ICCV)*, pp. 108–115.

Grossberg, M. D. and Nayar, S. K. (2004). Modeling the space of camera response functions. *IEEE Transactions on Pattern Analysis and Machine Intelligence*, 26(10):1272–1282.

Grosse, R. and Ba, J. (2019). University of Toronto csc 421/2516: Neural networks and deep learning. Slides and notes available at https://www.cs.toronto.edu/~rgrosse/courses/csc421_2019.

Groueix, T., Fisher, M., Kim, V. G., Russell, B. C., and Aubry, M. (2018). A papier-mâché approach to learning 3D surface generation. In *IEEE Conference on Computer Vision and Pattern Recognition (CVPR)*.

Grundmann, M., Kwatra, V., and Essa, I. (2011). Auto-directed video stabilization with robust L1 optimal camera paths. In *IEEE Computer Society Conference on Computer Vision and Pattern Recognition (CVPR)*.

Grundmann, M., Kwatra, V., Castro, D., and Essa, I. (2012). Calibration-free rolling shutter removal. In *International Conference on Computational Photography (ICCP)*, pp. 1–8.

Gu, C., Sun, C., Ross, D. A., Vondrick, C., Pantofaru, C., Li, Y., Vijayanarasimhan, S., Toderici, G., Ricco, S., Sukthankar, R., Schmid, C., and Malik, J. (2018). AVA: A video dataset of spatio-temporally localized atomic visual actions. In *IEEE Conference on Computer Vision and Pattern Recognition (CVPR)*.

Gu, C., Lim, J., Arbelaez, P., and Malik, J. (2009). Recognition using regions. In *IEEE Computer Society Conference on Computer Vision and Pattern Recognition (CVPR)*.

Gu, J., Wang, Z., Kuen, J., Ma, L., Shahroudy, A., Shuai, B., Liu, T., Wang, X., Wang, G., Cai, J., and Chen, T. (2018). Recent advances in convolutional neural networks. *Pattern Recognition*, 77:354–377.

Gu, S. and Timofte, R. (2019). A brief review of image denoising algorithms and beyond. In Escalera, S., Ayache, S., Wan, J., Madadi, M., Güçlü, U., and Baró, X. (eds), *Inpainting and Denoising Challenges*, pp. 1–21, Springer.

Gu, S., Zhang, L., Zuo, W., and Feng, X. (2014). Weighted nuclear norm minimization with application to image denoising. In *IEEE Conference on Computer Vision and Pattern Recognition (CVPR)*.

Gu, X., Gortler, S. J., and Hoppe, H. (2002). Geometry images. *ACM Transactions on Graphics*, 21(3):355–361.

Gu, X., Fan, Z., Zhu, S., Dai, Z., Tan, F., and Tan, P. (2020). Cascade cost volume for high-resolution multi-view stereo and stereo matching. In *IEEE/CVF Conference on Computer Vision and Pattern Recognition (CVPR)*.

Guan, P., Weiss, A., Bălan, A. O., and Black, M. J. (2009). Estimating human shape and pose from a single image. In *International Conference on Computer Vision (ICCV)*.

Guennebaud, G. and Gross, M. (2007). Algebraic point set surfaces. *ACM Transactions on Graphics*, 26(3).

Guennebaud, G., Germann, M., and Gross, M. (2008). Dynamic sampling and rendering of algebraic point set surfaces. *Computer Graphics Forum*, 27(2):653–662.

Guenter, B., Grimm, C., Wood, D., Malvar, H., and Pighin, F. (1998). Making faces. In *ACM SIGGRAPH Conference Proceedings*, pp. 55–66.

Guillard, B., Remelli, E., and Fua, P. (2020). UCLID-Net: Single view reconstruction in object space. In *Advances in Neural Information Processing Systems (NeurIPS)*.

Guizzo, E. (2008). Kiva systems: Three engineers, hundreds of robots, one warehouse. *IEEE Spectrum*, 45(7):26–34.

Gulbins, J. and Gulbins, R. (2009). *Photographic Multishot Techniques: High Dynamic Range, Super-Resolution, Extended Depth of Field, Stitching*. Rocky Nook.

Güler, R. A., Neverova, N., and Kokkinos, I. (2018). DensePose: Dense human pose estimation in the wild. In *IEEE Conference on Computer Vision and Pattern Recognition (CVPR)*.

Guo, C., Pleiss, G., Sun, Y., and Weinberger, K. Q. (2017). On calibration of modern neural networks. In *International Conference on Machine Learning (ICML)*, pp. 1321–1330.

Guo, Y., Wang, H., Hu, Q., Liu, H., Liu, L., and Bennamoun, M. (2020). Deep learning for 3D point clouds: A survey. *IEEE Transactions on Pattern Analysis and Machine Intelligence*.

Gupta, A., Dollár, P., and Girshick, R. (2019). LVIS: A dataset for large vocabulary instance segmentation. In *IEEE/CVF Conference on Computer Vision and Pattern Recognition (CVPR)*.

Gupta, S. and Malik, J. (2015). Visual semantic role labeling. *arXiv preprint arXiv:1505.04474*.

Habermann, M., Xu, W., Zollhöefer, M., Pons-Moll, G., and Theobalt, C. (2019). LiveCap: Real-time human performance capture from monocular video. *ACM Transactions On Graphics (Proc. SIGGRAPH)*, 38(2):1–17.

Hadsell, R., Chopra, S., and LeCun, Y. (2006). Dimensionality reduction by learning an invariant mapping. In *IEEE Computer Society Conference on Computer Vision and Pattern Recognition (CVPR)*, pp. 1735–1742.

Haefner, B., Ye, Z., Gao, M., Wu, T., Queau, Y., and Cremers, D. (2019). Variational uncalibrated photometric stereo under general lighting. In *IEEE/CVF International Conference on Computer Vision (ICCV)*.

Hagelskjær, F. and Buch, A. G. (2020). PointVoteNet: Accurate object detection and 6 DOF pose estimation in point clouds. In *2020 IEEE International Conference on Image Processing (ICIP)*, pp. 2641–2645.

Hager, G. D. and Belhumeur, P. N. (1998). Efficient region tracking with parametric models of geometry and illumination. *IEEE Transactions on Pattern Analysis and Machine Intelligence*, 20(10):1025–1039.

Hall, D., Dayoub, F., Skinner, J., Zhang, H., Miller, D., Corke, P., Carneiro, G., Angelova, A., and Sünderhauf, N. (2020). Probabilistic object detection: Definition and evaluation. In *IEEE Winter Conference on Applications of Computer Vision (WACV)*, pp. 1031–1040.

Hall, R. (1989). *Illumination and Color in Computer Generated Imagery*. Springer-Verlag, New York.

Haller, I. and Nedevschi, S. (2012). Design of interpolation functions for subpixel-accuracy stereo-vision systems. *IEEE Transactions on Image Processing*, 21(2):889–898.

Haller, M., Billinghurst, M., and Thomas, B. (2007). *Emerging Technologies of Augmented Reality: Interfaces and Design*. IGI Publishing.

Hampel, F. R., Ronchetti, E. M., Rousseeuw, P. J., and Stahel, W. A. (1986). *Robust Statistics : The Approach Based on Influence Functions*. Wiley, New York.

Han, F. and Zhu, S.-C. (2005). Bottom-up/top-down image parsing by attribute graph grammar. In *International Conference on Computer Vision (ICCV)*, pp. 1778–1785.

Han, J., Shao, L., Xu, D., and Shotton, J. (2013). Enhanced computer vision with Microsoft Kinect sensor: A review. *IEEE Transactions on Cybernetics*, 43(5):1318–1334.

Han, K., Wang, Y., Chen, H., Chen, X., Guo, J., Liu, Z., Tang, Y., Xiao, A., Xu, C., Xu, Y., Zhang1, Y., and Tao, D. (2020). A survey on visual transformer. *arXiv preprint arXiv:2012.12556.*

Han, S., Mao, H., and Dally, W. J. (2016). Deep compression: Compressing deep neural networks with pruning, trained quantization and Huffman coding. In *International Conference on Learning Representations (ICLR).*

Han, X.-F., Laga, H., and Bennamoun, M. (2021). Image-based 3D object reconstruction: State-of-the-art and trends in the deep learning era. *IEEE Transactions on Pattern Analysis and Machine Intelligence*, 43(5):1578–1604.

Handa, A., Whelan, T., McDonald, J., and Davison, A. J. (2014). A benchmark for RGB-D visual odometry, 3D reconstruction and SLAM. In *IEEE International Conference on Robotics and Automation (ICRA)*, pp. 1524–1531.

Häne, C., Zach, C., Cohen, A., Angst, R., and Pollefeys, M. (2013). Joint 3D scene reconstruction and class segmentation. In *IEEE Computer Society Conference on Computer Vision and Pattern Recognition (CVPR).*

Hanna, K. J. (1991). Direct multi-resolution estimation of ego-motion and structure from motion. In *IEEE Workshop on Visual Motion*, pp. 156–162.

Hannah, M. J. (1974). *Computer Matching of Areas in Stereo Images*. Ph.D. thesis, Stanford University.

Hannah, M. J. (1988). Test results from SRI's stereo system. In *Image Understanding Workshop*, pp. 740–744.

Hansard, M., Lee, S., Choi, O., and Horaud, R. P. (2012). *Time-of-flight cameras: principles, methods and applications.* Springer Science & Business Media.

Hansen, M., Anandan, P., Dana, K., van der Wal, G., and Burt, P. (1994). Real-time scene stabilization and mosaic construction. In *IEEE Workshop on Applications of Computer Vision (WACV)*, pp. 54–62.

Hanson, A. R. and Riseman, E. M. (eds). (1978). *Computer Vision Systems*, Academic Press, New York.

Haralick, R. M. and Shapiro, L. G. (1985). Image segmentation techniques. *Computer Vision, Graphics, and Image Processing*, 29(1):100–132.

Haralick, R. M. and Shapiro, L. G. (1992). *Computer and Robot Vision*. Addison-Wesley, Reading, MA.

Haralick, R. M., Lee, C.-N., Ottenberg, K., and Nölle, M. (1994). Review and analysis of solutions of the three point perspective pose estimation problem. *International Journal of Computer Vision*, 13(3):331–356.

Hardie, R. C., Barnard, K. J., and Armstrong, E. E. (1997). Joint MAP registration and high-resolution image estimation using a sequence of undersampled images. *IEEE Transactions on Image Processing*, 6(12):1621–1633.

Hare, S., Golodetz, S., Saffari, A., Vineet, V., Cheng, M.-M., Hicks, S. L., and Torr, P. H. S. (2015). Struck: Structured output tracking with kernels. *IEEE Transactions on Pattern Analysis and Machine Intelligence*, 38(10):2096–2109.

Hariharan, B., Arbeláez, P., Girshick, R., and Malik, J. (2014). Simultaneous detection and segmentation. In *European Conference on Computer Vision (ECCV)*, pp. 297–312.

Hariharan, B., Arbeláez, P., Girshick, R., and Malik, J. (2015). Hypercolumns for object segmentation and fine-grained localization. In *IEEE Conference on Computer Vision and Pattern Recognition (CVPR).*

Haritaoglu, I., Harwood, D., and Davis, L. S. (2000). W^4: Real-time surveillance of people and their activities. *IEEE Transactions on Pattern Analysis and Machine Intelligence*, 22(8):809–830.

Harker, M. and O'Leary, P. (2008). Least squares surface reconstruction from measured gradient fields. In *IEEE Computer Society Conference on Computer Vision and Pattern Recognition (CVPR).*

Härkönen, E., Hertzmann, A., Lehtinen, J., and Paris, S. (2020). GANSpace: Discovering interpretable gan controls. In *Advances in Neural Information Processing Systems (NeurIPS).*

Harris, C. and Stephens, M. J. (1988). A combined corner and edge detector. In *Alvey Vision Conference*, pp. 147–152.

Hartley, R. and Kang, S. B. (2007). Parameter-free radial distortion correction with center of distortion estimation. *IEEE Transactions on Pattern Analysis and Machine Intelligence*, 31(8):1309–1321.

Hartley, R., Gupta, R., and Chang, T. (1992). Estimation of relative camera positions for uncalibrated cameras. In *European Conference on Computer Vision (ECCV)*, pp. 579–587.

Hartley, R., Trumpf, J., Dai, Y., and Li, H. (2013). Rotation averaging. *International Journal of Computer Vision*, 103(3):267–305.

Hartley, R. I. (1994a). Projective reconstruction and invariants from multiple images. *IEEE Transactions on Pattern Analysis and Machine Intelligence*, 16(10):1036–1041.

Hartley, R. I. (1994b). Self-calibration from multiple views of a rotating camera. In *European Conference on Computer Vision (ECCV)*, pp. 471–478.

Hartley, R. I. (1997a). In defense of the 8-point algorithm. *IEEE Transactions on Pattern Analysis and Machine Intelligence*, 19(6):580–593.

Hartley, R. I. (1997b). Self-calibration of stationary cameras. *International Journal of Computer Vision*, 22(1):5–23.

Hartley, R. I. (1998). Chirality. *International Journal of Computer Vision*, 26(1):41–61.

Hartley, R. I. and Kang, S. B. (2005). Parameter-free radial distortion correction with centre of distortion estimation. In *International Conference on Computer Vision (ICCV)*, pp. 1834–1841.

Hartley, R. I. and Sturm, P. (1997). Triangulation. *Computer Vision and Image Understanding*, 68(2):146–157.

Hartley, R. I. and Zisserman, A. (2004). *Multiple View Geometry in Computer Vision*. Cambridge University Press, Cambridge, UK, 2nd edition.

Hartley, R. I., Hayman, E., de Agapito, L., and Reid, I. (2000). Camera calibration and the search for infinity. In *IEEE Computer Society Conference on Computer Vision and Pattern Recognition (CVPR)*, pp. 510–517.

Hasinoff, S. W. and Kutulakos, K. N. (2011). Light-efficient photography. *IEEE Transactions on Pattern Analysis and Machine Intelligence*, 33(11):2203–2214.

Hasinoff, S. W., Durand, F., and Freeman, W. T. (2010). Noise-optimal capture for high dynamic range photography. In *IEEE Computer Society Conference on Computer Vision and Pattern Recognition (CVPR)*.

Hasinoff, S. W., Kang, S. B., and Szeliski, R. (2006). Boundary matting for view synthesis. *Computer Vision and Image Understanding*, 103(1):22–32.

Hasinoff, S. W., Kutulakos, K. N., Durand, F., and Freeman, W. T. (2009). Time-constrained photography. In *International Conference on Computer Vision (ICCV)*.

Hasinoff, S. W., Sharlet, D., Geiss, R., Adams, A., Barron, J. T., Kainz, F., Chen, J., and Levoy, M. (2016). Burst photography for high dynamic range and low-light imaging on mobile cameras. *ACM Transactions on Graphics (Proc. SIGGRAPH Asia)*, 35(6):192:1–192:12.

Hasson, Y., Varol, G., Tzionas, D., Kalevatykh, I., Black, M. J., Laptev, I., and Schmid, C. (2019). Learning joint reconstruction of hands and manipulated objects. In *IEEE/CVF Conference on Computer Vision and Pattern Recognition (CVPR)*.

Hastie, T., Tibshirani, R., and Friedman, J. (2009). *The Elements of Statistical Learning: Data Mining, Inference, and Prediction*. Springer, New York, 2nd edition.

Havaei, M., Davy, A., Warde-Farley, D., Biard, A., Courville, A., Bengio, Y., Pal, C., Jodoin, P.-M., and Larochelle, H. (2017). Brain tumor segmentation with deep neural networks. *Medical Image Analysis*, 35:18–31.

Hayes, B. (2008). Computational photography. *American Scientist*, 96:94–99.

Hays, J. and Efros, A. A. (2007). Scene completion using millions of photographs. *ACM Transactions on Graphics*, 26(3).

Hays, J., Leordeanu, M., Efros, A. A., and Liu, Y. (2006). Discovering texture regularity as a higher-order correspondence problem. In *European Conference on Computer Vision (ECCV)*, pp. 522–535.

He, K., Girshick, R., and Dollár, P. (2019). Rethinking imagenet pre-training. In *IEEE/CVF International Conference on Computer Vision (ICCV)*.

He, K., Sun, J., and Tang, X. (2013). Guided image filtering. *IEEE Transactions on Pattern Analysis and Machine Intelligence*, 35(6):1397–1409.

He, K., Gkioxari, G., Dollár, P., and Girshick, R. (2017). Mask R-CNN. In *IEEE International Conference on Computer Vision (ICCV)*.

He, K., Zhang, X., Ren, S., and Sun, J. (2015). Delving deep into rectifiers: Surpassing human-level performance on imagenet classification. In *IEEE International Conference on Computer Vision (ICCV)*.

He, K., Zhang, X., Ren, S., and Sun, J. (2015). Spatial pyramid pooling in deep convolutional networks for visual recognition. *IEEE Transactions on Pattern Analysis and Machine Intelligence*, 37(9):1904–1916.

He, K., Zhang, X., Ren, S., and Sun, J. (2016a). Deep residual learning for image recognition. In *IEEE Conference on Computer Vision and Pattern Recognition (CVPR)*.

He, K., Zhang, X., Ren, S., and Sun, J. (2016b). Identity mappings in deep residual networks. In *European Conference on Computer Vision*, pp. 630–645.

He, K., Fan, H., Wu, Y., Xie, S., and Girshick, R. (2020). Momentum contrast for unsupervised visual representation learning. In *IEEE/CVF Conference on Computer Vision and Pattern Recognition (CVPR)*.

He, L., Ren, X., Gao, Q., Zhao, X., Yao, B., and Chao, Y. (2017). The connected-component labeling problem: A review of state-of-the-art algorithms. *Pattern Recognition*, 70:25–43.

He, L.-W. and Zhang, Z. (2005). Real-time whiteboard capture and processing using a video camera for teleconferencing. In *IEEE International Conference on Acoustics, Speech, and Signal Processing (ICASSP)*, pp. 1113–1116.

He, M., Liao, J., Sander, P. V., and Hoppe, H. (2017). Gigapixel panorama video loops. *ACM Transactions on Graphics (Proc. SIGGRAPH)*, 37(1):1–15.

He, X., Zemel, R. S., and Carreira-Perpiñán, M. A. (2004). Multiscale conditional random fields for image labeling. In *IEEE Computer Society Conference on Computer Vision and Pattern Recognition (CVPR)*, pp. 695–702.

He, X., Zemel, R. S., and Ray, D. (2006). Learning and incorporating top-down cues in image segmentation. In *European Conference on Computer Vision (ECCV)*, pp. 338–351.

Healey, G. E. and Kondepudy, R. (1994). Radiometric CCD camera calibration and noise estimation. *IEEE Transactions on Pattern Analysis and Machine Intelligence*, 16(3):267–276.

Healey, G. E. and Shafer, S. A. (1992). *Color. Physics-Based Vision: Principles and Practice*, Jones & Bartlett, Cambridge, MA.

Heath, M. D., Sarkar, S., Sanocki, T., and Bowyer, K. W. (1998). Comparison of edge detectors. *Computer Vision and Image Understanding*, 69(1):38–54.

Hebb, D. O. (1949). *The organization of behavior: a neuropsychological theory*. John Wiley & Sons Inc.

Hebert, M. (2000). Active and passive range sensing for robotics. In *IEEE International Conference on Robotics and Automation*, pp. 102–110.

Hecht, E. (2015). *Optics*. Pearson Addison Wesley, Reading, MA, 5th edition.

Heckbert, P. (1986). Survey of texture mapping. *IEEE Computer Graphics and Applications*, 6(11):56–67.

Heckbert, P. (1989). *Fundamentals of Texture Mapping and Image Warping*. Master's thesis, The University of California at Berkeley.

Hedborg, J., Forssén, P.-E., Felsberg, M., and Ringaby, E. (2012). Rolling shutter bundle adjustment. In *IEEE Computer Society Conference on Computer Vision and Pattern Recognition (CVPR)*.

Hedman, P. and Kopf, J. (2018). Instant 3D photography. *ACM Transactions on Graphics (Proc. SIGGRAPH)*, 37(4):101.

Hedman, P., Alsisan, S., Szeliski, R., and Kopf, J. (2017). Casual 3D photography. *ACM Transactions on Graphics (Proc. SIGGRAPH Asia)*, 36(6):234.

Hedman, P., Ritschel, T., Drettakis, G., and Brostow, G. (2016). Scalable inside-out image-based rendering. *ACM Transactions on Graphics (Proc. SIGGRAPH Asia)*, 35(6):1–11.

Hedman, P., Srinivasan, P. P., Mildenhall, B., Barron, J. T., and Debevec, P. (2021). Baking neural radiance fields for real-time view synthesis. *arXiv preprint arXiv:2103.14645*.

Hedman, P., Philip, J., Price, T., Frahm, J.-M., Drettakis, G., and Brostow, G. (2018). Deep blending for free-viewpoint image-based rendering. *ACM Transactions on Graphics (Proc. SIGGRAPH Asia)*, 37(6):257.

Heeger, D. J. (1988). Optical flow using spatiotemporal filters. *International Journal of Computer Vision*, 1(1):279–302.

Heeger, D. J. and Bergen, J. R. (1995). Pyramid-based texture analysis/synthesis. In *ACM SIGGRAPH Conference Proceedings*, pp. 229–238.

Heinly, J., Dunn, E., and Frahm, J.-M. (2014). Correcting for duplicate scene structure in sparse 3D reconstruction. In *European Conference on Computer Vision*, pp. 780–795.

Heinly, J., Schönberger, J. L., Dunn, E., and Frahm, J.-M. (2015). Reconstructing the world* in six days *(as captured by the Yahoo 100 million image dataset). In *IEEE Conference on Computer Vision and Pattern Recognition (CVPR)*.

Heisele, B., Serre, T., and Poggio, T. (2007). A component-based framework for face detection and identification. *International Journal of Computer Vision*, 74(2):167–181.

Heisele, B., Ho, P., Wu, J., and Poggio, T. (2003). Face recognition: component-based versus global approaches. *Computer Vision and Image Understanding*, 91(1–2):6–21.

Hendrycks, D., Basart, S., Mu, N., Kadavath, S., Wang, F., Dorundo, E., Desai, R., Zhu, T., Parajuli, S., Guo, M., Song, D., Steinhardt, J., and Gilmer, J. (2020). The many faces of robustness: A critical analysis of out-of-distribution generalization. *arXiv preprint arXiv:2006.16241*.

Hendrycks, D., Zhao, K., Basart, S., Steinhardt, J., and Song, D. (2021). Natural adversarial examples. In *IEEE/CVF Conference on Computer Vision and Pattern Recognition (CVPR)*.

Henriques, J. F., Caseiro, R., Martins, P., and Batista, J. (2014). High-speed tracking with kernelized correlation filters. *IEEE Transactions on Pattern Analysis and Machine Intelligence*, 37(3):583–596.

Herley, C. (2005). Automatic occlusion removal from minimum number of images. In *International Conference on Image Processing (ICIP)*, pp. 1046–1049–16.

Hernández, C. and Schmitt, F. (2004). Silhouette and stereo fusion for 3D object modeling. *Computer Vision and Image Understanding*, 96(3):367–392.

Hernández, C. and Vogiatzis, G. (2010). Self-calibrating a real-time monocular 3d facial capture system. In *International Symposium on 3D Data Processing, Visualization and Transmission (3DPVT)*.

Hernández, C., Vogiatzis, G., and Cipolla, R. (2008). Multiview photometric stereo. *IEEE Transactions on Pattern Analysis and Machine Intelligence*, 30(3):548–554.

Hernandez, C., Vogiatzis, G., Brostow, G. J., Stenger, B., and Cipolla, R. (2007). Non-rigid photometric stereo with colored lights. In *International Conference on Computer Vision (ICCV)*.

Herrmann, C., Wang, C., Strong Bowen, R., Keyder, E., and Zabih, R. (2018a). Object-centered image stitching. In *European Conference on Computer Vision (ECCV)*.

Herrmann, C., Wang, C., Strong Bowen, R., Keyder, E., Krainin, M., Liu, C., and Zabih, R. (2018b). Robust image stitching with multiple registrations. In *European Conference on Computer Vision (ECCV)*.

Hershberger, J. and Snoeyink, J. (1992). *Speeding Up the Douglas-Peucker Line-Simplification Algorithm*. Technical Report TR-92-07, Computer Science Department, The University of British Columbia.

Hertzmann, A., Jacobs, C. E., Oliver, N., Curless, B., and Salesin, D. H. (2001). Image analogies. In *ACM SIGGRAPH Conference Proceedings*, pp. 327–340.

Hidalgo, G., Raaj, Y., Idrees, H., Xiang, D., Joo, H., Simon, T., and Sheikh, Y. (2019). Single-network whole-body pose estimation. In *IEEE/CVF International Conference on Computer Vision (ICCV)*.

Hiep, V. H., Keriven, R., Pons, J.-P., and Labatut, P. (2009). Towards high-resolution large-scale multi-view stereo. In *IEEE Computer Society Conference on Computer Vision and Pattern Recognition (CVPR)*.

Hillman, P., Hannah, J., and Renshaw, D. (2001). Alpha channel estimation in high resolution images and image sequences. In *IEEE Computer Society Conference on Computer Vision and Pattern Recognition (CVPR)*, pp. 1063–1068.

Hilton, A., Fua, P., and Ronfard, R. (2006). Modeling people: Vision-based understanding of a person's shape, appearance, movement, and behaviour. *Computer Vision and Image Understanding*, 104(2–3):87–89.

Hilton, A., Stoddart, A. J., Illingworth, J., and Windeatt, T. (1996). Reliable surface reconstruction from multiple range images. In *European Conference on Computer Vision (ECCV)*, pp. 117–126.

Hinckley, K., Sinclair, M., Hanson, E., Szeliski, R., and Conway, M. (1999). The VideoMouse: a camera-based multi-degree-of-freedom input device. In *ACM Symposium on User Interface Software and Technology (UIST)*, pp. 103–112.

Hinterstoisser, S., Benhimane, S., Navab, N., Fua, P., and Lepetit, V. (2008). Online learning of patch perspective rectification for efficient object detection. In *IEEE Computer Society Conference on Computer Vision and Pattern Recognition (CVPR)*.

Hinton, G. (2012). Neural networks for machine learning lecture 6e. rmsprop: Divide the gradient by a running average of its recent magnitude. https://www.cs.toronto.edu/~hinton/coursera/lecture6/lec6.pdf.

Hinton, G., Deng, L., Yu, D., Dahl, G., Mohamed, A.-r., Jaitly, N., Senior, A., Vanhoucke, V., Nguyen, P., and Kingsbury, B. (2012). Deep neural networks for acoustic modeling in speech recognition. *IEEE Signal Processing Magazine*, 29.

Hinton, G., Vinyals, O., and Dean, J. (2015). Distilling the knowledge in a neural network. *arXiv preprint arXiv:1503.02531*.

Hinton, G. E. (1977). *Relaxation and its Role in Vision*. Ph.D. thesis, University of Edinburgh.

Hinton, G. E. and Salakhutdinov, R. R. (2006). Reducing the dimensionality of data with neural networks. *Science*, 313(5786):504–507.

Hinton, G. E. and Zemel, R. S. (1994). Autoencoders, minimum description length and helmholtz free energy. In *Advances in Neural Information Processing Systems (NeurIPS)*, pp. 3–10.

Hirschmüller, H. (2008). Stereo processing by semiglobal matching and mutual information. *IEEE Transactions on Pattern Analysis and Machine Intelligence*, 30(2):328–341.

Hirschmüller, H. and Scharstein, D. (2009). Evaluation of stereo matching costs on images with radiometric differences. *IEEE Transactions on Pattern Analysis and Machine Intelligence*, 31(9):1582–1599.

Hirshberg, D. A., Loper, M., Rachlin, E., and Black, M. J. (2012). Coregistration: Simultaneous alignment and modeling of articulated 3D shape. In *European Conference on Computer Vision (ECCV)*, pp. 242–255.

Hjaltason, G. R. and Samet, H. (2003). Index-driven similarity search in metric spaces. *ACM Transactions on Database Systems*, 28(4):517–580.

Hochreiter, S. and Schmidhuber, J. (1997). Long short-term memory. *Neural Computation*, 9(8):1735–1780.

Hodaň, T., Michel, F., Brachmann, E., Kehl, W., GlentBuch, A., Kraft, D., Drost, B., Vidal, J., Ihrke, S., Zabulis, X., Sahin, C., Manhardt, F., Tombari, F., Kim, T.-K., Matas, J., and Rother, C. (2018). BOP: Benchmark for 6D object pose estimation. In *European Conference on Computer Vision (ECCV)*.

Hoffer, E., Banner, R., Golan, I., and Soudry, D. (2018). Norm matters: efficient and accurate normalization schemes in deep networks. In *Advances in Neural Information Processing Systems (NeurIPS)*, pp. 2160–2170.

Hofinger, M., Rota Bulò, S., Porzi, L., Knapitsch, A., Pock, T., and Kontschieder, P. (2020). Improving optical flow on a pyramid level. In *European Conference on Computer Vision (ECCV)*.

Hofmann, T. (1999). Probabilistic latent semantic indexing. In *ACM SIGIR Conference on Research and Development in Informaion Retrieval*, pp. 50–57.

Hogg, D. (1983). Model-based vision: A program to see a walking person. *Image and Vision Computing*, 1(1):5–20.

Hoiem, D., Efros, A. A., and Hebert, M. (2005a). Automatic photo pop-up. *ACM Transactions on Graphics (Proc. SIGGRAPH)*, 24(3):577–584.

Hoiem, D., Efros, A. A., and Hebert, M. (2005b). Geometric context from a single image. In *International Conference on Computer Vision (ICCV)*, pp. 654–661.

Hoiem, D., Efros, A. A., and Hebert, M. (2008). Putting objects in perspective. *International Journal of Computer Vision*, 80(1):3–15.

Hoiem, D., Rother, C., and Winn, J. (2007). 3D LayoutCRF for multi-view object class recognition and segmentation. In *IEEE Computer Society Conference on Computer Vision and Pattern Recognition (CVPR)*.

Holynski, A., Curless, B., Seitz, S. M., and Szeliski, R. (2021). Animating pictures with Eulerian motion fields. In *IEEE/CVF Conference on Computer Vision and Pattern Recognition (CVPR)*.

Holynski, A., Geraghty, D., Frahm, J.-M., Sweeney, C., and Szeliski, R. (2020). Reducing drift in structure from motion using extended features. In *International Conference on 3D Vision (3DV)*, pp. 51–60.

Honauer, K., Johannsen, O., Kondermann, D., and Goldluecke, B. (2016). A dataset and evaluation methodology for depth estimation on 4D light fields. In *Asian Conference on Computer Vision*, pp. 19–34.

Hoover, A., Jean-Baptiste, G., Jiang, X., Flynn, P. J., Bunke, H., Goldgof, D. B., Bowyer, K., Eggert, D. W., Fitzgibbon, A., and Fisher, R. B. (1996). An experimental comparison of range image segmentation algorithms. *IEEE Transactions on Pattern Analysis and Machine Intelligence*, 18(7):673–689.

Hoppe, H. (1996). Progressive meshes. In *ACM SIGGRAPH Conference Proceedings*, pp. 99–108.

Hoppe, H., DeRose, T., Duchamp, T., McDonald, J., and Stuetzle, W. (1992). Surface reconstruction from unorganized points. *Computer Graphics (SIGGRAPH)*, 26(2):71–78.

Horn, B. K. P. (1974). Determining lightness from an image. *Computer Graphics and Image Processing*, 3(1):277–299.

Horn, B. K. P. (1975). Obtaining shape from shading information. In Winston, P. H. (ed.), *The Psychology of Computer Vision*, pp. 115–155, McGraw-Hill, New York.

Horn, B. K. P. (1977). Understanding image intensities. *Artificial Intelligence*, 8(2):201–231.

Horn, B. K. P. (1986). *Robot Vision.* MIT Press, Cambridge, Massachusetts.

Horn, B. K. P. (1987). Closed-form solution of absolute orientation using unit quaternions. *Journal of the Optical Society of America A*, 4(4):629–642.

Horn, B. K. P. (1990). Height and gradient from shading. *International Journal of Computer Vision*, 5(1):37–75.

Horn, B. K. P. and Brooks, M. J. (1986). The variational approach to shape from shading. *Computer Vision, Graphics, and Image Processing*, 33:174–208.

Horn, B. K. P. and Brooks, M. J. (eds). (1989). *Shape from Shading*, MIT Press, Cambridge, Massachusetts.

Horn, B. K. P. and Schunck, B. G. (1981). Determining optical flow. *Artificial Intelligence*, 17:185–203.

Horn, B. K. P. and Weldon Jr., E. J. (1988). Direct methods for recovering motion. *International Journal of Computer Vision*, 2(1):51–76.

Hornung, A., Wurm, K. M., Bennewitz, M., Stachniss, C., and Burgard, W. (2013). OctoMap: an efficient probabilistic 3D mapping framework based on octrees. *Autonomous Robots*, 34(3):189–206.

Horowitz, S. L. and Pavlidis, T. (1976). Picture segmentation by a tree traversal algorithm. *Journal of the ACM*, 23(2):368–388.

Horry, Y., Anjyo, K.-I., and Arai, K. (1997). Tour into the picture: Using a spidery mesh interface to make animation from a single image. In *ACM SIGGRAPH Conference Proceedings*, pp. 225–232.

Hosni, A., Rhemann, C., Bleyer, M., Rother, C., and Gelautz, M. (2013). Fast cost-volume filtering for visual correspondence and beyond. *IEEE Transactions on Pattern Analysis and Machine Intelligence*, 35(2):504–511.

Hough, P. V. C. (1962). Method and means for recognizing complex patterns. *U.S. Patent*, 3,069,654.

Houghton, J. (2013). Finding the no-parallax point. https://www.johnhpanos.com/epcalib.htm.

Houhou, N., Thiran, J.-P., and Bresson, X. (2008). Fast texture segmentation using the shape operator and active contour. In *IEEE Computer Society Conference on Computer Vision and Pattern Recognition (CVPR).*

Howard, A. G., Zhu, M., Chen, B., Kalenichenko, D., Wang, W., Weyand, T., Andreetto, M., and Adam, H. (2017). MobileNets: Efficient convolutional neural networks for mobile vision applications. *arXiv preprint arXiv:1704.04861.*

Howe, N. R., Leventon, M. E., and Freeman, W. T. (2000). Bayesian reconstruction of 3D human motion from single-camera video. In *Advances in Neural Information Processing Systems (NeurIPS).*

Hsieh, Y. C., McKeown, D., and Perlant, F. P. (1992). Performance evaluation of scene registration and stereo matching for cartographic feature extraction. *IEEE Transactions on Pattern Analysis and Machine Intelligence*, 14(2):214–238.

Hu, J., Shen, L., and Sun, G. (2018). Squeeze-and-excitation networks. In *IEEE Conference on Computer Vision and Pattern Recognition (CVPR).*

Hu, W., Tan, T., Wang, L., and Maybank, S. (2004). A survey on visual surveillance of object motion and behaviors. *IEEE Transactions on Systems, Man, and Cybernetics, Part C: Applications and Reviews*, 34(3):334–352.

Hu, X. and Mordohai, P. (2012). A quantitative evaluation of confidence measures for stereo vision. *IEEE Transactions on Pattern Analysis and Machine Intelligence*, 34(11):2121–2133.

Hu, Y., Fua, P., Wang, W., and Salzmann, M. (2020). Single-stage 6D object pose estimation. In *IEEE/CVF Conference on Computer Vision and Pattern Recognition (CVPR).*

Hu, Y., Hugonot, J., Fua, P., and Salzmann, M. (2019). Segmentation-driven 6d object pose estimation. In *IEEE/CVF Conference on Computer Vision and Pattern Recognition (CVPR).*

Hua, G., Brown, M., and Winder, S. (2007). Discriminant embedding for local image descriptors. In *International Conference on Computer Vision (ICCV).*

Huang, G., Liu, Z., van der Maaten, L., and Weinberger, K. Q. (2017). Densely connected convolutional networks. In *IEEE Conference on Computer Vision and Pattern Recognition (CVPR).*

Huang, G. B., Ramesh, M., Berg, T., and Learned-Miller, E. (2007). *Labeled Faces in the Wild: A Database for Studying Face Recognition in Unconstrained Environments.* Technical Report 07-49, University of Massachusetts, Amherst.

Huang, H.-P., Tseng, H.-Y., Lee, H.-Y., and Huang, J.-B. (2020). Semantic view synthesis. In *European Conference on Computer Vision (ECCV)*.

Huang, J., Rathod, V., Sun, C., Zhu, M., Korattikara, A., Fathi, A., Fischer, I., Wojna, Z., Song, Y., Guadarrama, S., and Murphy, K. (2017). Speed/accuracy trade-offs for modern convolutional object detectors. In *IEEE Conference on Computer Vision and Pattern Recognition (CVPR)*.

Huang, J., Dai, A., Guibas, L. J., and Nießner, M. (2017). 3DLite: Towards commodity 3D scanning for content creation. *ACM Transactions on Graphics (Proc. SIGGRAPH Asia)*, 36(6):203.

Huang, J., Zhou, Y., Funkhouser, T., and Guibas, L. J. (2019). FrameNet: Learning local canonical frames of 3D surfaces from a single RGB image. In *IEEE/CVF International Conference on Computer Vision (ICCV)*.

Huang, J., Zhu, Z., Guo, F., and Huang, G. (2020). The devil is in the details: Delving into unbiased data processing for human pose estimation. In *IEEE/CVF Conference on Computer Vision and Pattern Recognition (CVPR)*.

Huang, K., Wang, Y., Zhou, Z., Ding, T., Gao, S., and Ma, Y. (2018). Learning to parse wireframes in images of man-made environments. In *IEEE Conference on Computer Vision and Pattern Recognition (CVPR)*.

Huang, P.-H., Matzen, K., Kopf, J., Ahuja, N., and Huang, J.-B. (2018). DeepMVS: Learning multi-view stereopsis. In *IEEE Conference on Computer Vision and Pattern Recognition (CVPR)*.

Huang, S., Qin, F., Xiong, P., Ding, N., He, Y., and Liu, X. (2020). TP-LSD: Tri-points based line segment detector. In *European Conference on Computer Vision (ECCV)*.

Huang, T. S. (1981). *Image Sequence Analysis*. Springer-Verlag, Berlin, Heidelberg.

Huang, X. and Belongie, S. (2017). Arbitrary style transfer in real-time with adaptive instance normalization. In *IEEE International Conference on Computer Vision (ICCV)*.

Huang, X., Liu, M.-Y., Belongie, S., and Kautz, J. (2018). Multimodal unsupervised image-to-image translation. In *European Conference on Computer Vision (ECCV)*.

Huang, Y., Shen, P., Tai, Y., Li, S., Liu, X., Li, J., Huang, F., and Ji, R. (2020). Improving face recognition from hard samples via distribution distillation loss. In *European Conference on Computer Vision (ECCV)*.

Huang, Z., Zeng, Z., Liu, B., Fu, D., and Fu, J. (2020). Pixel-BERT: Aligning image pixels with text by deep multi-modal transformers. *arXiv preprint arXiv:2004.00849*.

Huber, P. J. (1981). *Robust Statistics*. John Wiley & Sons, New York.

Hudson, D. A. and Manning, C. D. (2019). GQA: A new dataset for real-world visual reasoning and compositional question answering. In *IEEE/CVF Conference on Computer Vision and Pattern Recognition (CVPR)*.

Hudson, D. A. and Zitnick, C. L. (2021). Generative adversarial transformers. *arXiv preprint arXiv:2103.01209*.

Huffman, D. A. (1971). Impossible objects and nonsense sentences. *Machine Intelligence*, 8:295–323.

Hughes, J. F., van Dam, A., McGuire, M., Sklar, D. F., Foley, J. D., Feiner, S. K., and Akeley, K. (2013). *Computer graphics: principles and practice (3rd ed.)*. Addison-Wesley Professional, Boston, MA, USA.

Huguet, F. and Devernay, F. (2007). A variational method for scene flow estimation from stereo sequences. In *International Conference on Computer Vision (ICCV)*.

Hui, T.-W. and Loy, C. C. (2020). LiteFlowNet3: Resolving correspondence ambiguity for more accurate optical flow estimation. In *European Conference on Computer Vision (ECCV)*.

Hui, T.-W., Tang, X., and Loy, C. C. (2021). A lightweight optical flow CNN — revisiting data fidelity and regularization. *IEEE Transactions on Pattern Analysis and Machine Intelligence*, 43(8):2555–2569.

Hur, J. and Roth, S. (2017). MirrorFlow: Exploiting symmetries in joint optical flow and occlusion estimation. In *IEEE International Conference on Computer Vision (ICCV)*.

Hur, J. and Roth, S. (2019). Iterative residual refinement for joint optical flow and occlusion estimation. In *IEEE/CVF Conference on Computer Vision and Pattern Recognition (CVPR)*.

Hur, J. and Roth, S. (2020). Optical flow estimation in the deep learning age. In Noceti, N., Sciutti, A., and Rea, F. (eds), *Modelling Human Motion*, pp. 119–140, Springer.

Huttenlocher, D. P., Klanderman, G., and Rucklidge, W. (1993). Comparing images using the Hausdorff distance. *IEEE Transactions on Pattern Analysis and Machine Intelligence*, 15(9):850–863.

Hutter, F., Kotthoff, L., and Vanschoren, J. (2019). *Automated machine learning: methods, systems, challenges*. Springer Nature.

Huynh, D. Q., Hartley, R., and Heyden, A. (2003). Outlier correcton in image sequences for the affine camera. In *International Conference on Computer Vision (ICCV)*, pp. 585–590.

Iandola, F. N., Han, S., Moskewicz, M. W., Ashraf, K., Dally, W. J., and Keutzer, K. (2016). SqueezeNet: AlexNet-level accuracy with 50x fewer parameters and¡ 0.5 MB model size. *arXiv preprint arXiv:1602.07360*.

Iddan, G. J. and Yahav, G. (2001). 3D imaging in the studio (and elsewhere...). In *Three-Dimensional Image Capture and Applications IV*, pp. 48–55.

Igarashi, T., Nishino, K., and Nayar, S. (2007). The appearance of human skin: A survey. *Foundations and Trends® in Computer Graphics and Computer Vision*, 3(1):1–95.

Ikeuchi, K. (1981). Shape from regular patterns. *Artificial Intelligence*, 22(1):49–75.

Ikeuchi, K. and Horn, B. K. P. (1981). Numerical shape from shading and occluding boundaries. *Artificial Intelligence*, 17:141–184.

Ikeuchi, K. and Miyazaki, D. (eds). (2007). *Digitally Archiving Cultural Objects*, Springer, Boston, MA.

Ikeuchi, K. and Sato, Y. (eds). (2001). *Modeling From Reality*, Kluwer Academic Publishers, Boston.

Ikeuchi, K., Matsushita, Y., Sagawa, R., Kawasaki, H., Mukaigawa, Y., Furukawa, R., and Miyazaki, D. (2020). *Active Lighting and Its Application for Computer Vision*. Springer.

Ilg, E., Saikia, T., Keuper, M., and Brox, T. (2018). Occlusions, motion and depth boundaries with a generic network for disparity, optical flow or scene flow estimation. In *European Conference on Computer Vision (ECCV)*, pp. 626–643.

Ilg, E., Mayer, N., Saikia, T., Keuper, M., Dosovitskiy, A., and Brox, T. (2017). Flownet 2.0: Evolution of optical flow estimation with deep networks. In *IEEE Conference on Computer Vision and Pattern Recognition (CVPR)*.

Ilg, E., Çiçek, O., Galesso, S., Klein, A., Makansi, O., Hutter, F., and Brox, T. (2018). Uncertainty estimates and multi-hypotheses networks for optical flow. In *European Conference on Computer Vision (ECCV)*.

Illingworth, J. and Kittler, J. (1988). A survey of the Hough transform. *Computer Vision, Graphics, and Image Processing*, 44:87–116.

Ilyas, A., Santurkar, S., Tsipras, D., Engstrom, L., Tran, B., and Madry, A. (2019). Adversarial examples are not bugs, they are features. In *NeurIPS*.

Innmann, M., Zollhöfer, M., Nießner, M., Theobalt, C., and Stamminger, M. (2016). VolumeDeform: Real-time volumetric non-rigid reconstruction. In *European Conference on Computer Vision (ECCV)*, pp. 362–379.

Intille, S. S. and Bobick, A. F. (1994). Disparity-space images and large occlusion stereo. In *European Conference on Computer Vision (ECCV)*.

Ioffe, S. and Szegedy, C. (2015). Batch normalization: Accelerating deep network training by reducing internal covariate shift. *arXiv preprint arXiv:1502.03167*.

Irani, M. and Anandan, P. (1998). Video indexing based on mosaic representations. *Proceedings of the IEEE*, 86(5):905–921.

Irani, M. and Anandan, P. (1999). About direct methods. In *International Workshop on Vision Algorithms*, pp. 267–277.

Irani, M. and Peleg, S. (1991). Improving resolution by image registration. *Graphical Models and Image Processing*, 53(3):231–239.

Irani, M., Hsu, S., and Anandan, P. (1995). Video compression using mosaic representations. *Signal Processing: Image Communication*, 7:529–552.

Irani, M., Rousso, B., and Peleg, S. (1994). Computing occluding and transparent motions. *International Journal of Computer Vision*, 12(1):5–16.

Irani, M., Rousso, B., and Peleg, S. (1997). Recovery of ego-motion using image stabilization. *IEEE Transactions on Pattern Analysis and Machine Intelligence*, 19(3):268–272.

Irschara, A., Zach, C., Frahm, J.-M., and Bischof, H. (2009). From structure-from-motion point clouds to fast location recognition. In *IEEE Computer Society Conference on Computer Vision and Pattern Recognition (CVPR)*.

Isaksen, A., McMillan, L., and Gortler, S. J. (2000). Dynamically reparameterized light fields. In *ACM SIGGRAPH Conference Proceedings*, pp. 297–306.

Isard, M. and Blake, A. (1998). CONDENSATION—conditional density propagation for visual tracking. *International Journal of Computer Vision*, 29(1):5–28.

Ishiguro, H., Yamamoto, M., and Tsuji, S. (1992). Omni-directional stereo. *IEEE Transactions on Pattern Analysis and Machine Intelligence*, 14(2):257–262.

Ishikawa, H. (2003). Exact optimization for Markov random fields with convex priors. *IEEE Transactions on Pattern Analysis and Machine Intelligence*, 25(10):1333–1336.

Isidoro, J. and Sclaroff, S. (2003). Stochastic refinement of the visual hull to satisfy photometric and silhouette consistency constraints. In *International Conference on Computer Vision (ICCV)*, pp. 1335–1342.

Isobe, T., Jia, X., Gu, S., Li, S., Wang, S., and Tian, Q. (2020). Video super-resolution with recurrent structure-detail network. In *European Conference on Computer Vision (ECCV)*.

Isola, P., Zhu, J.-Y., Zhou, T., and Efros, A. A. (2017). Image-to-image translation with conditional adversarial networks. In *IEEE Conference on Computer Vision and Pattern Recognition (CVPR)*.

Ivanchenko, V., Shen, H., and Coughlan, J. (2009). Elevation-based stereo implemented in real-time on a GPU. In *IEEE Workshop on Applications of Computer Vision (WACV)*.

Izadi, S., Kim, D., Hilliges, O., Molyneaux, D., Newcombe, R., Kohli, P., Shotton, J., Hodges, S., Freeman, D., and Davison, A. (2011). KinectFusion: real-time 3D reconstruction and interaction using a moving depth camera. In *ACM Symposium on User Interface Software and Technology (UIST)*, pp. 559–568.

Jacob, P., Picard, D., Histace, A., and Klein, E. (2019). Metric learning with HORDE: High-order regularizer for deep embeddings. In *IEEE/CVF International Conference on Computer Vision (ICCV)*.

Jacobs, C. E., Finkelstein, A., and Salesin, D. H. (1995). Fast multiresolution image querying. In *ACM SIGGRAPH Conference Proceedings*, pp. 277–286.

Jacquet, B., Hane, C., Koser, K., and Pollefeys, M. (2013). Real-world normal map capture for nearly flat reflective surfaces. In *International Conference on Computer Vision (ICCV)*.

Jähne, B. (1997). *Digital Image Processing*. Springer-Verlag, Berlin.

Jain, A. K. and Dubes, R. C. (1988). *Algorithms for Clustering Data*. Prentice Hall, Englewood Cliffs, New Jersey.

Jain, A. K., Bolle, R. M., and Pankanti, S. (eds). (1999). *Biometrics: Personal Identification in Networked Society*, Kluwer.

Jain, A. K., Duin, R. P. W., and Mao, J. (2000). Statistical pattern recognition: A review. *IEEE Transactions on Pattern Analysis and Machine Intelligence*, 22(1):4–37.

Jain, A. K., Topchy, A., Law, M. H. C., and Buhmann, J. M. (2004). Landscape of clustering algorithms. In *International Conference on Pattern Recognition (ICPR)*, pp. 260–263.

Jain, D. S., Xiong, B., and Grauman, K. (2017). FusionSeg: Learning to combine motion and appearance for fully automatic segmentation of generic objects in videos. In *IEEE Conference on Computer Vision and Pattern Recognition (CVPR)*.

Janai, J., Güney, F., Behl, A., and Geiger, A. (2020). Computer vision for autonomous vehicles: Problems, datasets and state of the art. *Foundations and Trends® in Computer Graphics and Vision*, 12(1–3):1–308.

Janai, J., Guney, F., Ranjan, A., Black, M., and Geiger, A. (2018). Unsupervised learning of multi-frame optical flow with occlusions. In *European Conference on Computer Vision (ECCV)*.

Jancosek, M. and Pajdla, T. (2011). Robust, accurate and weakly-supported-surfaces preserving multi-view reconstruction. In *IEEE Computer Society Conference on Computer Vision and Pattern Recognition (CVPR)*.

Jayaraman, D., Sha, F., and Grauman, K. (2014). Decorrelating semantic visual attributes by resisting the urge to share. In *IEEE Conference on Computer Vision and Pattern Recognition (CVPR)*.

Jégou, H., Douze, M., and Schmid, C. (2008). Hamming embedding and weak geometric consistency for large scale image search. In *European Conference on Computer Vision (ECCV)*, pp. 304–317.

Jegou, H., Douze, M., and Schmid, C. (2009). On the burstiness of visual elements. In *IEEE Computer Society Conference on Computer Vision and Pattern Recognition (CVPR)*.

Jégou, H., Douze, M., and Schmid, C. (2010). Product quantization for nearest neighbor search. *IEEE Transactions on Pattern Analysis and Machine Intelligence*, 33(1):117–128.

Jégou, H., Harzallah, H., and Schmid, C. (2007). A contextual dissimilarity measure for accurate and efficient image search. In *IEEE Computer Society Conference on Computer Vision and Pattern Recognition (CVPR)*.

Jégou, H., Perronnin, F., Douze, M., Sánchez, J., Pérez, P., and Schmid, C. (2012). Aggregating local image descriptors into compact codes. *IEEE Transactions on Pattern Analysis and Machine Intelligence*, 34(9):1704–1716.

Jenkin, M. R. M., Jepson, A. D., and Tsotsos, J. K. (1991). Techniques for disparity measurement. *CVGIP: Image Understanding*, 53(1):14–30.

Jensen, H. W., Marschner, S. R., Levoy, M., and Hanrahan, P. (2001). A practical model for subsurface light transport. In *ACM SIGGRAPH Conference Proceedings*, pp. 511–518.

Jensen, R., Dahl, A., Vogiatzis, G., Tola, E., and Aanaes, H. (2014). Large scale multi-view stereopsis evaluation. In *IEEE Conference on Computer Vision and Pattern Recognition (CVPR)*.

Jeong, Y., Nistér, D., Steedly, D., Szeliski, R., and Kweon, I.-S. (2010). Pushing the envelope of modern methods for bundle adjustment. In *IEEE Computer Society Conference on Computer Vision and Pattern Recognition (CVPR)*.

Jeong, Y., Nistér, D., Steedly, D., Szeliski, R., and Kweon, I.-S. (2012). Pushing the envelope of modern methods for bundle adjustment. *IEEE Transactions on Pattern Analysis and Machine Intelligence*, 34(8):1605–1617.

Ji, D., Dunn, E., and Frahm, J.-M. (2014). 3D reconstruction of dynamic textures in crowd sourced data. In *European Conference on Computer Vision*, pp. 143–158.

Ji, S., Xu, W., Yang, M., and Yu, K. (2013). 3D convolutional neural networks for human action recognition. *IEEE Transactions on Pattern Analysis and Machine Intelligence*, 35(1):221–231.

Ji, X., Henriques, J. F., and Vedaldi, A. (2019). Invariant information clustering for unsupervised image classification and segmentation. In *IEEE/CVF International Conference on Computer Vision (ICCV)*.

Jia, J. and Tang, C.-K. (2003). Image registration with global and local luminance alignment. In *International Conference on Computer Vision (ICCV)*, pp. 156–163.

Jia, J., Sun, J., Tang, C.-K., and Shum, H.-Y. (2006). Drag-and-drop pasting. *ACM Transactions on Graphics*, 25(3):631–636.

Jia, Y., Shelhamer, E., Donahue, J., Karayev, S., Long, J., Girshick, R., Guadarrama, S., and Darrell, T. (2014). Caffe: Convolutional architecture for fast feature embedding. In *ACM International Conference on Multimedia*, pp. 675–678.

Jiang, C. M., Sud, A., Makadia, A., Huang, J., Nießner, M., and Funkhouser, T. (2020). Local implicit grid representations for 3D scenes. In *IEEE/CVF Conference on Computer Vision and Pattern Recognition (CVPR)*.

Jiang, H., Misra, I., Rohrbach, M., Learned-Miller, E., and Chen, X. (2020). In defense of grid features for visual question answering. In *IEEE/CVF Conference on Computer Vision and Pattern Recognition (CVPR)*.

Jiang, H., Sun, D., Jampani, V., Lv, Z., Learned-Miller, E., and Kautz, J. (2019). SENSE: A shared encoder network for scene-flow estimation. In *IEEE/CVF International Conference on Computer Vision (ICCV)*.

Jiang, H., Sun, D., Jampani, V., Yang, M.-H., Learned-Miller, E., and Kautz, J. (2018). Super SloMo: High quality estimation of multiple intermediate frames for video interpolation. In *IEEE Conference on Computer Vision and Pattern Recognition (CVPR)*.

Jiang, N., Cui, Z., and Tan, P. (2013). A global linear method for camera pose registration. In *International Conference on Computer Vision (ICCV)*.

Jiang, S., Lu, Y., Li, H., and Hartley, R. (2021). Learning optical flow from a few matches. In *IEEE/CVF Conference on Computer Vision and Pattern Recognition (CVPR)*.

Jiang, S., Campbell, D., Lu, Y., Li, H., and Hartley, R. (2021). Learning to estimate hidden motions with global motion aggregation. *arXiv preprint arXiv:2104.02409*.

Jiang, W., Trulls, E., Hosang, J., Tagliasacchi, A., and Yi, K. M. (2021). COTR: Correspondence transformer for matching across images. *arXiv preprint arXiv:2103.14167*.

Jiang, Y., Chang, S., and Wang, Z. (2021). TransGAN: Two transformers can make one strong GAN. *arXiv preprint arXiv:2102.07074*.

Jiang, Z., Wong, T.-T., and Bao, H. (2003). Practical super-resolution from dynamic video sequences. In *IEEE Computer Society Conference on Computer Vision and Pattern Recognition (CVPR)*, pp. 549–554.

Jiao, L., Zhang, F., Liu, F., Yang, S., Li, L., Feng, Z., and Qu, R. (2019). A survey of deep learning-based object detection. *IEEE Access*, 7:128837–128868.

Jin, Q., Facciolo, G., and Morel, J.-M. (2020). A review of an old dilemma: Demosaicking first, or denoising first? In *IEEE/CVF Conference on Computer Vision and Pattern Recognition (CVPR) Workshops*.

Jin, Y., Mishkin, D., Mishchuk, A., Matas, J., Fua, P., Yi, K. M., and Trulls, E. (2021). Image matching across wide baselines: From paper to practice. *International Journal of Computer Vision*, 129(2):517–547.

Jing, L. and Tian, Y. (2020). Self-supervised visual feature learning with deep neural networks: A survey. *IEEE Transactions on Pattern Analysis and Machine Intelligence*, (accepted).

Johnson, A. E. and Hebert, M. (1999). Using spin images for efficient object recognition in cluttered 3D scenes. *IEEE Transactions on Pattern Analysis and Machine Intelligence*, 21(5):433–448.

Johnson, A. E. and Kang, S. B. (1997). Registration and integration of textured 3-D data. In *International Conference on Recent Advances in 3-D Digital Imaging and Modeling*, pp. 234–241.

Johnson, J. (2020). University of Michigan EECS 498-007 / 598-005 course: Deep learning for computer vision. Slides available on the Schedule page of https://web.eecs.umich.edu/~justincj/teaching/eecs498.

Johnson, J., Alahi, A., and Fei-Fei, L. (2016). Perceptual losses for real-time style transfer and super-resolution. In *European Conference on Computer Vision*, pp. 694–711.

Johnson, J., Douze, M., and Jégou, H. (2021). Billion-scale similarity search with GPUs. *IEEE Transactions on Big Data*, 7(3):535–547.

Johnson, J., Gupta, A., and Fei-Fei, L. (2018). Image generation from scene graphs. In *IEEE Conference on Computer Vision and Pattern Recognition (CVPR)*.

Johnson, J., Karpathy, A., and Fei-Fei, L. (2016). DenseCap: Fully convolutional localization networks for dense captioning. In *IEEE Conference on Computer Vision and Pattern Recognition (CVPR)*.

Johnson, S. and Everingham, M. (2011). Learning effective human pose estimation from inaccurate annotation. In *IEEE Computer Society Conference on Computer Vision and Pattern Recognition (CVPR)*.

Jojic, N. and Frey, B. J. (2001). Learning flexible sprites in video layers. In *IEEE Computer Society Conference on Computer Vision and Pattern Recognition (CVPR)*, pp. 199–206.

Jones, D. G. and Malik, J. (1992). A computational framework for determining stereo correspondence from a set of linear spatial filters. In *European Conference on Computer Vision (ECCV)*, pp. 397–410.

Jonschkowski, R., Stone, A., Barron, J. T., Gordon, A., Konolige, K., and Angelova, A. (2020). What matters in unsupervised optical flow. In *European Conference on Computer Vision (ECCV)*.

Joo, H., Simon, T., Li, X., Liu, H., Tan, L., Gui, L., Banerjee, S., Godisart, T. S., Nabbe, B., Matthews, I., Kanade, T., Nobuhara, S., and Sheikh, Y. (2019). Panoptic studio: A massively multiview system for social interaction capture. *IEEE Transactions on Pattern Analysis and Machine Intelligence*, 41(1):190–204.

Joo, H., Neverova, N., and Vedaldi, A. (2020). Exemplar fine-tuning for 3D human pose fitting towards in-the-wild 3D human pose estimation. *arXiv preprint arXiv:2004.03686*.

Joo, H., Simon, T., and Sheikh, Y. (2018). Total capture: A 3D deformation model for tracking faces, hands, and bodies. In *IEEE Conference on Computer Vision and Pattern Recognition (CVPR)*.

Joo, H., Liu, H., Tan, L., Gui, L., Nabbe, B., Matthews, I., Kanade, T., Nobuhara, S., and Sheikh, Y. (2015). Panoptic studio: A massively multiview system for social motion capture. In *IEEE International Conference on Computer Vision (ICCV)*.

Joshi, N., Matusik, W., and Avidan, S. (2006). Natural video matting using camera arrays. *ACM Transactions on Graphics*, 25(3):779–786.

Joshi, N., Szeliski, R., and Kriegman, D. J. (2008). PSF estimation using sharp edge prediction. In *IEEE Computer Society Conference on Computer Vision and Pattern Recognition (CVPR)*.

Joshi, N., Kang, S. B., Zitnick, C. L., and Szeliski, R. (2010). Image deblurring using inertial measurement sensors. *ACM Transactions on Graphics (Proc. SIGGRAPH)*, 29(3):30.

Joshi, N., Zitnick, C. L., Szeliski, R., and Kriegman, D. J. (2009). Image deblurring and denoising using color priors. In *IEEE Computer Society Conference on Computer Vision and Pattern Recognition (CVPR)*.

Joshi, N., Mehta, S., Drucker, S., Stollnitz, E., Hoppe, H., Uyttendaele, M., and Cohen, M. (2012). Cliplets: juxtaposing still and dynamic imagery. In *ACM Symposium on User Interface Software and Technology (UIST)*, pp. 251–260.

Ju, S. X., Black, M. J., and Jepson, A. D. (1996). Skin and bones: Multi-layer, locally affine, optical flow and regularization with transparency. In *IEEE Computer Society Conference on Computer Vision and Pattern Recognition (CVPR)*, pp. 307–314.

Ju, S. X., Black, M. J., and Yacoob, Y. (1996). Cardboard people: a parameterized model of articulated image motion. In *International Conference on Automatic Face and Gesture Recognition*, pp. 38–44.

Jung, S., Hwang, S., Shin, H., and Shim, D. H. (2018). Perception, guidance, and navigation for indoor autonomous drone racing using deep learning. *IEEE Robotics and Automation Letters*, 3(3):2539–2544.

Jurie, F. and Dhome, M. (2002). Hyperplane approximation for template matching. *IEEE Transactions on Pattern Analysis and Machine Intelligence*, 24(7):996–1000.

Kadir, T., Zisserman, A., and Brady, M. (2004). An affine invariant salient region detector. In *European Conference on Computer Vision (ECCV)*, pp. 228–241.

Kaehler, A. and Bradski, G. (2017). *Learning OpenCV 3: Computer Vision in C++ with the OpenCV Library*. O'Reilly Media.

Kaess, M., Ranganathan, A., and Dellaert, F. (2008). iSAM: Incremental smoothing and mapping. *IEEE Transactions on Robotics*, 24(6):1365–1378.

Kaess, M., Johannsson, H., Roberts, R., Ila, V., Leonard, J. J., and Dellaert, F. (2012). iSAM2: Incremental smoothing and mapping using the Bayes tree. *International Journal of Robotics Research*, 31(2):216–235.

Kaftory, R., Schechner, Y., and Zeevi, Y. (2007). Variational distance-dependent image restoration. In *IEEE Computer Society Conference on Computer Vision and Pattern Recognition (CVPR)*.

Kahl, F. and Hartley, R. (2008). Multiple-view geometry under the l_∞-norm. *IEEE Transactions on Pattern Analysis and Machine Intelligence*, 30(11):1603–1617.

Kakadiaris, I. and Metaxas, D. (2000). Model-based estimation of 3D human motion. *IEEE Transactions on Pattern Analysis and Machine Intelligence*, 22(12):1453–1459.

Kalantari, N. K. and Ramamoorthi, R. (2017). Deep high dynamic range imaging of dynamic scenes. *ACM Transactions On Graphics (Proc. SIGGRAPH)*, 36(4):144–1.

Kalantari, N. K., Wang, T.-C., and Ramamoorthi, R. (2016). Learning-based view synthesis for light field cameras. *ACM Transactions on Graphics (TOG)*, 35(6):1–10.

Kambhamettu, C., Goldgof, D., He, M., and Laskov, P. (2003). 3d nonrigid motion analysis under small deformations. *Image and Vision Computing*, 21(3):229–245.

Kambhamettu, C., Goldgof, D. B., Terzopoulos, D., and Huang, T. S. (1994). Nonrigid motion analysis. In *Handbook of Pattern Recognition and Image Processing: Computer Vision, Volume 2*, pp. 405–430, Academic Press.

Kaminsky, R. S., Snavely, N., Seitz, S. M., and Szeliski, R. (2009). Alignment of 3D point clouds to overhead images. In *Second IEEE Workshop on Internet Vision*.

Kamnitsas, K., Ferrante, E., Parisot, S., Ledig, C., Nori, A. V., Criminisi, A., Rueckert, D., and Glocker, B. (2016). DeepMedic for brain tumor segmentation. In *International MICCAI Brainlesion Workshop*, pp. 138–149.

Kamnitsas, K., Ledig, C., Newcombe, V. F. J., Simpson, J. P., Kane, A. D., Menon, D. K., Rueckert, D., and Glocker, B. (2017). Efficient multi-scale 3D CNN with fully connected CRF for accurate brain lesion segmentation. *Medical Image Analysis*, 36:61–78.

Kamvar, S. D., Klein, D., and Manning, C. D. (2002). Interpreting and extending classical agglomerative clustering algorithms using a model-based approach. In *International Conference on Machine Learning*, pp. 283–290.

Kanade, T. (1977). *Computer Recognition of Human Faces*. Birkhauser, Basel.

Kanade, T. (1980). A theory of the origami world. *Artificial Intelligence*, 13:279–311.

Kanade, T. (ed.). (1987). *Three-Dimensional Machine Vision*, Kluwer Academic Publishers, Boston.

Kanade, T. (1994). Development of a video-rate stereo machine. In *Image Understanding Workshop*, pp. 549–557.

Kanade, T. and Okutomi, M. (1994). A stereo matching algorithm with an adaptive window: Theory and experiment. *IEEE Transactions on Pattern Analysis and Machine Intelligence*, 16(9):920–932.

Kanade, T., Rander, P. W., and Narayanan, P. J. (1997). Virtualized reality: constructing virtual worlds from real scenes. *IEEE MultiMedia Magazine*, 4(1):34–47.

Kanade, T., Yoshida, A., Oda, K., Kano, H., and Tanaka, M. (1996). A stereo machine for video-rate dense depth mapping and its new applications. In *IEEE Computer Society Conference on Computer Vision and Pattern Recognition (CVPR)*, pp. 196–202.

Kanatani, K. (1998). Geometric information criterion for model selection. *International Journal of Computer Vision*, 26(3):171–189.

Kanatani, K. and Matsunaga, C. (2000). Closed-form expression for focal lengths from the fundamental matrix. In *Asian Conference on Computer Vision (ACCV)*, pp. 128–133.

Kanatani, K. and Morris, D. D. (2001). Gauges and gauge transformations for uncertainty description of geometric structure with indeterminacy. *IEEE Transactions on Information Theory*, 47(5):2017–2028.

Kang, J., Jo, Y., Oh, S. W., Vajda, P., and Kim, S. J. (2020). Deep space-time video upsampling networks. In *European Conference on Computer Vision (ECCV)*.

Kang, S. B. (1998). *Depth Painting for Image-based Rendering Applications*. Technical Report, Compaq Computer Corporation, Cambridge Research Lab.

Kang, S. B. (1999). A survey of image-based rendering techniques. In *Videometrics VI*, pp. 2–16.

Kang, S. B. (2001). Radial distortion snakes. *IEICE Trans. Inf. & Syst.*, E84-D(12):1603–1611.

Kang, S. B. and Jones, M. (2002). Appearance-based structure from motion using linear classes of 3-D models. *International Journal of Computer Vision*, 49(1):5–22.

Kang, S. B. and Szeliski, R. (1997). 3-D scene data recovery using omnidirectional multibaseline stereo. *International Journal of Computer Vision*, 25(2):167–183.

Kang, S. B. and Szeliski, R. (2004). Extracting view-dependent depth maps from a collection of images. *International Journal of Computer Vision*, 58(2):139–163.

Kang, S. B. and Weiss, R. (1997). Characterization of errors in compositing panoramic images. In *IEEE Computer Society Conference on Computer Vision and Pattern Recognition (CVPR)*, pp. 103–109.

Kang, S. B. and Weiss, R. (1999). Characterization of errors in compositing panoramic images. *Computer Vision and Image Understanding*, 73(2):269–280.

Kang, S. B. and Weiss, R. (2000). Can we calibrate a camera using an image of a flat, textureless Lambertian surface? In *European Conference on Computer Vision (ECCV)*, pp. 640–653.

Kang, S. B., Szeliski, R., and Anandan, P. (2000). The geometry-image representation tradeoff for rendering. In *International Conference on Image Processing (ICIP)*, pp. 13–16.

Kang, S. B., Szeliski, R., and Chai, J. (2001). Handling occlusions in dense multi-view stereo. In *IEEE Computer Society Conference on Computer Vision and Pattern Recognition (CVPR)*, pp. 103–110.

Kang, S. B., Szeliski, R., and Shum, H.-Y. (1997). A parallel feature tracker for extended image sequences. *Computer Vision and Image Understanding*, 67(3):296–310.

Kang, S. B., Szeliski, R., and Uyttendaele, M. (2004). *Seamless Stitching using Multi-Perspective Plane Sweep*. Technical Report MSR-TR-2004-48, Microsoft Research.

Kang, S. B., Li, Y., Tong, X., and Shum, H.-Y. (2006). Image-based rendering. *Foundations and Trends® in Computer Graphics and Computer Vision*, 2(3):173–258.

Kang, S. B., Uyttendaele, M., Winder, S., and Szeliski, R. (2003). High dynamic range video. *ACM Transactions on Graphics (Proc. SIGGRAPH)*, 22(3):319–325.

Kang, S. B., Webb, J., Zitnick, L., and Kanade, T. (1995). A multibaseline stereo system with active illumination and real-time image acquisition. In *International Conference on Computer Vision (ICCV)*, pp. 88–93.

Kannala, J., Rahtu, E., Brandt, S. S., and Heikkila, J. (2008). Object recognition and segmentation by non-rigid quasi-dense matching. In *IEEE Computer Society Conference on Computer Vision and Pattern Recognition (CVPR)*.

Kappeler, A., Yoo, S., Dai, Q., and Katsaggelos, A. K. (2016). Video super-resolution with convolutional neural networks. *IEEE Transactions on Computational Imaging*, 2(2):109–122.

Kappes, J. H., Andres, B., Hamprecht, F. A., Schnörr, C., Nowozin, S., Batra, D., Kim, S., Kausler, B. X., Kröger, T., Lellmann, J., Komodakis, N., Savchynskyy, B., and Rother, C. (2015). A comparative study of modern inference techniques for structured discrete energy minimization problems. *International Journal of Computer Vision*, 115(3):211–252. Code and benchmark at http://hciweb2.iwr.uni-heidelberg.de/opengm/index.php?l0=benchmark.

Karaimer, H. C. and Brown, M. S. (2016). A software platform for manipulating the camera imaging pipeline. In *European Conference on Computer Vision*, pp. 429–444. https://karaimer.github.io/camera-pipeline.

Karpathy, A. and Fei-Fei, L. (2015). Deep visual-semantic alignments for generating image descriptions. In *IEEE Conference on Computer Vision and Pattern Recognition (CVPR)*.

Karpathy, A., Joulin, A., and Fei-Fei, L. F. (2014). Deep fragment embeddings for bidirectional image sentence mapping. In *Advances in Neural Information Processing Systems (NeurIPS)*, pp. 1889–1897.

Karpathy, A., Toderici, G., Shetty, S., Leung, T., Sukthankar, R., and Fei-Fei, L. (2014). Large-scale video classification with convolutional neural networks. In *IEEE Conference on Computer Vision and Pattern Recognition (CVPR)*.

Karras, T., Laine, S., and Aila, T. (2019). A style-based generator architecture for generative adversarial networks. In *IEEE/CVF Conference on Computer Vision and Pattern Recognition (CVPR)*.

Karras, T., Aila, T., Laine, S., and Lehtinen, J. (2018). Progressive growing of gans for improved quality, stability, and variation. In *International Conference on Learning Representations (ICLR)*.

Karras, T., Laine, S., Aittala, M., Hellsten, J., Lehtinen, J., and Aila, T. (2020). Analyzing and improving the image quality of StyleGAN. In *IEEE/CVF Conference on Computer Vision and Pattern Recognition (CVPR)*.

Kass, M. (1988). Linear image features in stereopsis. *International Journal of Computer Vision*, 1(4):357–368.

Kass, M., Witkin, A., and Terzopoulos, D. (1988). Snakes: Active contour models. *International Journal of Computer Vision*, 1(4):321–331.

Kato, H., Billinghurst, M., Poupyrev, I., Imamoto, K., and Tachibana, K. (2000). Virtual object manipulation on a table-top AR environment. In *International Symposium on Augmented Reality (ISAR)*.

Kato, H., Beker, D., Morariu, M., Ando, T., Matsuoka, T., Kehl, W., and Gaidon, A. (2020). Differentiable rendering: A survey. *arXiv preprint arXiv:2006.12057*.

Kaucic, R., Hartley, R. I., and Dano, N. (2001). Plane-based projective reconstruction. In *International Conference on Computer Vision (ICCV)*, pp. 420–427.

Kaufman, L. and Rousseeuw, P. J. (1990). *Finding Groups in Data: An Introduction to Cluster Analysis*. John Wiley & Sons, Hoboken.

Kaufmann, E., Gehrig, M., Foehn, P., Ranftl, R., Dosovitskiy, A., Koltun, V., and Scaramuzza, D. (2019). Beauty and the beast: Optimal methods meet learning for drone racing. In *International Conference on Robotics and Automation (ICRA)*, pp. 690–696.

Kazhdan, M. and Hoppe, H. (2013). Screened Poisson surface reconstruction. *ACM Transactions on Graphics (ToG)*, 32(3):29.

Kazhdan, M., Bolitho, M., and Hoppe, H. (2006). Poisson surface reconstruction. In *Eurographics Symposium on Geometry Processing*, pp. 61–70.

Ke, Y. and Sukthankar, R. (2004). PCA-SIFT: a more distinctive representation for local image descriptors. In *IEEE Computer Society Conference on Computer Vision and Pattern Recognition (CVPR)*, pp. 506–513.

Kehl, R. and Van Gool, L. (2006). Markerless tracking of complex human motions from multiple views. *Computer Vision and Image Understanding*, 104(2–3):190–209.

Kellnhofer, P., Jebe, L. C., Jones, A., Spicer, R., Pulli, K., and Wetzstein, G. (2021). Neural Lumigraph rendering. In *IEEE/CVF Conference on Computer Vision and Pattern Recognition (CVPR)*.

Kemelmacher-Shlizerman, I., Seitz, S. M., Miller, D., and Brossard, E. (2016). The MegaFace benchmark: 1 million faces for recognition at scale. In *IEEE Conference on Computer Vision and Pattern Recognition (CVPR)*.

Kendall, A. and Cipolla, R. (2017). Geometric loss functions for camera pose regression with deep learning. In *IEEE Conference on Computer Vision and Pattern Recognition (CVPR)*.

Kendall, A., Grimes, M., and Cipolla, R. (2015). PoseNet: A convolutional network for real-time 6-dof camera relocalization. In *IEEE International Conference on Computer Vision (ICCV)*.

Kendall, A., Martirosyan, H., Dasgupta, S., and Henry, P. (2017). End-to-end learning of geometry and context for deep stereo regression. In *International Conference on Computer Vision (ICCV)*, pp. 66–75.

Kenney, C., Zuliani, M., and Manjunath, B. (2005). An axiomatic approach to corner detection. In *IEEE Computer Society Conference on Computer Vision and Pattern Recognition (CVPR)*, pp. 191–197.

Keren, D., Peleg, S., and Brada, R. (1988). Image sequence enhancement using sub-pixel displacements. In *IEEE Computer Society Conference on Computer Vision and Pattern Recognition (CVPR)*, pp. 742–746.

Khamis, S., Fanello, S., Rhemann, C., Kowdle, A., Valentin, J., and Izadi, S. (2018). StereoNet: Guided hierarchical refinement for real-time edge-aware depth prediction. In *European Conference on Computer Vision (ECCV)*.

Khan, S., Naseer, M., Hayat, M., Zamir, S. W., Khan, F. S., and Shah, M. (2021). Transformers in vision: A survey. *arXiv preprint arXiv:2101.01169*.

Khan, Z., Balch, T., and Dellaert, F. (2005). MCMC-based particle filtering for tracking a variable number of interacting targets. *IEEE Transactions on Pattern Analysis and Machine Intelligence*, 27(11):1805–1819.

Khoreva, A., Benenson, R., Hosang, J., Hein, M., and Schiele, B. (2017). Simple does it: Weakly supervised instance and semantic segmentation. In *IEEE Conference on Computer Vision and Pattern Recognition (CVPR)*.

Khoreva, A., Benenson, R., Ilg, E., Brox, T., and Schiele, B. (2019). Lucid data dreaming for video object segmentation. *International Journal of Computer Vision*, 127(9):1175–1197.

Kim, C., Zimmer, H., Pritch, Y., Sorkine-Hornung, A., and Gross, M. H. (2013). Scene reconstruction from high spatio-angular resolution light fields. *ACM Transactions on Graphics (Proc. SIGGRAPH)*, 32(4).

Kim, H., Garrido, P., Tewari, A., Xu, W., Thies, J., Nießner, M., Pérez, P., Richardt, C., Zollhöfer, M., and Theobalt, C. (2018). Deep video portraits. *ACM Transactions on Graphics (Proc. SIGGRAPH)*, 37(4):163.

Kim, H. J., Dunn, E., and Frahm, J.-M. (2017). Learned contextual feature reweighting for image geo-localization. In *IEEE Conference on Computer Vision and Pattern Recognition (CVPR)*.

Kim, H.-J., Lee, J. S., and Yang, H.-S. (2007). Human action recognition using a modified convolutional neural network. In *International Symposium on Neural Networks*, pp. 715–723.

Kim, J., Kolmogorov, V., and Zabih, R. (2003). Visual correspondence using energy minimization and mutual information. In *International Conference on Computer Vision (ICCV)*, pp. 1033–1040.

Kim, S. J., Lin, H. T., Lu, Z., Süsstrunk, S., Lin, S., and Brown, M. S. (2012). A new in-camera imaging model for color computer vision and its application. *IEEE Transactions on Pattern Analysis and Machine Intelligence*, 34(12):2289–2302.

Kim, T., Cha, M., Kim, H., Lee, J. K., and Kim, J. (2017). Learning to discover cross-domain relations with generative adversarial networks. In *International Conference on Machine Learning (ICML)*, pp. 1857–1865.

Kimmel, R. (1999). Demosaicing: image reconstruction from color CCD samples. *IEEE Transactions on Image Processing*, 8(9):1221–1228.

Kimura, S., Shinbo, T., Yamaguchi, H., Kawamura, E., and Nakano, K. (1999). A convolver-based real-time stereo machine (SAZAN). In *IEEE Computer Society Conference on Computer Vision and Pattern Recognition (CVPR)*, pp. 457–463.

Kindermann, R. and Snell, J. L. (1980). *Markov Random Fields and Their Applications*. American Mathematical Society.

King, D. (1997). *The Commissar Vanishes*. Henry Holt and Company.

Kingma, D. P. and Ba, J. (2015). Adam: A method for stochastic optimization. In *International Conference on Learning Representations (ICLR)*.

Kingma, D. P. and Welling, M. (2013). Auto-encoding variational Bayes. *arXiv preprint arXiv:1312.6114*.

Kingma, D. P. and Welling, M. (2019). An introduction to variational autoencoders. *Foundations and Trends® in Machine Learning*, 12(4):307–392.

Kirby, M. and Sirovich, L. (1990). Application of the Karhunen–Lòeve procedure for the characterization of human faces. *IEEE Transactions on Pattern Analysis and Machine Intelligence*, 12(1):103–108.

Kirillov, A., Lin, T.-Y., Cui, Y., Ronchi, M. R., Neverova, N., Khalidov, V., Girshick, R., Dollar, P., Gupta, A., and Girshick, R. (2020). COCO + LVIS joint recognition challenge. In *European Conference on Computer Vision (ECCV) Workshops*. https://cocodataset.org/workshop/coco-lvis-eccv-2020.html.

Kirillov, A., Girshick, R., He, K., and Dollár, P. (2019). Panoptic feature pyramid networks. In *IEEE/CVF Conference on Computer Vision and Pattern Recognition (CVPR)*.

Kirillov, A., He, K., Girshick, R., Rother, C., and Dollár, P. (2019). Panoptic segmentation. In *IEEE/CVF Conference on Computer Vision and Pattern Recognition (CVPR)*.

Kirkpatrick, S., Gelatt, C. D. J., and Vecchi, M. P. (1983). Optimization by simulated annealing. *Science*, 220:671–680.

Kirovski, D., Jojic, N., and Jancke, G. (2004). Tamper-resistant biometric IDs. In *ISSE 2004 - Securing Electronic Business Processes: Highlights of the Information Security Solutions Europe 2004 Conference*, pp. 160–175.

Kittler, J. and Föglein, J. (1984). Contextual classification of multispectral pixel data. *Image and Vision Computing*, 2(1):13–29.

Klare, B. F., Klein, B., Taborsky, E., Blanton, A., Cheney, J., Allen, K., Grother, P., Mah, A., and Jain, A. K. (2015). Pushing the frontiers of unconstrained face detection and recognition: IARPA Janus Benchmark A. In *IEEE Conference on Computer Vision and Pattern Recognition (CVPR)*.

Klaus, A., Sormann, M., and Karner, K. (2006). Segment-based stereo matching using belief propagation and a self-adapting dissimilarity measure. In *International Conference on Pattern Recognition (ICPR)*, pp. 15–18.

Klein, G. and Murray, D. (2007). Parallel tracking and mapping for small AR workspaces. In *International Symposium on Mixed and Augmented Reality (ISMAR)*.

Klein, G. and Murray, D. (2008). Improving the agility of keyframe-based SLAM. In *European Conference on Computer Vision (ECCV)*, pp. 802–815.

Klein, G. and Murray, D. (2009). Parallel tracking and mapping on a camera phone. In *International Symposium on Mixed and Augmented Reality (ISMAR)*.

Klein, S., Staring, M., and Pluim, J. P. W. (2007). Evaluation of optimization methods for nonrigid medical image registration using mutual information and B-splines. *IEEE Transactions on Image Processing*, 16(12):2879–2890.

Klingensmith, M., Dryanovski, I., Srinivasa, S. S., and Xiao, J. (2015). Chisel: Real time large scale 3D reconstruction onboard a mobile device using spatially-hashed signed distance fields. In *Robotics: Science and Systems*.

Klingner, B., Martin, D., and Roseborough, J. (2013). Street view motion-from-structure-from-motion. In *International Conference on Computer Vision (ICCV)*.

Klinker, G. J. (1993). *A Physical Approach to Color Image Understanding*. A K Peters, Wellesley, Massachusetts.

Klinker, G. J., Shafer, S. A., and Kanade, T. (1990). A physical approach to color image understanding. *International Journal of Computer Vision*, 4(1):7–38.

Kluger, F., Ackermann, H., Yang, M. Y., and Rosenhahn, B. (2017). Deep learning for vanishing point detection using an inverse gnomonic projection. In *German Conference on Pattern Recognition*, pp. 17–28.

Knapitsch, A., Park, J., Zhou, Q.-Y., and Koltun, V. (2017). Tanks and temples: Benchmarking large-scale scene reconstruction. *ACM Transactions on Graphics (Proc. SIGGRAPH)*, 36(4).

Kneip, L. and Furgale, P. (2014). OpenGV: A unified and generalized approach to real-time calibrated geometric vision. In *IEEE International Conference on Robotics and Automation (ICRA)*, pp. 1–8.

Knöbelreiter, P. and Pock, T. (2019). Learned collaborative stereo refinement. In *German Conference on Pattern Recognition (GCPR)*, pp. 3–17.

Knöbelreiter, P., Reinbacher, C., Shekhovtsov, A., and Pock, T. (2017). End-to-end training of hybrid CNN-CRF models for stereo. In *IEEE Conference on Computer Vision and Pattern Recognition (CVPR)*, pp. 1456–1465.

Knopp, J., Sivic, J., and Pajdla, T. (2010). Avoiding confusing features in place recognition. In *European Conference on Computer Vision (ECCV)*, pp. 748–761.

Kocabas, M., Athanasiou, N., and Black, M. J. (2020). VIBE: Video inference for human body pose and shape estimation. In *IEEE/CVF Conference on Computer Vision and Pattern Recognition (CVPR)*.

Koch, R., Pollefeys, M., and Van Gool, L. J. (2000). Realistic surface reconstruction of 3D scenes from uncalibrated image sequences. *Journal Visualization and Computer Animation*, 11:115–127.

Koenderink, J. J. (1990). *Solid Shape*. MIT Press, Cambridge, Massachusetts.

Koethe, U. (2003). Integrated edge and junction detection with the boundary tensor. In *International Conference on Computer Vision (ICCV)*, pp. 424–431.

Kohler, J., Daneshmand, H., Lucchi, A., Hofmann, T., Zhou, M., and Neymeyr, K. (2019). Exponential convergence rates for batch normalization: The power of length-direction decoupling in non-convex optimization. In *Proceedings of Machine Learning Research*, pp. 806–815.

Kohli, P. (2007). *Minimizing Dynamic and Higher Order Energy Functions using Graph Cuts*. Ph.D. thesis, Oxford Brookes University.

Kohli, P. and Torr, P. H. S. (2007). Dynamic graph cuts for efficient inference in markov random fields. *IEEE Transactions on Pattern Analysis and Machine Intelligence*, 29(12):2079–2088.

Kohli, P. and Torr, P. H. S. (2008). Measuring uncertainty in graph cut solutions. *Computer Vision and Image Understanding*, 112(1):30–38.

Kohli, P., Kumar, M. P., and Torr, P. H. S. (2009). $\mathcal{P}^3$ & beyond: Move making algorithms for solving higher order functions. *IEEE Transactions on Pattern Analysis and Machine Intelligence*, 31(9):1645–1656.

Kohli, P., Ladický, L., and Torr, P. H. S. (2009). Robust higher order potentials for enforcing label consistency. *International Journal of Computer Vision*, 82(3):302–324.

Kokaram, A. (2004). On missing data treatment for degraded video and film archives: a survey and a new Bayesian approach. *IEEE Transactions on Image Processing*, 13(3):397–415.

Kolesnikov, A., Zhai, X., and Beyer, L. (2019). Revisiting self-supervised visual representation learning. In *IEEE/CVF Conference on Computer Vision and Pattern Recognition (CVPR)*.

Kolesnikov, A., Beyer, L., Zhai, X., Puigcerver, J., Yung, J., Gelly, S., and Houlsby, N. (2020). Big transfer (BiT): General visual representation learning. In *European Conference on Computer Vision (ECCV)*.

Kolev, K. and Cremers, D. (2008). Integration of multiview stereo and silhouettes via convex functionals on convex domains. In *European Conference on Computer Vision (ECCV)*, pp. 752–765.

Kolev, K. and Cremers, D. (2009). Continuous ratio optimization via convex relaxation with applications to multiview 3D reconstruction. In *IEEE Computer Society Conference on Computer Vision and Pattern Recognition (CVPR)*.

Kolev, K., Klodt, M., Brox, T., and Cremers, D. (2009). Continuous global optimization in multiview 3D reconstruction. *International Journal of Computer Vision*, 84(1):80–96.

Koller, D. and Friedman, N. (2009). *Probabilistic Graphical Models: Principles and Techniques*. MIT Press, Cambridge, Massachusetts.

Kolmogorov, V. and Boykov, Y. (2005). What metrics can be approximated by geo-cuts, or global optimization of length/area and flux. In *International Conference on Computer Vision (ICCV)*, pp. 564–571.

Kolmogorov, V. and Zabih, R. (2002). Multi-camera scene reconstruction via graph cuts. In *European Conference on Computer Vision (ECCV)*, pp. 82–96.

Kolmogorov, V. and Zabih, R. (2004). What energy functions can be minimized via graph cuts? *IEEE Transactions on Pattern Analysis and Machine Intelligence*, 26(2):147–159.

Kolmogorov, V., Criminisi, A., Blake, A., Cross, G., and Rother, C. (2006). Probabilistic fusion of stereo with color and contrast for bi-layer segmentation. *IEEE Transactions on Pattern Analysis and Machine Intelligence*, 28(9):1480–1492.

Komodakis, N. and Tziritas, G. (2007). Image completion using efficient belief propagation via priority scheduling and dynamic pruning. *IEEE Transactions on Image Processing*, 29(11):2649–2661.

Komodakis, N., Tziritas, G., and Paragios, N. (2008). Performance vs computational efficiency for optimizing single and dynamic MRFs: Setting the state of the art with primal dual strategies. *Computer Vision and Image Understanding*, 112(1):14–29.

Kondermann, D., Nair, R., Honauer, K., Krispin, K., Andrulis, J., Brock, A., Gussefeld, B., Rahimimoghaddam, M., Hofmann, S., Brenner, C., and Jahne, B. (2016). The HCI benchmark suite: Stereo and flow ground truth with uncertainties for urban autonomous driving. In *IEEE Conference on Computer Vision and Pattern Recognition (CVPR) Workshops*.

Konolige, K. (1997). Small vision systems: Hardware and implementation. In *International Symposium on Robotics Research*, pp. 203–212.

Kononenko, D. and Lempitsky, V. (2015). Learning to look up: Realtime monocular gaze correction using machine learning. In *IEEE Conference on Computer Vision and Pattern Recognition (CVPR)*.

Kopf, J. (2016). 360 video stabilization. *Transactions on Graphics (Proc. SIGGRAPH Asia)*, 35(6):195.

Kopf, J., Alsisan, S., Ge, F., Chong, Y., Matzen, K., Quigley, O., Patterson, J., Tirado, J., Wu, S., and Cohen, M. F. (2019). Practical 3D photography. In *CVPR Workshop on Computer Vision for Augmented and Virtual Reality*.

Kopf, J., Matzen, K., Alsisan, S., Quigley, O., Ge, F., Chong, Y., Patterson, J., Frahm, J.-M., Wu, S., Yu, M., Zhang, P., He, Z., Vajda, P., Saraf, A., and Cohen, M. (2020). One shot 3D photography. *ACM Transactions on Graphics (Proc. SIGGRAPH)*, 39(4):76.

Kopf, J., Cohen, M., and Szeliski, R. (2014). First-person hyperlapse videos. *ACM Transactions on Graphics (Proc. SIGGRAPH)*, 33(4):63.

Kopf, J., Rong, X., and Huang, J.-B. (2021). Robust consistent video depth estimation. In *IEEE/CVF Conference on Computer Vision and Pattern Recognition (CVPR)*.

Kopf, J., Chen, B., Szeliski, R., and Cohen, M. F. (2010). Street slide: Browsing street level imagery. *ACM Transactions on Graphics (Proc. SIGGRAPH)*, 29(4):96.

Kopf, J., Cohen, M. F., Lischinski, D., and Uyttendaele, M. (2007). Joint bilateral upsampling. *ACM Transactions on Graphics*, 26(3).

Kopf, J., Uyttendaele, M., Deussen, O., and Cohen, M. F. (2007). Capturing and viewing gigapixel images. *ACM Transactions on Graphics*, 26(3).

Kopf, J., Langguth, F., Scharstein, D., Szeliski, R., and Goesele, M. (2013). Image-based rendering in the gradient domain. *ACM Transactions on Graphics (Proc. SIGGRAPH Asia)*, 32(6):199.

Kopf, J., Lischinski, D., Deussen, O., Cohen-Or, D., and Cohen, M. (2009). Locally adapted projections to reduce panorama distortions. *Computer Graphics Forum (EGSR)*, 28(4).

Korman, S. and Litman, R. (2018). Latent RANSAC. In *IEEE Conference on Computer Vision and Pattern Recognition (CVPR)*.

Köstinger, M., Hirzer, M., Wohlhart, P., Roth, P. M., and Bischof, H. (2012). Large scale metric learning from equivalence constraints. In *IEEE Computer Society Conference on Computer Vision and Pattern Recognition (CVPR)*.

Košecká, J. and Zhang, W. (2005). Extraction, matching and pose recovery based on dominant rectangular structures. *Computer Vision and Image Understanding*, 100(3):174–293.

Kotovenko, D., Sanakoyeu, A., Lang, S., and Ommer, B. (2019). Content and style disentanglement for artistic style transfer. In *IEEE/CVF International Conference on Computer Vision (ICCV)*.

Koutis, I. (2007). *Combinatorial and algebraic tools for optimal multilevel algorithms*. Ph.D. thesis, Carnegie Mellon University. Technical Report CMU-CS-07-131.

Koutis, I. and Miller, G. L. (2008). Graph partitioning into isolated, high conductance clusters: theory, computation and applications to preconditioning. In *Symposium on Parallel Algorithms and Architectures*, pp. 137–145.

Koutis, I., Miller, G. L., and Tolliver, D. (2009). Combinatorial preconditioners and multilevel solvers for problems in computer vision and image processing. In *International Symposium on Visual Computing (ISVC)*.

Kovar, L., Gleicher, M., and Pighin, F. (2002). Motion graphs. *ACM Transactions on Graphics*, 21(3):473–482.

Kovashka, A., Russakovsky, O., Fei-Fei, L., and Grauman, K. (2016). Crowdsourcing in computer vision. *Foundations and Trends® in Computer Graphics and Vision*, 10(3):177–243.

Krähenbühl, P. and Koltun, V. (2011). Efficient inference in fully connected CRFs with Gaussian edge potentials. In *Advances in Neural Information Processing Systems (NeurIPS)*, pp. 109–117.

Krähenbühl, P. and Koltun, V. (2013). Parameter learning and convergent inference for dense random fields. In *International Conference on Machine Learning (ICML)*, pp. 513–521.

Krähenbühl, P., Doersch, C., Donahue, J., and Darrell, T. (2016). Data-dependent initializations of convolutional neural networks. In *International Conference on Learning Representations (ICLR)*.

Kraus, K. (1997). *Photogrammetry*. Dümmler, Bonn.

Krause, J., Jin, H., Yang, J., and Fei-Fei, L. (2015). Fine-grained recognition without part annotations. In *IEEE Conference on Computer Vision and Pattern Recognition (CVPR)*.

Krishna, R., Zhu, Y., Groth, O., Johnson, J., Hata, K., Kravitz, J., Chen, S., Kalantidis, Y., Li, L.-J., Shamma, D. A., Bernstein, M. S., and Fei-Fei, L. (2017). Visual genome: Connecting language and vision using crowdsourced dense image annotations. *International Journal of Computer Vision*, 123(1):32–73.

Krishnan, D. and Fergus, R. (2009). Fast image deconvolution using hyper-Laplacian priors. In *Advances in Neural Information Processing Systems (NeurIPS)*.

Krishnan, D. and Szeliski, R. (2011). Multigrid and multilevel preconditioners for computational photography. *ACM Transactions on Graphics (Proc. SIGGRAPH Asia)*, 30(6):177.

Krishnan, D., Fattal, R., and Szeliski, R. (2013). Efficient preconditioning of Laplacian matrices for computer graphics. *ACM Transactions on Graphics (Proc. SIGGRAPH)*, 32(4):142.

Kristan, M., Leonardis, A., Matas, J., Felsberg, M. *et al.* (2020). The eighth visual object tracking VOT2020 challenge results. In *European Conference on Computer Vision (ECCV) Workshops*.

Kristan, M., Matas, J., Leonardis, A., Vojir, T., Pflugfelder, R., Fernandez, G., Nebehay, G., Porikli, F., and Čehovin, L. (2016). A novel performance evaluation methodology for single-target trackers. *IEEE Transactions on Pattern Analysis and Machine Intelligence*, 38(11):2137–2155.

Krizhevsky, A. (2009). *Learning multiple layers of features from tiny images*. Master's thesis, University of Toronto.

Krizhevsky, A., Sutskever, I., and Hinton, G. E. (2012). Imagenet classification with deep convolutional neural networks. In *Advances in Neural Information Processing Systems (NeurIPS)*, pp. 1097–1105.

Kroeger, T., Timofte, R., Dai, D., and Van Gool, L. (2016). Fast optical flow using dense inverse search. In *European Conference on Computer Vision*, pp. 471–488.

Kruskal, J. B. (1964a). Multidimensional scaling by optimizing goodness of fit to a nonmetric hypothesis. *Psychometrika*, 29(1):1–27.

Kruskal, J. B. (1964b). Nonmetric multidimensional scaling: a numerical method. *Psychometrika*, 29(2):115–129.

Kubota, T. (2019). Stanford's robotics legacy. *Stanford — News*. https://news.stanford.edu/2019/01/16/stanfords-robotics-legacy.

Kuehne, H., Jhuang, H., Garrote, E., Poggio, T., and Serre, T. (2011). HMDB: A large video database for human motion recognition. In *International Conference on Computer Vision (ICCV)*.

Kuglin, C. D. and Hines, D. C. (1975). The phase correlation image alignment method. In *IEEE Conference on Cybernetics and Society*, pp. 163–165.

Kuhn, A., Hirschmüller, H., Scharstein, D., and Mayer, H. (2017). A TV prior for high-quality scalable multi-view stereo reconstruction. *International Journal of Computer Vision*, 124(1):2–17.

Kuhn, A., Sormann, C., Rossi, M., Erdler, O., and Fraundorfer, F. (2020). DeepC-MVS: Deep confidence prediction for multi-view stereo reconstruction. In *International Conference on 3D Vision (3DV)*, pp. 404–413.

Kukelova, Z., Albl, C., Sugimoto, A., and Pajdla, T. (2018). Linear solution to the minimal absolute pose rolling shutter problem. In *Asian Conference on Computer Vision*, pp. 265–280.

Kukelova, Z., Albl, C., Sugimoto, A., Schindler, K., and Pajdla, T. (2020). Minimal rolling shutter absolute pose with unknown focal length and radial distortion. In *European Conference on Computer Vision (ECCV)*.

Kulis, B. (2013). Metric learning: A survey. *Foundations and Trends® in Machine Learning*, 5(4):287–364.

Kulis, B. and Grauman, K. (2009). Kernelized locality-sensitive hashing for scalable image search. In *International Conference on Computer Vision (ICCV)*.

Kulkarni, G., Premraj, V., Ordonez, V., Dhar, S., Li, S., Choi, Y., Berg, A. C., and Berg, T. L. (2013). BabyTalk: Understanding and generating simple image descriptions. *IEEE Transactions on Pattern Analysis and Machine Intelligence*, 35(12):2891–2903.

Kumar, M., Weissenborn, D., and Kalchbrenner, N. (2021). Colorization transformer. *arXiv preprint arXiv:2102.04432*.

Kumar, M. P. (2008). *Combinatorial and Convex Optimization for Probabilistic Models in Computer Vision*. Ph.D. thesis, Oxford Brookes University.

Kumar, M. P. and Torr, P. H. S. (2006). Fast memory-efficient generalized belief propagation. In *European Conference on Computer Vision (ECCV)*, pp. 451–463.

Kumar, M. P., Torr, P. H. S., and Zisserman, A. (2008). Learning layered motion segmentations of video. *International Journal of Computer Vision*, 76(3):301–319.

Kumar, M. P., Veksler, O., and Torr, P. H. S. (2011). Improved moves for truncated convex models. *Journal of Machine Learning Research*, 12:31–67.

Kumar, M. P., Zisserman, A., and Torr, P. H. (2009). Efficient discriminative learning of parts-based models. In *International Conference on Computer Vision (ICCV)*.

Kumar, R., Anandan, P., and Hanna, K. (1994). Direct recovery of shape from multiple views: A parallax based approach. In *International Conference on Pattern Recognition (ICPR)*, pp. 685–688.

Kumar, R., Anandan, P., Irani, M., Bergen, J., and Hanna, K. (1995). Representation of scenes from collections of images. In *IEEE Workshop on Representations of Visual Scenes*, pp. 10–17.

Kumar, S. and Hebert, M. (2003). Discriminative random fields: A discriminative framework for contextual interaction in classification. In *International Conference on Computer Vision (ICCV)*, pp. 1150–1157.

Kumar, S. and Hebert, M. (2006). Discriminative random fields. *International Journal of Computer Vision*, 68(2):179–202.

Kumawat, S., Verma, M., Nakashima, Y., and Raman, S. (2021). Depthwise spatio-temporal STFT convolutional neural networks for human action recognition. *IEEE Transactions on Pattern Analysis and Machine Intelligence*, ():1–1.

Kundur, D. and Hatzinakos, D. (1996). Blind image deconvolution. *IEEE Signal Processing Magazine*, 13(3):43–64.

Kupyn, O., Martyniuk, T., Wu, J., and Wang, Z. (2019). DeblurGAN-v2: Deblurring (orders-of-magnitude) faster and better. In *IEEE/CVF International Conference on Computer Vision (ICCV)*.

Kupyn, O., Budzan, V., Mykhailych, M., Mishkin, D., and Matas, J. (2018). DeblurGAN: Blind motion deblurring using conditional adversarial networks. In *IEEE Conference on Computer Vision and Pattern Recognition (CVPR)*.

Kurakin, A., Goodfellow, I., and Bengio, S. (2016). Adversarial examples in the physical world. *arXiv preprint arXiv:1607.02533*.

Kushal, A., Self, B., Furukawa, Y., Gallup, D., Hernández, C., Curless, B., and Seitz, S. M. (2012). Photo tours. In *International Conference on 3D Imaging, Modeling, Processing, Visualization and Transmission (3DIMPVT)*, pp. 57–64

Kuster, C., Popa, T., Bazin, J.-C., Gotsman, C., and Gross, M. (2012). Gaze correction for home video conferencing. *ACM Transactions on Graphics (Proc. SIGGRAPH Asia)*, 31(6):174.

Kutulakos, K. N. (2000). Approximate N-view stereo. In *European Conference on Computer Vision (ECCV)*, pp. 67–83.

Kutulakos, K. N. and Seitz, S. M. (2000). A theory of shape by space carving. *International Journal of Computer Vision*, 38(3):199–218.

Kuznetsova, A., Rom, H., Alldrin, N., Uijlings, J., Krasin, I., Pont-Tuset, J., Kamali, S., Popov, S., Malloci, M., Kolesnikov, A., Duerig, T., and Ferrari, V. (2020). The open images dataset V4: Unified image classification, object detection, and visual relationship detection at scale. *International Journal of Computer Vision*.

Kwatra, V., Essa, I., Bobick, A., and Kwatra, N. (2005). Texture optimization for example-based synthesis. *ACM Transactions on Graphics (Proc. SIGGRAPH)*, 24(5):795–802.

Kwatra, V., Schödl, A., Essa, I., Turk, G., and Bobick, A. (2003). Graphcut textures: Image and video synthesis using graph cuts. *ACM Transactions on Graphics (Proc. SIGGRAPH)*, 22(3):277–286.

Kybic, J. and Unser, M. (2003). Fast parametric elastic image registration. *IEEE Transactions on Image Processing*, 12(11):1427–1442.

Labatut, P., Pons, J.-P., and Keriven, R. (2007). Efficient multi-view reconstruction of large-scale scenes using interest points, delaunay triangulation and graph cuts. In *International Conference on Computer Vision (ICCV)*.

Labbé, Y., Carpentier, J., Aubry, M., and Sivic, J. (2020). CosyPose: Consistent multi-view multi-object 6d pose estimation. In *European Conference on Computer Vision (ECCV)*.

Lafferty, J., McCallum, A., and Pereira, F. (2001). Conditional random fields: Probabilistic models for segmenting and labeling sequence data. In *International Conference on Machine Learning*.

Lafortune, E. P. F., Foo, S.-C., Torrance, K. E., and Greenberg, D. P. (1997). Non-linear approximation of reflectance functions. In *ACM SIGGRAPH Conference Proceedings*, pp. 117–126.

Laga, H., Jospin, L. V., Boussaid, F., and Bennamoun, M. (2020). A survey on deep learning techniques for stereo-based depth estimation. *IEEE Transactions on Pattern Analysis and Machine Intelligence*.

Lai, S.-H. and Vemuri, B. C. (1997). Physically based adaptive preconditioning for early vision. *IEEE Transactions on Pattern Analysis and Machine Intelligence*, 19(6):594–607.

Lake, B. M., Salakhutdinov, R., and Tenenbaum, J. B. (2015). Human-level concept learning through probabilistic program induction. *Science*, 350(6266):1332–1338.

Lalonde, J.-F., Hoiem, D., Efros, A. A., Rother, C., Winn, J., and Criminisi, A. (2007). Photo clip art. *ACM Transactions on Graphics*, 26(3).

Lampert, C., Nickisch, H., and Harmeling, S. (2009). Learning to detect unseen object classes by between-class attribute transfer. In *IEEE Computer Society Conference on Computer Vision and Pattern Recognition (CVPR)*.

Lampert, C. H. (2008). Kernel methods in computer vision. *Foundations and Trends® in Computer Graphics and Computer Vision*, 4(3):193–285.

Lampert, C. H., Blaschko, M. B., and Hofmann, T. (2008). Beyond sliding windows: Object localization by efficient subwindow search. In *IEEE Computer Society Conference on Computer Vision and Pattern Recognition (CVPR)*.

Lampert, C. H., Nickisch, H., and Harmeling, S. (2014). Attribute-based classification for zero-shot visual object categorization. *IEEE Transactions on Pattern Analysis and Machine Intelligence*, 36(3):453–465.

Langer, M. S. and Zucker, S. W. (1994). Shape from shading on a cloudy day. *Journal Optical Society America, A*, 11(2):467–478.

Lanitis, A., Taylor, C. J., and Cootes, T. F. (1997). Automatic interpretation and coding of face images using flexible models. *IEEE Transactions on Pattern Analysis and Machine Intelligence*, 19(7):742–756.

Laptev, I., Marszalek, M., Schmid, C., and Rozenfeld, B. (2008). Learning realistic human actions from movies. In *IEEE Computer Society Conference on Computer Vision and Pattern Recognition (CVPR)*.

Lapuschkin, S., Wäldchen, S., Binder, A., Montavon, G., Samek, W., and Müller, K.-R. (2019). Unmasking Clever Hans predictors and assessing what machines really learn. *Nature Communications*, 10(1):1–8.

Larson, G. W. (1998). LogLuv encoding for full-gamut, high-dynamic range images. *Journal of Graphics Tools*, 3(1):15–31.

Larson, G. W., Rushmeier, H., and Piatko, C. (1997). A visibility matching tone reproduction operator for high dynamic range scenes. *IEEE Transactions on Visualization and Computer Graphics*, 3(4):291–306.

Laurentini, A. (1994). The visual hull concept for silhouette-based image understanding. *IEEE Transactions on Pattern Analysis and Machine Intelligence*, 16(2):150–162.

Lavallée, S. and Szeliski, R. (1995). Recovering the position and orientation of free-form objects from image contours using 3-D distance maps. *IEEE Transactions on Pattern Analysis and Machine Intelligence*, 17(4):378–390.

Laveau, S. and Faugeras, O. D. (1994). 3-D scene representation as a collection of images. In *International Conference on Pattern Recognition (ICPR)*, pp. 689–691.

Lazebnik, S., Schmid, C., and Ponce, J. (2005). A sparse texture representation using local affine regions. *IEEE Transactions on Pattern Analysis and Machine Intelligence*, 27(8):1265–1278.

Lazebnik, S., Schmid, C., and Ponce, J. (2006). Beyond bags of features: Spatial pyramid matching for recognizing natural scene categories. In *IEEE Computer Society Conference on Computer Vision and Pattern Recognition (CVPR)*, pp. 2169–2176.

Le Gall, D. (1991). MPEG: A video compression standard for multimedia applications. *Communications of the ACM*, 34(4):46–58.

Leal-Taixé, L. and Nießner, M. (2020). TUM IN2364: Advanced deep learning for computer vision (ADL4CV). Slides available at https://dvl.in.tum.de/teaching/adl4cv-ss20.

Leal-Taixé, L. and Nießner, M. (2021). TUM IN2346: Introduction to deep learning (I2DL). Slides and videos available at https://niessner.github.io/I2DL.

Leal-Taixé, L., Milan, A., Reid, I., Roth, S., and Schindler, K. (2015). MOTChallenge 2015: Towards a benchmark for multi-target tracking. *arXiv preprint arXiv:1504.01942*. arXiv: 1504.01942.

Learned-Miller, E., Huang, G. B., RoyChowdhury, A., Li, H., and Hua, G. (2016). Labeled faces in the wild: A survey. In *Advances in Face Detection and Facial Image Analysis,*, pp. 189–248.

Leclerc, Y. G. (1989). Constructing simple stable descriptions for image partitioning. *International Journal of Computer Vision*, 3(1):73–102.

LeCun, Y. and Bengio, Y. (2020). The future is self-supervised. In *International Conference on Learning Representations (ICLR)*. Keynote talk online at https://iclr.cc/virtual_2020/speaker_7.html.

LeCun, Y. and Canziani, A. (2020). NYU DS-GA 1008 course: Deep learning. Video and slides available at https://atcold.github.io/pytorch-Deep-Learning.

LeCun, Y., Bengio, Y., and Hinton, G. (2015). Deep learning. *Nature*, 521(7553):436–444.

LeCun, Y., Cortes, C., and Burges, C. J. C. (1998). The MNIST database. http://yann.lecun.com/exdb/mnist.

LeCun, Y., Bottou, L., Bengio, Y., and Haffner, P. (1998). Gradient-based learning applied to document recognition. *Proceedings of the IEEE*, 86(11):2278–2324.

Lee, A. W. F., Sweldens, W., Schröder, P., Cowsar, L., and Dobkin, D. (1998). MAPS: Multiresolution adaptive parameterization of surfaces. In *ACM SIGGRAPH Conference Proceedings*, pp. 95–104.

Lee, D., Ryu, S., Yeon, S., Lee, Y., Kim, D., Han, C., Cabon, Y., Weinzaepfel, P., Guerin, N., Csurka, G., and Humenberger, M. (2021). Large-scale localization datasets in crowded indoor spaces. In *IEEE/CVF Conference on Computer Vision and Pattern Recognition (CVPR)*.

Lee, H., Kim, T., Chung, T.-y., Pak, D., Ban, Y., and Lee, S. (2020). AdaCoF: Adaptive collaboration of flows for video frame interpolation. In *IEEE/CVF Conference on Computer Vision and Pattern Recognition (CVPR)*.

Lee, J., Chai, J., Reitsma, P. S. A., Hodgins, J. K., and Pollard, N. S. (2002). Interactive control of avatars animated with human motion data. *ACM Transactions on Graphics*, 21(3):491–500.

Lee, J. A. and Verleysen, M. (2007). *Nonlinear dimensionality reduction*. Springer.

Lee, J. H., Ha, H., Dong, Y., Tong, X., and Kim, M. H. (2020). TextureFusion: High-quality texture acquisition for real-time RGB-D scanning. In *IEEE/CVF Conference on Computer Vision and Pattern Recognition (CVPR)*.

Lee, K.-Y. and Sim, J.-Y. (2020). Warping residual based image stitching for large parallax. In *IEEE/CVF Conference on Computer Vision and Pattern Recognition (CVPR)*.

Lee, M.-C., Chen, W.-G., Lin, C.-L. B., Gu, C., Markoc, T., Zabinsky, S. I., and Szeliski, R. (1997). A layered video object coding system using sprite and affine motion model. *IEEE Transactions on Circuits and Systems for Video Technology*, 7(1):130–145.

Lee, M. E. and Redner, R. A. (1990). A note on the use of nonlinear filtering in computer graphics. *IEEE Computer Graphics and Applications*, 10(3):23–29.

Lee, M. W. and Cohen, I. (2006). A model-based approach for estimating human 3D poses in static images. *IEEE Transactions on Pattern Analysis and Machine Intelligence*, 28(6):905–916.

Lee, S., Wolberg, G., and Shin, S. Y. (1997). Data interpolation using multilevel b-splines. *IEEE Transactions on Visualization and Computer Graphics*, 3(3):228–244.

Lee, S., Wolberg, G., Chwa, K.-Y., and Shin, S. Y. (1996). Image metamorphosis with scattered feature constraints. *IEEE Transactions on Visualization and Computer Graphics*, 2(4):337–354.

Lee, Y. D., Terzopoulos, D., and Waters, K. (1995). Realistic facial modeling for animation. In *ACM SIGGRAPH Conference Proceedings*, pp. 55–62.

Lee, Y. J. and Grauman, K. (2010). Object-graphs for context-aware category discovery. In *IEEE Computer Society Conference on Computer Vision and Pattern Recognition (CVPR)*.

Lee, Y. J., Kim, J., and Grauman, K. (2011). Key-segments for video object segmentation. In *International Conference on Computer Vision (ICCV)*.

Lee, Y. J., Zitnick, C. L., and Cohen, M. F. (2011). ShadowDraw: real-time user guidance for freehand drawing. *ACM Transactions on Graphics (Proc. SIGGRAPH)*, 30(4):1–10.

Lei, C. and Yang, Y.-H. (2009). Optical flow estimation on coarse-to-fine region-trees using discrete optimization. In *International Conference on Computer Vision (ICCV)*.

Leibe, B., Leonardis, A., and Schiele, B. (2008). Robust object detection with interleaved categorization and segmentation. *International Journal of Computer Vision*, 77(1–3):259–289.

Leibe, B., Seemann, E., and Schiele, B. (2005). Pedestrian detection in crowded scenes. In *IEEE Computer Society Conference on Computer Vision and Pattern Recognition (CVPR)*, pp. 878–885.

Lempitsky, V. and Boykov, Y. (2007). Global optimization for shape fitting. In *IEEE Computer Society Conference on Computer Vision and Pattern Recognition (CVPR)*.

Lempitsky, V. and Ivanov, D. (2007). Seamless mosaicing of image-based texture maps. In *IEEE Computer Society Conference on Computer Vision and Pattern Recognition (CVPR)*.

Lempitsky, V., Blake, A., and Rother, C. (2008). Image segmentation by branch-and-mincut. In *European Conference on Computer Vision (ECCV)*, pp. 15–29.

Lempitsky, V., Roth, S., and Rother, C. (2008). FlowFusion: Discrete-continuous optimization for optical flow estimation. In *IEEE Computer Society Conference on Computer Vision and Pattern Recognition (CVPR)*.

Lempitsky, V., Rother, C., and Blake, A. (2007). LogCut — efficient graph cut optimization for Markov random fields. In *International Conference on Computer Vision (ICCV)*.

Lenc, K. and Vedaldi, A. (2016). Learning covariant feature detectors. In *European Conference on Computer Vision (ECCV)*, pp. 100–117.

Lengyel, J. and Snyder, J. (1997). Rendering with coherent layers. In *ACM SIGGRAPH Conference Proceedings*, pp. 233–242.

Lensch, H. P. A., Kautz, J., Goesele, M., Heidrich, W., and Seidel, H.-P. (2003). Image-based reconstruction of spatial appearance and geometric detail. *ACM Transactions on Graphics*, 22(2):234–257.

Leonardis, A., Jaklič, A., and Solina, F. (1997). Superquadrics for segmenting and modeling range data. *IEEE Transactions on Pattern Analysis and Machine Intelligence*, 19(11):1289–1295.

Lepetit, V. and Fua, P. (2005). Monocular model-based 3D tracking of rigid objects. *Foundations and Trends® in Computer Graphics and Computer Vision*, 1(1).

Lepetit, V. and Fua, P. (2006). Keypoint recognition using randomized trees. *IEEE Transactions on Pattern Analysis and Machine Intelligence*, 28(9):1465–1479.

Lepetit, V., Pilet, J., and Fua, P. (2004). Point matching as a classification problem for fast and robust object pose estimation. In *IEEE Computer Society Conference on Computer Vision and Pattern Recognition (CVPR)*, pp. 244–250.

Lepetit, V., Pilet, J., and Fua, P. (2006). Keypoint recognition using randomized trees. *IEEE Transactions on Pattern Analysis and Machine Intelligence*, 28(9):1465–1479.

Leung, T. K., Burl, M. C., and Perona, P. (1995). Finding faces in cluttered scenes using random labeled graph matching. In *International Conference on Computer Vision (ICCV)*, pp. 637–644.

Leutenegger, S., Chli, M., and Siegwart, R. (2011). BRISK: Binary robust invariant scalable keypoints. In *International Conference on Computer Vision (ICCV)*.

Levenberg, K. (1944). A method for the solution of certain problems in least squares. *Quarterly of Applied Mathematics*, 2:164–168.

Levin, A. (2006). Blind motion deblurring using image statistics. In *Advances in Neural Information Processing Systems (NeurIPS)*.

Levin, A. and Szeliski, R. (2004). Visual odometry and map correlation. In *IEEE Computer Society Conference on Computer Vision and Pattern Recognition (CVPR)*, pp. 611–618.

Levin, A. and Szeliski, R. (2006). *Motion Uncertainty and Field of View*. Technical Report MSR-TR-2006-37, Microsoft Research.

Levin, A. and Weiss, Y. (2007). User assisted separation of reflections from a single image using a sparsity prior. *IEEE Transactions on Pattern Analysis and Machine Intelligence*, 29(9):1647–1654.

Levin, A., Acha, A. R., and Lischinski, D. (2008). Spectral matting. *IEEE Transactions on Pattern Analysis and Machine Intelligence*, 30(10):1699–1712.

Levin, A., Lischinski, D., and Weiss, Y. (2004). Colorization using optimization. *ACM Transactions on Graphics*, 23(3):689–694.

Levin, A., Lischinski, D., and Weiss, Y. (2008). A closed form solution to natural image matting. *IEEE Transactions on Pattern Analysis and Machine Intelligence*, 30(2):228–242.

Levin, A., Zomet, A., and Weiss, Y. (2004). Separating reflections from a single image using local features. In *IEEE Computer Society Conference on Computer Vision and Pattern Recognition (CVPR)*, pp. 306–313.

Levin, A., Fergus, R., Durand, F., and Freeman, W. T. (2007). Image and depth from a conventional camera with a coded aperture. *ACM Transactions on Graphics*, 26(3).

Levin, A., Weiss, Y., Durand, F., and Freeman, B. (2009). Understanding and evaluating blind deconvolution algorithms. In *IEEE Computer Society Conference on Computer Vision and Pattern Recognition (CVPR)*.

Levin, A., Weiss, Y., Durand, F., and Freeman, B. (2011). Understanding blind deconvolution algorithms. *IEEE Transactions on Pattern Analysis and Machine Intelligence*, 33(12):2354–2367.

Levin, A., Zomet, A., Peleg, S., and Weiss, Y. (2004). Seamless image stitching in the gradient domain. In *European Conference on Computer Vision (ECCV)*, pp. 377–389.

Levoy, M. (1988). Display of surfaces from volume data. *IEEE Computer Graphics and Applications*, 8(3):29–37.

Levoy, M. (2006). Light fields and computational imaging. *Computer*, 39(8):46–55.

Levoy, M. (2010). Digital photography — Stanford CS 178. https://graphics.stanford.edu/courses/cs178-10.

Levoy, M., Pulli, K., Curless, B., Rusinkiewicz, S., Koller, D., Pereira, L., Ginzton, M., Anderson, S., Davis, J., Ginsberg, J., Shade, J., and Fulk, D. (2000). The digital Michelangelo project: 3D scanning of large statues. In *ACM SIGGRAPH Conference Proceedings*, pp. 131–144.

Levoy, M. and Hanrahan, P. (1996). Light field rendering. In *ACM SIGGRAPH Conference Proceedings*, pp. 31–42.

Levoy, M. and Whitted, T. (1985). *The Use of Points as a Display Primitive*. Technical Report 85-022, University of North Carolina at Chapel Hill.

Levoy, M., Ng, R., Adams, A., Footer, M., and Horowitz, M. (2006). Light field microscopy. *ACM Transactions on Graphics*, 25(3):924–934.

Levoy, M., Chen, B., Vaish, V., Horowitz, M., McDowall, I., and Bolas, M. (2004). Synthetic aperture confocal imaging. *ACM Transactions on Graphics*, 23(3):825–834.

Lew, M. S., Sebe, N., Djeraba, C., and Jain, R. (2006). Content-based multimedia information retrieval: State of the art and challenges. *ACM Transactions on Multimedia Computing, Communications and Applications*, 2(1):1–19.

Leyvand, T., Cohen-Or, D., Dror, G., and Lischinski, D. (2008). Data-driven enhancement of facial attractiveness. *ACM Transactions on Graphics*, 27(3):38.

Lhuillier, M. and Quan, L. (2002). Match propagation for image-based modeling and rendering. *IEEE Transactions on Pattern Analysis and Machine Intelligence*, 24(8):1140–1146.

Lhuillier, M. and Quan, L. (2005). A quasi-dense approach to surface reconstruction from uncalibrated images. *IEEE Transactions on Pattern Analysis and Machine Intelligence*, 27(3):418–433.

Li, A., Thotakuri, M., Ross, D. A., Carreira, J., Vostrikov, A., and Zisserman, A. (2020). The AVA-Kinetics localized human actions video dataset. *arXiv preprint arXiv:2005.00214*.

Li, B., Qi, X., Lukasiewicz, T., and Torr, P. H. S. (2019). Controllable text-to-image generation. In *Advances in Neural Information Processing Systems (NeurIPS)*.

Li, F.-F., Johnson, J., and Yeung, S. (2019). Stanford CS231n course: Convolutional neural networks for visual recognition. Slides available on the Detailed Syllabus page of http://cs231n.stanford.edu, with course notes at https://cs231n.github.io.

Li, H. and Hartley, R. (2007). The 3D–3D registration problem revisited. In *International Conference on Computer Vision (ICCV)*.

Li, H., Lin, Z., Shen, X., Brandt, J., and Hua, G. (2015). A convolutional neural network cascade for face detection. In *IEEE Conference on Computer Vision and Pattern Recognition (CVPR)*.

Li, H., Xu, Z., Taylor, G., Studer, C., and Goldstein, T. (2018). Visualizing the loss landscape of neural nets. In *Advances in Neural Information Processing Systems (NeurIPS)*.

Li, H., Yao, J., Bazin, J.-C., Lu, X., Xing, Y., and Liu, K. (2018). A monocular SLAM system leveraging structural regularity in Manhattan world. In *IEEE International Conference on Robotics and Automation (ICRA)*, pp. 2518–2525.

Li, H., Kim, P., Zhao, J., Joo, K., Cai, Z., Liu, Z., and Liu, Y.-H. (2020). Globally optimal and efficient vanishing point estimation in Atlanta world. In *European Conference on Computer Vision (ECCV)*.

Li, J. and Zhang, Y. (2013). Learning SURF cascade for fast and accurate object detection. In *IEEE Computer Society Conference on Computer Vision and Pattern Recognition (CVPR)*.

Li, J., Robles-Kelly, A., You, S., and Matsushita, Y. (2019). Learning to minify photometric stereo. In *IEEE/CVF Conference on Computer Vision and Pattern Recognition (CVPR)*.

Li, K., Wang, S., Zhang, X., Xu, Y., Xu, W., and Tu, Z. (2021). Pose recognition with cascade transformers. *arXiv preprint arXiv:2104.06976*.

Li, L., Yu, X., Zhang, S., Zhao, X., and Zhang, L. (2017). 3d cost aggregation with multiple minimum spanning trees for stereo matching. *Applied Optics*, 56(12):3411–3420.

Li, M. and Mourikis, A. I. (2013). High-precision, consistent EKF-based visual-inertial odometry. *International Journal of Robotics Research*, 32(6):690–711.

Li, S. (1995). *Markov Random Field Modeling in Computer Vision*. Springer-Verlag.

Li, S. and Deng, W. (2020). Deep facial expression recognition: A survey. *IEEE Transactions on Affective Computing*.

Li, S., Yao, Y., Fang, T., and Quan, L. (2018). Reconstructing thin structures of manifold surfaces by integrating spatial curves. In *IEEE Conference on Computer Vision and Pattern Recognition (CVPR)*.

Li, S. Z. and Jain, A. K. (eds). (2005). *Handbook of Face Recognition*. Springer.

Li, W., Tao, X., Guo, T., Qi, L., Lu, J., and Jia, J. (2020). MuCAN: Multi-correspondence aggregation network for video super-resolution. In *European Conference on Computer Vision (ECCV)*.

Li, X., Yin, X., Li, C., Zhang, P., Hu, X., Zhang, L., Wang, L., Hu, H., Dong, L., Wei, F., Choi, Y., and Gao, J. (2020). Oscar: Object-semantics aligned pre-training for vision-language tasks. In *European Conference on Computer Vision (ECCV)*.

Li, X., Gunturk, B., and Zhang, L. (2008). Image demosaicing: A systematic survey. In *Visual Communications and Image Processing*, p. 68221J.

Li, X., Wu, C., Zach, C., Lazebnik, S., and Frahm, J.-M. (2008). Modeling and recognition of landmark image collections using iconic scene graphs. In *European Conference on Computer Vision (ECCV)*, pp. 427–440.

Li, X., Zhang, Y., Liu, C., Shuai, B., Zhu, Y., Brattoli, B., Chen, H., Marsic, I., and Tighe, J. (2021). VidTr: Video transformer without convolutions. *arXiv preprint arXiv:2104.11746*.

Li, Y. and Huttenlocher, D. P. (2008). Learning for optical flow using stochastic optimization. In *European Conference on Computer Vision (ECCV)*, pp. 379–391.

Li, Y., Crandall, D. J., and Huttenlocher, D. P. (2009). Landmark classification in large-scale image collections. In *International Conference on Computer Vision (ICCV)*.

Li, Y., Snavely, N., and Huttenlocher, D. P. (2010). Location recognition using prioritized feature matching. In *European Conference on Computer Vision (ECCV)*, pp. 791–804.

Li, Y., Wang, T., and Shum, H.-Y. (2002). Motion texture: a two-level statistical model for character motion synthesis. *ACM Transactions on Graphics*, 21(3):465–472.

Li, Y., Shum, H.-Y., Tang, C.-K., and Szeliski, R. (2004). Stereo reconstruction from multiperspective panoramas. *IEEE Transactions on Pattern Analysis and Machine Intelligence*, 26(1):44–62.

Li, Y., Snavely, N., Huttenlocher, D., and Fua, P. (2012). Worldwide pose estimation using 3D point clouds. In *European Conference on Computer Vision (ECCV)*, pp. 15–29.

Li, Y., Sun, J., Tang, C.-K., and Shum, H.-Y. (2004). Lazy snapping. *ACM Transactions on Graphics (Proc. SIGGRAPH)*, 23(3):303–308.

Li, Y., Qi, H., Dai, J., Ji, X., and Wei, Y. (2017). Fully convolutional instance-aware semantic segmentation. In *IEEE Conference on Computer Vision and Pattern Recognition (CVPR)*.

Li, Z. and Snavely, N. (2018). MegaDepth: Learning single-view depth prediction from internet photos. In *IEEE Conference on Computer Vision and Pattern Recognition (CVPR)*.

Li, Z., Sunkavalli, K., and Chandraker, M. (2018). Materials for masses: SVBRDF acquisition with a single mobile phone image. In *European Conference on Computer Vision (ECCV)*.

Li, Z., Niklaus, S., Snavely, N., and Wang, O. (2021). Neural scene flow fields for space-time view synthesis of dynamic scenes. In *IEEE/CVF Conference on Computer Vision and Pattern Recognition (CVPR)*.

Li, Z., Xian, W., Davis, A., and Snavely, N. (2020). Crowdsampling the plenoptic function. In *European Conference on Computer Vision (ECCV)*.

Li, Z., Shafiei, M., Ramamoorthi, R., Sunkavalli, K., and Chandraker, M. (2020). Inverse rendering for complex indoor scenes: Shape, spatially-varying lighting and SVBRDF from a single image. In *IEEE/CVF Conference on Computer Vision and Pattern Recognition (CVPR)*.

Li, Z., Xu, Z., Ramamoorthi, R., Sunkavalli, K., and Chandraker, M. (2018). Learning to reconstruct shape and spatially-varying reflectance from a single image. *ACM Transactions on Graphics (Proc. SIGGRAPH Asia)*, 37(6):269.

Li, Z., Dekel, T., Cole, F., Tucker, R., Snavely, N., Liu, C., and Freeman, W. T. (2019). Learning the depths of moving people by watching frozen people. In *IEEE/CVF Conference on Computer Vision and Pattern Recognition (CVPR)*.

Liang, L., Xiao, R., Wen, F., and Sun, J. (2008). Face alignment via component-based discriminative search. In *European Conference on Computer Vision (ECCV)*, pp. 72–85.

Liang, L., Liu, C., Xu, Y.-Q., Guo, B., and Shum, H.-Y. (2001). Real-time texture synthesis by patch-based sampling. *ACM Transactions on Graphics*, 20(3):127–150.

Liang, Z., Feng, Y., Guo, Y., Liu, H., Chen, W., Qiao, L., Zhou, L., and Zhang, J. (2018). Learning for disparity estimation through feature constancy. In *IEEE Conference on Computer Vision and Pattern Recognition (CVPR)*, pp. 2811–2820.

Liao, J., Finch, M., and Hoppe, H. (2015). Fast computation of seamless video loops. *ACM Transactions on Graphics (Proc. SIGGRAPH Asia)*, 34(6):1–10.

Liao, Z., Joshi, N., and Hoppe, H. (2013). Automated video looping with progressive dynamism. *ACM Transactions on Graphics (ToG)*, 32(4):1–10.

Liba, O., Murthy, K., Tsai, Y.-T., Brooks, T., Xue, T., Karnad, N., He, Q., Barron, J. T., Sharlet, D., Geiss, R., Hasinoff, S. W., Pritch, Y., and Levoy, M. (2019). Handheld mobile photography in very low light. *ACM Transactions on Graphics (Proc. SIGGRAPH Asia)*, 38(6):164:1–164:16.

Liebowitz, D. and Zisserman, A. (1998). Metric rectification for perspective images of planes. In *IEEE Computer Society Conference on Computer Vision and Pattern Recognition (CVPR)*, pp. 482–488.

Lim, H., Sinha, S. N., Cohen, M. F., and Uyttendaele, M. (2012). Real-time image-based 6-DOF localization in large-scale environments. In *IEEE Computer Society Conference on Computer Vision and Pattern Recognition (CVPR)*.

Lim, J. J., Zitnick, C. L., and Dollár, P. (2013). Sketch tokens: A learned mid-level representation for contour and object detection. In *IEEE Computer Society Conference on Computer Vision and Pattern Recognition (CVPR)*.

Lim, J. J., Arbeláez, P., Gu, C., and Malik, J. (2009). Context by region ancestry. In *International Conference on Computer Vision (ICCV)*.

Lin, C.-C., Pankanti, S. U., Natesan Ramamurthy, K., and Aravkin, A. Y. (2015). Adaptive as-natural-as-possible image stitching. In *IEEE Conference on Computer Vision and Pattern Recognition (CVPR)*.

Lin, C. Y. (2004). ROUGE: A package for automatic evaluation of summaries. In *Text Summarization Branches Out*, pp. 74–81.

Lin, D., Kapoor, A., Hua, G., and Baker, S. (2010). Joint people, event, and location recognition in personal photo collections using cross-domain context. In *European Conference on Computer Vision (ECCV)*, pp. 243–256.

Lin, K., Jiang, N., Cheong, L.-F., Do, M., and Lu, J. (2016). SEAGULL: Seam-guided local alignment for parallax-tolerant image stitching. In *European Conference on Computer Vision (ECCV)*, pp. 370–385.

Lin, K.-E., Xu, Z., Mildenhall, B., Srinivasan, P. P., Hold-Geoffroy, Y., DiVerdi, S., Sun, Q., Sunkavalli, K., and Ramamoorthi, R. (2020). Deep multi depth panoramas for view synthesis. In *European Conference on Computer Vision (ECCV)*.

Lin, S., Ryabtsev, A., Sengupta, S., Curless, B. L., Seitz, S. M., and Kemelmacher-Shlizerman, I. (2021). Real-time high-resolution background matting. In *IEEE/CVF Conference on Computer Vision and Pattern Recognition (CVPR)*.

Lin, T.-Y., Goyal, P., Girshick, R., He, K., and Dollár, P. (2017). Focal loss for dense object detection. In *IEEE International Conference on Computer Vision (ICCV)*.

Lin, T.-Y., Dollár, P., Girshick, R., He, K., Hariharan, B., and Belongie, S. (2017). Feature pyramid networks for object detection. In *IEEE Conference on Computer Vision and Pattern Recognition (CVPR)*.

Lin, T.-Y., Maire, M., Belongie, S., Hays, J., Perona, P., Ramanan, D., Dollár, P., and Zitnick, C. L. (2014). Microsoft COCO: Common objects in context. In *European Conference on Computer Vision*, pp. 740–755.

Lin, W.-C., Hays, J., Wu, C., Kwatra, V., and Liu, Y. (2006). Quantitative evaluation of near regular texture synthesis algorithms. In *IEEE Computer Society Conference on Computer Vision and Pattern Recognition (CVPR)*, pp. 427–434.

Lin, Y., Pintea, S. L., and van Gemert, J. C. (2020). Deep Hough-transform line priors. In *European Conference on Computer Vision (ECCV)*.

Lindeberg, T. (1990). Scale-space for discrete signals. *IEEE Transactions on Pattern Analysis and Machine Intelligence*, 12(3):234–254.

Lindeberg, T. (1993). Detecting salient blob-like image structures and their scales with a scale-space primal sketch: a method for focus-of-attention. *International Journal of Computer Vision*, 11(3):283–318.

Lindeberg, T. (1994). Scale-space theory: A basic tool for analysing structures at different scales. *Journal of Applied Statistics*, 21(2):224–270.

Lindeberg, T. (1998a). Edge detection and ridge detection with automatic scale selection. *International Journal of Computer Vision*, 30(2):116–154.

Lindeberg, T. (1998b). Feature detection with automatic scale selection. *International Journal of Computer Vision*, 30(2):79–116.

Lindeberg, T. and Gårding, J. (1997). Shape-adapted smoothing in estimation of 3-D shape cues from affine deformations of local 2-D brightness structure. *Image and Vision Computing*, 15(6):415–434.

Lippman, A. (1980). Movie maps: An application of the optical videodisc to computer graphics. *Computer Graphics (SIGGRAPH)*, 14(3):32–43.

Lischinski, D., Farbman, Z., Uyttendaele, M., and Szeliski, R. (2006). Interactive local adjustment of tonal values. *ACM Transactions on Graphics (Proc. SIGGRAPH)*, 25(3):646–653.

Littlefield, R. (2006). Theory of the "no-parallax" point in panorama photography. https://www.janrik.net/PanoPostings/NoParallaxPoint/TheoryOfTheNoParallaxPoint.pdf.

Litvinov, A. and Schechner, Y. Y. (2005). Radiometric framework for image mosaicking. *Journal of the Optical Society of America A*, 22(5):839–848.

Litwiller, D. (2005). CMOS vs. CCD: Maturing technologies, maturing markets. *Photonics Spectra*, (8):54–59.

Litwinowicz, P. (1997). Processing images and video for an impressionist effect. In *ACM SIGGRAPH Conference Proceedings*, pp. 407–414.

Litwinowicz, P. and Williams, L. (1994). Animating images with drawings. In *ACM SIGGRAPH Conference Proceedings*, pp. 409–412.

Liu, C., Zoph, B., Neumann, M., Shlens, J., Hua, W., Li, L.-J., Fei-Fei, L., Yuille, A., Huang, J., and Murphy, K. (2018). Progressive neural architecture search. In *European Conference on Computer Vision (ECCV)*.

Liu, C. and Freeman, W. T. (2010). A high-quality video denoising algorithm based on reliable motion estimation. In *European Conference on Computer Vision (ECCV)*, pp. 706–719.

Liu, C. and Sun, D. (2013). On Bayesian adaptive video super resolution. *IEEE Transactions on Pattern Analysis and Machine Intelligence*, 36(2):346–360.

Liu, C., Yuen, J., and Torralba, A. (2009). Nonparametric scene parsing: Label transfer via dense scene alignment. In *IEEE Computer Society Conference on Computer Vision and Pattern Recognition (CVPR)*.

Liu, C., Szeliski, R., Kang, S. B., Zitnick, C. L., and Freeman, W. T. (2008). Automatic estimation and removal of noise from a single image. *IEEE Transactions on Pattern Analysis and Machine Intelligence*, 30(2):299–314.

Liu, C., Dollár, P., He, K., Girshick, R., Yuille, A., and Xie, S. (2020). Are labels necessary for neural architecture search? In *European Conference on Computer Vision (ECCV)*.

Liu, F., Gleicher, M., Jin, H., and Agarwala, A. (2009). Content-preserving warps for 3D video stabilization. *ACM Transactions on Graphics*, 28(3):44.

Liu, F., Gleicher, M., Wang, J., Jin, H., and Agarwala, A. (2011). Subspace video stabilization. *ACM Transactions on Graphics*, 30(1):Article 4.

Liu, G., Reda, F. A., Shih, K. J., Wang, T.-C., Tao, A., and Catanzaro, B. (2018). Image inpainting for irregular holes using partial convolutions. In *European Conference on Computer Vision (ECCV)*.

Liu, H., Simonyan, K., and Yang, Y. (2019). DARTS: Differentiable architecture search. In *International Conference on Learning Representations (ICLR)*.

Liu, H., Dai, Z., So, D. R., and Le, Q. V. (2021). Pay attention to MLPs. *arXiv preprint arXiv:2105.08050*.

Liu, L., Gu, J., Lin, K. Z., Chua, T.-S., and Theobalt, C. (2020). Neural sparse voxel fields. In *Advances in Neural Information Processing Systems (NeurIPS)*.

Liu, L., Chen, N., Ceylan, D., Theobalt, C., Wang, W., and Mitra, N. J. (2018). CurveFusion: Reconstructing thin structures from RGBD sequences. *ACM Transactions on Graphics (Proc. SIGGRAPH)*, 37(6):218.

Liu, M.-Y., Breuel, T., and Kautz, J. (2017). Unsupervised image-to-image translation networks. In *Advances in neural information processing systems (NeurIPS)*, pp. 700–708.

Liu, S., Yuan, L., Tan, P., and Sun, J. (2013). Bundled camera paths for video stabilization. *ACM Transactions on Graphics (Proc. SIGGRAPH)*, 32(4):78.

Liu, S., Qi, L., Qin, H., Shi, J., and Jia, J. (2018). Path aggregation network for instance segmentation. In *IEEE Conference on Computer Vision and Pattern Recognition (CVPR)*.

Liu, W., Anguelov, D., Erhan, D., Szegedy, C., Reed, S., Fu, C.-Y., and Berg, A. C. (2016). SSD: Single shot multibox detector. In *European Conference on Computer Vision (ECCV)*, pp. 21–37.

Liu, Y., Ott, M., Goyal, N., Du, J., Joshi, M., Chen, D., Levy, O., Lewis, M., Zettlemoyer, L., and Stoyanov, V. (2019). RoBERTa: A robustly optimized BERT pretraining approach. *arXiv preprint arXiv:1907.11692*.

Liu, Y., Collins, R. T., and Tsin, Y. (2004). A computational model for periodic pattern perception based on frieze and wallpaper groups. *IEEE Transactions on Pattern Analysis and Machine Intelligence*, 26(3):354–371.

Liu, Y., Lin, W.-C., and Hays, J. (2004). Near-regular texture analysis and manipulation. *ACM Transactions on Graphics*, 23(3):368–376.

Liu, Y.-L., Lai, W.-S., Yang, M.-H., Chuang, Y.-Y., and Huang, J.-B. (2020a). Learning to see through obstructions. In *IEEE/CVF Conference on Computer Vision and Pattern Recognition (CVPR)*.

Liu, Y.-L., Lai, W.-S., Chen, Y.-S., Kao, Y.-L., Yang, M.-H., Chuang, Y.-Y., and Huang, J.-B. (2020b). Single-image HDR reconstruction by learning to reverse the camera pipeline. In *IEEE/CVF Conference on Computer Vision and Pattern Recognition (CVPR)*.

Liu, Z., Yuan, L., Tang, X., Uyttendaele, M., and Sun, J. (2014). Fast burst images denoising. *ACM Transactions on Graphics (Proc. SIGGRAPH Asia)*, 33(6):1–9.

Liu, Z., Lin, Y., Cao, Y., Hu, H., Wei, Y., Zhang, Z., Lin, S., and Guo, B. (2021). Swin transformer: Hierarchical vision transformer using shifted windows. *arXiv preprint arXiv:2103.14030*.

Livingstone, M. (2008). *Vision and Art: The Biology of Seeing*. Abrams, New York.

Lobay, A. and Forsyth, D. A. (2006). Shape from texture without boundaries. *International Journal of Computer Vision*, 67(1):71–91.

Logothetis, F., Mecca, R., and Cipolla, R. (2019). A differential volumetric approach to multi-view photometric stereo. In *IEEE/CVF International Conference on Computer Vision (ICCV)*.

Lombardi, S., Saragih, J., Simon, T., and Sheikh, Y. (2018). Deep appearance models for face rendering. *ACM Transactions on Graphics (Proc. SIGGRAPH)*, 37(4):68.

Lombardi, S., Simon, T., Saragih, J., Schwartz, G., Lehrmann, A., and Sheikh, Y. (2019). Neural volumes: Learning dynamic renderable volumes from images. *ACM Transactions on Graphics (Proc. SIGGRAPH)*, 38(4):65.

Long, J., Shelhamer, E., and Darrell, T. (2015). Fully convolutional networks for semantic segmentation. In *IEEE Conference on Computer Vision and Pattern Recognition (CVPR)*.

Longere, P., Delahunt, P. B., Zhang, X., and Brainard, D. H. (2002). Perceptual assessment of demosaicing algorithm performance. *Proceedings of the IEEE*, 90(1):123–132.

Longuet-Higgins, H. C. (1981). A computer algorithm for reconstructing a scene from two projections. *Nature*, 293:133–135.

Loop, C. and Zhang, Z. (1999). Computing rectifying homographies for stereo vision. In *IEEE Computer Society Conference on Computer Vision and Pattern Recognition (CVPR)*, pp. 125–131.

Loper, M., Mahmood, N., Romero, J., Pons-Moll, G., and Black, M. J. (2015). SMPL: A skinned multi-person linear model. *ACM Transactions on Graphics (Proc. SIGGRAPH Asia)*, 34(6):1–16.

Lorensen, W. E. and Cline, H. E. (1987). Marching cubes: A high resolution 3D surface construction algorithm. *Computer Graphics (SIGGRAPH)*, 21(4):163–169.

Lorusso, A., Eggert, D., and Fisher, R. B. (1995). A comparison of four algorithms for estimating 3-D rigid transformations. In *British Machine Vision Conference (BMVC)*, pp. 237–246.

Loshchilov, I. and Hutter, F. (2019). Decoupled weight decay regularization. In *International Conference on Learning Representations (ICLR)*.

Lourakis, M. I. A. and Argyros, A. A. (2009). SBA: A software package for generic sparse bundle adjustment. *ACM Transactions on Mathematical Software*, 36(1):2.

Lowe, D. G. (1988). Organization of smooth image curves at multiple scales. In *International Conference on Computer Vision (ICCV)*, pp. 558–567.

Lowe, D. G. (1989). Organization of smooth image curves at multiple scales. *International Journal of Computer Vision*, 3(2):119–130.

Lowe, D. G. (1999). Object recognition from local scale-invariant features. In *International Conference on Computer Vision (ICCV)*, pp. 1150–1157.

Lowe, D. G. (2004). Distinctive image features from scale-invariant keypoints. *International Journal of Computer Vision*, 60(2):91–110.

Lowel, S. and Singer, W. (1992). Selection of intrinsic horizontal connections in the visual cortex by correlated neuronal activity. *Science*, 255(5041):209–212.

Lowry, S., Sünderhauf, N., Newman, P., Leonard, J. J., Cox, D., Corke, P., and Milford, M. J. (2015). Visual place recognition: A survey. *IEEE Transactions on Robotics*, 32(1):1–19.

Lu, E., Cole, F., Dekel, T., Zisserman, A., Freeman, W. T., and Rubinstein, M. (2021). Omnimatte: Associating objects and their effects in video. In *IEEE/CVF Conference on Computer Vision and Pattern Recognition (CVPR)*.

Lu, J., Batra, D., Parikh, D., and Lee, S. (2019). ViLBERT: Pretraining task-agnostic visiolinguistic representations for vision-and-language tasks. In *Advances in Neural Information Processing Systems (NeurIPS)*, pp. 13–23.

Lu, J., Xiong, C., Parikh, D., and Socher, R. (2017). Knowing when to look: Adaptive attention via a visual sentinel for image captioning. In *IEEE Conference on Computer Vision and Pattern Recognition (CVPR)*.

Lu, J., Yang, J., Batra, D., and Parikh, D. (2018). Neural baby talk. In *IEEE Conference on Computer Vision and Pattern Recognition (CVPR)*.

Lu, W.-L., Ting, J.-A., Little, J. J., and Murphy, K. P. (2013). Learning to track and identify players from broadcast sports videos. *IEEE Transactions on Pattern Analysis and Machine Intelligence*, 35(7):1704–1716.

Lucas, B. D. and Kanade, T. (1981). An iterative image registration technique with an application in stereo vision. In *International Joint Conference on Artificial Intelligence (IJCAI)*, pp. 674–679.

Luo, K., Guan, T., Ju, L., Huang, H., and Luo, Y. (2019). P-MVSNet: Learning patch-wise matching confidence aggregation for multi-view stereo. In *IEEE/CVF International Conference on Computer Vision (ICCV)*.

Luo, W., Schwing, A., and Urtasun, R. (2016). Efficient deep learning for stereo matching. In *IEEE Conference on Computer Vision and Pattern Recognition (CVPR)*, pp. 5695–5703.

Luo, W., Li, Y., Urtasun, R., and Zemel, R. (2016). Understanding the effective receptive field in deep convolutional neural networks. In *Advances in Neural Information Processing Systems (NeurIPS)*, pp. 4898–4906.

Luo, W., Xing, J., Milan, A., Zhang, X., Liu, W., and Kim, T.-K. (2021). Multiple object tracking: A literature review. *Artificial Intelligence*, 293:103448.

Luo, X., Huang, J.-B., Szeliski, R., Matzen, K., and Kopf, J. (2020). Consistent video depth estimation. *ACM Transactions on Graphics (Proc. SIGGRAPH)*, 39(4):71.

Luo, X., Kong, Y., Lawrence, J., Martin-Brualla, R., and Seitz, S. (2020). KeystoneDepth: History in 3D. In *International Conference on 3D Vision (3DV)*, pp. 463–472.

Luo, Z., Shen, T., Zhou, L., Zhang, J., Yao, Y., Li, S., Fang, T., and Quan, L. (2019). ContextDesc: Local descriptor augmentation with cross-modality context. In *IEEE/CVF Conference on Computer Vision and Pattern Recognition (CVPR)*.

Luo, Z., Shen, T., Zhou, L., Zhu, S., Zhang, R., Yao, Y., Fang, T., and Quan, L. (2018). GeoDesc: Learning local descriptors by integrating geometry constraints. In *European Conference on Computer Vision (ECCV)*.

Luo, Z., Zhou, L., Bai, X., Chen, H., Zhang, J., Yao, Y., Li, S., Fang, T., and Quan, L. (2020). ASLFeat: Learning local features of accurate shape and localization. In *IEEE/CVF Conference on Computer Vision and Pattern Recognition (CVPR)*.

Luong, Q.-T. and Faugeras, O. D. (1996). The fundamental matrix: Theory, algorithms, and stability analysis. *International Journal of Computer Vision*, 17(1):43–75.

Luong, Q.-T. and Viéville, T. (1996). Canonical representations for the geometries of multiple projective views. *Computer Vision and Image Understanding*, 64(2):193–229.

Lyu, S. and Simoncelli, E. (2008). Nonlinear image representation using divisive normalization. In *IEEE Computer Society Conference on Computer Vision and Pattern Recognition (CVPR)*.

Lyu, S. and Simoncelli, E. (2009). Modeling multiscale subbands of photographic images with fields of Gaussian scale mixtures. *IEEE Transactions on Pattern Analysis and Machine Intelligence*, 31(4):693–706.

Ma, L., Lin, Z., Barnes, C., Efros, A. A., and Lu, J. (2020). Unselfie: Translating selfies to neutral-pose portraits in the wild. In *European Conference on Computer Vision (ECCV)*.

Ma, N., Zhang, X., Zheng, H.-T., and Sun, J. (2018). ShuffleNet V2: Practical guidelines for efficient CNN architecture design. In *European Conference on Computer Vision (ECCV)*.

Ma, W.-C., Wang, S., Hu, R., Xiong, Y., and Urtasun, R. (2019). Deep rigid instance scene flow. In *IEEE/CVF Conference on Computer Vision and Pattern Recognition (CVPR)*.

Ma, W.-C., Hawkins, T., Peers, P., Chabert, C.-F., Weiss, M., and Debevec, P. (2007). Rapid acquisition of specular and diffuse normal maps from polarized spherical gradient illumination. In *Eurographics Symposium on Rendering*.

Ma, W.-C., Jones, A., Chiang, J.-Y., Hawkins, T., Frederiksen, S., Peers, P., Vukovic, M., Ouhyoung, M., and Debevec, P. (2008). Facial performance synthesis using deformation-driven polynomial displacement maps. *ACM Transactions on Graphics*, 27(5):121.

Ma, Y., Derksen, H., Hong, W., and Wright, J. (2007). Segmentation of multivariate mixed data via lossy data coding and compression. *IEEE Transactions on Pattern Analysis and Machine Intelligence*, 29(9):1546–1562.

Ma, Y., Soatto, S., Kosecká, J., and Sastry, S. S. (2012). *An Invitation to 3-D Vision: From Images to Geometric Models*. Springer.

MacDonald, L. (ed.). (2006). *Digital Heritage: Applying Digital Imaging to Cultural Heritage*, Butterworth-Heinemann.

MacKay, D. J. C. (2003). *Information Theory, Inference, and Learning Algorithms*. Cambridge University Press, Cambridge, UK.

Madry, A., Makelov, A., Schmidt, L., Tsipras, D., and Vladu, A. (2018). Towards deep learning models resistant to adversarial attacks. In *International Conference on Learning Representations (ICLR)*.

Madsen, K., Nielsen, H. B., and Tingleff, O. (2004). Methods for non-linear least squares problems. Informatics and Mathematical Modelling, Technical University of Denmark (DTU)

Maes, F., Collignon, A., Vandermeulen, D., Marchal, G., and Suetens, P. (1997). Multimodality image registration by maximization of mutual information. *IEEE Transactions on Medical Imaging*, 16(2):187–198.

Maggioni, M., Boracchi, G., Foi, A., and Egiazarian, K. (2012). Video denoising, deblocking, and enhancement through separable 4-d nonlocal spatiotemporal transforms. *IEEE Transactions on Image Processing*, 21(9):3952–3966.

Magnor, M. (2005). *Video-Based Rendering*. A. K. Peters, Wellesley, MA.

Magnor, M. and Girod, B. (2000). Data compression for light-field rendering. *IEEE Transactions on Circuits and Systems for Video Technology*, 10(3):338–343.

Magnor, M., Ramanathan, P., and Girod, B. (2003). Multi-view coding for image-based rendering using 3-D scene geometry. *IEEE Transactions on Circuits and Systems for Video Technology*, 13(11):1092–1106.

Mahajan, D., Huang, F.-C., Matusik, W., Ramamoorthi, R., and Belhumeur, P. (2009). Moving gradients: A path-based method for plausible image interpolation. *ACM Transactions on Graphics*, 28(3):42.

Mahajan, D., Girshick, R., Ramanathan, V., He, K., Paluri, M., Li, Y., Bharambe, A., and van der Maaten, L. (2018). Exploring the limits of weakly supervised pretraining. In *European Conference on Computer Vision (ECCV)*.

Mahmood, N., Ghorbani, N., Troje, N. F., Pons-Moll, G., and Black, M. J. (2019). AMASS: Archive of motion capture as surface shapes. In *IEEE/CVF International Conference on Computer Vision (ICCV)*.

Maimone, M., Cheng, Y., and Matthies, L. (2007). Two years of visual odometry on the Mars exploration rovers. *Journal of Field Robotics*, 24(3).

Maire, M., Arbelaez, P., Fowlkes, C., and Malik, J. (2008). Using contours to detect and localize junctions in natural images. In *IEEE Computer Society Conference on Computer Vision and Pattern Recognition (CVPR)*.

Maitre, M., Shinagawa, Y., and Do, M. N. (2008). Symmetric multi-view stereo reconstruction from planar camera arrays. In *IEEE Computer Society Conference on Computer Vision and Pattern Recognition (CVPR)*.

Maji, S. and Berg, A. C. (2009). Max-margin additive classifiers for detection. In *International Conference on Computer Vision (ICCV)*.

Malik, J. and Rosenholtz, R. (1997). Computing local surface orientation and shape from texture for curved surfaces. *International Journal of Computer Vision*, 23(2):149–168.

Malik, J., Belongie, S., Leung, T., and Shi, J. (2001). Contour and texture analysis for image segmentation. *International Journal of Computer Vision*, 43(1):7–27.

Malisiewicz, T. and Efros, A. A. (2008). Recognition by association via learning per-exemplar distances. In *IEEE Computer Society Conference on Computer Vision and Pattern Recognition (CVPR)*.

Malladi, R., Sethian, J. A., and Vemuri, B. C. (1995). Shape modeling with front propagation. *IEEE Transactions on Pattern Analysis and Machine Intelligence*, 17(2):158–176.

Mallat, S. G. (1989). A theory for multiresolution signal decomposition: the wavelet representation. *IEEE Transactions on Pattern Analysis and Machine Intelligence*, PAMI-11(7):674–693.

Malvar, H. S. (1990). Lapped transforms for efficient transform/subband coding. *IEEE Transactions on Acoustics, Speech, and Signal Processing*, 38(6):969–978.

Malvar, H. S. (1998). Biorthogonal and nonuniform lapped transforms for transform coding with reduced blocking and ringing artifacts. *IEEE Transactions on Signal Processing*, 46(4):1043–1053.

Malvar, H. S. (2000). Fast progressive image coding without wavelets. In *IEEE Data Compressions Conference*, pp. 243–252.

Malvar, H. S., He, L.-W., and Cutler, R. (2004). High-quality linear interpolation for demosaicing of Bayer-patterned color images. In *IEEE International Conference on Acoustics, Speech, and Signal Processing (ICASSP)*, pp. 485–488.

Mancini, T. A. and Wolff, L. B. (1992). 3D shape and light source location from depth and reflectance. In *IEEE Computer Society Conference on Computer Vision and Pattern Recognition (CVPR)*, pp. 707–709.

Manjunathi, B. S. and Ma, W. Y. (1996). Texture features for browsing and retrieval of image data. *IEEE Transactions on Pattern Analysis and Machine Intelligence*, 18(8):837–842.

Mann, S. and Picard, R. W. (1994). Virtual bellows: Constructing high-quality images from video. In *First IEEE International Conference on Image Processing (ICIP)*, pp. 363–367.

Mann, S. and Picard, R. W. (1995). On being 'undigital' with digital cameras: Extending dynamic range by combining differently exposed pictures. In *IS&T's Annual Conference*, pp. 422–428, Washington, D. C.

Manning, C. D., Raghavan, P., and Schütze, H. (2008). *Introduction to Information Retrieval*. Cambridge University Press.

Mansimov, E., Parisotto, E., Ba, J. L., and Salakhutdinov, R. (2016). Generating images from captions with attention. In *International Conference on Learning Representations (ICLR)*.

Marcus, G. (2020). The next decade in AI: four steps towards robust artificial intelligence. *arXiv preprint arXiv:2002.06177*.

Marpe, D., Schwarz, H., and Wiegand, T. (2003). Context-based adaptive binary arithmetic coding in the H.264/AVC video compression standard. *IEEE Transactions on Circuits and Systems for Video Technology*, 13(7):620–636.

Marquardt, D. W. (1963). An algorithm for least-squares estimation of nonlinear parameters. *Journal of the Society for Industrial and Applied Mathematics*, 11(2):431–441.

Marr, D. (1982). *Vision: A Computational Investigation into the Human Representation and Processing of Visual Information*. W. H. Freeman, San Francisco.

Marr, D. and Hildreth, E. (1980). Theory of edge detection. *Proceedings of the Royal Society of London*, B 207:187–217.

Marr, D. and Nishihara, H. K. (1978). Representation and recognition of the spatial organization of three-dimensional shapes. *Proc. Roy. Soc. London, B*, 200:269–294.

Marr, D. and Poggio, T. (1976). Cooperative computation of stereo disparity. *Science*, 194:283–287.

Marr, D. C. and Poggio, T. (1979). A computational theory of human stereo vision. *Proceedings of the Royal Society of London*, B 204:301–328.

Marroquin, J., Mitter, S., and Poggio, T. (1987). Probabilistic solution of ill-posed problems in computational vision. *Journal of the American Statistical Association*, 82(397):76–89.

Marroquin, J. L. (1983). *Design of Cooperative Networks*. Working Paper 253, Artificial Intelligence Laboratory, Massachusetts Institute of Technology.

Marschner, S. and Shirley, P. (2015). *Fundamentals of computer graphics*. A K Peters/CRC Press, 4th edition.

Marschner, S. R., Westin, S. H., Lafortune, E. P. F., and Torrance, K. E. (2000). Image-based bidirectional reflectance distribution function measurement. *Applied Optics*, 39(16):2592–2600.

Marszalek, M., Laptev, I., and Schmid, C. (2009). Actions in context. In *IEEE Computer Society Conference on Computer Vision and Pattern Recognition (CVPR)*.

Martin, D., Fowlkes, C., and Malik, J. (2004). Learning to detect natural image boundaries using local brightness, color, and texture cues. *IEEE Transactions on Pattern Analysis and Machine Intelligence*, 26(5):530–549.

Martin, D., Fowlkes, C., Tal, D., and Malik, J. (2001). A database of human segmented natural images and its application to evaluating segmentation algorithms and measuring ecological statistics. In *International Conference on Computer Vision (ICCV)*, pp. 416–423.

Martin, W. N. and Aggarwal, J. K. (1983). Volumetric description of objects from multiple views. *IEEE Transactions on Pattern Analysis and Machine Intelligence*, PAMI-5(2):150–158.

Martin-Brualla, R., Pandey, R., Yang, S., Pidlypenskyi, P., Taylor, J., Valentin, J., Khamis, S., Davidson, P., Tkach, A., Lincoln, P., Kowdle, A., Rhemann, C., Goldman, D. B., Keskin, C., Seitz, S., Izadi, S., and Fanello, S. (2018). LookinGood:

Enhancing performance capture with real-time neural re-rendering. *ACM Transactions on Graphics (Proc. SIGGRAPH Asia)*, 37(6):255.

Martin-Brualla, R., Pandey, R., Bouaziz, S., Brown, M., and Goldman, D. B. (2020). GeLaTO: Generative latent textured objects. In *European Conference on Computer Vision (ECCV)*.

Martin-Brualla, R., Radwan, N., Sajjadi, M. S. M., Barron, J. T., Dosovitskiy, A., and Duckworth, D. (2021). NeRF in the wild: Neural radiance fields for unconstrained photo collections. In *IEEE/CVF Conference on Computer Vision and Pattern Recognition (CVPR)*.

Martinec, D. and Pajdla, T. (2007). Robust rotation and translation estimation in multiview reconstruction. In *IEEE Computer Society Conference on Computer Vision and Pattern Recognition (CVPR)*.

Masi, I., Wu, Y., Hassner, T., and Natarajan, P. (2018). Deep face recognition: A survey. In *Conference on Graphics, Patterns and Images (SIBGRAPI)*, pp. 471–478.

Massey, M. and Bender, W. (1996). Salient stills: Process and practice. *IBM Systems Journal*, 35(3&4):557–573.

Matas, J., Chum, O., Urban, M., and Pajdla, T. (2004). Robust wide baseline stereo from maximally stable extremal regions. *Image and Vision Computing*, 22(10):761–767.

Matei, B. C. and Meer, P. (2006). Estimation of nonlinear errors-in-variables models for computer vision applications. *IEEE Transactions on Pattern Analysis and Machine Intelligence*, 28(10):1537–1552.

Mathias, M., Benenson, R., Pedersoli, M., and Van Gool, L. (2014). Face detection without bells and whistles. In *European Conference on Computer Vision*, pp. 720–735.

Matsushita, Y. and Lin, S. (2007a). A probabilistic intensity similarity measure based on noise distributions. In *IEEE Computer Society Conference on Computer Vision and Pattern Recognition (CVPR)*.

Matsushita, Y. and Lin, S. (2007b). Radiometric calibration from noise distributions. In *IEEE Computer Society Conference on Computer Vision and Pattern Recognition (CVPR)*.

Matsushita, Y., Ofek, E., Ge, W., Tang, X., and Shum, H.-Y. (2006). Full-frame video stabilization with motion inpainting. *IEEE Transactions on Pattern Analysis and Machine Intelligence*, 28(7):1150–1163.

Matthews, I. and Baker, S. (2004). Active appearance models revisited. *International Journal of Computer Vision*, 60(2):135–164.

Matthews, I., Xiao, J., and Baker, S. (2007). 2D vs. 3D deformable face models: Representational power, construction, and real-time fitting. *International Journal of Computer Vision*, 75(1):93–113.

Matthies, L., Kanade, T., and Szeliski, R. (1989). Kalman filter-based algorithms for estimating depth from image sequences. *International Journal of Computer Vision*, 3(3):209–236.

Maturana, D. and Scherer, S. (2015). VoxNet: A 3D convolutional neural network for real-time object recognition. In *EEE/RSJ International Conference on Intelligent Robots and Systems (IROS)*, pp. 922–928.

Matusik, W., Buehler, C., and McMillan, L. (2001). Polyhedral visual hulls for real-time rendering. In *Eurographics Workshop on Rendering Techniques*, pp. 115–126.

Matusik, W., Pfister, H., Brand, M., and McMillan, L. (2003). A data-driven reflectance model. *ACM Transactions on Graphics (Proc. SIGGRAPH)*, 22(3):759–769.

Matusik, W., Buehler, C., Raskar, R., Gortler, S. J., and McMillan, L. (2000). Image-based visual hulls. In *ACM SIGGRAPH Conference Proceedings*, pp. 369–374.

Matzen, K., Cohen, M., Evans, B., Kopf, J., and Szeliski, R. (2017). Low-cost 360 stereo photography and video capture. *ACM Transactions on Graphics (Proc. SIGGRAPH)*, 36(4):148.

Mayer, N., Ilg, E., Fischer, P., Hazirbas, C., Cremers, D., Dosovitskiy, A., and Brox, T. (2018). What makes good synthetic training data for learning disparity and optical flow estimation? *International Journal of Computer Vision*, 126(9):942–960.

Mayer, N., Ilg, E., Häusser, P., Fischer, P., Cremers, D., Dosovitskiy, A., and Brox, T. (2016). A large dataset to train convolutional networks for disparity, optical flow, and scene flow estimation. In *IEEE Conference on Computer Vision and Pattern Recognition (CVPR)*, pp. 4040–4048.

Mayhew, J. E. W. and Frisby, J. P. (1980). The computation of binocular edges. *Perception*, 9:69–87.

Maze, B., Adams, J., Duncan, J. A., Kalka, N., Miller, T., Otto, C., Jain, A. K., Niggel, W. T., Anderson, J., Cheney, J., and Grother, P. (2018). IARPA Janus Benchmark - C: Face dataset and protocol. In *International Conference on Biometrics (ICB)*, pp. 158–165.

Mazumdar, A., Alaghi, A., Barron, J. T., Gallup, D., Ceze, L., Oskin, M., and Seitz, S. M. (2017). A hardware-friendly bilateral solver for real-time virtual reality video. In *Proceedings of High Performance Graphics*, p. 13.

McAllester, D. (2020). TTIC 31230: Fundamentals of deep learning. Slides available at https://mcallester.github.io/ttic-31230/Fall2020.

McCamy, C. S., Marcus, H., and Davidson, J. G. (1976). A color-rendition chart. *Journal of Applied Photogrammetric Engineering*, 2(3):95–99.

McCane, B., Novins, K., Crannitch, D., and Galvin, B. (2001). On benchmarking optical flow. *Computer Vision and Image Understanding*, 84(1):126–143.

McClelland, J. L., Rumelhart, D. E., and PDP Research Group. (1987). *Parallel distributed processing*. Volume 2, MIT Press.

McGuire, M., Matusik, W., Pfister, H., Hughes, J. F., and Durand, F. (2005). Defocus video matting. *ACM Transactions on Graphics (Proc. SIGGRAPH)*, 24(3):567–576.

McInerney, T. and Terzopoulos, D. (1993). A finite element model for 3D shape reconstruction and nonrigid motion tracking. In *International Conference on Computer Vision (ICCV)*, pp. 518–523.

McInerney, T. and Terzopoulos, D. (1996). Deformable models in medical image analysis: A survey. *Medical Image Analysis*, 1(2):91–108.

McInerney, T. and Terzopoulos, D. (1999). Topology adaptive deformable surfaces for medical image volume segmentation. *IEEE Transactions on Medical Imaging*, 18(10):840–850.

McInerney, T. and Terzopoulos, D. (2000). T-snakes: Topology adaptive snakes. *Medical Image Analysis*, 4:73–91.

McInnes, L., Healy, J., and Melville, J. (2018). UMAP: Uniform manifold approximation and projection for dimension reduction. *arXiv preprint arXiv:1802.03426*.

McLauchlan, P. F. (2000). A batch/recursive algorithm for 3D scene reconstruction. In *IEEE Computer Society Conference on Computer Vision and Pattern Recognition (CVPR)*, pp. 738–743.

McLauchlan, P. F. and Jaenicke, A. (2002). Image mosaicing using sequential bundle adjustment. *Image and Vision Computing*, 20(9–10):751–759.

McLean, G. F. and Kotturi, D. (1995). Vanishing point detection by line clustering. *IEEE Transactions on Pattern Analysis and Machine Intelligence*, 17(11):1090–1095.

McMillan, L. and Bishop, G. (1995). Plenoptic modeling: An image-based rendering system. In *ACM SIGGRAPH Conference Proceedings*, pp. 39–46.

McMillan, L. and Gortler, S. (1999). Image-based rendering: A new interface between computer vision and computer graphics. *Computer Graphics*, 33(4):61–64.

McQueen, J., Meila, M., VanderPlas, J., and Zhang, Z. (2016). megaman: Manifold learning with millions of points. *arXiv preprint arXiv:1603.02763*.

Meehan, J. (1990). *Panoramic Photography*. Watson-Guptill.

Mehta, D., Sridhar, S., Sotnychenko, O., Rhodin, H., Shafiei, M., Seidel, H.-P., Xu, W., Casas, D., and Theobalt, C. (2017). VNect: Real-time 3D human pose estimation with a single RGB camera. *ACM Transactions on Graphics (Proc. SIGGRAPH)*, 36(4):1–14.

Mehta, S., Rastegari, M., Shapiro, L., and Hajishirzi, H. (2019). ESPNetv2: A light-weight, power efficient, and general purpose convolutional neural network. In *IEEE/CVF Conference on Computer Vision and Pattern Recognition (CVPR)*.

Mehta, S., Rastegari, M., Caspi, A., Shapiro, L., and Hajishirzi, H. (2018). ESPNet: Efficient spatial pyramid of dilated convolutions for semantic segmentation. In *European Conference on Computer Vision (ECCV)*.

Mei, X. and Ling, H. (2009). Robust visual tracking using l1 minimization. In *International Conference on Computer Vision (ICCV)*.

Meilă, M. and Shi, J. (2000). Learning segmentation by random walks. In *Advances in Neural Information Processing Systems (NeurIPS)*.

Meilă, M. and Shi, J. (2001). A random walks view of spectral segmentation. In *Workshop on Artificial Intelligence and Statistics*, pp. 177–182.

Meinhardt, T. and Leal-Taixé, L. (2020). Make one-shot video object segmentation efficient again. In *Advances in Neural Information Processing Systems (NeurIPS)*, pp. 10607–10619.

Meinhardt, T., Kirillov, A., Leal-Taixé, L., and Feichtenhofer, C. (2021). TrackFormer: Multi-object tracking with transformers. *arXiv preprint arXiv:2101.02702*.

Meister, S., Hur, J., and Roth, S. (2018). UnFlow: Unsupervised learning of optical flow with a bidirectional census loss. In *AAAI Conference on Artificial Intelligence (AAAI)*.

Meka, A., Haene, C., Pandey, R., Zollhöfer, M., Fanello, S., Fyffe, G., Kowdle, A., Yu, X., Busch, J., Dourgarian, J., Denny, P., Bouaziz, S., Lincoln, P., Whalen, M., Harvey, G., Taylor, J., Izadi, S., Tagliasacchi, A., Debevec, P., Theobalt, C., Valentin, J., and Rhemann, C. (2019). Deep reflectance fields: High-quality facial reflectance field inference from color gradient illumination. *ACM Transactions on Graphics (Proc. SIGGRAPH)*, 38(4).

Melekhov, I., Ylioinas, J., Kannala, J., and Rahtu, E. (2017). Relative camera pose estimation using convolutional neural networks. In *International Conference on Advanced Concepts for Intelligent Vision Systems (ACIVS)*, pp. 675–687.

Meltzer, J. and Soatto, S. (2008). Edge descriptors for robust wide-baseline correspondence. In *IEEE Computer Society Conference on Computer Vision and Pattern Recognition (CVPR)*.

Mémin, E. and Pérez, P. (2002). Hierarchical estimation and segmentation of dense motion fields. *International Journal of Computer Vision*, 44(2):129–155.

Menet, S., Saint-Marc, P., and Medioni, G. (1990a). Active contour models: overview, implementation and applications. In *IEEE International Conference on Systems, Man and Cybernetics*, pp. 194–199.

Menet, S., Saint-Marc, P., and Medioni, G. (1990b). B-snakes: implementation and applications to stereo. In *Image Understanding Workshop*, pp. 720–726.

Mentzer, F., Agustsson, E., Tschannen, M., Timofte, R., and Van Gool, L. (2019). Practical full resolution learned lossless image compression. In *IEEE/CVF Conference on Computer Vision and Pattern Recognition (CVPR)*.

Menze, M. and Geiger, A. (2015). Object scene flow for autonomous vehicles. In *IEEE Conference on Computer Vision and Pattern Recognition (CVPR)*.

Mertens, T., Kautz, J., and Reeth, F. V. (2007). Exposure fusion. In *Pacific Graphics*, pp. 382–390.

Mescheder, L., Oechsle, M., Niemeyer, M., Nowozin, S., and Geiger, A. (2019). Occupancy networks: Learning 3D reconstruction in function space. In *IEEE/CVF Conference on Computer Vision and Pattern Recognition (CVPR)*.

Meshry, M., Goldman, D. B., Khamis, S., Hoppe, H., Pandey, R., Snavely, N., and Martin-Brualla, R. (2019). Neural rerendering in the wild. In *IEEE/CVF Conference on Computer Vision and Pattern Recognition (CVPR)*.

Metaxas, D. and Terzopoulos, D. (2002). Dynamic deformation of solid primitives with constraints. *ACM Transactions on Graphics (Proc. SIGGRAPH)*, 21(3):309–312.

Metropolis, N., Rosenbluth, A. W., Rosenbluth, M. N., Teller, A. H., and Teller, E. (1953). Equations of state calculations by fast computing machines. *Journal of Chemical Physics*, 21:1087–1091.

Meyer, C. D. (2000). *Matrix Analysis and Applied Linear Algebra*. Society for Industrial and Applied Mathematics, Philadephia.

Meyer, Y. (1993). *Wavelets: Algorithms and Applications*. Society for Industrial and Applied Mathematics, Philadephia.

Micusik, B. and Wildenauer, H. (2017). Plane refined structure from motion. In *Scandinavian Conference on Image Analysis*, pp. 29–40.

Mičušík, B. and Košecká, J. (2009). Piecewise planar city 3D modeling from street view panoramic sequences. In *IEEE Computer Society Conference on Computer Vision and Pattern Recognition (CVPR)*.

Mičušìk, B., Wildenauer, H., and Košecká, J. (2008). Detection and matching of rectilinear structures. In *IEEE Computer Society Conference on Computer Vision and Pattern Recognition (CVPR)*.

Mihajlovic, M., Weder, S., Pollefeys, M., and Oswald, M. R. (2021). DeepSurfels: Learning online appearance fusion. In *IEEE/CVF Conference on Computer Vision and Pattern Recognition (CVPR)*.

Mikolajczyk, K. and Schmid, C. (2004). Scale & affine invariant interest point detectors. *International Journal of Computer Vision*, 60(1):63–86.

Mikolajczyk, K. and Schmid, C. (2005). A performance evaluation of local descriptors. *IEEE Transactions on Pattern Analysis and Machine Intelligence*, 27(10):1615–1630.

Mikolajczyk, K., Schmid, C., and Zisserman, A. (2004). Human detection based on a probabilistic assembly of robust part detectors. In *European Conference on Computer Vision (ECCV)*, pp. 69–82.

Mikolajczyk, K., Tuytelaars, T., Schmid, C., Zisserman, A., Matas, J., Schaffalitzky, F., Kadir, T., and Van Gool, L. J. (2005). A comparison of affine region detectors. *International Journal of Computer Vision*, 65(1–2):43–72.

Mikolov, T., Sutskever, I., Chen, K., Corrado, G., and Dean, J. (2013). Distributed representations of words and phrases and their compositionality. *arXiv preprint arXiv:1310.4546*.

Mikulik, A., Perdoch, M., Chum, O., and Matas, J. (2013). Learning vocabularies over a fine quantization. *International Journal of Computer Vision*, 103:163–175.

Mildenhall, B., Barron, J. T., Chen, J., Sharlet, D., Ng, R., and Carroll, R. (2018). Burst denoising with kernel prediction networks. In *IEEE Conference on Computer Vision and Pattern Recognition (CVPR)*.

Mildenhall, B., Srinivasan, P. P., Tancik, M., Barron, J. T., Ramamoorthi, R., and Ng, R. (2020). NeRF: Representing scenes as neural radiance fields for view synthesis. In *European Conference on Computer Vision (ECCV)*.

Mildenhall, B., Srinivasan, P. P., Ortiz-Cayon, R., Kalantari, N. K., Ramamoorthi, R., Ng, R., and Kar, A. (2019). Local light field fusion: Practical view synthesis with prescriptive sampling guidelines. *ACM Transactions on Graphics (Proc. SIGGRAPH)*, 38(4):29:1–29:14.

Milgram, D. L. (1975). Computer methods for creating photomosaics. *IEEE Transactions on Computers*, C-24(11):1113–1119.

Milgram, D. L. (1977). Adaptive techniques for photomosaicking. *IEEE Transactions on Computers*, C-26(11):1175–1180.

Mirza, M. and Osindero, S. (2014). Conditional generative adversarial nets. *arXiv preprint arXiv:1411.1784*.

Mishchuk, A., Mishkin, D., Radenović, F., and Matas, J. (2017). Working hard to know your neighbor's margins: Local descriptor learning loss. In *Advances in Neural Information Processing Systems (NeurIPS)*, pp. 4826–4837.

Mishkin, D. (2021). WxBS: Relaunching challenging benchmark for image matching. https://ducha-aiki.github.io/wide-baseline-stereo-blog/2021/07/30/Reviving-WxBS-benchmark.html.

Mishkin, D. and Matas, J. (2016). All you need is a good init. In *International Conference on Learning Representations (ICLR)*.

Mishkin, D., Matas, J., and Perdoch, M. (2015). MODS: Fast and robust method for two-view matching. *Computer Vision and Image Understanding*, 141:81–93.

Mishkin, D., Radenovic, F., and Matas, J. (2018). Repeatability is not enough: Learning affine regions via discriminability. In *European Conference on Computer Vision (ECCV)*.

Misra, I. and Maaten, L. v. d. (2020). Self-supervised learning of pretext-invariant representations. In *IEEE/CVF Conference on Computer Vision and Pattern Recognition (CVPR)*.

Misra, I., Zitnick, C. L., and Hebert, M. (2016). Shuffle and learn: unsupervised learning using temporal order verification. In *European Conference on Computer Vision (ECCV)*, pp. 527–544.

Mitiche, A. and Bouthemy, P. (1996). Computation and analysis of image motion: A synopsis of current problems and methods. *International Journal of Computer Vision*, 19(1):29–55.

Mitsunaga, T. and Nayar, S. K. (1999). Radiometric self calibration. In *IEEE Computer Society Conference on Computer Vision and Pattern Recognition (CVPR)*, pp. 374–380.

Mittal, A. and Davis, L. S. (2003). M_2 tracker: A multi-view approach to segmenting and tracking people in a cluttered scene. *International Journal of Computer Vision*, 51(3):189–203.

Mittal, A., Moorthy, A. K., and Bovik, A. C. (2012). No-reference image quality assessment in the spatial domain. *IEEE Transactions on Image Processing*, 21(12):4695–4708.

Miyato, T., Kataoka, T., Koyama, M., and Yoshida, Y. (2018). Spectral normalization for generative adversarial networks. *arXiv preprint arXiv:1802.05957*.

Moeslund, T. B. and Granum, E. (2001). A survey of computer vision-based human motion capture. *Computer Vision and Image Understanding*, 81(3):231–268.

Moeslund, T. B., Hilton, A., and Krüger, V. (2006). A survey of advances in vision-based human motion capture and analysis. *Computer Vision and Image Understanding*, 104(2–3):90–126.

Moezzi, S., Katkere, A., Kuramura, D., and Jain, R. (1996). Reality modeling and visualization from multiple video sequences. *IEEE Computer Graphics and Applications*, 16(6):58–63.

Mogadala, A., Kalimuthu, M., and Klakow, D. (2021). Trends in integration of vision and language research: A survey of tasks, datasets, and methods. *Journal of Artificial Intelligence Research*, 71:1183–1317.

Moghaddam, B. and Pentland, A. (1997). Probabilistic visual learning for object representation. *IEEE Transactions on Pattern Analysis and Machine Intelligence*, 19(7):696–710.

Moghaddam, B., Jebara, T., and Pentland, A. (2000). Bayesian face recognition. *Pattern Recognition*, 33(11):1771–1782.

Mohan, A., Papageorgiou, C., and Poggio, T. (2001). Example-based object detection in images by components. *IEEE Transactions on Pattern Analysis and Machine Intelligence*, 23(4):349–361.

Möller, K. D. (1988). *Optics*. University Science Books, Mill Valley, CA.

Montemerlo, M., Becker, J., Bhat, S., Dahlkamp, H., Dolgov, D. *et al.* (2008). Junior: The Stanford entry in the Urban Challenge. *Journal of Field Robotics*, 25(9):569–597.

Moon, G., Shiratori, T., and Lee, K. M. (2020). DeepHandMesh: A weakly-supervised deep encoder-decoder framework for high-fidelity hand mesh modeling. In *European Conference on Computer Vision (ECCV)*.

Moon, G., Yu, S.-I., Wen, H., Shiratori, T., and Lee, K. M. (2020). InterHand2.6M: A dataset and baseline for 3D interacting hand pose estimation from a single RGB image. In *European Conference on Computer Vision (ECCV)*.

Moon, P. and Spencer, D. E. (1981). *The Photic Field*. MIT Press, Cambridge, Massachusetts.

Moons, T., Van Gool, L., and Vergauwen, M. (2010). 3D reconstruction from multiple images. *Foundations and Trends® in Computer Graphics and Computer Vision*, 4(4).

Moosavi-Dezfooli, S.-M., Fawzi, A., and Frossard, P. (2016). DeepFool: A simple and accurate method to fool deep neural networks. In *IEEE Conference on Computer Vision and Pattern Recognition (CVPR)*.

Moosmann, F., Nowak, E., and Jurie, F. (2008). Randomized clustering forests for image classification. *IEEE Transactions on Pattern Analysis and Machine Intelligence*, 30(9):1632–1646.

Moravec, H. (1983). The Stanford cart and the CMU rover. *Proceedings of the IEEE*, 71(7):872–884.

Moravec, H. P. (1980). *Obstacle avoidance and navigation in the real world by a seeing robot rover*. Ph.D. thesis, Stanford University Department of Computer Science.

Moravec, H. P. (1983). The Stanford Cart and the CMU Rover. *Proceedings of the IEEE*, 71(7):872–884.

Mordvintsev, A., Olah, C., and Tyka, M. (2015). Inceptionism: Going deeper into neural networks. *Google AI Blog*. https://ai.googleblog.com/2015/06/inceptionism-going-deeper-into-neural.html.

Moreno-Noguer, F., Lepetit, V., and Fua, P. (2007). Accurate non-iterative $O(n)$ solution to the PnP problem. In *International Conference on Computer Vision (ICCV)*.

Mori, G. (2005). Guiding model search using segmentation. In *International Conference on Computer Vision (ICCV)*, pp. 1417–1423.

Mori, G., Ren, X., Efros, A., and Malik, J. (2004). Recovering human body configurations: Combining segmentation and recognition. In *IEEE Computer Society Conference on Computer Vision and Pattern Recognition (CVPR)*, pp. 326–333.

Morimoto, C. and Chellappa, R. (1997). Fast 3D stabilization and mosaic construction. In *IEEE Computer Society Conference on Computer Vision and Pattern Recognition (CVPR)*, pp. 660–665.

Morita, T. and Kanade, T. (1997). A sequential factorization method for recovering shape and motion from image streams. *IEEE Transactions on Pattern Analysis and Machine Intelligence*, 19(8):858–867.

Morris, D. D. and Kanade, T. (1998). A unified factorization algorithm for points, line segments and planes with uncertainty models. In *International Conference on Computer Vision (ICCV)*, pp. 696–702.

Morrone, M. and Burr, D. (1988). Feature detection in human vision: A phase dependent energy model. *Proceedings of the Royal Society of London B*, 235:221–245.

Mortensen, E. N. (1999). Vision-assisted image editing. *Computer Graphics*, 33(4):55–57.

Mortensen, E. N. and Barrett, W. A. (1995). Intelligent scissors for image composition. In *ACM SIGGRAPH Conference Proceedings*, pp. 191–198.

Mortensen, E. N. and Barrett, W. A. (1998). Interactive segmentation with intelligent scissors. *Graphical Models and Image Processing*, 60(5):349–384.

Mortensen, E. N. and Barrett, W. A. (1999). Toboggan-based intelligent scissors with a four parameter edge model. In *IEEE Computer Society Conference on Computer Vision and Pattern Recognition (CVPR)*, pp. 452–458.

Moulon, P., Monasse, P., and Marlet, R. (2013). Global fusion of relative motions for robust, accurate and scalable structure from motion. In *International Conference on Computer Vision (ICCV)*.

Moulon, P., Monasse, P., Perrot, R., and Marlet, R. (2016). OpenMVG: Open multiple view geometry. In *International Workshop on Reproducible Research in Pattern Recognition*, pp. 60–74.

Mourikis, A. I. and Roumeliotis, S. I. (2007). A multi-state constraint Kalman filter for vision-aided inertial navigation. In *IEEE International Conference on Robotics and Automation*, pp. 3565–3572.

Mueller, F., Bernard, F., Sotnychenko, O., Mehta, D., Sridhar, S., Casas, D., and Theobalt, C. (2018). GANerated hands for real-time 3D hand tracking from monocular RGB. In *IEEE Conference on Computer Vision and Pattern Recognition (CVPR)*.

Mueller, P., Zeng, G., Wonka, P., and Van Gool, L. (2007). Image-based procedural modeling of facades. *ACM Transactions on Graphics*, 26(3).

Mühlich, M. and Mester, R. (1998). The role of total least squares in motion analysis. In *European Conference on Computer Vision (ECCV)*, pp. 305–321.

Muja, M. and Lowe, D. G. (2009). Fast approximate nearest neighbors with automatic algorithm configuration. In *International Conference on Computer Vision Theory and Applications (VISAPP)*.

Muja, M. and Lowe, D. G. (2014). Scalable nearest neighbor algorithms for high dimensional data. *IEEE Transactions on Pattern Analysis and Machine Intelligence*, 36(11):2227–2240.

Mukherjee, D., Wu, Q. M. J., and Wang, G. (2015). A comparative experimental study of image feature detectors and descriptors. *Machine Vision and Applications*, 26(4):443–466.

Mumford, D. and Shah, J. (1989). Optimal approximations by piecewise smooth functions and variational problems. *Comm. Pure Appl. Math.*, XLII(5):577–685.

Mundy, J. L. (2006). Object recognition in the geometric era: A retrospective. In Ponce, J., Hebert, M., Schmid, C., and Zisserman, A. (eds), *Toward Category-Level Object Recognition*, pp. 3–28, Springer, New York.

Mundy, J. L. and Zisserman, A. (eds). (1992). *Geometric Invariance in Computer Vision*. MIT Press, Cambridge, Massachusetts.

Mur-Artal, R. and Tardós, J. D. (2017). ORB-SLAM2: An open-source SLAM system for monocular, stereo, and RGB-D cameras. *IEEE Transactions on Robotics*, 33(5):1255–1262.

Mur-Artal, R., Montiel, J. M. M., and Tardos, J. D. (2015). ORB-SLAM: a versatile and accurate monocular SLAM system. *IEEE Transactions on Robotics*, 31(5):1147–1163.

Murase, H. and Nayar, S. K. (1995). Visual learning and recognition of 3-D objects from appearance. *International Journal of Computer Vision*, 14(1):5–24.

Murphy, E. P. (2005). *A Testing Procedure to Characterize Color and Spatial Quality of Digital Cameras Used to Image Cultural Heritage*. Master's thesis, Rochester Institute of Technology.

Murphy, K., Torralba, A., and Freeman, W. T. (2003). Using the forest to see the trees: A graphical model relating features, objects, and scenes. In *Advances in Neural Information Processing Systems (NeurIPS)*.

Murphy, K. P. (2012). *Machine learning: a probabilistic perspective*. MIT Press.

Murphy-Chutorian, E. and Trivedi, M. M. (2009). Head pose estimation in computer vision: A survey. *IEEE Transactions on Pattern Analysis and Machine Intelligence*, 31(4):607–626.

Murray, R. M., Li, Z. X., and Sastry, S. S. (1994). *A Mathematical Introduction to Robotic Manipulation*. CRC Press.

Musgrave, K., Belongie, S., and Lim, S.-N. (2020). A metric learning reality check. In *European Conference on Computer Vision (ECCV)*.

Mutch, J. and Lowe, D. G. (2008). Object class recognition and localization using sparse features with limited receptive fields. *International Journal of Computer Vision*, 80(1):45–57.

Myszkowski, K., Mantiuk, R., and Krawczyk, G. (2008). High dynamic range video. *Synthesis Lectures on Computer Graphics and Animation*, 1(1):1–158.

Nagel, H. H. (1986). Image sequences—ten (octal) years—from phenomenology towards a theoretical foundation. In *International Conference on Pattern Recognition (ICPR)*, pp. 1174–1185.

Nagel, H.-H. and Enkelmann, W. (1986). An investigation of smoothness constraints for the estimation of displacement vector fields from image sequences. *IEEE Transactions on Pattern Analysis and Machine Intelligence*, PAMI-8(5):565–593.

Nah, S., Hyun Kim, T., and Mu Lee, K. (2017). Deep multi-scale convolutional neural network for dynamic scene deblurring. In *IEEE Conference on Computer Vision and Pattern Recognition (CVPR)*.

Nah, S., Timofte, R., Gu, S., Baik, S., Hong, S., Moon, G., Son, S., and Mu Lee, K. (2019). NTIRE 2019 challenge on video super-resolution: Methods and results. In *IEEE/CVF Conference on Computer Vision and Pattern Recognition (CVPR) Workshops*.

Nair, V. and Hinton, G. E. (2010). Rectified linear units improve restricted Boltzmann machines. In *International Conference on Machine Learning (ICML)*, pp. 807–814.

Nakamura, Y., Matsuura, T., Satoh, K., and Ohta, Y. (1996). Occlusion detectable stereo—occlusion patterns in camera matrix. In *IEEE Computer Society Conference on Computer Vision and Pattern Recognition (CVPR)*, pp. 371–378.

Nakao, T., Kashitani, A., and Kaneyoshi, A. (1998). Scanning a document with a small camera attached to a mouse. In *IEEE Workshop on Applications of Computer Vision (WACV)*, pp. 63–68.

Nalwa, V. S. (1987). Edge-detector resolution improvement by image interpolation. *IEEE Transactions on Pattern Analysis and Machine Intelligence*, PAMI-9(3):446–451.

Nalwa, V. S. (1993). *A Guided Tour of Computer Vision*. Addison-Wesley, Reading, MA.

Nalwa, V. S. and Binford, T. O. (1986). On detecting edges. *IEEE Transactions on Pattern Analysis and Machine Intelligence*, PAMI-8(6):699–714.

Nam, S., Hwang, Y., Matsushita, Y., and Kim, S. J. (2016). A holistic approach to cross-channel image noise modeling and its application to image denoising. In *IEEE Conference on Computer Vision and Pattern Recognition (CVPR)*.

Nandoriya, A., Elgharib, M., Kim, C., Hefeeda, M., and Matusik, W. (2017). Video reflection removal through spatio-temporal optimization. In *IEEE International Conference on Computer Vision (ICCV)*.

Narasimhan, S. G. and Nayar, S. K. (2005). Enhancing resolution along multiple imaging dimensions using assorted pixels. *IEEE Transactions on Pattern Analysis and Machine Intelligence*, 27(4):518–530.

Narayanan, P., Rander, P., and Kanade, T. (1998). Constructing virtual worlds using dense stereo. In *International Conference on Computer Vision (ICCV)*, pp. 3–10.

Nayar, S., Watanabe, M., and Noguchi, M. (1995). Real-time focus range sensor. In *International Conference on Computer Vision (ICCV)*, pp. 995–1001.

Nayar, S. K. (2006). Computational cameras: Redefining the image. *Computer*, 39(8):30–38.

Nayar, S. K. and Branzoi, V. (2003). Adaptive dynamic range imaging: Optical control of pixel exposures over space and time. In *International Conference on Computer Vision (ICCV)*, pp. 1168–1175.

Nayar, S. K. and Mitsunaga, T. (2000). High dynamic range imaging: Spatially varying pixel exposures. In *IEEE Computer Society Conference on Computer Vision and Pattern Recognition (CVPR)*, pp. 472–479.

Nayar, S. K. and Nakagawa, Y. (1994). Shape from focus. *IEEE Transactions on Pattern Analysis and Machine Intelligence*, 16(8):824–831.

Nayar, S. K., Ikeuchi, K., and Kanade, T. (1991). Shape from interreflections. *International Journal of Computer Vision*, 6(3):173–195.

Nayar, S. K., Watanabe, M., and Noguchi, M. (1996). Real-time focus range sensor. *IEEE Transactions on Pattern Analysis and Machine Intelligence*, 18(12):1186–1198.

Nazeri, K., Ng, E., Joseph, T., Qureshi, F., and Ebrahimi, M. (2019). EdgeConnect: Structure guided image inpainting using edge prediction. In *IEEE International Conference on Computer Vision (ICCV) Workshops*.

Nech, A. and Kemelmacher-Shlizerman, I. (2017). Level playing field for million scale face recognition. In *IEEE Conference on Computer Vision and Pattern Recognition (CVPR)*.

Negahdaripour, S. (1998). Revised definition of optical flow: Integration of radiometric and geometric cues for dynamic scene analysis. *IEEE Transactions on Pattern Analysis and Machine Intelligence*, 20(9):961–979.

Nehab, D. and Hoppe, H. (2014). A fresh look at generalized sampling. *Foundations and Trends® in Computer Graphics and Vision*, 8(1):1–84.

Nehab, D., Rusinkiewicz, S., Davis, J., and Ramamoorthi, R. (2005). Efficiently combining positions and normals for precise 3d geometry. *ACM Transactions on Graphics (Proc. SIGGRAPH)*, 24(3):536–543.

Nene, S. and Nayar, S. K. (1997). A simple algorithm for nearest neighbor search in high dimensions. *IEEE Transactions on Pattern Analysis and Machine Intelligence*, 19(9):989–1003.

Nene, S. A., Nayar, S. K., and Murase, H. (1996). *Columbia Object Image Library (COIL-100)*. Technical Report CUCS-006-96, Department of Computer Science, Columbia University.

Neoral, M., Šochman, J., and Matas, J. (2018). Continual occlusion and optical flow estimation. In *Asian Conference on Computer Vision*, pp. 159–174.

Netravali, A. and Robbins, J. (1979). Motion-compensated television coding: Part 1. *Bell System Tech.*, 58(3):631–670.

Neuhold, G., Ollmann, T., Rota Bulo, S., and Kontschieder, P. (2017). The mapillary vistas dataset for semantic understanding of street scenes. In *IEEE International Conference on Computer Vision (ICCV)*.

Nevatia, R. (1977). A color edge detector and its use in scene segmentation. *IEEE Transactions on Systems, Man, and Cybernetics*, SMC-7(11):820–826.

Nevatia, R. and Binford, T. (1977). Description and recognition of curved objects. *Artificial Intelligence*, 8:77–98.

Newcombe, R., Lovegrove, S., and Davison, A. (2011). DTAM: Dense tracking and mapping in real-time. In *International Conference on Computer Vision (ICCV)*.

Newcombe, R. A., Izadi, S., Hilliges, O., Molyneaux, D., Kim, D., Davison, A. J., Kohli, P., Shotton, J., Hodges, S., and Fitzgibbon, A. W. (2011). Kinectfusion: Real-time dense surface mapping and tracking. In *IEEE/ACM International Symposium on Mixed and Augmented Reality (ISMAR)*, pp. 127–136.

Newcombe, R. A., Fox, D., and Seitz, S. M. (2015). DynamicFusion: Reconstruction and tracking of non-rigid scenes in real-time. In *IEEE Conference on Computer Vision and Pattern Recognition (CVPR)*.

Newell, A. and Deng, J. (2020). How useful is self-supervised pretraining for visual tasks? In *IEEE/CVF Conference on Computer Vision and Pattern Recognition (CVPR)*.

Ng, A. Y., Jordan, M. I., and Weiss, Y. (2001). On spectral clustering: Analysis and an algorithm. In *Advances in Neural Information Processing Systems (NeurIPS)*, pp. 849–854.

Ng, R. (2005). Fourier slice photography. *ACM Transactions on Graphics (Proc. SIGGRAPH)*, 24(3):735–744.

Ng, R., Levoy, M., Bréedif, M., Duval, G., Horowitz, M., and Hanrahan, P. (2005). *Light Field Photography with a Hand-held Plenoptic Camera*. Technical Report CSTR 2005-02, Stanford University.

Ng, T., Balntas, V., Tian, Y., and Mikolajczyk, K. (2020). SOLAR: Second-order loss and attention for image retrieval. In *European Conference on Computer Vision (ECCV)*.

Ng, Y.-H. J., Hausknecht, M., Vijayanarasimhan, S., Vinyals, O., Monga, R., and Toderici, G. (2015). Beyond short snippets: Deep networks for video classification. In *IEEE Conference on Computer Vision and Pattern Recognition (CVPR)*.

Nguyen, A., Yosinski, J., and Clune, J. (2015). Deep neural networks are easily fooled: High confidence predictions for unrecognizable images. In *IEEE Conference on Computer Vision and Pattern Recognition (CVPR)*.

Nguyen, A., Yosinski, J., and Clune, J. (2016). Multifaceted feature visualization: Uncovering the different types of features learned by each neuron in deep neural networks. In *ICML Visualization for Deep Learning Workshop*.

Nguyen, M. H., Torresani, L., la Torre, F. D., and Rother, C. (2009). Weakly supervised discriminative localization and classification: A joint learning process. In *International Conference on Computer Vision (ICCV)*.

Nielsen, M., Florack, L. M. J., and Deriche, R. (1997). Regularization, scale-space, and edge-detection filters. *Journal of Mathematical Imaging and Vision*, 7(4):291–307.

Nielsen, M. A. (2015). *Neural Networks and Deep Learning*. Volume 25, Determination Press San Francisco, CA.

Nielson, G. M. (1993). Scattered data modeling. *IEEE Computer Graphics and Applications*, 13(1):60–70.

Niemeyer, M., Mescheder, L., Oechsle, M., and Geiger, A. (2019). Occupancy flow: 4d reconstruction by learning particle dynamics. In *IEEE/CVF International Conference on Computer Vision (ICCV)*.

Niemeyer, M., Mescheder, L., Oechsle, M., and Geiger, A. (2020). Differentiable volumetric rendering: Learning implicit 3D representations without 3D supervision. In *IEEE/CVF Conference on Computer Vision and Pattern Recognition (CVPR)*.

Nießner, M., Zollhöfer, M., Izadi, S., and Stamminger, M. (2013). Real-time 3D reconstruction at scale using voxel hashing. *ACM Transactions on Graphics (Proc. SIGGRAPH Asia)*, 32(6).

Niklaus, S. and Liu, F. (2018). Context-aware synthesis for video frame interpolation. In *IEEE Conference on Computer Vision and Pattern Recognition (CVPR)*.

Niklaus, S. and Liu, F. (2020). Softmax splatting for video frame interpolation. In *IEEE/CVF Conference on Computer Vision and Pattern Recognition (CVPR)*.

Niklaus, S., Mai, L., and Liu, F. (2017). Video frame interpolation via adaptive convolution. In *IEEE Conference on Computer Vision and Pattern Recognition (CVPR)*.

Niklaus, S., Mai, L., Yang, J., and Liu, F. (2019). 3D Ken Burns effect from a single image. *ACM Transactions on Graphics (Proc. SIGGRAPH Asia*, 38(6):184:1–184:15.

Nilsback, M.-E. and Zisserman, A. (2006). A visual vocabulary for flower classification. In *IEEE Computer Society Conference on Computer Vision and Pattern Recognition (CVPR)*, pp. 1447–1454.

Nir, T., Bruckstein, A. M., and Kimmel, R. (2008). Over-parameterized variational optical flow. *International Journal of Computer Vision*, 76(2):205–216.

Nishihara, H. K. (1984). Practical real-time imaging stereo matcher. *OptEng*, 23(5):536–545.

Nistér, D. (2003). Preemptive RANSAC for live structure and motion estimation. In *International Conference on Computer Vision (ICCV)*, pp. 199–206.

Nistér, D. (2004). An efficient solution to the five-point relative pose problem. *IEEE Transactions on Pattern Analysis and Machine Intelligence*, 26(6):756–777.

Nistér, D. and Stewénius, H. (2006). Scalable recognition with a vocabulary tree. In *IEEE Computer Society Conference on Computer Vision and Pattern Recognition (CVPR)*, pp. 2161–2168.

Nistér, D. and Stewénius, H. (2008). Linear time maximally stable extremal regions. In *European Conference on Computer Vision (ECCV)*, pp. 183–196.

Nistér, D., Naroditsky, O., and Bergen, J. (2006). Visual odometry for ground vehicle applications. *Journal of Field Robotics*, 23(1):3–20.

Noborio, H., Fukada, S., and Arimoto, S. (1988). Construction of the octree approximating three-dimensional objects by using multiple views. *IEEE Transactions on Pattern Analysis and Machine Intelligence*, PAMI-10(6):769–782.

Nocedal, J. and Wright, S. J. (2006). *Numerical Optimization*. Springer, New York, 2nd edition.

Noh, H., Hong, S., and Han, B. (2015). Learning deconvolution network for semantic segmentation. In *IEEE International Conference on Computer Vision (ICCV)*.

Noh, H., Araujo, A., Sim, J., Weyand, T., and Han, B. (2017). Large-scale image retrieval with attentive deep local features. In *IEEE International Conference on Computer Vision (ICCV)*.

Nomura, Y., Zhang, L., and Nayar, S. K. (2007). Scene collages and flexible camera arrays. In *Eurographics Symposium on Rendering*.

Nordström, N. (1990). Biased anisotropic diffusion: A unified regularization and diffusion approach to edge detection. *Image and Vision Computing*, 8(4):318–327.

Noroozi, M. and Favaro, P. (2016). Unsupervised learning of visual representations by solving jigsaw puzzles. In *European Conference on Computer Vision*, pp. 69–84.

Novotny, D., Larlus, D., and Vedaldi, A. (2017). AnchorNet: A weakly supervised network to learn geometry-sensitive features for semantic matching. In *IEEE Conference on Computer Vision and Pattern Recognition (CVPR)*.

Nowozin, S. and Lampert, C. H. (2011). Structured learning and prediction in computer vision. *Foundations and Trends® in Computer Graphics and Vision*, 6(3–4):185–365.

Ntavelis, E., Romero, A., Bigdeli, S., Timofte, R. *et al.* (2020a). AIM 2020 challenge on image extreme inpainting. In *European Conference on Computer Vision (ECCV) Workshops*.

Ntavelis, E., Romero, A., Kastanis, I., Van Gool, L., and Timofte, R. (2020b). SESAME: Semantic editing of scenes by adding, manipulating or erasing objects. In *European Conference on Computer Vision (ECCV)*.

Obdržálek, S. and Matas, J. (2006). Object recognition using local affine frames on maximally stable extremal regions. In Ponce, J., Hebert, M., Schmid, C., and Zisserman, A. (eds), *Toward Category-Level Object Recognition*, pp. 83–104, Springer, New York.

Oberweger, M., Rad, M., and Lepetit, V. (2018). Making deep heatmaps robust to partial occlusions for 3D object pose estimation. In *European Conference on Computer Vision (ECCV)*.

O'Brian, M. (2019). As robots take over warehousing, workers pushed to adapt. *Associated Press*. https://apnews.com/article/056b44f5bfff11208847aa9768f10757.

Oechsle, M., Mescheder, L., Niemeyer, M., Strauss, T., and Geiger, A. (2019). Texture fields: Learning texture representations in function space. In *IEEE/CVF International Conference on Computer Vision (ICCV)*.

Oh, B. M., Chen, M., Dorsey, J., and Durand, F. (2001). Image-based modeling and photo editing. In *ACM SIGGRAPH Conference Proceedings*, pp. 433–442.

Oh, T.-H., Joo, K., Joshi, N., Wang, B., So Kweon, I., and Bing Kang, S. (2017). Personalized cinemagraphs using semantic understanding and collaborative learning. In *IEEE International Conference on Computer Vision (ICCV)*.

Ohlander, R., Price, K., and Reddy, D. R. (1978). Picture segmentation using a recursive region splitting method. *Computer Graphics and Image Processing*, 8(3):313–333.

Ohta, Y. and Kanade, T. (1985). Stereo by intra- and inter-scanline search using dynamic programming. *IEEE Transactions on Pattern Analysis and Machine Intelligence*, PAMI-7(2):139–154.

Ohtake, Y., Belyaev, A., Alexa, M., Turk, G., and Seidel, H.-P. (2003). Multi-level partition of unity implicits. *ACM Transactions on Graphics (Proc. SIGGRAPH)*, 22(3):463–470.

Okutomi, M. and Kanade, T. (1992). A locally adaptive window for signal matching. *International Journal of Computer Vision*, 7(2):143–162.

Okutomi, M. and Kanade, T. (1993). A multiple baseline stereo. *IEEE Transactions on Pattern Analysis and Machine Intelligence*, 15(4):353–363.

Okutomi, M. and Kanade, T. (1994). A stereo matching algorithm with an adaptive window: Theory and experiment. *IEEE Transactions on Pattern Analysis and Machine Intelligence*, 16(9):920–932.

Olah, C., Mordvintsev, A., and Schubert, L. (2017). Feature visualization. *Distill*. https://distill.pub/2017/feature-visualization.

Olah, C., Satyanarayan, A., Johnson, I., Carter, S., Schubert, L., Ye, K., and Mordvintsev, A. (2018). The building blocks of interpretability. *Distill*. https://distill.pub/2018/building-blocks.

Oliensis, J. (2005). The least-squares error for structure from infinitesimal motion. *International Journal of Computer Vision*, 61(3):259–299.

Oliensis, J. and Hartley, R. (2007). Iterative extensions of the Sturm/Triggs algorithm: Convergence and nonconvergence. *IEEE Transactions on Pattern Analysis and Machine Intelligence*, 29(12):2217–2233.

Oliva, A. and Torralba, A. (2001). Modeling the shape of the scene: a holistic representation of the spatial envelope. *International Journal of Computer Vision*, 42(3):145–175.

Oliva, A. and Torralba, A. (2007). The role of context in object recognition. *Trends in Cognitive Sciences*, 11(12):520–527.

Omer, I. and Werman, M. (2004). Color lines: Image specific color representation. In *IEEE Computer Society Conference on Computer Vision and Pattern Recognition (CVPR)*, pp. 946–953.

Ong, E.-J., Micilotta, A. S., Bowden, R., and Hilton, A. (2006). Viewpoint invariant exemplar-based 3D human tracking. *Computer Vision and Image Understanding*, 104(2–3):178–189.

Ono, Y., Trulls, E., Fua, P., and Yi, K. M. (2018). LF-Net: learning local features from images. In *Advances in Neural Information Processing Systems (NeurIPS)*, pp. 6234–6244.

Oord, A. v. d., Li, Y., and Vinyals, O. (2018). Representation learning with contrastive predictive coding. *arXiv preprint arXiv:1807.03748*.

Opelt, A., Pinz, A., and Zisserman, A. (2006). A boundary-fragment-model for object detection. In *European Conference on Computer Vision (ECCV)*, pp. 575–588.

OpenGL-ARB. (1997). *OpenGL Reference Manual: The Official Reference Document to OpenGL, Version 1.1*. Addison-Wesley, Reading, MA, 2nd edition.

Oppenheim, A. V. and Schafer, A. S. (1996). *Signals and Systems*. Prentice Hall, Englewood Cliffs, New Jersey, 2nd edition.

Oppenheim, A. V., Schafer, R. W., and Buck, J. R. (1999). *Discrete-Time Signal Processing*. Prentice Hall, Englewood Cliffs, New Jersey, 2nd edition.

Ordonez, V., Kulkarni, G., and Berg, T. L. (2011). Im2Text: Describing images using 1 million captioned photographs. In *Advances in Neural Information Processing Systems (NeurIPS)*, pp. 1143–1151.

Oren, M. and Nayar, S. (1997). A theory of specular surface geometry. *International Journal of Computer Vision*, 24(2):105–124.

O'Rourke, J. and Badler, N. I. (1980). Model-based image analysis of human motion using constraint propagation. *IEEE Transactions on Pattern Analysis and Machine Intelligence*, 2(6):522–536.

Osher, S. and Paragios, N. (eds). (2003). *Geometric Level Set Methods in Imaging, Vision, and Graphics*, Springer.

Osman, A. A. A., Bolkart, T., and Black, M. J. (2020). STAR: Sparse trained articulated human body regressor. In *European Conference on Computer Vision (ECCV)*.

Osman Ulusoy, A., Black, M. J., and Geiger, A. (2017). Semantic multi-view stereo: Jointly estimating objects and voxels. In *IEEE Conference on Computer Vision and Pattern Recognition (CVPR)*.

Osuna, E., Freund, R., and Girosi, F. (1997). Training support vector machines: An application to face detection. In *IEEE Computer Society Conference on Computer Vision and Pattern Recognition (CVPR)*, pp. 130–136.

O'Toole, A. J., Jiang, F., Roark, D., and Abdi, H. (2006). Predicting human face recognition. In Zhao, W.-Y. and Chellappa, R. (eds), *Face Processing: Advanced Methods and Models*, Elsevier.

O'Toole, A. J., Phillips, P. J., Jiang, F., Ayyad, J., Pénard, N., and Abdi, H. (2009). Face recognition algorithms surpass humans matching faces over changes in illumination. *IEEE Transactions on Pattern Analysis and Machine Intelligence*, 29(9):1642–1646.

Ott, M., Lewis, J. P., and Cox, I. J. (1993). Teleconferencing eye contact using a virtual camera. In *INTERACT and CHI Conference Companion on Human Factors in Computing Systems*, pp. 109–110.

Otte, M. and Nagel, H.-H. (1994). Optical flow estimation: advances and comparisons. In *European Conference on Computer Vision (ECCV)*, pp. 51–60.

Ouyang, W. and Wang, X. (2013). Joint deep learning for pedestrian detection. In *International Conference on Computer Vision (ICCV)*.

Ovide, S. (2020). A case for banning facial recognition. *New York Times*.

Owens, A. and Efros, A. A. (2018). Audio-visual scene analysis with self-supervised multisensory features. In *European Conference on Computer Vision (ECCV)*.

Oztireli, C., Guennebaud, G., and Gross, M. (2008). Feature preserving point set surfaces. *Computer Graphics Forum*, 28(2):493–501.

Özuysal, M., Calonder, M., Lepetit, V., and Fua, P. (2010). Fast keypoint recognition using random ferns. *IEEE Transactions on Pattern Analysis and Machine Intelligence*, 32(3):448–461.

Özyeşil, O. and Singer, A. (2015). Robust camera location estimation by convex programming. In *IEEE Conference on Computer Vision and Pattern Recognition (CVPR)*.

Özyeşil, O., Voroninski, V., Basri, R., and Singer, A. (2017). A survey of structure from motion. *Acta Numerica*, 26:305–364.

Paglieroni, D. W. (1992). Distance transforms: Properties and machine vision applications. *Graphical Models and Image Processing*, 54(1):56–74.

Pal, C., Szeliski, R., Uyttendaele, M., and Jojic, N. (2004). Probability models for high dynamic range imaging. In *IEEE Computer Society Conference on Computer Vision and Pattern Recognition (CVPR)*, pp. 173–180.

Pal, C. J., Weinman, J. J., Tran, L. C., and Scharstein, D. (2012). On learning conditional random fields for stereo. *International Journal of Computer Vision*, 99(3):319–337.

Palmer, S. E. (1999). *Vision Science: Photons to Phenomenology*. The MIT Press, Cambridge, Massachusetts.

Pan, S. J. and Yang, Q. (2009). A survey on transfer learning. *IEEE Transactions on Knowledge and Data Engineering*, 22(10):1345–1359.

Pan, Z., Yu, W., Yi, X., Khan, A., Yuan, F., and Zheng, Y. (2019). Recent progress on generative adversarial networks (GANs): A survey. *IEEE Access*, 7:36322–36333.

Pandey, R., Tkach, A., Yang, S., Pidlypenskyi, P., Taylor, J., Martin-Brualla, R., Tagliasacchi, A., Papandreou, G., Davidson, P., Keskin, C., Izadi, S., and Fanello, S. (2019). Volumetric capture of humans with a single RGBD camera via semi-parametric learning. In *IEEE/CVF Conference on Computer Vision and Pattern Recognition (CVPR)*.

Pang, J., Sun, W., Ren, J., Yang, C., and Yan, Q. (2017). Cascade residual learning: A two-stage convolutional neural network for stereo matching. In *International Conference on Computer Vision (ICCV) Workshops*.

Papageorgiou, C. and Poggio, T. (2000). A trainable system for object detection. *International Journal of Computer Vision*, 38(1):15–33.

Papazoglou, A. and Ferrari, V. (2013). Fast object segmentation in unconstrained video. In *International Conference on Computer Vision (ICCV)*.

Papenberg, N., Bruhn, A., Brox, T., Didas, S., and Weickert, J. (2006). Highly accurate optic flow computation with theoretically justified warping. *International Journal of Computer Vision*, 67(2):141–158.

Papernot, N., Faghri, F., Carlini, N., Goodfellow, I., Feinman, R. *et al.* (2018). CleverHans v2.1.0: Adversarial examples library. *arXiv preprint arXiv:1610.00768*.

Papert, S. (1966). *The Summer Vision Project*. Technical Report AIM-100, Artificial Intelligence Group, Massachusetts Institute of Technology. https://hdl.handle.net/1721.1/6125.

Papineni, K., Roukos, S., Ward, T., and Zhu, W.-J. (2002). BLEU: a method for automatic evaluation of machine translation. In *Annual Meeting of the Association for Computational Linguistics (ACL)*, pp. 311–318.

Paragios, N. and Deriche, R. (2000). Geodesic active contours and level sets for the detection and tracking of moving objects. *IEEE Transactions on Pattern Analysis and Machine Intelligence*, 22(3):266–280.

Paragios, N. and Sgallari, F. (2009). Special issue on scale space and variational methods in computer vision. *International Journal of Computer Vision*, 84(2).

Paragios, N., Faugeras, O. D., Chan, T., and Schnörr, C. (eds). (2005). *International Workshop on Variational, Geometric, and Level Set Methods in Computer Vision (VLSM)*, Springer.

Parikh, D. and Grauman, K. (2011). Relative attributes. In *International Conference on Computer Vision (ICCV)*.

Paris, S. and Durand, F. (2007). A topological approach to hierarchical segmentation using mean shift. In *IEEE Computer Society Conference on Computer Vision and Pattern Recognition (CVPR)*.

Paris, S. and Durand, F. (2009). A fast approximation of the bilateral filter using a signal processing approach. *International Journal of Computer Vision*, 81(1):24–52.

Paris, S., Hasinoff, S. W., and Kautz, J. (2011). Local Laplacian filters: Edge-aware image processing with a laplacian pyramid. *ACM Transactions on Graphics (Proc. SIGGRAPH)*, 30(4):68.

Paris, S., Kornprobst, P., Tumblin, J., and Durand, F. (2008). Bilateral filtering: Theory and applications. *Foundations and Trends® in Computer Graphics and Computer Vision*, 4(1):1–73.

Park, H. and Lee, K. M. (2017). Look wider to match image patches with convolutional neural networks. *IEEE Signal Processing Letters*, 24(12):1788–1792.

Park, J., Ko, K., Lee, C., and Kim, C.-S. (2020). BMBC: Bilateral motion estimation with bilateral cost volume for video interpolation. In *European Conference on Computer Vision (ECCV)*.

Park, J., Sinha, S. N., Matsushita, Y., Tai, Y.-W., and Kweon, I. S. (2017). Robust multiview photometric stereo using planar mesh parameterization. *IEEE Transactions on Pattern Analysis and Machine Intelligence*, 39(8):1591–1604.

Park, J. J., Holynski, A., and Seitz, S. M. (2020). Seeing the world in a bag of chips. In *IEEE/CVF Conference on Computer Vision and Pattern Recognition (CVPR)*.

Park, J. J., Newcombe, R., and Seitz, S. (2018). Surface light field fusion. In *International Conference on 3D Vision (3DV)*, pp. 12–21.

Park, J. J., Florence, P., Straub, J., Newcombe, R., and Lovegrove, S. (2019). DeepSDF: Learning continuous signed distance functions for shape representation. In *IEEE/CVF Conference on Computer Vision and Pattern Recognition (CVPR)*.

Park, K., Sinha, U., Barron, J. T., Bouaziz, S., Goldman, D. B., Seitz, S. M., and Martin-Brualla, R. (2020). Deformable neural radiance fields. *arXiv preprint arXiv:2011.12948*.

Park, K., Sinha, U., Hedman, P., Barron, J. T., Bouaziz, S., Goldman, D. B., Martin-Brualla, R., and Seitz, S. M. (2021). HyperNeRF: A higher-dimensional representation for topologically varying neural radiance fields. *arXiv preprint arXiv:2106.13228*.

Park, M., Brocklehurst, K., Collins, R. T., and Liu, Y. (2009). Deformed lattice detection in real-world images using mean-shift belief propagation. *IEEE Transactions on Pattern Analysis and Machine Intelligence*, 31(10):1804–1816.

Park, S. C., Park, M. K., and Kang, M. G. (2003). Super-resolution image reconstruction: A technical overview. *IEEE Signal Processing Magazine*, 20:21–36.

Park, T., Liu, M.-Y., Wang, T.-C., and Zhu, J.-Y. (2019). Semantic image synthesis with spatially-adaptive normalization. In *IEEE/CVF Conference on Computer Vision and Pattern Recognition (CVPR)*.

Park, T., Zhu, J.-Y., Wang, O., Lu, J., Shechtman, E., Efros, A. A., and Zhang, R. (2020). Swapping autoencoder for deep image manipulation. In *Advances in Neural Information Processing Systems (NeurIPS)*.

Parke, F. I. and Waters, K. (1996). *Computer Facial Animation*. A K Peters, Wellesley, Massachusetts.

Parker, J. A., Kenyon, R. V., and Troxel, D. E. (1983). Comparison of interpolating methods for image resampling. *IEEE Transactions on Medical Imaging*, MI-2(1):31–39.

Parkhi, O. M., Vedaldi, A., and Zisserman, A. (2015). Deep face recognition. In *British Machine Vision Conference*, p. 6.

Parkhi, O. M., Vedaldi, A., Zisserman, A., and Jawahar, C. B. (2012). Cats and dogs. In *IEEE Computer Society Conference on Computer Vision and Pattern Recognition (CVPR)*.

Parmar, N., Vaswani, A., Uszkoreit, J., Kaiser, L., Shazeer, N., Ku, A., and Tran, D. (2018). Image transformer. In *International Conference on Machine Learning (ICML)*, pp. 4055–4064.

Parra Pozo, A., Toksvig, M., Filiba Schrager, T., Hsu, J., Mathur, U., Sorkine-Hornung, A., Szeliski, R., and Cabral, B. (2019). An integrated 6DoF video camera and system design. *ACM Transactions on Graphics (Proc. SIGGRAPH Asia)*, 38(6):216.

Pathak, D., Krahenbuhl, P., and Darrell, T. (2015). Constrained convolutional neural networks for weakly supervised segmentation. In *IEEE International Conference on Computer Vision (ICCV)*.

Pathak, D., Krahenbuhl, P., Donahue, J., Darrell, T., and Efros, A. A. (2016). Context encoders: Feature learning by inpainting. In *IEEE Conference on Computer Vision and Pattern Recognition (CVPR)*.

Patrick, M., Asano, Y. M., Fong, R., Henriques, J. F., Zweig, G., and Vedaldi, A. (2020). Multi-modal self-supervision from generalized data transformations. *arXiv preprint arXiv:2003.04298*.

Patron-Perez, A., Lovegrove, S., and Sibley, G. (2015). A spline-based trajectory representation for sensor fusion and rolling shutter cameras. *International Journal of Computer Vision*, 113(3):208–219.

Pattanaik, S. N., Ferwerda, J. A., Fairchild, M. D., and Greenberg, D. P. (1998). A multiscale model of adaptation and spatial vision for realistic image display. In *ACM SIGGRAPH Conference Proceedings*, pp. 287–298.

Pauly, M., Keiser, R., Kobbelt, L. P., and Gross, M. (2003). Shape modeling with point-sampled geometry. *ACM Transactions on Graphics (Proc. SIGGRAPH)*, 21(3):641–650.

Pauly, M., Mitra, N. J., Wallner, J., Pottmann, H., and Guibas, L. J. (2008). Discovering structural regularity in 3D geometry. *ACM Transactions on Graphics*, 27(3):43.

Pautrat, R., Larsson, V., Oswald, M. R., and Pollefeys, M. (2020). Online invariance selection for local feature descriptors. In *European Conference on Computer Vision (ECCV)*.

Pavlakos, G., Choutas, V., Ghorbani, N., Bolkart, T., Osman, A. A. A., Tzionas, D., and Black, M. J. (2019). Expressive body capture: 3D hands, face, and body from a single image. In *IEEE/CVF Conference on Computer Vision and Pattern Recognition (CVPR)*.

Pavlidis, T. (1977). *Structural Pattern Recognition*. Springer-Verlag, Berlin; New York.

Pavlidis, T. and Liow, Y.-T. (1990). Integrating region growing and edge detection. *IEEE Transactions on Pattern Analysis and Machine Intelligence*, 12(3):225–233.

Pavlović, V., Sharma, R., and Huang, T. S. (1997). Visual interpretation of hand gestures for human-computer interaction: A review. *IEEE Transactions on Pattern Analysis and Machine Intelligence*, 19(7):677–695.

Paysan, P., Knothe, R., Amberg, B., Romdhani, S., and Vetter, T. (2009). A 3D face model for pose and illumination invariant face recognition. In *IEEE International Conference on Advanced Video and Signal Based Surveillance*, pp. 296–301.

Pearl, J. (1988). *Probabilistic reasoning in intelligent systems: networks of plausible inference*. Morgan Kaufmann Publishers, Los Altos.

Peleg, R., Ben-Ezra, M., and Pritch, Y. (2001). Omnistereo: Panoramic stereo imaging. *IEEE Transactions on Pattern Analysis and Machine Intelligence*, 23(3):279–290.

Peleg, S. (1981). Elimination of seams from photomosaics. *Computer Vision, Graphics, and Image Processing*, 16(1):1206–1210.

Peleg, S. and Herman, J. (1997). Panoramic mosaics by manifold projection. In *IEEE Computer Society Conference on Computer Vision and Pattern Recognition (CVPR)*, pp. 338–343.

Peleg, S. and Rav-Acha, A. (2006). Lucas-Kanade without iterative warping. In *International Conference on Image Processing (ICIP)*, pp. 1097–1100.

Peleg, S., Rousso, B., Rav-Acha, A., and Zomet, A. (2000). Mosaicing on adaptive manifolds. *IEEE Transactions on Pattern Analysis and Machine Intelligence*, 22(10):1144–1154.

Penev, P. and Atick, J. (1996). Local feature analysis: A general statistical theory for object representation. *Network Computation and Neural Systems*, 7:477–500.

Peng, S., Liu, Y., Huang, Q., Zhou, X., and Bao, H. (2019). PVNet: Pixel-wise voting network for 6DoF pose estimation. In *IEEE/CVF Conference on Computer Vision and Pattern Recognition (CVPR)*.

Peng, S., Niemeyer, M., Mescheder, L., Pollefeys, M., and Geiger, A. (2020). Convolutional occupancy networks. In *European Conference on Computer Vision (ECCV)*.

Peng, S., Jiang, W., Pi, H., Li, X., Bao, H., and Zhou, X. (2020). Deep snake for real-time instance segmentation. In *IEEE/CVF Conference on Computer Vision and Pattern Recognition (CVPR)*.

Peng, S., Zhang, Y., Xu, Y., Wang, Q., Shuai, Q., Bao, H., and Zhou, X. (2021). Neural body: Implicit neural representations with structured latent codes for novel view synthesis of dynamic humans. In *IEEE/CVF Conference on Computer Vision and Pattern Recognition (CVPR)*.

Penner, E. and Zhang, L. (2017). Soft 3D reconstruction for view synthesis. *ACM Transactions on Graphics (Proc. SIGGRAPH Asia)*, 36(6):1–11.

Pentland, A. P. (1984). Local shading analysis. *IEEE Transactions on Pattern Analysis and Machine Intelligence*, PAMI-6(2):170–179.

Pentland, A. P. (1986). Perceptual organization and the representation of natural form. *Artificial Intelligence*, 28(3):293–331.

Pentland, A. P. (1987). A new sense for depth of field. *IEEE Transactions on Pattern Analysis and Machine Intelligence*, PAMI-9(4):523–531.

Pentland, A. P. (1994). Interpolation using wavelet bases. *IEEE Transactions on Pattern Analysis and Machine Intelligence*, 16(4):410–414.

Perazzi, F., Wang, O., Gross, M., and Sorkine-Hornung, A. (2015). Fully connected object proposals for video segmentation. In *IEEE International Conference on Computer Vision (ICCV)*.

Perazzi, F., Khoreva, A., Benenson, R., Schiele, B., and Sorkine-Hornung, A. (2017). Learning video object segmentation from static images. In *IEEE Conference on Computer Vision and Pattern Recognition (CVPR)*.

Perazzi, F., Pont-Tuset, J., McWilliams, B., Van Gool, L., Gross, M., and Sorkine-Hornung, A. (2016). A benchmark dataset and evaluation methodology for video object segmentation. In *IEEE Conference on Computer Vision and Pattern Recognition (CVPR)*.

Pérez, P., Blake, A., and Gangnet, M. (2001). JetStream: Probabilistic contour extraction with particles. In *International Conference on Computer Vision (ICCV)*, pp. 524–531.

Pérez, P., Gangnet, M., and Blake, A. (2003). Poisson image editing. *ACM Transactions on Graphics (Proc. SIGGRAPH)*, 22(3):313–318.

Perona, P. (1995). Deformable kernels for early vision. *IEEE Transactions on Pattern Analysis and Machine Intelligence*, 17(5):488–499.

Perona, P. and Malik, J. (1990a). Detecting and localizing edges composed of steps, peaks and roofs. In *International Conference on Computer Vision (ICCV)*, pp. 52–57.

Perona, P. and Malik, J. (1990b). Scale space and edge detection using anisotropic diffusion. *IEEE Transactions on Pattern Analysis and Machine Intelligence*, 12(7):629–639.

Perreault, S. and Hébert, P. (2007). Median filtering in constant time. *IEEE Transactions on Image Processing*, 16(9):2389–2394.

Persson, M. and Nordberg, K. (2018). Lambda twist: An accurate fast robust perspective three point (P3P) solver. In *European Conference on Computer Vision (ECCV)*.

Peters, J. and Reif, U. (2008). *Subdivision Surfaces*. Springer.

Petersen, K. B. and Pedersen, M. S. (2012). The matrix cookbook.

Petschnigg, G., Agrawala, M., Hoppe, H., Szeliski, R., Cohen, M., and Toyama, K. (2004). Digital photography with flash and no-flash image pairs. *ACM Transactions on Graphics (Proc. SIGGRAPH)*, 23(3):664–672.

Pfister, H., Zwicker, M., van Baar, J., and Gross, M. (2000). Surfels: Surface elements as rendering primitives. In *ACM SIGGRAPH Conference Proceedings*, pp. 335–342.

Pflugfelder, R. (2008). *Self-calibrating Cameras in Video Surveillance*. Ph.D. thesis, Graz University of Technology.

Pham, H., Guan, M. Y., Zoph, B., Le, Q. V., and Dean, J. (2018). Efficient neural architecture search via parameter sharing. In *International Conference on Machine Learning (ICML)*.

Philbin, J. and Zisserman, A. (2008). Object mining using a matching graph on very large image collections. In *Indian Conference on Computer Vision, Graphics and Image Processing*.

Philbin, J., Chum, O., Isard, M., Sivic, J., and Zisserman, A. (2007). Object retrieval with large vocabularies and fast spatial matching. In *IEEE Computer Society Conference on Computer Vision and Pattern Recognition (CVPR)*.

Philbin, J., Chum, O., Sivic, J., Isard, M., and Zisserman, A. (2008). Lost in quantization: Improving particular object retrieval in large scale image databases. In *IEEE Computer Society Conference on Computer Vision and Pattern Recognition (CVPR)*.

Philip, J., Gharbi, M., Zhou, T., Efros, A. A., and Drettakis, G. (2019). Multi-view relighting using a geometry-aware network. *ACM Transactions on Graphics (Proc. SIGGRAPH)*, 38(4):78:1–78:14.

Phillips, P. J., Moon, H., Rizvi, S. A., and Rauss, P. J. (2000). The FERET evaluation methodology for face recognition algorithms. *IEEE Transactions on Pattern Analysis and Machine Intelligence*, 22(10):1090–1104.

Phillips, P. J., Scruggs, W. T., O'Toole, A. J., Flynn, P. J., Bowyer, K. W., Schott, C. L., and Sharpe, M. (2010). FRVT 2006 and ICE 2006 large-scale experimental results. *IEEE Transactions on Pattern Analysis and Machine Intelligence*, 32(5):831–846.

Phong, B. T. (1975). Illumination for computer generated pictures. *Communications of the ACM*, 18(6):311–317.

Pickup, L. C. (2007). *Machine Learning in Multi-frame Image Super-resolution*. Ph.D. thesis, University of Oxford.

Pickup, L. C. and Zisserman, A. (2009). Automatic retrieval of visual continuity errors in movies. In *ACM International Conference on Image and Video Retrieval*.

Pickup, L. C., Capel, D. P., Roberts, S. J., and Zisserman, A. (2007). Overcoming registration uncertainty in image super-resolution: Maximize or marginalize? *EURASIP Journal on Advances in Signal Processing*, 2010(Article ID 23565).

Pickup, L. C., Capel, D. P., Roberts, S. J., and Zisserman, A. (2009). Bayesian methods for image super-resolution. *The Computer Journal*, 52.

Pighin, F., Szeliski, R., and Salesin, D. H. (2002). Modeling and animating realistic faces from images. *International Journal of Computer Vision*, 50(2):143–169.

Pighin, F., Hecker, J., Lischinski, D., Salesin, D. H., and Szeliski, R. (1998). Synthesizing realistic facial expressions from photographs. In *ACM SIGGRAPH Conference Proceedings*, pp. 75–84.

Pilet, J., Lepetit, V., and Fua, P. (2008). Fast non-rigid surface detection, registration, and realistic augmentation. *International Journal of Computer Vision*, 76(2).

Pinheiro, P. O., Lin, T.-Y., Collobert, R., and Dollár, P. (2016). Learning to refine object segments. In *European Conference on Computer Vision (ECCV)*, pp. 75–91.

Pintore, G., Mura, C., Ganovelli, F., Fuentes-Perez, L., Pajarola, R., and Gobbetti, E. (2020). State-of-the-art in automatic 3D reconstruction of structured indoor environments. In *Computer Graphics Forum*, pp. 667–699.

Pinz, A. (2005). Object categorization. *Foundations and Trends® in Computer Graphics and Computer Vision*, 1(4):255–353.

Pion, N., Humenberger, M., Csurka, G., Cabon, Y., and Sattler, T. (2020). Benchmarking image retrieval for visual localization. In *International Conference on 3D Vision (3DV)*, pp. 483–494.

Pishchulin, L., Andriluka, M., Gehler, P., and Schiele, B. (2013). Poselet conditioned pictorial structures. In *IEEE Computer Society Conference on Computer Vision and Pattern Recognition (CVPR)*.

Pishchulin, L., Insafutdinov, E., Tang, S., Andres, B., Andriluka, M., Gehler, P. V., and Schiele, B. (2016). DeepCut: Joint subset partition and labeling for multi person pose estimation. In *IEEE Conference on Computer Vision and Pattern Recognition (CVPR)*.

Pizer, S. M., Amburn, E. P., Austin, J. D., Cromartie, R., Geselowitz, A., Greer, T., Romeny, B. T. H., and Zimmerman, J. B. (1987). Adaptive histogram equalization and its variations. *Computer Vision, Graphics, and Image Processing*, 39(3):355–368.

Platel, B., Balmachnova, E., Florack, L., and ter Haar Romeny, B. (2006). Top-points as interest points for image matching. In *European Conference on Computer Vision (ECCV)*, pp. 418–429.

Platt, J. (1989). *Fast training of support vector machines using sequential minimal optimization*, pp. 185–208. MIT Press.

Platt, J. (2000a). Probabilities for support vector machines. In *Advances in Large Margin Classifiers*, pp. 61–74.

Platt, J. C. (2000b). Optimal filtering for patterned displays. *IEEE Signal Processing Letters*, 7(7):179–180.

Plötz, T. and Roth, S. (2017). Benchmarking denoising algorithms with real photographs. In *IEEE Conference on Computer Vision and Pattern Recognition (CVPR)*.

Plummer, B. A., Wang, L., Cervantes, C. M., Caicedo, J. C., Hockenmaier, J., and Lazebnik, S. (2017). Flickr30k Entities: Collecting region-to-phrase correspondences for richer image-to-sentence models. *International Journal of Computer Vision*, 123(1):74–93.

Pock, T., Schoenemann, T., Graber, G., Bischof, H., and Cremers, D. (2008). A convex formulation of continuous multi-label problems. In *European Conference on Computer Vision (ECCV)*, pp. 792–805.

Poelman, C. J. and Kanade, T. (1997). A paraperspective factorization method for shape and motion recovery. *IEEE Transactions on Pattern Analysis and Machine Intelligence*, 19(3):206–218.

Poggi, M., Tosi, F., Batsos, K., Mordohai, P., and Mattoccia, S. (2021). On the synergies between machine learning and binocular stereo for depth estimation from images: a survey. *IEEE Transactions on Pattern Analysis and Machine Intelligence*, (accepted)().

Poggi, M., Kim, S., Tosi, F., Kim, S., Aleotti, F., Min, D., Sohn, K., and Mattoccia, S. (2021). On the confidence of stereo matching in a deep-learning era: a quantitative evaluation. *IEEE Transactions on Pattern Analysis and Machine Intelligence*.

Poggio, T., Little, J., Gamble, E., Gillet, W., Geiger, D., Weinshall, D., Villalba, M., Larson, N., Cass, T., Bülthoff, H., Drumheller, M., Oppenheimer, P., Yang, W., and Hurlbert, A. (1988). The MIT vision machine. In *Image Understanding Workshop*, pp. 177–198.

Poggio, T. and Koch, C. (1985). Ill-posed problems in early vision: from computational theory to analogue networks. *Proceedings of the Royal Society of London*, B 226:303–323.

Poggio, T., Gamble, E., and Little, J. (1988). Parallel integration of vision modules. *Science*, 242(4877):436–440.

Poggio, T., Torre, V., and Koch, C. (1985). Computational vision and regularization theory. *Nature*, 317(6035):314–319.

Polana, R. and Nelson, R. C. (1997). Detection and recognition of periodic, nonrigid motion. *International Journal of Computer Vision*, 23(3):261–282.

Pollard, S. B., Mayhew, J. E. W., and Frisby, J. P. (1985). PMF: A stereo correspondence algorithm using a disparity gradient limit. *Perception*, 14:449–470.

Pollefeys, M., Nistér, D., Frahm, J.-M., Akbarzadeh, A., Mordohai, P., Clipp, B., Engels, C., Gallup, D., Kim, S.-J., Merrell, P., Salmi, C., Sinha, S., Talton, B., Wang, L., Yang, Q., Stewénius, H., Yang, R., Welch, G., and Towles, H. (2008). Detailed real-time urban 3D reconstruction from video. *International Journal of Computer Vision*, 78(2–3):143–167.

Pollefeys, M. and Van Gool, L. (2002). From images to 3D models. *Communications of the ACM*, 45(7):50–55.

Pollefeys, M., Koch, R., and Van Gool, L. (1999). Self-calibration and metric reconstruction in spite of varying and unknown internal camera parameters. *International Journal of Computer Vision*, 32(1):7–25.

Pollefeys, M., Van Gool, L., Vergauwen, M., Verbiest, F., Cornelis, K., Tops, J., and Koch, R. (2004). Visual modeling with a hand-held camera. *International Journal of Computer Vision*, 59(3):207–232.

Pomerleau, D. A. (1989). ALVINN: An autonomous land vehicle in a neural network. In *Advances in Neural Information Processing Systems (NeurIPS)*, pp. 305–313.

Pomerleau, F., Colas, F., and Siegwart, R. (2015). A review of point cloud registration algorithms for mobile robotics. *Foundations and Trends® in Robotics*, 4(1):1–104.

Ponce, J., Hebert, M., Schmid, C., and Zisserman, A. (eds). (2006). *Toward Category-Level Object Recognition*, Springer, New York.

Pons, J.-P., Keriven, R., and Faugeras, O. (2005). Modelling dynamic scenes by registering multi-view image sequences. In *IEEE Computer Society Conference on Computer Vision and Pattern Recognition (CVPR)*, pp. 822–827.

Pons, J.-P., Keriven, R., and Faugeras, O. (2007). Multi-view stereo reconstruction and scene flow estimation with a global image-based matching score. *International Journal of Computer Vision*, 72(2):179–193.

Pont-Tuset, J., Arbeláez, P., Barron, J. T., Marques, F., and Malik, J. (2017). Multiscale combinatorial grouping for image segmentation and object proposal generation. *IEEE Transactions on Pattern Analysis and Machine Intelligence*, 39(1):128–140.

Pont-Tuset, J., Perazzi, F., Caelles, S., Arbeláez, P., Sorkine-Hornung, A., and Van Gool, L. (2017). The 2017 DAVIS challenge on video object segmentation. *arXiv preprint arXiv:1704.00675*.

Poppe, R. (2010). A survey on vision-based human action recognition. *Image and Vision Computing*, 28(6):976–990.

Porter, T. and Duff, T. (1984). Compositing digital images. *Computer Graphics (SIGGRAPH)*, 18(3):253–259.

Portilla, J. and Simoncelli, E. P. (2000). A parametric texture model based on joint statistics of complex wavelet coefficients. *International Journal of Computer Vision*, 40(1):49–71.

Portilla, J., Strela, V., Wainwright, M., and Simoncelli, E. P. (2003). Image denoising using scale mixtures of Gaussians in the wavelet domain. *IEEE Transactions on Image Processing*, 12(11):1338–1351.

Potetz, B. and Lee, T. S. (2008). Efficient belief propagation for higher-order cliques using linear constraint nodes. *Computer Vision and Image Understanding*, 112(1):39–54.

Potmesil, M. (1987). Generating octree models of 3D objects from their silhouettes in a sequence of images. *Computer Vision, Graphics, and Image Processing*, 40:1–29.

Pratt, W. K. (2007). *Digital Image Processing*. Wiley-Interscience, Hoboken, NJ, 4th edition.

Prazdny, K. (1985). Detection of binocular disparities. *Biological Cybernetics*, 52:93–99.

Preparata, F. P. and Shamos, M. I. (1985). *Computational Geometry: An Introduction*. Academic Press, New York.

Prince, S. J. D. (2012). *Computer vision: models, learning, and inference*. Cambridge University Press.

Pritchett, P. and Zisserman, A. (1998). Wide baseline stereo matching. In *International Conference on Computer Vision (ICCV)*, pp. 754–760.

Proesmans, M., Van Gool, L., and Defoort, F. (1998). Reading between the lines – a method for extracting dynamic 3D with texture. In *International Conference on Computer Vision (ICCV)*, pp. 1081–1086.

Prusinkiewicz, P. and Lindenmayer, A. (1996). *The algorithmic beauty of plants*. Springer.

Pullen, K. and Bregler, C. (2002). Motion capture assisted animation: texturing and synthesis. *ACM Transactions on Graphics*, 21(3):501–508.

Pulli, K. (1999). Multiview registration for large datasets. In *International Conference on 3D Digital Imaging and Modeling (3DIM)*, pp. 160–168.

Pulli, K., Abi-Rached, H., Duchamp, T., Shapiro, L., and Stuetzle, W. (1998). Acquisition and visualization of colored 3D objects. In *International Conference on Pattern Recognition (ICPR)*, pp. 11–15.

Pumarola, A., Corona, E., Pons-Moll, G., and Moreno-Noguer, F. (2021). D-NeRF: Neural radiance fields for dynamic scenes. In *IEEE/CVF Conference on Computer Vision and Pattern Recognition (CVPR)*.

Punnappurath, A. and Brown, M. S. (2019). Reflection removal using a dual-pixel sensor. In *IEEE/CVF Conference on Computer Vision and Pattern Recognition (CVPR)*.

Qi, C. R., Su, H., Mo, K., and Guibas, L. J. (2017). PointNet: Deep learning on point sets for 3D classification and segmentation. In *IEEE Conference on Computer Vision and Pattern Recognition (CVPR)*.

Qiao, S., Wang, H., Liu, C., Shen, W., and Yuille, A. (2019a). Micro-batch training with batch-channel normalization and weight standardization. *arXiv preprint arXiv:1903.10520*.

Qiao, S., Wang, H., Liu, C., Shen, W., and Yuille, A. (2019b). Rethinking normalization and elimination singularity in neural networks. *arXiv preprint arXiv:1911.09738*.

Qin, Y., Frosst, N., Raffel, C., Cottrell, G., and Hinton, G. (2020a). Deflecting adversarial attacks. *arXiv preprint arXiv:2002.07405*.

Qin, Y., Frosst, N., Sabour, S., Raffel, C., Cottrell, G., and Hinton, G. (2020b). Detecting and diagnosing adversarial images with class-conditional capsule reconstructions. In *International Conference on Learning Representations (ICLR).*

Quack, T., Leibe, B., and Van Gool, L. (2008). World-scale mining of objects and events from community photo collections. In *Conference on Image and Video Retrieval*, pp. 47–56.

Quam, L. H. (1984). Hierarchical warp stereo. In *Image Understanding Workshop*, pp. 149–155.

Quan, L. and Lan, Z. (1999). Linear N-point camera pose determination. *IEEE Transactions on Pattern Analysis and Machine Intelligence*, 21(8):774–780.

Quan, L. and Mohr, R. (1989). Determining perspective structures using hierarchical Hough transform. *Pattern Recognition Letters*, 9(4):279–286.

Rabe, C., Müller, T., Wedel, A., and Franke, U. (2010). Dense, robust, and accurate motion field estimation from stereo image sequences in real-time. In *European Conference on Computer Vision (ECCV)*, pp. 582–595.

Rabinovich, A., Vedaldi, A., Galleguillos, C., Wiewiora, E., and Belongie, S. (2007). Objects in context. In *International Conference on Computer Vision (ICCV).*

Rademacher, P. and Bishop, G. (1998). Multiple-center-of-projection images. In *ACM SIGGRAPH Conference Proceedings*, pp. 199–206.

Radenović, F., Tolias, G., and Chum, O. (2019). Fine-tuning CNN image retrieval with no human annotation. *IEEE Transactions on Pattern Analysis and Machine Intelligence*, 41(7):1655–1668.

Radenović, F., Iscen, A., Tolias, G., Avrithis, Y., and Chum, O. (2018). Revisiting Oxford and Paris: Large-scale image retrieval benchmarking. In *IEEE Conference on Computer Vision and Pattern Recognition (CVPR).*

Radenović, F., Schönberger, J. L., Ji, D., Frahm, J.-M., Chum, O., and Matas, J. (2016). From dusk till dawn: Modeling in the dark. In *IEEE Conference on Computer Vision and Pattern Recognition (CVPR).*

Radford, A., Kim, J. W., Hallacy, C., Ramesh, A., Goh, G., Agarwal, S., Sastry, G., Askell, A., Mishkin, P., Clark, J., Krueger, G., and Sutskever, I. (2021). Learning transferable visual models from natural language supervision. *arXiv preprint arXiv:2103.00020.*

Radford, A., Metz, L., and Chintala, S. (2015). Unsupervised representation learning with deep convolutional generative adversarial networks. *arXiv preprint arXiv:1511.06434.* presented at ICLR 2016.

Radosavovic, I., Kosaraju, R. P., Girshick, R., He, K., and Dollár, P. (2020). Designing network design spaces. In *IEEE/CVF Conference on Computer Vision and Pattern Recognition (CVPR).*

Ragan-Kelley, J., Barnes, C., Adams, A., Paris, S., Durand, F., and Amarasinghe, S. (2013). Halide: a language and compiler for optimizing parallelism, locality, and recomputation in image processing pipelines. In *ACM SIGPLAN Notices*, pp. 519–530.

Raginsky, M. and Lazebnik, S. (2009). Locality-sensitive binary codes from shift-invariant kernels. In *Advances in Neural Information Processing Systems (NeurIPS).*

Raguram, R., Wu, C., Frahm, J.-M., and Lazebnik, S. (2011). Modeling and recognition of landmark image collections using iconic scene graphs. *International Journal of Computer Vision*, 95(3):213–239.

Raguram, R., Chum, O., Pollefeys, M., Matas, J., and Frahm, J.-M. (2012). USAC: a universal framework for random sample consensus. *IEEE Transactions on Pattern Analysis and Machine Intelligence*, 35(8):2022–2038.

Raman, S. and Chaudhuri, S. (2007). A matte-less, variational approach to automatic scene compositing. In *International Conference on Computer Vision (ICCV).*

Raman, S. and Chaudhuri, S. (2009). Bilateral filter based compositing for variable exposure photography. In *Eurographics*.

Ramanan, D. and Baker, S. (2009). Local distance functions: A taxonomy, new algorithms, and an evaluation. In *International Conference on Computer Vision (ICCV).*

Ramanan, D., Forsyth, D., and Zisserman, A. (2005). Strike a pose: Tracking people by finding stylized poses. In *IEEE Computer Society Conference on Computer Vision and Pattern Recognition (CVPR)*, pp. 271–278.

Ramanarayanan, G. and Bala, K. (2007). Constrained texture synthesis via energy minimization. *IEEE Transactions on Visualization and Computer Graphics*, 13(1):167–178.

Ramer, U. (1972). An iterative procedure for the polygonal approximation of plane curves. *Computer Graphics and Image Processing*, 1(3):244–256.

Ramesh, A., Pavlov, M., Goh, G., Gray, S., Voss, C., Radford, A., Chen, M., and Sutskever, I. (2021). Zero-shot text-to-image generation. *arXiv preprint arXiv:2102.12092*. Blog at https://openai.com/blog/dall-e.

Ramnath, K., Koterba, S., Xiao, J., Hu, C., Matthews, I., Baker, S., Cohn, J., and Kanade, T. (2008). Multi-view AAM fitting and construction. *International Journal of Computer Vision*, 76(2):183–204.

Ranftl, R., Lasinger, K., Hafner, D., Schindler, K., and Koltun, V. (2020). Towards robust monocular depth estimation: Mixing datasets for zero-shot cross-dataset transfer. *IEEE Transactions on Pattern Analysis and Machine Intelligence*, (accepted).

Ranjan, A. and Black, M. J. (2017). Optical flow estimation using a spatial pyramid network. In *IEEE Conference on Computer Vision and Pattern Recognition (CVPR)*.

Raskar, R. and Tumblin, J. (2010). *Computational Photography: Mastering New Techniques for Lenses, Lighting, and Sensors*. A K Peters, Wellesley, Massachusetts.

Raskar, R., Tan, K.-H., Feris, R., Yu, J., and Turk, M. (2004). Non-photorealistic camera: Depth edge detection and stylized rendering using multi-flash imaging. *ACM Transactions on Graphics*, 23(3):679–688.

Rastegari, M., Ordonez, V., Redmon, J., and Farhadi, A. (2016). XNOR-Net: Imagenet classification using binary convolutional neural networks. In *European Conference on Computer Vision*, pp. 525–542.

Rav-Acha, A., Kohli, P., Fitzgibbon, A., and Rother, C. (2008). Unwrap mosaics: A new representation for video editing. *ACM Transactions on Graphics*, 27(3):17.

Rav-Acha, A., Pritch, Y., Lischinski, D., and Peleg, S. (2005). Dynamosaics: Video mosaics with non-chronological time. In *IEEE Computer Society Conference on Computer Vision and Pattern Recognition (CVPR)*, pp. 58–65.

Rawat, W. and Wang, Z. (2017). Deep convolutional neural networks for image classification: A comprehensive review. *Neural Computation*, 29(9):2352–2449.

Ray, S. F. (2002). *Applied Photographic Optics*. Focal Press, Oxford, 3rd edition.

Razavi, A., van den Oord, A., and Vinyals, O. (2019). Generating diverse high-fidelity images with VQ-VAE-2. *arXiv preprint arXiv:1906.00446*.

Recht, B., Roelofs, R., Schmidt, L., and Shankar, V. (2019). Do ImageNet classifiers generalize to ImageNet? In *International Conference on Machine Learning (ICML)*, pp. 5389–5400.

Redmon, J. and Farhadi, A. (2017). YOLO9000: Better, faster, stronger. In *IEEE Conference on Computer Vision and Pattern Recognition (CVPR)*.

Redmon, J. and Farhadi, A. (2018). YOLOv3: An incremental improvement. *arXiv preprint arXiv:1804.02767*.

Redmon, J., Divvala, S., Girshick, R., and Farhadi, A. (2016). You only look once: Unified, real-time object detection. In *IEEE Conference on Computer Vision and Pattern Recognition (CVPR)*.

Reed, S., Akata, Z., Yan, X., Logeswaran, L., Schiele, B., and Lee, H. (2016). Generative adversarial text to image synthesis. In *International Conference on Machine Learning (ICML)*, pp. 1060–1069.

Rehg, J. and Kanade, T. (1994). Visual tracking of high DOF articulated structures: an application to human hand tracking. In *European Conference on Computer Vision (ECCV)*, pp. 35–46.

Rehg, J. and Witkin, A. (1991). Visual tracking with deformation models. In *IEEE International Conference on Robotics and Automation*, pp. 844–850.

Rehg, J., Morris, D. D., and Kanade, T. (2003). Ambiguities in visual tracking of articulated objects using two- and three-dimensional models. *International Journal of Robotics Research*, 22(6):393–418.

Reichenbach, S. E., Park, S. K., and Narayanswamy, R. (1991). Characterizing digital image acquisition devices. *Optical Engineering*, 30(2):170–177.

Reinhard, E., Stark, M., Shirley, P., and Ferwerda, J. (2002). Photographic tone reproduction for digital images. *ACM Transactions on Graphics (Proc. SIGGRAPH)*, 21(3):267–276.

Reinhard, E., Heidrich, W., Debevec, P., Pattanaik, S., Ward, G., and Myszkowski, K. (2010). *High dynamic range imaging: acquisition, display, and image-based lighting*. Morgan Kaufmann, 2nd edition.

Reiser, C., Peng, S., Liao, Y., and Geiger, A. (2021). KiloNeRF: Speeding up neural radiance fields with thousands of tiny mlps. *arXiv preprint arXiv:2103.13744*.

Reitmayr, G. and Drummond, T. W. (2006). Going out: robust model-based tracking for outdoor augmented reality. In *IEEE/ACM International Symposium on Mixed and Augmented Reality (ISMAR)*, pp. 109–118.

Remelli, E., Lukoianov, A., Richter, S., Guillard, B., Bagautdinov, T., Baque, P., and Fua, P. (2020). MeshSDF: Differentiable iso-surface extraction. In *Advances in Neural Information Processing Systems (NeurIPS)*, pp. 22468–22478.

Ren, J. (2020). Awesome self-supervised learning. online bibliography at https://github.com/jason718/awesome-self-supervised-learning.

Ren, M., Zeng, W., Yang, B., and Urtasun, R. (2018). Learning to reweight examples for robust deep learning. *arXiv preprint arXiv:1803.09050*.

Ren, S., Cao, X., Wei, Y., and Sun, J. (2014). Face alignment at 3000 fps via regressing local binary features. In *IEEE Conference on Computer Vision and Pattern Recognition (CVPR)*.

Ren, S., He, K., Girshick, R., and Sun, J. (2015). Faster R-CNN: Towards real-time object detection with region proposal networks. In *Advances in Neural Information Processing Systems (NeurIPS)*, pp. 91–99.

Ren, Y., Yu, X., Zhang, R., Li, T. H., Liu, S., and Li, G. (2019). StructureFlow: Image inpainting via structure-aware appearance flow. In *IEEE International Conference on Computer Vision (ICCV)*.

Ren, Z., Gallo, O., Sun, D., Yang, M.-H., Sudderth, E., and Kautz, J. (2019). A fusion approach for multi-frame optical flow estimation. In *IEEE Winter Conference on Applications of Computer Vision (WACV)*, pp. 2077–2086.

Revaud, J., Weinzaepfel, P., Harchaoui, Z., and Schmid, C. (2015). EpicFlow: Edge-preserving interpolation of correspondences for optical flow. In *IEEE Conference on Computer Vision and Pattern Recognition (CVPR)*.

Revaud, J., Weinzaepfel, P., Souza, C. D., and Humenberger, M. (2019). R2D2: repeatable and reliable detector and descriptor. In *Advances in Neural Information Processing Systems (NeurIPS)*.

Rezatofighi, H., Tsoi, N., Gwak, J., Sadeghian, A., Reid, I., and Savarese, S. (2019). Generalized intersection over union: A metric and a loss for bounding box regression. In *IEEE/CVF Conference on Computer Vision and Pattern Recognition (CVPR)*.

Rezende, D. J., Mohamed, S., and Wierstra, D. (2014). Stochastic backpropagation and approximate inference in deep generative models. In *International Conference on Machine Learning (ICML)*, pp. 1278–1286.

Rhemann, C., Rother, C., and Gelautz, M. (2008). Improving color modeling for alpha matting. In *British Machine Vision Conference (BMVC)*.

Rhemann, C., Rother, C., Rav-Acha, A., and Sharp, T. (2008). High resolution matting via interactive trimap segmentation. In *IEEE Computer Society Conference on Computer Vision and Pattern Recognition (CVPR)*.

Rhemann, C., Rother, C., Wang, J., Gelautz, M., Kohli, P., and Rott, P. (2009). A perceptually motivated online benchmark for image matting. In *IEEE Computer Society Conference on Computer Vision and Pattern Recognition (CVPR)*.

Richardson, I. E. G. (2003). *H.264 and MPEG-4 Video Compression: Video Coding for Next Generation Multimedia*. Wiley.

Richardt, C., Pritch, Y., Zimmer, H., and Sorkine-Hornung, A. (2013). Megastereo: Constructing high-resolution stereo panoramas. In *IEEE Computer Society Conference on Computer Vision and Pattern Recognition (CVPR)*.

Richter, S. R. and Roth, S. (2018). Matryoshka networks: Predicting 3D geometry via nested shape layers. In *IEEE Conference on Computer Vision and Pattern Recognition (CVPR)*.

Richter, S. R., Hayder, Z., and Koltun, V. (2017). Playing for benchmarks. In *IEEE International Conference on Computer Vision (ICCV)*.

Riegler, G. and Koltun, V. (2020a). Free view synthesis. In *European Conference on Computer Vision (ECCV)*.

Riegler, G. and Koltun, V. (2020b). Stable view synthesis. *arXiv preprint arXiv:2011.07233*.

Rioul, O. and Vetterli, M. (1991). Wavelets and signal processing. *IEEE Signal Processing Magazine*, 8(4):14–38.

Rioux, M. and Bird, T. (1993). White laser, synced scan. *IEEE Computer Graphics and Applications*, 13(3):15–17.

Rioux, M., Bechthold, G., Taylor, D., and Duggan, M. (1987). Design of a large depth of view three-dimensional camera for robot vision. *Optical Engineering*, 26(12):1245–1250.

Rippel, O. and Bourdev, L. (2017). Real-time adaptive image compression. In *International Conference on Machine Learning (ICML)*, p. 2922–2930.

Rippel, O., Nair, S., Lew, C., Branson, S., Anderson, A. G., and Bourdev, L. (2019). Learned video compression. In *IEEE/CVF International Conference on Computer Vision (ICCV)*.

Riseman, E. M. and Arbib, M. A. (1977). Computational techniques in the visual segmentation of static scenes. *Computer Graphics and Image Processing*, 6(3):221–276.

Ritter, G. X. and Wilson, J. N. (2000). *Handbook of Computer Vision Algorithms in Image Algebra.* CRC Press, Boca Raton, 2nd edition.

Robert, C. P. (2007). *The Bayesian Choice: From Decision-Theoretic Foundations to Computational Implementation.* Springer-Verlag, New York.

Roberts, L. G. (1965). Machine perception of three-dimensional solids. In Tippett, J. T., Borkowitz, D. A., Clapp, L. C., Koester, C. J., and Vanderburgh Jr., A. (eds), *Optical and Electro-Optical Information Processing*, pp. 159–197, MIT Press, Cambridge, Massachusetts.

Roberts, R., Sinha, S. N., Szeliski, R., and Steedly, D. (2011). Structure from motion for scenes with large duplicate structures. In *IEEE Computer Society Conference on Computer Vision and Pattern Recognition (CVPR)*, pp. 3137–3144.

Robertson, D. and Cipolla, R. (2004). An image-based system for urban navigation. In *British Machine Vision Conference*, pp. 656–665.

Robertson, D. P. and Cipolla, R. (2002). Building architectural models from many views using map constraints. In *European Conference on Computer Vision (ECCV)*, pp. 155–169.

Robertson, D. P. and Cipolla, R. (2009). Architectural modelling. In Varga, M. (ed.), *Practical Image Processing and Computer Vision*, John Wiley.

Robertson, N. and Reid, I. (2006). A general method for human activity recognition in video. *Computer Vision and Image Understanding*, 104(2–3):232–248.

Roble, D. (1999). Vision in film and special effects. *Computer Graphics*, 33(4):58–60.

Roble, D. and Zafar, N. B. (2009). Don't trust your eyes: cutting-edge visual effects. *Computer*, 42(7):35–41.

Rodriguez, M., Ahmed, J., and Shah, M. (2008). Action MACH: A spatio-temporal maximum average correlation height filter for action recognition. In *IEEE Computer Society Conference on Computer Vision and Pattern Recognition (CVPR).*

Rodríguez, R. G., Vazquez-Corral, J., and Bertalmío, M. (2019). Issues with common assumptions about the camera pipeline and their impact in HDR imaging from multiple exposures. *SIAM Journal on Imaging Sciences*, 12(4):1627–1642.

Rogez, G., Rihan, J., Ramalingam, S., Orrite, C., and Torr, P. H. S. (2008). Randomized trees for human pose detection. In *IEEE Computer Society Conference on Computer Vision and Pattern Recognition (CVPR).*

Rogmans, S., Lu, J., Bekaert, P., and Lafruit, G. (2009). Real-time stereo-based views synthesis algorithms: A unified framework and evaluation on commodity GPUs. *Signal Processing: Image Communication*, 24:49–64.

Rohr, K. (1994). Towards model-based recognition of human movements in image sequences. *Computer Vision, Graphics, and Image Processing*, 59(1):94–115.

Román, A. and Lensch, H. P. A. (2006). Automatic multiperspective images. In *Eurographics Symposium on Rendering*, pp. 83–92.

Román, A., Garg, G., and Levoy, M. (2004). Interactive design of multi-perspective images for visualizing urban landscapes. In *IEEE Visualization*, pp. 537–544.

Romdhani, S. and Vetter, T. (2003). Efficient, robust and accurate fitting of a 3D morphable model. In *International Conference on Computer Vision (ICCV)*, pp. 59–66.

Romdhani, S., Torr, P. H. S., Schölkopf, B., and Blake, A. (2001). Computationally efficient face detection. In *International Conference on Computer Vision (ICCV)*, pp. 695–700.

Romero, J., Tzionas, D., and Black, M. J. (2017). Embodied hands: Modeling and capturing hands and bodies together. *ACM Transactions on Graphics (Proc. SIGGRAPH Asia)*, 36(6):1–17.

Rong, X., Demandolx, D., Matzen, K., Chatterjee, P., and Tian, Y. (2020). Burst denoising via temporally shifted wavelet transforms. In *European Conference on Computer Vision (ECCV).*

Rong, Y., Shiratori, T., and Joo, H. (2020). FrankMocap: Fast monocular 3D hand and body motion capture by regression and integration. *arXiv preprint arXiv:2008.08324.*

Ronneberger, O., Fischer, P., and Brox, T. (2015). U-Net: Convolutional networks for biomedical image segmentation. In *International Conference on Medical Image Computing and Computer-Assisted Intervention – MICCAI*, pp. 234–241.

Rosales, R. and Sclaroff, S. (2000). Inferring body pose without tracking body parts. In *IEEE Computer Society Conference on Computer Vision and Pattern Recognition (CVPR)*, pp. 721–727.

Rosenblatt, F. (1958). The perceptron: a probabilistic model for information storage and organization in the brain. *Psychological Review*, 65(6):386.

Rosenfeld, A. (1980). Quadtrees and pyramids for pattern recognition and image processing. In *International Conference on Pattern Recognition (ICPR)*, pp. 802–809.

Rosenfeld, A. (ed.). (1984). *Multiresolution Image Processing and Analysis*, Springer-Verlag, New York.

Rosenfeld, A. and Davis, L. S. (1979). Image segmentation and image models. *Proceedings of the IEEE*, 67(5):764–772.

Rosenfeld, A. and Kak, A. C. (1976). *Digital Picture Processing*. Academic Press, New York.

Rosenfeld, A. and Pfaltz, J. L. (1966). Sequential operations in digital picture processing. *Journal of the ACM*, 13(4):471–494.

Rosenfeld, A., Hummel, R. A., and Zucker, S. W. (1976). Scene labeling by relaxation operations. *IEEE Transactions on Systems, Man, and Cybernetics*, SMC-6:420–433.

Ross, D. A., Lim, J., Lin, R.-S., and Yang, M.-H. (2008). Incremental learning for robust visual tracking. *International Journal of Computer Vision*, 77(1-3):125–141.

Rosten, E. and Drummond, T. (2005). Fusing points and lines for high performance tracking. In *International Conference on Computer Vision (ICCV)*, pp. 1508–1515.

Rosten, E., Porter, R., and Drummond, T. (2010). Faster and better: A machine learning approach to corner detection. *IEEE Transactions on Pattern Analysis and Machine Intelligence*, 32(1):105–119.

Roth, S. and Black, M. J. (2007a). On the spatial statistics of optical flow. *International Journal of Computer Vision*, 74(1):33–50.

Roth, S. and Black, M. J. (2007b). Steerable random fields. In *International Conference on Computer Vision (ICCV)*.

Roth, S. and Black, M. J. (2009). Fields of experts. *International Journal of Computer Vision*, 82(2):205–229.

Rother, C. (2002). A new approach for vanishing point detection in architectural environments. *Image and Vision Computing*, 20(9–10):647–656.

Rother, C. (2003). Linear multi-view reconstruction of points, lines, planes and cameras using a reference plane. In *International Conference on Computer Vision (ICCV)*, pp. 1210–1217.

Rother, C. and Carlsson, S. (2002). Linear multi view reconstruction and camera recovery using a reference plane. *International Journal of Computer Vision*, 49(2/3):117–141.

Rother, C., Kolmogorov, V., and Blake, A. (2004). "GrabCut"—interactive foreground extraction using iterated graph cuts. *ACM Transactions on Graphics (Proc. SIGGRAPH)*, 23(3):309–314.

Rother, C., Kohli, P., Feng, W., and Jia, J. (2009). Minimizing sparse higher order energy functions of discrete variables. In *IEEE Computer Society Conference on Computer Vision and Pattern Recognition (CVPR)*.

Rother, C., Kolmogorov, V., Lempitsky, V., and Szummer, M. (2007). Optimizing binary MRFs via extended roof duality. In *IEEE Computer Society Conference on Computer Vision and Pattern Recognition (CVPR)*.

Rothganger, F., Lazebnik, S., Schmid, C., and Ponce, J. (2006). 3D object modeling and recognition using local affine-invariant image descriptors and multi-view spatial constraints. *International Journal of Computer Vision*, 66(3):231–259.

Rousseeuw, P. J. (1984). Least median of squares regresssion. *Journal of the American Statistical Association*, 79:871–880.

Rousseeuw, P. J. and Leroy, A. M. (1987). *Robust Regression and Outlier Detection*. Wiley, New York.

Rousson, M. and Paragios, N. (2008). Prior knowledge, level set representations, and visual grouping. *International Journal of Computer Vision*, 76(3):231–243.

Roweis, S. T. and Saul, L. K. (2000). Nonlinear dimensionality reduction by locally linear embedding. *science*, 290(5500):2323–2326.

Rowland, D. A. and Perrett, D. I. (1995). Manipulating facial appearance through shape and color. *IEEE Computer Graphics and Applications*, 15(5):70–76.

Rowley, H. A., Baluja, S., and Kanade, T. (1998). Neural network-based face detection. *IEEE Transactions on Pattern Analysis and Machine Intelligence*, 20(1):23–38.

Roy, S. and Cox, I. J. (1998). A maximum-flow formulation of the N-camera stereo correspondence problem. In *International Conference on Computer Vision (ICCV)*, pp. 492–499.

Rozenfeld, S., Shimshoni, I., and Lindenbaum, M. (2011). Dense mirroring surface recovery from 1d homographies and sparse correspondences. *IEEE Transactions on Pattern Analysis and Machine Intelligence*, 33(2):325–327.

Rublee, E., Rabaud, V., Konolige, K., and Bradski, G. (2011). ORB: an efficient alternative to SIFT or SURF. In *International Conference on Computer Vision (ICCV)*.

Rudin, L. I., Osher, S., and Fatemi, E. (1992). Nonlinear total variation based noise removal algorithms. *Physica D: Nonlinear Phenomena*, 60(1–4):259–268.

Rumelhart, D. E., Hinton, G. E., and Williams, R. J. (1986a). Learning internal representations by error propagation. In Rumelhart, D. E., McClelland, J. L., and the PDP research group (eds), *Parallel Distributed Processing: Explorations in the Microstructure of Cognition*, pp. 318–362, Bradford Books, Cambridge, Massachusetts.

Rumelhart, D. E., Hinton, G. E., and Williams, R. J. (1986b). Learning internal representations by error propagation. *Nature*, 323:533–536.

Rusinkiewicz, S. (2019). A symmetric objective function for ICP. *ACM Transactions on Graphics (Proc. SIGGRAPH)*, 38(4):85:1–85:7.

Rusinkiewicz, S. and Levoy, M. (2000). QSplat: A multiresolution point rendering system for large meshes. In *ACM SIGGRAPH Conference Proceedings*, pp. 343–352.

Rusinkiewicz, S., Hall-Holt, O., and Levoy, M. (2002). Real-time 3D model acquisition. *ACM Transactions on Graphics*, 21(3):438–446.

Russ, J. C. (2007). *The Image Processing Handbook*. CRC Press, Boca Raton, 5th edition.

Russakovsky, O., Deng, J., Su, H., Krause, J., Satheesh, S., Ma, S., Huang, Z., Karpathy, A., Khosla, A., Bernstein, M., Berg, A. C., and Fei-Fei, L. (2015). ImageNet large scale visual recognition challenge. *International Journal of Computer Vision*, 115(3):211–252.

Russell, B., Efros, A., Sivic, J., Freeman, W., and Zisserman, A. (2006). Using multiple segmentations to discover objects and their extent in image collections. In *IEEE Computer Society Conference on Computer Vision and Pattern Recognition (CVPR)*, pp. 1605–1612.

Russell, B. C., Torralba, A., Murphy, K. P., and Freeman, W. T. (2008). LabelMe: A database and web-based tool for image annotation. *International Journal of Computer Vision*, 77(1–3):157–173.

Russell, B. C., Torralba, A., Liu, C., Fergus, R., and Freeman, W. T. (2007). Object recognition by scene alignment. In *Advances in Neural Information Processing Systems (NeurIPS)*.

Ruzon, M. A. and Tomasi, C. (2000). Alpha estimation in natural images. In *IEEE Computer Society Conference on Computer Vision and Pattern Recognition (CVPR)*, pp. 18–25.

Ruzon, M. A. and Tomasi, C. (2001). Edge, junction, and corner detection using color distributions. *IEEE Transactions on Pattern Analysis and Machine Intelligence*, 23(11):1281–1295.

Ryan, T. W., Gray, R. T., and Hunt, B. R. (1980). Prediction of correlation errors in stereo-pair images. *Optical Engineering*, 19(3):312–322.

Saad, Y. (2003). *Iterative Methods for Sparse Linear Systems*. Society for Industrial and Applied Mathematics, 2nd edition.

Saikia, T., Marrakchi, Y., Zela, A., Hutter, F., and Brox, T. (2019). AutoDispNet: Improving disparity estimation with AutoML. In *IEEE/CVF International Conference on Computer Vision (ICCV)*.

Saint-Marc, P., Chen, J. S., and Medioni, G. (1991). Adaptive smoothing: A general tool for early vision. *IEEE Transactions on Pattern Analysis and Machine Intelligence*, 13(6):514–529.

Saito, H. and Kanade, T. (1999). Shape reconstruction in projective grid space from large number of images. In *IEEE Computer Society Conference on Computer Vision and Pattern Recognition (CVPR)*, pp. 49–54.

Saito, S., Simon, T., Saragih, J., and Joo, H. (2020). PIFuHD: Multi-level pixel-aligned implicit function for high-resolution 3D human digitization. In *IEEE/CVF Conference on Computer Vision and Pattern Recognition (CVPR)*.

Saito, S., Huang, Z., Natsume, R., Morishima, S., Kanazawa, A., and Li, H. (2019). PIFu: Pixel-aligned implicit function for high-resolution clothed human digitization. In *IEEE/CVF International Conference on Computer Vision (ICCV)*.

Salakhutdinov, R. and Hinton, G. (2009). Deep boltzmann machines. In *Artificial Intelligence and Statistics*, pp. 448–455.

Salimans, T. and Kingma, D. P. (2016). Weight normalization: A simple reparameterization to accelerate training of deep neural networks. In *Advances in Neural Information Processing Systems (NeurIPS)*, pp. 901–909.

Salzmann, M. and Fua, P. (2010). Deformable surface 3d reconstruction from monocular images. *Synthesis Lectures on Computer Vision*, 2(1):1–113.

Samano, N., Zhou, M., and Calway, A. (2020). You are here: Geolocation by embedding maps and images. In *European Conference on Computer Vision (ECCV)*.

Samet, H. (1989). *The Design and Analysis of Spatial Data Structures*. Addison-Wesley, Reading, Massachusetts.

Sander, P. T. and Zucker, S. W. (1990). Inferring surface trace and differential structure from 3-D images. *IEEE Transactions on Pattern Analysis and Machine Intelligence*, 12(9):833–854.

Sandler, M., Howard, A., Zhu, M., Zhmoginov, A., and Chen, L.-C. (2018). Mobilenetv2: Inverted residuals and linear bottlenecks. In *IEEE Conference on Computer Vision and Pattern Recognition (CVPR)*.

Sang, S. and Chandraker, M. (2020). Single-shot neural relighting and SVBRDF estimation. In *European Conference on Computer Vision (ECCV)*.

Santo, H., Waechter, M., and Matsushita, Y. (2020). Deep near-light photometric stereo for spatially varying reflectances. In *European Conference on Computer Vision (ECCV)*.

Santurkar, S., Tsipras, D., Ilyas, A., and Madry, A. (2018). How does batch normalization help optimization? In *Advances in Neural Information Processing Systems (NeurIPS)*, pp. 2483–2493.

Sapiro, G. (2001). *Geometric Partial Differential Equations and Image Analysis*. Cambridge University Press.

Sapp, B. and Taskar, B. (2013). MODEC: Multimodal decomposable models for human pose estimation. In *IEEE Computer Society Conference on Computer Vision and Pattern Recognition (CVPR)*.

Sarlin, P.-E., Unagar, A., Larsson, M., Germain, H., Toft, C., Larsson, V., Pollefeys, M., Lepetit, V., Hammarstrand, L., Kahl, F., and Sattler, T. (2021). Back to the feature: Learning robust camera localization from pixels to pose. In *IEEE/CVF Conference on Computer Vision and Pattern Recognition (CVPR)*.

Sarlin, P.-E., Cadena, C., Siegwart, R., and Dymczyk, M. (2019). From coarse to fine: Robust hierarchical localization at large scale. In *IEEE/CVF Conference on Computer Vision and Pattern Recognition (CVPR)*.

Sarlin, P.-E., DeTone, D., Malisiewicz, T., and Rabinovich, A. (2020). SuperGlue: Learning feature matching with graph neural networks. In *IEEE/CVF Conference on Computer Vision and Pattern Recognition (CVPR)*.

Sato, Y. and Ikeuchi, K. (1996). Reflectance analysis for 3D computer graphics model generation. *Graphical Models and Image Processing*, 58(5):437–451.

Sato, Y., Wheeler, M., and Ikeuchi, K. (1997). Object shape and reflectance modeling from observation. In *ACM SIGGRAPH Conference Proceedings*, pp. 379–387.

Sattler, T., Maddern, W., Toft, C., Torii, A., Hammarstrand, L., Stenborg, E., Safari, D., Okutomi, M., Pollefeys, M., Sivic, J., Kahl, F., and Pajdla, T. (2018). Benchmarking 6DOF outdoor visual localization in changing conditions. In *IEEE Conference on Computer Vision and Pattern Recognition (CVPR)*.

Sattler, T., Leibe, B., and Kobbelt, L. (2011). Fast image-based localization using direct 2D-to-3D matching. In *International Conference on Computer Vision (ICCV)*.

Sattler, T., Leibe, B., and Kobbelt, L. (2017). Efficient & effective prioritized matching for large-scale image-based localization. *IEEE Transactions on Pattern Analysis and Machine Intelligence*, 39(9):1744–1756.

Sattler, T., Havlena, M., Schindler, K., and Pollefeys, M. (2016). Large-scale location recognition and the geometric burstiness problem. In *IEEE Conference on Computer Vision and Pattern Recognition (CVPR)*.

Sattler, T., Zhou, Q., Pollefeys, M., and Leal-Taixé, L. (2019). Understanding the limitations of CNN-based absolute camera pose regression. In *IEEE/CVF Conference on Computer Vision and Pattern Recognition (CVPR)*.

Savarese, S., Chen, M., and Perona, P. (2005). Local shape from mirror reflections. *International Journal of Computer Vision*, 64(1):31–67.

Savarese, S., Andreetto, M., Rushmeier, H. E., Bernardini, F., and Perona, P. (2007). 3D reconstruction by shadow carving: Theory and practical evaluation. *International Journal of Computer Vision*, 71(3):305–336.

Sawhney, H. S. (1994). Simplifying motion and structure analysis using planar parallax and image warping. In *International Conference on Pattern Recognition (ICPR)*, pp. 403–408.

Sawhney, H. S. and Ayer, S. (1996). Compact representation of videos through dominant multiple motion estimation. *IEEE Transactions on Pattern Analysis and Machine Intelligence*, 18(8):814–830.

Sawhney, H. S. and Hanson, A. R. (1991). Identification and 3D description of 'shallow' environmental structure over a sequence of images. In *IEEE Computer Society Conference on Computer Vision and Pattern Recognition (CVPR)*, pp. 179–185.

Sawhney, H. S. and Kumar, R. (1999). True multi-image alignment and its application to mosaicing and lens distortion correction. *IEEE Transactions on Pattern Analysis and Machine Intelligence*, 21(3):235–243.

Sawhney, H. S., Kumar, R., Gendel, G., Bergen, J., Dixon, D., and Paragano, V. (1998). VideoBrush: Experiences with consumer video mosaicing. In *IEEE Workshop on Applications of Computer Vision (WACV)*, pp. 56–62.

Sawhney, H. S., Arpa, A., Kumar, R., Samarasekera, S., Aggarwal, M., Hsu, S., Nister, D., and Hanna, K. (2002). Video flashlights: real time rendering of multiple videos for immersive model visualization. In *Eurographics Workshop on Rendering*, pp. 157–168.

Saxena, A., Sun, M., and Ng, A. Y. (2009). Make3D: Learning 3D scene structure from a single still image. *IEEE Transactions on Pattern Analysis and Machine Intelligence*, 31(5):824–840.

Schaffalitzky, F. and Zisserman, A. (2000). Planar grouping for automatic detection of vanishing lines and points. *Image and Vision Computing*, 18:647–658.

Schaffalitzky, F. and Zisserman, A. (2002). Multi-view matching for unordered image sets, or "How do I organize my holiday snaps?". In *European Conference on Computer Vision (ECCV)*, pp. 414–431.

Scharr, H., Black, M. J., and Haussecker, H. W. (2003). Image statistics and anisotropic diffusion. In *International Conference on Computer Vision (ICCV)*, pp. 840–847.

Scharstein, D. (1994). Matching images by comparing their gradient fields. In *International Conference on Pattern Recognition (ICPR)*, pp. 572–575.

Scharstein, D. (1999). *View Synthesis Using Stereo Vision*. Volume 1583, Springer-Verlag.

Scharstein, D. and Szeliski, R. (1998). Stereo matching with nonlinear diffusion. *International Journal of Computer Vision*, 28(2):155–174.

Scharstein, D. and Szeliski, R. (2002). A taxonomy and evaluation of dense two-frame stereo correspondence algorithms. *International Journal of Computer Vision*, 47(1):7–42.

Scharstein, D. and Szeliski, R. (2003). High-accuracy stereo depth maps using structured light. In *IEEE Computer Society Conference on Computer Vision and Pattern Recognition (CVPR)*, pp. 195–202.

Scharstein, D., Hirschmüller, H., Kitajima, Y., Krathwohl, G., Nešić, N., Wang, X., and Westling, P. (2014). High-resolution stereo datasets with subpixel-accurate ground truth. In *German Conference on Pattern Recognition*, pp. 31–42.

Schechner, Y. Y., Nayar, S. K., and Belhumeur, P. N. (2009). Multiplexing for optimal lighting. *IEEE Transactions on Pattern Analysis and Machine Intelligence*, 29(8):1339–1354.

Schilling, H., Gutsche, M., Brock, A., Spath, D., Rother, C., and Krispin, K. (2020). Mind the gap - a benchmark for dense depth prediction beyond lidar. In *IEEE/CVF Conference on Computer Vision and Pattern Recognition (CVPR) Workshops*.

Schindler, G. and Dellaert, F. (2004). Atlanta world: An expectation maximization framework for simultaneous low-level edge grouping and camera calibration in complex man-made environments. In *IEEE Computer Society Conference on Computer Vision and Pattern Recognition (CVPR)*.

Schindler, G., Brown, M., and Szeliski, R. (2007). City-scale location recognition. In *IEEE Computer Society Conference on Computer Vision and Pattern Recognition (CVPR)*.

Schindler, G., Krishnamurthy, P., Lublinerman, R., Liu, Y., and Dellaert, F. (2008). Detecting and matching repeated patterns for automatic geo-tagging in urban environments. In *IEEE Computer Society Conference on Computer Vision and Pattern Recognition (CVPR)*.

Schmid, C. and Mohr, R. (1997). Local grayvalue invariants for image retrieval. *IEEE Transactions on Pattern Analysis and Machine Intelligence*, 19(5):530–534.

Schmid, C. and Zisserman, A. (1997). Automatic line matching across views. In *IEEE Computer Society Conference on Computer Vision and Pattern Recognition (CVPR)*, pp. 666–671.

Schmid, C., Mohr, R., and Bauckhage, C. (2000). Evaluation of interest point detectors. *International Journal of Computer Vision*, 37(2):151–172.

Schmidt, U. and Roth, S. (2014). Shrinkage fields for effective image restoration. In *IEEE Conference on Computer Vision and Pattern Recognition (CVPR)*.

Schmitt, C., Donne, S., Riegler, G., Koltun, V., and Geiger, A. (2020). On joint estimation of pose, geometry and svBRDF from a handheld scanner. In *IEEE/CVF Conference on Computer Vision and Pattern Recognition (CVPR)*.

Schneiderman, H. and Kanade, T. (2004). Object detection using the statistics of parts. *International Journal of Computer Vision*, 56(3):151–177.

Schödl, A. and Essa, I. (2002). Controlled animation of video sprites. In *ACM Symposium on Computater Animation*.

Schödl, A., Szeliski, R., Salesin, D. H., and Essa, I. (2000). Video textures. In *ACM SIGGRAPH Conference Proceedings*, pp. 489–498.

Schoenemann, T. and Cremers, D. (2008). High resolution motion layer decomposition using dual-space graph cuts. In *IEEE Computer Society Conference on Computer Vision and Pattern Recognition (CVPR)*.

Schölkopf, B. and Smola, A. J. (2001). *Learning with Kernels: Support Vector Machines, Regularization, Optimization, and Beyond.* MIT Press.

Schönberger, J. L. and Frahm, J.-M. (2016). Structure-from-motion revisited. In *IEEE Conference on Computer Vision and Pattern Recognition (CVPR)*.

Schönberger, J. L., Sinha, S. N., and Pollefeys, M. (2018). Learning to fuse proposals from multiple scanline optimizations in semi-global matching. In *European Conference on Computer Vision (ECCV)*.

Schönberger, J. L., Zheng, E., Pollefeys, M., and Frahm, J.-M. (2016). Pixelwise view selection for unstructured multi-view stereo. In *European Conference on Computer Vision (ECCV)*.

Schöps, T., Sattler, T., and Pollefeys, M. (2020). SurfelMeshing: Online surfel-based mesh reconstruction. *IEEE Transactions on Pattern Analysis and Machine Intelligence*, 42(10):2494–2507.

Schöps, T., Sattler, T., and Pollefeys, M. (2019a). BAD SLAM: Bundle adjusted direct RGB-D slam. In *IEEE/CVF Conference on Computer Vision and Pattern Recognition (CVPR)*.

Schöps, T., Sattler, T., and Pollefeys, M. (2019b). SurfelMeshing: Online surfel-based mesh reconstruction. *IEEE Transactions on Pattern Analysis and Machine Intelligence*, 42(10):2494–2507.

Schöps, T., Schönberger, J. L., Galliani, S., Sattler, T., Schindler, K., Pollefeys, M., and Geiger, A. (2017). A multi-view stereo benchmark with high-resolution images and multi-camera videos. In *IEEE Conference on Computer Vision and Pattern Recognition (CVPR)*.

Schröder, P. and Sweldens, W. (1995). Spherical wavelets: Efficiently representing functions on the sphere. In *ACM SIGGRAPH Conference Proceedings*, pp. 161–172.

Schroff, F., Kalenichenko, D., and Philbin, J. (2015). FaceNet: A unified embedding for face recognition and clustering. In *IEEE Conference on Computer Vision and Pattern Recognition (CVPR)*.

Schubert, D., Demmel, N., Usenko, V., Stuckler, J., and Cremers, D. (2018). Direct sparse odometry with rolling shutter. In *European Conference on Computer Vision (ECCV)*.

Schubert, D., Goll, T., Demmel, N., Usenko, V., Stückler, J., and Cremers, D. (2018). The TUM VI benchmark for evaluating visual-inertial odometry. In *IEEE/RSJ International Conference on Intelligent Robots and Systems (IROS)*, pp. 1680–1687.

Schüldt, C., Laptev, I., and Caputo, B. (2004). Recognizing human actions: a local SVM approach. In *International Conference on Pattern Recognition (ICPR)*, pp. 32–36.

Schuler, C. J., Hirsch, M., Harmeling, S., and Schölkopf, B. (2016). Learning to deblur. *IEEE Transactions on Pattern Analysis and Machine Intelligence*, 38(7):1439–1451.

Schultz, R. R. and Stevenson, R. L. (1996). Extraction of high-resolution frames from video sequences. *IEEE Transactions on Image Processing*, 5(6):996–1011.

Se, S., Lowe, D., and Little, J. (2002). Global localization using distinctive visual features. In *IEEE/RSJ International Conference on Intelligent Robots and Systems (IROS)*, pp. 226–231.

Sebastian, T. B. and Kimia, B. B. (2005). Curves vs. skeletons in object recognition. *Signal Processing*, 85(2):246–263.

Sederberg, T. W. and Parry, S. R. (1986). Free-form deformations of solid geometric models. *Computer Graphics (SIGGRAPH)*, 20(4):151–160.

Sederberg, T. W., Gao, P., Wang, G., and Mu, H. (1993). 2D shape blending: An intrinsic solution to the vertex path problem. In *ACM SIGGRAPH Conference Proceedings*, pp. 15–18.

Seitz, P. (1989). Using local orientation information as image primitive for robust object recognition. In *SPIE Vol. 1199, Visual Communications and Image Processing IV*, pp. 1630–1639.

Seitz, S. (2001). The space of all stereo images. In *International Conference on Computer Vision (ICCV)*, pp. 26–33.

Seitz, S. and Szeliski, R. (1999). Applications of computer vision to computer graphics. *Computer Graphics*, 33(4):35–37. Guest Editors' introduction to the Special Issue.

Seitz, S., Curless, B., Diebel, J., Scharstein, D., and Szeliski, R. (2006). A comparison and evaluation of multi-view stereo reconstruction algorithms. In *IEEE Computer Society Conference on Computer Vision and Pattern Recognition (CVPR)*, pp. 519–526.

Seitz, S. M. and Baker, S. (2009). Filter flow. In *International Conference on Computer Vision (ICCV)*.

Seitz, S. M. and Dyer, C. M. (1996). View morphing. In *ACM SIGGRAPH Conference Proceedings*, pp. 21–30.

Seitz, S. M. and Dyer, C. M. (1997). Photorealistic scene reconstruction by voxel coloring. In *IEEE Computer Society Conference on Computer Vision and Pattern Recognition (CVPR)*, pp. 1067–1073.

Seitz, S. M. and Dyer, C. M. (1999). Photorealistic scene reconstruction by voxel coloring. *International Journal of Computer Vision*, 35(2):151–173.

Seitz, S. M. and Dyer, C. R. (1997). View invariant analysis of cyclic motion. *International Journal of Computer Vision*, 25(3):231–251.

Sejnowski, T. J. (2018). *The deep learning revolution*. MIT Press.

Seki, A. and Pollefeys, M. (2017). Sgm-nets: Semi-global matching with neural networks. In *IEEE Conference on Computer Vision and Pattern Recognition (CVPR)*.

Selvaraju, R. R., Cogswell, M., Das, A., Vedantam, R., Parikh, D., and Batra, D. (2017). Grad-cam: Visual explanations from deep networks via gradient-based localization. In *IEEE International Conference on Computer Vision (ICCV)*.

Sen, P., Kalantari, N. K., Yaesoubi, M., Darabi, S., Goldman, D. B., and Shechtman, E. (2012). Robust patch-based HDR reconstruction of dynamic scenes. *ACM Transactions On Graphics (Proc. SIGGRAPH)*, 31(6):203–1.

Sengupta, S., Jayaram, V., Curless, B., Seitz, S. M., and Kemelmacher-Shlizerman, I. (2020). Background matting: The world is your green screen. In *IEEE/CVF Conference on Computer Vision and Pattern Recognition (CVPR)*.

Sennrich, R., Haddow, B., and Birch, A. (2015). Neural machine translation of rare words with subword units. *arXiv preprint arXiv:1508.07909*.

Sermanet, P., Kavukcuoglu, K., Chintala, S., and Lecun, Y. (2013). Pedestrian detection with unsupervised multi-stage feature learning. In *IEEE Computer Society Conference on Computer Vision and Pattern Recognition (CVPR)*.

Serra, J. (1982). *Image Analysis and Mathematical Morphology*. Academic Press, New York.

Serra, J. and Vincent, L. (1992). An overview of morphological filtering. *Circuits, Systems and Signal Processing*, 11(1):47–108.

Serrano, A., Kim, I., Chen, Z., DiVerdi, S., Gutierrez, D., Hertzmann, A., and Masia, B. (2019). Motion parallax for 360 RGBD video. *IEEE Transactions on Visualization and Computer Graphics*, 25(5):1817–1827.

Serre, T., Wolf, L., and Poggio, T. (2005). Object recognition with features inspired by visual cortex. In *IEEE Computer Society Conference on Computer Vision and Pattern Recognition (CVPR)*, pp. 994–1000.

Sethian, J. (1999). *Level Set Methods and Fast Marching Methods*. Cambridge University Press, Cambridge, 2nd edition.

Sevilla-Lara, L., Sun, D., Jampani, V., and Black, M. J. (2016). Optical flow with semantic segmentation and localized layers. In *IEEE Conference on Computer Vision and Pattern Recognition (CVPR)*.

Shade, J., Gortler, S., He, L., and Szeliski, R. (1998). Layered depth images. In *ACM SIGGRAPH Conference Proceedings*, pp. 231–242.

Shade, J., Lischinski, D., Salesin, D., DeRose, T., and Snyder, J. (1996). Hierarchical images caching for accelerated walkthroughs of complex environments. In *ACM SIGGRAPH Conference Proceedings*, pp. 75–82.

Shafer, S. A. (1985). Using color to separate reflection components. *Color Research and Applications*, 10(4):210–218.

Shafer, S. A., Healey, G., and Wolff, L. (1992). *Physics-Based Vision: Principles and Practice*. Jones & Bartlett, Cambridge, MA.

Shafique, K. and Shah, M. (2005). A noniterative greedy algorithm for multiframe point correspondence. *IEEE Transactions on Pattern Analysis and Machine Intelligence*, 27(1):51–65.

Shah, J. (1993). A nonlinear diffusion model for discontinuous disparity and half-occlusion in stereo. In *IEEE Computer Society Conference on Computer Vision and Pattern Recognition (CVPR)*, pp. 34–40.

Shaham, T. R., Dekel, T., and Michaeli, T. (2019). SinGAN: Learning a generative model from a single natural image. In *IEEE/CVF International Conference on Computer Vision (ICCV)*.

Shakhnarovich, G., Darrell, T., and Indyk, P. (eds). (2006). *Nearest-Neighbor Methods in Learning and Vision: Theory and Practice*, MIT Press.

Shakhnarovich, G., Viola, P., and Darrell, T. (2003). Fast pose estimation with parameter-sensitive hashing. In *International Conference on Computer Vision (ICCV)*, pp. 750–757.

Shan, C., Gong, S., and McOwan, P. W. (2009). Facial expression recognition based on local binary patterns: A comprehensive study. *Image and Vision Computing*, 27(6):803–816.

Shan, D., Geng, J., Shu, M., and Fouhey, D. F. (2020). Understanding human hands in contact at internet scale. In *IEEE/CVF Conference on Computer Vision and Pattern Recognition (CVPR)*.

Shan, Q., Adams, R., Curless, B., Furukawa, Y., and Seitz, S. M. (2013). The visual Turing test for scene reconstruction. In *International Conference on 3D Vision*, pp. 25–32.

Shan, Q., Wu, C., Curless, B., Furukawa, Y., Hernández, C., and Seitz, S. M. (2014). Accurate geo-registration by ground-to-aerial image matching. In *International Conference on 3D Vision*, pp. 525–532.

Shan, Y., Liu, Z., and Zhang, Z. (2001). Model-based bundle adjustment with application to face modeling. In *International Conference on Computer Vision (ICCV)*, pp. 644–641.

Shane, J. (2019). *You Look Like a Thing and I Love You: How Artificial Intelligence Works and why It's Making the World a Weirder Place*. Voracious.

Shao, S., Li, Z., Zhang, T., Peng, C., Yu, G., Zhang, X., Li, J., and Sun, J. (2019). Objects365: A large-scale, high-quality dataset for object detection. In *IEEE/CVF International Conference on Computer Vision (ICCV)*.

Sharif Razavian, A., Azizpour, H., Sullivan, J., and Carlsson, S. (2014). CNN features off-the-shelf: an astounding baseline for recognition. In *IEEE Conference on Computer Vision and Pattern Recognition Workshops*, pp. 806–813.

Sharma, P., Ding, N., Goodman, S., and Soricut, R. (2018). Conceptual captions: A cleaned, hypernymed, image alt-text dataset for automatic image captioning. In *Annual Meeting of the Association for Computational Linguistics (ACL)*, pp. 2556–2565.

Sharon, E., Galun, M., Sharon, D., Basri, R., and Brandt, A. (2006). Hierarchy and adaptivity in segmenting visual scenes. *Nature*, 442(7104):810–813.

Sharp, T., Keskin, C., Robertson, D., Taylor, J., Shotton, J., Kim, D., Rhemann, C., Leichter, I., Vinnikov, A., Wei, Y., Freedman, D., Kohli, P., Krupka, E., Fitzgibbon, A., and Izadi, S. (2015). Accurate, robust, and flexible real-time hand tracking. In *ACM Conference on Human Factors in Computing Systems*, pp. 3633–3642.

Shashua, A. and Toelg, S. (1997). The quadric reference surface: Theory and applications. *International Journal of Computer Vision*, 23(2):185–198.

Shashua, A. and Wexler, Y. (2001). Q-warping: Direct computation of quadratic reference surfaces. *IEEE Transactions on Pattern Analysis and Machine Intelligence*, 23(8):920–925.

Shaw, D. and Barnes, N. (2006). Perspective rectangle detection. In *ECCV Workshop on Applications of Computer Vision*.

Shewchuk, J. R. (1994). An introduction to the conjugate gradient method without the agonizing pain. Unpublished manuscript, available on author's homepage (https://www.cs.berkeley.edu/~jrs). An earlier version appeared as a Carnegie Mellon University Technical Report, CMU-CS-94-125.

Shi, B., Mo, Z., Wu, Z., Duan, D., Yeung, S.-K., and Tan, P. (2019). A benchmark dataset and evaluation for non-Lambertian and uncalibrated photometric stereo. *IEEE Transactions on Pattern Analysis and Machine Intelligence*, 41(2):271–284.

Shi, J. and Malik, J. (2000). Normalized cuts and image segmentation. *IEEE Transactions on Pattern Analysis and Machine Intelligence*, 8(22):888–905.

Shi, J. and Tomasi, C. (1994). Good features to track. In *IEEE Computer Society Conference on Computer Vision and Pattern Recognition (CVPR)*, pp. 593–600.

Shi, W., Caballero, J., Huszar, F., Totz, J., Aitken, A. P., Bishop, R., Rueckert, D., and Wang, Z. (2016). Real-time single image and video super-resolution using an efficient sub-pixel convolutional neural network. In *IEEE Conference on Computer Vision and Pattern Recognition (CVPR)*.

Shih, M.-L., Su, S.-Y., Kopf, J., and Huang, J.-B. (2020). 3D photography using context-aware layered depth inpainting. In *IEEE/CVF Conference on Computer Vision and Pattern Recognition (CVPR)*.

Shimizu, M. and Okutomi, M. (2001). Precise sub-pixel estimation on area-based matching. In *International Conference on Computer Vision (ICCV)*, pp. 90–97.

Shin, C., Jeon, H.-G., Yoon, Y., So Kweon, I., and Joo Kim, S. (2018). EPINET: A fully-convolutional neural network using epipolar geometry for depth from light field images. In *IEEE Conference on Computer Vision and Pattern Recognition (CVPR)*.

Shizawa, M. and Mase, K. (1991). A unified computational theory of motion transparency and motion boundaries based on eigenenergy analysis. In *IEEE Computer Society Conference on Computer Vision and Pattern Recognition (CVPR)*, pp. 289–295.

Shlezinger, M. I. (1968). The interaction of learning and self-organization in pattern recognition. *Kibernetica*, 4(2):66–71.

Shocher, A., Gandelsman, Y., Mosseri, I., Yarom, M., Irani, M., Freeman, W. T., and Dekel, T. (2020). Semantic pyramid for image generation. In *IEEE/CVF Conference on Computer Vision and Pattern Recognition (CVPR)*.

Shoemake, K. (1985). Animating rotation with quaternion curves. *Computer Graphics (SIGGRAPH)*, 19(3):245–254.

Shotton, J., Girshick, R., Fitzgibbon, A., Sharp, T., Cook, M., Finocchio, M., Moore, R., Kohli, P., Criminisi, A., Kipman, A., and Blake, A. (2013). Efficient human pose estimation from single depth images. *IEEE Transactions on Pattern Analysis and Machine Intelligence*, 35(12):2821–2840.

Shotton, J., Blake, A., and Cipolla, R. (2005). Contour-based learning for object detection. In *International Conference on Computer Vision (ICCV)*, pp. 503–510.

Shotton, J., Johnson, M., and Cipolla, R. (2008). Semantic texton forests for image categorization and segmentation. In *IEEE Computer Society Conference on Computer Vision and Pattern Recognition (CVPR)*.

Shotton, J., Winn, J., Rother, C., and Criminisi, A. (2009). Textonboost for image understanding: Multi-class object recognition and segmentation by jointly modeling appearance, shape and context. *International Journal of Computer Vision*, 81(1):2–23.

Shotton, J., Glocker, B., Zach, C., Izadi, S., Criminisi, A., and Fitzgibbon, A. (2013). Scene coordinate regression forests for camera relocalization in RGB-D images. In *IEEE Computer Society Conference on Computer Vision and Pattern Recognition (CVPR)*.

Shotton, J., Fitzgibbon, A., Cook, M., Sharp, T., Finocchio, M., Moore, R., Kipman, A., and Blake, A. (2011). Real-time human pose recognition in parts from single depth images. In *IEEE Computer Society Conference on Computer Vision and Pattern Recognition (CVPR)*.

Shrivastava, A., Hudson, D., Murahari, V., Das, A., Kottur, S., Batra, D., Parikh, D., and Agrawal, A. (2020). CVPR 2020 visual question answering and dialog workshop. https://visualqa.org/workshop.html.

Shufelt, J. (1999). Performance evaluation and analysis of vanishing point detection techniques. *IEEE Transactions on Pattern Analysis and Machine Intelligence*, 21(3):282–288.

Shum, H.-Y. and He, L.-W. (1999). Rendering with concentric mosaics. In *ACM SIGGRAPH Conference Proceedings*, pp. 299–306.

Shum, H.-Y. and Szeliski, R. (1999). Stereo reconstruction from multiperspective panoramas. In *International Conference on Computer Vision (ICCV)*, pp. 14–21.

Shum, H.-Y. and Szeliski, R. (2000). Construction of panoramic mosaics with global and local alignment. *International Journal of Computer Vision*, 36(2):101–130. Erratum published July 2002, 48(2):151–152.

Shum, H.-Y., Chan, S.-C., and Kang, S. B. (2007). *Image-Based Rendering*. Springer, New York, NY.

Shum, H.-Y., Han, M., and Szeliski, R. (1998). Interactive construction of 3D models from panoramic mosaics. In *IEEE Computer Society Conference on Computer Vision and Pattern Recognition (CVPR)*, pp. 427–433.

Shum, H.-Y., Ikeuchi, K., and Reddy, R. (1995). Principal component analysis with missing data and its application to polyhedral modeling. *IEEE Transactions on Pattern Analysis and Machine Intelligence*, 17(9):854–867.

Shum, H.-Y., Kang, S. B., and Chan, S.-C. (2003). Survey of image-based representations and compression techniques. *IEEE Transactions on Circuits and Systems for Video Technology*, 13(11):1020–1037.

Shum, H.-Y., Wang, L., Chai, J.-X., and Tong, X. (2002). Rendering by manifold hopping. *International Journal of Computer Vision*, 50(2):185–201.

Shum, H.-Y., Sun, J., Yamazaki, S., Li, Y., and Tang, C.-K. (2004). Pop-up light field: An interactive image-based modeling and rendering system. *ACM Transactions on Graphics*, 23(2):143–162.

Sibley, D., Mei, C., Reid, I. D., and Newman, P. (2009). Adaptive relative bundle adjustment. In *Robotics: Science and Systems*, p. 33.

Sibley, G., Mei, C., Reid, I., and Newman, P. (2010). Vast-scale outdoor navigation using adaptive relative bundle adjustment. *International Journal of Robotics Research*, 29(8):958–980.

Sidenbladh, H. and Black, M. J. (2003). Learning the statistics of people in images and video. *International Journal of Computer Vision*, 54(1):189–209.

Sidenbladh, H., Black, M. J., and Fleet, D. J. (2000). Stochastic tracking of 3D human figures using 2D image motion. In *European Conference on Computer Vision (ECCV)*, pp. 702–718.

Sigal, L. and Black, M. (2006a). Measure locally, reason globally: Occlusion-sensitive articulated pose estimation. In *IEEE Computer Society Conference on Computer Vision and Pattern Recognition (CVPR)*, pp. 2041–2048.

Sigal, L. and Black, M. J. (2006b). Predicting 3D people from 2D pictures. In *AMDO 2006 - IV Conference on Articulated Motion and Deformable Objects*, pp. 185–195.

Sigal, L., Balan, A., and Black, M. J. (2010). HumanEva: Synchronized video and motion capture dataset and baseline algorithm for evaluation of articulated human motion. *International Journal of Computer Vision*, 87(1–2):4–27.

Sigal, L., Bhatia, S., Roth, S., Black, M. J., and Isard, M. (2004). Tracking loose-limbed people. In *IEEE Computer Society Conference on Computer Vision and Pattern Recognition (CVPR)*, pp. 421–428.

Sigurdsson, G. A., Varol, G., Wang, X., Farhadi, A., Laptev, I., and Gupta, A. (2016). Hollywood in homes: Crowdsourcing data collection for activity understanding. In *European Conference on Computer Vision*, pp. 510–526.

Silberman, N., Hoiem, D., Kohli, P., and Fergus, R. (2012). Indoor segmentation and support inference from RGBD images. In *European Conference on Computer Vision (ECCV)*, pp. 746–760.

Sillion, F. and Puech, C. (1994). *Radiosity and Global Illumination*. Morgan Kaufmann.

Silpa-Anan, C. and Hartley, R. (2008). Optimised kd-trees for fast image descriptor matching. In *IEEE Conference on Computer Vision and Pattern Recognition*, pp. 1–8.

Sim, T., Baker, S., and Bsat, M. (2003). The CMU pose, illumination, and expression database. *IEEE Transactions on Pattern Analysis and Machine Intelligence*, 25(12):1615–1618.

Simard, P. Y., Steinkraus, D., and Platt, J. C. (2003). Best practices for convolutional neural networks applied to visual document analysis. In *International Conference on Document Analysis and Recognition (ICDAR)*, pp. 958–962.

Simard, P. Y., Bottou, L., Haffner, P., and Cun, Y. L. (1998). Boxlets: a fast convolution algorithm for signal processing and neural networks. In *Advances in Neural Information Processing Systems (NeurIPS)*, pp. 571–577.

Simon, I. and Seitz, S. M. (2008). Scene segmentation using the wisdom of crowds. In *European Conference on Computer Vision (ECCV)*, pp. 541–553.

Simon, I., Snavely, N., and Seitz, S. M. (2007). Scene summarization for online image collections. In *International Conference on Computer Vision (ICCV)*.

Simoncelli, E. P. (1999). Bayesian denoising of visual images in the wavelet domain. In Müller, P. and Vidakovic, B. (eds), *Bayesian Inference in Wavelet Based Models*, pp. 291–308, Springer-Verlag, New York.

Simoncelli, E. P. and Adelson, E. H. (1990a). Non-separable extensions of quadrature mirror filters to multiple dimensions. *Proceedings of the IEEE*, 78(4):652–664.

Simoncelli, E. P. and Adelson, E. H. (1990b). Subband transforms. In Woods, J. (ed.), *Subband Coding*, pp. 143–191, Kluwer Academic Press, Norwell, MA.

Simoncelli, E. P., Adelson, E. H., and Heeger, D. J. (1991). Probability distributions of optic flow. In *IEEE Computer Society Conference on Computer Vision and Pattern Recognition (CVPR)*, pp. 310–315.

Simoncelli, E. P., Freeman, W. T., Adelson, E. H., and Heeger, D. J. (1992). Shiftable multiscale transforms. *IEEE Transactions on Information Theory*, 38(3):587–607.

Simonyan, K. and Zisserman, A. (2014a). Two-stream convolutional networks for action recognition in videos. In *Advances in Neural Information Processing Systems (NeurIPS)*, pp. 568–576.

Simonyan, K. and Zisserman, A. (2014b). Very deep convolutional networks for large-scale image recognition. *arXiv preprint arXiv:1409.1556*.

Simonyan, K., Vedaldi, A., and Zisserman, A. (2013). Deep inside convolutional networks: Visualising image classification models and saliency maps. *arXiv preprint arXiv:1312.6034*.

Singaraju, D., Grady, L., and Vidal, R. (2008). Interactive image segmentation via minimization of quadratic energies on directed graphs. In *IEEE Computer Society Conference on Computer Vision and Pattern Recognition (CVPR)*.

Singaraju, D., Rother, C., and Rhemann, C. (2009). New appearance models for natural image matting. In *IEEE Computer Society Conference on Computer Vision and Pattern Recognition (CVPR)*.

Singaraju, D., Grady, L., Sinop, A. K., and Vidal, R. (2011). Continuous-valued MRF for image segmentation. In Blake, A., Kohli, P., and Rother, C. (eds), *Markov Random Fields for Vision and Image Processing*, pp. 127–142, MIT Press.

Singh, K. K., Mahajan, D., Grauman, K., Lee, Y. J., Feiszli, M., and Ghadiyaram, D. (2020). Don't judge an object by its context: Learning to overcome contextual bias. In *IEEE/CVF Conference on Computer Vision and Pattern Recognition (CVPR)*.

Sinha, P., Balas, B., Ostrovsky, Y., and Russell, R. (2006). Face recognition by humans: Nineteen results all computer vision researchers should know about. *Proceedings of the IEEE*, 94(11):1948–1962.

Sinha, S. N. and Pollefeys, M. (2005). Multi-view reconstruction using photo-consistency and exact silhouette constraints: A maximum-flow formulation. In *International Conference on Computer Vision (ICCV)*, pp. 349–356.

Sinha, S. N., Scharstein, D., and Szeliski, R. (2014). Efficient high-resolution stereo matching using local plane sweeps. In *IEEE Computer Society Conference on Computer Vision and Pattern Recognition (CVPR)*.

Sinha, S. N., Steedly, D., and Szeliski, R. (2009). Piecewise planar stereo for image-based rendering. In *IEEE International Conference on Computer Vision (ICCV)*, pp. 1881–1888.

Sinha, S. N., Steedly, D., and Szeliski, R. (2010). A multi-stage linear approach to structure from motion. In *ECCV Workshop on Reconstruction and Modeling of Large-Scale 3D Virtual Environments*.

Sinha, S. N., Kopf, J., Goesele, M., Scharstein, D., and Szeliski, R. (2012). Image-based rendering for scenes with reflections. *ACM Transactions on Graphics (Proc. SIGGRAPH)*, 31(4):100.

Sinha, S. N., Steedly, D., Szeliski, R., Agrawala, M., and Pollefeys, M. (2008). Interactive 3D architectural modeling from unordered photo collections. *ACM Transactions on Graphics (Proc. SIGGRAPH Asia)*, 27(5):159.

Sinop, A. K. and Grady, L. (2007). A seeded image segmentation framework unifying graph cuts and random walker which yields a new algorithm. In *International Conference on Computer Vision (ICCV)*.

Sitzmann, V., Zollhöfer, M., and Wetzstein, G. (2019). Scene representation networks: Continuous 3D-structure-aware neural scene representations. In *Advances in Neural Information Processing Systems (NeurIPS)*, pp. 1119–1130.

Sitzmann, V., Chan, E. R., Tucker, R., Snavely, N., and Wetzstein, G. (2020). MetaSDF: Meta-learning signed distance functions. In *Advances in Neural Information Processing Systems (NeurIPS)*.

Sitzmann, V., Martel, J. N. P., Bergman, A. W., Lindell, D. B., and Wetzstein, G. (2020). Implicit neural representations with periodic activation functions. In *Advances in Neural Information Processing Systems (NeurIPS)*.

Sitzmann, V., Thies, J., Heide, F., Nießner, M., Wetzstein, G., and Zollhöfer, M. (2019). DeepVoxels: Learning persistent 3D feature embeddings. In *IEEE/CVF Conference on Computer Vision and Pattern Recognition (CVPR)*.

Sivic, J. and Zisserman, A. (2003). Video Google: A text retrieval approach to object matching in videos. In *International Conference on Computer Vision (ICCV)*, pp. 1470–1477.

Sivic, J. and Zisserman, A. (2009). Efficient visual search of videos cast as text retrieval. *IEEE Transactions on Pattern Analysis and Machine Intelligence*, 31(4):591–606.

Sivic, J., Zitnick, C. L., and Szeliski, R. (2006). Finding people in repeated shots of the same scene. In *British Machine Vision Conference (BMVC)*, pp. 909–918.

Sivic, J., Russell, B., Zisserman, A., Freeman, W. T., and Efros, A. A. (2008). Unsupervised discovery of visual object class hierarchies. In *IEEE Computer Society Conference on Computer Vision and Pattern Recognition (CVPR)*.

Slabaugh, G. G., Culbertson, W. B., Slabaugh, T. G., Culbertson, B., Malzbender, T., and Stevens, M. (2004). Methods for volumetric reconstruction of visual scenes. *International Journal of Computer Vision*, 57(3):179–199.

Slama, C. C. (ed.). (1980). *Manual of Photogrammetry*. American Society of Photogrammetry, Falls Church, Virginia, 4th edition.

Smelyanskiy, V. N., Cheeseman, P., Maluf, D. A., and Morris, R. D. (2000). Bayesian super-resolved surface reconstruction from images. In *IEEE Computer Society Conference on Computer Vision and Pattern Recognition (CVPR)*, pp. 375–382.

Smeulders, A. W. M., Worring, M., Santini, S., Gupta, A., and Jain, R. C. (2000). Content-based image retrieval at the end of the early years. *IEEE Transactions on Pattern Analysis and Machine Intelligence*, 22(12):477–490.

Smeulders, A. W. M., Chu, D. M., Cucchiara, R., Calderara, S., Dehghan, A., and Shah, M. (2014). Visual tracking: An experimental survey. *IEEE Transactions on Pattern Analysis and Machine Intelligence*, 36(7):1442–1468.

Sminchisescu, C. and Triggs, B. (2001). Covariance scaled sampling for monocular 3D body tracking. In *IEEE Computer Society Conference on Computer Vision and Pattern Recognition (CVPR)*, pp. 447–454.

Sminchisescu, C., Kanaujia, A., and Metaxas, D. (2006). Conditional models for contextual human motion recognition. *Computer Vision and Image Understanding*, 104(2–3):210–220.

Sminchisescu, C., Kanaujia, A., Li, Z., and Metaxas, D. (2005). Discriminative density propagation for 3D human motion estimation. In *IEEE Computer Society Conference on Computer Vision and Pattern Recognition (CVPR)*, pp. 390–397.

Smith, A. R. and Blinn, J. F. (1996). Blue screen matting. In *ACM SIGGRAPH Conference Proceedings*, pp. 259–268.

Smith, B. M., Zhang, L., Jin, H., and Agarwala, A. (2009). Light field video stabilization. In *International Conference on Computer Vision (ICCV)*.

Smith, S. M. and Brady, J. M. (1997). SUSAN—a new approach to low level image processing. *International Journal of Computer Vision*, 23(1):45–78.

Smola, A. and Li, M. (2019). Berkeley stat 157 course (spring 2019): Introduction to deep learning. Slides and Jupyter notebooks available at https://courses.d2l.ai/berkeley-stat-157.

Smolic, A. and Kauff, P. (2005). Interactive 3-D video representation and coding technologies. *Proceedings of the IEEE*, 93(1):98–110.

Snavely, N., Seitz, S. M., and Szeliski, R. (2006). Photo tourism: Exploring photo collections in 3D. *ACM Transactions on Graphics (Proc. SIGGRAPH)*, 25(3):835–846.

Snavely, N., Seitz, S. M., and Szeliski, R. (2008a). Modeling the world from internet photo collections. *International Journal of Computer Vision*, 80(2):189–210.

Snavely, N., Seitz, S. M., and Szeliski, R. (2008b). Skeletal graphs for efficient structure from motion. In *IEEE Computer Society Conference on Computer Vision and Pattern Recognition (CVPR)*.

Snavely, N., Garg, R., Seitz, S. M., and Szeliski, R. (2008). Finding paths through the world's photos. *ACM Transactions on Graphics (Proc. SIGGRAPH)*, 27(3).

Snavely, N., Simon, I., Goesele, M., Szeliski, R., and Seitz, S. M. (2010). Scene reconstruction and visualization from community photo collections. *Proceedings of the IEEE*, 98(8):1370–1390.

Sohn, K. (2016). Improved deep metric learning with multi-class n-pair loss objective. In *Advances in Neural Information Processing Systems (NeurIPS)*, pp. 1857–1865.

Sohoni, N. S., Aberger, C. R., Leszczynski, M., Zhang, J., and Ré, C. (2019). Low-memory neural network training: A technical report. https://web.stanford.edu/~nims/low_memory_training.pdf.

Soille, P. (2006). Morphological image compositing. *IEEE Transactions on Pattern Analysis and Machine Intelligence*, 28(5):673–683.

Solà, J., Deray, J., and Atchuthan, D. (2019). A micro lie theory for state estimation in robotics. *arXiv preprint arXiv:1812.01537*.

Solina, F. and Bajcsy, R. (1990). Recovery of parametric models from range images: The case for superquadrics with global deformations. *IEEE Transactions on Pattern Analysis and Machine Intelligence*, 12(2):131–147.

Song, H. O., Xiang, Y., Jegelka, S., and Savarese, S. (2016). Deep metric learning via lifted structured feature embedding. In *IEEE Conference on Computer Vision and Pattern Recognition (CVPR)*.

Song, J., Chen, X., and Hilliges, O. (2020). Human body model fitting by learned gradient descent. In *European Conference on Computer Vision (ECCV)*.

Song, X., Zhao, X., Fang, L., Hu, H., and Yu, Y. (2020). EdgeStereo: An effective multi-task learning network for stereo matching and edge detection. *International Journal of Computer Vision*, 128(4):910–930.

Song, X., Yang, G., Zhu, X., Zhou, H., Wang, Z., and Shi, J. (2021). AdaStereo: A simple and efficient approach for adaptive stereo matching. In *IEEE/CVF Conference on Computer Vision and Pattern Recognition (CVPR)*.

Soomro, K., Zamir, A. R., and Shah, M. (2012). *UCF101: A dataset of 101 human actions classes from videos in the wild.* Technical Report CRCV-TR-12-01, University of Central Florida.

Sotiras, A., Davatzikos, C., and Paragios, N. (2013). Deformable medical image registration: A survey. *IEEE Transactions on Medical Imaging*, 32(7):1153–1190.

Soucy, M. and Laurendeau, D. (1992). Multi-resolution surface modeling from multiple range views. In *IEEE Computer Society Conference on Computer Vision and Pattern Recognition (CVPR)*, pp. 348–353.

Springenberg, J., Dosovitskiy, A., Brox, T., and Riedmiller, M. (2015). Striving for simplicity: The all convolutional net. In *ICLR (workshop track)*.

Spurr, A., Iqbal, U., Molchanov, P., Hilliges, O., and Kautz, J. (2020). Weakly supervised 3D hand pose estimation via biomechanical constraints. In *European Conference on Computer Vision (ECCV)*.

Srinivas, A., Lin, T.-Y., Parmar, N., Shlens, J., Abbeel, P., and Vaswani, A. (2021). Bottleneck transformers for visual recognition. In *IEEE/CVF Conference on Computer Vision and Pattern Recognition (CVPR)*.

Srinivasan, P. P., Deng, B., Zhang, X., Tancik, M., Mildenhall, B., and Barron, J. T. (2021). NeRV: Neural reflectance and visibility fields for relighting and view synthesis. In *IEEE/CVF Conference on Computer Vision and Pattern Recognition (CVPR)*.

Srinivasan, P. P., Tucker, R., Barron, J. T., Ramamoorthi, R., Ng, R., and Snavely, N. (2019). Pushing the boundaries of view extrapolation with multiplane images. In *IEEE/CVF Conference on Computer Vision and Pattern Recognition (CVPR)*.

Srinivasan, S., Chellappa, R., Veeraraghavan, A., and Aggarwal, G. (2005). Electronic image stabilization and mosaicking algorithms. In Bovik, A. (ed.), *Handbook of Image and Video Processing*, Academic Press.

Srivasan, P., Liang, P., and Hackwood, S. (1990). Computational geometric methods in volumetric intersections for 3D reconstruction. *Pattern Recognition*, 23(8):843–857.

Srivastava, N., Mansimov, E., and Salakhudinov, R. (2015). Unsupervised learning of video representations using LSTMs. In *International Conference on Machine Learning (ICML)*, pp. 843–852.

Srivastava, N., Hinton, G., Krizhevsky, A., Sutskever, I., and Salakhutdinov, R. (2014). Dropout: a simple way to prevent neural networks from overfitting. *Journal of Machine Learning Research*, 15(1):1929–1958.

Stallkamp, J., Schlipsing, M., Salmen, J., and Igel, C. (2012). Man vs. computer: Benchmarking machine learning algorithms for traffic sign recognition. *Neural Networks*, 32:323–332.

Stamos, I., Liu, L., Chen, C., Wolberg, G., Yu, G., and Zokai, S. (2008). Integrating automated range registration with multiview geometry for the photorealistic modeling of large-scale scenes. *International Journal of Computer Vision*, 78(2–3):237–260.

Stark, J. A. (2000). Adaptive image contrast enhancement using generalizations of histogram equalization. *IEEE Transactions on Image Processing*, 9(5):889–896.

Stauffer, C. and Grimson, W. (1999). Adaptive background mixture models for real-time tracking. In *IEEE Computer Society Conference on Computer Vision and Pattern Recognition (CVPR)*, pp. 246–252.

Steedly, D. and Essa, I. (2001). Propagation of innovative information in non-linear least-squares structure from motion. In *International Conference on Computer Vision (ICCV)*, pp. 223–229.

Steedly, D., Essa, I., and Dellaert, F. (2003). Spectral partitioning for structure from motion. In *International Conference on Computer Vision (ICCV)*, pp. 996–1003.

Steedly, D., Pal, C., and Szeliski, R. (2005). Efficiently registering video into panoramic mosaics. In *International Conference on Computer Vision (ICCV)*, pp. 1300–1307.

Steele, R. and Jaynes, C. (2005). Feature uncertainty arising from covariant image noise. In *IEEE Computer Society Conference on Computer Vision and Pattern Recognition (CVPR)*, pp. 1063–1070.

Steele, R. M. and Jaynes, C. (2006). Overconstrained linear estimation of radial distortion and multi-view geometry. In *European Conference on Computer Vision (ECCV)*, pp. 253–264.

Stein, A., Hoiem, D., and Hebert, M. (2007). Learning to extract object boundaries using motion cues. In *International Conference on Computer Vision (ICCV)*.

Stein, F. and Medioni, G. (1992). Structural indexing: Efficient 3-D object recognition. *IEEE Transactions on Pattern Analysis and Machine Intelligence*, 14(2):125–145.

Stein, G. (1995). Accurate internal camera calibration using rotation, with analysis of sources of error. In *International Conference on Computer Vision (ICCV)*, pp. 230–236.

Stein, G. (1997). Lens distortion calibration using point correspondences. In *IEEE Computer Society Conference on Computer Vision and Pattern Recognition (CVPR)*, pp. 602–608.

Stein, S. (2020). See what iPhone 12 Pro's new lidar feature can do with this 3D-scanning app. *c—net*. https://www.cnet.com/news/what-lidar-feature-apple-iphone-12-pro-can-do-3d-scanning-app-canvas.

Stenborg, E., Sattler, T., and Hammarstrand, L. (2020). Using image sequences for long-term visual localization. In *International Conference on 3D Vision (3DV)*, pp. 938–948.

Stenger, B., Thayananthan, A., Torr, P. H. S., and Cipolla, R. (2006). Model-based hand tracking using a hierarchical bayesian filter. *IEEE Transactions on Pattern Analysis and Machine Intelligence*, 28(9):1372–1384.

Stewart, C. V. (1999). Robust parameter estimation in computer vision. *SIAM Reviews*, 41(3):513–537.

Stiller, C. and Konrad, J. (1999). Estimating motion in image sequences: A tutorial on modeling and computation of 2D motion. *IEEE Signal Processing Magazine*, 16(4):70–91.

Stollnitz, E. J., DeRose, T. D., and Salesin, D. H. (1996). *Wavelets for Computer Graphics: Theory and Applications*. Morgan Kaufmann, San Francisco.

Strang, G. (1988). *Linear Algebra and its Applications*. Harcourt, Brace, Jovanovich, Publishers, San Diego, 3rd edition.

Strang, G. (1989). Wavelets and dilation equations: A brief introduction. *SIAM Reviews*, 31(4):614–627.

Strecha, C., Tuytelaars, T., and Van Gool, L. (2003). Dense matching of multiple wide-baseline views. In *International Conference on Computer Vision (ICCV)*, pp. 1194–1201.

Strecha, C., von Hansen, W., Van Gool, L., Fua, P., and Thoennessen, U. (2008). On benchmarking camera calibration and multi-view stereo. In *IEEE Computer Society Conference on Computer Vision and Pattern Recognition (CVPR)*.

Stühmer, J., Nowozin, S., Fitzgibbon, A., Szeliski, R., Perry, T., Acharya, S., Cremers, D., and Shotton, J. (2015). Model-based tracking at 300Hz using raw time-of-flight observations. In *IEEE International Conference on Computer Vision (ICCV)*.

Sturm, J., Engelhard, N., Endres, F., Burgard, W., and Cremers, D. (2012). A benchmark for the evaluation of RGB-D SLAM systems. In *IEEE/RSJ International Conference on Intelligent Robots and Systems*, pp. 573–580.

Sturm, P. (2005). Multi-view geometry for general camera models. In *IEEE Computer Society Conference on Computer Vision and Pattern Recognition (CVPR)*, pp. 206–212.

Sturm, P. and Ramalingam, S. (2004). A generic concept for camera calibration. In *European Conference on Computer Vision (ECCV)*, pp. 1–13.

Sturm, P. and Triggs, W. (1996). A factorization based algorithm for multi-image projective structure and motion. In *European Conference on Computer Vision (ECCV)*, pp. 709–720.

Su, N. M. and Crandall, D. J. (2021). The affective growth of computer vision. In *IEEE/CVF Conference on Computer Vision and Pattern Recognition (CVPR)*.

Su, X. and Zhang, Q. (2010). Dynamic 3-d shape measurement method: a review. *Optics and Lasers in Engineering*, 48(2):191–204.

Subramanya, A. and Talukdar, P. P. (2014). Graph-based semi-supervised learning. *Synthesis Lectures on Artificial Intelligence and Machine Learning*, 8(4):1–125.

Sudderth, E. B., Torralba, A., Freeman, W. T., and Willsky, A. S. (2008). Describing visual scenes using transformed objects and parts. *International Journal of Computer Vision*, 77(1–3):291–330.

Sullivan, G. J., Ohm, J.-R., Han, W.-J., and Wiegand, T. (2012). Overview of the high efficiency video coding (hevc) standard. *IEEE Transactions on Circuits and Systems for Video Technology*, 22(12):1649–1668.

Sullivan, S. and Ponce, J. (1998). Automatic model construction and pose estimation from photographs using triangular splines. *IEEE Transactions on Pattern Analysis and Machine Intelligence*, 20(10):1091–1096.

Sun, C., Shrivastava, A., Singh, S., and Gupta, A. (2017). Revisiting unreasonable effectiveness of data in deep learning era. In *IEEE International Conference on Computer Vision (ICCV)*.

Sun, C., Myers, A., Vondrick, C., Murphy, K., and Schmid, C. (2019). VideoBERT: A joint model for video and language representation learning. In *IEEE/CVF International Conference on Computer Vision (ICCV)*.

Sun, D., Liu, C., and Pfister, H. (2014). Local layering for joint motion estimation and occlusion detection. In *IEEE Conference on Computer Vision and Pattern Recognition (CVPR)*.

Sun, D., Roth, S., and Black, M. J. (2010). Secrets of optical flow estimation and their principles. In *IEEE Computer Society Conference on Computer Vision and Pattern Recognition (CVPR)*.

Sun, D., Roth, S., and Black, M. J. (2014). A quantitative analysis of current practices in optical flow estimation and the principles behind them. *International Journal of Computer Vision*, 106(2):115–137.

Sun, D., Sudderth, E. B., and Black, M. J. (2012). Layered segmentation and optical flow estimation over time. In *IEEE Computer Society Conference on Computer Vision and Pattern Recognition (CVPR)*.

Sun, D., Sudderth, E. B., and Pfister, H. (2015). Layered RGBD scene flow estimation. In *IEEE Conference on Computer Vision and Pattern Recognition (CVPR)*.

Sun, D., Roth, S., Lewis, J. P., and Black, M. J. (2008). Learning optical flow. In *European Conference on Computer Vision (ECCV)*, pp. 83–97.

Sun, D., Yang, X., Liu, M.-Y., and Kautz, J. (2018). PWC-Net: CNNs for optical flow using pyramid, warping, and cost volume. In *IEEE Conference on Computer Vision and Pattern Recognition (CVPR)*.

Sun, D., Yang, X., Liu, M.-Y., and Kautz, J. (2019). Models matter, so does training: An empirical study of CNNs for optical flow estimation. *IEEE Transactions on Pattern Analysis and Machine Intelligence*, 42(6):1408–1423.

Sun, D., Wulff, J., Sudderth, E. B., Pfister, H., and Black, M. J. (2013). A fully-connected layered model of foreground and background flow. In *IEEE Computer Society Conference on Computer Vision and Pattern Recognition (CVPR)*.

Sun, D., Vlasic, D., Herrmann, C., Jampani, V., Krainin, M., Chang, H., Zabih, R., Freeman, W. T., and Liu, C. (2021). AutoFlow: Learning a better training set for optical flow. In *IEEE/CVF Conference on Computer Vision and Pattern Recognition (CVPR)*.

Sun, J., Zheng, N., and Shum, H. (2003). Stereo matching using belief propagation. *IEEE Transactions on Pattern Analysis and Machine Intelligence*, 25(7):787–800.

Sun, J., Cao, W., Xu, Z., and Ponce, J. (2015). Learning a convolutional neural network for non-uniform motion blur removal. In *IEEE Conference on Computer Vision and Pattern Recognition (CVPR)*.

Sun, J., Jia, J., Tang, C.-K., and Shum, H.-Y. (2004). Poisson matting. *ACM Transactions on Graphics (Proc. SIGGRAPH)*, 23(3):315–321.

Sun, J., Li, Y., Kang, S. B., and Shum, H.-Y. (2006). Flash matting. *ACM Transactions on Graphics*, 25(3):772–778.

Sun, J., Yuan, L., Jia, J., and Shum, H.-Y. (2004). Image completion with structure propagation. *ACM Transactions on Graphics (Proc. SIGGRAPH)*, 24(3):861–868.

Sun, J., Shen, Z., Wang, Y., Bao, H., and Zhou, X. (2021). LoFTR: Detector-free local feature matching with transformers. In *IEEE/CVF Conference on Computer Vision and Pattern Recognition (CVPR)*.

Sun, T., Barron, J. T., Tsai, Y.-T., Xu, Z., Yu, X., Fyffe, G., Rhemann, C., Busch, J., Debevec, P., and Ramamoorthi, R. (2019). Single image portrait relighting. *ACM Transactions on Graphics (Proc. SIGGRAPH)*, 38(4):79.

Sun, X., Xie, Y., Luo, P., and Wang, L. (2017). A dataset for benchmarking image-based localization. In *IEEE Conference on Computer Vision and Pattern Recognition (CVPR)*.

Sun, Y., Tzeng, E., Darrell, T., and Efros, A. A. (2019). Unsupervised domain adaptation through self-supervision. *arXiv preprint arXiv:1909.11825*.

Sung, K.-K. and Poggio, T. (1998). Example-based learning for view-based human face detection. *IEEE Transactions on Pattern Analysis and Machine Intelligence*, 20(1):39–51.

Sutherland, I. E. (1974). Three-dimensional data input by tablet. *Proceedings of the IEEE*, 62(4):453–461.

Sutton, R. (2019). The bitter lesson. http://www.incompleteideas.net/IncIdeas/BitterLesson.html.

Svoboda, J., Anoosheh, A., Osendorfer, C., and Masci, J. (2020). Two-stage peer-regularized feature recombination for arbitrary image style transfer. In *IEEE/CVF Conference on Computer Vision and Pattern Recognition (CVPR)*.

Swain, M. J. and Ballard, D. H. (1991). Color indexing. *International Journal of Computer Vision*, 7(1):11–32.

Swaminathan, R., Kang, S. B., Szeliski, R., Criminisi, A., and Nayar, S. K. (2002). On the motion and appearance of specularities in image sequences. In *European Conference on Computer Vision (ECCV)*, pp. 508–523.

Sweeney, C., Hollerer, T., and Turk, M. (2015). Theia: A fast and scalable structure-from-motion library. In *ACM International Conference on Multimedia*, pp. 693–696.

Sweeney, C., Kneip, L., Hollerer, T., and Turk, M. (2015). Computing similarity transformations from only image correspondences. In *IEEE Conference on Computer Vision and Pattern Recognition (CVPR)*.

Sweldens, W. (1996). Wavelets and the lifting scheme: A 5 minute tour. *Z. Angew. Math. Mech.*, 76 (Suppl. 2):41–44.

Sweldens, W. (1997). The lifting scheme: A construction of second generation wavelets. *SIAM J. Math. Anal.*, 29(2):511–546.

Swendsen, R. H. and Wang, J.-S. (1987). Nonuniversal critical dynamics in Monte Carlo simulations. *Physical Review Letters*, 58(2):86–88.

Sze, V., Chen, Y., Yang, T., and Emer, J. S. (2017). Efficient processing of deep neural networks: A tutorial and survey. *Proceedings of the IEEE*, 105(12):2295–2329.

Szegedy, C., Ioffe, S., Vanhoucke, V., and Alemi, A. A. (2017). Inception-v4, Inception-ResNet and the impact of residual connections on learning. In *Thirty-first AAAI Conference on Artificial Intelligence*.

Szegedy, C., Vanhoucke, V., Ioffe, S., Shlens, J., and Wojna, Z. (2016). Rethinking the inception architecture for computer vision. In *IEEE Conference on Computer Vision and Pattern Recognition (CVPR)*.

Szegedy, C., Zaremba, W., Sutskever, I., Bruna, J., Erhan, D., Goodfellow, I., and Fergus, R. (2013). Intriguing properties of neural networks. *arXiv preprint arXiv:1312.6199*.

Szegedy, C., Liu, W., Jia, Y., Sermanet, P., Reed, S., Anguelov, D., Erhan, D., Vanhoucke, V., and Rabinovich, A. (2015). Going deeper with convolutions. In *IEEE Conference on Computer Vision and Pattern Recognition (CVPR)*.

Szeliski, R. (1989). *Bayesian Modeling of Uncertainty in Low-Level Vision*. Kluwer Academic Publishers, Boston.

Szeliski, R. (1990a). Bayesian modeling of uncertainty in low-level vision. *International Journal of Computer Vision*, 5(3):271–301.

Szeliski, R. (1990b). Fast surface interpolation using hierarchical basis functions. *IEEE Transactions on Pattern Analysis and Machine Intelligence*, 12(6):513–528.

Szeliski, R. (1991a). Fast shape from shading. *CVGIP: Image Understanding*, 53(2):129–153.

Szeliski, R. (1991b). Shape from rotation. In *IEEE Computer Society Conference on Computer Vision and Pattern Recognition (CVPR)*, pp. 625–630.

Szeliski, R. (1993). Rapid octree construction from image sequences. *CVGIP: Image Understanding*, 58(1):23–32.

Szeliski, R. (1994). Image mosaicing for tele-reality applications. In *IEEE Workshop on Applications of Computer Vision (WACV)*, pp. 44–53.

Szeliski, R. (1996). Video mosaics for virtual environments. *IEEE Computer Graphics and Applications*, 16(2):22–30.

Szeliski, R. (1999a). A multi-view approach to motion and stereo. In *IEEE Computer Society Conference on Computer Vision and Pattern Recognition (CVPR)*, pp. 157–163.

Szeliski, R. (1999b). Prediction error as a quality metric for motion and stereo. In *International Conference on Computer Vision (ICCV)*, pp. 781–788.

Szeliski, R. (2005). ICCV2005 computer vision contest. Originally hosted at http://research.microsoft.com/iccv2005/Contest/ and now archived at https://szeliski.org/ICCV_2005_Contest/.

Szeliski, R. (2006a). Image alignment and stitching: A tutorial. *Foundations and Trends® in Computer Graphics and Computer Vision*, 2(1):1–104.

Szeliski, R. (2006b). Locally adapted hierarchical basis preconditioning. *ACM Transactions on Graphics (Proc. SIGGRAPH)*, 25(3):1135–1143.

Szeliski, R. (2010). *Computer Vision: Algorithms and Applications*. Springer, New York, 1st edition. A free PDF version can be found at https://szeliski.org/Book/1stEdition.htm.

Szeliski, R. and Coughlan, J. (1997). Spline-based image registration. *International Journal of Computer Vision*, 22(3):199–218.

Szeliski, R. and Golland, P. (1999). Stereo matching with transparency and matting. *International Journal of Computer Vision*, 32(1):45–61. Special Issue for Marr Prize papers.

Szeliski, R. and Hinton, G. (1985). Solving random-dot stereograms using the heat equation. In *IEEE Computer Society Conference on Computer Vision and Pattern Recognition (CVPR)*, pp. 284–288.

Szeliski, R. and Ito, M. R. (1986). New Hermite cubic interpolator for two-dimensional curve generation. *IEE Proceedings E*, 133(6):341–347.

Szeliski, R. and Kang, S. B. (1994). Recovering 3D shape and motion from image streams using nonlinear least squares. *Journal of Visual Communication and Image Representation*, 5(1):10–28.

Szeliski, R. and Kang, S. B. (1995). Direct methods for visual scene reconstruction. In *IEEE Workshop on Representations of Visual Scenes*, pp. 26–33.

Szeliski, R. and Kang, S. B. (1997). Shape ambiguities in structure from motion. *IEEE Transactions on Pattern Analysis and Machine Intelligence*, 19(5):506–512.

Szeliski, R. and Lavallée, S. (1996). Matching 3-D anatomical surfaces with non-rigid deformations using octree-splines. *International Journal of Computer Vision*, 18(2):171–186.

Szeliski, R. and Scharstein, D. (2004). Sampling the disparity space image. *IEEE Transactions on Pattern Analysis and Machine Intelligence*, 26(3):419–425.

Szeliski, R. and Shum, H.-Y. (1996). Motion estimation with quadtree splines. *IEEE Transactions on Pattern Analysis and Machine Intelligence*, 18(12):1199–1210.

Szeliski, R. and Shum, H.-Y. (1997). Creating full view panoramic image mosaics and environment maps. In *ACM SIGGRAPH Conference Proceedings*, pp. 251–258.

Szeliski, R. and Tonnesen, D. (1992). Surface modeling with oriented particle systems. *Computer Graphics (SIGGRAPH)*, 26(2):185–194.

Szeliski, R. and Torr, P. (1998). Geometrically constrained structure from motion: Points on planes. In *European Workshop on 3D Structure from Multiple Images of Large-Scale Environments (SMILE)*, pp. 171–186.

Szeliski, R. and Weiss, R. (1998). Robust shape recovery from occluding contours using a linear smoother. *International Journal of Computer Vision*, 28(1):27–44.

Szeliski, R., Avidan, S., and Anandan, P. (2000). Layer extraction from multiple images containing reflections and transparency. In *IEEE Computer Society Conference on Computer Vision and Pattern Recognition (CVPR)*, pp. 246–253.

Szeliski, R., Tonnesen, D., and Terzopoulos, D. (1993a). Curvature and continuity control in particle-based surface models. In *SPIE Vol. 2031, Geometric Methods in Computer Vision II*, pp. 172–181.

Szeliski, R., Tonnesen, D., and Terzopoulos, D. (1993b). Modeling surfaces of arbitrary topology with dynamic particles. In *IEEE Computer Society Conference on Computer Vision and Pattern Recognition (CVPR)*, pp. 82–87.

Szeliski, R., Uyttendaele, M., and Steedly, D. (2008). *Fast Poisson Blending using Multi-Splines*. Technical Report MSR-TR-2008-58, Microsoft Research.

Szeliski, R., Winder, S., and Uyttendaele, M. (2010). *High-quality multi-pass image resampling*. Technical Report MSR-TR-2010-10, Microsoft Research.

Szeliski, R., Zabih, R., Scharstein, D., Veksler, O., Kolmogorov, V., Agarwala, A., Tappen, M., and Rother, C. (2008). A comparative study of energy minimization methods for Markov random fields with smoothness-based priors. *IEEE Transactions on Pattern Analysis and Machine Intelligence*, 30(6):1068–1080.

Szummer, M. and Picard, R. W. (1996). Temporal texture modeling. In *IEEE International Conference on Image Processing (ICIP)*, pp. 823–826.

Tabb, M. and Ahuja, N. (1997). Multiscale image segmentation by integrated edge and region detection. *IEEE Transactions on Image Processing*, 6(5):642–655.

Taguchi, Y., Wilburn, B., and Zitnick, C. L. (2008). Stereo reconstruction with mixed pixels using adaptive over-segmentation. In *IEEE Computer Society Conference on Computer Vision and Pattern Recognition (CVPR)*.

Taheri, O., Ghorbani, N., Black, M. J., and Tzionas, D. (2020). GRAB: A dataset of whole-body human grasping of objects. In *European Conference on Computer Vision (ECCV)*.

Taigman, Y., Yang, M., Ranzato, M., and Wolf, L. (2014). DeepFace: Closing the gap to human-level performance in face verification. In *IEEE Conference on Computer Vision and Pattern Recognition (CVPR)*.

Taira, H., Okutomi, M., Sattler, T., Cimpoi, M., Pollefeys, M., Sivic, J., Pajdla, T., and Torii, A. (2018). InLoc: Indoor visual localization with dense matching and view synthesis. In *IEEE Conference on Computer Vision and Pattern Recognition (CVPR)*.

Taira, H., Rocco, I., Sedlar, J., Okutomi, M., Sivic, J., Pajdla, T., Sattler, T., and Torii, A. (2019). Is this the right place? Geometric semantic pose verification for indoor visual localization. In *IEEE/CVF International Conference on Computer Vision (ICCV)*.

Takamatsu, J., Matsushita, Y., and Ikeuchi, K. (2008). Estimating radiometric response functions from image noise variance. In *European Conference on Computer Vision (ECCV)*, pp. 623–637.

Takeda, H., Farsiu, S., and Milanfar, P. (2007). Kernel regression for image processing and reconstruction. *IEEE Transactions on Image Processing*, 16(2):349–366.

Takikawa, T., Litalien, J., Yin, K., Kreis, K., Loop, C., Nowrouzezahrai, D., Jacobson, A., McGuire, M., and Fidler, S. (2021). Neural geometric level of detail: Real-time rendering with implicit 3D shapes. In *IEEE/CVF Conference on Computer Vision and Pattern Recognition (CVPR)*.

Talebi, H. and Milanfar, P. (2018). NIMA: Neural image assessment. *IEEE Transactions on Image Processing*, 27(8):3998–4011.

Tam, G. K. L., Cheng, Z.-Q., Lai, Y.-K., Langbein, F. C., Liu, Y., Marshall, D., Martin, R. R., Sun, X.-F., and Rosin, P. L. (2012). Registration of 3D point clouds and meshes: A survey from rigid to nonrigid. *IEEE Transactions on Visualization and Computer Graphics*, 19(7):1199–1217.

Tan, M. and Le, Q. V. (2019). EfficientNet: Rethinking model scaling for convolutional neural networks. In *International Conference on Machine Learning (ICML)*.

Tan, M. and Le, Q. V. (2021). EfficientNetV2: Smaller models and faster training. *arXiv preprint arXiv:2104.00298*.

Tan, M., Pang, R., and Le, Q. V. (2020). EfficientDet: Scalable and efficient object detection. In *IEEE/CVF Conference on Computer Vision and Pattern Recognition (CVPR)*.

Tanaka, M. and Okutomi, M. (2008). Locally adaptive learning for translation-variant MRF image priors. In *IEEE Computer Society Conference on Computer Vision and Pattern Recognition (CVPR)*.

Tancik, M., Mildenhall, B., Wang, T., Schmidt, D., Srinivasan, P. P., Barron, J. T., and Ng, R. (2021). Learned initializations for optimizing coordinate-based neural representations. In *IEEE/CVF Conference on Computer Vision and Pattern Recognition (CVPR)*.

Tancik, M., Srinivasan, P. P., Mildenhall, B., Fridovich-Keil, S., Raghavan, N., Singhal, U., Ramamoorthi, R., Barron, J. T., and Ng, R. (2020). Fourier features let networks learn high frequency functions in low dimensional domains. In *Advances in Neural Information Processing Systems (NeurIPS)*.

Tang, D., Dou, M., Lincoln, P., Davidson, P., Guo, K., Taylor, J., Fanello, S., Keskin, C., Kowdle, A., Bouaziz, S., Izadi, S., and Tagliasacchi, A. (2018). Real-time compression and streaming of 4D performances. *ACM Transactions on Graphics (Proc. SIGGRAPH Asia)*, 37(6):256.

Tang, D., Singh, S., Chou, P. A., Hane, C., Dou, M., Fanello, S., Taylor, J., Davidson, P., Guleryuz, O. G., Zhang, Y., Izadi, S., Tagliasacchi, A., Bouaziz, S., and Keskin, C. (2020). Deep implicit volume compression. In *IEEE/CVF Conference on Computer Vision and Pattern Recognition (CVPR)*.

Tang, Y., Borisyuk, F., Malreddy, S., Li, Y., Liu, Y., and Kirshner, S. (2019). MSURU: Large scale e-commerce image classification with weakly supervised search data. In *ACM SIGKDD International Conference on Knowledge Discovery & Data Mining*, pp. 2518–2526.

Taniai, T., Matsushita, Y., Sato, Y., and Naemura, T. (2018). Continuous 3D label stereo matching using local expansion moves. *IEEE Transactions on Pattern Analysis and Machine Intelligence*, 40(11):2725–2739.

Tankovich, V., Hane, C., Zhang, Y., Kowdle, A., Fanello, S., and Bouaziz, S. (2021). HITNet: Hierarchical iterative tile refinement network for real-time stereo matching. In *IEEE/CVF Conference on Computer Vision and Pattern Recognition (CVPR)*.

Tao, H., Sawhney, H. S., and Kumar, R. (2001). A global matching framework for stereo computation. In *International Conference on Computer Vision (ICCV)*, pp. 532–539.

Tao, X., Gao, H., Liao, R., Wang, J., and Jia, J. (2017). Detail-revealing deep video super-resolution. In *IEEE International Conference on Computer Vision (ICCV)*.

Tao, X., Gao, H., Shen, X., Wang, J., and Jia, J. (2018). Scale-recurrent network for deep image deblurring. In *IEEE Conference on Computer Vision and Pattern Recognition (CVPR)*.

Tappen, M. F. (2007). Utilizing variational optimization to learn Markov random fields. In *IEEE Computer Society Conference on Computer Vision and Pattern Recognition (CVPR)*.

Tappen, M. F. and Freeman, W. T. (2003). Comparison of graph cuts with belief propagation for stereo, using identical MRF parameters. In *International Conference on Computer Vision (ICCV)*, pp. 900–907.

Tappen, M. F., Freeman, W. T., and Adelson, E. H. (2005). Recovering intrinsic images from a single image. *IEEE Transactions on Pattern Analysis and Machine Intelligence*, 27(9):1459–1472.

Tappen, M. F., Russell, B. C., and Freeman, W. T. (2003). Exploiting the sparse derivative prior for super-resolution and image demosaicing. In *International Workshop on Statistical and Computational Theories of Vision*.

Tappen, M. F., Liu, C., Freeman, W., and Adelson, E. (2007). Learning Gaussian conditional random fields for low-level vision. In *IEEE Computer Society Conference on Computer Vision and Pattern Recognition (CVPR)*.

Tardif, J.-P. (2009). Non-iterative approach for fast and accurate vanishing point detection. In *International Conference on Computer Vision (ICCV)*.

Tardif, J.-P., Sturm, P., and Roy, S. (2007). Plane-based self-calibration of radial distortion. In *International Conference on Computer Vision (ICCV)*.

Tardif, J.-P., Sturm, P., Trudeau, M., and Roy, S. (2009). Calibration of cameras with radially symmetric distortion. *IEEE Transactions on Pattern Analysis and Machine Intelligence*, 31(9):1552–1566.

Tariq, T., Tursun, O. T., Kim, M., and Didyk, P. (2020). Why are deep representations good perceptual quality features? In *European Conference on Computer Vision (ECCV)*.

Tassano, M., Delon, J., and Veit, T. (2019). DVDnet: A fast network for deep video denoising. In *IEEE International Conference on Image Processing (ICIP)*, pp. 1805–1809.

Tassano, M., Delon, J., and Veit, T. (2020). FastDVDnet: Towards real-time deep video denoising without flow estimation. In *IEEE/CVF Conference on Computer Vision and Pattern Recognition (CVPR)*.

Tatarchenko, M., Dosovitskiy, A., and Brox, T. (2017). Octree generating networks: Efficient convolutional architectures for high-resolution 3D outputs. In *IEEE International Conference on Computer Vision (ICCV)*.

Tatarchenko, M., Richter, S. R., Ranftl, R., Li, Z., Koltun, V., and Brox, T. (2019). What do single-view 3D reconstruction networks learn? In *IEEE/CVF Conference on Computer Vision and Pattern Recognition (CVPR)*.

Taubin, G. (1995). Curve and surface smoothing without shrinkage. In *International Conference on Computer Vision (ICCV)*, pp. 852–857.

Taubman, D. S. and Marcellin, M. W. (2002). JPEG2000: Standard for interactive imaging. *Proceedings of the IEEE*, 90(8):1336–1357.

Tay, Y., Dehghani, M., Bahri, D., and Metzler, D. (2020). Efficient transformers: A survey. *arXiv preprint arXiv:2009.06732*.

Taylor, C. J. (2003). Surface reconstruction from feature based stereo. In *International Conference on Computer Vision (ICCV)*, pp. 184–190.

Taylor, C. J., Debevec, P. E., and Malik, J. (1996). Reconstructing polyhedral models of architectural scenes from photographs. In *European Conference on Computer Vision (ECCV)*, pp. 659–668.

Taylor, C. J., Kriegman, D. J., and Anandan, P. (1991). Structure and motion in two dimensions from multiple images: A least squares approach. In *IEEE Workshop on Visual Motion*, pp. 242–248.

Taylor, J., Bordeaux, L., Cashman, T., Corish, B., Keskin, C., Sharp, T., Soto, E., Sweeney, D., Valentin, J., Luff, B., Topalian, A., Wood, E., Khamis, S., Kohli, P., Izadi, S., Banks, R., Fitzgibbon, A., and Shotton, J. (2016). Efficient and precise interactive hand tracking through joint, continuous optimization of pose and correspondences. *ACM Transactions on Graphics (Proc. SIGGRAPH)*, 35(4):Article 143.

Taylor, J., Shotton, J., Sharp, T., and Fitzgibbon, A. (2012). The Vitruvian manifold: Inferring dense correspondences for one-shot human pose estimation. In *IEEE Computer Society Conference on Computer Vision and Pattern Recognition (CVPR)*.

Taylor, P. (2009). *Text-to-Speech Synthesis*. Cambridge University Press, Cambridge.

Tech, G., Chen, Y., Müller, K., Ohm, J.-R., Vetro, A., and Wang, Y.-K. (2015). Overview of the multiview and 3D extensions of high efficiency video coding. *IEEE Transactions on Circuits and Systems for Video Technology*, 26(1):35–49.

Teed, Z. and Deng, J. (2020a). DeepV2D: Video to depth with differentiable structure from motion. In *International Conference on Learning Representations (ICLR)*.

Teed, Z. and Deng, J. (2020b). RAFT: Recurrent all-pairs field transforms for optical flow. In *European Conference on Computer Vision (ECCV)*.

Tek, K. and Kimia, B. B. (2003). Symmetry maps of free-form curve segments via wave propagation. *International Journal of Computer Vision*, 54(1–3):35–81.

Tekalp, M. (1995). *Digital Video Processing*. Prentice Hall, Upper Saddle River, NJ.

Telea, A. (2004). An image inpainting technique based on fast marching method. *Journal of Graphics Tools*, 9(1):23–34.

Teller, S., Antone, M., Bodnar, Z., Bosse, M., Coorg, S., Jethwa, M., and Master, N. (2003). Calibrated, registered images of an extended urban area. *International Journal of Computer Vision*, 53(1):93–107.

Tenenbaum, J. B., De Silva, V., and Langford, J. C. (2000). A global geometric framework for nonlinear dimensionality reduction. *science*, 290(5500):2319–2323.

Teodosio, L. and Bender, W. (1993). Salient video stills: Content and context preserved. In *ACM Multimedia*, pp. 39–46.

Terzakis, G. and Lourakis, M. (2020). A consistently fast and globally optimal solution to the perspective-n-point problem. In *European Conference on Computer Vision (ECCV)*.

Terzopoulos, D. (1983). Multilevel computational processes for visual surface reconstruction. *Computer Vision, Graphics, and Image Processing*, 24:52–96.

Terzopoulos, D. (1986a). Image analysis using multigrid relaxation methods. *IEEE Transactions on Pattern Analysis and Machine Intelligence*, PAMI-8(2):129–139.

Terzopoulos, D. (1986b). Regularization of inverse visual problems involving discontinuities. *IEEE Transactions on Pattern Analysis and Machine Intelligence*, PAMI-8(4):413–424.

Terzopoulos, D. (1988). The computation of visible-surface representations. *IEEE Transactions on Pattern Analysis and Machine Intelligence*, PAMI-10(4):417–438.

Terzopoulos, D. (1999). Visual modeling for computer animation: Graphics with a vision. *Computer Graphics*, 33(4):42–45.

Terzopoulos, D. and Fleischer, K. (1988). Deformable models. *The Visual Computer*, 4(6):306–331.

Terzopoulos, D. and Metaxas, D. (1991). Dynamic 3D models with local and global deformations: Deformable superquadrics. *IEEE Transactions on Pattern Analysis and Machine Intelligence*, 13(7):703–714.

Terzopoulos, D. and Szeliski, R. (1992). Tracking with Kalman snakes. In Blake, A. and Yuille, A. L. (eds), *Active Vision*, pp. 3–20, MIT Press, Cambridge, Massachusetts.

Terzopoulos, D. and Waters, K. (1990). Analysis of facial images using physical and anatomical models. In *International Conference on Computer Vision (ICCV)*, pp. 727–732.

Terzopoulos, D. and Witkin, A. (1988). Physically-based models with rigid and deformable components. *IEEE Computer Graphics and Applications*, 8(6):41–51.

Terzopoulos, D., Witkin, A., and Kass, M. (1987). Symmetry-seeking models and 3D object reconstruction. *International Journal of Computer Vision*, 1(3):211–221.

Terzopoulos, D., Witkin, A., and Kass, M. (1988). Constraints on deformable models: Recovering 3D shape and nonrigid motion. *Artificial Intelligence*, 36(1):91–123.

Tewari, A., Zollhöfer, M., Zhu, J.-Y., Park, T., Thies, J., Theobalt, C., Shysheya, A., Sitzmann, V., Meshry, M., Mildenhall, B., Xu, Z., Philip, J., Meka, A., Fanello, S., Pandey, R. K., Lombardi, S., and Fried, O. (2020). CVPR 2020 tutorial on neural rendering. https://www.neuralrender.com.

Tewari, A., Fried, O., Thies, J., Sitzmann, V., Lombardi, S., Sunkavalli, K., Martin-Brualla, R., Simon, T., Saragi, J., Nießner, M., Pandey, R., Fanello, S., Wetzstein, G., Zhu, J.-Y., Theobalt, C., Agrawala, M., Shechtman, E., Goldman, D. B., and Zollhöfer, M. (2020). State of the art on neural rendering. *Computer Graphics Forum (Eurographics)*, 39(2).

Thayananthan, A., Iwasaki, M., and Cipolla, R. (2008). Principled fusion of high-level model and low-level cues for motion segmentation. In *IEEE Computer Society Conference on Computer Vision and Pattern Recognition (CVPR)*.

Thies, J., Zollhöfer, M., and Nießner, M. (2019). Deferred neural rendering: Image synthesis using neural textures. *ACM Transactions on Graphics (Proc. SIGGRAPH)*, 38(4):66.

Thies, J., Zollhofer, M., Stamminger, M., Theobalt, C., and Nießner, M. (2016). Face2Face: Real-time face capture and reenactment of RGB videos. In *IEEE Conference on Computer Vision and Pattern Recognition (CVPR)*.

Thies, J., Zollhöfer, M., Theobalt, C., Stamminger, M., and Nießner, M. (2018). HeadOn: Real-time reenactment of human portrait videos. *ACM Transactions on Graphics (Proc. SIGGRAPH)*, 37(4):164.

Thirthala, S. and Pollefeys, M. (2005). The radial trifocal tensor: A tool for calibrating the radial distortion of wide-angle cameras. In *IEEE Computer Society Conference on Computer Vision and Pattern Recognition (CVPR)*, pp. 321–328.

Thomee, B., Shamma, D. A., Friedland, G., Elizalde, B., Ni, K., Poland, D., Borth, D., and Li, L.-J. (2016). YFCC100M: The new data in multimedia research. *Communications of the ACM*, 59(2):64–73.

Thorpe, C., Hebert, M. H., Kanade, T., and Shafer, S. A. (1988). Vision and navigation for the Carnegie-Mellon Navlab. *IEEE Transactions on Pattern Analysis and Machine Intelligence*, 10(3):362–373.

Thrun, S., Burgard, W., and Fox, D. (2005). *Probabilistic Robotics*. The MIT Press, Cambridge, Massachusetts.

Thrun, S., Montemerlo, M., Dahlkamp, H., Stavens, D., Aron, A. *et al.* (2006). Stanley: the robot that won the DARPA Grand Challenge. *Journal of Field Robotics*, 23(9):661–692.

Tian, Q. and Huhns, M. N. (1986). Algorithms for subpixel registration. *Computer Vision, Graphics, and Image Processing*, 35:220–233.

Tian, Y., Chen, X., and Ganguli, S. (2021). Understanding self-supervised learning dynamics without contrastive pairs. In *International Conference on Machine Learning (ICML)*.

Tian, Y., Fan, B., and Wu, F. (2017). L2-Net: Deep learning of discriminative patch descriptor in Euclidean space. In *IEEE Conference on Computer Vision and Pattern Recognition (CVPR)*.

Tian, Y., Luo, P., Wang, X., and Tang, X. (2015). Deep learning strong parts for pedestrian detection. In *IEEE International Conference on Computer Vision (ICCV)*.

Tian, Y., Balntas, V., Ng, T., Barroso-Laguna, A., Demiris, Y., and Mikolajczyk, K. (2020). D2D: Keypoint extraction with describe to detect approach. In *Asian Conference on Computer Vision (ACCV)*.

Tian, Y., Yu, X., Fan, B., Wu, F., Heijnen, H., and Balntas, V. (2019). SOSNet: Second order similarity regularization for local descriptor learning. In *IEEE/CVF Conference on Computer Vision and Pattern Recognition (CVPR)*.

Tibshirani, R. (1996). Regression shrinkage and selection via the lasso. *Journal of the Royal Statistical Society: Series B (Methodological)*, 58(1):267–288.

Tighe, J. and Lazebnik, S. (2013). Finding things: Image parsing with regions and per-exemplar detectors. In *IEEE Computer Society Conference on Computer Vision and Pattern Recognition (CVPR)*.

Tikhonov, A. N. and Arsenin, V. Y. (1977). *Solutions of Ill-Posed Problems*. V. H. Winston, Washington, D. C.

Tocci, M. D., Kiser, C., Tocci, N., and Sen, P. (2011). A versatile HDR video production system. *ACM Transactions on Graphics (Proc. SIGGRAPH)*, 30(4).

Toint, P. L. (1987). On large scale nonlinear least squares calculations. *SIAM J. Sci. Stat. Comput.*, 8(3):416–435.

Tola, E., Lepetit, V., and Fua, P. (2010). DAISY: An efficient dense descriptor applied to wide-baseline stereo. *IEEE Transactions on Pattern Analysis and Machine Intelligence*, 32(5):815–830.

Tolias, G., Jenicek, T., and Chum, O. (2020). Learning and aggregating deep local descriptors for instance-level recognition. In *European Conference on Computer Vision (ECCV)*.

Tolias, G., Sicre, R., and Jégou, H. (2016). Particular object retrieval with integral max-pooling of CNN activations. In *International Conference on Learning Representations (ICLR)*.

Tolliver, D. and Miller, G. (2006). Graph partitioning by spectral rounding: Applications in image segmentation and clustering. In *IEEE Computer Society Conference on Computer Vision and Pattern Recognition (CVPR)*, pp. 1053–1060.

Tolstikhin, I., Houlsby, N., Kolesnikov, A., Beyer, L., Zhai, X., Unterthiner, T., Yung, J., Keysers, D., Uszkoreit, J., Lucic, M., and Dosovitskiy, A. (2021). MLP-Mixer: An all-MLP architecture for vision. *arXiv preprint arXiv:2105.01601*.

Tomasi, C. and Kanade, T. (1992). Shape and motion from image streams under orthography: A factorization method. *International Journal of Computer Vision*, 9(2):137–154.

Tomasi, C. and Manduchi, R. (1998). Bilateral filtering for gray and color images. In *International Conference on Computer Vision (ICCV)*, pp. 839–846.

Tombari, F., Mattoccia, S., and Di Stefano, L. (2007). Segmentation-based adaptive support for accurate stereo correspondence. In *Pacific-Rim Symposium on Image and Video Technology*.

Tombari, F., Mattoccia, S., Di Stefano, L., and Addimanda, E. (2008). Classification and evaluation of cost aggregation methods for stereo correspondence. In *IEEE Computer Society Conference on Computer Vision and Pattern Recognition (CVPR)*.

Tommasini, T., Fusiello, A., Trucco, E., and Roberto, V. (1998). Making good features track better. In *IEEE Computer Society Conference on Computer Vision and Pattern Recognition (CVPR)*, pp. 178–183.

Tompson, J., Jain, A., LeCun, Y., and Bregler, C. (2014). Joint training of a convolutional network and a graphical model for human pose estimation. In *Advances in Neural Information Processing Systems (NeurIPS)*, pp. 1799–1807.

Tonioni, A., Tosi, F., Poggi, M., Mattoccia, S., and di Stefano, L. (2019). Real-time self-adaptive deep stereo. In *IEEE Conference on Computer Vision and Pattern Recognition (CVPR)*, pp. 195–204.

Torborg, J. and Kajiya, J. T. (1996). Talisman: Commodity realtime 3D graphics for the PC. In *ACM SIGGRAPH Conference Proceedings*, pp. 353–363.

Torii, A., Sivic, J., Pajdla, T., and Okutomi, M. (2013). Visual place recognition with repetitive structures. In *IEEE Computer Society Conference on Computer Vision and Pattern Recognition (CVPR)*.

Torii, A., Arandjelović, R., Sivic, J., Okutomi, M., and Pajdla, T. (2018). 24/7 place recognition by view synthesis. *IEEE Transactions on Pattern Analysis and Machine Intelligence*, 40(2):257–271.

Torr, P. (2002). Bayesian model estimation and selection for epipolar geometry and generic manifold fitting. *International Journal of Computer Vision*, 50(1):35–61.

Torr, P. H. S. and Fitzgibbon, A. W. (2004). Invariant fitting of two view geometry. *IEEE Transactions on Pattern Analysis and Machine Intelligence*, 26(5):648–650.

Torr, P. H. S. and Murray, D. (1997). The development and comparison of robust methods for estimating the fundamental matrix. *International Journal of Computer Vision*, 24(3):271–300.

Torr, P. H. S. and Zisserman, A. (2000). MLESAC: A new robust estimator with application to estimating image geometry. *Computer Vision and Image Understanding*, 78(1):138–156.

Torr, P. H. S., Szeliski, R., and Anandan, P. (1999). An integrated Bayesian approach to layer extraction from image sequences. In *International Conference on Computer Vision (ICCV)*, pp. 983–990.

Torr, P. H. S., Szeliski, R., and Anandan, P. (2001). An integrated Bayesian approach to layer extraction from image sequences. *IEEE Transactions on Pattern Analysis and Machine Intelligence*, 23(3):297–303.

Torralba, A. (2003). Contextual priming for object detection. *International Journal of Computer Vision*, 53(2):169–191.

Torralba, A. (2007). Classifier-based methods. In *CVPR Short Course on Recognizing and Learning Object Categories*. https://people.csail.mit.edu/torralba/shortCourseRLOC.

Torralba, A. (2008). Object recognition and scene understanding. MIT Course 6.870, http://people.csail.mit.edu/torralba/courses/6.870/6.870.recognition.htm.

Torralba, A. and Efros, A. (2011). Unbiased look at dataset bias. In *IEEE Computer Society Conference on Computer Vision and Pattern Recognition (CVPR)*.

Torralba, A., Freeman, W. T., and Fergus, R. (2008). 80 million tiny images: a large dataset for non-parametric object and scene recognition. *IEEE Transactions on Pattern Analysis and Machine Intelligence*, 30(11):1958–1970.

Torralba, A., Murphy, K. P., and Freeman, W. T. (2004). Contextual models for object detection using boosted random fields. In *Advances in Neural Information Processing Systems (NeurIPS)*.

Torralba, A., Weiss, Y., and Fergus, R. (2008). Small codes and large databases of images for object recognition. In *IEEE Computer Society Conference on Computer Vision and Pattern Recognition (CVPR)*.

Torralba, A., Murphy, K. P., Freeman, W. T., and Rubin, M. A. (2003). Context-based vision system for place and object recognition. In *International Conference on Computer Vision (ICCV)*, pp. 273–280.

Torrance, K. E. and Sparrow, E. M. (1967). Theory for off-specular reflection from roughened surfaces. *Journal of the Optical Society of America A*, 57(9):1105–1114.

Torresani, L. (2014). *Weakly Supervised Learning*, pp. 883–885. Springer, Boston, MA.

Torresani, L., Hertzmann, A., and Bregler, C. (2008). Non-rigid structure-from-motion: Estimating shape and motion with hierarchical priors. *IEEE Transactions on Pattern Analysis and Machine Intelligence*, 30(5):878–892.

Toshev, A. and Szegedy, C. (2014). DeepPose: Human pose estimation via deep neural networks. In *IEEE Conference on Computer Vision and Pattern Recognition (CVPR)*.

Tosi, F., Liao, Y., Schmitt, C., and Geiger, A. (2021). SMD-Nets: Stereo mixture density networks. In *IEEE/CVF Conference on Computer Vision and Pattern Recognition (CVPR)*.

Touvron, H., Bojanowski, P., Caron, M., Cord, M., El-Nouby, A., Grave, E., Joulin, A., Synnaeve, G., Verbeek, J., and Jégou, H. (2021). ResMLP: Feedforward networks for image classification with data-efficient training. *arXiv preprint arXiv:2105.03404*.

Touvron, H., Sablayrolles, A., Douze, M., Cord, M., and Jégou, H. (2020). Grafit: Learning fine-grained image representations with coarse labels. *arXiv preprint arXiv:2011.12982*.

Touvron, H., Cord, M., Douze, M., Massa, F., Sablayrolles, A., and Jégou, H. (2020). Training data-efficient image transformers & distillation through attention. *arXiv preprint arXiv:2012.12877*.

Toyama, K. (1998). *Prolegomena for Robust Face Tracking*. Technical Report MSR-TR-98-65, Microsoft Research.

Toyama, K., Krumm, J., Brumitt, B., and Meyers, B. (1999). Wallflower: Principles and practice of background maintenance. In *International Conference on Computer Vision (ICCV)*, pp. 255–261.

Tran, D., Wang, H., Torresani, L., and Feiszli, M. (2019). Video classification with channel-separated convolutional networks. In *IEEE/CVF International Conference on Computer Vision (ICCV)*.

Tran, D., Bourdev, L., Fergus, R., Torresani, L., and Paluri, M. (2015). Learning spatiotemporal features with 3D convolutional networks. In *IEEE International Conference on Computer Vision (ICCV)*.

Tran, D., Wang, H., Torresani, L., Ray, J., LeCun, Y., and Paluri, M. (2018). A closer look at spatiotemporal convolutions for action recognition. In *IEEE Conference on Computer Vision and Pattern Recognition (CVPR)*.

Trefethen, L. N. and Bau, D. (1997). *Numerical Linear Algebra*. SIAM.

Treisman, A. (1985). Preattentive processing in vision. *Computer Vision, Graphics, and Image Processing*, 31(2):156–177.

Triggs, B. (1996). Factorization methods for projective structure and motion. In *IEEE Computer Society Conference on Computer Vision and Pattern Recognition (CVPR)*, pp. 845–851.

Triggs, B. (2004). Detecting keypoints with stable position, orientation, and scale under illumination changes. In *European Conference on Computer Vision (ECCV)*, pp. 100–113.

Triggs, B., Zisserman, A., and Szeliski, R. (eds). (2000). *Vision Algorithms: Theory and Practice*, Springer, Berlin. International Workshop on Vision Algorithms, Corfu, Greece, September 1999.

Triggs, B., McLauchlan, P. F., Hartley, R. I., and Fitzgibbon, A. W. (1999). Bundle adjustment — a modern synthesis. In *International Workshop on Vision Algorithms*, pp. 298–372.

Trobin, W., Pock, T., Cremers, D., and Bischof, H. (2008). Continuous energy minimization via repeated binary fusion. In *European Conference on Computer Vision (ECCV)*, pp. 677–690.

Troccoli, A. and Allen, P. (2008). Building illumination coherent 3D models of large-scale outdoor scenes. *International Journal of Computer Vision*, 78(2–3):261–280.

Trottenberg, U., Oosterlee, C. W., and Schuller, A. (2000). *Multigrid*. Academic Press.

Trucco, E. and Verri, A. (1998). *Introductory Techniques for 3-D Computer Vision*. Prentice Hall, Upper Saddle River, NJ.

Truong, P., Danelljan, M., Van Gool, L., and Timofte, R. (2021). Learning accurate dense correspondences and when to trust them. In *IEEE/CVF Conference on Computer Vision and Pattern Recognition (CVPR)*.

Tsai, P. S. and Shah, M. (1994). Shape from shading using linear approximation. *Image and Vision Computing*, 12:487–498.

Tsai, R. Y. (1987). A versatile camera calibration technique for high-accuracy 3D machine vision metrology using off-the-shelf TV cameras and lenses. *IEEE Journal of Robotics and Automation*, RA-3(4):323–344.

Tschumperlé, D. (2006). Curvature-preserving regularization of multi-valued images using PDEs. In *European Conference on Computer Vision (ECCV)*, pp. 295–307.

Tschumperlé, D. and Deriche, R. (2005). Vector-valued image regularization with PDEs: A common framework for different applications. *IEEE Transactions on Pattern Analysis and Machine Intelligence*, 27:506–517.

Tseng, E., Yu, F., Yang, Y., Mannan, F., Arnaud, K. S., Nowrouzezahrai, D., Lalonde, J.-F., and Heide, F. (2019). Hyperparameter optimization in black-box image processing using differentiable proxies. *ACM Transactions on Graphics (Proc. SIGGRAPH)*, 38(4):27:1–27:14.

Tsin, Y., Kang, S. B., and Szeliski, R. (2006). Stereo matching with linear superposition of layers. *IEEE Transactions on Pattern Analysis and Machine Intelligence*, 28(2):290–301.

Tsin, Y., Ramesh, V., and Kanade, T. (2001). Statistical calibration of CCD imaging process. In *International Conference on Computer Vision (ICCV)*, pp. 480–487.

Tu, Z., Chen, X., Yuille, A. L., and Zhu, S.-C. (2005). Image parsing: Unifying segmentation, detection, and recognition. *International Journal of Computer Vision*, 63(2):113–140.

Tucker, R. and Snavely, N. (2020). Single-view view synthesis with multiplane images. In *IEEE/CVF Conference on Computer Vision and Pattern Recognition (CVPR)*.

Tulsiani, S., Gupta, S., Fouhey, D. F., Efros, A. A., and Malik, J. (2018). Factoring shape, pose, and layout from the 2D image of a 3D scene. In *IEEE Conference on Computer Vision and Pattern Recognition (CVPR)*.

Tumblin, J. and Rushmeier, H. E. (1993). Tone reproduction for realistic images. *IEEE Computer Graphics and Applications*, 13(6):42–48.

Tumblin, J. and Turk, G. (1999). LCIS: A boundary hierarchy for detail-preserving contrast reduction. In *ACM SIGGRAPH Conference Proceedings*, pp. 83–90.

Tumblin, J., Agrawal, A., and Raskar, R. (2005). Why I want a gradient camera. In *IEEE Computer Society Conference on Computer Vision and Pattern Recognition (CVPR)*, pp. 103–110.

Turcot, P. and Lowe, D. G. (2009). Better matching with fewer features: The selection of useful features in large database recognition problems. In *ICCV Workshop on Emergent Issues in Large Amounts of Visual Data (WS-LAVD)*.

Turk, G. and Levoy, M. (1994). Zippered polygonal meshes from range images. In *ACM SIGGRAPH Conference Proceedings*, pp. 311–318.

Turk, G. and O'Brien, J. (2002). Modelling with implicit surfaces that interpolate. *ACM Transactions on Graphics*, 21(4):855–873.

Turk, M. and Pentland, A. (1991). Eigenfaces for recognition. *Journal of Cognitive Neuroscience*, 3(1):71–86.

Tuytelaars, T. and Mikolajczyk, K. (2008). Local invariant feature detectors: A survey. *Foundations and Trends® in Computer Graphics and Computer Vision*, 3(3):177–280.

Tuytelaars, T. and Van Gool, L. (2004). Matching widely separated views based on affine invariant regions. *International Journal of Computer Vision*, 59(1):61–85.

Tuytelaars, T., Van Gool, L., and Proesmans, M. (1997). The cascaded Hough transform. In *International Conference on Image Processing (ICIP)*, pp. 736–739.

Uijlings, J. R. R., Van De Sande, K. E. A., Gevers, T., and Smeulders, A. W. M. (2013). Selective search for object recognition. *International Journal of Computer Vision*, 104(2):154–171.

Ullman, S. (1979). The interpretation of structure from motion. *Proceedings of the Royal Society of London*, B-203:405–426.

Ulyanov, D., Vedaldi, A., and Lempitsky, V. (2017). Improved texture networks: Maximizing quality and diversity in feed-forward stylization and texture synthesis. In *IEEE Conference on Computer Vision and Pattern Recognition (CVPR)*.

Unnikrishnan, R., Pantofaru, C., and Hebert, M. (2007). Toward objective evaluation of image segmentation algorithms. *IEEE Transactions on Pattern Analysis and Machine Intelligence*, 29(6):828–944.

Unser, M. (1999). Splines: A perfect fit for signal and image processing. *IEEE Signal Processing Magazine*, 16(6):22–38.

Urmson, C., Anhalt, J., Bagnell, D., Baker, C., Bittner, R. *et al.* (2008). Autonomous driving in urban environments: Boss and the urban challenge. *Journal of Field Robotics*, 25(8):425–466.

Urtasun, R., Fleet, D. J., and Fua, P. (2006). Temporal motion models for monocular and multiview 3D human body tracking. *Computer Vision and Image Understanding*, 104(2–3):157–177.

Uyttendaele, M., Eden, A., and Szeliski, R. (2001). Eliminating ghosting and exposure artifacts in image mosaics. In *IEEE Computer Society Conference on Computer Vision and Pattern Recognition (CVPR)*, pp. 509–516.

Uyttendaele, M., Criminisi, A., Kang, S. B., Winder, S., Hartley, R., and Szeliski, R. (2004). Image-based interactive exploration of real-world environments. *IEEE Computer Graphics and Applications*, 24(3):52–63.

Vaillant, R. and Faugeras, O. D. (1992). Using extremal boundaries for 3-D object modeling. *IEEE Transactions on Pattern Analysis and Machine Intelligence*, 14(2):157–173.

Vaish, V., Szeliski, R., Zitnick, C. L., Kang, S. B., and Levoy, M. (2006). Reconstructing occluded surfaces using synthetic apertures: Shape from focus vs. shape from stereo. In *IEEE Computer Society Conference on Computer Vision and Pattern Recognition (CVPR)*, pp. 2331–2338.

Valentin, J., Kowdle, A., Barron, J. T., Wadhwa, N., Dzitsiuk, M., Schoenberg, M., Verma, V., Csaszar, A., Turner, E., Dryanovski, I., Afonso, J., Pascoal, J., Tsotsos, K., Leung, M., Schmidt, M., Guleryuz, O., Khamis, S., Tankovitch, V., Fanello, S., Izadi, S., and Rhemann, C. (2018). Depth from motion for smartphone AR. *ACM Transactions on Graphics (Proc. SIGGRAPH Asia)*, 37(6):193:1–193:19.

van de Weijer, J. and Schmid, C. (2006). Coloring local feature extraction. In *European Conference on Computer Vision (ECCV)*, pp. 334–348.

van den Hengel, A., Dick, A., Thormählen, T., Ward, B., and Torr, P. H. S. (2007). VideoTrace: Rapid interactive scene modeling from video. *ACM Transactions on Graphics*, 26(3).

van den Oord, A., Vinyals, O., and Kavukcuoglu, K. (2017). Neural discrete representation learning. In *Advances in Neural Information Processing Systems (NeurIPS)*.

van den Oord, A., Kalchbrenner, N., Espeholt, L., kavukcuoglu, k., Vinyals, O., and Graves, A. (2016). Conditional image generation with PixelCNN decoders. In *Advances in Neural Information Processing Systems (NeurIPS)*.

van der Maaten, L. (2014). Accelerating t-SNE using tree-based algorithms. *Journal of Machine Learning Research*, 15(1):3221–3245.

van der Maaten, L. and Hinton, G. (2008). Visualizing data using t-SNE. *Journal of Machine Learning Research*, 9(Nov):2579–2605.

van Dijk, T. and de Croon, G. (2019). How do neural networks see depth in single images? In *IEEE/CVF International Conference on Computer Vision (ICCV)*.

Van Horn, G., Branson, S., Farrell, R., Haber, S., Barry, J., Ipeirotis, P., Perona, P., and Belongie, S. (2015). Building a bird recognition app and large scale dataset with citizen scientists: The fine print in fine-grained dataset collection. In *IEEE Conference on Computer Vision and Pattern Recognition (CVPR)*.

Van Horn, G., Mac Aodha, O., Song, Y., Cui, Y., Sun, C., Shepard, A., Adam, H., Perona, P., and Belongie, S. (2018). The iNaturalist species classification and detection dataset. In *IEEE Conference on Computer Vision and Pattern Recognition (CVPR)*.

Van Huffel, S. and Lemmerling, P. (eds). (2002). *Total Least Squares and Errors-in-Variables Modeling*, Springer.

Van Huffel, S. and Vandewalle, J. (1991). *The Total Least Squares Problem: Computational Aspects and Analysis*. Society for Industrial and Applied Mathematics, Philadephia.

van Ouwerkerk, J. D. (2006). Image super-resolution survey. *Image and Vision Computing*, 24(10):1039–1052.

Varol, G., Laptev, I., and Schmid, C. (2017). Long-term temporal convolutions for action recognition. *IEEE Transactions on Pattern Analysis and Machine Intelligence*, 40(6):1510–1517.

Vasconcelos, N. (2007). From pixels to semantic spaces: Advances in content-based image retrieval. *Computer*, 40(7):20–26.

Vasilescu, M. A. O. and Terzopoulos, D. (2007). Multilinear (tensor) image synthesis, analysis, and recognition. *IEEE Signal Processing Magazine*, 24(6):118–123.

Vaswani, A., Huang, A., and Manning, C. (2019). Stanford CS224n course: Natural language processing with deep learning. Video and slides available at https://web.stanford.edu/class/archive/cs/cs224n/cs224n.1194/index.html#schedule.

Vaswani, A., Shazeer, N., Parmar, N., Uszkoreit, J., Jones, L., Gomez, A. N., Kaiser, Ł., and Polosukhin, I. (2017). Attention is all you need. In *Advances in Neural Information Processing Systems (NeurIPS)*, pp. 5998–6008.

Vedantam, R., Lawrence Zitnick, C., and Parikh, D. (2015). CIDEr: Consensus-based image description evaluation. In *IEEE Conference on Computer Vision and Pattern Recognition (CVPR)*.

Vedula, S., Baker, S., and Kanade, T. (2005). Image-based spatio-temporal modeling and view interpolation of dynamic events. *ACM Transactions on Graphics*, 24(2):240–261.

Vedula, S., Baker, S., Rander, P., Collins, R., and Kanade, T. (2005). Three-dimensional scene flow. *IEEE Transactions on Pattern Analysis and Machine Intelligence*, 27(3):475–480.

Veksler, O. (1999). *Efficient Graph-based Energy Minimization Methods in Computer Vision*. Ph.D. thesis, Cornell University.

Veksler, O. (2001). Stereo matching by compact windows via minimum ratio cycle. In *International Conference on Computer Vision (ICCV)*, pp. 540–547.

Veksler, O. (2003). Fast variable window for stereo correspondence using integral images. In *IEEE Computer Society Conference on Computer Vision and Pattern Recognition (CVPR)*, pp. 556–561.

Veksler, O. (2005). Stereo correspondence by dynamic programming on a tree. In *IEEE Computer Society Conference on Computer Vision and Pattern Recognition (CVPR)*, pp. 384–390.

Verbin, D. and Zickler, T. (2020). Toward a universal model for shape from texture. In *IEEE/CVF Conference on Computer Vision and Pattern Recognition (CVPR)*.

Vergauwen, M. and Van Gool, L. (2006). Web-based 3D reconstruction service. *Machine Vision and Applications*, 17(2):321–329.

Vetter, T. and Poggio, T. (1997). Linear object classes and image synthesis from a single example image. *IEEE Transactions on Pattern Analysis and Machine Intelligence*, 19(7):733–742.

Viazovetskyi, Y., Ivashkin, V., and Kashin, E. (2020). Stylegan2 distillation for feed-forward image manipulation. In *European Conference on Computer Vision (ECCV)*.

Vicente, S., Kolmogorov, V., and Rother, C. (2008). Graph cut based image segmentation with connectivity priors. In *IEEE Computer Society Conference on Computer Vision and Pattern Recognition (CVPR)*.

Vidal, J., Lin, C.-Y., Lladó, X., and Martí, R. (2018). A method for 6d pose estimation of free-form rigid objects using point pair features on range data. *Sensors*, 18(8):2678.

Vidal, R., Ma, Y., and Sastry, S. S. (2016). *Generalized Principal Component Analysis*. Springer.

Viéville, T. and Faugeras, O. D. (1990). Feedforward recovery of motion and structure from a sequence of 2D-lines matches. In *International Conference on Computer Vision (ICCV)*, pp. 517–520.

Vincent, L. and Soille, P. (1991). Watersheds in digital spaces: An efficient algorithm based on immersion simulations. *IEEE Transactions on Pattern Analysis and Machine Intelligence*, 13(6):583–596.

Vineet, V., Warrell, J., and Torr, P. H. S. (2014). Filter-based mean-field inference for random fields with higher-order terms and product label-spaces. *International Journal of Computer Vision*, 110(3):290–307.

Vinyals, O., Toshev, A., Bengio, S., and Erhan, D. (2015). Show and tell: A neural image caption generator. In *IEEE Conference on Computer Vision and Pattern Recognition (CVPR)*.

Viola, P. and Wells III, W. (1997). Alignment by maximization of mutual information. *International Journal of Computer Vision*, 24(2):137–154.

Viola, P., Jones, M. J., and Snow, D. (2003). Detecting pedestrians using patterns of motion and appearance. In *International Conference on Computer Vision (ICCV)*, pp. 734–741.

Viola, P. A. and Jones, M. J. (2004). Robust real-time face detection. *International Journal of Computer Vision*, 57(2):137–154.

Vlasic, D., Baran, I., Matusik, W., and Popović, J. (2008). Articulated mesh animation from multi-view silhouettes. *ACM Transactions on Graphics*, 27(3):97.

Vlasic, D., Brand, M., Pfister, H., and Popović, J. (2005). Face transfer with multilinear models. *ACM Transactions on Graphics (Proc. SIGGRAPH)*, 24(3):426–433.

Vogel, C., Schindler, K., and Roth, S. (2015). 3D scene flow estimation with a piecewise rigid scene model. *International Journal of Computer Vision*, 115(1):1–28.

Vogiatzis, G., Torr, P., and Cipolla, R. (2005). Multi-view stereo via volumetric graph-cuts. In *IEEE Computer Society Conference on Computer Vision and Pattern Recognition (CVPR)*, pp. 391–398.

Vogiatzis, G., Hernández, C., Torr, P., and Cipolla, R. (2007). Multi-view stereo via volumetric graph-cuts and occlusion robust photo-consistency. *IEEE Transactions on Pattern Analysis and Machine Intelligence*, 29(12):2241–2246.

Voigtlaender, P., Krause, M., Ošep, A., Luiten, J., Sekar, B. B. G., Geiger, A., and Leibe, B. (2019). MOTS: Multi-object tracking and segmentation. In *IEEE/CVF Conference on Computer Vision and Pattern Recognition (CVPR)*.

von Marcard, T., Henschel, R., Black, M. J., Rosenhahn, B., and Pons-Moll, G. (2018). Recovering accurate 3D human pose in the wild using IMUs and a moving camera. In *European Conference on Computer Vision (ECCV)*.

Vondrick, C., Shrivastava, A., Fathi, A., Guadarrama, S., and Murphy, K. (2018). Tracking emerges by colorizing videos. In *European Conference on Computer Vision (ECCV)*.

Vu, H., Labatut, P., Pons, J., and Keriven, R. (2012). High accuracy and visibility-consistent dense multiview stereo. *IEEE Transactions on Pattern Analysis and Machine Intelligence*, 34(5):889–901.

Wadhwa, N., Garg, R., Jacobs, D. E., Feldman, B. E., Kanazawa, N., Carroll, R., Movshovitz-Attias, Y., Barron, J. T., Pritch, Y., and Levoy, M. (2018). Synthetic depth-of-field with a single-camera mobile phone. *ACM Transactions on Graphics (Proc. SIGGRAPH)*, 37(4):Article 64.

Waechter, M., Moehrle, N., and Goesele, M. (2014). Let there be color! Large-scale texturing of 3D reconstructions. In *European Conference on Computer Vision*, pp. 836–850.

Waechter, M., Beljan, M., Fuhrmann, S., Moehrle, N., Kopf, J., and Goesele, M. (2017). Virtual rephotography: Novel view prediction error for 3D reconstruction. *ACM Transactions on Graphics*, 36(1):8:1–8:11.

Wagner, D., Reitmayr, G., Mulloni, A., Drummond, T., and Schmalstieg, D. (2008). Pose tracking from natural features on mobile phones. In *IEEE/ACM International Symposium on Mixed and Augmented Reality (ISMAR)*, pp. 125–134.

Wah, C., Branson, S., Welinder, P., Perona, P., and Belongie, S. (2011). *The Caltech-UCSD Birds-200-2011 Dataset*. Technical Report CNS-TR-2011-001, California Institute of Technology.

Wahba, G. and Wendelberger, J. (1980). Some new mathematical methods for variational objective analysis using splines and cross validation. *Monthly Weather Review*, 108:1122–1143.

Wainwright, M. J. and Jordan, M. I. (2008). Graphical models, exponential families, and variational inference. *Foundations and Trends® in Machine Learning*, 1(1–2):1–305.

Waithe, P. and Ferrie, F. (1991). From uncertainty to visual exploration. *IEEE Transactions on Pattern Analysis and Machine Intelligence*, 13(10):1038–1049.

Walker, E. L. and Herman, M. (1988). Geometric reasoning for constructing 3D scene descriptions from images. *Artificial Intelligence*, 37:275–290.

Wallace, G. K. (1991). The JPEG still picture compression standard. *Communications of the ACM*, 34(4):30–44.

Wallace, J. R., Cohen, M. F., and Greenberg, D. P. (1987). A two-pass solution to the rendering equation: A synthesis of ray tracing and radiosity methods. *Computer Graphics (SIGGRAPH)*, 21(4):311–320.

Wallace, R., Stentz, A., Thorpe, C., Moravec, H., Whittaker, W., and Kanade, T. (1985). First results in robot road-following. In *International Joint Conference on Artificial Intelligence (IJCAI)*.

Waltz, D. L. (1975). Understanding line drawings of scenes with shadows. In Winston, P. H. (ed.), *The Psychology of Computer Vision*, McGraw-Hill, New York.

Wan, A., Dai, X., Zhang, P., He, Z., Tian, Y., Xie, S., Wu, B., Yu, M., Xu, T., Chen, K., Vajda, P., and Gonzalez, J. E. (2020). FBNetV2: Differentiable neural architecture search for spatial and channel dimensions. In *IEEE/CVF Conference on Computer Vision and Pattern Recognition (CVPR)*.

Wandell, B. A. (1995). *Foundations of Vision*. Sinauer Associates. https://foundationsofvision.stanford.edu.

Wang, F., Galliani, S., Vogel, C., Speciale, P., and Pollefeys, M. (2021). PatchmatchNet: Learned multi-view patchmatch stereo. In *IEEE/CVF Conference on Computer Vision and Pattern Recognition (CVPR)*.

Wang, H. and Oliensis, J. (2010). Shape matching by segmentation averaging. *IEEE Transactions on Pattern Analysis and Machine Intelligence*, 32(4):619–635.

Wang, H. and Schmid, C. (2013). Action recognition with improved trajectories. In *International Conference on Computer Vision (ICCV)*.

Wang, H., Ge, S., Lipton, Z., and Xing, E. P. (2019). Learning robust global representations by penalizing local predictive power. In *Advances in Neural Information Processing Systems (NeurIPS)*, pp. 10506–10518.

Wang, H., Li, Y., Wang, Y., Hu, H., and Yang, M.-H. (2020). Collaborative distillation for ultra-resolution universal style transfer. In *IEEE/CVF Conference on Computer Vision and Pattern Recognition (CVPR)*.

Wang, J., Sun, K., Cheng, T., Jiang, B., Deng, C., Zhao, Y., Liu, D., Mu, Y., Tan, M., Wang, X., Liu, W., and Xiao, B. (2020). Deep high-resolution representation learning for visual recognition. *IEEE Transactions on Pattern Analysis and Machine Intelligence*, (accepted).

Wang, J. and Cohen, M. F. (2005). An iterative optimization approach for unified image segmentation and matting. In *International Conference on Computer Vision (ICCV)*.

Wang, J. and Cohen, M. F. (2007a). Optimized color sampling for robust matting. In *IEEE Computer Society Conference on Computer Vision and Pattern Recognition (CVPR)*.

Wang, J. and Cohen, M. F. (2007b). Simultaneous matting and compositing. In *IEEE Computer Society Conference on Computer Vision and Pattern Recognition (CVPR)*.

Wang, J. and Cohen, M. F. (2009). Image and video matting: A survey. *Foundations and Trends® in Computer Graphics and Computer Vision*, 3(2).

Wang, J., Agrawala, M., and Cohen, M. F. (2007). Soft scissors: An interactive tool for realtime high quality matting. *ACM Transactions on Graphics*, 26(3).

Wang, J., Bhat, P., Colburn, R. A., Agrawala, M., and Cohen, M. F. (2005). Interactive video cutout. *ACM Transactions on Graphics (Proc. SIGGRAPH)*, 24(3):585–594.

Wang, J. Y. A. and Adelson, E. H. (1994). Representing moving images with layers. *IEEE Transactions on Image Processing*, 3(5):625–638.

Wang, L., Kang, S. B., Szeliski, R., and Shum, H.-Y. (2001). Optimal texture map reconstruction from multiple views. In *IEEE Computer Society Conference on Computer Vision and Pattern Recognition (CVPR)*, pp. 347–354.

Wang, L., Xiong, Y., Wang, Z., Qiao, Y., Lin, D., Tang, X., and Van Gool, L. (2019). Temporal segment networks for action recognition in videos. *IEEE Transactions on Pattern Analysis and Machine Intelligence*, 41(11):2740–2755.

Wang, N., Zhang, Y., Li, Z., Fu, Y., Liu, W., and Jiang, Y.-G. (2018). Pixel2Mesh: Generating 3D mesh models from single RGB images. In *European Conference on Computer Vision (ECCV)*.

Wang, P., Liu, L., Chen, N., Chu, H.-K., Theobalt, C., and Wang, W. (2020). Vid2Curve: Simultaneous camera motion estimation and thin structure reconstruction from an RGB video. *ACM Transactions on Graphics (Proc. SIGGRAPH)*, 39(4).

Wang, P., Liu, L., Liu, Y., Theobalt, C., Komura, T., and Wang, W. (2021). NeuS: Learning neural implicit surfaces by volume rendering for multi-view reconstruction. *arXiv preprint arXiv:2106.10689*.

Wang, P., Shen, X., Russell, B., Cohen, S., Price, B., and Yuille, A. L. (2016). SURGE: Surface regularized geometry estimation from a single image. In *Advances in Neural Information Processing Systems (NeurIPS)*, pp. 172–180.

Wang, Q., Zhou, X., Hariharan, B., and Snavely, N. (2020). Learning feature descriptors using camera pose supervision. In *European Conference on Computer Vision (ECCV)*.

Wang, Q., Zhang, L., Bertinetto, L., Hu, W., and Torr, P. H. (2019). Fast online object tracking and segmentation: A unifying approach. In *IEEE/CVF Conference on Computer Vision and Pattern Recognition (CVPR)*.

Wang, Q., Wang, Z., Genova, K., Srinivasan, P. P., Zhou, H., Barron, J. T., Martin-Brualla, R., Snavely, N., and Funkhouser, T. (2021). IBRNet: Learning multi-view image-based rendering. In *IEEE/CVF Conference on Computer Vision and Pattern Recognition (CVPR)*.

Wang, R., Geraghty, D., Matzen, K., Szeliski, R., and Frahm, J.-M. (2020). VPLNet: Deep single view normal estimation with vanishing points and lines. In *IEEE/CVF Conference on Computer Vision and Pattern Recognition (CVPR)*.

Wang, S., Fidler, S., and Urtasun, R. (2015). Lost shopping! monocular localization in large indoor spaces. In *IEEE International Conference on Computer Vision (ICCV)*.

Wang, T.-C., Liu, M.-Y., Zhu, J.-Y., Liu, G., Tao, A., Kautz, J., and Catanzaro, B. (2018). Video-to-video synthesis. In *Conference on Neural Information Processing Systems (NeurIPS)*.

Wang, W., Shen, J., and Porikli, F. (2015). Saliency-aware geodesic video object segmentation. In *IEEE Conference on Computer Vision and Pattern Recognition (CVPR)*.

Wang, X., Girshick, R., Gupta, A., and He, K. (2018). Non-local neural networks. In *IEEE Conference on Computer Vision and Pattern Recognition (CVPR)*.

Wang, Y. and Zhu, S.-C. (2003). Modeling textured motion: Particle, wave and sketch. In *International Conference on Computer Vision (ICCV)*, pp. 213–220.

Wang, Y., Curless, B. L., and Seitz, S. M. (2020). People as scene probes. In *European Conference on Computer Vision (ECCV)*.

Wang, Y., Sun, Y., Liu, Z., Sarma, S. E., Bronstein, M. M., and Solomon, J. M. (2019). Dynamic graph CNN for learning on point clouds. *ACM Transactions On Graphics*, 38(5):146.

Wang, Z., Bovik, A. C., and Simoncelli, E. P. (2005). Structural approaches to image quality assessment. In Bovik, A. C. (ed.), *Handbook of Image and Video Processing*, pp. 961–974, Elsevier Academic Press.

Wang, Z., Bovik, A. C., Sheikh, H. R., and Simoncelli, E. P. (2004). Image quality assessment: From error visibility to structural similarity. *IEEE Transactions on Image Processing*, 13(4):600–612.

Wang, Z.-F. and Zheng, Z.-G. (2008). A region based stereo matching algorithm using cooperative optimization. In *IEEE Computer Society Conference on Computer Vision and Pattern Recognition (CVPR)*.

Wanner, S. and Goldluecke, B. (2014). Variational light field analysis for disparity estimation and super-resolution. *IEEE Transactions on Pattern Analysis and Machine Intelligence*, 36(3):606–619.

Ward, G. (1992). Measuring and modeling anisotropic reflection. *Computer Graphics (SIGGRAPH)*, 26(4):265–272.

Ward, G. (1994). The radiance lighting simulation and rendering system. In *ACM SIGGRAPH Conference Proceedings*, pp. 459–472.

Ward, G. (2003). Fast, robust image registration for compositing high dynamic range photographs from hand-held exposures. *Journal of Graphics Tools*, 8(2):17–30.

Ward, G. (2004). High dynamic range image encodings. http://www.anyhere.com/gward/hdrenc/hdr_encodings.html.

Ware, C., Arthur, K., and Booth, K. S. (1993). Fish tank virtual reality. In *INTERCHI*, pp. 37–42.

Warren, J. and Weimer, H. (2001). *Subdivision Methods for Geometric Design: A Constructive Approach*. Morgan Kaufmann.

Watanabe, M. and Nayar, S. K. (1998). Rational filters for passive depth from defocus. *International Journal of Computer Vision*, 27(3):203–225.

Watt, A. (1995). *3D Computer Graphics*. Addison-Wesley, Harlow, England, 3rd edition.

Weber, J. and Malik, J. (1995). Robust computation of optical flow in a multi-scale differential framework. *International Journal of Computer Vision*, 14(1):67–81.

Weber, M., Welling, M., and Perona, P. (2000). Unsupervised learning of models for recognition. In *European Conference on Computer Vision (ECCV)*, pp. 18–32.

Wedel, A., Cremers, D., Pock, T., and Bischof, H. (2009). Structure- and motion-adaptive regularization for high accuracy optic flow. In *International Conference on Computer Vision (ICCV)*.

Wedel, A., Brox, T., Vaudrey, T., Rabe, C., Franke, U., and Cremers, D. (2011). Stereoscopic scene flow computation for 3D motion understanding. *International Journal of Computer Vision*, 95(1):29–51.

Wedel, A., Rabe, C., Vaudrey, T., Brox, T., Franke, U., and Cremers, D. (2008). Efficient dense scene flow from sparse or dense stereo data. In *European Conference on Computer Vision (ECCV)*, pp. 739–751.

Weder, S., Schonberger, J., Pollefeys, M., and Oswald, M. R. (2020). RoutedFusion: Learning real-time depth map fusion. In *IEEE/CVF Conference on Computer Vision and Pattern Recognition (CVPR)*.

Weder, S., Schonberger, J. L., Pollefeys, M., and Oswald, M. R. (2021). NeuralFusion: Online depth fusion in latent space. In *IEEE/CVF Conference on Computer Vision and Pattern Recognition (CVPR)*.

Wei, C. Y. and Quan, L. (2004). Region-based progressive stereo matching. In *IEEE Computer Society Conference on Computer Vision and Pattern Recognition (CVPR)*, pp. 106–113.

Wei, D., Lim, J. J., Zisserman, A., and Freeman, W. T. (2018). Learning and using the arrow of time. In *IEEE Conference on Computer Vision and Pattern Recognition (CVPR)*.

Wei, L.-Y. and Levoy, M. (2000). Fast texture synthesis using tree-structured vector quantization. In *ACM SIGGRAPH Conference Proceedings*, pp. 479–488.

Wei, L.-Y., Lefebvre, S., Kwatra, V., and Turk, G. (2009). State of the art in example-based texture synthesis. In *Eurographics 2009 - State of the Art Reports*, pp. 93–117.

Wei, S.-E., Saragih, J., Simon, T., Harley, A. W., Lombardi, S., Perdoch, M., Hypes, A., Wang, D., Badino, H., and Sheikh, Y. (2019). VR facial animation via multiview image translation. *ACM Transactions on Graphics (Proc. SIGGRAPH)*, 38(4):67.

Wei, S.-E., Ramakrishna, V., Kanade, T., and Sheikh, Y. (2016). Convolutional pose machines. In *IEEE Conference on Computer Vision and Pattern Recognition (CVPR)*.

Weickert, J. (1998). *Anisotropic Diffusion in Image Processing*. Tuebner, Stuttgart.

Weickert, J., ter Haar Romeny, B. M., and Viergever, M. A. (1998). Efficient and reliable schemes for nonlinear diffusion filtering. *IEEE Transactions on Image Processing*, 7(3):398–410.

Weinberger, K. Q. and Saul, L. K. (2009). Distance metric learning for large margin nearest neighbor classification. *Journal of Machine Learning Research*, 10(2).

Weinhaus, F. M. and Devarajan, V. (1997). Texture mapping 3d models of real-world scenes. *ACM Computing Surveys (CSUR)*, 29(4):325–365.

Weinland, D., Ronfard, R., and Boyer, E. (2006). Free viewpoint action recognition using motion history volumes. *Computer Vision and Image Understanding*, 104(2–3):249–257.

Weinland, D., Ronfard, R., and Boyer, E. (2011). A survey of vision-based methods for action representation, segmentation and recognition. *Computer Vision and Image Understanding*, 115(2):224–241.

Weinzaepfel, P., Revaud, J., Harchaoui, Z., and Schmid, C. (2013). DeepFlow: Large displacement optical flow with deep matching. In *International Conference on Computer Vision (ICCV)*.

Weise, T., Bouaziz, S., Li, H., and Pauly, M. (2011). Realtime performance-based facial animation. *ACM Transactions On Graphics (Proc. SIGGRAPH)*, 30(4):1–10.

Weiss, Y. (1997). Smoothness in layers: Motion segmentation using nonparametric mixture estimation. In *IEEE Computer Society Conference on Computer Vision and Pattern Recognition (CVPR)*, pp. 520–526.

Weiss, Y. (1999). Segmentation using eigenvectors: A unifying view. In *International Conference on Computer Vision (ICCV)*, pp. 975–982.

Weiss, Y. (2001). Deriving intrinsic images from image sequences. In *International Conference on Computer Vision (ICCV)*, pp. 7–14.

Weiss, Y. and Adelson, E. H. (1996). A unified mixture framework for motion segmentation: Incorporating spatial coherence and estimating the number of models. In *IEEE Computer Society Conference on Computer Vision and Pattern Recognition (CVPR)*, pp. 321–326.

Weiss, Y. and Freeman, B. (2007). What makes a good model of natural images? In *IEEE Computer Society Conference on Computer Vision and Pattern Recognition (CVPR)*.

Weiss, Y., Torralba, A., and Fergus, R. (2008). Spectral hashing. In *Advances in Neural Information Processing Systems (NeurIPS)*.

Wells, III, W. M. (1986). Efficient synthesis of Gaussian filters by cascaded uniform filters. *IEEE Transactions on Pattern Analysis and Machine Intelligence*, 8(2):234–239.

Wendland, H. (2004). *Scattered data approximation*. Cambridge University Press.

Weng, C.-Y., Curless, B., and Kemelmacher-Shlizerman, I. (2020). Vid2Actor: Free-viewpoint animatable person synthesis from video in the wild. *arXiv preprint arXiv:2012.12884*.

Weng, J., Ahuja, N., and Huang, T. S. (1993). Optimal motion and structure estimation. *IEEE Transactions on Pattern Analysis and Machine Intelligence*, 15(9):864–884.

Weng, L. (2019). Self-supervised representation learning. *lilianweng.github.io/lil-log*. https://lilianweng.github.io/lil-log/2019/11/10/self-supervised-learning.html.

Wenger, A., Gardner, A., Tchou, C., Unger, J., Hawkins, T., and Debevec, P. (2005). Performance relighting and reflectance transformation with time-multiplexed illumination. *ACM Transactions on Graphics (Proc. SIGGRAPH)*, 24(3):756–764.

Werlberger, M., Pock, T., and Bischof, H. (2010). Motion estimation with non-local total variation regularization. In *IEEE Computer Society Conference on Computer Vision and Pattern Recognition (CVPR)*.

Werlberger, M., Trobin, W., Pock, T., Bischof, H., Wedel, A., and Cremers, D. (2009). Anisotropic Huber-L1 optical flow. In *British Machine Vision Conference (BMVC)*.

Werner, T. and Zisserman, A. (2002). New techniques for automated architectural reconstruction from photographs. In *European Conference on Computer Vision (ECCV)*, pp. 541–555.

Westin, S. H., Arvo, J. R., and Torrance, K. E. (1992). Predicting reflectance functions from complex surfaces. *Computer Graphics (SIGGRAPH)*, 26(4):255–264.

Weston, J., Bengio, S., and Usunier, N. (2011). WSABIE: Scaling up to large vocabulary image annotation. In *International Joint Conference on Artificial Intelligence (IJCAI)*, p. 2764–2770.

Westover, L. (1989). Interactive volume rendering. In *Workshop on Volume Visualization*, pp. 9–16.

Wexler, Y., Fitzgibbon, A., and Zisserman, A. (2002). Bayesian estimation of layers from multiple images. In *European Conference on Computer Vision (ECCV)*, pp. 487–501.

Wexler, Y., Shechtman, E., and Irani, M. (2007). Space-time completion of video. *IEEE Transactions on Pattern Analysis and Machine Intelligence*, 29(3):463–476.

Weyrich, T., Matusik, W., Pfister, H., Bickel, B., Donner, C., Tu, C., McAndless, J., Lee, J., Ngan, A., Jansen, H. W., and Gross, M. (2006). Analysis of human faces using a measurement-based skin reflectance model. *ACM Transactions on Graphics*, 25(3):1013–1024.

Weyrich, T., Lawrence, J., Lensch, H. P. A., Rusinkiewicz, S., and Zickler, T. (2009). Principles of appearance acquisition and representation. *Foundations and Trends® in Computer Graphics and Computer Vision*, 4(2):75–191.

Wheeler, M. D., Sato, Y., and Ikeuchi, K. (1998). Consensus surfaces for modeling 3D objects from multiple range images. In *International Conference on Computer Vision (ICCV)*, pp. 917–924.

Whelan, T., Salas-Moreno, R. F., Glocker, B., Davison, A. J., and Leutenegger, S. (2016). ElasticFusion: Real-time dense SLAM and light source estimation. *International Journal of Robotics Research*, 35(14):1697–1716.

Whelan, T., Goesele, M., Lovegrove, S. J., Straub, J., Green, S., Szeliski, R., Butterfield, S., Verma, S., and Newcombe, R. (2018). Reconstructing scenes with mirror and glass surfaces. *ACM Transactions on Graphics (Proc. SIGGRAPH)*, 37(4):102.

White, R. and Forsyth, D. (2006). Combining cues: Shape from shading and texture. In *IEEE Computer Society Conference on Computer Vision and Pattern Recognition (CVPR)*, pp. 1809–1816.

White, R., Crane, K., and Forsyth, D. A. (2007). Capturing and animating occluded cloth. *ACM Transactions on Graphics*, 26(3).

Wickramasinghe, U., Fua, P., and Knott, G. (2021). Deep active surface models. In *IEEE/CVF Conference on Computer Vision and Pattern Recognition (CVPR)*.

Widrow, B. and Hoff, M. E. (1960). *Adaptive switching circuits*. Technical Report, Stanford Univ Ca Stanford Electronics Labs.

Wiejak, J. S., Buxton, H., and Buxton, B. F. (1985). Convolution with separable masks for early image processing. *Computer Vision, Graphics, and Image Processing*, 32(3):279–290.

Wightman, R. (2021). PyTorch image models. https://github.com/rwightman/pytorch-image-models.

Wilburn, B., Xu, H., and Matsushita, Y. (2008). Radiometric calibration using temporal irradiance mixtures. In *IEEE Computer Society Conference on Computer Vision and Pattern Recognition (CVPR)*.

Wilburn, B., Joshi, N., Vaish, V., Talvala, E.-V., Antunez, E., Barth, A., Adams, A., Horowitz, M., and Levoy, M. (2005). High performance imaging using large camera arrays. *ACM Transactions on Graphics (Proc. SIGGRAPH)*, 24(3):765–776.

Wilczkowiak, M., Brostow, G. J., Tordoff, B., and Cipolla, R. (2005). Hole filling through photomontage. In *British Machine Vision Conference (BMVC)*, pp. 492–501.

Wiles, O., Gkioxari, G., Szeliski, R., and Johnson, J. (2020). SynSin: End-to-end view synthesis from a single image. In *IEEE/CVF Conference on Computer Vision and Pattern Recognition (CVPR)*.

Williams, D. and Burns, P. D. (2001). Diagnostics for digital capture using MTF. In *IS&T PICS Conference*, pp. 227–232.

Williams, L. (1983). Pyramidal parametrics. *Computer Graphics (SIGGRAPH)*, 17(3):1–11.

Williams, L. (1990). Performace driven facial animation. *Computer Graphics (SIGGRAPH)*, 24(4):235–242.

Williams, O., Blake, A., and Cipolla, R. (2003). A sparse probabilistic learning algorithm for real-time tracking. In *International Conference on Computer Vision (ICCV)*, pp. 353–360.

Williams, T. L. (1999). *The Optical Transfer Function of Imaging Systems*. Institute of Physics Publishing, London.

Wilson, K. and Snavely, N. (2013). Network principles for SfM: Disambiguating repeated structures with local context. In *International Conference on Computer Vision (ICCV)*.

Wilson, K. and Snavely, N. (2014). Robust global translations with 1DSfM. In *European Conference on Computer Vision*, pp. 61–75.

Wilson, K. and Wehrwein, S. (2020). Visualizing spectral bundle adjustment uncertainty. In *International Conference on 3D Vision (3DV)*, pp. 663–671.

Winder, S. and Brown, M. (2007). Learning local image descriptors. In *IEEE Computer Society Conference on Computer Vision and Pattern Recognition (CVPR)*.

Wingfield, N. (2019). Inside Amazon Go, a store of the future. *New York Times*. Jan 21.

Winkenbach, G. and Salesin, D. H. (1994). Computer-generated pen-and-ink illustration. In *ACM SIGGRAPH Conference Proceedings*, pp. 91–100, Orlando, Florida.

Winn, J. and Shotton, J. (2006). The layout consistent random field for recognizing and segmenting partially occluded objects. In *IEEE Computer Society Conference on Computer Vision and Pattern Recognition (CVPR)*, pp. 37–44.

Winnemöller, H., Olsen, S. C., and Gooch, B. (2006). Real-time video abstraction. *ACM Transactions on Graphics*, 25(3):1221–1226.

Winston, P. H. (ed.). (1975). *The Psychology of Computer Vision*, McGraw-Hill, New York.

Wiskott, L., Fellous, J.-M., Krüger, N., and von der Malsburg, C. (1997). Face recognition by elastic bunch graph matching. *IEEE Transactions on Pattern Analysis and Machine Intelligence*, 19(7):775–779.

Witkin, A. (1981). Recovering surface shape and orientation from texture. *Artificial Intelligence*, 17(1–3):17–45.

Witkin, A. (1983). Scale-space filtering. In *International Joint Conference on Artificial Intelligence (IJCAI)*, pp. 1019–1022.

Witkin, A., Terzopoulos, D., and Kass, M. (1986). Signal matching through scale space. In *National Conference on Artificial Intelligence (AAAI)*, pp. 714–719.

Witkin, A., Terzopoulos, D., and Kass, M. (1987). Signal matching through scale space. *International Journal of Computer Vision*, 1:133–144.

Wizadwongsa, S., Phongthawee, P., Yenphraphai, J., and Suwajanakorn, S. (2021). NeX: Real-time view synthesis with neural basis expansion. In *IEEE/CVF Conference on Computer Vision and Pattern Recognition (CVPR)*.

Wolberg, G. (1990). *Digital Image Warping*. IEEE Computer Society Press, Los Alamitos.

Wolberg, G. and Pavlidis, T. (1985). Restoration of binary images using stochastic relaxation with annealing. *Pattern Recognition Letters*, 3:375–388.

Wolf, L., Hassner, T., and Maoz, I. (2011). Face recognition in unconstrained videos with matched background similarity. In *IEEE Computer Society Conference on Computer Vision and Pattern Recognition (CVPR)*.

Wolff, L. B., Shafer, S. A., and Healey, G. E. (eds). (1992a). *Radiometry. Physics-Based Vision: Principles and Practice*, Jones & Bartlett, Cambridge, MA.

Wolff, L. B., Shafer, S. A., and Healey, G. E. (eds). (1992b). *Shape Recovery. Physics-Based Vision: Principles and Practice*, Jones & Bartlett, Cambridge, MA.

Wood, D. N., Finkelstein, A., Hughes, J. F., Thayer, C. E., and Salesin, D. H. (1997). Multiperspective panoramas for cel animation. In *ACM SIGGRAPH Conference Proceedings*, pp. 243–250.

Wood, D. N., Azuma, D. I., Aldinger, K., Curless, B., Duchamp, T., Salesin, D. H., and Stuetzle, W. (2000). Surface light fields for 3D photography. In *ACM SIGGRAPH Conference Proceedings*, pp. 287–296.

Woodford, O., Reid, I., Torr, P. H., and Fitzgibbon, A. (2008). Global stereo reconstruction under second order smoothness priors. In *IEEE Computer Society Conference on Computer Vision and Pattern Recognition (CVPR)*.

Woodham, R. J. (1981). Analysing images of curved surfaces. *Artificial Intelligence*, 17:117–140.

Woodham, R. J. (1994). Gradient and curvature from photometric stereo including local confidence estimation. *Journal of the Optical Society of America, A*, 11:3050–3068.

Wren, C. R., Azarbayejani, A., Darrell, T., and Pentland, A. P. (1997). Pfinder: Real-time tracking of the human body. *IEEE Transactions on Pattern Analysis and Machine Intelligence*, 19(7):780–785.

Wright, S. (2006). *Digital Compositing for Film and Video*. Focal Press, 2nd edition.

Wronski, B., Garcia-Dorado, I., Ernst, M., Kelly, D., Krainin, M., Liang, C.-K., Levoy, M., and Milanfar, P. (2019). Handheld multi-frame super-resolution. *ACM Transactions on Graphics (Proc. SIGGRAPH)*, 38(4):28:1–28:18.

Wu, B., Dai, X., Zhang, P., Wang, Y., Sun, F., Wu, Y., Tian, Y., Vajda, P., Jia, Y., and Keutzer, K. (2019). FBNet: Hardware-aware efficient ConvNet design via differentiable neural architecture search. In *IEEE/CVF Conference on Computer Vision and Pattern Recognition (CVPR)*.

Wu, C. (2013). Towards linear-time incremental structure from motion. In *International Conference on 3D Vision (3DV)*.

Wu, C., Frahm, J.-M., and Pollefeys, M. (2010). Detecting large repetitive structures with salient boundaries. In *European Conference on Computer Vision (ECCV)*, pp. 142–155.

Wu, C., Agarwal, S., Curless, B., and Seitz, S. (2011). Multicore bundle adjustment. In *IEEE Computer Society Conference on Computer Vision and Pattern Recognition (CVPR)*.

Wu, C., Liu, Y., Dai, Q., and Wilburn, B. (2010). Fusing multiview and photometric stereo for 3D reconstruction under uncalibrated illumination. *IEEE Transactions on Visualization and Computer Graphics*, 17(8):1082–1095.

Wu, C., Clipp, B., Li, X., Frahm, J.-M., and Pollefeys, M. (2008). 3D model matching with viewpoint-invariant patches (VIP). In *IEEE Computer Society Conference on Computer Vision and Pattern Recognition (CVPR)*.

Wu, C.-Y., Manmatha, R., Smola, A. J., and Krahenbuhl, P. (2017). Sampling matters in deep embedding learning. In *IEEE International Conference on Computer Vision (ICCV)*.

Wu, C.-Y., Girshick, R., He, K., Feichtenhofer, C., and Krahenbuhl, P. (2020). A multigrid method for efficiently training video models. In *IEEE/CVF Conference on Computer Vision and Pattern Recognition (CVPR)*.

Wu, C.-Y., Feichtenhofer, C., Fan, H., He, K., Krahenbuhl, P., and Girshick, R. (2019). Long-term feature banks for detailed video understanding. In *IEEE/CVF Conference on Computer Vision and Pattern Recognition (CVPR)*.

Wu, G., Masia, B., Jarabo, A., Zhang, Y., Wang, L., Dai, Q., Chai, T., and Liu, Y. (2017). Light field image processing: An overview. *IEEE Journal of Selected Topics in Signal Processing*, 11(7):926–954.

Wu, H., Xiao, B., Codella, N., Liu, M., Dai, X., Yuan, L., and Zhang, L. (2021). CvT: Introducing convolutions to vision transformers. *arXiv preprint arXiv:2103.15808*.

Wu, Y. and He, K. (2018). Group normalization. In *European Conference on Computer Vision (ECCV)*.

Wu, Y., Lim, J., and Yang, M.-H. (2013). Online object tracking: A benchmark. In *IEEE Computer Society Conference on Computer Vision and Pattern Recognition (CVPR)*.

Wu, Z., Xiong, Y., Yu, S. X., and Lin, D. (2018). Unsupervised feature learning via non-parametric instance discrimination. In *IEEE Conference on Computer Vision and Pattern Recognition (CVPR)*.

Wu, Z., Song, S., Khosla, A., Yu, F., Zhang, L., Tang, X., and Xiao, J. (2015). 3D ShapeNets: A deep representation for volumetric shapes. In *IEEE Conference on Computer Vision and Pattern Recognition (CVPR)*.

Wug Oh, S., Lee, J.-Y., Sunkavalli, K., and Joo Kim, S. (2018). Fast video object segmentation by reference-guided mask propagation. In *IEEE Conference on Computer Vision and Pattern Recognition (CVPR)*.

Wulff, J. and Black, M. J. (2015). Efficient sparse-to-dense optical flow estimation using a learned basis and layers. In *IEEE Conference on Computer Vision and Pattern Recognition (CVPR)*.

Wyszecki, G. and Stiles, W. S. (2000). *Color Science: Concepts and Methods, Quantitative Data and Formulae*. John Wiley & Sons, New York, 2nd edition.

Xia, X., Zhang, M., Xue, T., Sun, Z., Fang, H., Kulis, B., and Chen, J. (2020). Joint bilateral learning for real-time universal photorealistic style transfer. In *European Conference on Computer Vision (ECCV)*.

Xian, K., Shen, C., Cao, Z., Lu, H., Xiao, Y., Li, R., and Luo, Z. (2018). Monocular relative depth perception with web stereo data supervision. In *IEEE Conference on Computer Vision and Pattern Recognition (CVPR)*.

Xian, W., Huang, J.-B., Kopf, J., and Kim, C. (2021). Space-time neural irradiance fields for free-viewpoint video. In *IEEE/CVF Conference on Computer Vision and Pattern Recognition (CVPR)*.

Xian, Y., Lampert, C. H., Schiele, B., and Akata, Z. (2019). Zero-shot learning—–a comprehensive evaluation of the good, the bad and the ugly. *IEEE Transactions on Pattern Analysis and Machine Intelligence*, 41(9):2251–2265.

Xiang, Y., Schmidt, T., Narayanan, V., and Fox, D. (2018). PoseCNN: A convolutional neural network for 6D object pose estimation in cluttered scenes. *Robotics: Science and Systems (RSS)*.

Xiao, F., Lee, Y. J., Grauman, K., Malik, J., and Feichtenhofer, C. (2020). Audiovisual SlowFast networks for video recognition. *arXiv preprint arXiv:2001.08740*.

Xiao, H., Rasul, K., and Vollgraf, R. (2017). Fashion-MNIST: a novel image dataset for benchmarking machine learning algorithms. *arXiv preprint arXiv:1708.07747*.

Xiao, J. and Shah, M. (2003). Two-frame wide baseline matching. In *International Conference on Computer Vision (ICCV)*, pp. 603–609.

Xiao, J. and Shah, M. (2005). Motion layer extraction in the presence of occlusion using graph cuts. *IEEE Transactions on Pattern Analysis and Machine Intelligence*, 27(10):1644–1659.

Xiao, T., Liu, Y., Zhou, B., Jiang, Y., and Sun, J. (2018). Unified perceptual parsing for scene understanding. In *European Conference on Computer Vision (ECCV)*.

Xiaofeng, R. and Bo, L. (2012). Discriminatively trained sparse code gradients for contour detection. In *Advances in Neural Information Processing Systems (NeurIPS)*, pp. 584–592.

Xie, C., Tan, M., Gong, B., Wang, J., Yuille, A. L., and Le, Q. V. (2020). Adversarial examples improve image recognition. In *IEEE/CVF Conference on Computer Vision and Pattern Recognition (CVPR)*.

Xie, Q., Luong, M.-T., Hovy, E., and Le, Q. V. (2020). Self-training with noisy student improves imagenet classification. In *IEEE/CVF Conference on Computer Vision and Pattern Recognition (CVPR)*.

Xie, S., Girshick, R., Kirillov, A., Wu, Y., Feichtenhofer, C., Fan, H., Gkioxari, G., Johnson, J., Ravi, N., Dollár, P., and Lo, W.-Y. (2019). ICCV 2019 tutorial on visual recognition for images, video, and 3D. https://alexander-kirillov.github.io/tutorials/visual-recognition-iccv19.

Xie, S., Girshick, R., Kirillov, A., Wu, Y., Feichtenhofer, C., Fan, H., Gkioxari, G., Johnson, J., Ravi, N., Dollár, P., and Lo, W.-Y. (2020). ECCV 2020 tutorial on visual recognition for images, video, and 3D. https://s9xie.github.io/Tutorials/ECCV2020.

Xie, S. and Tu, Z. (2015). Holistically-nested edge detection. In *IEEE International Conference on Computer Vision (ICCV)*.

Xie, S., Kirillov, A., Girshick, R., and He, K. (2019). Exploring randomly wired neural networks for image recognition. In *IEEE/CVF International Conference on Computer Vision (ICCV)*.

Xie, S., Girshick, R., Dollár, P., Tu, Z., and He, K. (2017). Aggregated residual transformations for deep neural networks. In *IEEE Conference on Computer Vision and Pattern Recognition (CVPR)*.

Xiong, Y. and Turkowski, K. (1997). Creating image-based VR using a self-calibrating fisheye lens. In *IEEE Computer Society Conference on Computer Vision and Pattern Recognition (CVPR)*, pp. 237–243.

Xiong, Y. and Turkowski, K. (1998). Registration, calibration and blending in creating high quality panoramas. In *IEEE Workshop on Applications of Computer Vision (WACV)*, pp. 69–74.

Xu, H., Bazavan, E. G., Zanfir, A., Freeman, W. T., Sukthankar, R., and Sminchisescu, C. (2020). GHUM & GHUML: Generative 3D human shape and articulated pose models. In *IEEE/CVF Conference on Computer Vision and Pattern Recognition (CVPR)*.

Xu, J., Ranftl, R., and Koltun, V. (2017). Accurate optical flow via direct cost volume processing. In *IEEE Conference on Computer Vision and Pattern Recognition (CVPR)*.

Xu, K., Ba, J., Kiros, R., Cho, K., Courville, A., Salakhudinov, R., Zemel, R., and Bengio, Y. (2015). Show, attend and tell: Neural image caption generation with visual attention. In *International Conference on Machine Learning (ICML)*, pp. 2048–2057.

Xu, L., Chen, J., and Jia, J. (2008). A segmentation based variational model for accurate optical flow estimation. In *European Conference on Computer Vision (ECCV)*, pp. 671–684.

Xu, N., Price, B., Cohen, S., and Huang, T. (2017). Deep image matting. In *IEEE Conference on Computer Vision and Pattern Recognition (CVPR)*.

Xu, N., Yang, L., Fan, Y., Yang, J., Yue, D., Liang, Y., Price, B., Cohen, S., and Huang, T. (2018). YouTube-VOS: Sequence-to-sequence video object segmentation. In *European Conference on Computer Vision (ECCV)*.

Xu, Q., Wang, W., Ceylan, D., Mech, R., and Neumann, U. (2019). DISN: Deep implicit surface network for high-quality single-view 3D reconstruction. In *Advances in Neural Information Processing Systems (NeurIPS)*, pp. 492–502.

Xu, R. and Wunsch, D. C. (2005). Survey of clustering algorithms. *IEEE Transactions on Neural Networks*, 16(3):645–678.

Xu, T., Zhang, P., Huang, Q., Zhang, H., Gan, Z., Huang, X., and He, X. (2018). AttnGAN: Fine-grained text to image generation with attentional generative adversarial networks. In *IEEE Conference on Computer Vision and Pattern Recognition (CVPR)*.

Xue, T., Rubinstein, M., Liu, C., and Freeman, W. T. (2015). A computational approach for obstruction-free photography. *ACM Transactions on Graphics (Proc. SIGGRAPH)*, 34(4):79.

Xue, T., Chen, B., Wu, J., Wei, D., and Freeman, W. T. (2019). Video enhancement with task-oriented flow. *International Journal of Computer Vision*, 127(8):1106–1125.

Xue, T., Owens, A., Scharstein, D., Goesele, M., and Szeliski, R. (2019). Multi-frame stereo matching with edges, planes, and superpixels. *Image and Vision Computing*, 91:103771.

Yamins, D. L. K. and DiCarlo, J. J. (2016). Using goal-driven deep learning models to understand sensory cortex. *Nature Neuroscience*, 19(3):356–365.

Yan, H., Liu, Y., and Furukawa, Y. (2017). Turning an urban scene video into a cinemagraph. In *IEEE Conference on Computer Vision and Pattern Recognition (CVPR)*.

Yan, J., Wei, Z., Yi, H., Ding, M., Zhang, R., Chen, Y., Wang, G., and Tai, Y.-W. (2020). Dense hybrid recurrent multi-view stereo net with dynamic consistency checking. In *European Conference on Computer Vision (ECCV)*.

Yan, X., Misra, I., Gupta, A., Ghadiyaram, D., and Mahajan, D. (2020). ClusterFit: Improving generalization of visual representations. In *IEEE/CVF Conference on Computer Vision and Pattern Recognition (CVPR)*.

Yang, B., Yan, J., Lei, Z., and Li, S. Z. (2014). Aggregate channel features for multi-view face detection. In *IEEE International Joint Conference on Biometrics*, pp. 1–8.

Yang, C., Lu, X., Lin, Z., Shechtman, E., Wang, O., and Li, H. (2017). High-resolution image inpainting using multi-scale neural patch synthesis. In *IEEE Conference on Computer Vision and Pattern Recognition (CVPR)*.

Yang, D., El Gamal, A., Fowler, B., and Tian, H. (1999). A 640x512 CMOS image sensor with ultra-wide dynamic range floating-point pixel level ADC. *IEEE Journal of Solid State Circuits*, 34(12):1821–1834.

Yang, G., Manela, J., Happold, M., and Ramanan, D. (2019). Hierarchical deep stereo matching on high-resolution images. In *IEEE Conference on Computer Vision and Pattern Recognition (CVPR)*, pp. 5515–5524.

Yang, G., Zhao, H., Shi, J., Deng, Z., and Jia, J. (2018). SegStereo: Exploiting semantic information for disparity estimation. In *European Conference on Computer Vision (ECCV)*, pp. 660–676.

Yang, H., Antonante, P., Tzoumas, V., and Carlone, L. (2020). Graduated non-convexity for robust spatial perception: From non-minimal solvers to global outlier rejection. *IEEE Robotics and Automation Letters*, 5(2):1127–1134.

Yang, J., Li, H., Dai, Y., and Tan, R. T. (2016). Robust optical flow estimation of double-layer images under transparency or reflection. In *IEEE Conference on Computer Vision and Pattern Recognition (CVPR)*.

Yang, J., Mao, W., Alvarez, J. M., and Liu, M. (2020). Cost volume pyramid based depth inference for multi-view stereo. In *IEEE/CVF Conference on Computer Vision and Pattern Recognition (CVPR)*.

Yang, L., Jin, R., Sukthankar, R., and Jurie, F. (2008). Unifying discriminative visual codebook generation with classifier training for object category recognition. In *IEEE Computer Society Conference on Computer Vision and Pattern Recognition (CVPR)*.

Yang, L., Luo, P., Change Loy, C., and Tang, X. (2015). A large-scale car dataset for fine-grained categorization and verification. In *IEEE Conference on Computer Vision and Pattern Recognition (CVPR)*.

Yang, M.-H., Ahuja, N., and Tabb, M. (2002). Extraction of 2D motion trajectories and its application to hand gesture recognition. *IEEE Transactions on Pattern Analysis and Machine Intelligence*, 24(8):1061–1074.

Yang, M.-H., Kriegman, D. J., and Ahuja, N. (2002). Detecting faces in images: A survey. *IEEE Transactions on Pattern Analysis and Machine Intelligence*, 24(1):34–58.

Yang, Q. (2015). Stereo matching using tree filtering. *IEEE Transactions on Pattern Analysis and Machine Intelligence*, 37(4):834–846.

Yang, Q., Wang, L., Yang, R., Stewénius, H., and Nistér, D. (2009). Stereo matching with color-weighted correlation, hierarchical belief propagation and occlusion handling. *IEEE Transactions on Pattern Analysis and Machine Intelligence*, 31(3):492–504.

Yang, S. and Scherer, S. (2019). Monocular object and plane SLAM in structured environments. *IEEE Robotics and Automation Letters*, 4(4):3145–3152.

Yang, S., Luo, P., Loy, C.-C., and Tang, X. (2015). From facial parts responses to face detection: A deep learning approach. In *IEEE International Conference on Computer Vision (ICCV)*.

Yang, S., Luo, P., Loy, C.-C., and Tang, X. (2016). Wider face: A face detection benchmark. In *IEEE Conference on Computer Vision and Pattern Recognition (CVPR)*.

Yang, S., Wang, Z., Wang, Z., Xu, N., Liu, J., and Guo, Z. (2019). Controllable artistic text style transfer via shape-matching GAN. In *IEEE/CVF International Conference on Computer Vision (ICCV)*.

Yang, T.-Y., Kien Nguyen, D., Heijnen, H., and Balntas, V. (2019). DAME WEB: DynAmic MEan with Whitening Ensemble Binarization for landmark retrieval without human annotation. In *IEEE International Conference on Computer Vision (ICCV) Workshops*.

Yang, Y. and Ramanan, D. (2011). Articulated pose estimation with flexible mixtures-of-parts. In *IEEE Computer Society Conference on Computer Vision and Pattern Recognition (CVPR)*.

Yang, Y., Yuille, A., and Lu, J. (1993). Local, global, and multilevel stereo matching. In *IEEE Computer Society Conference on Computer Vision and Pattern Recognition (CVPR)*, pp. 274–279.

Yang, Z., He, X., Gao, J., Deng, L., and Smola, A. (2016). Stacked attention networks for image question answering. In *IEEE Conference on Computer Vision and Pattern Recognition (CVPR)*.

Yao, B. and Fei-Fei, L. (2012). Recognizing human-object interactions in still images by modeling the mutual context of objects and human poses. *IEEE Transactions on Pattern Analysis and Machine Intelligence*, 34(9):1691–1703.

Yao, J., Fidler, S., and Urtasun, R. (2012). Describing the scene as a whole: Joint object detection, scene classification and semantic segmentation. In *IEEE Computer Society Conference on Computer Vision and Pattern Recognition (CVPR)*.

Yao, Y., Luo, Z., Li, S., Fang, T., and Quan, L. (2018). MVSNet: Depth inference for unstructured multi-view stereo. In *European Conference on Computer Vision (ECCV)*.

Yao, Y., Luo, Z., Li, S., Shen, T., Fang, T., and Quan, L. (2019). Recurrent mvsnet for high-resolution multi-view stereo depth inference. In *IEEE/CVF Conference on Computer Vision and Pattern Recognition (CVPR)*.

Yao, Y., Luo, Z., Li, S., Zhang, J., Ren, Y., Zhou, L., Fang, T., and Quan, L. (2020). BlendedMVS: A large-scale dataset for generalized multi-view stereo networks. In *IEEE/CVF Conference on Computer Vision and Pattern Recognition (CVPR)*.

Yaou, M.-H. and Chang, W.-T. (1994). Fast surface interpolation using multiresolution wavelets. *IEEE Transactions on Pattern Analysis and Machine Intelligence*, 16(7):673–689.

Yariv, L., Kasten, Y., Moran, D., Galun, M., Atzmon, M., Basri, R., and Lipman, Y. (2020). Multiview neural surface reconstruction with implicit lighting and material. In *Advances in Neural Information Processing Systems (NeurIPS)*.

Yatskar, M., Zettlemoyer, L., and Farhadi, A. (2016). Situation recognition: Visual semantic role labeling for image understanding. In *IEEE Conference on Computer Vision and Pattern Recognition (CVPR)*.

Yatziv, L. and Sapiro, G. (2006). Fast image and video colorization using chrominance blending. *IEEE Transactions on Image Processing*, 15(5):1120–1129.

Ye, W., Zhao, Y., and Vela, P. A. (2019). Characterizing SLAM benchmarks and methods for the robust perception age. *arXiv preprint arXiv:1905.07808*. Presented at the ICRA 2019 Workshop on Dataset Generation and Benchmarking of SLAM Algorithms for Robotics and VR/AR.

Yedidia, J. S., Freeman, W. T., and Weiss, Y. (2001). Understanding belief propagation and its generalization. In *International Joint Conference on Artificial Intelligence (IJCAI)*.

Yezzi, A. and Soatto, S. (2003). Stereoscopic segmentation. *International Journal of Computer Vision*, 53(1):31–43.

Yezzi, A. J., Kichenassamy, S., Kumar, A., Olver, P., and Tannenbaum, A. (1997). A geometric snake model for segmentation of medical imagery. *IEEE Transactions on Medical Imaging*, 16(2):199–209.

Yi, K. M., Trulls, E., Lepetit, V., and Fua, P. (2016). LIFT: Learned invariant feature transform. In *European Conference on Computer Vision*, pp. 467–483.

Yi, K. M., Verdie, Y., Fua, P., and Lepetit, V. (2016). Learning to assign orientations to feature points. In *IEEE Conference on Computer Vision and Pattern Recognition (CVPR)*.

Yi, Z., Zhang, H., Tan, P., and Gong, M. (2017). DualGAN: Unsupervised dual learning for image-to-image translation. In *IEEE International Conference on Computer Vision (ICCV)*.

Yi, Z., Tang, Q., Azizi, S., Jang, D., and Xu, Z. (2020). Contextual residual aggregation for ultra high-resolution image inpainting. In *IEEE/CVF Conference on Computer Vision and Pattern Recognition (CVPR)*.

Yilmaz, A. and Shah, M. (2006). Matching actions in presence of camera motion. *Computer Vision and Image Understanding*, 104(2–3):221–231.

Yilmaz, A., Javed, O., and Shah, M. (2006). Object tracking: A survey. *ACM Computing Surveys (CSUR)*, 38(4):13.

Yin, P., Criminisi, A., Winn, J., and Essa, I. (2007). Tree-based classifiers for bilayer video segmentation. In *IEEE Computer Society Conference on Computer Vision and Pattern Recognition (CVPR)*.

Yin, W., Liu, Y., Shen, C., and Yan, Y. (2019). Enforcing geometric constraints of virtual normal for depth prediction. In *IEEE/CVF International Conference on Computer Vision (ICCV)*.

Yin, Z., Darrell, T., and Yu, F. (2019). Hierarchical discrete distribution decomposition for match density estimation. In *IEEE/CVF Conference on Computer Vision and Pattern Recognition (CVPR)*, pp. 6044–6053.

Yoon, J. S., Kim, K., Gallo, O., Park, H. S., and Kautz, J. (2020). Novel view synthesis of dynamic scenes with globally coherent depths from a monocular camera. In *IEEE/CVF Conference on Computer Vision and Pattern Recognition (CVPR)*.

Yoon, K.-J. and Kweon, I.-S. (2006). Adaptive support-weight approach for correspondence search. *IEEE Transactions on Pattern Analysis and Machine Intelligence*, 28(4):650–656.

You, Q., Jin, H., Wang, Z., Fang, C., and Luo, J. (2016). Image captioning with semantic attention. In *IEEE Conference on Computer Vision and Pattern Recognition (CVPR)*.

Young, P., Lai, A., Hodosh, M., and Hockenmaier, J. (2014). From image descriptions to visual denotations: New similarity metrics for semantic inference over event descriptions. *Transactions of the Association for Computational Linguistics*, 2:67–78.

Yserentant, H. (1986). On the multi-level splitting of finite element spaces. *Numerische Mathematik*, 49:379–412.

Yu, A., Ye, V., Tancik, M., and Kanazawa, A. (2020). pixelNeRF: Neural radiance fields from one or few images. *arXiv preprint arXiv:2012.02190*.

Yu, A., Li, R., Tancik, M., Li, H., Ng, R., and Kanazawa, A. (2021). PlenOctrees for real-time rendering of neural radiance fields. *arXiv preprint arXiv:2103.14024*.

Yu, F. and Koltun, V. (2016). Multi-scale context aggregation by dilated convolutions. In *International Conference on Learning Representations (ICLR)*.

Yu, J. and Ramamoorthi, R. (2018). Selfie video stabilization. In *European Conference on Computer Vision (ECCV)*.

Yu, J. and Ramamoorthi, R. (2020). Learning video stabilization using optical flow. In *IEEE/CVF Conference on Computer Vision and Pattern Recognition (CVPR)*.

Yu, J., Ramamoorthi, R., Cheng, K., Sarkis, M., and Bi, N. (2021). Real-time selfie video stabilization. In *IEEE/CVF Conference on Computer Vision and Pattern Recognition (CVPR)*.

Yu, J., Lin, Z., Yang, J., Shen, X., Lu, X., and Huang, T. S. (2018). Generative image inpainting with contextual attention. In *IEEE Conference on Computer Vision and Pattern Recognition (CVPR)*.

Yu, J., Lin, Z., Yang, J., Shen, X., Lu, X., and Huang, T. S. (2019). Free-form image inpainting with gated convolution. In *IEEE/CVF International Conference on Computer Vision (ICCV)*.

Yu, L., Chen, Y.-C., and Li, L. (2020). CVPR 2020 tutorial on self-supervised learning for vision-and-language. https://rohit497.github.io/Recent-Advances-in-Vision-and-Language-Research.

Yu, S. X. and Shi, J. (2003). Multiclass spectral clustering. In *International Conference on Computer Vision (ICCV)*, pp. 313–319.

Yu, Y. and Malik, J. (1998). Recovering photometric properties of architectural scenes from photographs. In *ACM SIGGRAPH Conference Proceedings*, pp. 207–218.

Yu, Y., Debevec, P., Malik, J., and Hawkins, T. (1999). Inverse global illumination: Recovering reflectance models of real scenes from photographs. In *ACM SIGGRAPH Conference Proceedings*, pp. 215–224.

Yuan, K., Guo, S., Liu, Z., Zhou, A., Yu, F., and Wu, W. (2021). Incorporating convolution designs into visual transformers. *arXiv preprint arXiv:2103.11816*.

Yuan, L., Sun, J., Quan, L., and Shum, H.-Y. (2007). Image deblurring with blurred/noisy image pairs. *ACM Transactions on Graphics*, 26(3).

Yuan, L., Sun, J., Quan, L., and Shum, H.-Y. (2008). Progressive inter-scale and intra-scale non-blind image deconvolution. *ACM Transactions on Graphics*, 27(3):74.

Yuan, L., Wen, F., Liu, C., and Shum, H.-Y. (2004). Synthesizing dynamic texture with closed-loop linear dynamic system. In *European Conference on Computer Vision (ECCV)*, pp. 603–616.

Yücer, K., Kim, C., Sorkine-Hornung, A., and Sorkine-Hornung, O. (2016). Depth from gradients in dense light fields for object reconstruction. In *International Conference on 3D Vision (3DV)*, pp. 249–257.

Yücer, K., Sorkine-Hornung, A., Wang, O., and Sorkine-Hornung, O. (2016). Efficient 3D object segmentation from densely sampled light fields with applications to 3D reconstruction. *ACM Transactions on Graphics*, 35(3):22.

Yuille, A. (1991). Deformable templates for face recognition. *Journal of Cognitive Neuroscience*, 3(1):59–70.

Yuille, A. and Poggio, T. (1984). *A Generalized Ordering Constraint for Stereo Correspondence*. A. I. Memo 777, Artificial Intelligence Laboratory, Massachusetts Institute of Technology.

Yuille, A., Vincent, L., and Geiger, D. (1992). Statistical morphology and Bayesian reconstruction. *Journal of Mathematical Imaging and Vision*, 1(3):223–238.

Zabih, R. and Woodfill, J. (1994). Non-parametric local transforms for computing visual correspondence. In *European Conference on Computer Vision (ECCV)*, pp. 151–158.

Zach, C. (2008). Fast and high quality fusion of depth maps. In *International Symposium on 3D Data Processing, Visualization and Transmission (3DPVT)*.

Zach, C., Klopschitz, M., and Pollefeys, M. (2010). Disambiguating visual relations using loop constraints. In *IEEE Computer Society Conference on Computer Vision and Pattern Recognition (CVPR)*.

Zach, C., Pock, T., and Bischof, H. (2007a). A duality based approach for realtime TV-L1 optical flow. In *Pattern Recognition (DAGM)*.

Zach, C., Pock, T., and Bischof, H. (2007b). A globally optimal algorithm for robust TV-L^1 range image integration. In *International Conference on Computer Vision (ICCV)*.

Zafeiriou, S., Zhang, C., and Zhang, Z. (2015). A survey on face detection in the wild: past, present and future. *Computer Vision and Image Understanding*, 138:1–24.

Zaragoza, J., Chin, T.-J., Brown, M. S., and Suter, D. (2013). As-projective-as-possible image stitching with moving dlt. In *IEEE Computer Society Conference on Computer Vision and Pattern Recognition (CVPR)*.

Zebedin, L., Bauer, J., Karner, K., and Bischof, H. (2008). Fusion of feature- and area-based information for urban buildings modeling from aerial imagery. In *European Conference on Computer Vision (ECCV)*, pp. 873–886.

Zeiler, M. D. (2012). Adadelta: an adaptive learning rate method. *arXiv preprint arXiv:1212.5701*.

Zeiler, M. D. and Fergus, R. (2014). Visualizing and understanding convolutional networks. In *European Conference on Computer Vision*, pp. 818–833.

Zellers, R., Bisk, Y., Farhadi, A., and Choi, Y. (2019). From recognition to cognition: Visual commonsense reasoning. In *IEEE/CVF Conference on Computer Vision and Pattern Recognition (CVPR)*.

Zelnik-Manor, L. and Perona, P. (2007). Automating joiners. In *Symposium on Non Photorealistic Animation and Rendering*.

Zendel, O. *et al.* (2020). Robust vision challenge. In *European Conference on Computer Vision (ECCV) Workshops*. http://www.robustvision.net.

Zendel, O., Honauer, K., Murschitz, M., Steininger, D., and Fernandez Dominguez, G. (2018). WildDash - creating hazard-aware benchmarks. In *European Conference on Computer Vision (ECCV)*.

Zeng, Y., Fu, J., and Chao, H. (2020). Learning joint spatial-temporal transformations for video inpainting. In *European Conference on Computer Vision*, pp. 528–543.

Zeng, Y., Fu, J., Chao, H., and Guo, B. (2019). Learning pyramid-context encoder network for high-quality image inpainting. In *IEEE/CVF Conference on Computer Vision and Pattern Recognition (CVPR)*.

Zeng, Y., Lin, Z., Yang, J., Zhang, J., Shechtman, E., and Lu, H. (2020). High-resolution image inpainting with iterative confidence feedback and guided upsampling. *arXiv preprint arXiv:2005.11742*.

Zhai, A., Wu, H.-Y., Tzeng, E., Park, D. H., and Rosenberg, C. (2019). Learning a unified embedding for visual search at pinterest. In *ACM SIGKDD International Conference on Knowledge Discovery & Data Mining*, pp. 2412–2420.

Zhai, X., Oliver, A., Kolesnikov, A., and Beyer, L. (2019). S4L: Self-supervised semi-supervised learning. In *IEEE/CVF International Conference on Computer Vision (ICCV)*.

Zhang, A., Lipton, Z. C., Li, M., and Smola, A. J. (2021). Dive into deep learning. release 0.16.1, latest draft at https://d2l.ai.

Zhang, C. and Chen, T. (2004). A survey on image-based rendering - representation, sampling and compression. *EURASIP Signal Processing: Image Communication*, 19(1):1–28.

Zhang, F. and Liu, F. (2014). Parallax-tolerant image stitching. In *IEEE Conference on Computer Vision and Pattern Recognition (CVPR)*.

Zhang, F., Prisacariu, V. A., Yang, R., and Torr, P. (2019). GA-Net: Guided aggregation net for end-to-end stereo matching. In *IEEE Conference on Computer Vision and Pattern Recognition (CVPR)*, pp. 185–194.

Zhang, F., Qi, X., Yang, R., Prisacariu, V., Wah, B., and Torr, P. (2020). Domain-invariant stereo matching networks. In *European Conference on Computer Vision (ECCV)*.

Zhang, G., Jia, J., Wong, T.-T., and Bao, H. (2008). Recovering consistent video depth maps via bundle optimization. In *IEEE Computer Society Conference on Computer Vision and Pattern Recognition (CVPR)*.

Zhang, H., Dauphin, Y. N., and Ma, T. (2019). Fixup initialization: Residual learning without normalization. In *International Conference on Learning Representations (ICLR)*.

Zhang, H., Cisse, M., Dauphin, Y. N., and Lopez-Paz, D. (2018). *mixup*: Beyond empirical risk minimization. In *International Conference on Learning Representations (ICLR)*.

Zhang, H., Dai, Y., Li, H., and Koniusz, P. (2019). Deep stacked hierarchical multi-patch network for image deblurring. In *IEEE/CVF Conference on Computer Vision and Pattern Recognition (CVPR)*.

Zhang, H., Goodfellow, I., Metaxas, D., and Odena, A. (2019). Self-attention generative adversarial networks. In *International Conference on Machine Learning (ICML)*, pp. 7354–7363.

Zhang, H., Sciutto, C., Agrawala, M., and Fatahalian, K. (2021). Vid2Player: Controllable video sprites that behave and appear like professional tennis players. *ACM Transactions on Graphics (TOG)*, 40(3):1–16.

Zhang, H., Xu, T., Li, H., Zhang, S., Wang, X., Huang, X., and Metaxas, D. N. (2017). StackGAN: Text to photo-realistic image synthesis with stacked generative adversarial networks. In *IEEE International Conference on Computer Vision (ICCV)*.

Zhang, H., Xu, T., Li, H., Zhang, S., Wang, X., Huang, X., and Metaxas, D. N. (2018). StackGAN++: Realistic image synthesis with stacked generative adversarial networks. *IEEE Transactions on Pattern Analysis and Machine Intelligence*, 41(8):1947–1962.

Zhang, J., McMillan, L., and Yu, J. (2006). Robust tracking and stereo matching under variable illumination. In *IEEE Computer Society Conference on Computer Vision and Pattern Recognition (CVPR)*, pp. 871–878.

Zhang, J., Marszalek, M., Lazebnik, S., and Schmid, C. (2007). Local features and kernels for classification of texture and object categories: a comprehensive study. *International Journal of Computer Vision*, 73(2):213–238.

Zhang, J., Yao, Y., Li, S., Luo, Z., and Fang, T. (2020). Visibility-aware multi-view stereo network. In *British Machine Vision Conference (BMVC)*.

Zhang, K., Riegler, G., Snavely, N., and Koltun, V. (2020). NeRF++: Analyzing and improving neural radiance fields. *arXiv preprint arXiv:2010.07492*.

Zhang, K., Luan, F., Wang, Q., Bala, K., and Snavely, N. (2021). PhySG: Inverse rendering with spherical gaussians for physics-based material editing and relighting. In *IEEE/CVF Conference on Computer Vision and Pattern Recognition (CVPR)*.

Zhang, K., Zuo, W., Chen, Y., Meng, D., and Zhang, L. (2017). Beyond a Gaussian denoiser: Residual learning of deep CNN for image denoising. *IEEE Transactions on Image Processing*, 26(7):3142–3155.

Zhang, L. and Seitz, S. (2007). Estimating optimal parameters for MRF stereo from a single image pair. *IEEE Transactions on Pattern Analysis and Machine Intelligence*, 29(2):331–342.

Zhang, L., Curless, B., and Seitz, S. (2003). Spacetime stereo: Shape recovery for dynamic scenes. In *IEEE Computer Society Conference on Computer Vision and Pattern Recognition (CVPR)*, pp. 367–374.

Zhang, L., Dugas-Phocion, G., Samson, J.-S., and Seitz, S. M. (2002). Single view modeling of free-form scenes. *Journal of Visualization and Computer Animation*, 13(4):225–235.

Zhang, L., Lin, L., Liang, X., and He, K. (2016). Is faster R-CNN doing well for pedestrian detection? In *European Conference on Computer Vision*, pp. 443–457.

Zhang, L., Snavely, N., Curless, B., and Seitz, S. M. (2004). Spacetime faces: High resolution capture for modeling and animation. *ACM Transactions on Graphics*, 23(3):548–558.

Zhang, N., Donahue, J., Girshick, R., and Darrell, T. (2014). Part-based R-CNNs for fine-grained category detection. In *European Conference on Computer Vision (ECCV)*, pp. 834–849.

Zhang, R. (2019). Making convolutional networks shift-invariant again. In *International Conference on Machine Learning (ICML)*.

Zhang, R., Isola, P., and Efros, A. A. (2016). Colorful image colorization. In *European Conference on Computer Vision*, pp. 649–666.

Zhang, R., Tsai, P.-S., Cryer, J. E., and Shah, M. (1999). Shape from shading: A survey. *IEEE Transactions on Pattern Analysis and Machine Intelligence*, 21(8):690–706.

Zhang, R., Isola, P., Efros, A. A., Shechtman, E., and Wang, O. (2018). The unreasonable effectiveness of deep features as a perceptual metric. In *IEEE Conference on Computer Vision and Pattern Recognition (CVPR)*.

Zhang, R., Zhu, J.-Y., Isola, P., Geng, X., Lin, A. S., Yu, T., and Efros, A. A. (2017). Real-time user-guided image colorization with learned deep priors. *ACM Transactions on Graphics (Proc. SIGGRAPH)*, 9(4).

Zhang, S. (2018). High-speed 3D shape measurement with structured light methods: A review. *Optics and Lasers in Engineering*, 106:119–131.

Zhang, S., Benenson, R., and Schiele, B. (2017). CityPersons: A diverse dataset for pedestrian detection. In *IEEE Conference on Computer Vision and Pattern Recognition (CVPR)*.

Zhang, W. and Kosecka, J. (2006). Image based localization in urban environments. In *International Symposium on 3D Data Processing, Visualization and Transmission (3DPVT)*, pp. 33–40.

Zhang, X., Zhou, X., Lin, M., and Sun, J. (2018). ShuffleNet: An extremely efficient convolutional neural network for mobile devices. In *IEEE Conference on Computer Vision and Pattern Recognition (CVPR)*.

Zhang, X. C., Barron, J. T., Tsai, Y.-T., Pandey, R., Zhang, X., Ng, R., and Jacobs, D. E. (2020). Portrait shadow manipulation. *ACM Transactions on Graphics (Proc. SIGGRAPH)*, 39(4):78.

Zhang, Y. and Kambhamettu, C. (2003). On 3D scene flow and structure recovery from multiview image sequences. *IEEE Transactions on Systems, Man, and Cybernetics*, 33(4):592–606.

Zhang, Y., Wadhwa, N., Orts-Escolano, S., Häne, C., Fanello, S., and Garg, R. (2020). Du^{2}Net: Learning depth estimation from dual-cameras and dual-pixels. In *European Conference on Computer Vision (ECCV)*.

Zhang, Z. (1994). Iterative point matching for registration of free-form curves and surfaces. *International Journal of Computer Vision*, 13(2):119–152.

Zhang, Z. (1998a). Determining the epipolar geometry and its uncertainty: A review. *International Journal of Computer Vision*, 27(2):161–195.

Zhang, Z. (1998b). On the optimization criteria used in two-view motion analysis. *IEEE Transactions on Pattern Analysis and Machine Intelligence*, 20(7):717–729.

Zhang, Z. (2000). A flexible new technique for camera calibration. *IEEE Transactions on Pattern Analysis and Machine Intelligence*, 22(11):1330–1334.

Zhang, Z. (2012). Microsoft Kinect sensor and its effect. *IEEE MultiMedia*, 19(2):4–10.

Zhang, Z. and He, L.-W. (2007). Whiteboard scanning and image enhancement. *Digital Signal Processing*, 17(2):414–432.

Zhang, Z. and Shan, Y. (2000). A progressive scheme for stereo matching. In *European Workshop on 3D Structure from Multiple Images of Large-Scale Environments (SMILE)*, pp. 68–85.

Zhang, Z. and Wang, J. (2007). MLLE: Modified locally linear embedding using multiple weights. In *Advances in Neural Information Processing Systems (NeurIPS)*, pp. 1593–1600.

Zhang, Z. and Zha, H. (2004). Principal manifolds and nonlinear dimensionality reduction via tangent space alignment. *SIAM Journal on Scientific Computing*, 26(1):313–338.

Zhang, Z., Deriche, R., Faugeras, O., and Luong, Q. (1995). A robust technique for matching two uncalibrated images through the recovery of the unknown epipolar geometry. *Artificial Intelligence*, 78:87–119.

Zhang, Z., Li, Z., Bi, N., Zheng, J., Wang, J., Huang, K., Luo, W., Xu, Y., and Gao, S. (2019). PPGNet: Learning point-pair graph for line segment detection. In *IEEE/CVF Conference on Computer Vision and Pattern Recognition (CVPR)*.

Zhao, C., Sun, Q., Zhang, C., Tang, Y., and Qian, F. (2020). Monocular depth estimation based on deep learning: An overview. *arXiv preprint arXiv:2003.06620*.

Zhao, G. and Pietikäinen, M. (2007). Dynamic texture recognition using local binary patterns with an application to facial expressions. *IEEE Transactions on Pattern Analysis and Machine Intelligence*, 29(6):915–928.

Zhao, H., Shi, J., Qi, X., Wang, X., and Jia, J. (2017). Pyramid scene parsing network. In *IEEE Conference on Computer Vision and Pattern Recognition (CVPR)*.

Zhao, W., Chellappa, R., Phillips, P. J., and Rosenfeld, A. (2003). Face recognition: A literature survey. *ACM Computing Surveys (CSUR)*, 35(4):399–358.

Zhao, Y., Xu, S., Bu, S., Jiang, H., and Han, P. (2019). GSLAM: A general SLAM framework and benchmark. In *IEEE/CVF International Conference on Computer Vision (ICCV)*.

Zhao, Y., Huang, Z., Li, T., Chen, W., LeGendre, C., Ren, X., Shapiro, A., and Li, H. (2019). Learning perspective undistortion of portraits. In *IEEE/CVF International Conference on Computer Vision (ICCV)*.

Zhao, Z.-Q., Zheng, P., Xu, S.-t., and Wu, X. (2019). Object detection with deep learning: A review. *IEEE Transactions on Neural Networks and Learning Systems*, 30(11):3212–3232.

Zheng, E., Wang, K., Dunn, E., and Frahm, J.-M. (2014). Joint object class sequencing and trajectory triangulation (JOST). In *European Conference on Computer Vision*, pp. 599–614.

Zheng, J. Y. (1994). Acquiring 3-D models from sequences of contours. *IEEE Transactions on Pattern Analysis and Machine Intelligence*, 16(2):163–178.

Zheng, K. C., Kang, S. B., Cohen, M., and Szeliski, R. (2007). Layered depth panoramas. In *IEEE Computer Society Conference on Computer Vision and Pattern Recognition (CVPR)*.

Zheng, L., Yang, Y., and Tian, Q. (2018). SIFT meets CNN: A decade survey of instance retrieval. *IEEE Transactions on Pattern Analysis and Machine Intelligence*, 40(5):1224–1244.

Zheng, N. and Xue, J. (2009). Manifold learning. In *Statistical Learning and Pattern Analysis for Image and Video Processing*, pp. 87–119, Springer, London.

Zheng, S., Lu, J., Zhao, H., Zhu, X., Luo, Z., Wang, Y., Fu, Y., Feng, J., Xiang, T., Torr, P. H. S., and Zhang, L. (2020). Rethinking semantic segmentation from a sequence-to-sequence perspective with transformers. *arXiv preprint arXiv:2012.15840*.

Zheng, S., Jayasumana, S., Romera-Paredes, B., Vineet, V., Su, Z., Du, D., Huang, C., and Torr, P. H. S. (2015). Conditional random fields as recurrent neural networks. In *IEEE International Conference on Computer Vision (ICCV)*.

Zheng, Y., Lin, S., and Kang, S. B. (2006). Single-image vignetting correction. In *IEEE Computer Society Conference on Computer Vision and Pattern Recognition (CVPR)*, pp. 461–468.

Zheng, Y., Yu, J., Kang, S.-B., Lin, S., and Kambhamettu, C. (2008). Single-image vignetting correction using radial gradient symmetry. In *IEEE Computer Society Conference on Computer Vision and Pattern Recognition (CVPR)*.

Zheng, Y., Zhou, X. S., Georgescu, B., Zhou, S. K., and Comaniciu, D. (2006). Example based non-rigid shape detection. In *European Conference on Computer Vision (ECCV)*, pp. 423–436.

Zheng, Y.-T., Zhao, M., Song, Y., Adam, H., Buddemeier, U., Bissacco, A., Brucher, F., Chua, T.-S., and Neven, H. (2009). Tour the world: building a web-scale landmark recognition engine. In *IEEE Computer Society Conference on Computer Vision and Pattern Recognition (CVPR)*.

Zhong, J. and Sclaroff, S. (2003). Segmenting foreground objects from a dynamic, textured background via a robust Kalman filter. In *International Conference on Computer Vision (ICCV)*, pp. 44–50.

Zhou, B., Khosla, A., Lapedriza, A., Oliva, A., and Torralba, A. (2015). Object detectors emerge in deep scene CNNs. In *International Conference on Learning Representations (ICLR)*.

Zhou, B., Zhao, H., Puig, X., Xiao, T., Fidler, S., Barriuso, A., and Torralba, A. (2019). Semantic understanding of scenes through the ADE20K dataset. *International Journal of Computer Vision*, 127(3):302–321.

Zhou, C., Lin, S., and Nayar, S. (2009). Coded aperture pairs for depth from defocus. In *International Conference on Computer Vision (ICCV)*.

Zhou, H., Sattler, T., and Jacobs, D. W. (2016). Evaluating local features for day-night matching. In *European Conference on Computer Vision (ECCV) Workshops*, pp. 724–736.

Zhou, H., Hadap, S., Sunkavalli, K., and Jacobs, D. W. (2019). Deep single-image portrait relighting. In *IEEE/CVF International Conference on Computer Vision (ICCV)*.

Zhou, H., Zou, D., Pei, L., Ying, R., Liu, P., and Yu, W. (2015). StructSLAM: Visual SLAM with building structure lines. *IEEE Transactions on Vehicular Technology*, 64(4):1364–1375.

Zhou, L. (2020). CVPR 2020 tutorial on recent advances in visual captioning. https://rohit497.github.io/Recent-Advances-in-Vision-and-Language-Research.

Zhou, L., Kalantidis, Y., Chen, X., Corso, J. J., and Rohrbach, M. (2019). Grounded video description. In *IEEE/CVF Conference on Computer Vision and Pattern Recognition (CVPR)*.

Zhou, L., Palangi, H., Zhang, L., Hu, H., Corso, J. J., and Gao, J. (2020). Unified vision-language pre-training for image captioning and VQA. In *AAAI Conference on Artificial Intelligence (AAAI)*, pp. 13041–13049.

Zhou, Q., Sattler, T., and Leal-Taixé, L. (2021). Patch2Pix: Epipolar-guided pixel-level correspondences. In *IEEE/CVF Conference on Computer Vision and Pattern Recognition (CVPR)*.

Zhou, Q., Sattler, T., Pollefeys, M., and Leal-Taixé, L. (2019). To learn or not to learn: Visual localization from essential matrices. In *IEEE International Conference on Robotics and Automation (ICRA)*.

Zhou, Q.-Y. and Koltun, V. (2013). Dense scene reconstruction with points of interest. *ACM Transactions on Graphics (Proc. SIGGRAPH)*, 32(4):112.

Zhou, Q.-Y. and Koltun, V. (2014). Color map optimization for 3D reconstruction with consumer depth cameras. *ACM Transactions on Graphics (Proc. SIGGRAPH)*, 33(4):155.

Zhou, Q.-Y., Miller, S., and Koltun, V. (2013). Elastic fragments for dense scene reconstruction. In *International Conference on Computer Vision (ICCV)*.

Zhou, Q.-Y., Park, J., and Koltun, V. (2018). Open3D: A modern library for 3D data processing. *arXiv preprint arXiv:1801.09847*.

Zhou, T., Brown, M., Snavely, N., and Lowe, D. G. (2017). Unsupervised learning of depth and ego-motion from video. In *IEEE Conference on Computer Vision and Pattern Recognition (CVPR)*.

Zhou, T., Tucker, R., Flynn, J., Fyffe, G., and Snavely, N. (2018). Stereo magnification: Learning view synthesis using multiplane images. *ACM Transactions on Graphics (Proc. SIGGRAPH)*, 37(4):65:1–65:12.

Zhou, X., Koltun, V., and Krähenbühl, P. (2020). Tracking objects as points. In *European Conference on Computer Vision (ECCV)*.

Zhou, Y., Qi, H., and Ma, Y. (2019). End-to-end wireframe parsing. In *IEEE/CVF International Conference on Computer Vision (ICCV)*.

Zhou, Y., Qi, H., Huang, J., and Ma, Y. (2019a). NeurVPS: Neural vanishing point scanning via conic convolution. In *Advances in Neural Information Processing Systems (NeurIPS)*, pp. 866–875.

Zhou, Y., Qi, H., Zhai, Y., Sun, Q., Chen, Z., Wei, L.-Y., and Ma, Y. (2019b). Learning to reconstruct 3D Manhattan wireframes from a single image. In *IEEE/CVF International Conference on Computer Vision (ICCV)*.

Zhou, Z., Furukawa, Y., Ma, Y., Gao, S., Liu, C., Zhou, Y., Luo, L., Zheng, J., Zhang, J., and Tang, R. (2020). Holistic scene structures for 3D vision. In *European Conference on Computer Vision (ECCV) Workshops*. https://holistic-3d.github.io/eccv20.

Zhou, Z., Furukawa, Y., and Ma, Y. (2019). ICCV 2019 tutorial on holistic 3D reconstruction: Learning to reconstruct holistic 3D structures from sensorial data. https://holistic-3d.github.io/iccv19.

Zhou, Z.-H. (2018). A brief introduction to weakly supervised learning. *National Science Review*, 5(1):44–53.

Zhu, D. and Smith, W. A. P. (2020). Least squares surface reconstruction on arbitrary domains. In *European Conference on Computer Vision (ECCV)*.

Zhu, J.-Y. (2021). Carnegie Mellon University 16-726 course: Learning-based image synthesis. https://learning-image-synthesis.github.io.

Zhu, J.-Y., Krähenbühl, P., Shechtman, E., and Efros, A. A. (2016). Generative visual manipulation on the natural image manifold. In *European Conference on Computer Vision (ECCV)*, pp. 597–613.

Zhu, J.-Y., Park, T., Isola, P., and Efros, A. A. (2017). Unpaired image-to-image translation using cycle-consistent adversarial networks. In *IEEE International Conference on Computer Vision (ICCV)*.

Zhu, J.-Y., Zhang, R., Pathak, D., Darrell, T., Efros, A. A., Wang, O., and Shechtman, E. (2017). Toward multimodal image-to-image translation. In *Advances in Neural Information Processing Systems (NeurIPS)*, pp. 465–476.

Zhu, S.-C. and Mumford, D. (2006). A stochastic grammar of images. *Foundations and Trends® in Computer Graphics and Computer Vision*, 2(4).

Zhu, S. C. and Yuille, A. L. (1996). Region competition: Unifying snakes, region growing, and Bayes/MDL for multiband image segmentation. *IEEE Transactions on Pattern Analysis and Machine Intelligence*, 18(9):884–900.

Zhu, X. and Goldberg, A. B. (2009). Introduction to semi-supervised learning. *Synthesis Lectures on Artificial Intelligence and Machine Learning*, 3(1):1–130.

Zhu, Y., Li, X., Liu, C., Zolfaghari, M., Xiong, Y., Wu, C., Zhang, Z., Tighe, J., Manmatha, R., and Li, M. (2020). A comprehensive study of deep video action recognition. *arXiv preprint arXiv:2012.06567*.

Zhu, Z. and Kanade, T. (2008). Modeling and representations of large-scale 3D scenes. *International Journal of Computer Vision*, 78(2–3):119–120.

Zhuang, B. and Tran, Q.-H. (2020). Image stitching and rectification for hand-held cameras. In *European Conference on Computer Vision (ECCV)*.

Zhuang, C., Yan, S., Nayebi, A., Schrimpf, M., Frank, M., DiCarlo, J., and Yamins, D. (2020). Unsupervised neural network models of the ventral visual stream. *bioRxiv preprint 2020.06.16.155556*.

Zickler, T. E., Belhumeur, P. N., and Kriegman, D. J. (2002). Helmholtz stereopsis: Exploiting reciprocity for surface reconstruction. *International Journal of Computer Vision*, 49(2):215–227.

Zimmer, H., Bruhn, A., and Weickert, J. (2011). Optic flow in harmony. *International Journal of Computer Vision*, 93(3):368–388.

Zimmermann, C. and Brox, T. (2017). Learning to estimate 3D hand pose from single RGB images. In *IEEE International Conference on Computer Vision (ICCV)*.

Zisserman, A. (2018). Self-supervised learning. Slides from talk given at INRIA, available at https://project.inria.fr/paiss/files/2018/07/zisserman-self-supervised.pdf.

Zisserman, A., Giblin, P. J., and Blake, A. (1989). The information available to a moving observer from specularities. *Image and Vision Computing*, 7(1):38–42.

Zitnick, C. L. and Dollár, P. (2014). Edge boxes: Locating object proposals from edges. In *European Conference on Computer Vision*, pp. 391–405.

Zitnick, C. L. and Kanade, T. (2000). A cooperative algorithm for stereo matching and occlusion detection. *IEEE Transactions on Pattern Analysis and Machine Intelligence*, 22(7):675–684.

Zitnick, C. L. and Kang, S. B. (2007). Stereo for image-based rendering using image over-segmentation. *International Journal of Computer Vision*, 75(1):49–65.

Zitnick, C. L., Jojic, N., and Kang, S. B. (2005). Consistent segmentation for optical flow estimation. In *International Conference on Computer Vision (ICCV)*, pp. 1308–1315.

Zitnick, C. L., Kang, S. B., Uyttendaele, M., Winder, S., and Szeliski, R. (2004). High-quality video view interpolation using a layered representation. *ACM Transactions on Graphics (Proc. SIGGRAPH)*, 23(3):600–608.

Zitov'aa, B. and Flusser, J. (2003). Image registration methods: A survey. *Image and Vision Computing*, 21:997–1000.

Zoghlami, I., Faugeras, O., and Deriche, R. (1997). Using geometric corners to build a 2D mosaic from a set of images. In *IEEE Computer Society Conference on Computer Vision and Pattern Recognition (CVPR)*, pp. 420–425.

Zollhöfer, M., Stotko, P., Görlitz, A., Theobalt, C., Nießner, M., Klein, R., and Kolb, A. (2018). State of the art on 3D reconstruction with RGB-D cameras. *Computer Graphics Forum (Eurographics)*, 37(2):625–652.

Zollhöfer, M., Thies, J., Garrido, P., Bradley, D., Beeler, T., Pérez, P., Stamminger, M., Nießner, M., and Theobalt, C. (2018). State of the art on monocular 3D face reconstruction, tracking, and applications. *Computer Graphics Forum (Eurographics)*, 37(2):523–550.

Zongker, D. E., Werner, D. M., Curless, B., and Salesin, D. H. (1999). Environment matting and compositing. In *ACM SIGGRAPH Conference Proceedings*, pp. 205–214.

Zoph, B. and Le, Q. V. (2017). Neural architecture search with reinforcement learning. In *International Conference on Learning Representations (ICLR)*.

Zoph, B., Vasudevan, V., Shlens, J., and Le, Q. V. (2018). Learning transferable architectures for scalable image recognition. In *IEEE Conference on Computer Vision and Pattern Recognition (CVPR)*.

Zoran, D. and Weiss, Y. (2011). From learning models of natural image patches to whole image restoration. In *International Conference on Computer Vision (ICCV)*.

Zorin, D., Schröder, P., and Sweldens, W. (1996). Interpolating subdivision for meshes with arbitrary topology. In *ACM SIGGRAPH Conference Proceedings*, pp. 189–192.

Zuo, C., Huang, L., Zhang, M., Chen, Q., and Asundi, A. (2016). Temporal phase unwrapping algorithms for fringe projection profilometry: A comparative review. *Optics and Lasers in Engineering*, 85:84–103.

Žbontar, J. and LeCun, Y. (2016). Stereo matching by training a convolutional neural network to compare image patches. *Journal of Machine Learning Research*, 17(65):1–32.

Index

3D Rotations, *see* Rotations
3D alignment, 410
    absolute orientation, 410, 651
    orthogonal Procrustes, 411
3D convolutional neural networks, 252
3D model capture, 676
3D photography, 676, 690
3D scanning, 647
3D video, 706

Absolute orientation, 410, 651
Activation functions, 217
    rectified linear unit (ReLU), 217
    sigmoid, 217
Active appearance model (AAM), 291
Active contours, 373
Active illumination, 647
Active rangefinding, 647
Active shape model (ASM), 291, 376
Active stereo, 649
Activity recognition, 673
Adadelta, 230
AdaGrad, 230
Adam optimization algorithm, 231
Adaptive smoothing, 108
Adversarial examples, 248, 321
Affine transforms, 33, 36
Affinities (segmentation), 391
AlexNet neural network, 238
Algebraic multigrid, 388
Algorithms
    testing, viii
Aliasing, 67, 491
Alignment, *see* Image alignment
Alpha
    opacity, 91
    pre-multiplied, 91
Alpha matte, 91
Ambient illumination, 57
Analog to digital conversion (ADC), 66
Anisotropic diffusion, 108
Anisotropic filtering, 140
Anti-aliasing filter, 68, 491
Aperture, 60
Aperture problem, 453
Applications, 3
    3D model capture, 676
    3D model reconstruction, 576
    3D photography, 676, 690
    augmented reality, 555, 587
    automotive safety, 6
    background replacement, 617
    biometrics, 289
    colorization, 169
    digit classification, 237
    digital heritage, 654
    document scanning, 414
    edge editing, 372
    facial animation, 665
    flash photography, 506
    frame interpolation, 473
    gaze correction, 612
    head tracking, 611
    hole filling, 531
    image search, 287
    industrial, 6
    intelligent photo editing, 314
    Internet photos, 576
    location recognition, 555
    machine inspection, 6
    match move, 574
    medical imaging, 6, 310, 461
    morphing, 142
    mosaic-based video compression, 417
    non-photorealistic rendering, 532
    Optical character recognition (OCR), 3
    panography, 405
    performance-driven animation, 363

R. Szeliski, *Computer Vision*, Texts in Computer Science,
https://doi.org/10.1007/978-3-030-34372-9

photo pop-up, 314
Photo Tourism, 686
Photomontage, 435
planar pattern tracking, 555
rolling shutter wobble removal, 468
rotoscoping, 380
scene completion, 314
scratch removal, 531
segmentation, 181
self-driving vehicles, 6
single view reconstruction, 548
style transfer, 534
synthetic re-focusing, 698
tonal adjustment, 95
video denoising, 469
video stabilization, 457
video summarization, 417
video-based walkthroughs, 708
view morphing, 568
visual effects, 6
whiteboard scanning, 414
z-keying, 617
Arc length parameterization of a curve, 370
Architectural reconstruction, 660
Area statistics, 113
mean (centroid), 113
perimeter, 113
second moment (inertia), 113
Aspect ratio, 46, 47
Atrous convolution, 234
Attention, 257
Augmented reality, 439, 555, 587
Auto-calibration, 566
Autoencoder, 235, 261
Automatic gain control (AGC), 66
Average pooling, 234
Axis/angle representation of rotations, 37

B-snake, 375
B-spline, 141, 375, 377, 381, 461
cubic, 120
multilevel, 655
octree, 659
Backbone network, 237
Background plate, 528
Background subtraction (maintenance), 668
Backpropagation, 214, 226
gradient checkpointing, 228
guided, 245
Backside illumination (back-illuminated) sensor, 65
Backward convolution, 235
Bag of words (keypoints), 279
distance metrics, 281
Band-pass filter, 101
Bartlett filter, *see* Bilinear kernel
Barycentric coordinates, 156
Batch channel normalization, 222
Batch normalization, *see* Deep neural networks
Bayer pattern (RGB sensor mosaic), 74
demosaicing, 74, 515
Bayes' rule, 171, 750
MAP (maximum a posteriori) estimate, 750
posterior distribution, 750
Bayesian classification, 194
Bayesian modeling, 170, 750
MAP estimate, 171, 750
matting, 522
posterior distribution, 171, 750
prior distribution, 171, 750
uncertainty, 171
Belief propagation (BP), 175
Benchmarks, 756
Bias, 89, 447
Bias-variance tradeoff, 162
Bidirectional Reflectance Distribution Function, *see* BRDF
Bilateral filter, 107, 147
joint, 506
range kernel, 107
tone mapping, 502
Bilateral solver, 168
Bilinear blending, 94
Bilinear kernel, 100
Biometrics, 289
Bipartite problem, 572
Blind image deconvolution, 508
Block-based motion estimation (block matching), 448
Blocks world, 10
Blue screen matting, 91, 145, 519
Blur kernel, 60
estimation, 491, 540
Blur removal, 118, 147
Body color, 56
Boltzmann distribution, 171, 751
Boosting, 297
decision stump, 298
weak learner, 298
Border (boundary) effects, 98, 146
Boundary detection, 368
Box filter, 100
Boxlet, 104
BRDF, 54
anisotropic, 55
isotropic, 55

recovery, 675
spatially varying (SVBRDF), 675
Brightness, 89
Brightness constancy, 3, 446
constraint, 446, 452, 462
Bundle adjustment, 570

C3D network, 253
Calibration, *see* Camera calibration
Calibration matrix, 45
Camera calibration, 44, 83
accuracy, 590
aliasing, 491
extrinsic, 45, 552
intrinsic, 44, 545
optical blur, 491, 540
patterns, 546
photometric, 486
plumb-line method, 551, 591
point spread function, 491, 540
radial distortion, 550
radiometric, 486, 494, 538
rotational motion, 549, 590
slant edge, 491
vanishing points, 547
vignetting, 489
Camera matrix, 45, 48
Catadioptric optics, 62
Category-level recognition, 278
bag of words, 279
part-based, 282
surveys, 326
CCD, 63
blooming, 63
Central difference, 101
Chain rule
as used in backpropagation, 226
Chained transformations, 554, 571
Chamfer matching, 111
Channel-separated convolution, 234
Characteristic function, 113, 379, 653, 658
Characteristic polynomial, 731
cheirality, 559, 563
Cholesky factorization, 732
algorithm, 732
incomplete, 740
sparse, 738
Chromatic aberration, 61, 592
Chromaticity coordinates, 71
CIE L*a*b*, *see* Color
CIE L*u*v*, *see* Color
CIE XYZ, *see* Color
Circle of confusion, 60
CLAHE, *see* Histogram equalization
Classification, 190, 191
Bayesian, 194
CLIP, 251, 321
Closing, 111
Clustering, 205
agglomerative, 205, 387
cluster analysis, 386, 394
divisive, 205, 387
CMOS, 64
CNN stereo matching costs, 619
Co-vector, 34
Coefficient matrix, 167
Collineation, 36
Color, 69
balance, 75, 82, 144, 145
camera, 73
demosaicing, 74, 515
fringing, 516
hue, saturation, value (HSV), 77
L*a*b*, 72
L*u*v*, 72, 205, 389
primaries, 70
profile, 488
ratios, 77
RGB, 70
transform, 90
twist, 75, 90
XYZ, 70
YIQ, 76
YUV, 76
Color filter array (CFA), 74, 515
Color line model, 524
ColorChecker chart, 488
Colorization, 169
Compositing, 91, 143, 145
image stitching, 426
opacity, 91
over operator, 91
surface, 426
transparency, 91
Compression, 78
Computational photography, 483
active illumination, 507
flash and non-flash, 506
high dynamic range, 494
references, 485, 537
tone mapping, 499
Concentric mosaic, 418, 698
CONDENSATION, 377
Condition number, 739

Conditional batch normalization, 223
Conditional generative adversarial network, 264
Conditional random field (CRF), 177, 308, 613
  dense, 179
  fully connected, 179
Confidence calibration, 224
Confusion matrix (table), 353
Conic section, 30
Conjugate gradient descent (CG), 739
  algorithm, 739
  non-linear, 740
  preconditioned, 740
Connected components, 113
Constellation model, 283
Content-based image retrieval (CBIR), 287, 358
Content-preserving warps, 425, 458
Continuation method, 168
Contour
  arc length parameterization, 370
  chain code, 370
  detection, 368
  matching, 371, 397
  smoothing, 371
Contrast, 89
Contrastive (metric) learning, 251
Contrastive loss, 224, 225
Controlled-continuity spline, 165
Convolution, 96
  kernel, 96
  mask, 96
  superposition, 96
Convolutional neural networks
  1×1 convolutions, 233
Convolutional neural networks (CNNs), 231
Coring, 126, 149
Correlation, 96, 448
  windowed, 450
Correspondence map, 455
Cramer–Rao lower bound, 410, 454, 752
Cross-entropy loss, 198, 223
  multi-class, 198
Cross-validation, 162
Cube map
  Hough transform, 383
  image stitching, 429
Curve
  arc length parameterization, 370
  evolution, 371
  matching, 371
  smoothing, 371
Cylindrical coordinates, 418
DALL·E, 263
Data energy (term), 171, 751
Data fitting
  robust, 162
Data interpolation, 155
Dataset augmentation, *see* Deep neural networks
Dataset bias, 249
Datasets and test databases, 756
Decimation, 122
Decimation kernels
  bicubic, 123
  binomial, 122, 123
  QMF, 123
  windowed sinc, 122
Decision theory, 191
Decision trees and random forests, 202
Deconvolution network, 245
Deep learning, 189, 214
  courses, 267
  history, 267
  layers, 215
  surveys, 267
  textbooks, 267
Deep neural networks, 214
  3D, 252
  3D point clouds and meshes, 254
  activation functions, 217
  adversarial examples, 248
  AlexNet, 238
  architectures, 238
  attention, 256
  backbone, 237
  backbone (trunk), 249
  backpropagation, 226
  batch normalization, 220
  bottleneck, 235
  channels, 232
  convolutional neural networks, 231
  dataset augmentation, 219
  deconvolution, 245
  dropout, 220
  efficient (mobile) networks, 241
  feature maps, 232
  fine-tuning, 249
  fully convolutional, 237
  generative adversarial networks (GANs), 263
  GoogLeNet, 238
  group normalization, 222
  He initialization, 226
  head (branches), 249
  instance normalization, 222
  layer normalization, 222

learning rate, 229, 269
loss functions, 223
LSTMs, 255
minibatch stochastic gradient descent, 229
model zoo, 242
momentum, 230
neural architecture search (NAS), 243
optimization, 228
pre-training, 249
recurrent networks (RNNs), 255
regularization, 219
ResNet, 239
sequence modeling, 255
size and efficiency, 243
solvers, 228
spatio-temporal, 255
stereo matching, 618
stochastic gradient descent (SGD), 229
training, 228
transformer, 256
U-Net, 237
VGG, 238
visualization, 244
weight initialization, 225
weight sharing, 233
weights, 215
Delaunay triangulation, 156
Demosaicing (Bayer), 74, 515
Denoising
image, 513
video, 469
Dense captioning, 322
Dense conditional random field (CRF), 179
Depth estimation
monocular, 632
Depth from defocus, 646
Depth map, *see* Disparity map
Depth of field, 60, 81
Depth recovery, *see* Stereo
deep networks, 618
multi-view, 620
Depthwise convolution, 234
DETR, 260
Di-chromatic reflection model, 58
Difference matting (keying), 91, 145, 520, 668
Difference of Gaussians (DoG), 125
Difference of low-pass (DOLP), 127
Diffuse reflection, 56
Diffusion
anisotropic, 108
Digital camera, 63
color, 73
color filter array (CFA), 74
compression, 78
Dilation, 111
Dimensionality reduction
non-linear embeddings, 211
Direct current (DC), 79
Direct linear transform (DLT), 552
Direct sparse matrix techniques, 737
Directional derivative, 102
selectivity, 103
Discrete cosine transform (DCT), 79, 117
Discrete Fourier transform (DFT), 115
Discriminant analysis
Fisher, 197
linear, 197
quadratic, 197
Discriminative models
decision trees and random forests, 202
deep neural networks, 214
feedforward networks, 214, 219
logistic regression, 198
support vector machines, 199
Discriminative random field (DRF), 178
Disparity, 44, 601
Disparity map, 601
geometric consistency, 623
multiple, 622
Disparity space image (DSI), 601
generalized, 603
Displaced frame difference (DFD), 446
Displacement field, 141
Distance from face space (DFFS), 210
Distance in face space (DIFS), 210
Distance map, *see* Distance transform
Distance transform, 111
Euclidean, 112
image stitching, 430
Manhattan (city block), 111
signed, 112
Domain (of a function), 89
Domain adaptation, 249
Domain scaling law, 138
Downsampling, *see* Decimation
Downstream task, 249
Dropout, *see* Deep neural networks
DSAC, *see* RANSAC
Dynamic programming (DP), 614
monotonicity, 615
ordering constraint, 615
scanline optimization, 615
Dynamic snake, 376
Dynamic texture, 705

Edge detection, 364, 396
  boundary detection, 368
  Canny, 365
  chain code, 370
  color, 367
  Difference of Gaussian, 366
  edgel (edge element), 366
  hysteresis, 368
  Laplacian of Gaussian, 366
  linking, 368, 397
  marching cubes, 366
  scale selection, 366
  steerable filter, 366
  zero crossing, 366
Eigenface, 209
Eigenvalue decomposition, 209, 375, 729
Eigenvector, 729
Elastic deformations, 461
  image registration, 461
Elastic nets, 374
Elliptical weighted average (EWA), 139
Empirical risk minimization, 191
Energy functions
  regular, 173
  sub-modular, 173
Energy-based models, 153, 163
Environment map, 53, 696
Environment matte, 699
Epipolar constraint, 560
Epipolar geometry, 560, 599
  pure rotation, 564
  pure translation, 564
Epipolar line, 599
Epipolar plane, 600, 605
  image (EPI), 621, 694
Epipolar volume, 694
Epipole, 561, 599
Erosion, 111
Error rates
  accuracy (ACC), 354
  false negative (FN), 353
  false positive (FP), 353
  positive predictive value (PPV), 354
  precision, 354
  recall, 354
  ROC curve, 354
  true negative (TN), 353
  true positive (TP), 353
Errors-in-variable model, 422, 734
  heteroscedastic, 735
ESAC, *see* RANSAC
Essential matrix, 561
  5-point algorithm, 562
  eight-point algorithm, 561
  re-normalization, 562
  seven-point algorithm, 562
  twisted pair, 563
Estimation theory, 745
Euclidean transformation, 33, 35
Euler angles, 37
Expectation maximization (EM), 207
Exponential twist, 38
Exposure bracketing, 494
Exposure value (EV), 61, 486

F-number (stop), 60, 82
F-score, 303, 354
F-theta lens, 52
Face detection, 295
  boosting, 297
  cascade of classifiers, 299
  clustering and PCA, 297
  neural networks, 297
  support vector machines, 297
Face modeling, 663
Face recognition, 289
  active appearance model, 291
  eigenface, 209
  elastic bunch graph matching, 290
  local binary patterns (LBP), 327
  local feature analysis, 290
Face transfer, 702
Facial expression recognition, 290
Facial motion capture, 663, 668, 702
Factor graph, 171, 172, 584, 751
Factorization, 14, 568
  missing data, 569
  projective, 570
Fast Fourier transform (FFT), 115
Fast marching method (FMM), 379
Feature descriptor, 347, 396
  bias and gain normalization, 348
  GLOH, 349
  patch, 348
  PCA-SIFT, 349
  performance (evaluation), 350
  quantization, 281, 358
  RootSIFT, 349
  SIFT, 348
  steerable filter, 349
Feature detection, 335, 337, 395
  Adaptive non-maximal suppression, 341
  affine invariance, 346
  auto-correlation, 337

Förstner, 340
Harris, 340
Laplacian of Gaussian, 343
MSER, 346
region, 347
repeatability, 342
rotation invariance, 344
scale invariance, 343
Feature maps in deep neural networks, 232
Feature matching, 335, 352, 396
densification, 358
efficiency, 356
error rates, 353
hashing, 356
indexing structure, 356
k-d trees, 357
locality sensitive hashing, 356
nearest neighbor, 355
strategy, 353
verification, 358
Feature tracking, 361, 396
affine, 361
learning, 362
Feature tracks, 568, 576
Feature-based alignment, 403
2D, 403
3D, 410
iterative, 406
Jacobian, 404
least squares, 403
match verification, 277
RANSAC, 409
robust, 408
Field of Experts (FoE), 175
Fill factor, 65
Fill-in, 572, 737
Filter
adaptive, 108
band-pass, 101
bilateral, 107, 147
directional derivative, 102
edge-preserving, 106, 108
guided, 109, 147
Laplacian of Gaussian, 102
median, 105
moving average, 100
non-linear, 105
separable, 99, 146
steerable, 102, 147
Filter coefficients, 96
Filter kernel, *see* Kernel
Finding faces, *see* Face detection
Fine-grained category recognition, 285
Fine-tuning deep neural networks, 249
Finite element analysis, 166
stiffness matrix, 167
Finite impulse response (FIR) filter, 95, 104
Fisher discriminant analysis, 197
Fisher information matrix, 405, 410, 746, 752
Fisheye lens, 52
Flash and non-flash merging, 506
Flash matting, 528
Flip-book animation, 438
Flying spot scanner, 648
Focal length, 46, 47, 60
Focus, 60
shape-from, 646, 679
Focus of expansion (FOE), 564
Form factor, 59
Forward mapping, *see* Forward warping
Forward warping, 135, 150
Fourier transform, 113, 147
discrete, 115
magnitude (gain), 114
phase (shift), 114
power spectrum, 117
two-dimensional, 115
Fourier-based motion estimation, 449
Frame interpolation, 473
Free-viewpoint video, 706, 708
Fully connected (FC) layer, 216
Fully connected conditional random field (CRF), 179
Fully convolutional network, 237
Fundamental matrix, 565
estimation, *see* Essential matrix
Fundamental radiometric relation, 63

Gain, 89, 447
Gamma, 90
Gamma correction, 75, 82
Gap closing (image stitching), 416
Garbage matte, 528
Gated convolution, 234
Gaussian kernel, 100
Gaussian Markov random field (GMRF), 174, 179, 510
Gaussian mixture model, 374, 377
color model, 522
expectation maximization (EM), 207
mixing coefficient, 207
Gaussian mixture models, 206
Gaussian pyramid, 123
Gaussian scale mixtures (GSM), 175
Gaussians mixture model
soft assignment, 207

Gaze correction, 612
Geman–McClure function, 446
Generalized cylinders, 10, 650, 655
Generalized mean pooling (GeM), 235
Generative adversarial networks (GANs), 263
 conditional, 264
 discriminator, 263
 generator, 263
Generative models, 155, 170, 194, 261, 750
 probabilistic generative classification, 194
Geodesic active contour, 379
Geodesic distance (segmentation), 183
Geometric image formation, 29
Geometric lens aberrations, 61
Geometric primitives, 29
 homogeneous coordinates, 29
 lines, 30, 31
 normal vectors, 30, 31
 planes, 31
 points, 29, 31
Geometric transformations
 2D, 32, 135
 3D, 35
 3D perspective, 36
 3D rotations, 36
 affine, 33, 36
 bilinear, 35
 calibration matrix, 45
 collineation, 36
 Euclidean, 33, 35
 forward warping, 135, 150
 hierarchy, 34
 homography, 33, 36, 49, 413
 inverse warping, 136
 perspective, 33
 projections, 41
 projective, 33
 rigid body, 33, 35
 scaled rotation, 33, 36
 similarity, 33, 36
 translation, 32, 35
Geometry image, 656
Gesture recognition, 668
Gibbs distribution, 171, 751
Gimbal lock, 37
Gist (of a scene), 285, 314
Global illumination, 58
GooLeNet neural network, 238
Gradient checkpointing, 228
Gradient location-orientation histogram (GLOH), 349
Graduated non-convexity (GNC), 168
Graph cuts
 MRF inference, 173
 normalized cuts, 391
Graph-based segmentation, 388
Graphical models, 170, 171, 283
Grassfire transform, 112, 371, 430
Ground control points, 411, 562
Group normalization, 222
Guided image filter, 109, 147

Hammersley–Clifford theorem, 171, 751
Hand tracking, 670
Harris corner detector, *see* Feature detection
HDR imaging, *see* High dynamic range (HDR) imaging
He initialization rule for neural network weights, 226
Head tracking, 611
 active appearance model (AAM), 291
Helmholtz reciprocity, 55
Hessian, 167, 341, 405, 406, 410, 453, 456, 733
 eigenvalues, 454
 image, 453, 463
 inverse, 410, 454
 local, 462, 463
 patch-based, 456
 rank-deficient, 575
 reduced motion, 572
 sparse, 572, 593, 737
Heteroscedastic, 405, 735
Hidden Markov model (HMM), 703
Hierarchical motion estimation, 448
High dynamic range (HDR) imaging, 494
 formats, 499
 tone mapping, 499
 video, 499
Highest confidence first (HCF), 173
Hilbert transform pair, 103
Hinge loss, 201
Hinton diagrams, 244
Histogram equalization, 92, 145
 locally adaptive, 94, 146
Histogram intersection, 281
Histogram of oriented gradients (HOG), 300
History of computer vision, 9
Hole filling, 531
Holistic 3D reconstruction, 662
Homogeneous coordinates, 29, 559
Homography, 33, 49, 413
Hough transform, 381
 cascaded, 383
 cube map, 383
 generalized, 381
Human activity recognition, 316
Human body shape modeling, 671

Human motion tracking, 668
activity recognition, 673
adaptive shape modeling, 671
background subtraction, 668
flow-based, 669
initialization, 669
kinematic models, 669
particle filtering, 671
probabilistic models, 671
Hyper-Laplacian, 168, 174, 175
Hyperlapse videos, 711
Hyperparameters, 162, 231

Ideal points, 30
Ill-conditioned problems, 157
Ill-posed (ill-conditioned) problems, 163
Illusions, 3
Image alignment
feature-based, 403, 604
intensity-based, 445
Image analogies, 532
Image blending
feathering, 432
GIST, 437
gradient domain, 435
image stitching, 430
Poisson, 435
pyramid, 132, 435
Image center, 46
Image classification, 278
Image compositing, *see* Compositing
Image compression, 78
Image decimation, 122
Image deconvolution, *see* Blur removal
Image denoising, 147, 513
Image filtering, *see* Filter
Image formation
geometric, 29
photometric, 53
Image gradient, 102, 109, 452
constraint, 166
Image interpolation, 119
Image matting, 518, 541
Image processing, 87
textbooks, 87, 143, 184
Image pyramid, 119, 148
Image quality assessment, 119
Image resampling, 135, 147
test images, 148
Image restoration, 118
blur removal, 118, 147
deblocking, 186
denoising, 118, 147
noise removal, 151
Image sensing, *see* Sensing
Image statistics, 113
Image stitching, 411
automatic, 425
bundle adjustment, 421
compositing, 426
coordinate transformations, 429
cube map, 429
cylindrical, 418, 440
deghosting, 425, 433, 441
exposure compensation, 437
feathering, 432
gap closing, 416
global alignment, 421
homography, 413
motion models, 412
panography, 405
parallax removal, 424
photogrammetry, 411
pixel selection, 430
planar perspective motion, 413
recognizing panoramas, 425
rotational motion, 414
seam selection, 432
spherical, 419
up vector selection, 423
Image Transformer, 259
Image warping, 135, 149, 449
Image-based modeling, 685
Image-based rendering, 475, 681
concentric mosaic, 698
environment matte, 699
impostors, 688
layered depth image, 688
layers, 688
light field, 693
Lumigraph, 693
modeling vs. rendering continuum, 701
multiplane image (MPI), 689
reflections, 690
sprites, 688
surface light field, 696
unstructured Lumigraph, 696
view interpolation, 683
view-dependent texture maps, 685
Image-based visual hull, 631
Image to image translation, 264
Implicit surface, 658
Impostors, *see* Sprites
Impulse response, 96

Inception module, 239
Incremental refinement
    motion estimation, 449, 451
Incremental rotation, 40
Indexing structure, 356
Indicator function, 658
Inductive spatial locality bias, 257
Industrial applications, 6
Infinite impulse response (IIR) filter, 104
Influence function, 168, 408, 749
Information criteria (BIC, AIC), 162
Information matrix, 405, 410, 575, 746, 752
Inpainting, 531
Instance normalization, 222
Instance recognition, 276
    geometric alignment, 277
    inverted index, 358
    large-scale, 358
    match verification, 277
    query expansion, 359
    stop list, 359
    visual words, 359
Integrability constraint, 643
Integral image, 103
Integrating sphere, 488
Intelligent scissors, 378
Interaction potential, 171, 172, 751
Interactive computer vision, 677
Interactive segmentation, 181
International Color Consortium (ICC), 488
Internet photos, 576
Interpolation, 119
    scattered data, 155
Interpolation kernels
    bicubic, 121
    bilinear, 120
    binomial, 120
    sinc, 121
    spline, 121
Intrinsic camera calibration, 545
Intrinsic images, 10
Inverse kinematics (IK), 669
Inverse mapping, *see* Inverse warping
Inverse problems, 3, 163
Inverse warping, 136
ISO setting, 66
Isomap, 211
Iterated conditional modes (ICM), 173
Iterative back projection (IBP), 509
Iterative closest point (ICP), 374, 411, 651
Iterative feature-based alignment, 406
Iterative sparse matrix techniques, 738
    conjugate gradient, 739
Iteratively reweighted least squares, 199
    (IRLS), 408, 455, 554, 749
Iteratively reweighted least squares (IRLS), 162

Jacobian, 404, 452, 554, 571, 736
    image, 453
    motion, 456
    sparse, 572, 593, 737
Joint bilateral filter, 506
Joint domain (feature space), 389

K-d trees, 357
K-means, 206
K-nearest neighbors (kNN), 192
Kalman snakes, 376
Kanade–Lucas–Tomasi (KLT) tracker, 362
Karhunen–Loève transform, 117, 209
Kernel, 100
    bilinear, 100
    Gaussian, 100
    low-pass, 101
    Sobel operator, 101
    unsharp mask, 101
Kernel basis function, 166
Kernel density estimation, 158
Kernel functions, 157
Kernel methods, 157
Kernel regression, 157, 158, 201
Keypoint detection, *see* Feature detection
KinectFusion, 653
Kinematic model (chain), 669
Kruppa equations, 567

L*a*b*, *see* Color
L*u*v*, *see* Color
L_1 norm, 168, 446, 463, 659
L_∞ norm, 574
Lambertian reflection, 56
Laplacian Eigenmaps, 211
Laplacian matting, 525
Laplacian of Gaussian (LoG) filter, 102
Laplacian pyramid, 125
    blending, 132, 148, 435
    perfect reconstruction, 125
Lasso (least absolute shrinkage and selection operator), 158
Latent Dirichlet process (LDP), 284
Layered depth image (LDI), 688
Layered depth panorama, 698
Layered motion estimation, 470
    reflections, 474
    transparent, 474

Layers in image-based rendering, 688
Layout consistent random field, 308
Learning, 189
    classification, 191
        nearest neighbors, 192
    contrastive (metric), 251
    deep neural networks, 214
    regression, 191
    self-supervised, 249
    semi-supervised, 212, 250
    student-teacher, 252
    supervised, 190, 191
    test phase, 191
    training phase, 191
    unsupervised, 190, 205
    weak, 250
    weakly supervised, 214
Learning rate, *see* Deep neural networks
Least median of squares (LMS), 409
Least squares
    iterative solvers, 554, 738
    linear, 81, 403, 410, 445, 729, 733, 744, 747
    non-linear, 406, 554, 559, 736, 748
    robust, *see* Robust least squares
    sparse, 571, 737
    total, 734
    weighted, 170, 405, 506, 508
LeNet-5, 231
Lens
    compound, 61
    nodal point, 61
    thin, 60
Lens distortions, 51
    calibration, 550
    decentering, 52
    radial, 51
    spline-based, 52
    tangential, 52
Lens law, 60
Level of detail (LOD), 656
Level sets, 379, 380
    fast marching method, 379
    geodesic active contour, 379
Levenberg–Marquardt, 407, 576, 593, 736, 765
Lidar (Light Detection and Ranging), 647
Lifting, *see* Wavelets
Light field
    higher dimensional, 700
    light slab, 694
    ray space, 694
    rendering, 693
    surface, 696
Lightness, 72
Line at infinity, 30
Line detection, 381
    Hough transform, 381
    RANSAC, 383
    simplification, 381, 398
    successive approximation, 381, 398
Line equation, 30, 31
Line fitting, 81
    uncertainty, 398
Line hull, *see* Visual hull
Line labeling, 10
Line process, 184, 613, 751
Line segment detector (LSD), 384
Line spread function (LSF), 491
Line support regions, 384
Line-based structure from motion, 581
Linear algebra, 727
    least squares, 733
    matrix decompositions, 728
    references, 728
Linear blend, 89
Linear discriminant analysis (LDA), 197
Linear filtering, 95
Linear operator, 89
    superposition, 89
Linear shift invariant (LSI) filter, 96
Live-wire, 378
Local distance functions, 211
Local Laplacian, 127, 149
Local Linear Embedding (LLE), 211
Local operator, 95
Locality sensitive hashing (LSH), 356
Localization, 555
Locally adaptive histogram equalization, 94
Location recognition, 555
Log likelihood, 194
Log odds, 195
Logistic regression, 196, 198
Logistic sigmoid function (curve), 195
Logit, 195
Long short-term memory (LSTM), 255
Loopy belief propagation (LBP), 175
Loss function, 223
    ArcFace, 225
    contrastive, 224
    cross-entropy, 223
    perceptual, 224
    triplet, 224
Low-pass filter, 101
    sinc, 101
Lumigraph, 693

unstructured, 696
Luminance, 71
Lumisphere, 696

M-estimator, 162, 408, 446, 748
Machine learning, *see* Learning
textbooks, 267
Machine learning models
discriminative vs. generative, 198
MAGSAC, *see* RANSAC
Mahalanobis distance, 207, 211, 746
Manhattan world, 582
Manifold learning, 211
Manifold mosaic, 432, 718
Markov chain Monte Carlo (MCMC), 747
Markov random field, 170, 751
cliques, 172, 751
directed edges, 182
flux, 182
inference, *see* MRF inference
layout consistent, 308
learning parameters, 170
line process, 184, 613, 751
neighborhood, 172, 751
order, 172, 751
random walker, 183
stereo matching, 613
Marr's framework, 10
computational theory, 11
hardware implementation, 11
representations and algorithms, 11
Masked convolution, 234
Match move, 574
Matrix decompositions, 728
Cholesky, 732
eigenvalue (ED), 729
QR, 731
singular value (SVD), 728
square root, 732
Matte reflection, 56
Matting, 91, 518, 541
alpha matte, 91
Bayesian, 522
blue screen, 91, 145, 519
difference, 91, 145, 520, 668
flash, 528
GrabCut, 524
Laplacian, 525
natural, 521
optimization-based, 524
Poisson, 524
shadow, 528
smoke, 527
triangulation, 519, 528
trimap, 521
two screen, 519
video, 528
Max pooling, 234
Maximally stable extremal region (MSER), 346
Maximum *a posteriori* (MAP) estimate, 171, 750
Maximum margin classifier, 199
Mean absolute difference (MAD), 607
Mean average precision, 354
Mean shift, 389
Mean square error (MSE), 79, 607
Measurement equation (model), 558, 745
Measurement model, *see* Bayesian model
Medial axis transform (MAT), 112
Median absolute deviation (MAD), 446
Median filter, 105
Medical image registration, 461
Medical image segmentation, 310
Membrane, 165
Mesh-based warping, 140, 149
Metamer, 71
Metric learning, 211
Metric tree, 358
Minibatch stochastic gradient descent), 229
MIP-mapping, 138
trilinear, 139
Mixture of Gaussians, *see* Gaussian mixture model
MLESAC, *see* RANSAC
Model selection, 162, 413, 750
Model zoo, *see* Deep neural networks
Model-based reconstruction, 660
architecture, 660
heads and faces, 663
human body, 668
Model-based stereo, 661, 685
Models
Bayesian, 170, 750
forward, 3
physically based, 12
physics-based, 3
probabilistic, 3
Modulation transfer function (MTF), 69, 491
Momentum, *see* Deep neural networks
Monocular depth estimation, 632
Morphable model
body, 671
face, 665, 702
multidimensional, 702
Morphing, 142, 150, 683, 684
3D body, 671

3D face, 665
automated, 481
facial feature, 702
feature-based, 142, 150
flow-based, 481
video textures, 703
view morphing, 684, 719
Morphological operator, 110
Morphology, 110
Mosaic, *see* Image stitching
Mosaics
motion models, 412
video compression, 417
whiteboard and document scanning, 414
Motion compensated video compression, 448, 478
Motion compensation, 79
Motion estimation, 445
affine, 455
aperture problem, 453
compositional, 457
Fourier-based, 449
frame interpolation, 473
hierarchical, 448
incremental refinement, 451
layered, 470
learning, 457, 463
linear appearance variation, 454
optical flow, 461
parametric, 455
patch-based, 445, 456
phase correlation, 451
quadtree spline-based, 460
reflections, 474
rolling shutter, 468
spline-based, 459
translational, 445
transparent, 474
uncertainty modeling, 454
Motion field, 455
Motion models
learned, 457
Motion segmentation, 482
Motion stereo, 622
Moving least squares (MLS), 657
MRF inference, 173, 751
alpha expansion, 175
belief propagation, 175
expansion move, 175
graph cuts, 173
highest confidence first, 173
iterated conditional modes, 173
loopy belief propagation, 175
simulated annealing, 173
stochastic gradient descent, 173
swap move (alpha-beta), 175
Multi-frame motion estimation, 468
Multi-layer perceptron (MLP), 216
Multi-pass transforms, 140
Multi-perspective panoramas, 418, 698
Multi-perspective plane sweep (MPPS), 424
Multi-view stereo, 620
epipolar plane image, 621
evaluation, 630
initialization requirements, 630
point clouds, 625
reconstruction algorithm, 629
scene representation, 625
shape priors, 629
silhouettes, 630
space carving, 629
spatio-temporally shiftable window, 622
taxonomy, 624
visibility, 628
volumetric, 624, 626
voxel coloring, 629
Multidimensional scaling (MDS), 211
Multigrid, 741
algebraic (AMG), 388, 741
Multinomial logistic regression objective, 198
Multiplane image (MPI), 689
Multiple hypothesis tracking, 377
Multiple object tracking, 478
Multiple-center-of-projection images, 418, 698, 718
Multiresolution representation, 123
Mutual information, 448, 461

Naïve Bayes, 195
Natural image matting, 521
Nearest neighbor, 192
distance ratio (NNDR), 355
matching, *see* Feature matching
Negative posterior log likelihood, 171, 746, 750
Neighborhood operator, 95, 105
Neural architecture search (NAS), 243
Neural network
backbone, 242
backpropagation, 214
branches, 242
confidence calibration, 224
fine tuning, 242
for face detection, 297
head(s), 242
pre-trained, 242
trunk, 242

Neural network pooling
    average, 234
    generalized mean (GeM), 235
    max, 234
    unpooling, 235
Neural rendering, 711
    depth images and layers, 713
    implicit functions using MLPs, 715
    texture mapped meshes and models, 713
    voxel grids, 714
Neural textures, 713
Nodal point, 61, 414
Noise (sensor), 66, 488
Noise level function (NLF), 66, 82, 488, 539
Noise removal, *see* (enoising)118, 151
Non-linear filter, 105, 143
Non-linear least squares
    seeLeast squares, 406
Non-maximal suppression, *see* Feature detection
Non-parametric density modeling, 389
Non-photorealistic rendering (NPR), 532
Normal equations, 404, 452, 733, 736
Normal map (geometry image), 656
Normal vector, 31
Normalized cross-correlation (NCC), 448, 479, 607
Normalized cuts, 391
    intervening contour, 392
Normalized device coordinates (NDC), 43, 47
Normalized exponential, 194
Normalized sum of squared differences (NSSD), 448
Norms
    L_1, 168, 446, 463, 659
    L_∞, 574
Novel view synthesis (NVS), 683
Nyquist rate/frequency, 68

Object detection, 295
    car, 299, 327
    face, 295
    part-based, 300
    pedestrian, 294, 299
Object tracking, 477
Object-centered projection, 50
Obstruction-free photography, 475
Occluding contours, 604
Octree reconstruction, 631
Octree spline, 461
Omnidirectional vision systeghyms, 708
One-hot encoding, 198
Opacity, 91
OpenGV, 565, 765
Opening, 111
Operator linearity, 89
Optic flow, *see* Optical flow
Optical flow, 461
    anisotropic smoothness, 463
    benchmarks and datasets, 465
    coarse-to-fine, 465
    combinatorial optimization, 465
    constraint equation, 452
    deep learning, 466
    evaluation, 465
    fusion move, 465
    global and local, 463
    Large displacement, 465
    Markov random field, 465
    multi-frame, 468
    neural networks, 466
    normal flow, 453
    patch-based, 462
    region-based, 472
    regularization, 462
    robust regularization, 463
    smoothness, 462
    task-oriented, 468
    total variation, 463
    variational, 463
Optical illusions, 3
Optical transfer function (OTF), 69, 491
Optical triangulation, 647
Optics, 59
    chromatic aberration, 61
    Seidel aberrations, 61
    vignetting, 62, 539
Optimal motion estimation, 570
Oriented particles (points), 657
Orthogonal Procrustes, 411
Orthographic projection, 41
Osculating circle, 605
Over operator, 91
Overfitting, 159, 193
Overview, 18

Padding, 98, 146
Pairwise alignment, 403
Panography, 405, 439
Panorama, *see* Image stitching
Panorama with depth, 418, 603, 698
Para-perspective projection, 43
Parallel tracking and mapping (PTAM), 587
Parameter sensitive hashing, 356
Parametric motion estimation, 455
Parametric surface, 655

Parametric transformation, 135, 149
Part-based recognition, 282
    constellation model, 283
Partial convolution, 234
Particle filtering, 377, 671, 747
Partition function, 198
Parzen window, 158
Patch-based motion estimation, 445
PatchMatch, 530
PatchMatch Stereo, 614
Peak signal-to-noise Ratio (PSNR), 79, 118
Pedestrian detection, 299
Penumbra, 53
Perceptual loss, 119, 224, 535
Perceptual similarity metrics, 118, 119, 224
Performance-driven animation, 363, 667, 702
Perspective n-point problem (PnP), 553
Perspective projection, 43
Perspective transform (2D), 33
Phase correlation, 451, 479
Phong shading, 57
Photo pop-up, 314
Photo Tourism, 686
Photo-mosaic, 411
Photoconsistency, 602, 628
Photometric image formation, 53
    calibration, 486
    global illumination, 58
    lighting, 53
    optics, 59
    radiosity, 58
    reflectance, 54
    shading, 57
Photometric stereo, 643
Photometry, 53
Photomontage, 435
Physically based models, 12
Physics-based vision, 14
Pictorial structures, 10, 16, 282
Pixel transform, 89
Plücker coordinates, 32
Planar pattern tracking, 555
Plane at infinity, 31
Plane equation, 31
Plane plus parallax, 48, 459, 471, 602, 688
Plane sweep, 602, 636
Plane-based structure from motion, 582
Plenoptic function, 693
Plenoptic modeling, 684
Plumb-line calibration method, 551, 591
Point distribution model, 375
Point operator, 87
Point process, 87
Point spread function (PSF), 68
    estimation, 491, 540
Point-based representations, 657
Points at infinity, 30
Poisson
    blending, 435
    equations, 659
    matting, 524
    noise, 66
    surface reconstruction, 659
Polar coordinates, 30
Polar projection, 52, 420
Polyphase filter, 120
Pop-out effect, 4
Pose estimation, 552
    iterative, 554
    RGB-D data, 652
Power spectrum, 117
Pre-training deep neural networks, 249
Precision, *see* Error rates
    mean average, 354
Preconditioning, 740
Pretext task, 249
Principal component analysis (PCA), 209, 297, 375, 730, 746
    face modeling, 663
    generalized, 731
    missing data, 569, 731
Principal point, *see* Image center
Prior energy (term), 171, 751
Prior model, *see* Bayesian model
Probabilistic generative classification, 194
Profile curves, 604
Progressive mesh (PM), 656
Projections
    object-centered, 50
    orthographic, 41
    para-perspective, 43
    perspective, 43
Projective (uncalibrated) reconstruction, 565
Projective depth, 49, 602
Projective disparity, 49, 602
Projective space, 30
PROSAC (PROgressive SAmple Consensus), 409
PSNR, *see* Peak signal-to-noise ratio
Pull-push algorithm, 157
Pyramid, 119, 148
    blending, 132, 148
    Gaussian, 123
    half-octave, 127
    Laplacian, 125
    motion estimation, 449

octave, 123
radial frequency implementation, 132
steerable, 132
Pyramid match kernel, 281

QR factorization, 731
Quadratic discriminant analysis, 197
Quadratic form, 166
Quadrature mirror filter (QMF), 123
Quadric equation, 30, 32
Quadtree spline
motion estimation, 460
restricted, 460
Quaternions, 39
antipodal, 39
multiplication, 40
Query by image content (QBIC), 287
Query expansion, 359
Quincunx sampling, 127

Radial basis function, 141, 157, 166, 655
Radial distortion, 51
barrel, 51
calibration, 550
parameters, 51
pincushion, 51
Radiance map, 497
Radiometric image formation, 53
Radiometric response function, 486
Radiometry, 53
Radiosity, 59
Random forests, 202
Random walker, 183
Range (of a function), 89
Range data, *see* Range scan
Range image, *see* Range scan
Range scan
alignment, 651, 679
large scenes, 654
merging, 652
registration, 651, 679
segmentation, 651
volumetric, 653
Range sensing (rangefinding), 647
coded pattern, 648
light stripe, 647
shadow stripe, 648, 679
spacetime stereo, 650
stereo, 649
texture pattern (checkerboard), 648
time of flight, 649
Ranking loss, 224, 251
RANSAC
inliers, 409
preemptive, 409
progressive (PROSAC), 409
RANSAC (RANdom SAmple Consensus), 383, 409, 562, 749
RAW image format, 66
Ray space (light field), 694
Ray tracing, 58
Rayleigh quotient, 392
Recall, *see* Error rates
Receiver Operating Characteristic
area under the curve (AUC), 354
mean average precision, 354
ROC curve, 354, 396
Recognition, 273
category (class), 278
color similarity, 287
context, 284
contour-based, 327
face, 289
instance, 276
part-based, 282
scene understanding, 284
semantic segmentation, 307
shape context, 327
Rectangle detection, 386
Rectification, 600, 635
standard rectified geometry, 601
Rectified linear unit (ReLU), 217
Recurrent neural networks (RNNs), 255
Recursive filter, 104
Reference plane, 49
Reflectance, 54
Reflectance map, 642
Reflectance modeling, 674
Reflection
di-chromatic, 58
diffuse, 56
specular, 56
Reflection layers, 474, 690
Region merging and splitting, 205, 387
Registration, *see* Image Alignment
feature-based, 403
intensity-based, 445
medical image, 461
Regression, 155, 190, 191
Regularization, 158, 163, 460
neural network, 219
robust, 168
weight decay, 219
Regularization parameter, 165

Residual error, 162, 403, 408, 409, 446, 452, 456, 463, 558, 571, 733, 739
Residual network (ResNet), 239
RGB (red green blue), *see* Color
Ridge regression, 158
Rigid body transformation, 33, 35
Risk minimization, 191
RMSProp, 230
Robust data fitting, 162
Robust error metric, *see* Robust penalty function
Robust least squares, 385, 408, 446, 748
 iteratively reweighted, 408, 455, 554, 749
Robust loss function, 162
Robust penalty function, 168, 446, 454, 509, 603, 607, 608, 613, 749
Robust regularization, 168
Robust statistics, 446, 748
 inliers, 409
 loss function, 162
 M-estimator, 162, 408, 446, 748
Rodrigues' formula, 38
Rolling shutter wobble removal, 468
Root mean square error (RMS), 79, 447
Rotations, 36
 Euler angles, 37
 axis/angle, 37
 exponential twist, 38
 incremental, 40
 interpolation, 40
 quaternions, 39
 Rodrigues' formula, 38

Sampling, 67
Scale invariant feature transform (SIFT), 348
Scale-space, 12, 102, 126, 380
Scatter matrix, 209
Scattered data approximation, 155
 overfitting, 159
 underfitting, 159
Scattered data interpolation, 141, 155
Scene completion, 314
Scene flow, 624, 706
Scene understanding, 284
 gist, 285, 314
Schur complement, 572, 737
Scratch removal, 531
Seam selection in image stitching, 432
Second-order cone programming (SOCP), 574
Seed and grow
 stereo, 604
 structure from motion, 577
Segmentation
 active contours, 373
 affinities, 391
 binary MRF, 172, 181
 CONDENSATION, 377
 connected components, 113
 energy-based, 181
 Gaussian mixture model, 206
 geodesic active contour, 379
 geodesic distance, 183
 GrabCut, 182, 524
 graph cuts, 181
 graph-based, 388
 hierarchical, 388
 intelligent scissors, 378
 joint feature space, 389
 k-means, 206
 level sets, 379
 mean shift, 389
 medical image, 310
 merging, 205, 387
 minimum description length (MDL), 181
 Mumford–Shah, 181
 non-parametric, 389
 normalized cuts, 391
 probabilistic aggregation, 388
 random walker, 183
 snakes, 373
 splitting, 205, 387
 stereo matching, 616
 thresholding, 110
 tobogganing, 378, 388
 watershed, 387
 weighted aggregation (SWA), 392
Seidel aberrations, 61
Self-attention, 258
Self-calibration, 566
 bundle adjustment, 568
 Kruppa equations, 567
Self-supervised learning, 249
Semantic image synthesis, 264
Semantic segmentation, 307
Semi-global matching (SGM), 616, 619
Semi-supervised learning, 212, 250
 transductive vs. inductive, 212
Sensing, 63
 aliasing, 67, 491
 color, 69
 color balance, 75
 gamma, 75
 pipeline, 64, 487
 sampling, 67
 sampling pitch, 65

Sensor noise, 66, 488
- amplifier, 66
- dark current, 66
- fixed pattern, 66
- shot noise, 66

Separable filtering, 99, 146
Sequential minimal optimization (SMO), 202
Shading, 57
- equation, 56
- shape-from, 642

Shadow matting, 528
Shape context, 327, 372
Shape from
- focus, 646, 679
- photometric stereo, 643
- profiles, 604
- shading, 642
- silhouettes, 630
- specularities, 646
- stereo, 595
- texture, 645

Shape parameters, 291, 375
Shape-from-X, 12
- focus, 12
- photometric stereo, 12
- shading, 12
- texture, 12

Shift invariance, 96
Shiftable multi-scale transform, 132
Shutter speed, 64
Siamese network, 224
Sigmoid activation function, 217
Sigmoid function, 195
Signed distance function, 379, 651, 652, 657, 658
Silhouette-based reconstruction, 630
- octree, 631
- visual hull, 630

Similarity metrics
- perceptual, 119

Similarity metrics (perceptual), 118
Similarity transform, 33, 36
Simulated annealing, 173
Simultaneous localization and mapping (SLAM), 583
Sinc filter
- interpolation, 121
- low-pass, 101
- windowed, 121

Singe image depth estimation, 632
Single view metrology, 548, 591
Singular value decomposition (SVD), 728
Skeletal set, 574, 577
Skeleton, 112, 371
Skew, 45, 46
Slant edge calibration, 491
Slippery spring, 374
SlowFast neural network architecture, 253
Smoke matting, 527
Smoothness constraint, 165
Smoothness penalty, 165
Snakes, 373
- ballooning, 373
- dynamic, 376
- internal energy, 373
- Kalman, 376
- shape priors, 375
- slippery spring, 374

Soft assignment, 207
Softmax function, 194, 198
Software, 761
Space carving
- multi-view stereo, 629

Spacetime stereo, 650
Sparse flexible model, 283
Sparse matrices, 737
- compressed sparse row (CSR), 737
- skyline storage, 737

Sparse methods
- direct, 737
- iterative, 738

Spatial pyramid matching, 281
Spatially varying bidirectional reflectance distribution function (SVBRDF), 676
Spectral (weight) normalization, 223
Spectral response function, 73
Spectral sensitivity, 73
Specular flow, 646
Specular reflection, 56
Spherical coordinates, 31, 383, 419
Spherical linear interpolation, 40
Spin image, 651
Splatting, *see* Forward warping, 157
- volumetric, 657

Spline, 156
- controlled continuity, 165
- octree, 461
- quadtree, 460
- tensor product, 156
- thin plate, 165

Spline-based motion estimation, 459
Splining images, *see* Laplacian pyramid blending
Sprites
- image-based rendering, 688
- motion estimation, 470
- video, 705

video compression, 417
with depth, 688
Statistical decision theory, 745, 747
Statistical models: discriminative vs. generative, 198
Steerable filter, 102, 147
Steerable pyramid, 132
Steerable random field, 174
Stereo, 595
aggregation methods, 609, 636
coarse-to-fine, 614
cooperative algorithms, 613
correspondence, 597
curve-based, 604
deep networks, 618
dense correspondence, 606
depth map, 597
dynamic programming, 614
edge-based, 604
epipolar geometry, 599
feature-based, 604
global optimization, 612, 637
graph cut, 613
layers, 617
local methods, 609
model-based, 661, 685
multi-view, 620
non-parametric similarity measures, 607
photoconsistency, 602
plane sweep, 602, 636
rectification, 600, 635
region-based, 609
scanline optimization, 615
seed and grow, 604
segmentation-based, 609, 616
semi-global matching (SGM), 616, 619
shiftable window, 622
similarity measure, 607
spacetime, 650
sparse correspondence, 604
sub-pixel refinement, 610
support region, 609
taxonomy, 597, 606
uncertainty, 611
window-based, 609, 636
winner-take-all (WTA), 610
Stereo-based head tracking, 611
Stiffness matrix, 167
Stitching, *see* Image stitching
Stochastic gradient descent (SGD), 173, 229
Strided convolution, 234
Structural Similarity (SSIM) index, 119
Structure from motion, 545
bas-relief ambiguity, 575
bundle adjustment, 570
constrained, 580
factorization, 568
feature tracks, 576
iterative factorization, 570
line-based, 581
multi-frame, 568
orthographic, 569
plane-based, 570, 582
projective factorization, 570
seed and grow, 577
self-calibration, 566
skeletal set, 574, 577
two-frame, 560
uncertainty, 575
Student-teacher learning, 252
Style transfer, 534
Sub-modular energy functions, 173
Subdivision surface, 655
subdivision connectivity, 656
Subspace learning, 211
Sum of absolute differences (SAD), 446, 479, 607
Sum of squared differences (SSD), 445, 479, 607
bias and gain, 447
Fourier-based computation, 450
normalized, 448
surface, 338
weighted, 447
windowed, 447
Sum of sum of squared differences (SSSD), 621
Summed area table, 103
Super-resolution, 508, 540
example-based, 510
faces, 510
hallucination, 510
prior, 510
video, 513
Superposition principle, 89
Superquadric, 658
Supervised learning, 190, 191
SuperVision neural network, 238
Support vector machine (SVM), 199, 297, 300
Support vectors, 200
Surface element (surfel), 657
Surface interpolation, 655
Surface light field, 696
Surface representations, 654
non-parametric, 655
parametric, 655
point-based, 657
simplification, 656

splines, 655
subdivision surface, 655
symmetry-seeking, 655
triangle mesh, 655
Surface simplification, 656

t-distributed Stochastic Neighbor Embedding (t-SNE), 211
Telecentric lens, 41, 646
Temporal derivative, 452, 463
Temporal texture, 705
Testing algorithms, viii
TextonBoost, 307
Texture addressing mode, 98
Texture map
recovery, 674
view-dependent, 674, 685
Texture mapping
anisotropic filtering, 140
MIP-mapping, 138
multi-pass, 140
trilinear interpolation, 139
Texture synthesis, 529, 542
by numbers, 533
hole filling, 531
image quilting, 530
non-parametric, 529
transfer, 532
Texture, shape-from, 645
Thin lens, 60
Thin-plate spline, 165
Thresholding, 110
Through-the-lens camera control, 555, 575
Tobogganing, 378, 388
Tonal adjustment, 95, 145, 146
Tone mapping, 499
adaptive, 500
bilateral filter, 502
global, 499
gradient domain, 502
halos, 500
interactive, 503
local, 500
scale selection, 503
Total least squares (TLS), 455, 734
Total variation, 168, 463, 659
Tracking
feature, 361
head, 611
human motion, 668
multiple hypothesis, 377
multiple object, 478
object, 477
planar pattern, 555
PTAM, 587
Training error, 162
Transfer learning, 243, 249
Transformers, 257
Translational motion estimation, 445
bias and gain, 447
Transparency, 91
Transposed convolution, 235
Travelling salesman problem (TSP), 374
Tri-chromatic sensing, 70
Tri-stimulus values, 70, 74
Triangular irregular network (TIN), 156
Triangulation, 558
planar, 156
Trilinear interpolation, *see* MIP-mapping
Trimap (matting), 521
Triplet loss, 224
Truncated signed distance function (TSDF), 652, 657
Trust region method, 736
Two-dimensional Fourier transform, 115

U-Net, 237
Uncertainty
correspondence, 405
modeling, 410, 752
weighting, 405
Underfitting, 159, 193
Unpooling, 235
Unsharp mask, 101
Unsupervised learning, 190, 205
clustering, 205
principal component analysis, 209
Upsampling, *see* Interpolation

Validation error, 162
Validation set, 159
Vanishing point
detection, 384, 399
Hough, 385
modeling, 661
uncertainty, 399
Variable reordering, 737
minimum degree, 737
multi-frontal, 737
nested dissection, 737
Variable state dimension filter (VSDF), 573
Variational autoencoder (VAE), 262
Variational method, 164
Variational methods, 163
VGG neural network, 238
Video compression

motion compensated, 448
Video compression (coding), 478
Video denoising, 469
Video matting, 528
Video object segmentation, 476
Video objects (coding), 470
Video segmentation, 476
Video sprites, 705
Video stabilization, 457, 480
Video super-resolution, 513
Video texture, 703
Video understanding, 316
Video-based animation, 702
Video-based rendering, 701
3D video, 706
animating pictures, 705
sprites, 705
video texture, 703
virtual viewpoint video, 706
walkthroughs, 708
View correlation, 575
View interpolation, 568, 683, 719
View morphing, 568, 684, 703
View-dependent texture maps, 685
Vignetting, 62, 447, 489, 539
mechanical, 63
natural, 62
Virtual viewpoint video, 706
Vision Transformer, 259
Visual hull, 630
image-based, 631
Visual illusions, 3
Visual localization, 555
Visual object tracking, 477
Visual odometry, 583
Visual place recognition, 555
Visual search, 287
Visual similarity (search), 287
Visual words, 279, 358, 359
Visual-inertial odometry, 585
Vocabulary tree, 358
Volumetric 3D reconstruction, 624
Volumetric performance capture, 708
Volumetric range image processing (VRIP), 652
Volumetric representations, 658
Voronoi diagram, 432
Voxel coloring multi-view stereo, 629
VQ-VAE, 262

Watershed, 387, 389
basins, 387, 389
oriented, 388
Wavelets, 128, 149
compression, 149
lifting, 129
overcomplete, 128, 132
second generation, 131
self-inverting, 132
tight frame, 128
weighted, 131
Weak learning, 250
Weakly supervised learning, 214
Weaving wall, 605
Weight decay, 158, 219
Weight initialization, *see* Deep neural networks
Weight sharing, 233
Weight standardization, 222
Weighted least squares (WLS), 170, 503
Weighted prediction (bias and gain), 447
White balance, 75, 82
Wiener filter, 117
Wire removal, 531
Wrapping mode, 98

XYZ, *see* Color

Zippering, 652

Printed in the United States
by Baker & Taylor Publisher Services